Regulated Chemicals Directory 1995

REGULATED CHEMICALS DIRECTORY 1995

Compiled by ChemADVISOR®, Inc.
Pittsburgh, Pennsylvania

VNR SPRINGER SCIENCE+BUSINESS MEDIA, LLC

For more information contact:

Van Nostrand Reinhold
115 Fifth Avenue
New York, NY 10003

International Thomson Publishing GmbH
Königswinterer Str. 418
53227 Bonn
Germany

International Thomson Publishing Europe
Berkshire House,168-173
High Holborn, London WC1V 7AA
England

International Thomson Publishing Asia
221 Henderson Road #05-10
Henderson Building.
Singapore 0315

Thomas Nelson Australia
102 Dodds Street
South Melbourne 3205
Victoria, Australia

International Thomson Publishing Japan
Hirakawacho Kyowa Building, 3F
2-2-1 Hirakawacho
Chiyoda-ku, 102 Tokyo
Japan

Nelson Canada
1120 Birchmount Road
Scarborough, Ontario
Canada M1K 5G4

International Thomson Editores
Campos Eliseos 385, Piso 7
Col. Polanco
11560 Mexico D.F. Mexico

This edition of *Regulated Chemicals Directory* ™ was prepared in collaboration with ChemADVISOR®, Inc., Regulatory Compliance Products & Services, Pittsburgh, PA, from the ChemADVISOR LOLI™ database, using software developed by Automated Publishing/ Pre-press Services, Phoenix, AZ.

1 2 3 4 5 6 7 8 9 10 BBR 01 00 99 98 97 96 95

ISSN 1058 1707
ISBN 978-0-442-02124-5   ISBN 978-94-011-4910-5 (eBook)
DOI 10.1007/978-94-011-4910-5

Contents

Acknowledgments

The *Regulated Chemicals Directory*™ is prepared from a database of chemical regulatory information maintained by ChemADVISOR®, Inc. Maintenance of the database is currently under the supervision of Scott A. Amoroso. The electronic database is updated on almost a daily basis to keep pace with regulatory changes and additions. The technical content and structure of the database are under the supervision of Patricia Dsida.

The *Regulated Chemicals Directory*™ is the product of a close collaboration between Van Nostrand Reinhold and Chapman and Hall scientific publishers and ChemADVISOR®, Inc. We wish to thank Dr. Barbara Goldman, Editor at Chapman & Hall, who originally approached Patricia Dsida with the idea of developing a reference book based on the ChemADVISOR® database, and who worked closely with ChemADVISOR® to create the *RCD*™.

Values from the ACGIH publication, "Threshold Limit Values and Biological Exposure Indices for 1994-95," were used for the ChemADVISOR® database with the kind permission of the American Conference of Governmental Industrial Hygienists.

Preface

The *Regulated Chemicals Directory*™ is meant to be a convenient source of information for everyone who needs to keep up-to-date regarding the regulations and recommendations that pertain to chemical substances. The *RCD*™ is designed to be the first reference book to consult when beginning compliance efforts. Every regulatory or advisory list used in the *RCD*™ is keyed to its source, to help readers who need more detailed information on regulations, recommendations, or guidelines readily locate source documents.

Some organizations now center their compliance efforts on computerized information stored in cross-referenced databases. A unique feature of the *RCD*™ is the availability of an electronic version suitable for use on IBM-compatible personal computers, download onto mainframes and CD-ROM players. Both the print and electronic versions are updated with the same timeliness. For more information on the electronic versions of the *Regulated Chemicals Directory*™, contact ChemADVISOR®, Inc. directly (750 William Pitt Way, Pittsburgh, PA 15238, phone 1-800-466-3750).

Many companies working on product development need information on what may be regulated in the future. The *RCD*™ provides selected information on pending regulations and in-progress testing lists, which can provide a starting place for tracking future regulatory considerations.

Information for the *RCD*™ is continually gathered and updated. Suggestions from readers for information that should be added to the *RCD*™ or for other ways to improve the book are welcomed by Van Nostrand Reinhold.

—Patricia L. Dsida, Pres.
ChemADVISOR®, Inc.

Part A.
Chemical Lists and Indexes

Section 1.
Introduction

INTRODUCTION

Regulatory compliance has become a major concern for all industries that handle, manufacture, process, store, transport, or dispose of chemicals. In an effort to protect human health and the environment from present or future damage, numerous federal, state, and local agencies have issued regulations, advisories, and recommendations regarding chemical hazards. Each such publication or document defines its own area of concern and specific requirements. As a result, anyone who uses chemicals is confronted with the enormous job of staying current and in compliance with a complicated, overlapping, and constantly changing set of requirements.

Managing this much information is expensive and time-consuming, especially because regulatory requirements are published in many different places. The *Regulated Chemicals Directory*™ was developed to answer the need for a single, reliable publication that would integrate and condense many thousands of pages of regulations and guidelines. Designed as the *first* reference source to consult when beginning compliance efforts, the *RCD*™ provides rapid answers to key questions about the regulatory status of a particular chemical. The *RCD*™ is based on the database compiled and maintained by ChemADVISOR®, Inc., containing data on more than 10,000 chemicals gleaned from dozens of separate regulatory listings.

Selection of Chemical Lists

ChemADVISOR® selects chemical lists from U.S. government regulatory and advisory groups that review chemical materials, as well as from national associations that review health and/or environmental effects as part of their charters. Information from these lists is provided under the *RCD*™ headings "Health and Safety Lists" and "Environmental Lists." List sources include the major regulatory and advisory agencies in the United States, such as the Environmental Protection Agency (EPA), the Occupational Safety and Health Administration (OSHA), the National Institute of Occupational Safety and Health (NIOSH), and such national associations as the American Conference of Governmental Industrial Hygienists (ACGIH) and the American Industrial Hygiene Association (AIHA). Separate sub-lists are maintained for each type of data published in a source document, e.g., carcinogenicity, skin absorption, exposure limits. This facilitates comparison of a particular type of data across several regulatory or advisory groups. In electronic form, this feature allows a user to identify all materials having a certain property, i.e., all chemicals considered carcinogenic by one or more lists.

A major regulatory concern in the United States is the preparation of Material Safety Data Sheet (MSDS) documents, in compliance with the OSHA Hazard Communication Standard. Compliance requires that the following information be provided: (i) exposure limits and (ii) identification of carcinogenic materials, as defined and listed by OSHA, the National Toxicology Program (NTP), and the International Agency for Research on Cancer (IARC). The "Health and Safety Lists" were selected to provide accurate information for preparation and maintenance of MSDS documents.

In addition to the information required by the OSHA Standard, the *RCD*™ provides information on exposure limits identified by NIOSH, on chemicals currently being tested by NTP but for which limits have not yet been finalized, on workplace exposure limits set by the AIHA, and on substances designated "Hazardous Materials" by the U.S. Department of Transportation.

The MSDS has become an integral part of occupational materials information, and, as such, has taken on an additional role in the transmission of environmental information. Many suppliers now use the MSDS to provide information required by the Superfund Amendments and Reauthorization Act (SARA). The "Environmental Lists" include this required information, plus information on chemicals currently designated as hazardous wastes under the Resource Conservation and Recovery Act (RCRA), information also provided on the MSDS by many suppliers.

As trade becomes increasingly international, so do concerns about health, safety, and the environment. Gathering accurate and up-to-date information from countries outside the U.S. and Canada requires careful and lengthy review. The international lists included in the current edition of the *RCD*™ are those ChemADVISOR® judged to be the most reliable, accurate, and well-maintained. The "International Lists" component of the *RCD*™ database will grow rapidly as new lists become available for review.

Information provided by state governments under various Right-to-Know or similar regulatory statutes was included selectively, since there are a large number of such laws, many of which are redundant. State lists that merely reference another source of information (such as the OSHA Hazard Communication Standard or other Federal statutes) were not included. All unique lists of chemicals issued by state governments were considered for inclusion in the database, although not all were eventually selected. The "State Lists" section will expand in future editions to include additional lists requested by *RCD*™ users.

The section on "Proposed Regulations" does not hold the weight of law as yet. Selected proposed changes were included for the benefit of individuals who want to begin preparing for regulatory changes well before enactment.

Chemical Nomenclature and CAS Numbers

Since a particular chemical may be known by several names, Chemical Abstracts Service (CAS) of the American Chemical Society has assigned unique registry numbers to chemical compounds. The *RCD*™ database uses CAS Registry Numbers as the primary means of storing and sorting information. Before list information is entered into the database, ChemADVISOR® carefully reviews the list to assure its accuracy. Their working guidelines for review are to meet the goal of the regulatory agency, while being as consistent with reasonable chemical principles as possible.

Often errors are detected in the original publication from which a list has been derived, most commonly publication of an incorrect CAS number. Since the software used for the database incorporates the algorithm used by the ACS to reduce input errors ("check sum digit" analysis), CAS numbers that are incorrect or out of sequence usually are detected immediately. ChemADVISOR® then searches online databases to locate the correct CAS Registry Number corresponding to the chemical name.

If a regulatory document does not specify a CAS number for a compound, ChemADVISOR® assigns one where appropriate. In cases where listed substances have no specific CAS numbers (for example, "Cotton Dust [Raw]"), ChemADVISOR® assigns a chemical identification number that is structured like a CAS number but begins with the prefix "RR" (i.e. RR-00001-1).

In a few instances, Chemical Abstracts Services has assigned more than one CAS Registry Number to a unique chemical. In such cases, ChemADVISOR® includes *both* CAS numbers.

The CAS number is intended to refer to a unique chemical substance, but regulators may provide questionable chemical names for particular CAS numbers. The most common inconsistencies occur with metals. For example, the *RCD*™ Reference Name for CAS number 7440-48-4 is "Cobalt." However, different lists regulate this CAS number variously as "Cobalt and Compounds," "Cobalt Metal, Dust, and Fume," or "Cobalt, Elemental." Whenever a regulation assigns a specific name to a specific CAS number, the *RCD*™ includes that name, even if it is chemically inaccurate. However, if a regulation assigns *no* CAS number to a substance—for example "Cobalt Compounds"—the *RCD*™ assigns a chemical identification number, not a CAS number.

"See Also" Cross-reference to Generic and Element Categories

ChemADVISOR® has cross-referenced many specific CAS numbers to more general categories in the database, including such generic groups as hazardous waste streams, and isomers or other crystalline forms. Substances containing more than one major element also contain a cross-reference, for example "Lead" and "Arsenic" for the entry "Lead Arsenate." Such connections, while useful, are not complete, and the reader may find other relevant substances that should be looked up for a given CAS number.

A good example of the usefulness of the "See Also" feature of the *RCD*™ is the special case of the individual chemical components of a particular "hazardous waste" defined by the Resource Conservation and Recovery Act (RCRA). If the EPA has defined a hazardous waste as containing one or more chemicals, then the Regulatory Summaries entry for each chemical component includes a cross-reference to the appropriate hazardous waste number. The APPENDIX supplies the RCRA definitions for all F- and K-Series Hazardous Wastes.

Organization of the *RCD*™

All of the lists used to compile the *RCD*™ are summarized in the LIST SUMMARY TABLE that opens Part A, Section 2, "Chemical Lists and List Descriptions." The three-letter Synonym Codes, used throughout the book, refer the reader to sources that regulate a given compound under a particular chemical synonym. (A reference table identifying the Synonym Codes immediately precedes each index in Section 3.)

The LIST DESCRIPTIONS section provides essential information on the regulatory sources used to compile data for the *RCD*™. Each list description includes the source publication, the address of the issuing body, the background and rationale underlying the list, and an explanation of the categories and values specified. Since the *RCD*™ is designed as a convenient initial reference for beginning compliance efforts, the original source documents should be consulted for more details on the particular regulations, recommendations or guidelines.

Indexes

Part A, Section 3, "Cross-Reference Indexes of Chemical Names and Synonyms," contains two indexes designed to simplify the task of determining (i) whether a particular substance appears on one of the regulatory or advisory lists used in the *RCD*™ and (ii) the specific name under which it is included in the REGULATORY SUMMARIES section (Part B).

Since a substance with a unique CAS number may be regulated under several chemical names by different agencies, the *RCD*™ indexes substances both by CAS number and by chemical synonyms. The regulatory synonym most frequently used for a chemical substance—its Reference Name—is the name under which the substance's Regulatory Summary is listed in Part B. Reference Names are printed in bold type throughout the *RCD*™ to help the reader rapidly identify them.

The NUMERICAL INDEX BY CAS NUMBER lists all the chemical synonyms commonly associated with a particular CAS registry number. The substance's Reference Name is printed in bold type next to the CAS number. The three-letter Synonym Codes in parentheses identify which lists use a specific synonym (see "Chemical Nomenclature and CAS Numbers," above).

Some regulated substances have not been assigned CAS Registry Numbers by Chemical Abstracts Services. In the *RCD*™, these substances have been assigned chemical information numbers that begin with the prefix "RR" and are structured like CAS numbers, e.g., RR-00001-1 for "Cotton Dust (Raw)." Substances assigned such RR numbers are listed in numerical order at the end of the CAS number index.

The ALPHABETICAL INDEX BY CHEMICAL NAME includes both the exact chemical names used by the list sources and commonly used synonyms, with the identifying Synonym Codes in parentheses. Reference Names (printed in bold type in the alphabetical index) are given in the right-hand column of this index, to simplify the search for a particular Regulatory Summary.

Regulatory Summaries

The heart of the *RCD*™ is Part B of the book, the REGULATORY SUMMARIES. Each entry in the REGULATORY SUMMARIES section is listed alphabetically under the Reference Name, the most frequently used regulatory synonym. The CAS number is included in each Regulatory Summary to confirm the identity of the substance. Reference Names may be found by looking up a particular substance either by name in the ALPHABETICAL INDEX BY CHEMICAL NAME or by CAS number in the NUMERICAL INDEX BY CAS NUMBER.

Each entry in the REGULATORY SUMMARIES integrates data from health and safety, environmental, state, and international lists, and includes proposed changes to regulatory or advisory status. All the regulations and reports on which the substance appears, along with the applicable data values, are included.

Each data field contains information originally published as footnotes, table headers, or other constructs in the source document. Such features often are needed for a proper understanding of the regulatory intent and information in the entry.

Database Maintenance and *RCD*™ Updates

Once a list has been added to the database, ChemADVISOR® institutes a regular program of monitoring additions or corrections to list information, including daily review of the Federal Register for U.S. regulatory information. Selected state registers are regularly reviewed, and ChemADVISOR® subscribes to many publications and online data services. Any changes are thoroughly documented, and the indexes and list descriptions are modified whenever a change is made. The "Proposed Changes"

component requires particularly careful scrutiny and updating, since regulatory policy can change dramatically and rapidly. Each year, the entire *RCD*™ is completely reissued, incorporating all updates and corrections made to the database during the previous 12 months.

Section 2.
Chemical Lists and List Descriptions

All the lists used to compile the *RCD*™ are summarized in the LIST SUMMARY TABLE that opens this section. The LIST DESCRIPTIONS that follow provide essential information on the regulatory sources used to compile data in the *RCD*™.

Each list description includes the source publication, the address of the issuing body, the background and rationale underlying the list, and an explanation of the categories and values specified. The three letter "synonym codes" given in parentheses are abbreviations used throughout the book to identify the sources that regulate a given substance under a particular chemical synonym. (A key to the synonym codes is provided at the beginning of each index in Section 3).

Since the *RCD*™ is designed as a convenient initial reference for beginning compliance efforts, the original source documents should be consulted for more details on the particular regulations, recommendations or guidelines.

LIST SUMMARY TABLE

Health And Safety Lists

Environmental Lists

International Lists

REGULATORY SOURCE	LIST NAME	PAGE

State Lists

Proposed Regulations

LIST DESCRIPTIONS

HEALTH AND SAFETY LISTS

ACGIH 1995 - Biological Exposure Indices   (BEI)

SOURCE

American Conference of Governmental Industrial
    Hygienists (ACGIH)
"Threshold Limit Values and Biological Exposure
    Indices for 1994-95"

ACGIH
Kemper Woods Center
1330 Kemper Woods Center
Cincinnati, OH 45240
Requests for interpretations and other information:
Telephone (513) 772-2020
FAX (513) 772-3355

LIST DESCRIPTION

The American Conference of Governmental Industrial
Hygienists (ACGIH) is an organization devoted to the
administrative and technical aspects of occupational
and environmental health. The Conference is a
professional society, not a government agency. ACGIH
publishes the "TLV/BEI Booklet" each year.

The ACGIH has prefaced their publication with a
policy statement on the uses of TLVs and BEIs. That
policy statement is as follows:

The Threshold Limit Values (TLVs) and Biological Exposure Indices (BEIs) are developed as guidelines to assist in the control of health hazards. These
recommendations or guidelines are intended for use
in the practice of industrial hygiene, to be interpreted
and applied only by a person trained in this discipline. They are not developed for use as legal standards, and the ACGIH does not advocate their use
as such. However, it is recognized that in certain circumstances individuals or organizations may wish to
make use of these recommendations or guidelines as
a supplement to their occupational safety and health
program. The ACGIH will not oppose their use in
this manner, if the use of TLVs and BEIs in these
instances will contribute to the overall improvement
in worker protection. However, the user must recognize the constraints and limitations subject to their
proper use and bear the responsibility for such use.

This is a list of substances for which ACGIH has
established "Biological Exposure Indices" (BEIs).

Biological monitoring consists of an assessment of
overall exposure to chemicals which are present in
the workplace through measurement of the appropriate
determinant(s) in biological specimens collected from
the worker at the specified time. BEIs serve
as reference values intended as guidelines for the
evaluation of potential health hazards.

The data field indicates the determinant chemical
or material, the reference level for that determinant,
the sampling time for the material and "notations."
Notations and rationale for sampling times are defined
by ACGIH in the introduction to the BEIs. Definitions
of the Notations used in the data field are:

"Ns" Notation. This notation indicates that the
determinant is NONSPECIFIC since it is observed
after exposure to some other chemicals. These
nonspecific tests are preferred because they are easy
to use and usually offer a better correlation with
exposure than specific tests. In such instances,
a BEI for a specific, less quantitative biological
determinant is recommended as a confirmatory
test. The documentation should be consulted for
information on factors affecting interpretation of
these BEIs.

"Cf" Notation. This notation indicates that the
biological determinant is an indicator of exposure
to the chemical, but the quantitative interpretation
of the measurements is ambiguous. Their BEIs
should be applied cautiously. These biological
determinants should be used as confirmatory tests,
mainly for confirmation of exposures indicated by
measurements of a nonspecific determinant or as a
screening test if a quantitative test is not practical.

"Sc" Notation. This notation indicates that
an identifiable population group might have an
increased susceptibility to the effect of the chemical,
thus leaving it unprotected by the recommended BEI.
The specific BEI documentation should be consulted
for information.

"B" Notation. This notation indicates that
the determinant is usually present in a significant
amount in biological specimens collected from
subjects who have not been occupationally exposed.
Such background levels are included in the BEI
value. For information on background levels, consult
the specific documentation.

ACGIH 1995 - Short Term Exposure Limits   (TLV)

SOURCE

American Conference of Governmental Industrial
    Hygienists (ACGIH)
"1994-1995 Threshold Limit Values for Chemical
    Substances and Physical Agents and Biological
    Exposure Indices"

ACGIH
1330 Kemper Meadow Drive
Cincinnati, OH 45240
Requests for interpretations and other information:
Telephone (513) 742-2020
FAX (513) 742-3355

LIST DESCRIPTION

The American Conference of Governmental Industrial
Hygienists (ACGIH) is an organization devoted to the
administrative and technical aspects of occupational
and environmental health. The Conference is a
professional society, not a government agency. ACGIH
publishes the "TLV/BEI Booklet" each year. The TLVs,
as issued by the American Conference of Governmental
Industrial Hygienists, are recommendation and should
be used as guidelines for good practices. In spite of
the fact that serious injury is not believed likely as a
result of exposure to the threshold limit concentrations,

the best practice is to maintain concentrations of all atmospheric contaminants as low as is practical.

The ACGIH has prefaced their publication with a policy statement on the uses of TLVs and BEIs. That policy statement is as follows:

The Threshold Limit Values (TLVs) and Biological Exposure Indices (BEIs) are developed as guidelines to assist in the control of health hazards. These recommendations or guidelines are intended for use in the practice of industrial hygiene, to be interpreted and applied only by a person trained in this discipline. They are not developed for use as legal standards, and the ACGIH does not advocate their use as such. However, it is recognized that in certain circumstances individuals or organizations may wish to make use of these recommendations or guidelines as a supplement to their occupational safety and health program. The ACGIH will not oppose their use in this manner, if the use of TLVs and BEIs in these instances will contribute to the overall improvement in worker protection. However, the user must recognize the constraints and limitations subject to their proper use and bear the responsibility for such use.

There are three categories of Threshold Limit Values: TLV-TWA (Time Weighted Averages), TLV-STEL (Short Term Exposure Limits), and TLV-C (Ceiling Limit). This is a list of substances for which ACGIH has established a Threshold Limit Value - Short-Term Exposure Limit (TLV-STEL).

The TLV-STEL is defined as the concentration to which workers can be exposed continuously for a short period of time without suffering from 1) irritation, 2) chronic or irreversible tissue damage, or 3) narcosis of sufficient degree to increase the likelihood of accidental injury, impair self-rescue or materially reduce work efficiency. Provided that the daily TLV-TWA is not exceeded, it is not a separate independent exposure limit; rather, it supplements the time-weighted average limit where there are recognized acute effects from a substance whose toxic effects are primarily of a chronic nature. STELS are recommended only where toxic effects have been reported from high short-term exposures in either humans or animals.

A STEL is defined as a 15-minute TWA exposure which should not be exceeded at any time during a workday even if the 8-hour TWA is within the TLV-TWA. Exposures above the TLV-TWA up to the STEL should not be longer than 15 minutes and should not occur more than 4 times per day. There should be at least 60 minutes between successive exposures in this range. An average period other than 15 minutes may be recommended when this is warranted by the observed biological effects.

This list contains all substances for which ACGIH has recommended a TLV-STEL.

The data field contains the TLV-STEL value in ppm and/or mg/m3 followed by the letters "STEL" to denote a Short Term Exposure Limit plus any comments provided by ACGIH.

ACGIH 1995 - Ceiling Limits (TLV)

SOURCE

American Conference of Governmental Industrial Hygienists (ACGIH)
"1994-1995 Threshold Limit Values for Chemical Substances and Physical Agents and Biological Exposure Indices"

ACGIH
Kemper Woods Center
1330 Kemper Meadow Drive
Cincinnati, OH 45240
Requests for interpretations and other information:
Telephone (513) 742-2020
FAX (513) 742-3355

LIST DESCRIPTION

The American Conference of Governmental Industrial Hygienists (ACGIH) is an organization devoted to the administrative and technical aspects of occupational and environmental health. The Conference is a professional society, not a government agency. ACGIH publishes the "TLV/BEI Booklet" each year. The TLVs, as issued by the American Conference of Governmental Industrial Hygienists, are recommendation and should be used as guidelines for good practives. In spite of the fact that serious injury is not believed likely as a result of exposure to the threshold limit concentrations, the best practice is to maintain concentrations of all atmospheric contaminants as low as is practical.

The ACGIH has prefaced their publication with a policy statement on the uses of TLVs and BEIs. That policy statement is as follows:

The Threshold Limit Values (TLVs) and Biological Exposure Indices (BEIs) are developed as guidelines to assist in the control of health hazards. These recommendations or guidelines are intended for use in the practice of industrial hygiene, to be interpreted and applied only by a person trained in this discipline. They are not developed for use as legal standards, and the ACGIH does not advocate their use as such. However, it is recognized that in certain circumstances individuals or organizations may wish to make use of these recommendations or guidelines as a supplement to their occupational safety and health program. The ACGIH will not oppose their use in this manner, if the use of TLVs and BEIs in these instances will contribute to the overall improvement in worker protection. However, the user must recognize the constraints and limitations subject to their proper use and bear the responsibility for such use.

There are three categories of Threshold Limit Values: TLV-TWA (Time Weighted Averages), TLV-STEL (Short Term Exposure Limits), and TLV-C (Ceiling Limits).

This is a list of substances for which ACGIH has established a Threshold Limit Value-Ceiling (TLV-C). The TLV-C is defined as the concentration that should not be exceeded during any part of the working exposure. In conventional industrial hygiene practice, if instantaneous monitoring is not feasible, the TLV-C can be assessed by sampling over a 15-minute period except for those substances that may cause immediate irritation when exposure is short.

The data field contains the letter "C" plus the TLV-C value in ppm and/or mg/m3 plus any footnotes or comments provided by ACGIH.

ACGIH 1995 - Skin Designations (TLV)
SOURCE

American Conference of Governmental Industrial Hygienists (ACGIH)
"Threshold Limit Values and Biological Exposure Indices for 1994-95"
ACGIH
Kemper Woods Center
1330 Kemper Meadow Drive
Cincinnati, OH 45240
Requests for interpretations and other information:
Telephone (513) 742-2020
FAX (513) 742-3355

LIST DESCRIPTION

The American Conference of Governmental Industrial Hygienists (ACGIH) is an organization devoted to the administrative and technical aspects of occupational and environmental health. The Conference is a professional society, not a government agency. ACGIH publishes the "TLV/BEI Booklet" each year. The TLVs, as issued by the American Conference of Governmental Industrial Hygienists, are recommendation and should be used as guidelines for good practives.

The ACGIH has prefaced their publication with a policy statement on the uses of TLVs and BEIs. That policy statement is as follows:

The Threshold Limit Values (TLVs) and Biological Exposure Indices (BEIs) are developed as guidelines to assist in the control of health hazards. These recommendations or guidelines are intended for use in the practice of industrial hygiene, to be interpreted and applied only by a person trained in this discipline. They are not developed for use as legal standards, and the ACGIH does not advocate their use as such. However, it is recognized that in certain circumstances individuals or organizations may wish to make use of these recommendations or guidelines as a supplement to their occupational safety and health program. The ACGIH will not oppose their use in this manner, if the use of TLVs and BEIs in these instances will contribute to the overall improvement in worker protection. However, the user must recognize the constraints and limitations subject to their proper use and bear the responsibility for such use.

Some substances for which a TLV has been established contain a "skin" notation. Listed substances followed by the designation "skin" refer to the potential contribution to the overall exposure by the cutaneous route including mucous membranes and eye, either by airborne, or, more particularly, by direct contact with the substance. Vehicles can alter skin absorption. This attention-calling designation is intended to suggest

appropriate measures for the prevention of cutaneous absorption so that the TLV is not invalidated.

This list identifies all materials in the TLV/BEI Booklet which include the "skin" designation.

The data field contains the word "skin - potential for cutaneous absorption."

ACGIH 1995 - Carcinogens (TLV)
SOURCE

American Conference of Governmental Industrial Hygienists (ACGIH)
"1994-1995 Threshold Limit Values for Chemical Substances and Physical Agents and Biological Exposure Indices"
ACGIH
Kemper Woods Center
1330 Kemper Meadow Drive
Cincinnati, OH 45240
Requests for interpretations and other information:
Telephone (513) 742-2020
FAX (513) 742-3355

LIST DESCRIPTION

The American Conference of Governmental Industrial Hygienists (ACGIH) is an organization devoted to the administrative and technical aspects of occupational and environmental health. The Conference is a professional society, not a government agency. ACGIH publishes the "TLV/BEI Booklet" each year. ACGIH classifies certain substances found in the occupational environment as either confirmed or suspected human carcinogens.

The ACGIH has prefaced their publication with a policy statement on the uses of TLVs and BEIs. That policy statement is as follows:

The Threshold Limit Values (TLVs) and Biological Exposure Indices (BEIs) are developed as guidelines to assist in the control of health hazards. These recommendations or guidelines are intended for use in the practice of industrial hygiene, to be interpreted and applied only by a person trained in this discipline. They are not developed for use as legal standards, and the ACGIH does not advocate their use as such. However, it is recognized that in certain circumstances individuals or organizations may wish to make use of these recommendations or guidelines as a supplement to their occupational safety and health program. The ACGIH will not oppose their use in this manner, if the use of TLVs and BEIs in these instances will contribute to the overall improvement in worker protection. However, the user must recognize the constraints and limitations subject to their proper use and bear the responsibility for such use.

In order to recognize the qualitative difference in research results, five categories of carcinogens are designated by ACGIH:
NOTE: The adoption of a 5 category system was transferred to theadopted status in the 1992-93 publication.

A1 - Confirmed Human Carcinogen: The agent is carcinogenic to humans based on the weight of evidence from epidemiologic studies of, or convincing clinical evidence in, exposed humans.

A2 - Suspected Human Carcinogen: The agent is carcinogenic in experimental animals at dose levels, by route(s) of administration, at sites(s), of histologic type(s), or by mechanism(s) that are considered relevant to worker exposure. Available epidemiologic studies are conflicting or insufficient to confirm an increased risk of cancer in exposed humans.

A3 - Animal Carcinogen: The agent is carcinogenic in experimental animals at a relatively high dose, by route(s) of administration, at site(s), of histologic types(s), or by mechanism(s) that are not considered relevant to worker exposure. Available epidemiologic studies do not confirm an increased risk of cancer in exposed humans. Available evidence suggests that the agent is not likely to cause cancer in humans except under uncommon or unlikely routes or levels of exposure.

A4 - Not Classifiable as a Carcinogen: There are inadequate data on which to classify the agent in terms of its carcinogenicity in humans and/or animals.

A5 - Not Suspected as a Human Carcinogen: The agent is not suspected to be a human carcinogen on the basis of properly conducted epidemiologic studies in humans. These studies have sufficiently long follow-up, reliable exposure histories, sufficiently high dose, and adequate statistical power to conclude that exposure to the agent does not convey a significant risk of cancer to humans. Evidence suggesting a lack of carcinogenicity in experimental animals will be considered if it is supported by other relevant data.

This list contains all the entries for which ACGIH has identified one of these categories.

The Data field contains "A1-Confirmed Human Carcinogen", "A2-Suspected Human Carcinogen", "A3-Animal Carcinogen", "A4-Not Classifiable as a Carcinogen" or "Not Suspected as a Human Carcinogen".

ACGIH 1995 - Time Weighted Averages   (TLV)

SOURCE

American Conference of Governmental Industrial Hygienists (ACGIH)
"1994-1995 Threshold Limit Values for Chemical Substances and Physical Agents and Biological Exposure Indices"

ACGIH
Kemper Woods Center
1330 Kemper Meadow Drive
Cincinnati, OH 45240
Requests for interpretations and other information:
Telephone (513) 742-2020
FAX (513) 742-3355

LIST DESCRIPTION

The American Conference of Governmental Industrial Hygienists (ACGIH) is an organization devoted to the administrative and technical aspects of occupational and environmental health. The Conference is a professional society, not a government agency. ACGIH publishes the "TLV/BEI Booklet" each year. The TLVs, as issued by the American Conference of Governmental Industrial Hygienists, are recommendations and should be used as guidelines for good practices. In spite of the fact that serious injury is not believed likely as a result of exposure to the threshold limit concentrations, the best practice is to maintain concentrations of all atmospheric contaminants as low as is practical.

The ACGIH has prefaced their publication with a policy statement on the uses of TLVs and BEIs. That policy statement is as follows:

The Threshold Limit Values (TLVs) and Biological Exposure Indices (BEIs) are developed as guidelines to assist in the control of health hazards. These recommendations or guidelines are intended for use in the practice of industrial hygiene, to be interpreted and applied only by a person trained in this discipline. They are not developed for use as legal standards, and the ACGIH does not advocate their use as such. However, it is recognized that in certain circumstances individuals or organizations may wish to make use of these recommendations or guidelines as a supplement to their occupational safety and health program. The ACGIH will not oppose their use in this manner, if the use of TLVs and BEIs in these instances will contribute to the overall improvement in worker protection. However, the user must recognize the constraints and limitations subject to their proper use and bear the responsibility for such use.

There are three categories of Threshold Limit Values: TLV-TWA (Time Weighted Averages), TLV-STEL (Short Term Exposure Limits), and TLV-C (Ceiling Limit).

This is a list of substances for which ACGIH has established a Threshold Limit Value - Time Weighted Average (TLV-TWA). The TLV-TWA is defined as the time-weighted average concentration for a normal 8-hour workday and a 40-hour workweek, to which nearly all workers may be repeatedly exposed, day after day, without adverse effect.

The data field indicates the TLV-TWA in ppm and/or mg/m3 plus the letters "TWA" to denote a Time Weighted Average. Any comments or clarifications used by ACGIH are also included.

U.S.DOT - Appendix A Table 1 - Hazardous (HM1) Substances

SOURCE

U.S. Department of Transportation
Research and Special Programs Administration
49 CFR Subchapter C - Hazardous Materials Regulations
Part 172 Hazardous Materials Tables
Paragraph 172.101 Hazardous Materials Table Appendix A Table 1 - Hazardous Substances Other Than Radionuclides
Amended through June 20, 1994 (59 FR 31822)

LIST DESCRIPTION

The Hazardous Materials Transportation Act (HMTA), which became law in 1975, gives the Department of Transportation (DOT) authority to regulate the movement within the United States of substances that may pose a threat to health, safety, property or the environment when transported by air, highway, rail, or water.

Under HMTA, DOT has delegated authority for regulation of hazardous materials to the Research and Special Programs Administration (RSPA). This organization, with its Office of Hazardous Materials Transportation, drafts and issues hazardous materials regulations, exemptions, registration certificates, and packaging and container certification for all transport modes.

The Hazardous Materials Table (49 CFR 172.101) has several thousand entries applicable to various materials including explosives, flammables, oxidizing materials, organic peroxides, corrosives, gases, poisons, radioactive substances and etiologic agents. The Table is composed of the main table and two Appendices. Appendix A contains materials designated as hazardous substances and wastes and Appendix B contains a list of Marine Pollutants.

Appendix A lists materials which are listed or designated as hazardous substances under section 101(14) of the Comprehensive Environmental Response, Compensation, and Liability Act (CERCLA). DOT, by law, must add these to Table 101. Appendix A is divided into 2 tables which are entitled Table 1 - Hazardous Substances Other Than Radionuclides and Table 2 - Radionuclides. A material listed in this appendix is regulated as a hazardous material and a hazardous substance under this subchapter if it meets the definition of a hazardous substance in paragraph 171.8 of this subchapter.

This list contains all materials which are listed in Table 1 of Appendix A. ChemADVISOR has provided the Chemical Abstracts Service Numbers for these chemicals as used by the EPA (see CERCLA/SARA - Hazardous Substances and Their Reportable Quantities). Each chemical is listed by only one name. The name chosen is the name as used by the EPA.

The data field includes the Reportable Quantity in units of pounds and (kilograms).

Only chemical substances and chemical groups have been included: entries such as "articles," "explosives,""ammunition," etc. are not included here. Chemical substances may be listed one or more times in the Hazardous Materials Table depending on physical property qualifications used in the shipping name description. The chemical substance or group is listed here only once. YOU MUST CHECK THE HAZARDOUS MATERIALS TABLE TO FIND THE PROPER SHIPPING DESCRIPTION OR DESCRIPTIONS FOR EACH ENTRY.

The data field includes all the UN/NA identification numbers for entries based on that substance and the words "regulated by DOT." For mixtures of substances each separate substance includes the UN/NA number for the mixture.

U.S.DOT - Appendix A Table 2 - Radionuclides (HM1)

SOURCE

U.S. Department of Transportation
Research and Special Programs Administration
49 CFR Subchapter C - Hazardous Materials Regulations
Part 172 Hazardous Materials Tables
Paragraph 172.101 Hazardous Materials Table Appendix A Table 1 - Hazardous Substances Other Than Radionuclides
As published in 49 CFR revised as of October 1, 1993 (58 FR 51524)

LIST DESCRIPTION

The Hazardous Materials Transportation Act (HMTA), which became law in 1975, gives the Department of Transportation (DOT) authority to regulate the movement within the United States of substances that may pose a threat to health, safety, property or the environment when transported by air, highway, rail, or water.

Under HMTA, DOT has delegated authority for regulation of hazardous materials to the Research and Special Programs Administration (RSPA). This organization, with its Office of Hazardous Materials Transportation, drafts and issues hazardous materials regulations, exemptions, registration certificates, and packaging and container certification for all transport modes.

The Hazardous Materials Table (49 CFR 172.101) has several thousand entries applicable to various materials including explosives, flammables, oxidizing materials, organic peroxides, corrosives, gases, poisons, radioactive substances and etiologic agents. The Table is composed of the main table and two Appendices. Appendix A contains materials designated as hazardous substances and wastes and Appendix B contains a list of Marine Pollutants.

Appendix A lists materials which are listed or designated as hazardous substances under section 101(14) of the Comprehensive Environmental Response, Compensation, and Liability Act (CERCLA). DOT, by law, must add these to Table 101. Appendix A is divided into 2 tables which are entitled Table 1 - Hazardous Substances Other Than Radionuclides and Table 2 - Radionuclides. A material listed in this appendix is regulated as a hazardous material and a hazardous substance under this subchapter if it meets the definition of a hazardous substance in paragraph 171.8 of this subchapter.

This list contains all materials which are listed in Table 2 of Appendix A. ChemADVISOR has provided the Chemical Abstracts Service Numbers for these chemicals.

The data field includes the Reportable Quantity in units of Curies(Ci) and terabecquerels (TBq).

Only chemical substances and chemical groups have been included: entries such as "articles," "explosives,""ammunition," etc. are not included here. Chemical substances may be listed one or more times in the Hazardous Materials Table depending on physical property qualifications used in the shipping name description. The chemical substance or group is listed here only once. YOU MUST CHECK THE HAZARDOUS MATERIALS TABLE TO FIND THE PROPER SHIPPING DESCRIPTION OR DESCRIPTIONS FOR EACH ENTRY.

The data field includes all the UN/NA identification numbers for entries based on that substance and the words "regulated by DOT." For mixtures of substances each separate substance includes the UN/NA number for the mixture.

U.S.DOT - Organic Peroxides Table   (DOT)

SOURCE

U.S. Department of Transportation
Research and Special Programs Administration
49 CFR Subchapter C - Hazardous Materials Regulations
Part 172 Hazardous Materials Tables
Paragraph 173.225 Packaging requirements
Organic Peroxides Table - Columns 1 and 2 only.
ChemADVISOR has added Chemical Abstracts Services Registry Numbers where found.
NOTE: This list is based on the revisions made under Docket HM181

LIST DESCRIPTION

The Hazardous Materials Transportation Act (HMTA), which became law in 1975, gives the Department of Transportation (DOT) authority to regulate the movement within the United States of substances that may pose a threat to health, safety, property, or the environment when transported by air, highway, rail, or water.

Under HMTA, DOT has delegated authority for regulation of hazardous materials to the Research and Special Programs Administration (RSPA). This organization, with its Office of Hazardous Materials Transportation, drafts and issues hazardous materials regulations, exemptions, registration certificates, and packaging and container certification for all transport modes.

The Hazardous Materials Table (Table 101) designates the materials therein as hazardous materials for the purpose of transportation of those materials. When Table 101 specifies that an organic peroxide be packaged under Section 173.225, each package must conform to the general requirements of Subpart B of Part 173 and to the applicable requirements of Part 178 of Subchapter C. Non-bulk packagings must meet Packing Group II performance levels. Packing Group I and Packing Group III non-bulk packagings are not authorized. Organic peroxides which require temperature control are subject to the provision of 173.21(f).

The Organic Peroxides Table specifies, by technical name, those organic peroxides that are authorized for transportation and not subject to the approval provisions of 173.128. An organic peroxide identified by technical name in the Organic Peroxides Table is authorized for tranportation only if it conforms to all applicable provisions of the table.

The chemical names used in this list are taken from Column 1 ofthe Organic Peroxides Table. The Chemical Abstracts Services Registry Numbers were searched using the name given in column 1.

Column 2 lists the identification (ID) number which is used to identify the proper shipping name in Table 101. The ID number or numbers appear in the data field.

U.S.DOT - Appendix B - Marine Pollutants   (HM1)

SOURCE

U.S. Department of Transportation
Research and Special Programs Administration
49 CFR Subchapter C - Hazardous Materials Regulations
Part 172 Hazardous Materials Tables
Appendix B to Paragraph 172.101 Hazardous Materials Table
Final Rule Federal Register November 5, 1992 (57 FR 52930) effective date January 1, 1993.

For further information contact the Office of Hazardous Materials Standards, RSPA, 400 Seventh Street SW, Washington, DC 20590-0001.

LIST DESCRIPTION

The Hazardous Materials Transportation Act (HMTA), which became law in 1975, gives the Department of Transportation (DOT) authority to regulate the movement within the United States of substances that may pose a threat to health, safety, property, or the environment when transported by air, highway, rail, or water.

Under HMTA, DOT has delegated authority for regulation of hazardous materials to the Research and Special Programs Administration (RSPA). This organization, with its Office of Hazardous Materials Transportation, drafts and issues hazardous materials regulations, exemptions, registration certificates, and packaging and container certification for all transport modes.

The Hazardous Materials Table (49 CFR 172.101) has several thousand entries applicable to various materials including explosives, flammables, oxidizing materials, organic peroxides, corrosives, gases, poisons, radioactive substances and etiologic agents. The Table is composed of the main table and an Appendix. This amendment renames the current Appendix as Appendix A and adds a new Appendix B titled "List of Marine Pollutants".

The addition of Appendix B is part of amendments made to list and regulate, in all modes of transportation, those materials identified as marine pollutants by the International Maritime Organization (IMO). These changes implement the provisions of Annex III of the 1973 International Convention for the Prevention of Pollution from Ships, as modified by the Protocol of 1978 (MARPOL 73/78). Annex III sets forth general regulations for the transport of harmful packaged substances. Many of these substances, such as pesticides and herbicides, are known to kill or retard the growth of marine life and to bioaccumulate in marine organisms, causing potential danger to the food chain, including health risks to humans as well as to birds and other wildlife that eat fish or shellfish.

The regulations of Annex III apply to "harmful substances"identified as marine pollutants under the 1990 consolidated edition of the International Maritime Dangerous Goods (IMDG) Code. Marine pollutants are identified in the individual schedules and the General Index of the IMDG Code by the letters "P" or "PP". The letters "PP" identify those materials that are regulated as severe marine pollutants when in concentrations of 1% or more. The letter "P" identifies those materials that are marine pollutants when in concentrations of 10% or more.

In addition to adding regulations for marine pollutants transported by vessel, as required by Annex III, RSPA also has added regulations for transportation of marine pollutants by air, rail and highway.

This list contains all the entries in Appendix B. Chemical Abstracts Service Registry Numbers are not used by DOT and have been added by ChemADVISOR, Inc.

The data field indicates whether the substance is a marine pollutant (i.e. regulated at 10% or greater) or a severe marine pollutant (i.e. regulated at 1% or greater). Any additional qualifications to the entry are also included in the data field.

U.S.DOT - Substances from 49 CFR 172.101   (HM1)

SOURCE

U.S. Department of Transportation
Research and Special Programs Administration
49 CFR Subchapter C - Hazardous Materials Regulations
Part 172 Hazardous Materials Tables
Paragraph 172.101 Hazardous Materials Table
As published in 49 CFR with amendments through June 2, 1994 (59 FR 28487)

LIST DESCRIPTION

The Hazardous Materials Transportation Act (HMTA), which became law in 1975, gives the Department

of Transportation (DOT) authority to regulate the movement within the United States of substances that may pose a threat to health, safety, property or the environment when transported by air, highway, rail, or water.

Under HMTA, DOT has delegated authority for regulation of hazardous materials to the Research and Special Programs Administration (RSPA). This organization, with its Office of Hazardous Materials Transportation, drafts and issues hazardous materials regulations, exemptions, registration certificates, and packaging and container certification for all transport modes.

The Hazardous Materials Table (49 CFR 1 2.101) has several thousand entries applicable to various materials including explosives, flammables, oxidizing materials, organic peroxides, corrosives, gases, poisons, radioactive substances and etiologic agents. The Table is composed of the main table and two Appendices. Appendix A contains materials designated as hazardous substances and wastes (Table 1) and radionuclides (Table 2) and Appendix B contains a list of Marine Pollutants.

Appendix A lists materials which are listed or designated as hazardous substances under section 101(14) of the Comprehensive Environmental Response, Compensation, and Liability Act (CERCLA). DOT, by law, must add these to Table 101. ChemADVISOR provides Appendix A Table 1 and Table 2 and Appendix B (Marine Pollutants) as separate lists.

Only chemical substances and chemical groups have been included: entries such as "articles," "explosives,""ammunition," etc. are not included here. Chemical substances may be listed one or more times in the Hazardous Materials Table depending on physical property qualifications used in the shipping name description. The chemical substance or group is listed here only once. YOU MUST CHECK THE HAZARDOUS MATERIALS TABLE TO FIND THE PROPER SHIPPING DESCRIPTION OR DESCRIPTIONS FOR EACH ENTRY.

The data field includes all the UN/NA identification numbers for entries based on that substance and the words "regulated by DOT." For mixtures of substances each separate substance includes the UN/NA number for the mixture.

U.S.DOT - Hazard Classes   (HM1)

SOURCE

U.S. Department of Transportation
Research and Special Programs Administration
49 CFR Subchapter C - Hazardous Materials Regulations
Part 172 Hazardous Materials Tables
Paragraph 172.101 Hazardous Materials Table
As published in 49 CFR with amendments through June 2, 1994 (59 FR 28487)
NOTE: This list is based on the revisions made under Docket HM181 - hazard classes are provided as numbers rather than text.

LIST DESCRIPTION

The Hazardous Materials Transportation Act (HMTA), which became law in 1975, gives the Department of Transportation (DOT) authority to regulate the movement within the United States of substances that may pose a threat to health, safety, property, or the environment when transported by air, highway, rail, or water.

Under HMTA, DOT has delegated authority for regulation of hazardous materials to the Research and Special Programs Administration (RSPA). This organization, with its Office of Hazardous Materials Transportation, drafts and issues hazardous materials regulations, exemptions, registration certificates, and packaging and container certification for all transport modes.

The Hazardous Materials Table (Table 101) designates the materials therein as hazardous materials for the purpose of transportation of those materials. For each listed material, the Table identifies the hazard class or specifies that the material is forbidden in transportation. Column 3 of Table 101 identifies the Hazard class or Division corresponding to each PROPER SHIPPING NAME. Several proper shipping names may correspond to a single chemical - you MUST check 49 CFR 172.101 to determine the exact hazard class if you are shipping materials. ChemADVISOR provides this list only as a reference for screening or general information concerning these materials.

The following are the class or division names and a reference to the specific section of 49 CFR which contain definitions for classifying hazardous materials:

Class 1.1 - Explosives (with a mass explosion hazard)

49 CFR 173.21

Class 1.2 - Explosives (with a projection hazard)

49 CFR 173.53

Class 1.3 - Explosives (with predominately a fire hazard)

49 CFR 173.50

Class 1.4 - Explosives (with no significant blast hazard)

49 CFR 173.50

Class 1.5 - Very insensitive explosives; blasting agents

49 CFR 173.50

Class 1.6 - Extremely insensitive detonating substances

49 CFR 173.50

Class 2.1 - Flammable gas

49 CFR 173.115

Class 2.2 - Non-flammable compressed gas

49 CFR 173.115

Class 2.3 - Poisonous gas

49 CFR 173.115

Class 3 - Flammable and combustible liquid

49 CFR 173.120

Class 4.1 - Flammable Solid

49 CFR 173.124

Class 4.2 - Spontaneously combustible material

49 CFR 173.124

Class 4.3 - Dangerous when wet material

49 CFR 173.124

Class 5.1 - Oxidizer

49 CFR 173.128

Class 5.2 - Organic Peroxide

49 CFR 173.128

Class 6.1 - Poisonous materials

49 CFR 173.132

Class 6.2 - Infectious substance (etiologic agent)

49 CFR 173.134

Class 7 - Radioactive material

49 CFR 173.403

Class 8 - Corrosive material

49 CFR 173.136

Class 9 - Miscellaneous hazardous material

49 CFR 173.140

Only chemical substances, chemical groups or groups related to a chemical hazard have been included in this listing.

The data field includes multiple hazard classes and clarifying notes for chemical groups where this is applicable.

U.S.DOT - Substances which are Poisonous by (HM1) Inhalation

SOURCE

U.S. Department of Transportation
Research and Special Programs Administration (DHM-21)
Washington, DC 20590-0001
Attention: Dr. George E. Cushmac
List of Gaseous Hazardous Materials - Poisonous by Inhalation (GHM-PBI)(Rev/2) and List of Liquid Hazardous Materials - Poisonous by Inhalation (LHM-PBI)(Rev/2) March 1, 1993

LIST DESCRIPTION

The Hazardous Materials Transportation Act (HMTA), which became law in 1975, gives the Department of Transportation (DOT) authority to regulate the movement within the United States of substances that may pose a threat to health, safety, property, or the environment when transported by air, highway, rail, or water.

Under HMTA, DOT has delegated authority for regulation of hazardous materials to the Research and Special Programs Administration (RSPA). This organization, with its Office of Hazardous Materials Transportation, drafts and issues hazardous materials regulations, exemptions, registration certificates, and packaging and container certification for all transport modes.

The Hazardous Materials Table (49 CFR 172.101) has several thousand entries applicable to various materials including explosives, flammables, oxidizing materials, organic peroxides, corrosives, gases, poisons, radioactive substances and etiologic agents. The Table is composed of the main table and an Appendix. The appendix contains materials designated as hazardous substances and wastes.

RSPA has prepared a guidance document which provides assistance to the regulated community in identifying liquid and gaseous substances included in the Hazardous Materials Table which are considered to be "Poisonous by Inhalation". This document lists the substances by the technical or generic name given in Table 101 and provides additional information such as the LC50, Saturated Vapor Concentration and ERG Guide Number. This guidance document does not list all the requirements of 49 CFR pertaining to these chemicals, but does provide a convenient listing of these materials for those who must comply. You must check all the requirements of 49 CFR 172 before shipping these materials.

This list contains all the entries in both guidance documents (LHM-PBI and GHM-PBI).

The data field contains the identification number (UN/NA number), any qualifications for the material (such as "anhydrous", "stabilized", etc.), and identifies it as a gaseous orliquid hazardous material poisonous by inhalation.

IARC - Group 4 (probably not carcinogenic) (IAR)

SOURCE

International Agency for Research on Cancer (IARC)
"IARC Monographs on the Evaluation of Carcinogenic Risk to Humans - Overall Evaluations of Carcinogenicity: An Updating of IARC Monographs Volumes 1 to 42" - Supplement 7
IARC Monographs Volumes 43 to 57 - Summary of Final Evaluations from individual monographs
World Health Organization
Lyon, France

LIST DESCRIPTION

The International Agency for Research on Cancer (IARC), an Agency of the World Health Organization (WHO), has established a program with the objective of publishing in the form of monographs critical reviews of data on carcinogenicity for agents to which humans are known to be exposed, and on specific exposure situations; to evaluate these data in terms of human risk with the help of international working groups of experts in carcinogenesis and related fields; and to indicate where additional research efforts are needed. Evaluations of the strength of the evidence for carcinogenicity arising from human and experimental animal data are made using standard terms. The categorization of an agent is a matter of scientific judgment reflecting the strength of the evidence derived from the evaluated studies.

The term "carcinogenic risk" in the IARC Monographs series is taken to mean the probability that exposure to an agent will lead to cancer in humans.

Inclusion of an agent in the Monographs does not imply that it is a carcinogen, only that the published data have been examined. Equally, the fact that an agent has not yet been evaluated in a monograph does not mean that it is not carcinogenic.

The evaluations of carcinogenic risk are made by international working groups of independent scientists and are qualitative in nature. No recommendation is given for regulation or legislation.

The listings provided here are taken from the "Overall Evaluation" for each substance. Supplement 7 contained a summary of the overall evaluations for the substances presented in Volumes 1 to 42. The Overall Evaluation is a consideration of the body of evidence as a whole in order to reach an evaluation of the carcinogenicity to humans of an agent, mixture or circumstance of exposure. ChemADVISOR has included only the overall evaluations for chemical substances or groups of chemical substances.

IARC sometimes makes an evaluation for a group of chemicalcompounds. When supporting data indicate that other, related compounds for which there is no direct evidence of capacity to induce cancer in animals or in humans, may also be carcinogenic, a statement describing the rationale for this conclusion is added to the evaluation narrative; an additional evaluation may be made for this broader group of compounds if the strength of the evidence warrants it.

ChemADVISOR has indicated this differentiation by putting these related compounds in a list designated "Group Unspecified"and indicating the chemical group to which they are related.

This list identifies those agents with an overall evaluation of Group 4. Group 4 is defined in the Preamble to Supplement 7 as:

> Group 4 - The agent is probably not carcinogenic to humans. This category is used for agents for which there is evidence suggesting lack of carcinogenicity in humans together with evidence suggesting lack of carcinogenicity in experimental animals. In some circumstances, agents for which there is inadequate evidence of, or no data on carcinogenicity in humans but evidence suggesting lack of carcinogenicity in experimental animals, consistently and strongly supported by a broad range of other relevant data, may be classified in this group.

The data field contains the word "present" to indicate its presence on the list, plus any footnotes or comments used by IARC to clarify the classification.

IARC - Group 3 (not classifiable)   (IAR)

SOURCE

International Agency for Research on Cancer (IARC) "IARC Monographs on the Evaluation of Carcinogenic Risk to Humans - Overall Evaluations of Carcinogenicity: An Updating of IARC Monographs Volumes 1 to 42" - Supplement 7
IARC Monographs Volumes 43 to 57 - Summary of Final Evaluations from individual monographs
World Health Organization
Lyon, France

LIST DESCRIPTION

The International Agency for Research on Cancer (IARC), an Agency of the World Health Organization (WHO), has established a program with the objective of publishing in the form of monographs critical reviews of data on carcinogenicity for agents to which humans are known to be exposed, and on specific exposure situations; to evaluate these data in terms of human risk with the help of international working groups of experts in carcinogenesis and related fields; and to indicate where additional research efforts are needed. Evaluations of the strength of the evidence for carcinogenicity arising from human and experimental animal data are made using standard terms. The categorization of an agent is a matter of scientific judgment reflecting the strength of the evidence derived from the evaluated studies.

The term "carcinogenic risk" in the IARC Monographs series is taken to mean the probabiity that exposure to an agent will lead to cancer in humans.

Inclusion of an agent in the Monographs does not imply that it is a carcinogen, only that the pubished data have been examined. Equally, the fact that an agent has not yet been evaluated in a monograph does not mean that it is not carcinogenic.

The evaluations of carcinogenic risk are made by international working groups of independent scientists and are qualitative in nature. No recommendation is given for regulation or legislation.

The listings provided here are taken from the "Overall Evaluation" for each substance. Supplement 7 contained a summary of the overall evaluations for the substances presented in Volumes 1 to 42. The Overall Evaluation is a consideration of the body of evidence as a whole in order to reach an evaluation of the carcinogenicity to humans of an agent, mixture or circumstance of exposure. ChemADVISOR has included only the overall evaluations for chemical substances or groups of chemical substances.

IARC sometimes makes an evaluation for a group of chemicalcompounds. When supporting data indicate that other, related compounds for which there is no direct evidence of capacity to induce cancer in animals or in humans, may also be carcinogenic, a statement describing the rationale for this conclusion is added to the evaluation narrative; an additional evaluation may be made for this broader group of compounds if the strength of the evidence warrants it.

ChemADVISOR has indicated this differentiation by putting these related compounds in a list designated "Group Unspecified"and indicating the chemical group to which they are related.

This list identifies those agents with an overall evaluation of Group 3. Group 3 is defined in the Preamble to Supplement 7 as:

> Group 3 -The agent is not classifiable as to its carcinogenicity to humans. Agents are placed in this category when they do not fall into any other group.

The data field contains the word "present" to indicate its presence on the list, plus any footnotes or comments used by IARC to clarify the classification.

IARC - Group Unspecified   (IAR)

SOURCE

International Agency for Research on Cancer (IARC) "IARC Monographs on the Evaluation of Carcinogenic Risk to Humans - Overall Evaluations of Carcinogenicity: An Updating of IARC Monographs Volumes 1 to 42" - Supplement 7
IARC Monographs Volumes 43 to 57 - Summary of Final Evaluations from individual monographs
World Health Organization
Lyon, France

LIST DESCRIPTION

The International Agency for Research on Cancer (IARC), an Agency of the World Health Organization (WHO), has established a program with the objective of publishing in the form of monographs critical reviews of data on carcinogenicity for agents to which humans are known to be exposed, and on specific exposure situations; to evaluate these data in terms of human risk with the help of international working groups of experts in carcinogenesis and related fields; and to indicate where additional research efforts are needed. Evaluations of the strength of the evidence for carcinogenicity arising from human and experimental animal data are made using standard terms. The categorization of an agent is a matter of scientific judgment reflecting the strength of the evidence derived from the evaluated studies.

The term "carcinogenic risk" in the IARC Monographs series is taken to mean the probabiity that exposure to an agent will lead to cancer in humans.

Inclusion of an agent in the Monographs does not imply that it is a carcinogen, only that the pubished data have been examined. Equally, the fact that an agent has not yet been evaluated in a monograph does not mean that it is not carcinogenic.

The evaluations of carcinogenic risk are made by international working groups of independent scientists and are qualitative in nature. No recommendation is given for regulation or legislation.

The listings provided here are taken from the "Overall Evaluation" for each substance. Supplement 7 contained a summary of the overall evaluations for the substances presented in Volumes 1 to 42. The remaining evaluations are taken directly from the published monograph. The Overall Evaluation is a consideration of the body of evidence as a whole in order to reach an evaluation of the carcinogenicity to humans of an agent, mixture or circumstance of exposure. ChemADVISOR has included only the overall evaluations for chemical substances or groups of chemical substances.

IARC sometimes makes an evaluation for a group of chemical compounds. When supporting data indicate that other, related compounds for which there is no direct evidence of capacity to induce cancer in animals or in humans, may also be carcinogenic, a statement describing the rationale for this conclusion is added to the evaluation narrative; an additional evaluation may be made for this broader group of compounds if the strength of the evidence warrants it.

ChemADVISOR has indicated this differentiation by putting these related compounds in a list designated "Group Unspecified"and indicating the chemical group to which they are related.

This list contains all those substances mentioned within a category (e.g., "androgenic (anabolic) steroids") assigned overall classification of group 1, 2A or 2B, but for which IARC made no specific individual classification (e.g., oxymetholone).

The data field contains the word "present" and the generic group of chemicals under which it was identified in the IARC monograph.

IARC - Group 1 (carcinogenic to humans)   (IAR)

SOURCE

International Agency for Research on Cancer (IARC)
"IARC Monographs on the Evaluation of Carcinogenic Risk to Humans - Overall Evaluations of Carcinogenicity: An Updating of IARC Monographs Volumes 1 to 42" - Supplement 7
IARC Monographs Volumes 43 to 57 - Summary of Final Evaluations from individual monographs
World Health Organization
Lyon, France

LIST DESCRIPTION

The International Agency for Research on Cancer (IARC), an Agency of the World Health Organization (WHO), has established a program with the objective of publishing, in the form of monographs, critical reviews of data on carcinogenicity for agents to which humans are known to be exposed, and on specific exposure situations; to evaluate these data in terms of human risk with the help of international working groups of experts in carcinogenesis and related fields; and to indicate where additional research efforts are needed. Evaluations of the strength of the evidence for carcinogenicity arising from human and experimental animal data are made using standard terms. The categorization of an agent is a matter of scientific judgment reflecting the strength of the evidence derived from the evaluated studies.

The term "carcinogenic risk" in the IARC Monographs series is taken to mean the probabiity that exposure to an agent will lead to cancer in humans.

Inclusion of an agent in the Monographs does not imply that it is a carcinogen, only that the pubished data have been examined. Equally, the fact that an agent has not yet been evaluated in a monograph does not mean that it is not carcinogenic.

The evaluations of carcinogenic risk are made by international working groups of independent scientists and are qualitative in nature. No recommendation is given for regulation or legislation.

The listings provided here are taken from the "Overall Evaluation" for each substance. Supplement 7 contained a summary of the overall evaluations for the substances presented in Volumes 1 to 42. The Overall Evaluation is a consideration of the body of evidence as a whole in order to reach an evaluation of the carcinogenicity to humans of an agent, mixture or circumstance of exposure. ChemADVISOR has included only the overall evaluations for chemical substances or groups of chemical substances.

IARC sometimes makes an evaluation for a group of chemicalcompounds. When supporting data indicate that other, related compounds for which there is no direct evidence of capacity to induce cancer in animals or in humans, may also be carcinogenic, a statement describing the rationale for this conclusion is added to the evaluation narrative; an additional evaluation may be made for this broader group of compounds if the strength of the evidence warrants it.

ChemADVISOR has indicated this differentiation by putting these related compounds in a list designated "Group Unspecified"and indicating the chemical group to which they are related.

This list identifies those agents with an overall evaluation of Group 1. Group 1 is defined in the Preamble to Supplement 7 as:

Group 1 - Carcinogenic to humans. This category is used only when there is "sufficient evidence" of carcinogenicity in humans.

The data field contains the word "present" to indicate its presence on the list, plus any footnotes or comments used by IARC to clarify the entry.

IARC - Group 2B (sufficient animal data)   (IAR)
SOURCE

International Agency for Research on Cancer (IARC)
"IARC Monographs on the Evaluation of Carcinogenic Risk to Humans - Overall Evaluations of Carcinogenicity: An Updating of IARC Monographs Volumes 1 to 42" - Supplement 7
IARC Monographs Volumes 43 to 57 - Summary of Final Evaluations from individual monographs
World Health Organization
Lyon, France

LIST DESCRIPTION:

The International Agency for Research on Cancer (IARC), an Agency of the World Health Organization (WHO), has established a program with the objective of publishing in the form of monographs critical reviews of data on carcinogenicity for agents to which humans are known to be exposed, and on specific exposure situations; to evaluate these data in terms of human risk with the help of international working groups of experts in carcinogenesis and related fields; and to indicate where additional research efforts are needed. Evaluations of the strength of the evidence for carcinogenicity arising from human and experimental animal data are made using standard terms. The categorization of an agent is a matter of scientific judgment reflecting the strength of the evidence derived from the evaluated studies.

The term "carcinogenic risk" in the IARC Monographs series is taken to mean the probabiity that exposure to an agent will lead to cancer in humans.

Inclusion of an agent in the Monographs does not imply that it is a carcinogen, only that the pubished data have been examined. Equally, the fact that an agent has not yet been evaluated in a monograph does not mean that it is not carcinogenic.

The evaluations of carcinogenic risk are made by international working groups of independent scientists and are qualitative in nature. No recommendation is given for regulation or legislation.

The listings provided here are taken from the "Overall Evaluation" for each substance. Supplement 7 contained a summary of the overall evaluations for the substances presented in Volumes 1 to 42. The Overall Evaluation is a consideration of the body of evidence as a whole in order to reach an evaluation of the carcinogenicity to humans of an agent, mixture or circumstance of exposure. ChemADVISOR has included only the overall evaluations for chemical substances or groups of chemical substances.

IARC sometimes makes an evaluation for a group of chemicalcompounds. When supporting data indicate that other, related compounds for which there is no direct evidence of capacity to induce cancer in animals or in humans, may also be carcinogenic, a statement describing the rationale for this conclusion is added to the evaluation narrative; an additional evaluation may be made for this broader group of compounds if the strength of the evidence warrants it.

ChemADVISOR has indicated this differentiation by putting these related compounds in a list designated "Group Unspecified"and indicating the chemical group to which they are related.

This list identifies those agents with an overall evaluation of Group 2B. Group 2B is defined in the Preamble to Supplement 7 as:

Group 2B - Possibly carcinogenic to humans. This category is generally used for agents for which there is "limited evidence" in humans in the absence of "sufficient evidence" in experimental animals. It may also be used when there is "inadequate evidence" of carcinogenicity in humans or when human data are nonexistent but there is "sufficient evidence" of carcinogenicity in experimental animals. In some instances, an agent for which there is inadequate evidence or no data in humans but "limited evidence" of carcinogenicity in experimental animals together with supporting evidence from other relevant data may be placed in this group.

The data field contains the word "present" to indicate its presence on the list, plus any footnotes or comments used by IARC to clarify the classification.

IARC - Group 2A (limited human data)   (IAR)
SOURCE

International Agency for Research on Cancer (IARC)
"IARC Monographs on the Evaluation of Carcinogenic Risk to Humans - Overall Evaluations of Carcinogenicity: An Updating of IARC Monographs Volumes 1 to 42" - Supplement 7
IARC Monographs Volumes 43 to 57 - Summary of Final Evaluations from individual monographs
World Health Organization
Lyon, France

LIST DESCRIPTION

The International Agency for Research on Cancer (IARC), an Agency of the World Health Organization (WHO), has established a program with the objective of publishing in the form of monographs critical reviews of data on carcinogenicity for agents to which humans are known to be exposed, and on specific exposure situations; to evaluate these data in terms of human risk with the help of international working groups of experts in carcinogenesis and related fields; and to indicate where additional research efforts are needed. Evaluations of the strength of the evidence for carcinogenicity arising from human and experimental animal data are made using standard terms. The categorization of an agent is a matter of scientific judgment reflecting the strength of the evidence derived from the evaluated studies.

The term "carcinogenic risk" in the IARC Monographs series is taken to mean the probability that exposure to an agent will lead to cancer in humans.

Inclusion of an agent in the Monographs does not imply that it is a carcinogen, only that the published data have been examined. Equally, the fact that an agent has not yet been evaluated in a monograph does not mean that it is not carcinogenic.

The evaluations of carcinogenic risk are made by international working groups of independent scientists and are qualitative in nature. No recommendation is given for regulation or legislation.

The listings provided here are taken from the "Overall Evaluation" for each substance. Supplement 7 contained a summary of the overall evaluations for the substances presented in Volumes 1 to 42. The Overall Evaluation is a consideration of the body of evidence as a whole in order to reach an evaluation of the carcinogenicity to humans of an agent, mixture or circumstance of exposure. ChemADVISOR has included only the overall evaluations for chemical substances or groups of chemical substances.

IARC sometimes makes an evaluation for a group of chemicalcompounds. When supporting data indicate that other, related compounds for which there is no direct evidence of capacity to induce cancer in animals or in humans, may also be carcinogenic, a statement describing the rationale for this conclusion is added to the evaluation narrative; an additional evaluation may be made for this broader group of compounds if the strength of the evidence warrants it.

ChemADVISOR has indicated this differentiation by putting these related compounds in a list designated "Group Unspecified"and indicating the chemical group to which they are related.

This list identifies those agents with an overall evaluation of Group 2A. Group 2A is defined in the Preamble to Supplement 7 as:

Group 2A - Probably carcinogenic to humans. This category is used when there is "limited evidence" of carcinogenicity in humans and "sufficient evidence" of carcinogenicity in experimental animals. Exceptionally, an agent may be classified into this category solely on the basis of "limited evidence" of carcinogenicity in humans or of "sufficient evidence" of carcinogenicity in experimental animals strengthened by supporting evidence from other relevant data.

The data field contains the word "present" to indicate its presence on the list, plus any footnotes or comments used by IARC to clarify the classification.

NIOSH - Selected LD50s and LC50s (—)

SOURCE

Registry of Toxic Effects of Chemical Substances (RTECS)
Centers for Disease Control and Prevention
National Institute for Occupational Safety and Health
Division of Standards Development and Technology Transfer
4676 Columbia Parkway
Cincinnati, OH 45226-9966
Doris V. Sweet, editor
FAX (513) 533-8588

LIST DESCRIPTION

An LD50 is defined as the calculated dose of a substance which is expected to cause the death of 50% of an entire defined experimental animal population. It is determined from the exposure to the substance by any route (other than inhalation) of a significant number from that population.

An LC50 is defined as a calculated concentration of a substance in air, exposure to which for a specified length of time is expected to cause the death of 50% of an entire defined experimental animal population. It is determined from the exposure to the substance of a significant number from that population.

The units of dose measurement are expressed in terms of the quantity administered per unit body weight, or quantity per skin surface area, or quantity per unit volume of the respired. air. In addition, the duration of time over which the dose was administered is also listed, as needed.

Dose amounts are generally expressed as milligrams (one thousandth of a gram) per kilogram (mg/kg). In some cases, because of dose size grams per kiligram (gm/kg), micrograms (one millionth of a gram) per kilogram (ug/kg), or nanograms (one billionth of a gram) per kilogram (ng/kg) are used. Doses may also be reported as millilters/kilogram (ml/kg).

Concentrations of gaseous substance in air are generally listed as parts of vapor or gas per million parts of air by volume (ppm). However, parts per hundred (pph or per cent), parts per billion (ppb), or parts per trillion (ppt) may be used. If the substance is a solid or a liquid, the concentrations are listed preferably as milligrams per cubic meter (mg/m3) but may, as applicable, be listed as micrograms per cubic meter (ug/m3).

LD50 and LC50 data represent single dose exposures.

The last printed edition of RTECS, the 1985-86 edition, is now out of print and there are no plans for future editions. However, the Registry is maintained and updated electronically each quarter by the National Institute for Occupational Safety and Health (NIOSH). RTECS is maintained in compliance with the requirements of Section 20(a)(6) of the Occupational Safety and Health Act of 1970.

The database includes primary skin and eye irritation, mutagenic effects, reproductive effects, tumorigenic effects, and acute toxicity data. A recent addition is other toxic effects data from multiple dose studies. NIOSH policy is to record the lowest dose or lowest exposure concentration reported to cause the tabulated effect. It is also NIOSH policy NOT to evaluate the data, but to tabulate the values reported.

For the substances listed in the database, ChemADVISOR has queried the latest version of the RTECS database and selected the following pieces of data to compile this list:

Oral LD50s in rats and mice ONLY;

Dermal LD50s in rabbits ONLY;

LC50s in rats and mice or in selected other species where rats and mice are not available.

The data field contains the values obtained, relevant dosage units and exposure times where applicable.

NIOSH - Health Standards - Exposure Limits   (NHS)

SOURCE

"NIOSH Recommendations for Occupational Safety and Health Standards 1988"

U.S. Department of Health and Human Services
Public Health Service
National Institute for Occupational Safety and Health
Centers for Disease Control
Atlanta, Georgia 30333

Morbidity and Mortality Weekly Report Supplement
August 26, 1988
Volume 37 No. S-7

LIST DESCRIPTION

The National Institute of Occupational Safety and Health (NIOSH) develops and periodically revises recommendations for limits of exposure to potentially hazardous substances or conditions in the workplace. The recommendations are then published and transmitted to the Occupational Safety and Health Administration (OSHA) or the Mine Safety and Health Administration (MSHA) of the U.S. Department of Labor for use in promulgating legal standards.

NIOSH recommendations are published in a variety of documents. Criteria documents specify a NIOSH recommended exposure limit (REL) and appropriate preventive measures designed to reduce or eliminate adverse health effects.

Special hazard reviews, occupational hazard assessments, alerts, and technical guidelines are other types of NIOSH documents that complement the Institute's recommendations for standards. NIOSH periodically presents testimony before various Congressional committees and at regulatory hearings convened by OSHA or MSHA. The testimony always includes the current NIOSH policy concerning the hazard in question. NIOSH Current Intelligence Bulletins (CIBs) review and evaluate new and emerging information on occupational hazards.

This group of lists is taken from a supplement to the "Morbidity and Mortality Weekly Report" in which NIOSH presented a rapid reference to the most recent NIOSH REL or other recommendation for each potential hazard. Information in that publication is contained in this group of lists.

This list identifies the substances for which NIOSH has established a Recommended Exposure Limit (REL) or other recommendation for exposure control. The limit is given as parts per million or milligrams per cubic meter as a 10 hour Time Weighted Average (TWA). If the substance is listed as part of a group of chemicals this information is also included.

NOTE: These lists include only chemical compounds or groups.

NIOSH - Health Standards - Health Effects and   (NHS) Precautions

SOURCE

"NIOSH Recommendations for Occupational Safety and Health Standards 1988"

U.S. Department of Health and Human Services
Public Health Service
National Institute for Occupational Safety and Health
Centers for Disease Control
Atlanta, Georgia 30333

Morbidity and Mortality Weekly Report Supplement
August 26, 1988
Volume 37 No. S-7

LIST DESCRIPTION

The National Institute of Occupational Safety and Health (NIOSH) develops and periodically revises recommendations for limits of exposure to potentially hazardous substances or conditions in the workplace. The recommendations are then published and transmitted to the Occupational Safety and Health Administration (OSHA) or the Mine Safety and Health Administration (MSHA) of the U.S. Department of Labor for use in promulgating legal standards.

NIOSH recommendations are published in a variety of documents. Criteria documents specify a NIOSH recommended exposure limit (REL) and appropriate preventive measures designed to reduce or eliminate adverse health effects.

Special hazard reviews, occupational hazard assessments, alerts, and technical guidelines are other types of NIOSH documents that complement the Institute's recommendations for standards. NIOSH periodically presents testimony before various Congressional committees and at regulatory hearings convened by OSHA or MSHA. The testimony always includes the current NIOSH policy concerning the hazard in question. NIOSH Current Intelligence Bulletins (CIBs) review and evaluate new and emerging information on occupational hazards.

This group of lists is taken from a supplement to the "Morbidity and Mortality Weekly Report" in which NIOSH presented a rapid reference to the most recent NIOSH REL or other recommendations for each potential hazard. Information in that publication is contained in this group of lists.

This list identifies substances with information contained in the "Health Effect(s) Considered" column plus any precautions provided in the "comments" section. Health effects citedare for humans unless otherwise indicated. The data field contains the health effects cited and the precautions recommended.

NOTE: These lists include only chemical compounds or groups.

NIOSH - Health Standards - Carcinogenic (NHS) Chemicals

SOURCE

"NIOSH Recommendations for Occupational Safety and Health Standards 1988"

U.S. Department of Health and Human Services
Public Health Service
National Institute for Occupational Safety and Health
Centers for Disease Control
Atlanta, Georgia 30333

Morbidity and Mortality Weekly Report Supplement
August 26, 1988
Volume 37 No. S-7

LIST DESCRIPTION

The National Institute of Occupational Safety and Health (NIOSH) developes and periodically revises recommendations for limits of exposure to potentially hazardous substances or conditions in the workplace. The recommendations are then published and transmitted to the Occupational Safety and Health Administration (OSHA) or the Mine Safety and Health Administration (MSHA) of the U.S. Department of Labor for use in promulgating legal standards.

NIOSH recommendations are published in a variety of documents. Criteria documents specify a NIOSH recommended exposure limit (REL) and appropriate preventive measures designed to reduce or eliminate adverse health effects.

Special hazard reviews, occupational hazard assessments, alerts, and technical guidelines are other types of NIOSH documents that complement the Institute's recommendations for standards. NIOSH periodically presents testimony before various Congressional committees and at regulatory hearings convened by OSHA or MSHA. The testimony always includes the current NIOSH policy concerning the hazard in question. NIOSH Current Intelligence Bulletins (CIBs) review and evaluate new and emerging information on occupational hazards.

This group of lists is taken from a supplement to the "Morbidity and Mortality Weekly Report" in which NIOSH presented a rapid reference to the most recent NIOSH REL or other recommendation for each potential hazard. Each column heading (except the OSHA PEL) in that publication is contained in this group of lists.

This list identifies substances for which the Health Effects Considered includes a carcinogenic effect in animals and/or humans. The data field contains the statement "potential humancarcinogen" and, if applicable, any generic group under which NIOSH had included the substance.

NOTE: These lists include only chemical compounds or groups.

NIOSH 1990 - Pocket Guide - Carcinogens (NIO)

SOURCE

"NIOSH Pocket Guide to Chemical Hazards"
U.S. Department of Health and Human Services
Centers for Disease Control
National Institute for Occupational Safety and Health (NIOSH)
June 1990
DHHS (NIOSH) Publication No. 90-117

LIST DESCRIPTION

The National Institute For Occupational Safety and Health (NIOSH) publishes the "Pocket Guide to Chemical Hazards" which presents information taken, in part, from the NIOSH/OSHA Occupational Health Guidelines for Chemical Hazards, from National Institute for Occupational Safety and Health (NIOSH) criteria documents and Current Intelligence Bulletins, and from recognized references in the fields of industrial hygiene, occupational medicine, toxicology, and analytical chemistry. The Pocket Guide contains data on chemical structures or formulas, identification codes, synonyms, exposure limits, chemical and physical properties, incompatibilities and reactivities, measurement methods, respirator selections, signs and symptoms of exposure, and procedures for emergency treatment. ChemADVISOR has included only selected portions of all the data contained in the Pocket Guide.

The substances included in this list are those which NIOSH has identified as "Carcinogens" (Ca) according to the criteria specified in Appendix A of the Guide. This information is contained under the heading "Exposure limits" in the main body of the Guide.

Appendix A states:

NIOSH has identified numerous chemicals that should be treated as occupational carcinogens even though OSHA has not identified them as such. In determining their carcinogenicity, NIOSH uses a classification outlined in 29 CFR 1990.103, which states in part:

'Potential occupational carcinogen' means any substance, or combination or mixture of substances, which causes an increased incidence of benign and/or malignant neoplasms, or a substantial decrease in the latency period between exposure and onset of neoplasms in humans or in one or more experimental mammalian species as the resultof any oral, respiratory or dermal exposure, or any other exposure which results in the induction of tumors at a site other than the site of administration. This definition also includes any substance which is metabolized into one or more potential occupational carcinogens by mammals.

The data field contains the notation "occupational carcinogen" and any generic group (if applicable).

NIOSH 1990 - Pocket Guide - IDLHs   (NIO)
SOURCE

"NIOSH Pocket Guide to Chemical Hazards"
U.S. Department of Health and Human Services
Centers for Disease Control
National Institute for Occupational Safety and Health (NIOSH)
June 1990
DHHS (NIOSH) Publication No. 90-117

LIST DESCRIPTION

The National Institute For Occupational Safety and Health (NIOSH) publishes the "Pocket Guide to Chemical Hazards" which presents information taken, in part, from the NIOSH/OSHA Occupational Health Guidelines for Chemical Hazards, from National Institute for Occupational Safety and Health (NIOSH) criteria documents and Current Intelligence Bulletins, and from recognized references in the fields of industrial hygiene, occupational medicine, toxicology, and analytical chemistry. The Pocket Guide contains data on chemical structures or formulas, identification codes, synonyms, exposure limits, chemical and physical properties, incompatibilities and reactivities, measurement methods, respirator selections, signs and symptoms of exposure, and procedures for emergency treatment. ChemADVISOR has included only selected portions of all the data contained in the Pocket Guide.

This list contains all the substances for which NIOSH has established an "Immediately Dangerous to Life and Health" (IDLH) level. The IDLH level is defined for the purpose of respirator selection and represents the maximum concentration from which one could escape within 30 minutes without a respirator and without experiencing any escape-impairing (e.g., severe eye irritation) or irreversible health effects.

The notation "Ca" appears in this column for all substances that NIOSH considers to be potential human carcinogens. However, the IDLHs that were originally determined in the Standards Completion Program are shown in brackets following the "Ca"designation (note: carcinogenic effects were not considered in the Standards Completion Program).

The data field indicates the IDLH levels in ppm, mg/m3 and the notation "(not considering carcinogenic effects)" (if applicable). Substances with only a Ca notation (and no IDLH indicated) are not included in this list but in the "NIOSH 1990-Pocket Guide-Carcinogens" list.

NIOSH 1990 - Pocket Guide - Skin List   (NIO)
SOURCE

"NIOSH Pocket Guide to Chemical Hazards"
U.S. Department of Health and Human Services
Centers for Disease Control
National Institute for Occupational Safety and Health (NIOSH)
June 1990
DHHS (NIOSH) Publication No. 90-117

LIST DESCRIPTION

The National Institute For Occupational Safety and Health (NIOSH) publishes the "Pocket Guide to Chemical Hazards" which presents information taken, in part, from the NIOSH/OSHA Occupational Health Guidelines for Chemical Hazards, from National Institute for Occupational Safety and Health (NIOSH) criteria documents and Current Intelligence Bulletins, and from recognized references in the fields of industrial hygiene, occupational medicine, toxicology, and analytical chemistry. The Pocket Guide contains data on chemical structures or formulas, identification codes, synonyms, exposure limits, chemical and physical properties, incompatibilities and reactivities, measurement methods, respirator selections, signs and symptoms of exposure, and procedures for emergency treatment. ChemADVISOR has included only selected portions of all the data contained in the Pocket Guide.

This list contains all the substances for which NIOSH has indicated a potential for dermal absorption. Skin exposure should be prevented as necessary through the use of good work practices and gloves, coveralls, goggles, and other appropriate equipment.

The data field identifies these entries with the words "potential for dermal absorption." If the substance is listed as part of a group of substances, that group is also identified in the data field.

NIOSH 1990 - Pocket Guide - Target Organs   (NIO)
SOURCE

"NIOSH Pocket Guide to Chemical Hazards"
U.S. Department of Health and Human Services
Centers for. Disease Control
National Institute for Occupational Safety and Health (NIOSH)
June 1990
DHHS (NIOSH) Publication No. 90-117

LIST DESCRIPTION

The National Institute For Occupational Safety and Health (NIOSH) publishes the "Pocket Guide to Chemical Hazards" which presents information taken, in part, from the NIOSH/OSHA Occupational Health Guidelines for Chemical Hazards, from National Institute for Occupational Safety and Health (NIOSH) criteria documents and Current Intelligence Bulletins, and from recognized references in the fields of industrial hygiene, occupational medicine, toxicology, and analytical chemistry. The Pocket Guide contains data on chemical structures or formulas, identification codes, synonyms, exposure limits, chemical and physical properties, incompatibilities and reactivities, measurement methods, respirator selections, signs and symptoms of exposure, and procedures for emergency treatment. ChemADVISOR has included only selected portions of all the data contained in the Pocket Guide.

This list contains all the substances listed in the Pocket Guide since every entry identifies some affected organ system.

The organs that are affected by exposure to each substance are identified in the data field. CVS = Cerebrovascular system; CNS = Central Nervous System; GI = Gastrointestinal Tract.

NIOSH 1990 - Pocket Guide - RELs   (NIO)
SOURCE

"NIOSH Pocket Guide to Chemical Hazards"
U.S. Department of Health and Human Services
Centers for Disease Control
National Institute for Occupational Safety and Health (NIOSH)
June 1990
DHHS (NIOSH) Publication No. 90-117

LIST DESCRIPTION

The National Institute For Occupational Safety and Health (NIOSH) publishes the "Pocket Guide to Chemical Hazards" which presents information taken, in part, from the NIOSH/OSHA Occupational Health Guidelines for Chemical Hazards, from National Institute for Occupational Safety and Health (NIOSH) criteria documents and Current Intelligence Bulletins, and from recognized references in the fields of industrial hygiene, occupational medicine, toxicology, and analytical chemistry. The Pocket Guide contains data on chemical structures or formulas, identification codes, synonyms, exposure limits, chemical and physical properties, incompatibilities and reactivities, measurement methods, respirator selections, signs and symptoms of exposure, and procedures for emergency treatment. ChemADVISOR has included only selected portions of all the data contained in the Pocket Guide.

This list contains all substances for which NIOSH has established recommended exposure limits. NIOSH Recommended Exposure Limits (RELs) are: 10-hour Time Weighted Averages (TWA); 15-minute Short Term Exposure Limits (STEL); and Ceilings (C) which should not be exceeded at any time.

The data field contains the data value for the REL in ppm or mg/m3 followed by "TWA," Ceiling values are preceded by a "C," and Short Term Exposure Limits are followed by "STEL."

NIOSH also publishes additional information and recommendations in the "Morbidity and Mortality Weekly Report" (see NIOSH Health Standards lists).

NTP Chemical Status Reports - Evidence of (CSR) Carcinogenicity
SOURCE

National Toxicology Program
Division of Toxicology Research and Testing
Chemical Status Reports
Produced from NTP Chemtrack System
last report dated April 5, 1993
For additional information contact -
Central Data Management, Mail Drop A0-01
NIEHS
PO Box 12233
Research Triangle Park, NC 27709
telephone (919) 541-3419

LIST DESCRIPTION

The National Toxicology Program consists of relevant toxicology activities of the National Institutes of Health's National Institute of Environmental Health Sciences (NIH/NIEHS), the

Center for Disease Control's National Institute for Occupational Safety and Health (CDC/NIOSH), and the Food and Drug Administration's National Center for Toxicological Research (FDA/NCTR). The National Cancer Institute (NCI) was a charter member; however, its primary participating component, the Carcinogenesis Bioassay Program, was transferred to the NIEHS by the Secretary of Health and Human Services in 1981.

Toxicologic research and applied studies activities are divided into four major program areas: cellular and genetic toxicology; carcinogenesis; reproductive and developmental toxicology; and toxicologic characterization. The last area covers activities in cardiac, cutaneous, immunologic, neurobehavioral, and respiratory toxicologies, and chemical disposition and chemical pathology, as well as studies oriented to specific chemicals, chemical classes, or mixtures.

Chemicals studied by the National Toxicology Program are selected mainly on the basis of human exposure, production levels, chemical structure, and available toxicologic data.

This group of substances is taken from the public distribution copy of the National Toxicology Program, Division of Toxicology Research and Testing, Chemical Status Report. It is produced from the NTP Chemtrack System. The Chemical Status Report (CSR) identifies substances being tested by NTP, in various stages of testing and in

various stages of report generation as of the date of the last report.

This list identifies the chemicals included in the Chemical Status Report for which results of tests for carcinogenicity have already been published. Some chemicals have had more than one test performed.

The data field gives the summary of the carcinogenicity tests as provided in the Chemical Status Report. Some chemicals have had more than one test performed.

NTP Chemical Status Reports - Testing Status (CSR) and NTIS Number

SOURCE

National Toxicology Program
Division of Toxicology research and Testing
Chemical Status Reports
Produced from NTP Chemtrack System
latest revision dated October 5, 1994
For additional information contact -
Central Data Management, Mail Drop A0-01
NIEHS
PO Box 12233
Research Triangle Park, NC 27709
telephone (919) 541-3419

LIST DESCRIPTION

The National Toxicology Program consists of relevant toxicology activities of the National Institutes of Health's National Institute of Environmental Health Sciences (NIH/NIEHS), the Center for Disease Control's National Institute for Occupational Safety and Health (CDC/NIOSH), and the Food and Drug Administration's National Center for Toxicological Research (FDA/NCTR). The National Cancer Institute (NCI) was a charter member; however, its primary participating component, the Carcinogenesis Bioassay Program, was transferred to the NIEHS by the Secretary of Health and Human Services in 1981.

Toxicologic research and applied studies activities are divided into four major program areas: cellular and genetic toxicology; carcinogenesis; reproductive and developmental toxicology; and toxicologic characterization. The last area covers activities in cardiac, cutaneous, immunologic, neurobehavioral, and respiratory toxicologies, and chemical disposition and chemical pathology, as well as studies oriented to specific chemicals, chemical classes, or mixtures.

Chemicals studied by the National Toxicology Program are selected mainly on the basis of human exposure, production levels, chemical structure, and available toxicologic data.

This group of substances is taken from the public distribution copy of the National Toxicology Program, Division of Toxicology Research and Testing, Chemical Status Report. It is produced from the NTP Chemtrack System. The Chemical Status Report (CSR) identifies substances being tested by NTP, in various stages of testing and in various stages of report generation as of the date of the report.

This list contains all the substances listed in the latestManagement Status Report produced from NTP Chemtrack System, public distribution copy.

The data field indicates the progress of the study and the NTIS order number (if any).

NTP Seventh Report - Suspect Carcinogens (NTP)

SOURCE

Seventh Annual Report on Carcinogens 1994 - Summary
U.S. Department of Health and Human Services
Public Health Service
National Toxicology Program
For more information contact:
Public Health Service
National Toxicology Program
P.O. Box 12233, MD AO-10
Research Triangle Park, NC 27709

LIST DESCRIPTION

The National Toxicology Program consists of relevant toxicology activities of the National Institutes of Health's National institute of Environmental Health Sciences (NIH/NIEHS), the Center for Disease Control's National Institute for Occupational Safety and Health (CDC/NIOSH), and the Food and Drug Administration's National Center for Toxicological Research (FDA/NCTR). The National Cancer Institute (NCI) was a charter member; however, its primary participating component, the Carcinogenesis Bioassay Program, was transferred to the NIEHS by the Secretary of Health and Human Services in 1981.

The Seventh Annual Report on Carcinogens, prepared by the NTP is issued pursuant to Public Law 95-622 of November 9, 1978. This law requires the Secretary of the Department of Health and Human Services to publish an annual report that contains "a list of all substances (i) which either are known to be carcinogens or which may reasonably be anticipated to be carcinogens and (ii) to which a significant number of persons residing in the United States are exposed..."

For the purpose of this Report, "known carcinogens" are defined as those substances for which the evidence from human studies indicates that there is a causal relationship between exposure to the substance and human cancer. Substances "which may reasonably be anticipated to be carcinogens" are defined as "those for which there is limited evidence of carcinogenicity in humans or sufficient evidence of carcinogenicity in experimental animals."

The Seventh Annual Report contains most of the substances, groups of substances, and some of the technological processes that were listed in the Sixth Annual Report. The Seventh Annual Report presents information on 6 additional substances. Each has been chosen either from substances tested by the NCI Carcinogenesis Testing Program or the National Toxicology Program (NTP); from designations of the participating agencies; or from substancesevaluated by the IARC Working Groups. Other substances from the same sources will be added to subsequent Annual Reports.

The Seventh Annual Report contains entries on the carcinogenicity of seven metals (arsenic, beryllium, cadmium, chromium, lead, nickel, and thorium). The entries for the individual metals identify those compounds of the metal (and, where appropriate, the elemental metal itself) for which evidence of carcinogenicity in environmentally-exposed humans or experimental animals is sufficient. Relatively few of the many different forms (elemental, salts, complexes, chelates, etc.) of the metals have been fully evaluated for carcinogenicity. The various factors that can influence the carcinogenic potential of a given metal form that should be considered include route of exposure, absorption, distribution, valence state, metabolism, elimination, as well as potential for specific biochemical interactions in cells. However, in the absence of specific information a metal shown to be carcinogenic in one of its forms should be considered as being potentially carcinogenic in its other forms.

This list identifies all the substances or groups of substances, and medical treatments that are listed as reasonably anticipated to be carcinogens. CAS Numbers for substances included in groups are derived from individual reports.

The data field contains the words "suspect carcinogen" and the generic group in which the chemical was included (if applicable). NOTE: The words "reasonably anticipated to be carcinogens" has been shortened by ChemADVISOR to "suspect carcinogen" for ease in entry and tracking.

NTP Seventh Report - Known Carcinogens   (NTP)
SOURCE

Seventh Annual Report on Carcinogens 1994 - Summary
U.S. Department of Health and Human Services
Public Health Service
National Toxicology Program
For more information contact:
Public Health Service
National Toxicology Program
P.O. Box 12233, MD AO-10
Research Triangle Park, NC 27709

LIST DESCRIPTION

The National Toxicology Program consists of relevant toxicology activities of the National Institutes of Health's National institute of Environmental Health Sciences (NIH/NIEHS), the Center for Disease Control's National Institute for Occupational Safety and Health (CDC/NIOSH), and the Food and Drug Administration's National Center for Toxicological Research (FDA/NCTR). The National Cancer Institute (NCI) was a charter member; however, its primary participating component, the Carcinogenesis Bioassay Program, was transferred to the NIEHS by the Secretary of Health and Human Services in 1981.

The Seventh Annual Report on Carcinogens, prepared by the NTP is issued pursuant to Public Law 95-622 of November 9, 1978. This law requires the Secretary of the Department of Health and Human Services to publish an annual report that contains "a list of all substances (i) which either are known to be carcinogens or which may reasonably be anticipated to be carcinogens and (ii) to which a significant number of persons residing in the United States are exposed..."

For the purpose of this Report, "known carcinogens" are defined as those substances for which the evidence from human studies indicates that there is a causal relationship between exposure to the substance and human cancer. Substances "which may reasonably be anticipated to be carcinogens" are defined as "those for which there is limited evidence of carcinogenicity in humans or sufficient evidence of carcinogenicity in experimental animals."

The Seventh Annual Report contains most of the substances, groups of substances, and some of the technological processes that were listed in the Sixth Annual Report. The Seventh Annual Report present information on 6 additional substances. Each has been chosen either from substances tested by the NCI Carcinogenesis Testing Program or the National Toxicology Program (NTP); from designations of the participating agencies; or from substances evaluated by the IARC Working Groups. Other substances from the same sources will be added to subsequent Annual Reports.

The Seventh Annual Report contains entries on the carcinogenicity of seven metals (arsenic, beryllium, cadmium, chromium, lead, nickel, and thorium). The entries for the individual metals identify those compounds of the metal (and, where appropriate, the elemental metal itself) for which evidence of carcinogenicity in environmentally-exposed humans or experimental animals is sufficient. Relatively few of the many different forms (elemental, salts, complexes, chelates, etc.) of the metals have been fully evaluated for carcinogenicity. the various factors that can influence the carcinogenic potential of a given metal form that should be considered include route of exposure, absorption, distribution, valence state, metabolism, elimination, as well as potential for specific biochemical interactions in cells. However, in the absence of specific information a metal shown to be carcinogenic in one of its forms should be considered as being potentially carcinogenic in its other forms.

This list identifies all the substances or groups of substances, and medical treatments that are listed as reasonably anticipated to be carcinogens. CAS Numbers for substances are derived from individual reports.

The data field contains the words "known carcinogen" and the generic group in which the chemical was included (if applicable).

OSHA - Final PELs - Time Weighted Averages   (OS1)
SOURCE

Department of Labor
Occupational Safety and Health Administration
29 CFR Part 1910 Subpart Z 1910.1000 Table Z-1 Limits for Air Contaminants
as published in the Federal Register June 30, 1993 (58 FR 35338)
NOTE: These replace the Vacated 1989 Air Contaminant Limits

LIST DESCRIPTION

The Occupational Safety and Health Act was effective on April 28, 1971. This Act was established to assure "safe and healthful working conditions for working men and women." The Act authorizes development of standards to assure both safety and health, including 1) setting standards for exposure to various chemical substance in the workplace, 2) listing permissible exposure limits for airborne contaminants that are not subject to such standards, and 3) informing employees of the dangers posed by substances with which they work.

The implementing regulations pertaining to health standards are contained in 29 CFR 1910 Subpart Z (Toxic and Hazardous Substances). Substances for which a specific standard exists are included in 29 CFR 1910.1001 to 1910.1101. Permissible Exposure Limits for Air Contaminants are listed in 29 CFR 1910.1000. The Hazard Communication Standard is defined at 29 CFR 1910.1200 and Occupational Exposures to Hazardous Chemicals in Laboratories is contained at 29 CFR 1910.1450.

On July 7, 1992, the Eleventh Circuit Court of Appeals issued a decision stating that the revised air contaminants standard was vacated. The new (1989) limits remained in effect and enforced past July 7, 1992. On October 23, 1992 the Eleventh Circuit denied OSHA's request for rehearing. After consideration, the Government decided not to petition the Supreme Court for review. OSHA directed its compliance officers to cease enforcing the new exposure limits (i.e. the 1989 limits) starting March 23, 1993. As of that date OSHA resumed enforcing, and employers were required to comply with, the air contaminant exposure limits that were in effect prior to the issuance of the new limits on January 19, 1989. These are the exposure limits (both airborne limits and skin designations) which are located in the "Transitional Limits" columns of 29 CFR 1910.1000 Table Z-1-A.

The Court of Appeals decision does not directly affect the 25 States operating their own OSHA-approved occupational safety andhealth State plans as their standards were adopted under authority of applicable State law. The 25 State plan States are: Alaska, Arizona, California, Connecticut, Hawaii, Indiana, Iowa, Kentucky, Maryland, Michigan, Minnesota, New Mexico, New York, Nevada, North Carolina, Oregon, Puerto Rico, South Carolina, Tennessee, Utah, Vermont, Virgin Islands, Virginia, Washington and Wyoming (Connecticut and New York are public sector only plans).

Some States are continuing to enforce the 1989 Air Contaminants standard or an "at least as effective" equivalent.

This group of lists contains the information described in the Federal Register Final Rule published June 30, 1993.

The data field contains the PEL TWA (Time Weighted Average) in ppm, mg/m3 and comments relating to the entry.

OSHA - Final PELs - Ceiling Limits   (OS1)

SOURCE

Department of Labor
Occupational Safety and Health Administration
29 CFR Part 1910 Subpart Z 1910.1000 Table Z-1 Limits for Air Contaminants
as published in the Federal Register June 30, 1993 (58 FR 35338)
NOTE: These replace the Vacated 1989 Air Contaminant Limits

LIST DESCRIPTION

The Occupational Safety and Health Act was effective on April 28, 1971. This Act was established to assure "safe and healthful working conditions for working men and women." The Act authorizes development of standards to assure both safety and health, including 1) setting standards for exposure to various chemical substance in the workplace, 2) listing permissible exposure limits for airborne contaminants that are not subject to such standards, and 3) informing employees of the dangers posed by substances with which they work.

The implementing regulations pertaining to health standards are contained in 29 CFR 1910 Subpart Z (Toxic and Hazardous Substances). Substances for which a specific standard exists are included in 29 CFR 1910.1001 to 1910.1101. Permissible Exposure Limits for Air Contaminants are listed in 29 CFR 1910.1000. The Hazard Communication Standard is defined at 29 CFR 1910.1200 and Occupational Exposures to Hazardous Chemicals in Laboratories is contained at 29 CFR 1910.1450.

On July 7, 1992, the Eleventh Circuit Court of Appeals issued a decision stating that the revised air contaminants standard was vacated. The new (1989) limits remained in effect and enforced past July 7, 1992. On October 23, 1992 the Eleventh Circuit denied OSHA's request for rehearing. After consideration, the Government decided not to petition the Supreme Court for review. OSHA directed its compliance officers to cease enforcing the new exposure limits (i.e. the 1989 limits) starting March 23, 1993. As of that date OSHA resumed enforcing, and employers were required to comply with, the air contaminant exposure limits that were in effect prior to the issuance of the new limits on January 19, 1989. These are the exposure limits (both airborne limits and skin designations) which are located in the "Transitional Limits" columns of 29 CFR 1910.1000 Table Z-1-A.

The Court of Appeals decision does not directly affect the 25 States operating their own OSHA-approved occupational safety andhealth State plans as their standards were adopted under authority of applicable State law. The 25 State plan States are: Alaska, Arizona, California, Connecticut, Hawaii, Indiana, Iowa, Kentucky, Maryland, Michigan, Minnesota, New Mexico, New York, Nevada, North Carolina, Oregon, Puerto Rico, South Carolina, Tennessee, Utah, Vermont, Virgin Islands, Virginia,

Washington and Wyoming (Connecticut and New York are public sector only plans).

Some States are continuing to enforce the 1989 Air Contaminants standard or an "at least as effective" equivalent.

This group of lists contains the information described in the Federal Register Final Rule published June 30, 1993. This list identifies those substances for which a Ceiling Limit is established.

The data field contains the Ceiling Limit in ppm, mg/m3 and comments relating to the entry.

OSHA - Final PELs - Skin Notations   (OS1)

SOURCE

Department of Labor
Occupational Safety and Health Administration
29 CFR Part 1910 Subpart Z 1910.1000 Table Z-1
  Limits for Air Contaminants
as published in the Federal Register June 30, 1993 (58 FR 35338)
NOTE: These replace the Vacated 1989 Air Contaminant Limits

LIST DESCRIPTION

The Occupational Safety and Health Act was effective on April 28, 1971. This Act was established to assure "safe and healthful working conditions for working men and women." The Act authorizes development of standards to assure both safety and health, including 1) setting standards for exposure to various chemical substance in the workplace, 2) listing permissible exposure limits for airborne contaminants that are not subject to such standards, and 3) informing employees of the dangers posed by substances with which they work.

The implementing regulations pertaining to health standards are contained in 29 CFR 1910 Subpart Z (Toxic and Hazardous Substances). Substances for which a specific standard exists are included in 29 CFR 1910.1001 to 1910.1101. Permissible Exposure Limits for Air Contaminants are listed in 29 CFR 1910.1000. The Hazard Communication Standard is defined at 29 CFR 1910.1200 and Occupational Exposures to Hazardous Chemicals in Laboratories is contained at 29 CFR 1910.1450.

On July 7, 1992, the Eleventh Circuit Court of Appeals issued a decision stating that the revised air contaminants standard was vacated. The new (1989) limits remained in effect and enforced past July 7, 1992. On October 23, 1992 the Eleventh Circuit denied OSHA's request for rehearing. After consideration, the Government decided not to petition the Supreme Court for review. OSHA directed its compliance officers to cease enforcing the new exposure limits (i.e. the 1989 limits) starting March 23, 1993. As of that date OSHA resumed enforcing, and employers were required to comply with, the air contaminant exposure limits that were in effect prior to the issuance of the new limits on January 19, 1989. These are the exposure limits (both airborne limits and skin

designations) which are located in the "Transitional Limits" columns of 29 CFR 1910.1000 Table Z-1-A.

The Court of Appeals decision does not directly affect the 25 States operating their own OSHA-approved occupational safety andhealth State plans as their standards were adopted under authority of applicable State law. The 25 State plan States are: Alaska, Arizona, California, Connecticut, Hawaii, Indiana, Iowa, Kentucky, Maryland, Michigan, Minnesota, New Mexico, New York, Nevada, North Carolina, Oregon, Puerto Rico, South Carolina, Tennessee, Utah, Vermont, Virgin Islands, Virginia, Washington and Wyoming (Connecticut and New York are public sector only plans).

Some States are continuing to enforce the 1989 Air Contaminants standard or an "at least as effective" equivalent.

This group of lists contains the information described in the Federal Register Final Rule published June 30, 1993. This list identifies those substances which contain a (skin) notation.

The data field contains the words "prevent or reduce skin absorption".

OSHA - Select Carcinogens   (—)

SOURCE

Department of Labor
Occupational Safety and Health Administration
29 CFR Part 1910.1450
Includes substances from the 6th NTP Report and IARC Monographs 1 through 57

LIST DESCRIPTION

The Occupational Safety and Health Act was effective on April 28, 1971. This Act was established to assure "safe and healthful working conditions for working men and women." The Act authorizes development of standards to assure both safety and health, including 1) setting standards for exposure to various chemical substance in the workplace, 2) listing permissible exposure limits for airborne contaminants that are not subject to such standards, and 3) informing employees of the dangers posed by substances with which they work.

The implementing regulations pertaining to health standards are contained in 29 CFR 1910 Subpart Z (Toxic and Hazardous Substances). Substances for which a specific standard exists are included in 29 CFR 1910.1001 to 1910.1101. Permissible Exposure Limits for Air Contaminants are listed in 29 CFR 1910.1000. The Hazard Communication Standard is defined at 29 CFR 1910.1200 and Occupational Exposures to Hazardous Chemicals in Laboratories is contained at 29 CFR 1910.1450.

29 CFR 1910.1450 is often called the "Laboratory Standard", although its correct designation is "Occupational Exposures to Hazardous Chemicals in Laboratories." It applies to all employers engaged in the laboratory use of hazardous chemicals as defined in that standard. (These definitions are quite specific and should be referred to for further information). Para-

graph (b) of the Laboratory Standard defines a "select carcinogen" as follows:

"Select carcinogen" means any substance which meets one of the following criteria:

(i) it is regulated by OSHA as a carcinogen; or
(ii) it is listed under the category, "known to be carcinogens" in the Annual Report on Carcinogens published by the National Toxicology Program (NTP) (latesteditions); or
(iii) it is listed under Group 1 ("carcinogenic to humans") by the International Agency for Research on Cancer Monographs (IARC) (latest edition) or
(iv) it is listed in either Group 2A or 2B by IARC or under the category "reasonably anticipated to be carcinogens" by NTP, AND causes statistically significant tumor incidence in experimental animals in accordance with any of the following criteria:

(A) after inhalation exposure of 6-7 hours per day, 5 days per week, for a significant portion of a lifetime to dosages of less than 10 mg/m3;
(B) after repeated skin application of less than 300 mg/kg of body weight per week; or
(C) after oral dosages of less than 50 mg/kg of body weight per day.

Thus, any chemical identified in (i), (ii) or (iii) above is automatically considered a "select carcinogen." ChemADVISOR has merged (eliminating duplicate entries) the materials identified by the lists described in (i), (ii) and (iii) above. Those materials identified only in sub-paragraph (iv) meet the definition of "select carcinogen" if the material is contained on the identified lists and meets the experimental criteria as outlined.

ChemADVISOR has merged (eliminating duplicate entries) the materials identified by the lists described in subparagraph (iv) above and arbitrarily named these "possible select carcinogens" for the purposes of this database. This addition of the word "possible" only serves to distinguish those substances which must be contained on the specified lists and meet the additional criteria specified in sub-paragraph (iv).

This list contains all the substances from sources identified in sub-paragraphs (i), (ii) and (iii) of the "select carcinogens" definition in 1910.1450(b). The data field contains the word "present" to indicate that it meets the criteria for a select carcinogen.

OSHA - Possible Select Carcinogens   (—)

SOURCE

Department of Labor
Occupational Safety and Health Administration
29 CFR Part 1910.1450
Includes substances from the 6th NTP Report and IARC Monographs 1 through 57

LIST DESCRIPTION

The Occupational Safety and Health Act was effective on April 28, 1971. This Act was established to assure

"safe and healthful working conditions for working men and women." The Act authorizes development of standards to assure both safety and health, including 1) setting standards for exposure to various chemical substance in the workplace, 2) listing permissible exposure limits for airborne contaminants that are not subject to such standards, and 3) informing employees of the dangers posed by substances with which they work.

The implementing regulations pertaining to health standards are contained in 29 CFR 1910 Subpart Z (Toxic and Hazardous Substances). Substances for which a specific standard exists are included in 29 CFR 1910.1001 to 1910.1101. Permissible Exposure Limits for Air Contaminants are listed in 29 CFR 1910.1000. The Hazard Communication Standard is defined at 29 CFR 1910.1200 and Occupational Exposures to Hazardous Chemicals in Laboratories is contained at 29 CFR 1910.1450.

29 CFR 1910.1450 is often called the "Laboratory Standard," although its correct designation is "Occupational Exposures to Hazardous Chemicals in Laboratories". It applies to all employers engaged in the laboratory use of hazardous chemicals as defined in that standard. (These definitions are quite specific and should be referred to for further information). Paragraph (b) of the Laboratory Standard defines a "select carcinogen" as follows:

"Select carcinogen" means any substance which meets one of the following criteria:

(i) it is regulated by OSHA as a carcinogen; or
(ii) it is listed under the category, "known to be carcinogens" in the Annual Report on Carcinogens published by the National Toxicology Program (NTP) (latest editions); or
(iii) it is listed under Group 1 ("carcinogenic to humans") by the International Agency for Research on Cancer Monographs (IARC) (latest edition); or
(iv) it is listed in either Group 2A or 2B by IARC or under the category "reasonably anticipated to be carcinogens" by NTP, AND causes statistically significant tumor incidence in experimental animals in accordance with any of the following criteria:

(A) after inhalation exposure of 6-7 hours per day, 5 days per week, for a significant portion of a lifetime to dosages of less than 10 mg/m3; or
(B) after repeated skin application of less than 300 mg/kg of body weight per week; or
(C) after oral dosages of less than 50 mg/kg of body weight per day.

Thus, any chemical identified in (i), (ii) or (iii) above is considered a "select carcinogen" without further qualification. ChemADVISOR has merged (eliminating duplicate entries) the materials identified by the lists described in (i), (ii) and (iii) above. Those materials identified only in sub-paragraph (iv) meet the definition of "select carcinogen" if the material is contained on the identified lists and meets the experimental criteria as outlined.

ChemADVISOR has merged (eliminating duplicate entries) the materials identified by the lists described in subparagraph (iv) above and arbitrarily termed these "possible select carcinogens" for the purposes of this database. The addition of the word "possible" only serves to distinquish those substances which must be contained on the specified lists and meet the additional criteria specified in sub-paragraph (iv).

This list contains all the substances identified in subparagraph (iv) of the "select carcinogens" definition in 1910.1450(b). The data field contains the word "present"to indicate its status as a possible select carcinogen under this standard.

OSHA - List of Highly Hazardous Chemicals (OS3)

SOURCE

Occupational Safety and Health Administration
29 CFR Part 1910 Subpart H
Section 1910.119
Process safety management of highly hazardous chemicals
July 1, 1992 edition of CFR

LIST DESCRIPTION

This addition to 29 CFR Part 1910 Subpart H adds a new section on process safety management of highly hazardous chemicals. This section contains requirements intended to eliminate the incidence or mitigate the consequences of highly hazardous chemical releases, fires, and explosions. This regulation accomplishes its goal by requiring a comprehensive management program: A holistic approach that integrates technologies, procedures, and management practices.

Workplaces covered include any activity conducted by an employer that involves a highly hazardous chemical including any use, storage, manufacturing, handling, processing, or movement or any combination of these activities.

The highly hazardous chemicals covered include any chemical listed in the mandatory Appendix A which is present at or above the specified threshold quantity. Appendix A is a compilation of highly hazardous chemicals that can cause a serious chemical accident, by toxicity, or reactivity, and a consequent serious danger to the employees in a workplace. Appendix A is based on information drawn from a variety of sources including among others:

The Environmental Protection Agency
The Department of Transportation
The World Bank
The National Fire Protection Association
The Health and Safety Commission of the United Kingdom
The States of Delaware and New Jersey

Every chemical in Appendix A is on at least one list compiled by these agencies and organizations as warranting a high degree of management control due to its extremely hazardous nature. Most of the chemicals are on several lists.

OSHA realizes that these lists vary in chemicals as well as quantities. Based on a review of these sources, OSHA has included those toxics and reactives it believes are most significant in potentially becoming a catastrophic event. OSHA has also sought todevelop a reasonable listing of threshold quantities based on a review of the data available that sufficiently address potential catastrophic amounts of chemicals.

In addition to the chemicals specified in Appendix A, the proposed standard would cover:

1) Processes involving flammable liquids or gases in quantities of 10,000 pounds or more (with several exceptions)
2) Manufacture of explosives
3) Manufacture of pyrotechnics including fireworks and flares
4) Newly developed toxic chemicals which meet certain specified criteria (i.e. the Substance Hazard Index)

This list includes the substances specified in Appendix A.

The data field contains the recommended threshold quantity.

OSHA - 1989 Vacated PELs - Skin Designation (OS1)

SOURCE

Department of Labor
Occupational Safety and Health Administration
29 CFR Part 1910 Subpart Z -1910.1000 to end
CFR Revised as of July 1, 1992
These values have no official status. The only enforceable contaminant levels are those listed as "OSHA Final PELs".

LIST DESCRIPTION

The Occupational Safety and Health Act was effective on April 28, 1971. This Act was established to assure "safe and healthful working conditions for working men and women." The Act authorizes development of standards to assure both safety and health, including 1) setting standards for exposure to various chemical substance in the workplace, 2) listing permissible exposure limits for airborne contaminants that are not subject to such standards, and 3) informing employees of the dangers posed by substances with which they work.

The implementing regulations pertaining to health standards are contained in 29 CFR 1910 Subpart Z (Toxic and Hazardous Substances). Substances for which a specific standard exists are included in 29 CFR 1910.1001 to 1910.1101. Permissible Exposure Limits for Air Contaminants are listed in 29 CFR 1910.1000. The Hazard Communication Standard is defined at 29 CFR 1910.1200 and Occupational Exposures to Hazardous Chemicals in Laboratories is contained at 29 CFR 1910.1450.

This group of lists contains the information described in 29 CFR 1910.1000 - Table Z-1-A "Limits for Air Contaminants." This list contains substances for which OSHA has given a "skin" designation. The "skin" designation is defined in 1910.1000(a)(4) "to prevent or reduce skin absorption, an employee's skin exposure to substances listed in Table Z-1-A with an "X" in one or both of the Skin Designation columns following the substance name shall be prevented or reduced to the extent necessary in the circumstances through the use of gloves, coveralls, goggles, or other appropriate personal protective equipment, engineering controls or work practices." These values appear in Table Z-1-A under "Final Rule Limits - Skin Designation."

The data field contains the statement "prevent or reduce skin absorption."

OSHA also notes that employers in General Industry (i.e. thosecovered by 29 CFR 1910) may use any combination of controls to achieve these limits until December 31, 1992 as set forth in 29 CFR 1910.1000(f).

OSHA - 1989 Vacated PELs - Ceiling Limits   (OS1)
SOURCE

Department of Labor
Occupational Safety and Health Administration
29 CFR Part 1910 Subpart Z - 1910.1000
CFR Revised as of July 1, 1992
These values have no official status. The only enforceable contaminant levels are those listed as "OSHA Final PELs".

LIST DESCRIPTION

The Occupational Safety and Health Act was effective on April 28, 1971. This Act was established to assure "safe and healthful working conditions for working men and women." The Act authorizes development of standards to assure both safety and health, including 1) setting standards for exposure to various chemical substance in the workplace, 2) listing permissible exposure limits for airborne contaminants that are not subject to such standards, and 3) informing employees of the dangers posed by substances with which they work.

The implementing regulations pertaining to health standards are contained in 29 CFR 1910 Subpart Z (Toxic and Hazardous Substances). Substances for which a specific standard exists are included in 29 CFR 1910.1001 to 1910.1101. Permissible Exposure Limits for Air Contaminants are listed in 29 CFR 1910.1000. The Hazard Communication Standard is defined at 29 CFR 1910.1200 and Occupational Exposures to Hazardous Chemicals in Laboratories is contained at 29 CFR 1910.1450.

This group of lists contains the information described in 29 CFR 1910.1000 - Table Z-1-A "Limits for Air Contaminants." This list contains substances for which OSHA has established a Ceiling. A Ceiling is defined in 1910.1000(a)(5)(iii) as "an employee's exposure which shall not be exceeded during any part of the work day. If instantaneous monitoring is not feasible, then the ceiling shall be assessed as a 15-minute time weighted average exposure which shall not be exceeded at any time over a working day." These values appear in Table Z-1-A under "Final Rule Limits - Ceiling."

The data field contains the letter "C" to denote Ceiling, the ceiling value in ppm and mg/m3 and comments (if any) used by OSHA. OSHA defines these units as:

ppm = Parts of vapor or gas per million parts of contaminated air by volume at 25 deg. C and 760 torr.

mg/m3 = Approximate milligrams of substance per cubic meter of air.

OSHA also notes that employers in General Industry (i.e. those covered by 29 CFR 1910) may use any combination of controls to achieve these limits until December 31, 1992 as set forth in 29 CFR 1910.1000(f).

OSHA - 29 CFR 1910 Specifically Regulated   (OS1) Chemicals
SOURCE

Department of Labor
Occupational Safety and Health Administration
29 CFR Part 1910 Subpart Z - 1910.1001 to 1101
amended by 57 FR 24310 (Asbestos), 57 FR 35630 (4,4' - MDA), 57 FR 42102 (Cadmium)
NOTE: This portion of the OSHA regulations was not affected by the court decision vacating the 1989 Air Contaminant levels.

LIST DESCRIPTION

The Occupational Safety and Health Act was effective on April 28, 1971. This Act was established to assure "safe and healthful working conditions for working men and women". The Act authorizes development of standards to assure both safety and health, including 1) setting standards for exposure to various chemical substance in the workplace, 2) listing permissible exposure limits for airborne contaminants that are not subject to such standards, and 3) informing employees of the dangers posed by substances with which they work.

The implementing regulations pertaining to health standards are contained in 29 CFR 1910 Subpart Z (Toxic and Hazardous Substances). Substances for which a specific standard exists are included in 29 CFR 1910.1001 to 1910.1048. Permissible Exposure Limits for Air Contaminants are listed in 29 CFR 1910.1000. The Hazard Communication Standard is defined at 29 CFR 1910.1200 and Occupational Exposures to Hazardous Chemicals in Laboratories is contained at 29 CFR 1910.1450.

This list contains only those substances for which a substance specific regulation exists (i.e., 29 CFR 1910.1001 to 1048) through the date listed above.

Generally each specifically regulated substance includes:
1) A definition of the specific chemical or agent
2) A Permissible Exposure Limit and Action level

3) Requirements for Exposure Monitoring

4) Definition of Regulated Areas

5) Description of Methods of Compliance (i.e. engineering controls, work practices)

6) Requirements for Respiratory Protection

7) Requirements for Protective equipment and clothing

8) Handling Emergencies

9) Medical Surveillance

10) Hazard Communication (including labels)

The data field indicates the established Exposure Limit, Action Level, specific labeling which is applicable to the substance, and the CFR citation where additional requirements (such as exposure monitoring) are contained.

OSHA - 1989 Vacated PELs - Time Weighted (OS1) Averages

SOURCE

Department of Labor

Occupational Safety and Health Administration

29 CFR Part 1910 Subpart Z -1910.1000 to end

With amendments and corrections published in Federal Register July 1, 1992 (57 FR 29204)

These values have no official status. The only enforceable contaminant levels are those listed as "OSHA Final PELs".

LIST DESCRIPTION

The Occupational Safety and Health Act was effective on April 28, 1971. This Act was established to assure "safe and healthful working conditions for working men and women." The Act authorizes development of standards to assure both safety and health, including 1) setting standards for exposure to various chemical substance in the workplace, 2) listing permissible exposure limits for airborne contaminants that are not subject to such standards, and 3) informing employees of the dangers posed by substances with which they work.

The implementing regulations pertaining to health standards are contained in 29 CFR 1910 Subpart Z (Toxic and Hazardous Substances). Substances for which a specific standard exists are included in 29 CFR 1910.1001 to 1910.1101. Permissible Exposure Limits for Air Contaminants are listed in 29 CFR 1910.1000. The Hazard Communication Standard is defined at 29 CFR 1910.1200 and Occupational Exposures to Hazardous Chemicals in Laboratories is contained at 29 CFR 1910.1450.

On July 1, 1992 OSHA published a final rule correcting 29 CFR 1910.1000. The corrections made include:

1) correcting the nomenclature for the bivalent and trivalent chromium compounds back to that of the original standard.

2) corrections to the nomenclature for the forms of crystalline silica.

3) addition of a footnote clarifying that all inert or nuisance dusts, whether mineral, inorganic or organic are covered by the Particulate Not

Otherwise Regulated (PNOR) limit in Table Z-1-A.

4) correction of footnote to change date to December 31, 1993 for requirements of 29 CFR1910.1000(f)(2)(ii).

5) footnote b is restated to be clearer.

6) footnote added to the carbon monoxide ceiling entry reflecting OSHA's enforcement policy that it is appropriate to monitor the 200 ppm ceiling over a 5 minute period, with an instantaneous ceiling of 1500 ppm (the IDLH level).

This group of lists contains the information described in corrected 29 CFR 1910.1000 - Table Z-1-A "Limits for Air Contaminants". This list contains substances for which OSHA has established a Permissible Exposure Limit - 8-hour Time Weighted Average (TWA). These values appear in Table Z-1-A under "Final Rule Limits - TWA."

The data field contains the TWA value in ppm and mg/m3 and comments (if any) used by OSHA. OSHA defines these units as:

ppm = Parts of vapor or gas per million parts of contaminated air by volume at 25 deg. C and 760 torr.

mg/m3 = Approximate milligrams of substance per cubic meter of air.

OSHA also notes that employers in General Industry (i.e. those covered by 29 CFR 1910) may use any combination of controls to achieve these limits until December 31, 1993 (corrected per item 4 above) as set forth in 29 CFR 1910.1000(f).

OSHA - 1989 Vacated PELs - Short Term (OS1) Exposure Limits

SOURCE

Department of Labor

Occupational Safety and Health Administration

29 CFR Part 1910 Subpart Z -1910.1000 to end

with corrections published in Federal Register July 1, 1992 (57 FR 29204)

These values have no official status. The only enforceable contaminant levels are those listed as "OSHA Final PELs".

LIST DESCRIPTION

The Occupational Safety and Health Act was effective on April 28, 1971. This Act was established to assure "safe and healthful working conditions for working men and women." The Act authorizes development of standards to assure both safety and health, including 1) setting standards for exposure to various chemical substance in the workplace, 2) listing permissible exposure limits for airborne contaminants that are not subject to such standards, and 3) informing employees of the dangers posed by substances with which they work.

The implementing regulations pertaining to health standards are contained in 29 CFR 1910 Subpart Z (Toxic and Hazardous Substances). Substances for which a specific standard exists are included in 29 CFR 1910.1001 to 1910.1101. Permissible Exposure Limits for Air Contaminants are listed in 29 CFR 1910.1000. The Hazard Communication Standard is defined at 29 CFR 1910.1200 and Occupational Exposures to Hazardous Chemicals in Laboratories is contained at 29 CFR 1910.1450.

On July 1, 1992 OSHA published a final rule correcting 29 CFR 1910.1000. The corrections made include:

1) correcting the nomenclature for the bivalent and trivalent chromium compounds back to that of the original standard.
2) corrections to the nomenclature for the forms of crystalline silica.
3) addition of a footnote clarifying that all inert or nuisance dusts, whether mineral, inorganic or organic are covered by the Particulate Not Otherwise Regulated (PNOR) limit in Table Z-1-A.
4) correction of footnote to change date to December 31, 1993 for requirements of 29 CFR1910.1000(f)(2)(ii).
5) footnote b is restated to be clearer.
6) footnote added to the carbon monoxide ceiling entry reflecting OSHA's enforcement policy that it is appropriate to monitor the 200 ppm ceiling over a 5 minute period, with an instantaneous ceiling of 1500 ppm (the IDLH level).

This group of lists contains the information described in 29 CFR 1910.1000 - Table Z-1-A "Limits for Air Contaminants". This list contains substances for which OSHA has established a Short Term Exposure Limit - STEL (Duration is for 15 minutes unless otherwise noted). These values appear in Table Z-1-A under "Final Rule Limits - STEL."

The data field contains the STEL value in ppm and mg/m3 and comments (if any) by OSHA. OSHA defines these units as:

ppm = Parts of vapor or gas per million parts of contaminated air by volume at 25 deg. C and 760 torr.

mg/m3 = Approximate milligrams of substance per cubic meter of air.

OSHA also notes that employers in General Industry (i.e. those covered by 29 CFR 1910) may use any combination of controls to achieve these limits until December 31, 1993 as set forth in 29 CFR 1910.1000(f).

## AIHA - Odor Threshold Values						(OTV)
SOURCE

Odor Thresholds for Chemicals with Established Occupational Health Standards - 1989 report
American Industrial Hygiene Association
2700 Prosperity Avenue
Suite 250
Fairfax, Virginia 22031
NOTE: new address December 1994

LIST DESCRIPTION

This report was prepared by the American Industrial Hygiene Association, a non-profit professional association, as a necessary first step in developing better odor threshold reference information for chemicals with Threshold Limit Values (TLVs). Phase I of the project identified primary experimental odor threshold determinations in the literature; Phase II evaluated odor threshold methodology against a set of objective criteria. A best estimate either of detection or of recognition odor threshold then was developed using the geometric mean of odor threshold values from sources accepted in Phase II critique.

The geometric mean value or recommended best estimate for the odor threshold for each of the compounds for which there is a published Threshold Limit Value (1986-87) is given in Table 5.1 of the report. The Geometric means were computed for the mean odor threshold values. This is a common practice in sensory evaluation as it accounts for the wide range of response over several orders of magnitude. The means were rounded off to two significant digits. Where values were given as a range, the geometric mean of the two points was taken for the threshold.

In some cases, the mean value for detection is higher than the mean value for recognition. This is a result of pooling of several data sets for the geometric mean. Of the 182 compounds which were evaluated in the study only 110 compounds had odor threshold values that meet the evaluation criteria.

This list contains all of the geometric mean air odor thresholds, both detectable and recognizable, which are identified in Table 5.1 of the report.

The data field contains the mean value in ppm for recognizable and detectable levels. For those compounds which were evaluated but for which the data did not meet the evaluation criteria the statement "no geometric mean air odor threshold" is contained in the data field.

## AIHA - WEEL - Skin Absorption Designations	(WEL)
SOURCE

American Industrial Hygiene Association
2700 Prosperity Ave. Suite 250
Fairfax, Virginia 22031
Telephone (703) 849-8888

Prepared by the Workplace Environmental Exposure Levels Committee. Guides current as of December 22, 1993.

LIST DESCRIPTION

The American Industrial Hygiene Association is an organization of industrial hygienists whose professional interests include occupational and environmental health. The AIHA is a professional society, not a government agency.

Workplace Environmental Exposure Level (WEEL) Guides are developed by the AIHA WEEL Committee. This Committee investigates agents that have no current exposure guidelines established by other organizations. They represent the workplace exposure levels to which, it is believed, nearly all employees could be repeatedly exposed without adverse effects. All WEELs are expressed as either time-weighted average (TWA) concentrations or ceiling values; however, different time periods are specified depending on the properties of the agent.

An 8-hour TWA indicates a time-weighted average concentration for a normal 8-hour workday and 40-hour workweek. A ceiling limit should not be exceeded at any time. When it is believed that excursion levels should be more limited, a shorter duration TWA may be recommended, either in conjunction with or in place of an 8-hour TWA value. The time specified is relevant to exposure, not necessarily to sampling.

This list includes all the substances for which AIHA has assigned a "skin" designation. The addition of the word "skin" indicates that the material may be absorbed in toxicologically significant amounts through the skin. Skin contact can contribute to the overall exposure and invalidate the TWA exposure evaluations.

The data field indicates this evaluation with the words "skin absorber."

AIHA - WEEL - Ceilings or Short Term Time (WEL) Weighted Averages

SOURCE

American Industrial Hygiene Association
2700 Prosperity Avenue, Suite 250
Fairfax, Virginia 22031
Telephone (703) 849-8888

Prepared by the Workplace Environmental Exposure Levels Committee. Guides current as of December 22, 1993

LIST DESCRIPTION

The American Industrial Hygiene Association is an organization of industrial hygienists whose professional interests include occupational and environmental health. The AIHA is a professional society, not a government agency.

Workplace Environmental Exposure Level (WEEL) Guides are developed by the AIHA WEEL Committee. This Committee investigates agents that have no current exposure guidelines established by other organizations. They represent the workplace exposure levels to which, it is believed, nearly all employees could be repeatedly exposed without adverse effects. All WEELs are expressed as either time-weighted average (TWA) concentrations or ceiling values; however, different time periods are specified depending on the properties of the agent.

An 8-hour TWA indicates a time-weighted average concentration for a normal 8-hour workday and 40-hour workweek. A ceiling limit should not be exceeded at any time. When it is believed that excursion levels should be more limited, a shorter duration TWA may be recommended, either in conjunction with or in place of an 8-hour TWA value. The time specified is relevant to exposure, not necessarily to sampling.

This list includes all the substances for which AIHA has assigned a ceiling (C) or Short Term TWA (15 minutes).

The data field indicates the limit in ppm or mg/m3. Short Term Limits are for 15 minute exposures TWA plus any comments on the entry added by AIHA.

AIHA - WEEL - Time Weighted Averages (WEL)

SOURCE

American Industrial Hygiene Association
2700 Prosperity Ave., Suite 250
Fairfax, Virginia 22031
Telephone (703) 849-8888

Prepared by the Workplace Environmental Exposure Levels Committee. Guides current as of December, 1994 (set 17)

LIST DESCRIPTION

The American Industrial Hygiene Association is an organization of industrial hygienists whose professional interests include occupational and environmental health. The AIHA is a professional society, not a government agency.

Workplace Environmental Exposure Level (WEEL) Guides are developed by the AIHA WEEL Committee. This Committee investigates agents that have no current exposure guidelines established by other organizations. They represent the workplace exposure levels to which, it is believed, nearly all employees could be repeatedly exposed without adverse effects. All WEELs are expressed as either time-weighted average (TWA) concentrations or ceiling values; however, different time periods are specified depending on the properties of the agent.

An 8-hour TWA indicates a time-weighted average concentration for a normal 8-hour workday and 40-hour workweek. A ceiling limit should not be exceeded at any time. When it is believed that excursion levels should be more limited, a shorter duration TWA may be recommended, either in conjunction with or in place of an 8-hour TWA value. The time specified is relevant to exposure, not necessarily to sampling.

This list includes all the substances for which AIHA has assigned a 8-hour Time-Weighted Average workplace environmental exposure limit.

The data field indicates the limit in ppm and/or mg/m3 TWA plus any comments on the entry added by AIHA.

NFPA - National Fire Protection Association - (NFP) Flash Points

SOURCE

Fire Protection Guide on Hazardous Materials
Eleventh Edition
NFPA 325 Fire Hazard Properties of Flammable Liquids, Gases, Volatile Solids 1994, pages 325-10 to 325-94
National Fire Protection Association
1 Batterymarch Park
P.O. Box 9101
Quincy, MA 02269-9101

LIST DESCRIPTION

The National Fire Protection Association was organized in 1896 to promote the science and improve the methods of fire protection and fire prevention. It is a nonprofit, educational, voluntary-membership organization recognized internationally as a
clearinghouse for information on fire prevention, fire fighting procedures and means of fire protection. It is also recognized internationally as an authoritative source for fire-loss
experience.

The 1994 edition (eleventh edition) is an amended version of the 1991 edition.

The information in NFPA 325 is not a code, standard or recommended practice, as these terms are defined by NFPA. It is only a compilation of basic fire protection properties of various materials, prefaced by an explanation of the properties covered. the data contained have been collected from numerous authoritative sources, including the U.S.Bureau of Mines, Factory Mutual Research Corporation, and Underwriters Laboratories, as well as from the manufacturers of the materials. The originating source of the data is on file at NFPA headquarters and may be obtained upon request.

The values for any given property are representative figures deemed suitable for general use. Where differences exist in reference sources, the value selected for inclusion in this compilation is conservative. Slight differences are to be expected even between the most competent agencies owning to differences in the purity of the samples tested, and minor differences in technical manipulation and observation. Where there is difference of opinion as to the actual value of a property of a given material or where the validity of the data presented is questioned, further tests should then be conducted on representative samples of the specific material in question by a qualified testing laboratory.

Flash point of a liquid is the minimum temperature at which the liquid gives off sufficient vapor to form an ignitible mixture with the air near the surface of the liquid or within the test vessel used. By ignitable mixture is meant a mixture within the flammable range (between upper and lower limits) that is capable of the propagation of flame away from the source of ignition.

In general, there is little information in the literature on the degree of purity of the material tested and it is assumed that there is little industrial use of high purity

materials. The flash point figure represents closed cup tests with very few exceptions. For older data, there may be no indication as to the method used.

For further information on the flash point test procedures used, see NFPA 321, Standard on Basic Classification of Flammable and Combustible Liquids.

This list contains all the entries for which NFPA has provided some information on Flash Point. NFPA has not included Chemical Abstracts Service (CAS) Numbers with the 325-1994 Table. ChemADVISOR has searched RTECS and the TSCA Inventory to provide the CAS Numbers identified here. It should be understood that NFPA specifically does not include CAS Numbers with its list because it recognizes that industrial materials seldom are pure materials and that CAS Numbers are the tools of chemists, not fire protection personnel. We recognize that NFPA has designed their information specifically with fire protection personnel in mind, but the usefulness and reliability of the NFPA information has reached a much broader audience which now includes government regulators, industry regulatory compliance personnel, environmental assessment personnel, chemical laboratories (which in many cases may be using the high purity form of the chemical) and the general population of employers and employees who obtain information about chemicals from Material Safety Data Sheets. Much of this expansion is due to regulatory requirements which use CAS Numbers for the careful identification of materials. ChemADVISOR has supplied CAS Numbers to the NFPA listing to assist this group of users.

The data supplied in this list includes only Flash Point information (Hazard Identification information is provided in a separate list). This information is generally needed for compliance with various regulations, but must be used with the qualifications stated by NFPA in the above paragraphs. The data includes the Flash Point Temperature in degrees Fahrenheit and degrees Celsius. The Celsius degrees have been rounded off by NFPA to the nearest degree.

For entries containing the statement "* see list description" NFPA has provided two list entries for the same CAS Numbered chemical. Some of these entries have similar data and are not noted here. Where there are significant differences in the data we have noted the alternative data here.

CAS 584-03-2 Alpha Butylene Glycol: under 1,2 Butanediol the flash point is given as 104(40).

CAS 2050-60-4 Butyl Oxalate: under Dibutyl Oxalate the flash point in given as 220(104).

CAS 112-34-5 Diethylene Glycol Monobutyl Ether: under Diethylene Glycol n-Butyl Ether the flash point is given as 230(110).

CAS 103-23-1 Dioctyl Adipate: under Di-2-Ethylhexyl Adipate the flash point is given as 385(196).

CAS 111-15-9 Ethylene Glycol Monomethyl Ether Acetate: under 2-Ethoxyethyl Acetate the flash point is given as 117(47).

CAS 818-61-1 2-Hydroxyethyl Acrylate: under Ethylene Glycol Monoacrylate the flash point is given as 220 (104).

CAS 94-96-2 Octylene Glycol: under 2-Ethyl-1,3-Hexanediol the flash point is given as 260(127).

CAS 107-41-5 2-Methyl-2,4-Pentanediol: under Hexylene Glycol the flash point is given as 215(102).

CAS 8008-20-6 Fuel Oil No.1: under Jet Fuels JP-5 the flash point is given as 95-145(35-63) and under Ultrasene the flash point is given as 175(79).

CAS 110-49-6 Ethylene Glycol Monomethyl Ether Acetate: under Methyl Cellosolve Acetate and Methyl Glycol Acetate the flash point is given as 111(44).

CAS 6032-29-7 Sec-Amyl Alcohol: under Methyl Propyl Carbinol the flash point is given as 105(41).

CAS 122-99-6 Ethylene Glycol Phenyl Ether: under Phenoxy Ethyl Alcohol the flash point is given as 250(121).

CAS 112-70-9 Tridecanol: under Tridecyl alcohol the flash point is given as 180(82).

CAS 540-84-1 Isooctane: under Trimethylpentane the flash point is given as 10(-12).

NFPA - National Fire Protection Association - (NFP) Hazard Identification

SOURCE

Fire Protection Guide on Hazardous Materials
Eleventh Edition
NFPA 325 Fire Hazard Properties of Flammable Liquids, Gases, Volatile Solids 1994, pages 325-10 to 325-94
National Fire Protection Association
1 Batterymarch Park
P.O. Box 9101
Quincy, MA 02269-9101

LIST DESCRIPTION

The National Fire Protection Association was organized in 1896 to promote the science and improve the methods of fire protection and fire prevention. It is a nonprofit, educational, voluntary-membership organization recognized internationally as a clearinghouse for information on fire prevention, fire fighting procedures and means of fire protection. It is also recognized internationally as an authoritative source for fire-loss experience. The 1994 edition (eleventh edition) is an amended version of the 1991 edition.

The information in NFPA 325 is not a code, standard or recommended practice, as these terms are defined by NFPA. It is only a compilation of basic fire protection properties of various materials, prefaced by an explanation of the properties covered. the data contained have been collected from numerous authoritative sources, including the U.S.Bureau of Mines, Factory Mutual Research Corporation, and Underwriters Laboratories, as well as from the manufacturers of the materials. The originating source of the data is on file at NFPA headquarters and may be obtained upon request.

The increased use of chemicals, many of which introduce hazards other than flammability, led to the need for a simple hazard identification system that could be immediately recognized by emergency response personnel. This led to the development of NFPA 704 Hazard Identification System, otherwise known as the NFPA diamond. This system is completely described in NFPA 704 -Recommended System for the Identification of the Fire Hazard of Materials. This system provides simple, readily recognized and easily understood markings that give, at a glance, a general idea of the inherent hazards of the material and the order of severity of these hazards as they relate to fire protection, exposure, and control. The system's objectives are to provide an appropriate alert signal and on-the-spot information to safeguard the lives of both public and private emergency response personnel. Thesystem also assists in planning for effective fire fighting operations and may be used by plant design engineers and plant protection and safety personnel.

The system identifies the hazards of a material in terms of three categories - Health, Flammability, and Reactivity. It indicates the order of severity of these hazards by means of a numerical rating of 0, indicating no special hazard, to 4, indicating extreme hazard. The three hazard categories were selected after studying about 35 inherent and environmental hazards of materials that could affect fire fighting operations. The five degrees of hazard were decided upon as necessary to give the required information.

The following commentary on degrees of hazard are an interpretation of the information contained within NFPA 704 and are related specifically to the fire fighting aspects. Refer to NFPA 704 for a detailed discussion of the identification system.

NOTE: The hazard identification rating definitions below and the actual ratings in the text are based on definitions from the 1985 edition of NFPA 704. Materials have not been rated using the new definitions appearing in the 1990 edition of NFPA 704.

HEALTH HAZARD RATING

In general, health hazard in fire fighting is that of a single exposure which may vary from a few seconds up to an hour. Only hazards arising out of an inherent property of the material are considered. The following explanation is based upon protective equipment normally used by fire fighters.

4 Materials too dangerous to health to expose fire fighters. A few whiffs of the vapor could cause death or the vapor or liquid could be fatal on penetrating the fire fighter's normal full protective clothing. The normal full protective clothing and breathing apparatus available to the average fire department will not provide adequate protection against inhalation or skin contact with these materials.

3 Materials extremely hazardous to health but areas may be entered with extreme care. Full protective clothing, including self-contained breathing apparatus, coat, pants, gloves, boots, and bands around legs, arms and waist should be provided. No skin surface should be exposed.

2 Materials hazardous to health, but areas may be entered freely with full-faced mask self-contained breathing apparatus which provides eye protection.

1 Materials only slightly hazardous to health. It may be desirable to wear self-contained breathing apparatus.

0 Materials which on exposure under fire conditions would offer no hazard beyond that of ordinary combustible material.

FLAMMABILITY HAZARD RATING

Susceptibility to burning is the basis for assigning degrees within this category. The method of attacking the fire is influenced by this susceptibility factor.

4 Very flammable gases or very volatile flammable liquids. Shut off flow and keep cooling water streams on exposed tanks or containers.

3 Materials which can be ignited under almost all normal temperature conditions. Water may be ineffective because of the low flash point.

2 Materials which must be moderately heated before ignition will occur. Water spray may be used to extinguish the fire because the material can be cooled below its flash point.

1 Materials that must be preheated before ignition can occur. Water may cause frothing if it gets below the surface of the liquid and turns to steam. However, water fog gently applied to the surface will cause a frothing which will extinguish the fire.

0 Materials that will not burn.

REACTIVITY HAZARD RATING

The assignment of degrees in the reactivity category is based upon the susceptibility of materials to release energy either by themselves or in combination with water. Fire exposure was one of the factors considered along with conditions of shock and pressure.

4 Materials which (in themselves) are readily capable of detonation or of explosive decomposition or explosive reaction at normal temperatures and pressures. Includes materials which are sensitive to mechanical or localized thermal shock. If a chemical with this hazard rating is in an advanced or massive fire, the area should be evacuated.

3 Materials which (in themselves) are capable of detonation or of explosive decomposition or of explosive reaction but which require a strong initiating source or which must be heated under confinement before initiation. Includes materials which are sensitive to thermal or mechanical shock at elevated temperatures and pressure or which react explosively with water without requiring heat or confinement. Fire fighting should be done from an explosion-resistant location.

2 Materials which (in themselves) are normally unstable and readily undergo violent chemical change but do not detonate. Includes materials which can undergo chemical change with rapid release of energy at normal temperatures and pressures or which can undergo violent chemical change at elevated temperatures and pressures.

Also includes those materials which may react violently with water or which may form potentially explosive mixtures with water. In advanced or massive fires, fire fighting should be done from a safe distance or from a protected location.

1 Materials which (in themselves) are normally stable but which may become unstable at elevated temperatures and pressures or which may react with water with some release of energy but not violently. Caution must be used in approaching the fire and applying water.

0 Materials which (in themselves) are normally stable even under fire exposure conditions and which are not reactive with water. Normal fire fighting procedures may be used.

This list contains all the entries for which NFPA has provided some information on Hazard Identification (i.e. hazard ratings for one or more of the categories). NFPA has not included Chemical Abstracts Service (CAS) Numbers with the 325M-1984 Table.

ChemADVISOR has searched RTECS and the TSCA Inventory to provide the CAS Numbers identified here. It should be understood that NFPA specifically does not include CAS Numbers with its list because it recognizes that industrial materials seldom are pure materials and that CAS Numbers are the tools of chemists, not fire protection personnel. We recognize that NFPA has designed their information specifically with fire protection personnel in mind, but the usefulness and reliability of the NFPA information has reached a much broader audience which now includes government regulators, industry regulatory compliance personnel, environmental assessment personnel, chemical laboratories (which in many cases may be using the high purity form of the chemical) and the general population of employers and employees who obtain information about chemicals from Material Safety Data Sheets. Much of this expansion is due to regulatory requirements which use CAS Numbers for the careful identification of materials. ChemADVISOR has supplied CAS Numbers to the NFPA listing to assist this group of users.

The data supplied in this list includes the information contained in the columns under Suggested Hazard Identification. Where adequate information was not available to NFPA (i.e. spaces were left blank in the NFPA listing), no entry is made for that category.

For entries containing the statement "*See list description" NFPA has provided two list entries for the same CAS Numbered chemical. Some of these entries have similar data and are not noted here. Where there are significant differences in the data we have noted the alternative data here.

CAS 111-69-3 Adiponitrile: under Adipyldinitrile the hazard identification ratings are listed as health-4; flammability-2

CAS 622-08-2 Ethylene Glycol Monobenzyl Ether: under Glycol Benzyl Ether the hazard identification ratings are listed as health-0; flammability-1; reactivity-0

CAS 123-31-9 P-Dihydroxybenzene: under Hydroquinone the hazard identification ratings are listed as health-2; flammability-1; reactivity-0

FDA - Controlled Substances Act - Precursor (FDA) Chemicals

SOURCE

Food and Drug Administration
Controlled Substances Act, as amended and the Controlled Substances Import and Export Act
21 CFR Part 1310
21 CFR 1310.02(a) Listed Precursor Chemicals

LIST DESCRIPTION

The Food and Drug Administration has been given authority under the Controlled Substances Act and the Controlled Substances Import and Export Act to require certain records to be kept by regulated persons who participate in regulated transactions involving listed chemicals (precursors or essential chemicals).

Regulated persons include any individual, corporation, partnership, association, or other legal entity who manufactures, distributes, imports, or exports a listed chemical, a tableting machine, or an encapsulating machine.

Records of the transactions for precursors shall be maintained for 4 years and records of transactions for essential chemicals shall be kept for 2 years. Any regulated transaction involving an extraordinary quantity of a listed chemical, an uncommon method of payment or delivery, or any other circumstance that the regulated person believes may indicate that the listed chemical will be used in violation of this Act must be reported.

Also any unusual or excessive loss or disappearance of a listed chemical under the control of the regulated person must be reported. If a regulated material is lost in-transit the supplier must report the loss.

No transactions are permitted with a person whose description or other identifying characteristic the Drug Enforcement Administration (DEA) has previously provided to the regulated person.

Reports are to be made to the Special Agent in Charge of the DEA Divisional Office for the area in which the regulated person making the report is located.

This list contains all the chemicals in 21 CFR 1310.02(a). ChemADVISOR has added Chemical Abstracts Registry Numbers where applicable.

The data field contains the quantitative threshold or the cumulative amount for multiple transactions within a calendarmonth to be utilized in determining whether a receipt, sale, importation or exportation is a regulated transaction.

FDA - Controlled Substances Act - Essential (FDA) Chemicals

SOURCE

Food and Drug Administration
Controlled Substances Act, as amended and the Controlled Substances Import and Export Act
21 CFR Part 1310
21 CFR 1310.02(b) Listed Essential Chemicals

LIST DESCRIPTION

The Food and Drug Administration has been given authority under the Controlled Substances Act and the Controlled Substances Import and Export Act to require certain records to be kept by regulated persons who participate in regulated transactions involving listed chemicals (precursors or essential chemicals).

Regulated persons include any individual, corporation, partnership, association, or other legal entity who manufactures, distributes, imports, or exports a listed chemical, a tableting machine, or an encapsulating machine.

Records of the transactions for precursors shall be maintained for 4 years and records of transactions for essential chemicals shall be kept for 2 years. Any regulated transaction involving an extraordinary quantity of a listed chemical, an uncommon method of payment or delivery, or any other circumstance that the regulated person believes may indicate that the listed chemical will be used in violation of this Act must be reported.

Also any unusual or excessive loss or disappearance of a listed chemical under the control of the regulated person must be reported. If a regulated material is lost in-transit the supplier must report the loss.

No transactions are permitted with a person whose description or other identifying characteristic the Drug Enforcement Administration (DEA) has previously provided to the regulated person.

Reports are to be made to the Special Agent in Charge of the DEA Divisional Office for the area in which the regulated person making the report is located.

This list contains all the chemicals in 21 CFR 1310.02(b). ChemADVISOR has added Chemical Abstracts Registry Numbers where applicable.

The data field contains the quantitative threshold or the cumulative amount for multiple transactions within a calendarmonth to be utilized in determining whether a receipt, sale, importation or exportation is a regulated transaction for a domestic sale and for imports and exports.

ENVIRONMENTAL LISTS

**CERCLA/SARA - Section 302 Extremely Haz-   (302)
ardous Substances and TPQs**
SOURCE

Environmental Protection Agency
Comprehensive Environmental Response, Compensation, and Liability Act
Superfund Amendments and Reauthorization Act
40 CFR Subchapter J - Superfund, Emergency Planning, and Community Right-to-Know Programs.
Part 355 - Emergency Planning and Notification
Appendix A
40 CFR 355 Appendix A
with amendments through October 12, 1994 (59 FR
51816) which deleted CAS Numbers 1642-54-2
(diethylcarbamazine citrate), 122-14-5 (fenitrothion), 1314-56-3 (phosphorus pentoxide), 13494-80-
9 (tellurium) and changed TPQ of Isophorone Diisocyanate (CAS 4098-71-9) to 1,000 pounds.

LIST DESCRIPTION

The Comprehensive Environmental Response, Compensation, and Liability Act of 1980 (CERCLA), commonly known as "Superfund," provided the Congressional mandate to remove or "clean up" abandoned and
inactive hazardous waste sites as well as Federal assistance in cases of toxic emergency situations. Congress
made the Environmental Protection Agency (EPA) the
lead agency in implementation of CERCLA. The law
provides authority for EPA to collect the cost of cleaning up a release from the responsible parties. CERCLA
incorporates in one place the provisions for responding
to releases of hazardous substances into the environment, whether intentional or accidental and whether
one-time or continuing.

Under CERCLA, spills or discharges into the environment of certain amounts of substances designated
as hazardous must be reported immediately to the National Response Center, which was originally established under the Clean Water Act. The center will
notify all appropriate government agencies, which will
coordinate response to the discharge.

Major amendments and the reauthorization of
CERCLA occurred in 1986 with passage of the
Superfund Amendments and Reauthorization Act
(SARA). The amendments made major changes in the
original act by adding stricter cleanup standards, more
control over the settlement process and increased state
and public involvement. SARA also contains a self-
contained law designated as Title III – the Emergency
Planning and Community Right-to-Know Act (often
referred to as "SARA Title III" or EPCRA).

Part 355 deals with the Emergency Planning and
Notification requirements of Title III. The requirements
of this part apply to any facility at which there is present
an amount of any extremely hazardous substance (EHS)
equal to or in excess of its threshold planning quantity
(TPQ). The "amount of any extremely hazardous
substance" means the total amount of an extremely

hazardous substance present at any one time at a
facility at a concentration greater than the one percent
(1%) by weight, regardless of location, number of
containers, or method of storage.

Extremely hazardous substances that are solids are
subject to either of two threshold planning quantities as
designated in Appendix A. The lower quantity applies
only if the solid exists in powdered form and has a
particle size less than 100 microns; or is handled in
solution or in molten form; or meets the criteria for
a National Fire Protection Association (NFPA) rating
of 2, 3, or 4 for reactivity. If the solid does not meet
any of these criteria, it is subject to the upper threshold
planning quantity.

This list contains all the substances identified in
Appendix A to Part 355.

The data field contains the Threshold Planning
Quantity (lower limit and upper limits for solids) in
pounds.   Reportable Quantities are identified in a
separate list.

**CERCLA/SARA - Section 313 Emissions Re-   (313)
porting**
SOURCE

Environmental Protection Agency
Comprehensive Environmental Response, Compensation, and Liability Act
Superfund Amendments and Reauthorization Act
40 CFR Subchapter J - Superfund, Emergency Planning, and Community Right-to-Know Programs.
Part 372 - Toxic Chemical Release Reporting
Subpart D - Specific Toxic Chemical Listings
Paragraph 372.65 Chemicals and chemical categories
40 CFR 372.65
Amended through November 30, 1994 (59 FR
61432)(addition of 286 chemicals and chemical
categories)

LIST DESCRIPTION

The Comprehensive Environmental Response, Compensation, and Liability Act of 1980 (CERCLA), commonly known as "Superfund," provided the Congressional mandate to remove or "clean up" abandoned and
inactive hazardous waste sites as well as Federal assistance in cases of toxic emergency situations. Congress
made the Environmental Protection Agency (EPA) the
lead agency in implementation of CERCLA. The law
provides authority for EPA to collect the cost of cleaning up a release from the responsible parties. CERCLA
incorporates in one place the provisions for responding
to releases of hazardous substances into the environment, whether intentional or accidental and whether
one-time or continuing.

Under CERCLA, spills or discharges into the environment of certain amounts of substances designated
as hazardous must be reported immediately to the National Response Center, which was originally established under the Clean Water Act. The center will
notify all appropriate government agencies, which will
coordinate response to the discharge.

Major amendments and the reauthorization of CERCLA occurred in 1986 with passage of the Superfund Amendments and Reauthorization Act (SARA). The amendments made major changes in the original act by adding stricter cleanup standards, more control over the settlement process and increased state and public involvement. SARA also contains a self-contained law designated as Title III – the Emergency Planning and Community Right-to-Know Act (often referred to as "SARA Title III" or EPCRA).

Part 372 provides requirements for the submission of information relating to the release of chemicals designated as toxic. This part requires suppliers to notify persons to whom they distribute mixtures or trade name products containing toxic chemicals that they contain such chemicals. Paragraph 372.25 defines threshold amounts for purposes of reporting. The threshold amounts vary according to the activity of the facility. Facilities containing more than the threshold amounts must make their report to EPA using Reporting Form R. These reports are compiled into the Toxic Release Inventory, a publicly available database that contains information on estimated total environmental releases of the chemicals identified in this list.

Under the Pollution Prevention Act of 1990, Congress expanded the reporting requirement under this section to include gathering information on U.S. companies' pollution prevention efforts. EPA revised Form R for the 1991 reporting year to include the expanded reporting requirements. The revised Form Rs were made available on July 1, 1991 and include detailed instructions for completing the form. They were submitted to the Federal Register for publication but were not available at the time of this update.

The revised Form R contained a revision to the de minimis exemption - beneficiation activities are no longer excluded from this exemption. Under any circumstances, toxic chemicals received in mixtures or trade name products under the de minimis value of one percent, or 0.1 percent if carcinogenic, are exempted from threshold determinations and release calculations.

This list identifies all chemicals and chemical categories identified 40 CFR 372.65 and the de minimis levels as published in the July 1, 1991 EPA filing instructions for Form R.

The data field contains the de minimis reporting percent for Form R.

ATSDR - Priority List for Tox Profiles   (ATS)

SOURCE

1993 CERCLA Priority List Of Hazardous Substances That Will Be the Subject Of toxicological Profiles and Support Document
February 1994

U.S. Department of Health and Human Services
Public Health Service
Agency for Toxic Substances and Disease Registry
Quality Assurance Branch, Division of Toxicology
1600 Clifton Road, N.E.
Atlanta, GA 30333
(404) 639-6030

LIST DESCRIPTION

The Agency for Toxic Substances and Disease Registry (ATSDR) is part of the Public Health Service and is based in Atlanta, Georgia. It was created by Congress to implement the health-related sections of laws that protect the public from hazardous wastes or environmental spills of hazardous substances. ATSDR derives its authority from three separate acts of Congress:

The Comprehensive Environmental Response, Compensation, and Liability Act of 1980 (CERCLA), commonly known as "Superfund," provided the Congressional mandate to remove or "clean up" abandoned and inactive hazardous waste sites, and to provide Federal assistance in cases of toxic emergency situations. Congress made the Environmental Protection Agency (EPA) the lead agency in implementation of CERCLA and created ATSDR to implement the health-related sections.

In 1984, amendments to the Resource Conservation and Recovery Act of 1976 (RCRA), which provides for the management of legitimate hazardous waste storage or destruction facilities, charged ATSDR to conduct health assessments at these sites when requested by EPA, States, or private citizens, and to assist EPA in determining which substances should be regulated and the levels at which they may pose a threat to human health.

In 1986, amendments to CERCLA, known as the Superfund Amendments and Reauthorization Act of 1986 (SARA), broadened ATSDR's authority and layed the framework for additional programs.

Section 104(i)(2) of CERCLA required that ATSDR and EPA prepare a list, in order of priority, of at least 100 hazardous substances that are most commonly found at facilities on the National Priorities List (NPL) and which, in their sole discretion, are determined to pose the most significant potential threat to human health. CERCLA also required the agencies to revise the priority list to include 100 or more additional hazardous substances and to include at least 25 additional hazardous substances in each of the three successive years following the 1988 revision. CERCLA also requires that ATSDR and EPA shall revise this list not less than once every year to include additional hazardous substances which are determined to pose the most significant potential threat to human health. Each substance on the CERCLA priority list of hazardous substances is a candidate to become the subject of a toxicological profile prepared by ATSDR and the subsequent identification of priority data needs.

New candidate substances (substances which have been found at three or more NPL sites) were assigned a toxicity/environmental score (TES) using the EPA Reportable Quantity methodology, and were added to the pool of substances previously considered for the annual list. All substances were then evaluated together for consideration on the priority list.

The approach used to generate the revised priority list was summarized in the October 17, 1991 Federal Register publication. Using the same approach, and the same algorithm this year, over 600 candidate substances have been ranked to create the current list of 275 substances.

This list identifies all the substances on the 1993 Priority List of Hazardous Substances.

The data field contains the rank order of the substance. Rank order is assigned based on the total number of points (i.e. the highest TES score of 1690 points has rank 1 and the lowest score of 431 has rank 275).

U.S. DOT - Appendix A Table 1 - Hazardous (EPA) Substances

SOURCE

U.S. Department of Transportation
Research and Special Programs Administration
49 CFR Subchapter C - Hazardous Materials Regulations
Appendix A to 172.101 - List of Hazardous Substances and Reportable Quantities
Table 1 - Hazardous Substances Other than Radionuclides
As published in 49 CFR revised as of October 1, 1992 and updated through 57 FR 59308 December 15, 1992.
NOTE: This list is based on the revisions made under Docket HM181.

LIST DESCRIPTION

The Hazardous Materials Transportation Act (HMTA), which became law in 1975, gives the Department of Transportation (DOT) authority to regulate the movement within the United States of substances that may pose a threat to health, safety, property or the environment when transported by air, highway, rail, or water.

Under HMTA, DOT has delegated authority for regulation of hazardous materials to the Research and Special Programs Administration (RSPA). This organization, with its Office of Hazardous Materials Transportation, drafts and issues hazardous materials regulations, exemptions, registration certificates, and packaging and container certification for all transport modes.

The Hazardous Materials Table (49 CFR 172.101) has several thousand entries applicable to various

materials including explosives, flammables, oxidizing materials, organic peroxides, corrosives, gases, poisons, radioactive substances and etiologic agents. The Table is composed of the main table and two Appendices. Appendix A contains materials designated as hazardous substances and wastes and Appendix B contains a list of Marine Pollutants.

Appendix A lists materials which are listed or designated as hazardous substances under section 101(14) of the Comprehensive Environmental Response, Compensation, and Liability Act (CERCLA). DOT, by law, must add these to Table 101. Appendix A is divided into 2 tables which are entitled " Table 1 - Hazardous Substances Other Than Radionuclides" and " Table 2 - Radionuclides" . A material listed in this appendix is regulated as a hazardous material and a hazardous substance under this subchapter if it meets the definition of a hazardous substance in paragraph 171.8 of this subchapter.

This list contains all materials which are listed in Table 1 of Appendix A. ChemADVISOR has provided the Chemical AbstractsService Numbers for these chemicals as used by the EPA (see CERCLA/SARA - Hazardous Substances and Their Reportable Quantities). Each chemical is listed by only one name. The name chosen is the name as used by the EPA.

The data field includes the Reportable Quantity in units of pounds and (kilograms).

U.S. DOT - Appendix A Table 2 - Radionuclides (EPA)

SOURCE

U.S. Department of Transportation
Research and Special Programs Administration
49 CFR Subchapter C - Hazardous Materials Regulations
Appendix A to 172.101 - List of Hazardous Substances and Reportable Quantities
Table 2 - Radionuclides
As published in 49 CFR revised as of October 1, 1992 and updated through 57 FR 59308 December 15, 1992.
NOTE: This list is based on the revisions made under Docket HM181.

LIST DESCRIPTION

The Hazardous Materials Transportation Act (HMTA), which became law in 1975, gives the Department of Transportation (DOT) authority to regulate the movement within the United States of substances that may pose a threat to health, safety, property or the environment when transported by air, highway, rail, or water.

Under HMTA, DOT has delegated authority for regulation of hazardous materials to the Research and Special Programs Administration (RSPA). This organization, with its Office of Hazardous Materials Transportation, drafts and issues hazardous materials regulations, exemptions, registration certificates, and packaging and container certification for all transport modes.

The Hazardous Materials Table (49 CFR 172.101) has several thousand entries applicable to various materials including explosives, flammables, oxidizing materials, organic peroxides, corrosives, gases, poisons, radioactive substances and etiologic agents. The Table is composed of the main table and two Appendices. Appendix A contains materials designated as hazardous substances and wastes and Appendix B contains a list of Marine Pollutants.

Appendix A lists materials which are listed or designated as hazardous substances under section 101(14) of the Comprehensive Environmental Response, Compensation, and Liability Act (CERCLA). DOT, by law, must add these to Table 101. Appendix A is divided into 2 tables which are entitled " Table 1 - Hazardous Substances Other Than Radionuclides" and " Table 2 - Radionuclides" . A material listed in this appendix is regulated as a hazardous material and a hazardous substance under this subchapter if it meets the definition of a hazardous substance in paragraph 171.8 of this subchapter.

This list contains all materials which are listed in Table 2 of Appendix A. ChemADVISOR has provided the Chemical Abstracts Service Numbers for these chemicals.

The data field includes the Reportable Quantity in units of curies (Ci) and terabecquerels (TBq).

Clean Air Act (1990) - List of Hazardous Air (CAA) Contaminants

SOURCE

Public Law 101-549
101st Congress
Title III - Hazardous Air Pollutants
November 15, 1990

LIST DESCRIPTION

The Clean Air Act of 1990 was passed to amend the Clean Air Act in order to provide for attainment and maintenance of health protective national ambient air quality standards, and for other purposes.

The Law has 9 major section (titles):

Title I - Provisions for Attainment and Maintenance of National Ambient Air Quality Standards
Title II - Provisions Relating to Mobile Sources
Title III - Hazardous Air Pollutants
Title IV - Acid Deposition Control
Title V - Permits
Title VI - Stratospheric Ozone Protection
Title VII - Provisions Relating to Enforcement
Title VIII - Miscellaneous Provisions
Title IX - Clean Air Research

This Law will be under the jurisdiction of the Environmental Protection Agency. The Law is one of the most complex pieces of legislation ever passed by Congress and contains unusually technical language. EPA is still in the process of codifying the law into regulations.

Title III of the 1990 CAA amends Section 112 of the current Clean Air Act which defines Hazardous Air Pollutants. The law contains an initial list of Pollutants with provisions for revisions by the Administrator "as appropriate." Revisions to the list can add pollutants which present, or may present, through inhalation or other routes of exposure, a threat of adverse human health effects (including, but not limited to, substances which are known to be, or may reasonably be anticipated to be carcinogenic, mutagenic, teratogenic, neurotoxic, which cause reproductive dysfunction, or which are acutely or chronically toxic) or adverse environmental effects whether through ambient concentrations, bioaccumulation, deposition, or otherwise, but not including releases subject to regulation under subsection (r) as a result of emissions to the air.

This list contains the initial list of pollutants identified inthis law.

The data field contains the word "present" to indicate its presence on this list.

Class 1 Ozone Depletors   (CAA)

SOURCE

Environmental Protection Agency
Clean Air Act (1990) Public Law 101-549 Title VI (Stratospheric Ozone Protection)
40 CFR Part 82
Subpart A - Production and Consumption Controls
Appendix F to Subpart A - Listing of Ozone Depleting Chemicals amended through December 10, 1993 (58 FR 65018)

LIST DESCRIPTION

The Clean Air Act of 1990 was passed to amend the Clean Air Act in order to provide for attainment and maintenance of health protective national ambient air quality standards, and for other purposes. Title VI of the Clean Air Act Amendments of 1990 establish a comprehensive regime for phasing out ozone-depleting substances. Building on the Montreal Protocol on Substances that Deplete the Ozone Layer and the London Amendments of 1990 and the EPA regulations implementing the Protocol (see 53 FR 30566, August 12, 1988), title VI calls for the phase-out of most ozone-depleting substances by the year 2000 and the imposition of other controls designed to minimize the emissions of such substances prior to their elimination.

Section 602 of that Act specifies the creation of two lists. One list contains fully halogenated chlorofluorocarbons (CFCs), halons, carbon tetrachloride and methyl chloroform, collectively referred to as "Class I" substances. The second list contains hydrochlorofluorocarbons (HCFCs) termed "Class II"substances. Section 602 also requires EPA at the time it publishes the initial lists to assign each listed substance numerical values representing the substance's ozone-depletion potential (ODP), chlorine and bromine loading potentials, and atmospheric lifetime.

This section also requires that the initial list of class I and class II substance shall also include the isomers of the listed substances, other than 1,1,2-trichloroethane (an isomer of methyl chloroform). EPA does not currently have adequate data on the potential ozone-depletion effects of all possible isomers of the listed substances and reserves publications of their values for a later date.

Class I chemicals are those the Administrator finds cause or contribute significantly to harmful effects on the stratospheric ozone layer, including all chemicals that have an ozone depletion potential of 0.2 or greater.

Class II chemicals are those the Administrator finds are known or reasonably anticipated to cause or contribute to harmful effects on the stratospheric ozone layer.

On December 10, 1993, EPA published a notice of final rulemaking which added methyl bromide to the list of class I substances, in response to new scientific information, a petition submitted under section 602 of the Act, and the decision of the Protocol Parties to classify methyl bromide as a controlled substance with an ozone-depleting potential (ODP) of 0.7. EPA also added hydrobromofluorocarbons (HBFCs) to the list of class I substances. EPA makes final a freeze on the production and consumption of HBFCs starting January 1, 1994 at 1991 baseline levels. Only one HBFC is currently in production, HBFC-22B1, used as a fire suppressant with and ODP of 0.74. Use of this chemical is extremely limited, and it is only manufactured by one company.

This list contains all the substances identified in Appendix F to Subpart A of Part 82 published in the December 10th notice. CAS Numbers were obtained from Chemical Abstracts Services and were not published in the Federal Register Notice.

The data field contains the Ozone Depletion Potential.

Class 2 Ozone Depletors   (CAA)

SOURCE

Environmental Protection Agency
Clean Air Act (1990) Public Law 101-549 Title VI (Stratospheric Ozone Protection)
40 CFR Part 82
Federal Register Notice January 22, 1991 (56 FR 2420) listing Ozone-Depleting Substances. Related Notices of Proposed Rulemaking at 56 FR 49548 and 57 FR 19166.

LIST DESCRIPTION

The Clean Air Act of 1990 was passed to amend the Clean Air Act in order to provide for attainment and maintenance of health protective national ambient air quality standards, and for other purposes. Title VI of the Clean Air Act Amendments of 1990 establish a comprehensive regime for phasing out ozone-depleting substances. Building on the Montreal Protocol on Substances that Deplete the Ozone Layer and the London Amendments of 1990 and the EPA regulations implementing the Protocol (see 53 FR 30566, August 12, 1988), title VI calls for the phase-out of most ozone-depleting substances by the year 2000 and the

imposition of other controls designed to minimize the emissions of such substances prior to their elimination.

Section 602 of that Act specifies the creation of two lists. One list contains fully halogenated chlorofluorocarbons (CFCs), halons, carbon tetrachloride and methyl chloroform, collectively referred to as "Class I" substances. The second list contains hydrochlorofluorocarbons (HCFCs) termed "Class II" substances. Section 602 also requires EPA at the time it publishes the initial lists to assign each listed substance numerical values representing the substance's ozone-depletion potential (ODP), chlorine and bromine loading potentials, and atmospheric lifetime.

This section also requires that the initial list of class I and class II substance shall also include the isomers of the listed substances, other than 1,1,2-trichloroethane (an isomer of methyl chloroform). EPA does not currently have adequate data on the potential ozone-depletion effects of all possible isomers of the listed substances and reserves publications of their values for a later date.

Class I chemicals are those the Administrator finds cause or contribute significantly to harmful effects on the stratospheric ozone layer, including all chemicals that have an ozone depletion potential of 0.2 or greater.

Class II chemicals are those the Administrator finds are known or reasonably anticipated to cause or contribute to harmful effects on the stratospheric ozone layer.

On September 30, 1991 EPA published in the Federal Register a Notice of proposed rulemaking further implementing the Montreal Protocol. The list of ozone-depleting chemicals appeared as Appendix A to that proposal. The Ozone Depleting Potential (ODP) was identified in that publication as the Ozone Depletion Weight (ODW).

On May 4, 1992 (57 FR 19166) EPA published a notice of proposed rulemaking to require warning labels on containers of, and products containing or manufactured with, certain ozone-depleting substances pursuant to section 611 of the Clean Air Act, as amended. EPA also proposed to require permanent labels on products containing ozone-depleting substances that can be recovered or recycled pursuant to section 608 of the Clean Air Act, as amended. The substances affected by this proposal included both class I and class II chemicals.

This list contains all the substances identified as Class II substances (also called controlled substances) published in the September 30, 1991 FR proposal in Appendix A. CAS Numbers were obtained from Chemical Abstracts Services and were not published in the Federal Register Notice.

The data field contains the Ozone Depletion Weight (ODW) as published in the September 30, 1991 FR notice Appendix A.

CAA - Toxic Substances for Accidental Release (EP3) Prevention

SOURCE

Clean Air Act Section 112(r)(3)
40 CFR Part 68 - Chemical Accident Provisions
Subpart C - Regulated Substances for Accidental Release Prevention
40 CFR 68.130 Table 1 - List of Regulated Toxic Substances and Threshold Quantities for Accidental Release Prevention
Final Rule published in 59 FR 4478 (January 31, 1994). Rule effective March 2, 1994.

LIST DESCRIPTION

The Clean Air Act (CAA) of 1990 was passed to amend the Clean Air Act in order to provide for attainment and maintenance of health protective national ambient air quality standards, and for other purposes.

In the CAA Amendments Congress added subsection (r) to section 112 for the prevention of chemical accidents. The goals of the chemical accident prevention provisions are to focus on chemicals that pose a significant hazard to the community should an accident occur, to prevent their accidental release and to minimize the consequences of such releases.

Section 112(r) of the CAA has a number of provisions. Under section 112(r) owners and operators of stationary sources who produce, process, handle, or store substances listed under section 112(r)(3) or any other extremely hazardous substances have a general duty to initiate specific activities to prevent and mitigate accidental releases. The general duty requirements apply to stationary sources regardless of the quantity of substances managed at the facility. Activities such as identifying hazards which may result from accidental releases using appropriate hazard assessment techniques; designing, maintaining and operating a safe facility; and minimizing the consequences of accidental releases if they occur would be essential activities to be taken as necessary to satisfy the general duty requirements. As a matter of business practice, owners and operators of these stationary sources have a duty to conduct these activities under section 112(r) in the same manner and to the same extent as an employer's duties under OSHA's general duty clause in section 654 of title 20 of the United States Code.

Section 112(r)(3) of the CAA requires EPA to promulgate an initial list of at least 100 substances (regulated substances) that are known to cause, or may be reasonably anticipated to cause, death, injury, or serious adverse effects to human health or theenvironment if accidentally released. EPA is required to set threshold quantities for each listed substance.

Under CAA section 112(r)(7), the Act requires EPA to promulgate reasonable regulations and appropriate guidance to provide for the prevention and detection of accidental releases and for responses to such releases. The accident prevention regulations will apply to stationary sources that have present more than a threshold quantity of a regulated substance.

These regulations shall address, as appropriate, the use, operation, repair, and maintenance of equipment to monitor, detect, inspect, and control releases, including training of personnel in the use and maintenance of equipment or in the conduct of periodic inspections. The regulations shall include requirements for the development and submission of Risk Management Plans (RMPs) by regulated sources. The RMP shall include a hazard assessment, a prevention program, and an emergency response program.

In developing the list of substances EPA was required to consider the list of extremely hazardous substances (EHSs) promulgated under EPCRA (SARA Title III) section 302. In addition Congress listed 16 substances to be included in the initial list. Toxic substances were included on the list based on their toxicity, physical state, vapor pressure, production volume, and accident history. Toxicity criteria used to identify chemicals as extremely hazardous substances (EHSs) under EPCRA were used as criteria. Threshold quantities were set for toxic substances based on a ranking method that considers toxicity and volatility of the chemicals. EPA assigned identical thresholds to chemicals with similar ranking scores, ranging from 500 pounds to 10,000 pounds.

A threshold quantity of a regulated substance is present at a stationary source if the total quantity of the regulated substance contained in a process exceeds the threshold.

Explosives listed by DOT as Division 1.1 in 49 CFR 172.101 are covered under section 112(r) of the Clean Air Act. The threshold quantity for explosives is 5,000 pounds.

This list contains all the entries in Table 1 to paragraph 68.130 (Table 1 presents the list in alphabetical order and Table 2 presents the list in CAS Number order).

The data field contains the Threshold quantity in pounds.

CAA - Flammable Substances for Accidental (EP3) Release Prevention

SOURCE

Clean Air Act Section 112(r)(3)
40 CFR Part 68 - Chemical Accident Provisions
Subpart C - Regulated Substances for Accidental Release Prevention
40 CFR 68.130 Table 1 - List of Regulated Toxic Substances and Threshold Quantities for Accidental Release Prevention
Final Rule published in 59 FR 4478 (January 31, 1994). Rule effective March 2, 1994.

LIST DESCRIPTION

The Clean Air Act (CAA) of 1990 was passed to amend the Clean Air Act in order to provide for attainment and maintenance of health protective national ambient air quality standards, and for other purposes.

In the CAA Amendments Congress added subsection (r) to section 112 for the prevention of chemical accidents. The goals of the chemical accident prevention provisions are to focus on chemicals that pose a significant hazard to the community should an accident occur, to prevent their accidental release and to minimize the consequences of such releases.

Section 112(r) of the CAA has a number of provisions. Under section 112(r) owners and operators of stationary sources who produce, process, handle, or store substances listed under section 112(r)(3) or any other extremely hazardous substances have a general duty to initiate specific activities to prevent and mitigate accidental releases. The general duty requirements apply to stationary sources regardless of the quantity of substances managed at the facility. Activities such as identifying hazards which may result from accidental releases using appropriate hazard assessment techniques; designing, maintaining and operating a safe facility; and minimizing the consequences of accidental releases if they occur would be essential activities to be taken as necessary to satisfy the general duty requirements. As a matter of business practice, owners and operators of these stationary sources have a duty to conduct these activities under section 112(r) in the same manner and to the same extent as an employer's duties under OSHA's general duty clause in section 654 of title 20 of the United States Code.

Section 112(r)(3) of the CAA requires EPA to promulgate an initial list of at least 100 substances (regulated substances) that are known to cause, or may be reasonably anticipated to cause, death, injury, or serious adverse effects to human health or theenvironment if accidentally released. EPA is required to set threshold quantities for each listed substance.

Under CAA section 112(r)(7), the Act requires EPA to promulgate reasonable regulations and appropriate guidance to provide for the prevention and detection of accidental releases and for responses to such releases. The accident prevention regulations will apply to stationary sources that have present more than a threshold quantity of a regulated substance. These regulations shall address, as appropriate, the use, operation, repair, and maintenance of equipment to monitor, detect, inspect, and control releases, including training of personnel in the use and maintenance of equipment or in the conduct of periodic inspections. The regulations shall include requirements for the development and submission of Risk Management Plans (RMPs) by regulated sources. The RMP shall include a hazard assessment, a prevention program, and an emergency response program.

In developing the list of substances EPA was required to consider the list of extremely hazardous substances (EHSs) promulgated under EPCRA (SARA Title III) section 302. In addition Congress listed 16 substances to be included in the initial list.

Flammable gases and volatile flammable liquids were included on the list based on the flash point and boiling point criteria used by the National Fire Protection Association (NFPA) for its hightest flammability hazard ranking (flash point below 73 deg F and boiling point below 100 deg F. Only flammable

substances in commercial production were listed. The threshold quantity for flammable substances was set at 10,000 pounds based on the potential for a vapor cloud explosion.

A threshold quantity of a regulated substance is present at a stationary source if the total quantity of the regulated substance contained in a process exceeds the threshold.

Explosives listed by DOT as Division 1.1 in 49 CFR 172.101 are covered under section 112(r) of the Clean Air Act. They are not included in 40 CFR 68.130 but referenced the DOT regulation. The threshold quantity for explosives is 5,000 pounds.

This list contains all the entries in Table 1 to paragraph 68.130 (Table 1 presents the list in alphabetical order and Table 2 presents the list in CAS Number order).

The data field contains the Threshold quantity in pounds.

CAA - HON Rule - SOCMI Chemicals   (HON)

SOURCE

Environmental Protection Agency
Code of Federal Regulations Title 40
Part 63 - National Emission Standards for Hazardous Air Pollutants for Source Categories
Subpart F - National Emission Standards for Organic Hazardous Air Pollutants from the Symthetic Organic Chemical Manufacturing Industry
Table 1 to Subpart F - Synthetic Organic Chemical Manufacturing Industry Chemicals
Final Rule published in the Federal Register April 22, 1994 (59 FR 19402)

LIST DESCRIPTION

The Clean Air Act of 1990 was passed to amend the Clean Air Act in order to provide for attainment and maintenance of health protective national ambient air quality standards, and for other purposes.

Section 112 of the Act requires that the EPA establish regulations setting emission standards for categories of sources of HAP emissions. In addition, the Act sets out specific criteria for establishing a minimum level of control, and criteria to be considered in evaluating control options more stringent than the minimum control level.

On December 31, 1992, the EPA proposed to regulate the emissions of certain organic hazardous air pollutants from synthetic organic chemical manufacturing industry (SOCMI) production processes which are part of major sources under section 112 of the Clean Air Act as amended in 1990. EPA's final rule is referred to as the hazardous organic NESHAP or the HON rule.

The HON requires sources to achieve emission limits reflecting the application of the maximum achievable control technology consistent with section 112(d) and 112(b) of the Act. The rule regulates the emissions of 112 of the organic chemicals identified in the Act's list of 189 hazardous air pollutants at both new and existing SOCMI sources and from equipment leaks at sources in certain polymer and resin production processes, and certain miscellaneous processes.

The rule consist of four subparts in 40 CFR Part 63. Subpart F provides the applicability criteria for SOCMI sources, requires that owners and operators of SOCMI sources comply with subparts G and H, and specifies general recordkeeping and reporting requirements. The specific control, monitoring, reporting, and recordkeeping requirements are stated in subpart G for processvents, storage vessels, transfer racks, and wastewater streams, and in subpart H for equipment leaks. Subpart I provides the applicability criteria for the non-SOCMI processes subject to the negotiated regulation for equipment leaks and requires owners and operators to comply with subpart H.

The rule applies to chemical manufacturing process units that are: 1) Part of a major source as defined in section 112 of the Act; 2) produce as a primary product a SOCMI chemical listed in Table 1 of subpart F; and 3) use as a reactant or manufacture as a product, by-product, or co-product one or more of the organic HAP's listed in Table 2 of subpart F.

This list contains all the chemical substances in Table 1 to Subpart F of Part 63.

The data field contains the date that existing sources shall be in compliance with the rule.

CAA - HON Rule - Organic HAPs   (HON)
SOURCE

Environmental Protection Agency
Code of Federal Regulations Title 40
Part 63 - National Emission Standards for Hazardous Air Pollutants for Source Categories
Subpart F - National Emission Standards for Organic Hazardous Air Pollutants from the Symthetic Organic Chemical Manufacturing Industry
Table 2 to Subpart F - Organic Hazardous Air Pollutants
Final Rule published in the Federal Register April 22, 1994 (59 FR 19402)

LIST DESCRIPTION

The Clean Air Act of 1990 was passed to amend the Clean Air Act in order to provide for attainment and maintenance of health protective national ambient air quality standards, and for other purposes.

Section 112 of the Act requires that the EPA establish regulations setting emission standards for categories of sources of HAP emissions. In addition, the Act sets out specific criteria for establishing a minimum level of control, and criteria to be considered in evaluating control options more stringent than the minimum control level.

On December 31, 1992, the EPA proposed to regulate the emissions of certain organic hazardous air pollutants from synthetic organic chemical manufacturing industry (SOCMI) production processes which are part of major sources under section 112 of the Clean Air Act as amended in 1990. EPA's final rule is referred to as the hazardous organic NESHAP or the HON rule.

The HON requires sources to achieve emission limits reflecting the application of the maximum achievable control technology consistent with section 112(d) and 112(b) of the Act. The rule regulates the emissions of 112 of the organic chemicals identified in the Act's list of 189 hazardous air pollutants at both new and existing SOCMI sources and from equipment leaks at sources in certain polymer and resin production processes, and certain miscellaneous processes.

The rule consist of four subparts in 40 CFR Part 63. Subpart F provides the applicability criteria for SOCMI sources, requires that owners and operators of SOCMI sources comply with subparts G and H, and specifies general recordkeeping and reporting requirements. The specific control, monitoring, reporting, and recordkeeping requirements are stated in subpart G for processvents, storage vessels, transfer racks, and wastewater streams, and in subpart H for equipment leaks. Subpart I provides the applicability criteria for the non-SOCMI processes subject to the negotiated regulation for equipment leaks and requires owners and operators to comply with subpart H.

Subpart F lists 112 organic HAP's that the EPA has determined may be emitted from SOCMI processes because they are either produced as a product or used as a reactant. The emissions of these 112 organic chemicals are regulated by subparts F, G, and H.

The rule applies to chemical manufacturing process units that are: 1) Part of a major source as defined in section 112 of the Act; 2) produce as a primary product a SOCMI chemical listed in Table 1 of subpart F; and 3) use as a reactant or manufacture as a product, by-product, or co-product one or more of the organic HAP's listed in Table 2 of subpart F.

This list contains all the chemical substances in Table 2 to Subpart F of Part 63.

The data field contains only the word " present" and any footnotes which apply to Table 2.

Clean Water Act - Priority Pollutants   (CWA)
SOURCE

Environmental Protection Agency
Clean Water Act
Subchapter N - Effluent Guidelines and Standards
Part 423 Appendix A
CFR July 1992 Edition

LIST DESCRIPTION

The Clean Water Act regulates the discharge of non-toxic and toxic pollutants into waterways by municipal, industrial and other point sources and by non-point sources of pollution. Types of pollution include toxic substances, organic wastes, sediment washed from agricultural or construction operations, acid, bacteria and viruses, nutrients, heat, and oil and grease. The act gives authority to the federal government and covers all surface waters in the United States.

Major provisions of the act that deal with chemicals are the following:

Section 303: Water quality criteria and standards

Sections 301, 304, and 307: Effluent limitation and guidelines
Section 311: Control of discharges of oil and hazardous substances.

Section 307 requires the Administrator to establish pretreatment standards which will prevent the discharge of any pollutant into publicly owned treatment works which pollutant interferes with, passes through untreated, or with which it is otherwise incompatible. Regulations dealing with effluent guidelines and Standards are located in Subchapter N of 40 CFR. Under this authority, EPA has established a list of "Toxic Pollutants"in Part 400 along with a time frame by which EPA must implement the requirements of the law.

The list of 65 Toxic Pollutants includes many broad chemical groups with the potential to cover many hundreds of materials. EPA has established a list of "priority pollutants" to better define some of the broad categories. This list appears as an appendix to Part 423 and is cited in 423.17 as a compliance option for pretreatment standards for new sources (see 423.17(d)(2).

This list includes all entries in Appendix A - "126 Priority Pollutants."

The data field includes the word "present" to indicate its presence on this list.

Clean Water Act - Toxic Pollutants   (CWA)
SOURCE

Environmental Protection Agency
Clean Water Act
Subchapter N - Effluent Guidelines and Standards
Part 401 - General Provisions
40 CFR 401.15 - Toxic Pollutants
CFR July 1992 Edition

LIST DESCRIPTION

The Clean Water Act regulates the discharge of non-toxic and toxic pollutants into waterways by municipal, industrial and other point sources and by non-point sources of pollution. Types of pollution include toxic substances, organic wastes, sediment washed from agricultural or construction operations, acid, bacteria and viruses, nutrients, heat, and oil and grease. The act gives authority to the federal government and covers all surface waters in the United States.

Major provisions of the act that deal with chemicals are the following:

Section 303: Water quality criteria and standards
Sections 301, 304, and 307: Effluent limitation and guidelines
Section 311: Control of discharges of oil and hazardous substances.

Section 307 requires the Administrator to establish pretreatment standards which will prevent the discharge of any pollutant into publicly owned treatment works which pollutant interferes with, passes through untreated, or with which it is otherwise incompatible. Regulations dealing with effluent guidelines and

Standards are located in Subchapter N of 40 CFR. Under this authority, EPA has established a list of "Toxic Pollutants"in Part 400 along with a time frame by which EPA must implement the requirements of the law.

This list includes all entries identified as "Toxic Pollutants" at 40 CFR 401.15.

The data field includes the word "present" to indicate its presence on this list and any generic chemical category under which it is contained.

Clean Water Act - Hazardous Substances   (CWA)
SOURCE

Environmental Protection Agency
Clean Water Act
Subchapter D - Water Programs
Part 116 - Designation of Hazardous Substances
Table 116.4 - List of Hazardous Substances
40 CFR 116.4
CFR July 1992 Edition

LIST DESCRIPTION

The Clean Water Act regulates the discharge of non-toxic and toxic pollutants into waterways by municipal, industrial and other point sources and by non-point sources of pollution. Types of pollution include toxic substances, organic wastes, sediment washed from agricultural or construction operations, acid, bacteria and viruses, nutrients, heat, and oil and grease. The act gives authority to the federal government and covers all surface waters in the United States.

Major provisions of the act that deal with chemicals are the following:

Section 303: Water quality criteria and standards
Sections 301, 304, and 307: Effluent limitation and guidelines
Section 311: Control of discharges of oil and hazardous substances.

Section 311 of the act establishes mechanisms for cleaning up spills and other releases of oil and hazardous substances into navigable water. Regulations appear at 40 CFR 109 to 117. In Part 116 EPA has designated about 300 substances as hazardous when spilled or discharged into navigable waters and established "reportable quantities"- the minimum amount of substance that must be reported to the federal government's National Response Center when spilled.

This section of the CWA works closely with the cleanup provisions of the Comprehensive Environmental Response, Compensation and Liability Act and the cleanup provisions (including the establishment of the Reportable Quantity) are virtually identical in the two laws. In order to avoid confusion, the Reportable Quantities are provided ONLY with the CERCLA RQ Hazardous Materials List in this database.

This list contains all entries on Table 116.4A.

The data field identifies all entries with the word "present" and any chemical group to which EPA has identified the substance.

EPA - Carcinogen Hazard Ranking for RQ (EP2) Adjustment

SOURCE

Environmental Protection Agency
Office of Health and Environmental Assessment
Human Health Assessment Group (formerly Carcinogens Assessment Group)
Publication EPA/600/8-89/053
June 1988
"Methodology for Evaluating Potential Carcinogenicity in Support of Reportable Quantity Adjustments Pursuant to CERCLA Section 102"

LIST DESCRIPTION

This report describes the technical methodology the Agency has used in developing a hazard ranking for potential carcinogens in order to adjust reportable quantities (RQs) under Section 102 of the Comprehensive Environmental Response, Compensation, and Liability Act of 1980 (CERCLA). Under Section 102(a) the RQ of any hazardous substance designated in Section 101(14) is 1 pound unless a different RQ has been established pursuant to Section 311(b)(4) of the Federal Water Pollution Control Act. In Section 102(a) of CERCLA these statutory RQs may be adjusted by regulations establishing different quantities to be reported upon release of a hazardous substance. Section 102(a) also gives the EPA authority to establish a single RQ for each hazardous substance, regardless of the environmental medium into which the substance is released.

The RQ adjustment methodology is based, in part, on the methodology used to establish RQs under the Clean Water Act. The methodology begins with an evaluation of the intrinsic physical, chemical, and toxicological properties associated with each hazardous substance. The intrinsic properties evaluated, called primary criteria, are: aquatic toxicity, mammalian toxicity (oral, dermal, and inhalation), ignitability, reactivity, chronic toxicity, and potential carcinogenicity. The Agency ranks each intrinsic property (other than potential carcinogenicity) on a five-tier scale, associating a specific range of values on each scale with a particular RQ value (corresponding to RQs of 1, 10, 100, 1000, and 5000 pounds). The lowest of all of the tentative RQs becomes the primary criteria RQ for that hazardous substance. The Agency has determined that no potential carcinogen shall be assigned a primary criteria RQ above 100 pounds. The Agency has always regarded potential carcinogens with special concern and in its regulatory actions has sought to minimize carcinogenic risks.

As a consequence of the Agency's decision to adopt a 100 pound maximum RQ for potential carcinogens, the Human Health Assessment Group was requested to rank potential carcinogens on a three-tier scale (high, medium, and low) that corresponds to RQ levels of 1, 10 and 100 pounds.

This list includes all substances (or groups of substances) contained in the Appendix to the report for which a Hazard Ranking was made. Several substances listed in the Appendix were not ranked based on a variety of reasons explained in the report. The Appendix is titled "Hazard Ranking of Potential Carcinogens."

The data field contains the ranking (high, medium or low).

TSCA - Multichemical Test Rules - Neurotoxicity (TSC)

SOURCE

Environmental Protection Agency
Toxic Substance and Control Act
40 CFR Part 799 Identification of Specific Chemical Substance and Mixture Testing Requirements.
Subpart D - Multichemical Test Rules
40 CFR 799.5050
Administrative stay - published in Federal Register June 27, 1994 (59 FR 33184)

LIST DESCRIPTION

Under Section 4 of the Act EPA can require industry to test chemical substances already on the market or about to be produced. EPA is issuing a final rule under Section 4 requiring manufacturers and processors of 10 substances to conduct testing for neurotoxicity. These substances are related in that all are volatile solvents with high production volumes, occupational exposure, presence in and/or release to the environment, and, with the exception of 2-ethoxyethanol, consumer exposure. This rule requires cognitive function and screening level tests for neurotoxicity.

EPA has developed this rule to test a number of substances for a single toxicological endpoint, neurotoxicity. EPA believes that available data on the neurotoxic effects of many chemicals in commerce are insufficient to evaluate human health risk.

Initial selection of specific organic solvents by EPA was based on five criteria:

1) production level greater than 10 million pounds,
2) occupational exposure greater than 10,000 workers,
3) consumer exposure,
4) vapor pressure greater than 5 mmHg
5) presence in or release to the environment.

Health Effects testing includes:

Acute Neurotoxicity:
    Functional observational battery
    Motor activity
Subchronic neurotoxicity:
    Functional observational battery
    Motor activity
    Neuropathology
    Schedule-controlled operant behavior

On June 27, 1994 EPA announced its decision to stay the Multi-Substance Rule for the Testing of Neurotoxicity at 40 CFR 799.5050 pending final action on a proposed revocation of the final test rule. The proposed revocation of the final test rule was also published in the June 27 Federal Register.

The manufacturers of 7 or the 10 chemicals subject to the final test rule have agreed, subject to certain conditions set forth in the settlement agreement to conduct a set of neurotoxicity and pharmacokinetics testing under the enforceable consent agreements.

This list identifies the substances in 40 CFR 799.5050 Table 1.

The data field contains "administrative stay for neurotoxicity tests effective June 27, 1994".

EPA - Master Testing List   (OTS)

SOURCE

Environmental Protection Agency
Office of Pollution Prevention and Toxics
Existing Chemicals Program
For additional information contact:
Susan Hazen, Director
Environmental Assistance Division (TS-799)
U.S. EPA
401 M Street SW
Washington DC 20460
telephone: (202)554-1404

LIST DESCRIPTION

EPA has been using the MTL since 1990 to set its chemical testing agenda. Section 4 of the Toxic Substance Control Act (TSCA) gives EPA the authority to require chemical manufacturers and processors to test chemicals. Under Section 4, EPA has the authority to require testing after finding that (1) a chemical substance may present an unreasonable risk of injury to human health or the environment, or the chemical is produced in substantial quantities which could result in significant or substantial human or environmental exposure, and (2) available data to evaluate the chemical are inadequate, and (3) testing is needed to develop the necessary data. EPA's Chemical Testing Program also continues to work with industry to develop test data by way of consent orders and voluntary testing agreements.

The purposes of the MTL are to (1) identify chemical testing needs of the Federal Government (including EPA) and international programs of interest to the U.S., (2) focus limited EPA resources on the highest priority chemical testing needs, (3) identify and publicize EPA's testing priorities for industrial chemicals, (4) obtain broad public comment on EPA's Chemical Testing Program and its priorities, and (5) encourage initiatives by industry to provide EPA with the priority data needs identified on the MTL.

Existing test data, as well as any suggestions for subsequent versions of the MTL should be sent in triplicate to the TSCA Public Docket (TS-793), Attn - TSCA Section 4 Master Testing List, Office of Pollution Prevention and Toxics, U.S. EPA, 401 M St. SW, Washington, DC 20440.

EPA is making test results and the results of the Agency's review of test data available to the public through summaries that are added to TSCATS (TSCA Test Submissions), a publicly accessible computerized data base. In addition, information about testing decisions resulting from Risk Management meetings are contained inthe administrative record, a central collection point established by OPPT for materials on each chemical handled by OPPT's Existing Chemicals Program.

OPPT receives requests for test data and nominations for additional testing from other EPA offices, Federal and state agencies, the ITC, and other sources outside the program. OPPT screens these chemicals and others identified through its Existing Chemicals Program.

This list contains all chemicals identified in the "Index of Chemicals Sorted by CAS Number" from the MTL document cited.

The data field contains only the word "present". The source of the request for testing (i.e. other agencies or groups) has not been added.

List of Pesticide Product Inert Ingredients   (PST)

SOURCE

Environmental Protection Agency
Office of Pesticides and Toxic Substances
List of Pesticide Product Inert Ingredients

Public Response and Program Resources Branch
(703) 305-5805

LIST DESCRIPTION

The referenced document was obtained from the Freedom of Information Office, Office of Pesticide Programs, EPA and is dated January 15, 1992. The following is the introduction to the list provided by EPA:

The attached document is a comprehensive list of pesticide product inert ingredients and their associated Chemical Abstract Service (CAS) Registry Numbers. This list is inclusive of the previously published Lists 1,2, and 4, as well as those inerts which were to be compiled as List 3. This list represents OPP's most exhaustive and complete effort to date to identify and compile pesticide product inert ingredients.

Pesticide regulation is a dynamic process whereby products are continually being registered and cancelled, and product formulations frequently are revised or alternated. This inherent factor affecting the accuracy of any compilation of inert ingredients should be noted.

At the present time, OPP records do not readily permit the confirmation of the presence of any given listed inert in currently registered pesticide products. Therefore, this list should not be construed as a compilation of "acceptable" inert ingredients but rather as a candidate list of inert ingredients which may or may not be present in currently registered pesticide products.

Conversely, the fact that a particular chemical is not on the list does not mean that it is not present as an inert ingredient in a registered pesticide product; it may have been inadvertently omitted from the list. Any chemical not appearing on this list will generally be considered a new inert ingredient unless appropriate documentation is provided which establishes the presence of the chemical in currently registered pesticide products.

If you have any questions, please contact the Public Response and Program Resources Branch on (703) 305-5805.

The comprehensive list of chemicals attached to the above letter is provided here.

The list did not provide any toxicological classification (as in the previous lists). The data field contains only the word 'present'.

RCRA - Universal Treatment Standards (LDR)   (—)

SOURCE

Environmental Protection Agency
Resource Conservation and Recovery Act
40 CFR Subchapter I - Solid Wastes
Part 268 - Land Disposal Restrictions
Subpart D - Treatment Standards
Paragraph 268.48 - Universal Treatment Standards
Table UTS
Final Rule published at Federal Register September 19, 1994 (59 FR 47982). Effective December 19, 1994

LIST DESCRIPTION

The Resource Conservation and Recovery Act was passed by Congress in 1976 to deal with the control of all varieties of solid waste disposal, including the disposal of wastes that EPA lists as hazardous. RCRA defines hazardous waste, among other things, as solid waste that may pose a substantial present or potential hazard to human health and the environment when improperly treated, stored, transported, disposed, or otherwise managed.

The primary responsibility for determining whether wastes exhibit hazardous characteristics rests with generators, for whom standardization and availability of testing protocols are essential. Once a waste has been determined to be hazardous, those who generate, transport, or dispose of it must comply with the variety of notification, recordkeeping, permitting and monitoring requirements under RCRA.

EPA defines which solid wastes are hazardous by either identifying the characteristics of hazardous waste or listing particular hazardous wastes. Identifying characteristics of hazardous waste and listing hazardous wastes are distinct and fundamentally different mechanisms for defining hazardous wastes. The hazardous waste characteristics promulgated by EPA designate broad classes of wastes which are clearly hazardous by virtue of an inherent property which would result in harm to human health or the environment if a waste is mismanaged. Test methods and regulatory levels for each characteristic property are then established.

Part 268 identifies hazardous wastes that are restricted from land disposal and defines those limited circumstances under which an otherwise prohibited waste may continue to be land disposed. Subpart D covers Treatment Standards.

The September 19, 1994 amendments simplify and provide consistency in requirements. EPA has established a single set of requirements, referred to as universal treatment standards that apply to mosthazardous wastes. Universal treatment standards are established for each constituent in nonwastewater form and a single UTS for each constituent in wastewater form, regardless of the hazardous waste containing the constituent.

The universal treatment standards apply to all listed and characteristic wastes for which treatment standards have been promulgated, with two exceptions. The first exception is teh TC metal wastes (D004-D011). These metal wastes will be addressed in the future Phase IV LDR rule. The second exception is those for which the treatment standard is a specified method of treatment. Most of these wastes must continue to be treated using those required technologies.

In most cases (59%), UTS are the same as the previous treatment standards. Thirty three percent of the standards went up or down within a factor of ten of the original of the original standard, while 8 percent underwent larger changes (3 percent of the total number of UTS becoming significantly more stringent.

Table UTS identifies the hazardous constituents, along with the nonwastewater and wastewater treatment standard levels that are used to regulate most prohibited hazardous wastes with numerical limits. Wastewater standards are given in milligrams/liter based on analysis of composite samples. Nonwastewater standard are given as mg/kg unless noted as " mg/l TCLP".

RCRA - F Series Wastes   (—)

SOURCE

Environmental Protection Agency
Resource Conservation and Recovery Act
40 CFR Subchapter I - Solid Wastes
Part 261 - Identification and Listing of Hazardous Waste
Subpart D - Lists of Hazardous Wastes
Paragraphs 261.31 - Hazardous Wastes from non-specific sources
40 CFR July, 1992

LIST DESCRIPTION

The Resource Conservation and Recovery Act was passed by Congress in 1976 to deal with the control of all varieties of solid waste disposal - including the disposal of wastes that EPA lists as hazardous. RCRA defines hazardous waste, among other things, as solid waste that may pose a substantial present or potential hazard to human health and the environment when improperly treated, stored, transported, disposed, or otherwise managed.

The primary responsibility for determining whether wastes exhibit hazardous characteristics rests with generators, for whom standardization and availability of testing protocols are essential. Once a waste has been determined to be hazardous, those who generate, transport, or dispose of it must comply with the variety of notification, recordkeeping, permitting and monitoring requirements under RCRA.

EPA defines which solid wastes are hazardous by either identifying the characteristics of hazardous waste or listing particular hazardous wastes. Identifying characteristics of hazardous waste and listing hazardous wastes are distinct and fundamentally different mechanisms for defining hazardous wastes. The hazardous waste characteristics promulgated by EPA designate broad classes of wastes which are clearly hazardous by virtue of an inherent property which would result in harm to human health or the environment if a waste is mismanaged. Test methods and regulatory levels for each characteristic property are then established.

For those wastes defined as hazardous by virtue of their listing in the regulations, EPA must indicate the basis for listing. The criteria are defined in 40 CFR 261.11 and include the following:

Ignitability, Corrosivity, Reactivity as defined in Subpart C

Acutely Hazardous - oral LD50 (rat) less than 50 mg/kg inhalation LC50 (rat) less than 2 mg/L dermal LD50 (rabbit) less than 200 mg/kg or otherwise capable of causing or significantly contributing to an increase in serious irreversible, or incapacitating reversible, illness.

Toxic Wastes - it contains any of the toxic constituents listed in Appendix VIII and, after considering a variety of factors identified in 261.11(3), the Administrator concludes that the waste is capable of posing a substantial present or potential hazard to human health or the environment when improperly treated, stored, transported or disposed of, or otherwise managed.

Substances will be listed in Appendix VIII only if they have been shown in scientific studies to have toxic, carcinogenic, mutagenic or teratogenic effects on humans or other life forms.

This list identifies the Hazardous Wastes defined in 40 CFR 261.31 as wastes from non-specific sources. All contain the letter "F" in the Hazardous Waste number and may be called "F Series Wastes." The full description of each F series waste is provided in the Appendix. The identity of these wastes often include reference to specific chemicals. Any specific chemical mentioned as part of an F series waste include a reference to the F waste number in the "See Also" portion of the Regulatory Summary.

The data includes the basis for listing the waste (Toxic, Reactive, Ignitable, Corrosive, Acutely Hazardous).

RCRA - Substances Banned from Land Disposal   (—)

SOURCE

Environmental Protection Agency
Resource Conservation and Recovery Act
40 CFR Subchapter I - Solid Wastes
Part 268 - Land Disposal Restrictions
Subpart C - Prohibition on Land Disposal
Paragraph 268.35 - Third Third wastes
with amendments through June 20, 1994 (59 FR 31551)

LIST DESCRIPTION

The Resource Conservation and Recovery Act was passed by Congress in 1976 to deal with the control of all varieties of solid waste disposal, including the disposal of wastes that EPA lists as hazardous. RCRA defines hazardous waste, among other things, as solid waste that may pose a substantial present or potential hazard to human health and the environment when improperly treated, stored, transported, disposed, or otherwise managed.

The primary responsibility for determining whether wastes exhibit hazardous characteristics rests with generators, for whom standardization and availability of testing protocols are essential. Once a waste has been determined to be hazardous, those who generate, transport, or dispose of it must comply with the variety of notification, recordkeeping, permitting and monitoring requirements under RCRA.

EPA defines which solid wastes are hazardous by either identifying the characteristics of hazardous waste or listing particular hazardous wastes. Identifying characteristics of hazardous waste and listing hazardous wastes are distinct and fundamentally different mechanisms for defining hazardous wastes. The hazardous waste characteristics promulgated by EPA designate broad classes of wastes which are clearly hazardous by virtue of an inherent property which would result in harm to human health or the environment if a waste is mismanaged. Test methods and regulatory levels for each characteristic property are then established.

Part 268 identifies hazardous wastes that are restricted from land disposal and defines those limited circumstances under which an otherwise prohibited waste may continue to be land disposed. Subpart C identifies specific wastes and the time frame and other conditions under which prohibitions on land disposal are enacted.

The Federal Register Notice of June 26, 1992 adds an effective date of May 8, 1993 for D006 (lead materials stored before secondary smelting). It also requires that the owner or operator of each secondary lead smelting facility shall submit to EPA information as described in 40 CFR 268.35(k) on or before March 1, 1993.

This list includes all wastes for which the effective date for prohibition has already passed (as of October, 1992). It includes wastes identified in 268.30 (solvent wastes), 268.32 (California list wastes), 268.33 (First Third), 268. 34 (Second Third) and 268.35 (Third

Third) and the seven additional wastes added in the August 18, 1992 amendment.

The data field contains the word (present) and identifies if the material is listed for wastewaters.

RCRA - K Series Wastes   (—)

SOURCE

Environmental Protection Agency
Resource Conservation and Recovery Act
40 CFR Subchapter I - Solid Wastes
Part 261 - Identification and Listing of Hazardous Waste
Subpart D - Lists of Hazardous Wastes
Paragraphs 261.32 - Hazardous Wastes from specific sources
40 CFR July, 1992 with amendments through October 15, 1992 (57 FR 47376).

LIST DESCRIPTION

The Resource Conservation and Recovery Act was passed by Congress in 1976 to deal with the control of all varieties of solid waste disposal, including the disposal of wastes that EPA lists as hazardous. RCRA defines hazardous waste, among other things, as solid waste that may pose a substantial present or potential hazard to human health and the environment when improperly treated, stored, transported, disposed, or otherwise managed.

The primary responsibility for determining whether wastes exhibit hazardous characteristics rests with generators, for whom standardization and availability of testing protocols are essential. Once a waste has been determined to be hazardous, those who generate, transport, or dispose of it must comply with the variety of notification, recordkeeping, permitting and monitoring requirements under RCRA.

EPA defines which solid wastes are hazardous by either identifying the characteristics of hazardous waste or listing particular hazardous wastes. Identifying characteristics of hazardous waste and listing hazardous wastes are distinct and fundamentally different mechanisms for defining hazardous wastes. The hazardous waste characteristics promulgated by EPA designate broad classes of wastes which are clearly hazardous by virtue of an inherent property which would result in harm to human health or the environment if a waste is mismanaged. Test methods and regulatory levels for each characteristic property are then established.

For those wastes defined as hazardous by virtue of their listing in the regulations, EPA must indicate the basis for listing. The criteria are defined in 40 CFR 261.11 and include the following:

Ignitability, Corrosivity, Reactivity as defined in Subpart C
Acutely Hazardous - oral LD50 (rat) less than 50mg/kg; inhalation LC50 (rat) less than 2 mg/L; dermal LD50 (rabbit) less than 200 mg/kg; or otherwise capable of causing or significantly contributing to an increase in serious irreversible, or incapacitating reversible, illness.

Toxic Wastes - it contains any of the toxic constituents listed in Appendix VIII and, after considering a variety of factors identified in 261.11(3), the Administrator concludes that the waste is capable of posing a substantial present or potential hazard to human health or the environment when improperly treated, stored, transported or disposed of, or otherwise managed.

Substances will be listed in Appendix VIII only if they have been shown in scientific studies to have toxic, carcinogenic, mutagenic or teratogenic effects on humans or other life forms.

This list identifies the Hazardous Wastes from specific sources described in 40 CFR 261.32. All contain the letter "K" in the Hazardous Waste number and may be called "K Series Wastes." The identity of these wastes often include reference to specific chemicals. Any specific chemical mentioned as part of an K series waste include a reference to the K waste number in the "See Also" portion of the Regulatory Summary. The full description of each K series waste is provided in the Appendix.

The data includes the basis for listing the waste (Toxic, Reactive, Ignitable, Corrosive, Acutely Hazardous).

RCRA - TSD Facilities Groundwater Monitoring   (PQL)

SOURCE

Environmental Protection Agency
Resource Conservation and Recovery Act
40 CFR Subchapter I - Solid Wastes
Part 264 - Standards for Owners and Operators of
Hazardous Waste Treatment, Storage, and Disposal Facilities
Appendix IX - Groundwater Monitoring List
40 CFR July, 1992 edition

LIST DESCRIPTION

The Resource Conservation and Recovery Act was passed by Congress in 1976 to deal with the control of all varieties of solid waste disposal, including the disposal of wastes that EPA lists as hazardous. RCRA defines hazardous waste, among other things, as solid waste that may pose a substantial present or potential hazard to human health and the environment when improperly treated, stored, transported, disposed, or otherwise managed.

The primary responsibility for determining whether wastes exhibit hazardous characteristics rests with generators, for whom standardization and availability of testing protocols are essential. Once a waste has been determined to be hazardous, those who generate, transport, or dispose of it must comply with the variety of notification, recordkeeping, permitting and monitoring requirements under RCRA.

Part 264 establishes minimum national standards which define the acceptable management of hazardous wastes. The standards in Part 264 apply to owners and operators of all facilities which treat, store, or dispose of hazardous waste, except as specifically provided. Part 264 does not apply to the treatment or storage of hazardous waste before it is loaded onto an ocean vessel for incineration or disposal at sea, or of hazardous waste before it is injected underground. Other qualification to requirements of this Part are found in Subpart A.

Subpart F of Part 264 defines the requirements for releases from solid waste management units including general ground-water monitoring requirements (264.97), the detection monitoring program requirements (264.98) and the compliance monitoring program requirements (264.99). Detection monitoring must include sampling the ground water in all monitoring wells for the constituents in the list contained in Appendix IX to Part 264.

This list includes substances listed in Appendix IX (Groundwater Monitoring List) of Part 264.

The data field contains the suggested test method(s) as defined in EPA Report SW-846 and the Practical Quantitation Limit (PQL). The regulatory requirements apply to the list of substances ONLY. The test methods and PQLs are given for informational purposes only. The PQL is defined as the lowest concentration of analytes in ground waters that can be reliably determined within specified limits of precision and accuracy by the indicated methods under routine laboratory operating conditions. The PQLs listed are generally stated to one significant figure. CAUTION: The PQL values in many cases are based only on a general estimate for the method and not on a determination for individual compounds. PQLs are not a part of the regulation.

RCRA - D Series - Chronic Toxicity Reference (RC2) Levels

SOURCE

Environmental Protection Agency
Resource Conservation and Recovery Act
55 FR 11804 (March 29, 1990) Preamble to Toxicity
  Characteristics Revisions

LIST DESCRIPTION

The Resource Conservation and Recovery Act was passed by Congress in 1976 to deal with the control of all varieties of solid waste disposal - including the disposal of wastes that EPA lists as hazardous. RCRA defines hazardous waste, among other things, as solid waste that may pose a substantial present or potential hazard to human health and the environment when improperly treated, stored, transported, disposed, or otherwise managed.

The primary responsibility for determining whether wastes exhibit hazardous characteristics rests with generators, for whom standardization and availability of testing protocols are essential. Once a waste has been determined to be hazardous, those who generate, transport, or dispose of it must comply with the variety of notification, recordkeeping, permitting and monitoring requirements under RCRA.

EPA defines which solid wastes are hazardous by either identifying the characteristics of hazardous waste

or listing particular hazardous wastes. Identifying characteristics of hazardous waste and listing hazardous wastes are distinct and fundamentally different mechanisms for defining hazardous wastes. The hazardous waste characteristics promulgated by EPA designate broad classes of wastes which are clearly hazardous by virtue of an inherent property which would result in harm to human health or the environment if a waste is mismanaged. Test methods and regulatory levels for each characteristic property are then established.

The four hazardous waste characteristics are: ignitability, corrosivity, reactivity and toxicity. The characteristics of ignitability, corrosivity and reactivity are based on physical tests identified in 40 CFR 261.21, 261.22 and 261.23 respectively. The characteristic of toxicity is established if, by the using the Toxicity Characteristic Leachability Procedure (TCLP), the extract from a representative sample of the waste contains any of the contaminants listed in Table 1 at the concentration equal to or greater than the respective value given in that Table. The Toxicity Characteristic Leachability Procedure is designed to measure the potential for toxic constituents in the waste to leach out and contaminate ground water.

In establishing the Maximum Concentration of Contaminant for the Toxicity Characteristic, the Environmental Protection Agency developed "chronic toxicity reference levels." Chronic toxicity reference levels are those levels below which chronic exposure for individual toxicants in drinking water is considered safe or considered to pose minimal risk (in the case of carcinogens). EPA used Drinking Water Standards for some toxicants, Maximum Contaminant Level Goals (MCLGs) for some and for the remaining established chronic toxicity reference levels using Reference Doses (RfDs) for non-carcinogens and Risk-Specific Doses (RSDs) for carcinogens.

This list identifies the substances for which EPA has established Chronic Toxicity Reference Levels for use in establishing the Maximum Concentration of Contaminants for the Toxicity Characteristic (40 CFR 262.24 Table I). These were published in the March 29, 1990 Federal Register (page 11804) in the preamble to the final rule establishing the TCLP.

The data field contains chronic toxicity reference level in mg/L.

RCRA - D Series - Maximum Concentrations of (RC2) Contaminants

SOURCE

Environmental Protection Agency
Resource Conservation and Recovery Act
40 CFR Subchapter I - Solid Wastes
Part 261 - Identification and Listing of Hazardous
  Waste
Subpart C - Characteristics of Hazardous Waste
Paragraphs 261.21 to 261.24
40 CFR July 1992

LIST DESCRIPTION

The Resource Conservation and Recovery Act was passed by Congress in 1976 to deal with the control of all varieties of solid waste disposal, including the disposal of wastes that EPA lists as hazardous. RCRA defines hazardous waste, among other things, as solid waste that may pose a substantial present or potential hazard to human health and the environment when improperly treated, stored, transported, disposed, or otherwise managed.

The primary responsibility for determining whether wastes exhibit hazardous characteristics rests with generators, for whom standardization and availability of testing protocols are essential. Once a waste has been determined to be hazardous, those who generate, transport, or dispose of it must comply with the variety of notification, recordkeeping, permitting and monitoring requirements under RCRA.

EPA defines which solid wastes are hazardous by either identifying the characteristics of hazardous waste or listing particular hazardous wastes. Identifying characteristics of hazardous waste and listing hazardous wastes are distinct and fundamentally different mechanisms for defining hazardous wastes. The hazardous waste characteristics promulgated by EPA designate broad classes of wastes which are clearly hazardous by virtue of an inherent property which would result in harm to human health or the environment if a waste is mismanaged. Test methods and regulatory levels for each characteristic property are then established.

The four hazardous waste characteristics are: ignitability, corrosivity, reactivity and toxicity. The characteristics of ignitability, corrosivity and reactivity are based on physical tests identified in 40 CFR 261.21, 261.22 and 261.23 respectively. The characteristic of toxicity is established if, by the using the Toxicity Characteristic Leachability Procedure (TCLP), the extract from a representative sample of the waste contains any of the contaminants listed in Table 1 at the concentration equal to or greater than the respective value given in that Table. TheToxicity Characteristic Leachability Procedure is designed to measure the potential for toxic constituents in the waste to leach out and contaminate ground water. A solid waste that exhibits the characteristic of toxicity has the EPA Hazardous Waste Number specified in Table I which corresponds to the toxic contaminant causing it to be hazardous.

This list identifies the constituents listed in Table I of 40 CFR 261.24 as well as the Hazardous Waste Numbers for the broad classes of ignitability, corrosivity and reactivity.

The data field contains the maximum concentrations of contaminants (the regulatory level identified in Table I) below which the waste would not be considered hazardous and the hazardous waste number by which the waste must be identified. All wastes meeting these four broad categories are identified with the letter "D" and are often called "D Series" wastes.

RCRA - Basis for Listing - Appendix VII   (RCA)

SOURCE

Environmental Protection Agency
Resource Conservation and Recovery Act
40 CFR Subchapter I - Solid Wastes
Part 261 - Identification and Listing of Hazardous Waste
Appendix VII - Basis for Listing Hazardous Waste
with amendments through October 15, 1992 (57 FR 47376)

LIST DESCRIPTION

The Resource Conservation and Recovery Act was passed by Congress in 1976 to deal with the control of all varieties of solid waste disposal, including the disposal of wastes that EPA lists as hazardous. RCRA defines hazardous waste, among other things, as solid waste that may pose a substantial present or potential hazard to human health and the environment when improperly treated, stored, transported, disposed, or otherwise managed.

The primary responsibility for determining whether wastes exhibit hazardous characteristics rests with generators, for whom standardization and availability of testing protocols are essential. Once a waste has been determined to be hazardous, those who generate, transport, or dispose of it must comply with the variety of notification, recordkeeping, permitting and monitoring requirements under RCRA.

EPA defines which solid wastes are hazardous by either identifying the characteristics of hazardous waste or listing particular hazardous wastes. Identifying characteristics of hazardous waste and listing hazardous wastes are distinct and fundamentally different mechanisms for defining hazardous wastes. The hazardous waste characteristics promulgated by EPA designate broad classes of wastes which are clearly hazardous by virtue of an inherent property which would result in harm to human health or the environment if a waste is mismanaged. Test methods and regulatory levels for each characteristic property are then established.

For those wastes defined as hazardous by virtue of their listing in the regulations, EPA must indicate the basis for listing. The criteria are defined in 40 CFR 261.11 and include the following:

Ignitability, Corrosivity, Reactivity as defined in Subpart C

Acutely Hazardous - oral LD50 (rat) less than 50 mg/kg; inhalation LC50 (rat) less than 2 mg/L; dermal LD50 (rabbit) less than 200 mg/kg; or otherwise capable of causing or significantlycontributing to an increase in serious irreversible, or incapacitating reversible, illness.

Toxic Wastes - it contains any of the toxic constituents listed in Appendix VIII and, after considering a variety of factors identified in 261.11(3), the Administrator concludes that the waste is capable of posing a substantial present or potential hazard to human health or the environment when improperly treated, stored, transported or disposed of, or otherwise managed.

Substances will be listed in Appendix VIII only if they have been shown in scientific studies to have toxic, carcinogenic, mutagenic or teratogenic effects on humans or other life forms.

The wastes identified from non-specific sources (F Series wastes) and from Specific Sources (K Series wastes) may contain one or more hazardous constituents which contribute to the hazardous properties of the waste. Appendix VII to Part 261 identifies those constituents. Appendix III identifies chemical analysis test methods which are used to determine whether a sample contains a given Appendix VII or VIII toxic constituent.

The data includes the waste stream Hazardous Waste Identification Numbers in which that substance contributes to the hazard.

RCRA - U Series Wastes   (RCU)
SOURCE

Environmental Protection Agency
Resource Conservation and Recovery Act
40 CFR Subchapter I - Solid Wastes
Part 261 - Identification and Listing of Hazardous Waste
Subpart D - Lists of Hazardous Wastes
Paragraphs 261.33(f) - Discarded commercial chemical products, off-specification species, container residues, and spill residues thereof – Toxic Wastes
40 CFR July, 1992

LIST DESCRIPTION

The Resource Conservation and Recovery Act was passed by Congress in 1976 to deal with the control of all varieties of solid waste disposal, including the disposal of wastes that EPA lists as hazardous. RCRA defines hazardous waste, among other things, as solid waste that may pose a substantial present or potential hazard to human health and the environment when improperly treated, stored, transported, disposed, or otherwise managed.

The primary responsibility for determining whether wastes exhibit hazardous characteristics rests with generators, for whom standardization and availability of testing protocols are essential. Once a waste has been determined to be hazardous, those who generate, transport, or dispose of it must comply with the variety of notification, recordkeeping, permitting and monitoring requirements under RCRA.

EPA defines which solid wastes are hazardous by either identifying the characteristics of hazardous waste or listing particular hazardous wastes. Identifying characteristics of hazardous waste and listing hazardous wastes are distinct and fundamentally different mechanisms for defining hazardous wastes. The hazardous waste characteristics promulgated by EPA designate broad classes of wastes which are clearly hazardous by virtue of an inherent property which would result in harm to human health or the environment if a waste

is mismanaged. Test methods and regulatory levels for each characteristic property are then established.

For those waste defined as hazardous by virtue of their listing in the regulations, EPA must indicate the basis for listing. The criteria are defined in 40 CFR 261.11 and include the following:

Ignitability, Corrosivity, Reactivity as defined in Subpart C
Acutely Hazardous - oral LD50 (rat) less than 50mg/kg; inhalation LC50 (rat) less than 2 mg/L; dermal Ld50 (rabbit) less than 200 mg/kg; or otherwise capable of causing or significantly contributing to an increase in serious irreversible, or incapacitating reversible, illness.
Toxic Wastes - it contains any of the toxic constituents listed in Appendix VIII and, after considering a variety of factors identified in 261.11(3), the Administrator concludes that the waste is capable of posing a substantial present or potential hazard to human health or the environment when improperly treated, stored, transported or disposed of, or otherwise managed.

Substances will be listed in Appendix VIII only if they have been shown in scientific studies to have toxic, carcinogenic, mutagenic or teratogenic effects on humans or other life forms.

This list identifies the commercial chemical products, manufacturing chemical intermediates or off-specification commercial chemical products or manufacturing chemical intermediates referred to in paragraphs (a) through (d) of Section 261.33. All the substances listed have been determined by EPA to be Toxic Wastes (T). The Hazardous Waste Numbers for these wastes all begin with the letter "U."

The data includes the Hazardous Waste Number and any ADDITIONAL basis for listing (other than toxicity).

RCRA - Hazardous Constituents - Appendix (RCU) VIII
SOURCE

Environmental Protection Agency
Resource Conservation and Recovery Act
40 CFR Subchapter I - Solid Wastes
Part 261 - Identification and Listing of Hazardous Waste
Appendix VIII - Hazardous Wastes
with amendments through June 20, 1994 (59 FR 31551)

LIST DESCRIPTION

The Resource Conservation and Recovery Act was passed by Congress in 1976 to deal with the control of all varieties of solid waste disposal, including the disposal of wastes that EPA lists as hazardous. RCRA defines hazardous waste, among other things, as solid waste that may pose a substantial present or potential hazard to human health and the environment when improperly treated, stored, transported, disposed, or otherwise managed.

The primary responsibility for determining whether wastes exhibit hazardous characteristics rests with generators, for whom standardization and availability of testing protocols are essential. Once a waste has been determined to be hazardous, those who generate, transport, or dispose of it must comply with the variety of notification, recordkeeping, permitting and monitoring requirements under RCRA.

EPA defines which solid wastes are hazardous by either identifying the characteristics of hazardous waste or listing particular hazardous wastes. Identifying characteristics of hazardous waste and listing hazardous wastes are distinct and fundamentally different mechanisms for defining hazardous wastes. The hazardous waste characteristics promulgated by EPA designate broad classes of wastes which are clearly hazardous by virtue of an inherent property which would result in harm to human health or the environment if a waste is mismanaged. Test methods and regulatory levels for each characteristic property are then established.

For those wastes defined as hazardous by virtue of their listing in the regulations, EPA must indicate the basis for listing. The criteria are defined in 40 CFR 261.11 and include the following:

Ignitability, Corrosivity, Reactivity as defined in Subpart C

Acutely Hazardous - oral LD50 (rat) less than 50 mg/kg; inhalation LC50 (rat) less than 2 mg/L; dermal LD50 (rabbit) less than 200 mg/kg; or otherwise capable of causing or significantlycontributing to an increase in serious irreversible, or incapacitating reversible, illness.

Toxic Wastes - it contains any of the toxic constituents listed in Appendix VIII and, after considering a variety of factors identified in 261.11(3), the Administrator concludes that the waste is capable of posing a substantial present or potential hazard to human health or the environment when improperly treated, stored, transported or disposed of, or otherwise managed.

Substances will be listed in Appendix VIII only if they have been shown in scientific studies to have toxic, carcinogenic, mutagenic or teratogenic effects on humans or other life forms.

This list contains all entries in Appendix VIII.

The data field identifies the "P" Series or "U"Series waste Numbers. If the substance does not have a waste number listed, the words "hazardous constituent - no waste number" are provided.

RCRA - P Series Wastes   (RCU)

SOURCE

Environmental Protection Agency
Resource Conservation and Recovery Act
40 CFR Subchapter I - Solid Wastes
Part 261 - Identification and Listing of Hazardous Waste
Subpart D - Lists of Hazardous Wastes
Paragraphs 261.33(e) - Discarded commercial chemical products, off-specification species, container residues, and spill residues thereof – Acutely Hazardous Wastes
with amendments through June 20, 1994 (59 FR 31551)

LIST DESCRIPTION

The Resource Conservation and Recovery Act was passed by Congress in 1976 to deal with the control of all varieties of solid waste disposal - including the disposal of wastes that EPA lists as hazardous. RCRA defines hazardous waste, among other things, as solid waste that may pose a substantial present or potential hazard to human health and the environment when improperly treated, stored, transported, disposed, or otherwise managed.

The primary responsibility for determining whether wastes exhibit hazardous characteristics rests with generators, for whom standardization and availability of testing protocols are essential. Once a waste has been determined to be hazardous, those who generate, transport, or dispose of it must comply with the variety of notification, recordkeeping, permitting and monitoring requirements under RCRA.

EPA defines which solid wastes are hazardous by either identifying the characteristics of hazardous waste or listing particular hazardous wastes. Identifying characteristics of hazardous waste and listing hazardous wastes are distinct and fundamentally different mechanisms for defining hazardous wastes. The hazardous waste characteristics promulgated by EPA designate broad classes of wastes which are clearly hazardous by virtue of an inherent property which would result in harm to human health or the environment if a waste is mismanaged. Test methods and regulatory levels for each characteristic property are then established.

For those waste defined as hazardous by virtue of their listing in the regulations, EPA must indicate the basis for listing. The criteria are defined in 40 CFR 261.11 and include the following:

Ignitability, Corrosivity, Reactivity as defined in Subpart C

Acutely Hazardous - oral LD50 (rat) less than 50mg/kg inhalation LC50 (rat) less than 2 mg/L dermal LD50 (rabbit) less than 200 mg/kg or otherwise capable of causing or significantly contributing to an increase in serious irreversible, or incapacitating reversible, illness.

Toxic Wastes - it contains any of the toxic constituents listed in Appendix VIII and, after considering a variety of factors identified in

261.11(3), the Administrator concludes that the waste is capable of posing a substantial present or potential hazard to human health or the environment when improperly treated, stored, transported or disposed of, or otherwise managed.

Substances will be listed in Appendix VIII only if they have been shown in scientific studies to have toxic, carcinogenic, mutagenic or teratogenic effects on humans or other life forms.

This list identifies the commercial chemical products, manufacturing chemical intermediates or off-specification commercial chemical products or manufacturing chemical intermediates referred to in paragraphs (a) through (d) of Section 261.33. All the substances listed have been determined by EPA to be Acute Hazardous Wastes (H). The Hazardous Waste Numbers for these wastes all begin with the letter "P."

The data includes the Hazardous Waste Number and any ADDITIONAL basis for listing (other than acutely hazardous).

Safe Drinking Water Act - Monitoring   (SDW)
SOURCE

Environmental Protection Agency
Safe Drinking Water Act (SDWA)
Subchapter D - Water Programs
Part 141 - National Primary Drinking Water Regulations
Subpart E - Special Regulations, including Monitoring Regulations and Prohibition on Lead Use
40 CFR 141.40 Special monitoring for organic chemicals
141.40(e) - monitoring required; 141.40(j) - monitoring required at discretion of the State.
CFR July 1, 1992 edition

LIST DESCRIPTION

The Safe Drinking Water Act mandates establishment of uniform federal standards for drinking water quality, and sets up a system to regulate underground injection of wastes and other substances that could contaminate underground water sources. (Contamination of surface water is protected under the Clean Water Act).

Under the law, EPA sets two types of drinking water standards. Primary standards apply to substances which may have an adverse effect on health. These are enforced by the states. Compliance is mandatory. Secondary standards provide guidelines on substances or conditions that affect color, taste, smell, and other physical characteristics of drinking water. These standards are advisory, not mandatory. Drinking water obtained from underground sources must be tested to see that it meets the primary standards. Some states also require that water meet the secondary standards.

The law also bans underground injection of certain materials in or near an underground water source, and requires issuing permits, monitoring, and recordkeeping for underground injection that is allowable.

The Safe Drinking Water Act was amended in June 1986 to remove the EPA's discretion in deciding whether to set standards for contaminants in drinking water. EPA was required to issue binding standards for 83 drinking water contaminants within three years.

All community and non-transient, non-community water systems must monitor for the contaminants listed in 40 CFR 141.40(e). Monitoring for the substances listed in 40 CFR 141.40(j) is required at the discretion of the state.

This list identifies the substances in 40 CFR 141.40(e) and 141.40(j).

The data field indicates monitoring required or required at the discretion of the state.

Safe Drinking Water Act - MCLGs   (SDW)
SOURCE

Environmental Protection Agency
Safe Drinking Water Act (SDWA)
Subchapter D - Water Programs
Part 141 - National Primary Drinking Water Regulations
Subpart F - Maximum Contaminant Level Goals
40 CFR 141.50, 141.51 - MCLGs for organic chemicals, inorganic chemicals - Amendments through July 17, 1992 (57 FR 31776)

LIST DESCRIPTION

The Safe Drinking Water Act mandates establishment of uniform federal standards for drinking water quality, and set up a system to regulate underground injection of wastes and other substances that could contaminate underground water sources. (Contamination of surface water is protected under the Clean Water Act).

Under the law, EPA sets two types of drinking water standards. Primary standards apply to substances which may have an adverse effect on health. These are enforced by the states. Compliance is mandatory. Secondary standards provide guidelines on substances or conditions that affect color, taste, smell, and other physical characteristics of drinking water. These standards are advisory, not mandatory. Drinking water obtained from underground sources must be tested to see that it meets the primary standards. Some states also require that water meet the secondary standards.

The law also bans underground injection of certain materials in or near an underground water source, and requires issuing permits, monitoring, and recordkeeping for underground injection that is allowable.

The Safe Drinking Water Act was amended in June 1986 to remove the EPA's discretion in deciding whether to set standards for contaminants in drinking water. EPA was required to issue binding standards for 83 drinking water contaminants within three years.

Primary standards set limits on contaminants that may affect health and are enforced by the state. States must measure and monitor the amounts of the primary contaminants in public water supplies and report the results to EPA periodically. Municipal governments and private water companies that supply water to the public must test water quality to make certain that no contaminants exceed federal limits.

The Safe Drinking Water Act as amended in 1986 requires EPA to publish "maximum contaminant level goals" (MCLGs) for contaminants which, in the judgment of the Administrator, "mayhave any adverse effect on the health of persons and which are known or anticipated to occur in public water systems." MCLGs are to be set at a level at which "no known or anticipated adverse effects on the health of persons occur and which allow an adequate margin of safety."

MCLGs are non-enforceable goals. An MCL must be set as close to the MCLG as feasible using the best technology, treatment techniques and other means which the Administrator finds are available.

On January 30, 1991 a final rule was published in the Federal Register (56 FR 3526) which established MCLs and MCLGs for 26 synthetic organic chemicals and 7 inorganic chemicals. This finalized (and made several changes in) the Proposed Rule published on May 22, 1989 (note deletion of Proposed List from the LOLI/RCD database). On June 7, 1991 (56 FR 26460) EPA finalized the MCLGs for lead and copper, on July 1, 1991 (56 FR 30266) EPA finalized the MCLs and MCLGs for aldicarb, aldicarb sulfoxide, aldicarb sulfone, pentachlorophenol, and barium and on July 17, 1992 (57 FR 31776) EPA finalized

MCLGs and MCLs (and NPDWRs) for 18 synthetic organic chemicals and 5 inorganic chemicals.

This list includes all chemicals for which Maximum Contaminant Level Goals (MCLGs) have been established.

The data field provides the level in mg/L.

Safe Drinking Water Act - SMCLs (SDW)

SOURCE

Environmental Protection Agency
Safe Drinking Water Act (SDWA)
Subchapter D - Water Programs
Part 143 - National Secondary Drinking Water Regulations
40 CFR 143.3 - SMCLs
CFR July 1, 1992 edition

LIST DESCRIPTION

The Safe Drinking Water Act mandates establishment of uniform federal standards for drinking water quality, and sets up a system to regulate underground injection of wastes and other substances that could contaminate underground water sources. (Contamination of surface water is protected under the Clean Water Act).

Under the law, EPA sets two types of drinking water standards. Primary standards apply to substances which may have an adverse effect on health. These are enforced by the states. Compliance is mandatory. Secondary standards provide guidelines on substances or conditions that affect color, taste, smell, and other physical characteristics of drinking water. These standards are advisory, not mandatory. Drinking water obtained from underground sources must be tested to see that it meets the primary standards. Some states also require that water meet the secondary standards.

The law also bans underground injection of certain materials in or near an underground water source, and requires issuing permits, monitoring, and recordkeeping for underground injection that is allowable.

The Safe Drinking Water Act was amended in June 1986 to remove the EPA's discretion in deciding whether to set standards for contaminants in drinking water. EPA was required to issue binding standards for 83 drinking water contaminants within three years.

Primary standards set limits on contaminants that may affect health and are enforced by the state. States must measure and monitor the amounts of the primary contaminants in public water supplies and report the results to EPA periodically. Municipal governments and private water companies that supply water to the public must test water quality to make certain that no contaminants exceed federal limits.

National Secondary Drinking Water Regulations control contaminants in drinking water that primarily affect the esthetic qualities relating to the public acceptance of drinking water. At considerably higher concentrations of these contaminants, healthimplications may also exist as well as esthetic-degradation. The regulations are not federally enforceable, but are intended as guidelines for the states.

This list includes all substances (not physical qualities) for which SMCLs have been established.

The data field contains the recommended level in mg/L.

Safe Drinking Water Act - MCLs (SDW)

SOURCE

Environmental Protection Agency
Safe Drinking Water Act (SDWA)
Subchapter D - Water Programs
Part 141 - National Primary Drinking Water Regulations
Subpart B - Maximum Contaminant Levels
40 CFR 141.11, 141.12 - MCLs for inorganic chemicals, organic chemicals
Incorporates changes through July 17, 1992 (57 FR 31776)

LIST DESCRIPTION

The Safe Drinking Water Act mandates establishment of uniform federal standards for drinking water quality, and sets up a system to regulate underground injection of wastes and other substances that could contaminate underground water sources. (Contamination of surface water is protected under the Clean Water Act).

Under the law, EPA sets two types of drinking water standards. Primary standards apply to substances which may have an adverse effect on health. These are enforced by the states. Compliance is mandatory. Secondary standards provide guidelines on substances or conditions that affect color, taste, smell, and other physical characteristics of drinking water. These standards are advisory, not mandatory. Drinking water obtained from underground sources must be tested to see that it meets the primary standards. Some states also require that water meet the secondary standards.

The law also bans underground injection of certain materials in or near an underground water source, and requires issuing permits, monitoring, and recordkeeping for underground injection that is allowable.

The Safe Drinking Water Act was amended in June 1986 to remove the EPA's discretion in deciding whether to set standards for contaminants in drinking water. EPA was required to issue binding standards for 83 drinking water contaminants within three years.

Primary standards set limits on contaminants which may affect health and which are enforced by the state. States must measure and monitor the amounts of the primary contaminants in public water supplies and report the results to EPA periodically. Municipal governments and private water companies that supply water to the public must test water quality to make certain that no contaminants exceed federal limits.

On January 30, 1991 a final rule was published in the Federal Register (56 FR 3526) which established MCLs and MCLGs for 26synthetic organic chemicals and 7 inorganic chemicals. This finalized (and made several changes in) the Proposed Rule published on May 22, 1989 (note deletion of Proposed List from the LOLI/RCD database). On June 7, 1991 (56 FR 26460) EPA finalized the MCLGs for lead and copper, on July 1, 1991 (56 FR 30266) EPA finalized the MCLs and MCLGs for aldicarb, aldicarb sulfoxide, aldicarb sulfone, pentachlorophenol, and barium and on July 17, 1992 (57 FR 31776) EPA finalized

MCLGs and MCLs (and NPDWRs) for 18 synthetic organic chemicals and 5 inorganic chemicals.

This list includes all substances that have a Maximum Contaminant Level established at 40 CFR 141.11 (inorganic chemicals), 40 CFR 141.12 (organic chemicals) and 40 CFR 141.15 (Radium 226, and Radium 228) through the July 17, 1992 Federal Register modification.

The data field contains the Maximum Contaminant Level in mg/L or in PicoCuries/1.

TSCA - CAIR - Regulated Chemicals   (CAI)
SOURCE:

Environmental Protection Agency
Toxic Substances Control Act
40 CFR Subpart D - Comprehensive Assessment Information Rule Reporting and Recordkeeping Requirements
Paragraph 704.225 (Chemical substances matrix)
July 1, 1992 edition

LIST DESCRIPTION

In 1976 Congress passed the Toxic Substance Control Act as a method to control exposure and use of raw industrial chemicals that fall outside the jurisdiction of other environmental laws. TSCA was passed to assure that chemicals would be evaluated before use to make sure they pose no unnecessary risk to health or the environment. Under TSCA, EPA can require that the effects of an existing chemical on health and/or the environment be evaluated through a series of toxicity tests.

TSCA has the following four major purposes:

To screen new chemicals to see if they pose a risk (covered in Section 5 of the Act);

To require testing of chemicals identified as possible risks to health or the environment (covered in Section 4 of the Act);

To gather information on existing chemicals (covered in Section 8 of the Act);

To control chemicals proven to pose a risk (covered in Section 6 of the Act).

Section 8 of TSCA gives EPA authority to require chemical manufacturers and processors to collect, record, or submit certain information on chemicals. Among the types of information involved are data on chemical production, use, exposure, and disposal; records of allegations of significant adverse reactions to health or the environment; unpublished health and safety studies; and notification of previously unknown risks:

Section 8(a) General reporting (40 CFR 712)
Section 8(c) Allegation of adverse reactions (40 CFR 717)
Section 8(d) Health and safety studies (40 CFR 716)
Section 8(e) Substantial risk

The Comprehensive Assessment Information Rule (CAIR) standardizes certain Section 8(a) rules by providing a set of uniform questions, requiring the submission of information on a standard reporting form and establishing uniform reporting and recordkeeping provision that supplement those described elsewhere (Subpart A of part 704). CAIR reports must be submitted for the time period 2/8/87 to 2/5/89.

This list contains substances identified by EPA for reporting under the Comprehensive Assessment Information Rule (CAIR) at 40 CFR 704.225.

The data field identifies who must report.

TSCA - Chemical Test Rules   (CTR)
SOURCE

Environmental Protection Agency
Toxic Substances Control Act
40 CFR Part 799 Identification of Specific Chemical Substance and Mixture Testing Requirements.
Subpart B - Specific Chemical Test Rules
40 CFR 799.500 through 799.4440
through January 1, 1994

LIST DESCRIPTION

In 1976 Congress passed the Toxic Substance Control Act as a method to control exposure and use of raw industrial chemicals that fall outside the jurisdiction of other environmental laws. TSCA was passed to assure that chemicals would be evaluated before use to make sure they pose no unnecessary risk to health or the environment. Under TSCA, EPA can require that the effects of an existing chemical on health and/or the environment be evaluated through a series of toxicity tests.

TSCA has the following four major purposes:

To screen new chemicals to see if they pose a risk (covered in Section 5 of the Act);

To require testing of chemicals identified as possible risks to health or the environment (covered in Section 4 of the Act);

To gather information on existing chemicals (covered in Section 8 of the Act);

To control chemicals proven to pose a risk (covered in Section 6 of the Act).

Under Section 4 of the Act EPA can require industry to test chemical substances already on the market or about to be produced. The primary way chemicals are chosen for testing is from designation by the Interagency Testing Committee (ITC). Once the committee designates a chemical for testing, EPA has one year either to propose a test rule on the substance or to explain why testing of the substance is not needed.

Part 790 of 40 CFR defines procedures governing testing consent agreements and test rules on chemical substances and mixtures under section 4 of TSCA. In addition to the authority to develop rules requiring testing by manufacturers and processors of chemical substances and mixtures, EPA can enter into enforceable consent agreements requiring testing where they provide procedural safeguards equivalent to those that apply where testing is conducted by rule.

Part 799 of 40 CFR identifies the specific chemical substances (or mixtures) and the tests which must be conducted for those substances identified by test rules or under consent agreements.

This list includes those substances in 40 CFR 799 which are the subject of specific chemical test rules. Each substance has an individual citation which includes the specific testing required.

The data field contains information on who must do the testing, the specific citation in 40 CFR 799 which identifies the exact tests to be conducted and any generic group under which the substance has been classed by EPA.

TSCA - Health and Safety Reporting List   (TSC)
SOURCE

Environmental Protection Agency
Toxic Substances Control Act
40 CFR Part 712 - Health and Safety Data Reporting
Subpart B - Specific Chemical Listings
40 CFR 716.120(a) List of substances, (b) reserved, (c)
By category and (d) Listed members of categories
Amended through May 2, 1994 (59 FR 22519) which deleted diethyl phthalate (CAS 84-66-2) and methyl methacrylate (CAS 80-62-6).

LIST DESCRIPTION

In 1976 Congress passed the Toxic Substance Control Act as a method to control exposure and use of raw industrial chemicals that fall outside the jurisdiction of other environmental laws. TSCA was passed to assure that chemicals would be evaluated before use to make sure they pose no unnecessary risk to health or the environment. Under TSCA, EPA can require that the

effects of an existing chemical on health and/or the environment be evaluated through a series of toxicity tests.

TSCA has the following four major purposes:

To screen new chemicals to see if they pose a risk (covered in Section 5 of the Act);

To require testing of chemicals identified as possible risks to health or the environment (covered in Section 4 of the Act);

To gather information on existing chemicals (covered in Section 8 of the Act);

To control chemicals proven to pose a risk (covered in Section 6 of the Act).

Section 8 of TSCA gives EPA authority to require chemical manufacturers and processors to collect, record, or submit certain information on chemicals. Among the types of information involved are data on chemical production, use, exposure, and disposal; records of allegations of significant adverse reactions to health or the environment; unpublished health and safety studies; and notification of previously unknown risks:

Section 8(a) General reporting (40 CFR 712)
Section 8(c) Allegation of adverse reactions (40 CFR 717)
Section 8(d) Health and safety studies (40 CFR 716)
Section 8(e) Substantial risk

Part 716 establishes requirements for the submission of lists and copies of health and safety studies on chemical substances and mixtures selected for priority consideration for testing rules under section 4(a) and on other chemical substances and mixtures for which EPA requires health and safety information.

This list identifies the substances contained in 40 CFR 716.120(a) list of substances, 716.120(c) categories of substances and 716.120(d) listed members of categories. EPA provides examples of substances included within the categories identified in 716.120(c) but states that ALL substances within the category are subject to all the provisions of Part 716. Examples are not included in the listing here.

In a final rule published August 11, 1993 (58 FR 42675) EPA amended the sunset dates on 92 chemical substances, mixtures and categories. For these chemicals reporting under this rule is no longer required except for those studies which were initiated before the end of the sunset period or were ongoing at the end of the sunset period. The rule is effective November 9, 1993.

Section 716.120(d) identifies categories and states that only those chemical substances specifically listed within a category are subject to all the provisions of part 716 for the time period from the effective date of the rule until the sunset date (10 years from the effective date).

The data field contains the effective date and any comment or clarification used by EPA. The sunset date is 10 years from the effective date. Sunset dates other than 10 years from the effective date are identified in the data field.

TSCA - Chemicals with Significant New Use (TSC) Rules

SOURCE

Environmental Protection Agency
Toxic Substances Control Act
40 CFR Part 721 - Significant New Uses of Chemical Substances
Subpart E - Significant New uses for specific Chemical Substances
40 CFR 721.224 through 721.2585
Amended by Federal Register notices through June 6, 1994 (59 FR 29202 and 29255)

LIST DESCRIPTION

In 1976 Congress passed the Toxic Substance Control Act as a method to control exposure and use of raw industrial chemicals that fall outside the jurisdiction of other environmental laws. TSCA was passed to assure that chemicals would be evaluated before use to make sure they pose no unnecessary risk to health or the environment. Under TSCA, EPA can require that the effects of an existing chemical on health and/or the environment be evaluated through a series of toxicity tests.

TSCA has the following four major purposes:

To screen new chemicals to see if they pose a risk (covered in Section 5 of the Act);

To require testing of chemicals identified as possible risks to health or the environment (covered in Section 4 of the Act);

To gather information on existing chemicals (covered in Section 8 of the Act);

To control chemicals proven to pose a risk (covered in Section 6 of the Act).

Shortly after the act was passed, EPA began to compile an inventory of all existing chemicals and asked manufacturers to submit chemicals for inclusion. Once the initial inventory was completed, all chemicals not listed were considered new chemicals and subject to premanufacture review as of July 1, 1979. The requirements for premanufacture notification (PMN) are contained in Section 5 of the Act and covered by regulations in 40 CFR 720-723.

One way the Agency can follow-up on chemicals once they begin to be manufactured in order to monitor the actual condition of use, is to propose a "significant new use" rule (SNUR) under Section 5(a)(2) of TSCA. Part 721 of the regulations covers requirements for significant new uses including procedures for manufacturers, importers, and processors to report on those significant new uses.

Part 721 establishes categories of significant new uses in Supart B. These include areas such as protection in the workplace, hazard communication program, industrial, commercial and consumer activities, disposal, release to water and various recordkeeping requirements. These are referenced in the requirements for specific chemical substances listed in Subpart E.

This list contains the substances identified in Subpart E. The data field contains the Premanufacturing Noti-

fication number for substances not yet assigned CAS numbers (or held confidential by the manufacturer).

TSCA - HDD/HDF - Chemicals Required for (TSC) Testing

SOURCE

Environmental Protection Agency
Toxic Substances Control Act
40 CFR Part 766 - Dibenzo-paradioxins/dibenzofurans
Subpart B - Specific Chemical testing/reporting requirements
40 CFR 766.25
July 1, 1992 Edition

LIST DESCRIPTION

In 1976 Congress passed the Toxic Substance Control Act as a method to control exposure and use of raw industrial chemicals that fall outside the jurisdiction of other environmental laws. TSCA was passed to assure that chemicals would be evaluated before use to make sure they pose no unnecessary risk to health or the environment. Under TSCA, EPA can require that the effects of an existing chemical on health and/or the environment be evaluated through a series of toxicity tests.

TSCA has the following four major purposes:

To screen new chemicals to see if they pose a risk (covered in Section 5 of the Act);

To require testing of chemicals identified as possible risks to health or the environment (covered in Section 4 of the Act);

To gather information on existing chemicals (covered in Section 8 of the Act);

To control chemicals proven to pose a risk (covered in Section 6 of the Act).

The Agency has a variety of options for controlling chemicals including the following:

use of hazard warning labels
ban on manufacture, use, or import
requirements for specific controls during use or manufacture
court action to protect the public health or the environment

The actions taken so far under Section 6 (40 CFR 760, 761, 762, 763, 775) include control of polychlorinated biphenyls (PCBs), chlorofluorocarbons (CFCs), friable asbestos in schools, and dibenzo-para-dioxins/dibenzofurans. Each of the substances was handled differently.

Part 766 defines the requirements EPA has established to ascertain whether certain specified chemical substances may be contaminated with halogenated dibenzodioxins (HDDs)/dibenzofurans (HDFs). Manufacturers and processors of chemical substances identified here must submit test data showing the results of their analysis plus other information defined under this part.

This list includes the chemical substances for testing identified in 40 CFR 766.25.

The data field contains the word "present" to indicate its presence on this list.

TSCA - HDD/HDF - Precursors Required for (TSC) Reporting

SOURCE

Environmental Protection Agency
Toxic Substances Control Act
40 CFR Part 766 - Dibenzo-paradioxins/dibenzofurans
Subpart B - Specific Chemical testing/reporting
  requirements
40 CFR 766.38
July 1, 1992 Edition

LIST DESCRIPTION

In 1976 Congress passed the Toxic Substance Control Act as a method to control exposure and use of raw industrial chemicals that fall outside the jurisdiction of other environmental laws. TSCA was passed to assure that chemicals would be evaluated before use to make sure they pose no unnecessary risk to health or the environment. Under TSCA, EPA can require that the effects of an existing chemical on health and/or the environment be evaluated through a series of toxicity tests.

TSCA has the following four major purposes:

To screen new chemicals to see if they pose a risk (covered in Section 5 of the Act);
To require testing of chemicals identified as possible risks to health or the environment (covered in Section 4 of the Act);
To gather information on existing chemicals (covered in Section 8 of the Act);
To control chemicals proven to pose a risk (covered in Section 6 of the Act).

The Agency has a variety of options for controlling chemicals including the following:

use of hazard warning labels
ban on manufacture, use, or import
requirements for specific controls during use or
  manufacture
court action to protect the public health or the
  environment

The actions taken so far under Section 6 (40 CFR 760, 761, 762, 763, 775) include control of polychlorinated biphenyls (PCBs), chlorofluorocarbons (CFCs), friable asbestos in schools, and dibenzo-para-dioxins/dibenzofurans. Each of the substances was handled differently.

Part 766 defines the requirements EPA has established to ascertain whether certain specified chemical substances may be contaminatedwith halogenated dibenzodioxins (HDDs)/dibenzofurans (HDFs).

Precursor chemical substances are produced under conditions that will not yield HDDs and HDFs, but their molecular structure is conducive to HDD/HDF formation under favorable reaction conditions when they are used to produce other chemicals or products. All persons who manufacture or import a chemical product produced using any of the chemical substances listed here as feedstocks or intermediates must report (unless less than 100 kilograms per year is used for research and development purposes).

This list contains the substances included in 40 CFR 766.38(a).

The data field contains the word "present" to indicate its presence on this list.

TSCA - Substances Subject to Testing Consent (TSC) Orders

SOURCE

Environmental Protection Agency
Toxic Substances Control Act
40 CFR Part 799 Identification of Specific Chemical
  Substance and Mixture Testing Requirements.
Subpart C - Testing Consent Orders
40 CFR 799.5000 Amended through August 1, 1994
  59 FR 38917 (addition of bisphenol A diglycidyl
  ether)

LIST DESCRIPTION

In 1976 Congress passed the Toxic Substance Control Act as a method to control exposure and use of raw industrial chemicals that fall outside the jurisdiction of other environmental laws. TSCA was passed to assure that chemicals would be evaluated before use to make sure they pose no unnecessary risk to health or the environment. Under TSCA, EPA can require that the effects of an existing chemical on health and/or the environment be evaluated through a series of toxicity tests.

TSCA has the following four major purposes:

To screen new chemicals to see if they pose a risk (covered in Section 5 of the Act);
To require testing of chemicals identified as possible risks to health or the environment (covered in Section 4 of the Act);
To gather information on existing chemicals (covered in Section 8 of the Act);
To control chemicals proven to pose a risk (covered in Section 6 of the Act).

Under Section 4 of the Act EPA can require industry to test chemical substances already on the market or about to be produced. The primary way chemicals are chosen for testing is from designation by the Interagency Testing Committee (ITC). Once the committee designates a chemical for testing, EPA has one year either to propose a test rule on the substance or to explain why testing of the substance is not needed.

Part 790 of 40 CFR defines procedures governing testing consent agreements and test rules on chemical substances and mixtures under section 4 of TSCA. In addition to the authority to develope rules requiring testing by manufacturers and processors of chemical substances and mixtures, EPA can enter into enforceable consent agreements requiring testing where they provide proceduralsafeguards equivalent to those that apply where testing is conducted by rule.

Part 799 of 40 CFR identifies the specific chemical substances (or mixtures) and the tests which must be conducted for those substances identified by test rules or under consent agreements.

This list includes those substances in 40 CFR 799.5000 which are the subject of testing consent orders adopted under 40 CFR part 790.

The data field contains the identification of the area for which the chemical or mixture must be tested. (The exact nature of the tests in these areas is described in 40 CFR parts 796 - Chemical Fate, 797 - Environmental Effects and 798 - Health Effects).

TSCA - Multichemical Test Rules - Waste (TSC) Constituents
SOURCE

Environmental Protection Agency
Toxic Substances Control Act
40 CFR Part 799 Identification of Specific Chemical Substance and Mixture Testing Requirements.
Subpart D - Multichemical Test Rules
40 CFR 799.5055
July 1, 1992 Edition

LIST DESCRIPTION

In 1976 Congress passed the Toxic Substance Control Act as a method to control exposure and use of raw industrial chemicals that fall outside the jurisdiction of other environmental laws. TSCA was passed to assure that chemicals would be evaluated before use to make sure they pose no unnecessary risk to health or the environment. Under TSCA, EPA can require that the effects of an existing chemical on health and/or the environment be evaluated through a series of toxicity tests.

TSCA has the following four major purposes:

To screen new chemicals to see if they pose a risk (covered in Section 5 of the Act);

To require testing of chemicals identified as possible risks to health or the environment (covered in Section 4 of the Act);

To gather information on existing chemicals (covered in Section 8 of the Act);

To control chemicals proven to pose a risk (covered in Section 6 of the Act).

Under Section 4 of the Act EPA can require industry to test chemical substances already on the market or about to be produced. The primary way chemicals are chosen for testing is from designation by the Interagency Testing Committee (ITC). Once the committee designates a chemical for testing, EPA has one year either to propose a test rule on the substance or to explain why testing of the substance is not needed.

Part 790 of 40 CFR defines procedures governing testing consent agreements and test rules on chemical substances and mixtures under section 4 of TSCA. In addition to the authority to develope rules requiring testing by manufacturers and processors of chemical substances and mixtures, EPA can enter into enforceable consent agreements requiring testing where

they provide procedural safeguards equivalent to those that apply where testing is conducted by rule.

Part 799 of 40 CFR identifies the specific chemical substances (or mixtures) and the tests which must be conducted for those substances identified by test rules or under consent agreements.

This list includes the hazardous waste constituents identified in 40 CFR 799.5055 which are the subject of multichemical test rules.

The data field contains the identification of the type of test which must be conducted for each waste constituent. The full description of the tests is provided in various subparagraphs to paragraph 40 CFR 799.5055.

TSCA - PAIR - Reporting List (TSC)
SOURCE

Environmental Protection Agency
Toxic Substances Control Act
40 CFR Part 712 - Chemical Information Rules
Subpart B - Manufacturers Reporting - Preliminary Assessment Information 40 CFR 712.30(d) through (x)
Amended through May 2, 1994 (59 FR 22519) which deleted diethyl phthalate (CAS 84-66-2)and methyl methacrylate (CAS 80-62-6).

LIST DESCRIPTION

In 1976 Congress passed the Toxic Substance Control Act as a method to control exposure and use of raw industrial chemicals that fall outside the jurisdiction of other environmental laws. TSCA was passed to assure that chemicals would be evaluated before use to make sure they pose no unnecessary risk to health or the environment. Under TSCA, EPA can require that the effects of an existing chemical on health and/or the environment be evaluated through a series of toxicity tests.

TSCA has the following four major purposes:

To screen new chemicals to see if they pose a risk (covered in Section 5 of the Act);

To require testing of chemicals identified as possible risks to health or the environment (covered in Section 4 of the Act);

To gather information on existing chemicals (covered in Section 8 of the Act);

To control chemicals proven to pose a risk (covered in Section 6 of the Act).

Under Section 4 of the Act EPA can require industry to test chemical substances already on the market or about to be produced. The primary way chemicals are chosen for testing is from designation by the Interagency Testing Committee (ITC). Once the

committee designates a chemical for testing, EPA has one year either to propose a test rule on the substance or to explain why testing of the substance is not needed.

Section 8 of TSCA gives EPA authority to require chemical manufacturers and processors to collect, record, or submit certain information on chemicals.

Among the types of information involved are data on chemical production, use, exposure, and disposal;

records of allegations of significant adverse reactions to health or the environment; unpublished health and safety studies; and notification of previously unknown risks:

Section 8(a) General reporting (40 CFR 712)
Section 8(c) Allegation of adverse reactions (40 CFR 717)
Section 8(d) Health and safety studies (40 CFR 716)
Section 8(e) Substantial risk

Part 712 establishes procedures for chemical manufacturers and processors to report production, use, and exposure-related information on listed chemical substances. The chemical substances, mixtures, and categories of substances or mixtures included in the PAIR reports are those which have been recommended by the ITC for testing consideration but not designated for Agency response within 12 months. The information in each manufacturer's report must cover the latest complete corporate fiscal year as of

the effective date. The effective date will be 30 days after the Federal Register publishes a rule amendment. For full information on specific reporting requirements, see Subpart A of Part 704.

This list contains the substances and mixtures identified at 40 CFR 712.30(a) through (x).

The data field contains the reporting date (which is 60 days after the effective date).

TSCA - Code of Federal Regulation Citations   (TSC)
SOURCE

Environmental Protection Agency
Office of Prevention, Pesticides and Toxic Substances
Toxic Substances CAS Number - Chemical Index.
Prepared by EPA for information only and printed on pages 731 to 748 of 40 CFR Parts 790 to End, revised as of July 1, 1992

LIST DESCRIPTION

In 1976 Congress passed the Toxic Substance Control Act as a method to control exposure and use of raw industrial chemicals that fall outside the jurisdiction of other environmental laws. TSCA was passed to assure that chemicals would be evaluated before use to make sure they pose no unnecessary risk to health or the environment. Under TSCA, EPA can require that the effects of an existing chemical on health and/or the environment be evaluated through a series of toxicity tests.

TSCA has the following four major purposes:

To screen new chemicals to see if they pose a risk (covered in Section 5 of the Act);
To require testing of chemicals identified as possible risks to health or the environment (covered in Section 4 of the Act);
To gather information on existing chemicals (covered in Section 8 of the Act);
To control chemicals proven to pose a risk (covered in Section 6 of the Act).

Shortly after the act was passed, EPA began to compile an inventory of all existing chemicals and asked manufacturers to submit chemicals for inclusion. Once the initial inventory was completed, all chemicals not listed were considered new chemicals and subject to premanufacture review as of July 1, 1979. The requirements for premanufacture notification (PMN) are contained in Section 5 of the Act and covered by regulations in 40 CFR 720-723.

One way the Agency can follow-up on chemicals once they begin to be manufactured in order to monitor the actual condition of use is to propose a "significant new use" rule (SNUR) under Section 5(a)(2) of TSCA (40 CFR 721).

Under Section 4 of the Act EPA can require industry to test chemical substances already on the market or about to be produced. The primary way chemicals are chosen for testing is from designation by the Interagency Testing Committee (ITC). Once the committee designates a chemical for testing, EPA has one year either to propose a test rule on the substance or to explain why testing of the substance is not needed.

The Agency has a variety of options for controlling chemicals including:

use of hazard warning labels
ban on manufacture, use, or import
requirements for specific controls during use or manufacture
court action to protect the public health or the environment

The actions taken so far under Section 6 (40 CFR 760, 761, 762, 763, 775) include control of polychlorinated biphenyls (PCBs), chlorofluorocarbons (CFCs), friable asbestos in schools, and dibenzo-paradioxins/dibenzofurans. Each of the substances was handled differently.

Section 8 of TSCA gives EPA authority to require chemical manufacturers and processors to collect, record, or submit certain information on chemicals. Among the types of information involved are data on chemical production, use, exposure, and disposal; records of allegations of significant adverse reactions to health or the environment; unpublished health and safety studies; and notification of previously unknown risks:

Section 8(a) General reporting (40 CFR 712)
Section 8(c) Allegation of adverse reactions (40 CFR 717)
Section 8(d) Health and safety studies (40 CFR 716)
Section 8(e) Substantial risk

The Office of Prevention, Pesticides and Toxic Substances, EPA, has developed an Index relating the Toxic Substance CAS Number to a citation or citations in 40 CFR where information concerning that chemical can be found. This listing is provided by EPA for information purposes only.

EPA identifies the contact person for information on this list as:

Joan Sutter
Federal Register Staff
  Office of Prevention, Pesticides and Toxic Substances (OPPTS)
EPA
401 M St. SW TS-788B
Washington, DC 20460
(202) 382-3405
The data field contains the CFR citation(s).

TSCA - Section 12(b) - Export Notification   (TSE)

SOURCE

Environmental Protection Agency
Toxic Substances Control Act
Document from the TSCA Assistance Information Service
Export Division
(202) 488-3821
Using publication updated through June 30, 1994
  NOTE: ChemADVISOR added Part Two to this document on January 1, 1995 (non-CAS numbered entries)

LIST DESCRIPTION

Under Section 12(b) of the Toxic Substances Control Act (TSCA), exporters must notify the U.S. Environmental Protection Agency if they export or intend to export a chemical substance or mixture:

1) for which submission of data is required under Section 4 or 5(b);
2) for which an order has been issued under Section 5;
3) for which a rule has been proposed or promulgated under Section 5 or 6;
4) or with respect to which an action is pending or relief has been granted under Section 5 or 7

Beginning January 1, 1994, substances subject to section 4 regulations require a one-time notification only (see 58 FR 40238). Annual notification still applies for regulations under sections 5 and 6.

Notifications must be submitted in writing and need only include: the exporter's name and address, the name of the regulated chemical, the date(s) of export or intended export, the importing country (copuntries), and the section of TSCA under which EPA has taken action. Notifications should be sent to the following address:

USEPA
TSCA Document Processing Center (7407)
Room G-099
401 M Street SW
Washington, DC 20460
The Export Division of the TSCA Assistance Information Service provides a document intended to be an information resource. Thedocument provides a listing of chemicals that trigger 12(b) export notification. The chemicals on the list are identified by Chemical Abstracts Service Registry numbers or Premanufacture Notice (PMN) numbers. The list also includes appropriate citations from the Federal Register and Code of Federal Regulations.

EPA makes the following note to the document:
This document is intended to be an information resource. It may contain errors, omissions, etc., and therefore should not be used in lieu of the Federal Register or Code of Federal Regulations for the purposes of compliance. Updates will be made quarterly. For further assistance concerning information contained in this document, please contact the TSCA Assistance Information Service Export Division at (202) 488-3821. For all other information concerning TSCA, contact the TSCA Assistance Information Service Hotline (202) 554-1404.

The list prepared by ChemADVISOR, Inc. contains, with a few exceptions, just those entries identified by a valid Chemical Abstracts Services Registry Number. Most entries identified only by PMN Number and Generic Name are not included.

The data field contains the statement "export notification required", identification of any chemical group designated by EPA to which that material may belong and the Section of TSCA under which the chemical data is required. The data field also indicates if a material is listed because of a PROPOSED reporting requirement.

INTERNATIONAL LISTS

Australian Exposure Standards - Time Weighted (AUS) Averages

SOURCE

Australian National Occupational Health and Safety Commission

Exposure Standards for Atmospheric Contaminants in the Occupational Environment - 2nd Edition

Guidance Note on the Interpretation of Exposure Standards for Atmospheric Contaminants in the Occupational Environment [NOHSC:3008(1991)]

Adopted National Exposure Standards for Atmospheric Contaminants in the Occupational Environment [NOHSC:1003(1991)]

October 1991

LIST DESCRIPTION

The National Occupational Health and Safety Commission (NOHSC) is a tripartite body established by the Commonwealth Government to develop, facilitate and implement a national occupational health and safety strategy.

NOHSC has prepared the following policy statement on the use of exposure Standards:

These exposure standards are guides to be used in the control of occupational health hazards. They should not be used as fine dividing lines between safe and dangerous concentrations of chemicals. They are not a measure of relative toxicity and should not be applied in the control of community air pollution. Interpretation of the exposure standards should be undertaken by an appropriately qualified and experienced person.

Guidance notes are advisory technical documents issued by the National Commission. In contrast to national standards and national codes of practice, guidance notes have no legal standing as documents declared by the National Commission under s.38(1) of the National Occupational Health and Safety Commission Act, and may not be suitable for reference in Commonwealth, State or Territory legislation.

The National Commission, having considered the recommendations prepared by the Exposure Standards Working Group, now declares "National Exposure Standards for Atmospheric Contaminants in the Occupational Environment" pursuant to section 38(1) of the "National Occupational Health and Safety Commission Act 1985 (Cwlth)."

Except where modified by consideration of excursion limits, exposure standards apply to long term exposure to a substance over an eight-hour day, for a five-day working week, over an entire working life. The exposure standards only consider absorption via inhalation and are valid only on the condition that significant skin absorption cannot occur.

For some rapidly acting substances and irritants, the averaging of the airborne concentration over an eight-hour period is inappropriate. These substances may induce acute effects after relatively brief exposure to high concentrations and so the exposure standard for these substances represents a maximum or peak concentration to which workers may be exposed. Although it is recognized that there are analytical limitations to the measurement of some substances, compliance with these "peak limitation" exposure standards should be determined over the shortest analytically practicable period of time, but under no circumstances should a single determination exceed 15 minutes.

This list contains all entries in NOHSC:1003(1991) section 4 (Exposure Standards for Atmospheric Contaminants in the Occupational Environment) for which NOHSC has identified an 8 hour time-weighted average exposure (TWA) and/or a Peak Limitation.

The data field contains the exposure limit in mg/m3 or ppm, any Peak Limitation if specified, asphyxiant notation and any chemical group to which the substance may be included.

Australian Exposure Standards - Short Term (AUS) Exposure Limit

SOURCE

Australian National Occupational Health and Safety Commission

Exposure Standards for Atmospheric Contaminants in the Occupational Environment - 2nd Edition

Guidance Note on the Interpretation of Exposure Standards for Atmospheric Contaminants in the Occupational Environment [NOHSC:3008(1991)]

Adopted National Exposure Standards for Atmospheric Contaminants in the Occupational Environment [NOHSC:1003(1991)]

October 1991

LIST DESCRIPTION

The National Occupational Health and Safety Commission (NOHSC) is a tripartite body established by the Commonwealth Government to develop, facilitate and implement a national occupational health and safety strategy.

NOHSC has prepared the following policy statement on the use of exposure Standards:

These exposure standards are guides to be used in the control of occupational health hazards. They should not be used as fine dividing lines between safe and dangerous concentrations of chemicals. They are not a measure of relative toxicity and should not be applied in the control of community air pollution. Interpretation of the exposure standards should be undertaken by an appropriately qualified and experienced person.

Guidance notes are advisory technical documents issued by the National Commission. In contrast to national standards and national codes of practice, guidance notes have no legal standing as documents declared by the National Commission under s.38(1) of the National Occupational Health and Safety

Commission Act, and may not be suitable for reference in Commonwealth, State or Territory legislation.

The National Commission, having considered the recommendations prepared by the Exposure Standards Working Group, now declares "National Exposure Standards for Atmospheric Contaminants in the Occupational Environment" pursuant to section 38(1) of the "National Occupational Health and Safety Commission Act 1985 (Cwlth)."

This list contains all entries for which NOHSC has identified aShort Term Exposure Limit. The Short Term Exposure Limit is defined as the concentration to which workers can be exposed continuously for a short period of time without suffering from 1) irritation, 2) chronic or irreversible tissue damage, or 3) narcosis of sufficient degree to increase the likelihood of accidental injury, impair self-rescue or materially reduce work efficiency, and provided that the daily TWA is not exceeded, it is not a separate independent exposure limit; rather, it supplements the time-weighted average limit where there are recognized acute effects from a substance whose toxic effects are primarily of a chronic nature. STELS are recommended only where toxic effects have been reported from high short-term exposures in either humans or animals.

A STEL is defined as a 15-minute TWA exposure which should not be exceeded at any time during a workday even if the 8-hour TWA is within the TWA. Exposures above the TWA up to the STEL should not be longer than 15 minutes and should not occur more than four times per day. There should be at least 60 minutes between successive exposures in this range. An average period other than 15 minutes may be recommended when this is warranted by observed biological effects.

This list contains all entries in NOHSC:1003(1991) section 4 (Exposure Standards for Atmospheric Contaminants in the Occupational Environment) for which NOHSC has identified a STEL.

The data field contains the exposure limit in mg/m3 or ppm and the notation "STEL."

Australian Exposure Standards - Skin Effects   (AUS)

SOURCE

Australian National Occupational Health and Safety
Commission
Exposure Standards for Atmospheric Contaminants in
the Occupational Environment - 2nd Edition
Guidance Note on the Interpretation of Exposure
Standards for Atmospheric Contaminants in the
Occupational Environment [NOHSC:3008(1991)]
Adopted National Exposure Standards for Atmospheric
Contaminants in the Occupational Environment
[NOHSC:1003(1991)]
October 1991

LIST DESCRIPTION

The National Occupational Health and Safety Commission (NOHSC) is a tripartite body established by the Commonwealth Government to develop, facilitate and implement a national occupational health and safety strategy.

NOHSC has prepared the following policy statement on the use of exposure Standards:

These exposure standards are guides to be used in the control of occupational health hazards. They should not be used as fine dividing lines between safe and dangerous concentrations of chemicals. They are not a measure of relative toxicity and should not be applied in the control of community air pollution. Interpretation of the exposure standards should be undertaken by an appropriately qualified and experienced person.

Guidance notes are advisory technical documents issued by the National Commission. In contrast to national standards and national codes of practice, guidance notes have no legal standing as documents declared by the National Commission under s.38(1) of the National Occupational Health and Safety Commission Act, and may not be suitable for reference in Commonwealth, State or Territory legislation.

The National Commission, having considered the recommendations prepared by the Exposure Standards Working Group, now declares "National Exposure Standards for Atmospheric Contaminants in the Occupational Environment" pursuant to section 38(1) of the "National Occupational Health and Safety Commission Act 1985 (Cwlth)."

Certain substances can readily penetrate the intact skin and thusbecome absorbed into the body. Frequently, there will be no accompanying skin damage. In some instances this dermal absorption can pose a far greater danger than inhalation exposure.

Some substances can cause a specific immune response in some people. Such substances are called sensitizers and the development of a specific immune response is termed "sensitization." Exposure to a sensitizer, once sensitization has occurred, may manifest itself as a skin rash or inflammation or as an asthmatic condition, and in some individuals this reaction can be extremely severe.

Following the induction of a sensitized state, an affected individual may subsequently react to exposure to minute levels of that substance. Although low values have been assigned to strong sensitizing agents, compliance with the recommended exposure standard may not provide adequate protection for a hypersensitive individual.

This list contains all entries in NOHSC:1003(1991) section 4 (Exposure Standards for Atmospheric Contaminants in the Occupational Environment) for which NOHSC has identified the potential for skin sensitization or potential for skin absorption.

The data field identifies the potential skin effect(s).

Australian Exposure Standards - Carcinogens (AUS)

SOURCE

Australian National Occupational Health and Safety Commission

Exposure Standards for Atmospheric Contaminants in the Occupational Environment - 2nd Edition

Guidance Note on the Interpretation of Exposure Standards for Atmospheric Contaminants in the Occupational Environment [NOHSC:3008(1991)]

Adopted National Exposure Standards for Atmospheric Contaminants in the Occupational Environment [NOHSC:1003(1991)]

October 1991

LIST DESCRIPTION

The National Occupational Health and Safety Commission (NOHSC) is a tripartite body established by the Commonwealth Government to develop, facilitate and implement a national occupational health and safety strategy.

NOHSC has prepared the following policy statement on the use of exposure Standards:

These exposure standards are guides to be used in the control of occupational health hazards. They should not be used as fine dividing lines between safe and dangerous concentrations of chemicals. They are not a measure of relative toxicity and should not be applied in the control of community air pollution. Interpretation of the exposure standards should be undertaken by an appropriately qualified and experienced person.

Guidance notes are advisory technical documents issued by the National Commission. In contrast to national standards and national codes of practice, guidance notes have no legal standing as documents declared by the National Commission under s.38(1) of the National Occupational Health and Safety Commission Act, and may not be suitable for reference in Commonwealth, State or Territory legislation.

The National Commission, having considered the recommendations prepared by the Exposure Standards Working Group, now declares "National Exposure Standards for Atmospheric Contaminants in the Occupational Environment" pursuant to section 38(1) of the National Occupational Health and Safety Commission Act 1985 (Cwlth).

*Chemical substances which have been identified as suspected or established carcinogens, or substances associated with industrial processes which have been identified as suspected or established carcinogens, have been highlighted in the list of adopted exposure standards. The Commission of the European Communities (EEC) system of classification of carcinogenic substances (i.e. European Chemical Industry Ecology and Toxicology Centre, "A Guide to the Classification of Carcinogens, Mutagens and Teratogens under the Sixth Amendment," Technical Report No. 21, Brussels, February 1986) is used to indicate the strength of the causal association between these substances and the development of cancer.

Category 1:

Established human carcinogens are those substances known to be carcinogenic to humans. There is sufficient evidence to establish a causal association between human exposure to these substances and the development of cancer.

Category 2:

Probable human carcinogens are those substances for which there is sufficient evidence to provide a strong presumption that human exposure may result in the development of cancer. This evidence is generally based on appropriate long term animal studies, limited epidemiological evidence or other relevant information.

Category 3:

Substances suspected of having carcinogenic potential are those substances which have possible carcinogenic effects on humans but in respect of which the available information is not adequate for making a satisfactory assessment. There is some evidence from appropriate animal or epidemiological studies, but this is insufficient to place the substance in Category 2.

This list contains all entries in NOHSC:1003(1991) section 4 (Exposure Standards for Atmospheric Contaminants in the Occupational Environment) for which NOHSC has identified the potential for carcinogenicity.

The data field identifies the NOHSC classification as "confirmed carcinogen (category 1)," "probable carcinogen" (category 2) and "suspect carcinogen"(category 3).

Australian Exposure Standards - Under Review (AUS)

SOURCE

Australian National Occupational Health and Safety Commission

Exposure Standards for Atmospheric Contaminants in the Occupational Environment - 2nd Edition

Guidance Note on the Interpretation of Exposure Standards for Atmospheric Contaminants in the Occupational Environment [NOHSC:3008(1991)]

Adopted National Exposure Standards for Atmospheric Contaminants in the Occupational Environment [NOHSC:1003(1991)]

October 1991

LIST DESCRIPTION

The National Occupational Health and Safety Commission (NOHSC) is a tripartite body established by the Commonwealth Government to develop, facilitate and implement a national occupational health and safety strategy.

NOHSC has prepared the following policy statement on the use of exposure Standards:

These exposure standards are guides to be used in the control of occupational health hazards. They should not be used as fine dividing lines between safe and dangerous concentrations of chemicals. They are not a measure of relative toxicity and should not be applied in the control of community air pollution. Interpretation of the exposure standards should be undertaken by an appropriately qualified and experienced person.

Guidance notes are advisory technical documents issued by the National Commission. In contrast to national standards and national codes of practice, guidance notes have no legal standing as documents declared by the National Commission under s.38(1) of the National Occupational Health and Safety Commission Act, and may not be suitable for reference in Commonwealth, State or Territory legislation.

The National Commission, having considered the recommendations prepared by the Exposure Standards Working Group, now declares "National Exposure Standards for Atmospheric Contaminants in the Occupational Environment" pursuant to section 38(1) of the "National Occupational Health and Safety Commission Act 1985 (Cwlth)."

Appendix 3 to the report identifies "Issues and Substances Under Review." The substances listed in this Appendix havebeen identified by the Exposure Standards Working group as requiring further review.

The Exposure Standards Working Group will be collecting and evaluating relevant occupational health data on these substances. Draft documentation will be prepared.

This list identifies all the substances for which exposure limits are currently under review by NOHSC. (Issues under review are not included).

The data field contains the statement "exposure limits under review."

Canada - WHMIS: Ingredient Disclosure List   (WHM)
SOURCE

Registration SOR/88-64,31 December 1987
Hazardous Products Act
Canada Gazette, Part II, Volume 122, No. 2

LIST DESCRIPTION

Canada has established a "Workplace Hazardous Materials Information System" (WHMIS) through the Occupational Safety and Health Branch of Labour Canada. This system establishes requirements for information on Material Safety Data Sheets and Labels. As part of the criteria for establishing coverage under WHMIS, the supplier or employer must consider the criteria established for hazardous materials in the Hazardous Products Act.

WHMIS is implemented by a combination of federal and provincial legislation. The federal Hazardous

Products Act as amended by Bill C-70 (Chapter 30 [1987] of the Statutes of Canada) requires suppliers of hazardous materials called controlled products to provide adequate labels and MSDSs as a condition of sale and importation. The Controlled Products Regulations, issued under the authority of the Hazardous Products Act, contain requirements which specify the form and content of supplier labels, the types and arrangement of information on material safety data sheets, conditions of exemption and the details of the criteria that define a controlled product.

The Ingredient Disclosure List issued pursuant to subsection 17(1) of the Hazardous Products Act contains the names of chemicals which must be identified on material safety data sheets if they are present in controlled products above a specified concentration.

All substances contained on the IDL are included in this list.

The data field indicates the concentration above which disclosure is required and the item number of that substance on the IDL printed in French.

German (DFG) - MAK Values   (MAK)
SOURCE

Deutsche Forschungsgemeinschaft (DFG)
MAK and BAT Values 1994
Report No. 30 of the Commission for the Investigation of Health Hazards of Chemical Compounds in the Work Area
Deutsche Forschungsgemeinschaft
Kennedyallee 40
D-5300 Bonn 2, Federal Republic of Germany
Telephone (0228) 885-1, Telex 17228312 Telefax (0228) 885-2221

LIST DESCRIPTION

The MAK value (maximum concentration value in the workplace) is defined as the maximum permissible concentration of a chemical compound present in the air within a working area (as gas, vapor or particulate matter) which, according to current knowledge, generally does not impair the health of the employee or cause any undue annoyance. Under these conditions, exposure can be safely repeated over a daily period of eight hours, constituting an average work week of 40 hours. As a rule, the MAK value is integrated as an average concentration over periods of up to one workday or one shift. In establishing the MAK values, primarily the effects of the compounds have been taken into account; however, where possible, practical criteria posed by the working procedures or the patterns of exposure which they determine have also been considered. Scientifically based criteria for health protection, rather than their technical or economical feasibility, are employed.

The main reasons for setting a MAK value at a certain concentration are presented in a series of publications by the Commission for Investigation of Health Hazards of Chemical Compounds in the Work Area entitled "Toxikologisch-arbeitsmedizinische Begrundung von MAK-Werten." In particular, it contains critical evaluations of the present state of knowledge. Annual supplements of "Begrundungen" are in preparation.

MAK values promote the protection of health in the workplace. They provide a basis for judgement of the toxic potential or safety of a prevailing concentration in the work area. However, they do not provide constants from which the presence or absence of a health hazard after longer or shorter periods of exposure can be computed; nor can, in an isolated case, a proven or suspected damage to health be deduced from MAK values or from the classification of a substance as carcinogenic. Such deductions can be made only on the basis of medical findings, taking into consideration all the circumstances of the particular case. Therefore, on principle, statements in the MAK List are not to be seen as a priority judgements in individual cases. In addition to exposure via the respiratory system, a number of other factors may determine the kind and degree of damaging effects: sensitizing potential, skin absorption, corrosiveness, combustibility, vapor pressure, etc. Adherence to MAK values, on principle, does not eliminate the necessity for medical health surveillance of the exposed individuals.

This list contains all entries from the current Report which have an MAK value assigned.

The data field contains the MAK value in ppm (ml/m3) and/or in mg/m3.

German (DFG) - Peak Limitations   (MAK)
SOURCE

Deutsche Forschungsgemeinschaft (DFG)
MAK and BAT values 1994
Report No. 30 of the Commission for the Investigation of Health Hazards of Chemical Compounds in the Work Area

Deutsche Forschungsgemeinschaft
Kennedyallee 40
D-5300 Bonn 2, Federal Republic of Germany
Telephone (0228) 885-1, Telex 17228312 Telefax (0228) 885-2221

LIST DESCRIPTION

MAK values have always been conceived and applied as 8 hour time weighted averages. In practice, however, the actual concentrations of chemical compounds in the workplace air fluctuate frequently and to a considerable extent. Excursions above the time weighted average must be restricted for many substances in order to avoid impairment of health. As the degree of risk to health from such peak concentrations depends on the particular behavior and effects of the substance in question, a system has been developed in which as many as possible of the substances in the MAK Values List have been classified into distinct categories whose definitions take into account both toxicological considerations and analytical practicability.

The 8 hour time weighted average must in any case be observed. Under these conditions the following categories apply for limitation of excursions above the 8 hour time weighted average:

I = Local irritants

2 times MAK, 5 minutes, momentary value 8 times per shift

II = Substances with systemic effects - onset less than or equal to 2 hours.

Half-life less than 2 hours 2 times MAK, 30 minute, average value 4 times per shift
Half-life 2 hours up to shift length 5 times MAK, 30 minute, average value 2 times per shift

III = Substances with systemic effects - onset of effect greater than 2 hours

Half-life greater than shift length (strongly cumulative)
10 times MAK, 30 minutes, average value 1 time per shift

IV = Substance eliciting very weak effects

MAK greater than 500 ppm
2 times MAK, 60 minutes, momentary value 3 times per shift

V = Substances having intensive odor

2 times MAK, 10 minutes, momentary value 4 times per shift

This list contains all entries from the 26th Report which have a Peak Limitation category assigned.

The data field contains the recommendation for limiting peak exposures (categories are expressed in words rather than category numerals).

German (DFG) - Skin/Sensitizers   (MAK)
SOURCE

Deutsche Forschungsgemeinschaft (DFG)
MAK and BAT Values 1994
Report No. 30 of Commission for the Investigation of Health Hazards of Chemical Compounds in the Work Area
Deutsche Forschungsgemeinschaft
Kennedyallee 40
D-5300 Bonn 2, Federal Republic of Germany
Telephone (0228) 885-1, Telex 17228312 Telefax (0228) 885-2221

LIST DESCRIPTION

The allergies caused by substances at the workplace affect mostly the skin (contact eczema, contact urticaria), the respiratory passages (rhinitis, bronchitis, asthma, alveolitis) and the conjunctiva (blepharoconjunctivitis). The kind of allergy is determined by the route of uptake but also by the chemical properties of the substance and its aggregation state. A number of allergens can cause both skin lesions and respiratory symptoms.

The object of identification here is to list those substances which, under workplace conditions, have an allergenic potential strong enough to make it necessary to designate the substance as an allergen. In this context it is pointed out that the sensitization capacity or sensitization potential of a substance is not the same as the frequency of sensitizations which it causes, that is, the frequency of allergies detected. Whereas the sensitization potential is a property resulting from the chemistry of the substance, the sensitization frequency is determined not only by the sensitization potential but also and very markedly by the distribution of the substance and the possibililities for contact with it.

Following sensitization (e.g. of the skin or the respiratory system), individually disposed allergic effects can emerge with varying degrees of rapidity and severity as a result of exposure to different compounds. Even adherence to MAK values cannot provide protection against such reactions.

Chemical compounds causing a noticeably higher than normal number of sensitivities (exemplified as hypersensitivity of an allergic nature) are designated by an "S" in the relevant column in the list of MAK values. Although this classification could be subject to discussion, this designation does indicate that caution should be exercised in industrial use of the compound. Those having a danger of photo-contact sensitization are designated by "S(P)".

Certain compounds can penetrate the epidermis easily; absorption of such compounds through the skin can pose, in effect, a larger danger of toxicity than their inhalation. Therefore, potentially fatal poisonings, frequently without warning symptoms, can result from aniline, nitrobenzene, nitroglycol, phenols, certain pesticides, etc. Such chemical compounds are specified by an "H" in the relevant column in the list of MAK values. In handling such compounds, meticulous cleanliness of the skin, hair and clothing is imperative for health protection. The letter "H," however, does not indicate a potential danger of skin irritation.

This list contains all entries in the current MAK table which have the designation "S", "S(P)" or "H."

The data field contains the words "danger of cutaneous absorption" for those substances with the "H"designation and "danger of sensitization (skin or respiratory)" for those with the "S" designation and "dange or photo-contact sensitization" for those with the "S(P)" designation.

German (DFG) - Carcinogens   (MAK)

SOURCE

Deutsche Forschungsgemeinschaft (DFG)
MAK and BAT Values 1994
Report No. 30 of the Commission for the Investigation of Health Hazards of Chemical Compounds in the Work Area

Deutsche Forschungsgemeinschaft
Kennedyallee 40
D-5300 Bonn 2, Federal Republic of Germany
Telephone (0228) 885-1, Telex 17228312 Telefax (0228) 885-2221

LIST DESCRIPTION

The handling of proven or potential carcinogenic compounds requires extraordinary caution and protective measure for health care. Therefore, these substances are presented separately and are categorized as follows:

A: Working materials which have been unequivo-cally proven carcinogenic

    A1: Compounds capable of inducing malignant tumors as shown by experience with humans

    A2: Compounds which in the Commission's opinion have proven so far to be unmistakably carcinogenic in animal experimentation only; namely under conditions which are comparable to those for possible exposure of a human being at the workplace, or from which such comparability can be deduced

Compounds in Category A, whose effects, according to current knowledge, manifest a distinct cancer risk for humans, have no concentration value listed in Section IIa (the main MAK table) since no values have been established for a safe concentration range. For some of these compounds, even their uptake through intact skin poses a great danger.

B: Compounds which are justifiably suspected of having carcinogenic potential

Recent results of cancer research call for the consideration of additional materials for which a noteworthy carcinogenic potential can be suspected and which urgently need further clarification. Where MAK values already exist for the compounds listed in category B, they are tentatively retained.

Although compounds in Category B are not subject to the strict handling regulations of Category A, the health surveillance of employees using these substances must be intensified, with the goal of minimizing the exposure as much as possible and of proving or ruling out causal relationships between the effects of the compound and cancer. In addition, the branches of industry which produce and process such compounds are requested - as are all relevantly involved research laboratories - to participate in the effort to shed light on the cancer-correlation question and, where necessary, to search for harmless alternative substances.

Also included in the category of carcinogenic working materials are specific groups of substances.

Among the specific groups the Commission has identified "Fibrous Dusts". Not only certain kinds of asbestos, but also the fibrous zeolite, erionite, is considered to produce tumors in man. In addition, a number of fibrous dusts have been shown to produce tumors in experimental animals after inhalative, intratracheal or direct local administration into the ches (intrapleura) or abdominal (intraperitoneal) cavity. In principle, all kinds of elongated dust particles have the potential, like asbestos fibers, to cause tumors if they are sufficiently long, thin and durable in vivo.

According to the internationally accepted convention only the particles with a ratio of length to diameter greater than 3 to 1 and which are longer than 5 micrometers and have a diameter less than 3 micrometer are counted. The term fibrous dust is used here for fibers of these dimensions. With such fibrous dists in animal studies, the number of fibers has been shown to correlate positively with the tumor incidence.

After intraperitoneal or intrapleural administration, carcinogenic effects were demonstrated for almost all inorganic fibers. Therefore, in the interests of health protection, all inorganic fibers are currently suspected of having carcinogenic potunless other evidence of carcinogenic effects is available, are classified as suspected carcinogens.

This list contains all the entries which the Commission has identified as unequivocally proven (category A) or justifiably suspect (catgory B) carcinogenic working materials.

The data field has summarized the categories above using the following wording:

A1 = "proven carcinogen"
A2 = "animal evidence of carcinogenicity"
B = "suspected carcinogen"

German (DFG) - Pregnancy (MAK)

SOURCE

Deutsche Forschungsgemeinschaft (DFG)
MAK and BAT Values 1994
Report No. 30 of Commission for the Investigation of Health Hazards of Chemical Compounds in the Work Area
Deutsche Forschungsgemeinschaft
Kennedyallee 40
D-5300 Bonn 2, Federal Republic of Germany
Telephone (0228) 885-1, Telex 17228312 Telefax (0228) 885-2221

LIST DESCRIPTION

Maximum concentration values in the workplace (MAK values) are established for healthy persons of working age. Epidemiological and animal-experimental findings which point to embryotoxic and/or fetotoxic effects from occupational chemicals are taken into account in the "Toxikologisch-arbeitsmedizinische Begrundung von MAK-Werten" and "von BAT-Werten." The unconditional adoption of MAK values and BAT values as guidelines during pregnancy is not possible because adherence to these values does not guarantee, in every case, that the unborn child is reliably protected from the embryotoxic and/or fetotoxic effects of workplace substances.

Numerous occupational chemicals have not yet been investigated or have not been sufficiently tested for embryotoxic and/or fetotoxic effects. The currently available animal-experimental studies of embryotoxic and/or fetotoxic effects of workplace substances were not only carried out with diverse methods, but also with varying degrees of thoroughness. It is usually not safe to justify or quantify an embryotoxic and/or fetotoxic risk for a human on the basis of these studies because, in the individual case, the risk for a human can exist even if animal tests are negative or if the dose is significantly lower than the threshold dose as determined in animal experimentation.

The Commission intends to test those health hazardous occupational chemicals already included in the MAK list to determine whether an embryotoxic and/or fetotoxic risk can be excluded by adherence to MAK values and to BAT values, whether such a risk has been reliably proved, or if it must be assumed as probable according to the existing information.

Detailed discussion of the fundamentals and possibilities of classification of embryotoxic and/or fetotoxic working materials has led to a division of the substances in the MAK values list into the following groups:

Group A: A risk of damage to the developing embryo or fetus has been unequivocally demonstrated. Exposure of pregnant women can lead to damage to the developing organism even when MAK and BAT values are adhered to.
Group B: According to currently available information, a risk of damage to the developing embryo or fetus must be considered to be probable. Damage to the developing organism cannot be excluded when pregnant women are exposed even when MAK and BAT values are adhered to.
Group C: There is no reason to fear a risk of damage to the developing embryo or fetus when MAK and BAT values are adhered to.
Group D: Classification in one of the groups A to C is not yet possible because although the data available may indicate a trend they are not sufficient for a final evaluation.

This list contains all entries from the current Report which are provided with a pregnancy group as described above.

The data field describes each group as follows:

Group A: "risk to embryo/fetus unequivocal"
Group B: "risk to embryo/fetus probable"
Group C: "no risk to embryo/fetus if exposure limits adhered to"
Group D: "classification not yet possible"

Israel - Time Weighted Averages   (—)

SOURCE

Occupational Exposure Limits for Chemical Substances
in Israel
Dr. Leon Y Naim, M.D., M.P.H.
Dr. Curt Lemesch, M.D., M.P.H.
Medical Inspectors, Ministry of Labour and Welfare of
Israel.
1990

LIST DESCRIPTION

A basic element in the control of occupational hazards
is the setting of threshold levels of exposure to
hazardous chemical agents, below which the exposed
worker can cope successfully with the stress with no
significant threat to his or her health.

Israel has followed in this respect the ILO policy
aimed at the definition of operational levels which can
be incorporated into occupational health legislation and
fully implemented at the workplace.

In 1983 Israel adopted, as legally binding, the
Threshold Limit Values (TLV) recommended by the
ACGIH, which are updated automatically with each
new yearly issue of the TLV booklet.

The Ministry of Labour and Welfare of Israel has
also adopted the definitions of TWA, STEL and Ceiling
recommended by the ACGIH, and defined, in addition,
the ACTION LEVEL (AL) as half the TWA for all
the chemical substances and a quarter of the TWA for
hazardous dust causing Pneumoconiosis. Nevertheless,
for some substances Israel has adopted different TLVs
than those recommended by the ACGIH, as of 1990.

This list contains all the substances for which a time-
weighted average (8 hour) exposure limit is established.

The data field contains the TWA exposure limit in
mg/m3 and/or ppm. Values in parentheses () are under
review and subject to change.

Israel - Short Term Exposure Limits   (—)

SOURCE

Occupational Exposure Limits for Chemical Substances
in Israel
Dr. Leon Y Naim, M.D., M.P.H.
Dr. Curt Lemesch, M.D., M.P.H.
Medical Inspectors, Ministry of Labour and Welfare of
Israel.
1990

LIST DESCRIPTION

A basic element in the control of occupational hazards
is the setting of threshold levels of exposure to
hazardous chemical agents, below which the exposed
worker can cope successfully with the stress with no
significant threat to his or her health.

Israel has followed in this respect the ILO policy
aimed at the definition of operational levels which can
be incorporated into occupational health legislation and
fully implemented at the workplace.

In 1983 Israel adopted, as legally binding, the
Threshold Limit Values (TLV) recommended by the

ACGIH, which are updated automatically with each
new yearly issue of the TLV booklet.

The Ministry of Labour and Welfare of Israel has
also adopted the definitions of TWA, STEL and Ceiling
recommended by the ACGIH, and defined, in addition,
the ACTION LEVEL (AL) as half the TWA for all
the chemical substances and a quarter of the TWA for
hazardous dust causing Pneumoconiosis. Nevertheless,
for some substances Israel has adopted different TLVs
than those recommended by the ACGIH, as of 1990.

This list contains all the substances for which
a short term exposure limit (15 minutes) has been
established. A short Term Exposure Limit is defined
as the concentration to which workers can be exposed
continuously for a short period of time without suffering
from 1) irritation, 2) chronic or irreversible tissue
damage, or 3) narcosis of sufficient degree to increase
the likelihood of accidental injury, impair self-rescue
or materially reduce work efficiency. STELs are
recommended only where toxic effects have been
reported from high short-term exposures in either
humans or animals.

A STEL is defined as a 15 minute TWA exposure
which should not be exceeded at any time during a
work day.

The data field contains the TWA exposure limit in
mg/m3 and/or ppm and the letters STEL to indicate a
short term exposure limit. Values in parentheses () are
under review and subject to change.

Israel - Ceiling Exposure Limits   (—)

SOURCE

Occupational Exposure Limits for Chemical Substances
in Israel
Dr. Leon Y Naim, M.D., M.P.H.
Dr. Curt Lemesch, M.D., M.P.H.
Medical Inspectors, Ministry of Labour and Welfare of
Israel.
1990

LIST DESCRIPTION

A basic element in the control of occupational hazards
is the setting of threshold levels of exposure to
hazardous chemical agents, below which the exposed
worker can cope successfully with the stress with no
significant threat to his or her health.

Israel has followed in this respect the ILO policy
aimed at the definition of operational levels which can
be incorporated into occupational health legislation and
fully implemented at the workplace.

In 1983 Israel adopted, as legally binding, the
Threshold Limit Values (TLV) recommended by the
ACGIH, which are updated automatically with each
new yearly issue of the TLV booklet.

The Ministry of Labour and Welfare of Israel has
also adopted the definitions of TWA, STEL and Ceiling
recommended by the ACGIH, and defined, in addition,
the ACTION LEVEL (AL) as half the TWA for all
the chemical substances and a quarter of the TWA for
hazardous dust causing Pneumoconiosis. Nevertheless,

for some substances Israel has adopted different TLVs than those recommended by the ACGIH, as of 1990.

This list contains all the substances for which a Ceiling Exposure Limit has been established. A Ceiling Exposure is defined as the concentration that should not be exceeded during any part of the working exposure. In conventional industrial hygiene practice if instantaneous monitoring is not feasible, then the Ceiling can be assessed by sampling over a 15-minute period except for those substances which may cause immediate irritation when exposures are short.

The data field contains the Ceiling exposure limit in mg/m3 and/or ppm. Values in parentheses () are under review and subject to change.

Israel - Action Levels   (—)
SOURCE

Occupational Exposure Limits for Chemical Substances
  in Israel
Dr. Leon Y Naim, M.D., M.P.H.
Dr. Curt Lemesch, M.D., M.P.H.
Medical Inspectors, Ministry of Labour and Welfare of
  Israel.
1990

LIST DESCRIPTION

A basic element in the control of occupational hazards is the setting of threshold levels of exposure to hazardous chemical agents, below which the exposed worker can cope successfully with the stress with no significant threat to his or health.

Israel has followed in this respect the ILO policy aimed at the definition of operational levels which can be incorporated into occupational health legislation and fully implemented at the workplace.

In 1983 Israel adopted, as legally binding, the Threshold Limit Values (TLV) recommended by the ACGIH, which are updated automatically with each new yearly issue of the TLV booklet.

The Ministry of Labour and Welfare of Israel has also adopted the definitions of TWA, STEL and Ceiling recommended by the ACGIH, and defined, in addition, the ACTION LEVEL (AL) as half the TWA for all the chemical substances and a quarter of the TWA for hazardous dust causing Pneumoconiosis. Nevertheless, for some substances Israel has adopted different TLVs than those recommended by the ACGIH, as of 1990.

This list contains all the substances for which Israel has established an Action Level.

The data field contains the Action Level in mg/m3 and/or ppm.

United Kingdom - Maximum Exposure Limits -   (GBR)
TWAs
SOURCE

EH40/92 Occupational Exposure Limits 1992
Health and Safety Executive
HMSO Publications Centre
PO Box 276, London SW8 5DT
Telephone (071)873 9090

LIST DESCRIPTION

EH40/92 contains the lists of maximum exposure limits and occupational exposure standards for use with The Control of Substances Hazardous to Health (COSHH) regulations (1988) established by the Health and Safety Executive of the UK.

The advice in EH40 should be taken in the context of the requirements of the COSHH Regulations, especially Regulation 6 (Assessment), Regulation 7 (Control of exposure), Regulations 8 and 9 (Use and maintenance of control measures) and regulation 10 (Monitoring of exposure). Substances hazardous to health are defined in regulations 2 and 5. Additional guidance may be found in the COSHH General Approved Code of Practice (ACOP) and in the Control of Carcinogenic Substances ACOP.

There are two types of occupational exposure limit, defined in Regulation 2 of the COSHH Regulations and applied in Regulation 7. These are maximum exposure limits (MELs), (which are listed in schedule 1 of COSHH and are therefore part of the Regulations), and occupational exposure standards (OESs). The key difference between the two types of limit is that an OES is set at a level at which there is no indication of risk to health; for a MEL, a residual risk may exist and the level set takes socio-economic factors into account.

A MEL is the maximum concentration of an airborne substance, averaged over a reference period, to which employees may be exposed by inhalation under any circumstances and is specified, together with the appropriate reference period, in Schedule 1 of the COSHH Regulations.

An OES is the concentration of an airborne substance, averaged over a reference period, at which, according to current knowledge, there is no evidence that it is likely to be injurious to employees if they are exposed by inhalation, day after day, to that concentration. The OESs are specified in a list approved by the Health and Safety Commission.

Regulation 7 of the COSHH Regulations lays down the requirements for the use of MELs and OESs for the purposes of achieving adequatecontrol. Regulation 7(4) requires that where there is exposure to a substance for which a MEL is specified in schedule 1, the control of exposures shall, so far as inhalation of that substance is concerned, only be treated as being adequate if the level of exposure is reduced so far as is reasonably practicable and in any case below the MEL.

Regulation 7(5) of the COSHH regulations requires that, without prejudice to the generality of regulation 7(1), where there is exposure to a substance for which an OES has been approved, the control of exposure shall, so far as the inhalation of that substance is concerned, be treated as being adequate if:

(a) that OES is not exceeded; or

(b) where that OES is exceeded, the employer identifies the reasons for the standard being exceeded and takes appropriate action to remedy the situation as soon as is reasonably practicable.

The lists of occupational exposure limits given in EH40/92, unless otherwise stated, relate to personal exposure to substances hazardous to health in the air of the workplace.

The units of measurement in occupational exposure limits, concentrations of gases and vapours in air are usually expressed in parts per million (ppm), a measure of concentration by volume, as well as in milligrams per cubic metre of air (mg/ m3), a measure of concentration by mass. In converting from ppm to mg/m3 a temperature of 25 deg. C and an atmospheric pressure of one bar are used. Concentrations of airborne particles (fume, dust, etc) are usually expressed in mg/m3. In the case of dusts the limits in the tables refer to the "total inhalable" fraction unless specifically indicated as referring to the "respirable"fraction. Exceptionally the limits for man-made mineral fibre can be expressed as either mg/m3 or as fibres per millilitre of air (fibres per ml).

This list contains the entries from Table 1: List of maximum exposure limits for which a "long-term exposure limit (8 hour Time Weighted Average reference period)" has been established. The MELs of the dusts included in the list refer to the total inhalable dust fraction, unless otherwise stated.

The data field contains the 8-hour Time Weighted Averages (TWAs) expressed as ppm and/or mg/m3.

United Kingdom - Occupational Exposure Standards - Notes   (GBR)

SOURCE

EH40/92 Occupational Exposure Limits 1992
Health and Safety Executive
HMSO Publications Centre
PO Box 276, London SW8 5DT
Telephone (071)873 9090

LIST DESCRIPTION

EH40/92 contains the lists of maximum exposure limits and occupational exposure standards for use with The Control of Substances Hazardous to Health (COSHH) regulations (1988) established by the Health and Safety Executive of the UK.

The advice in EH40 should be taken in the context of the requirements of the COSHH Regulations, especially Regulation 6 (Assessment), Regulation 7 (Control of exposure), Regulations 8 and 9 (Use and maintenance of control measures) and regulation 10 (Monitoring of exposure). Substances hazardous to health are defined in regulations 2 and 5. Additional guidance may be found in the COSHH General Approved Code of

Practice (ACOP) and in the Control of Carcinogenic Substances ACOP.

There are two types of occupational exposure limit, defined in Regulation 2 of the COSHH Regulations and applied in Regulation 7. These are maximum exposure limits (MELs), (which are listed in schedule 1 of COSHH and are therefore part of the Regulations), and occupational exposure standards (OESs). The key difference between the two types of limit is that an OES is set at a level at which there is no indication of risk to health; for a MEL, a residual risk may exist and the level set takes socio-economic factors into account.

A MEL is the maximum concentration of an airborne substance, averaged over a reference period, to which employees may be exposed by inhalation under any circumstances and is specified, together with the appropriate reference period, in Schedule 1 of the COSHH Regulations.

An OES is the concentration of an airborne substance, averaged over a reference period, at which, according to current knowledge, there is no evidence that it is likely to be injurious to employees if they are exposed by inhalation, day after day, to that concentration. The OESs are specified in a list approved by the Health and Safety Commission.

Regulation 7 of the COSHH Regulations lays down the requirements for the use of MELs and OESs for the purposes of achieving adequatecontrol. Regulation 7(4) requires that where there is exposure to a substance for which a MEL is specified in schedule 1, the control of exposures shall, so far as inhalation of that substance is concerned, only be treated as being adequate if the level of exposure is reduced so far as is reasonably practicable and in any case below the MEL.

Regulation 7(5) of the COSHH regulations requires that, without prejudice to the generality of regulation 7(1), where there is exposure to a substance for which an OES has been approved, the control of exposure shall, so far as the inhalation of that substance is concerned, be treated as being adequate if:

(a) that OES is not exceeded; or

(b) where that OES is exceeded, the employer identifies the reasons for the standard being exceeded and takes appropriate action to remedy the situation as soon as is reasonably practicable.

In general, for most substances the main route of entry into the body is by inhalation and the exposure limits given in EH40/92 relate to exposure by this route. Certain substances, such as phenol, aniline and certain pesticides, which are marked in the Tables in EH40/92 with an 'Sk' notation, have the ability to penetrate the intact skin and thus become absorbed into the body. Absorption through the skin can result from localised contamination, for example from a splash on the skin or clothing, or in certain cases from exposure to high atmospheric concentrations of vapour. Serious effects can result with little or no worning and it is necessary to take special precautions to prevent skin contact when handling these substances. Where the properties of the substances and the methods of use provide a potential exposure route via skin absorption, then these factors should be taken into account in determining the adequacy of the control measures. Further information is given in paragraph 30 of the COSHH General Approved Code of Practice and the COSHH Pesticides Approved Code of Practice.

Certain substances may cause sensitisation of the respiratory tract if inhaled, or of the skin if contact occurs. Respiratory sensitisers can cause asthma, rhinitis, or extrinsic allergic alveolitis. Skin sensitisers cause allergic contact dermatitis. Substances which cause skin sensitisation are not necessarily respiratory sensitisers or vice-versa. Only a proportion of the exposed population will become sensitised, and those who do could not have been identified in advance. Individuals who have become sensitised may produce symptoms of ill-health after exposure even to minute concentrations of the sensitiser.

Wherever it is reasonably practicable exposure to sensitisers should be prevented. Where this cannot be achieved exposure should be kept as low as is reasonably practicable and activities giving rise to short-term peak-cvoncentrations should receive particularattention. As with other substances the spread of contamination by sensitisers to other working areas should also be prevented, as far as is reasonably practicable.

This list contains all the entries from EH40/92 which have a 'Sk' or 'Sen' notation in the "Notes" column of Table 2 (List of approved occupational exposure standards). It is noted that the entries in this column are NOT a part of the approved occupational exposure standards.

The data field contains the words 'can be absorbed through skin' for those substances with a 'Sk' notation and the words 'capable of causing respiratory sensitisation' for those substances with a 'Sen' notation.

United Kingdom - Occupational Exposure Standards - STELs   (GBR)

SOURCE

EH40/92 Occupational Exposure Limits 1992
Health and Safety Executive
HMSO Publications Centre
PO Box 276, London SW8 5DT
Telephone (071)873 9090

LIST DESCRIPTION

EH40/92 contains the lists of maximum exposure limits and occupational exposure standards for use with The Control of Substances Hazardous to Health (COSHH) regulations (1988) established by the Health and Safety Executive of the UK.

The advice in EH40 should be taken in the context of the requirements of the COSHH Regulations, especially Regulation 6 (Assessment), Regulation 7 (Control of exposure), Regulations 8 and 9 (Use and maintenance of control measures) and regulation 10 (Monitoring of exposure). Substances hazardous to health are defined in regulations 2 and 5. Additional guidance may be found in the COSHH General Approved Code of Practice (ACOP) and in the Control of Carcinogenic Substances ACOP.

There are two types of occupational exposure limit, defined in Regulation 2 of the COSHH Regulations and applied in Regulation 7. These are maximum exposure limits (MELs), (which are listed in schedule 1 of COSHH and are therefore part of the Regulations), and occupational exposure standards (OESs). The key difference between the two types of limit is that an OES is set at a level at which there is no indication of risk to health; for a MEL, a residual risk may exist and the level set takes socio-economic factors into account.

A MEL is the maximum concentration of an airborne substance, averaged over a reference period, to which employees may be exposed by inhalation under any circumstances and is specified, together with the appropriate reference period, in Schedule 1 of the COSHH Regulations.

An OES is the concentration of an airborne substance, averaged over a reference period, at which, according to current knowledge, there is no evidence that it is likely to be injurious to employees if they are exposed by inhalation, day after day, to that concentration. The OESs are specified in a list approved by the Health and Safety Commission.

Regulation 7 of the COSHH Regulations lays down the requirementsfor the use of MELs and OESs for the purposes of achieving adequate control. Regulation 7(4) requires that where there is exposure to a substance for which a MEL is specified in schedule 1, the control of exposures shall, so far as inhalation of that substance is concerned, only be treated as being adequate if the level of exposure is reduced so far as is reasonably practicable and in any case below the MEL.

Regulation 7(5) of the COSHH regulations requires that, without prejudice to the generality of regulation 7(1), where there is exposure to a substance for which an OES has been approved, the control of exposure shall, so far as the inhalation of that substance is concerned, be treated as being adequate if:

(a) that OES is not exceeded; or
(b) where that OES is exceeded, the employer identifies the reasons for the standard being exceeded and takes appropriate action to remedy the situation as soon as is reasonably practicable.

The lists of occupational exposure limits given in EH40/92, unless otherwise stated, relate to personal exposure to substances hazardous to health in the air of the workplace.

The units of measurement in occupational exposure limits, concentrations of gases and vapours in air are usually expressed in parts per million (ppm), a measure of concentration by volume, as well as in milligrams per cubic metre of air (mg/ m3), a measure of concentration by mass. In converting from ppm to mg/m3 a temperature of 25 deg. C and an atmospheric pressure of one bar are used. Concentrations of airborne particles (fume, dust, etc) are usually expressed in mg/m3. In the case of dusts the limits in the tables refer to the "total inhalable" fraction unless specifically indicated as referring to the "respirable"fraction. Exceptionally the limits for man-made mineral fibre can be expressed as either mg/m3 or as fibres per millilitre of air (fibres per ml).

This list contains the entries from Table 2: List of approved occupational exposure standards for which a "short-term exposure limit (10-minute reference period)" has been established. The OESs of the dusts included in the list refer to the total inhalable dust fraction, unless otherwise stated.

The data field contains the short-term exposure limit (10-minute reference period) expressed as ppm and/or mg/m3.

United Kingdom - Occupational Exposure Stan- (GBR) dards - TWAs
SOURCE

EH40/92 Occupational Exposure Limits 1992
Health and Safety Executive
HMSO Publications Centre
PO Box 276, London SW8 5DT
Telephone (071)873 9090

LIST DESCRIPTION

EH40/92 contains the lists of maximum exposure limits and occupational exposure standards for use with The Control of Substances Hazardous to Health (COSHH) regulations (1988) established by the Health and Safety Executive of the UK.

The advice in EH40 should be taken in the context of the requirements of the COSHH Regulations, especially Regulation 6 (Assessment), Regulation 7 (Control of exposure), Regulations 8 and 9 (Use and maintenance of control measures) and regulation 10 (Monitoring of exposure). Substances hazardous to health are defined in regulations 2 and 5. Additional guidance may be found in the COSHH General Approved Code of Practice (ACOP) and in the Control of Carcinogenic Substances ACOP.

There are two types of occupational exposure limit, defined in Regulation 2 of the COSHH Regulations and applied in Regulation 7. These are maximum exposure limits (MELs), (which are listed in schedule 1 of COSHH and are therefore part of the Regulations), and occupational exposure standards (OESs). The key difference between the two types of limit is that an OES is set at a level at which there is no indication of risk to health; for a MEL, a residual risk may exist and the level set takes socio-economic factors into account.

A MEL is the maximum concentration of an airborne substance, averaged over a reference period, to which employees may be exposed by inhalation under any circumstances and is specified, together with the appropriate reference period, in Schedule 1 of the COSHH Regulations.

An OES is the concentration of an airborne substance, averaged over a reference period, at which, according to current knowledge, there is no evidence that it is likely to be injurious to employees if they are exposed by inhalation, day after day, to that concentration. The OESs are specified in a list approved by the Health and Safety Commission.

Regulation 7 of the COSHH Regulations lays down the requirements for the use of MELs and OESs for the purposes of achieving adequatecontrol. Regulation 7(4) requires that where there is exposure to a substance for which a MEL is specified in schedule 1, the control of exposures shall, so far as inhalation of that substance is concerned, only be treated as being adequate if the level of exposure is reduced so far as is reasonably practicable and in any case below the MEL.

Regulation 7(5) of the COSHH regulations requires that, without prejudice to the generality of regulation 7(1), where there is exposure to a substance for which an OES has been approved, the control of exposure shall, so far as the inhalation of that substance is concerned, be treated as being adequate if:

(a) that OES is not exceeded; or

(b) where that OES is exceeded, the employer identifies the reasons for the standard being exceeded and takes appropriate action to remedy the situation as soon as is reasonably practicable.

The lists of occupational exposure limits given in EH40/92, unless otherwise stated, relate to personal exposure to substances hazardous to health in the air of the workplace.

The units of measurement in occupational exposure limits, concentrations of gases and vapours in air are usually expressed in parts per million (ppm), a measure of concentration by volume, as well as in milligrams per cubic metre of air (mg/ m3), a measure of concentration by mass. In converting from ppm to mg/m3 a temperature of 25 deg. C and an atmospheric pressure of one bar are used. Concentrations of airborne particles (fume, dust, etc) are usually expressed in mg/m3. In the case of dusts the limits in the tables refer to the "total inhalable" fraction unless specifically indicated as referring to the "respirable"fraction. Exceptionally the limits for man-made mineral fibre can be expressed as either mg/m3 or as fibres per millilitre of air (fibres per ml).

This list contains the entries from Table 2: List of approved occupational exposure standards for which a "long-term exposure limit (8 hour Time Weighted Average reference period)" has been established. The OESs of the dusts included in the list refer to the total inhalable dust fraction, unless otherwise stated.

The data field contains the 8-hour Time Weighted Averages (TWAs) expressed as ppm and/or mg/m3.

United Kingdom - Maximum Exposure Limits - (GBR) Notes
SOURCE

EH40/92 Occupational Exposure Limits 1992
Health and Safety Executive
HMSO Publications Centre
PO Box 276, London SW8 5DT
Telephone (071)873 9090

LIST DESCRIPTION

EH40/92 contains the lists of maximum exposure limits and occupational exposure standards for use with The Control of Substances Hazardous to Health (COSHH) regulations (1988) established by the Health and Safety Executive of the UK.

The advice in EH40 should be taken in the context of the requirements of the COSHH Regulations, especially Regulation 6 (Assessment), Regulation 7 (Control of exposure), Regulations 8 and 9 (Use and maintenance of control measures) and regulation 10 (Monitoring of exposure). Substances hazardous to health are defined in regulations 2 and 5. Additional guidance may be found in the COSHH General Approved Code of Practice (ACOP) and in the Control of Carcinogenic Substances ACOP.

There are two types of occupational exposure limit, defined in Regulation 2 of the COSHH Regulations and applied in Regulation 7. These are maximum exposure limits (MELs), (which are listed in schedule 1 of COSHH and are therefore part of the Regulations), and occupational exposure standards (OESs). The key difference between the two types of limit is that an OES is set at a level at which there is no indication of risk to health; for a MEL, a residual risk may exist and the level set takes socio-economic factors into account.

A MEL is the maximum concentration of an airborne substance, averaged over a reference period, to which employees may be exposed by inhalation under any circumstances and is specified, together with the appropriate reference period, in Schedule 1 of the COSHH Regulations.

An OES is the concentration of an airborne substance, averaged over a reference period, at which, according to current knowledge, there is no evidence that it is likely to be injurious to employees if they are exposed by inhalation, day after day, to that concentration. The OESs are specified in a list approved by the Health and Safety Commission.

Regulation 7 of the COSHH Regulations lays down the requirements for the use of MELs and OESs for the purposes of achieving adequatecontrol. Regulation 7(4) requires that where there is exposure to a substance for which a MEL is specified in schedule 1, the control of exposures shall, so far as inhalation of that substance is concerned, only be treated as being adequate if the level of exposure is reduced so far as is reasonably practicable and in any case below the MEL.

Regulation 7(5) of the COSHH regulations requires that, without prejudice to the generality of regulation 7(1), where there is exposure to a substance for which an OES has been approved, the control of exposure shall, so far as the inhalation of that substance is concerned, be treated as being adequate if:

(a) that OES is not exceeded; or
(b) where that OES is exceeded, the employer identifies the reasons for the standard being exceeded and takes appropriate action to remedy the situation as soon as is reasonably practicable.

In general, for most substances the main route of entry into the body is by inhalation and the exposure limits given in EH40/92 relate to exposure by this route. Certain substances, such as phenol, aniline and certain pesticides, which are marked in the Tables in EH40/92 with an 'Sk' notation, have the ability to penetrate the intact skin and thus become absorbed into the body. Absorption through the skin can result from localized contamination, for example from a splash on the skin or clothing, or in certain cases from exposure to high atmospheric concentrations of vapour. Serious effects can result with little or no warning and it is necessary to take special precautions to prevent skin contact when handling these substances. Where the properties of the substances and the methods of use provide a potential exposure route via skin absorption, then these factors should be taken into account in determining the adequacy of the control measures. Further information is given in paragraph 30 of the COSHH General Approved Code of Practice and the COSHH Pesticides Approved Code of Practice.

Certain substances may cause sensitization of the respiratory tract if inhaled, or of the skin if contact occurs. Respiratory sensitizers can cause asthma, rhinitis, or extrinsic allergic alveolitis. Skin sensitizers cause allergic contact dermatitis. Substances which cause skin sensitization are not necessarily respiratory sensitizers or vice-versa. Only a proportion of the exposed population will become sensitized, and those who do could not have been identified in advance. Individuals who have become sensitized may produce symptoms of ill-health after exposure even to minute concentrations of the sensitizer.

Wherever it is reasonably practicable exposure to sensitizers should be prevented. Where this cannot be achieved exposure should be kept as low as is reasonably practicable and activities giving rise to short-term peak-concentrations should receive particularattention. As with other substances the spread of contamination by sensitizers to other working areas should also be prevented, as far as is reasonably practicable.

This list contains all the entries from EH40/92 which have a 'Sk' or 'Sen' notation in the "Notes" column of Table 1 (List of maximum exposure limits). It is noted that the entries in this column are NOT a part of Schedule 1 of COSHH.

The data field contains the words "can be absorbed through skin" for those substances with a 'Sk' notation and the words "capable of causing respiratory sensitization" for those substances with a 'Sen' notation.

United Kingdom - Maximum Exposure Limits - (GBR) STELs

SOURCE

EH40/92 Occupational Exposure Limits 1992
Health and Safety Executive
HMSO Publications Centre
PO Box 276, London SW8 5DT
Telephone (071)873 9090

LIST DESCRIPTION

EH40/92 contains the lists of maximum exposure limits and occupational exposure standards for use with The Control of Substances Hazardous to Health (COSHH) regulations (1988) established by the Health and Safety Executive of the UK.

The advice in EH40 should be taken in the context of the requirements of the COSHH Regulations, especially Regulation 6 (Assessment), Regulation 7 (Control of exposure), Regulations 8 and 9 (Use and maintenance of control measures) and regulation 10 (Monitoring of exposure). Substances hazardous to health are defined in regulations 2 and 5. Additional guidance may be found in the COSHH General Approved Code of Practice (ACOP) and in the Control of Carcinogenic Substances ACOP.

There are two types of occupational exposure limit, defined in Regulation 2 of the COSHH Regulations and applied in Regulation 7. These are maximum exposure limits (MELs), (which are listed in schedule 1 of COSHH and are therefore part of the Regulations), and occupational exposure standards (OESs). The key difference between the two types of limit is that an OES is set at a level at which there is no indication of risk to health; for a MEL, a residual risk may exist and the level set takes socio-economic factors into account.

A MEL is the maximum concentration of an airborne substance, averaged over a reference period, to which employees may be exposed by inhalation under any circumstances and is specified, together with the appropriate reference period, in Schedule 1 of the COSHH Regulations.

An OES is the concentration of an airborne substance, averaged over a reference period, at which, according to current knowledge, there is no evidence that it is likely to be injurious to employees if they are exposed by inhalation, day after day, to that concentration. The OESs are specified in a list approved by the Health and Safety Commission.

Regulation 7 of the COSHH Regulations lays down the requirements for the use of MELs and OESs for the purposes of achieving adequate control. Regulation 7(4) requires that where there is exposure to a substance for which a MEL is specified in schedule 1, the control of exposures shall, so far as inhalation of that substance is concerned, only be treated as being adequate if the level of exposure is reduced so far as is reasonably practicable and in any case below the MEL.

Regulation 7(5) of the COSHH regulations requires that, without prejudice to the generality of regulation 7(1), where there is exposure to a substance for which an OES has been approved, the control of exposure shall, so far as the inhalation of that substance is concerned, be treated as being adequate if:

(a) that OES is not exceeded; or

(b) where that OES is exceeded, the employer identifies the reasons for the standard being exceeded and takes appropriate action to remedy the situation as soon as is reasonably practicable.

The lists of occupational exposure limits given in EH40/92, unless otherwise stated, relate to personal exposure to substances hazardous to health in the air of the workplace.

The units of measurement in occupational exposure limits, concentrations of gases and vapours in air are usually expressed in parts per million (ppm), a measure of concentration by volume, as well as in milligrams per cubic metre of air (mg/m3), a measure of concentration by mass. In converting from ppm to mg/m3 a temperature of 25 deg. C and an atmospheric pressure of one bar are used. Concentrations of airborne particles (fume, dust, etc) are usually expressed in mg/m3. In the case of dusts the limits in the tables refer to the "total inhalable" fraction unless specifically indicated as referring to the "respirable" fraction. Exceptionally the limits for man-made mineral fibre can be expressed as either mg/m3 or as fibres per millilitre of air (fibres per ml).

This list contains the entries from Table 1: List of maximum exposure limits for which a "short-term exposure limit (10-minute reference period)" has been established. For substances assigned a short-term MEL (e.g. a 10-minute reference period), the MEL should never be exceeded.

The MELs of the dusts included in the list refer to the total inhalable dust fraction, unless otherwise stated.

The data field contains the short-term exposure limit (10-minute reference period) expressed as ppm and/or mg/m3.

Canada - NPRI (National Pollutant Release (CAN) Inventory)

SOURCE

Department of the Environment
Canadian Environmental Protection Act
subsection 16(1)
Extract Canada Gazette, Part I
March 27, 1993
Notice with Respect to Substances in the National Pollutant Release Inventory
Schedule I - National Pollutant Release Inventory Substances

LIST DESCRIPTION

The March 27, 1993 Gazette gave notice that any person who owns or operates a facility described in Schedule II of the notice, and who was engaged, in 1993, in any activity described in Schedule III to the notice and who possesses or who may reasonably be expected to have access to the types of information described in Schedule IV with respect to the substances listed in Schedule I must provide that information to the Minister no later than June 1, 1994.

Schedule II excludes the manufacturing, processing or otherwise using any substance listed in Schedule I to this notice

1) at concentrations of less than one percent by weight.

2) using less than 10 tonnes of substances during 1993

Also excluded from reporting is any facility or any part thereof which is used exclusively for

3) Education or training of students, such as universities, colleges, and schools

4) Research on, or testing of, substances listed in Schedule I

5) The maintenance and repair of transportation vehicles, such as automobiles, trucks, locomotives, ships or aircraft

6) The distribution, storage, or retail sale of fuels

7) The wholesale or retail sale of articles or products which contain substances listed in Schedule I to this notice, as long as thesubstances are not released to the environment during normal use at the facility

8) The retail sale of substances listed in Schedule I to this notice

9) Growing, harvesting, or management of renewable natural resources, such as fisheries, forestry or agriculture, as opposed to facilities which process or otherwise use their products

10) Mining of materials which contain substances listed in Schedule I to this notice, but not those facilities engaged in further processing of these mined materials

11) Drilling or operating wells to obtain oil and gas products which contain substances listed in Schedule I to this notice, but not those facilities engaged in further processing of these oil and gas products.

The types of information which are subject to notice are described in Schedule IV to this Notice. Facilities which are required to provide information in response to this notice must do so in the manner specified in the guidance document " Reporting to the 1993 National Pollutant Release Inventory" , which will be mailed to reporting facilities in mid-1993.

This list contains all entries in Schedule I.

The data field contains any clarification of the entry, if applicable, and the word "present".

Canada - CEPA Schedule II Part II - Toxic (CN2) Substances (Export)

SOURCE

Department of the Environment
Canadian Environmental Protection Act
paragraph 42(1)(a)
Canada Gazette, Part II
12 November 1992
Toxic Substances Export Notification Regulations - List of Toxic Substances Requiring Export Notification in Part II of Schedule II of CEPA

LIST DESCRIPTION

The Toxic Substances Export Notification Regulations address the exchange of information with exporting countries about toxic substances in international trade.

These regulations are promulgated and enforced under the Canadian Environmental Protection Act (CEPA) and represent the best method of ensuring the private sector maintain compliance with Canada's international commitments in regard to the export of substantially restricted substances.

A company is affected by the regulations when it proposes to export for the first time to a country a substantially restricted substance listed on the List of Toxic Substances Requiring Export Notification in Part II of Schedule II of CEPA. These regulations require that the exporter send to the designated national authority an export notification containing the information prescribed in Schedule I and II of the regulations. A copy of this export notification is to be sent simultaneously to Environment Canada. This consignment is required because it is essential for the authority receiving these substances to have pertinent information on the substance they will have to deal with.

The federal government has published (12 November 1992) a List of Toxic Substance Requiring Export Notification under Part II of Schedule II of CEPA.

This list contains all the substances identified in the List of Toxic Substances Requiring Export Notification.

The data field contains the word "present".

Canada - National Air Quality Objectives - (CN3) Schedule I

SOURCE

Department of the Environment
Canadian Environmental Protection Act
subsection 8(1)
Canada Gazette, Part I
August 12, 1989
National Ambient Air Quality Objectives for Air Contaminants - Schedule I

LIST DESCRIPTION

Subsection 8(1) of the Canadian Environmental Protection Act states that the Minister of the Environment shall formulate environmental quality objectives specifying goals or purposes toward which an environmental effort is directed, including goals or purposes stated in quantitative or qualitative terms.

The annexed National Ambient Air Quality Objectives for air contaminants, prescribed between 1974 and 1978 under the now repealed Clean Air Act, have been consolidated and incorporated under the Canadian Environmental Protection Act.

The Objectives are undergoing review by the Canadian Council of Ministers of the Environment. The results of this review will be made public, and any changes or additions will be published in the Canada Gazette.

Three ranges of ambient air quality objective have been formulated - "desirable", "acceptable" and "tolerable". The maximum desirable level is the long-term goal for air quality and provides a basis for an antidegradation policy for unpolluted parts of the country, and for continuing development of pollution control technology.

The maximum acceptable level is intended to provide adequate protection against effects on soil, water, vegetation, materials, animals, visibility, personal comfort, and well-being.

The maximum tolerable level denotes time-based concentration of air contaminants beyond which, due to a diminishing margin of safety, appropriate action is required without delay to protect the health of the general population.

This list includes all air contaminants listed in Column I of Schedule I.

The data field contains the Concentrations identified in Column II and the Range of Quality indicated in Column III of Schedule I.

Canada - CEPA - Priority Substances List   (CN4)

SOURCE

Department of the Environment
Canadian Environmental Protection Act
subsection 12(1)
Canada Gazette, Part I
February 11, 1989
Priority Substances List
For more information contact Mr. G. Allard, Director, Commercial Chemicals Branch, Conservation and Protection, Department of the Environment, 351 St. Joseph Blvd., Hull, Quebec K1A 0H3

LIST DESCRIPTION

The Canadian Environmental Protection Act (CEPA) authorizes the Ministers of National Health and Welfare and of the Environment to conduct research and collect information on a wide variety of substances that may contaminate the environment and cause adverse effects on human health or the environment. It is not possible to assess simultaneously all the substances that may pose a threat to health or the environment. Therefore it was necessary to select a manageable number that should be given priority for assessment, as required by subsection 12(1) of CEPA.

Substances that appear on the Priority Substances List must be assessed to determine whether they are toxic according to the definition specified in section 11 of the Act.

In preparing the CEPA Priority Substances List for the first time, the Ministers of the Environment and of National Health and Welfare gave consideration to the advice provided in various reports submitted by the Canadian Environmental Advisory Council. A substance was selected for the List if it met at least one of the following three criteria:

1. The substance causes or has the potential to cause adverse effects on human health or the environment.
2. The substance accumulates or could accumulate to signifcant concentrations in air, water, soil, sediment or tissue.

3. The substance is released or may be released into the environment in significant quantities or concentrations.

Substances that appear on Schedule I of CEPA (the List of ToxicSubstances) have already been found to be toxic as defined under CEPA Section 11 and are subject to regulations. These substances are not included in the Priority Substances List.

Substances that are not on Schedule I and for which control options are being developed are also not on the List. These substances include carbon dioxide, Halons, oxides of nitrogen, oxides of sulphur, volatile organic compounds (including transportation fuels) and chlorophenate releases resulting from wood protection and preservation practices. Following the preparation of these control options, the Ministers will consider proposing regulations for these substances.

This list contains the substances identified in the February 11th Gazette publication with additional information provided by the Commercial Chemicals Branch on substances removed from that list since its publication. The following 8 substances were indicated as being removed from the published list:

Methyl tertiary butyl ether
Polychlorinated dibenzodioxins
Polychlorinated dibenzofurans
Toluene
bis(2-Chloroethyl)ether
bis(Chloromethyl)ether
Chloromethyl methyl ether
Methyl methacrylate

The data field indicates the estimated time for completion of the assessment reports for the substances. In collecting information to carry out the assessment of a particular priority substance, the opportunity may be taken to collect information on the "class of substances" (CEPA Section 3) of which the priority substance is a member.

Canada - CEPA Schedule I - Toxic Substances   (CN5)

SOURCE

Department of the Environment
Canadian Environmental Protection Act
Sections 13, 33 to 37
Schedule I
May 21, 1993
Unofficial copy - M. Menard, Regulatory Affairs. This document summarizes all regulatory information printed in the Canada Gazette through May 21, 1993.

LIST DESCRIPTION

The Canadian Environmental Protection Act (CEPA) authorizes the Ministers of National Health and Welfare and of the Environment to conduct research and collect information on a wide variety of substances that may contaminate the environment and cause adverse effects on human health or the environment. It is not possible to assess simultaneously all the substances that may pose a threat to health or the environment. Therefore it was necessary to select a manageable number that should be given priority for assessment, as required by subsection 12(1) of CEPA. Substances selected for priority consideration are listed on the Priority Substances List.

Section 13 of CEPA requires the Ministers to review the report of the assessment and publish in the Canada Gazette a summary of the report including a statement of whether the Ministers intend to recommend that the substance be added under subsection 33(1) to the List of Toxic Substances in Schedule I. The Governor in Council if satisfied that a substance is toxic, on the recommendation of the Ministers, make an order adding the substance to the List of Toxic Substances in Schedule I.

The order adding a substance to the List of Toxic Substances in Schedule I is effective on the coming into force of regulations made under subsection 34(1) with respect to the substance.

This list contains all entries provided on the unofficial copy provided by the Regulatory Affairs department. This list represents a summation of various Gazette entries through the May 21, 1993 date. One entry has been modified for assignment of a shorter name. The name provided in the database has been shorten to " Fuel containing toxic substances that are dangerous goods" . The complete entry is as follows:

> Fuel containing toxic substances that are dangerous goods within the meaning of section 2 of the Transportation of Dangerous Goods Act and that (a) are neither normal components of the fuel nor additives designed to improve the characteristicsor the performance of the fuel; or (b) are normal components of the fuel or additives designed to improve the characteristics or performance of the fuel, but are present in quantities or concentrations greater than those generally accepted by industry standards.

The data field represents the type of regulations which are applicable to that substance as specified in subsection 34(1) of the Act.

Canada - CEPA Schedule III Part I - Prohibited (CN6) Substances (Ocean Dumping)

SOURCE

Department of the Environment
Canadian Environmental Protection Act
Sections 71 and 72 (Part VI - Ocean Dumping)
Schedule III Part 1
Canada Gazette 28 June 1988

LIST DESCRIPTION

Part VI of the Canadian Environmental Protection Act (Sections 66 to 86) covers responsibility for protection of the environment from deliberate disposal at sea from ships, aircraft, platforms or other structures. Sections 71 and 72 grant the Minister authority to grant permits for controlled dumping under certain conditions.

As stated in Section 71(3)(c) no permit may be granted unless, in the opinion of the Minister, the dumping or disposal of a certain quantity of the substances is necessary to avert an emergency that poses an unacceptable risk relating to human health and admits of no other feasible solution.

If such conditions as described in 71(3)(c) exist in relation to a substance specified in Schedule III, no permit may be granted in respect of the substance unless:

1. consultation has, if practicable, taken place with any foreign state that is likely to be affected by the proposed dumping; and
2. notification of the proposed dumping has been given to the organization responsible under the Convention for secretariat duties.

Part I of Schedule III includes a list of prohibited substances and Part II contains a list of restricted substances.

This list contains the entries from Schedule III Part I.

The data field contains the word " present" .

Canada - CEPA Schedule III Part II - Restricted (CN7) Substances (Ocean Dumping)

SOURCE

Department of the Environment
Canadian Environmental Protection Act
Sections 71 and 72 (Part VI - Ocean Dumping)
Schedule III Part II
Canada Gazette 28 June 1988

LIST DESCRIPTION

Part VI of the Canadian Environmental Protection Act (Sections 66 to 86) covers responsibility for protection of the environment from deliberate disposal at sea from ships, aircraft, platforms or other structures. Sections 71 and 72 grant the Minister authority to grant permits for controlled dumping under certain conditions.

As stated in Section 71(3)(c) no permit may be granted unless, in the opinion of the Minister, the dumping or disposal of a certain quantity of the substances is necessary to avert an emergency that poses an unacceptable risk relating to human health and admits of no other feasible solution.

If such conditions as described in 71(3)(c) exist in relation to a substance specified in Schedule III, no permit may be granted in respect of the substance unless:

1. consultation has, if practicable, taken place with any foreign state that is likely to be affected by the proposed dumping; and
2. notification of the proposed dumping has been given to the organization responsible under the Convention for secretariat duties.

Part I of Schedule III includes a list of prohibited substances and Part II contains a list of restricted substances.

This list contains the entries from Schedule III Part II.

The data field contains the word " present"

Canada - CEPA Schedule II Part I - Prohibited Substances (Export) (CN8)

SOURCE

Department of the Environment
Canadian Environmental Protection Act
paragraph 41(2)
Canada Gazette, Part II
12 November 1992
Toxic Substances Export Notification Regulations - List of Toxic Substances Requiring Export Notification in Part II of Schedule II of CEPA

LIST DESCRIPTION

The Toxic Substances Export Notification Regulations address the exchange of information with exporting countries about toxic substances in international trade. These regulations are promulgated and enforced under the Canadian Environmental Protection Act (CEPA) and represent the best method of ensuring the private sector maintain compliance with Canada's international commitments in regard to the export of substantially restricted substances.

Section 41 of the Canadian Environmental Protection Act provides that the Governor in Council may, on the recommendation of the Ministers, make an order adding to the List of Prohibited Substances in Part I of Schedule II any toxic substance the use of which is prohibited in Canada by or under an Act of Parliament and may, in the same manner, delete any toxic substance from that List.

No person shall export any toxic substance specified on the List of Prohibited Substances in Part I of Schedule II except for the purpose of destroying the substance or complying with a direction under subparagraph 40(b)(iii).

The federal government has published (12 November 1992) a List of Prohibited Substances under Part I of Schedule II of CEPA.

This list contains all the substances identified in the List of Prohibited Substances in Schedule II Part I.

The data field contains the word "present".

Canada - Drinking Water Quality - AOs (CD1)

SOURCE

Health and Welfare Canada
Federal-Provincial Advisory Committee on Environmental and Occupational Health
Federal-Provincial Subcommittee on Drinking Water
Guidelines for Canadian Drinking Water Quality - Fifth Edition 1993
Available by mail from Canada Communication Group - Publishing, Ottawa, Canada K1A 0S9
Catalog No. H48-10/1993E

LIST DESCRIPTION

"Guidelines for Canadian Drinking Water Quality", Fifth Edition, includes the guidelines approved by the Conference of Deputy Ministers of Health as of June 1992. In the period since the fourth edition was published (1989), no evidence has been presented that would warrant changing the values proposed in that edition. These values have, therefore, been adopted and are presented in this edition as confirmed guidelines.

The guidelines and recommendations listed in the Fifth Edition are intended to apply to all drinking water supplies, public and private. However, they should not be regarded as legally enforceable standards unless promulgated as such by the appropriate provincial, territorial or federal agency. Judicious use of the guidelines will result in the provision of drinking water that is both wholesome and protective of public health.

The Fifth Edition provides guidelines as "Maximum Acceptable Concentration" (MAC), "Interim Maximum Acceptable Concentration" (IMAC) and "Aesthetic Objective" (AO).

Maximum acceptable concentrations have been established for certain substances that are known or suspected to cause adverse effects on health. Each maximum acceptable concentration has been derived to safeguard health assuming lifelong consumption of drinking water containing the substance at that concentration. When the maximum acceptable concentration for a substance is exceeded, however, the minimum action required is immediate resampling. If the maximum acceptable concentration continues to be exceeded, the local authority responsible for drinking water supplies should be consulted concerning appropriate corrective action.

For those substances for which there are insufficient toxicological data to derive a maximum acceptable concentration with reasonable certainty, interim values have been recommended, taking into account the available health-related data, but employing a larger safety factory to compensate for the additional uncertaintiesinvolved. Interim maximum acceptable concentrations were also established for those substances for which estimated lifetime risks of cancer associated with the guideline (the lowest concentration in drinking water that is practicably achievable using available analytical or treatment methods) are greater than those deemed to be essentially negligible.

Aesthetic objectives apply to certain substances or characteristics of drinking water that can affect its acceptance by consumers or interfere with practices for supplying good quality water. For certain parameters, both aesthetic objectives and health-related guidelines have been derived. Where only aesthetic objectives are specified, these values are below those considered to constitute a health hazard. However, if a concentration in drinking water is well above an aesthetic objective, there is a possibility of a health hazard.

These lists contain those entries related to chemical substances (i.e. color, turbidity, etc. are not included).

This list contains all chemicals for which an Aesthetic Objective has been established.

The data field contains the Aesthetic Objective as less than or equal to an amount in mg/Liter.

Canada - Drinking Water Quality - MACs   (CD1)

SOURCE

Health and Welfare Canada
Federal-Provincial Advisory Committee on Environmental and Occupational Health
Federal-Provincial Subcommittee on Drinking Water
Guidelines for Canadian Drinking Water Quality - Fifth Edition 1993
Available by mail from Canada Communication Group - Publishing, Ottawa, Canada K1A 0S9
Catalog No. H48-10/1993E

LIST DESCRIPTION

"Guidelines for Canadian Drinking Water Quality", Fifth Edition, includes the guidelines approved by the Conference of Deputy Ministers of Health as of June 1992. In the period since the fourth edition was published (1989), no evidence has been presented that would warrant changing the values proposed in that edition. These values have, therefore, been adopted and are presented in this edition as confirmed guidelines.

The guidelines and recommendations listed in the Fifth Edition are intended to apply to all drinking water supplies, public and private. However, they should not be regarded as legally enforceable standards unless promulgated as such by the appropriate provincial, territorial or federal agency. Judicious use of the guidelines will result in the provision of drinking water that is both wholesome and protective of public health.

The Fifth Edition provides guidelines as "Maximum Acceptable Concentration" (MAC), "Interim Maximum Acceptable Concentration" (IMAC) and "Aesthetic Objective" (AO).

Maximum acceptable concentrations have been established for certain substances that are known or suspected to cause adverse effects on health. Each maximum acceptable concentration has been derived to safeguard health assuming lifelong consumption of drinking water containing the substance at that concentration. When the maximum acceptable concentration for a substance is exceeded, however, the minimum action required is immediate resampling. If the maximum acceptable concentration continues to be exceeded, the local authority responsible for drinking water supplies should be consulted concerning appropriate corrective action.

For those substances for which there are insufficient toxicological data to derive a maximum acceptable concentration with reasonable certainty, interim values have been recommended, taking into account the available health-related data, but employing a larger safety factor to compensate for the additional uncertainties involved. Interim maximum acceptable concentrations were also established for those substances for which estimated lifetime risks of cancer associated with the guideline (the lowest concentration in drinking water that is practicably achievable using available analytical or treatment methods) are greater than those deemed to be essentially negligible.

Aesthetic objectives apply to certain substances or characteristics of drinking water that can affect its acceptance by consumers or interfere with practices for supplying good quality water. For certain parameters, both aesthetic objectives and health-related guidelines have been derived. Where only aesthetic objectives are specified, these values are below those considered to constitute a health hazard. However, if a concentration in drinking water is well above an aesthetic objective, there is a possibility of a health hazard.

These lists contain those entries related to chemical substances (i.e. color, turbidity, etc. are not included).

This list contains all chemicals for which a Maximum Acceptable Concentration has been established.

The data field contains the Maximum Acceptable Concentration (MAC) in mg/Liter.

Canada - Drinking Water Quality - IMACs   (CD1)

SOURCE

Health and Welfare Canada
Federal-Provincial Advisory Committee on Environmental and Occupational Health
Federal-Provincial Subcommittee on Drinking Water
Guidelines for Canadian Drinking Water Quality - Fifth Edition 1993
Available by mail from Canada Communication Group - Publishing, Ottawa, Canada K1A 0S9
Catalog No. H48-10/1993E

LIST DESCRIPTION

"Guidelines for Canadian Drinking Water Quality", Fifth Edition, includes the guidelines approved by the Conference of Deputy Ministers of Health as of June 1992. In the period since the fourth edition was published (1989), no evidence has been presented that would warrant changing the values proposed in that edition. These values have, therefore, been adopted and are presented in this edition as confirmed guidelines.

The guidelines and recommendations listed in the Fifth Edition are intended to apply to all drinking water supplies, public and private. However, they should not be regarded as legally enforceable standards unless promulgated as such by the appropriate provincial, territorial or federal agency. Judicious use of the guidelines will result in the provision of drinking water that is both wholesome and protective of public health.

The Fifth Edition provides guidelines as "Maximum Acceptable Concentration" (MAC), "Interim Maximum Acceptable Concentration" (IMAC) and "Aesthetic Objective" (AO).

Maximum acceptable concentrations have been established for certain substances that are known or suspected to cause adverse effects on health. Each maximum acceptable concentration has been derived to safeguard health assuming lifelong consumption of drinking water containing the substance at that concentration. When the maximum acceptable concentration for a substance is exceeded, however, the minimum action required is immediate resampling. If the maximum acceptable concentration continues to be exceeded, the local authority responsible for drinking water supplies should be consulted concerning appropriate corrective action.

For those substances for which there are insufficient toxicological data to derive a maximum acceptable concentration with reasonable certainty, interim values have been recommended, taking into account the available health-related data, but employing a larger safety factory to compensate for the additional uncertaintiesinvolved. Interim maximum acceptable concentrations were also established for those substances for which estimated lifetime risks of cancer associated with the guideline (the lowest concentration in drinking water that is practicably achievable using available analytical or treatment methods) are greater than those deemed to be essentially negligible.

Aesthetic objectives apply to certain substances or characteristics of drinking water that can affect its acceptance by consumers or interfere with practices for supplying good quality water. For certain parameters, both aesthetic objectives and health-related guidelines have been derived. Where only aesthetic objectives are specified, these values are below those considered to constitute a health hazard. However, if a concentration in drinking water is well above an aesthetic objective, there is a possibility of a health hazard.

These lists contain those entries related to chemical substances (i.e. color, turbidity, etc. are not included).

This list contains all chemicals for which an Interim Maximum Acceptable Concentration has been established.

The data field contains the Interim Maximum Acceptable Concentration (MAC) in mg/Liter.

Canada - Ontario - OHSA - CEVs   (ONT)

SOURCE

Ontario Occupational Health and Safety Act
Ontario Regulation 833 (formerly 654/86)
As last amended by Ontario Regulation 513/92
Control of Exposure to Biological or Chemical Agents
This list also includes exposure limits established by
Ontarios Regulations:

Reg 835 Acrylonitrile
Reg 836 Arsenic
Reg 837 Asbestos
Reg 839 Benzene
Reg 840 Coke Oven Emissions
Reg 841 Ethylene Oxide
Reg 842 Isocyanates
Reg 843 Lead
Reg 844 Mercury
Reg 845 Silica
Reg 846 Vinyl Chloride

LIST DESCRIPTION

The Minister of Ontario has established regulations under the Ontario Occupational Health and Safety Act to control the exposure of workers to biological or chemical agents. Every employer shall take all measures reasonably necessary in the circumstances

to protect workers from exposure to a hazardous biological or chemical agent because of the storage, handling, processing or use of such agent in the work place.

Such measures shall include the provision and use of engineering controls, work practices, hygiene facilities and practices and personal protective equipment.

Every employer shall take measures to limit the daily and weekly exposure of workers to a biological or chemical agent listed here to the concentration limit expressed as a time-weighted average exposure value (TWAEV), a short-term exposure value (STEV) or a ceiling exposure value (CEV).

In determining the exposure of workers to a hazardous biological or chemical agent no regard shall be had to the wearing and use of personal protective equipment. An employer shall protect workers from exposure to a hazardous biological or chemical agent without requiring the workers to wear and use personal protective equipment.

Where engineering controls are not in existence or are not obtainable, are not reasonable or not practical, are renderedineffective because of a temporary breakdown of such controls, or are ineffective to prevent, control or limit exposure because of an emergency, the employer shall provide, and workers shall wear and use, personal protective equipment appropriate in the circumstances to protect the workers from exposure to a hazardous biological or chemical agent.

This list contains all chemical and biological agents for which a Ceiling Exposure Value has been established under Regulation 833 (Part 5 and 11) or the Designated Substance Regulations (DSR) cited above. For substances for which a Designated Substance Regulation exists there can be other requirements. The specific regulations should be consulted for these substances.

The data field contains the CEV in parts per million (ppm) and/or in milligrams per cubic meter of air (mg/m3) and any clarification statements (physical form, footnotes, etc.). The title of parts other than part 4 where this value is cited (i.e. values applying to workplaces to which the DSR does not apply) are also included.

Canada - Ontario - OHSA - STEVs   (ONT)
SOURCE

Ontario Occupational Health and Safety Act
Ontario Regulation 833
As last amended by Ontario Regulation 597/94 made
   on September 15, 1994, filed September 16, 1994.
   Effective October 31, 1994.
Control of Exposure to Biological or Chemical Agents
This list also includes exposure limits established by
   Ontarios Regulations:

Reg 835 Acrylonitrile
Reg 836 Arsenic
Reg 837 Asbestos
Reg 839 Benzene
Reg 840 Coke Oven Emissions
Reg 841 Ethylene Oxide
Reg 842 Isocyanates
Reg 843 Lead
Reg 844 Mercury
Reg 845 Silica
Reg 846 Vinyl Chloride

LIST DESCRIPTION

The Minister of Ontario has established regulations under the Ontario Occupational Health and Safety Act to control the exposure of workers to biological or chemical agents. Every employer shall take all measures reasonably necessary in the circumstances to protect workers from exposure to a hazardous biological or chemical agent because of the storage, handling, processing or use of such agent in the work place.

Such measures shall include the provision and use of engineering controls, work practices, hygiene facilities and practices and personal protective equipment.

Every employer shall take measures to limit the daily and weekly exposure of workers to a biological or chemical agent listed here to the concentration limit expressed as a time-weighted average exposure value (TWAEV), a short-term exposure value (STEV) or a ceiling exposure value (CEV).

In determining the exposure of workers to a hazardous biological or chemical agent no regard shall be had to the wearing and use of personal protective equipment. An employer shall protect workers from exposure to a hazardous biological or chemical agent without requiring the workers to wear and use personal protective equipment.

Where engineering controls are not in existence or are not obtainable, are not reasonable or not practical, are rendered ineffective because of a temporary breakdown of such controls, or are ineffective to prevent, control or limit exposure beacuse of an emergency, the employer shall provide, and workers shall wear and use, personal protective equipment appropriate in the circumstances to protect the workers from exposure to a hazardous biological or chemical agent.

This list contains all chemical and biological agents for which a Short Term Exposure Value has

been established under Regulation 833 (Parts 4,6,8, and 11) or the Designated Substance Regulations (DSR) cited above. For substances for which a Designated Substance Regulation exists there can be other requirements. The specific regulations should be consulted for these substances.

The data field contains the STEV in parts per million (ppm) and/or in milligrams per cubic meter of air (mg/m3) and any clarification statements (physical form, footnotes, etc.). The title of parts other than part 4 where this value is cited (i.e. mineral dusts, nuisance particulates, agents of variable composition, simple asphyxiants and values applying to workplaces to which the DSR does not apply) are also included.

Canada - Ontario - OHSA - TWAEVs   (ONT)
SOURCE

Ontario Occupational Health and Safety Act
Ontario Regulation 833
As last amended by Ontario Regulation 597/94 made
   September 15, 1994, filed September 16, 1994.
   Effective October 31, 1994
Control of Exposure to Biological or Chemical Agents
This list also includes exposure limits established by
   Ontarios Regulations:

Reg 835 Acrylonitrile
Reg 836 Arsenic
Reg 837 Asbestos
Reg 839 Benzene
Reg 840 Coke Oven Emissions
Reg 841 Ethylene Oxide
Reg 842 Isocyanates
Reg 843 Lead
Reg 844 Mercury
Reg 845 Silica
Reg 846 Vinyl Chloride

LIST DESCRIPTION

The Minister of Ontario has established regulations under the Ontario Occupational Health and Safety Act to control the exposure of workers to biological or chemical agents. Every employer shall take all measures reasonably necessary in the circumstances to protect workers from exposure to a hazardous biological or chemical agent because of the storage, handling, processing or use of such agent in the work place.

Such measures shall include the provision and use of engineering controls, work practices, hygiene facilities and practices and personal protective equipment.

Every employer shall take measures to limit the daily and weekly exposure of workers to a biological or chemical agent listed here to the concentration limit expressed as a time-weighted average exposure value (TWAEV), a short-term exposure value (STEV) or a ceiling exposure value (CEV).

In determining the exposure of workers to a hazardous biological or chemical agent no regard shall be had to the wearing and use of personal protective equipment. An employer shall protect workers from exposure to a hazardous biological or chemical agent without requiring the workers to wear and use personal protective equipment.

Where engineering controls are not in existence or are not obtainable, are not reasonable or not practical, are rendered ineffective because of a temporary breakdown of such controls, or are ineffective to prevent, control or limit exposure beacuse of an emergency, the employer shall provide, and workers shall wear and use, personal protective equipment appropriate in the circumstances to protect the workers from exposure to a hazardous biological or chemical agent.

This list contains all chemical and biological agents for which a Time Weighted Average Exposure Value has been established under Regulation 833 (Parts 4,6,7,8,9 and 11) or the Designated Substance Regulations (DSR) cited above. For substances for which a Designated Substance Regulation exists there can be other requirements. The specific regulations should be consulted for these substances.

The data field contains the TWAEV in parts per million (ppm) and/or in milligrams per cubic meter of air (mg/m3) and any clarification statements (physical form, footnotes, etc.). The title of parts other than part 4 where this value is cited (i.e. mineral dusts, nuisance particulates, agents of variable composition, simple asphyxiants and values applying to workplaces to which the DSR does not apply) are also included.

Canada - Ontario - OHSA - Designated (ONT) Substances

SOURCE

Ontario Occupational Health and Safety Act
Revised Statutes of Ontario, 1980
Chapter 321

LIST DESCRIPTION

The Minister of Ontario has established specific regulations for certain designated substances. These regulations include the requirement for an emergency program or a control program which may include an assessment to be made, in writing, of the exposure or likelihood of exposure in a workplace to the substance. Such a program can include a code for repiratory equipment,a code for measuring airborne levels of the substance, includingvarious testing and sampling methods, a code for medicalsurveillance of exposed workers and training requirements.

Each designated substance includes requirements for the employer to take all necessary measures and procedures by means of engineering controls, work practices and hygiene practices and facilities to ensure that the time-weighted average exposure of a worker to the designated substance shall not exceed the designated levels.

Such compliance shall be achieved without requiring a worker to wear and use respiratory equipment. The time-weighted average exposure of a worker to the substance shall be calculated in accordance with

the Code as established in the designated substance regulation.

This list identifies all chemicals or groups of chemicals whichhave been determined to be designated substances.

The data field includes the Time-Weighted Average Exposure Value (TWAEV) designated in the specific regulation for that substance and the Regulation number of the specific regulation. The TWAEV is given in parts per million (ppm) or milligrams per cubic meter (mg/m3). Limits for isocyanates are given in ppm of isocyanates and micromoles of isocyanates per cubic meter of air.

Canada - Ontario - OHSA - Skin Notation   (ONT)

SOURCE

Ontario Occupational Health and Safety Act
Ontario Regulation 833 (formerly 654/86)
As last amended by Ontario Regulation 513/92
Control of Exposure to Biological or Chemical Agents

LIST DESCRIPTION

The Minister of Ontario has established regulations under the Ontario Occupational Health and Safety Act to control the exposure of workers to biological or chemical agents. Every employer shall take all measures reasonably necessary in the circumstances to protect workers from exposure to a hazardous biological or chemical agent because of the storage, handling, processing or use of such agent in the work place.

Such measures shall include the provision and use of engineering controls, work practices, hygiene facilities and practices and personal protective equipment.

Every employer shall take measures to limit the daily and weekly exposure of workers to a biological or chemical agent listed here to the concentration limit expressed as a time-weighted average exposure value (TWAEV), a short-term exposure value (STEV) or a ceiling exposure value (CEV).

In determining the exposure of workers to a hazardous biological or chemical agent no regard shall be had to the wearing and use of personal protective equipment. An employer shall protect workers from exposure to a hazardous biological or chemical agent without requiring the workers to wear and use personal protective equipment.

Where engineering controls are not in existence or are not obtainable, are not reasonable or not practical, are rendered ineffective because of a temporary breakdown of such controls, or are ineffective to prevent, control or limit exposure beacuse of an emergency, the employer shall provide, and workers shall wear and use, personal protective equipment appropriate in the circumstances to protect the workers from exposure to a hazardous biological or chemical agent.

Some substances for which an airborne exposure limit has been established contain a "skin" notation. Listed substances followed by the designation "skin" refer to the potential contribution to the overall exposure by the cutaneous route including mucous membranes and eye, either by airborne, or, more particularly, by direct contact with the substance. Vehicles can alter skin absorption. This designation is intended to suggestappropriate measures for the prevention of cutaneous absorption so that the exposure limit is not invalidated.

This list contains all chemical and biological agents contained in Part 4 and Part 8 (agents of variable composition) of Regulation 833 which such a skin notation.

The data field contains the words " absorption through skin, eyes, or mucous membranes" , any clarification statements (physical form, footnotes, etc.). If information was taken from other than part 4, that part is identified by name in parenthesis.

Canada - British Columbia - Ceiling Limits   (BC1)
SOURCE

Industrial Health and Safety Regulations, Sections 12, 13, 35, 36, 76 and Appendices A and B
Pursuant to Section 71 of the Worker's Compensation Act
With revisions effective November 1, 1993
Workers' Compensation Board of British Columbia
6951 Westminster Highway
Richmond, British Columbia, Canada V7C 1C6
Telephone 276-3100

LIST DESCRIPTION

These regulations have been promulgated after public hearings and publication in the Gazette. They represent minimum requirements and apply to all persons working in or contributing to the production of those industries coming within the scope of the Workers' Compensation Act of British Columbia, except for those industries for which industrial health and safety regulations have been specifically provided under some other Act of Canada or the Province of British Columbia.

Under Section 13 of the Industrial Health and Safety Regulations, a worker's exposure to airborne contaminants shall be limited to the stated permissible concentrations as specified in Appendix A and Appendix B and the preambles thereto.

Permissible concentrations refer to airborne concentrations of substances and represent conditions under which it is believed that nearly all workers may be repeatedly exposed day after day without adverse effect. Because of wide variation in individual susceptibility, however, a small percentage of workers may experience discomfort from some substances at concentrations at or below the permissible concentrations; at concentrations at or below the permissible concentration; a smaller percentage may be affected more seriously by aggravation of a pre-existing condition or by development of an occupational illness.

The categories of permissible concentrations are specified as:

a) 8 hour Exposure Limit - The time weighted average concentration for a normal 8 hour workday or a 40 hour work week;

b) 15 minute Exposure Limit - The maximal concentration to which workers can be exposed for a period up to 15 minutes without suffering from irritation, chronic or irreversible tissue change, or narcosis of sufficient degree to increase accident proneness, impair self-rescue, or materially reduce work efficiency, providedthat not more than four excursions per day are permitted with at least 60 minutes between exposure periods and provided that the 8 hour exposure limit is not exceeded. The 15 minute exposure limit is considered a maximal allowable concentration or ceiling not to be exceeded at any time during the 15 minute excursion period;

c) Ceiling (C) - The concentration of a chemical substance that should not be exceeded at any time;

d) (K) - Substances used in industry which have proven carcinogenic in man or have induced cancer in animals under appropriate experimental conditions, (see Appendix B), or for which there are no permissible concentrations; and

e) "Skin" Notation. Substances which carry the notation of "Skin" refer to the potential contribution to the overall exposure by skin absorption including mucous membranes and eye, either by airborne, or more particularly, by direct contact with the substance.

The list contains all the entries in Appendix A which are prefixed by the letter (C).

The data field contains the Permissible Exposure Limit in parts per million (ppm) and/or milligrams per cubic meter of air (mg/m3).

Canada - British Columbia - 15 Minute Exposure   (BC1)
Limits
SOURCE

Industrial Health and Safety Regulations, Sections 12, 13, 35, 36, 76 and Appendices A and B
Pursuant to Section 71 of the Worker's Compensation Act
With revisions effective November 1, 1993
Workers' Compensation Board of British Columbia
6951 Westminster Highway
Richmond, British Columbia, Canada V7C 1C6
Telephone 276-3100

LIST DESCRIPTION

These regulations have been promulgated after public hearings and publication in the Gazette. They represent minimum requirements and apply to all persons working in or contributing to the production of those industries coming within the scope of the Workers' Compensation Act of British Columbia, except for those industries for which industrial health and safety regulations have been specifically provided under some other Act of Canada or the Province of British Columbia.

Under Section 13 of the Industrial Health and Safety Regulations, a worker's exposure to airborne contaminants shall be limited to the stated permissible concentrations as specified in Appendix A and Appendix B and the preambles thereto.

Permissible concentrations refer to airborne concentrations of substances and represent conditions under which it is believed that nearly all workers may be repeatedly exposed day after day without adverse effect. Because of wide variation in individual susceptibility, however, a small percentage of workers may experience discomfort from some substances at concentrations at or below the permissible concentrations; at concentrations at or below the permissible concentration; a smaller percentage may be affected more seriously by aggravation of a pre-existing condition or by development of an occupational illness.

The categories of permissible concentrations are specified as:

a) 8 hour Exposure Limit - The time weighted average concentration for a normal 8 hour workday or a 40 hour work week;

b) 15 minute Exposure Limit - The maximal concentration to which workers can be exposed for a period up to 15 minutes without suffering from irritation, chronic or irreversible tissue change, or narcosis of sufficient degree to increase accident proneness, impair self-rescue, or materially reduce work efficiency, providedthat not more than four excursions per day are permitted with at least 60 minutes between exposure periods and provided that the 8 hour exposure limit is not exceeded. The 15 minute exposure limit is considered a maximal allowable concentration or ceiling not to be exceeded at any time during the 15 minute excursion period;

c) Ceiling (C) - The concentration of a chemical substance that should not be exceeded at any time;

d) (K) - Substances used in industry which have proven carcinogenic in man or have induced cancer in animals under appropriate experimental conditions, (see Appendix B), or for which there are no permissible concentrations; and

e) "Skin" Notation. Substances which carry the notation of "Skin" refer to the potential contribution to the overall exposure by skin absorption including mucous membranes and eye, either by airborne, or more particularly, by direct contact with the substance.

The list contains all the entries in Appendix A for which a 15 Minute Permissible Concentration Limit is defined.

The data field contains the 15 Minute Permissible Exposure Limit in parts per million (ppm) and/or milligrams per cubic meter of air (mg/m3) followed by the designation STEL (for Short Term Exposure Limit).

Canada - British Columbia - Skin Notations   (BC1)

SOURCE

Industrial Health and Safety Regulations, Sections 12, 13, 35, 36, 76 and Appendices A and B
Pursuant to Section 71 of the Worker's Compensation Act
With revisions effective November 1, 1993
Workers' Compensation Board of British Columbia
6951 Westminster Highway
Richmond, British Columbia, Canada V7C 1C6
Telephone 276-3100

LIST DESCRIPTION

These regulations have been promulgated after public hearings and publication in the Gazette. They represent minimum requirements and apply to all persons working in or contributing to the production of those industries coming within the scope of the Workers' Compensation Act of British Columbia, except for those industries for which industrial health and safety regulations have been specifically provided under some other Act of Canada or the Province of British Columbia.

Under Section 13 of the Industrial Health and Safety Regulations, a worker's exposure to airborne contaminants shall be limited to the stated permissible concentrations as specified in Appendix A and Appendix B and the preambles thereto.

Permissible concentrations refer to airborne concentrations of substances and represent conditions under which it is believed that nearly all workers may be repeatedly exposed day after day without adverse effect. Because of wide variation in individual susceptibility, however, a small percentage of workers may experience discomfort from some substances at concentrations at or below the permissible concentrations; at concentrations at or below the permissible concentration; a smaller percentage may be affected more seriously by aggravation of a pre-existing condition or by development of an occupational illness.

The categories of permissible concentrations are specified as:

a) 8 hour Exposure Limit - The time weighted average concentration for a normal 8 hour workday or a 40 hour work week;

b) 15 minute Exposure Limit - The maximal concentration to which workers can be exposed for a period up to 15 minutes without suffering from irritation, chronic or irreversible tissue change, or narcosis of sufficient degree to increase accident proneness, impair self-rescue, or materially reduce work efficiency, providedthat not more than four excursions per day are permitted with at least 60 minutes between exposure periods and provided that the 8 hour exposure limit is not exceeded. The 15 minute exposure limit is considered a maximal allowable concentration or ceiling not to be exceeded at any time during the 15 minute excursion period;

c) Ceiling (C) - The concentration of a chemical substance that should not be exceeded at any time;

d) (K) - Substances used in industry which have proven carcinogenic in man or have induced cancer in animals under appropriate experimental conditions, (see Appendix B), or for which there are no permissible concentrations; and

e) "Skin" Notation. Substances which carry the notation of "Skin" refer to the potential contribution to the overall exposure by skin absorption including mucous membranes and eye, either by airborne, or more particularly, by direct contact with the substance.

The list contains all the entries in Appendix A which are followed by the word Skin.

The data field contains the words "skin - potential for skin absorption".

Canada - British Columbia - Carcinogens (BC1)

SOURCE

Industrial Health and Safety Regulations, Sections 12, 13, 35, 36, 76 and Appendices A and B
Pursuant to Section 71 of the Worker's Compensation Act
With revisions effective November 1, 1993
Workers' Compensation Board of British Columbia
6951 Westminster Highway
Richmond, British Columbia, Canada V7C 1C6
Telephone 276-3100

LIST DESCRIPTION

These regulations have been promulgated after public hearings and publication in the Gazette. They represent minimum requirements and apply to all persons working in or contributing to the production of those industries coming within the scope of the Workers' Compensation Act of British Columbia, except for those industries for which industrial health and safety regulations have been specifically provided under some other Act of Canada or the Province of British Columbia.

Under Section 13 of the Industrial Health and Safety Regulations, a worker's exposure to airborne contaminants shall be limited to the stated permissible concentrations as specified in Appendix A and Appendix B and the preambles thereto.

Permissible concentrations refer to airborne concentrations of substances and represent conditions under which it is believed that nearly all workers may be repeatedly exposed day after day without adverse effect. Because of wide variation in individual susceptibility, however, a small percentage of workers may experience discomfort from some substances at concentrations at or below the permissible concentrations; at concentrations at or below the permissible concentration; a smaller percentage may be affected more seriously by aggravation of a pre-existing condition or by development of an occupational illness.

The categories of permissible concentrations are specified as:

a) 8 hour Exposure Limit - The time weighted average concentration for a normal 8 hour workday or a 40 hour work week;

b) 15 minute Exposure Limit - The maximal concentration to which workers can be exposed for a period up to 15 minutes without suffering from irritation, chronic or irreversible tissue change, or narcosis of sufficient degree to increase accident proneness, impair self-rescue, or materially reduce work efficiency, providedthat not more than four excursions per day are permitted with at least 60 minutes between exposure periods and provided that the 8 hour exposure limit is not exceeded. The 15 minute exposure limit is considered a maximal allowable concentration or ceiling not to be exceeded at any time during the 15 minute excursion period;

c) Ceiling (C) - The concentration of a chemical substance that should not be exceeded at any time;

d) (K) - Substances used in industry which have proven carcinogenic in man or have induced cancer in animals under appropriate experimental conditions, (see Appendix B), or for which there are no permissible concentrations; and

e) "Skin" Notation. Substances which carry the notation of "Skin" refer to the potential contribution to the overall exposure by skin absorption including mucous membranes and eye, either by airborne, or more particularly, by direct contact with the substance.

The list contains all the entries in Appendix A which are preceded by the letter "K" and all the entries in Appendix B (carcinogens which have established limits are listed both in Appendix A and Appendix B).

The data field contains the word "carcinogens" and the designated exposure limit, if any. Substances without an exposure limit and those for which no contact by any route is permitted are indicated in words.

Canada - British Columbia - 8 Hour Exposure Limits (BC1)

SOURCE

Industrial Health and Safety Regulations, Sections 12, 13, 35, 36, 76 and Appendices A and B
Pursuant to Section 71 of the Worker's Compensation Act
With revisions effective November 1, 1993
Workers' Compensation Board of British Columbia
6951 Westminster Highway
Richmond, British Columbia, Canada V7C 1C6
Telephone 276-3100

LIST DESCRIPTION

These regulations have been promulgated after public hearings and publication in the Gazette. They represent minimum requirements and apply to all persons working in or contributing to the production of those industries coming within the scope of the Workers' Compensation Act of British Columbia, except for those industries for which industrial health and safety regulations have been specifically provided under some other Act of Canada or the Province of British Columbia.

Under Section 13 of the Industrial Health and Safety Regulations, a worker's exposure to airborne contaminants shall be limited to the stated permissible concentrations as specified in Appendix A and Appendix B and the preambles thereto.

Permissible concentrations refer to airborne concentrations of substances and represent conditions under which it is believed that nearly all workers may be repeatedly exposed day after day without adverse effect. Because of wide variation in individual susceptibility, however, a small percentage of workers may experience discomfort from some substances at concentrations at or below the permissible concentrations; at concentrations at or below the permissible concentration; a smaller percentage may be affected more seriously by aggravation of a pre-existing condition or by development of an occupational illness.

The categories of permissible concentrations are specified as:

a) 8 hour Exposure Limit - The time weighted average concentration for a normal 8 hour workday or a 40 hour work week;

b) 15 minute Exposure Limit - The maximal concentration to which workers can be exposed for a period up to 15 minutes without suffering from irritation, chronic or irreversible tissue change, or narcosis of sufficient degree to increase accident proneness, impair self-rescue, or materially reduce work efficiency, providedthat not more than four excursions per day are permitted with at least 60 minutes between exposure periods and provided that the 8 hour exposure limit is not exceeded. The 15 minute exposure limit is considered a maximal allowable concentration or ceiling not to be exceeded at any time during the 15 minute excursion period;

c) Ceiling (C) - The concentration of a chemical substance that should not be exceeded at any time;

d) (K) - Substances used in industry which have proven carcinogenic in man or have induced cancer in animals under appropriate experimental conditions, (see Appendix B), or for which there are no permissible concentrations; and

e) "Skin" Notation. Substances which carry the notation of "Skin" refer to the potential contribution to the overall exposure by skin absorption including mucous membranes and eye, either by airborne, or more particularly, by direct contact with the substance.

The list contains all the entries in Appendix A which provide an 8 hour Time Weighted Average (TWA) Permissible Concentration.

The data field contains the 8 Hour Time Weighted Average Permissible Exposure Limit in parts per million (ppm) and/or milligrams per cubic meter of air (mg/m3).

Mexico - Instruction No. 10 - TWAs   (MEX)

SOURCE

Mexican Laws and Regulations Governing Occupational Safety and Health

General Regulation Governing Workplace Safety and Hygiene

Instruction No. 10 - Safety and Hygiene Conditions at Workplaces Where Chemical Substances Capable of Causing Contamination of the Work Environment are Produced, Stored, or Handled

Published (in English) by Occupational Safety and Health Administration, U. S. Department of Labor

January, 1993

LIST DESCRIPTION

The Occupational Safety and Health Administration (OSHA) has compiled and translated into English a selection of laws and regulations governing workplace safety and hygiene in Mexico. OSHA has undertaken this unofficial translation to make these documents more widely available. Other laws and regulations may be in effect in Mexico that are applicable to certain activities. Some of the documents published here, which were current as of April 1992, may subsequently have been revised or superseded. OSHA has published these documents for informational purposes - they should not be relied on as legal documents. The Spanish versions of these documents, which are published by the Mexican Secretariat of Labor and Social Welfare, are the definitive texts and should be referred to for legal purposes.

Title Eight of the General Regulations on Workplace Safety and Hygiene, article 135 states:

Workplace atmospheric contaminants are physical agents and chemical or biological elements or compounds capable of altering atmospheric conditions in the workplace and which, because of their properties, concentrations, level, and duration can affect workers' health. the Secretary of Labor and Social Welfare shall, in the applicable instruction, determine the maximum permissible levels of exposure to such agents in the workplace.

The maximum permissible levels as required under article 135 of Title 8 are provided in Instruction No. 10, titled " Safety and Hygiene Conditions at Workplaces Where Chemical Substances Capable of Causing Contamination of the Work Environment are Produced, Stored, or Handled".

This instruction is mandatory and is intended to spell out measures designed to improve safety and hygiene conditions at workplaces that produce, store, or handle chemical substances that, by virtue of their properties, concentration levels, and exposure durations, are capable of contaminating the work environment and adversely affecting the health of workers, as well as to establish the maximum permissible concentration levels of said substances, in accordance with the type of exposure. Employers must compile, keep, maintain up-to-date, and show to the competent authorities records of the concentration levels of chemical substances mentioned in this Instruction, in accordance with the information provided in Annex I of said Instruction.

The maximum permissible concentration levels to which workers may be exposed are shown in Table I, which shows three different categories of exposure limits:

a) The time-weighted average (TWA) permissible exposure limit (PEL) for eight hours of daily exposure, to which nearly all workers may be repeatedly exposed without experiencing any adverse health effects.

b) The short-term exposure limit (STEL), which shall not be exceeded in any 15 minute period of the working day. Exposures at the STEL shall not occur more than 4 times per day and shall be separated by intervals of at least 60 minutes between successive STEL exposures. In any case, exposures at the STEL may not exceed the 8-hour TWA PEL.

c) The ceiling (or peak) exposure limit. Exposures at the ceiling may not be exceeded during any part of the working exposure, even for a moment.

Table I also includes Appendices A,B,C and D and Appendix I and II. Appendix II defines the term "Skin" as:

This term, added to some of the substances in the list, indicates that the element or compound is capable of being absorbed through the skin by means of simple skin contact. This effect must be taken into account because, when this occurs, the maximum permissible concentration level proposed can be invalidated by the contribution of the entry through the skin, including membranes, mucosa, and eyes, to the worker's overall exposure.

Appendix A lists carcinogens. Three types of carcinogens are defined:

A1 - potential carcinogenic contaminants

A2 - potential carcinogens for humans based onlimited epidemiological evidence.

A3 - Carcinogens in humans. Substances associated with industrial processes, recognized as being potential carcinogens although they have not been assigned a maximum permissible exposure value. Workers may not be exposed in any way or form to these substances, which is why specific control methods must be used.

Appendix B provides definitions and limits for Mineral Dust.

Appendix C provides a definition and identification of pure asphyxiants.

Appendix D defines a method for calculating the maximum permissible concentrations for mixtures of contaminants.

Annex I to Instruction 10 identifies the contents of the evaluation report which must be provided under the general provisions of Instruction 10 which states:

Employers shall compile, keep, maintain up-to-date, and show to the competent authorities records of the concentration levels of chemical substances mentioned in this Instruction, in accordance with the information provided in Annex I of said Instruction.

This is done to facilitate the adoption of safety and hygiene measures aimed at controlling exposure to said substances.

Annex II provides definitions of technical terms used in the instruction.

This list identifies all the substances in Table I which have an established 8-hour Time Weighted Average (TWA) Permissible Exposure Limit (PEL) including the simple asphyxiants identified in Appendix C and the mineral dusts and other particulate contaminants listed in Appendix B.

The data field provides the limit in parts per million (ppm) or milligrams per cubic meter (mg/m3).

Mexico - Instruction No. 10 - STELs (MEX)

SOURCE

Mexican Laws and Regulations Governing Occupational Safety and Health

General Regulation Governing Workplace Safety and Hygiene

Instruction No. 10 - Safety and Hygiene Conditions at Workplaces Where Chemical Substances Capable of Causing Contamination of the Work Environment are Produced, Stored, or Handled

Published (in English) by Occupational Safety and Health Administration, U. S. Department of Labor

January, 1993

LIST DESCRIPTION

The Occupational Safety and Health Administration (OSHA) has compiled and translated into English a selection of laws and regulations governing workplace safety and hygiene in Mexico. OSHA has undertaken this unofficial translation to make these documents more widely available. Other laws and regulations may be in effect in Mexico that are applicable to certain activities. Some of the documents published here, which were current as of April 1992, may subsequently have been revised or superseded. OSHA has published these documents for informational purposes - they should not be relied on as legal documents. The Spanish versions of these documents, which are published by the Mexican Secretariat of Labor and Social Welfare, are the definitive texts and should be referred to for legal purposes.

Title Eight of the General Regulations on Workplace Safety and Hygiene, article 135 states:

Workplace atmospheric contaminants are physical agents and chemical or biological elements or compounds capable of altering atmospheric conditions in the workplace and which, because of their properties, concentrations, level, and duration can affect workers' health. the Secretary of Labor and Social Welfare shall, in the applicable instruction, determine the maximum permissible levels of exposure to such agents in the workplace.

The maximum permissible levels as required under article 135 of Title 8 are provided in Instruction No. 10, titled " Safety and Hygiene Conditions at Workplaces Where Chemical Substances Capable of

Causing Contamination of the Work Environment are Produced, Stored, or Handled".

This instruction is mandatory and is intended to spell out measures designed to improve safety and hygiene conditions at workplaces that produce, store, or handle chemical substances that, by virtue of their properties, concentration levels, and exposure durations, are capable of contaminating the work environment and adversely affecting the health of workers, as well as to establish the maximum permissible concentration levels of said substances, in accordance with the type of exposure. Employers must compile, keep, maintain up-to-date, and show to the competent authorities records of the concentration levels of chemical substances mentioned in this Instruction, in accordance with the information provided in Annex I of said Instruction.

The maximum permissible concentration levels to which workers may be exposed are shown in Table I, which shows three different categories of exposure limits:

a) The time-weighted average (TWA) permissible exposure limit (PEL) for eight hours of daily exposure, to which nearly all workers may be repeatedly exposed without experiencing any adverse health effects.

b) The short-term exposure limit (STEL), which shall not be exceeded in any 15 minute period of the working day. Exposures at the STEL shall not occur more than 4 times per day and shall be separated by intervals of at least 60 minutes between successive STEL exposures. In any case, exposures at the STEL may not exceed the 8-hour TWA PEL.

c) The ceiling (or peak) exposure limit. Exposures at the ceiling may not be exceeded during any part of the working exposure, even for a moment.

Table I also includes Appendices A,B,C and D and Appendix I and II. Appendix II defines the term "Skin" as:

This term, added to some of the substances in the list, indicates that the element or compound is capable of being absorbed through the skin by means of simple skin contact. This effect must be taken into account because, when this occurs, the maximum permissible concentration level proposed can be invalidated by the contribution of the entry through the skin, including membranes, mucosa, and eyes, to the worker's overall exposure.

Appendix A lists carcinogens. Three types of carcinogens are defined:

A1 - potential carcinogenic contaminants

A2 - potential carcinogens for humans based on limited epidemiological evidence.

A3 - Carcinogens in humans. Substances associated with industrial processes, recognized as being potential carcinogens although they have not been assigned a maximum permissible exposure value. Workers may not be exposed in any way or form to these substances, which is why specific control methods must be used.

Appendix B provides definitions and limits for Mineral Dust.

Appendix C provides a definition and identification of pure asphyxiants.

Appendix D defines a method for calculating the maximum permissible concentrations for mixtures of contaminants.

Annex I to Instruction 10 identifies the contents of the evaluation report which must be provided under the general provisions of Instruction 10 which states:

Employers shall compile, keep, maintain up-to-date, and show to the competent authorities records of the concentration levels of chemical substances mentioned in this Instruction, in accordance with the information provided in Annex I of said Instruction. This is done to facilitate the adoption of safety and hygiene measures aimed at controlling exposure to said substances.

Annex II provides definitions of technical terms used in the instruction.

This list identifies all the substances in Table I which have an established Short Term Exposure Limit (STEL).

The data field provides the limit in parts per million (ppm) or milligrams per cubic meter (mg/m3).

Mexico - Instruction No. 10 - Carcinogens   (MEX)

SOURCE

Mexican Laws and Regulations Governing Occupational Safety and Health

General Regulation Governing Workplace Safety and Hygiene

Instruction No. 10 - Safety and Hygiene Conditions at Workplaces Where Chemical Substances Capable of Causing Contamination of the Work Environment are Produced, Stored, or Handled

Published (in English) by Occupational Safety and Health Administration, U. S. Department of Labor

January, 1993

LIST DESCRIPTION

The Occupational Safety and Health Administration (OSHA) has compiled and translated into English a selection of laws and regulations governing workplace safety and hygiene in Mexico. OSHA has undertaken this unofficial translation to make these documents more widely available. Other laws and regulations may be in effect in Mexico that are applicable to certain activities. Some of the documents published here, which were current as of April 1992, may subsequently have been revised or superseded. OSHA has published these documents for informational purposes - they should not be relied on as legal documents. The Spanish versions of these documents, which are published by the Mexican Secretariat of Labor and Social Welfare, are the definitive texts and should be referred to for legal purposes.

Title Eight of the General Regulations on Workplace Safety and Hygiene, article 135 states:

Workplace atmospheric contaminants are physical agents and chemical or biological elements or compounds capable of altering atmospheric conditions in the workplace and which, because of their properties, concentrations, level, and duration can affect workers' health. the Secretary of Labor and Social Welfare shall, in the applicable instruction, determine the maximum permissible levels of exposure to such agents in the workplace.

The maximum permissible levels as required under article 135 of Title 8 are provided in Instruction No. 10, titled " Safety and Hygiene Conditions at Workplaces Where Chemical Substances Capable of Causing Contamination of the Work Environment are Produced, Stored, or Handled".

This instruction is mandatory and is intended to spell out measures designed to improve safety and hygiene conditions at workplaces that produce, store, or handle chemical substances that, by virtue of their properties, concentration levels, and exposure durations, are capable of contaminating the work environment and adversely affecting the health of workers, as well as to establish the maximum permissible concentration levels of said substances, in accordance with the type of exposure. Employers must compile, keep, maintain up-to-date, and show to the competent authorities records of the concentration levels of chemical substances mentioned in this Instruction, in accordance with the information provided in Annex I of said Instruction.

The maximum permissible concentration levels to which workers may be exposed are shown in Table I, which shows three different categories of exposure limits:

a) The time-weighted average (TWA) permissible exposure limit (PEL) for eight hours of daily exposure, to which nearly all workers may be repeatedly exposed without experiencing any adverse health effects.

b) The short-term exposure limit (STEL), which shall not be exceeded in any 15 minute period of the working day. Exposures at the STEL shall not occur more than 4 times per day and shall be separated by intervals of at least 60 minutes between successive STEL exposures. In any case, exposures at the STEL may not exceed the 8-hour TWA PEL.

c) The ceiling (or peak) exposure limit. Exposures at the ceiling may not be exceeded during any part of the working exposure, even for a moment.

Table I also includes Appendices A,B,C and D and Appendix I and II. Appendix II defines the term "Skin" as:

This term, added to some of the substances in the list, indicates that the element or compound is capable of being absorbed through the skin by means of simple skin contact. This effect must be taken into account because, when this occurs, the maximum permissible concentration level proposed can be invalidated by the contribution of the entry through the skin, including membranes, mucosa, and eyes, to the worker's overall exposure.

Appendix A lists carcinogens. Three types of carcinogens are defined:

A1 - potential carcinogenic contaminants

A2 - potential carcinogens for humans based onlimited epidemiological evidence.

A3 - Carcinogens in humans. Substances associated with industrial processes, recognized as being potential carcinogens although they have not been assigned a maximum permissible exposure value. Workers may not be exposed in any way or form to these substances, which is why specific control methods must be used.

Appendix B provides definitions and limits for Mineral Dust.

Appendix C provides a definition and identification of pure asphyxiants.

Appendix D defines a method for calculating the maximum permissible concentrations for mixtures of contaminants.

Annex I to Instruction 10 identifies the contents of the evaluation report which must be provided under the general provisions of Instruction 10 which states:

Employers shall compile, keep, maintain up-to-date, and show to the competent authorities records of the concentration levels of chemical substances mentioned in this Instruction, in accordance with the information provided in Annex I of said Instruction. This is done to facilitate the adoption of safety and hygiene measures aimed at controlling exposure to said substances.

Annex II provides definitions of technical terms used in the instruction.

This list identifies all the substances in Appendix A to Table I.

The data field provides the following words to identify the 3 classes of carcinogens:

carcinogen in humans (class A3)

potential carcinogen in humans - limited epidemiological evidence (class A2)

potentially carcinogenic contaminant (class A1)

Mexico - Instruction No. 10 - Skin Designation   (MEX)

SOURCE

Mexican Laws and Regulations Governing Occupational Safety and Health

General Regulation Governing Workplace Safety and Hygiene

Instruction No. 10 - Safety and Hygiene Conditions at Workplaces Where Chemical Substances Capable of Causing Contamination of the Work Environment are Produced, Stored, or Handled

Published (in English) by Occupational Safety and Health Administration, U. S. Department of Labor

January, 1993

LIST DESCRIPTION

The Occupational Safety and Health Administration (OSHA) has compiled and translated into English a selection of laws and regulations governing workplace safety and hygiene in Mexico. OSHA has undertaken this unofficial translation to make these documents more widely available. Other laws and regulations may be in effect in Mexico that are applicable to certain activities. Some of the documents published here, which were current as of April 1992, may subsequently have been revised or superseded. OSHA has published these documents for informational purposes - they should not be relied on as legal documents. The Spanish versions of these documents, which are published by the Mexican Secretariat of Labor and Social Welfare, are the definitive texts and should be referred to for legal purposes.

Title Eight of the General Regulations on Workplace Safety and Hygiene, article 135 states:

Workplace atmospheric contaminants are physical agents and chemical or biological elements or compounds capable of altering atmospheric conditions in the workplace and which, because of their properties, concentrations, level, and duration can affect workers' health. the Secretary of Labor and Social Welfare shall, in the applicable instruction, determine the maximum permissible levels of exposure to such agents in the workplace.

The maximum permissible levels as required under article 135 of Title 8 are provided in Instruction No. 10, titled " Safety and Hygiene Conditions at Workplaces Where Chemical Substances Capable of Causing Contamination of the Work Environment are Produced, Stored, or Handled".

This instruction is mandatory and is intended to spell out measures designed to improve safety and hygiene conditions at workplaces that produce, store, or handle chemical substances that, by virtue of their properties, concentration levels, and exposure durations, are capable of contaminating the work environment and adversely affecting the health of workers, as well as to establish the maximum permissible concentration levels of said substances, in accordance with the type of exposure. Employers must compile, keep, maintain up-to-date, and show to the competent authorities records of the concentration levels of chemical substances mentioned in this Instruction, in accordance with the information provided in Annex I of said Instruction.

The maximum permissible concentration levels to which workers may be exposed are shown in Table I, which shows three different categories of exposure limits:

a) The time-weighted average (TWA) permissible exposure limit (PEL) for eight hours of daily exposure, to which nearly all workers may be repeatedly exposed without experiencing any adverse health effects.

b) The short-term exposure limit (STEL), which shall not be exceeded in any 15 minute period of the working day. Exposures at the STEL shall not occur more than 4 times per day and shall be separated by intervals of at least 60 minutes between successive STEL exposures. In any case, exposures at the STEL may not exceed the 8-hour TWA PEL.

c) The ceiling (or peak) exposure limit. Exposures at the ceiling may not be exceeded during any part of the working exposure, even for a moment.

Table I also includes Appendices A,B,C and D and Appendix I and II. Appendix II defines the term "Skin" as:

This term, added to some of the substances in the list, indicates that the element or compound is capable of being absorbed through the skin by means of simple skin contact. This effect must be taken into account because, when this occurs, the maximum permissible concentration level proposed can be invalidated by the contribution of the entry through the skin, including membranes, mucosa, and eyes, to the worker's overall exposure.

Appendix A lists carcinogens. Three types of carcinogens are defined:

A1 - potential carcinogenic contaminants

A2 - potential carcinogens for humans based on limited epidemiological evidence.

A3 - Carcinogens in humans. Substances associated with industrial processes, recognized as being potential carcinogens although they have not been assigned a maximum permissible exposure value. Workers may not be exposed in any way or form to these substances, which is why specific control methods must be used.

Appendix B provides definitions and limits for Mineral Dust.

Appendix C provides a definition and identification of pure asphyxiants.

Appendix D defines a method for calculating the maximum permissible concentrations for mixtures of contaminants.

Annex I to Instruction 10 identifies the contents of the evaluation report which must be provided under the general provisions of Instruction 10 which states:

Employers shall compile, keep, maintain up-to-date, and show to the competent authorities records of the concentration levels of chemical substances mentioned in this Instruction, in accordance with the information provided in Annex I of said Instruction. This is done to facilitate the adoption of safety and hygiene measures aimed at controlling exposure to said substances.

Annex II provides definitions of technical terms used in the instruction.

This list identifies all the substances in Table I which have an X under the "Skin" column. The definition of a skin designation is taken from Annex II to Table I.

The data field provides the words "skin - potential for cutaneous absorption".

Mexico - Instruction No. 10 - Ceiling Limits (MEX)

SOURCE

Mexican Laws and Regulations Governing Occupational Safety and Health

General Regulation Governing Workplace Safety and Hygiene

Instruction No. 10 - Safety and Hygiene Conditions at Workplaces Where Chemical Substances Capable of Causing Contamination of the Work Environment are Produced, Stored, or Handled

Published (in English) by Occupational Safety and Health Administration, U. S. Department of Labor

January, 1993

LIST DESCRIPTION

The Occupational Safety and Health Administration (OSHA) has compiled and translated into English a selection of laws and regulations governing workplace safety and hygiene in Mexico. OSHA has undertaken this unofficial translation to make these documents more widely available. Other laws and regulations may be in effect in Mexico that are applicable to certain activities. Some of the documents published here, which were current as of April 1992, may subsequently have been revised or superseded. OSHA has published these documents for informational purposes - they should not be relied on as legal documents. The Spanish versions of these documents, which are published by the Mexican Secretariat of Labor and Social Welfare, are the definitive texts and should be referred to for legal purposes.

Title Eight of the General Regulations on Workplace Safety and Hygiene, article 135 states:

Workplace atmospheric contaminants are physical agents and chemical or biological elements or compounds capable of altering atmospheric conditions in the workplace and which, because of their properties, concentrations, level, and duration can affect workers' health. the Secretary of Labor and Social Welfare shall, in the applicable instruction, determine the maximum permissible levels of exposure to such agents in the workplace.

The maximum permissible levels as required under article 135 of Title 8 are provided in Instruction No. 10, titled " Safety and Hygiene Conditions at Workplaces Where Chemical Substances Capable of Causing Contamination of the Work Environment are Produced, Stored, or Handled".

This instruction is mandatory and is intended to spell out measures designed to improve safety and hygiene conditions at workplaces that produce, store, or handle chemical substances that, by virtue of their properties, concentration levels, and exposure durations, are capable of contaminating the work environment and adversely affecting the health of workers, as well as to establish the maximum permissible concentration levels of said substances, in accordance with the type of exposure. Employers must compile, keep, maintain up-to-date, and show to the competent authorities records of the concentration levels of chemical substances mentioned in this Instruction, in accordance with the information provided in Annex I of said Instruction.

The maximum permissible concentration levels to which workers may be exposed are shown in Table I, which shows three different categories of exposure limits:

a) The time-weighted average (TWA) permissible exposure limit (PEL) for eight hours of daily exposure, to which nearly all workers may be repeatedly exposed without experiencing any adverse health effects.

b) The short-term exposure limit (STEL), which shall not be exceeded in any 15 minute period of the working day. Exposures at the STEL shall not occur more than 4 times per day and shall be separated by intervals of at least 60 minutes between successive STEL exposures. In any case, exposures at the STEL may not exceed the 8-hour TWA PEL.

c) The ceiling (or peak) exposure limit. Exposures at the ceiling may not be exceeded during any part of the working exposure, even for a moment.

Table I also includes Appendices A,B,C and D and Appendix I and II. Appendix II defines the term "Skin" as:

This term, added to some of the substances in the list, indicates that the element or compound is capable of being absorbed through the skin by means of simple skin contact. This effect must be taken into account because, when this occurs, the maximum permissible concentration level proposed can be invalidated by the contribution of the entry through the skin, including membranes, mucosa, and eyes, to the worker's overall exposure.

Appendix A lists carcinogens. Three types of carcinogens are defined:

A1 - potential carcinogenic contaminants

A2 - potential carcinogens for humans based onlimited epidemiological evidence.

A3 - Carcinogens in humans. Substances associated with industrial processes, recognized as being potential carcinogens although they have not been assigned a maximum permissible exposure value. Workers may not be exposed in any way or form to these substances, which is why specific control methods must be used.

Appendix B provides definitions and limits for Mineral Dust.

Appendix C provides a definition and identification of pure asphyxiants.

Appendix D defines a method for calculating the maximum permissible concentrations for mixtures of contaminants.

Annex I to Instruction 10 identifies the contents of the evaluation report which must be provided under the general provisions of Instruction 10 which states:

Employers shall compile, keep, maintain up-to-date, and show to the competent authorities records of the concentration levels of chemical substances mentioned in this Instruction, in accordance with the information provided in Annex I of said Instruction. This is done to facilitate the adoption of safety and hygiene measures aimed at controlling exposure to said substances.

Annex II provides definitions of technical terms used in the instruction.

This list identifies all the substances in Table I which have an established Ceiling exposure limit. These substances are identified in the list with the letter (P) for Peak.

The data field provides the limit in parts per million (ppm) or milligrams per cubic meter (mg/m3).

Mexico - Wastewater - Organic Toxic Pollutants (MX1) and Heavy Metals

SOURCE

United Mexican States (Mexico)
General Law on Ecological Equilibrium and Environmental Protection
National Water Commission (CNA)
Official Mexican Norm (NOM) NOM-CCA-ECOL/1993
Electric Power Plant Wastewater Discharge Limits
Organic Toxic Pollutants and Heavy Metals in Annex A
dated 10/18/93

LIST DESCRIPTION

According to the Federal Law on Weights and Measures and Standardization, the President of the National Advisory Committee on Standardization for Environmental Protection is responsible for overseeing the establishment of an Official Mexican Norm NOM-PA-CCA-001/93. This norm should contain the designated "maximum allowable limits of pollutants in wastewater discharges from conventional steam electric power plants to receiving waterbodies."

This said norm was then published and distributed so that interested parties could comment on its overall effectiveness. The comments were reviewed by the National Advisory Committee on Standardization so that the appropriate alterations could be made. Through the National Institute of Ecology, the Secretariat of Social Development responded to these comments in the Ecological Gazette, Volume V, Special Edition in October of 1993.

After review and agreement on the content of the proposed norm, by the Department of Agriculture and Water Resources and the National Advisory Committee on Standardization for Environmental Protection the Official Mexican Norm NOM-PA-CCA-001-ECOL/1993 was established.

In part it contained the maximum allowable limits in the form of a table identifying parameters and their daily average and instantaneous maximum allowable limits. Complying with these established limits became mandatory in October 1993.

In addition, a subsection of the norm includes those situations where although the maximum allowable limits are followed, the effects of the pollutants on the waterbody are still detrimental. In accordance with these situations the Secretariat of Agriculture and Water Resources was given the duty of creating stricter maximum allowable limits for a potential list of parameters. These parameters include the organic toxic pollutants and heavy metals listed in Annex A of the norm.

This data field contains the potential pollutants whose maximum allowable limits are to be acted upon (if necessary), followed by their "[subgroup name]."

Mexico - Drinking Water - Ecological Criteria (MX2)

SOURCE

United Mexican States (Mexico)
General Law on Ecological Equilibrium and Environmental Protection
Secretariat of Social Development (SEDESOL)
Resolution CE-CCA-001/89 - Ecological Criteria for Water Quality
Table 1 - Ecological Criteria for Water Quality - Maximum levels
dated 12/02/89

LIST DESCRIPTION

Under the General Law of Ecological Balance and Environmental Protection, ecological criteria for water quality must be established and observed. Ecological criteria for water quality were defined by setting certain levels of parameters or substances that may be maintained by a body of water in order for the said water source to be a suitable means for 1)a drinking water supply, 2)recreational activities with direct water contact, 3)agricultural irrigation, 4)livestock feed, 5)aquaculture use, and 6)for the preservation and development of present aquatic life.

The purpose of identifying these limits will become apparent when these levels are compared to the quality of water in its present situation. Many programs will be implemented in order to meet these criteria to prevent and control water pollution. This task can be accomplished by restoring the water quality of chemically-polluted and impoverished systems and by enforcing a means of protection for those systems that have already met the criteria. Once all systems are in check, a means of continual monitoring will be required.

This list covers the ecological criteria for water quality as it affects the supply source for drinking water. Based on the physical and chemical properties of the substances, appropriate levels of their presence were established. Since the levels are often determined to be very low, it was necessary to pool some of the data by extrapolating the information from a mathematic model. Some of the substances have shown evidence of accumulation or have even revealed a cancer risk. Please note: The aforementioned "parameters constitute the minimum quality required for the use and exploitation of water." Exceeding these levels will endanger both human and aquatic prosperity.

The data field contains the established levels of each individual substance in mg/l. Other health hazards are included in the footnote.

Canada - Quebec - Ceiling Limits   (QBC)

SOURCE

Revised Regulations of Quebec
Supplement to the Revised Regulations
Gazette officielle du Quebec, Part 2
Amendments through 17 January 1990 (effective February 15, 1990)
Environment Quality Act (R.S.Q., c.Q-2)
An Act respecting occupational health and safety (R.S.Q., c.S-2.1)

LIST DESCRIPTION

The purpose of this portion of the Regulation is to govern the presence of dusts, gases, fumes, vapours and mists.

Subject to Section 8 of the Regulation, any establishment whose operation could cause the emission of gases, dusts, fumes, vapours and mists into the work area must be operated so that the concentration of any gas, dust, fume, vapour or mist does not exceed, in the respiratory zone of the workers, the standards

provided for in Schedule A for any time period specified in Schedule A.

Unless the Commission de la sante et de la securite du travail uses its power to issue an order in accordance with section 66 of the Act respecting occupational health and safety (R.S.Q., c. S-2.1), the use of crocidolite or amosite or of a product containing either of those materials is prohibited, except where their replacement is

not reasonable or practicable.

Dusts, gases, vapours and mists present in the work environment must be sampled, measured and analyzed so as to obtain a degree of accuracy equal to that obtained in accordance with the methods described in the Guide d'echantillonnage des contaminants de l'air en milieu de travail, published in 1989 by the Direction des

laboratoires de l'Institut de recherche en sante et en securite du travail du Quebec.

This list is taken from Schedule A (ss. 5, 11 and 21) - Permissible concentrations of gases, dusts, fumes, vapours or mists in the work environment (Part I - General Table) and includes those entries which are designated with the letter P in the Remarks column indicating that the levels given are Ceiling Limits that must never

be exceeded for any time period whatsoever.

The term mppcf means million particles per cubic foot of air, and mg/m3 means milligram per cubic metre of air.

The data field contains the Ceiling Limit concentration in parts per million (ppm) and/or milligrams per cubic meter of air (mg/m3) taken from the General Table in Part I.

Canada - Quebec - Short Term Exposure Values   (QBC)

SOURCE

Revised Regulations of Quebec
Supplement to the Revised Regulations
Gazette officielle du Quebec, Part 2
Amendments through 17 January 1994 (published September 7, 1994 O.C. 1248-94). Effective September 22, 1994 and September 22, 1995 (Sections 3 and 8 and paragraph 3 of Section 9 and Section 10)
An Act respecting occupational health and safety (R.S.Q., c.S-2.1)

LIST DESCRIPTION

The purpose of this portion of the Regulation is to govern the presence of dusts, gases, fumes, vapours and mists. Subject to Section 8 of the Regulation, any establishment whose operation could cause the emission of gases, dusts, fumes, vapours and mists into the work area must be operated so that the concentration of any gas, dust, fume, vapour or mist does not exceed, in the respiratory zone of the workers, the standards provided for in Schedule A for any time period specified in Schedule A.

Unless the Commission de la sante et de la securite du travail uses its power to issue an order in accordance with section 66 of the Act respecting occupational health and safety (R.S.Q., c. S-2.1), the use of crocidolite or amosite or of a product containing either of those materials is prohibited, except where their replacement is not reasonable or practicable.

Dusts, gases, vapours and mists present in the work environment must be sampled, measured and analyzed so as to obtain a degree of accuracy equal to that obtained in accordance with the methods described in the Guide d'echantillonnage des contaminants de l'air en milieu de travail, published in 1989 by the Direction des laboratoires de l'Institut de recherche en sante et en securite du travail du Quebec.

This list is taken from Schedule A (ss. 5, 11 and 21) - Permissible concentrations of gases, dusts, fumes, vapours or mists in the work environment and includes entries from Parts I and IV.

The 1994 amendments (O.C. 1248-94) added after paragrph p of Section 1 the following:

"respiratory zone": a hemisphere having a 300 mm radius extending in front of the face and measured from the mid point of an imaginary line joining the ears.

The Time Weighted Average Exposure Value is the time weighted average concentration for an 8 hour work day and a 40 hour work week of a chemical substance (in the form of gases, dusts, fumes, vapours or mists) present in the air in a worker's respiratory zone.

The Short Term Exposure Value is the 15 minute time-weighted average concentration for exposure to a chemical substance (in the form of gases, dusts, fumes, vapours or mists) present in the air in a worker's respiratory zone which should not be exceeded at any time during the work day, even if the time weighted average exposure value is not exceeded.

The data field contains the Short Term Exposure Value in parts per million (ppm) and/or milligrams per cubic meter of air (mg/m3) taken from the General Table in Part I and Part IV.

Canada - Quebec - Time-Weighted Average (QBC) Exposure Values

SOURCE

Revised Regulations of Quebec
Supplement to the Revised Regulations
Gazette officielle du Quebec, Part 2
Amendments through 17 January 1994 (published September 7, 1994 O.C. 1248-94). Effective September 22, 1994 and September 22, 1995 (Sections 3 and 8 and paragraph 3 of Section 9 and Section 10)
An Act respecting occupational health and safety (R.S.Q., c.S-2.1)

LIST DESCRIPTION

The purpose of this portion of the Regulation is to govern the presence of dusts, gases, fumes, vapours and mists. Subject to Section 8 of the Regulation, any establishment whose operation could cause the emission of gases, dusts, fumes, vapours and mists into the work area must be operated so that the concentration of any gas, dust, fume, vapour or mist does not exceed, in the respiratory zone of the workers, the standards provided for in Schedule A for any time period specified in
Schedule A.

Unless the Commission de la sante et de la securite du travail uses its power to issue an order in accordance with section 66 of the Act respecting occupational health and safety (R.S.Q., c. S-2.1), the use of crocidolite or amosite or of a product containing either of those materials is prohibited, except where their replacement is
not reasonable or practicable.

Dusts, gases, vapours and mists present in the work environment must be sampled, measured and analyzed so as to obtain a degree of accuracy equal to that obtained in accordance with the methods described in the Guide d'echantillonnage des contaminants de l'air en milieu de travail, published in 1989 by the Direction des laboratoires of the Institut de recherche en sante et en securite du travail du Quebec.

This list is taken from Schedule A (ss. 5, 11 and 21) - Permissible concentrations of gases, dusts, fumes, vapours or mists in the work environment and includes entries from Parts I, II and V.

The 1994 amendments (O.C. 1248-94) added after paragrph p of Section 1 the following:
"respiratory zone": a hemisphere having a 300 mm radiumextending in front of the face and measured from the mid point of an imaginary line joining the ears.

The Time Weighted Average Exposure Value is the time weighted average concentration for an 8 hour work day and a 40 hour work week of a chemical substance (in the form of gases, dusts, fumes, vapours or mists) present in the air in a worker's respiratory zone.

The data field contains the 8 Hour Time Weighted Average Exposure Value in parts per million (ppm) and/or milligrams per cubic meter of air (mg/m3) taken from the General Table in Part I and the list of Substances of Which The Recirculation is Prohibited in Part IV.

Canada - Quebec - Skin Designations   (QBC)

SOURCE

Revised Regulations of Quebec
Supplement to the Revised Regulations
Gazette officielle du Quebec, Part 2
Amendments through 17 January 1994 (published September 7, 1994 O.C. 1248-94). Effective September 22, 1994 and September 22, 1995 (Sections 3 and 8 and paragraph 3 of Section 9 and Section 10)
An Act respecting occupational health and safety (R.S.Q., c.S-2.1)

LIST DESCRIPTION

The purpose of this portion of the Regulation is to govern the presence of dusts, gases, fumes, vapours and mists.

Subject to Section 8 of the Regulation, any establishment whose operation could cause the emission of gases, dusts, fumes, vapours and mists into the work area must be operated so that the concentration of any gas, dust, fume, vapour or mist does not exceed, in the respiratory zone of the workers, the standards provided for in Schedule A for any time period specified in Schedule A.

Unless the Commission de la sante et de la securite du travail uses its power to issue an order in accordance with section 66 of the Act respecting occupational health and safety (R.S.Q., c. S-2.1), the use of crocidolite or amosite or of a product containing either of those materials is prohibited, except where their replacement is not reasonable or practicable.

Dusts, gases, vapours and mists present in the work environment must be sampled, measured and analyzed so as to obtain a degree of accuracy equal to that obtained in accordance with the methods described in the Guide d'echantillonnage des contaminants de l'air en milieu de travail, published in 1989 by the Direction des laboratoires de the Institut de recherche en sante et en securite du travail du Quebec.

This list is taken from Schedule A (ss. 5, 11 and 21) - Permissible concentrations of gases, dusts, fumes, vapours or mists in the work environment and (Part I - General Table) and includes those entries which are designated with the letter X in the Remarks column indicating a substance whose toxicity is percutaneous, i.e. that the worker absorbs first through his skin.

The data field contains the words "absorbed through the skin".

Canada - Quebec - Carcinogens   (QBC)

SOURCE

Revised Regulations of Quebec
Supplement to the Revised Regulations
Gazette officielle du Quebec, Part 2
Amendments through 17 January 1994 (published September 7, 1994 O.C. 1248-94). Effective September 22, 1994 and September 22, 1995 (Sections 3 and 8 and paragraph 3 of Section 9 and Section 10)
An Act respecting occupational health and safety (R.S.Q., c.S-2.1)

LIST DESCRIPTION

The purpose of this portion of the Regulation is to govern the presence of dusts, gases, fumes, vapours and mists.

Subject to Section 8 of the Regulation, any establishment whose operation could cause the emission of gases, dusts, fumes, vapours and mists into the work area must be operated so that the concentration of any gas, dust, fume, vapour or mist does not exceed, in the respiratory zone of the workers, the standards provided for in Schedule A for any time period specified in Schedule A.

Unless the Commission de la sante et de la securite du travail uses its power to issue an order in accordance with section 66 of the Act respecting occupational health and safety (R.S.Q., c. S-2.1), the use of crocidolite or amosite or of a product containing either of those materials is prohibited, except where their replacement is not reasonable or practicable.

Dusts, gases, vapours and mists present in the work environment must be sampled, measured and analyzed so as to obtain a degree of accuracy equal to that obtained in accordance with the methods described in the Guide d'echantillonnage des contaminants de l'air en milieu de travail, published in 1989 by the Direction des laboratoires de l'Institut de recherche en sante et en securite du travail du Quebec.

This list is taken from Schedule A (ss. 5, 11 and 21) - Permissible concentrations of gases, dusts, fumes, vapours or mists in the work environment and (Part I - General Table) and includes those entries which are designated with C1, C2 or C3 in the Remarks - Carcinogen column.

The data field contains the following:
C1 carcinogen: effect detected in humans
C2 carcinogen: effect suspected in humans
C3 carcinogen: effect detected in animals

Canada - Alberta - Designated Substances   (ALB)

SOURCE

Province of Alberta
Occupational Health and Safety Act
Chemical Hazards Regulation
Publication Services
11510 Kingsway
Edmonton, T5G 2Y5
Telephone 427-4952, FAX 452-0668

LIST DESCRIPTION

The Province of Alberta, under the Occupational Health and Safety Act has provided Regulations for certain chemical hazards. Included in the Chemical Hazards Regulations are:

Provisions of General Application - Part 1
Controlled Products - Part 2
Asbestos - Part 3
Coal Dust - Part 4
Silica - Part 5
General - Part 6

Part I includes in Section 2(1) the employers obligation to ensure that each worker's exposure by inhalation to any substance listed in Schedule 1 is kept as low as is reasonably practicable, and does not exceed its Occupational Exposure Limit. Where no Occupational Exposure Limit has been established for a harmful substance present at a work site, an employer must ensure that all steps are taken to keep each worker's exposure to that harmful substance as low as is reasonably practicable.

Measurement of the airborne concentration of a harmful substance shall use methods in accordance with the NIOSH Manual (amended through 4th supplement, August 15, 1990).

Where a worker may become contaminated by a harmful substance, the employer must provide suitable showers, change rooms or other appropriate means to enable workers to remove the contamination prior to leaving the work site.

Schedule 2 to Part I identifies substances which, if present as a pure substance in an amount exceeding 10 kilograms or in a mixture at a concentration equal to or greater than 0.1% by weight, must have a code of practice governing the storage, handling, use and disposal of the substance. The employer must establish such a code of practice including procedures to be used to prevent the uncontrolled release of the substances and procedures to befollowed in the event of a release. The employer must also establish a code of practice governing the operation of the process. Substances listed in Schedule 2 are considered "designated substances" for the purposes of Section 24 of the Act.

The categories identified in Schedule 1 are defined as:

a) "8 hour Occupational Exposure Limit" - means the time weighted average concentration of an airborne substance listed in Schedule 1 for an 8 hour period

b) "15 minute Occupational Exposure Limit" - means the time-weighted average concentration of an airborne substance listed in Schedule 1 for a 15 minute period

c) "ceiling Occupational Exposure Limit" - means the maximum concentration at any point in time of an airborne substance listed in Schedule 1.

d) "Skin" Notation means that the substance can be absorbed through the intact skin.

e) "fibre" means a particulate material having a diameter equal to or less than 3 micrometres, a length equal to or greater than 5 micrometres, and an aspect (length/diameter) ratio equal to or greater than 3:1

This list contains all the entries in Schedule 2 (substances referred to in sections 9 and 12.

The data field contains the words "designated substance - requires code of practice".

Canada - Alberta - Skin Designation   (ALB)

SOURCE

Province of Alberta
Occupational Health and Safety Act
Chemical Hazards Regulation
Publication Services
11510 Kingsway
Edmonton, T5G 2Y5
Telephone 427-4952, FAX 452-0668

LIST DESCRIPTION

The Province of Alberta, under the Occupational Health and Safety Act has provided Regulations for certain chemical hazards. Included in the Chemical Hazards Regulations are:

Provisions of General Application - Part 1
Controlled Products - Part 2
Asbestos - Part 3
Coal Dust - Part 4
Silica - Part 5
General - Part 6

Part I includes in Section 2(1) the employers obligation to ensure that each worker's exposure by inhalation to any substance listed in Schedule 1 is kept as low as is reasonably practicable, and does not exceed its Occupational Exposure Limit. Where no Occupational Exposure Limit has been established for a harmful substance present at a work site, an employer must ensure that all steps are taken to keep each worker's exposure to that harmful substance as low as is reasonably practicable.

Measurement of the airborne concentration of a harmful substance shall use methods in accordance with the NIOSH Manual (amended through 4th supplement, August 15, 1990).

Where a worker may become contaminated by a harmful substance, the employer must provide suitable showers, change rooms or other appropriate means to enable workers to remove the contamination prior to leaving the work site.

Schedule 2 to Part I identifies substances which, if present as a pure substance in an amount exceeding 10 kilograms or in a mixture at a concentration equal to or greater than 0.1% by weight, must have a code of practice governing the storage, handling, use and disposal of the substance. The employer must establish such a code of practice including procedures to be used to prevent the uncontrolled release of the substances and procedures to befollowed in the event of a release. The employer must also establish a code of practice governing the operation of the process. Substances listed in Schedule 2 are considered

"designated substances" for the purposes of Section 24 of the Act.

The categories identified in Schedule 1 are defined as:

a) "8 hour Occupational Exposure Limit" - means the time weighted average concentration of an airborne substance listed in Schedule 1 for an 8 hour period

b) "15 minute Occupational Exposure Limit" - means the time-weighted average concentration of an airborne substance listed in Schedule 1 for a 15 minute period

c) "ceiling Occupational Exposure Limit" - means the maximum concentration at any point in time of an airborne substance listed in Schedule 1.

d) "Skin" Notation means that the substance can be absorbed through the intact skin.

e) "fibre" means a particulate material having a diameter equal to or less than 3 micrometres, a length equal to or greater than 5 micrometres, and an aspect (length/diameter) ratio equal to or greater than 3:1

This list contains all the entries in Schedule 1 (including Tables 1, 2 and 3) which provide a "skin" notation.

The data field contains the words "can be absorbed through the intact skin".

Canada - Alberta - Ceiling Occupational Expo-   (ALB)
sure Limits

SOURCE

Province of Alberta
Occupational Health and Safety Act
Chemical Hazards Regulation
Publication Services
11510 Kingsway
Edmonton, T5G 2Y5
Telephone 427-4952, FAX 452-0668

LIST DESCRIPTION

The Province of Alberta, under the Occupational Health and Safety Act has provided Regulations for certain chemical hazards. Included in the Chemical Hazards Regulations are:

Provisions of General Application - Part 1
Controlled Products - Part 2
Asbestos - Part 3
Coal Dust - Part 4
Silica - Part 5
General - Part 6

Part I includes in Section 2(1) the employers obligation to ensure that each worker's exposure by inhalation to any substance listed in Schedule 1 is kept as low as is reasonably practicable, and does not exceed its Occupational Exposure Limit. Where no Occupational Exposure Limit has been established for a harmful substance present at a work site, an employer must ensure that all steps are taken to keep each worker's exposure to that harmful substance as low as is reasonably practicable.

Measurement of the airborne concentration of a harmful substance shall use methods in accordance with the NIOSH Manual (amended through 4th supplement, August 15, 1990).

Where a worker may become contaminated by a harmful substance, the employer must provide suitable showers, change rooms or other appropriate means to enable workers to remove the contamination prior to leaving the work site.

Schedule 2 to Part I identifies substances which, if present as a pure substance in an amount exceeding 10 kilograms or in a mixture at a concentration equal to or greater than 0.1% by weight, must have a code of practice governing the storage, handling, use and disposal of the substance. The employer must establish such a code of practice including procedures to be used to prevent theuncontrolled release of the substances and procedures to be followed in the event of a release. The employer must also establish a code of practice governing the operation of the process. Substances listed in Schedule 2 are considered "designated substances" for the purposes of Section 24 of the Act.

The categories identified in Schedule 1 are defined as:

a) "8 hour Occupational Exposure Limit" - means the time weighted average concentration of an airborne substance listed in Schedule 1 for an 8 hour period

b) "15 minute Occupational Exposure Limit" - means the time-weighted average concentration of an airborne substance listed in Schedule 1 for a 15 minute period

c) "ceiling Occupational Exposure Limit" - means the maximum concentration at any point in time of an airborne substance listed in Schedule 1.

d) "Skin" Notation means that the substance can be absorbed through the intact skin.

e) "fibre" means a particulate material having a diameter equal to or less than 3 micrometres, a length equal to or greater than 5 micrometres, and an aspect (length/diameter) ratio equal to or greater than 3:1

This list contains all the entries in Schedule 1 (including Tables 1,2 and 3) which establish a Ceiling Occupational Exposure Limit.

The data field contains the Ceiling Occupational Exposure Limit in parts per million (ppm) and/or milligrams per cubic meter of air (mg/m3). The value is preceded by the letter "C" to denote a ceiling limit.

Canada - Alberta - 15 Minute Occupational (ALB) Exposure Limit

SOURCE

Province of Alberta
Occupational Health and Safety Act
Chemical Hazards Regulation
Publication Services
11510 Kingsway
Edmonton, T5G 2Y5
Telephone 427-4952, FAX 452-0668

LIST DESCRIPTION

The Province of Alberta, under the Occupational Health and Safety Act has provided Regulations for certain chemical hazards. Included in the Chemical Hazards Regulations are:

Provisions of General Application - Part 1
Controlled Products - Part 2
Asbestos - Part 3
Coal Dust - Part 4
Silica - Part 5
General - Part 6

Part I includes in Section 2(1) the employers obligation to ensure that each worker's exposure by inhalation to any substance listed in Schedule 1 is kept as low as is reasonably practicable, and does not exceed its Occupational Exposure Limit. Where no Occupational Exposure Limit has been established for a harmful substance present at a work site, an employer must ensure that all steps are taken to keep each worker's exposure to that harmful substance as low as is reasonably practicable.

Measurement of the airborne concentration of a harmful substance shall use methods in accordance with the NIOSH Manual (amended through 4th supplement, August 15, 1990).

Where a worker may become contaminated by a harmful substance, the employer must provide suitable showers, change rooms or other appropriate means to enable workers to remove the contamination prior to leaving the work site.

Schedule 2 to Part I identifies substances which, if present as a pure substance in an amount exceeding 10 kilograms or in a mixture at a concentration equal to or greater than 0.1% by weight, must have a code of practice governing the storage, handling, use and disposal of the substance. The employer must establish such a code of practice including procedures to be used to prevent theuncontrolled release of the substances and procedures to be followed in the event of a release. The employer must also establish a code of practice governing the operation of the process. Substances listed in Schedule 2 are considered "designated substances" for the purposes of Section 24 of the Act.

The categories identified in Schedule 1 are defined as:

a) "8 hour Occupational Exposure Limit" - means the time weighted average concentration of an airborne substance listed in Schedule 1 for an 8 hour period

b) "15 minute Occupational Exposure Limit" - means the time-weighted average concentration of an airborne substance listed in Schedule 1 for a 15 minute period

c) "ceiling Occupational Exposure Limit" - means the maximum concentration at any point in time of an airborne substance listed in Schedule 1.

d) "Skin" Notation means that the substance can be absorbed through the intact skin.

e) "fibre" means a particulate material having a diameter equal to or less than 3 micrometres, a length equal to or greater than 5 micrometres, and an aspect (length/diameter) ratio equal to or greater than 3:1

This list contains all the entries in Schedule 1 (including Tables 1,2 and 3) which establish a 15 minute Time Weighted Average (TWA) Occupational Exposure Limit.

The data field contains the 15 Minute Time Weighted Average (TWA) Occupational Exposure Limit in parts per million (ppm) and/or milligrams per cubic meter of air (mg/m3). To distinguish these 15 minute TWAs from the 8-hour TWAs these are followed by the letters STEL which stand for "Short Term Exposure Limit".

Canada - Alberta - 8 Hour Occupational (ALB) Exposure Limits

SOURCE

Province of Alberta
Occupational Health and Safety Act
Chemical Hazards Regulation
Publication Services
11510 Kingsway
Edmonton, T5G 2Y5
Telephone 427-4952, FAX 452-0668

LIST DESCRIPTION

The Province of Alberta, under the Occupational Health and Safety Act has provided Regulations for certain chemical hazards. Included in the Chemical Hazards Regulations are:

Provisions of General Application - Part 1
Controlled Products - Part 2
Asbestos - Part 3
Coal Dust - Part 4
Silica - Part 5
General - Part 6

Part I includes in Section 2(1) the employers obligation to ensure that each worker's exposure by inhalation to any substance listed in Schedule 1 is kept as low as is reasonably practicable, and does not exceed its Occupational Exposure Limit. Where no Occupational Exposure Limit has been established for a harmful substance present at a work site, an employer must ensure that all steps are taken to keep each worker's exposure to that harmful substance as low as is reasonably practicable.

Measurement of the airborne concentration of a harmful substance shall use methods in accordance with the NIOSH Manual (amended through 4th supplement, August 15, 1990).

Where a worker may become contaminated by a harmful substance, the employer must provide suitable showers, change rooms or other appropriate means to enable workers to remove the contamination prior to leaving the work site.

Schedule 2 to Part I identifies substances which, if present as a pure substance in an amount exceeding 10 kilograms or in a mixture at a concentration equal to or greater than 0.1% by weight, must have a code of practice governing the storage, handling, use and disposal of the substance. The employer must establish such a code of practice including procedures to be used to prevent the uncontrolled release of the substances and procedures to befollowed in the event of a release. The employer must also establish a code of practice governing the operation of the process. Substances listed in Schedule 2 are considered "designated substances" for the purposes of Section 24 of the Act.

The categories identified in Schedule 1 are defined as:

a) "8 hour Occupational Exposure Limit" - means the time weighted average concentration of an airborne substance listed in Schedule 1 for an 8 hour period

b) "15 minute Occupational Exposure Limit" - means the time-weighted average concentration of an airborne substance listed in Schedule 1 for a 15 minute period

c) "ceiling Occupational Exposure Limit" - means the maximum concentration at any point in time of an airborne substance listed in Schedule 1.

d) "Skin" Notation means that the substance can be absorbed through the intact skin.

e) "fibre" means a particulate material having a diameter equal to or less than 3 micrometres, a length equal to or greater than 5 micrometres, and an aspect (length/diameter) ratio equal to or greater than 3:1

This list contains all the entries in Schedule 1 (including Tables 1,2 and 3) which establish an 8 hour Time Weighted Average (TWA) Occupational Exposure Limit.

The data field contains the 8 Hour Time Weighted Average Occupational Exposure Limit in parts per million (ppm) and/or milligrams per cubic meter of air (mg/m3) followed by the letters TWA.

STATE LISTS

California - Prop. 65 - No Significant Risk (C65) Levels

SOURCE

State of California
California Environmental Protection Agency
Office of Environmental Health Hazard Assessment
601 North 7th Street
P.O. Box 942732
Sacramento, CA 94234-7320
Title 22 California Code of Regulations, Section 12705
Chapter 3. Safe Drinking Water and Toxic Enforcement Act of 1986
Article 7. No Significant Risk Levels
Implementing regulations Division 2 of Title 22
Additions and deletions made by Regulatory Action Nos. 92-0915-04S (effective 11/08/92)

LIST DESCRIPTION

The Safe Drinking Water and Toxic Enforcement Act of 1986 (Proposition 65) was passed in November, 1986 and became effective January 1, 1987. The Act requires the Governor to publish a list of chemicals known to the State to cause cancer or reproductive toxicity and provide updates and revisions to the list no less frequently than annually. For chemicals so listed, warnings are required for knowing an intentional exposures, and knowing discharges to the State's drinking water sources.

No warning is required if daily exposure to a chemical is deemed to pose no significant risk within the meaning of the California Health and Safety Code Section 25249.10(c). California Code of Regulations (CCR) Title 22, Chapter 3, Article 7 Section 12705, subsections (b),(c) and (d) set forth levels deemed to pose no significant risk and the methods by which the State has made such determinations.

Subsection (b) of Section 12705 specifies no significant risk levels based on risk assessment conducted or reviewed by the lead agency (CalEPA Office of Environmental Health Hazard Assessment).

Subsection (c) of Section 12705 specifies no significant risk levels based on state or federal risk assessments. Any interested party may request the lead agency to reevaluate a level established in this subsection based on scientific considerations that indicate the need for the lead agency to develop its own risk assessment or to conduct a detailed review of the risk assessment used to derive the level in question.

Subsection (d) specifies no significant risk levels determined bythe lead agency using an expedited method consistent with the procedures specified in Section 12703. Any interested party may request the lead agency to reevaluate a level established in this subsection based on scientific considerations that indicate the need for the lead agency to develop its own risk assessment or to conduct a detailed review of the risk assessment used to derive the level in question.

This list contains all substances identified in Section 12705 subsections (b), (c) and (d) for which "no significant risk levels" are established.

The data field contains the no significant risk levels.

California - Prop. 65 - Reproductive - Male (C65)

SOURCE

State of California
California Environmental Protection Agency
Office of Environmental Health Hazard Assessment
601 North 7th Street
Sacramento, CA 94234-7320
Title 22 California Code of Regulations
Implementing regulations Division 2 of Title 22
Update - Entries dated through October 1, 1994

LIST DESCRIPTION

The Safe Drinking Water and Toxic Enforcement Act of 1986 (Proposition 65) was passed in November, 1986 and became effective January 1, 1987. The Act requires the Governor to publish a list of chemicals known to the State to cause cancer or reproductive toxicity and provide updates and revisions to the list no less frequently than annually. For chemicals so listed, warnings are required for knowing and intentional exposures, and knowing discharges to the State's drinking water sources.

The Act contains several exemptions

1) First, no warning is required if exposures to listed carcinogens would result in a risk lower than the level of "no significan risk" (defined as one excess case of cancer per 100,000 individuals exposed over a 70 year lifetime) or if exposures to listed reproductive toxicants are less than one one-thousandth of the no observable effect level (NOEL). Similarly, discharges to the State's drinking water sources are not prohibited if they pose no significant risk or if they do not exceed one one-thousandth of the NOEL.

2) Second, the Act is not applicable to businesses employing fewer than 10 employees.

3) Third, the Act is not applicable to government agencies.

4) Fourth, the Act is not applicable to drinking water utilities.

Enforcement is via the attorney general, district attorneys, certain city attorneys, and private citizens. The burden of proof is affected by the Proposition; the plaintiff is required to show that an exposure or a discharge occurred and then the burden shifts to the defendant to show that such action did not result in exposures or discharges greater than those allowed by the Act.

No warning is required if daily exposure to a chemical is deemed to pose no significant risk within the meaning of the California Health and Safety Code Section 25249.10(c). California Code of Regulations (CCR) Title 22, Chapter 3, Article 7 Section 12705, subsections (b),(c) and (d) set forth levels deemed to pose no significant risk and the methods by which the State has made such determinations.

On July 17, 1991 the lead agency for this law was transferred from the Health and Welfare Agency to the Environmental Protection Agency's Office of Environmental Health Hazard Assessment by Executive Order W-15-91.

The Act provides three mechanisms by which a chemical is listed:

1) if, in the opinion of the state's qualified experts, it has been clearly shown, through scientifically valid testing according to generally accepted principles, to cause cancer, or reproductive toxicity. A Scientific Advisory Board serves as the state's qualified experts for the purpose of Proposition 65. The Developmental and Reproductive Toxicant (DART) and the Cancer Identification Committees have been established to address the listing of chemicals under Proposition 65.

2) if a body considered to be authoritative by the state's qualified experts has formally identified it as causing cancer or reproductive toxicity. The U.S. Environmental Protection Agency, the U.S. Food and Drug Administration, the International Agency for Research on Cancer, the National Institute for Occupational Safety and Health, and the National Toxicology Program have been designated as "authoritative bodies" for purposes of the Act.

3) if an agency of the state or federal government has formally required it to be labeled or identified as causing cancer or reproductive toxicity. Title 22, California Code of Regulations (22 CCR), Sections 12306 and 12902, respectively, address the last two listing mechanisms.

This list contains all those substances which the State has designated as known to cause reproductive toxicity in males.

The data field includes the words "male reproductive toxicity" and the date the substance was initially listed.

California - Prop. 65 - Cancer List   (C65)
SOURCE

State of California
California Environmental Protection Agency
Office of Environmental Health Hazard Assessment
601 North 7th Street
P.O. Box 942732
Sacramento CA 94234-7320
Title 22 California Code of Regulations
Implementing regulations Division 2 of Title 22
Update - Entries dated through October 1, 1994

LIST DESCRIPTION

The Safe Drinking Water and Toxic Enforcement Act of 1986 (Proposition 65) was passed in November, 1986 and became effective January 1, 1987. The Act requires the Governor to publish a list of chemicals known to the State to cause cancer or reproductive toxicity and provide updates and revisions to the list no less frequently than annually. For chemicals so listed, warnings are required for knowing and intentional exposures, and knowing discharges to the State's drinking water sources.

The Act contains several exemptions

1) First, no warning is required if exposures to listed carcinogens would result in a risk lower than the level of "no significan risk" (defined as one excess case of cancer per 100,000 individuals exposed over a 70 year lifetime) or if exposures to listed reproductive toxicants are less than one one-thousandth of the no observable effect level (NOEL). Similarly, discharges to the State's drinking water sources are not prohibited if they pose no significant risk or if they do not exceed one one-thousandth of the NOEL.

2) Second, the Act is not applicable to businesses employing fewer than 10 employees.

3) Third, the Act is not applicable to government agencies.

4) Fourth, the Act is not applicable to drinking water utilities.

Enforcement is via the attorney general, district attorneys, certain city attorneys, and private citizens. The burden of proof is affected by the Proposition; the plaintiff is required to show that an exposure or a discharge occurred and then the burden shifts to the defendant to show that such action did not result in exposures or discharges greater than those allowed by the Act.

No warning is required if daily exposure to a chemical is deemed to pose no significant risk within the meaning of the California Health and Safety Code Section 25249.10(c). California Code of Regulations (CCR) Title 22, Chapter 3, Article 7 Section 12705, subsections (b),(c) and (d) set forth levels deemed to pose no significant risk and the methods by which the State has made such determinations.

On July 17, 1991 the lead agency for this law was transferred from the Health and Welfare Agency to the Environmental Protection Agency's Office of Environmental Health Hazard Assessment by Executive Order W-15-91.

The Act provides three mechanisms by which a chemical is listed:

1) if, in the opinion of the state's qualified experts, it has been clearly shown, through scientifically valid testing according to generally accepted principles, to cause cancer, or reproductive toxicity. A Scientific Advisory Board serves as the state's qualified experts for the purpose of Proposition 65. The Developmental and Reproductive Toxicant (DART) and the Cancer Identification Committees have been established to address the listing of chemicals under Proposition 65.

2) if a body considered to be authoritative by the state's qualified experts has formally identified it as causing cancer or reproductive toxicity. The U.S. Environmental Protection Agency, the U.S. Food and Drug Administration, the International Agency for Research on Cancer, the National Institute for Occupational Safety and Health, and the National Toxicology Program have been

designated as "authoritative bodies" for purposes of the Act.

3) if an agency of the state or federal government has formally required it to be labeled or identified as causing cancer or reproductive toxicity. Title 22, California Code of Regulations (22 CCR), Sections 12306 and 12902, respectively, address the last two listing mechanisms.

This list contains all those substances which the State has designated as known to cause cancer.

The data field includes the word "carcinogen" and the date on which the substance was initially listed.

California - Prop. 65 - Developmental Toxicity   (C65)

SOURCE

State of California
California Environmental Protection Agency
Office of Environmental Health Hazard Assessment
601 North 7th Street
P.O. Box 942732
Sacramento CA 94234-7320
Title 22 California Code of Regulations
Implementing regulations Division 2 of Title 22
Update - Entries dated through October 1, 1994

LIST DESCRIPTION

The Safe Drinking Water and Toxic Enforcement Act of 1986 (Proposition 65) was passed in November, 1986 and became effective January 1, 1987. The Act requires the Governor to publish a list of chemicals known to the State to cause cancer or reproductive toxicity and provide updates and revisions to the list no less frequently than annually. For chemicals so listed, warnings are required for knowing and intentional exposures, and knowing discharges to the State's drinking water sources.

The Act contains several exemptions

1) First, no warning is required if exposures to listed carcinogens would result in a risk lower than the level of "no significan risk" (defined as one excess case of cancer per 100,000 individuals exposed over a 70 year lifetime) or if exposures to listed reproductive toxicants are less than one one-thousandth of the no observable effect level (NOEL). Similarly, discharges to the State's drinking water sources are not prohibited if they pose no significant risk or if they do not exceed one one-thousandth of the NOEL.

2) Second, the Act is not applicable to businesses employing fewer than 10 employees.

3) Third, the Act is not applicable to government agencies.

4) Fourth, the Act is not applicable to drinking water utilities.

Enforcement is via the attorney general, district attorneys, certain city attorneys, and private citizens. The burden of proof is affected by the Proposition; the plaintiff is required to show that an exposure or a discharge occurred and then the burden shifts to the defendant to show that such action did not result inexposures or discharges greater than those allowed by the Act.

No warning is required if daily exposure to a chemical is deemed to pose no significant risk within the meaning of the California Health and Safety Code Section 25249.10(c). California Code of Regulations (CCR) Title 22, Chapter 3, Article 7 Section 12705, subsections (b),(c) and (d) set forth levels deemed to pose no significant risk and the methods by which the State has made such determinations.

On July 17, 1991 the lead agency for this law was transferred from the Health and Welfare Agency to the Environmental Protection Agency's Office of Environmental Health Hazard Assessment by Executive Order W-15-91.

The Act provides three mechanisms by which a chemical is listed:

1) if, in the opinion of the state's qualified experts, it has been clearly shown, through scientifically valid testing according to generally accepted principles, to cause cancer, or reproductive toxicity. A Scientific Advisory Board serves as the state's qualified experts for the purpose of Proposition 65. The Developmental and Reproductive Toxicant (DART) and the Cancer Identification Committees have been established to address the listing of chemicals under Proposition 65.

2) if a body considered to be authoritative by the state's qualified experts has formally identified it as causing cancer or reproductive toxicity. The U.S. Environmental Protection Agency, the U.S. Food and Drug Administration, the International Agency for Research on Cancer, the National Institute for Occupational Safety and Health, and the National Toxicology Program have been designated as "authoritative bodies" for purposes of the Act.

3) if an agency of the state or federal government has formally required it to be labeled or identified as causing cancer or reproductive toxicity. Title 22, California Code of Regulations (22 CCR), Sections 12306 and 12902, respectively, address the last two listing mechanisms.

This list contains all those substances which the State has designated as known to cause developmental toxicity.

The data field includes the words "developmental toxicity."and the date the substance was originally listed.

California - Prop. 65 - Reproductive - Female   (C65)

SOURCE

State of California
California Environmental Protection Agency
Office of Environmental Health Hazard Assessment
601 North 7th Street
P.O. Box 942732
Sacramento, CA 94234-7320
Title 22 California Code of Regulations
Implementing regulations Division 2 of Title 22
Update - Entries dated through October 1, 1994

LIST DESCRIPTION

The Safe Drinking Water and Toxic Enforcement Act of 1986 (Proposition 65) was passed in November, 1986 and became effective January 1, 1987. The Act requires the Governor to publish a list of chemicals known to the State to cause cancer or reproductive toxicity and provide updates and revisions to the list no less frequently than annually. For chemicals so listed, warnings are required for knowing and intentional exposures, and knowing discharges to the State's drinking water sources.

The Act contains several exemptions

1) First, no warning is required if exposures to listed carcinogens would result in a risk lower than the level of "no significan risk" (defined as one excess case of cancer per 100,000 individuals exposed over a 70 year lifetime) or if exposures to listed reproductive toxicants are less than one one-thousandth of the no observable effect level (NOEL). Similarly, discharges to the State's drinking water sources are not prohibited if they pose no significant risk or if they do not exceed one one-thousandth of the NOEL.

2) Second, the Act is not applicable to businesses employing fewer than 10 employees.

3) Third, the Act is not applicable to government agencies.

4) Fourth, the Act is not applicable to drinking water utilities.

Enforcement is via the attorney general, district attorneys, certain city attorneys, and private citizens. The burden of proof is affected by the Proposition; the plaintiff is required to show that an exposure or a discharge occurred and then the burden shifts to the defendant to show that such action did not result in exposures or discharges greater than those allowed by the Act.

No warning is required if daily exposure to a chemical is deemed to pose no significant risk within the meaning of the California Health and Safety Code Section 25249.10(c). California Code of Regulations (CCR) Title 22, Chapter 3, Article 7 Section 12705, subsections (b),(c) and (d) set forth levels deemed to pose no significant risk and the methods by which the State has made such determinations.

On July 17, 1991 the lead agency for this law was transferred from the Health and Welfare Agency to the Environmental Protection Agency's Office of Environmental Health Hazard Assessment by Executive Order W-15-91.

The Act provides three mechanisms by which a chemical is listed:

1) if, in the opinion of the state's qualified experts, it has been clearly shown, through scientifically valid testing according to generally accepted principles, to cause cancer, or reproductive toxicity. A Scientific Advisory Board serves as the state's qualified experts for the purpose of Proposition 65. The Developmental and Reproductive Toxicant (DART) and the Cancer Identification Committees have been established to address the listing of chemicals under Proposition 65.

2) if a body considered to be authoritative by the state's qualified experts has formally identified it as causing cancer or reproductive toxicity. The U.S. Environmental Protection Agency, the U.S. Food and Drug Administration, the International Agency for Research on Cancer, the National Institute for Occupational Safety and Health, and the National Toxicology Program have been designated as "authoritative bodies" for purposes of the Act.

3) if an agency of the state or federal government has formally required it to be labeled or identified as causing cancer or reproductive toxicity. Title 22, California Code of Regulations (22 CCR), Sections 12306 and 12902, respectively, address the last two listing mechanisms.

This list contains all those substances which the State has designated as known to cause reproductive toxicity in females.

The data field includes the words "female reproductive toxicity" and the date the substance was initially listed.

California Air Bill 2588 Appendix A - II   (CAB)

SOURCE

Air Toxics "Hot Spots" Information and Assessment Act of 1987
California Code of Regulations Title 17, Chapter 7.6 Section 93300 and Title 26, Part 6 Section 44300
List of substances in Appendix A of Title 17 California Code of Regulations, Sections 90700 through 90704 and in Appendices A-I and A-II
"Emission Inventory Criteria and Guidelines Regulation Pursuant to the Air Toxics Hot Spots Information and Assessment Act of 1987" - as amended June 1990, September 1990, June 1991 and June 1993. The latest amendments became effective on January 31, 1994.
For more information contact:
Richard Bode, Manager of the Special Pollutants Emission Inventory Section
California Air Resources Board
(916)322-3807

LIST DESCRIPTION

In September 1987, Governor Deukmejian signed into law Assembly Bill 2588, the Air Toxic "Hot Spots" Information and Assessment Act. Under this law stationary sources are required to report the type and quantity of certain substances which their facilities routinely release into the air. The goals of the Act are to collect emission data, to identify facilities having localized impacts, to ascertain health risks, and to notify nearby residents of significant risks. Air releases of interest are those that result from the routine operation of a facility or that are predictable, including but not limited to continuous and intermittent releases and process upsets or leaks.

The process established by the Act requires owners or operators of facilities to prepare and submit to air pollution control districts an air toxics emission inventory plan, a subsequent emission inventory, and for high priority facilities, a health risk assessment. The risk assessment must be reviewed by the Office of Environmental Health Hazard Assessment (OEHHA) and approved by the local air pollution control district (APCD). If the APCD judges that significant health risks are associated with emissions from the facility, operators must notify all exposed individuals.

The Air Resources Board (ARB) is required to develop a program to make the emission data collected under the Air Toxics Hot Spots Program available to the public. If requested, districts must make health risk assessments available for public review. Districts must also publish annual reports which summarize the health risk assessment program and rank facilities according to the cancer risk posed, identify facilities posing noncancer health risks and describe the status of development of control measures, if any.

The Air Toxic Hot Spots program will complement the ARB's existing air toxics control program by locating sources of substances not currently under evaluation and by providing exposure information necessary for establishment of priorities and regulatory action.

The Act requires the ARB to compile and maintain a list of substances posing a chronic or acute health threat when present in the air. The Act identifies by reference currently over 700 substances which are required to be included on the list. The list (which is updated annually) is used in determining which facilities are covered by the law's requirements and is used by the ARB to develop reporting guidelines for various types of facilities. The ARB may remove substances from the list if criteria outlined in the law are met. A facility is subject to the Act if it (1) manufactures, formulates, uses, or releases a listed substance (or a substance which reacts to form a listed substance) and emits 10 tons or more per year of total organic gases, particulate matter, nitrogen oxides or sulfur oxides, (2) is listed in any existing toxics use of toxics air emission survey, inventory, or report released or compiled by an APCD, or (3) if it manufactures, formulates, uses, or releases a listed substance (or substance which reacts to form a listed substance) and emits less than 10 tons per year of a

criteria pollutant and is subject to emission inventory requirements.

The Act requires facilities meeting the applicability criteria to prepare air toxics emission inventory plans and subsequently, emission inventory reports. Facility operators must first submit to the APCD a proposed emission inventory plan indicating how emissions will be measured or calculated. The district must approve, modify or return the inventory plan to the operator for revisions with 120 days. The Emission inventory Criteria and Guidelines Regulation, which were first approved by the ARB in April 1989, were developed by ARB staff in consultation with APCDs.

Once a district approves a plan, the operator must implement the plan and submit the emission data to the district within 180 days. Emission inventories must be updated biennially according to procedures developed by the ARB. For certain classes of facilities, districts must prepare industrywide inventories; individual facility reports are not required. Districts will determine which facilities will be covered by industrywide inventories based on conditions such as economic hardship and small business status.

Emission reports are due within 180 days of district approval of the emission inventory plan. After reviewing emission inventory data, districts must rank facilities for purposes of risk assessment into high, intermediate, and low priority categories. Facilities which districts designate as high priority must submit a risk assessment to the district for approval. In addition, the district may require facilities beyond those designated as high priority to submit a health risk assessment to the district. Risk assessments are reviewed by districts and by the OEHHA. Facility operators must notify all exposed persons of the risk assessment results if the district determines that there is a significant health risk associated with emissions from the facility.

In establishing priorities, the district is to consider the potency, toxicity, quantity and volume of hazardous materials released from the facility, the proximity of the facility to potential receptors and any other factors that the district determines may indicate that the facility may pose a significant risk. The district is required to hold a public hearing prior to the final establishment of priorities and categories.

This list contains the substances identified in Appendix A - II and titled "List of Substances for Which Production, Use, or Other Presence Must be Reported"

The data field contains the date the Board approved addition of the substance to the original list. The original list (for which no date is listed in the data field) was approved by the Board in July 1988. The data field also identifies substances which must be treated as a human carcinogen or potential human carcinogen for purposes of this section. Those substances so designated contain the words "known or potential carcinogen".

California Air Bill 2588 Appendix A - I   (CAB)

SOURCE

Air Toxics "Hot Spots" Information and Assessment Act of 1987

California Code of Regulations Title 17, Chapter 7.6 Section 93300 and Title 26, Part 6 Section 44300

List of substances in Appendix A of Title 17 California Code of Regulations, Sections 90700 through 90704 and in Appendices A-I and A-II

"Emission Inventory Criteria and Guidelines Regulation Pursuant to the Air Toxics Hot Spots Information and Assessment Act of 1987" - as amended June 1990, September 1990, June 1991 and June 1993. The latest amendments became effective on January 31, 1994.

For more information contact:

Richard Bode, Manager of the Special Pollutants Emission Inventory Section

California Air Resources Board

(916)322-3807

LIST DESCRIPTION

In September 1987, Governor Deukmejian signed into law Assembly Bill 2588, the Air Toxic "Hot Spots" Information and Assessment Act. Under this law stationary sources are required to report the type and quantity of certain substances which their facilities routinely release into the air. The goals of the Act are to collect emission data, to identify facilities having localized impacts, to ascertain health risks, and to notify nearby residents of significant risks. Air releases of interest are those that result from the routine operation of a facility or that are predictable, including but not limited to continuous and intermittent releases and process upsets or leaks.

The process established by the Act requires owners or operators of facilities to prepare and submit to air pollution control districts an air toxics emission inventory plan, a subsequent emission inventory, and for high priority facilities, a health risk assessment. The risk assessment must be reviewed by the Office of Environmental Health Hazard Assessment (OEHHA) and approved by the local air pollution control district (APCD). If the APCD judges that significant health risks are associated with emissions from the facility, operators must notify all exposed individuals.

The Air Resources Board (ARB) is required to develop a program to make the emission data collected under the Air Toxics Hot Spots Program available to the public. If requested, districts must make health risk assessments available for public review. Districts must also publish annual reports which summarize the health risk assessment program and rank facilities according to the cancer risk posed, identify facilities posing noncancer health risks and describe the status of development of control measures, if any.

The Air Toxic Hot Spots program will complement the ARB's existing air toxics control program by locating sources of substances not currently under

evaluation and by providing exposure information necessary for establishment of priorities and regulatory action.

The Act requires the ARB to compile and maintain a list of substances posing a chronic or acute health threat when present in the air. The Act identifies by reference currently over 700 substances which are required to be included on the list. The list (which is updated annually) is used in determining which facilities are covered by the law's requirements and is used by the ARB to develop reporting guidelines for various types of facilities. The ARB may remove substances from the list if criteria outlined in the law are met. A facility is subject to the Act if it (1) manufactures, formulates, uses, or releases a listed substance (or a substance which reacts to form a listed substance) and emits 10 tons or more per year of total organic gases, particulate matter, nitrogen oxides or sulfur oxides, (2) is listed in any existing toxics use of toxics air emission survey, inventory, or report released or compiled by an APCD, or (3) if it manufactures, formulates, uses, or releases a listed substance (or substance which reacts to form a listed substance) and emits less than 10 tons per year of a criteria pollutant and is subject to emission inventory requirements.

The Act requires facilities meeting the applicability criteria to prepare air toxics emission inventory plans and subsequently, emission inventory reports. Facility operators must first submit to the APCD a proposed emission inventory plan indicating how emissions will be measured or calculated. The district must approve, modify or return the inventory plan to the operator for revisions with 120 days. The Emission inventory Criteria and Guidelines Regulation, which were first approved by the ARB in April 1989, were developed by ARB staff in consultation with APCDs.

Once a district approves a plan, the operator must implement the plan and submit the emission data to the district within 180 days. Emission inventories must be updated biennially according to procedures developed by the ARB. For certain classes of facilities, districts must prepare industrywide inventories; individual facility reports are not required. Districts will determine which facilities will be covered by industrywide inventories based on conditions such as economic hardship and small business status.

Emission reports are due within 180 days of district approval of the emission inventory plan. After reviewing emission inventory data, districts must rank facilities for purposes of risk assessment into high, intermediate, and low priority categories. Facilities which districts designate as high priority must submit a risk assessment to the district for approval. In addition, the district may require facilities beyond those designated as high priority to submit a health risk assessment to the district. Risk assessments are reviewed by districts and by the OEHHA. Facility operators must notify all exposed persons of the risk assessment results if the district determines that there is a significant health risk associated with emissions from the facility.

In establishing priorities, the district is to consider the potency, toxicity, quantity and volume of hazardous materials released from the facility, the proximity of the facility to potential receptors and any other factors that the district determines may indicate that the facility may pose a significant risk. The district is required to hold a public hearing prior to the final establishment of priorities and categories.

This list contains the substances identified in Appendix A - I and titled "List of Substances for Which Emissions Must be Quantified"

The data field contains the date the Board approved addition of the substance to the original list. The original list (for which no date is listed in the data field) was approved by the Board in July 1988. The data field also identifies substances which must be treated as a human carcinogen or potential human carcinogen for purposes of this section. Those substances so designated contain the words "known or potential carcinogen".

California - Exposure Limits - PELs   (CEX)
SOURCE

State of California
California Code of Regulations
Title 8. Industrial Relations
Division 1. Department of Industrial Relations
Chapter 4. Division of Industrial Safety
Subchapter 7. General Industry Safety Orders
Group 16. Control of Hazardous Substances
Article 107. Dusts, Fumes, Mists, Vapors and Gases
Section 5155. Airborne Contaminants (revised 4-3-92) Table AC-1 Permissible Exposure Limits for Chemical Contaminants

LIST DESCRIPTION

The State of California, under the California Code of Regulations has established requirements for controlling employee exposure to airborne contaminants and skin contact with those substances which are readily absorbed through the skin and are designated by the "S" notation in Table AC-1 at all places of employment in the State.

Table AC-1 presents concentration limits for airborne contaminants to which nearly all workers may be exposed daily during a 40-hour work week for a working lifetime without adverse effect. Because of some variation in individual susceptibility, an occasional worker may suffer discomfort, aggravation of a pre-existing condition, or occupational disease upon exposure to concentrations even below the values specified in these table.

A Permissible Exposure Limit (PEL) is defined as the maximum permitted 8-hour time-weighted average concentration of an airborne contaminant. An employee exposure to an airborne contaminant in a workday, expressed as an 8-hour TWA concentration, shall not exceed the PEL specified for the substance in Table AC-1.

This list contains all the substances contained in Table AC-1 for which a PEL has been established, and/or for which a reference to the section number in Article 110 "Regulated Carcinogens"of Title 8 is provided.

The data field contains the PEL in ppm (parts per million) and/or mg/cubic meter. If a substance is also covered under Article 110, the Action Level in ppm or mg/cubic meter and reference to the specific section in Article 110 is given.

California - Exposure Limits - STELs   (CEX)
SOURCE

State of California
California Code of Regulations
Title 8. Industrial Relations
Division 1. Department of Industrial Relations
Chapter 4. Division of Industrial Safety
Subchapter 7. General Industry Safety Orders
Group 16. Control of Hazardous Substances
Article 107. Dusts, Fumes, Mists, Vapors and Gases
Section 5155. Airborne Contaminants (revised 4-3-92) Table AC-1 Permissible Exposure Limits for Chemical Contaminants

LIST DESCRIPTION

The State of California, under the California Code of Regulations has established requirements for controlling employee exposure to airborne contaminants and skin contact with those substances which are readily absorbed through the skin and are designated by the "S" notation in Table AC-1 at all places of employment in the State.

Table AC-1 presents concentration limits for airborne contaminants to which nearly all workers may be exposed daily during a 40-hour work week for a working lifetime without adverse effect. Because of some variation in individual susceptibiity, an occasional worker may suffer discomfort, aggravation of a pre-existing condition, or occupational disease upon exposure to concentrations even below the values specified in these table.

A Short Term Exposure Limit (STEL) is defined as a 15-minute time-weighted average exposure which is not to be exceeded at any time during a workday even if the 8-hour time-weighted average is below the PEL. An averaging period other than 15 minutes may be specified in the footnotes.

An employee exposure to an airborne contaminant, expressed as a 15 minute time-weighted average concentration, shall not exceed the STEL specified for the substance in Table AC-1 at any time during the workday. If another averaging period is indicated in the footnotes to Table AC-1, the time-weighted average exposure over that time period shall not exceed the specified STEL at any time during the workday.

This list includes all substances for which a Short Term Exposure Limit is indicated in Table AC-1.

The data field indicates the STEL in ppm (parts per million) and/or in mg/cubic meter.

California - Exposure Limits - Ceilings   (CEX)

SOURCE

State of California
California Code of Regulations
Title 8. Industrial Relations
Division 1. Department of Industrial Relations
Chapter 4. Division of Industrial Safety
Subchapter 7. General Industry Safety Orders
Group 16. Control of Hazardous Substances
Article 107. Dusts, Fumes, Mists, Vapors and Gases
Section 5155. Airborne Contaminants (revised 4-3-92) Table AC-1 Permissible Exposure Limits for Chemical Contaminants

LIST DESCRIPTION

The State of California, under the California Code of Regulations has established requirements for controlling employee exposure to airborne contaminants and skin contact with those substances which are readily absorbed through the skin and are designated by the "S" notation in Table AC-1 at all places of employment in the State.

Table AC-1 presents concentration limits for airborne contaminants to which nearly all workers may be exposed daily during a 40-hour work week for a working lifetime without adverse effect. Because of some variation in individual susceptibiity, an occasional worker may suffer discomfort, aggravation of a pre-existing condition, or occupational disease upon exposure to concentrations even below the values specified in these table.

A Ceiling Limit is defined as the maximum concentration of an airborne contaminant to which an employee may be exposed at any time. Employee exposure shall be controlled such that the applicable ceiling limit specified in Table AC-1 for any airborne contaminant is not exceeded at any time.

This list contains all substances on Table AC-1 which have an established Ceiling Limit.

The data field contains the Ceiling Limit in ppm (parts per million) and/or in mg/cubic meter.

California - Exposure Limits - Skin Notation   (CEX)

SOURCE

State of California
California Code of Regulations
Title 8. Industrial Relations
Division 1. Department of Industrial Relations
Chapter 4. Division of Industrial Safety
Subchapter 7. General Industry Safety Orders
Group 16. Control of Hazardous Substances
Article 107. Dusts, Fumes, Mists, Vapors and Gases
Section 5155. Airborne Contaminants (revised 4-3-92) Table AC-1 Permissible Exposure Limits for Chemical Contaminants

LIST DESCRIPTION

The State of California, under the California Code of Regulations has established requirements for controlling employee exposure to airborne contaminants and skin contact with those substances which are readily absorbed through the skin and are designated by the "S" notation in Table AC-1 at all places of employment in the State.

Table AC-1 presents concentration limits for airborne contaminants to which nearly all workers may be exposed daily during a 40-hour work week for a working lifetime without adverse effect. Because of some variation in individual susceptibiity, an occasional worker may suffer discomfort, aggravation of a pre-existing condition, or occupational disease upon exposure to concentrations even below the values specified in these table.

The substances designated by "S" in the skin notation column of Table AC-1 may be absorbed into the bloodstream through the skin, the mucous membranes and/or the eye, and contribute to the overall exposure. Appropriate protective clothing shall be provided for and used by employees as necessary to prevent skin absorption. This requirement does not remove the employer's responsibility to provide appropriate protection from corrosive or skin irritating materials which may not bear the "S" notation.

This list contains all the substances in Table AC-1 which have the "S" notation.

The data field contains the words "material may be absorbed through the skin, eyes or mucous membrane".

California - Exposure Limits - Carcinogens   (CEX)

SOURCE

State of California
California Code of Regulations
Title 8. Industrial Relations
Division 1. Department of Industrial Relations
Chapter 4. Division of Industrial Safety
Subchapter 7. General Industry Safety Orders
Group 16. Control of Hazardous Substances
Article 110. Regulated Carcinogens
Section 5509. Carcinogens

LIST DESCRIPTION

The State of California, under the California Code of Regulations has established requirements for controlling employee exposure to carcinogens. Article 110 of the Code identifies materials and processes which are regulated as carcinogens by California Occupational Safety and Health Administration. Section 5209 of Article 110 identifies 13 materials which are not listed in Table AC-1. There are no action levels or exposure limits established in this Section, but specific requirements for handling in closed systems.

Other materials covered under Article 110 are included in Table AC-1 with a reference to the Section Number in Article 110, the Permissible Exposure Limit, the Action Level and any clarifying remarks.

It should be noted that the State of California also identifies certain substances as carcinogenic under Title 22 of the Code of Regulations (Safe Drinking Water and Toxic Enforcement Act which is also identified as Proposition 65). These substances are incorporated into Title 8 of the Code by reference in Section 5194

(Hazard Communication) Appendix E. They are not referenced under Article 110 of Title 8.

This list identifies all the substances covered under Section 5209.

The data field contains the words which are required for identification of regulated areas "Cancer-Suspect Agent"and the percentage of solid or liquid mixtures below which these regulations do not apply.

California - Directors List of Hazardous Sub- (CAL) stances (8 CCR 339)
SOURCE:

Director of the Department of Industrial Relations
Title 8, California Code of Regulations
Chapter 3.2, Group 1, Sections 330, et.seq.
Section 339 The Hazardous Substance List
Proposed December 23, 1991, made final January 1, 1993.

LIST DESCRIPTION

Existing law (Labor Code Section 6360, et. seq.) and Title 8 of the California Code of Regulations, Section 337, require the Director of the Department of Industrial Relations to establish and at least every two years review a list of the substances which the Director has concluded to be potentially hazardous when present occupationally. The Director's List of Hazardous Substances (Title 8 of the California Code of Regulations, Section 339) was initially adopted by the Director on August 25, 1982, based on specific criteria set forth in Labor Code Section 6382. The List was submitted to the California Occupational Safety and Health Standards Board for modification and approval. Pursuant to Labor Code Section 6399.2, the provisions of the Act became operative on February 21, 1983.

Labor Code Section 6382 and Title 8 of the California Code of Regulations, Section 337, require that the Director at least every two years review and revise the existing List and submit it to the Occupational Safety and Health Standards Board for possible modifications and approval. The law requires the Director to review the five source listings from which the List is drawn, and to revise the List to include new substances or to exclude substances no longer on such listings. Further, the Director shall consider petitions submitted to date by any member of the public to modify the List or the concentration requirements.

The Director has completed the biennial review and has reviewed petitions to modify the List. The revised list was proposed on December 23, 1991 and made final in December 1992. The Director, as a result of consideration of petitions, proposed two new source reference numbers (the Governor's List of Chemicals Known to Cause Cancer or Reproductive Toxicity and petitions to add individual substance to the Director's list which are not listed in any of the five listings in Labor Code 6382(b). As a result of the hearings held, the final list does NOT contain these two sources.

The finalized source list is as follows:

1. International Agency for Research on Can-cer(IARC)
2. Environmental Protection Agency lists pursuant to the Clean Air and Clean Water Acts
3. General Industry Safety Order Section 5155
4. California Department of Food and Agriculture's list of Restricted Materials
5. Information Alerts put out by the Hazard Evaluation and Information Service pursuant to Labor Code Section 147.2

Inquiries may be directed to:

Ron E. Medeiros
Staff Counsel, DOSH
Department of Industrial Relations
Division of Occupational Safety and Health
Legal Unit
Post Office Box 420603
San Francisco, CA 94142
(415) 703-4361

This list contains all the entries on the Director's List as finalized.

The data field contains only the footnotes as supplied. The reference code(s) for the source of the entry is not included.

Florida Hazardous Substance List   (FLA)
SOURCE

Florida Department of Labor and Employment Security
Division of Safety
2002 Old St., Augustine Road
Building E, Suite 45
Tallahassee, Florida 32399-0663
1-800-367-4378 or (904)488-3044
Florida Substance List
(Substances covered under Chapter 442, Florida Statutes)
Chapter 381-30 (formerly Chapter 38F-41)
Toxic Substance in the Workplace
With revisions effective March 13, 1992

LIST DESCRIPTION

The Legislature has determined that exposure to toxic substances encountered in the course of employment may create a danger to the health and safety of employees and their families in the State of Florida. In Chapter 442, Florida Statutes, the Legislative has mandated that employees have a right to know about toxic substances in their workplace and that employees be given information concerning the nature of toxic substances with which they are working.

The Toxic Substance Advisory Council created by section 442.105, Florida Statutes, has submitted recommendations on the Florida Substance List to the Secretary of Labor and Employment Security. The Secretary has determined that the substances set out in this rule should be considered toxic substances subject to the provision of Chapter 442, Florida Statutes.

This list contains all the entries on the Florida Substance List (38F-41.03).

The data field indicates the presence on this list with the word "present."

Massachusetts Right to Know List   (MSL)
SOURCE
The Commonwealth of Massachusetts
Department of Public Health
Department of Environmental Quality Engineering
Division of Industrial Safety
Regulation Chapter Number and Heading:
105 CMR 670.000 Administrative Bulletin Concerning Massachusetts Substance List for "Right to Know" Law,
Massachusetts General Law Chapter 111F
Appendix A Effective May 26, 1989

LIST DESCRIPTION
The Massachusetts Substance List is contained in Appendix A of 105 CMR 670.00. The Massachusetts Substance List contains those toxic or hazardous substances to which the provisions of M.G. L. c. 111F apply.

Any substance which appears on any of the following source lists shall be added to the Massachusetts Substance List:

1) Substances found to have at least sufficient evidence of carcinogenicity in animals as indicated in monographs published by the International Agency for Research on Cancer (IARC).
2) Substances designated as toxic or hazardous substances by the United States Occupational Safety and Health Administration and identified in 29 CFR 1910.1000 (Subpart Z) and substances identified in "Occupational Health Guidelines for Chemical Hazards" published by the National Institute for Occupational Safety and Health.
3) Substances listed in the most recent edition of the "Annual Report on Carcinogens" published by the National Toxicology Program of the United States Public Health Service.
4) Substances for which a threshold limit Value (TLV) has been established by the American Conference of Governmental Industrial Hygienists.
5) Substances listed by the National Fire Protection Association in "Hazardous Chemicals Data" (NFPA 325M).
6) Substances listed by the National Fire Protection Association and rated 2 through 4 as health hazards or rated 3 through 6 as flammability or reactivity hazards in "Fire Hazard Properties of Flammable Liquids, Gases, Volatile Solids" (NFPA 325M).
7) Substances listed as carcinogens by the Carcinogen Assessment Group of the United States Environmental Protection Agency.
8) Pesticides which have been classified for restricted use by the United States Environmental Protection Agency pursuant to 40 CFR 162.30.
9) Substances listed in a review by the National Cancer Institute scientists published in the Journal of Toxicology and Environmental Health, 8:251-280, Tables 3 through 6, and in subsequent published reviews by the National Cancer Institute scientists of substances which meet the criteria of the National Toxicology Program for significant carcinogenic effect.

Substances are designated as Carcinogens, Mutagens, Teratogens or Neurotoxins when they are designated as such on one of the source lists above or if the Commissioner determines, based upon a preponderance of the substantial and valid scientific evidence reasonably available to the Commissioner, that exposure to the substance may pose a risk.

An Extraordinarily Hazardous Substance is a substance which is designated a carcinogen, or for which there exists valid and substantial evidence that the substance has an oral LD50 of 25 milligrams or less per kilogram in one or more species of test animals, or an LC50 of 0.5 milligrams per liter in one or more species of test animals exposed for a period of up to 8 hours. A listing of an LD50 or LC50 in the Registry of Toxic Effects of Chemical Substances (RTECS) published by the National Institute for Occupational Safety and Health, June 1983, or any amendment thereto, shall be deemed to establish that valid and substantial scientific evidence exists.

This list contains all entries from the Massachusetts Substance List.

The data field identifies Carcinogens, Mutagens, Teratogens, Neurotoxins and Extraordinarily Hazardous substances.

Minnesota Hazardous Substance List   (MIN)
SOURCE
Minnesota Department of Labor and Industry
Occupational Safety and Health (OSHA)
Chapter 5206 - Employee Right-to-Know Standards
Part 5206.0400 Hazardous Substances
Includes amendments through May 14, 1990

The Office of Revisor of Statutes
7th Floor, State Office Building
St. Paul, Minnesota 55155

LIST DESCRIPTION
The Commissioner has determined that the list of hazardous substances in subpart 4 shall be covered by the provisions of this chapter. The hazardous substance list includes the majority of hazardous substances that will be encountered in Minnesota; it does not include all hazardous substances and will not always be current. Employers shall exercise reasonable diligence in evaluating their workplace for the presence of other recognized hazardous substances and shall assure that employees are provided with the rights stated in this chapter.

This list includes all substances identified in 5206.0400 subpart 5.

The data field identifies substances designated by Minnesota to be carcinogens (carcinogens) and those that have potential for skin absorption (skin). All substances on the list are noted as "present."

NJ Right to Know

NJL

SOURCE

New Jersey Worker and Community Right to Know Act
P.L. 1983, C.315, N.J.S.A. 34:5A-1
New Jersey Department of Environmental Protection
   and Energy Community Right to Know Program
Bureau of Hazardous Substances Information
CN 405
Trenton, NJ 08625-0405
telephone (609)292-6714

LIST DESCRIPTION

The State of New Jersey, under the New Jersey Worker and Community Right to Know Act (N.J.S.A. 34:5A as amended), requires annual reporting of inventories of hazardous substances which are stored, produced or used at a place where business is conducted in the State of New Jersey. This annual reporting is required by state and federal laws. The information collected is available to the public and emergency responders such as police and fire

departments. It is also used to supplement other regulatory programs within the state and to allow proper planning for a response to an emergency at a facility which may threaten the surrounding community or environment.

The state and federal Community Right to Know (CRTK) laws have similar requirements. They include the reporting of hazardous substance inventories and releases of hazardous substances to the environment. New Jersey employers who are engaged in certain types of business activities specified by the New Jersey Worker and Community Right to Know Act are required to complete and return a survey for each of their facility locations. The nature and scope of this reporting is detailed in an annual publication titled Community Right to Know Survey. If you believe you might be covered by the regulations you must receive a copy of this document for full reporting instructions.

Contained in the Survey are two tables listing substances which must be reported. Table A contains the Environmental Hazardous Substance List/Hazardous Materials Table. This table consists of substances identified as hazardous substances by the State of New Jersey and substances identified as Hazardous Materials by the U.S. Department of Transportation (listed at 49 CFR 172.101). These substances must be reported if they were present at any quantity in a covered facility during the reporting year. Table D contains U.S. EPA's Extremely Hazardous Substance List. These substances must be reported if they were present at the facility at the threshold planning quantity or 500 pounds, whichever is less.

Table D is identical to CERCLA/SARA - Extremely Hazardous
Substances and is not repeated here (any changes in this portion of NJ's list are made by EPA and will be

included in ChemADVISOR's CERCLA/SARA listing as they are made). ChemADVISOR has included in this list all substances identified in Table A.

The data field includes only the New Jersey substance number. All DOT identification numbers (i.e. UN/NA numbers) are maintained by DOT in the Hazardous Materials Table (49 CFR 172.101) and are found in ChemADVISOR's DOT listings. They will be updated as U.S. DOT updates that list.

NJ Special Hazardous Substances   **(NJL)**

SOURCE

New Jersey Worker and Community Right to Know
   Act
P.L. 1983, C.315, N.J.S.A. 34:5A-1
New Jersey Department of Environmental Protection
Division of Environmental Quality
Bureau of Hazardous Substances Information
CN 405
   Trenton, NJ 08625-0405
   telephone (609)292-6714
New Jersey Department of Health
Division of Occupational and Environmental Health
Right To Know Program
CN 368
Trenton, NJ 08625-0368
Right To Know Hazardous Substance List -
   December 1989

LIST DESCRIPTION

The State of New Jersey, under the New Jersey Worker and Community Right to Know Act (N.J.S.A. 34:5A as amended), requires annual reporting of inventories of hazardous substances which are stored, produced or used at a place where business is conducted in the State of New Jersey. This annual reporting is required by state and federal laws. The information collected is available to the public and emergency responders such as police and fire

departments. It is also used to supplement other regulatory programs within the state and to allow proper planning for a response to an emergency at a facility which may threaten the surrounding community or environment.

The state and federal Community Right to Know (CRTK) laws have similar requirements. They include the reporting of hazardous substance inventories and releases of hazardous substances to the environment. New Jersey employers who are engaged in certain types of business activities specified by the New Jersey Worker and Community Right to Know Act are required to complete and return a survey for each of their facility locations. The nature and scope of this reporting is detailed in an annual publication titled Community Right to Know Survey. If you believe you might be covered by the regulations you must receive a copy of this document for full reporting instructions.

"Special Health Hazards" are designated by the NJ Department of Health and are identified in the December 1989 list only (not in the annual NJDEP survey). They are identified in the publication by a pound sign preceeding the chemical name. Special Health Hazards are substances which contain one or moreof the following hazard categories:

CA - Carcinogen
MU - Mutagen
TE - Teratogen
CO - Corrosive
F4 - Flammable - Fourth Degree
F3 - Flammable - Third Degree
R4 - Reactive - Fourth Degree
R3 - Reactive - Third Degree
R2 - Reactive - Second Degree

The carcinogenic, mutagenic and teratogenic substances are special health hazard substances when present as pure substances or in mixtures at a concentration of on-tenth of one percent (0.1%) or greater. The flammable, reactive/explosive, and corrosive substances are special health hazard substances when present as pure substances or present at a concentration of one percent or greater in a mixture which meets the hazard criteria as defined in N.J.A.C. 8:59-10.2(A).

This list contains all the substances identified as a special health hazard by the presence of a pound sign in the 1989 list.

The data field contains the category or categories for which the substance was listed as a special health hazard.

Pennsylvania Right to Know List   (PAL)

SOURCE

Pennsylvania Department of Labor and Industry
34 PA. Code CH.323
Pennsylvania Bulletin
Volume 19 Number 32
Saturday, August 12,1989
Hazardous Substance List
Effective January 1, 1990
For more information contact:
James H. Tinney
Director, Bureau of Worker and Community Right-to-
    Know
Room 1503 Labor and Industry Building
Harrisburg, PA 17120
(717) 783-2071

LIST DESCRIPTION

The Department of Labor and Industry, under the authority contained in section 3 of the Worker and Community Right-to-Know Act (35 P.S. paragraph 7303) maintains a list of substances with probable adverse human and/or environmental effects. This list is utilized as the basis upon which employers, employees and the general public may become informed about hazardous substances introduced into the workplace and into the general environment by employers subject to the act.

Section 3(a) of the act requires that the Hazardous Substance List be composed of, but not limited to, substances found on 11 recognized lists of environmental and occupational hazards. These lists are as follows:

1) The EPA list of toxic pollutants and hazardous substance prepared under sections 307 and 311 of the Federal Clean Water Act
2) The EPA list of hazardous air pollutants prepared under section 112 of the Federal Clean Air Act
3) the EPA list of restricted use pesticides found at 40 CFR 162.30 (relating to optional procedures for classification of pesticide uses by regulation).
4) The EPA Carcinogen Assessment Group's List of Carcinogens
5) The OSHA list of toxic and hazardous substances-found in 29 CFR 1910, Subpart Z
6) The International Agency for Research on Cancer sublist, entitled "Substances Found to Have at Least Sufficient Evidence of Carcinogenicity in Animals"
7) The National Toxicology Program's list of substances published in the latest Annual Report on Carcinogens
8) The National Fire-Protection Association list found in "Hazardous Chemicals Data" (NFPA 49)
9) The National Fire Protection Association list found in Fire Hazard Properties of Flammable Liquids, Gases, Volatile Solids (NFPA 325M), but only those substances found on sublists for health items, categories 2, 3 and 4; sublists for reactivity items, categories 3 and 4; sublists for flammability, categories 3 and 4
10) The American Conference of Governmental Industrial Hygienists' list found in "Threshold Limit Value for Chemical Substances and Physical Agents in the Workplace."
11) The National Cancer Institute sublist, entitled "Carcinogens Bioassays With at Least Evidence Suggestive of Carcinogenic Effect,"but including only those substances which satisfy criteria of the National Toxicology Program indicating significant carcinogenic effect.

An Environmental Hazard is defined as "any substance, emission or discharge determined by the Department to be a hazardous substance and which, because of its particular or extreme properties, poses a danger if released into the environment."

A Special Hazardous Substance is defined as any hazardous substance which, because of its particular toxicity, tumorigenicity, mutagenicity, reproductive toxicity, flammability, explosiveness, corrosivity or reactivity, poses a special hazard to health and safety. These substances will be subject to disclosure whenever they constitute 0.01% of any mixture.

This list identifies entries using the nomenclature of the International Union of Pure and Applied Chemists.

The data field includes the notation established by Pennsylvania for Environmental Hazard and Special Hazardous Substance.

**Pennsylvania RTK - Special Hazardous Sub-   (PAL)
stances**

SOURCE

Pennsylvania Department of Labor and Industry
34 PA. Code CH.323
Pennsylvania Bulletin
Volume 19 Number 32
Saturday, August 12,1989
Hazardous Substance List
Effective January 1, 1990
For more information contact:
James H. Tinney
Director, Bureau of Worker and Community Right-to-
  Know
Room 1503 Labor and Industry Building
Harrisburg, PA 17120
(717) 783-2071

LIST DESCRIPTION

The Department of Labor and Industry, under the
authority contained in section 3 of the Worker and
Community Right-to-Know Act (35 P.S. paragraph
7303) maintains a list of substances with probable
adverse human and/or environmental effects. This
list is utilized as the basis upon which employers,
employees and the general public may become
informed about hazardous substances introduced into
the workplace and into the general environment by
employers subject to the act.

Section 3(a) of the act requires that the Hazardous
Substance List be composed of, but not limited
to, substances found on 11 recognized lists of
environmental and occupational hazards. These lists
are as follows:

1) The EPA list of toxic pollutants and hazardous
   substance prepared under sections 307 and 311
   of the Federal Clean Water Act
2) The EPA list of hazardous air pollutants prepared
   under section 112 of the Federal Clean Air Act
3) the EPA list of restricted use pesticides found at
   40 CFR 162.30 (relating to optional procedures
   for classification of pesticide uses by regulation).
4) The EPA Carcinogen Assessment Group's List of
   Carcinogens
5) The OSHA list of toxic and hazardous substances-
   found in 29 CFR 1910, Subpart Z
6) The International Agency for Research on Cancer
   sublist, entitled "Substances Found to Have at
   Least Sufficient Evidence of Carcinogenicity in
   Animals"
7) The National Toxicology Program's list of
   substances published in the latest Annual Report
   on Carcinogens
8) The National Fire Protection Association list
   found in "Hazardous Chemicals Data" (NFPA 49)
9) The National Fire Protection Association list
   found in Fire Hazard Properties of Flammable
   Liquids, Gases, Volatile Solids (NFPA 325M),
   but only those substances found on sublists for
   health items, categories 2, 3 and 4; sublists for

reactivity items, categories 3 and 4; sublists for
flammability, categories 3 and 4
10) The American Conference of Governmental
    Industrial Hygienists' list found in "Threshold
    Limit Value for Chemical Substances and
    Physical Agents in the Workplace."
11) The National Cancer Institute sublist, entitled
    "Carcinogens Bioassays With at Least Evidence
    Suggestive of Carcinogenic Effect,"but including
    only those substances which satisfy criteria of
    the National Toxicology Program indicating
    significant carcinogenic effect.

An Environmental Hazard is defined as "any
substance, emission or discharge determined by the
Department to be a hazardous substance and which,
because of its particular or extreme properties, poses a
danger if released into the environment."

A Special Hazardous Substance is defined as any
hazardous substance which, because of its particular
toxicity, tumorigenicity, mutagenicity, reproductive
toxicity, flammability, explosiveness, corrosivity or
reactivity, poses a special hazard to health and safety.
These substances will be subject to disclosure whenever
they constitute 0.01% of any mixture.

Pennsylvania has chosen to use the chemical names
assigned by the International Union of Pure and
Applied Chemists (IUPAC). This list contains all the
substances which have an "S" designation, that is have
been determined by the State to be "Special Hazardous
Substances".

The data field contains the word "present".

PROPOSED REGULATIONS

ACGIH 1995 - Proposed Biological Exposure (BEI) Indices

SOURCE

ACGIH "Threshold Limit Values and Biological Exposure Indices for 1994-95"

ACGIH
Kemper Woods Center
1330 Kemper Meadow Drive
Cincinnati, OH 45240
Requests for interpretations and other information:
Telephone (513) 772-2020
FAX (513) 772-3355

LIST DESCRIPTION

The American Conference of Governmental Industrial Hygienists (ACGIH) is an organization devoted to the administrative and technical aspects of occupational and environmental health. The Conference is a professional society, not a government agency. ACGIH publishes the "TLV/BEI Booklet" each year. The TLVs, as issued by the American Conference of Governmental Industrial Hygienists, are recommendations and should be used as guidelines for good practices. In spite of the fact that serious injury is not believed likely as a result of exposure to the threshold limit concentrations, the best practice is to maintain concentrations of all atmospheric contaminants as low as is practical.

This portion of the TLV/BEI lists is a list of substances for which ACGIH has published a notice of intent to establish biological exposure indices.

The data field indicates the determinant chemical or material, the reference level for that determinant, the sampling time for the material and "notations." Notations and rationale for sampling times are defined by ACGIH in the introduction to the BEIs. Definitions of the Notations used in the data field are:

"Ns" Notation. This notation indicates that the determinant is NONSPECIFIC since it is observed after exposure to some other chemicals. These nonspecific tests are preferred because they are easy to use and usually offer a better correlation with exposure than specific tests. In such instances, a BEI for a specific, less quantitative biological determinant is recommended as a confirmatory test. The documentation should be consulted for information on factors affecting interpretation of these BEIs.

"Cf" Notation. This notation indicated that the biological determinant is an indicator of exposure to the chemical, but the quantitative interpretation of the measurements is ambiguous. Their BEIs should be applied cautiously. These biological determinants should be used as confirmatory tests, mainly for confirmation of exposures indicated by measurements of a nonspecific determinant or as a screening test if a quantitative test is not practical.

"Sc" Notation. This notation indicates that an identifiable population group might have an increased susceptibility to the effect of the chemical, thus leaving it unprotected by the recommended BEI. The specific BEI documentation should be consulted for information.

"B" Notation. This notation indicates that the determinant is usually present in a significant amount in biological specimens collected from subjects who have not been occupationally exposed. Such background levels are included in the BEI value. For information on background levels, consult the specific documentation.

ACGIH 1995 - Notice of Intended Changes (TLV)

SOURCE:

ACGIH "Threshold Limit Values and Biological Exposure Indices for 1994-95"

ACGIH
Kemper Woods Center
1330 Kemper Meadow Drive
Cincinnati, OH 45240
Requests for interpretations and other information:
Telephone (513) 772-2020
FAX (513) 772-3355

LIST DESCRIPTION

The American Conference of Governmental Industrial Hygienists (ACGIH) is an organization devoted to the administrative and technical aspects of occupational and environmental health. The Conference is a professional society, not a government agency. ACGIH publishes the "TLV/BEI Booklet" each year. The TLVs, as issued by the American Conference of Governmental Industrial Hygienists, are recommendations and should be used as guidelines for good practices. In spite of the fact that serious injury is not believed likely as a result of exposure to the threshold limit concentrations, the best practice is to maintain concentrations of all atmospheric contaminants as low as is practical.

This portion of the TLV/BEI lists is a list of substances for which a limit has been proposed for the first time, or for which a change in the adopted listing has been proposed. These are considered trial limits and will remain in the listing for 1 year. If, after one year, no evidence comes to light that questions the appropriateness of the values, the values will be reconsidered to the "Adopted" list.

The data field indicates the TLV-TWA in ppm and/or mg/m3 plus the letters "TWA" to denote a Time Weighted Average. If the material has potential for absorption through the skin, it is noted by the addition of the word "skin". Any comments or clarifications used by ACGIH are also included.

CERCLA - Proposed Hazardous Substance (EPA) Additions

SOURCE

Environmental Protection Agency
Comprehensive Environmental Response, Compensation, and Liability Act
Superfund Amendments and Reauthorization Act
40 CFR Subchapter J - Superfund, Emergency Planning, and Community Right-to-Know Programs.
Part 302 - Designation, Reportable Quantities, and Notification
Paragraph 302.4 - Designation of Hazardous Substances
PROPOSED ADDITION TO 40 CFR 302.4 Table 302.4
54 Federal Register 3388 (January 23, 1989)
Docket Number 102RQ-232EHS

LIST DESCRIPTION

The Environmental Protection Agency (EPA) proposed to designate 232 extremely hazardous substances listed under Title III of the Superfund Amendments and Reauthorization Act (SARA) as hazardous substances under the Comprehensive Environmental Response, Compensation, and Liability Act (CERCLA). 134 of the substances listed as Extremely Hazardous Substances (EHS) are already designated as CERCLA hazardous substance.

This proposed rule does not include RQ adjustments for the 232 substances proposed to be designated as hazardous under CERCLA section 102. (See separate listing in this database).

This list includes all the substances identified in this proposal.

The data field contains the statutory RQ (set at 1 pound).

CERCLA - 1989 Proposed RQ Adjustments (EPA)

SOURCE

Environmental Protection Agency
Comprehensive Environmental Response, Compensation, and Liability Act
Superfund Amendments and Reauthorization Act
40 CFR Subchapter J - Superfund, Emergency Planning, and Community Right-to-Know Programs.
Part 302 - Designation, Reportable Quantities, and Notification
Paragraph 302.4 - Designation of Hazardous Substances
PROPOSED amendment to 40 CFR 302.4 Table 302.4
54 Federal Register 35988 (August 30, 1989)
Docket Number 102RQ-251EHS

LIST DESCRIPTION

On January 23, 1989, the Environmental Protection Agency (EPA) proposed to designate 232 extremely hazardous substances listed under Title III of the Superfund Amendments and Reauthorization Act (SARA) as hazardous substances under the Comprehensive Environmental Response, Compensation, and Liability Act

(CERCLA). 134 of the substances listed as Extremely Hazardous Substances (EHS) are already designated as CERCLA hazardous substances. In that proposal EPA set the Reportable Quantities under the statutory requirements of CERCLA Section 102 (i.e. 1 pound).

EPA proposes to adjust the RQs of 251 substances by amending Table 302.4 of 40 CFR 302.4 and Appendices A and B of 40 CFR Part 355. Of these 251 substances, 232 are EHSs that were proposed for designation as CERCLA hazardous substances on January 23, 1989. The remaining 19 substances are EHSs that were already listed as CERCLA hazardous substances, but had RQs greater than their TPQs.

The list includes all substances identified in the proposal which are to be added to Table 302.4 and substances already included on Table 302.4 for which only the RQ is changed.

The data field indicates the proposed adjusted RQ in pounds (kilograms).

NTP - Proposed Additions to Annual Report on (NTP) Carcinogens

SOURCE

Additions to 8th Annual Report - 56 Federal Register 37366
Additions to 9th Annual Report - 57 Federal Register 24806
Comments on the actions proposed for the Annual Report on Carcinogens should be sent to the National Toxicology Program Public Information Office, MD B2-04, P.O. Box 12233, Research Triangle Park, North Carolina 27709

LIST DESCRIPTION

The National Toxicology Program consists of relevant toxicology activities of the National Institutes of Health's National institute of Environmental Health Sciences (NIH/NIEHS), the Center for Disease Control's National Institute for Occupational Safety and Health (CDC/NIOSH), and the Food and Drug Administration's National Center for Toxicological Research (FDA/NCTR). The National Cancer Institute (NCI) was a charter member; however, its primary participating component, the Carcinogenesis Bioassay Program, was transferred to the NIEHS by the Secretary of Health & Human Services in 1981.

The Annual Report on Carcinogens, prepared by the NTP is issued pursuant to Public Law 95-622 of November 9, 1978. This law requires the Secretary of the Department of Health and Human Services to publish an annual report that contains "a list of all substances (i) which either are known to be carcinogens or which may reasonably be anticipated to be carcinogens and (ii) to which a significant number of persons residing in the United States are exposed..."

For the purpose of this Report, "known carcinogens" are defined as those substances for which the evidence from human studies indicates that there is a causal relationship between exposure to the substance and human cancer. Substances "which may reasonably be anticipated to be carcinogens"are defined as "those for which there is limited evidence of carcinogenicity in humans or sufficient evidence of carcinogenicity in experimental animals."

The latest "Annual Report" issued in final form is the 7th Annual Report. NTP does publish in the Federal Register a "Call for Public Comments" on substances which the Program plans to add to the next Annual Report.

This list consists of the Proposed additions to the Eighth and Ninth Annual Reports as published in the Call for Public Comments.

The data field contains the classification for each substance.

OSHA - Proposed Amendments to OSHA (OS1) Specifically Regulated Chemicals

SOURCE

Department of Labor
Occupational Safety and Health Administration
29 CFR Part 1910 Subpart Z
Additions proposed at 56 FR 57129 (Nov. 7, 1991 - Methylene Chloride) and 58 FR 15526 (March 23, 1993 - 4 specific glycol ethers)

LIST DESCRIPTION

The Occupational Safety and Health Act became effective on April 28, 1971. This Act was established to assure safe and healthful working conditions for working men and women. The Act authorizes development of standards to assure both safety and health, including 1) setting standards for exposure to various chemical substance in the workplace, 2) listing permissible exposure limits for airborne contaminants that are not subject to such standards, and 3) informing employees of the dangers posed by substances with which they work.

The implementing regulations pertaining to health standards are contained in 29 CFR 1910 Subpart Z (Toxic and Hazardous Substances). Substances for which a specific standard exists are included in 29 CFR 1910.1001 to 1910.1048. Permissible Exposure Limits for Air Contaminants are listed in 29 CFR 1910.1000. The Hazard Communication Standard is defined at 29 CFR 1910.1200 and Occupational Exposures to Hazardous Chemicals in Laboratories is contained at 29 CFR 1910.1450.

This list contains substances for which OSHA has PROPOSED to establish specific standards.

Safe Drinking Water Act - Priority List (SD4)

SOURCE

Environmental Protection Agency
Safe Drinking Water Act (SDWA)
Priority List of Substances Which May Require Regulation Under the Safe Drinking Water Act Notice
56 Federal Register 1470 January 14, 1991
Docket Number - Not assigned.

Contact Jitendra Saxena, Ph.D.
Office of Drinking Water (WH-550D)
(202) 475-9579

LIST DESCRIPTION

The Safe Drinking Water Act as amended in 1986 requires EPA to publish a triennial list of contaminants which are known or anticipated to occur in drinking water and which may require regulation under the Act. The Drinking Water Priority List (DWPL) serves as a list of candidate contaminants for regulation under the Act. EPA published the DWPL containing 53 contaminants/contaminant groups on January 22, 1988. The present notice establishes a revised DWPL (1991 version) or "candidates" for future regulations. The list is comprised of 50 substances carried over from the 1988 DWPL and 27 new substances. The total number of contaminants/contaminant groups on the revised DWPL is 77.

This list contains all substances included in Table 5 - "Priority List (1991 Version) of Contaminants Which May Require Regulation Under the Safe Drinking Water Act."

The data field contains the word "present" to indicate their presence on this list.

TSCA - Proposed Testing Rule for Glycidyls (EPA)

SOURCE:

40 CFR Parts 704 and 799
Glycidol and its Derivative Category; Proposed Test Rule With Reporting and Recordkeeping Requirements
Federal Register November 7, 1991
56 FR 57144
Document Control Number OPTS-42051A

For further information:
David Kling
Director, Environmental Assistance Division (TS-799)
Office of Toxic Substance, Room E0543B
401 M Street SW
Washington, DC 20460
(202) 544-1404

LIST DESCRIPTION

EPA, under Section 4 of the Toxic Substances Control Act (TSCA), is proposing that manufacturers and processors of chemical substances listed on the public or confidential portions of the TSCA section 8(b) Chemical Substance Inventory which belong to the "Category of glycidol and its derivatives" be required to perform health effects testing. EPA is also proposing under TSCA section 8(a) that manufacturers and importers of glycidyls be required to report to EPA the volume of manufacture and importation of the substances in accordance with 40 CFR part 704 to allow EPA to determine when certain tests are to be performed.

For purposes of this proposed rule, EPA has defined "glycidyls" as glycidol itself and any of its esters or ethers which are currently listed on, or are subsequently listed on, the public or confidential portions of the TSCA section 8(b) Inventory of Chemical Substances.

EPA is proposing a testing scheme based on subcategories of related substances as well as individual substances. EPA would select one substance within a subcategory for testing which would be paid for by manufacturers and processors of all the substances within the subcategory. The mutagenicity and oncogenicity data from the representative member would then be used for risk assessment for all members of the subcategory.

The Interagency Testing Committee (ITC) designated the category of "glycidol and its derivatives" for health effects testing. The definition of this category and the reasons for this designation are discussed in the Federal Register of October 30, 1978 (43 FR 50630).

This list identifies the 66 category members listed on the publicportion of the TSCA Chemical Substance Inventory.

The data field indicates to which subcategory that substance belongs and any specific testing which EPA has identified for that substance. As indicated in the proposal results of testing will apply to all members of the subcategory.

TSCA - ITC 32nd Report Priority Testing List (TSC)
SOURCE

Environmental Protection Agency
Toxic Substances Control Act
Thirty Second Report of the Interagency Testing Committee
58 Federal Register 38490 (July 16, 1993)
Notice
Docket Number OPPTS-41039

LIST DESCRIPTION

The Interagency Testing Committee (ITC), established under section 4(e) of the Toxic Substances Control Act (TSCA), was created by the U.S. Congress to recommend TSCA regulable chemicals and chemical groups to the Administrator of the EPA for Priority Testing consideration and to facilitate coordination of chemical testing sponsored or required by the U.S. Government organizations represented on the Committee. Congress directed the Committee to (1) organize their recommendations as the Priority testing List, (2) revise the Priority Testing List at least every 6 months and (3) transmit these revisions to the EPA Administrator for publication in the Federal Register.

The ITC transmitted its Thirty Second Report to the Administrator of EPA on June 2, 1993. The Committee

revised the Priority Testing List by adding one group of 34 chemicals to the list for priority consideration by the EPA Administrator for promulgation of test rules under section 4(a) of the Act. These chemicals are designated for response within 12 months. Therefore, in response to the ITC's designation, EPA will either initiate rulemaking under section 4(a) of TSCA, enter into a testing consent agreement, or publish a Federal Register notice explaining the reasons for not initiating such rulemaking within 12 months.

The ITC is also removing four designated chemicals and two recommended chemicals and eight recommended chemical groups added in the Twenty-Eighth report. There are no recommended with intent-to-designate chemicals or chemical groups in the Thirty-Second Report.

The entries on the Priority Testing List are designated, recommended with intent-to-designate or recommended by the Committee.

This list contains all the substances and groups of substances contained in the 32nd Report.

The data field contains the Committee's recommendation (designated, recommended with intent-to-designate, or recommended).

TSCA - ITC 33rd Report Priority Testing List (T33)
SOURCE

Environmental Protection Agency
Toxic Substances Control Act
Thirty Third Report of the Interagency Testing Committee
59 Federal Register 3764 (January 26, 1994)
Notice
Docket Number OPPTS-41040

LIST DESCRIPTION

The Interagency Testing Committee (ITC), established under section 4(e) of the Toxic Substances Control Act (TSCA), was created by the U.S. Congress to recommend TSCA regulable chemicals and chemical groups to the Administrator of the EPA for Priority Testing consideration and to facilitate coordination of chemical testing sponsored or required by the U.S. Government organizations represented on the Committee. Congress directed the Committee to (1) organize their recommendations as the Priority testing List, (2) revise the Priority Testing List at least every 6 months and (3) transmit these revisions to the EPA Administrator for publication in the Federal Register.

The ITC transmitted its Thirty Third Report to the Administrator of EPA on November 29, 1993. The Committee revised the Priority Testing List by removing 2 chemicals and 4 chemical groups from the List of priority coonsideration by the EPA Administrator for promulgation of test rules under section 4(a) of the Act. The chemicals being removed are 2,4-dinitrophenol and 3,4-dimethylphenol. The groups being removed are imidazolium and ethoxylated quaternary ammonium compounds (8 chemicals), brominated flame retardants (23 chemicals) and alkyl phosphates (20 chemicals). The report indicates that removal of these chemicals and chemical groups from the List should be interpreted only as a reordering of priorities and not a statement that the testing recommended earlier has been completed or is not needed. Prior to removing chemicals from the List, the ITC reviewed about 80 TSCA section 8(a) reports and 850 TSCA section 8(d) studies. The ITC is sharing summaries of these reviews with appropriate EPA programs offices and other concerned Federal agencies to assist them in risk assessment and management activities.

There are no additions to the Priority Testing List.

The chemical substances and chemical groups recommended for deletion from the Priority List were previously added to the TSCA section 8(d) Health and Safety Data Reporting Rule (40 CFR Part 716) in response to the ITC testing recommendations. These substances will be considered for removal from the 8(d) rule duringthe next 8(d) rule "biennial review" scheduled to occur during 1995. See 40 CFR 716.65(b).

The entries on the Priority Testing List are designated, recommended with intent-to-designate or recommended by the Committee.

This list contains all the substances and chemical groups of substances contained in the 33rd Report. Two groups, "OSHA chemicals with no dermal toxicity data" and "OSHA chemicals with insufficient dermal absorption data" are not assigned a generic number since these designations do not meet ChemADVISOR's criteria for chemical groups. The individual chemicals are included in the 31st and 32nd reports.

The data field contains the Committee's recommendation (designated, recommended with intent-to-designate, or recommended).

TSCA - ITC 34th Report Priority Testing List (T34)

SOURCE

Environmental Protection Agency
Toxic Substances Control Act
Thirty Fourth Report of the Interagency Testing Committee
59 Federal Register 35720 (July 13, 1994)
Notice
Docket Number OPPTS-41041

LIST DESCRIPTION

The Interagency Testing Committee (ITC), established under section 4(e) of the Toxic Substances Control Act (TSCA), was created by the U.S. Congress to recommend TSCA regulable chemicals and chemical groups to the Administrator of the EPA for Priority Testing consideration and to facilitate coordination of chemical testing sponsored or required by the U.S. Government organizations represented on the Committee. Congress directed the Committee to (1) organize their recommendations as the Priority testing List, (2) revise the Priority Testing List at least every 6 months and (3) transmit these revisions to the EPA Administrator for publication in the Federal Register.

The ITC transmitted its Thirty Fourth Report to the Administrator of EPA on May 17, 1994. The ITC revised the Priority Testing List by:
(1)changing a recommendation for one chemical, white phosphorus, to a designation
(2) recommending two chemicals, ethyl tert-butyl ether and tert-amyl methyl ether, and
(3) removing eight chemicals from the Priority List.
The eight chemicals being removed from the List are:
(1) methyl methacrylate - REMOVED
(2) diethyl phthalate - REMOVED
(3) N-phenyl-1-naphthylamine - REMOVED
(4) acetophenone - REMOVED
(5) phenol - REMOVED
(6) N,N-dimethylaniline - REMOVED
(7) ethyl acetate - REMOVED
(8) 2,6-dimethylphenol

The Report states the reasons for the removal of these chemicals from the List. Written comments on the 34th report must be submitted by August 12, 1994.

A notice will be published at a later date in the Federal Registeradding the substances recommended in the ITC's 34th Report to the TSCA section 8(d) Health and Safety Data Reporting Rule (40 CFR part 716), which requires the reporting of unpublished health and safety studies on the listed chemicals.

The current TSCA section 4(e) Priority Testing List contains 12 chemicals and 12 chemical groups, with 2 chemical groups and 3 chemicals designated for testing.

This list contains the TSCA section 4(e) Priority Testing List as of May 1994. The entries on the Priority Testing List are designated, recommended with intent-to-designate or recommended by the Committee. Three groups, "OSHA chemicals with no dermal toxicity data", "OSHA chemicals with insufficient dermal absorption data" and "Substantially produced chemicals in need of subchronic toxicity testing" are not assigned a generic number and included in ChemADVISOR's listing since these designations do not meet ChemADVISOR's criteria for chemical groups (i.e. they have no chemically related designation).

The data field contains the Committee's recommendation (designated, recommended with intent-to-designate, or recommended).

TSCA - ITC 31st Report Priority Testing List (TSC)

SOURCE

Environmental Protection Agency
Toxic Substances Control Act
Thirty First Report of the Interagency Testing Committee
58 Federal Register 26898 (May 5, 1993)
Notice
Docket Number OPPTS-41038

LIST DESCRIPTION

The Interagency Testing Committee (ITC), established under section 4(e) of the Toxic Substances Control Act (TSCA), was created by the U.S. Congress to recommend TSCA regulable chemicals and chemical groups to the Administrator of the EPA for Priority Testing consideration and to facilitate coordination of chemical testing sponsored or required by the U.S. Government organizations represented on the Committee. Congress directed the Committee to (1) organize their recommendations as the Priority testing List, (2) revise the Priority Testing List at least every 6 months and (3) transmit these revisions to the EPA Administrator for publication in the Federal Register.

The ITC transmitted its Thirty First Report to the Administrator of EPA on January 28, 1993. The Committee revised the Priority Testing List by adding one group of 24 chemicals to the list for priority consideration by the EPA Administrator for promulgation of test rules under section 4(a) of the Act. These chemicals are designated for response within 12 months. Therefore, in response to the ITC's designation, EPA will either initiate rulemaking under section 4(a) of TSCA, enter into a testing consent agreement, or publish a Federal Register notice explaining the reasons for not initiating such rulemaking within 12 months.

The ITC is also revising the recommendations for two chemical groups that were recommended in the ITC's Twenty Eighth Report. The propylene glycol ethers and esters group is revised by the addition of 2 chemicals and the removal of 29. The methyl ethylene glycol ethers and esters group is modified by the removal of eight chemicals. The ITC's reasons for revising these chemical groups are stated in the Thirty First Report. There are no recommended with intent-to-designate chemicals or chemical groups in the Thirty First Report.

The entries on the Priority Testing List are designated, recommended with intent-to-designate or recommended by the Committee.

This list contains all the substances and groups of substancescontained in the 31th Report.

The data field contains the Committee's recommendation (designated, recommended with intent-to-designate, or recommended).

TSCA - Proposed Substances for Developmen- (TSC) tal/Reproductive Testing

SOURCE

Environmental Protection Agency
Toxic Substance and Control Act
40 CFR Part 799 Identification of Specific Chemical
 Substance and Mixture Testing Requirements.
Subpart B - Specific Chemical Test Rules
PROPOSED addition of 40 CFR 799.5050
56 FR 9092 (March 4, 1991)
Docket Number OPTS-42123

LIST DESCRIPTION

Under Section 4 of the Act EPA can require industry to test chemical substances already on the market or about to be produced. The substances selected as candidates for this rule met one or more of the following criteria:

1) EPA received a TSCA section 8(e) notice of substantial risk

2) Available screening level data or other data on the substances provide suggestive evidence that the substance may be toxic and more definitive data are needed to adequately assess risk.

3) Available data on structurally related substances provide suggestive evidence that the substance may be toxic.

4) Adequate developmental toxicity data on one mammalian species are available, but testing in an additional mammalian species is needed to adequately assess risk.

The available studies are not acceptable to EPA because they do not conform to the guidelines for state-of-the-art methodology provided in 40 CFR parts 795 through 798.

This lists identifies the substances proposed for addition to 40 CFR 799.5050.

The data field contains which type of testing must be conducted and the route of administration considered appropriate.

Canada - Ontario - Proposed Occupational (ONT) TWAEVs

SOURCE

Ontario Occupational Health and Safety Act
Ontario Regulation 833 (formerly 654/86)
As last amended by Ontario Regulation 597/94
Control of Exposure to Biological or Chemical Agents
Notice of Proposed Changes to Occupational Exposure
 Limits of 101 Substances. Published in the August
 1, 1992 issue of The Ontario Gazette.
Supplementary Notice Regarding Proposed Changes to
 Occupational Exposure Limits of 101 Substances.
 Published in The Ontario Gazette October 31, 1992.
Deletion of entries made final by 597/94

LIST DESCRIPTION

The Minister of Labour, Ontario, has proposed changes to the Occupational Exposure Limits for 101 substances currently listed in Ontario Regulation 833 (formerly 654/86) and in certain desinated substance regulation made under the Occupational Health and Safety Act. The Joint Steering Committee on Hazardous Substance in the Workplace was established by the Minister of Labour to develop and review regulations designed to control worker exposure to hazardous substances.

The Occupational Exposure Limits Task Force of the Joint Steering Committee, consisting of labour and management representatives, reviewed exposure limits in five jurisdictions which follow a comprehensive process for the setting of exposure limits based on labour-management consultation. The five jurisdictions are the United Kingdom, Germany, Sweden, Norway and the Netherlands. Information on the relevant documentation for the proposed limit and information on the other limits from the five jurisdictions for the substances listed in the proposal is available from the Ministry, on request.

This list contains the proposed changes in the Time Weighted Average Exposure Values (TWAEVs) as published in the October 31, 1992 publication in The Ontario Gazette.

The data field contains the proposed TWAEV in parts per million (ppm) and/or milligrams per cubic meter of air (mg/m3).

Canada - Ontario - Proposed Occupational (ONT) STEVs

SOURCE

Ontario Occupational Health and Safety Act
Ontario Regulation 833
As last amended by Ontario Regulation 597/94
Control of Exposure to Biological or Chemical Agents
Notice of Proposed Changes to Occupational Exposure Limits of 101 Substances. Published in the August 1, 1992 issue of The Ontario Gazette.
Supplementary Notice Regarding Proposed Changes to Occupational Exposure Limits of 101 Substances. Published in The Ontario Gazette October 31, 1992.
Deletion of substances made final by 597/94

LIST DESCRIPTION

The Minister of Labour, Ontario, has proposed changes to the Occupational Exposure Limits for 101 substances currently listed in Ontario Regulation 833 (formerly 654/86) and in certain desinated substance regulation made under the Occupational Health and Safety Act. The Joint Steering Committee on Hazardous Substance in the Workplace was established by the Minister of Labour to develop and review regulations designed to control worker exposure to hazardous substances.

The Occupational Exposure Limits Task Force of the Joint Steering Committee, consisting of labour and management representatives, reviewed exposure limits in five jurisdictions which follow a comprehensive process for the setting of exposure limits based on labour-management consultation. The five jurisdictions are the United Kingdom, Germany, Sweden, Norway and the Netherlands. Information on the relevant documentation for the proposed limit and information on the other limits from the five jurisdictions for the substances listed in the proposal is available from the Ministry, on request.

This list contains the proposed changes in the Short Term Exposure Values (STEVs) as published in the October 31, 1992 publication in The Ontario Gazette.

The data field contains the proposed STEV in parts per million (ppm) and/or milligrams per cubic meter of air (mg/m3).

Canada - Ontario - Proposed Occupational (ONT) CEVs

SOURCE

Ontario Occupational Health and Safety Act
Ontario Regulation 833 (formerly 654/86)
As last amended by Ontario Regulation 513/92
Control of Exposure to Biological or Chemical Agents
Notice of Proposed Changes to Occupational Exposure Limits of 101 Substances. Published in the August 1, 1992 issue of The Ontario Gazette.
Supplementary Notice Regarding Proposed Changes to Occupational Exposure Limits of 101 Substances. Published in The Ontario Gazette October 31, 1992.

LIST DESCRIPTION

The Minister of Labour, Ontario, has proposed changes to the Occupational Exposure Limits for 101 substances currently listed in Ontario Regulation 833 (formerly 654/86) and in certain desinated substance regulation made under the Occupational Health and Safety Act. The Joint Steering Committee on Hazardous Substance in the Workplace was established by the Minister of Labour to develop and review regulations designed to control worker exposure to hazardous substances.

The Occupational Exposure Limits Task Force of the Joint Steering Committee, consisting of labour and management representatives, reviewed exposure limits in five jurisdictions which follow a comprehensive process for the setting of exposure limits based on labour-management consultation. The five jurisdictions are the United Kingdom, Germany, Sweden, Norway and the Netherlands. Information on the relevant documentation for the proposed limit and information on the other limits from the five jurisdictions for the substances listed in the proposal is available from the Ministry, on request.

This list contains the proposed changes in the Ceiling Exposure Values (CEVs) as published in the October 31, 1992 publication in The Ontario Gazette.

The data field contains the proposed CEV in parts per million (ppm) and/or milligrams per cubic meter of air (mg/m3).

Section 3.
Cross-Reference Indexes of Chemical Names and Synonyms

This section contains two indexes designed to simplify the task of determining (i) whether a particular substance appears on one of the regulatory or advisory lists used in the *RCD*™ and (ii) the chemical name under which it is included in the REGULATORY SUMMARIES section of the *RCD*™ (Part B).

Since a substance with a unique CAS number may be regulated under several chemical names by different agencies, the *RCD*™ indexes substances both by CAS number and by chemical synonyms. The regulatory synonym most frequently used for a chemical substance has been designated its "Reference Name," and it is the name under which the the substance's Regulatory Summary is listed. Reference Names are printed in bold type throughout the *RCD*™ to help the reader identify them.

The NUMERICAL INDEX BY CAS NUMBER lists all the chemical synonyms commonly associated with a particular CAS registry number. The substance's Reference Name is printed in bold type next to the CAS number. The 3-letter Synonym Codes in parentheses identify which sources use that specific synonym; a key to these Synonym Codes appears on the next page (See also Section 2 of the *RCD*™, "Lists and List Descriptions").

Some regulated substances have not been assigned CAS registry numbers by Chemical Abstracts Services. In the *RCD*™, these substances have been given chemical information numbers that begin with the prefix "RR" and are structured like CAS numbers, e.g., RR-00001-1, Cotton Dust (Raw). Substances assigned such RR numbers are listed in numerical order at the end of the CAS number index.

The ALPHABETICAL INDEX BY CHEMICAL NAME includes both the exact chemical names used by the list sources and commonly used synonyms. Reference Names are given in the right-hand column of this index.

SYNONYM REFERENCE CODES

302	CERCLA/SARA - Superfund (CERCLA) and Superfund Amendments (SARA)
313	CERCLA/SARA - Superfund (CERCLA) and Superfund Amendments (SARA)
ALB	Alberta Occupational Health and Safety Act
ATS	CERCLA/SARA - Superfund (CERCLA) and Superfund Amendments (SARA)
AUS	Australian National Occupational Health and Safety Commission (NOHSC)
BC1	Workers' Compensation Board of British Columbia
BEI	American Conference of Governmental Industrial Hygienists (ACGIH) 1994-95 TLV Booklet
C65	State of California
CAA	Clean Air Act
CAB	State of California
CAI	Toxic Substance Control Act (TSCA)
CAN	Canadian Environmental Protection Agency
CD1	Health and Welfare Canada
CEX	State of California
CN2	Canadian Environmental Protection Agency
CN3	Canadian Environmental Protection Agency
CN4	Canadian Environmental Protection Agency
CN5	Canadian Environmental Protection Agency
CN6	Canadian Environmental Protection Agency
CN7	Canadian Environmental Protection Agency
CN8	Canadian Environmental Protection Agency
CSR	National Toxicology Program (NTP)
CTR	Toxic Substance Control Act (TSCA)
CW2	Clean Water Act

CW3	Clean Water Act
CWA	Clean Water Act
EP2	Environmental Protection Agency (EPA)
EPA	CERCLA/SARA - Superfund (CERCLA) and Superfund Amendments (SARA)
FDA	U.S. Food and Drug Administration
FLA	State of Florida
GBR	United Kingdom Health and Safety Executive
HM1	Department of Transportation (49 CFR 172)
IAR	International Agency for Research on Cancer
MAK	Deutsche Forschungsgemeinschaft (DFG) of Germany
MEX	Mexican Laws and Regulations Governing Occupational Safety and Health and Environment
MIN	State of Minnesota
MSL	State of Massachusetts
MX1	Mexican Laws and Regulations Governing Occupational Safety and Health and Environment
MX2	Mexican Laws and Regulations Governing Occupational Safety and Health and Environment
NEU	Environmental Protection Agency (EPA)
NFP	National Fire Protection Association - 325M Eleventh Edition
NHS	National Institute for Occupational Safety and Health (NIOSH) 1990 Pocket Guide
NIO	National Institute for Occupational Safety and Health (NIOSH) 1990 Pocket Guide
NJL	State of New Jersey
NTP	National Toxicology Program (NTP)
PAL	State of Pennsylvania

ONT	Ontario Occupational Health and Safety Act
OS1	Occupational Safety and Health Administration (OSHA) 29 CFR 1910
OTS	Environmental Protection Agency (EPA)
OTV	American Industrial Hygiene Association (AIHA)
PQL	Resource Conservation and Recovery Act (RCRA)
PST	Office of Pesticides and Toxic Substances, U.S. EPA
QBC	Quebec Environment Quality Act
RC2	Resource Conservation and Recovery Act (RCRA)
RCA	Resource Conservation and Recovery Act (RCRA)
RCU	Resource Conservation and Recovery Act (RCRA)
SD4	Safety Drinking Water Act (SDWA)
SDW	Safety Drinking Water Act (SDWA)
T32	Toxic Substance Control Act (TSCA)
T33	Toxic Substance Control Act (TSCA)
T34	Toxic Substance Control Act (TSCA)
TLV	American Conference of Governmental Industrial Hygienists (ACGIH) 1994-95 TLV Booklet
TSC	Toxic Substance Control Act (TSCA)
TSE	Toxic Substance Control Act (TSCA)
WEL	American Industrial Hygiene Association (AIHA)
WHM	Canadian Occupational Safety and Health Workplace Hazardous Materials Information System (WHMIS)

Chemical Name	Reference Name
2698-41-1 O-CHLOROBENZYLIDENE MALONONI-TRILE [MEX, MEX]	O-CHLOROBENZYLIDENE MALONITRILE
92-67-1 4-AMINODIPHENYL [MEX]	4-AMINOBIPHENYL
92-93-3 4-NITRODIPHENYL [MEX]	4-NITROBIPHENYL
71751-41-2 ABAMECTIN [AVERMECTIN B1]	ABAMECTIN [AVERMECTIN B1]-PHENOXY]PROPANOIC ACID, BUTYL ESTER
3383-96-8 ABATE [BC1, BC1]	TEMEPHOS
3383-96-8 ABATE (TEMEPHOS) [QBC, ALB, ALB]	TEMEPHOS
8021-27-0 ABIES ALBA OIL	ABIES ALBA OIL
514-10-3 ABIETIC ACID	ABIETIC ACID
10107-99-0 ABIETIC ACID, DIETHYLENE GLYCOL ESTER	ABIETIC ACID, DIETHYLENE GLYCOL ESTER
14351-66-7 ABIETIC ACIDS, SODIUM SALT	ABIETIC ACIDS, SODIUM SALT
6798-76-1 ABIETIC ACID, ZINC SALT	ABIETIC ACID, ZINC SALT
9003-56-9 ABS RESIN	ABS RESIN
83-32-9 ACENAPHTHENE	ACENAPHTHENE
208-96-8 ACENAPHTHYLENE	ACENAPHTHYLENE
83-32-9 ACENAPHTHYLENE, 1,2-DIHYDRO- [PAL]	ACENAPHTHENE
602-87-9 ACENAPHTHYLENE, 1,2-DIHYDRO-5-NI-TRO [PAL, PAL]	5-NITROACENAPHTHENE
30560-19-1 ACEPHATE	ACEPHATEL)-O,O-DIMETHYLPHOSPHOROTHIOATE]
30560-19-1 ACEPHATE (ACETYLPHOSPHORAMIDEOTH-IOIC ACID O,S-DIMETHYL ESTER [313]	ACEPHATEL)-O,O-DIMETHYLPHOSPHOROTHIOATE]
105-57-7 ACETAL	ACETAL
75-07-0 ACETALDEHYDE	ACETALDEHYDE
75-39-8 ACETALDEHYDE AMMONIA	ACETALDEHYDE AMMONIA
107-20-0 ACETALDEHYDE, CHLORO- [PAL, TSC, TSC, TSC]	CHLOROACETALDEHYDE
84-83-3 ACETALDEHYDE, (1,3-DIHYDRO-1,3,3-TRIMETHYL-2H-INDOL-2-YLIDENE)	ACETALDEHYDE, (1,3-DIHYDRO-1,3,3-TRIMETHYL-2H-INDOL-2-YLIDEN
9002-91-9 ACETALDEHYDE, HOMOPOLYMER	ACETALDEHYDE, HOMOPOLYMER
107-29-9 ACETALDEHYDE OXIME	ACETALDEHYDE OXIME
75-87-6 ACETALDEHYDE, TRICHLORO [NJL]	CHLORAL
75-87-6 ACETALDEHYDE, TRICHLORO- [PAL, TSC, TSC, TSC]	CHLORAL
107-89-1 ACETALDOL [FLA, HON]	ALDOL
107-29-9 ACETALDOXIME [NJL]	ACETALDEHYDE OXIME
60-35-5 ACETAMIDE	ACETAMIDE
60-35-5 ACETAMIDE (ETHANAMIDE) [CAI, TSC]	ACETAMIDE
591-08-2 ACETAMIDE, N-(AMINOTHIOXOMETHYL)- [MSL, PAL, TSC, TSC]	1-ACETYL-2-THIOUREA
613-35-4 ACETAMIDE, N,N'-[1,1'BIPHENYL]-4,4'-DIYLBIS- [PAL, PAL]	N,N'-DIACETYLBENZIDINE
3956-55-6 ACETAMIDE, N-[5-[BIS(2-(ACETYLOXY)ETHYL) AMINO]-2-[(2-BROMO-4,6-DINITROPHENYL) AZO]-4-ETHOXYPHENYL]-	ACETAMIDE, N-[5-[BIS(2-(ACETYLOXY)ETHYL)AMINO]-2-[(2-BROMO-
3618-72-2 ACETAMIDE, N-[5-[BIS[2-(ACETYLOXY)ETHYL] AMINO]-2-[(2-BROMO-4,6-DINITROPHENYL) AZO]-4-METHOXYPHENYL]- [TSC, TSC, TSC, TSC]	C.I. DISPERSE BLUE 79:1 ACETAMIDE, N-[5-[BIS[2-(ACETYLOXYL)
3618-73-3 ACETAMIDE, N-(5-(BIS(2-(ACETYLOXY)ETHYL) AMINO)-2-((2-CHLORO-4,6-DINITROPHENYL) AZO)-4-METHOXYPHENYL)-	ACETAMIDE, N-(5-(BIS(2-(ACETYLOXY)ETHYL) AMINO)-2-((2-CHLORO-4,6-DINITROPHENYL)AZO]-4-METHOXYPHENYL]-

Chemical Name	Reference Name
21429-43-6 ACETAMIDE, N-[5-[BIS[2-(ACETYLOXY)ETHYL] AMINO]-2-[(2-CHLORO-4,6-DINITROPHENYL) AZO]-4-ETHOXYPHENYL]-	ACETAMIDE, N-[5-[BIS[2-(ACETYLOXY)ETHYL]AMINO]-2-[(2-CHLORO-
1119-49-9 ACETAMIDE, N-BUTYL- [PAL]	N-BUTYL ACETAMIDE
91-49-6 ACETAMIDE, N-BUTYL-N-PHENYL- [PAL]	N-BUTYLACETANILIDE
56-75-7 ACETAMIDE, 2,2-DICHLORO-N-[2-HYDROXY-1-(HYDROXYMETHYL)-2-(NITROPHENYL)ETHYL] -,[R-(R*,R*)]- [PAL, PAL]	CHLORAMPHENICOLBENZOPYRAN-7-YL) O,O-DIETHYL ESTER
79660-25-6 ACETAMIDE, 2,2-DICHLORO-N-(1,3-DIOXOLAN-2-YLMETHYL)-N-2-PROPENYL-	ACETAMIDE, 2,2-DICHLORO-N-(1,3-DIOXOLAN-2-YL-METHYL)-N-2-PROP
685-91-6 ACETAMIDE, N,N-DIETHYL-	ACETAMIDE, N,N-DIETHYL-
127-19-5 ACETAMIDE, N,N-DIMETHYL- [PAL]	DIMETHYL ACETAMIDE
62-44-2 ACETAMIDE, N-(4-ETHOXYPHENYL)- [PAL, PAL]	PHENACETIN
53-96-3 ACETAMIDE, N-FLUOREN-2-YL [EP2]	2-ACETYLAMINOFLUORENE
53-96-3 ACETAMIDE, N-9H-FLUOREN-2-YL- [PAL, PAL]	2-ACETYLAMINOFLUORENE
28314-03-6 ACETAMIDE, N-9H-FLUOREN-1-YL-	ACETAMIDE, N-9H-FLUOREN-1-YL-O2:O3,O4]DI-, DIPOTASSIUM, TRIHYDRATE, STEREOISOMER
640-19-7 ACETAMIDE, 2-FLUORO- [PAL, RCU, RCU, TSC, TSC]	FLUOROACETAMIDE
640-19-7 ACETAMIDE, 2-FLUORO [TSE]	FLUOROACETAMIDE
103-89-9 ACETAMIDE, N-(4-METHYLPHENYL)- [PAL]	P-ACETOTOLUIDIDE
531-82-8 ACETAMIDE, N-[4-(5-NITRO-2-FURANYL)-2-THIAZOLYL]- [PAL, PAL]	N-[4-(5-NITRO-2-FURYL)-2-THIAZOLYL]ACETAMIDE
RR-01650-2 ACETAMIDE, N-[4-(PENTYLOXY)PHENYL] - ACETAMIDE, N-[2-NITRO-4-(PENTYLOXY) PHENYL]-, AND ACETAMIDE, N-[2-AMINO-4-(PENTYLOXY)PHENYL]	ACETAMIDE, N-[4-(PENTYLOXY)PHENYL]- ACETAMIDE, N-[2-NITRO-4
103-84-4 ACETAMIDE, N-PHENYL-	ACETAMIDE, N-PHENYL-
138-31-8 P-ACETAMIDOBENZENESTIBONIC ACID, SODIUM SALT	P-ACETAMIDOBENZENESTIBONIC ACID, SODIUM SALT
103-84-4 ACETANILIDE [FLA, HON, MSL, NFP, NFP]	ACETAMIDE, N-PHENYL-
1068-57-1 ACETHYDRAZIDE	ACETHYDRAZIDE
64-19-7 ACETIC ACID	ACETIC ACID
631-61-8 ACETIC ACID, AMMONIUM SALT [PAL]	AMMONIUM ACETATE
108-24-7 ACETIC ACID, ANHYDRIDE [PAL]	ACETIC ANHYDRIDE
123-86-4 ACETIC ACID, BUTYL ESTER [PAL, PST]	N-BUTYL ACETATE
540-88-5 ACETIC ACID-TERT-BUTYL ESTER [NJL]	TERT-BUTYL ACETATE
543-90-8 ACETIC ACID, CADMIUM SALT [PAL]	CADMIUM ACETATE
79-11-8 ACETIC ACID, CHLORO- [PAL]	CHLOROACETIC ACID
105-39-5 ACETIC ACID, CHLORO-, ETHYL ESTER [PAL]	ETHYL CHLOROACETATE
96-34-4 ACETIC ACID, CHLORO-, METHYL ESTER [PAL]	METHYL CHLOROACETATE
3926-62-3 ACETIC ACID, CHLORO-, SODIUM SALT [OTS]	SODIUM CHLOROACETATE
1066-30-4 ACETIC ACID, CHROMIUM(3+) SALT [PAL]	CHROMIC ACETATEO-1,5,5A,6,9,9A-HEXAHYDRO-, 3,3-DIOXIDE
142-71-2 ACETIC ACID, COPPER(2+) SALT [PAL]	CUPRIC ACETATE
372-09-8 ACETIC ACID, CYANO-	ACETIC ACID, CYANO-MONOHYDROCHLORIDE
105-56-6 ACETIC ACID, CYANO-, ETHYL ESTER [PAL]	ETHYL CYANOACETATE
25168-24-5 ACETIC ACID, 2,2'-[(DIBUTYLSTANNYLENE)BIS (THIO)]BIS-, DIISOOCTYL ESTER [TSC, TSC]	DIBUTYLTIN BIS(ISOOCTYL MERCAPTOACETATE)-, DIISOOCTYL ESTER (Z,Z)-
94-75-7 ACETIC ACID (2,4-DICHLOROPHENOXY)- [EPA]	2,4-D
94-75-7 ACETIC ACID, (2,4-DICHLOROPHENOXY)- [PAL]	2,4-D

136   © Van Nostrand Reinhold 1995

Chemical Name	Reference Name
1929-73-3 ACETIC ACID, (2,4-DICHLOROPHENOXY)-, 2-BUTOXYETHYL ESTER [MSL, PAL]	2,4-D BUTOXYETHYL ESTER
1320-18-9 ACETIC ACID, (2,4-DICHLOROPHENOXY)-, 2-BUTOXYMETHYLETHYL ESTER [MSL]	2,4-D PROPYLENE GLYCOL BUTYL ETHER ESTER
1320-18-9 ACETIC ACID, (2,4-DICHLOROPHENOXY)-, 2-BUTOXYMETHYLETHYLESTER [PAL]	2,4-D PROPYLENE GLYCOL BUTYL ETHER ESTER
94-80-4 ACETIC ACID, (2,4-DICHLOROPHENOXY)-, BUTYL ESTER [MSL, PAL]	N-BUTYL 2,4-D ESTER
2971-38-2 ACETIC ACID, (2,4-DICHLOROPHENOXY)-, 4-CHLORO-2-BUTENYL ESTER [MSL]	2,4-D CHLOROCROTYL ESTER
2971-38-2 ACETIC ACID, (2,4-DICHLOROPHENOXY)-,4-CHLORO-2-BUTENYL ESTER [PAL]	2,4-D CHLOROCROTYL ESTER
25168-26-7 ACETIC ACID, (2,4-DICHLOROPHENOXY)-, ISOOCTYL ESTER [MSL, PAL]	2,4-D ISOOCTYL ESTER
94-11-1 ACETIC ACID, (2,4-DICHLOROPHENOXY)-, ISOPROPYL ESTER [MSL]	2,4-D ESTERS
1928-38-7 ACETIC ACID, (2,4-DICHLOROPHENOXY)-, METHYL ESTER [MSL, PAL]	2,4-D METHYL ESTER
94-11-1 ACETIC ACID, (2,4-DICHLOROPHENOXY)-, 1-METHYLETHYL ESTER [PAL]	2,4-D ESTERS
94-79-1 ACETIC ACID, (2,4-DICHLOROPHENOXY)-, 1-METHYLPROPYL ESTER [PAL]	SEC-BUTYL 2,4-D ESTER
1928-61-6 ACETIC ACID, (2,4-DICHLOROPHENOXY)-, PROPYL ESTER [MSL, PAL]	2,4-D PROPYL ESTER
540-88-5 ACETIC ACID, 1,1-DIMETHYLETHYL ESTER [PAL]	TERT-BUTYL ACETATE
26636-01-1 ACETIC ACID, 2,2'-[(DIMETHYLSTANNYLENE) BIS(THIO)]BIS-, DIISOOCTYL ESTER [TSC, TSC]	DIBUTYLTIN S,S'-BIS(ISOOCTYL MERCAPTOACETATE) ETHYLPHENYL)-
2016-56-0 ACETIC ACID, DODECYLAMINE SALT	ACETIC ACID, DODECYLAMINE SALT
108-05-4 ACETIC ACID ETHENYL ESTER [PAL]	VINYL ACETATE
10031-87-5 ACETIC ACID, 2-ETHYLBUTYL ESTER	ACETIC ACID, 2-ETHYLBUTYL ESTER
141-78-6 ACETIC ACID ETHYL ESTER [PAL]	ETHYL ACETATEDIMETHYL ESTER, (E)-
103-09-3 ACETIC ACID, 2-ETHYLHEXYL ESTER [PAL]	2-ETHYLHEXYL ACETATE
62-74-8 ACETIC ACID, FLUORO-, SODIUM SALT [PAL, RCU, RCU, TSC, TSC]	SODIUM FLUOROACETATE
64-19-7 ACETIC ACID, GLACIAL [NFP, NFP]	ACETIC ACID
110-19-0 ACETIC ACID, ISOBUTYL ESTER [PST]	ISOBUTYL ACETATE
2949-22-6 ACETIC ACID, ISOCYANATO-, ETHYL ESTER	ACETIC ACID, ISOCYANATO-, ETHYL ESTER
301-04-2 ACETIC ACID, LEAD(2+) SALT [PAL, PAL]	LEAD ACETATER
68-11-1 ACETIC ACID, MERCAPTO- [PAL]	THIOGLYCOLIC ACID
123-81-9 ACETIC ACID, MERCAPTO-, 1,2-ETHANEDIYL ESTER [PAL]	GLYCOL DIMERCAPTOACETATE
79-20-9 ACETIC ACID, METHYL ESTER [PAL]	METHYL ACETATE
108-21-4 ACETIC ACID, 1-METHYLETHYL ESTER [PAL]	ISOPROPYL ACETATE
54849-38-6 ACETIC ACID, 2,2',2"-[(METHYLSTANNYLIDYNE)TRIS(THIO)]TRIS-, TRIISOOCTYL ESTER [TSC, TSC]	MONOMETHYLTIN TRIS(ISOOOCTYL MERCAPTOACETATE)
105-46-4 ACETIC ACID, 1-METHYLPROPYL ESTER [PAL]	SEC-BUTYL ACETATE
110-19-0 ACETIC ACID, 2-METHYLPROPYL ESTER [PAL]	ISOBUTYL ACETATE
112-14-1 ACETIC ACID, OCTYL ESTER	ACETIC ACID, OCTYL ESTER
628-63-7 ACETIC ACID, PENTYL ESTER [PAL]	N-AMYL ACETATE
53496-15-4 ACETIC ACID, SEC-PENTYL ESTER	ACETIC ACID, SEC-PENTYL ESTERXY)ACETYL]-.OMEGA.-BUTOXY-

Chemical Name	Reference Name
114-83-0 ACETIC ACID, 2-PHENYLHYDRAZIDE	ACETIC ACID, 2-PHENYLHYDRAZIDE
591-87-7 ACETIC ACID, 2-PROPENYL ESTER [PAL]	ALLYL ACETATE
109-60-4 ACETIC ACID, PROPYL ESTER [PAL]	N-PROPYL ACETATE
563-68-8 ACETIC ACID, THALLIUM(I) SALT [MSL, PAL]	THALLIUM(I) ACETATE
76-03-9 ACETIC ACID, TRICHLORO- [PAL]	TRICHLOROACETIC ACID
140-41-0 ACETIC ACID TRICHLORO-, COMPD. WITH 3-(P-CHLOROPHENYL)-1,	ACETIC ACID TRICHLORO-, COMPD. WITH 3-(P-CHLOROPHENYL)-1,
93-76-5 ACETIC ACID, (2,4,5-TRICHLOROPHENOXY)- [PAL]	2,4,5-T
2545-59-7 ACETIC ACID, (2,4,5-TRICHLOROPHENOXY)-, 2-BUTOXYETHYL ESTER	ACETIC ACID, (2,4,5-TRICHLOROPHENOXY)-, 2-BU-TOXYETHYL ESTER
93-79-8 ACETIC ACID, (2,4,5-TRICHLOROPHENOXY)-, BUTYL ESTER	ACETIC ACID, (2,4,5-TRICHLOROPHENOXY)-, BUTYL ESTER
1319-72-8 ACETIC ACID, (2,4,5-TRICHLOROPHENOXY)-, COMPOUND WITH 1-AMINO-2-PROPANOL(1:1)	ACETIC ACID, (2,4,5-TRICHLOROPHENOXY)-, COMPOUND WITH 1-AMIN
2008-46-0 ACETIC ACID, (2,4,5-TRICHLOROPHENOXY)-, COMPOUND WITH N,N-DIETHYLETHANAMINE	ACETIC ACID, (2,4,5-TRICHLOROPHENOXY)-, COMPOUND WITH N,N-
3813-14-7 ACETIC ACID, (2,4,5-TRICHLOROPHENOXY)-, COMPOUND WITH 2,2',2"-NITROTRIS (ETHANOL) (1:1)	ACETIC ACID, (2,4,5-TRICHLOROPHENOXY)-, COMPOUND WITH 2,2',2
6369-97-7 ACETIC ACID, (2,4,5-TRICHLOROPHENOXY)-, COMPOUND WITH N-METHYLMETHANAMINE	ACETIC ACID, (2,4,5-TRICHLOROPHENOXY)-, COMPOUND WITH N-METH
6369-96-6 ACETIC ACID, (2,4,5-TRICHLOROPHENOXY)-, COMPOUND WITH TRIMETHYLAMINE	ACETIC ACID, (2,4,5-TRICHLOROPHENOXY)-, COMPOUND WITH TRIMET
1928-47-8 ACETIC ACID, (2,4,5-TRICHLOROPHENOXY)-, 2-ETHYLHEXYL ESTER	ACETIC ACID, (2,4,5-TRICHLOROPHENOXY)-, 2-ETHYL-HEXYL ESTER
25168-15-4 ACETIC ACID, (2,4,5-TRICHLOROPHENOXY)-, ISOOCTYL ESTER	ACETIC ACID, (2,4,5-TRICHLOROPHENOXY)-, ISOOCTYL ESTER
61792-07-2 ACETIC ACID, (2,4,5-TRICHLOROPHENOXY)-, 1-METHYL PROPYL ESTER	ACETIC ACID, (2,4,5-TRICHLOROPHENOXY)-, 1-METHYL PROPYL ESTE
13560-99-1 ACETIC ACID, (2,4,5-TRICHLOROPHENOXY)-, SODIUM SALT [PAL]	2,4,5-T SALTS
RR-00780-7 ACETIC ACID, WATER SOLUTIONS	ACETIC ACID, WATER SOLUTIONS
557-34-6 ACETIC ACID, ZINC SALT [PAL]	ZINC ACETATE
108-24-7 ACETIC ANHYDRIDE	ACETIC ANHYDRIDE
102-01-2 ACETOACETANILIDE	ACETOACETANILIDE
92-15-9 O-ACETOACETANISIDIDE	O-ACETOACETANISIDIDE
92-15-9 O-ACETOACET ANISIDIDE [NFP, NFP]	O-ACETOACETANISIDIDE
93-68-5 O-ACETOACET0TOLUIDIDE [FLA]	ACETOACET-ORTHO-TOLUIDIDE
122-82-7 ACETOACET-P-PHENETIDE [FLA, MSL]	BUTANAMIDE, N-(4-ETHOXYPHENYL)-3-OXO-
122-82-7 ACETOACET-PARA-PHENETIDE [NFP, NFP]	BUTANAMIDE, N-(4-ETHOXYPHENYL)-3-OXO-
93-68-5 ACETOACET-ORTHO-TOLUIDIDE	ACETOACET-ORTHO-TOLUIDIDE
93-68-5 ACETOACET-O-TOLUIDIDE [MSL]	ACETOACET-ORTHO-TOLUIDIDE
97-36-9 M-ACETOACET XYLIDIDE	M-ACETOACET XYLIDIDE
92-15-9 ACETOACETYL-O-ANISIDINE [FLA]	O-ACETOACETANISIDIDE
122-80-5 P-ACETOAMINOANILINE	P-ACETOAMINOANILINE
34256-82-1 ACETOCHLOR	ACETOCHLOR
968-81-0 ACETOHEXAMIDE	ACETOHEXAMIDE
546-88-3 ACETOHYDROXAMIC ACID	ACETOHYDROXAMIC ACID
50-78-2 ACETOL (2) [FLA]	ACETYLSALICYLIC ACID (ASPIRIN)
103-90-2 ACETOMINOPHEN (4-HYDROXYACETANILIDE)	ACETOMINOPHEN (4-HYDROXYACETANILIDE)
93-08-3 2-ACETONAPHTHONE	2-ACETONAPHTHONE

Chemical Name	Reference Name
941-98-0 1-ACETONAPHTHONE	1-ACETONAPHTHONE
67-64-1 ACETONE	ACETONE
116-09-6 ACETONE ALCOHOL	ACETONE ALCOHOL
75-86-5 ACETONE CYANOHYDRIN	ACETONE CYANOHYDRIN
RR-00099-7 ACETONE OIL	ACETONE OIL
RR-00099-7 ACETONE OILS [HM1, HM1]	ACETONE OIL
1752-30-3 ACETONE THIOSEMICARBAZIDE	ACETONE THIOSEMICARBAZIDE
75-05-8 ACETONITRILE	ACETONITRILE
110-13-4 ACETONYL ACETONE [NFP, NFP]	2,5-HEXANEDIONE
81-81-2 3-(ALPHA-ACETONYLBENZYL)-4-HYDROXY-COUMARIN [FLA]	WARFARIN
62-44-2 P-ACETOPHENETIDIDE [FLA]	PHENACETIN
98-86-2 ACETOPHENONE	ACETOPHENONE
103-89-9 P-ACETOTOLUIDIDE	P-ACETOTOLUIDIDE
828-00-2 6-ACETOXY-2,4-DIMETHYL-M-DIOXANE [PST]	DIMETHOXANE
62-38-4 ACETOXYPHENYLMERCURY [FLA]	PHENYLMERCURIC ACETATE
900-95-8 ACETOXYTRIPHENYL STANNONE [MSL]	STANNANE, ACETOXYTRIPHENYL-
79-27-6 ACETYLENE TETRABROMIDE (1,1,2,2-TETRA-BROMOETHANE) [QBC]	ACETYLENE TETRABROMIDE
123-54-6 ACETYL ACETONE [FLA]	2,4-PENTANEDIONE
123-54-6 ACETYLACETONE [NJL]	2,4-PENTANEDIONE
37187-22-7 ACETYL ACETONE PEROXIDE	ACETYL ACETONE PEROXIDE
121-60-8 4-(ACETYLAMINO) BENZENESULFONYL CHLORIDE	4-(ACETYLAMINO) BENZENESULFONYL CHLORIDE
4075-79-0 4-ACETYLAMINOBIPHENYL	4-ACETYLAMINOBIPHENYL
53-96-3 2-ACETYLAMINOFLUORENE	2-ACETYLAMINOFLUORENE
89-52-1 N-ACETYLANTHRANILIC ACID	N-ACETYLANTHRANILIC ACID
RR-01776-5 N-ACETYLANTHRANILIC ACID SALTS	N-ACETYLANTHRANILIC ACID SALTS
61788-48-5 ACETYLATED LANOLIN	ACETYLATED LANOLIN
6178-49-0 ACETYLATED LANOLIN ALCOHOL	ACETYLATED LANOLIN ALCOHOL
644-31-5 ACETYL BENZOYL PEROXIDE	ACETYL BENZOYL PEROXIDE
506-96-7 ACETYL BROMIDE	ACETYL BROMIDE
598-21-0 ACETYL BROMIDE, BROMO- [TSC, TSC]	BROMOACETYL BROMIDE
75-36-5 ACETYL CHLORIDE	ACETYL CHLORIDE
79-36-7 ACETYL CHLORIDE [CAL, CWA]	DICHLOROACETYL CHLORIDE
79-04-9 ACETYL CHLORIDE, CHLORO- [PAL]	CHLOROACETYL CHLORIDE
79-36-7 ACETYL CHLORIDE, DICHLORO- [PAL]	DICHLOROACETYL CHLORIDE
3179-56-4 ACETYL CYCLOHEXANE SULFONYL PEROXIDE	ACETYL CYCLOHEXANE SULFONYL PEROXIDE
74-86-2 ACETYLENE	ACETYLENE
540-49-8 ACETYLENE DIBROMIDE [HM1]	1,2-DIBROMOETHENE
540-59-0 ACETYLENE DICHLORIDE [FLA, CAL]	1,2-DICHLOROETHYLENE
540-59-0 TRANS-ACETYLENE, DICHLORIDE [PAL]	1,2-DICHLOROETHYLENE
540-59-0 ACETYLENE DICHLORIDE (1,2-DICHLOROETHYLENE) [ALB, ALB]	1,2-DICHLOROETHYLENE
RR-01390-1 ACETYLENE SILVER NITRATE	ACETYLENE SILVER NITRATE
79-27-6 ACETYLENE TETRABROMIDE	ACETYLENE TETRABROMIDE
79-34-5 ACETYLENE TETRACHLORIDE [CAL, FLA, HM1, HM1]	1,1,2,2-TETRACHLOROETHANE
79-01-6 ACETYLENE TRICHLORIDE [FLA]	TRICHLOROETHYLENE
74-86-2 ACETYLENE [ETHYNE] [EP3]	ACETYLENE

Chemical Name	Reference Name
142-26-7 N-ACETYL ETHANOLAMINE	N-ACETYL ETHANOLAMINE
507-02-8 ACETYL IODIDE	ACETYL IODIDE
513-86-0 ACETYL METHYL CARBINOL	ACETYL METHYL CARBINOL
1696-20-4 4-ACETYL MORPHOLINE [FLA]	MORPHOLINE, 4-ACETYL-
1696-20-4 N-ACETYL MORPHOLINE [MSL, NFP, NFP]	MORPHOLINE, 4-ACETYL-
50-78-2 2-(ACETYLOXY) BENZOIC ACID [ONT]	ACETYLSALICYLIC ACID (ASPIRIN)
110-22-5 ACETYL PEROXIDE [MSL, NFP, NJL, NJL]	DIACETYL PEROXIDE
1071-73-4 3-ACETYL-1-PROPANOL	3-ACETYL-1-PROPANOL
89-84-9 4-ACETYLRESORCINOL	4-ACETYLRESORCINOL
50-78-2 ACETYLSALICYLIC ACID [AUS, CAL]	ACETYLSALICYLIC ACID (ASPIRIN)
50-78-2 O-ACETYLSALICYLIC ACID [GBR]	ACETYLSALICYLIC ACID (ASPIRIN)
50-78-2 ACETYLSALICYLIC ACID [MIN, MSL, NJL]	ACETYLSALICYLIC ACID (ASPIRIN)
50-78-2 ACETYLSALICYLIC ACID (ASPIRIN)	ACETYLSALICYLIC ACID (ASPIRIN)
591-08-2 1-ACETYL-2-THIOUREA	1-ACETYL-2-THIOUREA
77-89-4 ACETYL TRIETHYL CITRATE	ACETYL TRIETHYL CITRATE
RR-00238-0 ACID MODIFIED ACRYLATED EPOXIDE	ACID MODIFIED ACRYLATED EPOXIDE
633-96-5 ACID ORANGE 7	ACID ORANGE 7
1936-15-8 ACID ORANGE 10	ACID ORANGE 10
17372-87-1 ACID RED 87	ACID RED 87
64742-24-1 ACID SLUDGE [NJL]	SLUDGE ACID
64742-14-9 ACID TREATED LIGHT DISTILLATE (PETROLEUM)	ACID TREATED LIGHT DISTILLATE (PETROLEUM)
64742-17-2 ACID TREATED RESIDUAL OIL (PETROLEUM)	ACID TREATED RESIDUAL OIL (PETROLEUM) SOLVENT
62476-59-9 ACIFLUORFEN	ACIFLUORFEN
62476-59-9 ACIFLUORFEN, SODIUM SALT [5-(2-CHLORO-4-(TRIFLUOROMETHYL)PHENOXY)-2-NITRO-BENZOIC ACID, SODIUM SALT] [313]	ACIFLUORFENAN-2-YL]-METHYL-1H-1,2,4,-TRIAZOLE]
94-75-7 ACITIC ACID, (2,4-DICHLOROPHENOXY)- [TSC, TSC]	2,4-D
260-94-6 ACRIDINE	ACRIDINE
494-38-2 ACRIDINE ORANGE	ACRIDINE ORANGE
8048-52-0 ACRIFLAVINIUM CHLORIDE	ACRIFLAVINIUM CHLORIDE
107-02-8 ACROLEIN	ACROLEIN
107-02-8 ACROLEIN (2-PROPENAL) [OS3]	ACROLEIN
100-73-2 ACROLEIN DIMER	ACROLEIN DIMER
107-02-8 ACROLEINE [QBC, QBC]	ACROLEIN
107-02-8 ACROLEIN [2-PROPENAL] [EP3]	ACROLEIN
7008-42-6 ACRONYCINE	ACRONYCINE
107-02-8 ACRYLALDEHYDE [GBR, GBR]	ACROLEIN
79-06-1 ACRYLAMIDE	ACRYLAMIDE
9003-06-9 ACRYLAMIDE-ACRYLIC ACID POLYMER	ACRYLAMIDE-ACRYLIC ACID POLYMER
9003-06-9 ACRYLAMIDE-ACRYLIC ACID RESIN [PST]	ACRYLAMIDE-ACRYLIC ACID POLYMER
RR-01205-5 ACRYLAMIDE, POLYMERS WITH TETRAALKYL AMMONIUM SALT AND POLYALKYL, AMINO ALKYL METHACRYLAMIDE SALT	ACRYLAMIDE, POLYMERS WITH TETRAALKYL AMMONIUM SALT AND POLYA
RR-00905-2 ACRYLAMIDE, POLYMER WITH SUBSTITUTED ALKYLACRYLAMIDE SALT	ACRYLAMIDE, POLYMER WITH SUBSTITUTED ALKYLACRYLAMIDE SALT
25085-02-3 ACRYLAMIDE-SODIUM ACRYLATE POLYMER	ACRYLAMIDE-SODIUM ACRYLATE POLYMER
RR-01672-8 ACRYLAMIDE-SUBSTITUTED EPOXY	ACRYLAMIDE-SUBSTITUTED EPOXY
RR-01765-2 ACRYLATE ESTERS	ACRYLATE ESTERS
RR-01669-3 ACRYLATED EPOXY PHENOLIC RESIN	ACRYLATED EPOXY PHENOLIC RESIN

Chemical Name	Reference Name
RR-01575-8 ACRYLATES OF ALIPHATIC POLYOL	ACRYLATES OF ALIPHATIC POLYOL
RR-01258-8 ACRYLATE SUBSTITUTED SILOXANES AND SILICONES	ACRYLATE SUBSTITUTED SILOXANES AND SILICONES
79-10-7 ACRYLIC ACID	ACRYLIC ACID
141-32-2 ACRYLIC ACID, N-BUTYL ESTER [MAK, MAK, MAK, MAK]	BUTYL ACRYLATE
25750-84-9 ACRYLIC ACID, BUTYL ESTER, POLYMER WITH ETHYLENE	ACRYLIC ACID, BUTYL ESTER, POLYMER WITH ETHYLENE
140-88-5 ACRYLIC ACID, ETHYL ESTER [MAK, MAK, MAK, MAK]	ETHYL ACRYLATE
79-10-7 ACRYLIC ACID (GLACIAL) [NFP, NFP]	ACRYLIC ACID
96-33-3 ACRYLIC ACID, METHYL ESTER [MAK, MAK, MAK]	METHYL ACRYLATE
24968-79-4 ACRYLIC ACID METHYL ESTER, POLYMER WITH ACRYLONITRILE AND 1,3-BUTADIENE	ACRYLIC ACID METHYL ESTER, POLYMER WITH ACRYLONITRILE AND 1,
9003-01-4 ACRYLIC ACID POLYMER [PST]	ACRYLIC RESIN
9003-04-7 ACRYLIC ACID POLYMER, SODIUM SALT [PST]	SODIUM POLYACRYLATE
9010-77-9 ACRYLIC ACID POLYMER WITH ETHYLENE	ACRYLIC ACID POLYMER WITH ETHYLENE
25987-30-8 ACRYLIC ACID, POLYMER WITH ACRYLAMIDE, SODIUM SALT [PST]	2-PROPENOIC ACID, POLYMER WITH 2-PROPENAMIDE, SODIUM SALT-
RR-01217-9 ACRYLIC ACID, POLYMER WITH SUBSTITUTED ETHENE	ACRYLIC ACID, POLYMER WITH SUBSTITUTED ETHENE
9007-16-3 ACRYLIC ACID-SUCROSE POLYALLYL ETHER POLYMER	ACRYLIC ACID-SUCROSE POLYALLYL ETHER POLYMER
RR-01524-7 ACRYLIC FIBRES	ACRYLIC FIBRES
9003-01-4 ACRYLIC RESIN	ACRYLIC RESIN
107-13-1 ACRYLONITRILE	ACRYLONITRILE
107-13-1 ACRYLONITRILE (VINYL CYANIDE) [ALB, ALB, ALB, ALB]	ACRYLONITRILE
9003-56-9 ACRYLONITRILE-BUTADIENE-STYRENE COPOLYMERS [IAR]	ABS RESIN
25014-41-9 ACRYLONITRILE POLYMER	ACRYLONITRILE POLYMER
107-13-1 ACRYLONITRILE [2-PROPENENITRILE] [EP3]	ACRYLONITRILE
814-68-6 ACRYLYL CHLORIDE	ACRYLYL CHLORIDE3-YL]IMINO]BIS-
814-68-6 ACRYLYL CHLORIDE [2-PROPENOYL CHLORIDE] [EP3]	ACRYLYL CHLORIDE
15755-98-3 ACTINIUM 224	ACTINIUM 224
14265-85-1 ACTINIUM 225	ACTINIUM 225
20379-10-6 ACTINIUM 226	ACTINIUM 226
14952-40-0 ACTINIUM 227	ACTINIUM 227
14331-83-0 ACTINIUM 228	ACTINIUM 228
12172-67-7 ACTINOLITE	ACTINOLITE
13768-00-8 ACTINOLITE [QBC, QBC]	ACTINOLITE, NON-ASBESTIFORM
77536-66-4 ACTINOLITE ASBESTOS [MSL]	ASBESTOS, ACTINOLITE
13768-00-8 ACTINOLITE, NON-ASBESTIFORM	ACTINOLITE, NON-ASBESTIFORM
50-76-0 ACTINOMYCIN D	ACTINOMYCIN D
1406-16-2 ACTIVATED ERGOSTEROL	ACTIVATED ERGOSTEROL
73-24-5 ADENINE (6-AMINOPURINE)	ADENINE (6-AMINOPURINE)
RR-01322-9 ADHESIVES	ADHESIVES
628-94-4 ADIPAMIDE	ADIPAMIDE
124-04-9 ADIPIC ACID	ADIPIC ACID
123-79-5 ADIPIC ACID, DIOCTYL ESTER	ADIPIC ACID, DIOCTYL ESTER

Chemical Name	Reference Name
626-86-8 ADIPIC ACID MONOMETHYL ESTER	ADIPIC ACID MONOMETHYL ESTER
111-69-3 ADIPONITRILE	ADIPONITRILE
111-50-2 ADIPOYL CHLORIDE	ADIPOYL CHLORIDE
23214-92-8 ADRIAMYCIN	ADRIAMYCIN
23214-92-8 ADRIAMYCIN (DOXORUBICIN HYDROCHLO-RIDE) [C65]	ADRIAMYCINIHYDROCHLORIDE)
RR-01514-5 AEROSOL DISPENSERS	AEROSOL DISPENSERS
RR-01323-0 AEROSOLS	AEROSOLS
3688-53-7 AF-2 [MIN]	FURYLFURAMIDE (AF-2)
51004-61-6 AF 2(FOAMING AGENT)	AF 2(FOAMING AGENT)
3688-53-7 AF-2; [2-(2-FURYL)-3-(5-NITRO-2-FURYL)] ACRYLAMIDE [C65, C65]	FURYLFURAMIDE (AF-2)
3688-53-7 AF-2 (2-(2-FURYL)-3-(5-NITRO-2-FURYL)ACRY-LAMIDE) [IAR]	FURYLFURAMIDE (AF-2)
3688-53-7 AF-2 [2-(2-FURYL)-3-(5-NITRO-2-FURYL)ACRY-LAMIDE] [CAL]	FURYLFURAMIDE (AF-2)OS)
1162-65-8 AFLATOXIN B1	AFLATOXIN B1
7220-81-7 AFLATOXIN B2	AFLATOXIN B2
1165-39-5 AFLATOXIN G1	AFLATOXIN G1
7241-98-7 AFLATOXIN G2	AFLATOXIN G2
6795-23-9 AFLATOXIN M1	AFLATOXIN M1
1402-68-2 AFLATOXINS	AFLATOXINS
1402-68-2 AFLATOXINS, NATURALLY OCCURRING MIX-TURES OF [IAR]	AFLATOXINS
9002-18-0 AGAR	AGAR
2757-90-6 AGARITINE	AGARITINE
RR-01325-2 AIR	AIR
15972-60-8 ALACHLOR	ALACHLOR
15972-60-8 ALACHLOR (2-CHLORO-2',6'-DIETHYL-N-METHOXYMETHYL ACETANILIDE) [CN8]	ALACHLORTHYL PHOSPHATE)
RR-00966-5 ALANINE, N-(2-CARBOXYETHYL)-N-ALKYL, SALT	ALANINE, N-(2-CARBOXYETHYL)-N-ALKYL, SALT
9006-50-2 ALBUMIN EGG	ALBUMIN EGG
RR-00964-3 ALCOHOL, ALKALI METAL SALT	ALCOHOL, ALKALI METAL SALT
RR-01271-5 ALCOHOL C-13 - C-15 POLY(1-3) ETHOXYLATE	ALCOHOL C-13 - C-15 POLY(1-3) ETHOXYLATE
RR-01272-6 ALCOHOL C-6 - C-17 (SECONDARY)POLY(3-6) ETHOXYLATE	ALCOHOL C-6 - C-17 (SECONDARY)POLY(3-6) ETHOXY-LATE
RR-00113-8 ALCOHOL, DENATURED	ALCOHOL, DENATURED
RR-00111-6 ALCOHOLIC BEVERAGES [C65, C65, IAR]	ALCOHOLS
RR-00111-6 ALCOHOL, N.O.S. [NJL]	ALCOHOLS
RR-00111-6 ALCOHOLS	ALCOHOLS
68439-45-2 ALCOHOLS, C6-12, ETHOXYLATED	ALCOHOLS, C6-12, ETHOXYLATED
69013-18-9 ALCOHOLS, C8-18, ETHOXYLATED PROPOXY-LATED	ALCOHOLS, C8-18, ETHOXYLATED PROPOXYLATED
69013-19-0 ALCOHOLS, C8-22, ETHOXYLATED	ALCOHOLS, C8-22, ETHOXYLATED
97043-91-9 ALCOHOLS, C9-16, ETHOXYLATED	ALCOHOLS, C9-16, ETHOXYLATED
69227-22-1 ALCOHOLS, C10-16, ETHOXYLATED PROPOXY-LATED	ALCOHOLS, C10-16, ETHOXYLATED PROPOXYLATED
68131-40-8 ALCOHOLS, C11-15-SECONDARY, ETHOXY-LATED	ALCOHOLS, C11-15-SECONDARY, ETHOXYLATED
68439-50-9 ALCOHOLS, C12-14, ETHOXYLATED	ALCOHOLS, C12-14, ETHOXYLATED

Chemical Name	Reference Name
69227-21-0 ALCOHOLS, C12-18, ETHOXYLATED PROPOXYLATED	ALCOHOLS, C12-18, ETHOXYLATED PROPOXYLATED
68526-94-3 ALCOHOLS, C12-20, ETHOXYLATED	ALCOHOLS, C12-20, ETHOXYLATED
68439-49-6 ALCOHOLS, C16-18, ETHOXYLATED	ALCOHOLS, C16-18, ETHOXYLATED
68131-39-5 ALCOHOLS, C-12 - C-15, ETHOXYLATED	ALCOHOLS, C-12 - C-15, ETHOXYLATED
68131-39-5 ALCOHOL C12 - C15, POLY(1-3) ETHOXYLATE [HM1]	ALCOHOLS, C-12 - C-15, ETHOXYLATED
68213-23-0 ALCOHOLS, C12-18, ETHOXYLATED	ALCOHOLS, C12-18, ETHOXYLATED
RR-00111-6 ALCOHOLS, N.O.S. [HM1, HM1]	ALCOHOLS
61791-28-4 ALCOHOLS, TALLOW, ETHOXYLATED	ALCOHOLS, TALLOW, ETHOXYLATED
RR-00114-9 ALDEHYDES	ALDEHYDES
RR-00114-9 ALDEHYDES, N.O.S. [HM1, HM1, NJL]	ALDEHYDES
116-06-3 ALDICARB	ALDICARB
1646-88-4 ALDICARB SULFONE	ALDICARB SULFONE
1646-87-3 ALDICARB SULFOXIDE	ALDICARB SULFOXIDE
107-89-1 ALDOL	ALDOL
309-00-2 ALDRIN	ALDRIN
309-00-2 ALDRIN[1,4:5,8-DIMETHANONAPHTHALENE,1,2,3,4,10,10-HEXACHLORO-1,4,4A,5,8,8A,HEXAHYDRO-(1ALPHA,4ALPHA,4BETA,5ALPHA, 8ALPHA,8BETA)-] [313]	ALDRIN
309-00-2 ALDRIN ((1R,4S,4AS,5S,8R,8AR)1,2,3,4,10,10-HEXACHLORO-1,4,4A,5,8,8A-HEXAHYDRO-1,4:5,8-DIMETHANONAPHTHALENE) [CN2]	ALDRIN
309-00-2 ALDRIN (ISO) [GBR, GBR, GBR]	ALDRIN
RR-01614-8 ALDRIN AND DIELDRIN	ALDRIN AND DIELDRIN
RR-01028-6 ALFALFA	ALFALFA
9005-38-3 ALGENIC ACID, SODIUM SALT	ALGENIC ACID, SODIUM SALT
9049-05-2 ALGIN GUM	ALGIN GUM
RR-01681-9 ALIPHATIC DICARBOXYLIC ACID SALT	ALIPHATIC DICARBOXYLIC ACID SALT
RR-01680-8 ALIPHATIC DIFUNCTIONAL ACRYLIC ACID ESTER	ALIPHATIC DIFUNCTIONAL ACRYLIC ACID ESTER
RR-00239-1 ALIPHATIC DIURETHANE ACRYLATE ESTER	ALIPHATIC DIURETHANE ACRYLATE ESTER
RR-01733-4 ALIPHATIC ETHER	ALIPHATIC ETHER
8032-32-4 ALIPHATIC NAPHTHA [NJL]	LIGROINE
RR-01019-5 ALIPHATIC POLYGLYCIDYL ETHER	ALIPHATIC POLYGLYCIDYL ETHER
72-48-0 ALIZARIN [HON]	1,2-DIHYDROXYANTHRAQUINONE
RR-00116-1 ALKALI METAL ALLOYS	ALKALI METAL ALLOYS
RR-00117-2 ALKALI METAL AMALGAM, N.O.S.	ALKALI METAL AMALGAM, N.O.S.
RR-00117-2 ALKALI METAL AMALGAMS [HM1, HM1]	ALKALI METAL AMALGAM, N.O.S.
RR-00118-3 ALKALI METAL AMIDES [HM1, HM1]	ALKALI METAL AMIDES, N.O.S.
RR-00118-3 ALKALI METAL AMIDES, N.O.S.	ALKALI METAL AMIDES, N.O.S.
RR-00119-4 ALKALI METAL DISPERSIONS, N.O.S.	ALKALI METAL DISPERSIONS, N.O.S.
RR-00119-4 ALKALI METAL DISPERSIONS [HM1, HM1]	ALKALI METAL DISPERSIONS, N.O.S.
RR-01264-6 ALKALI METAL NITRITES	ALKALI METAL NITRITESIC PENTAERYTHRITOL TETRAESTER
RR-00121-8 ALKALINE CORROSIVE LIQUID, N.O.S.	ALKALINE CORROSIVE LIQUID, N.O.S.
RR-00122-9 ALKALINE EARTH METAL ALLOYS, N.O.S.	ALKALINE EARTH METAL ALLOYS, N.O.S.
RR-00123-0 ALKALINE EARTH METAL AMALGAMS, N.O.S.	ALKALINE EARTH METAL AMALGAMS, N.O.S.
RR-00123-0 ALKALINE EARTH METAL AMALGAMS [HM1, HM1]	ALKALINE EARTH METAL AMALGAMS, N.O.S.

Chemical Name	Reference Name
RR-00124-1 ALKALINE EARTH METAL DISPERSIONS, N.O.S.	ALKALINE EARTH METAL DISPERSIONS, N.O.S.
RR-01339-8 ALKALOIDS	ALKALOIDS
RR-00125-2 ALKALOID, SALTS, N.O.S.	ALKALOID, SALTS, N.O.S.
RR-01229-3 ALKANAMINIUM, POLYALKYL-[(2-METHYL-1-OXO-2-PROPENYL)OXY]SALT, POLYMER WITH ACRYLAMIDE AND SUBSTITUTED ALKYL METHACRYLATE	ALKANAMINIUM, POLYALKYL-[(2-METHYL-1-OXO-2-PROPENYL)OXY]SALT
68551-17-7 ALKANES [MIN]	ALKANES, C10-13-ISO-
68920-70-7 ALKANES, C6-18, CHLORO-	ALKANES, C6-18, CHLORO-ETRAKIS-, OCTAAMMONIUM SALT
68920-70-7 ALKANES, C(6-18), CHLORO- [TSC]	ALKANES, C6-18, CHLORO-ETRAKIS-, OCTAAMMONIUM SALT
68551-17-7 ALKANES, C10-13-ISO-	ALKANES, C10-13-ISO-
68955-41-9 ALKANES, C(10-18)-BROMOCHLORO-	ALKANES, C(10-18)-BROMOCHLORO-
61788-76-9 ALKANES, CHLORO- [TSC]	CHLOROALKANES
75-75-2 ALKANE SULFONIC ACID	ALKANE SULFONIC ACID
RR-00938-1 ALKANOIC ACID, BUTANEDIOL AND CYCLO-HEXANEALKANOL POLYMER	ALKANOIC ACID, BUTANEDIOL AND CYCLOHEX-ANEALKANOL POLYMER
RR-00215-3 ALKEHYLDICARBOXYLIC ACIDS, POLYMERS WITH ALKANEPOLYOL ANDTDI, ALKANOL BLOCKED, ACRYLATE	ALKEHYLDICARBOXYLIC ACIDS, POLYMERS WITH ALKANEPOLYOL ANDALPHA-[2-AMINOETHYLETHYL]-X-(2-AMINOETHYLETHOXY)
RR-00221-1 ALKENOIC ACID, TRISUBSTITUTED BENZYL-DISUBSTITUTED PHENYLESTER	ALKENOIC ACID, TRISUBSTITUTED BENZYL-DISUB-STITUTED PHENYLTONE TRIOL AND ALKOXYLATED ALKANEPOLYOL, HYDROXYALKYL METHACRYLATE ESTER
RR-00222-2 ALKENOIC ACID, TRISUBSTITUTED PHENY-LALKYL DISUBSTITUTEDPHENYL ESTER	ALKENOIC ACID, TRISUBSTITUTED PHENYLALKYL DISUBSTITUTEDESTER
RR-01690-0 ALKENYL ETHER OF ALKANETRIOL POLYMER	ALKENYL ETHER OF ALKANETRIOL POLYMER
RR-00243-7 ALKOXYLATED ALKANE POLYOL, POLYACRY-LATE ESTER	ALKOXYLATED ALKANE POLYOL, POLYACRYLATE ESTER
RR-01212-4 ALKYOXYLATED DIALKYLDIETHYLENETRI-AMINE, ALKYL SULFATE SALT	ALKYOXYLATED DIALKYLDIETHYLENETRIAMINE, ALKYL SULFATE SALTLAMINE
RR-01746-9 ALKYL ANTHRAQUINONES	ALKYL ANTHRAQUINONES
RR-00946-1 ALKYL ALKENOATE, AZOBIS-	ALKYL ALKENOATE, AZOBIS-
RR-00022-6 ALKYLALUMINUMS [OS3]	ALUMINUM, ALKYLS (NOC)
RR-00126-3 ALKYLAMINES, N.O.S. [HM1, HM1]	ALKYLAMINES OR POLYALKYLAMINES
RR-00126-3 ALKYLAMINES OR POLYALKYLAMINES	ALKYLAMINES OR POLYALKYLAMINES
RR-01195-0 ALKYLAMINE TETRACHLOROPHENATE	ALKYLAMINE TETRACHLOROPHENATE) ETHANOL
RR-00199-0 3-ALKYL-2-(2-ANILINO)VINYLTHIAZOLINIUM SALT	3-ALKYL-2-(2-ANILINO)VINYLTHIAZOLINIUM SALT
RR-00129-6 ALKYL, ARYL, OR TOLUENE SULFONIC ACID	ALKYL, ARYL, OR TOLUENE SULFONIC ACID
RR-01178-9 ALKYLARYL SUBSTITUTED PHOSPHITE	ALKYLARYL SUBSTITUTED PHOSPHITE
RR-00179-6 ALKYLATED DIARYLAMINE, SULFURIZED	ALKYLATED DIARYLAMINE, SULFURIZEDESTERS
RR-00174-1 ALKYLATED DIPHENYL OXIDE	ALKYLATED DIPHENYL OXIDE
RR-01570-3 ALKYLATED DIPHENYLS	ALKYLATED DIPHENYLS
RR-01688-6 ALKYLATED SULFONATED DIPHENYL OXIDE, ALKALI AND AMINE SALTS	ALKYLATED SULFONATED DIPHENYL OXIDE, ALKALI AND AMINE SALTS
RR-00992-7 ALKYLBENZENE SULFONATE, AMINE SALT	ALKYLBENZENE SULFONATE, AMINE SALT
RR-01233-9 ALKYLBENZENESULFONIC ACIDS AND SODIUM SALTS	ALKYLBENZENESULFONIC ACIDS AND SODIUM SALT-SALOGEN ACID SALTS
68391-01-5 ALKYLBENZYLDIMETHYLAMMONIUM CHLO-RIDE	ALKYLBENZYLDIMETHYLAMMONIUM CHLORIDEYL-NOR, HYDROXIDE, MONOSODIUM SALT
RR-00930-3 ALKYLBISOXYALKYL (SUBSTITUTED-1,1-DIMETHYLETHYLPHENYL) BENZOTRIAZOLE	ALKYLBISOXYALKYL (SUBSTITUTED-1,1-DIMETHYLETHYLPHENYL) BENZO

Chemical Name	Reference Name
RR-01712-9 ALKYL-, BROMO-, CHLORO-, HYDROX-YMETHYL DIARYL ETHERS	ALKYL-, BROMO-, CHLORO-, HYDROXYMETHYL DI-ARYL ETHERS
68891-29-2 ALPHA-ALKYL(C8-C10)-OMEGA-HYDROXY-POLY(OXYETHYLENE) AMMONIUMSULFATE	ALPHA-ALKYL(C8-C10)-OMEGA-HYDROXYPOLY (OXYETHYLENE) AMMONIUM
RR-01032-2 ALKYL(C8-10) POLYETHOXYPOLYPROPOXY-BENZENE ETHER	ALKYL(C8-10) POLYETHOXYPOLYPROPOXYBENZENE ETHER
68987-80-4 ALKYL (C8-C12) GLYCIDYL ETHER [OTS]	OXIRANE, MONO[C(6-12)-ALKYLOXY)METHYL]DERIVA-TIVES
RR-01033-3 ALPHA-ALKYL(C8-C14)-OMEGA-HYDROXY-POLY(OXYPROPYLENE) BLOCK COPOLYMER WITH POLYOXYETHYLENE	ALPHA-ALKYL(C8-C14)-OMEGA-HYDROXYPOLY (OXYPROPYLENE) BLOCK CO
68585-36-4 ALKYL(C10-14) OXYPOLY(ETHYLENEOXY) ETHYL PHOSPHATE	ALKYL(C10-14) OXYPOLY(ETHYLENEOXY)ETHYL PHOS-PHATESODIUM SALT
RR-01030-0 ALKYL(C10-14) POLY(OXYETHYLENE)POLY (OXYPROPYLENE) CONDENSATE WITH MONOSTEARYL ACID PHOSPHATE	ALKYL(C10-14) POLY(OXYETHYLENE)POLY(OXYPROPY-LENE) CONDENSATE
70592-80-2 ALKYL(C10-16) DIMETHYLAMINE OXIDE	ALKYL(C10-16) DIMETHYLAMINE OXIDEHANOLAMINE SALTS
68585-34-2 ALPHA-ALKYL(C10-16)-OMEGA-HYDROXYPOLY (OXYETHYLENE) SULFATE,SODIUM SALT	ALPHA-ALKYL(C10-16)-OMEGA-HYDROXYPOLY (OXYETHYLENE) SULFATE,LTS
69227-22-1 ALPHA-ALKYL(C10-16)-OMEGA-HYDROXY-POLYOXYETHYLENE POLYOXYPROPYLENE POLYOXYETHYLENE [PST]	ALCOHOLS, C10-16, ETHOXYLATED PROPOXYLATED
68081-84-5 ALKYL (C10 - C16) GLYCIDYL ETHER [EPA]	OXIRANE, MONO[(C10-16-ALKYLOXY)METHYL] DERIVATIVES
68081-84-5 ALKYL (C10-C16) GLYCIDYL ETHER [OTS]	OXIRANE, MONO[(C10-16-ALKYLOXY)METHYL] DERIVATIVES
RR-01031-1 ALKYL(C11-15) PHENOXYPOLY(OXYETHY-LENE) ETHANOL	ALKYL(C11-15) PHENOXYPOLY(OXYETHYLENE) ETHANOL WITH MONOSTEARYL ACID PHOSPHATE
68955-55-5 ALKYL(C12-14) DIMETHYL AMINE OXIDE	ALKYL(C12-14) DIMETHYL AMINE OXIDENTAMINE
68424-85-1 ALKYL (C12-16)DIMETHYLBENZYLAMMONIUM CHLORIDE	ALKYL (C12-16)DIMETHYLBENZYLAMMONIUM CHLO-RIDE
68424-85-1 ALKYL(C12-16) DIMETHYL BENZYL AMMO-NIUM CHLORIDE [PST]	ALKYL (C12-16)DIMETHYLBENZYLAMMONIUM CHLO-RIDE
68609-97-2 ALKYL (C12-C14) GLYCIDYL ETHER	ALKYL (C12-C14) GLYCIDYL ETHER
68908-63-4 ALPHA-ALKYL(C12-C15)-OMEGA-HYDROXY-POLYOXYETHYLENE	ALPHA-ALKYL(C12-C15)-OMEGA-HYDROXYPOLY-OXYETHYLENE
68987-80-4 ALKYL (C6 - C12) GLYCIDYL ETHER [EPA]	OXIRANE, MONO[C(6-12)-ALKYLOXY)METHYL]DERIVA-TIVES
68609-96-1 ALKYL (C8-C10) GLYCIDYL ETHER	ALKYL (C8-C10) GLYCIDYL ETHER
RR-01248-6 ALKYLCARBAMIC ACID, ALKYNYL ESTER	ALKYLCARBAMIC ACID, ALKYNYL ESTER
8001-54-5 N-ALKYLDIMETHYLBENZYL AMMONIUM CHLORIDE	N-ALKYLDIMETHYLBENZYL AMMONIUM CHLORIDE
RR-00178-5 ALKYLENE GLYCOL TEREPHTHALATE AND SUBSTITUTED BENZOATEESTERS	ALKYLENE GLYCOL TEREPHTHALATE AND SUBSTI-TUTED BENZOATE
RR-00984-7 ALKYLENEBIS(SUBSTITUTED CARBOMONO-CYCLE), EPICHLOROHYDRIN, DISUBSTITUTED HETERMONOCYCLE, ACRYLATE POLYMER	ALKYLENEBIS(SUBSTITUTED CARBOMONOCYCLE), EPICHLOROHYDRIN, DI-1,3-PROPANEDIYL(BIS[3-(OXI-RANYLMETHOXY)-
RR-01689-7 ALKYLENEDIOALKYL ETHER	ALKYLENEDIOALKYL ETHER
RR-00285-7 ALKYL EPOXIDES	ALKYL EPOXIDES
RR-00923-4 ALKYL ESTER	ALKYL ESTER
RR-01029-7 ALKYL (FATTY ACIDS OF COCONUT OIL) DIMETHYLAMMONIUM BETAINE	ALKYL (FATTY ACIDS OF COCONUT OIL) DIMETHY-LAMMONIUM BETAINE
RR-00925-6 ALKYL (HETEROCYCLICYL) PHENYLAZO-HETERO MONOCYCLIC POLYONE	ALKYL (HETEROCYCLICYL) PHENYLAZOHETERO MONOCYCLIC POLYONE

Chemical Name	Reference Name
RR-00927-8 ALKYL (HETEROCYCLICYL) PHENYLAZO-HETERO MONOCYCLIC POLYONE, [(ALKYLIM-IDAZOLYL)METHYL] DERIVATIVE	ALKYL (HETEROCYCLICYL) PHENYLAZOHETERO MONOCYCLIC POLYONE, [
70206-24-5 ALKYL IMIDAZOLINIUM METHYL SULFATE (DERIVED FROM OLEIC ACID)	ALKYL IMIDAZOLINIUM METHYL SULFATE (DERIVED FROM OLEIC ACID)
RR-00945-0 ALKYL PEROXY-2-ETHYL HEXANOATE	ALKYL PEROXY-2-ETHYL HEXANOATE
RR-00133-2 ALKYL PHENOL, N.O.S.	ALKYL PHENOL, N.O.S.
RR-00133-2 ALKYLPHENOLS, N.O.S. [HM1, HM1, HM1]	ALKYL PHENOL, N.O.S.
RR-01652-4 ALKYLPHENOXYPOLY(OXYETHYLENE SULFU-RIC ACID ESTER, SUBSTITUTED AMINE SALT	ALKYLPHENOXYPOLY(OXYETHYLENE SULFURIC ACID ESTER, SUBSTITUTE
RR-00192-3 ALKYLPHENOXYPOLYALKOXYAMINE	ALKYLPHENOXYPOLYALKOXYAMINE
RR-01659-1 ALKYL PHOSPHONATE AMMONIUM SALTS	ALKYL PHOSPHONATE AMMONIUM SALTS
RR-00284-6 ALKYL PHTHALATES	ALKYL PHTHALATES
RR-00973-4 ALKYL POLYETHYLENE GLYCOL PHOSPHATE, POTASSIUM SALT	ALKYL POLYETHYLENE GLYCOL PHOSPHATE, POTASSIUM SALT
RR-01219-1 ALKYL SUBSTITUTED DIAROMATIC HYDRO-CARBONS	ALKYL SUBSTITUTED DIAROMATIC HYDROCARBONS
RR-01726-5 ALKYLSULFONIUM SALT	ALKYLSULFONIUM SALT
RR-00283-5 ALKYLTIN COMPOUNDS	ALKYLTIN COMPOUNDS
97-59-6 ALLANTOIN	ALLANTOIN
584-79-2 ALLETHRIN	ALLETHRIN
28057-48-9 D-TRANS-ALLETHRIN [D-TRANS-CHRYSAN-THEMIC ACID OF D-ALLETHRONE]	D-TRANS-ALLETHRIN [D-TRANS-CHRYSANTHEMIC ACID OF D-ALLETHRONHENYL]-3(2H)-PYRIDAZINONE]
591-87-7 ALLYL ACETATE	ALLYL ACETATE
107-18-6 ALLYL ALCOHOL	ALLYL ALCOHOL
107-18-6 ALLYL ALCOHOL (2-PROPEN-1-OL) [CN2]	ALLYL ALCOHOL
107-18-6 ALLYL ALCOHOL [2-PROPEN-1-OL] [EP3]	ALLYL ALCOHOL
107-11-9 ALLYLAMINE	ALLYLAMINE
107-11-9 ALLYL AMINE [NJL, NJL]	ALLYLAMINE
107-11-9 ALLYLAMINE [2-PROPEN-1-AMINE] [EP3]	ALLYLAMINE
106-95-6 ALLYL BROMIDE	ALLYL BROMIDE
123-68-2 ALLYL CAPROATE	ALLYL CAPROATE
107-05-1 ALLYL CHLORIDE	ALLYL CHLORIDE
107-05-1 ALLYL CHLORIDE (3-CHLOROPROPENE) [QBC, QBC, RCA]	ALLYL CHLORIDE
2937-50-0 ALLYL CHLOROCARBONATE	ALLYL CHLOROCARBONATE
2937-50-0 ALLYL CHLOROFORMATE [HM1, HM1, HM1, MSL, NFP, NFP, WHM]	ALLYL CHLOROCARBONATE
109-75-1 ALLYL CYANIDE	ALLYL CYANIDE
13361-32-5 ALLYL CYANOACETATE	ALLYL CYANOACETATE
106-92-3 ALLYL 2,3-EPOXYPROPYL ETHER [GBR, GBR, GBR]	ALLYL GLYCIDYL ETHER
557-40-4 ALLYL ETHER [MSL, NFP, NFP]	DIALLYL ETHER
25327-89-3 ALLYL ETHER OF TETRABROMOBISPHENOL-A	ALLYL ETHER OF TETRABROMOBISPHENOL-A
557-31-3 ALLYL ETHYL ETHER	ALLYL ETHYL ETHER
1838-59-1 ALLYL FORMATE	ALLYL FORMATE
106-92-3 ALLYL GLYCIDYL ETHER	ALLYL GLYCIDYL ETHER
106-92-3 ALLYL GLYCIDYL ETHER (AGE) [MSL, NHS, NHS, QBC, QBC, TLV, TLV, WHM, AUS, AUS, AUS, BC1, BC1, BC1]	ALLYL GLYCIDYL ETHER
869-29-4 ALLYLIDENE DIACETATE	ALLYLIDENE DIACETATE
556-56-9 ALLYL IODIDE	ALLYL IODIDE
57-06-7 ALLYL ISOTHIOCYANATE	ALLYL ISOTHIOCYANATE

Chemical Name	Reference Name
2835-39-4 ALLYL ISOVALERATE	ALLYL ISOVALERATE
96-05-9 ALLYL METHACRYLATE	ALLYL METHACRYLATE
94-59-7 4-ALLYL-1,2-(METHYLENEDIOXY)-BEN-ZENE [FLA]	SAFROLE
12012-95-2 ALLYL PALLADIUM CHLORIDE DIMER	ALLYL PALLADIUM CHLORIDE DIMER
2179-59-1 ALLYL PROPYL DISULFIDE	ALLYL PROPYL DISULFIDE
96-18-4 ALLYL TRICHLORIDE [FLA]	1,2,3-TRICHLOROPROPANE
107-37-9 ALLYL TRICHLOROSILANE	ALLYL TRICHLOROSILANE
RR-01034-4 ALMOND HULLS	ALMOND HULLSPOLYMER WITH POLYOXYETHYLENE
8001-97-6 ALOE	ALOE
8001-97-6 ALOE VERA GEL [PST]	ALOE
1344-28-1 ALPHA-ALUMINA [MEX]	ALUMINUM OXIDE
532-27-4 ALPHA-CHLOROACETOPHENONE [MEX]	ALPHA-CHLOROACETOPHENONE
26148-68-5 A-ALPHA-C	A-ALPHA-C
26148-68-5 A-ALPHA-C(2-AMINO-9H-PYRIDO[2,3-B]IN-DOLE) [IAR]	A-ALPHA-C
26148-68-5 A-ALPHA-C (2-AMINO-9H-PYRIDO[2,3-B]IN-DOLE) [C65, C65]	A-ALPHA-C
12587-46-1 ALPHA RADIATION	ALPHA RADIATION
28981-97-7 ALPRAZOLAM	ALPRAZOLAM
1344-28-1 A-ALUMINA [MIN]	ALUMINUM OXIDE
1344-28-1 ALPHA-ALUMINA [OS1, OS1, WHM, ONT]	ALUMINUM OXIDE
16853-85-3 ALUMINATE(1-), TETRAHYDRO-, LITHIUM,(T-4)- [PAL]	LITHIUM ALUMINUM HYDRIDE
7429-90-5 ALUMINUM	ALUMINUM
14682-66-7 ALUMINUM 26	ALUMINUM 26
7429-90-5 ALUMINUM METAL AND OXIDE [ALB, ALB, MEX, MEX]	ALUMINUM
RR-00135-4 ALUMINUM ALKYL CHLORIDE	ALUMINUM ALKYL CHLORIDE
RR-00022-6 ALUMINUM ALKYL COMPOUNDS [WHM, GBR]	ALUMINUM, ALKYLS (NOC)
RR-00136-5 ALUMINUM ALKYL HALIDES	ALUMINUM ALKYL HALIDES
RR-01361-6 ALUMINUM ALKYL HYDRIDES	ALUMINUM ALKYL HYDRIDES
RR-00022-6 ALUMINUM, ALKYLS [CAL]	ALUMINUM, ALKYLS (NOC)
RR-00022-6 ALUMINUM ALKYLS [NJL]	ALUMINUM, ALKYLS (NOC)
RR-00022-6 ALUMINUM, ALKYLS [OS1]	ALUMINUM, ALKYLS (NOC)
RR-00022-6 ALUMINUM ALKYLS [PAL, QBC]	ALUMINUM, ALKYLS (NOC)
RR-00022-6 ALUMINUM, ALKYLS (NOC)	ALUMINUM, ALKYLS (NOC)
16962-07-5 ALUMINUM BOROHYDRIDE	ALUMINUM BOROHYDRIDE
7727-15-3 ALUMINUM BROMIDE	ALUMINUM BROMIDE
12656-43-8 ALUMINUM CARBIDE	ALUMINUM CARBIDE
7446-70-0 ALUMINUM CHLORIDE	ALUMINUM CHLORIDE
7446-70-0 ALUMINUM CHLORIDE (ALCL3) [PAL]	ALUMINUM CHLORIDE
96-10-6 ALUMINUM, CHLORODIETHYL- [PAL]	DIETHYLALUMINUM CHLORIDE
563-43-9 ALUMINUM, DICHLOROETHYL- [PAL]	ETHYLALUMINUM DICHLORIDEESTER
871-27-2 ALUMINUM, DIETHYLHYDRO- [PAL]	DIETHYLALUMINUM HYDRIDE
300-92-5 ALUMINUM DISTEARATE [CEX]	ALUMINUM, HYDROXYBIS(STEARATO)-
69011-71-8 ALUMINUM DROSS	ALUMINUM DROSS
7429-90-5 ALUMINUM, ELEMENTAL [WHM]	ALUMINUM
12003-41-7 ALUMINUM FERROSILICON	ALUMINUM FERROSILICON
73680-58-7 ALUMINUM FLUOROSULFATE, HYDRATE	ALUMINUM FLUOROSULFATE, HYDRATE

Chemical Name	Reference Name
7784-21-6 ALUMINUM HYDRIDE	ALUMINUM HYDRIDE
1191-15-7 ALUMINUM, HYDROBIS(2-METHYLPROPYL)- [PAL]	DIISOBUTYLALUMINUM HYDRIDE
2036-15-9 ALUMINUM, HYDRODIPROPYL- [PAL]	DIPROPYLALUMINUM HYDRIDE
21645-51-2 ALUMINUM HYDROXIDE	ALUMINUM HYDROXIDE
300-92-5 ALUMINUM, HYDROXYBIS(STEARATO)-	ALUMINUM, HYDROXYBIS(STEARATO)-
555-31-7 ALUMINUM ISO-PROPOXIDE	ALUMINUM ISO-PROPOXIDE
RR-00621-3 ALUMINUM MAGNESIUM PHOSPHIDE	ALUMINUM MAGNESIUM PHOSPHIDE
1327-43-1 ALUMINUM - MAGNESIUM SILICATE [PST]	MAGNESIUM ALUMINUM SILICATE
RR-01035-5 ALUMINUM - MAGNESIUM STEARATE	ALUMINUM - MAGNESIUM STEARATE
7429-90-5 ALUMINUM METAL [CEX, OS1, OS1]	ALUMINUM
7429-90-5 ALUMINUM METAL AND OXIDE [CAL]	ALUMINUMTOPHOS)
13473-90-0 ALUMINUM NITRATE	ALUMINUM NITRATE
6028-57-5 ALUMINUM OCTANOATE	ALUMINUM OCTANOATE
1302-74-5 ALUMINUM OXIDE [ALB, MAK]	CORUNDUM
1344-28-1 ALUMINUM OXIDE	ALUMINUM OXIDE
1344-28-1 ALUMINUM OXIDE [MAK, MAK, MAK]	ALUMINUM OXIDE
1344-28-1 ALUMINUM OXIDE (AL2O3) [PAL]	ALUMINUM OXIDE
1302-74-5 ALUMINUM OXIDE (ALUNDUM, CORUNDUM, ALUMINE) [QBC]	CORUNDUM
7784-30-7 ALUMINUM PHOSPHATE	ALUMINUM PHOSPHATE
20859-73-8 ALUMINUM PHOSPHIDE	ALUMINUM PHOSPHIDE
20859-73-8 ALUMINUM PHOSPHIDE (ALP) [PAL]	ALUMINUM PHOSPHIDE.ALPHA.-L-LYXO-HEXOPY-RANOSYL)OXY]-7,8,9,10-TETRAHYDRO-6,8,11-TRI-HYDROXY-8-(HYDROXYACETYL)-1-METHOXY-, HY-DROCHLORIDE, (8S-CIS)-
RR-00139-8 ALUMINUM PHOSPHIDE PESTICIDES	ALUMINUM PHOSPHIDE PESTICIDES
7429-90-5 ALUMINUM POWDER [PST]	ALUMINUM
RR-00141-2 ALUMINUM POWDER	ALUMINUM POWDER
RR-00546-9 ALUMINUM PRODUCTION	ALUMINUM PRODUCTION
RR-00019-1 ALUMINUM, PYRO POWDERS	ALUMINUM, PYRO POWDERS
RR-00019-1 ALUMINUM PYRO POWDERS [CAL, PAL, QBC]	ALUMINUM, PYRO POWDERS
61789-65-9 ALUMINUM RESINATE	ALUMINUM RESINATE
RR-00021-5 ALUMINUM SALTS, SOLUBLE [GBR]	ALUMINUM, SOLUBLE SALTS
1327-36-2 ALUMINUM SILICATE	ALUMINUM SILICATE
1335-30-4 ALUMINUM SILICATE, HYDRATE	ALUMINUM SILICATE, HYDRATE
1335-30-4 ALUMINUM SILICATE, HYDRATED [PST]	ALUMINUM SILICATE, HYDRATE
12042-55-6 ALUMINUM SILICON [HM1, HM1]	ALUMINUM SILICON (AL SI)
57485-31-1 ALUMINUM SILICON	ALUMINUM SILICON
12042-55-6 ALUMINUM SILICON (AL SI)	ALUMINUM SILICON (AL SI)
50810-25-8 ALUMINUM SILICON (AL SI5)	ALUMINUM SILICON (AL SI5)ETHYL)O-ETHYL ESTER
15096-52-3 ALUMINUM SODIUM FLUORIDE [WHM]	SODIUM ALUMINUM FLUORIDE
11138-49-1 ALUMINUM SODIUM OXIDE	ALUMINUM SODIUM OXIDE
RR-00021-5 ALUMINUM, SOLUBLE SALTS	ALUMINUM, SOLUBLE SALTS
RR-00021-5 ALUMINUM SOLUBLE SALTS [PAL]	ALUMINUM, SOLUBLE SALTS
7047-84-9 ALUMINUM STEARATE	ALUMINUM STEARATE
10043-01-3 ALUMINUM SULFATE	ALUMINUM SULFATE
12263-85-3 ALUMINUM, TRIBROMOTRIMETHYLDI- [PAL]	METHYLALUMINUM SESQUIBROMIDE
1116-70-7 ALUMINUM, TRIBUTYL- [PAL]	TRIBUTYL ALUMINUM
12075-68-2 ALUMINUM, TRICHLOROTRIETHYLDI- [PAL]	ETHYL ALUMINUM SESQUICHLORIDE

Chemical Name	Reference Name
12542-85-7 ALUMINUM, TRICHLOROTRIMETHYLDI- [PAL]	METHYL ALUMINUM SESQUICHLORIDE
97-93-8 ALUMINUM, TRIETHYL- [PAL]	TRIETHYLALUMINUM
75-24-1 ALUMINUM, TRIMETHYL- [PAL]	TRIMETHYLALUMINIUM
102-67-0 ALUMINUM, TRIPROPYL- [PAL]	TRIPROPYL ALUMINUM
100-99-2 ALUMINUM, TRIS(2-METHYLPROPYL)- [PAL]	TRIISOBUTYL ALUMINUM
637-12-7 ALUMINUM TRISTEARATE [CEX]	ALUMINUM STEARATE
RR-00021-5 ALUMINUM, WATER-SOLUBLE SALTS, N.O.S. [WHM]	ALUMINUM, SOLUBLE SALTS
RR-00020-4 ALUMINUM, WELDING FUMES	ALUMINUM, WELDING FUMES
RR-00020-4 ALUMINUM WELDING FUMES [PAL]	ALUMINUM, WELDING FUMES
1344-28-1 ALUNDUM (AL2O3) [BC1, BC1]	ALUMINUM OXIDE
915-67-3 AMARANTH [IAR]	C.I. ACID RED 27, TRISODIUM SALT
29492-78-2 AMERICIUM 237	AMERICIUM 237
18233-96-0 AMERICIUM 238	AMERICIUM 238
16652-10-1 AMERICIUM 239	AMERICIUM 239
15116-95-7 AMERICIUM 240	AMERICIUM 240
14596-10-2 AMERICIUM 241	AMERICIUM 241
13981-54-9 AMERICIUM 242	AMERICIUM 242
RR-00495-5 AMERICIUM 242M	AMERICIUM 242M
14993-75-0 AMERICIUM 243	AMERICIUM 243
15756-26-0 AMERICIUM 244	AMERICIUM 244
RR-00493-3 AMERICIUM 244M	AMERICIUM 244M
16415-43-3 AMERICIUM 245	AMERICIUM 245
15776-16-6 AMERICIUM 246	AMERICIUM 246
RR-00492-2 AMERICIUM 246M	AMERICIUM 246M
834-12-8 AMETRYN (N-ETHYL-N'-(1-METHYLETHYL)-6-(METHYLTHIO)-1,3,5,-TRIAZINE-2,4-DIAMINE)	AMETRYN (N-ETHYL-N'-(1-METHYLETHYL)-6-(METHYLTHIO)-1,3,5,-TR
68603-42-9 AMIDES, COCO, N,N-BIS(2-HYDROXYETHYL) [PST]	COCONUT DIETHANOLAMIDE
68425-44-5 AMIDES, COCO, N-(HYDROXYETHYL), ETHOXYLATED	AMIDES, COCO, N-(HYDROXYETHYL), ETHOXYLATED
RR-01209-9 AMIDINODITHIOPROPIONIC ACID HY-DROCHLORIDE	AMIDINODITHIOPROPIONIC ACID HYDROCHLORIDE
70904-61-9 AMIDOSULFOSUCCINATE	AMIDOSULFOSUCCINATE
39831-55-5 AMIKACIN SULFATE	AMIKACIN SULFATE
68155-33-9 AMINES, C14-18-ALKYL, ETHOXYLATED	AMINES, C14-18-ALKYL, ETHOXYLATED
61791-24-0 AMINES, SOYA ALKYL, ETHOXYLATED	AMINES, SOYA ALKYL, ETHOXYLATED
68153-99-1 AMINES, N-TALLOW ALKYLTRIMETHYLENEDI-, DIOLEATES	AMINES, N-TALLOW ALKYLTRIMETHYLENEDI-, DI-OLEATESLDEHYDE POLYMER
4657-93-6 5-AMINOACENAPHTHENE	5-AMINOACENAPHTHENE
99-03-6 M-AMINOACETOPHENONE	M-AMINOACETOPHENONE
99-92-3 P-AMINOACETOPHENONE	P-AMINOACETOPHENONE
134-50-9 9-AMINOACRIDINE HYDROCHLORIDE	9-AMINOACRIDINE HYDROCHLORIDE
RR-00241-5 AMINO ACRYLATE MONOMER	AMINO ACRYLATE MONOMER
82-45-1 1-AMINOANTHRAQUINONE	1-AMINOANTHRAQUINONE
117-79-3 2-AMINOANTHRAQUINONE	2-AMINOANTHRAQUINONE
117-79-3 2-AMINO-ANTHRAQUINONE [FLA]	2-AMINOANTHRAQUINONE
83-07-8 4-AMINOANTIPYRINE	4-AMINOANTIPYRINE
60-09-3 AMINOAZOBENZENE [WHM]	4-AMINOAZOBENZENE
60-09-3 4-AMINOAZOBENZENE	4-AMINOAZOBENZENE

Chemical Name	Reference Name
60-09-3 P-AMINOAZOBENZENE [C65]	4-AMINOAZOBENZENE
60-09-3 PARA-AMINOAZOBENZENE [CAL, IAR, MIN]	4-AMINOAZOBENZENE
101-50-8 4-AMINOAZOBENZENE-3,4'-DISULFONIC ACID	4-AMINOAZOBENZENE-3,4'-DISULFONIC ACID
97-56-3 O-AMINOAZOTOLUENE	O-AMINOAZOTOLUENE
97-56-3 ORTHO-AMINOAZOTOLUENE [C65, C65, IAR, MIN]	O-AMINOAZOTOLUENE
97-56-3 AMINOAZOTOLUENE [WHM]	O-AMINOAZOTOLUENE
98-44-2 2-AMINO-P-BENZENEDISULFONIC ACID	2-AMINO-P-BENZENEDISULFONIC ACID
1126-34-7 M-AMINOBENZENESULFONIC ACID, SODIUM SALT	M-AMINOBENZENESULFONIC ACID, SODIUM SALT
150-13-0 P-AMINOBENZOIC ACID	P-AMINOBENZOIC ACID
150-13-0 PARA-AMINOBENZOIC ACID [IAR, WEL]	P-AMINOBENZOIC ACID
1137-41-3 4-AMINOBENZOPHENONE	4-AMINOBENZOPHENONE
92-67-1 4-AMINOBIPHENYL	4-AMINOBIPHENYL
92-67-1 4-AMINOBIPHENYL (4-AMINODIPHENYL) [C65, C65]	4-AMINOBIPHENYL
92-67-1 4-AMINODIPHENYLE [QBC, QBC, QBC]	4-AMINOBIPHENYL
96-20-8 2-AMINO-1-BUTANOL	2-AMINO-1-BUTANOL
2032-59-9 AMINOCARB	AMINOCARB
95-85-2 2-AMINO-4-CHLOROPHENOL	2-AMINO-4-CHLOROPHENOL
81-49-2 1-AMINO-2,4-DIBROMOANTHRAQUINONE	1-AMINO-2,4-DIBROMOANTHRAQUINONE
140-80-7 2-AMINO-5-DIETHYL AMINOPENTANE [NJL]	2-AMINO-5-DIETHYLAMINOPENTANE
140-80-7 2-AMINO-5-DIETHYLAMINOPENTANE	2-AMINO-5-DIETHYLAMINOPENTANE
72243-90-4 3-[[4-AMINO-9,10-DIHYDRO-9,10-DIOXO-3-[SULFO-4-(1,1,3,3-TETRAMETHYLBUTYL)PHENOXY]-1-ANTHRACENYL]AMINO]-2,4,6-TRIMETHYL BENZENESULFONIC ACID DIS-ODIUM SALT	3-[[4-AMINO-9,10-DIHYDRO-9,10-DIOXO-3-[SULFO-4-(1,1,3,3-TETR
27277-00-5 2-AMINO-4,5-DIHYDRO-6-METHYL-4-PROPYL-5-TRIAZOLO-(1,5-C)-PYRAMIDIN-5-ONE	2-AMINO-4,5-DIHYDRO-6-METHYL-4-PROPYL-5-TRIAZOLO-(1,5-C)-PYR
109-55-7 1-AMINO-3-DIMETHYLAMINO PROPANE [OTS]	3-(DIMETHYLAMINO)-PROPYLAMINE
RR-00503-8 2-AMINO-2,3-DIMETHYLBUTYRONITRILE	2-AMINO-2,3-DIMETHYLBUTYRONITRILE
68298-46-4 7-AMINO-2,2-DIMETHYL-2,3-DIHYDROBENZO-FURAN	7-AMINO-2,2-DIMETHYL-2,3-DIHYDROBENZOFURAN
68808-54-8 3-AMINO-1,4-DIMETHYL-5H-PYRIDO (4,3-B) INDOLE ACETATE	3-AMINO-1,4-DIMETHYL-5H-PYRIDO (4,3-B) INDOLE ACETATE
92-67-1 4-AMINODIPHENYL [ALB, AUS, AUS, AUS, BC1, BC1, BC1, CAL, CEX, CEX, MIN, NHS, NHS, NHS, NIO, NIO, OS1, TLV, TLV, WHM]	4-AMINOBIPHENYL
101-54-2 P-AMINODIPHENYLAMINE	P-AMINODIPHENYLAMINE
67730-10-3 2-AMINODIPYRIDO(1,2-A:3',2'-D)IMIDA-ZOLE [MSL]	GLU-P-2
141-43-5 2-AMINOETHANOL [GBR, GBR, MAK, MAK, ONT, ONT, PST]	ETHANOLAMINE
141-43-5 2-AMINOETHANOL (ETHANOLAMINE) [QBC, QBC, ALB, ALB]	ETHANOLAMINE
17026-81-2 3-AMINO-4-ETHOXYACETANILIDE	3-AMINO-4-ETHOXYACETANILIDE
110-76-9 AMINOETHOXYETHANOL	AMINOETHOXYETHANOL
110-76-9 2-(2-AMINOETHOXY)ETHANOL [HM1, HM1]	AMINOETHOXYETHANOL
929-06-6 2-(2-AMINOETHOXY)ETHANOL	2-(2-AMINOETHOXY)ETHANOL
929-06-6 2-(2-AMINOETHOXY)-ETHANOL [TSC, TSC, TSC]	2-(2-AMINOETHOXY)ETHANOL

Chemical Name	Reference Name
118-28-5 5-AMINO-6-ETHOXY-2-NAPHTHALENESUL-FONIC ACID	5-AMINO-6-ETHOXY-2-NAPHTHALENESULFONIC ACID
132-32-1 3-AMINO-9-ETHYLCARBAZOLE	3-AMINO-9-ETHYLCARBAZOLE
6109-97-3 3-AMINO-9-ETHYLCARBAZOLE HCL	3-AMINO-9-ETHYLCARBAZOLE HCL
6109-97-3 3-AMINO-9-ETHYLCARBAZOLE HYDROCHLO-RIDE [C65, C65]	3-AMINO-9-ETHYLCARBAZOLE HCL
6109-97-3 3-AMINO-9-ETHYLCARBAZOLE, HYDROCHLO-RIDE [MSL]	3-AMINO-9-ETHYLCARBAZOLE HCL
111-41-1 AMINOETHYLETHANOLAMINE	AMINOETHYLETHANOLAMINE
111-41-1 (2-AMINOETHYL) ETHANOLAMINE [FLA, MSL, NFP, NFP]	AMINOETHYLETHANOLAMINE
111-41-1 AMINOETHYL ETHANOLAMINE [WHM]	AMINOETHYLETHANOLAMINE
RR-01642-2 AMINOETHYLETHYLENE UREA METHACRY-LAMIDE	AMINOETHYLETHYLENE UREA METHACRYLAMIDE
2038-03-1 4-(2-AMINOETHYL)-MORPHOLINE	4-(2-AMINOETHYL)-MORPHOLINE
140-31-8 1-(2-AMINOETHYL) PIPERAZINE	1-(2-AMINOETHYL) PIPERAZINE
140-31-8 N-AMINOETHYLPIPERAZINE [HM1, HM1, NJL, NJL]	1-(2-AMINOETHYL) PIPERAZINE
140-31-8 N-(2-AMINOETHYL)PIPERAZINE [WHM]	1-(2-AMINOETHYL) PIPERAZINE
125-84-8 AMINOGLUTETHIMIDE	AMINOGLUTETHIMIDE
RR-01471-1 AMINOGLYCOSIDES	AMINOGLYCOSIDES
548-93-6 2-AMINO-3-HYDROXYBENZOIC ACID	2-AMINO-3-HYDROXYBENZOIC ACID
17601-96-6 2-AMINO-4-[(2-HYDROXYETHYL)SULFONYL] PHENOL	2-AMINO-4-[(2-HYDROXYETHYL)SULFONYL]PHENOL
87-02-5 7-AMINO-4-HYDROXY-2-NAPHTHALENESUL-FONIC ACID	7-AMINO-4-HYDROXY-2-NAPHTHALENESULFONIC ACID
82-28-0 1-AMINO-2-METHYLANTHRAQUINONE	1-AMINO-2-METHYLANTHRAQUINONE
4475-95-0 2-AMINO-2-METHYLBUTANENITRILE	2-AMINO-2-METHYLBUTANENITRILE
67730-11-4 2-AMINO-6-METHYLDIPYRIDO[1,2-A:3',2'-D] IMIDAZOLE [MSL]	GLU-P-1
76180-96-6 2-AMINO-3-METHYLIMIDAZO(4,5-F)QUINO-LINE [MSL]	IQ
2763-96-4 5-(AMINOMETHYL)-3-ISOXAZOLOL [RCU, RCU]	MUSCIMOL
19355-69-2 2-AMINO-2-METHYLPROPANENITRILE	2-AMINO-2-METHYLPROPANENITRILE
124-68-5 2-AMINO-2-METHYL-1-PROPANOL	2-AMINO-2-METHYL-1-PROPANOL
2854-16-2 1-AMINO-2-METHYL-2-PROPANOL	1-AMINO-2-METHYL-2-PROPANOL
3731-51-9 2-AMINO-4-METHYLPYRIDINE	2-AMINO-4-METHYLPYRIDINE
68006-83-7 2-AMINO-3-METHYL-9H-PYRIDO(2,3,-B)IN-DOLE [MSL]	2-AMINO-3-METHYL-9H-PYRIDO(2,3,-B)INDOLE (METHYL A-ALPHA-C)
72254-58-1 3-AMINO-1-METHYL-5H-PYRIDO (4,3-B) INDOLE ACETATE	3-AMINO-1-METHYL-5H-PYRIDO (4,3-B) INDOLE AC-ETATE
68006-83-7 2-AMINO-3-METHYL-9H-PYRIDO(2,3,-B)INDOLE (METHYL A-ALPHA-C)	2-AMINO-3-METHYL-9H-PYRIDO(2,3,-B)INDOLE (METHYL A-ALPHA-C)E]
98-30-6 2-AMINO-4-(METHYLSULONYL)PHENOL	2-AMINO-4-(METHYLSULONYL)PHENOL
131-27-1 3-AMINO-1,5-NAPHTHALENEDISULFONIC ACID	3-AMINO-1,5-NAPHTHALENEDISULFONIC ACID
118-33-2 6-AMINO-1,3-NAPHTHALENEDISULFONIC ACID	6-AMINO-1,3-NAPHTHALENEDISULFONIC ACID
81-16-3 2-AMINO-1-NAPHTHALENESULFONIC ACID	2-AMINO-1-NAPHTHALENESULFONIC ACID
52218-35-6 2-[(6-AMINO-2-NAPHTHALENYL)SULFONYL] ETHANOL	2-[(6-AMINO-2-NAPHTHALENYL)SULFONYL] ETHANOL
117-62-4 2-AMINO-1,5-NAPHTHALINEDISULFONIC ACID	2-AMINO-1,5-NAPHTHALINEDISULFONIC ACID
99-56-9 2-AMINO-4-NITROANILINE	2-AMINO-4-NITROANILINE
5307-14-2 4-AMINO-2-NITROANILINE [WHM]	2-NITRO-P-PHENYLENEDIAMINE

Chemical Name	Reference Name
59716-87-9 2-AMINO-5-(5-NITRO-2-FURYL)-1,3,4-THIADIA-ZOLE	2-AMINO-5-(5-NITRO-2-FURYL)-1,3,4-THIADIAZOLE
99-57-0 2-AMINO-4-NITROPHENOL	2-AMINO-4-NITROPHENOL
119-34-6 4-AMINO-2-NITROPHENOL [CSR, CSR, IAR, MSL]	4-HYDROXY-3-NITROANILINE
121-88-0 2-AMINO-5-NITROPHENOL	2-AMINO-5-NITROPHENOL
121-66-4 2-AMINO-5-NITROTHIAZOLE	2-AMINO-5-NITROTHIAZOLE
95-55-6 O-AMINOPHENOL	O-AMINOPHENOL
123-30-8 P-AMINOPHENOL	P-AMINOPHENOL
123-30-8 AMINOPHENOL, P- [OTS]	P-AMINOPHENOL
591-27-5 M-AMINOPHENOL	M-AMINOPHENOL
27598-85-2 AMINOPHENOL	AMINOPHENOL
51-78-5 P-AMINOPHENOL HYDROCHLORIDE	P-AMINOPHENOL HYDROCHLORIDE
27598-85-2 AMINOPHENOLS [HM1, HM1, NJL]	AMINOPHENOL
RR-01747-0 AMINOPHENOL SULFONIC ACID	AMINOPHENOL SULFONIC ACID
130-17-6 2-(4-AMINOPHENYL)-6-METHYL-7-BENZOTHIA-ZOLE SULFONIC ACID	2-(4-AMINOPHENYL)-6-METHYL-7-BENZOTHIAZOLE SULFONIC ACID
2454-37-7 (M-AMINOPHENYL) METHYL CARBINOL [FLA]	BENZENEMETHANOL, 3-AMINO-.ALPHA.-METHYL-
2454-37-7 (M-AMINOPHENYL)METHYL CARBINOL [MSL]	BENZENEMETHANOL, 3-AMINO-.ALPHA.-METHYL-
2454-37-7 (M-AMINOPHENYL) METHYL CARBINOL [NFP, NFP]	BENZENEMETHANOL, 3-AMINO-.ALPHA.-METHYL-
599-61-1 3-AMINOPHENYL SULFONE	3-AMINOPHENYL SULFONE
5246-57-1 2-[(3-AMINOPHENYL)SULFONYL]ETHANOL	2-[(3-AMINOPHENYL)SULFONYL]ETHANOL
78-96-6 1-AMINO-2-PROPANOL [FLA, MSL, NFP, NFP]	ISOPROPANOLAMINE
156-87-6 3-AMINOPROPANOL [FLA, MSL, NFP, NFP]	PROPANOLAMINE
70-69-9 4-AMINOPROPIOPHENONE [MSL]	PROPIOPHENONE, 4-AMINO-
3312-60-5 N-(3-AMINOPROPYL) CYCLOHEXYLAMINE	N-(3-AMINOPROPYL) CYCLOHEXYLAMINE
4985-85-7 AMINOPROPYLDIETHANOLAMINE	AMINOPROPYLDIETHANOLAMINE
123-00-2 N-AMINOPROPYLMORPHOLINE	N-AMINOPROPYLMORPHOLINE
123-00-2 4-AMINOPROPYL MORPHOLINE [FLA]	N-AMINOPROPYLMORPHOLINE
123-00-2 N-(3-AMINOPROPYL) MORPHOLINE [MSL, NFP, NFP]	N-AMINOPROPYLMORPHOLINE
123-00-2 AMINOPROPYLMORPHOLINE [NJL, NJL]	N-AMINOPROPYLMORPHOLINE
54-62-6 AMINOPTERIN	AMINOPTERIN
1606-67-3 1-AMINO PYRENE	1-AMINO PYRENE
462-08-8 3-AMINO PYRIDINE	3-AMINO PYRIDINE
462-08-8 3-AMINOPYRIDINE [NJL]	3-AMINO PYRIDINE
504-24-5 4-AMINOPYRIDINE	4-AMINOPYRIDINE
504-29-0 2-AMINOPYRIDINE	2-AMINOPYRIDINE
RR-01354-7 AMINOPYRIDINES	AMINOPYRIDINES
26148-68-5 A-C (2-AMINO-9H-PYRIDO[2,3-B]INDOLE) [CAL]	A-ALPHA-C
26148-68-5 2-AMINO-9H-PYRIDO(2,3-B)INDOLE [MSL]	A-ALPHA-C
58-15-1 1-AMINOPYRINE	1-AMINOPYRINE
65-49-6 4-AMINOSALICYLIC ACID	4-AMINOSALICYLIC ACID
36768-62-4 4-AMINO-2,2,6,6-TETRAMETHYLPIPERIDINE	4-AMINO-2,2,6,6-TETRAMETHYLPIPERIDINE
79-19-6 1-AMINO-2-THIOUREA [WHM]	THIOSEMICARBAZIDE
61-82-5 3-AMINO-1H-1,2,4-TRIAZOLE [ONT]	AMITROLE
61-82-5 3-AMINO 1,2,4- TRIAZOLE (AMITROLE) [QBC, QBC]	AMITROLE
61-82-5 3-AMINO-1,2,4-TRIAZOLE (AMITROLE) [ALB, ALB]	AMITROLE

Chemical Name	Reference Name
1918-02-1 4-AMINO-3,5,6-TRICHLORO-2-PYRIDINECAR-BOXYLIC ACID [ONT, ONT]	PICLORAM
139-13-9 AMINOTRIETHANOIC ACID [PST]	NITRILOTRIACETIC ACID (NTA)
6419-19-8 AMINOTRI(METHYLENEPHOSPHONIC ACID)	AMINOTRI(METHYLENEPHOSPHONIC ACID)
2432-99-7 11-AMINOUNDECANOIC ACID	11-AMINOUNDECANOIC ACIDL]AZO][1,1'-BIPHENYL]-4-YL]AZO]-2-HYDROXY-, DISODIUM SALT
78-53-5 AMITON	AMITON
3734-97-2 AMITON OXALATE	AMITON OXALATE
33089-61-1 AMITRAZ	AMITRAZ
61-82-5 AMITROL [MAK]	AMITROLE
61-82-5 AMITROLE	AMITROLE
7664-41-7 AMMONIA	AMMONIA
7664-41-7 AMMONIA, ANHYDROUS [NFP, NFP]	AMMONIA
12125-02-9 AMMONIUM CHLORIDE FUME [MEX, MEX]	AMMONIUM CHLORIDE
631-61-8 AMMONIUM ACETATE	AMMONIUM ACETATE
7784-25-0 AMMONIUM ALUM	AMMONIUM ALUM
7784-44-3 AMMONIUM ARSENATE	AMMONIUM ARSENATE
12164-94-2 AMMONIUM AZIDE	AMMONIUM AZIDE
1863-63-4 AMMONIUM BENZOATE	AMMONIUM BENZOATE
1066-33-7 AMMONIUM BICARBONATE	AMMONIUM BICARBONATE
7789-09-5 AMMONIUM BICHROMATE	AMMONIUM BICHROMATE
1341-49-7 AMMONIUM BIFLUORIDE	AMMONIUM BIFLUORIDE
7803-63-6 AMMONIUM BISULFATE [PST]	AMMONIUM HYDROGEN SULFATE
10192-30-0 AMMONIUM BISULFITE	AMMONIUM BISULFITE
13843-59-9 AMMONIUM BROMATE	AMMONIUM BROMATE
12124-97-9 AMMONIUM BROMIDE	AMMONIUM BROMIDECYCLOPENTADIEN-1-YL]-
12124-97-9 AMMONIUM BROMIDE ((NH4)BR) [PAL]	AMMONIUM BROMIDECYCLOPENTADIEN-1-YL]-
68037-05-8 AMMONIUM C6-10-ALKYL POLYOXYETHY-LENE SULFATE	AMMONIUM C6-10-ALKYL POLYOXYETHYLENE SUL-FATE
1111-78-0 AMMONIUM CARBAMATE	AMMONIUM CARBAMATE
10361-29-2 AMMONIUM CARBONATE	AMMONIUM CARBONATE
9005-42-9 AMMONIUM CASEINATE	AMMONIUM CASEINATE
10192-29-7 AMMONIUM CHLORATE	AMMONIUM CHLORATE
12125-02-9 AMMONIUM CHLORIDE	AMMONIUM CHLORIDE
12125-02-9 AMMONIUM CHLORIDE ((NH4)CL) [PAL]	AMMONIUM CHLORIDE
12125-02-9 AMMONIUM CHLORIDE (FUME) [QBC, QBC]	AMMONIUM CHLORIDE
12125-02-9 AMMONIUM CHLORIDE FUME [CAL, ONT, ONT, OS1, OS1, TLV, TLV]	AMMONIUM CHLORIDE
19168-23-1 AMMONIUM CHLOROPALLADATE	AMMONIUM CHLOROPALLADATE
13820-40-1 AMMONIUM CHLOROPALLADITE	AMMONIUM CHLOROPALLADITE
16919-58-7 AMMONIUM CHLOROPLATINATE	AMMONIUM CHLOROPLATINATE
7788-98-9 AMMONIUM CHROMATE	AMMONIUM CHROMATE
3012-65-5 AMMONIUM CITRATE, DIBASIC	AMMONIUM CITRATE, DIBASIC
3012-65-5 AMMONIUM CITRATE DIBASIC [CAL, CWA]	AMMONIUM CITRATE, DIBASIC
7789-09-5 AMMONIUM DICHROMATE [FLA, HM1, HM1, HM1, MSL, NJL, WHM]	AMMONIUM BICHROMATE
129-17-9 AMMONIUM, (4-(ALPHA-(P-(DIETHYLAMINO) PHENYL)-2,4-DISULFOBENZYLIDENE)-2,5-CY-CLOHEXADIEN-1-YLIDENE)DIETHYL-, HY-DROXIDE, MONOSODIUM SALT	AMMONIUM, (4-(ALPHA-(P-(DIETHYLAMINO)PHENYL)-2,4-
94313-89-0 AMMONIUM DIISODECYL SULFOSUCCINATE	AMMONIUM DIISODECYL SULFOSUCCINATE

Chemical Name	Reference Name
29595-25-3 AMMONIUM DINITRO-O-CRESOLATE	AMMONIUM DINITRO-O-CRESOLATE
RR-01036-6 AMMONIUM DODECYL ALCOHOL POLY-OXYETHYLENE PHOSPHATE	AMMONIUM DODECYL ALCOHOL POLYOXYETHYLENE PHOSPHATE
32612-48-9 AMMONIUM DODECYL POLYOXYETHYLENE SULFATE [PST]	POLYETHYLENE GLYCOL MONOLAURYL ETHER SULFATE AMMONIUM SALT
25954-13-6 AMMONIUM ETHYL CARBAMOYLPHOSPHO-NATE	AMMONIUM ETHYL CARBAMOYLPHOSPHONATE
13826-83-0 AMMONIUM FLUOBORATE	AMMONIUM FLUOBORATE
12125-01-8 AMMONIUM FLUORIDE	AMMONIUM FLUORIDE
1341-49-7 AMMONIUM FLUORIDE ((NH4)(HF2)) [PAL]	AMMONIUM BIFLUORIDE
12125-01-8 AMMONIUM FLUORIDE ((NH4)F) [PAL]	AMMONIUM FLUORIDE
13826-83-0 AMMONIUM FLUOROBORATE [NJL]	AMMONIUM FLUOBORATE
16919-19-0 AMMONIUM FLUOROSILICATE [PST]	AMMONIUM SILICOFLUORIDE
1309-32-6 AMMONIUM FLUOSILICATE	AMMONIUM FLUOSILICATE
540-69-2 AMMONIUM FORMATE	AMMONIUM FORMATE
RR-01391-2 AMMONIUM FULMINATE	AMMONIUM FULMINATE
7783-20-2 AMMONIUM HYDROGEN SULFATE [WHM]	AMMONIUM SULFATE
7803-63-6 AMMONIUM HYDROGEN SULFATE	AMMONIUM HYDROGEN SULFATE
12124-99-1 AMMONIUM HYDROSULFIDE SOLUTION	AMMONIUM HYDROSULFIDE SOLUTION
1336-21-6 AMMONIUM HYDROXIDE	AMMONIUM HYDROXIDE
1336-21-6 AMMONIUM HYDROXIDE ((NH4)(OH)) [PAL]	AMMONIUM HYDROXIDE
515-98-0 AMMONIUM LACTATE	AMMONIUM LACTATE
2235-54-3 AMMONIUM LAURYL SULFATE	AMMONIUM LAURYL SULFATE
7803-55-6 AMMONIUM METAVANADATE [HM1, HM1, HM1, NJL, WHM]	AMMONIUM VANADATE
13106-76-8 AMMONIUM MOLYBDATE	AMMONIUM MOLYBDATE
6484-52-2 AMMONIUM NITRATE	AMMONIUM NITRATE
RR-00142-3 AMMONIUM NITRATE FERTILIZERS	AMMONIUM NITRATE FERTILIZERS
RR-00148-9 AMMONIUM NITRATE-FUEL OIL MIXTURES	AMMONIUM NITRATE-FUEL OIL MIXTURES
57608-40-9 AMMONIUM NITRATE PHOSPHATE	AMMONIUM NITRATE PHOSPHATE
RR-00149-0 AMMONIUM NITRATE-SULFATE MIXTURES	AMMONIUM NITRATE-SULFATE MIXTURES
9051-57-4 AMMONIUM NONYLPHENYL POLYOXYETHY-LENE SULFATE	AMMONIUM NONYLPHENYL POLYOXYETHYLENE SULFATE
544-60-5 AMMONIUM OLEATE	AMMONIUM OLEATE
1113-38-8 AMMONIUM OXALATE	AMMONIUM OXALATE
14258-49-2 AMMONIUM OXALATE [CAL, EPA]	OXALIC ACID, AMMONIUM SALT
6009-70-7 AMMONIUM OXALATE, MONOHYDRATE	AMMONIUM OXALATE, MONOHYDRATE
5972-73-6 AMMONIUM OXALATE, UNSPECIFIED HY-DRATE	AMMONIUM OXALATE, UNSPECIFIED HYDRATE
7790-98-9 AMMONIUM PERCHLORATE	AMMONIUM PERCHLORATE
3825-26-1 AMMONIUM PERFLUOROOCTANOATE	AMMONIUM PERFLUOROOCTANOATE
7787-36-2 AMMONIUM PERMANGANATE [OS3]	BARIUM PERMANGANATE
13446-10-1 AMMONIUM PERMANGANATE	AMMONIUM PERMANGANATE
7727-54-0 AMMONIUM PERSULFATE	AMMONIUM PERSULFATE
7722-76-1 AMMONIUM PHOSPHATE (MONOBASIC) [PST]	MONOAMMONIUM PHOSPHATE
131-74-8 AMMONIUM PICRATE	AMMONIUM PICRATE
12259-92-6 AMMONIUM POLYSULFIDE	AMMONIUM POLYSULFIDE
RR-00150-3 AMMONIUM POLYVANADATE	AMMONIUM POLYVANADATE
16919-19-0 AMMONIUM SILICOFLUORIDE	AMMONIUM SILICOFLUORIDE
1002-89-7 AMMONIUM STEARATE	AMMONIUM STEARATE
7773-06-0 AMMONIUM SULFAMATE	AMMONIUM SULFAMATE

Chemical Name	Reference Name
7773-06-0 AMMONIUM SULFAMATE (AMMATE) [MSL]	AMMONIUM SULFAMATE
7773-06-0 AMMONIUM SULFAMATE (AMATE) [QBC]	AMMONIUM SULFAMATE
7773-06-0 AMMONIUM SULFAMATE (AMMATE) [FLA, BC1, BC1]	AMMONIUM SULFAMATE
7783-20-2 AMMONIUM SULFATE	AMMONIUM SULFATE
12135-76-1 AMMONIUM SULFIDE	AMMONIUM SULFIDE
12135-76-1 AMMONIUM SULFIDE ((NH4)2S) [PAL]	AMMONIUM SULFIDE
10196-04-0 AMMONIUM SULFITE	AMMONIUM SULFITE
7773-06-0 AMMONIUM SULPHAMATE [AUS]	AMMONIUM SULFAMATE
7773-06-0 AMMONIUM SULPHAMIDATE [GBR, GBR]	AMMONIUM SULFAMATE
7783-20-2 AMMONIUM SULPHATE, SOLUTION [NJL]	AMMONIUM SULFATE
3164-29-2 AMMONIUM TARTRATE [CAL, EPA, NJL]	AMMONIUM TARTRATE, DIAMMONIUM SALT
14307-43-8 AMMONIUM TARTRATE	AMMONIUM TARTRATE
3164-29-2 AMMONIUM TARTRATE, DIAMMONIUM SALT	AMMONIUM TARTRATE, DIAMMONIUM SALT
13453-06-0 AMMONIUM TELLURATE	AMMONIUM TELLURATE
1762-95-4 AMMONIUM THIOCYANATE	AMMONIUM THIOCYANATE
7783-18-8 AMMONIUM THIOSULFATE	AMMONIUM THIOSULFATE
11115-67-6 AMMONIUM VANADATE	AMMONIUM VANADATE
52628-25-8 AMMONIUM ZINC CHLORIDE [PAL]	ZINC AMMONIUM CHLORIDE [VAN]
7631-86-9 AMORPHOUS SILICA [FLA, MSL]	SILICA, AMORPHOUS
12172-73-5 AMOSITE [QBC, QBC, QBC]	ASBESTOS, AMOSITE
12172-73-5 AMOSITE ASBESTOS [MSL, WHM]	ASBESTOS, AMOSITE
300-62-9 AMPHETAMINE	AMPHETAMINE
60-13-9 DL-AMPHETAMINE SULFATE	DL-AMPHETAMINE SULFATE
69-53-4 AMPICILLIN	AMPICILLIN
7177-48-2 AMPICILLIN TRIHYDRATE	AMPICILLIN TRIHYDRATE
628-63-7 AMYL ACETATE [PST]	N-AMYL ACETATE
123-92-2 ISO-AMYL ACETATE [CWA]	ISOAMYL ACETATE
625-16-1 TERT-AMYL ACETATE	TERT-AMYL ACETATE
626-38-0 SEC-AMYL ACETATE	SEC-AMYL ACETATE
628-63-7 N-AMYL ACETATE	N-AMYL ACETATE
628-63-7 AMYL ACETATE [EPA, HM1, HM1, HM1, MAK, NFP, NFP]	N-AMYL ACETATE
628-63-7 AMYL ACETATE, N- [OTS]	N-AMYL ACETATE
12789-46-7 AMYL ACID PHOSPHATE	AMYL ACID PHOSPHATE
71-41-0 AMYL ALCOHOL	AMYL ALCOHOL
71-41-0 N-AMYL ALCOHOL [WHM]	AMYL ALCOHOL
6032-29-7 SEC-AMYL ALCOHOL	SEC-AMYL ALCOHOL
RR-00108-1 AMYL ALCOHOLS	AMYL ALCOHOLS
6032-29-7 SEC-AMYL ALCOHOL [NFP, NFP]	SEC-AMYL ALCOHOL
110-58-7 AMYL AMINE	AMYL AMINE
110-58-7 AMYLAMINE [FLA, MSL, NFP, NFP]	AMYL AMINE
625-30-9 SEC-AMYLAMINE	SEC-AMYLAMINE
2049-92-5 P-TERT-AMYLANILINE	P-TERT-AMYLANILINE
538-68-1 AMYLBENZENE [NFP, NFP]	BENZENE, PENTYL-
110-53-2 AMYL BROMIDE [NFP, NFP]	1-BROMOPENTANE
540-18-1 AMYL BUTYRATE	AMYL BUTYRATE
RR-00145-6 AMYL BUTYRATES	AMYL BUTYRATES
543-59-9 AMYL CHLORIDE	AMYL CHLORIDE

Chemical Name	Reference Name
594-36-5 TERT-AMYL CHLORIDE	TERT-AMYL CHLORIDE
RR-00063-5 AMYL CHLORIDES [HM1, HM1]	AMYL CHLORIDES (MIXED)
RR-00063-5 AMYL CHLORIDES (MIXED)	AMYL CHLORIDES (MIXED)
4292-92-6 AMYLCYCLOHEXANE	AMYLCYCLOHEXANE
627-20-3 BETA-AMYLENE-CIS	BETA-AMYLENE-CIS
646-04-8 BETA-AMYLENE-TRANS	BETA-AMYLENE-TRANS
25377-72-4 N-AMYLENE [HM1, HM1]	AMYLENE, NORMAL
25377-72-4 AMYLENE, NORMAL	AMYLENE, NORMAL
693-65-2 AMYL ETHER	AMYL ETHER
638-49-3 AMYL FORMATE	AMYL FORMATE
RR-00144-5 AMYL FORMATES	AMYL FORMATES
3425-61-4 TERT-AMYL HYDROPEROXIDE	TERT-AMYL HYDROPEROXIDE
RR-00620-2 AMYL LACTATE	AMYL LACTATE
RR-00807-1 AMYL LAURATE	AMYL LAURATE
RR-00513-0 AMYL MALEATE	AMYL MALEATE
110-66-7 AMYL MERCAPTAN	AMYL MERCAPTAN
RR-01196-1 AMYL MERCAPTANS	AMYL MERCAPTANS
994-05-8 TERT-AMYL METHYL ETHER	TERT-AMYL METHYL ETHER
110-43-0 AMYL METHYL KETONE [HM1, HM1]	METHYL N-AMYL KETONE
RR-00336-1 AMYL NAPHTHALENE	AMYL NAPHTHALENE
1002-16-0 AMYL NITRATE	AMYL NITRATE
110-46-3 AMYL NITRITE	AMYL NITRITE
9005-84-9 AMYLODEXTRIN	AMYLODEXTRIN
RR-00601-9 AMYL OLEATE	AMYL OLEATE
RR-00824-2 AMYL OXALATE	AMYL OXALATE
RR-00924-5 TERT-AMYL PEROXY ALKYLENE ESTER	TERT-AMYL PEROXY ALKYLENE ESTER
RR-01749-2 TERT-AMYL PEROXY-2-EHTYLHEXANOATE	TERT-AMYL PEROXY-2-EHTYLHEXANOATE
RR-01750-5 TERT-AMYLPEROXY-3,5,5-TRIMETHYLHEX-ANOATE	TERT-AMYLPEROXY-3,5,5-TRIMETHYLHEXANOATE
136-81-2 O-AMYL PHENOL	O-AMYL PHENOLSALT
25735-67-5 P-SEC-AMYLPHENOL	P-SEC-AMYLPHENOL
6382-07-6 2-(P-TERT-AMYLPHENOXY) ETHANOL	2-(P-TERT-AMYLPHENOXY) ETHANOL
RR-00762-5 2-(P-TERT-AMYLPHENOXY) ETHYL LAURATE	2-(P-TERT-AMYLPHENOXY) ETHYL LAURATE
5137-52-0 P-TERT-AMYLPHENYL ACETATE	P-TERT-AMYLPHENYL ACETATE
RR-00876-4 P-TERT-AMYLPHENYL BUTYL ETHER	P-TERT-AMYLPHENYL BUTYL ETHER
2050-04-6 AMYL PHENYL ETHER	AMYL PHENYL ETHER
1320-05-4 P-TERT-AMYLPHENYL METHYL ETHER	P-TERT-AMYLPHENYL METHYL ETHER
624-54-4 AMYL PROPIONATE	AMYL PROPIONATE
2050-08-0 AMYL SALICYLATE	AMYL SALICYLATE
6382-13-4 AMYL STEARATE	AMYL STEARATE
RR-00083-9 AMYL SULFIDES, MIXED	AMYL SULFIDES, MIXED
1320-01-0 AMYL TOLUENE [FLA, MSL, NFP, NFP]	BENZENE, METHYLPENTYL-
107-72-2 AMYL TRICHLOROSILANE	AMYL TRICHLOROSILANE
107-72-2 AMYLTRICHLOROSILANE [WHM]	AMYL TRICHLOROSILANE
1320-21-4 AMYL XYLYL ETHER [MSL, NFP, NFP]	BENZENE, DIMETHYL(PENTYLOXY)-
RR-00057-7 ANABOLIC STEROIDS [C65, C65]	ANDROGENIC (ANABOLIC) STEROIDS
RR-00057-7 ANABOLIC STEROIDS (ANDROGENIC STEROIDS) [CAL]	ANDROGENIC (ANABOLIC) STEROIDS
RR-01525-8 ANAESTHETICS, VOLATILE	ANAESTHETICS, VOLATILE

Chemical Name	Reference Name
RR-00055-5 ANALGESIC MIXTURE CONTAINING PHENACETIN [MIN]	ANALGESIC MIXTURES CONTAINING PHENACETIN
RR-00055-5 ANALGESIC MIXTURES CONTAINING PHENACETIN	ANALGESIC MIXTURES CONTAINING PHENACETIN
1317-70-0 ANATASE (TIO2)	ANATASE (TIO2)
RR-00057-7 ANDROGENIC (ANABOLIC) STEROIDS	ANDROGENIC (ANABOLIC) STEROIDS
434-07-1 ANDROSTAN-3-ONE, 17-HYDROXY-2-(HYDROXYMETHYLENE)-17-METHYL-,(5.ALPHA., 17.BETA.)- [PAL, PAL]	OXYMETHOLONEMONOHYDROCHLORIDE
58-22-0 ANDROST-4-EN-3-ONE, 17-HYDROXY-, (17.BETA.)- [PAL, PAL]	TESTOSTERONE
104-46-1 P-ANETHOLE [PST]	BENZENE, 1-METHOXY-4-(1-PROPENYL)-
RR-01526-9 ANGELICIN PLUS ULTRAVIOLET A RADIATION	ANGELICIN PLUS ULTRAVIOLET A RADIATION
RR-01472-2 ANGIOTENSIN CONVERTING ENZYME (ACE)	ANGIOTENSIN CONVERTING ENZYME (ACE)
RR-01472-2 ANGIOTENSIN CONVERTING ENZYME (ACE) INHIBITORS [C65]	ANGIOTENSIN CONVERTING ENZYME (ACE)
101-05-3 ANILAZINE	ANILAZINE
101-05-3 ANILAZINE [4,6-DICHLORO-N-(2-CHLOROPHENYL)-1,3,5-TRIAZIN-2-AMINE] [313]	ANILAZINE
62-53-3 ANILINE	ANILINE
62-53-3 ANILINE AND HOMOLOGUES [ALB, ALB, ALB, MEX, MEX, MEX]	ANILINE
RR-00282-4 ANILINE AND CHLORO-, BROMO-, AND/OR NITROANILINES	ANILINE AND CHLORO-, BROMO-, AND/OR NITROANILINES
62-53-3 ANILINE AND HOMOLOGUES [ONT, ONT, ONT, ONT]	ANILINE
62-53-3 ANILINE AND HOMOLOGS [NIO, NIO, NIO, NIO, NIO, OS1, OS1, OS1, OS1]	ANILINE
62-53-3 ANILINE AND HOMOLOGUES [AUS, AUS, MIN, TLV, TLV]	ANILINE
142-04-1 ANILINE HYDROCHLORIDE	ANILINE HYDROCHLORIDE
104-94-9 ANILINE, 4-METHOXY- [OTS]	P-ANISIDINE
88-05-1 ANILINE, 2,4,6-TRIMETHYL-	ANILINE, 2,4,6-TRIMETHYL-
122-98-5 2-ANILINOETHANOL [NFP, NFP]	N-PHENYL ETHANOLAMINE
122-37-2 4-ANILINOPHENOL	4-ANILINOPHENOL
123-11-5 P-ANISALDEHYDE	P-ANISALDEHYDE
135-02-4 O-ANISALDEHYDE	O-ANISALDEHYDE
90-04-0 O-ANISIDINE	O-ANISIDINE
90-04-0 ORTHO-ANISIDINE [C65, C65, IAR]	O-ANISIDINE
104-94-9 P-ANISIDINE	P-ANISIDINE
104-94-9 PARA-ANISIDINE [IAR]	P-ANISIDINE
536-90-3 M-ANISIDINE	M-ANISIDINE
29191-52-4 ANISIDINE [FLA]	ANISIDINE (O-, P- ISOMERS)
29191-52-4 ANISIDINE (ORTHO AND PARA ISOMERS) [CEX, CEX]	ANISIDINE (O-, P- ISOMERS)
64070-14-0 O-ANISIDINE ANTIMONYL TARTRATE	O-ANISIDINE ANTIMONYL TARTRATE
20265-97-8 P-ANISIDINE HYDROCHLORIDE	P-ANISIDINE HYDROCHLORIDE
134-29-2 O-ANISIDINE HYDROCHLORIDE	O-ANISIDINE HYDROCHLORIDE
134-29-2 ORTHO-ANISIDINE HYDROCHLORIDE [C65, C65]	O-ANISIDINE HYDROCHLORIDE
29191-52-4 ANISIDINE (O- AND P- ISOMERS) [MSL]	ANISIDINE (O-, P- ISOMERS)
29191-52-4 ANISIDINE (O-, P- ISOMERS)	ANISIDINE (O-, P- ISOMERS)

Chemical Name	Reference Name
29191-52-4 ANISIDINE (O-P ISOMERS) [MIN]	ANISIDINE (O-, P- ISOMERS)
29191-52-4 ANISIDINES [HM1, HM1]	ANISIDINE (O-, P- ISOMERS)
RR-01197-2 ORTHO-ANISIDINES	ORTHO-ANISIDINES
29191-52-4 ANISIDINE (SUM OF O-, P- ISOMERS) [ONT, ONT]	ANISIDINE (O-, P- ISOMERS)
117-37-3 ANISINDIONE	ANISINDIONE
100-66-3 ANISOLE	ANISOLE
1300-64-7 ANISOYL CHLORIDE	ANISOYL CHLORIDE
191-26-4 ANTHANTHRENE	ANTHANTHRENE
77536-67-5 ANTHOPHYLITE [QBC, QBC]	ASBESTOS, ANTHOPHYLITE
77536-67-5 ANTHOPHYLITE ASBESTOS [IAR]	ASBESTOS, ANTHOPHYLITE
17068-78-9 ANTHOPHYLLITE [PAL, PAL]	ANTHOPHYLLITE, NON-ASBESTIFORM
77536-67-5 ANTHOPHYLLITE ASBESTOS [MSL]	ASBESTOS, ANTHOPHYLITE
17068-78-9 ANTHOPHYLLITE, NON-ASBESTIFORM	ANTHOPHYLLITE, NON-ASBESTIFORM
120-20-7 ANTHRACENE	ANTHRACENE
117-79-3 9,10-ANTHRACENEDIONE, 2-AMINO- [PAL, PAL]	2-AMINOANTHRAQUINONE
17418-58-5 9,10-ANTHRACENEDIONE, 1-AMINO-4-HYDROXY-2-PHENOXY-	9,10-ANTHRACENEDIONE, 1-AMINO-4-HYDROXY-2-PHENOXY-
82-28-0 9,10-ANTHRACENEDIONE, 1-AMINO-2-METHYL- [PAL, PAL]	1-AMINO-2-METHYLANTHRAQUINONE
12217-79-7 9,10-ANTHRACENEDIONE, 1,5-DIAMINOCHLORO-4,8-DIHYDROXY-	9,10-ANTHRACENEDIONE, 1,5-DIAMINOCHLORO-4,8-DIHYDROXY-
129-15-7 9,10-ANTHRACENEDIONE, 2-METHYL-1-NITRO- [PAL, PAL]	2-METHYL-1-NITROANTHRAQUINONE
128-86-9 2,6-ANTHRACENEDISULFONIC ACID, 4,8-DIAMINO-9,10-DIHYDRO-1,5-DIHYDROXY-9,10-DIOXO-	2,6-ANTHRACENEDISULFONIC ACID, 4,8-DIAMINO-9,10-DIHYDRO-1,5-
2861-02-1 2,6-ANTHRACENEDISULFONIC ACID, 4,8-DIAMINO-9,10-DIHYDRO-1,5-DIHYDROXY-9,10-DIOXO-, DISODIUM SALT [TSC, TSC, TSC]	ANTHRAQUINONE BLUE
RR-00781-8 ANTHRACENE OILS [MIN]	ANTHRACENE OILS (COAL TAR DERIVED PRODUCT)
RR-00781-8 ANTHRACENE OILS (COAL TAR DERIVED PRODUCT)	ANTHRACENE OILS (COAL TAR DERIVED PRODUCT)
6247-34-3 2-ANTHRACENESULFONIC ACID, 4-((4-(ACETYLAMINO)PHENYL)AMINO)-1-AMINO-9,10-DIHYDRO-9,10-DIOXO-	2-ANTHRACENESULFONIC ACID, 4-((4-(ACETYLAMINO)PHENYL)AMINO)-
6424-85-7 2-ANTHRACENESULFONIC ACID, 4-[[4-(ACETYLAMINO)PHENYL]AMINO]-1-AMINO-9,10-DIHYDRO-9,10-DIOXO-, MONOSODIUM SALT	2-ANTHRACENESULFONIC ACID, 4-[[4-(ACETYLAMINO)PHENYL]AMINO]-
118-92-3 O-ANTHRANILIC ACID	O-ANTHRANILIC ACID
118-92-3 ANTHRANILIC ACID [FDA, IAR]	O-ANTHRANILIC ACID
87-25-2 ANTHRANILIC ACID, ETHYL ESTER	ANTHRANILIC ACID, ETHYL ESTER
RR-01772-1 ANTHRANILIC ACID SALTS	ANTHRANILIC ACID SALTS
84-65-1 ANTHRAQUINONE	ANTHRAQUINONE
2861-02-1 ANTHRAQUINONE BLUE	ANTHRAQUINONE BLUE
131-08-8 2-ANTHRAQUINONE SULFONIC ACID, SODIUM SALT	2-ANTHRAQUINONE SULFONIC ACID, SODIUM SALT
28300-74-5 ANTIMONATE(2-), BIS[.MU.-[2,3-DIHYDROXYBUTANEDIOATO(4-)-O1,O2:O3,O4]DI-, DIPOTASSIUM, TRIHYDRATE, STEREOISOMER [PAL]	ANTIMONY POTASSIUM TARTRATE
7440-36-0 ANTIMONY	ANTIMONY
17620-10-9 ANTIMONY 115	ANTIMONY 115
15755-27-8 ANTIMONY 116	ANTIMONY 116

Chemical Name	Reference Name
RR-00488-6 ANTIMONY 116M	ANTIMONY 116M
15755-18-7 ANTIMONY 117	ANTIMONY 117
RR-00487-5 ANTIMONY 118M	ANTIMONY 118M
14914-68-2 ANTIMONY 119	ANTIMONY 119
14391-68-5 ANTIMONY 120	ANTIMONY 120
14374-79-9 ANTIMONY 122	ANTIMONY 122
14683-10-4 ANTIMONY 124	ANTIMONY 124
RR-00486-4 ANTIMONY 124M	ANTIMONY 124M
14234-35-6 ANTIMONY 125	ANTIMONY 125
15756-32-8 ANTIMONY 126	ANTIMONY 126
RR-00484-2 ANTIMONY 126M	ANTIMONY 126M
13968-50-8 ANTIMONY 127	ANTIMONY 127
15756-34-0 ANTIMONY 128	ANTIMONY 128
14331-88-5 ANTIMONY 129	ANTIMONY 129
15756-35-1 ANTIMONY 130	ANTIMONY 130
15756-29-3 ANTIMONY 131	ANTIMONY 131
7440-36-0 ANTIMONY AND COMPOUNDS [ALB, ALB, MEX]	ANTIMONY
1309-64-4 ANTIMONY TRIOXIDE, HANDLING AND USE [ALB, ALB]	ANTIMONY TRIOXIDE
6923-52-0 ANTIMONY(III) ACETATE	ANTIMONY(III) ACETATE
7440-36-0 ANTIMONY AND COMPOUNDS [AUS, BC1]	ANTIMONY
7440-36-0 ANITMONY AND COMPOUNDS [MIN]	ANTIMONY
RR-00585-6 ANTIMONY AND COMPOUNDS [NIO, NIO, NIO, OS1, OS1, TLV, CEX, GBR]	ANTIMONY COMPOUNDS
7647-18-9 ANTIMONY CHLORIDE (SBCL5) [PAL]	ANTIMONY PENTACHLORIDE
RR-00585-6 ANTIMONY COMPOUNDS	ANTIMONY COMPOUNDS
RR-00585-6 ANTIMONY COMPOUNDS [QBC]	ANTIMONY COMPOUNDS
RR-00554-9 ANTIMONY COMPOUNDS, INORGANIC, N.O.S. [HM1, HM1]	ANTIMONY, INORGANIC COMPOUNDS
RR-00585-6 ANTIMONY COMPOUNDS, N.O.S. [RCU, WHM]	ANTIMONY COMPOUNDS
7440-36-0 ANTIMONY, ELEMENTAL [WHM]	ANTIMONY
7783-70-2 ANTIMONY FLUORIDE (SBF5) [PAL]	ANTIMONY PENTAFLUORIDE
7803-52-3 ANTIMONY HYDRIDE [ONT]	STIBINE
RR-00554-9 ANTIMONY, INORGANIC COMPOUNDS	ANTIMONY, INORGANIC COMPOUNDS
58164-88-8 ANTIMONY LACTATE	ANTIMONY LACTATE
1327-33-9 ANTIMONY OXIDE	ANTIMONY OXIDE
1309-64-4 ANTIMONY OXIDE (ANTIMONY TRIOXIDE) [C65]	ANTIMONY TRIOXIDE
1309-64-4 ANTIMONY OXIDE (SB2O3) [PAL]	ANTIMONY TRIOXIDE
7647-18-9 ANTIMONY PENTACHLORIDE	ANTIMONY PENTACHLORIDE
7783-70-2 ANTIMONY(V) PENTAFLUORIDE [WHM]	ANTIMONY PENTAFLUORIDE
7783-70-2 ANTIMONY PENTAFLUORIDE	ANTIMONY PENTAFLUORIDE
1315-04-4 ANTIMONY PENTASULFIDE	ANTIMONY PENTASULFIDE
28300-74-5 ANTIMONY POTASSIUM TARTRATE	ANTIMONY POTASSIUM TARTRATE
1345-04-6 ANTIMONY SULFIDE	ANTIMONY SULFIDE
1315-04-4 ANTIMONY SULFIDE (SB2S5) [PAL]	ANTIMONY PENTASULFIDE
7789-61-9 ANTIMONY TRIBROMIDE	ANTIMONY TRIBROMIDE
10025-91-9 ANTIMONY TRICHLORIDE	ANTIMONY TRICHLORIDE
10025-91-9 ANTIMONY(III) TRICHLORIDE [WHM]	ANTIMONY TRICHLORIDE

Chemical Name	Reference Name
7783-56-4 ANTIMONY TRIFLUORIDE	ANTIMONY TRIFLUORIDE
1309-64-4 ANTIMONY TRIOXIDE	ANTIMONY TRIOXIDE
1309-64-4 ANTIMONY TRIOXIDE [QBC, QBC]	ANTIMONY TRIOXIDE
RR-01639-7 ANTIMONY TRIOXIDE PRODUCTION	ANTIMONY TRIOXIDE PRODUCTION
1345-04-6 ANTIMONY TRISULFIDE [TSC, TSC, TSC]	ANTIMONY SULFIDE
1397-94-0 ANTIMYCIN [NJL]	ANTIMYCIN A
1397-94-0 ANTIMYCIN A	ANTIMYCIN A
86-88-4 ANTU	ANTU
86-88-4 ANTU (ALPHA NAPHTHYLTHIOUREA) [OS1, OS1]	ANTU
86-88-4 ANTU (ALPHA-NAPHTHYL THIOUREA) [BC1, BC1, ALB, ALB]	ANTU
52-46-0 APHOLATE	APHOLATE
RR-01037-7 APPLE POMACE	APPLE POMACE
26125-61-1 P-ARAMIDE	P-ARAMIDE
140-57-8 ARAMITE	ARAMITE
1119-34-2 L-ARGININE HCL	L-ARGININE HCL
7440-37-1 ARGON	ARGON
25729-41-3 ARGON 39	ARGON 39
14163-25-8 ARGON 41	ARGON 41
37324-23-5 AROCHLOR 1262	AROCHLOR 1262
11100-14-4 AROCHLOR 1268	AROCHLOR 1268
11120-29-9 AROCHLOR 4465	AROCHLOR 4465
12674-11-2 AROCLOR 1016	AROCLOR 1016
11104-28-2 AROCLOR 1221	AROCLOR 1221
11141-16-5 AROCLOR 1232	AROCLOR 1232
71328-89-7 AROCLOR 1240	AROCLOR 1240
53469-21-9 AROCLOR 1242	AROCLOR 1242
12672-29-6 AROCLOR 1248	AROCLOR 1248
11097-69-1 AROCLOR 1254 [ATS, CSR, CSR, EP2, EPA, NTP, PAL, PAL, RCA]	CHLORODIPHENYL (54% CHLORINE)
11097-69-1 AROCLOR 1254 (PCB) [WHM]	CHLORODIPHENYL (54% CHLORINE)
11096-82-5 AROCLOR 1260	AROCLOR 1260
RR-01009-3 AROMATIC AMINE COMPOUND	AROMATIC AMINE COMPOUND
RR-01686-4 AROMATIC AMINE POLYOLS	AROMATIC AMINE POLYOLS
RR-01668-2 AROMATIC AMINO ETHER	AROMATIC AMINO ETHER
RR-00286-8 AROMATIC C9 FRACTION FROM PETROLEUM REFINING	AROMATIC C9 FRACTION FROM PETROLEUM REFINING
RR-00164-9 AROMATIC DIAMINES	AROMATIC DIAMINESDIGLYCERYLETHER AND ALKYLENEPOLYOLS POLYGLYCIDYLETHERS
RR-01766-3 C9 AROMATIC HYDROCARBON FRACTION	C9 AROMATIC HYDROCARBON FRACTION
RR-01010-6 AROMATIC NITRO COMPOUND	AROMATIC NITRO COMPOUND
68477-30-5 AROMATIC PETROLEUM DERIVATIVE SOLVENT [PST]	DISTILLATES (PETROLEUM), CATALYTIC REFORMER FRACTIONATOR
68477-31-6 AROMATIC PETROLEUM HYDROCARBON SOLVENT [PST]	PETROLEUM DISTILLATES, CATALYTIC, REFORMER FRACTIONATOR RESI
RR-01660-4 AROMATIC SULFONIC ACID COMPOUND WITH AMINE	AROMATIC SULFONIC ACID COMPOUND WITH AMINE
RR-01200-0 ARSENATES, N.O.S.	ARSENATES, N.O.S.
7440-38-2 ARSENIC	ARSENIC
14809-44-0 ARSENIC 69	ARSENIC 69
14809-45-1 ARSENIC 70	ARSENIC 70

Chemical Name	Reference Name
16685-55-5 ARSENIC 71	ARSENIC 71
15755-33-6 ARSENIC 72	ARSENIC 72
15422-59-0 ARSENIC 73	ARSENIC 73
14304-78-0 ARSENIC 74	ARSENIC 74
15575-20-9 ARSENIC 76	ARSENIC 76
14687-61-7 ARSENIC 77	ARSENIC 77
15755-35-8 ARSENIC 78	ARSENIC 78
7440-38-2 ARSENIC AND SOLUBLE COMPOUNDS [ALB, ALB, MEX]	ARSENIC
1327-53-3 ARSENIC TRIOXIDE (PRODUCTION) [MEX, MEX]	ARSENIC TRIOXIDE
7778-39-4 ARSENIC ACID	ARSENIC ACID
1327-52-2 ARSENIC ACID H3AS04 [EPA]	ARSENIC ACID
7778-39-4 ARSENIC ACID (H3ASO4) [RCU, RCU]	ARSENIC ACID
7778-44-1 ARSENIC ACID (H3ASO4), CALCIUM SALT (2:3) [PAL]	CALCIUM ARSENATE
7778-43-0 ARSENIC ACID, DISODIUM SALT [MSL, FLA]	DISODIUM HYDROGEN ARSENATE
53404-12-9 ARSENIC ACID, LEAD (4+) SALT	ARSENIC ACID, LEAD (4+) SALT
7645-25-2 ARSENIC ACID, LEAD SALT [MSL]	LEAD ARSENATE, UNSPECIFIED
7645-25-2 ARSENIC ACID (H3ASO4), LEAD SALT [PAL]	LEAD ARSENATE, UNSPECIFIED
7784-40-9 ARSENIC ACID (H3ASO4), LEAD (2+)SALT (1:1) [PAL]	LEAD ARSENATE
10102-48-4 ARSENIC ACID (H3ASO4), LEAD(4+) SALT (3:2) [PAL]	LEAD ARSENATE (PB3(ASS04)2)
7784-41-0 ARSENIC ACID (H3ASO4), MONOPOTASSIUM SALT [PAL, PAL]	POTASSIUM ARSENATE
7631-89-2 ARSENIC ACID, SODIUM SALT [MSL]	SODIUM ARSENATE
7631-89-2 ARSENIC ACID (H3ASO4), SODIUM SALT [PAL, PAL]	SODIUM ARSENATE
13464-38-5 ARSENIC ACID, TRISODIUM SALT	ARSENIC ACID, TRISODIUM SALT
RR-00153-6 ARSENICAL DUST [HM1, HM1]	ARSENICAL DUST OR FLUE DUST
RR-00153-6 ARSENICAL DUST OR FLUE DUST	ARSENICAL DUST OR FLUE DUST
RR-00154-7 ARSENICAL PESTICIDES, N.O.S.	ARSENICAL PESTICIDES, N.O.S.
RR-00625-7 ARSENIC AND ARSENIC COMPOUNDS [IAR]	ARSENIC COMPOUNDS, N.O.S.
RR-00625-7 ARSENIC AND COMPOUNDS [GBR]	ARSENIC COMPOUNDS, N.O.S.
7440-38-2 ARSENIC AND SOLUBLE COMPOUNDS [AUS, AUS]	ARSENIC
7784-42-1 ARSENIC AND SOLUBLE COMPOUNDS INCLUDING ARSINE [BEI]	ARSINE
7784-33-0 ARSENIC BROMIDE	ARSENIC BROMIDE
7784-34-1 ARSENIC CHLORIDE [FLA, MSL]	ARSENOUS TRICHLORIDE
RR-00625-7 ARSENIC COMPOUNDS [313, BC1, CAL, CN4, CN7, CW3, QBC]	ARSENIC COMPOUNDS, N.O.S.
RR-00625-7 ARSENIC COMPOUNDS (INORGANIC INCLUDING ARSINE) [CAA]	ARSENIC COMPOUNDS, N.O.S.
RR-00625-7 ARSENIC COMPOUNDS, N.O.S.	ARSENIC COMPOUNDS, N.O.S.
56320-22-0 ARSENIC DISULFIDE	ARSENIC DISULFIDE
7440-38-2 ARSENIC, ELEMENTAL [WHM]	ARSENIC
RR-00065-7 ARSENIC, INORGANIC [NHS, NHS, NHS]	INORGANIC ARSENIC
RR-00065-7 ARSENIC (INORGANIC ARSENIC COMPOUNDS) [C65]	INORGANIC ARSENIC

Chemical Name	Reference Name
RR-00065-7 ARSENIC, INORGANIC COMPOUNDS [CAB, NJL]	INORGANIC ARSENIC
7440-38-2 ARSENIC (INORGANIC COMPOUNDS) [NIO, NIO, NIO, NIO]	ARSENIC
7784-45-4 ARSENIC IODIDE(ASI3) [NJL]	ARSENOUS TRIIODIDE
RR-00035-1 ARSENIC, ORGANIC COMPOUNDS [MIN]	ARSENIC ORGANIC COMPOUNDS
RR-00035-1 ARSENIC ORGANIC COMPOUNDS	ARSENIC ORGANIC COMPOUNDS
1327-53-3 ARSENIC OXIDE (AS2O3) [PAL, PAL]	ARSENIC TRIOXIDE
1303-28-2 ARSENIC OXIDE (AS2O5) [PAL, PAL]	ARSENIC PENTOXIDE
1303-28-2 ARSENIC PENTOXIDE	ARSENIC PENTOXIDE
1303-33-9 ARSENIC SULFIDE (AS2S3) [PAL]	ARSENIC TRISULFIDE
7784-34-1 ARSENIC TRICHLORIDE [CWA, EP2, EPA, HM1, HM1, HM1, HM1, HM1, NJL]	ARSENOUS TRICHLORIDE
7784-34-1 ARSENIC(III) TRICHLORIDE [WHM]	ARSENOUS TRICHLORIDE
60646-36-8 ARSENIC TRICHLORIDE	ARSENIC TRICHLORIDE
1327-53-3 ARSENIC TRIOXIDE	ARSENIC TRIOXIDE
1327-53-3 ARSENIC TRIOXIDE (PRODUCTION) [QBC, QBC, ALB, ALB, ALB]	ARSENIC TRIOXIDE
1327-53-3 ARSENIC TRIOXIDE PRODUCTION [BC1, BC1, BC1, MIN]	ARSENIC TRIOXIDE
1303-33-9 ARSENIC TRISULFIDE	ARSENIC TRISULFIDE
RR-00586-7 ARSENIC, WATER-SOLUBLE COMPOUNDS, N.O.S.	ARSENIC, WATER-SOLUBLE COMPOUNDS, N.O.S.
1327-53-3 ARSENOUS OXIDE [FLA]	ARSENIC TRIOXIDE
13464-37-4 ARSENOUS ACID, TRISODIUM SALT	ARSENOUS ACID, TRISODIUM SALT
1327-53-3 ARSENOUS OXIDE [302]	ARSENIC TRIOXIDE
7784-34-1 ARSENOUS TRICHLORIDE	ARSENOUS TRICHLORIDE
7784-45-4 ARSENOUS TRIIODIDE	ARSENOUS TRIIODIDE
7784-42-1 ARSINE	ARSINE
696-28-6 ARSINE,DICHLOROPHENYL [MSL]	DICHLOROPHENYLARSINE
692-42-2 ARSINE, DIETHYL-	ARSINE, DIETHYL-
12044-79-0 ARSINO, THIOXO-	ARSINO, THIOXO-
52740-16-6 ARSONIC ACID, CALCIUM SALT (1:1) [PAL]	CALCIUM ARSENITE
10124-50-2 ARSONIC ACID, POTASSIUM SALT [PAL]	POTASSIUM ARSENITE
696-28-6 ARSONOUS DICHLORIDE, PHENYL- [TSC, TSC]	DICHLOROPHENYLARSINE
RR-00281-3 ARYL PHOSPHATES	ARYL PHOSPHATES
RR-01218-0 ARYL SULFONATE OF A FATTY ACID MIXTURE, POLYAMINE CONDENSATE	ARYL SULFONATE OF A FATTY ACID MIXTURE, POLYAMINE CONDENSATE
17068-78-9 ASBESTIFORM MINERAL(S) [TSC, TSC, TSC, TSC, TSC]	ANTHOPHYLLITE, NON-ASBESTIFORM
1332-21-4 ASBESTOS	ASBESTOS
77536-66-4 ASBESTOS, ACTINOLITE	ASBESTOS, ACTINOLITE
1332-21-4 ASBESTOS (ALL FORMS) [BC1]	ASBESTOS
1332-21-4 ASBESTOS, ALL FORMS [TLV, TLV, TLV]	ASBESTOS
12172-73-5 ASBESTOS, AMOSITE	ASBESTOS, AMOSITE
12172-73-5 ASBESTOS-AMOSITE [TLV, TLV]	ASBESTOS, AMOSITE
77536-67-5 ASBESTOS, ANTHOPHYLITE	ASBESTOS, ANTHOPHYLITE
12001-29-5 ASBESTOS, CHRYSOTILE	ASBESTOS, CHRYSOTILE
12001-29-5 ASBESTOS-CHRYSOTILE [TLV, TLV]	ASBESTOS, CHRYSOTILE
12001-28-4 ASBESTOS, CROCIDOLITE	ASBESTOS, CROCIDOLITE

Chemical Name	Reference Name
12001-28-4 ASBESTOS-CROCIDOLITE [TLV, TLV]	ASBESTOS, CROCIDOLITE
1332-21-4 ASBESTOS DUST [FLA]	ASBESTOS
12172-73-5 ASBESTOS FIBRES, AMOSITE [ONT, ONT, ONT, ONT]	ASBESTOS, AMOSITE
12001-28-4 ASBESTOS FIBRES, CROCIDOLITE [ONT, ONT, ONT, ONT]	ASBESTOS, CROCIDOLITE
1332-21-4 ASBESTOS FIBRES, OTHER [ONT, ONT, ONT, ONT]	ASBESTOS
12172-73-5 ASBESTOS, GRUNERITE [PAL, PAL]	ASBESTOS, AMOSITE
1332-21-4 ASBESTOS, OTHER FORMS [MIN]	ASBESTOS
14567-73-8 ASBESTOS, TREMOLITE [CSR, CSR, BC1, BC1, IAR]	TREMOLITEYL-CCNU)
77536-68-6 ASBESTOS, TREMOLITE	ASBESTOS, TREMOLITE
512-85-6 ASCARIDOLE	ASCARIDOLE
50-81-7 L-ASCORBIC ACID	L-ASCORBIC ACID
137-66-6 ASCORBYL PALMITATE	ASCORBYL PALMITATE
9015-68-3 ASPARAGINASE	ASPARAGINASE
38916-42-6 DL-ASPARTIC ACID, N-(3-CARBOXY-1-OXO-3-SULFOPROPYL)-N-OCTADECYL-, TETRA-SODIUM SALT	DL-ASPARTIC ACID, N-(3-CARBOXY-1-OXO-3-SULFO-PROPYL)-N-OCTADE
8052-42-4 ASPHALT	ASPHALT
8052-42-4 ASPHALT (PETROLEUM) FUMES [MEX, MEX, BC1, BC1, QBC]	ASPHALT
8052-42-4 ASPHALT (PETROLEUM FUMES) [ALB, ALB]	ASPHALT
8052-42-4 ASPHALT FUMES [FLA, MSL, NHS, NHS, NJL, ONT]	ASPHALT
8052-42-4 ASPHALT (PETROLEUM) FUMES [MIN, CAL, CEX, TLV]	ASPHALT
50-78-2 ASPIRIN [C65, C65]	ACETYLSALICYLIC ACID (ASPIRIN)
8003-03-0 ASPIRIN, PHENACETIN, AND CAFFIENE	ASPIRIN, PHENACETIN, AND CAFFIENE
20601-76-7 ASTATINE 207	ASTATINE 207
15755-39-2 ASTATINE 211	ASTATINE 211
3337-71-1 ASULAM	ASULAM
1912-24-9 ATRAZINE	ATRAZINE
1912-24-9 ATRAZINE (6-CHLORO-N-ETHYL-N'-(1-METHYLETHYL)-1,3,5-TRIAZINE-2,4-DIAMINE) [313]	ATRAZINE
12174-11-7 ATTAPULGITE	ATTAPULGITE
8031-18-3 ATTAPULGITE/PALYGORSKITE [MAK]	FULLERS EARTH
2465-27-2 AURAMINE	AURAMINE
RR-00525-4 AURAMINE MANUFACTURE [PAL, PAL]	AURAMINE, MANUFACTURE OF
RR-00525-4 AURAMINE, MANUFACTURE OF	AURAMINE, MANUFACTURE OF
2465-27-2 AURAMINE (TECHNICAL GRADE) [MIN, IAR]	AURAMINE
12192-57-3 AUROTHIOGLUCOSE	AUROTHIOGLUCOSE
320-67-2 AZACITIDINE [C65, IAR]	5-AZACYTIDINE
320-67-2 5-AZACYTIDINE	5-AZACYTIDINE
115-02-6 AZASERINE	AZASERINE
446-86-6 AZATHIOPRINE	AZATHIOPRINE
54-25-1 6-AZAURIDINE	6-AZAURIDINE
RR-01392-3 AZAUROLIC ACID	AZAUROLIC ACID
320-67-2 AZCYTIDINE [NTP]	5-AZACYTIDINE

Chemical Name	Reference Name
105-60-2 2H-AZEPIN-2-ONE, HEXAHYDRO- [PAL, TSC, TSC]	CAPROLACTAM
RR-01633-1 3'-AZIDO-3'-DEOXYTHYMIDINE + 2',3'-DIDEOXYINOSINE (AIDS INITIATIVE)	3'-AZIDO-3'-DEOXYTHYMIDINE + 2',3'-DIDEOXYINOSINE (AIDS INITATIVE)
30516-87-1 3'-AZIDO-3'-DEOXYTHYMIDINE (AIDS INITIA-TIVE)	3'-AZIDO-3'-DEOXYTHYMIDINE (AIDS INITIATIVE)
RR-01632-0 3'-AZIDO-3'-DEOXYTHYMIDINE/2',3'-DIDEOXY-CYTIDINE (AIDS INITIATIVE)	3'-AZIDO-3'-DEOXYTHYMIDINE/2',3'-DIDEOXYCYTIDINE (AIDS INITI
4472-06-4 AZIDODITHIOCARBONIC ACID	AZIDODITHIOCARBONIC ACID
53422-49-4 AZIDOETHYL NITRATE	AZIDOETHYL NITRATE
RR-01393-4 AZIDO GUANIDINE PICRATE	AZIDO GUANIDINE PICRATE
RR-01388-7 5-AZIDO-1-HYDROXY TETRAZOLE	5-AZIDO-1-HYDROXY TETRAZOLE
RR-01394-5 AZIDO HYDROXY TETRAZOLE	AZIDO HYDROXY TETRAZOLE
RR-01386-5 3-AZIDO-1,2-PROPYLENE GLYCOL DINITRATE	3-AZIDO-1,2-PROPYLENE GLYCOL DINITRATE
2642-71-9 AZINPHOS-ETHYL	AZINPHOS-ETHYL
86-50-0 AZINPHOS-METHYL	AZINPHOS-METHYL
86-50-0 AZINPHOSMETHYL [CSR, CSR]	AZINPHOS-METHYL
86-50-0 AZINPHOS-METHYL (GUTHION) [ALB, ALB, ALB]	AZINPHOS-METHYL
86-50-0 AZINPHOS-METHYL (ISO) [GBR, GBR, GBR]	AZINPHOS-METHYL
151-56-4 AZIRIDINE [EP2, EPA, IAR, PAL, RCU, RCU]	ETHYLENEIMINE
1072-52-2 1-AZIRIDINEETHANOL	1-AZIRIDINEETHANOL
75-55-8 AZIRIDINE, 2-METHTL- [PAL, PAL]	PROPYLENEIMINE
52-24-4 AZIRIDINE, 1,1',1"-PHOSPHINOTHIOYLI-DYNETRIS- [PAL, PAL]	TRIS(1-AZIRIDINYL)PHOSPHINE SULFIDE
1072-52-2 2-(1-AZIRIDINYL)ETHANOL [CAL, IAR]	1-AZIRIDINEETHANOL
545-55-1 1-AZIRIDINYL PHOSPHINE OXIDE (TRIS) [NJL]	TRIS(1-AZIRIDINYL) PHOSPHINE OXIDE
800-24-8 AZIRIDYL BENZOQUINONE	AZIRIDYL BENZOQUINONE
50-07-7 AZIRINO[2',3':3,4]PYRROLO[1,2-A]INDOLE-4,7-DIONE,6-AMINO-8-[[(AMINOCARBONYL)OXY]METHYL]-1,1A,2,8,8A,8B-HEXAHYDRO-8A-METHOXY-5-MeTHYL-,[1AS-(A.ALPHA.,8.BETA., 8A.ALPHA., 8B.ALPHA.)]- [PAL, PAL]	MITOMYCIN C
103-33-3 AZOBENZENE	AZOBENZENE
78-67-1 AZOBISISOBUTYRONITRILE	AZOBISISOBUTYRONITRILE
78-67-1 2,2'-AZOBIS(2-METHYL PROPIONITRILE [FLA]	AZOBISISOBUTYRONITRILE
123-77-3 AZODICARBONAMIDE	AZODICARBONAMIDE
123-77-3 AZODICARBOXAMIDE [PST]	AZODICARBONAMIDE
15545-97-8 2,2'-AZODI-(2,4-DIMETHYL-4-METHOXY-VALERONITRILE)	2,2'-AZODI-(2,4-DIMETHYL-4-METHOXYVALERONI-TRILE)
28604-91-3 2,2'-AZODI-(2,4-DIMETHYLVALERONITRILE)	2,2'-AZODI-(2,4-DIMETHYLVALERONITRILE)
25551-14-8 AZODI-(1,1'-HEXAHYDROBENZONITRILE)	AZODI-(1,1'-HEXAHYDROBENZONITRILE)
25551-14-8 1,1'-AZODI-(HEXAHYDROBENZONITRILE) [HM1, HM1]	AZODI-(1,1'-HEXAHYDROBENZONITRILE)
78-67-1 AZODIISOBUTYRONITRILE [HM1, HM1, NJL]	AZOBISISOBUTYRONITRILE
RR-00768-1 AZODI(2-METHYLBUTYRONITRITE)	AZODI(2-METHYLBUTYRONITRITE)
RR-00768-1 2,2'-AZODI-(2-METHYL-BUTYRONITRILE) [HM1, HM1]	AZODI(2-METHYLBUTYRONITRITE)
RR-01395-6 AZOTETRAZOLE	AZOTETRAZOLE
495-48-7 AZOXYBENZENE	AZOXYBENZENE
1405-87-4 BACITRACIN	BACITRACIN
51142-18-8 BACTOPEPTONE	BACTOPEPTONE

Chemical Name	Reference Name
RR-01038-8 BANCROFT CLAY	BANCROFT CLAY
101-27-9 BARBAN	BARBAN
83-79-4 BARBASCO [PST]	ROTENONE (COMMERCIAL)
RR-01473-3 BARBITURATES	BARBITURATES
7440-39-3 BARIUM	BARIUM
15229-36-4 BARIUM 126	BARIUM 126
15741-25-0 BARIUM 128	BARIUM 128
14914-75-1 BARIUM 131	BARIUM 131
RR-00483-1 BARIUM 131M	BARIUM 131M
13981-41-4 BARIUM 133	BARIUM 133
RR-00482-0 BARIUM 133M	BARIUM 133M
RR-00481-9 BARIUM 135M	BARIUM 135M
14378-25-7 BARIUM 139	BARIUM 139
14798-08-4 BARIUM 140	BARIUM 140
15741-29-4 BARIUM 141	BARIUM 141
18879-37-3 BARIUM 142	BARIUM 142
7440-39-3 BARIUM (SOLUBLE COMPOUNDS) [ALB, ALB]	BARIUM
7440-39-3 BARIUM, SOLUBLE COMPOUNDS [MEX]	BARIUM
543-80-6 BARIUM ACETATE	BARIUM ACETATE
RR-00161-6 BARIUM ALLOYS	BARIUM ALLOYS
18810-58-7 BARIUM AZIDE	BARIUM AZIDE
13967-90-3 BARIUM BROMATE	BARIUM BROMATE
513-77-9 BARIUM CARBONATE	BARIUM CARBONATE
13477-00-4 BARIUM CHLORATE	BARIUM CHLORATE
10361-37-2 BARIUM CHLORIDE	BARIUM CHLORIDE
10326-27-9 BARIUM CHLORIDE DIHYDRATE	BARIUM CHLORIDE DIHYDRATE
10294-40-3 BARIUM CHROMATE	BARIUM CHROMATE
RR-00555-0 BARIUM COMPOUNDS [MAK, MAK, 313]	BARIUM COMPOUNDS, N.O.S.
RR-00555-0 BARIUM COMPOUNDS, N.O.S.	BARIUM COMPOUNDS, N.O.S.
RR-00049-7 BARIUM COMPOUNDS, SOLUBLE [GBR]	BARIUM SOLUBLE COMPOUNDS
RR-00555-0 BARIUM COMPOUNDS, SOLUBLE, N.O.S. [HM1, HM1, HM1]	BARIUM COMPOUNDS, N.O.S.
542-62-1 BARIUM CYANIDE	BARIUM CYANIDE
542-62-1 BARIUM CYANIDE (BA(CN)2) [PAL]	BARIUM CYANIDE
10031-16-0 BARIUM DICHROMATE	BARIUM DICHROMATE
7787-32-8 BARIUM FLUORIDE	BARIUM FLUORIDE
13477-10-6 BARIUM HYPOCHLORITE	BARIUM HYPOCHLORITE
13701-59-2 BARIUM METABORATE	BARIUM METABORATE
10022-31-8 BARIUM NITRATE	BARIUM NITRATE
28987-17-9 BARIUM NONYLPHENATE	BARIUM NONYLPHENATE
1304-28-5 BARIUM OXIDE	BARIUM OXIDE
13465-95-7 BARIUM PERCHLORATE	BARIUM PERCHLORATE
7787-36-2 BARIUM PERMANGANATE	BARIUM PERMANGANATE
1304-29-6 BARIUM PEROXIDE	BARIUM PEROXIDE
1304-29-6 BARIUM PEROXIDE (BA(O2)) [PAL]	BARIUM PEROXIDE
7787-41-9 BARIUM SELENATE	BARIUM SELENATE
13718-59-7 BARIUM SELENITE	BARIUM SELENITE
17125-80-3 BARIUM SILICOFLUORIDE	BARIUM SILICOFLUORIDE
7440-39-3 BARIUM, SOLUBLE COMPOUNDS [MIN, OS1, OS1]	BARIUM

Chemical Name	Reference Name
RR-00049-7 BARIUM SOLUBLE COMPOUNDS	BARIUM SOLUBLE COMPOUNDS
RR-00049-7 BARIUM, SOLUBLE COMPOUNDS [BC1, CAL]	BARIUM SOLUBLE COMPOUNDS
RR-00049-7 BARIUM (SOLUBLE COMPOUNDS) [NIO, NIO, NIO, QBC]	BARIUM SOLUBLE COMPOUNDS
6865-35-6 BARIUM STEARATE	BARIUM STEARATE
7727-43-7 BARIUM SULFATE	BARIUM SULFATE
7727-43-7 BARIUM, SULFATE [MIN]	BARIUM SULFATE
7727-43-7 BARIUM SULFATE (1:1) [PST]	BARIUM SULFATE
7727-43-7 BARIUM SULPHATE [AUS, GBR]	BARIUM SULFATE
RR-00049-7 BARIUM, WATER-SOLUBLE COMPOUNDS, N.O.S. [WHM]	BARIUM SOLUBLE COMPOUNDS
RR-01039-9 BARIUM ZINC METHYLBENZOATE-2-ETHYL-HEXANOATE	BARIUM ZINC METHYLBENZOATE-2-ETHYLHEX-ANOATE
12009-21-1 BARIUM ZIRCONATE	BARIUM ZIRCONATE
8002-48-0 BARLEY, MALT	BARLEY, MALT
142-03-0 BASIC ALUMINUM ACETATE	BASIC ALUMINUM ACETATE
50922-29-7 BASIC ZINC CHROMATE	BASIC ZINC CHROMATE
RR-00166-1 BATTERY FLUID, ACID	BATTERY FLUID, ACID
RR-00167-2 BATTERY FLUID, ALKALI	BATTERY FLUID, ALKALI
1318-16-7 BAUXITE (AL2O3.XH2O)	BAUXITE (AL2O3.XH2O)
114-26-1 BAYGON [MIN]	PROPOXUR
114-26-1 BAYGON (PROPOXUR) [ALB, ALB, BC1, BC1]	PROPOXUR
55-38-9 BAYTEX (FENTHION) [ALB, ALB, ALB, ALB]	FENTHION
53469-21-9 PCB-1242 (AROCHLOR 1242) [CW2]	AROCLOR 1242
8012-89-3 BEESWAX [PST]	WHITE BEESWAX
8007-93-0 BELLADONNALEAF	BELLADONNALEAF
22781-23-3 BENDIOCARB	BENDIOCARB
22781-23-3 BENDIOCARB [2,2-DIMETHYL-1,3-BENZODI-OXOL-4-OL METHYLCARBAMATE] [313]	BENDIOCARBINE-3,5-DIONE]
1861-40-1 BENFLURALIN (N-BUTYL-N-ETHYL-2,6-DINITRO-4-(TRIFLUOROMETHYL)BEN-ZENAMINE [313]	N-BUTYL-N-ETHYL-2,6-DINITRO-4-(TRIFLUO-ROMETHYL)BENZENAMINE
17804-35-2 BENOMYL	BENOMYL
17804-35-2 BENOMYL [MEX, MEX]	BENOMYL
17804-35-2 BENOMYL (ISO) [GBR, GBR]	BENOMYL
495-73-8 BENQUINOX	BENQUINOX
25057-89-0 BENTAZON	BENTAZON
1302-78-9 BENTONITE	BENTONITE
70131-50-9 BENTONITE, ACID-LEACHED	BENTONITE, ACID-LEACHED
225-11-6 BENZ[A]ACRIDINE	BENZ[A]ACRIDINE
56-55-3 BENZ(A)ANTHRACENE	BENZ(A)ANTHRACENE
57-97-6 BENZ(A)ANTHRACENE, 7,12-DIMETHYL- [EPA, NJL, PAL, PAL]	7,12-DIMETHYLBENZ(A)ANTHRACENE
50-32-8 BENZ(A)PYRENE [QBC, QBC]	BENZO(A)PYRENE
225-51-4 BENZ(C)ACRIDINE	BENZ(C)ACRIDINE
51787-44-1 BENZ(C)ACRIDINE [FLA]	BENZ(C)ACRIDINE, 7,8,9,11-TETRAMETHYL-
51787-44-1 BENZ(C)ACRIDINE, 7,8,9,11-TETRAMETHYL-	BENZ(C)ACRIDINE, 7,8,9,11-TETRAMETHYL-M SALT, DIHYDRATE
205-99-2 BENZ(E)ACEPHENANTHRYLENE [PAL, PAL]	BENZO(B)FLUORANTHENE

Chemical Name	Reference Name
56-49-5 BENZ(J)ACEANTHRYLENE, 1,2-DIHYRO-3-METHYL- [PAL, PAL]	3-METHYLCHOLANTHRENE
98-87-3 BENZAL CHLORIDE	BENZAL CHLORIDE
100-52-7 BENZALDEHYDE	BENZALDEHYDE
3132-99-8 BENZALDEHYDE, 3-BROMO-	BENZALDEHYDE, 3-BROMO-
1200-14-2 BENZALDEHYDE, 4-BUTYL-	BENZALDEHYDE, 4-BUTYL-
89-98-5 BENZALDEHYDE, 2-CHLORO-	BENZALDEHYDE, 2-CHLORO-
104-88-1 BENZALDEHYDE, 4-CHLORO- [PAL, TSC, TSC, TSC]	P-CHLOROBENZALDEHYDE
120-21-8 BENZALDEHYDE, 4-(DIETHYLAMINO)-	BENZALDEHYDE, 4-(DIETHYLAMINO)-
17754-90-4 BENZALDEHYDE, 4-(DIETHYLAMINO)-2-HYDROXY-	BENZALDEHYDE, 4-(DIETHYLAMINO)-2-HYDROXY-1,1-DIOXIDE-
95-01-2 BENZALDEHYDE, 2,4-DIHYDROXY-	BENZALDEHYDE, 2,4-DIHYDROXY-
100-10-7 BENZALDEHYDE, 4-(DIMETHYLAMINO)-	BENZALDEHYDE, 4-(DIMETHYLAMINO)-
93-02-7 BENZALDEHYDE, 2,5-DIMETHOXY-	BENZALDEHYDE, 2,5-DIMETHOXY-
120-14-9 BENZALDEHYDE, 3,4-DIMETHOXY- [TSC, TSC, TSC]	VERATRALDEHYDE
28602-27-9 BENZALDEHYDE, (DIMETHYLAMINO)-	BENZALDEHYDE, (DIMETHYLAMINO)-
939-97-9 BENZALDEHYDE, 4-(1,1-DIMETHYLETHYL)-	BENZALDEHYDE, 4-(1,1-DIMETHYLETHYL)-
10031-82-0 BENZALDEHYDE, 4-ETHOXY-	BENZALDEHYDE, 4-ETHOXY-
121-32-4 BENZALDEHYDE, 3-ETHOXY-4-HYDROXY-	BENZALDEHYDE, 3-ETHOXY-4-HYDROXY-
90-02-8 BENZALDEHYDE, 2-HYDROXY- [TSC, TSC, TSC]	SALICYLALDEHYDE
123-08-0 BENZALDEHYDE, 4-HYDROXY-	BENZALDEHYDE, 4-HYDROXY-
121-33-5 BENZALDEHYDE, 4-HYDROXY-3-METHOXY- [TSC, TSC, TSC]	VANILLIN
97-51-8 BENZALDEHYDE, 2-HYDROXY-5-NITRO-	BENZALDEHYDE, 2-HYDROXY-5-NITRO-
123-11-5 BENZALDEHYDE, 4-METHOXY- [TSC, TSC, TSC]	P-ANISALDEHYDE
135-02-4 BENZALDEHYDE, 2-METHOXY- [PAL, TSC, TSC, TSC]	O-ANISALDEHYDE
104-87-0 BENZALDEHYDE, 4-METHYL-	BENZALDEHYDE, 4-METHYL-
1334-78-7 BENZALDEHYDE, METHYL- [TSC, TSC, TSC]	TOLYL ALDEHYDE
552-89-6 BENZALDEHYDE, 2-NITRO-	BENZALDEHYDE, 2-NITRO-
39515-51-0 BENZALDEHYDE, 3-PHENOXY-	BENZALDEHYDE, 3-PHENOXY-
455-19-6 BENZALDEHYDE, 4-(TRIFLUOROMETHYL)-	BENZALDEHYDE, 4-(TRIFLUOROMETHYL)-
55-21-0 BENZAMIDE	BENZAMIDE
1420-04-8 BENZAMIDE, 5-CHLORO-N-(2-CHLORO-4-NITROPHENYL)-2-HYDROXY-,COMPD. WITH 2-AMINOETHANOL (1:1) [PAL]	CLONITRALIDLINE ENZYME)
23950-58-5 BENZAMIDE, 3,5-DICHLORO-N-(1,1-DIMETHYL-2-PROPYNYL)- [PAL]	PRONAMIDE-LYXO-HEXOPYRANOSYL)OXY]-7,8,9,10-TETRAHYDRO-6,8,11-TRIHYDROXYACETYL)-1-METHOXY, (8S-CIS)-
87-10-5 BENZAMIDE, 3,5-DIBROMO-N-(4-BROMOPHENYL)-2-HYDROXY- [TSC, TSC, TSC, TSC]	3,4',5-TRIBROMOSALICYLANILIDE
148-01-6 BENZAMIDE, 2-METHYL-3,5-DINITRO- [PAL]	DINITOLMIDE
366-70-1 BENZAMIDE, N-(1-METHYLETHYL)-4-[(2-METHYLHYDRAZINO)METHYL]-,MONOHYDROCHLORIDE [PAL, PAL]	PROCARBAZINE HYDROCHLORIDE3.ALPHA.,4.BETA., 5.ALPHA.,6.BETA.)-
671-16-9 BENZAMIDE, N-(1-METHYLETHYL)-4-[(2-METHYLHYDRAZINO)METHYL]- [PAL, PAL]	PROCARBAZINE
56-55-3 1,2-BENZANTHRACENE (BENZO(A) ANTHRACENE) [CW2]	BENZ(A)ANTHRACENE
82-05-3 BENZANTHRONE	BENZANTHRONE

Chemical Name	Reference Name
768-52-5 BENZEDRINE [NFP, NFP]	N-ISOPROPYLANILINE
62-53-3 BENZENAMINE [PAL]	ANILINE
62-53-3 BENZENAMINE (ANILINE) [TSC, TSC]	ANILINE
1208-52-2 BENZENAMINE, 2-[(4-AMINOPHENYL)METHYL]	BENZENAMINE, 2-[(4-AMINOPHENYL)METHYL]-
106-40-1 BENZENAMINE, 4-BROMO-	BENZENAMINE, 4-BROMO-
99-29-6 BENZENAMINE, 2-BROMO-6-CHLORO-4-NITRO-	BENZENAMINE, 2-BROMO-6-CHLORO-4-NITRO-
1817-73-8 BENZENAMINE, 2-BROMO-4,6-DINITRO-	BENZENAMINE, 2-BROMO-4,6-DINITRO-
1126-78-9 BENZENAMINE, N-BUTYL- [PAL]	N-BUTYLANILINE
2465-27-2 BENZENAMINE, 4-4'-CARBONIMIDOYLBIS[N, N-DIMETHYL-,MONOHYDROCHLORIDE [PAL, PAL]	AURAMINE
95-51-2 BENZENAMINE, 2-CHLORO- [TSC, TSC, TSC]	2-CHLOROANILINE
106-47-8 BENZENAMINE, 4-CHLORO- [PAL, TSC, TSC, TSC]	P-CHLOROANILINE
108-42-9 BENZENAMINE, 3-CHLORO- [TSC, TSC]	M-CHLOROANILINE
3531-19-9 BENZENAMINE, 2-CHLORO-4,6-DINITRO-	BENZENAMINE, 2-CHLORO-4,6-DINITRO-[1,1'-BIPHENYL]-4-YL]AZO]-7-HYDROXY-, DISODIUM SALT
5388-62-5 BENZENAMINE, 4-CHLORO-2,6-DINITRO-	BENZENAMINE, 4-CHLORO-2,6-DINITRO-
92-49-9 BENZENAMINE, N-(2-CHLOROETHYL)-N-ETHYL-	BENZENAMINE, N-(2-CHLOROETHYL)-N-ETHYL-
141-85-5 BENZENAMINE, 3-CHLORO-, HYDROCHLORIDE	BENZENAMINE, 3-CHLORO-, HYDROCHLORIDE
87-63-8 BENZENAMINE, 2-CHLORO-6-METHYL- [TSC, TSC]	6-CHLORO-2-METHYL ANILINE
95-69-2 BENZENAMINE, 4-CHLORO-2-METHYL- [PAL, PAL, TSC, TSC]	4-CHLORO-O-TOLUIDINE
87-63-8 BENZENAMINE, 2-CHLORO-6-METHYL [TSE]	6-CHLORO-2-METHYL ANILINE
95-69-2 BENZENAMINE, 4-CHLORO-2-METHYL [TSE]	4-CHLORO-O-TOLUIDINE
3165-93-3 BENZENAMINE, 4-CHLORO-2-METHYL-, HYDROCHLORIDE	BENZENAMINE, 4-CHLORO-2-METHYL-, HYDROCHLORIDE
89-63-4 BENZENAMINE, 4-CHLORO-2-NITRO-	BENZENAMINE, 4-CHLORO-2-NITRO-ESTER
121-87-9 BENZENAMINE, 2-CHLORO-4-NITRO-	BENZENAMINE, 2-CHLORO-4-NITRO-
635-22-3 BENZENAMINE, 4-CHLORO-3-NITRO-	BENZENAMINE, 4-CHLORO-3-NITRO-
6283-25-6 BENZENAMINE, 2-CHLORO-5-NITRO-	BENZENAMINE, 2-CHLORO-5-NITRO-1-AMINO-9,10-DIHYDRO-9,10-DIOXO-
827-94-1 BENZENAMINE, 2,6-DIBROMO-4-NITRO-	BENZENAMINE, 2,6-DIBROMO-4-NITRO-
130169-66-3 BENZENAMINE, 2,5-DIBUTOXY-4-(4-MORPHOLINYL)-, SULFATE	BENZENAMINE, 2,5-DIBUTOXY-4-(4-MORPHOLINYL)-, SULFATEMETHOXY-
613-29-6 BENZENAMINE, N,N-DIBUTYL-	BENZENAMINE, N,N-DIBUTYL-
95-76-1 BENZENAMINE, 3,4-DICHLORO- [PAL, TSC, TSC, TSC]	3,4-DICHLOROANILINE
95-82-9 BENZENAMINE, 2,5-DICHLORO-	BENZENAMINE, 2,5-DICHLORO-
554-00-7 BENZENAMINE, 2,4-DICHLORO-	BENZENAMINE, 2,4-DICHLORO-
608-27-5 BENZENAMINE, 2,3-DICHLORO-	BENZENAMINE, 2,3-DICHLORO-
626-43-7 BENZENAMINE, 3,5-DICHLORO-	BENZENAMINE, 3,5-DICHLORO-
99-30-9 BENZENAMINE, 2,6-DICHLORO-4-NITRO- [TSC, TSC, TSC, TSC]	2,6-DICHLORO-4-NITROANILINE
91-66-7 BENZENAMINE, N,N-DIETHYL- [PAL]	N,N-DIETHYLANILINE
102-56-7 BENZENAMINE, 2,5-DIMETHOXY- [PAL]	2,5-DIMETHOXYANILINE
87-59-2 BENZENAMINE, 2,3-DIMETHYL-	BENZENAMINE, 2,3-DIMETHYL-
87-62-7 BENZENAMINE, 2,6-DIMETHYL- [PAL]	2,6-XYLIDENE

Chemical Name	Reference Name
121-69-7 BENZENAMINE, N,N-DIMETHYL- [PAL]	DIMETHYLANILINE
60-11-7 BENZENAMINE, N,N-DIMETHYL-4-(PHENY-LAZO)- [PAL, PAL]	4-DIMETHYLAMINOAZOBENZENE
2049-92-5 BENZENAMINE, 4-(1,1-DIMETHYLPROPYL)-[PAL]	P-TERT-AMYLANILINE
97-02-9 BENZENAMINE, 2,4-DINITRO- [PAL, TSC, TSC, TSC]	2,4-DINITROANILINE
603-34-9 BENZENAMINE, N,N-DIPHENYL [PAL]	TRIPHENYL AMINE
94-70-2 BENZENAMINE, 2-ETHOXY- [PAL]	O-PHENETIDINE
156-43-4 BENZENAMINE, 4-ETHOXY- [OTS, PAL]	P-PHENETIDINE
13410-72-5 BENZENAMINE, 4-ETHOXY-N-[(5-NITRO-2-FURANYL)METHYLENE]-	BENZENAMINE, 4-ETHOXY-N-[(5-NITRO-2-FURANYL)METHYLENE]-
103-69-5 BENZENAMINE, N-ETHYL- [PAL]	N-ETHYLANILINE
10137-80-1 BENZENAMINE, N-(2-ETHYLHEXYL)-	BENZENAMINE, N-(2-ETHYLHEXYL)-
142-04-1 BENZENAMINE, HYDROCHLORIDE [PAL, TSC, TSC]	ANILINE HYDROCHLORIDE
135-20-6 BENZENAMINE, N-HYDROXY-N-NITROSO-, AMMONIUM SALT [PAL, PAL]	CUPFERRON
126505-35-9 BENZENAMINE, 4-ISOCYANATO-N,N-BIS(4-ISOCYANATOPHENYL)-2,5-DIMETHOXY- [TSC]	·2,4,8,10-TETRAOXA-3,9-DIPHOSPHASPIRO[5,5] UNDE-CANE, 3,9-BIS[
90-04-0 BENZENAMINE, 2-METHOXY- [PAL, PAL]	O-ANISIDINE
104-94-9 BENZENAMINE, 4-METHOXY- [PAL]	P-ANISIDINE
536-90-3 BENZENAMINE, 3-METHOXY- [OTS]	M-ANISIDINE
29191-52-4 BENZENAMINE, AR-METHOXY- [PAL]	ANISIDINE (O-, P- ISOMERS)
134-29-2 BENZENAMINE, 2-METHOXY-, HYDROCHLO-RIDE [PAL, PAL]	O-ANISIDINE HYDROCHLORIDEROMETHYL)THIO]-
120-71-8 BENZENAMINE, 2-METHOXY-5-METHYL- [PAL, PAL]	P-CRESIDINE
99-59-2 BENZENAMINE, 2-METHOXY-5-NITRO- [PAL, PAL]	5-NITRO-O-ANISIDINE
95-53-4 BENZENAMINE, 2-METHYL- [PAL, PAL, TSC, TSC]	O-TOLUIDINE
100-61-8 BENZENAMINE, N-METHYL [PAL]	N-METHYL ANILINE
106-49-0 BENZENAMINE, 4-METHYL- [PAL, TSC, TSC]	P-TOLUIDINE
RR-01237-3 BENZENAMINE, 4-(1-METHYLBUTOXY)-, HY-DROCHLORIDE	BENZENAMINE, 4-(1-METHYLBUTOXY)-, HYDROCHLO-RIDEUCT
101-77-9 BENZENAMINE, 4-4'-METHYLENEBIS- [PAL, TSC, TSC, TSC]	4,4'-METHYLENEDIANILINE
101-14-4 BENZENAMINE, 4-4'-METHYLENEBIS-(2-CHLORO)- [PAL, PAL]	4,4'-METHYLENEBIS(2-CHLOROANILINE) (MBOCA)
101-14-4 BENZENAMINE, 4,4'-METHYLENEBIS[2-CHLORO-](MBOCA) [CAI]	4,4'-METHYLENEBIS(2-CHLOROANILINE) (MBOCA)
101-61-1 BENZENAMINE, 4-4'-METHYLENEBIS-(N,N-DIMETHYL)- [PAL, PAL]	4,4'-METHYLENEBIS(N,N-DIMETHYL) BENZENAMINE
838-88-0 BENZENAMINE, 4-4'-METHYLENEBIS-(2-METHYL)- [PAL, PAL]	4,4'-METHYLENE BIS(2-METHYLANILINE)3-YL]IMINO]BIS-
768-52-5 BENZENAMINE, N-(1-METHYLETHYL)- [PAL]	N-ISOPROPYLANILINE
636-21-5 BENZENAMINE, 2-METHYL-, HYDROCHLO-RIDE [PAL, PAL]	O-TOLUIDINE HYDROCHLORIDE
97-56-3 BENZENAMINE, 2-METHYL-4-[(2-METHYLPHENYL)AZO]- [PAL, PAL]	O-AMINOAZOTOLUENE
89-62-3 BENZENAMINE, 4-METHYL-2-NITRO- [PAL]	2-NITRO-P-TOLUIDINE

Chemical Name	Reference Name
99-55-8 BENZENAMINE, 2-METHYL-5-NITRO- [EPA, PAL]	5-NITRO-O-TOLUIDINE
479-45-8 BENZENAMINE, N-METHYL-N,2,4,6-TETRANI-TRO- [PAL]	TETRYLDIETHYL ESTER
88-74-4 BENZENAMINE, 2-NITRO- [TSC, TSC, TSC]	O-NITROANILINE
99-09-2 BENZENAMINE, 3-NITRO- [TSC, TSC]	M-NITROANILINE
100-01-6 BENZENAMINE, 4-NITRO- [PAL, TSC, TSC, TSC, TSC]	P-NITROANILINE
836-30-6 BENZENAMINE, 4-NITRO-N-PHENYL- [OTS]	P-NITRODIPHENYLAMINE
86-30-6 BENZENAMINE, N-NITROSO-N-PHENYL- [PAL]	DIPHENYLNITROSAMINEER
156-10-5 BENZENAMINE, 4-NITROSO-N-PHENYL- [PAL, PAL]	P-NITROSODIPHENYLAMINEYDRO-, (2R-TRANS)-
101-80-4 BENZENAMINE, 4-4'-OXYBIS- [PAL, PAL]	4,4'-DIAMINODIPHENYL ETHER
28434-86-8 BENZENAMINE, 4-4'-OXYBIS(2-CHLORO)- [PAL, PAL]	3,3'-DICHLORO-4,4'-DIAMINODIPHENYL ETHER
122-39-4 BENZENAMINE, N-PHENYL- [PAL]	DIPHENYLAMINE
2716-10-1 BENZENAMINE, 4,4'-[1,4-PHENYLENEBIS[1-METHYLETHYLIDENE]BIS[2,6-DIMETHYL-,	BENZENAMINE, 4,4'-[1,4-PHENYLENEBIS[1-METHYLETHYLIDENE]BIS[
2716-12-3 BENZENAMINE, 4,4'-[1,3-PHENYLENEBIS[1-METHYLETHYLIDENE]BIS[2,6-DIMETHYL-,	BENZENAMINE, 4,4'-[1,3-PHENYLENEBIS[1-METHYLETHYLIDENE]BIS[2,6-DIMETHYL-,
139-65-1 BENZENAMINE, 4-4'-THIOBIS- [PAL, PAL]	4,4'-THIODIANILINE
147-82-0 BENZENAMINE, 2,4,6-TRIBROMO-	BENZENAMINE, 2,4,6-TRIBROMO-
634-93-5 BENZENAMINE, 2,4,6-TRICHLORO-	BENZENAMINE, 2,4,6-TRICHLORO-
98-16-8 BENZENAMINE, 3-(TRIFLUOROMETHYL)-	BENZENAMINE, 3-(TRIFLUOROMETHYL)-
71-43-2 BENZENE	BENZENE
71-43-2 BENZENE [NFP, NFP]	BENZENE
122-78-1 BENZENEACETALDEHYDE	BENZENEACETALDEHYDE
104-09-6 BENZENEACETALDEHYDE, 4-METHYL-	BENZENEACETALDEHYDE, 4-METHYL-
93-53-8 BENZENEACETALDEHYDE, .ALPHA.-METHYL-	BENZENEACETALDEHYDE, .ALPHA.-METHYL-
510-15-6 BENZENEACETIC ACID, 4-CHLORO-.ALPHA.-(4-CHLOROPHENYL)-.ALPHA.-HYDROXY-,ETHYL ESTER [PAL]	CHLOROBENZILATE
114-49-8 BENZENEACETIC ACID, .ALPHA.-(HY-DROXYMETHYL)-, 9-METHYL-3-OXA-9-AZATRICYCLO[3.3.1.02,4]NON-7-YL ESTER, HYDROBROMIDE, [7(S)-(1.AL-PHA.2.BETA.4.BETA.5.ALPHA.7.BETA)]- [TSC, TSC]	SCOPOLAMINE HYDROBROMIDE
6106-46-3 BENZENEACETIC ACID, ALPHA-(HYDROX-YMETHYL)-, 9-METHYL-3-OXA-9-AZATRICYCLO [3.3.1.02,4]NON-7-YL ESTER, [7(S)-(1-ALPHA,2-BETA,4-BETA,5-ALPHA,7-BETA)]-, COMPOUND WITH METHYL NITRATE (1:1)	BENZENEACETIC ACID, ALPHA-(HYDROXYMETHYL)-, 9-METHYL-3-OXA-)-, (Z)-9-OCTADECENOATE (SALT)
6106-81-6 BENZENEACETIC ACID, ALPHA-(HYDROX-YMETHYL)-, 9-METHYL-3-OXA-9-AZATRICYCLO [3.3.1.02,4]NON-7-YL ESTER, N-OXIDE, HYDRO-BROMI DE, [7(S)-(1-ALPHA,2-BETA,4-BETA,5-AL-PHA,7-BETA)]-	BENZENEACETIC ACID, ALPHA-(HYDROXYMETHYL)-, 9-METHYL-3-OXA-9PHENYLPROPOXY)-9,9-DIMETHYL-, [7 (S)-(1A,2B,4B,5A,7B)-, NITRATE (SALT)
140-29-4 BENZENEACETONITRILE [PAL]	BENZYL CYANIDE
16532-79-9 BENZENE ACETONITRILE, 4-BROMO- [TSC, TSC]	4-BROMOBENZYLCYANIDEDIYL)BIS(AZO)BIS[4-AMINO-5-HYDROXY-1,3-NAPHTHALENE DISULFONATO](8-)]DI-, TETRASODIUM
98-05-5 BENZENEARSONIC ACID	BENZENEARSONIC ACID
28347-13-9 BENZENE, BIS(CHLOROMETHYL)- [MSL]	XYLYLENE DICHLORIDE
25854-16-4 BENZENE, BIS(ISOCYANATOMETHYL)-	BENZENE, BIS(ISOCYANATOMETHYL)-

Chemical Name	Reference Name
2778-42-9 BENZENE, 1,3-BIS(1-ISOCYANATO-1-METHYLETHYL-	BENZENE, 1,3-BIS(1-ISOCYANATO-1-METHYLETHYL-
108-86-1 BENZENE, BROMO- [PAL, TSC, TSC, TSC]	BROMOBENZENE
35884-77-6 BENZENE, BROMODIMETHYL	BENZENE, BROMODIMETHYL
2493-02-9 BENZENE, 1-BROMO-4-ISOCYANATO-	BENZENE, 1-BROMO-4-ISOCYANATO-TER
95-46-5 BENZENE, 1-BROMO-2-METHYL [PAL]	O-BROMOTOLUENE
106-38-7 BENZENE, 1-BROMO-4-METHYL- [PAL]	P-BROMOTOLUENE
51632-16-7 BENZENE, 1-(BROMOMETHYL)-3-PHENOXY-	BENZENE, 1-(BROMOMETHYL)-3-PHENOXY-
101-55-3 BENZENE, 1-BROMO-4-PHENOXY	BENZENE, 1-BROMO-4-PHENOXY
305-03-3 BENZENEBUTANOIC ACID, 4-[BIS(2-CHLOROETHYL)AMINO]- [PAL, PAL]	CHLORAMBUCILHYL)-3-METHYL-1-OXOBUTOXY] METHYL]-2,3,5,7A-TETRAHYDRO-1H-PYRROLIZIN-1-YL ESTER, [1S[1.ALPHA.(Z),7(2S*,3R*),7A.ALPHA.]-
104-51-8 BENZENE, BUTYL- [PAL, TSC, TSC]	BUTYL BENZENE
614-45-9 BENZENECARBOPEROXOIC ACID, 1,1-DIMETHYLETHYL ESTER [PAL]	TERT-BUTYL PEROXYBENZOATE
108-90-7 BENZENE, CHLORO- [PAL, TSC, TSC, TSC, TSC]	CHLOROBENZENE
2100-42-7 BENZENE, 2-CHLORO-1,4-DIMETHOXY- [PAL]	2,5-DIMETHOXYCHLOROBENZENE
97-00-7 BENZENE, 1-CHLORO-2,4-DINITRO- [PAL]	1-CHLORO-2,4-DINITROBENZENE
25567-67-3 BENZENE, CHLORODINITRO- [PAL]	DINITROCHLOROBENZENE
2039-87-4 BENZENE, 1-CHLORO-2-ETHENYL- [PAL]	O-CHLOROSTYRENE
622-86-6 BENZENE, (2-CHLOROETHOXY)-	BENZENE, (2-CHLOROETHOXY)-
104-12-1 BENZENE, 1-CHLORO-4-ISOCYANATO-	BENZENE, 1-CHLORO-4-ISOCYANATO-
2909-38-8 BENZENE, 1-CHLORO-3-ISOCYANATO-	BENZENE, 1-CHLORO-3-ISOCYANATO-
95-49-8 BENZENE, 1-CHLORO-2-METHYL- [PAL, TSC, TSC, TSC]	O-CHLOROTOLUENE
100-44-7 BENZENE, (CHLOROMETHYL)- [PAL]	BENZYL CHLORIDE
100-44-7 BENZENE, CHLOROMETHYL- [TSC, TSC, TSC]	BENZYL CHLORIDE
106-43-4 BENZENE, 1-CHLORO-4-METHYL- [TSC, TSC]	P-CHLOROTOLUENE
25168-05-2 BENZENE, CHLOROMETHYL- [PAL]	CHLOROTOLUENES
30030-25-2 BENZENE, (CHLOROMETHYL)ETHENYL-	BENZENE, (CHLOROMETHYL)ETHENYL-
100-14-1 BENZENE, 1-(CHLOROMETHYL)-4-NITRO-	BENZENE, 1-(CHLOROMETHYL)-4-NITRO-
88-73-3 BENZENE, 1-CHLORO-2-NITRO- [PAL]	O-NITROCHLOROBENZENE
100-00-5 BENZENE, 1-CHLORO-4-NITRO- [PAL]	P-NITROCHLOROBENZENE
121-73-3 BENZENE, 1-CHLORO-3-NITRO- [PAL]	M-NITROCHLOROBENZENE
25167-93-5 BENZENE, CHLORONITRO- [PAL]	CHLORONITROBENZENE
777-37-7 BENZENE, 1-CHLORO-4-NITRO-2-(TRIFLUO-ROMETHYL)- [PAL]	4-NITRO-1-CHLOROBENZO-2-TRIFLUORIDE
7005-72-3 BENZENE, 1-CHLORO-4-PHENOXY- [PAL]	4-CHLOROPHENYL PHENYL ETHERPROPENYL ESTER, (E)-
5216-25-1 BENZENE, 1-CHLORO-4-(TRICHLOROMETHYL)- [TSC, TSC]	4-CHLOROBENZOTRICHLORIDE
88-16-4 BENZENE, 1-CHLORO-2-(TRIFLUOROMETHYL)-	BENZENE, 1-CHLORO-2-(TRIFLUOROMETHYL)-
98-56-6 BENZENE, 1-CHLORO-4-TRIFLUOROMETHYL- [TSC, TSC, TSC]	P-CHLORO-A,A,A-TRIFLUOROTOLUENE
98-56-6 BENZENE, 1-CHLORO-4-(TRIFLUORMETHYL)- [OTS]	P-CHLORO-A,A,A-TRIFLUOROTOLUENE
827-52-1 BENZENE, CYCLOHEXYL- [PAL]	CYCLOHEXYLBENZENE
104-72-3 BENZENE, DECYL- [PAL]	DECYLBENZENE
95-54-5 1,2-BENZENEDIAMINE [TSC, TSC]	O-PHENYLENEDIAMINE
106-50-3 1,4-BENZENEDIAMINE [PAL]	P-PHENYLENE DIAMINE

Chemical Name	Reference Name
108-45-2 1,3-BENZENEDIAMINE [TSC, TSC]	M-PHENYLENEDIAMINE
106-50-3 1,4-BENZENEDIAMINE (P-PHENYLENEDI-AMINE) [TSC, TSC, TSC]	P-PHENYLENE DIAMINE
3081-14-9 1,4-BENZENEDIAMINE, N,N'-BIS(1,4-DIMETHYLPENTYL)-	1,4-BENZENEDIAMINE, N,N'-BIS(1,4-DIMETHYLPENTYL)-
101-96-2 1,4-BENZENEDIAMINE, N,N'-BIS(1-METHYL-PROPYL)-	1,4-BENZENEDIAMINE, N,N'-BIS(1-METHYLPROPYL)-
3663-23-8 1,2-BENZENEDIAMINE, 4-BUTYL-	1,2-BENZENEDIAMINE, 4-BUTYL-ROXYPHENYL)AZO] [1,1'-BIPHENYL]-4-YL]-AZO]-6-(PHENYLAZO)-, DISODIUM SALT
95-83-0 1,2-BENZENEDIAMINE, 4-CHLORO- [PAL, PAL, TSC, TSC]	4-CHLORO-O-PHENYLENEDIAMINE
5131-60-2 1,3-BENZENEDIAMINE, 4-CHLORO- [TSC, TSC]	4-CHLORO-M-PHENYLENEDIAMINE
615-46-3 1,4-BENZENEDIAMINE, 2-CHLORO-, DIHY-DROCHLORIDE	1,4-BENZENEDIAMINE, 2-CHLORO-, DIHYDROCHLORIDE
42389-30-0 1,2-BENZENEDIAMINE, 5-CHLORO-3-NITRO-	1,2-BENZENEDIAMINE, 5-CHLORO-3-NITRO-
6219-71-2 1,4-BENZENEDIAMINE, 2-CHLORO-, SULFATE	1,4-BENZENEDIAMINE, 2-CHLORO-, SULFATE
68239-80-5 1,3-BENZENEDIAMINE, 4-CHLORO-, SULFATE (1:1)	1,3-BENZENEDIAMINE, 4-CHLORO-, SULFATE (1:1)
68459-98-3 1,2-BENZENEDIAMINE, 4-CHLORO-, SULFATE (1:1)	1,2-BENZENEDIAMINE, 4-CHLORO-, SULFATE (1:1))METHYLAMMONIO]METHYLETHYL]-OMEGA-HY-DROXY-, N,N'-DITALLOW ACYL DERIVATIVES, METHYL SULFATES (SALTS)
20103-09-7 1,4-BENZENEDIAMINE, 2,5-DICHLORO-	1,4-BENZENEDIAMINE, 2,5-DICHLORO-
541-69-5 1,3-BENZENEDIAMINE, DIHYDROCHLORIDE	1,3-BENZENEDIAMINE, DIHYDROCHLORIDE
615-28-1 1,2-BENZENEDIAMINE, DIHYDROCHLORIDE	1,2-BENZENEDIAMINE, DIHYDROCHLORIDE
624-18-0 1,4-BENZENEDIAMINE, DIHYDROCHLO-RIDE [TSC, TSC]	P-PHENYLENEDIAMINE DIHYDROCHLORIDE
793-24-8 1,4-BENZENEDIAMINE, N-(1,3-DIMETHYL-BUTYL-	1,4-BENZENEDIAMINE, N-(1,3-DIMETHYLBUTYL-
RR-00246-0 1,3-BENZENEDIAMINE, 4-(1,1-DIMETHYLETHYL)-AR-METHYL	1,3-BENZENEDIAMINE, 4-(1,1-DIMETHYLETHYL)-AR-METHYL
62654-17-5 1,4-BENZENEDIAMINE, ETHANEDIOATE (1:1)	1,4-BENZENEDIAMINE, ETHANEDIOATE (1:1)
1197-37-1 1,2-BENZENEDIAMINE, 4-ETHOXY-	1,2-BENZENEDIAMINE, 4-ETHOXY-
67801-06-3 1,3-BENZENEDIAMINE, 4-ETHOXY-, DIHY-DROCHLORIDE	1,3-BENZENEDIAMINE, 4-ETHOXY-, DIHYDROCHLORIDE
RR-00259-5 1,2-BENZENEDIAMINE, 4-ETHOXY-, SULFATE	1,2-BENZENEDIAMINE, 4-ETHOXY-, SULFATE
68015-98-5 1,3-BENZENEDIAMINE, 4-ETHOXY-, SULFATE (1:1)	1,3-BENZENEDIAMINE, 4-ETHOXY-, SULFATE (1:1)TE-TRAKIS[PHOSPHONATO](8-)N,N',O,O'',O'''',O''''']-, PEN-TAPOTASSIUM HYDROGEN, (OC-6-21)-
68966-84-7 1,3-BENZENEDIAMINE, AR-ETHYL-AR-METHYL-	1,3-BENZENEDIAMINE, AR-ETHYL-AR-METHYL-TE-TRAKIS[PHOSPHONATO](8-)], PENTASODIUM HYDRO-GEN, (OC-6-21)-
615-05-4 1,3-BENZENEDIAMINE, 4-METHOXY- [TSC, TSC]	2,4-DIAMINOANISOLE
5307-02-8 1,4-BENZENEDIAMINE, 2-METHOXY- [TSC, TSC]	2-METHOXY-1,4-BENZENEDIAMINE
614-94-8 1,3-BENZENEDIAMINE, 4-METHOXY-, DIHY-DROCHLORIDE	1,3-BENZENEDIAMINE, 4-METHOXY-, DIHYDROCHLO-RIDE
6219-67-6 1,3-BENZENEDIAMINE, 4-METHOXY-, SULFATE	1,3-BENZENEDIAMINE, 4-METHOXY-, SULFATE
39156-41-7 1,3-BENZENEDIAMINE, 4-METHOXY-, SULFATE (1:1) [PAL, PAL, TSC, TSC]	2,4-DIAMINOANISOLE SULFATE
95-70-5 1,4-BENZENEDIAMINE, 2-METHYL- [TSC, TSC]	2,5-DIAMINOTOLUENE
95-80-7 1,3-BENZENEDIAMINE, 4-METHYL- [PAL, PAL, TSC, TSC, TSC]	TOLUENE 2,4-DIAMINE

Chemical Name	Reference Name
108-71-4 1,3-BENZENEDIAMINE, 5-METHYL- [TSC, TSC]	TOLUENE-3,5-DIAMINE
496-72-0 1,2-BENZENEDIAMINE, 4-METHYL- [TSC, TSC]	DIAMINOTOLUENE
823-40-5 1,3-BENZENEDIAMINE, 2-METHYL- [TSC, TSC]	2,6-DIAMINOTOLUENE
2687-25-4 1,2-BENZENEDIAMINE, 3-METHYL-	1,2-BENZENEDIAMINE, 3-METHYL-[1,1'-BIPHENYL]-4, 4'-DIYL)BIS(AZO)]BIS[4-AMINO-5-HYDROXY-, TETRA-SODIUM SALT
25376-45-8 BENZENEDIAMINE, AR-METHYL- [TSC, TSC]	DIAMINOTOLUENE (MIXED ISOMERS)NYLOXY)-
104983-85-9 1,3-BENZENEDIAMINE, 2-METHYL-4,6-BIS (METHYLTHIO)-	1,3-BENZENEDIAMINE, 2-METHYL-4,6-BIS(METHYLTHIO)-T
615-45-2 1,4-BENZENEDIAMINE, 2-METHYL-, DIHY-DROCHLORIDE	1,4-BENZENEDIAMINE, 2-METHYL-, DIHYDROCHLORIDE
101-72-4 BENZENEDIAMINE, N-(1-METHYLETHYL)-N'-PHENYL-, 1,4- [OTS]	N-ISOPROPYL-N'-PHENYL-P-PHENYLENE-DIAMINE
6369-59-1 1,4-BENZENEDIAMINE, 2-METHYL-, SUL-FATE [TSC, TSC]	2,5-TOLUENEDIAMINE SULFATESULFOPHENYL)AZO] PHENYL]AZO][1,1'-BIPHENYL]-4-YL]AZO]-2- HYDROXY-3-METHYL-, DISODIUM SALT
615-50-9 1,4-BENZENEDIAMINE, 2-METHYL-, SULFATE (1:1)	1,4-BENZENEDIAMINE, 2-METHYL-, SULFATE (1:1)
99-56-9 1,2-BENZENEDIAMINE, 4-NITRO- [TSC, TSC]	2-AMINO-4-NITROANILINE
5042-55-7 1,3-BENZENEDIAMINE, 5-NITRO-	1,3-BENZENEDIAMINE, 5-NITRO-
5131-58-8 1,3-BENZENEDIAMINE, 4-NITRO-	1,3-BENZENEDIAMINE, 4-NITRO-
5307-14-2 1,4-BENZENEDIAMINE, 2-NITRO- [TSC, TSC]	2-NITRO-P-PHENYLENEDIAMINE
6219-77-8 1,2-BENZENEDIAMINE, 4-NITRO-, DIHY-DROCHLORIDE	1,2-BENZENEDIAMINE, 4-NITRO-, DIHYDROCHLORIDE
18266-52-9 1,4-BENZENEDIAMINE, 2-NITRO-, DIHY-DROCHLORIDE	1,4-BENZENEDIAMINE, 2-NITRO-, DIHYDROCHLORIDE
68239-82-7 1,2-BENZENEDIAMINE, 4-NITRO-, SULFATE (1:1)	1,2-BENZENEDIAMINE, 4-NITRO-, SULFATE (1:1)
68239-83-8 1,4-BENZENEDIAMINE, 2-NITRO-, SULFATE (1:1)	1,4-BENZENEDIAMINE, 2-NITRO-, SULFATE (1:1)
101-54-2 1,4-BENZENEDIAMINE, N-PHENYL- [OTS]	P-AMINODIPHENYLAMINE
541-70-8 1,3-BENZENEDIAMINE, SULFATE (1:1)	1,3-BENZENEDIAMINE, SULFATE (1:1)
16245-77-5 1,4-BENZENEDIAMINE, SULFATE (1:1)	1,4-BENZENEDIAMINE, SULFATE (1:1)
124737-31-1 BENZENEDIAZONIUM, 4-(DIMETHYLAMINO)-, SALT WITH 2-HYDROXY-5-SULFOBENZOIC ACID (1:1)	BENZENEDIAZONIUM, 4-(DIMETHYLAMINO)-, SALT WITH 2-HYDROXY-5-
619-97-6 BENZENE DIAZONIUM NITRATE	BENZENE DIAZONIUM NITRATE
626-17-5 1,3-BENZENEDICARBONITRILE [PAL]	M-PHTHALODINITRILE
100-20-9 1,4-BENZENEDICARBONYL DICHLORIDE [PAL]	TEREPHTHALOYL CHLORIDE
68584-15-6 1,3-BENZENEDICARBOXYLIC ACID, POLYMER WITH 5-AMINO-1,3,5-TRIMETHYLCYCLOHEXY-LAMINE, MODIFIED	1,3-BENZENEDICARBOXYLIC ACID, POLYMER WITH 5-AMINO-1,3,5-TRIESQUIOXANES AND POLYETHYLENE-POYPROPYLENE GLYCOL MONOBUTYL ETHER
26761-40-0 1,2-BENZENEDICARBOXYLIC ACID, DIISODE-CYL ESTER [TSE]	DIISODECYL PHTHALATE
68515-42-4 1,2-BENZENEDICARBOXYLIC ACID, DI-C(7-11)-BRANCHED AND LINEARALKYL ESTERS [TSE]	PHTHALIC ACID, DIALKYL (C7-C11) ESTERINE
117-81-7 1,2-BENZENEDICARBOXYLIC ACID, BIS(2-ETHYLHEXYL) ESTER [ONT, ONT, PAL, PAL, TSC, TSC, TSC]	DI(2-ETHYLHEXYL)PHTHALATE
6422-86-2 1,4-BENZENEDICARBOXYLIC ACID, BIS(2-ETHYLHEXYL) ESTER	1,4-BENZENEDICARBOXYLIC ACID, BIS(2-ETHYLHEXYL) ESTER
131-15-7 1,2-BENZENEDICARBOXYLIC ACID, BIS(1-METHYLHEPTYL) ESTER	1,2-BENZENEDICARBOXYLIC ACID, BIS(1-METHYLHEP-TYL) ESTER
7195-45-1 1,2-BENZENEDICARBOXYLIC ACID, BIS(OXI-RANYLMETHYL) ESTER	1,2-BENZENEDICARBOXYLIC ACID, BIS(OXIRANYL-METHYL) ESTER

Chemical Name	Reference Name
84-69-5 1,2-BENZENEDICARBOXYLIC ACID, BIS(2-METHYLPROPYL) ESTER	1,2-BENZENEDICARBOXYLIC ACID, BIS(2-METHYL-PROPYL) ESTER
85-70-1 1,2-BENZENEDICARBOXYLIC ACID, 2-BUTOXY-2-OXYETHYL BUTYLESTER	1,2-BENZENEDICARBOXYLIC ACID, 2-BUTOXY-2-OXYETHYL BUTYL
84-64-0 1,2-BENZENEDICARBOXYLIC ACID, BUTYL CYCLOHEXYL ESTER	1,2-BENZENEDICARBOXYLIC ACID, BUTYL CYCLO-HEXYL ESTER
85-69-8 1,2-BENZENEDICARBOXYLIC ACID, BUTYL 2-ETHYLHEXYL ESTER	1,2-BENZENEDICARBOXYLIC ACID, BUTYL 2-ETHYL-HEXYL ESTER
84-78-6 1,2-BENZENEDICARBOXYLIC ACID, BUTYL OCTYL ESTER	1,2-BENZENEDICARBOXYLIC ACID, BUTYL OCTYL ESTER
85-68-7 1,2-BENZENEDICARBOXYLIC ACID, BUTYL PHENYLMETHYL ESTER [PAL, TSC, TSC, TSC]	BUTYL BENZYL PHTHALATE
25724-58-7 1,2-BENZENEDICARBOXYLIC ACID, DECYL HEXYL ESTER	1,2-BENZENEDICARBOXYLIC ACID, DECYL HEXYL ESTER
119-07-3 1,2-BENZENEDICARBOXYLIC ACID, DECYL OCTYL ESTER	1,2-BENZENEDICARBOXYLIC ACID, DECYL OCTYL ESTER
131-17-9 1,2-BENZENEDICARBOXYLIC ACID, DI-2-PROPENYL ESTER [PAL]	DIALLYL PHTHALATE
84-74-2 1,2-BENZENEDICARBOXYLIC ACID, DIBUTYL ESTER [PAL, TSC, TSC, TSC]	DIBUTYL PHTHALATE
84-61-7 1,2-BENZENEDICARBOXYLIC ACID, DICYCLO-HEXYL ESTER	1,2-BENZENEDICARBOXYLIC ACID, DICYCLOHEXYL ESTER
84-66-2 1,2-BENZENEDICARBOXYLIC ACID, DIETHYL ESTER [PAL, TSC]	DIETHYL PHTHALATE
84-75-3 1,2-BENZENEDICARBOXYLIC ACID, DIHEXYL ESTER	1,2-BENZENEDICARBOXYLIC ACID, DIHEXYL ESTER
26761-40-0 1,2-BENZENEDICARBOXYLIC ACID, DIISODE-CYL ESTER [TSC, TSC, TSC]	DIISODECYL PHTHALATEOCTYL ESTER
28553-12-0 1,2-BENZENEDICARBOXYLIC ACID, DI-ISONONYL ESTER [TSC]	DIISONONYL PHTHALATE
27554-26-3 1,2-BENZENEDICARBOXYLIC ACID, DI-ISOOCTYL ESTER [TSC]	DIISOOCTYL PHTHALATE
131-11-3 1,2-BENZENEDICARBOXYLIC ACID, DIMETHYL ESTER [PAL, TSC, TSC, TSC]	DIMETHYL PHTHALATEDIHYDROXY-9,10-DIOXO-
84-76-4 1,2-BENZENEDICARBOXYLIC ACID, DINONYL ESTER	1,2-BENZENEDICARBOXYLIC ACID, DINONYL ESTER
117-84-0 1,2-BENZENEDICARBOXYLIC ACID, DIOCTYL ESTER [PAL, TSC, TSC]	DI-N-OCTYL PHTHALATE
119-06-2 1,2-BENZENEDICARBOXYLIC ACID, DITRIDE-CYL ESTER	1,2-BENZENEDICARBOXYLIC ACID, DITRIDECYL ESTER
3648-20-2 1,2-BENZENEDICARBOXYLIC ACID, DIUNDE-CYL ESTER	1,2-BENZENEDICARBOXYLIC ACID, DIUNDECYL ES-TER4,6-DINITROPHENYL)AZO]-4-METHOXYPHENYL]-
84-72-0 1,2-BENZENEDICARBOXYLIC ACID, 2-ETHOXY-2-OXOETHYL ETHYL ESTER	1,2-BENZENEDICARBOXYLIC ACID, 2-ETHOXY-2-OX-OETHYL ETHYL ESTE
85-71-2 1,2-BENZENEDICARBOXYLIC ACID, 2-ETHOXY-2-OXOETHYL METHYL ESTER [PAL]	METHYL PHTHALYL ETHYL GLYCOLATE
89-13-4 1,2-BENZENEDICARBOXYLIC ACID, 2-ETHYL-HEXYL-8-METHYLNONYLESTER	1,2-BENZENEDICARBOXYLIC ACID, 2-ETHYLHEXYL-8-METHYLNONYL
61702-81-6 1,2-BENZENEDICARBOXYLIC ACID, HEXYL ISODECYL ESTER	1,2-BENZENEDICARBOXYLIC ACID, HEXYL ISODECYL ESTER
131-17-9 1,2-BENZENEDICARBOXYLIC ACID, DI-2-PROPENYL ESTER [TSC, TSC, TSC]	DIALLYL PHTHALATE
61886-60-0 1,2-BENZENEDICARBOXYLIC ACID, ISODECYL TRIDECYL ESTER	1,2-BENZENEDICARBOXYLIC ACID, ISODECYL TRIDE-CYL ESTER
68515-49-1 1,2-BENZENEDICAROBOXIC ACID, DI-C(9-11)-BRANCHED ALKYL ESTERS, C(10-RICH [(DIISODECYL PHTHALATE (MIXED ISOMERS)] [TSC, TSC]	DIISODECYL PHTHALATE (MIXED ISOMERS)RS, C(13)-RICH [DITRIDCEYL PHTHALATE (MIXED ISOMERS)]

Chemical Name	Reference Name
68515-42-4 1,2-BENZENEDICAROBOXIC ACID, DI-C(7-11) -BRANCHED AND LINEARALKYL ESTERS [DI (HEPTYL-, NONYL-, UNDECYL) PHTHALATE (MIXED ISOMERS)] [TSC]	PHTHALIC ACID, DIALKYL (C7-C11) ESTER
68515-47-9 1,2-BENZENEDICAROBOXIC ACID, DI-C(11-14)-BRANCHED ALKYL ESTERS, C(13)-RICH [DITRIDCEYL PHTHALATE (MIXED ISOMERS)] [TSC, TSC]	DITRIDECYL PHTHALATE (MIXED ISOMERS)
68515-50-4 1,2-BENZENEDICAROBOXIC ACID, DIHEXYL ESTER, BRANCHED AND LINEAR [DIHEXYL PHTHALATE (MIXED ISOMERS)] [TSC, TSC]	DIHEXYL PHTHALATE (MIXED ISOMERS)S, C(10-RICH [(DIISODECYL PHTHALATE (MIXED ISOMERS)]
95-50-1 BENZENE, 1,2-DICHLORO- [PAL, TSC, TSC, TSC, TSC]	O-DICHLOROBENZENE
106-46-7 BENZENE, 1,4-DICHLORO- [PAL, PAL, TSC, TSC, TSC]	P-DICHLOROBENZENE
541-73-1 BENZENE, 1,3-DICHLORO- [PAL, TSC, TSC]	1,3-DICHLOROBENZENE
25321-22-6 BENZENE, DICHLORO- [PAL]	DICHLOROBENZENE (MIXED ISOMERS)
6607-45-0 BENZENE, (1,2-DICHLOROETHENYL)-	BENZENE, (1,2-DICHLOROETHENYL)-
72-55-9 BENZENE, 1,1'-(DICHLOROETHENYLIDENE)BIS [4-CHLORO- [PAL]	DDE
72-54-8 BENZENE, 1,1'-(2,2-DICHLOROETHYLIDENE)BIS [4-CHLORO- [PAL]	DDD
102-36-3 BENZENE, 1,2-DICHLORO-4-ISOCYANATO- [MSL, TSC, TSC, TSC]	ISOCYANIC ACID, 3,4-DICHLOROPHENYL ESTER
34893-92-0 BENZENE, 1,3-DICHLORO-5-ISOCYANATO- [TSC, TSC, TSC]	1,3-DICHLORO-5-PHENYL ISOCYANATEESTER
98-87-3 BENZENE, DICHLOROMETHYL- [TSC, TSC]	BENZAL CHLORIDE
89-61-2 BENZENE, 1,4-DICHLORO-2-NITRO- [OTS]	2,5-DICHLORONITROBENZENE
611-06-3 BENZENE, 2,4-DICHLORO-1-NITRO-	BENZENE, 2,4-DICHLORO-1-NITRO-
3209-22-1 BENZENE, 1,2-DICHLORO-3-NITRO-	BENZENE, 1,2-DICHLORO-3-NITRO-
1836-75-5 BENZENE, 2,4-DICHLORO-1-(4-NITROPHENOXY)- [PAL, PAL]	NITROFEN
328-84-7 BENZENE, 1,2-DICHLORO-4-(TRIFLUO-ROMETHYL)-3,4-DICHLOROBENZOTRIFLUO-RIDE [TSC, TSC, TSC, TSC]	3,4-DICHLOROBENZOTRIFLUORIDE
108-57-6 BENZENE, 1,3-DIETHENYL- [PAL]	DIVINYL BENZENE
1321-74-0 BENZENE, DIETHENYL- [PAL]	DIVINYL BENZENE
105-05-5 BENZENE, 1,4-DIETHYL- [OTS, PAL]	P-DIETHYL BENZENE
135-01-3 BENZENE, 1,2-DIETHYL- [PAL]	O-DIETHYL BENZENE
141-93-5 BENZENE, 1,3-DIETHYL- [PAL]	M-DIETHYL BENZENE
104-49-4 BENZENE, 1,4-DIISOCYANATO-	BENZENE, 1,4-DIISOCYANATO-
123-61-5 BENZENE, 1,3-DIISOCYANATO-	BENZENE, 1,3-DIISOCYANATO-
91-08-7 BENZENE, 1,3-DIISOCYANATO-2-METHYL- (2,6-TOLUENE DIISOCYANATE) [CAI]	TOLUENE-2,6-DIISOCYANATE
91-08-7 BENZENE, 1,3-DIISOCYANATO-2-METHYL- [TSC, TSC]	TOLUENE-2,6-DIISOCYANATE
584-84-9 BENZENE, 2,4-DIISOCYANATO-1-METHYL- [PAL, PAL, TSC, TSC]	TOLUENE 2,4-DIISOCYANATE
1321-38-6 BENZENE, DIISOCYANATOMETHYL-	BENZENE, DIISOCYANATOMETHYL-
26471-62-5 BENZENE,2,4-DIISOCYANATOMETHYL- [PAL]	TOLUENE DIISOCYANATE)METHYLENE)AMINO)-
26471-62-5 BENZENE, 1,3-DIISOCYANATOMETHYL- [TSC, TSC]	TOLUENE DIISOCYANATE
584-84-9 BENZENE, 2,4-DIISOCYANATO-1-METHYL- (2,4-TOLUENE DIISOCYANATE) [CAI]	TOLUENE 2,4-DIISOCYANATECHLORIDE)

Chemical Name	Reference Name
1321-38-6 BENZENE, DIISOCYANATOMETHYL- (UNSPECIFIED TOLUENE DIISOCYANATE) [CAI]	BENZENE, DIISOCYANATOMETHYL-
26471-62-5 BENZENE, 1,3-DIISOCYANATOMETHYL-(TOLUENE DIISOCYANATE) [CAI]	TOLUENE DIISOCYANATE
10031-75-1 BENZENE, 1,1'-(DIISOCYANATOMETHYLENE) BIS-	BENZENE, 1,1'-(DIISOCYANATOMETHYLENE)BIS-
1477-55-0 1,3-BENZENEDIMETHANAMINE [PAL, TSC, TSC]	M-XYLENE-ALPHA, ALPHA'-DIAMINE
95-47-6 BENZENE, 1,2-DIMETHYL- [PAL, TSC, TSC, TSC]	O-XYLENER
106-42-3 BENZENE, 1,4-DIMETHYL- [PAL, TSC, TSC, TSC, TSC]	P-XYLENE
108-38-3 BENZENE, 1,3-DIMETHYL- [PAL, TSC, TSC, TSC]	M-XYLENE
1330-20-7 BENZENE, DIMETHYL- [PAL, TSC, TSC]	XYLENES (O-, M-, P- ISOMERS)
98-06-6 BENZENE, (1,1-DIMETHYLETHYL)- [PAL, TSC, TSC]	TERT-BUTYLBENZENE
98-51-1 BENZENE, 1-(1,1-DIMETHYLETHYL)-4-METHYL- [PAL]	P-TERT-BUTYLTOLUENE
1320-21-4 BENZENE, DIMETHYL(PENTYLOXY)-	BENZENE, DIMETHYL(PENTYLOXY)-ESTER
RR-00967-6 BENZENE, 1,2-DIMETHYL-, POLYPROPENE DERIVATIVES, SULFONATED,POTASSIUM SALTS	BENZENE, 1,2-DIMETHYL-, POLYPROPENE DERIVATIVES, SULFONATED,
99-65-0 BENZENE, 1,3-DINITRO- [PAL]	M-DINITROBENZENE
100-25-4 BENZENE, 1,4-DINITRO- [PAL]	P-DINITROBENZENE
528-29-0 BENZENE, 1,2-DINITRO- [PAL]	O-DINITROBENZENE
25154-54-5 BENZENE, DINITRO- [PAL]	DINITROBENZENE (MIXED)
12385-08-9 BENZENEDIOL	BENZENEDIOL
108-46-3 1,3-BENZENEDIOL [PAL]	RESORCINOL
120-80-9 1,2-BENZENEDIOL [PAL, PST]	CATECHOL
123-31-9 1,4-BENZENEDIOL (HYDROQUINONE) [TSC, TSC, TSC]	HYDROQUINONE
98-29-3 1,2-BENZENEDIOL, 4-(1,1-DIMETHYLETHYL)- [PAL]	4-TERT-BUTYL CATECHOL
51-43-4 1,2-BENZENEDIOL, 4-[1-HYDROXY-2-(METHYLAMINO)ETHYL]	1,2-BENZENEDIOL, 4-[1-HYDROXY-2-(METHYLAMINO) ETHYL]
51-43-4 1,2-BENZENEDIOL,4-[1-HYDROXY-2-(METHYLAMINO)ETHYL]- [MSL]	1,2-BENZENEDIOL, 4-[1-HYDROXY-2-(METHYLAMINO) ETHYL]
15245-44-0 1,3-BENZENEDIOL, 2,4,6-TRINITRO-, LEAD SALT	1,3-BENZENEDIOL, 2,4,6-TRINITRO-, LEAD SALT
RR-01358-1 BENZENE-1,3-DISULFOHYDRAZIDE	BENZENE-1,3-DISULFOHYDRAZIDE
98-48-6 BENZENEDISULFONIC ACID [HON]	1,3-BENZENEDISULFONIC ACID
98-48-6 1,3-BENZENEDISULFONIC ACID	1,3-BENZENEDISULFONIC ACID
41098-56-0 1,4-BENZENEDISULFONIC ACID, 2,2'-[1,2-ETHENEDIYLBIS[(3-SULFO-4,1-PHENYLENE) IMINO[6-(DIETHYLAMINO)-1,3,5-TRIAZINE-4, 2-DIYL]IMINO]BIS-, HEXASODIUM SALT	1,4-BENZENEDISULFONIC ACID, 2,2'-[1,2-ETHENEDIYL-BIS[(3-SULFO
122-09-8 BENZENEETHANAMINE, ALPHA,ALPHA-DIMETHYL- [TSC, TSC]	ALPHA,ALPHA-DIMETHYLPHENETHYLAMINE
60-15-1 BENZENEETHANAMINE, .ALPHA.-METHYL-	BENZENEETHANAMINE, .ALPHA.-METHYL-
37853-59-1 BENZENE, 1,1'-[1,2-ETHANEDIYLBIS(OXY)]BIS[2, 4,6-TRIBROMO- [TSC, TSC, TSC, TSC]	FIREMASTER 680
100-42-5 BENZENE, ETHENYL- [PAL, TSC, TSC]	STYRENE
88497-56-7 BENZENE, ETHENYL-, HOMOPOLYMER, BROMINATED	BENZENE, ETHENYL-, HOMOPOLYMER, BROMINATED
25013-15-4 BENZENE, ETHENYLMETHYL- [PAL]	VINYL TOLUENE
1319-73-9 BENZENE, ETHENYL-, MONOMETHYL DERIV.	BENZENE, ETHENYL-, MONOMETHYL DERIV.

Chemical Name	Reference Name
100-41-4 BENZENE, ETHYL- [PAL, TSC, TSC, TSC]	ETHYLBENZENE
611-14-3 BENZENE, 1-ETHYL-2-METHYL-	BENZENE, 1-ETHYL-2-METHYL-
25550-14-5 BENZENE, ETHYLMETHYL-	BENZENE, ETHYLMETHYL-
462-06-6 BENZENE, FLUORO- [PAL]	FLUOROBENZENE
608-73-1 BENZENE HEXACHLORIDE [FLA]	HEXACHLOROCYCLOHEXANE (MIXED ISOMERS)
118-74-1 BENZENE, HEXACHLORO- [PAL, PAL]	HEXACHLOROBENZENE
103-71-9 BENZENE, ISOCYANATO- [TSC, TSC, TSC]	PHENYL ISOCYANATE
28178-42-9 BENZENE, 2-ISOCYANATO-1,3-BIS(1-METHYLETHYL)-	BENZENE, 2-ISOCYANATO-1,3-BIS(1-METHYLETHYL)-
28556-81-2 BENZENE, 2-ISOCYANATO-1,3-DIMETHYL-	BENZENE, 2-ISOCYANATO-1,3-DIMETHYL-
5873-54-1 BENZENE, 1-ISOCYANATO-2-(4-ISO-CYANATOPHENYL)METHYL- [TSC, TSC, TSC]	2,4'-DIPHENYLMETHANE DIISOCYANATE
75790-84-0 BENZENE, 2-ISOCYANATO-4-[(4-ISOCYANATO PHENYL)METHYL]-1-METHYL-	BENZENE, 2-ISOCYANATO-4-[(4-ISOCYANATO PHENYL) METHYL]-1-METHTHYLCYCLOHEXYL]METHYL]AMINO] CARBONYL]OXY]ETHYL ESTER
75790-87-3 BENZENE, 1-ISOCYANATO-2-[(4-ISO-CYANATOPHENYL)THIO]-	BENZENE, 1-ISOCYANATO-2-[(4-ISOCYANATOPHENYL) THIO]-YL-
614-68-6 BENZENE, 1-ISOCYANATO-2-METHYL-	BENZENE, 1-ISOCYANATO-2-METHYL-
622-58-2 BENZENE, 1-ISOCYANATO-4-METHYL-	BENZENE, 1-ISOCYANATO-4-METHYL-
100-28-7 BENZENE, 1-ISOCYANATO-4-NITRO-	BENZENE, 1-ISOCYANATO-4-NITRO-
329-01-1 BENZENE, 1-ISOCYANATO-3-(TRIFLUO-ROMETHYL)-	BENZENE, 1-ISOCYANATO-3-(TRIFLUOROMETHYL)-RIFLUORIDE
63-92-3 BENZENEMETHANAMINE, N-(2-CHLOROETHYL)-N-(1-METHYL-2-PHE-NOXYETHYL)-,HYDROCHLORIDE [PAL, PAL, PAL, PAL]	PHENOXYBENZAMINE HYDROCHLORIDE
772-54-3 BENZENEMETHANAMINE, N,N-DIETHYL-	BENZENEMETHANAMINE, N,N-DIETHYL-
92-59-1 BENZENEMETHANAMINE, N-ETHYL-N-PHENYL- [PAL]	ETHYLBENZYLANILINE
98-84-0 BENZENEMETHANAMINE,.ALPHA.-METHYL- [PAL]	ALPHA-METHYLBENZYLAMINE
2449-49-2 BENZENEMETHANAMINE, N,N,.ALPHA.-TRIMETHYL- [PAL]	ALPHA-METHYLBENZYL DIMETHYL AMINE
1694-09-3 BENZENEMETHANAMINIUM, N-[4-[[4-(DIMETHYLAMINO)PHENYL][4-[ETHYL[(3-SUL-FOPHENYL)METHYL]AMINO]PHENYL]METHY-LENE]-2,5- CYCLOHEXADIEN-1-YLIDENE]-N-ETHYL-SULFO-,HYDROXIDE, INNER SALT, SODIUM SALT [PAL, PAL]	BENZYL VIOLET 4B
100-53-8 BENZENEMETHANETHIOL [PAL]	BENZYL MERCAPTAN
100-51-6 BENZENEMETHANOL [PAL]	BENZYL ALCOHOL
7568-93-6 BENZENEMETHANOL,.ALPHA.-(AMINOMETHYL)-	BENZENEMETHANOL,.ALPHA.-(AMINOMETHYL)-
2454-37-7 BENZENEMETHANOL, 3-AMINO-.ALPHA.-METHYL-	BENZENEMETHANOL, 3-AMINO-.ALPHA.-METHYL-
13826-35-2 BENZENEMETHANOL, 3-PHENOXY-	BENZENEMETHANOL, 3-PHENOXY-
50789-44-1 BENZENEMETHANOL, 3-PHENOXY-, ACETATE	BENZENEMETHANOL, 3-PHENOXY-, ACETATE
104-46-1 BENZENE, 1-METHOXY-4-(1-PROPENYL)-	BENZENE, 1-METHOXY-4-(1-PROPENYL)-
106-44-5 BENZENE, 4-METHYL- [TSC, TSC, TSC]	P-CRESOL
108-88-3 BENZENE, METHYL- [PAL, TSC, TSC, TSC]	TOLUENE
RR-01238-4 BENZENE, 1-(1-METHYLBUTOXY)-4-NITRO-,	BENZENE, 1-(1-METHYLBUTOXY)-4-NITRO-,
121-14-2 BENZENE, 1-METHYL-2,4-DINITRO- [OTS, PAL]	2,4-DINITROTOLUENE
606-20-2 BENZENE, 2-METHYL-1,3-DINITRO- [PAL]	2,6-DINITROTOLUENE

Chemical Name	Reference Name
610-39-9 BENZENE, 4-METHYL-1,2-DINITRO- [PAL]	3,4-DINITROTOLUENE
25321-14-6 BENZENE, METHYLDINITRO- [PAL]	DINITROTOLUENE (MIXED ISOMERS)-LYXO-HEXOPY-RANOSYL)OXY]-7,8,9,10-TETRAHYDRO-6,8,11-TRI-HYDROXY-8-(HYDROXYACETYL)-1-METHOXY-, HY-DROCHLORIDE, (8S-CIS)-
101-68-8 BENZENE, 1,1'-METHYLENEBIS[4-ISOCYANATO-[PAL]	METHYLENE BISPHENOL ISOCYANATE (MDI)
101-68-8 BENZENE, 1,1'-METHYLENEBIS(4-ISOCYANATO-[TSC, TSC, TSC]	METHYLENE BISPHENOL ISOCYANATE (MDI)
2536-05-2 BENZENE, 1,1'-METHYLENEBIS(2-ISOCYANATO-	BENZENE, 1,1'-METHYLENEBIS(2-ISOCYANATO-
26447-40-5 BENZENE, 1,1-METHYLENEBIS[ISOCYANATO-[TSC, TSC, TSC]	1,1'-METHYLENEBIS (ISOCYANATO-) BENZENE
139-25-3 BENZENE, 1,1'-METHYLENEBIS[4-ISOCYANATO-3-METHYL-	BENZENE, 1,1'-METHYLENEBIS[4-ISOCYANATO-3-METHYL-
98-83-9 BENZENE, (1-METHYLETHENYL)- [TSC, TSC]	ALPHA-METHYL STYRENE
98-82-8 BENZENE, (1-METHYLETHYL)- [PAL]	CUMENE
25327-89-3 BENZENE, 1,1'-(1-METHYLETHYLIDENE)BIS[3,5-DIBROMO-4-(2-PROPENYLOXY)- [TSC, TSC, TSC, TSC]	ALLYL ETHER OF TETRABROMOBISPHENOL-A
77851-17-3 BENZENE, (1-METHYLETHYL)(2-PHENYLETHYL)-	BENZENE, (1-METHYLETHYL)(2-PHENYLETHYL)-
2422-91-5 BENZENE, 1,1',1"-METHYLIDYENETRIS(4-ISO-CYANATO-	BENZENE, 1,1',1"-METHYLIDYENETRIS(4-ISOCYANATO-
99-87-6 BENZENE, 1-METHYL-4-(1-METHYLETHYL)-[PAL, TSC]	P-CYMENE
88-72-2 BENZENE, 1-METHYL-2-NITRO- [PAL]	O-NITROTOLUENE
99-08-1 BENZENE, 1-METHYL-3-NITRO- [PAL]	M-NITROTOLUENE
99-99-0 BENZENE, 1-METHYL-4-NITRO- [PAL]	P-NITROTOLUENE
1321-12-6 BENZENE, METHYLNITRO- [PAL]	NITROTOLUENE (MIXED ISOMERS)
1320-01-0 BENZENE, METHYLPENTYL-	BENZENE, METHYLPENTYL-
3586-14-9 BENZENE, 1-METHYL-3-PHENOXY-	BENZENE, 1-METHYL-3-PHENOXY-
13414-54-5 BENZENE, 1-[(2-METHYL-2-PROPENYL)OXY]-2-NITRO-	BENZENE, 1-[(2-METHYL-2-PROPENYL)OXY]-2-NITRO-
135-98-8 BENZENE, (1-METHYLPROPYL)- [PAL, TSC, TSC]	SEC-BUTYLBENZENE
538-93-2 BENZENE, (2-METHYLPROPYL)- [PAL]	ISOBUTYLBENZENE
118-96-7 BENZENE, 2-METHYL-1,3,5-TRINITRO- [PAL]	2,4,6-TRINITROTOLUENE
98-95-3 BENZENE, NITRO- [PAL, TSC, TSC, TSC]	NITROBENZENE
101-84-8 BENZENE, 1,1'-OXYBIS- [PAL, TSC, TSC, TSC]	PHENYL ETHER
69834-19-1 BENZENE, 1,1'-OXYBIS[DODECYL-	BENZENE, 1,1'-OXYBIS[DODECYL-
28299-41-4 BENZENE, 1,1'-OXYBIS[METHYL-	BENZENE, 1,1'-OXYBIS[METHYL-
61702-88-3 BENZENE, 1,1'-OXYBIS[(1,1,3,3-TETRAMETHYL-BUTYL)-	BENZENE, 1,1'-OXYBIS[(1,1,3,3-TETRAMETHYLBUTYL)-
93-96-9 BENZENE, 1,1'-(OXYDIETHYLIDENE)BIS-	BENZENE, 1,1'-(OXYDIETHYLIDENE)BIS-
85-22-3 BENZENE, PENTABROMOETHYL- [TSC, TSC, TSC, TSC, TSC]	PENTABROMOETHYLBENZENE
87-83-2 BENZENE, PENTABROMOMETHYL-	BENZENE, PENTABROMOMETHYL-
608-93-5 BENZENE, PENTACHLORO- [PAL, TSC, TSC, TSC, TSC]	PENTACHLOROBENZENE
82-68-8 BENZENE, PENTACHLORONITRO- [PAL]	PENTACHLORONITROBENZENE
538-68-1 BENZENE, PENTYL-	BENZENE, PENTYL-
14684-25-4 BENZENE PHOSPHOROUS THIODICHLO-RIDE [NJL, NJL]	BENZENE PHOSPHORUS THIODICHLORIDE

Chemical Name	Reference Name
644-97-3 BENZENE PHOSPHORUS DICHLORIDE [NJL, NJL]	PHENYL PHOSPHORUS DICHLORIDE
14684-25-4 BENZENE PHOSPHORUS THIODICHLORIDE	BENZENE PHOSPHORUS THIODICHLORIDE
80-54-6 BENZENEPROPANAL, 4-(1,1-DIMETHYLETHYL)-.ALPHA.-METHYL-	BENZENEPROPANAL, 4-(1,1-DIMETHYLETHYL)-.ALPHA.-METHYL-
103-95-7 BENZENEPROPANAL, .ALPHA.-METHYL-4-(1-METHYLETHYL)-	BENZENEPROPANAL, .ALPHA.-METHYL-4-(1-METHYLETHYL)-
127519-17-9 BENZENEPROPANOIC ACID, 3-(2H-BENZO-TRIAZOL-2-YL)-5-(1,1-DIMETHYLETHYL)-4-HYDROXY-, C(7-9) BRANCHED AND LINEAR ALKYL ESTERS	BENZENEPROPANOIC ACID, 3-(2H-BENZOTRIAZOL-2-YL)-5-(1,1-DIMETROXY-, C(10-16)-ALKYL ETHERS
6386-38-5 BENZENEPROPANOIC ACID, 3,5-BIS(1,1-DIMETHYLETHYL)-4-HYDROXY-, METHYL ESTER	BENZENEPROPANOIC ACID, 3,5-BIS(1,1-DIMETHYLETHYL)-4-HYDROXY-
6386-38-5 BENZENEPROPANOIC ACID, 3,5-BIS(1,1-DIMETHYLETHYL)- [OTS]	BENZENEPROPANOIC ACID, 3,5-BIS(1,1-DIMETHYLETHYL)-4-HYDROXY-
103-65-1 BENZENE, PROPYL- [PAL]	PROPYL BENZENE
RR-00903-0 BENZENE, SUBSTITUTED, ALKYL ACRYLATE DERIVATIVE	BENZENE, SUBSTITUTED, ALKYL ACRYLATE DERIVATIVE
80-17-1 BENZENESULFOHYDRAZINE [HM1, HM1]	BENZENESULFONIC ACID, HYDRAZIDE
98-10-2 BENZENESULFONAMIDE	BENZENESULFONAMIDE
741-58-2 BENZENESULFONAMIDE, N-(2-MERCAP-TOETHYL)-, S-ESTER WITH O,O-DIISOPROPY-LPHOSPHORODITHIOATE	BENZENESULFONAMIDE, N-(2-MERCAPTOETHYL)-, S-ESTER WITH O,O-D
70-55-3 BENZENESULFONAMIDE, 4-METHYL-	BENZENESULFONAMIDE, 4-METHYL-
98-11-3 BENZENESULFONIC ACID [HON]	PHENOLSULPHONIC ACID
147170-38-5 BENZENESULFONIC ACID, 4-METHYL-, REAC-TION PRODUCTS WITH OXIRANE MONO[(C10, 16-ALKYLOXY)METHYL] DERIVATIVES AND 2, 2,4 (OR 2,4,4)-TRIMETHYL-1,6-HEXANEDIAMINE	BENZENESULFONIC ACID, 4-METHYL-, REACTION PRODUCTS WITH OXIR
121-47-1 BENZENESULFONIC ACID, 3-AMINO- [TSC, TSC]	METANILIC ACID
121-57-3 BENZENESULFONIC ACID, 4-AMINO-	BENZENESULFONIC ACID, 4-AMINO-
RR-01040-2 BENZENESULFONIC ACID, C8-24-ALKYL DERIVATIVES	BENZENESULFONIC ACID, C8-24-ALKYL DERIVATIVES
RR-01041-3 BENZENESULFONIC ACID, C8-24-ALKYL DERIVATIVES, AMMONIUM, MAGNESIUM, POTASSIUM OR ZINC SALT	BENZENESULFONIC ACID, C8-24-ALKYL DERIVATIVES, AMMONIUM, MAG
68411-30-3 BENZENESULFONIC ACID, C10-13-ALKYL DERIVATIVES SODIUM SALT	BENZENESULFONIC ACID, C10-13-ALKYL DERIVATIVES SODIUM SALT
68584-22-5 BENZENESULFONIC ACID, C10-16-ALKYL DERIVATIVES	BENZENESULFONIC ACID, C10-16-ALKYL DERIVATIVES-METHYLCYCLOHEXYLAMINE, MODIFIED
68584-23-6 BENZENESULFONIC ACID, C10-16-ALKYL DERIVATIVES, CALCIUM SALTS	BENZENESULFONIC ACID, C10-16-ALKYL DERIVATIVES, CALCIUM SALT
68584-25-8 BENZENESULFONIC ACID, C10-16-ALKYL DERIVATIVES, COMPOUNDS WITH TRI-ETHANOLAMINE	BENZENESULFONIC ACID, C10-16-ALKYL DERIVATIVES, COMPOUNDS WITH 2-PROPANAMINE
68584-26-9 BENZENESULFONIC ACID, C10-16-ALKYL DERIVATIVES, MAGNESIUM SALTS	BENZENESULFONIC ACID, C10-16-ALKYL DERIVATIVES, MAGNESIUM SATH TRIETHANOLAMINE
68584-27-0 BENZENESULFONIC ACID, C10-16-ALKYL DERIVATIVES, POTASSIUM SALTS	BENZENESULFONIC ACID, C10-16-ALKYL DERIVATIVES, POTASSIUM SALTS
68081-81-2 BENZENESULFONIC ACID, C10-16-ALKYL DERIVATIVES, SODIUM SALTS	BENZENESULFONIC ACID, C10-16-ALKYL DERIVATIVES, SODIUM SALTSD, GLYCEROL AND PHTHALIC ANHY-DRIDE
16110-89-7 BENZENESULFONIC ACID, 4-[(4,6-DICHLORO-1, 3,5-TRIAZIN-2-YL)AMINO]-	BENZENESULFONIC ACID, 4-[(4,6-DICHLORO-1,3,5-TRI-AZIN-2-YL)PHENYL]AZO]PHENYL]AZO][1,1'-BIPHENYL]-4-YL]AZO]-2-HYDROXY BENZOATO(4-)]-, DISODIUM

Chemical Name	Reference Name
27176-87-0 BENZENESULFONIC ACID, DODECYL- [PAL]	DODECYLBENZENESULFONIC ACID
26264-06-2 BENZENESULFONIC ACID, DODECYL-, CALCIUM SALT [PAL]	CALCIUM DODECYLBENZENE SULFONATE
54590-52-2 BENZENESULFONIC ACID, 4-DODECYL-, COMPD. WITH 1-AMINO-2-PROPANOL (1:1) [PAL]	ISOPROPANOLAMINE DODECYLBENZENESULFONATE (1:1)
27323-41-7 BENZENESULFONIC ACID, DODECYL-, COMPD. WITH 2,2',2"-NITRILOTRIS[ETHANOL](1:1) [PAL]	TRIETHANOLAMINE DODECYLBENZOSULFONATE
25155-30-0 BENZENESULFONIC ACID, DODECYL-, SODIUM SALT [PAL]	SODIUM DODECYLBENZENESULFONATE
81-11-8 BENZENESULFONIC ACID, 2,2'-(1,2-ETHANEDI- [OTS]	4,4'-DIAMINO-2,2'-STILBENEDISULFONIC ACID, DISODIUM SALT
80-17-1 BENZENESULFONIC ACID, HYDRAZIDE	BENZENESULFONIC ACID, HYDRAZIDE
127-82-2 BENZENESULFONIC ACID, 4-HYDROXY-, ZINC SALT (2:1) [PAL]	ZINC PHENOLSULFONATE
68411-30-3 BENZENESULFONIC ACID, LINEAR ALKYL, SODIUM SALT [WHM]	BENZENESULFONIC ACID, C10-13-ALKYL DERIVATIVES SODIUM SALT
104-15-4 BENZENESULFONIC ACID, 4-METHYL- [PAL]	P-TOLUENESULFONIC ACID
80-48-8 BENZENESULFONIC ACID, 4-METHYL-, METHYL ESTER [PAL]	METHYL-P-TOLUENESULFONATE
68648-98-6 BENZENESULFONIC ACID, MONO-C9-17-BRANCHED ALKYL DERIVATIVES	BENZENESULFONIC ACID, MONO-C9-17-BRANCHED ALKYL DERIVATIVES
68649-00-3 BENZENESULFONIC ACID, MONO-C9-17-BRANCHED ALKYL DERIVATIVES, ISOPROPYLAMINE SALTS	BENZENESULFONIC ACID, MONO-C9-17-BRANCHED ALKYL DERIVATIVES,
25167-32-2 BENZENESULFONIC ACID, OXYBIS[DODECYL-, DISODIUM SALT]	BENZENESULFONIC ACID, OXYBIS[DODECYL-,DISODIUM SALT]
98-09-9 BENZENESULFONYL CHLORIDE	BENZENESULFONYL CHLORIDE
98-09-9 BENZENE SULFONYL CHLORIDE [NJL, NJL]	BENZENESULFONYL CHLORIDE
95-94-3 BENZENE, 1,2,4,5-TETRACHLORO- [EPA, PAL, TSC, TSC, TSC, TSC, TSC]	1,2,4,5-TETRACHLOROBENZENE
634-66-2 BENZENE, 1,2,3,4-TETRACHLORO- [TSC, TSC]	1,2,3,4-TETRACHLOROBENZENE
634-90-2 BENZENE, 1,2,3,5-TETRACHLORO-	BENZENE, 1,2,3,5-TETRACHLORO-
108-98-5 BENZENETHIOL	BENZENETHIOL
108-98-5 BENZENETHIOL [TSC, TSC, TSC]	BENZENETHIOL
552-30-7 BENZENE-1,2,4-TRICARBOXYLIC ACID 1,2-ANHYDRIDE [GBR, GBR]	TRIMELLITIC ANHYDRIDE
3319-31-1 1,2,4-BENZENETRICARBOXYLIC ACID, TRIS (2-ETHYLHEXYL) ESTER	1,2,4-BENZENETRICARBOXYLIC ACID, TRIS (2-ETHYLHEXYL) ESTER,
87-61-6 BENZENE, 1,2,3-TRICHLORO- [TSC, TSC, TSC]	1,2,3-TRICHLOROBENZENE
108-70-3 BENZENE, 1,3,5-TRICHLORO- [TSC, TSC, TSC]	1,3,5-TRICHLOROBENZENE
120-82-1 BENZENE, 1,2,4-TRICHLORO- [PAL, TSC, TSC, TSC]	1,2,4-TRICHLOROBENZENE
50-29-3 BENZENE, 1,1'-(2,2,2-TRICHLOROETHYLIDENE) BIS[4-CHLORO- [PAL, PAL]	DDT
72-43-5 BENZENE, 1,1'-(2,2,2-TRICHLOROETHYLIDENE) BIS[4-METHOXY- [PAL]	METHOXYCHLORPHA.)-
98-07-7 BENZENE, (TRICHLOROMETHYL)- [PAL, PAL]	BENZOTRICHLORIDE
98-07-7 BENZENE, TRICHLOROMETHYL- [TSC, TSC]	BENZOTRICHLORIDE
98-08-8 BENZENE, (TRIFLUOROMETHYL)- [PAL]	BENZOTRIFLUORIDE
95-63-6 BENZENE, 1,2,4-TRIMETHYL- [TSC, TSC, TSC]	PSEUDOCUMENE
108-67-8 BENZENE, 1,3,5-TRIMETHYL- [TSC, TSC, TSC]	1,3,5-TRIMETHYLBENZENE
526-73-8 BENZENE, 1,2,3-TRIMETHYL- [TSC, TSC, TSC]	1,2,3-TRIMETHYLBENZENE

Chemical Name	Reference Name
25551-13-7 BENZENE, TRIMETHYL- [PAL, TSC, TSC, TSC]	TRIMETHYL BENZENE
99-35-4 BENZENE, 1,3,5-TRINITRO- [PAL]	TRINITROBENZENE
RR-01396-7 BENZENE TRIOZONIDE	BENZENE TRIOZONIDE
6742-54-7 BENZENE, UNDECYL-	BENZENE, UNDECYL-
121-54-0 BENZETHONIUM CHLORIDE	BENZETHONIUM CHLORIDE
91-01-0 BENZHYDROL	BENZHYDROL
92-87-5 BENZIDINE	BENZIDINE
92-87-5 BENZIDINE AND ITS SALTS [EP2, MAK, MAK]	BENZIDINE
RR-00532-3 BENZIDINE BASED DYES	BENZIDINE BASED DYES
RR-00532-3 BENZIDINE-BASED DYES [MIN, NHS, NHS, NHS]	BENZIDINE BASED DYES
531-85-1 BENZIDINE DIHYDROCHLORIDE [CSR]	[1,1'-BIPHENYL]-4,4'-DIAMINE, DIHYDROCHLORIDE
92-87-5 BENZIDINE PRODUCTION [BC1, BC1, BC1, QBC, QBC, QBC]	BENZIDINE
531-86-2 BENZIDINE SALT	BENZIDINE SALT
134-81-6 BENZIL	BENZIL
76-93-7 BENZILIC ACID	BENZILIC ACID
51-17-2 1-BENZIMIDAZOLE	1-BENZIMIDAZOLE
3615-21-2 BENZIMIDAZOLE, 4,5-DICHLORO-2-(TRIFLUO-ROMETHYL)-	BENZIMIDAZOLE, 4,5-DICHLORO-2-(TRIFLUO-ROMETHYL)-
8030-30-6 BENZIN	BENZIN
8030-30-6 BENZINE [NJL]	BENZIN
52821-24-6 1H-BENZ(DE)ISOQUINOLINE-1,3(2H)-DIONE, 2-(3-HYDROXYPROPYL)-6-[(3-HYDROXYPROPYL)AMINO]-	1H-BENZ(DE)ISOQUINOLINE-1,3(2H)-DIONE, 2-(3-HY-DROXYPROPYL)-6
2634-33-5 1,2-BENZISOTHIAZOLIN-3-ONE	1,2-BENZISOTHIAZOLIN-3-ONE
81-07-2 1,2-BENZISOTHIAZOL-3(2H)-ONE, 1,1-DIOX-IDE [PAL, PAL]	SACCHARIN
128-44-9 1,2-BENZISOTHIAZOL-3(2H)-ONE, 1,1-DIOXIDE, SODIUM SALT [PAL, PAL]	SODIUM SACCHARIN
56-55-3 BENZO(A)ANTHRACENE [ATS, PQL]	BENZ(A)ANTHRACENE
203-33-8 BENZO(A)FLUORANTHENE	BENZO(A)FLUORANTHENE
238-84-6 BENZO(A)FLUORENE	BENZO(A)FLUORENE
50-32-8 BENZO(A)PYRENE	BENZO(A)PYRENE
50-32-8 BENZO(A)PYRENE (3,4-BENZO-PYRENE) [CW2]	BENZO(A)PYRENE
50-32-8 BENZO(ALPHA)PYRENE [BC1, BC1]	BENZO(A)PYRENE
205-99-2 BENZO(B)FLUORANTHENE	BENZO(B)FLUORANTHENE
205-99-2 BENZO(B)FLUORANTHENE [UTS]	BENZO(B)FLUORANTHENE
89-32-7 1H,3H-BENZO[1,2-C:4,5-C']DIFURAN-1,3,5,7-TETRONE	1H,3H-BENZO[1,2-C:4,5-C']DIFURAN-1,3,5,7-TETRONE
RR-01527-0 BENZO[C]FLUORENE	BENZO[C]FLUORENE
195-19-7 BENZO[C]PHENANTHRENE	BENZO[C]PHENANTHRENE
192-97-2 BENZO(E)PYRENE	BENZO(E)PYRENE
191-24-2 BENZO(G)PERYLENE (1,12-BENZOPERYLENE) [MX1]	BENZO(GHI)PERYLENE
203-12-3 BENZO[GHI]FLUORANTHENE	BENZO[GHI]FLUORANTHENE
191-24-2 BENZO(GHI)PERYLENE	BENZO(GHI)PERYLENE
191-24-2 BENZO(G,H,I)PERYLENE [ATS, UTS]	BENZO(GHI)PERYLENE
205-82-3 BENZO(J)FLUORANTHENE	BENZO(J)FLUORANTHENE
206-44-0 BENZO(J,K)FLUORENE [EPA, MSL]	FLUORANTHENE
207-08-9 BENZO(K)FLUORANTHENE	BENZO(K)FLUORANTHENE

Chemical Name	**Reference Name**
207-08-9 BENZO(K)FLUORANTHENE (11,12-BENZOFLUO-RANTHENE) [MX1]	BENZO(K)FLUORANTHENE
189-55-9 BENZO(RST)PENTAPHENE [313, PAL, PAL]	DIBENZO(A,I)PYRENE
56-55-3 BENZO(A)ANTHRACENE (1,2-BENZOAN-THRACENE) [MX1]	BENZ(A)ANTHRACENE
218-01-9 BENZO(A)PHENANTHRENE [313]	CHRYSENE
50-32-8 BENZO(A)PYRENE (3,4-BENZOPYRENE) [MX1]	BENZO(A)PYRENE
RR-00974-5 BENZOATE ESTER	BENZOATE ESTER
94-09-7 BENZOCAINE [PST]	ETHYL P-AMINOBENZOATE
RR-01161-0 BENZOCYCLOBUTENE	BENZOCYCLOBUTENE
RR-01474-4 BENZODIAZEPINES	BENZODIAZEPINES
120-57-0 1,3-BENZODIOXOLE-5-CARBOXALDEHYDE	1,3-BENZODIOXOLE-5-CARBOXALDEHYDE
5780-07-4 1,3-BENZODIOXOLE-5-CARBOXALDEHYDE, 7-METHOXY- [TSC, TSC, TSC]	MYRISTIC ALDEHYDE
94-59-7 1,3-BENZODIOXOLE, 5-(2-PROPENYL)- [PAL, PAL]	SAFROLE
120-58-1 1,3-BENZODIOXOLE, 5-(1-PROPENYL)- [PAL, PAL]	ISOSAFROLE
94-58-6 1,3-BENZODIOXOLE, 5-PROPYL- [PAL, PAL]	DIHYDROSAFROLE
205-99-2 3,4-BENZOFLUORANTHENE [MX1]	BENZO(B)FLUORANTHENE
56832-73-6 BENZOFLUORANTHENE	BENZOFLUORANTHENE
205-99-2 3,4-BENZOFLUORANTHENE (BENZO(B) FLUO-RANTHENE) [CW2]	BENZO(B)FLUORANTHENE
207-08-9 11,12-BENZOFLUORANTHENE (BENZO(K)FLUO-RANTHENE) [CW2]	BENZO(K)FLUORANTHENE
271-89-6 BENZOFURAN	BENZOFURAN
68298-46-4 7-BENZOFURANAMINE, 2,3-DIHYDRO-2,2-DIMETHYL- [TSC, TSC, TSC]	7-AMINO-2,2-DIMETHYL-2,3-DIHYDROBENZOFURAN
13414-55-6 BENZOFURAN, 2,3-DIHYDRO-2,2-DIMETHYL-7-NITRO-	BENZOFURAN, 2,3-DIHYDRO-2,2-DIMETHYL-7-NITRO-
1563-66-2 7-BENZOFURANOL, 2,3-DIHYDRO-2,2-DIMETHYL-, METHYLCARBAMATE [PAL]	CARBOFURAN
65-85-0 BENZOIC ACID	BENZOIC ACID
50-78-2 BENZOIC ACID, 2-(ACETYLOXY)- [PAL]	ACETYLSALICYLIC ACID (ASPIRIN)ETHOXYBENZOYL)OXY]-, METHYL ESTER, (3.BETA.,16.BETA., 17.ALPHA., 18.BETA.,20.ALPHA.)-
6739-62-4 BENZOIC ACID, 2-[[2-AMINO-6-[[4'-[(3-CARBOXY-4-HYDROXYPHENYL)AZO]-3,3'-DIMETHOXY[1,1'-BIPHENYL]-4-YL]AZO]-5-HY-DROXY-7-S ULFO-1-NAPHTHALENYL]AZO]-5-NITRO-, TRISODIUM SALT [TSC]	C.I. DIRECT BLACK 91, TRISODIUM SALTAMINO)-3-SULFO-2-NAPHTHALENYL]AZO][1,1'-BIPHENYL]-4-YL]AZO] -8-HYDROXY-1,6-NAPHTHALENEDISULFONATO(7-)]DI-, TRISODIUM
2429-82-5 BENZOIC ACID, 5-[[4'-[7-AMINO-1-HYDROXY-3-SULFO-2-NAPHTHALENYL]AZO][1,1'-BIPHENYL]-4-YL]AZO]-2-HYDROXY-, DISODIUM SALT [TSC]	C.I. DIRECT BROWN 2, DISODIUM SALTDISULFO-7-[(4-SULFO-1-NAPHTHALENYL)AZO]-2-NAPHTHALENYL]AZO] -5-METHYLPHENYL]AZO] [1,1'-BIPHENYL]-4-YL]AZO]-2-HYDROXY-, TETRASODIUM SALT
2429-84-7 BENZOIC ACID, 5-[[4'-[2-AMINO-8-HYDROXY-6-SULFO-1-NAPTHALENYL]AZO][1,1'-BIPHENYL]-4-YL]AZO]-2-HYDROXY-, DISODIUM SALT [TSC]	C.I. DIRECT RED 1, DISODIUM SALT-5-METHYLPHENYL)AZO] [1,1'-BIPHENYL]-4-YL]AZO] 5-HYDROXY-6-(PHENYLAZO)-, DISODIUM SALT
2429-79-0 BENZOIC ACID, 5-[[4'-[1-AMINO-4-SULFO-2-NAPHTHALENYL]AZO][1,1'-BIPHENYL]-4-YL]AZO]-2-HYDROXY-, DISODIUM SALT	BENZOIC ACID, 5-[[4'-[1-AMINO-4-SULFO-2-NAPH-THALENYL]AZO][1,1'-BIPHENYL]-4,4'-DIYL)BIS(AZO)]BIS[5-AMINO-4-HYDROXY-, TETRASODIUM SALT
1863-63-4 BENZOIC ACID, AMMONIUM SALT [PAL]	AMMONIUM BENZOATE
136-60-7 BENZOIC ACID, BUTYL ESTER [PAL]	BUTYL BENZOATE
2905-65-9 BENZOIC ACID, 3-CHLORO-, METHYL ESTER	BENZOIC ACID, 3-CHLORO-, METHYL ESTER[1,1'-BIPHENYL]-4-YL]AZO]-2-HYDROXY-, DISODIUM SALT

Chemical Name	Reference Name
63734-62-3 BENZOIC ACID, 3-[2-CHLORO-4-(TRIFLUO-ROMETHYL)PHENOXY]-	BENZOIC ACID, 3-[2-CHLORO-4-(TRIFLUOROMETHYL) PHENOXY]-
72252-48-3 BENZOIC ACID, 3-[2-CHLORO-4-(TRIFLUO-ROMETHYL)PHENOXY]-, POTASSIUM SALT	BENZOIC ACID, 3-[2-CHLORO-4-(TRIFLUOROMETHYL) PHENOXY]-, POTA
2429-81-4 BENZOIC ACID, 5-[[4'-[[2,6-DIAMINO-3-[[8-HY-DROXY-3,6-DISULFO-7-[(4-SULFO-1-NAPH-THALENYL)AZO]-2-NAPHTHALENYL]AZO] -5-METHYLPHENYL]AZO] [1,1'-BIPHENYL]-4-YL]AZO]-2-HYDROXY-, TETRASODIUM SALT [TSC]	C.I. DIRECT BROWN 31, TETRASODIUM SALT[1,1'-BIPHENYL]-4-YL]AZO]-2-HYDROXY-, DISODIUM SALT
2586-58-5 BENZOIC ACID, 5-[[4'-[[2,6-DIAMINO-3-METHYL-5[(SULFOPHENYL)AZO]PHENYL]AZO][1,1'-BIPHENYL]-4-YL]AZO]-2-HYDROXY-, DISODIUM SALT	BENZOIC ACID, 5-[[4'-[[2,6-DIAMINO-3-METHYL-5[(SUL-FOPHENYL)A2-HYDROXY-1-NAPHTHALENYL)AZO]-3, 3'-DIMETHOXY[1,1'-BIPHENYL]-4-YL]AZO]-, DISODIUM SALT
6360-54-9 BENZOIC ACID, 5-[[4'-[[2,6-DIAMINO-3-METHYL-5-[(4-SULFOPHENYL)AZO]PHENYL] AZO][1,1'-BIPHENYL]-4-YL]AZO]-2- HYDROXY-3-METHYL-, DISODIUM SALT [TSC]	C.I. DIRECT BROWN 154-3,3'-DIMETHYL] [1,1'-BIPHENYL] -4-YL]AZO]-7-HYDROXY-, DISODIUM SALT
6637-88-3 BENZOIC ACID, 5-[[4'-[(2,6-DIAMINO-3-METHYL-5-SULFOPHENYL)AZO]-3,3'-DIMETHYL[1,1'-BIPHENYL]-4-YL]AZO]-2-HY-DROXY-, DISODIUM SALT [TSC]	C.I. DIRECT ORANGE 6, DISODIUM SALTULFO-2-NAPH-THALENYL)AZO]-3,3'-DIMETHOXY[1,1'-BIPHENYL]-4-YL] AZO]-4-HYDROXY-, DISODIUM SALT
1918-00-9 BENZOIC ACID, 3,6-DICHLORO-2-METHOXY-[PAL]	DICAMBAYL)-
2893-80-3 BENZOIC ACID, 5-[[4'-[[2,4-DIHYDROXY-3-[(4-SULFOPHENYL)AZO][1,1'-BIPHENYL]-4-YL]AZO]-2-HYDROXY-, DISODIUM SALT [TSC]	C.I. DIRECT BROWN 6, DISODIUM SALT
8014-91-3 BENZOIC ACID, 3,3'-[(3,7-DISULFO-1,5-NAPH-THALENE-DIYL)BIS[AZO(6-HYDROXY-3,1-PHENYLENE)AZO[6(OR 7)-SULFO-4,1- NAPH-THALENEDIYL]AZO[1,1'-BIPHENYL]-4,4'-DIYLAZO]BIS[6- HYDROXY-, HEXASODIUM SALT [TSC]	C.I. DIRECT BROWN
93-89-0 BENZOIC ACID, ETHYL ESTER	BENZOIC ACID, ETHYL ESTER
119-36-8 BENZOIC ACID, 2-HYDROXY-, METHYL ES-TER [PAL]	METHYL SALICYLATE
553-70-8 BENZOIC ACID, MAGNESIUM SALT	BENZOIC ACID, MAGNESIUM SALT
118-90-1 BENZOIC ACID, 2-METHYL-	BENZOIC ACID, 2-METHYL-
68092-47-7 BENZOIC ACID, 3-METHYL-, BARIUM SALT	BENZOIC ACID, 3-METHYL-, BARIUM SALTLT
61386-02-5 BENZOIC ACID, 3,3'-METHYLENEBIS[6-AMINO-DI-2-PROPENYL] ESTER	BENZOIC ACID, 3,3'-METHYLENEBIS[6-AMINO-DI-2-PROPENYL] ESTER
RR-01663-7 BENZOIC ACID, 2-(3-PHENYLBUTYLIDENE)AMINO-, METHYL ESTER	BENZOIC ACID, 2-(3-PHENYLBUTYLIDENE)AMINO-, METHYL ESTER
RR-00168-3 BENZOIC DERIVATIVE PESTICIDES, N.O.S.	BENZOIC DERIVATIVE PESTICIDES, N.O.S.
98-07-7 BENZOIC TRICHLORIDE (BENZOTRICHLORIDE) [313]	BENZOTRICHLORIDE
119-53-9 BENZOIN	BENZOIN
441-38-3 ALPHA-BENZOIN OXIME	ALPHA-BENZOIN OXIME
RR-01762-9 BENZOL DILUENT	BENZOL DILUENT
100-47-0 BENZONITRILE	BENZONITRILE
1194-65-6 BENZONITRILE, 2,6-DICHLORO- [PAL]	DICHLOBENIL
191-24-2 1,12-BENZOPERYLENE [CAL]	BENZO(GHI)PERYLENE
191-24-2 1,12-BENZOPERYLENE (BENZO(GHI)PERYLENE) [CW2]	BENZO(GHI)PERYLENE
119-61-9 BENZOPHENONE	BENZOPHENONE
5411-22-3 BENZOPHETAMINE HYDROCHLORIDE	BENZOPHETAMINE HYDROCHLORIDE

Chemical Name	Reference Name
83-79-4 [1]BENZOPYRANO[3,4-B]FURO[2,3-H][1]BEN-ZOPYRAN-6(6AH)-ONE,1,212,12A-TETRAHY-DRO-8,9-DIMETHOXY-2-(1-METHYLETHENYL)-, [2R-(2.ALPHA.,6A.ALPHA.,12A.ALPHA.)]- [PAL]	ROTENONE (COMMERCIAL)
91681-63-9 4H-BENZOPYRAN-4-ONE, 6-(2,3-DIHYDROXY-3-METHYLBUTYL)-3-(2,4-DIHYDROXYPHENYL)-7-HYDROXY-	4H-BENZOPYRAN-4-ONE, 6-(2,3-DIHYDROXY-3-METHYL-BUTYL)-3-(2,4-
81-81-2 2H-1-BENZOPYRAN-2-ONE, 4-HYDROXY-3-(3-OXO)-1-PHENYLBUTYL)- [PAL]	WARFARIN
154-23-4 2H-1-BENZOPYRAN-3,5,7-TRIOL, 2-(3,4-DIHY-DROXYPHENYL)-3,4-DIHYDRO-, (2R-TRANS)-	2H-1-BENZOPYRAN-3,5,7-TRIOL, 2-(3,4-DIHYDROX-YPHENYL)-3,4-DIH
106-51-4 P-BENZOQUINONE [EPA, GBR, GBR]	QUINONE
106-51-4 BENZOQUINONE [HM1, HM1, HM1]	QUINONE
106-51-4 P-BENZOQUINONE [NFP, NFP, NJL, RCU, RCU, WHM]	QUINONE
583-63-1 O-BENZOQUINONE	O-BENZOQUINONE
106-51-4 P-BENZOQUINONE (QUINONE) [ALB, ALB, QBC]	QUINONE
105-11-3 P-BENZOQUINONE DIOXIME	P-BENZOQUINONE DIOXIME
105-11-3 PARA-BENZOQUINONE DIOXIME [CAL, IAR]	P-BENZOQUINONE DIOXIME
95-16-9 BENZOTHIAZOLE	BENZOTHIAZOLE
95-14-7 1,2,3-BENZOTRIAZOLE	1,2,3-BENZOTRIAZOLE
95-14-7 1H-BENZOTRIAZOLE [MSL, TSC, TSC, TSC]	1,2,3-BENZOTRIAZOLE
29385-43-1 1H-BENZOTRIAZOLE, METHYL- [TSC, TSC, TSC]	TOLYL TRIAZOLE
133145-29-6 1H-BENZOTRIAZOLE, 5-(PENTYLOXY)-	1H-BENZOTRIAZOLE, 5-(PENTYLOXY)-]BIS[N-OXI-RANYLMETHYL]-
RR-01561-2 1H-BENZOTRIAZOLE, 5-(PENTYLOXY)-, POTAS-SIUM SALT	1H-BENZOTRIAZOLE, 5-(PENTYLOXY)-, POTASSIUM SALT
RR-01560-1 1H-BENZOTRIAZOLE, 5-(PENTYLOXY)-, SODIUM SALT	1H-BENZOTRIAZOLE, 5-(PENTYLOXY)-, SODIUM SALT
2440-22-4 2-BENZOTRIAZOLYL-4-METHYLPHENOL	2-BENZOTRIAZOLYL-4-METHYLPHENOL
2440-22-4 2-(2H-BENZOTRIAZOL-2-YL)-4-METHYLPHE-NOL [PST]	2-BENZOTRIAZOLYL-4-METHYLPHENOL
98-07-7 BENZOTRICHLORIDE	BENZOTRICHLORIDE
98-08-8 BENZOTRIFLUORIDE	BENZOTRIFLUORIDE
98730-04-2 2H-1,4-BENZOXAZINE, 4-(DICHLOROACETYL)-3, 4-DIHYDRO-3-METHYL-	2H-1,4-BENZOXAZINE, 4-(DICHLOROACETYL)-3,4-DIHY-DRO-3-METHYL-
273-53-0 BENZOXAZOLE	BENZOXAZOLE
RR-01397-8 BENZOXIDIAZOLES	BENZOXIDIAZOLES
582-61-6 BENZOYL AZIDE	BENZOYL AZIDE
98-88-4 BENZOYL CHLORIDE	BENZOYL CHLORIDE
94-36-0 BENZOYL PEROXIDE	BENZOYL PEROXIDE
5411-22-3 BENZPHETAMINE HYDROCHLORIDE [C65]	BENZOPHETAMINE HYDROCHLORIDE
120-78-5 BENZTHIAZOLE DISULFIDE [OTS]	2,2'-DIBENZOTHIAZYL DISULFIDE
98-87-3 BENZYL DICHLORIDE [HON]	BENZAL CHLORIDE
140-11-4 BENZYL ACETATE	BENZYL ACETATE
100-51-6 BENZYL ALCOHOL	BENZYL ALCOHOL
120-51-4 BENZYL BENZOATE	BENZYL BENZOATE
61789-73-9 BENZYL BIS(HYDROGENATED TALLOW ALKYL) METHYL AMMONIUM CHLORIDE	BENZYL BIS(HYDROGENATED TALLOW ALKYL) METHYL AMMONIUM CHLORI
100-39-0 BENZYL BROMIDE	BENZYL BROMIDE
85-68-7 BENZYL BUTYL PHTHALATE [NFP, NFP, GBR]	BUTYL BENZYL PHTHALATE
100-44-7 BENZYL CHLORIDE	BENZYL CHLORIDE

Chemical Name	Reference Name
100-44-7 BENZYL CHLORIDE (CHLOROTOLUENE) [QBC]	BENZYL CHLORIDE
501-53-1 BENZYL CHLOROFORMATE	BENZYL CHLOROFORMATE
120-32-1 O-BENZYL-P-CHLOROPHENOL	O-BENZYL-P-CHLOROPHENOL
120-32-1 2-BENZYL-4-CHLOROPHENOL [PST, TSC, TSC, TSC]	O-BENZYL-P-CHLOROPHENOL
103-41-3 BENZYL CINNAMATE	BENZYL CINNAMATE
140-29-4 BENZYL CYANIDE	BENZYL CYANIDE
64503-07-7 BENZYL DIBROMOACETATE	BENZYL DIBROMOACETATEHACRYLATE, STYRENE, HEXADEXYL METHACRYLATE AND TETRADECYL METHACRYLATE
98-87-3 BENZYL DICHLORIDE [WHM]	BENZAL CHLORIDE
772-54-3 N-BENZYLDIETHYLAMINE [FLA, MSL, NFP, NFP]	BENZENEMETHANAMINE, N,N-DIETHYL-
3734-33-6 BENZYL DIETHYL 2,6-XYLYLCARBAMOYL-METHYL AMMONIUM BENZOATE [PST]	BITREX
103-83-3 BENZYL DIMETHYLAMINE	BENZYL DIMETHYLAMINE
103-83-3 BENZYLDIMETHYLAMINE [WHM]	BENZYL DIMETHYLAMINE
61789-75-1 BENZYL DIMETHYL (TALLOW ALKYL) AMMO-NIUM CHLORIDE	BENZYL DIMETHYL (TALLOW ALKYL) AMMONIUM CHLORIDEDE
139-08-2 BENZYLDIMETHYLTETRADECYLAMMONIUM CHLORIDE	BENZYLDIMETHYLTETRADECYLAMMONIUM CHLO-RIDE
103-50-4 BENZYL ETHER	BENZYL ETHER
60864-33-7 BENZYL ETHER OF OCTYLPHENOXY-POLYETHOXY ETHANOL [PST]	TRITON CF 10
RR-00769-2 4-(BENZYL(ETHYL)AMINO)-3-ETHOXYBEN-ZENEDIAZONIUM ZINCCHLORIDE	4-(BENZYL(ETHYL)AMINO)-3-ETHOXYBENZENEDIAZO-NIUM ZINC
122-57-6 BENZYLIDENE ACETONE	BENZYLIDENE ACETONE
98-87-3 BENZYLIDENE CHLORIDE [HM1, HM1, HM1]	BENZAL CHLORIDE
620-05-3 BENZYL IODIDE	BENZYL IODIDE
100-53-8 BENZYL MERCAPTAN	BENZYL MERCAPTAN
RR-00770-5 4-(BENZYL(METHYL)AMINO)-3-ETHOXYBEN-ZENEDIAZONIUM ZINCCHLORIDE	4-(BENZYL(METHYL)AMINO)-3-ETHOXYBENZENEDIA-ZONIUM ZINCCHLORIDE
61-33-6 BENZYLPENICILLIN	BENZYLPENICILLIN
118-58-1 BENZYL SALICILATE	BENZYL SALICILATE
56-93-9 BENZYLTRIMETHYL AMMONIUM CHLORIDE	BENZYLTRIMETHYL AMMONIUM CHLORIDE
1694-09-3 BENZYL VIOLET 4B	BENZYL VIOLET 4B
56-55-3 BENZ[A]ANTHRACENE [TLV]	BENZ(A)ANTHRACENE
8007-75-8 BERGAMOT OIL	BERGAMOT OIL
16652-07-6 BERKELIUM 245	BERKELIUM 245
15715-02-3 BERKELIUM 246	BERKELIUM 246
15752-38-2 BERKELIUM 247	BERKELIUM 247
14900-25-5 BERKELIUM 249	BERKELIUM 249
15755-53-0 BERKELIUM 250	BERKELIUM 250
12161-82-9 BERTRONDITE	BERTRONDITE
1302-52-9 BERYL	BERYL
1302-52-9 BERYL (AL2BE3(SIO3)6) [PAL, PAL]	BERYL
7440-41-7 BERYLLIUM	BERYLLIUM
13966-02-4 BERYLLIUM 7	BERYLLIUM 7
14390-89-7 BERYLLIUM 10	BERYLLIUM 10
12770-50-2 BERYLLIUM ALUMINUM ALLOY	BERYLLIUM ALUMINUM ALLOY
12770-50-2 BERYLLIUM-ALUMINUM ALLOY [MSL]	BERYLLIUM ALUMINUM ALLOY

Chemical Name	Reference Name
RR-00557-2 BERYLLIUM AND BERYLLIUM COMPOUNDS [OS1, OS1, OS1, OS1, OS1, IAR, CEX, CEX, CEX, C65]	BERYLLIUM COMPOUNDS, N.O.S.
7440-41-7 BERYLLIUM AND COMPOUNDS [AUS, AUS, NIO, NIO, NIO, NIO, TLV, TLV]	BERYLLIUM
7440-41-7 BERYLLIUM AND ITS COMPOUNDS [MAK]	BERYLLIUM
66104-24-3 BERYLLIUM CARBONATE	BERYLLIUM CARBONATE
7787-47-5 BERYLLIUM CHLORIDE	BERYLLIUM CHLORIDE
7787-47-5 BERYLLIUM CHLORIDE (BECL2) [PAL, PAL]	BERYLLIUM CHLORIDE
RR-00557-2 BERYLLIUM COMPOUNDS [313, CAA, CAL, CN7, CW3, ONT]	BERYLLIUM COMPOUNDS, N.O.S.
RR-00557-2 BERYLLIUM COMPOUNDS, N.O.S.	BERYLLIUM COMPOUNDS, N.O.S.
7440-41-7 BERYLLIUM, ELEMENTAL [WHM]	BERYLLIUM
7787-49-7 BERYLLIUM FLUORIDE	BERYLLIUM FLUORIDE
7787-49-7 BERYLLIUM FLUORIDE (BEF2) [PAL, PAL]	BERYLLIUM FLUORIDE
13327-32-7 BERYLLIUM HYDROXIDE	BERYLLIUM HYDROXIDE
13327-32-7 BERYLLIUM HYDROXIDE (BE(OH)2) [PAL, PAL]	BERYLLIUM HYDROXIDE
13597-99-4 BERYLLIUM NITRATE	BERYLLIUM NITRATE
7787-55-5 BERYLLIUM NITRATE (HYDRATED) [MSL]	BERYLLIUM NITRATE TRIHYDRATE
7787-55-5 BERYLLIUM NITRATE TRIHYDRATE	BERYLLIUM NITRATE TRIHYDRATE
1304-56-9 BERYLLIUM OXIDE	BERYLLIUM OXIDE
1304-56-9 BERYLLIUM OXIDE (BEO) [PAL, PAL]	BERYLLIUM OXIDE
35089-00-0 BERYLLIUM PHOSPHATE	BERYLLIUM PHOSPHATE
7440-41-7 BERYLLIUM POWDER [EPA, RCU, RCU]	BERYLLIUM
15191-85-2 BERYLLIUM SILICATE [FLA, MSL]	SILICIC ACID (H4SIO4), BERYLLIUM SALT (1:2)
13510-49-1 BERYLLIUM SULFATE	BERYLLIUM SULFATE
7787-56-6 BERYLLIUM SULFATE TETRAHYDRATE	BERYLLIUM SULFATE TETRAHYDRATE
13510-49-1 BERYLLIUM SULPHATE [FLA, MSL]	BERYLLIUM SULFATE
39413-47-3 BERYLLIUM ZINC SILICATE [NTP]	ZINC BERYLLIUM SILICATE
1302-52-9 BERYL ORE [FLA, MSL, NTP]	BERYL
12587-47-2 BETA RADIATION	BETA RADIATION
RR-00027-1 BETEL QUID WITH TOBACCO	BETEL QUID WITH TOBACCO
58-89-9 GAMMA-BHC [RCA]	LINDANE
319-84-6 ALPHA-BHC	ALPHA-BHC
319-85-7 BETA-BHC	BETA-BHC3.BETA.,4.ALPHA.,5.BETA.,6.BETA.)-
319-86-8 DELTA-BHC	DELTA-BHC
608-73-1 BHC [MX2]	HEXACHLOROCYCLOHEXANE (MIXED ISOMERS)
58-89-9 GAMMA-BHC (LINDANE) [CW2, EP2]	LINDANE
319-86-8 DELTA-BHC (PCB-POLYCHLORINATED BIPHENYLS) [CW2]	DELTA-BHC
58-89-9 GAMMA-BHC (ISO) [GBR, GBR, GBR]	LINDANE
319-84-6 ALPHA-BHC-ALPHA [MX1]	ALPHA-BHC
319-85-7 BETA-BHC-BETA [MX1]	BETA-BHC
58-89-9 GAMMA-BHC [HM1, HM1]	LINDANE
58-89-9 GAMMA-BHC(LINDANE)-GAMMA [MX1]	LINDANE
RR-00172-9 BHUSA	BHUSA
RR-00782-9 BICHROMATES	BICHROMATES
121-46-0 BICYCLO(2.2.1)HEPTA-2,5-DIENE	BICYCLO(2.2.1)HEPTA-2,5-DIENE

Chemical Name	Reference Name
15271-41-7 BICYCLO[2.2.1]HEPTANE-2-CARBONITRILE, 5-CHLORO-6- [NJL]	BICYCLO[2.2.1]HEPTANE-2-CARBONITRILE, 5-CHLORO-6-((((METHYLA
15271-41-7 **BICYCLO[2.2.1]HEPTANE-2-CARBONITRILE, 5-CHLORO-6-((((METHYLAMINO)CARBONYL) OXY)IMINO)-,(1ST-(1-ALPHA, 2-BETA, 4-ALPHA, 5-ALPHA, 6E))-**	BICYCLO[2.2.1]HEPTANE-2-CARBONITRILE, 5-CHLORO-6-((((METHYLA
507-70-0 BICYCLO(2.2.1)HEPTAN-2-OL, 1,7,7-TRIMETHYL-, ENDO- [PAL]	BORNEOL
76-22-2 BICYCLO(2.2.1)HEPTAN-2-ONE, 1,7,7-TRIMETHYL- [PAL]	CAMPHOR
115-28-6 BICYCLO[2.2.1]HEPT-5-ENE-2,3-DICARBOXYLIC ACID, 1,4,5,6,7,7-HEXACHLORO- [TSC, TSC, TSC]	CHLORENDIC ACID-9-AZATRICYCLO[3.3.1.02,4] NON-7-YL ESTER, HYDROBROMIDE, [7(S)-(1.AL-PHA.2.BETA.4.BETA.5.ALPHA.7.BETA)]-
826-62-0 **BICYCLO [2,2,1]5-HEPTENE-2,3-DICARBOXYLIC ANHYDRIDE**	BICYCLO [2,2,1]5-HEPTENE-2,3-DICARBOXYLIC ANHYDRIDE
16219-75-3 BICYCLO(2.2.1)HEPT-2-ENE, 5-ETHYLIDENE- [PAL]	ETHYLIDENE NORBORNENEENYL)AZO]PHENYL]AZO] 1,1'-BIPHENYL]4-YL]AZO-2-HYDROXY BENZOATO(4-)]-, DISODIUM
80-56-8 BICYCLO(3.1.1)HEPT-2-ENE, 2,6,6-TRIMETHYL- [PAL]	ALPHA-PINENE
92-51-3 **BICYCLOHEXYL**	BICYCLOHEXYL
90-42-6 **[1,1'-BICYCLOHEXYL]-2-ONE**	[1,1'-BICYCLOHEXYL]-2-ONE
82657-04-3 **BIFENTHRIN**	BIFENTHRINETHOXY-1-METHYL-2-OXOETHYL ESTER]
18130-74-0 **BIFLUORIDE, N.O.S.**	BIFLUORIDE, N.O.S.
RR-01343-4 **BIFLUORIDES, N.O.S.**	BIFLUORIDES, N.O.S.
485-31-4 **BINAPACRYL**	BINAPACRYL
137-40-6 **BIOBAN-S**	BIOBAN-S
RR-01607-9 **BIOLOGICAL WARFARE SUBSTANCES**	BIOLOGICAL WARFARE SUBSTANCES
1464-53-5 2,2'-BIOXIRANE [EPA, PAL, PAL, TSC, TSC, TSC]	DIEPOXYBUTANETETRAKIS- (EDTMPA)
92-52-4 **BIPHENYL**	BIPHENYL
92-52-4 1,1'-BIPHENYL [PAL, TSC, TSC, TSC]	BIPHENYL
92-52-4 1,1-BIPHENYL [TSE]	BIPHENYL
92-52-4 BIPHENYL (DIPHENYL) [ATS, MEX, MEX, ALB, ALB]	BIPHENYL
90-41-5 **2-BIPHENYLAMINE**	2-BIPHENYLAMINE
90-41-5 [1,1'-BIPHENYL]-2-AMINE [PAL]	2-BIPHENYLAMINE
92-67-1 [1,1'-BIPHENYL]-4-AMINE [PAL, PAL]	4-AMINOBIPHENYL
2185-92-4 **2-BIPHENYLAMINE HYDROCHLORIDE**	2-BIPHENYLAMINE HYDROCHLORIDE
20282-70-6 [1,1'-BIPHENYL]-4,4'-BIS(DIAZONIUM), 3,3'-DIMETHOXY-	[1,1'-BIPHENYL]-4,4'-BIS(DIAZONIUM), 3,3'-DIMETHOXY-
92-66-0 1,1'-BIPHENYL, 4-BROMO- [PAL, TSC, TSC]	4-BROMODIPHENYL
2052-07-5 **1,1'-BIPHENYL, 2-BROMO-**	1,1'-BIPHENYL, 2-BROMO-
2113-57-7 **1,1'-BIPHENYL, 3-BROMO-**	1,1'-BIPHENYL, 3-BROMO-
1336-36-3 1,1-BIPHENYL, CHLORO DERIVATIVES [PAL, PAL]	POLYCHLORINATED BIPHENYLS
13654-09-6 1,1'-BIPHENYL, 2,2',3,3',4,4',5,5',6,6'-DE-CABROMO- [TSC, TSC]	DECARBROMOBIPHENYL
92-87-5 [1,1'-BIPHENYL]-4,4'-DIAMINE [PAL, PAL]	BENZIDINE
92-87-5 (1,1'-BIPHENYL)-4,4'-DIAMINE [TSC, TSC]	BENZIDINE
91-94-1 [1,1'-BIPHENYL]-4,4'-DIAMINE, 3-3'-DICHLORO- [PAL, PAL]	3,3'-DICHLOROBENZIDINE
531-85-1 **[1,1'-BIPHENYL]-4,4'-DIAMINE, DIHYDROCHLO-RIDE**	[1,1'-BIPHENYL]-4,4'-DIAMINE, DIHYDROCHLORIDE

Chemical Name	Reference Name
119-90-4 [1,1'-BIPHENYL]-4,4'-DIAMINE, 3-3'-DIMETHOXY- [PAL, PAL]	3,3'-DIMETHOXYBENZIDINE
119-93-7 [1,1'-BIPHENYL]-4,4'-DIAMINE, 3-3'-DIMETHYL- [PAL, PAL]	3,3'-DIMETHYLBENZIDINE
1331-47-1 [1,1'-BIPHENYL]-4,4'-DIAMINO, DICHLORO-	[1,1'-BIPHENYL]-4,4'-DIAMINO, DICHLORO-
92-86-4 1,1'-BIPHENYL, 4,4'-DIBROMO-	1,1'-BIPHENYL, 4,4'-DIBROMO-
91-97-4 1,1'-BIPHENYL, 4,4'-DIISOCYANATO-3,3'-DIMETHYL-	1,1'-BIPHENYL, 4,4'-DIISOCYANATO-3,3'-DIMETHYL-BIS [3-OXO-
36355-01-8 1,1'-BIPHENYL, HEXABROMO- [TSC, TSC]	HEXABROMOBIPHENYL
643-58-3 1,1'-BIPHENYL, 2-METHYL- [PAL]	2-METHYLBIPHENYL
92-93-3 1,1'-BIPHENYL, 4-NITRO- [PAL]	4-NITROBIPHENYL
28984-85-2 1,1'-BIPHENYL, NITRO- [PAL]	NITROBIPHENYL
27753-52-2 1,1'-BIPHENYL, NONABROMO-	1,1'-BIPHENYL, NONABROMO-
27858-07-7 1,1'-BIPHENYL, OCTABROMO- [TSC, TSC]	OCTABROMOBIPHENYL
92-69-3 (1,1'-BIPHENYL)-4-OL	(1,1'-BIPHENYL)-4-OL
580-51-8 [1,1'-BIPHENYL]-3-OL	[1,1'-BIPHENYL]-3-OLIS(AZO)]BIS[(4-AMINO-, DISODIUM SALT
92-04-6 [1,1'-BIPHENYL]-4-OL, 3-CHLORO-	[1,1'-BIPHENYL]-4-OL, 3-CHLORO-
28984-89-6 1,1'-BIPHENYL, PHENOXY-	1,1'-BIPHENYL, PHENOXY-
RR-01273-7 BIPHENYL PHENYL ETHER AND DIPHENYL OXIDE, MIXTURES	BIPHENYL PHENYL ETHER AND DIPHENYL OXIDE, MIXTURES
RR-01398-9 BIPHENYL TRIOZONIDE	BIPHENYL TRIOZONIDE
RR-00173-0 BIPYRIDILIUM PESTICIDES, N.O.S.	BIPYRIDILIUM PESTICIDES, N.O.S.
366-18-7 2,2'-BIPYRIDINE	2,2'-BIPYRIDINE
4685-14-7 4,4'-BIPYRIDINIUM, 1,1'-DIMETHYL- [PAL]	PARAQUAT
2074-50-2 4,4'-BIPYRIDINIUM, 1,1'-DIMETHYL-, BIS (METHYL SULFATE) [PAL]	PARAQUAT METHOSULFATE
1910-42-5 4,4'-BIPYRIDINIUM, 1,1'-DIMETHYL-, DICHLORIDE [PAL]	PARAQUAT
542-88-1 BIS (2-CHLOROMETHYL) ETHER [NJL, NJL]	BIS(CHLOROMETHYL) ETHER
597-71-7 2,2-BIS[(ACETYLOXY)METHYL]-1,3-PROPANEDIOL DIACETATE	2,2-BIS[(ACETYLOXY)METHYL]-1,3-PROPANEDIOL DIACETATE
RR-00994-9 BISALKYLATED FATTY ALKYL AMINE OXIDE	BISALKYLATED FATTY ALKYL AMINE OXIDE
RR-00211-9 N,N'-BIS[2-[2-(3-ALKYL)THIAZOLINE]VINYL]-1,4-PHENYLENEDIAMINE METHYL SULFATE, DOUBLE SALT	N,N'-BIS[2-[2-(3-ALKYL)THIAZOLINE]VINYL]-1,4-PHENYLENE
128-87-0 BIS(4-AMINO-1-ANTHRAQUINONYL)AMINE	BIS(4-AMINO-1-ANTHRAQUINONYL)AMINE
2549-93-1 1,4-BIS(AMINOMETHYL)CYCLOHEXANE	1,4-BIS(AMINOMETHYL)CYCLOHEXANE
56-18-8 BIS(AMINOPROPYL)AMINE [NJL, NJL]	3,3'-IMINOBISPROPYLAMINE
105-83-9 N,N-BIS(3-AMINOPROPYL)METHYLAMINE	N,N-BIS(3-AMINOPROPYL)METHYLAMINE
7209-38-3 BIS(AMINOPROPYL)PIPERAZINE	BIS(AMINOPROPYL)PIPERAZINE
2168-68-5 BIS(1-AZIRIDINYL)MORPHOLINOPHOSPHINE SULPHIDE	BIS(1-AZIRIDINYL)MORPHOLINOPHOSPHINE SULPHIDE
RR-00280-2 BISAZOBIPHENYL DYES	BISAZOBIPHENYL DYES
3296-90-0 2,2-BIS(BROMOMETHYL)-1,3-PROPANEDIOL	2,2-BIS(BROMOMETHYL)-1,3-PROPANEDIOL
117-83-9 BIS[2-N-BUTOXYETHYL] PHTHALATE	BIS[2-N-BUTOXYETHYL] PHTHALATE
17354-14-2 1,4-BIS(BUTYLAMINO)-9,10-ANTHRAQUINONE	1,4-BIS(BUTYLAMINO)-9,10-ANTHRAQUINONE
RR-00820-8 BIS(P-TERT-BUTYLPHENYL) PHENYL PHOSPHATE	BIS(P-TERT-BUTYLPHENYL) PHENYL PHOSPHATE
638-56-2 BIS(2-(2-CHLOROETHOXY)ETHYL)ETHER	BIS(2-(2-CHLOROETHOXY)ETHYL)ETHER
111-91-1 BIS(2-CHLOROETHOXY)METHANE	BIS(2-CHLOROETHOXY)METHANE

Chemical Name	Reference Name
66-75-1 5-(BIS(2-CHLOROETHYL)AMINO) URACIL [NJL]	URACIL MUSTARD
111-44-4 BIS(2-CHLOROETHYL) ETHER	BIS(2-CHLOROETHYL) ETHER
111-91-1 BIS(2-CHLOROETHYL) FORMAL [FLA, MSL, NFP, NFP]	BIS(2-CHLOROETHOXY)METHANE
494-03-1 N,N-BIS(2-CHLOROETHYL)-2-NAPHTHYLAMINE	N,N-BIS(2-CHLOROETHYL)-2-NAPHTHYLAMINE
494-03-1 N,N-BIS(2-CHLOROETHYL)-2-NAPHTHYLAMINE (CHLORNAPHAZINE) [IAR, C65]	N,N-BIS(2-CHLOROETHYL)-2-NAPHTHYLAMINE
154-93-8 N,N-BIS(2-CHLOROETHYL)-N-NITROSOUREA [FLA]	BISCHLOROETHYL NITROSOUREA
154-93-8 BISCHLOROETHYL NITROSOUREA	BISCHLOROETHYL NITROSOUREA
154-93-8 1,3-BIS(2-CHLOROETHYL)-1-NITROSOUREA [CAL]	BISCHLOROETHYL NITROSOUREA
154-93-8 1,3-BIS(CHLOROETHYL)-1-NITROSOUREA [CSR]	BISCHLOROETHYL NITROSOUREA
154-93-8 BISCHLOROETHYL NITROSOUREA (BCNU) [IAR]	BISCHLOROETHYL NITROSOUREA
154-93-8 BISCHLOROETHYL NITROSOUREA (BCNU) (CARMUSTINE) [C65, C65]	BISCHLOROETHYL NITROSOUREA
108-60-1 BIS(2-CHLOROISOPROPYL) ETHER	BIS(2-CHLOROISOPROPYL) ETHER
39638-32-9 BIS(2-CHLOROISOPROPYL)ETHER	BIS(2-CHLOROISOPROPYL)ETHER
13483-18-6 1,2-BIS(CHLOROMETHOXY)ETHANE	1,2-BIS(CHLOROMETHOXY)ETHANE
56894-91-8 1,4-BIS(CHLOROMETHOXYMETHYL)BENZENE	1,4-BIS(CHLOROMETHOXYMETHYL)BENZENE
542-88-1 BIS-CHLOROMETHYL ETHER [BC1, BC1]	BIS(CHLOROMETHYL) ETHER
542-88-1 BIS(CHLOROMETHYL) ETHER [CAB]	BIS(CHLOROMETHYL) ETHER
542-88-1 BIS-CHLOROMETHYL ETHER [CEX, CEX]	BIS(CHLOROMETHYL) ETHER
542-88-1 BISCHLOROMETHYL ETHER [MAK]	BIS(CHLOROMETHYL) ETHER
542-88-1 BIS-CHLOROMETHYL ETHER [NHS, NHS, NHS, NIO, NIO]	BIS(CHLOROMETHYL) ETHER
108-60-1 BIS(2-CHLORO-METHYLETHYL) ETHER [313]	BIS(2-CHLOROISOPROPYL) ETHER
108-60-1 BIS(2-CHLORO-1-METHYLETHYL)ETHER [IAR]	BIS(2-CHLOROISOPROPYL) ETHER
108-60-1 BIS(2-CHLORO-1-METHYLETHYL) ETHER [CSR, CSR]	BIS(2-CHLOROISOPROPYL) ETHER
108-60-1 BIS(2-CHLORO-METHYLETHYL) ETHER [PQL]	BIS(2-CHLOROISOPROPYL) ETHER
534-07-6 BIS(CHLOROMETHYL)KETONE	BIS(CHLOROMETHYL)KETONE
38051-10-4 2,2-BIS(CHLOROMETHYL)-1,3-PROPANEDIYL TETRAKIS(2-CHLOROETHYL) PHOSPHATE	2,2-BIS(CHLOROMETHYL)-1,3-PROPANEDIYL TETRAKIS (2-CHLOROETHYL
115-32-2 1,1-BIS(P-CHLOROPHENYL)-2,2,2-TRICHLOROETHANOL (DICOFOL) [CAL]	DICOFOL
102-54-5 BIS(ETA-CYCLOPENTADIENYL) IRON [ONT]	DICYCLOPENTADIENYL IRON
64047-28-5 BIS(DI-(BETA-CHLOROETHYL)SULFIDE)PALLADOUS CHLORIDE	BIS(DI-(BETA-CHLOROETHYL)SULFIDE)PALLADOUS CHLORIDE
RR-00821-9 BIS-DIETHYLENE GLYCOL MONOETHYL ETHER PHTHALATE	BIS-DIETHYLENE GLYCOL MONOETHYL ETHER PHTHALATE
110-18-9 1,2-BIS(DIMETHYLAMINO)ETHANE	1,2-BIS(DIMETHYLAMINO)ETHANE
RR-00818-4 BIS(2,4-DIMETHYLBUTYL) MALEATE	BIS(2,4-DIMETHYLBUTYL) MALEATE
128-37-0 2,6-BIS(1,1-DIMETHYLETHYL)-4-METHYLPHENOL [ONT]	2,6-DI-TERT-BUTYL-P-CRESOL
38304-52-8 1,3-BIS(5,5-DIMETHYL-1-GLYCIDYLHYDANTOIN-3-YL)-2-GLYCIDYLOXYPROPANE [EPA]	2,4-IMIDAZOLIDINEDIONE, 3,3'-[2-(OXIRANYLMETHOXY)-1,3-PROPAN
38304-52-8 1,3-BIS(5,5-DIMETHYL-1-GLYCIDYL-HYDANTOIN-3-YL)-2-GLYCIDYL [OTS]	2,4-IMIDAZOLIDINEDIONE, 3,3'-[2-(OXIRANYLMETHOXY)-1,3-PROPAN
3081-14-9 N,N'-BIS-(1,4-DIMETHYL-PENTYL)-P-PHENYLENEDIAMINE [FLA]	1,4-BENZENEDIAMINE, N,N'-BIS(1,4-DIMETHYLPENTYL)-

Chemical Name	Reference Name
3081-14-9 N,N'-BIS-(1,4-DIMETHYLPENTYL) P-PHENYLENEDIAMINE [NFP, NFP]	1,4-BENZENEDIAMINE, N,N'-BIS(1,4-DIMETHYLPENTYL)-
3081-14-9 N,N'-BIS(1,4-DIMETHYLPENTYL)-P-PHENYLENEDIAMINE [MSL]	1,4-BENZENEDIAMINE, N,N'-BIS(1,4-DIMETHYLPENTYL)-
137-26-8 BIS(DIMETHYLTHIOCARBAMOYL) DISULFIDE [CAL]	THIRAM
137-26-8 BIS(DIMETHYLTHIOCARBONYL) DISULFIDE [ONT]	THIRAM
3130-19-6 BIS[(3,4-EPOXYCYCLOHEXYL)METHYL] ADIPATE	BIS[(3,4-EPOXYCYCLOHEXYL)METHYL] ADIPATE
2386-90-5 BIS(2,3-EPOXYCYCLOPENTYL)ETHER	BIS(2,3-EPOXYCYCLOPENTYL)ETHER
126-80-7 1,3-BIS[3-(2,3-EPOXYPROPOXY)PROPYL]TETRAMETHYLDISILOXANE [EPA]	DISILOXANE, 1,1,3,3-TETRAMETHYL-1,3-BIS[3-OXIRANYLMETHOXY)
126-80-7 1,3-BIS[3-(2,3-EPOXYPROPOXY)-PROPYL)TETRAMETHYLDISILOXANE [OTS]	DISILOXANE, 1,1,3,3-TETRAMETHYL-1,3-BIS[3-OXIRANYLMETHOXY)
2238-07-5 BIS(2,3-EPOXYPROPYL) ETHER [GBR]	DIGLYCIDYL ETHER (DGE)
103-23-1 BIS (2-ETHYLHEXYL) ADIPATE	BIS (2-ETHYLHEXYL) ADIPATE
106-20-7 BIS(2-ETHYLHEXYL)AMINE [FLA, MSL]	DI(2-ETHYLHEXYL)AMINE
106-20-7 BIS(2-ETHYLHEXYL) AMINE [NFP, NFP]	DI(2-ETHYLHEXYL)AMINE
103-24-2 BIS(2-ETHYLHEXYL) AZELATE [PST]	DI-2-ETHYL HEXYL AZELATE
142-16-5 BIS(2-ETHYLHEXYL)-2-BUTENEDIOATE	BIS(2-ETHYLHEXYL)-2-BUTENEDIOATE
19074-24-9 BIS(2-ETHYLHEXYL) DODECANEDIOATE	BIS(2-ETHYLHEXYL) DODECANEDIOATE
RR-00819-5 BIS(2-ETHYLHEXYL)-ETHANOLAMINE	BIS(2-ETHYLHEXYL)-ETHANOLAMINE
137-89-3 BIS(2-ETHYLHEXYL) ISOPHTHALATE	BIS(2-ETHYLHEXYL) ISOPHTHALATE
142-16-5 BIS(2-ETHYLHEXYL) MALEATE [NFP, NFP]	BIS(2-ETHYLHEXYL)-2-BUTENEDIOATE
25168-24-5 BIS(2-ETHYLHEXYLOXYCARBONYLMETHYLTHIO)DIBUTYLSTANNANE [WHM]	DIBUTYLTIN BIS(ISOOCTYL MERCAPTOACETATE)
15546-12-0 BIS((2-ETHYLHEXYLOXY)MALEOLOXY) DIBUTYLSTANNANE	BIS((2-ETHYLHEXYLOXY)MALEOLOXY) DIBUTYLSTANNANE
298-07-7 BIS(2-ETHYLHEXYL) PHOSPHORIC ACID [NFP]	DI(2-ETHYLHEXYL)PHOSPHORIC ACID
117-81-7 BIS(2-ETHYLHEXYL) PHTHALATE [CAN, CW2, EP2]	DI(2-ETHYLHEXYL)PHTHALATE
117-81-7 BIS(2-ETHYLHEXYL)PHTHALATE [EPA, GBR, GBR]	DI(2-ETHYLHEXYL)PHTHALATE
117-81-7 BIS (2-ETHYLHEXYL) PHTHALATE [MX1]	DI(2-ETHYLHEXYL)PHTHALATE
117-81-7 BIS(2-ETHYLHEXYL)PHTHALATE [NJL, NJL, PQL]	DI(2-ETHYLHEXYL)PHTHALATE
117-81-7 BIS(2-ETHYLHEXYL) PHTHALATE [PST, RCA]	DI(2-ETHYLHEXYL)PHTHALATE
117-81-7 BIS(2-ETHYLHEXYL)PHTHALATE(DEHP) [CAA]	DI(2-ETHYLHEXYL)PHTHALATE
2915-57-3 BIS(2-ETHYLHEXYL) SUCCINATE	BIS(2-ETHYLHEXYL) SUCCINATE
71033-08-4 2,2-BIS[P-2-GLYCIDYLOXY-3-BUTOXYPROPYLOXY)-PHENYL]PROPANE [EPA]	OXIRANE, 2,2'-[(1-METHYLETHYLIDINE)BIS[4,1-PHENYLENEOXY[1-(BRASILOXANE
71033-08-4 2,2-BIS(P-GLYCIDYLOXY-3-BUTOXYPROPYLOXY)-PHENYL)PROPANE [OTS]	OXIRANE, 2,2'-[(1-METHYLETHYLIDINE)BIS[4,1-PHENYLENEOXY[1-(B
14228-73-0 BIS(GLYCIDYLOXYMETHYL) CYCLOHEXANE, 1,4- [OTS]	OXIRANE, 2,2'-[1,4-CYCLOHEXANEDILBIS (METHYLENEOXYMETHYLENE)
14228-73-0 1,4-BIS(GLYCIDYLOXYMETHYL) CYCLOHEXANE [EPA]	OXIRANE, 2,2'-[1,4-CYCLOHEXANEDILBIS (METHYLENEOXYMETHYLENE)
67786-03-2 [BIS(4-GLYCIDYLOXYPHENYL)]-(2-GLYCIDYLOXYPHENYL)METHANE [EPA]	OXIRANE, 2,2'-[[[(2-OXIRANYLMETHOXY)PHENYL]METHYLENE]BIS(4,1
67786-03-2 [BIS(4-GLYCIDYLOXYPHENYL)]-(2-GLYCIDYLOXYPHENYL)METHANE [OTS]	OXIRANE, 2,2'-[[[(2-OXIRANYLMETHOXY)PHENYL]METHYLENE]BIS(4,1

Chemical Name	Reference Name
61789-81-9 BIS(HYDROGENATED TALLOW ALKYL) DIMETHYL QUATERNARY COMPOUNDS, METHYL SULFATES	BIS(HYDROGENATED TALLOW ALKYL)DIMETHYL QUA-TERNARY COMPOUNDS,
71786-60-2 N,N-BIS(2-HYDROXYETHYL)-C12-18-ALKY-LAMINE	N,N-BIS(2-HYDROXYETHYL)-C12-18-ALKY-LAMINETHANOLAMINE SALTS
61791-47-7 BIS(2-HYDROXYETHYL) COCOAMINE OXIDE	BIS(2-HYDROXYETHYL) COCOAMINE OXIDE
77500-13-1 BIS(2-HYDROXYETHYL)-3-(DECYLOXY)PROPY-LAMINE OXIDE	BIS(2-HYDROXYETHYL)-3-(DECYLOXY)PROPYLAMINE OXIDE
120-40-1 N,N-BIS(2-HYDROXYETHYL)DODE-CANAMIDE [PST]	LAURIC ACID DIETHANOLAMIDE CONDENSATE
93-83-4 N,N-BIS(2-HYDROXYETHYL)-CIS-9-OCTADECE-NAMIDE [PST]	OLEIC ACID DIETHANOLAMIDE
140-95-4 N,N'-BIS(HYDROXYMETHYL) UREA [PST]	DIMETHYLOLUREA
117-82-8 BIS(2-METHOXYETHYL)PHTHALATE	BIS(2-METHOXYETHYL)PHTHALATE
93-96-9 BIS(ALPHA-METHYLBENZYL) ETHER [FLA]	BENZENE, 1,1'-(OXYDIETHYLIDENE)BIS-
37853-61-5 BISMETHYLETHER OF TETRABROMOBISPHE-NOL-A	BISMETHYLETHER OF TETRABROMOBISPHENOL-A
5012-62-4 2,6-BIS(1-METHYLHEPTADECYL)-P-CRESOL	2,6-BIS(1-METHYLHEPTADECYL)-P-CRESOL
1330-76-3 BIS(1-METHYLHEPTYL)-2-BUTENEDIOATE	BIS(1-METHYLHEPTYL)-2-BUTENEDIOATE
RR-00871-9 N,N-BIS(1-METHYLHEPTYL) ETHYLENEDI-AMINE	N,N-BIS(1-METHYLHEPTYL) ETHYLENEDIAMINE
10347-54-3 1,4-BIS(METHYLISOCYANATE)CYCLOHEX-ANE [313]	CYCLOHEXANE, 1,4-BIS(ISOCYANATOMETHYL)-
38661-72-2 1,3-BIS(METHYLISOCYANATE)CYCLOHEX-ANE [313]	CYCLOHEXANE, 1,3-BIS(ISOCYANATOMETHYL)-
7440-69-9 BISMUTH	BISMUTH
17239-85-9 BISMUTH 200	BISMUTH 200
14280-38-7 BISMUTH 201	BISMUTH 201
14687-50-4 BISMUTH 202	BISMUTH 202
24383-94-6 BISMUTH 203	BISMUTH 203
14333-38-1 BISMUTH 205	BISMUTH 205
15776-19-9 BISMUTH 206	BISMUTH 206
13982-38-2 BISMUTH 207	BISMUTH 207
14331-79-4 BISMUTH 210	BISMUTH 210
RR-00480-8 BISMUTH 210M	BISMUTH 210M
14913-49-6 BISMUTH 212	BISMUTH 212
15776-20-2 BISMUTH 213	BISMUTH 213
14733-03-0 BISMUTH 214	BISMUTH 214
61204-26-0 BISMUTH CHROMATE	BISMUTH CHROMATE
37293-14-4 BISMUTH TELLURIDE	BISMUTH TELLURIDE
1304-82-1 BISMUTH TELLURIDE (BI2TE3) [PAL]	BISMUTH TELLURIDE
37293-14-4 BISMUTH TELLURIDE (SE DOPED) [ALB, ALB]	BISMUTH TELLURIDE
37293-14-4 BISMUTH TELLURIDE, SE DOPED [OS1]	BISMUTH TELLURIDE
37293-14-4 BISMUTH TELLURIDE, SELENIUM-DOPED [BC1, BC1]	BISMUTH TELLURIDE
1304-82-1 BISMUTH TELLURIDE, UNDOPED [OS1, OS1]	BISMUTH TELLURIDE
41556-26-7 BIS (1,2,2,6,6-PENTAMETHYL-4-PIPERIDINYL) DECANEDIOATE [PST]	BIS (1,2,2,6,6-PENTAMETHYL-4-PIPERIDINYL) SEBACATE
41556-26-7 BIS (1,2,2,6,6-PENTAMETHYL-4-PIPERIDINYL) SEBACATE	BIS (1,2,2,6,6-PENTAMETHYL-4-PIPERIDINYL) SEBACATE
RR-01576-9 BIS(1,2,2,6,6-PENTAMETHYL-4-PIPERIDIN-4-OL) ESTER OF CYCLOALIPHATIC SPIROKETAL	BIS(1,2,2,6,6-PENTAMETHYL-4-PIPERIDIN-4-OL) ESTER OF CYCLOAL
80-05-7 BISPHENOL A	BISPHENOL A

Chemical Name	Reference Name
1675-54-3 BISPHENOL A DIGLYCIDYL ETHER	BISPHENOL A DIGLYCIDYL ETHER
RR-00231-3 BISPHENOL A, EPICHLOROHYDRIN, METHYLENEBIS(SUBSTITUTEDCARBOMONO-CYCLE), POLYALKYLENE GLYCOL, ALKANOL, METHACRYLATE POLYMER	BISPHENOL A, EPICHLOROHYDRIN, METHYLENEBIS (SUBSTITUTED
25068-38-6 BISPHENOL A-EPICHLOROHYDRIN POLYMER	BISPHENOL A-EPICHLOROHYDRIN POLYMER
RR-00947-2 BISPHENOL A, EPICHLOROHYDRIN, POLYALKYLENEPOLYOL AND POLYISO-CYANATO DERIVATIVE	BISPHENOL A, EPICHLOROHYDRIN, POLYALKYLENEPOLYOL AND POLYISO
RR-01566-7 BISPHENOL DERIVATIVE	BISPHENOL DERIVATIVE
25068-38-6 BISPHENOL A - EPICHLOROHYDRIN CONDENSATE [PST]	BISPHENOL A-EPICHLOROHYDRIN POLYMER
54208-63-8 BISPHENOL F DIGLYCIDYL ETHER [OTS, EPA]	OXIRANE, 2,2'-[METHYLENEBIS(2,1-PHENYLE-NEOXYMETHYLENE)]BIS-
RR-00913-2 BIS(SUBSTITUTED)CARBOMONOCYCLIC AZO-CARBOMONOCYLICOL	BIS(SUBSTITUTED)CARBOMONOCYCLIC AZOCAR-BOMONOCYLICOLSTITUTED CARBOPOLYCYCLES
52829-07-9 BIS(2,2,6,6-TETRAMETHYL-4-PIPERIDINYL) SEBACETATE [PST]	TINUVIN 770 (AMINE LIGHT STABILIZER)-[(3-HYDROX-YPROPYL)AMINO]-
RR-00953-0 BIS(2,2,6,6-TETRAMETHYLPIPERIDINYL) ESTER OF CYCLOALKYL SPIROKETAL	BIS(2,2,6,6-TETRAMETHYLPIPERIDINYL) ESTER OF CY-CLOALKYL SPIR
128-80-3 1,4-BIS(P-TOLYLAMINO)ANTHRAQUINONE	1,4-BIS(P-TOLYLAMINO)ANTHRAQUINONE
37853-59-1 1,2-BIS(TRIBROMOPHENOXY)-ETHANE [TSE]	FIREMASTER 680
56-35-9 BIS(TRIBUTYLTIN)OXIDE	BIS(TRIBUTYLTIN)OXIDE
56-35-9 BIS(TRIBUTYLTIN) OXIDE [313]	BIS(TRIBUTYLTIN)OXIDEER, PHOSPHOROTHIOIC ACID]
3064-70-8 BIS(TRICHLOROMETHYL)SULFONE	BIS(TRICHLOROMETHYL)SULFONE
2673-22-5 BIS(TRIDECYL) SODIUM SULFOSUCCINATE	BIS(TRIDECYL) SODIUM SULFOSUCCINATE
RR-00817-3 BIS(2,2,4-TRIMETHYLPENTANEDIOLISOBU-TYRATE) DIGLYCOLATE	BIS(2,2,4-TRIMETHYLPENTANEDIOLISOBUTYRATE) DIGLYCOLATE
7422-52-8 3-[BIS(TRIMETHYLSILOXY)METHYL] PROPYL GLYCIDYL ETHER [EPA]	TRISILOXANE, 1,1,1,3,5,5,5-HEPTAMETHYL-3-[3-(OXI-RANYLMETHOXY
7422-52-8 3-(BIS(TRIMETHYLSILOXY)METHYL)-PROPYL GLYCIDYL ETHER [OTS]	TRISILOXANE, 1,1,1,3,5,5,5-HEPTAMETHYL-3-[3-(OXI-RANYLMETHOXY
13356-08-6 BIS(TRIS(2-METHYL-2-PHENYLPROPYL)TIN) OXIDE	BIS(TRIS(2-METHYL-2-PHENYLPROPYL)TIN)OXIDE
RR-00177-4 BISULFITES	BISULFITES
RR-00177-4 BISULFITES, INORGANIC, AQUEOUS SOLUTION, N.O.S. [HM1, HM1, NJL]	BISULFITES
97-18-7 BITHIONOL [MAK]	PHENOL, 2,2'-THIOBIS(4,6-DICHLORO)-
4044-65-9 BITOSCANATE	BITOSCANATE
3734-33-6 BITREX	BITREX
8052-42-4 BITUMEN [MAK]	ASPHALT
8052-42-4 BITUMEN FUMES [AUS]	ASPHALT
RR-00061-3 BITUMENS [IAR, IAR, MIN]	BITUMENS, EXTRACTS OF STEAM-REFINED AND AIR-REFINED
RR-00061-3 BITUMENS, EXTRACTS OF STEAM-REFINED AND AIR-REFINED	BITUMENS, EXTRACTS OF STEAM-REFINED AND AIR-REFINED
RR-00340-7 BLACK NEWSPAPER INK	BLACK NEWSPAPER INK
21725-46-2 BLADEX [NJL]	CYANAZINE
11056-06-7 BLEOMYCIN	BLEOMYCIN
11056-06-7 BLEOMYCINS [CAL, IAR, MIN]	BLEOMYCIN
9003-11-6 BLOCK POLYMER OF POLYPROPYLENE OXIDE AND POLYETHYLENE OXIDE [PST]	POLYETHYLENE-POLYPROPYLENE GLYCOL
12001-28-4 BLUE ASBESTOS [HM1, HM1]	ASBESTOS, CROCIDOLITE

Chemical Name	Reference Name
129-17-9 BLUE VRS [CAL, IAR]	AMMONIUM, (4-(ALPHA-(P-(DIETHYLAMINO)PHENYL)-2, 4-
28249-77-6 BOLERO (THIOBENCARB)	BOLERO (THIOBENCARB)E]
35400-43-2 BOLSTAR (SULPROFOS) [ALB, ALB]	SULPROFOS
68409-75-6 BONE MEAL	BONE MEALCHLORIDES
RR-00539-0 BOOT AND SHOE MANUFACTURE AND REPAIR	BOOT AND SHOE MANUFACTURE AND REPAIR
RR-00539-0 BOOT AND SHOE MANUFACTURE AND REPAIR (CERTAIN EXPOSURES) [PAL, PAL]	BOOT AND SHOE MANUFACTURE AND REPAIR
10294-33-4 BORANE, TRIBROMO- [PAL]	BORON TRIBROMIDE
97-94-9 BORANE, TRIETHYL- [PAL]	TRIETHYLBORANE
7637-07-2 BORANE, TRIFLUORO- [PAL]	BORON TRIFLUORIDE
RR-00180-9 BORATE AND CHLORATE MIXTURES	BORATE AND CHLORATE MIXTURES
1303-96-4 BORATES, TETRA, SODIUM SALTS [TLV]	BORATES, TETRA, SODIUM SALTS, DECAHYDRATE
1330-43-4 BORATES, TETRA, SODIUM SALTS, ANHYDROUS	BORATES, TETRA, SODIUM SALTS, ANHYDROUS
1330-43-4 BORATES, TETRA, SODIUM SALTS ANHYDROUS [AUS, ONT]	BORATES, TETRA, SODIUM SALTS, ANHYDROUS
1303-96-4 BORATES, TETRA, SODIUM SALTS, DECAHYDRATE	BORATES, TETRA, SODIUM SALTS, DECAHYDRATE
1303-96-4 BORATES, TETRA, SODIUM SALTS DECAHYDRATE [AUS]	BORATES, TETRA, SODIUM SALTS, DECAHYDRATE
12179-04-3 BORATES, TETRA, SODIUM SALTS, PENTAHYDRATE	BORATES, TETRA, SODIUM SALTS, PENTAHYDRATE
12179-04-3 BORATES, TETRA, SODIUM SALTS PENTAHYDRATE [AUS]	BORATES, TETRA, SODIUM SALTS, PENTAHYDRATE
25481-93-0 BORATES, TETRA, SODIUM SALTS, PENTAHYDRATE	BORATES, TETRA, SODIUM SALTS, PENTAHYDRATE
13826-83-0 BORATE(1-), TETRAFLUORO-, AMMONIUM [PAL]	AMMONIUM FLUOBORATE
13814-96-5 BORATE(1-), TETRAFLUORO-, LEAD (2+) (2:1) [PAL]	LEAD FLUOBORATE
283-56-7 BORATRAN	BORATRAN
1303-96-4 BORAX [PST]	BORATES, TETRA, SODIUM SALTS, DECAHYDRATE
1303-96-4 BORAX (B4NA2O7.10H2O) [PAL]	BORATES, TETRA, SODIUM SALTS, DECAHYDRATE
RR-01042-4 BORAX PENTAHYDRATE	BORAX PENTAHYDRATENESIUM, POTASSIUM OR ZINC SALT
RR-00181-0 BORDEAUX ARSENITE [NJL]	BORDEAUX ARSENITE, LIQUID OR SOLID
RR-00181-0 BORDEAUX ARSENITE, LIQUID OR SOLID	BORDEAUX ARSENITE, LIQUID OR SOLID
10043-35-3 BORIC ACID [CSR, CSR]	BORIC ACID (H3BO3)
11113-50-1 BORIC ACID	BORIC ACIDER
10043-35-3 BORIC ACID (H3BO3)	BORIC ACID (H3BO3)
RR-00160-5 BORIC ACID, ALKYL AND SUBSTITUTED ALKYL ESTERS	BORIC ACID, ALKYL AND SUBSTITUTED ALKYL ESTERS
51845-86-4 BORIC ACID, ETHYL ESTER	BORIC ACID, ETHYL ESTER
688-74-4 BORIC ACID (H3BO3), TRIBUTYL ESTER [PAL]	TRIBUTYL BORATE
121-43-7 BORIC ACID (H3BO3), TRIMETHYL ESTER [PAL]	TRIMETHYL BORATE1-PROPENYL)-2,2-DIMETHYL-,2-METHYL-4-OXO-3-(2,4-PENTA DIENYL)-2-CYCLOPENTEN-1-YL ESTER, [R-[1.ALPHA.[3*(Z)], 3.BETA.(E)]-
5419-55-6 BORIC ACID (H3BO3), TRIS(1-METHYLETHYL) ESTER [PAL]	TRIISOPROPYL BORATE
13195-76-1 BORIC ACID (H3BO3), TRIS(2-METHYLPROPYL) ESTER	BORIC ACID (H3BO3), TRIS(2-METHYLPROPYL) ESTER
1332-07-6 BORIC ACID, ZINC SALT [PAL]	ZINC BORATE

Chemical Name	Reference Name
1303-86-2 BORIC ANHYDRIDE [WHM]	BORON OXIDE
76-22-2 BORNAN-2-ONE [GBR, GBR]	CAMPHOR
507-70-0 BORNEOL	BORNEOL
7440-42-8 BORON	BORON
54566-73-3 BORON OXIDE	BORON OXIDE
1303-86-2 BORON OXIDE (B2O3) [PAL]	BORON OXIDE
12008-41-2 BORON SODIUM OXIDE (B8NA2O13)	BORON SODIUM OXIDE (B8NA2O13)
1330-43-4 BORON SODIUM OXIDE (B4NA2O7) [PAL]	BORATES, TETRA, SODIUM SALTS, ANHYDROUS
10294-33-4 BORON TRIBROMIDE	BORON TRIBROMIDE
10294-34-5 BORON TRICHLORIDE	BORON TRICHLORIDE
10294-34-5 BORON TRICHLORIDE [BORANE, TRICHLORO-] [EP3]	BORON TRICHLORIDE
7637-07-2 BORON TRIFLUORIDE	BORON TRIFLUORIDE
753-53-7 BORON TRIFLUORIDE ACETIC ACID	BORON TRIFLUORIDE ACETIC ACID
753-53-7 BORON TRIFLUORIDE ACETIC ACID COMPLEX [NJL]	BORON TRIFLUORIDE ACETIC ACID
353-42-4 BORON TRIFLUORIDE COMPOUND WITH METHYL ETHER (1:1) [BORON, TRIFLUORO [OXYBIS[METANE]-, T-4- [EP3]	BORON TRIFLUORIDE COMPOUND WITH METHYL ETHER (1:1)
353-42-4 BORON TRIFLUORIDE COMPOUND WITH METHYL ETHER [MSL]	BORON TRIFLUORIDE COMPOUND WITH METHYL ETHER (1:1)
353-42-4 BORON TRIFLUORIDE COMPOUND WITH METHYL ETHER (1:1)	BORON TRIFLUORIDE COMPOUND WITH METHYL ETHER (1:1)
109-63-7 BORON TRIFLUORIDE DIETHYL ETHER- ATE [NJL, NJL]	BORON TRIFLUORIDE DIETHYLETHERATE
109-63-7 BORON TRIFLUORIDE DIETHYLETHERATE	BORON TRIFLUORIDE DIETHYLETHERATE
13319-75-0 BORON TRIFLUORIDE DIHYDRATE	BORON TRIFLUORIDE DIHYDRATE
353-42-4 BORON TRIFLUORIDE DIMETHYL ETHER- ATE [HM1, HM1, NJL]	BORON TRIFLUORIDE COMPOUND WITH METHYL ETHER (1:1)
109-63-7 BORON TRIFLUORIDE ETHERATE [FLA, NFP, NFP]	BORON TRIFLUORIDE DIETHYLETHERATE
109-63-7 BORON TRIFLOURIDE ETHERATE [MSL]	BORON TRIFLUORIDE DIETHYLETHERATE
RR-01344-5 BORON TRIFLUORIDE PROPIONIC ACID COMPLEX	BORON TRIFLUORIDE PROPIONIC ACID COMPLEX
7637-07-2 BORON TRIFLUORIDE [BORANE, TRIFLUORO-] [EP3]	BORON TRIFLUORIDE
109-63-7 BORON, TRIFLUORO[1,1'-OXYBIS[ETHANE]-, (T-4)- [PAL]	BORON TRIFLUORIDE DIETHYLETHERATE
67859-60-3 BOROXIN, TRIS((2-ETHYLHEXYL)OXY)-	BOROXIN, TRIS((2-ETHYLHEXYL)OXY)-
RR-00032-8 BRACKEN FERN	BRACKEN FERN
RR-00032-8 BRACKERN FERN [MSL]	BRACKEN FERN
RR-01043-5 BRAN	BRAN
RR-01044-6 BREAD CRUMBS	BREAD CRUMBS
3844-45-9 BRILLIANT BLUE FCF [CAL, IAR]	C.I. ACID BLUE 9, DISODIUM SALT
6104-58-1 BRILLIANT BLUE G (ACID BLUE 90)	BRILLIANT BLUE G (ACID BLUE 90)
6104-59-2 BRILLIANT BLUE R (ACID BLUE 83)	BRILLIANT BLUE R (ACID BLUE 83)
56073-10-0 BRODIFACOUM	BRODIFACOUM
314-40-9 BROMACIL	BROMACIL
314-40-9 BROMACIL (ISO) [GBR, GBR]	BROMACIL
314-40-9 BROMACIL (5-BROMO-6-METHYL-3-(1-METHYLPROPYL)-2,4-(1H,3H)-PYRIMIDINE-DIONE [313]	BROMACIL-1,4,4A,5,8,8A,HEXAHYDRO-(1ALPHA,4ALPHA, 4BETA,5ALPHA, 8ALPHA,8BETA)-]

Chemical Name	Reference Name
53404-19-6 BROMACIL, LITHIUM SALT [2,4-(1H,3H)-PYRIMIDINEDIONE, 5-BROMO-6-METHYL-3-(1-METHYLPROPYL), LITHIUM SALT]	BROMACIL, LITHIUM SALT [2,4-(1H,3H)-PYRIM-IDINEDIONE, 5-BROMOCARBOXYLIC ACID, (3-PHE-NOXYPHENYL)METHYL ESTER]
28772-56-7 BROMADIOLONE	BROMADIOLONE
116-81-4 BROMAMINE ACID	BROMAMINE ACID
RR-00183-2 BROMATES, INORGANIC, N.O.S.	BROMATES, INORGANIC, N.O.S.
7758-01-2 BROMIC ACID, POTASSIUM SALT [PAL]	POTASSIUM BROMATE
RR-00212-0 BROMINATED AROMATIC COMPOUND	BROMINATED AROMATIC COMPOUNDDIAMINE METHYL SULFATE, DOUBLE SALT
RR-00489-7 BROMINATED FLUOROCARBONS	BROMINATED FLUOROCARBONS
RR-01253-3 BROMINATED TRIAZINE DERIVATIVE	BROMINATED TRIAZINE DERIVATIVE1,2]TRIS(2-PROPENOATO-O-)
7726-95-6 BROMINE	BROMINE
15720-26-0 BROMINE 74	BROMINE 74
RR-00478-4 BROMINE 74M	BROMINE 74M
14809-47-3 BROMINE 75	BROMINE 75
15765-38-5 BROMINE 76	BROMINE 76
15765-39-6 BROMINE 77	BROMINE 77
14391-61-8 BROMINE 80	BROMINE 80
RR-00477-3 BROMINE 80M	BROMINE 80M
14686-69-2 BROMINE 82	BROMINE 82
14687-62-8 BROMINE 83	BROMINE 83
14331-90-9 BROMINE 84	BROMINE 84
RR-01399-0 BROMINE AZIDE	BROMINE AZIDE
13863-41-7 BROMINE CHLORIDE	BROMINE CHLORIDE
RR-00331-6 BROMINE COMPOUNDS (INORGANIC)	BROMINE COMPOUNDS (INORGANIC)
506-68-3 BROMINE CYANIDE [HM1, HM1, HM1, HM1, HM1]	CYANOGEN BROMIDE
7787-71-5 BROMINE FLUORIDE (BRF3) [PAL]	BROMINE TRIFLUORIDE
7789-30-2 BROMINE FLUORIDE (BRF5) [PAL]	BROMINE PENTAFLUORIDE
7789-30-2 BROMINE PENTAFLUORIDE	BROMINE PENTAFLUORIDE
7787-71-5 BROMINE TRIFLUORIDE	BROMINE TRIFLUORIDE
79-08-3 BROMOACETIC ACID	BROMOACETIC ACID
598-31-2 BROMOACETONE	BROMOACETONE
598-21-0 BROMOACETYL BROMIDE	BROMOACETYL BROMIDE
RR-01274-8 BROMOALLYLENE	BROMOALLYLENE
108-86-1 BROMOBENZENE	BROMOBENZENE
5798-79-8 BROMOBENZYL CYANIDE	BROMOBENZYL CYANIDE
16532-79-9 4-BROMOBENZYLCYANIDE	4-BROMOBENZYLCYANIDE
19472-24-3 ORTHO-BROMOBENZYL CYANIDE	ORTHO-BROMOBENZYL CYANIDE
RR-00085-1 BROMOBENZYL CYANIDE, ALL ISOMERS	BROMOBENZYL CYANIDE, ALL ISOMERS
RR-00085-1 BROMOBENZYL CYANIDES [HM1, HM1]	BROMOBENZYL CYANIDE, ALL ISOMERS
92-66-0 4-BROMO-1,1-BIPHENYL [TSE]	4-BROMODIPHENYL
2052-07-5 2-BROMO-1,1-BIPHENYL [TSE]	1,1'-BIPHENYL, 2-BROMO-
2113-57-7 3-BROMO-1,1-BIPHENYL [TSE]	1,1'-BIPHENYL, 3-BROMO-
35691-65-7 2-BROMO-2-(BROMOETHYL)PENTANEDINI-TRILE [PST]	1,2-DIBROMO-2,4-DICYANOBUTANE
35691-65-7 1-BROMO-1-(BROMOMETHYL)-1,3-PROPANEDI-CARBONITRILE [313]	1,2-DIBROMO-2,4-DICYANOBUTANE1H-IMIDAZOLE]
78-76-2 2-BROMOBUTANE	2-BROMOBUTANE
83463-62-1 BROMOCHLOROACETONITRILE	BROMOCHLOROACETONITRILE

Chemical Name	Reference Name
353-59-3 BROMOCHLORODIFLUOROMETHANE [CN5]	CHLORODIFLUOROBROMO-METHANE
353-59-3 BROMOCHLORODIFLUOROMETHANE (HALON 1211) [313]	CHLORODIFLUOROBROMO-METHANE
353-59-3 BROMOCHLORODIFLUOROMETHANE [CN2]	CHLORODIFLUOROBROMO-METHANE,5,8,8A-HEXAHY-DRO-1,4:5,8-DIMETHANONAPHTHALENE)
126-06-7 3-BROMO-1-CHLORO-5,5-DIMETHYL-2,4-IMIDA-ZOLIDINEDIONE	3-BROMO-1-CHLORO-5,5-DIMETHYL-2,4-IMIDAZO-LIDINEDIONE
107-04-0 1-BROMO-2-CHLOROETHANE	1-BROMO-2-CHLOROETHANE
74-97-5 BROMOCHLOROMETHANE [CAL, GBR, GBR, HM1, HM1, MAK, MAK, ONT, ONT, OTS, SDW]	CHLOROBROMOMETHANE
74-97-5 BROMOCHLOROMETHANE (CHLOROBRO-MOMETHANE) [ALB, ALB]	CHLOROBROMOMETHANE
74-97-5 BROMOCHLOROMETHANE/ CHLOROBRO-MOMETHANE [BC1, BC1]	CHLOROBROMOMETHANE
112-29-8 1-BROMODECANE	1-BROMODECANE
59-14-3 5-BROMO-2'-DEOXYURIDINE	5-BROMO-2'-DEOXYURIDINE
683-53-4 BROMODICHLOROETHANE	BROMODICHLOROETHANE
75-27-4 BROMODICHLOROMETHANE [C65, C65, CAL, CSR, CSR, IAR, NTP, PQL, RCA, SD4, SDW, UTS]	DICHLOROBROMOMETHANE
1940-42-7 4-BROMO-2,5-DICHLOROPHENOL	4-BROMO-2,5-DICHLOROPHENOL
610-38-8 4-BROMO-1,2-DINITROBENZENE	4-BROMO-1,2-DINITROBENZENE
92-66-0 4-BROMODIPHENYL	4-BROMODIPHENYL
74-96-4 BROMOETHANE [CAN, GBR, GBR, IAR, MAK]	ETHYL BROMIDE
74-96-4 BROMOETHANE (ETHYL BROMIDE) [CSR, CSR]	ETHYL BROMIDE
592-55-2 BROMOETHYL ETHYL ETHER	BROMOETHYL ETHYL ETHER
592-55-2 2-BROMOETHYL ETHYL ETHER [HM1, HM1]	BROMOETHYL ETHYL ETHER
RR-01349-0 BROMOETHYLPROPANES	BROMOETHYLPROPANES
75-25-2 BROMOFORM	BROMOFORM
75-25-2 BROMOFORM (TRIBROMOMETHANE) [CW2, 313, QBC, QBC]	BROMOFORM
111-25-1 1-BROMOHEXANE	1-BROMOHEXANE
74-83-9 BROMOMETHANE [GBR, GBR, GBR, SDW]	METHYL BROMIDE (BROMOMETHANE)
74-83-9 METHYL BROMIDE (BROMOMETHANE) [UTS]	METHYL BROMIDE (BROMOMETHANE)
74-96-4 BROMOMETHANE (ETHYL BROMIDE) [QBC, QBC]	ETHYL BROMIDE
74-83-9 BROMOMETHANE (METHYL BROMIDE) [313, QBC, QBC]	METHYL BROMIDE (BROMOMETHANE)
95-46-5 1-BROMO-2-METHYL BENZENE [FLA]	O-BROMOTOLUENE
107-82-4 BROMOMETHYLBUTANE	BROMOMETHYLBUTANE
107-82-4 1-BROMO-3-METHYLBUTANE [HM1, HM1]	BROMOMETHYLBUTANE
314-42-1 5-BROMO-6-METHYL-3-(1-METHYLETHYL)-2,4 (1H,3H)-PYRIMIDINEDIONE	5-BROMO-6-METHYL-3-(1-METHYLETHYL)-2,4(1H,3H)-PYRIMIDINEDION
78-77-3 1-BROMO-2-METHYLPROPANE	1-BROMO-2-METHYLPROPANE
507-19-7 2-BROMO-2-METHYLPROPANE	2-BROMO-2-METHYLPROPANE
30007-47-7 5-BROMO-5-NITRO-1,3-DIOXANE	5-BROMO-5-NITRO-1,3-DIOXANE
52-51-7 2-BROMO-2-NITRO-1,3-PROPANEDIOL	2-BROMO-2-NITRO-1,3-PROPANEDIOL
52-51-7 2-BROMO-2-NITROPROPANE-1,3-DIOL (BRONOPOL) [313]	2-BROMO-2-NITRO-1,3-PROPANEDIOL
7166-19-0 BETA-BROMO-BETA-NITROSTYRENE	BETA-BROMO-BETA-NITROSTYRENE
111-83-1 1-BROMOOCTANE	1-BROMOOCTANE
107-81-3 2-BROMOPENTANE	2-BROMOPENTANE
110-53-2 1-BROMOPENTANE	1-BROMOPENTANE

Chemical Name	Reference Name
29756-38-5 BROMOPENTANE	BROMOPENTANE
95-56-7 O-BROMOPHENOL	O-BROMOPHENOL
101-55-3 4-BROMOPHENYL PHENYL ETHER [MX2, UTS, CAL, CW2, MX1, PQL, RCA, RCU, RCU, TSC, TSE]	BENZENE, 1-BROMO-4-PHENOXY
4824-78-6 BROMOPHOS-ETHYL	BROMOPHOS-ETHYL
75-26-3 2-BROMOPROPANE	2-BROMOPROPANE
26446-77-5 BROMOPROPANE	BROMOPROPANE
106-95-6 3-BROMOPROPENE [HM1, HM1, HM1]	ALLYL BROMIDE
106-96-7 3-BROMOPROPYNE	3-BROMOPROPYNE
13465-73-1 BROMOSILANE	BROMOSILANE
95-46-5 O-BROMOTOLUENE	O-BROMOTOLUENE
106-38-7 P-BROMOTOLUENE	P-BROMOTOLUENE
598-73-2 BROMOTRIFLUOROETHYLENE	BROMOTRIFLUOROETHYLENE
75-63-8 BROMOTRIFLUOROMETHANE [CAL, CN2, CN5, GBR, GBR, MAK, MAK, ONT, QBC]	TRIFLUOROBROMOMETHANE
75-63-8 BROMOTRIFLUOROMETHANE (HALON 1301) [313]	TRIFLUOROBROMOMETHANE
598-73-2 BROMOTRILFUOROETHYLENE [ETHYLENE, BROMOTRIFLUORO-] [EP3]	BROMOTRIFLUOROETHYLENE
1689-84-5 BROMOXYNIL	BROMOXYNIL
1689-84-5 BROMOXYNIL (3,5-DIBROMO-4-HYDROXYBEN-ZONITRILE) [313]	BROMOXYNIL
1689-99-2 BROMOXYNIL OCTANOATE (OCTANOIC ACID, 2,6-DIBROMO-4-CYANOPHENYL ESTER)	BROMOXYNIL OCTANOATE (OCTANOIC ACID,2,6-DI-BROMO-4-CYANOPHENY
RR-00783-0 BRONZING LIQUID	BRONZING LIQUID
RR-00515-2 BROWN COAL TAR	BROWN COAL TAR
357-57-3 BRUCINE	BRUCINE
55-98-1 BUSULFAN	BUSULFAN
143-81-7 BUTABARBITAL SODIUM	BUTABARBITAL SODIUM
106-99-0 1,3-BUTADIENE	1,3-BUTADIENE
106-99-0 BUTA-1,3-DIENE [GBR]	1,3-BUTADIENE
106-99-0 BUTADIENE [HM1, HM1, MIN, WHM]	1,3-BUTADIENE
106-99-0 BUTADIENE (1,3-BUTADIENE) [OS1, OS1, BC1, BC1]	1,3-BUTADIENE
126-99-8 1,3-BUTADIENE, 2-CHLORO- [PAL, TSC, TSC, TSC]	2-CHLORO-1,3-BUTADIENE (CHLOROPRENE)
1653-19-6 1,3-BUTADIENE, 2,3-DICHLORO- [PAL]	2,3-DICHLOROBUTADIENE-1,3
87-68-3 1,3-BUTADIENE, 1,1,2,3,4,4-HEXACHLORO- [PAL, TSC, TSC, TSC]	HEXACHLOROBUTADIENE
78-79-5 1,3-BUTADIENE, 2-METHYL- [PAL]	ISOPRENE
930-22-3 BUTADIENE MONOXIDE	BUTADIENE MONOXIDE
9003-17-2 BUTADIENE RESIN [PST]	POLYBUTADIENE RESIN
9003-55-8 BUTADIENE-STYRENE COPOLYMER [PST]	STYRENE-BUTADIENE POLYMER
431-03-8 2,3-BUTADIONE	2,3-BUTADIONE
123-72-8 BUTANAL [PAL, TSC, TSC, TSC]	BUTYRALDEHYDE
97-96-1 BUTANAL, 2-ETHYL- [PAL]	2-ETHYLBUTYRALDEHYDE
107-89-1 BUTANAL, 3-HYDROXY- [PAL]	ALDOL
96-17-3 BUTANAL, 2-METHYL-	BUTANAL, 2-METHYL-
590-86-3 BUTANAL, 3-METHYL- [OTS, PAL, TSC, TSC, TSC]	ISOPENTALDEHYDE

Chemical Name	Reference Name
110-69-0 BUTANAL OXIME [CSR]	BUTYRALDOXIME
110-69-0 BUTANAL, OXIME [PAL]	BUTYRALDOXIME
91-96-3 BUTANAMIDE, N,N'-(3,3'-DIMETHYL [1,1'-BIPHENYL]-4,4'-DIYL)BIS[3-OXO- [TSC]	C.I. AZOIC COUPLING COMPONENT 5-4,4'-DIYL)BIS[3-HYDROXY-
97-36-9 BUTANAMIDE, N-(2,4-DIMETHYLPHENYL)-3-OXO- [PAL]	M-ACETOACET XYLIDIDE
122-82-7 BUTANAMIDE, N-(4-ETHOXYPHENYL)-3-OXO-	BUTANAMIDE, N-(4-ETHOXYPHENYL)-3-OXO-
92-15-9 BUTANAMIDE, N-(2-METHOXYPHENYL)-3-OXO- [PAL]	O-ACETOACETANISIDIDE
102-01-2 BUTANAMIDE, 3-OXO-N-PHENYL- [PAL]	ACETOACETANILIDE
109-73-9 1-BUTANAMINE [PAL]	BUTYLAMINE (N-)
13952-84-6 2-BUTANAMINE [PAL]	SEC-BUTYLAMINE
111-92-2 1-BUTANAMINE, N-BUTYL- [PAL, TSC, TSC]	DI-(N-BUTYL)AMINE
924-16-3 1-BUTANAMINE, N-BUTYL-N-NITROSO- [PAL, PAL]	N-NITROSODI-N-BUTYLAMINE
102-82-9 1-BUTANAMINE, N,N-DIBUTYL- [PAL]	TRIBUTYLAMINE
13360-63-9 1-BUTANAMINE, N-ETHYL- [PAL]	ETHYLBUTYLAMINE
110-68-9 1-BUTANAMINE, N-METHYL- [PAL]	N-METHYLBUTYLAMINE
626-23-3 2-BUTANAMINE, N-(1-METHYLPROPYL)- [PAL]	DI-SEC-BUTYLAMINE
106-97-8 BUTANE	BUTANE
106-97-8 N-BUTANE [MAK, MAK, PST]	BUTANE
109-65-9 BUTANE, 1-BROMO- [PAL]	BUTYL BROMIDE
78-86-4 BUTANE, 2-CHLORO-	BUTANE, 2-CHLORO-
109-69-3 BUTANE, 1-CHLORO- [PAL]	BUTYL CHLORIDE
107-84-6 BUTANE, 1-CHLORO-3-METHYL- [PAL]	ISOAMYL CHLORIDE
594-36-5 BUTANE, 2-CHLORO-2-METHYL- [PAL]	TERT-AMYL CHLORIDE
590-88-5 1,3-BUTANEDIAMINE	1,3-BUTANEDIAMINE
32280-46-9 1,3-BUTANEDIAMINE, N,N'-DIETHYL- [PAL]	N,N-DIETHYL-1,3-BUTANEDIAMINE
110-56-5 BUTANE, 1,4-DICHLORO- [PAL]	1,4-DICHLOROBUTANE
616-21-7 BUTANE, 1,2-DICHLORO- [PAL]	1,2-DICHLOROBUTANE
7581-97-7 BUTANE, 2,3-DICHLORO- [PAL]	2,3-DICHLOROBUTANE
75-83-2 BUTANE, 2,2-DIMETHYL- [PAL]	NEOHEXANE
79-29-8 BUTANE, 2,3-DIMETHYL- [PAL]	2,3-DIMETHYLBUTANE
3333-52-6 BUTANEDINITRILE, TETRAMETHYL- [PAL]	TETRAMETHYL SUCCINONITRILE
815-82-7 BUTANEDIOIC ACID, 2,3-DIHYDROXY-[R-(R*, R*)]-, COPPER(2+)SALT (1:1) [PAL]	CUPRIC TARTRATE
3164-29-2 BUTANEDIOIC ACID, 2,3-DIHYDROXY-[R-(R*, R*)]-, DIAMMONIUMSALT [PAL]	AMMONIUM TARTRATE, DIAMMONIUM SALT
97-65-4 BUTANEDIOIC ACID, METHYLENE-	BUTANEDIOIC ACID, METHYLENE-
107-88-0 1,3-BUTANEDIOL	1,3-BUTANEDIOL
107-88-0 1,3-BUTYLENE GLYCOL [HON]	1,3-BUTANEDIOL
110-63-4 1,4-BUTANEDIOL	1,4-BUTANEDIOL
513-85-9 2,3-BUTANEDIOL	2,3-BUTANEDIOL
584-03-2 1,2-BUTANEDIOL	1,2-BUTANEDIOL
584-03-2 BUTANEDIOL, 1,2- [OTS]	1,2-BUTANEDIOL
2425-79-8 1,4-BUTANEDIOL DIGLYCIDYL ETHER [EPA]	OXIRANE, 2,2'-[1,4-BUTANEDIYLBIS(OXYMETHYLENE)]BIS-
2425-79-8 BUTANEDIOL DIGLYCIDYL ETHER, 1,4- [OTS]	OXIRANE, 2,2'-[1,4-BUTANEDIYLBIS(OXYMETHYLENE)]BIS-

Chemical Name	Reference Name
2082-81-7 1,4-BUTANEDIOL DIMETHACRYLATE	1,4-BUTANEDIOL DIMETHACRYLATE
55-98-1 1,4-BUTANEDIOL DIMETHANESUL-FONATE [CAB]	BUSULFAN
55-98-1 1,4-BUTANEDIOL DIMETHANESULFONATE (BUSULFAN) [CAL]	BUSULFAN
55-98-1 1,4-BUTANEDIOL, DIMETHANESUL-FONATE [PAL, PAL]	BUSULFANHYDROCHLORIDE
55-98-1 1,4-BUTANEDIOL DIMETHANESULFONATE (BUSULFAN) [C65, C65]	BUSULFANE)
55-98-1 1,4-BUTANEDIOL DIMETHANE-SULPHONATE [FLA]	BUSULFAN
55-98-1 1,4-BUTANEDIOL DIMETHANE SULPHONATE [MIN, MSL]	BUSULFAN
55-98-1 1,4-BUTANEDIOL DIMETHANESULPHONATE (MYLERAN) [IAR]	BUSULFAN
55-98-1 1,4-BUTANEDIOL DIMETHYLSULFONATE (MYLERAN) [NTP]	BUSULFAN
431-03-8 2,3-BUTANEDIONE [FLA]	2,3-BUTADIONE
431-03-8 BUTANEDIONE [HM1, HM1]	2,3-BUTADIONE
431-03-8 2,3-BUTANEDIONE [NFP, NFP]	2,3-BUTADIONE
431-03-8 BUTANEDIONE [NJL]	2,3-BUTADIONE
431-03-8 2,3-BUTANEDIONE [PAL, PST]	2,3-BUTADIONE
111-34-2 BUTANE, 1-(ETHENYLOXY)- [PAL]	BUTYL VINYL ETHER
628-81-9 BUTANE, 1-ETHOXY- [PAL]	ETHYL BUTYL ETHER
111-36-4 BUTANE, 1-ISOCYANATO- [PAL, TSC, TSC, TSC]	N-BUTYL ISOCYANATE
78-78-4 BUTANE, 2-METHYL- [PAL]	ISOPENTANE
109-74-0 BUTANENITRILE [PAL]	BUTYRONITRILE
142-96-1 BUTANE, 1,1'-OXYBIS- [PAL]	BUTYL ETHER
1121-03-5 2,4-BUTANE SULTONE	2,4-BUTANE SULTONE
1633-83-6 1,4-BUTANESULTONE	1,4-BUTANESULTONE
1633-83-6 1,4-BUTANE SULTONE [MAK]	1,4-BUTANESULTONE
109-79-5 BUTANETHIOL [MAK, MAK]	BUTYL MERCAPTAN
109-79-5 1-BUTANETHIOL [NFP, NFP, NHS, NHS, ONT, PAL]	BUTYL MERCAPTAN
513-53-1 2-BUTANETHIOL	2-BUTANETHIOL
109-79-5 BUTANETHIOL (BUTYL MERCAPTAN) [ALB, ALB]	BUTYL MERCAPTAN
2084-18-6 2-BUTANETHIOL, 3-METHYL- [PAL]	3-METHYL-2-BUTANETHIOL
37971-36-1 BUTANETRICARBOXYLIC ACID, 1,2,4- [OTS]	2-PHOSPHONO-1,2-4-BUTANETRICARBOXYLIC ACID
40372-66-5 1,2,4-BUTANETRICARBOXYLIC ACID, 2-PHOS-PHONO-, SODIUM SALT [PST]	2-PHOSPHONO-1,2,4-BUTANETRICARBOXYLIC ACID, SODIUM SALT
464-06-2 BUTANE, 2,2,3-TRIMETHYL-	BUTANE, 2,2,3-TRIMETHYL-
3068-00-6 1,2,4-BUTANETRIOL	1,2,4-BUTANETRIOL
299-75-2 1,2,3,4-BUTANETROL, 1,4-DIMETHANESUL-FONATE, [S(R*,R*)]- [PAL, PAL]	TREOSULPHANESTER
107-92-6 BUTANOIC ACID [PAL]	BUTYRIC ACID
109-21-7 BUTANOIC ACID, BUTYL ESTER [PAL]	BUTYL BUTYRATE
123-20-6 BUTANOIC ACID, ETHENYL ESTER [PAL]	VINYL BUTYRATE
88-09-5 BUTANOIC ACID, 2-ETHYL- [PAL]	2-ETHYLBUTYRIC ACID
105-54-4 BUTANOIC ACID, ETHYL ESTER [PAL]	ETHYL BUTYRATE

Chemical Name	Reference Name
623-42-7 BUTANOIC ACID, METHYL ESTER [PAL]	METHYL BUTYRATE
141-97-9 BUTANOIC ACID, 3-OXO-, ETHYL ESTER [PAL]	ETHYL ACETOACETATE
105-45-3 BUTANOIC ACID, 3-OXO-, METHYL ESTER [PAL]	METHYLACETOACETATE
105-66-8 BUTANOIC ACID, PROPYL ESTER	BUTANOIC ACID, PROPYL ESTER
71-36-3 BUTAN-1-OL [GBR, GBR]	N-BUTYL ALCOHOL
71-36-3 BUTANOL [HM1, HM1, HM1]	N-BUTYL ALCOHOL
71-36-3 1-BUTANOL [NEU]	N-BUTYL ALCOHOL
71-36-3 BUTANOL, 1- [OTS]	N-BUTYL ALCOHOL
71-36-3 1-BUTANOL [PAL, PST]	N-BUTYL ALCOHOL
71-36-3 N-BUTANOL [RCU, RCU]	N-BUTYL ALCOHOL
75-65-0 TERT-BUTANOL [PST, TLV, TLV, WHM]	TERT-BUTYL ALCOHOL
78-92-2 BUTAN-2-OL [GBR, GBR]	SEC-BUTYL ALCOHOL
78-92-2 2-BUTANOL [PAL]	SEC-BUTYL ALCOHOL
78-92-2 SEC-BUTANOL [PST, WHM]	SEC-BUTYL ALCOHOL
2269-22-9 2-BUTANOL, ALUMINUM SALT	2-BUTANOL, ALUMINUM SALT
96-20-8 1-BUTANOL, 2-AMINO- [PAL]	2-AMINO-1-BUTANOL
682-09-7 1-BUTANOL, 2,2-BIS[(2-PROPENYLOXY) METHYL]-	1-BUTANOL, 2,2-BIS[(2-PROPENYLOXY)METHYL]-
97-95-0 1-BUTANOL, 2-ETHYL- [PAL]	SEC-HEXYL ALCOHOL
75-85-4 2-BUTANOL, 2-METHYL- [PAL]	2-METHYL-2-BUTANOL
123-51-3 1-BUTANOL, 3-METHYL- [PAL]	ISOAMYL ALCOHOL
137-32-6 1-BUTANOL, 2-METHYL- [PAL]	2-METHYL-1-BUTANOL
598-75-4 2-BUTANOL, 3-METHYL- [PAL]	3-METHYL-2-BUTANOL
123-92-2 1-BUTANOL, 3-METHYL-, ACETATE [PAL]	ISOAMYL ACETATE
625-16-1 2-BUTANOL, 2-METHYL-, ACETATE [PAL]	TERT-AMYL ACETATE
78-93-3 2-BUTANONE [ATS, BC1, BC1, CAL, EPA, FDA]	METHYL ETHYL KETONE
78-93-3 BUTAN-2-ONE [GBR, GBR]	METHYL ETHYL KETONE
78-93-3 2-BUTANONE [MAK, MAK, MAK, NIO, NIO, NIO, ONT, ONT, ONT, ONT, PAL, PST]	METHYL ETHYL KETONE
78-93-3 2-BUTANONE (METHYL ETHYL KETONE) [OS1, OS1, OS1, ALB, ALB, TSC, TSC, TSC]	METHYL ETHYL KETONE
563-80-4 2-BUTANONE, 3-METHYL- [PAL]	METHYL ISOPROPYL KETONE
96-29-7 2-BUTANONE, OXIME [TSC, TSC, TSC]	METHYL ETHYL KETOXIME
1338-23-4 2-BUTANONE, PEROXIDE [PAL]	METHYL ETHYL KETONE PEROXIDE
4170-30-3 2-BUTENAL [PAL]	CROTONALDEHYDEPHENYL) ESTER
4170-30-3 2-BUTENAL (CROTONALDEHYDE) [TSC, TSC, TSC, TSC]	CROTONALDEHYDEYLENE)OXY]BIS-
4170-30-3 2-BUTENAL, INHIBITED [HM1, HM1, HM1, HM1, HM1]	CROTONALDEHYDE
106-98-9 BUTENE, 1- [OTS]	1-BUTENE
107-01-7 2-BUTENE	2-BUTENE
107-01-7 BUTENE, 2- [OTS]	2-BUTENE
624-64-6 2-BUTENE-TRANS [MSL, NFP, NFP]	2-BUTENE, (E)-
25167-67-3 BUTENE [EP3]	BUTYLENE
106-98-9 1-BUTENE	1-BUTENE
624-64-6 2-BUTENE, (E)-	2-BUTENE, (E)-
624-64-6 2-BUTENE-(E) [FLA]	2-BUTENE, (E)-
590-18-1 2-BUTENE, (Z)- [PAL]	2-BUTENE-CIS

Chemical Name	Reference Name
4784-77-4 2-BUTENE, 1-BROMO- [PAL]	1-CROTYL BROMIDE
591-97-9 2-BUTENE, 1-CHLORO- [PAL]	1-CROTYL CHLORIDE
4461-41-0 2-BUTENE, 2-CHLORO- [PAL]	2-CHLOROBUTENE-2
590-18-1 2-BUTENE-CIS	2-BUTENE-CIS
764-41-0 2-BUTENE, 1,4-DICHLORO- [EPA, MSL, PAL]	1,4-DICHLORO-2-BUTENE
926-57-8 2-BUTENE, 1,3-DICHLORO- [PAL]	1,3-DICHLOROBUTENE-2
64037-54-3 1-BUTENE, 3,4-DICHLORO-,(.+-.)-1 [PAL]	3,4-DICHLOROBUTENE-1
563-78-0 1-BUTENE, 2,3-DIMETHYL- [PAL]	2,3-DIMETHYL-1-BUTENE
563-79-1 2-BUTENE, 2,3-DIMETHYL- [PAL]	2,3-DIMETHYL-2-BUTENE
110-17-8 2-BUTENEDIOIC ACID, (E)- [PAL]	FUMARIC ACID
110-16-7 2-BUTENEDIOIC ACID, (Z)- [PAL]	MALEIC ACID
623-91-6 2-BUTENEDIOIC ACID (E)-, DIETHYL ESTER [OTS]	FUMARIC ACID, DIETHYL ESTER
RR-01227-1 2-BUTENEDIOIC ACID (Z), MONO(2-((1-OXO-PROPENYLOXY)ETHYL)ESTER	2-BUTENEDIOIC ACID (Z), MONO(2-((1-OXO-PROPENY-LOXY)ETHYL)EST
19201-36-6 2-BUTENEDIOIC ACID (Z-)-, MONO [2-[(1-OXO-2-PROPENYL)OXY]ETHYL]-ESTER	2-BUTENEDIOIC ACID (Z-)-, MONO [2-[(1-OXO-2-PROPENYL)OXY]ETH
110-64-5 2-BUTENE-1,4-DIOL	2-BUTENE-1,4-DIOL
29733-86-6 BUTENEDIOL	BUTENEDIOL
513-35-9 2-BUTENE, 2-METHYL- [PAL]	2-METHYL-2-BUTENE.ALPHA.-HYDROXY-,ETHYL ESTER
563-45-1 1-BUTENE, 3-METHYL- [PAL]	3-METHYL-1-BUTENE
563-46-2 1-BUTENE, 2-METHYL- [PAL]	2-METHYL-1-BUTENE (TECHNICAL)
26249-20-7 BUTENE OXIDE	BUTENE OXIDE
25167-67-3 BUTENES [PAL]	BUTYLENE
624-64-6 2-BUTENE-TRANS [2-BUTENE, (E)] [EP3]	2-BUTENE, (E)-
2431-50-7 1-BUTENE, 2,3,4-TRICHLORO-	1-BUTENE, 2,3,4-TRICHLORO-
2431-50-7 BUTENE, 2,3,4-TRICHLORO-, 1- [OTS]	1-BUTENE, 2,3,4-TRICHLORO-
594-56-9 1-BUTENE, 2,3,3-TRIMETHYL-	1-BUTENE, 2,3,3-TRIMETHYL-
3724-65-0 2-BUTENOIC ACID [PAL]	CROTONIC ACID
25168-21-2 2-BUTENOIC ACID, 4,4'-[(DIBUTYLSTANNY-LENE)BIS(OXY)]BIS(4-OXO-, DIISOOCTYL ESTER (Z,Z)- [TSC, TSC, TSC]	DIBUTYLTIN BIS (ISOOCTYL MALEATE)-2-BUTENOIC ACID, 4,4'-[(DI
7786-34-7 2-BUTENOIC ACID, 3-[(DIMETHOXYPHOS-PHINYL)OXY]-, METHYL ESTER [PAL]	MEVINPHOS
14861-06-4 2-BUTENOIC ACID, ETHENYL ESTER	2-BUTENOIC ACID, ETHENYL ESTER
10544-63-5 2-BUTENOIC ACID, ETHYL ESTER [PAL]	ETHYL CROTONATE
303-34-4 2-BUTENOIC ACID, 2-METHYL-, 7-[[2,3-DIHY-DROXY-2-(1-METHOXYETHYL)-3-METHYL-1-OXOBUTOXY]METHYL]-2,3,5,7A-TETRAHYDRO-1H-PYRROLIZIN-1-YL ESTER, [1S[1.ALPHA.(Z),7 (2S*,3R*),7A.ALPHA.]- [PAL, PAL]	LASIOCARPINEHYDROCHLORIDE
6117-91-5 2-BUTEN-1-OL [PAL]	CROTONYL ALCOHOL
115-18-4 3-BUTEN-2-OL, 2-METHYL-	3-BUTEN-2-OL, 2-METHYL-
115-19-5 3-BUTEN-2-OL, 2-METHYL- [PAL]	3-METHYL BUTYNOL
78-94-4 3-BUTEN-2-ONE [FLA, PAL]	METHYL VINYL KETONE
814-78-8 3-BUTEN-2-ONE, 3-METHYL- [PAL]	METHYL ISOPROPENYL KETONE
689-97-4 1-BUTEN-3-YNE	1-BUTEN-3-YNE
78-80-8 1-BUTEN-3-YNE, 2-METHYL- [PAL]	ISOPROPENYL ACETYLENE
1070-19-5 TERT-BUTOXYCARBONYL AZIDE	TERT-BUTOXYCARBONYL AZIDE
111-76-2 2-BUTOXYETHANOL	2-BUTOXYETHANOL

Chemical Name	Reference Name
111-76-2 2-BUTOXY ETHANOL [FLA, NJL]	2-BUTOXYETHANOL
111-76-2 2-BUTOXYETHANOL (EGBE) [TLV, TLV]	2-BUTOXYETHANOL
111-76-2 2-BUTOXYETHANOL (BUTYL CELLOSOLVE) [ALB, ALB, ALB, BC1, BC1, BC1]	2-BUTOXYETHANOL
124-17-4 2-(2-BUTOXYETHOXY)ETHYL ACETATE	2-(2-BUTOXYETHOXY)ETHYL ACETATE
1120-23-6 2-BETA-BUTOXYETHOXYETHYL CHLORIDE	2-BETA-BUTOXYETHOXYETHYL CHLORIDE
1120-23-6 2,BETA-BUTOXYETHOXYETHYL CHLORIDE [NFP, NFP]	2-BETA-BUTOXYETHOXYETHYL CHLORIDE
124-16-3 1-(BUTOXYETHOXY)-2-PROPANOL [FLA, MSL, NFP, NFP]	2-PROPANOL, 1-(2-BUTOXYETHOXY)-
124-16-3 BUTOXYETHOXYPROPANOL [PST]	2-PROPANOL, 1-(2-BUTOXYETHOXY)-
64051-23-6 2-BUTOXYETHYL DIHYDROGENPHOSPHATE, DIETHYLAMINE SALT	2-BUTOXYETHYL DIHYDROGENPHOSPHATE, DIETHYLAMINE SALT
RR-00791-0 BETA-BUTOXYETHYL SALICYLATE	BETA-BUTOXYETHYL SALICYLATE
4435-53-4 BUTOXYL	BUTOXYL
2426-08-6 (BUTOXYMETHYL)OXIRANE [ONT, ONT, ONT]	N-BUTYL GLYCIDYL ETHER (BGE)
9003-13-8 BUTOXYPOLYPROPYLENE GLYCOL [PST]	POLYPROPYLENE GLYCOL, MONOBUTYL ETHER
5131-66-8 1-BUTOXY-2-PROPANOL	1-BUTOXY-2-PROPANOL
RR-01674-0 BUTOXY-SUBSTITUTED ETHER ALKANE	BUTOXY-SUBSTITUTED ETHER ALKANE
RR-01045-7 BUTOXYTRIETHYLENE GLYCOL PHOSPHATE	BUTOXYTRIETHYLENE GLYCOL PHOSPHATE
88-85-7 2-SEC-BUTYL-4,6-DINITROPHENOL (DINOSEB) [UTS]	2-SEC-BUTYL-4,6-DINITROPHENOL (DINOSEB)
1119-49-9 N-BUTYL ACETAMIDE	N-BUTYL ACETAMIDE
91-49-6 N-BUTYLACETANILIDE	N-BUTYLACETANILIDE
105-46-4 SEC-BUTYL ACETATE	SEC-BUTYL ACETATE
123-86-4 N-BUTYL ACETATE	N-BUTYL ACETATE
123-86-4 BUTYL ACETATE [EPA, GBR, GBR, MSL, NFP, NFP]	N-BUTYL ACETATE
123-86-4 BUTYL ACETATE, N- [OTS]	N-BUTYL ACETATE
540-88-5 TERT-BUTYL ACETATE	TERT-BUTYL ACETATE
RR-00147-8 BUTYL ACETATES	BUTYL ACETATES
591-60-6 BUTYL ACETOACETATE	BUTYL ACETOACETATE
140-04-5 BUTYL ACETYL RICINOLEATE	BUTYL ACETYL RICINOLEATENYL)METHYLENE] AMINO]-
12788-93-1 BUTYL ACID PHOSPHATE [HM1, HM1, NJL, NJL]	BUTYL PHOSPHORIC ACID
12788-93-1 N-BUTYL ACID PHOSPHATE [PST]	BUTYL PHOSPHORIC ACID
107-58-4 N-TERT-BUTYLACRYLAMIDE	N-TERT-BUTYLACRYLAMIDE
141-32-2 BUTYL ACRYLATE	BUTYL ACRYLATE
141-32-2 N-BUTYLACRYLATE [HON]	BUTYL ACRYLATE
141-32-2 N-BUTYL ACRYLATE [IAR, ONT, TLV]	BUTYL ACRYLATE
25951-38-6 BUTYL ACRYLATE-HYDROXYETHYL ACRYLATE-METHYL METHACRYLATE COPOLYMER	BUTYL ACRYLATE-HYDROXYETHYL ACRYLATE-METHYL METHACRYLATE COP
RR-01251-1 BUTYL ACRYLATE, POLYMER WITH SUBSTITUTED METHYL STYRENE, METHACRYLATE, AND SUBSTITUTED SILANE	BUTYL ACRYLATE, POLYMER WITH SUBSTITUTED METHYL STYRENE, METYL SUBSTITUTED
71-36-3 N-BUTYL ALCOHOL	N-BUTYL ALCOHOL
71-36-3 BUTYL ALCOHOL [NFP, NFP]	N-BUTYL ALCOHOL
75-65-0 TERT-BUTYL ALCOHOL	TERT-BUTYL ALCOHOL
78-92-2 SEC-BUTYL ALCOHOL	SEC-BUTYL ALCOHOL
110-69-0 BUTYL ALDEHYDE OXIME [NJL]	BUTYRALDOXIME
75-64-9 TERT-BUTYLAMINE	TERT-BUTYLAMINE

Chemical Name	Reference Name
109-73-9 BUTYLAMINE [ALB, ALB, BC1, BC1, CEX, CEX]	BUTYLAMINE (N-)
109-73-9 N-BUTYLAMINE [CWA, GBR, GBR, MAK, MAK, MAK]	BUTYLAMINE (N-)
109-73-9 BUTYLAMINE [MEX, MEX, NFP, NFP, NJL, NJL]	BUTYLAMINE (N-)
109-73-9 N-BUTYLAMINE [OTV]	BUTYLAMINE (N-)
109-73-9 BUTYLAMINE [QBC, QBC]	BUTYLAMINE (N-)
109-73-9 N-BUTYLAMINE [TLV, TLV]	BUTYLAMINE (N-)
513-49-5 SEC-BUTYLAMINE	SEC-BUTYLAMINE
513-49-5 SEC-BUTYLAMINE (S-) [MSL]	SEC-BUTYLAMINE
13952-84-6 SEC-BUTYLAMINE	SEC-BUTYLAMINE
13952-84-6 (RS)-SEC-BUTYLAMINE [CWA]	SEC-BUTYLAMINE
109-73-9 BUTYLAMINE (N-)	BUTYLAMINE (N-)
109-73-9 N-BUTYLAMINE [ONT, ONT]	BUTYLAMINE (N-)
26094-13-3 BUTYLAMINE OLEATE	BUTYLAMINE OLEATE
3775-90-4 TERT-BUTYLAMINOETHYL METHACRYLATE	TERT-BUTYLAMINOETHYL METHACRYLATEENYL) PHENOXY]
1126-78-9 N-BUTYLANILINE	N-BUTYLANILINE
1126-78-9 BUTYLANILINE [NJL]	N-BUTYLANILINE
1126-78-9 N-(N-BUTYL)ANILINE [WHM]	N-BUTYLANILINE
RR-01577-0 TERT-BUTYLARSINE	TERT-BUTYLARSINE
489-01-0 BUTYLATED HYDROXYANISOLE	BUTYLATED HYDROXYANISOLE
25013-16-5 BUTYLATED HYDROXYANISOLE (BHA) [IAR, WHM]	BUTYLATED HYDROXYANISOLE
128-37-0 BUTYLATED HYDROXYTOLUENE [CSR, CSR]	2,6-DI-TERT-BUTYL-P-CRESOL
128-37-0 BUTYLATED HYDROXYTOLUENE (BHT) [MSL]	2,6-DI-TERT-BUTYL-P-CRESOL
128-37-0 BUTYLATED HYDROXYTOLUENE [PST]	2,6-DI-TERT-BUTYL-P-CRESOL
128-37-0 BUTYLATED HYDROXYTOLUENE (BHT) [WHM, FLA, IAR]	2,6-DI-TERT-BUTYL-P-CRESOL
26160-96-3 BUTYLATED POLYVINYLPYRROLIDONE	BUTYLATED POLYVINYLPYRROLIDONE
98-06-6 TERT-BUTYLBENZENE	TERT-BUTYLBENZENE
104-51-8 BUTYL BENZENE	BUTYL BENZENE
104-51-8 BUTYLBENZENE [FLA, MSL, NFP, NFP]	BUTYL BENZENE
104-51-8 N-BUTYLBENZENE [SDW]	BUTYL BENZENE
135-98-8 SEC-BUTYLBENZENE	SEC-BUTYLBENZENE
RR-00151-4 BUTYL BENZENES	BUTYL BENZENES
3622-84-2 N-BUTYLBENZENESULFONAMIDE	N-BUTYLBENZENESULFONAMIDE
136-60-7 BUTYL BENZOATE	BUTYL BENZOATE
98-73-7 P-TERT-BUTYLBENZOIC ACID	P-TERT-BUTYLBENZOIC ACID
98-73-7 4-TERT-BUTYL BENZOIC ACID [PST]	P-TERT-BUTYLBENZOIC ACID
583-03-9 ALPHA-BUTYL-BENZYL ALCOHOL	ALPHA-BUTYL-BENZYL ALCOHOL
85-68-7 BUTYL BENZYL PHTHALATE	BUTYL BENZYL PHTHALATE
54532-97-7 2-BUTYLBIPHENYL	2-BUTYLBIPHENYL
109-65-9 BUTYL BROMIDE	BUTYL BROMIDE
109-65-9 N-BUTYL BROMIDE [HM1, HM1, WHM]	BUTYL BROMIDE
109-21-7 BUTYL BUTYRATE	BUTYL BUTYRATE
75-84-3 TERT-BUTYL CARBINOL [FLA, MSL, NFP, NFP]	1-PROPANOL, 2,2-DIMETHYL-
98-29-3 4-TERT-BUTYL CATECHOL	4-TERT-BUTYL CATECHOL
98-29-3 P-TERT-BUTYLCATECHOL [CSR]	4-TERT-BUTYL CATECHOL
98-29-3 4-TERT BUTYLCATECHOL [WEL, WEL]	4-TERT-BUTYL CATECHOL

Chemical Name	Reference Name
111-76-2 BUTYL CELLOSOLVE [PST]	2-BUTOXYETHANOL
78-86-4 SEC-BUTYL CHLORIDE [MSL, NFP, NFP]	BUTANE, 2-CHLORO-
109-69-3 BUTYL CHLORIDE	BUTYL CHLORIDE
109-69-3 N-BUTYL CHLORIDE [CSR, CSR]	BUTYL CHLORIDE
507-20-0 TERT-BUTYL CHLORIDE	TERT-BUTYL CHLORIDE
592-34-7 N-BUTYL CHLOROFORMATE	N-BUTYL CHLOROFORMATE
592-34-7 BUTYL CHLOROFORMATE [NJL, NJL]	N-BUTYL CHLOROFORMATE
17462-58-7 SEC-BUTYL CHLOROFORMATE	SEC-BUTYL CHLOROFORMATE
98-28-2 4-TERT-BUTYL-2-CHLOROPHENOL	4-TERT-BUTYL-2-CHLOROPHENOL
299-86-5 4-TERT-BUTYL-2-CHLOROPHENYLMETHYL METHYLPHOSPHORAMIDATE [CAL]	CRUFOMATE
1189-85-1 TERT-BUTYL CHROMATE	TERT-BUTYL CHROMATE
98-27-1 P-TERT-BUTYL-O-CRESOL	P-TERT-BUTYL-O-CRESOL
1333-13-7 TERT-BUTYL-M-CRESOL	TERT-BUTYL-M-CRESOL
3457-61-2 TERT-BUTYL CUMYL PEROXIDE	TERT-BUTYL CUMYL PEROXIDE
1678-93-9 BUTYLCYCLOHEXANE	BUTYLCYCLOHEXANE
3178-22-1 TERT-BUTYLCYCLOHEXANE	TERT-BUTYLCYCLOHEXANE
RR-00882-2 SEC-BUTYLCYCLOHEXANE	SEC-BUTYLCYCLOHEXANE
98-52-2 4-TERT-BUTYLCYCLOHEXANOL	4-TERT-BUTYLCYCLOHEXANOL
10108-56-2 N-BUTYLCYCLOHEXYLAMINE	N-BUTYLCYCLOHEXYLAMINE
70042-58-9 TERT-BUTYLCYCLOHEXYLCHLOROFOR-MATE [NJL]	TERT-BUTYLCYCLOHEXYL CHLOROFORMATE
70042-58-9 TERT-BUTYLCYCLOHEXYL CHLOROFORMATE	TERT-BUTYLCYCLOHEXYL CHLOROFORMATE
RR-00822-0 BUTYLCYCLOPENTANE	BUTYLCYCLOPENTANE
RR-00823-1 BUTYLDECALIN	BUTYLDECALIN
RR-00884-4 TERT-BUTYLDECALIN	TERT-BUTYLDECALIN
94-79-1 SEC-BUTYL 2,4-D ESTER	SEC-BUTYL 2,4-D ESTER
94-80-4 N-BUTYL 2,4-D ESTER	N-BUTYL 2,4-D ESTER
995-33-5 N-BUTYL-4,4-DI(TERT-BUTYL-PEROXY) VALER-ATE [NJL]	N-BUTYL-4,4-DI(TERT-BUTYL-PEROXY)VALERATE
995-33-5 N-BUTYL-4,4-DI(TERT-BUTYL-PEROXY)VALER-ATE	N-BUTYL-4,4-DI(TERT-BUTYL-PEROXY)VALERATE
102-79-4 N-BUTYLDIETHANOLAMINE	N-BUTYLDIETHANOLAMINE
2160-93-2 TERT-BUTYLDIETHANOLAMINE	TERT-BUTYLDIETHANOLAMINE
88-85-7 2-SEC-BUTYL-4,6-DINITROPHENOL [RCA]	2-SEC-BUTYL-4,6-DINITROPHENOL (DINOSEB)
25167-67-3 BUTYLENE	BUTYLENE
107-88-0 BETA-BUTYLENE GLYCOL [NFP, NFP]	1,3-BUTANEDIOL
513-85-9 BUTYLENE GLYCOL (PSEUDO) [NFP, NFP]	2,3-BUTANEDIOL
584-03-2 ALPHA-BUTYLENE GLYCOL [NFP, NFP]	1,2-BUTANEDIOL
106-88-7 1,2-BUTYLENE OXIDE	1,2-BUTYLENE OXIDE
106-88-7 BUTYLENE OXIDE [PST, WEL]	1,2-BUTYLENE OXIDE
3266-23-7 2,3-BUTYLENE OXIDE [NFP, NFP]	OXIRANE, 2,3-DIMETHYL-
RR-01610-4 2,3,4,5-BIS(2-BUTYLENE)TETRAHYDRO-2-FUR-FURAL	2,3,4,5-BIS(2-BUTYLENE)TETRAHYDRO-2-FURFURAL
111-75-1 N-BUTYLETHANOLAMINE	N-BUTYLETHANOLAMINE
111-75-1 N-BUTYL ETHANOLAMINE [NFP, NFP]	N-BUTYLETHANOLAMINE
142-96-1 BUTYL ETHER	BUTYL ETHER
1861-40-1 N-BUTYL-N-ETHYL-2,6-DINITRO-4-(TRIFLUO-ROMETHYL)BENZENAMINE	N-BUTYL-N-ETHYL-2,6-DINITRO-4-(TRIFLUO-ROMETHYL)BENZENAMINE
592-84-7 BUTYL FORMATE	BUTYL FORMATE

Chemical Name	Reference Name
592-84-7 N-BUTYL FORMATE [HM1, HM1]	BUTYL FORMATE
2426-08-6 N-BUTYL GLYCIDYL ETHER [CAL, CEX, EPA, MAK, MAK, MIN, NIO, NIO, NIO]	N-BUTYL GLYCIDYL ETHER (BGE)
2426-08-6 BUTYL GLYCIDYL ETHER, N- [OTS]	N-BUTYL GLYCIDYL ETHER (BGE)
7665-72-7 TERT-BUTYL GLYCIDYL ETHER	TERT-BUTYL GLYCIDYL ETHER
7665-72-7 BUTYL GLYCIDYL ETHER, TERT- [OTS]	TERT-BUTYL GLYCIDYL ETHER
2426-08-6 N-BUTYL GLYCIDYL ETHER (BGE)	N-BUTYL GLYCIDYL ETHER (BGE)
2426-08-6 BUTYL GLYCIDYL ETHER (BGE) [WHM]	N-BUTYL GLYCIDYL ETHER (BGE)
2426-08-6 N-BUTYL GLYCIDYL OXIDE (ETHER) [QBC]	N-BUTYL GLYCIDYL ETHER (BGE)
7397-62-8 BUTYL GLYCOLATE	BUTYL GLYCOLATE
75-91-2 TERT-BUTYL HYDROPEROXIDE	TERT-BUTYL HYDROPEROXIDE
75-91-2 TERT-BUTYLHYDROPEROXIDE [MAK]	TERT-BUTYL HYDROPEROXIDE
75-91-2 BUTYL HYDROPEROXIDE (TERTIARY) [OS3]	TERT-BUTYL HYDROPEROXIDE
1948-33-0 T-BUTYLHYDROQUINONE	T-BUTYLHYDROQUINONE
1948-33-0 TERT-BUTYLHYDROQUINONE [PST]	T-BUTYLHYDROQUINONE
4316-42-1 N-(N-BUTYL)IMIDAZOLE	N-(N-BUTYL)IMIDAZOLE
4316-42-1 N-N-BUTYL-IMIDAZOLE [NJL]	N-(N-BUTYL)IMIDAZOLE
513-48-4 SEC-BUTYL IODIDE	SEC-BUTYL IODIDE
111-36-4 N-BUTYL ISOCYANATE	N-BUTYL ISOCYANATE
111-36-4 N-BUTYLISOCYANATE [NJL]	N-BUTYL ISOCYANATE
1609-86-5 TERT-BUTYL ISOCYANATE	TERT-BUTYL ISOCYANATE
1609-86-5 TERT-BUTYLISOCYANATE [NJL]	TERT-BUTYL ISOCYANATE
30026-92-7 TERT-BUTYL ISOPROPYL BENZENE HY-DROPEROXIDE	TERT-BUTYL ISOPROPYL BENZENE HYDROPEROXIDE
109-19-3 BUTYL ISOVALERATE	BUTYL ISOVALERATE
138-22-7 N-BUTYL LACTATE	N-BUTYL LACTATE
138-22-7 BUTYL LACTATE [GBR, NFP, NFP]	N-BUTYL LACTATE
109-72-8 BUTYL LITHIUM	BUTYL LITHIUM
109-79-5 BUTYL MERCAPTAN	BUTYL MERCAPTAN
109-79-5 N-BUTYL MERCAPTAN [AUS, CAL, CEX]	BUTYL MERCAPTAN
97-88-1 BUTYL METHACRYLATE	BUTYL METHACRYLATE
97-88-1 N-BUTYL METHACRYLATE [NJL]	BUTYL METHACRYLATE
32458-06-3 BUTYL METHACRYLATE, METHYL METHACRYLATE, 2-HYDROXYETHYL METHACRYLATE AND STYRENE COPOLYMER	BUTYL METHACRYLATE, METHYL METHACRYLATE, 2-HYDROXYETHYL METH
9003-63-8 N-BUTYL METHACRYLATE, POLYMER-IZED [PST]	METHACRYLIC ACID, BUTYL ESTER, POLYMER
628-28-4 BUTYL METHYL ETHER	BUTYL METHYL ETHER
1931-62-0 TERT-BUTYL MONOPEROXYMALEATE	TERT-BUTYL MONOPEROXYMALEATE
31711-50-9 BUTYL NAPHTHALENE	BUTYL NAPHTHALENE
31711-50-9 BUTYLNAPHTHALENE [PST]	BUTYL NAPHTHALENE
25638-17-9 BUTYLNAPHTHALENESULFONIC ACID, SODIUM SALT	BUTYLNAPHTHALENESULFONIC ACID, SODIUM SALT
928-45-0 BUTYL NITRATE	BUTYL NITRATE
544-16-1 BUTYL NITRITE	BUTYL NITRITE
3913-02-8 2-BUTYLOCTANOL	2-BUTYLOCTANOL
142-77-8 BUTYL OLEATE	BUTYL OLEATE
2050-60-4 BUTYL OXALATE	BUTYL OXALATE
94-26-8 BUTYL PARABAN	BUTYL PARABAN
107-71-1 TERT-BUTYL PERACETATE [MAK, NFP, NFP]	TERT-BUTYL PEROXYACETATE

Chemical Name	Reference Name
614-45-9 TERT-BUTYL PERBENZOATE [CSR, FLA, MSL, NFP, NFP, PST]	TERT-BUTYL PEROXYBENZOATE
614-45-9 BUTYL PERBENZOATE (TERTIARY) [OS3]	TERT-BUTYL PEROXYBENZOATE
110-05-4 TERT-BUTYL PEROXIDE [PST]	DI-TERT-BUTYL PEROXIDE
107-71-1 TERT-BUTYL PEROXYACETATE	TERT-BUTYL PEROXYACETATE
614-45-9 TERT-BUTYL PEROXYBENZOATE	TERT-BUTYL PEROXYBENZOATE
614-45-9 TERT-BUTYL PEROXYBENZOATE [NJL, NJL]	TERT-BUTYL PEROXYBENZOATE
23474-91-1 TERT-BUTYL PEROXYCROTONATE	TERT-BUTYL PEROXYCROTONATE
16215-49-9 N-BUTYL PEROXYDICARBONATE [HM1]	BUTYL PEROXYDICARBONATE
16215-49-9 BUTYL PEROXYDICARBONATE	BUTYL PEROXYDICARBONATE
2550-33-6 TERT-BUTYL PEROXYDIETHYL-ACETATE	TERT-BUTYL PEROXYDIETHYL-ACETATE
RR-00185-4 T-BUTYL PEROXYDIETHYL-ACETATE WITH T-BUTYL PEROXYBENZOATE	T-BUTYL PEROXYDIETHYL-ACETATE WITH T-BUTYL PEROXYBENZOATE
3006-82-4 TERT-BUTYL PEROXY(2-ETHYL)-HEXANOATE	TERT-BUTYL PEROXY(2-ETHYL)-HEXANOATE
109-13-7 TERT-BUTYL PEROXYISOBUTYRATE	TERT-BUTYL PEROXYISOBUTYRATE
RR-00771-6 TERT-BUTYL PEROXYISONONANOATE	TERT-BUTYL PEROXYISONONANOATE
2372-21-6 TERT-BUTYL PEROXYISOPROPYL CARBONATE	TERT-BUTYL PEROXYISOPROPYL CARBONATE
26748-41-4 TERT-BUTYL PEROXYNEODECANOATE	TERT-BUTYL PEROXYNEODECANOATE
25251-51-8 3-TERT-BUTYL PEROXY-3-PHENYLPHTHALIDE	3-TERT-BUTYL PEROXY-3-PHENYLPHTHALIDE
RR-00772-7 TERT-BUTYL PEROXYPHTHALATE	TERT-BUTYL PEROXYPHTHALATE
927-07-1 TERT-BUTYL PEROXYPIVALATE	TERT-BUTYL PEROXYPIVALATE
RR-01751-6 TERT-BUTYLPEROXY STEARYLCARBONATE	TERT-BUTYLPEROXY STEARYLCARBONATE
13122-18-4 TERT-BUTYL PEROXY-3,5,5-TRIMETHYLHEX-ANOATE	TERT-BUTYL PEROXY-3,5,5-TRIMETHYLHEXANOATE
89-72-5 O-SEC-BUTYLPHENOL	O-SEC-BUTYLPHENOL
89-72-5 2-SEC-BUTYLPHENOL [GBR, GBR]	O-SEC-BUTYLPHENOL
98-54-4 P-TERT-BUTYL PHENOL	P-TERT-BUTYL PHENOL
98-54-4 P-TERT-BUTYLPHENOL [PST]	P-TERT-BUTYL PHENOL
98-54-4 4-TERT-BUTYLPHENOL [WHM]	P-TERT-BUTYL PHENOL
1638-22-8 P-BUTYLPHENOL	P-BUTYLPHENOL
3180-09-4 O-BUTYLPHENOL	O-BUTYLPHENOL
4074-43-5 M-BUTYLPHENOL	M-BUTYLPHENOL
28805-86-9 BUTYL PHENOL	BUTYL PHENOL
68332-64-9 P-TERT-BUTYLPHENOL, ETHOXYLATED, PHOS-PHATED	P-TERT-BUTYLPHENOL, ETHOXYLATED, PHOSPHATED
28805-86-9 BUTYLPHENOLS [HM1, HM1, HM1]	BUTYL PHENOL
713-46-2 BETA(P-TERT-BUTYLPHENOXY) ETHANOL	BETA(P-TERT-BUTYLPHENOXY) ETHANOL
RR-00794-3 BETA-(P-TERT-BUTYLPHENOXY) ETHYL AC-ETATE	BETA-(P-TERT-BUTYLPHENOXY) ETHYL ACETATE
140-57-8 2-(P-TERT-BUTYLPHENOXY)ISOPROPYL-2-CHLOROETHYL SULFITE [CAL]	ARAMITE
56803-37-3 TERT-BUTYLPHENYL DIPHENYL PHOSPHATE	TERT-BUTYLPHENYL DIPHENYL PHOSPHATE
1126-79-0 BUTYL PHENYL ETHER	BUTYL PHENYL ETHER
3101-60-8 P-TERT-BUTYL PHENYL GLYCIDYL ETHER	P-TERT-BUTYL PHENYL GLYCIDYL ETHER
3101-60-8 BUTYLPHENYL GLYCIDYL ETHER, P-TERT- [OTS]	P-TERT-BUTYL PHENYL GLYCIDYL ETHER
42479-87-8 4-TERT-BUTYL-2-PHENYLPHENOL	4-TERT-BUTYL-2-PHENYLPHENOL
126-73-8 BUTYL PHOSPHATE [NFP, NFP]	TRIBUTYL PHOSPHATE
12788-93-1 BUTYL PHOSPHORIC ACID	BUTYL PHOSPHORIC ACID
85-70-1 BUTYL PHTHALYL BUTYL GLYCOLATE [NFP, NFP]	1,2-BENZENEDICARBOXYLIC ACID, 2-BUTOXY-2-OXYETHYL BUTYL

Chemical Name	Reference Name
50769-39-6 BUTYLPOLYETHOXYETHANOL ESTERS OF PHOSPHORIC ACID	BUTYLPOLYETHOXYETHANOL ESTERS OF PHOSPHORIC ACIDODIUM SALT
590-01-2 BUTYL PROPIONATE	BUTYL PROPIONATE
151-13-3 BUTYL RICINOLEATE	BUTYL RICINOLEATE
109-43-3 BUTYL SEBACATE [NFP, NFP]	DI-N-BUTYL SEBACATE
123-95-5 BUTYL STEARATE	BUTYL STEARATE
25338-51-6 TERT-BUTYL STYRENE	TERT-BUTYL STYRENE
25338-51-6 TERT-BUTYLSTYRENE [NFP, NFP, PAL]	TERT-BUTYL STYRENE
73090-68-3 TERT-BUTYL TETRALIN	TERT-BUTYL TETRALIN
5593-70-4 BUTYL TITANATE	BUTYL TITANATE
98-51-1 P-TERT-BUTYLTOLUENE	P-TERT-BUTYLTOLUENE
98-51-1 P-TERT-BUTYL TOLUENE [MAK, MAK, TLV]	P-TERT-BUTYLTOLUENE
1595-04-6 M-BUTYLTOLUENE	M-BUTYLTOLUENE
1595-05-7 P-BUTYLTOLUENE	P-BUTYLTOLUENE
1595-11-5 O-BUTYLTOLUENE	O-BUTYLTOLUENE
98-51-1 PARA-TERTIARY-BUTYLTOLUENE [HM1]	P-TERT-BUTYLTOLUENE
RR-01353-6 BUTYLTOLUENES	BUTYLTOLUENES
7521-80-4 BUTYLTRICHLOROSILANE	BUTYLTRICHLOROSILANE
7521-80-4 BUTYL TRICHLOROSILANE [FLA, NFP, NFP, NJL, NJL]	BUTYLTRICHLOROSILANE
1118-46-3 BUTYLTRICHLOROSTANNANE	BUTYLTRICHLOROSTANNANE
81-15-2 5-TERT-BUTYL-2,4,6-TRINITRO-M-XY-LENE [HM1, HM1]	MUSK XYLENE
591-62-8 N-BUTYLURETHANE	N-BUTYLURETHANE
111-34-2 BUTYL VINYL ETHER	BUTYL VINYL ETHER
1879-09-0 6-TERT-BUTYL-2,4-XYLENOL [OTS]	2,4-DIMETHYL-6-TERT-BUTYLPHENOL
503-17-3 2-BUTYNE	2-BUTYNE
110-65-6 1,4-BUTYNEDIOL	1,4-BUTYNEDIOL
1606-85-5 2,2'-[2-BUTYNE-1,4-DIYL(OXY)] BISETHANOL [PST]	ETHANOL, 2,2'-[2-BUTYNE-1,4-DIYLBIS(OXY)]BIS-
115-19-5 3-BUTYN-2-OL, 2-METHYL- [OTS]	3-METHYL BUTYNOL
1423-60-5 3-BUTYN-2-ONE	3-BUTYN-2-ONE
123-72-8 BUTYRALDEHYDE	BUTYRALDEHYDE
123-72-8 N-BUTYRALDEHYDE [FLA, MSL]	BUTYRALDEHYDE
496-03-7 BUTYRALDOL	BUTYRALDOL
110-69-0 BUTYRALDOXIME	BUTYRALDOXIME
107-92-6 BUTYRIC ACID	BUTYRIC ACID
107-92-6 N-BUTYRIC ACID [CWA]	BUTYRIC ACID
106-31-0 BUTYRIC ANHYDRIDE	BUTYRIC ANHYDRIDE
96-48-0 BUTYROLACETONE [HON]	GAMMA-BUTYROLACTONE
96-48-0 GAMMA-BUTYROLACTONE	GAMMA-BUTYROLACTONE
96-48-0 BUTYROLACTONE [NFP, NFP]	GAMMA-BUTYROLACTONE
3068-88-0 BETA-BUTYROLACTONE	BETA-BUTYROLACTONE
3068-88-0 B-BUTYROLACTONE [IAR]	BETA-BUTYROLACTONE
109-74-0 BUTYRONITRILE	BUTYRONITRILE
109-74-0 N-BUTYRONITRILE [MIN, NHS, NHS]	BUTYRONITRILE
141-75-3 BUTYRYL CHLORIDE	BUTYRYL CHLORIDE
1330-38-7 C.I. 74180	C.I. 74180
1064-48-8 C.I. ACID BLACK 1	C.I. ACID BLACK 1
8005-03-6 C.I. ACID BLACK 2	C.I. ACID BLACK 2

Chemical Name	Reference Name
129-17-9 C.I. ACID BLUE 1, SODIUM SALT [PST]	AMMONIUM, (4-(ALPHA-(P-(DIETHYLAMINO)PHENYL)-2, 4-
3486-30-4 C.I. ACID BLUE 7, SODIUM SALT	C.I. ACID BLUE 7, SODIUM SALT
2650-18-2 C.I. ACID BLUE 9, DIAMMONIUM SALT	C.I. ACID BLUE 9, DIAMMONIUM SALT
3844-45-9 C.I. ACID BLUE 9, DISODIUM SALT	C.I. ACID BLUE 9, DISODIUM SALT
28631-66-5 C.I. ACID BLUE 22	C.I. ACID BLUE 22
6408-78-2 C.I. ACID BLUE 25 [PST]	FD&C BLUE NO.2
6424-85-7 C.I. ACID BLUE 40 [PST]	2-ANTHRACENESULFONIC ACID, 4-[[4-(ACETYLAMINO)PHENYL]AMINO]-
4474-24-2 C.I. ACID BLUE 80	C.I. ACID BLUE 80
28983-56-4 C.I. ACID BLUE 93	C.I. ACID BLUE 93
6408-80-6 C.I. ACID BLUE 145	C.I. ACID BLUE 145
72152-54-6 C.I. ACID BLUE 182	C.I. ACID BLUE 182
19381-50-1 C.I. ACID GREEN 1	C.I. ACID GREEN 1
4680-78-8 C.I. ACID GREEN 3	C.I. ACID GREEN 3
12768-78-4 C.I. ACID GREEN 16	C.I. ACID GREEN 16
4403-90-1 C.I. ACID GREEN 25 [PST]	D AND C GREEN NO. 5
71927-89-4 C.I. ACID GREEN 28	C.I. ACID GREEN 28
6373-74-6 C.I. ACID ORANGE 3	C.I. ACID ORANGE 3
633-96-5 C.I. ACID ORANGE 7, MONOSODIUM SALT [PST]	ACID ORANGE 7
1936-15-8 C.I. ACID ORANGE 10 [CSR, CSR, PST]	ACID ORANGE 10
2429-80-3 C.I. ACID ORANGE 45	C.I. ACID ORANGE 45
547-58-0 C.I. ACID ORANGE 52	C.I. ACID ORANGE 52
3734-67-6 C.I. ACID RED 1	C.I. ACID RED 1
3567-69-9 C.I. ACID RED 14	C.I. ACID RED 14ROCHLORIDE)
3567-69-9 C.I. ACID RED 14, DISODIUM SALT [PST]	C.I. ACID RED 14
5858-33-3 C.I. ACID RED 17, DISODIUM SALT	C.I. ACID RED 17, DISODIUM SALT
915-67-3 C.I. ACID RED 27 [PST]	C.I. ACID RED 27, TRISODIUM SALT
915-67-3 C.I. ACID RED 27, TRISODIUM SALT	C.I. ACID RED 27, TRISODIUM SALT
3567-66-6 C.I. ACID RED 33, DISODIUM SALT	C.I. ACID RED 33, DISODIUM SALT
3520-42-1 C.I. ACID RED 52	C.I. ACID RED 52
5413-75-2 C.I. ACID RED 73, DISODIUM SALT	C.I. ACID RED 73, DISODIUM SALT
3567-65-5 C.I. ACID RED 85	C.I. ACID RED 85
6459-94-5 C.I. ACID RED 114 [313]	C.I. ACID RED 114, DISODIUM SALT-PYRIMIDINEDIONE]
6844-74-2 C.I. ACID RED 101, DISODIUM SALT	C.I. ACID RED 101, DISODIUM SALT
6459-94-5 C.I. ACID RED 114 [C65, CSR, CSR]	C.I. ACID RED 114, DISODIUM SALT
6459-94-5 C.I. ACID RED 114, DISODIUM SALT	C.I. ACID RED 114, DISODIUM SALT
4321-69-1 C.I. ACID VIOLET 7, DISODIUM SALT	C.I. ACID VIOLET 7, DISODIUM SALT
6625-46-3 C.I. ACID VIOLET 12, DISODIUM SALT	C.I. ACID VIOLET 12, DISODIUM SALT
6359-82-6 C.I. ACID YELLOW 11, SODIUM SALT	C.I. ACID YELLOW 11, SODIUM SALT
6359-98-4 C.I. ACID YELLOW 17, DISODIUM SALT	C.I. ACID YELLOW 17, DISODIUM SALT
1934-21-0 C.I. ACID YELLOW 23, TRISODIUM SALT [PST]	FD&C YELLOW 5
6359-97-3 C.I. ACID YELLOW 34	C.I. ACID YELLOW 34
587-98-4 C.I. ACID YELLOW 36, MONOSODIUM SALT [PST]	METANIL YELLOW
518-47-8 C.I. ACID YELLOW 73 [PST]	SODIUM FLUORESCEIN
71873-51-3 C.I. ACID YELLOW 218	C.I. ACID YELLOW 218
91-92-9 C.I. AZOIC COUPLING COMPONENT	C.I. AZOIC COUPLING COMPONENT
91-96-3 C.I. AZOIC COUPLING COMPONENT 5	C.I. AZOIC COUPLING COMPONENT 5

Chemical Name	Reference Name
61-73-4 C.I. BASIC BLUE 9 [PST]	METHYLENE BLUE
633-03-4 C.I. BASIC GREEN 1	C.I. BASIC GREEN 1
569-64-2 C.I. BASIC GREEN 4 [313, CAN, MSL, NJL, PAL]	MALACHITE GREEN
18015-76-4 C.I. BASIC GREEN 4, OXALATE	C.I. BASIC GREEN 4, OXALATE
12768-82-0 C.I. BASIC ORANGE 15	C.I. BASIC ORANGE 15
989-38-8 C.I. BASIC RED 1	C.I. BASIC RED 1
569-61-9 C.I. BASIC RED 9 MONOHYDROCHLORIDE	C.I. BASIC RED 9 MONOHYDROCHLORIDE
65122-06-7 C.I. BASIC RED 14	C.I. BASIC RED 14
548-62-9 C.I. BASIC VIOLET 1	C.I. BASIC VIOLET 1
8004-87-3 C.I. BASIC VIOLET 1 [PST]	METHYL VIOLET
2390-63-8 C.I. BASIC VIOLET 11	C.I. BASIC VIOLET 11
6359-45-1 C.I. BASIC VIOLET 16	C.I. BASIC VIOLET 16
2465-27-2 C.I. BASIC YELLOW 2 [PST]	AURAMINE
2465-27-2 C.I. BASIC YELLOW 2, MONOHYDROCHLO-RIDE [NJL, NJL]	AURAMINE
25156-49-4 C.I. DIRECT BLACK 4	C.I. DIRECT BLACK 4
2429-83-6 C.I. DIRECT BLACK 4, DISODIUM SALT	C.I. DIRECT BLACK 4, DISODIUM SALT
1937-37-7 C.I. DIRECT BLACK 38	C.I. DIRECT BLACK 38
6739-62-4 C.I. DIRECT BLACK 91, TRISODIUM SALT	C.I. DIRECT BLACK 91, TRISODIUM SALT
61703-05-7 C.I. DIRECT BLACK 114	C.I. DIRECT BLACK 114
2610-05-1 C.I. DIRECT BLUE 1	C.I. DIRECT BLUE 1
25180-19-2 C.I. DIRECT BLUE 2	C.I. DIRECT BLUE 2
2429-73-4 C.I. DIRECT BLUE 2, TRISODIUM SALT	C.I. DIRECT BLUE 2, TRISODIUM SALT
2602-46-2 C.I. DIRECT BLUE 6	C.I. DIRECT BLUE 6
2429-71-2 C.I. DIRECT BLUE 8, DISODIUM SALT	C.I. DIRECT BLUE 8, DISODIUM SALT
72-57-1 C.I. DIRECT BLUE 14, TETRASODIUM SALT [NJL, NJL]	TRYPAN BLUE
2429-74-5 C.I. DIRECT BLUE 15	C.I. DIRECT BLUE 15
2586-57-4 C.I. DIRECT BLUE 22, DISODIUM SALT	C.I. DIRECT BLUE 22, DISODIUM SALT
25180-27-2 C.I. DIRECT BLUE 25	C.I. DIRECT BLUE 25
2150-54-1 C.I. DIRECT BLUE 25, TETRASODIUM SALT	C.I. DIRECT BLUE 25, TETRASODIUM SALT
1330-38-7 C.I. DIRECT BLUE 86 [PST]	C.I. 74180
12222-04-7 C.I. DIRECT BLUE 199	C.I. DIRECT BLUE 199
28407-37-6 C.I. DIRECT BLUE 218	C.I. DIRECT BLUE 218
8014-91-3 C.I. DIRECT BROWN	C.I. DIRECT BROWN
3811-71-0 C.I. DIRECT BROWN 1	C.I. DIRECT BROWN 1
25255-06-5 C.I. DIRECT BROWN 2	C.I. DIRECT BROWN 2
2429-82-5 C.I. DIRECT BROWN 2, DISODIUM SALT	C.I. DIRECT BROWN 2, DISODIUM SALT
25180-39-6 C.I. DIRECT BROWN 6	C.I. DIRECT BROWN 6
2893-80-3 C.I. DIRECT BROWN 6, DISODIUM SALT	C.I. DIRECT BROWN 6, DISODIUM SALT
25180-41-0 C.I. DIRECT BROWN 31	C.I. DIRECT BROWN 31
2429-81-4 C.I. DIRECT BROWN 31, TETRASODIUM SALT	C.I. DIRECT BROWN 31, TETRASODIUM SALT
6247-51-4 C.I. DIRECT BROWN 59	C.I. DIRECT BROWN 59
3476-90-2 C.I. DIRECT BROWN 59, DISODIUM SALT	C.I. DIRECT BROWN 59, DISODIUM SALT
16071-86-6 C.I. DIRECT BROWN 95	C.I. DIRECT BROWN 95
12222-20-7 C.I. DIRECT BROWN 111	C.I. DIRECT BROWN 111
6360-54-9 C.I. DIRECT BROWN 154	C.I. DIRECT BROWN 154
8014-91-3 C.I. DIRECT BROWN 74 [NJL, NJL]	C.I. DIRECT BROWN
25180-45-4 C.I. DIRECT GREEN 1	C.I. DIRECT GREEN 1

Chemical Name	Reference Name
3626-28-6 C.I. DIRECT GREEN 1, DISODIUM SALT	C.I. DIRECT GREEN 1, DISODIUM SALT
25180-46-5 C.I. DIRECT GREEN 6	C.I. DIRECT GREEN 6
4335-09-5 C.I. DIRECT GREEN 6, DISODIUM SALT	C.I. DIRECT GREEN 6, DISODIUM SALT
25180-47-6 C.I. DIRECT GREEN 8	C.I. DIRECT GREEN 8
5422-17-3 C.I. DIRECT GREEN 8, TRISODIUM SALT	C.I. DIRECT GREEN 8, TRISODIUM SALT
54579-28-1 C.I. DIRECT ORANGE 1	C.I. DIRECT ORANGE 1
6637-88-3 C.I. DIRECT ORANGE 6, DISODIUM SALT	C.I. DIRECT ORANGE 6, DISODIUM SALT
64083-59-6 C.I. DIRECT ORANGE 8	C.I. DIRECT ORANGE 8
25188-24-3 C.I. DIRECT RED 1	C.I. DIRECT RED 1
2429-84-7 C.I. DIRECT RED 1, DISODIUM SALT	C.I. DIRECT RED 1, DISODIUM SALT
992-59-6 C.I. DIRECT RED 2, DISODIUM SALT	C.I. DIRECT RED 2, DISODIUM SALT
25188-29-8 C.I. DIRECT RED 10	C.I. DIRECT RED 10
2429-70-1 C.I. DIRECT RED 10, DISODIUM SALT	C.I. DIRECT RED 10, DISODIUM SALT
25188-30-1 C.I. DIRECT RED 13	C.I. DIRECT RED 13
1937-35-5 C.I. DIRECT RED 13, DISODIUM SALT	C.I. DIRECT RED 13, DISODIUM SALT
573-58-0 C.I. DIRECT RED 28	C.I. DIRECT RED 28
3530-19-6 C.I. DIRECT RED 37	C.I. DIRECT RED 37
6358-29-8 C.I. DIRECT RED 39, DISODIUM SALT	C.I. DIRECT RED 39, DISODIUM SALT
25188-42-5 C.I. DIRECT RED 81	C.I. DIRECT RED 81
2610-11-9 C.I. DIRECT RED 81, DISODIUM SALT	C.I. DIRECT RED 81, DISODIUM SALT
75768-93-3 C.I. DIRECT RED 81, TRIETHANOLAMINE SALT	C.I. DIRECT RED 81, TRIETHANOLAMINE SALT
25188-44-7 C.I. DIRECT VIOLET 1	C.I. DIRECT VIOLET 1
2586-60-9 C.I. DIRECT VIOLET 1, DISODIUM SALT	C.I. DIRECT VIOLET 1, DISODIUM SALT
6227-14-1 C.I. DIRECT VIOLET 9, DISODIUM SALT	C.I. DIRECT VIOLET 9, DISODIUM SALT
25329-82-2 C.I. DIRECT VIOLET 22	C.I. DIRECT VIOLET 22
6426-67-1 C.I. DIRECT VIOLET 22, TRISODIUM SALT	C.I. DIRECT VIOLET 22, TRISODIUM SALT
6426-62-6 C.I. DIRECT YELLOW 20	C.I. DIRECT YELLOW 20
3214-47-9 C.I. DIRECT YELLOW 50, TETRASODIUM SALT	C.I. DIRECT YELLOW 50, TETRASODIUM SALT
2475-45-8 C.I. DISPERSE BLUE 1	C.I. DISPERSE BLUE 1
2475-46-9 C.I. DISPERSE BLUE 3 [PST]	1-METHYLAMINO-4-ETHANOLAMINOANTHRAQUINONE
3618-72-2 C.I. DISPERSE BLUE 79:1 [TSE]	C.I. DISPERSE BLUE 79:1 ACETAMIDE, N-[5-[BIS[2-(ACETYLOXYL)
3618-72-2 C.I. DISPERSE BLUE 79:1 ACETAMIDE, N-[5-[BIS[2-(ACETYLOXYL)ETHYL]AMINO]-2-[(2-BROMO-4,6-DINITROPHENYL)AZO]-4-METHOXY-PHENYL]-	C.I. DISPERSE BLUE 79:1 ACETAMIDE, N-[5-[BIS[2-(ACETYLOXYL)LENE)]BIS-
2832-40-8 C.I. DISPERSE YELLOW 3	C.I. DISPERSE YELLOW 3
6416-68-8 C.I. FLUORESCENT BRIGHTENER 46	C.I. FLUORESCENT BRIGHTENER 46
2353-45-9 C.I. FOOD GREEN 3, DISODIUM SALT [PST]	FAST GREEN FCF
4548-53-2 C.I. FOOD RED 1	C.I. FOOD RED 1
3761-53-3 CI FOOD RED 5 [313]	PONCEAU MXDIUM SALT)
3761-53-3 C.I. FOOD RED 5 [NJL]	PONCEAU MX
81-88-9 C.I. FOOD RED 15	C.I. FOOD RED 15
81-88-9 CI FOOD RED 15 [313]	C.I. FOOD RED 15
25956-17-6 C.I. FOOD RED 17 [PST]	FD&C RED 40OLYMER
104-23-4 C.I. FOOD YELLOW 6	C.I. FOOD YELLOW 6
147-14-8 C.I. PIGMENT BLUE 15 [OTS]	PHTHALOCYANINE BLUE
14038-43-8 C.I. PIGMENT BLUE 27 [PST]	FERRIC FERROCYANIDE
57455-37-5 C.I. PIGMENT BLUE 29 [PST]	ULTRAMARINE BLUE

Chemical Name	Reference Name
12002-03-8 C.I. PIGMENT GREEN 21 [PAL]	COPPER ACETOARSENITE
2425-85-6 C.I. PIGMENT RED 3	C.I. PIGMENT RED 3
6410-41-9 C.I. PIGMENT RED 5	C.I. PIGMENT RED 5
6471-49-4 C.I. PIGMENT RED 23	C.I. PIGMENT RED 23
7023-61-2 C.I. PIGMENT RED 48, CALCIUM SALT	C.I. PIGMENT RED 48, CALCIUM SALT
3564-21-4 C.I. PIGMENT RED 48, DISODIUM SALT [PST]	PIGMENT RED 28 (PERMANENT RED)
27757-95-5 C.I. PIGMENT RED 52	C.I. PIGMENT RED 52
5858-82-2 C.I. PIGMENT RED 52, DISODIUM SALT [PST]	PIGMENT RED 48, CI NO. 15865
12656-85-8 C.I. PIGMENT RED 104 [PST]	MOLYBDATE ORANGE (LEAD CHROMATE PIGMENT)
5979-28-2 C.I. PIGMENT YELLOW 16 [PST]	PIGMENT YELLOW 16
51274-00-1 C.I. PIGMENT YELLOW 42 [PST]	IRON OXIDE YELLOW
13515-40-7 C.I. PIGMENT YELLOW 73	C.I. PIGMENT YELLOW 73UM METHYL SULFATE
12225-18-2 C.I. PIGMENT YELLOW 97 [PST]	PIGMENT YELLOW 97, CI 11767
15790-07-5 C.I. PIGMENT YELLOW 104	C.I. PIGMENT YELLOW 104
5590-18-1 C.I. PIGMENT YELLOW 110 [PST]	1H-ISOINDOL-1-ONE
68187-51-9 C.I. PIGMENT YELLOW 119 [PST]	ZINC IRON YELLOW
30125-47-4 C.I. PIGMENT YELLOW 138	C.I. PIGMENT YELLOW 138
6358-31-2 C.I. PIGMENT YELLOW 74 [PST]	PIGMENT YELLOW 74
74204-30-1 C.I. REACTIVE BLUE 52	C.I. REACTIVE BLUE 52
72152-45-5 C.I. REACTIVE GREEN 12	C.I. REACTIVE GREEN 12
72139-15-2 C.I. REACTIVE RED 56	C.I. REACTIVE RED 56
17354-14-2 C.I. SOLVENT BLUE 35 [PST]	1,4-BIS(BUTYLAMINO)-9,10-ANTHRAQUINONE
14233-37-5 C.I. SOLVENT BLUE 36	C.I. SOLVENT BLUE 36
61969-42-4 C.I. SOLVENT BLUE 53	C.I. SOLVENT BLUE 53
4395-65-7 C.I. SOLVENT BLUE 68	C.I. SOLVENT BLUE 68
71819-49-3 C.I. SOLVENT BLUE 98	C.I. SOLVENT BLUE 98
128-80-3 C.I. SOLVENT GREEN 3E [PST]	1,4-BIS(P-TOLYLAMINO)ANTHRAQUINONE
495-54-5 C.I. SOLVENT ORANGE 3	C.I. SOLVENT ORANGE 3
3118-97-6 C.I. SOLVENT ORANGE 7	C.I. SOLVENT ORANGE 7
71819-51-7 C.I. SOLVENT RED 164	C.I. SOLVENT RED 164
27354-18-3 C.I. SOLVENT RED 169	C.I. SOLVENT RED 169
81-48-1 C.I. SOLVENT VIOLET 13	C.I. SOLVENT VIOLET 13
492-80-8 C.I. SOLVENT YELLOW (AURAMINE) [NJL]	AURAMINE
60-09-3 C.I. SOLVENT YELLOW 1 [NJL]	4-AMINOAZOBENZENE
97-56-3 CI SOLVENT YELLOW 3 [313]	O-AMINOAZOTOLUENE
97-56-3 C.I. SOLVENT YELLOW 3 [NJL]	O-AMINOAZOTOLUENE
842-07-9 C.I. SOLVENT YELLOW 14	C.I. SOLVENT YELLOW 14
8003-22-3 C.I. SOLVENT YELLOW 33 [PST]	D & C YELLOW NO. 11
492-80-8 CI SOLVENT YELLOW 34 (AURAMINE) [313]	AURAMINE
2321-07-5 C.I. SOLVENT YELLOW 94 [PST]	FLUORESCEIN
67990-27-6 C.I. SOLVENT YELLOW 107	C.I. SOLVENT YELLOW 107TE AND TUNG OIL
2379-81-9 C.I. VAT BLACK 27	C.I. VAT BLACK 27
6424-76-6 C.I. VAT BLUE 16	C.I. VAT BLUE 16
6373-20-2 C.I. VAT BLUE 22	C.I. VAT BLUE 22
2475-33-4 C.I. VAT BROWN 1	C.I. VAT BROWN 1
25704-81-8 C.I. VAT GREEN 2	C.I. VAT GREEN 2
128-66-5 C.I. VAT YELLOW 4	C.I. VAT YELLOW 4
68783-08-4 C7-C35 REFINED PETROLEUM OIL	C7-C35 REFINED PETROLEUM OIL
RR-00502-7 C9 AROMATIC HYDROCARBON FRACTION	C9 AROMATIC HYDROCARBON FRACTION

Chemical Name	Reference Name
68459-31-4 C9-19 FATTY ACID ESTER PHTHALIC ALKYL	C9-19 FATTY ACID ESTER PHTHALIC ALKYL
497-18-7 CABAZIDE [HM1]	CARBONIC DIHYDRAZIDE
75-60-5 CACODYLIC ACID	CACODYLIC ACID
68309-98-8 CADMATE (6-), [[[1,2-ETHANEDIYLBIS[NITRILO-BIS(METHYLENE)]TETRAKIS[PHOSPHONATO] (8-)-N,N',O,O",O"",O""""]-, PENTAPOTASSIUM HYDROGEN, (OC-6-21)-	CADMATE (6-), [[[1,2-ETHANEDIYLBIS[NITRILOBIS (METHYLENE)]
7440-43-9 CADMIUM	CADMIUM
30905-38-5 CADMIUM 104	CADMIUM 104
14709-52-5 CADMIUM 107	CADMIUM 107
14109-32-1 CADMIUM 109	CADMIUM 109
14336-66-4 CADMIUM 113	CADMIUM 113
RR-00474-0 CADMIUM 113M	CADMIUM 113M
14336-68-6 CADMIUM 115	CADMIUM 115
RR-00473-9 CADMIUM 115M	CADMIUM 115M
15139-70-5 CADMIUM 117	CADMIUM 117
RR-00471-7 CADMIUM 117M	CADMIUM 117M
1306-19-0 CADMIUM OXIDE FUME [ALB, ALB, MEX]	CADMIUM OXIDE
RR-01638-6 CADMIUM OXIDE PRODUCTION (MEASURED AS CADMIUM) [MEX]	CADMIUM OXIDE PRODUCTION
7440-43-9 CADMIUM, DUSTS AND SALTS [MEX, MEX]	CADMIUM
543-90-8 CADMIUM ACETATE	CADMIUM ACETATE
RR-00559-4 CADMIUM AND CADMIUM COMPOUNDS [IAR, GBR]	CADMIUM COMPOUNDS
7440-43-9 CADMIUM AND ITS COMPOUNDS [MAK]	CADMIUM
7789-42-6 CADMIUM BROMIDE	CADMIUM BROMIDE
7789-42-6 CADMIUM BROMIDE (CDBR2) [PAL]	CADMIUM BROMIDE
513-78-0 CADMIUM CARBONATE	CADMIUM CARBONATE
10108-64-2 CADMIUM CHLORIDE	CADMIUM CHLORIDE
10108-64-2 CADMIUM CHLORIDE (CDCL2) [PAL, PAL]	CADMIUM CHLORIDE
RR-00559-4 CADMIUM COMPOUNDS	CADMIUM COMPOUNDS
RR-00559-4 CADMIUM COMPOUNDS, N.O.S. [RCU, NJL]	CADMIUM COMPOUNDS
14239-68-0 CADMIUM DIETHYLDITHIOCARBAMATE	CADMIUM DIETHYLDITHIOCARBAMATE
7440-43-9 CADMIUM DUST [NIO, NIO, NIO, NIO]	CADMIUM
7440-43-9 CADMIUM, DUST & SALTS [BC1, BC1]	CADMIUM
7440-43-9 CADMIUM, DUST & SALTS (AS CD) [ALB, ALB]	CADMIUM
7440-43-9 CADMIUM DUST AND SALTS [QBC, QBC]	CADMIUM
7440-43-9 CADMIUM, ELEMENTAL [WHM]	CADMIUM
14486-19-2 CADMIUM FLUOBORATE	CADMIUM FLUOBORATE
1306-19-0 CADMIUM FUME [NIO, NIO, NIO, NIO, OS1, OS1, OS1]	CADMIUM OXIDE
RR-00559-4 CADMIUM INORGANIC COMPOUNDS [BEI]	CADMIUM COMPOUNDS
7790-80-9 CADMIUM IODIDE	CADMIUM IODIDE
10325-94-7 CADMIUM NITRATE	CADMIUM NITRATE
12656-57-4 CADMIUM ORANGE PIGMENT	CADMIUM ORANGE PIGMENT
1306-19-0 CADMIUM OXIDE	CADMIUM OXIDE
1306-19-0 CADMIUM OXIDE (CDO) [PAL, PAL]	CADMIUM OXIDE
1306-19-0 CADMIUM OXIDE FUME [BC1, BC1, QBC, QBC]	CADMIUM OXIDE
1306-19-0 CADMIUM OXIDE, PRODUCTION [ONT, ONT]	CADMIUM OXIDE
RR-01638-6 CADMIUM OXIDE PRODUCTION	CADMIUM OXIDE PRODUCTION

Chemical Name	Reference Name
RR-00559-4 CADMIUM SALTS, N.O.S. [WHM]	CADMIUM COMPOUNDS
2223-93-0 CADMIUM STEARATE	CADMIUM STEARATE
141-00-4 CADMIUM SUCCINATE	CADMIUM SUCCINATE
10124-36-4 CADMIUM SULFATE	CADMIUM SULFATE
1306-23-6 CADMIUM SULFIDE	CADMIUM SULFIDE
1306-23-6 CADMIUM SULFIDE (CDS) [PAL, PAL]	CADMIUM SULFIDE
10124-36-4 CADMIUM SULPHATE [FLA, MSL]	CADMIUM SULFATE
1306-23-6 CADMIUM SULPHIDE [FLA, GBR, HM1, MSL]	CADMIUM SULFIDE
21351-79-1 CAESIUM HYDROXIDE [AUS, GBR]	CESIUM HYDROXIDE
331-39-5 CAFFEIC ACID	CAFFEIC ACID
58-08-2 CAFFEINE	CAFFEINE
7440-70-2 CALCIUM	CALCIUM
14092-95-6 CALCIUM 41	CALCIUM 41
13966-05-7 CALCIUM 45	CALCIUM 45
14391-99-2 CALCIUM 47	CALCIUM 47
1317-65-3 CACIUM CARBONATE (LIMESTONE) [QBC]	CALCIUM CARBONATE
471-34-1 CALCIUM CARBONATE [MEX, MEX]	CARBONIC ACID, CALCIUM SALT (1:1)
13463-98-4 CALCIUM ABIETATE	CALCIUM ABIETATE
5743-26-0 CALCIUM ACETATE, MONOHYDRATE	CALCIUM ACETATE, MONOHYDRATE
RR-01046-8 CALCIUM ALKYL(C8-24)BENZENESULFONATE	CALCIUM ALKYL(C8-24)BENZENESULFONATE
RR-01047-9 CALCIUM AND SODIUM SALTS OF SUGAR DERIVED ACIDS	CALCIUM AND SODIUM SALTS OF SUGAR DERIVED ACIDS
10103-62-5 CALCIUM ARSENATE	CALCIUM ARSENATE
RR-01275-9 CALCIUM ARSENATE AND CALCIUM ARSENITE, MIXTURES, SOLID	CALCIUM ARSENATE AND CALCIUM ARSENITE, MIXTURES, SOLID
52740-16-6 CALCIUM ARSENITE	CALCIUM ARSENITE
15194-98-6 CALCIUM ARSENITE (2:1)	CALCIUM ARSENITE (2:1)
27152-57-4 CALCIUM ARSENITE (2:3)	CALCIUM ARSENITE (2:3)
2090-05-3 CALCIUM BENZOATE	CALCIUM BENZOATE
14307-33-6 CALCIUM BICHROMATE [FLA]	CHROMIC ACID (H2CR2O7), CALCIUM SALT (1:1)
19372-44-2 CALCIUM, BIS(2,4-PENTANEDIONATO-O,O')	CALCIUM, BIS(2,4-PENTANEDIONATO-O,O')ZENESULFONATE)
13780-03-5 CALCIUM BISULFITE	CALCIUM BISULFITE
12007-56-6 CALCIUM BORATE	CALCIUM BORATE
75-20-7 CALCIUM CARBIDE	CALCIUM CARBIDE
75-20-7 CALCIUM CARBIDE (CAC2) [PAL]	CALCIUM CARBIDE
471-34-1 CALCIUM CARBONATE [ALB, GBR]	CARBONIC ACID, CALCIUM SALT (1:1)
471-34-1 CALCIUM CARBONATE, INCLUDING MARBLE [ONT]	CARBONIC ACID, CALCIUM SALT (1:1)
471-34-1 CALCIUM CARBONATE [PST]	CARBONIC ACID, CALCIUM SALT (1:1)
1317-65-3 CALCIUM CARBONATE	CALCIUM CARBONATE
471-34-1 CALCIUM CARBONATE/MARBLE [BC1, BC1, QBC]	CARBONIC ACID, CALCIUM SALT (1:1)
68442-82-0 CALCIUM CARBONATE DIMETHYLHEXANOATE	CALCIUM CARBONATE DIMETHYLHEXANOATEALDEHYDE
10137-74-3 CALCIUM CHLORATE	CALCIUM CHLORATE
10043-52-4 CALCIUM CHLORIDE	CALCIUM CHLORIDE
64175-94-6 CALCIUM CHLORIDE HYDROXIDE HYPOCHLORITE, DIHYDRATE	CALCIUM CHLORIDE HYDROXIDE HYPOCHLORITE, DIHYDRATE
14674-72-7 CALCIUM CHLORITE	CALCIUM CHLORITE
13765-19-0 CALCIUM CHROMATE	CALCIUM CHROMATE

Chemical Name	Reference Name
813-94-5 CALCIUM CITRATE	CALCIUM CITRATE
156-62-7 CALCIUM CYANAMIDE	CALCIUM CYANAMIDE
592-01-8 CALCIUM CYANIDE	CALCIUM CYANIDE
592-01-8 CALCIUM CYANIDE (CA(CN)2) [PAL]	CALCIUM CYANIDE
14307-33-6 CALCIUM DICHROMATE [MSL]	CHROMIC ACID (H2CR2O7), CALCIUM SALT (1:1)
62-33-9 CALCIUM DISODIUM ETHYLENEDIAMINETE-TRAACETATE [PST]	DISODIUM CALCIUM EDTA
13846-18-9 CALCIUM DITHIONITE	CALCIUM DITHIONITE
26264-06-2 CALCIUM DODECYLBENZENE SULFONATE	CALCIUM DODECYLBENZENE SULFONATE
26264-06-2 CALCIUM DODECYLBENZENESUL-FONATE [CAL]	CALCIUM DODECYLBENZENE SULFONATE
136-51-6 CALCIUM 2-ETHYLHEXANOATE	CALCIUM 2-ETHYLHEXANOATE
68478-54-6 CALCIUM 2-ETHYLHEXANOATE/ISONONANOATE COM-PLEXES	CALCIUM 2-ETHYLHEXANOATE/ISONONANOATE COM-PLEXES
7789-75-5 CALCIUM FLUORIDE (CAF2)	CALCIUM FLUORIDE (CAF2)
544-17-2 CALCIUM FORMATE	CALCIUM FORMATE
57308-10-8 CALCIUM HYDRIDE	CALCIUM HYDRIDE
13780-03-5 CALCIUM HYDROGEN SULFITE [NJL, NJL]	CALCIUM BISULFITE
13846-18-9 CALCIUM HYDROSULFITE [NJL]	CALCIUM DITHIONITE
1305-62-0 CALCIUM HYDROXIDE	CALCIUM HYDROXIDE
1305-62-0 CALCIUM HYDROXIDE (CA(OH)2) [PAL]	CALCIUM HYDROXIDE
7778-54-3 CALCIUM HYPOCHLORITE	CALCIUM HYPOCHLORITE
7789-80-2 CALCIUM IODATE	CALCIUM IODATE
53988-05-9 CALCIUM ISONONANOATE	CALCIUM ISONONANOATE
8061-52-7 CALCIUM LIGNOSULFONATE	CALCIUM LIGNOSULFONATE
13573-15-4 CALCIUM MANGANESE SILICON	CALCIUM MANGANESE SILICON
7789-82-4 CALCIUM MOLYBDATE	CALCIUM MOLYBDATE
61789-36-4 CALCIUM NAPHTHENATE	CALCIUM NAPHTHENATE
27253-33-4 CALCIUM NEODECANOATE	CALCIUM NEODECANOATE
10124-37-5 CALCIUM NITRATE	CALCIUM NITRATE
13477-34-4 CALCIUM NITRATE.4H20	CALCIUM NITRATE.4H20
6107-56-8 CALCIUM OCTANOATE	CALCIUM OCTANOATE
1305-78-8 CALCIUM OXIDE	CALCIUM OXIDE
1305-78-8 CALCIUM OXIDE (CAO) [PAL]	CALCIUM OXIDE
13477-36-6 CALCIUM PERCHLORATE	CALCIUM PERCHLORATE
10118-76-0 CALCIUM PERMANGANATE	CALCIUM PERMANGANATE
1305-79-9 CALCIUM PEROXIDE	CALCIUM PEROXIDE
61789-86-4 CALCIUM PETROLEUM SULFONATE	CALCIUM PETROLEUM SULFONATE METHYL SULFATES
10103-46-5 CALCIUM PHOSPHATE	CALCIUM PHOSPHATE
1305-99-3 CALCIUM PHOSPHIDE	CALCIUM PHOSPHIDE
25987-55-7 CALCIUM POLYACRYLATE	CALCIUM POLYACRYLATE
4075-81-4 CALCIUM PROPIONATE	CALCIUM PROPIONATE
9007-13-0 CALCIUM RESINATE	CALCIUM RESINATE
9007-13-0 CALCIUM SALT OF TALL-OIL ROSIN [PST]	CALCIUM RESINATE
61789-86-4 CALCIUM SALTS OF PETROLEUM SULFONIC ACIDS [PST]	CALCIUM PETROLEUM SULFONATE METHYL SULFATES
68187-71-3 CALCIUM SALTS OF TALL-OIL FATTY ACIDS	CALCIUM SALTS OF TALL-OIL FATTY ACIDSIS(HY-DROXYETHYL) METHYL, ETHOXYLATED, CHLORIDES
14019-91-1 CALCIUM SELENATE	CALCIUM SELENATE
1344-95-2 CALCIUM SILICATE	CALCIUM SILICATE

Chemical Name	Reference Name
1344-95-2 CALCIUM SILICATE (SYNTHETIC) [TLV]	CALCIUM SILICATE
12737-18-7 CALCIUM SILICIDE	CALCIUM SILICIDE
12013-56-8 CALCIUM SILICON	CALCIUM SILICON
23209-59-8 CALCIUM SODIUM METAPHOSPHATE	CALCIUM SODIUM METAPHOSPHATE
1592-23-0 CALCIUM STEARATE	CALCIUM STEARATE
7778-18-9 CALCIUM SULFATE	CALCIUM SULFATE
10101-41-4 CALCIUM SULFATE DIHYDRATE	CALCIUM SULFATE DIHYDRATE
10101-41-4 CALCIUM SULFATE, DIHYDRATE (GYPSUM) [PST]	CALCIUM SULFATE DIHYDRATE
10034-76-1 CALCIUM SULFATE HEMIHYDRATE	CALCIUM SULFATE HEMIHYDRATE
10101-41-4 CALCIUM SULFATE, INCLUDING PLASTER OF PARIS [ONT]	CALCIUM SULFATE DIHYDRATE
7778-18-9 CALCIUM SULPHATE [AUS]	CALCIUM SULFATE
10124-41-1 CALCIUM THIOSULFATE	CALCIUM THIOSULFATE
29887-08-9 CALCO OIL BLUE	CALCO OIL BLUE
16044-16-9 CALIFORNIUM 244	CALIFORNIUM 244
15117-45-0 CALIFORNIUM 246	CALIFORNIUM 246
15758-24-4 CALIFORNIUM 248	CALIFORNIUM 248
15237-97-5 CALIFORNIUM 249	CALIFORNIUM 249
13982-11-1 CALIFORNIUM 250	CALIFORNIUM 250
15765-19-2 CALIFORNIUM 251	CALIFORNIUM 251
13981-17-4 CALIFORNIUM 252	CALIFORNIUM 252
15720-29-3 CALIFORNIUM 253	CALIFORNIUM 253
22095-76-7 CALIFORNIUM 254	CALIFORNIUM 254
8001-35-2 CAMPHECHLOR [302]	TOXAPHENE
8001-35-2 CAMPHECLOR [HM1, HM1]	TOXAPHENE
79-92-5 CAMPHENE	CAMPHENE
76-22-2 CAMPHOR	CAMPHOR
464-49-3 (+)-CAMPHOR	(+)-CAMPHOR
21368-68-3 DL-CAMPHOR	DL-CAMPHOR
76-22-2 CAMPHOR, SYNTHETIC [MEX]	CAMPHOR
8008-51-3 CAMPHOR OIL	CAMPHOR OIL
8008-51-3 CAMPHOR OIL (LIGHT) [FLA, MSL, NFP, NFP]	CAMPHOR OIL
76-22-2 CAMPHOR, SYNTHETIC [ALB, ALB, AUS, AUS, BC1, BC1]	CAMPHOR
76-22-2 CAMPHOR (SYNTHETIC) [CEX, NIO, NIO, NIO]	CAMPHOR
76-22-2 CAMPHOR, SYNTHETIC [OS1, OS1, TLV, TLV]	CAMPHOR
76-22-2 CAMPHOUR, SYNTHETIC [QBC, QBC]	CAMPHOR
RR-01048-0 CANARY SEED	CANARY SEED
68476-78-8 CANE SYRUP [PST]	MOLASSES
56-25-7 CANTHARIDIN	CANTHARIDIN
RR-01664-8 CAPPED ALIPHATIC ISOCYANATE	CAPPED ALIPHATIC ISOCYANATE
334-48-5 CAPRIC ACID	CAPRIC ACID
142-62-1 CAPROIC ACID	CAPROIC ACID
105-60-2 CAPROLACTAM	CAPROLACTAM
105-60-2 SIGMA-CAPROLACTAM [MAK, MAK]	CAPROLACTAM
105-60-2 CAPROLACTAM [QBC, QBC]	CAPROLACTAM
105-60-2 CAPROLACTAM DUST AND VAPOR [MSL]	CAPROLACTAM
105-60-2 CAPROLACTAM, VAPOUR AND AEROSOL [ONT]	CAPROLACTAM

Chemical Name	Reference Name
502-44-3 CAPROLACTONE	CAPROLACTONE
RR-00993-8 CAPROLACTONE MODIFIED ACRYLATE MONOMER	CAPROLACTONE MODIFIED ACRYLATE MONOMER
RR-00217-5 CAPROLACTONE, POLYMER WITH HEXAM-ETHYLENE DIISOCYANATE, HYDROXYALKYL ACRYLATE ESTER, REACTION PRODUCTS WITH SUBSTITUTED ALKANOIC ACID AND METAL HETEROMONOCYCLE	CAPROLACTONE, POLYMER WITH HEXAMETHYLENE DIISOCYANATE, HYDROTDI, ALKANOL BLOCKED, ACRYLATE
124-13-0 CAPRYLALDEHYDE	CAPRYLALDEHYDE
26402-26-6 CAPRYLIC ACID MONOGLYCERIDE	CAPRYLIC ACID MONOGLYCERIDE
37332-31-3 CAPRYLIC/CAPRIC TRIGLYCERIDE	CAPRYLIC/CAPRIC TRIGLYCERIDE
111-64-8 CAPRYLYL CHLORIDE	CAPRYLYL CHLORIDE
7530-07-6 CAPRYLYL PEROXIDE [NJL]	CAPRYLYL PEROXIDE (N-OCTANOYL PEROXIDE)
7530-07-6 CAPRYLYL PEROXIDE (N-OCTANOYL PEROX-IDE)	CAPRYLYL PEROXIDE (N-OCTANOYL PEROXIDE)
2425-06-1 CAPTAFOL	CAPTAFOL
2939-80-2 CAPTAFOL, CIS-	CAPTAFOL, CIS-
2425-06-1 CAPTAFOL (DIFOLATAN) [QBC, QBC, OS1, ALB, ALB, ALB, BC1, BC1]	CAPTAFOL
2425-06-1 CAPTAFOL (ISO) [GBR, GBR]	CAPTAFOL
133-06-2 CAPTAN	CAPTAN
133-06-2 CAPTAN [1H-ISOINDOLE-1,3(2H)-DIONE,3A,4, 7,7A-TETRAHYDRO-2-[(TRICHLOROMETHYL) THIO]-] [313]	CAPTAN
133-06-2 CAPTAN (ISO) [GBR, GBR]	CAPTAN
8028-89-5 CARAMEL	CARAMEL
51-83-2 CARBACHOL CHLORIDE	CARBACHOL CHLORIDE
RR-00186-5 CARBAMATE PESTICIDES, N.O.S.	CARBAMATE PESTICIDES, N.O.S.
17804-35-2 CARBAMIC ACID, [1-[(BUTYLAMINO)CAR-BONYL]-1H-BENZIMIDAZOL-2-YL]-, METHYL ESTER [PAL]	BENOMYL
51-79-6 CARBAMIC ACID, ETHYL ESTER [EPA, MAK, PAL, PAL]	URETHANE
78812-39-2 CARBAMIC ACID, MANGANESE SALT	CARBAMIC ACID, MANGANESE SALT
26419-73-8 CARBAMIC ACID, METHYL-, O-(((2,4-DIMETHYL-1,3-DITHIOLAN-2-YL)METHYLENE) AMINO)-	CARBAMIC ACID, METHYL-, O-(((2,4-DIMETHYL-1,3-DITHIOLAN-2-YL
615-53-2 CARBAMIC ACID, METHYLNITROSO-, ETHYL ESTER [PAL, PAL]	N-NITROSO-N-METHYLURETHANE
1111-78-0 CARBAMIC ACID, MONOAMMONIUM SALT [PAL]	AMMONIUM CARBAMATE
RR-00225-5 CARBAMIC ACID, (TRIALKYLOXYSILYALKYL)-SUBSTITUTED ACRYLATEESTER	CARBAMIC ACID, (TRIALKYLOXYSILYALKYL)-SUBSTI-TUTED ACRYLATE
88-10-8 CARBAMIC CHLORIDE, DIETHYL- [PAL]	DIETHYLCARBAMOYL CHLORIDE
79-44-7 CARBAMIC CHLORIDE, DIMETHYL- [PAL, PAL]	DIMETHYLCARBAMOYL CHLORIDE
630-10-4 CARBAMIMIDOSELENOIC ACID	CARBAMIMIDOSELENOIC ACID
95-06-7 CARBAMODITHIOIC ACID, DIETHYL-, 2-CHLORO-2-PROPENYL ESTER [PAL, PAL]	SULFALLATE
111-54-6 CARBAMODITHIOIC ACID, 1,2-ETHANEDIYLBIS, SALTS AND ESTERS [EPA]	1,2-ETHANEDIYLBISCARBAMODITHIOIC ACID
34731-32-3 CARBAMODITHIOIC ACID, 1,2-ETHANEDIYL ESTER	CARBAMODITHIOIC ACID, 1,2-ETHANEDIYL ESTERTRI-AZINE-TRIONE
2303-16-4 CARBAMOTHIOIC ACID, BIS(1-METHYLETHYL)-, S-(2,3-DICHLORO-2-PROPENYL) ESTER [PAL]	DIALLATESIUM SALT
63-25-2 CARBARYL	CARBARYL

Chemical Name	Reference Name
63-25-2 CARBARYL [1-NAPHTHALENOL, METHYLCAR-BAMATE] [313]	CARBARYLSTER)
63-25-2 CARBARYL (SEVIN) [NIO, NIO, NIO, OS1, OS1, QBC]	CARBARYL
63-25-2 CARBARYL (ISO) [GBR, GBR]	CARBARYL
63-25-2 CARBARYL (SEVIN) [ALB, ALB, BC1, BC1]	CARBARYL
86-74-8 9H-CARBAZOLE	9H-CARBAZOLEESTER
86-74-8 CARBAZOLE [ATS, CAL, IAR]	9H-CARBAZOLE
RR-01528-1 3-CARBETHOXYPSORALEN	3-CARBETHOXYPSORALEN
112-15-2 CARBINOL ACETATE	CARBINOL ACETATE
1563-66-2 CARBOFURAN	CARBOFURAN
1563-66-2 CARBOFURAN (FURADAN) [OS1, QBC, ALB, ALB, BC1]	CARBOFURAN
1563-66-2 CARBOFURAN (ISO) [GBR]	CARBOFURAN
7786-34-7 ALPHA-2-CARBOMETHOXY-1-METHYLVINYL DIMETHYL PHOSPHATE (MEVINPHOS) [CAL]	MEVINPHOS
7440-44-0 CARBON	CARBON
14333-33-6 CARBON 11	CARBON 11
14762-75-5 CARBON 14	CARBON 14
56-23-5 CARBON TETRACHLORIDE [QBC, QBC, QBC]	CARBON TETRACHLORIDE
75-73-0 CARBON TETRAFLUORIDE [HON]	TETRAFLUOROMETHANE
75-15-0 CARBON BISULPHIDE [HM1, HM1, HM1, HM1]	CARBON DISULFIDE
1333-86-4 CARBON BLACK	CARBON BLACK
7440-44-0 CARBON BLACK [CEX]	CARBON
1333-86-4 CARBON BLACK - EXTRACTS [CAL]	CARBON BLACK
RR-00060-2 CARBON BLACK EXTRACTS	CARBON BLACK EXTRACTS
RR-00060-2 CARBON-BLACK EXTRACTS [C65, IAR]	CARBON BLACK EXTRACTS
1333-86-4 CARBON BLACKS [IAR]	CARBON BLACK
124-38-9 CARBON DIOXIDE	CARBON DIOXIDE
8070-50-6 CARBON DIOXIDE AND ETHYLENE OXIDE MIXTURES	CARBON DIOXIDE AND ETHYLENE OXIDE MIXTURES
75-15-0 CARBON DISULFIDE	CARBON DISULFIDE
75-15-0 CARBON DISULPHIDE [AUS, AUS, AUS, GBR, GBR]	CARBON DISULFIDE
10361-29-2 CARBONIC ACID, AMMONIUM SALT [PAL]	AMMONIUM CARBONATE
13106-47-3 CARBONIC ACID, BERYLLIUM SALT (1:1) [PAL, PAL]	BERYLLIUM CARBONATE
30714-78-4 CARBONIC ACID, BUTYL ETHYL ESTER [PAL]	ETHYL BUTYL CARBONATE
471-34-1 CARBONIC ACID, CALCIUM SALT (1:1)	CARBONIC ACID, CALCIUM SALT (1:1)
506-87-6 CARBONIC ACID, DIAMMONIUM SALT [PAL]	AMMONIUM CARBONATE
105-58-8 CARBONIC ACID, DIETHYL ESTER [PAL]	DIETHYL CARBONATE
616-38-6 CARBONIC ACID, DIMETHYL ESTER [PAL]	DIMETHYL CARBONATE
1066-33-7 CARBONIC ACID, MONOAMMONIUM SALT [PAL]	AMMONIUM BICARBONATE
16337-84-1 CARBONIC ACID, NICKEL SALT	CARBONIC ACID, NICKEL SALT
3333-67-3 CARBONIC ACID, NICKEL(2+) SALT (1:1) [PAL, PAL]	NICKEL CARBONATE
17237-93-3 CARBONIC ACID, NICKEL(2+) SALT (2:1)	CARBONIC ACID, NICKEL(2+) SALT (2:1)
3486-35-9 CARBONIC ACID, ZINC SALT (1:1) [PAL]	ZINC CARBONATE
124-38-9 CARBONIC ANHYDRIDE (CARBON DIOXIDE) [QBC, QBC]	CARBON DIOXIDE

Chemical Name	Reference Name
75-44-5 CARBONIC DICHLORIDE [PAL]	PHOSGENE
353-50-4 CARBONIC DIFLUORIDE [PAL, RCU, RCU]	CARBONYL FLUORIDE
497-18-7 CARBONIC DIHYDRAZIDE	CARBONIC DIHYDRAZIDE
630-08-0 CARBON MONOXIDE	CARBON MONOXIDE
541-41-3 CARBONOCHLORIDIC ACID, ETHYL ESTER [PAL]	ETHYL CHLOROFORMATE
2937-50-0 CARBONOCHLORIDIC ACID, 2-PROPENYL ESTER [PAL]	ALLYL CHLOROCARBONATE
13889-92-4 CARBONOCHLORIDOTHIOIC ACID, S-PROPYL ESTER	CARBONOCHLORIDOTHIOIC ACID, S-PROPYL ESTER
140-89-6 CARBONODITHIOIC ACID, O-ETHYL ESTER, POTASSIUM SALT	CARBONODITHIOIC ACID, O-ETHYL ESTER, POTASSIUM SALT
4452-58-8 CARBONOPEROXOIC ACID, DISODIUM SALT [NJL]	SODIUM PERCARBONATE
463-58-1 CARBON OXIDE SULFIDE (COS) [PAL]	CARBONYL SULFIDE
353-50-4 CARBON OXYFLUORIDE (CARBONIC DIFLUORIDE) [TSC, TSE]	CARBONYL FLUORIDE
463-58-1 CARBON OXYSULFIDE [FLA, MSL, NFP, NFP]	CARBONYL SULFIDE
463-58-1 CARBON OXYSULFIDE [CARBON OXIDE SULFIDE (COS)] [EP3]	CARBONYL SULFIDE
558-13-4 CARBON TETRABROMIDE	CARBON TETRABROMIDE
558-13-4 CARBON TETRABROMIDE (TETRABROMOMETHANE) [QBC, QBC]	CARBON TETRABROMIDE
56-23-5 CARBON TETRACHLORIDE	CARBON TETRACHLORIDE
56-23-5 CARBON TETRACHLORIDE (TETRACHLOROMETHANE) [CW2, CN2]	CARBON TETRACHLORIDE
75-44-5 CARBONYL CHLORIDE [ONT]	PHOSGENE
75-44-5 CARBONYL CHLORIDE (PHOSGENE) [BC1, QBC, ALB, ALB]	PHOSGENE
353-50-4 CARBONYL FLUORIDE	CARBONYL FLUORIDE
463-58-1 CARBONYL SULFIDE	CARBONYL SULFIDE
786-19-6 CARBOPHENOTHION	CARBOPHENOTHION
41575-94-4 CARBOPLATIN	CARBOPLATIN
RR-01232-8 CARBOPOLYCYCLICOL AZOALKYLAMINOALKYLCARBOMONOCYCLIC ESTER, HALOGEN ACID SALTS	CARBOPOLYCYCLICOL AZOALKYLAMINOALKYLCARBOMONOCYCLIC ESTER, HE COPOLYMER
5234-68-4 CARBOXIN (5,6-DIHYDRO-2-METHYL-N-PHENYL-1,4-OXATHIIN-3-CARBOXAMIDE)	CARBOXIN (5,6-DIHYDRO-2-METHYL-N-PHENYL-1,4-OXATHIIN-3-CARBO
68630-89-7 6-CARBOXY-4-HEXYL-2-CYCLOHEXENE-1-OCTANOIC ACID, MONOPOTASSIUM SALT	6-CARBOXY-4-HEXYL-2-CYCLOHEXENE-1-OCTANOIC ACID, MONOPOTASSI
RR-01721-0 CARBOXYLIC ACID GLYCIDYL ESTERS	CARBOXYLIC ACID GLYCIDYL ESTERS
9000-11-7 CARBOXYMETHYL CELLULOSE	CARBOXYMETHYL CELLULOSE
68649-45-6 CARBOXYPOLYMETHYLENE RESIN	CARBOXYPOLYMETHYLENE RESIN ISOPROPYLAMINE SALTS
77-65-6 CARBROMAL	CARBROMAL
RR-01049-1 CARDBOARD	CARDBOARD
78-44-4 CARISOPRODOL	CARISOPRODOL
3567-69-9 CARMOISINE [IAR]	C.I. ACID RED 14
8015-86-9 CARNAUBA WAX	CARNAUBA WAX
RR-00544-7 CARPENTRY AND JOINERY	CARPENTRY AND JOINERY
9000-07-1 CARRAGEENAN	CARRAGEENAN
9000-07-1 CARRAGEENAN, DEGRADED [IAR]	CARRAGEENAN
9000-07-1 CARRAGEENAN (DEGRADED) [MSL]	CARRAGEENAN

Chemical Name	Reference Name
RR-01529-2 CARRAGEENAN, NATIVE	CARRAGEENAN, NATIVE
RR-01050-4 CARROTS	CARROTS
15263-52-2 CARTAP HYDROCHLORIDE	CARTAP HYDROCHLORIDE
499-75-2 CARVACROL	CARVACROL
2244-16-8 CARVONE	CARVONE
2244-16-8 D-CARVONE [CSR, CSR]	CARVONE
6485-40-1 (-)-CARVONE	(-)-CARVONE
9000-71-9 CASEIN	CASEIN
RR-00861-7 CASINGHEAD GASOLINE [NJL]	GASOLINE (CASINGHEAD)
8001-79-4 CASTOR OIL	CASTOR OIL
8001-78-3 CASTOR OIL (HYDROGENATED) [NFP]	HYDROGENATED CASTOR OIL
68071-54-5 CASTOR OIL, DEHYDRATED, POLYMER WITH P-TERT-BUTYLBENZOIC ACID, GLYCEROL AND PHTHALIC ANHYDRIDE	CASTOR OIL, DEHYDRATED, POLYMER WITH P-TERT-BUTYLBENZOIC ACI PRODUCTS WITH POLYETHYLENE-POYPROPYLENE GLYCOL MONOACETATE ALLYL ETHER
105839-17-6 CASTOR OIL, EPOXIDIZED	CASTOR OIL, EPOXIDIZED-5-(1,1-DIMETHYLETHYL)-4-HYDROXYPHENYL)-1-OXOPROPYL)-OMEGA- HY-DROXY-
8001-78-3 CASTOR OIL, HYDROGENATED [PST]	HYDROGENATED CASTOR OIL
RR-01051-5 CASTOR OIL, MALEIC ANHYDRIDE, AND POLYETHYLENE GLYCOL COPOLYMER	CASTOR OIL, MALEIC ANHYDRIDE, AND POLYETHY-LENE GLYCOL COPOLY
68187-84-8 CASTOR OIL, OXIDIZED	CASTOR OIL, OXIDIZED
61791-12-6 CASTOR OIL, POLYOXYETHYLATED [PST]	ETHOXYLATED CASTOR OILTHYL, ETHOXYLATED, CHLORIDES
8002-33-3 CASTOR OIL, SULFATED	CASTOR OIL, SULFATED
68187-76-8 CASTOR OIL, SULFATED, SODIUM SALT [PST]	SULFONATED CASTOR OIL, SODIUM SALT
535-89-7 CASTRIX [NJL]	CRIMIDINE
120-80-9 CATECHOL	CATECHOL
120-80-9 CATECHOL (PYROCATECHOL) [FLA, OS1, OS1, QBC, QBC, ALB, ALB, BC1]	CATECHOL
RR-01052-6 CAT FOOD	CAT FOODMER
RR-00193-4 CAUSTIC ALKALI LIQUIDS, N.O.S.	CAUSTIC ALKALI LIQUIDS, N.O.S.
RR-00801-5 CAUSTIC ARSENIC OIL	CAUSTIC ARSENIC OIL
8000-27-9 CEDAR WOOD OIL	CEDAR WOOD OIL
469-61-4 ALPHA-CEDRENE	ALPHA-CEDRENE
67874-81-1 CEDROL METHYL ETHER	CEDROL METHYL ETHER
61790-53-2 CELITE [PST]	DIATOMACEOUS EARTH
111-15-9 CELLOSOLVE ACETATE [PST]	2-ETHOXYETHYL ACETATE
111-15-9 CELLOSOLVE ACETATE (2-ETHOXYETHYLAC-ETATE) [ALB, ALB, ALB]	2-ETHOXYETHYL ACETATE
RR-00195-6 CELLULOID	CELLULOID
9004-34-6 CELLULOSE	CELLULOSE
9004-34-6 CELLULOSE (PAPER FIBER) [MEX, MEX, ALB, BC1, BC1]	CELLULOSE
9004-36-8 CELLULOSE ACETATE BUTYRATE [PST]	CELLULOSE ACETATE BUTYRATE
9004-36-8 CELLULOSE ACETATE BUTYRATE	CELLULOSE ACETATE BUTYRATE
9000-11-7 CELLULOSE, CARBOXYMETHYL ETHER [PST]	CARBOXYMETHYL CELLULOSE
9004-32-4 CELLULOSE CARBOXYMETHYL ETHER, SODIUM SALT [PST]	SODIUM CARBOXYMETHYL CELLULOSE
68610-92-4 CELLULOSE, OMEGA-ETHER WITH ETHOXY-LATED 2-HYDROXY-3-(TRIMETHYLAMMONIO) PROPANOL, CHLORIDE	CELLULOSE, OMEGA-ETHER WITH ETHOXYLATED 2-HYDROXY-3-(TRIMETH

Chemical Name	Reference Name
9004-57-3 CELLULOSE, ETHYL ETHER [PST]	ETHYLCELLULOSE
9004-58-4 CELLULOSE, ETHYL 2-HYDROXYETHYL ETHER [PST]	ETHYL HYDROXYETHYL CELLULOSE
9004-62-0 CELLULOSE, 2-HYDROXYETHYL ETHER [PST]	HYDROXYETHYL CELLULOSE
9004-65-3 CELLULOSE HYDROXYPROPYL METHYL ETHER	CELLULOSE HYDROXYPROPYL METHYL ETHER
9004-64-2 CELLULOSE, 2-HYDROXYPROPYL ETHER [PST]	HYDROXYPROPYL ETHER OF CELLULOSE
9004-70-0 CELLULOSE NITRATE [OS3]	NITROCELLULOSE
9004-70-0 CELLULOSE, NITRATE [PAL]	NITROCELLULOSE
RR-01760-7 CELLULOSE NITRATE, WET WITH ALCOHOL	CELLULOSE NITRATE, WET WITH ALCOHOL
9004-34-6 CELLULOSE (PAPER FIBER) [QBC]	CELLULOSE
9004-34-6 CELLULOSE (PAPER FIBRE) [AUS]	CELLULOSE
68442-85-3 CELLULOSE, REGENERATED	CELLULOSE, REGENERATED
65997-15-1 CEMENT, PORTLAND, CHEMICALS [PAL]	PORTLAND CEMENT
142844-00-6 CERAMIC FIBERS [C65, MAK]	REFRACTORY CERAMIC FIBERS
142844-00-6 CERAMIC FIBRES [AUS, IAR]	REFRACTORY CERAMIC FIBERS AVERAGE DEGREE OF CHLORINATION 60%
142844-00-6 CERAMIC FIBRES OF RESPIRABLE SIZE [NTP]	REFRACTORY CERAMIC FIBERS
8001-75-0 CERESIN WAX	CERESIN WAX
7440-45-1 CERIUM	CERIUM
15055-11-5 CERIUM 134	CERIUM 134
15757-94-5 CERIUM 135	CERIUM 135
13968-49-5 CERIUM 137	CERIUM 137
RR-00470-6 CERIUM 137M	CERIUM 137M
13982-30-4 CERIUM 139	CERIUM 139
13967-74-3 CERIUM 141	CERIUM 141
14119-19-8 CERIUM 143	CERIUM 143
14762-78-8 CERIUM 144	CERIUM 144
56797-01-4 CERIUM 2-ETHYLHEXOATE	CERIUM 2-ETHYLHEXOATE
10108-73-3 CERIUM NITRATE	CERIUM NITRATE
RR-00081-7 CERTAIN COMBINED CHEMOTHERAPY FOR LYMPHOMAS [C65, MIN]	COMBINED CHEMOTHERAPY FOR LYMPHOMAS
7440-46-2 CESIUM	CESIUM
15758-27-7 CESIUM 125	CESIUM 125
15720-35-1 CESIUM 127	CESIUM 127
15047-05-9 CESIUM 129	CESIUM 129
15066-92-9 CESIUM 130	CESIUM 130
14914-76-2 CESIUM 131	CESIUM 131
15758-03-9 CESIUM 132	CESIUM 132
13967-70-9 CESIUM 134	CESIUM 134
RR-00469-3 CESIUM 134M	CESIUM 134M
15726-30-4 CESIUM 135	CESIUM 135
RR-00468-2 CESIUM 135M	CESIUM 135M
14234-29-8 CESIUM 136	CESIUM 136
10045-97-3 CESIUM 137	CESIUM 137
15758-29-9 CESIUM 138	CESIUM 138
534-17-8 CESIUM CARBONATE	CESIUM CARBONATE
7647-17-8 CESIUM CHLORIDE	CESIUM CHLORIDE
13454-78-9 CESIUM CHROMATE	CESIUM CHROMATE
21351-79-1 CESIUM HYDROXIDE	CESIUM HYDROXIDE

Chemical Name	Reference Name
21351-79-1 CESIUM HYDROXIDE (CS(OH)) [PAL]	CESIUM HYDROXIDETHYLTHIO)-
7789-17-5 CESIUM IODIDE	CESIUM IODIDE
7789-18-6 CESIUM NITRATE	CESIUM NITRATE
10294-54-9 CESIUM SULFATE	CESIUM SULFATE
36653-82-4 CETYL ALCOHOL	CETYL ALCOHOL
7128-91-8 CETYLDIMETHYLAMINE OXIDE	CETYLDIMETHYLAMINE OXIDE
29710-31-4 CETYL OCTANOATE	CETYL OCTANOATE
75-69-4 CFC-11 [CAA]	TRICHLOROFLUOROMETHANE
354-56-3 CFC-111 [CAA]	PENTACHLOROFLUOROETHANE
76-12-0 CFC-112 [CAA]	1,1,2,2-TETRACHLORO-1,2-DIFLUOROETHANE
76-13-1 CFC-113 [CAA]	1,1,2-TRICHLORO-1,2,2-TRIFLUOROETHANE
76-14-2 CFC-114 [CAA]	DICHLOROTETRAFLUOROETHANE
76-15-3 CFC-115 [CAA]	CHLOROPENTAFLUOROETHANE
75-71-8 CFC-12 [CAA]	DICHLORODIFLUOROMETHANE
75-72-9 CFC-13 [CAA]	CHLOROTRIFLUOROMETHANE
135401-87-5 CFC-211 [CAA]	HEPTACHLOROFLUOROPROPANE
134452-44-1 CFC-212 [CAA]	HEXACHLORODIFLUOROPROPANE
134237-31-3 CFC-213 [CAA]	PENTACHLOROTRIFLUOROPROPANE
29255-31-0 CFC-214 [CAA]	TETRACHLOROTETRAFLUOROPROPANE
4259-43-2 CFC-215 [CAA]	1,1,1-TRICHLORO-2,2,3,3,3,3-PENTAFLUOROPROPANE
662-01-1 CFC-216 [CAA]	1,3-DICHLORO-1,1,2,2,3,3,-HEXAFLUOROPROPANE
125426-39-3 CFC-217 [CAA]	2-CHLORO-1,1,1,2,3,3,3-HEPTAFLUOROPROPANE
42509-80-8 CGA-12223 [MSL]	PHOSPHOROTHIOIC ACID, O-(5-CHLORO-1-(1-METHYLETHYL)-1H-1,2,4
RR-01053-7 CHEESE	CHEESE
RR-01608-0 CHEMICAL WARFARE SUBSTANCES	CHEMICAL WARFARE SUBSTANCES
474-25-9 CHENODIOL	CHENODIOL
63449-39-8 CHLORINATED HYDROCARBONS (CHORI-NATED PARAFFINS)	CHLORINATED HYDROCARBONS (CHORINATED PARAFFINS)
63449-39-8 CHLORINATED PARAFFIN WAXES [CN4]	CHLORINATED HYDROCARBONS (CHORINATED PARAFFINS)
2439-01-2 CHINOMETHIONAT [6-METHYL-1,3-DITHIOLO [4,5-B]QUINOXALIN-2-ONE]	CHINOMETHIONAT [6-METHYL-1,3-DITHIOLO[4,5-B] QUINOXALIN-2-ONE
75-87-6 CHLORAL	CHLORAL
302-17-0 CHLORAL HYDRATE [CSR]	1,1-ETHANEDIOL, 2,2,2-TRICHLORO-
133-90-4 CHLORAMBEN	CHLORAMBEN
133-90-4 CHLORAMBEN [BENZOIC ACID, 3-AMINO-2,5-DICHLORO-] [313]	CHLORAMBENTRICHLOROMETHYL)THIO]-]
305-03-3 CHLORAMBUCIL	CHLORAMBUCIL
RR-01022-0 CHLORAMINATED WATER	CHLORAMINATED WATER
10599-90-3 CHLORAMINE	CHLORAMINE
127-65-1 CHLORAMINE-T	CHLORAMINE-T
10599-90-3 CHLORAMINES [SD4]	CHLORAMINE
56-75-7 CHLORAMPHENICOL	CHLORAMPHENICOL
982-57-0 CHLORAMPHENICOL SODIUM SUCCINATE	CHLORAMPHENICOL SODIUM SUCCINATE
118-75-2 CHLORANIL [OTS]	2,3,5,6-TETRACHLORO-2,5-CYCLOHEXADIENE-1,4-DIONE
14866-68-3 CHLORATE	CHLORATE
RR-00180-9 CHLORATE AND BORATE MIXTURES [NJL, HM1, HM1]	BORATE AND CHLORATE MIXTURES

Chemical Name	Reference Name
RR-00200-6 CHLORATE AND MAGNESIUM CHLORIDE MIXTURES	CHLORATE AND MAGNESIUM CHLORIDE MIXTURES
RR-00201-7 CHLORATES, INORGANIC, N.O.S.	CHLORATES, INORGANIC, N.O.S.
14362-31-3 CHLORCYCLIZINE HYDROCHLORIDE	CHLORCYCLIZINE HYDROCHLORIDE
57-74-9 CHLORDANE	CHLORDANE
57-74-9 CHLORDANE (ALPHA AND GAMMA ISOMERS) [UTS]	CHLORDANE
5103-74-2 TRANS-CHLORDANE [ATS]	GAMMA CHLORDANE
5103-71-9 ALPHA CHLORDANE	ALPHA CHLORDANE
5103-74-2 GAMMA CHLORDANE	GAMMA CHLORDANE
57-74-9 CHLORDANE (4,7-METHANOINDAN, 1,2,4,5,6,7, 8,8-OCTACHLORO-2,3,3A,4,7,7A-HEXAHYDRO-] [313]	CHLORDANE
57-74-9 CHLORDANE (1,2,4,5,6,7,8,8-OCTACHLORO-2,3, 3A,4,7,7A-HEXAHYDRO-4,7-METHANOINDENE) [CN2]	CHLORDANENO-6H,14H-INDOLO[3,2,1-IJ]OXEPINO[2,3,4-DE]PYRROLO[2,3-H]QUINOLIN-14-ONE)
57-74-9 CHLORDANE, ALPHA & GAMMA ISO-MERS [RCU, RCU]	CHLORDANE
57-74-9 CHLORDANE (ANALYTICAL GRADE) [CSR, CSR]	CHLORDANE
RR-01530-5 CHLORDANE/HEPTACHLOR	CHLORDANE/HEPTACHLOR
57-74-9 CHLORDANE (TECHNICAL MIXTURE AND METABOLITES) [CW2, CW3]	CHLORDANE
143-50-0 CHLORDECONE [CAB, CAL, MIN]	KEPONE
143-50-0 CHLORDECONE (KEPONE) [C65, C65, C65, CSR]	KEPONE
3734-48-3 CHLORDENE	CHLORDENE
58-25-3 CHLORDIAZEPOXIDE	CHLORDIAZEPOXIDE
438-41-5 CHLORDIAZEPOXIDE HYDROCHLORIDE	CHLORDIAZEPOXIDE HYDROCHLORIDE
6164-98-3 CHLORDIMEFORM	CHLORDIMEFORM
115-28-6 CHLORENDIC ACID	CHLORENDIC ACID
470-90-6 CHLORFENVINFOS [302]	CHLORFENVINPHOS
470-90-6 CHLORFENVINPHOS	CHLORFENVINPHOS
56-95-1 CHLORHEXIDINE DIACETATE	CHLORHEXIDINE DIACETATE
7790-93-4 CHLORIC ACID	CHLORIC ACID
13477-00-4 CHLORIC ACID, BARIUM SALT [PAL]	BARIUM CHLORATE
10137-74-3 CHLORIC ACID, CALCIUM SALT [PAL]	CALCIUM CHLORATE
3811-04-9 CHLORIC ACID, POTASSIUM SALT [PAL]	POTASSIUM CHLORATEYL ESTER
7775-09-9 CHLORIC ACID, SODIUM SALT [PAL]	SODIUM CHLORATE
10361-95-2 CHLORIC ACID, ZINC SALT [PAL]	ZINC CHLORATE
16887-00-6 CHLORIDE	CHLORIDE
90982-32-4 CHLORIMURON ETHYL [ETHYL-2-[[[(4-CHLORO-6-METHOXYPYRIMIDIN-2-YL) -CAR-BONYL]-AMINO]SULFONYL]BENZOATE]	CHLORIMURON ETHYL [ETHYL-2-[[[(4-CHLORO-6-METHOXYPYRIMIDIN-2
8001-35-2 CHLORINATED CAMPHENE [MEX, MEX]	TOXAPHENE
RR-00077-1 CHLORINATED BENZENES	CHLORINATED BENZENES
55720-99-5 CHLORINATED BIPHENYL OXIDE [MAK, MAK]	CHLORINATED DIPHENYL OXIDE
53469-21-9 CHLORINATED BIPHENYLS (42% CHLORINE) [MAK, MAK, MAK, MAK, MAK]	AROCLOR 1242
11097-69-1 CHLORINATED BIPHENYLS (54% CHLORINE) [MAK, MAK, MAK, MAK, MAK]	CHLORODIPHENYL (54% CHLORINE)

Chemical Name	Reference Name
8001-35-2 CHLORINATED CAMPHENE [ALB, ALB, ALB, AUS, AUS, AUS, BC1, BC1, BC1, CEX, CEX, CEX, MAK, MAK, MAK, MIN, NIO, NIO, NIO, NIO, OS1, OS1, OS1, OS1, OS1]	TOXAPHENE
8001-35-2 CHLORINATED CAMPHENE (TOXAPHENE) [TLV, TLV, TLV, QBC, QBC, QBC, QBC]	TOXAPHENE
RR-01024-2 CHLORINATED CRESOLS	CHLORINATED CRESOLS
RR-01531-6 CHLORINATED DIBENZODIOXINS	CHLORINATED DIBENZODIOXINS
31242-93-0 CHLORINATED DIPHENYL OXIDE	CHLORINATED DIPHENYL OXIDE
31242-93-0 CHLORINATED DIPHENYL OXIDE (ETHER) [QBC]	CHLORINATED DIPHENYL OXIDE
55720-99-5 CHLORINATED DIPHENYL OXIDE	CHLORINATED DIPHENYL OXIDE
RR-01624-0 CHLORINATED DIPHENYL OXIDES	CHLORINATED DIPHENYL OXIDES
RR-01023-1 CHLORINATED DRINKING WATER [IAR]	CHLORINATED WATER
RR-00202-8 CHLORINATED ETHANES	CHLORINATED ETHANES
76-13-1 CHLORINATED FLUOROCARBON (CFC-113) [CAB]	1,1,2-TRICHLORO-1,2,2-TRIFLUOROETHANE
RR-00747-6 CHLORINATED FLUOROCARBONS	CHLORINATED FLUOROCARBONS
70776-03-3 CHLORINATED NAPHTHALENE [CW3]	NAPHTHALENE, CHLORO DERIVATIVES
70776-03-3 CHLORINATED NAPHTHALENE, N.O.S. [RCU]	NAPHTHALENE, CHLORO DERIVATIVES
70776-03-3 CHLORINATED NAPHTHALENES [CAL]	NAPHTHALENE, CHLORO DERIVATIVES
RR-00279-9 CHLORINATED NAPHTHALENES	CHLORINATED NAPHTHALENES
61788-76-9 CHLORINATED PARAFFINS (C10 - C13) [HM1]	CHLOROALKANES
108171-26-2 CHLORINATED PARAFFINS [C65]	CHLORINATED PARAFFINS: C12, 60% CHLORINE
RR-00278-8 CHLORINATED PARAFFINS	CHLORINATED PARAFFINS
108171-26-2 CHLORINATED PARAFFINS: C12, 60% CHLORINE	CHLORINATED PARAFFINS: C12, 60% CHLORINE
108171-26-2 CHLORINATED PARAFFINS (C12, 60% CHLORINE) [NTP]	CHLORINATED PARAFFINS: C12, 60% CHLORINE
108171-27-3 CHLORINATED PARAFFINS: C23, 43% CHLORINE	CHLORINATED PARAFFINS: C23, 43% CHLORINE
108171-26-2 CHLORINATED PARAFFINS OF AVERAGE CARBON-CHAIN LENGTH C12 AND AVERAGE DEGREE OF CHLORINATION 60% [IAR]	CHLORINATED PARAFFINS: C12, 60% CHLORINE
108171-26-2 CHLORINATED PARAFFINS [CAL]	CHLORINATED PARAFFINS: C12, 60% CHLORINE
RR-00204-0 CHLORINATED PHENOLS	CHLORINATED PHENOLS
64754-90-1 CHLORINATED POLYETHYLENE	CHLORINATED POLYETHYLENE
9006-03-5 CHLORINATED RUBBER	CHLORINATED RUBBER
RR-00534-5 ALPHA-CHLORINATED TOLUENES	ALPHA-CHLORINATED TOLUENES
56802-99-4 CHLORINATED TRISODIUM PHOSPHATE	CHLORINATED TRISODIUM PHOSPHATE
RR-01613-7 CHLORINATED WASTEWATER EFFLUENTS	CHLORINATED WASTEWATER EFFLUENTS
RR-01023-1 CHLORINATED WATER	CHLORINATED WATER
63449-39-8 CHLORINATED WAX [PST]	CHLORINATED HYDROCARBONS (CHORINATED PARAFFINS)
RR-00112-7 CHLORINATION/CHLORAMINATION BY PRODUCTS (MISC.)	CHLORINATION/CHLORAMINATION BY PRODUCTS (MISC.)
7782-50-5 CHLORINE	CHLORINE
13981-43-6 CHLORINE 36	CHLORINE 36
14158-34-0 CHLORINE 38	CHLORINE 38
15585-26-9 CHLORINE 39	CHLORINE 39
13973-88-1 CHLORINE AZIDE	CHLORINE AZIDE
506-77-4 CHLORINE CYANIDE [HM1, HM1, HM1, HM1, HM1]	CYANOGEN CHLORIDE

Chemical Name	Reference Name
10049-04-4 CHLORINE DIOXIDE	CHLORINE DIOXIDE
7790-93-4 CHLORINE DIOXIDE HYDRATE [NJL]	CHLORIC ACID
10049-04-4 CHLORINE DIOXIDE [CHLORINE OXIDE (CLO2)] [EP3]	CHLORINE DIOXIDE
7790-91-2 CHLORINE FLUORIDE (CLF3) [PAL]	CHLORINE TRIFLUORIDE
7791-21-1 CHLORINE MONOXIDE [FLA, MSL, NFP, NFP]	CHLORINE OXIDE (CL2O)
7791-21-1 CHLORINE MONOXIDE [CHLORINE OXIDE] [EP3]	CHLORINE OXIDE (CL2O)
7791-21-1 CHLORINE OXIDE (CL2O)	CHLORINE OXIDE (CL2O)
10049-04-4 CHLORINE OXIDE (CLO2) [PAL]	CHLORINE DIOXIDE-HYDROXY-6-METHOXY, (3AR-CIS)-
13637-63-3 CHLORINE PENTAFLUORIDE	CHLORINE PENTAFLUORIDE
7790-91-2 CHLORINE TRIFLUORIDE	CHLORINE TRIFLUORIDE
14998-27-7 CHLORITE	CHLORITE
RR-00205-1 CHLORITE, INORGANIC, N.O.S.	CHLORITE, INORGANIC, N.O.S.
RR-00205-1 CHLORITES, INORGANIC, N.O.S. [HM1, HM1]	CHLORITE, INORGANIC, N.O.S.
RR-00101-4 CHLORMADINONE ACETATE AND OESTRO-GENS	CHLORMADINONE ACETATE AND OESTROGENS
24934-91-6 CHLORMEPHOS	CHLORMEPHOS
999-81-5 CHLORMEQUAT CHLORIDE	CHLORMEQUAT CHLORIDE
494-03-1 CHLORNAPHAZINE [EP2, EPA, NJL, NJL, RCU, RCU]	N,N-BIS(2-CHLOROETHYL)-2-NAPHTHYLAMINE
107-20-0 CHLOROACETALDEHYDE	CHLOROACETALDEHYDE
79-07-2 2-CHLOROACETAMIDE	2-CHLOROACETAMIDE
532-28-5 CHLOROACETHOPHENONE (PHENACYL CHLO-RIDE) [QBC]	MANDELONITRILE
96-34-4 CHLOROACETIC ACID METHYL ESTER [MAK, MAK, MAK, MAK]	METHYL CHLOROACETATE
79-11-8 CHLOROACETIC ACID	CHLOROACETIC ACID
78-95-5 CHLOROACETONE	CHLOROACETONE
107-14-2 CHLOROACETONITRILE	CHLOROACETONITRILE
532-27-4 ALPHA-CHLOROACETOPHENONE	ALPHA-CHLOROACETOPHENONE
532-27-4 2-CHLOROACETOPHENONE [313]	ALPHA-CHLOROACETOPHENONEOPHENYL)-ALPHA,-HYDROXY-, ETHYL ESTER]
532-27-4 ALPHA-CHLOROACETOPHENONE (PHENACYL CHLORIDE) [ALB, ALB]	ALPHA-CHLOROACETOPHENONE
532-27-4 A-CHLOROACETOPHENONE [AUS]	ALPHA-CHLOROACETOPHENONE
532-27-4 2-CHLOROACETOPHENONE [CAA, CAB, GBR, HON, HON, NJL]	ALPHA-CHLOROACETOPHENONE
532-27-4 A-CHLOROACETOPHENONE [TSC, TSC, TSC]	ALPHA-CHLOROACETOPHENONE
532-27-4 CHLOROACETOPHENONE [WHM]	ALPHA-CHLOROACETOPHENONE
1341-24-8 CHLOROACETOPHENONE	CHLOROACETOPHENONE
1341-24-8 CHLOROACETO PHENONE [NFP, NFP]	CHLOROACETOPHENONE
532-27-4 A-CHLOROACETOPHENONE (PHENACYL CHLORIDE) [OS1, OS1]	ALPHA-CHLOROACETOPHENONE
532-27-4 2-CHLOROACETOPHENONE (CN) [CSR, CSR]	ALPHA-CHLOROACETOPHENONE
532-27-4 ALPHA-CHLOROACETOPHENONE (PHENACYL CHLORIDE) [BC1]	ALPHA-CHLOROACETOPHENONE
140-49-8 4-(CHLOROACETYL)ACETANILIDE	4-(CHLOROACETYL)ACETANILIDE
79-04-9 CHLOROACETYL CHLORIDE	CHLOROACETYL CHLORIDE
61788-76-9 CHLOROALKANES	CHLOROALKANES
RR-00206-2 CHLOROALKYL ETHERS	CHLOROALKYL ETHERS

Chemical Name	Reference Name
RR-00206-2 CHLOROALKYL ETHERS, N.O.S. [RCU]	CHLOROALKYL ETHERS
RR-01714-1 CHLOROALKYL PHOSPHATES	CHLOROALKYL PHOSPHATES
4080-31-3 1-(3-CHLOROALLYL)-3,5,7-TRIAZA-1-AZONI-AADAMANTANE CHLORIDE [313, PST]	3,5,7-TRIAZA-1-AZONIATRICYCLODECANE-1-(3-CHLORO-2-PROPENYL)-
615-66-7 3-CHLORO-4-AMINOANILINE	3-CHLORO-4-AMINOANILINE
73090-69-4 CHLORO-4-TERT-AMYLPHENOL	CHLORO-4-TERT-AMYLPHENOL
RR-00629-1 2-CHLORO-4-TERT-AMYLPHENYL METHYL ETHER	2-CHLORO-4-TERT-AMYLPHENYL METHYL ETHER
95-51-2 2-CHLOROANILINE	2-CHLOROANILINE
95-51-2 O-CHLOROANILINE [CSR]	2-CHLOROANILINE
106-47-8 P-CHLOROANILINE	P-CHLOROANILINE
106-47-8 PARA-CHLOROANILINE [IAR]	P-CHLOROANILINE
106-47-8 4-CHLOROANILINE [TSE]	P-CHLOROANILINE
108-42-9 M-CHLOROANILINE	M-CHLOROANILINE
20265-96-7 P-CHLOROANILINE HYDROCHLORIDE	P-CHLOROANILINE HYDROCHLORIDE
27134-26-5 CHLOROANILINES	CHLOROANILINES
93-50-5 P-CHLORO-O-ANISIDINE	P-CHLORO-O-ANISIDINE
RR-01347-8 CHLOROANISIDINES	CHLOROANISIDINES
104-88-1 P-CHLOROBENZALDEHYDE	P-CHLOROBENZALDEHYDE
2698-41-1 O-CHLOROBENZALMALONONITRILE (CS) [CSR, CSR]	O-CHLOROBENZYLIDENE MALONITRILE
108-90-7 CHLOROBENZENE	CHLOROBENZENE
108-90-7 CHLOROBENZENE (MONO) [QBC]	CHLOROBENZENE
108-90-7 CHLOROBENZENE (MONOCHLOROBENZENE) [ALB, ALB, BC1]	CHLOROBENZENE
95-83-0 4-CHLORO-1,2-BENZENEDIAMINE [FLA]	4-CHLORO-O-PHENYLENEDIAMINE
RR-01737-8 CHLOROBENZENES, N.O.S.	CHLOROBENZENES, N.O.S.
5138-90-9 P-CHLOROBENZENESULFONIC ACID, SODIUM SALT	P-CHLOROBENZENESULFONIC ACID, SODIUM SALT
510-15-6 CHLOROBENZILATE	CHLOROBENZILATE
510-15-6 CHLOROBENZILATE [BENZENEACETIC ACID, 4-CHLORO-ALPHA-(4-CHLOROPHENYL)-ALPHA,-HYDROXY-, ETHYL ESTER] [313]	CHLOROBENZILATE
118-91-2 O-CHLOROBENZOIC ACID	O-CHLOROBENZOIC ACID
5216-25-1 4-CHLOROBENZOTRICHLORIDE	4-CHLOROBENZOTRICHLORIDE
88-16-4 O-CHLOROBENZOTRIFLUORIDE [FLA, MSL, NFP, NFP]	BENZENE, 1-CHLORO-2-(TRIFLUOROMETHYL)-
52181-51-8 CHLOROBENZOTRIFLUORIDE	CHLOROBENZOTRIFLUORIDE
52181-51-8 CHLOROBENZOTRIFLUORIDES [HM1, HM1]	CHLOROBENZOTRIFLUORIDE
94-17-7 P-CHLOROBENZOYL PEROXIDE	P-CHLOROBENZOYL PEROXIDE
2698-41-1 O-CHLOROBENZYLIDENE MALONONITRILE [TLV, TLV]	O-CHLOROBENZYLIDENE MALONITRILE
104-83-6 P-CHLOROBENZYL CHLORIDE	P-CHLOROBENZYL CHLORIDE
104-83-6 PARA-CHLOROBENZYL CHLORIDE [HM1]	P-CHLOROBENZYL CHLORIDE
RR-01318-3 CHLOROBENZYLCHLORIDES	CHLOROBENZYLCHLORIDES
2698-41-1 O-CHLOROBENZYLIDENE MALONITRILE	O-CHLOROBENZYLIDENE MALONITRILE
2698-41-1 O-CHLOROBENZYLIDENE MALONONITRILE (OCBM) [FLA, MSL, WHM]	O-CHLOROBENZYLIDENE MALONITRILE
27323-18-8 CHLOROBIPHENYL	CHLOROBIPHENYL
92-04-6 3-CHLORO-4-BIPHENYLOL [FLA]	[1,1'-BIPHENYL]-4-OL, 3-CHLORO-
RR-01555-4 CHLOROBIPHENYLS	CHLOROBIPHENYLS

Chemical Name	Reference Name
RR-01719-6 2-CHLORO-4,6-BIS(SUBSTITUTED)-1,3,5-TRI-AZINE,DIHYDROCHLORIDE	2-CHLORO-4,6-BIS(SUBSTITUTED)-1,3,5-TRIAZINE,DIHY-DROCHLORIDE
107-04-0 1-CHLORO-2-BROMOETHANE [TSE]	1-BROMO-2-CHLOROETHANE
74-97-5 CHLOROBROMOMETHANE	CHLOROBROMOMETHANE
74-97-5 CHLOROBROMOMETHANE [QBC]	CHLOROBROMOMETHANE
109-70-6 1-CHLORO-3-BROMOPROPANE	1-CHLORO-3-BROMOPROPANE
126-99-8 2-CHLOROBUTA-1,3-DIENE [GBR, GBR]	2-CHLORO-1,3-BUTADIENE (CHLOROPRENE)
126-99-8 2-CHLORO-1,3-BUTADIENE [NFP, NFP, ONT, ONT, ONT, ONT, RCA]	2-CHLORO-1,3-BUTADIENE (CHLOROPRENE)
126-99-8 2-CHLORO-1,3-BUTADIENE (BETA-CHLORO-PRENE) [ALB, ALB, ALB]	2-CHLORO-1,3-BUTADIENE (CHLOROPRENE)
78-86-4 2-CHLOROBUTANE [FLA]	BUTANE, 2-CHLORO-
109-69-3 CHLOROBUTANE, 1- [OTS]	BUTYL CHLORIDE
25154-42-1 CHLOROBUTANE	CHLOROBUTANE
25154-42-1 CHLOROBUTANES [HM1, HM1]	CHLOROBUTANE
4461-41-0 2-CHLOROBUTENE-2	2-CHLOROBUTENE-2
59-50-7 4-CHLORO-M-CRESOL	4-CHLORO-M-CRESOL
59-50-7 PARA-CHLORO-META-CRESOL [CAL]	4-CHLORO-M-CRESOL
59-50-7 P-CHLORO-M-CRESOL [PQL, RCA, RCU, RCU, UTS]	4-CHLORO-M-CRESOL
1321-10-4 CHLOROCRESOL	CHLOROCRESOL
RR-00588-9 CHLORO-M-CRESOL, ALL ISOMERS	CHLORO-M-CRESOL, ALL ISOMERS
RR-00589-0 CHLORO-O-CRESOL, ALL ISOMERS	CHLORO-O-CRESOL, ALL ISOMERS
RR-00590-3 CHLORO-P-CRESOL, ALL ISOMERS	CHLORO-P-CRESOL, ALL ISOMERS
1321-10-4 CHLOROCRESOLS [HM1, HM1, NJL]	CHLOROCRESOL
13347-42-7 4-CHLORO-2-CYCLOPENTYLPHENOL	4-CHLORO-2-CYCLOPENTYLPHENOL
143-50-0 CHLORODECONE (KEPONE) [IAR]	KEPONE
42350-99-2 2-CHLORO-4,6-DI-TERT-AMYLPHENOL	2-CHLORO-4,6-DI-TERT-AMYLPHENOL
124-48-1 CHLORODIBROMOMETHANE	CHLORODIBROMOMETHANE
3380-34-5 5-CHLORO-2-(2,4-DICHLOROPHENOXY)PHENOL	5-CHLORO-2-(2,4-DICHLOROPHENOXY)PHENOL
96-10-6 CHLORODIETHYLALUMINUM [OS3]	DIETHYLALUMINUM CHLORIDE
RR-00775-0 3-CHLORO-4-DIETHYLAMINO-BENZENEDIAZO-NIUM ZINC CHLORIDE	3-CHLORO-4-DIETHYLAMINO-BENZENEDIAZONIUM ZINC CHLORIDECHLORIDE
1609-19-4 CHLORODIETHYLSILANE	CHLORODIETHYLSILANE
353-59-3 CHLORODIFLUOROBROMO-METHANE	CHLORODIFLUOROBROMO-METHANE
353-59-3 CHLORODIFLUOROBROMOMETHANE [NJL]	CHLORODIFLUOROBROMO-METHANE
75-68-3 1-CHLORO-1,1-DIFLUOROETHANE	1-CHLORO-1,1-DIFLUOROETHANE
75-68-3 1-CHLORO-1,1-DIFLUOROETHANE (HCFC-142B) [313]	1-CHLORO-1,1-DIFLUOROETHANE
75-68-3 CHLORODIFLUOROETHANE [NJL]	1-CHLORO-1,1-DIFLUOROETHANE
75-68-3 CHLORODIFLUOROETHANE (FC 142) [WHM]	1-CHLORO-1,1-DIFLUOROETHANE
27497-51-4 CHLORODIFLUOROETHANES	CHLORODIFLUOROETHANES
75-45-6 CHLORODIFLUOROMETHANE	CHLORODIFLUOROMETHANE
75-45-6 CHLORODIFLUOROMETHANE (FC 22) [WHM]	CHLORODIFLUOROMETHANE
75-45-6 CHLORODIFLUOROMETHANE (HCFC-22) [313]	CHLORODIFLUOROMETHANE
75-45-6 CHLORODIFLUOROMETHANE (FC-22) [CAL]	CHLORODIFLUOROMETHANE
75-45-6 CHLORODIFLUOROMETHANE (FREON 22) [QBC]	CHLORODIFLUOROMETHANE
134190-53-7 CHLORODIFLUOROPROPANE	CHLORODIFLUOROPROPANE
96-24-2 3-CHLORO-1,2-DIHYDROXYPROPANE	3-CHLORO-1,2-DIHYDROXYPROPANE

226

Chemical Name	Reference Name
97-50-7 5-CHLORO-2,4-DIMETHOXYANILINE	5-CHLORO-2,4-DIMETHOXYANILINE
88-04-0 4-CHLORO-3,5-DIMETHYLPHENOL [PST]	P-CHLORO-M-XYLENOL
97-00-7 1-CHLORO-2,4-DINITROBENZENE	1-CHLORO-2,4-DINITROBENZENE
RR-00587-8 CHLORODINITROBENZENE, ALL ISOMERS	CHLORODINITROBENZENE, ALL ISOMERS
RR-00587-8 CHLORODINITROBENZENES [HM1, HM1, HM1]	CHLORODINITROBENZENE, ALL ISOMERS
53469-21-9 CHLORODIPHENYL (42% CHLORINE) [FLA, TLV, TLV, MIN, MSL, NIO, NIO, NIO, NIO]	AROCLOR 1242
53469-21-9 CHLORODIPHENYL (42% CHLORINE) (PCB) [OS1, OS1, OS1, OS1]	AROCLOR 1242
53469-21-9 CHLORODIPHENYL (42% CHLORINE) PCB [WHM]	AROCLOR 1242
11097-69-1 CHLORODIPHENYL (54% CHLORINE) (PCB) [OS1, OS1, OS1, OS1]	CHLORODIPHENYL (54% CHLORINE)
53469-21-9 CHLORODIPHENYL 42% [MEX, MEX, MEX]	AROCLOR 1242
11097-69-1 CHLORODIPHENYL 54% [MEX, MEX, MEX]	CHLORODIPHENYL (54% CHLORINE)
53469-21-9 CHLORODIPHENYL (42% CHLORINE) [ALB, ALB, ALB, ALB, BC1, BC1, BC1, CEX, CEX]	AROCLOR 1242
53469-21-9 CHLORODIPHENYL (42% CL) [QBC, QBC, QBC]	AROCLOR 1242
11097-69-1 CHLORODIPHENYL (54% CHLORINE)	CHLORODIPHENYL (54% CHLORINE)
11097-69-1 CHLORODIPHENYL (54% CL) [QBC, QBC, QBC]	CHLORODIPHENYL (54% CHLORINE)
106-89-8 1-CHLORO-2,3-EPOXY-PROPANE (EPICHLORO-HYDRIN) [ALB, ALB, ALB]	EPICHLOROHYDRIN
106-89-8 1-CHLORO-2,3-EPOXYPROPANE [ONT, ONT]	EPICHLOROHYDRIN
106-89-8 N-CHLORO-2,3-EPOXYPROPANE [RCU, RCU]	EPICHLOROHYDRIN
106-89-8 1-CHLORO-2,3-EPOXYPROPANE (EPICHLORHY-DRIN) [QBC, QBC, QBC]	EPICHLOROHYDRIN
106-89-8 1-CHLORO-2,3-EPOXY PROPANE (EPICHLORO-HYDRIN) [BC1, BC1, BC1, BC1]	EPICHLOROHYDRIN
75-00-3 CHLOROETHANE [ATS, CAN, CSR, CSR, CW2, EPA, GBR, GBR, IAR, MAK, MX1, NHS, NHS, OTS, PST, RCA, SD4, SDW, TSE, UTS]	ETHYL CHLORIDE
75-00-3 CHLOROETHANE (ETHYL CHLORIDE) [313, C65]	ETHYL CHLORIDE
107-07-3 CHLOROETHANOL [302, EPA, EPA]	ETHYLENE CHLOROHYDRIN
107-07-3 2-CHLOROETHANOL [MAK, MAK, MAK, MAK, NFP, NFP, ONT, ONT]	ETHYLENE CHLOROHYDRIN
107-07-3 2-CHLOROETHANOL (ETHYLENE CHLOROHY-DRIN) [CSR, CSR, BC1, BC1, BC1]	ETHYLENE CHLOROHYDRIN
542-58-5 2-CHLOROETHANOL ACETATE [FLA]	ETHANOL, 2-CHLORO-, ACETATE
75-01-4 CHLOROETHENE [PST]	VINYL CHLORIDE
111-91-1 BIS(2-CHLOROETHOXY)METHANE [UTS, UTS]	BIS(2-CHLOROETHOXY)METHANE
542-58-5 2-CHLOROETHYL ACETATE [MSL, NFP, NFP]	ETHANOL, 2-CHLORO-, ACETATE
RR-00826-4 CHLOROETHYL ACETATE	CHLOROETHYL ACETATE
RR-00825-3 CHLORO-4-ETHYLBENZENE	CHLORO-4-ETHYLBENZENE
627-11-2 CHLOROETHYL CHLOROFORMATE	CHLOROETHYL CHLOROFORMATE
627-11-2 CHLOROETHYLCHLOROFORMATE [NJL]	CHLOROETHYL CHLOROFORMATE
13010-47-4 1-(2-CHLORO ETHYL)-3-CYCLOHEXYL-1-NI-TROSOUREA [MIN]	1-(2-CHLOROETHYL)-3-CYCLOHEXYL-1-NITROSOUREA
13010-47-4 1-(2-CHLOROETHYL)-3-CYCLOHEXYL-1-NI-TROSOUREA	1-(2-CHLOROETHYL)-3-CYCLOHEXYL-1-NITROSOUREA
13010-47-4 1-(2-CHLOROETHYL)-3-CYCLOHEXYL-1-NI-TROSOUREA (CCNU) (LOMUSTINE) [C65, C65]	1-(2-CHLOROETHYL)-3-CYCLOHEXYL-1-NITROSOUREA

Chemical Name	Reference Name
13010-47-4 1-(2-CHLOROETHYL)-3-CYCLOHEXYL-1-NI-TROSOUREA (CCNU) [IAR, NTP]	1-(2-CHLOROETHYL)-3-CYCLOHEXYL-1-NITROSOUREA
75-01-4 CHLOROETHYLENE (VINYL CHLORIDE) [BC1, BC1, QBC, QBC, QBC, ALB, ALB, ALB]	VINYL CHLORIDE
693-07-2 2-CHLOROETHYL ETHYL SULFIDE	2-CHLOROETHYL ETHYL SULFIDE
107-27-7 CHLOROETHYLMERCURY [FLA]	ETHYLMERCURIC CHLORIDE
13909-09-6 1-(2-CHLORO ETHYL)-3-(4-METHYLCYCLO-HEXYL)-1-NITROSOUREA [MIN]	1-(2-CHLOROETHYL)-3-(4-METHYLCYCLOHEXYL)-1-NITROSOUREA
13909-09-6 1-(2-CHLOROETHYL)-3-(4-METHYL-CYCLO-HEXYL)-1-NITROSOUREA (ME-CCNU) [NTP]	1-(2-CHLOROETHYL)-3-(4-METHYLCYCLOHEXYL)-1-NITROSOUREA
13909-09-6 1-(2-CHLOROETHYL)-3-(4-METHYLCYCLO-HEXYL)-1-NITROSOUREA	1-(2-CHLOROETHYL)-3-(4-METHYLCYCLOHEXYL)-1-NITROSOUREA
13909-09-6 1-(2-CHLOROETHYL)-3-(4-METHYLCYCLO-HEXYL)-1-NITROSOUREA (METHYL-CCNU) [C65, IAR, CAL]	1-(2-CHLOROETHYL)-3-(4-METHYLCYCLOHEXYL)-1-NITROSOUREA
999-81-5 2-CHLOROETHYLTRIMETHYLAMMONIUM CHLORIDE [CSR, CSR]	CHLORMEQUAT CHLORIDE
110-75-8 2-CHLOROETHYL VINYL ETHER	2-CHLOROETHYL VINYL ETHER
110-75-8 2-CHLOROETHYL VINYL ETHER (MIXED) [CW2]	2-CHLOROETHYL VINYL ETHER
110-75-8 2-CHLOROETHYL VINYL ETHER (MIXTURE) [MX1]	2-CHLOROETHYL VINYL ETHER
RR-00558-3 2-CHLOROETHYL-2-XENYL ETHER	2-CHLOROETHYL-2-XENYL ETHER
RR-01767-4 CHLOROFLUOROALKANES, FULLY HALO-GENATED	CHLOROFLUOROALKANES, FULLY HALOGENATED
348-51-6 O-CHLOROFLUOROBENZENE	O-CHLOROFLUOROBENZENE
RR-00078-2 CHLOROFLUOROCARBON	CHLOROFLUOROCARBON
593-70-4 CHLOROFLUOROMETHANE	CHLOROFLUOROMETHANE
350-30-1 3-CHLORO-4-FLUORONITROBENZENE	3-CHLORO-4-FLUORONITROBENZENE
134190-54-8 CHLOROFLUOROPROPANE	CHLOROFLUOROPROPANE
67-66-3 CHLOROFORM	CHLOROFORM
67-66-3 CHLOROFORM (TRICHLOROMETHANE) [BC1, BC1, CW2, QBC, QBC, OS1, OS1, ALB, ALB]	CHLOROFORM
RR-00207-3 CHLOROFORMATE COMPOUNDS, N.O.S. [WHM]	CHLOROFORMATE, N.O.S.
RR-00207-3 CHLOROFORMATE, N.O.S.	CHLOROFORMATE, N.O.S.
RR-00207-3 CHLOROFORMATES, N.O.S. [HM1, HM1, HM1]	CHLOROFORMATE, N.O.S.
15159-40-7 N-CHLOROFORMYL MORPHOLINE	N-CHLOROFORMYL MORPHOLINE
67-66-3 CHLOROFORM [METHANE, TRICHLORO-] [EP3]	CHLOROFORM
125426-39-3 2-CHLORO-1,1,1,2,3,3,3-HEPTAFLUORO-PROPANE	2-CHLORO-1,1,1,2,3,3,3-HEPTAFLUOROPROPANE
629-06-1 1-CHLOROHEPTANE	1-CHLOROHEPTANE
134308-72-8 CHLOROHEXAFLUOROPROPANE	CHLOROHEXAFLUOROPROPANE
544-10-5 1-CHLOROHEXANE	1-CHLOROHEXANE
18472-51-0 CHLOROHEXIDINE DIGLUCONATE	CHLOROHEXIDINE DIGLUCONATE
615-67-8 CHLOROHYDROQUINONE	CHLOROHYDROQUINONE
2832-19-1 2-CHLORO-N-HYDROXYMETHYLACETAMIDE	2-CHLORO-N-HYDROXYMETHYLACETAMIDE
90-03-9 CHLORO(O-HYDROXYPHENYL)MERCURY	CHLORO(O-HYDROXYPHENYL)MERCURY
13057-78-8 CHLOROISOCYANURIC ACID	CHLOROISOCYANURIC ACID
108-60-1 BIS(2-CHLOROISOPROPYL) ETHER [UTS]	BIS(2-CHLOROISOPROPYL) ETHER
74-87-3 CHLOROMETHANE [ATS, CAN, GBR, GBR, OTS, SD4, SDW]	METHYL CHLORIDE
74-87-3 CHLOROMETHANE (METHYL CHLORIDE) [313, UTS]	METHYL CHLORIDE
6806-86-6 CHLOROMETHYL [FLA]	CHLOROMETHYL (OXIRANE)

Chemical Name	Reference Name
6806-86-6 CHLOROMETHYL (OXIRANE)	CHLOROMETHYL (OXIRANE)
87-60-5 3-CHLORO-2-METHYL ANILINE	3-CHLORO-2-METHYL ANILINE
87-63-8 6-CHLORO-2-METHYL ANILINE	6-CHLORO-2-METHYL ANILINE
95-74-9 3-CHLORO-4-METHYL ANILINE [NJL]	3-CHLORO-P-TOLUIDINE
95-79-4 5-CHLORO-2-METHYL ANILINE [NJL]	5-CHLORO-O-TOLUIDINE
95-81-8 2-CHLORO-5-METHYL ANILINE	2-CHLORO-5-METHYL ANILINE
615-65-6 2-CHLORO-4-METHYL ANILINE	2-CHLORO-4-METHYL ANILINE
7149-75-9 4-CHLORO-3-METHYL ANILINE	4-CHLORO-3-METHYL ANILINE
29027-17-6 2-CHLORO-3-METHYL ANILINE	2-CHLORO-3-METHYL ANILINE
95-49-8 1-CHLORO-2-METHYL BENZENE [FLA]	O-CHLOROTOLUENE
22128-62-7 CHLOROMETHYL CHLOROFORMATE [WHM]	CHLOROMETHYLCHLOROFORMATE
22128-62-7 CHLOROMETHYLCHLOROFORMATE	CHLOROMETHYLCHLOROFORMATE
26172-55-4 5-CHLORO-2-METHYL-2,3-DIHYDROISOTHIA-ZOL-3-ONE [MAK, MAK, MAK]	5-CHLORO-2-METHYL-3-ISOTHIAZOLONE
542-88-1 BIS(CHLOROMETHYL) ETHER	BIS(CHLOROMETHYL) ETHER
542-88-1 CHLOROMETHYL ETHER [302]	BIS(CHLOROMETHYL) ETHER
542-88-1 CHLOROMETHYL BIS ETHER [QBC, QBC]	BIS(CHLOROMETHYL) ETHER
542-88-1 CHLOROMETHYL ETHER [METHANE, OXYBIS [CHLORO-] [EP3]	BIS(CHLOROMETHYL) ETHER
76649-15-5 2-CHLORO-1-METHYLETHYL BIS(2-CHLORO-PROPYL) PHOSPHATE	2-CHLORO-1-METHYLETHYL BIS(2-CHLOROPROPYL) PHOSPHATESSIUM SALT
3188-13-4 CHLOROMETHYL ETHYL ETHER	CHLOROMETHYL ETHYL ETHER
26172-55-4 5-CHLORO-2-METHYL-3-ISOTHIAZOLONE	5-CHLORO-2-METHYL-3-ISOTHIAZOLONE
107-30-2 CHLOROMETHYL METHYL ETHER [METHANE, CHLOROMETHOXY-] [EP3]	CHLOROMETHYL METHYL ETHER
107-30-2 CHLOROMETHYL METHYL ETHER	CHLOROMETHYL METHYL ETHER
107-30-2 CHLOROMETHYL METHYL ETHER (TECHNI-CAL GRADE) [C65, C65, EP2, IAR]	CHLOROMETHYL METHYL ETHER
42874-96-4 2-CHLORO-1-(3-METHYLPHENOXY)-4-(TRIFLU-OROMETHYL)BENZENE	2-CHLORO-1-(3-METHYLPHENOXY)-4-(TRIFLUO-ROMETHYL)BENZENE) PHOSPHATE
28479-22-3 3-CHLORO-4-METHYL PHENYL ISO-CYANATE [NJL]	3-CHLORO-4-METHYLPHENYL ISOCYANATE
15545-48-9 N'-(3-CHLORO-4-METHYLPHENYL)-N,N-DIMETHYLUREA	N'-(3-CHLORO-4-METHYLPHENYL)-N,N-DIMETHYLUREA
28479-22-3 3-CHLORO-4-METHYLPHENYL ISOCYANATE	3-CHLORO-4-METHYLPHENYL ISOCYANATE
563-47-3 3-CHLORO-2-METHYLPROPENE	3-CHLORO-2-METHYLPROPENE
563-47-3 3-CHLORO-2-METHYL-1-PROPENE [313]	3-CHLORO-2-METHYLPROPENE
6959-47-3 2-CHLOROMETHYLPYRIDINE HYDROCHLO-RIDE	2-CHLOROMETHYLPYRIDINE HYDROCHLORIDE
6959-48-4 3-CHLOROMETHYLPYRIDINE HYDROCHLO-RIDE	3-CHLOROMETHYLPYRIDINE HYDROCHLORIDE
6959-48-4 3-(CHLOROMETHYL)PYRIDINE HYDROCHLO-RIDE [MSL]	3-CHLOROMETHYLPYRIDINE HYDROCHLORIDE
RR-00919-8 2-CHLORO-N-METHYL-N-SUBSTITUTED AC-ETAMIDE	2-CHLORO-N-METHYL-N-SUBSTITUTED ACETAMIDE-TRIAZINE, REACTION PRODUCTS WITH N-BUTYL-2,2,6,6-TETRAMETHYL-4-PIPERIDINAMINE
90-13-1 1-CHLORONAPHTHALENE [NFP, NFP]	NAPHTHALENE, 1-CHLORO-
91-58-7 BETA-CHLORONAPHTHALENE	BETA-CHLORONAPHTHALENE
91-58-7 2-CHLORONAPHTHALENE [CAL, CW2, NJL, PQL, RCA, UTS]	BETA-CHLORONAPHTHALENE
25586-43-0 CHLORONAPHTHATLENE [HON]	NAPHTHALENE, CHLORO-
25586-43-0 CHLORONAPHTHALENE [MX1]	NAPHTHALENE, CHLORO-

Chemical Name	Reference Name
25586-43-0 CHLORONAPHTHALENES [MX2]	NAPHTHALENE, CHLORO-
89-63-4 4-CHLORO-2-NITROANILINE [CSR]	BENZENAMINE, 4-CHLORO-2-NITRO-
41587-36-4 CHLORONITROANILINE	CHLORONITROANILINE
41587-36-4 CHLORONITROANILINE, ALL ISOMERS [WHM]	CHLORONITROANILINE
41587-36-4 CHLORONITROANILINES [HM1, HM1, HM1]	CHLORONITROANILINE
88-73-3 2-CHLORONITROBENZENE [CSR]	O-NITROCHLOROBENZENE
88-73-3 1-CHLORO-2-NITROBENZENE [FLA, MAK, MAK]	O-NITROCHLOROBENZENE
88-73-3 O-CHLORONITROBENZENE [MSL]	O-NITROCHLOROBENZENE
100-00-5 4-CHLORONITROBENZENE [CSR]	P-NITROCHLOROBENZENE
100-00-5 1-CHLORO-4-NITROBENZENE [GBR, GBR, GBR]	P-NITROCHLOROBENZENE
100-00-5 P-CHLORONITROBENZENE [HON]	P-NITROCHLOROBENZENE
100-00-5 1-CHLORO-4-NITROBENZENE [MAK, MAK, MAK, MAK]	P-NITROCHLOROBENZENE
121-73-3 M-CHLORONITROBENZENE [FLA, HON, MSL]	M-NITROCHLOROBENZENE
25167-93-5 CHLORONITROBENZENE	CHLORONITROBENZENE
100-00-5 P-CHLORONITROBENZENE [ONT]	P-NITROCHLOROBENZENE
99-60-5 2-CHLORO-4-NITROBENZOIC ACID	2-CHLORO-4-NITROBENZOIC ACID
777-37-7 2-CHLORO-5-NITROBENZOTRIFLUORIDE [MSL, FLA, NFP, NFP]	4-NITRO-1-CHLOROBENZO-2-TRIFLUORIDE
598-92-5 1-CHLORO-1-NITROETHANE	1-CHLORO-1-NITROETHANE
89-64-5 4-CHLORO-2-NITROPHENOL	4-CHLORO-2-NITROPHENOL
594-71-8 2-CHLORO-2-NITROPROPANE	2-CHLORO-2-NITROPROPANE
600-25-9 1-CHLORO-1-NITROPROPANE	1-CHLORO-1-NITROPROPANE
89-60-1 4-CHLORO-3-NITROTOLUENE	4-CHLORO-3-NITROTOLUENE
25567-68-4 CHLORONITROTOLUENE	CHLORONITROTOLUENE
25567-68-4 CHLORONITROTOLUENES [HM1, HM1, HM1]	CHLORONITROTOLUENE
111-85-3 1-CHLOROOCTANE [HM1]	OCTYL CHLORIDE
87-60-5 P-CHLORO-O-TOLUIDINE [CAB]	3-CHLORO-2-METHYL ANILINE
63449-39-8 CHLOROPARRAFFINS [MAK]	CHLORINATED HYDROCARBONS (CHORINATED PARAFFINS)
76-15-3 CHLOROPENTAFLUOROETHANE	CHLOROPENTAFLUOROETHANE
76-15-3 CHLOROPENTAFLUOROETHANE (FREON 115) [QBC]	CHLOROPENTAFLUOROETHANE
76-15-3 CHLOROPENTAFLUOROETHANE (FC 115) [WHM]	CHLOROPENTAFLUOROETHANE
134237-41-5 CHLOROPENTAFLUOROPROPANE	CHLOROPENTAFLUOROPROPANE
543-59-9 1-CHLOROPENTANE [FLA, MSL]	AMYL CHLORIDE
937-14-4 3-CHLOROPEROXYBENZOIC ACID	3-CHLOROPEROXYBENZOIC ACID
3691-35-8 CHLOROPHACINONE	CHLOROPHACINONE
3691-35-8 CHLOROPHACINONE (ROZOL) [EPA, EPA]	CHLOROPHACINONE
RR-01276-0 CHLOROPHENATES	CHLOROPHENATES
RR-01276-0 CHLOROPHENATES, N.O.S. [HM1, HM1, HM1]	CHLOROPHENATES
RR-00816-2 BETA-CHLOROPHENETOLE	BETA-CHLOROPHENETOLE
95-57-8 2-CHLOROPHENOL	2-CHLOROPHENOL
95-57-8 O-CHLOROPHENOL [HON, MSL, NFP, NFP]	2-CHLOROPHENOL
95-57-8 2-CHLOROPHENOL [RCA]	2-CHLOROPHENOL
95-57-8 O-CHLOROPHENOL [RCU, RCU, TSC, WHM]	2-CHLOROPHENOL
106-48-9 P-CHLOROPHENOL	P-CHLOROPHENOL
106-48-9 4-CHLOROPHENOL [NJL, NJL]	P-CHLOROPHENOL

Chemical Name	Reference Name
108-43-0 M-CHLOROPHENOL	M-CHLOROPHENOL
108-43-0 3-CHLOROPHENOL [NJL, NJL]	M-CHLOROPHENOL
26982-03-6 CHLOROPHENOL (MIXED ISOMERS)	CHLOROPHENOL (MIXED ISOMERS)
26982-03-6 CHLOROPHENOLS [313, CAB, CAL, HM1, HM1, HM1, IAR, PAL, PAL]	CHLOROPHENOL (MIXED ISOMERS)
26982-03-6 CHLOROPHENOLS, N.O.S. [NJL]	CHLOROPHENOL (MIXED ISOMERS)
122-88-3 P-CHLOROPHENOXYACETIC ACID	P-CHLOROPHENOXYACETIC ACID
RR-00499-9 CHLOROPHENOXY DERIVATIVE ACIDS, ESTER, ETHERS, AMINES AND OTHER SALTS	CHLOROPHENOXY DERIVATIVE ACIDS, ESTER, ETHERS, AMINES AND OT
RR-00059-9 CHLOROPHENOXY HERBICIDES	CHLOROPHENOXY HERBICIDES
7203-90-9 1-(4-CHLOROPHENYL)-3,3-DIMETHYL TRIAZINE	1-(4-CHLOROPHENYL)-3,3-DIMETHYL TRIAZINE
95-83-0 4-CHLORO-O-PHENYLENEDIAMINE	4-CHLORO-O-PHENYLENEDIAMINE
95-83-0 4-CHLORO-ORTHO-PHENYLENEDIAMINE [C65, C65]	4-CHLORO-O-PHENYLENEDIAMINE
5131-60-2 4-CHLORO-M-PHENYLENEDIAMINE	4-CHLORO-M-PHENYLENEDIAMINE
61702-44-1 2-CHLORO-P-PHENYLENEDIAMINE SULFATE	2-CHLORO-P-PHENYLENEDIAMINE SULFATE
532-27-4 2-CHLORO-1-PHENYLETHANONE [ONT]	ALPHA-CHLOROACETOPHENONE
104-12-1 P-CHLOROPHENYL ISOCYANATE [313, HM1]	BENZENE, 1-CHLORO-4-ISOCYANATO-
92-04-6 2-CHLORO-4-PHENYLPHENOL [MSL, NFP, NFP, TSC]	[1,1'-BIPHENYL]-4-OL, 3-CHLORO-
7005-72-3 4-CHLOROPHENYL PHENYL ETHER	4-CHLOROPHENYL PHENYL ETHERPROPENYL ESTER, (E)-
5344-82-1 1-(O-CHLOROPHENYL)THIOUREA [RCU, RCU]	THIOUREA, (2-CHLOROPHENYL)-
26571-79-9 CHLOROPHENYL TRICHLOROSILANE [NJL]	CHLOROPHENYLTRICHLOROSILANE
26571-79-9 CHLOROPHENYLTRICHLOROSILANE	CHLOROPHENYLTRICHLOROSILANE
52-68-6 CHLOROPHOS [MSL]	TRICHLORFON
479-61-8 CHLOROPHYLL	CHLOROPHYLL
76-06-2 CHLOROPICRIN	CHLOROPICRIN
8004-09-9 CHLOROPICRIN AND METHYL BROMIDE MIXTURE [OS3]	CHLOROPICRIN/METHYL BROMIDE
RR-00806-0 CHLOROPICRIN AND METHYL CHLORIDE MIXTURE [OS3]	CHLOROPICRIN/METHYL CHLORIDE
76-06-2 CHLOROPICRINE (NITROCHLOROMETHANE) [QBC]	CHLOROPICRIN
8004-09-9 CHLOROPICRIN/METHYL BROMIDE	CHLOROPICRIN/METHYL BROMIDE
RR-00806-0 CHLOROPICRIN/METHYL CHLORIDE	CHLOROPICRIN/METHYL CHLORIDERTZ FRACTION
4300-97-4 CHLOROPIVALOYL CHLORIDE	CHLOROPIVALOYL CHLORIDE
16941-12-1 CHLOROPLATINIC ACID	CHLOROPLATINIC ACID
126-99-8 2-CHLORO-1,3-BUTADIENE (CHLOROPRENE)	2-CHLORO-1,3-BUTADIENE (CHLOROPRENE)
126-99-8 CHLOROPRENE [313]	2-CHLORO-1,3-BUTADIENE (CHLOROPRENE)
126-99-8 B-CHLOROPRENE [AUS, AUS]	2-CHLORO-1,3-BUTADIENE (CHLOROPRENE)
126-99-8 BETA-CHLOROPRENE [MAK, MAK, MAK, MAK, MEX, MEX, MIN, NIO, NIO, NIO, NIO, OS1, OS1, OS1, OS1, TLV, TLV, BC1, BC1, BC1, OTV, T32]	2-CHLORO-1,3-BUTADIENE (CHLOROPRENE)
126-99-8 BETA-CHLOROPRENE (2-CHLORO-1,3-BUTADIENE) [QBC, QBC]	2-CHLORO-1,3-BUTADIENE (CHLOROPRENE)
75-29-6 2-CHLOROPROPANE	2-CHLOROPROPANE
78-89-7 2-CHLORO-1-PROPANOL [FLA, MSL, NFP, NFP]	PROPYLENE CHLOROHYDRIN
127-00-4 1-CHLORO-2-PROPANOL	1-CHLORO-2-PROPANOL
627-30-5 3-CHLORO-1-PROPANOL	3-CHLORO-1-PROPANOL
627-30-5 3-CHLOROPROPANOL-1 [HM1, HM1]	3-CHLORO-1-PROPANOL

Chemical Name	Reference Name
627-30-5 3-CHLOROPROPANOL [NJL]	3-CHLORO-1-PROPANOL
557-98-2 2-CHLOROPROPENE	2-CHLOROPROPENE
101-21-3 CHLOROPROPHAM [IAR]	CIPC (ISOPROPYL N-(3-CHLOROPHENYL) CARBAMATE)
598-78-7 2-CHLOROPROPIONIC ACID	2-CHLOROPROPIONIC ACID
598-78-7 ALPHA-CHLOROPROPIONIC ACID [HM1, HM1, NFP, NFP]	2-CHLOROPROPIONIC ACID
28554-00-9 CHLOROPROPIONIC ACID	CHLOROPROPIONIC ACID
542-76-7 3-CHLOROPROPIONITRILE	3-CHLOROPROPIONITRILE
7623-09-8 2-CHLOROPROPIONYL CHLORIDE	2-CHLOROPROPIONYL CHLORIDE
71808-64-5 3-(CHLOROPROPYL)DIMETHOXY-[3-(OXI-RANYLMETHOXY)PROPYL] SILANE [EPA]	SILANE, [(3-CHLOROPROPYL)DIMETHOXY[3-(OXI-RANYLMETHOXY)PROPYL
107-05-1 3-CHLOROPROPYLENE [UTS]	ALLYL CHLORIDE
557-98-2 2-CHLOROPROPYLENE [HM1, HM1, HM1, MSL]	2-CHLOROPROPENE
557-98-2 2-CHLORO PROPYLENE [NFP, NFP]	2-CHLOROPROPENE
590-21-6 1-CHLOROPROPYLENE	1-CHLOROPROPYLENE
557-98-2 2-CHLOROPROPYLENE [FLA]	2-CHLOROPROPENE
RR-01277-1 ALPHA-CHLOROPROPYLENE	ALPHA-CHLOROPROPYLENE
557-98-2 2-CHLOROPROPYLENE [1-PROPENE, 2-CHLORO-] [EP3]	2-CHLOROPROPENE
590-21-6 1-CHLOROPROPYLENE [1-PROPENE, 1-CHLORO-] [EP3]	1-CHLOROPROPYLENE
109-09-1 2-CHLOROPYRIDINE	2-CHLOROPYRIDINE
2921-88-2 CHLOROPYRIFOS [MIN]	CHLORPYRIFOS
54-05-7 CHLOROQUINE	CHLOROQUINE
95-88-5 4-CHLORORESORCINOL	4-CHLORORESORCINOL
RR-00208-4 CHLOROSILANES, N.O.S.	CHLOROSILANES, N.O.S.
1331-28-8 CHLOROSTYRENE	CHLOROSTYRENE
1331-28-8 0-CHLOROSTYRENE [ALB, ALB]	CHLOROSTYRENE
1331-28-8 O-CHLOROSTYRENE [CAL, CEX, CEX, MEX, MEX, QBC, QBC]	CHLOROSTYRENE
2039-87-4 O-CHLOROSTYRENE	O-CHLOROSTYRENE
7790-94-5 CHLOROSULFONIC ACID	CHLOROSULFONIC ACID
7790-94-5 CHLOROSULFURIC ACID [FLA, MSL, PAL]	CHLOROSULFONIC ACID
7790-94-5 CHLOROSULPHONIC ACID [GBR]	CHLOROSULFONIC ACID
354-25-6 1-CHLORO-1,1,2,2-TETRAFLUOROETHANE (HCFC-124A)	1-CHLORO-1,1,2,2-TETRAFLUOROETHANE (HCFC-124A)
2837-89-0 2-CHLORO-1,1,1,2-TETRAFLUOROETHANE (HCFC-124) [313]	ETHANE, 2-CHLORO-1,1,1,2-TETRAFLUORO-
63938-10-3 CHLOROTETRAFLUOROETHANE	CHLOROTETRAFLUOROETHANE
679-85-6 3-CHLORO-1,1,2,2-TETRAFLUOROPROPANE	3-CHLORO-1,1,2,2-TETRAFLUOROPROPANE
1897-45-6 CHLOROTHALONIL	CHLOROTHALONIL
1897-45-6 CHLOROTHALONIL [1,3-BENZENEDICARBONI-TRILE,2,4,5,6-TETRACHLORO-] [313]	CHLOROTHALONILBENZENAMINE
106-54-7 P-CHLOROTHIOPHENOL	P-CHLOROTHIOPHENOL
94-74-6 4-CHLORO-O-TOLOXY ACETIC ACID	4-CHLORO-O-TOLOXY ACETIC ACID
95-49-8 O-CHLOROTOLUENE	O-CHLOROTOLUENE
95-49-8 2-CHLOROTOLUENE [GBR, PST]	O-CHLOROTOLUENE
106-43-4 P-CHLOROTOLUENE	P-CHLOROTOLUENE
108-41-8 M-CHLOROTOLUENE	M-CHLOROTOLUENE
25168-05-2 CHLOROTOLUENE [ATS, NFP, NFP]	CHLOROTOLUENES
25168-05-2 CHLOROTOLUENES	CHLOROTOLUENES

Chemical Name	Reference Name
87-60-5 PARA-CHLORO-ORTHO-TOLUIDINE [MSL]	3-CHLORO-2-METHYL ANILINE
95-69-2 4-CHLORO-O-TOLUIDINE	4-CHLORO-O-TOLUIDINE
95-69-2 P-CHLORO-O-TOLUIDINE [313, C65, C65, CAL]	4-CHLORO-O-TOLUIDINE
95-69-2 PARA-CHLORO-ORTHO-TOLUIDINE [IAR, MIN]	4-CHLORO-O-TOLUIDINE
95-69-2 P-CHLORO-O-TOLUIDINE [NTP]	4-CHLORO-O-TOLUIDINE
95-74-9 3-CHLORO-P-TOLUIDINE	3-CHLORO-P-TOLUIDINE
95-79-4 5-CHLORO-O-TOLUIDINE	5-CHLORO-O-TOLUIDINE
RR-00592-5 CHLORO-M-TOLUIDINE, ALL ISOMERS	CHLORO-M-TOLUIDINE, ALL ISOMERS
RR-00593-6 CHLORO-O-TOLUIDINE, ALL ISOMERS, N.O.S.	CHLORO-O-TOLUIDINE, ALL ISOMERS, N.O.S.
RR-00594-7 CHLORO-P-TOLUIDINE, ALL ISOMERS	CHLORO-P-TOLUIDINE, ALL ISOMERS
RR-00593-6 PARA-CHLORO-ORTHO-TOLUIDINE AND ITS STRONG ACID SALTS [IAR]	CHLORO-O-TOLUIDINE, ALL ISOMERS, N.O.S.
7790-94-5 PARA-CHLORO-ORTHO-TOLUIDINE HY-DROCHLORIDE [IAR]	CHLOROSULFONIC ACID
3165-93-3 4-CHLORO-O-TOLUIDINE, HYDROCHLO-RIDE [EP2]	BENZENAMINE, 4-CHLORO-2-METHYL-, HYDROCHLO-RIDE
3165-93-3 4-CHLORO-O-TOLUIDINE HYDROCHLO-RIDE [HM1, HM1, HM1, MSL, NJL, WHM]	BENZENAMINE, 4-CHLORO-2-METHYL-, HYDROCHLO-RIDE
7790-94-5 4-CHLORO-O-TOLUIDINE, HYDROCHLO-RIDE [EPA]	CHLOROSULFONIC ACID
7745-89-3 3-CHLORO-P-TOLUIDINE HYDROCHLO-RIDE [CAL]	STARLICIDE
3165-93-3 4-CHLORO-O-TOLUIDINE HYDROCHLO-RIDE [CSR, CSR]	BENZENAMINE, 4-CHLORO-2-METHYL-, HYDROCHLO-RIDE
RR-00115-0 CHLOROTOLUIDINES	CHLOROTOLUIDINES
569-57-3 CHLOROTRIANISENE	CHLOROTRIANISENE
1929-82-4 2-CHLORO-6-(TRICHLOROMETHYL) PYRIDINE (N-SERVE) [ALB, ALB, BC1, BC1]	NITRAPYRIN
1929-82-4 2-CHLORO-6-(TRICHLOROMETHYL)PYRI-DINE [CAL, GBR, GBR, ONT, ONT]	NITRAPYRIN
1929-82-4 2-CHLORO-6-TRICHLORO-METHYL PYRI-DINE [OS1, OS1]	NITRAPYRIN
75-88-7 2-CHLORO-1,1,1-TRIFLUOROETHANE (HCFC-133A) [313]	2-CHLORO-1,1,1-TRIFLUOROETHANE
75-88-7 2-CHLORO-1,1,1-TRIFLUOROETHANE	2-CHLORO-1,1,1-TRIFLUOROETHANE
75-88-7 CHLOROTRIFLUOROETHANE [HM1, HM1]	2-CHLORO-1,1,1-TRIFLUOROETHANE
1330-45-6 CHLOROTRIFLUOROETHANE	CHLOROTRIFLUOROETHANE
79-38-9 CHLOROTRIFLUOROETHYLENE	CHLOROTRIFLUOROETHYLENE
75-72-9 CHLOROTRIFLUOROMETHANE	CHLOROTRIFLUOROMETHANE
75-72-9 CHLOROTRIFLUOROMETHANE (CFC-13) [313]	CHLOROTRIFLUOROMETHANE
460-35-5 3-CHLORO-1,1,1-TRIFLUORO-PROPANE (HCFC-253FB)	3-CHLORO-1,1,1-TRIFLUORO-PROPANE (HCFC-253FB)
134237-44-8 CHLOROTRIFLUOROPROPANE	CHLOROTRIFLUOROPROPANE
RR-01167-6 CHLOROTRIFLUOROPYRIDINE	CHLOROTRIFLUOROPYRIDINE
98-56-6 P-CHLORO-A,A,A-TRIFLUOROTOLUENE	P-CHLORO-A,A,A-TRIFLUOROTOLUENE
7758-19-2 CHLOROUS ACID, SODIUM SALT [PAL]	SODIUM CHLORITE
1982-47-4 CHLOROXURON	CHLOROXURON
88-04-0 P-CHLORO-M-XYLENOL	P-CHLORO-M-XYLENOL
54749-90-5 CHLOROZOTOCIN	CHLOROZOTOCIN
113-92-8 CHLORPHENIRAMINE MALEATE	CHLORPHENIRAMINE MALEATE
50-53-3 CHLORPROMAZINE (2-CHLORO-10-(3-DIMETHYLAMINOPROPYL) PHENOTHIAZINE)	CHLORPROMAZINE (2-CHLORO-10-(3-DIMETHY-LAMINOPROPYL) PHENOTHI

Chemical Name	Reference Name
94-20-2 **CHLORPROPAMIDE**	CHLORPROPAMIDE
2921-88-2 **CHLORPYRIFOS**	CHLORPYRIFOS
2921-88-2 CHLORPYRIFOS (DURSBAN) [BC1, BC1, BC1]	CHLORPYRIFOS
2921-88-2 CHLORPYRIFOS-ETHYL (DURSBAN) [QBC, QBC]	CHLORPYRIFOS
5598-13-0 **CHLORPYRIFOS METHYL [O,O-DIMETHYL-O-(3,5,6-TRICHLORO-2-PYRIDYL)PHOSPHOROTH-IOATE**	CHLORPYRIFOS METHYL [O,O-DIMETHYL-O-(3,5,6-TRICHLORO-2-PYRID
64902-72-3 **CHLORSULFURON [2-CHLORO-N-[[4-METHOXY-6-METHYL-1,3,5-TRIAZIN-2-YL) AMINO]CAR-BONYL]BENZENESULFONAMIDE**	CHLORSULFURON [2-CHLORO-N-[[4-METHOXY-6-METHYL-1,3,5-TRIAZINNOXY)-2-NITRO-BENZOIC ACID, SODIUM SALT]
500-28-7 **CHLORTHION**	CHLORTHION
21923-23-9 **CHLORTHIOPHOS**	CHLORTHIOPHOS
57-88-5 **CHOLESTEROL**	CHOLESTEROL
62-49-7 **CHOLINE**	CHOLINE
67-48-1 **CHOLINE CHLORIDE**	CHOLINE CHLORIDE
RR-01554-3 **CHROMATES**	CHROMATES
RR-01554-3 CHROMATES (WATER INSOLUBLE CHROMIUM (IV) COMPOUNDS [ALB]	CHROMATES
1308-38-9 CHROME OXIDE (CR2O3) [PST]	CHROMIUM (III) OXIDE
1066-30-4 **CHROMIC ACETATE**	CHROMIC ACETATE
11115-74-5 **CHROMIC ACID**	CHROMIC ACID
13530-68-2 **CHROMIC ACID (H2CR2O7)**	CHROMIC ACID (H2CR2O7)
7738-94-5 CHROMIC (VI) ACID (H2CRO4) [OS1, OS1]	CHROMIC ACID
7738-94-5 CHROMIC ACID (H2CRO4) [WHM]	CHROMIC ACID
7738-94-5 CHROMIC ACID AND CHROMATES [BC1, NIO, NIO, NIO, NIO]	CHROMIC ACID
1189-85-1 CHROMIC ACID (H2CRO4), BIS(1,1-DIMETHYLETHYL) ESTER [PAL]	TERT-BUTYL CHROMATE
13765-19-0 CHROMIC ACID (H2CRO4), CALCIUM SALT (1:1) [PAL, PAL]	CALCIUM CHROMATE
14307-33-6 **CHROMIC ACID (H2CR2O7), CALCIUM SALT (1:1)**	CHROMIC ACID (H2CR2O7), CALCIUM SALT (1:1)
7788-98-9 CHROMIC ACID (H2CRO4), DIAMMONIUM SALT [PAL]	AMMONIUM CHROMATE
7789-09-5 CHROMIC ACID (H2CR2O7), DIAMMONIUM SALT [PAL]	AMMONIUM BICHROMATE
14307-35-8 CHROMIC ACID (H2CRO4), DILITHIUM SALT [PAL]	LITHIUM CHROMATE
7778-50-9 CHROMIC ACID (H2CR2O7), DIPOTASSIUM SALT [PAL]	POTASSIUM DICHROMATE
7789-00-6 CHROMIC ACID (H2CRO4), DIPOTASSIUM SALT [PAL]	POTASSIUM CHROMATE
7775-11-3 CHROMIC ACID (H2CRO4), DISODIUM SALT [PAL]	SODIUM CHROMATE
10588-01-9 CHROMIC ACID (H2CR2O7), DISODIUM SALT [PAL]	SODIUM DICHROMATE
11115-74-5 CHROMIC ACID ESTER [EPA]	CHROMIC ACID
7758-97-6 CHROMIC ACID (H2CRO4), LEAD(2+) SALT (1:1) [PAL, PAL]	LEAD CHROMATE
13423-61-5 **CHROMIC ACID (H2CRO4), MAGNESIUM SALT (1:1)**	CHROMIC ACID (H2CRO4), MAGNESIUM SALT (1:1)
7789-06-2 CHROMIC ACID (H2CRO4), STRONTIUM SALT (1:1) [PAL, PAL]	STRONTIUM CHROMATE

Chemical Name	Reference Name
13530-65-9 CHROMIC ACID (H2CRO4), ZINC SALT (1:1) [PAL, PAL]	ZINC CHROMATE
14018-95-2 CHROMIC ACID (H2CR2O7), ZINC SALT (1:1) [PAL]	ZINC DICHROMATE
10025-73-7 CHROMIC CHLORIDE	CHROMIC CHLORIDE
10025-73-7 CHROMIC (III) CHLORIDE [MSL]	CHROMIC CHLORIDE
10049-05-5 CHROMIC(II) CHLORIDE	CHROMIC(II) CHLORIDE
7788-97-8 CHROMIC FLUORIDE	CHROMIC FLUORIDE
10101-53-8 CHROMIC SULFATE	CHROMIC SULFATE
1308-31-2 CHROMITE	CHROMITE
1308-31-2 CHROMITE ORE [MSL]	CHROMITE
RR-00070-4 CHROMITE ORE PROCESSING	CHROMITE ORE PROCESSING
RR-00070-4 CHROMITE ORE PROCESSING (CHROMATE) [CEX]	CHROMITE ORE PROCESSING
RR-00070-4 CHROMITE ORE PROCESSING (CHROMATES) [QBC, QBC]	CHROMITE ORE PROCESSING
RR-00070-4 CHROMITE ORE PROCESSING (CHROMATE) [PAL, TLV, TLV]	CHROMITE ORE PROCESSING
7440-47-3 CHROMIUM	CHROMIUM
7440-47-3 CHROMIUM (TOTAL) [UTS]	CHROMIUM
RR-01791-4 CHROMIUM (VI) COMPOUNDS (CERTAIN WATER INSOLUBLE FORMS)	CHROMIUM (VI) COMPOUNDS (CERTAIN WATER INSOLUBLE FORMS)
14833-09-1 CHROMIUM 48	CHROMIUM 48
15758-14-2 CHROMIUM 49	CHROMIUM 49
14392-02-0 CHROMIUM 51	CHROMIUM 51
22541-79-3 CHROMIUM (II)	CHROMIUM (II)
16065-83-1 CHROMIUM (III) [SD4]	CHROMIUM, ION (CR3+)
18540-29-9 CHROMIUM (VI)	CHROMIUM (VI)
7440-47-3 CHROMIUM, METAL [MEX]	CHROMIUM
1066-30-4 CHROMIUM (III) ACETATE [WHM]	CHROMIC ACETATE
21679-31-2 CHROMIUM ACETYLACETONATE	CHROMIUM ACETYLACETONATE
1333-82-0 CHROMIUM ANHYDRIDE [MSL]	CHROMIUM TRIOXIDE (CRO3)
29689-14-3 CHROMIUM CARBONATE	CHROMIUM CARBONATE
13007-92-6 CHROMIUM CARBONYL	CHROMIUM CARBONYL
10025-73-7 CHROMIUM(III) CHLORIDE [WHM]	CHROMIC CHLORIDE
10049-05-5 CHROMIUM CHLORIDE (CRCL2) [PAL]	CHROMIC(II) CHLORIDE
24613-89-6 CHROMIUM(III) CHROMATE	CHROMIUM(III) CHROMATE
24613-89-6 CHROMIUM CHROMATE [MSL]	CHROMIUM(III) CHROMATE
RR-00023-7 CHROMIUM (II) COMPOUNDS	CHROMIUM (II) COMPOUNDS
RR-00024-8 CHROMIUM (III) COMPOUNDS	CHROMIUM (III) COMPOUNDS
RR-00026-0 CHROMIUM (VI) COMPOUNDS	CHROMIUM (VI) COMPOUNDS
RR-00026-0 CHROMIUM COMPOUNDS, HEXAVALENT [PAL, PAL]	CHROMIUM (VI) COMPOUNDS
RR-00634-8 CHROMIUM COMPOUNDS	CHROMIUM COMPOUNDS
RR-00023-7 CHROMIUM (II) COMPOUNDS, N.O.S. [WHM]	CHROMIUM (II) COMPOUNDS
RR-00024-8 CHROMIUM (III) COMPOUNDS, N.O.S. [WHM]	CHROMIUM (III) COMPOUNDS
RR-00026-0 CHROMIUM (VI) COMPOUNDS, N.O.S. [WHM]	CHROMIUM (VI) COMPOUNDS
RR-00634-8 CHROMIUM COMPOUNDS, N.O.S. [RCU, NJL]	CHROMIUM COMPOUNDS
RR-00024-8 CHROMIUM COMPOUNDS, TRIVALENT [IAR]	CHROMIUM (III) COMPOUNDS

Chemical Name	Reference Name
RR-00025-9 CHROMIUM (VI) COMPOUNDS- WATER SOLU-BLE	CHROMIUM (VI) COMPOUNDS- WATER SOLUBLE
14977-61-8 CHROMIUM, DICHLORODIOXO-, (T-4)- [PAL]	CHROMYL CHLORIDE
7440-47-3 CHROMIUM, ELEMENTAL [WHM]	CHROMIUM
18540-29-9 CHROMIUM, HEXAVALENT [ATS]	CHROMIUM (VI)
RR-00026-0 CHROMIUM (HEXAVALENT COMPOUNDS) [C65, C65]	CHROMIUM (VI) COMPOUNDS
RR-00024-8 CHROMIUM III COMPOUNDS [ALB, ALB]	CHROMIUM (III) COMPOUNDS
RR-00560-7 CHROMIUM, INORGANIC COMPOUNDS	CHROMIUM, INORGANIC COMPOUNDS
RR-00560-7 CHROMIUM (VI) INORGANIC COM-POUNDS [GBR]	CHROMIUM, INORGANIC COMPOUNDS
22541-79-3 CHROMIUM, ION (CR2+) [PAL]	CHROMIUM (II)(METHYLTHIO)PHENYL ESTER
16065-83-1 CHROMIUM, ION (CR3+)	CHROMIUM, ION (CR3+)
18540-29-9 CHROMIUM, ION (CR6+) [PAL]	CHROMIUM (VI)
11119-70-3 CHROMIUM LEAD OXIDE	CHROMIUM LEAD OXIDE
7440-47-3 CHROMIUM (METAL) [AUS]	CHROMIUM
7440-47-3 CHROMIUM METAL [CEX, IAR, MIN, NIO, NIO, ONT, TLV, TLV]	CHROMIUM
7789-02-8 CHROMIUM (III) NITRATE	CHROMIUM (III) NITRATE
7789-02-8 CHROMIUM NITRATE [HM1, HM1]	CHROMIUM (III) NITRATE
13548-38-4 CHROMIUM NITRATE	CHROMIUM NITRATE
1308-38-9 CHROMIUM (III) OXIDE	CHROMIUM (III) OXIDE
1333-82-0 CHROMIUM(VI) OXIDE(1:3) [FLA]	CHROMIUM TRIOXIDE (CRO3)
1333-82-0 CHROMIUM OXIDE [NJL, NJL]	CHROMIUM TRIOXIDE (CRO3)
1333-82-0 CHROMIUM(VI) OXIDE [WHM]	CHROMIUM TRIOXIDE (CRO3)
1333-82-0 CHROMIUM OXIDE (CRO3) [PAL, PAL]	CHROMIUM TRIOXIDE (CRO3)
1308-38-9 CHROMIUM (III) OXIDE(2:3) [NJL]	CHROMIUM (III) OXIDE
14977-61-8 CHROMIUM OXYCHLORIDE [HM1, HM1, HM1, NJL, NJL]	CHROMYL CHLORIDE
7789-04-0 CHROMIUM PHOSPHATE	CHROMIUM PHOSPHATE
10141-00-1 CHROMIUM POTASSIUM SULFATE [MSL]	CHROMIUM(III) POTASSIUM SULFATE
10141-00-1 CHROMIUM(III) POTASSIUM SULFATE	CHROMIUM(III) POTASSIUM SULFATE
12680-48-7 CHROMIUM SODIUM OXIDE	CHROMIUM SODIUM OXIDE
RR-00024-8 CHROMIUM, SOLUBLE CHROMIC SALTS [BC1]	CHROMIUM (III) COMPOUNDS
RR-00023-7 CHROMIUM, SOLUBLE CHROMOUS SALTS [BC1]	CHROMIUM (II) COMPOUNDS
10101-53-8 CHROMIUM SULFATE [MSL]	CHROMIC SULFATE
10101-53-8 CHROMIUM(III) SULFATE [WHM]	CHROMIC SULFATE
1333-82-0 CHROMIUM TRIOXIDE [NTP]	CHROMIUM TRIOXIDE (CRO3)
1333-82-0 CHROMIUM TRIOXIDE (CRO3)	CHROMIUM TRIOXIDE (CRO3)
1333-82-0 CHROMIUM TRIOXIDE, ANHYDROUS [HM1, HM1]	CHROMIUM TRIOXIDE (CRO3)
RR-00023-7 CHROMIUM VI COMPOUNDS, WATER SOLU-BLE [QBC]	CHROMIUM (II) COMPOUNDS
RR-01554-3 CHROMIUM VI COMPOUNDS, CERTAIN INSOL-UBLE SALTS [QBC]	CHROMATES
7440-47-3 CHROMIUM (VI), WATER SOLUBLE FUME [BEI]	CHROMIUM
12018-19-8 CHROMIUM ZINC OXIDE	CHROMIUM ZINC OXIDE
7059-24-7 CHROMOMYCIN A3	CHROMOMYCIN A3
64093-79-4 CHROMOSULFURIC ACID	CHROMOSULFURIC ACID
10049-05-5 CHROMOUS CHLORIDE [CWA, EPA, NJL]	CHROMIC(II) CHLORIDE

Chemical Name	Reference Name
13825-86-0 CHROMOUS (II) SULFATE	CHROMOUS (II) SULFATE
14977-61-8 CHROMYL CHLORIDE	CHROMYL CHLORIDE
218-01-9 CHRYSENE	CHRYSENE
3697-24-3 CHRYSENE, 5-METHYL- [PAL, PAL]	5-METHYLCHRYSENE
532-82-1 CHRYSOIDINE	CHRYSOIDINE
12001-29-5 CHRYSOTILE [ATS, QBC, QBC, QBC]	ASBESTOS, CHRYSOTILE
12001-29-5 CHRYSOTILE (MG3H2(SIO4)2.H2O) [PAL, PAL]	ASBESTOS, CHRYSOTILE
12001-29-5 CHRYSOTILE ASBESTOS [MSL, WHM]	ASBESTOS, CHRYSOTILE
12001-29-5 CHRYSOTILE DUST [FLA]	ASBESTOS, CHRYSOTILE
6373-74-6 CI ACID ORANGE 3 [IAR]	C.I. ACID ORANGE 3
6459-94-5 CI ACID RED 114 [IAR]	C.I. ACID RED 114, DISODIUM SALT
569-64-2 C.I. BASIC GREEN 4 [CAB]	MALACHITE GREEN
79217-60-0 CICLOSPORIN [IAR]	CYCLOSPORINE
8003-69-8 C.I. DIRECT BLACK 80	C.I. DIRECT BLACK 80
2429-74-5 CI DIRECT BLUE 15 [IAR]	C.I. DIRECT BLUE 15
51481-61-9 CIMETIDINE	CIMETIDINE
470-82-6 1,8-CINEOL	1,8-CINEOL
25402-06-6 CINERIN I	CINERIN I
104-55-2 CINNAMALDEHYDE	CINNAMALDEHYDE
140-10-3 TRANS-CINNAMIC ACID	TRANS-CINNAMIC ACID
104-54-1 CINNAMIC ALCOHOL [PST]	CINNAMYL ALCOHOL
8007-80-5 CINNAMON BARK OIL	CINNAMON BARK OIL
104-54-1 CINNAMYL ALCOHOL	CINNAMYL ALCOHOL
87-29-6 CINNAMYL ANTHRANILATE	CINNAMYL ANTHRANILATE
101-21-3 CIPC (ISOPROPYL N-(3-CHLOROPHENYL) CAR-BAMATE)	CIPC (ISOPROPYL N-(3-CHLOROPHENYL) CARBAMATE)
2425-85-6 CI PIGMENT RED 3 [IAR]	C.I. PIGMENT RED 3
5103-71-9 CIS-CHLORDANE [ATS]	ALPHA CHLORDANE
15663-27-1 CISPLATIN	CISPLATIN
616-02-4 CITRACONIC ANHYDRIDE	CITRACONIC ANHYDRIDE
5392-40-5 CITRAL	CITRAL
77-92-9 CITRIC ACID	CITRIC ACID
52217-48-8 CITRIC ACID, TRIS(DIMETHYLAMINE) SALT	CITRIC ACID, TRIS(DIMETHYLAMINE) SALT
518-75-2 CITRININ	CITRININ
8000-29-1 CITRONELLA OIL	CITRONELLA OIL
RR-00827-5 CITRONELLEL	CITRONELLEL
106-22-9 CITRONELLOL	CITRONELLOL
68514-76-1 CITRUS PULP, ORANGE	CITRUS PULP, ORANGE
6358-53-8 CITRUS RED NO. 2	CITRUS RED NO. 2
70131-50-9 CLAY [PST]	BENTONITE, ACID-LEACHED
RR-00103-6 CLEANING SOLVENTS, 140 (60) CLASS	CLEANING SOLVENTS, 140 (60) CLASS
8052-41-3 CLEANING SOLVENT, STODDARD SOL-VENT [NFP, NFP]	STODDARD SOLVENT
637-07-0 CLOFIBRATE	CLOFIBRATE
50-41-9 CLOMIPHENE CITRATE	CLOMIPHENE CITRATE
1420-04-8 CLONITRALID	CLONITRALIDLINE ENZYME)
2971-90-6 CLOPIDOL	CLOPIDOL
2971-90-6 CLOPIDOL (COYDEN) [BC1, BC1, ALB, ALB]	CLOPIDOL
133-90-4 CLORAMBEN [NJL]	CHLORAMBEN

Chemical Name	Reference Name
57109-90-7 **CLORAZEPATE DIPOTASSIUM**	CLORAZEPATE DIPOTASSIUM
RR-00011-3 COAL [BC1, MSL]	COAL DUST
65996-93-2 COAL TAR PITCH VOLATILES [MEX, MEX]	COAL TAR PITCHES
RR-00011-3 COAL (BITUMINOUS) DUST [CEX]	COAL DUST
RR-01400-6 COAL BRIQUETTES, HOT	COAL BRIQUETTES, HOT
RR-00011-3 COAL DUST	COAL DUST
RR-00011-3 COAL (BITUMINOUS) DUST [CAL]	COAL DUST
RR-00805-9 COAL DUST (GREATER THAN OR EQUAL TO 5% SIO2), RESPIRABLE QUARTZ FRACTION [OS1, OS1]	COAL DUST, > 5% QUARTZ
RR-00805-9 COAL DUST, > 5% QUARTZ	COAL DUST, > 5% QUARTZ
RR-00216-4 COAL GAS	COAL GAS
RR-00535-6 COAL GASIFICATION	COAL GASIFICATION
RR-00535-6 COAL GASSIFICATION PROCESS (OLDER) [PAL, PAL]	COAL GASIFICATION
RR-00786-3 COAL SOOT	COAL SOOT
8007-45-2 COAL TAR	COAL TAR
8007-45-2 COAL TAR (COAL TAR PITCH) [NJL]	COAL TAR
8001-58-9 COAL TAR CREOSOTE [NJL, NJL]	CREOSOTE
65996-92-1 COAL TAR DISTILLATE	COAL TAR DISTILLATE
65996-92-1 COAL TAR DISTILLATES, FLAMMABLE [HM1, HM1]	COAL TAR DISTILLATE
65996-91-0 COAL TAR LIGHT OIL	COAL TAR LIGHT OIL
65996-79-4 COAL TAR NAPHTHA [NJL]	NAPHTHA (COAL TAR)
65996-91-0 COAL TAR OIL [MAK]	COAL TAR LIGHT OIL
65996-93-2 COAL TAR PITCH [ATS, MSL, NFP, NFP, NHS, NHS, NHS]	COAL TAR PITCHES
65996-93-2 COAL TAR PITCHES	COAL TAR PITCHES
65996-93-2 COAL-TAR PITCHES [IAR]	COAL TAR PITCHES
65996-93-2 COAL TAR PITCH VOLATILES [CAL, ONT, ALB, ALB, ALB, AUS, AUS, CEX, OS1, OS1, TLV, TLV, WHM]	COAL TAR PITCHES
65996-93-2 COAL TAR PITCH VOLATILES (BENZENE-SOLUBLE FRACTION) [NIO, NIO, NIO, NIO]	COAL TAR PITCHES
65996-93-2 COAL TAR PITCH VOLATILES (BENZENE SOLUBLE FRACTION) [QBC, QBC]	COAL TAR PITCHES
8007-45-2 COAL-TARS [IAR]	COAL TAR
8007-45-2 COAL TARS [MIN]	COAL TAR
RR-00787-4 COAL TARS (DURING DESTRUCTIVE DISTILLATION)	COAL TARS (DURING DESTRUCTIVE DISTILLATION)
7440-48-4 **COBALT**	COBALT
13982-25-7 **COBALT 55**	COBALT 55
14093-03-9 **COBALT 56**	COBALT 56
13981-50-5 **COBALT 57**	COBALT 57
13981-38-9 **COBALT 58**	COBALT 58
RR-00467-1 **COBALT 58M**	COBALT 58M
10198-40-0 **COBALT 60**	COBALT 60
RR-00466-0 **COBALT 60M**	COBALT 60M
13981-83-4 **COBALT 61**	COBALT 61
RR-00465-9 **COBALT 62M**	COBALT 62M
7440-48-4 COBALT, METAL, DUST AND FUME [MEX]	COBALT

Chemical Name	Reference Name
71-48-7 COBALT(II) ACETATE	COBALT(II) ACETATE
11114-92-4 COBALT ALLOY, CO, CR [FLA]	COBALT CHROMIUM ALLOY
7440-48-4 COBALT AND BIOAVAILABLE COBALT COMPOUNDS [MAK, MAK]	COBALT
RR-00107-0 COBALT AND COBALT COMPOUNDS [IAR]	COBALT COMPOUNDS
67969-67-9 COBALTATE (6-), [[[1,2-ETHANEDIYLBIS[NITRILOBIS(METHYLENE)]TETRAKIS[PHOSPHONATO](8-)-N,N',O,O',O'''',O''''']-, PENTASODIUM HYDROGEN, (OC-6-21)-	COBALTATE (6-), [[[1,2-ETHANEDIYLBIS[NITRILOBIS(METHYLENE)]TETRAKIS[PHOSPHONATO](8-)-N,N',O,O'',O'''',O''''']-, PENTAPOTASSIUM HYDROGEN, (OC-6-21)-
7789-43-7 COBALT(II) BROMIDE [WHM]	COBALTOUS BROMIDE
7789-43-7 COBALT BROMIDE (COBR2) [PAL]	COBALTOUS BROMIDE
513-79-1 COBALT CARBONATE	COBALT CARBONATE
10210-68-1 COBALT CARBONYL	COBALT CARBONYL
37264-96-3 COBALT CARBONYL [CEX, PAL]	COBALT CARBONYLS
10210-68-1 COBALT CARBONYL (CO2(CO)8) [EPA, EPA, WHM]	COBALT CARBONYL
37264-96-3 COBALT CARBONYLS	COBALT CARBONYLS
7646-79-9 COBALT(II) CHLORIDE [WHM]	COBALTOUS CHLORIDE
13455-25-9 COBALT CHROMATE	COBALT CHROMATE
11114-92-4 COBALT CHROMIUM ALLOY	COBALT CHROMIUM ALLOY
11114-92-4 COBALT-CHROMIUM ALLOY [MSL]	COBALT CHROMIUM ALLOY
RR-00107-0 COBALT COMPOUNDS	COBALT COMPOUNDS
RR-00107-0 COBALT COMPOUNDS, N.O.S. [NJL]	COBALT COMPOUNDS
12013-10-4 COBALT DISULFIDE	COBALT DISULFIDE
7440-48-4 COBALT DUST AND FUME [ALB, ALB]	COBALT
7440-48-4 COBALT, ELEMENTAL [WHM]	COBALT
62207-76-5 COBALT, ((2,2'-(1,2-ETHANEDIYLBIS(NITRILOMETHYLIDYNE))BIS(6-FLUOROPHENOLATO))(2)-	COBALT, ((2,2'-(1,2-ETHANEDIYLBIS(NITRILOMETHYLIDYNE))BIS(6-
62207-76-5 COBALT(II),N,N'-ETHYLENEBIS (3-FLUOROSALICYLIDENEIMINATO)- [MSL]	COBALT, ((2,2'-(1,2-ETHANEDIYLBIS(NITRILOMETHYLIDYNE))BIS(6-
62207-76-5 COBALT(II), N,N'-ETHYLENEBIS(3-FLUOROSALICYLIDENEIMINATO)- [WHM]	COBALT, ((2,2'-(1,2-ETHANEDIYLBIS(NITRILOMETHYLIDYNE))BIS(6-
26490-63-1 COBALT(II) FLUOBORATE	COBALT(II) FLUOBORATE
10026-17-2 COBALT(II) FLUORIDE	COBALT(II) FLUORIDE
7440-48-4 COBALT FUME AND DUST [QBC]	COBALT
16842-03-8 COBALT HYDROCARBONYL	COBALT HYDROCARBONYL
21041-93-0 COBALT HYDROXIDE	COBALT HYDROXIDE
7440-48-4 COBALT METAL, DUST, AND FUME [OS1, OS1]	COBALT
7440-48-4 COBALT METAL, DUST AND FUME [BC1, BC1, BC1]	COBALT
7440-48-4 COBALT METAL, DUST, AND FUME [NIO, NIO, NIO]	COBALT
7440-48-4 COBALT METAL POWDER [C65, IAR]	COBALT
13762-14-6 COBALT(II) MOLYBDATE	COBALT(II) MOLYBDATE
61789-51-3 COBALT NAPHTHA [NFP, NFP]	COBALT NAPHTHENATE
61789-51-3 COBALT NAPHTHENATE	COBALT NAPHTHENATE
61789-51-3 COBALT NAPHTHENATES [HM1, HM1]	COBALT NAPHTHENATE
27253-31-2 COBALT NEODECANOATE	COBALT NEODECANOATE
10026-22-9 COBALT NITRATE	COBALT NITRATE
10141-05-6 COBALT(II) NITRATE	COBALT(II) NITRATE

Chemical Name	Reference Name
13586-82-8 COBALT OCTOATE	COBALT OCTOATE
7789-43-7 COBALTOUS BROMIDE	COBALTOUS BROMIDE
7646-79-9 COBALTOUS CHLORIDE	COBALTOUS CHLORIDE
544-18-3 COBALTOUS FORMATE	COBALTOUS FORMATE
10141-05-6 COBALTOUS NITRATE [FLA]	COBALT(II) NITRATE
14017-41-5 COBALTOUS SULFAMATE	COBALTOUS SULFAMATE
1307-96-6 COBALT(II) OXIDE	COBALT(II) OXIDE
13455-36-2 COBALT(II) PHOSPHATE	COBALT(II) PHOSPHATE
1560-69-6 COBALT PROPIONATE	COBALT PROPIONATE
RR-00219-7 COBALT RESINATE, PRECIPITATED	COBALT RESINATE, PRECIPITATED
10124-43-3 COBALT SULFATE	COBALT SULFATE
10124-43-3 COBALT(II) SULFATE [WHM]	COBALT SULFATE
10026-24-1 COBALT(II) SULFATE (1:1), HEPTAHYDRATE	COBALT(II) SULFATE (1:1), HEPTAHYDRATE
10026-24-1 COBALT SULFATE HEPTAHYDRATE [CSR]	COBALT(II) SULFATE (1:1), HEPTAHYDRATE
1317-42-6 COBALT(II) SULFIDE	COBALT(II) SULFIDE
61789-52-4 COBALT TALLATE	COBALT TALLATE
10210-68-1 COBALT TETRACARBONYL [QBC]	COBALT CARBONYL
16842-03-8 COBALT, TETRACARBONYLHYDRO- [PAL]	COBALT HYDROCARBONYLMETHYL ESTER
50-36-2 COCAINE	COCAINE
124-87-8 COCCULUS [HM1, HM1, HM1, NJL]	PICROTOXIN
8002-31-1 COCOA	COCOA
RR-01226-0 COCO ACID TRIAMINE CONDENSATE, POLY-CARBOXYLIC ACID SALTS	COCO ACID TRIAMINE CONDENSATE, POLYCAR-BOXYLIC ACID SALTS
61789-40-0 N-(COCO ALKYL) AMIDO PROPYL DIMETHYL BETAINE [PST]	COCOAMIDO(PROPYLBETAINE)
61788-93-0 COCO ALKYLDIMETHYLAMINES	COCO ALKYLDIMETHYLAMINES
61788-90-7 COCO ALKYLDIMETHYLAMINES, N-OX-IDES [PST]	N,N-DIMETHYL (COCONUT OIL ALKYL) AMINE OXIDE
68424-94-2 (COCO ALKYL)DIMETHYL BETAINES	(COCO ALKYL)DIMETHYL BETAINES
61789-18-2 COCO ALKYLTRIMETHYL QUATERNARY AMMONIUM CHLORIDES	COCO ALKYLTRIMETHYL QUATERNARY AMMONIUM CHLORIDES
61789-40-0 COCOAMIDO(PROPYLBETAINE)	COCOAMIDO(PROPYLBETAINE)
RR-01055-9 COCOAMINO PROPIONIC ACID	COCOAMINO PROPIONIC ACID
RR-01056-0 COCOAMPHOCARBOXYGLYCINATE, DISODIUM SALT	COCOAMPHOCARBOXYGLYCINATE, DISODIUM SALT
142-78-9 COCOMONOETHANOLAMIDE	COCOMONOETHANOLAMIDE
68603-42-9 COCONUT DIETHANOLAMIDE	COCONUT DIETHANOLAMIDE
61789-30-8 COCONUT FATTY ACIDS, POTASSIUM SALT	COCONUT FATTY ACIDS, POTASSIUM SALT
68140-00-1 COCONUT MONOETHANOLAMIDE	COCONUT MONOETHANOLAMIDE
8001-31-8 COCONUT OIL [NFP, NFP, PST]	COPRA
61790-63-4 COCONUT OIL ACID, DIETHANOLAMINE SALT	COCONUT OIL ACID, DIETHANOLAMINE SALT
68603-42-9 COCONUT OIL ACID DIETHANOLAMIDE CON-DENSATE [CSR]	COCONUT DIETHANOLAMIDE
61788-47-4 COCONUT OIL FATTY ACIDS	COCONUT OIL FATTY ACIDS
RR-01057-1 COCONUT OIL, POLYMER WITH ISOPH-THALIC ACID, TRIMELITTIC ANHYDRIDE AND TRIMETHYLOLPROPANE	COCONUT OIL, POLYMER WITH ISOPHTHALIC ACID, TRIMELITTIC ANHY
61789-31-9 COCONUT OIL SOAP [PST]	POTASSIUM COCONUT FATTY OIL SOAP
RR-01054-8 COCO SHELL FLOUR	COCO SHELL FLOUR
76-57-3 CODEINE	CODEINE
8001-69-2 COD-LIVER OIL	COD-LIVER OIL

Chemical Name	Reference Name
8001-69-2 COD LIVER OIL [NFP, NFP]	COD-LIVER OIL
97553-00-9 COD OIL, SULFONATED	COD OIL, SULFONATED
RR-01058-2 COFFEE [IAR]	COFFEE GROUNDS
RR-01058-2 COFFEE GROUNDS	COFFEE GROUNDS
65996-77-2 COKE (COAL)	COKE (COAL)
RR-00528-7 COKE OVEN EMISSIONS	COKE OVEN EMISSIONS
RR-00536-7 COKE PRODUCTION	COKE PRODUCTION
64-86-8 COLCHICINE	COLCHICINE
9004-70-0 COLLODION [FLA, MSL, NFP, NFP]	NITROCELLULOSE
RR-00081-7 COMBINED CHEMOTHERAPY FOR LYM-PHOMAS	COMBINED CHEMOTHERAPY FOR LYMPHOMAS
RR-00543-6 COMBINED ORAL CONTRACEPTIVES	COMBINED ORAL CONTRACEPTIVES
RR-00220-0 COMBUSTIBLE LIQUID, N.O.S.	COMBUSTIBLE LIQUID, N.O.S.
RR-00152-5 COMMERCIAL HEXANE	COMMERCIAL HEXANE
RR-00224-4 COMPRESSED OR LIQUIFIED GASES, N.O.S.	COMPRESSED OR LIQUIFIED GASES, N.O.S.
RR-01059-3 CONDENSATION PRODUCT OF SORBITOL EPICHLOROHYDRIN WITH THE OLEIC ACID DIAMIDE OF DIETHYLENETRIAMINE	CONDENSATION PRODUCT OF SORBITOL EPICHLORO-HYDRIN WITH THE OLDRIDE AND TRIMETHYLOL-PROPANE
RR-01744-7 CONJUGATED ESTROGENS	CONJUGATED ESTROGENS
7280-37-7 CONJUGATED ESTROGENS: PIPERAZINE ESTRONE SULFATE [PAL, PAL]	PIPERAZINE ESTRONE SULFATE
16680-47-0 CONJUGATED ESTROGENS: SODIUM EQUILIN SULFATE [PAL, PAL]	SODIUM EQUILIN SULFATE
438-67-5 CONJUGATED ESTROGENS: SODIUM ESTRONE SULFATE [PAL, PAL]	SODIUM ESTRONE SULFATE(5.ALPHA.,17.BETA.)-
RR-00082-8 CONJUGATED OESTROGENS [IAR, MIN, PAL, PAL]	CONJUGATED ESTROGENS
7440-50-8 COPPER	COPPER
13982-06-4 COPPER 60	COPPER 60
15128-03-7 COPPER 61	COPPER 61
13981-25-4 COPPER 64	COPPER 64
15757-86-5 COPPER 67	COPPER 67
7440-50-8 COPPER, FUME, DUSTS AND MISTS [MEX, MEX]	COPPER
142-71-2 COPPER ACETATE [PST]	CUPRIC ACETATE
12002-03-8 COPPER ACETOARSENITE	COPPER ACETOARSENITE
12540-13-5 COPPER ACETYLIDE	COPPER ACETYLIDE
RR-01401-7 COPPER AMINE AZIDE	COPPER AMINE AZIDE
10290-12-7 COPPER ARSENITE	COPPER ARSENITE
RR-00228-8 COPPER BASED PESTICIDES, N.O.S.	COPPER BASED PESTICIDES, N.O.S.
12069-69-1 COPPER(II) CARBONATE HYDROXIDE	COPPER(II) CARBONATE HYDROXIDE
26506-47-8 COPPER CHLORATE	COPPER CHLORATE
1344-67-8 COPPER CHLORIDE	COPPER CHLORIDE
1344-67-8 COPPER(II) CHLORIDE [WHM]	COPPER CHLORIDE
7758-89-6 COPPER(I) CHLORIDE	COPPER(I) CHLORIDE
7447-39-4 COPPER CHLORIDE (CUCL2) [PAL]	CUPRIC CHLORIDE
13548-42-0 COPPER CHROMATE	COPPER CHROMATE
RR-00595-8 COPPER COMPOUNDS [313, CAL, CN7, CW3]	COPPER COMPOUNDS, N.O.S.
RR-00595-8 COPPER COMPOUNDS, N.O.S.	COPPER COMPOUNDS, N.O.S.
544-92-3 COPPER CYANIDE	COPPER CYANIDE
39377-49-6 COPPER CYANIDE [NJL, HM1, HM1]	COPPER CYANIDE (VAN)

Chemical Name	Reference Name
39377-49-6 COPPER CYANIDE (VAN)	COPPER CYANIDE (VAN)
6046-93-1 COPPER DIACETATE MONOHYDRATE	COPPER DIACETATE MONOHYDRATE
137-29-1 COPPER(II) DIMETHYLDITHIOCARBAMATE	COPPER(II) DIMETHYLDITHIOCARBAMATE
7440-50-8 COPPER, ELEMENTAL [WHM]	COPPER
20427-59-2 COPPER HYDROXIDE	COPPER HYDROXIDE
10380-28-6 COPPER(II) 8-HYDROXYQUINOLINATE	COPPER(II) 8-HYDROXYQUINOLINATE
10380-28-6 COPPER 8-HYDROXYQUINOLINE [IAR]	COPPER(II) 8-HYDROXYQUINOLINATE
RR-00561-8 COPPER, INORGANIC COMPOUNDS	COPPER, INORGANIC COMPOUNDS
1338-02-9 COPPER NAPHTHENATE	COPPER NAPHTHENATE
3251-23-8 COPPER NITRATE [PST]	CUPRIC NITRATE
3251-23-8 COPPER(II) NITRATE [WHM]	CUPRIC NITRATE
10031-43-3 COPPER(II) NITRATE, TRIHYDRATE (1:2:3)	COPPER(II) NITRATE, TRIHYDRATE (1:2:3)
814-91-5 COPPER OXALATE (CUC204)	COPPER OXALATE (CUC204)
1317-39-1 COPPER(+1) OXIDE	COPPER(+1) OXIDE
1317-39-1 COPPER(I) OXIDE [WHM]	COPPER(+1) OXIDE
147-14-8 COPPER PHTHALOCYANIDE BLUE [PST]	PHTHALOCYANINE BLUE
81457-65-0 COPPER, [29H, 31H-PHTHALOCYANINATO(2-)-N29,N30,N31,N32-, [[3-(METHYLETHOXY)-, PROPYLAMINOSULFONYL DERIVATIVES	COPPER, [29H, 31H-PHTHALOCYANINATO(2-)-N29,N30, N31,N32-, [[3
15123-69-0 COPPER SELENATE	COPPER SELENATE
10214-40-1 COPPER SELENITE	COPPER SELENITE
14264-31-4 COPPER SODIUM CYANIDE	COPPER SODIUM CYANIDE
7758-98-7 COPPER SULFATE [CWA, PST]	CUPRIC SULFATE
7758-98-7 COPPER(II) SULFATE [WHM]	CUPRIC SULFATE
7758-99-8 COPPER (II) SULFATE PENTAHYDRATE (1:1:5)	COPPER (II) SULFATE PENTAHYDRATE (1:1:5)
1317-40-4 COPPER(II) SULFIDE	COPPER(II) SULFIDE
22205-45-4 COPPER(I) SULFIDE	COPPER(I) SULFIDE
10380-29-7 COPPER(2+), TETRAAMINE-, SULFATE (1:1), MONOHYDRATE [PAL]	CUPRIC SULFATE AMMONIATED
RR-01402-8 COPPER TETRAMINE NITRATE	COPPER TETRAMINE NITRATE
8001-31-8 COPRA	COPRA
61789-98-8 CORK	CORK
RR-01060-6 CORN	CORNEIC ACID DIAMIDE OF DIETHYLENETRIAMINE
8001-30-7 CORN OIL	CORN OIL
9005-25-8 CORNSTARCH [PST]	STARCH
8029-43-4 CORN SYRUP	CORN SYRUP
191-07-1 CORONENE	CORONENE
RR-00233-5 CORROSIVE LIQUIDS, N.O.S.	CORROSIVE LIQUIDS, N.O.S.
RR-00236-8 CORROSIVE SOLIDS, N.O.S.	CORROSIVE SOLIDS, N.O.S.
1302-74-5 CORUNDUM	CORUNDUM
RR-00001-1 COTTON [HM1, HM1, PST]	COTTON DUST (RAW)
RR-00001-1 COTTON DUST [CAL, CEX, GBR, MSL, NHS, NHS]	COTTON DUST (RAW)
RR-00001-1 COTTON DUST (RAW)	COTTON DUST (RAW)
RR-00001-1 COTTON DUST, RAW [AUS]	COTTON DUST (RAW)
RR-00001-1 COTTON DUST RAW [BC1, BC1]	COTTON DUST (RAW)
RR-00001-1 COTTON DUST, RAW [MIN, ONT, PAL, QBC, TLV]	COTTON DUST (RAW)
68424-10-2 COTTONSEED MEAL	COTTONSEED MEAL

Chemical Name	Reference Name
8001-29-4 COTTONSEED OIL	COTTONSEED OIL
8001-29-4 COTTONSEED OIL REFINED [NFP, NFP]	COTTONSEED OIL
81-82-3 COUMACHLOR	COUMACHLOR
117-52-2 COUMAFURYL	COUMAFURYL
56-72-4 COUMAPHOS	COUMAPHOS
91-64-5 COUMARIN	COUMARIN
RR-00244-8 COUMARIN DERIVATIVE PESTICIDES, N.O.S.	COUMARIN DERIVATIVE PESTICIDES, N.O.S.
91-44-1 COUMARIN, 7-(DIETHYLAMINO)-4-METHYL-	COUMARIN, 7-(DIETHYLAMINO)-4-METHYL-
63393-89-5 COUMARONE-INDENE RESIN	COUMARONE-INDENE RESIN
5836-29-3 COUMATETRALYL	COUMATETRALYL
RR-01061-7 CRACKED OATS AND WHEAT	CRACKED OATS AND WHEAT
68476-87-9 CRACKED RESIDUES DERIVED FROM THE REFINING OF CRUDE OIL [IAR]	PETROLEUM REFINING RESIDUES, POLYMERIZED
136-78-7 CRAG HERBICIDE [BC1, BC1, FLA, MSL, NIO, NIO, NIO]	SESONE
136-78-7 CRAG HERBICIDE (SESONE) [OS1, OS1]	SESONE
136-78-7 CRAG HERBICIDE (SODIUM 2,4-DICHLOROPHE-NOXYETHYL SULPHATE) [ALB, ALB]	SESONE
8001-58-9 CREOSOTE	CREOSOTE
8001-58-9 CREOSOTE (COAL) [NTP]	CREOSOTE
8001-58-9 CREOSOTE (COAL TAR) [HM1, HM1]	CREOSOTE
8001-58-9 CREOSOTE, COAL TAR [MSL]	CREOSOTE
RR-01601-3 CREOSOTE IMPREGNATED WASTE MATERI-ALS	CREOSOTE IMPREGNATED WASTE MATERIALS
8001-58-9 CREOSOTE OIL [NFP, NFP]	CREOSOTE
61789-28-4 CREOSOTE OIL	CREOSOTE OIL
61789-28-4 CREOSOTE, OIL [MSL]	CREOSOTE OIL
8001-58-9 CREOSOTES [CAL]	CREOSOTE
RR-00091-9 CREOSOTES	CREOSOTES
8021-39-4 CREOSOTE, WOOD	CREOSOTE, WOOD
8021-39-4 CREOSOTE (WOOD TAR) [HM1]	CREOSOTE, WOOD
8021-39-4 CREOSOTE (WOOD) [NTP]	CREOSOTE, WOOD
102-50-1 M-CRESIDINE	M-CRESIDINE
102-50-1 META-CRESIDINE [IAR]	M-CRESIDINE
120-71-8 P-CRESIDINE	P-CRESIDINE
120-71-8 PARA-CRESIDINE [C65, C65, IAR]	P-CRESIDINE
95-48-7 O-CRESOL	O-CRESOL
95-48-7 CRESOL, O- [302]	O-CRESOL
95-48-7 CRESOL, ORTHO- [ATS]	O-CRESOL
95-48-7 ORTHO-CRESOL [CTR]	O-CRESOL
95-48-7 O-CRESOL [PQL, RC2, RC2]	O-CRESOL
106-44-5 P-CRESOL	P-CRESOL
106-44-5 CRESOL, PARA- [ATS]	P-CRESOL
106-44-5 PARA-CRESOL [CTR]	P-CRESOL
108-39-4 M-CRESOL	M-CRESOL
108-39-4 META-CRESOL [CTR]	M-CRESOL
1319-77-3 CRESOL	CRESOL
1319-77-3 M- OR P- CRESOL [NFP, NFP]	CRESOL
1319-77-3 CRESOL (ALL ISOMERS) [QBC, QBC]	CRESOL

Chemical Name	Reference Name
1319-77-3 CRESOL (CRESYLIC ACID) [RCU, RCU]	CRESOL
1319-77-3 CRESOL AND CRESYLIC ACIDS (MIXED) [HON, HON]	CRESOL
1319-77-3 CRESOL ISOMERS [MEX, MEX]	CRESOL
1319-77-3 CRESOL (SUM OF O-, M-, AND P- ISOMERS) [ONT, ONT]	CRESOL
1319-77-3 CRESOL, ALL ISOMERS [ALB, ALB, ALB, AUS, AUS, BC1, BC1]	CRESOL
1319-77-3 CRESOL (ALL ISOMERS) [CEX, CEX]	CRESOL
1319-77-3 CRESOL, ALL ISOMERS [MIN]	CRESOL
1319-77-3 CRESOL (ALL ISOMERS) [NIO, NIO, NIO]	CRESOL
1319-77-3 CRESOL, ALL ISOMERS [OS1, OS1, OS1, OS1, TLV, TLV]	CRESOL
95-48-7 O-CRESOL AND CRESYLIC ACID [HON, HON]	O-CRESOL
106-44-5 P-CRESOL AND CRESYLIC ACID [HON, HON]	P-CRESOL
108-39-4 M-CRESOL AND CRESYLIC ACID [HON, HON]	M-CRESOL
1319-77-3 CRESOL (MIXED ISOMERS) [313, CAN, CSR]	CRESOL
122436-67-3 M-CRESOL, POLYMER WITH FORMALDEHYDE AND SULFANILIC ACID	M-CRESOL, POLYMER WITH FORMALDEHYDE AND SULFANILIC ACIDCYL METHACRYLATE, 2-MORPHOLI-NOETHYL METHACRYLATE COPOLYMER
1319-77-3 CRESOLS [EPA]	CRESOL
1319-77-3 CRESOLS (CRESYLIC ACID) [RCA]	CRESOL
1319-77-3 CRESOLS, ALL ISOMERS [GBR, GBR]	CRESOL
1319-77-3 CRESOLS/CRESYLIC ACID (ISOMERS AND MIXTURE) [CAA]	CRESOL
1319-77-3 CRESOLS (O-; M-; P-) [HM1, HM1, HM1, HM1]	CRESOL
140-39-6 P-CRESYL ACETATE [NFP, NFP]	P-TOLYL ACETATE
10303-47-6 CRESYL DIPHENYL PHOSPHATE	CRESYL DIPHENYL PHOSPHATE
2210-79-9 O-CRESYL GLYCIDYL ETHER [EPA]	OXIRANE, [(2-METHYLPHENOXY)METHYL]-
2210-79-9 CRESYL GLYCIDYL ETHER, O- [OTS]	OXIRANE, [(2-METHYLPHENOXY)METHYL]-
26447-14-3 CRESYL GLYCIDYL ETHER	CRESYL GLYCIDYL ETHER
26447-14-3 CRESYL GLYCIDYL ETHER (MIXED ISOMERS) [EPA, OTS]	CRESYL GLYCIDYL ETHER
1319-77-3 CRESYLIC ACID [NJL, PST]	CRESOL
12002-51-6 CRESYLIC ACID, POTASSIUM SALT	CRESYLIC ACID, POTASSIUM SALT
34689-46-8 CRESYLIC ACID SODIUM SALT	CRESYLIC ACID SODIUM SALT
535-89-7 CRIMIDINE	CRIMIDINE
14464-46-1 CRISTOBALITE (SIO2) [PAL]	SILICA, CRISTOBALITE
14464-46-1 CRISTOBALITE DUST [FLA, MSL]	SILICA, CRISTOBALITE
12001-28-4 CROCIDOLITE [PAL, PAL, QBC, QBC, QBC]	ASBESTOS, CROCIDOLITE
61105-31-5 CROCIDOLITE (FE2MG3NA2(SIO3)8)	CROCIDOLITE (FE2MG3NA2(SIO3)8)
53799-46-5 CROCIDOLITE (FE5NA2(SIO3)8)	CROCIDOLITE (FE5NA2(SIO3)8)
12001-28-4 CROCIDOLITE ASBESTOS [MSL, WHM]	ASBESTOS, CROCIDOLITE
12001-28-4 CROCIDOLITE DUST [FLA]	ASBESTOS, CROCIDOLITE
74811-65-7 CROSCARMELLOSE SODIUM	CROSCARMELLOSE SODIUM
RR-01062-8 CROSS LINKED POLYMER OF SEBACRYL CHLORIDE & POLYMETHYLENEPOLYPHENYL ISOCYANATE, WITH ETHYLENEDIAMINE AND DIETHYLENETRIA MINE	CROSS LINKED POLYMER OF SEBACRYL CHLORIDE & POLYMETHYLENEPOL
123-73-9 CROTONALDEHYDE	CROTONALDEHYDE
123-73-9 (E)-CROTONALDEHYDE [WHM]	CROTONALDEHYDE

Chemical Name	Reference Name
4170-30-3 CROTONALDEHYDE	CROTONALDEHYDE
4170-30-3 CROTONALDEHYDE [CSR, ONT]	CROTONALDEHYDE
123-73-9 CROTONALDEHYDE, (E)- [302]	CROTONALDEHYDE
123-73-9 CROTONALDEHYDE (E) [MSL]	CROTONALDEHYDE
123-73-9 CROTONALDEHYDE, (E)- [OS1, OS1]	CROTONALDEHYDE
123-73-9 CROTONALDEHYDE, (E)- [2-BUTENAL, (E)-] [EP3]	CROTONALDEHYDE
4170-30-3 CROTONALDEHYDE, INHIBITED [NJL]	CROTONALDEHYDE
4170-30-3 CROTONALDEHYDE [2-BUTENAL] [EP3]	CROTONALDEHYDE
3724-65-0 CROTONIC ACID	CROTONIC ACID
4786-20-3 CROTONONITRILE	CROTONONITRILE
6117-91-5 CROTONYL ALCOHOL	CROTONYL ALCOHOL
503-17-3 CROTONYLENE [HM1, HM1, NJL]	2-BUTYNE
7700-17-6 CROTOXYPHOS	CROTOXYPHOS
4784-77-4 1-CROTYL BROMIDE	1-CROTYL BROMIDE
591-97-9 1-CROTYL CHLORIDE	1-CROTYL CHLORIDE
8002-05-9 CRUDE OIL [CN6]	PETROLEUM DISTILLATES (NAPHTHA)
RR-01606-8 CRUDE OIL WASTES	CRUDE OIL WASTES
299-86-5 CRUFOMATE	CRUFOMATE
299-86-5 CRUFOMATE (R) [QBC]	CRUFOMATE
1320-37-2 CRYOFLUORANE (INN) [GBR, GBR]	DICHLOROTETRAFLUOROETHANE
RR-00109-2 CRYPTOSPORIDUM	CRYPTOSPORIDUM
RR-00087-3 CRYSTALLINE SILICA [IAR]	SILICA, CRYSTALLINE (GENERAL FORM)
98-82-8 CUMENE	CUMENE
98-82-8 CUMENE (1-METHYLETHYLBENZENE) [TSC, TSC, TSC]	CUMENE
80-15-9 CUMENE HYDROPEROXIDE	CUMENE HYDROPEROXIDE
536-60-7 CUMINIC ALCOHOL	CUMINIC ALCOHOL
23383-59-7 CUMYL PEROXYPIVALATE	CUMYL PEROXYPIVALATE
80-15-9 CUMYL HYDROPEROXIDE [DOT]	CUMENE HYDROPEROXIDE
26748-47-0 CUMYL PEROXYNEODECANOATE	CUMYL PEROXYNEODECANOATE
RR-00580-1 CUMYL PEROXYPIVALATE	CUMYL PEROXYPIVALATE
61578-04-9 P-CUMYLPHENYL GLYCIDYL ETHER [EPA]	OXIRANE, [[4-(1-METHYL-1-PHENYLETHYL)PHENOXY]METHYL]-
61578-04-9 CUMYLPHENYL GLYCIDYL ETHER, P- [OTS]	OXIRANE, [[4-(1-METHYL-1-PHENYLETHYL)PHENOXY]METHYL]-
135-20-6 CUPFERRON	CUPFERRON
135-20-6 CUPFERRON [BENZENEAMINE, N-HYDROXY-N-NITROSO, AMMONIUM SALT] [313]	CUPFERRON
16143-79-6 CUPRATE(4-), [U-[[6,6'-[3,3'-DIHYDROXY[1,1'-BIPHENYL]-4,4'-DIYL)BIS(AZO)BIS[4-AMINO-5-HYDROXY-1,3-NAPHTHALENE DISULFONATO](8-)]DI-, TETRASODIUM	CUPRATE(4-), [U-[[6,6'-[3,3'-DIHYDROXY[1,1'-BIPHENYL]-4,4'-AMINO]-
6656-03-7 CUPRATE(3-), [U-[7-[[3,3'-DIHYDROXY-4'-[[1-HYDROXY-6-(PHENYLAMINO)-3-SULFO-2-NAPHTHALENYL]AZO][1,1'-BIPHENYL]-4-YL]AZO] -8-HYDROXY-1,6-NAPHTHALENEDISULFONATO(7-)]DI-, TRISODIUM	CUPRATE(3-), [U-[7-[[3,3'-DIHYDROXY-4'-[[1-HYDROXY-6-(PHENYLAZO]-3,3'-DIMETHYL[1,1'-BIPHENYL]-4-YL]AZO]-2-HYDROXY-, DISODIUM SALT
16071-86-6 CUPRATE(2-), [5-[[4'-[[2,6-DIHYDROXY-3-[(2-HYDROXY-5-SULFOPHENYL)AZO]PHENYL]AZO] 1,1'-BIPHENYL]4-YL]AZO-2-HYDROXY BENZOATO(4-)]-, DISODIUM [PAL, PAL]	C.I. DIRECT BROWN 95

Chemical Name	Reference Name
16071-86-6 CUPRATE(2-), [5-[[4'-[[2,6-DIHYDROXY-3-[(2-HYDROXY-5-SULFOPHENYL)AZO]PHENYL] AZO][1,1'-BIPHENYL]-4-YL]AZO]-2-HYDROXY BENZOATO(4-)]-, DISODIUM [TSC]	C.I. DIRECT BROWN 95
67989-89-3 CUPRATE(6-), [[[1,2-ETHANEDIYLBIS[NITRILO-BIS(METHYLENE)]TETRAKIS[PHOSPHONATO] (8-)N,N',O,O'',O''',O'''']-, PENTAPOTASSIUM HYDROGEN, (OC-6-21)-	CUPRATE(6-), [[[1,2-ETHANEDIYLBIS[NITRILOBIS (METHYLENE)]TETRAKIS[PHOSPHONATO](8-)-N,N',O,O', O'',O'''']-, PENTASODIUM HYDROGEN, (OC-6-21)-
14025-15-1 CUPRATE(2-), ((ETHYLENEDINITRILO)TE-TRAACETATO)-, DISODIUM	CUPRATE(2-), ((ETHYLENEDINITRILO)TETRAACETATO)-, DISODIUM
142-71-2 CUPRIC ACETATE	CUPRIC ACETATE
12002-03-8 CUPRIC ACETOARSENITE [CWA, EP2, EPA]	COPPER ACETOARSENITE
7447-39-4 CUPRIC CHLORIDE	CUPRIC CHLORIDE
14763-77-0 CUPRIC CYANIDE	CUPRIC CYANIDE
527-09-3 CUPRIC GLUCONATE	CUPRIC GLUCONATE
3251-23-8 CUPRIC NITRATE	CUPRIC NITRATE
814-91-5 CUPRIC OXALATE [NJL]	COPPER OXALATE (CUC2O4)
5893-66-3 CUPRIC OXALATE	CUPRIC OXALATE
7758-98-7 CUPRIC SULFATE	CUPRIC SULFATE
7758-99-8 CUPRIC SULFATE [CSR]	COPPER (II) SULFATE PENTAHYDRATE (1:1:5)
10380-29-7 CUPRIC SULFATE AMMONIATED	CUPRIC SULFATE AMMONIATED
10380-29-7 CUPRIC SULFATE, AMMONIATED [NJL]	CUPRIC SULFATE AMMONIATED
815-82-7 CUPRIC TARTRATE	CUPRIC TARTRATE
13426-91-0 CUPRIETHYLENEDIAMINE	CUPRIETHYLENEDIAMINE
7758-89-6 CUPROUS CHLORIDE [HM1]	COPPER(I) CHLORIDE
1317-39-1 CUPROUS OXIDE [PST]	COPPER(+1) OXIDE
41198-08-7 CURACRON	CURACRON
30989-40-3 CURIUM 238	CURIUM 238
15411-90-2 CURIUM 240	CURIUM 240
15411-91-3 CURIUM 241	CURIUM 241
15510-73-3 CURIUM 242	CURIUM 242
15757-87-6 CURIUM 243	CURIUM 243
13981-15-2 CURIUM 244	CURIUM 244
15621-76-8 CURIUM 245	CURIUM 245
15757-90-1 CURIUM 246	CURIUM 246
15758-32-4 CURIUM 247	CURIUM 247
15758-33-5 CURIUM 248	CURIUM 248
15701-07-2 CURIUM 249	CURIUM 249
420-04-2 CYANAMIDE	CYANAMIDE
156-62-7 CYANAMIDE, CALCIUM SALT (1:1) [PAL]	CALCIUM CYANAMIDE
1467-79-4 CYANAMIDE, DIMETHYL- [PAL]	DIMETHYLCYANAMIDE
21725-46-2 CYANAZINE	CYANAZINE
74-90-8 CYANHYDRIC ACID (HYDROGEN CYANIDE) [QBC, QBC]	HYDROGEN CYANIDE
57-12-5 CYANIDE [ATS, CD1, CEX, CEX, NJL, PAL, PQL, QBC, QBC, SDW, SDW]	CYANIDE ANION
RR-00573-2 CYANIDE, ALL INORGANIC COMPOUNDS [NJL]	CYANIDE, INORGANIC COMPOUNDS
57-12-5 CYANIDE ANION	CYANIDE ANION
57-12-5 CYANIDE (ANION) [MSL]	CYANIDE ANION
57-12-5 CYANIDE (AS CN) [BC1, BC1]	CYANIDE ANION
57-12-5 CYANIDE (COMPLEXED) [RCA]	CYANIDE ANION

Chemical Name	Reference Name
57-12-5 CYANIDE COMPOUNDS [CAA]	CYANIDE ANION
RR-00812-8 CYANIDE COMPOUNDS	CYANIDE COMPOUNDS
RR-00573-2 CYANIDE COMPOUNDS, INORGANIC, N.O.S. [WHM]	CYANIDE, INORGANIC COMPOUNDS
RR-00812-8 CYANIDE COMPOUNDS, N.O.S. [NJL]	CYANIDE COMPOUNDS
RR-00573-2 CYANIDE, INORGANIC COMPOUNDS	CYANIDE, INORGANIC COMPOUNDS
57-12-5 CYANIDE MIXTURES OR SOLUTIONS [HM1, HM1, HM1, HM1]	CYANIDE ANION
57-12-5 CYANIDE, POTASSIUM AND SODIUM [ONT, ONT]	CYANIDE ANION
57-12-5 CYANIDES [ALB, ALB, ALB, AUS, AUS, CAN, CN7, CW3, EPA, GBR, GBR, MAK, MAK, MAK, MEX, MEX, OS1, OS1, UTS]	CYANIDE ANION
RR-00500-5 CYANIDE (SALTS)	CYANIDE (SALTS)
RR-00500-5 CYANIDE SALTS [NHS, NHS]	CYANIDE (SALTS)
RR-00573-2 CYANIDES, INORGANIC, N.O.S. [HM1, HM1, HM1]	CYANIDE, INORGANIC COMPOUNDS
RR-00573-2 CYANIDES, INORGANIC SALTS [CAL]	CYANIDE, INORGANIC COMPOUNDS
RR-00500-5 CYANIDES (SOLUBLE CYANIDE SALTS), NOT OTHERWISE SPECIFIED [RCU, RCU]	CYANIDE (SALTS)
57-12-5 CYANIDE, TOTAL [CW2]	CYANIDE ANION
107-91-5 2-CYANOACETAMIDE	2-CYANOACETAMIDE
372-09-8 CYANOACETIC ACID [HON, FLA, MSL]	ACETIC ACID, CYANO-
RR-01711-8 CYANOACRYLATES	CYANOACRYLATES
106-71-8 2-CYANOETHYL ACRYLATE	2-CYANOETHYL ACRYLATE
702-03-4 N-(2-CYANOETHYL)CYCLOHEXYLAMINE	N-(2-CYANOETHYL)CYCLOHEXYLAMINE
2074-87-5 CYANOGEN	CYANOGEN
506-68-3 CYANOGEN BROMIDE	CYANOGEN BROMIDE
506-68-3 CYANOGEN BROMIDE (CN)BR [RCU, RCU]	CYANOGEN BROMIDE
506-77-4 CYANOGEN CHLORIDE	CYANOGEN CHLORIDE
2074-87-5 CYANOGENE [QBC]	CYANOGEN
506-78-5 CYANOGEN IODIDE	CYANOGEN IODIDE
460-19-5 CYANOGEN [ETHANEDINITRILE] [EP3]	CYANOGEN
2636-26-2 CYANOPHOS	CYANOPHOS
108-80-5 CYANURIC ACID [PST]	ISOCYANURIC ACID
108-77-0 CYANURIC CHLORIDE	CYANURIC CHLORIDE
675-14-9 CYANURIC FLUORIDE	CYANURIC FLUORIDE
5637-83-2 CYANURIC TRIAZIDE	CYANURIC TRIAZIDE
108-77-0 CYANURIC TRICHLORIDE [NJL, NJL]	CYANURIC CHLORIDE
14901-08-7 CYCASIN	CYCASIN
RR-01025-3 CYCLAMATES	CYCLAMATES
103-95-7 CYCLAMEN ALDEHYDE [NFP, NFP]	BENZENEPROPANAL, .ALPHA.-METHYL-4-(1-METHYLETHYL)-
100-88-9 CYCLAMIC ACID	CYCLAMIC ACID
RR-01223-7 CYCLIC AMIDE	CYCLIC AMIDE
RR-01679-5 CYCLIC PHOSPHAZENE, METHACRYLATE DERIVATIVE	CYCLIC PHOSPHAZENE, METHACRYLATE DERIVATIVE
1134-23-2 CYCLOATE	CYCLOATE
287-23-0 CYCLOBUTANE	CYCLOBUTANE
4806-61-5 CYCLOBUTANE, ETHYL- [PAL]	ETHYL CYCLOBUTANE
81228-87-7 CYCLOBUTYL CHLOROFORMATE	CYCLOBUTYL CHLOROFORMATE

Chemical Name	Reference Name
81228-87-7 CYCLOBUTYLCHLOROFORMATE [NJL, NJL]	CYCLOBUTYL CHLOROFORMATE
12663-46-6 CYCLOCHLOROTINE	CYCLOCHLOROTINE
294-62-2 CYCLODODECANE	CYCLODODECANE
3194-55-6 CYCLODODECANE, 1,2,5,6,9,10-HEXABROMO- [TSC, TSC, TSC]	HEXABROMOCYCLODODECANE
25637-99-4 CYCLODODECANE, HEXABROMO-	CYCLODODECANE, HEXABROMO-
4904-61-4 1,5,9-CYCLODODECATRIENE	1,5,9-CYCLODODECATRIENE
7585-39-9 CYCLOHEPTAAMYLOSE	CYCLOHEPTAAMYLOSE
291-64-5 CYCLOHEPTANE	CYCLOHEPTANE
544-25-2 CYCLOHEPTATRIENE	CYCLOHEPTATRIENE
628-92-2 CYCLOHEPTENE	CYCLOHEPTENE
106-51-4 2,5-CYCLOHEXADIENE, 1,4-DIONE- [PAL]	QUINONE
106-51-4 2,5-CYCLOHEXADIENE-1,4-DIONE (QUINONE) [TSC, TSC, TSC]	QUINONE
68-76-8 2,5-CYCLOHEXADIENE, 1,4-DIONE, 2,3,5-TRIS(1-AZIRIDINYL)- [PAL, PAL]	TRIS(AZIRIDINYL)-P-BENZOQUINONE
108-91-8 CYCLOHEXANAMINE [PAL]	CYCLOHEXYLAMINE
10108-56-2 CYCLOHEXANAMINE, N-BUTYL- [PAL]	N-BUTYLCYCLOHEXYLAMINE
101-83-7 CYCLOHEXANAMINE, N-CYCLOHEXYL- [PAL]	DICYCLOHEXYLAMINE
5459-93-8 CYCLOHEXANAMINE, N-ETHYL- [PAL]	N-ETHYLCYCLOHEXYLAMINE
5432-61-1 CYCLOHEXANAMINE, N-(2-ETHYLHEXYL)- [PAL]	N-2-(ETHYLHEXYL)-CYCLOHEXYLAMINE
1195-42-2 CYCLOHEXANAMINE, N-(1-METHYLETHYL)- [PAL]	ISOPROPYL CYCLOHEXYLAMINE
110-82-7 CYCLOHEXANE	CYCLOHEXANE
14228-73-0 CYCLOHEXANE, 1,4-BIS[2,3-EPOXYPROPOXY) METHYL]- [TSC]	OXIRANE, 2,2'-[1,4-CYCLOHEXANEDILBIS (METHYLENEOXYMETHYLENE)
10347-54-3 CYCLOHEXANE, 1,4-BIS(ISOCYANATOMETHYL)-	CYCLOHEXANE, 1,4-BIS(ISOCYANATOMETHYL)-
38661-72-2 CYCLOHEXANE, 1,3-BIS(ISOCYANATOMETHYL)-	CYCLOHEXANE, 1,3-BIS(ISOCYANATOMETHYL)-
7027-11-4 CYCLOHEXANECARBONITRILE, 1,3,3-TRIMETHYL-5-OXO-,	CYCLOHEXANECARBONITRILE, 1,3,3-TRIMETHYL-5-OXO-,
98-89-5 CYCLOHEXANE CARBOXYLIC ACID	CYCLOHEXANE CARBOXYLIC ACID
542-18-7 CYCLOHEXANE, CHLORO- [PAL]	CYCLOHEXYL CHLORIDE
2615-25-0 CIS- AND TRANS-1,4-CYCLOHEXANEDIAMINE	CIS- AND TRANS-1,4-CYCLOHEXANEDIAMINE
2615-25-0 (TRANS) 1,4-CYCLOHEXANEDIAMINE [TSE]	CIS- AND TRANS-1,4-CYCLOHEXANEDIAMINE
15827-56-2 (CIS) 1,4-CYLCOHEXANEDIAMINE	(CIS) 1,4-CYLCOHEXANEDIAMINE
3322-93-8 CYCLOHEXANE, 1,2-DIBROMO-4-(1,2-DIBROMOETHYL)- [TSC, TSC, TSC]	1,2-DIBROMO-4-(1,2-DIBROMOETHYL) CYCLOHEXANE
5493-45-8 1,2-CYCLOHEXANEDICARBOXYLIC ACID, BIS (OXIRANYLMETHYL) ESTER	1,2-CYCLOHEXANEDICARBOXYLIC ACID, BIS(OXIRANYLMETHYL) ESTER
RR-01653-5 1,2-CYCLOHEXANEDICARBOXYLIC ACID, 2,2-BIS[[[[2-[(OXIRANYLMETHOXY)CARBONYL] CYCLOHEXYLCARBONYL]OXY]METHYL]-1,3-PROPANEDIYL BIS(OXIRANYLMETHYL) ESTER	1,2-CYCLOHEXANEDICARBOXYLIC ACID, 2,2-BIS[[[[2-[(OXIRANYLMETD AMINE SALT
1331-43-7 CYCLOHEXANE, DIETHYL- [PAL]	DIETHYLCYCLOHEXANE
2556-36-7 CYCLOHEXANE, 1,4-DIISOCYANATO-	CYCLOHEXANE, 1,4-DIISOCYANATO-
2556-36-7 1,4-CYCLOHEXANE DIISOCYANATE [313]	CYCLOHEXANE, 1,4-DIISOCYANATO-
105-08-8 1,4-CYCLOHEXANE DIMETHANOL	1,4-CYCLOHEXANE DIMETHANOL
589-90-2 CYCLOHEXANE, 1,4-DIMETHYL- [PAL]	1,4-DIMETHYLCYCLOHEXANE
591-21-9 CYCLOHEXANE, 1,3-DIMETHYL- [PAL]	1,3-DIMETHYL CYCLOHEXANE

Chemical Name	Reference Name
624-29-3 CYCLOHEXANE, 1,4-DIMETHYL-, CIS- [PAL]	1,4-DIMETHYLCYCLOHEXANE-CIS
2207-04-7 CYCLOHEXANE, 1,4-DIMETHYL-, TRANS- [PAL]	1,4-DIMETHYLCYCLOHEXANE-TRANS
1678-91-7 CYCLOHEXANE, ETHYL- [PAL]	ETHYL CYCLOHEXANE
68239-06-5 CYCLOHEXANE, 2-HEPTYL-3,4-BIS(9-ISO-CYANATONONYL)-1-PENTYL-	CYCLOHEXANE, 2-HEPTYL-3,4-BIS(9-ISO-CYANATONONYL)-1-PENTYL-TETRAKIS-, TETRAPOTASSIUM SALT
608-73-1 CYCLOHEXANE, 1,2,3,4,5,6-HEXACHLORO- [PAL, PAL]	HEXACHLOROCYCLOHEXANE (MIXED ISOMERS)
319-84-6 CYCLOHEXANE, 1,2,3,4,5,6-HEXACHLORO-, (1.ALPHA.,2.ALPHA.,3.BETA.,4.ALPHA.,5.BETA.,6.BETA.)- [PAL, PAL, PAL, PAL]	ALPHA-BHC, (13.ALPHA.,14.ALPHA.)-
319-85-7 CYCLOHEXANE, 1,2,3,4,5,6-HEXACHLORO-, (1.ALPHA.,2.BETA.,3.ALPHA.,4.BETA.,5.ALPHA.,6.BETA.)- [PAL, PAL]	BETA-BHC3.BETA.,4.ALPHA.,5.BETA.,6.BETA.)-
319-86-8 CYCLOHEXANE, 1,2,3,4,5,6-HEXACHLORO-, (1.ALPHA.,2.ALPHA.,3.ALPHA.,4.BETA.,5.ALPHA.,6.BETA.)- [PAL]	DELTA-BHC3.ALPHA.,4.BETA.,5.ALPHA.,6.BETA.)-
3173-53-3 CYCLOHEXANE, ISOCYANATO- [TSC, TSC, TSC]	CYCLOHEXYL ISOCYANATE
4098-71-9 CYCLOHEXANE, 5-ISOCYANATO-1-(ISO-CYANATOMETHYL)-1,3,3-TRIMETHYL- [PAL]	ISOPHORONE DIISOCYANATE
4098-71-9 CYCLOHEXANE, 5-ISOCYANATO-1-(ISO-CYANATOMETHYL)-1,3,3-TRIMETHYL- [TSC, TSC, TSC]	ISOPHORONE DIISOCYANATE
108-87-2 CYCLOHEXANE, METHYL- [PAL]	METHYLCYCLOHEXANE
5124-30-1 CYCLOHEXANE, 1,1'-METHYLENEBIS[4-ISO-CYANATO- [PAL]	METHYLENE BIS(4-CYCLOHEXYLISOCYANATE)
5124-30-1 CYCLOHEXANE, 1,1'-METHYLENEBIS(4-ISO-CYANATO- [TSC, TSC, TSC]	METHYLENE BIS(4-CYCLOHEXYLISOCYANATE)
1122-60-7 CYCLOHEXANE, NITRO- [PAL]	NITROCYCLOHEXANE
87-84-3 CYCLOHEXANE, 1,2,3,4,5-PENTABROMO-6-CHLORO- [TSC, TSC, TSC, TSC]	1,2,3,4,5-PENTABROMO-6-CHLORO-CYCLOHEXANE
30554-72-4 CYCLOHEXANE, TETRABROMODICHLORO-	CYCLOHEXANE, TETRABROMODICHLORO-
1569-69-3 CYCLOHEXANETHIOL	CYCLOHEXANETHIOL
30554-73-5 CYCLOHEXANE, TRIBROMOCHLORO-	CYCLOHEXANE, TRIBROMOCHLORO-
30554-73-5 CYCLOHEXANE, TRIBROMOTRICHLORO- [TSC, TSC, TSC]	CYCLOHEXANE, TRIBROMOCHLORO-
108-93-0 CYCLOHEXANOL	CYCLOHEXANOL
25639-42-3 CYCLOHEXANOL, METHYL- [PAL]	METHYLCYCLOHEXANOL
116-02-9 CYCLOHEXANOL, 3,3,5-TRIMETHYL- [PAL]	3,3,5-TRIMETHYLCYCLOHEXANOLPHENYL] ESTER
1321-60-4 CYCLOHEXANOL, TRIMETHYL- [PAL]	TRIMETHYLCYCLOHEXANOL
108-94-1 CYCLOHEXANONE	CYCLOHEXANONE
1867-66-9 CYCLOHEXANONE, 2-(O-CHLOROPHENYL)-2-(METHYLAMINE)-, HYDROCHLORIDE	CYCLOHEXANONE, 2-(O-CHLOROPHENYL)-2-(METHYLAMINE)-, HYDROCHL
583-60-8 CYCLOHEXANONE, 2-METHYL- [PAL]	O-METHYLCYCLOHEXANONE
100-64-1 CYCLOHEXANONE OXIME	CYCLOHEXANONE OXIME
12262-58-7 CYCLOHEXANONE PEROXIDE	CYCLOHEXANONE PEROXIDE
110-83-8 CYCLOHEXENE	CYCLOHEXENE
100-50-5 3-CYCLOHEXENE-1-CARBOXALDEHYDE [PAL, TSC, TSC, TSC]	1,2,3,6-TETRAHYDROBENZALDEHYDE
27939-60-2 3-CYCLOHEXENE-1-CARBOXALDEHYDE, DIMETHYL-	3-CYCLOHEXENE-1-CARBOXALDEHYDE, DIMETHYL-
31906-04-4 3-CYCLOHEXENE-1-CARBOXALDEHYDE, 4-(4-HYDROXY-4-METHYLPENTYL)-	3-CYCLOHEXENE-1-CARBOXALDEHYDE, 4-(4-HYDROXY-4-METHYLPENTYL)

Chemical Name	Reference Name
52475-86-2 3-CYCLOHEXENE-1-CARBOXALDEHYDE, 1-METHYL-4-(4-METHYL-3-PENTENYL)-	3-CYCLOHEXENE-1-CARBOXALDEHYDE, 1-METHYL-4-(4-METHYL-3-PENTEWAMINE OXIDE, PHOSPHATE
66327-54-6 3-CYCLOHEXENE-1-CARBOXALDEHYDE, 1-METHYL-4-(4-METHYLPENTYL)-	3-CYCLOHEXENE-1-CARBOXALDEHYDE, 1-METHYL-4-(4-METHYLPENTYL)-CHLORO-1-(CHLOROMETHYL) ETHYL] ESTER
37677-14-8 3-CYCLOHEXENE-1-CARBOXALDEHYDE, 4-(4-METHYL-3-PENTENYL)-	3-CYCLOHEXENE-1-CARBOXALDEHYDE, 4-(4-METHYL-3-PENTENYL)-
1423-46-7 3-CYCLOHEXENE-1-CARBOXALDEHYDE, 2,4,6-TRIMETHYL-	3-CYCLOHEXENE-1-CARBOXALDEHYDE, 2,4,6-TRIMETHYL-
1321-16-0 CYCLOHEXENECARBOXALDEHYDE [PAL]	TETRAHYDROBENZALDEHYDE
100-40-3 CYCLOHEXENE, 4-ETHENYL- [PAL]	4-VINYLCYCLOHEXENE
591-47-9 CYCLOHEXENE, 4-METHYL- [PAL]	4-METHYLCYCLOHEXENE
53980-88-4 2-CYCLOHEXENE-1-OCTANOIC ACID, 5(OR 6)-CARBOXY-4-HEXYL-	2-CYCLOHEXENE-1-OCTANOIC ACID, 5(OR 6)-CARBOXY-4-HEXYL-
930-68-7 2-CYCLOHEXENE-1-ONE	2-CYCLOHEXENE-1-ONE
286-20-4 CYCLOHEXENE OXIDE [CSR]	7-OXABICYCLO[4.1.0]HEPTANE
930-68-7 CYCLOHEXENONE [NFP, NFP]	2-CYCLOHEXENE-1-ONE
78-59-1 2-CYCLOHEXEN-1-ONE, 3,5,5-TRIMETHYL- [PAL, TSC, TSC, TSC]	ISOPHORONE
10137-69-6 CYCLOHEXENYL TRICHLOROSILANE	CYCLOHEXENYL TRICHLOROSILANE
66-81-9 CYCLOHEXIMIDE	CYCLOHEXIMIDE
622-45-7 CYCLOHEXYL ACETATE	CYCLOHEXYL ACETATE
108-91-8 CYCLOHEXYLAMINE	CYCLOHEXYLAMINE
108-91-8 CYCLOHEXYLAMINE [CYCLOHEXANAMINE] [EP3]	CYCLOHEXYLAMINE
827-52-1 CYCLOHEXYLBENZENE	CYCLOHEXYLBENZENE
95-33-0 N-CYCLOHEXYL-2-BENZOTHIAZOLESULFE-NAMIDE	N-CYCLOHEXYL-2-BENZOTHIAZOLESULFENAMIDE
542-18-7 CYCLOHEXYL CHLORIDE	CYCLOHEXYL CHLORIDE
6531-86-8 CYCLOHEXYLCYCLOHEXANOL	CYCLOHEXYLCYCLOHEXANOL
131-89-5 2-CYCLOHEXYL-4,6-DINITROPHENOL [EPA, RCU, RCU]	4,6-DINITRO-O-CYCLOHEXYLPHENOL
RR-00828-6 CYCLOHEXYL FORMATE	CYCLOHEXYL FORMATE
3173-53-3 CYCLOHEXYL ISOCYANATE	CYCLOHEXYL ISOCYANATE
1569-69-3 CYCLOHEXYL MERCAPTAN [HM1, HM1, NJL]	CYCLOHEXANETHIOL
119239-21-3 CYCLOHEXYL METHACRYLATE, 2-HYDROXYLETHYL METHACRYLATE, ISODECYL METHACRYLATE, 2-MORPHOLINOETHYL METHACRYLATE COPOLYMER	CYCLOHEXYL METHACRYLATE, 2-HYDROXYLETHYL METHACRYLATE, ISODE
119-42-6 O-CYCLOHEXYLPHENOL	O-CYCLOHEXYLPHENOL
98-12-4 CYCLOHEXYL TRICHLOROSILANE [NJL, NJL]	CYCLOHEXYLTRICHLOROSILANE
98-12-4 CYCLOHEXYLTRICHLOROSILANE	CYCLOHEXYLTRICHLOROSILANE
121-82-4 CYCLONITE	CYCLONITE
121-82-4 CYCLONITE [QBC, QBC]	CYCLONITE
1552-12-1 1,5-CYCLOOCTADIENE	1,5-CYCLOOCTADIENE
29965-97-7 CYCLOOCTADIENE	CYCLOOCTADIENE
29965-97-7 CYCLOOCTADIENES [HM1, HM1]	CYCLOOCTADIENE
629-20-9 CYCLOOCTATETRAENE	CYCLOOCTATETRAENE
27208-37-3 CYCLOPENTA[C,D]PYRENE	CYCLOPENTA[C,D]PYRENE
542-92-7 CYCLOPENTADIENE	CYCLOPENTADIENE
542-92-7 1,3-CYCLOPENTADIENE [FLA, PAL, WHM]	CYCLOPENTADIENE
77-47-4 1,3-CYCLOPENTADIENE, 1,2,3,4,5,5-HEXACHLORO- [TSC, TSC, TSC, PAL]	HEXACHLOROCYCLOPENTADIENEETRAHYDRO-

Chemical Name	Reference Name
287-92-3 CYCLOPENTANE	CYCLOPENTANE
1640-89-7 CYCLOPENTANE, ETHYL- [PAL]	ETHYL CYCLOPENTANE
745-65-3 CYCLOPENTANEHEPTANOIC ACID, 3-.ALPHA.-HYDROXY-2-(3-HYDROXY-1-OCTENYL)-5-OXO-	CYCLOPENTANEHEPTANOIC ACID, 3-.ALPHA.-HYDROXY-2-(3-HYDROXY-1IISOPROPYLPHOSPHO-RODITHIOATE
96-37-7 CYCLOPENTANE, METHYL- [PAL]	METHYLCYCLOPENTANE
96-41-3 CYCLOPENTANOL	CYCLOPENTANOL
120-92-3 CYCLOPENTANONE	CYCLOPENTANONE
142-29-0 CYCLOPENTENE	CYCLOPENTENE
4501-58-0 3-CYCLOPENTENE-1-ACETALDEHYDE, 2,2,3-TRIMETHYL-	3-CYCLOPENTENE-1-ACETALDEHYDE, 2,2,3-TRIMETHYL-HYDROXYPHENYL)AZO] [1,1'-BIPHENYL]-4-YL]-AZO]-3-[(4-NITRO PHENYL)AZO]-, DISODIUM SALT
50-18-0 CYCLOPHOSPHAMIDE	CYCLOPHOSPHAMIDE
50-18-0 CYCLOPHOSPHAMIDE (ANHYDROUS) [C65, C65, C65, C65, C65]	CYCLOPHOSPHAMIDE
6055-19-2 CYCLOPHOSPHAMIDE C	CYCLOPHOSPHAMIDE C
6055-19-2 CYCLOPHOSPHAMIDE (HYDRATED) [C65, C65, C65, C65, C65]	CYCLOPHOSPHAMIDE C
69430-24-6 CYCLOPOLYDIMETHYLSILOXANE [TSC, TSC]	CYCLOSILANES, DIMETHYL-)METHYLAMMONIO] METHYLETHYL]-OMEGA-HYDROXY-, N,N'-DITALLOW ACYL DERIVATIVES, METHYL SULFATES (SALTS)
75-19-4 CYCLOPROPANE	CYCLOPROPANE
121-21-1 CYCLOPROPANECARBOXYLIC ACID, 2,2-DIMETHYL-3-(2-METHYL-1-PROPENYL)-,2-METHYL-4-OXO-3-(2,4-PENTADIENYL)-2-CYCLOPENTEN-1-YL ESTER, [1R[1.ALPHA.[S*()], 3.BETA.]- [PAL]	PYRETHRIN I
121-29-9 CYCLOPROPANECARBOXYLIC ACID, 3-(3-METHOXY-2-METHYL-3-OXO-1-PROPENYL)-2,2-DIMETHYL-,2-METHYL-4-OXO-3-(2,4-PENTA DIENYL)-2-CYCLOPENTEN-1-YL ESTER, [R-[1.ALPHA.[3*(Z)], 3.BETA.(E)]- [PAL]	PYRETHRIN IIPROPENYL)-,2-METHYL-4-OXO-3-(2,4-PENTADIENYL)-2-CYCLOPENTEN-1-YL ESTER, [1R [1.ALPHA.[S*()],3.BETA.]-
69430-24-6 CYCLOSILANES, DIMETHYL-	CYCLOSILANES, DIMETHYL-
59865-13-3 CYCLOSPORIN [NTP]	CYCLOSPORIN A
59865-13-3 CYCLOSPORIN A	CYCLOSPORIN A
79217-60-0 CYCLOSPORINE	CYCLOSPORINE
2691-41-0 CYCLOTETRAMETHYLENETETRANITRAMINE	CYCLOTETRAMETHYLENETETRANITRAMINEER
121-82-4 CYCLOTRIMETHYLENE-TRINITRAMINE [CAL]	CYCLONITE
121-82-4 CYCLOTRIMETHYLENETRINITRAMINE (RDX) [ATS]	CYCLONITE
68359-37-5 CYFLUTHRIN [3-(2,2-DICHLOROETHENYL)-2, 2-DIMETHYLCYCLOPROPANECARBOXYLIC ACID, CYANO(4-FLUORO-3-PHENOXYPHENYL) METHYL ESTER]	CYFLUTHRIN [3-(2,2-DICHLOROETHENYL)-2, 2-DIMETHYLCYCLOPROPANEETHYLCYCLO-PROPANECARBOXYLIC ACID CYANO(3-PHENOXYPHENYL) METHYL ESTER]
68085-85-8 CYHALOTHRIN [3-(2-CHLORO-3,3,3-TRIFLUORO-1-PROPENYL)-2,2-DIMETHYLCYCLOPROPANECARBOXYLIC ACID CYANO(3-PHENOXYPHENYL) METHYL ESTER]	CYHALOTHRIN [3-(2-CHLORO-3,3,3-TRIFLUORO-1-PROPENYL)-2,2-DIM-[4-(TRIFLUOROMETHYL)PHENYL]-1-[2-[4-(TRIFLUOROMETHYL)PHENYL]ETHENYL]-2-PROPENYLIDENE]HYDRAZONE]
13121-70-5 CYHEXATIN	CYHEXATIN
13121-70-5 CYHEXATIN [QBC]	CYHEXATIN
13121-70-5 CYHEXATIN (ISO) [GBR, GBR]	CYHEXATIN
13121-70-5 CYHEXATIN (TRICYCLOHEXYLTIN HYDROX-IDE) [CN8]	CYHEXATIN
99-87-6 P-CYMENE	P-CYMENE
25155-15-1 CYMENE	CYMENE
25155-15-1 CYMENES [HM1, HM1]	CYMENE

Chemical Name	Reference Name
52315-07-8 CYPERMETHRIN	CYPERMETHRIN
66215-27-8 CYROMAZINE	CYROMAZINE
52-90-4 L-CYSTEINE	L-CYSTEINE
147-94-4 CYSTOSINE ARABINOSIDE	CYSTOSINE ARABINOSIDE
147-94-4 CYTARABINE [C65, CAB, MSL]	CYSTOSINE ARABINOSIDE
147-94-4 CYTARBINE [CSR]	CYSTOSINE ARABINOSIDE
21739-91-3 CYTEMBENA	CYTEMBENA
14930-96-2 CYTOCHALASIN B	CYTOCHALASIN B
4465-94-5 CYTOXAL ALCOHOL	CYTOXAL ALCOHOL
94-75-7 2,4-D	2,4-D
94-75-7 2,4-D (2,4-DICHLOROPHENOXYACETIC ACID) [UTS]	2,4-D
94-75-7 2,4-D [ACETIC ACID, (2,4-DICHLOROPHENOXY) -] [313]	2,4-D
94-75-7 2,4-D (2,4-DICHLOROPHENOXYACETIC ACID) [BC1, BC1]	2,4-D
94-75-7 2,4-D (DICHLORYLPHENOXYACETIC ACID) [OS1, OS1]	2,4-D
94-75-7 2,4-D ACID [CWA]	2,4-D
1929-73-3 2,4-D BUTOXYETHYL ESTER	2,4-D BUTOXYETHYL ESTER
94-80-4 2,4-D BUTYL ESTER [313, CWA]	N-BUTYL 2,4-D ESTER
2971-38-2 2,4-D CHLOROCROTYL ESTER	2,4-D CHLOROCROTYL ESTER
94-79-1 2,4-D ESTER [CWA]	SEC-BUTYL 2,4-D ESTER
25168-26-7 2,4-D ESTER (2,4-D ISOOCTYL ESTER) [CWA]	2,4-D ISOOCTYL ESTER
94-11-1 2,4-D ESTERS	2,4-D ESTERS
25168-26-7 2,4-D ESTERS [MSL, NJL, NJL, NJL, NJL, NJL, NJL, NJL, NJL]	2,4-D ISOOCTYL ESTER
53467-11-1 2,4-D ESTERS	2,4-D ESTERS
25168-26-7 2,4-D ISOOCTYL ESTER	2,4-D ISOOCTYL ESTER
1928-38-7 2,4-D METHYL ESTER	2,4-D METHYL ESTER
1928-61-6 2,4-D PROPYL ESTER	2,4-D PROPYL ESTER
1320-18-9 2,4-D PROPYLENE GLYCOL BUTYL ETHER ESTER	2,4-D PROPYLENE GLYCOL BUTYL ETHER ESTER
94-75-7 2,4-D, SALTS AND ESTERS [CAA]	2,4-D
94-75-7 2,4-D, SALTS & ESTERS [RCU, RCU]	2,4-D
4403-90-1 D AND C GREEN NO. 5	D AND C GREEN NO. 5
3468-63-1 D&C ORANGE NO. 17	D&C ORANGE NO. 17
596-03-2 D&C ORANGE NO. 5	D&C ORANGE NO. 5
2092-56-0 D&C RED NO. 8	D&C RED NO. 8
5160-02-1 D & C RED NO. 9	D & C RED NO. 9
5160-02-1 D&C RED NO. 9 [C65, C65]	D & C RED NO. 9
5160-02-1 D & C RED NO. 9 (CI PIGMENT RED 53:1) [IAR]	D & C RED NO. 9
81-88-9 D&C RED NO. 19 [C65]	C.I. FOOD RED 15
8003-22-3 D & C YELLOW NO. 11	D & C YELLOW NO. 11
94-75-7 2,4-D (ISO) [GBR, GBR]	2,4-D
RR-00631-5 D001-IGNITABLE UNLISTED HAZARDOUS WASTES	D001-IGNITABLE UNLISTED HAZARDOUS WASTES
RR-00632-6 D002-CORROSIVE UNLISTED HAZARDOUS WASTES	D002-CORROSIVE UNLISTED HAZARDOUS WASTES
RR-00633-7 D003-REACTIVE UNLISTED HAZARDOUS WASTES	D003-REACTIVE UNLISTED HAZARDOUS WASTES

Chemical Name	Reference Name
4342-03-4 DACARBAZINE	DACARBAZINE
94-75-7 2,4-D ACID [ATS]	2,4-D
75-99-0 DALAPON [SDW, SDW]	2,2-DICHLOROPROPIONIC ACID
1596-84-5 DAMINOZIDE	DAMINOZIDE
17230-88-5 DANAZOL	DANAZOL
81-88-9 D AND C RED NO. 19 [CAB]	C.I. FOOD RED 15
5281-04-9 D AND C RED NO 7 [OTS]	LITHOL RUBINE
117-10-2 DANTRON (CHRYSAZIN; 1,8-DIHYDROXYAN-THRAQUINONE) [C65, C65, IAR]	1,8-DIHYDROXYANTHRAQUINONE
80-08-0 DAPSONE	DAPSONE
20830-81-3 DAUNOMYCIN	DAUNOMYCIN
23541-50-6 DAUNORUBICIN HYDROCHLORIDE	DAUNORUBICIN HYDROCHLORIDE
12011-76-6 DAWSONITE	DAWSONITE
533-74-4 DAZOMET (TETRAHYDRO-3,5-DIMETHYL-2H-1,3,5-THIADIAZINE-2-THIONE) [313]	TETRAHYDRO-3,5-DIMETHYL-2H-1,3,5-THIADIAZINE-2-THIONE
53404-60-7 DAZOMET, SODIUM SALT [TETRAHYDRO-3,5-DIMETHYL-2H-1,3,5-THIADIAZINE-2-THIONE, ION(1-), SODIUM]	DAZOMET, SODIUM SALT [TETRAHYDRO-3,5-DIMETHYL-2H-1,3,5-THIAD
94-82-6 2,4-DB [313]	2,4-DB (2,4-DICHLORO-PHENOXYBUTYRIC ACID)
94-82-6 2,4-DB (2,4-DICHLORO-PHENOXYBUTYRIC ACID)	2,4-DB (2,4-DICHLORO-PHENOXYBUTYRIC ACID)
96-12-8 DBCP [NJL, NJL]	1,2-DIBROMO-3-CHLOROPROPANE
96-12-8 DBCP (1,2-DIBROMO-3-CHLOROPROPANE) [CN2]	1,2-DIBROMO-3-CHLOROPROPANE
1861-32-1 DCPA (AND ITS ACID METABOLITES) [SD4]	TEREPHTHALIC ACID, TETRACHLORO-, DIMETHYL ESTER
53-19-0 O,P'-DDD	O,P'-DDD
72-54-8 DDD	DDD
72-54-8 4-4'-DDD (P-P'-TDE) [MX1]	DDD
72-54-8 4,4'-DDD [PQL]	DDD
72-54-8 P,P'-DDD [RCA, UTS]	DDD
72-54-8 4,4'-DDD (P,P-TDE) [CW2]	DDD
72-54-8 DDD, P,P'- [ATS]	DDD
72-54-8 DDD (DICHLORODIPHENYLDICHLORO-ETHANE) [C65]	DDD
72-55-9 DDE	DDE
72-55-9 P,P'-DDE [MSL]	DDE
72-55-9 4-4'-DDE (P-P'-DDX) [MX1]	DDE
72-55-9 4,4'-DDE [PQL]	DDE
72-55-9 P,P'-DDE [RCA, UTS]	DDE
3424-82-6 O,P'-DDE	O,P'-DDE
3547-04-4 DDE	DDE
72-55-9 4,4'-DDE (P,P-DDX) [CW2]	DDE
72-55-9 DDE, P,P'- [ATS]	DDE
72-55-9 DDE (DICHLORODIPHENYLDICHLORO-ETHYLENE) [C65]	DDE
50-29-3 DDT	DDT
50-29-3 4,4'-DDT [CW2]	DDT
50-29-3 4-4'-DDT [MX1]	DDT
50-29-3 4,4'-DDT [PQL]	DDT

Chemical Name	Reference Name
50-29-3 P,P'-DDT [RCA, UTS]	DDT
789-02-6 O,P'-DDT	O,P'-DDT
50-29-3 DDT (DICHLORODIPHENYLTRICHLOROSILANE) [TLV]	DDT
50-29-3 DDT (1,1,1-TRICHLORO-2,2-BIS(P-CHLOR-PHENYL) ETHANE) [CAL]	DDT
50-29-3 DDT (ZEIDANE) [QBC, QBC]	DDT
50-29-3 DDT, P,P'- [ATS]	DDT
50-29-3 DDT (DICHLORODIPHENYLTRICHLOROETHANE) [AUS, BC1, BC1, TSC, C65]	DDT
50-29-3 DDT (1,1,1-TRICHLORO-2,2-BIS (4-CHLOROPHENYL) ETHANE) [CN2]	DDT
50-29-3 DDT AND METABOLITES [CW3]	DDT
RR-00080-6 DDT, DDE, AND DDD (IN COMBINATION)	DDT, DDE, AND DDD (IN COMBINATION)
RR-00628-0 DDT METABOLITES	DDT METABOLITES
62-73-7 DDVP (DICHLORVOS) [C65, C65]	DICHLORVOS
114-26-1 DDVP [FLA, MSL]	PROPOXUR
62-73-7 DDVP (DICHLORVOS) [ALB, ALB, ALB]	DICHLORVOS
17702-41-9 DECABORANE	DECABORANE
17702-41-9 DECABORANE (14) [302]	DECABORANE
17702-41-9 DECABORANE(14) [PAL]	DECABORANE
1163-19-5 DECABROMOBIPHENYL ETHER [NJL]	DECABROMODIPHENYL OXIDE
1163-19-5 DECABROMODIPHENYL ETHER [OTS, TSC, TSC, TSC, TSC]	DECABROMODIPHENYL OXIDEMETHYLENE-, [1R-(1R*, 4R*,6R*,10S*)]-
1163-19-5 DECABROMODIPHENYL OXIDE	DECABROMODIPHENYL OXIDE
25152-84-5 2,4-DECADIENAL	2,4-DECADIENAL
91-17-8 DECAHYDRONAPHTHALENE	DECAHYDRONAPHTHALENE
91-17-8 DECALIN [CSR]	DECAHYDRONAPHTHALENE
541-02-6 DECAMETHYLCYCLOPENTASILOXANE	DECAMETHYLCYCLOPENTASILOXANE
141-62-8 DECAMETHYLTETRASILOXANE	DECAMETHYLTETRASILOXANE
112-31-2 DECANAL	DECANAL
2016-57-1 1-DECANAMINE [PAL]	DECYLAMINEPHENYL)AZO][1,1'-BIPHENYL]-4-YL]AZO]-5-HYDROXY-6-(PHENYLAZO)-,DISODIUM SALT
100545-50-4 1-DECANAMINE, N-DECYL-N-METHYL-N-OXIDE	1-DECANAMINE, N-DECYL-N-METHYL-N-OXIDE-OMEGA-((TETRAHYDRO-2-FURANYL)METHOXY)-
124-18-5 DECANE	DECANE
124-18-5 N-DECANE [HM1, HM1, OTS]	DECANE
111-20-6 DECANEDIOIC ACID	DECANEDIOIC ACID
143-10-2 1-DECANETHIOL	1-DECANETHIOL
30174-58-4 TERT-DECANETHIOL	TERT-DECANETHIOL
122-62-3 DECANOIC ACID, BIS(2-ETHYLHEXYL) ESTER	DECANOIC ACID, BIS(2-ETHYLHEXYL) ESTER
112-30-1 1-DECANOL	1-DECANOL
112-30-1 DECANOL [NFP, NFP]	1-DECANOL
693-54-9 2-DECANONE	2-DECANONE
762-12-9 DECANOYL PEROXIDE	DECANOYL PEROXIDE
13654-09-6 DECARBROMOBIPHENYL	DECARBROMOBIPHENYL
872-05-9 1-DECENE	1-DECENE
872-05-9 DECENE, N- [OTS]	1-DECENE
2156-96-9 DECYL ACRYLATE	DECYL ACRYLATE

Chemical Name	Reference Name
68526-90-9 DECYL ALCOHOL BOTTOMS	DECYL ALCOHOL BOTTOMS
26183-52-8 DECYL ALCOHOL, ETHOXYLATED	DECYL ALCOHOL, ETHOXYLATED
2016-57-1 DECYLAMINE	DECYLAMINE
104-72-3 DECYLBENZENE	DECYLBENZENE
1322-98-1 DECYLBENZENESULFONIC ACID, SODIUM SALT [PST]	SODIUM DECYLBENZENESULFONATE
41444-55-7 DECYL GLUCOSIDE	DECYL GLUCOSIDE
30174-58-4 TERT-DECYLMERCAPTAN [FLA, MSL, NFP, NFP]	TERT-DECANETHIOL
RR-00830-0 DECYLNAPHTHALENE	DECYLNAPHTHALENE
RR-00829-7 DECYL NITRATE	DECYL NITRATE
9038-29-3 (DECYLOXY)POLY(OXYETHYLENE)POLY(OXYPROPYLENE) [PST]	OXIRANE, METHYL-, POLYMER WITH OXIRANE, DECYL ETHER
18760-44-6 3-(DECYLOXY)TETRAHYDROTHIOPHENE 1,1-DIOXIDE	3-(DECYLOXY)TETRAHYDROTHIOPHENE 1,1-DIOXIDE
70191-75-2 DECYL PHENOXYBENZENEDISULFONIC ACID	DECYL PHENOXYBENZENEDISULFONIC ACID
36445-71-3 DECYL PHENOXYBENZENEDISULFONIC ACID, DISODIUM SALT	DECYL PHENOXYBENZENEDISULFONIC ACID, DIS-ODIUM SALT
84-77-5 DECYL PHTHALATE	DECYL PHTHALATE
126-86-3 5-DECYNE-4,7-DIOL, 2,4,7,9-TETRAMETHYL-	5-DECYNE-4,7-DIOL, 2,4,7,9-TETRAMETHYL-
78-48-8 DEF	DEF7-METHANO-1H-INDENE]
80584-98-1 DEHYDROABIETYLAMINE, ETHOXYLATED	DEHYDROABIETYLAMINE, ETHOXYLATEDENYL-
51344-62-8 DEHYDROABIETYLAMINE-ETHYLENE OXIDE ADDUCT	DEHYDROABIETYLAMINE-ETHYLENE OXIDE ADDUCT
51344-62-8 DEHYDROABIETYLAMINE-ETHYLENE OXIDE CONDENSATE [PST]	DEHYDROABIETYLAMINE-ETHYLENE OXIDE ADDUCT
520-45-6 DEHYDROACETIC ACID	DEHYDROACETIC ACID
84-17-3 DEHYDROSTILBESTROL [FLA]	DIENOESTROL
319-86-8 DELTA-BHC-DELTA [MX1]	DELTA-BHC
52918-63-5 DELTAMETHRIN	DELTAMETHRIN
64-73-3 DEMECLOCYCLINE HYDROCHLORIDE	DEMECLOCYCLINE HYDROCHLORIDE
8022-00-2 DEMETHON-METHYL [MAK, MAK, MAK]	METHYL DEMETON
8065-48-3 DEMETON	DEMETON
8065-48-3 DEMETON (SYSTOX) [BC1, BC1, BC1]	DEMETON
298-03-3 DEMETON-O	DEMETON-O
126-75-0 DEMETON-S	DEMETON-S
8022-00-2 DEMETON-METHYL, MIXTURE OF O AND S [QBC, QBC]	METHYL DEMETON
919-86-8 DEMETON-S-METHYL	DEMETON-S-METHYL
8065-48-3 DEMETON, MIXTURE OF O AND S (SYSTOX) (R) [QBC, QBC]	DEMETON
8065-48-3 DEMETON (SYSTOX) [OS1, OS1, OS1, OS1]	DEMETON
RR-00113-8 DENATURED ALCOHOL [HM1, HM1, NFP, NFP, PAL]	ALCOHOL, DENATURED
83-44-3 DEOXYCHOLIC ACID	DEOXYCHOLIC ACID
51481-10-8 DEOXYNIVALENOL	DEOXYNIVALENOL
RR-00249-3 DERIVATIVE OF TETRACHLOROETHYLENE	DERIVATIVE OF TETRACHLOROETHYLENE
13684-56-5 DESMEDIPHAM	DESMEDIPHAM
1928-43-4 2,4-D 2-ETHYLHEXYL ESTER	2,4-D 2-ETHYLHEXYL ESTER
53404-37-8 2,4-D 2-ETHYL-4-METHYLPENTYL ESTER	2,4-D 2-ETHYL-4-METHYLPENTYL ESTER-6-METHYL-3-(1-METHYLPROPYL), LITHIUM SALT]
7782-39-0 DEUTERIUM	DEUTERIUM

Chemical Name	Reference Name
9004-53-9 DEXTRIN	DEXTRIN
50-99-7 DEXTROSE [PST]	GLUCOSE
84-74-2 DI-N-BUTYL PHTHALATE [UTS]	DIBUTYL PHTHALATE
621-64-7 DI-N-PROPYLNITROSAMINE [UTS]	N-NITROSODI-N-PROPYLAMINE
123-42-2 DIACETONE ALCOHOL	DIACETONE ALCOHOL
123-42-2 DIACETONE ALCOHOL (4-HYDROXY-4-METHYL-2-PENTANONE) [BC1, BC1, OS1, OS1]	DIACETONE ALCOHOL
123-42-2 DIACETONE ALCOHOL (4-HYDROXY-4-METHYL-2-PENTANONE) [ALB, ALB]	DIACETONE ALCOHOL
54693-46-8 DIACETONE ALCOHOL PEROXIDE	DIACETONE ALCOHOL PEROXIDE
25260-60-0 1,4-DIACETOXYBUT-2-ENE (1,4)	1,4-DIACETOXYBUT-2-ENE (1,4)
83-63-6 DIACETYLAMINOAZOTOLUENE	DIACETYLAMINOAZOTOLUENE
613-35-4 N,N'-DIACETYLBENZIDINE	N,N'-DIACETYLBENZIDINE
110-22-5 DIACETYL PEROXIDE	DIACETYL PEROXIDE
110-22-5 DI-ACETYL PEROXIDE [FLA]	DIACETYL PEROXIDE
RR-01064-0 DIACETYL TARTARIC ACID ESTERS OF MONO AND DIGLYCERIDES OF EDIBLE FATS	DIACETYL TARTARIC ACID ESTERS OF MONO AND DIGLYCERIDES OF ED FATTY ACIDS
RR-01063-9 DI ACETYL TARTARIC ESTERS OF MONO AND DIGLYCERIDES OF EDIBLE FATTY ACIDS	DI ACETYL TARTARIC ESTERS OF MONO AND DIGLYCERIDES OF EDIBLEYPHENYL ISOCYANATE, WITH ETHYLENEDIAMINE AND DIETHYLENETRIA MINE
10311-84-9 DIALIFOR [302, CAL]	DIALIFOS
10311-84-9 DIALIFOS	DIALIFOS
RR-00978-9 DI(ALKANEPOLYOL) ETHER, POLYACRYLATE	DI(ALKANEPOLYOL) ETHER, POLYACRYLATEACRYLATE
RR-00191-2 DIALKENYLAMIDE	DIALKENYLAMIDE
83968-18-7 DIALKYL 79 PHTHALATE	DIALKYL 79 PHTHALATE
RR-00971-2 DIALKYLAMINO ALKANOATE, METAL SALT	DIALKYLAMINO ALKANOATE, METAL SALT, AND BUTADIENE
RR-01245-3 DIALKYLDITHIOPHOSPHORIC ACID, ALIPHATIC AMINE SALT	DIALKYLDITHIOPHOSPHORIC ACID, ALIPHATIC AMINE SALT
RR-00414-8 DIALKYLNITROSAMINES	DIALKYLNITROSAMINES
RR-00961-0 DIALKYL PHOSPHORODITHIOATE PHOSPHATE COMPOUNDS, 2-PROPENAMIDE, N-[3-(DIMETHYLAMINO)PROPYL]-	DIALKYL PHOSPHORODITHIOATE PHOSPHATE COMPOUNDS, 2-PROPENAMIDYANATO-2,2,4(OR 2,4,4)-TRIMETHYLHEXANE, 2-HYDROXYETHYL-ACRYLATED-BLOCKED
2303-16-4 DIALLATE	DIALLATE
2303-16-4 DI-ALLATE [HM1, HM1]	DIALLATE
2303-16-4 DIALLATE [CARBAMOTHIOIC ACID, BIS (1-METHYLETHYL)-,S-(2,3-DICHLORO-2-PROPENYL)ESTER] [313]	DIALLATE
2998-04-1 DIALLYL ADIPATE	DIALLYL ADIPATE
124-02-7 DIALLYLAMINE	DIALLYLAMINE
37764-25-3 N,N-DIALLYL-2,3-DICHLOROACETAMIDE	N,N-DIALLYL-2,3-DICHLOROACETAMIDE
557-40-4 DIALLYL ETHER	DIALLYL ETHER
557-40-4 DIALLYLETHER [FLA, NJL, NJL]	DIALLYL ETHER
999-21-3 DIALLYL MALEATE	DIALLYL MALEATE
131-17-9 DIALLYL PHTHALATE	DIALLYL PHTHALATE
615-05-4 2,4-DIAMINOANISOLE	2,4-DIAMINOANISOLE
615-05-4 2,4-DIAMINOANISOLE AND ITS SALTS [NHS, NHS, NHS]	2,4-DIAMINOANISOLE
39156-41-7 2,4-DIAMINOANISOLE SULFATE	2,4-DIAMINOANISOLE SULFATE
39156-41-7 2,4-DIAMINOANISOLE SULPHATE [FLA]	2,4-DIAMINOANISOLE SULFATE
128-95-0 1,4-DIAMINOANTHRAQUINONE	1,4-DIAMINOANTHRAQUINONE

Chemical Name	Reference Name
91-95-2 3,3'-DIAMINOBENZIDINE	3,3'-DIAMINOBENZIDINE
7411-49-6 3,3'-DIAMINOBENZIDINE TETRAHYDROCHLO-RIDE	3,3'-DIAMINOBENZIDINE TETRAHYDROCHLORIDE
101-80-4 4,4'-DIAMINODIPHENYL ETHER	4,4'-DIAMINODIPHENYL ETHER
101-80-4 4,4'-DIAMINODIPHENYL ETHER (4,4'-OXYDIAN-ILINE) [C65, C65]	4,4'-DIAMINODIPHENYL ETHER
101-77-9 4,4'-DIAMINODIPHENYL METHANE [HM1, HM1]	4,4'-METHYLENEDIANILINE
101-77-9 4,4'-DIAMINODIPHENYLMETHANE [MAK, MAK]	4,4'-METHYLENEDIANILINE
101-77-9 4,4'-DIAMINODIPHENYLMETHANE (MDA) [WHM]	4,4'-METHYLENEDIANILINE
80-08-0 4,4'-DIAMINODIPHENYL SULFONE [TSC, TSC, TSC]	DAPSONE
107-15-3 1,2-DIAMINOETHANE [GBR, ONT]	ETHYLENEDIAMINE
107-15-3 1,2 DIAMINOETHANE (ETHYLENEDIAMINE) [ALB, ALB]	ETHYLENEDIAMINE
5700-49-2 1,2-DIAMINOETHANE, DIHYDROIODIDE	1,2-DIAMINOETHANE, DIHYDROIODIDE
99-56-9 1,2-DIAMINO-4-NITROBENZENE [IAR]	2-AMINO-4-NITROANILINE
5307-14-2 1,4-DIAMINO-2-NITROBENZENE (2-NITRO-PARA-PHENYLENEDIAMINE) [IAR]	2-NITRO-P-PHENYLENEDIAMINE
137-09-7 DIAMINOPHENOL HYDROCHLORIDE [HON]	2,4-DIAMINOPHENOL DIHYDROCHLORIDE
137-09-7 2,4-DIAMINOPHENOL DIHYDROCHLORIDE	2,4-DIAMINOPHENOL DIHYDROCHLORIDE
109-76-2 1,3-DIAMINOPROPANE [WHM]	1,3-PROPANEDIAMINE
616-29-5 1,3-DIAMINO-2-PROPANOL	1,3-DIAMINO-2-PROPANOL
7336-20-1 4,4'-DIAMINO-2,2'-STILBENEDISULFONIC ACID, DISODIUM SALT	4,4'-DIAMINO-2,2'-STILBENEDISULFONIC ACID, DISODIUM SALT
95-70-5 2,5-DIAMINOTOLUENE	2,5-DIAMINOTOLUENE
95-80-7 DIAMINOTOLUENE (MIXED) [EP2]	TOLUENE 2,4-DIAMINE
95-80-7 2,4-DIAMINOTOLUENE [NTP]	TOLUENE 2,4-DIAMINE
95-80-7 DIAMINOTOLUENE, 2,4- [OTS]	TOLUENE 2,4-DIAMINE
496-72-0 DIAMINOTOLUENE	DIAMINOTOLUENE
496-72-0 3,4-DIAMINOTOLUENE [MSL, NJL]	DIAMINOTOLUENE
823-40-5 2,6-DIAMINOTOLUENE	2,6-DIAMINOTOLUENE
25376-45-8 DIAMINOTOLUENE [PAL, MSL]	DIAMINOTOLUENE (MIXED ISOMERS)
25376-45-8 DIAMINOTOLUENE (MIXED) [C65, CAL]	DIAMINOTOLUENE (MIXED ISOMERS)
25376-45-8 DIAMINOTOLUENE, MIXED ISOMERS [NJL]	DIAMINOTOLUENE (MIXED ISOMERS)
25376-45-8 DIAMINOTOLUENE (MIXED ISOMERS)	DIAMINOTOLUENE (MIXED ISOMERS)
14323-43-4 DIAMMINEDICHLOROPALLADIUM	DIAMMINEDICHLOROPALLADIUM
15684-18-1 CIS-DIAMMINEDICHLOROPALLADIUM(II)	CIS-DIAMMINEDICHLOROPALLADIUM(II)
15663-27-1 CIS-DIAMMINEDICHLOROPLATINUM(II) [WHM]	CISPLATIN
20824-56-0 DIAMMONIUM ETHYLENEDIAMINETETRAAC-ETATE	DIAMMONIUM ETHYLENEDIAMINETETRAACETATE
6009-70-7 DIAMMONIUM OXALATE [MSL]	AMMONIUM OXALATE, MONOHYDRATE
7727-54-0 DIAMMONIUM PEROXYDISULPHATE [GBR]	AMMONIUM PERSULFATE
7783-28-0 DIAMMONIUM PHOSPHATE	DIAMMONIUM PHOSPHATE
95-80-7 2,4-DIAMOTOLUENE [313]	TOLUENE 2,4-DIAMINE
2050-92-2 DI-N-AMYLAMINE	DI-N-AMYLAMINE
2050-92-2 DIAMYLAMINE [FLA, HM1, HM1, HM1, MSL, NFP, NFP, NJL]	DI-N-AMYLAMINE
RR-00836-6 DIAMYLBENZENE	DIAMYLBENZENE
RR-00837-7 DIAMYLBIPHENYL	DIAMYLBIPHENYL

Chemical Name	Reference Name
RR-00833-3 DI-TERT-AMYLCYCLOHEXANOL	DI-TERT-AMYLCYCLOHEXANOL
RR-00839-9 DIAMYLENE	DIAMYLENE
10099-71-5 DIAMYL MALEATE	DIAMYL MALEATE
50696-42-9 DIAMYL NAPHTHALENE	DIAMYL NAPHTHALENE
138-00-1 2,4-DIAMYLPHENOL [FLA, MSL, NFP, NFP]	PHENOL, 2,4-DIPENTYL-
RR-00834-4 DI-TERT-AMYLPHENOXY ETHANOL	DI-TERT-AMYLPHENOXY ETHANOL
131-18-0 DIAMYL PHTHALATE	DIAMYL PHTHALATE
872-10-6 DIAMYL SULFIDE [FLA, MSL, NFP, NFP]	PENTANE, 1,1'-THIOBIS-
119-90-4 O-DIANISIDINE [MAK, NFP, NFP]	3,3'-DIMETHOXYBENZIDINE
RR-00062-4 O-DIANISIDINE-BASED DYES	O-DIANISIDINE-BASED DYES
82-22-4 1,1-DIANTHRIMIDE	1,1-DIANTHRIMIDE
6358-85-6 DIARYLANILIDE YELLOW	DIARYLANILIDE YELLOW
61790-53-2 DIATOMACEOUS EARTH	DIATOMACEOUS EARTH
68855-54-9 DIATOMACEOUS EARTH, UNCALCINED [ONT]	SILICA, AMORPHOUS, DIATOMACEOUS EARTH
61790-53-2 DIATOMACEOUS EARTH, NATURAL [GBR]	DIATOMACEOUS EARTH
280-57-9 1,4-DIAZABICYCLO[2,2,2]OCTANE	1,4-DIAZABICYCLO[2,2,2]OCTANE
123-77-3 DIAZENEDICARBOXAMIDE [OTS]	AZODICARBONAMIDE
439-14-5 DIAZEPAM	DIAZEPAM
26747-90-0 1,3-DIAZETIDINE-2,4-DIONE, 1,3-BIS(3-ISO-CYANATOMETHYLPHENYL)-	1,3-DIAZETIDINE-2,4-DIONE, 1,3-BIS(3-ISO-CYANATOMETHYLPHENYL)OCTYL ESTER
2294-47-5 P-DIAZIDOBENZENE	P-DIAZIDOBENZENE
RR-01365-0 1,2-DIAZIDOETHANE	1,2-DIAZIDOETHANE
333-41-5 DIAZINON	DIAZINON
333-41-5 DIAZINON (ISO) [GBR, GBR, GBR]	DIAZINON
RR-01407-3 DIAZOAMINOTETRAZOLE	DIAZOAMINOTETRAZOLE
87-31-0 DIAZODINTIROPHENOL	DIAZODINTIROPHENOL
4682-03-5 DIAZODINITROPHENOL	DIAZODINITROPHENOL
883-40-9 DIAZODIPHENYLMETHANE	DIAZODIPHENYLMETHANE
78491-02-8 DIAZOLIDINYL UREA	DIAZOLIDINYL UREADIYL) C11-14 ISOALKYL ETHERS, (C13-RICH)
334-88-3 DIAZOMETHANE	DIAZOMETHANE
RR-00763-6 2-DIAZO-1-NAPHTHOL-4-SULFOCHLORIDE	2-DIAZO-1-NAPHTHOL-4-SULFOCHLORIDE
RR-00764-7 2-DIAZO-1-NAPHTHOL-5-SULFOCHLORIDE	2-DIAZO-1-NAPHTHOL-5-SULFOCHLORIDE
RR-00763-6 2-DIAZO-1-NAPHTHOL-4-SULPHO-CHLO-RIDE [HM1, HM1]	2-DIAZO-1-NAPHTHOL-4-SULFOCHLORIDE
RR-00764-7 2-DIAZO-1-NAPHTHOL-5-SULPHO-CHLO-RIDE [HM1, HM1]	2-DIAZO-1-NAPHTHOL-5-SULFOCHLORIDE
RR-01408-4 DIAZONIUM NITRATES	DIAZONIUM NITRATES
RR-01409-5 DIAZONIUM PERCHLORATES	DIAZONIUM PERCHLORATES
5239-06-5 1,3-DIAZOPROPANE	1,3-DIAZOPROPANE
15845-52-0 DIBASIC LEAD PHOSPHATE	DIBASIC LEAD PHOSPHATE
56189-09-4 DIBASIC LEAD STEARATE	DIBASIC LEAD STEARATE
215-58-7 DIBENZ[A,C]ANTHRACENE	DIBENZ[A,C]ANTHRACENE
226-36-8 DIBENZ(A,H)ACRIDINE	DIBENZ(A,H)ACRIDINE
53-70-3 DIBENZ[A,H]ANTHRACENE [IAR]	DIBENZO(A,H)ANTHRACENE
189-55-9 DIBENZ(A,I)PYRENE [EPA]	DIBENZO(A,I)PYRENE
224-42-0 DIBENZ(A,J)ACRIDINE	DIBENZ(A,J)ACRIDINE
224-41-9 DIBENZ[A,J]ANTHRACENE	DIBENZ[A,J]ANTHRACENE
128-79-0 4,4'-DIBENZAMIDO-1,1'-DIANTHRIMIDE	4,4'-DIBENZAMIDO-1,1'-DIANTHRIMIDE

Chemical Name	Reference Name
53-70-3 1,2,5,6-DIBENZANTHRACENE (DIBENZO(A,H) ANTHRACENE) [CW2]	DIBENZO(A,H)ANTHRACENE
116-90-5 4,4'-DIBENZANTHRONYL	4,4'-DIBENZANTHRONYL
5385-75-1 DIBENZO[A,E]FLUORANTHENE	DIBENZO[A,E]FLUORANTHENEXAMIDE)
192-65-4 DIBENZO(A,E)PYRENE	DIBENZO(A,E)PYRENE
53-70-3 DIBENZO(A,H)ANTHRACENE	DIBENZO(A,H)ANTHRACENE
53-70-3 DIBENZO(A,H)ANTHRACENE [ATS]	DIBENZO(A,H)ANTHRACENE
189-64-0 DIBENZO(A,H)PYRENE	DIBENZO(A,H)PYRENE
189-55-9 DIBENZO(A,I)PYRENE	DIBENZO(A,I)PYRENE
191-30-0 DIBENZO(A,L)PYRENE	DIBENZO(A,L)PYRENE
1746-01-6 DIBENZO[B,E][1,4]DIOXIN, 2,3,7,8-TETRA-CHLORO- [PAL, PAL]	2,3,7,8-TETRACHLORODIBENZO-P-DIOXIN (TCDD)
189-64-0 DIBENZO(B,DEF)CHRYSENE [PAL, PAL]	DIBENZO(A,H)PYRENE
194-59-2 7H-DIBENZO(C,G)CARBAZOLE	7H-DIBENZO(C,G)CARBAZOLE
194-59-2 DIBENZO(C,G)CARBAZOLE [NTP]	7H-DIBENZO(C,G)CARBAZOLE
192-47-2 DIBENZO[H,RST]PENTAPHENE	DIBENZO[H,RST]PENTAPHENE
14187-32-7 DIBENZO-18-CROWN-6	DIBENZO-18-CROWN-6
262-12-4 DIBENZO-P-DIOXIN	DIBENZO-P-DIOXIN
262-12-4 DIBENZO-PARA-DIOXIN [CN5]	DIBENZO-P-DIOXIN
132-64-9 DIBENZOFURAN	DIBENZOFURAN
126-15-8 4A(4H)-DIBENZOFURANCARBOXALDEHYDE, 1, 5A,6,9,9B,-HEXAHYDRO-	4A(4H)-DIBENZOFURANCARBOXALDEHYDE, 1,5A,6,9,9A, 9B,-HEXAHYDRO
132-64-9 DIBENZOFURANS [CAA]	DIBENZOFURAN
189-55-9 1,2:7,8-DIBENZOPYRENE [EP2]	DIBENZO(A,I)PYRENE
120-78-5 2,2'-DIBENZOTHIAZYL DISULFIDE	2,2'-DIBENZOTHIAZYL DISULFIDE
132-65-0 DIBENZOTHIOPHENE	DIBENZOTHIOPHENE
RR-00563-0 DIBENZOYL CHLORIDE	DIBENZOYL CHLORIDE
94-36-0 DIBENZOYL PEROXIDE [GBR, MAK, MAK, MAK, OS3]	BENZOYL PEROXIDE
18414-36-3 DIBENZYLDICHLOROSILANE	DIBENZYLDICHLOROSILANE
18414-36-3 DIBENZYL DICHLOROSILANE [NJL, NJL]	DIBENZYLDICHLOROSILANE
150-60-7 DIBENZYL DISULFIDE	DIBENZYL DISULFIDE
103-50-4 DIBENZYL ETHER [NFP, NFP]	BENZYL ETHER
2144-45-8 DIBENZYL PEROXYDICARBONATE	DIBENZYL PEROXYDICARBONATE
1304-82-1 DIBISMUTH TRITELLURIDE [GBR, GBR]	BISMUTH TELLURIDE
19287-45-7 DIBORAN [QBC]	DIBORANE
19287-45-7 DIBORANE	DIBORANE
19287-45-7 DIBORANE(6) [PAL]	DIBORANE
1303-86-2 DIBORON TRIOXIDE [GBR, GBR]	BORON OXIDE
300-76-5 DIBROM [BC1, BC1, MAK, MAK]	NALED
300-76-5 DIBROM (NALED) [ALB, ALB]	NALED
3252-43-5 DIBROMOACETONITRILE	DIBROMOACETONITRILE
624-61-3 DIBROMOACETYLENE	DIBROMOACETYLENE
106-37-6 P-DIBROMOBENZENE	P-DIBROMOBENZENE
26249-12-7 DIBROMOBENZENE	DIBROMOBENZENE
92-86-4 4,4-DIBROMO-1,1-BIPHENYL [TSE]	1,1'-BIPHENYL, 4,4'-DIBROMO-
3479-86-5 DIBROMOBUTANONE	DIBROMOBUTANONE
25109-57-3 1,2-DIBROMO-3-BUTANONE	1,2-DIBROMO-3-BUTANONE
25109-57-3 1,2-DIBROMOBUTAN-3-ONE [HM1, HM1]	1,2-DIBROMO-3-BUTANONE

Chemical Name	Reference Name
124-48-1 DIBROMOCHLOROMETHANE [PQL, SD4]	CHLORODIBROMOMETHANE
96-12-8 1,2-DIBROMO-3-CHLOROPROPANE	1,2-DIBROMO-3-CHLOROPROPANE
96-12-8 DIBROMOCHLOROPROPANE [HM1, HM1, HM1, NHS, NHS]	1,2-DIBROMO-3-CHLOROPROPANE
67708-83-2 DIBROMOCHLOROPROPANE	DIBROMOCHLOROPROPANE
96-12-8 1,2-DIBROMO-3-CHLOROPROPANE (DBCP) [313, C65, C65, C65, MSL, SDW, SDW, SDW, WHM]	1,2-DIBROMO-3-CHLOROPROPANE
3322-93-8 1,2-DIBROMO-4-(1,2-DIBROMOETHYL) CYCLO-HEXANE	1,2-DIBROMO-4-(1,2-DIBROMOETHYL) CYCLOHEXANE
35691-65-7 1,2-DIBROMO-2,4-DICYANOBUTANE	1,2-DIBROMO-2,4-DICYANOBUTANE1H-IMIDAZOLE]
75-61-6 DIBROMODIFLUOROMETHANE [BC1, BC1, CAL, GBR, GBR, HM1, HM1, MAK, MAK, ONT]	DIFLUORODIBROMOMETHANE
75-61-6 DIBROMODIFLUOROMETHANE (FREON 12B2) [QBC]	DIFLUORODIBROMOMETHANE
10318-26-0 DIBROMODULCITOL	DIBROMODULCITOL
106-93-4 1,2-DIBROMOETHANE [ATS, CSR, CSR, HM1, HM1, HM1, HM1, HM1, MAK, MAK]	ETHYLENE DIBROMIDE (1,2-DIBROMOETHANE)
106-93-4 1,2-DIBROMOETHANE (ETHYLENE DIBROMIDE) [313, BC1, BC1, BC1, BC1, QBC, QBC, QBC, NTP]	ETHYLENE DIBROMIDE (1,2-DIBROMOETHANE)
106-93-4 1,2 DIBROMOETHANE (ETHYLENE DIBROMIDE) [ALB, ALB, ALB, ALB]	ETHYLENE DIBROMIDE (1,2-DIBROMOETHANE)
106-93-4 1,2-DIBROMOETHANE (ETHYLENE DIBROMIDE) [GBR, GBR]	ETHYLENE DIBROMIDE (1,2-DIBROMOETHANE)
540-49-8 1,2-DIBROMOETHENE	1,2-DIBROMOETHENE
35243-89-1 1,2-DIBROMOPROPYL GLYCIDYL ETHER [EPA]	OXIRANE, [(1,2-DIBROMOPROPOXY)METHYL]-
488-41-5 DIBROMOMANNITOL	DIBROMOMANNITOL
74-95-3 DIBROMOMETHANE [HM1, HM1, HM1, HM1, HON, MSL, RCA, SD4, SDW, TSE, UTS, WHM]	METHYLENE BROMIDE
22421-59-6 2,6-DIBROMO-4-METHYLPHENYL GYCIDYL ETHER	2,6-DIBROMO-4-METHYLPHENYL GYCIDYL ETHER
75150-13-9 2,4-DIBROMO-6-METHYLPHENYL GYCIDYL ETHER	2,4-DIBROMO-6-METHYLPHENYL GYCIDYL ETHERDE-CANOIC ACID
22421-59-6 DIBROMO-4-METHYLPHENYL GLYCIDYL ETHER, 2,6- [OTS]	2,6-DIBROMO-4-METHYLPHENYL GYCIDYL ETHER
10222-01-2 2,2-DIBROMO-3-NITRILOPROPIONAMIDE	2,2-DIBROMO-3-NITRILOPROPIONAMIDE
99-28-5 2,6-DIBROMO-4-NITROPHENOL	2,6-DIBROMO-4-NITROPHENOL
615-58-7 2,4-DIBROMOPHENOL	2,4-DIBROMOPHENOL
20217-01-0 2,4-DIBROMOPHENYL GLYCIDYL ETHER [EPA]	OXIRANE, [(2,4-DIBROMOPHENOXY)METHYL]-
20217-01-0 DIBROMOPHENYL GLYCIDYL ETHER, 2,4- [OTS]	OXIRANE, [(2,4-DIBROMOPHENOXY)METHYL]-
78-75-1 1,2-DIBROMOPROPANE	1,2-DIBROMOPROPANE
96-13-9 2,3-DIBROMO-1-PROPANOL	2,3-DIBROMO-1-PROPANOL
35243-89-1 DIBROMOPROPYL GLYCIDYL ETHER, 1,2- [OTS]	OXIRANE, [(1,2-DIBROMOPROPOXY)METHYL]-
2577-72-2 3,5-DIBROMOSALICYLANILIDE	3,5-DIBROMOSALICYLANILIDE
124-73-2 DIBROMOTETRAFLUOROETHANE [CN2]	DIBROMOTETRAFLUOROETHANE (HALON 2402)
124-73-2 DIBROMOTETRAFLUOROETHANE (HALON 2402)	DIBROMOTETRAFLUOROETHANE (HALON 2402)
124-73-2 DIBROMOTETRAFLUOROETHANE [CN5]	DIBROMOTETRAFLUOROETHANE (HALON 2402)
117-83-9 DIBUTOXY ETHYL PHTHALATE [NFP, NFP]	BIS[2-N-BUTOXYETHYL] PHTHALATE
2568-90-3 DIBUTOXYMETHANE	DIBUTOXYMETHANE
112-98-1 DIBUTOXY TETRAGLYCOL	DIBUTOXY TETRAGLYCOL
1563-90-2 N,N-DIBUTYLACETAMIDE	N,N-DIBUTYLACETAMIDE

Chemical Name	Reference Name
105-99-7 DIBUTYL ADIPATE	DIBUTYL ADIPATE
105-99-7 DI-BUTYL ADIPATE [OTS]	DIBUTYL ADIPATE
111-92-2 DI-(N-BUTYL)AMINE	DI-(N-BUTYL)AMINE
111-92-2 DIBUTYLAMINE [FLA, MSL, NFP, NFP, NJL, NJL]	DI-(N-BUTYL)AMINE
111-92-2 DI-N-BUTYLAMINE [PST]	DI-(N-BUTYL)AMINE
626-23-3 DI-SEC-BUTYLAMINE	DI-SEC-BUTYLAMINE
102-81-8 2-N-DIBUTYLAMINOETHANOL	2-N-DIBUTYLAMINOETHANOL
102-81-8 2-(DIBUTYLAMINO)ETHANOL [CAL]	2-N-DIBUTYLAMINOETHANOL
102-81-8 DIBUTYLAMINOETHANOL [HM1, HM1]	2-N-DIBUTYLAMINOETHANOL
102-81-8 DIBYUTYLAMINOETHANOL [NFP, NFP]	2-N-DIBUTYLAMINOETHANOL
102-81-8 2-(DIBUTYLAMINO)ETHANOL [ONT, ONT]	2-N-DIBUTYLAMINOETHANOL
102-81-8 N,N-DI-N-BUTYLAMINOETHANOL [WHM]	2-N-DIBUTYLAMINOETHANOL
RR-00552-7 1-DIBUTYLAMINO-2-PROPANOL	1-DIBUTYLAMINO-2-PROPANOL
613-29-6 N,N-DIBUTYLANILINE [FLA, MSL, NFP, NFP]	BENZENAMINE, N,N-DIBUTYL-
128-37-0 2,6-DI-TERT-BUTYL-P-CRESOL	2,6-DI-TERT-BUTYL-P-CRESOL
128-37-0 2,6-DITERBUTYL-P-CRESOL [QBC]	2,6-DI-TERT-BUTYL-P-CRESOL
25377-21-3 DI-TERT-BUTYL-P-CRESOL	DI-TERT-BUTYL-P-CRESOL
15520-11-3 DI(4-TERT-BUTYLCYCLOHEXYL)-PEROXYDI-CARBONATE	DI(4-TERT-BUTYLCYCLOHEXYL)-PEROXYDICARBON-ATE
136-23-2 DIBUTYLDITHIOCARBAMIC ACID, ZINC SALT [PST]	ZINC DIBUTYL DITHIOCARBAMATE
142-96-1 DIBUTYL ETHER [FLA, HM1, HM1, MSL, NFP, NFP]	BUTYL ETHER
105-75-9 DI-N-BUTYL FUMARATE	DI-N-BUTYL FUMARATE
107-66-4 DIBUTYL HYDROGEN PHOSPHATE [GBR, GBR]	DIBUTYL PHOSPHATE
88-58-4 2,5-DI-TERT-BUTYLHYDROQUINONE	2,5-DI-TERT-BUTYLHYDROQUINONE
88-58-4 2,5-DI(TERT-BUTYL)HYDROQUINONE [PST]	2,5-DI-TERT-BUTYLHYDROQUINONE
3126-90-7 DIBUTYL ISOPHTHALATE	DIBUTYL ISOPHTHALATE
2109-64-0 DIBUTYLISOPROPANOLAMINE	DIBUTYLISOPROPANOLAMINEESTER
105-76-0 DIBUTYL MALEATE [NFP, NFP]	MALEIC ACID, DIBUTYL ESTER
25417-20-3 DIBUTYLNAPHTHALENESULFONIC ACID, SODIUM SALT	DIBUTYLNAPHTHALENESULFONIC ACID, SODIUM SALT
2050-60-4 DIBUTYL OXALATE [PST]	BUTYL OXALATE
128-37-0 2,6-DI-TERT-BUTYL-P-CRESOL [ALB, ALB, MEX, MEX]	2,6-DI-TERT-BUTYL-P-CRESOL
110-05-4 DI-TERT-BUTYL PEROXIDE	DI-TERT-BUTYL PEROXIDE
110-05-4 DIBUTYL PEROXIDE (TERTIARY) [OS3]	DI-TERT-BUTYL PEROXIDE
2167-23-9 2,2-DI(TERT-BUTYLPEROXY)-BUTANE	2,2-DI(TERT-BUTYLPEROXY)-BUTANE
3006-86-8 1,1-DI(TERT-BUTYLPEROXY)-CYCLOHEXANE	1,1-DI(TERT-BUTYLPEROXY)-CYCLOHEXANE
RR-01372-9 2,2-DI-(4,4-DI-TERT-BUTYLPEROXYCYCLO-HEXYL) PROPANE	2,2-DI-(4,4-DI-TERT-BUTYLPEROXYCYCLOHEXYL) PROPANE
RR-01406-2 DI-N-BUTYL PEROXYDICARBONATE	DI-N-BUTYL PEROXYDICARBONATE
19910-65-7 DI-SEC-BUTYL PEROXYDICARBONATE	DI-SEC-BUTYL PEROXYDICARBONATE
RR-00250-6 1,4-DI-(2-T-BUTYLPEROXY ISOPROPYL)BEN-ZENE	1,4-DI-(2-T-BUTYLPEROXY ISOPROPYL)BENZENE
RR-01752-7 DI-(2-TERT-BUTYLPEROXYISOPROPYL)-BEN-ZENE	DI-(2-TERT-BUTYLPEROXYISOPROPYL)-BENZENE
2155-71-7 DI-TERT-BUTYLPEROXYPHTHALATE	DI-TERT-BUTYLPEROXYPHTHALATE
RR-01405-1 DI-(TERT-BUTYLPEROXY) PHTHALATE	DI-(TERT-BUTYLPEROXY) PHTHALATE
RR-00253-9 2,2-DI(T-BUTYLPEROXY)PROPANE	2,2-DI(T-BUTYLPEROXY)PROPANE

Chemical Name	Reference Name
6731-36-8 1,1-DI(TERT-BUTYLPEROXY)-3,3,5-TRIMETHYL-CYCLOHEXANE [NJL]	1,1-DI(TERT-BUTYLPEROXY)-3,3,5-TRIMETHYLCYCLO-HEXANE PEROXIDE
6731-36-8 1,1-DI(TERT-BUTYLPEROXY)-3,3,5-TRIMETHYL-CYCLOHEXANE PEROXIDE	1,1-DI(TERT-BUTYLPEROXY)-3,3,5-TRIMETHYLCYCLO-HEXANE PEROXIDE
128-39-2 2,6-DI-T-BUTYLPHENOL	2,6-DI-T-BUTYLPHENOL
128-39-2 DI-TERT-BUTYLPHENOL [OTS]	2,6-DI-T-BUTYLPHENOL
128-39-2 2,6-DI-TERT-BUTYL PHENOL [TSC, TSC, TSC]	2,6-DI-T-BUTYLPHENOL
101-96-2 N,N'-DI-SEC-BUTYL-P-PHENYLENEDI-AMINE [NFP, NFP]	1,4-BENZENEDIAMINE, N,N'-BIS(1-METHYLPROPYL)-
101-96-2 N,N-DI-SEC-BUTYL-P-PHENYLENEDI-AMINE [FLA]	1,4-BENZENEDIAMINE, N,N'-BIS(1-METHYLPROPYL)-
101-96-2 N,N'-DI-SEC-BUTYL-P-PHENYLENEDI-AMINE [MSL]	1,4-BENZENEDIAMINE, N,N'-BIS(1-METHYLPROPYL)-
2528-36-1 DIBUTYL PHENYL PHOSPHATE	DIBUTYL PHENYL PHOSPHATE
2528-36-1 DIBUTYLPHENYLPHOSPHATE [MIN]	DIBUTYL PHENYL PHOSPHATE
2528-36-1 DI(N-BUTYL) PHENYL PHOSPHATE [OTS]	DIBUTYL PHENYL PHOSPHATE
107-66-4 DIBUTYL PHOSPHATE	DIBUTYL PHOSPHATE
1809-19-4 DIBUTYL PHOSPHITE	DIBUTYL PHOSPHITE
84-74-2 DIBUTYL PHTHALATE	DIBUTYL PHTHALATE
84-74-2 DI-N-BUTYL PHTHALATE [ATS]	DIBUTYL PHTHALATE
84-74-2 DIBUTYLPHTHALATE [CAA]	DIBUTYL PHTHALATE
84-74-2 DI-N-BUTYL PHTHALATE [CW2, CWA, HM1, HM1, MX1]	DIBUTYL PHTHALATE
84-74-2 DIBUTYLPHTHALATE [NIO, NIO, NIO]	DIBUTYL PHTHALATE
84-74-2 DI-N-BUTYL PHTHALATE [PQL, RCA]	DIBUTYL PHTHALATE
84-74-2 DIBUTYL PHTHALATE [TSE]	DIBUTYL PHTHALATE
109-43-3 DI-N-BUTYL SEBACATE	DI-N-BUTYL SEBACATE
5831-88-9 N.N-DIBUTYL STEARAMIDE	N.N-DIBUTYL STEARAMIDE
5831-88-9 N,N-DIBUTYL STEARAMIDE [NFP, NFP]	N.N-DIBUTYL STEARAMIDE
141-03-7 DI-N-BUTYL SUCCINATE	DI-N-BUTYL SUCCINATE
87-92-3 N-DIBUTYL TARTRATE	N-DIBUTYL TARTRATE
96-69-5 6,6'-DI-TERT-BUTYL-4,4'-THIODI-M-CRESOL [GBR, GBR]	4,4'-THIOBIS(6-TERT-BUTYL-M-CRESOL)
109-46-6 DIBUTYL THIOUREA	DIBUTYL THIOUREA
109-46-6 N.N'-DIBUTYL THIOUREA [PST]	DIBUTYL THIOUREA
25168-21-2 DIBUTYLTIN BIS (ISOOCTYL MALEATE)-2-BUTENOIC ACID, 4,4'-[(DIBUTYLSTANNYLENE)BIS(OXY)]BIS[4-OXO-, DIISOOCTYL ESTER, (Z,Z)-	DIBUTYLTIN BIS (ISOOCTYL MALEATE)-2-BUTENOIC ACID, 4,4'-[(DI
25168-24-5 DIBUTYLTIN BIS(ISOOCTYL MERCAPTOAC-ETATE)	DIBUTYLTIN BIS(ISOOCTYL MERCAPTOACETATE)
26636-01-1 DIBUTYLTIN S,S'-BIS(ISOOCTYL MERCAP-TOACETATE)	DIBUTYLTIN S,S'-BIS(ISOOCTYL MERCAPTOACETATE) ETHYLPHENYL)-
1185-81-5 DIBUTYLTIN BIS(LAURYL MERCAPTIDE)	DIBUTYLTIN BIS(LAURYL MERCAPTIDE)
15546-11-9 DI-N-BUTYLTIN BIS(METHYL MALEATE)	DI-N-BUTYLTIN BIS(METHYL MALEATE)
1067-33-0 DIBUTYLTIN DIACETATE	DIBUTYLTIN DIACETATE
683-18-1 DIBUTYLTIN DICHLORIDE	DIBUTYLTIN DICHLORIDE
2781-10-4 DI-N-BUTYLTIN DI-2-ETHYLHEXANOATE	DI-N-BUTYLTIN DI-2-ETHYLHEXANOATE
77-58-7 DIBUTYLTIN DILAURATE	DIBUTYLTIN DILAURATE
15546-16-4 DI-N-BUTYLTIN DI(MONOBUTYL)MALEATE	DI-N-BUTYLTIN DI(MONOBUTYL)MALEATE
5847-55-2 DIBUTYLTIN DISTEARATE	DIBUTYLTIN DISTEARATE
78-04-6 DIBUTYLTIN MALEATE	DIBUTYLTIN MALEATE

Chemical Name	Reference Name
15546-11-9 DIBUTYLTIN METHYL MALEATE [WHM]	DI-N-BUTYLTIN BIS(METHYL MALEATE)
818-08-6 DIBUTYLTIN OXIDE	DIBUTYLTIN OXIDE
4253-22-9 DIBUTYLTIN SULFIDE	DIBUTYLTIN SULFIDE
RR-00872-0 N,N-DIBUTYLTOLUENESULFONAMIDE	N,N-DIBUTYLTOLUENESULFONAMIDE
105-77-1 DIBUTYL XANTHOGEN DISULFIDE	DIBUTYL XANTHOGEN DISULFIDE
73398-64-8 DI(C8-18)ALKYL DIMETHYL AMMONIUM CHLORIDE	DI(C8-18)ALKYL DIMETHYL AMMONIUM CHLORIDE
68153-33-3 DI-C10-16-ALKYL DIMETHYL AMMONIUM CHLORIDE	DI-C10-16-ALKYL DIMETHYL AMMONIUM CHLORIDE
1918-00-9 DICAMBA	DICAMBA
1918-00-9 DICAMBA (3,6-DICHLORO-2-METHOXYBEN-ZOIC ACID) [313]	DICAMBA-2,4-DIAMINE)
131-15-7 DICAPRYL PHTHALATE [NFP, NFP]	1,2-BENZENEDICARBOXYLIC ACID, BIS(1-METHYLHEP-TYL) ESTER
RR-00258-4 DICARBOXYLIC ACID MONOESTER	DICARBOXYLIC ACID MONOESTER
26322-14-5 DICETYL PEROXYDICARBONATE	DICETYL PEROXYDICARBONATE
1194-65-6 DICHLOBENIL	DICHLOBENIL
97-17-6 DICHLOFENTHION	DICHLOFENTHION
117-80-6 DICHLONE	DICHLONE
99-30-9 DICHLORAN (2,6-DICHLORO-4-NITROANILINE) [313]	2,6-DICHLORO-4-NITROANILINE
RR-01446-0 N,N'-DICHLORAZODICARBONAMIDINE	N,N'-DICHLORAZODICARBONAMIDINE
79-02-7 DICHLOROACETALDEHYDE	DICHLOROACETALDEHYDE
79-43-6 DICHLOROACETIC ACID	DICHLOROACETIC ACID
534-07-6 1,3-DICHLOROACETONE [HM1, HM1, WHM]	BIS(CHLOROMETHYL)KETONE
3018-12-0 DICHLOROACETONITRILE	DICHLOROACETONITRILE
79-36-7 DICHLOROACETYL CHLORIDE	DICHLOROACETYL CHLORIDE
7572-29-4 DICHLOROACETYLENE	DICHLOROACETYLENE
7572-29-4 DICHLORO ACETYLENE [OS3]	DICHLOROACETYLENE
95-76-1 3,4-DICHLOROANILINE	3,4-DICHLOROANILINE
27134-27-6 DICHLOROANILINE	DICHLOROANILINE
27134-27-6 DICHLOROANILINE (MIXED ISOMERS) [HON]	DICHLOROANILINE
27134-27-6 DICHLOROANILINE, ALL ISOMERS [WHM]	DICHLOROANILINE
RR-01278-2 DICHLOROANILINES	DICHLOROANILINES
8023-53-8 DICHLOROBENZALKONIUM CHLORIDE	DICHLOROBENZALKONIUM CHLORIDE
95-76-1 3,4-DICHLOROBENZENAMINE [FLA]	3,4-DICHLOROANILINE
95-50-1 O-DICHLOROBENZENE	O-DICHLOROBENZENE
95-50-1 1,2-DICHLOROBENZENE [313, ATS, CAB, CAN, CTR, CW2, CW3, FLA, GBR]	O-DICHLOROBENZENE
95-50-1 ORTHO-DICHLOROBENZENE [IAR]	O-DICHLOROBENZENE
95-50-1 1,2-DICHLOROBENZENE [MAK, MAK, MAK, MAK, MX1, NJL]	O-DICHLOROBENZENE
95-50-1 ORTHO-DICHLOROBENZENE [RCA]	O-DICHLOROBENZENE
95-50-1 1,2-DICHLOROBENZENE [TSE]	O-DICHLOROBENZENE
106-46-7 P-DICHLOROBENZENE	P-DICHLOROBENZENE
106-46-7 1,4-DICHLOROBENZENE [313, ATS, CTR, CW2, CW3, GBR, GBR]	P-DICHLOROBENZENE
106-46-7 PARA-DICHLOROBENZENE [IAR]	P-DICHLOROBENZENE
106-46-7 1,4-DICHLOROBENZENE [MAK, MAK, MAK, MX1, NJL, NTP, RC2, RC2]	P-DICHLOROBENZENE
106-46-7 PARA-DICHLOROBENZENE [SDW, SDW]	P-DICHLOROBENZENE

Chemical Name	Reference Name
106-46-7 1,4-DICHLOROBENZENE [TSE]	P-DICHLOROBENZENE
541-73-1 1,3-DICHLOROBENZENE	1,3-DICHLOROBENZENE
541-73-1 M-DICHLOROBENZENE [CAL, HON, PQL, RCA, RCU, RCU, SDW, UTS, WHM]	1,3-DICHLOROBENZENE
25321-22-6 DICHLOROBENZENE [ATS]	DICHLOROBENZENE (MIXED ISOMERS)
95-50-1 1,2-DICHLOROBENZENE (O-DICHLOROBENZENE) [CSR, CSR]	O-DICHLOROBENZENE
106-46-7 1,4-DICHLOROBENZENE(P) [CAA]	P-DICHLOROBENZENE
106-46-7 1,4-DICHLOROBENZENE (P-DICHLOROBENZENE) [CSR, CSR]	P-DICHLOROBENZENE
95-50-1 DICHLOROBENZENE, 1,2- [CD1, CD1]	O-DICHLOROBENZENE
95-50-1 1,2-DICHLOROBENZENE [CN4, HM1, HM1, HM1, HM1]	O-DICHLOROBENZENE
106-46-7 1,4-DICHLOROBENZENE [CAN]	P-DICHLOROBENZENE
106-46-7 DICHLOROBENZENE, 1,4- [CD1, CD1]	P-DICHLOROBENZENE
106-46-7 1,4-DICHLOROBENZENE [CN4]	P-DICHLOROBENZENE
25321-22-6 DICHLOROBENZENE (META; ORTHO; PARA) [HM1, HM1]	DICHLOROBENZENE (MIXED ISOMERS)
25321-22-6 DICHLOROBENZENE (MIXED) [EPA, MSL]	DICHLOROBENZENE (MIXED ISOMERS)
25321-22-6 DICHLOROBENZENE (MIXED ISOMERS)	DICHLOROBENZENE (MIXED ISOMERS)
25321-22-6 DICHLOROBENZENE, N.O.S. [RCU]	DICHLOROBENZENE (MIXED ISOMERS)
25321-22-6 DICHLOROBENZENES [MX2, RCA]	DICHLOROBENZENE (MIXED ISOMERS)
25321-22-6 DICHLOROBENZENES, MIXED ISOMERS [NJL]	DICHLOROBENZENE (MIXED ISOMERS)
91-94-1 3,3-DICHLOROBENZIDENE [CW2]	3,3'-DICHLOROBENZIDINE
91-94-1 DICHLOROBENZIDINE [WHM]	3,3'-DICHLOROBENZIDINE
91-94-1 3,3'-DICHLOROBENZIDINE	3,3'-DICHLOROBENZIDINE
91-94-1 3,3-DICHLOROBENZIDINE [313, BC1, BC1, BC1, CAA]	3,3'-DICHLOROBENZIDINE
91-94-1 DICHLOROBENZIDINE [CAL]	3,3'-DICHLOROBENZIDINE
91-94-1 3,3'-DICHLOROBENZIDINE [QBC, QBC, QBC]	3,3'-DICHLOROBENZIDINE
91-94-1 3,3'-DICHLOROBENZIDINE (AND ITS SALTS) [NIO, NIO]	3,3'-DICHLOROBENZIDINE
612-83-9 3,3'-DICHLOROBENZIDINE DIHYDROCHLORIDE	3,3'-DICHLOROBENZIDINE DIHYDROCHLORIDE
64969-34-2 3,3'-DICHLOROBENZIDINE SULFATE	3,3'-DICHLOROBENZIDINE SULFATE-2-YL) AMINO] CARBONYL]BENZENESULFONAMIDE
24072-75-1 5,6-DICHLORO-2-BENZOTHIAZOLAMINE	5,6-DICHLORO-2-BENZOTHIAZOLAMINE
328-84-7 3,4-DICHLOROBENZOTRIFLUORIDE	3,4-DICHLOROBENZOTRIFLUORIDE
133-14-2 2,4-DICHLOROBENZOYL PEROXIDE	2,4-DICHLOROBENZOYL PEROXIDE
RR-01753-8 DI-4-CHLOROBENZOYL PEROXIDE	DI-4-CHLOROBENZOYL PEROXIDE
72-54-8 1,1-DICHLORO-2,2-BIS-(P-CHLOROPHENYL) ETHANE [NJL]	DDD
72-55-9 1,1-DICHLORO-2,2-BIS(P-CHLOROPHENYL) ETHYLENE [WHM, CAL]	DDE
73506-91-9 DICHLOROBROMOETHANE	DICHLOROBROMOETHANE
75-27-4 DICHLOROBROMOMETHANE	DICHLOROBROMOMETHANE
1653-19-6 2,3-DICHLOROBUTADIENE-1,3	2,3-DICHLOROBUTADIENE-1,3
110-56-5 1,4-DICHLOROBUTANE	1,4-DICHLOROBUTANE
616-21-7 1,2-DICHLOROBUTANE	1,2-DICHLOROBUTANE
7581-97-7 2,3-DICHLOROBUTANE	2,3-DICHLOROBUTANE
110-57-6 TRANS-1,4-DICHLOROBUTENE	TRANS-1,4-DICHLOROBUTENE

Chemical Name	Reference Name
110-57-6 TRANS-1,4-DICHLORO-2-BUTENE [313, MSL, PQL]	TRANS-1,4-DICHLOROBUTENE
760-23-6 3,4-DICHLORO-1-BUTENE	3,4-DICHLORO-1-BUTENE
760-23-6 3,4-DICHLOROBUTENE [TSC, TSC, TSC]	3,4-DICHLORO-1-BUTENE
764-41-0 1,4-DICHLORO-2-BUTENE	1,4-DICHLORO-2-BUTENE
764-41-0 1,4-DICHLOROBUTENE-2 [MAK]	1,4-DICHLORO-2-BUTENE
926-57-8 1,3-DICHLOROBUTENE-2	1,3-DICHLOROBUTENE-2
7415-31-8 1,3-DICHLORO-2-BUTENE	1,3-DICHLORO-2-BUTENE
11069-19-5 DICHLOROBUTENE	DICHLOROBUTENE
64037-54-3 3,4-DICHLOROBUTENE-1	3,4-DICHLOROBUTENE-1
28434-86-8 3,3'-DICHLORO-4,4'-DIAMINODIPHENYL ETHER	3,3'-DICHLORO-4,4'-DIAMINODIPHENYL ETHER
33857-26-0 2,7-DICHLORODIBENZO-P-DIOXIN	2,7-DICHLORODIBENZO-P-DIOXIN
111-44-4 2,2-DICHLORODIETHYL ETHER [HM1, HM1, HM1, HM1]	BIS(2-CHLOROETHYL) ETHER
111-44-4 2,2'-DICHLORODIETHYL ETHER [MAK, MAK, MAK]	BIS(2-CHLOROETHYL) ETHER
1649-08-7 1,2-DICHLORO-1,1-DIFLUOROETHANE	1,2-DICHLORO-1,1-DIFLUOROETHANE
1649-08-7 1,2-DICHLORO-1,1-DIFLUOROETHANE (HCFC-132B) [313]	1,2-DICHLORO-1,1-DIFLUOROETHANEUOROMETHYL)-]
27156-03-2 DICHLORODIFLUOROETHYLENE	DICHLORODIFLUOROETHYLENE
75-71-8 DICHLORODIFLUOROMETHANE	DICHLORODIFLUOROMETHANE
75-71-8 DICHLORODIFLUOROMETHANE (CFC-12) [313]	DICHLORODIFLUOROMETHANE
75-71-8 DICHLORODIFLUOROMETHANE (FC-12) [CAL]	DICHLORODIFLUOROMETHANE
75-71-8 DICHLORODIFLUOROMETHANE (FREON-12) [QBC]	DICHLORODIFLUOROMETHANE
75-71-8 DICHLORODIFLUOROMETHANE (FC 12) [WHM]	DICHLORODIFLUOROMETHANE
134190-52-6 DICHLORODIFLUOROPROPANE	DICHLORODIFLUOROPROPANE
118-52-5 1,3-DICHLORO-5,5-DIMETHYL HYDANTOIN	1,3-DICHLORO-5,5-DIMETHYL HYDANTOIN
118-52-5 1,3-DICHLORO-5,5-DIMETHYLHYDAN-TOIN [CAL]	1,3-DICHLORO-5,5-DIMETHYL HYDANTOIN
118-52-5 3-DICHLORO 5,5-DIMETHYL/HYDANTOIN [QBC, QBC]	1,3-DICHLORO-5,5-DIMETHYL HYDANTOIN
118-52-5 1,3-DICHLORO-5,5-DIMETHYLHYDAN-TOIN [NIO, NIO, WHM]	1,3-DICHLORO-5,5-DIMETHYL HYDANTOIN
72-55-9 P,P'-DICHLORODIPHENOLDICHLOROETHY-LENE [CSR, CSR]	DDE
80-10-4 DICHLORODIPHENYLSILANE [FLA]	DIPHENYL DICHLOROSILANE
80-07-9 P,P'-DICHLORODIPHENYL SULFONE [CSR]	SULFONYL BIS-(4-CHLOROBENZENE)
50-29-3 DICHLORO DIPHENYL TRICHLOROETHANE (D.D.T.) [FLA]	DDT
50-29-3 DICHLORODIPHENYLTRICHLOROETHANE (DDT) [CSR, CSR, OS1, OS1, OS1, OS1]	DDT
75-35-4 1,1-DICHLOROETHYLENE [MX1]	VINYLIDENE CHLORIDE
75-34-3 1,1-DICHLOROETHANE	1,1-DICHLOROETHANE
75-34-3 DICHLOROETHANE, 1,1- [OTS]	1,1-DICHLOROETHANE
107-06-2 1,2-DICHLOROETHANE [ATS, BC1, BC1, CAN, CN4, CSR, CSR, CW2, EP2, IAR, MAK, MX1, MX2, NJL, NJL, NTP, ONT, ONT, ONT, QBC, QBC, QBC, RC2, RC2, RCA, SDW, SDW, UTS]	ETHYLENE DICHLORIDE
1300-21-6 DICHLOROETHANE [ATS]	ETHANE DICHLORIDE
107-06-2 1,2-DICHLOROETHANE (ETHYLENE DICHLO-RIDE) [313]	ETHYLENE DICHLORIDE

Chemical Name	Reference Name
75-34-3 1,1-DICHLOROETHANE (ETHYLIDENE CHLORIDE) [ALB, ALB]	1,1-DICHLOROETHANE
107-06-2 1,2-DICHLOROETHANE (ETHYLENE DICHLORIDE) [ALB, ALB]	ETHYLENE DICHLORIDE
107-06-2 1,2-DICHLOROETHANE (ETHYLENEDICHLORIDE) (EDC) [HON, HON]	ETHYLENE DICHLORIDE
107-06-2 DICHLOROETHANE, 1,2- [CD1]	ETHYLENE DICHLORIDE
75-35-4 1,1-DICHLOROETHENE [ATS, ONT, ONT]	VINYLIDENE CHLORIDE
156-59-2 1,2-DICHLOROETHENE, CIS- [ATS]	CIS-1,2-DICHLOROETHYLENE
156-60-5 1,2-DICHLOROETHENE, TRANS- [ATS]	1,2-TRANS-DICHLOROETHYLENE
540-59-0 1,2-DICHLOROETHENE [ONT, ONT]	1,2-DICHLOROETHYLENE
RR-01279-3 DICHLOROETHER	DICHLOROETHER
111-44-4 DICHLOROETHYL ETHER [FLA]	BIS(2-CHLOROETHYL) ETHER
10140-87-1 1,2-DICHLOROETHYL ACETATE [MSL]	ETHANOL, 1,2-DICHLORO-, ACETATE
75-35-4 1,1-DICHLOROETHYLENE [CW2, CW3, CWA, EPA]	VINYLIDENE CHLORIDE
75-35-4 1,1 DICHLOROETHYLENE [FLA]	VINYLIDENE CHLORIDE
75-35-4 1,1-DICHLOROETHYLENE [HM1, HM1, HM1, HM1, MX2, OTS, QBC, RC2, RC2, RCA, RCU, RCU, SDW, SDW, UTS]	VINYLIDENE CHLORIDE
156-59-2 CIS-1,2-DICHLOROETHYLENE	CIS-1,2-DICHLOROETHYLENE
156-59-2 DICHLOROETHYLENE-CIS [FLA, MSL]	CIS-1,2-DICHLOROETHYLENE
156-60-5 1,2-TRANS-DICHLOROETHYLENE	1,2-TRANS-DICHLOROETHYLENE
156-60-5 TRANS-1,2-DICHLOROETHYLENE [CSR]	1,2-TRANS-DICHLOROETHYLENE
156-60-5 DICHLOROETHYLENE-TRANS [FLA]	1,2-TRANS-DICHLOROETHYLENE
156-60-5 1,2 DICHLOROETHYLENE [MEX, MEX]	1,2-TRANS-DICHLOROETHYLENE
156-60-5 DICHLOROETHYLENE-TRANS [MSL]	1,2-TRANS-DICHLOROETHYLENE
156-60-5 1,2-DICHLOROETHYLENE [MX2]	1,2-TRANS-DICHLOROETHYLENE
156-60-5 TRANS-1,2-DICHLOROETHYLENE [PQL, RCA]	1,2-TRANS-DICHLOROETHYLENE
156-60-5 1,2-DICHLOROETHYLENE [RCU, RCU]	1,2-TRANS-DICHLOROETHYLENE
156-60-5 TRANS-1,2-DICHLOROETHYLENE [SDW, SDW, SDW]	1,2-TRANS-DICHLOROETHYLENE
540-59-0 1,2-DICHLOROETHYLENE	1,2-DICHLOROETHYLENE
540-59-0 CIS & TRANS 1,2-DICHLOROETHYLENE [CSR]	1,2-DICHLOROETHYLENE
540-59-0 1,2 DICHLOROETHYLENE [QBC]	1,2-DICHLOROETHYLENE
540-59-0 SYM-DICHLOROETHYLENE [WHM]	1,2-DICHLOROETHYLENE
25323-30-2 DICHLOROETHYLENE [HM1, HM1]	DICHLOROETHYLENES
75-35-4 1,1-DICHLOROETHYLENE (VINYLIDENE CHLORIDE) [EP2]	VINYLIDENE CHLORIDE
75-35-4 1,1 DICHLOROETHYLENE (VINYLIDENE DICHLORIDE) [ALB, ALB]	VINYLIDENE CHLORIDE
25323-30-2 DICHLOROETHYLENE, N.O.S. [RCU]	DICHLOROETHYLENES
25323-30-2 DICHLOROETHYLENES	DICHLOROETHYLENES
111-44-4 DICHLOROETHYL ETHER [302, ALB, ALB, ALB, AUS, AUS, AUS, BC1, BC1, BC1, CEX, CEX, CEX]	BIS(2-CHLOROETHYL) ETHER
111-44-4 DICHLOROETHYLETHER (BIS(2-CHLOROETHYL) ETHER) [HON, HON]	BIS(2-CHLOROETHYL) ETHER
111-44-4 DICHLOROETHYL ETHER [MIN]	BIS(2-CHLOROETHYL) ETHER
111-44-4 2,2'-DICHLOROETHYL ETHER [NFP, NFP]	BIS(2-CHLOROETHYL) ETHER

Chemical Name	Reference Name
111-44-4 DICHLOROETHYL ETHER [NIO, NIO, NIO, NIO, NIO, OS1, OS1, OS1, OS1, OS1, QBC, QBC, QBC, TLV, TLV, TLV, WHM]	BIS(2-CHLOROETHYL) ETHER
111-44-4 DICHLOROETHYL ETHER (BIS(2-CHLOROETHYL)ETHER) [CAA]	BIS(2-CHLOROETHYL) ETHER
RR-01280-6 DICHLOROETHYL OXIDE	DICHLOROETHYL OXIDE
505-60-2 DICHLOROETHYL SULFIDE [HM1]	MUSTARD GAS
505-60-2 2,2'-DICHLOROETHYL SULFIDE [MAK]	MUSTARD GAS
RR-01410-8 DICHLOROETHYL SULFIDE	DICHLOROETHYL SULFIDE
7572-29-4 DICHLOROETHYNE [ONT]	DICHLOROACETYLENE
1717-00-6 1,1-DICHLORO-1-FLUOROETHANE (HCFC-141B) [313]	ETHANE, 1,1-DICHLORO-1-FLUORO-L ESTER)
75-43-4 DICHLOROFLUOROMETHANE	DICHLOROFLUOROMETHANE
75-43-4 DICHLOROFLUOROMETHANE [AUS, CEX, FLA, GBR, MAK, MAK, MSL, NJL, ONT]	DICHLOROFLUOROMETHANE
75-43-4 DICHLOROFLUOROMETHANE (FREON 21) [QBC]	DICHLOROFLUOROMETHANE
75-43-4 DICHLOROFLUOROMETHANE [TLV]	DICHLOROFLUOROMETHANE
134237-45-9 DICHLOROFLUOROPROPANE	DICHLOROFLUOROPROPANE
RR-00562-9 1,3-DICHLORO-2,4-HEXADIENE	1,3-DICHLORO-2,4-HEXADIENE
662-01-1 1,3-DICHLORO-1,1,2,2,3,3,-HEXAFLUORO-PROPANE	1,3-DICHLORO-1,1,2,2,3,3,-HEXAFLUOROPROPANE
2163-00-0 1,6-DICHLOROHEXANE	1,6-DICHLOROHEXANE
2782-57-2 DICHLOROISOCYANURIC ACID	DICHLOROISOCYANURIC ACID
2244-21-5 DICHLOROISOCYANURIC ACID, POTASSIUM SALT [WHM]	POTASSIUM DICHLOROISOCYANURATE
2893-78-9 DICHLOROISOCYANURIC ACID, SODIUM SALT	DICHLOROISOCYANURIC ACID, SODIUM SALT
108-60-1 DICHLOROISOPROPYL ETHER [EPA]	BIS(2-CHLOROISOPROPYL) ETHER
108-60-1 2,2-DICHLORO ISOPROPYL ETHER [FLA]	BIS(2-CHLOROISOPROPYL) ETHER
108-60-1 DICHLOROISOPROPYL ETHER [HM1, HM1, HM1]	BIS(2-CHLOROISOPROPYL) ETHER
108-60-1 2,2-DICHLORO ISOPROPYL ETHER [MSL, NFP, NFP]	BIS(2-CHLOROISOPROPYL) ETHER
3188-13-4 DICHLOROISOPROPYL ETHER [WHM, NFP, NFP]	CHLOROMETHYL ETHYL ETHER
13674-87-8 1,3-DICHLORO ISOPROPYL PHOSPHATE	1,3-DICHLORO ISOPROPYL PHOSPHATE
108-60-1 DICHLOROISOPROPYL ETHER [RCU, RCU]	BIS(2-CHLOROISOPROPYL) ETHER
75-09-2 DICHLOROMETHANE [CAN, CD1, CN4, GBR, HM1, HM1, HM1, HM1, IAR, MAK, MAK, MAK, MAK, ONT, SDW, SDW, SDW]	METHYLENE CHLORIDE
75-09-2 DICHLOROMETHANE (METHYLENE CHLO-RIDE) [NTP, 313, C65, C65, ALB, ALB]	METHYLENE CHLORIDE
528-74-5 DICHLOROMETHOTREXATE	DICHLOROMETHOTREXATE
111-91-1 DICHLOROMETHOXY ETHANE [RCU, RCU]	BIS(2-CHLOROETHOXY)METHANE
41683-62-9 1,2-DICHLOROMETHOXYETHANE	1,2-DICHLOROMETHOXYETHANE
101-14-4 2,2'-DICHLORO-4,4'-METHYLENEDIANILINE (MBOCA) [GBR, GBR]	4,4'-METHYLENEBIS(2-CHLOROANILINE) (MBOCA)
542-88-1 DICHLOROMETHYL ETHER [EPA, RCU, RCU]	BIS(CHLOROMETHYL) ETHER
149-74-6 DICHLOROMETHYLPHENYLSILANE [MSL, PAL, FLA]	METHYLPHENYLDICHLOROSILANE
149-74-6 DICHLOROMETHYPHENYLSILANE [302, EPA, EPA]	METHYLPHENYLDICHLOROSILANE

Chemical Name	Reference Name
75-43-4 DICHLOROMONOFLUOROMETHANE (HCFC-21) [313]	DICHLOROFLUOROMETHANE
75-43-4 DICHLOROMONOFLUOROMETHANE (FC 21) [WHM]	DICHLOROFLUOROMETHANE
75-43-4 DICHLOROMONOFLUOROMETHANE (FC-21) [CAL]	DICHLOROFLUOROMETHANE
99-30-9 2,6-DICHLORO-4-NITROANILINE	2,6-DICHLORO-4-NITROANILINE
89-61-2 2,5-DICHLORONITROBENZENE	2,5-DICHLORONITROBENZENE
89-61-2 1,4-DICHLORO-2-NITROBENZENE [TSC]	2,5-DICHLORONITROBENZENE
99-54-7 3,4-DICHLORONITROBENZENE	3,4-DICHLORONITROBENZENE
594-72-9 1,1-DICHLORO-1-NITROETHANE	1,1-DICHLORO-1-NITROETHANE
594-72-9 1,1-DICHLORO-1-NITRO ETHANE [NFP, NFP]	1,1-DICHLORO-1-NITROETHANE
17741-62-7 4-[4-[(2,6-DICHLORO-4-NITROPHENYL)AZO] PHENYL]THIOMORPHOLINE,1,1-DIOXIDE-	4-[4-[(2,6-DICHLORO-4-NITROPHENYL)AZO]PHENYL] THIOMORPHOLINE,
595-44-8 1,1-DICHLORO-1-NITRO PROPANE [FLA, MSL, NFP, NFP]	PROPANE, 1,1-DICHLORO-1-NITRO-
127564-92-5 DICHLOROPENTAFLUOROPROPANE	DICHLOROPENTAFLUOROPROPANELORIDE)
13474-88-9 1,1-DICHLORO-1,2,2,3,3-PENTAFLUORO-PROPANE (HCFC-225CC)	1,1-DICHLORO-1,2,2,3,3-PENTAFLUOROPROPANE (HCFC-225CC)XANE)
111512-56-2 1,1-DICHLORO-1,2,3,3,3-PENTAFLUORO-PROPANE (HCFC-225EB)	1,1-DICHLORO-1,2,3,3,3-PENTAFLUOROPROPANE (HCFC-225EB)-YL)-METHYLAMINO)CARBONYL)AMINO)SUL-FONYL)-, METHYL ESTER]
422-44-6 1,2-DICHLORO-1,1,2,3,3-PENTAFLUORO-PROPANE (HCFC-225BB)	1,2-DICHLORO-1,1,2,3,3-PENTAFLUOROPROPANE (HCFC-225BB)
431-86-7 1,2-DICHLORO-1,1,3,3,3-PENTAFLUORO-PROPANE (HCFC-225DA)	1,2-DICHLORO-1,1,3,3,3-PENTAFLUOROPROPANE (HCFC-225DA)
136013-79-1 1,3-DICHLORO-1,1,2,3,3-PENTAFLUORO-PROPANE (HCFC-225EA)	1,3-DICHLORO-1,1,2,3,3-PENTAFLUOROPROPANE (HCFC-225EA)
128903-21-9 2,2-DICHLORO-1,1,1,3,3-PENTAFLUORO-PROPANE (HCFC-225AA)	2,2-DICHLORO-1,1,1,3,3-PENTAFLUOROPROPANE (HCFC-225AA)
422-48-0 2,3-DICHLORO-1,1,1,2,3-PENTAFLUORO-PROPANE (HCFC-225BA)	2,3-DICHLORO-1,1,1,2,3-PENTAFLUOROPROPANE (HCFC-225BA)
507-55-1 1,3-DICHLORO-1,1,2,2,3-PENTAFLUORO-PROPANE	1,3-DICHLORO-1,1,2,2,3-PENTAFLUOROPROPANE
507-55-1 1,3-DICHLORO-1,1,2,2,3-PENTAFLUORO-PROPANE (HCFC-225CB) [313]	1,3-DICHLORO-1,1,2,2,3-PENTAFLUOROPROPANE
422-56-0 3,3-DICHLORO-1,1,1,2,2-PENTAFLUORO-PROPANE	3,3-DICHLORO-1,1,1,2,2-PENTAFLUOROPROPANE
422-56-0 3,3-DICHLORO-1,1,1,2,2-PENTAFLUORO-PROPANE (HCFC-225CA) [313]	3,3-DICHLORO-1,1,1,2,2-PENTAFLUOROPROPANE
628-76-2 1,5-DICHLOROPENTANE	1,5-DICHLOROPENTANE
30586-10-8 DICHLOROPENTANE	DICHLOROPENTANE
30586-10-8 DICHLOROPENTANES [NFP, NFP, PAL]	DICHLOROPENTANE
97-23-4 DICHLOROPHEN	DICHLOROPHEN
97-23-4 DICHLOROPHENE [PST]	DICHLOROPHEN
97-23-4 DICHLOROPHENE [2,2'-METHYLENE-BIS(4-CHLOROPHENOL)] [313]	DICHLOROPHEN
87-65-0 2,6-DICHLOROPHENOL	2,6-DICHLOROPHENOL
95-77-2 3,4-DICHLOROPHENOL	3,4-DICHLOROPHENOL
120-83-2 2,4-DICHLOROPHENOL	2,4-DICHLOROPHENOL
576-24-9 2,3-DICHLOROPHENOL	2,3-DICHLOROPHENOL
583-78-8 2,5-DICHLOROPHENOL	2,5-DICHLOROPHENOL
120-83-2 DICHLOROPHENOL, 2,4- [CD1, CD1]	2,4-DICHLOROPHENOL

Chemical Name	Reference Name
RR-01281-7 DICHLOROPHENOLS	DICHLOROPHENOLS
94-11-1 2,4-DICHLOROPHENOXYACETIC ACID, ISO-PROPYL ESTER [NJL]	2,4-D ESTERS
94-75-7 2,4-DICHLOROPHENOXYACETATE ACID [MX2]	2,4-D
53467-11-1 2,4-DICHLOROPHENOXYACETIC ACID, ESTER [NJL]	2,4-D ESTERS
94-75-7 2,4-DICHLOROPHENOXYACETIC ACID [FLA]	2,4-D
94-75-7 2,4-DICHLOROPHENOXY ACETIC ACID [MAK, MAK, MAK, MAK]	2,4-D
94-75-7 2,4-DICHLOROPHENOXYACETIC ACID [MSL]	2,4-D
94-75-7 (2,4-DICHLOROPHENOXY) ACETIC ACID [ONT]	2,4-D
94-75-7 DICHLOROPHENOXYACETIC ACID, 2,4- (2,4-D) [CD1]	2,4-D
94-75-7 2,4-DICHLOROPHENOXYACETIC ACID (2,4-D) [RCA]	2,4-D
RR-01282-8 2,4-DICHLOROPHENOXYACETIC ACID DI-ETHANOLAMINE SALT	2,4-DICHLOROPHENOXYACETIC ACID DI-ETHANOLAMINE SALT
RR-01283-9 2,4-DICHLOROPHENOXYACETIC ACID DIMETHYLAMINE SALT	2,4-DICHLOROPHENOXYACETIC ACID DIMETHY-LAMINE SALT
RR-01284-0 2,4-DICHLOROPHENOXYACETIC ACID TRIISO-PROPYLAMINE SALT	2,4-DICHLOROPHENOXYACETIC ACID TRIISOPROPY-LAMINE SALT
94-75-7 2,4-DICHLOROPHENOXYACETIC ACID [TSE]	2,4-D
6341-97-5 2,4-DICHLOROPHENOXYACETIC ACID ESTER	2,4-DICHLOROPHENOXYACETIC ACID ESTER
28165-71-1 DICHLOROPHENOXYACETIC ACID ESTER	DICHLOROPHENOXYACETIC ACID ESTER
28165-71-1 2,6-DICHLOROPHENOXYACETIC ACID ESTER [NJL]	DICHLOROPHENOXYACETIC ACID ESTER
94-11-1 (2,4-DICHLOROPHENOXY) ACETIC ACID ESTERS [ONT]	2,4-D ESTERS
2702-72-9 2,4-DICHLOROPHENOXYACETIC ACID SODIUM SALT	2,4-DICHLOROPHENOXYACETIC ACID SODIUM SALT
136-78-7 2-(2,4-DICHLOROPHENOXY)ETHANOL HYDROGEN SULFATE SODIUM SALT [ONT]	SESONE
120-36-5 2[2,4-(DICHLOROPHENOXY)]-PROPANOIC ACID [TSE]	2[2,4-(DICHLOROPHENOXY)]-PROPIONIC ACID
120-36-5 2[2,4-(DICHLOROPHENOXY)]-PROPIONIC ACID	2[2,4-(DICHLOROPHENOXY)]-PROPIONIC ACID
696-28-6 DICHLOROPHENYLARSINE	DICHLOROPHENYLARSINE
609-20-1 2,6-DICHLORO-P-PHENYLENEDIAMINE	2,6-DICHLORO-P-PHENYLENEDIAMINE
609-20-1 2,6-DICHLORO-PARA-PHENYLENEDI-AMINE [CAL, IAR]	2,6-DICHLORO-P-PHENYLENEDIAMINE
2612-57-9 2,4-DICHLORO-1-PHENYL ISOCYANATE	2,4-DICHLORO-1-PHENYL ISOCYANATE
RR-00584-5 DICHLOROPHENYL ISOCYANATE, ALL ISOMERS	DICHLOROPHENYL ISOCYANATE, ALL ISOMERS
RR-00584-5 DICHLOROPHENYL ISOCYANATES [HM1, HM1]	DICHLOROPHENYL ISOCYANATE, ALL ISOMERS
102-36-3 1,2-DICHLORO-4-PHENYL ISOCYANATE [NJL]	ISOCYANIC ACID, 3,4-DICHLOROPHENYL ESTER
5392-82-5 1,4-DICHLORO-2-PHENYL ISOCYANATE	1,4-DICHLORO-2-PHENYL ISOCYANATE
34893-92-0 1,3-DICHLORO-5-PHENYL ISOCYANATE	1,3-DICHLORO-5-PHENYL ISOCYANATE
39920-37-1 1,3-DICHLORO-2-PHENYL ISOCYANATE	1,3-DICHLORO-2-PHENYL ISOCYANATE
41195-90-8 1,2-DICHLORO-3-PHENYL ISOCYANATE	1,2-DICHLORO-3-PHENYL ISOCYANATE
1836-75-5 2,4-DICHLOROPHENYL P-NITROPHENYL ETHER [CAL]	NITROFEN
50-29-3 DICHLOROPHENYLTRICHLOROETHANE (DDT) + METABOLITES [CD1]	DDT

Chemical Name	Reference Name
27137-85-5 DICHLOROPHENYL TRICHLOROSILANE [NJL, NJL]	TRICHLORO(DICHLOROPHENYL)SILANE
27137-85-5 DICHLOROPHENYLTRICHLOROSILANE [HM1, HM1, HM1, WHM]	TRICHLORO(DICHLOROPHENYL)SILANE
78-87-5 1,2-DICHLOROPROPANE	1,2-DICHLOROPROPANE
78-99-9 1,1-DICHLOROPROPANE	1,1-DICHLOROPROPANE
142-28-9 1,3-DICHLOROPROPANE	1,3-DICHLOROPROPANE
594-20-7 2,2-DICHLOROPROPANE [MSL, SD4, SDW]	PROPANE, 2,2-DICHLORO-
26638-19-7 DICHLOROPROPANE	DICHLOROPROPANE
78-87-5 1,2-DICHLOROPROPANE (PROPYLENE DICHLORIDE) [CSR, CSR, ALB, ALB]	1,2-DICHLOROPROPANE
8003-19-8 DICHLOROPROPANE-DICHLOROPROPENE (MIXTURE)	DICHLOROPROPANE-DICHLOROPROPENE (MIXTURE)
26952-23-8 DICHLOROPROPANE, N.O.S. [RCU, RCU]	DICHLOROPROPENE
26638-19-7 DICHLOROPROPANES [CAL]	DICHLOROPROPANE
96-23-1 1,3-DICHLORO-2-PROPANOL	1,3-DICHLORO-2-PROPANOL
96-23-1 1,3-DICHLOROPROPANOL-2 [HM1, HM1]	1,3-DICHLORO-2-PROPANOL
96-23-1 1,3-DICHLOROPROPANOL [NJL, TSC, TSC]	1,3-DICHLORO-2-PROPANOL
96-23-1 1,3-DICHLORO-2-PROPANOL [TSE]	1,3-DICHLORO-2-PROPANOL
26545-73-3 DICHLOROPROPANOL, N.O.S. [RCU]	DICHLOROPROPANOLS
26545-73-3 DICHLOROPROPANOLS	DICHLOROPROPANOLS
78-88-6 2,3-DICHLOROPROPENE	2,3-DICHLOROPROPENE
78-88-6 2,3-DICHLORO-1-PROPENE [PST]	2,3-DICHLOROPROPENE
542-75-6 1,3-DICHLOROPROPENE	1,3-DICHLOROPROPENE
542-75-6 DICHLOROPROPENE [AUS, AUS, AUS, CEX, CEX]	1,3-DICHLOROPROPENE
542-75-6 TRANS-1,3-DICHLOROPROPENE [NJL, NJL]	1,3-DICHLOROPROPENE
542-75-6 1,3-DICHLOROPROPENE (CIS AND TRANS ISOMERS) [QBC, QBC, QBC]	1,3-DICHLOROPROPENE
542-75-6 1,3-DICHLOROPROPENE [TLV, TLV]	1,3-DICHLOROPROPENE
542-75-6 DICHLOROPROPENE [WHM]	1,3-DICHLOROPROPENE
563-58-6 1,1-DICHLOROPROPENE	1,1-DICHLOROPROPENE
10061-01-5 CIS-1,3-DICHLOROPROPENE	CIS-1,3-DICHLOROPROPENE
10061-01-5 1,3-DICHLOROPROPENE, CIS- [ATS]	CIS-1,3-DICHLOROPROPENE
10061-02-6 TRANS-1,3-DICHLOROPROPENE	TRANS-1,3-DICHLOROPROPENE
26952-23-8 DICHLOROPROPENE	DICHLOROPROPENE
542-75-6 1,3-DICHLOROPROPENE (TELONE II) [CSR, CSR]	1,3-DICHLOROPROPENE
26952-23-8 DICHLOROPROPENES [CAL]	DICHLOROPROPENE
542-75-6 1,3-DICHLOROPROPENE (TECHNICAL GRADE) [IAR, MIN, NTP]	1,3-DICHLOROPROPENE
75-99-0 2,2-DICHLOROPROPIONIC ACID	2,2-DICHLOROPROPIONIC ACID
127-20-8 2,2-DICHLOROPROPIONIC ACID, SODIUM SALT	2,2-DICHLOROPROPIONIC ACID, SODIUM SALT
78-88-6 2,3-DICHLOROPROPYLENE [CSR]	2,3-DICHLOROPROPENE
542-75-6 1,3-DICHLOROPROPYLENE [313]	1,3-DICHLOROPROPENE
542-75-6 1,2-DICHLOROPROPYLENE [MX1, MX2]	1,3-DICHLOROPROPENE
542-75-6 1,2-DICHLOROPROPYLENE (1,3-DICHLOROPROPENE) [CW2]	1,3-DICHLOROPROPENE
320-72-9 3,5-DICHLOROSALICYLIC ACID	3,5-DICHLOROSALICYLIC ACID

Chemical Name	Reference Name
4109-96-0 DICHLOROSILANE	DICHLOROSILANE
4109-96-0 DICHLOROSILANE [SILANE, DICHLORO-] [EP3]	DICHLOROSILANE
6607-45-0 ALPHA,BETA-DICHLOROSTRYENE [FLA, NFP, NFP]	BENZENE, (1,2-DICHLOROETHENYL)-
6607-45-0 ALPHA,BETA-DICHLOROSTYRENE [MSL]	BENZENE, (1,2-DICHLOROETHENYL)-
76-14-2 1,2-DICHLOROTETRAFLUOROETHANE [CEX]	DICHLOROTETRAFLUOROETHANE
76-14-2 1,2-DICHLORO-1,1,2,2-TETRAFLUO-ROETHANE [MAK, MAK, WHM]	DICHLOROTETRAFLUOROETHANE
76-14-2 DICHLOROTETRAFLUOROETHANE	DICHLOROTETRAFLUOROETHANE
76-14-2 DICHLOROTETRAFLUOROETHANE (CFC-114) [313]	DICHLOROTETRAFLUOROETHANE
76-14-2 1,2-DICHLORO-1,1,2,2-TETRAFLUOROETHANE (FC-114) [CAL]	DICHLOROTETRAFLUOROETHANE
76-14-2 1,2-DICHLORO-1,1,2,2-TETRAFLUO-ROETHANE [ONT]	DICHLOROTETRAFLUOROETHANE
76-14-2 DICHLOROTETRAFLUOROETHANE (FREON 114) [QBC]	DICHLOROTETRAFLUOROETHANE
1320-37-2 DICHLOROTETRAFLUOROETHANE	DICHLOROTETRAFLUOROETHANE
127564-83-4 DICHLOROTETRAFLUOROPROPANE	DICHLOROTETRAFLUOROPROPANE
95-73-8 2,4-DICHLOROTOLUENE	2,4-DICHLOROTOLUENE
118-69-4 2,6-DICHLOROTOLUENE	2,6-DICHLOROTOLUENE
2782-57-2 DICHLORO-S-TRIAZINETRIONE [FLA, MSL]	DICHLOROISOCYANURIC ACID
RR-00255-1 DICHLOROTRIAZINETRIONE AND ITS SALTS	DICHLOROTRIAZINETRIONE AND ITS SALTS
812-04-4 1,1-DICHLORO-1,2,2-TRIFLUOROETHANE (HCFC-123B)	1,1-DICHLORO-1,2,2-TRIFLUOROETHANE (HCFC-123B)
306-83-2 2,2-DICHLORO-1,1,1-TRIFLUOROETHANE (HCFC-123) [313]	ETHANE, 2,2-DICHLORO-1,1,1-TRIFLUORO-STER PHOS-PHOROTHIOIC ACID]
306-83-2 2,2-DICHLORO-1,1,1-TRIFLUOROETHANE [MAK]	ETHANE, 2,2-DICHLORO-1,1,1-TRIFLUORO-
34077-87-7 DICHLOROTRIFLUOROETHANE	DICHLOROTRIFLUOROETHANE-N,N'-DIMETHYLUREA]
90454-18-5 DICHLORO-1,1,2-TRIFLUOROETHANE	DICHLORO-1,1,2-TRIFLUOROETHANE4-TRIAZOLE-1-PROPANETRILE]
354-23-4 1,2-DICHLORO-1,1,2-TRIFLUOROETHANE (HCFC-123A)	1,2-DICHLORO-1,1,2-TRIFLUOROETHANE (HCFC-123A)
3615-21-2 4,5-DICHLORO-2-(TRIFLUOROMETHYL)-BENZ-IMIDAZOLE [MSL]	BENZIMIDAZOLE, 4,5-DICHLORO-2-(TRIFLUO-ROMETHYL)-
134237-43-7 DICHLOROTRIFLUOROPROPANE	DICHLOROTRIFLUOROPROPANE
1737-93-5 3,5-DICHLORO-2,4,6-TRIFLUOROPYRIDINE	3,5-DICHLORO-2,4,6-TRIFLUOROPYRIDINE
594-31-0 DICHLOROTRIPHENYLANTIMONY	DICHLOROTRIPHENYLANTIMONY
RR-01411-9 DICHLOROVINYLCHLOROARSINE	DICHLOROVINYLCHLOROARSINE
623-25-6 2,2'-DICHLORO-P-XYLENE	2,2'-DICHLORO-P-XYLENE
62-73-7 DICHLORVOS	DICHLORVOS
62-73-7 DICHLORVOS (DDVP) [AUS, AUS, BC1, BC1, BC1, MAK, MAK, MAK, MAK, OS1, OS1, OS1, OS1]	DICHLORVOS
62-73-7 DICHLORVOS (PHOSPHORIC ACID, 2,2,-DICHLOROETHENYL DIMETHYL ESTER) [313]	DICHLORVOS3BETA,4ALPHA,5ALPHA,6BETA)-]
62-73-7 DICHLORVOS (DDVP) [CEX, CEX]	DICHLORVOS
62-73-7 DICHLORVOS (ISO) [GBR, GBR, GBR]	DICHLORVOS
RR-01620-6 DICHROMATES	DICHROMATES
51338-27-3 DICLOFOP-METHYL [CD1]	HOELON
51338-27-3 DICLOFOP-METHYL [2-[4-(2,4-DICHLOROPHE-NOXY)PHENOXY] PROPANOIC ACID, METHYL ESTER] [313]	HOELON

Chemical Name	Reference Name
102-30-7 DICLORAN	DICLORAN
61789-77-3 DI(COCO ALKYL) DIMETHYL AMMONIUM CHLORIDE	DI(COCO ALKYL) DIMETHYL AMMONIUM CHLORIDE
115-32-2 DICOFOL	DICOFOL
115-32-2 DICOFOL [BENZENEMETHANOL,4-CHLORO-ALPHA-(4-CHLOROPHENYL)-ALPHA-(TRICHLORMETHYL)-] [313]	DICOFOL
141-66-2 DICROTOPHOS	DICROTOPHOS
141-66-2 DICROTOPHOS (BIDRIN) [BC1, BC1]	DICROTOPHOS
141-66-2 DICROTOPHOS (DIDRIN) [QBC, QBC]	DICROTOPHOS
141-66-2 DICROTOPHOS (BIDRIN) [ALB, ALB, ALB]	DICROTOPHOS
66-76-2 DICUMAROL	DICUMAROL
80-43-3 DICUMYL PEROXIDE	DICUMYL PEROXIDE
626-17-5 1,3-DICYANOBENZENE [TSC, TSC, TSC]	M-PHTHALODINITRILE
111-69-3 1,4-DICYANOBUTANE [OTS]	ADIPONITRILE
461-58-5 DICYANODIAMIDE [PST]	GUANIDINE, CYANO
74849-88-0 DICYANOETHYL DIETHYLENETRIAMINE	DICYANOETHYL DIETHYLENETRIAMINE
101-83-7 DICYCLOHEXYLAMINE	DICYCLOHEXYLAMINE
3129-91-7 DICYCLOHEXYLAMMONIUM NITRITE	DICYCLOHEXYLAMMONIUM NITRITE
3882-06-2 DICYCLOHEXYLAMMONIUM NITRATE	DICYCLOHEXYLAMMONIUM NITRATE
4979-32-2 N,N-DICYCLOHEXYL-2-BENZOTHIAZOLE SULFENAMIDE	N,N-DICYCLOHEXYL-2-BENZOTHIAZOLE SULFENAMIDE
538-75-0 DICYCLOHEXYLCARBODIIMIDE	DICYCLOHEXYLCARBODIIMIDE
5124-30-1 DICYCLOHEXYLMETHANE-4,4'-DIISO-CYANATE [CAL]	METHYLENE BIS(4-CYCLOHEXYLISOCYANATE)
5124-30-1 DICYCLOHEXYLMETHANE 4,4'-DIISOCYANATE (HYDROGENATED MDI) [NHS, NHS]	METHYLENE BIS(4-CYCLOHEXYLISOCYANATE)
1758-61-8 DICYCLOHEXYL PEROXIDE [MAK]	DI(1-HYDROXYCYCLO-HEXYL)PEROXIDE
1561-49-5 DICYCLOHEXYL PEROXY-DICARBONATE	DICYCLOHEXYL PEROXY-DICARBONATE
84-61-7 DICYCLOHEXYL PHTHALATE [GBR]	1,2-BENZENEDICARBOXYLIC ACID, DICYCLOHEXYL ESTER
1212-29-9 N,N'-DICYCLOHEXYLTHIOUREA	N,N'-DICYCLOHEXYLTHIOUREA
77-73-6 DICYCLOPENTADIENE	DICYCLOPENTADIENE
RR-01065-1 DICYCLOPENTADIENE, POLYMER WITH (MIXED STYRENE AND ALPHA-METHYL-STYRENE), (MIXED INDENE AND METHYL IN-DENE), AND VINYL TOLUENE	DICYCLOPENTADIENE, POLYMER WITH (MIXED STYRENE AND ALPHA-METIBLE FATS
102-54-5 DICYCLOPENTADIENYL IRON	DICYCLOPENTADIENYL IRON
849-99-0 DICYCLOHEXYL ADIPATE	DICYCLOHEXYL ADIPATE
762-12-9 DIDECANOYL PEROXIDE [DOT]	DECANOYL PEROXIDE
7173-51-5 DIDECYLDIMETHYLAMMONIUM CHLORIDE	DIDECYLDIMETHYLAMMONIUM CHLORIDE
2456-28-2 DIDECYL ETHER	DIDECYL ETHER
7481-89-2 2',3'-DIDEOXYCYTIDINE (AIDS INITIATIVE)	2',3'-DIDEOXYCYTIDINE (AIDS INITIATIVE)
69655-05-6 2',3'-DIDEOXYINOSINE (AIDS INITIATIVE)	2',3'-DIDEOXYINOSINE (AIDS INITIATIVE)
RR-00256-2 2,2-DI(4,4-DI-T-BUTYL-PEROXY CYCLOHEXYL)-PROPANE 42%	2,2-DI(4,4-DI-T-BUTYL-PEROXY CYCLOHEXYL)-PROPANE 42%
RR-00257-3 DI-2,4-DICHLOROBENZOYL PEROXIDE	DI-2,4-DICHLOROBENZOYL PEROXIDE
133-14-2 DI-2,4-DICHLOROBENZOYL PEROXIDE [HM1]	2,4-DICHLOROBENZOYL PEROXIDE
110-18-9 1,2-DI-(DIMETHYLAMINO)ETHANE [HM1, HM1]	1,2-BIS(DIMETHYLAMINO)ETHANE
110-18-9 1,2-DI(DIMETHYLAMINO) ETHANE [NJL]	1,2-BIS(DIMETHYLAMINO)ETHANE
RR-00581-2 DIDYMIUM NITRATE	DIDYMIUM NITRATE
60-57-1 DIELDRIN	DIELDRIN

Chemical Name	Reference Name
60-57-1 DIELDRIN ((1R,4S,4AS,5R,6R,7S,8S,8AR)1,2,3,4,10, 10-HEXACHLORO-1,4,4A,5,6,7,8,8A-OCTAHY-DRO-6,7-EPOXY-1,4:5,8-DIMETHANONAPHTHA-LENE) [CN2]	DIELDRIN
60-57-1 DIELDRIN (ISO) [GBR, GBR, GBR]	DIELDRIN
60-57-1 DIELDRINE [QBC, QBC]	DIELDRIN
84-17-3 DIENESTROL [C65, CAL]	DIENOESTROL
84-17-3 DIENOESTROL	DIENOESTROL
298-18-0 DL-DIEPOXYBUTANE	DL-DIEPOXYBUTANE
1464-53-5 DIEPOXYBUTANE	DIEPOXYBUTANE
1464-53-5 1,2:3,4-DIEPOXYBUTANE [EP2]	DIEPOXYBUTANE
1464-53-5 1,2,3,4-DIEPOXYBUTANE [NJL, NJL]	DIEPOXYBUTANE
RR-00270-0 DIESEL ENGINE EMISSIONS [MAK]	DIESEL ENGINE EXHAUST
RR-00270-0 DIESEL ENGINE EXHAUST	DIESEL ENGINE EXHAUST
RR-00270-0 DIESEL EXHAUST [NHS, NHS, NHS]	DIESEL ENGINE EXHAUST
68334-30-5 DIESEL FUEL [NJL]	DIESEL OIL (PETROLEUM)
68334-30-5 DIESEL FUEL (MARINE) [CAB]	DIESEL OIL (PETROLEUM)
68334-30-5 DIESEL FUEL MARINE [CSR]	DIESEL OIL (PETROLEUM)
68334-30-5 DIESEL FUEL OIL [NFP, NFP, PAL]	DIESEL OIL (PETROLEUM)
68334-30-5 DIESEL OIL (PETROLEUM)	DIESEL OIL (PETROLEUM)
56863-02-6 DIETHANOLAMIDE OF LINOLEIC ACID	DIETHANOLAMIDE OF LINOLEIC ACID
111-42-2 DIETHANOLAMINE	DIETHANOLAMINE
111-42-2 DIETHANOLAMINE (2,2'-IMINODIETHANOL) [HON, HON]	DIETHANOLAMINE
111-42-2 DIETHANOLAMINE (ETHANOL, 2,2'-IMINOBIS-) [TSC, TSC, TSC]	DIETHANOLAMINE
68133-37-9 DIETHANOLAMINE ETHYLENEDIAMINETE-TRAACETATE	DIETHANOLAMINE ETHYLENEDIAMINETETRAAC-ETATE
7487-79-8 DIETHANOLAMINE LAURATE	DIETHANOLAMINE LAURATE
13961-86-9 DIETHANOLAMINE OLEATE [PST]	OLEIC ACID DIETHANOLAMINE
143-00-0 DIETHANOLAMMONIUM DODECYL SULFATE	DIETHANOLAMMONIUM DODECYL SULFATE
RR-01412-0 DIETHANOL NITROSAMINE DINITRATE	DIETHANOL NITROSAMINE DINITRATE
38727-55-8 DIETHATYL ETHYL	DIETHATYL ETHYL
6175-45-7 2,2-DIETHOXYACETOPHENONE	2,2-DIETHOXYACETOPHENONE
462-95-3 DIETHOXYMETHANE	DIETHOXYMETHANE
RR-01360-5 2,5-DIETHOXY-4-MORPHOLINOBENZENEDIA-ZONIUM ZINC CHLORIDE	2,5-DIETHOXY-4-MORPHOLINOBENZENEDIAZONIUM ZINC CHLORIDELORIDE
3054-95-3 3,3-DIETHOXYPROPENE	3,3-DIETHOXYPROPENE
2235-46-3 N,N-DIETHYLACETOACETAMIDE	N,N-DIETHYLACETOACETAMIDE
1619-57-4 DIETHYL ACETOACETATE	DIETHYL ACETOACETATE
141-28-6 DIETHYL ADIPATE	DIETHYL ADIPATE
96-10-6 DIETHYLALUMINUM CHLORIDE	DIETHYLALUMINUM CHLORIDE
871-27-2 DIETHYLALUMINUM HYDRIDE	DIETHYLALUMINUM HYDRIDE
109-89-7 DIETHYLAMINE	DIETHYLAMINE
100-37-8 2-DIETHYLAMINOETHANOL	2-DIETHYLAMINOETHANOL
100-37-8 DIETHYLAMINOETHANOL [ALB, ALB, ALB, BC1, BC1]	2-DIETHYLAMINOETHANOL
100-37-8 2-(DIETHYLAMINO)ETHANOL [CAL, CEX, CEX]	2-DIETHYLAMINOETHANOL
100-37-8 DIETHYLAMINOETHANOL [MEX, MEX, NJL]	2-DIETHYLAMINOETHANOL
100-37-8 2-(DIETHYLAMINO)ETHANOL [ONT, ONT]	2-DIETHYLAMINOETHANOL
100-37-8 DIETHYLAMINOETHANOL [QBC, QBC, WHM]	2-DIETHYLAMINOETHANOL

Chemical Name	Reference Name
58145-14-5 DIETHYLAMINO ETHANOLAMINE	DIETHYLAMINO ETHANOLAMINE
2426-54-2 2-(DIETHYLAMINO)ETHYL ACRYLATE	2-(DIETHYLAMINO)ETHYL ACRYLATE
105-16-8 2-(N,N-DIETHYLAMINO)ETHYL METHACRY-LATE	2-(N,N-DIETHYLAMINO)ETHYL METHACRYLATE
64399-38-8 2-(DIETHYLAMINO)ETHYL METHACRYLATE, POLYMER WITH DODECYL METHACRYLATE, STYRENE, HEXADEXYL METHACRYLATE AND TETRADECYL METHACRYLATE	2-(DIETHYLAMINO)ETHYL METHACRYLATE, POLYMER WITH DODECYL METFURANCARBOXYLIC ACID AND TRIMETHYLOLPROPANE, LINOLEATE
104-78-9 3-(DIETHYLAMINO)PROPYLAMINE	3-(DIETHYLAMINO)PROPYLAMINE
104-78-9 DIETHYLAMINOPROPYLAMINE [HM1, HM1]	3-(DIETHYLAMINO)PROPYLAMINE
91-66-7 N,N-DIETHYLANILINE	N,N-DIETHYLANILINE
91-66-7 DIETHYL ANILINE [NJL]	N,N-DIETHYLANILINE
121-69-7 N,N-DIETHYL ANILINE (N,N-DIMETHYLANI-LINE) [CAA]	DIMETHYLANILINE
579-66-8 2,6-DIETHYLANILINE	2,6-DIETHYLANILINE
692-42-2 DIETHYLARSINE [MSL]	ARSINE, DIETHYL-
105-05-5 P-DIETHYL BENZENE	P-DIETHYL BENZENE
135-01-3 O-DIETHYL BENZENE	O-DIETHYL BENZENE
141-93-5 M-DIETHYL BENZENE	M-DIETHYL BENZENE
25340-17-4 DIETHYLBENZENE	DIETHYLBENZENE
32280-46-9 N,N-DIETHYL-1,3-BUTANEDIAMINE	N,N-DIETHYL-1,3-BUTANEDIAMINE
RR-00832-2 DI-2-ETHYLBUTYL PHTHALATE	DI-2-ETHYLBUTYL PHTHALATE
20624-25-3 DIETHYLCARBAMODITHIOIC ACID SODIUM SALT TRIHYDRATE	DIETHYLCARBAMODITHIOIC ACID SODIUM SALT TRI-HYDRATE
1642-54-2 DIETHYLCARBAMAZINE CITRATE	DIETHYLCARBAMAZINE CITRATE
88-10-8 DIETHYLCARBAMOYL CHLORIDE	DIETHYLCARBAMOYL CHLORIDE
88-10-8 DIETHYL CARBAMOYL CHLORIDE [FLA]	DIETHYLCARBAMOYL CHLORIDE
88-10-8 DIETHYL CARBAMYL CHLORIDE [MSL, NFP, NFP]	DIETHYLCARBAMOYL CHLORIDE
584-02-1 DIETHYLCARBINOL [NJL, NJL]	3-PENTANOL
584-02-1 DIETHYL CARBINOL [WHM]	3-PENTANOL
105-58-8 DIETHYL CARBONATE	DIETHYL CARBONATE
814-49-3 DIETHYL CHLOROPHOSPHATE	DIETHYL CHLOROPHOSPHATE
814-49-3 DIETHYLCHLOROPHOSPHATE [MSL]	DIETHYL CHLOROPHOSPHATE
95-06-7 DIETHYL-2-CHLORO-2-PROPANYL ESTER CARBAMODITHIOIC ACID [FLA]	SULFALLATE
2524-04-1 DIETHYL CHLOROTHIOPHOSPHATE [OTS]	DIETHYLTHIOPHOSPHORYL CHLORIDE
1331-43-7 DIETHYLCYCLOHEXANE	DIETHYLCYCLOHEXANE
1719-53-5 DIETHYLDICHLOROSILANE	DIETHYLDICHLOROSILANE
1719-53-5 DIETHYL DICHLOROSILANE [NJL, NJL]	DIETHYLDICHLOROSILANE
134190-37-7 DIETHYLDIISOCYANATOBENZENE	DIETHYLDIISOCYANATOBENZENE
85-98-3 1,3-DIETHYL-1,3-DIPHENYL UREA [NFP, NFP]	ETHYL CENTRALITE
124-17-4 DIETHYLENE GLYCOL MONOBUTYL ETHER ACETATE [HON]	2-(2-BUTOXYETHOXY)ETHYL ACETATE
112-15-2 DIETHYLENE GLYCOL MONOETHYL ETHER ACETATE [HON]	CARBINOL ACETATE
RR-01162-1 DIETHYLENEDIAMINE	DIETHYLENEDIAMINE
RR-01162-1 N,N-DIETHYLENEDIAMINE [HM1, HM1]	DIETHYLENEDIAMINE
111-46-6 DIETHYLENE GLYCOL	DIETHYLENE GLYCOL
142-22-3 DIETHYLENE GLYCOL BIS (ALLYLCARBON-ATE)	DIETHYLENE GLYCOL BIS (ALLYLCARBONATE)

Chemical Name	Reference Name
4246-51-9 DIETHYLENE GLYCOL BIS(3-AMINOPROPYL) ETHER	DIETHYLENE GLYCOL BIS(3-AMINOPROPYL) ETHER
RR-00840-2 DIETHYLENE GLYCOL BIS (2-BUTOXYETHYL CARBONATE)	DIETHYLENE GLYCOL BIS (2-BUTOXYETHYL CARBONATE)
RR-00841-3 DIETHYLENE GLYCOL BIS (BUTYL CARBONATE)	DIETHYLENE GLYCOL BIS (BUTYL CARBONATE)
RR-00842-4 DIETHYLENE GLYCOL BIS (PHENYLCARBONATE)	DIETHYLENE GLYCOL BIS (PHENYLCARBONATE)
112-34-5 DIETHYLENE GLYCOL BUTYL ETHER [CTR, TSC]	DIETHYLENE GLYCOL MONOBUTYL ETHER
124-17-4 DIETHYLENE GLYCOL BUTYL ETHER ACETATE [CTR]	2-(2-BUTOXYETHOXY)ETHYL ACETATE
112-34-5 DIETHYLENE GLYCOL BUTYL ETHER [TSE]	DIETHYLENE GLYCOL MONOBUTYL ETHER
124-17-4 DIETHYLENE GLYCOL BUTYL ETHER ACETATE [TSE]	2-(2-BUTOXYETHOXY)ETHYL ACETATE
628-68-2 DIETHYLENE GLYCOL DIACETATE	DIETHYLENE GLYCOL DIACETATE
120-55-8 DIETHYLENE GLYCOL DIBENZOATE	DIETHYLENE GLYCOL DIBENZOATE
112-73-2 DIETHYLENE GLYCOL DIBUTYL ETHER	DIETHYLENE GLYCOL DIBUTYL ETHER
112-36-7 DIETHYLENE GLYCOL DIETHYL ETHER	DIETHYLENE GLYCOL DIETHYL ETHER
RR-00843-5 DIETHYLENE GLYCOL DIETHYL LEVULINATE	DIETHYLENE GLYCOL DIETHYL LEVULINATE
2358-84-1 DIETHYLENE GLYCOL DIMETHACRYLATE	DIETHYLENE GLYCOL DIMETHACRYLATE
111-96-6 DIETHYLENE GLYCOL DIMETHYL ETHER	DIETHYLENE GLYCOL DIMETHYL ETHER
111-96-6 DIETHYLENE GLYCOL DIMETHYL ETHER (DGDDME) [CAL]	DIETHYLENE GLYCOL DIMETHYL ETHER
693-21-0 DIETHYLENEGLYCOL DINITRATE	DIETHYLENEGLYCOL DINITRATE
RR-00844-6 DIETHYLENE GLYCOL DIPROPIONATE	DIETHYLENE GLYCOL DIPROPIONATE
111-90-0 DIETHYLENE GLYCOL ETHYL ETHER [NFP, NFP]	DIETHYLENE GLYCOL MONOETHYL ETHER
RR-00845-7 DIETHYLENE GLYCOL ETHYL ETHER PHTHALATE	DIETHYLENE GLYCOL ETHYL ETHER PHTHALATE
111-77-3 DIETHYLENE GLYCOL METHYL ETHER [FLA, MSL, NFP, NFP]	DIETHYLENE GLYCOL MONOMETHYL ETHER
629-38-9 DIETHYLENE GLYCOL METHYL ETHER ACETATE	DIETHYLENE GLYCOL METHYL ETHER ACETATE
112-34-5 DIETHYLENE GLYCOL MONOBUTYL ETHER	DIETHYLENE GLYCOL MONOBUTYL ETHER
112-34-5 DIETHYLENE GLYCOL MONOBUTYL ETHER [NFP, NFP]	DIETHYLENE GLYCOL MONOBUTYL ETHER
124-17-4 DIETHYLENE GLYCOL MONOBUTYL ETHER ACETATE [NFP, NFP]	2-(2-BUTOXYETHOXY)ETHYL ACETATE
111-90-0 DIETHYLENE GLYCOL MONOETHYL ETHER	DIETHYLENE GLYCOL MONOETHYL ETHER
RR-01164-3 DIETHYLENE GLYCOL MONOETHYL ETHER ACETATE	DIETHYLENE GLYCOL MONOETHYL ETHER ACETATE
112-59-4 DIETHYLENE GLYCOL MONOHEXYL ETHER	DIETHYLENE GLYCOL MONOHEXYL ETHER
18912-80-6 DIETHYLENE GLYCOL MONOISOBUTYL ETHER [NFP, NFP]	ETHANOL, 2-[2-(2-METHYLPROPOXY)ETHOXY]-
111-77-3 DIETHYLENE GLYCOL MONOMETHYL ETHER	DIETHYLENE GLYCOL MONOMETHYL ETHER
5405-88-9 DIETHYLENE GLYCOL MONOMETHYL ETHER FORMAL	DIETHYLENE GLYCOL MONOMETHYL ETHER FORMAL
104-68-7 DIETHYLENE GLYCOL MONOPHENYL ETHER	DIETHYLENE GLYCOL MONOPHENYL ETHER
RR-00846-8 DIETHYLENE GLYCOL PHTHALATE	DIETHYLENE GLYCOL PHTHALATE
111-40-0 DIETHYLENE TRIAMINE	DIETHYLENE TRIAMINE
111-40-0 DIETHYLENETRIAMINE [CAL, CEX, CEX, ONT, ONT, OTS, PST, QBC, QBC, TSE, WHM]	DIETHYLENE TRIAMINE
111-40-0 DIETHYLENETRIAMINE (DTA) [CTR]	DIETHYLENE TRIAMINE

Chemical Name	Reference Name
19529-38-5 DIETHYLENETRIAMINEPENTAACETIC ACID, DISODIUM IRON(III) SALT	DIETHYLENETRIAMINEPENTAACETIC ACID, DISODIUM IRON(III) SALT
35365-94-7 N,N-DIETHYLETHANAMINE PHOSPHATE	N,N-DIETHYLETHANAMINE PHOSPHATE
100-37-8 N,N-DIETHYLETHANOLAMINE [FLA, MSL, NFP, NFP]	2-DIETHYLAMINOETHANOL
100-37-8 DIETHYLETHANOLAMINE [PST]	2-DIETHYLAMINOETHANOL
60-29-7 DIETHYL ETHER [GBR, GBR, HM1, HM1, HM1, NEU, NJL, NJL, OTS, TSC, TSC, TSC]	ETHYL ETHER
60-29-7 DIETHYL ETHER (ETHYL ETHER) [ALB, ALB]	ETHYL ETHER
100-36-7 DIETHYLETHYLENE DIAMINE	DIETHYLETHYLENE DIAMINE
100-36-7 N,N-DIETHYLETHYLENE-DIAMINE [FLA]	DIETHYLETHYLENE DIAMINE
100-36-7 N,N-DIETHYLETHYLENEDIAMINE [MSL]	DIETHYLETHYLENE DIAMINE
100-36-7 N,N-DIETHYLETHYLENE-DIAMINE [NFP, NFP]	DIETHYLETHYLENE DIAMINE
623-91-6 DIETHYL FUMARATE [NFP, NFP]	FUMARIC ACID, DIETHYL ESTER
16484-86-9 DIETHYL GLYCOL	DIETHYL GLYCOL
26645-10-3 DIETHYLGOLD BROMIDE	DIETHYLGOLD BROMIDE
103-23-1 DI(2-ETHYLHEXYL)ADIPATE [CAL, CSR, CSR, IAR]	BIS (2-ETHYLHEXYL) ADIPATE
103-23-1 DI(2-ETHYLHEXYL) ADIPATE [OTS]	BIS (2-ETHYLHEXYL) ADIPATE
103-23-1 DI(2-ETHYLHEXYL)ADIPATE [SDW, SDW]	BIS (2-ETHYLHEXYL) ADIPATE
106-20-7 DI(2-ETHYLHEXYL)AMINE	DI(2-ETHYLHEXYL)AMINE
103-24-2 DI-2-ETHYL HEXYL AZELATE	DI-2-ETHYL HEXYL AZELATE
10143-60-9 DI-2-ETHYLHEXYL ETHER	DI-2-ETHYLHEXYL ETHER
141-02-6 DI(2-ETHYLHEXYL)FUMARATE	DI(2-ETHYLHEXYL)FUMARATE
16111-62-9 DI(2-ETHYLHEXYL)PEROXY-DICARBONATE	DI(2-ETHYLHEXYL)PEROXY-DICARBONATE
298-07-7 DI(2-ETHYLHEXYL)PHOSPHORIC ACID	DI(2-ETHYLHEXYL)PHOSPHORIC ACID
117-81-7 DI(2-ETHYLHEXYL)PHTHALATE	DI(2-ETHYLHEXYL)PHTHALATE
117-81-7 DI-2-ETHYLHEXYL PHTHALATE (DEHP) [NHS, NHS, NHS]	DI(2-ETHYLHEXYL)PHTHALATE
117-81-7 DIETHYLHEXYL PHTHALATE [RCU, RCU]	DI(2-ETHYLHEXYL)PHTHALATE
117-81-7 DI(2-ETHYLHEXYL)PHTHALATE (DEHP) [WHM]	DI(2-ETHYLHEXYL)PHTHALATE
1615-80-1 1,2-DIETHYLHYDRAZINE	1,2-DIETHYLHYDRAZINE
1615-80-1 N,N'-DIETHYLHYDRAZINE [EPA]	1,2-DIETHYLHYDRAZINE
1615-80-1 DIETHYL HYDRAZINE [MIN]	1,2-DIETHYLHYDRAZINE
1615-80-1 1,2-DIETHYL HYDRAZINE [NJL, NJL]	1,2-DIETHYLHYDRAZINE
1615-80-1 N,N'-DIETHYLHYDRAZINE [RCU, RCU]	1,2-DIETHYLHYDRAZINE
3710-84-7 DIETHYL HYDROXYLAMINE	DIETHYL HYDROXYLAMINE
3710-84-7 N,N-DIETHYLHYDROXYLAMINE [PST]	DIETHYL HYDROXYLAMINE
96-22-0 DIETHYL KETONE	DIETHYL KETONE
3352-87-2 N,N-DIETHYLLAURAMIDE	N,N-DIETHYLLAURAMIDE
557-18-6 DIETHYLMAGNESIUM	DIETHYLMAGNESIUM
557-18-6 DIETHYL MAGNESIUM [NJL]	DIETHYLMAGNESIUM
141-05-9 DIETHYL MALEATE	DIETHYL MALEATE
105-53-3 DIETHYL MALONATE	DIETHYL MALONATE
627-44-1 DIETHYLMERCURY	DIETHYLMERCURY
3288-58-2 O,O-DIETHYL S-METHYL DITHIOPHOSPHATE	O,O-DIETHYL S-METHYL DITHIOPHOSPHATE
115-90-2 O,O-DIETHYL O-[4-METHYLSULFINYL)PHENYL] PHOSPHOROTHIOATE (FENSULFOTHION) [CAL]	FENSULFOTHION
115-91-3 O,O-DIETHYL O-[4-METHYLSULFINYL)PHENYL] PHOSPHOROTHIOATE (FENSULFOTHION)	O,O-DIETHYL O-[4-METHYLSULFINYL)PHENYL]PHOS-PHOROTHIOATE (FENSULFOTHION)

Chemical Name	Reference Name
311-45-5 DIETHYL-P-NITROPHENYL PHOSPHATE	DIETHYL-P-NITROPHENYL PHOSPHATE
60-29-7 DIETHYL OXIDE (ETHER) [QBC, QBC]	ETHYL ETHER
1067-20-5 3,3-DIETHYLPENTANE	3,3-DIETHYLPENTANE
628-37-5 DIETHYL PEROXIDE	DIETHYL PEROXIDE
14666-78-5 DIETHYL PEROXYDICARBONATE	DIETHYL PEROXYDICARBONATE
72-56-0 DI(P-ETHYLPHENYL)DICHLOROETHANE	DI(P-ETHYLPHENYL)DICHLOROETHANE
93-05-0 DIETHYL-P-PHENYLENEDIAMINE [MSL]	N,N-DIETHYL-P-PHOSPHORIC ACID
762-04-9 DIETHYL PHOSPHITE	DIETHYL PHOSPHITE
93-05-0 N,N-DIETHYL-P-PHOSPHORIC ACID	N,N-DIETHYL-P-PHOSPHORIC ACID
84-66-2 DIETHYL PHTHALATE	DIETHYL PHTHALATE
115-76-4 2,2-DIETHYL-1,3-PROPANEDIOL	2,2-DIETHYL-1,3-PROPANEDIOL
297-97-2 O,O-DIETHYL O-2-PYRAZINYL PHOSPHOROTH-IOATE [PQL]	O,O-DIETHYL O-PYRAZINYL PHOSPHOROTHIATE
297-97-2 O,O-DIETHYL O-PYRAZINYL PHOSPHOROTH-IATE	O,O-DIETHYL O-PYRAZINYL PHOSPHOROTHIATE
1609-47-8 DIETHYL PYROCARBONATE	DIETHYL PYROCARBONATE
110-40-7 DIETHYL SEBACATE	DIETHYL SEBACATE
627-53-2 DIETHYL SELENIDE [FLA, MSL, NFP]	ETHANE, 1,1'-SELENOBIS-
RR-00873-1 N,N-DIETHYLSTEARAMIDE	N,N-DIETHYLSTEARAMIDE
56-53-1 DIETHYLSTILBESTROL	DIETHYLSTILBESTROL
56-53-1 DIETHYLSTILBESTROL (DES) [C65, C65, C65]	DIETHYLSTILBESTROL
123-25-1 DIETHYL SUCCINATE	DIETHYL SUCCINATE
64-67-5 DIETHYL SULFATE	DIETHYL SULFATE
352-93-2 DIETHYL SULFIDE	DIETHYL SULFIDE
64-67-5 DIETHYL SULPHATE [IAR]	DIETHYL SULFATE
87-91-2 DIETHYL TARTRATE	DIETHYL TARTRATE
636-09-9 DIETHYL TEREPHTHALATE	DIETHYL TEREPHTHALATE
2524-04-1 DIETHYLTHIOPHOSPHORYL CHLORIDE	DIETHYLTHIOPHOSPHORYL CHLORIDE
105-55-5 N,N'-DIETHYLTHIOUREA	N,N'-DIETHYLTHIOUREA
134-62-3 N,N-DIETHYL-M-TOLUAMIDE	N,N-DIETHYL-M-TOLUAMIDE
557-20-0 DIETHYLZINC	DIETHYLZINC
557-20-0 DIETHYL ZINC [FLA, MSL]	DIETHYLZINC
35367-38-5 DIFLUBENZURON	DIFLUBENZURON
75-68-3 DIFLUORO-1-CHLOROETHANE [FLA, MSL, NFP, NFP]	1-CHLORO-1,1-DIFLUOROETHANE
75-61-6 DIFLUORODIBROMOMETHANE	DIFLUORODIBROMOMETHANE
75-37-6 1,1-DIFLUOROETHANE	1,1-DIFLUOROETHANE
75-37-6 DIFLUOROETHANE [HM1, HM1, MSL]	1,1-DIFLUOROETHANE
75-37-6 DIFLUOROETHANE [ETHANE, 1,1-DIFLUORO-] [EP3]	1,1-DIFLUOROETHANE
75-38-7 1,1-DIFLUOROETHYLENE [HM1, HM1, MAK]	VINYLIDENE FLUORIDE
13779-41-4 DIFLUOROPHOSPHORIC ACID	DIFLUOROPHOSPHORIC ACID
76-12-0 1,2-DIFLUORO-1,1,2,2-TETRACHLOROETHANE [CSR]	1,1,2,2-TETRACHLORO-1,2-DIFLUOROETHANE
71-63-6 DIGITOXIN	DIGITOXIN
1323-83-7 DIGYLCERYL STEARATE	DIGYLCERYL STEARATE
5026-74-4 4-(DIGLYCIDYLAMINO)PHENYL GLYCIDYL ETHER [EPA, OTS]	OXIRANEMETHANAMINE, N-[4-(OXIRANYLMETHOXY) PHENYL]-N-
5493-45-8 DIGLYCIDYL ESTER OF HEXAHYDROPH-THALIC ACID [EPA]	1,2-CYCLOHEXANEDICARBOXYLIC ACID, BIS(OXI-RANYLMETHYL) ESTER

Chemical Name	Reference Name
5493-45-8 DIGLYCIDYL ESTER OF HEXAHYDRO-PH-THALIC ACID [OTS]	1,2-CYCLOHEXANEDICARBOXYLIC ACID, BIS(OXIRANYLMETHYL) ESTER
7195-45-1 DIGLYCIDYL ESTER OF PHTHALIC ACID [EPA, OTS]	1,2-BENZENEDICARBOXYLIC ACID, BIS(OXIRANYL-METHYL) ESTER
2238-07-5 DIGLYCIDYL ETHER [302, CAL, CEX, EPA, EPA, EPA, FLA, MSL, NIO, NIO, NIO, NIO, NJL, OTS]	DIGLYCIDYL ETHER (DGE)
2238-07-5 DIGLYCIDYL ETHER (DGE)	DIGLYCIDYL ETHER (DGE)
1675-54-3 DIGLYCIDYL ETHER OF BISPHENOL A [PST]	BISPHENOL A DIGLYCIDYL ETHERPROPANAMINIUM IODIDE
RR-00935-8 DIGLYCIDYL ETHER OF DISUBSTITUTED CAR-BOPOLYCYCLE	DIGLYCIDYL ETHER OF DISUBSTITUTED CARBOPOLY-CYCLE
2238-07-5 DIGLYCIDYL OXIDE (ETHER) [QBC]	DIGLYCIDYL ETHER (DGE)
13561-08-5 2,6-DIGLYCIDYLPHENYL GLYCIDYL ETHER [EPA]	OXIRANE, 2,2'-(OXIRANYLMETHOXY)-1,3-PHENYLENE] BIS(METHYLENE)
13561-08-5 DIGLYCIDYLPHENYL GLYCIDYL ETHER, 2,6-[OTS]	OXIRANE, 2,2'-(OXIRANYLMETHOXY)-1,3-PHENYLENE] BIS(METHYLENE)
101-90-6 DIGLYCIDYL RESORCINOL ETHER	DIGLYCIDYL RESORCINOL ETHER
2238-07-5 DIGLYCIDYL RESORCINOL ETHER [MIN]	DIGLYCIDYL ETHER (DGE)
101-90-6 DIGLYCIDYL RESORCINOL ETHER (DGRE) [C65, C65, CSR, CSR]	DIGLYCIDYL RESORCINOL ETHER
RR-00847-9 DIGLYCOL CHLOROFORMATE	DIGLYCOL CHLOROFORMATE
6288-89-7 DIGLYCOL CHLOROHYDRIN	DIGLYCOL CHLOROHYDRIN
RR-00848-0 DIGLYCOL DIACETATE	DIGLYCOL DIACETATE
RR-00849-1 DIGLYCOL DILEVULINATE	DIGLYCOL DILEVULINATE
110-99-6 DIGLYCOLIC ACID	DIGLYCOLIC ACID
141-20-8 DIGLYCOL LAURATE	DIGLYCOL LAURATE
106-11-6 DIGLYCOL MONOSTEARATE	DIGLYCOL MONOSTEARATE
20830-75-5 DIGOXIN	DIGOXIN
111381-89-6 DI(HEPTYL-, NONYL-) PHTHALATE (BRANCHED AND LINEAR ISOMERS)	DI(HEPTYL-, NONYL-) PHTHALATE (BRANCHED AND LINEAR ISOMERS)WITH 2-ETHYL-2-(HYDROXYMETHYL)-1,3-PROPANEDIOL (3:1) DI-2-PROPENOATE, METHYL ETHER
RR-00277-7 DI[HEPTYL, NONYL, UNDECYL] PHTHALATE (D711P)	DI[HEPTYL, NONYL, UNDECYL] PHTHALATE (D711P) EAR [DIHEXYL PHTHALATE (MIXED ISOMERS)]
3648-21-3 DIHEPTYL PHTHALATE	DIHEPTYL PHTHALATE
68515-44-6 DIHEPTYL PHTHALATE (BRANCHED AND LINEAR ISOMERS) [TSC]	PHTHALIC ACID, DIALKYL (C7) ESTERALKYL ESTERS [DI(HEPTYL-, NONYL-, UNDECYL) PHTHALATE (MIXED ISOMERS)]
111381-90-9 DI(HEPTYL-, UNDECYL-) PHTHALATE (BRANCHED AND LINEAR ISOMERS)	DI(HEPTYL-, UNDECYL-) PHTHALATE (BRANCHED AND LINEAR ISOMERS
143-16-8 DIHEXYLAMINE	DIHEXYLAMINE
109-31-9 DI-N-HEXYL AZELATE	DI-N-HEXYL AZELATE
84-75-3 DIHEXYL PHTHALATE [TSE]	1,2-BENZENEDICARBOXYLIC ACID, DIHEXYL ESTER
68515-50-4 DIHEXYL PHTHALATE (MIXED ISOMERS)	DIHEXYL PHTHALATE (MIXED ISOMERS)S, C(10-RICH [(DIISODECYL PHTHALATE (MIXED ISOMERS)]
119-84-6 3,4-DIHYDROCOUMARIN	3,4-DIHYDROCOUMARIN
1563-66-2 2,3-DIHYDRO-2,2-DIMETHYL-7-BENZO-FU-RANYL METHYLCARBAMATE (CARBOFURAN) [CAL]	CARBOFURAN
7778-77-0 DIHYDROGEN POTASSIUM PHOSPHATE	DIHYDROGEN POTASSIUM PHOSPHATE
422-05-9 1,1-DIHYDROPERFLUOROPROPANOL	1,1-DIHYDROPERFLUOROPROPANOL
2614-76-8 2,2-DIHYDROPEROXY PROPANE	2,2-DIHYDROPEROXY PROPANE
110-87-2 DIHYDROPYRAN	DIHYDROPYRAN
25512-65-6 DIHYDRO-2H-PYRAN	DIHYDRO-2H-PYRAN

Chemical Name	Reference Name
123-33-1 1,2-DIHYDRO-3,6-PYRIDAZINEDIONE	1,2-DIHYDRO-3,6-PYRIDAZINEDIONE
94-58-6 DIHYDROSAFROLE	DIHYDROSAFROLE
128-46-1 DIHYDROSTREPTOMYCIN	DIHYDROSTREPTOMYCIN
147-47-7 1,2-DIHYDRO-2,2,4-TRIMETHYLQUINOLIN (MONOMER)	1,2-DIHYDRO-2,2,4-TRIMETHYLQUINOLIN (MONOMER)
26780-96-1 1,2-DIHYDRO-2,2,4-TRIMETHYLQUINOLINE POLYMER	1,2-DIHYDRO-2,2,4-TRIMETHYLQUINOLINE POLYMER
26780-96-1 1,2-DIHYDRO-2,2,4-TRIMETHYLQUINOLINE (POLYMER) [CSR]	1,2-DIHYDRO-2,2,4-TRIMETHYLQUINOLINE POLYMER
81-64-1 1,4-DIHYDROXY-9,10-ANTHRACENEDIONE	1,4-DIHYDROXY-9,10-ANTHRACENEDIONE
72-48-0 1,2-DIHYDROXYANTHRAQUINONE	1,2-DIHYDROXYANTHRAQUINONE
117-10-2 1,8-DIHYDROXYANTHRAQUINONE	1,8-DIHYDROXYANTHRAQUINONE
120-80-9 O-DIHYDROXYBENZENE [NFP, NFP]	CATECHOL
120-80-9 1,2-DIHYDROXYBENZENE [NJL, ONT]	CATECHOL
123-31-9 P-DIHYDROXYBENZENE [NFP, NFP]	HYDROQUINONE
123-31-9 1,4-DIHYDROXYBENZENE [ONT, ONT, ONT]	HYDROQUINONE
123-31-9 DIHYDROXYBENZENE (HYDROQUINONE) [ALB, ALB]	HYDROQUINONE
27138-57-4 DIHYDROXYBENZOIC ACID (RESORCYLIC ACID)	DIHYDROXYBENZOIC ACID (RESORCYLIC ACID)
131-56-6 2,4-DIHYDROXYBENZOPHENONE	2,4-DIHYDROXYBENZOPHENONE
1758-61-8 DI(1-HYDROXYCYCLO-HEXYL)PEROXIDE	DI(1-HYDROXYCYCLO-HEXYL)PEROXIDE
137-08-6 N-(2,4-DIHYDROXY-3,3-DIMETHYL-1-OXOBUTYL)-.BETA.-ALANINE, CALCIUM SALT (2:1), (R)-	N-(2,4-DIHYDROXY-3,3-DIMETHYL-1-OXOBUTYL)-.BETA.-ALANINE, CA
150-25-4 N,N-(2-DIHYDROXYETHYL)GLYCINE	N,N-(2-DIHYDROXYETHYL)GLYCINE
794-93-4 DIHYDROXYMETHYLFURATRIZINE [CAL]	PANFURAN S
67859-56-7 2,3-DIHYDROXYPROPYL-3'-(HEXYLTHIO)PRO-PIONATE	2,3-DIHYDROXYPROPYL-3'-(HEXYLTHIO)PROPIONATE
517-92-0 1,8-DIHYDROXY-2,4,5,7-TETRANITROAN-THRAQUINONE	1,8-DIHYDROXY-2,4,5,7-TETRANITROANTHRAQUINONE
RR-01403-9 DI-(1-HYDROXYTETRAZOLE)	DI-(1-HYDROXYTETRAZOLE)
RR-01413-1 DIIDOACETYLENE	DIIDOACETYLENE
20018-09-1 1-(DIIODOMETHYL)SULFONYL-4-METHYL BENZENE	1-(DIIODOMETHYL)SULFONYL-4-METHYL BENZENE
3437-84-1 DI-ISOBUTRYL PEROXIDE	DI-ISOBUTRYL PEROXIDE
141-04-8 DIISOBUTYL ADIPATE	DIISOBUTYL ADIPATE
1191-15-7 DIISOBUTYLALUMINUM HYDRIDE	DIISOBUTYLALUMINUM HYDRIDE
110-96-3 DI-ISOBUTYLAMINE	DI-ISOBUTYLAMINE
110-96-3 DIISOBUTYLAMINE [FLA, MSL, NFP, NFP]	DI-ISOBUTYLAMINE
108-82-7 DIISOBUTYL CARBINOL [NFP, NFP]	DIISOBUTYL CARBITOL
108-82-7 DIISOBUTYL CARBITOL	DIISOBUTYL CARBITOL
25167-70-8 DIISOBUTYLENE	DIISOBUTYLENE
25167-70-8 DI-ISOBUTYLENE [NJL, NJL]	DIISOBUTYLENE
108-83-8 DIISOBUTYL KETONE	DIISOBUTYL KETONE
108-83-8 DIISOBUTYLKETONE [QBC]	DIISOBUTYL KETONE
27213-90-7 DIISOBUTYLNAPHTHALENESULFONIC ACID, SODIUM SALT	DIISOBUTYLNAPHTHALENESULFONIC ACID, SODIUM SALT
121-54-0 P-DIISOBUTYLPHENOXYETHOXYETHYL DIMETHYL BENZYL AMMONIUM CHLO-RIDE [PST]	BENZETHONIUM CHLORIDE

Chemical Name	Reference Name
84-69-5 DIISOBUTYL PHTHALATE [GBR, NFP, NFP]	1,2-BENZENEDICARBOXYLIC ACID, BIS(2-METHYL-PROPYL) ESTER
RR-00547-0 DIISOCYANATES	DIISOCYANATES
4128-73-8 4,4'-DIISOCYANATODIPHENYL ETHER	4,4'-DIISOCYANATODIPHENYL ETHER
75790-87-3 2,4'-DIISOCYANATODIPHENYL SULFIDE [313]	BENZENE, 1-ISOCYANATO-2-[(4-ISOCYANATOPHENYL) THIO]-
26471-62-5 2,4-DIISOCYANATOMETHYLBENZENE [MSL]	TOLUENE DIISOCYANATEMOYL) OXIME
27178-16-1 DIISODECYL ADIPATE	DIISODECYL ADIPATE
25550-98-5 DIISODECYLPHENYL PHOSPHITE [PST]	PHOSPHOROUS ACID, (DIISODECYL)PHENYL ETHER
25550-98-5 DIISODECYL PHENYL PHOSPHITE [TSE]	PHOSPHOROUS ACID, (DIISODECYL)PHENYL ETHER
26761-40-0 DIISODECYL PHTHALATE	DIISODECYL PHTHALATE
68515-49-1 DIISODECYL PHTHALATE (MIXED ISOMERS)	DIISODECYL PHTHALATE (MIXED ISOMERS)RS, C(13)-RICH [DITRIDCEYL PHTHALATE (MIXED ISOMERS)]
28553-12-0 DIISONONYL PHTHALATE	DIISONONYL PHTHALATE
28553-12-0 DIISONONYLPHTHALATE [PST]	DIISONONYL PHTHALATE
27215-10-7 DIISOOCTYL ACID PHOSPHATE	DIISOOCTYL ACID PHOSPHATE
3658-48-8 DIISOOCTYL PHOSPHITE	DIISOOCTYL PHOSPHITE
27554-26-3 DIISOOCTYL PHTHALATE	DIISOOCTYL PHTHALATE
110-97-4 DIISOPROPANOLAMINE	DIISOPROPANOLAMINE
108-18-9 DIISOPROPYLAMINE	DIISOPROPYLAMINE
99-62-7 1,3-DIISOPROPYLBENZENE	1,3-DIISOPROPYLBENZENE
25321-09-9 DIISOPROPYL BENZENE	DIISOPROPYL BENZENE
26762-93-6 DIISOPROPYLBENZENE HYDROPEROXIDE	DIISOPROPYLBENZENE HYDROPEROXIDE
25321-09-9 DIISOPROPYL BENZENES [HM1]	DIISOPROPYL BENZENE
69009-90-1 DIISOPROPYL BIPHENYL	DIISOPROPYL BIPHENYL
69009-90-1 DIISOPROPYLBIPHENYL [PST]	DIISOPROPYL BIPHENYL
693-13-0 DIISOPROPYLCARBODIIMIDE	DIISOPROPYLCARBODIIMIDE
96-80-0 N,N-DIISOPROPYL ETHANOLAMINE	N,N-DIISOPROPYL ETHANOLAMINE
96-80-0 N,N-DIISOPROPYLETHANOLAMINE [NFP, NFP]	N,N-DIISOPROPYL ETHANOLAMINE
96-80-0 DIISOPROPYLETHANOLAMINE [NJL]	N,N-DIISOPROPYL ETHANOLAMINE
108-20-3 DIISOPROPYL ETHER [GBR, GBR, HM1, HM1, NJL, NJL, QBC, QBC]	ISOPROPYL ETHER
55-91-4 DIISOPROPYL FLUOROPHOSPHATE [EPA]	ISOFLUORPHATE
55-91-4 DIISOPROPYLFLUOROPHOSPHATE (DFP) [RCU, RCU]	ISOFLUORPHATE
10099-70-4 DIISOPROPYL MALEATE	DIISOPROPYL MALEATE
105-64-6 DIISOPROPYL PEROXYDICARBONATE [MSL]	ISOPROPYL PEROXYDICARBONATE
105-64-6 DI-ISOPROPYL PEROXYDICARBONATE [NJL, NJL]	ISOPROPYL PEROXYDICARBONATE
105-64-6 DIISOPROPYL PEROXYDICARBONATE [FLA, NFP, NFP, OS3, DOT]	ISOPROPYL PEROXYDICARBONATE
27923-56-4 DIISOPROPYLPHENOLS	DIISOPROPYLPHENOLS
RR-01553-2 DIISOPROPYL SULFATE	DIISOPROPYL SULFATE
RR-01516-7 DI-ISOTRIDECYL PEROXYDICARBONATE	DI-ISOTRIDECYL PEROXYDICARBONATE
82065-80-3 DIISOTRIDECYL PEROXYDICARBONATE, TECHNICALLY PURE	DIISOTRIDECYL PEROXYDICARBONATE, TECHNICALLY PURE
82065-80-3 DIISOTRIDECYL PEROXYDICARBONATE [DOT]	DIISOTRIDECYL PEROXYDICARBONATE, TECHNICALLY PURE
674-82-8 DIKETENE	DIKETENE
105-74-8 DILAUROYL PEROXIDE [DOT, FLA, MAK, MSL, NJL, OS3]	LAUROYL PEROXIDE

Chemical Name	Reference Name
123-28-4 DILAURYL BETA-THIODIPROPIONATE	DILAURYL BETA-THIODIPROPIONATE
123-28-4 DILAURYL THIODIPROPIONATE [PST]	DILAURYL BETA-THIODIPROPIONATE
RR-01192-7 DI-LINEAR 79 PHTHALATE	DI-LINEAR 79 PHTHALATE
6144-28-1 DILINOLEIC ACID	DILINOLEIC ACID
115-26-4 DIMEFOX	DIMEFOX
RR-00163-8 DIMER ACIDS, POLYMER WITH POLYALKY-LENE GLYCOL, BISPHENOL A-DIGLYC-ERYLETHER AND ALKYLENEPOLYOLS POLYG-LYCIDYLETHERS	DIMER ACIDS, POLYMER WITH POLYALKYLENE GLY-COL, BISPHENOL A-
RR-00909-6 2,5-DIMERCAPTO-1,3,4-THIADIAZOLE, ALKYL POLYCARBOXYLATE	2,5-DIMERCAPTO-1,3,4-THIADIAZOLE, ALKYL POLY-CARBOXYLATE
541-47-9 BETA, BETA-DIMETHACRYLIC ACID	BETA, BETA-DIMETHACRYLIC ACID
309-00-2 1,4:5,8-DIMETHANONAPHTHALENE, 1,2,3,4,10, 10-HEXACHLORO-1,4,4A,5,8,8A-HEXAHYDRO-, (1.ALPHA.,4.ALPHA.,4A.BETA.,5.ALPHA., 8.AL-PHA.,8A.BETA.)- [PAL]	ALDRIN
72-20-8 2,7:3,6-DIMETHANONAPHTH[2,3-B]OXIRENE, 3,4,5,6,9,9-HEXACHLORO-1A,2,2A,3,6,6A,7,7A-OCTAHYDRO-,(1A.ALPHA.,2.BETA.,2A.BET A., 2A.BETA.,3.ALPHA.,6.ALPHA.,6A.BETA.,7.BETA., 7A.ALPHA.)- [PAL, PAL]	ENDRIN
55290-64-7 DIMETHIPIN [2,3,-DIHYDRO-5,6-DIMETHYL-1,4-DITHIIN-1,1,4,4-TETRAOXIDE]	DIMETHIPIN [2,3,-DIHYDRO-5,6-DIMETHYL-1,4-DITHIIN-1,1,4,4-TEIAZINE-2-THIONE, ION(1-), SODIUM]
79-64-1 DIMETHISTERONE	DIMETHISTERONE
RR-00102-5 DIMETHISTERONE AND OESTROGENS	DIMETHISTERONE AND OESTROGENS
60-51-5 DIMETHOATE	DIMETHOATE
828-00-2 DIMETHOXANE	DIMETHOXANE
102-56-7 2,5-DIMETHOXYANILINE	2,5-DIMETHOXYANILINE
54150-69-5 2,4-DIMETHOXYANILINE HYDROCHLORIDE	2,4-DIMETHOXYANILINE HYDROCHLORIDE
119-90-4 3,3'-DIMETHOXYBENZIDINE	3,3'-DIMETHOXYBENZIDINE
20325-40-0 3,3'-DIMETHOXYBENZIDINE DIHYDROCHLO-RIDE (O-DIANISIDINE DIHYDROCHLORIDE) [313]	3,3'-DIMETHOXYBENZIDINE DIHYDROCHLO-RIDEDIMETHYLETHYL)-1,3,4-OXADIAZOL-2(3H)-ONE]
111984-09-9 3,3'-DIMETHOXYBENZIDINE HYDROCHLORIDE (O-DIANISIDINE HYDROCHLORIDE)	3,3'-DIMETHOXYBENZIDINE HYDROCHLORIDE (O-DIANISIDINE HYDROCH
119-90-4 3,3'-DIMETHOXYBENZIDINE (ORTHO-TOLI-DINE) [IAR]	3,3'-DIMETHOXYBENZIDINE
119-90-4 3,3'-DIMETHOXYBENZIDINE (O-DIANISIDINE) [RCU, RCU]	3,3'-DIMETHOXYBENZIDINE
119-90-4 3,3'-DIMETHOXYBENZIDINE (ORTHO-DIANI-SIDINE) [C65]	3,3'-DIMETHOXYBENZIDINE
20325-40-0 3,3'-DIMETHOXYBENZIDINE DIHYDROCHLO-RIDE	3,3'-DIMETHOXYBENZIDINE DIHYDROCHLORIDE
20325-40-0 3,3'-DIMETHOXYBENZIDINE DIHYDROCHLO-RIDE (ORTHO-DIANISIDINE DIHYDROCHLO-RIDE) [C65]	3,3'-DIMETHOXYBENZIDINE DIHYDROCHLORIDE
91-93-0 3,3'-DIMETHOXYBENZIDINE-4,4'-DIISO-CYANATE	3,3'-DIMETHOXYBENZIDINE-4,4'-DIISOCYANATE
2100-42-7 2,5-DIMETHOXYCHLOROBENZENE	2,5-DIMETHOXYCHLOROBENZENE
110-71-4 1,2-DIMETHOXYETHANE [HM1, HM1]	ETHYLENE GLYCOL DIMETHYL ETHER
534-15-6 1,1-DIMETHOXYETHANE	1,1-DIMETHOXYETHANE
117-82-8 DIMETHOXYETHYL PHTHALATE [NFP, NFP]	BIS(2-METHOXYETHYL)PHTHALATE
109-87-5 DIMETHOXYMETHANE [GBR, GBR, MAK, ONT]	METHYLAL
109-87-5 DIMETHOXYMETHANE (METHYLAL) [QBC, ALB, ALB]	METHYLAL

Chemical Name	Reference Name
6923-22-4 3-(DIMETHOXYPHOSPHINYLOXY)-N-METHYL-CIS-CROTONAMIDE (MONOCROTOPHOS) [CAL]	MONOCROTOPHOS
RR-01163-2 1,1-DIMETHOXY-2-PROPENE	1,1-DIMETHOXY-2-PROPENE
523-80-8 4,7-DIMETHOXY-5-(2-PROPENYL)-1,3-BENZODI-OXOLE	4,7-DIMETHOXY-5-(2-PROPENYL)-1,3-BENZODIOXOLE
71808-64-5 DIMETHOXYSILANE, (3-GLYCIDOXY-PROPYL) (3-CHLOROPROPYL)- [OTS]	SILANE, [(3-CHLOROPROPYL)DIMETHOXY[3-(OXI-RANYLMETHOXY)PROPYL
143-24-8 DIMETHOXY TETRAGLYCOL [NFP, NFP]	2,5,8,11,14-PENTAOXAPENTADECANE
99-98-9 DIMETHYL-P-PHENYLENEDIAMINE	DIMETHYL-P-PHENYLENEDIAMINE
9006-65-9 DIMETHICONE [TSC]	DIMETHYL SILICONE
120-61-6 DIMETHYL TEREPHTHALATE [HON]	DIMETHYL P-PHTHALATE
127-19-5 DIMETHYL ACETAMIDE	DIMETHYL ACETAMIDE
127-19-5 N,N-DIMETHYL ACETAMIDE [BEI]	DIMETHYL ACETAMIDE
127-19-5 N,N-DIMETHYLACETAMIDE [CAL]	DIMETHYL ACETAMIDE
127-19-5 DIMETHYLACETAMIDE [CEX, CEX]	DIMETHYL ACETAMIDE
127-19-5 N,N-DIMETHYL ACETAMIDE [FLA]	DIMETHYL ACETAMIDE
127-19-5 N,N-DIMETHYLACETAMIDE [GBR, GBR, GBR]	DIMETHYL ACETAMIDE
127-19-5 N,N-DIMETHYL ACETAMIDE [MAK, MAK, MAK, MAK]	DIMETHYL ACETAMIDE
127-19-5 DIMETHYLACETAMIDE [NFP, NFP]	DIMETHYL ACETAMIDE
127-19-5 N,N-DIMETHYLACETAMIDE [ONT, ONT]	DIMETHYL ACETAMIDE
127-19-5 DIMETHYLACETAMIDE [OTS, QBC, QBC]	DIMETHYL ACETAMIDE
127-19-5 N,N-DIMETHYL ACETAMIDE [TLV, TLV]	DIMETHYL ACETAMIDE
127-19-5 N,N-DIMETHYLACETAMIDE [WHM]	DIMETHYL ACETAMIDE
124-40-3 DIMETHYLAMINE	DIMETHYLAMINE
2300-66-5 DIMETHYLAMINE DICAMBA	DIMETHYLAMINE DICAMBASTER)
2008-39-1 DIMETHYLAMINE (2,4-DICHLOROPHENOXY) ACETATE	DIMETHYLAMINE (2,4-DICHLOROPHENOXY)ACETATE
37452-11-2 DIMETHYLAMINE DODECYLBENZENESUL-FONATE	DIMETHYLAMINE DODECYLBENZENESULFONATE
73455-30-8 DIMETHYLAMINE ETHYLENEDIAMINETE-TRAACETATE	DIMETHYLAMINE ETHYLENEDIAMINETETRAACETATE
506-59-2 DIMETHYLAMINE HYDROCHLORIDE	DIMETHYLAMINE HYDROCHLORIDE
124-40-3 DIMETHYLAMINE [METHANAMINE, N-METHYL-] [EP3]	DIMETHYLAMINE
926-64-7 2-DIMETHYLAMINOACETONITRILE [HM1, HM1]	DIMETHYLAMINOACETONITRILE
926-64-7 2-DIMETHYLAMINOACETO-NITRILE [NJL]	DIMETHYLAMINOACETONITRILE
926-64-7 DIMETHYLAMINOACETONITRILE	DIMETHYLAMINOACETONITRILE
60-11-7 DIMETHYLAMINOAZOBENZENE [EP2]	4-DIMETHYLAMINOAZOBENZENE
60-11-7 4-DIMETHYLAMINOAZOBENZENE	4-DIMETHYLAMINOAZOBENZENE
60-11-7 DIMETHYL AMINOAZOBENZENE [CAA]	4-DIMETHYLAMINOAZOBENZENE
60-11-7 P-(DIMETHYLAMINO)AZOBENZENE [PQL]	4-DIMETHYLAMINOAZOBENZENE
60-11-7 P-DIMETHYLAMINOAZOBENZENE [RCA, UTS, EPA, FLA]	4-DIMETHYLAMINOAZOBENZENE
60-11-7 PARA-DIMETHYLAMINOAZOBENZENE [IAR]	4-DIMETHYLAMINOAZOBENZENE
60-11-7 P-DIMETHYLAMINOAZOBENZENE [MSL]	4-DIMETHYLAMINOAZOBENZENE
RR-01532-7 PARA-DIMETHYLAMINOAZOBENZENEDIAZO SODIUM SULPHONATE	PARA-DIMETHYLAMINOAZOBENZENEDIAZO SODIUM SULPHONATE
1300-73-8 DIMETHYLAMINOBENZENE (XYLIDENE) [ALB, ALB, ALB]	XYLIDINE

Chemical Name	Reference Name
21245-02-3 4-(DIMETHYLAMINO)BENZOIC ACID, 2-ETHYL-HEXYL ESTER	4-(DIMETHYLAMINO)BENZOIC ACID, 2-ETHYLHEXYL ESTER
RR-00765-8 4-DIMETHYLAMINO-6-(2-DIMETHY-LAMINOETHOXY)TOLUENE-2-DIAZONIUMZINC CHLORIDE	4-DIMETHYLAMINO-6-(2-DIMETHYLAMINOETHOXY) TOLUENE-2-DIAZONIUM
17268-47-2 3-DIMETHYLAMINO-N,N-DIMETHYLPROPI-ONAMIDE	3-DIMETHYLAMINO-N,N-DIMETHYLPROPIONAMIDE
108-01-0 2-(DIMETHYLAMINO) ETHANOL	2-(DIMETHYLAMINO) ETHANOL
108-01-0 DIMETHYLAMINOETHANOL [OTS]	2-(DIMETHYLAMINO) ETHANOL
1704-62-7 DIMETHYLAMINOETHOXYETHANOL	DIMETHYLAMINOETHOXYETHANOL
2867-47-2 2-(DIMETHYLAMINO) ETHYL METHACRY-LATE [MSL, FLA, NFP, NFP]	N,N-DIMETHYLAMINOETHYL METHACRYLATE
2867-47-2 DIMETHYLAMINOETHYL METHACRY-LATE [HM1, HM1, NJL]	N,N-DIMETHYLAMINOETHYL METHACRYLATE
2867-47-2 N,N-DIMETHYLAMINOETHYL METHACRYLATE	N,N-DIMETHYLAMINOETHYL METHACRYLATE
3030-47-5 N-[2-(DIMETHYLAMINO)ETHYL]-N,N'.N'-TRIMETHYL-1,2-ETHANEDIAMINE	N-[2-(DIMETHYLAMINO)ETHYL]-N,N'.N'-TRIMETHYL-1,2-ETHANEDIAMI
55738-54-0 TRANS-2-((DIMETHYLAMINO)METHYLIMINO)-5-(2-(5-NITRO-2-FURYL)VINYL)-1,3,4-OXADIA-ZOLE	TRANS-2-((DIMETHYLAMINO)METHYLIMINO)-5-(2-(5-NITRO-2-FURYL)
25338-55-0 DIMETHYLAMINOMETHYLPHENOL	DIMETHYLAMINOMETHYLPHENOL
1738-25-6 BETA-DIMETHYLAMINOPROPIONITRILE	BETA-DIMETHYLAMINOPROPIONITRILE
1738-25-6 3-(DIMETHYLAMINO)-PROPIONITRILE [NFP, NFP]	BETA-DIMETHYLAMINOPROPIONITRILE
109-55-7 3-(DIMETHYLAMINO)-PROPYLAMINE	3-(DIMETHYLAMINO)-PROPYLAMINE
71113-21-8 [[[3-(DIMETHYLAMINO)PROPYL]IMINO]BIS(METHYLENE)]BISPHOSPHONIC ACID, MONO-HYDROCHLORIDE	[[[3-(DIMETHYLAMINO)PROPYL]IMINO]BIS(METHYLENE)]BISPHOSPHONER
22975-76-4 4,4'-DIMETHYLANGELICIN PLUS ULTRAVIO-LET A RADIATION	4,4'-DIMETHYLANGELICIN PLUS ULTRAVIOLET A RADI-ATION
RR-01533-8 4,5'-DIMETHYLANGELICIN PLUS ULTRAVIO-LET A RADIATION	4,5'-DIMETHYLANGELICIN PLUS ULTRAVIOLET A RADI-ATION
121-69-7 DIMETHYLANILINE	DIMETHYLANILINE
121-69-7 N,N-DIMETHYLANILINE [313, ALB, ALB, ALB, AUS, AUS, AUS, CAB, CAL, CAN, CEX, CEX, CEX, CSR, CSR, EPA, GBR, GBR, GBR, HM1, HM1, IAR, MAK, MAK, MAK, MAK, NFP, NFP, ONT, ONT, ONT]	DIMETHYLANILINE
121-69-7 DIMETHYLANILINE, N,N- [OTS]	DIMETHYLANILINE
121-69-7 N,N-DIMETHYLANILINE [T33, WHM]	DIMETHYLANILINE
121-69-7 DIMETHYLANILINE (N,N-DIMETHYLANILINE) [BC1, BC1, BC1, TLV, TLV, TLV]	DIMETHYLANILINE
121-69-7 DIMETHYLANILINE (N,N) [HON, HON]	DIMETHYLANILINE
121-69-7 DIMETHYLANILINE (N,N-DIMETHYLANILINE) [OS1, OS1, OS1, OS1, OS1]	DIMETHYLANILINE
108-69-0 3,5-DIMETHYLANILINE	3,5-DIMETHYLANILINE
121-69-7 DIMETHYLANILINE N- [QBC, QBC, QBC]	DIMETHYLANILINE
87-62-7 2,6-DIMETHYLANILINE (2,6-XYLIDENE) [IAR]	2,6-XYLIDENE
85-91-6 DIMETHYL ANTHRANILATE	DIMETHYL ANTHRANILATE
75-60-5 DIMETHYLARSINIC ACID [ATS, HM1, HM1, HM1, HM1]	CACODYLIC ACID
57-97-6 7,12-DIMETHYLBENZ(A)ANTHRACENE	7,12-DIMETHYLBENZ(A)ANTHRACENE
95-47-6 1,2-DIMETHYL BENZENE [FLA]	O-XYLENE
1330-20-7 DIMETHYLBENZENE [QBC, QBC]	XYLENES (O-, M-, P- ISOMERS)

Chemical Name	Reference Name
1330-20-7 DIMETHYLBENZENE (SUM OF O-, M-, AND P- ISOMERS) [ONT, ONT]	XYLENES (O-, M-, P- ISOMERS)
1300-73-8 AR,AR-DIMETHYLBENZENEAMINE (SUM OF ALL ISOMERS) [ONT, ONT]	XYLIDINE
119-93-7 3,3'-DIMETHYLBENZIDINE	3,3'-DIMETHYLBENZIDINE
119-93-7 3,3'-DIMETHYL BENZIDINE [CAA]	3,3'-DIMETHYLBENZIDINE
119-93-7 3,3'-DIMETHYLBENZIDINE (O-TOLIDINE) [313]	3,3'-DIMETHYLBENZIDINEHA-(TRICHLORMETHYL)-]
119-93-7 3,3'-DIMETHYLBENZIDINE (ORTHO-TOLIDINE) [IAR]	3,3'-DIMETHYLBENZIDINE
119-93-7 3,3'-DIMETHYLBENZIDINE (ORTHO-TOLUIDINE) [C65]	3,3'-DIMETHYLBENZIDINE
41766-75-0 3,3'-DIMETHYLBENZIDINE DIHYDROFLUORIDE (O-TOLUIDINE DIHYDROFLUORIDE)	3,3'-DIMETHYLBENZIDINE DIHYDROFLUORIDE (O-TOLUIDINE DIHYDROFPHOROTHIOATE]
41766-75-0 3,3'-DIMETHYLBENZIDINE DIHYDROFLUORIDE (O-TOLIDINE DIHYDROFLUORIDE) [313]	3,3'-DIMETHYLBENZIDINE DIHYDROFLUORIDE (O-TOLUIDINE DIHYDROFPHOROTHIOATE]
612-82-8 3,3'-DIMETHYLBENZIDINE DIHYDROCHLORIDE (O-TOLIDINE DIHYDROCHLORIDE) [313]	3,3'-DIMETHYLBENZIDINE DIHYDROCHLORIDE
612-82-8 3,3'-DIMETHYLBENZIDINE DIHYDROCHLORIDE	3,3'-DIMETHYLBENZIDINE DIHYDROCHLORIDE
3034-79-5 DI(2-METHYLBENZOYL)PEROXIDE	DI(2-METHYLBENZOYL)PEROXIDE
617-94-7 2,2-DIMETHYLBENZYL ALCOHOL	2,2-DIMETHYLBENZYL ALCOHOL
151-05-3 DIMETHYLBENZYLCARBINYL ACETATE	DIMETHYLBENZYLCARBINYL ACETATE
61789-72-8 DIMETHYL BENZYL HYDROGENATED TALLOW AMMONIUM CHLORIDE	DIMETHYL BENZYL HYDROGENATED TALLOW AMMONIUM CHLORIDE
80-15-9 ALPHA,ALPHA-DIMETHYLBENZYL HYDROPEROXIDE [MAK]	CUMENE HYDROPEROXIDE
80-15-9 ALPHA, ALPHA-DIMETHYLBENZYLHYDROPEROXIDE [RCU, RCU]	CUMENE HYDROPEROXIDE
122-19-0 DIMETHYLBENZYLOCTADECYLAMMONIUM CHLORIDE	DIMETHYLBENZYLOCTADECYLAMMONIUM CHLORIDE
75-83-2 2,2-DIMETHYLBUTANE [FLA, MSL, NFP, NFP]	NEOHEXANE
79-29-8 2,3-DIMETHYLBUTANE	2,3-DIMETHYLBUTANE
563-78-0 2,3-DIMETHYL-1-BUTENE	2,3-DIMETHYL-1-BUTENE
563-79-1 2,3-DIMETHYL-2-BUTENE	2,3-DIMETHYL-2-BUTENE
108-84-9 1,3-DIMETHYLBUTYL ACETATE [GBR, GBR, NFP, NFP]	SEC-HEXYL ACETATE
108-09-8 1,3-DIMETHYLBUTYLAMINE	1,3-DIMETHYLBUTYLAMINE
68134-06-5 1,3-DIMETHYLBUTYL GLYCIDYL ETHER	1,3-DIMETHYLBUTYL GLYCIDYL ETHER
68134-06-5 DIMETHYLBUTYL GLYCIDYL ETHER, 1,3- [OTS]	1,3-DIMETHYLBUTYL GLYCIDYL ETHER
1879-09-0 2,4-DIMETHYL-6-TERT-BUTYLPHENOL	2,4-DIMETHYL-6-TERT-BUTYLPHENOLORIDE
14433-76-2 N,N-DIMETHYLCAPRAMIDE	N,N-DIMETHYLCAPRAMIDE
1118-92-9 N,N-DIMETHYLCAPRYLAMIDE	N,N-DIMETHYLCAPRYLAMIDE
79-44-7 DIMETHYLCARBAMOL CHLORIDE [MIN]	DIMETHYLCARBAMOYL CHLORIDE
79-44-7 DIMETHYL CARBAMOYL CHLORIDE [AUS, AUS, CAA, TLV]	DIMETHYLCARBAMOYL CHLORIDE
79-44-7 DIMETHYLCARBAMOYL CHLORIDE	DIMETHYLCARBAMOYL CHLORIDE
79-44-7 DIMETHYLCARBAMOYL CHLORIDE [NJL, NJL]	DIMETHYLCARBAMOYL CHLORIDE
79-44-7 N,N-DIMETHYLCARBAMOYL CHLORIDE [WHM]	DIMETHYLCARBAMOYL CHLORIDE
79-44-7 DIMETHYLCARBAMYL CHLORIDE [313]	DIMETHYLCARBAMOYL CHLORIDE
79-44-7 DIMETHYL CARBAMYL CHLORIDE [BC1, BC1]	DIMETHYLCARBAMOYL CHLORIDE
79-44-7 DIMETHYLCARBAMYL CHLORIDE [EPA, QBC, QBC, C65, C65]	DIMETHYLCARBAMOYL CHLORIDE

Chemical Name	Reference Name
616-38-6 DIMETHYL CARBONATE	DIMETHYL CARBONATE
97-97-2 DIMETHYL CHLORACETAL [FLA, NFP, NFP]	DIMETHYL DICHLORACETAL
75-78-5 DIMETHYLCHLOROSILANE [FLA]	DIMETHYLDICHLOROSILANE
2524-03-0 DIMETHYL CHLOROTHIOPHOSPHATE [313, OTS]	DIMETHYL PHOSPHOROCHLORIDOTHIOATE
61788-90-7 N,N-DIMETHYL (COCONUT OIL ALKYL) AMINE OXIDE	N,N-DIMETHYL (COCONUT OIL ALKYL) AMINE OXIDE
1467-79-4 DIMETHYLCYANAMIDE	DIMETHYLCYANAMIDE
98-94-2 N,N-DIMETHYLCYCLOHEXANAMINE [PST]	N,N-DIMETHYLCYCLOHEXYLAMINE
583-57-3 1,2-DIMETHYLCYCLOHEXANE	1,2-DIMETHYLCYCLOHEXANE
589-90-2 1,4-DIMETHYLCYCLOHEXANE	1,4-DIMETHYLCYCLOHEXANE
591-21-9 1,3-DIMETHYL CYCLOHEXANE	1,3-DIMETHYL CYCLOHEXANE
591-21-9 1,3-DIMETHYLCYCLOHEXANE [NFP, NFP]	1,3-DIMETHYL CYCLOHEXANE
624-29-3 1,4-DIMETHYLCYCLOHEXANE-CIS	1,4-DIMETHYLCYCLOHEXANE-CIS
2207-04-7 1,4-DIMETHYLCYCLOHEXANE-TRANS	1,4-DIMETHYLCYCLOHEXANE-TRANS
RR-01348-9 DIMETHYLCYCLOHEXANES	DIMETHYLCYCLOHEXANES
98-94-2 DIMETHYL CYCLOHEXYLAMINE [HM1, HM1]	N,N-DIMETHYLCYCLOHEXYLAMINE
98-94-2 N,N-DIMETHYLCYCLOHEXYLAMINE	N,N-DIMETHYLCYCLOHEXYLAMINE
98-94-2 N,N-DIMETHYLCYCLOHEXYL AMINE [NJL, NJL]	N,N-DIMETHYLCYCLOHEXYLAMINE
RR-00850-4 DIMETHYL DECALIN	DIMETHYL DECALIN
2605-79-0 N,N-DIMETHYLDECYLAMINE OXIDE	N,N-DIMETHYLDECYLAMINE OXIDE
13052-09-0 2,5-DIMETHYL-2,5-DI(2-ETHYLHEXANOYLPER-OXY)-HEXANE	2,5-DIMETHYL-2,5-DI(2-ETHYLHEXANOYLPEROXY)-HEXANE
RR-01066-2 DIMETHYL DIALKYL AMMONIUM CHLORIDE POWDER	DIMETHYL DIALKYL AMMONIUM CHLORIDE POWDER-HYLSTYRENE), (MIXED INDENE AND METHYL INDENE), AND VINYL TOLUENE
2618-77-1 2,5-DIMETHYL-2,5-DI-(BENZOYLPEROXY)HEX-ANE	2,5-DIMETHYL-2,5-DI-(BENZOYLPEROXY)HEXANE
300-76-5 DIMETHYL-1,2-DIBROMO-2,2-DICHLOROETHYL PHOSPHATE [NIO, NIO, NIO, NIO, OS1, OS1, OS1]	NALED
300-76-5 DIMETHYL-1,2-DIBROMO-2,2-DICHLOROETHYL PHOSPHATE (NALED) [NJL]	NALED
300-76-5 O,O-DIMETHYL O-(1,2-DIBROMO-2,2-DICHLOROETHYL) PHOSPHATE (NALED) [CAL]	NALED
78-63-7 2,5-DIMETHYL-2,5-DI(TERT-BUTYLPEROXY) HEXANE	2,5-DIMETHYL-2,5-DI(TERT-BUTYLPEROXY)HEXANE
1068-27-5 2,5-DIMETHYL-2,5-DI(TERT-BUTYLPEROXY) HEXYNE-3	2,5-DIMETHYL-2,5-DI(TERT-BUTYLPEROXY)HEXYNE-3
97-97-2 DIMETHYL DICHLORACETAL	DIMETHYL DICHLORACETAL
56343-50-1 N,N-DIMETHYL DICHLOROACETAMIDE	N,N-DIMETHYL DICHLOROACETAMIDE
75-78-5 DIMETHYLDICHLOROSILANE	DIMETHYLDICHLOROSILANE
75-78-5 DIMETHYL DICHLOROSILANE [NJL, NJL]	DIMETHYLDICHLOROSILANE
75-78-5 DIMETHYLDICHLOROSILANE [SILANE, DICHLORODIMETHYL-] [EP3]	DIMETHYLDICHLOROSILANE
62-73-7 DIMETHYL-O,O-DICHLOROVINYL-2,2-PHOS-PHATE (TECHNICAL) [NFP, NFP]	DICHLORVOS
78-62-6 DIMETHYLDIETHOXYSILANE	DIMETHYLDIETHOXYSILANE
RR-01754-9 2,5-DIMETHYL-2,5-DI-(2-ETHYLHEXANOYLPER-OXY) HEXANE	2,5-DIMETHYL-2,5-DI-(2-ETHYLHEXANOYLPEROXY) HEXANE
3025-88-5 2,5-DIMETHYL-2,5-DIHYDRO-PEROXYHEXANE	2,5-DIMETHYL-2,5-DIHYDRO-PEROXYHEXANE

Chemical Name	Reference Name
3025-88-5 2,5-DIMETHYL-2,5-DIHYDROPEROXY HEXANE [HM1]	2,5-DIMETHYL-2,5-DIHYDRO-PEROXYHEXANE
15414-89-8 1,3-DIMETHYL-5,5-DIMETHYLHYDANTOIN	1,3-DIMETHYL-5,5-DIMETHYLHYDANTOIN
107-64-2 DIMETHYL DIOCTADECYL AMMONIUM CHLORIDE [PST]	DISTEARYLDIMETHYLAMMONIUM CHLORIDE
73138-28-0 DIMETHYLDIOCTADECYLAMMONIUM BENTONITE	DIMETHYLDIOCTADECYLAMMONIUM BENTONITE
25136-55-4 DIMETHYLDIOXANE	DIMETHYLDIOXANE
25136-55-4 DIMETHYLDIOXANES [HM1, HM1]	DIMETHYLDIOXANE
RR-00785-2 1,3-DIMETHYL-1,3-DIPHENYLCYCLOBUTANE	1,3-DIMETHYL-1,3-DIPHENYLCYCLOBUTANE
91-97-4 3,3'-DIMETHYL-4,4'-DIPHENYLENE DIISOCYANATE [313]	1,1'-BIPHENYL, 4,4'-DIISOCYANATO-3,3'-DIMETHYL-
139-25-3 3,3'-DIMETHYLDIPHENYLMETHANE-4,4'-DIISOCYANATE [313]	BENZENE, 1,1'-METHYLENEBIS[4-ISOCYANATO-3-METHYL-
68083-14-7 DIMETHYLDIPHENYLSILOXANE [TSC, TSC]	POLYDIMETHYLDIPHENYL SILOXANE COPOLYMER
624-92-0 DIMETHYLDISULFIDE	DIMETHYLDISULFIDE
624-92-0 DIMETHYL DISULFIDE [HM1, HM1, TSC]	DIMETHYLDISULFIDE
112-18-5 N,N-DIMETHYLDODCEYLAMINE [OTS]	LAURYLDIMETHYLAMINE
1643-20-5 N,N-DIMETHYLDODECYLAMINE OXIDE [PST]	LAURYLDIMETHYLAMINE OXIDE
108-01-0 DIMETHYLETHANOLAMINE [HM1, HM1]	2-(DIMETHYLAMINO) ETHANOL
115-10-6 DIMETHYL ETHER	DIMETHYL ETHER
14857-34-2 DIMETHYLETHOXYSILOXANE	DIMETHYLETHOXYSILOXANE
17963-04-1 3-(DIMETHYLETHOXYSILYL)PROPYL GLYCIDYL ETHER [EPA, OTS]	SILANE, ETHOXYDIMETHYL[3-(OXIRANYL-METHOXY)PROPYL]-
598-56-1 N,N-DIMETHYL ETHYLAMINE	N,N-DIMETHYL ETHYLAMINE
598-56-1 N,N-DIMETHYLETHYLAMINE [GBR, GBR]	N,N-DIMETHYL ETHYLAMINE
1068-87-7 2,4-DIMETHYL-3-ETHYL PENTANE	2,4-DIMETHYL-3-ETHYL PENTANE
1068-87-7 2,4-DIMETHYL-3-ETHYLPENTANE [NFP, NFP]	2,4-DIMETHYL-3-ETHYL PENTANE
68-12-2 DIMETHYLFORMAMIDE	DIMETHYLFORMAMIDE
68-12-2 N,N-DIMETHYLFORMAMIDE [313]	DIMETHYLFORMAMIDE
68-12-2 DIMETHYL FORMAMIDE [CAA]	DIMETHYLFORMAMIDE
68-12-2 N,N-DIMETHYLFORMAMIDE [CAL, HM1, HM1, NFP, NFP, ONT, ONT]	DIMETHYLFORMAMIDE
68-12-2 DIMETHYL FORMAMIDE [OTV]	DIMETHYLFORMAMIDE
68-12-2 N,N-DIMETHYLFORMAMIDE [PST]	DIMETHYLFORMAMIDE
68-12-2 DIMETHYL FORMAMIDE [TSC, TSC, TSC]	DIMETHYLFORMAMIDE
68-12-2 N,N-DIMETHYLFORMAMIDE [WHM]	DIMETHYLFORMAMIDE
68-12-2 N,N-DIMETHYLFORMAMIDE (DMF) [BEI, BEI]	DIMETHYLFORMAMIDE
68-12-2 DIMETHYL FORMAMIDE (DMF) [EPA]	DIMETHYLFORMAMIDE
68-12-2 DIMETHYLFORMAMIDE (N,N-) [HON, HON]	DIMETHYLFORMAMIDE
625-86-5 2,5-DIMETHYLFURAN	2,5-DIMETHYLFURAN
RR-00851-5 DIMETHYL GLYCOL PHTHALATE	DIMETHYL GLYCOL PHTHALATE
4032-86-4 3,3-DIMETHYLHEPTANE	3,3-DIMETHYLHEPTANE
108-83-8 2,6-DIMETHYL-4-HEPTANONE (DIISOBUTYL KETONE) [ALB, ALB]	DIISOBUTYL KETONE
108-83-8 2,6-DIMETHYLHEPTAN-4-ONE [GBR]	DIISOBUTYL KETONE
108-83-8 2,6-DIMETHYLHEPTANONE [NJL]	DIISOBUTYL KETONE
108-83-8 2,6-DIMETHYL-4-HEPTANONE [ONT]	DIISOBUTYL KETONE
112-69-6 N,N-DIMETHYL-1-HEXADECYLAMINE	N,N-DIMETHYL-1-HEXADECYLAMINE
584-94-1 2,3-DIMETHYLHEXANE	2,3-DIMETHYLHEXANE

Chemical Name	Reference Name
589-43-5 2,4-DIMETHYLHEXANE	2,4-DIMETHYLHEXANE
107-54-0 DIMETHYL HEXYNOL [NFP, NFP]	1-HEXYN-3-OL, 3,5-DIMETHYL-
107-54-0 3,5-DIMETHYL-1-HEXYN-3-OL [PST]	1-HEXYN-3-OL, 3,5-DIMETHYL-
77-71-4 5,5-DIMETHYLHYDANTOIN	5,5-DIMETHYLHYDANTOIN
57-14-7 1,1-DIMETHYLHYDRAZINE	1,1-DIMETHYLHYDRAZINE
57-14-7 DIMETHYLHYDRAZINE [302]	1,1-DIMETHYLHYDRAZINE
57-14-7 1,1-DIMETHYL HYDRAZINE [313, CAA]	1,1-DIMETHYLHYDRAZINE
57-14-7 1,1-DIMETHYLHYDRAZINE [NIO, NIO, NIO, NIO]	1,1-DIMETHYLHYDRAZINE
57-14-7 DIMETHYLHYDRAZINE, 1,1- [OS3]	1,1-DIMETHYLHYDRAZINE
540-73-8 1,2-DIMETHYLHYDRAZINE	1,2-DIMETHYLHYDRAZINE
57-14-7 1,1-DIMETHYLHYDRAZINE (UDMH) [C65, RCA]	1,1-DIMETHYLHYDRAZINE
306-37-6 SYM-DIMETHYLHYDRAZINE DIHYDROCHLORIDE	SYM-DIMETHYLHYDRAZINE DIHYDROCHLORIDE
540-73-8 DIMETHYLHYDRAZINE, SYMMETRICAL [HM1, HM1, HM1]	1,2-DIMETHYLHYDRAZINE
57-14-7 DIMETHYLHYDRAZINE, UNSYMMETRI-CAL [HM1, HM1, HM1, HM1]	1,1-DIMETHYLHYDRAZINE
57-14-7 1,1-DIMETHYLHYDRAZINE [HYDRAZINE, 1,1-DIMETHYL-] [EP3]	1,1-DIMETHYLHYDRAZINE
868-85-9 DIMETHYL HYDROGEN PHOSPHITE	DIMETHYL HYDROGEN PHOSPHITE
68037-59-2 DIMETHYLHYDROPOLYSILOXANE	DIMETHYLHYDROPOLYSILOXANE
125972-19-2 N,N-DIMETHYLISOOCTADECANAMINE, N-OXIDE	N,N-DIMETHYLISOOCTADECANAMINE, N-OXIDE
1459-93-4 DIMETHYL ISOPHTHALATE	DIMETHYL ISOPHTHALATE
1459-93-4 DIMETHYLISOPHTHALATE [NFP, NFP]	DIMETHYL ISOPHTHALATE
108-16-7 N,N-DIMETHYLISOPROPANOLAMINE	N,N-DIMETHYLISOPROPANOLAMINE
2999-74-8 DIMETHYLMAGNESIUM	DIMETHYLMAGNESIUM
624-48-6 DIMETHYL MALEATE	DIMETHYL MALEATE
108-59-8 DIMETHYL MALONATE	DIMETHYL MALONATE
593-74-8 DIMETHYLMERCURY	DIMETHYLMERCURY
593-74-8 DIMETHYL MERCURY [NJL, NJL]	DIMETHYLMERCURY
756-79-6 DIMETHYL METHYLPHOSPHONATE	DIMETHYL METHYLPHOSPHONATE
28434-00-6 2,2-DIMETHYL-3-(2-METHYL-1-PROPENYL)CY-CLOPROPANECARBOXYLIC ACID, 2-METHYL-4-OXO-3-(2-PROPENYL)-2-CYCLOPENTEN-1-YL ESTER	2,2-DIMETHYL-3-(2-METHYL-1-PROPENYL)CYCLO-PROPANECARBOXYLIC A
115361-68-7 DIMETHYLMETHYL 3,3,3-TRIFLUOROPROPYL SILOXANE	DIMETHYLMETHYL 3,3,3-TRIFLUOROPROPYL SILOXANE
67762-94-1 DIMETHYLMETHYLVINYLSILOXANE	DIMETHYLMETHYLVINYLSILOXANEICA
141-91-3 2,6-DIMETHYLMORPHOLINE	2,6-DIMETHYLMORPHOLINE
597-25-1 DIMETHYL MORPHOLINOPHOSPHORAMIDATE	DIMETHYL MORPHOLINOPHOSPHORAMIDATE
27178-87-6 DIMETHYLNAPHTHALENESULFONIC ACID, SODIUM SALT	DIMETHYLNAPHTHALENESULFONIC ACID, SODIUM SALT
2050-99-9 2,8-DIMETHYL-5-NONANONE	2,8-DIMETHYL-5-NONANONE
2571-88-2 N,N-DIMETHYLOCTADECYLAMINE OXIDE	N,N-DIMETHYLOCTADECYLAMINE OXIDE
122-19-0 DIMETHYL OCTADECYL BENZYL AMMONIUM CHLORIDE [PST]	DIMETHYLBENZYLOCTADECYLAMMONIUM CHLORIDE
15869-92-8 3,4-DIMETHYLOCTANE	3,4-DIMETHYLOCTANE
RR-00531-2 2,3-DIMETHYLOCTANE	2,3-DIMETHYLOCTANE
7378-99-6 N,N-DIMETHYLOCTYLAMINE	N,N-DIMETHYLOCTYLAMINE

Chemical Name	Reference Name
78-66-0 3,6-DIMETHYL-4-OCTYNE-3,6-DIOL [PST]	4-OCTYNE-3,6-DIOL, 3,6-DIMETHYL-
2664-42-8 N,N-DIMETHYLOLEAMIDE	N,N-DIMETHYLOLEAMIDE
14351-50-9 N,N-DIMETHYLOLEYLAMINE OXIDE	N,N-DIMETHYLOLEYLAMINE OXIDE
140-95-4 DIMETHYLOLUREA	DIMETHYLOLUREA
62073-57-8 DIMETHYLOL UREA - FORMALDEHYDE - MONOMETHYLOL UREA POLYMER	DIMETHYLOL UREA - FORMALDEHYDE - MONOMETHYLOL UREA POLYMER
51200-87-4 4,4-DIMETHYLOXAZOLIDINE	4,4-DIMETHYLOXAZOLIDINE
86-50-0 O,O-DIMETHYL S-(4-OXO-BENZOTRI-AZINO-3-METHYL) PHOSPHORODITHIOATE (AZINPHOS METHYL) [CAL]	AZINPHOS-METHYL
83-26-1 2-(2,2-DIMETHYL-1-OXOPROPYL)-H-INDENE-1,3 (2H)-DIONE [ONT]	PINDONE
32749-94-3 2,3-DIMETHYL PENTALDEHYDE [MSL]	PENTANAL, 2,3-DIMETHYL-
RR-00784-1 2,3-DIMETHYLPENTALDEHYDE	2,3-DIMETHYLPENTALDEHYDE
108-08-7 2,4-DIMETHYLPENTANE	2,4-DIMETHYLPENTANE
565-59-3 2,3-DIMETHYLPENTANE	2,3-DIMETHYLPENTANE
600-36-2 2,4-DIMETHYL-3-PENTANOL	2,4-DIMETHYL-3-PENTANOL
22349-59-3 1,4-DIMETHYLPHENANTHRENE	1,4-DIMETHYLPHENANTHRENE
122-09-8 ALPHA,ALPHA-DIMETHYL PHENETHY-LAMINE [NJL]	ALPHA,ALPHA-DIMETHYLPHENETHYLAMINE
95-65-8 3,4-DIMETHYLPHENOL	3,4-DIMETHYLPHENOL
105-67-9 2,4-DIMETHYLPHENOL	2,4-DIMETHYLPHENOL
576-26-1 2,6-DIMETHYLPHENOL	2,6-DIMETHYLPHENOL
576-26-1 DIMETHYLPHENOL, 2,6- [OTS]	2,6-DIMETHYLPHENOL
RR-01285-1 DIMETHYLPHENOLS	DIMETHYLPHENOLS
122-09-8 ALPHA,ALPHA-DIMETHYLPHENETHYLAMINE	ALPHA,ALPHA-DIMETHYLPHENETHYLAMINE
7227-91-0 3,3-DIMETHYL-1-PHENYLTRIAZENE	3,3-DIMETHYL-1-PHENYLTRIAZENE
10265-92-6 O,S-DIMETHYL PHOSPHORAMIDOTH-IOATE [CAL]	METHAMIDOPHOS
2524-03-0 DIMETHYL PHOSPHOROCHLORIDOTHIOATE	DIMETHYL PHOSPHOROCHLORIDOTHIOATE
2524-03-0 O,O-DIMETHYL PHOSPHOROCHLORIDOTH-IOATE [WHM]	DIMETHYL PHOSPHOROCHLORIDOTHIOATE
950-37-8 O,O-DIMETHYL PHOSPHORODITHIOATE, S-ESTER WITH 4-(MERCAPTOMETHYL)-2-METHOXY-O2-1,3,4-THIADIAZOLIN-5-ONE [CAL]	METHIDATHION
131-11-3 DIMETHYLPHTHALATE [AUS, BC1, BC1]	DIMETHYL PHTHALATE
120-61-6 DIMETHYL P-PHTHALATE	DIMETHYL P-PHTHALATE
131-11-3 DIMETHYL PHTHALATE	DIMETHYL PHTHALATE
131-11-3 N,N-DIMETHYLPHTHALATE [FLA]	DIMETHYL PHTHALATE
131-11-3 DIMETHYLPHTHALATE [MIN, MSL, NIO, NIO, NIO, OS1, OS1, QBC, TLV]	DIMETHYL PHTHALATE
131-11-3 DIMETHYL PHTHALATE (DMP) [WHM]	DIMETHYL PHTHALATE
106-55-8 DIMETHYLPIPERAZINE, CIS-	DIMETHYLPIPERAZINE, CIS-
106-55-8 DIMETHYLPIPERAZINE-CIS [NFP, NFP]	DIMETHYLPIPERAZINE, CIS-
6284-84-0 DIMETHYL PIPERAZINE-CIS	DIMETHYL PIPERAZINE-CIS
68037-74-1 DIMETHYLPOLYSILOXANES	DIMETHYLPOLYSILOXANES
463-82-1 2,2-DIMETHYLPROPANE	2,2-DIMETHYLPROPANE
463-82-1 DIMETHYLPROPANE [NJL, NJL]	2,2-DIMETHYLPROPANE
75-98-9 2,2-DIMETHYLPROPANOIC ACID	2,2-DIMETHYLPROPANOIC ACID
926-63-6 DIMETHYLPROPYLAMINE	DIMETHYLPROPYLAMINE

Chemical Name	Reference Name
926-63-6 DIMETHYL-N-PROPYLAMINE [HM1, HM1]	DIMETHYLPROPYLAMINE
RR-00169-4 1,1-DIMETHYLPROPYL PEROXYESTER	1,1-DIMETHYLPROPYL PEROXYESTER
123-32-0 2,5-DIMETHYLPYRAZINE	2,5-DIMETHYLPYRAZINE
106-79-6 DIMETHYL SEBACATE	DIMETHYL SEBACATE
9006-65-9 DIMETHYL SILICONE	DIMETHYL SILICONE
63148-62-9 DIMETHYL SILICONES AND SILOXANE [TSC, TSC]	POLY(DIMETHYLSILOXANE)
67762-90-7 DIMETHYL SILICONES AND SILOXANES, REACTION PRODUCTS WITH SILICA [TSC, TSC]	DIMETHYL SILICONE POLYMER WITH SILICA-PROPENYL]OXY]-N,N,N-TRIMETHYL-, CHLORIDE
67762-90-7 DIMETHYL SILICONE POLYMER WITH SILICA	DIMETHYL SILICONE POLYMER WITH SILICA-PROPENYL]OXY]-N,N,N-TRIMETHYL-, CHLORIDE
RR-01571-4 DIMETHYL-3-SUBSTITUTED HETEROMONOCYCLE	DIMETHYL-3-SUBSTITUTED HETEROMONOCYCLE
RR-01572-5 DIMETHYL-3-SUBSTITUTED HETEROMONOCYCLE AMINE	DIMETHYL-3-SUBSTITUTED HETEROMONOCYCLE AMINE
13360-57-1 DIMETHYLSULFAMOYL CHLORIDE	DIMETHYLSULFAMOYL CHLORIDE
77-78-1 DIMETHYL SULFATE	DIMETHYL SULFATE
75-18-3 DIMETHYL SULFIDE	DIMETHYL SULFIDE
1003-78-7 DIMETHYL SULFOLANE	DIMETHYL SULFOLANE
67-71-0 DIMETHYLSULFONE	DIMETHYLSULFONE
67-68-5 DIMETHYL SULFOXIDE	DIMETHYL SULFOXIDE
77-78-1 DIMETHYL SULPHATE [AUS, AUS, AUS, IAR]	DIMETHYL SULFATE
68308-74-7 N,N-DIMETHYL TALL-OIL FATTY AMIDES	N,N-DIMETHYL TALL-OIL FATTY AMIDES
120-61-6 DIMETHYL TEREPHTHALATE [CSR, CSR, MIN, NFP, NFP, OTS]	DIMETHYL P-PHTHALATE
3332-27-2 N,N-DIMETHYLTETRADECYLAMINE OXIDE	N,N-DIMETHYLTETRADECYLAMINE OXIDE
993-12-4 DIMETHYL THIOPHOSPHORYL CHLORIDE	DIMETHYL THIOPHOSPHORYL CHLORIDE
2524-03-0 DIMETHYL THIOPHOSPHORYL CHLORIDE [HM1, HM1, NJL, NJL]	DIMETHYL PHOSPHOROCHLORIDOTHIOATE
RR-01755-0 2,5-DIMETHYL-2,5-DI-(3,5,5-TRI-METHYLHEXANOYLPEROXY) HEXANE	2,5-DIMETHYL-2,5-DI-(3,5,5-TRI-METHYLHEXANOYLPEROXY) HEXANE
32749-94-3 2,3-DIMETHYL VALERALDEHYDE [FLA]	PENTANAL, 2,3-DIMETHYL-
513-37-1 DIMETHYLVINYLCHLORIDE [C65, C65]	DIMETHYLVINYLCHLORIDE (DMVC)
513-37-1 DIMETHYLVINYL CHLORIDE [NTP]	DIMETHYLVINYLCHLORIDE (DMVC)
513-37-1 DIMETHYLVINYLCHLORIDE (DMVC)	DIMETHYLVINYLCHLORIDE (DMVC)
513-37-1 DIMETHYLVINYL CHLORIDE (DMVC) [CSR, CSR]	DIMETHYLVINYLCHLORIDE (DMVC)
544-97-8 DIMETHYLZINC	DIMETHYLZINC
644-64-4 DIMETILAN	DIMETILAN
RR-01239-5 DIMETRIDAZOLE	DIMETRIDAZOLE
53220-22-7 DIMYRISTYL PEROXYDI-CARBONATE	DIMYRISTYL PEROXYDI-CARBONATE
53220-22-7 DIMYRISTYL PEROXYDICARBONATE [NJL]	DIMYRISTYL PEROXYDI-CARBONATE
29903-04-6 DI-(1-NAPHTHOYL) PEROXIDE	DI-(1-NAPHTHOYL) PEROXIDE
93-46-9 N,N'-DI-BETA-NAPHTHYL-P-PHENYLENE-DIAMINE	N,N'-DI-BETA-NAPHTHYL-P-PHENYLENE-DIAMINE
16215-49-9 DI-N-BUTYL PEROXYDICARBONATE [DOT]	BUTYL PEROXYDICARBONATE
131-89-5 DINEX [NJL]	4,6-DINITRO-O-CYCLOHEXYLPHENOL
148-01-6 DINITOLMIDE	DINITOLMIDE
148-01-6 DINITOLMIDE [QBC]	DINITOLMIDE
148-01-6 DINITOLMIDE (3,5-DINITRO-O-TOLUAMIDE) [OS1]	DINITOLMIDE

Chemical Name	Reference Name
534-52-1 DINITRO-O-CRESOL [MEX, MEX, MEX]	4,6-DINITRO-O-CRESOL
97-02-9 2,4-DINITROANILINE	2,4-DINITROANILINE
26471-56-7 DINITROANILINE	DINITROANILINE
26471-56-7 DINITROANILINE, ALL ISOMERS [WHM]	DINITROANILINE
99-65-0 M-DINITROBENZENE	M-DINITROBENZENE
99-65-0 1,3,-DINITROBENZENE [ATS]	M-DINITROBENZENE
99-65-0 META-DINITROBENZENE [RCA]	M-DINITROBENZENE
100-25-4 P-DINITROBENZENE	P-DINITROBENZENE
100-25-4 1,4-DINITROBENZENE [RCA, UTS]	P-DINITROBENZENE
528-29-0 O-DINITROBENZENE	O-DINITROBENZENE
25154-54-5 DINITROBENZENE [MAK, MAK, MEX, MEX, MEX, NJL]	DINITROBENZENE (MIXED)
528-29-0 DINITROBENZENE (ALPHA-) [OS1, OS1, OS1, OS1]	O-DINITROBENZENE
99-65-0 DINITROBENZENE (META-) [OS1, OS1]	M-DINITROBENZENE
100-25-4 DINITROBENZENE (PARA-) [OS1, OS1]	P-DINITROBENZENE
25154-54-5 DINITROBENZENE (SUM OF M-, O-, AND P-ISOMERS) [ONT, ONT]	DINITROBENZENE (MIXED)
25154-54-5 DINITROBENZENE, ALL ISOMERS [ALB, ALB, ALB, GBR, GBR, GBR]	DINITROBENZENE (MIXED)
25154-54-5 DINITROBENZENE (ALL ISOMERS) [QBC, QBC, BC1, BC1, BC1, NIO, NIO, NIO, NIO, OS1, OS1]	DINITROBENZENE (MIXED)
25154-54-5 DINITROBENZENE (MIXED)	DINITROBENZENE (MIXED)
25154-54-5 DINITROBENZENE, N.O.S. [RCU]	DINITROBENZENE (MIXED)
25154-54-5 DINITROBENZENES [HM1, HM1, HM1]	DINITROBENZENE (MIXED)
25154-54-5 DINITROBENZENES (NOS) [HON]	DINITROBENZENE (MIXED)
528-29-0 1,2-DINITRO BENZOL [NFP, NFP]	O-DINITROBENZENE
88-85-7 DINITROBUTYL PHENOL (DINOSEB) [313]	2-SEC-BUTYL-4,6-DINITROPHENOL (DINOSEB)
25567-67-3 DINITROCHLOROBENZENE	DINITROCHLOROBENZENE
25567-67-3 DINITROCHLOROBENZENES [HM1]	DINITROCHLOROBENZENE
534-52-1 4,6-DINITRO-O-CRESOL	4,6-DINITRO-O-CRESOL
534-52-1 DINITROCRESOL [302]	4,6-DINITRO-O-CRESOL
534-52-1 DINITRO-O-CRESOL [AUS, AUS, BC1, BC1, BC1, FLA, MIN, MSL]	4,6-DINITRO-O-CRESOL
534-52-1 DINITRO-ORTHO-CRESOL [NHS, NHS]	4,6-DINITRO-O-CRESOL
534-52-1 DINITRO-O-CRESOL [NIO, NIO, NIO, NIO, ONT, ONT, OS1, OS1, OS1, OS1, QBC, QBC, TLV, TLV]	4,6-DINITRO-O-CRESOL
534-52-1 4,6-DINITRO-O-CRESOL, AND SALTS [CAA]	4,6-DINITRO-O-CRESOL
534-52-1 4,6-DINITRO-O-CRESOL, & SALTS [RCU, RCU]	4,6-DINITRO-O-CRESOL
131-89-5 4,6-DINITRO-O-CYCLOHEXYLPHENOL	4,6-DINITRO-O-CYCLOHEXYLPHENOL
RR-01414-2 DINITRO-7,8-DIMETHYLGLYCOLURIL	DINITRO-7,8-DIMETHYLGLYCOLURIL
RR-01367-2 1,3-DINITRO-5,5-DIMETHYL HYDANTOIN	1,3-DINITRO-5,5-DIMETHYL HYDANTOIN
RR-01366-1 1,3-DINITRO-4,5-DINITROSOBENZENE	1,3-DINITRO-4,5-DINITROSOBENZENE
600-40-8 1,1-DINITROETHANE	1,1-DINITROETHANE
7570-26-5 1,2-DINITROETHANE	1,2-DINITROETHANE
22506-53-2 3,9-DINITROFLUORANTHENE	3,9-DINITROFLUORANTHENE
105735-71-5 3,7-DINITROFLUORANTHENE	3,7-DINITROFLUORANTHENE
RR-01333-2 DINITROGLYCOLURIL	DINITROGLYCOLURIL
625-76-3 DINITROMETHANE	DINITROMETHANE

Chemical Name	Reference Name
27478-34-8 DINITRONAPHTHALENES	DINITRONAPHTHALENES
534-52-1 DINITRO-O-CRESOL [ALB, ALB, ALB, MX2]	4,6-DINITRO-O-CRESOL
534-52-1 4,6-DINITRO-ORTO-CRESOL [MX1]	4,6-DINITRO-O-CRESOL
148-01-6 3,5-DINITRO-O-TOLUAMIDE (ZOALENE) [ALB, ALB]	DINITOLMIDE
RR-00262-0 DINITROPHENATES ALKALI METALS, DRY OR WETTED	DINITROPHENATES ALKALI METALS, DRY OR WETTED
51-28-5 2,4-DINITROPHENOL	2,4-DINITROPHENOL
329-71-5 2,5-DINITROPHENOL	2,5-DINITROPHENOL
573-56-8 2,6-DINITROPHENOL	2,6-DINITROPHENOL
25550-58-7 DINITROPHENOL	DINITROPHENOL
25550-58-7 DINITROPHENOL, ALL ISOMERS [WHM]	DINITROPHENOL
RR-00263-1 DINITROPHENOLATES	DINITROPHENOLATES
119-26-6 2,4-DINITROPHENYLHYDRAZINE	2,4-DINITROPHENYLHYDRAZINE
RR-01415-3 DINITROPROPYLENE GLYCOL	DINITROPROPYLENE GLYCOL
42397-64-8 1,6-DINITROPYRENE	1,6-DINITROPYRENE
42397-65-9 1,8-DINITROPYRENE	1,8-DINITROPYRENE
75321-20-9 1,3-DINITROPYRENE	1,3-DINITROPYRENE
519-44-8 2,4-DINITRORESORCINOL	2,4-DINITRORESORCINOL
616-74-0 4,6-DINITRORESORCINOL	4,6-DINITRORESORCINOL
35860-51-6 DINITRORESORCINOL	DINITRORESORCINOL
RR-01385-4 2,4-DINITRORESORCINOL	2,4-DINITRORESORCINOL
RR-01387-6 4,6-DINITRORESORCINOL	4,6-DINITRORESORCINOL
25550-55-4 DINITROSOBENZENE	DINITROSOBENZENE
RR-01416-4 DINITROSOBENZYLAMIDINE AND SALTS	DINITROSOBENZYLAMIDINE AND SALTS
133-55-1 N,N'-DINITROSO-N,N'-DIMETHYL TEREPHTHA-LAMIDE [HM1, HM1]	TEREPHTHALAMIDE, N,N'-DIMETHYL-N,N'-DINITROSO-
101-25-7 DINITROSOPENTAMETHYLENETETRAMINE	DINITROSOPENTAMETHYLENETETRAMINE
101-25-7 N,N-DINITROSOPENTAMETHYLENETE-TRAMINE [HM1, HM1]	DINITROSOPENTAMETHYLENETETRAMINE
101-25-7 N,N'-DINITROSOPENTAMETHYLENETE-TRAMINE [PST]	DINITROSOPENTAMETHYLENETETRAMINE
140-79-4 1,4-DINITROSOPIPERAZINE	1,4-DINITROSOPIPERAZINE
RR-01373-0 2,2-DINITROSTILBENE	2,2-DINITROSTILBENE
RR-01368-3 1,4-DINITRO-1,1,4,4-TETRAMETHYLOLBU-TANETETRANITRATE	1,4-DINITRO-1,1,4,4-TETRAMETHYLOLBUTANETETRANI-TRATE
148-01-6 3,5-DINITRO-O-TOLUAMIDE [CAL, MIN]	DINITOLMIDE
148-01-6 3,5-DINITRO-O-TOLUAMIDE (ZOALENE) [BC1, BC1]	DINITOLMIDE
121-14-2 2,4-DINITROTOLUENE	2,4-DINITROTOLUENE
121-14-2 DINITROTOLUENE [ALB, ALB, ALB, AUS, AUS, FLA, MEX, MEX, MEX, MIN]	2,4-DINITROTOLUENE
606-20-2 2,6-DINITROTOLUENE	2,6-DINITROTOLUENE
610-39-9 3,4-DINITROTOLUENE	3,4-DINITROTOLUENE
25321-14-6 DINITROTOLUENE [ATS, BC1, BC1, BC1, CW3, EPA, MSL, NIO, NIO, NIO, NIO, NIO, NJL, OS1, OS1, OS1, OS1, OS1, QBC, QBC, TLV, TLV, TLV]	DINITROTOLUENE (MIXED ISOMERS)
25321-14-6 DINITROTOLUENE (SUM OF ALL ISOMERS) [ONT, ONT]	DINITROTOLUENE (MIXED ISOMERS)
25321-14-6 DINITROTOLUENE (MIXED) [EP2]	DINITROTOLUENE (MIXED ISOMERS)
25321-14-6 DINITROTOLUENE (MIXED ISOMERS)	DINITROTOLUENE (MIXED ISOMERS)

Chemical Name	Reference Name
25321-14-6 DINITROTOLUENES [MAK, MAK, NHS, NHS, NHS]	DINITROTOLUENE (MIXED ISOMERS)
608-50-4 2,4-DINITRO-1,3,5-TRIMETHYLBENZENE	2,4-DINITRO-1,3,5-TRIMETHYLBENZENE
RR-01404-0 DI-(BETA-NITROXYETHYL) AMMONIUM NITRATE	DI-(BETA-NITROXYETHYL) AMMONIUM NITRATE
33453-96-2 A,A'-DI-(NITROXY) METHYLETHER	A,A'-DI-(NITROXY) METHYLETHER
RR-01370-7 1,9-DINITROXY PENTAMETHYLENE-2,4,6,8-TETRAMINE	1,9-DINITROXY PENTAMETHYLENE-2,4,6,8-TETRAMINE
973-21-7 DINOBUTON	DINOBUTON
39300-45-3 DINOCAP	DINOCAP
762-13-0 DI-N-NONANOYL PEROXIDE	DI-N-NONANOYL PEROXIDE
39464-64-7 DINONYLPHENOL, ETHOXYLATED, PHOSPHATED	DINONYLPHENOL, ETHOXYLATED, PHOSPHATED
RR-01067-3 O,P-DINONYLPHENOL, ETHOXYLATED, PHOSPHATED, AMMONIUM SALT	O,P-DINONYLPHENOL, ETHOXYLATED, PHOSPHATED, AMMONIUM SALT
RR-01068-4 O,P-DINONYLPHENOL, ETHOXYLATED, PHOSPHATED, CALCIUM SALT	O,P-DINONYLPHENOL, ETHOXYLATED, PHOSPHATED, CALCIUM SALT
72067-21-1 O,P-DINONYLPHENOL, ETHOXYLATED, PHOSPHATED, POTASSIUM SALT	O,P-DINONYLPHENOL, ETHOXYLATED, PHOSPHATED, POTASSIUM SALT
70903-62-7 O,P-DINONYLPHENOL, ETHOXYLATED, PHOSPHATED, SODIUM SALT	O,P-DINONYLPHENOL, ETHOXYLATED, PHOSPHATED, SODIUM SALT
RR-01069-5 O,P-DINONYLPHENOL, ETHOXYLATED, PHOSPHATED, ZINC SALT	O,P-DINONYLPHENOL, ETHOXYLATED, PHOSPHATED, ZINC SALT
84-76-4 DINONYL PHTHALATE [GBR]	1,2-BENZENEDICARBOXYLIC ACID, DINONYL ESTER
68515-45-7 DINONYL PHTHALATE (BRANCHED AND LINEAR ISOMERS) [TSC]	PHTHALIC ACID, DIALKYL (C9) ESTER
111381-91-0 DI(NONYL-, UNDECYL-) PHTHALATE (BRANCHED AND LINEAR ISOMERS)	DI(NONYL-, UNDECYL-) PHTHALATE (BRANCHED AND LINEAR ISOMERS))
1420-07-1 DINOTERB	DINOTERB
61789-76-2 DIOCO ALKYLAMINE	DIOCO ALKYLAMINE
762-16-3 DI-N-OCTANOYL PEROXIDE	DI-N-OCTANOYL PEROXIDE
103-23-1 DIOCTYL ADIPATE [NFP, NFP]	BIS (2-ETHYLHEXYL) ADIPATE
103-24-2 DIOCTYL AZELATE [NFP, NFP]	DI-2-ETHYL HEXYL AZELATE
101-67-7 P,P'-DIOCTYLDIPHENYLAMINE	P,P'-DIOCTYLDIPHENYLAMINE
629-82-3 DIOCTYL ETHER	DIOCTYL ETHER
2915-53-9 DIOCTYL MALEATE	DIOCTYL MALEATE
117-81-7 DI-SEC-OCTYL PHTHALATE [AUS, AUS]	DI(2-ETHYLHEXYL)PHTHALATE
117-81-7 DI-SEC OCTYL PHTHALATE (DI-2-ETHYL-HEXYL-PHTHALATE) [BC1, BC1]	DI(2-ETHYLHEXYL)PHTHALATE
117-81-7 DI-SEC-OCTYL PHTHALATE [CEX, CEX]	DI(2-ETHYLHEXYL)PHTHALATE
117-81-7 DI-SEC-OCTYL-PHTHALATE [MIN]	DI(2-ETHYLHEXYL)PHTHALATE
117-81-7 DI-SEC OCTYL PHTHALATE [MSL, NIO, NIO, NIO]	DI(2-ETHYLHEXYL)PHTHALATE
117-81-7 DI-SEC-OCTYL PHTHALATE [TLV, TLV]	DI(2-ETHYLHEXYL)PHTHALATE
117-84-0 DI-N-OCTYL PHTHALATE	DI-N-OCTYL PHTHALATE
117-84-0 DIOCTYL PHTHALATE [MSL, NFP, NFP, PST, WHM]	DI-N-OCTYL PHTHALATE
117-81-7 DI-SEC OCTYL PHTHALATE (DI-2-ETHYL-HEXYL-PHTHALATE) [OS1, OS1, OS1]	DI(2-ETHYLHEXYL)PHTHALATE
20727-33-7 DIOCTYL SODIUM SULFOSUCCINATE	DIOCTYL SODIUM SULFOSUCCINATE
4654-26-6 DIOCTYL TEREPHTHALATE	DIOCTYL TEREPHTHALATE
26401-97-8 DIOCTYL TIN	DIOCTYL TIN
16091-18-2 DIOCTYLTIN MALEATE	DIOCTYLTIN MALEATE

Chemical Name	Reference Name
15571-58-1 DIOCTYLTIN BIS(2-ETHYLHEXYL THIOGLYCO-LATE)	DIOCTYLTIN BIS(2-ETHYLHEXYL THIOGLYCOLATE)
33568-99-9 DIOCTYLTIN BIS(ISOOCTYL MALEATE)	DIOCTYLTIN BIS(ISOOCTYL MALEATE)
26401-97-8 DIOCTYLTIN BIS(ISOOCTYL THIOGLYCOLATE) [MAK, MAK, MAK]	DIOCTYL TIN
3542-36-7 DIOCTYLTIN DICHLORIDE	DIOCTYLTIN DICHLORIDE
870-08-6 DIOCTYLTIN OXIDE	DIOCTYLTIN OXIDE
6988-21-2 DIOXACARB	DIOXACARB
123-91-1 1,4-DIOXANE	1,4-DIOXANE
123-91-1 DIOXANE [AUS, AUS]	1,4-DIOXANE
123-91-1 P-DIOXANE [CAL]	1,4-DIOXANE
123-91-1 DIOXANE [HM1, HM1, HM1]	1,4-DIOXANE
123-91-1 1,4-DIOXANE (1,4-DIETHYLENEOXIDE) [HON, HON]	1,4-DIOXANE
123-91-1 DIOXANE [MIN]	1,4-DIOXANE
123-91-1 P-DIOXANE [NFP, NFP]	1,4-DIOXANE
123-91-1 DIOXANE [NHS, NHS, NHS, NIO, NIO, NIO, NIO, OTV, TLV, TLV]	1,4-DIOXANE
123-91-1 1,4-DIOXANE (1,4-DIETHYLENEOXIDE) [CAA]	1,4-DIOXANE
123-91-1 DIOXANE (DIETHYLENE DIOXIDE) [OS1, OS1, OS1, OS1]	1,4-DIOXANE
123-91-1 DIOXANE (TECH. GRADE) [QBC, QBC, QBC]	1,4-DIOXANE
123-91-1 DIOXANE, TECH. GRADE [MEX, MEX, MEX]	1,4-DIOXANE
25136-55-4 1,4-DIOXANE, DIMETHYL- [PAL]	DIMETHYLDIOXANE
78-34-2 2,3-P-DIOXANEDITHIOL S,S-BIS(O,O-DI-ETHYLPHOSPHORODITHIOATE)(DIOXATHION) [CAL]	DIOXATHION
123-91-1 DIOXANE, TECH. GRADE [ALB, ALB, ALB, BC1, BC1]	1,4-DIOXANE
123-91-1 P-DIOXANE, TECH. GRADE [CEX, CEX]	1,4-DIOXANE
123-91-1 1,4-DIOXANE, TECHNICAL GRADE [GBR, GBR, GBR]	1,4-DIOXANE
828-00-2 1,3-DIOXAN-4-OL, 2,6-DIMETHYL-, AC-ETATE [TSC, TSC]	DIMETHOXANE
110843-98-6 1,5-DIOXASPIRO [5.5] UNDECANE-3,3-DICAR-BOXYLIC ACID BIS (1,2,2,6,6-PENTAMETHYL-4-PIPERIDINYL) ESTER	1,5-DIOXASPIRO [5.5] UNDECANE-3,3-DICARBOXYLIC ACID BIS (1,2EAMINE
78-34-2 DIOXATHION	DIOXATHION
78-34-2 DIOXATHION (DELNAV) [BC1, BC1, OS1, OS1]	DIOXATHION
78-34-2 DIOXATHION (DELNAV) (R) [QBC, QBC]	DIOXATHION
78-34-2 DIOXATHION (DELNAV) [ALB, ALB, ALB]	DIOXATHION
78-34-2 DIOXATHION (ISO) [GBR, GBR]	DIOXATHION
RR-01476-6 DIOXINS, POLYHALOGENATED DIBENZO-P-	DIOXINS, POLYHALOGENATED DIBENZO-P-
100-79-8 DIOXOLANE	DIOXOLANE
646-06-0 1,3-DIOXOLANE	1,3-DIOXOLANE
646-06-0 DIOXOLANE [HM1, HM1, MSL, NJL, NJL]	1,3-DIOXOLANE
126-39-6 1,3-DIOXOLANE, 2-ETHYL-2-METHYL- [PAL]	2-METHYL-2-ETHYL-1,3-DIOXOLANE
96-49-1 1,3-DIOXOLAN-2-ONE [PAL]	ETHYLENE CARBONATE
138-86-3 DIPENTENE	DIPENTENE
1941-79-3 DIPEROXYAZELAIC ACID	DIPEROXYAZELAIC ACID
RR-01756-1 DIPEROXY DODECANE DIACID	DIPEROXY DODECANE DIACID

Chemical Name	Reference Name
RR-01655-7 DIPEROXY KETAL	DIPEROXY KETAL
82-66-6 DIPHACINONE	DIPHACINONE
957-51-7 DIPHENAMID [313]	DIPHENAMIDE
957-51-7 DIPHENAMIDE	DIPHENAMIDE
147-24-0 DIPHENHYDRAMINE HYDROCHLORIDE	DIPHENHYDRAMINE HYDROCHLORIDE
94-75-7 DIPHENOXYACID-2,4 (2,4-D) [QBC, QBC]	2,4-D
3061-36-7 1,4-DIPHENOXYBENZENE	1,4-DIPHENOXYBENZENEDIHYDROXY-9,10-DIOXO-, DISODIUM SALT
RR-01757-2 DI-(2 PHENOXYETHYL) PEROXYDICARBONATE	DI-(2 PHENOXYETHYL) PEROXYDICARBONATE
92-52-4 DIPHENYL [BC1, BC1, HM1, NIO, NIO, NIO, NJL, WHM]	BIPHENYL
92-52-4 DIPHENYL (BIPHENYL) [OS1, OS1]	BIPHENYL
101-84-8 DIPHENYL OXIDE [HON]	PHENYL ETHER
104-15-4 TOLUENE SULFONIC ACIDS [HON]	P-TOLUENESULFONIC ACID
634-93-5 TRICHLOROANILINE (2,4,6-) [HON]	BENZENAMINE, 2,4,6-TRICHLORO-
26471-62-5 TOLUENE DIISOCYANATES (MIXTURE) [HON]	TOLUENE DIISOCYANATE
117-34-0 DIPHENYLACETIC ACID	DIPHENYLACETIC ACID
122-39-4 DIPHENYLAMINE	DIPHENYLAMINE
122-39-4 N,N-DIPHENYLAMINE [ONT, WHM]	DIPHENYLAMINE
578-94-9 DIPHENYLAMINECHLORO-ARSINE	DIPHENYLAMINECHLORO-ARSINE
578-94-9 DIPHENYLAMINE CHLOROARSINE [HM1, HM1, HM1]	DIPHENYLAMINECHLORO-ARSINE
RR-00120-7 1,1-DIPHENYLBUTANE	1,1-DIPHENYLBUTANE
712-48-1 DIPHENYLCHLOROARSINE	DIPHENYLCHLOROARSINE
3287-06-7 DIPHENYL DECYL PHOSPHITE	DIPHENYL DECYL PHOSPHITE
492-17-1 2,4'-DIPHENYLDIAMINE	2,4'-DIPHENYLDIAMINE
80-10-4 DIPHENYL DICHLOROSILANE	DIPHENYL DICHLOROSILANE
80-10-4 DIPHENYLDICHLOROSILANE [MSL, NFP, NFP, WHM]	DIPHENYL DICHLOROSILANE
RR-00853-7 DIPHENYLDODECYL PHOSPHITE	DIPHENYLDODECYL PHOSPHITE
103-29-7 1,2-DIPHENYLETHANE (SYM)	1,2-DIPHENYLETHANE (SYM)
38888-98-1 1,1-DIPHENYLETHANE (UNS)	1,1-DIPHENYLETHANE (UNS)
101-84-8 DIPHENYL ETHER [GBR, HM1, MAK, WHM]	PHENYL ETHER
RR-00209-5 DIPHENYL ETHER/BIPHENYL MIXTURE	DIPHENYL ETHER/BIPHENYL MIXTURE
102-06-7 1,3-DIPHENYLGUANIDINE	1,3-DIPHENYLGUANIDINE
57-41-0 DIPHENYLHYDANTOIN (PHENYTOIN) [C65, C65, CAL, CSR]	PHENYTOIN
630-93-3 DIPHENYLHYDANTOIN, SODIUM SALT	DIPHENYLHYDANTOIN, SODIUM SALT
630-93-3 DIPHENYLHYDANTOIN (PHENYTOIN), SODIUM SALT [C65, CAL]	DIPHENYLHYDANTOIN, SODIUM SALT
122-66-7 1,2-DIPHENYLHYDRAZINE [ATS, CAA, CW2]	HYDRAZOBENZENE
122-66-7 DIPHENYLHYDRAZINE [CW3]	HYDRAZOBENZENE
122-66-7 1,2-DIPHENYLHYDRAZINE [EP2, EPA, HON, MX1, MX2]	HYDRAZOBENZENE
122-66-7 1,2-DIPHENYL HYDRAZINE [NJL, NJL]	HYDRAZOBENZENE
122-66-7 1,2-DIPHENYLHYDRAZINE [RCU, RCU, SD4, UTS]	HYDRAZOBENZENE
55299-18-8 DIPHENYLHYDRAZINE	DIPHENYLHYDRAZINE
122-66-7 1,2-DIPHENYLHYDRAZINE (HYDRAZOBEN-ZENE) [313]	HYDRAZOBENZENE
26544-23-0 DIPHENYL ISODECYL PHOSPHITE	DIPHENYL ISODECYL PHOSPHITE

Chemical Name	Reference Name
587-85-9 DIPHENYLMERCURY	DIPHENYLMERCURY
101-81-5 DIPHENYLMETHANE	DIPHENYLMETHANE
101-68-8 DIPHENYLMETHANE DIISOCYANATE (METHY-LENE BISPHENYLISOCYANATE) [ALB, ALB]	METHYLENE BISPHENOL ISOCYANATE (MDI)
101-68-8 DIPHENYLMETHANE-4,4'-DIISO-CYANATE [HM1, HM1, MAK, MAK, MAK, MAK]	METHYLENE BISPHENOL ISOCYANATE (MDI)
101-68-8 DIPHENYLMETHANE DIISOCYANATE (MDI) [NHS, NHS]	METHYLENE BISPHENOL ISOCYANATE (MDI)
101-68-8 4,4'-DIPHENYLMETHANE DIISO-CYANATE [WHM]	METHYLENE BISPHENOL ISOCYANATE (MDI)
5873-54-1 2,4'-DIPHENYLMETHANE DIISOCYANATE	2,4'-DIPHENYLMETHANE DIISOCYANATE
RR-01224-8 DIPHENYLMETHANE DIISOCYANATE (MDI) MODIFIED	DIPHENYLMETHANE DIISOCYANATE (MDI) MODIFIED
776-74-9 DIPHENYL METHYL BROMIDE	DIPHENYL METHYL BROMIDE
776-74-9 DIPHENYLMETHYL BROMIDE [NJL, NJL]	DIPHENYL METHYL BROMIDE
86-30-6 DIPHENYLNITROSAMINE	DIPHENYLNITROSAMINE
101-84-8 DIPHENYL OXIDE [NFP, NFP]	PHENYL ETHER
RR-01286-2 DIPHENYL OXIDE AND BIPHENYL PHENYL ETHER MIXTURES	DIPHENYL OXIDE AND BIPHENYL PHENYL ETHER MIXTURES
80-51-3 DIPHENYLOXIDE-4,4'-DISULFOHYDRAZIDE	DIPHENYLOXIDE-4,4'-DISULFOHYDRAZIDE
RR-00774-9 1,1-DIPHENYLPENTANE	1,1-DIPHENYLPENTANE
74-31-7 N,N'-DIPHENYL-P-PHENYLENEDIAMINE	N,N'-DIPHENYL-P-PHENYLENEDIAMINE
4712-55-4 DIPHENYL PHOSPHITE	DIPHENYL PHOSPHITE
84-62-8 DIPHENYL PHTHALATE	DIPHENYL PHTHALATE
RR-00627-9 1,1-DIPHENYLPROPANE	1,1-DIPHENYLPROPANE
127-63-9 DIPHENYLSULFONE	DIPHENYLSULFONE
75980-60-8 DIPHENYL-2,4,6-TRIMETHYLBENZOYL PHOS-PHINE OXIDE	DIPHENYL-2,4,6-TRIMETHYLBENZOYL PHOSPHINE OXIDE
RR-00852-6 DIPHENYL (O-XENYL) PHOSPHATE	DIPHENYL (O-XENYL) PHOSPHATE
152-16-9 DIPHOSPHORAMIDE, OCTAMETHYL- [FLA, 302, NJL, PAL]	OCTAMETHYLPYROPHOSPHORAMIDE
7758-16-9 DIPHOSPHORIC ACID, DISODIUM SALT	DIPHOSPHORIC ACID, DISODIUM SALT
107-49-3 DIPHOSPHORIC ACID, TETRAETHYL ES-TER [EPA, PAL, RCU, RCU]	TEPP
7722-88-5 DIPHOSPHORIC ACID, TETRASODIUM SALT [PAL, PST]	TETRASODIUM PYROPHOSPHATE
1314-80-3 DIPHOSPHORUS PENTASULPHIDE [GBR, GBR]	PHOSPHORUS PENTASULFIDE
2217-06-3 DIPICRYL SULFIDE	DIPICRYL SULFIDE
2164-07-0 DIPOTASSIUM ENDOTHALL [7-OXABICYCLO (2.2.1)HEPTANE-2,3-DICARBOXYLIC ACID, DIPOTASSIUM SALT]	DIPOTASSIUM ENDOTHALL [7-OXABICYCLO(2.2.1)HEP-TANE-2,3-DICARB
16920-93-7 DIPOTASSIUM HEXABROMOPLATINATE	DIPOTASSIUM HEXABROMOPLATINATE
16919-73-6 DIPOTASSIUM HEXACHLOROPALLADATE	DIPOTASSIUM HEXACHLOROPALLADATE
7727-21-1 DIPOTASSIUM PEROXYDISULPHATE [GBR]	POTASSIUM PERSULFATE
7758-11-4 DIPOTASSIUM PHOSPHATE	DIPOTASSIUM PHOSPHATE
53404-44-7 DIPOTASSIUM 12-SULFATO-9-OCTADE-CENOATE	DIPOTASSIUM 12-SULFATO-9-OCTADECENOATE
13826-93-2 DIPOTASSIUM TETRABROMOPALLADATE	DIPOTASSIUM TETRABROMOPALLADATE
3248-28-0 DIPROPIONYL PEROXIDE [DOT, HM1, NJL]	PROPIONYL PEROXIDE
2036-15-9 DIPROPYLALUMINUM HYDRIDE	DIPROPYLALUMINUM HYDRIDE
142-84-7 DIPROPYLAMINE	DIPROPYLAMINE
RR-00766-9 4-DIPROPYLAMINOBENZENEDIAZONIUM ZINC CHLORIDE	4-DIPROPYLAMINOBENZENEDIAZONIUM ZINC CHLO-RIDEZINC CHLORIDE

Chemical Name	Reference Name
110-98-5 DIPROPYLENE GLYCOL [HON, TSC, TSC, TSC]	2-PROPANOL, 1,1'-OXYBIS
25265-71-8 DIPROPYLENE GLYCOL	DIPROPYLENE GLYCOL
29911-28-2 DIPROPYLENE GLYCOL BUTYL ETHER [TSC, TSC, TSC]	2-PROPANOL, 1-(2-BUTOXY-1-METHYLETHOXY)-
94-51-9 DIPROPYLENE GLYCOL DIBENZOATE	DIPROPYLENE GLYCOL DIBENZOATE
34590-94-8 DIPROPYLENE GLYCOL METHYL ETHER [BC1, BC1, BC1]	DIPROPYLENE GLYCOL MONOMETHYL ETHER
20324-32-7 DIPROPYLENE GLYCOL METHYL ETHER	DIPROPYLENE GLYCOL METHYL ETHER
34590-94-8 DIPROPYLENE GLYCOL, METHYL ETHER [AUS, AUS, AUS]	DIPROPYLENE GLYCOL MONOMETHYL ETHER
34590-94-8 DIPROPYLENE GLYCOL METHYL ETHER [MIN, MSL, NIO, NIO, NIO, OS1, OS1, OS1, TLV, TLV, TLV, WHM, CEX, CEX, CEX]	DIPROPYLENE GLYCOL MONOMETHYL ETHER
30025-38-8 DIPROPYLENE GLYCOL MONOETHYL ETHER	DIPROPYLENE GLYCOL MONOETHYL ETHER
34590-94-8 DIPROPYLENE GLYCOL MONOMETHYL ETHER	DIPROPYLENE GLYCOL MONOMETHYL ETHER
34590-94-8 DIPROPYLENE GLYCOL, MONOMETHYL ETHER [MAK, MAK]	DIPROPYLENE GLYCOL MONOMETHYL ETHER
88917-22-0 DIPROPYLENE GLYCOL MONOMETHYL ETHER ACETATE [TSC, TSC, TSC]	PROPANOL 1 (OR 2)-2-METHOXYMETHYLETHOXY, ACETATE
RR-01070-8 DIPROPYLENETRIAMINE AMIDE OF TALL-OIL FATTY ACID REACTED WITH TALL-OIL FATTY ACID-POLYETHYLENE GLYCOL ESTER AND MALEIC ANHYDRIDE	DIPROPYLENETRIAMINE AMIDE OF TALL-OIL FATTY ACID REACTED WIT
111-43-3 DIPROPYL ETHER	DIPROPYL ETHER
136-45-8 DIPROPYL ISOCINCHOMERONATE	DIPROPYL ISOCINCHOMERONATE
123-19-3 DIPROPYL KETONE	DIPROPYL KETONE
621-64-7 DI-N-PROPYLNITROSAMINE [EPA, PQL, RCA]	N-NITROSODI-N-PROPYLAMINE
621-64-7 DI-N-PROPYLNITROSOAMINE [RCU, RCU]	N-NITROSODI-N-PROPYLAMINE
16066-38-9 DI-N-PROPYL PEROXYDICARBONATE	DI-N-PROPYL PEROXYDICARBONATE
85-00-7 DIPYRIDO[1,2-A:2',1'-C]PYRAZINEDIIUM, 6,7-DIHYDRO-,DIBROMIDE [PAL]	DIQUAT DIBROMIDE
2764-72-9 DIPYRIDO[1,2-A:2',1'-C]PYRAZINEDIIUM, 6,7-DIHYDRO- [PAL]	DIQUAT
85-00-7 DIQUAT [ALB, ALB, AUS, CEX, CWA, FLA, MEX, MEX, MIN, ONT, OS1, SDW, SDW]	DIQUAT DIBROMIDE
2764-72-9 DIQUAT	DIQUAT
2764-72-9 DIQUAT (REGLONE) [BC1, BC1]	DIQUAT
85-00-7 DIQUAT DIBROMIDE	DIQUAT DIBROMIDE
2764-72-9 DIQUAT DIBROMIDE [CWA]	DIQUAT
2764-72-9 DIQUAT, DIBROMIDE [NJL]	DIQUAT
85-00-7 DIQUAT DIBROMIDE (ISO) [GBR, GBR]	DIQUAT DIBROMIDE
1937-37-7 DIRECT BLACK 38 [FLA, MSL, NTP]	C.I. DIRECT BLACK 38
1937-37-7 DIRECT BLACK 38 (TECHNICAL GRADE) [C65, C65, IAR, MIN]	C.I. DIRECT BLACK 38
1937-37-7 DIRECT BLACK META [WHM]	C.I. DIRECT BLACK 38
2602-46-2 DIRECT BLUE 6 [FLA, MSL, NTP]	C.I. DIRECT BLUE 6
2602-46-2 DIRECT BLUE 6 (TECHNICAL GRADE) [C65, C65]	C.I. DIRECT BLUE 6
2602-46-2 DIRECT BLUE 6 (TECHNICAL-GRADE) [IAR]	C.I. DIRECT BLUE 6
2602-46-2 DIRECT BLUE 6 (TECHNICAL GRADE) [MIN, WHM]	C.I. DIRECT BLUE 6
314-13-6 DIRECT BLUE 53	DIRECT BLUE 53

Chemical Name	Reference Name
16071-86-6 DIRECT BROWN 95 [FLA, MSL]	C.I. DIRECT BROWN 95
16071-86-6 DIRECT BROWN 95 (TECHNICAL GRADE) [MIN, WHM, C65, C65, IAR]	C.I. DIRECT BROWN 95
1937-37-7 DIRECT BLACK 38 [CAB]	C.I. DIRECT BLACK 38
1937-37-7 DIRECT BLACK 38 (TECHNICAL GRADE) [CAL]	C.I. DIRECT BLACK 38
2602-46-2 DIRECT BLUE 6 [CAB]	C.I. DIRECT BLUE 6
2602-46-2 DIRECT BLUE 6 (TECHNICAL GRADE) [CAL]	C.I. DIRECT BLUE 6
16071-86-6 DIRECT BROWN 95 (TECH. GRADE) [CAB]	C.I. DIRECT BROWN 95
16071-86-6 DIRECT BROWN 95 [CAL]	C.I. DIRECT BROWN 95
94-91-7 N,N'-DISALICYLIDENE-1,2-DIAMINO-PROPANE [PST]	N,N'-DISALICYLIDENE-1,2-PROPANEDIAMINE
94-91-7 N,N'-DISALICYLIDENE-1,2-PROPANEDIAMINE	N,N'-DISALICYLIDENE-1,2-PROPANEDIAMINE
117-81-7 DI-SEC-OCTYL PHTHALATE [ALB, ALB, MEX, MEX]	DI(2-ETHYLHEXYL)PHTHALATE
126-80-7 DISILOXANE, 1,1,3,3-TETRAMETHYL-1,3-BIS[3-OXIRANYLMETHOXY)PROPYL]-	DISILOXANE, 1,1,3,3-TETRAMETHYL-1,3-BIS[3-OXI-RANYLMETHOXY)
RR-00264-2 DISINFECTANTS, N.O.S.	DISINFECTANTS, N.O.S.
27344-41-8 DISODIUM 4,4'-BIS(2-SULFOSTYRYL) BIPHENYL [WHM]	TINOPAL CBS
62-33-9 DISODIUM CALCIUM EDTA	DISODIUM CALCIUM EDTA
14025-15-1 DISODIUM CUPRIC ETHYLENEDIAMINETE-TRAACETATE [PST]	CUPRATE(2-), ((ETHYLENEDINITRILO)TETRAACETATO)-, DISODIUM
138-93-2 DISODIUM CYANODITHIOIMIDOCARBONATE	DISODIUM CYANODITHIOIMIDOCARBONATE
7681-57-4 DISODIUM DISULPHITE [GBR]	SODIUM METABISULFITE
28519-02-0 DISODIUM DODECYLDIPHENYL ETHER DISUL-FONATE	DISODIUM DODECYLDIPHENYL ETHER DISULFONATE
RR-01071-9 DISODIUM DODECYLIMIDAZOLINIUM DICAR-BOXYLATE	DISODIUM DODECYLIMIDAZOLINIUM DICARBOXY-LATEH TALL-OIL FATTY ACID-POLYETHYLENE GLY-COL ESTER AND MALEIC ANHYDRIDE
7575-62-4 DISODIUM 4-DODECYL-2,4'-OXYDIBENZENE-SULFONATE	DISODIUM 4-DODECYL-2,4'-OXYDIBENZENESULFONATE
12068-17-6 DISODIUM AR-(DODECYLPHENOXY)BENZENE DISULFONATE	DISODIUM AR-(DODECYLPHENOXY)BENZENE DISUL-FONATE
139-33-3 DISODIUM EDTA	DISODIUM EDTA
139-33-3 DISODIUM ETHYLENEDIAMINETETRAAC-ETATE [PST]	DISODIUM EDTA
14025-21-9 DISODIUM [(ETHYLENEDINITRILO)TETRAAC-ETATO] ZINC	DISODIUM [(ETHYLENEDINITRILO)TETRAACETATO] ZINC
10048-95-0 DISODIUM HYDROGEN ARSENATE	DISODIUM HYDROGEN ARSENATE
14729-89-6 DISODIUM IRON(II) ETHYLENEDIAMINETE-TRAACETATE	DISODIUM IRON(II) ETHYLENEDIAMINETETRAACETATE
15375-84-5 DISODIUM MANGANESE ETHYLENEDI-AMINETETRAACETATE	DISODIUM MANGANESE ETHYLENEDIAMINETETRAAC-ETATE
12008-41-2 DISODIUM OCTABORATE [PST]	BORON SODIUM OXIDE (B8NA2O13)
7782-85-6 DISODIUM ORTHOPHOSPHATE HEPTAHY-DRATE	DISODIUM ORTHOPHOSPHATE HEPTAHYDRATE
25167-32-2 DISODIUM 2,2'-OXYBIS(4-DODECYLBENZENE-SULFONATE) [PST]	BENZENESULFONIC ACID, OXYBIS[DODECYL-,DIS-ODIUM SALT]
7775-27-1 DISODIUM PEROXODISULPHATE [GBR]	SODIUM PERSULFATE
7558-79-4 DISODIUM PHOSPHATE [PST]	SODIUM PHOSPHATE DIBASIC
39354-45-5 DISODIUM SALT OF LAURYL ALCOHOL POLYETHYLENE GLYCOL ETHER SULFOSUC-CINATE	DISODIUM SALT OF LAURYL ALCOHOL POLYETHY-LENE GLYCOL ETHER SU
1330-43-4 DISODIUM TETRABORATE [GBR]	BORATES, TETRA, SODIUM SALTS, ANHYDROUS

Chemical Name	Reference Name
94-11-1 2,4-D ISOPROPYL ESTERS [313]	2,4-D ESTERS
RR-00267-5 DISPERSANT GAS, N.O.S.	DISPERSANT GAS, N.O.S.
2475-45-8 DISPERSE BLUE 1 [C65, C65, CAB, CAL, IAR]	C.I. DISPERSE BLUE 1
2832-40-8 DISPERSE YELLOW 3 [IAR]	C.I. DISPERSE YELLOW 3
107-64-2 DISTEARYLDIMETHYLAMMONIUM CHLORIDE	DISTEARYLDIMETHYLAMMONIUM CHLORIDE
52326-66-6 DISTEARYL PEROXYDI-CARBONATE	DISTEARYL PEROXYDI-CARBONATE
52326-66-6 DISTEARYL PEROXY DI-CARBONATE [NJL]	DISTEARYL PEROXYDI-CARBONATE
RR-01550-9 DISTILLATE (LIGHT) FUELS	DISTILLATE (LIGHT) FUELS
65996-91-0 DISTILLATES (COAL TAR), UPPER [PAL]	COAL TAR LIGHT OIL
64742-14-9 DISTILLATES (PETROLEUM), ACID TREATED LIGHT [PST]	ACID TREATED LIGHT DISTILLATE (PETROLEUM)
RR-00939-2 DISTILLATES (PETROLEUM), C(3-6), POLYMERS WITH STYRENE AND MIXED TERPENES	DISTILLATES (PETROLEUM), C(3-6), POLYMERS WITH STYRENE AND M
68477-30-5 DISTILLATES (PETROLEUM), CATALYTIC RE-FORMER FRACTIONATORRESIDUE, INTERME-DIATE BOILING	DISTILLATES (PETROLEUM), CATALYTIC REFORMER FRACTIONATOR
64741-53-3 DISTILLATES (PETROLEUM), HEAVY NAPTHENIC [PST]	PETROLEUM DISTILLATES, HEAVY NAPHTHENIC
64741-51-1 DISTILLATES (PETROLEUM), HEAVY PARAF-FINIC [PST]	PETROLEUM DISTILLATES, HEAVY PARAFFINIC
64741-52-2 DISTILLATES (PETROLEUM), LIGHT NAPH-THENIC [PST]	LIGHT NAPHTHENIC OIL
64741-50-0 DISTILLATES (PETROLEUM), LIGHT PARAF-FINIC [PST]	PETROLEUM DISTILLATES, LIGHT PARAFFINIC
64742-64-9 DISTILLATES (PETROLEUM), SOLVENT-DE-WAXED LIGHT NAPHTHENIC	DISTILLATES (PETROLEUM), SOLVENT-DEWAXED LIGHT NAPHTHENICPARAFFINIC
64741-97-5 DISTILLATES (PETROLEUM), SOLVENT RE-FINED LIGHT NAPHTHENIC	DISTILLATES (PETROLEUM), SOLVENT REFINED LIGHT NAPHTHENICPARAFFINIC
64741-44-2 DISTILLATES (PETROLEUM), STRAIGHT-RUN MIDDLE	DISTILLATES (PETROLEUM), STRAIGHT-RUN MIDDLE
RR-00196-7 DISUBSTITUTED ALKYL TRIAZINES	DISUBSTITUTED ALKYL TRIAZINES
RR-00265-3 DISUBSTITUTED DIAMINO ANISOLE	DISUBSTITUTED DIAMINO ANISOLE
RR-01675-1 DISUBSTITUTED DIPHENYLSULFONE	DISUBSTITUTED DIPHENYLSULFONE
RR-00921-2 DISUBSTITUTED NITROBENZENE	DISUBSTITUTED NITROBENZENE
RR-01206-6 DISUBSTITUTED PHENOXAZINE, CHLOROMET-ALATE SALT	DISUBSTITUTED PHENOXAZINE, CHLOROMETALATE SALTLKYL, AMINO ALKYL METHACRYLAMIDE SALT
RR-01207-7 FATTY ACID POLYAMINE CONDENSATE, PH-SOPHORIC ACID ESTER SALT	FATTY ACID POLYAMINE CONDENSATE, PHSOPHORIC ACID ESTER SALT
RR-01213-5 DISUBSTITUTED PHENYLAZO TRISUBSTI-TUTED NAPHTHALENE	DISUBSTITUTED PHENYLAZO TRISUBSTITUTED NAPH-THALENE
RR-00268-6 DISUCCINIC ACID PEROXIDE, 72% IN WATER	DISUCCINIC ACID PEROXIDE, 72% IN WATER
123-23-9 DISUCCINIC ACID PEROXIDE [DOT]	SUCCINIC ACID PEROXIDE
624-92-0 DISULFIDE, DIMETHYL (DIMETHYL DISULFIDE) [CAI]	DIMETHYLDISULFIDEE)
2179-59-1 DISULFIDE, 2-PROPENYL PROPYL [PAL]	ALLYL PROPYL DISULFIDE
97-77-8 DISULFIRAM	DISULFIRAM
RR-00906-3 DISULFONIC ACID ROSIN AMINE SALT OF A BENZIDINE DERIVATIVE	DISULFONIC ACID ROSIN AMINE SALT OF A BENZIDINE DERIVATIVE
298-04-4 DISULFOTON	DISULFOTON
298-04-4 DISULFOTON [QBC]	DISULFOTON
298-04-4 DISULFOTON (ISO) [GBR, GBR]	DISULFOTON
7681-57-4 DISULFUROUS ACID, DISODIUM SALT [PAL]	SODIUM METABISULFITE
7757-74-6 DISULFUROUS ACID, DISODIUM SALT	DISULFUROUS ACID, DISODIUM SALT

Chemical Name	Reference Name
7681-57-4 DISULFUROUS ACID, DISODIUM SALT [PST]	SODIUM METABISULFITE
5714-22-7 DISULPHUR DECAFLUORIDE [GBR, GBR]	SULFUR PENTAFLUORIDE
10025-67-9 DISULPHUR DICHLORIDE [GBR]	SULFUR MONOCHLORIDE
298-04-4 DISYSTON [BC1, BC1, BC1]	DISULFOTON
68783-78-8 DITALLOW DIMETHYL AMMONIUM CHLORIDE	DITALLOW DIMETHYL AMMONIUM CHLORIDE
68410-69-5 N,N'-DITALLOW N"-METHYL N""-POLYETHOXYAMIDO AMMONIUM METHOSULFATE [PST]	POLY(OXY-1,2-ETHANEDIYL), ALPHA-[2-[BIS(2-AMINOETHYL)METHYL
128-37-0 2,6-DITERT. BUTYL-P-CRESOL [BC1, BC1]	2,6-DI-TERT-BUTYL-P-CRESOL
25377-21-3 2,6-DITERTIARY-BUTYL-PARA-CRESOL [GBR]	DI-TERT-BUTYL-P-CRESOL
514-73-8 DITHIAZANINE IODIDE	DITHIAZANINE IODIDE
638-17-5 4H-1,3,5-DITHIAZINE, DIHYDRO-2,4,6-TRIMETHYL-, (2.ALPHA.,4.ALPHA.,6.ALPHA.)-[PAL]	THIALDINE
142-46-1 2,5-DITHIOBIUREA	2,5-DITHIOBIUREA
541-53-7 DITHIOBIURET	DITHIOBIURET
541-53-7 2,4-DITHIOBIURET [313]	DITHIOBIURETNE)
RR-00269-7 DITHIOCARBAMATE PESTICIDES, N.O.S.	DITHIOCARBAMATE PESTICIDES, N.O.S.
103-34-4 4,4'-DITHIODIMORPHOLINE	4,4'-DITHIODIMORPHOLINE
26419-73-8 1,3-DITHIOLANE-2-CARBOXYALDEHYDE, 2,4-DIMETHYL,O-(METHYLCARBAMOYL) OXIME [MSL]	CARBAMIC ACID, METHYL-, O-(((2,4-DIMETHYL-1,3-DITHIOLAN-2-YL
7775-14-6 DITHIONOUS ACID, DISODIUM SALT [PAL]	SODIUM DITHIONITE
7779-86-4 DITHIONOUS ACID, ZINC SALT (1:1) [PAL]	ZINC HYDROSULFITE
480-22-8 DITHRANOL	DITHRANOL
97-39-2 1,3-DI-O-TOLYLGUANIDINE	1,3-DI-O-TOLYLGUANIDINE
119-06-2 DITRIDECYL PHTHALATE [NFP, NFP, TSE]	1,2-BENZENEDICARBOXYLIC ACID, DITRIDECYL ESTER
68515-47-9 DITRIDECYL PHTHALATE (MIXED ISOMERS)	DITRIDECYL PHTHALATE (MIXED ISOMERS)
RR-00273-3 DI(3,5,5-TRIMETHYL-1,2-DIOXOLANYL-3) PEROXIDE, PASTE	DI(3,5,5-TRIMETHYL-1,2-DIOXOLANYL-3) PEROXIDE, PASTE
3851-87-4 DI(3,5,5-TRIMETHYLHEXANOYL) PEROXIDE	DI(3,5,5-TRIMETHYLHEXANOYL) PEROXIDE
3648-20-2 DIUNDECYL PHTHALATE [TSE]	1,2-BENZENEDICARBOXYLIC ACID, DIUNDECYL ESTER
330-54-1 DIURON	DIURON
330-54-1 DIURON (ISO) [GBR]	DIURON
22541-79-3 DIVALENT CHROMIUM COMPOUNDS [ONT]	CHROMIUM (II)
1314-62-1 DIVANADIUM PENTAOXIDE [GBR]	VANADIUM PENTOXIDE
821-08-9 DIVINYL ACETYLENE	DIVINYL ACETYLENE
108-57-6 DIVINYL BENZENE	DIVINYL BENZENE
108-57-6 DIVINYLBENZENE [GBR]	DIVINYL BENZENE
108-57-6 M-DIVINYL BENZENE [MSL]	DIVINYL BENZENE
108-57-6 DIVINYLBENZENE [NFP, NFP]	DIVINYL BENZENE
108-57-6 M-DIVINYLBENZENE [ONT, WHM]	DIVINYL BENZENE
1321-74-0 DIVINYL BENZENE	DIVINYL BENZENE
1321-74-0 DIVINYLBENZENE [CSR]	DIVINYL BENZENE
109-93-3 DIVINYL ETHER	DIVINYL ETHER
RR-00831-1 DI(O-XENYL) PHENYL PHOSPHATE	DI(O-XENYL) PHENYL PHOSPHATE
534-52-1 DNOC [HM1, HM1, HM1, HM1]	4,6-DINITRO-O-CRESOL
18766-38-6 DOCOSAMETHYLCYLCLOUNDECASILOXANE	DOCOSAMETHYLCYLCLOUNDECASILOXANE
556-70-7 DOCOSAMETHYLDECASILOXANE	DOCOSAMETHYLDECASILOXANE

Chemical Name	Reference Name
2385-85-5 DODECACHLOROPENTACYCLO [5.3.0.0.0.0] DECANE [CN5]	MIREX
540-97-6 DODECAMETHYLCYCLOHEXASILOXANE	DODECAMETHYLCYCLOHEXASILOXANE
141-63-9 DODECAMETHYLPENTASILOXANE	DODECAMETHYLPENTASILOXANE
112-54-9 DODECANAL [TSC, TSC, TSC]	LAURYL ALDEHYDE
112-40-3 DODECANE	DODECANE
693-23-2 DODECANEDIOIC ACID	DODECANEDIOIC ACID
112-55-0 1-DODECANETHIOL	1-DODECANETHIOL
142-18-7 DODECANOIC ACID, 2,3-DIHYDROXYPROPYL ESTER	DODECANOIC ACID, 2,3-DIHYDROXYPROPYL ESTER
111-82-0 DODECANOIC ACID, METHYL ESTER	DODECANOIC ACID, METHYL ESTER
5350-03-8 DODECANOIC ACID, PENTYL ESTER	DODECANOIC ACID, PENTYL ESTER
112-53-8 1-DODECANOL [NFP, NFP]	LAURYL ALCOHOL
112-53-8 DODECANOL, 1- [OTS]	LAURYL ALCOHOL
112-53-8 1-DODECANOL [PST]	LAURYL ALCOHOL
25103-58-6 TERT-DODECANTHIOL [PAL]	TERT-DODECYL MERCAPTAN
112-41-4 1-DODECENE	1-DODECENE
85081-53-4 DODECENYLSUCCINIC ACID, MONOTRIDECYL ESTER	DODECENYLSUCCINIC ACID, MONOTRIDECYL ESTER
19780-11-1 DODECENYLSUCCINIC ANHYDRIDE	DODECENYLSUCCINIC ANHYDRIDE
123-01-3 DODECYL BENZENE (N-) [HON]	DODECYL BENZENE (CRUDE)
121158-58-5 DODECYL PHENOL (BRANCHED)	DODECYL PHENOL (BRANCHED)
9002-92-0 DODECYL ALCOHOL, ETHOXYLATED [PST]	LAURYL POLYETHYLENE GLYCOL ETHER
26183-44-8 DODECYL ALCOHOL, ETHOXYLATED AND SULFATED	DODECYL ALCOHOL, ETHOXYLATED AND SULFATED
124-22-1 DODECYLAMINE	DODECYLAMINE
70955-37-2 DODECYL AND HIGHER ALIPHATIC KETONES	DODECYL AND HIGHER ALIPHATIC KETONES
28675-17-4 DODECYLANILINE	DODECYLANILINE
123-01-3 DODECYLBENZENE [OTS]	DODECYL BENZENE (CRUDE)
123-01-3 DODECYL BENZENE (CRUDE)	DODECYL BENZENE (CRUDE)
1886-81-3 DODECYL BENZENESULFONATE	DODECYL BENZENESULFONATE
27176-87-0 DODECYLBENZENESULFONIC ACID	DODECYLBENZENESULFONIC ACID
27176-87-0 DODECYLBENZENE SULFONIC ACID [NJL, NJL]	DODECYLBENZENESULFONIC ACID
68084-55-9 DODECYLBENZENESULFONIC ACID, N-(2-AMINOETHYL)ETHANOLAMINE SALT	DODECYLBENZENESULFONIC ACID, N-(2-AMINOETHYL) ETHANOLAMINE SA
1331-61-9 DODECYLBENZENESULFONIC ACID, AMMO-NIUM SALT	DODECYLBENZENESULFONIC ACID, AMMONIUM SALT
12068-09-6 DODECYLBENZENESULFONIC ACID, BUTY-LAMINE SALT	DODECYLBENZENESULFONIC ACID, BUTYLAMINE SALT
26264-06-2 DODECYLBENZENESULFONIC ACID, CALCIUM SALT [PST]	CALCIUM DODECYLBENZENE SULFONATE
26545-53-9 DODECYLBENZENESULFONIC ACID, DI-ETHANOLAMINE SALT	DODECYLBENZENESULFONIC ACID, DIETHANOLAMINE SALT
29061-61-8 DODECYLBENZENESULFONIC ACID, DIISO-PROPYLAMINE SALT	DODECYLBENZENESULFONIC ACID, DIISOPROPY-LAMINE SALT
60816-39-9 DODECYLBENZENESULFONIC ACID, N,N-DIMETHYL-1,3-PROPANEDIAMINESALT	DODECYLBENZENESULFONIC ACID, N,N-DIMETHYL-1,3-PROPANEDIAMINE
67952-66-3 DODECYLBENZENESULFONIC ACID, ETHYLENEDIAMINE SALT	DODECYLBENZENESULFONIC ACID, ETHYLENEDI-AMINE SALT
26264-05-1 DODECYLBENZENESULFONIC ACID, ISO-PROPYLAMINE SALT	DODECYLBENZENESULFONIC ACID, ISOPROPYLAMINE SALT
12068-08-5 DODECYLBENZENESULFONIC ACID, MORPHO-LINE SALT	DODECYLBENZENESULFONIC ACID, MORPHOLINE SALT

Chemical Name	Reference Name
60816-37-7 DODECYLBENZENESULFONIC ACID, 1,3-PROPANEDIAMINE SALT	DODECYLBENZENESULFONIC ACID, 1,3-PROPANEDI-AMINE SALT
25155-30-0 DODECYLBENZENESULFONIC ACID, SODIUM SALT [PST]	SODIUM DODECYLBENZENESULFONATE
RR-01073-1 DODECYLBENZENESULFONIC ACID, 1,1,2,3-TETRAMETHYLBUTYLAMINE SALT	DODECYLBENZENESULFONIC ACID, 1,1,2,3-TETRAM-ETHYLBUTYLAMINE S
12068-15-4 DODECYLBENZENESULFONIC ACID, STRON-TIUM SALT	DODECYLBENZENESULFONIC ACID, STRONTIUM SALT
27323-41-7 DODECYLBENZENESULFONIC ACID, TRI-ETHANOLAMINE SALT [PST]	TRIETHANOLAMINE DODECYLBENZOSULFONATE
29061-63-0 DODECYLBENZENESULFONIC ACID, TRIETHY-LAMINE SALT	DODECYLBENZENESULFONIC ACID, TRIETHYLAMINE SALT
12068-16-5 DODECYLBENZENESULFONIC ACID, ZINC SALT	DODECYLBENZENESULFONIC ACID, ZINC SALT
683-10-3 DODECYLBETAINE	DODECYLBETAINE
59227-89-3 N-DODECYLCAPROLACTAM	N-DODECYLCAPROLACTAM
6843-97-6 DODECYLDI(AMINOETHYL)GLYCINE	DODECYLDI(AMINOETHYL)GLYCINE
RR-01072-0 DODECYL DIMETHYL BENZYL AMMONIUM NAPHTHENATE	DODECYL DIMETHYL BENZYL AMMONIUM NAPHTHEN-ATE
6842-15-5 DODECYLENE (ALPHA) [NFP, NFP]	PROPYLENE TETRAMER
2439-10-3 DODECYLGUANIDINE MONOACETATE	DODECYLGUANIDINE MONOACETATE]
3655-00-3 3,3'-(DODECYLIMINO)DIPROPIONIC ACID, DISODIUM SALT	3,3'-(DODECYLIMINO)DIPROPIONIC ACID, DISODIUM SALT
112-55-0 DODECYL MERCAPTAN [PST]	1-DODECANETHIOL
25103-58-6 TERT-DODECYL MERCAPTAN	TERT-DODECYL MERCAPTAN
25719-52-2 DODECYL 2-METHYLACRYLATE POLYMER	DODECYL 2-METHYLACRYLATE POLYMER
2985-59-3 4-DODECYLOXY-2-HYDROXYBENZOPHENONE	4-DODECYLOXY-2-HYDROXYBENZOPHENONE
9004-82-4 DODECYLOXYPOLY(ETHYLENEOXY) ETHYL SULFATE, SODIUM SALT	DODECYLOXYPOLY(ETHYLENEOXY) ETHYL SULFATE, SODIUM SALT
27193-86-8 DODECYLPHENOL	DODECYLPHENOL
27193-86-8 DODECYL PHENOL [NFP, NFP]	DODECYLPHENOL
67993-50-4 DODECYLPHENOXYBENZENE DISULFONIC ACID	DODECYLPHENOXYBENZENE DISULFONIC ACID
9014-92-0 ALPHA-(DODECYLPHENYL)-OMEGA-HYDROXY-POLY(OXY-1,2-ETHANEDIYL)	ALPHA-(DODECYLPHENYL)-OMEGA-HYDROXY-POLY (OXY-1,2-ETHANEDIYL)
7631-98-3 N-DODECYLSARCOSINE, SODIUM SALT	N-DODECYLSARCOSINE, SODIUM SALT
2235-54-3 DODECYL SULFATE AMMONIUM SALT [PST]	AMMONIUM LAURYL SULFATE
3097-08-3 DODECYL SULFATE, MAGNESIUM SALT	DODECYL SULFATE, MAGNESIUM SALT
151-21-3 DODECYL SULFATE, SODIUM SALT [PST]	SODIUM LAURYL SULFATE
4706-78-9 DODECYL SULFURIC ACID, POTASSIUM SALT	DODECYL SULFURIC ACID, POTASSIUM SALT
3614-12-8 N-DODECYL-N-TETRADECYL BETA-ALANINE	N-DODECYL-N-TETRADECYL BETA-ALANINE
4484-72-4 DODECYL TRICHLOROSILANE	DODECYL TRICHLOROSILANE
2439-10-3 DODINE [DODECYLGUANIDINE MONOAC-ETATE] [313]	DODECYLGUANIDINE MONOACETATE]
16389-88-1 DOLOMITE [PST]	DOLOMITE [CAMG(CO3)2]
16389-88-1 DOLOMITE [CAMG(CO3)2]	DOLOMITE [CAMG(CO3)2]
2471-11-6 DOTRIACONTAMETHYLPENTADECASILOXANE	DOTRIACONTAMETHYLPENTADECASILOXANE
RR-01074-2 DOUGLAS FIR BARK	DOUGLAS FIR BARKALT
564-25-0 DOXYCYCLINE	DOXYCYCLINE
94088-85-4 DOXYCYCLINE CALCIUM	DOXYCYCLINE CALCIUM
24390-14-5 DOXYCYCLINE HYCLATE	DOXYCYCLINE HYCLATE
17086-28-1 DOXYCYCLINE MONOHYDRATE	DOXYCYCLINE MONOHYDRATE
469-21-6 DOXYLAMINE	DOXYLAMINE

Chemical Name	Reference Name
120-36-5 2,4,-DP [313]	2[2,4-(DICHLOROPHENOXY)]-PROPIONIC ACID
120-36-5 2,4-DP (2,4-DICHLORPHENOXY-PROPIONIC ACID) [CAL]	2[2,4-(DICHLOROPHENOXY)]-PROPIONIC ACID
5707-69-7 DRAZOXOLON	DRAZOXOLON
68911-49-9 DRIED BLOOD	DRIED BLOOD
2702-72-9 2,4-D SODIUM SALT [313]	2,4-DICHLOROPHENOXYACETIC ACID SODIUM SALT
150-69-6 DULCIN	DULCIN
944-22-9 DYFONATE [BC1]	FONOFOS
944-22-9 DYFONATE (FONOFOS) [ALB, ALB, ALB]	FONOFOS
1322-90-3 DYPNONE	DYPNONE
14982-00-4 DYSPROSIUM 155	DYSPROSIUM 155
14981-97-6 DYSPROSIUM 157	DYSPROSIUM 157
14280-34-3 DYSPROSIUM 159	DYSPROSIUM 159
13967-64-1 DYSPROSIUM 165	DYSPROSIUM 165
15840-01-4 DYSPROSIUM 166	DYSPROSIUM 166
5289-74-7 ECDYSTERONE	ECDYSTERONE
17109-49-8 EDIFENPHOS	EDIFENPHOS
60-00-4 EDTA [CAL]	ETHYLENEDIAMINE TETRAACETIC ACID (EDTA)
9006-50-2 EGG WHITE [PST]	ALBUMIN EGG
18772-36-6 EICOSAMETHYLCYCLODECASILOXANE	EICOSAMETHYLCYCLODECASILOXANE
2652-13-3 EICOSAMETHYLNONASILOXANE	EICOSAMETHYLNONASILOXANE
112-95-8 EICOSANE	EICOSANE
26885-07-4 N-EICOSANOYL-N-METHYLTAURINE, SODIUM SALT	N-EICOSANOYL-N-METHYLTAURINE, SODIUM SALT
26636-39-5 EICOSYLOXYPOLY(ETHYLENEOXY) ETHANOL	EICOSYLOXYPOLY(ETHYLENEOXY) ETHANOL
26150-38-9 EINSTEINIUM 250	EINSTEINIUM 250
26250-43-1 EINSTEINIUM 251	EINSTEINIUM 251
15840-02-5 EINSTEINIUM 253	EINSTEINIUM 253
15840-03-6 EINSTEINIUM 254	EINSTEINIUM 254
RR-00464-8 EINSTEINIUM 254M	EINSTEINIUM 254M
37319-17-8 ELMIRON (SODIUM PENTOSANPOLYSULFATE)	ELMIRON (SODIUM PENTOSANPOLYSULFATE)
112-62-9 EMERY	EMERY
1302-74-5 EMERY [TLV]	CORUNDUM
57407-26-8 EMERY	EMERY
316-42-7 EMETINE DIHYDROCHLORIDE	EMETINE DIHYDROCHLORIDE
316-42-7 EMETINE, DIHYDROCHLORIDE [NJL, PAL]	EMETINE DIHYDROCHLORIDE, (13.ALPHA.,14.ALPHA.)-
316-42-7 EMETINE, HYDROCHLORIDE [302]	EMETINE DIHYDROCHLORIDE
316-42-7 EMETINE HYDROCHLORIDE [CSR, CSR]	EMETINE DIHYDROCHLORIDE
518-82-1 EMODIN	EMODIN
115-29-7 ENDOSULFAN	ENDOSULFAN
959-98-8 ALPHA-ENDOSULFAN	ALPHA-ENDOSULFAN
959-98-8 ENDOSULFAN, ALPHA [ATS]	ALPHA-ENDOSULFAN
959-98-8 ALPHA - ENDOSULFAN [MSL]	ALPHA-ENDOSULFAN
959-98-8 ALPHA-ENDOSULFAN-ALPHA [MX1]	ALPHA-ENDOSULFAN
33213-65-9 BETA-ENDOSULFAN	BETA-ENDOSULFAN
33213-65-9 BETA - ENDOSULFAN [MSL]	BETA-ENDOSULFAN
33213-65-9 BETA-ENDOSULFAN-BETA [MX1]	BETA-ENDOSULFAN
115-29-7 ENDOSULFAN (THIODAN) [BC1, BC1, BC1]	ENDOSULFAN
115-29-7 ENDOSULFAN (THIODAN) (R) [QBC, QBC]	ENDOSULFAN

Chemical Name	Reference Name
33213-65-9 ENDOSULFAN, BETA- [ATS]	BETA-ENDOSULFAN
959-98-8 ENDOSULFAN I [PQL, RCA, UTS]	ALPHA-ENDOSULFAN
33213-65-9 ENDOSULFAN II [PQL, RCA, UTS]	BETA-ENDOSULFAN
115-29-7 ENDOSULFAN (ISO) [GBR, GBR, GBR]	ENDOSULFAN
115-29-7 ENDOSULFAN AND METABOLITES [CW3]	ENDOSULFAN
RR-00564-1 ENDOSULFAN METABOLITES	ENDOSULFAN METABOLITES
1031-07-8 ENDOSULFAN SULFATE	ENDOSULFAN SULFATE
145-73-3 ENDOTHALL	ENDOTHALL
2778-04-3 ENDOTHION	ENDOTHION
72-20-8 ENDRIN	ENDRIN
72-20-8 ENDRIN (1,2,3,4,10,10-HEXACHLORO-6,7-EPOXY-1,4,4A,5,6,7,8,8A-OCTAHYDRO-EXO-5,8-DIMETHANONAPHTHALENE) [CN2]	ENDRINO-1,4,4A,5,6,7,8,8A-OCTAHYDRO-6,7-EPOXY-1,4:5,8-DIMETHANONAPHTHALENE)
72-20-8 ENDRIN (ISO) [GBR, GBR, GBR]	ENDRIN
7421-93-4 ENDRIN ALDEHYDE	ENDRIN ALDEHYDE
72-20-8 ENDRIN AND METABOLITES [CW3]	ENDRIN
53494-70-5 ENDRIN KETONE	ENDRIN KETONE
RR-00565-2 ENDRIN METABOLITES	ENDRIN METABOLITES
13838-16-9 ENFLURANE	ENFLURANE
RR-01427-7 ENGINE STARTING FLUID, WITH FLAMMABLE GAS [NJL]	HYDRAZINE CHLORATE
RR-00030-6 ENVIRONMENTAL TOBACCO SMOKE [CAB]	TOBACCO SMOKE
15086-94-9 EOSIN	EOSIN
299-42-3 EPHEDRINE	EPHEDRINE
RR-01773-2 EPHEDRINE SALTS, OPTICAL ISOMERS, AND SALTS OF OPTICAL ISOMERS	EPHEDRINE SALTS, OPTICAL ISOMERS, AND SALTS OF OPTICAL ISOME
134-72-5 EPHEDRINE SULFATE	EPHEDRINE SULFATE
3132-64-7 EPIBROMOHYDRIN	EPIBROMOHYDRIN
106-89-8 EPICHLOROHYDRIN	EPICHLOROHYDRIN
106-89-8 EPICHLOROHYDRIN (1-CHLORO-2,3-EPOXYPROPANE) [CAA, HON, HON]	EPICHLOROHYDRIN
26658-42-4 EPICHLOROHYDRIN, POLYMER WITH TE-TRAETHYLENEPENTAMINE	EPICHLOROHYDRIN, POLYMER WITH TE-TRAETHYLENEPENTAMINE
106-89-8 EPICHLOROHYDRIN [OXIRANE, (CHLOROMETHYL)-] [EP3]	EPICHLOROHYDRIN-]
51-43-4 EPINEPHRINE [EPA, RCU, RCU]	1,2-BENZENEDIOL, 4-[1-HYDROXY-2-(METHYLAMINO) ETHYL]
55-31-2 EPINEPHRINE HYDROCHLORIDE	EPINEPHRINE HYDROCHLORIDE
2104-64-5 EPN	EPN
RR-01255-5 EPOXIDIZED COPOLYMER OF PHENOL AND SUBSTITUTED PHENOL	EPOXIDIZED COPOLYMER OF PHENOL AND SUBSTITUTED PHENOLSOCYANATE, AND ALKYLENE GLYCOLS, HYDROXYALKYL ACRYLATE ESTER
8016-11-3 EPOXIDIZED LINSEED OIL	EPOXIDIZED LINSEED OIL
61788-72-5 EPOXIDIZED OCTYL TALLATE [PST]	OCTYL EPOXYTALLATES
RR-00907-4 EPOXIDIZED POLYBUTENE	EPOXIDIZED POLYBUTENE
8013-07-8 EPOXIDIZED SOYBEAN OIL	EPOXIDIZED SOYBEAN OIL
RR-01075-3 EPOXIDIZED TALL-OIL FATTY ACIDS WITH TALL-OIL ROSIN	EPOXIDIZED TALL-OIL FATTY ACIDS WITH TALL-OIL ROSIN
94-70-2 2-EPOXYBENZAMINE [FLA]	O-PHENETIDINE
106-88-7 1,2-EPOXYBUTANE [CAA, CAB, CSR, CSR, IAR]	1,2-BUTYLENE OXIDE
2386-87-0 3,4-EPOXYCYCLOHEXANECARBOXYLIC ACID (3,4-EPOXYCYCLOHEXYLMETHYL) ESTER	3,4-EPOXYCYCLOHEXANECARBOXYLIC ACID (3,4-EPOXYCYCLOHEXYLMETH

Chemical Name	Reference Name
106-87-6 1,2-EPOXY-4-(EPOXYETHYL)-CYCLOHEX-ANE [ONT, ONT]	VINYL CYCLOHEXENE DIOXIDE
4016-11-9 1,2-EPOXY-3-ETHOXYPROPANE [HM1, HM1]	EPOXY ETHYLOXY PROPANE
106-87-6 1-EPOXYETHYL-3,4-EPOXYCYCLOHEX-ANE [IAR]	VINYL CYCLOHEXENE DIOXIDE
4016-11-9 EPOXY ETHYLOXY PROPANE	EPOXY ETHYLOXY PROPANE
141-37-7 3,4-EPOXY-6-METHYLCYCLOHEXYLMETHYL-3,4-EPOXY-6-METHYLCYCLOHEXANE CARBOXY-LATE	3,4-EPOXY-6-METHYLCYCLOHEXYLMETHYL-3,4-EPOXY-6-METHYLCYCLOHE
75-56-9 1,2-EPOXYPROPANE [FLA, ONT, ONT]	PROPYLENE OXIDE
75-56-9 1,2-EPOXYPROPANE (PROPYLENE OXIDE) [ALB, ALB]	PROPYLENE OXIDE
556-52-5 2,3-EPOXY-1-PROPANOL (GLYCIDOL) [ALB, ALB]	GLYCIDOL
4016-14-2 2,3-EPOXYPROPYL ISOPROPYL ETHER [GBR, GBR]	ISOPROPYL GLYCIDYL ETHER (IGE)
RR-01000-4 EPOXY RESIN	EPOXY RESIN
556-52-5 2,3-EPOXY-1-PROPANOL [NJL, ONT]	GLYCIDOL
556-52-5 2,3-EPOXY-1-PROPANOL (GLYCIDOL) [QBC]	GLYCIDOL
2443-39-2 CIS-9,10-EPOXYSTEARIC ACID	CIS-9,10-EPOXYSTEARIC ACID
14967-67-0 ERBIUM 161	ERBIUM 161
14041-43-1 ERBIUM 165	ERBIUM 165
15840-13-8 ERBIUM 169	ERBIUM 169
14391-45-8 ERBIUM 171	ERBIUM 171
15840-14-9 ERBIUM 172	ERBIUM 172
50-14-6 ERGOCALCIFEROL	ERGOCALCIFEROL
60-79-7 ERGONOVINE	ERGONOVINE
RR-01774-3 ERGONOVINE SALTS	ERGONOVINE SALTSRS
113-15-5 ERGOTAMINE	ERGOTAMINE
RR-01775-4 ERGOTAMINE SALTS	ERGOTAMINE SALTS
379-79-3 ERGOTAMINE TARTRATE	ERGOTAMINE TARTRATE
66733-21-9 ERIONITE	ERIONITE
66733-21-9 ERIONITE FIBER [TSC, TSE, TSC]	ERIONITE
89-65-6 ERYTHORBIC ACID	ERYTHORBIC ACID
114-07-8 ERYTHROMYCIN	ERYTHROMYCIN
643-22-1 ERYTHROMYCIN STEARATE	ERYTHROMYCIN STEARATE
8022-96-6 ESSENTIAL OILS	ESSENTIAL OILS
50-28-2 ESTRADIOL [NJL, NJL]	ESTRADIOL-17B
50-28-2 ESTRADIOL-17B	ESTRADIOL-17B
50-28-2 ESTRADIOL 17B [C65, C65]	ESTRADIOL-17B
22966-79-6 ESTRADIOL MUSTARD	ESTRADIOL MUSTARD
50-28-2 ESTRA-1,3,5(10)-TRIENE-3,17-DIOL(17.BETA.)-[PAL, PAL]	ESTRADIOL-17BAHYDRO-, 2-OXIDE
53-16-7 ESTRA-1,3,5(10)-TRIEN-17-ONE, 3-HYDROXY-[PAL, PAL]	ESTRONE
RR-01742-5 ESTROGENS, NON-STEROIDAL	ESTROGENS, NON-STEROIDAL
RR-01743-6 ESTROGENS, STEROIDAL	ESTROGENS, STEROIDAL
53-16-7 ESTRONE	ESTRONE
75-04-7 ETHANAMINE [PAL, TSC, TSC]	ETHYLAMINE
55-86-7 ETHANAMINE, 2-CHLORO-N-(2-CHLOROETHYL)-N-METHYL-,HYDROCHLO-RIDE [PAL, PAL]	NITROGEN MUSTARD HYDROCHLORIDE

Chemical Name	Reference Name
302-70-5 ETHANAMINE, 2-CHLORO-N-(2-CHLOROETHYL)-N-METHYL-, N-OXIDE,HY-DROCHLORIDE [PAL, PAL]	NITROGEN MUSTARD N-OXIDE HYDROCHLORIDE
121-44-8 ETHANAMINE, N,N-DIETHYL- [PAL, TSC, TSC]	TRIETHYLAMINE
122-07-6 ETHANAMINE, 2,2-DIMETHOXY-N-METHYL-	ETHANAMINE, 2,2-DIMETHOXY-N-METHYL-
109-89-7 ETHANAMINE, N-ETHYL- [PAL, TSC, TSC]	DIETHYLAMINE
55-18-5 ETHANAMINE, N-ETHYL-N-NITROSO- [PAL, PAL]	N-NITROSODIETHYLAMINE
10595-95-6 ETHANAMINE, N-METHYL-N-NITROSO- [PAL, PAL]	N-NITROSOMETHYLETHYLAMINE
68153-35-5 ETHANAMINIUM, 2-AMINO-N-(2-AMINOETHYL)-N-(2-HYDROXYETHYL)-N-METHYL-, N,N'-DITALLOW ACYL DERIVATIVES, METHYL SULFATES (SALTS)	ETHANAMINIUM, 2-AMINO-N-(2-AMINOETHYL)-N-(2-HY-DROXYETHYL)-N-ALKYL-1-(2-TALLOW AMIDOETHYL), METHYL SULFATES
64992-16-1 ETHANAMINIUM, 2-[[2-CYANO-3-[4-(DIETHY-LAMINO)PHENYL]-1-OXO-2-PROPENYL]OXY]-N,N,N-TRIMETHYL-, CHLORIDE	ETHANAMINIUM, 2-[[2-CYANO-3-[4-(DIETHYLAMINO) PHENYL]-1-OXO-2
74-84-0 ETHANE	ETHANE
112-26-5 ETHANE, 1,2-BIS(2-CHLOROETHOXY)-	ETHANE, 1,2-BIS(2-CHLOROETHOXY)-
37853-59-1 ETHANE, 1,2-BIS(2,4,6-TRIBROMOPHENOXY)-[OTS]	FIREMASTER 680
74-96-4 ETHANE, BROMO- [PAL]	ETHYL BROMIDE
107-04-0 ETHANE, 1-BROMO-2-CHLORO- [TSC, TSC]	1-BROMO-2-CHLOROETHANE
75-00-3 ETHANE, CHLORO- [PAL, TSC, TSC, TSC]	ETHYL CHLORIDE
75-68-3 ETHANE, 1-CHLORO-1,1-DIFLUORO- [PAL, TSC, TSC, TSC]	1-CHLORO-1,1-DIFLUOROETHANE
97-97-2 ETHANE, 2-CHLORO-1,1-DIMETHOXY- [PAL]	DIMETHYL DICHLORACETAL
598-92-5 ETHANE, 1-CHLORO-1-NITRO- [PAL]	1-CHLORO-1-NITROETHANE
76-15-3 ETHANE, CHLOROPENTAFLUORO- [PAL]	CHLOROPENTAFLUOROETHANE
2837-89-0 ETHANE, 2-CHLORO-1,1,1,2-TETRAFLUORO-	ETHANE, 2-CHLORO-1,1,1,2-TETRAFLUORO-
75-88-7 ETHANE, 2-CHLORO-1,1,1-TRIFLUORO- [TSC]	2-CHLORO-1,1,1-TRIFLUOROETHANE
107-22-2 ETHANEDIAL [TSC, TSC, TSC]	GLYOXAL
107-15-3 1,2-ETHANEDIAMINE [PAL]	ETHYLENEDIAMINE
111-40-0 1,2-ETHANEDIAMINE, N-(2-AMINOETHYL)-[PAL, TSC, TSC, TSC]	DIETHYLENE TRIAMINE
112-57-2 1,2-ETHANEDIAMINE, N-(2-AMINOETHYL)-N'-[2-[(2-AMINOETHYL)AMINO]ETHYL]- [PAL]	TETRAETHYLENEPENTAMINE
112-24-3 1,2-ETHANEDIAMINE, N,N-BIS(2-AMINOETHYL)-[PAL]	TRIETHYLENE TETRAMINE
100-36-7 1,2-ETHANEDIAMINE, N,N-DIETHYL- [PAL]	DIETHYLETHYLENE DIAMINE
91-80-5 1,2-ETHANEDIAMINE, N,N-DIETHYL-N'-2-PYRIDINYL-N'-(2-THIENYLMETHYL)- [PAL]	METHAPYRILENE
106-93-4 ETHANE, 1,2-DIBROMO- [PAL, PAL]	ETHYLENE DIBROMIDE (1,2-DIBROMOETHANE)
1300-21-6 ETHANE DICHLORIDE	ETHANE DICHLORIDE
75-34-3 ETHANE, 1,1-DICHLORO- [PAL, TSC, TSC, TSC]	1,1-DICHLOROETHANE
107-06-2 ETHANE, 1,2-DICHLORO- [PAL, PAL, TSC, TSC]	ETHYLENE DICHLORIDE
1649-08-7 ETHANE, 1,2-DICHLORO-1,1-DIFLUORO- [TSC, TSC]	1,2-DICHLORO-1,1-DIFLUOROETHANE
1717-00-6 ETHANE, 1,1-DICHLORO-1-FLUORO-	ETHANE, 1,1-DICHLORO-1-FLUORO-LENE)]BIS-
594-72-9 ETHANE, 1,1-DICHLORO-1-NITRO- [PAL]	1,1-DICHLORO-1-NITROETHANE
1320-37-2 ETHANE, DICHLOROTETRAFLUORO- [PAL]	DICHLOROTETRAFLUOROETHANE
306-83-2 ETHANE, 2,2-DICHLORO-1,1,1-TRIFLUORO-	ETHANE, 2,2-DICHLORO-1,1,1-TRIFLUORO-

Chemical Name	Reference Name
105-57-7 ETHANE, 1,1-DIETHOXY- [PAL]	ACETAL
629-14-1 ETHANE, 1,2-DIETHOXY- [PAL]	ETHYLENE GLYCOL DIETHYL ETHER
75-37-6 ETHANE, 1,1-DIFLUORO- [TSC, TSC, TSC]	1,1-DIFLUOROETHANE
110-71-4 ETHANE, 1,2-DIMETHOXY- [PAL]	ETHYLENE GLYCOL DIMETHYL ETHER
534-15-6 ETHANE, 1,1-DIMETHOXY- [TSC, TSC]	1,1-DIMETHOXYETHANE
RR-01242-0 ETHANEDIMIDIC ACID	ETHANEDIMIDIC ACID
460-19-5 ETHANEDINITRILE [PAL]	CYANOGEN
144-62-7 ETHANEDIOIC ACID [PAL]	OXALIC ACID5,5A,5B,6-DECACHLOROOCTAHYDRO-
2944-67-4 ETHANEDIOIC ACID, AMMONIUM IRON (3+) SALT [MSL]	OXALIC ACID, AMMONIUM IRON (3+) SALT (3:3:1)
2944-67-4 ETHANEDIOIC ACID, AMMONIUM IRON(3+) SALT (3:3:1) [PAL]	OXALIC ACID, AMMONIUM IRON (3+) SALT (3:3:1)
55488-87-4 ETHANEDIOIC ACID, AMMONIUM IRON SALT [MSL]	FERRIC AMMONIUM OXALATE, UNSPECIFIED HYDRATE
14258-49-2 ETHANEDIOIC ACID, AMMONIUM SALT [MSL, PAL]	OXALIC ACID, AMMONIUM SALT
6009-70-7 ETHANEDIOIC ACID, DIAMMONIUM SALT, MONOHYDRATE [PAL]	AMMONIUM OXALATE, MONOHYDRATE
6153-56-6 ETHANEDIOIC ACID, DIHYDRATE	ETHANEDIOIC ACID, DIHYDRATE
5972-73-6 ETHANEDIOIC ACID, MONOAMMONIUM SALT, MONOHYDRATE [MSL, PAL]	AMMONIUM OXALATE, UNSPECIFIED HYDRATE
107-21-1 ETHANE-1,2-DIOL [GBR, GBR]	ETHYLENE GLYCOL
107-21-1 1,2-ETHANEDIOL [PAL, PST]	ETHYLENE GLYCOL
6315-52-2 1,2-ETHANEDIOL BIS(4-METHYLBENZENESUL-FONATE)	1,2-ETHANEDIOL BIS(4-METHYLBENZENESULFONATE)
111-55-7 1,2-ETHANEDIOL, DIACETATE	1,2-ETHANEDIOL, DIACETATE
629-15-2 1,2-ETHANEDIOL DIFORMATE [NFP, NFP]	ETHYLENE GLYCOL DIFORMATE
628-96-6 1,2-ETHANEDIOL, DINITRATE [PAL]	ETHYLENE GLYCOL DINITRATE
302-17-0 1,1-ETHANEDIOL, 2,2,2-TRICHLORO-	1,1-ETHANEDIOL, 2,2,2-TRICHLORO-
79-40-3 ETHANEDITHIOAMIDE	ETHANEDITHIOAMIDE
111-54-6 1,2-ETHANEDIYLBISCARBAMODITHIOIC ACID	1,2-ETHANEDIYLBISCARBAMODITHIOIC ACID
RR-00918-7 N,N'-1,2-ETHANEDIYLBIS-, POLYMER WITH 2,4,6-TRICHLORO-1,3,5-TRIAZINE, REACTION PRODUCTS WITH N-BUTYL-2,2,6,6-TETRAM-ETHYL-4-PIPERIDINAMINE	N,N'-1,2-ETHANEDIYLBIS-, POLYMER WITH 2,4,6-TRICHLORO-1,3,5-
34621-99-3 1,2-ETHANEDIYL TETRAKIS(2-CHLORO-1-METHYLETHYLENE) PHOSPHATE	1,2-ETHANEDIYL TETRAKIS(2-CHLORO-1-METHYLETHYLENE) PHOSPHATEESTER
353-36-6 ETHANE, FLUORO- [PAL]	ETHYL FLUORIDE
67-72-1 ETHANE, HEXACHLORO- [PAL, TSC, TSC, TSC]	HEXACHLOROETHANE
109-90-0 ETHANE, ISOCYANATO- [TSC, TSC, TSC]	ETHYL ISOCYANATE
540-67-0 ETHANE, METHOXY- [PAL]	METHYL ETHYL ETHER
111-91-1 ETHANE, 1,1'-[METHYLENEBIS(OXY)]BIS[2-CHLORO- [PAL, TSC, TSC, TSC]	BIS(2-CHLOROETHOXY)METHANE
122-51-0 ETHANE, 1,1',1"-[METHYLIDYNETRIS(OXY)]TRIS- [PAL]	ETHYL ORTHOFORMATE
79-24-3 ETHANE, NITRO- [PAL, TSC, TSC, TSC]	NITROETHANE
60-29-7 ETHANE, 1,1'-OXYBIS- [PAL]	ETHYL ETHER
111-44-4 ETHANE, 1,1'-OXYBIS[2-CHLORO- [PAL]	BIS(2-CHLOROETHYL) ETHER
354-33-6 ETHANE, 1,1,1,2,2-PENTAFLUORO- [TSC, TSC, TSC]	ETHENE, PENTAFLUORO-
354-33-6 ETHANE, 1,1,1,2,2-PENTAFLUORO [TSE]	ETHENE, PENTAFLUORO-

Chemical Name	Reference Name
79-21-0 ETHANEPEROXOIC ACID [PAL]	PEROXYACETIC ACID
107-71-1 ETHANEPEROXOIC ACID, 1,1-DIMETHYLETHYL ESTER [PAL]	TERT-BUTYL PEROXYACETATE
* 627-53-2 ETHANE, 1,1'-SELENOBIS-	ETHANE, 1,1'-SELENOBIS-
3039-83-6 ETHANESULFONIC ACID, SODIUM SALT	ETHANESULFONIC ACID, SODIUM SALT
2917-94-4 ETHANESULFONIC ACID, 2-[2-[2-[4-(1,1,3,3-TETRAMETHYLBUTYL)PHENOXY]ETHOXY]ETHOXY]-, SODIUM SALT	ETHANESULFONIC ACID, 2-[2-[2-[4-(1,1,3,3-TETRAM-ETHYLBUTYL)PH
1622-32-8 ETHANESULFONYL CHLORIDE, 2-CHLORO-	ETHANESULFONYL CHLORIDE, 2-CHLORO-
79-27-6 ETHANE,1,1,2,2-TETRABROMO- [CAI]	ACETYLENE TETRABROMIDE
79-27-6 ETHANE, 1,1,2,2-TETRABROMO- [PAL, TSC]	ACETYLENE TETRABROMIDE
79-34-5 ETHANE, 1,1,2,2-TETRACHLORO- [PAL]	1,1,2,2-TETRACHLOROETHANE
630-20-6 ETHANE, 1,1,1,2-TETRACHLORO- [PAL, TSC, TSC]	1,1,1,2-TETRACHLOROETHANE
79-34-5 ETHANE, 1,1,2,2-TETRACHLORO- [TSC]	1,1,2,2-TETRACHLOROETHANE
76-11-9 ETHANE, 1,1,1,2-TETRACHLORO-2,2-DIFLUORO- [PAL]	1,1,1,2-TETRACHLORO-2,2-DIFLUOROETHANE
76-12-0 ETHANE, 1,1,2,2-TETRACHLORO-1,2-DIFLUORO- [PAL]	1,1,2,2-TETRACHLORO-1,2-DIFLUOROETHANE
811-97-2 ETHANE, 1,1,1,2-TETRAFLUORO- [TSC, TSC]	1,1,1,2-TETRAFLUOROETHANE
62-55-5 ETHANETHIOAMIDE [PAL, PAL]	THIOACETAMIDE
505-60-2 ETHANE, 1,1'-THIOBIS[2-CHLORO- [PAL, PAL]	MUSTARD GAS
75-08-1 ETHANETHIOL [GBR, GBR, MAK, MAK]	ETHYL MERCAPTAN
75-08-1 1-ETHANETHIOL [NHS, NHS]	ETHYL MERCAPTAN
75-08-1 ETHANETHIOL [ONT, PAL]	ETHYL MERCAPTAN
75-08-1 ETHANETHIOL (ETHYL MERCAPTAN) [ALB, ALB]	ETHYL MERCAPTAN
71-55-6 ETHANE, 1,1,1-TRICHLORO- [PAL, TSC, TSC, TSC, TSC]	1,1,1-TRICHLOROETHANE
79-00-5 ETHANE, 1,1,2-TRICHLORO- [PAL, TSC, TSC, TSC]	1,1,2-TRICHLOROETHANE
354-21-2 ETHANE, 1,2,2-TRICHLORODIFLUORO- [TSC]	1,2,2-TRICHLORO-1,1-DIFLUOROETHANE
76-13-1 ETHANE, 1,1,2-TRICHLORO-1,2,2-TRIFLUORO- [PAL, TSC, TSC]	1,1,2-TRICHLORO-1,2,2-TRIFLUOROETHANE
420-46-2 ETHANE, 1,1,1-TRIFLUORO	ETHANE, 1,1,1-TRIFLUORO
16752-77-5 ETHANIMIDOTHIOIC ACID, N-[[(METHYLAMINO)CARBONYL]OXY]-,METHYL ESTER [PAL]	METHOMYL
64-17-5 ETHANOL [GBR, HM1, HM1, MAK, MAK, MAK, ONT, PAL, PST, TLV, WHM]	ETHYL ALCOHOL
124029-00-1 ETHANOL,2-[1-[[2-[2-[[(4-METHYLPHENYL)SULFONYL]OXY]ETHOXY]ETHOXY]METHYL]-2-(PROPENYLOXY)ETHOXY]-,4-METHYLBENZENESULFONATE	ETHANOL,2-[1-[[2-[2-[[(4-METHYLPHENYL)SULFONYL]OXY]ETHOXY]ETOXY]-3-(2-PROPENYLOXY)-,4-METHYL-BENZENESULFONATE
141-43-5 ETHANOLAMINE	ETHANOLAMINE
RR-01417-5 ETHANOL AMINE DINITRATE	ETHANOL AMINE DINITRATE
141-43-5 ETHANOL, 2-AMINO- [PAL]	ETHANOLAMINE
929-06-6 2-ETHANOL, (2-AMINOETHOXY) [MSL]	2-(2-AMINOETHOXY)ETHANOL
111-41-1 ETHANOL, 2-[(2-AMINOETHYL)AMINO]- [PAL]	AMINOETHYLETHANOLAMINE
10138-74-6 ETHANOL, 2-[(2-AMINO-1-METHYLETHYL)AMINO]- [PAL]	N-(2-HYDROXYETHYL) PROPYLENE DIAMINE
111-76-2 ETHANOL, 2-BUTOXY- [PAL, TSC, TSC, TSC]	2-BUTOXYETHANOL

Chemical Name	Reference Name
7332-46-9 ETHANOL, 2-(2-BUTOXYETHOXY)-, PHOSPHATE (3:1)	ETHANOL, 2-(2-BUTOXYETHOXY)-, PHOSPHATE (3:1)
78-51-3 ETHANOL, 2-BUTOXY-, PHOSPHATE (3:1) [TSC, TSC, TSC]	TRIBUTYOXYETHYL PHOSPHATE
102-79-4 ETHANOL, 2,2'-(BUTYLIMINO)BIS- [PAL]	N-BUTYLDIETHANOLAMINE
1606-85-5 ETHANOL, 2,2'-[2-BUTYNE-1,4-DIYLBIS(OXY)] BIS-	ETHANOL, 2,2'-[2-BUTYNE-1,4-DIYLBIS(OXY)]BIS-
107-07-3 ETHANOL, 2-CHLORO- [PAL]	ETHYLENE CHLOROHYDRIN
542-58-5 ETHANOL, 2-CHLORO-, ACETATE	ETHANOL, 2-CHLORO-, ACETATE
115-96-8 ETHANOL, 2-CHLORO-, PHOSPHATE (3:1)	ETHANOL, 2-CHLORO-, PHOSPHATE (3:1)
115-96-8 ETHANOL, 2-CHLORO-, PHOSPHATE (3:1) (TRIS (2-CHLOROETHYL) PHOSPHATE) [CAI]	ETHANOL, 2-CHLORO-, PHOSPHATE (3:1)
140-08-9 ETHANOL, 2-CHLORO-, PHOSPHITE (3:1)	ETHANOL, 2-CHLORO-, PHOSPHITE (3:1)
2842-38-8 ETHANOL, 2-(CYCLOHEXYLAMINO)-	ETHANOL, 2-(CYCLOHEXYLAMINO)-
66422-95-5 ETHANOL, 2-(2,4-DIAMINOPHENOXY)-, DIHY-DROCHLORIDE	ETHANOL, 2-(2,4-DIAMINOPHENOXY)-, DIHYDROCHLO-RIDE
24442-57-7 ETHANOL, 1,2-DIBROMO-, ACETATE	ETHANOL, 1,2-DIBROMO-, ACETATE
102-81-8 ETHANOL, 2-(DIBUTYLAMINO)- [PAL]	2-N-DIBUTYLAMINOETHANOL
10140-87-1 ETHANOL, 1,2-DICHLORO-, ACETATE	ETHANOL, 1,2-DICHLORO-, ACETATE
136-78-7 ETHANOL, 2-(2,4-DICHLOROPHENOXY)-, HY-DROGEN SULFATE, SODIUMSALT [PAL]	SESONE
100-37-8 ETHANOL, 2-(DIETHYLAMINO)- [PAL]	2-DIETHYLAMINOETHANOL
108-01-0 ETHANOL, 2-(DIMETHYLAMINO)- [PAL]	2-(DIMETHYLAMINO) ETHANOL
2160-93-2 ETHANOL, 2,2'-[(1,1-DIMETHYLETHYL)IMINO] BIS- [PAL]	TERT-BUTYLDIETHANOLAMINE
112-27-6 ETHANOL, 2,2'-[1,2-ETHANEDIYLBIS(OXY)]BIS- [PAL]	TRIETHYLENE GLYCOL
110-80-5 ETHANOL, 2-ETHOXY- [PAL]	2-ETHOXYETHANOL
111-15-9 ETHANOL, 2-ETHOXY-, ACETATE [PAL]	2-ETHOXYETHYL ACETATE
112-15-2 ETHANOL, 2-(2-ETHOXYETHOXY)-, AC-ETATE [TSC, TSC, TSC]	CARBINOL ACETATE
139-87-7 ETHANOL, 2,2'-(ETHYLIMINO)BIS- [PAL]	N-ETHYLDIETHANOLAMINE
92-50-2 ETHANOL, 2-(ETHYLPHENYLAMINO)-	ETHANOL, 2-(ETHYLPHENYLAMINO)-
6752-33-6 ETHANOL, 2,2'-(HEXYLAMINO)BIS-,	ETHANOL, 2,2'-(HEXYLAMINO)BIS-,
6752-33-6 ETHANOL, 2,2'-(HEXYLAMINO)BIS- [TSE]	ETHANOL, 2,2'-(HEXYLAMINO)BIS-,
111-42-2. ETHANOL, 2,2'-IMINOBIS- [PAL]	DIETHANOLAMINE
60-24-2 ETHANOL, 2-MERCAPTO- [PAL]	2-MERCAPTOETHANOL
109-86-4 ETHANOL, 2-METHOXY- [PAL]	2-METHOXYETHANOL
110-49-6 ETHANOL, 2-METHOXY-, ACETATE [PAL]	2-METHOXYETHYL ACETATE
111-77-3 ETHANOL, 2-(2-METHOXYETHOXY)- [PAL]	DIETHYLENE GLYCOL MONOMETHYL ETHER
112-35-6 ETHANOL, 2-(2-(2-METHOXYETHOXY) ETHOXY)- [OTS]	TRIETHYLENE GLYCOL MONOMETHYL ETHER
109-83-1 ETHANOL, 2-(METHYLAMINO)- [PAL]	N-METHYLETHANOLAMINE
109-59-1 ETHANOL, 2-(1-METHYLETHOXY)- [PAL]	ISOPROPOXYETHANOL
4162-45-2 ETHANOL, 2,2'-[(1-METHYLETHYLIDENE)BIS [(2,6-DIBROMO-4,1-PHENYLENE)OXY]BIS- [TSC, TSC, TSC, TSC]	TETRABROMOBISPHENOL-A-BISETHOXYLATEYL-
93-90-3 ETHANOL, 2-(METHYLPHENYLAMINO)-	ETHANOL, 2-(METHYLPHENYLAMINO)-
91-99-6 ETHANOL, 2,2'-[(3-METHYLPHENYL)IMINO]BIS- [PAL]	M-TOLYDIETHANOLAMINE
4439-24-1 ETHANOL, 2-(2-METHYLPROPOXY)- [PAL]	ETHYLENE GLYCOL MONOISOBUTYL ETHER

Chemical Name	Reference Name
18912-80-6 ETHANOL, 2-[2-(2-METHYLPROPOXY)ETHOXY]-	ETHANOL, 2-[2-(2-METHYLPROPOXY)ETHOXY]-
102-71-6 ETHANOL, 2,2',2''-NITRILOBIS- [TSC, TSC, TSC]	TRIETHANOLAMINE
102-71-6 ETHANOL, 2,2',2''-NITRILOTRIS- [PAL]	TRIETHANOLAMINE
1116-54-7 ETHANOL, 2,2'-(NITROSOIMINO)BIS- [PAL, PAL]	N-NITROSODIETHANOLAMINE
111-46-6 ETHANOL, 2,2'-OXYBIS- [PAL]	DIETHYLENE GLYCOL
19249-03-7 ETHANOL,2,2'-[OXYBIS(2,1-ETHANEDIYLOXY)]BIS-,BIS(4-METHYLBENZENESULFONATE)	ETHANOL,2,2'-[OXYBIS(2,1-ETHANEDIYLOXY)]BIS-,BIS(4-METHYLBEN
37860-51-8 ETHANOL,2,2'-[OXYBIS(2,1-ETHANEDIYLOXY)]BIS,BIS(4-METHYLBENZENESULFONATE))	ETHANOL,2,2'-[OXYBIS(2,1-ETHANEDIYLOXY)]BIS,BIS(4-METHYLBENZ
122-98-5 ETHANOL, 2-(PHENYLAMINO)- [PAL]	N-PHENYL ETHANOLAMINE
622-08-2 ETHANOL, 2-(PHENYLMETHOXY)- [PAL]	ETHYLENE GLYCOL MONOBENZYL ETHER
114719-15-2 ETHANOL,2,2'-[[1-[(2-PROPENYLOXY)METHYL]-1,2-ETHANEDIYL]BIS(OXY)]BIS-,BIS(4-METHYL-BENZENESULFONATE)	ETHANOL,2,2'-[[1-[(2-PROPENYLOXY)METHYL]-1,2-ETHANEDIYL]BIS(-DIMETHYL-
111-48-8 ETHANOL, 2,2'-THIOBIS-	ETHANOL, 2,2'-THIOBIS-
532-27-4 ETHANONE, 2-CHLORO-1-PHENYL- [PAL]	ALPHA-CHLOROACETOPHENONE
98-86-2 ETHANONE, 1-PHENYL- [PAL]	ACETOPHENONE
1341-24-8 ETHANONE, 1-PHENYL-, MONOCHLORO DE-RIV. [PAL]	CHLOROACETOPHENONE
4549-40-0 ETHENAMINE, N-METHYL-N-NITROSO- [PAL, PAL]	N-NITROSOMETHYLVINYLAMINE
74-85-1 ETHENE [FLA, PAL]	ETHYLENE
593-60-2 ETHENE, BROMO- [PAL, TSC, TSC, TSC]	VINYL BROMIDE
75-01-4 ETHENE, CHLORO- [PAL, PAL]	VINYL CHLORIDE
110-75-8 ETHENE, (2-CHLOROETHOXY)- [PAL, TSC, TSC]	2-CHLOROETHYL VINYL ETHER
79-38-9 ETHENE, CHLOROTRIFLUORO- [PAL]	CHLOROTRIFLUOROETHYLENE
75-35-4 ETHENE, 1,1-DICHLORO- [PAL]	VINYLIDENE CHLORIDE
156-59-2 ETHENE, 1,2-DICHLORO-, (Z)- [PAL]	CIS-1,2-DICHLOROETHYLENE
156-60-5 ETHENE, 1,2-DICHLORO-, (E)- [PAL]	1,2-TRANS-DICHLOROETHYLENE
75-38-7 ETHENE, 1,1-DIFLUORO- [PAL, TSC, TSC, TSC]	VINYLIDENE FLUORIDE
109-92-2 ETHENE, ETHOXY- [PAL]	VINYL ETHYL ETHER
75-02-5 ETHENE, FLUORO- [PAL, TSC, TSC, TSC]	VINYL FLUORIDE
68441-17-8 ETHENE, HOMOPOLYMER, OXIDIZED	ETHENE, HOMOPOLYMER, OXIDIZED
107-25-5 ETHENE, METHOXY- [PAL]	VINYL METHYL ETHER
1663-35-0 ETHENE, (2-METHOXYETHOXY)-	ETHENE, (2-METHOXYETHOXY)-
109-93-3 ETHENE, 1,1'-OXYBIS- [PAL]	DIVINYL ETHER
354-33-6 ETHENE, PENTAFLUORO-	ETHENE, PENTAFLUORO-
127-18-4 ETHENE, TETRACHLORO- [PAL, PAL, TSC, TSC]	TETRACHLOROETHYLENE
116-14-3 ETHENE, TETRAFLUORO- [PAL, TSC, TSC]	TETRAFLUOROETHYLENE
9002-84-0 ETHENE, TETRAFLUORO-, HOMOPOLY-MER [PAL]	POLYTETRAFLUOROETHYLENE
79-01-6 ETHENE, TRICHLORO- [PAL]	TRICHLOROETHYLENE
359-11-5 ETHENE, TRIFLUORO-	ETHENE, TRIFLUORO-
557-75-5 ETHENOL	ETHENOL
463-51-4 ETHENONE [PAL]	KETENE
57-63-6 ETHINYLESTRADIOL	ETHINYLESTRADIOL
57-63-6 ETHINYLOESTRADIOL [CAL, FLA, MIN, MSL, NJL, NJL]	ETHINYLESTRADIOL
563-12-2 ETHION	ETHION

Chemical Name	Reference Name
563-12-2 ETHION (NIALATE) [BC1, BC1]	ETHION
563-12-2 ETHION (NIALATE) (R) [QBC, QBC]	ETHION
563-12-2 ETHION (NIALATE) [ALB, ALB, ALB]	ETHION
536-33-4 ETHIONAMIDE	ETHIONAMIDE
13194-48-4 ETHOPROP [MSL]	ETHOPROPHOSZOLIDINONE HYDROCHLORIDE
13194-48-4 ETHOPROPHOS	ETHOPROPHOS
13194-48-4 ETHOPROP [PHOSPHORODITHIOIC ACID, O-ETHYL S,S-DIPROPYL ESTER] [313]	ETHOPROPHOSOMPLEX]
110-80-5 2-ETHOXYACETYLENE [FLA]	2-ETHOXYETHANOL
927-80-0 ETHOXYACETYLENE [FLA, MSL, NFP, NFP]	ETHYNE, ETHOXY-
103-73-1 ETHOXYBENZENE [NFP, NFP]	PHENETOLE
RR-01241-9 ETHOXYBENZOTHIAZOLE DISULFIDE	ETHOXYBENZOTHIAZOLE DISULFIDEYL ACRYLATE
103-75-3 2-ETHOXY-3,4-DIHYDRO-2H-PYRAN	2-ETHOXY-3,4-DIHYDRO-2H-PYRAN
103-75-3 2-ETHOXY-3,4-DIHYDRO-2-PYRAN [MSL, NFP, NFP]	2-ETHOXY-3,4-DIHYDRO-2H-PYRAN
103-75-3 2-ETHOXY-3,4-DI-HYDRO-2-PYRAN [PAL]	2-ETHOXY-3,4-DIHYDRO-2H-PYRAN
60-29-7 ETHOXYETHANE [ONT, ONT]	ETHYL ETHER
110-80-5 2-ETHOXYETHANOL	2-ETHOXYETHANOL
110-80-5 ETHOXYETHANOL, 2- [OTS]	2-ETHOXYETHANOL
111-15-9 2-ETHOXYETHANOL [FLA]	2-ETHOXYETHYL ACETATE
110-80-5 2-ETHOXYETHANOL (EGEE) [TLV, TLV]	2-ETHOXYETHANOL
110-80-5 2-ETHOXYETHANOL (EEGE) [BEI]	2-ETHOXYETHANOL
110-80-5 2-ETHOXYETHANOL (GLYCOL MONOETHYL ETHER) [ALB, ALB, ALB]	2-ETHOXYETHANOLACETATE)
111-15-9 2-ETHOXYETHYL ACETATE	2-ETHOXYETHYL ACETATE
111-15-9 2-ETHOXYETHYLACETATE [NJL]	2-ETHOXYETHYL ACETATE
111-15-9 2-ETHOXYETHYL ACETATE (CELLOSOLVE ACETATE) [OS1, OS1, OS1, OS1, OS1, BC1, BC1, BC1]	2-ETHOXYETHYL ACETATE
111-15-9 2,2-ETHOXYETHYL ACETATE (CELLOSOLVE ACETATE) [QBC, QBC]	2-ETHOXYETHYL ACETATE
111-15-9 2-ETHOXYETHYL ACETATE (EGEEA) [TLV, TLV, BEI]	2-ETHOXYETHYL ACETATE
51344-60-6 ETHOXYLATED ABIETYLAMINE	ETHOXYLATED ABIETYLAMINE
61791-12-6 ETHOXYLATED CASTOR OIL	ETHOXYLATED CASTOR OILTHYL, ETHOXYLATED, CHLORIDES
61791-14-8 ETHOXYLATED COCO ALKYL AMINES [PST]	ETHOXYLATED COCONUT OIL ALKYL AMINE
61791-14-8 ETHOXYLATED COCONUT OIL ALKYL AMINE	ETHOXYLATED COCONUT OIL ALKYL AMINE
61524-98-9 ETHOXYLATED HYDROABIETHYL ALCOHOL	ETHOXYLATED HYDROABIETHYL ALCOHOL
61790-81-6 ETHOXYLATED LANOLIN [PST]	LANOLIN, ETHOXYLATED
9002-92-0 ETHOXYLATED LAURYL ALCOHOL [WHM]	LAURYL POLYETHYLENE GLYCOL ETHER
68511-39-7 ETHOXYLATED LINEAR C12-15-SEC-ALCOHOL SULFATE	ETHOXYLATED LINEAR C12-15-SEC-ALCOHOL SULFATE
68239-42-9 ETHOXYLATED METHYL GLUCOSIDE	ETHOXYLATED METHYL GLUCOSIDE
11096-42-7 ETHOXYLATED NONYLPHENOL COMPLEX WITH IODINE [PST]	POLYETHOXY POLYPROPOXY OLYETHOXY ETHANOL-IODINE COMPLEX
26027-38-3 ETHOXYLATED P-NONYL PHENOL	ETHOXYLATED P-NONYL PHENOLIDE
9036-19-5 ETHOXYLATED OCTYLPHENOL	ETHOXYLATED OCTYLPHENOL
RR-01076-4 ETHOXYLATED SORBITAN POLYSORBATE	ETHOXYLATED SORBITAN POLYSORBATE
61790-90-7 ETHOXYLATED SORBITOL HEXAESTER OF TALL-OIL ACIDS	ETHOXYLATED SORBITOL HEXAESTER OF TALL-OIL ACIDS

Chemical Name	Reference Name
RR-01077-5 ETHOXYLATED SORBITOL PENTAESTER OF TALL-OIL ACIDS	ETHOXYLATED SORBITOL PENTAESTER OF TALL-OIL ACIDS
RR-01665-9 ETHOXYLATED SUBSTITUTED NAPHTHOL	ETHOXYLATED SUBSTITUTED NAPHTHOL
61790-85-0 ETHOXYLATED N-(TALLOW ALKYL) TRIMETHYLENE DIAMINES	ETHOXYLATED N-(TALLOW ALKYL)TRIMETHYLENE DIAMINES
61791-26-2 ETHOXYLATED TALLOWAMINE [PST]	TALLOW AMINE, ETHOXYLATED
68132-78-5 ETHOXYLATED TALLOW AMINE HYDROCHLORIDE	ETHOXYLATED TALLOW AMINE HYDROCHLORIDE
100-29-8 4-ETHOXYNITROBENZENE [TSC, TSC, TSC]	4-NITROPHENETOLE
2806-85-1 3-ETHOXYPROPANOL [NFP, NFP]	3-ETHOXYPROPIONALDEHYDE
2806-85-1 3-ETHOXYPROPIONALDEHYDE	3-ETHOXYPROPIONALDEHYDE
1331-11-9 3-ETHOXYPROPIONIC ACID	3-ETHOXYPROPIONIC ACID
91-53-2 ETHOXYQUIN	ETHOXYQUIN
112-50-5 ETHOXYTRIGLYCOL [NFP, NFP]	TRIETHYLENE GLYCOL MONOETHYL ETHER
100-41-4 ETHYL BENZENE [ATS]	ETHYLBENZENE
75-00-3 ETHYL CHLORIDE (CHLOROETHANE) [HON, HON]	ETHYL CHLORIDE
107-12-0 ETHYL CYANIDE (PROPIONITRILE) [UTS]	PROPIONITRILE
631-71-0 ETHYL ABIETATE	ETHYL ABIETATE
625-50-3 N-ETHYLACETAMIDE	N-ETHYLACETAMIDE
529-65-7 N-ETHYL ACETANILIDE	N-ETHYL ACETANILIDE
141-78-6 ETHYL ACETATE	ETHYL ACETATE
141-78-6 ETHYLACETATE [T33]	ETHYL ACETATE
141-97-9 ETHYL ACETOACETATE	ETHYL ACETOACETATE
107-00-6 ETHYL ACETYLENE	ETHYL ACETYLENE
107-00-6 ETHYL ACETYLENE [1-BUTYNE] [EP3]	ETHYL ACETYLENE
RR-00855-9 ETHYL ACETYL GLYCOLATE	ETHYL ACETYL GLYCOLATE
RR-01165-4 ETHYLACROLEIN	ETHYLACROLEIN
140-88-5 ETHYL ACRYLATE	ETHYL ACRYLATE
26376-86-3 ETHYL ACRYLATE AND 2-ETHYLHEXYL ACRYLATE COPOLYMER	ETHYL ACRYLATE AND 2-ETHYLHEXYL ACRYLATE COPOLYMER
9010-88-2 ETHYL ACRYLATE, COPOLYMER WITH METHYL METHACRYLATE [PST]	ETHYL ACRYLATE-METHYL METHACRYLATE POLYMER
25035-68-1 ETHYL ACRYLATE-METHACRYLIC ACID-STYRENE POLYMER	ETHYL ACRYLATE-METHACRYLIC ACID-STYRENE POLYMER
9010-88-2 ETHYL ACRYLATE-METHYL METHACRYLATE POLYMER	ETHYL ACRYLATE-METHYL METHACRYLATE POLYMER
9003-32-1 ETHYL ACRYLATE POLYMER	ETHYL ACRYLATE POLYMER
64-17-5 ETHYL ALCOHOL	ETHYL ALCOHOL
64-17-5 ETHYL ALCOHOL (ETHANOL) [BC1, OS1, OS1, QBC, ALB, ALB]	ETHYL ALCOHOL
RR-00802-6 ETHYL ALCOHOL AND WATER	ETHYL ALCOHOL AND WATER
563-43-9 ETHYLALUMINUM DICHLORIDE	ETHYLALUMINUM DICHLORIDE
563-43-9 ETHYL ALUMINUM DICHLORIDE [NJL, NJL]	ETHYLALUMINUM DICHLORIDE
12075-68-2 ETHYL ALUMINUM SESQUICHLORIDE	ETHYL ALUMINUM SESQUICHLORIDE
12075-68-2 ETHYLALUMINUM SESQUICHLORIDE [MSL, NFP, NFP, FLA]	ETHYL ALUMINUM SESQUICHLORIDE
75-04-7 ETHYLAMINE	ETHYLAMINE
79171-09-8 ETHYLAMINE, DISTILLATION PRODUCTS	ETHYLAMINE, DISTILLATION PRODUCTS
79771-09-8 ETHYLAMINE, DISTILLATION RESIDUES	ETHYLAMINE, DISTILLATION RESIDUES
79771-09-8 ETHYLAMINE DISTILLATION RESIDUES [TSE]	ETHYLAMINE, DISTILLATION RESIDUES
RR-01783-4 ETHYLAMINE SALTS	ETHYLAMINE SALTS

311

Chemical Name	Reference Name
75-04-7 ETHYLAMINE [ETHANAMINE] [EP3]	ETHYLAMINE
94-09-7 ETHYL P-AMINOBENZOATE	ETHYL P-AMINOBENZOATE
110-73-6 2-ETHYLAMINOETHANOL	2-ETHYLAMINOETHANOL
106-68-3 ETHYL AMYL KETONE	ETHYL AMYL KETONE
106-68-3 ETHYL N-AMYL KETONE [WHM]	ETHYL AMYL KETONE
541-85-5 ETHYL AMYL KETONE	ETHYL AMYL KETONE
541-85-5 ETHYL SEC-AMYL KETONE (4-METHYL-3-HEPTANONE) [BC1]	ETHYL AMYL KETONE
541-85-5 ETHYL SEC-AMYL KETONE [CAL, CEX, FLA, WHM]	ETHYL AMYL KETONE
106-68-3 ETHYL AMYL KETONE (5-METHYL-3-HEPTANONE) [OS1, OS1, ALB, ALB]	ETHYL AMYL KETONE
103-69-5 N-ETHYLANILINE	N-ETHYLANILINE
103-69-5 ETHYLANILINE [NFP, NFP, NJL]	N-ETHYLANILINE
578-54-1 2-ETHYLANILINE	2-ETHYLANILINE
578-54-1 O-ETHYLANILINE [HON]	2-ETHYLANILINE
84-51-5 2-ETHYLANTHRAQUINONE	2-ETHYLANTHRAQUINONE
100-41-4 ETHYLBENZENE	ETHYLBENZENE
100-41-4 ETHYL BENZENE [AUS, AUS, AUS, BC1, BC1, BEI, CAA, FLA, MIN, MSL, NIO, NIO, NIO, NJL, NJL, OS1, OS1, OS1, OTV, RCA, TLV, TLV]	ETHYLBENZENE
93-89-0 ETHYL BENZOATE [NFP, NFP]	BENZOIC ACID, ETHYL ESTER
94-02-0 ETHYL BENZOYLACETATE	ETHYL BENZOYLACETATE
92-59-1 ETHYLBENZYLANILINE	ETHYLBENZYLANILINE
92-59-1 N-ETHYL-N-BENZYLANILINE [HM1, HM1, WHM]	ETHYLBENZYLANILINE
119-94-8 ETHYLBENZYLTOLUIDINE	ETHYLBENZYLTOLUIDINE
RR-00623-5 N-ETHYLBENZYLTOLUIDINE, ALL ISOMERS	N-ETHYLBENZYLTOLUIDINE, ALL ISOMERS
RR-00623-5 N-ETHYLBENZYLTOLUIDINES [HM1, HM1]	N-ETHYLBENZYLTOLUIDINE, ALL ISOMERS
538-07-8 ETHYLBIS(2-CHLOROETHYL)AMINE	ETHYLBIS(2-CHLOROETHYL)AMINE
6895-43-8 ETHYL BIXIN	ETHYL BIXIN3-PROPANEDIYL ESTER
34099-73-5 ETHYL BORATE	ETHYL BORATE
51845-86-4 ETHYL BORATE [HM1, HM1]	BORIC ACID, ETHYL ESTER
74-96-4 ETHYL BROMIDE	ETHYL BROMIDE
105-36-2 ETHYL BROMOACETATE	ETHYL BROMOACETATE
97-95-0 2-ETHYLBUTANOL [HM1, HM1]	SEC-HEXYL ALCOHOL
97-95-0 ETHYLBUTANOL [NJL]	SEC-HEXYL ALCOHOL
760-21-4 2-ETHYL-1-BUTENE	2-ETHYL-1-BUTENE
4468-93-3 ETHYL BUTOXY ETHANOL	ETHYL BUTOXY ETHANOL
10213-74-8 3-(2-ETHYLBUTOXY) PROPIONIC ACID	3-(2-ETHYLBUTOXY) PROPIONIC ACID
123-66-0 ETHYL BUTYL ACETATE [NJL]	ETHYL HEXANOATE
10031-87-5 2-ETHYLBUTYL ACETATE [NFP, NFP]	ACETIC ACID, 2-ETHYLBUTYL ESTER
3953-10-4 2-ETHYLBUTYL ACRYLATE	2-ETHYLBUTYL ACRYLATE
97-95-0 2-ETHYLBUTYL ALCOHOL [NFP, NFP]	SEC-HEXYL ALCOHOL
13360-63-9 ETHYLBUTYLAMINE	ETHYLBUTYLAMINE
30714-78-4 ETHYL BUTYL CARBONATE	ETHYL BUTYL CARBONATE
628-81-9 ETHYL BUTYL ETHER	ETHYL BUTYL ETHER
628-81-9 ETHYLBUTYL ETHER [NJL, NJL]	ETHYL BUTYL ETHER
637-92-3 ETHYL TERT-BUTYL ETHER	ETHYL TERT-BUTYL ETHER
4468-93-3 2-ETHYL BUTYL GLYCOL [NFP, NFP]	ETHYL BUTOXY ETHANOL

Chemical Name	Reference Name
106-35-4 ETHYL BUTYL KETONE	ETHYL BUTYL KETONE
106-35-4 ETHYL BUTYL KETONE (3-HEPTANONE) [WHM]	ETHYL BUTYL KETONE
106-35-4 ETHYLBUTYL KETONE (3-HEPTANONE) [BC1, BC1]	ETHYL BUTYL KETONE
106-35-4 ETHYL BUTYL KETONE (3-HEPTANONE) [OS1, OS1]	ETHYL BUTYL KETONE
106-35-4 ETHYLBUTYLKETONE (3-HEPTANONE) [QBC]	ETHYL BUTYL KETONE
106-35-4 ETHYL BUTYL KETONE (3-HEPTANONE) [ALB, ALB]	ETHYL BUTYL KETONE
115-84-4 2-ETHYL-2-BUTYL-1,3-PROPANEDIOL	2-ETHYL-2-BUTYL-1,3-PROPANEDIOL
97-96-1 2-ETHYLBUTYRALDEHYDE	2-ETHYLBUTYRALDEHYDE
97-96-1 ETHYLBUTYRALDEHYDE [NJL, NJL]	2-ETHYLBUTYRALDEHYDE
105-54-4 ETHYL BUTYRATE	ETHYL BUTYRATE
88-09-5 2-ETHYLBUTYRIC ACID	2-ETHYLBUTYRIC ACID
123-66-0 ETHYL CAPROATE [FLA, MSL, NFP, NFP]	ETHYL HEXANOATE
106-32-1 ETHYL CAPRYLATE	ETHYL CAPRYLATE
51-79-6 ETHYL CARBAMATE (URETHANE) [CAA, EP2, RCU, RCU]	URETHANE
9004-57-3 ETHYLCELLULOSE	ETHYLCELLULOSE
85-98-3 ETHYL CENTRALITE	ETHYL CENTRALITE
75-00-3 ETHYL CHLORIDE	ETHYL CHLORIDE
75-00-3 ETHYL CHLORIDE (CHLOROETHANE) [CAA, QBC]	ETHYL CHLORIDE
75-00-3 ETHYL CHLORIDE [ETHANE, CHLORO-] [EP3]	ETHYL CHLORIDE
105-39-5 ETHYL CHLOROACETATE	ETHYL CHLOROACETATE
541-41-3 ETHYL CHLOROFORMATE	ETHYL CHLOROFORMATE
535-13-7 ETHYL-2-CHLOROPROPIONATE	ETHYL-2-CHLOROPROPIONATE
2812-73-9 ETHYL CHLOROTHIOFORMATE	ETHYL CHLOROTHIOFORMATE
2941-64-2 S-ETHYL CHLOROTHIOFORMATE	S-ETHYL CHLOROTHIOFORMATE
2941-64-2 ETHYL CHLOROTHIOFORMATE [HM1, HM1]	S-ETHYL CHLOROTHIOFORMATE
103-36-6 ETHYL CINNAMATE	ETHYL CINNAMATE
10544-63-5 ETHYL CROTONATE	ETHYL CROTONATE
107-12-0 ETHYL CYANIDE [RCU, RCU, WHM]	PROPIONITRILE
105-56-6 ETHYL CYANOACETATE	ETHYL CYANOACETATE
7085-85-0 ETHYL CYANOACRYLATE	ETHYL CYANOACRYLATE
4806-61-5 ETHYL CYCLOBUTANE	ETHYL CYCLOBUTANE
4806-61-5 ETHYLCYCLOBUTANE [NFP, NFP]	ETHYL CYCLOBUTANE
1678-91-7 ETHYL CYCLOHEXANE	ETHYL CYCLOHEXANE
1678-91-7 ETHYLCYCLOHEXANE [NFP, NFP]	ETHYL CYCLOHEXANE
5459-93-8 N-ETHYLCYCLOHEXYLAMINE	N-ETHYLCYCLOHEXYLAMINE
1640-89-7 ETHYL CYCLOPENTANE	ETHYL CYCLOPENTANE
1640-89-7 ETHYLCYCLOPENTANE [NFP, NFP]	ETHYL CYCLOPENTANE
72-56-0 P,P-ETHYL DDD (PERTHANE) [MSL]	DI(P-ETHYLPHENYL)DICHLOROETHANE
110-38-3 ETHYL DECANOATE	ETHYL DECANOATE
RR-00287-9 ETHYL-3,3-DI(TERT-BUTYL-PEROXY)BU-TYRATE	ETHYL-3,3-DI(TERT-BUTYL-PEROXY)BUTYRATE
598-14-1 ETHYL DICHLOROARSINE	ETHYL DICHLOROARSINE
598-14-1 ETHYLDICHLOROARSINE [HM1, HM1, HM1, HM1]	ETHYL DICHLOROARSINE

Chemical Name	Reference Name
510-15-6 ETHYL 4,4'-DICHLOROBENZILATE [C65, C65, EP2, NJL, NJL]	CHLOROBENZILATE
1789-58-8 ETHYL DICHLOROSILANE	ETHYL DICHLOROSILANE
139-87-7 N-ETHYLDIETHANOLAMINE	N-ETHYLDIETHANOLAMINE
638-10-8 ETHYL 3,3-DIMETHACRYLATE	ETHYL 3,3-DIMETHACRYLATE
3006-13-1 N-ETHYL-N,N-DIMETHYL-1-DODECAMINIUM ETHYL SULFATE	N-ETHYL-N,N-DIMETHYL-1-DODECAMINIUM ETHYL SULFATEENOXY]ETHOXY]ETHOXY]-, SODIUM SALT
13194-48-4 O-ETHYL S,S-DIPROPYL PHOSPHO-RODITHIOATE (ETHOPROP) [CAL]	ETHOPROPHOS
759-94-4 ETHYL DIPROPYLTHIOCARBAMATE [EPTC]	ETHYL DIPROPYLTHIOCARBAMATE [EPTC]
74-85-1 ETHYLENE	ETHYLENE
107-15-3 ETHYLENE DIAMINE [QBC]	ETHYLENEDIAMINE
111-55-7 ETHYLENE GLYCOL DIACETATE [HON]	1,2-ETHANEDIOL, DIACETATE
629-14-1 ETHYLENE GLYCOL DIETHYL ETHER (1,2-DIETHOXYETHANE) [HON]	ETHYLENE GLYCOL DIETHYL ETHER
542-59-6 ETHYLENE GLYCOL MONOACETATE [HON]	ETHYLENE GLYCOL ACETATE
111-76-2 ETHYLENE GLYCOL MONOBUTYL ETHER [HON]	2-BUTOXYETHANOL
10020-43-6 ETHYLENE GLYCOL MONOCTYL ETHER	ETHYLENE GLYCOL MONOCTYL ETHER
110-80-5 ETHYLENE GLYCOL MONOETHYL ETHER [HON]	2-ETHOXYETHANOL
111-15-9 ETHYLENE GLYCOL MONOETHYL ETHER ACETATE [HON]	2-ETHOXYETHYL ACETATE
109-86-4 ETHYLENE GLYCOL MONOMETHYL ETHER ACETATE [HON]	2-METHOXYETHANOL
9010-77-9 ETHYLENE - ACRYLIC ACID COPOLY-MER [PST]	ACRYLIC ACID POLYMER WITH ETHYLENE
52907-07-0 ETHYLENE BIS(5,6-DIBROMONORBORNANE-2,3-DICARBOXIMIDE)	ETHYLENE BIS(5,6-DIBROMONORBORNANC-2,3-DICAR-BOXIMIDE)
142-59-6 ETHYLENE BIS DITHIOCARBAMATE	ETHYLENE BIS DITHIOCARBAMATE
111-54-6 ETHYLENEBISDITHIOCARBAMIC ACID, SALTS AND ESTERS [313]	1,2-ETHANEDIYLBISCARBAMODITHIOIC ACID
111-54-6 ETHYLENEBISDITHIOCARBAMIC ACID, SALTS & ESTERS [RCU, RCU]	1,2-ETHANEDIYLBISCARBAMODITHIOIC ACID
123-26-2 N,N'-ETHYLENEBIS(12-HYDROXYOCTADE-CANAMIDE)	N,N'-ETHYLENEBIS(12-HYDROXYOCTADECANAMIDE)
111-21-7 ETHYLENE BIS(OXYETHYLENE) DIACETATE	ETHYLENE BIS(OXYETHYLENE) DIACETATE
61262-53-1 ETHYLENE BIS(PENTABROMOPHENOXIDE)	ETHYLENE BIS(PENTABROMOPHENOXIDE)
110-30-5 N,N'-ETHYLENEBIS(STEARAMIDE)	N,N'-ETHYLENEBIS(STEARAMIDE)
32588-76-4 ETHYLENE BIS(TETRABROMOPHTHALIMIDE)	ETHYLENE BIS(TETRABROMOPHTHALIMIDE)ROPYL]-1-(OXIRANYLMETHYL)-
96-49-1 ETHYLENE CARBONATE	ETHYLENE CARBONATE
107-06-2 ETHYLENE CHLORIDE [HM1, HM1, HM1, HM1]	ETHYLENE DICHLORIDE
107-07-3 ETHYLENE CHLOROHYDRIN	ETHYLENE CHLOROHYDRIN
107-07-3 ETHYLENE CHLOROHYDRIN (2-CHLOROETHANOL) [ALB, ALB]	ETHYLENE CHLOROHYDRIN
107-07-3 ETHYLENE CHLOROHYDRINE [QBC, QBC]	ETHYLENE CHLOROHYDRIN
109-78-4 ETHYLENE CYANOHYDRIN	ETHYLENE CYANOHYDRIN
107-15-3 ETHYLENEDIAMINE	ETHYLENEDIAMINE
107-15-3 ETHYLENE DIAMINE [WHM]	ETHYLENEDIAMINE
15718-71-5 ETHYLENE DIAMINE DIPERCHLORATE	ETHYLENE DIAMINE DIPERCHLORATE
60-00-4 ETHYLENEDIAMINE TETRA-ACETIC ACID [NJL]	ETHYLENEDIAMINE TETRAACETIC ACID (EDTA)

Chemical Name	Reference Name
60-00-4 ETHYLENEDIAMINETETRAACETIC ACID [PST]	ETHYLENEDIAMINE TETRAACETIC ACID (EDTA)
60-00-4 ETHYLENEDIAMINE TETRAACETIC ACID (EDTA)	ETHYLENEDIAMINE TETRAACETIC ACID (EDTA)
17099-81-9 ETHYLENEDIAMINETETRAACETIC ACID, FE (III) CHELATE [PST]	FERRIC EDTA
60-00-4 ETHYLENEDIAMINETETRAACETIC ACID (EDTA) [CWA]	ETHYLENEDIAMINE TETRAACETIC ACID (EDTA)
64-02-8 ETHYLENEDIAMINETETRAACETIC ACID, TETRASODIUM SALT [PST]	TETRASODIUM EDTA
7379-27-3 ETHYLENEDIAMINETETRAACETIC ACID, POTASSIUM SALT	ETHYLENEDIAMINETETRAACETIC ACID, POTASSIUM SALT
60816-63-9 ETHYLENEDIAMINETETRAACETIC ACID, TRIETHYLAMINE SALT	ETHYLENEDIAMINETETRAACETIC ACID, TRIETHYLAMINE SALTSALT
17572-97-3 ETHYLENEDIAMINETETRAACETIC ACID, TRIPOTASSIUM SALT	ETHYLENEDIAMINETETRAACETIC ACID, TRIPOTASSIUM SALT
1429-50-1 ETHYLENEDIAMINETETRA(METHYLENEPHOSPHONIC) ACID	ETHYLENEDIAMINETETRA(METHYLENEPHOSPHONIC) ACID
107-15-3 ETHYLENEDIAMINE [1,2-ETHANEDIAMINE] [EP3]	ETHYLENEDIAMINE
106-93-4 ETHYLENE DIBROMIDE (1,2-DIBROMOETHANE)	ETHYLENE DIBROMIDE (1,2-DIBROMOETHANE)
106-93-4 ETHYLENE DIBROMIDE (DIBROMOETHANE) [HON, HON]	ETHYLENE DIBROMIDE (1,2-DIBROMOETHANE)
106-93-4 ETHYLENE DIBROMIDE [TLV, TLV]	ETHYLENE DIBROMIDE (1,2-DIBROMOETHANE)
106-93-4 ETHYLENE DIBROMIDE (DIBROMOETHANE) [CAA]	ETHYLENE DIBROMIDE (1,2-DIBROMOETHANE)
106-93-4 ETHYLENE DIBROMIDE (EDB) [SDW, SDW, SDW]	ETHYLENE DIBROMIDE (1,2-DIBROMOETHANE)
106-93-4 ETHYLENE DIBROMIDE (1,2-DIBROMO-ETHANE) [WHM]	ETHYLENE DIBROMIDE (1,2-DIBROMOETHANE)
106-93-4 ETHYLENE DIBROMIDE (1,2-DIBROMOETHANE) [CN2]	ETHYLENE DIBROMIDE (1,2-DIBROMOETHANE)
RR-01288-4 ETHYLENE DIBROMIDE AND METHYL BROMIDE MIXTURES, LIQUID	ETHYLENE DIBROMIDE AND METHYL BROMIDE MIXTURES, LIQUID
41291-34-3 ETHYLENE(5,6-DIBROMONORBORNANE-2,3-DICARBOXIMIDE)	ETHYLENE(5,6-DIBROMONORBORNANE-2,3-DICARBOXIMIDE)
75-34-3 ETHYLENE DICHLORIDE [RCU, RCU]	1,1-DICHLOROETHANE
107-06-2 ETHYLENE DICHLORIDE	ETHYLENE DICHLORIDE
107-06-2 ETHYLENE DICHLORIDE (1,2-DICHLORO-ETHANE) [FLA, MSL, WHM]	ETHYLENE DICHLORIDE
107-06-2 ETHYLENE DICHLORIDE (1,2-DICHLOROETHANE) [CAA, C65, C65, CN2]	ETHYLENE DICHLORIDE
628-96-6 ETHYLENE DINITRATE [GBR, GBR, GBR]	ETHYLENE GLYCOL DINITRATE
371-62-0 ETHYLENE FLUOROHYDRIN	ETHYLENE FLUOROHYDRIN
107-21-1 ETHYLENE GLYCOL	ETHYLENE GLYCOL
107-21-1 ETHYLENE GLYCOL (VAPOUR, MIST) [QBC]	ETHYLENE GLYCOL
542-59-6 ETHYLENE GLYCOL ACETATE	ETHYLENE GLYCOL ACETATE
3586-55-8 ETHYLENE GLYCOL BIS(SEMIFORMAL)	ETHYLENE GLYCOL BIS(SEMIFORMAL)
111-76-2 ETHYLENE GLYCOL N-BUTYL ETHER [NFP, NFP]	2-BUTOXYETHANOL
112-48-1 ETHYLENE GLYCOL DIBUTYL ETHER	ETHYLENE GLYCOL DIBUTYL ETHER
629-14-1 ETHYLENE GLYCOL DIETHYL ETHER	ETHYLENE GLYCOL DIETHYL ETHER
629-15-2 ETHYLENE GLYCOL DIFORMATE	ETHYLENE GLYCOL DIFORMATE

Chemical Name	Reference Name
2224-15-9 ETHYLENE GLYCOL DIGLYCIDYL ETHER [EPA, OTS]	OXIRANE, 2,2'-[1,2-ETHANEDIYLBIS (OXYMETHYLENE)] BIS-
110-71-4 ETHYLENE GLYCOL DIMETHYL ETHER	ETHYLENE GLYCOL DIMETHYL ETHER
628-96-6 ETHYLENE GLYCOL DINITRATE	ETHYLENE GLYCOL DINITRATE
628-96-6 ETHYLENE GLYCOL DINITRATE AND/OR NITROGLYCERIN [BC1, BC1]	ETHYLENE GLYCOL DINITRATE
628-96-6 ETHYLENE GLYCOL DINITRATE (EGDN) [NHS, NHS]	ETHYLENE GLYCOL DINITRATE
627-83-8 ETHYLENE GLYCOL DISTEARATE	ETHYLENE GLYCOL DISTEARATE
53404-49-2 ETHYLENE GLYCOL ETHER OF PINENE	ETHYLENE GLYCOL ETHER OF PINENE
RR-00856-0 ETHYLENE GLYCOL ETHYLBUTYL ETHER	ETHYLENE GLYCOL ETHYLBUTYL ETHER
RR-00857-1 ETHYLENE GLYCOL ETHYLHEXYL ETHER	ETHYLENE GLYCOL ETHYLHEXYL ETHER
109-59-1 ETHYLENE GLYCOL ISOPROPYL ETHER [FLA, MSL, NFP, NFP, WHM]	ISOPROPOXYETHANOL
110-49-6 ETHYLENE GLYCOL METHYL ETHER AC-ETATE (METHYL CELLOSOLVEACETATE) [ALB, ALB, ALB]	2-METHOXYETHYL ACETATE
110-49-6 ETHYLENE GLYCOL METHYL ETHER AC-ETATE [MIN]	2-METHOXYETHYL ACETATE
542-59-6 ETHYLENE GLYCOL MONOACETATE [NFP, NFP]	ETHYLENE GLYCOL ACETATE
818-61-1 ETHYLENE GLYCOL MONOACRYLATE	ETHYLENE GLYCOL MONOACRYLATE
622-08-2 ETHYLENE GLYCOL MONOBENZYL ETHER	ETHYLENE GLYCOL MONOBENZYL ETHER
111-76-2 ETHYLENE GLYCOL MONOBUTYL ETHER [CAB, HM1, HM1]	2-BUTOXYETHANOL
111-76-2 ETHYLENE GLYCOL, MONOBUTYL ETHER [MAK, MAK, MAK, MAK]	2-BUTOXYETHANOL
111-76-2 ETHYLENE GLYCOL MONOBUTYL ETHER [WHM, CAL]	2-BUTOXYETHANOL
111-76-2 ETHYLENE GLYCOL MONOBUTYL ETHER (EGMBE) [CSR]	2-BUTOXYETHANOL
112-07-2 ETHYLENE GLYCOL, MONOBUTYL ETHER ACETATE	ETHYLENE GLYCOL, MONOBUTYL ETHER ACETATE
112-07-2 ETHYLENE GLYCOL MONOBUTYL ETHER ACETATE [NFP, NFP]	ETHYLENE GLYCOL, MONOBUTYL ETHER ACETATE
110-80-5 ETHYLENE GLYCOL MONOETHYL ETHER [C65, C65, EPA, HM1, HM1, HM1]	2-ETHOXYETHANOL
110-80-5 ETHYLENE GLYCOL, MONOETHYL ETHER [MAK, MAK, MAK, MAK]	2-ETHOXYETHANOL
110-80-5 ETHYLENE GLYCOL MONOETHYL ETHER [NFP, NFP, RCU, RCU]	2-ETHOXYETHANOL
110-80-5 ETHYLENE GLYCOL MONOETHYL ETHER (EGMEE) [CSR]	2-ETHOXYETHANOL
111-15-9 ETHYLENE GLYCOL MONOETHYL ETHER ACETATE [CAB]	2-ETHOXYETHYL ACETATE
110-80-5 ETHYLENE GLYCOL MONOETHYL ETHER [CAL]	2-ETHOXYETHANOL
111-15-9 ETHYLENE GLYCOL MONOETHYL ETHER ACETATE [HM1, HM1]	2-ETHOXYETHYL ACETATE
111-15-9 ETHYLENE GLYCOL, MONOETHYL ETHER ACETATE [MAK, MAK, MAK, MAK]	2-ETHOXYETHYL ACETATE
106-74-1 ETHYLENE GLYCOL MONOETHYL ETHER ACRYLATE	ETHYLENE GLYCOL MONOETHYL ETHER ACRYLATE
111-15-9 ETHYLENE GLYCOL MONOETHYL ETHER ACETATE [C65, C65, CAL, NFP, NFP]	2-ETHOXYETHYL ACETATE

Chemical Name	Reference Name
112-25-4 ETHYLENE GLYCOL MONOHEXYL ETHER	ETHYLENE GLYCOL MONOHEXYL ETHER
4439-24-1 ETHYLENE GLYCOL MONOISOBUTYL ETHER	ETHYLENE GLYCOL MONOISOBUTYL ETHER
110-49-6 ETHYLENE GLYCOL MONOMETHYL ETHER ACETATE (METHYL CELLOSOLVEACETATE) [BC1, BC1, BC1]	2-METHOXYETHYL ACETATE
109-86-4 ETHYLENE GLYCOL MONOMETHYL ETHER [C65, C65, CAB, HM1, HM1]	2-METHOXYETHANOL
109-86-4 ETHYLENE GLYCOL, MONOMETHYL ETHER [MAK, MAK, MAK, MAK]	2-METHOXYETHANOL
109-86-4 ETHYLENE GLYCOL MONOMETHYL ETHER [NFP, NFP]	2-METHOXYETHANOL
110-49-6 ETHYLENE GLYCOL MONOMETHYL ETHER ACETATE [CAB]	2-METHOXYETHYL ACETATE
109-86-4 ETHYLENE GLYCOL MONOMETHYL ETHER [CAL]	2-METHOXYETHANOL
109-86-4 ETHYLENE GLYCOL MONOMETHYL ETHER (EGMME) [CSR]	2-METHOXYETHANOL
110-49-6 ETHYLENE GLYCOL MONOMETHYL ETHER ACETATE [C65, C65, CAL]	2-METHOXYETHYL ACETATE
110-49-6 ETHYLENE GLYCOL, MONOMETHYL ETHER ACETATE [MAK, MAK, MAK, MAK]	2-METHOXYETHYL ACETATE
110-49-6 ETHYLENE GLYCOL MONOMETHYL ETHER ACETATE [NFP, NFP]	2-METHOXYETHYL ACETATE
3121-61-7 ETHYLENE GLYCOL MONOMETHYL ETHER ACRYLATE [TSC, TSC, TSC]	2-METHOXYETHYL ACRYLATE
RR-00858-2 ETHYLENE GLYCOL MONOMETHYL ETHER ACETAL	ETHYLENE GLYCOL MONOMETHYL ETHER ACETAL
RR-00859-3 ETHYLENE GLYCOL MONOMETHYL ETHER FORMAL	ETHYLENE GLYCOL MONOMETHYL ETHER FORMAL
122-99-6 ETHYLENE GLYCOL MONOPHENYL ETHER	ETHYLENE GLYCOL MONOPHENYL ETHER
2807-30-9 ETHYLENE GLYCOL MONOPROPYL ETHER	ETHYLENE GLYCOL MONOPROPYL ETHER
111-60-4 ETHYLENE GLYCOL MONOSTEARATE [PST]	GLYCOL STEARATE
107-21-1 ETHYLENE GLYCOL PARTICULATE AND VAPOR [FLA]	ETHYLENE GLYCOL
122-99-6 ETHYLENE GLYCOL PHENYL ETHER [NFP, NFP]	ETHYLENE GLYCOL MONOPHENYL ETHER
68152-55-6 ETHYLENE GLYCOL POLYMER WITH FUMARIC ACID & ROSIN	ETHYLENE GLYCOL POLYMER WITH FUMARIC ACID & ROSIN,N',N'-TETRAMETHYL-1,2-ETHANEDIAMINE
107-21-1 ETHYLENE GLYCOL, VAPOR [ONT]	ETHYLENE GLYCOL
151-56-4 ETHYLENEIMINE	ETHYLENEIMINE
151-56-4 ETHYLENE IMINE (AZIRIDINE) [CAA]	ETHYLENEIMINE
151-56-4 ETHYLENEIMINE (AZIRIDINE) [313]	ETHYLENEIMINE
151-56-4 ETHYLENEIMINE [AZIRIDINE] [EP3]	ETHYLENEIMINE
75-21-8 ETHYLENE OXIDE	ETHYLENE OXIDE
75-21-8 ETHYLENE OXIDE (ETHER) [QBC, QBC]	ETHYLENE OXIDE
RR-01214-6 ETHYLENE OXIDE ADDUCT OF FATTY ACID ESTER WITH PENTAERYTHRITOL	ETHYLENE OXIDE ADDUCT OF FATTY ACID ESTER WITH PENTAERYTHRIT
9016-45-9 ETHYLENE OXIDE-NONYLPHENOL POLYMER	ETHYLENE OXIDE-NONYLPHENOL POLYMER
9038-95-3 ETHYLENE OXIDE-PROPYLENE OXIDE COPOLYMER MONOBUTYL ETHER	ETHYLENE OXIDE-PROPYLENE OXIDE COPOLYMER MONOBUTYL ETHER
11111-34-5 ETHYLENE OXIDE-PROPYLENE OXIDE COPOLYMER ETHYLENEDIAMINE ETHER	ETHYLENE OXIDE-PROPYLENE OXIDE COPOLYMER ETHYLENEDIAMINE ETH
75-21-8 ETHYLENE OXIDE [OXIRANE] [EP3]	ETHYLENE OXIDE

Chemical Name	Reference Name
24937-78-8 ETHYLENE, POLYMER WITH VINYL ACETATE [PST]	ETHYLENE-VINYL ACETATE POLYMER
9010-79-1 ETHYLENE-PROPYLENE POLYMER	ETHYLENE-PROPYLENE POLYMER
420-12-2 ETHYLENE SULFIDE	ETHYLENE SULFIDE
420-12-2 ETHYLENE SULPHIDE [CAL, IAR]	ETHYLENE SULFIDE
96-45-7 ETHYLENE THIOUREA	ETHYLENE THIOUREA
96-45-7 ETHYLENETHIOUREA [CAL]	ETHYLENE THIOUREA
96-45-7 ETHYLENE THIOUREA (ETU) [CSR, CSR]	ETHYLENE THIOUREA
96-45-7 ETHYLENETHIOUREA [EP2, EPA, RCU, RCU, SD4]	ETHYLENE THIOUREA
24937-78-8 ETHYLENE-VINYL ACETATE POLYMER	ETHYLENE-VINYL ACETATE POLYMER
74-85-1 ETHYLENE [ETHENE] [EP3]	ETHYLENE
151-56-4 ETHYLENIMINE [AUS, AUS, AUS, BC1, BC1, MAK, MAK, OTV, TLV, TLV]	ETHYLENEIMINE
151-56-4 ETHYLENIMINE (AZIRIDINE) [HON]	ETHYLENEIMINE
RR-01789-0 N-ETHYLEPHEDRINE	N-ETHYLEPHEDRINEOPTICAL ISOMERS
RR-01786-7 N-ETHYLEPHEDRINE SALTS, OPTICAL ISOMERS, AND SALTS OF OPTICAL ISOMERS	N-ETHYLEPHEDRINE SALTS, OPTICAL ISOMERS, AND SALTS OF OPTICAAL ISOMERS
110-73-6 N-ETHYLETHANOLAMINE [NFP, NFP]	2-ETHYLAMINOETHANOL
110-73-6 ETHYL ETHANOLAMINE [PST]	2-ETHYLAMINOETHANOL
60-29-7 ETHYL ETHER	ETHYL ETHER
60-29-7 ETHYL ETHER [ETHANE, 1,1'-OXYBIS-] [EP3]	ETHYL ETHER
763-69-9 ETHYL 3-ETHOXYPROPANOATE	ETHYL 3-ETHOXYPROPANOATE
RR-01287-3 ETHYL FLUID	ETHYL FLUID
353-36-6 ETHYL FLUORIDE	ETHYL FLUORIDE
109-94-4 ETHYL FORMATE	ETHYL FORMATE
122-51-0 ETHYL FORMATE (ORTHO) [NFP, NFP, MSL]	ETHYL ORTHOFORMATE
4016-11-9 ETHYL GLYCIDYL ETHER [EPA, OTS]	EPOXY ETHYLOXY PROPANE
124-03-8 ETHYLHEXADECYLDIMETHYL-AMMONIUM BROMIDE	ETHYLHEXADECYLDIMETHYL-AMMONIUM BROMIDE
78-21-7 4-ETHYL-4-HEXADECYLMORPHOLINIUM, ETHYL SULFATE	4-ETHYL-4-HEXADECYLMORPHOLINIUM, ETHYL SULFATE
123-05-7 2-ETHYLHEXALDEHYDE	2-ETHYLHEXALDEHYDE
123-05-7 ETHYL HEXALDEHYDE [NJL]	2-ETHYLHEXALDEHYDE
123-05-7 2-ETHYLHEXANAL [FLA, MSL, NFP, NFP]	2-ETHYLHEXALDEHYDE
94-96-2 2-ETHYL-1,3-HEXANEDIOL	2-ETHYL-1,3-HEXANEDIOL
123-66-0 ETHYL HEXANOATE	ETHYL HEXANOATE
149-57-5 2-ETHYLHEXANOIC ACID	2-ETHYLHEXANOIC ACID
149-57-5 ETHYL HEXANOIC ACID, 2- [OTS]	2-ETHYLHEXANOIC ACID
15956-58-8 2-ETHYLHEXANOIC ACID, MANGANESE SALT	2-ETHYLHEXANOIC ACID, MANGANESE SALT
7580-31-6 2-ETHYLHEXANOIC ACID, NICKEL SALT	2-ETHYLHEXANOIC ACID, NICKEL SALT
94-04-2 2-ETHYLHEXANOIC ACID, VINYL ESTER [FLA]	HEXANOIC ACID, 2-ETHYL-, ETHENYL ESTER
22464-99-9 2-ETHYLHEXANOIC ACID, ZIRCONIUM SALT [PST]	ZIRCONIUM ETHYL HEXOATE
104-76-7 2-ETHYLHEXANOL	2-ETHYLHEXANOL
104-76-7 ETHYLHEXANOL, 2- [OTS]	2-ETHYLHEXANOL
104-76-7 2-ETHYL-1-HEXANOL [PST, TSE]	2-ETHYLHEXANOL
26266-68-2 2-ETHYLHEXENAL [HM1]	HEXENAL, 2-ETHYL-
149-57-5 2-ETHYLHEXOIC ACID [WHM]	2-ETHYLHEXANOIC ACID
103-09-3 2-ETHYLHEXYL ACETATE	2-ETHYLHEXYL ACETATE

Chemical Name	Reference Name
103-11-7 2-ETHYLHEXYL ACRYLATE	2-ETHYLHEXYL ACRYLATE
103-11-7 2-ETHYL HEXYL ACRYLATE [PST]	2-ETHYLHEXYL ACRYLATE
60381-61-5 2-ETHYLHEXYL ACRYLATE, POLYMER WITH VINYL TOLUENE	2-ETHYLHEXYL ACRYLATE, POLYMER WITH VINYL TOLUENE
104-76-7 2-ETHYLHEXYL ALCOHOL [WHM]	2-ETHYLHEXANOL
104-75-6 2-ETHYLHEXYLAMINE	2-ETHYLHEXYLAMINE
104-75-6 2-ETHYL HEXYLAMINE [NJL]	2-ETHYLHEXYLAMINE
10137-80-1 N-2-(ETHYLHEXYL) ANILINE [FLA, MSL, NFP, NFP]	BENZENAMINE, N-(2-ETHYLHEXYL)-
123-04-6 2-ETHYLHEXYL CHLORIDE	2-ETHYLHEXYL CHLORIDE
24468-13-1 2-ETHYLHEXYL CHLOROFORMATE	2-ETHYLHEXYL CHLOROFORMATE
24468-13-1 ETHYL HEXYLCHLOROFORMATE [NJL, NJL]	2-ETHYLHEXYL CHLOROFORMATE
6197-30-4 2-ETHYLHEXYL 2-CYANO-3,3-DIPHENYLACRYLATE [PST]	2-PROPENOIC ACID, 2-CYANO-3,3-DIPHENYL-, 2-ETHYLHEXYL ESTER
5432-61-1 N-2-(ETHYLHEXYL)-CYCLOHEXYLAMINE	N-2-(ETHYLHEXYL)-CYCLOHEXYLAMINE
1070-03-7 2-ETHYLHEXYL DIHYDROGEN PHOSPHATE [PST]	PHOSPHORIC ACID, MONO(2-ETHYLHEXYL) ESTER
1241-94-7 ETHYLHEXYL DIPHENYL PHOSPHATE, 2- [OTS]	PHOSPHORIC ACID, 2-ETHYLHEXYL DIPHENYL ESTER
5756-43-4 2-ETHYLHEXYL ETHER	2-ETHYLHEXYL ETHER
2461-15-6 2-ETHYLHEXYL GLYCIDYL ETHER [EPA]	OXIRANE, [[(2-ETHYLHEXYL)OXY]METHYL]-
2461-15-6 ETHYLHEXYL GLYCIDYL ETHER, 2- [OTS]	OXIRANE, [[(2-ETHYLHEXYL)OXY]METHYL]-
688-84-6 2-ETHYLHEXYL METHACRYLATE	2-ETHYLHEXYL METHACRYLATE
5466-77-3 2-ETHYLHEXYL-P-METHOXYCINNAMATE	2-ETHYLHEXYL-P-METHOXYCINNAMATE
1559-35-9 2-(2-ETHYLHEXYLOXY)ETHANOL	2-(2-ETHYLHEXYLOXY)ETHANOL
117-81-7 BIS(2-ETHYLHEXYL)PHTHALATE [CN4]	DI(2-ETHYLHEXYL)PHTHALATE
118-60-5 2-ETHYLHEXYL SALICYLATE	2-ETHYLHEXYL SALICYLATE
126-92-1 2-ETHYLHEXYL SODIUM SULFATE	2-ETHYLHEXYL SODIUM SULFATE
72214-01-8 2-ETHYLHEXYL SULFATE	2-ETHYLHEXYL SULFATE
3031-74-1 ETHYL HYDROPEROXIDE	ETHYL HYDROPEROXIDE
120-47-8 ETHYL P-HYDROXYBENZOATE	ETHYL P-HYDROXYBENZOATE
9004-58-4 ETHYL HYDROXYETHYL CELLULOSE	ETHYL HYDROXYETHYL CELLULOSE
75-34-3 ETHYLIDENE CHLORIDE [CAL]	1,1-DICHLOROETHANE
50-14-6 1,2-ETHYLIDENE DICHLORIDE [PAL, NFP, NFP]	ERGOCALCIFEROL
75-34-3 ETHYLIDENE DICHLORIDE [313, HM1, HM1, HM1, HM1]	1,1-DICHLOROETHANE
75-34-3 ETHYLIDENE DICHLORIDE (1,1-DICHLOROETHANE) [HON]	1,1-DICHLOROETHANE
75-34-3 1,1-ETHYLIDENE DICHLORIDE [NFP, NFP]	1,1-DICHLOROETHANE
75-34-3 ETHYLIDENE DICHLORIDE (1,1-DICHLOROETHANE) [CAA]	1,1-DICHLOROETHANE
16219-75-3 ETHYLIDENE NORBORNENE	ETHYLIDENE NORBORNENE
16219-75-3 5-ETHYLIDENE-2-NORBORNENE [ONT, WHM]	ETHYLIDENE NORBORNENE
139-87-7 2,2'-(ETHYLIMINO)DIETHANOL [PST]	N-ETHYLDIETHANOLAMINE
75-03-6 ETHYL IODIDE	ETHYL IODIDE
97-62-1 ETHYL ISOBUTYRATE	ETHYL ISOBUTYRATE
109-90-0 ETHYL ISOCYANATE	ETHYL ISOCYANATE
106-67-2 2-ETHYLISOHEXANOL [NFP, NFP]	2-ETHYL-4-METHYL-1-PENTANOL
97-64-3 ETHYL LACTATE	ETHYL LACTATE
75-08-1 ETHYL MERCAPTAN	ETHYL MERCAPTAN
75-08-1 ETHYL MERCAPTAN [QBC]	ETHYL MERCAPTAN

Chemical Name	Reference Name
75-08-1 ETHYL MERCAPTAN [ETHANETHIOL] [EP3]	ETHYL MERCAPTAN
623-51-8 ETHYL MERCAPTOACETATE	ETHYL MERCAPTOACETATE
107-27-7 ETHYLMERCURIC CHLORIDE	ETHYLMERCURIC CHLORIDE
107-27-7 ETHYL MERCURIC CHLORIDE [NJL, NJL]	ETHYLMERCURIC CHLORIDE
2235-25-8 ETHYLMERCURIC PHOSPHATE	ETHYLMERCURIC PHOSPHATE
517-16-8 N-(ETHYLMERCURIC)-P-TOLUENESULPHO-NANNILIDE	N-(ETHYLMERCURIC)-P-TOLUENESULPHONANNILIDE
97-63-2 ETHYL METHACRYLATE	ETHYL METHACRYLATE
62-50-0 ETHYL METHANESULFONATE	ETHYL METHANESULFONATE
62-50-0 ETHYL METHANESULPHONATE [CAL, IAR, MIN]	ETHYL METHANESULFONATE
540-67-0 ETHYL METHYL ETHER [HM1, HM1]	METHYL ETHYL ETHER
RR-00622-4 7-ETHYL-2-METHYL-4-HENDECANOL	7-ETHYL-2-METHYL-4-HENDECANOL
78-93-3 ETHYL METHYL KETONE [HM1, HM1, HM1]	METHYL ETHYL KETONE
22224-92-6 ETHYL 3-METHYL-4-(METHYLTHIO)-PHENYL (1-METHYLETHYL) PHOS-PHORAMIDATE [CAL]	FENAMINPHOS
106-67-2 2-ETHYL-4-METHYL-1-PENTANOL	2-ETHYL-4-METHYL-1-PENTANOL
16588-67-3 3-[N-ETHYL-4-[[6-(METHYLSULFONYL)-2-BEN-ZOTHIAZOLYL]AZO]-M-TOLUIDINO]-PROPI-ONITRILE	3-[N-ETHYL-4-[[6-(METHYLSULFONYL)-2-BENZOTHIA-ZOLYL]AZO]-M-TO
139-88-8 7-ETHYL-2-METHYL-4-UNDECANOLSULFATE, SODIUM SALT [WHM]	TERGITOL NO. 4
100-74-3 N-ETHYLMORPHOLINE	N-ETHYLMORPHOLINE
100-74-3 N-ETHYL MORPHOLINE [FLA]	N-ETHYLMORPHOLINE
100-74-3 4-ETHYLMORPHOLINE [GBR, GBR, GBR]	N-ETHYLMORPHOLINE
100-74-3 N-ETHYL MORPHOLINE [MSL]	N-ETHYLMORPHOLINE
100-74-3 4-ETHYLMORPHOLINE [NFP, NFP]	N-ETHYLMORPHOLINE
100-74-3 ETHYLMORPHOLINE [QBC, QBC]	N-ETHYLMORPHOLINEOPHENYLMETHANE)
1127-76-0 1-ETHYLNAPHTHALENE	1-ETHYLNAPHTHALENE
625-58-1 ETHYL NITRATE	ETHYL NITRATE
109-95-5 ETHYL NITRITE	ETHYL NITRITE
109-95-5 ETHYL NITRITE [NITROUS ACID, ETHYL ES-TER] [EP3]	ETHYL NITRITE
2104-64-5 O-ETHYL O-P-NITROPHENYL PHENYLPHOS-PHONOTHIOATE [ONT, ONT]	EPN
759-73-9 N-ETHYL-N-NITROSUREA [IAR]	N-NITROSO-N-ETHYLUREA
1854-23-5 4,4'-(2-ETHYL-2-NITROTRIMETHYLENE)-DI-MORPHOLINE	4,4'-(2-ETHYL-2-NITROTRIMETHYLENE)-DIMORPHOLINE
123-29-5 ETHYL NONANOATE	ETHYL NONANOATE
5881-17-4 3-ETHYLOCTANE	3-ETHYLOCTANE
15869-86-0 4-ETHYLOCTANE	4-ETHYLOCTANE
122-51-0 ETHYL ORTHOFORMATE	ETHYL ORTHOFORMATE
95-92-1 ETHYL OXALATE	ETHYL OXALATE
22750-93-2 ETHYL PERCHLORATE	ETHYL PERCHLORATE
90-00-6 O-ETHYLPHENOL	O-ETHYLPHENOL
123-07-9 P-ETHYLPHENOL [FLA, MSL, NFP, NFP]	PHENOL, 4-ETHYL-
93-55-0 ETHYL PHENYL KETONE	ETHYL PHENYL KETONE
101-97-3 ETHYL PHENYLACETATE	ETHYL PHENYLACETATE
1125-27-5 ETHYL PHENYL DICHLOROSILANE	ETHYL PHENYL DICHLOROSILANE
993-43-1 ETHYLPHOSPHONOTHIOIC-DICHLORIDE	ETHYLPHOSPHONOTHIOIC-DICHLORIDE
993-43-1 ETHYL PHOSPHONOTHIOIC DICHLORIDE [HM1, HM1, HM1]	ETHYLPHOSPHONOTHIOIC-DICHLORIDE

Chemical Name	Reference Name
1498-40-4 ETHYL PHOSPHONOUS DICHLORIDE	ETHYL PHOSPHONOUS DICHLORIDE
1498-51-7 ETHYL PHOSPHORODICHLORIDATE	ETHYL PHOSPHORODICHLORIDATE
84-72-0 ETHYL PHTHALYL ETHYL GLYCOLATE [NFP, NFP]	1,2-BENZENEDICARBOXYLIC ACID, 2-ETHOXY-2-OX-OETHYL ETHYL ESTE
RR-01319-4 5-ETHYL-2-PICOLINE	5-ETHYL-2-PICOLINE
766-09-6 1-ETHYL PIPERDINE	1-ETHYL PIPERDINE
RR-01171-2 ETHYL PROPENYL ETHER	ETHYL PROPENYL ETHER
105-37-3 ETHYL PROPIONATE	ETHYL PROPIONATE
645-62-5 2-ETHYL-3-PROPYLACROLEIN	2-ETHYL-3-PROPYLACROLEIN
5309-52-4 2-ETHYL-3-PROPYLACRYLIC ACID	2-ETHYL-3-PROPYLACRYLIC ACID
628-32-0 ETHYL PROPYL ETHER	ETHYL PROPYL ETHER
RR-01790-3 N-ETHYLPSEUDOEPHEDRINE	N-ETHYLPSEUDOEPHEDRINE
RR-01788-9 N-ETHYLPSEUDOEPHEDRINE SALTS, OPTICAL ISOMERS, AND SALTS OFOPTICAL ISOMERS	N-ETHYLPSEUDOEPHEDRINE SALTS, OPTICAL ISOMERS, AND SALTS OFOPTICAL ISOMERS
118-61-6 ETHYL SALICYLATE	ETHYL SALICYLATE
5456-28-0 ETHYL SELENAC	ETHYL SELENAC
78-10-4 ETHYL SILICATE	ETHYL SILICATE
540-82-9 ETHYL SULFURIC ACID [NJL, NJL]	ETHYLSULPHURIC ACID
540-82-9 ETHYLSULPHURIC ACID	ETHYLSULPHURIC ACID
20941-65-5 ETHYL TELLURAC	ETHYL TELLURAC
542-90-5 ETHYL THIOCYANATE	ETHYL THIOCYANATE
542-90-5 ETHYLTHIOCYANATE [302]	ETHYL THIOCYANATE
110-77-0 2-(ETHYLTHIO)ETHANOL	2-(ETHYLTHIO)ETHANOL
611-14-3 O-ETHYLTOLUENE [NFP]	BENZENE, 1-ETHYL-2-METHYL-
620-14-4 M-ETHYLTOLUENE	M-ETHYLTOLUENE
622-96-8 P-ETHYLTOLUENE	P-ETHYLTOLUENE
620-14-4 M-ETHYLTOLUENE (BENZENE, 1-ETHYL-3-METHYL-) [TSC, TSC]	M-ETHYLTOLUENE
80-39-7 ETHYL P-TOLUENE SULFONAMIDE	ETHYL P-TOLUENE SULFONAMIDE
8047-99-2 N-ETHYLTOLUENESULFONAMIDE	N-ETHYLTOLUENESULFONAMIDE
80-40-0 ETHYL P-TOLUENE SULFONATE	ETHYL P-TOLUENE SULFONATE
94-68-8 N-ETHYL-O-TOLUIDINE	N-ETHYL-O-TOLUIDINE
102-27-2 N-ETHYL-M-TOLUIDINE	N-ETHYL-M-TOLUIDINE
622-57-1 N-ETHYL-P-TOLUIDINE	N-ETHYL-P-TOLUIDINE
622-57-1 ETHYL TOLUIDINE [NJL]	N-ETHYL-P-TOLUIDINE
RR-01356-9 N-ETHYLTOLUIDINES	N-ETHYLTOLUIDINES
80-39-7 N-ETHYL-P-TOLYLSULFONAMIDE [PST]	ETHYL P-TOLUENE SULFONAMIDE
115-21-9 ETHYLTRICHLORO SILANE [NFP, NFP]	TRICHLOROETHYLSILANE
115-21-9 ETHYLTRICHLOROSILANE [NJL, NJL]	TRICHLOROETHYLSILANE
1629-58-9 ETHYL VINYL KETONE	ETHYL VINYL KETONE
74-86-2 ETHYNE [PAL]	ACETYLENE
7572-29-4 ETHYNE, DICHLORO- [PAL]	DICHLOROACETYLENE
927-80-0 ETHYNE, ETHOXY-	ETHYNE, ETHOXY-
297-76-7 ETHYNODIOL DIACETATE	ETHYNODIOL DIACETATE
RR-00100-3 ETHYNODIOL DIACETATE AND OESTROGENS	ETHYNODIOL DIACETATE AND OESTROGENS
7414-83-7 ETIDRONATE DISODIUM	ETIDRONATE DISODIUM
33419-42-0 ETOPOSIDE	ETOPOSIDE
54350-48-0 ETRETINATE	ETRETINATE
97-53-0 EUGENOL	EUGENOL
14981-86-3 EUROPIUM 145	EUROPIUM 145

Chemical Name	Reference Name
14907-88-1 EUROPIUM 146	EUROPIUM 146
14191-78-7 EUROPIUM 147	EUROPIUM 147
15840-15-0 EUROPIUM 148	EUROPIUM 148
14907-89-2 EUROPIUM 149	EUROPIUM 149
15840-16-1 EUROPIUM 150	EUROPIUM 150
14683-23-9 EUROPIUM 152	EUROPIUM 152
RR-00463-7 EUROPIUM 152M	EUROPIUM 152M
15585-10-1 EUROPIUM 154	EUROPIUM 154
14391-16-3 EUROPIUM 155	EUROPIUM 155
14280-35-4 EUROPIUM 156	EUROPIUM 156
14280-36-5 EUROPIUM 157	EUROPIUM 157
14041-40-8 EUROPIUM 158	EUROPIUM 158
10025-76-0 EUROPIUM CHLORIDE	EUROPIUM CHLORIDE
314-13-6 EVANS BLUE [CAL, IAR]	DIRECT BLUE 53
15271-41-7 2-EXO-CHLORO-5-ENDO-CYANO-2-NOR-BORNANONE O-(METHYLCARBAMOLY) OXIME [MSL]	BICYCLO[2.2.1]HEPTANE-2-CARBONITRILE, 5-CHLORO-6-((((METHYLA
6252-76-2 EXT D & C RED NO. 3	EXT D & C RED NO. 3
64742-06-9 EXTRACTS (PETROLEUM), MIDDLE DISTIL-LATE SOLVENT	EXTRACTS (PETROLEUM), MIDDLE DISTILLATE SOLVENT
RR-00635-9 F001-HAZARDOUS WASTES	F001-HAZARDOUS WASTES
RR-00636-0 F002-HAZARDOUS WASTES	F002-HAZARDOUS WASTES
RR-00637-1 F003-HAZARDOUS WASTES	F003-HAZARDOUS WASTES
RR-00638-2 F004-HAZARDOUS WASTES	F004-HAZARDOUS WASTES
RR-00639-3 F005-HAZARDOUS WASTES	F005-HAZARDOUS WASTES
RR-00640-6 F006-HAZARDOUS WASTES	F006-HAZARDOUS WASTES
RR-00641-7 F007-HAZARDOUS WASTES	F007-HAZARDOUS WASTES
RR-00642-8 F008-HAZARDOUS WASTES	F008-HAZARDOUS WASTES
RR-00643-9 F009-HAZARDOUS WASTES	F009-HAZARDOUS WASTES
RR-00644-0 F010-HAZARDOUS WASTES	F010-HAZARDOUS WASTES
RR-00645-1 F011-HAZARDOUS WASTES	F011-HAZARDOUS WASTES
RR-00646-2 F012-HAZARDOUS WASTES	F012-HAZARDOUS WASTES
RR-00647-3 F019-HAZARDOUS WASTES	F019-HAZARDOUS WASTES
RR-00648-4 F020-HAZARDOUS WASTES	F020-HAZARDOUS WASTES
RR-00649-5 F021-HAZARDOUS WASTES	F021-HAZARDOUS WASTES
RR-00650-8 F022-HAZARDOUS WASTES	F022-HAZARDOUS WASTES
RR-00651-9 F023-HAZARDOUS WASTES	F023-HAZARDOUS WASTES
RR-00652-0 F024-HAZARDOUS WASTES	F024-HAZARDOUS WASTES
RR-00755-6 F025-HAZARDOUS WASTES	F025-HAZARDOUS WASTES
RR-00653-1 F026-HAZARDOUS WASTES	F026-HAZARDOUS WASTES
RR-00654-2 F027-HAZARDOUS WASTES	F027-HAZARDOUS WASTES
RR-00655-3 F028-HAZARDOUS WASTES	F028-HAZARDOUS WASTES
RR-00106-9 F032-HAZARDOUS WASTES	F032-HAZARDOUS WASTES
RR-00105-8 F034-HAZARDOUS WASTES	F034-HAZARDOUS WASTES
RR-00104-7 F035-HAZARDOUS WASTES	F035-HAZARDOUS WASTES
RR-00292-6 F037-HAZARDOUS WASTES	F037-HAZARDOUS WASTES
RR-00293-7 F038-HAZARDOUS WASTES	F038-HAZARDOUS WASTES
RR-00291-5 F039-HAZARDOUS WASTES	F039-HAZARDOUS WASTES
52-85-7 FAMPHUR	FAMPHUR
2353-45-9 FAST GREEN FCF	FAST GREEN FCF

Chemical Name	Reference Name
RR-01225-9 FATTY ACID AMINE CONDENSATE, POLYCAR-BOXYLIC ACID SALTS	FATTY ACID AMINE CONDENSATE, POLYCARBOXYLIC ACID SALTS
RR-00914-3 FATTY ACID, AMINE SALT	FATTY ACID, AMINE SALT
RR-01236-2 FATTY ACID, ESTER WITH STYRENATED PHE-NOL, ETHYLENE OXIDE ADDUCT	FATTY ACID, ESTER WITH STYRENATED PHENOL, ETHYLENE OXIDE ADDWITH STYRENATED PHENOL, ETHYLENE OXIDE ADDUCT
68551-42-8 FATTY ACIDS, C6-19-BRANCHED, MANGANESE SALTS	FATTY ACIDS, C6-19-BRANCHED, MANGANESE SALTS
67701-05-7 FATTY ACIDS, C8-18 AND C18-UNSATURATED	FATTY ACIDS, C8-18 AND C18-UNSATURATEDEROL
61788-89-4 FATTY ACIDS, C18-UNSATURATED, DIMERS	FATTY ACIDS, C18-UNSATURATED, DIMERS
61790-63-4 FATTY ACIDS, COCO, COMPOUNDS WITH DIETHANOLAMINE [PST]	COCONUT OIL ACID, DIETHANOLAMINE SALT
68938-15-8 FATTY ACIDS, COCO, HYDROGENATED	FATTY ACIDS, COCO, HYDROGENATED
61788-59-8 FATTY ACIDS, COCO, METHYL ESTERS	FATTY ACIDS, COCO, METHYL ESTERS
68154-36-9 FATTY ACIDS, COCO, MONOESTERS WITH SORBITAN	FATTY ACIDS, COCO, MONOESTERS WITH SORBITAN
RR-01078-6 FATTY ACIDS, COCO, REACTION PRODUCTS WITH 2-[(2-AMINOETHYL)AMINO] ETHANOL, ALKYLATION PRODUCTS WITH METHYL ACRYLATE, SODIUM SALTS	FATTY ACIDS, COCO, REACTION PRODUCTS WITH 2-[(2-AMINOETHYL)A
61789-32-0 FATTY ACIDS, COCO, 2-SULFOETHYL ESTERS, SODIUM SALTS	FATTY ACIDS, COCO, 2-SULFOETHYL ESTERS, SODIUM SALTS
67701-05-7 FATTY ACIDS CONTAINING LAURIC, MYRIS-TIC, PALMITIC, STEARIC,OLEIC, AND/OR LINOLEIC [PST]	FATTY ACIDS, C8-18 AND C18-UNSATURATEDEROL
RR-01079-7 FATTY ACIDS, DEHYDRATED CASTOR-OIL, POLYMERS WITH MALEIC ANHYDRIDE, TRI-ETHANOLAMINE SALTS	FATTY ACIDS, DEHYDRATED CASTOR-OIL, POLYMERS WITH MALEIC ANHMINO] ETHANOL, ALKYLATION PRODUCTS WITH METHYL ACRYLATE, SODIUM SALTS
61790-12-3 FATTY ACIDS, TALL-OIL [PST]	TALL OIL FATTY ACIDS
61789-01-3 FATTY ACIDS, TALL OIL, EPOXIDIZED, 2-ETHYLHEXYL ESTERS	FATTY ACIDS, TALL OIL, EPOXIDIZED, 2-ETHYLHEXYL ESTERS
68187-85-9 FATTY ACIDS, TALL-OIL, ESTERS WITH ETHY-LENE GLYCOL	FATTY ACIDS, TALL-OIL, ESTERS WITH ETHYLENE GLYCOL
68188-27-2 FATTY ACIDS, TALL-OIL, ESTERS WITH PEN-TAERYTHRITOL	FATTY ACIDS, TALL-OIL, ESTERS WITH PENTAERY-THRITOL
61791-00-2 FATTY ACIDS, TALL-OIL, ETHOXLATED [PST]	TALL OIL FATTY ACIDS, ETHOXYLATED
68650-09-9 FATTY ACIDS, TALL-OIL, MIXED ESTERS WITH GLYCEROL AND POLYETHYLENE GLYCOL	FATTY ACIDS, TALL-OIL, MIXED ESTERS WITH GLYC-EROL AND POLYET
8052-48-0 FATTY ACIDS, TALLOW, SODIUM SALTS	FATTY ACIDS, TALLOW, SODIUM SALTS
RR-01208-8 FATTY AMIDE	FATTY AMIDE
6408-78-2 FD&C BLUE NO.2	FD&C BLUE NO.2
25956-17-6 FD&C RED 40	FD&C RED 40OLYMER
1934-21-0 FD&C YELLOW 5	FD&C YELLOW 5
2783-94-0 FD & C YELLOW NO. 6	FD & C YELLOW NO. 6
15792-67-3 FD&C BLUE NO. 1 (ALUMINUM SALT)	FD&C BLUE NO. 1 (ALUMINUM SALT)
860-22-0 FD&C BLUE NO. 2 [PST]	1H-INDOLE-5-SULFONIC ACID,2-(1,3-DIHYDRO-3-OXO-5-SULFO-2H
1694-09-3 FD&C VIOLET NO. 1 [PST]	BENZYL VIOLET 4B
68476-25-5 FELDSPAR	FELDSPAR
140-56-7 FENAMINOSULF (LESAN)	FENAMINOSULF (LESAN)
22224-92-6 FENAMINPHOS	FENAMINPHOS
22224-92-6 FENAMIPHOS [ONT, ONT]	FENAMINPHOS
22224-92-6 FENAMIPHOS (NEMACUR) [ALB, ALB, ALB, ALB]	FENAMINPHOS

Chemical Name	Reference Name
60168-88-9 FENARIMOL [.ALPHA.-(2-CHLOROPHENYL)-.ALPHA.-4-CHLOROPHENYL)-5-PYRIM-IDINEMETHANOL]	FENARIMOL [.ALPHA.-(2-CHLOROPHENYL)-.ALPHA.-4-CHLOROPHENYL)-
13356-08-6 FENBUTATIN OXIDE (HEXAKIS(2-METHYL-2-PHENYL-PROPYL)DISTANNOXANE) [313]	BIS(TRIS(2-METHYL-2-PHENYLPROPYL)TIN)OXIDE]
299-84-3 FENCHLORPHOS (ISO) [GBR]	RONNEL
1195-79-5 FENCHONE	FENCHONE
512-13-0 ALPHA-FENCHYL ALCOHOL	ALPHA-FENCHYL ALCOHOL
122-14-5 FENITROTHION	FENITROTHION
66441-23-4 FENOXAPROP ETHYL [2-(4-((6-CHLORO-2-BEN-ZOXAZOLYLEN)OXY)PHENOXY) PROPANOIC ACID, ETHYL ESTER]	FENOXAPROP ETHYL [2-(4-((6-CHLORO-2-BENZOXA-ZOLYLEN)OXY)PHENO
72490-01-8 FENOXYCARB [2-(4-PHENOXYPHENOXY)ETHYL]CARBAMIC ACID ETHYL ESTER	FENOXYCARB [2-(4-PHENOXYPHENOXY)ETHYL]CAR-BAMIC ACID ETHYL ESLFONYL)-2-NITROBENZAMIDE]
39515-41-8 FENPROPATHRIN	FENPROPATHRIN
39515-41-8 FENPROPATHRIN [2,2,3,3-TETRAMETHYLCY-CLOPROPANE CARBOXYLIC ACID CYANO(3-PHENOXY-PHENYL) METHYL ESTER] [313]	FENPROPATHRIN
115-90-2 FENSULFOTHION	FENSULFOTHION
115-90-2 FENSULFOTHION (DASANIT) [BC1, QBC, OS1]	FENSULFOTHION
115-90-2 FENSULFOTHION (DANSANIT) [ALB, ALB]	FENSULFOTHION
55-38-9 FENTHION	FENTHION
55-38-9 FENTHION [O,O-DIMETHYL O-[3-METHYL-4-(METHYLTHIO)PHENYL] ESTER, PHOSPHO-ROTHIOIC ACID] [313]	FENTHIONL)-DIMETHYL ESTER]
900-95-8 FENTIN ACETATE [HM1]	STANNANE, ACETOXYTRIPHENYL-
76-87-9 FENTIN HYDROXIDE [HM1]	TRIPHENYLTIN HYDROXIDE
101-42-8 FENURON [CAL]	3-PHENYL-1,1-DIMETHYLUREA
4482-55-7 FENURON-TCA	FENURON-TCA
51630-58-1 FENVALERATE	FENVALERATE
51630-58-1 FENVALERATE [4-CHLORO-ALPHA-(1-METHYLETHYL)BENZENEACETIC ACID CYANO(3-PHENOXYPHENYL)METHYL ESTER] [313]	FENVALERATEC ACID, METHYL ESTER]
14484-64-1 FERBAM	FERBAM
14484-64-1 FERBAM (ISO) [GBR, GBR]	FERBAM
14484-64-1 FERBAME [QBC]	FERBAM
14484-64-1 FERBAM [TRIS(DIMETHYLCARBAMO-DITHIOATO-S,S') IRON [313]	FERBAM
15756-90-8 FERMIUM 252	FERMIUM 252
18396-20-8 FERMIUM 253	FERMIUM 253
15750-23-9 FERMIUM 254	FERMIUM 254
15750-24-0 FERMIUM 255	FERMIUM 255
15750-26-2 FERMIUM 257	FERMIUM 257
1185-57-5 FERRIC AMMONIUM CITRATE	FERRIC AMMONIUM CITRATE
2944-67-4 FERRIC AMMONIUM OXALATE [CWA, EPA]	OXALIC ACID, AMMONIUM IRON (3+) SALT (3:3:1)
14221-47-7 FERRIC AMMONIUM OXALATE	FERRIC AMMONIUM OXALATE
55488-87-4 FERRIC AMMONIUM OXALATE [PAL]	FERRIC AMMONIUM OXALATE, UNSPECIFIED HYDRATE
55488-87-4 FERRIC AMMONIUM OXALATE, UNSPECIFIED HYDRATE	FERRIC AMMONIUM OXALATE, UNSPECIFIED HYDRATE
10138-04-2 FERRIC AMMONIUM SULFATE	FERRIC AMMONIUM SULFATE
10102-49-5 FERRIC ARSENATE	FERRIC ARSENATE
63989-69-5 FERRIC ARSENITE	FERRIC ARSENITE

Chemical Name	Reference Name
7705-08-0 FERRIC CHLORIDE	FERRIC CHLORIDE
17099-81-9 FERRIC EDTA	FERRIC EDTA
14038-43-8 FERRIC FERROCYANIDE	FERRIC FERROCYANIDE
7783-50-8 FERRIC FLUORIDE	FERRIC FLUORIDE
7782-61-8 FERRIC NITRATE.9H20	FERRIC NITRATE.9H20
10421-48-4 FERRIC NITRATE	FERRIC NITRATE
1309-37-1 FERRIC OXIDE [IAR, PST, WHM]	IRON OXIDE
10028-22-5 FERRIC SULFATE	FERRIC SULFATE
102-54-5 FERROCENE [CSR, FLA, GBR, GBR, PAL, WHM]	DICYCLOPENTADIENYL IRON
69523-06-4 FERROCERIUM	FERROCERIUM
12604-53-4 FERROMANGANESE	FERROMANGANESE
8049-17-0 FERROSILICON	FERROSILICON
7783-85-9 FERROUS AMMONIUM SULFATE.6H20	FERROUS AMMONIUM SULFATE.6H20
10045-89-3 FERROUS AMMONIUM SULFATE	FERROUS AMMONIUM SULFATE
10102-50-8 FERROUS ARSENATE	FERROUS ARSENATE
7758-94-3 FERROUS CHLORIDE	FERROUS CHLORIDE
13478-10-9 FERROUS IRON (II) CHLORIDE.4H20	FERROUS IRON (II) CHLORIDE.4H20
5905-52-2 FERROUS LACTATE	FERROUS LACTATE
1345-25-1 FERROUS OXIDE	FERROUS OXIDE
7720-78-7 FERROUS SULFATE	FERROUS SULFATE
7782-63-0 FERROUS SULFATE [PAL]	FERROUS SULFATE HEPTAHYDRATE
7782-63-0 FERROUS SULFATE HEPTAHYDRATE	FERROUS SULFATE HEPTAHYDRATE
7782-63-0 FERROUS SULFATE (HEPAHYDRATE) [MSL]	FERROUS SULFATE HEPTAHYDRATE
7720-78-7 FERROUS SULPHATE [WHM]	FERROUS SULFATE
12604-58-9 FERROVANADIUM	FERROVANADIUM
12604-58-9 FERROVANADIUM (DUST) [QBC, QBC]	FERROVANADIUM
12604-58-9 FERROVANADIUM DUST [ALB, ALB, MEX, MEX, AUS, AUS, BC1, BC1, CAL, FLA, MSL, NIO, NIO, ONT, ONT, OS1, OS1, OS1, TLV, TLV]	FERROVANADIUM
RR-01080-0 FERTILIZER	FERTILIZERYDRIDE, TRIETHANOLAMINE SALTS
RR-00297-1 FERTILIZER AMMONIATING SOLUTION	FERTILIZER AMMONIATING SOLUTION
RR-00298-2 FIBER, ANIMAL OR VEGETABLE	FIBER, ANIMAL OR VEGETABLE
RR-00002-2 FIBER GLASS, FIBROUS OR DUST [QBC]	FIBROUS GLASS
RR-00298-2 FIBERS, ANIMAL OR VEGETABLE [HM1, HM1]	FIBER, ANIMAL OR VEGETABLE
RR-01808-6 FIBROUS GLASS, MICROFIBRES	FIBROUS GLASS, MICROFIBRES
RR-00002-2 FIBROUS GLASS	FIBROUS GLASS
RR-00002-2 FIBROUS GLASS DUST [PAL, TLV]	FIBROUS GLASS
RR-00089-5 FINE DUST CONTAINING GREATER THAN 1% QUARTZ	FINE DUST CONTAINING GREATER THAN 1% QUARTZ
RR-00015-7 FINE MINERAL FIBERS [CAA]	MINERAL WOOL FIBER
37853-59-1 FIREMASTER 680	FIREMASTER 680
59536-65-1 FIREMASTER BP 6 [PAL, PAL]	FIREMASTER BP-6
59536-65-1 FIREMASTER BP-6	FIREMASTER BP-6
67774-32-7 FIREMASTER FF-1	FIREMASTER FF-1
RR-01081-1 FISH MEAL	FISH MEAL
8016-13-5 FISH OIL	FISH OIL
RR-00306-5 FLAMMABLE GAS, N.O.S.	FLAMMABLE GAS, N.O.S.
RR-00309-8 FLAMMABLE LIQUIDS, N.O.S.	FLAMMABLE LIQUIDS, N.O.S.
RR-00312-3 FLAMMABLE SOLIDS, N.O.S.	FLAMMABLE SOLIDS, N.O.S.
RR-01082-2 FLOUR	FLOUR

Chemical Name	Reference Name
69806-50-4 FLUAZIFOP-BUTYL [2-[4-[[5-(TRIFLUO-ROMETHYL)-2-PYRIDINYL]OXY]-PHENOXY] PROPANOIC ACID, BUTYL ESTER	FLUAZIFOP-BUTYL [2-[4-[[5-(TRIFLUOROMETHYL)-2-PYRIDINYL]OXY]E(+)-CYANO (3-PHENOXYPHENYL) METHYL ESTER
70124-77-5 FLUCYTHRINATE	FLUCYTHRINATE
67711-90-4 FLUE DUST, POISONOUS	FLUE DUST, POISONOUS
65996-68-1 FLUE GASSES, FERROUS METAL, BLAST FUR-NACE	FLUE GASSES, FERROUS METAL, BLAST FURNACE
4301-50-2 FLUENETIL	FLUENETIL
RR-00598-1 FLUOBORATE COMPOUNDS, N.O.S.	FLUOBORATE COMPOUNDS, N.O.S.
2164-17-2 FLUOMETURON	FLUOMETURON
2164-17-2 FLUOMETURON [UREA, N,N-DIMETHYL-N'-[3-(TRIFLUOROMETHYL)PHENYL]-] [313]	FLUOMETURONOXYLIC ACID, DIPOTASSIUM SALT]
206-44-0 FLUORANTHENE	FLUORANTHENE
1306-05-4 FLUORAPATITE	FLUORAPATITE
86-73-7 FLUORENE	FLUORENEIAZIN-3(4H)-YL)METHYL] ESTER
86-73-7 9H-FLUORENE [PAL]	FLUORENEIAZIN-3(4H)-YL)METHYL] ESTER
RR-01247-5 FLUORENE SUBSTITUTED AROMATIC AMINE	FLUORENE SUBSTITUTED AROMATIC AMINE
53-96-3 N-FLUOREN-2-YLACETAMIDE [FLA]	2-ACETYLAMINOFLUORENE
RR-00934-7 4,4'-(9H-FLUOREN-9-YLIDENE)BIS	4,4'-(9H-FLUOREN-9-YLIDENE)BIS
2321-07-5 FLUORESCEIN	FLUORESCEIN
3570-80-7 FLUORESCEIN MERCURIC ACETATE	FLUORESCEIN MERCURIC ACETATE
17372-87-1 FLUORESCEIN, 2',4',5',7'-TETRABROMO-, DIS-ODIUM SALT [PST]	ACID RED 87
16423-68-0 FLUORESCEIN, 2',4',5',7'-TETRAIODO, DIS-ODIUM SALT	FLUORESCEIN, 2',4',5',7'-TETRAIODO, DISODIUM SALT
16984-48-8 FLUORIDE [ALB, ALB, ATS, BC1, CAL, CD1, CEX, GBR, QBC, RCA, SDW, SDW, SDW]	FLUORIDES
RR-00599-2 FLUORIDE COMPOUNDS, INORGANIC, N.O.S. [WHM]	FLUORIDES, INORGANIC
16984-48-8 FLUORIDES	FLUORIDES
RR-00599-2 FLUORIDES, INORGANIC	FLUORIDES, INORGANIC
7782-41-4 FLUORINE	FLUORINE
13981-56-1 FLUORINE 18	FLUORINE 18
640-19-7 FLUOROACETAMIDE	FLUOROACETAMIDE
640-19-7 FLUOROACETAMINE [FLA]	FLUOROACETAMIDE
640-19-7 FLUOROACETAMIDE/1081 [CAL, MSL]	FLUOROACETAMIDE
144-49-0 FLUOROACETIC ACID	FLUOROACETIC ACID
62-74-8 FLUOROACETIC ACID, SODIUM SALT [EPA]	SODIUM FLUOROACETATE
359-06-8 FLUOROACETYL CHLORIDE	FLUOROACETYL CHLORIDE
RR-00276-6 FLUOROALKENES	FLUOROALKENES
348-54-9 2-FLUOROANILINE	2-FLUOROANILINE
371-40-4 4-FLUOROANILINE	4-FLUOROANILINE
RR-01357-0 FLUOROANILINES	FLUOROANILINES
462-06-6 FLUOROBENZENE	FLUOROBENZENE
16872-11-0 FLUOROBORIC ACID	FLUOROBORIC ACID
1072-85-1 1-FLUORO-2-BROMOBENZENE	1-FLUORO-2-BROMOBENZENE
1073-06-9 1-FLUORO-3-BROMOBENZENE	1-FLUORO-3-BROMOBENZENE
RR-00073-7 FLUOROCARBON, POLYMERS [MIN]	POLYTETRAFLUORETHYLENE DECOMPOSITION PRODUCTS
RR-00073-7 FLUOROCARBON POLYMERS, DECOMPOSITION PRODUCTS OF [NHS, NHS]	POLYTETRAFLUORETHYLENE DECOMPOSITION PRODUCTS
13537-32-1 FLUOROPHOSPHORIC ACID [NJL, NJL]	MONOFLUOROPHOSPHORIC ACID

Chemical Name	Reference Name
13478-20-1 FLUOROPHOSPHORIC ACID, ANHYDROUS [HM1, HM1]	PHOSPHORYL FLUORIDE
RR-00314-5 FLUOROSILICATES, N.O.S.	FLUOROSILICATES, N.O.S.
16961-83-4 FLUOROSILICIC ACID	FLUOROSILICIC ACID
7789-21-1 FLUOROSULFONIC ACID	FLUOROSULFONIC ACID
7789-21-1 FLUOROSULPHONIC ACID [NJL, NJL]	FLUOROSULFONIC ACID
95-52-3 O-FLUOROTOLUENE	O-FLUOROTOLUENE
25496-08-6 FLUOROTOLUENE	FLUOROTOLUENE
25496-08-6 FLUOROTOLUENES [HM1, HM1]	FLUOROTOLUENE
75-69-4 FLUOROTRICHLOROMETHANE [BC1, BC1, CAL, NIO, NIO, NIO, OTS, QBC, SD4, SDW]	TRICHLOROFLUOROMETHANE
75-69-4 FLUOROTRICHLOROMETHANE (TRICHLOROFLUOROMETHANE) [OS1, OS1]	TRICHLOROFLUOROMETHANE
51-21-8 FLUOROURACIL	FLUOROURACIL
51-21-8 5-FLUOROURACIL [IAR, NJL, NJL]	FLUOROURACIL
51-21-8 FLUOROURACIL (5-FLUOROURACIL) [313]	FLUOROURACIL
12003-38-2 FLUORPHLOGOPITE (MG3K[ALF2O(SIO3)3])	FLUORPHLOGOPITE (MG3K[ALF2O(SIO3)3])
76-43-7 FLUOXYMESTERONE	FLUOXYMESTERONE
17617-23-1 FLURAZEPAM	FLURAZEPAM
1172-18-5 FLURAZEPAM HYDROCHLORIDE	FLURAZEPAM HYDROCHLORIDE
13311-84-7 FLUTAMIDE	FLUTAMIDENE)
69409-94-5 FLUVALINATE [N-[2-CHLORO-4-(TRIFLUOROMETHYL)PHENYL]-DL-VALINE(+)-CYANO (3-PHENOXYPHENYL)METHYL ESTER	FLUVALINATE [N-[2-CHLORO-4-(TRIFLUOROMETHYL) PHENYL]-DL-VALINCARBOXYLIC ACID, CYANO(4-FLUORO-3-PHENOXYPHENYL)METHYL ESTER]
RR-00064-6 FOAMING AGENTS	FOAMING AGENTS
133-07-3 FOLPET	FOLPET
72178-02-0 FOMESAFEN	FOMESAFEN
72178-02-0 FOMESAFEN [5-(2-CHLORO-4-TRIFLUOROMETHYL)PHENOXY)-N-METHYLSULFONYL)-2-NITROBENZAMIDE] [313]	FOMESAFEN
944-22-9 FONOFOS	FONOFOS
944-22-9 FONOFOS [QBC, QBC]	FONOFOS
2783-94-0 FOOD YELLOW #3 [PST]	FD & C YELLOW NO. 6ATE
8004-92-0 FOOD YELLOW #13 [PST]	QUINOLINE YELLOW
50-00-0 FORMALDEHYDE	FORMALDEHYDE
50-00-0 FORMALDEHYDE (SOLUTION) [EP3]	FORMALDEHYDE
RR-01235-1 FORMALDEHYDE, CONDENSATED POLYOXYETHYLENE FATTY ACID, ESTERWITH STYRENATED PHENOL, ETHYLENE OXIDE ADDUCT	FORMALDEHYDE, CONDENSATED POLYOXYETHYLENE FATTY ACID, ESTER
107-16-4 FORMALDEHYDE CYANOHYDRIN	FORMALDEHYDE CYANOHYDRIN
50-00-0 FORMALDEHYDE (GAS) [C65, C65, NTP]	FORMALDEHYDE
9084-06-4 FORMALDEHYDE-NAPHTHALENESULFONIC ACID POLYMER SODIUM SALT	FORMALDEHYDE-NAPHTHALENESULFONIC ACID POLYMER SODIUM SALT
RR-00995-0 FORMALDEHYDE, POLYMER WITH (CHLOROMETHYL)OXIRANE, 4,4'-(1-METHYL ETHYLIDENE) BIS(2,6-DIBROMOPHENOL) AND PHENOL, 2-METHYL-2-PROPENOATE	FORMALDEHYDE, POLYMER WITH (CHLOROMETHYL) OXIRANE, 4,4'-(1-ME
55845-06-2 FORMALDEHYDE, POLYMER WITH NONYLPHENOL AND OXIRANE	FORMALDEHYDE, POLYMER WITH NONYLPHENOL AND OXIRANE
75-12-7 FORMAMIDE	FORMAMIDE
68-12-2 FORMAMIDE, N,N-DIMETHYL- [PAL]	DIMETHYLFORMAMIDE
23422-53-9 FORMETANATE	FORMETANATE

Chemical Name	Reference Name
23422-53-9 FORMETANATE HYDROCHLORIDE [302, EPA, EPA, MSL]	FORMETANATE
64-18-6 FORMIC ACID	FORMIC ACID
592-84-7 FORMIC ACID, BUTYL ESTER [PAL]	BUTYL FORMATE
544-18-3 FORMIC ACID, COBALT(2+) SALT [PAL]	COBALTOUS FORMATE
109-94-4 FORMIC ACID, ETHYL ESTER [MAK, MAK, MAK, PAL]	ETHYL FORMATE
107-31-3 FORMIC ACID, METHYL ESTER [MAK, MAK, MAK, PAL]	METHYL FORMATE
625-55-8 FORMIC ACID, 1-METHYLETHYL ESTER [PAL]	ISOPROPYL FORMATE
542-55-2 FORMIC ACID, 2-METHYLPROPYL ESTER [PAL]	ISOBUTYL FORMATE
638-49-3 FORMIC ACID, PENTYL ESTER [PAL]	AMYL FORMATE
110-74-7 FORMIC ACID, PROPYL ESTER [PAL]	PROPYL FORMATE
557-41-5 FORMIC ACID, ZINC SALT [PAL]	ZINC FORMATE
2540-82-1 FORMOTHION	FORMOTHION
17702-57-7 FORMPARANATE	FORMPARANATE
140-56-7 FORMULATED FENAMINOSULF [CSR, CSR]	FENAMINOSULF (LESAN)
3570-75-0 2-(2-FORMYLHYDRAZINO)-4-(5-NITRO-2-FURYL)THIAZOLE	2-(2-FORMYLHYDRAZINO)-4-(5-NITRO-2-FURYL)THIAZOLE
21548-32-3 FOSTHIETAN	FOSTHIETAN
36840-25-2 FRANCIUM 222	FRANCIUM 222
15756-98-6 FRANCIUM 223	FRANCIUM 223
76-13-1 FREON 113 [313]	1,1,2-TRICHLORO-1,2,2-TRIFLUOROETHANE
RR-01083-3 FRIANITE	FRIANITE
3878-19-1 FUBERIDAZOLE	FUBERIDAZOLE
RR-00315-6 FUEL, AVIATION, TURBINE ENGINE	FUEL, AVIATION, TURBINE ENGINE
RR-01602-4 FUEL CONTAINING TOXIC SUBSTANCES THAT ARE DANGEROUS GOODS	FUEL CONTAINING TOXIC SUBSTANCES THAT ARE DANGEROUS GOODS
68476-26-6 FUEL GASES	FUEL GASES
8006-20-0 FUEL GASES, PRODUCER GAS [PAL]	GAS, PRODUCER
8021-92-9 FUEL GASES, WATER GAS	FUEL GASES, WATER GAS
68476-26-6 FUEL OIL [HM1, HM1]	FUEL GASES
68476-33-5 FUEL OIL	FUEL OIL
8008-20-6 FUEL OIL NO. 1 [NFP, NFP]	KEROSENE
68476-30-2 FUEL OIL NO. 2	FUEL OIL NO. 2
68476-31-3 FUEL OIL NO. 4	FUEL OIL NO. 4
RR-00860-6 FUEL OIL NO. 5	FUEL OIL NO. 5
68553-00-4 FUEL OIL NO. 6	FUEL OIL NO. 6
RR-00317-8 FUEL, PYROPHORIC, N.O.S.	FUEL, PYROPHORIC, N.O.S.
8031-18-3 FULLERS EARTH	FULLERS EARTH
RR-01418-6 FULMINATING GOLD	FULMINATING GOLD
628-86-4 FULMINATING MERCURY [HM1, HM1, HM1]	MERCURY FULMINATE
RR-01419-7 FULMINATING PLATINUM	FULMINATING PLATINUM
5610-59-3 FULMINATING SILVER	FULMINATING SILVER
628-86-4 FULMINIC ACID, MERCURY(II)SALT [MSL, PAL]	MERCURY FULMINATE
110-17-8 FUMARIC ACID	FUMARIC ACID
623-91-6 FUMARIC ACID, DIETHYL ESTER	FUMARIC ACID, DIETHYL ESTER
117-52-2 FUMARIN [NJL]	COUMAFURYL
764-42-1 FUMARONITRILE	FUMARONITRILE
627-63-4 FUMARYL CHLORIDE	FUMARYL CHLORIDE

Chemical Name	Reference Name
116355-83-0 FUMONISIN B1	FUMONISIN B1
116355-84-1 FUMONISIN B2	FUMONISIN B2
RR-00318-9 FUNGICIDES, N.O.S.	FUNGICIDES, N.O.S.
9006-26-2 2,5-FURADIONE POLYMER WITH ETHENE	2,5-FURADIONE POLYMER WITH ETHENE
98-01-1 2-FURALDEHYDE (FURFURAL) [GBR, GBR, GBR]	FURFURAL
110-00-9 FURAN	FURAN
3688-53-7 2-FURANACETAMIDE, .ALPHA.-[(5-NITRO-2-FURANYL)METHYLENE]- [PAL, PAL]	FURYLFURAMIDE (AF-2)2-THIAZOLYL]-
98-01-1 2-FURANCARBOXALDEHYDE [PAL, TSC, TSC, TSC]	FURFURAL
625-86-5 FURAN, 2,5-DIMETHYL- [PAL]	2,5-DIMETHYLFURAN
108-31-6 2,5-FURANDIONE [PAL]	MALEIC ANHYDRIDE
108-31-6 2,5-FURANDIONE (MALEIC ANHYDRIDE) [TSC, TSC, TSC]	MALEIC ANHYDRIDE
617-89-0 2-FURANMETHANAMINE [PAL]	FURFURYLAMINE
98-00-0 2-FURANMETHANOL [PAL]	FURFURYL ALCOHOL
97-99-4 2-FURANMETHANOL, TETRAHYDRO- [PAL]	TETRAHYDROFURFURYL ALCOHOL
534-22-5 FURAN, 2-METHYL- [PAL]	2-METHYLFURAN
RR-01477-7 FURANS, POLYHALOGENATED DIBENZO-	FURANS, POLYHALOGENATED DIBENZO-
109-99-9 FURAN, TETRAHYDRO- [PAL]	TETRAHYDROFURAN
96-47-9 FURAN, TETRAHYDRO-2-METHYL- [PAL]	METHYLTETRAHYDROFURAN
67-45-8 FURAZOLIDONE	FURAZOLIDONE
98-01-1 FURFURAL	FURFURAL
98-01-1 FURFURAL (FURFURAL ALDEHYDE) [QBC, QBC]	FURFURAL
623-17-6 FURFURYL ACETATE	FURFURYL ACETATE
98-00-0 FURFURYL ALCOHOL	FURFURYL ALCOHOL
98-00-0 FURFURYL ALCOHOL (FURFURAL) [QBC, QBC, QBC]	FURFURYL ALCOHOL
617-89-0 FURFURYLAMINE	FURFURYLAMINE
60568-05-0 FURMECYCLOX	FURMECYCLOX
RR-00545-8 FURNITURE AND CABINET MAKING [PAL, PAL]	WOOD INDUSTRIES - FURNITURE & CABINET MAKING
10048-13-2 7H-FURO[3',2':4,5]FURO[2,3-C]XANTHEN-7-ONE, 3A,12C-DIHYDRO-8-HYDROXY-6-METHOXY, (3AR-CIS)- [PAL, PAL]	STERIGMATOCYSTIN
26447-28-9 2-FUROIC ACID	2-FUROIC ACID
54-31-9 FUROSEMIDE	FUROSEMIDE
54-31-9 FUROSEMIDE (FRUSEMIDE) [IAR]	FUROSEMIDE
3688-53-7 FURYLFURAMIDE (AF-2)	FURYLFURAMIDE (AF-2)
3688-53-7 2-(2-FURYL)-3-(5-NITRO-2-FURYL)ACRY-LAMIDE [MSL]	FURYLFURAMIDE (AF-2)
23255-69-8 FUSARENON-X	FUSARENON-X
23255-69-8 FUSARENONE X [IAR]	FUSARENON-X
79748-81-5 FUSARIN C	FUSARIN C
7699-41-4 FUSED SILICA	FUSED SILICA
8013-75-0 FUSEL OIL	FUSEL OIL
23315-89-1 GADOLINIUM 145	GADOLINIUM 145
14952-32-0 GADOLINIUM 146	GADOLINIUM 146
14952-31-9 GADOLINIUM 147	GADOLINIUM 147
14119-21-2 GADOLINIUM 148	GADOLINIUM 148

Chemical Name	Reference Name
14937-16-7 GADOLINIUM 149	GADOLINIUM 149
14937-17-8 GADOLINIUM 151	GADOLINIUM 151
14867-54-0 GADOLINIUM 152	GADOLINIUM 152
14276-65-4 GADOLINIUM 153	GADOLINIUM 153
14041-42-0 GADOLINIUM 159	GADOLINIUM 159
10168-81-7 GADOLINIUM NITRATE	GADOLINIUM NITRATE
10138-52-0 GADOLINIUM TRICHLORIDE	GADOLINIUM TRICHLORIDE
RR-01420-0 GALACTSAN TRINITRATE	GALACTSAN TRINITRATE
149-91-7 GALLIC ACID	GALLIC ACID
121-79-9 GALLIC ACID, PROPYL ESTER [PST]	PROPYL GALLATERIDE
7440-55-3 GALLIUM	GALLIUM
16922-44-4 GALLIUM 65	GALLIUM 65
14119-08-5 GALLIUM 66	GALLIUM 66
14119-09-6 GALLIUM 67	GALLIUM 67
15757-14-9 GALLIUM 68	GALLIUM 68
14391-74-3 GALLIUM 70	GALLIUM 70
13982-22-4 GALLIUM 72	GALLIUM 72
15034-51-2 GALLIUM 73	GALLIUM 73
1303-00-0 GALLIUM ARSENIDE	GALLIUM ARSENIDE
13450-90-3 GALLIUM(III) CHLORIDE [WHM]	GALLIUM TRICHLORIDE
7440-55-3 GALLIUM, ELEMENTAL [WHM]	GALLIUM
13494-90-1 GALLIUM(III) NITRATE	GALLIUM(III) NITRATE
12024-21-4 GALLIUM OXIDE	GALLIUM OXIDE
12063-98-8 GALLIUM PHOSPHIDE	GALLIUM PHOSPHIDE
13450-90-3 GALLIUM TRICHLORIDE	GALLIUM TRICHLORIDE
58-89-9 GAMMA-BHC [UTS]	LINDANE
65996-68-1 GAS, BLAST FURNACE [MSL, NFP]	FLUE GASSES, FERROUS METAL, BLAST FURNACE
RR-00216-4 GAS, COAL GAS [NFP]	COAL GAS
RR-00528-7 GAS, COKE-OVEN [NFP]	COKE OVEN EMISSIONS
8006-14-2 GAS, NATURAL [MSL, NFP]	NATURAL GAS
RR-00322-5 GASOHOL	GASOHOL
64741-44-2 GAS OIL [HM1, HM1, NFP, NFP, NJL]	DISTILLATES (PETROLEUM), STRAIGHT-RUN MIDDLE
68476-26-6 GAS, OIL GAS [NFP]	FUEL GASES
8006-61-9 GASOLINE	GASOLINE
86290-81-5 GASOLINE [PAL]	GASOLINE, MOTOR FUEL
RR-00861-7 GASOLINE (CASINGHEAD)	GASOLINE (CASINGHEAD)
RR-00266-4 GASOLINE ENGINE EXHAUST (CONDEN-SATES/EXTRACTS)	GASOLINE ENGINE EXHAUST (CONDEN-SATES/EXTRACTS)
RR-01738-9 GASOLINE ENGINE EXHAUST	GASOLINE ENGINE EXHAUST
8006-61-9 GASOLINE, LEADED [HM1, HM1, HM1]	GASOLINE
86290-81-5 GASOLINE, MOTOR FUEL	GASOLINE, MOTOR FUEL
68425-31-0 GASOLINE, NATURAL	GASOLINE, NATURAL
68425-31-0 GASOLINE (NATURAL GAS), NATURAL [PAL]	GASOLINE, NATURAL
8006-20-0 GAS, PRODUCER	GAS, PRODUCER
RR-00803-7 GAS, WATER (CARBURETED) [NFP]	WATER GAS (CARBURETED)
8006-61-9 GAZOLINE [QBC, QBC, QBC]	GASOLINE
9000-70-8 GELATIN	GELATIN
106-24-1 GERANIOL	GERANIOL
105-87-3 GERANYL ACETATE	GERANYL ACETATE

Chemical Name	Reference Name
106-29-6 GERANYL BUTYRATE	GERANYL BUTYRATE
105-86-2 GERANYL FORMATE	GERANYL FORMATE
105-90-8 GERANYL PROPIONATE	GERANYL PROPIONATE
7782-65-2 GERMANE [GBR, GBR, HM1, HM1, HM1, PAL]	GERMANIUM TETRAHYDRIDE
7440-56-4 GERMANIUM	GERMANIUM
15756-84-0 GERMANIUM 66	GERMANIUM 66
15756-76-0 GERMANIUM 67	GERMANIUM 67
15756-77-1 GERMANIUM 68	GERMANIUM 68
15034-49-8 GERMANIUM 69	GERMANIUM 69
14374-81-3 GERMANIUM 71	GERMANIUM 71
14687-40-2 GERMANIUM 75	GERMANIUM 75
14687-59-3 GERMANIUM 77	GERMANIUM 77
15756-83-9 GERMANIUM 78	GERMANIUM 78
7782-65-2 GERMANIUM HYDRIDE [NJL]	GERMANIUM TETRAHYDRIDE
1310-53-8 GERMANIUM OXIDE	GERMANIUM OXIDE
10038-98-9 GERMANIUM TETRACHLORIDE	GERMANIUM TETRACHLORIDE
7782-65-2 GERMANIUM TETRAHYDRIDE	GERMANIUM TETRAHYDRIDE
77-06-5 GIBBERELLIC ACID	GIBBERELLIC ACID
12002-43-6 GILSONITE	GILSONITE
65997-17-3 GLASS [MIN]	GLASS, OXIDE
RR-00002-2 GLASS FIBERS [MAK]	FIBROUS GLASS
RR-00002-2 GLASS, FIBROUS OR DUST [BC1, CAL, ONT]	FIBROUS GLASS
RR-01545-2 GLASS FILAMENTS	GLASS FILAMENTS
65997-17-3 GLASS, OXIDE	GLASS, OXIDE
RR-00015-7 GLASSWOOL [IAR]	MINERAL WOOL FIBER
RR-00015-7 GLASS WOOL (RESPIRABLE SIZE) [NTP]	MINERAL WOOL FIBER
RR-00015-7 GLASSWOOL FIBERS [C65, CAB]	MINERAL WOOL FIBER
RR-00015-7 GLASSWOOL (INCLUDING SUPERFINE GLASS-FIBRE) [AUS]	MINERAL WOOL FIBER
67730-11-4 GLU-P-1	GLU-P-1
67730-11-4 GLU-P-1 (2-AMINO-6-METHYLDIPYRIDO[1,2-ALPHA:3',2'-D]IMIDAZOLE] [C65, C65, CAL, IAR]	GLU-P-1
67730-10-3 GLU-P-2	GLU-P-2
67730-10-3 GLU-P-2 (2-AMINOPYRIDO[1,2-ALPHA:3',2'-D]IMIDAZOLE) [C65, C65, CAL]	GLU-P-2
67730-10-3 GLU-P-2 (2-AMINOPYRIDO[1,2-ALPHA:3',2'-D]IMIDAZOLE] [IAR]	GLU-P-2
50-70-4 D-GLUCITOL	D-GLUCITOLPROPANEDIYLOXY-4,1-PHENYLENE(1-METHYLETHYLIDENE)-4,1- PHENYLENEOXYMETHY-LENE]BIS(OXIRANE)
10094-62-9 ALPHA,DELTA-GLUCOHEPTONIC ACID, SODIUM SALT, DIHYDRATE	ALPHA,DELTA-GLUCOHEPTONIC ACID, SODIUM SALT, DIHYDRATE
526-95-4 GLUCONIC ACID	GLUCONIC ACID
526-95-4 D-GLUCONIC ACID [PST]	GLUCONIC ACID
299-27-4 GLUCONIC ACID, MONOPOTASSIUM SALT	GLUCONIC ACID, MONOPOTASSIUM SALT
299-27-4 D-GLUCONIC ACID, MONOPOTASSIUM SALT [PST]	GLUCONIC ACID, MONOPOTASSIUM SALT
35087-77-5 D-GLUCONIC ACID, POTASSIUM SALT	D-GLUCONIC ACID, POTASSIUM SALT
527-07-1 GLUCONIC ACID, SODIUM SALT [PST]	SODIUM GLUCONATE
90-80-2 DELTA-GLUCONOLACTONE	DELTA-GLUCONOLACTONE

Chemical Name	Reference Name
57-50-1 .ALPHA.-D-GLUCOPYRANOSIDE,.BETA.-D-FRUCTOFURANOSYL [PAL]	SUCROSE
14901-08-7 .BETA.-D-GLUCOPYRANOSIDE,(METHYL-ONN-AZOXY)METHYL [PAL, PAL]	CYCASIN
29836-26-8 BETA-D-GLUCOPYRANOSIDE, 1-O-OCTYL	BETA-D-GLUCOPYRANOSIDE, 1-O-OCTYL
50-99-7 GLUCOSE	GLUCOSE
18883-66-4 D-GLUCOSE, 2-DEOXY-2-[[(METHYLNITROSOAMINO)CARBONYL]AMINO]- [PAL, PAL]	STREPTOZOTOCIN
RR-00862-8 GLUCOSE PENTAPROPIONATE	GLUCOSE PENTAPROPIONATE
68476-37-9 GLUE (AS DEPOLYMERIZED ANIMAL COLLAGEN)	GLUE (AS DEPOLYMERIZED ANIMAL COLLAGEN)
56-86-0 ALPHA-GLUTAMIC ACID	ALPHA-GLUTAMIC ACID
142-47-2 GLUTAMIC ACID, SODIUM SALT	GLUTAMIC ACID, SODIUM SALT
111-30-8 GLUTARALDEHYDE	GLUTARALDEHYDE
111-30-8 GLUTARALDEHYDE, ACTIVATED OR UNACTIVATED [BC1]	GLUTARALDEHYDE
110-94-1 GLUTARIC ACID	GLUTARIC ACID
108-55-4 GLUTARIC ANHYDRIDE	GLUTARIC ANHYDRIDE
70-18-8 GLUTATHIONE	GLUTATHIONE
367-47-5 GLYCERALDEHYDE	GLYCERALDEHYDE
56-81-5 GLYCERIN	GLYCERIN
56-81-5 GLYCERIN MIST [MEX]	GLYCERIN
616-23-9 ALPHA,BETA-GLYCERIN DICHLOROHYDRIN [NFP, NFP]	1-PROPANOL, 2,3-DICHLORO-
616-23-9 ALPHA, BETA-GLYCERIN DICHLOROHYDRIN [PAL]	1-PROPANOL, 2,3-DICHLORO-
56-81-5 GLYCERINE [NFP, NFP]	GLYCERIN
37220-82-9 GLYCERINE OLEATE [PST]	OLEIC ACID ESTER WITH 1,2,3-PROPANETRIOLM SALT
56-81-5 GLYCERIN MIST [ALB, ALB, AUS, BC1, MIN, ONT]	GLYCERIN
56-81-5 GLYCERIN (MIST) [OS1, OS1]	GLYCERIN
56-81-5 GLYCERIN MIST [QBC, TLV]	GLYCERIN
56-81-5 GLYCEROL [GBR, HON]	GLYCERIN
25791-96-2 GLYCEROL TRI(POLYOXYPROPYLENE) ETHER [HON]	POLY(OXYPROPYLENE) TRIOL
3568-29-4 GLYCEROL 1,3-DIGLYCIDYL ETHER	GLYCEROL 1,3-DIGLYCIDYL ETHER
623-87-0 GLYCEROL-1,3-DINITRATE	GLYCEROL-1,3-DINITRATE
RR-01421-1 GLYCEROL GLUCONATE TRINITRATE	GLYCEROL GLUCONATE TRINITRATE
RR-01422-2 GLYCEROL LACTATE TRINITRATE	GLYCEROL LACTATE TRINITRATE
96-24-2 GLYCEROL ALPHA-MONOCHLOROHYDRIN [HM1, HM1]	3-CHLORO-1,2-DIHYDROXYPROPANE
96-24-2 GLYCEROL-ALPHA-MONOCHLOROHYDRIN [NJL]	3-CHLORO-1,2-DIHYDROXYPROPANE
25496-72-4 GLYCEROL MONOOLEATE	GLYCEROL MONOOLEATE
102-76-1 GLYCEROL TRIACETATE	GLYCEROL TRIACETATE
13236-02-7 GLYCEROL TRIGLYCIDYL ETHER [EPA, OTS]	OXIRANE, 2,2',2"-[1,2,3-PROPANETRIYL TRIS (OXYMETHYLENE)]
55-63-0 GLYCEROL TRINITRATE [GBR, GBR, GBR]	NITROGLYCERIN
136-44-7 GLYCERYL P-AMINOBENZOATE	GLYCERYL P-AMINOBENZOATE
25637-84-7 GLYCERYL DIOLEATE	GLYCERYL DIOLEATE
26446-35-5 GLYCERYL MONOACETATE	GLYCERYL MONOACETATE
25496-72-4 GLYCERYL MONOOLEATE [PST]	GLYCEROL MONOOLEATE

Chemical Name	Reference Name
1323-38-2 GLYCERYL MONORICINOLEATE	GLYCERYL MONORICINOLEATE
31566-31-1 GLYCERYL MONOSTEARATE	GLYCERYL MONOSTEARATE
30618-84-9 GLYCERYL MONOTHIOGLYCOLATE	GLYCERYL MONOTHIOGLYCOLATE
123-94-4 GLYCERYL STEARATE	GLYCERYL STEARATE
102-76-1 GLYCERYL TRIACETATE [NFP, NFP, PST]	GLYCEROL TRIACETATE
60-01-5 GLYCERYL TRIBUTYRATE	GLYCERYL TRIBUTYRATE
139-45-7 GLYCERYL TRIPROPIONATE	GLYCERYL TRIPROPIONATE
139-44-6 GLYCERYL TRIS(12-HYDROXYSTEARATE)	GLYCERYL TRIS(12-HYDROXYSTEARATE)
765-34-4 GLYCIDALDEHYDE [C65, CAL, FLA, IAR, MIN, MSL, NJL, NJL, WHM]	GLYCIDYLALDEHYDE
556-52-5 GLYCIDOL	GLYCIDOL
556-52-5 GLYCIDOL (2,3-EPOXY-1-PROPANOL) [BC1, BC1]	GLYCIDOL
RR-00275-5 GLYCIDOL (OXIRANEMETHANOL) AND ITS DERIVATIVES	GLYCIDOL (OXIRANEMETHANOL) AND ITS DERIVATIVES
2530-83-8 GLYCIDOXYPROPYLTRIMETHOXYSILANE	GLYCIDOXYPROPYLTRIMETHOXYSILANE
2530-83-8 GLYCIDOXYPROPYLTRIMETHOXYSILANE, GAMMA- [OTS]	GLYCIDOXYPROPYLTRIMETHOXYSILANE
106-90-1 GLYCIDYL ACRYLATE	GLYCIDYL ACRYLATE
765-34-4 GLYCIDYLALDEHYDE	GLYCIDYLALDEHYDE
26761-45-5 GLYCIDYL ESTER OF NEODECANOIC ACID [EPA, OTS]	NEODECANOIC ACID, OXIRANYLMETHYL ESTER
RR-00551-6 GLYCIDYL ETHERS	GLYCIDYL ETHERS
106-91-2 GLYCIDYL METHACRYLATE	GLYCIDYL METHACRYLATE
5431-33-4 GLYCIDYL OLEATE	GLYCIDYL OLEATE
32568-89-1 3-(2-GLYCIDYLOXYPROPYL)-1-GLYCIDOL-5,5-DIMETHYL-HYDANTOIN [EPA, OTS]	2,4-IMIDAZOLIDINEDIONE, 5,5-DIMETHYL-3-[2-(OXIRANYLMETHOXY)
7460-84-6 GLYCIDYL STEARATE	GLYCIDYL STEARATE
3033-77-0 GLYCIDYL TRIMETHYL AMMONIUM CHLORIDE	GLYCIDYL TRIMETHYL AMMONIUM CHLORIDE
56-40-6 GLYCINE	GLYCINE
139-13-9 GLYCINE, N,N-BIS(CARBOXYMETHYL)- [PAL, PAL]	NITRILOTRIACETIC ACID (NTA)
60-00-4 GLYCINE, N,N'-1,2-ETHANEDIYLBIS[N-(CARBOXYMETHYL)- [PAL]	ETHYLENEDIAMINE TETRAACETIC ACID (EDTA)PHENOXYETHYL)-
13256-22-9 GLYCINE, N-METHYL-N-NITROSO- [PAL, PAL]	N-NITROSOSARCOSINE
RR-01084-4 GLYCINE OF TALL-OIL FATTY ACIDS	GLYCINE OF TALL-OIL FATTY ACIDS
1071-83-6 GLYCINE, N-(PHOSPHONOMETHYL)-	GLYCINE, N-(PHOSPHONOMETHYL)-
111-55-7 GLYCOL DIACETATE [NFP, NFP]	1,2-ETHANEDIOL, DIACETATE
123-81-9 GLYCOL DIMERCAPTOACETATE	GLYCOL DIMERCAPTOACETATE
RR-00067-9 GLYCOL ETHERS	GLYCOL ETHERS
79-14-1 GLYCOLIC ACID [PST]	HYDROXYACETIC ACID
2836-32-0 GLYCOLIC ACID, SODIUM SALT	GLYCOLIC ACID, SODIUM SALT
RR-00956-3 GLYCOL MONOBENZOATE	GLYCOL MONOBENZOATE
107-16-4 GLYCOLONITRILE [MIN, NHS, NHS]	FORMALDEHYDE CYANOHYDRIN
RR-00958-5 GLYCOL, POLYETHYLENE, 3-SULFO-2-HYDROXYPROPYL-P-(1,1,3,3-TETRAMETHYL-BUTYL)PHENYL ETHER, SODIUM SALT	GLYCOL, POLYETHYLENE, 3-SULFO-2-HYDROXYPROPYL-P-(1,1,3,3-TET
111-60-4 GLYCOL STEARATE	GLYCOL STEARATE
496-46-8 GLYCOURIL	GLYCOURIL
107-22-2 GLYOXAL	GLYOXAL
107-22-2 GLYOXAL DIHYDRATE [CSR]	GLYOXAL

333

Chemical Name	Reference Name
1071-83-6 GLYPHOSATE [CD1, CSR, SDW, SDW]	GLYCINE, N-(PHOSPHONOMETHYL)-
13982-20-2 GOLD 193	GOLD 193
15756-89-5 GOLD 194	GOLD 194
14320-93-5 GOLD 195	GOLD 195
10043-49-9 GOLD 198	GOLD 198
RR-00462-6 GOLD 198M	GOLD 198M
14391-11-8 GOLD 199	GOLD 199
20091-45-6 GOLD 200	GOLD 200
RR-00461-5 GOLD 200M	GOLD 200M
23238-59-7 GOLD 201	GOLD 201
10233-88-2 GOLD SODIUM THIOSULFATE	GOLD SODIUM THIOSULFATE
13453-07-1 GOLD TRICHLORIDE	GOLD TRICHLORIDE
RR-00014-6 GRAIN DUST	GRAIN DUST
RR-00014-6 GRAIN DUST (OAT, WHEAT, BARLEY) [QBC, CEX]	GRAIN DUST
RR-00014-6 GRAIN DUST (OATS, WHEAT, BARLEY) [AUS, ONT]	GRAIN DUST
RR-00014-6 GRAIN DUST (OAT, WHEAT, BARLEY) [OS1, OS1, PAL, TLV]	GRAIN DUST
RR-01085-5 GRAPES (POMACE)	GRAPES (POMACE)
7782-42-5 GRAPHITE	GRAPHITE
7782-42-5 GRAPHITE (NATURAL) [QBC]	GRAPHITE
7782-42-5 GRAPHITE (SYNTHETIC) [BC1]	GRAPHITE
7782-42-5 GRAPHITE (ALL FORMS) [AUS]	GRAPHITE
7782-42-5 GRAPHITE-CONTAINING DUSTS [MAK, MAK]	GRAPHITE
7440-44-0 GRAPHITE (NATURAL) [NIO]	CARBON
7782-42-5 GRAPHITE, NATURAL [ALB]	GRAPHITE
7782-42-5 GRAPHITE (NATURAL) [NIO]	GRAPHITE
7782-42-5 GRAPHITE, NATURAL [ONT, OS1, OS1]	GRAPHITE
7782-42-5 GRAPHITE (NATURAL) DUST [FLA, MSL]	GRAPHITE
RR-00012-4 GRAPHITE, SYNTHETIC	GRAPHITE, SYNTHETIC
RR-00012-4 GRAPHITE (SYNTHETIC) [PAL]	GRAPHITE, SYNTHETIC
RR-01086-6 GRAVEL	GRAVEL
126-07-8 GRISEOFULVIN	GRISEOFULVIN
RR-01768-5 GROUTS	GROUTS
9000-29-7 GUAIAC GUM	GUAIAC GUM
90-05-1 GUAIACOL	GUAIACOL
150-19-6 M-GUAIACOL	M-GUAIACOL
1455-77-2 GUANAZOLE	GUANAZOLE
461-58-5 GUANIDINE, CYANO	GUANIDINE, CYANO
50-01-1 GUANIDINE HYDROCHLORIDE	GUANIDINE HYDROCHLORIDE
70-25-7 GUANIDINE, N-METHYL-N'-NITRO-N-NITROSO- [EPA, PAL, PAL, RCU, RCU]	N-METHYL-N'-NITRO-N-NITROSOGUANIDINE
506-93-4 GUANIDINE MONONITRATE	GUANIDINE MONONITRATE
52470-25-4 GUANIDINE NITRATE [NJL, PST, HM1, HM1]	GUANIDINE NITRATE (VAN)
52470-25-4 GUANIDINE NITRATE (VAN)	GUANIDINE NITRATE (VAN)
RR-01424-4 GUANYL NITROSOAMINOGUANYLIDENE HY-DRAZINE	GUANYL NITROSOAMINOGUANYLIDENE HYDRAZINE
109-27-3 GUANYL NITROSOAMINOGUANYLTE-TRAZENE [HM1]	1-TETRAZENE-1-CARBOXIMIDIC ACID, 4-(AMINOMETHYL)-, 2-NITROSO

Chemical Name	Reference Name
RR-01328-5 GUANYL NITROSOAMINOGUANYLTETRAZENE	GUANYL NITROSOAMINOGUANYLTETRAZENE
9000-30-0 GUAR GUM	GUAR GUM
4680-78-8 GUINEA GREEN B [CAL, IAR]	C.I. ACID GREEN 3
6706-59-8 L-GULITOL	L-GULITOL
9000-01-5 GUM ARABIC	GUM ARABIC
9000-28-6 GUM GHATTI	GUM GHATTI
8050-09-7 GUM ROSIN [PST]	ROSIN
8050-07-5 GUM THUS	GUM THUS
9000-65-1 GUM TRAGACANTH	GUM TRAGACANTH
86-50-0 GUTHION [ATS, CWA, EPA, NJL]	AZINPHOS-METHYL
RR-00767-0 GUTTA-PERCHA SOLUTION	GUTTA-PERCHA SOLUTION
13397-24-5 GYPSUM [ALB, BC1, BC1, GBR, MEX, OS1, OS1]	GYPSUM (CA(SO4).2H2O)
13397-24-5 GYPSUM (CA(SO4).2H2O)	GYPSUM (CA(SO4).2H2O)
16568-02-8 GYROMITRIN	GYROMITRIN
16568-02-8 GYROMITRIN (ACETALDEHYDE METHYL-FORMYLHYDRAZONE) [CAL, C65, C65]	GYROMITRIN
1317-60-8 HAEMATITE [IAR]	HEMATITE
7440-58-6 HAFNIUM	HAFNIUM
14922-51-1 HAFNIUM 170	HAFNIUM 170
14093-11-9 HAFNIUM 172	HAFNIUM 172
15757-23-0 HAFNIUM 173	HAFNIUM 173
15750-13-7 HAFNIUM 175	HAFNIUM 175
RR-00460-4 HAFNIUM 177M	HAFNIUM 177M
RR-00458-0 HAFNIUM 178M	HAFNIUM 178M
RR-00451-3 HAFNIUM 179M	HAFNIUM 179M
RR-00449-9 HAFNIUM 180M	HAFNIUM 180M
14900-21-1 HAFNIUM 181	HAFNIUM 181
29492-85-1 HAFNIUM 182	HAFNIUM 182
RR-00448-8 HAFNIUM 182M	HAFNIUM 182M
15832-40-3 HAFNIUM 183	HAFNIUM 183
29687-28-3 HAFNIUM 184	HAFNIUM 184
7440-58-6 HAFNIUM AND COMPOUNDS [NIO, NIO]	HAFNIUM
7440-58-6 HAFNIUM, ELEMENTAL [WHM]	HAFNIUM
RR-01701-6 HAIR COLORANTS, PERSONAL USE OF	HAIR COLORANTS, PERSONAL USE OF
RR-01702-7 HAIRDRESSER OR BARBER, OCCUPATIONAL EXPOSURES AS	HAIRDRESSER OR BARBER, OCCUPATIONAL EXPOSURES AS
23092-17-3 HALAZEPAM	HALAZEPAM
12298-43-0 HALLOYSITE	HALLOYSITE
428-25-1 HALOALKYL EPOXIDE	HALOALKYL EPOXIDE
RR-00171-8 HALOALKYL SUBSTITUTED CYCLIC ETHERS	HALOALKYL SUBSTITUTED CYCLIC ETHERS
RR-00325-8 HALOETHERS	HALOETHERS
RR-00972-3 HALOGENATED ACRYLONITRILE	HALOGENATED ACRYLONITRILE
RR-00274-4 HALOGENATED ALKYL EPOXIDES	HALOGENATED ALKYL EPOXIDES
RR-01246-4 HALOGENATED BIPHENYL GLYCIDYL ETHERS	HALOGENATED BIPHENYL GLYCIDYL ETHERS
RR-01636-4 HALOGENATED ETHANES	HALOGENATED ETHANES
RR-01634-2 HALOGENATED ETHANES CLASS STUDY	HALOGENATED ETHANES CLASS STUDYIATIVE)
RR-01667-1 HALOGENATED PHENYL ALKANE	HALOGENATED PHENYL ALKANE
RR-01641-1 HALOGENATED PHOSPHATE ESTERS	HALOGENATED PHOSPHATE ESTERSIPHATIC SPIROKE-TAL
RR-00327-0 HALOMETHANES	HALOMETHANES

Chemical Name	Reference Name
RR-00327-0 HALOMETHANES, N.O.S. [RCU]	HALOMETHANES
353-59-3 HALON 1211 [CAA]	CHLORODIFLUOROBROMO-METHANE
75-63-8 HALON 1301 [CAA]	TRIFLUOROBROMOMETHANE
124-73-2 HALON 2402 [CAA]	DIBROMOTETRAFLUOROETHANE (HALON 2402)
RR-00929-0 HALONITROBENZOIC ACID, SUBSTITUTED	HALONITROBENZOIC ACID, SUBSTITUTED
RR-01176-7 HALOPHENYL SULFONAMIDE SALT	HALOPHENYL SULFONAMIDE SALT
151-67-7 HALOTHANE	HALOTHANE
RR-00328-1 HAY	HAY
2784-94-3 HC BLUE 1	HC BLUE 1
33229-34-4 HC BLUE 2	HC BLUE 2
2784-94-3 HC BLUE NO. 1 [IAR]	HC BLUE 1
33229-34-4 HC BLUE NO. 2 [IAR]	HC BLUE 2
134237-32-4 HCFC-121 [CAA]	TETRACHLOROFLUOROETHANE
354-21-2 HCFC-122 [CAA]	1,2,2-TRICHLORO-1,1-DIFLUOROETHANE
306-83-2 HCFC-123 [CAA]	ETHANE, 2,2-DICHLORO-1,1,1-TRIFLUORO-
2837-89-0 HCFC-124 [CAA]	ETHANE, 2-CHLORO-1,1,1,2-TETRAFLUORO-
134237-34-6 HCFC-131 [CAA]	TRICHLOROFLUOROETHANE
1649-08-7 HCFC-132B [CAA]	1,2-DICHLORO-1,1-DIFLUOROETHANE
75-88-7 HCFC-133A [CAA]	2-CHLORO-1,1,1-TRIFLUOROETHANE
1717-00-6 HCFC-141B [CAA]	ETHANE, 1,1-DICHLORO-1-FLUORO-
75-68-3 HCFC-142B [CAA]	1-CHLORO-1,1-DIFLUOROETHANE
75-43-4 HCFC-21 [CAA]	DICHLOROFLUOROMETHANE
75-45-6 HCFC-22 [CAA]	CHLORODIFLUOROMETHANE
134237-35-7 HCFC-221 [CAA]	HEXACHLOROFLUOROPROPANE
134237-36-8 HCFC-222 [CAA]	PENTACHLORODIFLUOROPROPANE
134237-37-9 HCFC-223 [CAA]	TETRACHLOROTRIFLUOROPROPANE
134237-38-0 HCFC-224 [CAA]	TRICHLOROTETRAFLUOROPROPANE
422-56-0 HCFC-225CA [CAA]	3,3-DICHLORO-1,1,1,2,2-PENTAFLUOROPROPANE
507-55-1 HCFC-225CB [CAA]	1,3-DICHLORO-1,1,2,2,3-PENTAFLUOROPROPANE
134308-72-8 HCFC-226 [CAA]	CHLOROHEXAFLUOROPROPANE
134190-48-0 HCFC-231 [CAA]	PENTACHLOROFLUOROPROPANE
134237-39-1 HCFC-232 [CAA]	TETRACHLORODIFLUOROPROPANE
134237-40-4 HCFC-233 [CAA]	TRICHLOROTRIFLUOROPROPANE
127564-83-4 HCFC-234 [CAA]	DICHLOROTETRAFLUOROPROPANE
134237-41-5 HCFC-235 [CAA]	CHLOROPENTAFLUOROPROPANE
134190-49-1 HCFC-241 [CAA]	TETRACHLOROFLUOROPROPANE
134237-42-6 HCFC-242 [CAA]	TRICHLORODIFLUOROPROPANE
134237-43-7 HCFC-243 [CAA]	DICHLOROTRIFLUOROPROPANE
679-85-6 HCFC-244 [CAA]	3-CHLORO-1,1,2,2-TETRAFLUOROPROPANE
134190-51-5 HCFC-251 [CAA]	TRICHLOROFLUOROPROPANE
134190-52-6 HCFC-252 [CAA]	DICHLORODIFLUOROPROPANE
134237-44-8 HCFC-253 [CAA]	CHLOROTRIFLUOROPROPANE
134237-45-9 HCFC-261 [CAA]	DICHLOROFLUOROPROPANE
134190-53-7 HCFC-262 [CAA]	CHLORODIFLUOROPROPANE
134190-54-8 HCFC-271 [CAA]	CHLOROFLUOROPROPANE
593-70-4 HCFC-31 [CAA]	CHLOROFLUOROMETHANE
319-84-6 ALPHA-HCH [IAR]	ALPHA-BHC

Chemical Name	Reference Name
319-85-7 BETA-HCH [IAR]	BETA-BHC
608-73-1 TECHNICAL HCH [MSL]	HEXACHLOROCYCLOHEXANE (MIXED ISOMERS)
58-89-9 GAMMA-HCH (LINDANE) [IAR]	LINDANE
2871-01-4 HC RED 3 [CSR, CSR]	HC RED NO. 3
2871-01-4 HC RED NO. 3	HC RED NO. 3
59820-43-8 HC YELLOW 4	HC YELLOW 4
59820-43-8 HC YELLOW NO. 4 [IAR]	HC YELLOW 4
64741-67-9 HEAVY AROMATIC BOTTOMS	HEAVY AROMATIC BOTTOMS
67891-79-6 HEAVY AROMATIC DISTILLATE (PETROLEUM)	HEAVY AROMATIC DISTILLATE (PETROLEUM)
64741-61-3 HEAVY CATALYTICALLY CRACKED DISTILLATES [IAR]	PETROLEUM DISTILLATES, HEAVY CATALYTIC CRACKED
64742-11-6 HEAVY NAPHTHENIC DISTILLATE SOLVENT EXTRACT	HEAVY NAPHTHENIC DISTILLATE SOLVENT EXTRACT
70592-78-8 HEAVY VACUUM DISTILLATES [IAR]	PETROLEUM DISTILLATES, VACUUM
12173-47-6 HECTORITE	HECTORITE
7440-59-7 HELIUM	HELIUM
1317-60-8 HEMATITE	HEMATITE
1120-21-4 HENDECANE [NFP, NFP]	UNDECANE
76-44-8 HEPTACHLOR	HEPTACHLOR
76-44-8 HEPTACHLOR[1,4,5,6,7,8,8-HEPTACHLORO-3A,4,7,7A-TETRAHYDRO-4,7-METHANO-1H-INDENE] [313]	HEPTACHLOR
76-44-8 HEPTACHLOR (AND ITS EPOXIDE) [RC2, RC2]	HEPTACHLOR
76-44-8 HEPTACHLOR AND METABOLITES [CW3]	HEPTACHLOR
1024-57-3 HEPTACHLOR EPOXIDE	HEPTACHLOR EPOXIDE)
1024-57-3 HEPTACHLOR EPOXIDE (BHC-HEXACHLOROCYCLOHEXANE) [CW2]	HEPTACHLOR EPOXIDE
76-44-8 HEPTACHLOR + HEPTACHLOR EPOXIDE [CD1]	HEPTACHLOR
RR-00566-3 HEPTACHLOR METABOLITES	HEPTACHLOR METABOLITES
37871-00-4 HEPTACHLOROBENZO-P-DIOXIN	HEPTACHLOROBENZO-P-DIOXIN
35822-46-9 1,2,3,4,6,7,8-HEPTACHLORODIBENZO-P-DIOXIN	1,2,3,4,6,7,8-HEPTACHLORODIBENZO-P-DIOXIN
RR-01174-5 HEPTACHLORODIBENZO-P-DIOXINS	HEPTACHLORODIBENZO-P-DIOXINS
38998-75-3 1,2,3,4,7,8,9-HEPTACHLORO DIBENZOFURAN	1,2,3,4,7,8,9-HEPTACHLORO DIBENZOFURAN
38998-75-3 HEPTACHLORODIBENZOFURAN [ATS]	1,2,3,4,7,8,9-HEPTACHLORO DIBENZOFURAN
67562-39-4 1,2,3,4,6,7,8-HEPTACHLORODIBENZOFURAN	1,2,3,4,6,7,8-HEPTACHLORODIBENZOFURAN
RR-01173-4 HEPTACHLORODIBENZOFURANS	HEPTACHLORODIBENZOFURANS
135401-87-5 HEPTACHLOROFLUOROPROPANE	HEPTACHLOROFLUOROPROPANE
1652-63-7 3-(((HEPTADECAFLUOROOCTYL)SULFONYL)AMINO)-N,N,N-TRIMETHYL-1-PROPANAMINIUM IODIDE	3-(((HEPTADECAFLUOROOCTYL)SULFONYL)AMINO)-N,N,N-TRIMETHYL-1-
52783-44-5 HEPTADECANOL	HEPTADECANOL
95-38-5 2-(8-HEPTADECENYL)-4,5-DIHYDRO-1H-IMIDAZOLE-1-ETHANOL [PST]	1H-IMIDAZOLE-1-ETHANOL, 2-(8-HEPTADACENYL)-4,5-DIHYDRO-
62449-33-6 2-(8-HEPTADECENYL)-4,5-DIHYDRO-1H-IMIDAZOLE-1-ETHANOL, MONOHYDROCHLORIDE, (Z)-	2-(8-HEPTADECENYL)-4,5-DIHYDRO-1H-IMIDAZOLE-1-ETHANOL, MONOH
21652-27-7 (Z)-2-(8-HEPTADECENYL)-2-IMIDAZOLINE-1-ETHANOL [PST]	OLEYL BASED IMIDAZOLINE
14408-42-5 2-(8-HEPTADECENYL)-4-METHYL-2-OXAZOLINE-4-METHANOL	2-(8-HEPTADECENYL)-4-METHYL-2-OXAZOLINE-4-METHANOL
95-19-2 HEPTADECYL HYDROXYETHYL IMIDAZOLINE [PST]	1-(HYDROXYETHYL)-2-(HEPTADECYL)IMIDAZOLINE

Chemical Name	Reference Name
105-28-2 2-HEPTADECYL-2-IMIDAZOLINE	2-HEPTADECYL-2-IMIDAZOLINE
13470-50-3 2-HEPTADECYL-1-METHYL-1-(2-STEAROY-LAMIDO)ETHYL-2-IMIDAZOLINIUM METHYL SULFATE	2-HEPTADECYL-1-METHYL-1-(2-STEAROYLAMIDO) ETHYL-2-IMIDAZOLINI
5910-79-2 HEPTADECYL SULFATE, SODIUM SALT	HEPTADECYL SULFATE, SODIUM SALT
504-20-1 2,5-HEPTADIEN-4-ONE, 2,6-DIMETHYL- [PAL]	PHORONE
111-71-7 HEPTANAL	HEPTANAL
122-40-7 HEPTANAL, 2-(PHENYLMETHYLENE)-	HEPTANAL, 2-(PHENYLMETHYLENE)-
111-68-2 1-HEPTANAMINE	1-HEPTANAMINE
142-82-5 N-HEPTANE [CAL, CEX, CEX, GBR, GBR]	HEPTANE (N-)
142-82-5 HEPTANE [HM1, HM1, MEX, MEX, MEX, NFP, NFP]	HEPTANE (N-)
142-82-5 N-HEPTANE [NIO, NIO, NIO, NJL, NJL, ONT, ONT, ONT, ONT]	HEPTANE (N-)
142-82-5 HEPTANE [OTV]	HEPTANE (N-)
142-82-5 N-HEPTANE [QBC, QBC, TSC, TSC, TSC]	HEPTANE (N-)
142-82-5 HEPTANE (N-HEPTANE) [BC1, BC1, FLA, OS1, OS1, OS1, AUS, AUS, MSL, TLV, TLV]	HEPTANE (N-)
123-04-6 HEPTANE, 3-(CHLOROMETHYL)- [PAL]	2-ETHYLHEXYL CHLORIDE
4032-86-4 HEPTANE, 3,3-DIMETHYL- [PAL]	3,3-DIMETHYLHEPTANE
103-44-6 HEPTANE, 3-[(ETHENYLOXY)METHYL]- [PAL]	VINYL 2-ETHYLHEXYL ETHER
142-82-5 HEPTANE (N-)	HEPTANE (N-)
1639-09-4 1-HEPTANETHIOL	1-HEPTANETHIOL
111-14-8 HEPTANOIC ACID	HEPTANOIC ACID
543-49-7 2-HEPTANOL	2-HEPTANOL
589-82-2 3-HEPTANOL	3-HEPTANOL
106-35-4 HEPTAN-3-ONE [GBR, GBR]	ETHYL BUTYL KETONE
106-35-4 3-HEPTANONE [ONT, PAL]	ETHYL BUTYL KETONE
110-43-0 HEPTAN-2-ONE [GBR]	METHYL N-AMYL KETONE
110-43-0 2-HEPTANONE [ONT, PAL, PST]	METHYL N-AMYL KETONE
123-19-3 4-HEPTANONE [NFP, NFP, ONT, PAL]	DIPROPYL KETONE
108-83-8 HEPTANONE, 2,6-DIMETHYL-, 4- [OTS]	DIISOBUTYL KETONE
108-83-8 4-HEPTANONE, 2,6-DIMETHYL- [PAL]	DIISOBUTYL KETONE
541-85-5 3-HEPTANONE, 5-METHYL- [PAL]	ETHYL AMYL KETONE
104503-68-6 3,6,9,12,15,18,21-HEPTAOXATETRATRIAOC-TANOIC ACID, SODIUM SALT	3,6,9,12,15,18,21-HEPTAOXATETRATRIAOCTANOIC ACID, SODIUM SAL
106-72-9 5-HEPTENAL, 2,6-DIMETHYL-	5-HEPTENAL, 2,6-DIMETHYL-
592-76-7 1-HEPTENE	1-HEPTENE
25339-56-4 HEPTENE	HEPTENE
25339-56-4 N-HEPTENE [HM1, HM1]	HEPTENE
81624-04-6 HEPTENE	HEPTENE
14686-13-6 2-HEPTENE, (E)- [PAL]	HEPTYLENE-2-TRANS
RR-00519-6 3-HEPTENE (MIXED CIS AND TRANS) [NFP, NFP]	3-HEPTENE (MIXED CIS- AND TRANS- ISOMERS)
RR-00519-6 3-HEPTENE (MIXED CIS- AND TRANS- ISO-MERS)	3-HEPTENE (MIXED CIS- AND TRANS- ISOMERS)
23560-59-0 HEPTENOPHOS	HEPTENOPHOS
111-70-6 N-HEPTYL ALCOHOL	N-HEPTYL ALCOHOL
111-68-2 HEPTYLAMINE [FLA, MSL, NFP, NFP]	1-HEPTANAMINE
629-06-1 NORMAL-HEPTYL CHLORIDE [HM1]	1-CHLOROHEPTANE

Chemical Name	Reference Name
14686-13-6 HEPTYLENE-2-TRANS	HEPTYLENE-2-TRANS
25339-56-4 HEPTYLENE [FLA, MSL, NFP, NFP]	HEPTENE
561-27-3 HEROIN	HEROIN
RR-00959-6 HETEROCYCLIC ALDEHYDE IMINE	HETEROCYCLIC ALDEHYDE IMINERAMETHYLBUTYL) PHENYL ETHER, SODIUM SALT
12027-67-7 HEXAAMMONIUM MOLYBDATE	HEXAAMMONIUM MOLYBDATE
36355-01-8 HEXABROMOBIPHENYL	HEXABROMOBIPHENYL
36355-01-8 HEXABROMO-1,1-BIPHENYL [TSE]	HEXABROMOBIPHENYL
59080-40-9 2,2',4,4',5,5'-HEXABROMO-1,1'-BIPHENYL	2,2',4,4',5,5'-HEXABROMO-1,1'-BIPHENYL
67774-32-7 HEXABROMOBIPHENYL (FIREMASTER FF-1) [NTP]	FIREMASTER FF-1
3194-55-6 HEXABROMOCYCLODODECANE	HEXABROMOCYCLODODECANE
116-16-5 HEXACHLOROACETONE	HEXACHLOROACETONE
118-74-1 HEXACHLOROBENZENE	HEXACHLOROBENZENE
87-68-3 HEXACHLOROBUTADIENE	HEXACHLOROBUTADIENE
87-68-3 HEXACHLORO-1,3-BUTADIENE [313, CSR, MAK, MAK, ONT, ONT, RC2, RC2, RCA, WHM]	HEXACHLOROBUTADIENE
608-73-1 HEXACHLOROCYCLOHEXANE [CW3, NTP, RCA]	HEXACHLOROCYCLOHEXANE (MIXED ISOMERS)
319-84-6 ALPHA-HEXACHLOROCYCLOHEXANE [313, FLA, MSL, NJL, NTP]	ALPHA-BHC
319-85-7 BETA-HEXACHLOROCYCLOHEXANE [NJL, NTP, FLA, MSL]	BETA-BHC
58-89-9 HEXACHLOROCYCLOHEXANE, GAMMA ISO-MER [C65, C65]	LINDANE
58-89-9 GAMMA-HEXACHLOROCYCLOHEXANE [MAK, MAK, MAK]	LINDANE
319-84-6 HEXACHLOROCYCLOHEXANE, ALPHA ISO-MER [C65]	ALPHA-BHC
319-85-7 HEXACHLOROCYCLOHEXANE, BETA ISO-MER [C65]	BETA-BHC
608-73-1 HEXACHLOROCYCLOHEXANE [ATS, CAL]	HEXACHLOROCYCLOHEXANE (MIXED ISOMERS)
608-73-1 1,2,3,4,5,6-HEXACHLOROCYCYCLOHEX-ANE [MAK, MAK]	HEXACHLOROCYCLOHEXANE (MIXED ISOMERS)
319-84-6 HEXAHCLOROCYCLOHEXANE, ALPHA- [ATS]	ALPHA-BHC
319-85-7 HEXACHOLOCYCLOHEXANE, BETA- [ATS]	BETA-BHC
319-86-8 HEXACHLOROCYCLOHEXANE, DELTA- [ATS]	DELTA-BHC
58-89-9 HEXACHLOROCYCLOHEXANE, GAMMA- [ATS]	LINDANE
608-73-1 HEXACHLOROCYCLOHEXANE ISOMERS [MIN]	HEXACHLOROCYCLOHEXANE (MIXED ISOMERS)
608-73-1 HEXACHLOROCYCLOHEXANE (MIXED ISO-MERS)	HEXACHLOROCYCLOHEXANE (MIXED ISOMERS)
77-47-4 HEXACHLOROCYCLO-PENTADIENE [OS1]	HEXACHLOROCYCLOPENTADIENE
77-47-4 HEXACHLOROCYCLOPENTADIENE	HEXACHLOROCYCLOPENTADIENE
19408-74-3 1,2,3,7,8,9-HEXACHLORODIBENZO-P-DIOXIN	1,2,3,7,8,9-HEXACHLORODIBENZO-P-DIOXIN
34465-46-8 HEXACHLORODIBENZODIOXIN	HEXACHLORODIBENZODIOXIN
34465-46-8 HEXACHLORODIBENZO-P-DIOXIN [ATS]	HEXACHLORODIBENZODIOXIN
39227-28-6 1,2,3,4,7,8-HEXACHLORODIBENZO-P-DIOXIN	1,2,3,4,7,8-HEXACHLORODIBENZO-P-DIOXIN
57653-85-7 1,2,3,6,7,8-HEXACHLORODIBENZO-P-DIOXIN	1,2,3,6,7,8-HEXACHLORODIBENZO-P-DIOXIN
RR-00509-4 HEXACHLORODIBENZO-P-DIOXINS	HEXACHLORODIBENZO-P-DIOXINS
RR-00509-4 HXCDDS (ALL HEXACHLORODIBENZO-P-DIOXINS) [UTS]	HEXACHLORODIBENZO-P-DIOXINS
55673-89-7 1,2,3,4,7,8,9-HEXACHLORODIBENZOFURAN	1,2,3,4,7,8,9-HEXACHLORODIBENZOFURAN

Chemical Name	Reference Name
55684-94-1 HEXACHLORODIBENZOFURAN	HEXACHLORODIBENZOFURAN
57117-44-9 1,2,3,6,7,8-HEXACHLORO DIBENZOFURAN	1,2,3,6,7,8-HEXACHLORO DIBENZOFURAN
70648-26-9 1,2,3,4,7,8-HEXACHLORO DIBENZOFURAN	1,2,3,4,7,8-HEXACHLORO DIBENZOFURAN
72918-21-9 1,2,3,7,8,9-HEXACHLORO DIBENZOFURAN	1,2,3,7,8,9-HEXACHLORO DIBENZOFURAN
60851-34-5 2,3,4,6,7,8-HEXACHLORO DIBENZOFURANS	2,3,4,6,7,8-HEXACHLORO DIBENZOFURANS
RR-00505-0 HEXACHLORODIBENZOFURANS	HEXACHLORODIBENZOFURANSHER SALTS
RR-00505-0 HXCDFS (ALL HEXACHLORODIBENZOFURANS) [UTS]	HEXACHLORODIBENZOFURANS
134452-44-1 HEXACHLORODIFLUOROPROPANE	HEXACHLORODIFLUOROPROPANE
31242-93-0 HEXACHLORO DIPHENYL OXIDE [NFP, PAL]	CHLORINATED DIPHENYL OXIDE
67-72-1 HEXACHLOROETHANE	HEXACHLOROETHANE
134237-35-7 HEXACHLOROFLUOROPROPANE	HEXACHLOROFLUOROPROPANE
1335-87-1 HEXACHLORONAPHTHALENE	HEXACHLORONAPHTHALENE
3389-71-7 1,2,3,4,7,7-HEXACHLORONORBORNADIENE	1,2,3,4,7,7-HEXACHLORONORBORNADIENE
70-30-4 HEXACHLOROPHENE	HEXACHLOROPHENE
70-30-4 HEXACHLOROPHENE (HCP) [CAL]	HEXACHLOROPHENE
1888-71-7 HEXACHLOROPROPENE	HEXACHLOROPROPENE
23732-94-7 HEXACOSAMETHYLCYCLOTRIDECASILOXANE	HEXACOSAMETHYLCYCLOTRIDECASILOXANE
2471-08-1 HEXACOSAMETHYLDODECASILOXANE	HEXACOSAMETHYLDODECASILOXANE
556-68-3 HEXADECAMETHYLCYCLOOCTASILOXANE	HEXADECAMETHYLCYCLOOCTASILOXANE
541-01-5 HEXADECAMETHYLHEPTASILOXANE	HEXADECAMETHYLHEPTASILOXANE
544-76-3 HEXADECANE	HEXADECANE
2917-26-2 1-HEXADECANETHIOL	1-HEXADECANETHIOL
25360-09-2 TERT-HEXADECANETHIOL	TERT-HEXADECANETHIOL
57-10-3 HEXADECANOIC ACID [OTS, PST, TSC]	PALMITIC ACID
629-73-2 HEXADECYLENE-1	HEXADECYLENE-1
15965-99-8 HEXADECYL GLYCIDYL ETHER [EPA, OTS]	OXIRANE, [(HEXADECYLOXY)METHYL]-
1120-01-0 HEXADECYL SULFATE, SODIUM SALT	HEXADECYL SULFATE, SODIUM SALT
5894-60-0 HEXADECYLTRICHLOROSILANE	HEXADECYLTRICHLOROSILANE
57-09-0 HEXADECYLTRIMETHYLAMMONIUM BROMIDE	HEXADECYLTRIMETHYLAMMONIUM BROMIDE
142-83-6 2,4-HEXADIENAL [CSR, FLA, MSL]	SORBALDEHYDE
80466-34-8 2,4-HEXADIENAL	2,4-HEXADIENAL
592-42-7 HEXADIENE	HEXADIENE
592-45-0 1,4-HEXADIENE	1,4-HEXADIENE
42296-74-2 HEXADIENE	HEXADIENE
RR-01352-5 HEXADIENES	HEXADIENES
821-08-9 1,5-HEXADIEN-3-YNE [PAL]	DIVINYL ACETYLENE
757-58-4 HEXAETHYL TETRAPHOSPHATE	HEXAETHYL TETRAPHOSPHATE
757-58-4 HEXAETHYLTETRAPHOSPHATE [HM1, HM1, HM1, HM1, HM1]	HEXAETHYL TETRAPHOSPHATE
684-16-2 HEXAFLUOROACETONE	HEXAFLUOROACETONE
10543-95-0 HEXAFLUOROACETONE HYDRATE	HEXAFLUOROACETONE HYDRATE
392-56-3 HEXAFLUOROBENZENE	HEXAFLUOROBENZENE
76-16-4 HEXAFLUOROETHANE	HEXAFLUOROETHANE
16940-81-1 HEXAFLUOROPHOSPHORIC ACID	HEXAFLUOROPHOSPHORIC ACID
920-66-1 1,1,1,3,3,3-HEXAFLUORO-2-PROPANOL	1,1,1,3,3,3-HEXAFLUORO-2-PROPANOL
116-15-4 HEXAFLUOROPROPENE [CTR, OTS]	HEXAFLUOROPROPYLENE
116-15-4 HEXAFLUOROPROPYLENE	HEXAFLUOROPROPYLENE
428-59-1 HEXAFLUOROPROPYLENE OXIDE	HEXAFLUOROPROPYLENE OXIDE

Chemical Name	Reference Name
105-60-2 HEXAHYDRO-2H-AZETIN-2-ONE [FLA]	CAPROLACTAM
7779-27-3 HEXAHYDRO-1,3,5-TRIETHYL-S-TRIAZINE	HEXAHYDRO-1,3,5-TRIETHYL-S-TRIAZINE
121-82-4 HEXAHYDRO-1,3,5-TRINITRO-1,3,5-TRI-AZINE [GBR, GBR, GBR, ONT, ONT]	CYCLONITE
3089-11-0 HEXAKIS(METHOXYMETHYL)MELAMINE	HEXAKIS(METHOXYMETHYL)MELAMINE
66-25-1 HEXALDEHYDE	HEXALDEHYDE
3089-11-0 HEXA(METHOXYMETHYL) MELAMINE [TSC, TSC, TSC]	HEXAKIS(METHOXYMETHYL)MELAMINE
541-05-9 HEXAMETHYLCYCLOTRISILOXANE [TSC, TSC]	TRIALKOXY SILANE
999-97-3 HEXAMETHYLDISILIZANE	HEXAMETHYLDISILIZANE
107-46-0 HEXAMETHYLDISILOXANE	HEXAMETHYLDISILOXANE
124-09-4 HEXAMETHYLENE DIAMINE [NJL, NJL]	HEXAMETHYLENEDIAMINE
124-09-4 HEXAMETHYLENEDIAMINE	HEXAMETHYLENEDIAMINE
4835-11-4 HEXAMETHYLENEDIAMINE, N,N'-DIBUTYL-	HEXAMETHYLENEDIAMINE, N,N'-DIBUTYL-
822-06-0 HEXAMETHYLENE-1,6-DIISOCYANATE [CAB]	HEXAMETHYLENE DIISOCYANATE
822-06-0 HEXAMETHYLENE DIISOCYANATE	HEXAMETHYLENE DIISOCYANATE
822-06-0 HEXAMETHYLENE-1,6-DIISOCYANATE [313, CAA]	HEXAMETHYLENE DIISOCYANATE
822-06-0 HEXAMETHYLENE-1,6-DIISOCYANATE (HDI) [ONT, ONT, ONT]	HEXAMETHYLENE DIISOCYANATE
822-06-0 HEXAMETHYLENE DIISOCYANATE (HDI) [NHS, NHS]	HEXAMETHYLENE DIISOCYANATE
822-06-0 HEXAMETHYLENE DIISOCYANATE, 1,6- [OTS]	HEXAMETHYLENE DIISOCYANATE
822-06-0 HEXAMETHYLENEDIISOCYANATE [WHM]	HEXAMETHYLENE DIISOCYANATE
4035-89-6 HEXAMETHYLENE DIISOCYANATE BIURET	HEXAMETHYLENE DIISOCYANATE BIURET
822-06-0 1,6-HEXAMETHYLENEDIISOCYANATE [MAK, MAK, MAK]	HEXAMETHYLENE DIISOCYANATE
28182-81-2 HEXAMETHYLENE DIISOCYANATE HO-MOPOLYMER	HEXAMETHYLENE DIISOCYANATE HOMOPOLYMER
629-11-8 HEXAMETHYLENE GLYCOL [OTS]	1,6-HEXANEDIOL
111-49-9 HEXAMETHYLENEIMINE	HEXAMETHYLENEIMINE
100-97-0 HEXAMETHYLENETETRAMINE [HON, PST]	METHENAMINE
24360-05-2 HEXAMETHYLENETETRAMINE HYDROGEN CHLORIDE	HEXAMETHYLENETETRAMINE HYDROGEN CHLORIDE
283-66-9 HEXAMETHYLENE TRIPEROXIDE DIAMINE	HEXAMETHYLENE TRIPEROXIDE DIAMINE
105554-30-1 HEXAMETHYLOL BENZENE HEXANITRATE	HEXAMETHYLOL BENZENE HEXANITRATE
531-18-0 HEXAMETHYLOLMELAMINE	HEXAMETHYLOLMELAMINE
680-31-9 HEXAMETHYL PHOSPHORAMIDE	HEXAMETHYL PHOSPHORAMIDE
680-31-9 HEXAMETHYLPHOSPHORAMIDE [313, C65, C65, CAA, CAL, IAR, MIN, NTP, TSC, TSC, TSE]	HEXAMETHYL PHOSPHORAMIDE
680-31-9 HEXAMETHYLPHOSPHORIC ACID TRI-AMDE [MAK]	HEXAMETHYL PHOSPHORAMIDE
548-62-9 HEXAMETHYL-P-ROSANILINE CHLORIDE [CSR]	C.I. BASIC VIOLET 1
22397-33-7 3,3,6,6,9,9-HEXAMETHYL-1,2,4,5-TETRAOXOCY-CLONONANE	3,3,6,6,9,9-HEXAMETHYL-1,2,4,5-TETRAOXOCY-CLONONANE
100-97-0 HEXAMINE [HM1, HM1, NJL]	METHENAMINE
66-25-1 HEXANAL [FLA, MSL, NFP, NFP, PAL]	HEXALDEHYDE
123-05-7 HEXANAL, 2-ETHYL- [PAL, TSC, TSC, TSC]	2-ETHYLHEXALDEHYDE
496-03-7 HEXANAL, 2-ETHYL-3-HYDROXY- [PAL]	BUTYRALDOL
5435-64-3 HEXANAL, 3,5,5-TRIMETHYL-	HEXANAL, 3,5,5-TRIMETHYL-
111-26-2 1-HEXANAMINE [PAL]	N-HEXYLAMINE

Chemical Name	Reference Name
104-75-6 1-HEXANAMINE, 2-ETHYL- [PAL]	2-ETHYLHEXYLAMINE
106-20-7 1-HEXANAMINE, 2-ETHYL-N-(2-ETHYLHEXYL)- [PAL]	DI(2-ETHYLHEXYL)AMINE
143-16-8 1-HEXANAMINE, N-HEXYL- [PAL]	DIHEXYLAMINE
110-54-3 HEXANE	HEXANE
110-54-3 N-HEXANE [313, BEI, CSR, GBR, NIO, NIO, NIO, NJL, NJL, ONT, ONT, OS1, OS1, TSC, WHM]	HEXANE
110-54-3 HEXANE (N-HEXANE) [BC1, BC1, AUS, TLV]	HEXANE
110-54-3 HEXANE (N-HIXANE) [QBC]	HEXANE
RR-01809-7 HEXANE (OTHER ISOMERS)	HEXANE (OTHER ISOMERS)
RR-01809-7 HEXANE (ISOMERS OTHER THAN "N") [QBC, QBC]	HEXANE (OTHER ISOMERS)
RR-00003-3 HEXANE (ALL ISOMERS) [CAL]	HEXANE ISOMERS
RR-00003-3 HEXANE, ALL ISOMERS [GBR, GBR, WHM]	HEXANE ISOMERS
544-10-5 HEXANE, 1-CHLORO- [PAL]	1-CHLOROHEXANE
25495-90-3 HEXANE, CHLORO-	HEXANE, CHLORO-
124-09-4 1,6-HEXANEDIAMINE [OTS, TLV]	HEXAMETHYLENEDIAMINE
4835-11-4 1,6-HEXANEDIAMINE,N,N'-DIBUTYL [MSL]	HEXAMETHYLENEDIAMINE, N,N'-DIBUTYL-
4835-11-4 1,6-HEXANEDIAMINE, N,N'-DIBUTYL- [TSC, TSC]	HEXAMETHYLENEDIAMINE, N,N'-DIBUTYL-
6055-52-3 1,6-HEXANEDIAMINE, DIHYDROCHLORIDE	1,6-HEXANEDIAMINE, DIHYDROCHLORIDE
822-06-0 HEXANE, 1,6-DIISOCYANATO- [TSC, TSC, TSC]	HEXAMETHYLENE DIISOCYANATE
15646-96-5 HEXANE, 1,6-DIISOCYANATO-2,4,4-TRIMETHYL-	HEXANE, 1,6-DIISOCYANATO-2,4,4-TRIMETHYL-TE-TRAKIS-, HEXASODIUM SALT
16938-22-0 HEXANE, 1,6-DIISOCYANATO-2,2,4-TRIMETHYL-	HEXANE, 1,6-DIISOCYANATO-2,2,4-TRIMETHYL-ETHYL ESTER
584-94-1 HEXANE, 2,3-DIMETHYL- [PAL]	2,3-DIMETHYLHEXANE
589-43-5 HEXANE, 2,4-DIMETHYL- [PAL]	2,4-DIMETHYLHEXANE
111-69-3 HEXANEDINITRILE [PAL, TSC, TSC]	ADIPONITRILE
68955-64-6 HEXANEDINITRILE, HYDROGENATED, HIGH-BOILING FRACTION, PHOSPHONOMETHY-LATED	HEXANEDINITRILE, HYDROGENATED, HIGH-BOILING FRACTION, PHOSPH
124-04-9 HEXANEDIOIC ACID [PAL]	ADIPIC ACID
RR-01644-4 HEXANEDIOIC ACID, DIETHENYL ESTER	HEXANEDIOIC ACID, DIETHENYL ESTER
RR-00960-9 HEXANEDIOIC ACID, POLYMER WITH 1,2-ETHANEDIOL AND 1,7-DIISOCYANATO-2,2,4(OR 2,4,4)-TRIMETHYLHEXANE, 2-HYDROXYETHYL-ACRYLATED-BLOCKED	HEXANEDIOIC ACID, POLYMER WITH 1,2-ETHANEDIOL AND 1,7-DIISOC
629-11-8 1,6-HEXANEDIOL	1,6-HEXANEDIOL
2935-44-6 2,5-HEXANEDIOL	2,5-HEXANEDIOLPYRIDINYL) ESTER
13048-33-4 1,6-HEXANEDIOL DIACRYLATE	1,6-HEXANEDIOL DIACRYLATE
13048-33-4 HEXANEDIOL DIACRYLATE [WEL]	1,6-HEXANEDIOL DIACRYLATE
110-13-4 2,5-HEXANEDIONE	2,5-HEXANEDIONE
111-50-2 HEXANEDIOYL DICHLORIDE [PAL]	ADIPOYL CHLORIDE
3074-75-7 HEXANE, 4-ETHYL-2-METHYL- [PAL]	2-METHYL-4-ETHYLHEXANE
3074-77-9 HEXANE, 3-ETHYL-4-METHYL- [PAL]	3-METHYL-4-ETHYLHEXANE
RR-00003-3 HEXANE ISOMERS	HEXANE ISOMERS
589-34-4 HEXANE, 3-METHYL- [PAL]	3-METHYLHEXANE
591-76-4 HEXANE, 2-METHYL- [PAL]	ISOHEPTANE
RR-00003-3 HEXANE, OTHER ISOMERS [AUS, AUS]	HEXANE ISOMERS

Chemical Name	Reference Name
RR-00003-3 HEXANE, OTHER ISOMERS OF [ONT, ONT]	HEXANE ISOMERS
RR-00003-3 HEXANE, OTHER ISOMERS [TLV, TLV]	HEXANE ISOMERS
112-58-3 HEXANE, 1,1'-OXYBIS- [PAL]	N-HEXYL ETHERNO]ETHYL]-
RR-00003-3 HEXANES [HM1, HM1]	HEXANE ISOMERS
111-31-9 1-HEXANETHIOL	1-HEXANETHIOL
3522-94-9 HEXANE, 2,2,5-TRIMETHYL-	HEXANE, 2,2,5-TRIMETHYL-
106-69-4 1,2,6-HEXANETRIOL	1,2,6-HEXANETRIOL
68959-23-9 1,2,6-HEXANETRIOL TRIGLYCIDYL ETHER [EPA]	OXIRANE, 2,2',2"-[1,2,6-HEXANETRIYLTRIS-(OXYMETHYLENE)]
68959-23-9 HEXANETRIOL TRIGLYCIDYL ETHER, 1,2,6-[OTS]	OXIRANE, 2,2',2"-[1,2,6-HEXANETRIYLTRIS-(OXYMETHYLENE)]
RR-01425-5 HEXANITROAZOXY BENZENE	HEXANITROAZOXY BENZENE
RR-01371-8 2,2',4,4',6,6'-HEXANITRO-3,3'-DIHYDROXYAZOBENZENE	2,2',4,4',6,6'-HEXANITRO-3,3'-DIHYDROXYAZOBENZENE
35860-31-2 HEXANITRODIPHENYLAMINE	HEXANITRODIPHENYLAMINE
RR-01374-1 2,3',4,4',6,6'-HEXANITRODIPHENYLETHER	2,3',4,4',6,6'-HEXANITRODIPHENYLETHER
RR-01445-9 N,N'-(HEXANITRODIPHENYL) ETHYLENE DINITRAMINE	N,N'-(HEXANITRODIPHENYL) ETHYLENE DINITRAMINE
RR-01426-6 HEXANITRODIPHENYL UREA	HEXANITRODIPHENYL UREA
918-37-6 HEXANITROETHANE	HEXANITROETHANE
29135-62-4 HEXANITROOXANILIDE	HEXANITROOXANILIDE
20062-22-0 HEXANITROSTILBENE [NJL]	HEXANITROSTILBINE
20062-22-0 HEXANITROSTILBINE	HEXANITROSTILBINE
142-62-1 HEXANOIC ACID [PAL, WHM]	CAPROIC ACID
149-57-5 HEXANOIC ACID, 2-ETHYL- [TSC, TSC, TSC]	2-ETHYLHEXANOIC ACID
RR-01648-8 HEXANOIC ACID, 2-ETHYL-, ETHENYL ESTER	HEXANOIC ACID, 2-ETHYL-, ETHENYL ESTEROXO-2-PROPENYL)OXY]ETHYL ESTER
18268-70-7 HEXANOIC ACID, 2-ETHYL-, DIESTER WITH TETRAETHYLENE GLYCOL	HEXANOIC ACID, 2-ETHYL-, DIESTER WITH TETRAETHYLENE GLYCOL
123-66-0 HEXANOIC ACID, ETHYL ESTER [PAL]	ETHYL HEXANOATE
94-04-2 HEXANOIC ACID, 2-ETHYL-, ETHENYL ESTER	HEXANOIC ACID, 2-ETHYL-, ETHENYL ESTER
111-27-3 N-HEXANOL	N-HEXANOL
111-27-3 1-HEXANOL [PAL]	N-HEXANOL
25917-35-5 HEXANOL	HEXANOL
105-60-2 1,6-HEXANOLACTAM [GBR, GBR]	CAPROLACTAM
104-76-7 1-HEXANOL, 2-ETHYL- [PAL, TSC, TSC, TSC]	2-ETHYLHEXANOL
25917-35-5 HEXANOLS [HM1, HM1]	HEXANOL
3452-97-9 1-HEXANOL, 3,5,5-TRIMETHYL-	1-HEXANOL, 3,5,5-TRIMETHYL-TETRAMETHYL ESTER
589-38-8 3-HEXANONE	3-HEXANONE
591-78-6 2-HEXANONE	2-HEXANONE
591-78-6 2-HEXANONE [BC1, BC1]	2-HEXANONE
591-78-6 HEXAN-2-ONE [GBR, GBR]	2-HEXANONE
591-78-6 2-HEXANONE (METHYL N-BUTYL KETONE) [OS1, OS1, ALB, ALB, ALB]	2-HEXANONE
110-12-3 2-HEXANONE, 5-METHYL- [PAL]	METHYL ISOAMYL KETONE
84812-04-4 1,4,7,10,13,16-HEXAOXACYCLOOCTADECANE,2-[(2-PROPENYLOXY)METHYL]	1,4,7,10,13,16-HEXAOXACYCLOOCTADECANE,2-[(2-PROPENYLOXY)METH
23523-12-8 HEXATRIACONTAMETHYLCYCLOOCTADECASILOXANE	HEXATRIACONTAMETHYLCYCLOOCTADECASILOXANE
18844-04-7 HEXATRIACONTAMETHYLHEPTADECASILOXANE	HEXATRIACONTAMETHYLHEPTADECASILOXANE
18540-29-9 HEXAVALENT CHROMIUM [MX2, RCA]	CHROMIUM (VI)

Chemical Name	Reference Name
RR-01769-6 HEXAVALENT CHROMIUM CHEMICALS	HEXAVALENT CHROMIUM CHEMICALS
RR-00026-0 HEXAVALENT CHROMIUM COMPOUNDS [IAR, ONT, ONT]	CHROMIUM (VI) COMPOUNDS
51235-04-2 HEXAZINONE	HEXAZINONEXAZOLIDINEDIONE]
505-57-7 2-HEXENAL	2-HEXENAL
645-62-5 2-HEXENAL, 2-ETHYL- [PAL]	2-ETHYL-3-PROPYLACROLEIN
26266-68-2 HEXENAL, 2-ETHYL-	HEXENAL, 2-ETHYL-
592-41-6 1-HEXENE	1-HEXENE
7688-21-3 2-HEXENE-CIS	2-HEXENE-CIS
592-43-8 2-HEXENE(MIXED CIS & TRANS) [MSL]	2-HEXENE (MIXED CIS & TRANS ISOMERS)
592-43-8 2-HEXENE (MIXED CIS & TRANS ISOMERS)	2-HEXENE (MIXED CIS & TRANS ISOMERS)
592-43-8 2-HEXENE (MIXED CIS- AND TRANS- ISOMERS) [PAL]	2-HEXENE (MIXED CIS & TRANS ISOMERS)
5309-52-4 2-HEXENOIC ACID, 2-ETHYL- [PAL]	2-ETHYL-3-PROPYLACRYLIC ACID
928-96-1 3-HEXEN-1-OL, (Z)-	3-HEXEN-1-OL, (Z)-
928-96-1 3-HEXENOL-CIS [NFP, NFP]	3-HEXEN-1-OL, (Z)-
84-16-2 HEXOESTROL	HEXOESTROL
RR-01329-6 HEXOLITE	HEXOLITE
108-10-1 HEXONE [MAK, MAK, MIN, NIO, NIO, NIO]	METHYL ISOBUTYL KETONE
108-10-1 HEXONE (METHYL ISOBUTYL KETONE) [OS1, OS1, OS1, BC1, BC1, BC1, QBC, QBC, ALB, ALB, ALB]	METHYL ISOBUTYL KETONE
108-84-9 SEC-HEXYL ACETATE	SEC-HEXYL ACETATE
142-92-7 N-HEXYL ACETATE	N-HEXYL ACETATE
142-92-7 SEC-HEXYL ACETATE [FLA]	N-HEXYL ACETATE
142-92-7 HEXYL ACETATE [NFP, NFP]	N-HEXYL ACETATE
142-92-7 SEC-HEXYL ACETATE [ONT]	N-HEXYL ACETATE
97-95-0 SEC-HEXYL ALCOHOL	SEC-HEXYL ALCOHOL
111-27-3 HEXYL ALCOHOL [NFP, NFP, PST]	N-HEXANOL
111-27-3 N-HEXYL ALCOHOL [WHM]	N-HEXANOL
111-26-2 N-HEXYLAMINE	N-HEXYLAMINE
111-26-2 HEXYLAMINE [FLA, MSL, NFP, NFP]	N-HEXYLAMINE
544-10-5 NORMAL-HEXYL CHLORIDE [HM1]	1-CHLOROHEXANE
101-86-0 HEXYL CINNAMIC ALDEHYDE [NFP, NFP]	OCTANAL, 2-(PHENYLMETHYLENE)-
107-41-5 HEXYLENE GLYCOL	HEXYLENE GLYCOL
112-58-3 N-HEXYL ETHER	N-HEXYL ETHER
112-58-3 HEXYL ETHER [FLA, MSL, NFP, NFP]	N-HEXYL ETHER
142-09-6 HEXYLMETHACRYLATE	HEXYLMETHACRYLATE
142-09-6 HEXYL METHACRYLATE [NFP, NFP]	HEXYLMETHACRYLATE
5434-57-1 HEXYL NEOPENTANOATE	HEXYL NEOPENTANOATE
136-77-6 4-HEXYLRESORCINOL	4-HEXYLRESORCINOL
928-65-4 HEXYLTRICHLOROSILANE	HEXYLTRICHLOROSILANE
928-65-4 HEXYL TRICHLOROSILANE [NJL, NJL]	HEXYLTRICHLOROSILANE
764-35-2 2-HEXYNE [PAL]	METHYL PROPYL ACETYLENE
105-31-7 1-HEXYN-3-OL	1-HEXYN-3-OL
107-54-0 1-HEXYN-3-OL, 3,5-DIMETHYL-	1-HEXYN-3-OL, 3,5-DIMETHYL-
RR-00459-1 HIGH-LEVEL RADIOACTIVE MATTER (INCLUDING RADIOACTIVE WASTES) [CN6]	RADIOACTIVE MATERIALS, N.O.S.
65996-89-6 HIGH-TEMPERATURE COAL TARS [NTP]	TAR, COAL, HIGH-TEMP

Chemical Name	Reference Name
680-31-9 HMPA (HEXAMETHYLPHOSPHORAMIDE) [QBC, QBC, QBC]	HEXAMETHYL PHOSPHORAMIDE
51338-27-3 HOELON	HOELON
15125-75-4 HOLMIUM 155	HOLMIUM 155
15832-34-5 HOLMIUM 157	HOLMIUM 157
15750-02-4 HOLMIUM 159	HOLMIUM 159
14391-20-9 HOLMIUM 161	HOLMIUM 161
15700-49-9 HOLMIUM 162	HOLMIUM 162
RR-00447-7 HOLMIUM 162M	HOLMIUM 162M
15749-97-0 HOLMIUM 164	HOLMIUM 164
RR-00446-6 HOLMIUM 164M	HOLMIUM 164M
RR-00444-4 HOLMIUM 166M	HOLMIUM 166M
15750-04-6 HOLMIUM 167	HOLMIUM-167
8028-66-8 HONEY	HONEY
23255-93-8 HYCANTHONE MESYLATE	HYCANTHONE MESYLATE
86-54-4 HYDRALAZINE	HYDRALAZINEIOATE (AZINPHOS METHYL)
67485-29-4 HYDRAMETHYLNON [TETRAHYDRO-5,5-DIMETHYL-2(1H)-PYRIMIDINONE[3-[4-(TRI-FLUOROMETHYL)PHENYL]-1-[2-[4-(TRIFLUO-ROMETHYL)PHENYL]ETHENYL]-2-PROPENYLI-DENE]HYDRAZONE]	HYDRAMETHYLNON [TETRAHYDRO-5,5-DIMETHYL-2(1H)-PYRIMIDINONE[3XY) PROPANOIC ACID, ETHYL ESTER]
10279-57-9 HYDRATED AMORPHOUS SILICA [PST]	HYDRATED SILICA
10279-57-9 HYDRATED SILICA	HYDRATED SILICA
302-01-2 HYDRAZINE	HYDRAZINE
302-01-2 HYDRAZINE (ANHYDROUS) [NFP, NFP]	HYDRAZINE
14546-44-2 HYDRAZINE AZIDE	HYDRAZINE AZIDE
79-19-6 HYDRAZINE CARBOTHIOAMIDE [RCU, RCU]	THIOSEMICARBAZIDE
16568-02-8 HYDRAZINECARBOXALDEHYDE, ETHYLIDEN-EMETHYL- [PAL, PAL]	GYROMITRIN
3570-75-0 HYDRAZINECARBOXALDEHYDE, 2-[4-(5-NI-TRO-2-FURANYL)-2-THIAZOLYL]- [PAL, PAL]	2-(2-FORMYLHYDRAZINO)-4-(5-NITRO-2-FURYL)THIA-ZOLE
57-56-7 HYDRAZINECARBOXAMIDE	HYDRAZINECARBOXAMIDE
57-56-7 HYDRAZINE CARBOXAMIDE [TSC]	HYDRAZINECARBOXAMIDE
RR-00965-4 HYDRAZINECARBOXAMIDE, N,N'-1,6-HEX-AENDIYLBIS [2,2-DIMETHYL-	HYDRAZINECARBOXAMIDE, N,N'-1,6-HEXAENDIYLBIS [2,2-DIMETHYL-
85095-61-0 HYDRAZINECARBOXAMIDE, N,N'-(METHY-LENE-4,1-PHENYLENE)BIS[2,2'-DIMETHYL-[TSC, TSC]	N,N'-(METHYLENEDI-4,1-PHENYLENE)BIS[2,2-DIMETHYL-YL]
563-41-7 HYDRAZINECARBOXAMIDE, MONOHY-DROCHLORIDE (SEMICARBAZIDE HY-DROCHLORIDE) [CAI]	SEMICARBAZIDE HYDROCHLORIDE
85095-61-0 HYDRAZINECARBOXYMIDE, N,N'-(METHYLENEDI-4,1-PHENYLENE)BIS[2,2-DIMETHYL] [TSE]	N,N'-(METHYLENEDI-4,1-PHENYLENE)BIS[2,2-DIMETHYL-
RR-01427-7 HYDRAZINE CHLORATE	HYDRAZINE CHLORATE
67880-17-5 HYDRAZINE DICARBONIC ACID DIAZIDE	HYDRAZINE DICARBONIC ACID DIAZIDE
1615-80-1 HYDRAZINE, 1,2-DIETHYL- [PAL, PAL]	1,2-DIETHYLHYDRAZINE
57-14-7 HYDRAZINE, 1,1-DIMETHYL- [PAL, PAL]	1,1-DIMETHYLHYDRAZINE
540-73-8 HYDRAZINE, 1,2-DIMETHYL- [PAL, PAL]	1,2-DIMETHYLHYDRAZINE
122-66-7 HYDRAZINE, 1,2-DIPHENYL- [PAL, PAL, TSC, TSC]	HYDRAZOBENZENE
530-50-7 HYDRAZINE, 1,1-DIPHENYL	HYDRAZINE, 1,1-DIPHENYL
7803-57-8 HYDRAZINE HYDRATE [NJL]	HYDRAZINE MONOHYDRATE

Chemical Name	Reference Name
60-34-4 HYDRAZINE, METHYL- [PAL]	METHYL HYDRAZINE
124993-63-1 HYDRAZINE, [4-[1-METHYLBUTOXY]PHENYL], MONOHYDROCHLORIDE	HYDRAZINE, [4-[1-METHYLBUTOXY]PHENYL], MONO-HYDROCHLORIDEPROPENYLOXY)-4-METHYLBENZENE-SULFONATE
7803-57-8 HYDRAZINE MONOHYDRATE	HYDRAZINE MONOHYDRATE
27978-54-7 HYDRAZINE PERCHLORATE	HYDRAZINE PERCHLORATE
100-63-0 HYDRAZINE, PHENYL- [PAL, TSC]	PHENYLHYDRAZINE
73506-32-8 HYDRAZINE SELENATE	HYDRAZINE SELENATE
10034-93-2 HYDRAZINE SULFATE	HYDRAZINE SULFATE
10034-93-2 HYDRAZINE, SULFATE (1:1) [PAL, PAL]	HYDRAZINE SULFATE
122-66-7 HYDRAZOBENZENE	HYDRAZOBENZENE
122-66-7 HYDRAZOBENZENE (1,2-DIPHENYLHY-DRAZINE) [C65, C65]	HYDRAZOBENZENE
7782-79-8 HYDRAZOIC ACID	HYDRAZOIC ACID
RR-00333-8 HYDRIDES, METAL, N.O.S.	HYDRIDES, METAL, N.O.S.
496-10-6 HYDRINDANE	HYDRINDANE
10034-85-2 HYDRIODIC ACID	HYDRIODIC ACID
10034-85-2 HYDRIOTIC ACID [FDA]	HYDRIODIC ACID
10035-10-6 HYDROBROMIC ACID [HM1, HM1, HM1, PAL]	HYDROGEN BROMIDE
RR-01708-3 HYDROBROMOFLUOROCARBONS	HYDROBROMOFLUOROCARBONS
RR-01708-3 HYDROBROMOFLUOROCARBONS (EXCEPT HBFC-22B1) [CAA]	HYDROBROMOFLUOROCARBONS
RR-00334-9 HYDROCARBON GASES, N.O.S.	HYDROCARBON GASES, N.O.S.
8020-83-5 HYDROCARBON OILS [PAL, PAL]	HYDROCARBON OILS, WHITE MINERAL OIL
8020-83-5 HYDROCARBON OILS, WHITE MINERAL OIL	HYDROCARBON OILS, WHITE MINERAL OIL
68476-40-4 HYDROCARBON PROPELLANT	HYDROCARBON PROPELLANT
64742-42-3 HYDROCARBON WAXES (PETROLEUM), CLAY-TREATED MICROCRYSTALLINE	HYDROCARBON WAXES (PETROLEUM), CLAY-TREATED MICROCRYSTALLINE
7647-01-0 HYDROCHLORIC ACID [313, CAA, CAB, CAN, CWA, EP3, IAR, OS3, PAL]	HYDROGEN CHLORIDE
75-88-7 HYDROCHLOROFLUOROCARBON (HCFC 133A) [TSE]	2-CHLORO-1,1,1-TRIFLUOROETHANE
1649-08-7 HYDROCHLOROFLUOROCARBON (HCFC 132B) [TSE]	1,2-DICHLORO-1,1-DIFLUOROETHANE
58-93-5 HYDROCHLOROTHIAZIDE	HYDROCHLOROTHIAZIDE
74-90-8 HYDROCYANIC ACID [302, CAB, EP3, HM1, HM1, HM1, HM1, HM1, NFP, NFP, PAL, RCA, TSC, TSC]	HYDROGEN CYANIDE
RR-01428-8 HYDROCYANIC ACID (PRUSSIC)	HYDROCYANIC ACID (PRUSSIC)
7664-39-3 HYDROFLUORIC ACID [CWA, OS3, PAL]	HYDROGEN FLUORIDE
16961-83-4 HYDROFLUOSILICIC ACID [MSL]	FLUOROSILICIC ACID
1333-74-0 HYDROGEN	HYDROGEN
RR-00957-4 HYDROGENATED ARYLATED POLYDECENE	HYDROGENATED ARYLATED POLYDECENE
8001-78-3 HYDROGENATED CASTOR OIL	HYDROGENATED CASTOR OIL
68334-00-9 HYDROGENATED COTTONSEED OIL	HYDROGENATED COTTONSEED OILSIN
30968-45-7 HYDROGENATED METHYL ABIETE	HYDROGENATED METHYL ABIETEAND ACRYLIC ACID
8016-70-4 HYDROGENATED SOYBEAN OIL	HYDROGENATED SOYBEAN OIL
61790-59-8 HYDROGENATED TALLOW ALKYL AMINE ACETATE	HYDROGENATED TALLOW ALKYL AMINE ACETATE
61788-32-7 HYDROGENATED TERPHENYLS	HYDROGENATED TERPHENYLS
10035-10-6 HYDROGEN BROMIDE	HYDROGEN BROMIDE
7647-01-0 HYDROGEN CHLORIDE	HYDROGEN CHLORIDE

Chemical Name	Reference Name
420-04-2 HYDROGEN CYANAMIDE [PST]	CYANAMIDE
74-90-8 HYDROGEN CYANIDE	HYDROGEN CYANIDE
7664-39-3 HYDROGEN FLUORIDE	HYDROGEN FLUORIDE
7664-39-3 HYDROGEN FLUORIDE (HYDROFLUORIC ACID) [CAA, TSC, TSC]	HYDROGEN FLUORIDE
7664-39-3 HYDROGEN FLUORIDE/HYDROFLUORIC ACID [EP3]	HYDROGEN FLUORIDE
10034-85-2 HYDROGEN IODIDE [HM1, HM1, HM1, WHM]	HYDRIODIC ACID
7722-84-1 HYDROGEN PEROXIDE	HYDROGEN PEROXIDE
7722-84-1 HYDROGEN PEROXIDE (90%) [ONT]	HYDROGEN PEROXIDE
7722-84-1 HYDROGEN PEROXIDE (CONC >52%) [PAL]	HYDROGEN PEROXIDE
7783-07-5 HYDROGEN SELENIDE	HYDROGEN SELENIDE
7783-07-5 HYDROGEN SELENIDE (H2SE) [PAL]	HYDROGEN SELENIDE
7783-06-4 HYDROGEN SULFIDE	HYDROGEN SULFIDE
7783-06-4 HYDROGEN SULFIDE (H2S) [PAL]	HYDROGEN SULFIDE
7783-06-4 HYDROGEN SULPHIDE [AUS, AUS, GBR, GBR]	HYDROGEN SULFIDE
25214-69-1 HYDROLYZED POLYACRYLONITRILE	HYDROLYZED POLYACRYLONITRILE
9015-54-7 HYDROLYZED PROTEIN	HYDROLYZED PROTEIN
75-91-2 HYDROPEROXIDE, 1,1-DIMETHYLETHYL- [OTS]	TERT-BUTYL HYDROPEROXIDE
75-91-2 HYDROPEROXIDE, 1,1-DIMETHYLETHYL [PAL]	TERT-BUTYL HYDROPEROXIDE
80-15-9 HYDROPEROXIDE, 1-METHYL-1-PHENYLETHYL- [EPA]	CUMENE HYDROPEROXIDE
80-15-9 HYDROPEROXIDE, 1-METHYL-1-PHENYLETHYL [PAL]	CUMENE HYDROPEROXIDE
80-15-9 HYDROPEROXIDE, 1-METHYL-1-PHENYLETHYL- [TSC, TSC]	CUMENE HYDROPEROXIDE
123-31-9 HYDROQUINONE	HYDROQUINONE
2425-01-6 HYDROQUINONE DIGLYCIDYL ETHER [EPA, OTS]	OXIRANE, 2,2'-[1,4-PHENYLENEBIS (OXYMETHYLENE)] BIS-
104-38-1 HYDROQUINONE DI-(BETA-HYDROXYETHYL) ETHER	HYDROQUINONE DI-(BETA-HYDROXYETHYL) ETHER
150-78-7 HYDROQUINONE, DIMETHYL ETHER	HYDROQUINONE, DIMETHYL ETHER
150-78-7 HYDROQUINONE DIMETHYL ETHER [PST]	HYDROQUINONE, DIMETHYL ETHER
150-76-5 HYDROQUINONE MONOMETHYL ETHER [NFP, NFP]	4-METHOXYPHENOL
79-14-1 HYDROXYACETIC ACID	HYDROXYACETIC ACID
99-93-4 P-HYDROXYACETOPHENONE	P-HYDROXYACETOPHENONE
141-31-1 HYDROXYADIPALDEHYDE	HYDROXYADIPALDEHYDE
RR-00223-3 HYDROXYALKYL METHACRYLATE, ALKYL ESTER	HYDROXYALKYL METHACRYLATE, ALKYL ESTER-PHENYL ESTER
150-76-5 P-HYDROXYANISOLE [PST]	4-METHOXYPHENOL
1689-82-3 4-HYDROXYAZOBENZENE	4-HYDROXYAZOBENZENE
99-06-9 M-HYDROXYBENZOIC ACID	M-HYDROXYBENZOIC ACID
99-96-7 P-HYDROXYBENZOIC ACID	P-HYDROXYBENZOIC ACID
RR-01691-1 1,4-BIS(3-HYDROXY-4-BENZOYLPHENOXY) BUTANE	1,4-BIS(3-HYDROXY-4-BENZOYLPHENOXY)BUTANE
RR-01013-9 2-(2-HYDROXY-3-TERT-BUTYL-5-METHYL-BENZYL)-4-METHYL-6-TERT-BUTYLPHENYL METHACRYLATE	2-(2-HYDROXY-3-TERT-BUTYL-5-METHYLBENZYL)-4-METHYL-6-TERT-BUERIVATIVE
107-75-5 HYDROXYCINTRONELLAL [NFP, NFP]	OCTANAL, 7-HYDROXY-3,7-DIMETHYL-
825-51-4 2-HYDROXYDECALIN	2-HYDROXYDECALIN
706-14-9 HYDROXYDECANOIC ACID, GAMMA-LACTONE	HYDROXYDECANOIC ACID, GAMMA-LACTONE

Chemical Name	Reference Name
RR-01289-5 HYDROXYDIMETHYLBENZENES	HYDROXYDIMETHYLBENZENES
141-66-2 3-HYDROXY-N,N-DIMETHYL-CIS-CROTON-AMIDE DIMETHYL PHOSPHATE (DICRO-TOPHOS) [CAL]	DICROTOPHOSXANE CARBOXYLATE
2809-21-4 1-HYDROXYETHANE-1,1-DIPHOSPHONIC ACID	1-HYDROXYETHANE-1,1-DIPHOSPHONIC ACID
RR-01359-2 3-(2-HYDROXYETHOXY)-4-PYRROLIDIN-1-YLBENZENEDAZONIUM ZINC CHLORIDE	3-(2-HYDROXYETHOXY)-4-PYRROLIDIN-1-YLBEN-ZENEDAZONIUM ZINC CH
818-61-1 2-HYDROXYETHYL ACRYLATE [NFP, NFP]	ETHYLENE GLYCOL MONOACRYLATE
818-61-1 HYDROXYETHYL ACRYLATE [WHM]	ETHYLENE GLYCOL MONOACRYLATE
5395-01-7 BETA-HYDROXYETHYLCARBAMATE	BETA-HYDROXYETHYLCARBAMATE
9004-62-0 HYDROXYETHYL CELLULOSE	HYDROXYETHYL CELLULOSE
2842-38-8 N-(2-HYDROXYETHYL) CYCLOHEXY-LAMINE [FLA, NFP, NFP]	ETHANOL, 2-(CYCLOHEXYLAMINO)-
2842-38-8 N-(2-HYDROXYETHYL)CYCLOHEXY-LAMINE [MSL]	ETHANOL, 2-(CYCLOHEXYLAMINO)-
1965-29-3 N-(HYDROXYETHYL)DIETHYLENETRIAMINE	N-(HYDROXYETHYL)DIETHYLENETRIAMINE
2764-13-8 2-HYDROXYETHYL DIMETHYL 3-OCTADE-CANAMIDOPROPYL AMMONIUM NITRATE	2-HYDROXYETHYL DIMETHYL 3-OCTADECANAMIDO-PROPYL AMMONIUM NITR
142-78-9 N-(2-HYDROXYETHYL)DODECANAMIDE [PST]	COCOMONOETHANOLAMIDE
139-89-9 N-HYDROXYETHYLETHYLENEDIAMINETRI-ACETIC ACID TRISODIUM SALT	N-HYDROXYETHYLETHYLENEDIAMINETRIACETIC ACID TRISODIUM SALT
27136-73-8 1-HYDROXYETHYL-2-HEPTADECENYL GLYOX-ALIDINE [PST]	SUBSTITUTED IMIDAZOLE
95-19-2 1-(HYDROXYETHYL)-2-(HEPTADECYL)IMIDA-ZOLINE	1-(HYDROXYETHYL)-2-(HEPTADECYL)IMIDAZOLINE
544-31-0 N-(2-HYDROXYETHYL)HEXADECANAMIDE	N-(2-HYDROXYETHYL)HEXADECANAMIDE
2809-21-4 1-HYDROXYETHYLIDENE-1,1-DIPHOSPHONIC ACID [PST]	1-HYDROXYETHANE-1,1-DIPHOSPHONIC ACID
868-77-9 2-HYDROXYETHYL METHACRYLATE	2-HYDROXYETHYL METHACRYLATE
622-40-2 4-(2-HYDROXYETHYL) MORPHOLINE [FLA, MSL, NFP, NFP]	MORPHOLINE ETHANOL
111-57-9 N-(2-HYDROXYETHYL)OCTADECANAMIDE	N-(2-HYDROXYETHYL)OCTADECANAMIDE
103-76-4 1-(2-HYDROXYETHYL) PIPERAZINE [NFP, NFP]	1-PIPERAZINEETHANOL
10138-74-6 N-(2-HYDROXYETHYL) PROPYLENE DIAMINE	N-(2-HYDROXYETHYL) PROPYLENE DIAMINE
10138-74-6 N-(2-HYDROXYETHYL) PROPYLENEDI-AMINE [NFP, NFP]	N-(2-HYDROXYETHYL) PROPYLENE DIAMINE
142-58-5 N-(2-HYDROXYETHYL)TETRADECANAMIDE	N-(2-HYDROXYETHYL)TETRADECANAMIDE
105-21-5 4-HYDROXYHEPTANOIC ACID, GAMMA-LAC-TONE	4-HYDROXYHEPTANOIC ACID, GAMMA-LACTONE
78-18-2 1-HYDROXY-1'-HYDROPEROXY-DICYCLO-HEXYL PEROXIDE [MAK]	CYCLOHEXANONE PEROXIDE
78-18-2 1-HYDROXY-1'-HYDROPEROXY DICYCLO-HEXYL PEROXIDE [NJL]	CYCLOHEXANONE PEROXIDE
7803-49-8 HYDROXYLAMINE	HYDROXYLAMINE
7803-49-8 HYDROXYLAMINE (OXAMMONIUM) [CAI]	HYDROXYLAMINE
5470-11-1 HYDROXYLAMINE, HYDROCHLORIDE	HYDROXYLAMINE, HYDROCHLORIDETE)
5470-11-1 HYDROXYLAMINE, HYDROCHLORIDE (HY-DROXYLAMMONIUM CHLORIDE) [CAI]	HYDROXYLAMINE, HYDROCHLORIDETE)
5470-11-1 HYDROXYLAMINE HYDROCHLORIDE (HY-DROXYLAMMONIUM CHLORIDE) [TSC]	HYDROXYLAMINE, HYDROCHLORIDE
59917-23-6 HYDROXYL AMINE IODIDE	HYDROXYL AMINE IODIDE
10039-54-0 HYDROXYLAMINE SULFATE [HM1, HM1, NJL, NJL, PST, WHM]	HYDROXYLAMINE SULFATE (2:1)

Chemical Name	Reference Name
10046-00-1 HYDROXYLAMINE SULFATE (1:1)	HYDROXYLAMINE SULFATE (1:1)
10046-00-1 HYDROXYLAMINE, SULFATE (1:1) (HYDROXY-LAMINE ACID SULFATE) [CAI]	HYDROXYLAMINE SULFATE (1:1)
10039-54-0 HYDROXYLAMINE SULFATE (2:1) (HYDROXY-LAMMONIUM) [CAI]	HYDROXYLAMINE SULFATE (2:1)
10039-54-0 HYDROXYLAMINE SULFATE (2:1)	HYDROXYLAMINE SULFATE (2:1)
7803-49-8 HYDROXYLAMMONIUM (OXAMMONIUM) [TSC]	HYDROXYLAMINE
131-57-7 2-HYDROXY-4-METHOXYBENZOPHENONE	2-HYDROXY-4-METHOXYBENZOPHENONE
131-57-7 2-HYDROXY-4-METHOXY BENZOPHE-NONE [PST]	2-HYDROXY-4-METHOXYBENZOPHENONE
4065-45-6 2-HYDROXY-4-METHOXYBENZOPHENONE-5-SULFONIC ACID	2-HYDROXY-4-METHOXYBENZOPHENONE-5-SULFONIC ACID
52299-20-4 2-[(HYDROXYMETHYL)AMINO]-2-METHYL PROPANOL	2-[(HYDROXYMETHYL)AMINO]-2-METHYL PROPANOL SALT
123-42-2 4-HYDROXY-4-METHYL-2-PENTANONE [CAL]	DIACETONE ALCOHOL
123-42-2 4-HYDROXY-4-METHYL-PENTAN-2-ONE [GBR, GBR]	DIACETONE ALCOHOL
123-42-2 4-HYDROXY-4-METHYL-2-PENTANONE [ONT, ONT]	DIACETONE ALCOHOL
481-39-0 5-HYDROXY-1,4-NAPHTHAQUINONE	5-HYDROXY-1,4-NAPHTHAQUINONE
92-70-6 3-HYDROXY-2-NAPHTHOIC ACID	3-HYDROXY-2-NAPHTHOIC ACID
92-70-6 2-HYDROXY-3-NAPHTHOIC ACID [OTS]	3-HYDROXY-2-NAPHTHOIC ACID
119-34-6 4-HYDROXY-3-NITROANILINE	4-HYDROXY-3-NITROANILINE
42808-36-6 9-HYDROXYOCTADECANOIC ACID, BUTYL ESTER, HYDROGEN SULFATE, SODIUM SALT	9-HYDROXYOCTADECANOIC ACID, BUTYL ESTER, HYDROGEN SULFATE, SETHYLSILYL)-OMEGA-[(TRIMETHYLSILYL)OXY]-
104-50-7 4-HYDROXYOCTANOIC ACID LACTONE	4-HYDROXYOCTANOIC ACID LACTONE
1843-05-6 2-HYDROXY-4-N-OCTOXYBENZOPHENONE	2-HYDROXY-4-N-OCTOXYBENZOPHENONE
3147-75-9 2-(2-HYDROXY-5-TERT-OCTYLPHENYL)BENZO-TRIAZOLE	2-(2-HYDROXY-5-TERT-OCTYLPHENYL)BENZOTRIA-ZOLE
120-80-9 O-HYDROXYPHENOL [EPA]	CATECHOL
120-80-9 HYDROXYPHENOL, O- [OTS]	CATECHOL
120-80-9 O-HYDROXYPHENOL [TSC]	CATECHOL
630-56-8 17A-HYDROXYPROGESTERONE CAPROATE	17A-HYDROXYPROGESTERONE CAPROATE
999-61-1 2-HYDROXYPROPYL ACRYLATE	2-HYDROXYPROPYL ACRYLATE
999-61-1 HYDROXYPROPYL ACRYLATE [CAL]	2-HYDROXYPROPYL ACRYLATE
7373-11-7 2-HYDROXYPROPYLAMINE NITRITE	2-HYDROXYPROPYLAMINE NITRITE
23054-61-7 N-(2-HYDROXYPROPYL)DECANAMIDE	N-(2-HYDROXYPROPYL)DECANAMIDE
9004-64-2 HYDROXYPROPYL ETHER OF CELLULOSE	HYDROXYPROPYL ETHER OF CELLULOSE
39421-75-5 HYDROXYPROPYL GUAR GUM	HYDROXYPROPYL GUAR GUM
9004-65-3 2-HYDROXYPROPYL METHYL CELLU-LOSE [PST]	CELLULOSE HYDROXYPROPYL METHYL ETHER
23054-60-6 N-(2-HYDROXYPROPYL)OCTANAMIDE	N-(2-HYDROXYPROPYL)OCTANAMIDE
109-00-2 3-HYDROXYPYRIDINE	3-HYDROXYPYRIDINE
148-24-3 8-HYDROXYQUINOLINE	8-HYDROXYQUINOLINE
134-31-6 8-HYDROXYQUINOLINE SULFATE	8-HYDROXYQUINOLINE SULFATE
26782-43-4 HYDROXYSENKIRKINE	HYDROXYSENKIRKINE
127-07-1 HYDROXYUREA	HYDROXYUREA
3761-53-3 3-HYDROXY-4-(2,4-XYLYLAZO)-2,7-NAPH-THALENEDISULFONIC ACID,DISODIUM SALT [WHM]	PONCEAU MX
14380-61-1 HYPOCHLORITE ION	HYPOCHLORITE ION

Chemical Name	Reference Name
RR-01547-4 HYPOCHLORITE SALTS	HYPOCHLORITE SALTS
RR-00337-2 HYPOCHLORITE SOLUTIONS	HYPOCHLORITE SOLUTIONS
7778-54-3 HYPOCHLOROUS ACID, CALCIUM SALT [PAL]	CALCIUM HYPOCHLORITE
7681-52-9 HYPOCHLOROUS ACID, SODIUM SALT [PAL]	SODIUM HYPOCHLORITE
10022-70-5 HYPOCHLOROUS ACID, SODIUM SALT, PENTAHYDRATE [PAL]	SODIUM HYPOCHLORITE (PENTAHYDRATE)
14448-38-5 HYPONITROUS ACID	HYPONITROUS ACID
6303-21-5 HYPOPHOSPHORUS ACID	HYPOPHOSPHORUS ACID
6737-68-4 HYTHERM BLUE E	HYTHERM BLUE E
108-84-9 HYXYL ACETATE - SEC [QBC]	SEC-HEXYL ACETATE
21416-87-5 ICRF-159	ICRF-159
3778-73-2 IFOSFAMIDE [C65, CAB]	ISOPHOSPHAMIDE
35554-44-0 IMAZALIL [1-[2-(2,4-DICHLOROPHENYL)-2-(2-PROPENYLOXY)ETHYL]-1H-IMIDAZOLE]	IMAZALIL [1-[2-(2,4-DICHLOROPHENYL)-2-(2-PROPENYLOXY)ETHYL]- ACID S-PROPYL ESTER]
118-52-5 2,4-IMIDAXOLIDINEDIONE, 1,3-DICHLORO-5,5-DIMETHYL- [PAL]	1,3-DICHLORO-5,5-DIMETHYL HYDANTOIN
288-32-4 1-IMIDAZOLE	1-IMIDAZOLE
4342-03-4 1H-IMIDAZOLE-4-CARBOXAMIDE, 5-(3,3-DIMETHYL-1-TRIAZENYL)- [PAL, PAL]	DACARBAZINE
95-38-5 1H-IMIDAZOLE-1-ETHANOL, 2-(8-HEPTADACENYL)-4,5-DIHYDRO-	1H-IMIDAZOLE-1-ETHANOL, 2-(8-HEPTADACENYL)-4,5-DIHYDRO-
443-48-1 1H-IMIDAZOLE-1-ETHANOL, 2-METHYL-5-NITRO- [PAL, PAL]	METRONIDAZOLE
693-98-1 1H-IMIDAZOLE, 2-METHYL- [PST]	2-METHYLIMIDAZOLE
68630-92-2 1H-IMIDAZOLE-1-PROPANOIC ACID, 3-[2-(2-CARBOXYETHYL)ETHYL]-2-HEPTYL-2,3-DIHYDRO-, DISODIUM SALT	1H-IMIDAZOLE-1-PROPANOIC ACID, 3-[2-(2-CARBOXYETHYL)ETHYL]-UM SALT
32568-89-1 2,4-IMIDAZOLIDINEDIONE, 5,5-DIMETHYL-3-[2-(OXIRANYLMETHOXY)PROPYL]-1-(OXIRANYLMETHYL)-	2,4-IMIDAZOLIDINEDIONE, 5,5-DIMETHYL-3-[2-(OXIRANYLMETHOXY)
32568-89-1 2,4-IMIDAZOLIDINEDIONE, 5,5-DIMETHYL-3-[2-(OXIRANYLMETHOXY)PROPYL]-1-(OXIRANYLMETHYL)- [TSC]	2,4-IMIDAZOLIDINEDIONE, 5,5-DIMETHYL-3-[2-(OXIRANYLMETHOXY)
57-41-0 2,4-IMIDAZOLIDINEDIONE, 5,5-DIPHENYL- [PAL, PAL]	PHENYTOIN
38304-52-8 2,4-IMIDAZOLIDINEDIONE, 3,3'-[2-(OXIRANYLMETHOXY)-1,3-PROPANEDIYL]BIS[5,5-DIMETHYL-1-(OXIRANYLMETHYL)-	2,4-IMIDAZOLIDINEDIONE, 3,3'-[2-(OXIRANYLMETHOXY)-1,3-PROPAN
96-45-7 2-IMIDAZOLIDINETHIONE [FLA, PAL, PAL]	ETHYLENE THIOUREA
1854-26-8 2-IMIDAZOLIDINONE, 4,5-DIHYDROXY-1,3-BIS	2-IMIDAZOLIDINONE, 4,5-DIHYDROXY-1,3-BIS
555-84-0 2-IMIDAZOLIDINONE, 1-[[(5-NITRO-2-FURANYL)METHYLENE]AMINO]- [PAL, PAL]	1-((5-NITROFURFURYLIDENE)AMINO)-2-IMIDAZOLIDINONE
61-57-4 2-IMIDAZOLIDINONE, 1-(5-NITRO-2-THIAZOLYL)- [PAL, PAL]	NIRIDAZOLE
39236-46-9 IMIDAZOLINIDYL UREA	IMIDAZOLINIDYL UREA
68390-66-9 2-IMIDAZOLINIUM, 1-(CARBOXYMETHYL)-1-(2-HYDROXYETHYL)-2-COCOYLNOR, HYDROXIDE, MONOSODIUM SALT	2-IMIDAZOLINIUM, 1-(CARBOXYMETHYL)-1-(2-HYDROXYETHYL)-2-COCO
68527-99-1 1H-IMIDAZOLIUM, 1,3-BIS(CARBOXYMETHYL)-4,5-DIHYDRO-1-(2-HYDROXYETHYL)-2-UNDECYL-, DIHYDROXIDE, DISODIUM SALT	1H-IMIDAZOLIUM, 1,3-BIS(CARBOXYMETHYL)-4,5-DIHYDRO-1-(2-HYDR
61702-73-6 1H-IMIDAZOLIUM, 1,1-BIS(CARBOXYMETHYL)-4,5-DIHYDRO-2-UNDECYL-, HYDROXIDE, DISODIUM SALT	1H-IMIDAZOLIUM, 1,1-BIS(CARBOXYMETHYL)-4,5-DIHYDRO-2-UNDECYL

Chemical Name	Reference Name
68630-96-6 1H-IMIDAZOLIUM, 1-(2-CARBOXYETHYL)-4, 5-DIHYDRO-3-(2-HYDROXYETHYL)-2-ISOHEP-TADECYL-, HYDROXIDE, INNER SALT	1H-IMIDAZOLIUM, 1-(2-CARBOXYETHYL)-4,5-DIHYDRO-3-(2-HYDROXYE2-HEPTYL-2,3-DIHYDRO-, DISODIUM SALT
RR-01088-8 IMIDAZOLIUM COMPOUNDS, 2-(C16-18 ALKYL) -1-(2-(C16-18 AMIDO)ETHYL)-4,5-DIHYDRO-1-METHYL, METHYL SULFATES	IMIDAZOLIUM COMPOUNDS, 2-(C16-18 ALKYL)-1-(2-(C16-18 AMIDO)EYLNOR, CHLORIDE, MONOSODIUM SALT
68650-39-5 IMIDAZOLIUM COMPOUNDS, 1-[2-(CAR-BOXYMETHOXY)ETHYL]-1-CARBOXYMETHYL) -4,5-DIHYDRO-2-NORCOCO ALKYL, HYDROX-IDES, SODIUM SALTS	IMIDAZOLIUM COMPOUNDS, 1-[2-(CARBOXYMETHOXY) ETHYL]-1-CARBOXYHYLENE GLYCOL
68122-86-1 IMIDAZOLIUM COMPOUNDS, 4,5-DIHYDRO-1-METHYL-2-NORTALLOWALKYL-1-(2-TALLOW AMIDOETHYL), METHYL SULFATES	IMIDAZOLIUM COMPOUNDS, 4,5-DIHYDRO-1-METHYL-2-NORTALLOW
RR-01089-9 IMIDAZOLIUM COMPOUNDS, 2-HEPTADECYL-4,5-DIHYDRO-1-METHYL-1-(2-TALLOW AMI-DOETHYL), METHYL SULFATES	IMIDAZOLIUM COMPOUNDS, 2-HEPTADECYL-4,5-DI-HYDRO-1-METHYL-1-(THYL)-4,5-DIHYDRO-1-METHYL, METHYL SULFATES
4035-89-6 IMIDODICARBONIC DIAMIDE, N,N'-2-TRIS(6-ISOCYANATOHEXYL)- [TSC, TSC, TSC]	HEXAMETHYLENE DIISOCYANATE BIURET
56-18-8 3,3'-IMINOBISPROPYLAMINE	3,3'-IMINOBISPROPYLAMINE
111-42-2 2,2'-IMINODIETHANOL [GBR]	DIETHANOLAMINE
111-40-0 2,2'-IMINODI(ETHYLAMINE) [GBR, GBR]	DIETHYLENE TRIAMINE
111-94-4 IMINODIPROPIONITRILE	IMINODIPROPIONITRILE
56-18-8 3,3'-IMINODIPROPYLAMINE [HM1, HM1]	3,3'-IMINOBISPROPYLAMINE
RR-01210-2 2-IMINO-1,3-THIAZIN-4-ONE-5,6-DIHY-DROMONOHYDROCHLORIDE	2-IMINO-1,3-THIAZIN-4-ONE-5,6-DIHYDROMONOHY-DROCHLORIDE
496-11-7 INDAN	INDAN
95-13-6 INDENE	INDENE
95-13-6 1H-INDENE [FLA, PAL]	INDENE
83-26-1 1H-INDENE-1,3-(2H)-DIONE, 2-(2,2-DIMETHYL-1-OXOPROPYL)- [PAL]	PINDONE
129-00-0 INDENO (1,2,3-CD) PYRENE [NJL]	PYRENE
193-39-5 INDENO(1,2,3-CD)PYRENE (2,3-O-PHENYLENEPYRENE)	INDENO(1,2,3-CD)PYRENE (2,3-O-PHENYLENEPYRENE)
193-39-5 INDENO(1,2,3-CD)PYRENE (2,3-O-PHENYLENE PYRENE) [CW2]	INDENO(1,2,3-CD)PYRENE (2,3-O-PHENYLENEPYRENE)
7440-74-6 INDIUM	INDIUM
14833-35-3 INDIUM 109	INDIUM 109
14133-75-6 INDIUM 110	INDIUM 110
15750-15-9 INDIUM 111	INDIUM 111
14391-66-3 INDIUM 112	INDIUM 112
RR-00443-3 INDIUM 113M	INDIUM 113M
RR-00442-2 INDIUM 114M	INDIUM 114M
14191-71-0 INDIUM 115	INDIUM 115
RR-00441-1 INDIUM 115M	INDIUM 115M
RR-00439-7 INDIUM 116M	INDIUM 116M
14914-66-0 INDIUM 117	INDIUM 117
RR-00438-6 INDIUM 117M	INDIUM 117M
RR-00437-5 INDIUM 119M	INDIUM 119M
RR-00600-8 INDIUM AND COMPOUNDS [QBC, ALB, ALB, AUS, BC1, BC1, OS1, TLV, GBR, GBR]	INDIUM COMPOUNDS, N.O.S.
RR-00600-8 INDIUM COMPOUNDS [CAL, CEX, ONT]	INDIUM COMPOUNDS, N.O.S.
RR-00600-8 INDIUM COMPOUNDS, N.O.S.	INDIUM COMPOUNDS, N.O.S.
7440-74-6 INDIUM, ELEMENTAL [WHM]	INDIUM

Chemical Name	Reference Name
13770-61-1 INDIUM NITRATE	INDIUM NITRATE
22398-80-7 INDIUM PHOSPHIDE	INDIUM PHOSPHIDE
120-72-9 INDOLE	INDOLE
120-72-9 1H-INDOLE [PST]	INDOLE
133-32-4 3-INDOLEBUTYRIC ACID	3-INDOLEBUTYRIC ACID
860-22-0 1H-INDOLE-5-SULFONIC ACID,2-(1,3-DIHYDRO-3-OXO-5-SULFO-2H	1H-INDOLE-5-SULFONIC ACID,2-(1,3-DIHYDRO-3-OXO-5-SULFO-2H
482-89-3 3H-INDOL-3-ONE, 2-(1,3-DIHYDRO-3-OXO-2H-	3H-INDOL-3-ONE, 2-(1,3-DIHYDRO-3-OXO-2H-
53-86-1 INDOMETHACIN	INDOMETHACIN
RR-00341-8 INK [HM1, HM1]	INKS
RR-00341-8 INKS	INKS
7440-38-2 INORGANIC ARSENIC [MIN]	ARSENIC
RR-00065-7 INORGANIC ARSENIC	INORGANIC ARSENIC
RR-00599-2 INORGANIC FLUORIDE COMPOUNDS [CAL]	FLUORIDES, INORGANIC
RR-00599-2 INORGANIC FLUORIDES [CN4]	FLUORIDES, INORGANIC
RR-00538-9 INORGANIC LEAD	INORGANIC LEAD
4691-65-0 INOSINE-5-MONOPHOSPHORIC ACID	INOSINE-5-MONOPHOSPHORIC ACID
RR-01429-9 INOSITOL HEXANITRATE	INOSITOL HEXANITRATE
RR-00342-9 INSECTICIDE, DRY, N.O.S.	INSECTICIDE, DRY, N.O.S.
RR-00343-0 INSECTICIDE GASES, N.O.S.	INSECTICIDE GASES, N.O.S.
RR-00345-2 INSECTICIDE, LIQUID, N.O.S.	INSECTICIDE, LIQUID, N.O.S.
RR-00007-7 INSOLUBLE TUNGSTEN [NHS, NHS]	TUNGSTEN, INSOLUBLE COMPOUNDS
RR-00015-7 INSULATION WOOL FIBRES, GLASS WOOL [QBC, QBC]	MINERAL WOOL FIBER
RR-01201-1 INSULATION WOOL FIBRES, ROCK WOOL [QBC, QBC]	ROCKWOOL
RR-01202-2 INSULATION WOOL FIBRES, SLAG WOOL [QBC, QBC]	SLAGWOOL
76543-88-9 INTERFERON A (AIDS INITIATIVE)	INTERFERON A (AIDS INITIATIVE)
RR-01430-2 INULIN TRINITRATE	INULIN TRINITRATE
8013-17-0 INVERT SUGAR	INVERT SUGAR
RR-00013-5 INVESTIGATIONAL ORAL CONTRACEPTIVES	INVESTIGATIONAL ORAL CONTRACEPTIVES
64-69-7 IODACETIC ACID	IODACETIC ACID
5634-39-9 IODINATED GLYCEROL	IODINATED GLYCEROL
7553-56-2 IODINE	IODINE
15480-34-9 IODINE 120	IODINE 120
RR-00436-4 IODINE 120M	IODINE 120M
15755-17-6 IODINE 121	IODINE 121
15715-08-9 IODINE 123	IODINE 123
14158-30-6 IODINE 124	IODINE 124
14158-31-7 IODINE 125	IODINE 125
14158-32-8 IODINE 126	IODINE 126
14391-72-1 IODINE 128	IODINE 128
15046-84-1 IODINE 129	IODINE 129
14914-02-4 IODINE 130	IODINE 130
10043-66-0 IODINE 131	IODINE 131
24267-56-9 IODINE-131	IODINE-131
14683-16-0 IODINE 132	IODINE 132
RR-00435-3 IODINE 132M	IODINE 132M
14834-67-4 IODINE 133	IODINE 133

Chemical Name	Reference Name
14914-27-3 IODINE 134	IODINE 134
14834-68-5 IODINE 135	IODINE 135
14696-82-3 IODINE AZIDE	IODINE AZIDE
7790-99-0 IODINE MONOCHLORIDE	IODINE MONOCHLORIDE
7783-66-6 IODINE PENTAFLUORIDE	IODINE PENTAFLUORIDE
144-48-9 2-IODOACETAMIDE	2-IODOACETAMIDE
591-50-4 IODOBENZENE	IODOBENZENE
513-48-4 2-IODOBUTANE [HM1, HM1, NJL]	SEC-BUTYL IODIDE
542-69-8 1-IODOBUTANE	1-IODOBUTANE
75-47-8 IODOFORM	IODOFORM
75-47-8 IODOFORM (TRIIODOMETHANE) [QBC]	IODOFORM
74-88-4 IODOMETHANE [RCA, UTS]	METHYL IODIDE
74-88-4 IODOMETHANE (METHYL IODIDE) [QBC, QBC, QBC]	METHYL IODIDE
513-38-2 IODO METHYLPROPANE	IODO METHYLPROPANE
513-38-2 1-IODO-2-METHYLPROPANE [NJL]	IODO METHYLPROPANE
558-17-8 2-IODO-2-METHYLPROPANE	2-IODO-2-METHYLPROPANE
RR-00146-7 IODOMETHYLPROPANES	IODOMETHYLPROPANES
75-30-9 2-IODOPROPANE	2-IODOPROPANE
107-08-4 1-IODOPROPANE	1-IODOPROPANE
26914-02-3 IODO PROPANE	IODO PROPANE
26914-02-3 IODOPROPANES [HM1, HM1]	IODO PROPANE
55406-53-6 3-IODO-2-PROPYNYL BUTYLCARBAMATE	3-IODO-2-PROPYNYL BUTYLCARBAMATETRAOXIDE]
55406-53-6 3-IODO-2-PROPYNYL BUTYL CARBA-MATE [PST]	3-IODO-2-PROPYNYL BUTYLCARBAMATE
RR-01431-3 IODOXY COMPOUNDS	IODOXY COMPOUNDS
RR-00158-1 IONIZING RADIATION	IONIZING RADIATION
127-41-3 IONONE, ALPHA (ALPHA-IONONE)	IONONE, ALPHA (ALPHA-IONONE)
8013-90-9 IONONE	IONONE
14901-07-6 IONONE, BETA (BETA-IONONE)	IONONE, BETA (BETA-IONONE)
618-76-8 IOXYNIL	IOXYNIL
3458-22-8 IPD	IPD
3458-22-8 IPD (3,3'-IMINOBIS-1-PROPANOL DIMETHANE-SULFONATE (ESTER) HYDROCHLORIDE) [CSR, CSR]	IPD
76180-96-6 IQ	IQ
76180-96-6 IQ (2-AMINO-3-METHYLIMIDAZO[4,5-F]QUINO-LINE) [C65, C65]	IQ
76180-96-6 IQ; (2-AMINO-3-METHYLIMIDAZO[4,5-F] QUINOLINE) [CAL]	IQ
76180-96-6 IQ (2-AMINO-3-METHYLIMIDAZO[4,5-F]QUINO-LINE) [IAR]	IQ
29054-62-4 IRIDIUM 182	IRIDIUM 182
27742-26-3 IRIDIUM 184	IRIDIUM 184
29054-43-1 IRIDIUM 185	IRIDIUM 185
24447-13-0 IRIDIUM 186	IRIDIUM 186
14834-71-0 IRIDIUM 187	IRIDIUM 187
15752-22-4 IRIDIUM 188	IRIDIUM 188
14265-84-0 IRIDIUM 189	IRIDIUM 189
14981-91-0 IRIDIUM 190	IRIDIUM 190
RR-00432-0 IRIDIUM 190M	IRIDIUM 190M

Chemical Name	Reference Name
14694-69-0 IRIDIUM 192	IRIDIUM 192
RR-00431-9 IRIDIUM 192M	IRIDIUM 192M
14158-35-1 IRIDIUM 194	IRIDIUM 194
RR-00430-8 IRIDIUM 194M	IRIDIUM 194M
15816-99-6 IRIDIUM 195	IRIDIUM 195
RR-00427-3 IRIDIUM 195M	IRIDIUM 195M
RR-01432-4 IRIDIUM NITRATOPENTAMINE IRIDIUM NITRATE	IRIDIUM NITRATOPENTAMINE IRIDIUM NITRATE
10025-97-5 IRIDIUM TETRACHLORIDE	IRIDIUM TETRACHLORIDE
7439-89-6 IRON	IRON
14093-04-0 IRON 52	IRON 52
14681-59-5 IRON 55	IRON 55
14596-12-4 IRON 59	IRON 59
32020-21-6 IRON 60	IRON 60
1309-37-1 IRON OXIDE FUME [MEX, MEX]	IRON OXIDE
7439-89-6 IRON (FE) [PST]	IRON
RR-00537-8 IRON AND STEEL FOUNDING	IRON AND STEEL FOUNDING
85763-69-5 IRON, C3-13-CARBOXYLATE NAPHTHENATE COMPLEXES	IRON, C3-13-CARBOXYLATE NAPHTHENATE COMPLEXES
13463-40-6 IRON CARBONYL [NFP, NFP]	IRON PENTACARBONYL
13463-40-6 IRON CARBONYL (FE(CO)5), (TB-5-11)- [PAL]	IRON PENTACARBONYL
7705-08-0 IRON CHLORIDE [NJL, NJL]	FERRIC CHLORIDE
7758-94-3 IRON CHLORIDE (FECL2) [PAL]	FERROUS CHLORIDE
7705-08-0 IRON CHLORIDE (FECL3) [PAL]	FERRIC CHLORIDE
9004-66-4 IRON DEXTRAN [CAL, FLA, MSL, PAL, PAL]	IRON DEXTRAN COMPLEX
9004-66-4 IRON DEXTRAN COMPLEX	IRON DEXTRAN COMPLEX
9004-66-4 IRON-DEXTRAN COMPLEX [IAR]	IRON DEXTRAN COMPLEX
8050-93-9 IRON DEXTRIN COMPLEX	IRON DEXTRIN COMPLEX
10294-53-8 IRON (III) DICHROMATE	IRON (III) DICHROMATE
7783-50-8 IRON FLUORIDE (FEF3) [PAL]	FERRIC FLUORIDE
22830-45-1 IRON (II) GLUCONATE	IRON (II) GLUCONATE
12645-49-7 IRON MANGANESE ZINC OXIDE	IRON MANGANESE ZINC OXIDE
12645-50-0 IRON NICKEL ZINC OXIDE	IRON NICKEL ZINC OXIDE
1332-37-2 IRON OXIDE	IRON OXIDE
1309-37-1 IRON OXIDE (FE2O3) [PAL]	IRON OXIDE
1309-37-1 IRON OXIDE DUST AND FUME [NIO, NIO]	IRON OXIDE
1309-37-1 IRON OXIDE FUME [ALB, ALB, BC1, BC1, CAL, FLA, MSL, NJL, OS1, OS1]	IRON OXIDE
1309-37-1 IRON OXIDE FUME (FE2O3) [TLV]	IRON OXIDE
12259-21-1 IRON OXIDE (FE2O3), HYDRATE	IRON OXIDE (FE2O3), HYDRATE
1332-37-2 IRON OXIDE, SPENT [HM1, HM1]	IRON OXIDE
51274-00-1 IRON OXIDE YELLOW	IRON OXIDE YELLOW
13463-40-6 IRON PENTACARBONYL	IRON PENTACARBONYL
13463-40-6 IRON, PENTACRABONYL [302]	IRON PENTACARBONYL
13463-40-6 IRON, PENTACARBONYL- [OS3]	IRON PENTACARBONYL
13463-40-6 IRON, PENTACARBONYL- [IRON CARBONYL (FE(CO)5), (TB-5-11)-] [EP3]	IRON PENTACARBONYL
RR-00521-0 IRON SALTS [GBR, GBR, PAL]	IRON SALTS (SOLUBLE)
RR-00521-0 IRON SALTS (SOLUBLE)	IRON SALTS (SOLUBLE)

Chemical Name	Reference Name
RR-00521-0 IRON SALTS, SOLUBLE [BC1, BC1, CAL, CEX, QBC, TLV]	IRON SALTS (SOLUBLE)
RR-00521-0 IRON SALTS, (WATER-SOLUBLE) [ONT]	IRON SALTS (SOLUBLE)
RR-01534-9 IRON SORBITOL-CITRIC ACID COMPLEX	IRON SORBITOL-CITRIC ACID COMPLEX
1309-37-1 IRON TRIOXIDE DUST AND FUME [QBC]	IRON OXIDE
14484-64-1 IRON, TRIS(DIMETHYLCARBAMODITHIOATO-S, S')-, (OC-6-11)- [PAL]	FERBAM
RR-00521-0 IRON, WATER-SOLUBLE SALTS, N.O.S. [WHM]	IRON SALTS (SOLUBLE)
8001-86-3 ISANO OIL	ISANO OIL
15503-86-3 ISATIDINE	ISATIDINE
142-15-4 ISETHIONIC ACID, OLEATE, SODIUM SALT	ISETHIONIC ACID, OLEATE, SODIUM SALT
123-92-2 ISOAMYL ACETATE	ISOAMYL ACETATE
123-92-2 ISO-AMYL ACETATE [EPA]	ISOAMYL ACETATE
123-51-3 ISOAMYL ALCOHOL	ISOAMYL ALCOHOL
123-51-3 ISOAMYL ALCOHOL (PRIMARY) [NIO, NIO, NIO]	ISOAMYL ALCOHOL
123-51-3 ISOAMYL ALCOHOL (PRIMARY AND SEC-ONDARY) [OS1, OS1, OS1]	ISOAMYL ALCOHOL
528-75-6 ISOAMYL ALCOHOL (SECONDARY)	ISOAMYL ALCOHOL (SECONDARY)
106-27-4 ISOAMYL BUTYRATE	ISOAMYL BUTYRATE
107-84-6 ISOAMYL CHLORIDE	ISOAMYL CHLORIDE
541-31-1 ISOAMYL MERCAPTAN	ISOAMYL MERCAPTAN
89-65-6 D-ISOASCORBIC ACID [PST]	ERYTHORBIC ACID
297-78-9 ISOBENZAN	ISOBENZAN
297-78-9 ISOBENZEN [302]	ISOBENZAN
552-30-7 5-ISOBENZOFURANCARBOXYLIC ACID, 1,3-DIHYDRO-1,3-DIOXO- [PAL]	TRIMELLITIC ANHYDRIDE
85-44-9 1,3-ISOBENZOFURANDIONE [PAL]	PHTHALIC ANHYDRIDEDIBROMIDE
632-79-1 1,3-ISOBENZOFURANDIONE, 4,5,6,7-TETRA-BROMO-	1,3-ISOBENZOFURANDIONE, 4,5,6,7-TETRABROMO-
125-12-2 ISOBORNYL ACETATE	ISOBORNYL ACETATE
115-31-1 ISOBORNYL THIOCYANOACETATE	ISOBORNYL THIOCYANOACETATE
75-28-5 ISOBUTANE	ISOBUTANE
75-28-5 ISOBUTANE [PROPANE, 2-METHYL-] [EP3]	ISOBUTANE
78-83-1 ISOBUTANOL [HM1, HM1, HM1, RCA, WHM]	ISOBUTYL ALCOHOL
115-11-7 ISOBUTENE [CSR]	ISOBUTYLENE
110-19-0 ISOBUTYL ACETATE	ISOBUTYL ACETATE
110-19-0 ISO-BUTYL ACETATE [CWA, EPA]	ISOBUTYL ACETATE
110-19-0 ISOBUTYL ACETATE (2-METHYLPROPYL AC-ETATE) [CAL]	ISOBUTYL ACETATE
106-63-8 ISOBUTYL ACRYLATE	ISOBUTYL ACRYLATE
78-83-1 ISOBUTYL ALCOHOL	ISOBUTYL ALCOHOL
78-81-9 ISOBUTYLAMINE	ISOBUTYLAMINE
78-81-9 ISO-BUTYLAMINE [CWA, EPA]	ISOBUTYLAMINE
78-81-9 ISOBUTYL AMINE [MAK, MAK, MAK]	ISOBUTYLAMINE
68457-75-0 ISOBUTYLATED, STYRENEATED CRESOLS	ISOBUTYLATED, STYRENEATED CRESOLS
538-93-2 ISOBUTYLBENZENE	ISOBUTYLBENZENE
539-90-2 ISOBUTYL BUTYRATE	ISOBUTYL BUTYRATE
513-36-0 ISOBUTYL CHLORIDE	ISOBUTYL CHLORIDE
543-27-1 ISOBUTYL CHLOROFORMATE	ISOBUTYL CHLOROFORMATE
RR-00863-9 ISOBUTYLCYCLOHEXANE	ISOBUTYLCYCLOHEXANE

Chemical Name	Reference Name
115-11-7 ISOBUTYLENE	ISOBUTYLENE
542-55-2 ISOBUTYL FORMATE	ISOBUTYL FORMATE
123-18-2 ISOBUTYL HEPTYL KETONE	ISOBUTYL HEPTYL KETONE
97-85-8 ISOBUTYL ISOBUTYRATE	ISOBUTYL ISOBUTYRATE
1873-29-6 ISOBUTYL ISOCYANATE	ISOBUTYL ISOCYANATE
97-86-9 ISOBUTYL METHACRYLATE	ISOBUTYL METHACRYLATE
97-86-9 ISOBUTYLMETHACRYLATE [NJL]	ISOBUTYL METHACRYLATE
28348-65-4 ISOBUTYLNAPTHALENESULFONIC ACID, SODIUM SALT	ISOBUTYLNAPTHALENESULFONIC ACID, SODIUM SALT
542-56-3 ISOBUTYL NITRITE	ISOBUTYL NITRITE
102-13-6 ISOBUTYL PHENYLACETATE	ISOBUTYL PHENYLACETATE
126-71-6 ISOBUTYL PHOSPHATE	ISOBUTYL PHOSPHATE-
540-42-1 ISOBUTYL PROPIONATE	ISOBUTYL PROPIONATE
78-84-2 ISOBUTYRALDEHYDE	ISOBUTYRALDEHYDE
79-31-2 ISOBUTYRIC ACID	ISOBUTYRIC ACID
79-31-2 ISO-BUTYRIC ACID [CWA, EPA, MSL]	ISOBUTYRIC ACID
97-72-3 ISOBUTYRIC ANHYDRIDE	ISOBUTYRIC ANHYDRIDE
78-82-0 ISOBUTYRONITRILE	ISOBUTYRONITRILE
78-82-0 ISOBUTYRONITRILE [PROPANENITRILE, 2-METHYL-] [EP3]	ISOBUTYRONITRILE
79-30-1 ISOBUTYRYL CHLORIDE	ISOBUTYRYL CHLORIDE
79-30-1 ISOBUTYRYLCHLORIDE [NJL]	ISOBUTYRYL CHLORIDE
RR-00294-8 ISOCYANATES [T33]	ISOCYANATES, ALL
RR-00294-8 ISOCYANATES, ALL	ISOCYANATES, ALL
RR-00294-8 ISOCYANATES, N.O.S. [HM1, HM1, HM1, NJL]	ISOCYANATES, ALL
RR-00955-2 ISOCYANATE TERMINATED POLYOLS	ISOCYANATE TERMINATED POLYOLS
71121-36-3 ISOCYANATOBENZOTRIFLUORIDE	ISOCYANATOBENZOTRIFLUORIDE
71121-36-3 ISOCYANATOBENZOTRIFLUORIDES [HM1, HM1, HM1]	ISOCYANATOBENZOTRIFLUORIDE
71121-36-3 ISOCYANATOBENZO-TRIFLUORIDE [NJL]	ISOCYANATOBENZOTRIFLUORIDE
71121-36-3 ISOCYANATOBENZOTRIFLUORIDES [WHM]	ISOCYANATOBENZOTRIFLUORIDE
106790-31-2 4-ISO-CYANATO-N,N-BIS(4-ISOCYANATOPHENYL)-2,5-DIMETHOXYBENZENEAMINE	4-ISOCYANATO-N,N-BIS(4-ISOCYANATOPHENYL)-2,5-DIMETHOXYBENZENAND 1,3-BENZENEDIAMINE, 4-METHYL-2,6-BIS(METHYLTHIO)
102-36-3 ISOCYANIC ACID, 3,4-DICHLOROPHENYL ESTER	ISOCYANIC ACID, 3,4-DICHLOROPHENYL ESTER
9016-87-9 ISOCYANIC ACID, POLY-METHYLENEPOLYPHENYLENE ESTER [TSC, TSC]	POLYMETHYLENE POLYPHENYLENE ISOCYANATE[AZO (6-HYDROXY-3,1-PHENYLENE)AZO[6(OR 7)-SULFO-4,1-NAPHTHALENEDIYL]AZO[1,1'-BIPHENYL]-4,4'-DIYLAZO] BIS[6- HYDROXY-, HEXASODIUM SALT
32052-51-0 ISOCYANIC ACID, TRIMETHYLCYCLOHEXYL ESTER	ISOCYANIC ACID, TRIMETHYLCYCLOHEXYL ESTER-
108-80-5 ISOCYANURIC ACID	ISOCYANURIC ACID
RR-00864-0 ISODECALDEHYDE	ISODECALDEHYDE
RR-00854-8 ISODECANE	ISODECANE
26403-17-8 ISODECANOIC ACID	ISODECANOIC ACID
25339-17-7 ISODECANOL	ISODECANOL
25339-17-7 ISODECANOL, MIXED ISOMERS [NFP, NFP]	ISODECANOL
1330-61-6 ISODECYL ACRYLATE [HM1]	2-PROPENOIC ACID, ISODECYL ESTER
25339-17-7 ISODECYL ALCOHOL [PST]	ISODECANOL
29761-21-5 ISODECYL DIPHENYL PHOSPHATE [HM1, OTS]	PHOSPHATE, ISODECYL DIPHENYL
60209-82-7 ISODECYL NEOPENTANOATE	ISODECYL NEOPENTANOATE

356

Chemical Name	Reference Name
31807-55-3 **ISODODECANE**	ISODODECANE
465-73-6 **ISODRIN**	ISODRIN
97-54-1 **ISOEUGENOL**	ISOEUGENOL
RR-01290-8 **ISOFENPHOS**	ISOFENPHOS
25311-71-1 **ISOFENPHOS [2-[[ETHOXYL[(1-METHYLETHYL) AMINO]PHOSPHINOTHIOYL]OXY]BENZOIC ACID 1-METHYLETHYL ESTER]**	ISOFENPHOS [2-[[ETHOXYL[(1-METHYLETHYL)AMINO] PHOSPHINOTHIOYLCARBAMIC ACID DIETHYL ESTER]
55-91-4 **ISOFLUORPHATE**	ISOFLUORPHATE
26675-46-7 **ISOFLURANE**	ISOFLURANE
31394-54-4 **ISOHEPTANE**	ISOHEPTANE
68975-47-3 **ISOHEPTENE**	ISOHEPTENE
36311-34-9 **ISOHEXADECANOL**	ISOHEXADECANOL
107-83-5 **ISOHEXANE**	ISOHEXANE
27236-46-0 **ISOHEXENE**	ISOHEXENE
77-74-7 **TERT-ISOHEXYL ALCOHOL**	TERT-ISOHEXYL ALCOHOL
5455-98-1 **1H-ISOINDOLE-1,3(2H)-DIONE, 2-(OXIRANYL-METHYL)-**	1H-ISOINDOLE-1,3(2H)-DIONE, 2-(OXIRANYLMETHYL)-
2425-06-1 **1H-ISOINDOLE-1,3(2H)-DIONE, 3A,4,7,7A-TE-TRAHYDRO-2-[(1,1,2,2-TETRACHLOROETHYL) THIO]- [PAL]**	CAPTAFOL5,5A,5B,6-DODECACHLOROOCTAHYDRO-
133-06-2 **1H-ISOINDOLE-1,3(2H)-DIONE, 3A,4,7,7A-TE-TRAHYDRO-2-[(TRICHLOROMETHYL)THIO]- [PAL]**	CAPTAN
3468-11-9 **ISOINDOLINE, 1,3-DIIMINO**	ISOINDOLINE, 1,3-DIIMINO
5590-18-1 **1H-ISOINDOL-1-ONE**	1H-ISOINDOL-1-ONE
55-22-1 **ISONICOTINIC ACID**	ISONICOTINIC ACID
54-85-3 **ISONICOTINIC ACID HYDRAZINE**	ISONICOTINIC ACID HYDRAZINE
54-85-3 **ISONICOTINIC ACID HYDRAZINE (ISONIAZID) [CAL, IAR]**	ISONICOTINIC ACID HYDRAZINE
58449-37-9 **ISONONANOYL PEROXIDE**	ISONONANOYL PEROXIDEIMINO)PENTANENITRILE)-, (T-4)-
30399-84-9 **ISOOCTADECANOIC ACID**	ISOOCTADECANOIC ACID
27458-93-1 **ISOOCTADECANOL**	ISOOCTADECANOL
540-84-1 **ISOOCTANE**	ISOOCTANE
26635-64-3 **ISOOCTANE [PAL]**	ISOOCTANE (VAN)
26635-64-3 **ISOOCTANE (VAN)**	ISOOCTANE (VAN)
25103-52-0 **ISOOCTANOIC ACID**	ISOOCTANOIC ACID
26952-21-6 **ISOOCTANOL [PAL]**	ISOOCTYL ALCOHOL
11071-47-9 **ISOOCTENE**	ISOOCTENE
11071-47-9 **ISOOCTENES [HM1, HM1, NFP, NFP, PAL]**	ISOOCTENE
29590-42-9 **ISO-OCTYL ACRYLATE [OTS]**	2-PROPENOIC ACID, ISOOCTYL ESTER
26952-21-6 **ISOOCTYL ALCOHOL**	ISOOCTYL ALCOHOL
26952-21-6 **ISOOCTYL ALCOHOL (ISOOCTANOL) [WHM]**	ISOOCTYL ALCOHOL
26952-21-6 **ISOOCTYL ALCOHOL (MIXED ISOMERS) [GBR]**	ISOOCTYL ALCOHOL
26401-27-4 **ISOOCTYL DIPHENYL PHOSPHITE**	ISOOCTYL DIPHENYL PHOSPHITE
RR-00865-1 **ISOOCTYL NITRATE**	ISOOCTYL NITRATE
4016-14-2 **ISOOPROPYL GLYCIDYL ETHER [OTS]**	ISOPROPYL GLYCIDYL ETHER (IGE)
64365-06-6 **ISOPARAFFINIC PETROLEUM HYDROCARBONS**	ISOPARAFFINIC PETROLEUM HYDROCARBONS
590-86-3 **ISOPENTALDEHYDE**	ISOPENTALDEHYDE
78-78-4 **ISOPENTANE**	ISOPENTANE
78-78-4 **ISO-PENTANE [MAK, MAK]**	ISOPENTANE
78-78-4 **ISOPENTANE [BUTANE, 2-METHYL-] [EP3]**	ISOPENTANE

Chemical Name	Reference Name
503-74-2 ISOPENTANOIC ACID	ISOPENTANOIC ACID
RR-00349-6 ISOPENTENES	ISOPENTENES
123-92-2 ISOPENTYL ACETATE [GBR, GBR]	ISOAMYL ACETATE
78-59-1 ISOPHORONE	ISOPHORONE
7027-11-4 ISOPHORONE NITRILE [HON]	CYCLOHEXANECARBONITRILE, 1,3,3-TRIMETHYL-5-OXO-,
2855-13-2 ISOPHORONE DIAMINE	ISOPHORONE DIAMINE
2855-13-2 ISOPHORONEDIAMINE [NJL, NJL]	ISOPHORONE DIAMINE
4098-71-9 ISOPHORONE DIISOCYANATE	ISOPHORONE DIISOCYANATE
4098-71-9 ISOPHORONEDIISOCYANATE [MAK, MAK, MAK]	ISOPHORONE DIISOCYANATE
4098-71-9 ISOPHORONE DIISOCYANATE (IPDI) [ONT, ONT, ONT, NHS, NHS]	ISOPHORONE DIISOCYANATE
3778-73-2 ISOPHOSPHAMIDE	ISOPHOSPHAMIDE
121-91-5 ISOPHTHALIC ACID	ISOPHTHALIC ACID
4891-67-2 ISOPHTHALIC ANHYDRIDE	ISOPHTHALIC ANHYDRIDE
99-63-8 ISOPHTHALOYL CHLORIDE	ISOPHTHALOYL CHLORIDE
78-79-5 ISOPRENE	ISOPRENE
78-79-5 ISOPRENE [1,3-BUTADIENE, 2-METHYL-] [EP3]	ISOPRENE
2631-40-5 ISOPROCARB	ISOPROCARB
67-63-0 ISOPROPANOL [ATS, CTR, HM1, HM1, OTS, TSE, WHM]	ISOPROPYL ALCOHOL
78-96-6 ISOPROPANOLAMINE	ISOPROPANOLAMINE
42504-46-1 ISOPROPANOLAMINE DODECYLBENZENE SULFONATE	ISOPROPANOLAMINE DODECYLBENZENE SULFONATE
42504-46-1 ISOPROPANOLAMINE DODECYL-BENZENESUL-FONATE [CAL]	ISOPROPANOLAMINE DODECYLBENZENE SULFONATE
54590-52-2 ISOPROPANOLAMINE DODECYLBENZENESUL-FONATE [CWA, EPA, NJL]	ISOPROPANOLAMINE DODECYLBENZENESULFONATE (1:1)
54590-52-2 ISOPROPANOLAMINE DODECYLBENZENESUL-FONATE (1:1)	ISOPROPANOLAMINE DODECYLBENZENESULFONATE (1:1)
108-22-5 ISOPROPENYL ACETATE	ISOPROPENYL ACETATE
78-80-8 ISOPROPENYL ACETYLENE	ISOPROPENYL ACETYLENE
98-83-9 ISOPROPENYLBENZENE [HM1, HM1, HM1]	ALPHA-METHYL STYRENE
98-83-9 ISOPROPENYL BENZENE [NJL]	ALPHA-METHYL STYRENE
10471-78-0 2-ISOPROPENYL-2-OXAZOLINE	2-ISOPROPENYL-2-OXAZOLINE
109-59-1 ISOPROPOXYETHANOL	ISOPROPOXYETHANOL
109-59-1 2-ISOPROPOXYETHANOL [MAK, MAK, MAK, MAK, ONT, ONT, OS1]	ISOPROPOXYETHANOL
114-26-1 2-ISOPROPOXYPHENYL N-METHYLCARBA-MATE (PROPOXUR) [CAL]	PROPOXUR
29387-84-6 3-ISOPROPOXY-1-PROPANOL [PST]	1,2-PROPANEDIOL, MONOISOPROPYL ETHER
110-47-4 3-ISOPROPOXYPROPIONITRILE	3-ISOPROPOXYPROPIONITRILE
108-21-4 ISOPROPYL ACETATE	ISOPROPYL ACETATE
1623-24-1 ISOPROPYL ACID PHOSPHATE [HM1, HM1, NJL, NJL]	ISOPROPYL PHOSPHORIC ACID
67-63-0 ISOPROPYL ALCOHOL	ISOPROPYL ALCOHOL
RR-00068-0 ISOPROPYL ALCOHOL MANUFACTURE (STRONG-ACID PROCESS)	ISOPROPYL ALCOHOL MANUFACTURE (STRONG-ACID PROCESS)
75-31-0 ISOPROPYLAMINE	ISOPROPYLAMINE
79771-08-7 ISOPROPYLAMINE, DISTILLATION RESIDUES	ISOPROPYLAMINE, DISTILLATION RESIDUES
79771-08-7 ISOPROPYLAMINE DISTILLATION RESIDUES [TSE]	ISOPROPYLAMINE, DISTILLATION RESIDUES

Chemical Name	Reference Name
12068-04-1 ISOPROPYLAMINE METHYLNAPHTHALENE-SULFONATE	ISOPROPYLAMINE METHYLNAPHTHALENESULFONATE
RR-01090-2 ISOPROPYLAMINE SALT OF OLEOYLISO-PROPANOLAMIDE DERIVATIVE OFSULFOSUC-CINIC ACID	ISOPROPYLAMINE SALT OF OLEOYLISO-PROPANOLAMIDE DERIVATIVE OF2-TALLOW AMI-DOETHYL), METHYL SULFATES
RR-01091-3 ISOPROPYLAMINE SALT OF STEARYLISO-PROPANOLAMIDE DERIVATIVE OF SULFOS-UCCINIC ACID	ISOPROPYLAMINE SALT OF STEARYLISO-PROPANOLAMIDE DERIVATIVE OFSULFOSUCCINIC ACID
26118-67-2 ISOPROPYLAMINE SULFONATE	ISOPROPYLAMINE SULFONATE
75-31-0 ISOPROPYLAMINE [2-PROPANAMINE] [EP3]	ISOPROPYLAMINE
643-28-7 N-ISOPROPYLANILINE	N-ISOPROPYLANILINE
643-28-7 O-ISOPROPYLANILINE [MSL]	N-ISOPROPYLANILINE
643-28-7 2-ISOPROPYL ANILINE [WHM]	N-ISOPROPYLANILINE
768-52-5 N-ISOPROPYLANILINE	N-ISOPROPYLANILINE
68987-86-0 ISOPROPYLATED CRESOL	ISOPROPYLATED CRESOLONOMETHYLATED
98-82-8 ISOPROPYLBENZENE [HM1, HM1, HM1, HM1, SDW, WHM]	CUMENE
98-82-8 ISOPROPYLBENZENE (CUMENE) [QBC, QBC]	CUMENE
939-48-0 ISOPROPYL BENZOATE	ISOPROPYL BENZOATE
RR-00866-2 ISOPROPYL BICYCLOHEXYL	ISOPROPYL BICYCLOHEXYL
25640-78-2 ISOPROPYL BIPHENYL	ISOPROPYL BIPHENYL
25640-78-2 2-ISOPROPYLBIPHENYL [NFP, NFP]	ISOPROPYL BIPHENYL
638-11-9 ISOPROPYL BUTYRATE	ISOPROPYL BUTYRATE
75-29-6 ISOPROPYL CHLORIDE [FLA, HM1, HM1, HM1, MSL, NFP, NFP]	2-CHLOROPROPANE
75-29-6 ISOPROPYL CHLORIDE [PROPANE, 2-CHLORO-] [EP3]	2-CHLOROPROPANE
105-48-6 ISOPROPYL CHLOROACETATE	ISOPROPYL CHLOROACETATE
108-23-6 ISOPROPYL CHLOROFORMATE	ISOPROPYL CHLOROFORMATE
40058-87-5 ISOPROPYL-2-CHLOROPROPIONATE	ISOPROPYL-2-CHLOROPROPIONATE
RR-01433-5 ISOPROPYLCUMYL HYDROPEROXIDE	ISOPROPYLCUMYL HYDROPEROXIDE
696-29-7 ISOPROPYLCYCLOHEXANE	ISOPROPYLCYCLOHEXANE
1195-42-2 ISOPROPYL CYCLOHEXYLAMINE	ISOPROPYL CYCLOHEXYLAMINE
1195-42-2 ISOPROPYLCYCLOHEXYLAMINE [NFP, NFP]	ISOPROPYL CYCLOHEXYLAMINE
109-56-8 ISOPROPYL ETHANOLAMINE	ISOPROPYL ETHANOLAMINE
108-20-3 ISOPROPYL ETHER	ISOPROPYL ETHER
625-55-8 ISOPROPYL FORMATE	ISOPROPYL FORMATE
4016-14-2 ISOPROPYL GLYCIDYL ETHER [CAL, CEX, CEX, EPA, MIN, NIO, NIO, NIO]	ISOPROPYL GLYCIDYL ETHER (IGE)
4016-14-2 ISOPROPYL GLYCIDYL ETHER (IGE)	ISOPROPYL GLYCIDYL ETHER (IGE)
52896-87-4 4-ISOPROPYLHEPTANE	4-ISOPROPYLHEPTANE
RR-01228-2 ISOPROPYLIDENE, BIS(1,1-DIMETHYLPROPYL) DERIVATIVE	ISOPROPYLIDENE, BIS(1,1-DIMETHYLPROPYL) DERIVA-TIVEER
80-05-7 4,4'-ISOPROPYLIDENEDIPHENOL [313, CAB, MSL, NJL]	BISPHENOL A
80-05-7 4,4-ISOPROPYLIDENEDIPHENOL [CAN]	BISPHENOL A
80-05-7 4,4'-ISOPROPYLIDENEDIPHENOL [PAL, PST]	BISPHENOL A
RR-01092-4 4,4'-ISOPROPYLIDENE DIPHENOL ALKYL (C12-15) PHOSPHITES	4,4'-ISOPROPYLIDENE DIPHENOL ALKYL (C12-15) PHOS-PHITES SULFOSUCCINIC ACID
617-50-5 ISOPROPYL ISOBUTYRATE	ISOPROPYL ISOBUTYRATE
1795-48-8 ISOPROPYL ISOCYANATE	ISOPROPYL ISOCYANATE
617-51-6 ISOPROPYL LACTATE	ISOPROPYL LACTATE

Chemical Name	Reference Name
63393-93-1 ISOPROPYL LANOLIN	ISOPROPYL LANOLIN
75-33-2 ISOPROPYL MERCAPTAN	ISOPROPYL MERCAPTAN
926-06-7 ISOPROPYL METHANE SULPHONATE	ISOPROPYL METHANE SULPHONATE
671-16-9 N-ISOPROPYL-ALPHA-(2-METHYLHYDRAZINO)-P-TOLUAMIDE [NJL]	PROCARBAZINE
119-38-0 ISOPROPYLMETHYLPYRAZOLYL DIMETHYL-CARBAMATE	ISOPROPYLMETHYLPYRAZOLYL DIMETHYLCARBA-MATE
110-27-0 ISOPROPYL MYRISTATE	ISOPROPYL MYRISTATE
28348-64-3 ISOPROPYLNAPTHALENESULFONIC ACID, SODIUM SALT	ISOPROPYLNAPTHALENESULFONIC ACID, SODIUM SALT
1712-64-7 ISOPROPYL NITRATE	ISOPROPYL NITRATE
RR-00813-9 ISOPROPYL OIL	ISOPROPYL OIL
RR-00813-9 ISOPROPYL OILS [IAR]	ISOPROPYL OIL
142-91-6 ISOPROPYL PALMITATE	ISOPROPYL PALMITATE
105-64-6 ISOPROPYL PEROXYDICARBONATE	ISOPROPYL PEROXYDICARBONATE
25168-06-3 ISOPROPYL PHENOL	ISOPROPYL PHENOL
28108-99-8 ISOPROPYLPHENYL DIPHENYL PHOS-PHATE [OTS]	PHOSPHATE, ISOPROPYLPHENYL DIPHENYL
101-72-4 N-ISOPROPYL-N'-PHENYL-P-PHENYLENE-DIAMINE	N-ISOPROPYL-N'-PHENYL-P-PHENYLENE-DIAMINE
1623-24-1 ISOPROPYL PHOSPHORIC ACID	ISOPROPYL PHOSPHORIC ACID
637-78-5 ISOPROPYL PROPIONATE	ISOPROPYL PROPIONATE
112-10-7 ISOPROPYL STEARATE	ISOPROPYL STEARATE
99-87-6 P-ISOPROPYLTOLUENE [SDW]	P-CYMENE
119-65-3 ISOQUINOLINE	ISOQUINOLINE
120-58-1 ISOSAFROLE	ISOSAFROLE
87-33-2 ISOSORBIDE DINITRATE MIXTURE	ISOSORBIDE DINITRATE MIXTURE
4759-48-2 ISOTERTINATE	ISOTERTINATE
26172-55-4 3(2H)-ISOTHIAZOLONE, 5-CHLORO-2-METHYL-[PST]	5-CHLORO-2-METHYL-3-ISOTHIAZOLONE
2682-20-4 3(2H)-ISOTHIAZOLONE, 2-METHYL- [PST]	2-METHYL-3-ISOTHIAZOLONE
26530-20-1 3(2H)-ISOTHIAZOLONE, 2-OCTYL-	3(2H)-ISOTHIAZOLONE, 2-OCTYL-
3129-90-6 ISOTHIOCYANIC ACID	ISOTHIOCYANIC ACID
4759-48-2 ISOTRETINOIN [C65, CAB]	ISOTERTINATE
27458-92-0 ISOTRIDECYL ALCOHOL	ISOTRIDECYL ALCOHOL
RR-01291-9 ISOXATHION	ISOXATHION
2763-96-4 3(2H)-ISOXAZOLONE, 5-(AMINOMETHYL)- [TSC, TSC]	MUSCIMOL
6870-67-3 JACOBINE	JACOBINE
61789-91-1 JAJOBA BEAN OIL	JAJOBA BEAN OIL
4466-14-2 JASMOLIN I	JASMOLIN I
1172-63-0 JASMOLIN II	JASMOLIN II
488-10-8 JASMONE	JASMONE
RR-01549-6 JET FUEL	JET FUEL
RR-00130-9 JET FUELS JET A AND JET A-1	JET FUELS JET A AND JET A-1
RR-00788-5 JET FUELS JET B	JET FUELS JET B
RR-00789-6 JET FUELS JP-4	JET FUELS JP-4
RR-00790-9 JET FUELS JP-6	JET FUELS JP-6
RR-00656-4 K001-HAZARDOUS WASTES	K001-HAZARDOUS WASTES
RR-00657-5 K002-HAZARDOUS WASTES	K002-HAZARDOUS WASTES
RR-00658-6 K003-HAZARDOUS WASTES	K003-HAZARDOUS WASTES

Chemical Name	Reference Name
RR-00659-7 K004-HAZARDOUS WASTES	K004-HAZARDOUS WASTES
RR-00660-0 K005-HAZARDOUS WASTES	K005-HAZARDOUS WASTES
RR-00661-1 K006-HAZARDOUS WASTES	K006-HAZARDOUS WASTES
RR-00662-2 K007-HAZARDOUS WASTES	K007-HAZARDOUS WASTES
RR-00663-3 K008-HAZARDOUS WASTES	K008-HAZARDOUS WASTES
RR-00664-4 K009-HAZARDOUS WASTES	K009-HAZARDOUS WASTES
RR-00665-5 K010-HAZARDOUS WASTES	K010-HAZARDOUS WASTES
RR-00666-6 K011-HAZARDOUS WASTES	K011-HAZARDOUS WASTES
RR-00667-7 K013-HAZARDOUS WASTES	K013-HAZARDOUS WASTES
RR-00668-8 K014-HAZARDOUS WASTES	K014-HAZARDOUS WASTES
RR-00669-9 K015-HAZARDOUS WASTES	K015-HAZARDOUS WASTES
RR-00670-2 K016-HAZARDOUS WASTES	K016-HAZARDOUS WASTES
RR-00671-3 K017-HAZARDOUS WASTES	K017-HAZARDOUS WASTES
RR-00672-4 K018-HAZARDOUS WASTES	K018-HAZARDOUS WASTES
RR-00673-5 K019-HAZARDOUS WASTES	K019-HAZARDOUS WASTES
RR-00674-6 K020-HAZARDOUS WASTES	K020-HAZARDOUS WASTES
RR-00675-7 K021-HAZARDOUS WASTES	K021-HAZARDOUS WASTES
RR-00676-8 K022-HAZARDOUS WASTES	K022-HAZARDOUS WASTES
RR-00677-9 K023-HAZARDOUS WASTES	K023-HAZARDOUS WASTES
RR-00678-0 K024-HAZARDOUS WASTES	K024-HAZARDOUS WASTES
RR-00679-1 K025-HAZARDOUS WASTES	K025-HAZARDOUS WASTES
RR-00680-4 K026-HAZARDOUS WASTES	K026-HAZARDOUS WASTES
RR-00681-5 K027-HAZARDOUS WASTES	K027-HAZARDOUS WASTES
RR-00682-6 K028-HAZARDOUS WASTES	K028-HAZARDOUS WASTES
RR-00683-7 K029-HAZARDOUS WASTES	K029-HAZARDOUS WASTES
RR-00684-8 K030-HAZARDOUS WASTES	K030-HAZARDOUS WASTES
RR-00685-9 K031-HAZARDOUS WASTES	K031-HAZARDOUS WASTES
RR-00686-0 K032-HAZARDOUS WASTES	K032-HAZARDOUS WASTES
RR-00687-1 K033-HAZARDOUS WASTES	K033-HAZARDOUS WASTES
RR-00688-2 K034-HAZARDOUS WASTES	K034-HAZARDOUS WASTES
RR-00689-3 K035-HAZARDOUS WASTES	K035-HAZARDOUS WASTES
RR-00690-6 K036-HAZARDOUS WASTES	K036-HAZARDOUS WASTES
RR-00691-7 K037-HAZARDOUS WASTES	K037-HAZARDOUS WASTES
RR-00692-8 K038-HAZARDOUS WASTES	K038-HAZARDOUS WASTES
RR-00693-9 K039-HAZARDOUS WASTES	K039-HAZARDOUS WASTES
RR-00694-0 K040-HAZARDOUS WASTES	K040-HAZARDOUS WASTES
RR-00695-1 K041-HAZARDOUS WASTES	K041-HAZARDOUS WASTES
RR-00696-2 K042-HAZARDOUS WASTES	K042-HAZARDOUS WASTES
RR-00697-3 K043-HAZARDOUS WASTES	K043-HAZARDOUS WASTES
RR-00698-4 K044-HAZARDOUS WASTES	K044-HAZARDOUS WASTES
RR-00699-5 K045-HAZARDOUS WASTES	K045-HAZARDOUS WASTES
RR-00700-1 K046-HAZARDOUS WASTES	K046-HAZARDOUS WASTES
RR-00701-2 K047-HAZARDOUS WASTES	K047-HAZARDOUS WASTES
RR-00702-3 K048-HAZARDOUS WASTES	K048-HAZARDOUS WASTES
RR-00703-4 K049-HAZARDOUS WASTES	K049-HAZARDOUS WASTES
RR-00704-5 K050-HAZARDOUS WASTES	K050-HAZARDOUS WASTES
RR-00705-6 K051-HAZARDOUS WASTES	K051-HAZARDOUS WASTES
RR-00706-7 K052-HAZARDOUS WASTES	K052-HAZARDOUS WASTES
RR-00707-8 K060-HAZARDOUS WASTES	K060-HAZARDOUS WASTES
RR-00708-9 K061-HAZARDOUS WASTES	K061-HAZARDOUS WASTES
RR-00709-0 K062-HAZARDOUS WASTES	K062-HAZARDOUS WASTES

Chemical Name	Reference Name
RR-00710-3 K064-HAZARDOUS WASTES	K064-HAZARDOUS WASTES
RR-00711-4 K065-HAZARDOUS WASTES	K065-HAZARDOUS WASTES
RR-00712-5 K066-HAZARDOUS WASTES	K066-HAZARDOUS WASTES
RR-00713-6 K069-HAZARDOUS WASTES	K069-HAZARDOUS WASTES
RR-00714-7 K071-HAZARDOUS WASTES	K071-HAZARDOUS WASTES
RR-00715-8 K073-HAZARDOUS WASTES	K073-HAZARDOUS WASTES
RR-00716-9 K083-HAZARDOUS WASTES	K083-HAZARDOUS WASTES
RR-00717-0 K084-HAZARDOUS WASTES	K084-HAZARDOUS WASTES
RR-00718-1 K085-HAZARDOUS WASTES	K085-HAZARDOUS WASTES
RR-00719-2 K086-HAZARDOUS WASTES	K086-HAZARDOUS WASTES
RR-00720-5 K087-HAZARDOUS WASTES	K087-HAZARDOUS WASTES
RR-00721-6 K088-HAZARDOUS WASTES	K088-HAZARDOUS WASTES
RR-00722-7 K090-HAZARDOUS WASTES	K090-HAZARDOUS WASTES
RR-00723-8 K091-HAZARDOUS WASTES	K091-HAZARDOUS WASTES
RR-00724-9 K093-HAZARDOUS WASTES	K093-HAZARDOUS WASTES
RR-00725-0 K094-HAZARDOUS WASTES	K094-HAZARDOUS WASTES
RR-00726-1 K095-HAZARDOUS WASTES	K095-HAZARDOUS WASTES
RR-00727-2 K096-HAZARDOUS WASTES	K096-HAZARDOUS WASTES
RR-00728-3 K097-HAZARDOUS WASTES	K097-HAZARDOUS WASTES
RR-00729-4 K098-HAZARDOUS WASTES	K098-HAZARDOUS WASTES
RR-00730-7 K099-HAZARDOUS WASTES	K099-HAZARDOUS WASTES
RR-00731-8 K100-HAZARDOUS WASTES	K100-HAZARDOUS WASTES
RR-00732-9 K101-HAZARDOUS WASTES	K101-HAZARDOUS WASTES
RR-00733-0 K102-HAZARDOUS WASTES	K102-HAZARDOUS WASTES
RR-00734-1 K103-HAZARDOUS WASTES	K103-HAZARDOUS WASTES
RR-00735-2 K104-HAZARDOUS WASTES	K104-HAZARDOUS WASTES
RR-00736-3 K105-HAZARDOUS WASTES	K105-HAZARDOUS WASTES
RR-00737-4 K106-HAZARDOUS WASTES	K106-HAZARDOUS WASTES
RR-00808-2 K107-HAZARDOUS WASTES	K107-HAZARDOUS WASTES
RR-00810-6 K108-HAZARDOUS WASTES	K108-HAZARDOUS WASTES
RR-00811-7 K109-HAZARDOUS WASTES	K109-HAZARDOUS WASTES
RR-00804-8 K110-HAZARDOUS WASTES	K110-HAZARDOUS WASTES
RR-00738-5 K111-HAZARDOUS WASTES	K111-HAZARDOUS WASTES
RR-00739-6 K112-HAZARDOUS WASTES	K112-HAZARDOUS WASTES
RR-00740-9 K113-HAZARDOUS WASTES	K113-HAZARDOUS WASTES
RR-00741-0 K114-HAZARDOUS WASTES	K114-HAZARDOUS WASTES
RR-00742-1 K115-HAZARDOUS WASTES	K115-HAZARDOUS WASTES
RR-00743-2 K116-HAZARDOUS WASTES	K116-HAZARDOUS WASTES
RR-00744-3 K117-HAZARDOUS WASTES	K117-HAZARDOUS WASTES
RR-00745-4 K118-HAZARDOUS WASTES	K118-HAZARDOUS WASTES
RR-00754-5 K123-HAZARDOUS WASTES	K123-HAZARDOUS WASTES
RR-00751-2 K124-HAZARDOUS WASTES	K124-HAZARDOUS WASTES
RR-00619-9 K125-HAZARDOUS WASTES	K125-HAZARDOUS WASTES
RR-00597-0 K126-HAZARDOUS WASTES	K126-HAZARDOUS WASTES
RR-00567-4 K131-HAZARDOUS WASTES	K131-HAZARDOUS WASTES
RR-00596-9 K132-HAZARDOUS WASTES	K132-HAZARDOUS WASTES
RR-00746-5 K136-HAZARDOUS WASTES	K136-HAZARDOUS WASTES
RR-00888-8 K141-HAZARDOUS WASTES	K141-HAZARDOUS WASTES
RR-00889-9 K142-HAZARDOUS WASTES	K142-HAZARDOUS WASTES
RR-00890-2 K143-HAZARDOUS WASTES	K143-HAZARDOUS WASTES
RR-00891-3 K144-HAZARDOUS WASTES	K144-HAZARDOUS WASTES

Chemical Name	Reference Name
RR-00892-4 K145-HAZARDOUS WASTES	K145-HAZARDOUS WASTES
RR-00893-5 K147-HAZARDOUS WASTES	K147-HAZARDOUS WASTES
RR-00894-6 K148-HAZARDOUS WASTES	K148-HAZARDOUS WASTES
RR-00895-7 K149-HAZARDOUS WASTES	K149-HAZARDOUS WASTES
RR-00896-8 K150-HAZARDOUS WASTES	K150-HAZARDOUS WASTES
RR-00897-9 K151-HAZARDOUS WASTES	K151-HAZARDOUS WASTES
520-18-3 KAEMPFEROL	KAEMPFEROL
59-01-8 KANAMYCIN	KANAMYCIN
61788-33-8 KANECHLOR 500	KANECHLOR 500
1332-58-7 KAOLIN	KAOLIN
9000-36-6 KARAYA	KARAYA
115-32-2 KELTHANE [CWA, PAL]	DICOFOLO-1,5,5A,6,9,9A-HEXAHYDRO-, 3-OXIDE
143-50-0 KEPONE	KEPONE
143-50-0 KEPONE (CHLORDECONE) [NTP]	KEPONE
8008-20-6 KEROSENE	KEROSENE
8008-20-6 KEROSENE (DEODERIZED) [PST]	KEROSENE
8008-20-6 KEROSINE [FLA, MSL]	KEROSENE
8008-20-6 KEROSINE (PETROLEUM) [PAL]	KEROSENE
463-51-4 KETENE	KETENE
RR-00350-9 KETONES, LIQUID, N.O.S.	KETONES, LIQUID, N.O.S.
7439-90-9 KRYPTON	KRYPTON
28522-15-8 KRYPTON 74	KRYPTON 74
28522-17-0 KRYPTON 76	KRYPTON 76
14983-72-3 KRYPTON 77	KRYPTON 77
15478-11-2 KRYPTON 79	KRYPTON 79
15678-91-8 KRYPTON 81	KRYPTON 81
RR-00426-2 KRYPTON 83M	KRYPTON 83M
13983-27-2 KRYPTON 85	KRYPTON 85
RR-00425-1 KRYPTON 85M	KRYPTON 85M
14809-68-8 KRYPTON 87	KRYPTON 87
14995-61-0 KRYPTON 88	KRYPTON 88
RR-01500-9 LACQUER	LACQUER
50-21-5 LACTIC ACID	LACTIC ACID
77501-63-4 LACTOFEN	LACTOFEN
77501-63-4 LACTOFEN [5-(2-CHLORO-4-(TRIFLUO-ROMETHYL)PHENOXY)-2-NITRO-2-ETHOXY-1-METHYL-2-OXOETHYL ESTER] [313]	LACTOFEN]PROPANOIC ACID ETHYL ESTER]
78-97-7 LACTONITRILE	LACTONITRILE
63-42-3 BETA-D-LACTOSE	BETA-D-LACTOSE
8006-54-0 LANOLIN	LANOLIN
61790-81-6 LANOLIN, ETHOXYLATED	LANOLIN, ETHOXYLATED
68458-88-8 LANOLIN, ETHOXYLATED PROPOXYLATED	LANOLIN, ETHOXYLATED PROPOXYLATEDENYL ETHER
15715-04-5 LANTHANUM 131	LANTHANUM 131
15066-93-0 LANTHANUM 132	LANTHANUM 132
15816-85-0 LANTHANUM 135	LANTHANUM 135
14834-69-6 LANTHANUM 137	LANTHANUM 137
15816-87-2 LANTHANUM 138	LANTHANUM 138
13981-28-7 LANTHANUM 140	LANTHANUM 140
15816-88-3 LANTHANUM 141	LANTHANUM 141
15816-89-4 LANTHANUM 142	LANTHANUM 142

Chemical Name	Reference Name
16729-61-6 LANTHANUM 143	LANTHANUM 143
10099-58-8 LANTHANUM CHLORIDE	LANTHANUM CHLORIDE
10099-59-9 LANTHANUM NITRATE	LANTHANUM NITRATE
61789-99-9 LARD	LARD
8016-28-2 LARD OIL [NFP, NFP]	OILS, LARD
303-34-4 LASIOCARPINE	LASIOCARPINE
RR-01517-8 LATEX, PAINTS	LATEX, PAINTS
68425-13-8 LATEX, POLYMER OF 2-METHYL-1,3-BUTADI-ENE	LATEX, POLYMER OF 2-METHYL-1,3-BUTADIENE
143-07-7 LAURIC ACID	LAURIC ACID
120-40-1 LAURIC ACID DIETHANOLAMIDE CONDEN-SATE	LAURIC ACID DIETHANOLAMIDE CONDENSATE
6272-74-8 N-LAUROYL ESTER OF COLAMINOFORMYL-METHYLPYRIDINIUM CHLORIDE	N-LAUROYL ESTER OF COLAMINOFORMYL-METHYLPYRIDINIUM CHLORIDE
4337-75-1 N-LAUROYL-N-METHYLTAURINE, SODIUM SALT	N-LAUROYL-N-METHYLTAURINE, SODIUM SALT
105-74-8 LAUROYL PEROXIDE	LAUROYL PEROXIDE
112-53-8 LAURYL ALCOHOL	LAURYL ALCOHOL
112-54-9 LAURYL ALDEHYDE	LAURYL ALDEHYDE
143-15-7 LAURYL BROMIDE	LAURYL BROMIDE
112-18-5 LAURYLDIMETHYLAMINE	LAURYLDIMETHYLAMINE
1643-20-5 LAURYLDIMETHYLAMINE OXIDE	LAURYLDIMETHYLAMINE OXIDE
2461-18-9 LAURYL GLYCIDYL ETHER [EPA, OTS]	OXIRANE, [(DODECYLOXY)METHYL]-
142-90-5 LAURYL METHACRYLATE	LAURYL METHACRYLATE
9002-92-0 LAURYL POLYETHYLENE GLYCOL ETHER	LAURYL POLYETHYLENE GLYCOL ETHER
151-41-7 LAURYL SULFATE	LAURYL SULFATE
7439-92-1 LEAD	LEAD
RR-00421-7 LEAD 195M	LEAD 195M
16646-00-7 LEAD 198	LEAD 198
27486-00-6 LEAD 199	LEAD 199
16645-99-1 LEAD 200	LEAD 200
17239-87-1 LEAD 201	LEAD 201
15752-86-0 LEAD 202	LEAD 202
RR-00420-6 LEAD 202M	LEAD 202M
14687-25-3 LEAD 203	LEAD 203
14119-28-9 LEAD 205	LEAD 205
14119-30-3 LEAD 209	LEAD 209
14255-04-0 LEAD 210	LEAD 210
15816-77-0 LEAD 211	LEAD 211
15092-94-1 LEAD 212	LEAD 212
15067-28-4 LEAD 214	LEAD 214
7439-92-1 LEAD, INORGANIC, DUST AND FUME [MEX, MEX]	LEAD
301-04-2 LEAD ACETATE	LEAD ACETATE
301-04-2 LEAD(2+) ACETATE [CSR]	LEAD ACETATE
546-67-8 LEAD (IV) ACETATE	LEAD (IV) ACETATE
1335-32-6 LEAD ACETATE, BASIC [WHM]	LEAD SUBACETATE
RR-00538-9 LEAD, ALL INORGANIC COMPOUNDS [NJL]	INORGANIC LEAD
RR-00630-4 LEAD AND COMPOUNDS [GBR]	LEAD COMPOUNDS
13510-89-9 LEAD ANTIMONATE	LEAD ANTIMONATE
12266-38-5 LEAD ANTIMONIDE	LEAD ANTIMONIDE

Chemical Name	Reference Name
3687-31-8 LEAD ARSENATE	LEAD ARSENATE
7645-25-2 LEAD ARSENATE [CN2]	LEAD ARSENATE, UNSPECIFIED
7784-40-9 LEAD ARSENATE	LEAD ARSENATE
7784-40-9 LEAD (II) ARSENATE [MSL]	LEAD ARSENATE
10102-48-4 LEAD ARSENATE [FLA, MIN]	LEAD ARSENATE (PB3(ASS04)2)
10102-48-4 LEAD (IV) ARSENATE [MSL]	LEAD ARSENATE (PB3(ASS04)2)
10102-48-4 LEAD ARSENATE (PB3(ASS04)2)	LEAD ARSENATE (PB3(ASS04)2)
RR-00132-1 LEAD ARSENATES	LEAD ARSENATES
7645-25-2 LEAD ARSENATE, UNSPECIFIED	LEAD ARSENATE, UNSPECIFIED
10031-13-7 LEAD ARSENITE	LEAD ARSENITE
RR-01292-0 LEAD ARSENITES	LEAD ARSENITES
13424-46-9 LEAD AZIDE	LEAD AZIDE
1335-32-6 LEAD, BIS(ACETATO-O)TETRAHYDROXYTRI- [PAL, PAL]	LEAD SUBACETATE
19010-66-3 LEAD BIS (DIMETHYLDITHIOCARBAMATE) [MSL]	LEAD DIMETHYLDITHIOCARBAMATE
56189-09-4 LEAD, BIS(OCTADECANOATO)DIOXODI- [PAL]	DIBASIC LEAD STEARATE
14720-53-7 LEAD BORATE	LEAD BORATE
13814-96-5 LEAD BORON FLUORIDE [MSL]	LEAD FLUOBORATE
15696-43-2 LEAD CAPRYLATE [MSL]	LEAD OCTOATE
598-63-0 LEAD CARBONATE	LEAD CARBONATE
1319-46-6 LEAD CARBONATE HYDROXIDE	LEAD CARBONATE HYDROXIDE
7758-95-4 LEAD CHLORIDE	LEAD CHLORIDE
7758-95-4 LEAD CHLORIDE (PBCL2) [PAL]	LEAD CHLORIDE
39390-00-6 LEAD CHLORIDE SILICATE	LEAD CHLORIDE SILICATE
12612-47-4 LEAD CHLORIDE (VAN)	LEAD CHLORIDE (VAN)
12612-47-4 LEAD CHLORIDE [PAL]	LEAD CHLORIDE (VAN)
7758-97-6 LEAD CHROMATE	LEAD CHROMATE
18454-12-1 LEAD CHROMATE [FLA]	LEAD CHROMATE OXIDE
18454-12-1 LEAD CHROMATE OXIDE	LEAD CHROMATE OXIDE
18454-12-1 LEAD CHROMATE OXIDE (PB2(CRO4)O) [PAL, PAL]	LEAD CHROMATE OXIDE
11113-70-5 LEAD CHROMATE SILICATE	LEAD CHROMATE SILICATE
RR-00630-4 LEAD COMPOUNDS	LEAD COMPOUNDS
RR-00630-4 LEAD COMPOUNDS, N.O.S. [HM1, HM1, HM1, NJL, RCU]	LEAD COMPOUNDS
20837-86-9 LEAD CYANAMIDE	LEAD CYANAMIDE
592-05-2 LEAD CYANIDE	LEAD CYANIDE
19010-66-3 LEAD DIMETHYLDITHIOCARBAMATE	LEAD DIMETHYLDITHIOCARBAMATE
1309-60-0 LEAD DIOXIDE [HM1, HM1, MSL, NJL]	LEAD PEROXIDE
7439-92-1 LEAD, ELEMENTAL [WHM]	LEAD
16996-40-0 LEAD 2-ETHYLHEXOATE	LEAD 2-ETHYLHEXOATE
13814-96-5 LEAD FLUOBORATE	LEAD FLUOBORATE
7783-46-2 LEAD FLUORIDE	LEAD FLUORIDE
7783-46-2 LEAD FLUORIDE (PBF2) [PAL]	LEAD FLUORIDE
25808-74-6 LEAD FLUOROSILICATE	LEAD FLUOROSILICATE
811-54-1 LEAD FORMATE	LEAD FORMATE
15845-52-0 LEAD HYDROGEN PHOSPHATE [MSL]	DIBASIC LEAD PHOSPHATE
19783-14-3 LEAD HYDROXIDE	LEAD HYDROXIDE
87903-39-7 LEAD HYDROXYSALICYLATE	LEAD HYDROXYSALICYLATE

Chemical Name	Reference Name
RR-00538-9 LEAD, INORGANIC [AUS, AUS, IAR, NHS, NHS]	INORGANIC LEAD
RR-00538-9 LEAD INORGANIC COMPOUNDS [TLV]	INORGANIC LEAD
RR-00538-9 LEAD, INORGANIC COMPOUNDS [CAB]	INORGANIC LEAD
RR-00538-9 LEAD, INORGANIC COMPOUNDS, N.O.S. [WHM]	INORGANIC LEAD
7439-92-1 LEAD, INORGANIC, DUST AND FUMES [FLA]	LEAD
RR-00538-9 LEAD, INORGANIC FUMES AND DUSTS [ALB]	INORGANIC LEAD
RR-00538-9 LEAD, INORGANIC SALTS , FUMES AND DUST [QBC]	INORGANIC LEAD
RR-00538-9 LEAD, INORGANIC, FUMES AND DUST [BC1, BC1, BC1]	INORGANIC LEAD
25659-31-8 LEAD IODATE	LEAD IODATE
10101-63-0 LEAD IODIDE	LEAD IODIDE
10101-63-0 LEAD IODIDE (PBI2) [PAL]	LEAD IODIDE
16996-51-3 LEAD LINOLEATE	LEAD LINOLEATE
7439-92-1 LEAD METAL [EPA]	LEAD
1068-61-7 LEAD METHACRYLATE	LEAD METHACRYLATE
35029-96-0 LEAD (II) METHYLTHIOLATE	LEAD (II) METHYLTHIOLATE
10190-55-3 LEAD MOLYBDATE	LEAD MOLYBDATE
1317-36-8 LEAD MONOXIDE	LEAD MONOXIDE
20403-41-2 LEAD MYRISTATE	LEAD MYRISTATE
50825-29-1 LEAD NAPHTHALATE	LEAD NAPHTHALATE
61790-14-5 LEAD NAPHTHENATE	LEAD NAPHTHENATE
12034-88-7 LEAD NEOBATE	LEAD NEOBATE
27253-28-7 LEAD NEODECANOATE	LEAD NEODECANOATE
10099-74-8 LEAD NITRATE	LEAD NITRATE
51317-24-9 LEAD NITRORESORCINATE	LEAD NITRORESORCINATE
15696-43-2 LEAD OCTOATE	LEAD OCTOATE
1120-46-3 LEAD OLEATE	LEAD OLEATE
RR-01021-9 LEAD ORES	LEAD ORES
814-93-7 LEAD OXALATE	LEAD OXALATE
1317-36-8 LEAD OXIDE [CSR, MSL]	LEAD MONOXIDE
1344-40-7 LEAD OXIDE PHOSPHONATE, HEMIHYDRATE	LEAD OXIDE PHOSPHONATE, HEMIHYDRATE
13637-76-8 LEAD PERCHLORATE	LEAD PERCHLORATE
1309-60-0 LEAD PEROXIDE	LEAD PEROXIDE
7446-27-7 LEAD PHOSPHATE	LEAD PHOSPHATE
15845-52-0 LEAD PHOSPHATE, DIBASIC [NJL]	DIBASIC LEAD PHOSPHATE
1344-40-7 LEAD PHOSPHITE, DIBASIC [HM1, HM1]	LEAD OXIDE PHOSPHONATE, HEMIHYDRATE
6838-85-3 LEAD PHTHALATE	LEAD PHTHALATE
25721-38-4 LEAD PICRATE	LEAD PICRATE
13453-66-2 LEAD PYROPHOSPHATE	LEAD PYROPHOSPHATE
41453-50-3 LEAD B-RESORCYLATE	LEAD B-RESORCYLATE
29473-77-6 LEAD SEBACATE	LEAD SEBACATE
7446-15-3 LEAD SELENATE	LEAD SELENATE
12069-00-0 LEAD SELENIDE	LEAD SELENIDE
7488-51-9 LEAD SELENITE	LEAD SELENITE
11120-22-2 LEAD SILICATE	LEAD SILICATE
67711-86-8 LEAD SILICATE SULFATE	LEAD SILICATE SULFATE
25808-74-6 LEAD SILICON FLUORIDE [MSL]	LEAD FLUOROSILICATE
7428-48-0 LEAD STEARATE	LEAD STEARATE
52652-59-2 LEAD STEARATE [EPA]	LEAD STEARATE DIBASIC

Chemical Name	Reference Name
52652-59-2 LEAD STEARATE DIBASIC	LEAD STEARATE DIBASIC
63918-97-8 LEAD STYPHNATE	LEAD STYPHNATE
1335-32-6 LEAD SUBACETATE	LEAD SUBACETATE
15739-80-7 LEAD SULFATE	LEAD SULFATE
1314-87-0 LEAD SULFIDE [CSR, CWA, EPA, MSL, NJL]	LEAD SULFIDE (PBS)
1314-87-0 LEAD SULFIDE (PBS)	LEAD SULFIDE (PBS)
15739-80-7 LEAD SULPHATE [NJL, NJL, NJL]	LEAD SULFATE
12065-68-8 LEAD TANTALATE	LEAD TANTALATE
815-84-9 LEAD TARTRATE [MSL]	TARTARIC ACID, LEAD(2+) SALT
1314-91-6 LEAD TELLURIDE	LEAD TELLURIDE
13845-35-7 LEAD TELLURITE	LEAD TELLURITE
78-00-2 LEAD TETRAETHYL [HM1, HM1, HM1, HM1, QBC, QBC]	TETRAETHYL LEAD
75-74-1 LEAD TETRAMETHYL [QBC, QBC]	TETRAMETHYL LEAD
1314-41-6 LEAD TETRAOXIDE	LEAD TETRAOXIDE
592-87-0 LEAD THIOCYANATE	LEAD THIOCYANATE
13478-50-7 LEAD THIOSULFATE	LEAD THIOSULFATE
12060-00-3 LEAD TITANIUM OXIDE (PB.TI.O3)	LEAD TITANIUM OXIDE (PB.TI.O3)
12060-00-3 LEAD TITANIUM TRIOXIDE [MSL]	LEAD TITANIUM OXIDE (PB.TI.O3)
15245-44-0 LEAD TRINITRORESORCINATE [MSL]	1,3-BENZENEDIOL, 2,4,6-TRINITRO-, LEAD SALT
1314-27-8 LEAD TRIOXIDE	LEAD TRIOXIDE
7759-01-5 LEAD TUNGSTEN OXIDE	LEAD TUNGSTEN OXIDE
10099-79-3 LEAD VANADATE	LEAD VANADATE
12060-01-4 LEAD ZIRCONATE	LEAD ZIRCONATE
8002-43-5 LECITHIN [PST]	SOYBEAN LECITHIN
RR-01722-1 LECITHINS, PHOSPHOLIPASE A2-HYDROLYZED	LECITHINS, PHOSPHOLIPASE A2-HYDROLYZED
8007-02-1 LEMONGRASS OIL	LEMONGRASS OIL
8008-56-8 LEMON OIL	LEMON OIL
2164-08-1 LENACIL	LENACIL
21609-90-5 LEPTOPHOS	LEPTOPHOS
21609-90-5 LEPTOPHOS (O-(4-BROMO-2,5-DICHLOROPHENYL) O-METHYL-PHENYLPHOSPHONOTHIOATE [CN8]	LEPTOPHOS)
112-56-1 LETHANE	LETHANE
123-76-2 LEVULINIC ACID	LEVULINIC ACID
541-25-3 LEWISITE	LEWISITE
67891-80-9 LIGHT AROMATIC DISTILLATE (PETROLEUM)	LIGHT AROMATIC DISTILLATE (PETROLEUM)
64741-59-9 LIGHT CATALYTICALLY CRACKED DISTIL-LATES [IAR]	PETROLEUM DISTILLATES, LIGHT CATALYTIC CRACKED
64742-03-6 LIGHT DISTILLATE SOLVENT EXTRACT (PETROLEUM)	LIGHT DISTILLATE SOLVENT EXTRACT (PETROLEUM) NAPHTHENIC
5141-20-8 LIGHT GREEN SF	LIGHT GREEN SF
5141-20-8 LIGHT GREEN SF (C.I. ACID GREEN 5) [NJL]	LIGHT GREEN SF
64741-52-2 LIGHT NAPHTHENIC OIL	LIGHT NAPHTHENIC OIL
64742-05-8 LIGHT PARAFFINIC DISTILLATE SOLVENT EXTRACT (PETROLEUM)	LIGHT PARAFFINIC DISTILLATE SOLVENT EXTRACT (PETROLEUM)SOLVENT
64741-58-8 LIGHT VACUUM DISTILLATES [IAR]	LIGHT VACUUM GAS OIL (PETROLEUM)
64741-58-8 LIGHT VACUUM GAS OIL (PETROLEUM)	LIGHT VACUUM GAS OIL (PETROLEUM)
105859-97-0 LIGNIN, ALKALI, REACTION PRODUCTS WITH DISODIUM SULFITE ANDFORMALDEHYDE	LIGNIN, ALKALI, REACTION PRODUCTS WITH DISODIUM SULFITE AND
8068-05-1 LIGNIN SULPHATE	LIGNIN SULPHATE

Chemical Name	Reference Name
8062-15-5 LIGNOSULFONATE	LIGNOSULFONATE
8062-15-5 LIGNOSULFONIC ACID [PST]	LIGNOSULFONATE
8061-53-8 LIGNOSULFONIC ACID, AMMONIUM SALT	LIGNOSULFONIC ACID, AMMONIUM SALT
8061-52-7 LIGNOSULFONIC ACID, CALCIUM SALT [PST]	CALCIUM LIGNOSULFONATE
68611-14-3 LIGNOSULFONIC ACID, ETHOXYLATED, SODIUM SALTS	LIGNOSULFONIC ACID, ETHOXYLATED, SODIUM SALT-SYLAMMONIO)PROPANOL, CHLORIDE
8061-54-9 LIGNOSULFONIC ACID, MAGNESIUM SALT	LIGNOSULFONIC ACID, MAGNESIUM SALT
8061-51-6 LIGNOSULFONIC ACID, SODIUM SALT [PST]	SODIUM LIGNO SULFONATE
8032-32-4 LIGROINE	LIGROINE
RR-01093-5 LIME GREEN PIGMENT	LIME GREEN PIGMENT
1317-65-3 LIMESTONE [ALB, BC1, BC1, GBR, MEX, ONT, PAL, PST]	CALCIUM CARBONATE
138-86-3 ALPHA-LIMONENE [PST]	DIPENTENE
5989-27-5 D-LIMONENE	D-LIMONENE
5989-54-8 L-LIMONENE	L-LIMONENE
78-70-6 LINALOOL (EX BOIS DE ROSE; SYNTHETIC)	LINALOOL (EX BOIS DE ROSE; SYNTHETIC)
58-89-9 LINDANE	LINDANE
58-89-9 LINDANE [CYCLOHEXANE, 1,2,3,4,5,6-HEXACHLORO-(1ALPHA,2ALPHA,3BETA,4ALPHA,5ALPHA,6BETA)-] [313]	LINDANE3A,4,7,7A-HEXAHYDRO-]
58-89-9 LINDANE (GAMMA-HEXACHLOROCYCLOHEXANE) [NTP]	LINDANE
58-89-9 LINDANE (1,2,3,4,5,6-HEXACHLOROCYCLOHEXANE OR GAMMA-BHC) [CN2]	LINDANEO-4,7-METHANOINDENE)
58-89-9 LINDANE (ALL ISOMERS) [CAA]	LINDANE
608-73-1 LINDANE AND OTHER HEXACHLOROCYCLOHEXANE ISOMERS [CAL, C65, C65]	HEXACHLOROCYCLOHEXANE (MIXED ISOMERS)
68555-09-9 LINEAR ALKYLBENZENE (C-20 TO C-48)	LINEAR ALKYLBENZENE (C-20 TO C-48)
3999-01-7 LINOLEAMIDE	LINOLEAMIDE
60-33-3 LINOLEIC ACID	LINOLEIC ACID
37189-83-6 LINOLEIC ACID DIMER-DIETHYLENETRIAMINE POLYMER	LINOLEIC ACID DIMER-DIETHYLENETRIAMINE POLYMER
8001-26-1 LINSEED OIL	LINSEED OIL
8001-26-1 LINSEED OIL, RAW [NFP, NFP]	LINSEED OIL
68153-88-8 LINSEED OIL, TUNG OIL, 4-(T-BUTYL)PHENOL, BISPHENOL A, FORMALDEHYDE POLYMER	LINSEED OIL, TUNG OIL, 4-(T-BUTYL)PHENOL, BISPHENOL A, FORMA
330-55-2 LINURON	LINURON
68476-85-7 LIQUEFIED PETROLEUM GAS [NJL]	L.P.G. (LIQUIFIED PETROLEUM GAS)
68476-85-7 LIQUEFIED PETROLEUM GAS (L.P.G.) [MSL]	L.P.G. (LIQUIFIED PETROLEUM GAS)
68476-85-7 LIQUEFIED PETROLEUM GAS (LPG) [GBR, GBR]	L.P.G. (LIQUIFIED PETROLEUM GAS)
RR-00375-8 LIQUIFIED NATURAL GAS	LIQUIFIED NATURAL GAS
7439-93-2 LITHIUM	LITHIUM
50475-76-8 LITHIUM ACETYLIDE ETHYLENEDIAMINE	LITHIUM ACETYLIDE ETHYLENEDIAMINE
RR-01350-3 LITHIUM ALKYLS	LITHIUM ALKYLS
16853-85-3 LITHIUM ALUMINUM HYDRIDE	LITHIUM ALUMINUM HYDRIDE
17476-04-9 LITHIUM ALUMINUM TRI-TERT-BUTOXY-HYDRIDE	LITHIUM ALUMINUM TRI-TERT-BUTOXY-HYDRIDE
7782-89-0 LITHIUM AMIDE	LITHIUM AMIDE
16949-15-8 LITHIUM BOROHYDRIDE	LITHIUM BOROHYDRIDE
554-13-2 LITHIUM CARBONATE	LITHIUM CARBONATE
13453-71-9 LITHIUM CHLORATE	LITHIUM CHLORATE

Chemical Name	Reference Name
7447-41-8 LITHIUM CHLORIDE	LITHIUM CHLORIDE
14307-35-8 LITHIUM CHROMATE	LITHIUM CHROMATE
919-16-4 LITHIUM CITRATE	LITHIUM CITRATE
13843-81-7 LITHIUM DICHROMATE	LITHIUM DICHROMATE
64082-35-5 LITHIUM FERROSILICON	LITHIUM FERROSILICON
7789-24-4 LITHIUM FLUORIDE	LITHIUM FLUORIDE
7580-67-8 LITHIUM HYDRIDE	LITHIUM HYDRIDE
7580-67-8 LITHIUM HYDRIDE (LIH) [PAL]	LITHIUM HYDRIDE
1310-65-2 LITHIUM HYDROXIDE	LITHIUM HYDROXIDE
1310-66-3 LITHIUM HYDROXIDE [MIN]	LITHIUM HYDROXIDE MONOHYDRATE
1310-66-3 LITHIUM HYDROXIDE MONOHYDRATE	LITHIUM HYDROXIDE MONOHYDRATE
13840-33-0 LITHIUM HYPOCHLORITE	LITHIUM HYPOCHLORITE
27253-30-1 LITHIUM NEODECANOATE	LITHIUM NEODECANOATE
7790-69-4 LITHIUM NITRATE	LITHIUM NITRATE
26134-62-3 LITHIUM NITRIDE	LITHIUM NITRIDE
12057-24-8 LITHIUM OXIDE	LITHIUM OXIDE
12031-80-0 LITHIUM PEROXIDE	LITHIUM PEROXIDE
68848-64-6 LITHIUM SILICON	LITHIUM SILICON
4485-12-5 LITHIUM STEARATE	LITHIUM STEARATE
10377-48-7 LITHIUM SULFATE	LITHIUM SULFATE
16853-85-3 LITHIUM TETRAHYDROALUMINATE [FLA, MSL]	LITHIUM ALUMINUM HYDRIDE
434-13-9 LITHOCHOLIC ACID	LITHOCHOLIC ACID
5281-04-9 LITHOL RUBINE	LITHOL RUBINE
9000-40-2 LOCUST BEAN GUM	LOCUST BEAN GUM
13010-47-4 LOMUSTINE [CSR]	1-(2-CHLOROETHYL)-3-CYCLOHEXYL-1-NITROSOUREA
8012-74-6 LONDON PURPLE	LONDON PURPLE
846-49-1 LORAZEPAM	LORAZEPAM
75330-75-5 LOVASTATIN	LOVASTATIN
68476-85-7 L.P.G. [CEX, MIN, NIO, NIO, NIO]	L.P.G. (LIQUIFIED PETROLEUM GAS)
68476-85-7 L.P.G. (LIQUIFIED PETROLEUM GAS)	L.P.G. (LIQUIFIED PETROLEUM GAS)
RR-00384-9 LUBRICANT BASE OILS AND DERIVED PRODUCTS [CAB]	LUBRICATING OILS
8012-45-1 LUBRICATING OIL, MINERAL [NFP, NFP]	PARAFFIN OIL
RR-00384-9 LUBRICATING OILS	LUBRICATING OILS
21884-44-6 LUTEOSKYRIN	LUTEOSKYRIN
15715-05-6 LUTETIUM 169	LUTETIUM 169
15741-32-9 LUTETIUM 170	LUTETIUM 170
15752-27-9 LUTETIUM 171	LUTETIUM 171
14093-12-0 LUTETIUM 172	LUTETIUM 172
14391-24-3 LUTETIUM 173	LUTETIUM 173
14914-12-6 LUTETIUM 174	LUTETIUM 174
RR-00419-3 LUTETIUM 174M	LUTETIUM 174M
14452-47-2 LUTETIUM 176	LUTETIUM 176
RR-00415-9 LUTETIUM 176M	LUTETIUM 176M
14265-75-9 LUTETIUM 177	LUTETIUM 177
RR-00410-4 LUTETIUM 177M	LUTETIUM 177M
14683-30-8 LUTETIUM 178	LUTETIUM 178
RR-00409-1 LUTETIUM 178M	LUTETIUM 178M
15755-89-2 LUTETIUM 179	LUTETIUM 179

Chemical Name	Reference Name
115-95-7 LYNALYL ACETATE (EX BOIS DE ROSE; SYNTHETIC)	LYNALYL ACETATE (EX BOIS DE ROSE; SYNTHETIC)
52-76-6 LYNOESTRENOL	LYNOESTRENOL
RR-00098-6 LYNOESTRENOL AND OESTROGENS	LYNOESTRENOL AND OESTROGENS
82-58-8 D-LYSERGIC ACID	D-LYSERGIC ACID
RR-01784-5 D-LYSERGIC ACID SALTS, OPTICAL ISOMERS, AND SALTS OF OPTICALISOMERS	D-LYSERGIC ACID SALTS, OPTICAL ISOMERS, AND SALTS OF OPTICAL
657-27-2 L-LYSINE HCL	L-LYSINE HCL
632-99-5 MAGENTA	MAGENTA
632-99-5 MAGENTA I [IAR, IAR]	MAGENTA
26261-57-4 MAGENTA II	MAGENTA II
3248-91-7 MAGENTA III	MAGENTA III
RR-00540-3 MAGENTA-MANUFACTURE OF	MAGENTA-MANUFACTURE OF
546-93-0 MAGNESITE [ALB, AUS, BC1, BC1, GBR, MEX, MEX, ONT, OS1, OS1, QBC, TLV]	MAGNESIUM CARBONATE
13717-00-5 MAGNESITE (MG(CO3))	MAGNESITE (MG(CO3))
7439-95-4 MAGNESIUM	MAGNESIUM
15092-71-4 MAGNESIUM 28	MAGNESIUM 28
1309-48-4 MAGNESIUM OXIDE FUME [MEX]	MAGNESIUM OXIDE
RR-00568-5 MAGNESIUM ALKYLS	MAGNESIUM ALKYLS
RR-00621-3 MAGNESIUM ALUMINUM PHOSPHIDE [HM1, HM1, NJL]	ALUMINUM MAGNESIUM PHOSPHIDE
1327-43-1 MAGNESIUM ALUMINUM SILICATE	MAGNESIUM ALUMINUM SILICATE
10103-50-1 MAGNESIUM ARSENATE	MAGNESIUM ARSENATE
7789-36-8 MAGNESIUM BROMATE	MAGNESIUM BROMATE
546-93-0 MAGNESIUM CARBONATE	MAGNESIUM CARBONATE
10326-21-3 MAGNESIUM CHLORATE	MAGNESIUM CHLORATE
7786-30-3 MAGNESIUM CHLORIDE	MAGNESIUM CHLORIDE
7791-19-7 MAGNESIUM CHLORIDE.6H20	MAGNESIUM CHLORIDE.6H20
7791-18-6 MAGNESIUM CHLORIDE, HEXAHYDRATE	MAGNESIUM CHLORIDE, HEXAHYDRATE
13423-61-5 MAGNESIUM CHROMATE [MSL]	CHROMIC ACID (H2CRO4), MAGNESIUM SALT (1:1)
7803-54-5 MAGNESIUM DIAMIDE	MAGNESIUM DIAMIDE
13423-61-5 MAGNESIUM DICHROMATE [FLA]	CHROMIC ACID (H2CRO4), MAGNESIUM SALT (1:1)
14104-85-9 MAGNESIUM DICHROMATE	MAGNESIUM DICHROMATE
555-54-4 MAGNESIUM DIPHENYL	MAGNESIUM DIPHENYL
7783-40-6 MAGNESIUM FLUORIDE	MAGNESIUM FLUORIDE
18972-56-0 MAGNESIUM FLUOROSILICATE [HM1, HM1]	MAGNESIUM SILICOFLUORIDE
60616-74-2 MAGNESIUM HYDRIDE	MAGNESIUM HYDRIDE
1309-42-8 MAGNESIUM HYDROXIDE	MAGNESIUM HYDROXIDE
18917-93-6 MAGNESIUM LACTATE	MAGNESIUM LACTATE
10377-60-3 MAGNESIUM NITRATE	MAGNESIUM NITRATE
1309-48-4 MAGNESIUM OXIDE	MAGNESIUM OXIDE
1309-48-4 MAGNESIUM OXIDE (MGO) [PAL]	MAGNESIUM OXIDE
1309-48-4 MAGNESIUM OXIDE FUME [ALB, ALB, BC1, FLA, MSL, NIO, NIO, ONT, ONT, ONT, OS1, OS1, TLV]	MAGNESIUM OXIDE
1309-48-4 MAGNESIUM OXIDE FUMES [QBC]	MAGNESIUM OXIDE
12286-12-3 MAGNESIUM OXIDE SULFATE	MAGNESIUM OXIDE SULFATE
10034-81-8 MAGNESIUM PERCHLORATE	MAGNESIUM PERCHLORATE
14452-57-4 MAGNESIUM PEROXIDE	MAGNESIUM PEROXIDE
12057-74-8 MAGNESIUM PHOSPHIDE	MAGNESIUM PHOSPHIDE

Chemical Name	Reference Name
1343-88-0 MAGNESIUM SILICATE	MAGNESIUM SILICATE
1343-90-4 MAGNESIUM SILICATE, HYDRATE	MAGNESIUM SILICATE, HYDRATE
39404-03-0 MAGNESIUM SILICIDE	MAGNESIUM SILICIDE
18972-56-0 MAGNESIUM SILICOFLUORIDE	MAGNESIUM SILICOFLUORIDE
557-04-0 MAGNESIUM STEARATE	MAGNESIUM STEARATE
7487-88-9 MAGNESIUM SULFATE	MAGNESIUM SULFATE
10034-99-8 MAGNESIUM SULFATE HEPTAHYDRATE	MAGNESIUM SULFATE HEPTAHYDRATE
569-64-2 MALACHITE GREEN	MALACHITE GREEN
569-64-2 MALACHITE GREEN CHLORIDE [CSR]	MALACHITE GREEN
2437-29-8 MALACHITE GREEN OXALATE	MALACHITE GREEN OXALATE
1634-78-2 MALAOXON	MALAOXON
121-75-5 MALATHION	MALATHION
121-75-5 MALATHION (ISO) [GBR, GBR]	MALATHION
123-33-1 MALEIC HYDRAZIDE [HON]	1,2-DIHYDRO-3,6-PYRIDAZINEDIONE
110-16-7 MALEIC ACID	MALEIC ACID
108-31-6 MALEIC ACID ANHYDRIDE [MAK, MAK, MAK]	MALEIC ANHYDRIDE
105-76-0 MALEIC ACID, DIBUTYL ESTER	MALEIC ACID, DIBUTYL ESTER
128-53-0 MALEIC ACID N-ETHYLIMIDE	MALEIC ACID N-ETHYLIMIDE
108-31-6 MALEIC ANHYDRIDE	MALEIC ANHYDRIDE
68139-89-9 MALEIC ANHYDRIDE, COMPOUND WITH TALL-OIL FATTY ACIDS	MALEIC ANHYDRIDE, COMPOUND WITH TALL-OIL FATTY ACIDSLFO-, N-COCO ACYL DERIVATIVES, HYDROXIDES, INNER SALTS
37199-81-8 MALEIC ANHYDRIDE-DIISOBUTYLENE POLYMER, SODIUM SALT	MALEIC ANHYDRIDE-DIISOBUTYLENE POLYMER, SODIUM SALT
9011-16-9 MALEIC ANHYDRIDE - METHYLVINYL ETHER COPOLYMER	MALEIC ANHYDRIDE - METHYLVINYL ETHER COPOLYMER
113221-69-5 MALEIC ANHYDRIDE, POLYMER WITH ETHYL ACRYLATE AND VINYL ACETATE, HYDROLYZED	MALEIC ANHYDRIDE, POLYMER WITH ETHYL ACRYLATE AND VINYL ACETT
37199-81-8 MALEIC ANHYDRIDE, POLYMER WITH 2,4,4-TRIMETHYLPENTENE, SODIUM SALT [PST]	MALEIC ANHYDRIDE-DIISOBUTYLENE POLYMER, SODIUM SALT
9011-13-6 MALEIC ANHYDRIDE-STYRENE POLYMER	MALEIC ANHYDRIDE-STYRENE POLYMER
123-33-1 MALEIC HYDRAZIDE [EPA, IAR, RCU, RCU, TSC, TSC, TSE]	1,2-DIHYDRO-3,6-PYRIDAZINEDIONE
28330-26-9 MALEIC HYDRAZIDE SODIUM SALT	MALEIC HYDRAZIDE SODIUM SALT
6915-15-7 MALIC ACID	MALIC ACID
542-78-9 MALONALDEHYDE	MALONALDEHYDE
24382-04-5 MALONALDEHYDE, SODIUM SALT	MALONALDEHYDE, SODIUM SALT
141-82-2 MALONIC ACID	MALONIC ACID
109-77-3 MALONONITRILE	MALONONITRILE
69-79-4 MALTOSE	MALTOSE
8018-01-7 MANCOZEB	MANCOZEB
532-28-5 MANDELONITRILE	MANDELONITRILE
12427-38-2 MANEB	MANEB
12427-38-2 MANEB [CARBAMODITHIOIC ACID, 1,2-ETHANEDIYLBIS-, MANGANESE COMPLEX] [313]	MANEBX]
12427-38-2 MANEB OR MANEB PREPARATIONS [HM1, HM1, HM1]	MANEB
7439-96-5 MANGANESE	MANGANESE
14392-03-1 MANGANESE 51	MANGANESE 51
14092-99-0 MANGANESE 52	MANGANESE 52

Chemical Name	Reference Name
RR-00408-0 MANGANESE 52M	MANGANESE 52M
14999-33-8 MANGANESE 53	MANGANESE 53
13966-31-9 MANGANESE 54	MANGANESE 54
14681-52-8 MANGANESE 56	MANGANESE 56
7439-96-5 MANGANESE AND COMPOUNDS [ALB]	MANGANESE
RR-00602-0 MANGANESE COMPOUNDS [MEX]	MANGANESE COMPOUNDS, N.O.S.
1344-43-0 MANGANESE FUME [MEX, MEX]	MANGANESE MONOXIDE
638-38-0 MANGANESE(II) ACETATE	MANGANESE(II) ACETATE
7439-96-5 MANGANESE AND COMPOUNDS [BC1]	MANGANESE
68442-99-9 MANGANESE BORON NEODECANOATE	MANGANESE BORON NEODECANOATE
598-62-9 MANGANESE CARBONATE	MANGANESE CARBONATE
7773-01-5 MANGANESE(II) CHLORIDE	MANGANESE(II) CHLORIDE
7773-01-5 MANGANESE CHLORIDE [PST]	MANGANESE(II) CHLORIDE
RR-00602-0 MANGANESE COMPOUNDS [NIO, NIO, 313, CAA, CAL, CEX, ONT]	MANGANESE COMPOUNDS, N.O.S.
RR-00602-0 MANGANESE COMPOUNDS, N.O.S.	MANGANESE COMPOUNDS, N.O.S.
12079-65-1 MANGANESE CYCLOPENTADIENYL TRICARBONYL	MANGANESE CYCLOPENTADIENYL TRICARBONYL
12079-65-1 MANGANESE CYCLOPENTADIENYLTRICARBONYL [CAL, ONT, ONT]	MANGANESE CYCLOPENTADIENYL TRICARBONYL
12079-65-1 MANGANESE CYCLOPENTA-DIENYL TRICARBONYL [OS1, OS1]	MANGANESE CYCLOPENTADIENYL TRICARBONYL
12079-65-1 MANGANESE, CYCLOPENTADIENYL-TRICARBONYL [CEX, CEX]	MANGANESE CYCLOPENTADIENYL TRICARBONYL
12079-65-1 MANGANESE CYCLOPENTADIENYLTRICARBONYL [MIN]	MANGANESE CYCLOPENTADIENYL TRICARBONYL
1313-13-9 MANGANESE DIOXIDE	MANGANESE DIOXIDE
7439-96-5 MANGANESE, ELEMENTAL [WHM]	MANGANESE
68609-86-9 MANGANESE, 2-ETHYLHEXANOATE NAPHTHENATE COMPLEXES	MANGANESE, 2-ETHYLHEXANOATE NAPHTHENATE COMPLEXES
1344-43-0 MANGANESE FUME [ALB, ALB, OS1, QBC, QBC]	MANGANESE MONOXIDE
RR-00602-0 MANGANESE INORGANIC COMPOUNDS [TLV]	MANGANESE COMPOUNDS, N.O.S.
37449-19-7 MANGANESE ISOOCTANOATE	MANGANESE ISOOCTANOATE
RR-01699-9 MANGANESE METHYLCYCLOPENTADIENYL TRICARBONYL	MANGANESE METHYLCYCLOPENTADIENYL TRICARBONYL
12108-13-3 MANGANESE METHYLCYCLOPENTADIENYL TRICARBONYL [QBC]	METHYLCYCLOPENTADIENYL MANGANESE TRICARBONYL
1344-43-0 MANGANESE MONOXIDE	MANGANESE MONOXIDE
1336-93-2 MANGANESE NAPHTHENATE	MANGANESE NAPHTHENATE
93918-16-2 MANGANESE NEONONOATE	MANGANESE NEONONOATE
10377-66-9 MANGANESE NITRATE	MANGANESE NITRATE
1317-35-7 MANGANESE OXIDE [FLA]	MANGANESE TETROXIDE
21129-18-0 MANGANESE PROPIONATE	MANGANESE PROPIONATE
9008-34-8 MANGANESE RESINATE	MANGANESE RESINATE
7785-87-7 MANGANESE(II) SULFATE	MANGANESE(II) SULFATE
7785-87-7 MANGANESE SULFATE [PST]	MANGANESE(II) SULFATE
7785-87-7 MANGANESE (II) SULFATE (1:1) [TSC, TSC]	MANGANESE(II) SULFATE
10034-96-5 MANGANESE SULFATE MONOHYDRATE	MANGANESE SULFATE MONOHYDRATE
1317-35-7 MANGANESE TETROXIDE	MANGANESE TETROXIDE
12079-65-1 MANGANESE, TRICARBONYL(.ETA.5-2,4-CYCLOPENTADIEN-1-YL)- [PAL]	MANGANESE CYCLOPENTADIENYL TRICARBONYL

Chemical Name	Reference Name
12108-13-3 MANGANESE, TRICARBONYL[(1,2,3,4,5-.ETA.)-1-METHYL-2,4-CYCLOPENTADIEN-1-YL]- [PAL]	METHYLCYCLOPENTADIENYL MANGANESE TRICARBONYL
12108-13-3 MANGANESE, TRICARBONYL METHYLCYCLOPENTADIENYL [302, EPA, EPA]	METHYLCYCLOPENTADIENYL MANGANESE TRICARBONYL
13446-34-9 MANGANOUS CHLORIDE	MANGANOUS CHLORIDE
1317-35-7 MANGANOUS-MANGANIC OXIDE [MAK, MAK]	MANGANESE TETROXIDE
1344-43-0 MANGANOUS OXIDE [PST]	MANGANESE MONOXIDE
RR-00015-7 MAN-MADE MINERAL FIBERS [MAK]	MINERAL WOOL FIBER
RR-00015-7 MAN-MADE MINERAL FIBRE [GBR]	MINERAL WOOL FIBER
91031-95-7 MANNITAN, COCONUT OIL ESTER	MANNITAN, COCONUT OIL ESTER
RR-01436-8 MANNITAN TETRANITRATE	MANNITAN TETRANITRATE
69-65-8 D-MANNITOL	D-MANNITOL
15825-70-4 MANNITOL HEXANITRATE	MANNITOL HEXANITRATE
15825-70-4 MANNITOL HEXANITRATE (NITROMANNITE) [HM1, HM1]	MANNITOL HEXANITRATE
RR-01656-8 MANNICH-BASED ADDUCT	MANNICH-BASED ADDUCT
551-74-6 MANNOMUSTINE	MANNOMUSTINE
RR-00540-3 MANUFACTURE OF MAGENTA [IAR, PAL, PAL]	MAGENTA-MANUFACTURE OF
RR-01094-6 MANURE	MANURE
59355-75-8 MAPP	MAPP
RR-01193-8 MARBLE	MARBLE
RR-01193-8 MARBLE / CALCIUM CARBONATE [MEX, MEX]	MARBLE
RR-01193-8 MARBLE/CALCIUM CARBONATE [BC1, BC1]	MARBLE
68334-30-5 MARINE DIESEL FUEL [IAR]	DIESEL OIL (PETROLEUM)
68916-96-1 MATE	MATE
94-74-6 MCPA [IAR]	4-CHLORO-O-TOLOXY ACETIC ACID
94-74-6 MCPA (2-METHYL-4-CHLORO-PHENOXY-ACETIC ACID) [CAL]	4-CHLORO-O-TOLOXY ACETIC ACID
68006-83-7 MEA-A-C [MIN]	2-AMINO-3-METHYL-9H-PYRIDO(2,3,-B)INDOLE (METHYL A-ALPHA-C)
68006-83-7 ME-A-ALPHA-C (2-AMINO-3-METHYL-9H-PYRIDO[2,3-B]INDOLE) [C65, C65, CAL]	2-AMINO-3-METHYL-9H-PYRIDO(2,3,-B)INDOLE (METHYL A-ALPHA-C)E)
68006-83-7 MEA-ALPHA-C (2-AMINO-3-METHYL-9H-PYRIDO[2,3-B]INDOLE) [IAR]	2-AMINO-3-METHYL-9H-PYRIDO(2,3,-B)INDOLE (METHYL A-ALPHA-C)E]
68131-12-4 MEAT MEAL	MEAT MEAL
51-75-2 MECHLORETHAMINE [302, EPA, EPA, PAL]	NITROGEN MUSTARD
93-65-2 MECOPROP [313]	2-(2-METHYL-4-CHLOROPHENOXY) PROPIONIC ACID
13045-94-8 MEDPHALAN	MEDPHALAN
71-58-9 MEDROXYPROGESTERONE ACETATE	MEDROXYPROGESTERONE ACETATE
595-33-5 MEGESTROL ACETATE	MEGESTROL ACETATE
RR-00096-4 MEGESTROL ACETATE AND OESTROGENS	MEGESTROL ACETATE AND OESTROGENS
77094-11-2 MEIQ (2-AMINO-3,4-DIMETHYLIMIDAZO[4,5-F]QUINOLINE)	MEIQ (2-AMINO-3,4-DIMETHYLIMIDAZO[4,5-F]QUINOLINE)
77500-04-0 MEIQX (2-AMINO-3,8-DIMETHYLIMIDAZO[4,5-F]QUINOXALINE)	MEIQX (2-AMINO-3,8-DIMETHYLIMIDAZO[4,5-F]QUINOXALINE)
108-78-1 MELAMINE	MELAMINE
148-82-3 MELPHALAN	MELPHALAN
25134-21-8 MEMTETRAHYDROPHTHALIC ANHYDRIDE	MEMTETRAHYDROPHTHALIC ANHYDRIDE
15752-34-8 MENDELEVIUM 257	MENDELEVIUM 257
29665-18-7 MENDELEVIUM 258	MENDELEVIUM 258
8002-50-4 MENHADEN OIL	MENHADEN OIL

Chemical Name	Reference Name
9002-68-0 MENOTROPINS	MENOTROPINS
80-52-4 P-MENTHANE-1,8-DIAMINE	P-MENTHANE-1,8-DIAMINE
80-47-7 P-MENTHANE HYDROPEROXIDE	P-MENTHANE HYDROPEROXIDE
1490-04-6 MENTHOL	MENTHOL
15356-70-4 DL-MENTHOL	DL-MENTHOL
950-10-7 MEPHOSFOLAN	MEPHOSFOLAN
57-53-4 MEPROBAMATE	MEPROBAMATE
150-76-5 MEQUINOL (INN) (P-METHOXYPHENOL) [GBR]	4-METHOXYPHENOL
156-57-0 MERCAPTAMINE HYDROCHLORIDE	MERCAPTAMINE HYDROCHLORIDE
RR-00360-1 MERCAPTANS, LIQUID, N.O.S.	MERCAPTANS, LIQUID, N.O.S.
68-11-1 MERCAPTOACETIC ACID [GBR]	THIOGLYCOLIC ACID
583-39-1 2-MERCAPTOBENZIMIDAZOLE	2-MERCAPTOBENZIMIDAZOLE
149-30-4 2-MERCAPTOBENZOTHIAZOLE	2-MERCAPTOBENZOTHIAZOLE
149-30-4 2-MERCAPTOBENZOTHIAZOLE (MBT) [313]	2-MERCAPTOBENZOTHIAZOLE
149-30-4 MERCAPTOBENZOTHIAZOLE [TSC, TSC, TSC]	2-MERCAPTOBENZOTHIAZOLE
2492-26-4 2-MERCAPTOBENZOTHIAZOLE, SODIUM SALT [PST]	SODIUM MERCAPTOBENZOTHIAZOLE
2032-65-7 MERCAPTODIMETHUR [CAL, EPA, HM1, HM1, NJL]	METHIOCARB
60-24-2 2-MERCAPTOETHANOL	2-MERCAPTOETHANOL
60-24-2 MERCAPTOETHANOL [WEL, WEL]	2-MERCAPTOETHANOL
60-23-1 B-MERCAPTOETHYLAMINE HCL	B-MERCAPTOETHYLAMINE HCL
4420-74-0 3-MERCAPTOPROPYLTRIMETHOXYSILANE	3-MERCAPTOPROPYLTRIMETHOXYSILANE
50-44-2 6-MERCAPTOPURINE	6-MERCAPTOPURINE
6112-76-1 MERCAPTOPURINE	MERCAPTOPURINE
70-49-5 MERCAPTOSUCCINIC ACID	MERCAPTOSUCCINIC ACID
RR-01332-1 5-MERCAPTOTETRAZOL-1-ACETIC ACID	5-MERCAPTOTETRAZOL-1-ACETIC ACID
RR-01293-1 MERCARBAM	MERCARBAM
1600-27-7 MERCURIC ACETATE	MERCURIC ACETATE
7784-37-4 MERCURIC ARSENATE	MERCURIC ARSENATE
583-15-3 MERCURIC BENZOATE	MERCURIC BENZOATE
7789-47-1 MERCURIC BROMIDE	MERCURIC BROMIDE
7487-94-7 MERCURIC CHLORIDE	MERCURIC CHLORIDE
592-04-1 MERCURIC CYANIDE	MERCURIC CYANIDE
RR-01295-3 MERCURIC GLUCONATE	MERCURIC GLUCONATE
7774-29-0 MERCURIC IODIDE	MERCURIC IODIDE
10045-94-0 MERCURIC NITRATE	MERCURIC NITRATE
21908-53-2 MERCURIC OXIDE	MERCURIC OXIDE
1335-31-5 MERCURIC OXYCYANIDE	MERCURIC OXYCYANIDE
591-89-9 MERCURIC POTASSIUM CYANIDE	MERCURIC POTASSIUM CYANIDE
1312-03-4 MERCURIC SUBSULFATE	MERCURIC SUBSULFATE
7783-35-9 MERCURIC SULFATE	MERCURIC SULFATE
7783-35-9 MERCURIC SULPHATE [HM1, HM1]	MERCURIC SULFATE
1344-48-5 MERCURIC SULPHIDE [HM1]	MERCURY SULFIDE
592-85-8 MERCURIC THIOCYANATE	MERCURIC THIOCYANATE
12002-19-6 MERCUROL	MERCUROL
631-60-7 MERCUROUS ACETATE	MERCUROUS ACETATE
38232-63-2 MERCUROUS AZIDE	MERCUROUS AZIDE
RR-01296-4 MERCUROUS BISULPHATE	MERCUROUS BISULPHATE
15385-58-7 MERCUROUS BROMIDE [HM1]	MERCURY BROMIDE

Chemical Name	Reference Name
7546-30-7 MERCUROUS CHLORIDE	MERCUROUS CHLORIDE
7783-30-4 MERCUROUS IODIDE	MERCUROUS IODIDE
10415-75-5 MERCUROUS NITRATE	MERCUROUS NITRATE
7782-86-7 MERCUROUS NITRATE (MONOHYDRATE) [MSL]	MERCUROUS NITRATE
15829-53-5 MERCUROUS OXIDE	MERCUROUS OXIDE
7783-36-0 MERCUROUS SULFATE	MERCUROUS SULFATE
7783-36-0 MERCUROUS SULPHATE [HM1]	MERCUROUS SULFATE
7439-97-6 MERCURY	MERCURY
7439-97-6 MERCURY (VAPOR) [MEX]	MERCURY
15116-82-2 MERCURY 193	MERCURY 193
RR-00407-9 MERCURY 193M	MERCURY 193M
15064-97-8 MERCURY 194	MERCURY 194
15756-15-7 MERCURY 195	MERCURY 195
RR-00406-8 MERCURY 195M	MERCURY 195M
13981-51-6 MERCURY 197	MERCURY 197
RR-00405-7 MERCURY 197M	MERCURY 197M
RR-00404-6 MERCURY 199M	MERCURY 199M
13982-78-0 MERCURY 203	MERCURY 203
7439-97-6 MERCURY VAPOR [QBC]	MERCURY
1600-27-7 MERCURY ACETATE [HM1, HM1, HM1]	MERCURIC ACETATE
RR-01297-5 MERCURY ACETATES	MERCURY ACETATES
62-38-4 MERCURY, (ACETATO-O)PHENYL- [PAL]	PHENYLMERCURIC ACETATE
68833-55-6 MERCURY ACETYLIDE	MERCURY ACETYLIDE
RR-00004-4 MERCURY, ALKYL COMPOUNDS	MERCURY, ALKYL COMPOUNDS
RR-00004-4 MERCURY ALKYLS [GBR, GBR, GBR]	MERCURY, ALKYL COMPOUNDS
RR-00138-7 MERCURY, ALL FORMS EXCEPT ALKYL [BC1, BC1, QBC, QBC]	MERCURY COMPOUNDS
10124-48-8 MERCURY AMMONIUM CHLORIDE	MERCURY AMMONIUM CHLORIDE
RR-00138-7 MERCURY AND MERCURY COMPOUNDS [C65]	MERCURY COMPOUNDS
RR-00005-5 MERCURY, ARYL AND INORGANIC COMPOUNDS	MERCURY, ARYL AND INORGANIC COMPOUNDS
RR-00361-2 MERCURY BASED PESTICIDES, N.O.S.	MERCURY BASED PESTICIDES, N.O.S.
583-15-3 MERCURY BENZOATE [HM1, HM1, HM1]	MERCURIC BENZOATE
RR-01294-2 MERCURY BISULPHATES	MERCURY BISULPHATES
10031-18-2 MERCURY BROMIDE [NJL]	MERCURY BROMIDE (HGBR)
15385-58-7 MERCURY BROMIDE	MERCURY BROMIDE
10031-18-2 MERCURY BROMIDE (HGBR)	MERCURY BROMIDE (HGBR)
15385-58-7 MERCURY (I) BROMIDE (1:1) [NJL]	MERCURY BROMIDE
RR-01342-3 MERCURY BROMIDES	MERCURY BROMIDES
10112-91-1 MERCURY CHLORIDE	MERCURY CHLORIDE
13444-75-2 MERCURY (II) CHROMATE	MERCURY (II) CHROMATE
13465-34-4 MERCURY (I) CHROMATE	MERCURY (I) CHROMATE
RR-00138-7 MERCURY COMPOUNDS	MERCURY COMPOUNDS
RR-00138-7 MERCURY COMPOUNDS, N.O.S. [HM1, HM1, HM1, NJL, RCU]	MERCURY COMPOUNDS
592-04-1 MERCURY CYANIDE [HM1, HM1, HM1, HM1]	MERCURIC CYANIDE
592-04-1 MERCURY CYANIDE (HG(CN)2) [PAL]	MERCURIC CYANIDE
7439-97-6 MERCURY, ELEMENTAL [WHM]	MERCURY
628-86-4 MERCURY FULMINATE	MERCURY FULMINATE

Chemical Name	Reference Name
63937-14-4 MERCURY GLUCONATE	MERCURY GLUCONATE
RR-00569-6 MERCURY, INORGANIC [NHS, NHS]	MERCURY, INORGANIC COMPOUNDS
RR-00569-6 MERCURY, INORGANIC COMPOUNDS	MERCURY, INORGANIC COMPOUNDS
37320-91-5 MERCURY IODIDE [HM1, HM1, HM1]	MERCURY IODIDE (VAN)
RR-01437-9 MERCURY IODIDE AQUABASIC AMMONOBASIC (IODIDE OF MILTON'S BASE)	MERCURY IODIDE AQUABASIC AMMONOBASIC (IODIDE OF MILTON'S BAS
37320-91-5 MERCURY IODIDE (VAN)	MERCURY IODIDE (VAN)
7439-97-6 MERCURY, METALLIC [ATS]	MERCURY
115-09-3 MERCURYMETHYLCHLORIDE	MERCURYMETHYLCHLORIDE
12136-15-1 MERCURY NITRIDE	MERCURY NITRIDE
12002-19-6 MERCURY NUCLEATE [HM1, HM1, HM1]	MERCUROL
1191-80-6 MERCURY OLEATE	MERCURY OLEATE
RR-00004-4 MERCURY (ORGANO) ALKYL COMPOUNDS [NIO, NIO, NIO, NIO]	MERCURY, ALKYL COMPOUNDS
12653-71-3 MERCURY OXIDE [HM1, HM1, HM1]	MERCURY OXIDE (VAN)
12653-71-3 MERCURY OXIDE (VAN)	MERCURY OXIDE (VAN)
1335-31-5 MERCURYOXYCYANIDE [HM1, HM1, HM1]	MERCURIC OXYCYANIDE
RR-01298-6 MERCURY POTASSIUM CYANIDE	MERCURY POTASSIUM CYANIDE
7783-33-7 MERCURY POTASSIUM IODIDE	MERCURY POTASSIUM IODIDE
5970-32-1 MERCURY SALICYLATE	MERCURY SALICYLATE
RR-01299-7 MERCURY SULFATES	MERCURY SULFATES
1344-48-5 MERCURY SULFIDE	MERCURY SULFIDE
53408-91-6 MERCURY THIOCYANATE [NJL, HM1, HM1]	MERCURY THIOCYANATE (VAN)
53408-91-6 MERCURY THIOCYANATE (VAN)	MERCURY THIOCYANATE (VAN)
7439-97-6 MERCURY VAPOR [ALB, ALB, NIO, NIO, NIO, NIO]	MERCURY
531-76-0 MERPHALAN	MERPHALAN
150-50-5 MERPHOS [313]	S,S,S-TRIBUTYL PHOSPHOROTRITHIOITE
108-67-8 MESITYLENE [MSL]	1,3,5-TRIMETHYLBENZENE
141-79-7 MESITYL OXIDE	MESITYL OXIDE
141-79-7 MESTIYL OXIDE (MO) [CTR]	MESITYL OXIDE
141-79-7 MESITYL OXIDE (ETHER) [QBC]	MESITYL OXIDE
72-33-3 MESTRANOL	MESTRANOL
2032-65-7 MESUROL [MSL]	METHIOCARB
10102-53-1 METAARSENIC ACID	METAARSENIC ACID
RR-01551-0 METABISULFITES	METABISULFITES
RR-00367-8 METAL ALKYL HALIDES, N.O.S.	METAL ALKYL HALIDES, N.O.S.
RR-00368-9 METAL ALKYL HYDRIDES, N.O.S.	METAL ALKYL HYDRIDES, N.O.S.
RR-00369-0 METAL ALKYLS, N.O.S.	METAL ALKYLS, N.O.S.
RR-00370-3 METAL ALKYL SOLUTION, N.O.S.	METAL ALKYL SOLUTION, N.O.S.
RR-00931-4 METALATED ALKYLPHENOL COPOLYMER	METALATED ALKYLPHENOL COPOLYMERTRIAZOLE
57837-19-1 METALAXYL	METALAXYL
108-62-3 METALDEHYDE	METALDEHYDE
37273-91-9 METALDEHYDE [HM1, HM1, PAL]	METALDEHYDE (VAN)
37273-91-9 METALDEHYDE (VAN)	METALDEHYDE (VAN)
7440-02-0 METALLIC NICKEL [IAR]	NICKEL
7440-62-2 METALLIC VANADIUM [NHS, NHS]	VANADIUM
RR-00917-6 METAL SALT OF A COMPLEX INORGANIC OXYACID	METAL SALT OF A COMPLEX INORGANIC OXYACID

Chemical Name	Reference Name
RR-01438-0 METAL SALTS OF METHYL NITRAMINE	METAL SALTS OF METHYL NITRAMINEE)
RR-00814-0 METAL WORKING FLUIDS	METAL WORKING FLUIDS
137-42-8 METAM-SODIUM	METAM-SODIUM
121-47-1 METANILIC ACID	METANILIC ACID
587-98-4 METANIL YELLOW	METANIL YELLOW
10124-56-8 METAPHOSPHORIC ACID, HEXASODIUM SALT [MSL]	SODIUM HEXAMETAPHOSPHATE
10124-56-8 METAPHOSPHORIC ACID (H6P6O18), HEXASODIUM SALT [PAL]	SODIUM HEXAMETAPHOSPHATE
7785-84-4 METAPHOSPHORIC ACID, TRISODIUM SALT	METAPHOSPHORIC ACID, TRISODIUM SALT
7785-84-4 METAPHOSPHORIC ACID (H3P3O9), TRISODIUM SALT [PAL]	METAPHOSPHORIC ACID, TRISODIUM SALT
62-51-1 METHACHOLINE CHLORIDE	METHACHOLINE CHLORIDE
10476-95-6 METHACROLEIN DIACETATE	METHACROLEIN DIACETATE
78-85-3 METHACRYLALDEHYDE	METHACRYLALDEHYDE
79-41-4 METHACRYLIC ACID	METHACRYLIC ACID
9003-63-8 METHACRYLIC ACID, BUTYL ESTER, POLYMER	METHACRYLIC ACID, BUTYL ESTER, POLYMER
97-63-2 METHACRYLIC ACID, ETHYL ESTER [WHM]	ETHYL METHACRYLATE
80-62-6 METHACRYLIC ACID, METHYL ESTER [MAK, MAK, MAK, MAK]	METHYL METHACRYLATE
25035-69-2 METHACRYLIC ACID, POLYMER WITH BUTYL ACRYLATE AND METHYL METHACRYLATE [PST]	METHYL METHACRYLATE-BUTYL ACRYLATE-METHACRYLIC ACID POLYMER
25035-68-1 METHACRYLIC ACID, POLYMER WITH STYRENE AND ETHYL ACRYLATE [PST]	ETHYL ACRYLATE-METHACRYLIC ACID-STYRENE POLYMER
5536-61-8 METHACRYLIC ACID, SODIUM SALT [PST]	2-PROPENOIC ACID, 2-METHYL-, SODIUM SALT
2530-85-0 METHACRYLIC ACID, 3-(TRIMETHOXYSILYL)PROPYL ESTER	METHACRYLIC ACID, 3-(TRIMETHOXYSILYL)PROPYL ESTER
760-93-0 METHACRYLIC ANHYDRIDE	METHACRYLIC ANHYDRIDE
RR-01011-7 METHACRYLIC ESTER	METHACRYLIC ESTER
126-98-7 METHACRYLONITRILE [302, CSR, GBR, GBR, HM1, HM1, HM1, HM1, MSL, NFP, NFP, PQL, RCA, RCU, RCU]	METHYLACRYLONITRILE
2530-85-0 GAMMA-METHACRYLOXYPROPYLTRIMETHOXY-SILANE [WHM]	METHACRYLIC ACID, 3-(TRIMETHOXYSILYL)PROPYL ESTER
920-46-7 METHACRYLOYL CHLORIDE	METHACRYLOYL CHLORIDE
30674-80-7 METHACRYLOYLOXYETHYL ISOCYANATE	METHACRYLOYLOXYETHYL ISOCYANATE
3963-95-9 METHACYCLINE HYDROCHLORIDE	METHACYCLINE HYDROCHLORIDE
513-42-8 METHALLYL ALCOHOL	METHALLYL ALCOHOL
563-47-3 METHALLYL CHLORIDE [FLA, MSL, NFP, NFP]	3-CHLORO-2-METHYLPROPENE
10265-92-6 METHAMIDOPHOS	METHAMIDOPHOS
74-89-5 METHANAMINE [PAL]	METHYLAMINE
75-50-3 METHANAMINE, N,N-DIMETHYL- [PAL]	TRIMETHYLAMINE
124-40-3 METHANAMINE, N-METHYL- [PAL]	DIMETHYLAMINE
62-75-9 METHANAMINE, N-METHYL-N-NITROSO- [PAL, PAL]	N-NITROSODIMETHYLAMINE
74-82-8 METHANE	METHANE
74-83-9 METHANE, BROMO- [PAL, TSC, TSC, TSC]	METHYL BROMIDE (BROMOMETHANE)[1,1'-DIPHENYL]-4,4'-DIYL)BIS(AZO)]BIS[5-AMINO-4-HYDROXY-, TETRASODIUM SALT
74-97-5 METHANE, BROMOCHLORO- [PAL, TSC, TSC, TSC, TSC]	CHLOROBROMOMETHANE

Chemical Name	Reference Name
75-27-4 METHANE, BROMODICHLORO- [PAL, TSC, TSC]	DICHLOROBROMOMETHANE
1511-62-2 METHANE, BROMODIFLUORO-	METHANE, BROMODIFLUORO-
1511-62-2 METHANE, BROMODIFLUORO [TSE]	METHANE, BROMODIFLUORO-
75-63-8 METHANE, BROMOTRIFLUORO- [PAL]	TRIFLUOROBROMOMETHANE
74-87-3 METHANE, CHLORO- [PAL, TSC, TSC, TSC, TSC]	METHYL CHLORIDE
75-45-6 METHANE, CHLORODIFLUORO- [PAL, TSC, TSC, TSC]	CHLORODIFLUOROMETHANE
107-30-2 METHANE, CHLOROMETHOXY- [PAL, PAL]	CHLOROMETHYL METHYL ETHER
334-88-3 METHANE, DIAZO- [PAL]	DIAZOMETHANEETHYL)-4-PYRIMIDINYL] ESTER
74-95-3 METHANE, DIBROMO- [PAL, TSC, TSC, TSC]	METHYLENE BROMIDE
124-48-1 METHANE, DIBROMOCHLORO- [PAL, TSC, TSC]	CHLORODIBROMOMETHANE
75-61-6 METHANE, DIBROMODIFLUORO- [PAL]	DIFLUORODIBROMOMETHANE
75-09-2 METHANE, DICHLORO- [PAL, PAL, TSC, TSC, TSC]	METHYLENE CHLORIDE
75-71-8 METHANE, DICHLORODIFLUORO- [PAL]	DICHLORODIFLUOROMETHANE
75-43-4 METHANE, DICHLOROFLUORO- [PAL]	DICHLOROFLUOROMETHANE
109-87-5 METHANE, DIMETHOXY- [PAL, TSC, TSC]	METHYLAL
52061-60-6 P-METHANE HYDROPEROXIDE	P-METHANE HYDROPEROXIDE
74-88-4 METHANE, IODO- [PAL, PAL]	METHYL IODIDEPHENYL]-4,4'-DIYL)BIS(AZO)]BIS[5-AMINO-4-HYDROXY-, TETRASODIUM SALT
624-83-9 METHANE, ISOCYANATO- [PAL, TSC, TSC, TSC]	METHYL ISOCYANATE
75-52-5 METHANE, NITRO- [PAL, TSC, TSC, TSC]	NITROMETHANE
115-10-6 METHANE, OXYBIS- [PAL]	DIMETHYL ETHER
542-88-1 METHANE, OXYBIS[CHLORO- [PAL, PAL]	BIS(CHLOROMETHYL) ETHER
594-42-3 METHANESULFENYL CHLORIDE, TRICHLORO- [PAL, TSC, TSC, TSC]	TRICHLOROMETHANE SULPHURYL CHLORIDE
1758-73-2 METHANESULFONIC ACID, AMINOIMINO-	METHANESULFONIC ACID, AMINOIMINO-
62-50-0 METHANESULFONIC ACID, ETHYL ESTER [NJL, PAL, PAL]	ETHYL METHANESULFONATE
66-27-3 METHANESULFONIC ACID, METHYL ESTER [NJL, PAL, PAL]	METHYL METHANESULFONATE
124-63-0 METHANE SULFONYL CHLORIDE	METHANE SULFONYL CHLORIDE
124-63-0 METHANESULFONYL CHLORIDE [HM1]	METHANE SULFONYL CHLORIDE
558-25-8 METHANESULFONYL FLUORIDE	METHANESULFONYL FLUORIDE
558-13-4 METHANE, TETRABROMO- [PAL]	CARBON TETRABROMIDE
56-23-5 METHANE, TETRACHLORO- [PAL, PAL]	CARBON TETRACHLORIDE
509-14-8 METHANE, TETRANITRO- [PAL]	TETRANITROMETHANE
75-18-3 METHANE, THIOBIS- [PAL]	DIMETHYL SULFIDE
74-93-1 METHANETHIOL [GBR, HM1, HM1, HM1, HM1, HM1]	METHYL MERCAPTAN
74-93-1 1-METHANETHIOL [NHS, NHS]	METHYL MERCAPTAN
74-93-1 METHANETHIOL [ONT, PAL, RCU, RCU, TSC, TSC, TSE]	METHYL MERCAPTAN
74-93-1 METHANETHIOL (METHYL MERCAPTAN) [ALB, ALB]	METHYL MERCAPTAN
75-25-2 METHANE, TRIBROMO- [PAL, TSC, TSC, TSC]	BROMOFORM
67-66-3 METHANE, TRICHLORO- [PAL, PAL, TSC, TSC]	CHLOROFORM
75-69-4 METHANE, TRICHLOROFLUORO- [PAL]	TRICHLOROFLUOROMETHANE
76-06-2 METHANE, TRICHLORONITRO- [PAL]	CHLOROPICRIN

Chemical Name	Reference Name
75-47-8 METHANE, TRIIODO- [PAL]	IODOFORM
55738-54-0 METHANIMIDAMIDE, N,N-DIMETHYL-N'-[5-[2-(5-NITRO-2-FURANYL)ETHENYL]-1,3,4-OXADIA-ZOL-2-YL]-,(E)- [PAL, PAL]	TRANS-2-((DIMETHYLAMINO)METHYLIMINO)-5-(2-(5-NITRO-2-FURYL)
33213-65-9 6,9-METHANO-2,4,3-BENZODIOXATHIEPIN, 6,7,8,9,10,10-HEXACHLORO-1,5,5A,6,9,9A-HEXAHY-DRO-, 3-OXIDE, (3.ALPHA.,5A.ALPHA.,6.BETA.,9.BETA.,9A.ALPHA.)- [PAL, PAL, PAL, PAL]	BETA-ENDOSULFAN
76-44-8 4,7-METHANO-1H-INDENE, 1,4,5,6,7,8,8-HEP-TACHLORO-3A,4,7,7A-TETRAHYDRO- [PAL]	HEPTACHLOR
57-74-9 4,7-METHANO-1H-INDENE, 1,2,4,5,6,7,8,8-OC-TACHLORO-2,3,3A,4,7,7A-HEXAHYDRO- [PAL]	CHLORDANE
77-73-6 4,7-METHANO-1H-INDENE, 3A,4,7,7A-TETRAHY-DRO- [PAL]	DICYCLOPENTADIENE
81-21-0 2,4-METHANO-2H-INDENO [1,2-B:5,6-B'] BISOXIRENE, OCTAHYDRO-	2,4-METHANO-2H-INDENO [1,2-B:5,6-B'] BISOXIRENE, OCTAHYDRO-
1024-57-3 2,5-METHANO-2H-INDENO[1,2-B]OXIRENE, 2,3,4,5,6,7,7-HEPTACHLORO-,(1A.ALPHA.,1B.BETA.,2.ALPHA.,5A.ALPHA.,5A.BETA.,6.BETA., A.ALPHA.)- [PAL]	HEPTACHLOR EPOXIDE
115-27-5 4,7-METHANOISOBENZOFURAN-1,3-DIONE, 4,5,6,7,8,8-HEXACHLORO-3A,4,7,7A-TETRAHYDRO-	4,7-METHANOISOBENZOFURAN-1,3-DIONE, 4,5,6,7,8,8-HEXACHLORO--9-AZATRICYCLO[3.3.1.02,4]NON-7-YL ESTER, HYDROBROMIDE, [7(S)-(1.ALPHA.2.BETA.4.BETA.5.ALPHA.7.BETA)]-
25134-21-8 4,7-METHANOISOBENZOFURAN-1,3-DIONE, 3A,4,7,7A-TETRAHYDROMETHYL- [TSC, TSC]	MEMTETRAHYDROPHTHALIC ANHYDRIDELENE)]BIS-, HOMOPOLYMER
67-56-1 METHANOL [313, BEI, BEI, CAA, CAN, CSR, EPA, GBR, GBR, GBR, HM1, HM1, HM1, HON, HON, MAK, MAK, MAK, MAK, MSL, ONT, ONT, ONT, PAL, TLV, TLV, TLV, UTS]	METHYL ALCOHOL
2216-51-5 1-METHANOL	1-METHANOL
590-96-5 METHANOL, (METHYL-ONN-AZOXY)- [PAL, PAL]	METHYLAZOXYMETHANOL
592-62-1 METHANOL, (METHYL-ONN-AZOXY)-, AC-ETATE (ESTER) [PAL, PAL]	METHYLAZOXYMETHYL ACETATE
794-93-4 METHANOL,[[6-[2-(5-NITRO-2-FURANYL)ETHENYL]-1,2,4-TRIAZIN-3-YL]IMINO]BIS- [PAL, PAL]	PANFURAN S
124-41-4 METHANOL, SODIUM SALT [PAL]	SODIUM METHYLATE
32315-10-9 METHANOL, TRICHLORO-, CARBONATE (2:1)	METHANOL, TRICHLORO-, CARBONATE (2:1)
90-94-8 METHANONE, BIS[4-(DIMETHY-LAMINO0PHENYL]- [PAL, PAL]	MICHLER'S KETONE
135-23-9 METHAPYRIDINE HYDROCHLORIDE	METHAPYRIDINE HYDROCHLORIDE
91-80-5 METHAPYRILENE	METHAPYRILENE
135-23-9 METHAPYRILENE HYDROCHLORIDE [CSR]	METHAPYRIDINE HYDROCHLORIDE
RR-01439-1 METHAZOIC ACID	METHAZOIC ACID
20354-26-1 METHAZOLE [2-(3,4-DICHLOROPHENYL)-4-METHYL-1,2,4-OXADIAZOLIDINE-3,5-DIONE]	METHAZOLE [2-(3,4-DICHLOROPHENYL)-4-METHYL-1,2,4-OXADIAZOLIDROCHLORIDE)
1982-37-2 METHDILAZINE	METHDILAZINE
100-97-0 METHENAMINE	METHENAMINE
2385-85-5 1,3,4-METHENO-1H-CYCLOBUTA[CD]PEN-TALENE, 1,1A,2,2,3,3A,4,5,5,5A,5B,6-DODE-CACHLOROOCTAHYDRO- [PAL, PAL]	MIREXPHENYL)AZO][1,1'-BIPHENYL]-4-YL]AZO]-5-HY-DROXY-6-(PHENYLAZO)-,DISODIUM SALT
143-50-0 1,3,4-METHENO-2H-CYCLOBUTA[CD]PEN-TALEN-2-ONE, 1,1A,3,3A,4,5,5,5A,5B,6-DE-CACHLOROOCTAHYDRO- [PAL, PAL]	KEPONE

Chemical Name	Reference Name
7421-93-4 1,2,4-METHENOCYCLOPENTA[CD]PENTA-LENE-5-CARBOXALDEHYDE, 2,2A,3,3,4,7-HEXACHLORODECAHYDRO-,(1.ALPHA., 2.BETA.,2A.BETA., 4.BETA.,4A.BETA.,5.BETA., 6A.BETA.6B.BETA.,7R*)- [PAL]	ENDRIN ALDEHYDE
950-37-8 METHIDATHION	METHIDATHION
60-56-0 METHIMAZOLE	METHIMAZOLE
2032-65-7 METHIOCARB	METHIOCARB
63-68-3 METHIONINE	METHIONINE
16752-77-5 METHOMYL	METHOMYL
16752-77-5 METHOMYL (LANNATE) [OS1, BC1, BC1]	METHOMYL
16752-77-5 METHOMYL (LANNATE) (R) [QBC]	METHOMYL
16752-77-5 METHOMYL (ISO) [GBR, GBR]	METHOMYL
16752-77-5 METHOMYL (LANNATE) [ALB, ALB, ALB]	METHOMYL
59-05-2 METHOTREXATE	METHOTREXATE
15475-56-6 METHOTREXATE SODIUM	METHOTREXATE SODIUM
94-74-6 METHOXONE (4-CHLORO-2-METHYLPHE-NOXY) ACETIC ACID (MCPA)) [313]	4-CHLORO-O-TOLOXY ACETIC ACID
3653-48-3 METHOXONE-SODIUM SALT ((4-CHLORO-2-METHYLPHENOXY) ACETATE SODIUM SALT)	METHOXONE-SODIUM SALT ((4-CHLORO-2-METHYLPHENOXY) ACETATE SO
298-81-7 METHOXSALEN [CAL]	8-METHOXYPSORALEN
RR-00056-6 METHOXSALEN WITH ULTRA-VIOLET [MIN]	8-METHOXYPSORALEN PLUS ULTRAVIOLET RADIA-TION
RR-00056-6 METHOXSALEN WITH ULTRA-VIOLET A THER-APY [PAL, PAL]	8-METHOXYPSORALEN PLUS ULTRAVIOLET RADIA-TION
RR-00056-6 METHOXSALEN WITH ULTRA-VIOLET A THER-APY (PUVA) [NTP]	8-METHOXYPSORALEN PLUS ULTRAVIOLET RADIA-TION
90-04-0 ORTHO-METHOXYANILINE [HM1]	O-ANISIDINE
135-02-4 O-METHOXYBENZALDEHYDE [FLA, MSL, NFP, NFP]	O-ANISALDEHYDE
5307-02-8 2-METHOXY-1,4-BENZENEDIAMINE	2-METHOXY-1,4-BENZENEDIAMINE
1747-60-0 6-METHOXY-2-BENZOTHIAZOLAMINE	6-METHOXY-2-BENZOTHIAZOLAMINE
105-13-5 4-METHOXYBENZYL ALCOHOL	4-METHOXYBENZYL ALCOHOL
2517-43-3 3-METHOXYBUTANOL	3-METHOXYBUTANOL
4435-53-4 3-METHOXYBUTYL ACETATE [NFP, NFP]	BUTOXYL
5281-76-5 3-METHOXYBUTYRALDEHYDE	3-METHOXYBUTYRALDEHYDE
72-43-5 METHOXYCHLOR	METHOXYCHLOR
72-43-5 METHOXYCHLOR [BENZENE, 1,1'-(2,2,2-TRICHLOROETHYLIDENE)BIS [4-METHOXY]-] [313]	METHOXYCHLOR
72-43-5 METHOXYCHLOR (ISO) [GBR]	METHOXYCHLOR
921-20-0 3-METHOXY-2,4-DIHYDROXYPENTANE	3-METHOXY-2,4-DIHYDROXYPENTANE
109-86-4 2-METHOXYETHANOL	2-METHOXYETHANOL
109-86-4 2-METHOXYETHANOL [QBC, QBC]	2-METHOXYETHANOL
109-86-4 2-METHOXYETHANOL (EGME) [TLV, TLV]	2-METHOXYETHANOL
109-86-4 2-METHOXYETHANOL (METHYL CELLOSOLVE) [BC1, BC1, BC1, ALB, ALB, ALB]	2-METHOXYETHANOL
110-49-6 2-METHOXYETHYL ACETATE	2-METHOXYETHYL ACETATE
110-49-6 2-METHOXYETHYL ACETATE [QBC, QBC]	2-METHOXYETHYL ACETATE
110-49-6 2-METHOXYETHYL ACETATE (EGMEA) [TLV, TLV]	2-METHOXYETHYL ACETATE
3121-61-7 2-METHOXYETHYL ACRYLATE	2-METHOXYETHYL ACRYLATE

Chemical Name	Reference Name
10232-93-6 METHOXYETHYL HYDROGEN MALEATE	METHOXYETHYL HYDROGEN MALEATE
151-38-2 METHOXYETHYLMERCURIC ACETATE	METHOXYETHYLMERCURIC ACETATE
16501-01-2 METHOXY ETHYL PHTHALATE	METHOXY ETHYL PHTHALATE
76-38-0 METHOXYFLURANE	METHOXYFLURANE
4461-52-3 METHOXYMETHANOL	METHOXYMETHANOL
107-98-2 2-METHOXY-1-METHYLETHANOL [PST]	PROPYLENE GLYCOL MONOMETHYL ETHER
25498-49-1 (2-(2-METHOXYMETHYLETHOXY) METHYLETHOXY) PROPANOL [PST]	TRIPROPYLENE GLYCOL MONOMETHYL ETHER
6427-21-0 METHOXYMETHYL ISOCYANATE	METHOXYMETHYL ISOCYANATE
107-70-0 4-METHOXY-4-METHYLPENTAN-2-ONE	4-METHOXY-4-METHYLPENTAN-2-ONE
107-70-0 4-METHOXY-4-METHYL-2-PENTANONE [PST]	4-METHOXY-4-METHYLPENTAN-2-ONE
150-76-5 4-METHOXYPHENOL	4-METHOXYPHENOL
150-76-5 P-METHOXYPHENOL [T32, WHM]	4-METHOXYPHENOL
107-98-2 1-METHOXYPROPAN-2-OL [GBR, GBR, GBR]	PROPYLENE GLYCOL MONOMETHYL ETHER
107-98-2 1-METHOXY-2-PROPANOL [ONT, ONT]	PROPYLENE GLYCOL MONOMETHYL ETHER
1589-47-5 2-METHOXY-1-PROPANOL	2-METHOXY-1-PROPANOL
1589-49-7 3-METHOXYPROPANOL	3-METHOXYPROPANOL
28677-93-2 METHOXY-1-PROPANOL [TSC, TSC, TSC]	1-PROPANOL, METHOXY-
108-65-6 1-METHOXY-2-PROPANOL ACETATE [PST]	PROPYLENE GLYCOL MONOMETHYL ETHER ACETATE
110-67-8 3-METHOXYPROPIONITRILE [FLA, MSL, NFP, NFP]	PROPANENITRILE, 3-METHOXY-
107-98-2 1-METHOXY-2-PROPONAL [HM1, HM1]	PROPYLENE GLYCOL MONOMETHYL ETHER
10213-77-1 2-[2-(2-METHOXYPROPOXY)PROPOXY]-1-PROPANOL	2-[2-(2-METHOXYPROPOXY)PROPOXY]-1-PROPANOL
70657-70-4 2-METHOXYPROPYL-1-ACETATE	2-METHOXYPROPYL-1-ACETATE
5332-73-0 3-METHOXYPROPYLAMINE	3-METHOXYPROPYLAMINE
RR-01159-6 METHOXYPROPYLAMINE DODECYLBENZENE-SULFONATE	METHOXYPROPYLAMINE DODECYLBENZENESUL-FONATE
104-45-0 1-METHOXY-4-PROPYLBENZENE	1-METHOXY-4-PROPYLBENZENE
298-81-7 8-METHOXYPSORALEN	8-METHOXYPSORALEN
484-20-8 5-METHOXYPSORALEN	5-METHOXYPSORALEN
RR-00056-6 8-METHOXYPSORALEN PLUS ULTRAVIOLET RADIATION	8-METHOXYPSORALEN PLUS ULTRAVIOLET RADIA-TION
RR-00056-6 8-METHOXYPSORALEN (METHOXSALEN) PLUS ULTRAVIOLET RADIATION [IAR]	8-METHOXYPSORALEN PLUS ULTRAVIOLET RADIA-TION
298-81-7 8-METHOXYPSORALEN WITH ULTRAVIOLET A THERAPY [C65]	8-METHOXYPSORALEN
484-20-8 5-METHOXYPSORALEN WITH ULTRAVIOLET A THERAPY [C65]	5-METHOXYPSORALEN
112-35-6 METHOXY TRIGLYCOL [NFP, NFP]	TRIETHYLENE GLYCOL MONOMETHYL ETHER
3610-27-3 METHOXYTRIGLYCOL ACETATE	METHOXYTRIGLYCOL ACETATE
59355-75-8 METHYL ACETYLENE - PROPADIENE MIX-TURE [MEX, MEX]	MAPP
74-83-9 METHYL BROMIDE (BROMOMETHANE) [HON, HON]	METHYL BROMIDE (BROMOMETHANE)
74-87-3 METHYL CHLORIDE (CHLOROMETHANE) [HON, HON]	METHYL CHLORIDE
71-55-6 METHYL CHLOROFORM [MEX, MEX]	1,1,1-TRICHLOROETHANE
78-93-3 METHYL ETHYL KETONE (2-BUTANONE) [HON, HON]	METHYL ETHYL KETONE
127-25-3 METHYL ABIETATE	METHYL ABIETATE
79-16-3 N-METHYLACETAMIDE	N-METHYLACETAMIDE

Chemical Name	Reference Name
79-20-9 METHYL ACETATE	METHYL ACETATE
105-45-3 METHYLACETOACETATE	METHYLACETOACETATE
105-45-3 METHYL ACETOACETATE [FLA, MSL, NFP, NFP]	METHYLACETOACETATE
122-00-9 P-METHYLACETOPHENONE	P-METHYLACETOPHENONE
122-00-9 P-METHYL ACETOPHENONE [NFP, NFP]	P-METHYLACETOPHENONE
74-99-7 METHYL ACETYLENE	METHYL ACETYLENE
74-99-7 METHYL ACETYLENE (PROPYNE) [BC1, BC1, OS1, OS1, WHM, QBC]	METHYL ACETYLENE
59355-75-8 METHYL ACETYLENE - PROPADIENE MIXTURE (MAPP) [ALB, ALB]	MAPP
59355-75-8 METHYL ACETYLENEPROPADIENE MIXTURE [CAL]	MAPP
59355-75-8 METHYL ACETYLENE-PROPADIENE MIXTURE [HM1, HM1, MIN, NIO, NIO, NIO]	MAPP
59355-75-8 METHYL ACETYLENE-PROPADIENE MIXTURE (MAPP) [AUS, AUS, BC1, BC1, FLA, ONT, ONT, OS1, OS1, OS1]	MAPP
59355-75-8 METHYLACETYLENE-PROPADIENE (MAPP) [QBC, QBC]	MAPP
59355-75-8 METHYL ACETYLENE-PROPADIENE MIXTURE (MAPP) [TLV, TLV]	MAPP
59355-75-8 METHYL ACETYLENE-PROPADIENE MIXTURE [CEX, CEX]	MAPP
123-73-9 3-METHYLACROLEINE [HM1, HM1]	CROTONALDEHYDE
79-39-0 2-METHYLACRYLAMIDE	2-METHYLACRYLAMIDE
1187-59-3 N-METHYLACRYLAMIDE	N-METHYLACRYLAMIDE
96-33-3 METHYL ACRYLATE	METHYL ACRYLATE
126-98-7 METHYLACRYLONITRILE	METHYLACRYLONITRILE
126-98-7 ALPHA-METHYLACRYLONITRILE [CAL, CEX, CEX]	METHYLACRYLONITRILE
126-98-7 METHYL ACRYLONITRILE [OS3, WHM]	METHYLACRYLONITRILE
126-98-7 METHYLACRYLONITRILE [2-PROPENENITRILE, 2-METHYL-] [EP3]	METHYLACRYLONITRILE
109-87-5 METHYLAL	METHYLAL
109-87-5 METHYLAL (DIMETHOXYMETHANE) [BC1]	METHYLAL
109-87-5 METHYLAL (DIMETHOXY-METHANE) [OS1, OS1]	METHYLAL
67-56-1 METHYL ALCOHOL	METHYL ALCOHOL
67-56-1 METHYL ALCOHOL (METHANOL) [QBC, QBC, QBC, BC1, BC1, BC1, WHM, ALB, ALB, ALB]	METHYL ALCOHOL
563-47-3 METHYL ALLYL CHLORIDE [HM1, HM1, NJL]	3-CHLORO-2-METHYLPROPENE
12263-85-3 METHYLALUMINUM SESQUIBROMIDE	METHYLALUMINUM SESQUIBROMIDE
12263-85-3 METHYL ALUMINUM SESQUIBROMIDE [NJL]	METHYLALUMINUM SESQUIBROMIDE
12542-85-7 METHYL ALUMINUM SESQUICHLORIDE	METHYL ALUMINUM SESQUICHLORIDE
12542-85-7 METHYLALUMINUM SESQUICHLORIDE [MSL, FLA, NFP, NFP]	METHYL ALUMINUM SESQUICHLORIDE
74-89-5 METHYLAMINE	METHYLAMINE
74-89-5 METHYL AMINE [MIN]	METHYLAMINE
RR-01443-7 METHYLAMINE DINITRAMINE AND SALTS	METHYLAMINE DINITRAMINE AND SALTS
14147-71-8 METHYLAMINE NITROFORM	METHYLAMINE NITROFORM
RR-01444-8 METHYLAMINE PERCHLORATE	METHYLAMINE PERCHLORATE

Chemical Name	Reference Name
RR-01782-3 METHYLAMINE SALTS	METHYLAMINE SALTSISOMERS
74-89-5 METHYLAMINE [METHANAMINE] [EP3]	METHYLAMINE
107-68-6 2-(METHYLAMINO)ETHANESULFONIC ACID	2-(METHYLAMINO)ETHANESULFONIC ACID
2475-46-9 1-METHYLAMINO-4-ETHANOLAMINOAN-THRAQUINONE	1-METHYLAMINO-4-ETHANOLAMINOANTHRAQUINONE
21160-95-2 METHYLAMMONIUM N-METHYLDITHIOCAR-BAMATE	METHYLAMMONIUM N-METHYLDITHIOCARBAMATE
108-84-9 METHYLAMYL ACETATE [HM1, HM1]	SEC-HEXYL ACETATE
108-84-9 METHYL AMYL ACETATE [NJL]	SEC-HEXYL ACETATE
7789-99-3 METHYLAMYL ACETATE	METHYLAMYL ACETATE
108-11-2 METHYL AMYL ALCOHOL [NJL]	METHYL ISOBUTYL CARBINOL
108-11-2 METHYL AMYL ALCOHOL (METHYL ISOBUTYL CARBINOL) [QBC, QBC, QBC, ALB, ALB, ALB]	METHYL ISOBUTYL CARBINOL
110-43-0 METHYL N-AMYL KETONE	METHYL N-AMYL KETONE
110-43-0 METHYL (N-AMYL) KETONE [FLA]	METHYL N-AMYL KETONE
110-43-0 METHYL AMYL KETONE [NFP, NFP]	METHYL N-AMYL KETONE
110-43-0 METHYL (N-AMYL) KETONE [NIO, NIO, NIO, NJL]	METHYL N-AMYL KETONE
110-43-0 METHYL N-AMYL KETONE (2-HEPTANONE) [QBC]	METHYL N-AMYL KETONE
110-43-0 METHYL AMYL KETONE [WHM]	METHYL N-AMYL KETONE
110-43-0 METHYL N-AMYL KETONE (2-HEPTANONE) [BC1, BC1]	METHYL N-AMYL KETONE
73459-03-7 5-METHYLANGELICIN	5-METHYLANGELICIN
RR-01535-0 5-METHYLANGELICIN PLUS ULTRAVIOLET A RADIATION	5-METHYLANGELICIN PLUS ULTRAVIOLET A RADIATION
95-53-4 O-METHYLANILINE [CAL]	O-TOLUIDINE
100-61-8 N-METHYL ANILINE	N-METHYL ANILINE
100-61-8 N-METHYLANILINE [CAL, CEX, CEX, GBR, GBR, MAK, MAK, MAK]	N-METHYL ANILINE
100-61-8 METHYLANILINE [NJL, QBC, QBC]	N-METHYL ANILINE
100-61-8 N-METHYLANILINE [WHM]	N-METHYL ANILINE
93-90-3 2-(N-METHYLANILINO)-ETHANOL [FLA]	ETHANOL, 2-(METHYLPHENYLAMINO)-
120-71-8 5-METHYL-O-ANISIDINE [MAK]	P-CRESIDINE
134-20-3 METHYL ANTHRANILATE	METHYL ANTHRANILATE
1321-94-4 METHYLATED NAPHTHALENES [PST]	METHYLNAPHTHALENE
75-55-8 2-METHYLAZIRIDINE [CAB, CAL, IAR, MIN]	PROPYLENEIMINE
75-55-8 2-METHYLAZIRIDINE (PROPYLENEIMINE) [NTP, C65]	PROPYLENEIMINE
590-96-5 METHYLAZOXYMETHANOL	METHYLAZOXYMETHANOL
592-62-1 METHYLAZOXYMETHANOL ACETATE [FLA, MSL]	METHYLAZOXYMETHYL ACETATE
592-62-1 METHYLAZOXYMETHYL ACETATE	METHYLAZOXYMETHYL ACETATE
108-88-3 METHYLBENZENE [ONT, ONT]	TOLUENE
100-61-8 N-METHYLBENZENEAMINE [ONT, ONT]	N-METHYL ANILINE
95-80-7 4-METHYL-1,3-BENZENEDIAMINE [FLA]	TOLUENE 2,4-DIAMINE
104-15-4 4-METHYLBENZENESULFONIC ACID [PST]	P-TOLUENESULFONIC ACID
93-58-3 METHYLBENZOATE	METHYLBENZOATE
93-58-3 METHYL BENZOATE [NFP, NFP]	METHYLBENZOATE
553-97-9 2-METHYL-1,4-BENZOQUINONE	2-METHYL-1,4-BENZOQUINONE

Chemical Name	Reference Name
92-36-4 4-(6-METHYL-2-BENZOTHIAZOLYL)-BEN-ZENAMINE	4-(6-METHYL-2-BENZOTHIAZOLYL)-BENZENAMINE
29385-43-1 METHYL-1H-BENZOTRIAZOLE [PST]	TOLYL TRIAZOLE
98-85-1 METHYL PHENYL CARBINOL	METHYL PHENYL CARBINOL
589-18-4 P-METHYLBENZYL ALCOHOL	P-METHYLBENZYL ALCOHOL
98-84-0 ALPHA-METHYLBENZYLAMINE	ALPHA-METHYLBENZYLAMINE
2449-49-2 ALPHA-METHYLBENZYL DIMETHYL AMINE	ALPHA-METHYLBENZYL DIMETHYL AMINE
93-96-9 ALPHA-METHYLBENZYL ETHER [MSL, NFP, NFP]	BENZENE, 1,1'-(OXYDIETHYLIDENE)BIS-
10309-79-2 1-METHYL-2-BENZYLHYDRAZINE	1-METHYL-2-BENZYLHYDRAZINE
643-58-3 2-METHYLBIPHENYL	2-METHYLBIPHENYL
644-08-6 4-METHYL BIPHENYL	4-METHYL BIPHENYL
51-75-2 N-METHYL-BIS(2-CHLOROETHYL)AMINE [MAK, MAK, WHM]	NITROGEN MUSTARD
102093-68-5 4-METHYL-2,6-BIS(METHYLTHIO)-	4-METHYL-2,6-BIS(METHYLTHIO)-YDROXYPHENYL] ETHYL ESTER
121-43-7 METHYL BORATE [FLA, MSL, NFP, NFP]	TRIMETHYL BORATE
74-83-9 METHYL BROMIDE [MX1]	METHYL BROMIDE (BROMOMETHANE)
74-83-9 METHYL BROMIDE (BROMOMETHANE) [CAA, CAA, CW2]	METHYL BROMIDE (BROMOMETHANE)
74-83-9 METHYL BROMIDE (BROMOMETHANE)	METHYL BROMIDE (BROMOMETHANE)
RR-01300-3 METHYL BROMIDE AND ETHYLENE DIBRO-MIDE MIXTURES, LIQUID	METHYL BROMIDE AND ETHYLENE DIBROMIDE MIX-TURES, LIQUID
96-32-2 METHYL BROMOACETATE	METHYL BROMOACETATE
504-60-9 1-METHYLBUTADIENE	1-METHYLBUTADIENE
96-17-3 2-METHYLBUTANAL [FLA]	BUTANAL, 2-METHYL-
78-78-4 2-METHYLBUTANE [FLA]	ISOPENTANE
2084-18-6 3-METHYL-2-BUTANETHIOL	3-METHYL-2-BUTANETHIOL
75-85-4 2-METHYL-2-BUTANOL	2-METHYL-2-BUTANOL
123-51-3 3-METHYLBUTAN-1-OL [GBR, GBR]	ISOAMYL ALCOHOL
137-32-6 2-METHYL-1-BUTANOL	2-METHYL-1-BUTANOL
598-75-4 3-METHYL-2-BUTANOL	3-METHYL-2-BUTANOL
563-80-4 3-METHYLBUTAN-2-ONE [HM1, HM1]	METHYL ISOPROPYL KETONE
563-80-4 3-METHYL-2-BUTANONE [ONT, WHM]	METHYL ISOPROPYL KETONE
513-35-9 2-METHYL-2-BUTENE	2-METHYL-2-BUTENE.ALPHA.-HYDROXY-,ETHYL ESTER
563-45-1 3-METHYL-1-BUTENE	3-METHYL-1-BUTENE
563-46-2 2-METHYL-1-BUTENE [EP3, HM1, HM1]	2-METHYL-1-BUTENE (TECHNICAL)
26760-64-5 METHYL BUTENE	METHYL BUTENE
26760-64-5 METHYLBUTENE [NJL, NJL]	METHYL BUTENE
563-46-2 2-METHYL-1-BUTENE (TECHNICAL)	2-METHYL-1-BUTENE (TECHNICAL)
563-46-2 2-METHYL-1-BUTENE (TECHNICAL GRADE) [NFP, NFP]	2-METHYL-1-BUTENE (TECHNICAL)
626-38-0 1-METHYLBUTYL ACETATE [GBR]	SEC-AMYL ACETATE
110-68-9 N-METHYLBUTYLAMINE	N-METHYLBUTYLAMINE
1634-04-4 METHYL TERT-BUTYL ETHER	METHYL TERT-BUTYL ETHER
1634-04-4 METHYL-T-BUTYL-ETHER [ATS]	METHYL TERT-BUTYL ETHER
1634-04-4 METHYL-T-BUTYL ETHER [SD4]	METHYL TERT-BUTYL ETHER
591-78-6 METHYL N-BUTYL KETONE [AUS, AUS, CAL]	2-HEXANONE
591-78-6 METHYL BUTYL KETONE [CEX, CEX, FLA, MSL, NFP, NFP]	2-HEXANONE
591-78-6 METHYL N-BUTYL KETONE [NHS, NHS]	2-HEXANONE

Chemical Name	Reference Name
591-78-6 METHYL-N-BUTYL KETONE [NJL]	2-HEXANONE
591-78-6 METHYL N-BUTYL KETONE [OTV, TLV, TLV, TSE]	2-HEXANONE
591-78-6 METHYL N-BUTYL KETONE (2-HEXANONE) [TSC, TSC]	2-HEXANONE
2409-55-4 4-METHYL-2-T-BUTYLPHENOL	4-METHYL-2-T-BUTYLPHENOL
115-19-5 3-METHYL BUTYNOL	3-METHYL BUTYNOL
115-19-5 2-METHYL-3-BUTYN-2-OL [PST]	3-METHYL BUTYNOL
96-17-3 2-METHYLBUTYRALDEHYDE [MSL, NFP, NFP]	BUTANAL, 2-METHYL-
623-42-7 METHYL BUTYRATE	METHYL BUTYRATE
598-55-0 METHYL CARBAMATE	METHYL CARBAMATE
616-38-6 METHYL CARBONATE [FLA, MSL, NFP, NFP]	DIMETHYL CARBONATE
13909-09-6 METHYL CCNU [CSR]	1-(2-CHLOROETHYL)-3-(4-METHYLCYCLOHEXYL)-1-NITROSOUREA
109-86-4 METHYL CELLOSOLVE [FLA, MIN, MSL, NIO, NIO, NIO]	2-METHOXYETHANOL
109-86-4 METHYL CELLOSOLVE (2-METHOXYETHANOL) [OS1, OS1, OS1, OS1, OS1]	2-METHOXYETHANOL
110-49-6 METHYL CELLOSOLVE ACETATE [NIO, NIO, NIO]	2-METHOXYETHYL ACETATE
110-49-6 METHYL CELLOSOLVE ACETATE (2-METHOXYETHYL ACETATE) [OS1, OS1, OS1, OS1, OS1]	2-METHOXYETHYL ACETATE
9004-67-5 METHYLCELLULOSE	METHYLCELLULOSE
9004-67-5 METHYL CELLULOSE [PST]	METHYLCELLULOSE
74-87-3 METHYL CHLORIDE	METHYL CHLORIDE
74-87-3 METHYL CHLORIDE (CHLOROMETHANE) [CAA, QBC, QBC, QBC, CW2]	METHYL CHLORIDE
74-87-3 METHYL CHLORIDE [METHANE, CHLORO-] [EP3]	METHYL CHLORIDE
96-34-4 METHYL CHLOROACETATE	METHYL CHLOROACETATE
80-63-7 METHYL-2-CHLOROACRYLATE	METHYL-2-CHLOROACRYLATE
80-63-7 METHYL 2-CHLOROACRYLATE [302]	METHYL-2-CHLOROACRYLATE
RR-01301-4 METHYLCHLOROBENZENES	METHYLCHLOROBENZENES
79-22-1 METHYL CHLOROCARBONATE [313, EPA, RCU, RCU, WHM]	METHYL CHLOROFORMATE
71-55-6 METHYL CHLOROFORM [BEI, CAB, CAL, CEX, CEX, CEX]	1,1,1-TRICHLOROETHANE
71-55-6 METHYLCHLOROFORM [HM1, HM1, HM1, HM1]	1,1,1-TRICHLOROETHANE
71-55-6 METHYL CHLOROFORM [MIN, NIO, NIO, NIO, NJL]	1,1,1-TRICHLOROETHANE
71-55-6 METHYLCHLOROFORM [OTV]	1,1,1-TRICHLOROETHANE
71-55-6 METHYL CHLOROFORM [RCU, RCU, TLV, TLV]	1,1,1-TRICHLOROETHANE
71-55-6 METHYL CHLOROFORM (1,1,1-TRICHLOROETHANE) [CAA, CAA, OS1, OS1, OS1, BC1, BC1, QBC, QBC]	1,1,1-TRICHLOROETHANE
79-22-1 METHYL CHLOROFORMATE	METHYL CHLOROFORMATE
79-22-1 METHYL CHLOROFORMATE [CARBONOCHLO-RIDIC ACID, METHYL ESTER] [EP3]	METHYL CHLOROFORMATE
71-55-6 METHYL CHLOROFORM 1,1,1-TRICHLOROETHANE [EPA]	1,1,1-TRICHLOROETHANE
107-30-2 METHYL CHLOROMETHYL ETHER [CAL, CEX, HM1, HM1, HM1, HM1, MIN, NHS, NHS, NHS, OS1, WHM]	CHLOROMETHYL METHYL ETHER

Chemical Name	Reference Name
94-81-5 4-(2-METHYL-4-CHLOROPHENOXY) BUTYRIC ACID	4-(2-METHYL-4-CHLOROPHENOXY) BUTYRIC ACID
93-65-2 2-(2-METHYL-4-CHLOROPHENOXY) PROPIONIC ACID	2-(2-METHYL-4-CHLOROPHENOXY) PROPIONIC ACID
17639-93-9 METHYL-2-CHLOROPROPIONATE	METHYL-2-CHLOROPROPIONATE
993-00-0 METHYLCHLOROSILANE	METHYLCHLOROSILANE
993-00-0 METHYL CHLOROSILANE [NJL]	METHYLCHLOROSILANE
56-49-5 3-METHYLCHOLANTHRENE	3-METHYLCHOLANTHRENE
1705-85-7 6-METHYLCHRYSENE	6-METHYLCHRYSENERBOFURAN)
3351-28-8 1-METHYLCHRYSENE	1-METHYLCHRYSENE
3351-30-2 4-METHYLCHRYSENE	4-METHYLCHRYSENEOBISISOBUTYRONITRILE)
3351-31-3 3-METHYLCHRYSENE	3-METHYLCHRYSENE
3351-32-4 2-METHYLCHRYSENE	2-METHYLCHRYSENE
3697-24-3 5-METHYLCHRYSENE	5-METHYLCHRYSENE
3697-24-3 5-METHYL CHRYSENE [MSL]	5-METHYLCHRYSENE
103-26-4 METHYL CINNAMATE	METHYL CINNAMATE
92-48-8 6-METHYLCOUMARIN	6-METHYLCOUMARIN
92-48-8 METHYL COUMARIN [CSR]	6-METHYLCOUMARIN
104-93-8 METHYL PARA-CRESOL	METHYL PARA-CRESOL
75-05-8 METHYL CYANIDE [HM1, HM1, HM1]	ACETONITRILE
137-05-3 METHYL 2-CYANOACRYLATE	METHYL 2-CYANOACRYLATE
137-05-3 METHYL-2-CYANOACRYLATE [MIN]	METHYL 2-CYANOACRYLATE
137-05-3 METHYL CYANOACRYLATE [WHM]	METHYL 2-CYANOACRYLATE
108-87-2 METHYLCYCLOHEXANE	METHYLCYCLOHEXANE
108-87-2 METHYL CYCLOHEXANE [OS1, OS1]	METHYLCYCLOHEXANE
583-59-5 2-METHYLCYCLOHEXANOL	2-METHYLCYCLOHEXANOL
589-91-3 4-METHYLCYCLOHEXANOL	4-METHYLCYCLOHEXANOL
591-23-1 3-METHYLCYCLOHEXANOL	3-METHYLCYCLOHEXANOL
25639-42-3 METHYLCYCLOHEXANOL	METHYLCYCLOHEXANOL
25639-42-3 METHYL CYCLOHEXANOL [NJL]	METHYLCYCLOHEXANOL
25639-42-3 METHYLCYCLOHEXANOL (META- AND PARA-ISOMER MIXTURE) [CEX]	METHYLCYCLOHEXANOL
590-67-0 METHYL CYCLOHEXANOLS	METHYL CYCLOHEXANOLS
108-87-2 METHYL CYCLOHEXANONE [HM1, HM1]	METHYLCYCLOHEXANE
583-60-8 O-METHYLCYCLOHEXANONE	O-METHYLCYCLOHEXANONE
583-60-8 2-METHYLCYCLOHEXANONE [GBR, GBR, GBR]	O-METHYLCYCLOHEXANONE
583-60-8 1-METHYLCYCLOHEXAN-2-ONE [MAK, MAK, MAK]	O-METHYLCYCLOHEXANONE
583-60-8 2-METHYLCYCLOHEXANONE [MSL, ONT, ONT, ONT]	O-METHYLCYCLOHEXANONE
583-60-8 METHYLCYCLOHEXANONE-O [QBC, QBC, QBC]	O-METHYLCYCLOHEXANONE
583-60-8 2-METHYLCYCLOHEXANONE [WHM]	O-METHYLCYCLOHEXANONE
1331-22-2 METHYLCYCLOHEXANONE	METHYLCYCLOHEXANONE
RR-01758-3 METHYLCYCLOHEXANONE PEROXIDE	METHYLCYCLOHEXANONE PEROXIDE
591-47-9 4-METHYLCYCLOHEXENE	4-METHYLCYCLOHEXENE
100-60-7 N-METHYL CYCLOHEXYLAMINE	N-METHYL CYCLOHEXYLAMINE
30232-11-2 METHYLCYCLOHEXYL ACETATE	METHYLCYCLOHEXYL ACETATE
26472-00-4 METHYL CYCLOPENTADIENE	METHYL CYCLOPENTADIENE
12108-13-3 METHYL CYCLOPENTADIENYL MANGANESE TRICARBONYL [WHM]	METHYLCYCLOPENTADIENYL MANGANESE TRICARBONYL

Chemical Name	Reference Name
12108-13-3 **METHYLCYCLOPENTADIENYL MANGANESE TRICARBONYL**	METHYLCYCLOPENTADIENYL MANGANESE TRICARBONYL
12108-13-3 2-METHYLCYCLOPENTADIENYL MANGANESE TRICARBONYL [CAL, MIN, ONT, ONT, TLV, TLV]	METHYLCYCLOPENTADIENYL MANGANESE TRICARBONYL
96-37-7 **METHYLCYCLOPENTANE**	METHYLCYCLOPENTANE
96-37-7 METHYL CYCLOPENTANE [NJL, NJL]	METHYLCYCLOPENTANE
6975-98-0 **2-METHYLDECANE**	2-METHYLDECANE
919-86-8 METHYL-S-DEMETON [NJL]	DEMETON-S-METHYL
8022-00-2 **METHYL DEMETON**	METHYL DEMETON
116-54-1 **METHYL DICHLOROACETATE**	METHYL DICHLOROACETATE
593-89-5 **METHYL DICHLOROARSINE**	METHYL DICHLOROARSINE
75-54-7 **METHYL DICHLOROSILANE**	METHYL DICHLOROSILANE
75-54-7 METHYLDICHLOROSILANE [FLA, MSL, NFP, NFP]	METHYL DICHLOROSILANE
105-59-9 **METHYLDIETHANOLAMINE**	METHYLDIETHANOLAMINE
105-59-9 N-METHYLDIETHANOLAMINE [NFP, NFP]	METHYLDIETHANOLAMINE
2897-60-1 3-(METHYLDIETHOXYSILYL)PROPYL GLYCIDYL ETHER [EPA, OTS]	SILANE, DIETHOXYMETHYL[3-(OXIRANYLMETHOXY) PROPYL]-
2050-24-0 **1-METHYL-3,5-DIETHYLBENZENE**	1-METHYL-3,5-DIETHYLBENZENE
RR-00867-3 **METHYL DIHYDROABIETATE**	METHYL DIHYDROABIETATE
148-01-6 2-METHYL-3,5-DINITROBENZAMIDE [ONT, ONT]	DINITOLMIDE
606-20-2 1-METHYL-2,6-DINITROBENZENE [MSL]	2,6-DINITROTOLUENE
534-52-1 2-METHYL-4,6-DINITROPHENOL [GBR, GBR, GBR]	4,6-DINITRO-O-CRESOL
99-80-9 **N-METHYL-N,4-DINITROSOANILINE**	N-METHYL-N,4-DINITROSOANILINE
108-32-7 4-METHYL-1,3-DIOXOLANE-2-ONE [PST]	PROPYLENE CARBONATE
75790-84-0 4-METHYLDIPHENYLMETHANE-3,4-DIISOCYANATE [313]	BENZENE, 2-ISOCYANATO-4-[(4-ISOCYANATO PHENYL) METHYL]-1-METH-3-HYDROXY-2-CYCLOHEXEN-1-ONE]
624-92-0 METHYL DISULFIDE [EPA, EPA, MSL, PAL]	DIMETHYLDISULFIDE
555-30-6 **ALPHA-METHYLDOPA**	ALPHA-METHYLDOPA
41372-08-1 **METHYLDOPA SESQUIHYDRATE**	METHYLDOPA SESQUIHYDRATE
4016-14-2 ((1-METHYLETHOXY)METHYL)-OXIRANE [ONT, ONT]	ISOPROPYL GLYCIDYL ETHER (IGE)
75-09-2 METHYLENE CHLORIDE [QBC, QBC]	METHYLENE CHLORIDE
101-68-8 METHYLENE DIPHENYL DIISOCYANATE (4,4') (MDI) [HON, HON]	METHYLENE BISPHENOL ISOCYANATE (MDI)
101-68-8 METHYLENE-BIS (4-PHENYL ISOCYANATE) [QBC]	METHYLENE BISPHENOL ISOCYANATE (MDI)
74-99-7 METHYLENE ACETYLENE (PROPYNE) [ALB, ALB]	METHYL ACETYLENE
101-14-4 4,4'-METHYLENEBIS (2-CHLOROANILINE) (MBOCA) [ALB, ALB, ALB, ALB]	4,4'-METHYLENEBIS(2-CHLOROANILINE) (MBOCA)
101-14-4 4,4'-METHYLENEBIS (2-CHLOROANILINE) [CAN]	4,4'-METHYLENEBIS(2-CHLOROANILINE) (MBOCA)
101-68-8 METHYLENEBIS (PHENYLISOCYANATE) [CAN]	METHYLENE BISPHENOL ISOCYANATE (MDI)
119-47-1 2,2'-METHYLENEBIS(6-TERT-BUTYL-P-CRESOL) [PST]	2,2'-METHYLENEBIS(4-METHYL-6-TERT-BUTYLPHENOL)
101-14-4 4,4'-METHYLENE-BIS-(2-CHLOROANILINE) [ATS]	4,4'-METHYLENEBIS(2-CHLOROANILINE) (MBOCA)
101-14-4 4,4'-METHYLENE BIS(2-CHLOROANILINE) [AUS, AUS, AUS, C65, C65]	4,4'-METHYLENEBIS(2-CHLOROANILINE) (MBOCA)
101-14-4 4,4-METHYLENE BIS(2-CHLOROANILINE) [CAA]	4,4'-METHYLENEBIS(2-CHLOROANILINE) (MBOCA)

Chemical Name	Reference Name
101-14-4 4,4'-METHYLENEBIS(2-CHLOROANILINE) [CAL, CEX, CEX, EP2, EPA]	4,4'-METHYLENEBIS(2-CHLOROANILINE) (MBOCA)
101-14-4 4,4'-METHYLENE BIS(2-CHLOROANILINE) [FLA, MAK, MAK]	4,4'-METHYLENEBIS(2-CHLOROANILINE) (MBOCA)
101-14-4 4,4'-METHYLENEBIS(2-CHLOROANILINE) [MIN]	4,4'-METHYLENEBIS(2-CHLOROANILINE) (MBOCA)
101-14-4 4,4'-METHYLENE BIS(2-CHLOROANILINE) [MSL]	4,4'-METHYLENEBIS(2-CHLOROANILINE) (MBOCA)
101-14-4 4,4'-METHYLENEBIS (2-CHLOROANILINE) [NJL, NJL]	4,4'-METHYLENEBIS(2-CHLOROANILINE) (MBOCA)
101-14-4 4,4'-METHYLENEBIS(2-CHLOROANILINE) [RCA, RCU, RCU]	4,4'-METHYLENEBIS(2-CHLOROANILINE) (MBOCA)
101-14-4 4,4'-METHYLENEBIS(2-CHLOROANILINE) (MBOCA)	4,4'-METHYLENEBIS(2-CHLOROANILINE) (MBOCA)
101-14-4 4,4'-METHYLENEBIS(2-CHLOROANILINE) [BC1, BC1, BC1]	4,4'-METHYLENEBIS(2-CHLOROANILINE) (MBOCA)
101-14-4 4,4'-METHYLENE BIS(2-CHLOROANILINE) (MOCA) [IAR]	4,4'-METHYLENEBIS(2-CHLOROANILINE) (MBOCA)
101-14-4 4,4'-METHYLENE BIS (2-CHLOROANILINE) (MOCA) [NHS, NHS, NHS]	4,4'-METHYLENEBIS(2-CHLOROANILINE) (MBOCA)
101-14-4 4,4'-METHYLENEBIS(2-CHLOROANILINE) [ONT, ONT]	4,4'-METHYLENEBIS(2-CHLOROANILINE) (MBOCA)
101-14-4 4,4'-METHYLENE BIS (2-CHLOROANILINE) (MBOCA) [OS1, OS1]	4,4'-METHYLENEBIS(2-CHLOROANILINE) (MBOCA)
101-14-4 METHYLENE 4,4-BIS (2-CHLOROANILINE) (3, 3-DICHLORO-4,4-DIAMINOPHENYLMETHANE) [QBC, QBC, QBC]	4,4'-METHYLENEBIS(2-CHLOROANILINE) (MBOCA)
101-14-4 4,4'-METHYLENE BIS(2-CHLOROANILINE) [MOCA] [TLV, TLV, TLV]	4,4'-METHYLENEBIS(2-CHLOROANILINE) (MBOCA)
101-14-4 4,4'-METHYLENEBIS(2-CHLOROANILINE) [UTS]	4,4'-METHYLENEBIS(2-CHLOROANILINE) (MBOCA)
5124-30-1 METHYLENE BIS(4-CYCLOHEXYLISOCYANATE)	METHYLENE BIS(4-CYCLOHEXYLISOCYANATE)
5124-30-1 METHYLENE BIS-(4-CYCLOHEXYLISO-CYANATE) [MIN]	METHYLENE BIS(4-CYCLOHEXYLISOCYANATE)
101-61-1 4,4'-METHYLENE BIS(N,N-DIMETHYLANILINE) [MAK]	4,4'-METHYLENEBIS(N,N-DIMETHYL) BENZENAMINE
101-61-1 4,4'-METHYLENEBIS(N,N-DIMETHYLANILINE) [NJL]	4,4'-METHYLENEBIS(N,N-DIMETHYL) BENZENAMINE
101-61-1 4,4'-METHYLENEBIS(N,N-DIMETHYL) BENZENAMINE	4,4'-METHYLENEBIS(N,N-DIMETHYL) BENZENAMINE
101-61-1 4,4'-METHYLENE BIS(N,N-DIMETHYL)BENZENAMINE [CAL, FLA]	4,4'-METHYLENEBIS(N,N-DIMETHYL) BENZENAMINE
88-24-4 2,2'-METHYLENEBIS(4-ETHYL-6-TERT-BUTYLPHENOL)	2,2'-METHYLENEBIS(4-ETHYL-6-TERT-BUTYLPHENOL)
5124-30-1 1,1'-METHYLENE BIS(4-ISOCYANATOCYCLO-HEXANE) [313]	METHYLENE BIS(4-CYCLOHEXYLISOCYANATE)
26447-40-5 1,1'-METHYLENEBIS (ISOCYANATO-) BENZENE	1,1'-METHYLENEBIS (ISOCYANATO-) BENZENE
RR-00218-6 METHYLENEBIS(4-ISOCYANATOBENZENE), POLYMER WITH POLYCAPROLACTONE TRIOL AND ALKOXYLATED ALKANEPOLYOL, HYDROXYALKYL METHACRYLATE ESTER	METHYLENEBIS(4-ISOCYANATOBENZENE), POLYMER WITH POLYCAPROLACXYALKYL ACRYLATE ESTER, REACTION PRODUCTS WITH SUBSTITUTED ALKANOIC ACID AND METAL HETEROMONOCYCLE
838-88-0 4,4'-METHYLENE BIS(2-METHYLANILINE)	4,4'-METHYLENE BIS(2-METHYLANILINE)
838-88-0 4,4'-METHYLENEBIS(2-METHYLANILINE) [MIN]	4,4'-METHYLENE BIS(2-METHYLANILINE)
119-47-1 2,2'-METHYLENEBIS(4-METHYL-6-TERT-BUTYLPHENOL)	2,2'-METHYLENEBIS(4-METHYL-6-TERT-BUTYLPHENOL)
101-68-8 METHYLENE BISPHENOL ISOCYANATE (MDI)	METHYLENE BISPHENOL ISOCYANATE (MDI)
39817-09-9 2,2'-[METHYLENEBIS(PHENYLE-NEOXYMETHYL)]BISOXIRANE [TSC]	OXIRANE, 2,2'-[METHYLENEBIS(PHENYLE-NEOXYMETHYLENE)]BIS-

Chemical Name	Reference Name
101-68-8 METHYLENE BISPHENYL ISOCYANATE [MIN, NIO, NIO, NIO, NJL]	METHYLENE BISPHENOL ISOCYANATE (MDI)
101-68-8 METHYLENE-BIS(4-PHENYL ISOCYANATE) [CAL]	METHYLENE BISPHENOL ISOCYANATE (MDI)
101-68-8 METHYLENE BISPHENYL ISOCYANATE (MBI) [FLA]	METHYLENE BISPHENOL ISOCYANATE (MDI)
101-68-8 METHYLENE BISPHENYL ISOCYANATE (MDI) [TLV]	METHYLENE BISPHENOL ISOCYANATE (MDI)
101-68-8 METHYLENEBIS(PHENYLISOCYANATE) (MDI) [313]	METHYLENE BISPHENOL ISOCYANATE (MDI)MINE]
101-68-8 METHYLENE BIS(PHENYLISOCYANATE) [CEX]	METHYLENE BISPHENOL ISOCYANATE (MDI)
101-68-8 METHYLENE BISPHENYL ISOCYANATE [OTV]	METHYLENE BISPHENOL ISOCYANATE (MDI)
41123-59-5 1,1'-[METHYLENEBIS(SULFONYL)] BIS-2-CHLOROETHANE	1,1'-[METHYLENEBIS(SULFONYL)] BIS-2-CHLOROETHANE
41123-69-7 2,2'-[METHYLENEBIS(SULFONYL)] BISETHANOL	2,2'-[METHYLENEBIS(SULFONYL)] BISETHANOL
3278-22-6 1,1'-[METHYLENE BIS(SULFONYL)]BISETHENE	1,1'-[METHYLENE BIS(SULFONYL)]BISETHENE
6317-18-6 METHYLENE BIS(THIOCYANATE)	METHYLENE BIS(THIOCYANATE)
RR-00904-1 METHYLENEBISTRISUBSTITUTED ANILINE	METHYLENEBISTRISUBSTITUTED ANILINE
61-73-4 METHYLENE BLUE	METHYLENE BLUE
7220-79-3 METHYLENE BLUE TRIHYDRATE	METHYLENE BLUE TRIHYDRATE
74-95-3 METHYLENE BROMIDE	METHYLENE BROMIDE
75-09-2 METHYLENE CHLORIDE	METHYLENE CHLORIDE
75-09-2 METHYLENE CHLORIDE (DICHLOROMETHANE) [HON, HON, TLV, TLV, BC1, CAA, CW2]	METHYLENE CHLORIDE
110-26-9 METHYLENE DIACRYLAMIDE	METHYLENE DIACRYLAMIDE
101-77-9 4,4'-METHYLENEDIANILINE	4,4'-METHYLENEDIANILINE
101-77-9 4,4'-METHYLENE DIANILINE [AUS, AUS, AUS, CEX, CEX]	4,4'-METHYLENEDIANILINE
101-77-9 4,4-METHYLENE DIANILINE [FLA, MIN]	4,4'-METHYLENEDIANILINE
101-77-9 4,4'-METHYLENE DIANILINE [MSL]	4,4'-METHYLENEDIANILINE
101-77-9 METHYLENEDIANILINE [NFP, NFP]	4,4'-METHYLENEDIANILINE
101-77-9 4,4'-METHYLENE DIANILINE [NJL, OS1, TLV, TLV, TLV]	4,4'-METHYLENEDIANILINE
13552-44-8 4,4-METHYLENEDIANILINE AND ITS DIHYDROCHLORIDE [PAL, PAL]	4,4'-METHYLENEDIANILINE DIHYDROCHLORIDE
13552-44-8 4,4'-METHYLENEDIANILINE DIHYDROCHLORIDE	4,4'-METHYLENEDIANILINE DIHYDROCHLORIDE
RR-00870-8 METHYLENE DIISOCYANATE	METHYLENE DIISOCYANATE
4676-39-5 3,4-METHYLENEDIOXYPHENYL-2-PROPANONE	3,4-METHYLENEDIOXYPHENYL-2-PROPANONE
120-58-1 1,2-METHYLENEDIOXY-4-PROPENYL BENZENE [NJL]	ISOSAFROLE
94-58-6 1,2-METHYLENEDIOXY-4-PROPYL BENZENE [FLA, NJL]	DIHYDROSAFROLE
101-68-8 METHYLENE DIPHENYL DIISOCYANATE (MDI) [CAA]	METHYLENE BISPHENOL ISOCYANATE (MDI)
101-68-8 METHYLENEDIPHENYL DIISOCYANATE, 4,4'- [OTS]	METHYLENE BISPHENOL ISOCYANATE (MDI)
101-68-8 4,4'-METHYLENEDIPHENYL DIISOCYANATE [IAR]	METHYLENE BISPHENOL ISOCYANATE (MDI)
85095-61-0 N,N'-(METHYLENEDI-4,1-PHENYLENE)BIS[2,2-DIMETHYL-	N,N'-(METHYLENEDI-4,1-PHENYLENE)BIS[2,2-DIMETHYL-HYL ESTER
38483-28-2 METHYLENE GLYCOL DINITRATE	METHYLENE GLYCOL DINITRATE

Chemical Name	Reference Name
552-79-4 N-METHYLEPHEDRINE	N-METHYLEPHEDRINE
RR-01785-6 N-METHYLEPHEDRINE SALTS, OPTICAL ISO-MERS, AND SALTS OF OPTICAL ISOMERS	N-METHYLEPHEDRINE SALTS, OPTICAL ISOMERS, AND SALTS OF OPTICISOMERS
68298-14-6 METHYL EPOXYSTEARATE AND TETRAETHY-LENE PENTAMINE(TEPA), REACTION PROD-UCTS	METHYL EPOXYSTEARATE AND TETRAETHYLENE PENTAMINE(TEPA), REAC
79-20-9 METHYL ESTER ACETIC ACID [FLA]	METHYL ACETATE
96-34-4 METHYL ESTER CHLOROACETIC ACID [FLA]	METHYL CHLOROACETATE
111-11-5 METHYL ESTER OCTANOIC ACID	METHYL ESTER OCTANOIC ACID
8050-13-3 METHYL ESTER OF ROSIN, PARTIALLY HY-DROGENATED	METHYL ESTER OF ROSIN, PARTIALLY HYDRO-GENATED
61788-60-1 METHYL ESTERS OF COTTONSEED OIL	METHYL ESTERS OF COTTONSEED OIL
80-48-8 METHYL ESTER P-TOLUENE SULFONIC ACID [FLA]	METHYL-P-TOLUENESULFONATE
109-83-1 N-METHYLETHANOLAMINE	N-METHYLETHANOLAMINE
115-10-6 METHYL ETHER [FLA, MSL, NFP, NFP]	DIMETHYL ETHER
115-10-6 METHYL ETHER [METHANE, OXYBIS-] [EP3]	DIMETHYL ETHER
126-39-6 2-METHYL-2-ETHYL-1,3-DIOXOLANE	2-METHYL-2-ETHYL-1,3-DIOXOLANE
RR-01716-3 METHYL ETHYLENE GLYCOL ETHERS AND ESTERS	METHYL ETHYLENE GLYCOL ETHERS AND ESTERS
540-67-0 METHYL ETHYL ETHER	METHYL ETHYL ETHER
3074-75-7 2-METHYL-4-ETHYLHEXANE	2-METHYL-4-ETHYLHEXANE
3074-77-9 3-METHYL-4-ETHYLHEXANE	3-METHYL-4-ETHYLHEXANE
72319-24-5 2,2'-[(1-METHYLETHYLIDENE)BIS(4,1-PHENYLE-NEOXY-3,1-PROPANEDIOXY [OTS]	OXIRANE, 2,2'-[(1-METHYLETHYLIDINE)BIS[4,1-PHENYLENEOXY-3,1-
RR-01221-5 2,2'-[(1-METHYLETHYLIDENE)BIS[4,1-PHENY-LOXY[1-(BUTOXYMETHYL)-[2,1-ETHANEDIYL]OXYMETHYLENE]BISOXIRANE, REACTION PRODUCT WITH A DIAMINE	2,2'-[(1-METHYLETHYLIDENE)BIS[4,1-PHENYLOXY[1-(BUTOXYMETHYL)
72319-24-5 2,2'-[(1-METHYLETHYLIDINE)BIS[4,1-PHENYLE-NEOXY-3,1-PROPANEDIYLOXY-4,1-PHENY-LENE(1-METHYLETHYLIDENE)-4,1- PHENYLE-NEOXYMETHYLENE]BIS(OXIRANE) [EPA]	OXIRANE, 2,2'-[(1-METHYLETHYLIDINE)BIS[4,1-PHENYLENEOXY-3,1-
78-93-3 METHYL ETHYL KETONE	METHYL ETHYL KETONE
78-93-3 METHYL ETHYL KETONE (2-BUTANONE) [CAA]	METHYL ETHYL KETONE
78-93-3 METHYL ETHYL KETONE (MEK) [FLA, MSL, QBC, QBC, RCU, RCU, TLV, TLV, WHM, AUS, AUS]	METHYL ETHYL KETONE
1338-23-4 METHYL ETHYL KETONE PEROXIDE	METHYL ETHYL KETONE PEROXIDE
1338-23-4 METHYL ETHYL KETONE PEROXIDE (MEKP) [OS1]	METHYL ETHYL KETONE PEROXIDE
1338-23-4 METHYL ETHYL KETONE PEROXIDES (MEKP) [GBR]	METHYL ETHYL KETONE PEROXIDE
1338-23-4 METHYL ETHYL KETONE PEROXIDES [ONT]	METHYL ETHYL KETONE PEROXIDE
96-29-7 METHYL ETHYL KETOXIME	METHYL ETHYL KETOXIME
609-26-7 2-METHYL-3-ETHYLPENTANE	2-METHYL-3-ETHYLPENTANE
104-89-2 2-METHYL-5-ETHYLPIPERIDINE	2-METHYL-5-ETHYLPIPERIDINE
104-90-5 2-METHYL-5-ETHYLPYRIDINE	2-METHYL-5-ETHYLPYRIDINE
93-15-2 METHYLEUGENOL	METHYLEUGENOL
93-15-2 METHYL EUGENOL [NFP, NFP]	METHYLEUGENOL
1706-01-0 3-METHYLFLUORANTHENE	3-METHYLFLUORANTHENE
33543-31-6 2-METHYLFLUORANTHENE	2-METHYLFLUORANTHENEISOOCTYL ESTER
593-53-3 METHYL FLUORIDE	METHYL FLUORIDE
453-18-9 METHYL FLUOROACETATE	METHYL FLUOROACETATE

Chemical Name	Reference Name
421-20-5 METHYL FLUOROSULFATE	METHYL FLUOROSULFATE
123-39-7 N-METHYLFORMAMIDE	N-METHYLFORMAMIDE
107-31-3 METHYL FORMATE	METHYL FORMATE
107-31-3 METHYL FORMATE [FORMIC ACID, METHYL ESTER] [EP3]	METHYL FORMATE
534-22-5 2-METHYLFURAN	2-METHYLFURAN
534-22-5 2-METHYL FURAN [FLA]	2-METHYLFURAN
534-22-5 METHYL FURAN [WHM]	2-METHYLFURAN
611-13-2 METHYL-2-FUROATE	METHYL-2-FUROATE
13225-10-0 A-METHYLGLUCOSIDE TETRANITRATE	A-METHYLGLUCOSIDE TETRANITRATE
84002-64-2 A-METHYLGLYCEROL TRINITRATE	A-METHYLGLYCEROL TRINITRATE
930-37-0 METHYL GLYCIDYL ETHER [EPA, OTS]	OXIRANE, METHOXYMETHYL-
4316-73-8 N-METHYLGLYCINE, SODIUM SALT	N-METHYLGLYCINE, SODIUM SALT
78-98-8 METHYLGLYOXAL [IAR]	PYRUVIC ALDEHYDE
2724-58-5 16-METHYLHEPTADECANOIC ACID	16-METHYLHEPTADECANOIC ACID
RR-00868-4 METHYL HEPTADECYL KETONE	METHYL HEPTADECYL KETONE
541-85-5 5-METHYLHELPTAN-3-ONE [GBR]	ETHYL AMYL KETONE
541-85-5 5-METHYL-3-HEPTANONE [NIO, NIO, NIO, ONT]	ETHYL AMYL KETONE
409-02-9 METHYLHEPTENONE	METHYLHEPTENONE
111-12-6 METHYL HEPTINE CARBONATE	METHYL HEPTINE CARBONATE
68134-07-6 6-METHYLHEPTYL GLYCIDYL ETHER	6-METHYLHEPTYL GLYCIDYL ETHER
68134-07-6 METHYLHEPTYL GLYCIDYL ETHER, 6- [OTS]	6-METHYLHEPTYL GLYCIDYL ETHER
821-55-6 METHYL HEPTYL KETONE [NFP, NFP]	2-NONANONE
589-34-4 3-METHYLHEXANE	3-METHYLHEXANE
591-76-4 2-METHYLHEXANE [NFP, NFP]	ISOHEPTANE
106-70-7 METHYL HEXANOATE	METHYL HEXANOATE
110-12-3 5-METHYLHEXAN-2-ONE [GBR, GBR, HM1, HM1]	METHYL ISOAMYL KETONE
110-12-3 5-METHYL-2-HEXANONE [ONT, PST]	METHYL ISOAMYL KETONE
110-12-3 5-METHYLHEXAN-2-ONE [WHM]	METHYL ISOAMYL KETONE
111-13-7 METHYL HEXYL KETONE [NFP, NFP]	2-OCTANONE
60-34-4 METHYL HYDRAZINE	METHYL HYDRAZINE
60-34-4 METHYLHYDRAZINE [C65, HM1, HM1, HM1, HM1]	METHYL HYDRAZINE
60-34-4 METHYL HYDRAZINE [MEX, MEX, MEX, MEX]	METHYL HYDRAZINE
502-39-6 METHYLHYDRAZINE [MSL, NFP, NFP, NHS, NHS, NHS, ONT, ONT, WHM, HM1, HM1]	METHYL MERCURY DICYANDIAMIDE
60-34-4 METHYL HYDRAZINE (MONOMETHYL HYDRAZINE) [OS1, OS1, OS1, OS1]	METHYL HYDRAZINE
60-34-4 METHYL HYDRAZINE [HYDRAZINE, METHYL-] [EP3]	METHYL HYDRAZINE
99-76-3 METHYL P-HYDROXYBENZOATE	METHYL P-HYDROXYBENZOATE
1487-49-6 METHYL-3-HYDROXYBUTYRATE	METHYL-3-HYDROXYBUTYRATE
616-47-7 1-METHYLIMIDAZOLE	1-METHYLIMIDAZOLE
693-98-1 2-METHYLIMIDAZOLE	2-METHYLIMIDAZOLE
822-36-6 4-METHYLIMIDAZOLE	4-METHYLIMIDAZOLE
68140-76-1 1,1'-(METHYLIMINO)BIS(3-CHLORO-2-PROPANOL), POLYMER WITH N,N,N',N'-TETRAMETHYL-1,2-ETHANEDIAMINE	1,1'-(METHYLIMINO)BIS(3-CHLORO-2-PROPANOL), POLYMER WITH N,N
83-34-1 3-METHYLINDOLE	3-METHYLINDOLE

Chemical Name	Reference Name
74-88-4 METHYL IODIDE	METHYL IODIDE
74-88-4 METHYL IODIDE (IODOMETHANE) [CAA]	METHYL IODIDE
74-88-4 METHYL IODINE [FLA]	METHYL IODIDE
1335-94-0 METHYL IONONE	METHYL IONONE
79-69-6 METHYLIONONES (A-)	METHYLIONONES (A-)
110-12-3 METHYL ISOAMYL KETONE	METHYL ISOAMYL KETONE
108-11-2 METHYL ISOBUTYL CARBINOL	METHYL ISOBUTYL CARBINOL
108-11-2 METHYL ISOBUTYL CARBINOL (MIBC) [WHM]	METHYL ISOBUTYL CARBINOL
108-10-1 METHYL ISOBUTYL KETONE	METHYL ISOBUTYL KETONE
108-10-1 METHYL ISOBUTYL KETONE (MIBK) [BEI]	METHYL ISOBUTYL KETONE
108-10-1 METHYL ISOBUTYL KETONE (HEXONE) [HON, HON]	METHYL ISOBUTYL KETONE
108-10-1 METHYLISOBUTYL KETONE [MSL]	METHYL ISOBUTYL KETONE
108-10-1 METHYL ISOBUTYL KETONE (HEXONE) [CAA]	METHYL ISOBUTYL KETONE
37206-20-5 METHYL ISOBUTYL KETONE PEROXIDE	METHYL ISOBUTYL KETONE PEROXIDE
547-63-7 METHYL ISOBUTYRATE	METHYL ISOBUTYRATE
624-83-9 METHYL ISOCYANATE	METHYL ISOCYANATE
624-83-9 METHYL ISOCYANATE (M.I.C.) [FLA]	METHYL ISOCYANATE
624-83-9 METHYL ISOCYANATE [METHANE, ISO-CYANATO-] [EP3]	METHYL ISOCYANATEO-]
93-16-3 METHYL ISO EUGENOL	METHYL ISO EUGENOL
814-78-8 METHYL ISOPROPENYL KETONE	METHYL ISOPROPENYL KETONE
814-78-8 METHYL ISOPROPENYL KETONE, INHIB-ITED [NJL]	METHYL ISOPROPENYL KETONE
563-80-4 METHYL ISOPROPYL KETONE	METHYL ISOPROPYL KETONE
2682-20-4 2-METHYL-3-ISOTHIAZOLONE	2-METHYL-3-ISOTHIAZOLONE
556-61-6 METHYL ISOTHIOCYANATE	METHYL ISOTHIOCYANATE
556-61-6 METHYL ISOTHIOCYANATE [ISOTHIO-CYANATOMETHANE] [313]	METHYL ISOTHIOCYANATE
556-24-1 METHYL ISOVALERATE	METHYL ISOVALERATE
547-64-8 METHYL LACTATE	METHYL LACTATE
75-86-5 2-METHYLLACTONITRILE [313, RCU, RCU]	ACETONE CYANOHYDRIN
75-16-1 METHYL MAGNESIUM BROMIDE	METHYL MAGNESIUM BROMIDE
74-93-1 METHYL MERCAPTAN	METHYL MERCAPTAN
74-93-1 METHYLMERCAPTAN [EPA]	METHYL MERCAPTAN
74-93-1 METHYL MERCAPTAN [METHANETHIOL] [EP3]	METHYL MERCAPTAN
3268-49-3 BETA-METHYL MERCAPTOPROPIONALDE-HYDE [NFP, NFP]	4-THIAPENTANAL
342-69-8 6-METHYLMERCAPTOPURINE RIBONUCLEO-SIDE	6-METHYLMERCAPTOPURINE RIBONUCLEOSIDE
502-39-6 METHYLMERCURIC DICYANAMIDE [302, MSL, PAL]	METHYL MERCURY DICYANDIAMIDE
502-39-6 METHYLMERCURIC DICYANDIAMIDE [FLA]	METHYL MERCURY DICYANDIAMIDE
593-74-8 METHYL MERCURY [CAB]	DIMETHYLMERCURY
22967-92-6 METHYLMERCURY	METHYLMERCURY
22967-92-6 METHYL MERCURY [C65, MAK, MAK, MAK, MAK]	METHYLMERCURY
502-39-6 METHYL MERCURY DICYANDIAMIDE	METHYL MERCURY DICYANDIAMIDE
80-62-6 METHYL METHACRYLATE	METHYL METHACRYLATE
25035-69-2 METHYL METHACRYLATE-BUTYL ACRYLATE-METHACRYLIC ACID POLYMER	METHYL METHACRYLATE-BUTYL ACRYLATE-METHACRYLIC ACID POLYMER

Chemical Name	Reference Name
9011-14-7 METHYL METHACRYLATE POLYMER	METHYL METHACRYLATE POLYMER
66-27-3 METHYL METHANESULFONATE	METHYL METHANESULFONATE
66-27-3 METHYL METHANE SULFONATE [WHM]	METHYL METHANESULFONATE
66-27-3 METHYL METHANESULPHONATE [IAR, MIN, NTP]	METHYL METHANESULFONATE
16752-77-5 S-METHYL N-[(METHYLCARBAMOYL)OXY]-THIOACETAMIDATE (METHOMYL) [CAL]	METHOMYL
109-02-4 METHYLMORPHOLINE	METHYLMORPHOLINE
109-02-4 4-METHYLMORPHOLINE [FLA, MSL, NFP, NFP]	METHYLMORPHOLINE
90-12-0 1-METHYLNAPHTHALENE	1-METHYLNAPHTHALENE
91-57-6 2-METHYLNAPHTHALENE	2-METHYLNAPHTHALENE
1321-94-4 METHYLNAPHTHALENE	METHYLNAPHTHALENE
1321-94-4 METHYLNAPHTHALENES [HM1]	METHYLNAPHTHALENE
26264-58-4 METHYLNAPHTHALENESULFONIC ACID, SODIUM SALT [PST]	NAPHTHALENESULFONIC ACID, METHYL-, SODIUM SALT
591-78-6 METHYL-N-BUTYL KETONE [QBC, QBC]	2-HEXANONE
RR-01440-4 METHYL NITRAMINE	METHYL NITRAMINE
598-58-3 METHYLNITRATE	METHYLNITRATE
624-91-9 METHYL NITRITE	METHYL NITRITE
129-15-7 2-METHYL-1-NITROANTHRAQUINONE	2-METHYL-1-NITROANTHRAQUINONE
129-15-7 2-METHYL-1-NITROANTHRAQUINONE (UNCERTAIN PURITY) [IAR]	2-METHYL-1-NITROANTHRAQUINONE
70-25-7 N-METHYL-N'-NITRO-N-NITROSOGUANIDINE	N-METHYL-N'-NITRO-N-NITROSOGUANIDINE
70-25-7 N-METHYL-N'-NITRO-N-NITROSO-GUANIDINE [CAL]	N-METHYL-N'-NITRO-N-NITROSOGUANIDINE
70-25-7 1-METHYL-3-NITRO-1-NITROSO-GUANIDINE [NJL, NJL]	N-METHYL-N'-NITRO-N-NITROSOGUANIDINE
70-25-7 N-METHYL-N'-NITRO-N-NITROSOGUANIDINE (MNNG) [IAR]	N-METHYL-N'-NITRO-N-NITROSOGUANIDINE
119-33-5 4-METHYL-2-NITRO-PHENOL	4-METHYL-2-NITRO-PHENOL
119-33-5 4-METHYL-2-NITROPHENOL [TSC, TSC, TSC]	4-METHYL-2-NITRO-PHENOLHEXACHLORO-
2581-34-2 3-METHYL-4-NITROPHENOL	3-METHYL-4-NITROPHENOL
RR-01302-5 METHYLNITROPHENOLS	METHYLNITROPHENOLS
60153-49-3 3-METHYLNITROSAMINOPROPIONITRILE [PAL, PAL]	3-METHYLNITROSOAMINOPROPIONITRILE
60153-49-3 3-METHYLNITROSAMINO PROPIONITRILE [MSL]	3-METHYLNITROSOAMINOPROPIONITRILE
64091-91-4 4-(METHYLNITROSAMINO)-1-(3-PYRIDYL)-1-BUTANONE [MSL]	4-(N-NITROSOMETHYLAMINO)-1-(3-PYRIDYL)-1-BUTANONE
60153-49-3 3-METHYLNITROSOAMINOPROPIONITRILE	3-METHYLNITROSOAMINOPROPIONITRILE
80-11-5 N-METHYL-N-NITROSO-P-TOLUENESULFONAMIDE	N-METHYL-N-NITROSO-P-TOLUENESULFONAMIDE
684-93-5 N-METHYL-N-NITROSOUREA [IAR]	N-NITROSO-N-METHYLUREA
615-53-2 N-METHYL-N-NITROSOURETHANE [IAR, MIN]	N-NITROSO-N-METHYLURETHANE
112-12-9 METHYL NONYL KETONE [NFP, NFP]	2-UNDECANONE
924-42-5 N-METHYLOACRYLAMIDE [C65]	N-METHYLOLACRYLAMIDE
2216-33-3 3-METHYLOCTANE	3-METHYLOCTANE
2216-34-4 4-METHYLOCTANE	4-METHYLOCTANE
3221-61-2 2-METHYLOCTANE	2-METHYLOCTANE
5340-36-3 3-METHYL-3-OCTANOL	3-METHYL-3-OCTANOL
924-42-5 N-METHYLOLACRYLAMIDE	N-METHYLOLACRYLAMIDE
112-62-9 METHYL OLEATE [PST]	EMERY

Chemical Name	Reference Name
110-25-8 N-METHYL-N-OLEOYLGLYCINE [PST]	OLEOYL SARCOSINE
137-20-2 N-METHYL-N-OLEOYLTAURINE, SODIUM SALT [PST, TSC, TSC, TSC]	SODIUM N-METHYL-N-OLEOYLTAURINE
3188-83-8 2-METHYLOL-4,4'-ISOPROPYLIDENEDIPHENOL DIGLYCIDYL ETHER	2-METHYLOL-4,4'-ISOPROPYLIDENEDIPHENOL DIGLY-CIDYL ETHER
1000-82-4 METHYLOLUREA	METHYLOLUREA
1000-82-4 METHYLOL UREA [OTS]	METHYLOLUREA
1000-82-4 METHYLOLUREA (HYDROXYMETHYL UREA) [TSC, TSC, TSC]	METHYLOLUREABIPHENYL]-4,4'-DIYL)BIS(AZO)]BIS[4-AMINO-, DISODIUM SALT
149-73-5 METHYL ORTHOFORMATE	METHYL ORTHOFORMATE
681-84-5 METHYL ORTHOSILICATE [HM1, HM1, HM1]	METHYL SILICATE
37311-02-7 METHYL OXIRANE POLYMER WITH OXIRANE, MONOOCTYL ETHER	METHYL OXIRANE POLYMER WITH OXIRANE, MONOOCTYL ETHER
61827-84-7 METHYL OXIRANE POLYMER WITH OXIRANE, OCTYL ETHER	METHYL OXIRANE POLYMER WITH OXIRANE, OCTYL ETHER
112-39-0 METHYL PALMITATE	METHYL PALMITATE
3737-55-1 N-METHYL-N-PALMITOYLTAURINE, SODIUM SALT	N-METHYL-N-PALMITOYLTAURINE, SODIUM SALT
298-00-0 METHYL PARATHION	METHYL PARATHION
RR-00869-5 METHYL PENTADECYL KETONE	METHYL PENTADECYL KETONE
926-56-7 4-METHYL-1,3-PENTADIENE	4-METHYL-1,3-PENTADIENE
1118-58-7 2-METHYL-1,3-PENTADIENE	2-METHYL-1,3-PENTADIENE
54363-49-4 METHYLPENTADIENE	METHYLPENTADIENE
39382-31-5 METHYLPENTADIENES	METHYLPENTADIENES
73513-30-1 METHYLPENTALDEHYDE	METHYLPENTALDEHYDE
96-14-0 3-METHYLPENTANE	3-METHYLPENTANE
107-83-5 2-METHYLPENTANE [NFP, NFP, WHM]	ISOHEXANE
43133-95-5 METHYLPENTANE	METHYLPENTANE
107-41-5 2-METHYLPENTANE-2,4-DIOL [GBR, GBR]	HEXYLENE GLYCOL
107-41-5 2-METHYL-2,4-PENTANEDIOL [NFP, NFP]	HEXYLENE GLYCOL
149-31-5 2-METHYL-1,3-PENTANEDIOL	2-METHYL-1,3-PENTANEDIOL
97-61-0 2-METHYLPENTANOIC ACID	2-METHYLPENTANOIC ACID
105-30-6 2-METHYL-1-PENTANOL	2-METHYL-1-PENTANOL
108-11-2 4-METHYLPENTAN-2-OL [GBR, GBR, GBR]	METHYL ISOBUTYL CARBINOL
108-11-2 4-METHYL-2-PENTANOL [MAK, MAK, MAK, ONT, ONT, ONT]	METHYL ISOBUTYL CARBINOL
590-36-3 2-METHYL-2-PENTANOL	2-METHYL-2-PENTANOL
590-36-3 2-METHYLPENTAN-2-OL [HM1, HM1]	2-METHYL-2-PENTANOL
1320-98-5 4-METHYL-2-PENTANOL	4-METHYL-2-PENTANOL
108-10-1 4-METHYLPENTAN-2-ONE [GBR, GBR, GBR]	METHYL ISOBUTYL KETONE
108-10-1 4-METHYL-2-PENTANONE [ONT, ONT]	METHYL ISOBUTYL KETONE
625-27-4 2-METHYL-2-PENTENE	2-METHYL-2-PENTENE
691-37-2 4-METHYL-1-PENTENE	4-METHYL-1-PENTENE
763-29-1 2-METHYL-1-PENTENE	2-METHYL-1-PENTENE
4461-48-7 4-METHYL-2-PENTENE	4-METHYL-2-PENTENE
141-79-7 4-METHYLPENT-3-EN-2-ONE [GBR, GBR]	MESITYL OXIDE
77-75-8 METHYLPENTYLNOL [HON]	1-PENTYN-3-OL, 3-METHYL-
77-75-8 3-METHYL-1-PENTYNOL [NFP, NFP]	1-PENTYN-3-OL, 3-METHYL-
832-69-9 1-METHYLPHENANTHRENE	1-METHYLPHENANTHRENE
298-59-9 METHYLPHENIDATE HYDROCHLORIDE	METHYLPHENIDATE HYDROCHLORIDE
3735-23-7 METHYL PHENKAPTON	METHYL PHENKAPTON

Chemical Name	Reference Name
95-48-7 2-METHYL PHENOL [FLA]	O-CRESOL
101-41-7 METHYL PHENYLACETATE	METHYL PHENYLACETATE
98-85-1 METHYLPHENYL CARBINOL [NFP, NFP]	METHYL PHENYL CARBINOL
50373-55-2 METHYL PHENYL CARBINYL ACETATE	METHYL PHENYL CARBINYL ACETATE
149-74-6 METHYLPHENYLDICHLOROSILANE	METHYLPHENYLDICHLOROSILANE
28005-74-5 2,2'-[(2-METHYLPHENYL)IMINO]BISETHANOL	2,2'-[(2-METHYLPHENYL)IMINO]BISETHANOL
98-06-6 2-METHYL-2-PHENYLPROPANE [HM1]	TERT-BUTYLBENZENE
89-25-8 3-METHYL-1-PHENYL-2-PYRAZOLIN-5-ONE	3-METHYL-1-PHENYL-2-PYRAZOLIN-5-ONE
676-97-1 METHYL PHOSPHONIC DICHLORIDE	METHYL PHOSPHONIC DICHLORIDE
676-97-1 METHYLPHOSPHONIC DICHLORIDE [EPA, EPA]	METHYL PHOSPHONIC DICHLORIDE
676-99-3 METHYL PHOSPHONIC DIFLUORIDE	METHYL PHOSPHONIC DIFLUORIDE
676-98-2 METHYL PHOSPHONOTHIOIC DICHLORIDE	METHYL PHOSPHONOTHIOIC DICHLORIDE
676-83-5 METHYL PHOSPHONOUS DICHLORIDE	METHYL PHOSPHONOUS DICHLORIDE
85-71-2 METHYL PHTHALYL ETHYL GLYCOLATE	METHYL PHTHALYL ETHYL GLYCOLATE
109-01-3 1-METHYL PIPERAZINE	1-METHYL PIPERAZINE
626-67-5 1-METHYLPIPERIDINE	1-METHYLPIPERIDINE
RR-01573-6 METHYLPOLYCHLORO ALIPHATIC KETONE	METHYLPOLYCHLORO ALIPHATIC KETONE
9004-73-3 METHYLPOLYSILOXANE	METHYLPOLYSILOXANE
78-82-0 2-METHYLPROPANENITRILE [FLA]	ISOBUTYRONITRILE
75-66-1 2-METHYL-2-PROPANETHIOL	2-METHYL-2-PROPANETHIOL
79-31-2 METHYLPROPANOIC ACID, 2- [OTS]	ISOBUTYRIC ACID
79-31-2 2-METHYLPROPANOIC ACID [TSC]	ISOBUTYRIC ACID
75-65-0 2-METHYLPROPAN-2-OL [GBR, GBR]	TERT-BUTYL ALCOHOL
78-83-1 2-METHYLPROPAN-1-OL [GBR, GBR]	ISOBUTYL ALCOHOL
78-85-3 2-METHYLPROPENAL [MSL, NFP, NFP]	METHACRYLALDEHYDE
115-11-7 2-METHYLPROPENE [FLA, MSL, NFP, NFP, OTS]	ISOBUTYLENE
10476-95-6 2-METHYL-2-PROPENE-1,1-DIOL DIACETATE [WHM]	METHACROLEIN DIACETATE
126-98-7 2-METHYL-2-PROPENENITRILE [ONT, ONT]	METHYLACRYLONITRILE
115-11-7 2-METHYLPROPENE [1-PROPENE, 2-METHYL-] [EP3]	ISOBUTYLENE
554-12-1 METHYL PROPIONATE	METHYL PROPIONATE
7795-91-7 2-(1-METHYLPROPOXY) ETHANOL	2-(1-METHYLPROPOXY) ETHANOL
764-35-2 METHYL PROPYL ACETYLENE	METHYL PROPYL ACETYLENE
557-17-5 METHYL PROPYL ETHER	METHYL PROPYL ETHER
557-17-5 METHYL-N-PROPYL ETHER [FLA, MSL]	METHYL PROPYL ETHER
557-17-5 METHYL N-PROPYL ETHER [NFP, NFP]	METHYL PROPYL ETHER
107-87-9 METHYL PROPYL KETONE	METHYL PROPYL KETONE
107-87-9 METHYL N-PROPYL KETONE [NHS, NHS, PST, WHM]	METHYL PROPYL KETONE
107-87-9 METHYL PROPYL KETONE (2-PENTANONE) [ALB, ALB]	METHYL PROPYL KETONE
14222-20-9 N-METHYLPSEUDOEPHEDRINE	N-METHYLPSEUDOEPHEDRINE
RR-01787-8 N-METHYLPSEUDOEPHEDRINE SALTS, OPTICAL ISOMERS, AND SALTS OFOPTICAL ISOMERS	N-METHYLPSEUDOEPHEDRINE SALTS, OPTICAL ISOMERS, AND SALTS OFL ISOMERS
109-08-0 2-METHYLPYRAZINE	2-METHYLPYRAZINE
108-99-6 3-METHYLPYRIDINE	3-METHYLPYRIDINE
108-99-6 3-METHYL PYRIDINE [HM1]	3-METHYLPYRIDINE
109-06-8 2-METHYLPYRIDINE [313]	2-PICOLINE

Chemical Name	Reference Name
RR-01536-1 7-METHYLPYRIDO[3,4-C]PSORALEN	7-METHYLPYRIDO[3,4-C]PSORALEN
96-54-8 1-METHYLPYRROLE	1-METHYLPYRROLE
96-54-8 METHYLPYRROLE [MSL, NFP, NFP]	1-METHYLPYRROLE
120-94-5 METHYLPYRROLIDINE	METHYLPYRROLIDINE
120-94-5 N-METHYLPYRROLIDINE [PST]	METHYLPYRROLIDINE
872-50-4 N-METHYL-2-PYRROLIDINONE [313, PST]	1-METHYL-2-PYRROLIDONE
872-50-4 N-METHYLPYRROLIDINONE [TSC, TSE]	1-METHYL-2-PYRROLIDONE
872-50-4 1-METHYL-2-PYRROLIDONE	1-METHYL-2-PYRROLIDONE
872-50-4 1-METHYL-2-PYRROLIDONE [MSL, NFP, NFP]	1-METHYL-2-PYRROLIDONE
872-50-4 METHYLPYRROLIDONE, N- [OTS]	1-METHYL-2-PYRROLIDONE
493-52-7 METHYL RED	METHYL RED
119-36-8 METHYL SALICYLATE	METHYL SALICYLATE
144-34-3 METHYL SELENAC	METHYL SELENAC
593-79-3 METHYL SELENIDE	METHYL SELENIDE
681-84-5 METHYL SILICATE	METHYL SILICATE
112-61-8 METHYL STEARATE	METHYL STEARATE
98-83-9 ALPHA-METHYL STYRENE	ALPHA-METHYL STYRENE
98-83-9 A-METHYL STYRENE [AUS, AUS]	ALPHA-METHYL STYRENE
98-83-9 ALPHA-METHYLSTYRENE [CAL, CEX, CEX, MSL, NFP, NFP, ONT, ONT, WHM]	ALPHA-METHYL STYRENE
25013-15-4 METHYL STYRENE (ALL ISOMERS) [MAK, MAK]	VINYL TOLUENE
25013-15-4 METHYLSTYRENES [HM1, HM1, HM1]	VINYL TOLUENE
25013-15-4 METHYLSTYRENES, ALL ISOMERS [GBR, GBR]	VINYL TOLUENE
512-42-5 METHYL SULFATE, SODIUM SALT	METHYL SULFATE, SODIUM SALT
17557-67-4 6-(METHYLSULFONYL)-2-BENZOTHIAZO-LAMINE	6-(METHYLSULFONYL)-2-BENZOTHIAZOLAMINE
74499-22-2 METHYL TALLATE	METHYL TALLATE
68413-04-7 METHYL, TALLOW DIETHYLENETRIAMINE CONDENSATE, PROPOXYLATED,METHYL SULFATE	METHYL, TALLOW DIETHYLENETRIAMINE CONDEN-SATE, PROPOXYLATED,AMMONIO]ETHYL]-OMEGA-HY-DROXY-, N,N'-DITALLOW ACYL DERIVATIVES, METHYL SULFATES (SALTS)
1634-04-4 METHYLTERTBUTYL ETHER [HM1, HM1]	METHYL TERT-BUTYL ETHER
58-18-4 METHYLTESTOSTERONE	METHYLTESTOSTERONE
58-18-4 METHYL TESTOSTERONE [CAL]	METHYLTESTOSTERONE
124-10-7 METHYL TETRADECANOATE	METHYL TETRADECANOATE
96-47-9 METHYLTETRAHYDROFURAN	METHYLTETRAHYDROFURAN
96-47-9 2-METHYLTETRAHYDROFURAN [FLA, MSL, NFP, NFP]	METHYLTETRAHYDROFURAN
25265-68-3 METHYL TETRAHYDROFURAN	METHYL TETRAHYDROFURAN
479-45-8 N-METHYL-N,2,4,6-TETRANITROANILINE [GBR, GBR, GBR]	TETRYL
556-64-9 METHYL THIOCYANATE	METHYL THIOCYANATE
556-64-9 METHYL THIOCYANATE [THIOCYANIC ACID, METHYL ESTER] [EP3]	METHYL THIOCYANATE
56-04-2 METHYLTHIOURACIL	METHYLTHIOURACIL
56-04-2 6-METHYL-2-THIOURACIL [NJL, NJL]	METHYLTHIOURACIL
80-48-8 METHYL-P-TOLUENESULFONATE	METHYL-P-TOLUENESULFONATE
80-48-8 METHYL TOLUENE SULFONATE [MSL, NFP, NFP]	METHYL-P-TOLUENESULFONATE
5137-55-3 METHYLTRICAPRYLAMMONIUM CHLORIDE	METHYLTRICAPRYLAMMONIUM CHLORIDE
598-99-2 METHYL TRICHLOROACETATE	METHYL TRICHLOROACETATE

Chemical Name	Reference Name
75-79-6 METHYL TRICHLOROSILANE	METHYL TRICHLOROSILANE
75-79-6 METHYLTRICHLOROSILANE [302, 313, FLA, MSL, NFP, NFP, OS3, WEL, WHM]	METHYL TRICHLOROSILANE
75-79-6 METHYLTRICHLOROSILANE [SILANE, TRICHLOROMETHYL-] [EP3]	METHYL TRICHLOROSILANE
2031-67-6 METHYLTRIETHOXYSILANE	METHYLTRIETHOXYSILANE
1185-55-3 METHYLTRIMETHOXYSILANE	METHYLTRIMETHOXYSILANE
RR-01442-6 METHYL TRIMETHYLOL METHANE TRINITRATE	METHYL TRIMETHYLOL METHANE TRINITRATE
118-96-7 1-METHYL-2,4,6-TRINITROBENZENE [ONT, ONT]	2,4,6-TRINITROTOLUENE
7332-32-3 METHYLTRITHION	METHYLTRITHION
5760-50-9 METHYL UNDECYLENATE	METHYL UNDECYLENATE
593-08-8 METHYL UNDECYL KETONE	METHYL UNDECYL KETONE
615-53-2 METHYL URETHANE [NJL]	N-NITROSO-N-METHYLURETHANE
123-15-9 2-METHYL VALERALDEHYDE	2-METHYL VALERALDEHYDE
123-15-9 2-METHYLVALERALDEHYDE [FLA]	2-METHYL VALERALDEHYDE
123-15-9 ALPHA-METHYLVALERALDEHYDE [HM1, HM1]	2-METHYL VALERALDEHYDE
123-15-9 2-METHYLVALERALDEHYDE [MSL, NFP, NFP]	2-METHYL VALERALDEHYDE
123-15-9 METHYL VALERALDEHYDE [NJL, NJL]	2-METHYL VALERALDEHYDE
1119-16-0 4-METHYL VALERALDEHYDE	4-METHYL VALERALDEHYDE
624-24-8 METHYL VALERATE	METHYL VALERATE
RR-01303-6 METHYLVINYLBENZENES	METHYLVINYLBENZENES
2554-06-5 METHYLVINYLCYCLOSILOXANE	METHYLVINYLCYCLOSILOXANE
78-94-4 METHYL VINYL KETONE	METHYL VINYL KETONE
140-76-1 2-METHYL-5-VINYLPYRIDINE [WHM]	PYRIDINE, 2-METHYL-5-VINYL-
8004-87-3 METHYL VIOLET	METHYL VIOLET
9006-42-2 METIRAM	METIRAM
51218-45-2 METOLACHLOR	METOLACHLOR
1129-41-5 METOLCARB	METOLCARB
21087-64-9 METRIBUZIN	METRIBUZIN
21087-64-9 METRIBUZIN (SENCOR) [ALB, ALB]	METRIBUZIN
443-48-1 METRONIDAZOLE	METRONIDAZOLE
7786-34-7 MEVINPHOS	MEVINPHOS
7786-34-7 MEVINPHOS (PHOSDRIN) [QBC, QBC, QBC, MEX, MEX, MEX]	MEVINPHOS
7786-34-7 MEVINPHOS (ISO) [GBR, GBR, GBR]	MEVINPHOS
7786-34-7 PHOSDRIN (MEVINPHOS) [ALB, ALB, ALB]	MEVINPHOS
315-18-4 MEXACARBATE	MEXACARBATE
12001-26-2 MICA	MICA
12003-38-2 MICA [PST]	FLUORPHLOGOPITE (MG3K[ALF2O(SIO3)3])
12001-26-2 MICA (CONTAINING LESS THAN 1% QUARTZ) [NIO, NIO]	MICA
12001-26-2 MICA DUST [FLA, MSL]	MICA
12001-26-2 MICA GROUP MINERALS [PAL]	MICA
90-94-8 MICHLER'S KETONE	MICHLER'S KETONE
63231-60-7 MICROCRYSTALLINE WAX	MICROCRYSTALLINE WAX
59467-96-8 MIDAZOLAM HYDROCHLORIDE	MIDAZOLAM HYDROCHLORIDE
137-30-4 MILBAN [MSL]	ZIRAM
68514-61-4 MILK, HYDROLYZED	MILK, HYDROLYZED
RR-01095-7 MILLET SEED	MILLET SEED

Chemical Name	Reference Name
RR-01096-8 MILO	MILO
8049-99-8 MILORGANITE	MILORGANITE
RR-01501-0 MINERAL FIBERS	MINERAL FIBERS
RR-01739-0 MINERAL FIBERS (OTHER THAN MANMADE)	MINERAL FIBERS (OTHER THAN MANMADE)
RR-01501-0 MINERAL FIBRES [CN4]	MINERAL FIBERS
8012-95-1 MINERAL OIL [NFP, NFP]	OIL MIST, MINERAL
8012-95-1 MINERAL OIL MIST [PAL, QBC, QBC]	OIL MIST, MINERAL
64741-49-7 MINERAL OIL, PETROLEUM CONDENSATES, VACUUM TOWER [MSL]	VACUUM TOWER CONDENSATE (PETROLEUM)
64742-20-7 MINERAL OIL, PETROLEUM DISTILLATES, ACID-TREATED HEAVYPARAFFINIC [MSL]	PETROLEUM DISTILLATES, ACID TREATED, HEAVY PARAFFINICNAPHTHENIC
64742-21-8 MINERAL OIL, PETROLEUM DISTILLATES, ACID-TREATED LIGHTPARAFFINIC [MSL]	PETROLEUM DISTILLATES, ACID TREATED, LIGHT PARAFFINICPARAFFINIC
64741-53-3 MINERAL OIL, PETROLEUM DISTILLATES, HEAVY NAPHTHENIC [MSL]	PETROLEUM DISTILLATES, HEAVY NAPHTHENIC
64741-51-1 MINERAL OIL, PETROLEUM DISTILLATES, HEAVY PARAFFINIC [MSL]	PETROLEUM DISTILLATES, HEAVY PARAFFINIC
64742-55-8 MINERAL OIL, PETROLEUM DISTILLATES, HYDROTREATED LIGHTPARAFFINIC [MSL, MSL]	PETROLEUM DISTILLATES, HYDROTREATED LIGHT PARAFFINICNAPHTHENIC
64741-52-2 MINERAL OIL, PETROLEUM DISTILLATES, LIGHT NAPHTHENIC [MSL]	LIGHT NAPHTHENIC OIL
64741-50-0 MINERAL OIL, PETROLEUM DISTILLATES, LIGHT PARAFFINIC [MSL]	PETROLEUM DISTILLATES, LIGHT PARAFFINIC
64742-56-9 MINERAL OIL, PETROLEUM DISTILLATES, SOLVENT-DEWAXED LIGHTPARAFFINIC [MSL, MSL]	PETROLEUM DISTILLATES, SOLVENT DEWAXED LIGHT PARAFFINICPARAFFINIC
64741-89-5 MINERAL OIL, PETROLEUM DISTILLATES, SOLVENT-REFINED LIGHTPARAFFINIC [MSL, MSL]	PETROLEUM DISTILLATES, SOLVENT-REFINED LIGHT PARAFFINIC
64742-11-6 MINERAL OIL, PETROLEUM EXTRACTS, HEAVY NAPHTHENIC DISTILLATE SOLVENT [MSL]	HEAVY NAPHTHENIC DISTILLATE SOLVENT EXTRACT
64742-04-7 MINERAL OIL, PETROLEUM EXTRACTS, HEAVY PARAFFINIC DISTILLATESOLVENT [MSL]	PARAFFINIC DISTILLATE SOLVENT EXTRACTATE SOLVENT
64742-03-6 MINERAL OIL, PETROLEUM EXTRACTS, LIGHT NAPHTHENIC DISTILLATE SOLVENT [MSL]	LIGHT DISTILLATE SOLVENT EXTRACT (PETROLEUM) NAPHTHENIC
64742-05-8 MINERAL OIL, PETROLEUM EXTRACTS, LIGHT PARAFFINIC DISTILLATESOLVENT [MSL]	LIGHT PARAFFINIC DISTILLATE SOLVENT EXTRACT (PETROLEUM)SOLVENT
64742-10-5 MINERAL OIL, PETROLEUM EXTRACTS, RESIDUAL OIL SOLVENT [MSL]	RESIDUAL OIL SOVENT EXTRACT (PETROLEUM)SOLVENT
64742-69-4 MINERAL OIL, PETROLEUM NAPHTHENIC OILS, CATALYTIC DEWAXEDLIGHT [MSL, MSL]	NAPHTHENIC OILS (PETROLEUM), CATALYTIC DEWAXED LIGHTHEAVY
64742-70-7 MINERAL OIL, PETROLEUM PARAFFIN OILS, CATALYTIC DEWAXED HEAVY [MSL]	PARAFFIN OILS (PETROLEUM), CATALYTIC DEWAXED HEAVYLIGHT
64742-71-8 MINERAL OIL, PETROLEUM PARAFFIN OILS, CATALYTIC DEWAXED LIGHT [MSL]	PARAFFIN OILS, CATALYTIC DEWAXED LIGHTY
64742-17-2 MINERAL OIL, PETROLEUM RESIDUAL OILS, ACID-TREATED [MSL]	ACID TREATED RESIDUAL OIL (PETROLEUM) SOLVENT
64742-18-3 MINERAL OIL, PETROLEUM RESIDUAL OILS, ACID-TREATED HEAVYNAPHTHENIC [MSL]	PETROLEUM DISTILLATES, ACID TREATED, HEAVY NAPHTHENIC
64742-19-4 MINERAL OIL, PETROLEUM RESIDUAL OILS, ACID-TREATED LIGHTNAPHTHENIC [MSL]	PETROLEUM DISTILLATES, ACID TREATED, LIGHT NAPHTHENICNAPHTHENIC
8002-05-9 MINERAL OILS [CAB, MIN]	PETROLEUM DISTILLATES (NAPHTHA)
64742-06-9 MINERAL SEAL OIL TYPICAL [NFP, NFP]	EXTRACTS (PETROLEUM), MIDDLE DISTILLATE SOLVENT

Chemical Name	Reference Name
8032-32-4 MINERAL SPIRITS [NFP, NFP]	LIGROINE
9005-90-7 MINERAL TURPENTINE [AUS, AUS]	TURPENTINE (RESIN FROM PINUS SPECIES PARTICULARLY PALUSTRIS)
RR-00015-7 MINERAL WOOL FIBER	MINERAL WOOL FIBER
RR-00015-7 MINERAL WOOL FIBRE [BC1]	MINERAL WOOL FIBER
RR-01502-1 MINING REAGENT, LIQUID	MINING REAGENT, LIQUID
8023-74-3 MINK OIL	MINK OIL
13614-98-7 MINOCYCLINE HYDROCHLORIDE	MINOCYCLINE HYDROCHLORIDE
2385-85-5 MIREX	MIREX
2385-85-5 MIREX (DODECACHLOROPENTACYCLO [5.3.0.0.0.0] DECANE) [CN8]	MIREX
62015-39-8 MISOPROSTOL	MISOPROSTOL
50-07-7 MITOMYCIN C	MITOMYCIN C
70476-82-3 MITOXANTRONE HYDROCHLORIDE	MITOXANTRONE HYDROCHLORIDE
68611-55-2 MIXED C10-16-ALKYL SULFATES	MIXED C10-16-ALKYL SULFATES
67254-79-9 MIXED FATTY ACIDS	MIXED FATTY ACIDSE AND N,N,N',N'-TETRAMETHYL-2-BUTENE-1,4-DIAMINE
RR-01097-9 MIXED FATTY AND ROSIN ACIDS	MIXED FATTY AND ROSIN ACIDS
106264-79-3 MIXTURE OF 1,3-BENZENEDIAMINE, 2-METHYL-4,6-BIS(METHYLTHIO)AND 1,3-BENZENEDIAMINE, 4-METHYL-2,6-BIS(METHYLTHIO)	MIXTURE OF 1,3-BENZENEDIAMINE, 2-METHYL-4,6-BIS (METHYLTHIO)
70-25-7 MNNG (N-METHYL-N'-NITRO-N-NITROSOGUANIDINE) [TSC, TSE]	N-METHYL-N'-NITRO-N-NITROSOGUANIDINE
526-99-8 MOCIC ACID	MOCIC ACID
RR-01537-2 MODACRYLIC FIBRES	MODACRYLIC FIBRES
RR-00928-9 MODIFIED ACRYLIC ESTER	MODIFIED ACRYLIC ESTER(ALKYLIMIDAZOLYL) METHYL] DERIVATIVE
RR-01262-4 MODIFIED HYDROCARBON RESIN	MODIFIED HYDROCARBON RESIN
68476-78-8 MOLASSES	MOLASSES
2212-67-1 MOLINATE [CAL]	ORDRAM (MOLINATE)
2212-67-1 MOLINATE (1H-AZEPINE-1-CARBOTHIOIC ACID, HEXAHYDRO-S-ETHYL ESTER) [313]	ORDRAM (MOLINATE)L]-]
12656-57-4 MOLYBDATE ORANGE [PST]	CADMIUM ORANGE PIGMENT
12656-85-8 MOLYBDATE ORANGE (LEAD CHROMATE PIGMENT)	MOLYBDATE ORANGE (LEAD CHROMATE PIGMENT)
RR-01621-7 MOLYBDATES	MOLYBDATES
7439-98-7 MOLYBDENUM	MOLYBDENUM
15690-77-4 MOLYBDENUM 90	MOLYBDENUM 90
14119-13-2 MOLYBDENUM 93	MOLYBDENUM 93
RR-00402-4 MOLYBDENUM 93M	MOLYBDENUM 93M
14119-15-4 MOLYBDENUM 99	MOLYBDENUM 99
14191-83-4 MOLYBDENUM 101	MOLYBDENUM 101
RR-00603-1 MOYLBDENUM COMPOUNDS [CAL]	MOLYBDENUM COMPOUNDS, N.O.S.
RR-00603-1 MOLYBDENUM COMPOUNDS [GBR, GBR]	MOLYBDENUM COMPOUNDS, N.O.S.
RR-00603-1 MOLYBDENUM COMPOUNDS, N.O.S.	MOLYBDENUM COMPOUNDS, N.O.S.
18868-43-4 MOLYBDENUM DIOXIDE [ONT]	MOLYBDENUM OXIDE
1317-33-5 MOLYBDENUM DISULFIDE [PST, ONT]	MOLYBDENUM (IV) SULFIDE
7439-98-7 MOLYBDENUM, ELEMENTAL [WHM]	MOLYBDENUM
7439-98-7 MOLYBDENUM (INSOLUBLE COMPOUNDS) [BC1, BC1, NIO, NIO]	MOLYBDENUM
RR-00037-3 MOLYBDENUM INSOLUBLE COMPOUNDS	MOLYBDENUM INSOLUBLE COMPOUNDS

Chemical Name	Reference Name
RR-00037-3 MOLYBDENUM, INSOLUBLE COMPOUNDS [QBC]	MOLYBDENUM INSOLUBLE COMPOUNDS
7439-98-7 MOLYBDENUM METAL [ONT]	MOLYBDENUM
18868-43-4 MOLYBDENUM OXIDE	MOLYBDENUM OXIDE
10241-05-1 MOLYBDENUM PENTACHLORIDE	MOLYBDENUM PENTACHLORIDE
RR-00036-2 MOLYBDENUM SOLUBLE COMPOUNDS	MOLYBDENUM SOLUBLE COMPOUNDS
RR-00036-2 MOLYBDENUM (SOLUBLE COMPOUNDS) [NIO, NIO]	MOLYBDENUM SOLUBLE COMPOUNDS
RR-00036-2 MOLYBDENUM, SOLUBLE COMPOUNDS [QBC]	MOLYBDENUM SOLUBLE COMPOUNDS
1317-33-5 MOLYBDENUM (IV) SULFIDE	MOLYBDENUM (IV) SULFIDE
1313-27-5 MOLYBDENUM TRIOXIDE	MOLYBDENUM TRIOXIDE
RR-01001-5 MONOACRYLATE	MONOACRYLATE
7722-76-1 MONOAMMONIUM PHOSPHATE	MONOAMMONIUM PHOSPHATE
1118-46-3 MONOBUTYLTIN TRICHLORIDE [WHM]	BUTYLTRICHLOROSTANNANE
25852-70-4 MONOBUTYLTIN TRIS(ISOOCTYL) MERCAPTOACETATE [TSC, TSC, TSC]	MONOBUTYLTIN TRIS (ISOOCTYL) MERCAPTO-ACETATE
25852-70-4 MONOBUTYLTIN TRIS (ISOOCTYL) MERCAPTO-ACETATE	MONOBUTYLTIN TRIS (ISOOCTYL) MERCAPTO-ACETATE
79-11-8 MONOCHLOROACETIC ACID [CSR, CSR, GBR, GBR, MIN, PST, WEL, WEL, WEL]	CHLOROACETIC ACID
78-95-5 MONOCHLOROACETONE [MSL, WHM]	CHLOROACETONE
108-90-7 MONOCHLOROBENZENE [CD1, CD1, CTR, SDW, SDW, SDW, TSE]	CHLOROBENZENE
75-45-6 MONOCHLORODIFLUOROMETHANE [PST]	CHLORODIFLUOROMETHANE
107-30-2 MONOCHLORODIMETHYL ETHER [MAK]	CHLOROMETHYL METHYL ETHER
76-15-3 (MONO)CHLOROPENTAFLUOROETHANE (CFC-115) [313]	CHLOROPENTAFLUOROETHANE
315-22-0 MONOCROTALINE	MONOCROTALINE
6923-22-4 MONOCROTOPHOS	MONOCROTOPHOS
6923-22-4 MONOCROTOPHOS (AZODRIN) [OS1, BC1]	MONOCROTOPHOS
6923-22-4 MONOCROTOPHOS (AZODRIN (R)) [QBC, QBC]	MONOCROTOPHOS
6923-22-4 MONOCROTOPHOS (AZODRIN) [ALB, ALB]	MONOCROTOPHOS
3921-30-0 MONODECYL ACID PHOSPHATE	MONODECYL ACID PHOSPHATE PHOSPHATE
141-43-5 MONOETHANOLAMINE [WHM]	ETHANOLAMINE
RR-01098-0 MONOETHANOLAMINE ALKYL(C8-C24) BENZENE SULFONATE	MONOETHANOLAMINE ALKYL(C8-C24) BENZENE SULFONATE
75-04-7 MONOETHYLAMINE [CWA, EPA]	ETHYLAMINE
13537-32-1 MONOFLUOROPHOSPHORIC ACID	MONOFLUOROPHOSPHORIC ACID
78-96-6 MONOISOPROPANOLAMINE [PST]	ISOPROPANOLAMINE
RR-00980-3 MONOMETHOXY NEOPENTYL GLYCOL PROPOXYLATE MONOACRYLATE	MONOMETHOXY NEOPENTYL GLYCOL PROPOXYLATE MONOACRYLATEOXY]ALKOXY] CARBONYLAMINO] SUBSTITUTED] AMINOCARBONYL]OXY-
74-89-5 MONOMETHYLAMINE [CWA, EPA]	METHYLAMINE
100-61-8 MONOMETHYL ANILINE [BC1, BC1, BC1, NIO, NIO, NIO, NIO, OS1, OS1, OS1, OS1]	N-METHYL ANILINE
100-61-8 MONOMETHYL ANILINE (N-METHYLANILINE) [ALB, ALB, ALB]	N-METHYL ANILINE
60-34-4 MONOMETHYL HYDRAZINE [BC1, BC1, MIN]	METHYL HYDRAZINE
54849-38-6 MONOMETHYLTIN TRIS(ISOOOCTYL MERCAPTOACETATE)	MONOMETHYLTIN TRIS(ISOOOCTYL MERCAPTOACETATE)DI-4,1-PHENYLENE]BIS[3-(OXIRANYLMETHOXY)
RR-00604-2 MONONITRO-M-TOLUIDINE	MONONITRO-M-TOLUIDINE

Chemical Name	Reference Name
RR-00605-3 MONONITRO-O-TOLUIDINE	MONONITRO-O-TOLUIDINE
RR-00606-4 MONONITRO-P-TOLUIDINE	MONONITRO-P-TOLUIDINE
13356-20-2 MONOOCTYLTIN OXIDE	MONOOCTYLTIN OXIDE
3091-25-6 MONOOCTYLTIN TRICHLORIDE	MONOOCTYLTIN TRICHLORIDE
27107-89-7 MONOOCTYLTIN TRIS(2-ETHYLHEXYL THIO-GLYCOLATE)	MONOOCTYLTIN TRIS(2-ETHYLHEXYL THIOGLYCO-LATE)
26401-86-5 MONOOCTYLTIN TRIS(ISOOCTYL THIOGLYCO-LATE)	MONOOCTYLTIN TRIS(ISOOCTYL THIOGLYCOLATE)
10058-23-8 MONOPOTASSIUM PEROXYMONOSULFATE	MONOPOTASSIUM PEROXYMONOSULFATE
7558-80-7 MONOSODIUM O-PHOSPHATE [PST]	SODIUM ACID PHOSPHATE
RR-00162-7 MONOSUBSTITUTED ALKOXYAMINOTRIZINES	MONOSUBSTITUTED ALKOXYAMINOTRIZINES
34651-95-1 MONO-(TRICHLORO)TETRA(MONOPOTASSIUM DICHLORO)-PENTA-S-TRIAZINE-TRIONE [PAL]	MONO-(TRICHLORO)TETRA(MONO-POTASSIUM DICHLORO)-PENTA-S-
34651-95-1 MONO-(TRICHLORO)TETRA(MONO-POTAS-SIUM DICHLORO)-PENTA-S-TRIAZINE-TRIONE	MONO-(TRICHLORO)TETRA(MONO-POTASSIUM DICHLORO)-PENTA-S-
150-68-5 MONURON	MONURON
140-41-0 MONURON-TCA [CAL]	ACETIC ACID TRICHLORO-, COMPD. WITH 3-(P-CHLOROPHENYL)-1,
RR-00053-3 MOPP	MOPP
RR-00053-3 MOPP AND OTHER COMBINED CHEMOTHER-APY INCLUDING ALKYLATING AGENTS [IAR]	MOPP
RR-00053-3 MOPP(COMBINED THERAPY W/NITROGEN MUSTARD,VINCRISTINE,PROCAR. [PAL, PAL]	MOPP
6033-05-2 MORPHINAN-3,6-DIOL, 7,8-DIDEHYDRO-4,5-EPOXY-17-METHYL-(5A,6A)-, (Z)-9-OCTADE-CENOATE (SALT)	MORPHINAN-3,6-DIOL, 7,8-DIDEHYDRO-4,5-EPOXY-17-METHYL-(5A,6A
110-91-8 MORPHOLINE	MORPHOLINE
1696-20-4 MORPHOLINE, 4-ACETYL-	MORPHOLINE, 4-ACETYL-[ETHYL[(3-SULFOPHENYL) METHYL]AMINO]PHENYL]METHYLENE]-2,5- CYCLO-HEXADIEN-1-YLIDENE]-N-ETHYL-SULFO-,HYDROXIDE, INNER SALT, SODIUM SALT
141-91-3 MORPHOLINE, 2,6-DIMETHYL- [PAL]	2,6-DIMETHYLMORPHOLINE
2038-03-1 4-MORPHOLINEETHANAMINE [PAL]	4-(2-AMINOETHYL)-MORPHOLINE
622-40-2 MORPHOLINE ETHANOL	MORPHOLINE ETHANOL
622-40-2 4-MORPHOLINEETHANOL [PAL]	MORPHOLINE ETHANOL
100-74-3 MORPHOLINE, 4-ETHYL- [PAL]	N-ETHYLMORPHOLINE
109-02-4 MORPHOLINE, 4-METHYL- [PAL]	METHYLMORPHOLINE
59-89-2 MORPHOLINE, 4-NITROSO- [PAL, PAL]	N-NITROSOMORPHOLINE
1095-66-5 MORPHOLINE OLEATE	MORPHOLINE OLEATE
92-53-5 MORPHOLINE, 4-PHENYL- [PAL]	4-PHENYLMORPHOLINE
123-00-2 4-MORPHOLINEPROPANAMINE [PAL]	N-AMINOPROPYLMORPHOLINE
139-91-3 5-(MORPHOLINOMETHYL)-3-[5-NITRO-FUR-FURYLIDENE)-AMINO]-2-OXAZOLIDINONE	5-(MORPHOLINOMETHYL)-3-[(5-NITRO-FURFURYLI-DENE)-AMINO]-2-OXA
3759-92-0 5-MORPHOLINOMETHYL-3[(5-NITROFUR-FURYLIDENE)AMINO]-2-OXAZOLIDINONE [IAR]	5-MORPHOLINOMETHYL-3((5-NITROFURFURYLIDENE) AMINO)-2-
139-91-3 5-(MORPHOLINOMETHYL)-3-(5-(NITRO-FURFURYLIDENE)AMINO)-2-OXAZOLIDI-NONE [FLA]	5-(MORPHOLINOMETHYL)-3-[(5-NITRO-FURFURYLI-DENE)-AMINO]-2-OXA
139-91-3 5-(MORPHOLINOMETHYL)-3-((5-NITRO-FURFURYLIDENE)AMINO)-2-OXAZOLIDI-NONE [MSL]	5-(MORPHOLINOMETHYL)-3-[(5-NITRO-FURFURYLI-DENE)-AMINO]-2-OXA
3759-92-0 5-MORPHOLINOMETHYL-3((5-NITROFUR-FURYLIDENE)AMINO)-2-OXAZOLIDINONE HY-DROCHLORIDE	5-MORPHOLINOMETHYL-3((5-NITROFURFURYLIDENE) AMINO)-2-

Chemical Name	Reference Name
3795-88-8 5-(MORPHOLINOMETHYL)-3-((5-NITROFUR-FURYLIDENE)AMINO)-2-OXAZOLIDINONE	5-(MORPHOLINOMETHYL)-3-((5-NITROFURFURYLIDENE)AMINO)-2-OXAZO
13146-28-6 DI-5-(MORPHOLINOMETHYL)-3((5-NITROFUR-FURYLIDENE)AMINO)-2-OXAZOLIDINONE HY-DROCHLORIDE	DI-5-(MORPHOLINOMETHYL)-3((5-NITROFURFURYLI-DENE)AMINO)-2-OXA
95-32-9 2-(4-MORPHOLINYLDITHIO)-BENZOTHIAZOLE	2-(4-MORPHOLINYLDITHIO)-BENZOTHIAZOLE
1318-93-0 MORTMORILLONITE	MORTMORILLONITE
86290-81-5 MOTOR FUEL, N.O.S. [NJL]	GASOLINE, MOTOR FUEL
RR-00071-5 MULTI SOURCE LEACHATE	MULTI SOURCE LEACHATE
2763-96-4 MUSCIMOL	MUSCIMOL
81-15-2 MUSK XYLENE	MUSK XYLENE
505-60-2 MUSTARD GAS	MUSTARD GAS
505-60-2 MUSTARD GAS (BIS(2-CHLOROETHYL)-SUL-FIDE) [WHM]	MUSTARD GAS
505-60-2 MUSTARD GAS [ETHANE, 1,1'-THIOBIS[2-CHLORO-] [313]	MUSTARD GAS
505-60-2 MUSTARD GAS (SULPHUR MUSTARD) [IAR]	MUSTARD GAS
57-06-7 MUSTARD OIL [MSL, NFP, NFP]	ALLYL ISOTHIOCYANATE
88671-89-0 MYCLOBUTANIL [.ALPHA.-BUTYL-.ALPHA.-(4-CHLOROPHENYL)-1H-1,2,4-TRIAZOLE-1-PROPANETRILE]	MYCLOBUTANIL [.ALPHA.-BUTYL-.ALPHA.-(4-CHLOROPHENYL)-1H-1,2,
25038-59-9 MYLAR	MYLAR
123-35-3 MYRCENE	MYRCENE
544-63-8 MYRISTIC ACID	MYRISTIC ACID
112-72-1 MYRISTIC ALCOHOL	MYRISTIC ALCOHOL
5780-07-4 MYRISTIC ALDEHYDE	MYRISTIC ALDEHYDE
110-54-3 N-HEXANE [MEX]	HEXANE
142-59-6 NABAM [313, HM1]	ETHYLENE BIS DITHIOCARBAMATE
3771-19-5 NAFENOPIN	NAFENOPIN
3794-64-7 NAFENOPIN [MIN]	SILVER HEPTAFLUOROBUTYRATE
86220-42-0 NAFRELIN ACETATE	NAFRELIN ACETATE
300-76-5 NALED	NALED
300-76-5 NALED [QBC, QBC]	NALED
300-76-5 NALED (ISO) [GBR, GBR]	NALED
389-08-2 NALIDIXIC ACID	NALIDIXIC ACID
8030-30-6 NAPHTHA [HM1, HM1, HM1, PAL, PST]	BENZIN
8030-31-7 NAPHTHA 49 DEGREE BE-COAL TAR TYPE	NAPHTHA 49 DEGREE BE-COAL TAR TYPE
20830-81-3 5,12-NAPHTHACENEDIONE, 8-ACETYL-10-[(3-AMINO-2,3,6-TRIDEOXY-.ALPHA.-L-LYXO-HEX-OPYRANOSYL)OXY]-7,8,9,10-TETRAHYDRO-6,8,11-TRIHYDROXY-8-(HYDROXYACETYL)-1-METHOXY-, HYDROCHLORIDE, (8S-CIS)- [PAL, PAL]	DAUNOMYCIN
23214-92-8 5,12-NAPHTHACENEDIONE, 10-[(3-AMINO-2,3,6-TRIDEOXY-.ALPHA.-L-LYXO-HEXOPYRA-NOSYL)OXY]-7,8,9,10-TETRAHYDRO-6,8,11-TRIHYDROXYACETYL)-1-METHOXY, (8S-CIS)-[PAL, PAL]	ADRIAMYCIN
25316-40-9 5,12-NAPHTHACENEDIONE, 10-[(3-AMINO-2,3,6-TRIDEOXY-.ALPHA.-L-LYXO-HEXOPYRA-NOSYL)OXY]-7,8,9,10-TETRAHYDRO-6,8,11-TRI-HYDROXY-8-(HYDROXYACETYL)-1-METHOXY-, HYDROCHLORIDE, (8S-CIS)-	5,12-NAPHTHACENEDIONE, 10-[(3-AMINO-2,3,6-TRIDEOXY-.ALPHA.-L
8030-30-6 NAPHTHA (COAL TAR) [NIO, NIO, NIO, OS1, OS1]	BENZIN

402

Chemical Name	Reference Name
8030-31-7 NAPHTHA, COAL TAR [CAL, CEX]	NAPHTHA 49 DEGREE BE-COAL TAR TYPE
65996-79-4 NAPHTHA (COAL TAR)	NAPHTHA (COAL TAR)
91-59-8 2-NAPHTHALENAMINE [PAL, PAL, RCU, RCU]	BETA-NAPHTHYLAMINE
134-32-7 1-NAPHTHALENAMINE [PAL, RCU, RCU]	ALPHA-NAPHTHYLAMINE
494-03-1 2-NAPHTHALENAMINE, N,N-BIS(2-CHLOROETHYL)- [PAL, PAL]	N,N-BIS(2-CHLOROETHYL)-2-NAPHTHYLAMINE
135-88-6 2-NAPHTHALENAMINE, N-PHENYL- [PAL]	N-PHENYL-BETA-NAPHTHYLAMINE
RR-01304-7 NAPHTHALENE	NAPHTHALENE
85-47-2 NAPHTHALENE SULFONIC ACID (A-)	NAPHTHALENE SULFONIC ACID (A-)
120-18-3 NAPHTHALENE SULFONIC ACID (B-) [HON]	2-NAPHTHALENESULFONIC ACID
86-87-3 1-NAPHTHALENEACETIC ACID	1-NAPHTHALENEACETIC ACID
66-77-3 1-NAPHTHALENECARBOXALDEHYDE	1-NAPHTHALENECARBOXALDEHYDE
91-92-9 2-NAPHTHALENECARBOXAMIDE, N,N'-(3,3'-DIMETHOXY[1,1'BIPHENYL]-4,4'-DIYL)BIS[3-HYDROXY- [TSC]	C.I. AZOIC COUPLING COMPONENT
90-13-1 NAPHTHALENE, 1-CHLORO-	NAPHTHALENE, 1-CHLORO-
91-58-7 NAPHTHALENE, 2-CHLORO- [PAL, TSC, TSC]	BETA-CHLORONAPHTHALENE
25586-43-0 NAPHTHALENE, CHLORO-	NAPHTHALENE, CHLORO-
70776-03-3 NAPHTHALENE, CHLORO DERIVATIVES	NAPHTHALENE, CHLORO DERIVATIVES
91-17-8 NAPHTHALENE, DECAHYDRO- [PAL]	DECAHYDRONAPHTHALENE
2243-62-1 1,5-NAPHTHALENEDIAMINE	1,5-NAPHTHALENEDIAMINE
81-84-5 NAPHTHALENEDICARBOXYLIC ANHYDRIDE	NAPHTHALENEDICARBOXYLIC ANHYDRIDE
1825-30-5 NAPHTHALENE, 1,5-DICHLORO-	NAPHTHALENE, 1,5-DICHLORO-
1825-31-6 NAPHTHALENE, 1,4-DICHLORO-	NAPHTHALENE, 1,4-DICHLORO-
2050-69-3 NAPHTHALENE, 1,2-DICHLORO-	NAPHTHALENE, 1,2-DICHLORO-PHENYL)AZO] [1,1'-BIPHENYL]-4-YL]AZO]-5-HYDROXY-6- (PHENYLAZO)-, DISODIUM SALT
2050-72-8 NAPHTHALENE, 1,6-DICHLORO-	NAPHTHALENE, 1,6-DICHLORO-
2050-73-9 NAPHTHALENE, 1,7-DICHLORO-	NAPHTHALENE, 1,7-DICHLORO-
2050-74-0 NAPHTHALENE, 1,8-DICHLORO-	NAPHTHALENE, 1,8-DICHLORO-
2050-75-1 NAPHTHALENE, 2,3-DICHLORO-	NAPHTHALENE, 2,3-DICHLORO-
2065-70-5 NAPHTHALENE, 2,6-DICHLORO-	NAPHTHALENE, 2,6-DICHLORO-
2198-75-6 NAPHTHALENE, 1,3-DICHLORO-	NAPHTHALENE, 1,3-DICHLORO-
2198-77-8 NAPHTHALENE, 2,7-DICHLORO-	NAPHTHALENE, 2,7-DICHLORO-
3173-72-6 1,5-NAPHTHALENE DIISOCYANATE [IAR]	NAPHTHALENE, 1,5-DIISOCYANATO-
25551-28-4 NAPHTHALENE DIISOCYANATE	NAPHTHALENE DIISOCYANATE
25551-28-4 NAPHTHALENE DIISOCYANATE (NDI) [NHS, NHS]	NAPHTHALENE DIISOCYANATE
3173-72-6 NAPHTHALENE, 1,5-DIISOCYANATO-	NAPHTHALENE, 1,5-DIISOCYANATO-
73090-68-3 NAPHTHALENE, (1,1-DIMETHYLETHYL)-1,2,3,4-TETRAHYDRO- [PAL]	TERT-BUTYL TETRALIN
130-15-4 1,4-NAPHTHALENEDIONE	1,4-NAPHTHALENEDIONE
117-80-6 1,4-NAPHTHALENEDIONE, 2,3-DICHLORO-[PAL]	DICHLONE
RR-01448-2 NAPHTHALENE DIOZONIDE	NAPHTHALENE DIOZONIDE
2429-73-4 2,7-NAPHTHALENEDISULFONIC ACID, 5-AMINO-3-[[4'-[(7-AMINO-1-HYDROXY-3-SULFO-2-NAPHTHALENYL)-AZO] [1,1'-BIPHENYL]-4-YL]AZO]-4-HYDROXY-, TRISODIUM SALT [TSC]	C.I. DIRECT BLUE 2, TRISODIUM SALTBIPHENYL]-4,4'-DIYL)BIS(AZO)]BIS[4-HYDROXY-, DISODIUM SALT

Chemical Name	Reference Name
1937-37-7 2,7-NAPHTHALENEDISULFONIC ACID, 4-AMINO-3-[[4'-[(2,4-DIAMINOPHENYL)AZO][1,1'-BIPHENYL]-4-YL]AZO]-5-HYDROXY-6-(PHENY-LAZO)-,DISODIUM SALT [TSC, TSC, TSC, PAL, PAL]	C.I. DIRECT BLACK 38
3567-65-5 1,3-NAPHTHALENEDISULFONIC ACID, 7-HY-DROXY-8-[[4'-[[4-METHYLPHENYL) SULFONYL]OXY]PHENYL]AZO][1,1'-BIPHENYL]-4-YL]AZO]-, DISODIUM SALT [TSC]	C.I. ACID RED 85
2586-57-4 1,3-NAPHTHALENEDISULFONIC ACID, 4-AMINO-5-HYDROXY-6-[[4'-[(2-HYDROXY-1-NAPHTHALENYL)AZO]-3,3'-DIMETHOXY[1,1'-BIPHENYL]-4-YL]AZO]-, DISODIUM SALT [TSC]	C.I. DIRECT BLUE 22, DISODIUM SALT
3626-28-6 2,7-NAPHTHALENEDISULFONIC ACID, 4-AMINO-5-HYDROXY-3-[[4'-HYDROXYPHENYL)AZO] [1,1'-BIPHENYL]-4-YL]-AZO]-6-(PHENY-LAZO)-, DISODIUM SALT [TSC]	C.I. DIRECT GREEN 1, DISODIUM SALT4,6-DINITRO-PHENYL)AZO)-4-METHOXYPHENYL)-
4335-09-5 2,7-NAPHTHALENEDISULFONIC ACID, 4-AMINO-5-HYDROXY-6[[4'-[(4-HYDROX-YPHENYL)AZO] [1,1'-BIPHENYL]-4-YL]-AZO]-3-[(4-NITRO PHENYL)AZO]-, DISODIUM SALT [TSC]	C.I. DIRECT GREEN 6, DISODIUM SALT
6449-35-0 1-NAPHTHALENEDISULFONIC ACID, 3-[[4'-[(6-AMINO-1-HYDROXY-3-SULFO-2-NAPH-THALENYL)AZO]-3,3'-DIMETHOXY[1,1'-BIPHENYL]-4-YL]AZO]-4-HYDROXY-, DIS-ODIUM SALT	1-NAPHTHALENEDISULFONIC ACID, 3-[[4'-[(6-AMINO-1-HYDROXY-3-S1-AMINO-9,10-DIHYDRO-9,10-DIOXO-, MONOSODIUM SALT
2602-46-2 2,7-NAPHTHALENEDISULFONIC ACID, 3,3'-[[1,1'-BIPHENYL]4,4'-DIYL)BIS(AZO)]BIS[5-AMINO-4-HYDROXY-, TETRASODIUM SALT [PAL, PAL]	C.I. DIRECT BLUE 6MONOHYDROCHLORIDE
2602-46-2 2,7-NAPHTHALENEDISULFONIC ACID, 3,3'-[[1,1'-BIPHENYL]-4,4'-DIYLBIS(AZO)]BIS[5-AMINO-4-HYDROXY-, TETRASODIUM SALT [TSC, TSC]	C.I. DIRECT BLUE 6
2429-74-5 2,7-NAPHTHALENEDISULFONIC ACID, 3,3'-[(3,3'-DIMETHOXY[1,1'-BIPHENYL]-4,4'-DIYL)BIS(AZO)]BIS[5-AMINO-4-HYDROXY-, TETRA-SODIUM SALT [TSC]	C.I. DIRECT BLUE 15HYDROXY-3-SULFO-2-NAPH-THALENYL)-AZO] [1,1'-BIPHENYL]-4- YL]AZO]-4-HY-DROXY-, TRISODIUM SALT
2610-05-1 1,3-NAPHTHALENEDISULFONIC ACID, 6,6'-[(3,3'-DIMETHOXY[1,1'-BIPHENYL]-4,4'-DIYL)BIS(AZO)]BIS[4-AMINO-5-HYDROXY-, TETRA-SODIUM SALT [TSC, TSC]	C.I. DIRECT BLUE 1DIYLBIS(AZO)]BIS[5-AMINO-4-HY-DROXY-, TETRASODIUM SALT
72-57-1 2,7-NAPHTHALENEDISULFONIC ACID, 3,3'-[(3,3'-DIMETHYL-[1,1'-DIPHENYL]-4,4'-DIYL)BIS(AZO)]BIS[5-AMINO-4-HYDROXY-, TETRA-SODIUM SALT [TSC, TSC]	TRYPAN BLUE
2150-54-1 2,7-NAPHTHALENEDISULFONIC ACID, 3,3'-[(3,3'-DIMETHYL[1,1'-BIPHENYL]-4,4'-DIYL)BIS(AZO)]BIS-[4,5-DIHYDROXY-, TETRASODIUM SALT [PAL, PAL, TSC]	C.I. DIRECT BLUE 25, TETRASODIUM SALT
3761-53-3 2,7-NAPHTHALENEDISULFONIC ACID, 4-[(2,4-DIMETHYLPHENYL)AZO]-3-HYDROXY-, DIS-ODIUM SALT [PAL, PAL]	PONCEAU MX
6358-29-8 1,3-NAPHTHALENEDISULFONIC ACID, 8-[[4'-[(4-ETHOXYPHENYL)AZO]-3,3'-DIMETHYL] [1,1'-BIPHENYL]-4-YL]AZO]-7-HYDROXY-, DISODIUM SALT [TSC, TSC]	C.I. DIRECT RED 39, DISODIUM SALT
3564-09-8 2,7-NAPHTHALENEDISULFONIC ACID, 3-HY-DROXY-4-[(2,4,5-TRIMETHYLPHENYL)AZO]-, DISODIUM SALT [PAL, PAL]	PONCEAU 3R
32241-08-0 NAPHTHALENE, HEPTACHLORO-	NAPHTHALENE, HEPTACHLORO-

Chemical Name	Reference Name
1335-87-1 NAPHTHALENE, HEXACHLORO- [PAL, TSC, TSC, TSC]	HEXACHLORONAPHTHALENE
90-12-0 NAPHTHALENE, 1-METHYL- [PAL]	1-METHYLNAPHTHALENE
2234-13-1 NAPHTHALENE, OCTACHLORO- [PAL, TSC, TSC, TSC]	OCTACHLORONAPHTHALENE
1321-64-8 NAPHTHALENE, PENTACHLORO- [PAL, TSC, TSC, TSC]	PENTACHLORONAPHTHALENE
1320-27-0 NAPHTHALENE, PENTYL-	NAPHTHALENE, PENTYL-
120-18-3 2-NAPHTHALENESULFONIC ACID	2-NAPHTHALENESULFONIC ACID
93-00-5 2-NAPHTHALENESULFONIC ACID, 6-AMINO-	2-NAPHTHALENESULFONIC ACID, 6-AMINO-E)
93-00-5 2-NAPHTHALENESULFONIC ACID, 6-AMINO- (BROENNER'S ACID) [CAI]	2-NAPHTHALENESULFONIC ACID, 6-AMINO-E)
573-58-0 1-NAPHTHALENESULFONIC ACID, 3,3'-[[1,1'-BIPHENYL]-4,4'-DIYLBIS(AZO)]BIS[(4-AMINO-, DISODIUM SALT [TSC]	C.I. DIRECT RED 28
2429-71-2 1-NAPHTHALENESULFONIC ACID, 3,3'-[3,3'-DIMETHOXY-[1,1'-BIPHENYL]-4,4'-DIYL)BIS (AZO)]BIS[4-HYDROXY-, DISODIUM SALT [TSC]	C.I. DIRECT BLUE 8, DISODIUM SALT
992-59-6 1-NAPHTHALENESULFONIC ACID, 3,3'-[(3,3'-DIMETHYL-[1,1'-BIPHENYL]-4,4'-DIYL)BIS(AZO)] BIS[4-AMINO-, DISODIUM SALT [TSC]	C.I. DIRECT RED 2, DISODIUM SALT
26264-58-4 NAPHTHALENESULFONIC ACID, METHYL-, SODIUM SALT	NAPHTHALENESULFONIC ACID, METHYL-, SODIUM SALT
26545-58-4 NAPHTHALENESULFONIC ACID, METH-LENEBIS-, SODIUM SALT	NAPHTHALENESULFONIC ACID, METHLENEBIS-, SODIUM SALT
RR-01099-1 NAPHTHALENESULFONIC ACID/PARAMETALDEHYDE CONDENSATE, AM-MONIUM SALT	NAPHTHALENESULFONIC ACID/PARAMETALDEHYDE CONDENSATE, AMMONIU
9017-33-8 NAPHTHALENESULFONIC ACID, POLYMER WITH FORMALDEHYDE	NAPHTHALENESULFONIC ACID, POLYMER WITH FORMALDEHYDE
9084-06-4 NAPHTHALENESULFONIC ACID, POLYMER WITH FORMALDEHYDE, SOIDUMSALT [PST]	FORMALDEHYDE-NAPHTHALENESULFONIC ACID POLYMER SODIUM SALT
532-02-5 2-NAPHTHALENESULFONIC ACID, SODIUM SALT	2-NAPHTHALENESULFONIC ACID, SODIUM SALT
81-30-1 1,4,5,8-NAPHTHALENE TETRACARBOXYLIC DIANHYDRIDE	1,4,5,8-NAPHTHALENE TETRACARBOXYLIC DIANHY-DRIDE
1335-88-2 NAPHTHALENE, TETRACHLORO- [PAL, TSC, TSC, TSC]	TETRACHLORONAPHTHALENE
119-64-2 NAPHTHALENE, 1,2,3,4-TETRAHYDRO- [PAL]	TETRAHYDRONAPHTHALENE
RR-00165-0 NAPHTHALENE, 1,2,3,4-TETRAHYDRO(1-PHENYLETHYL)	NAPHTHALENE, 1,2,3,4-TETRAHYDRO(1-PHENYLETHYL)
91-60-1 2-NAPHTHALENETHIOL	2-NAPHTHALENETHIOL
1321-65-9 NAPHTHALENE, TRICHLORO- [PAL, TSC, TSC, TSC]	TRICHLORONAPHTHALENE
6358-53-8 2-NAPHTHALENOL, 1-[(2,5-DIMETHOXYPHENYL)AZO]- [PAL, PAL]	CITRUS RED NO. 2
63-25-2 1-NAPHTHALENOL, METHYLCARBA-MATE [PAL]	CARBARYL
85-83-6 2-NAPHTHALENOL, 1-[[2-METHYL-4-[(2-METHYLPHENYL)AZO]PHENY	2-NAPHTHALENOL, 1-[[2-METHYL-4-[(2-METHYLPHENYL)AZO]PHENY
2646-17-5 2-NAPHTHALENOL, 1-[(2-METHYLPHENYL) AZO]- [PAL, PAL]	OIL ORANGE SSDIYL)BIS(AZO)]BIS[5-AMINO-4-HY-DROXY-, TETRASODIUM SALT
64742-94-5 NAPHTHA (PETROLEUM), HEAVY AROMATIC	NAPHTHA (PETROLEUM), HEAVY AROMATIC
68527-23-1 NAPHTHA (PETROLEUM), LIGHT STEAM-CRACKED AROMATIC	NAPHTHA (PETROLEUM), LIGHT STEAM-CRACKED AROMATIC
RR-00076-0 NAPHTHA (RUBBER SOLVENT) [MIN]	RUBBER SOLVENT (NAPHTHA)

Chemical Name	Reference Name
64742-89-8 NAPHTHA, SOLVENT [NJL]	SOLVENT NAPHTHA (PETROLEUM), LIGHT ALIPHATIC
8030-30-6 NAPHTHA VM&P [MSL]	BENZIN
8030-30-6 NAPHTHA V.M. & P. [NFP, NFP]	BENZIN
8030-30-6 NAPHTHA (VM&P NAPHTHA) [MIN]	BENZIN
1338-24-5 NAPHTHENIC ACIDS [HM1, HM1, PAL]	NAPTHENIC ACID
RR-01100-7 NAPHTHENIC ACID SOAP OF N-ALKYL (C16-C18)TRIMETHYLENEDIAMINE	NAPHTHENIC ACID SOAP OF N-ALKYL (C16-C18) TRIMETHYLENEDIAMINEM SALT
12001-85-3 NAPHTHENIC ACIDS, ZINC SALTS	NAPHTHENIC ACIDS, ZINC SALTS
64742-68-3 NAPHTHENIC OILS (PETROLEUM), CATALYTIC DEWAXED HEAVY	NAPHTHENIC OILS (PETROLEUM), CATALYTIC DE-WAXED HEAVYNAPHTHENIC
64742-69-4 NAPHTHENIC OILS (PETROLEUM), CATALYTIC DEWAXED LIGHT	NAPHTHENIC OILS (PETROLEUM), CATALYTIC DE-WAXED LIGHTHEAVY
192-65-4 NAPHTHO[1,2,3,4-DEF]CHRYSENE [PAL, PAL]	DIBENZO(A,E)PYRENE
90-15-3 ALPHA-NAPHTHOL	ALPHA-NAPHTHOL
90-15-3 1-NAPHTHOL [TSC, TSC, TSC]	ALPHA-NAPHTHOL
135-19-3 2-NAPHTHOL	2-NAPHTHOL
135-19-3 NAPHTHOL (B-) [HON]	2-NAPHTHOL
135-19-3 BETA-NAPHTHOL [NFP, NFP]	2-NAPHTHOL
93-18-5 BETA-NAPHTHOL ETHYL ETHER	BETA-NAPHTHOL ETHYL ETHER
567-18-0 NAPHTHOLSULFONIC ACID (1-)	NAPHTHOLSULFONIC ACID (1-)
130-15-4 1,4-NAPHTHOQUINONE [PQL, RCA]	1,4-NAPHTHALENEDIONE
91-59-8 BETA-NAPHTHYLAMINE	BETA-NAPHTHYLAMINE
91-59-8 2-NAPHTHYLAMINE [AUS, AUS, C65, C65, CAB, EP2, FLA, IAR, MAK, MAK, MSL, NTP, PQL, RCA, UTS]	BETA-NAPHTHYLAMINE
134-32-7 ALPHA-NAPHTHYLAMINE	ALPHA-NAPHTHYLAMINE
134-32-7 1-NAPHTHYLAMINE [C65, CAB, EP2, IAR, MSL, NFP, NFP, NJL, NJL, PQL]	ALPHA-NAPHTHYLAMINE
81-16-3 NAPHTHYLAMINE SULFONIC ACID (2,1-) [HON]	2-AMINO-1-NAPHTHALENESULFONIC ACID
84-86-6 NAPHTHYLAMINE SULFONIC ACID (1,4-)	NAPHTHYLAMINE SULFONIC ACID (1,4-)
91-23-6 O-NITROANISOLE [HON]	2-NITROANISOLE
91-59-8 NAPHTHYLAMINE (2-) [HON]	BETA-NAPHTHYLAMINE
100-17-4 P-NITROANISOLE [HON]	4-NITROANISOLE
119-79-9 1-NAPHTHYLAMINE-6-SULFONIC ACID	1-NAPHTHYLAMINE-6-SULFONIC ACID
134-32-7 NAPHTHYLAMINE (1-) [HON]	ALPHA-NAPHTHYLAMINE
RR-01449-3 NAPHTHYL AMINEPERCHLORATE	NAPHTHYL AMINEPERCHLORATE
3173-72-6 1,5-NAPHTHYLENE DIISOCYANATE [MAK, MAK, MAK]	NAPHTHALENE, 1,5-DIISOCYANATO-
1465-25-4 N-(1-NAPHTHYL)ETHYLENEDIAMINE DIHYDROCHLORIDE	N-(1-NAPHTHYL)ETHYLENEDIAMINE DIHYDROCHLORIDE
63-25-2 1-NAPHTHYL N-METHYL CARBAMATE [ONT]	CARBARYL
7090-25-7 1-NAPHTHYL METHYLNITROSOCARBONATE	1-NAPHTHYL METHYLNITROSOCARBONATE
17518-47-7 2-(1-NAPHTHYL)THIOACETAMIDE	2-(1-NAPHTHYL)THIOACETAMIDE
86-88-4 1-(1-NAPHTHYL)-2-THIOUREA [CAL]	ANTU
86-88-4 ALPHA-NAPHTHYLTHIOUREA [EPA]	ANTU
86-88-4 ALPHA NAPHTHYL THIOUREA (A.N.T.U.) [FLA]	ANTU
86-88-4 1-NAPHTYLTHIOUREA (ANTU) [IAR]	ANTU
86-88-4 1-NAPHTHYLTHIOUREA [MAK, MAK]	ANTU
86-88-4 ALPHA-NAPHTHYLTHIOUREA [ONT, RCU, RCU]	ANTU
30553-04-9 NAPHTHYLTHIOUREA	NAPHTHYLTHIOUREA

Chemical Name	Reference Name
6950-84-1 1-NAPHTHYLUREA	1-NAPHTHYLUREA
6950-84-1 NAPHTHYLUREA [HM1, HM1]	1-NAPHTHYLUREA
13114-62-0 2-NAPHTHYLUREA	2-NAPHTHYLUREA
3173-72-6 1,5-NAPHTHALENE DIISOCYANATE [313]	NAPHTHALENE, 1,5-DIISOCYANATO-
1338-24-5 NAPTHENIC ACID	NAPTHENIC ACID
8006-14-2 NATURAL GAS	NATURAL GAS
68425-31-0 NATURAL GASOLINE [NJL, NJL, HM1, HM1]	GASOLINE, NATURAL
9006-04-6 NATURAL RUBBERS	NATURAL RUBBERS
117-39-5 NATURAL YELLOW 10	NATURAL YELLOW 10
7786-17-6 NAUGAUHITE	NAUGAUHITE
8008-20-6 NAVY FUELS JP-5 [CSR, CSR]	KEROSENE
8002-64-0 NEATSFOOT OIL	NEATSFOOT OIL
1317-43-7 NEMALITE/BRUCITE	NEMALITE/BRUCITE
64093-79-4 NEOCHROMIUM [NJL]	CHROMOSULFURIC ACID
51240-95-0 NEODECANEPEROXOIC ACID, 1,1,3,3-TETRAM-ETHYLBUTYL ESTER	NEODECANEPEROXOIC ACID, 1,1,3,3-TETRAMETHYL-BUTYL ESTEROPENYL)OXY]ETHOXY]CARBONYL]AMINO]CYCLOHEXYL]METHYL]AMINO] CARBONYL]OXY]ETHYL ESTER
26896-20-8 NEODECANOIC ACID	NEODECANOIC ACID
26761-45-5 NEODECANOIC ACID, 2,3-EPOXYPROPYL ES-TER [TSC, TSC]	NEODECANOIC ACID, OXIRANYLMETHYL ESTER
26761-45-5 NEODECANOIC ACID, OXIRANYLMETHYL ESTER	NEODECANOIC ACID, OXIRANYLMETHYL ESTER
39049-04-2 NEODECANOIC ACID, ZIRCONIUM SALT	NEODECANOIC ACID, ZIRCONIUM SALTATE
22095-52-9 NEODYMIUM 136	NEODYMIUM 136
15700-34-2 NEODYMIUM 138	NEODYMIUM 138
18411-36-4 NEODYMIUM 139	NEODYMIUM 139
RR-00401-3 NEODYMIUM 139M	NEODYMIUM 139M
14877-64-6 NEODYMIUM 141	NEODYMIUM 141
14269-74-0 NEODYMIUM 147	NEODYMIUM 147
15749-81-2 NEODYMIUM 149	NEODYMIUM 149
15690-82-1 NEODYMIUM 151	NEODYMIUM 151
10024-93-8 NEODYMIUM CHLORIDE	NEODYMIUM CHLORIDE
75-83-2 NEOHEXANE	NEOHEXANE
1405-10-3 NEOMYCIN SULFATE	NEOMYCIN SULFATE
7440-01-9 NEON	NEON
RR-01651-3 NEONONANOIC ACID, ETHENYL ESTER	NEONONANOIC ACID, ETHENYL ESTER-(PENTYLOXY)PHENYL]-, AND ACETAMIDE, N-[2-AMINO-4-(PENTY-LOXY)PHENYL]
126-30-7 NEOPENTYL GLYCOL [NFP, NFP, PST]	1,3-PROPANEDIOL, 2,2-DIMETHYL-
2223-82-7 NEOPENTYL GLYCOL DIACRYLATE	NEOPENTYL GLYCOL DIACRYLATE
17557-23-2 NEOPENTYL GLYCOL DIGLYCIDYL ETHER [EPA, OTS]	OXIRANE, 2,2'-[(2,2-DIMETHYL-1,3-PROPANEDIYL)BIS(OXYMETHYLEN
9010-98-4 NEOPRENE RUBBER	NEOPRENE RUBBER
7696-12-0 NEOPYNAMIN	NEOPYNAMIN
29687-52-3 NEPTUNIUM 232	NEPTUNIUM 232
15832-46-9 NEPTUNIUM 233	NEPTUNIUM 233
15116-90-2 NEPTUNIUM 234	NEPTUNIUM 234
15700-37-5 NEPTUNIUM 235	NEPTUNIUM 235
15700-36-4 NEPTUNIUM 236	NEPTUNIUM 236
13994-20-2 NEPTUNIUM 237	NEPTUNIUM 237
15766-25-3 NEPTUNIUM 238	NEPTUNIUM 238

Chemical Name	Reference Name
13968-59-7 NEPTUNIUM 239	NEPTUNIUM 239
15690-84-3 NEPTUNIUM 240	NEPTUNIUM 240
56391-57-2 NETILMICIN SULFATE	NETILMICIN SULFATE
3033-62-3 NIAX A 1	NIAX A 1NE
62765-93-9 NIAX CATALYST ESN	NIAX CATALYST ESN
7440-02-0 NICKEL	NICKEL
7440-02-0 NICKEL (METAL) [QBC]	NICKEL
14932-64-0 NICKEL 56	NICKEL 56
13981-99-2 NICKEL 57	NICKEL 57
14336-70-0 NICKEL 59	NICKEL 59
13981-37-8 NICKEL 63	NICKEL 63
14833-49-9 NICKEL 65	NICKEL 65
15766-33-3 NICKEL 66	NICKEL 66
7440-02-0 NICKEL METAL [MEX]	NICKEL
RR-00798-7 NICKEL SULFIDE, ROASTING, FUME AND DUST [MEX, MEX]	NICKEL SULFIDE ROASTING
373-02-4 NICKEL(II) ACETATE	NICKEL(II) ACETATE
373-02-4 NICKEL ACETATE [NTP]	NICKEL(II) ACETATE
RR-01007-1 NICKEL ACRYLATE COMPLEX	NICKEL ACRYLATE COMPLEX
15699-18-0 NICKEL AMMONIUM SULFATE	NICKEL AMMONIUM SULFATE
15699-18-0 NICKEL(II) AMMONIUM SULFATE [WHM]	NICKEL AMMONIUM SULFATE
RR-00038-4 NICKEL AND CERTAIN NICKEL COMPOUNDS [C65]	NICKEL SOLUBLE COMPOUNDS
68958-88-3 NICKELATE(6-), [[[1,2-ETHANEDIYLBIS[NITRILOBIS(METHYLENE)]TETRAKIS[PHOSPHONATO](8-)], PENTASODIUM HYDROGEN, (OC-6-21)-	NICKELATE(6-), [[[1,2-ETHANEDIYLBIS[NITRILOBIS(METHYLENE)]TETRAKIS[PHOSPHONATO](8-)], PENTAPOTASSIUM HYDROGEN, (OC-6-21)-
3333-67-3 NICKEL CARBONATE	NICKEL CARBONATE
3333-67-3 NICKEL(II) CARBONATE [WHM]	NICKEL CARBONATE
12612-55-4 NICKEL CARBONYL [BC1, HM1, HM1, HM1, HM1, QBC]	NICKEL CARBONYL (VAN)
13463-39-3 NICKEL CARBONYL	NICKEL CARBONYL
13463-39-3 NICKEL CARBONYL (NICKEL TETRACARBONYL) [OS3]	NICKEL CARBONYL
13463-39-3 NICKEL CARBONYL NI(CO)4), (T-4)- [PAL, PAL]	NICKEL CARBONYL
12612-55-4 NICKEL CARBONYL (VAN)	NICKEL CARBONYL (VAN)
RR-00376-9 NICKEL CATALYST	NICKEL CATALYST
7718-54-9 NICKEL(II) CHLORIDE	NICKEL(II) CHLORIDE
7718-54-9 NICKEL CHLORIDE [EP2, EPA, NJL]	NICKEL(II) CHLORIDE
37211-05-5 NICKEL CHLORIDE	NICKEL CHLORIDE
7718-54-9 NICKEL CHLORIDE (NICL2) [PAL]	NICKEL(II) CHLORIDE
7791-20-0 NICKEL(II) CHLORIDE HEXAHYDRATE (1:2:6)	NICKEL(II) CHLORIDE HEXAHYDRATE (1:2:6)
RR-01101-8 NICKEL COMPLEX OF DIMETHYLAMINO METHYL OCTYL PHENOL	NICKEL COMPLEX OF DIMETHYLAMINO METHYL OCTYL PHENOL
RR-01673-9 NICKEL SALT OF AN ORGANO COMPOUND CONTAINING NITROGEN	NICKEL SALT OF AN ORGANO COMPOUND CONTAINING NITROGEN
RR-00800-4 NICKEL COMPOUNDS	NICKEL COMPOUNDS
RR-00800-4 NICKEL COMPOUNDS, N.O.S. [NJL, RCU]	NICKEL COMPOUNDS
557-19-7 NICKEL CYANIDE	NICKEL CYANIDE
15521-65-0 NICKEL DIMETHYLDITHIOCARBAMATE	NICKEL DIMETHYLDITHIOCARBAMATE
7440-02-0 NICKEL, ELEMENTAL [WHM]	NICKEL

Chemical Name	Reference Name
14708-14-6 NICKEL(II) FLUOBORATE	NICKEL(II) FLUOBORATE
26043-11-8 NICKEL(II) FLUOSILICATE	NICKEL(II) FLUOSILICATE
15843-02-4 NICKEL FORMATE	NICKEL FORMATE
11113-74-9 NICKEL HYDROXIDE [PAL]	NICKEL HYDROXIDE (VAN)
12054-48-7 NICKEL HYDROXIDE	NICKEL HYDROXIDE
12054-48-7 NICKEL(II) HYDROXIDE [WHM]	NICKEL HYDROXIDE
12125-56-3 NICKEL HYDROXIDE	NICKEL HYDROXIDE
11113-74-9 NICKEL HYDROXIDE (VAN)	NICKEL HYDROXIDE (VAN)
RR-00571-0 NICKEL, INORGANIC COMPOUNDS	NICKEL, INORGANIC COMPOUNDS
RR-00522-1 NICKEL INSOLUBLE COMPOUNDS	NICKEL INSOLUBLE COMPOUNDS
13462-90-3 NICKEL(II) IODIDE	NICKEL(II) IODIDE
7440-02-0 NICKEL METAL [BC1, BC1, BC1, ONT, ONT]	NICKEL
7440-02-0 NICKEL METAL AND OTHER COMPOUNDS [NIO, NIO, NIO]	NICKEL
7440-02-0 NICKEL, METAL, AND SOLUBLE COMPOUNDS [MIN]	NICKEL
1313-99-1 NICKEL MONOXIDE [IAR]	NICKEL OXIDE
13138-45-9 NICKEL NITRATE [FLA, HM1, HM1, NJL]	NICKEL NITRATE (2+ SALT)
13138-45-9 NICKEL(II) NITRATE [WHM]	NICKEL NITRATE (2+ SALT)
14216-75-2 NICKEL NITRATE	NICKEL NITRATE
13478-00-7 NICKEL(II) NITRATE, HEXAHYDRATE (1:2:6)	NICKEL(II) NITRATE, HEXAHYDRATE (1:2:6)
13138-45-9 NICKEL NITRATE (2+ SALT)	NICKEL NITRATE (2+ SALT)
17861-62-0 NICKEL NITRITE	NICKEL NITRITE
1271-28-9 NICKELOCENE	NICKELOCENE
1313-99-1 NICKEL OXIDE	NICKEL OXIDE
1313-99-1 NICKEL (II) OXIDE [CSR, ONT, ONT]	NICKEL OXIDE
1313-99-1 NICKEL(II) OXIDE [WHM]	NICKEL OXIDE
1314-06-3 NICKEL(III) OXIDE [WHM]	NICKEL TRIOXIDE
11099-02-8 NICKEL OXIDE	NICKEL OXIDE
1314-06-3 NICKEL OXIDE (NI2O3) [PAL, PAL]	NICKEL TRIOXIDE
1313-99-1 NICKEL OXIDE (NIO) [PAL, PAL]	NICKEL OXIDE
1314-06-3 NICKEL (III) OXIDE [ONT, ONT]	NICKEL TRIOXIDE
13637-71-3 NICKEL PERCHLORATE	NICKEL PERCHLORATE
RR-01450-6 NICKEL PICRATE	NICKEL PICRATE
14220-17-8 NICKEL POTASSIUM CYANIDE	NICKEL POTASSIUM CYANIDE
RR-00084-0 NICKEL REFINERY DUST FROM THE PYROMETALLURGICAL PROCESS	NICKEL REFINERY DUST FROM THE PYROMETALLURGICAL PROCESS
RR-00084-0 NICKEL REFINERY DUST FROM THE PYROMETALLURGICAL PROCESS [C65, C65]	NICKEL REFINERY DUST FROM THE PYROMETALLURGICAL PROCESS
RR-00084-0 NICKEL REFINING [PAL, PAL]	NICKEL REFINERY DUST FROM THE PYROMETALLURGICAL PROCESS
RR-00038-4 NICKEL SOLUBLE COMPOUNDS	NICKEL SOLUBLE COMPOUNDS
RR-00038-4 NICKEL, SOLUBLE COMPOUNDS [BC1, BC1, BC1, QBC]	NICKEL SOLUBLE COMPOUNDS
12035-72-2 NICKEL SUBSULFIDE	NICKEL SUBSULFIDE
12035-72-2 NICKEL SUBSULFIDE (NI3S2) [WHM]	NICKEL SUBSULFIDE
12035-72-2 NICKEL SUBSULPHIDE [FLA, MSL]	NICKEL SUBSULFIDE
7786-81-4 NICKEL SULFATE	NICKEL SULFATE
7786-81-4 NICKEL(II) SULFATE [WHM]	NICKEL SULFATE
10101-97-0 NICKEL(II) SULFATE HEXAHYDRATE (1:1:6)	NICKEL(II) SULFATE HEXAHYDRATE (1:1:6)

Chemical Name	Reference Name
10101-97-0 NICKEL SULFATE HEXAHYDRATE [CSR]	NICKEL(II) SULFATE HEXAHYDRATE (1:1:6)
11113-75-0 NICKEL(II) SULFIDE	NICKEL(II) SULFIDE
12035-72-2 NICKEL SULFIDE (NI3S2) [PAL, PAL]	NICKEL SUBSULFIDE
11113-75-0 NICKEL (II) SULFIDE [ONT, ONT]	NICKEL(II) SULFIDE
12035-72-2 NICKEL (III) SULFIDE [ONT, ONT]	NICKEL SUBSULFIDE
RR-00798-7 NICKEL SULFIDE ROASTING	NICKEL SULFIDE ROASTING
RR-00798-7 NICKEL SULFIDE ROASTING, FUME AND DUST [ALB, ALB, ALB, BC1, BC1, QBC, QBC, TLV, TLV]	NICKEL SULFIDE ROASTING
RR-01204-4 NICKEL SULFIDES, CRYSTALLINE	NICKEL SULFIDES, CRYSTALLINE
12142-88-0 NICKEL TELLURIDE	NICKEL TELLURIDE
13463-39-3 NICKEL TETRACARBONYL [HM1, HM1]	NICKEL CARBONYL
12035-39-1 NICKEL TITANIUM OXIDE	NICKEL TITANIUM OXIDE
1314-06-3 NICKEL TRIOXIDE	NICKEL TRIOXIDE
RR-00522-1 NICKEL, WATER-INSOLUBLE COMPOUNDS, N.O.S. [WHM]	NICKEL INSOLUBLE COMPOUNDS
RR-00038-4 NICKEL, WATER-SOLUBLE COMPOUNDS [ONT]	NICKEL SOLUBLE COMPOUNDS
RR-00038-4 NICKEL, WATER-SOLUBLE INORGANIC COMPOUNDS, N.O.S. [WHM]	NICKEL SOLUBLE COMPOUNDS
98-92-0 NICOTINAMIDE	NICOTINAMIDE
54-11-5 NICOTINE	NICOTINE
54-11-5 NICOTINE AND SALTS [EPA]	NICOTINE
RR-00378-1 NICOTINE, COMPOUNDS AND PREPARATIONS, N.O.S.	NICOTINE, COMPOUNDS AND PREPARATIONS, N.O.S.
RR-00378-1 NICOTINE COMPOUNDS, N.O.S. [HM1, HM1]	NICOTINE, COMPOUNDS AND PREPARATIONS, N.O.S.
2820-51-1 NICOTINE HYDROCHLORIDE	NICOTINE HYDROCHLORIDE
29790-52-1 NICOTINE SALICYLATE	NICOTINE SALICYLATE
54-11-5 NICOTINE, & SALTS [RCU, RCU]	NICOTINE
RR-00378-1 NICOTINE SALTS [313]	NICOTINE, COMPOUNDS AND PREPARATIONS, N.O.S.
6505-86-8 NICOTINE SULFATE	NICOTINE SULFATE
3275-73-8 NICOTINE TARTRATE	NICOTINE TARTRATE
59-67-6 NICOTINIC ACID [PST]	3-PYRIDINECARBOXYLIC ACID
7440-03-1 NIOBIUM	NIOBIUM
14681-74-4 NIOBIUM 88	NIOBIUM 88
15700-40-0 NIOBIUM 89	NIOBIUM 89
14681-65-3 NIOBIUM 90	NIOBIUM 90
RR-00400-2 NIOBIUM 93M	NIOBIUM 93M
14681-63-1 NIOBIUM 94	NIOBIUM 94
13967-76-5 NIOBIUM 95	NIOBIUM 95
RR-00398-5 NIOBIUM 95M	NIOBIUM 95M
15832-32-3 NIOBIUM 96	NIOBIUM 96
18496-04-3 NIOBIUM 97	NIOBIUM 97
15700-41-1 NIOBIUM 98	NIOBIUM 98
10026-12-7 NIOBIUM CHLORIDE	NIOBIUM CHLORIDE
61-57-4 NIRIDAZOLE	NIRIDAZOLE
139-94-6 NITHIAZIDE	NITHIAZIDELIDINONE
25168-04-1 NITROXYLENE [HON]	NITROXYLOL
479-45-8 NITRAMINE [ONT, ONT]	TETRYL
1929-82-4 NITRAPYRIN	NITRAPYRIN
1929-82-4 NITRAPYRIN [QBC, QBC]	NITRAPYRIN

Chemical Name	Reference Name
1929-82-4 NITRAPYRIN (2-CHLORO-6-(TRICHLOROMETHYL)PYRIDINE) [313]	NITRAPYRIN
14797-55-8 NITRATE	NITRATE
RR-01770-9 NITRATE COMPOUNDS	NITRATE COMPOUNDS
RR-01451-7 NITRATED PAPER	NITRATED PAPER
RR-00936-9 NITRATE POLYETHER POLYOL	NITRATE POLYETHER POLYOL
14797-55-8 NITRATES [MX2]	NITRATE
RR-00379-2 NITRATES, INORGANIC, N.O.S.	NITRATES, INORGANIC, N.O.S.
RR-01345-6 NITRATING ACID MIXTURES	NITRATING ACID MIXTURES
RR-01345-6 NITRATING ACID, MIXTURES [NJL]	NITRATING ACID MIXTURES
7697-37-2 NITRIC ACID	NITRIC ACID
6484-52-2 NITRIC ACID AMMONIUM SALT [PAL]	AMMONIUM NITRATE
10022-31-8 NITRIC ACID, BARIUM SALT [PAL]	BARIUM NITRATE
13597-99-4 NITRIC ACID, BERYLLIUM SALT [PAL]	BERYLLIUM NITRATE
7787-55-5 NITRIC ACID, BERYLLIUM SALT, TRIHYDRATE [PAL]	BERYLLIUM NITRATE TRIHYDRATE
928-45-0 NITRIC ACID, BUTYL ESTER [PAL]	BUTYL NITRATE
10141-05-6 NITRIC ACID, COBALT(2+) SALT [PAL]	COBALT(II) NITRATE
3251-23-8 NITRIC ACID, COPPER(2+) SALT [PAL]	CUPRIC NITRATE
625-58-1 NITRIC ACID, ETHYL ESTER [PAL]	ETHYL NITRATE
10421-48-4 NITRIC ACID, IRON(3+) SALT [PAL]	FERRIC NITRATE
10099-74-8 NITRIC ACID, LEAD (2+) SALT [MSL]	LEAD NITRATE
10099-74-8 NITRIC ACID, LEAD(2+) SALT [PAL]	LEAD NITRATE
10377-60-3 NITRIC ACID, MAGNESIUM SALT [PAL]	MAGNESIUM NITRATE
10045-94-0 NITRIC ACID, MERCURY(2+) SALT [PAL]	MERCURIC NITRATE
10415-75-5 NITRIC ACID, MERCURY(1+) SALT [PAL]	MERCUROUS NITRATE
7782-86-7 NITRIC ACID, MERCURY(1+) SALT, MONOHYDRATE [PAL]	MERCUROUS NITRATE
13138-45-9 NITRIC ACID, NICKEL (2+) SALT [MSL]	NICKEL NITRATE (2+ SALT)
13138-45-9 NITRIC ACID, NICKEL(2+) SALT [PAL]	NICKEL NITRATE (2+ SALT)
14216-75-2 NITRIC ACID, NICKEL SALT [MSL, PAL]	NICKEL NITRATE
1002-16-0 NITRIC ACID, PENTYL ESTER [PAL]	AMYL NITRATE
7757-79-1 NITRIC ACID POTASSIUM SALT [PAL]	POTASSIUM NITRATE
627-13-4 NITRIC ACID, PROPYL ESTER [PAL]	N-PROPYL NITRATE
7761-88-8 NITRIC ACID SILVER(1+) SALT [PAL]	SILVER NITRATE
7631-99-4 NITRIC ACID SODIUM SALT [PAL]	SODIUM NITRATE
10042-76-9 NITRIC ACID, STRONTIUM SALT [PAL]	STRONTIUM NITRATE
13823-29-5 NITRIC ACID, THORIUM(4+) SALT [PAL]	THORIUM NITRATE
15905-86-9 NITRIC ACID, URANIUM SALT	NITRIC ACID, URANIUM SALT
7779-88-6 NITRIC ACID, ZINC SALT [PAL]	ZINC NITRATE
13746-89-9 NITRIC ACID, ZIRCONIUM(4+) SALT [PAL]	ZIRCONIUM NITRATE
10102-43-9 NITRIC OXIDE	NITRIC OXIDE
10102-43-9 NITRIC OXIDE (NO) [QBC]	NITRIC OXIDE
10102-43-9 NITRIC OXIDE [NITROGEN OXIDE (NO)] [EP3]	NITRIC OXIDE
139-13-9 NITRILOTRIACETIC ACID [313, C65, C65, CAL, CAN, IAR, MIN, NJL, NJL, NTP]	NITRILOTRIACETIC ACID (NTA)
139-13-9 NITRILOTRIACETIC ACID (NTA)	NITRILOTRIACETIC ACID (NTA)
RR-01745-8 NITRILOTRIACETIC ACID (SALTS)	NITRILOTRIACETIC ACID (SALTS)

Chemical Name	Reference Name
15467-20-6 NITRILOTRIACETIC ACID DISODIUM SALT	NITRILOTRIACETIC ACID DISODIUM SALT
5064-31-3 NITRILOTRIACETIC ACID (NTA) TRISODIUM [MSL]	NITRILOTRIACETIC ACID TRISODIUM SALT
18662-53-8 NITRILOTRIACETIC ACID, TRISODIUM SALT MONOHYDRATE [CAB]	TRISODIUM NITRILOTRIACETATE MONOHYDRATE
18662-53-8 NITRILOTRIACETIC ACID TRISODIUM MONO-HYDRATE [CSR, CSR]	TRISODIUM NITRILOTRIACETATE MONOHYDRATE
5064-31-3 NITRILOTRIACETIC ACID TRISODIUM SALT	NITRILOTRIACETIC ACID TRISODIUM SALT
18662-53-8 NITRILOTRIACETIC ACID, TRISODIUM SALT MONOHYDRATE [C65, C65]	TRISODIUM NITRILOTRIACETATE MONOHYDRATE
66507-71-9 2,2',2"-NITRILOTRISETHANOL POLYMER WITH 1,4-DICHLORO-2-BUTENE AND N,N,N',N'-TE-TRAMETHYL-2-BUTENE-1,4-DIAMINE	2,2',2"-NITRILOTRISETHANOL POLYMER WITH 1,4-DICHLORO-2-BUTEN
6419-19-8 NITRILOTRIS(METHYLENE)TRISPHOSPHONIC ACID [PST]	AMINOTRI(METHYLENEPHOSPHONIC ACID)
14797-65-0 NITRITE	NITRITE
14797-65-0 NITRITES [ATS, MX2]	NITRITE
RR-00380-5 NITRITES, INORGANIC, N.O.S.	NITRITES, INORGANIC, N.O.S.
602-87-9 5-NITROACENAPHTHENE	5-NITROACENAPHTHENE
602-87-9 NITROACENAPHTHENE [FLA]	5-NITROACENAPHTHENE
1777-84-0 3-NITRO-P-ACETOPHENETIDE	3-NITRO-P-ACETOPHENETIDE
1777-84-0 3'-NITRO-P-ACETOPHENETIDE [MSL]	3-NITRO-P-ACETOPHENETIDE
121-89-1 3'-NITROACETOPHENONE	3'-NITROACETOPHENONE
119-34-6 2-NITRO-4-AMINOPHENOL [MAK, MAK]	4-HYDROXY-3-NITROANILINE
99-55-8 4-NITRO-2-AMINOTOLUENE [MAK]	5-NITRO-O-TOLUIDINE
88-74-4 O-NITROANILINE [UTS]	O-NITROANILINE
99-09-2 M-NITROANILINE	M-NITROANILINE
99-09-2 NITROANILINE, 3- [OTS]	M-NITROANILINE
100-01-6 P-NITROANILINE	P-NITROANILINE
100-01-6 4-NITROANILINE [GBR, GBR, MAK, MAK]	P-NITROANILINE
29757-24-2 N-NITROANILINE	N-NITROANILINE
29757-24-2 O-NITROANILINE [NJL, NJL]	N-NITROANILINE
100-01-6 NITROANILINE (PARA NITROANILINE) [OS3]	P-NITROANILINE
29757-24-2 NITROANILINES (O-; M-; P-) [HM1, HM1]	N-NITROANILINE
99-59-2 5-NITRO-O-ANISIDINE	5-NITRO-O-ANISIDINE
99-59-2 5-NITRO-ORTHO-ANISIDINE [IAR]	5-NITRO-O-ANISIDINE
91-23-6 2-NITROANISOLE	2-NITROANISOLE
91-23-6 O-NITROANISOLE [C65, CSR, NTP]	2-NITROANISOLE
100-17-4 4-NITROANISOLE	4-NITROANISOLE
100-17-4 NITROANISOLE [HM1, HM1]	4-NITROANISOLE
100-17-4 P-NITROANISOLE [NJL]	4-NITROANISOLE
555-03-3 3-NITROANISOLE	3-NITROANISOLE
602-60-8 9-NITROANTHRACENE	9-NITROANTHRACENE
619-17-0 4-NITROANTHRANILIC ACID	4-NITROANTHRANILIC ACID
555-16-8 P-NITROBENZALDEHYDE	P-NITROBENZALDEHYDE
98-95-3 NITROBENZENE	NITROBENZENE
22751-24-2 M-NITROBENZENE DIAZONIUM PERCHLO-RATE	M-NITROBENZENE DIAZONIUM PERCHLORATE
98-47-5 M-NITROBENZENESULFONIC ACID	M-NITROBENZENESULFONIC ACID
31212-28-9 NITROBENZENESULFONIC ACID [NJL, NJL]	NITROBENZENESULPHONIC ACID

Chemical Name	Reference Name
127-68-4 M-NITROBENZENESULFONIC ACID, SODIUM SALT	M-NITROBENZENESULFONIC ACID, SODIUM SALT
31212-28-9 NITROBENZENESULPHONIC ACID	NITROBENZENESULPHONIC ACID
94-52-0 6-NITROBENZIMIDAZOLE	6-NITROBENZIMIDAZOLE
20268-51-3 7-NITROBENZO[A]ANTHRACENE	7-NITROBENZO[A]ANTHRACENE
63041-90-7 6-NITROBENZO(A)PYRENE	6-NITROBENZO(A)PYRENE
62-23-7 P-NITROBENZOIC ACID	P-NITROBENZOIC ACID
121-92-6 M-NITROBENZOIC ACID	M-NITROBENZOIC ACID
RR-01687-5 NITROBENZOIC ACID OCTYL ESTER	NITROBENZOIC ACID OCTYL ESTER
RR-00198-9 6-NITRO-2(3H)-BENZOOXAZOLONE	6-NITRO-2(3H)-BENZOOXAZOLONE
2338-12-7 5-NITROBENZOTRIAZOL	5-NITROBENZOTRIAZOL
98-46-4 3-NITROBENZOTRIFLUORIDE	3-NITROBENZOTRIFLUORIDE
98-46-4 1,3-NITROBENZOTRIFLUORIDE [NFP, NFP]	3-NITROBENZOTRIFLUORIDE
98-46-4 M-NITROBENZOTRIFLUORIDE [WHM]	3-NITROBENZOTRIFLUORIDE
384-22-5 O-NITROBENZOTRIFLUORIDE	O-NITROBENZOTRIFLUORIDE
402-54-0 P-NITROBENZOTRIFLUORIDE	P-NITROBENZOTRIFLUORIDE
RR-01305-8 NITROBENZOTRIFLUORIDES	NITROBENZOTRIFLUORIDES
86-00-0 2-NITROBIPHENYL	2-NITROBIPHENYL
92-93-3 4-NITROBIPHENYL	4-NITROBIPHENYL
28984-85-2 NITROBIPHENYL	NITROBIPHENYL
577-19-5 O-NITROBROMOBENZENE	O-NITROBROMOBENZENE
585-79-5 M-NITROBROMOBENZENE	M-NITROBROMOBENZENE
586-78-7 P-NITROBROMOBENZENE	P-NITROBROMOBENZENE
586-78-7 NITROBROMOBENZENE [NJL]	P-NITROBROMOBENZENE
RR-01355-8 NITROBROMOBENZENES	NITROBROMOBENZENES
2224-44-4 4-(2-NITROBUTYL)MORPHOLINE	4-(2-NITROBUTYL)MORPHOLINE
9004-70-0 NITROCELLULOSE	NITROCELLULOSE
88-73-3 O-NITROCHLOROBENZENE	O-NITROCHLOROBENZENE
100-00-5 P-NITROCHLOROBENZENE	P-NITROCHLOROBENZENE
121-73-3 M-NITROCHLOROBENZENE	M-NITROCHLOROBENZENE
25167-93-5 NITROCHLOROBENZENE [FLA, MSL, NFP, NFP]	CHLORONITROBENZENE
118-83-2 1-NITRO-4-CHLOROBENZO-2-TRIFLUORIDE	1-NITRO-4-CHLOROBENZO-2-TRIFLUORIDE
121-17-5 2-NITRO-1-CHLOROBENZO-4-TRIFLUORIDE	2-NITRO-1-CHLOROBENZO-4-TRIFLUORIDE
121-17-5 3-NITRO-4-CHLOROBENZOTRIFLUORIDE [HM1, HM1, HM1, WHM]	2-NITRO-1-CHLOROBENZO-4-TRIFLUORIDE
777-37-7 4-NITRO-1-CHLOROBENZO-2-TRIFLUORIDE	4-NITRO-1-CHLOROBENZO-2-TRIFLUORIDE
25889-38-7 2-NITRO-4-CHLOROBENZO-1-TRIFLUORIDE	2-NITRO-4-CHLOROBENZO-1-TRIFLUORIDE
39974-35-1 1-NITRO-2-CHLOROBENZO-3-TRIFLUORIDE	1-NITRO-2-CHLOROBENZO-3-TRIFLUORIDE
7496-02-8 6-NITROCHRYSENE	6-NITROCHRYSENE
12167-20-3 NITROCRESOLS	NITROCRESOLS
1122-60-7 NITROCYCLOHEXANE	NITROCYCLOHEXANE
RR-01389-8 6-NITRO-4-DIAZOTOLUENE-3-SULFONIC ACID	6-NITRO-4-DIAZOTOLUENE-3-SULFONIC ACID
92-93-3 4-NITRODIPHENYL [AUS, AUS, BC1, BC1, CAL, MIN, MSL, QBC, QBC, QBC, TLV, TLV]	4-NITROBIPHENYL
836-30-6 P-NITRODIPHENYLAMINE	P-NITRODIPHENYLAMINE
79-24-3 NITROETHANE	NITROETHANE
26618-70-2 NITROETHYLENE POLYMER	NITROETHYLENE POLYMER
4528-34-1 NITROETHYL NITRATE	NITROETHYL NITRATE
1836-75-5 NITROFEN	NITROFEN
1836-75-5 NITROFEN [BENZENE, 2,4-DICHLORO-1-(4-NITROPHENOXY)-] [313]	NITROFEN

413

Chemical Name	Reference Name
1836-75-5 NITROFEN (TECHNICAL GRADE) [C65, C65, IAR]	NITROFEN
892-21-7 3-NITROFLUORANTHENE	3-NITROFLUORANTHENE
607-57-8 2-NITROFLUORENE	2-NITROFLUORENE
59-87-0 NITROFURAL (NITROFURAZONE) [IAR]	NITROFURAZONE
67-20-9 NITROFURANTOIN	NITROFURANTOINHYDROXYETHYL)-4-METHYLCHLO-RIDE, MONOHYDROCHLORIDE
59-87-0 NITROFURAZONE	NITROFURAZONE
555-84-0 1-((5-NITROFURFURYLIDENE)AMINO)-2-IMIDA-ZOLIDINONE	1-((5-NITROFURFURYLIDENE)AMINO)-2-IMIDAZOLIDI-NONE
555-84-0 1-((5-NITROFURFURYLIDENE)-AMINO)-2-IMI-DAZOLIDINONE [C65, C65]	1-((5-NITROFURFURYLIDENE)AMINO)-2-IMIDAZOLIDI-NONE
531-82-8 N-[4-(5-NITRO-2-FURYL)-2-THIAZOLYL]AC-ETAMIDE	N-[4-(5-NITRO-2-FURYL)-2-THIAZOLYL]ACETAMIDE
7727-37-9 NITROGEN	NITROGEN
RR-01026-4 NITROGEN COMPOUNDS, INORGANIC	NITROGEN COMPOUNDS, INORGANIC
10102-44-0 NITROGEN DIOXIDE	NITROGEN DIOXIDE
10544-72-6 NITROGEN DIOXIDE [EPA]	NITROGEN TETROXIDE
RR-00381-6 NITROGEN FERTILIZER SOLUTION	NITROGEN FERTILIZER SOLUTION
RR-01102-9 NITROGEN FIXING BACTERIA	NITROGEN FIXING BACTERIA
7783-54-2 NITROGEN FLUORIDE (NF3) [PAL]	NITROGEN TRIFLUORIDE
RR-01168-7 NITROGEN FLUORIDE OXIDE	NITROGEN FLUORIDE OXIDE
7727-37-9 NITROGEN (LIQUIFIED) [FLA, MSL]	NITROGEN
10024-97-2 NITROGEN MONOXIDE [GBR, GBR]	NITROUS OXIDEZYME)
51-75-2 NITROGEN MUSTARD	NITROGEN MUSTARD
51-75-2 NITROGEN MUSTARD [2-CHLORO-N-(2-CHLOROETHYL)-N-METHYLETHANAMINE] [313]	NITROGEN MUSTARD
51-75-2 NITROGEN MUSTARD (MECHLORETHAMINE) [C65, C65]	NITROGEN MUSTARD
55-86-7 NITROGEN MUSTARD HYDROCHLORIDE	NITROGEN MUSTARD HYDROCHLORIDE
55-86-7 NITROGEN MUSTARD HYDROCHLORIDE (MECHLORETHAMINE HYDROCHLORIDE) [C65, C65]	NITROGEN MUSTARD HYDROCHLORIDE
302-70-5 NITROGEN MUSTARD N-OXIDE [CAB]	NITROGEN MUSTARD N-OXIDE HYDROCHLORIDE
126-85-2 NITROGEN MUSTARD N-OXIDE	NITROGEN MUSTARD N-OXIDE
302-70-5 NITROGEN MUSTARD N-OXIDE HYDROCHLO-RIDE	NITROGEN MUSTARD N-OXIDE HYDROCHLORIDE
10024-97-2 NITROGEN OXIDE [NHS, NHS]	NITROUS OXIDE
10544-73-7 NITROGEN OXIDE (N2O3) [PAL]	NITROGEN TRIOXIDE
10544-72-6 NITROGEN OXIDE (N2O4) [PAL]	NITROGEN TETROXIDE
10102-43-9 NITROGEN OXIDE (NO) [PAL, RCU, RCU]	NITRIC OXIDE
10102-44-0 NITROGEN OXIDE (NO2) [PAL]	NITROGEN DIOXIDE
12033-49-7 NITROGEN OXIDE (NO3) [PAL]	NITROGEN TRIOXIDE
10544-72-6 NITROGEN TETROXIDE	NITROGEN TETROXIDE
10025-85-1 NITROGEN TRICHLORIDE	NITROGEN TRICHLORIDE
7783-54-2 NITROGEN TRIFLUORIDE	NITROGEN TRIFLUORIDE
13444-85-4 NITROGEN TRIIODIDE	NITROGEN TRIIODIDE
RR-01452-8 NITROGEN TRIIODIDE MONOAMINE	NITROGEN TRIIODIDE MONOAMINE
12033-49-7 NITROGEN TRIOXIDE	NITROGEN TRIOXIDE
55-63-0 NITROGLYCERIN	NITROGLYCERIN

Chemical Name	Reference Name
55-63-0 NITROGLYCERIN (NG) [AUS, AUS, NJL, NJL, TLV, TLV]	NITROGLYCERIN
55-63-0 NITROGLYCERINE [EPA, NFP, NFP, NIO, NIO, NIO, NIO]	NITROGLYCERIN
556-88-7 1-NITROGUANIDINE	1-NITROGUANIDINE
556-88-7 NITROGUANIDINE [HM1, HM1, NJL]	1-NITROGUANIDINE
RR-01453-9 NITROGUANIDINE NITRATE	NITROGUANIDINE NITRATE
2825-15-2 1-NITROHYDANTOIN	1-NITROHYDANTOIN
8007-56-5 NITROHYDROCHLORIC ACID [HM1, HM1, NJL, NJL]	NITROHYDROCHLORIC ACID (AQUA REGIA)
8007-56-5 NITROHYDROCHLORIC ACID (AQUA REGIA)	NITROHYDROCHLORIC ACID (AQUA REGIA)
20820-44-4 NITRO ISOBUTANE TRIOL TRINITRATE	NITRO ISOBUTANE TRIOL TRINITRATE
75-52-5 NITROMETHANE	NITROMETHANE
RR-01447-1 N-NITRO-N-METHYLGLYCOLAMIDE NITRATE	N-NITRO-N-METHYLGLYCOLAMIDE NITRATE
24884-69-3 2-NITRO-2-METHYLPROPANOL NITRATE	2-NITRO-2-METHYLPROPANOL NITRATE
86-57-7 1-NITRONAPHTHALENE	1-NITRONAPHTHALENE
581-89-5 2-NITRONAPHTHALENE	2-NITRONAPHTHALENE
27254-36-0 NITRONAPHTHALENE	NITRONAPHTHALENE
20539-63-3 3-NITROPERYLENE	3-NITROPERYLENE
100-29-8 4-NITROPHENETOLE	4-NITROPHENETOLE
88-75-5 O-NITROPHENOL	O-NITROPHENOL
88-75-5 2-NITROPHENOL [313, ATS, CAB, CW2, MX1, MX2, NJL]	O-NITROPHENOL
100-02-7 P-NITROPHENOL	P-NITROPHENOL
100-02-7 4-NITROPHENOL [313, ATS, CAA, CW2, MX1, MX2, NJL, PST, RCA]	P-NITROPHENOL
554-84-7 M-NITROPHENOL	M-NITROPHENOL
554-84-7 3-NITROPHENOL [NJL]	M-NITROPHENOL
25154-55-6 NITROPHENOL (MIXED) [CWA, EPA, MSL]	NITROPHENOL (MIXED ISOMERS)
25154-55-6 NITROPHENOL (MIXED ISOMERS)	NITROPHENOL (MIXED ISOMERS)
25154-55-6 NITROPHENOL, MIXED ISOMERS [NJL]	NITROPHENOL (MIXED ISOMERS)
25154-55-6 NITROPHENOLS [CW3, PAL]	NITROPHENOL (MIXED ISOMERS)
RR-00932-5 NITROPHENOXYALKANOIC ACID SUBSTITUTED THIAZINO HYDRAZIDE	NITROPHENOXYALKANOIC ACID SUBSTITUTED THIAZINO HYDRAZIDE
RR-01434-6 M-NITROPHENYLDINITRO METHANE	M-NITROPHENYLDINITRO METHANE
99-56-9 4-NITRO-O-PHENYLENEDIAMINE [CSR, CSR]	2-AMINO-4-NITROANILINE
5307-14-2 2-NITRO-P-PHENYLENEDIAMINE	2-NITRO-P-PHENYLENEDIAMINE
5255-75-4 P-NITROPHENYL GLYCIDYL ETHER [EPA]	OXIRANE, [(4-NITROPHENOXY)METHYL]-
5255-75-4 NITROPHENYL GLYCIDYL ETHER, P- [OTS]	OXIRANE, [(4-NITROPHENOXY)METHYL]-
41687-30-3 2-[(3-NITROPHENYL)SULFONYL] ETHANOL	2-[(3-NITROPHENYL)SULFONYL] ETHANOL
79-46-9 2-NITROPROPANE	2-NITROPROPANE
108-03-2 1-NITROPROPANE	1-NITROPROPANE
25322-01-4 NITROPROPANE	NITROPROPANE
25322-01-4 NITROPROPANES [HM1, HM1]	NITROPROPANE
504-88-1 3-NITROPROPIONIC ACID	3-NITROPROPIONIC ACID
789-07-1 2-NITROPYRENE	2-NITROPYRENE
5522-43-0 1-NITROPYRENE	1-NITROPYRENE
57835-92-4 4-NITROPYRENE	4-NITROPYRENEVINYL)-1,3,4-OXADIAZOLE
RR-00815-1 NITROPYRENES (MONO-, DI-, TRI-, AND TETRA-)	NITROPYRENES (MONO-, DI-, TRI-, AND TETRA-)
3565-26-2 4-NITROQUINOLINE-1-OXIDE	4-NITROQUINOLINE-1-OXIDE

Chemical Name	Reference Name
RR-00382-7 NITROSAMINES	NITROSAMINES
RR-00382-7 NITROSAMINES, N.O.S. [RCU]	NITROSAMINES
1133-64-8 N'-NITROSOANABASINE	N'-NITROSOANABASINE
71267-22-6 N'-NITROSOANATABINE	N'-NITROSOANATABINE
38252-74-3 N-NITROSO-N-BUTYL-N-(3-CARBOXYPROPYL) AMINE	N-NITROSO-N-BUTYL-N-(3-CARBOXYPROPYL)AMINE
3817-11-6 N-NITROSO-N-BUTYL-N-(4-HYDROXYBUTYL) AMINE	N-NITROSO-N-BUTYL-N-(4-HYDROXYBUTYL)AMINE
924-16-3 N-NITROSODI-N-BUTYLAMINE	N-NITROSODI-N-BUTYLAMINE
924-16-3 N-NITROSO-DI-N-BUTYLAMINE [CAL, RCA]	N-NITROSODI-N-BUTYLAMINE
924-16-3 N-NITROSODIBUTYLAMINE [WHM]	N-NITROSODI-N-BUTYLAMINE
1116-54-7 N-NITROSODIETHANOLAMINE	N-NITROSODIETHANOLAMINE
55-18-5 N-NITROSODIETHYLAMINE	N-NITROSODIETHYLAMINE
62-75-9 N-NITROSODIMETHYLAMINE	N-NITROSODIMETHYLAMINE
62-75-9 NITROSODIMETHYLAMINE [302]	N-NITROSODIMETHYLAMINE
62-75-9 N-NITROSODIMETHYLAMINE (DIMETHYLNI-TROSOAMINE) [BC1, BC1, BC1]	N-NITROSODIMETHYLAMINE
62-75-9 N-NITROSODIMETHYLAMINE (DIMETHYL-NITROSAMINE) [ALB, ALB]	N-NITROSODIMETHYLAMINE
138-89-6 P-NITROSODIMETHYLANILINE	P-NITROSODIMETHYLANILINE
86-30-6 N-NITROSODIPHENYLAMINE [313, ATS, MX1, MX2]	DIPHENYLNITROSAMINE
156-10-5 P-NITROSODIPHENYLAMINE	P-NITROSODIPHENYLAMINE
156-10-5 PARA-NITROSODIPHENYLAMINE [IAR]	P-NITROSODIPHENYLAMINE
601-77-4 N-NITROSODI-I-PROPYLAMINE	N-NITROSODI-I-PROPYLAMINE
621-64-7 N-NITROSODI-N-PROPYLAMINE	N-NITROSODI-N-PROPYLAMINE
621-64-7 N-NITROSO-DI-N-PROPYLAMINE [CAL]	N-NITROSODI-N-PROPYLAMINE
621-64-7 N-NITROSODIPROPYLAMINE [MX2, WHM]	N-NITROSODI-N-PROPYLAMINE
612-64-6 N-NITROSOETHYLPHENYLAMINE	N-NITROSOETHYLPHENYLAMINE
759-73-9 N-NITROSO-N-ETHYLUREA	N-NITROSO-N-ETHYLUREA
29291-35-8 N-NITROSOFOLIC ACID	N-NITROSOFOLIC ACID
70-25-7 NITROSOGUANIDINE [HM1, HM1, HM1]	N-METHYL-N'-NITRO-N-NITROSOGUANIDINE
55557-01-2 N-NITROSOGUVACINE	N-NITROSOGUVACINE
55557-02-3 N-NITROSOGUVACOLINE	N-NITROSOGUVACOLINE
30310-80-6 N-NITROSOHYDROXYPROLINE	N-NITROSOHYDROXYPROLINE
60153-49-3 3-(N-NITROSOMETHYLAMINO)PROPIONI-TRILE [C65, CAL, MIN]	3-METHYLNITROSOAMINOPROPIONITRILE
85502-23-4 3-(N-NITROSOMETHYLAMINO)PROPIONALDE-HDYE	3-(N-NITROSOMETHYLAMINO)PROPIONALDEHDYE
64091-91-4 4-(N-NITROSOMETHYLAMINO)-1-(3-PYRIDYL)-1-BUTANONE	4-(N-NITROSOMETHYLAMINO)-1-(3-PYRIDYL)-1-BU-TANONE
64091-91-4 4-(N-NITROSOMETHYLAMINO)-1-(3-PYRIDYL)-1-BUTANONE (NNK) [NTP, IAR]	4-(N-NITROSOMETHYLAMINO)-1-(3-PYRIDYL)-1-BU-TANONE
64091-90-3 3-(N-NITROSOMETHYLAMINO)-4-(3-PYRIDYL)-1-BUTANAL (NNA)	3-(N-NITROSOMETHYLAMINO)-4-(3-PYRIDYL)-1-BU-TANAL (NNA)
64091-91-4 4-(N-NITROSOMETHYLAMINO)-1-(3-PYRIDYL)-1-BUTANONE (NNK) [CAL]	4-(N-NITROSOMETHYLAMINO)-1-(3-PYRIDYL)-1-BU-TANONE
10595-95-6 N-NITROSOMETHYLETHYLAMINE	N-NITROSOMETHYLETHYLAMINE
614-00-6 N-NITROSOMETHYLPHENYLAMINE	N-NITROSOMETHYLPHENYLAMINE
684-93-5 N-NITROSO-N-METHYLUREA	N-NITROSO-N-METHYLUREA
684-93-5 N-NITROSO-N-METHYLUREA CAR-BAMIDE [MSL]	N-NITROSO-N-METHYLUREA

Chemical Name	Reference Name
615-53-2 N-NITROSO-N-METHYLURETHANE	N-NITROSO-N-METHYLURETHANE
615-53-2 NITROSOMETHYLURETHANE [WHM]	N-NITROSO-N-METHYLURETHANE
4549-40-0 N-NITROSOMETHYLVINYLAMINE	N-NITROSOMETHYLVINYLAMINE
59-89-2 N-NITROSOMORPHOLINE	N-NITROSOMORPHOLINE
16543-55-8 N-NITROSONORNICOTINE	N-NITROSONORNICOTINE
80508-23-2 N'-NITROSONORNICOTINE [IAR, WHM, MSL]	PYRIDINE, 3-(1-NITROSO-2-PYRROLIDINYL)
100-75-4 N-NITROSOPIPERIDINE	N-NITROSOPIPERIDINE
RR-01543-0 N-NITROSOPROLINE	N-NITROSOPROLINE
930-55-2 N-NITROSOPYRROLIDINE	N-NITROSOPYRROLIDINE
13256-22-9 N-NITROSOSARCOSINE	N-NITROSOSARCOSINE
9056-38-6 NITROSTARCH	NITROSTARCH
102-96-5 BETA-NITROSTYRENE	BETA-NITROSTYRENE
RR-01454-0 NITROSUGARS	NITROSUGARS
2696-92-6 NITROSYL CHLORIDE	NITROSYL CHLORIDE
7782-78-7 NITROSYLSULFURIC ACID [NJL, NJL]	NITROSYLSULPHURIC ACID
7782-78-7 NITROSYLSULPHURIC ACID	NITROSYLSULPHURIC ACID
88-72-2 O-NITROTOLUENE	O-NITROTOLUENE
88-72-2 2-NITROTOLUENE [MAK, MAK]	O-NITROTOLUENE
88-72-2 NITROTOLUENE, 2- [OTS]	O-NITROTOLUENE
99-08-1 M-NITROTOLUENE	M-NITROTOLUENE
99-08-1 NITROTOLUENE [MIN]	M-NITROTOLUENE
99-99-0 P-NITROTOLUENE	P-NITROTOLUENE
1321-12-6 NITROTOLUENE [BC1, BC1, BC1, EPA, MAK, MAK, MAK, MEX, MEX, MSL, NJL, NJL]	NITROTOLUENE (MIXED ISOMERS)
1321-12-6 NITROTOLUENE (ALL ISOMERS) [HON, QBC, QBC]	NITROTOLUENE (MIXED ISOMERS)
1321-12-6 NITROTOLUENE (SUM OF M-, O-, AND P- ISOMERS) [ONT, ONT, ONT]	NITROTOLUENE (MIXED ISOMERS)
1321-12-6 NITROTOLUENE, ALL ISOMERS [GBR, GBR, GBR]	NITROTOLUENE (MIXED ISOMERS)
1321-12-6 NITROTOLUENE (MIXED ISOMERS)	NITROTOLUENE (MIXED ISOMERS)
1321-12-6 NITROTOLUENES [HM1, HM1, HM1]	NITROTOLUENE (MIXED ISOMERS)
89-62-3 2-NITRO-P-TOLUIDINE	2-NITRO-P-TOLUIDINE
99-55-8 5-NITRO-O-TOLUIDINE	5-NITRO-O-TOLUIDINE
99-55-8 5-NITRO-ORTHO-TOLUIDINE [IAR]	5-NITRO-O-TOLUIDINE
RR-01334-3 NITROTRIAZOLONE	NITROTRIAZOLONE
76-06-2 NITROTRICHLOROMETHANE (CHLOROPICRIN) [ALB, ALB]	CHLOROPICRIN
556-89-8 NITRO UREA	NITRO UREA
109-95-5 NITROUS ACID, ETHYL ESTER [PAL]	ETHYL NITRITE
13826-65-8 NITROUS ACID, LEAD (2+) SALT	NITROUS ACID, LEAD (2+) SALT
7632-00-0 NITROUS ACID, SODIUM SALT [PAL]	SODIUM NITRITE
10024-97-2 NITROUS OXIDE	NITROUS OXIDE
804-36-4 NITROVIN	NITROVIN
99-51-4 M-NITROXYLENE	M-NITROXYLENE
RR-00607-5 NITROXYLENE, ALL ISOMERS	NITROXYLENE, ALL ISOMERS
RR-00607-5 NITROXYLENES (O-; M-; P-) [HM1, HM1, HM1]	NITROXYLENE, ALL ISOMERS
25168-04-1 NITROXYLOL	NITROXYLOL
2608-48-2 5-(4-NITROPHENYL)-2,4-PENTADIEN-1-AL	5-(4-NITROPHENYL)-2,4-PENTADIEN-1-AL

Chemical Name	Reference Name
60153-49-3 3-(N-NITROSOMETHYLAMINO)PROPIONI-TRILE [IAR]	3-METHYLNITROSOAMINOPROPIONITRILE
23282-20-4 NIVALENOL	NIVALENOL
27753-52-2 NONABROMO-1,1'-BIPHENYL [TSE]	1,1'-BIPHENYL, NONABROMO-
18435-45-5 NONADECANE	NONADECANE
124-19-6 NONANAL	NONANAL
111-84-2 NONANE	NONANE
3221-61-2 NONANE (ISO) [NFP, NFP, NFP]	2-METHYLOCTANE
1455-21-6 1-NONANETHIOL	1-NONANETHIOL
25360-10-5 TERT-NONANETHIOL [PAL]	TERT-NONYL MERCAPTAN
112-05-0 NONANOIC ACID [PST]	PELARGONIC ACID
143-08-8 1-NONANOL	1-NONANOL
123-17-1 4-NONANOL, 2,6,8-TRIMETHYL-	4-NONANOL, 2,6,8-TRIMETHYL-
821-55-6 2-NONANONE	2-NONANONE
123-18-2 4-NONANONE, 2,6,8-TRIMETHYL- [PAL]	ISOBUTYL HEPTYL KETONE
27215-95-8 NONENE	NONENE
RR-00383-8 NONFLAMMABLE GAS, N.O.S.	NONFLAMMABLE GAS, N.O.S.
RR-00051-1 NONSTEROIDAL OESTROGENS [IAR]	OESTROGENS, NONSTEROIDAL
143-13-5 NONYL ACETATE	NONYL ACETATE
1081-77-2 NONYLBENZENE	NONYLBENZENE
1081-77-2 NONYLBENZENE (BRANCHED) [HON]	NONYLBENZENE
25360-10-5 TERT-NONYL MERCAPTAN	TERT-NONYL MERCAPTAN
27193-93-7 NONYLNAPHTHALENE	NONYLNAPHTHALENE
68877-34-9 N-(NONYLOXYPROPYL)-1,3-PROPANEDIAMINE	N-(NONYLOXYPROPYL)-1,3-PROPANEDIAMINE
25154-52-3 NONYLPHENOL	NONYLPHENOL
25154-52-3 NONYL PHENOL [WHM]	NONYLPHENOL
84852-15-3 4-NONYLPHENOL, BRANCHED	4-NONYLPHENOL, BRANCHED
84852-15-3 NONYLPHENOL, 4-BRANCHED [OTS]	4-NONYLPHENOL, BRANCHED
26027-38-3 P-NONYLPHENOL ETHOXYLATED [PST]	ETHOXYLATED P-NONYL PHENOLIDE
51811-79-1 NONYLPHENOL, ETHOXYLATED AND PHOS-PHATED	NONYLPHENOL, ETHOXYLATED AND PHOSPHATED)
51609-41-7 P-NONYLPHENOL, ETHOXYLATED AND PHOS-PHATED	P-NONYLPHENOL, ETHOXYLATED AND PHOSPHATED
9081-17-8 NONYLPHENOL, ETHOXYLATED AND SUL-FATED	NONYLPHENOL, ETHOXYLATED AND SULFATED
RR-01103-0 P-NONYLPHENOL, ETHOXYLATED, PHOS-PHATED, CALCIUM SALT	P-NONYLPHENOL, ETHOXYLATED, PHOSPHATED, CAL-CIUM SALT
106151-63-7 P-NONYLPHENOL, ETHOXYLATED, PHOS-PHATED, DIPOTASSIUM SALT	P-NONYLPHENOL, ETHOXYLATED, PHOSPHATED, DIPOTASSIUM SALTFORMALDEHYDE
67922-57-0 NONYLPHENOL, ETHOXYLATED, PHOS-PHATED, MAGNESIUM SALT	NONYLPHENOL, ETHOXYLATED, PHOSPHATED, MAG-NESIUM SALT
59139-23-0 NONYLPHENOL, ETHOXYLATED, PHOS-PHATED, MONOETHANOLAMINE SALT	NONYLPHENOL, ETHOXYLATED, PHOSPHATED, MO-NOETHANOLAMINE SALT
52503-15-8 NONYLPHENOL, ETHOXYLATED, PHOS-PHATED, POTASSIUM SALT	NONYLPHENOL, ETHOXYLATED, PHOSPHATED, POTAS-SIUM SALT
9014-90-8 NONYLPHENOL, ETHOXYLATED, SULFATED, SODIUM SALT [PST]	POLYETHYLENE GLYCOL NONYLPHENYL ETHER SODIUM SULFATE
57451-03-3 NONYLPHENOL, ETHOXYLATED, SULFATED, TRIETHANOLAMINE SALT	NONYLPHENOL, ETHOXYLATED, SULFATED, TRI-ETHANOLAMINE SALT
30526-26-2 NONYLPHENOL PHOSPHATE	NONYLPHENOL PHOSPHATE
37340-60-6 NONYLPHENOXYPOLY(ETHYLENEOXY) ETHYL PHOSPHATE, SODIUM SALT	NONYLPHENOXYPOLY(ETHYLENEOXY) ETHYL PHOS-PHATE, SODIUM SALT

Chemical Name	Reference Name
6178-32-1 P-NONYLPHENYL GLYCIDYL ETHER [EPA, OTS]	OXIRANE, [(4-NONYLPHENOXY)METHYL]-
54612-36-1 NONYLPHENYLPOLYOXYETHYLENE SULFOS-UCCINATE	NONYLPHENYLPOLYOXYETHYLENE SULFOSUCCINATE
5283-67-0 NONYLTRICHLOROSILANE	NONYLTRICHLOROSILANE
5283-67-0 NONYL TRICHLOROSILANE [NJL, NJL]	NONYLTRICHLOROSILANE
991-42-4 NORBORMIDE	NORBORMIDE
121-46-0 NORBORNADIENE [FLA]	BICYCLO(2.2.1)HEPTA-2,5-DIENE
121-46-0 2,5-NORBORNADIENE [MSL, NFP, NFP]	BICYCLO(2.2.1)HEPTA-2,5-DIENE
121-46-0 2,5-NORBORNADIENE (DICYCLOHEPTADIENE) [HM1, HM1]	BICYCLO(2.2.1)HEPTA-2,5-DIENE
315-22-0 20-NORCROTALANAN-11,15-DIONE, 14,19-DIHYDRO-12,13-DIHYDROXY-, (13.ALPHA., 14.ALPHA.)- [PAL, PAL]	MONOCROTALINE
68-22-4 NORETHISTERONE	NORETHISTERONE
68-22-4 NORETHISTERONE (NORETHINDRONE) [C65, C65]	NORETHISTERONE
51-98-9 NORETHISTERONE ACETATE [CAL]	NORETHISTERONE ACETATE (NORETHINDRONE AC-ETATE)
51-98-9 NORETHISTERONE ACETATE (NORETHIN-DRONE ACETATE)	NORETHISTERONE ACETATE (NORETHINDRONE AC-ETATE)
RR-00039-5 NORETHISTERONE AND OESTROGENS	NORETHISTERONE AND OESTROGENS
68-23-5 NORETHYNODREL	NORETHYNODREL
RR-00550-5 NORETHYNODREL AND OESTROGENS	NORETHYNODREL AND OESTROGENS
27314-13-2 NORFLURAZON [4-CHLORO-5-METHYLAMINO) -2-[3-(TRIFLUOROMETHYL)PHENYL]-3(2H)-PYRIDAZINONE]	NORFLURAZON [4-CHLORO-5-METHYLAMINO)-2-[3-(TRIFLUOROMETHYL)P
6533-00-2 NORGESTREL	NORGESTREL
RR-00034-0 NORGESTREL AND OESTROGENS	NORGESTREL AND OESTROGENS
116712-07-3 NORPOR 10	NORPOR 10ATE, HYDROLYZED
57-63-6 19-NORPREGNA-1,3,5(10)-TRIEN-20-YNE-3,17-DIOL, (17.ALPHA.)- [PAL, PAL]	ETHINYLESTRADIOL
72-33-3 19-NORPREGNA-1,3,5(10)-TRIEN-20-YN-17-OL, 3-METHOXY-, (17.ALPHA.)- [PAL, PAL]	MESTRANOL
68-22-4 19-NORPREGN-4-EN-20-YN-3-ONE, 17-HY-DROXY-, (17.ALPHA.)- [PAL, PAL]	NORETHISTERONE
36393-56-3 NORPSEUDOEPHEDRINE	NORPSEUDOEPHEDRINE
RR-01777-6 NORPSEUDOEPHEDRINE SALTS, OPTICAL ISOMERS, AND SALTS OF OPTICAL ISOMERS	NORPSEUDOEPHEDRINE SALTS, OPTICAL ISOMERS, AND SALTS OF OPTI
303-81-1 NOVOBIOCIN	NOVOBIOCIN
18662-53-8 NTA TRISODIUM SALT.H2O [MSL]	TRISODIUM NITRILOTRIACETATE MONOHYDRATE
RR-00072-6 NUISANCE PARTICULATES	NUISANCE PARTICULATES
9008-75-7 NYLON	NYLON
25038-54-4 NYLON 6	NYLON 6HACRYLATE
88-74-4 O-NITROANILINE	O-NITROANILINE
RR-01104-1 OATS AND OATMEAL	OATS AND OATMEAL
303-47-9 OCHRATOXIN A	OCHRATOXIN A
68917-09-9 OCOTEA CYMBARUM OIL	OCOTEA CYMBARUM OIL
68917-09-9 OCOTEA OIL [PST]	OCOTEA CYMBARUM OIL
27858-07-7 OCTABROMOBIPHENYL	OCTABROMOBIPHENYL
32536-52-0 OCTABROMODIPHENYL ETHER [OTS, TSC, TSC, TSC, TSC]	OCTABROMODIPHENYLOXIDE
32536-52-0 OCTABROMODIPHENYLOXIDE	OCTABROMODIPHENYLOXIDE

Chemical Name	Reference Name
3268-87-9 OCTACHLORODIBENZO-P-DIOXIN	OCTACHLORODIBENZO-P-DIOXIN
39001-02-0 OCTACHLORODIBENZOFURAN	OCTACHLORODIBENZOFURAN
2234-13-1 OCTACHLORONAPHTHALENE	OCTACHLORONAPHTHALENE
2471-09-2 OCTACOSAMETHYLTRIDECASILOXANE	OCTACOSAMETHYLTRIDECASILOXANE
556-71-8 OCTADECAMETHYLCYCLONONASILOXANE	OCTADECAMETHYLCYCLONONASILOXANE
556-69-4 OCTADECAMETHYLOCTASILOXANE	OCTADECAMETHYLOCTASILOXANE
110-30-5 OCTADECANAMIDE, N,N'-1,2-ETHANEDIYLBIS- [OTS, PST]	N,N'-ETHYLENEBIS(STEARAMIDE)
107-64-2 1-OCTADECANAMINIUM, N,N-DIMETHYL-N-OCTAD [OTS]	DISTEARYLDIMETHYLAMMONIUM CHLORIDE
593-45-3 OCTADECANE	OCTADECANE
112-96-9 OCTADECANE, 1-ISOCYANATO-	OCTADECANE, 1-ISOCYANATO-
2885-00-9 1-OCTADECANETHIOL	1-OCTADECANETHIOL
1072-35-1 OCTADECANOIC ACID, LEAD(2+) SALT [PAL]	STEARIC ACID, LEAD (2+) SALT
7428-48-0 OCTADECANOIC ACID, LEAD SALT [PAL]	LEAD STEARATE3,3,4,7-HEXACHLORODECAHYDRO-, (1.ALPHA.,2.BETA.,2A.BETA., 4.BETA.,4A.BETA.,5.BETA., 6A.BETA.6B.BETA.,7R*)-
1330-80-9 9-OCTADECANOIC ACID (Z)-, MONOESTER WITH 1,2-PROPANEDIOL	9-OCTADECANOIC ACID (Z)-, MONOESTER WITH 1,2-PROPANEDIOL
68443-05-0 9-OCTADECENOIC ACID (Z)-, SULFONATED, SODIUM SALT	9-OCTADECENOIC ACID (Z)-, SULFONATED, SODIUM SALT
557-05-1 OCTADECANOIC ACID, ZINC SALT [PAL]	ZINC STEARATE
112-92-5 1-OCTADECANOL	1-OCTADECANOL
112-92-5 OCTADECANOL, 1- [OTS]	1-OCTADECANOL
112-90-3 9-OCTADECEN-1-AMINE, (Z)- [OTS]	OLEYLAMINE
112-80-1 9-OCTADECENOIC ACID (Z)- [PAL]	OLEIC ACID
140-04-5 9-OCTADECENOIC ACID, 12-(ACETYLOXY)-, BUTYL ESTER, [R-(Z)]- [PAL]	BUTYL ACETYL RICINOLEATENYL)METHYLENE] AMINO]-
26094-13-3 9-OCTADECENOIC ACID (Z)-, COMPD. WITH 1-BUTANAMINE (1:1) [PAL]	BUTYLAMINE OLEATE
68988-76-1 9-OCTADECENOIC ACID (Z)-, SULFONATED	9-OCTADECENOIC ACID (Z)-, SULFONATED
143-28-2 CIS-9-OCTADECEN-1-OL	CIS-9-OCTADECEN-1-OL
7173-62-8 N-CIS-9-OCTADECENYL-1,3-PROPANEDIAMINE	N-CIS-9-OCTADECENYL-1,3-PROPANEDIAMINE
1847-55-8 CIS-9-OCTADECENYL SULFATE, SODIUM SALT	CIS-9-OCTADECENYL SULFATE, SODIUM SALT
124-30-1 N-OCTADECYLAMINE	N-OCTADECYLAMINE
2190-04-7 OCTADECYLAMINE ACETATE	OCTADECYLAMINE ACETATE
2082-79-3 OCTADECYL 3-(3',5'-DI-TERT-BUTYL-4'-HY-DROXYPHENYL)PROPIONATE	OCTADECYL 3-(3',5'-DI-TERT-BUTYL-4'-HYDROX-YPHENYL)PROPIONATE
112-88-9 OCTADECYLENE, ALPHA-	OCTADECYLENE, ALPHA-
16245-97-9 N-OCTADECYL GLYCIDYL ETHER [EPA]	OXIRANE, [(OCTADECYLOXY)METHYL]-
16245-97-9 OCTADECYL GLYCIDYL ETHER, N- [OTS]	OXIRANE, [(OCTADECYLOXY)METHYL]-
51617-79-9 ALPHA-(OCTADECYLPHENYL)-OMEGA-HY-DROXYPOLY(OXY-1,2-ETHANEDIYL)	ALPHA-(OCTADECYLPHENYL)-OMEGA-HYDROXYPOLY (OXY-1,2-ETHANEDIYL
1120-04-3 OCTADECYL SULFATE, SODIUM SALT	OCTADECYL SULFATE, SODIUM SALT
112-04-9 OCTADECYL TRICHLOROSILANE [NJL, NJL]	OCTADECYLTRICHLOROSILANE
112-04-9 OCTADECYLTRICHLOROSILANE	OCTADECYLTRICHLOROSILANE
112-04-9 OCTADECYL TRICHLOROSILANE [FLA]	OCTADECYLTRICHLOROSILANE
106-26-3 2,6-OCTADIENAL, 3,7-DIMETHYL-, (Z)-	2,6-OCTADIENAL, 3,7-DIMETHYL-, (Z)-
141-27-5 2,6-OCTADIENAL, 3,7-DIMETHYL-,(E)-	2,6-OCTADIENAL, 3,7-DIMETHYL-,(E)-
63597-41-1 OCTADIENE	OCTADIENE

Chemical Name	Reference Name
RR-01369-4 1,7-OCTADINE-3,5-DIYNE-1,8-DIMETHOXY-9-OCTADECYNOIC ACID	1,7-OCTADINE-3,5-DIYNE-1,8-DIMETHOXY-9-OCTADE-CYNOIC ACID
360-89-4 OCTAFLUOROBUT-2-ENE	OCTAFLUOROBUT-2-ENE
115-25-3 OCTAFLUOROCYCLOBUTANE	OCTAFLUOROCYCLOBUTANE
76-19-7 OCTAFLUOROPROPANE	OCTAFLUOROPROPANE
69155-42-6 1,1,1,3,5,7,77-OCTAMETHYL-3,5-BIS(6,7-EPOXY-4-OXAHEPTYL) TETRASILOXANE [EPA]	TETRASILOXANE, 1,1,1,3,5,7,7-OCTAMETHYL-3,5-BIS[3-(OXIRANYLM
69155-42-6 1,1,1,3,5,7,7,7-OCTAMETHYL-3,5-BIS(6,7-EPOXY-4-OXAHEPTYL)- [OTS]	TETRASILOXANE, 1,1,1,3,5,7,7-OCTAMETHYL-3,5-BIS[3-(OXIRANYLM
556-67-2 OCTAMETHYLCYCLOTETRASILOXANE	OCTAMETHYLCYCLOTETRASILOXANE
152-16-9 OCTAMETHYLPYROPHOSPHORAMIDE	OCTAMETHYLPYROPHOSPHORAMIDE
107-51-7 OCTAMETHYLTRISILOXANE	OCTAMETHYLTRISILOXANE
124-13-0 OCTANAL [TSC, TSC, TSC]	CAPRYLALDEHYDE
5988-91-0 OCTANAL, 3,7-DIMETHYL-	OCTANAL, 3,7-DIMETHYL-
107-75-5 OCTANAL, 7-HYDROXY-3,7-DIMETHYL-	OCTANAL, 7-HYDROXY-3,7-DIMETHYL-
3613-30-7 OCTANAL, 7-METHOXY-3,7-DIMETHYL-	OCTANAL, 7-METHOXY-3,7-DIMETHYL-HENYL) SULFONYL]OXY]PHENYL]AZO][1,1'-BIPHENYL]-4-YL]AZO]-, DISODIUM SALT
101-86-0 OCTANAL, 2-(PHENYLMETHYLENE)-	OCTANAL, 2-(PHENYLMETHYLENE)-
111-86-4 1-OCTANAMINE	1-OCTANAMINE
1120-48-5 1-OCTANAMINE, N-OCTYL-	1-OCTANAMINE, N-OCTYL-
111-65-9 OCTANE	OCTANE
111-65-9 N-OCTANE [GBR, GBR]	OCTANE
111-65-9 OCTANE (ALL ISOMERS) [MAK, MAK]	OCTANE
2216-33-3 OCTANE, 3-METHYL [PAL]	3-METHYLOCTANE
2216-34-4 OCTANE, 4-METHYL [PAL]	4-METHYLOCTANE
3221-61-2 OCTANE, 2-METHYL [PAL]	2-METHYLOCTANE
111-88-6 1-OCTANETHIOL	1-OCTANETHIOL
124-07-2 OCTANOIC ACID	OCTANOIC ACID
2191-10-8 OCTANOIC ACID, CADMIUM SALT (2:1)	OCTANOIC ACID, CADMIUM SALT (2:1)
106-32-1 OCTANOIC ACID, ETHYL ESTER [PAL]	ETHYL CAPRYLATE
6304-39-8 OCTANOIC ACID, HYDRAZIDE	OCTANOIC ACID, HYDRAZIDE
111-11-5 OCTANOIC ACID, METHYL ESTER [OTS]	METHYL ESTER OCTANOIC ACID
557-09-5 OCTANOIC ACID, ZINC SALT [PST]	ZINC OCTOATE
18312-04-4 OCTANOIC ACID, ZIRCONIUM SALT	OCTANOIC ACID, ZIRCONIUM SALT
111-87-5 1-OCTANOL	1-OCTANOL
589-98-0 3-OCTANOL	3-OCTANOL
5978-70-1 2-OCTANOL	2-OCTANOL
106-68-3 3-OCTANONE [PAL]	ETHYL AMYL KETONE
111-13-7 2-OCTANONE	2-OCTANONE
111-64-8 OCTANOYL CHLORIDE [PAL]	CAPRYLYL CHLORIDE
546-56-5 OCTAPHENYLCYCLOTETRASILOXANE	OCTAPHENYLCYCLOTETRASILOXANE
36938-52-0 OCTATRIACONTAMETHYLOCTADECASILOXANE	OCTATRIACONTAMETHYLOCTADECASILOXANE
106-23-0 6-OCTENAL, 3,7-DIMETHYL-	6-OCTENAL, 3,7-DIMETHYL-
5949-05-3 6-OCTENAL, 3,7-DIMETHYL-, (S)-	6-OCTENAL, 3,7-DIMETHYL-, (S)-
111-66-0 1-OCTENE	1-OCTENE
111-67-1 2-OCTENE	2-OCTENE
25377-83-7 2-OCTENE (MIXED CIS- AND TRANS- ISOMERS) [NFP, NFP]	OCTENE (MIXED ISOMERS)
25377-83-7 OCTENE (MIXED ISOMERS)	OCTENE (MIXED ISOMERS)

Chemical Name	Reference Name
29171-20-8 6-OCTEN-1-YN-3-OL, 3,7-DIMETHYL-	6-OCTEN-1-YN-3-OL, 3,7-DIMETHYL-
RR-01330-9 OCTOLITE	OCTOLITE
111-87-5 OCTYL ALCOHOL [NFP, NFP, PST]	1-OCTANOL
68526-82-9 OCTYL ALCOHOL BOTTOMS	OCTYL ALCOHOL BOTTOMS
RR-01335-4 OCTYL ALDEHYDES	OCTYL ALDEHYDES
107-45-9 TERT-OCTYLAMINE	TERT-OCTYLAMINE
111-86-4 OCTYLAMINE [FLA, MSL, NFP, NFP]	1-OCTANAMINE
113-48-4 N-OCTYL BICYCLOHEPTENEDICARBOXIMIDE	N-OCTYL BICYCLOHEPTENEDICARBOXIMIDE
2305-05-7 GAMMA-N-OCTYL-GAMMA-N-BUTYROLAC-TONE	GAMMA-N-OCTYL-GAMMA-N-BUTYROLACTONE
111-85-3 OCTYL CHLORIDE	OCTYL CHLORIDE
4175-37-5 OCTYL DIPHENYLAMINE	OCTYL DIPHENYLAMINE
4175-37-5 OCTYLDIPHENYLAMINE [PST]	OCTYL DIPHENYLAMINE
533-42-6 2-OCTYLDODECANOL	2-OCTYLDODECANOL
94-96-2 OCTYLENE GLYCOL [NFP, NFP]	2-ETHYL-1,3-HEXANEDIOL
61788-72-5 OCTYL EPOXYTALLATES	OCTYL EPOXYTALLATES
RR-01522-5 TERT-OCTYL HYDROPEROXIDE	TERT-OCTYL HYDROPEROXIDE
29710-25-6 OCTYL 12-HYDROXYSTEARATE	OCTYL 12-HYDROXYSTEARATE
141-59-3 TERT-OCTYL MERCAPTAN	TERT-OCTYL MERCAPTAN
141-59-3 TERT-OCTYLMERCAPTAN [HM1, HM1, HM1]	TERT-OCTYL MERCAPTAN
31800-88-1 OCTYLOXYPOLY(ETHYLENEOXY)ETHYL PHOSPHATE	OCTYLOXYPOLY(ETHYLENEOXY)ETHYL PHOSPHATE
RR-00750-1 TERT-OCTYL PEROXY-2-ETHYLHEXANOATE	TERT-OCTYL PEROXY-2-ETHYLHEXANOATE
27193-28-8 OCTYLPHENOL	OCTYLPHENOL
1322-97-0 2-(OCTYLPHENOXY)-ETHANOL	2-(OCTYLPHENOXY)-ETHANOL
2553-08-4 P-OCTYLPHENYL SALICYLATE	P-OCTYLPHENYL SALICYLATE
117-81-7 OCTYL PHTHALATE, DI-SEC [QBC, QBC, QBC]	DI(2-ETHYLHEXYL)PHTHALATE
6969-49-9 N-OCTYL SALICYLATE	N-OCTYL SALICYLATE
142-31-4 OCTYL SULFATE, SODIUM SALT [PST]	SODIUM OCTYL SULFATE
5283-66-9 OCTYL TRICHLOROSILANE	OCTYL TRICHLOROSILANE
78-66-0 4-OCTYNE-3,6-DIOL, 3,6-DIMETHYL-	4-OCTYNE-3,6-DIOL, 3,6-DIMETHYL-
50-28-2 OESTRADIOL-17 BETA [FLA]	ESTRADIOL-17B
50-28-2 OESTRADIOL-17B [CAL]	ESTRADIOL-17B
RR-00549-2 OESTRADIOL-17B AND ESTERS	OESTRADIOL-17B AND ESTERS
22966-79-6 OESTRADIOL MUSTARD [CAL, IAR]	ESTRADIOL MUSTARD
RR-00542-5 OESTROGEN-PROGESTIN COMBIN., SEQUENTIAL ORAL CONTRACEPTIVES	OESTROGEN-PROGESTIN COMBIN., SEQUENTIAL ORAL CONTRACEPTIVES
RR-01544-1 OESTROGEN-PROGESTIN REPLACEMENT THERAPY	OESTROGEN-PROGESTIN REPLACEMENT THERAPY
RR-00541-4 OESTROGEN REPLACEMENT THERAPY	OESTROGEN REPLACEMENT THERAPY
RR-00051-1 OESTROGENS, NONSTEROIDAL	OESTROGENS, NONSTEROIDAL
RR-00052-2 OESTROGENS, STEROIDAL	OESTROGENS, STEROIDAL
53-16-7 OESTRONE [FLA, IAR, MIN, MSL]	ESTRONE
68476-26-6 OIL GAS [NJL]	FUEL GASES
8012-95-1 OIL, MINERAL [WHM]	OIL MIST, MINERAL
8012-95-1 OIL (MINERAL) MIST [CEX]	OIL MIST, MINERAL
8012-95-1 OIL, MINERAL - MIST [ONT, ONT, ONT, ONT]	OIL MIST, MINERAL
8012-95-1 OIL MIST [CAL]	OIL MIST, MINERAL
8012-95-1 OIL MIST, MINERAL	OIL MIST, MINERAL

Chemical Name	Reference Name
8012-95-1 OIL MIST (MINERAL) [NIO, NIO]	OIL MIST, MINERAL
8007-70-3 OIL OF ANISE	OIL OF ANISE
8013-76-1 OIL OF BITTER ALMOND	OIL OF BITTER ALMOND
8000-27-9 OIL OF CEDARWOOD [PST]	CEDAR WOOD OIL
8000-29-1 OIL OF CITRONELLA [PST]	CITRONELLA OIL
8008-56-8 OIL OF LEMON [PST]	LEMON OIL
8007-02-1 OIL OF LEMONGRASS [PST]	LEMONGRASS OIL
2646-17-5 OIL ORANGE SS	OIL ORANGE SS
8001-86-3 OILS, BOLEKO [PAL]	ISANO OIL
8008-51-3 OILS, CAMPHOR [PAL]	CAMPHOR OIL
8016-28-2 OILS, LARD	OILS, LARD
68213-57-0 OILS, MENHADEN, POLYMERIZED	OILS, MENHADEN, POLYMERIZED
8007-40-7 OILS, MUSTARD	OILS, MUSTARD
8000-26-8 OILS, PINE NEEDLE	OILS, PINE NEEDLE
8037-19-2 OILS, TOBACCO	OILS, TOBACCO
68334-28-1 OILS, VEGETABLE, HYDROGENATED	OILS, VEGETABLE, HYDROGENATED
23696-28-8 OLAQUINDOX (N-(2-HYDROXYETHYL)-3-METHYL-2-QUINOXALINECARBOXAMIDE 1,4-DIOXIDE)	OLAQUINDOX (N-(2-HYDROXYETHYL)-3-METHYL-2-QUINOXALINECARBOXA
69898-00-6 ALPHA-OLEFINS	ALPHA-OLEFINS
RR-00963-2 ALPHA-OLEFIN SULFONATE, POTASSIUM SALT	ALPHA-OLEFIN SULFONATE, POTASSIUM SALTALKYLENEPOLYOL, DISUBSTITUTED ALKANES AND HYDROXYETHYL ACRYLATE
RR-01234-0 ALPHA-OLEFIN SULFONATE, SODIUM SALTS	ALPHA-OLEFIN SULFONATE, SODIUM SALTS
112-80-1 OLEIC ACID	OLEIC ACID
93-83-4 OLEIC ACID DIETHANOLAMIDE	OLEIC ACID DIETHANOLAMIDE
13961-86-9 OLEIC ACID DIETHANOLAMINE	OLEIC ACID DIETHANOLAMINE
93-83-4 OLEIC ACID DIETHANOLAMINE CONDENSATE [CSR]	OLEIC ACID DIETHANOLAMIDE
RR-01105-2 OLEIC ACID ESTER OF TETRA(HYDROXYETHYL)ETHYLENEDIAMINE	OLEIC ACID ESTER OF TETRA(HYDROXYETHYL) ETHYLENEDIAMINE
37220-82-9 OLEIC ACID ESTER WITH 1,2,3-PROPANETRIOL	OLEIC ACID ESTER WITH 1,2,3-PROPANETRIOLM SALT
71012-10-7 OLEIC ACID, 2-(2-(2-(2-HYDROXYETHOXY)ETHOXY)ETHOXY)ETHYL ESTER	OLEIC ACID, 2-(2-(2-(2-HYDROXYETHOXY)ETHOXY)ETHOXY)ETHYL EST
2717-15-9 OLEIC ACID, TRIETHANOLAMINE SALT [PST]	TRIETHANOLAMINE OLEATE
7347-29-7 OLEOYL IMIDAZOLINE	OLEOYL IMIDAZOLINE
RR-00792-1 OLEO OIL	OLEO OIL
110-25-8 OLEOYL SARCOSINE	OLEOYL SARCOSINE
8014-95-7 OLEUM [OS3]	SULFURIC ACID
8014-95-7 OLEUM (FUMING SULFURIC ACID) [FUMING SULFURIC ACID, MIXTUREWITH SULFUR TRIOXIDE] [EP3]	SULFURIC ACID
112-90-3 OLEYLAMINE	OLEYLAMINE
112-90-3 OLEYLAMINE (Z-9-OCTADECEN-1-AMINE) [TSC, TSC, TSC]	OLEYLAMINE
21652-27-7 OLEYL BASED IMIDAZOLINE	OLEYL BASED IMIDAZOLINE
7722-71-6 OLEYL ETHER PHOSPHATE (ACID)	OLEYL ETHER PHOSPHATE (ACID)
37310-83-1 OLEYL ETHER PHOSPHATE (NEUTRAL)	OLEYL ETHER PHOSPHATE (NEUTRAL)
60501-41-9 OLEYL GLYCIDYL ETHER [EPA, OTS]	OXIRANE, [(9-OCTADECENYLOXY)METHYL]-, (Z)-

Chemical Name	Reference Name
RR-01106-3 OLIGOESTER DERIVFED BY CONDENSATION OF ADIPIC ACID, PHTHALIC ANHYDRIDE, ETHYLENE GLYCOL, N-OCTYL ALCOHOL AND N-DECYL ALCOHOL	OLIGOESTER DERIVFED BY CONDENSATION OF ADIPIC ACID, PHTHALIC
RR-01211-3 OLIGOMETRIC SILICIC ACID ESTER COMPOUND WITH AN HYDROXYLALKYLAMINE	OLIGOMETRIC SILICIC ACID ESTER COMPOUND WITH AN HYDROXYLALKY
8001-25-0 OLIVE OIL	OLIVE OIL
1317-71-1 OLIVINE	OLIVINE
1317-71-1 OLIVINE SAND [BC1]	OLIVINE
1121-30-8 OMADINE	OMADINE
8002-72-0 ONIONS, OIL	ONIONS, OIL
RR-00543-6 ORAL CONTRACEPTIVES [CAL]	COMBINED ORAL CONTRACEPTIVES
RR-00543-6 ORAL CONTRACEPTIVES, COMBINED [C65]	COMBINED ORAL CONTRACEPTIVES
RR-00542-5 ORAL CONTRACEPTIVES, SEQUENTIAL [C65]	OESTROGEN-PROGESTIN COMBIN., SEQUENTIAL ORAL CONTRACEPTIVES
1936-15-8 ORANGE G [IAR]	ACID ORANGE 10
523-44-4 ORANGE I	ORANGE I
8008-57-9 ORANGE OIL	ORANGE OIL
2212-67-1 ORDRAM (MOLINATE)	ORDRAM (MOLINATE)L]-]
RR-01807-5 ORGANIC SYNTHETIC FIBRES (CARBON AND GRAPHITE FIBRES)	ORGANIC SYNTHETIC FIBRES (CARBON AND GRAPHITE FIBRES)
RR-00385-0 ORGANIC PEROXIDES, N.O.S.	ORGANIC PEROXIDES, N.O.S.
14265-44-2 ORGANIC PHOSPHATE [HM1, HM1, HM1]	PHOSPHATE
14265-44-2 ORGANIC PHOSPHATE COMPOUND, SOLID (POISON B) [NJL]	PHOSPHATE
RR-00390-7 ORGANIC PHOSPHORUS COMPOUND, MIXED WITH COMPRESSED GAS	ORGANIC PHOSPHORUS COMPOUND, MIXED WITH COMPRESSED GAS
RR-00004-4 ORGANO (ALKYL) MERCURY [MIN]	MERCURY, ALKYL COMPOUNDS
RR-00004-4 ORGANO(ALKYL)MERCURY [MSL]	MERCURY, ALKYL COMPOUNDS
RR-00391-8 ORGANOCHLORINE PESTICIDES, N.O.S.	ORGANOCHLORINE PESTICIDES, N.O.S.
RR-01603-5 ORGANOHALOGEN COMPOUNDS	ORGANOHALOGEN COMPOUNDS
RR-00395-2 ORGANOPHOSPHOROUS PESTICIDES, N.O.S.	ORGANOPHOSPHOROUS PESTICIDES, N.O.S.
RR-00069-1 ORGANORHODIUM COMPLEX	ORGANORHODIUM COMPLEX
RR-00069-1 ORGANORHODIUM COMPLEX (PMN-82-147) [PAL]	ORGANORHODIUM COMPLEX
RR-01609-1 ORGANOSILICON COMPOUNDS	ORGANOSILICON COMPOUNDS
RR-01661-5 ORGANOTIN CATALYSTS	ORGANOTIN CATALYSTS
RR-00134-3 ORGANOTIN COMPOUNDS (NON-PESTICIDAL USES) [CN4]	ORGANOTIN COMPOUNDS, N.O.S.
RR-00134-3 ORGANOTIN COMPOUNDS, N.O.S.	ORGANOTIN COMPOUNDS, N.O.S.
RR-01725-4 ORGANOTIN LITHIUM COMPOUNDS	ORGANOTIN LITHIUM COMPOUNDS
RR-00399-6 ORGANOTIN PESTICIDES, N.O.S.	ORGANOTIN PESTICIDES, N.O.S.
3184-13-2 L-ORNITHINE HCL	L-ORNITHINE HCL
65-86-1 OROTIC ACID	OROTIC ACID
RR-01306-9 ORTHOARSENIC ACID	ORTHOARSENIC ACID
7664-38-2 ORTHOPHOSPHORIC ACID [GBR, GBR]	PHOSPHORIC ACID
19044-88-3 ORYZALIN [4-(DIPROPYLAMINO)-3,5-DINITROBENZENESULFONAMIDE]	ORYZALIN [4-(DIPROPYLAMINO)-3,5-DINITROBENZENE-SULFONAMIDE]
14993-35-2 OSMIUM 180	OSMIUM 180
14993-64-7 OSMIUM 181	OSMIUM 181
14993-36-3 OSMIUM 182	OSMIUM 182
15766-50-4 OSMIUM 185	OSMIUM 185

Chemical Name	Reference Name
RR-00397-4 OSMIUM 189M	OSMIUM 189M
14119-24-5 OSMIUM 191	OSMIUM 191
RR-00396-3 OSMIUM 191M	OSMIUM 191M
16057-77-5 OSMIUM 193	OSMIUM 193
15766-57-1 OSMIUM 194	OSMIUM 194
20816-12-0 OSMIUM OXIDE (OSO4), (T-4)- [PAL]	OSMIUM TETROXIDE
RR-00608-6 OSMIUM, TETRAAMINETETRACHLORO-	OSMIUM, TETRAAMINETETRACHLORO-
20816-12-0 OSMIUM TETRAOXIDE [GBR, GBR]	OSMIUM TETROXIDE
20816-12-0 OSMIUM TETROXIDE	OSMIUM TETROXIDE
630-60-4 OUABAIN	OUABAIN
71526-07-3 1-OXA-4-AZASPIRO(4,5)DECANE, 4-(DICHLOROACETYL)-	1-OXA-4-AZASPIRO(4,5)DECANE, 4-(DICHLOROACETYL)-
71526-07-3 1-OXA-4-AZASPIRO[4,5]DECANE, 4-(DICHLOROACETYL)- [TSE]	1-OXA-4-AZASPIRO(4,5)DECANE, 4-(DICHLOROACETYL)- ALKYL ESTERS
155-41-9 3-OXA-9-AZONIATRICYCLO[3.3.1.02,4]NONANE, 7-(3-HYDROXY-1-OXO-2-PHENYLPROPOXY)-9,9-DIMETHYL-, BROMIDE, [7(S)-(1-ALPHA, 2-BETA, 4-BETA,5-ALPHA,7-BETA)]-	3-OXA-9-AZONIATRICYCLO[3.3.1.02,4]NONANE, 7-(3-HYDROXY-1-OXO
6106-46-3 3-OXA-9-AZONIATRICYCLO[3.3.1.0]NONANE, 7-(3-HYDROXY-1-OXO-2-PHENYLPROPOXY)-9,9-DIMETHYL-, [7(S)-(1A,2B,4B,5A,7B)-, NITRATE (SALT) [TSC, TSC]	BENZENEACETIC ACID, ALPHA-(HYDROXYMETHYL)-, 9-METHYL-3-OXA-
286-20-4 7-OXABICYCLO[4.1.0]HEPTANE	7-OXABICYCLO[4.1.0]HEPTANE
145-73-3 7-OXABICYCLO[2.2.1]HEPTANE-2,3-DICARBOXYLIC ACID [TSC, TSC]	ENDOTHALL
RR-01015-1 OXABICYCLO[4.1.0]HEPTANE, 3-ETHENYL, HOMOPOLYMER, ETHER WITH2-ETHYL-2-(HYDROXYMETHYL)-1,3-PROPANEDIOL (3:1), EPOXIZED	OXABICYCLO[4.1.0]HEPTANE, 3-ETHENYL, HOMOPOLYMER, ETHER WITHHEXYL ACRYLATE, METHACRYLIC ACID AND SUBSTITUTED BISBENZENE
470-67-7 7-OXABICYCLO[2.2.1]HEPTANE, 1-METHYL-4-(1-METHYLETHYL)-	7-OXABICYCLO[2.2.1]HEPTANE, 1-METHYL-4-(1-METHYLETHYL)-
96-08-2 7-OXABICYCLO[4.1.0]HEPTANE, 1-METHYL-4-(2-METHYLOXIRANYL)-	7-OXABICYCLO[4.1.0]HEPTANE, 1-METHYL-4-(2-METHYLOXIRANYL)-
106-87-6 7-OXABICYCLO[4.1.0]HEPTANE, 3-OXIRANYL- [PAL, TSC, TSC]	VINYL CYCLOHEXENE DIOXIDE
285-67-6 6-OXABICYCLO[3.1.0]HEXANE	6-OXABICYCLO[3.1.0]HEXANE-2-PHENYLPROPOXY)-9,9-DIMETHYL-, BROMIDE, [7(S)-(1-ALPHA, 2-BETA,4-BETA, 5-ALPHA,7-BETA)]-
19666-30-9 OXADIAZON	OXADIAZON
RR-01351-4 OXALATES	OXALATES
144-62-7 OXALIC ACID	OXALIC ACID
2944-67-4 OXALIC ACID, AMMONIUM IRON (3+) SALT (3:3:1)	OXALIC ACID, AMMONIUM IRON (3+) SALT (3:3:1)
14258-49-2 OXALIC ACID, AMMONIUM SALT	OXALIC ACID, AMMONIUM SALT
460-19-5 OXALONITRILE [GBR]	CYANOGEN
471-46-5 OXAMIDE	OXAMIDE
23135-22-0 OXAMYL	OXAMYL
23135-22-0 OXAMYL (VYDATE) [SDW, SDW]	OXAMYL
15980-15-1 1,4-OXATHIANE	1,4-OXATHIANE
1120-71-4 1-2-OXATHIOLANE 2,2-DIOXIDE [NJL]	1,3-PROPANE SULTONE
1120-71-4 1,2-OXATHIOLANE, 2,2-DIOXIDE [PAL, PAL]	1,3-PROPANE SULTONE
1139-30-6 5-OXATRICYCLO[8.2.0.04,6]DODECANE, 4,12,12-TRIMETHYL-9-METHYLENE-, [1R-(1R*,4R*,6R*, 10S*)]-	5-OXATRICYCLO[8.2.0.04,6]DODECANE, 4,12,12-TRIMETHYL-9-

Chemical Name	Reference Name
34314-63-1 **3-OXAURACIL**	3-OXAURACIL
50-18-0 **2H-1,3,2-OXAZAPHOSPHORIN-2-AMINE, N, N-BIS(2-CHLOROETHYL)TETRAHYDRO-, 2-OXIDE [PAL, PAL]**	CYCLOPHOSPHAMIDE
604-75-1 **OXAZEPAM**	OXAZEPAM
129-20-4 **OXAZOLODINE-1,3**	OXAZOLODINE-1,3
139-91-3 **2-OXAZOLIDINONE, 5-(4-MORPHOLINYL-METHYL)-3-[[(5-NITRO-2-FURANYL)METHY-LENE]AMINO]- [PAL, PAL]**	5-(MORPHOLINOMETHYL)-3-[(5-NITRO-FURFURYLI-DENE)-AMINO]-2-OXA
6542-37-6 **1H,3H,5H-OXAZOLO(3,4-C)OXAZOLE-7A,(7H)-METHANOL**	1H,3H,5H-OXAZOLO(3,4-C)OXAZOLE-7A,(7H)-METHANOL
59720-42-2 **1H,3H,5H-OXAZOLO(3,4-C)OXAZOLE, METHANOL DERIVATIVE**	1H,3H,5H-OXAZOLO(3,4-C)OXAZOLE, METHANOL DERIVATIVE
56709-13-8 **1H,3H,5H-OXAZOLO(3,4-C)OXAZOLE, POLY (OXYMETHYLENE)DERIVATIVE**	1H,3H,5H-OXAZOLO(3,4-C)OXAZOLE, POLY(OXYMETHY-LENE)DERIVATIVE
RR-00915-4 **2-OXEPANONE, POLYMER WITH 4,4'-(1-METHYLETHYLIDENE)BISPHENOLAND 2,2-[(1-METHYLETHYLIDENE)BIS(4,1-PHENYLE-NEOXYMETHYLENE)]BISOXIRANE, GRAFT**	2-OXEPANONE, POLYMER WITH 4,4'-(1-METHYLETHYLI-DENE)BISPHENOL
78-71-7 **OXETANE, 3,3-BIS(CHLOROMETHYL)-**	OXETANE, 3,3-BIS(CHLOROMETHYL)-
57-57-8 **2-OXETANONE [PAL, PAL]**	BETA-PROPIOLACTONE
3068-88-0 **2-OXETANONE, 4-METHYL- [PAL, PAL]**	BETA-BUTYROLACTONE
674-82-8 **2-OXETANONE, 4-METHYLENE- [PAL]**	DIKETENE
RR-01556-5 **OXIDANTS (OZONE)**	OXIDANTS (OZONE)
68441-17-8 **OXIDIZED POLYETHYLENE [PST]**	ETHENE, HOMOPOLYMER, OXIDIZED
RR-00403-5 **OXIDIZERS, N.O.S.**	OXIDIZERS, N.O.S.
RR-00403-5 **OXIDIZING SUBSTANCES, N.O.S. [HM1, HM1]**	OXIDIZERS, N.O.S.
75-21-8 **OXIRANE [PAL, PAL, TSC, TSC, TSC]**	ETHYLENE OXIDE
3132-64-7 **OXIRANE, BROMOMETHYL- [TSC, TSC, TSC]**	EPIBROMOHYDRIN
2425-79-8 **OXIRANE, 2,2'-[1,4-BUTANEDIYLBIS (OXYMETHYLENE)]BIS-**	OXIRANE, 2,2'-[1,4-BUTANEDIYLBIS(OXYMETHYLENE)] BIS-
2426-08-6 **OXIRANE, (BUTOXYMETHYL)- [PAL]**	N-BUTYL GLYCIDYL ETHER (BGE)-TETRA-CHLOROETHYL)THIO]-
2426-08-6 **OXIRANE, BUTOXYMETHYL- [TSC, TSC]**	N-BUTYL GLYCIDYL ETHER (BGE)
25085-99-8 **OXIRANE,2,2'-4-BUTYLIDENEBISPHENYLE-NEOXYMETHYLENE (DGEBA)**	OXIRANE,2,2'-4-BUTYLIDENEBISPHENYLE-NEOXYMETHYLENE (DGEBA)
77-83-8 **OXIRANECARBOXYLIC ACID, 3-METHYL-3-PHENYL-, ETHYL ESTER**	OXIRANECARBOXYLIC ACID, 3-METHYL-3-PHENYL-, ETHYL ESTER
121-39-1 **OXIRANECARBOXYLIC ACID, 3-PHENYL-, ETHYL ESTER**	OXIRANECARBOXYLIC ACID, 3-PHENYL-, ETHYL ESTER
106-89-8 **OXIRANE, (CHLOROMETHYL)- [PAL, PAL]**	EPICHLOROHYDRIN
106-89-8 **OXIRANE, CHLOROMETHYL- [TSC, TSC, TSC]**	EPICHLOROHYDRIN
14228-73-0 **OXIRANE, 2,2'-[1,4-CYCLOHEXANEDILBIS (METHYLENEOXYMETHYLENE)]BIS-**	OXIRANE, 2,2'-[1,4-CYCLOHEXANEDILBIS (METHYLE-NEOXYMETHYLENE)
2855-19-8 **OXIRANE, DECYL-**	OXIRANE, DECYL-
20217-01-0 **OXIRANE, [(2,4-DIBROMOPHENOXY)METHYL]-**	OXIRANE, [(2,4-DIBROMOPHENOXY)METHYL]-
35243-89-1 **OXIRANE, [(1,2-DIBROMOPROPOXY)METHYL]-**	OXIRANE, [(1,2-DIBROMOPROPOXY)METHYL]-
558-30-5 **OXIRANE, 2,2-DIMETHYL-**	OXIRANE, 2,2-DIMETHYL-
3266-23-7 **OXIRANE, 2,3-DIMETHYL-**	OXIRANE, 2,3-DIMETHYL-
7665-72-7 **OXIRANE, [(1,1-DIMETHYLETHOXY)METHYL]- [TSC]**	TERT-BUTYL GLYCIDYL ETHER
3101-60-8 **OXIRANE, [[4-(1,1-DIMETHYLETHYL)PHENOXY] METHYL]- [TSC]**	P-TERT-BUTYL PHENYL GLYCIDYL ETHER

Chemical Name	Reference Name
17557-23-2 OXIRANE, 2,2'-[(2,2-DIMETHYL-1,3-PROPANEDIYL)BIS(OXYMETHYLENE)]BIS-	OXIRANE, 2,2'-[(2,2-DIMETHYL-1,3-PROPANEDIYL)BIS (OXYMETHYLEN
3234-28-4 OXIRANE, DODECYL-	OXIRANE, DODECYL-
2461-18-9 OXIRANE, [(DODECYLOXY)METHYL]-	OXIRANE, [(DODECYLOXY)METHYL]-
2224-15-9 OXIRANE, 2,2'-[1,2-ETHANEDIYLBIS (OXYMETHYLENE)]BIS-	OXIRANE, 2,2'-[1,2-ETHANEDIYLBIS (OXYMETHYLENE)] BIS-
7328-97-4 OXIRANE, 2,2',2",2"'-[1,2-ETHANEDIYLI-DENETETRAKIS-(4,1-PHENYLENEOXYMETHY-LENE)]TETRAKIS-	OXIRANE, 2,2',2",2"'-[1,2-ETHANEDIYLIDENETETRAKIS-
930-22-3 OXIRANE, ETHENYL- [PAL, TSC, TSC]	BUTADIENE MONOXIDE
4016-11-9 OXIRANE, ETHOXYMETHYL- [TSC, TSC]	EPOXY ETHYLOXY PROPANE
106-88-7 OXIRANE, ETHYL- [PAL, TSC, TSC, TSC]	1,2-BUTYLENE OXIDE
2461-15-6 OXIRANE, [[(2-ETHYLHEXYL)OXY]METHYL]-	OXIRANE, [[(2-ETHYLHEXYL)OXY]METHYL]-
67860-04-2 OXIRANE, HEPTADECYL-	OXIRANE, HEPTADECYL-
7390-81-0 OXIRANE, HEXADECYL-	OXIRANE, HEXADECYL-
15965-99-8 OXIRANE, [(HEXADECYLOXY)METHYL]-	OXIRANE, [(HEXADECYLOXY)METHYL]-
RR-00982-5 OXIRANE, 2,2'-(1,6-HEXANEDIYLBIS (OXYMETHYLENE)) BIS-	OXIRANE, 2,2'-(1,6-HEXANEDIYLBIS(OXYMETHYLENE)) BIS-TYLAMIONO) ETHANOL
68959-23-9 OXIRANE, 2,2',2"-[1,2,6-HEXANETRIYLTRIS-(OXYMETHYLENE)]TRIS-	OXIRANE, 2,2',2"-[1,2,6-HEXANETRIYLTRIS-(OXYMETHY-LENE)]
130728-76-6 OXIRANEMETHANAMINE, N,N'-[METHYLENEBIS[2-ETHYL-4,1-PHENYLENE]BIS [N-OXIRANYLMETHYL]-	OXIRANEMETHANAMINE, N,N'-[METHYLENEBIS[2-ETHYL-4,1-PHENYLENEHEXYLIMINO]-2,1-ETHANEDIYL] TRIS-[3,3,4,5,5-PENTAMETHYL]-,
5026-74-4 OXIRANEMETHANAMINE, N-[4-(OXIRANYL-METHOXY)PHENYL]-N-(OXIRANYLMETHYL)-	OXIRANEMETHANAMINE, N-[4-(OXIRANYLMETHOXY) PHENYL]-N-
556-52-5 OXIRANEMETHANOL [PAL]	GLYCIDOL
556-52-5 OXIRANEMETHANOL (GLYCIDOL) [TSC, TSC]	GLYCIDOL
930-37-0 OXIRANE, METHOXYMETHYL-	OXIRANE, METHOXYMETHYL-
75-56-9 OXIRANE, METHYL- [PAL, PAL, TSC, TSC, TSC]	PROPYLENE OXIDE
39817-09-9 OXIRANE, 2,2'-[METHYLENEBIS(PHENYLE-NEOXYMETHYLENE)]BIS-	OXIRANE, 2,2'-[METHYLENEBIS(PHENYLE-NEOXYMETHYLENE)]BIS-
54208-63-8 OXIRANE, 2,2'-[METHYLENEBIS(2,1-PHENYLE-NEOXYMETHYLENE)]BIS-	OXIRANE, 2,2'-[METHYLENEBIS(2,1-PHENYLE-NEOXYMETHYLENE)]BIS-
4016-14-2 OXIRANE, [(1-METHYLETHOXY)METHYL]- [PAL, TSC, TSC]	ISOPROPYL GLYCIDYL ETHER (IGE)
71033-08-4 OXIRANE, 2,2'-[(1-METHYLETHYLIDINE)BIS[4, 1-PHENYLENEOXY[1-(BUTOXYMETHYL)-2,1-ETHANEDIYL]OXYMETHYLENE]BIS-	OXIRANE, 2,2'-[(1-METHYLETHYLIDINE)BIS[4,1-PHENYLENEOXY[1-(B
25085-99-8 OXIRANE, 2,2'-[(1-METHYLETHYLIDINE)BIS (4,1-PHENYLENEOXYMETHYLENE)]BIS-, HO-MOPOLYMER [TSC, TSC, TSC, TSC]	OXIRANE,2,2'-4-BUTYLIDENEBISPHENYLE-NEOXYMETHYLENE (DGEBA)
72319-24-5 OXIRANE, 2,2'-[(1-METHYLETHYLIDINE)BIS[4, 1-PHENYLENEOXY-3,1-PROPANEDIYLOXY-4, 1-PHENYLENE(1-METHYLETHYLIDENE)-4,1-PHENYLENEOXYMETHYLENE]BIS-	OXIRANE, 2,2'-[(1-METHYLETHYLIDINE)BIS[4,1-PHENYLENEOXY-3,1-
2210-79-9 OXIRANE, [(2-METHYLPHENOXY)METHYL]-	OXIRANE, [(2-METHYLPHENOXY)METHYL]-
26447-14-3 OXIRANE, [(METHYLPHENOXY)METHYL]- [TSC, TSC]	CRESYL GLYCIDYL ETHER
61578-04-9 OXIRANE, [[4-(1-METHYL-1-PHENYLETHYL) PHENOXY]METHYL]-	OXIRANE, [[4-(1-METHYL-1-PHENYLETHYL)PHENOXY] METHYL]-
9038-29-3 OXIRANE, METHYL-, POLYMER WITH OXI-RANE, DECYL ETHER	OXIRANE, METHYL-, POLYMER WITH OXIRANE, DECYL ETHER

Chemical Name	Reference Name
61725-89-1 OXIRANE METHYL-, POLYMER WITH OXIRANE, TRIDECYL ETHER	OXIRANE METHYL-, POLYMER WITH OXIRANE, TRIDECYL ETHER-, HYDROXIDE, DISODIUM SALT
68987-80-4 OXIRANE, MONO[C(6-12)-ALKYLOXY)METHYL] DERIVATIVES	OXIRANE, MONO[C(6-12)-ALKYLOXY)METHYL]DERIVATIVES
68609-96-1 OXIRANE, MONO[C(8-10)-ALKYLOXY)METHYL] DERIVATIVES [TSC]	ALKYL (C8-C10) GLYCIDYL ETHERAMMONIO]-2-HYDROXYPROPYL]-OMEGA-HYDROXY-, N-COCO ACYL DERIVATIVES, METHYL SULFATES (SALTS)
68081-84-5 OXIRANE, MONO[(C10-16-ALKYLOXY)METHYL] DERIVATIVES	OXIRANE, MONO[(C10-16-ALKYLOXY)METHYL] DERIVATIVES
68081-84-5 OXIRANE, MONO[(C10-C16-ALKYLOXY) METHYL]DERIVATIVES [TSC, TSC]	OXIRANE, MONO[(C10-16-ALKYLOXY)METHYL] DERIVATIVESTETRAKIS[PHOSPHONATO](6-)-N,N',O,O", O"",O""']-, PENTAAMMONIUM HYDROGEN, (OC-6-21)-
68609-97-2 OXIRANE, MONO[(C(12-14)-ALKYLOXY) METHYL]DERIVATIVES [TSC]	ALKYL (C12-C14) GLYCIDYL ETHER
5255-75-4 OXIRANE, [(4-NITROPHENOXY)METHYL]-	OXIRANE, [(4-NITROPHENOXY)METHYL]-
6178-32-1 OXIRANE, [(4-NONYLPHENOXY)METHYL]-	OXIRANE, [(4-NONYLPHENOXY)METHYL]-
60501-41-9 OXIRANE, [(9-OCTADECENYLOXY)METHYL]-, (Z)-	OXIRANE, [(9-OCTADECENYLOXY)METHYL]-, (Z)-
16245-97-9 OXIRANE, [(OCTADECYLOXY)METHYL]-	OXIRANE, [(OCTADECYLOXY)METHYL]-
106-83-2 OXIRANEOCTANOIC ACID, 3-OCTYL-, BUTYL ESTER	OXIRANEOCTANOIC ACID, 3-OCTYL-, BUTYL ESTER
141-38-8 OXIRANEOCTANOIC ACID, 3-OCTYL-, 2-ETHYL-HEXYL ESTER	OXIRANEOCTANOIC ACID, 3-OCTYL-, 2-ETHYLHEXYL ESTER
106-84-3 OXIRANEOCTANOIC ACID, 3-OCTYL-, OCTYL ESTER	OXIRANEOCTANOIC ACID, 3-OCTYL-, OCTYL ESTER
2404-44-6 OXIRANE, OCTYL-	OXIRANE, OCTYL-
13561-08-5 OXIRANE, 2,2'-(OXIRANYLMETHOXY)-1,3-PHENYLENE]BIS(METHYLENE)]BIS-	OXIRANE, 2,2'-(OXIRANYLMETHOXY)-1,3-PHENYLENE]BIS(METHYLENE)
67786-03-2 OXIRANE, 2,2'-[[[(2-OXIRANYLMETHOXY) PHENYL]METHYLENE]BIS(4,1-PHENYLE-NEOXYMETHYLENE)]BIS-	OXIRANE, 2,2'-[[[(2-OXIRANYLMETHOXY)PHENYL] METHYLENE]BIS(4,1
2238-07-5 OXIRANE, 2,2'-[OXYBIS(METHYLENE)]BIS- [PAL, TSC, TSC]	DIGLYCIDYL ETHER (DGE)
22092-38-2 OXIRANE, PENTADECYL-	OXIRANE, PENTADECYL-4,6-DINITROPHENYL)AZO]-4-ETHOXYPHENYL]-
122-60-1 OXIRANE, (PHENOXYMETHYL)- [PAL, TSC, TSC]	PHENYL GLYCIDYL ETHER
96-09-3 OXIRANE, PHENYL- [PAL, TSC, TSC]	STYRENE OXIDE
101-90-6 OXIRANE, 2,2'-[1,3-PHENYLENEBIS (OXYMETHYLENE)]BIS- [TSC, TSC]	DIGLYCIDYL RESORCINOL ETHER
2425-01-6 OXIRANE, 2,2'-[1,4-PHENYLENEBIS (OXYMETHYLENE)]BIS-	OXIRANE, 2,2'-[1,4-PHENYLENEBIS (OXYMETHYLENE)] BIS-
13236-02-7 OXIRANE, 2,2',2"-[1,2,3-PROPANETRIYL TRIS (OXYMETHYLENE)]TRIS-	OXIRANE, 2,2',2"-[1,2,3-PROPANETRIYL TRIS (OXYMETHYLENE)]
106-92-3 OXIRANE, [(2-PROPENYLOXY)METHYL]- [PAL, TSC, TSC]	ALLYL GLYCIDYL ETHER
68517-02-2 OXIRANE, 2,2',2"-[PROPYLIDYNETRIS(4,1-PHENYLENEOXYMETHYLENE)]TRIS-	OXIRANE, 2,2',2"-[PROPYLIDYNETRIS(4,1-PHENYLE-NEOXYMETHYLENE
7320-37-8 OXIRANE, TETRADECYL-	OXIRANE, TETRADECYL-
38954-75-5 OXIRANE, [(TETRADECYLOXY)METHYL]-	OXIRANE, [(TETRADECYLOXY)METHYL]-
3083-25-8 OXIRANE, (2,2,2-TRICHLOROETHYL)- [TSC, TSC, TSC, TSC]	TRICHLOROBUTYLENE OXIDE
38565-52-5 OXIRANE, (2,2,3,3,4,4,5,5,6,6,7,7,7-TRIDECAFLU-OROHEPTYL)-	OXIRANE, (2,2,3,3,4,4,5,5,6,6,7,7,7-TRIDECAFLUOROHEP-TYL)-
18633-25-5 OXIRANE, TRIDECYL-	OXIRANE, TRIDECYL-

Chemical Name	Reference Name
428-59-1 OXIRANE, TRIFLUORO(TRIFLUOROMETHYL)- [TSC, TSC, TSC, TSC]	HEXAFLUOROPROPYLENE OXIDE
68551-07-5 OXO ALCOHOL STILL BOTTOMS (C8-18 ALCO- HOLS)	OXO ALCOHOL STILL BOTTOMS (C8-18 ALCOHOLS) OXYETHYL)-2-UNDECYL-, DIHYDROXIDE, DISODIUM SALT
68130-43-8 OXO ALCOHOL STILL BOTTOMS, SULFATED, SODIUM SALT	OXO ALCOHOL STILL BOTTOMS, SULFATED, SODIUM SALT
RR-01568-9 OXOSUBSTITUTED AMINOALKANOIC ACID DERIVATIVE	OXOSUBSTITUTED AMINOALKANOIC ACID DERIVATIVE
RR-00985-8 OXYALKANEPOLYOL POLYACRYLATE	OXYALKANEPOLYOL POLYACRYLATESUBSTITUTED HETERMONOCYCLE, ACRYLATE POLYMER
101-84-8 1,1'-OXYBISBENZENE [ONT, ONT]	PHENYL ETHER
80-51-3 P,P'-OXYBIS(BENZENESULFONYLHYDRAZIDE) [TSC, TSC, TSC]	DIPHENYLOXIDE-4,4'-DISULFOHYDRAZIDE
111-44-4 1,1'-OXYBIS(2-CHLOROETHANE) [ONT, ONT, ONT]	BIS(2-CHLOROETHYL) ETHER
542-88-1 OXYBIS (CHLOROMETHANE) [ONT]	BIS(CHLOROMETHYL) ETHER
74007-80-0 OXYBIS(DIBUTYL(2,4,5-TRICHLOROPHENOXY) TIN)	OXYBIS(DIBUTYL(2,4,5-TRICHLOROPHENOXY)TIN)
7460-82-4 2,2-OXYBISETHANE BIS(4-METHYLBENZENE- SULFONATE)	2,2-OXYBISETHANE BIS(4-METHYLBENZENESUL- FONATE)
2238-07-5 2,2'-(OXYBIS(METHYLENE))BISOXIRANE [ONT]	DIGLYCIDYL ETHER (DGE)
53061-10-2 1,1'-[OXYBIS(METHYLENESULFONYL)] BIS-2- CHLOROETHANE	1,1'-[OXYBIS(METHYLENESULFONYL)] BIS-2- CHLOROETHANEETHER WITH 2-ETHYL-2-(HYDROX- YMETHYL)-1,3-PROPANEDIOL (3:1)
26750-50-5 1,1'-[OXYBIS(METHYLENESULFONYL)] BISETHENE	1,1'-[OXYBIS(METHYLENESULFONYL)] BISETHENE-
36724-43-3 2,2'-[OXYBIS(METHYLENESULFONYL)] BISETHANOL	2,2'-[OXYBIS(METHYLENESULFONYL)] BISETHANOL
58-36-6 10,10'-OXYBISPHENOXARSINE	10,10'-OXYBISPHENOXARSINE
108-20-3 2,2'-OXYBIS(PROPANE) [ONT, ONT]	ISOPROPYL ETHER
27304-13-8 OXYCHLORDANE	OXYCHLORDANE
301-12-2 OXYDEMETON-METHYL	OXYDEMETON-METHYL
301-12-2 OXYDEMETONMETHYL [CAL]	OXYDEMETON-METHYLLED)
301-12-2 OXYDEMETON METHYL [MSL]	OXYDEMETON-METHYL
301-12-2 OXYDEMETON METHYL [S-(2-(ETHYL- SULFINYL)ETHYL) O,O-DIMETHYL ESTER PHOSPHOROTHIOIC ACID] [313]	OXYDEMETON-METHYL
101-80-4 4,4'-OXYDIANILINE [CSR, CSR]	4,4'-DIAMINODIPHENYL ETHER
101-80-4 4,4-OXYDIANILINE [FLA]	4,4'-DIAMINODIPHENYL ETHER
101-80-4 4,4'-OXYDIANILINE [MAK, MAK, NTP]	4,4'-DIAMINODIPHENYL ETHER
19666-30-9 OXYDIAZON [3-[2,4-DICHLORO-5-(1- METHYLETHOXY)PHENYL]-5-(1,1- DIMETHYLETHYL)-1,3,4-OXADIAZOL-2(3H)- ONE] [313]	OXADIAZON
53461-82-8 OXYDI-2,1-ETHANEDIYL TETRAKIS(2- CHLOROETHYL) PHOSPHATE	OXYDI-2,1-ETHANEDIYL TETRAKIS(2-CHLOROETHYL) PHOSPHATE
111-46-6 2,2'-OXYDIETHANOL [GBR]	DIETHYLENE GLYCOL
102-77-2 N-(OXYDIETHYLENE)BENZOTHIAZOLE-2- SULFENAMIDE	N-(OXYDIETHYLENE)BENZOTHIAZOLE-2-SULFENAMIDE
1656-48-0 3,3'-OXYDIPROPIONITRILE	3,3'-OXYDIPROPIONITRILE
2497-07-6 OXYDISULFOTON	OXYDISULFOTON
31866-76-9 1-OXYETHYL-2-STEARIC IMIDAZOLINE	1-OXYETHYL-2-STEARIC IMIDAZOLINE
42874-03-3 OXYFLUORFEN	OXYFLUORFENUORIDE)

Chemical Name	Reference Name
7782-44-7 OXYGEN	OXYGEN
7782-44-7 OXYGEN (DISSOLVED) [MX2]	OXYGEN
7783-41-7 OXYGEN DIFLUORIDE	OXYGEN DIFLUORIDE
7783-41-7 OXYGEN DIFLUORIDE (FLUORINE MONOXIDE) [OS3]	OXYGEN DIFLUORIDE
7783-41-7 OXYGEN FLUORIDE (OF2) [PAL]	OXYGEN DIFLUORIDE
7782-44-7 OXYGEN (LIQUID) [FLA, MSL]	OXYGEN
434-07-1 OXYMETHOLONE	OXYMETHOLONE
129-20-4 OXYPHENBUTAZONE [IAR]	OXAZOLODINE-1,3
79-57-2 OXYTETRACYCLINE	OXYTETRACYCLINE
2058-46-0 OXYTETRACYCLINE HYDROCHLORIDE	OXYTETRACYCLINE HYDROCHLORIDE
RR-01107-4 OYSTER SHELLS	OYSTER SHELLS ANHYDRIDE, ETHYLENE GLYCOL, N-OCTYL ALCOHOL AND N-DECYL ALCOHOL
10028-15-6 OZONE	OZONE
RR-00110-5 OZONE BY-PRODUCTS	OZONE BY-PRODUCTS
RR-01518-9 PAINT	PAINT
RR-00316-7 PAINT MANUFACTURE AND PAINTING	PAINT MANUFACTURE AND PAINTING
7440-05-3 PALLADIUM	PALLADIUM
15690-69-4 PALLADIUM 100	PALLADIUM 100
15749-54-9 PALLADIUM 101	PALLADIUM 101
14967-68-1 PALLADIUM 103	PALLADIUM 103
17637-99-9 PALLADIUM 107	PALLADIUM 107
14981-64-7 PALLADIUM 109	PALLADIUM 109
7647-10-1 PALLADIUM(II) CHLORIDE	PALLADIUM(II) CHLORIDE
10102-05-3 PALLADIUM DINITRATE	PALLADIUM DINITRATE
57-10-3 PALMITIC ACID	PALMITIC ACID
8023-79-8 PALM KERNEL OIL	PALM KERNEL OIL
8002-75-3 PALM OIL	PALM OIL
8049-47-6 PANCREATIN	PANCREATIN
794-93-4 PANFURAN S	PANFURAN S
794-93-4 PANFURAN S (CONTAINING DIHYDROX-YMETHYLFURATRIZINE) [IAR]	PANFURAN S
9001-73-4 PAPAIN	PAPAIN
RR-01108-5 PAPER	PAPER
RR-01503-2 PAPER, UNSATURATED OIL TREATED, INCOM-PLETELY DRY	PAPER, UNSATURATED OIL TREATED, INCOMPLETELY DRY
68991-42-4 PAPRIKA	PAPRIKA
RR-01805-3 PARA-ARAMIDE FIBRES (KEVLAR, TWARON)	PARA-ARAMIDE FIBRES (KEVLAR, TWARON)
103-90-2 PARACETAMOL (ACETAMINOPHEN) [IAR]	ACETOMINOPHEN (4-HYDROXYACETANILIDE)
59-50-7 PARACHLOROMETA CRESOL [CW2]	4-CHLORO-M-CRESOL
59-50-7 PARA-CHLORO-META-CRESOL [MX1]	4-CHLORO-M-CRESOL
8002-74-2 PARAFFIN WAX FUME [MEX, MEX]	PARAFFIN WAXES AND HYDROCARBON WAXES
64742-04-7 PARAFFINIC DISTILLATE SOLVENT EXTRACT	PARAFFINIC DISTILLATE SOLVENT EXTRACTATE SOLVENT
8012-45-1 PARAFFIN OIL	PARAFFIN OIL
64742-71-8 PARAFFIN OILS, CATALYTIC DEWAXED LIGHT	PARAFFIN OILS, CATALYTIC DEWAXED LIGHTY
64742-70-7 PARAFFIN OILS (PETROLEUM), CATALYTIC DEWAXED HEAVY	PARAFFIN OILS (PETROLEUM), CATALYTIC DEWAXED HEAVYLIGHT
8002-74-2 PARAFFIN WAX [CEX, GBR, GBR, MIN, PST]	PARAFFIN WAXES AND HYDROCARBON WAXES
8002-74-2 PARAFFIN WAXES AND HYDROCARBON WAXES	PARAFFIN WAXES AND HYDROCARBON WAXES

Chemical Name	Reference Name
63449-39-8 PARAFFIN WAXES AND HYDROCARBON WAXES, CHLORINATED [TSC]	CHLORINATED HYDROCARBONS (CHORINATED PARAFFINS)
8002-74-2 PARAFFIN WAX FUME [ALB, ALB]	PARAFFIN WAXES AND HYDROCARBON WAXES
8002-74-2 PARAFFIN WAX (FUME) [AUS]	PARAFFIN WAXES AND HYDROCARBON WAXES
8002-74-2 PARAFFIN WAX FUME [BC1, BC1, CAL, FLA, MSL, ONT, OS1, QBC, TLV]	PARAFFIN WAXES AND HYDROCARBON WAXES
30525-89-4 PARAFORMALDEHYDE	PARAFORMALDEHYDE
30525-89-4 PARAFORMALDEHYDE (POLYMER) [MSL]	PARAFORMALDEHYDE
123-63-7 PARALDEHYDE	PARALDEHYDE
115-67-3 PARAMETHADIONE	PARAMETHADIONE
52061-60-6 PARAMETHANE HYDROPEROXIDE [NJL]	P-METHANE HYDROPEROXIDE
311-45-5 PARAOXON [HM1, HM1, NJL]	DIETHYL-P-NITROPHENYL PHOSPHATE
1910-42-5 PARAQUAT	PARAQUAT
1910-42-5 PARAQUAT [302, CAL, MAK, MAK, MAK, MIN, NIO, NIO, NIO, NIO, NJL]	PARAQUAT
4685-14-7 PARAQUAT	PARAQUAT
1910-42-5 PARAQUAT (GRAMOXONE) [BC1, BC1]	PARAQUAT
1910-42-5 PARAQUAT RESPIRABLE DUST [MEX]	PARAQUAT
2074-50-2 PARAQUAT BIS(METHYL SULFATE) [MSL]	PARAQUAT METHOSULFATE
1910-42-5 PARAQUAT DICHLORIDE [313]	PARAQUATO-]
1910-42-5 PARAQUAT DICHLORIDE (ISO) [GBR]	PARAQUAT
2074-50-2 PARAQUAT METHOSULFATE	PARAQUAT METHOSULFATE
1910-42-5 PARAQUAT RESPIRABLE SIZES [QBC, ALB, ALB]	PARAQUAT
569-61-9 PARAROSANILINE HYDROCHLORIDE [WHM]	C.I. BASIC RED 9 MONOHYDROCHLORIDE
10048-32-5 PARASORBIC ACID	PARASORBIC ACID
56-38-2 PARATHION	PARATHION
56-38-2 PARATHION (PHOSPHOROTHIOIC ACID, 0,0-DIETHYL-0-(4-NITROPHENYL)ESTER) [313]	PARATHION
56-38-2 PARATHION (ISO) [GBR, GBR, GBR]	PARATHION
RR-00411-5 PARATHION AND COMPRESSED GAS MIXTURE	PARATHION AND COMPRESSED GAS MIXTURE
100-02-7 PARATHION DEGRADATION PRODUCT (4-NITROPHENOL) [SD4]	P-NITROPHENOL
298-00-0 PARATHION-METHYL [302, HM1, HM1, HM1, HM1, HM1]	METHYL PARATHION
298-00-0 PARATHION-METHYL (ISO) [GBR, GBR, GBR]	METHYL PARATHION
12002-03-8 PARIS GREEN [FLA, 302, MSL]	COPPER ACETOARSENITE
8000-68-8 PARSLEY APIOLE	PARSLEY APIOLE
RR-00079-3 PARTICULATE POLYCYCLIC AROMATIC HYDROCARBONS (PPAH) [BC1, BC1]	POLYCYCLIC ORGANIC MATTER
RR-00079-3 PARTICULATE POLYCYCLIC AROMATIC HYDROCARBONS [PAL]	POLYCYCLIC ORGANIC MATTER
RR-00072-6 PARTICULATES NOT OTHERWISE REGULATED [OS1, OS1]	NUISANCE PARTICULATES
RR-00072-6 PARTICULATES NOT OTHERWISE CLASSIFIED (PNOC) [TLV, TLV]	NUISANCE PARTICULATES
65996-93-2 PARTICULATE POLYCYCLIC AROMATIC HYDROCARBONS [BC1]	COAL TAR PITCHES
149-29-1 PATULIN	PATULIN
12674-11-2 PCB-1016 (AROCHLOR 1016) [CW2]	AROCLOR 1016
11104-28-2 PCB-1221 (AROCHLOR 1221) [CW2]	AROCLOR 1221
11141-16-5 PCB-1232 (AROCHLOR 1232) [CW2]	AROCLOR 1232

Chemical Name	Reference Name
12672-29-6 PCB-1248 (AROCHLOR 1248) [CW2]	AROCLOR 1248
11097-69-1 PCB-1254 (AROCHLOR 1254) [CW2]	CHLORODIPHENYL (54% CHLORINE)
11096-82-5 PCB-1260 (AROCHLOR 1260) [CW2]	AROCLOR 1260
12674-11-2 PCB-1016 (AROCHLOR 1016) [MX1]	AROCLOR 1016
11104-28-2 PCB-1221 (AROCHLOR 1221) [MX1]	AROCLOR 1221
11141-16-5 PCB-1232 (AROCHLOR 1232) [MX1]	AROCLOR 1232
53469-21-9 PCB-1242 (AROCHLOR 1242) [MX1]	AROCLOR 1242
12672-29-6 PCB-1248 (AROCHLOR 1248) [MX1]	AROCLOR 1248
11097-69-1 PCB-1254 (AROCHLOR 1254) [MX1]	CHLORODIPHENYL (54% CHLORINE)
11096-82-5 PCB-1260 (AROCHLOR 1260) [MX1]	AROCLOR 1260
53469-21-9 PCBS (42% CHLORINE) [AUS, AUS, AUS, AUS, AUS]	AROCLOR 1242
11097-69-1 PCBS (54% CHLORINE) [AUS, AUS, AUS, AUS, AUS]	CHLORODIPHENYL (54% CHLORINE)
1336-36-3 TOTAL PCBS (SUM OF ALL PCB ISOMERS, OR ALL AROCLORS) [UTS]	POLYCHLORINATED BIPHENYLS
RR-01109-6 PEANUT BUTTER	PEANUT BUTTER
8002-03-7 PEANUT OIL	PEANUT OIL
68476-82-4 PEANUTS	PEANUTS
RR-01110-9 PEANUT SHELLS	PEANUT SHELLS
RR-01111-0 PEAT MOSS	PEAT MOSS
1114-71-2 PEBULATE [BUTYLETHYLCARBAMOTHIOIC ACID S-PROPYL ESTER]	PEBULATE [BUTYLETHYLCARBAMOTHIOIC ACID S-PROPYL ESTER]OROPHENYL)ETHENYL DIMETHYL ESTER]
RR-01112-1 PECAN SHELL FLOUR	PECAN SHELL FLOUR
9000-69-5 PECTIN	PECTIN
9005-07-6 PEG (400) DIOLEATE	PEG (400) DIOLEATE
112-05-0 PELARGONIC ACID	PELARGONIC ACID
40487-42-1 PENDIMETHALIN [N-(1-ETHYLPROPYL)-3,4-DIMETHYL-2,6-DINITROBENZENEAMINE]	PENDIMETHALIN [N-(1-ETHYLPROPYL)-3,4-DIMETHYL-2,6-DINITROBENID CYANO(3-PHENOXY-PHENYL) METHYL ESTER]
52-67-5 D-PENICILLAMINE	D-PENICILLAMINE
52-67-5 PENICILLAMINE [C65]	D-PENICILLAMINE
90-65-3 PENICILLIC ACID	PENICILLIC ACID
132-98-9 PENICILLIN VK	PENICILLIN VK
19624-22-7 PENTABORANE	PENTABORANE
19624-22-7 PENTABORANE(9) [PAL]	PENTABORANE
87-84-3 1,2,3,4,5-PENTABROMO-6-CHLORO-CYCLOHEX-ANE	1,2,3,4,5-PENTABROMO-6-CHLORO-CYCLOHEXANE
32534-81-9 PENTABROMODIPHENYL ETHER [OTS, TSC, TSC, TSC, TSC]	PENTABROMODIPHENYLOXIDE
32534-81-9 PENTABROMODIPHENYLOXIDE	PENTABROMODIPHENYLOXIDE
32534-81-9 PENTABROMODIPHENYL OXIDE [TSE]	PENTABROMODIPHENYLOXIDEXO-2-PROPENYL)OXY] ETHYL ESTER
75-95-6 PENTABROMOETHANE	PENTABROMOETHANE
85-22-3 PENTABROMOETHYLBENZENE	PENTABROMOETHYLBENZENE
608-71-9 PENTABROMOPHENOL	PENTABROMOPHENOL
13463-40-6 PENTACARBONYLIRON [GBR]	IRON PENTACARBONYL
67-43-6 PENTACARBOXYMETHYL DIETHYLENETRI-AMINE	PENTACARBOXYMETHYL DIETHYLENETRIAMINE
RR-01761-8 PENT-ACETATE	PENT-ACETATE
1825-21-4 PENTACHLOROANISOLE	PENTACHLOROANISOLE

Chemical Name	Reference Name
608-93-5 PENTACHLOROBENZENE	PENTACHLOROBENZENE
36088-22-9 1,2,3,7,8-PENTACHLORODIBENZO-P-DIOXIN	1,2,3,7,8-PENTACHLORODIBENZO-P-DIOXIN
36088-22-9 PENTACHLORODIBENZO-P-DIOXIN [ATS]	1,2,3,7,8-PENTACHLORODIBENZO-P-DIOXIN
40321-76-4 1,2,3,7,8-PENTACHLORODIBENZO-P-DIOXIN	1,2,3,7,8-PENTACHLORODIBENZO-P-DIOXIN
RR-00511-8 PENTACHLORODIBENZO-P-DIOXINS	PENTACHLORODIBENZO-P-DIOXINS
RR-00511-8 PECDDS (ALL PENTACHLORODIBENZO-P-DIOXINS) [UTS]	PENTACHLORODIBENZO-P-DIOXINS
30402-15-4 PENTACHLORODIBENZOFURAN	PENTACHLORODIBENZOFURAN
57117-41-6 1,2,3,7,8-PENTACHLORO DIBENZOFURAN	1,2,3,7,8-PENTACHLORO DIBENZOFURAN
57117-31-4 2,3,4,7,8-PENTACHLORO DIBENZOFURANS	2,3,4,7,8-PENTACHLORO DIBENZOFURANS
RR-00507-2 PENTACHLORODIBENZOFURANS	PENTACHLORODIBENZOFURANS
RR-00507-2 PECDFS (ALL PENTACHLORODIBENZOFU-RANS) [UTS]	PENTACHLORODIBENZOFURANS
134237-36-8 PENTACHLORODIFLUOROPROPANE	PENTACHLORODIFLUOROPROPANE
76-01-7 PENTACHLOROETHANE	PENTACHLOROETHANE
354-56-3 PENTACHLOROFLUOROETHANE	PENTACHLOROFLUOROETHANE
134190-48-0 PENTACHLOROFLUOROPROPANE	PENTACHLOROFLUOROPROPANE
1321-64-8 PENTACHLORONAPHTHALENE	PENTACHLORONAPHTHALENE
82-68-8 PENTACHLORONITROBENZENE	PENTACHLORONITROBENZENE
82-68-8 PENTACHLORONITROBENZENE (DCNB) [EPA]	PENTACHLORONITROBENZENE
82-68-8 PENTACHLORONITROBENZENE (PCNB) [RCU, RCU]	PENTACHLORONITROBENZENE
82-68-8 PENTACHLORONITROBENZENE (QUINTOBEN-ZENE) [CAA]	PENTACHLORONITROBENZENE
79-21-0 PERACETIC ACID [HON]	PEROXYACETIC ACID
87-86-5 PENTACHLOROPHENOL	PENTACHLOROPHENOL
594-42-3 PERCHLOROMETHYLMERCAPTAN [HON]	TRICHLOROMETHANE SULPHURYL CHLORIDE
87-86-5 PENTACHLOROPHENOL (PCP) [313]	PENTACHLOROPHENOL
RR-00512-9 PENTACHLOROPHENOL DERIVATIVES	PENTACHLOROPHENOL DERIVATIVES
3772-94-9 PENTACHLOROPHENYL LAURATE	PENTACHLOROPHENYL LAURATE
133-49-3 PENTACHLOROTHIOPHENOL	PENTACHLOROTHIOPHENOL
134237-31-3 PENTACHLOROTRIFLUOROPROPANE	PENTACHLOROTRIFLUOROPROPANE
2570-26-5 PENTADECYLAMINE	PENTADECYLAMINE
504-60-9 1,3-PENTADIENE [EP3, EPA]	1-METHYLBUTADIENE
504-60-9 PENTADIENE, 1,3- [OTS]	1-METHYLBUTADIENE
504-60-9 1,3-PENTADIENE [PAL]	1-METHYLBUTADIENE
504-60-9 1,3-PENTADIENE (CIS & TRANS MIXED) [MSL]	1-METHYLBUTADIENE
504-60-9 1,3-PENTADIENE (CIS AND TRANS MIX) [NFP, NFP]	1-METHYLBUTADIENE
926-56-7 1,3-PENTADIENE, 4-METHYL- [PAL]	4-METHYL-1,3-PENTADIENE
1118-58-7 1,3-PENTADIENE, 2-METHYL- [PAL]	2-METHYL-1,3-PENTADIENE
504-60-9 1,3-PENTADIENE (MIXED CIS & TRANS) [FLA]	1-METHYLBUTADIENE
78-11-5 PENTAERYTHRITE TETRANITRATE	PENTAERYTHRITE TETRANITRATE
115-77-5 PENTAERYTHRITOL	PENTAERYTHRITOL
68333-69-7 PENTAERYTHRITOL ESTER OF MALEIC ANHY-DRIDE - MODIFIED WOOD ROSIN [PST]	ROSIN, MALEATED, POLYMER WITH PENTAERYTHRI-TOL
RR-01260-2 PENTAERYTHRITOL, MIXED ESTERS WITH CARBOXYLIC ACIDS	PENTAERYTHRITOL, MIXED ESTERS WITH CAR-BOXYLIC ACIDSH FATTY ACID OILS AND ESTERS, AND GLYCERIDE TRIESTERS
68554-37-0 PENTAERYTHRITOL, PHTHALIC ANHYDRIDE AND TALL-OIL POLYMER WITH ROSIN	PENTAERYTHRITOL, PHTHALIC ANHYDRIDE AND TALL-OIL POLYMER WIT

Chemical Name	Reference Name
4196-86-5 PENTAERYTHRITOL TETRABENZOATE	PENTAERYTHRITOL TETRABENZOATE
78-11-5 PENTAERYTHRITOL TETRANITRATE [CSR, CSR]	PENTAERYTHRITE TETRANITRATE
3524-68-3 PENTAERYTHRITOL TRIACRYLATE	PENTAERYTHRITOL TRIACRYLATE
28188-24-1 PENTAERYTHRITOL TRISTEARATE	PENTAERYTHRITOL TRISTEARATE
700-12-9 1,2,3,4,5-PENTAMETHYL BENZENE	1,2,3,4,5-PENTAMETHYL BENZENE
142-68-7 PENTAMETHYLENE OXIDE	PENTAMETHYLENE OXIDE
30586-18-6 PENTAMETHYLHEPTANE	PENTAMETHYLHEPTANE
41444-43-3 SEC-PENTAMINE	SEC-PENTAMINE
110-62-3 PENTANAL [PAL, TSC, TSC, TSC]	VALERALDEHYDE
32749-94-3 PENTANAL, 2,3-DIMETHYL-	PENTANAL, 2,3-DIMETHYL-
123-15-9 PENTANAL, 2-METHYL- [PAL]	2-METHYL VALERALDEHYDE
73513-30-1 PENTANAL, METHYL- [PAL]	METHYLPENTALDEHYDE
110-58-7 1-PENTANAMINE [PAL]	AMYL AMINE
625-30-9 2-PENTANAMINE [PAL]	SEC-AMYLAMINE
41444-43-3 SEC-PENTANAMINE [PAL]	SEC-PENTAMINE
621-77-2 1-PENTANAMINE, N,N-DIPENTYL-	1-PENTANAMINE, N,N-DIPENTYL-
108-09-8 2-PENTANAMINE, 4-METHYL- [PAL]	1,3-DIMETHYLBUTYLAMINE
2050-92-2 1-PENTANAMINE, N-PENTYL- [PAL]	DI-N-AMYLAMINE
107-45-9 2-PENTANAMINE, 2,4,4-TRIMETHYL- [PAL]	TERT-OCTYLAMINE
109-66-0 PENTANE	PENTANE
109-66-0 N-PENTANE [MAK, MAK, NIO, NIO, NIO, PST]	PENTANE
463-82-1 TERT-PENTANE [MAK, MAK]	2,2-DIMETHYLPROPANE
109-66-0 PENTANE, ALL ISOMERS [GBR, GBR]	PENTANE
110-53-2 PENTANE, 1-BROMO- [PAL]	1-BROMOPENTANE
543-59-9 PENTANE, 1-CHLORO- [PAL]	AMYL CHLORIDE
111-30-8 PENTANEDIAL [PAL, TSC, TSC, TSC]	GLUTARALDEHYDE
628-76-2 PENTANE, 1,5-DICHLORO- [PAL]	1,5-DICHLOROPENTANE
1067-20-5 PENTANE, 3,3-DIETHYL- [PAL]	3,3-DIETHYLPENTANE
108-08-7 PENTANE, 2,4-DIMETHYL- [PAL]	2,4-DIMETHYLPENTANE
565-59-3 PENTANE, 2,3-DIMETHYL- [PAL]	2,3-DIMETHYLPENTANE
111-29-5 1,5-PENTANEDIOL	1,5-PENTANEDIOL
625-69-4 2,4-PENTANEDIOL	2,4-PENTANEDIOL
149-31-5 1,3-PENTANEDIOLE, 2-METHYL- [PAL]	2-METHYL-1,3-PENTANEDIOL
107-41-5 2,4-PENTANEDIOL, 2-METHYL- [PAL]	HEXYLENE GLYCOL
123-54-6 2,4-PENTANEDIONE	2,4-PENTANEDIONE
123-54-6 PENTANE-2,4-DIONE [HM1, HM1]	2,4-PENTANEDIONE
600-14-6 2,3-PENTANEDIONE	2,3-PENTANEDIONE
1068-87-7 PENTANE, 3-ETHYL-2,4-DIMETHYL- [PAL]	2,4-DIMETHYL-3-ETHYL PENTANE
609-26-7 PENTANE, 3-ETHYL-2-METHYL- [PAL]	2-METHYL-3-ETHYLPENTANE
96-14-0 PENTANE, 3-METHYL- [PAL]	3-METHYLPENTANE
107-83-5 PENTANE, 2-METHYL- [PAL]	ISOHEXANE
760-21-4 PENTANE, 3-METHYLENE- [PAL]	2-ETHYL-1-BUTENE
75405-06-0 PENTANENITRILE, 3-AMINO-	PENTANENITRILE, 3-AMINO-
109-66-0 N-PENTANES [HM1, HM1]	PENTANE
1186-53-4 PENTANE, 2,2,3,4-TETRAMETHYL-	PENTANE, 2,2,3,4-TETRAMETHYL-+) SALT
7154-79-2 PENTANE, 2,2,3,3-TETRAMETHYL-	PENTANE, 2,2,3,3-TETRAMETHYL-

Chemical Name	Reference Name
872-10-6 PENTANE, 1,1'-THIOBIS-	PENTANE, 1,1'-THIOBIS-
110-66-7 1-PENTANETHIOL [NHS, NHS, PAL]	AMYL MERCAPTAN
RR-01307-0 PENTANETHIOLS	PENTANETHIOLS
141-59-3 2-PENTANETHIOL, 2,4,4-TRIMETHYL- [PAL]	TERT-OCTYL MERCAPTAN
540-84-1 PENTANE, 2,2,4-TRIMETHYL- [PAL, TSC, TSC]	ISOOCTANE
560-21-4 PENTANE, 2,3,3-TRIMETHYL-	PENTANE, 2,3,3-TRIMETHYL-
564-02-3 PENTANE, 2,2,3-TRIMETHYL-	PENTANE, 2,2,3-TRIMETHYL-
21985-87-5 PENTANITROANILINE	PENTANITROANILINE
109-52-4 PENTANOIC ACID [FLA, MSL, NFP, NFP, PAL]	VALERIC ACID
71-41-0 1-PENTANOL [PAL]	AMYL ALCOHOL
584-02-1 3-PENTANOL	3-PENTANOL
6032-29-7 2-PENTANOL [MSL, PAL]	SEC-AMYL ALCOHOL
626-38-0 2-PENTANOL, ACETATE [PAL]	SEC-AMYL ACETATE
6032-29-7 2-PENTANOLE [FLA]	SEC-AMYL ALCOHOL
108-11-2 2-PENTANOL, 4-METHYL [PAL]	METHYL ISOBUTYL CARBINOL
54972-97-3 1-PENTANOL, METHYL-	1-PENTANOL, METHYL-PROPANOL (1:1)
96-22-0 PENTAN-3-ONE [GBR, GBR]	DIETHYL KETONE
96-22-0 3-PENTANONE [ONT, PAL]	DIETHYL KETONE
107-87-9 2-PENTANONE [BC1, BC1, CAL]	METHYL PROPYL KETONE
107-87-9 PENTAN-2-ONE [GBR, GBR]	METHYL PROPYL KETONE
107-87-9 2-PENTANONE [MAK, MAK, NIO, NIO, NIO, ONT, ONT, PAL]	METHYL PROPYL KETONE
107-87-9 2-PENTANONE (METHYL PROPYL KETONE) [OS1, OS1, OS1]	METHYL PROPYL KETONE
123-42-2 2-PENTANONE, 4-HYDROXY-4-METHYL- [PAL]	DIACETONE ALCOHOL
108-10-1 2-PENTANONE, 4-METHYL- [PAL, TSC, TSC, TSC]	METHYL ISOBUTYL KETONE
112-98-1 5,8,11,14,17-PENTAOXAHENEICOSANE [PAL]	DIBUTOXY TETRAGLYCOL
31206-94-7 2,5,8,10,13-PENTAOXAHEXADEC-15-ENOIC ACID,9,14-DIOXO-2-[(1-OXO-2-PROPENYL)OXY] ETHYL ESTER	2,5,8,10,13-PENTAOXAHEXADEC-15-ENOIC ACID,9,14-DIOXO-2-[(1-O
RR-01647-7 2,5,8,10,13-PENTAOXAHEXADEC-15-ENOIC ACID, 9,14-DIOXO-2-[(1-OXO-2-PROPENYL)OXY] ETHYL ESTER	2,5,8,10,13-PENTAOXAHEXADEC-15-ENOIC ACID, 9,14-DIOXO-2-[(1-ESTER
RR-01729-8 3,6,9,12,16-PENTAOXANONADEC-18-ENE-1,14-DIOL,BIS(4-METHYLBENZENSULFONATE)	3,6,9,12,16-PENTAOXANONADEC-18-ENE-1,14-DIOL,BIS (4-METHYLBENESULFONATE)
143-24-8 2,5,8,11,14-PENTAOXAPENTADECANE	2,5,8,11,14-PENTAOXAPENTADECANE
80-46-6 PENTAPHEN [MSL, NFP, NFP]	P-TERT-PENTYLPHENOL
7758-29-4 PENTASODIUM TRIPHOSPHATE	PENTASODIUM TRIPHOSPHATE
109-67-1 1-PENTENE	1-PENTENE
646-04-8 2-PENTENE, (E)- [EP3, PAL]	BETA-AMYLENE-TRANS
627-20-3 2-PENTENE, (Z)- [EP3, PAL]	BETA-AMYLENE-CIS
625-27-4 2-PENTENE, 2-METHYL- [PAL]	2-METHYL-2-PENTENE
691-37-2 1-PENTENE, 4-METHYL- [PAL]	4-METHYL-1-PENTENE
763-29-1 1-PENTENE, 2-METHYL- [PAL]	2-METHYL-1-PENTENE
4461-48-7 2-PENTENE, 4-METHYL- [PAL]	4-METHYL-2-PENTENE
107-39-1 1-PENTENE, 2,4,4-TRIMETHYL-	1-PENTENE, 2,4,4-TRIMETHYL-
107-40-4 2-PENTENE, 2,4,4-TRIMETHYL-	2-PENTENE, 2,4,4-TRIMETHYL-
565-76-4 1-PENTENE, 2,3,4-TRIMETHYL-	1-PENTENE, 2,3,4-TRIMETHYL-
598-96-9 2-PENTENE, 3,4,4-TRIMETHYL- [PAL]	3,4,4-TRIMETHYL-2-PENTENE

Chemical Name	Reference Name
25167-70-8 PENTENE, 2,4,4-TRIMETHYL- [PAL]	DIISOBUTYLENE
141-79-7 3-PENTEN-2-ONE, 4-METHYL- [PAL]	MESITYL OXIDE
57-33-0 PENTOBARBITAL SODIUM	PENTOBARBITAL SODIUM
12772-47-3 PENTOL	PENTOL
12772-47-3 1-PENTOL [HM1, HM1]	PENTOL
8066-33-9 PENTOLITE	PENTOLITE
628-63-7 PENTYL ACETATE [GBR, GBR, WHM]	N-AMYL ACETATE
71-41-0 PENTYL ALCOHOL [PST]	AMYL ALCOHOL
80-46-6 P-TERT-PENTYLPHENOL	P-TERT-PENTYLPHENOL
69867-71-6 PENTYL PHENYL ACID PHOSPHATE	PENTYL PHENYL ACID PHOSPHATE
707-19-7 PENTYL PROPARGYL ALCOHOL [ALB, ALB, ALB]	PROPARGYL ALCOHOL
1320-21-4 PENTYL XYLYL ETHER [FLA]	BENZENE, DIMETHYL(PENTYLOXY)-
627-19-0 1-PENTYNE	1-PENTYNE
77-75-8 1-PENTYN-3-OL, 3-METHYL-	1-PENTYN-3 OL, 3-METHYL-
79-21-0 PERACETIC ACID [302, 313, CAN, EPA, EPA, MAK, MAK, NFP, NFP, OS3]	PEROXYACETIC ACID
RR-00797-6 PERACETIC ACID DILUTED WITH 60% OF ACETIC ACID	PERACETIC ACID DILUTED WITH 60% OF ACETIC ACID
79-21-0 PERACETIC ACID [ETHANEPEROXIC ACID] [EP3]	PEROXYACETIC ACID
7632-04-4 PERBORIC ACID, SODIUM SALT [PST]	SODIUM PERBORATE
RR-00609-7 PERCHLORATE COMPOUNDS, N.O.S.	PERCHLORATE COMPOUNDS, N.O.S.
RR-00413-7 PERCHLORATES, INORGANIC, N.O.S.	PERCHLORATES, INORGANIC, N.O.S.
7601-90-3 PERCHLORIC ACID	PERCHLORIC ACID
7790-98-9 PERCHLORIC ACID, AMMONIUM SALT [PAL]	AMMONIUM PERCHLORATE
10034-81-8 PERCHLORIC ACID, MAGNESIUM SALT [PAL]	MAGNESIUM PERCHLORATE
7778-74-7 PERCHLORIC ACID, POTASSIUM SALT [PAL]	POTASSIUM PERCHLORATE
7601-89-0 PERCHLORIC ACID, SODIUM SALT [PAL]	SODIUM PERCHLORATE
127-18-4 PERCHLOROETHYLENE [AUS, AUS, AUS, BC1, BC1, BC1, BC1, BEI, BEI, CAL, CEX, CEX, HM1, HM1, HM1, HM1, MEX, MEX, MEX, MIN, NFP, NFP, OTV, QBC, QBC, QBC]	TETRACHLOROETHYLENE
127-18-4 PERCHLOROETHYLENE (TETRACHLOROETHYLENE) [TLV, TLV, TLV]	TETRACHLOROETHYLENE
127-18-4 PERCHLOROETHYLENE [WHM]	TETRACHLOROETHYLENE
127-18-4 PERCHLOROETHYLENE (TETRACHLOROETHYLENE) [OS1, OS1, OS1, ALB, ALB, ALB]	TETRACHLOROETHYLENE
594-42-3 PERCHLOROMETHYL MERCAPTAN [MEX, 313, ALB, ALB, CEX]	TRICHLOROMETHANE SULPHURYL CHLORIDE
594-42-3 PERCHLOROMETHYLMERCAPTAN [302, BC1]	TRICHLOROMETHANE SULPHURYL CHLORIDE
594-42-3 PERCHLOROMETHYLMERCAPTAN [METHANE-SULFENYL CHLORIDE, TRICHLORO-] [EP3]	TRICHLOROMETHANE SULPHURYL CHLORIDE-]
594-42-3 PERCHLOROMETHYLMERCAPTAN [QBC]	TRICHLOROMETHANE SULPHURYL CHLORIDE
594-42-3 PERCHLOROMETHYL MERCAPTAN [TLV]	TRICHLOROMETHANE SULPHURYL CHLORIDE
7616-94-6 PERCHLORYL FLUORIDE	PERCHLORYL FLUORIDE
RR-01231-7 PERFLUOROALKYL AROMATIC CARBAMATE MODIFIED ALKYL METHACRYLATE COPOLYMER	PERFLUOROALKYL AROMATIC CARBAMATE MODIFIED ALKYL METHACRYLAT
RR-00190-1 PERFLUOROALKYL EPOXIDE	PERFLUOROALKYL EPOXIDESALT
RR-01363-8 PERFLUOROETHYLVINYL ETHER	PERFLUOROETHYLVINYL ETHER
355-42-0 PERFLUORO-N-HEXANE	PERFLUORO-N-HEXANE

Chemical Name	Reference Name
382-21-8 PERFLUOROISOBUTYLENE	PERFLUOROISOBUTYLENE
RR-01362-7 PERFLUOROMETHYLVINYL ETHER	PERFLUOROMETHYLVINYL ETHER
311-89-7 PERFLUOROTRIBUTYLAMINE	PERFLUOROTRIBUTYLAMINE
RR-01519-0 PERFUMERY PRODUCTS, WITH FLAMMABLE SOLVENT	PERFUMERY PRODUCTS, WITH FLAMMABLE SOLVENT
8024-43-9 PERFUMES AND ESSENCES, JASMIN	PERFUMES AND ESSENCES, JASMIN
8024-43-9 PERFUMES, JASMIN [PST]	PERFUMES AND ESSENCES, JASMIN
RR-00943-8 PERHALO ALKOXY ETHER	PERHALO ALKOXY ETHER
5743-97-5 PERHYDROPHENANTHRENE	PERHYDROPHENANTHRENE
536-59-4 PERILLA ALCOHOL	PERILLA ALCOHOL
68132-21-8 PERILLA OIL	PERILLA OIL
130885-09-5 PERLITE	PERLITEARBONYLIMINO-1,3-PROPANEDIYL (DIMETHYLIMINO)-1,2-ETHANEDIYL DICHLORIDE], AL-PHA-(2-CHLOROETHYL)-OMEGA-(2-CHLOROETHOXY)
93763-70-3 PERLITE DUST [AUS, PAL]	PERLITE DIHYDROXYPHENYL)-7-HYDROXY-
93763-70-3 PERLITE, EXPANDED [PST]	PERLITE
RR-00416-0 PERMANGANATES, INORGANIC, N.O.S.	PERMANGANATES, INORGANIC, N.O.S.
13446-10-1 PERMANGANIC ACID (HMNO4), AMMONIUM SALT [PAL]	AMMONIUM PERMANGANATE
7722-64-7 PERMANGANIC ACID (HMNO4), POTASSIUM SALT [PAL]	POTASSIUM PERMANGANATE
10101-50-5 PERMANGANIC ACID, SODIUM SALT [PST]	SODIUM PERMANGANATE
52645-53-1 PERMETHRIN	PERMETHRIN
52645-53-1 PERMETHRIN [3-(2,2-DICHLOROETHENYL)-2, 2-DIMETHYLCYCLOPROPANECARBOXYLIC ACID, (3-PHENOXYPHENYL)METHYL ESTER] [313]	PERMETHRIND CYANO(3-PHENOXYPHENYL)METHYL ESTER]
110-05-4 PEROXIDE, BIS(1,1-DIMETHYLETHYL) [PAL]	DI-TERT-BUTYL PEROXIDE
105-74-8 PEROXIDE, BIS(1-OXODODECYL) [PAL]	LAUROYL PEROXIDE
110-22-5 PEROXIDE, DIACETYL [PAL]	DIACETYL PEROXIDE
94-36-0 PEROXIDE, DIBENZOYL [PAL]	BENZOYL PEROXIDE
628-37-5 PEROXIDE, DIETHYL [PAL]	DIETHYL PEROXIDE
RR-00417-1 PEROXIDES, INORGANIC, N.O.S.	PEROXIDES, INORGANIC, N.O.S.
25812-30-0 PEROXISOME PROJECT (GEMFIBROZIL)	PEROXISOME PROJECT (GEMFIBROZIL)
50892-23-4 PEROXISOME PROJECT (WY-14643)	PEROXISOME PROJECT (WY-14643)
79-21-0 PEROXYACETIC ACID	PEROXYACETIC ACID
79-21-0 PEROXYACETIC ACID [NJL, NJL]	PEROXYACETIC ACID
105-64-6 PEROXYDICARBONIC ACID, BIS(1-METHYLETHYL) ESTER [PAL]	ISOPROPYL PEROXYDICARBONATE
7727-21-1 PEROXYDISULFURIC ACID ([(HO)S(0)2]2O2), DIPOTASSIUM SALT [PAL]	POTASSIUM PERSULFATE
RR-01604-6 PERSISTENT SYNTHETIC MATERIALS (IN-CLUDING PERSISTENT PLASTICS)	PERSISTENT SYNTHETIC MATERIALS (INCLUDING PERSISTENT PLASTIC
RR-00610-0 PERSULFATE COMPOUNDS	PERSULFATE COMPOUNDS
RR-00610-0 PERSULFATES (ALKALI METALS) [ALB, ALB]	PERSULFATE COMPOUNDS
RR-00610-0 PERSULFATES, ALKALI METAL [ONT, ONT]	PERSULFATE COMPOUNDS
72-56-0 PERTHANE [CAL]	DI(P-ETHYLPHENYL)DICHLOROETHANE
198-55-0 PERYLENE	PERYLENE
RR-00418-2 PESTICIDES AND THEIR BY-PRODUCTS [CN7]	PESTICIDES, N.O.S.
RR-00418-2 PESTICIDES, N.O.S.	PESTICIDES, N.O.S.
60102-37-6 PETASITENINE	PETASITENINE

Chemical Name	Reference Name
86290-81-5 PETROL (GASOLINE) [AUS, AUS]	GASOLINE, MOTOR FUEL
8009-03-8 PETROLATUM	PETROLATUM
8002-05-9 PETROLEUM [FLA, IAR, IAR, PAL]	PETROLEUM DISTILLATES (NAPHTHA)
RR-01763-0 PETROLEUM ETHER	PETROLEUM ETHER
8002-05-9 PETROLEUM CRUDE [MSL]	PETROLEUM DISTILLATES (NAPHTHA)
8002-05-9 PETROLEUM, CRUDE [NFP, NFP]	PETROLEUM DISTILLATES (NAPHTHA)
RR-01605-7 PETROLEUM DISTILLATE RESIDUES	PETROLEUM DISTILLATE RESIDUESS)
64742-18-3 PETROLEUM DISTILLATES, ACID TREATED, HEAVY NAPHTHENIC	PETROLEUM DISTILLATES, ACID TREATED, HEAVY NAPHTHENIC
64742-20-7 PETROLEUM DISTILLATES, ACID TREATED, HEAVY PARAFFINIC	PETROLEUM DISTILLATES, ACID TREATED, HEAVY PARAFFINICNAPHTHENIC
64742-19-4 PETROLEUM DISTILLATES, ACID TREATED, LIGHT NAPHTHENIC	PETROLEUM DISTILLATES, ACID TREATED, LIGHT NAPHTHENICNAPHTHENIC
64742-21-8 PETROLEUM DISTILLATES, ACID TREATED, LIGHT PARAFFINIC	PETROLEUM DISTILLATES, ACID TREATED, LIGHT PARAFFINICPARAFFINIC
68477-31-6 PETROLEUM DISTILLATES, CATALYTIC, RE-FORMER FRACTIONATOR RESIDUE, LOW-BOILING	PETROLEUM DISTILLATES, CATALYTIC, REFORMER FRACTIONATOR RESI
64741-61-3 PETROLEUM DISTILLATES, HEAVY CAT-ALYTIC CRACKED	PETROLEUM DISTILLATES, HEAVY CATALYTIC CRACKED
64741-53-3 PETROLEUM DISTILLATES, HEAVY NAPH-THENIC	PETROLEUM DISTILLATES, HEAVY NAPHTHENIC
64741-51-1 PETROLEUM DISTILLATES, HEAVY PARAF-FINIC	PETROLEUM DISTILLATES, HEAVY PARAFFINIC
64742-53-6 PETROLEUM DISTILLATES, HYDROTREATED LIGHT NAPHTHENIC	PETROLEUM DISTILLATES, HYDROTREATED LIGHT NAPHTHENICPARAFFINIC
64742-55-8 PETROLEUM DISTILLATES, HYDROTREATED LIGHT PARAFFINIC	PETROLEUM DISTILLATES, HYDROTREATED LIGHT PARAFFINICNAPHTHENIC
64741-59-9 PETROLEUM DISTILLATES, LIGHT CATALYTIC CRACKED	PETROLEUM DISTILLATES, LIGHT CATALYTIC CRACKED
64741-50-0 PETROLEUM DISTILLATES, LIGHT PARAF-FINIC	PETROLEUM DISTILLATES, LIGHT PARAFFINIC
8002-05-9 PETROLEUM DISTILLATES (NAPHTHA)	PETROLEUM DISTILLATES (NAPHTHA)
8002-05-9 PETROLEUM DISTILLATES, N.O.S. [NJL]	PETROLEUM DISTILLATES (NAPHTHA)
8002-05-9 PETROLEUM DISTILLATES (NAPHTHA) (RUB-BER SOLVENT) [OS1, OS1]	PETROLEUM DISTILLATES (NAPHTHA)
RR-00422-8 PETROLEUM DISTILLATES, N.O.S.	PETROLEUM DISTILLATES, N.O.S.
64742-56-9 PETROLEUM DISTILLATES, SOLVENT DE-WAXED LIGHT PARAFFINIC	PETROLEUM DISTILLATES, SOLVENT DEWAXED LIGHT PARAFFINICPARAFFINIC
64741-89-5 PETROLEUM DISTILLATES, SOLVENT-REFINED LIGHT PARAFFINIC	PETROLEUM DISTILLATES, SOLVENT-REFINED LIGHT PARAFFINIC
70592-78-8 PETROLEUM DISTILLATES, VACUUM	PETROLEUM DISTILLATES, VACUUM
68476-85-7 PETROLEUM GASES, LIQUEFIED [PAL]	L.P.G. (LIQUIFIED PETROLEUM GAS)
68476-86-8 PETROLEUM GASES, LIQUIFIED [HM1, HM1]	PETROLEUM GASES, LIQUIFIED, SWEETENED
68476-86-8 PETROLEUM GASES, LIQUIFIED, SWEETENED	PETROLEUM GASES, LIQUIFIED, SWEETENED
64742-95-6 PETROLEUM NAPHTHA, LIGHT AROMATIC	PETROLEUM NAPHTHA, LIGHT AROMATIC
RR-00423-9 PETROLEUM OIL	PETROLEUM OIL
68476-87-9 PETROLEUM REFINING RESIDUES, POLYMER-IZED	PETROLEUM REFINING RESIDUES, POLYMERIZED
64742-16-1 PETROLEUM RESINS	PETROLEUM RESINS
RR-01546-3 PETROLEUM SOLVENTS	PETROLEUM SOLVENTS
RR-00877-5 PETROLEUM SULFONATE	PETROLEUM SULFONATE
64742-42-3 PETROLEUM WAX, CLAY-TREATED, MICRO-CRYSTALLINE [PST]	HYDROCARBON WAXES (PETROLEUM), CLAY-TREATED MICROCRYSTALLINE

Chemical Name	Reference Name
86290-81-5 PETROL, LEADED [HM1]	GASOLINE, MOTOR FUEL
12751-36-9 PHARMAMEDIA	PHARMAMEDIA
555-10-2 BETA-PHELLANDRENE	BETA-PHELLANDRENE
63-98-9 PHENACEMIDE	PHENACEMIDE
62-44-2 PHENACETIN	PHENACETIN
70-11-1 PHENACYL BROMIDE	PHENACYL BROMIDE
85-01-8 PHENANTHRENE	PHENANTHRENE
66-71-7 1,10-PHENANTHROLINE	1,10-PHENANTHROLINE
58-36-6 PHENARSAZINE OXIDE [MSL, NJL]	10,10'-OXYBISPHENOXARSINE
94-78-0 PHENAZOPYRIDINE	PHENAZOPYRIDINE
136-40-3 PHENAZOPYRIDINE HYDROCHLORIDE	PHENAZOPYRIDINE HYDROCHLORIDE
156-51-4 PHENELZINE SULPHATE	PHENELZINE SULPHATE
3546-10-9 PHENESTERIN	PHENESTERIN
60-12-8 PHENETHYL ALCOHOL	PHENETHYL ALCOHOL
64-04-0 PHENETHYLAMINE	PHENETHYLAMINE
94-70-2 O-PHENETIDINE	O-PHENETIDINE
156-43-4 P-PHENETIDINE	P-PHENETIDINE
1321-31-9 PHENETIDINE	PHENETIDINE
1321-31-9 PHENETIDINES [HM1, HM1]	PHENETIDINE
103-73-1 PHENETOLE	PHENETOLE
834-28-6 PHENFORMIN HYDROCHLORIDE	PHENFORMIN HYDROCHLORIDE
103-03-7 PHENICARBAZIDE	PHENICARBAZIDE
94-78-0 3-(PHENOAZO)-2-6-PYRADINEDIAMINE [FLA]	PHENAZOPYRIDINE
50-06-6 PHENOBARBITAL	PHENOBARBITAL
50-06-6 PHENOBARBITOL [MIN]	PHENOBARBITAL
108-95-2 PHENOL	PHENOL
53894-28-3 PHENOL, [[(2-AMINOETHYL)AMINO]METHYL]-	PHENOL, [[(2-AMINOETHYL)AMINO]METHYL]-
25973-55-1 PHENOL, 2-(2H-BENZOTRIAZOL-2-YL)-4,6-BIS(1,1-DIMETHYLPROPYL)- [PST]	TINUVIN 328 (BENZOTRIAZOLE UV ABSORBER)
128-37-0 PHENOL, 2,6-BIS(1,1-DIMETHYLETHYL)-4-METHYL- [PAL]	2,6-DI-TERT-BUTYL-P-CRESOL
95-57-8 PHENOL, 2-CHLORO- [PAL]	2-CHLOROPHENOL
106-48-9 PHENOL, 4-CHLORO- [PAL]	P-CHLOROPHENOL
42350-99-2 PHENOL, 2-CHLORO-4,6-BIS(1,1-DIMETHYL-PROPYL)- [PAL]	2-CHLORO-4,6-DI-TERT-AMYLPHENOL
88-04-0 PHENOL, 4-CHLORO-3,5-DIMETHYL- [TSC, TSC, TSC]	P-CHLORO-M-XYLENOL
98-28-2 PHENOL, 2-CHLORO-4-(1,1-DIMETHYLETHYL)- [PAL]	4-TERT-BUTYL-2-CHLOROPHENOL
73090-69-4 PHENOL, CHLORO-4-(1,1-DIMETHYLPROPYL)- [PAL]	CHLORO-4-TERT-AMYLPHENOL
59-50-7 PHENOL, 4-CHLORO-3-METHYL- [PAL]	4-CHLORO-M-CRESOL
50594-77-9 PHENOL, 3-[2-CHLORO-4-(TRIFLUOROMETHYL)PHENOXY]-, ACETATE	PHENOL, 3-[2-CHLORO-4-(TRIFLUOROMETHYL)PHE-NOXY]-, ACETATE1-ETHANEDIYL)]ESTER
119-42-6 PHENOL, 2-CYCLOHEXYL- [PAL]	O-CYCLOHEXYLPHENOL
137-09-7 PHENOL, 2,4-DIAMINO-, DIHYDROCHLO-RIDE [TSC, TSC]	2,4-DIAMINOPHENOL DIHYDROCHLORIDE
15872-73-8 PHENOL, 2,4-DIAMINO-6-METHYL-	PHENOL, 2,4-DIAMINO-6-METHYL-
65879-44-9 PHENOL, 2,4-DIAMINO-6-METHYL-, HY-DROCHLORIDE	PHENOL, 2,4-DIAMINO-6-METHYL-, HYDROCHLORIDE
615-58-7 PHENOL, 2,4-DIBROMO- [TSC, TSC, TSC, TSC]	2,4-DIBROMOPHENOL

439

Chemical Name	Reference Name
69882-11-7 PHENOL, 2,4(OR 2,6)-DIBROMO-, HOMOPOLY-MER	PHENOL, 2,4(OR 2,6)-DIBROMO-, HOMOPOLYMER
120-83-2 PHENOL, 2,4-DICHLORO- [PAL]	2,4-DICHLOROPHENOL
56-53-1 PHENOL, 4,4'-(1,2-DIETHYL-1,2-ETHENEDIYL) BIS-, (E)- [PAL, PAL]	DIETHYLSTILBESTROL
472-41-3 PHENOL, 4-(3,4-DIHYDRO-2,2,4-TRIMETHYL-2H-1-BENZOPYRAN-4-YL)-	PHENOL, 4-(3,4-DIHYDRO-2,2,4-TRIMETHYL-2H-1-BEN-ZOPYRAN-4-YL)
105-67-9 PHENOL, 2,4-DIMETHYL- [PAL]	2,4-DIMETHYLPHENOL
1300-71-6 PHENOL, DIMETHYL- [PAL, TSC, TSC]	XYLENOL
315-18-4 PHENOL, 4-(DIMETHYLAMINO)-3,5-DIMETHYL-, METHYLCARBAMATE(ESTER) [PAL]	MEXACARBATEYL)-
98-27-1 PHENOL, 4-(1,1-DIMETHYLETHYL)-2-METHYL- [PAL]	P-TERT-BUTYL-O-CRESOL
1333-13-7 PHENOL, (1,1-DIMETHYLETHYL)-3-METHYL- [PAL]	TERT-BUTYL-M-CRESOL
78-33-1 PHENOL, 4-(1,1-DIMETHYLETHYL)-, PHOS-PHATE (3:1)	PHENOL, 4-(1.!-DIMETHYLETHYL)-, PHOSPHATE (3:1)
2032-65-7 PHENOL, 3,5-DIMETHYL-4-(METHYLTHIO)-, METHYLCARBAMATE [PAL]	METHIOCARB
25155-23-1 PHENOL, DIMETHYL-, PHOSPHATE (3:1) [TSC, TSC]	PHOSPHATE, TRIXYLYLYL-
80-46-6 PHENOL, 4-(1,1-DIMETHYLPROPYL)- [PAL]	P-TERT-PENTYLPHENOL
51-28-5 PHENOL, 2,4-DINITRO- [PAL]	2,4-DINITROPHENOL
329-71-5 PHENOL, 2,5-DINITRO- [PAL]	2,5-DINITROPHENOL3.ALPHA.,4.BETA.,5.ALPHA.,6.BETA.)-
573-56-8 PHENOL, 2,6-DINITRO- [PAL]	2,6-DINITROPHENOL
25550-58-7 PHENOL, DINITRO- [PAL]	DINITROPHENOL
138-00-1 PHENOL, 2,4-DIPENTYL-	PHENOL, 2,4-DIPENTYL-
123-07-9 PHENOL, 4-ETHYL-	PHENOL, 4-ETHYL-
RR-00498-8 PHENOLIC COMPOUNDS	PHENOLIC COMPOUNDS
RR-00498-8 PHENOLIC COMPOUNDS (4AAP) [CAL]	PHENOLIC COMPOUNDS
4151-51-3 PHENOL, 4-ISOCYANATO-, PHOSPHOROTH-IOATE (3:1) (ESTER)	PHENOL, 4-ISOCYANATO-, PHOSPHOROTHIOATE (3:1) (ESTER)YL-
68937-41-7 PHENOL ISOPROPYLATED PHOSPHATE [OTS]	TRIS ISOPROPYLATED PHENYL PHOSPHATE
150-76-5 PHENOL, 4-METHOXY- [PAL]	4-METHOXYPHENOL
95-48-7 PHENOL, 2-METHYL- [PAL, TSC, TSC, TSC]	O-CRESOL
106-44-5 PHENOL, 4-METHYL- [PAL]	P-CRESOL
108-39-4 PHENOL, 3-METHYL- [PAL, TSC, TSC, TSC]	M-CRESOL
1319-77-3 PHENOL, METHYL- [PAL, TSC, TSC]	CRESOL
534-52-1 PHENOL, 2-METHYL-4,6-DINITRO- [PAL]	4,6-DINITRO-O-CRESOL
97-23-4 PHENOL, 2,2'-METHYLENEBIS(4-CHLORO- [TSC, TSC]	DICHLOROPHEN
5384-21-4 PHENOL, 4,4'-METHYLENEBIS(2,6-DIMETHYL)-	PHENOL, 4,4'-METHYLENEBIS(2,6-DIMETHYL)-
93589-69-6 PHENOL, 4,4'-[METHYLENEBIS(OXY-2,1-ETHANEDIYLTHIO)]BIS-	PHENOL, 4,4'-[METHYLENEBIS(OXY-2,1-ETHANEDIYLTHIO)]BIS-
114-26-1 PHENOL, 2-(1-METHYLETHOXY)-, METHYL-CARBAMATE [PAL]	PROPOXUR
80-05-7 PHENOL, 4,4'-(1-METHYLETHYLIDENE)BIS- [TSC, TSC, TSC]	BISPHENOL A
64-00-6 PHENOL, 3-(1-METHYLETHYL)-, METHYLCAR-BAMATE	PHENOL, 3-(1-METHYLETHYL)-, METHYLCARBAMATE
2581-34-2 PHENOL, 3-METHYL-4-NITRO- [OTS]	3-METHYL-4-NITROPHENOL
89-72-5 PHENOL, 2-(1-METHYLPROPYL)- [PAL]	O-SEC-BUTYLPHENOL

Chemical Name	Reference Name
88-75-5 PHENOL, 2-NITRO- [PAL]	O-NITROPHENOL
100-02-7 PHENOL, 4-NITRO- [PAL]	P-NITROPHENOL
554-84-7 PHENOL, 3-NITRO- [PAL]	M-NITROPHENOL
25154-52-3 PHENOL, NONYL- [PAL]	NONYLPHENOL
26523-78-4 PHENOL, NONYL-, PHOSPHITE (3:1) [PST]	TRI[NONYLPHENYL] PHOSPHITE
90884-29-0 PHENOL, 4,4'-(OXYBIS(2,1-ETHANEDIYLTHIO)) BIS-	PHENOL, 4,4'-(OXYBIS(2,1-ETHANEDIYLTHIO))BIS-
87-86-5 PHENOL, PENTACHLORO- [PAL]	PENTACHLOROPHENOL
136-81-2 PHENOL, 2-PENTYL- [PAL]	O-AMYL PHENOLSALT
77-09-8 PHENOLPHTHALEIN	PHENOLPHTHALEIN
143-74-8 PHENOL RED	PHENOL RED
91-15-6 PHTHALONITRILE [HON]	O-PHTHALODINITRILE
108-99-6 PICOLINE (B-) [HON]	3-METHYLPYRIDINE
1333-39-7 PHENOLSULFONIC ACIDS (ALL ISOMERS) [HON]	PHENOLSULPHONIC ACID
80-09-1 PHENOL, 4,4'-SULFONYLBIS- [TSC, TSC, TSC]	4,4'-SULFONYLDIPHENOL
1333-39-7 PHENOLSULPHONIC ACID	PHENOLSULPHONIC ACID
58-90-2 PHENOL, 2,3,4,6-TETRACHLORO- [PAL]	2,3,4,6-TETRACHLOROPHENOL3.BETA.,4.ALPHA.,5.ALPHA.,6.BETA.)-
140-66-9 PHENOL, 4-(1,1,3,3-TETRAMETHYLBUTYL)-	PHENOL, 4-(1,1,3,3-TETRAMETHYLBUTYL)-
4418-66-0 PHENOL, 2,2'-THIOBIS(4-CHLORO-6-METHYL)-	PHENOL, 2,2'-THIOBIS(4-CHLORO-6-METHYL)-
97-18-7 PHENOL, 2,2'-THIOBIS(4,6-DICHLORO)-	PHENOL, 2,2'-THIOBIS(4,6-DICHLORO)-
96-69-5 PHENOL, 4,4'-THIOBIS[2-(1,1-DIMETHYLETHYL)-5-METHYL- [PAL]	4,4'-THIOBIS(6-TERT-BUTYL-M-CRESOL)
118-79-6 PHENOL, 2,4,6-TRIBROMO-	PHENOL, 2,4,6-TRIBROMO-
88-06-2 PHENOL, 2,4,6-TRICHLORO- [PAL, PAL]	2,4,6-TRICHLOROPHENOL
95-95-4 PHENOL, 2,4,5-TRICHLORO- [PAL]	2,4,5-TRICHLOROPHENOL
609-19-8 PHENOL, 3,4,5-TRICHLORO- [PAL]	3,4,5-TRICHLOROPHENOL
933-75-5 PHENOL, 2,3,6-TRICHLORO- [PAL]	2,3,6-TRICHLOROPHENOL
933-78-8 PHENOL, 2,3,5-TRICHLORO- [PAL]	2,3,5-TRICHLOROPHENOL
15950-66-0 PHENOL, 2,3,4-TRICHLORO- [PAL]	2,3,4-TRICHLOROPHENOL
25167-82-2 PHENOL, TRICHLORO- [PAL]	TRICHLOROPHENOL
527-60-6 PHENOL, 2,4,6-TRIMETHYL-	PHENOL, 2,4,6-TRIMETHYL-
88-89-1 PHENOL, 2,4,6-TRINITRO- [PAL]	PICRIC ACID
52277-29-9 PHENOLSULFONIC ACID - FORMALDEHYDE - UREA CONDENSATE, SODIUM SALT	PHENOLSULFONIC ACID - FORMALDEHYDE - UREA CONDENSATE, SODIUM
92-84-2 PHENOTHIAZINE	PHENOTHIAZINE
92-84-2 10H-PHENOTHIAZINE [PAL, TSC, TSC]	PHENOTHIAZINE
26002-80-2 PHENOTHRIN [2,2-DIMETHYL-3-(2-METHYL-1-PROPENYL)CYCLOPROPANECARBOXYLIC ACID (3-PHENOXYPHENYL)METHYL ESTER]	PHENOTHRIN [2,2-DIMETHYL-3-(2-METHYL-1-PROPENYL)CYCLOPROPANE]OXY]BENZOIC ACID 1-METHYLETHYL ESTER]
58-36-6 PHENOXARSINE, 10,10'-OXYDI- [302, PAL]	10,10'-OXYBISPHENOXARSINE
122-59-8 PHENOXYACETIC ACID	PHENOXYACETIC ACID
RR-00776-1 PHENOXYACETIC ACID HERBICIDES	PHENOXYACETIC ACID HERBICIDES
59-96-1 PHENOXYBENZAMINE	PHENOXYBENZAMINE
63-92-3 PHENOXYBENZAMINE HYDROCHLORIDE	PHENOXYBENZAMINE HYDROCHLORIDE
122-99-6 2-PHENOXYETHANOL [TSC, TSC, TSC]	ETHYLENE GLYCOL MONOPHENYL ETHER
RR-00874-2 N-(2-PHENOXYETHYL) ANILINE	N-(2-PHENOXYETHYL) ANILINE
122-60-1 (PHENOXYMETHYL) OXIRANE [ONT]	PHENYL GLYCIDYL ETHER

Chemical Name	Reference Name
RR-00424-0 PHENOXY PESTICIDES, N.O.S.	PHENOXY PESTICIDES, N.O.S.
770-35-4 1-PHENOXY-2-PROPANOL [TSC, TSC, TSC]	2-PROPANOL, 1-PHENOXY-
435-97-2 PHENPROCOUMON	PHENPROCOUMON
RR-01308-1 PHENTHOATE	PHENTHOATE
696-28-6 PHENYL DICHLOROARSINE [FLA]	DICHLOROPHENYLARSINE
101-84-8 PHENYL ETHER VAPOR [MEX, MEX]	PHENYL ETHER
101-84-8 PHENYL ETHER, VAPOUR [QBC, QBC]	PHENYL ETHER
108-98-5 PHENYL MERCAPTAN [MEX]	BENZENETHIOL
RR-01654-6 PHENYL(DISUBSTITUTEDPOLYCYCLIC)	PHENYL(DISUBSTITUTEDPOLYCYCLIC)HOXY)CARBONYL]CYCLOHEXYLCARBONYL]OXY]METHYL]-1,3-PROPANEDIYL BIS(OXIRANYLMETHYL) ESTER
122-78-1 PHENYLACETALDEHYDE [NFP, NFP]	BENZENEACETALDEHYDE
122-79-2 PHENYL ACETATE	PHENYL ACETATE
103-82-2 PHENYLACETIC ACID	PHENYLACETIC ACID
RR-01778-7 PHENYLACETIC ACID SALTS	PHENYLACETIC ACID SALTSCAL ISOMERS
140-29-4 PHENYLACETONITRILE [HM1, HM1, NJL, WHM]	BENZYL CYANIDE
103-80-0 PHENYLACETYL CHLORIDE	PHENYLACETYL CHLORIDE
148-82-3 L-PHENYLALANINE, 4-[BIS(2-CHLOROETHYL)AMINO]- [PAL, PAL]	MELPHALAN5,5A,5B,6-DECACHLOROOCTAHYDRO-
531-76-0 DL-PHENYLAMINE, 4-[BIS(2-CHLOROETHYL)AMINO]- [PAL, PAL]	MERPHALAN
91-40-7 PHENYL ANTHRANILIC ACID (ALL ISOMERS)	PHENYL ANTHRANILIC ACID (ALL ISOMERS)
98-05-5 PHENYLARSONIC ACID [WHM]	BENZENEARSONIC ACID
93-99-2 PHENYL BENZOATE	PHENYL BENZOATE
104-51-8 1-PHENYLBUTANE [HM1]	BUTYL BENZENE
135-98-8 2-PHENYLBUTANE [HM1]	SEC-BUTYLBENZENE
50-33-9 PHENYLBUTAZONE	PHENYLBUTAZONE
1560-06-1 1-PHENYL-2-BUTENE	1-PHENYL-2-BUTENE
622-44-6 PHENYLCARBYLAMINE CHLORIDE	PHENYLCARBYLAMINE CHLORIDE
1885-14-9 PHENYLCHLOROFORMATE	PHENYLCHLOROFORMATE
1885-14-9 PHENYL CHLOROFORMATE [WHM]	PHENYLCHLOROFORMATE
696-28-6 PHENYL DICHLOROARSINE [302]	DICHLOROPHENYLARSINE
696-28-6 PHENYLDICHLOROARSINE [NJL]	DICHLOROPHENYLARSINE
696-28-6 PHENYL DICHLOROARSINE [PAL]	DICHLOROPHENYLARSINE
1254-78-0 PHENYL DIDECYL PHOSPHITE	PHENYL DIDECYL PHOSPHITE
120-07-0 N-PHENYL DIETHANOLAMINE	N-PHENYL DIETHANOLAMINE
120-07-0 N-PHENYLDIETHANOLAMINE [NFP, NFP]	N-PHENYL DIETHANOLAMINE
101-42-8 3-PHENYL-1,1-DIMETHYLUREA	3-PHENYL-1,1-DIMETHYLUREA
1666-13-3 PHENYL DISELENIDE	PHENYL DISELENIDE
RR-00878-6 PHENYL DI-O-XENYL PHOSPHATE	PHENYL DI-O-XENYL PHOSPHATE
95-54-5 O-PHENYLENEDIAMINE	O-PHENYLENEDIAMINE
95-54-5 1,2-PHENYLENEDIAMINE [313]	O-PHENYLENEDIAMINE
95-54-5 ORTHO-PHENYLENEDIAMINE [CTR]	O-PHENYLENEDIAMINE
95-54-5 PHENYLENEDIAMINE, ORTHO- [OTS]	O-PHENYLENEDIAMINE
95-54-5 ORTHO-PHENYLENEDIAMINE [TSE]	O-PHENYLENEDIAMINE
106-50-3 P-PHENYLENE DIAMINE	P-PHENYLENE DIAMINE
106-50-3 P-PHENYLENEDIAMINE [313, AUS, AUS, CAA, CAL, CAN, CEX, CEX]	P-PHENYLENE DIAMINE
106-50-3 PARA-PHENYLENEDIAMINE [CTR]	P-PHENYLENE DIAMINE

Chemical Name	Reference Name
106-50-3 P-PHENYLENEDIAMINE [GBR, GBR]	P-PHENYLENE DIAMINE
106-50-3 PARA-PHENYLENEDIAMINE [IAR]	P-PHENYLENE DIAMINE
106-50-3 P-PHENYLENEDIAMINE [MAK, MAK, MAK, MAK, NJL, ONT, ONT]	P-PHENYLENE DIAMINE
106-50-3 PHENYLENEDIAMINE, PARA- [OTS]	P-PHENYLENE DIAMINE
106-50-3 P-PHENYLENEDIAMINE [PQL, TLV]	P-PHENYLENE DIAMINE
106-50-3 PARA-PHENYLENEDIAMINE [TSE]	P-PHENYLENE DIAMINE
108-45-2 M-PHENYLENEDIAMINE	M-PHENYLENEDIAMINE
108-45-2 1,3-PHENYLENEDIAMINE [313]	M-PHENYLENEDIAMINE
108-45-2 META-PHENYLENEDIAMINE [CTR, IAR]	M-PHENYLENEDIAMINE
108-45-2 PHENYLENEDIAMINE, META- [OTS]	M-PHENYLENEDIAMINE
108-45-2 META-PHENYLENEDIAMINE [TSE]	M-PHENYLENEDIAMINE
25265-76-3 PHENYLENEDIAMINE [QBC, QBC, RCU]	PHENYLENEDIAMINES
615-28-1 1,2-PHENYLENEDIAMINE DIHYDROCHLO-RIDE [313]	1,2-BENZENEDIAMINE, DIHYDROCHLORIDELORIDE)
624-18-0 P-PHENYLENEDIAMINE DIHYDROCHLORIDE	P-PHENYLENEDIAMINE DIHYDROCHLORIDE
624-18-0 1,4-PHENYLENEDIAMINE DIHYDROCHLO-RIDE [313]	P-PHENYLENEDIAMINE DIHYDROCHLORIDE
RR-01435-7 M-PHENYLENE DIAMINEDIPERCHLORATE	M-PHENYLENE DIAMINEDIPERCHLORATE
25265-76-3 PHENYLENEDIAMINES	PHENYLENEDIAMINES
25265-76-3 PHENYLENEDIAMINES (BENZENEDIAMINES) [TSC]	PHENYLENEDIAMINES-, DIISOOCTYL ESTER (Z,Z)-
16245-77-5 PARA-PHENYLENEDIAMINE, SULFATE SALT [CTR]	1,4-BENZENEDIAMINE, SULFATE (1:1)
54-17-1 M-PHENYLENEDIAMINE, SULFATE SALT	M-PHENYLENEDIAMINE, SULFATE SALT
541-70-8 META-PHENYLENEDIAMINE, SULFATE SALT [CTR]	1,3-BENZENEDIAMINE, SULFATE (1:1)
104-49-4 1,4-PHENYLENE DIISOCYANATE [313]	BENZENE, 1,4-DIISOCYANATO-
123-61-5 1,3-PHENYLENE DIISOCYANATE [313]	BENZENE, 1,3-DIISOCYANATO-
61-76-7 PHENYLEPHRINE HYDROCHLORIDE	PHENYLEPHRINE HYDROCHLORIDE
122-60-1 PHENYL-2,3-EPOXYPROPYL ETHER [GBR]	PHENYL GLYCIDYL ETHER
122-98-5 N-PHENYL ETHANOLAMINE	N-PHENYL ETHANOLAMINE
122-98-5 N-PHENYLETHANOLAMINE [FLA, MSL]	N-PHENYL ETHANOLAMINE
101-84-8 PHENYL ETHER	PHENYL ETHER
101-84-8 PHENYL ETHER (VAPOR) [ALB, ALB]	PHENYL ETHER
8004-13-5 PHENYL ETHER-BIPHENYL MIXTURE [MIN, PAL]	PHENYL ETHER-BIPHENYL MIXTURE VAPOR
8004-13-5 PHENYL ETHER-BIPHENYL MIXTURE VAPOR	PHENYL ETHER-BIPHENYL MIXTURE VAPOR
8004-13-5 PHENYL ETHER-BIPHENYL MIXTURE (VAPOR) [NIO, NIO]	PHENYL ETHER-BIPHENYL MIXTURE VAPOR
8004-13-5 PHENYL ETHER-BIPHENYL MIXTURE, VA-POR [OS1, OS1]	PHENYL ETHER-BIPHENYL MIXTURE VAPOR
8004-13-5 PHENYL ETHER-DIPHENYL MIXTURE VA-POR [ALB, ALB]	PHENYL ETHER-BIPHENYL MIXTURE VAPOR
8004-13-5 PHENYL ETHER-DIPHENYL MIXTURE (VAPOR) [BC1, BC1]	PHENYL ETHER-BIPHENYL MIXTURE VAPOR
101-84-8 PHENYL ETHER, VAPOR [CAL]	PHENYL ETHER
101-84-8 PHENYL ETHER VAPOR [FLA, MSL]	PHENYL ETHER
101-84-8 PHENYL ETHER (VAPOR) [NIO, NIO]	PHENYL ETHER
101-84-8 PHENYL ETHER (VAPOUR) [BC1, BC1]	PHENYL ETHER
103-45-7 2-PHYENYLETHYL ACETATE	2-PHYENYLETHYL ACETATE

Chemical Name	Reference Name
103-45-7 PHENYLETHYL ACETATE (BETA) [NFP, NFP]	2-PHYENYLETHYL ACETATE
100-42-5 PHENYLETHYLENE (STYRENE) [ALB, ALB, ALB]	STYRENE
92-50-2 N-PHENYL-N-ETHYLETHANOLAMINE [FLA, NFP, NFP, MSL]	ETHANOL, 2-(ETHYLPHENYLAMINO)-
122-60-1 PHENYL GLYCIDYL ETHER	PHENYL GLYCIDYL ETHER
122-60-1 PHENYL GLYCIDYL ETHER (PGE) [FLA, MSL, OS1, OS1, TLV, TLV, TLV, WHM, AUS, AUS, BC1, BC1, QBC, QBC]	PHENYL GLYCIDYL ETHER
93-56-1 PHENYL GLYCOL ETHER	PHENYL GLYCOL ETHER
100-63-0 PHENYLHYDRAZINE	PHENYLHYDRAZINE
59-88-1 PHENYLHYDRAZINE HYDROCHLORIDE	PHENYLHYDRAZINE HYDROCHLORIDE
103-71-9 PHENYL ISOCYANATE	PHENYL ISOCYANATE
108-98-5 PHENYL MERCAPTAN [OTV, ALB, ALB, AUS, BC1, CAL, CEX, FLA, HM1, HM1, HM1, HM1, MSL, OS1]	BENZENETHIOL
108-98-5 PHENYLMERCAPTAN [QBC]	BENZENETHIOL
108-98-5 PHENYL MERCAPTAN [TLV, WHM]	BENZENETHIOL
62-38-4 PHENYLMERCURIC ACETATE	PHENYLMERCURIC ACETATE
RR-00428-4 PHENYLMERCURIC COMPOUNDS, N.O.S. [HM1, HM1, HM1]	PHENYLMERCURIC COMPOUND, SOLID, N.O.S.
RR-00428-4 PHENYLMERCURIC COMPOUND, SOLID, N.O.S.	PHENYLMERCURIC COMPOUND, SOLID, N.O.S.
100-57-2 PHENYLMERCURIC HYDROXIDE	PHENYLMERCURIC HYDROXIDE
55-68-5 PHENYLMERCURIC NITRATE	PHENYLMERCURIC NITRATE
62-38-4 PHENYLMERCURY ACETATE [302, EPA, RCU, RCU]	PHENYLMERCURIC ACETATE
63134-33-8 4-[[4-(PHENYLMETHOXY)PHENYL]SULFONYL] PHENOL	4-[[4-(PHENYLMETHOXY)PHENYL]SULFONYL] PHENOL
93-90-3 PHENYLMETHYL ETHANOL AMINE [MSL, NFP, NFP]	ETHANOL, 2-(METHYLPHENYLAMINO)-
89-25-8 1-PHENYL-3-METHYL-5-PYRAZOLONE [CSR, CSR]	3-METHYL-1-PHENYL-2-PYRAZOLIN-5-ONE
92-53-5 4-PHENYLMORPHOLINE	4-PHENYLMORPHOLINE
90-30-2 N-PHENYL-1-NAPHTHYLAMINE	N-PHENYL-1-NAPHTHYLAMINE
135-88-6 N-PHENYL-BETA-NAPHTHYLAMINE	N-PHENYL-BETA-NAPHTHYLAMINE
135-88-6 N-PHENYL-2-NAPHTHYLAMINE [CAL, CSR, CSR, IAR, MAK]	N-PHENYL-BETA-NAPHTHYLAMINE
135-88-6 PHENYL-BETA-NAPHTHYLAMINE [NHS, NHS, NHS]	N-PHENYL-BETA-NAPHTHYLAMINE
96-09-3 PHENYL-OXIRANE [FLA]	STYRENE OXIDE
132-27-4 O-PHENYLPHENATE, SODIUM [C65, C65]	SODIUM O-PHENYLPHENOL
80-46-6 P-(TERT-PHENYL) PHENOL [FLA]	P-TERT-PENTYLPHENOL
90-43-7 2-PHENYLPHENOL	2-PHENYLPHENOL
90-43-7 O-PHENYLPHENOL [CAN, CSR, CSR]	2-PHENYLPHENOL
90-43-7 ORTHO-PHENYLPHENOL [IAR]	2-PHENYLPHENOL
90-43-7 O-PHENYLPHENOL [NFP, NFP, NJL]	2-PHENYLPHENOL
90-43-7 O-PHENYL PHENOL [WHM]	2-PHENYLPHENOL
132-27-4 O-PHENYLPHENOL, SODIUM SALT [PST]	SODIUM O-PHENYLPHENOL
101-54-2 N-PHENYL-P-PHENYLENEDIAMINE [CSR, CSR]	P-AMINODIPHENYLAMINE
638-21-1 PHENYLPHOSPHINE	PHENYLPHOSPHINE
644-97-3 PHENYL PHOSPHORUS DICHLORIDE	PHENYL PHOSPHORUS DICHLORIDE
3497-00-5 PHENYL PHOSPHORUS THIODICHLORIDE	PHENYL PHOSPHORUS THIODICHLORIDE
122-97-4 3-PHENYL-1-PROPANOL	3-PHENYL-1-PROPANOL

Chemical Name	Reference Name
14838-15-4 PHENYLPROPANOLAMINE	PHENYLPROPANOLAMINE
RR-01779-8 PHENYLPROPANOLAMINE SALTS, OPTICAL ISOMERS, AND SALTS OF OPTICAL ISOMERS	PHENYLPROPANOLAMINE SALTS, OPTICAL ISOMERS, AND SALTS OF OPT
98-83-9 2-PHENYLPROPENE [GBR]	ALPHA-METHYL STYRENE
122-97-4 PHENYLPROPYL ALCOHOL [NFP, NFP]	3-PHENYL-1-PROPANOL
1335-10-0 PHENYL PROPYL ALDEHYDE	PHENYL PROPYL ALDEHYDE
118-55-8 PHENYL SALICYLATE	PHENYL SALICYLATE
2097-19-0 PHENYLSILATRANE	PHENYLSILATRANE
139-66-2 PHENYL SULFIDE	PHENYL SULFIDE
17688-68-5 4-PHENYLTHIOMORPHOLINE, 1,1-DIOXIDE-	4-PHENYLTHIOMORPHOLINE, 1,1-DIOXIDE-
103-85-5 PHENYLTHIOUREA	PHENYLTHIOUREA
103-85-5 1-PHENYL-2-THIOUREA [CSR, CSR, WHM]	PHENYLTHIOUREA
28652-72-4 PHENYLTOLUENE	PHENYLTOLUENE
28652-72-4 PHENYLTOLUENE O [NFP, NFP]	PHENYLTOLUENE
98-13-5 PHENYL TRICHLOROSILANE	PHENYL TRICHLOROSILANE
98-13-5 PHENYLTRICHLOROSILANE [MSL]	PHENYL TRICHLOROSILANE
98-13-5 PHENYL TRICHLORO SILANE [NFP, NFP]	PHENYL TRICHLOROSILANE
2116-84-9 PHENYL TRIMETHYL SILOXANE	PHENYL TRIMETHYL SILOXANE
64-10-8 PHENYLUREA	PHENYLUREA
RR-00429-5 PHENYL UREA PESTICIDES, N.O.S.	PHENYL UREA PESTICIDES, N.O.S.
57-41-0 PHENYTOIN	PHENYTOIN
105650-23-5 PHIP (2-AMINO-1-METHYL-6-PHENYLIMIDAZO-[4,5-B]PYRIDINE)	PHIP (2-AMINO-1-METHYL-6-PHENYLIMIDAZO-[4,5-B] PYRIDINE)
108-73-6 PHLOROGLUCINOL	PHLOROGLUCINOL
298-02-2 PHORATE	PHORATE
298-02-2 PHORATE (THEMET (R)) [QBC, QBC, QBC]	PHORATE
298-02-2 PHORATE (THIMET) [BC1, BC1, BC1]	PHORATE
298-02-2 PHORATE (ISO) [GBR, GBR, GBR]	PHORATE
298-02-2 PHORATE (THIMET) [ALB, ALB, ALB]	PHORATE
504-20-1 PHORONE	PHORONE
4104-14-7 PHOSACETIM	PHOSACETIM
2310-17-0 PHOSALONE	PHOSALONE
7786-34-7 PHOSDRIN [NIO, NIO, NIO, NIO, NJL]	MEVINPHOS
7786-34-7 PHOSDRIN (MEVINPHOS) [BC1, BC1, BC1, OS1, OS1, OS1, OS1, OS1]	MEVINPHOS
947-02-4 PHOSFOLAN	PHOSFOLAN
75-44-5 PHOSGENE	PHOSGENE
75-44-5 PHOSGENE (CARBONYL CHLORIDE) [OS1, OS1, WHM]	PHOSGENE
75-44-5 PHOSGENE (CARBON DICHLORIDE) [EP3]	PHOSGENE
732-11-6 PHOSMET	PHOSMET
13396-80-0 9-PHOSPHABICYCLONONANE	9-PHOSPHABICYCLONONANE
13396-80-0 9-PHOSPHABICYCLONONANES (CYCLOOCTA-DIENE PHOSPHINES) [HM1, HM1]	9-PHOSPHABICYCLONONANE
13171-21-6 PHOSPHAMIDON	PHOSPHAMIDON
13171-21-6 PHOSPHAMIDON (2-CHLORO-2-DIETHYLCAR-BAMOYL-1-METHYLVINYL DEMETHYL PHOS-PHATE) [CN8]	PHOSPHAMIDON
14265-44-2 PHOSPHATE	PHOSPHATE
65652-41-7 PHOSPHATE, BIS(TERT-BUTYLPHENYL) PHENYL	PHOSPHATE, BIS(TERT-BUTYLPHENYL) PHENYL

Chemical Name	Reference Name
56803-37-3 PHOSPHATE, TERT-BUTYLPHENYL DIPHENYL [OTS]	TERT-BUTYLPHENYL DIPHENYL PHOSPHATE
51363-64-5 PHOSPHATE, DIISODECYL PHENYL	PHOSPHATE, DIISODECYL PHENYL
RR-01731-2 PHOSPHATED POLYARYLPHENOL ETHOXYLATE, POTASSIUM SALT	PHOSPHATED POLYARYLPHENOL ETHOXYLATE, POTASSIUM SALTHYL]-,BIS(4-METHYLBENZENESUL-FONATE)
29761-21-5 PHOSPHATE, ISODECYL DIPHENYL	PHOSPHATE, ISODECYL DIPHENYL
28108-99-8 PHOSPHATE, ISOPROPYLPHENYL DIPHENYL	PHOSPHATE, ISOPROPYLPHENYL DIPHENYL
14265-44-2 PHOSPHATES [MX2]	PHOSPHATE
78-33-1 PHOSPHATE, TRIS(TERT-BUTYLPHENYL) [OTS]	PHENOL, 4-(1,1-DIMETHYLETHYL)-, PHOSPHATE (3:1)
26967-76-0 PHOSPHATE, TRIS(ISOPROPYLPHENYL)	PHOSPHATE, TRIS(ISOPROPYLPHENYL)
25155-23-1 PHOSPHATE, TRIXYLYL	PHOSPHATE, TRIXYLYL
7803-51-2 PHOSPHINE	PHOSPHINE
7803-51-2 PHOSPHINE (HYDROGEN PHOSPHIDE) [OS3]	PHOSPHINE
638-21-1 PHOSPHINE, PHENYL- [PAL]	PHENYLPHOSPHINELPHA.,6.ALPHA.)-
603-35-0 PHOSPHINE, TRIPHENYL-	PHOSPHINE, TRIPHENYL-
947-02-4 PHOSPHOLAN [EPA, EPA]	PHOSFOLAN
1809-19-4 PHOSPHONIC ACID, DIBUTYL ESTER [PAL]	DIBUTYL PHOSPHITE
1429-50-1 PHOSPHONIC ACID, (1,2-ETHANEDIYL-BIS(NI-TRILOBIS(METHYLENE)))TETRAKIS- (EDTMPA) [TSC]	ETHYLENEDIAMINETETRA(METHYLENEPHOSPHONIC) ACID
34274-30-1 PHOSPHONIC ACID, [1,2-ETHANEDIYLBIS[NI-TRILOBIS(METHYLENE)]TETRAKIS-, POTAS-SIUM SALT [TSC]	PHOSPHORIC ACID, [1,2-ETHANEDIYLBIS[NITRILOBIS (METHYLENE)]INDENYL ESTER
68901-17-7 PHOSPHONIC ACID, [1,2-ETHANEDIYLBIS[NI-TRILOBIS(METHYLENE)]TETRAKIS-, OCTAAM-MONIUM SALT	PHOSPHONIC ACID, [1,2-ETHANEDIYLBIS[NITRILOBIS (METHYLENE)]T
14860-53-8 PHOSPHONIC ACID, (1-HYDROXYETHYLIDINE) BIS-, TETRAPOTASSIUMSALT	PHOSPHONIC ACID, (1-HYDROXYETHYLIDINE)BIS-, TETRAPOTASSIUM
3794-83-0 PHOSPHONIC ACID, (1-HYDROXYETHYLIDINE) BIS-, TETRASODIUM SALT	PHOSPHONIC ACID, (1-HYDROXYETHYLIDINE)BIS-, TETRASODIUM SALTE
6419-19-8 PHOSPHONIC ACID, [NITRILOTRIS(METHY-LENE)]TRIS- [OTS]	AMINOTRI(METHYLENEPHOSPHONIC ACID)
52-68-6 PHOSPHONIC ACID, (2,2,2-TRICHLORO-1-HY-DROXYETHYL)-, DIMETHYLESTER [PAL]	TRICHLORFON
RR-00175-2 PHOSPHONIUM SALT	PHOSPHONIUM SALT
37971-36-1 2-PHOSPHONO-1,2-4-BUTANETRICARBOXYLIC ACID	2-PHOSPHONO-1,2-4-BUTANETRICARBOXYLIC ACID
40372-66-5 2-PHOSPHONO-1,2,4-BUTANETRICARBOXYLIC ACID, SODIUM SALT	2-PHOSPHONO-1,2,4-BUTANETRICARBOXYLIC ACID, SODIUM SALT
37971-36-1 2-PHOSPHONOBUTANE-1,2,4-TRICARBOXYLIC ACID [PST]	2-PHOSPHONO-1,2-4-BUTANETRICARBOXYLIC ACID
RR-01658-0 PHOSPHONOCARBOXYLATE SALTS	PHOSPHONOCARBOXYLATE SALTS
944-22-9 PHOSPHONODITHIOIC ACID, ETHYL-, O-ETHYL S-PHENYL ESTER [PAL]	FONOFOS
50782-69-9 PHOSPHONOTHIOIC ACID, METHYL-, S-[2-[BIS (1-METHYLETHYL)AMINOETHYL)O-ETHYL ESTER	PHOSPHONOTHIOIC ACID, METHYL-, S-[2-[BIS(1-METHYLETHYL)AMINO
2703-13-1 PHOSPHONOTHIOIC ACID, METHYL-,O-ETHYL O-(4-(METHYLTHIO)PHENYL) ESTER	PHOSPHONOTHIOIC ACID, METHYL-,O-ETHYL O-(4-(METHYLTHIO)PHENY
2665-30-7 PHOSPHONOTHIOIC ACID,METHYL-,O-(4-NI-TROPHENYL)O-PHENYL)ESTER [MSL]	PHOSPHONOTHIOIC ACID, METHYL-,O-(4-NITRO-PHENYL) O-PHENYL EST
2665-30-7 PHOSPHONOTHIOIC ACID, METHYL-,O-(4-NITROPHENYL) O-PHENYL ESTER	PHOSPHONOTHIOIC ACID, METHYL-,O-(4-NITRO-PHENYL) O-PHENYL EST

Chemical Name	Reference Name
2104-64-5 PHOSPHONOTHIOIC ACID, PHENYL-, O-ETHYL O-(4-NITROPHENYL)ESTER [PAL]	EPN
26915-70-8 ALPHA-PHOSPHONO-OMEGA-(TRIDECYLOXY) POLY(OXY-1,2-ETHANEDIYL)	ALPHA-PHOSPHONO-OMEGA-(TRIDECYLOXY) POLY (OXY-1,2-ETHANEDIYL)
RR-01666-0 PHOSPHORAMIDE	PHOSPHORAMIDE
299-86-5 PHOSPHORAMIDIC ACID, METHYL-, 2-CHLORO-4-(1,1-DIMETHYLETHYL)PHENYL METHYL ESTER [PAL]	CRUFOMATEESTER
22224-92-6 PHOSPHORAMIDIC ACID, (1-METHYLETHYL)-, ETHYL-3-METHYL-4-(METHYLTHIO)PHENYL ESTER [PAL]	FENAMINPHOS
10265-92-6 PHOSPHORAMIDOTHIOIC ACID, O,S-DIMETHYL ESTER [PAL]	METHAMIDOPHOS
4104-14-7 PHOSPHORAMIDOTHIOIC ACID, (1-IMI-NOETHYL)-, O,O-BIS(4-CHLOROPHENYL) ESTER [PAL]	PHOSACETIMTRIMETHYL-
10026-13-8 PHOSPHORANE, PENTACHLORO- [PAL]	PHOSPHORUS PENTACHLORIDE
7664-38-2 PHOSPHORIC ACID	PHOSPHORIC ACID
35089-00-0 PHOSPHORIC ACID, BERYLLIUM SALT [PAL, PAL]	BERYLLIUM PHOSPHATE
13598-15-7 PHOSPHORIC ACID, BERYLLIUM SALT (1:1) [PAL, PAL]	BERYLLIUM PHOSPHATE
13598-26-0 PHOSPHORIC ACID, BERYLLIUM SALT (2:3)	PHOSPHORIC ACID, BERYLLIUM SALT (2:3)
66108-37-0 PHOSPHORIC ACID, 2,2-BIS(BROMOMETHYL)-3-CHLOROPROPYL BIS[2-CHLORO-1-(CHLOROMETHYL)ETHYL] ESTER	PHOSPHORIC ACID, 2,2-BIS(BROMOMETHYL)-3-CHLORO-PROPYL BIS[2-
141-65-1 PHOSPHORIC ACID, BIS(2-ETHYLHEXYL)ESTER, SODIUM SALT	PHOSPHORIC ACID, BIS(2-ETHYLHEXYL)ESTER, SODIUM SALT
16368-97-1 PHOSPHORIC ACID, BIS(2-ETHYLHEXYL) PHENYL ESTER	PHOSPHORIC ACID, BIS(2-ETHYLHEXYL) PHENYL ESTER
298-07-7 PHOSPHORIC ACID, BIS(2-ETHYLHEXYL) ESTER [TSC, TSC, TSC]	DI(2-ETHYLHEXYL)PHOSPHORIC ACID
69011-04-7 PHOSPHORIC ACID, BUTYL ESTER, MAN-GANESE(2+) SALT	PHOSPHORIC ACID, BUTYL ESTER, MANGANESE(2+) SALT
129733-59-1 PHOSPHORIC ACID, C(6-12)-ALKYL ESTERS, COMPOUND WITH 2-(DIBUTYLAMIONO) ETHANOL	PHOSPHORIC ACID, C(6-12)-ALKYL ESTERS, COM-POUND WITH 2-(DIBUHYLETHYL)-4-HYDROXY-, C(7-9) BRANCHED AND LINEAR ALKYL ESTERS
129733-59-1 PHOSPHORIC ACID, ALKYL ESTERS, COMPOUND WITH 2-(DIBUTYLAMINO) ETHANOL [TSE]	PHOSPHORIC ACID, C(6-12)-ALKYL ESTERS, COMPOUND WITH 2-(DIBU,2,6,6-PENTAMETHYL-4-PIPERIDINYL) ESTER
470-90-6 PHOSPHORIC ACID, 2-CHLORO-1-(2,4-DICHLOROPHENYL)ETHENYLDIETHYL ESTER [PAL]	CHLORFENVINPHOS
13171-21-6 PHOSPHORIC ACID, 2-CHLORO-3-(DIETHY-LAMINO)-1-METHYL-3-OXO-1-PROPENYL DIMETHYL ESTER [PAL]	PHOSPHAMIDON
300-76-5 PHOSPHORIC ACID, 1,2-DIBROMO-2,2-DICHLOROETHYL DIMETHYL ESTER [PAL]	NALEDPHENYL METHYL ESTER
107-66-4 PHOSPHORIC ACID, DIBUTYL ESTER [OTS, PAL, TSC, TSC, TSC]	DIBUTYL PHOSPHATE
2528-36-1 PHOSPHORIC ACID, DIBUTYL PHENYL ESTER [TSC, TSC]	DIBUTYL PHENYL PHOSPHATE
62-73-7 PHOSPHORIC ACID, 2,2-DICHLOROETHENYL DIMETHYL ESTER [PAL]	DICHLORVOS

Chemical Name	Reference Name
7057-92-3 PHOSPHORIC ACID, DIDODECYL ESTER	PHOSPHORIC ACID, DIDODECYL ESTER)AZO]-3,3'-DIMETHOXY[1,1'-BIPHENYL]-4-YL]AZO]-5-HYDROXY-7-SULFO-1-NAPHTHALENYL]AZO]-5-NITRO-, TRISODIUM SALT
51363-64-5 PHOSPHORIC ACID, DIISODECYL PHENYL ESTER [TSC, TSC]	PHOSPHATE, DIISODECYL PHENYL
27215-10-7 PHOSPHORIC ACID, DIISOOCTYL ESTER [TSC, TSC, TSC]	DIISOOCTYL ACID PHOSPHATE
141-66-2 PHOSPHORIC ACID, 3-(DIMETHYLAMINO)-1-METHYL-3-OXO-1-PROPENYLDIMETHYL ESTER, (E)- [PAL]	DICROTOPHOS
6923-22-4 PHOSPHORIC ACID, DIMETHYL 1-METHYL-3-(METHYLAMINO)-3-OXO-1-PROPENYL ESTER, (E)- [PAL]	MONOCROTOPHOS
3254-63-5 PHOSPHORIC ACID, DIMETHYL 4-(METHYLTHIO)PHENYL ESTER	PHOSPHORIC ACID, DIMETHYL 4-(METHYLTHIO) PHENYL ESTER
56803-37-3 PHOSPHORIC ACID, (1,1-DIMETHYLETHYL) PHENYL DIPHENYL ESTER [TSC, TSC]	TERT-BUTYLPHENYL DIPHENYL PHOSPHATEENYLENE) ESTER
7558-79-4 PHOSPHORIC ACID, DISODIUM SALT [MSL, PAL]	SODIUM PHOSPHATE DIBASIC
10039-32-4 PHOSPHORIC ACID, DISODIUM SALT, DECAHYDRATE [MSL]	PHOSPHORIC ACID, DISODIUM SALT, DODECAHYDRATE
10039-32-4 PHOSPHORIC ACID, DISODIUM SALT, DODECAHYDRATE	PHOSPHORIC ACID, DISODIUM SALT, DODECAHYDRATE
10140-65-5 PHOSPHORIC ACID, DISODIUM SALT, HYDRATE	PHOSPHORIC ACID, DISODIUM SALT, HYDRATE
12751-23-4 PHOSPHORIC ACID, DODECYL ESTER	PHOSPHORIC ACID, DODECYL ESTER
68188-96-5 PHOSPHORIC ACID, [1,2-ETHANEDIYLBIS[NITRILOBIS(METHYLENE)]TETRAKIS-, TETRAPOTASSIUM SALT	PHOSPHORIC ACID, [1,2-ETHANEDIYLBIS[NITRILO-BIS(METHYLENE)]METHYL-, N,N'-DITALLOW ACYL DERIVATIVES, METHYL SULFATES (SALTS)
RR-00901-8 PHOSPHORIC ACID, 1,2-ETHANEDIYL TETRAKIS(2-CHLORO-1-METHYLETHYL) ESTER	PHOSPHORIC ACID, 1,2-ETHANEDIYL TETRAKIS(2-CHLORO-1-METHYLET
33125-86-9 PHOSPHORIC ACID, 1,2-ETHANEDIYL TETRAKIS (2-CHLOROETHYL)ESTER [TSC, TSC, TSC]	TETRAKIS(2-CHLOROETHYL) ETHYLENE DIPHOSPHATE
1241-94-7 PHOSPHORIC ACID, 2-ETHYLHEXYL DIPHENYL ESTER	PHOSPHORIC ACID, 2-ETHYLHEXYL DIPHENYL ESTER
12645-31-7 PHOSPHORIC ACID, 2-ETHYLHEXYL ESTER	PHOSPHORIC ACID, 2-ETHYLHEXYL ESTER
29761-21-5 PHOSPHORIC ACID, ISODECYL DIPHENYL ESTER [TSC, TSC]	PHOSPHATE, ISODECYL DIPHENYL
7446-27-7 PHOSPHORIC ACID, LEAD(2+) SALT (2:3) [PAL, PAL]	LEAD PHOSPHATE
28108-99-8 PHOSPHORIC ACID, (1-METHYLETHYL)PHENYL DIPHENYL ESTER [TSC]	PHOSPHATE, ISOPROPYLPHENYL DIPHENYL
26444-49-5 PHOSPHORIC ACID, METHYLPHENYL DIPHENYL ESTER	PHOSPHORIC ACID, METHYLPHENYL DIPHENYL ESTER
26444-49-5 PHOSPHORIC ACID, METHYLPHENYLDIPHENYLS [OTS]	PHOSPHORIC ACID, METHYLPHENYL DIPHENYL ESTER
34364-42-6 PHOSPHORIC ACID, (1-METHYL-1-PHENYLETHYL)PHENYL DIPHENYLESTER	PHOSPHORIC ACID, (1-METHYL-1-PHENYLETHYL) PHENYL DIPHENYLTETRAKIS-, POTASSIUM SALT
1623-15-0 PHOSPHORIC ACID, MONOBUTYL ESTER	PHOSPHORIC ACID, MONOBUTYL ESTER
68909-59-1 PHOSPHORIC ACID MONO(C8-C10)ALKYL SODIUM SALTS	PHOSPHORIC ACID MONO(C8-C10)ALKYL SODIUM SALTS
1070-03-7 PHOSPHORIC ACID, MONO(2-ETHYLHEXYL) ESTER	PHOSPHORIC ACID, MONO(2-ETHYLHEXYL) ESTER
3900-04-7 PHOSPHORIC ACID, MONOHEXYL ESTER	PHOSPHORIC ACID, MONOHEXYL ESTERYL ESTER

Chemical Name	Reference Name
18351-85-4 PHOSPHORIC ACID, MONO(2-METHYLPHENYL) ESTER	PHOSPHORIC ACID, MONO(2-METHYLPHENYL) ESTER2-YL]-, METHYL ESTER
812-00-0 PHOSPHORIC ACID, MONOMETHYL ESTER	PHOSPHORIC ACID, MONOMETHYL ESTER
1623-24-1 PHOSPHORIC ACID, MONO(1-METHYLETHYL) ESTER [TSC, TSC, TSC]	ISOPROPYL PHOSPHORIC ACID
2958-09-0 PHOSPHORIC ACID, MONOOCTADECYL ESTER	PHOSPHORIC ACID, MONOOCTADECYL ESTER
3991-73-9 PHOSPHORIC ACID, MONOOCTYL ESTER	PHOSPHORIC ACID, MONOOCTYL ESTER4,6-DINITRO-PHENYL)AZO]-4-ETHOXYPHENYL]-
126-73-8 PHOSPHORIC ACID TRIBUTYL ESTER [PAL]	TRIBUTYL PHOSPHATE
126-73-8 PHOSPHORIC ACID, TRIBUTYL ESTER [TSC, TSC, TSC]	TRIBUTYL PHOSPHATE
78-40-0 PHOSPHORIC ACID, TRIETHYL ESTER [TSC, TSC, TSC, PST]	TRIETHYL PHOSPHATE
512-56-1 PHOSPHORIC ACID, TRIMETHYL ESTER [OTS]	TRIMETHYLPHOSPHATE
115-86-6 PHOSPHORIC ACID, TRIPHENYL ESTER [PAL, TSC, TSC]	TRIPHENYL PHOSPHATEHEXACHLORO-
78-42-2 PHOSPHORIC ACID, TRIS(ETHYLHEXYL) ESTER [TSC, TSC, TSC]	TRIS(2-ETHYLHEXYL)PHOSPHATE
78-30-8 PHOSPHORIC ACID, TRIS(2-METHYLPHENYL) ESTER [PAL, TSC, TSC]	TRIORTHOCRESYL PHOSPHATE
78-32-0 PHOSPHORIC ACID, TRIS(4-METHYLPHENYL) ESTER	PHOSPHORIC ACID, TRIS(4-METHYLPHENYL) ESTER
126-71-6 PHOSPHORIC ACID, TRIS(2-METHYLPROPYL) ESTER [TSC, TSC, TSC]	ISOBUTYL PHOSPHATE-
563-04-2 PHOSPHORIC ACID, TRIS(3-METHYLPHENYL) ESTER	PHOSPHORIC ACID, TRIS(3-METHYLPHENYL) ESTER
1330-78-5 PHOSPHORIC ACID, TRIS(METHYLPHENYL) ESTER [TSC, TSC]	TRICRESYL PHOSPHATE
7601-54-9 PHOSPHORIC ACID, TRISODIUM SALT [MSL, PAL]	TRISODIUM PHOSPHATE
10361-89-4 PHOSPHORIC ACID, TRISODIUM SALT, DEC-AHYDRATE	PHOSPHORIC ACID, TRISODIUM SALT, DECAHYDRATE
10101-89-0 PHOSPHORIC ACID, TRISODIUM SALT, DODEC-AHYDRATE [MSL, PAL]	TRIBASIC SODIUM PHOSPHATE DODECAHYDRATE
1314-56-3 PHOSPHORIC ANHYDRIDE [NJL, NJL, WHM]	PHOSPHORUS PENTOXIDE
680-31-9 PHOSPHORIC TRIAMIDE, HEXAMETHYL- [PAL, PAL]	HEXAMETHYL PHOSPHORAMIDE
1498-51-7 PHOSPHORODICHLORIDIC ACID, ETHYL ESTER [TSC, TSC, TSC]	ETHYL PHOSPHORODICHLORIDATE
298-04-4 PHOSPHORODITHIOIC ACID, O,O-DIETHYL S-[2-(ETHYLTHIO)ETHYL]ESTER [PAL]	DISULFOTONESTER
298-02-2 PHOSPHORODITHIOIC ACID, O,O-DIETHYL S-[(ETHYLTHIO)METHYL]ESTER [PAL]	PHORATE
3288-58-2 PHOSPHORODITHIOIC ACID, O,O-DIETHYL-S-METHYL ESTER [TSC, TSC]	O,O-DIETHYL S-METHYL DITHIOPHOSPHATE
86-50-0 PHOSPHORODITHIOIC ACID, O,O-DIMETHYL S-[(4-OXO-1,2,3-BENZOTRIAZIN-3(4H)-YL) METHYL] ESTER [PAL]	AZINPHOS-METHYL
78-34-2 PHOSPHORODITHIOIC ACID, S,S'-1,4-DIOXANE-2,3-DIYL O,O,O',O'-TETRAETHYL ESTER [PAL]	DIOXATHION
13194-48-4 PHOSPHORODITHIOIC ACID, O-ETHYL S,S-DIPROPYL ESTER [PAL]	ETHOPROPHOS1-PROPENYL DIMETHYL ESTER
35400-43-2 PHOSPHORODITHIOIC ACID, O-ETHYL O-[4-(METHYLTHIO)PHENYL]S-PROPYL ESTER [PAL]	SULPROFOS

Chemical Name	Reference Name
950-37-8 PHOSPHORODITHIOIC ACID, S-[(5-METHOXY-2-OXO-1,3,4-THIADIAZOL-3(2H)-YL)METHYL] O,O-DIMETHYL ESTER [PAL]	METHIDATHION
563-12-2 PHOSPHORODITHIOIC ACID, S,S'-METHYLENE O,O,O',O'-TETRAETHYLESTER [PAL]	ETHION
4259-15-8 PHOSPHORODITHIOIC ACID, O,O-BIS(2-ETHY-LENE- [OTS]	ZINC O,O-BIS(2-ETHYLHEXYL) PHOSPHORODITHIOATE
28629-66-5 PHOSPHORODITHIOIC ACID, O,O-DIISOOCTYL	PHOSPHORODITHIOIC ACID, O,O-DIISOOCTYL
RR-00497-7 PHOSPHORODITHIOIC AND PHOSPHOTHIOIC ACID ESTERS	PHOSPHORODITHIOIC AND PHOSPHOTHIOIC ACID ESTERS
42509-80-8 PHOSPHOROTHIOIC ACID, O-(5-CHLORO-1-(1-METHYLETHYL)-1H-1,2,4-TRIAZOL-3-YL) O,O-DIETHYL ESTER	PHOSPHOROTHIOIC ACID, O-(5-CHLORO-1-(1-METHYLETHYL)-1H-1,2,4
56-72-4 PHOSPHOROTHIOIC ACID, O-(3-CHLORO-4-METHYL-2-OXO-2H-1-BENZOPYRAN-7-YL) O,O-DIETHYL ESTER [PAL]	COUMAPHOS
8065-48-3 PHOSPHOROTHIOIC ACID, O,O-DIETHYL O-[2-(ETHYLTHIO)ETHYL)ESTER, MIXT. WITH O,O-DIETHYL S-[2-(ETHYLTHIO)ETHYL] PHOSPHO-ROTHIOATE [PAL]	DEMETON
333-41-5 PHOSPHOROTHIOIC ACID, O,O-DIETHYL O-[6-METHYL-2-(1-METHYLETHYL)-4-PYRIMIDINYL] ESTER [PAL]	DIAZINON
115-90-2 PHOSPHOROTHIOIC ACID, O,O-DIETHYL O-[4-(METHYLSULFINYL)PHENYL] ESTER [PAL]	FENSULFOTHION
56-38-2 PHOSPHOROTHIOIC ACID, O,O-DIETHYL O-(4-NITROPHENYL) ESTER [PAL]	PARATHION
2921-88-2 PHOSPHOROTHIOIC ACID, O,O-DIETHYL O-(3,5,6-TRICHLORO-2-PYRIDINYL) ESTER [PAL]	CHLORPYRIFOSM SALT
55-38-9 PHOSPHOROTHIOIC ACID, O,O-DIMETHYL O-[3-METHYL-4(METHYLTHIO)PHENYL] ES-TER [PAL]	FENTHION
2587-90-8 PHOSPHOROTHIOIC ACID, O,O-DIMETHYL-S-(2-(METHYLTHIO)ETHYL)ESTER [MSL]	PHOSPHOROTHIOIC ACID, O,O-DIMETHYL-S-(2-METHYLTHIO)ETHYL EST
2587-90-8 PHOSPHOROTHIOIC ACID, O,O-DIMETHYL-S-(2-METHYLTHIO)ETHYLESTER [PAL]	PHOSPHOROTHIOIC ACID, O,O-DIMETHYL-S-(2-METHYLTHIO)ETHYL EST
2587-90-8 PHOSPHOROTHIOIC ACID, O,O-DIMETHYL-S-(2-METHYLTHIO)ETHYL ESTER	PHOSPHOROTHIOIC ACID, O,O-DIMETHYL-S-(2-METHYLTHIO)ETHYL EST
298-00-0 PHOSPHOROTHIOIC ACID, O,O-DIMETHYL O-(4-NITROPHENYL) ESTER [PAL]	METHYL PARATHION
299-84-3 PHOSPHOROTHIOIC ACID, O,O-DIMETHYL O-(2,4,5-TRICHLOROPHENYL)ESTER [PAL]	RONNEL
8022-00-2 PHOSPHOROTHIOIC ACID, O-[2-(ETHYLTHIO)ETHYL] O,O-DIMETHYLESTER, MIXT. WITH S-[2-(ETHYLTHIO)ETHYL] O,O-DIMETHYL PHOSPHOROTHIOATE [PAL]	METHYL DEMETON
3383-96-8 PHOSPHOROTHIOIC ACID, O,O'-(THIODI-4,1-PHENYLENE) O,O,O',O'-TETRAMETHYL ES-TER [PAL]	TEMEPHOS
333-41-5 PHOSPHOROTHIONIC ACID, O,O-DIETHYL O-(6-METHYL-2-(1-METHYLETHYL)-4-PYRIMIDINYL) ESTER [ONT, ONT]	DIAZINON
10294-56-1 PHOSPHOROUS ACID [PST]	PHOSPHOROUS ACID, ORTHO
13598-36-2 PHOSPHOROUS ACID	PHOSPHOROUS ACID
10294-56-1 PHOSPHOROUS ACID, ORTHO	PHOSPHOROUS ACID, ORTHO
13598-36-2 PHOSPHOROUS ACID, ORTHO [NJL]	PHOSPHOROUS ACID
109-47-7 PHOSPHOROUS ACID, DIBUTYL ESTER	PHOSPHOROUS ACID, DIBUTYL ESTER

Chemical Name	Reference Name
25550-98-5 PHOSPHOROUS ACID, (DIISODECYL)PHENYL ETHER	PHOSPHOROUS ACID, (DIISODECYL)PHENYL ETHER
102-85-2 PHOSPHOROUS ACID, TRIBUTYL ESTER	PHOSPHOROUS ACID, TRIBUTYL ESTER
121-45-9 PHOSPHOROUS ACID, TRIMETHYL ESTER [PAL]	TRIMETHYL PHOSPHITE
101-02-0 PHOSPHOROUS ACID, TRIPHENYL ESTER [PST]	TRIPHENYL PHOSPHITE
7719-12-2 PHOSPHOROUS TRICHLORIDE [PAL]	PHOSPHORUS TRICHLORIDE
7723-14-0 PHOSPHORUS	PHOSPHORUS
7723-14-0 PHOSPHORUS (YELLOW) [QBC, MEX, MEX]	PHOSPHORUS
14596-37-3 PHOSPHORUS 32	PHOSPHORUS 32
15749-66-3 PHOSPHORUS 33	PHOSPHORUS 33
7723-14-0 PHOSPHORUS (YELLOW)) [ALB, ALB]	PHOSPHORUS
7723-14-0 PHOSPHORUS (YELLOW OR WHITE) [ONT]	PHOSPHORUS
121-45-9 PHOSPHORUS ACID, TRIMETHYL ESTER [TSC, TSC]	TRIMETHYL PHOSPHITE
RR-01336-5 PHOSPHORUS, AMORPHOUS	PHOSPHORUS, AMORPHOUS
RR-01027-5 PHOSPHORUS COMPOUNDS [CAB]	PHOSPHORUS COMPOUNDS, INORGANIC
RR-01027-5 PHOSPHORUS COMPOUNDS, INORGANIC	PHOSPHORUS COMPOUNDS, INORGANIC
12037-82-0 PHOSPHORUS HEPTASULPHIDE	PHOSPHORUS HEPTASULPHIDE
7789-59-5 PHOSPHORUS OXYBROMIDE	PHOSPHORUS OXYBROMIDE
10025-87-3 PHOSPHORUS OXYCHLORIDE	PHOSPHORUS OXYCHLORIDE
10025-87-3 PHOSPHORUS OXYCHLORIDE [PHOSPHORYL CHLORIDE] [EP3]	PHOSPHORUS OXYCHLORIDEWITH SULFUR TRIOXIDE]
7789-69-7 PHOSPHORUS PENTABROMIDE	PHOSPHORUS PENTABROMIDE
10026-13-8 PHOSPHORUS PENTACHLORIDE	PHOSPHORUS PENTACHLORIDE
7647-19-0 PHOSPHORUS PENTAFLUORIDE	PHOSPHORUS PENTAFLUORIDE
1314-80-3 PHOSPHORUS PENTASULFIDE	PHOSPHORUS PENTASULFIDE
1314-80-3 PHOSPHORUS PENTASULPHIDE [AUS, AUS]	PHOSPHORUS PENTASULFIDE
1314-56-3 PHOSPHORUS PENTOXIDE	PHOSPHORUS PENTOXIDE
1314-85-8 PHOSPHORUS SESQUISULFIDE	PHOSPHORUS SESQUISULFIDE
1314-80-3 PHOSPHORUS SULFIDE [RCU, RCU]	PHOSPHORUS PENTASULFIDE
1314-80-3 PHOSPHORUS SULFIDE (P2S5) [PAL]	PHOSPHORUS PENTASULFIDE
1314-85-8 PHOSPHORUS SULFIDE (P4S3) [PAL]	PHOSPHORUS SESQUISULFIDE
7789-60-8 PHOSPHORUS TRIBROMIDE	PHOSPHORUS TRIBROMIDE
7719-12-2 PHOSPHORUS TRICHLORIDE	PHOSPHORUS TRICHLORIDE
7719-12-2 PHOSPHORUS TRICHLORIDE [PHOSPHOROUS TRICHLORIDE] [EP3]	PHOSPHORUS TRICHLORIDE
7783-55-3 PHOSPHORUS TRIFLUORIDE	PHOSPHORUS TRIFLUORIDE
1314-24-5 PHOSPHORUS TRIOXIDE	PHOSPHORUS TRIOXIDE
12165-69-4 PHOSPHORUS TRISULFIDE	PHOSPHORUS TRISULFIDE
7723-14-0 PHOSPHORUS, WHITE OR YELLOW [HM1, HM1, HM1, HM1]	PHOSPHORUS
7723-14-0 PHOSPHORUS (YELLOW) [AUS, BC1, BC1]	PHOSPHORUS
7723-14-0 PHOSPHORUS, YELLOW [CEX]	PHOSPHORUS
7723-14-0 PHOSPHORUS (YELLOW) [FLA]	PHOSPHORUS
7723-14-0 PHOSPHORUS, YELLOW [GBR, GBR, MAK, MAK, MAK]	PHOSPHORUS
7723-14-0 PHOSPHORUS (YELLOW) [MSL, NIO, NIO, NJL, NJL, OS1, OS1, TLV]	PHOSPHORUS
7723-14-0 PHOSPHORUS (YELLOW OR WHITE) [313, CAN]	PHOSPHORUS

Chemical Name	Reference Name
RR-00941-6 PHOSPHORYLATED CAPROLACTONE, ALKY-LOXOHETEROMONO-CYCLE AND POLYALKY-LENE POLYOL ALKYL ETHER	PHOSPHORYLATED CAPROLACTONE, ALKYLOXO-HETEROMONO-CYCLE AND POR
RR-00940-5 PHOSPHORYLATED OXOHETEROMONOCYCLE POLYOXYETHYLENE ALKYL ETHER	PHOSPHORYLATED OXOHETEROMONOCYCLE POLY-OXYETHYLENE ALKYL ETHEIXED TERPENES
10025-87-3 PHOSPHORYL CHLORIDE [PAL]	PHOSPHORUS OXYCHLORIDE
13478-20-1 PHOSPHORYL FLUORIDE	PHOSPHORYL FLUORIDE
10025-87-3 PHOSPHORYL TRICHLORIDE [GBR, GBR]	PHOSPHORUS OXYCHLORIDE
RR-01637-5 PHOSPHOTUNGSTIC ACIDS	PHOSPHOTUNGSTIC ACIDS
13366-73-9 PHOTODIELDRIN	PHOTODIELDRIN
88-96-0 PHTHALAMIDE	PHTHALAMIDE
RR-00433-1 PHTHALATE ESTERS	PHTHALATE ESTERS
88-99-3 PHTHALIC ACID	PHTHALIC ACID
100-21-0 PHTHALIC ACID [UTS]	TEREPHTHALIC ACID
68515-44-6 PHTHALIC ACID, DIALKYL (C7) ESTER	PHTHALIC ACID, DIALKYL (C7) ESTERALKYL ESTERS [DI(HEPTYL-, NONYL-, UNDECYL) PHTHALATE (MIXED ISOMERS)]
68515-42-4 PHTHALIC ACID, DIALKYL (C7-C11) ESTER	PHTHALIC ACID, DIALKYL (C7-C11) ESTER
68515-45-7 PHTHALIC ACID, DIALKYL (C9) ESTER	PHTHALIC ACID, DIALKYL (C9) ESTER
64382-04-3 PHTHALIC ACID, POLYMER WITH 1,3-DI-HYDRO-1,3-DIOXO-5-ISOBENZOFURANCAR-BOXYLIC ACID AND TRIMETHYLOLPROPANE, LINOLEATE	PHTHALIC ACID, POLYMER WITH 1,3-DIHYDRO-1,3-DIOXO-5-ISOBENZO
85-44-9 PHTHALIC ANHYDRIDE	PHTHALIC ANHYDRIDE
3006-93-7 PHTHALIC DIMALEIMIDE	PHTHALIC DIMALEIMIDE
85-41-6 PHTHALIMIDE	PHTHALIMIDE
RR-00434-2 PHTHALIMIDE DERIVATIVE PESTICIDES, N.O.S.	PHTHALIMIDE DERIVATIVE PESTICIDES, N.O.S.
147-14-8 PHTHALOCYANINE BLUE	PHTHALOCYANINE BLUE
91-15-6 O-PHTHALODINITRILE	O-PHTHALODINITRILE
626-17-5 M-PHTHALODINITRILE	M-PHTHALODINITRILE
84-80-0 PHYLLOQUINONE	PHYLLOQUINONE
84-80-0 PHYLLOQUINONE (VITAMIN K1) [MSL]	PHYLLOQUINONE
57-47-6 PHYSOSTIGMINE	PHYSOSTIGMINE
57-64-7 PHYSOSTIGMINE,SALICYLATE [MSL]	PHYSOSTIGMINE SALICYLATE (1:1)
57-64-7 PHYSOSTIGMINE SALICYLATE (1:1)	PHYSOSTIGMINE SALICYLATE (1:1)
57-64-7 PHYSOSTIGMINE, SALICYLATE (1:1) [302, PAL]	PHYSOSTIGMINE SALICYLATE (1:1)
RR-01705-0 PICKLED VEGETABLES, TRADITIONAL ASIAN	PICKLED VEGETABLES, TRADITIONAL ASIAN
1918-02-1 PICLORAM	PICLORAM
1918-02-1 PICLORAM (TORDON) [BC1, BC1]	PICLORAM
1918-02-1 PICLORAM (ISO) [GBR, GBR]	PICLORAM
1918-02-1 PICLORAM (TORDON) [ALB, ALB]	PICLORAM
1918-02-1 PICLORAME (TORDON (R)) [QBC]	PICLORAM
108-89-4 P-PICOLINE	P-PICOLINE
108-89-4 4-PICOLINE [FLA, MSL, NFP, NFP, WEL, WEL, WEL]	P-PICOLINE
108-99-6 3-PICOLINE [WEL, WEL, WEL]	3-METHYLPYRIDINE
109-06-8 2-PICOLINE	2-PICOLINE
109-06-8 O-PICOLINE [WHM]	2-PICOLINE
1333-41-1 PICOLINE [NJL]	PICOLINES
104-90-5 2-PICOLINE, 5-ETHYL- [OTS]	2-METHYL-5-ETHYLPYRIDINE

Chemical Name	Reference Name
1333-41-1 PICOLINES	PICOLINES
98-98-6 PICOLINIC ACID	PICOLINIC ACID
88-89-1 PICRIC ACID	PICRIC ACID
3324-58-1 PICRIC ACID, SODIUM SALT	PICRIC ACID, SODIUM SALT
RR-01520-3 PICRITE	PICRITE
124-87-8 PICROTOXIN	PICROTOXIN
3564-21-4 PIGMENT RED 28 (PERMANENT RED)	PIGMENT RED 28 (PERMANENT RED)
5858-82-2 PIGMENT RED 48, CI NO. 15865	PIGMENT RED 48, CI NO. 15865
5979-28-2 PIGMENT YELLOW 16	PIGMENT YELLOW 16
6358-31-2 PIGMENT YELLOW 74	PIGMENT YELLOW 74
12225-18-2 PIGMENT YELLOW 97, CI 11767	PIGMENT YELLOW 97, CI 11767
7681-93-8 PIMARICIN	PIMARICIN
76-09-5 PINACOL	PINACOL
473-55-2 PINANE	PINANE
28324-52-9 PINANE HYDROPEROXIDE	PINANE HYDROPEROXIDE
28324-52-9 PINANYL HYDROPEROXIDE [DOT]	PINANE HYDROPEROXIDE
83-26-1 PINDONE	PINDONE
83-26-1 PINDONE [QBC]	PINDONE
83-26-1 PINDONE (2-PIVALYL-1,3-INDANDIONE) [OS1, OS1]	PINDONE
80-56-8 ALPHA-PINENE	ALPHA-PINENE
127-91-3 BETA-PINENE	BETA-PINENE
1330-16-1 PINENE	PINENE
25719-60-2 BETA-PINENE POLYMER	BETA-PINENE POLYMER
8002-09-3 PINE OIL	PINE OIL
RR-00879-7 PINE PITCH	PINE PITCH
8011-48-1 PINE TAR	PINE TAR
RR-00777-2 PINE TAR OIL [NFP, NFP, PAL]	TAR-OIL, WOOD
110-85-0 PIPERAZINE	PIPERAZINE
142-64-3 PIPERAZINE DIHYDRO-CHLORIDE [OS1]	PIPERAZINE DIHYDROCHLORIDE
142-64-3 PIPERAZINE, DIHYDROCHLORIDE [PAL]	PIPERAZINE DIHYDROCHLORIDE
142-64-3 PIPERAZINE DIHYDROCHLORIDE	PIPERAZINE DIHYDROCHLORIDE
7280-37-7 PIPERAZINE ESTRONE SULFATE	PIPERAZINE ESTRONE SULFATE
140-31-8 1-PIPERAZINEETHANAMINE [PAL]	1-(2-AMINOETHYL) PIPERAZINE
103-76-4 1-PIPERAZINEETHANOL	1-PIPERAZINEETHANOL
142-64-3 PIPERAZINE HYDROCHLORIDE [WHM]	PIPERAZINE DIHYDROCHLORIDE
109-01-3 PIPERAZINE, 1-METHYL- [PAL]	1-METHYL PIPERAZINE
130277-45-1 PIPERAZINONE, 1,1',1"-[1,3,5-TRIAZINE-2, 4,6-TRIYLTRIS[[CYCLOHEXYLIMINO]-2,1-ETHANEDIYL]TRIS-[3,3,4,5,5-PENTAMETHYL]-,	PIPERAZINONE, 1,1',1"-[1,3,5-TRIAZINE-2,4,6-TRIYLTRIS [[CYCLO
110-89-4 PIPERIDINE	PIPERIDINE
2591-86-8 1-PIPERIDINECARBOXALDEHYDE	1-PIPERIDINECARBOXALDEHYDEZO]PHENYL]AZO][1,1'-BIPHENYL]-4-YL]AZO]-2-HYDROXY-, DISODIUM SALT
66-81-9 2,6-PIPERIDINEDIONE, 4-[2-(3,5-DIMETHYL-2-OXOCYCLOHEXYL)-2-HYDROXYETHYL]-, [1S-[1.ALPHA.(S*),3.ALPHA.,5.BETA.]- [PAL]	CYCLOHEXIMIDE
104-89-2 PIPERIDINE, 5-ETHYL-2-METHYL- [PAL]	2-METHYL-5-ETHYLPIPERIDINE
100-75-4 PIPERIDINE, 1-NITROSO- [PAL, PAL]	N-NITROSOPIPERIDINE
RR-01780-1 PIPERIDINE SALTS	PIPERIDINE SALTSICAL ISOMERS
120-57-0 PIPERONAL [FDA]	1,3-BENZODIOXOLE-5-CARBOXALDEHYDE

Chemical Name	Reference Name
51-03-6 PIPERONYL BUTOXIDE	PIPERONYL BUTOXIDE
120-62-7 PIPERONYL SULFOXIDE	PIPERONYL SULFOXIDE
54-91-1 PIPOBROMANE	PIPOBROMANE
5281-13-0 PIPROTAL	PIPROTAL
23505-41-1 PIRIMFOS-ETHYL [MSL]	PIRIMPHOS-ETHYL
23103-98-2 PIRIMICARB	PIRIMICARB
RR-01309-2 PIRIMIPHOS-ETHYL	PIRIMIPHOS-ETHYL
29232-93-7 PIRIMIPHOS METHYL [O-(2-(DIETHYLAMINO)-6-METHYL-4-PYRIMIDINYL)-O,O-DIMETHYLPHOSPHOROTHIOATE]	PIRIMIPHOS METHYL [O-(2-(DIETHYLAMINO)-6-METHYL-4-PYRIMIDINY]
23505-41-1 PIRIMPHOS-ETHYL	PIRIMPHOS-ETHYL
83-26-1 PIVAL [MSL, NJL]	PINDONE
83-26-1 PIVAL (2-PIVALYL-1,3-INDANDIONE) (PINDONE) [ALB, ALB]	PINDONE
1955-45-9 PIVALOLACTONE	PIVALOLACTONE
3282-30-2 PIVALOYL CHLORIDE	PIVALOYL CHLORIDE
83-26-1 2-PIVALOYL-1,3-INDANDIONE [FLA]	PINDONE
83-26-1 PIVAL (2-PIVALOYL-1,3-INDANDIONE) [BC1, BC1]	PINDONE
83-26-1 2-PIVALYL-1,3-INDANDIONE (PINDONE) [CAL]	PINDONE
26499-65-0 PLASTER OF PARIS	PLASTER OF PARIS
10025-65-7 PLATINIUM(II) CHLORIDE [WHM]	PLATINOUS CHLORIDE
7440-06-4 PLATINIUM, ELEMENTAL [WHM]	PLATINUM
53231-79-1 PLATINIUM(II) SULFATE	PLATINIUM(II) SULFATE
10025-65-7 PLATINOUS CHLORIDE	PLATINOUS CHLORIDE
15663-27-1 CIS-PLATINOUS DIAMMINE DICHLORIDE [FLA]	CISPLATIN
7440-06-4 PLATINUM	PLATINUM
RR-00046-4 PLATINUM (SOLUBLE SALTS) [QBC]	PLATINUM SOLUBLE SALTS
14993-39-6 PLATINUM 186	PLATINUM 186
14922-70-4 PLATINUM 188	PLATINUM 188
15055-30-8 PLATINUM 189	PLATINUM 189
15706-36-2 PLATINUM 191	PLATINUM 191
15735-70-3 PLATINUM 193	PLATINUM 193
RR-00394-1 PLATINUM 193M	PLATINUM 193M
RR-00393-0 PLATINUM 195M	PLATINUM 195M
15735-74-7 PLATINUM 197	PLATINUM 197
RR-00392-9 PLATINUM 197M	PLATINUM 197M
15706-54-4 PLATINUM 199	PLATINUM 199
29687-31-8 PLATINUM 200	PLATINUM 200
7440-06-4 PLATINUM METAL [MEX]	PLATINUM
7440-06-4 PLATINUM COMPOUNDS [MAK, MAK]	PLATINUM
15663-27-1 PLATINUM, DIAMMINEDICHLORO-, (SP-4-2)-[PAL, PAL]	CISPLATIN
7440-06-4 PLATINUM, METAL [CAL]	PLATINUM
7440-06-4 PLATINUM METAL [GBR]	PLATINUM
RR-00046-4 PLATINUM SALTS, SOLUBLE [GBR, GBR]	PLATINUM SOLUBLE SALTS
RR-00046-4 PLATINUM SOLUBLE SALTS	PLATINUM SOLUBLE SALTS
RR-00046-4 PLATINUM, SOLUBLE SALTS [BC1]	PLATINUM SOLUBLE SALTS
RR-00046-4 PLATINUM (SOLUBLE SALTS) [NIO, NIO]	PLATINUM SOLUBLE SALTS
13454-96-1 PLATINUM(IV) TETRACHLORIDE [WHM]	PLATINUM TETRACHLORIDE

Chemical Name	Reference Name
13454-96-1 PLATINUM TETRACHLORIDE	PLATINUM TETRACHLORIDE
RR-00046-4 PLATINUM, WATER-SOLUBLE COMPOUNDS [ONT]	PLATINUM SOLUBLE SALTS
RR-00046-4 PLATINUM, WATER-SOLUBLE SALTS, N.O.S. [WHM]	PLATINUM SOLUBLE SALTS
18378-89-7 PLICAMYCIN	PLICAMYCIN
78-00-2 PLUMBANE, TETRAETHYL- [PAL]	TETRAETHYL LEAD
75-74-1 PLUMBANE, TETRAMETHYL- [PAL]	TETRAMETHYL LEAD
7440-07-5 PLUTONIUM	PLUTONIUM
34018-47-8 PLUTONIUM 234	PLUTONIUM 234
14928-39-3 PLUTONIUM 235	PLUTONIUM 235
15411-92-4 PLUTONIUM 236	PLUTONIUM 236
15411-93-5 PLUTONIUM 237	PLUTONIUM 237
13981-16-3 PLUTONIUM 238	PLUTONIUM 238
15117-48-3 PLUTONIUM 239	PLUTONIUM 239
14119-33-6 PLUTONIUM 240	PLUTONIUM 240
14119-32-5 PLUTONIUM 241	PLUTONIUM 241
13982-10-0 PLUTONIUM 242	PLUTONIUM 242
15706-37-3 PLUTONIUM 243	PLUTONIUM 243
14119-34-7 PLUTONIUM 244	PLUTONIUM 244
18784-52-6 PLUTONIUM 245	PLUTONIUM 245
RR-01113-2 POE SORBITAN DITALLATE	POE SORBITAN DITALLATE
RR-00440-0 POISONOUS LIQUIDS, N.O.S.	POISONOUS LIQUIDS, N.O.S.
RR-00445-5 POISONOUS SOLIDS, N.O.S.	POISONOUS SOLIDS, N.O.S.
RR-01504-3 POLISH	POLISH
16729-74-1 POLONIUM 203	POLONIUM 203
16729-76-3 POLONIUM 205	POLONIUM 205
15720-45-3 POLONIUM 207	POLONIUM 207
13981-52-7 POLONIUM 210	POLONIUM 210
9003-05-8 POLYACRYLAMIDE [PST]	POLYACRYLAMIDES
9003-05-8 POLYACRYLAMIDES	POLYACRYLAMIDES
9003-01-4 POLYACRYLIC ACID [IAR]	ACRYLIC RESIN
25014-41-9 POLYACRYLONITRILE [PST]	ACRYLONITRILE POLYMER
68908-35-0 POLYACRYLONITRILE, OXIDIZED	POLYACRYLONITRILE, OXIDIZEDSULFATE
RR-00981-4 POLYALKYLENE GLYCOL ALKYL ETHER ACRYLATE	POLYALKYLENE GLYCOL ALKYL ETHER ACRYLATE
RR-01261-3 POLYALKYLENE GLYCOL SUBSTITUTED ACETATE	POLYALKYLENE GLYCOL SUBSTITUTED ACETATE
RR-01559-8 POLYALKYLENE POLYAMINE	POLYALKYLENE POLYAMINE
RR-00916-5 POLYALKYLENEPOLYOL ALKYLAMINE	POLYALKYLENEPOLYOL ALKYLAMINEAND 2,2-[(1-METHYLETHYLIDENE)BIS(4,1-PHENYLENEOXYMETHYLENE)]BISOXIRANE, GRAFT
RR-00226-6 POLYALKYLPOLYSILAZANE, BIS(SUBSTITUTED ACRYLATE)	POLYALKYLPOLYSILAZANE, BIS(SUBSTITUTED ACRYLATE)ESTER
25038-54-4 POLYAMIDE RESINS [PST]	NYLON 6HACRYLATE
RR-01574-7 POLYAMINE DITHIOCARBAMATE	POLYAMINE DITHIOCARBAMATE
RR-01565-6 POLYAMINOPOLYACID	POLYAMINOPOLYACID
RR-00140-1 POLYAMYL NAPTHALENE MIXTURE OF POLYMERS	POLYAMYL NAPTHALENE MIXTURE OF POLYMERS
67774-32-7 POLYBROMINATED BIPHENYL MIXTURE (FIREMASTER FF-1) [CSR, CSR]	FIREMASTER FF-1

Chemical Name	Reference Name
67774-32-7 POLYBROMINATED BIPHENYLS [CAL, NJL, IAR, MIN, NTP, FLA, NJL, ATS, NJL]	FIREMASTER FF-1
RR-00086-2 POLYBROMINATED BIPHENYLS (PBB) (GENERIC)	POLYBROMINATED BIPHENYLS (PBB)(GENERIC)
RR-00086-2 POLYBROMINATED BIPHENYLS [C65, C65, C65, CN5, CN8]	POLYBROMINATED BIPHENYLS (PBB)(GENERIC)SPHONOTHIOATE
59536-65-1 POLYBROMINATED BIPHENYLS (PBB'S) [MSL]	FIREMASTER BP-6
RR-00086-2 POLYBROMINATED BIPHENYLS (PBBS) [313]	POLYBROMINATED BIPHENYLS (PBB)(GENERIC)
RR-00086-2 POLYBROMINATED BIPHENYLS (PBB) [PAL]	POLYBROMINATED BIPHENYLS (PBB)(GENERIC)
67774-32-7 POLYBROMINATED BIPHENYLS (2,4,5,2',4',5'-HEXABROMO BIPHENYL) [WHM]	FIREMASTER FF-1
9003-17-2 POLYBUTADIENE RESIN	POLYBUTADIENE RESIN
9003-29-6 POLYBUTENE	POLYBUTENE
9003-49-0 POLYBUTYL ACRYLATE	POLYBUTYL ACRYLATE
9003-28-5 POLYBUTYLENE	POLYBUTYLENE
RR-01771-0 POLYCHLORINATED ALKANES	POLYCHLORINATED ALKANES
1336-36-3 POLYCHLORINATED BIPHENYLS	POLYCHLORINATED BIPHENYLS
1336-36-3 POLYCHLORINATED BIPHENYLS (PCBS) [313]	POLYCHLORINATED BIPHENYLS
1336-36-3 POLYCHLORINATED BIPHENYLS (AROCLORS) [CAA]	POLYCHLORINATED BIPHENYLS
1336-36-3 POLYCHLORINATED BIPHENYLS (PCBS) [CW3, EP2, EPA]	POLYCHLORINATED BIPHENYLS
1336-36-3 POLYCHLORINATED BIPHENYLS (PCB'S) [MSL]	POLYCHLORINATED BIPHENYLS
1336-36-3 POLYCHLORINATED BIPHENYLS (PCBS) [ONT, ONT, ONT, ONT]	POLYCHLORINATED BIPHENYLS
1336-36-3 POLYCHLORINATED BIPHENYLS, N.O.S. [RCU]	POLYCHLORINATED BIPHENYLS
1336-36-3 POLYCHLORINATED BIPHENYLS (PCB) [WHM]	POLYCHLORINATED BIPHENYLS
11096-82-5 POLYCHLORINATED BIPHENYLS (CONTAINING 60% OR MORE CHLORINE) [C65]	AROCLOR 1260
RR-00296-0 POLYCHLORINATED DIBENZO-P-DIOXINS	POLYCHLORINATED DIBENZO-P-DIOXINS
RR-00296-0 POLYCHLORINATED DIBENZO-PARA-DIOXINS [CN5]	POLYCHLORINATED DIBENZO-P-DIOXINS
RR-00295-9 POLYCHLORINATED DIBENZOFURANS	POLYCHLORINATED DIBENZOFURANS
11126-42-4 POLYCHLORINATED O-TERPHENYL	POLYCHLORINATED O-TERPHENYL
RR-01611-5 POLYCHLORINATED TERPHENYLS	POLYCHLORINATED TERPHENYLS
12642-23-8 POLYCHLORINATED TRIPHENYLS	POLYCHLORINATED TRIPHENYLS
12642-23-8 POLYCHLORINATED TRIPHENYLS (AROCLOR 5442) [NJL]	POLYCHLORINATED TRIPHENYLS
1336-36-3 POLYCHLOROBIPHENYLS [CAL]	POLYCHLORINATED BIPHENYLS
9010-98-4 POLYCHLOROPRENE [IAR]	NEOPRENE RUBBER
RR-01523-6 POLYCYCLIC AROMATIC HYDROCARBONS (BENZENE-SOLUBLE FRACTION) [MEX, MEX]	POLYCYCLIC AROMATIC HYDROCARBONS
RR-01740-3 POLYCYCLIC AROMATIC HYDROCARBON DERIVATIVES	POLYCYCLIC AROMATIC HYDROCARBON DERIVATIVES
RR-01523-6 POLYCYCLIC AROMATIC HYDROCARBONS	POLYCYCLIC AROMATIC HYDROCARBONS
RR-00079-3 POLYCYCLIC ORGANIC MATTER	POLYCYCLIC ORGANIC MATTER
68083-14-7 POLYDIMETHYLDIPHENYL SILOXANE COPOLYMER	POLYDIMETHYLDIPHENYL SILOXANE COPOLYMER
9016-00-6 POLYDIMETHYLSILOXANE [TSC, TSC]	POLY[OXY(DIMETHYLSILYLENE)]
63148-62-9 POLY(DIMETHYLSILOXANE)	POLY(DIMETHYLSILOXANE)
RR-01723-2 POLYEPOXYPOLYOL	POLYEPOXYPOLYOL
RR-01692-2 POLYESTER POLYURETHANE ACRYLATE	POLYESTER POLYURETHANE ACRYLATE

Chemical Name	Reference Name
25667-42-9 POLYETHER-SULPHONE	POLYETHER-SULPHONE
37271-20-8 POLYETHOXYAMINE HK	POLYETHOXYAMINE HK
69013-18-9 POLYETHOXYLATED POLYPROPOXYLATED C8-C18 ALCOHOLS [PST]	ALCOHOLS, C8-18, ETHOXYLATED PROPOXYLATED
68551-13-3 POLYETHOXYLATED POLYPROPOXYLATED C12-15 ALCOHOLS	POLYETHOXYLATED POLYPROPOXYLATED C12-15 ALCOHOLS
68755-33-9 POLYETHOXYLATED PRIMARY AMINE (C14-18)	POLYETHOXYLATED PRIMARY AMINE (C14-18) METHYL)-4,5-DIHYDRO-2-NORCOCO ALKYL, HYDROX-IDES, SODIUM SALTS
11096-42-7 POLYETHOXY POLYPROPOXY OLYETHOXY ETHANOL-IODINE COMPLEX	POLYETHOXY POLYPROPOXY OLYETHOXY ETHANOL-IODINE COMPLEX
9002-88-4 POLYETHYLENE	POLYETHYLENE
25322-68-3 POLYETHYLENE GLYCOL	POLYETHYLENE GLYCOL
68412-54-4 POLYETHYLENE GLYCOL BRANCHED NONYLPHENYL ETHER	POLYETHYLENE GLYCOL BRANCHED NONYLPHENYL ETHERSULFATE
9004-74-4 POLYETHYLENE GLYCOL METHYL ETHER	POLYETHYLENE GLYCOL METHYL ETHER
9004-77-7 POLYETHYLENE GLYCOL, MONOBUTYL ETHER	POLYETHYLENE GLYCOL, MONOBUTYL ETHER
9004-81-3 POLYETHYLENE GLYCOL MONOLAURATE	POLYETHYLENE GLYCOL MONOLAURATE
32612-48-9 POLYETHYLENE GLYCOL MONOLAURYL ETHER SULFATE AMMONIUM SALT	POLYETHYLENE GLYCOL MONOLAURYL ETHER SUL-FATE AMMONIUM SALT
9004-96-0 POLYETHYLENE GLYCOL MONOOLEATE	POLYETHYLENE GLYCOL MONOOLEATE
9004-98-2 POLYETHYLENE GLYCOL MONOOLEYL ETHER	POLYETHYLENE GLYCOL MONOOLEYL ETHER
26027-37-2 POLYETHYLENE GLYCOL MONOETHER WITH OLEIC ACID MONO-ETHANOLAMIDE	POLYETHYLENE GLYCOL MONOETHER WITH OLEIC ACID MONO-ETHANOLAM
9014-90-8 POLYETHYLENE GLYCOL NONYLPHENYL ETHER SODIUM SULFATE	POLYETHYLENE GLYCOL NONYLPHENYL ETHER SODIUM SULFATE
9002-93-1 POLYETHYLENE GLYCOL OCTYLPHENOL ETHER	POLYETHYLENE GLYCOL OCTYLPHENOL ETHER
9036-19-5 POLYETHYLENE GLYCOL OCTYLPHENYL ETHER [WHM]	ETHOXYLATED OCTYLPHENOL
55069-68-6 POLYETHYLENE GLYCOL, OLEIC ACID DI-ESTER	POLYETHYLENE GLYCOL, OLEIC ACID DIESTER
25322-68-3 POLYETHYLENE GLYCOLS [MIN, NFP, NFP]	POLYETHYLENE GLYCOL
9004-99-3 POLYETHYLENE GLYCOL STEARATE	POLYETHYLENE GLYCOL STEARATE
9002-98-6 POLYETHYLENEIMINE	POLYETHYLENEIMINE
9003-11-6 POLYETHYLENE-POLYPROPYLENE GLYCOL	POLYETHYLENE-POLYPROPYLENE GLYCOL
9038-95-3 POLYETHYLENE-POLYPROPYLENE GLYCOLS MONOBUTYL ETHER [PST]	ETHYLENE OXIDE-PROPYLENE OXIDE COPOLYMER MONOBUTYL ETHER
24938-04-3 POLYETHYLENE TEREPHTHALATE - POLYETHYLENE ISOPHTHALATE FILM	POLYETHYLENE TEREPHTHALATE - POLYETHYLENE ISOPHTHALATE FILM
RR-00976-7 POLYFLUOROSULFONIC ACID SALT	POLYFLUOROSULFONIC ACID SALTCOMPLEX
53973-98-1 POLYGEENAN	POLYGEENAN
25618-55-7 POLYGLYCERINE	POLYGLYCERINE
9007-48-1 POLYGLYCEROL ESTER OF OLEIC ACID	POLYGLYCEROL ESTER OF OLEIC ACID
66070-87-9 POLYGLYCERYL PHTHALATE ESTER OF CO-CONUT OIL FATTY ACID	POLYGLYCERYL PHTHALATE ESTER OF COCONUT OIL FATTY ACID
25722-70-7 POLYGLYCIDOL	POLYGLYCIDOL
RR-01310-5 POLYHALOGENATED BIPHENYLS	POLYHALOGENATED BIPHENYLS
RR-01008-2 POLYMER	POLYMER
9016-87-9 POLYMERIC DIPHENYLMETHANE DIISO-CYANATE [313]	POLYMETHYLENE POLYPHENYLENE ISOCYANATE
9016-87-9 POLYMERIC MDI [MAK]	POLYMETHYLENE POLYPHENYLENE ISOCYANATE
9003-49-0 POLYMERIZED BUTYL ACRYLATE [PST]	POLYBUTYL ACRYLATE

Chemical Name	Reference Name
RR-01117-6 POLYMERIZED FATTY ACID PHOSPHATE (100% C12-C18)	POLYMERIZED FATTY ACID PHOSPHATE (100% C12-C18)C ACID AND AMINOPROPYL METHACRYLATE
RR-01017-3 POLYMER OF ADIPIC ACID, ALKANEPOLYOL, ALKYLDIISOCYANATOCARBOMONOCYCLE, HYDROXYALKYL ACRYLATE ESTER	POLYMER OF ADIPIC ACID, ALKANEPOLYOL, ALKYLDI-ISOCYANATOCARBO
RR-01254-4 POLYMER OF ALKANEDIOIC ACID, METHYLENEBISCARBOMONOCYCLIC DIISO-CYANATE, AND ALKYLENE GLYCOLS, HY-DROXYALKYL ACRYLATE ESTER	POLYMER OF ALKANEDIOIC ACID, METHYLENEBIS-CARBOMONOCYCLIC DII
RR-00188-7 POLYMER OF ALKANEPOLYOL AND POLYALKYLPOLYISOCYANATOCARBOMONO-CYCLE, ACETONE OXIME-BLOCKED	POLYMER OF ALKANEPOLYOL AND POLYALKYLPOLY-ISOCYANATOCARBOMONO
RR-00189-8 POLYMER OF ALKENOIC ACID, SUBSTITUTED ALKYLACRYLATE, SODIUMSALT	POLYMER OF ALKENOIC ACID, SUBSTITUTED ALKY-LACRYLATE, SODIUMCYCLE, ACETONE OXIME-BLOCKED
RR-00227-7 POLYMER OF ALKYL CARBOMONOCYCLE DI-ISOCYANATE WITH ALKANEPOLYOL POLY-ACRYLATES	POLYMER OF ALKYL CARBOMONOCYCLE DIISO-CYANATE WITH ALKANE
RR-00970-1 POLYMER OF BISPHENOL A DIGLYCIDYL ETHER, SUBSTITUTED ALKENES, AND BUTA-DIENE	POLYMER OF BISPHENOL A DIGLYCIDYL ETHER, SUB-STITUTED ALKENES
RR-01116-5 POLYMER OF N-BUTYL ACRYLATE, METHYL METHACRYLATE, METHACRYLIC ACID AND AMINOPROPYL METHACRYLATE	POLYMER OF N-BUTYL ACRYLATE, METHYL METHACRYLATE, METHACRYLI
RR-01257-7 POLYMER OF DISODIUM MALEATE, ALLYL ETHER, AND ETHYLENE OXIDE	POLYMER OF DISODIUM MALEATE, ALLYL ETHER, AND ETHYLENE OXIDE
RR-01179-0 POLYMER OF DISUBSTITUTED PHTHA-LATE, DIOXOHETEROPOLYCYCLE, AND METHACRYLIC ACID	POLYMER OF DISUBSTITUTED PHTHALATE, DIOXO-HETEROPOLYCYCLE, AN
RR-00999-4 POLYMER OF HYDROXYETHYL ACRYLATE AND POLYISOCYANATE	POLYMER OF HYDROXYETHYL ACRYLATE AND POLY-ISOCYANATE
RR-00962-1 POLYMER OF ISOPHORONE DIISO-CYANATE, TRIMETHYLOLPROPANE, POLYALKYLENEPOLYOL, DISUBSTITUTED ALKANES AND HYDROXYETHYL ACRYLATE	POLYMER OF ISOPHORONE DIISOCYANATE, TRIMETHY-LOLPROPANE, POLYE, N-[3-(DIMETHYLAMINO)PROPYL]
RR-00949-4 POLYMER OF POLYETHYLENEPOLYAMINE AND ALKANEDIOL DIGLYCIDYL ETHER	POLYMER OF POLYETHYLENEPOLYAMINE AND ALKA-NEDIOL DIGLYCIDYL E
RR-01014-0 POLYMER OF STYRENE, SUBSTITUTED ALKYL METHACRYLATES, 2-ETHYLHEXYL ACRY-LATE, METHACRYLIC ACID AND SUBSTI-TUTED BISBENZENE	POLYMER OF STYRENE, SUBSTITUTED ALKYL METHACRYLATES, 2-ETHYLTYLPHENYL METHACRY-LATE
RR-00911-0 POLYMER OF SUBSTITUTED ALKYLPHENOL FORMALDEHYDE AND PHTHALICANHYDRIDE, ACRYLATE	POLYMER OF SUBSTITUTED ALKYLPHENOL FORMALDEHYDE AND PHTHALIC
RR-01175-6 POLYMER OF SUBSTITUTED ARYL OLEFIN	POLYMER OF SUBSTITUTED ARYL OLEFIN
RR-00987-0 POLYMER OF SUBSTITUTED PHENOL, FORMALDEHYDE, EPICHLOROHYDRIN, AND DISUBSTITUTED BENZENE	POLYMER OF SUBSTITUTED PHENOL, FORMALDE-HYDE, EPICHLOROHYDRINALKYL ACRYLATE, AND ISOPHORONE DIISOCYANATE
30938-41-1 POLYMER OF VINYL ACETATE, N-BUTYL ACRYLATE, VINYL CHLORIDE,AND ACRYLIC ACID	POLYMER OF VINYL ACETATE, N-BUTYL ACRYLATE, VINYL CHLORIDE,
RR-01240-8 POLYMETHYLCARBOMONOCYCLE, REACTION PRODUCT WITH 2-HYDROXYETHYL ACRY-LATE	POLYMETHYLCARBOMONOCYCLE, REACTION PROD-UCT WITH 2-HYDROXYETH
9016-87-9 POLYMETHYLENE POLYPHENYLENE ISO-CYANATE	POLYMETHYLENE POLYPHENYLENE ISOCYANATE
9016-87-9 POLYMETHYLENEPOLYPHENYLENE ISO-CYANATE [PST]	POLYMETHYLENE POLYPHENYLENE ISOCYANATE
9016-87-9 POLYMETHYLENE POLYPHENYL ISO-CYANATE [IAR]	POLYMETHYLENE POLYPHENYLENE ISOCYANATE
9011-14-7 POLYMETHYL METHACRYLATE [IAR, PST]	METHYL METHACRYLATE POLYMER

458

Chemical Name	Reference Name
RR-01190-5 POLYMETHYLOCTADECYLSILOXANE	POLYMETHYLOCTADECYLSILOXANE
RR-01810-0 POLYNUCLEAR AROMATIC HYDROCARBIDES	POLYNUCLEAR AROMATIC HYDROCARBIDES
RR-00450-2 POLYNUCLEAR AROMATIC HYDROCARBONS	POLYNUCLEAR AROMATIC HYDROCARBONS
RR-00245-9 POLYOL CARBOXYLATE ESTER	POLYOL CARBOXYLATE ESTER
RR-01806-4 POLYOLEFINES FIBRES	POLYOLEFINES FIBRES
RR-01115-4 POLY-OXO ALUMINUM STEARATE	POLY-OXO ALUMINUM STEARATEONIO)ETHYL-OMEGA-HYDROXY-, CHLORIDE
RR-01118-7 POLYOXYALKYLENE GLYCOL	POLYOXYALKYLENE GLYCOL
RR-01180-3 POLYOXY ALKYLENE GLYCOL AMINE	POLYOXY ALKYLENE GLYCOL AMINED METHACRYLIC ACID
RR-01119-8 POLYOXYALKYLENE SILOXANE	POLYOXYALKYLENE SILOXANE
RR-01677-3 POLYOXYALKYLENE SUBSTITUTED ARO-MATIC AZO COLORANT	POLYOXYALKYLENE SUBSTITUTED AROMATIC AZO COLORANT
52277-33-5 POLY(OXY-1,4-BUTANEDIYL), ALPHA-(1-OXO-2-(HYDROXYETHYL)TALLOWAMINE OXIDE, PHOSPHATE [TSC, TSC]	POLY(OXY-1,4-BUTANEDIYL), ALPHA-(1-OXO-2-PROPENYL)-X-[(1-
52277-33-5 POLY(OXY-1,4-BUTANEDIYL), ALPHA-(1-OXO-2-PROPENYL)-X-[(1-OXO-2-PROPENYL)OXY]-	POLY(OXY-1,4-BUTANEDIYL), ALPHA-(1-OXO-2-PROPENYL)-X-[(1-
9016-00-6 POLY[OXY(DIMETHYLSILYLENE)]	POLY[OXY(DIMETHYLSILYLENE)]
104810-48-2 POLY(OXY-1,2-ETHANEDIYL), ALPHA-(3-(3-(2H-BENZOTRIAZOL-2-YL)-5-(1,1-DIMETHYLETHYL)-4-HYDROXYPHENYL)-1-OXOPROPYL)-OMEGA-HYDROXY-	POLY(OXY-1,2-ETHANEDIYL), ALPHA-(3-(3-(2H-BENZO-TRIAZOL-2-YL)-(TRICHLOROMETHYL)-, ETHYL ESTER
68410-69-5 POLY(OXY-1,2-ETHANEDIYL), ALPHA-[2-[BIS(2-AMINOETHYL)METHYLAMMONIO]ETHYL]-OMEGA-HYDROXY-, N,N'-DITALLOW ACYL DERIVATIVES, METHYL SULFATES (SALTS)	POLY(OXY-1,2-ETHANEDIYL), ALPHA-[2-[BIS(2-AMINOETHYL)METHYLAMMONIO]ETHYL-OMEGA]-HY-DROXY-, N,N'-BIS(HYDROGENATED TALLOW ACYL) DERIVATIVES, METHYL SULFATES (SALTS)
68554-06-3 POLY(OXY-1,2-ETHANEDIYL), ALPHA-[3-[BIS(2-AMINOETHYL)METHYLAMMONIO]-2-HYDROX-YPROPYL]-OMEGA-HYDROXY-, N-COCO ACYL DERIVATIVES, METHYL SULFATES (SALTS)	POLY(OXY-1,2-ETHANEDIYL), ALPHA-[3-[BIS(2-AMINOETHYL)METHYLEAR [DIHEXYL PHTHALATE (MIXED ISOMERS)]
70914-09-9 POLY(OXY-1,2-ETHANEDIYL), ALPHA-[2-[BIS(2-AMINOETHYL)METHYLAMMONIO]ETHYL]-OMEGA-HYDROXY-, N,N'-DI-C(14-18) ACYL DERIVAT IVES, METHYL SULFATES (SALTS)	POLY(OXY-1,2-ETHANEDIYL), ALPHA-[2-[BIS(2-AMINOETHYL)METHYLAOMEGA-[[(4-OXIRANYL-METHOXY)BENZOYL]OXY]-
70632-06-3 POLY(OXY-1,2-ETHANEDIYL), ALPHA-(CAR-BOXYMETHYL)-OMEGA-HYDROXY-, C12-15-ALKYL ETHERS, SODIUM SALTS	POLY(OXY-1,2-ETHANEDIYL), ALPHA-(CAR-BOXYMETHYL)-OMEGA-HYDROX
130547-87-4 POLY[OXY-1,2-ETHANEDIYL(DIMETHYLIM-INO)-1,3-PROPANEDIYLIMINOCARBONYLIM-INO-1,3-PROPANEDIYL(DIMETHYLIMINO)-1,2-ETHANEDIYL DICHLORIDE], ALPHA-(2-CHLOROETHYL)-OMEGA-(2-CHLOROETHOXY)-	POLY[OXY-1,2-ETHANEDIYL(DIMETHYLIMINO)-1,3-PROPANEDIYLIMINOC
106158-22-9 POLY(OXY-1,2-ETHANEDIYL), .ALPHA.-HYDRO-.W.-HYDROXY-, ETHERWITH 2-ETHYL-2-(HY-DROXYMETHYL)-1,3-PROPANEDIOL (3:1) DI-2-PROPENOATE, METHYL ETHER	POLY(OXY-1,2-ETHANEDIYL), .ALPHA.-HYDRO-.W.-HYDROXY-, ETHER
52495-71-3 POLY(OXY-1,2-ETHANEDIYL), ALPHA-HYDRO-3-(OXIRANYLMETHYL)-1,3-PROPANEDIOL (3:1)	POLY(OXY-1,2-ETHANEDIYL), ALPHA-HYDRO-3-(OXI-RANYLMETHYL)-1,3NYL)-
52495-71-3 POLY(OXY-1,2-ETHANEDIYL), ALPHA-HY-DRO-3-(OXIRANYLMETHOXY)-,ETHER WITH 2-ETHYL-2-(HYDROXYMETHYL)-1,3-PROPANE-DIOL (3:1) [TSC, TSC]	POLY(OXY-1,2-ETHANEDIYL), ALPHA-HYDRO-3-(OXI-RANYLMETHYL)-1,3WAMINE OXIDE, PHOSPHATE
RR-00235-7 POLY(OXY-1,2-ETHANEDIYL), ALPHA-(2-METHYL-1-OXO-2-PROPENYL)-X-HYDROXY-, C10-16-ALKYL ETHERS	POLY(OXY-1,2-ETHANEDIYL), ALPHA-(2-METHYL-1-OXO-2-PROPENYL)-
54140-64-6 POLY[OXY-1,2-ETHANEDIYL), ALPHA,ALPHA'-[(1-METHYLETHYLIDENE)DI-4,1-PHENYLENE]BIS[3-(OXIRANYLMETHOXY)- [TSC, TSC]	POLY[OXY(METHYL-1,2-ETHANEDIYL)], ALPHA,ALPHA'-[(1-METHYLETHETHER WITH 2-ETHYL-2-(HYDROX-YMETHYL)-1,3-PROPANEDIOL (3:1)

Chemical Name	Reference Name
69943-75-5 POLY(OXY-1,2-ETHANEDIYL), ALPHA-[4-OXI-RANYLMETHOXY)BENZOYL]-OMEGA-[[(4-OXI-RANYLMETHOXY)BENZOYL]OXY]-	POLY(OXY-1,2-ETHANEDIYL), ALPHA-[4-OXIRANYL-METHOXY)BENZOYL]-
125304-11-2 POLY(OXY-1,2-ETHANEDIYL), ALPHA-(1-OXO-2-PROPENYL)-OMEGA-HYDROXY-, C(10-16)-ALKYL ETHERS	POLY(OXY-1,2-ETHANEDIYL), ALPHA-(1-OXO-2-PROPENYL)-OMEGA-HYDSULFOBENZOIC ACID (1:1)
RR-01120-1 POLYOXYETHYLENE AMYLPHENOL - FORMALDEHYDE RESIN	POLYOXYETHYLENE AMYLPHENOL - FORMALDEHYDE RESIN
68412-54-4 POLYOXYETHYLENE BRANCHED-C9-ALKYLPHENOL [PST]	POLYETHYLENE GLYCOL BRANCHED NONYLPHENYL ETHERSULFATE
30704-63-3 POLYOXYETHYLENE P-TERT-BUTYLPHENOL - FORMALDEHYDE RESIN	POLYOXYETHYLENE P-TERT-BUTYLPHENOL - FORMALDEHYDE RESIN
70247-86-8 POLYOXYETHYLENE C4-18-ALKYL ESTER OF PHOSPHORIC ACID, MONOETHANOLAMINE SALTS	POLYOXYETHYLENE C4-18-ALKYL ESTER OF PHOS-PHORIC ACID, MONOET
68439-72-5 POLYOXYETHYLENE C8-18 AND C18 UNSATU-RATED, ALKYLAMINES	POLYOXYETHYLENE C8-18 AND C18 UNSATURATED, ALKYLAMINES
71549-82-1 POLYOXYETHYLENE C12-13-ALKYL ESTER OF PHOSPHORIC ACID, MONOETHANOLAMINE SALTS	POLYOXYETHYLENE C12-13-ALKYL ESTER OF PHOS-PHORIC ACID, MONOE
68155-01-1 POLYOXYETHYLENE CETYL AND OLEYL ALCOHOLS	POLYOXYETHYLENE CETYL AND OLEYL ALCOHOLS
31512-74-0 POLY[OXYETHYLENE (DIMETHYLIMINO) ETHYLENE (DIMETHYLIMINO) ETHYLENE DICHLORIDE]	POLY[OXYETHYLENE (DIMETHYLIMINO) ETHYLENE (DIMETHYLIMINO) ET
9014-93-1 POLYOXYETHYLENE DINONYLPHENOL	POLYOXYETHYLENE DINONYLPHENOL
9005-07-6 POLYOXYETHYLENE DIOLEATE [PST]	PEG (400) DIOLEATE
9005-08-7 POLYOXYETHYLENE DISTEARATE	POLYOXYETHYLENE DISTEARATE
26636-40-8 POLYOXYETHYLENE DOCOSYL ETHER	POLYOXYETHYLENE DOCOSYL ETHER
13081-34-0 POLY(OXYETHYLENE) 10-DODECYLMERCAP-TAN	POLY(OXYETHYLENE) 10-DODECYLMERCAPTAN
9014-92-0 POLYOXYETHYLENE DODECYLPHENOL [PST]	ALPHA-(DODECYLPHENYL)-OMEGA-HYDROXY-POLY(OXY-1,2-ETHANEDIYL)
8050-33-7 POLYOXYETHYLENE ESTER OF ROSIN	POLYOXYETHYLENE ESTER OF ROSIN
RR-01121-2 POLYOXYETHYLENE ESTERS OF MONO & DICARBOXYLIC ACID & OIL SOLUBLE SUL-FONATES	POLYOXYETHYLENE ESTERS OF MONO & DICAR-BOXYLIC ACID & OIL SOL
32492-61-8 POLYOXYETHYLENE ISOPROPYLIDENEDIPHE-NOL	POLYOXYETHYLENE ISOPROPYLIDENEDIPHENO-LACRYLATE AND STYRENE COPOLYMER
68648-38-4 POLYOXYETHYLENE LANOLIN ALCOHOL	POLYOXYETHYLENE LANOLIN ALCOHOL
9002-92-0 POLYOXYETHYLENE LAURYL ETHER [NFP, NFP]	LAURYL POLYETHYLENE GLYCOL ETHER
71243-46-4 POLYOXYETHYLENE LINEAR PRIMARY C8-16 ALCOHOLS	POLYOXYETHYLENE LINEAR PRIMARY C8-16 ALCO-HOLSIC ACID, MONOHYDROCHLORIDE
60874-89-7 POLY(OXYETHYLENE)METHYLENEBIS(DI-AMYLPHENOL)	POLY(OXYETHYLENE)METHYLENEBIS(DIAMYLPHENOL)
41928-09-0 POLY(OXYETHYLENE)METHYLENEBIS (OCTYLPHENOL)	POLY(OXYETHYLENE)METHYLENEBIS(OCTYLPHENOL)
9004-77-7 POLYOXYETHYLENE MONOBUTYL ETHER [PST]	POLYETHYLENE GLYCOL, MONOBUTYL ETHER
9004-95-9 POLYOXYETHYLENE MONOHEXADECYL ETHER	POLYOXYETHYLENE MONOHEXADECYL ETHER
9004-81-3 POLYOXYETHYLENE MONOLAURATE [PST]	POLYETHYLENE GLYCOL MONOLAURATE
9004-96-0 POLYOXYETHYLENE MONOLEATE [PST]	POLYETHYLENE GLYCOL MONOOLEATE
9004-98-2 POLYOXYETHYLENE MONO(CIS-9-OCTADE-CENYL) ETHER [PST]	POLYETHYLENE GLYCOL MONOOLEYL ETHER

Chemical Name	Reference Name
9005-00-9 POLYOXYETHYLENE MONOOCTADECYL ETHER	POLYOXYETHYLENE MONOOCTADECYL ETHER
9036-19-5 POLYOXYETHYLENE MONOOCTYLPHENYL ETHER [PST]	ETHOXYLATED OCTYLPHENOL
9004-99-3 POLYOXYETHYLENE MONOSTEARATE [PST]	POLYETHYLENE GLYCOL STEARATE
27306-79-2 POLYOXYETHYLENE MONOTETRADECYL ETHER	POLYOXYETHYLENE MONOTETRADECYL ETHERAMIDIN-5-ONE
9016-45-9 POLYOXYETHYLENE NONYLPHENOL [PST]	ETHYLENE OXIDE-NONYLPHENOL POLYMER
69029-39-6 POLYOXYETHYLENE POLYOXYPROPYLENE DI-SEC-BUTYLPHENOL	POLYOXYETHYLENE POLYOXYPROPYLENE DI-SEC-BUTYLPHENOL
RR-01123-4 POLYOXYETHYLENE POLYOXYPROPYLENE TERT-C12-13-ALKYL AMINE	POLYOXYETHYLENE POLYOXYPROPYLENE TERT-C12-13-ALKYL AMINEED CAPRYLIC AND CAPRIC ACIS
RR-01122-3 POLYOXYETHYLENE POLYOXYPROPY-LENE MONOISOPROPANOLAMIDE OF MIXED CAPRYLIC AND CAPRIC ACIS	POLYOXYETHYLENE POLYOXYPROPYLENE MONOISO-PROPANOLAMIDE OF MIXUBLE SULFONATES
37251-69-7 POLYOXYETHYLENE POLYOXYPROPYLENE NONYLPHENOL	POLYOXYETHYLENE POLYOXYPROPYLENE NONYLPHENOL
37280-82-3 POLYOXYETHYLENE POLYOXYPROPYLENE PHOSPHATE	POLYOXYETHYLENE POLYOXYPROPYLENE PHOS-PHATE
9005-65-6 POLYOXYETHYLENE 20 SORBITAN MONOOLEATE	POLYOXYETHYLENE 20 SORBITAN MONOOLEATE
54846-79-6 POLYOXYETHYLENE SORBITAN HEP-TAOLEATE	POLYOXYETHYLENE SORBITAN HEPTAOLEATE
9005-64-5 POLYOXYETHYLENE SORBITAN MONOLAU-RATE [PST]	POLYSORBATE 20
9005-65-6 POLYOXYETHYLENE SORBITAN MONOLEATE [PST]	POLYOXYETHYLENE 20 SORBITAN MONOOLEATE
9005-66-7 POLYOXYETHYLENE SORBITAN MONOPALMI-TATE	POLYOXYETHYLENE SORBITAN MONOPALMITATE
9005-67-8 POLYOXYETHYLENE SORBITAN MONOS-TEARATE	POLYOXYETHYLENE SORBITAN MONOSTEARATE
61790-86-1 POLYOXYETHYLENE SORBITAN MONOTAL-LATE	POLYOXYETHYLENE SORBITAN MONOTALLATE
RR-01124-5 POLYOXYETHYLENE SORBITAN TETRATAL-LATE	POLYOXYETHYLENE SORBITAN TETRATALLATE
9005-70-3 POLYOXYETHYLENE SORBITAN TRIOLEATE	POLYOXYETHYLENE SORBITAN TRIOLEATE
9005-71-4 POLYOXYETHYLENE SORBITAN TRISTEARATE	POLYOXYETHYLENE SORBITAN TRISTEARATE
RR-01125-6 POLYOXYETHYLENE SORBITOL	POLYOXYETHYLENE SORBITOL
57171-56-9 POLYOXYETHYLENE SORBITOL HEXAOLEATE	POLYOXYETHYLENE SORBITOL HEXAOLEATE
63089-86-1 POLYOXYETHYLENE SORBITOL TE-TRAOLEATE	POLYOXYETHYLENE SORBITOL TETRAOLEATEY-DROCHLORIDE, (Z)-
61791-23-9 POLYOXYETHYLENE SOYA ACID ESTERS	POLYOXYETHYLENE SOYA ACID ESTERS
61791-24-0 POLYOXYETHYLENE SOYA ALKYL AMINES [PST]	AMINES, SOYA ALKYL, ETHOXYLATED
9014-85-1 POLYOXYETHYLENE 2,4,7,9-TETRAMETHYL-5-DECYNE-4,7-DIOL	POLYOXYETHYLENE 2,4,7,9-TETRAMETHYL-5-DECYNE-4,7-DIOL
24938-91-8 POLYOXYETHYLENE TRIDECYL ALCO-HOL [PST]	TRIDECYL ALCOHOL, ETHOXYLATED
69011-36-5 POLYOXYETHYLENE TRIMETHYLDECYL ALCOHOL	POLYOXYETHYLENE TRIMETHYLDECYL ALCOHOL
9002-81-7 POLY(OXYMETHYLENE)	POLY(OXYMETHYLENE)ESTER, MIXT. WITH O,O-DI-ETHYL S-[2-(ETHYLTHIO)ETHYL] PHOSPHOROTHIOATE
9002-81-7 POLYOXYMETHYLENE [PST]	POLY(OXYMETHYLENE)
9015-98-9 POLY(OXYMETHYLENE), .ALPHA.-HYDRO-.OMEGA.-HYDROXY-	POLY(OXYMETHYLENE), .ALPHA.-HYDRO-.OMEGA.-HYDROXY-

Chemical Name	Reference Name
68413-04-7 POLY[OXY(METHYL-1,2-ETHANEDIYL)], AL-PHA-[2-[BIS(2-AMINOETHYL)METHYLAMMO-NIO]METHYLETHYL]-OMEGA-HYDROXY-, N,N'-DITALLOW ACYL DERIVATIVES, METHYL SULFATES (SALTS) [TSC, TSC]	METHYL, TALLOW DIETHYLENETRIAMINE CONDEN-SATE, PROPOXYLATED,AMMONIO]ETHYL]-OMEGA-HY-DROXY-, N,N'-DITALLOW ACYL DERIVATIVES, METHYL SULFATES (SALTS)
53467-11-1 POLY(OXY(METHYL-1,2-ETANEDIYL)),A-[(2,4-DICHLOROPHENOXY)ACETYL]-OMEGA-BUTOXY- [MSL]	2,4-D ESTERS
53467-11-1 POLY[OXY(METHYL-1,2-ETHANEDIYL)],.AL-PHA.-[(2,4-DICHLOROPHENOXY)ACETYL]-.OMEGA.-BUTOXY- [PAL]	2,4-D ESTERS
RR-00983-6 POLY[OXY(METHYL-1,2-ETHANEDIYL)], AL-PHA,ALPHA'-(2,2-DIMETHYL-1,3-PROPANEDIYL (BIS[3-(OXIRANYLMETHOXY)-	POLY[OXY(METHYL-1,2-ETHANEDIYL)], ALPHA,ALPHA'-(2,2-DIMETHYL
RR-01114-3 POLY(OXY(METHYL-1,2-ETHANEDIYL)), AL-PHA-(2-(DIETHYLMETHYLAMMONIO)ETHYL-OMEGA-HYDROXY-, CHLORIDE	POLY(OXY(METHYL-1,2-ETHANEDIYL)), ALPHA-(2-(DIETHYLMETHYLAMM
54140-64-6 POLY[OXY(METHYL-1,2-ETHANEDIYL)], AL-PHA,ALPHA'-[(1-METHYLETHYLIDENE)DI-4,1-PHENYLENE]BIS[3-(OXIRANYLMETHOXY)-	POLY[OXY(METHYL-1,2-ETHANEDIYL)], ALPHA,ALPHA'-[(1-METHYLETH
42557-13-1 POLY[OXY[METHYL(3,3,3-TRIFLUOROPROPYL) SILYLENE], ALPHA-(TRIMETHYLSILYL)-OMEGA-[(TRIMETHYLSILYL)OXY]-	POLY[OXY[METHYL(3,3,3-TRIFLUOROPROPYL)SILY-LENE], ALPHA-(TRIM
89800-10-2 POLYOXY(1-OXO-1,6-HEXANEDIYL), AL-PHA-HYDRO-OMEGA-HYDROXY-ESTER WITH 2,2'-(OXYBIS(METHYLENE)BIS(2-HY-DROXYMETHYL)-1,3-PROPANEDIOL)-2-PROPENOATE, 2-PROP	POLYOXY(1-OXO-1,6-HEXANEDIYL), ALPHA-HYDRO-OMEGA-HYDROXY-ESTETHYL]-5-ETHYL-1,3-DIOXAN-5-YL)METHYL ESTER
96915-50-3 POLYOXY[OXY(1-OXO-1,6 HEXANEDIYL)], AL-PHA-(1-OXO-2-PROPENYL)-OMEGA-((TETRAHY-DRO-2-FURANYL)METHOXY)-	POLYOXY[OXY(1-OXO-1,6 HEXANEDIYL)], ALPHA-(1-OXO-2-PROPENYL)ER WITH 3-HYDROXY-2,2-DIMETHYLPROPYL 3-HYDROXY-2,2-DIMETHYL-PROPANOATE (2:1), DI-2-[PROPANOATE
68648-12-4 POLYOXYPROPYLENE DITALL-OIL ESTER	POLYOXYPROPYLENE DITALL-OIL ESTERTHYL)-2-ISOHEPTADECYL-, HYDROXIDE, INNER SALT
25791-96-2 POLY(OXYPROPYLENE) GLYCEROL TRI-ETHER [PST]	POLY(OXYPROPYLENE) TRIOL
37281-78-0 POLYOXYPROPYLENE OLEATE BUTYL ETHER	POLYOXYPROPYLENE OLEATE BUTYL ETHER
25791-96-2 POLY(OXYPROPYLENE) TRIOL	POLY(OXYPROPYLENE) TRIOL
68458-49-1 POLYPHOSPHORIC ACIDS, ESTERS WITH POLYETHYLENEGLYCOL-NONYLPHENYL ETHER	POLYPHOSPHORIC ACIDS, ESTERS WITH POLYETHYLENEGLYCOL-NONYLPH
RR-01718-5 POLYPIPERIDINOL-ACRYLATE METHACRY-LATE	POLYPIPERIDINOL-ACRYLATE METHACRYLATE
9003-07-0 POLYPROPYLENE	POLYPROPYLENE
25322-69-4 POLYPROPYLENE GLYCOL	POLYPROPYLENE GLYCOL
37286-64-9 POLYPROPYLENE GLYCOL METHYL ETHER	POLYPROPYLENE GLYCOL METHYL ETHER
52673-60-6 POLYPROPYLENE GLYCOL ALPHA-METHYL GLUCOSIDE ETHER (4:1)	POLYPROPYLENE GLYCOL ALPHA-METHYL GLUCO-SIDE ETHER (4:1)
61849-72-7 POLYPROPYLENE GLYCOL BETA-METHYL GLUCOSIDE ETHER (4:1)	POLYPROPYLENE GLYCOL BETA-METHYL GLUCOSIDE ETHER (4:1)
9003-13-8 POLYPROPYLENE GLYCOL, MONOBUTYL ETHER	POLYPROPYLENE GLYCOL, MONOBUTYL ETHER
25322-69-4 POLYPROPYLENE GLYCOLS [MIN, NFP, NFP]	POLYPROPYLENE GLYCOL
25231-21-4 POLYPROPYLENE STEARYL ETHER	POLYPROPYLENE STEARYL ETHER
9005-64-5 POLYSORBATE 20	POLYSORBATE 20
9005-65-6 POLYSORBATE 80 (GLYCOL) [CSR, CSR]	POLYOXYETHYLENE 20 SORBITAN MONOOLEATE
9003-53-6 POLYSTYRENE	POLYSTYRENE

Chemical Name	Reference Name
9003-53-6 POLYSTYRENE BEADS, EXPANDED, MIXTURE WITH FLAMMABLE LIQUID [NJL]	POLYSTYRENE
9003-53-6 POLYSTYRENE RESIN [PST]	POLYSTYRENE
RR-01657-9 POLYSUBSTITUTED PHENYLAZOPOLYSUBSTITUTED PHENYL DYE	POLYSUBSTITUTED PHENYLAZOPOLYSUBSTITUTED PHENYL DYE
RR-00242-6 POLYSUBSTITUTED POLYOL	POLYSUBSTITUTED POLYOL
RR-00933-6 POLY(SUBSTITUTED TRIAZINYL) PIPERAZINE	POLY(SUBSTITUTED TRIAZINYL) PIPERAZINE
70750-53-7 POLYTERPENE RESINS, SYNTHETIC	POLYTERPENE RESINS, SYNTHETIC-, C12-15-ALKYL ETHERS, SODIUM SALTS
9002-84-0 POLYTETRAFLUOROETHYLENE	POLYTETRAFLUOROETHYLENE
9002-84-0 POLYTETRAFLUOROETHYLENE DECOMPOSITION PRODUCTS [BC1]	POLYTETRAFLUOROETHYLENE
RR-00073-7 POLYTETRAFLUORETHYLENE DECOMPOSITION PRODUCTS	POLYTETRAFLUORETHYLENE DECOMPOSITION PRODUCTS
68988-56-7 POLYTRIMETHYLHYDROSILYSILICONE	POLYTRIMETHYLHYDROSILYSILICONE
68400-67-9 POLYURETHANE	POLYURETHANE
RR-01538-3 POLYVINYL ACETATE	POLYVINYL ACETATE
9002-89-5 POLYVINYL ALCOHOL	POLYVINYL ALCOHOL
63148-65-2 POLYVINYL BUTYRAL	POLYVINYL BUTYRAL
63148-65-2 POLYVINYL BUTYRAL RESIN [PST]	POLYVINYL BUTYRAL
24991-31-9 POLYVINYL BUTYRATE	POLYVINYL BUTYRATE3-BUTADIENE
9002-86-2 POLYVINYL CHLORIDE	POLYVINYL CHLORIDE
9002-86-2 POLYVINYL CHLORIDE (PVC) [GBR]	POLYVINYL CHLORIDE
9002-86-2 POLYVINYL CHLORIDE RESIN [PST]	POLYVINYL CHLORIDE
25104-37-4 POLY(VINYL ETHYL ETHER)	POLY(VINYL ETHYL ETHER)
9003-39-8 POLYVINYL PYRROLIDONE	POLYVINYL PYRROLIDONE
9003-39-8 POLYVINYLPYRROLIDONE [PST]	POLYVINYL PYRROLIDONE
25086-89-9 POLYVINYLPYRROLIDONE - VINYL ACETATE COPOLYMER	POLYVINYLPYRROLIDONE - VINYL ACETATE COPOLYMER
3564-09-8 PONCEAU 3R	PONCEAU 3R
3761-53-3 PONCEAU MX	PONCEAU MXOXAZOLIDINONE HYDROCHLORIDE
4548-53-2 PONCEAU SX [IAR]	C.I. FOOD RED 1
37523-33-4 POP NONYLPHENOL - FORMALDEHYDE RESIN	POP NONYLPHENOL - FORMALDEHYDE RESIN
RR-00880-0 POPPY SEED OIL	POPPY SEED OIL
65997-15-1 PORTLAND CEMENT	PORTLAND CEMENT
7440-09-7 POTASSIUM	POTASSIUM
13966-00-2 POTASSIUM 40	POTASSIUM 40
14378-21-3 POTASSIUM 42	POTASSIUM 42
14903-02-7 POTASSIUM 43	POTASSIUM 43
14378-22-4 POTASSIUM 44	POTASSIUM 44
15706-41-9 POTASSIUM 45	POTASSIUM 45
127-08-2 POTASSIUM ACETATE	POTASSIUM ACETATE
11135-81-2 POTASSIUM ALLOY, NONBASE (SODIUM-POTASSIUM ALLOYS) [PAL]	POTASSIUM-SODIUM ALLOY
10043-67-1 POTASSIUM ALUMINUM SULFATE	POTASSIUM ALUMINUM SULFATE
7784-41-0 POTASSIUM ARSENATE	POTASSIUM ARSENATE
13464-35-2 POTASSIUM ARSENITE	POTASSIUM ARSENITE
10124-50-2 POTASSIUM ARSONITE [EPA]	POTASSIUM ARSENITE
298-14-6 POTASSIUM BICARBONATE	POTASSIUM BICARBONATE
7778-50-9 POTASSIUM BICHROMATE [CWA, EP2, EPA]	POTASSIUM DICHROMATE
7789-29-9 POTASSIUM BIFLUORIDE	POTASSIUM BIFLUORIDE

Chemical Name	Reference Name
23746-34-1 POTASSIUM BIS(2-HYDROXYETHYL)DITHIO-CARBAMATE	POTASSIUM BIS(2-HYDROXYETHYL)DITHIOCARBAMATE
1332-77-0 POTASSIUM BORATE	POTASSIUM BORATE
12045-78-2 POTASSIUM BORATE TETRAHYDRATE	POTASSIUM BORATE TETRAHYDRATE
13762-51-1 POTASSIUM BOROHYDRIDE	POTASSIUM BOROHYDRIDE
7758-01-2 POTASSIUM BROMATE	POTASSIUM BROMATE
7758-02-3 POTASSIUM BROMIDE	POTASSIUM BROMIDE
584-08-7 POTASSIUM CARBONATE	POTASSIUM CARBONATE
12397-35-2 POTASSIUM CARBONYL	POTASSIUM CARBONYL
3811-04-9 POTASSIUM CHLORATE	POTASSIUM CHLORATE
7447-40-7 POTASSIUM CHLORIDE	POTASSIUM CHLORIDE
7789-00-6 POTASSIUM CHROMATE	POTASSIUM CHROMATE
85712-26-1 POTASSIUM N,N-BIS (HYDROXYETHYL) CO-COAMINE OXIDE PHOSPHATE	POTASSIUM N,N-BIS (HYDROXYETHYL) COCOAMINE OXIDE PHOSPHATE-DIMETHYL-
85712-26-1 POTASSIUM N,N-BIS (HYDROXYETHYL) CO-COAMINE OXIDE PHOSPHATE [TSE]	POTASSIUM N,N-BIS (HYDROXYETHYL) COCOAMINE OXIDE PHOSPHATE2-DIMETHYL]
RR-01127-8 POTASSIUM N-COCO-N-HYDROXYETHYL AMINO-3-ETHOXY PROPANE SULFONATE	POTASSIUM N-COCO-N-HYDROXYETHYL AMINO-3-ETHOXY PROPANE SULFO
61789-31-9 POTASSIUM COCONUT FATTY OIL SOAP	POTASSIUM COCONUT FATTY OIL SOAP
61789-30-8 POTASSIUM COCONUT OIL SOAP [PST]	COCONUT FATTY ACIDS, POTASSIUM SALT
RR-01126-7 POTASSIUM CRESYL E5 PHOSPHATE	POTASSIUM CRESYL E5 PHOSPHATE
13682-73-0 POTASSIUM CUPROCYANIDE	POTASSIUM CUPROCYANIDE
590-28-3 POTASSIUM CYANATE	POTASSIUM CYANATE
151-50-8 POTASSIUM CYANIDE	POTASSIUM CYANIDE
151-50-8 POTASSIUM CYANIDE (K(CN)) [PAL]	POTASSIUM CYANIDE
2244-21-5 POTASSIUM DICHLOROISOCYANURATE	POTASSIUM DICHLOROISOCYANURATE
2244-21-5 POTASSIUM DICHLORO-S TRIAZINETRI-ONE [FLA]	POTASSIUM DICHLOROISOCYANURATE
2244-21-5 POTASSIUM DICHLORO-S-TRIAZINETRI-ONE [MSL]	POTASSIUM DICHLOROISOCYANURATE
7778-50-9 POTASSIUM DICHROMATE	POTASSIUM DICHROMATE
7784-41-0 POTASSIUM DIHYDROGEN ARSENATE [HM1, HM1, HM1, HM1, WHM]	POTASSIUM ARSENATE
128-03-0 POTASSIUM DIMETHYLDITHIOCARBAMATE	POTASSIUM DIMETHYLDITHIOCARBAMATE
RR-00752-3 POTASSIUM DITHIONITE	POTASSIUM DITHIONITE
7379-27-3 POTASSIUM ETHYLENEDIAMINETETRAAC-ETATE [PST]	ETHYLENEDIAMINETETRAACETIC ACID, POTASSIUM SALT
3164-85-0 POTASSIUM 2-ETHYLHEXANOATE	POTASSIUM 2-ETHYLHEXANOATE
7789-23-3 POTASSIUM FLUORIDE	POTASSIUM FLUORIDE
23745-86-0 POTASSIUM FLUOROACETATE	POTASSIUM FLUOROACETATE
16871-90-2 POTASSIUM FLUOROSILICATE [HM1, HM1]	POTASSIUM SILICOFLUORIDE
16924-00-8 POTASSIUM HEPTAFLUOROTANTALATE	POTASSIUM HEPTAFLUOROTANTALATE
16921-30-5 POTASSIUM HEXACHLOROPLATINATE(IV)	POTASSIUM HEXACHLOROPLATINATE(IV)
7789-29-9 POTASSIUM HYDROGEN FLUORIDE [NJL, NJL]	POTASSIUM BIFLUORIDE
7646-93-7 POTASSIUM HYDROGEN SULFATE	POTASSIUM HYDROGEN SULFATE
14293-73-3 POTASSIUM HYDROSULFITE	POTASSIUM HYDROSULFITE
1310-58-3 POTASSIUM HYDROXIDE	POTASSIUM HYDROXIDE
1310-58-3 POTASSIUM HYDROXIDE (K(OH)) [PAL]	POTASSIUM HYDROXIDE
7778-66-7 POTASSIUM HYPOCHLORITE	POTASSIUM HYPOCHLORITE
7681-11-0 POTASSIUM IODIDE	POTASSIUM IODIDE
16731-55-8 POTASSIUM METABISULFITE	POTASSIUM METABISULFITE
RR-01337-6 POTASSIUM, METAL ALLOYS	POTASSIUM, METAL ALLOYS

Chemical Name	Reference Name
13769-43-2 POTASSIUM METAVANADATE	POTASSIUM METAVANADATE
137-41-7 POTASSIUM N-METHYLDITHIOCARBAMATE	POTASSIUM N-METHYLDITHIOCARBAMATE
12136-45-7 POTASSIUM MONOXIDE [HM1, HM1]	POTASSIUM OXIDE
7757-79-1 POTASSIUM NITRATE	POTASSIUM NITRATE
RR-00452-4 POTASSIUM NITRATE AND SODIUM NITRITE, MIXTURE	POTASSIUM NITRATE AND SODIUM NITRITE, MIXTURE
RR-00452-4 POTASSIUM NITRATE AND SODIUM NITRITE MIXTURES [HM1, HM1]	POTASSIUM NITRATE AND SODIUM NITRITE, MIXTURE
7758-09-0 POTASSIUM NITRITE	POTASSIUM NITRITE
143-18-0 POTASSIUM OLEATE	POTASSIUM OLEATE
12136-45-7 POTASSIUM OXIDE	POTASSIUM OXIDE
10025-98-6 POTASSIUM PALLADIUM CHLORIDE	POTASSIUM PALLADIUM CHLORIDE
7778-73-6 POTASSIUM PENTACHLOROPHENATE	POTASSIUM PENTACHLOROPHENATE
7778-74-7 POTASSIUM PERCHLORATE	POTASSIUM PERCHLORATE
7722-64-7 POTASSIUM PERMANGANATE	POTASSIUM PERMANGANATE
17014-71-0 POTASSIUM PEROXIDE	POTASSIUM PEROXIDE
17014-71-0 POTASSIUM PEROXIDE (K2(O2)) [PAL]	POTASSIUM PEROXIDE
7727-21-1 POTASSIUM PERSULFATE	POTASSIUM PERSULFATE
7758-11-4 POTASSIUM PHOSPHATE (DIBASIC) [PST]	DIPOTASSIUM PHOSPHATE
7778-77-0 POTASSIUM PHOSPHATE, MONOBASIC [PST]	DIHYDROGEN POTASSIUM PHOSPHATE
20770-41-6 POTASSIUM PHOSPHIDE	POTASSIUM PHOSPHIDE
7320-34-5 POTASSIUM PYROPHOSPHATE	POTASSIUM PYROPHOSPHATE
7790-62-7 POTASSIUM PYROSULFATE	POTASSIUM PYROSULFATE
7492-30-0 POTASSIUM RICINOLEATE	POTASSIUM RICINOLEATE
RR-01128-9 POTASSIUM SALT OF NAPHTHALENE SULFONIC ACID - FORMALDEHYDE CONDENSATE	POTASSIUM SALT OF NAPHTHALENE SULFONIC ACID - FORMALDEHYDE CNATE
61790-50-9 POTASSIUM SALT OF WOOD ROSIN ACIDS [PST]	RESIN ACIDS AND ROSIN ACIDS, POTASSIUM SALT
RR-00453-5 POTASSIUM SALTS OF AROMATIC NITRO-DERIVATIVES [HM1, HM1]	POTASSIUM SALTS OF NITRO-AROMATIC DERIVATIVES, EXPLOSIVE
69669-25-6 POTASSIUM SALTS OF FATTY ACIDS (C12-C20)	POTASSIUM SALTS OF FATTY ACIDS (C12-C20)PYLENE POLYOXYETHYLENE
RR-00453-5 POTASSIUM SALTS OF NITRO-AROMATIC DERIVATIVES, EXPLOSIVE	POTASSIUM SALTS OF NITRO-AROMATIC DERIVATIVES, EXPLOSIVE
7790-59-2 POTASSIUM SELENATE	POTASSIUM SELENATE
1312-76-1 POTASSIUM SILICATE	POTASSIUM SILICATE
16871-90-2 POTASSIUM SILICOFLUORIDE	POTASSIUM SILICOFLUORIDE
506-61-6 POTASSIUM SILVER CYANIDE	POTASSIUM SILVER CYANIDE
11135-81-2 POTASSIUM-SODIUM ALLOY	POTASSIUM-SODIUM ALLOY
24634-61-5 POTASSIUM SORBATE	POTASSIUM SORBATE
593-29-3 POTASSIUM STEARATE	POTASSIUM STEARATE
7778-80-5 POTASSIUM SULFATE [PST]	SULFURIC ACID DIPOTASSIUM SALT
67785-93-7 POTASSIUM SULFATORICINOLEATE	POTASSIUM SULFATORICINOLEATE
1312-73-8 POTASSIUM SULFIDE	POTASSIUM SULFIDE
37248-34-3 POTASSIUM SULFIDE [PAL]	POTASSIUM SULFIDE (VAN)
1312-73-8 POTASSIUM SULFIDE (K2S) [PAL]	POTASSIUM SULFIDE
37248-34-3 POTASSIUM SULFIDE (VAN)	POTASSIUM SULFIDE (VAN)
10117-38-1 POTASSIUM SULFITE	POTASSIUM SULFITE
12030-88-5 POTASSIUM SUPEROXIDE	POTASSIUM SUPEROXIDE
85712-27-2 POTASSIUM N,N'-BIS(HYDROXYETHYL) TALLOWAMINE OXIDE PHOSPHATE	POTASSIUM N,N'-BIS(HYDROXYETHYL) TALLOWAMINE OXIDE PHOSPHATE
1332-77-0 POTASSIUM TETRABORATE [PST]	POTASSIUM BORATE

Chemical Name	Reference Name
53535-27-6 POTASSIUM TETRACHLOROPHENATE	POTASSIUM TETRACHLOROPHENATE
10025-99-7 POTASSIUM TETRACHLOROPLATINATE(II)	POTASSIUM TETRACHLOROPLATINATE(II)
333-20-0 POTASSIUM THIOCYANATE	POTASSIUM THIOCYANATE
12030-97-6 POTASSIUM TITANATE	POTASSIUM TITANATE
12030-97-6 POTASSIUM TITANATE (K2TIO3) [MAK]	POTASSIUM TITANATE
12056-46-1 POTASSIUM TITANATE (K2TI2O5)	POTASSIUM TITANATE (K2TI2O5)
12056-49-4 POTASSIUM TITANATE (K2TI4O9)	POTASSIUM TITANATE (K2TI4O9)
12056-51-8 POTASSIUM TITANATE (K2TI6O13) [MAK]	TITANATE [TI6O13(2-)], DIPOTASSIUM
59766-31-3 POTASSIUM TITANATE (K2TI8O17)	POTASSIUM TITANATE (K2TI8O17)
30526-22-8 POTASSIUM TOLUENESULFONATE	POTASSIUM TOLUENESULFONATE
13845-36-8 POTASSIUM TRIPOLYPHOSPHATE	POTASSIUM TRIPOLYPHOSPHATE
140-89-6 POTASSIUM XANTHATE [FLA, MSL, NFP, NFP]	CARBONODITHIOIC ACID, O-ETHYL ESTER, POTASSIUM SALT
RR-01129-0 POTATOES	POTATOESONDENSATE
22095-53-0 PRASEODYMIUM 136	PRASEODYMIUM 136
15125-66-3 PRASEODYMIUM 137	PRASEODYMIUM 137
RR-00389-4 PRASEODYMIUM 138M	PRASEODYMIUM 138M
14191-76-5 PRASEODYMIUM 139	PRASEODYMIUM 139
14191-64-1 PRASEODYMIUM 142	PRASEODYMIUM 142
RR-00388-3 PRASEODYMIUM 142M	PRASEODYMIUM 142M
14981-79-4 PRASEODYMIUM 143	PRASEODYMIUM 143
14119-05-2 PRASEODYMIUM 144	PRASEODYMIUM 144
15765-23-8 PRASEODYMIUM 145	PRASEODYMIUM 145
15765-24-9 PRASEODYMIUM 147	PRASEODYMIUM 147
10361-79-2 PRASEODYMIUM CHLORIDE	PRASEODYMIUM CHLORIDE
10361-80-5 PRASEODYMIUM NITRATE	PRASEODYMIUM NITRATE
2955-38-6 PRAZEPAM	PRAZEPAM
112926-00-8 PRECIPITATED SILICA [ONT, PAL]	SILICA, AMORPHOUS, PRECIPITATED AND GEL
29069-24-7 PREDNIMUSTINE	PREDNIMUSTINE
50-24-8 PREDNISOLONE	PREDNISOLONE
53-03-2 PREDNISONE	PREDNISONE
57-83-0 PREGN-4-ENE-3,20-DIONE [PAL, PAL]	PROGESTERONE
125-33-7 PRIMACLONE	PRIMACLONE
57-66-9 PROBENECID	PROBENECID
671-16-9 PROCARBAZINE	PROCARBAZINE
366-70-1 PROCARBAZINE HYDROCHLORIDE	PROCARBAZINE HYDROCHLORIDE
32809-16-8 PROCYMIDONE	PROCYMIDONE
41198-08-7 PROFENOFOS [O-(4-BROMO-2-CHLOROPHENYL)-O-ETHYL-S-PROPYL PHOSPHOROTHIOATE] [313]	CURACRONZENEAMINE]
952-23-8 PROFLAVINE [MSL]	PROFLAVIN HYDROCHLORIDE
RR-01539-4 PROFLAVINE SALTS	PROFLAVINE SALTS
952-23-8 PROFLAVIN HYDROCHLORIDE	PROFLAVIN HYDROCHLORIDE
57-83-0 PROGESTERONE	PROGESTERONE
RR-00018-0 PROGESTERONE AND OESTROGENS	PROGESTERONE AND OESTROGENS
RR-00033-9 PROGESTINS	PROGESTINS
2631-37-0 PROMECARB	PROMECARB
58-33-3 PROMETHAZINE HYDROCHLORIDE	PROMETHAZINE HYDROCHLORIDE
14952-27-3 PROMETHIUM 141	PROMETHIUM 141
14834-72-1 PROMETHIUM 143	PROMETHIUM 143
14834-73-2 PROMETHIUM 144	PROMETHIUM 144

Chemical Name	Reference Name
15706-44-2 PROMETHIUM 145	PROMETHIUM 145
14834-74-3 PROMETHIUM 146	PROMETHIUM 146
14380-75-7 PROMETHIUM 147	PROMETHIUM 147
14683-19-3 PROMETHIUM 148	PROMETHIUM 148
RR-00387-2 PROMETHIUM 148M	PROMETHIUM 148M
15765-31-8 PROMETHIUM 149	PROMETHIUM 149
15720-47-5 PROMETHIUM 150	PROMETHIUM 150
15766-03-7 PROMETHIUM 151	PROMETHIUM 151
1610-18-0 PROMETON	PROMETON
7287-19-6 PROMETRYN	PROMETRYN
7287-19-6 PROMETRYN [N,N'-BIS(1-METHYLETHYL)-6-METHYLTHIO-1,3,5-TRIAZINE-2,4-DIAMINE] [313]	PROMETRYN
23950-58-5 PRONAMIDE	PRONAMIDE
51-02-5 PRONETALOL HYDROCHLORIDE	PRONETALOL HYDROCHLORIDE
50-34-0 PROPANTHELINE BROMIDE	PROPANTHELINE BROMIDE
1918-16-7 PROPACHLOR	PROPACHLOR
1918-16-7 PROPACHLOR [2-CHLORO-N-(1-METHYLETHYL)-N-PHENYLACETAMINDE] [313]	PROPACHLOR
463-49-0 PROPADIENE	PROPADIENE
463-49-0 PROPADIENE [1,2-PROPADIENE] [EP3]	PROPADIENE
123-38-6 PROPANAL [NFP, NFP, OTS, PAL, TSC, TSC, TSC]	PROPIONALDEHYDE
2806-85-1 PROPANAL, 3-ETHOXY- [PAL]	3-ETHOXYPROPIONALDEHYDE
597-31-9 PROPANAL, 3-HYDROXY-2,2-DIMETHYL-	PROPANAL, 3-HYDROXY-2,2-DIMETHYL-
78-84-2 PROPANAL, 2-METHYL- [OTS, PAL, TSC, TSC, TSC]	ISOBUTYRALDEHYDE
116-06-3 PROPANAL, 2-METHYL-2-(METHYLTHIO)-,O-[(METHYLAMINO)CARBONYL]OXIME [PAL]	ALDICARB
3268-49-3 PROPANAL, 3-(METHYLTHIO)- [TSC, TSC, TSC]	4-THIAPENTANAL
75-31-0 2-PROPANAMINE [PAL]	ISOPROPYLAMINE
107-10-8 1-PROPANAMINE	1-PROPANAMINE
17256-39-2 2-PROPANAMINE, 1-CHLORO-N,N-DIMETHYL-, HYDROCHLORIDE	2-PROPANAMINE, 1-CHLORO-N,N-DIMETHYL-, HYDROCHLORIDE
102-69-2 1-PROPANAMINE, N,N-DIPROPYL- [PAL]	TRIPROPYLAMINE
5332-73-0 1-PROPANAMINE, 3-METHOXY- [PAL]	3-METHOXYPROPYLAMINE
75-64-9 2-PROPANAMINE, 2-METHYL- [PAL]	TERT-BUTYLAMINE
78-81-9 1-PROPANAMINE, 2-METHYL- [PAL]	ISOBUTYLAMINE
108-18-9 2-PROPANAMINE, N-(1-METHYLETHYL)- [PAL]	DIISOPROPYLAMINE
110-96-3 1-PROPANAMINE, 2-METHYL-N-(2-METHYL-PROPYL)- [PAL]	DI-ISOBUTYLAMINE
621-64-7 1-PROPANAMINE, N-NITROSO-N-PROPYL- [PAL, PAL]	N-NITROSODI-N-PROPYLAMINE
142-84-7 1-PROPANAMINE, N-PROPYL- [PAL, TSC, TSC]	DIPROPYLAMINE
68139-30-0 PROPANAMINIUM, N-(3-AMINOPROPYL)-2-HYDROXY-N,N-DIMETHYL-3-SULFO-, N-COCO ACYL DERIVATIVES, HYDROXIDES, INNER SALTS	PROPANAMINIUM, N-(3-AMINOPROPYL)-2-HYDROXY-N, N-DIMETHYL-3-SU
74-98-6 PROPANE	PROPANE
106-94-5 PROPANE, 1-BROMO- [PAL]	N-PROPYL BROMIDE
75-29-6 PROPANE, 2-CHLORO- [PAL, TSC, TSC]	2-CHLOROPROPANE
540-54-5 PROPANE, 1-CHLORO- [PAL, TSC, TSC]	PROPYL CHLORIDE

Chemical Name	Reference Name
507-20-0 PROPANE, 2-CHLORO-2-METHYL- [PAL]	TERT-BUTYL CHLORIDE
513-36-0 PROPANE, 1-CHLORO-2-METHYL- [PAL]	ISOBUTYL CHLORIDE
594-71-8 PROPANE, 2-CHLORO-2-NITRO- [PAL]	2-CHLORO-2-NITROPROPANE
600-25-9 PROPANE, 1-CHLORO-1-NITRO- [PAL]	1-CHLORO-1-NITROPROPANE
78-90-0 1,2-PROPANEDIAMINE [FLA, PAL]	PROPYLENE DIAMINE
109-76-2 1,3-PROPANEDIAMINE	1,3-PROPANEDIAMINE
3312-60-5 1,3-PROPANEDIAMINE, N-CYCLOHEXYL- [PAL]	N-(3-AMINOPROPYL) CYCLOHEXYLAMINE
104-78-9 1,3-PROPANEDIAMINE, N,N-DIETHYL- [PAL]	3-(DIETHYLAMINO)PROPYLAMINE
109-55-7 1,3-PROPANEDIAMINE, N,N-DIMETHYL- [PAL]	3-(DIMETHYLAMINO)-PROPYLAMINE
96-12-8 PROPANE, 1,2-DIBROMO-3-CHLORO- [PAL, PAL, TSC, TSC]	1,2-DIBROMO-3-CHLOROPROPANE
78-87-5 PROPANE, 1,2-DICHLORO- [PAL, TSC, TSC, TSC]	1,2-DICHLOROPROPANE
78-99-9 PROPANE, 1,1-DICHLORO- [PAL, TSC, TSC]	1,1-DICHLOROPROPANE
142-28-9 PROPANE, 1,3-DICHLORO- [PAL, TSC, TSC, TSC]	1,3-DICHLOROPROPANE
594-20-7 PROPANE, 2,2-DICHLORO-	PROPANE, 2,2-DICHLORO-
595-44-8 PROPANE, 1,1-DICHLORO-1-NITRO-	PROPANE, 1,1-DICHLORO-1-NITRO-
463-82-1 PROPANE, 2,2-DIMETHYL- [PAL]	2,2-DIMETHYLPROPANE
109-77-3 PROPANEDINITRILE [TSC, TSC, TSC]	MALONONITRILE
2698-41-1 PROPANEDINITRILE, [(2-CHLOROPHENYL) METHYLENE]- [PAL]	O-CHLOROBENZYLIDENE MALONITRILE
57-55-6 PROPANE-1,2-DIOL [GBR]	1,2-PROPYLENE GLYCOL
57-55-6 1,2-PROPANEDIOL [PAL]	1,2-PROPYLENE GLYCOL
504-63-2 1,3-PROPANEDIOL	1,3-PROPANEDIOL
3296-90-0 1,3-PROPANEDIOL, 2,2-BIS(BROMOETHYL)- DIBROMONEOPENTYL GLYCOL [TSC, TSC, TSC]	2,2-BIS(BROMOMETHYL)-1,3-PROPANEDIOL
115-77-5 1,3-PROPANEDIOL, 2,2-BIS(HYDROXYMETHYL)- [PAL]	PENTAERYTHRITOL
115-84-4 1,3-PROPANEDIOL, 2-BUTYL-2-ETHYL- [PAL]	2-ETHYL-2-BUTYL-1,3-PROPANEDIOL
623-84-7 1,2-PROPANEDIOL DIACETATE	1,2-PROPANEDIOL DIACETATE
115-76-4 1,3-PROPANEDIOL, 2,2-DIETHYL- [PAL]	2,2-DIETHYL-1,3-PROPANEDIOL
126-30-7 1,3-PROPANEDIOL, 2,2-DIMETHYL-	1,3-PROPANEDIOL, 2,2-DIMETHYL-
126-30-7 PROPANEDIOL, 2,2-DIMETHYL-, 1,3- [OTS]	1,3-PROPANEDIOL, 2,2-DIMETHYL-
6423-43-4 1,2-PROPANEDIOL, DINITRATE [PAL]	PROPYLENE GLYCOL DINITRATE
77-99-6 1,3-PROPANEDIOL, 2-ETHYL-2-(HYDROX-YMETHYL)-	1,3-PROPANEDIOL, 2-ETHYL-2-(HYDROXYMETHYL)-
77-99-6 PROPANEDIOL, 2-ETHYL-2-(HYDROX-YMETHYL)-, 1,3- [OTS]	1,3-PROPANEDIOL, 2-ETHYL-2-(HYDROXYMETHYL)-
126-11-4 1,3-PROPANEDIOL, 2-HYDROXYMETHYL-2-NITRO-	1,3-PROPANEDIOL, 2-HYDROXYMETHYL-2-NITRO-
29387-84-6 1,2-PROPANEDIOL, MONOISOPROPYL ETHER	1,2-PROPANEDIOL, MONOISOPROPYL ETHER
126-58-9 1,3-PROPANEDIOL, 2,2'-[OXYBIS(METHYLENE)]	1,3-PROPANEDIOL, 2,2'-[OXYBIS(METHYLENE)]
114719-19-6 1,2-PROPANEDIOL,3-(2-PROPENYLOXY)-,BIS(4-METHYLBENZENESULFONATE)	1,2-PROPANEDIOL,3-(2-PROPENYLOXY)-,BIS(4-METHYL-BENZENESULFONOXY)]BIS-,BIS(4-METHYLBENZENESUL-FONATE)
6178-32-1 PROPANE, 1,2-EPOXY-3-(P-NONYLPHENOXY)- [TSC]	OXIRANE, [(4-NONYLPHENOXY)METHYL]--AZATRICY-CLO[3.3.1.02,4]NON-7-YL ESTER, N-OXIDE, HYDROBROMI DE, [7(S)-(1-ALPHA,2-BETA,4-BETA,5-ALPHA,7-BETA)]-
926-65-8 PROPANE, 2-(ETHENYLOXY)-	PROPANE, 2-(ETHENYLOXY)-
109-53-5 PROPANE, 1-(ETHENYLOXY)-2-METHYL- [PAL]	VINYL ISOBUTYL ETHER

Chemical Name	Reference Name
628-32-0 PROPANE, 1-ETHOXY- [PAL]	ETHYL PROPYL ETHER
431-89-0 PROPANE, 1,1,1,2,3,3,3-HEPTAFLUORO-	PROPANE, 1,1,1,2,3,3,3-HEPTAFLUORO-
110-78-1 PROPANE, 1-ISOCYANATO- [TSC, TSC, TSC]	N-PROPYL ISOCYANATE
557-17-5 PROPANE, 1-METHOXY- [PAL]	METHYL PROPYL ETHER
1634-04-4 PROPANE, 2-METHOXY-2-METHYL- [TSC, TSC, TSC, TSC]	METHYL TERT-BUTYL ETHER
75-28-5 PROPANE, 2-METHYL- [PAL]	ISOBUTANE
107-12-0 PROPANENITRILE [PAL]	PROPIONITRILE
107-12-0 PROPANENITRILE (ETHYL CYANIDE) [RCA]	PROPIONITRILE
78-67-1 PROPANENITRILE, 2,2'-AZOBIS[2-METHYL- [PAL]	AZOBISISOBUTYRONITRILE
702-03-4 PROPANENITRILE, 3-(CYCLOHEXYLAMINO)- [PAL]	N-(2-CYANOETHYL)CYCLOHEXYLAMINE
78-97-7 PROPANENITRILE, 2-HYDROXY- [OTS, PAL, TSC, TSC]	LACTONITRILE
109-78-4 PROPANENITRILE, 3-HYDROXY- [PAL]	ETHYLENE CYANOHYDRIN
75-86-5 PROPANENITRILE, 2-HYDROXY-2-METHYL- [PAL, TSC, TSC]	ACETONE CYANOHYDRIN
110-67-8 PROPANENITRILE, 3-METHOXY-	PROPANENITRILE, 3-METHOXY-
78-82-0 PROPANENITRILE, 2-METHYL- [PAL]	ISOBUTYRONITRILE
79-46-9 PROPANE, 2-NITRO- [PAL, PAL]	2-NITROPROPANE
108-03-2 PROPANE, 1-NITRO- [PAL]	1-NITROPROPANE
108-20-3 PROPANE, 2,2'-OXYBIS- [PAL]	ISOPROPYL ETHER
111-43-3 PROPANE, 1,1'-OXYBIS- [PAL]	DIPROPYL ETHER
108-60-1 PROPANE, 2,2'-OXYBIS(1-CHLORO- [TSC, TSC, TSC]	BIS(2-CHLOROISOPROPYL) ETHER
39638-32-9 PROPANE, 2,2'-OXYBIS[2-CHLORO- [PAL]	BIS(2-CHLOROISOPROPYL)ETHER
63283-80-7 PROPANE, 2,2'-OXYBIS[DICHLORO-	PROPANE, 2,2'-OXYBIS[DICHLORO-
927-07-1 PROPANEPEROXOIC ACID, 2,2-DIMETHYL-, 1,1-DIMETHYLETHYL ESTER [PAL]	TERT-BUTYL PEROXYPIVALATE
1120-71-4 1,3-PROPANE SULTONE	1,3-PROPANE SULTONE
1120-71-4 PROPANE SULTONE [313, AUS, AUS, MIN, QBC, QBC, TLV, WHM]	1,3-PROPANE SULTONE
812-03-3 PROPANE, 1,1,1,2-TETRACHLORO-	PROPANE, 1,1,1,2-TETRACHLORO-
1070-78-6 PROPANE, 1,1,1,3-TETRACHLORO-	PROPANE, 1,1,1,3-TETRACHLORO-
18495-30-2 PROPANE, 1,1,2,3-TETRACHLORO-	PROPANE, 1,1,2,3-TETRACHLORO-
107-03-9 1-PROPANETHIOL	1-PROPANETHIOL
79869-58-2 PROPANETHIOL	PROPANETHIOL
75-66-1 2-PROPANETHIOL, 2-METHYL- [PAL]	2-METHYL-2-PROPANETHIOL
79869-58-2 PROPANETHIOLS [HM1, HM1]	PROPANETHIOL
1185-57-5 1,2,3-PROPANETRICARBOXYLIC ACID, 2-HYDROXY-, AMMONIUM IRON(3+) SALT [PAL]	FERRIC AMMONIUM CITRATE
3012-65-5 1,2,3-PROPANETRICARBOXYLIC ACID, 2-HYDROXY-, DIAMMONIUM SALT [PAL]	AMMONIUM CITRATE, DIBASIC
77-90-7 1,2,3-PROPANETRICARBOXYLIC ACID, 2-(ACETYLOXY)-, TRIBUTYL ESTER	1,2,3-PROPANETRICARBOXYLIC ACID, 2-(ACETYLOXY)-, TRIBUTYL ES
96-18-4 PROPANE, 1,2,3-TRICHLORO- [PAL, TSC, TSC]	1,2,3-TRICHLOROPROPANE
56-81-5 1,2,3-PROPANETRIOL [PAL]	GLYCERIN2-(NITROPHENYL)ETHYL]-,[R-(R*,R*)]-
55-63-0 1,2,3-PROPANETRIOL, TRINITRATE [PAL]	NITROGLYCERINPHENYL] ESTER
74398-71-3 1,2,3-PROPANETRIYL ESTER OF 12-(OXIRANYL-METHOXY)-9-OCTADECANOIC ACID	1,2,3-PROPANETRIYL ESTER OF 12-(OXIRANYL-METHOXY)-9-OCTAOXY

469

Chemical Name	Reference Name
709-98-8 **PROPANIL**	PROPANIL
709-98-8 PROPANIL [N-(3,4-DICHLOROPHENYL) PROPANAMIDE] [313]	PROPANIL
79-09-4 PROPANOIC ACID [PAL]	PROPIONIC ACID
123-62-6 PROPANOIC ACID, ANHYDRIDE [PAL]	PROPIONIC ANHYDRIDE
590-01-2 PROPANOIC ACID, BUTYL ESTER [PAL]	BUTYL PROPIONATE
75-99-0 PROPANOIC ACID, 2,2-DICHLORO- [PAL]	2,2-DICHLOROPROPIONIC ACID
RR-01643-3 **PROPANOIC ACID, 2,2-DIMETHYL-, ETHENYL ESTER**	PROPANOIC ACID, 2,2-DIMETHYL-, ETHENYL ESTER
105-38-4 **PROPANOIC ACID, ETHENYL ESTER**	PROPANOIC ACID, ETHENYL ESTER
4324-38-3 **PROPANOIC ACID, 3-ETHOXY-**	PROPANOIC ACID, 3-ETHOXY-
763-69-9 PROPANOIC ACID, 3-ETHOXY-, ETHYL ESTER [PST]	ETHYL 3-ETHOXYPROPANOATE
10213-74-8 PROPANOIC ACID, 3-(2-ETHYLBUTOXY)- [PAL]	3-(2-ETHYLBUTOXY) PROPIONIC ACID
105-37-3 PROPANOIC ACID, ETHYL ESTER [PAL]	ETHYL PROPIONATE
138-22-7 PROPANOIC ACID, 2-HYDROXY-, BUTYL ESTER [PAL]	N-BUTYL LACTATE
97-64-3 PROPANOIC ACID, 2-HYDROXY-, ETHYL ESTER [PAL]	ETHYL LACTATE
617-51-6 PROPANOIC ACID, 2-HYDROXY-, 1-METHYLETHYL ESTER [PAL]	ISOPROPYL LACTATE
79-31-2 PROPANOIC ACID, 2-METHYL- [PAL]	ISOBUTYRIC ACID
6846-50-0 PROPANOIC ACID, 2-METHYL-, 2,2-DIMETHYL-1-(1-METHYLETHYL)-1,3-PROPANEDIYL ESTER [PST]	2,2,4-TRIMETHYL-1,3-PENTANEDIOL DIISOBUTYRATE
96-33-3 2-PROPANOIC ACID, METHYL ESTER [FLA]	METHYL ACRYLATE
554-12-1 PROPANOIC ACID, METHYL ESTER [PAL]	METHYL PROPIONATE
42978-66-5 2-PROPANOIC ACID, (1-METHYL-1,2-ETHANEDIYL)BIS[OXY(METHYL-2,1-ETHANEDIYL)]ESTER [TSC, TSC, TSC]	TRIPROPYLENE GLYCOL DIACRYLATE
97-62-1 PROPANOIC ACID, 2-METHYL-, ETHYL ESTER [PAL]	ETHYL ISOBUTYRATE
25265-77-4 PROPANOIC ACID, 2-METHYL-, MONOESTER [OTS]	2,2,4-TRIMETHYLPENTANE-1,3-DIOL MONOISOBUTYRATE
3771-19-5 PROPANOIC ACID, 2-METHYL-2-[4-(1,2,3,4-TETRAHYDRO-1-NAPHTHALENYL)PHENOXY] [PAL, PAL]	NAFENOPIN3-HYDROXY-, DISODIUM SALT
106-36-5 PROPANOIC ACID, PROPYL ESTER [PAL]	PROPYL PROPIONATE
93-72-1 PROPANOIC ACID, 2-(2,4,5-TRICHLOROPHENOXY)- [PAL]	SILVEX (2,4,5-TP)
32534-95-5 PROPANOIC ACID, 2-(2,4,5-TRICHLOROPHENOXY)-, ISOOCTYL ESTER [PAL]	2,4,5-TP ACID ESTERS
123-62-6 PROPANOIC ANHYDRIDE [TSC, TSC, TSC]	PROPIONIC ANHYDRIDE
67-63-0 PROPAN-2-OL [GBR, GBR, GBR]	ISOPROPYL ALCOHOL
67-63-0 2-PROPANOL [PAL]	ISOPROPYL ALCOHOL
71-23-8 PROPAN-1-OL [GBR, GBR, GBR]	N-PROPYL ALCOHOL
71-23-8 N-PROPANOL [HM1, HM1]	N-PROPYL ALCOHOL
71-23-8 1-PROPANOL [PAL]	N-PROPYL ALCOHOL
71-23-8 N-PROPANOL [PST]	N-PROPYL ALCOHOL
124028-99-5 **2-PROPANOL,1-[2-[2-[[(4-METHYLPHENYL)SULFONYL]OXY]ETHOXY]ETHOXY]-3-(2-PROPENYLOXY)-,4-METHYLBENZENESULFONATE**	2-PROPANOL,1-[2-[2-[[(4-METHYLPHENYL)SULFONYL]OXY]ETHOXY]ETH
67-63-0 2-PROPANOL (ISOPROPANOL) [TSC, TSC, TSC]	ISOPROPYL ALCOHOL

Chemical Name	Reference Name
555-31-7 2-PROPANOL, ALUMINUM SALT [PST]	ALUMINUM ISO-PROPOXIDE
156-87-6 PROPANOLAMINE	PROPANOLAMINE
78-96-6 2-PROPANOL, 1-AMINO- [PAL]	ISOPROPANOLAMINE
156-87-6 1-PROPANOL, 3-AMINO- [PAL]	PROPANOLAMINE
124-68-5 1-PROPANOL, 2-AMINO-2-METHYL- [PAL]	2-AMINO-2-METHYL-1-PROPANOL
5131-66-8 2-PROPANOL, 1-BUTOXY- [TSC, TSC, TSC]	1-BUTOXY-2-PROPANOL
124-16-3 2-PROPANOL, 1-(2-BUTOXYETHOXY)-	2-PROPANOL, 1-(2-BUTOXYETHOXY)-
29911-28-2 2-PROPANOL, 1-(2-BUTOXY-1-METHYLETHOXY)-	2-PROPANOL, 1-(2-BUTOXY-1-METHYLETHOXY)-
78-89-7 1-PROPANOL, 2-CHLORO- [PAL]	PROPYLENE CHLOROHYDRIN
127-00-4 2-PROPANOL, 1-CHLORO- [PAL]	1-CHLORO-2-PROPANOL
6145-73-9 1-PROPANOL, 2-CHLORO-, PHOSPHATE (3:1)	1-PROPANOL, 2-CHLORO-, PHOSPHATE (3:1)
13674-84-5 2-PROPANOL, 1-CHLORO-, PHOSPHATE (3:1)	2-PROPANOL, 1-CHLORO-, PHOSPHATE (3:1)
13674-84-5 2-PROPANOL, 1-CHLORO-, PHOSPHATE (3:1) [OTS]	2-PROPANOL, 1-CHLORO-, PHOSPHATE (3:1)
616-29-5 2-PROPANOL, 1,3-DIAMINO- [PAL]	1,3-DIAMINO-2-PROPANOL
96-13-9 1-PROPANOL, 2,3-DIBROMO- [TSC, TSC, TSC]	2,3-DIBROMO-1-PROPANOL
126-72-7 1-PROPANOL, 2,3-DIBROMO-, PHOSPHATE (3:1) [PAL, PAL]	TRIS(2,3-DIBROMOPROPYL)PHOSPHATE
2109-64-0 2-PROPANOL, 1-(DIBUTYLAMINO)- [PAL]	DIBUTYLISOPROPANOLAMINEESTER
616-23-9 1-PROPANOL, 2,3-DICHLORO-	1-PROPANOL, 2,3-DICHLORO-
13674-87-8 2-PROPANOL, 1,3-DICHLORO-, PHOSPHATE (3:1) [TSC, TSC, TSC]	1,3-DICHLORO ISOPROPYL PHOSPHATE
75-84-3 1-PROPANOL, 2,2-DIMETHYL-	1-PROPANOL, 2,2-DIMETHYL-
108-16-7 2-PROPANOL, 1-(DIMETHYLAMINO)- [PAL]	N,N-DIMETHYLISOPROPANOLAMINE
57018-52-7 2-PROPANOL, 1-(1,1-DIMETHYLETHOXY)- [TSC, TSC, TSC]	PROPYLENE GLYCOL T-BUTYL ETHER
36483-57-5 1-PROPANOL, 2,2-DIMETHYL-, TRIBROMO DERIVATIVE	1-PROPANOL, 2,2-DIMETHYL-, TRIBROMO DERIVATIVE
110-97-4 2-PROPANOL, 1,1'-IMINOBIS- [PAL]	DIISOPROPANOLAMINE
19721-22-3 1-PROPANOL, 3-MERCAPTO	1-PROPANOL, 3-MERCAPTOYL]-ESTER
124213-39-4 2-PROPANOL,1-[2-[[(4-METHYLPHENYL)SULFONYL]OXY]ETHOXY]-3-(2-PROPENYLOXY)-4-METHYLBENZENESULFONATE	2-PROPANOL,1-[2-[[(4-METHYLPHENYL)SULFONYL]OXY]ETHOXY]-3-(2-HOXY]METHYL]-2-(PROPENYLOXY)ETHOXY]-,4-METHYLBENZENESULFONATE
107-98-2 2-PROPANOL, 1-METHOXY- [PAL, TSC, TSC, TSC]	PROPYLENE GLYCOL MONOMETHYL ETHER
28677-93-2 1-PROPANOL, METHOXY-	1-PROPANOL, METHOXY-
108-65-6 2-PROPANOL, 1-METHOXY-, ACETATE [TSC, TSC, TSC]	PROPYLENE GLYCOL MONOMETHYL ETHER ACETATE
20324-32-7 2-PROPANOL, 1-(2-METHOXY-1-METHYLETHOXY)- [TSC, TSC, TSC]	DIPROPYLENE GLYCOL METHYL ETHER
34590-94-8 PROPANOL, (2-METHOXYMETHYLETHOXY)- [PAL]	DIPROPYLENE GLYCOL MONOMETHYL ETHERO-1,5,5A,6,9,9A-HEXAHYDRO-, 3-OXIDE, (3.ALPHA.,5A.ALPHA.,6.BETA.,9.BETA.,9A.ALPHA.)-
88917-22-0 PROPANOL 1 (OR 2)-2-METHOXYMETHYLETHOXY, ACETATE	PROPANOL 1 (OR 2)-2-METHOXYMETHYLETHOXY, ACETATE
25498-49-1 PROPANOL, [2-(2-METHOXYMETHYLETHOXY)METHYLETHOXY- [TSC, TSC, TSC]	TRIPROPYLENE GLYCOL MONOMETHYL ETHER
10213-77-1 1-PROPANOL, 2-[2-(2-METHOXYPROPOXY)PROPOXY]- [TSC]	2-[2-(2-METHOXYPROPOXY)PROPOXY]-1-PROPANOL
75-65-0 2-PROPANOL, 2-METHYL- [PAL]	TERT-BUTYL ALCOHOL
78-83-1 1-PROPANOL, 2-METHYL- [PAL, TSC, TSC]	ISOBUTYL ALCOHOL

Chemical Name	Reference Name
24800-44-0 **PROPANOL, [(1-METHYL-1,2-ETHANEDIYL)BIS (OXY)]BIS-**	PROPANOL, [(1-METHYL-1,2-ETHANEDIYL)BIS(OXY)]BIS-
23436-19-3 **2-PROPANOL, 1-(2-METHYLPROPOXY)-**	2-PROPANOL, 1-(2-METHYLPROPOXY)-
122-20-3 **2-PROPANOL, 1,1',1"-NITROTRIS- [PAL]**	TRIISOPROPANOLAMINE
110-98-5 **2-PROPANOL, 1,1'-OXYBIS**	2-PROPANOL, 1,1'-OXYBIS
25265-71-8 PROPANOL, OXYBIS- [PAL]	DIPROPYLENE GLYCOL
RR-01230-6 **1-PROPANOL, 3,3'-OXYBIS[2,3-BIS(BRO-MOMETHYL)]-**	1-PROPANOL, 3,3'-OXYBIS[2,3-BIS(BROMOMETHYL)]-, POLYMER WITH ACRYLAMIDE AND SUBSTITUTED ALKYL METHACRYLATE
770-35-4 **2-PROPANOL, 1-PHENOXY-**	2-PROPANOL, 1-PHENOXY-
1569-01-3 **2-PROPANOL, 1-PROPOXY-**	2-PROPANOL, 1-PROPOXY-
67-64-1 **2-PROPANONE [PAL]**	ACETONE
598-31-2 2-PROPANONE, 1-BROMO- [TSC, TSC]	BROMOACETONE
534-07-6 2-PROPANONE, 1,3-DICHLORO- [TSC, TSC]	BIS(CHLOROMETHYL)KETONE
684-16-2 2-PROPANONE, 1,1,1,3,3,3-HEXAFLUORO- [PAL]	HEXAFLUOROACETONE
79-03-8 PROPANOYL CHLORIDE [PAL]	PROPIONYL CHLORIDE
7292-16-2 **PROPAPHOS**	PROPAPHOS
2312-35-8 **PROPARGITE**	PROPARGITE
707-19-7 **PROPARGYL ALCOHOL**	PROPARGYL ALCOHOL
106-96-7 PROPARGYL BROMIDE [302, EPA, EPA, NFP, NFP]	3-BROMOPROPYNE
107-02-8 2-PROPENAL [PAL, TSC, TSC, TSC]	ACROLEIN
13586-68-0 **2-PROPENAL, 3,4-(1,1-DIMETHYLETHYL) PHENYL-2-METHYL-**	2-PROPENAL, 3,4-(1,1-DIMETHYLETHYL)PHENYL-2-METHYL-
1504-74-1 **2-PROPENAL, 3-(2-METHOXYPHENYL)-**	2-PROPENAL, 3-(2-METHOXYPHENYL)-
78-85-3 2-PROPENAL, 2-METHYL- [PAL, TSC, TSC, TSC]	METHACRYLALDEHYDE
101-39-3 **2-PROPENAL, 2-METHYL-3-PHENYL-**	2-PROPENAL, 2-METHYL-3-PHENYL-
104-55-2 2-PROPENAL, 3-PHENYL- [TSC, TSC, TSC]	CINNAMALDEHYDE
1331-92-6 **2-PROPENAL, 3-PHENYL, MONOPENTYL DERIVATIVE**	2-PROPENAL, 3-PHENYL-, MONOPENTYL DERIVATIVE
79-06-1 2-PROPENAMIDE [PAL]	ACRYLAMIDE
79-06-1 2-PROPENAMIDE (ACRYLAMIDE) [TSC, TSC, TSC]	ACRYLAMIDE
RR-00910-9 **2-PROPENAMIDE, N-[3-(DIMETHYLAMINO) PROPYL]-**	2-PROPENAMIDE, N-[3-(DIMETHYLAMINO)PROPYL]-
107-11-9 2-PROPEN-1-AMINE [PAL]	ALLYLAMINE
115-07-1 1-PROPENE [PAL]	PROPYLENE
106-95-6 1-PROPENE, 3-BROMO- [PAL]	ALLYL BROMIDE
107-05-1 1-PROPENE, 3-CHLORO- [PAL]	ALLYL CHLORIDE
557-98-2 1-PROPENE, 2-CHLORO- [PAL]	2-CHLOROPROPENE
590-21-6 1-PROPENE, 1-CHLORO- [PAL]	1-CHLOROPROPYLENE
563-47-3 1-PROPENE, 3-CHLORO-2-METHYL- [PAL, PAL]	3-CHLORO-2-METHYLPROPENE
78-88-6 1-PROPENE, 2,3-DICHLORO- [PAL, TSC, TSC]	2,3-DICHLOROPROPENE
542-75-6 1-PROPENE, 1,3-DICHLORO- [PAL, PAL, TSC, TSC]	1,3-DICHLOROPROPENE
563-54-2 **1-PROPENE, 1,2-DICHLORO-**	1-PROPENE, 1,2-DICHLORO-CHLORIDE)
563-58-6 1-PROPENE, 1,1-DICHLORO- [TSC, TSC]	1,1-DICHLOROPROPENE
26952-23-8 1-PROPENE, DICHLORO- [PAL, TSC, TSC]	DICHLOROPROPENE
8003-19-8 1-PROPENE, 1,3-DICHLORO-, MIXT. WITH 1,2-DICHLOROPROPANE [PAL]	DICHLOROPROPANE-DICHLOROPROPENE (MIXTURE)

Chemical Name	Reference Name
869-29-4 2-PROPENE-1,1-DIOL, DIACETATE [PAL]	ALLYLIDENE DIACETATE
3917-15-5 1-PROPENE, 3-(ETHENYLOXY)-	1-PROPENE, 3-(ETHENYLOXY)-
928-55-2 1-PROPENE, 1-ETHOXY-	1-PROPENE, 1-ETHOXY-
1888-71-7 1-PROPENE, 1,1,2,3,3,3-HEXACHLORO- [TSC, TSC, TSC]	HEXACHLOROPROPENE
116-15-4 1-PROPENE, 1,1,2,3,3,3-HEXAFLUORO- [TSC, TSC]	HEXAFLUOROPROPYLENE
1476-23-9 PROPENE, 3-ISOCYANATO-	PROPENE, 3-ISOCYANATO-
1476-23-9 PROPENE, 3-ISOCYANATO- [TSC, TSC, TSC]	PROPENE, 3-ISOCYANATO-
57-06-7 1-PROPENE, 3-ISOTHIOCYANATO- [PAL]	ALLYL ISOTHIOCYANATE
115-11-7 1-PROPENE, 2-METHYL- [PAL]	ISOBUTYLENE
107-13-1 2-PROPENENITRILE [PAL, PAL]	ACRYLONITRILE
126-98-7 2-PROPENENITRILE, 2-METHYL- [PAL]	METHYLACRYLONITRILE
9003-54-7 2-PROPENENITRILE POLYMER WITH ETHENYLBENZENE	2-PROPENENITRILE POLYMER WITH ETHENYLBENZENE
RR-00996-1 2-PROPENENITRILE, POLYMER WITH 1,3-BU-TADIENE, 3-CARBOXY-1-CYANO-1-METHYL-PROPYL-TERMINATED, POLYMERS WITH EPICHLOROHYDRIN, FORMALDEHYDE, 4,4'-(1-METHYL ETHYLIDENE)BIS(2,6-DI	2-PROPENENITRILE, POLYMER WITH 1,3-BUTADIENE, 3-CARBOXY-1-CYTHYL ETHYLIDENE) BIS(2,6-DIBRO-MOPHENOL) AND PHENOL, 2-METHYL-2-PROPENOATE
557-40-4 1-PROPENE, 3,3'-OXYBIS- [PAL]	DIALLYL ETHER
10436-39-2 1-PROPENE, 1,1,2,3-TETRACHLORO- [TSC, TSC]	1,1,2,3-TETRACHLOROPROPENE
677-21-4 1-PROPENE, 3,3,3-TRIFLUORO-	1-PROPENE, 3,3,3-TRIFLUORO-
13987-01-4 1-PROPENE, TRIMER [PAL]	TRIPROPYLENE
RR-01177-8 PROPENOATE-TERMINATED ALKYL SUBSTI-TUTED SILYL ESTER	PROPENOATE-TERMINATED ALKYL SUBSTITUTED SILYL ESTER
79-10-7 2-PROPENOIC ACID [PAL]	ACRYLIC ACID
RR-01732-3 2-PROPENOIC ACID 3-(TRIMETHOXYSILYL) PROPYL ESTER	2-PROPENOIC ACID 3-(TRIMETHOXYSILYL) PROPYL ESTER
95-39-6 2-PROPENOIC ACID, BICYCLO[2.2.1]HEPT-5-EN-2-YL-, METHYL ESTER	2-PROPENOIC ACID, BICYCLO[2.2.1]HEPT-5-EN-2-YL-, METHYL ESTE
141-32-2 2-PROPENOIC ACID, BUTYL ESTER [PAL, TSC, TSC]	BUTYL ACRYLATE
RR-01685-3 2-PROPENOIC ACID, C-18-26 AND C>20 ALKYL ESTERS	2-PROPENOIC ACID, C-18-26 AND C>20 ALKYL ESTERS
6606-65-1 2-PROPENOIC ACID, 2-CYANO-, BUTYL ESTER	2-PROPENOIC ACID, 2-CYANO-, BUTYL ESTER
6197-30-4 2-PROPENOIC ACID, 2-CYANO-3,3-DIPHENYL-, 2-ETHYLHEXYL ESTER	2-PROPENOIC ACID, 2-CYANO-3,3-DIPHENYL-, 2-ETHYL-HEXYL ESTER
21982-43-4 2-PROPENOIC ACID, 2-CYANO-, ETHOXY ETHYL ESTER	2-PROPENOIC ACID, 2-CYANO-, ETHOXY ETHYL ESTER
106-71-8 2-PROPENOIC ACID, 2-CYANOETHYL ES-TER [PAL, TSC, TSC]	2-CYANOETHYL ACRYLATE
7085-85-0 2-PROPENOIC ACID, 2-CYANO-, ETHYL ES-TER [TSC, TSC]	ETHYL CYANOACRYLATE
1069-55-2 2-PROPENOIC ACID, 2-CYANO-, ISOBUTYL ESTER	2-PROPENOIC ACID, 2-CYANO-, ISOBUTYL ESTER
27816-23-5 2-PROPENOIC ACID, 2-CYANO-, 2-METHOXYETHYL ESTER	2-PROPENOIC ACID, 2-CYANO-, 2-METHOXYETHYL ESTER
137-05-3 2-PROPENOIC ACID, 2-CYANO-, METHYL ES-TER [PAL, TSC, TSC]	METHYL 2-CYANOACRYLATE
10586-17-1 2-PROPENOIC ACID, 2-CYANO-, 1-METHYLETHYL ESTER	2-PROPENOIC ACID, 2-CYANO-, 1-METHYLETHYL ES-TER
7324-02-9 2-PROPENOIC ACID, 2-CYANO-, 2-PROPENYL ESTER	2-PROPENOIC ACID, 2-CYANO-, 2-PROPENYL ESTER

Chemical Name	Reference Name
23023-91-8 2-PROPENOIC ACID, 2-CYANO-, 2,2,2-TRIFLUO-ROMETHYL ESTER	2-PROPENOIC ACID, 2-CYANO-, 2,2,2-TRIFLUO-ROMETHYL ESTER
2156-96-9 2-PROPENOIC ACID, DECYL ESTER [PAL, TSC, TSC]	DECYL ACRYLATEPHENYL]-4,4'-DIYL)BIS(AZO)]BIS-[4,5-DIHYDROXY-, TETRASODIUM SALT
19660-16-3 2-PROPENOIC ACID, 2,3-DIBROMOPROPYL ESTER	2-PROPENOIC ACID, 2,3-DIBROMOPROPYL ESTER
2426-54-2 2-PROPENOIC ACID, 2-(DIETHYLAMINO)ETHYL ESTER [PAL, TSC, TSC]	2-(DIETHYLAMINO)ETHYL ACRYLATE
87320-05-6 2-PROPENOIC ACID, [2-[1,1-DIMETHYL-2-[(1-OXO-2-PROPENYL)OXY]ETHYL]-5-ETHYL-1,3-DIOXAN-5-YL)METHYL ESTER	2-PROPENOIC ACID, [2-[1,1-DIMETHYL-2-[(1-OXO-2-PROPENYL)OXY]
1663-39-4 2-PROPENOIC ACID, 1,1-DIMETHYLETHYL ESTER	2-PROPENOIC ACID, 1,1-DIMETHYLETHYL ESTER
2223-82-7 2-PROPENOIC ACID, 2,2-DIMETHYL-1,3-PROPANEDIYL ESTER [TSC, TSC]	NEOPENTYL GLYCOL DIACRYLATE
17977-09-2 2-PROPENOIC ACID, 2,2-DINITROPROPYL ESTER	2-PROPENOIC ACID, 2,2-DINITROPROPYL ESTER
RR-01684-2 2-PROPENOIC ACID, DOCOSYL ESTER	2-PROPENOIC ACID, DOCOSYL ESTERLKYL TRIOL
106-74-1 2-PROPENOIC ACID, 2-ETHOXYETHYL ESTER [TSC, TSC]	ETHYLENE GLYCOL MONOETHYL ETHER ACRYLATE
3953-10-4 2-PROPENOIC ACID, 2-ETHYLBUTYL ESTER [PAL, TSC, TSC]	2-ETHYLBUTYL ACRYLATEYL ESTER
140-88-5 2-PROPENOIC ACID, ETHYL ESTER [PAL, PAL, TSC, TSC, TSC]	ETHYL ACRYLATE
103-11-7 2-PROPENOIC ACID, 2-ETHYLHEXYL ESTER [PAL, TSC, TSC]	2-ETHYLHEXYL ACRYLATE
33791-58-1 2-PROPENOIC ACID, 3A,4,5,6,7,7A-HEXAHYDRO-4,7-METHANO-1H-INDENYL ESTER	2-PROPENOIC ACID, 3A,4,5,6,7,7A-HEXAHYDRO-4,7-METHANO-1H-ESTER
13048-33-4 2-PROPENOIC ACID, 1,6-HEXANEDIYL ESTER [TSC, TSC]	1,6-HEXANEDIOL DIACRYLATE
2499-95-8 2-PROPENOIC ACID, HEXYL ESTER	2-PROPENOIC ACID, HEXYL ESTER
2421-27-4 2-PROPENOIC ACID, 2-HYDROXYBUTYL ESTER	2-PROPENOIC ACID, 2-HYDROXYBUTYL ESTER
818-61-1 2-PROPENOIC ACID, 2-HYDROXYETHYL ESTER [PAL, TSC, TSC]	ETHYLENE GLYCOL MONOACRYLATE
RR-00234-6 2-PROPENOIC ACID, 1-[HYDROXYMETHYL]PROPYL ESTER	2-PROPENOIC ACID, 1-[HYDROXYMETHYL]PROPYL ESTERX-HYDROXY METHACRYLATE
1330-61-6 2-PROPENOIC ACID, ISODECYL ESTER	2-PROPENOIC ACID, ISODECYL ESTER
29590-42-9 2-PROPENOIC ACID, ISOOCTYL ESTER	2-PROPENOIC ACID, ISOOCTYL ESTER
79-41-4 2-PROPENOIC ACID, 2-METHYL- [PAL]	METHACRYLIC ACID
RR-01004-8 2-PROPENOIC ACID, 3-(DIMETHYLAMINO)-2,2-DIMETHYLPROPYL ESTER	2-PROPENOIC ACID, 3-(DIMETHYLAMINO)-2,2-DIMETHYLPROPYL ESTER-DIOXO-5,12-DIAZAHEXADE-CANE, 1,16-DIYL ESTER
96478-09-0 2-PROPENOIC ACID, 2-METHYL-, 2-[3-(2H-BENZOTRIAZOL-2-YL)-4-HYDROXYPHENYL]ETHYL ESTER	2-PROPENOIC ACID, 2-METHYL-, 2-[3-(2H-BENZOTRIA-ZOL-2-YL)-4-H
16715-83-6 2-PROPENOIC ACID, 2-METHYL-, 2-[BIS(1-METHYLETHYL)AMINO]ETHYL ESTER	2-PROPENOIC ACID, 2-METHYL-, 2-[BIS(1-METHYLETHYL)AMINO]LUIDINO]-PROPIONITRILE
2082-81-7 2-PROPENOIC ACID, 2-METHYL-, 1,4-BUTANEDIYL ESTER [TSC, TSC]	1,4-BUTANEDIOL DIMETHACRYLATE
97-88-1 2-PROPENOIC ACID, 2-METHYL-, BUTYL ESTER [TSC, TSC, TSC, PAL]	BUTYL METHACRYLATE
101-43-9 2-PROPENOIC ACID, 2-METHYL-, CYCLOHEXYL ESTER	2-PROPENOIC ACID, 2-METHYL-, CYCLOHEXYL ESTER
105-16-8 2-PROPENOIC ACID, 2-METHYL-, 2-(DIETHYLAMINO)ETHYL ESTER [TSC, TSC]	2-(N,N-DIETHYLAMINO)ETHYL METHACRYLATE

Chemical Name	Reference Name
2867-47-2 2-PROPENOIC ACID, 2-METHYL-, 2-(DIMETHY-LAMINO)ETHYL ESTER [PAL, TSC, TSC]	N,N-DIMETHYLAMINOETHYL METHACRYLATEDIHY-DROXY-9,10-DIOXO-, DISODIUM SALT
3775-90-4 2-PROPENOIC ACID, 2-METHYL-, 2-[(1,1-DIMETHYLETHYL)AMINO]ETHYL ESTER [PAL, TSC, TSC]	TERT-BUTYLAMINOETHYL METHACRYLATE4,6-DINI-TROPHENYL)AZO]-4-METHOXYPHENYL]-
585-07-9 2-PROPENOIC ACID, 2-METHYL-, 1,1-DIMETHYLETHYL ESTER	2-PROPENOIC ACID, 2-METHYL-, 1,1-DIMETHYLETHYL ESTER
142-90-5 2-PROPENOIC ACID, 2-METHYL-, DODECYL ESTER [TSC, TSC]	LAURYL METHACRYLATE
96-33-3 2-PROPENOIC ACID, METHYL ESTER [PAL, TSC, TSC]	METHYL ACRYLATE
97-90-5 2-PROPENOIC ACID, 2-METHYL-, 1,2-ETHANEDIYL ESTER	2-PROPENOIC ACID, 2-METHYL-, 1,2-ETHANEDIYL ESTER
109-16-0 2-PROPENOIC ACID, 2-METHYL-, 1,2-ETHANEDIYLBIS(OXY-2,1-ETHANEDIYL) ES-TER [TSC, TSC]	TRIETHYLENE GLYCOL DIMETHACRYLATE
97-63-2 2-PROPENOIC ACID, 2-METHYL-, ETHYL ES-TER [PAL, TSC, TSC, TSC]	ETHYL METHACRYLATE
688-84-6 2-PROPENOIC ACID, 2-METHYL-, 2-ETHYL-HEXYL ESTER [TSC, TSC]	2-ETHYLHEXYL METHACRYLATE
689-12-3 2-PROPENOIC ACID, 1-METHYLETHYL ESTER	2-PROPENOIC ACID, 1-METHYLETHYL ESTER
55205-38-4 2-PROPENOIC ACID, (1-METHYLETHYLIDENE)BIS(2,6-DIBROMO-4,1-PHENYLENE) ESTER [TSC, TSC, TSC, TSC]	TETRABROMOBISPHENOL-A DIACRYLATE
142-09-6 2-PROPENOIC ACID, 2-METHYL-, HEXYL ES-TER [TSC, TSC]	HEXYLMETHACRYLATE
868-77-9 2-PROPENOIC ACID, 2-METHYL-, 2-HYDROX-YETHYL ESTER [TSC, TSC]	2-HYDROXYETHYL METHACRYLATE
923-26-2 2-PROPENOIC ACID, 2-METHYL-, 2-HYDROX-YPROPYL ESTER	2-PROPENOIC ACID, 2-METHYL-, 2-HYDROXYPROPYL ESTER
30674-80-7 2-PROPENOIC ACID, 2-METHYL-2-ISOCYANA-TOETHYL ESTER [TSC, TSC, TSC]	METHACRYLOYLOXYETHYL ISOCYANATE
73597-26-9 2-PROPENOIC ACID, 2-METHYL-, 2-[[[[[5-ISO-CYANATO-1,3,3-TRIMETHYLCYCLOHEXYL]METHYL]AMINO]CARBONYL]OXY]ETHYL ES-TER	2-PROPENOIC ACID, 2-METHYL-, 2-[[[[[5-ISOCYANATO-1,3,3-TRIMEMMONIO]ETHYL]-OMEGA-HYDROXY-, N,N'-DI-C(14-18) ACYL DERIVAT IVES, METHYL SULFATES (SALTS)
80-62-6 2-PROPENOIC ACID, 2-METHYL-, METHYL ESTER [TSC, PAL]	METHYL METHACRYLATE
97-86-9 2-PROPENOIC ACID, 2-METHYL-, 2-METHYL-PROPYL ESTER [TSC, TSC]	ISOBUTYL METHACRYLATE
32360-05-7 2-PROPENOIC ACID, 2-METHYL-, OCTADECYL ESTER	2-PROPENOIC ACID, 2-METHYL-, OCTADECYL ESTER
82428-30-6 2-PROPENOIC ACID, 2-METHYL-, 7-OXABICY-CLO[4.1.0]HEPT-3-YLMETHYL ESTER	2-PROPENOIC ACID, 2-METHYL-, 7-OXABICYCLO[4.1.0]HEPT-3-YLMET
106-91-2 2-PROPENOIC ACID, 2-METHYL-, OXIRANYL-METHYL ESTER [TSC, TSC]	GLYCIDYL METHACRYLATE
2358-84-1 2-PROPENOIC ACID, 2-METHYL-, OXYDI-2,1-ETHANEDIYL ESTER [TSC, TSC]	DIETHYLENE GLYCOL DIMETHACRYLATE
106-63-8 2-PROPENOIC ACID, 2-METHYLPROPYL ES-TER [PAL, TSC, TSC, TSC]	ISOBUTYL ACRYLATE
96-05-9 2-PROPENOIC ACID, 2-METHYL-, 2-PROPENYL ESTER [TSC, TSC]	ALLYL METHACRYLATE
2210-28-8 2-PROPENOIC ACID, 2-METHYL-, PROPYL ESTER	2-PROPENOIC ACID, 2-METHYL-, PROPYL ESTER
5536-61-8 2-PROPENOIC ACID, 2-METHYL-, SODIUM SALT	2-PROPENOIC ACID, 2-METHYL-, SODIUM SALT
2455-24-5 2-PROPENOIC ACID, 2-METHYL-, (TETRAHY-DRO-2-FURANYL)METHYL ESTER [TSC, TSC]	TETRAHYDROFURFURYL METHACRYLATE

Chemical Name	Reference Name
2530-85-0 2-PROPENOIC ACID, 2-METHYL-, 3-(TRIMETHOXYSILYL)PROPYL ESTER [TSC, TSC]	METHACRYLIC ACID, 3-(TRIMETHOXYSILYL)PROPYL ESTER
7534-94-3 2-PROPENOIC ACID, 2-METHYL-, 1,7,7-TRIMETHYLBICYCLO[2.2.1]HEPT-2-YL ESTER, EXO-	2-PROPENOIC ACID, 2-METHYL-, 1,7,7-TRIMETHYLBICYCLO[2.2.1]
RR-01003-7 2-PROPENOIC ACID, 2-METHYL-, 7,7,9-TRIMETHYL-4,13-DIOXO-3,14-DIOXO-5,12-DI-AZAHEXADECANE, 1,16-DIYL ESTER	2-PROPENOIC ACID, 2-METHYL-, 7,7,9-TRIMETHYL-4,13-DIOXO-3,14
RR-01005-9 2-PROPENOIC ACID, 2-METHYL-, 3,3,5-TRIMETHYL CYCLOHEXYL ESTER	2-PROPENOIC ACID, 2-METHYL-, 3,3,5-TRIMETHYL CYCLOHEXYL ESTE
25584-83-2 2-PROPENOIC ACID, MONOESTER WITH 1,2-PROPANEDIOL	2-PROPENOIC ACID, MONOESTER WITH 1,2-PROPANE-DIOL
RR-01646-6 2-PROPENOIC ACID, 2-[[(1-METHYLETHOXY) CARBONYL]AMINO]ETHYLESTER	2-PROPENOIC ACID, 2-[[(1-METHYLETHOXY) CAR-BONYL]AMINO]ETHYL
79637-74-4 2-PROPENOIC ACID, OCTAHYDRO-4,7-METHANO-1H-INDENYL ESTER	2-PROPENOIC ACID, OCTAHYDRO-4,7-METHANO-1H-INDENYL ESTER
RR-00989-2 2-PROPENOIC ACID [OCTAHYDRO-4,7-METHANO-1H-INDENE-1,5(1,6 OR2,5)-DIYL]BIS (METHYLENE) ESTER	2-PROPENOIC ACID [OCTAHYDRO-4,7-METHANO-1H-INDENE-1,5(1,6 OR, AND DISUBSTITUTED BENZENE
2499-59-4 2-PROPENOIC ACID, OCTYL ESTER	2-PROPENOIC ACID, OCTYL ESTERMONOHYDROCHLO-RIDE
64630-63-3 2-PROPENOIC ACID, 7-OXABICYCLO[4.1.0] HEPT-3-YLMETHYL ESTER	2-PROPENOIC ACID, 7-OXABICYCLO[4.1.0]HEPT-3-YL-METHYL ESTER
106-90-1 2-PROPENOIC ACID, OXIRANYLMETHYL ES-TER [TSC, TSC]	GLYCIDYL ACRYLATE
115965-75-8 2-PROPENOIC ACID, 2-(2-OXO-3-OXAZO-LIDINYL) ETHYL ESTER	2-PROPENOIC ACID, 2-(2-OXO-3-OXAZOLIDINYL) ETHYL ESTERATE)
25987-30-8 2-PROPENOIC ACID, POLYMER WITH 2-PROPE-NAMIDE, SODIUM SALT	2-PROPENOIC ACID, POLYMER WITH 2-PROPENAMIDE, SODIUM SALT-
925-60-0 2-PROPENOIC ACID, PROPYL ESTER	2-PROPENOIC ACID, PROPYL ESTER
RR-01683-1 2-PROPENOIC ACID, REACTION PRODUCT WITH 2-OXEPROPANONE AND ALKYL TRIOL	2-PROPENOIC ACID, REACTION PRODUCT WITH 2-OXEPROPANONE AND A
7446-81-3 2-PROPENOIC ACID, SODIUM SALT [TSC, TSC]	SODIUM ACRYLATE
3076-04-8 2-PROPENOIC ACID, TRIDECYL ESTER	2-PROPENOIC ACID, TRIDECYL ESTER
RR-01006-0 2-PROPENOIC ACID, 3,3,5-TRIMETHYLCYCLO-HEXYL ESTER	2-PROPENOIC ACID, 3,3,5-TRIMETHYLCYCLOHEXYL ESTERR
42404-50-2 2-PROPENOIC ACID, 2-[[[[[1,3,3-TRIMETHYL-5-[[[2-[(1-OXO-2-PROPENYL)OXY]ETHOXY] CARBONYL]AMINO]CYCLOHEXYL]METHYL] AMINO] CARBONYL]OXY]ETHYL ESTER	2-PROPENOIC ACID, 2-[[[[[1,3,3-TRIMETHYL-5-[[[2-[(1-OXO-2-PRLFONATE)
107-18-6 2-PROPEN-1-OL [PAL]	ALLYL ALCOHOL
108-22-5 1-PROPEN-2-OL, ACETATE [PAL]	ISOPROPENYL ACETATE
513-42-8 2-PROPEN-1-OL, 2-METHYL- [PAL]	METHALLYL ALCOHOL
107-05-1 PROPENYL CHLORIDE (CIS-; TRANS-) [HM1, HM1, HM1, HM1]	ALLYL CHLORIDE
928-55-2 PROPENYL ETHYL ETHER [FLA, MSL, NFP, NFP]	1-PROPENE, 1-ETHOXY-
106-92-3 ((2-PROPENYLOXY)METHYL)OXIRANE [FLA, ONT, ONT, ONT]	ALLYL GLYCIDYL ETHER
31218-83-4 PROPETAMPHOS	PROPETAMPHOS
31218-83-4 PROPETAMPHOS [3-[[(ETHYLAMINO) METHOXYPHOSPHINOTHIOYL]OXY]-2-BUTENOIC ACID, 1-METHYLETHYL ESTER] [313]	PROPETAMPHOS
122-42-9 PROPHAM	PROPHAM

Chemical Name	Reference Name
60207-90-1 PROPICONAZOLE [1-[2-(2,4-DICHLOROPHENYL) -4-PROPYL-1,3-DIOXOLAN-2-YL]-METHYL-1H-1, 2,4,-TRIAZOLE]	PROPICONAZOLE [1-[2-(2,4-DICHLOROPHENYL)-4- PROPYL-1,3-DIOXOL5-PYRIMIDINEMETHANOL]
57-57-8 BETA-PROPIOLACTONE	BETA-PROPIOLACTONE
57-57-8 PROPIOLACTONE, BETA [302]	BETA-PROPIOLACTONE
123-38-6 PROPIONALDEHYDE	PROPIONALDEHYDE
79-09-4 PROPIONIC ACID	PROPIONIC ACID
73826-29-6 PROPIONIC ACID, 2-(2,4,5-TRICHLOROPHE- NOXY)-, P-CHLOROPHENACYL ESTER	PROPIONIC ACID, 2-(2,4,5-TRICHLOROPHENOXY)-, P- CHLOROPHENACY
123-62-6 PROPIONIC ANHYDRIDE	PROPIONIC ANHYDRIDE
107-12-0 PROPIONIC NITRILE [MSL, NFP, NFP]	PROPIONITRILE
107-12-0 PROPIONITRILE	PROPIONITRILE
542-76-7 PROPIONITRILE, 3-CHLORO- [FLA, 302]	3-CHLOROPROPIONITRILE
542-76-7 PROPIONITRILE, 3-CHLORO [MSL]	3-CHLOROPROPIONITRILE
542-76-7 PROPIONITRILE, 3-CHLORO- [PAL]	3-CHLOROPROPIONITRILE
107-12-0 PROPIONITRILE [PROPANENITRILE] [EP3]	PROPIONITRILE
79-03-8 PROPIONYL CHLORIDE	PROPIONYL CHLORIDE
3248-28-0 PROPIONYL PEROXIDE	PROPIONYL PEROXIDE
70-69-9 PROPIOPHENONE, 4-AMINO-	PROPIOPHENONE, 4-AMINO-
70-69-9 PROPIOPHENONE, 4'-AMINO- [PAL]	PROPIOPHENONE, 4-AMINO-
114-26-1 PROPOXUR	PROPOXUR
114-26-1 PROPOXUR (BAYGON) [CAA, OS1, QBC]	PROPOXUR
114-26-1 PROPOXUR (ISO) [GBR, GBR]	PROPOXUR
114-26-1 PROPOXUR [PHENOL, 2-(1-METHYLETHOXY)-, METHYLCARBAMATE] [313]	PROPOXUR
23305-64-8 2-[2-(PROPOXYETHOXY)ETHOXY]-ETHANOL	2-[2-(PROPOXYETHOXY)ETHOXY]-ETHANOL
1569-01-3 N-PROPOXY-2-PROPANOL [PST]	2-PROPANOL, 1-PROPOXY-
71-23-8 PROPYL ALCOHOL [MEX, MEX]	N-PROPYL ALCOHOL
109-60-4 N-PROPYL ACETATE	N-PROPYL ACETATE
109-60-4 PROPYL ACETATE [MAK, MAK, NFP, NFP, WHM]	N-PROPYL ACETATE
71-23-8 N-PROPYL ALCOHOL	N-PROPYL ALCOHOL
71-23-8 PROPYL ALCOHOL [AUS, AUS, AUS, BC1, BC1, BC1, MIN, MSL, NFP, NFP, NJL, NJL, OTV]	N-PROPYL ALCOHOL
71-23-8 PROPYL ALCOHOL (PROPANOL) [QBC, QBC, QBC, FLA]	N-PROPYL ALCOHOL
107-10-8 PROPYLAMINE [FLA, HM1, HM1, HM1, MSL, NFP, NFP, NJL, NJL, WHM]	1-PROPANAMINE
103-65-1 PROPYL BENZENE	PROPYL BENZENE
103-65-1 PROPYLBENZENE [FLA]	PROPYL BENZENE
103-65-1 N-PROPYL BENZENE [HM1, HM1]	PROPYL BENZENE
103-65-1 PROPYLBENZENE [MSL, NFP, NFP]	PROPYL BENZENE
103-65-1 PROPYLBENZENE, N- [OTS]	PROPYL BENZENE
103-65-1 N-PROPYLBENZENE [SDW, TSE]	PROPYL BENZENE
RR-00261-9 2-PROPYLBIPHENYL	2-PROPYLBIPHENYL
106-94-5 N-PROPYL BROMIDE	N-PROPYL BROMIDE
105-66-8 N-PROPYL BUTYRATE [FLA, MSL, NFP, NFP]	BUTANOIC ACID, PROPYL ESTER
627-12-3 N-PROPYL CARBAMATE	N-PROPYL CARBAMATE
540-54-5 PROPYL CHLORIDE	PROPYL CHLORIDE
109-61-5 PROPYLCHLOROFORMATE	PROPYLCHLOROFORMATE

Chemical Name	Reference Name
109-61-5 PROPYL CHLOROFORMATE [302]	PROPYLCHLOROFORMATE
109-61-5 N-PROPYL CHLOROFORMATE [HM1, HM1, HM1]	PROPYLCHLOROFORMATE
109-61-5 PROPYL CHLOROFORMATE [MSL, PAL]	PROPYLCHLOROFORMATE
109-61-5 N-PROPYL CHLOROFORMATE [WHM]	PROPYLCHLOROFORMATE
109-61-5 PROPYL CHLOROFORMATE [CARBONOCHLO-RIDIC ACID, PROPYLESTER] [EP3]	PROPYLCHLOROFORMATE
13889-92-4 PROPYL CHLOROTHIOFORMATE [FLA, MSL, NFP, NFP]	CARBONOCHLORIDOTHIOIC ACID, S-PROPYL ESTER
RR-00881-1 PROPYLCYCLOHEXANE	PROPYLCYCLOHEXANE
2040-96-2 PROPYLCYCLOPENTANE	PROPYLCYCLOPENTANE
115-07-1 PROPYLENE	PROPYLENE
115-07-1 PROPYLENE (PROPENE) [313, MSL]	PROPYLENE
78-87-5 PROPYLENE DICHLORIDE [MEX, MEX]	1,2-DICHLOROPROPANE
115-07-1 PROPYLENE (1-PROPENE) [EP3]	PROPYLENE
108-32-7 PROPYLENE CARBONATE	PROPYLENE CARBONATE
78-89-7 PROPYLENE CHLOROHYDRIN	PROPYLENE CHLOROHYDRIN
78-90-0 PROPYLENE DIAMINE	PROPYLENE DIAMINE
78-90-0 1,2-PROPYLENEDIAMINE [HM1, HM1]	PROPYLENE DIAMINE
78-90-0 PROPYLENEDIAMINE [MSL, NFP, NFP, NJL, NJL]	PROPYLENE DIAMINE
78-87-5 PROPYLENE DICHLORIDE [OTV, AUS, AUS]	1,2-DICHLOROPROPANE
78-87-5 PROPYLENE DICHLORIDE (1,2-DICHLORO-PROPANE) [BC1, BC1, CAA]	1,2-DICHLOROPROPANE
78-87-5 PROPYLENE DICHLORIDE [CEX, CEX, FLA, HM1, HM1, HM1, HM1, MIN, MSL, NFP, NFP, NIO, NIO, NIO, OS1, OS1, OS1, RCU, RCU, TLV, TLV, WHM]	1,2-DICHLOROPROPANE
78-87-5 PROPYLENE DICHLORIDE (1,2-DICHLORO-PROPANE) [HON, HON]	1,2-DICHLOROPROPANE
6423-43-4 PROPYLENE DINITRATE [GBR, GBR, GBR]	PROPYLENE GLYCOL DINITRATE
RR-01131-4 PROPYLENE-ETHYLENE THIOETHER	PROPYLENE-ETHYLENE THIOETHER
57-55-6 1,2-PROPYLENE GLYCOL	1,2-PROPYLENE GLYCOL
57-55-6 PROPYLENE GLYCOL [HON, MIN, NFP, NFP, PST]	1,2-PROPYLENE GLYCOL
9005-37-2 PROPYLENE GLYCOL ALGINATE	PROPYLENE GLYCOL ALGINATE
1331-17-5 PROPYLENE GLYCOL, ALLYL ETHER	PROPYLENE GLYCOL, ALLYL ETHER
57018-52-7 PROPYLENE GLYCOL T-BUTYL ETHER	PROPYLENE GLYCOL T-BUTYL ETHER
6423-43-4 PROPYLENE GLYCOL DINITRATE	PROPYLENE GLYCOL DINITRATE
6423-43-4 1,2-PROPYLENE GLYCOL DINITRATE [ONT, ONT, WHM]	PROPYLENE GLYCOL DINITRATE
RR-01715-2 PROPYLENE GLYCOL ETHERS AND ESTERS	PROPYLENE GLYCOL ETHERS AND ESTERS
1569-02-4 PROPYLENE GLYCOL ETHYL ETHER	PROPYLENE GLYCOL ETHYL ETHER
RR-01130-3 PROPYLENE GLYCOL ISOBUTYL AND HIGHER HOMOLOGS	PROPYLENE GLYCOL ISOBUTYL AND HIGHER HO-MOLOGS
RR-01169-8 PROPYLENE GLYCOL ISOPROPYL ETHER	PROPYLENE GLYCOL ISOPROPYL ETHER
107-98-2 PROPYLENE GLYCOL METHYL ETHER [FLA, MSL, NFP, NFP]	PROPYLENE GLYCOL MONOMETHYL ETHER
108-65-6 PROPYLENE GLYCOL 1-METHYL ETHER-2-ACETATE [MAK, MAK, MAK]	PROPYLENE GLYCOL MONOMETHYL ETHER ACETATE
107-98-2 PROPYLENE GLYCOL 1-METHYL ETHER [MAK, MAK, MAK]	PROPYLENE GLYCOL MONOMETHYL ETHER

Chemical Name	Reference Name
108-65-6 PROPYLENE GLYCOL METHYL ETHER ACETATE (99% PURE) [NFP, NFP]	PROPYLENE GLYCOL MONOMETHYL ETHER ACETATE
999-61-1 PROPYLENE GLYCOL MONOACRYLATE [NFP, NFP, PAL]	2-HYDROXYPROPYL ACRYLATE
29387-86-8 PROPYLENE GLYCOL MONOBUTYL ETHER	PROPYLENE GLYCOL MONOBUTYL ETHER
5131-66-8 PROPYLENE GLYCOL MONOBUTYL ETHER [PST]	1-BUTOXY-2-PROPANOL
107-98-2 PROPYLENE GLYCOL MONOMETHYL ETHER	PROPYLENE GLYCOL MONOMETHYL ETHER
84540-57-8 PROPYLENE GLYCOL MONOMETHYL ETHER ACETATE	PROPYLENE GLYCOL MONOMETHYL ETHER ACETATE-(METHYLETHOXY)-, PROPYLAMINOSULFONYL DERIVATIVES
1330-80-9 PROPYLENE GLYCOL MONOOLEATE [PST]	9-OCTADECANOIC ACID (Z)-, MONOESTER WITH 1,2-PROPANEDIOL
30136-13-1 PROPYLENE GLYCOL MONOPROPYL ETHER	PROPYLENE GLYCOL MONOPROPYL ETHER
75-55-8 PROPYLENEIMINE	PROPYLENEIMINE
75-55-8 PROPYLENE IMINE [AUS, AUS, AUS, BC1, BC1]	PROPYLENEIMINE
75-55-8 1,2-PROPYLENEIMINE [EP2, EPA]	PROPYLENEIMINE
75-55-8 PROPYLENE IMINE [FLA, MAK, MAK, MSL, NIO, NIO, NIO, NIO, NIO, OS1, OS1, OS1, OS1, QBC, QBC, QBC]	PROPYLENEIMINE
75-55-8 1,2-PROPYLENEIMINE [RCU, RCU]	PROPYLENEIMINE
75-55-8 PROPYLENE IMINE [TLV, TLV, TLV]	PROPYLENEIMINE
75-55-8 PROPYLENEIMINE [AZIRIDINE, 2-METHYL-] [EP3]	PROPYLENEIMINE
75-56-9 PROPYLENE OXIDE	PROPYLENE OXIDE
75-56-9 1,2-PROPYLENE OXIDE [CSR, CSR]	PROPYLENE OXIDE
75-56-9 PROPYLENE OXIDE [OXIRANE, METHYL-] [EP3]	PROPYLENE OXIDE
1072-43-1 PROPYLENE SULFIDE	PROPYLENE SULFIDE
6842-15-5 PROPYLENE TETRAMER	PROPYLENE TETRAMER
75-55-8 1,2-PROPYLENIMINE (2-METHYL AZIRIDINE) [CAA]	PROPYLENEIMINE
111-43-3 PROPYL ETHER [FLA]	DIPROPYL ETHER
111-43-3 N-PROPYL ETHER [MSL, NFP, NFP]	DIPROPYL ETHER
110-74-7 PROPYL FORMATE	PROPYL FORMATE
RR-00137-6 PROPYL FORMATES	PROPYL FORMATES
121-79-9 PROPYL GALLATE	PROPYL GALLATE
2785-87-7 4-PROPYLGUAIACOL	4-PROPYLGUAIACOL
94-13-3 PROPYL-P-HYDROXYBENZOATE	PROPYL-P-HYDROXYBENZOATE
RR-01311-6 PROPYLIDENE DICHLORIDE	PROPYLIDENE DICHLORIDE
110-78-1 N-PROPYL ISOCYANATE	N-PROPYL ISOCYANATE
110-78-1 PROPYL ISOCYANATE [NJL]	N-PROPYL ISOCYANATE
107-03-9 N-PROPYL MERCAPTAN [MIN]	1-PROPANETHIOL
107-03-9 PROPYL MERCAPTAN [MSL, NJL]	1-PROPANETHIOL
627-13-4 N-PROPYL NITRATE	N-PROPYL NITRATE
627-13-4 PROPYL NITRATE [NFP, NFP, OS3]	N-PROPYL NITRATE
645-56-7 P-PROPYLPHENOL	P-PROPYLPHENOL
106-36-5 PROPYL PROPIONATE	PROPYL PROPIONATE
51-52-5 PROPYLTHIOURACIL	PROPYLTHIOURACIL
51-52-5 6-PROPYL-2-THIOURACIL [NJL]	PROPYLTHIOURACIL
141-57-1 PROPYL TRICHLOROSILANE	PROPYL TRICHLOROSILANE
141-57-1 PROPYLTRICHLOROSILANE [FLA, MSL, NFP, NFP]	PROPYL TRICHLOROSILANE

Chemical Name	Reference Name
74-99-7 PROPYNE [FLA, MSL, NFP, NFP, ONT, ONT]	METHYL ACETYLENE
74-99-7 1-PROPYNE [PAL]	METHYL ACETYLENE
106-96-7 1-PROPYNE, 3-BROMO- [PAL]	3-BROMOPROPYNE
74-99-7 PROPYNE [1-PROPYNE] [EP3]	METHYL ACETYLENE
107-19-7 PROP-2-YN-1-OL [GBR, GBR, GBR]	PROPARGYL ALCOHOL
107-19-7 2-PROPYN-1-OL [PAL, TSC, TSC]	PROPARGYL ALCOHOL
29901-97-1 PROTACTINIUM 227	PROTACTINIUM 227
15766-09-3 PROTACTINIUM 228	PROTACTINIUM 228
15766-10-6 PROTACTINIUM 230	PROTACTINIUM 230
14331-85-2 PROTACTINIUM 231	PROTACTINIUM 231
15766-06-0 PROTACTINIUM 232	PROTACTINIUM 232
13981-14-1 PROTACTINIUM 233	PROTACTINIUM 233
15100-28-4 PROTACTINIUM 234	PROTACTINIUM 234
RR-01132-5 PROTEIN COLLOID	PROTEIN COLLOID
2275-18-5 PROTHOATE	PROTHOATE
95-63-6 PSEUDOCUMENE	PSEUDOCUMENE
90-82-4 PSEUDOEPHEDRINE	PSEUDOEPHEDRINE
RR-01781-2 PSEUDOEPHEDRINE SALTS, OPTICAL ISOMERS, AND SALTS OF OPTICALISOMERS	PSEUDOEPHEDRINE SALTS, OPTICAL ISOMERS, AND SALTS OF OPTICAL
87625-62-5 PTAQUILOSIDE	PTAQUILOSIDE
98-51-1 P-TERT-BUTYLTOLUENE [HON]	P-TERT-BUTYLTOLUENE
1332-09-8 PUMICE	PUMICE
446-86-6 1H-PURINE, 6-[(1-METHYL-4-NITRO-1H-IMIDA-ZOL-5-YL)THIO]- [PAL, PAL]	AZATHIOPRINE
9002-86-2 PVC [MAK]	POLYVINYL CHLORIDE
100-73-2 2H-PYRAN-2-CARBOXALDEHYDE, 3,4-DIHY-DRO- [PAL]	ACROLEIN DIMER
110-87-2 2H-PYRAN, 3,4-DIHYDRO- [PAL]	DIHYDROPYRAN
3174-74-1 2H-PYRAN, 3,6-DIHYDRO-	2H-PYRAN, 3,6-DIHYDRO-SALT
25512-65-6 2H-PYRAN, DIHYDRO- [PAL]	DIHYDRO-2H-PYRAN
142-68-7 2H-PYRAN, TETRAHYDRO- [PAL]	PENTAMETHYLENE OXIDE
98-96-4 PYRAZINAMIDE	PYRAZINAMIDE
109-08-0 PYRAZINE, METHYL- [PAL]	2-METHYLPYRAZINE
92-43-3 3-PYRAZOLIDINONE, 1-PHENYL-	3-PYRAZOLIDINONE, 1-PHENYL-
13457-18-6 PYRAZOPHOS	PYRAZOPHOS
129-00-0 PYRENE	PYRENE
129-00-0 PYRENE (BENZO[DEF]PHENATHRENE) [CAI]	PYRENESPHATE)
121-21-1 PYRETHRIN I	PYRETHRIN I
121-29-9 PYRETHRIN II	PYRETHRIN II
121-29-9 PYRETHRINS [NJL, NJL]	PYRETHRIN II
8003-34-7 PYRETHRINS (ISO) [GBR, GBR]	PYRETHRUM
8003-34-7 PYRETHRINS AND PYRETHROIDS [PAL]	PYRETHRUM
8003-34-7 PYRETHRUM	PYRETHRUM
504-24-5 4-PYRIDINAMINE [MSL]	4-AMINOPYRIDINE
504-29-0 2-PYRIDINAMINE [PAL]	2-AMINOPYRIDINE
1918-02-1 2-PYRIDINCARBOXYLIC ACID, 4-AMINO-3,5,6-TRICHLORO- [PAL]	PICLORAM
110-86-1 PYRIDINE	PYRIDINE
504-24-5 PYRIDINE, 4-AMINO [FLA]	4-AMINOPYRIDINE

Chemical Name	Reference Name
504-24-5 PYRIDINE, 4-AMINO- [302, PAL]	4-AMINOPYRIDINE
100-48-1 4-PYRIDINECARBONITRILE	4-PYRIDINECARBONITRILE
100-54-9 3-PYRIDINECARBONITRILE	3-PYRIDINECARBONITRILE
100-70-9 2-PYRIDINECARBONITRILE	2-PYRIDINECARBONITRILE
1121-60-4 2-PYRIDINECARBOXALDEHYDE	2-PYRIDINECARBOXALDEHYDE
59-67-6 3-PYRIDINECARBOXYLIC ACID	3-PYRIDINECARBOXYLIC ACID
59-67-6 PYRIDINECARBOXYLIC ACID, 3- [OTS]	3-PYRIDINECARBOXYLIC ACID
1929-82-4 PYRIDINE, 2-CHLORO-6-(TRICHLOROMETHYL)- [PAL]	NITRAPYRIN
94-78-0 2,6-PYRIDINEDIAMINE, 3-(PHENYLAZO)- [PAL, PAL]	PHENAZOPYRIDINE
136-40-3 2,6-PYRIDINEDIAMINE, 3-(PHENYLAZO)-, MONOHYDROCHLORIDE [PAL, PAL]	PHENAZOPYRIDINE HYDROCHLORIDE
499-83-2 2,6-PYRIDINEDICARBOXYLIC ACID	2,6-PYRIDINEDICARBOXYLIC ACID
5408-74-2 PYRIDINE, 2-ETHENYL-5-ETHYL-	PYRIDINE, 2-ETHENYL-5-ETHYL-
104-90-5 PYRIDINE, 5-ETHYL-2-METHYL- [PAL]	2-METHYL-5-ETHYLPYRIDINE
108-89-4 PYRIDINE, 4-METHYL- [OTS, PAL, TSC, TSC]	P-PICOLINE
108-99-6 PYRIDINE, 3-METHYL- [OTS, TSC, TSC]	3-METHYLPYRIDINE
109-06-8 PYRIDINE, 2-METHYL- [EPA, OTS, PAL, TSC, TSC]	2-PICOLINE
1333-41-1 PYRIDINE, METHYL- [TSC, TSC]	PICOLINES
54-11-5 PYRIDINE, 3-(1-METHYL-2-PYRROLIDINYL)-, (S)- [PAL]	NICOTINE
140-76-1 PYRIDINE, 2-METHYL-5-VINYL-	PYRIDINE, 2-METHYL-5-VINYL-
1124-33-0 PYRIDINE, 4-NITRO-, 1-OXIDE-	PYRIDINE, 4-NITRO-, 1-OXIDE-
80508-23-2 PYRIDINE, 3-(1-NITROSO-2-PYRROLIDINYL)	PYRIDINE, 3-(1-NITROSO-2-PYRROLIDINYL)
16543-55-8 PYRIDINE, 3-(1-NITROSO-2-PYRROLIDINYL)-, (S)- [PAL, PAL]	N-NITROSONORNICOTINEENYL)AZO]PHENYL]AZO] 1, 1'-BIPHENYL]4-YL]AZO-2-HYDROXY BENZOATO(4-)]-, DISODIUM
15598-34-2 PYRIDINE PERCHLORATE	PYRIDINE PERCHLORATE
2402-79-1 PYRIDINE, 2,3,5,6-TETRACHLORO-	PYRIDINE, 2,3,5,6-TETRACHLORO-
2971-90-6 4-PYRIDINOL, 3,5-DICHLORO-2,6-DIMETHYL- [PAL]	CLOPIDOL
62450-06-0 5H-PYRIDO[4,3-B]INDOL-3-AMINE, 1,4-DIMETHYL- [PAL, PAL]	TRP-P-1
62450-07-1 5H-PYRIDO[4,3-B]INDOL-3-AMINE, 1-METHYL- [PAL, PAL]	TRP-P-2
RR-01540-7 PYRIDO[3,4-C]PSORALEN	PYRIDO[3,4-C]PSORALEN
41468-25-1 PYRIDOXAL-S'-PHOSPHATE	PYRIDOXAL-S'-PHOSPHATE-4,1-PHENYLENE)IMINO[6-(DIETHYLAMINO)-1,3,5-TRIAZINE-4,2-DIYL]IMINO]BIS-, HEXASODIUM SALT
504-29-0 2-PYRIDYLAMINE [GBR, GBR]	2-AMINOPYRIDINE
91-84-9 PYRILAMINE	PYRILAMINE
58-14-0 PYRIMETHAMINE	PYRIMETHAMINE
66-75-1 2,4(1H,3H)-PYRIMIDINEDIONE, 5-[BIS(2-CHLOROETHYL)AMINO]- [PAL, PAL]	URACIL MUSTARD
314-40-9 2,4(1H,3H)-PYRIMIDINEDIONE, 5-BROMO-6-METHYL-3-(1-METHYLPROPYL)- [PAL]	BROMACIL4A,5,8,8A-HEXAHYDRO-, (1.ALPHA.,4.ALPHA., 4A.BETA.,5.ALPHA., 8.ALPHA.,8A.BETA.)-
56-04-2 4(1H)-PYRIMIDINONE, 2,3-DIHYDRO-6-METHYL-2-THIOXO- [PAL, PAL]	METHYLTHIOURACIL
51-52-5 4(1H)-PYRIMIDINONE, 2,3-DIHYDRO-6-PROPYL-2-THIOXO- [PAL, PAL]	PROPYLTHIOURACIL

Chemical Name	Reference Name
53558-25-1 PYRIMINIL	PYRIMINIL
120-80-9 PYROCATECHOL [GBR]	CATECHOL
87-66-1 PYROGALLOL	PYROGALLOL
59789-51-4 1H-PYROLE-2,5-DIONE, 1-(2,4,6-TRIBRO-MOPHENYL)-	1H-PYROLE-2,5-DIONE, 1-(2,4,6-TRIBROMOPHENYL)-DI-4, 1-PHENYLENE]BIS[3-(OXIRANYLMETHOXY)-
RR-00454-6 PYROPHORIC LIQUIDS, N.O.S.	PYROPHORIC LIQUIDS, N.O.S.
RR-00455-7 PYROPHORIC METALS OR ALLOYS, N.O.S.	PYROPHORIC METALS OR ALLOYS, N.O.S.
RR-00456-8 PYROPHORIC SOLIDS, N.O.S.	PYROPHORIC SOLIDS, N.O.S.
12269-78-2 PYROPHYLLITE	PYROPHYLLITE
RR-00457-9 PYROSULFURIC ACID	PYROSULFURIC ACID
7791-27-7 PYROSULFURYL CHLORIDE	PYROSULFURYL CHLORIDE
7791-27-7 PYROSULPHURYL CHLORIDE [NJL, NJL]	PYROSULFURYL CHLORIDE
RR-01160-9 PYROXYLIN SOLUTION	PYROXYLIN SOLUTION
109-97-7 1H-PYRROLE	1H-PYRROLE
109-97-7 PYRROLE [FLA, MSL, NFP, NFP]	1H-PYRROLE
27417-39-6 1H-PYRROLE, METHYL-	1H-PYRROLE, METHYL-NITRILOTRIS[ETHANOL](1:1)
123-75-1 PYRROLIDINE	PYRROLIDINE
120-94-5 PYRROLIDINE, 1-METHYL- [PAL]	METHYLPYRROLIDINE
930-55-2 PYRROLIDINE, 1-NITROSO- [PAL, PAL]	N-NITROSOPYRROLIDINE
616-45-5 2-PYRROLIDINONE	2-PYRROLIDINONE
872-50-4 2-PYRROLIDINONE, 1-METHYL- [PAL]	1-METHYL-2-PYRROLIDONE
616-45-5 2-PYRROLIDONE [NFP, NFP]	2-PYRROLIDINONE
78-98-8 PYRUVIC ALDEHYDE	PYRUVIC ALDEHYDE
14808-60-7 QUARTZ	QUARTZ
14808-60-7 QUARTZ (SIO2) [PAL]	QUARTZ
14808-60-7 QUARTZ-CONTAINING FINE DUSTS (> 1% QUARTZ) [MAK, MAK]	QUARTZ
14464-46-1 QUARTZ, CRISTOBALITE [MAK, MAK]	SILICA, CRISTOBALITE
14808-60-7 QUARTZ DUST [FLA]	QUARTZ
15468-32-3 QUARTZ, TRIDYMITE [MAK, MAK]	SILICA, TRIDYMITE
68915-32-2 QUASSIA, EXTRACT	QUASSIA, EXTRACT
68391-01-5 QUATERNARY AMMONIUM COMPOUNDS, BENZYL-C12-18-ALKYLDIMETHYL,CHLO-RIDES [PST]	ALKYLBENZYLDIMETHYLAMMONIUM CHLORIDEYL-NOR, HYDROXIDE, MONOSODIUM SALT
61791-10-4 QUATERNARY AMMONIUM COMPOUNDS, COCO ALKYLBIS(HYDROXYETHYL)METHYL, ETHOXYLATED, CHLORIDES	QUATERNARY AMMONIUM COMPOUNDS, COCO ALKYLBIS(HYDROXYETHYL)ME
68187-69-9 QUATERNARY AMMONIUM COMPOUNDS, (HY-DROGENATED TALLOW ALKYL) BIS(HYDROX-YETHYL) METHYL, ETHOXYLATED, CHLO-RIDES	QUATERNARY AMMONIUM COMPOUNDS, (HYDRO-GENATED TALLOW ALKYL) B
RR-01567-8 QUATERNARY AMMONIUM SALT OF FLUORI-NATED ALKYLARYL AMIDE	QUATERNARY AMMONIUM SALT OF FLUORINATED ALKYLARYL AMIDE
RR-01457-3 QUEBRACHITOL PENTANITRATE	QUEBRACHITOL PENTANITRATE
RR-00793-2 QUENCHING OIL	QUENCHING OIL
117-39-5 QUERCETIN [CAL, CSR, CSR, IAR]	NATURAL YELLOW 10
1393-03-9 QUILLAJA (SAPONIN)	QUILLAJA (SAPONIN)
1047-16-1 QUINACRIDONE	QUINACRIDONE
69-05-6 QUINACRINE HYDROCHLORIDE	QUINACRINE HYDROCHLORIDE
91-63-4 QUINALDINE	QUINALDINE
13593-03-8 QUINALPHOS	QUINALPHOS

Chemical Name	Reference Name
130-95-0 QUININE	QUININE
91-22-5 QUINOLINE	QUINOLINE
56-57-5 QUINOLINE, 4-NITRO, 1-OXIDE [NJL, NJL]	4-NITROQUINOLINE-1-OXIDE
56-57-5 QUINOLINE, 4-NITRO-, 1-OXIDE [PAL]	4-NITROQUINOLINE-1-OXIDE
8004-92-0 QUINOLINE YELLOW	QUINOLINE YELLOW
106-51-4 QUINONE	QUINONE
106-51-4 P-QUINONE [CAN]	QUINONE
106-51-4 PARA-QUINONE [IAR]	QUINONE
105-11-3 P-QUINONE DIOXIME [MSL]	P-BENZOQUINONE DIOXIME
59-40-5 N-(2-QUINOXALINYL)-SULFANILIDE	N-(2-QUINOXALINYL)-SULFANILIDE
82-68-8 QUINTOZENE [NJL, NJL]	PENTACHLORONITROBENZENE
82-68-8 QUINTOZENE (PENTACHLORONITROBENZENE) [CAL, IAR]	PENTACHLORONITROBENZENE
82-68-8 QUINTOZENE [PENTACHLORONITROBENZENE] [313]	PENTACHLORONITROBENZENE
76578-14-8 QUIZALOFOP-ETHYL [2-[4-[(6-CHLORO-2-QUINOXALINYL)OXY]PHENOXY]PROPANOIC ACID ETHYL ESTER]	QUIZALOFOP-ETHYL [2-[4-[(6-CHLORO-2-QUINOX-ALINYL)OXY]PHENOXY
RR-00459-1 RADIOACTIVE MATERIALS, N.O.S.	RADIOACTIVE MATERIALS, N.O.S.
RR-00066-8 RADIONUCLIDES	RADIONUCLIDES
RR-00066-8 RADIONUCLIDES (INCLUDING RADON) [CAA]	RADIONUCLIDES
7440-14-4 RADIUM	RADIUM
15623-45-7 RADIUM 223	RADIUM 223
13233-32-4 RADIUM 224	RADIUM 224
13981-53-8 RADIUM 225	RADIUM 225
13982-63-3 RADIUM 226	RADIUM 226
15743-84-7 RADIUM 227	RADIUM 227
15262-20-1 RADIUM 228	RADIUM 228
10043-92-2 RADON	RADON
22481-48-7 RADON 220	RADON 220
14859-67-7 RADON 222	RADON 222
RR-01133-6 RAISINS	RAISINS
8002-13-9 RAPESEED OIL	RAPESEED OIL
8002-13-9 RAPE SEED OIL [NFP, NFP]	RAPESEED OIL
RR-01505-4 RARE GASES, MIXTURES	RARE GASES, MIXTURES
12713-03-0 RAW UMBER	RAW UMBER
121-82-4 RDX [BC1, BC1, BC1]	CYCLONITE
RR-00948-3 REACTION PRODUCT OF ALKANEDIOL AND EPICHLOROHYDRIN	REACTION PRODUCT OF ALKANEDIOL AND EPICHLOROHYDRINCYANATO DERIVATIVE
RR-00986-9 REACTION PRODUCT OF ALKYL CARBOXYLIC ACIDS, ALKANE POLYOLS,ALKYL ACRYLATE, AND ISOPHORONE DIISOCYANATE	REACTION PRODUCT OF ALKYL CARBOXYLIC ACIDS, ALKANE POLYOLS,
RR-00975-6 REACTION PRODUCT OF ALKYLPHENOL, TETRALKYL TITANATE, AND TINCOMPLEX	REACTION PRODUCT OF ALKYLPHENOL, TETRALKYL TITANATE, AND TIN
RR-01134-7 REACTION PRODUCT OF AMINOKETONES AND FORMALDEHYDE	REACTION PRODUCT OF AMINOKETONES AND FORMALDEHYDE
RR-00232-4 REACTION PRODUCT OF A MONOALKYL SUCCINIC ANHYDRIDE WITH ANX-HYDROXY METHACRYLATE	REACTION PRODUCT OF A MONOALKYL SUCCINIC AN-HYDRIDE WITH ANCARBOMONOCYCLE), POLYALKY-LENE GLYCOL, ALKANOL, METHACRYLATE POLYMER
60857-97-8 REACTION PRODUCT OF HYDROXYETHYL ACRYLATE AND METHYL OXIRANE	REACTION PRODUCT OF HYDROXYETHYL ACRYLATE AND METHYL OXIRANE

Chemical Name	Reference Name
RR-01263-5 REACTION PRODUCT OF ETHOXYLATED FATTY ACID OILS AND A PHENOLIC PENTAERYTHRITOL TETRAESTER	REACTION PRODUCT OF ETHOXYLATED FATTY ACID OILS AND A PHENOL
RR-01259-9 REACTION PRODUCTS OF PHENOLIC PENTAERYTHRITOL TETRAESTER WITH FATTY ACID OILS AND ESTERS, AND GLYCERIDE TRIESTERS	REACTION PRODUCTS OF PHENOLIC PENTAERYTHRITOL TETRAESTER WIT
RR-01249-7 REACTION PRODUCTS OF SUBSTITUTED HYDROXYLALKANES AND POLYALKYLPOLYISOCYANATOCARBOMONOCYCLE	REACTION PRODUCTS OF SUBSTITUTED HYDROXYLALKANES AND POLYALK
RR-01256-6 RECOVERED METAL HYDROXIDE	RECOVERED METAL HYDROXIDE
RR-00753-4 RED SQUILL	RED SQUILL
RR-01506-5 REDUCING LIQUID	REDUCING LIQUID
68783-08-4 REFINED PETROLEUM PRODUCTS [CN6]	C7-C35 REFINED PETROLEUM OIL
RR-01792-5 REFRACTORY FIBRES (CERAMIC OR OTHERS)	REFRACTORY FIBRES (CERAMIC OR OTHERS)
142844-00-6 REFRACTORY CERAMIC FIBERS	REFRACTORY CERAMIC FIBERS AVERAGE DEGREE OF CHLORINATION 60%
RR-00472-8 REFRIGERANT GASES, N.O.S.	REFRIGERANT GASES, N.O.S.
50-55-5 RESERPINE	RESERPINE
RR-01741-4 RESIDUAL (HEAVY) FUEL OILS	RESIDUAL (HEAVY) FUEL OILS
64741-71-5 RESIDUAL (HEAVY) FUEL OILS [C65, IAR]	RESIDUAL HEAVY POLYMER (PETROLEUM)
64741-71-5 RESIDUAL HEAVY POLYMER (PETROLEUM)	RESIDUAL HEAVY POLYMER (PETROLEUM)
64742-10-5 RESIDUAL OIL SOVENT EXTRACT (PETROLEUM)	RESIDUAL OIL SOVENT EXTRACT (PETROLEUM)SOLVENT
61790-50-9 RESIN ACIDS AND ROSIN ACIDS, POTASSIUM SALT	RESIN ACIDS AND ROSIN ACIDS, POTASSIUM SALT
61790-51-0 RESIN ACIDS AND ROSIN ACIDS, SODIUM SALTS	RESIN ACIDS AND ROSIN ACIDS, SODIUM SALTS
9010-69-9 RESIN AND ROSIN ACIDS, ZINC SALTS [PST]	ZINC RESINATE
RR-01507-6 RESIN FAST BLACK WP	RESIN FAST BLACK WP
RR-00475-1 RESIN OIL	RESIN OIL
RR-00476-2 RESIN SOLUTION	RESIN SOLUTION
10453-86-8 RESMETHRIN	RESMETHRIN
10453-86-8 RESMETHRIN [[5-(PHENYLMETHYL)-3-FURANYL]METHYL 2,2-DIMETHYL-3-(2-METHYL-1-PROPENYL)CYCLOPROPANECARBOXYLATE] [313]	RESMETHRIN
108-46-3 RESORCINOL	RESORCINOL
101-90-6 RESORCINOL DIGLYCIDYL ETHER [OTS, WHM, EPA]	DIGLYCIDYL RESORCINOL ETHER
RR-01018-4 RESORCINOL, FORMALDEHYDE, SUBSTITUTED CARBOMONOCYCLE RESIN	RESORCINOL, FORMALDEHYDE, SUBSTITUTED CARBOMONOCYCLE RESINMONOCYCLE, HYDROXYALKYL ACRYLATE ESTER
302-79-4 ALL-TRANS RETINOIC ACID	ALL-TRANS RETINOIC ACID
4759-48-2 RETINOIC ACID [MSL]	ISOTERTINATE
65646-68-6 RETINOID PROJECT 2 (4-HYDROXYPHENYL)	RETINOID PROJECT 2 (4-HYDROXYPHENYL)
68-26-8 RETINOL/RETINYL ESTERS [CAB]	VITAMIN A
127-47-9 RETINOL, ACETATE	RETINOL, ACETATE
68-26-8 RETINOL/RETINYL ESTERS [CAL, C65]	VITAMIN A
480-54-6 RETRORSINE	RETRORSINE
96949-21-2 RHAMSAN GUM	RHAMSAN GUM
18853-09-3 RHENIUM 177	RHENIUM 177
18853-08-2 RHENIUM 178	RHENIUM 178
14993-65-8 RHENIUM 181	RHENIUM 181

484

Chemical Name	Reference Name
21459-71-2 RHENIUM 182	RHENIUM 182
14983-46-1 RHENIUM 184	RHENIUM 184
RR-00386-1 RHENIUM 184M	RHENIUM 184M
14998-63-1 RHENIUM 186	RHENIUM 186
RR-00377-0 RHENIUM 186M	RHENIUM 186M
14391-29-8 RHENIUM 187	RHENIUM 187
14378-26-8 RHENIUM 188	RHENIUM 188
RR-00374-7 RHENIUM 188M	RHENIUM 188M
15765-78-3 RHENIUM 189	RHENIUM 189
989-38-8 RHODAMINE 6G [CSR, CSR, IAR]	C.I. BASIC RED 1
81-88-9 RHODAMINE B [IAR, WHM]	C.I. FOOD RED 15
989-38-8 RHODAMINE 6G [CAL]	C.I. BASIC RED 1
81-88-9 RHODAMINE B [CAL, PST]	C.I. FOOD RED 15
37299-86-8 RHODAMINE WT	RHODAMINE WT
141-84-4 RHODANINE	RHODANINE
141-25-3 RHODINOL	RHODINOL
7440-16-6 RHODIUM	RHODIUM
15765-79-4 RHODIUM 99	RHODIUM 99
RR-00373-6 RHODIUM 99M	RHODIUM 99M
15765-80-7 RHODIUM 100	RHODIUM 100
14378-53-1 RHODIUM 101	RHODIUM 101
RR-00372-5 RHODIUM 101M	RHODIUM 101M
15765-82-9 RHODIUM 102	RHODIUM 102
RR-00371-4 RHODIUM 102M	RHODIUM 102M
RR-00366-7 RHODIUM 103M	RHODIUM 103M
14913-89-4 RHODIUM 105	RHODIUM 105
RR-00365-6 RHODIUM 106M	RHODIUM 106M
15706-50-0 RHODIUM 107	RHODIUM 107
7440-16-6 RHODIUM, METAL [MEX]	RHODIUM
RR-00040-8 RHODIUM, SOLUBLE SALTS [MEX]	RHODIUM SOLUBLE COMPOUNDS
10049-07-7 RHODIUM CHLORIDE [ONT, ONT, ONT]	RHODIUM TRICHLORIDE
RR-00611-1 RHODIUM COMPOUNDS [CAL]	RHODIUM COMPOUNDS, N.O.S.
RR-00611-1 RHODIUM COMPOUNDS, N.O.S.	RHODIUM COMPOUNDS, N.O.S.
7440-16-6 RHODIUM, ELEMENTAL [WHM]	RHODIUM
RR-00047-5 RHODIUM INSOLUBLE COMPOUNDS	RHODIUM INSOLUBLE COMPOUNDS
7440-16-6 RHODIUM METAL [ONT, ONT, ONT]	RHODIUM
7440-16-6 RHODIUM, METAL AND INSOLUBLE COMPOUNDS [ALB, ALB]	RHODIUM
7440-16-6 RHODIUM, METAL FUME AND DUSTS [BC1, BC1, FLA, QBC]	RHODIUM
7440-16-6 RHODIUM (METAL FUME AND INSOLUBLE COMPOUNDS) [NIO, NIO]	RHODIUM
10139-58-9 RHODIUM NITRATE	RHODIUM NITRATE
RR-00040-8 RHODIUM SOLUBLE COMPOUNDS	RHODIUM SOLUBLE COMPOUNDS
RR-00040-8 RHODIUM (SOLUBLE COMPOUNDS) [NIO, NIO]	RHODIUM SOLUBLE COMPOUNDS
RR-00040-8 RHODIUM, SOLUBLE SALTS [BC1, BC1, QBC]	RHODIUM SOLUBLE COMPOUNDS
10489-46-0 RHODIUM SULFATE	RHODIUM SULFATE
10049-07-7 RHODIUM TRICHLORIDE	RHODIUM TRICHLORIDE
RR-00047-5 RHODIUM, WATER INSOLUBLE COMPOUNDS [ONT, ONT, ONT]	RHODIUM INSOLUBLE COMPOUNDS

Chemical Name	Reference Name
RR-00040-8 RHODIUM WATER-SOLUBLE COM-POUNDS [ONT, ONT, ONT]	RHODIUM SOLUBLE COMPOUNDS
36791-04-5 RIBAVIRIN	RIBAVIRIN
RR-01135-8 RICE	RICE
68553-81-1 RICE BRAN OIL	RICE BRAN OIL
9009-86-3 RICIN	RICIN
23246-96-0 RIDDELLINE	RIDDELLINE
13292-46-1 RIFAMPICIN	RIFAMPICIN
RR-01201-1 ROCKWOOL	ROCKWOOL
RR-01201-1 ROCK WOOL [MAK]	ROCKWOOL
RR-01201-1 ROCKWOOL FIBERS [CAB]	ROCKWOOL
RR-00479-5 RODENTICIDES, LIQUID OR SOLID, N.O.S.	RODENTICIDES, LIQUID OR SOLID, N.O.S.
299-84-3 RONNEL	RONNEL
8050-09-7 ROSIN	ROSIN
RR-00074-8 ROSIN CORE SOLDER PYROLYSIS PRODUCTS, AS FORMALDEHYDE	ROSIN CORE SOLDER PYROLYSIS PRODUCTS, AS FORMALDEHYDE
RR-00074-8 ROSIN CORE SOLDER PYROLYSIS PRODUCTS (AS FORMALDEHYDE) [QBC]	ROSIN CORE SOLDER PYROLYSIS PRODUCTS, AS FORMALDEHYDE
68333-69-7 ROSIN, MALEATED, POLYMER WITH PEN-TAERYTHRITOL	ROSIN, MALEATED, POLYMER WITH PENTAERYTHRI-TOL
8002-16-2 ROSIN OIL	ROSIN OIL
65997-05-9 ROSIN, PARTIALLY DIMERIZED [PST]	ROSIN POLYMERS
65997-06-0 ROSIN, PARTIALLY HYDROGENATED	ROSIN, PARTIALLY HYDROGENATED
65997-05-9 ROSIN POLYMERS	ROSIN POLYMERS
67700-49-6 ROSIN POLYMER WITH P-TERT-BUTYLPHE-NOL, FORMALDEHYDE AND GLYCEROL	ROSIN POLYMER WITH P-TERT-BUTYLPHENOL, FORMALDEHYDE AND GLYC
83-79-4 ROTENONE [CSR, CSR, HM1, MAK, MEX, MEX, NIO, NIO, NJL, NJL, OS1, OS1]	ROTENONE (COMMERCIAL)
83-79-4 ROTENONE (ISO) [GBR, GBR]	ROTENONE (COMMERCIAL)
83-79-4 ROTENONE (COMMERCIAL)	ROTENONE (COMMERCIAL)
83-79-4 ROTENONE, COMMERCIAL [CAL, CEX]	ROTENONE (COMMERCIAL)
1309-37-1 ROUGE [ONT]	IRON OXIDE
RR-00006-6 ROUGE	ROUGE
RR-00006-6 ROUGE DUST [AUS]	ROUGE
121-19-7 ROXARSONE	ROXARSONE
9006-04-6 RUBBER [AUS, GBR, HM1, HM1, NJL, PST]	NATURAL RUBBERS
RR-00548-1 RUBBER INDUSTRY	RUBBER INDUSTRY
RR-00548-1 RUBBER INDUSTRY (MANUFACTURE) [PAL, PAL]	RUBBER INDUSTRY
RR-00548-1 RUBBER PROCESS [GBR]	RUBBER INDUSTRY
RR-00076-0 RUBBER SOLVENT [ONT, PAL]	RUBBER SOLVENT (NAPHTHA)
8030-30-6 RUBBER SOLVENT (NAPHTHA) [TLV]	BENZIN
RR-00076-0 RUBBER SOLVENT (NAPHTHA)	RUBBER SOLVENT (NAPHTHA)
RR-00076-0 RUBBER SOLVENT, NAPHTHA [NHS, NHS]	RUBBER SOLVENT (NAPHTHA)
RR-00076-0 RUBBER SOLVENT (NAPHTHA) [QBC]	RUBBER SOLVENT (NAPHTHA)
7440-17-7 RUBIDIUM	RUBIDIUM
14809-48-4 RUBIDIUM 79	RUBIDIUM 79
18268-34-3 RUBIDIUM 81	RUBIDIUM 81
RR-00364-5 RUBIDIUM 81M	RUBIDIUM 81M
RR-00363-4 RUBIDIUM 82M	RUBIDIUM 82M
17056-36-9 RUBIDIUM 83	RUBIDIUM 83

Chemical Name	Reference Name
15765-86-3 RUBIDIUM 84	RUBIDIUM 84
14932-53-7 RUBIDIUM 86	RUBIDIUM 86
13982-13-3 RUBIDIUM 87	RUBIDIUM 87
14928-36-0 RUBIDIUM 88	RUBIDIUM 88
14191-65-2 RUBIDIUM 89	RUBIDIUM 89
7791-11-9 RUBIDIUM CHLORIDE	RUBIDIUM CHLORIDE
13446-72-5 RUBIDIUM CHROMATE	RUBIDIUM CHROMATE
13446-73-6 RUBIDIUM DICHROMATE	RUBIDIUM DICHROMATE
1310-82-3 RUBIDIUM HYDROXIDE	RUBIDIUM HYDROXIDE
7790-29-6 RUBIDIUM IODIDE	RUBIDIUM IODIDE
23537-16-8 RUGULOSIN	RUGULOSIN
15125-02-7 RUTHENIUM 94	RUTHENIUM 94
15758-35-7 RUTHENIUM 97	RUTHENIUM 97
13968-53-1 RUTHENIUM 103	RUTHENIUM 103
14331-95-4 RUTHENIUM 105	RUTHENIUM 105
13967-48-1 RUTHENIUM 106	RUTHENIUM 106
16845-29-7 RUTHENIUM CHLORIDE HYDROXIDE	RUTHENIUM CHLORIDE HYDROXIDE
12036-10-1 RUTHENIUM OXIDE	RUTHENIUM OXIDE
1317-80-2 RUTILE (TIO2)	RUTILE (TIO2)
15662-33-6 RYANODINE	RYANODINE
RR-01136-9 RYE FLOUR	RYE FLOUR
8047-67-4 SACCHARATED IRON OXIDE	SACCHARATED IRON OXIDE
81-07-2 SACCHARIN	SACCHARIN
81-07-2 SACCHARIN AND SALTS [EPA]	SACCHARIN
81-07-2 SACCHARIN, & SALTS [RCU, RCU]	SACCHARIN
81-07-2 SACCHARIN [1,2-BENZISOTHIAZOL-3(2H)-ONE, 1,1-DIOXIDE] [313]	SACCHARIN
128-44-9 SACCHARIN, SODIUM [C65]	SODIUM SACCHARIN
128-44-9 SACCHARIN SODIUM SALT [PST]	SODIUM SACCHARIN
8001-23-8 SAFFLOWER OIL	SAFFLOWER OIL
94-59-7 SAFROLE	SAFROLE
14167-18-1 SALCOMINE	SALCOMINE
90-02-8 SALICYLALDEHYDE	SALICYLALDEHYDE
65-45-2 SALICYLAMIDE	SALICYLAMIDE
599-79-1 SALICYLAZOSULFAPYRIDINE	SALICYLAZOSULFAPYRIDINE
69-72-7 SALICYLIC ACID	SALICYLIC ACID
3811-49-2 SALITHION	SALITHION
RR-01707-2 SALTED FISH, CHINESE STYLE	SALTED FISH, CHINESE STYLE
RR-01220-4 SALT OF CYCLODIAMINE AND MINERAL ACID	SALT OF CYCLODIAMINE AND MINERAL ACID
14877-67-9 SAMARIUM 141	SAMARIUM 141
RR-00362-3 SAMARIUM 141M	SAMARIUM 141M
15701-12-9 SAMARIUM 142	SAMARIUM 142
15065-02-8 SAMARIUM 145	SAMARIUM 145
14280-31-0 SAMARIUM 146	SAMARIUM 146
14392-33-7 SAMARIUM 147	SAMARIUM 147
15715-94-3 SAMARIUM 151	SAMARIUM 151
15766-00-4 SAMARIUM 153	SAMARIUM 153
14391-31-2 SAMARIUM 155	SAMARIUM 155
15759-70-3 SAMARIUM 156	SAMARIUM 156
10361-83-8 SAMARIUM NITRATE	SAMARIUM NITRATE

Chemical Name	Reference Name
10361-82-7 SAMARIUM TRICHLORIDE	SAMARIUM TRICHLORIDE
115-71-9 SANTALOL	SANTALOL
39393-37-8 SANTICIZER 711	SANTICIZER 711LFOSUCCINATE
107-44-8 SARIN	SARIN
RR-01137-0 SAWDUST	SAWDUST
14276-61-0 SCANDIUM 43	SCANDIUM 43
14391-94-7 SCANDIUM 44	SCANDIUM 44
RR-00359-8 SCANDIUM 44M	SCANDIUM 44M
13967-63-0 SCANDIUM 46	SCANDIUM 46
14391-96-9 SCANDIUM 47	SCANDIUM 47
14391-86-7 SCANDIUM 48	SCANDIUM 48
14391-97-0 SCANDIUM 49	SCANDIUM 49
85-83-6 SCARLET RED [IAR]	2-NAPHTHALENOL, 1-[[2-METHYL-4-[(2-METHYLPHENYL)AZO]PHENY
152-16-9 SCHRADAN [CAL, MSL]	OCTAMETHYLPYROPHOSPHORAMIDE
114-49-8 SCOPOLAMINE HYDROBROMIDE	SCOPOLAMINE HYDROBROMIDE
6533-68-2 SCOPOLAMINE HYDROBROMIDE TRIHYDRATE	SCOPOLAMINE HYDROBROMIDE TRIHYDRATE
RR-01138-1 SEAWEED	SEAWEED
78-92-2 SEC-BUTANOL [TLV]	SEC-BUTYL ALCOHOL
88-85-7 2-SEC-BUTYL-4,6-DINITROPHENOL (DINOSEB)	2-SEC-BUTYL-4,6-DINITROPHENOL (DINOSEB)
309-43-3 SECOBARBITAL SODIUM	SECOBARBITAL SODIUM
RR-00485-3 SELENATES, N.O.S.	SELENATES, N.O.S.
7783-08-6 SELENIC ACID	SELENIC ACID
7783-00-8 SELENIOUS ACID [302, EPA, WHM]	SELENOUS ACID
7782-82-3 SELENIOUS ACID, MONOSODIUM SALT [MSL]	SODIUM SELENITE
7782-49-2 SELENIUM	SELENIUM
7783-79-1 SELENIUM (HEXAFLUORIDE) [QBC]	SELENIUM HEXAFLUORIDE
19869-93-3 SELENIUM 70	SELENIUM 70
15422-57-8 SELENIUM 73	SELENIUM 73
RR-00358-7 SELENIUM 73M	SELENIUM 73M
14265-71-5 SELENIUM 75	SELENIUM 75
15758-45-9 SELENIUM 79	SELENIUM 79
15422-58-9 SELENIUM 81	SELENIUM 81
RR-00357-6 SELENIUM 81M	SELENIUM 81M
14687-60-6 SELENIUM 83	SELENIUM 83
7782-49-2 SELENIUM COMPOUNDS [MEX]	SELENIUM
RR-00612-2 SELENIUM AND COMPOUNDS [TLV, GBR]	SELENIUM COMPOUNDS
7782-49-2 SELENIUM COMPOUNDS [ALB, ALB, AUS, BC1, MAK, MAK, MIN, NIO, NIO, OS1, OS1]	SELENIUM
RR-00612-2 SELENIUM COMPOUNDS	SELENIUM COMPOUNDS
RR-00612-2 SELENIUM COMPOUNDS, N.O.S. [RCU, WHM, NJL]	SELENIUM COMPOUNDS
7446-08-4 SELENIUM DIOXIDE	SELENIUM DIOXIDE
7446-08-4 SELENIUM(IV) DIOXIDE [WHM]	SELENIUM DIOXIDE
7488-56-4 SELENIUM DISULFIDE [EPA, MSL, NJL, PAL]	SELENIUM SULFIDE
7488-56-4 SELENIUM(IV) DISULFIDE [WHM]	SELENIUM SULFIDE
7782-49-2 SELENIUM, ELEMENTAL [WHM]	SELENIUM
7783-79-1 SELENIUM FLUORIDE (SEF6), (OC-6-11)- [PAL]	SELENIUM HEXAFLUORIDE
7783-79-1 SELENIUM HEXAFLUORIDE	SELENIUM HEXAFLUORIDE
12033-59-9 SELENIUM NITRIDE	SELENIUM NITRIDE

Chemical Name	Reference Name
12640-89-0 SELENIUM OXIDE [NJL, HM1, HM1]	SELENIUM OXIDE (VAN)
7446-08-4 SELENIUM OXIDE (SEO2) [PAL]	SELENIUM DIOXIDE
12640-89-0 SELENIUM OXIDE (VAN)	SELENIUM OXIDE (VAN)
7791-23-3 SELENIUM OXYCHLORIDE	SELENIUM OXYCHLORIDE
56093-45-9 SELENIUM SULFIDE	SELENIUM SULFIDEETHENYL]-1,3,4-OXADIAZOL-2-YL]-, (E)-
7446-34-6 SELENIUM SULFIDE (SES) [PAL, PAL]	SELENIUM SULFIDE
1132-39-4 1,1'-SELENOBISBENZENE	1,1'-SELENOBISBENZENE
1464-42-2 SELENOMETHIONINE	SELENOMETHIONINE
630-10-4 SELENOUREA [EPA, RCU, RCU]	CARBAMIMIDOSELENOIC ACID
7783-00-8 SELENIOUS ACID [FLA]	SELENOUS ACID
7783-00-8 SELENOUS ACID	SELENOUS ACID
RR-00504-9 SELSUN	SELSUN
563-41-7 SEMICARBAZIDE HYDROCHLORIDE	SEMICARBAZIDE HYDROCHLORIDE
563-41-7 SEMICARBAZIDE HYDROCHLORIDE (HYDRAZINECARBOXAMIDE, MONOHYDROCHLORIDE) [TSC]	SEMICARBAZIDE HYDROCHLORIDE
480-81-9 SENECIPHYLLINE	SENECIPHYLLINE
2318-18-5 SENKIRKINE	SENKIRKINE
18307-23-8 SEPIOLITE	SEPIOLITE
RR-00542-5 SEQUENTIAL ORAL CONTRACEPTIVES [IAR, MIN]	OESTROGEN-PROGESTIN COMBIN., SEQUENTIAL ORAL CONTRACEPTIVES
115-02-6 L-SERINE, DIAZOACETATE (ESTER) [PAL, PAL]	AZASERINE
8008-74-0 SESAME OIL	SESAME OIL
136-78-7 SESONE	SESONE
136-78-7 SESONE (CRAG HERBICIDE) [MEX]	SESONE
136-78-7 SESONE (CRAG) [QBC]	SESONE
74051-80-2 SETHOXYDIM [2-[1-(ETHOXYIMINO)BUTYL]-5-[2-(ETHYLTHIO)PROPYL]-3-HYDROXY-2-CYCLOHEXEN-1-ONE]	SETHOXYDIM [2-[1-(ETHOXYIMINO)BUTYL]-5-[2-(ETHYLTHIO)PROPYL]TER
68476-95-9 SHALE	SHALE
68308-34-9 SHALE OIL [NJL]	SHALE OILS
68308-34-9 SHALE OILS	SHALE OILS
68308-34-9 SHALE-OILS [C65]	SHALE OILS
9000-59-3 SHELLAC	SHELLAC
138-59-0 SHIKIMIC ACID	SHIKIMIC ACID
1982-49-6 SIDURON [CAL]	UREA, 1-(2-METHYLCYCLOHEXYL)-3-PHENYL-
7803-62-5 SILANE	SILANE
7803-62-5 SILANE (SILICON TETRAHYDRIDE) [QBC]	SILANE
3037-72-7 SILANE, (4-AMINOBUTYL)DIETHOXYMETHYL-	SILANE, (4-AMINOBUTYL)DIETHOXYMETHYL-
7521-80-4 SILANE, BUTYLTRICHLORO- [PAL]	BUTYLTRICHLOROSILANE
1609-19-4 SILANE, CHLORODIETHYL- [PAL]	CHLORODIETHYLSILANE
71808-64-5 SILANE, [(3-CHLOROPROPYL)DIMETHOXY[3-(OXIRANYLMETHOXY)PROPYL]-	SILANE, [(3-CHLOROPROPYL)DIMETHOXY[3-(OXIRANYLMETHOXY)PROPYL
75-77-4 SILANE, CHLOROTRIMETHYL- [EPA, EPA, OTS, PAL]	TRIMETHYLCHLOROSILANE
75-78-5 SILANE, DICHLORODIMETHYL- [EPA, EPA, OTS, PAL]	DIMETHYLDICHLOROSILANE
80-10-4 SILANE, DICHLORODIPHENYL- [PAL]	DIPHENYL DICHLOROSILANE
1789-58-8 SILANE, DICHLOROETHYL- [PAL]	ETHYL DICHLOROSILANE
75-54-7 SILANE, DICHLOROMETHYL- [OTS, PAL]	METHYL DICHLOROSILANE

Chemical Name	Reference Name
2897-60-1 SILANE, DIETHOXYMETHYL[3-(OXIRANYL-METHOXY)PROPYL]-	SILANE, DIETHOXYMETHYL[3-(OXIRANYLMETHOXY)PROPYL]-
RR-00950-7 SILANE, (1,1-DIMETHYLETHOXY)DIMETHOXY (2-METHYLPROPYL)-	SILANE, (1,1-DIMETHYLETHOXY)DIMETHOXY (2-METHYLPROPYL)-THER
78-08-0 SILANE, ETHENYLTRIETHOXY-	SILANE, ETHENYLTRIETHOXY-
17963-04-1 SILANE, ETHOXYDIMETHYL[3-(OXIRANYL-METHOXY)PROPYL]-	SILANE, ETHOXYDIMETHYL[3-(OXIRANYL-METHOXY)PROPYL]-
10025-78-2 SILANE, TRICHLORO- [PAL]	TRICHLOROSILANE
98-12-4 SILANE, TRICHLOROCYCLOHEXYL- [PAL]	CYCLOHEXYLTRICHLOROSILANE
75-94-5 SILANE, TRICHLOROETHENYL- [PAL]	VINYLTRICHLOROSILANE
115-21-9 SILANE, TRICHLOROETHYL- [PAL]	TRICHLOROETHYLSILANE
5894-60-0 SILANE, TRICHLOROHEXADECYL- [PAL]	HEXADECYLTRICHLOROSILANE
75-79-6 SILANE, TRICHLOROMETHYL- [EPA, EPA, OTS, PAL]	METHYL TRICHLOROSILANE
112-04-9 SILANE, TRICHLOROOCTADECYL- [PAL]	OCTADECYLTRICHLOROSILANE
107-72-2 SILANE, TRICHLOROPENTYL- [PAL]	AMYL TRICHLOROSILANE
98-13-5 SILANE, TRICHLOROPHENYL- [PAL]	PHENYL TRICHLOROSILANE
107-37-9 SILANE, TRICHLORO-2-PROPENYL- [PAL]	ALLYL TRICHLOROSILANE
141-57-1 SILANE, TRICHLOROPROPYL- [PAL]	PROPYL TRICHLOROSILANE
78-07-9 SILANE, TRIETHOXYETHYL- [PST]	TRIETHOXYETHYL SILANE
2031-67-6 SILANE, TRIETHOXYMETHYL- [PST]	METHYLTRIETHOXYSILANE
2761-24-2 SILANE, TRIETHOXYPENTYL-	SILANE, TRIETHOXYPENTYL-
3388-04-3 SILANE, TRIMETHOXY[2-(7-OXABICYCLO[4.1.0]HEPT-3-YL)ETHYL]	SILANE, TRIMETHOXY[2-(7-OXABICYCLO[4.1.0]HEPT-3-YL)ETHYL]
2530-83-8 SILANE, TRIMETHOXY[3-(OXIRANYL-METHOXY)PROPYL]- [TSC]	GLYCIDOXYPROPYLTRIMETHOXYSILANE
4253-34-3 SILANETRIOL, METHYL-, TRIACETATE	SILANETRIOL, METHYL-, TRIACETATE
7631-86-9 SILICA [CAL, MIN, ONT, ONT, ONT, PAL]	SILICA, AMORPHOUS
14808-60-7 SILICA (QUARTZ) [QBC, QBC]	QUARTZ
7631-86-9 SILICA, AMORPHOUS (INCLUDING NATURAL DIATOMACEOUS EARTH) [QBC]	SILICA, AMORPHOUS
15468-32-3 SILICA-CRYSTALLINE, TRIDYMITE [QBC]	SILICA, TRIDYMITE
1317-95-9 SILICA-CRYSTALLINE, TRIPOLI [QBC]	SILICA-TRIPOLI
7631-86-9 SILICA, AMORPHOUS	SILICA, AMORPHOUS
68855-54-9 SILICA, AMORPHOUS [ALB]	SILICA, AMORPHOUS, DIATOMACEOUS EARTH
68855-54-9 SILICA, AMORPHOUS, DIATOMACEOUS EARTH	SILICA, AMORPHOUS, DIATOMACEOUS EARTH
68855-54-9 SILICA-AMORPHOUS DIATOMACEUS EARTH [MIN]	SILICA, AMORPHOUS, DIATOMACEOUS EARTH
61790-53-2 SILICA, AMORPHOUS, DIATOMACEOUS EARTH, CONTAINING LESS THAN1% CRYSTALLINE SILICA [OS1, OS1]	DIATOMACEOUS EARTH
68855-54-9 SILICA, AMORPHOUS, DIATOMACEOUS EARTH (CALCINED) [MAK, MAK]	SILICA, AMORPHOUS, DIATOMACEOUS EARTH
61790-53-2 SILICA-AMORPHOUS, DIATOMACEOUS EARTH (UNCALCINED) [AUS]	DIATOMACEOUS EARTH
61790-53-2 SILICA, AMORPHOUS, DIATOMACEOUS EARTH [CEX]	DIATOMACEOUS EARTH
61790-53-2 SILICA, AMORPHOUS, DIATOMACEOUS EARTH (UNCALCINED) [MAK, MAK]	DIATOMACEOUS EARTH
61790-53-2 SILICA - AMORPHOUS, DIATOMACEOUS EARTH (UNCALCINED) [TLV, TLV]	DIATOMACEOUS EARTH

Chemical Name	Reference Name
68855-54-9 SILICA AMORPHOUS DIATOMACEOUS EARTH (UNCALCINED) [PAL]	SILICA, AMORPHOUS, DIATOMACEOUS EARTH
69012-64-2 SILICA - AMORPHOUS, FUME [TLV]	SILICA FUME (AMORPHOUS)
7631-86-9 SILICA, AMORPHOUS, FUMED [WHM]	SILICA, AMORPHOUS
7631-86-9 SILICA-AMORPHOUS, FUMED SILICA [AUS]	SILICA, AMORPHOUS
60676-86-0 SILICA, AMORPHOUS, FUSED [WHM]	SILICA, FUSED
7699-41-4 SILICA, AMORPHOUS, FUSED SILICA [MAK, MAK]	FUSED SILICA
63231-67-4 SILICA, AMORPHOUS GEL [QBC]	SILICA GEL
112926-00-8 SILICA, AMORPHOUS, PRECIPITATED AND GEL	SILICA, AMORPHOUS, PRECIPITATED AND GEL
112926-00-8 SILICA-AMORPHOUS, PRECIPITATED SILICA [AUS]	SILICA, AMORPHOUS, PRECIPITATED AND GEL
1343-98-2 SILICA, AMORPHOUS PRECIPITATED [QBC]	SILICIC ACID
112926-00-8 SILICA - AMORPHOUS, PRECIPITATED SILICA AND SILICA GEL [TLV]	SILICA, AMORPHOUS, PRECIPITATED AND GEL
60676-86-0 SILICA, AMORPHOUS, QUARTZ GLASS [MAK, MAK]	SILICA, FUSED
60676-86-0 SILICA - AMORPHOUS, SILICA, FUSED [TLV]	SILICA, FUSED
63231-67-4 SILICA-AMORPHOUS, SILICA GEL [AUS]	SILICA GEL
14464-46-1 SILICA, CRISTOBALITE	SILICA, CRISTOBALITE
RR-00087-3 SILICA, CRYSTALLINE [GBR, NIO, NIO, NIO, C65, NHS, NHS]	SILICA, CRYSTALLINE (GENERAL FORM)
14464-46-1 SILICA-CRYSTALLINE, CRISTOBALITE [MIN]	SILICA, CRISTOBALITE
14464-46-1 SILICA CRYSTALLINE CRISTOBALITE [OS1, OS1]	SILICA, CRISTOBALITE
14464-46-1 SILICA - CRYSTALLINE, CRISTOBALITE [TLV]	SILICA, CRISTOBALITE
14464-46-1 SILICA, CRYSTALLINE, CRISTOBALITE [WHM, CEX]	SILICA, CRISTOBALITE
RR-00087-3 SILICA, CRYSTALLINE (GENERAL FORM)	SILICA, CRYSTALLINE (GENERAL FORM)
14808-60-7 SILICA-CRYSTALLINE, QUARTZ [MIN]	QUARTZ
14808-60-7 SILICA, CRYSTALLINE QUARTZ [OS1, OS1]	QUARTZ
14808-60-7 SILICA - CRYSTALLINE, QUARTZ [TLV]	QUARTZ
14808-60-7 SILICA, CRYSTALLINE, QUARTZ [WHM, CEX]	QUARTZ
60676-86-0 SILICA-CRYSTALLINE, SILICA [MIN]	SILICA, FUSED
60676-86-0 SILICA, CRYSTALLINE, SILICA, FUSED [CEX]	SILICA, FUSED
15468-32-3 SILICA, CRYSTALLINE, TRIDYMITE [CEX]	SILICA, TRIDYMITE
15468-32-3 SILICA-CRYSTALLINE, TRIDYMITE [MIN]	SILICA, TRIDYMITE
15468-32-3 SILICA, CRYSTALLINE TRIDYMITE [OS1, OS1]	SILICA, TRIDYMITE
15468-32-3 SILICA - CRYSTALLINE, TRIDYMITE [TLV]	SILICA, TRIDYMITE
15468-32-3 SILICA, CRYSTALLINE, TRIDYMITE [WHM]	SILICA, TRIDYMITE
1317-95-9 SILICA, CRYSTALLINE, TRIPOLI [CEX]	SILICA-TRIPOLI
1317-95-9 SILICA-CRYSTALLINE, TRIPOLI [MIN]	SILICA-TRIPOLI
1317-95-9 SILICA, CRYSTALLINE TRIPOLI [OS1, OS1]	SILICA-TRIPOLI
1317-95-9 SILICA - CRYSTALLINE, TRIPOLI [TLV]	SILICA-TRIPOLI
1317-95-9 SILICA, CRYSTALLINE, TRIPOLI [WHM]	SILICA-TRIPOLI
7631-86-9 SILICA FLOUR [ALB]	SILICA, AMORPHOUS
69012-64-2 SILICA FUME (AMORPHOUS)	SILICA FUME (AMORPHOUS)
60676-86-0 SILICA, FUSED	SILICA, FUSED
60676-86-0 SILICA, FUSED CRYSTALLINE [QBC]	SILICA, FUSED

Chemical Name	Reference Name
60676-86-0 SILICA, FUSED, DUST [FLA, MSL]	SILICA, FUSED
63231-67-4 SILICA GEL	SILICA GEL
7631-86-9 SILICA SAND [BC1]	SILICA, AMORPHOUS
14807-96-6 SILICA, TALC, NON-ASBESTOS FORM [FLA]	TALC
16919-19-0 SILICATE(2-), HEXAFLUORO-, DIAMMONIUM [PAL]	AMMONIUM SILICOFLUORIDE
16871-71-9 SILICATE(2-), HEXAFLUORO-, ZINC (1:1) [PAL]	ZINC SILICOFLUORIDE
RR-00010-2 SILICATE SOAPSTONE DUST [FLA]	SOAPSTONE
15468-32-3 SILICA, TRIDYMITE	SILICA, TRIDYMITE
1317-95-9 SILICA-TRIPOLI	SILICA-TRIPOLI
1317-95-9 SILICA, TRIPOLI [NJL]	SILICA-TRIPOLI
60676-86-0 SILICA, VITREOUS [PST]	SILICA, FUSED
1343-98-2 SILICIC ACID	SILICIC ACID
1344-00-9 SILICIC ACID, ALUMINUM SODIUM SALT (1:1:1) [PST]	SODIUM ALUMINUM SILICATE
15191-85-2 SILICIC ACID (H4SIO4), BERYLLIUM SALT (1:2)	SILICIC ACID (H4SIO4), BERYLLIUM SALT (1:2)
58500-38-2 SILICIC ACID, BERYLLIUM SALT	SILICIC ACID, BERYLLIUM SALT
39413-47-3 SILICIC ACID, BERYLLIUM ZINC SALT [PAL, PAL]	ZINC BERYLLIUM SILICATE
1344-95-2 SILICIC ACID, CALCIUM SALT [PAL]	CALCIUM SILICATE
75364-04-4 SILICIC ACID (H6SI2O7), COBALT(2+) MAGNESIUM SALT (1:2:1)	SILICIC ACID (H6SI2O7), COBALT(2+) MAGNESIUM SALT (1:2:1)
6834-92-0 SILICIC ACID, DISODIUM SALT, PENTAHYDRATE [PST]	SODIUM METASILICATE
11099-06-2 SILICIC ACID, ETHYL ESTER	SILICIC ACID, ETHYL ESTER
2157-42-8 SILICIC ACID (H6SI2O7), HEXAETHYL ESTER	SILICIC ACID (H6SI2O7), HEXAETHYL ESTER
12002-26-5 SILICIC ACID, METHYL ESTER	SILICIC ACID, METHYL ESTER
4521-94-2 SILICIC ACID (H8SI3O10), OCTAETHYL ESTER	SILICIC ACID (H8SI3O10), OCTAETHYL ESTER
4421-95-8 SILICIC ACID (H8SI3O10), OCTAMETHYL ESTER	SILICIC ACID (H8SI3O10), OCTAMETHYL ESTER
78-10-4 SILICIC ACID (H4SIO4), TETRAETHYL ESTER [PAL]	ETHYL SILICATE
78-10-4 SILICIC ACID, TETRAETHYL ESTER [MAK, MAK]	ETHYL SILICATE
78-13-7 SILICIC ACID (H4SIO4), TETRAKIS(2-ETHYLBUTYL) ESTER	SILICIC ACID (H4SIO4), TETRAKIS(2-ETHYLBUTYL) ESTER
681-84-5 SILICIC ACID (H4SIO4), TETRAMETHYL ESTER [PAL]	METHYL SILICATE
16961-83-4 SILICOFLUORIC ACID [NJL]	FLUOROSILICIC ACID
RR-00613-3 SILICOFLUORIDE COMPOUNDS, N.O.S.	SILICOFLUORIDE COMPOUNDS, N.O.S.
RR-00613-3 SILICOFLUORIDES, N.O.S. [NJL]	SILICOFLUORIDE COMPOUNDS, N.O.S.
7440-21-3 SILICON	SILICON
14276-49-4 SILICON 31	SILICON 31
15092-72-5 SILICON 32	SILICON 32
409-21-2 SILICON CARBIDE	SILICON CARBIDE
12327-32-1 SILICON CARBIDE (SI2C3)	SILICON CARBIDE (SI2C3)
409-21-2 SILICON CARBIDE (SIC) [PAL]	SILICON CARBIDE
7631-86-9 SILICON DIOXIDE [PST]	SILICA, AMORPHOUS
7803-62-5 SILICONE TETRAHYDRIDE [MEX]	SILANE
RR-01662-6 SILICONE ESTER POLYACRYLATE	SILICONE ESTER POLYACRYLATE
63148-62-9 SILICONES AND SILOXANES, DIMETHYL [PST]	POLY(DIMETHYLSILOXANE)

Chemical Name	Reference Name
7440-21-3 SILICON POWDER [NJL]	SILICON
10026-04-7 SILICON TETRACHLORIDE	SILICON TETRACHLORIDE
7783-61-1 SILICON TETRAFLUORIDE	SILICON TETRAFLUORIDE
7803-62-5 SILICON TETRAHYDRIDE [AUS]	SILANE
7803-62-5 SILICON TETRAHYDRIDE (SILANE) [BC1, BC1]	SILANE
7803-62-5 SILICON TETRAHYDRIDE [CEX, OS1, TLV]	SILANE
7803-62-5 SILICON TETRAHYDRIDE (SILANE) [ALB, ALB]	SILANE
RR-01139-2 SILKWORM PUPAE	SILKWORM PUPAE
RR-01713-0 SILOXANES	SILOXANES
70131-67-8 SILOXANES AND SILICONES, DI-ME, HY-DROXY-TERMINATED	SILOXANES AND SILICONES, DI-ME, HYDROXY-TERMINATED
68937-54-2 SILOXANES AND SILICONES, DIMETHYL, 3-HYDROXYPROPYL METHYL, ETHOXYLATED	SILOXANES AND SILICONES, DIMETHYL, 3-HYDROXYPROPYL METHYL, EETHACRYLATE POLYMER
68037-64-9 SILOXANES AND SILICONES, DIMETHYL, METHYL HYDROGEN, REACTION PRODUCTS WITH POLYETHYLENE-POYPROPYLENE GLYCOL MONOACETATE ALLYL ETHER	SILOXANES AND SILICONES, DIMETHYL, METHYL HYDROGEN, REACTION
68554-65-4 SILOXANES AND SILICONES, DIMETHYL, POLYMERS WITH METHYL SILSESQUIOXANES AND POLYETHYLENE-POYPROPYLENE GLYCOL MONOBUTYL ETHER	SILOXANES AND SILICONES, DIMETHYL, POLYMERS WITH METHYL SILSH ROSIN
63148-56-1 SILOXANES AND SILICONES, METHYL 3,3,3-TRIFLUOROPROPYL	SILOXANES AND SILICONES, METHYL 3,3,3-TRIFLUOROPROPYL
7440-22-4 SILVER	SILVER
14833-32-0 SILVER 102	SILVER 102
14967-69-2 SILVER 103	SILVER 103
15116-79-7 SILVER 104	SILVER 104
RR-00356-5 SILVER 104M	SILVER 104M
14928-14-4 SILVER 105	SILVER 105
14333-39-2 SILVER 106	SILVER 106
RR-00290-4 SILVER 106M	SILVER 106M
RR-00355-4 SILVER 108M	SILVER 108M
RR-00354-3 SILVER 110M	SILVER 110M
15760-04-0 SILVER 111	SILVER 111
14331-86-3 SILVER 112	SILVER 112
15760-07-3 SILVER 115	SILVER 115
7440-22-4 SILVER, METAL [MEX]	SILVER
13092-75-6 SILVER ACETYLIDE	SILVER ACETYLIDE
23606-32-8 SILVER AMMONIUM NITRATE	SILVER AMMONIUM NITRATE
7784-08-9 SILVER ARSENITE	SILVER ARSENITE
13863-88-2 SILVER AZIDE	SILVER AZIDE
7783-91-7 SILVER CHLORITE	SILVER CHLORITE
7784-01-2 SILVER CHROMATE	SILVER CHROMATE
RR-00574-3 SILVER COMPOUNDS	SILVER COMPOUNDS
RR-00574-3 SILVER COMPOUNDS, N.O.S. [NJL, RCU]	SILVER COMPOUNDS
506-64-9 SILVER CYANIDE	SILVER CYANIDE
7440-22-4 SILVER, ELEMENTAL [WHM]	SILVER
3794-64-7 SILVER HEPTAFLUOROBUTYRATE	SILVER HEPTAFLUOROBUTYRATESODIUM SALT
7440-22-4 SILVER, METAL [ONT]	SILVER
7440-22-4 SILVER, METAL AND SOLUBLE COMPOUNDS [BC1, BC1, OS1, OS1]	SILVER
7440-22-4 SILVER (METAL DUST AND SOLUBLE COMPOUNDS) [NIO, NIO]	SILVER

Chemical Name	Reference Name
7761-88-8 SILVER NITRATE	SILVER NITRATE
RR-01312-7 SILVER ORTHOARSENITE	SILVER ORTHOARSENITE
533-51-7 SILVER OXALATE	SILVER OXALATE
20667-12-3 SILVER OXIDE	SILVER OXIDE
25455-73-6 SILVER OXIDE (AG2O2)	SILVER OXIDE (AG2O2)
509-09-1 SILVER PENTAFLUOROPROPIONATE	SILVER PENTAFLUOROPROPIONATE
146-84-9 SILVER PICRATE	SILVER PICRATE
506-61-6 SILVER POTASSIUM CYANIDE [WHM]	POTASSIUM SILVER CYANIDE
RR-00041-9 SILVER SOLUBLE COMPOUNDS	SILVER SOLUBLE COMPOUNDS
RR-00041-9 SILVER, WATER SOLUBLE COMPOUNDS [ONT]	SILVER SOLUBLE COMPOUNDS
RR-00041-9 SILVER, WATER-SOLUBLE COMPOUNDS, N.O.S. [WHM]	SILVER SOLUBLE COMPOUNDS
93-72-1 SILVEX (2,4,5-TP)	SILVEX (2,4,5-TP)
93-72-1 SILVEX [PQL]	SILVEX (2,4,5-TP)
93-72-1 SILVEX (2,4,5-TP) [RCA]	SILVEX (2,4,5-TP)
122-34-9 SIMAZINE	SIMAZINE
8050-81-5 SIMETHICONE	SIMETHICONE
64742-61-6 SLACK WAX [PST]	SLACK WAX (PETROLEUM)
64742-61-6 SLACK WAX (PETROLEUM)	SLACK WAX (PETROLEUM)
RR-01202-2 SLAGWOOL	SLAGWOOL
RR-01202-2 SLAG WOOL [MAK]	SLAGWOOL
RR-01202-2 SLAGWOOL FIBERS [CAB]	SLAGWOOL
64742-24-1 SLUDGE ACID	SLUDGE ACID
12199-37-0 SMECTITE	SMECTITE
12199-37-0 SMECTITE-GROUP MINERALS [PST]	SMECTITE
RR-00010-2 SOAP [PST]	SOAPSTONE
RR-00010-2 SOAPSTONE	SOAPSTONE
RR-00010-2 SOAPSTONE (CONTAINING LESS THAN 1% QUARTZ) [NIO, NIO]	SOAPSTONE
RR-00010-2 SOAPSTONE DUST [PAL]	SOAPSTONE
8006-28-8 SODA LIME	SODA LIME
7440-23-5 SODIUM	SODIUM
13966-32-0 SODIUM 22	SODIUM 22
13982-04-2 SODIUM 24	SODIUM 24
127-09-3 SODIUM ACETATE	SODIUM ACETATE
7558-80-7 SODIUM ACID PHOSPHATE	SODIUM ACID PHOSPHATE
7758-16-9 SODIUM ACID PYROPHOSPHATE [PST]	DIPHOSPHORIC ACID, DISODIUM SALT
7446-81-3 SODIUM ACRYLATE	SODIUM ACRYLATE
25085-02-3 SODIUM ACRYLATE, POLYMER WITH ACRY-LAMIDE [PST]	ACRYLAMIDE-SODIUM ACRYLATE POLYMER
9005-38-3 SODIUM ALGINATE [PST]	ALGENIC ACID, SODIUM SALT
11138-49-1 SODIUM ALUMINATE [HM1, HM1, PST, NJL, NJL]	ALUMINUM SODIUM OXIDE
1302-42-7 SODIUM ALUMINATE (NAALO2)	SODIUM ALUMINATE (NAALO2)
15096-52-3 SODIUM ALUMINUM FLUORIDE	SODIUM ALUMINUM FLUORIDE
13770-96-2 SODIUM ALUMINUM HYDRIDE	SODIUM ALUMINUM HYDRIDE
7785-88-8 SODIUM ALUMINUM PHOSPHATE	SODIUM ALUMINUM PHOSPHATE
1344-00-9 SODIUM ALUMINUM SILICATE	SODIUM ALUMINUM SILICATE
11110-52-4 SODIUM AMALGAM	SODIUM AMALGAM
7782-92-5 SODIUM AMIDE	SODIUM AMIDE

Chemical Name	Reference Name
12055-09-3 SODIUM AMMONIUM VANADATE	SODIUM AMMONIUM VANADATE
128-56-3 SODIUM ANTHRAQUINONE-1-SULFONATE	SODIUM ANTHRAQUINONE-1-SULFONATE
11112-10-0 SODIUM ANTIMONATE	SODIUM ANTIMONATE
127-85-5 SODIUM ARSANILATE	SODIUM ARSANILATE
7631-89-2 SODIUM ARSENATE	SODIUM ARSENATE
7784-46-5 SODIUM ARSENITE	SODIUM ARSENITE
134-03-2 SODIUM ASCORBATE	SODIUM ASCORBATE
26628-22-8 SODIUM AZIDE	SODIUM AZIDE
26628-22-8 SODIUM AZIDE (NA(N3)) [302, PAL]	SODIUM AZIDE
532-32-1 SODIUM BENZOATE	SODIUM BENZOATE
144-55-8 SODIUM BICARBONATE	SODIUM BICARBONATE
10588-01-9 SODIUM BICHROMATE [CWA, EP2, EPA, FLA, MSL]	SODIUM DICHROMATE
1333-83-1 SODIUM BIFLUORIDE	SODIUM BIFLUORIDE
577-11-7 SODIUM BIS(2-ETHYLHEXYL) SULFOSUCCI-NATE [PST]	DIOCTYL SODIUM SULFOSUCCINATE
7681-38-1 SODIUM BISULFATE	SODIUM BISULFATE
7631-90-5 SODIUM BISULFITE	SODIUM BISULFITE
7631-90-5 SODIUM BISULPHITE [AUS]	SODIUM BISULFITE
1330-43-4 SODIUM BORATE [MSL, WHM]	BORATES, TETRA, SODIUM SALTS, ANHYDROUS
1303-96-4 SODIUM BORATES [NJL]	BORATES, TETRA, SODIUM SALTS, DECAHYDRATE
16940-66-2 SODIUM BOROHYDRIDE	SODIUM BOROHYDRIDE
7789-38-0 SODIUM BROMATE	SODIUM BROMATE
7647-15-6 SODIUM BROMIDE	SODIUM BROMIDE
124-65-2 SODIUM CACODYLATE	SODIUM CACODYLATE
1984-06-1 SODIUM CAPRYLATE	SODIUM CAPRYLATE
497-19-8 SODIUM CARBONATE	SODIUM CARBONATE
9004-32-4 SODIUM CARBOXYMETHYL CELLULOSE	SODIUM CARBOXYMETHYL CELLULOSE
9005-46-3 SODIUM CASEINATE	SODIUM CASEINATE
7775-09-9 SODIUM CHLORATE	SODIUM CHLORATE
7647-14-5 SODIUM CHLORIDE	SODIUM CHLORIDE
7758-19-2 SODIUM CHLORITE	SODIUM CHLORITE
3926-62-3 SODIUM CHLOROACETATE	SODIUM CHLOROACETATE
7775-11-3 SODIUM CHROMATE	SODIUM CHROMATE
13517-17-4 SODIUM CHROMATE DECAHYDRATE	SODIUM CHROMATE DECAHYDRATE
68-04-2 SODIUM CITRATE	SODIUM CITRATE
RR-01313-8 SODIUM COPPER CYANIDE	SODIUM COPPER CYANIDE
28348-53-0 SODIUM CUMENESULFONATE	SODIUM CUMENESULFONATE
14264-31-4 SODIUM CUPROCYANIDE [HM1, HM1, HM1, NJL]	COPPER SODIUM CYANIDE
143-33-9 SODIUM CYANIDE	SODIUM CYANIDE
143-33-9 SODIUM CYANIDE (NA(CN)) [302, PAL]	SODIUM CYANIDE
139-05-9 SODIUM CYCLAMATE	SODIUM CYCLAMATE
1322-98-1 SODIUM DECYLBENZENESULFONATE	SODIUM DECYLBENZENESULFONATE
142-87-0 SODIUM DECYL SULFATE [PST]	SULFURIC ACID, MONODECYL ESTER, SODIUM SALT
4418-26-2 SODIUM DEHYDROACETATE	SODIUM DEHYDROACETATE
RR-00760-3 SODIUM 2-DIAZO-1-NAPHTHOL-4-SULFONATE	SODIUM 2-DIAZO-1-NAPHTHOL-4-SULFONATE
RR-00761-4 SODIUM 2-DIAZO-1-NAPHTHOL-5-SULFONATE	SODIUM 2-DIAZO-1-NAPHTHOL-5-SULFONATE
1982-69-0 SODIUM DICAMBA [3,6-DICHLORO-2-METHOXYBENZOIC ACID, SODIUM SALT]	SODIUM DICAMBA [3,6-DICHLORO-2-METHOXYBEN-ZOIC ACID, SODIUM S
2156-56-1 SODIUM DICHLOROACETATE	SODIUM DICHLOROACETATE

Chemical Name	**Reference Name**
2893-78-9 SODIUM DICHLORO-ISOCYANATE [NJL]	DICHLOROISOCYANURIC ACID, SODIUM SALT
2893-78-9 SODIUM DICHLORO ISOCYANURATE [FLA]	DICHLOROISOCYANURIC ACID, SODIUM SALT
51580-86-0 SODIUM DICHLOROISOCYANURATE DIHY-DRATE	SODIUM DICHLOROISOCYANURATE DIHYDRATE
136-78-7 SODIUM 2-(2,4-DICHLOROPHENOXY)ETHYL SULFATE [CAL]	SESONE)
136-78-7 SODIUM 2-(2,4-DICHLOROPHENOXY)ETHYL SULPHATE [GBR, GBR]	SESONE
2893-78-9 SODIUM DICHLORO-S-TRIAZINETRIONE [MSL]	DICHLOROISOCYANURIC ACID, SODIUM SALT
51580-86-0 SODIUM DICHLORO-S-TRIAZINE TRIONEDIHY-DRATE [MSL]	SODIUM DICHLOROISOCYANURATE DIHYDRATE
10588-01-9 SODIUM DICHROMATE	SODIUM DICHROMATE
148-18-5 SODIUM DIETHYLDITHIOCARBAMATE	SODIUM DIETHYLDITHIOCARBAMATE
3006-15-3 SODIUM 1,4-DIHEXYL SULFOSUCCINATE	SODIUM 1,4-DIHEXYL SULFOSUCCINATE
1322-93-6 SODIUM DIISOPROPYLNAPHTHALENESUL-FONATE	SODIUM DIISOPROPYLNAPHTHALENESULFONATE
128-04-1 SODIUM DIMETHYLDITHIOCARBAMATE	SODIUM DIMETHYLDITHIOCARBAMATE
2312-76-7 SODIUM DINITRO-ORTHO-CRESOLATE	SODIUM DINITRO-ORTHO-CRESOLATE
2312-76-7 SODIUM DINITRO-O-CRESOLATE [HM1, HM1, HM1]	SODIUM DINITRO-ORTHO-CRESOLATE
25641-53-6 SODIUM DINITRO-O-CRESOLATE	SODIUM DINITRO-O-CRESOLATE
3246-20-6 SODIUM DINONYL SULFOSUCCINATE	SODIUM DINONYL SULFOSUCCINATE
1639-66-3 SODIUM DIOCTYL SULFOSUCCINATE [PST]	DIOCTYL SODIUM SULFOSUCCINATE
7631-90-5 SODIUM DISULFITE [FLA]	SODIUM BISULFITE
7775-14-6 SODIUM DITHIONITE	SODIUM DITHIONITE
2673-22-5 SODIUM DITRIDECYLSULFOSUCCINATE [PST]	BIS(TRIDECYL) SODIUM SULFOSUCCINATE
25155-30-0 SODIUM DODECYLBENZENE SUL-FONATE [MSL, NJL]	SODIUM DODECYLBENZENESULFONATE
25155-30-0 SODIUM DODECYLBENZENESULFONATE	SODIUM DODECYLBENZENESULFONATE
25155-30-0 SODIUM DODECYLBENZENE-SUL-FONATE [CAL]	SODIUM DODECYLBENZENESULFONATE
53467-00-8 SODIUM DODECYL DIPHENYL OXIDE SUL-FONATE	SODIUM DODECYL DIPHENYL OXIDE SULFONATE
RR-01140-5 SODIUM DODECYLPHENYL POLYOXYETHY-LENE PHOSPHATES	SODIUM DODECYLPHENYL POLYOXYETHYLENE PHOS-PHATES
16680-47-0 SODIUM EQUILIN SULFATE	SODIUM EQUILIN SULFATE
6381-77-7 SODIUM ERYTHORBATE	SODIUM ERYTHORBATE
438-67-5 SODIUM ESTRONE SULFATE	SODIUM ESTRONE SULFATE
17421-79-3 SODIUM ETHYLENEDIAMINETETRAACETATE	SODIUM ETHYLENEDIAMINETETRAACETATE
126-92-1 SODIUM 2-ETHYLHEXYL SULFATE [PST]	2-ETHYLHEXYL SODIUM SULFATE
54-64-8 SODIUM O-(ETHYLMERCURITHIO)BENZOATE	SODIUM O-(ETHYLMERCURITHIO)BENZOATE
62-74-8 SODIUM FLUORACETATE [QBC, QBC, QBC]	SODIUM FLUOROACETATE
518-47-8 SODIUM FLUORESCEIN	SODIUM FLUORESCEIN
7681-49-4 SODIUM FLUORIDE	SODIUM FLUORIDE
1333-83-1 SODIUM FLUORIDE (NA(HF2)) [PAL]	SODIUM BIFLUORIDE
7681-49-4 SODIUM FLUORIDE (NAF) [PAL]	SODIUM FLUORIDE
RR-00490-0 SODIUM FLUORIDES	SODIUM FLUORIDES
62-74-8 SODIUM FLUOROACETATE	SODIUM FLUOROACETATE
62-74-8 SODIUM FLUOROACETATE (1080) [BC1, BC1, BC1]	SODIUM FLUOROACETATE
62-74-8 SODIUM FLUOROACETATE [TLV, TLV]	SODIUM FLUOROACETATE
16893-85-9 SODIUM FLUOROSILICATE [NJL]	SODIUM SILICOFLUORIDE

Chemical Name	Reference Name
149-44-0 SODIUM FORMALDEHYDE SULFOXYLATE	SODIUM FORMALDEHYDE SULFOXYLATE
141-53-7 SODIUM FORMATE	SODIUM FORMATE
527-07-1 SODIUM GLUCONATE	SODIUM GLUCONATE
10124-56-8 SODIUM HEXAMETAPHOSPHATE	SODIUM HEXAMETAPHOSPHATE
7646-69-7 SODIUM HYDRIDE	SODIUM HYDRIDE
7646-69-7 SODIUM HYDRIDE (NAH) [PAL]	SODIUM HYDRIDE
1333-83-1 SODIUM HYDROGEN FLUORIDE [HM1, HM1, HM1, NJL, NJL]	SODIUM BIFLUORIDE
7681-38-1 SODIUM HYDROGEN SULFATE [HM1, HM1, NJL, NJL]	SODIUM BISULFATE
7631-90-5 SODIUM HYDROGENSULPHITE [GBR]	SODIUM BISULFITE
16721-80-5 SODIUM HYDROSULFIDE	SODIUM HYDROSULFIDE
7775-14-6 SODIUM HYDROSULFITE [FLA, MSL]	SODIUM DITHIONITE
16721-80-5 SODIUM HYDROSULPHIDE [WHM]	SODIUM HYDROSULFIDE
1310-73-2 SODIUM HYDROXIDE	SODIUM HYDROXIDE
1310-73-2 SODIUM HYDROXIDE (NA(OH)) [PAL]	SODIUM HYDROXIDE
7681-52-9 SODIUM HYPOCHLORITE	SODIUM HYPOCHLORITE
10022-70-5 SODIUM HYPOCHLORITE [EPA, NJL]	SODIUM HYPOCHLORITE (PENTAHYDRATE)
10022-70-5 SODIUM HYPOCHLORITE (PENTAHYDRATE)	SODIUM HYPOCHLORITE (PENTAHYDRATE)
13721-43-2 SODIUM HYPOPHOSPHATE	SODIUM HYPOPHOSPHATE
7681-53-0 SODIUM HYPOPHOSPHITE	SODIUM HYPOPHOSPHITE
7681-82-5 SODIUM IODIDE	SODIUM IODIDE
15708-41-5 SODIUM IRON(III) ETHYLENETETRAACETATE	SODIUM IRON(III) ETHYLENETETRAACETATE
60874-90-0 SODIUM ISOPROPYL ISOBUTYL NAPHTHALE-NESULFONATE	SODIUM ISOPROPYL ISOBUTYL NAPHTHALENESUL-FONATE
32612-48-9 SODIUM LAURYL ETHER SULFATE [WHM]	POLYETHYLENE GLYCOL MONOLAURYL ETHER SUL-FATE AMMONIUM SALT
151-21-3 SODIUM LAURYL SULFATE	SODIUM LAURYL SULFATE
13150-00-0 SODIUM LAURYL TRIETHOXY SULFATE	SODIUM LAURYL TRIETHOXY SULFATE
8061-51-6 SODIUM LIGNO SULFONATE	SODIUM LIGNO SULFONATE
2492-26-4 SODIUM MERCAPTOBENZOTHIAZOLE	SODIUM MERCAPTOBENZOTHIAZOLE
7784-46-5 SODIUM METAARSENITE [HM1, HM1, HM1, HM1]	SODIUM ARSENITE
7681-57-4 SODIUM METABISULFITE	SODIUM METABISULFITE
7681-57-4 SODIUM METABISULPHITE [AUS]	SODIUM METABISULFITE
7775-19-1 SODIUM METABORATE	SODIUM METABORATE
6834-92-0 SODIUM METASILICATE	SODIUM METASILICATE
13718-26-8 SODIUM METAVANADATE [WHM]	THIRAM
124-41-4 SODIUM METHOXIDE [HON]	SODIUM METHYLATE
124-41-4 SODIUM METHYLATE	SODIUM METHYLATE
137-20-2 SODIUM N-METHYL-N-OLEOYLTAURINE	SODIUM N-METHYL-N-OLEOYLTAURINE
7346-80-7 SODIUM N-METHYL-N-OLEYLTAURATE	SODIUM N-METHYL-N-OLEYLTAURATE
7631-95-0 SODIUM MOLYBDATE	SODIUM MOLYBDATE
RR-01142-7 SODIUM MONO, DI AND TRIBUTYLNAPH-THALENESULFONATE	SODIUM MONO, DI AND TRIBUTYLNAPHTHALENESUL-FONATE
RR-01141-6 SODIUM MONO AND DI-C8-13-ALKYL PHE-NOXY BENZENE DISULFONATES	SODIUM MONO AND DI-C8-13-ALKYL PHENOXY BEN-ZENE DISULFONATES
12401-86-4 SODIUM MONOXIDE	SODIUM MONOXIDE
7631-99-4 SODIUM NITRATE	SODIUM NITRATE
7632-00-0 SODIUM NITRITE	SODIUM NITRITE
14402-89-2 SODIUM NITROFERRICYANIDE	SODIUM NITROFERRICYANIDE

Chemical Name	Reference Name
RR-01143-8 SODIUM N-NONYLDIPHENYL ETHER SUL-FONATE	SODIUM N-NONYLDIPHENYL ETHER SULFONATE
60883-84-3 SODIUM NONYLMETHYLNAPHTHALENESUL-FONATE	SODIUM NONYLMETHYLNAPHTHALENESULFONATE
5324-84-5 SODIUM 1-OCTANESULFONATE	SODIUM 1-OCTANESULFONATE
RR-01144-9 SODIUM OCTYL PHENOXY DIETHOXYETHYL SULFATE	SODIUM OCTYL PHENOXY DIETHOXYETHYL SULFATE
142-31-4 SODIUM OCTYL SULFATE	SODIUM OCTYL SULFATE
143-19-1 SODIUM OLEATE	SODIUM OLEATE
68439-56-5 SODIUM ALPHA-OLEFINSULFONATE	SODIUM ALPHA-OLEFINSULFONATE
29169-69-5 SODIUM N-OLEYL TAURINE	SODIUM N-OLEYL TAURINE
15922-78-8 SODIUM OMADINE	SODIUM OMADINEN"-HEXAMTHYL-
132-27-4 SODIUM O-PHENYLPHENOXIDE [313]	SODIUM O-PHENYLPHENOL
13464-38-5 SODIUM ORTHOARSENATE [HM1]	ARSENIC ACID, TRISODIUM SALT
12007-92-0 SODIUM PENTABORATE	SODIUM PENTABORATE
131-52-2 SODIUM PENTACHLOROPHENATE	SODIUM PENTACHLOROPHENATE
13393-71-0 SODIUM PENTADECYLSULFATE	SODIUM PENTADECYLSULFATE
7632-04-4 SODIUM PERBORATE	SODIUM PERBORATE
15630-89-4 SODIUM PERCARBONATE	SODIUM PERCARBONATE
15630-89-4 SODIUM PERCARBONATE COMPOUND WITH HYDROGEN PEROXIDE [NJL]	SODIUM PERCARBONATE
7601-89-0 SODIUM PERCHLORATE	SODIUM PERCHLORATE
10101-50-5 SODIUM PERMANGANATE	SODIUM PERMANGANATE
1313-60-6 SODIUM PEROXIDE	SODIUM PEROXIDE
1313-60-6 SODIUM PEROXIDE (NA2(O2)) [PAL]	SODIUM PEROXIDE
7775-27-1 SODIUM PERSULFATE	SODIUM PERSULFATE
139-02-6 SODIUM PHENOLATE	SODIUM PHENOLATE
132-27-4 SODIUM ORTHO-PHENYLPHENATE [CAL, IAR, MIN]	SODIUM O-PHENYLPHENOL
132-27-4 SODIUM O-PHENYLPHENOL	SODIUM O-PHENYLPHENOL
7632-05-5 SODIUM PHOSPHATE	SODIUM PHOSPHATE
10140-65-5 SODIUM PHOSPHATE, DIBASIC [NJL, NJL]	PHOSPHORIC ACID, DISODIUM SALT, HYDRATE
7558-79-4 SODIUM PHOSPHATE DIBASIC	SODIUM PHOSPHATE DIBASIC
7558-79-4 SODIUM PHOSPHATE, DIBASIC [NJL]	SODIUM PHOSPHATE DIBASIC
7758-29-4 SODIUM PHOSPHATE, TRIBASIC [NJL]	PENTASODIUM TRIPHOSPHATE
7765-84-6 SODIUM PHOSPHATE, TRIBASIC	SODIUM PHOSPHATE, TRIBASIC
10361-89-4 SODIUM PHOSPHATE, TRIBASIC [NJL, NJL, NJL, NJL]	PHOSPHORIC ACID, TRISODIUM SALT, DECAHYDRATE
10124-56-8 SODIUM PHOSPHATE, TRIBASIC (SODIUM HEX-AMETAPHOSPHATE) [CWA]	SODIUM HEXAMETAPHOSPHATE
10361-89-4 SODIUM PHOSPHATE, TRIBASIC, DECAHY-DRATE [CWA]	PHOSPHORIC ACID, TRISODIUM SALT, DECAHYDRATE
14986-84-6 SODIUM PHOSPHATE TRIBISINE	SODIUM PHOSPHATE TRIBISINE
12058-85-4 SODIUM PHOSPHIDE	SODIUM PHOSPHIDE
831-52-7 SODIUM PICRAMATE	SODIUM PICRAMATE
RR-01458-4 SODIUM PICRYL PEROXIDE	SODIUM PICRYL PEROXIDE
9003-04-7 SODIUM POLYACRYLATE	SODIUM POLYACRYLATE
67785-61-9 SODIUM POLYMETHACRYLATE	SODIUM POLYMETHACRYLATEOLEIC, AND/OR LINOLEIC
25704-18-1 SODIUM POLYSTYRENE SULFONATE	SODIUM POLYSTYRENE SULFONATE
11135-81-2 SODIUM POTASSIUM ALLOYS [FLA, MSL]	POTASSIUM-SODIUM ALLOY
12736-96-8 SODIUM POTASSIUM ALUMINUM SILICATE	SODIUM POTASSIUM ALUMINUM SILICATE

Chemical Name	Reference Name
137-40-6 SODIUM PROPIONATE [PST]	BIOBAN-S
3811-73-2 SODIUM PYRIDITHIONE	SODIUM PYRIDITHIONE
3811-73-2 SODIUM PYRITHIONE [MAK, MAK, MAK, MAK]	SODIUM PYRIDITHIONE
5323-95-5 SODIUM RICINOLEATE	SODIUM RICINOLEATE
128-44-9 SODIUM SACCHARIN	SODIUM SACCHARIN
54-21-7 SODIUM SALICYLATE	SODIUM SALICYLATE
RR-00213-1 SODIUM SALT OF AN ALKYLATED, SULFONATED AROMATIC	SODIUM SALT OF AN ALKYLATED, SULFONATED AROMATIC
12788-84-0 SODIUM SALT OF N-COCO BETA AMINO BUTYRIC ACID	SODIUM SALT OF N-COCO BETA AMINO BUTYRIC ACID
68441-84-9 SODIUM SALT OF CRESOL SULFONIC ACID CONDENSED WITH UREA FORMALDEHYDE	SODIUM SALT OF CRESOL SULFONIC ACID CONDENSED WITH UREA FORM
61790-51-0 SODIUM SALT OF HYDROCARBON INSOLUBLE FRACTION OF ROSIN [PST]	RESIN ACIDS AND ROSIN ACIDS, SODIUM SALTS
RR-01145-0 SODIUM SALT OF OLEIC ACID AMIDE OF PROTEIN HYDROLYSATE	SODIUM SALT OF OLEIC ACID AMIDE OF PROTEIN HYDROLYSATE
630-93-3 SODIUM SALT OF PHENYTOIN [PAL, PAL]	DIPHENYLHYDANTOIN, SODIUM SALT
RR-00491-1 SODIUM SALTS OF AROMATIC NITRO-DERIVATIVES, N.O.S. [HM1, HM1]	SODIUM SALTS OF NITRO-AROMATIC DERIVATIVES, N.O.S.
RR-00491-1 SODIUM SALTS OF NITRO-AROMATIC DERIVATIVES, N.O.S.	SODIUM SALTS OF NITRO-AROMATIC DERIVATIVES, N.O.S.
68608-26-4 SODIUM SALTS OF PETROLEUM SULFONIC ACIDS [PST]	SULFONIC ACIDS, PETROLEUM, SODIUM SALTSIS(HYDROXYETHYL) METHYL, CHLORIDES
78330-30-0 SODIUM SALTS OF ALPHA-SULFO-OMEGA-HYDROXYPOLY(OXY-1,2-ETHANEDIYL) C11-14 ISOALKYL ETHERS, (C13-RICH)	SODIUM SALTS OF ALPHA-SULFO-OMEGA-HYDROXY-POLY(OXY-1,2-ETHANE
13410-01-0 SODIUM SELENATE	SODIUM SELENATE
10102-18-8 SODIUM SELENITE	SODIUM SELENITE
533-96-0 SODIUM SESQUICARBONATE	SODIUM SESQUICARBONATE
16893-85-9 SODIUM SILICA FLUORIDE [MSL]	SODIUM SILICOFLUORIDE
1344-09-8 SODIUM SILICATE	SODIUM SILICATE
16893-85-9 SODIUM SILICOFLUORIDE	SODIUM SILICOFLUORIDE
822-16-2 SODIUM STEARATE	SODIUM STEARATE
13845-18-6 SODIUM SULFAMATE	SODIUM SULFAMATE
7757-82-6 SODIUM SULFATE [PST]	SODIUM SULFATE (SOLUTION)
7757-82-6 SODIUM SULFATE (SOLUTION)	SODIUM SULFATE (SOLUTION)
1344-08-7 SODIUM SULFIDE	SODIUM SULFIDE
16721-80-5 SODIUM SULFIDE [FLA]	SODIUM HYDROSULFIDE
16721-80-5 SODIUM SULFIDE (NA(SH)) [PAL]	SODIUM HYDROSULFIDE
7757-83-7 SODIUM SULFITE	SODIUM SULFITE
68443-05-0 SODIUM SULFONATED OLEATE [PST]	9-OCTADECENOIC ACID (Z)-, SULFONATED, SODIUM SALT
20526-58-3 SODIUM SULFOSUCCINATE	SODIUM SULFOSUCCINATE
12034-12-7 SODIUM SUPEROXIDE	SODIUM SUPEROXIDE
61791-41-1 SODIUM N-(TALL-OIL ALKYL)-N-METHYLTAURINE	SODIUM N-(TALL-OIL ALKYL)-N-METHYLTAURINE
8052-48-0 SODIUM TALLOW SOAP [PST]	FATTY ACIDS, TALLOW, SODIUM SALTS
868-18-8 SODIUM TARTRATE	SODIUM TARTRATE
10102-20-2 SODIUM TELLURITE	SODIUM TELLURITE
1330-43-4 SODIUM TETRABORATE [WHM, PST]	BORATES, TETRA, SODIUM SALTS, ANHYDROUS
1191-50-0 SODIUM TETRADECYL SULFATE	SODIUM TETRADECYL SULFATE
RR-01459-5 SODIUM TETRANITRATE	SODIUM TETRANITRATE

Chemical Name	Reference Name
540-72-7 SODIUM THIOCYANATE	SODIUM THIOCYANATE
7772-98-7 SODIUM THIOSULFATE	SODIUM THIOSULFATE
12068-03-0 SODIUM TOLUENESULFONATE	SODIUM TOLUENESULFONATE
64665-10-7 SODIUM TRIBUTYLNAPHTHALENESULFONATE	SODIUM TRIBUTYLNAPHTHALENESULFONATE
650-51-1 SODIUM TRICHLOROACETATE	SODIUM TRICHLOROACETATE
136-32-3 SODIUM-2,4,5-TRICHLOROPHENATE	SODIUM-2,4,5-TRICHLOROPHENATE
54116-08-4 SODIUM TRIDECYLPOLY(OXYETHYLENE) SULFATE	SODIUM TRIDECYLPOLY(OXYETHYLENE) SULFATE
3026-63-9 SODIUM TRIDECYL SULFATE	SODIUM TRIDECYL SULFATE
1323-19-9 SODIUM TRIISOPROPYLNAPHTHALENESUL-FONATE	SODIUM TRIISOPROPYLNAPHTHALENESULFONATE
7758-29-4 SODIUM TRIPOLYPHOSPHATE [PST]	PENTASODIUM TRIPHOSPHATE
13472-45-2 SODIUM TUNGSTATE	SODIUM TUNGSTATE
63182-08-1 SODIUM VINYLBENZENESULFONATE, POLY-MER WITH DIVINYLBENZENE	SODIUM VINYLBENZENESULFONATE, POLYMER WITH DIVINYLBENZENE
1300-72-7 SODIUM XYLENE SULFONATE	SODIUM XYLENE SULFONATE
1300-72-7 SODIUM XYLENESULFONATE [CSR]	SODIUM XYLENE SULFONATE
126-17-0 SOLASODINE	SOLASODINE
RR-00008-8 SOLUBLE TUNGSTEN [NHS, NHS]	TUNGSTEN, SOLUBLE COMPOUNDS
RR-00060-2 SOLVENT (BENZENE) EXTRACTS OF MOST CARBON BLACKS [PAL, PAL]	CARBON BLACK EXTRACTS
64742-96-7 SOLVENT NAPHTHA (PETROLEUM), HEAVY ALIPHATIC	SOLVENT NAPHTHA (PETROLEUM), HEAVY ALIPHATIC
64742-94-5 SOLVENT NAPHTHA (PETROLEUM), HEAVY AROMATIC [PST]	NAPHTHA (PETROLEUM), HEAVY AROMATIC
64742-89-8 SOLVENT NAPHTHA (PETROLEUM), LIGHT ALIPHATIC	SOLVENT NAPHTHA (PETROLEUM), LIGHT ALIPHATIC
64742-95-6 SOLVENT NAPHTHA (PETROLEUM), LIGHT AROMATIC [PST, TSC, TSC]	PETROLEUM NAPHTHA, LIGHT AROMATIC
64742-88-7 SOLVENT NAPHTHA (PETROLEUM), MEDIUM ALIPHATIC	SOLVENT NAPHTHA (PETROLEUM), MEDIUM ALIPHATIC
4645-07-2 SOLVENT YELLOW #72	SOLVENT YELLOW #72
RR-00028-2 SOOT [PAL, PAL]	SOOTS
RR-00028-2 SOOTS	SOOTS
RR-00795-4 SOOTS, TARS, AND CERTAIN MINERAL OILS [CAL]	SOOTS, TARS, AND MINERAL OILS
RR-00795-4 SOOTS, TARS, AND MINERAL OILS	SOOTS, TARS, AND MINERAL OILS
142-83-6 SORBALDEHYDE	SORBALDEHYDE
110-44-1 SORBIC ACID	SORBIC ACID
590-00-1 SORBIC ACID, POTASSIUM SALT	SORBIC ACID, POTASSIUM SALT
29116-98-1 SORBITAN DIOLEATE	SORBITAN DIOLEATE
26266-57-9 SORBITAN MONOHEXADECANOATE	SORBITAN MONOHEXADECANOATE
1338-39-2 SORBITAN MONOLAURATE	SORBITAN MONOLAURATE
1338-41-6 SORBITAN MONOOCTADECANOATE	SORBITAN MONOOCTADECANOATE
1338-43-8 SORBITAN MONOOLEATE	SORBITAN MONOOLEATE
1338-41-6 SORBITAN MONOSTEARATE [PST]	SORBITAN MONOOCTADECANOATE
8007-43-0 SORBITAN, 9-OCTADECENOATE	SORBITAN, 9-OCTADECENOATE
8007-43-0 SORBITAN SESQUIOLEATE [PST]	SORBITAN, 9-OCTADECENOATE
61790-88-3 SORBITAN, TALL-OIL FATTY ACID TRIESTERS, ETHOXYLATED	SORBITAN, TALL-OIL FATTY ACID TRIESTERS, ETHOXYLATED
26658-19-5 SORBITAN, TRIOCTADECANOATE	SORBITAN, TRIOCTADECANOATE
26266-58-0 SORBITAN TRIOLEATE [PST]	(Z,Z,Z)-TRI-9-OCTADECENOATE SORBITAN

Chemical Name	Reference Name
50-70-4 SORBITOL [PST]	D-GLUCITOLPROPANEDIYLOXY-4,1-PHENYLENE(1-METHYLETHYLIDENE)-4,1- PHENYLENEOXYMETHY-LENE]BIS(OXIRANE)
68648-20-4 SORBITOL, TALL-OIL FATTY ACID SESQUIESTERS, ETHOXYLATED	SORBITOL, TALL-OIL FATTY ACID SESQUIESTERS, ETHOXYLATED
61791-34-2 N-(SOYA ALKYL)-N-ETHYLMORPHOLINIUM ETHYLSULFATE	N-(SOYA ALKYL)-N-ETHYLMORPHOLINIUM ETHYLSUL-FATE
RR-01146-1 N-(SOYA ALKYL)-N-METHYLMORPHOLINIUM SULFATE	N-(SOYA ALKYL)-N-METHYLMORPHOLINIUM SULFATE
63148-69-6 SOYA ALKYL RESIN	SOYA ALKYL RESIN
68308-53-2 SOYA FATTY ACIDS	SOYA FATTY ACIDS
RR-01147-2 SOYBEAN HULLS	SOYBEAN HULLS
8002-43-5 SOYBEAN LECITHIN	SOYBEAN LECITHIN
68308-36-1 SOYBEAN MEAL	SOYBEAN MEALTION PRODUCTS
8001-22-7 SOYBEAN OIL	SOYBEAN OIL
8001-22-7 SOY BEAN OIL [NFP, NFF]	SOYBEAN OIL
68122-64-5 SOYBEAN OIL POLYMERIZED	SOYBEAN OIL POLYMERIZED
67989-28-0 SOYBEAN OIL, POLYMER WITH PENTAERY-THRITOL, TOLUENEDIISOCYANATE AND TUNG OIL	SOYBEAN OIL, POLYMER WITH PENTAERYTHRITOL, TOLUENEDIISOCYANA
68513-95-1 SOY FLOUR	SOY FLOUR
8008-79-5 SPEARMINT OIL	SPEARMINT OIL
8002-24-2 SPERM OIL	SPERM OIL
8002-24-2 SPERM OIL NO. 2 [PAL]	SPERM OIL
8025-81-8 SPIRAMYCIN	SPIRAMYCIN
52-01-7 SPIRONOLACTONE	SPIRONOLACTONE
900-95-8 STANNANE, ACETOXYTRIPHENYL-	STANNANE, ACETOXYTRIPHENYL-
1185-81-5 STANNANE, DIBUTYLBIS(DODECYLTHIO)- [TSC, TSC, TSC]	DIBUTYLTIN BIS(LAURYL MERCAPTIDE)
77-58-7 STANNANE, DIBUTYLBIS[(1-OXODODECYL) OXY]- [TSC, TSC, TSC]	DIBUTYLTIN DILAURATE
1983-10-4 STANNANE, FLUOROTRIBUTYL-	STANNANE, FLUOROTRIBUTYL-
7646-78-8 STANNANE, TETRACHLORO- [PAL]	STANNIC CHLORIDE
594-27-4 STANNANE, TETRAMETHYL-	STANNANE, TETRAMETHYL-
595-90-4 STANNANE, TETRAPHENYL-	STANNANE, TETRAPHENYL-
688-73-3 STANNANE, TRI-N-BUTYL-, HYDRIDE	STANNANE, TRI-N-BUTYL-, HYDRIDE
13121-70-5 STANNANE, TRICYCLOHEXYLHYDROXY- [PAL]	CYHEXATIN
7646-78-8 STANNIC CHLORIDE	STANNIC CHLORIDE
10026-06-9 STANNIC CHLORIDE, HYDRATED	STANNIC CHLORIDE, HYDRATED
18282-10-5 STANNIC OXIDE	STANNIC OXIDE
12440-42-5 STANNIC PHOSPHIDE	STANNIC PHOSPHIDE
25324-56-5 STANNIC PHOSPHIDE [NJL]	TIN PHOSPHIDE
7772-99-8 STANNOUS CHLORIDE	STANNOUS CHLORIDE
10025-69-1 STANNOUS CHLORIDE DIHYDRATE	STANNOUS CHLORIDE DIHYDRATE
7783-47-3 STANNOUS FLUORIDE	STANNOUS FLUORIDE
814-94-8 STANNOUS OXALATE	STANNOUS OXALATE
9005-25-8 STARCH	STARCH
9005-84-9 STARCH [GBR]	AMYLODEXTRIN
53124-00-8 STARCH, HYDROGEN PHOSPHATE, 2-HYDROX-YPROPYLETHER	STARCH, HYDROGEN PHOSPHATE, 2-HYDROXYPROPYL-LETHER
9049-76-7 STARCH HYDROXYPROPYL ETHER	STARCH HYDROXYPROPYL ETHER
7745-89-3 STARLICIDE	STARLICIDE

Chemical Name	Reference Name
3758-54-1 STEARAMIDOPROPYL DIMETHYL 2-HYDROX-YETHYL AMMONIUM DIHYDROGEN PHOS-PHATE [PST]	STEARAMINOPROPYLDIMETHYL-.BETA.-HYDROX-YETHYLAMMONIUM PHOSPAT
3758-54-1 STEARAMINOPROPYLDIMETHYL-.BETA.-HY-DROXYETHYLAMMONIUM PHOSPATE	STEARAMINOPROPYLDIMETHYL-.BETA.-HYDROX-YETHYLAMMONIUM PHOSPAT
RR-00075-9 STEARATES	STEARATES
57-11-4 STEARIC ACID	STEARIC ACID
1072-35-1 STEARIC ACID, LEAD (2+) SALT	STEARIC ACID, LEAD (2+) SALT
7428-48-0 STEARIC ACID, LEAD SALT [MSL]	LEAD STEARATE
56189-09-4 STEARIC ACID, LEAD SALT, DIBASIC [MSL, MSL]	DIBASIC LEAD STEARATE
112-92-5 STEARYL ALCOHOL [NFP]	1-OCTADECANOL
18312-31-7 STEARYL OCTANOATE	STEARYL OCTANOATE
10048-13-2 STERIGMATOCYSTIN	STERIGMATOCYSTIN
RR-00052-2 STEROIDAL OESTROGENS [IAR]	OESTROGENS, STEROIDAL
6543-62-0 STIBAMINE	STIBAMINE
554-76-7 STIBANILIC ACID	STIBANILIC ACID
7803-52-3 STIBINE	STIBINE
7803-52-3 STIBINE (ANTIMONY HYDRIDE) [OS3]	STIBINE
7789-61-9 STIBINE, TRIBROMO- [PAL]	ANTIMONY TRIBROMIDE
10025-91-9 STIBINE, TRICHLORO- [PAL]	ANTIMONY TRICHLORIDE
7783-56-4 STIBINE, TRIFLUORO- [PAL]	ANTIMONY TRIFLUORIDE
103-30-0 TRANS-STILBENE	TRANS-STILBENE
588-59-0 STILBENE	STILBENE
8052-41-3 STODDARD SOLVENT	STODDARD SOLVENT
RR-00883-3 STRAW OIL	STRAW OIL
57-92-1 STREPTOMYCIN	STREPTOMYCIN
3810-74-0 STREPTOMYCIN SULFATE	STREPTOMYCIN SULFATE
18883-66-4 STREPTOZOCIN [NJL, NJL]	STREPTOZOTOCIN
18883-66-4 STREPTOZOTOCIN	STREPTOZOTOCIN
7440-24-6 STRONTIUM	STRONTIUM
15701-15-2 STRONTIUM 80	STRONTIUM 80
14809-49-5 STRONTIUM 81	STRONTIUM 81
14809-51-9 STRONTIUM 83	STRONTIUM 83
13967-73-2 STRONTIUM 85	STRONTIUM 85
RR-00353-2 STRONTIUM 85M	STRONTIUM 85M
RR-00352-1 STRONTIUM 87M	STRONTIUM 87M
14158-27-1 STRONTIUM 89	STRONTIUM 89
10098-97-2 STRONTIUM 90	STRONTIUM 90
14331-91-0 STRONTIUM 91	STRONTIUM 91
14928-29-1 STRONTIUM 92	STRONTIUM 92
15195-06-9 STRONTIUM ARSENITE	STRONTIUM ARSENITE
7791-10-8 STRONTIUM CHLORATE	STRONTIUM CHLORATE
10476-85-4 STRONTIUM CHLORIDE (SRCL2)	STRONTIUM CHLORIDE (SRCL2)
7789-06-2 STRONTIUM CHROMATE	STRONTIUM CHROMATE
RR-00556-1 STRONTIUM DICHROMATE	STRONTIUM DICHROMATE
13814-98-7 STRONTIUM FLUOBORATE	STRONTIUM FLUOBORATE
7783-48-4 STRONTIUM FLUORIDE	STRONTIUM FLUORIDE
10042-76-9 STRONTIUM NITRATE	STRONTIUM NITRATE
15195-06-9 STRONTIUM ORTHOARSENITE [HM1, HM1, HM1]	STRONTIUM ARSENITE

Chemical Name	Reference Name
13450-97-0 STRONTIUM PERCHLORATE	STRONTIUM PERCHLORATE
1314-18-7 STRONTIUM PEROXIDE	STRONTIUM PEROXIDE
1314-18-7 STRONTIUM PEROXIDE (SR(O2)) [PAL]	STRONTIUM PEROXIDE
12504-13-1 STRONTIUM PHOSPHIDE	STRONTIUM PHOSPHIDE
1314-96-1 STRONTIUM SULFIDE	STRONTIUM SULFIDE
57-24-9 STRYCHNIDIN-10-ONE [PAL]	STRYCHNINE
357-57-3 STRYCHNIDIN-10-ONE, 2,3-DIMETHOXY- [TSC, TSC]	BRUCINE
57-24-9 STRYCHNINE	STRYCHNINE
57-24-9 STRYCHNINE (2,4A,5,5A,7,8,15,15A,15B,15C, DECAHYDRO-4,6-METHANO-6H,14H-INDOLO [3,2,1-IJ]OXEPINO[2,3,4-DE]PYRROLO[2,3-H] QUINOLIN-14-ONE) [CN2]	STRYCHNINE
57-24-9 STRYCHNINE AND SALTS [EPA]	STRYCHNINE
57-24-9 STRYCHNINE, & SALTS [RCU, RCU]	STRYCHNINE
RR-01709-4 STRYCHNINE SALTS	STRYCHNINE SALTS
60-41-3 STRYCHNINE, SULFATE	STRYCHNINE, SULFATE
60-41-3 STRYCHNINE SULFATE [302, MSL]	STRYCHNINE, SULFATE
100-42-5 STYRENE	STYRENE
100-42-5 STYRENE, MONOMER [OTV, MEX, MEX]	STYRENE
25085-34-1 STYRENE ACRYLATE COPOLYMER	STYRENE ACRYLATE COPOLYMER
25085-34-1 STYRENE ACRYLIC ACID COPOLYMER [PST]	STYRENE ACRYLATE COPOLYMER
9003-54-7 STYRENE-ACRYLONITRILE COPOLYMERS [IAR]	2-PROPENENITRILE POLYMER WITH ETHENYLBENZENE
9003-55-8 STYRENE-BUTADIENE COPOLYMERS [IAR]	STYRENE-BUTADIENE POLYMER
9003-55-8 STYRENE-BUTADIENE POLYMER	STYRENE-BUTADIENE POLYMER
9003-70-7 STYRENE-DIVINYLBENZENE COPOLYMER	STYRENE-DIVINYLBENZENE COPOLYMER
9003-70-7 STYRENE-DIVINYL BENZENE COPOLYMER RESIN MATRIX [PST]	STYRENE-DIVINYLBENZENE COPOLYMER
9011-13-6 STYRENE-MALEIC ANHYDRIDE RESIN [PST]	MALEIC ANHYDRIDE-STYRENE POLYMER
100-42-5 STYRENE, MONOMER [AUS, AUS, CAL]	STYRENE
100-42-5 STYRENE (MONOMER) [CEX, CEX, CEX]	STYRENE
100-42-5 STYRENE MONOMER [HM1, HM1, HM1, HM1]	STYRENE
100-42-5 STYRENE, MONOMER [MIN]	STYRENE
100-42-5 STYRENE MONOMER [NJL, NJL]	STYRENE
100-42-5 STYRENE, MONOMER [TLV, TLV, TLV]	STYRENE
100-42-5 STYRENE, MONOMER (PHENYLETHYLENE) [QBC, QBC, QBC, QBC, BC1, BC1, BC1, BC1]	STYRENE
96-09-3 STYRENE OXIDE	STYRENE OXIDE
96-09-3 STYRENE, OXIDE [MIN]	STYRENE OXIDE
1395-21-7 SUBSTILISINS [MIN]	SUBTILISINS (PROTEOLYTIC ENZYMES AS 100% PURE ENZYME)
RR-01734-5 SUBSTITUTED 2-NITROBENZENESULFON-AMIDE	SUBSTITUTED 2-NITROBENZENESULFONAMIDE
RR-01735-6 SUBSTITUTED 2-AMINOBENZENESULFON-AMIDE	SUBSTITUTED 2-AMINOBENZENESULFONAMIDE
RR-01645-5 SUBSTITUTED ACRYLAMIDE	SUBSTITUTED ACRYLAMIDE
RR-00237-9 SUBSTITUTED ACRYLATED ALKOXYLATED ALIPHATIC POLYOL	SUBSTITUTED ACRYLATED ALKOXYLATED ALIPHATIC POLYOLX-HYDROXY-, C10-16 ALKYL ETHERS
RR-00920-1 SUBSTITUTED ALIPHATIC ACID HALIDE	SUBSTITUTED ALIPHATIC ACID HALIDE
RR-00942-7 SUBSTITUTED ALKYL HALIDE	SUBSTITUTED ALKYL HALIDELYALKYLENE POLYOL ALKYL ETHER

Chemical Name	Reference Name
RR-00159-2 SUBSTITUTED ALKYL PEROXYHEXANE CARBOXYLATE (MIXED ISOMERS)	SUBSTITUTED ALKYL PEROXYHEXANE CARBOXYLATE (MIXED ISOMERS)
RR-00922-3 SUBSTITUTED AMINOBENZOIC ACID ESTER	SUBSTITUTED AMINOBENZOIC ACID ESTER
RR-00210-8 SUBSTITUTED AROMATIC	SUBSTITUTED AROMATIC
RR-01693-3 SUBSTITUTED BENZENEDIAZONIUM	SUBSTITUTED BENZENEDIAZONIUM
RR-01724-3 SUBSTITUTED BENZENEDICARBOXYLIC ACID ESTER	SUBSTITUTED BENZENEDICARBOXYLIC ACID ESTER
RR-01012-8 SUBSTITUTED BENZENEDICARBOXYLIC ACID, POLY(ALKYL ACRYLATE) DERIVATIVE	SUBSTITUTED BENZENEDICARBOXYLIC ACID, POLY (ALKYL ACRYLATE) D
RR-00968-7 SUBSTITUTED BENZENESULFONIC ACID, ALKALI METAL SALT	SUBSTITUTED BENZENESULFONIC ACID, ALKALI METAL SALTPOTASSIUM SALTS
RR-00990-5 2-SUBSTITUTED BENZOTRIAZOLE	2-SUBSTITUTED BENZOTRIAZOLE2,5)-DIYL]BIS(METHYLENE) ESTER
RR-00977-8 SUBSTITUTED BIS(HYDROXYALKANE) POLYMER WITH EPICHLOROHYDRIN,ACRYLATE	SUBSTITUTED BIS(HYDROXYALKANE) POLYMER WITH EPICHLOROHYDRIN,
RR-00991-6 SUBSTITUTED CARBOHETEROCYCLIC BUTANE TETRACARBOXYLATE	SUBSTITUTED CARBOHETEROCYCLIC BUTANE TETRACARBOXYLATE
RR-01222-6 SUBSTITUTED CYCLOHEXYLDIAMINO ETHYL ESTER	SUBSTITUTED CYCLOHEXYLDIAMINO ETHYL ESTER-[2, 1-ETHANEDIYL]OXYMETHYLENE]BISOXIRANE, REACTION PRODUCT WITH A DIAMINE
RR-01569-0 SUBSTITUTED DIACRYLATE	SUBSTITUTED DIACRYLATE
RR-00902-9 SUBSTITUTED DIALKYL OXAZOLONE	SUBSTITUTED DIALKYL OXAZOLONEHYL) ESTER
RR-01649-9 SUBSTITUTED DICHLOROBENZOTHIAZOLES	SUBSTITUTED DICHLOROBENZOTHIAZOLES
RR-01216-8 SUBSTITUTED ETHANOLAMINE	SUBSTITUTED ETHANOLAMINE
RR-01558-7 SUBSTITUTED ETHYL ALKENAMIDE	SUBSTITUTED ETHYL ALKENAMIDE
RR-00979-0 SUBSTITUTED HYDROXYALKYL ALKNENOATE, [[[[[(1-OXO-2-PROPENYL)OXY]ALKOXY] CARBONYLAMINO]SUBSTITUTED] AMINOCARBONYL]OXY-	SUBSTITUTED HYDROXYALKYL ALKNENOATE, [[[[[(1-OXO-2-PROPENYL)
RR-00944-9 SUBSTITUTED HYDROXYLAMINE	SUBSTITUTED HYDROXYLAMINE
27136-73-8 SUBSTITUTED IMIDAZOLE	SUBSTITUTED IMIDAZOLE
RR-00254-0 SUBSTITUTED METHYLPYRIDINE	SUBSTITUTED METHYLPYRIDINE
RR-00898-0 SUBSTITUTED NITRILE	SUBSTITUTED NITRILE
RR-01676-2 SUBSTITUTED NITROBENZENE	SUBSTITUTED NITROBENZENE
RR-00496-6 SUBSTITUTED NITROPHENOL PESTICIDES, N.O.S.	SUBSTITUTED NITROPHENOL PESTICIDES, N.O.S.
RR-00229-9 SUBSTITUTED OXIDE-ALKYLENE POLYMER, METHACRYLATE	SUBSTITUTED OXIDE-ALKYLENE POLYMER, METHACRYLATEPOLYOL POLYACRYLATES
RR-00131-0 SUBSTITUTED OXIRANE	SUBSTITUTED OXIRANEANE MONO[(C10,16-ALKYLOXY)METHYL] DERIVATIVES AND 2,2,4 (OR 2,4,4)-TRIMETHYL-1,6-HEXANEDIAMINE
RR-00252-8 SUBSTITUTED 2-PHENOXYPYRIDINE	SUBSTITUTED 2-PHENOXYPYRIDINE
RR-01671-7 SUBSTITUTED PHENYL AZO SUBSTITUED BENZENEDIAZONIUM SALT	SUBSTITUTED PHENYL AZO SUBSTITUED BENZENEDIAZONIUM SALT
RR-01215-7 SUBSTITUTED PHENYLIMINO CARBAMATE DERIVATIVE	SUBSTITUTED PHENYLIMINO CARBAMATE DERIVATIVEOL
RR-00197-8 SUBSTITUTED PHOSPHATE ESTER	SUBSTITUTED PHOSPHATE ESTER
RR-00260-8 SUBSTITUTED POLYGLYCIDYL BENZENEAMINE	SUBSTITUTED POLYGLYCIDYL BENZENEAMINE
RR-01727-6 SUBSTITUTED QUINOLINE	SUBSTITUTED QUINOLINE
RR-01182-5 SUBSTITUTED SPIRO OXAZINE	SUBSTITUTED SPIRO OXAZINE
RR-00926-7 SUBSTITUTED TRIAZINE ISOCYANURATE	SUBSTITUTED TRIAZINE ISOCYANURATE
RR-01243-1 SUBSTITUTED TRIAZOLE	SUBSTITUTED TRIAZOLE
RR-00952-9 SUBSTITUTED TRIPHENYLMETHANE	SUBSTITUTED TRIPHENYLMETHANE
9014-01-1 SUBTILISIN [PST]	SUBTILISINS (PROTEOLYTIC ENZYMES)

Chemical Name	Reference Name
1395-21-7 SUBTILISIN BPN [WHM]	SUBTILISINS (PROTEOLYTIC ENZYMES AS 100% PURE ENZYME)
9014-01-1 SUBTILISINS [CEX]	SUBTILISINS (PROTEOLYTIC ENZYMES)
9014-01-1 SUBTILISINS (PROTEOLYTIC ENZYMES)	SUBTILISINS (PROTEOLYTIC ENZYMES)
9014-01-1 SUBTILISINS (PROTEOLYTIC ENZYMES AS 100% PURE CRYSTALLINE ENZYME) [ONT]	SUBTILISINS (PROTEOLYTIC ENZYMES)
1395-21-7 SUBTILISINS (PROTEOLYTIC ENZYMES AS 100 PERCENT PURE CRYSTALLINE ENZYME) [PAL]	SUBTILISINS (PROTEOLYTIC ENZYMES AS 100% PURE ENZYME)
1395-21-7 SUBTILISINS (PROTEOLYTIC ENZYMES AS 100% PURE CRYSTALLINE ENZYME) [TLV]	SUBTILISINS (PROTEOLYTIC ENZYMES AS 100% PURE ENZYME)
1395-21-7 SUBTILISINS (PROTEOLYTIC ENZYMES AS 100% PURE ENZYME)	SUBTILISINS (PROTEOLYTIC ENZYMES AS 100% PURE ENZYME)
1395-21-7 SUBTILISINS (100% PURE CRYSTALLINE EN-ZYME) [QBC]	SUBTILISINS (PROTEOLYTIC ENZYMES AS 100% PURE ENZYME)
9014-01-1 SUBTILISINS (PROTEOLYTIC ENZYMES AS 100% PURE CRYSTALLINE ENZYME) [GBR, GBR]	SUBTILISINS (PROTEOLYTIC ENZYMES)
110-15-6 SUCCINIC ACID	SUCCINIC ACID
123-23-9 SUCCINIC ACID PEROXIDE	SUCCINIC ACID PEROXIDE
108-30-5 SUCCINIC ANHYDRIDE	SUCCINIC ANHYDRIDE
123-56-8 SUCCINIMIDE	SUCCINIMIDE
110-61-2 SUCCINONITRILE	SUCCINONITRILE
57-50-1 SUCROSE	SUCROSE
126-14-7 SUCROSE OCTAACETATE	SUCROSE OCTAACETATE
30236-29-4 SUCROSE OCTANITRATE	SUCROSE OCTANITRATE
85-86-9 SUDAN G	SUDAN G
842-07-9 SUDAN I [IAR]	C.I. SOLVENT YELLOW 14
3118-97-6 SUDAN II [IAR]	C.I. SOLVENT ORANGE 7
85-86-9 SUDAN III [IAR]	SUDAN G
6416-57-5 SUDAN BROWN RR	SUDAN BROWN RR
842-07-9 SUDAN I [CAL]	C.I. SOLVENT YELLOW 14
3118-97-6 SUDAN II [CAL]	C.I. SOLVENT ORANGE 7
6368-72-5 SUDAN RED 7B	SUDAN RED 7B
57-50-1 SUGAR [PST]	SUCROSE
127-69-5 SULFAFURAZOLE (SULPHISOXASOLE) [IAR]	SULFISOXAZOLE
95-06-7 SULFALLATE	SULFALLATE
57-68-1 SULFAMETHAZINE	SULFAMETHAZINE
723-46-6 SULFAMETHOXAZOLE	SULFAMETHOXAZOLE
5329-14-6 SULFAMIC ACID	SULFAMIC ACID
14017-41-5 SULFAMIC ACID, COBALT(2+) SALT (2:1) [PAL]	COBALTOUS SULFAMATE
7773-06-0 SULFAMIC ACID, MONOAMMONIUM SALT [PAL]	AMMONIUM SULFAMATE
121-57-3 SULFANILIC ACID [HON]	BENZENESULFONIC ACID, 4-AMINO-
RR-00518-5 SULFATE	SULFATE
68422-69-5 SULFATED BUTYL OLEATE	SULFATED BUTYL OLEATE
8002-33-3 SULFATED CASTOR OIL [PST]	CASTOR OIL, SULFATED
541-70-8 SULFATE SALTS OF META-PHENYLENEDI-AMINE [TSE]	1,3-BENZENEDIAMINE, SULFATE (1:1)
16245-77-5 SULFATE SALTS OF PARA-PHENYLENEDI-AMINE [TSE]	1,4-BENZENEDIAMINE, SULFATE (1:1)
72-14-0 SULFATHIAZOLE	SULFATHIAZOLE
18496-25-8 SULFIDE	SULFIDE

Chemical Name	Reference Name
127-69-5 SULFISOXAZOLE	SULFISOXAZOLE
RR-01552-1 SULFITES	SULFITES
123-43-3 SULFOACETIC ACID	SULFOACETIC ACID
126-33-0 SULFOLANE	SULFOLANE
77-79-2 3-SULFOLENE	3-SULFOLENE
77-79-2 SULFOLENE [TSC, TSC, TSC]	3-SULFOLENE
74222-97-2 SULFFOMETURON METHYL	SULFFOMETURON METHYL
74222-97-2 SULFOMETURON METHYL [TLV, TLV]	SULFFOMETURON METHYL
RR-01244-2 SULFONAMIDE	SULFONAMIDE
68187-76-8 SULFONATED CASTOR OIL, SODIUM SALT	SULFONATED CASTOR OIL, SODIUM SALT
RR-01710-7 SULFONES	SULFONES
68608-26-4 SULFONIC ACIDS, PETROLEUM, SODIUM SALTS	SULFONIC ACIDS, PETROLEUM, SODIUM SALTSIS(HYDROXYETHYL) METHYL, CHLORIDES
80-07-9 SULFONYL BIS-(4-CHLOROBENZENE)	SULFONYL BIS-(4-CHLOROBENZENE)
2580-77-0 2,2'-SULFONYL BIS-ETHANOL	2,2'-SULFONYL BIS-ETHANOL
80-08-0 4,4'-SULFONYLDIANILINE (DAPSONE) [CSR, CSR]	DAPSONE
80-09-1 4,4'-SULFONYLDIPHENOL	4,4'-SULFONYLDIPHENOL
97-05-2 5-SULFOSALICYLIC ACID	5-SULFOSALICYLIC ACID
3689-24-5 SULFOTEP	SULFOTEP
3689-24-5 SULFOTEP [QBC, QBC]	SULFOTEP
3689-24-5 SULFOTEP (TEDP) [MEX, MEX, MEX]	SULFOTEP
3689-24-5 SULFOTEP (ISO) [GBR, GBR]	SULFOTEP
3689-24-5 SULFOTEPP [FLA, NJL, PQL]	SULFOTEP
3569-57-1 SULFOXIDE, 3-CHLOROPROPYL OCTYL	SULFOXIDE, 3-CHLOROPROPYL OCTYLLPHENYL)AZO]-, DISODIUM SALT
7704-34-9 SULFUR	SULFUR
15117-53-0 SULFUR 35	SULFUR 35
10545-99-0 SULFUR CHLORIDE [NJL, NJL, PST, WHM, HM1]	SULFUR DICHLORIDE
12771-08-3 SULFUR CHLORIDE	SULFUR CHLORIDE
10025-67-9 SULFUR CHLORIDE (S2CL2) [PAL]	SULFUR MONOCHLORIDE
10545-99-0 SULFUR CHLORIDE (SCL2) [PAL]	SULFUR DICHLORIDE
RR-01166-5 SULFUR CHLORIDE AND CARBON TETRA-CHLORIDE MIXTURES	SULFUR CHLORIDE AND CARBON TETRACHLORIDE MIXTURES
RR-01170-1 SULFUR CHLORIDE PENTAFLUORIDE	SULFUR CHLORIDE PENTAFLUORIDE
RR-01346-7 SULFUR CHLORIDES	SULFUR CHLORIDES
RR-01148-3 SULFUR COATED UREA	SULFUR COATED UREA
10545-99-0 SULFUR DICHLORIDE	SULFUR DICHLORIDE
7446-09-5 SULFUR DIOXIDE	SULFUR DIOXIDE
7446-09-5 SULFUR DIOXIDE (ANHYDROUS) [EP3]	SULFUR DIOXIDE
7783-60-0 SULFUR FLUORIDE (SF4), (T-4)- [PAL]	SULFUR TETRAFLUORIDE
10546-01-7 SULFUR FLUORIDE (SF5)	SULFUR FLUORIDE (SF5)
2551-62-4 SULFUR FLUORIDE (SF6), (OC-6-11)- [PAL]	SULFUR HEXAFLUORIDE
2551-62-4 SULFUR HEXAFLUORIDE	SULFUR HEXAFLUORIDE
8014-95-7 SULFURIC ACID	SULFURIC ACID
10043-01-3 SULFURIC ACID, ALUMINUM SALT (3:2) [PAL]	ALUMINUM SULFATE
10045-89-3 SULFURIC ACID, AMMONIUM IRON(2+) SALT (2:2:1) [PAL]	FERROUS AMMONIUM SULFATE
15699-18-0 SULFURIC ACID, AMMONIUM NICKEL(2+) SALT (2:2:1) [PAL]	NICKEL AMMONIUM SULFATE

Chemical Name	Reference Name
13510-49-1 SULFURIC ACID, BERYLLIUM SALT (1:1) [PAL, PAL]	BERYLLIUM SULFATE
RR-01149-4 SULFURIC ACID, C8-10-ALKYL ESTERS, COMPOUNDS WITH ISOPROPANOLAMINE	SULFURIC ACID, C8-10-ALKYL ESTERS, COMPOUNDS WITH ISOPROPANO
10124-36-4 SULFURIC ACID, CADMIUM SALT (1:1) [PAL, PAL]	CADMIUM SULFATE
10101-53-8 SULFURIC ACID, CHROMIUM(3+) SALT (3:2) [PAL]	CHROMIC SULFATE
10124-43-3 SULFURIC ACID, COBALT (2+) SALT (1:1) [TSC, TSC]	COBALT SULFATE
7758-98-7 SULFURIC ACID COPPER(2+) SALT (1:1) [PAL]	CUPRIC SULFATE
7783-20-2 SULFURIC ACID DIAMMONIUM SALT [PAL]	AMMONIUM SULFATE
64-67-5 SULFURIC ACID, DIETHYL ESTER [TSC, TSC, PAL, PAL]	DIETHYL SULFATEPHENOXYETHYL)-,HYDROCHLORIDE
77-78-1 SULFURIC ACID, DIMETHYL ESTER [FLA, PAL, PAL, TSC, TSC, TSC]	DIMETHYL SULFATE
7778-80-5 SULFURIC ACID DIPOTASSIUM SALT	SULFURIC ACID DIPOTASSIUM SALT
7446-18-6 SULFURIC ACID, DITHALLIUM(1+) SALT [PAL]	THALLOUS SULFATE
8014-95-7 SULFURIC ACID, FUMING [NJL]	SULFURIC ACID
7720-78-7 SULFURIC ACID, IRON(2+) SALT (1:1) [PAL]	FERROUS SULFATE
10028-22-5 SULFURIC ACID, IRON(3+) SALT (3:2) [PAL]	FERRIC SULFATE
7446-14-2 SULFURIC ACID, LEAD (2+) SALT [MSL]	LEAD SULFATE
7446-14-2 SULFURIC ACID, LEAD(2+) SALT (1:1) [PAL]	LEAD SULFATE
15739-80-7 SULFURIC ACID, LEAD SALT [MSL]	LEAD SULFATE
7783-35-9 SULFURIC ACID, MERCURY(2+) SALT (1:1) [PAL]	MERCURIC SULFATE
8014-95-7 SULFURIC ACID (MIXTURE WITH SULFUR TRIOXIDE) [MSL]	SULFURIC ACID
142-87-0 SULFURIC ACID, MONODECYL ESTER, SODIUM SALT	SULFURIC ACID, MONODECYL ESTER, SODIUM SALT
7786-81-4 SULFURIC ACID, NICKEL(2+) SALT (1:1) [PAL]	NICKEL SULFATE
10031-59-1 SULFURIC ACID, THALLIUM SALT [PAL]	THALLIUM SULFATE
7733-02-0 SULFURIC ACID, ZINC SALT (1:1) [PAL]	ZINC SULFATE
14644-61-2 SULFURIC ACID, ZIRCONIUM(4+) SALT (2:1) [PAL]	ZIRCONIUM SULFATE
RR-01016-2 SULFURIZED ALKYLPHENOLS	SULFURIZED ALKYLPHENOLS2-ETHYL-2-(HYDROXYMETHYL)-1,3-PROPANEDIOL (3:1), EPOXIZED
10025-67-9 SULFUR MONOCHLORIDE	SULFUR MONOCHLORIDE
12771-08-3 SULFUR MONOCHLORIDE [EPA, NJL]	SULFUR CHLORIDE
7782-99-2 SULFUROUS ACID	SULFUROUS ACID
140-57-8 SULFUROUS ACID, 2-CHLOROETHYL 2-[4-(1,1-DIMETHYLETHYL)PHENOXY]-1-METHYLETHYL ESTER [PAL, PAL]	ARAMITENYL)METHYLENE]AMINO]-
10196-04-0 SULFUROUS ACID, DIAMMONIUM SALT [PAL]	AMMONIUM SULFITE
2312-35-8 SULFUROUS ACID, 2-[4-(1,1-DIMETHYLETHYL)PHENOXY]CYCLOHEXYL2-PROPYNYL ESTER [PAL]	PROPARGITEPROPENYL) ESTER
10192-30-0 SULFUROUS ACID, MONOAMMONIUM SALT [PAL]	AMMONIUM BISULFITE
7631-90-5 SULFUROUS ACID, MONOSODIUM SALT [PAL]	SODIUM BISULFITE
7782-99-2 SULFUROUS ACID SOLUTION [NJL, NJL]	SULFUROUS ACID
5714-22-7 SULFUR PENTAFLUORIDE	SULFUR PENTAFLUORIDE
7783-60-0 SULFUR TETRAFLUORIDE	SULFUR TETRAFLUORIDE

Chemical Name	Reference Name
7783-60-0 SULFUR TETRAFLUORIDE [SULFUR FLUORIDE (SF4), (T-4)-] [EP3]	SULFUR TETRAFLUORIDE
7446-11-9 SULFUR TRIOXIDE	SULFUR TRIOXIDE
7791-25-5 SULFURYL CHLORIDE	SULFURYL CHLORIDE
2699-79-8 SULFURYL FLUORIDE	SULFURYL FLUORIDE
2699-79-8 SULFURYL FLUORIDE [VIKANE] [313]	SULFURYL FLUORIDE
5329-14-6 SULPHAMIC ACID [NJL]	SULFAMIC ACID
RR-00518-5 SULPHATE [CD1]	SULFATE
RR-00518-5 SULPHATES [MX2]	SULFATE
18496-25-8 SULPHIDE (AS H2S) [CD1]	SULFIDE
18496-25-8 SULPHIDES [MX2]	SULFIDE
7704-34-9 SULPHUR [PST]	SULFUR
RR-00937-0 SULPHUR-BRIDGED SUBSTITUTED PHENOLS	SULPHUR-BRIDGED SUBSTITUTED PHENOLS
7446-09-5 SULPHUR DIOXIDE [AUS, AUS, CN3, GBR, GBR]	SULFUR DIOXIDE
2551-62-4 SULPHUR HEXAFLUORIDE [AUS, GBR, GBR]	SULFUR HEXAFLUORIDE
7664-93-9 SULPHURIC ACID [AUS, AUS, GBR]	SULFURIC ACID
10025-67-9 SULPHUR MONOCHLORIDE [AUS]	SULFUR MONOCHLORIDE
5714-22-7 SULPHUR PENTAFLUORIDE [AUS]	SULFUR PENTAFLUORIDE
7783-60-0 SULPHUR TETRAFLUORIDE [AUS, GBR, GBR]	SULFUR TETRAFLUORIDE
2699-79-8 SULPHURYL DIFLUORIDE [GBR, GBR]	SULFURYL FLUORIDE
2699-79-8 SULPHURYL FLUORIDE [AUS, AUS]	SULFURYL FLUORIDE
35400-43-2 SULPROFOS	SULPROFOS
35400-43-2 SULPROFOS [O-ETHYL O-[4-(METHYLTHIO) PHENYL]PHOSPHORODITHIOIC ACID S-PROPYL ESTER] [313]	SULPROFOS
35400-43-2 SULPROPHOS [HM1]	SULPROFOS
68937-99-5 SUNFLOWER SEEDS	SUNFLOWER SEEDSTHOXYLATED
2783-94-0 SUNSET YELLOW FCF [IAR]	FD & C YELLOW NO. 6
RR-01557-6 SUSPENDED PARTICULATE MATTER	SUSPENDED PARTICULATE MATTER
122-10-1 SWAT	SWAT
1918-18-9 SWEP	SWEP
22571-95-5 SYMPHYTINE	SYMPHYTINE
93-76-5 2,4,5-T	2,4,5-T
93-76-5 2,4,5-T (2,4,5-TRICHLOROPHENOXYACETIC ACID) [UTS]	2,4,5-T
93-76-5 2,4,5-T ACID [CWA]	2,4,5-T
32534-95-5 2,4,5-T ACID ESTERS [MSL]	2,4,5-TP ACID ESTERS
6369-97-7 2,4,5-T AMINE [NJL, NJL, NJL, NJL, NJL]	ACETIC ACID, (2,4,5-TRICHLOROPHENOXY)-, COMPOUND WITH N-METH
42589-07-1 2,4,5-T AMINE	2,4,5-T AMINE
93-79-8 2,4,5-T BUTYL ESTER [EPA]	ACETIC ACID, (2,4,5-TRICHLOROPHENOXY)-, BUTYL ESTER
61792-07-2 2,4,5-T ESTER [NJL, NJL, NJL, NJL]	ACETIC ACID, (2,4,5-TRICHLOROPHENOXY)-, 1-METHYL PROPYL ESTE
25168-15-4 2,4,5-T ESTERS [CWA, PAL]	ACETIC ACID, (2,4,5-TRICHLOROPHENOXY)-, ISOOCTYL ESTER
13560-99-1 2,4,5-T SALT; ACETIC ACID 2,4,5-TRICHLOROPHENOXY-SODIUM SALT [CAL]	2,4,5-T SALTS
13560-99-1 2,4,5-T SALTS	2,4,5-T SALTS
93-76-5 2,4,5-T; 2,4,5-TRICHLORO-PHENOXYACETIC ACID [CAL]	2,4,5-T

Chemical Name	Reference Name
93-76-5 2,4,5-T (ISO) [GBR, GBR]	2,4,5-T
77-81-6 TABUN	TABUN
RR-01150-7 TACKS	TACKSLAMINE
14807-96-6 TALC	TALC
14807-96-6 TALC (FIBROUS) [BC1, BC1]	TALC
14807-96-6 TALC (MG3H2(SIO3)4) [PAL]	TALC
14807-96-6 TALC (NON-FIBROUS) [QBC, QBC]	TALC
14807-96-6 TALC (CONTAINING NO ASBESTOS FIBERS) [ONT]	TALC
RR-00029-3 TALC (FIBROUS) [ALB, ALB, QBC]	TALC CONTAINING ASBESTOS FIBERS
14807-96-6 TALC (NON-ASBESTIFORM) [ALB]	TALC
RR-00029-3 TALC CONTAINING ASBESTIFORM FIBERS [C65, IAR]	TALC CONTAINING ASBESTOS FIBERS
RR-00029-3 TALC CONTAINING ASBESTOS FIBERS	TALC CONTAINING ASBESTOS FIBERS
RR-00029-3 TALC (CONTAINING ASBESTOS FIBERS) [MIN, TLV]	TALC CONTAINING ASBESTOS FIBERS
14807-96-6 TALC (CONTAINING NO ASBESTOS) [OS1, OS1]	TALC
14807-96-6 TALC (CONTAINING NO ASBESTOS FIBERS) [TLV]	TALC
14807-96-6 TALC (CONTAINING NO ASBESTOS AND LESS THAN 1% QUARTZ) [NIO, NIO]	TALC
14807-96-6 TALC (NONASBESTIFORM, RESP. AND FIBROUS) [MIN]	TALC
14807-96-6 TALC NOT CONTAINING ASBESTIFORM FIBRES [IAR]	TALC
8002-26-4 TALL OIL	TALL OIL
61789-01-3 TALL OIL, EPOXIDIZED, 2-ETHYLHEXYL ESTERS [PST]	FATTY ACIDS, TALL OIL, EPOXIDIZED, 2-ETHYLHEXYL ESTERS
65071-95-6 TALL OIL, ETHOXYLATED	TALL OIL, ETHOXYLATED
61790-12-3 TALL OIL FATTY ACIDS	TALL OIL FATTY ACIDS
61791-00-2 TALL OIL FATTY ACIDS, ETHOXYLATED	TALL OIL FATTY ACIDS, ETHOXYLATED
RR-01151-8 TALL OIL FATTY ACID SOAP N-ALKYL(C16-C18)TRIMETHYLENEDIAMINE	TALL OIL FATTY ACID SOAP N-ALKYL(C16-C18) TRIMETHYLENEDIAMINE
68953-36-6 TALL OIL FATTY ACIDS, REACTION PRODUCT WITH TETRAETHYLENE PENTAMINE	TALL OIL FATTY ACIDS, REACTION PRODUCT WITH TETRAETHYLENE PE
RR-01250-0 TALL OIL FATTY ACIDS, REACTION PRODUCTS WITH POLYAMINES, ALKYL SUBSTITUTED	TALL OIL FATTY ACIDS, REACTION PRODUCTS WITH POLYAMINES, ALKYLPOLYISOCYANATOCARBOMONO-CYCLE
8052-10-6 TALL OIL ROSIN	TALL OIL ROSIN
61789-97-7 TALLOW [NFP, NFP, PST]	TALLOW OIL
61791-53-5 N-TALLOW ALKYLTRIMETHYLENEDIAMINE OLEATES [PST]	TALLOW DIAMINE
61791-26-2 TALLOW AMINE, ETHOXYLATED	TALLOW AMINE, ETHOXYLATED
RR-01152-9 TALLOW BIS-HYDROXYETHYL GLYCINATE	TALLOW BIS-HYDROXYETHYL GLYCINATE
61791-53-5 TALLOW DIAMINE	TALLOW DIAMINE
68122-86-1 TALLOW IMIDAZOLINIUM METHYL SULFATE [PST]	IMIDAZOLIUM COMPOUNDS, 4,5-DIHYDRO-1-METHYL-2-NORTALLOW
61789-97-7 TALLOW OIL	TALLOW OIL
54965-24-1 TAMOXIFEN CITRATE	TAMOXIFEN CITRATE
1401-55-4 TANNIC ACID	TANNIC ACID
1401-55-4 TANNIC ACID AND TANNINS [CAL]	TANNIC ACID
1401-55-4 TANNINS [PST]	TANNIC ACID

Chemical Name	Reference Name
7440-25-7 TANTALUM	TANTALUM
15759-26-9 TANTALUM 172	TANTALUM 172
22095-77-8 TANTALUM 173	TANTALUM 173
15758-54-0 TANTALUM 174	TANTALUM 174
15759-28-1 TANTALUM 175	TANTALUM 175
15758-55-1 TANTALUM 176	TANTALUM 176
15759-27-0 TANTALUM 177	TANTALUM 177
RR-00289-1 TANTALUM 178	TANTALUM 178
14391-27-6 TANTALUM 179	TANTALUM 179
15759-29-2 TANTALUM 180	TANTALUM 180
RR-00351-0 TANTALUM 180M	TANTALUM 180M
13982-00-8 TANTALUM 182	TANTALUM 182
RR-00348-5 TANTALUM 182M	TANTALUM 182M
14683-36-4 TANTALUM 183	TANTALUM 183
15701-21-0 TANTALUM 184	TANTALUM 184
15701-22-1 TANTALUM 185	TANTALUM 185
15701-16-3 TANTALUM 186	TANTALUM 186
7440-25-7 TANTALUM, ELEMENTAL [WHM]	TANTALUM
7440-25-7 TANTALUM METAL [ONT, ONT]	TANTALUM
7440-25-7 TANTALUM, METAL [TLV]	TANTALUM
7440-25-7 TANTALUM (METAL AND OXIDE DUST) [NIO, NIO]	TANTALUM
7440-25-7 TANTALUM, METAL AND OXIDE DUST [OS1, OS1]	TANTALUM
1314-61-0 TANTALUM OXIDE	TANTALUM OXIDE
1314-61-0 TANTALUM, OXIDE DUSTS [TLV]	TANTALUM OXIDE
39300-88-4 TARA GUM	TARA GUM
8007-45-2 TAR, COAL [PAL, PAL]	COAL TAR
65996-89-6 TAR, COAL, HIGH-TEMP	TAR, COAL, HIGH-TEMP
65996-90-9 TAR, COAL, LOW-TEMP	TAR, COAL, LOW-TEMP
8002-29-7 TAR OILS	TAR OILS
RR-00777-2 TAR-OIL, WOOD	TAR-OIL, WOOD
RR-00516-3 TARS (POLYCYCLIC AROMATIC HYDROCAR-BONS)	TARS (POLYCYCLIC AROMATIC HYDROCARBONS)
127-18-4 TETRACHLOROETHYLENE (PERCHLOROETHY-LENE) [HON, HON]	TETRACHLOROETHYLENE
526-83-0 TARTARIC ACID	TARTARIC ACID
632-79-1 TETRABROMOPHTHALIC ANHYDRIDE [HON]	1,3-ISOBENZOFURANDIONE, 4,5,6,7-TETRABROMO-
87-69-4 TARTARIC ACID [PST]	TARTARIC ACID (D, I)
87-69-4 TARTARIC ACID (D, I)	TARTARIC ACID (D, I)
14307-43-8 TARTARIC ACID, AMMONIUM SALT [MSL]	AMMONIUM TARTRATE
3164-29-2 TARTARIC ACID, DIAMMONIUM SALT [MSL]	AMMONIUM TARTRATE, DIAMMONIUM SALT
815-84-9 TARTARIC ACID, LEAD(2+) SALT	TARTARIC ACID, LEAD(2+) SALT
50-31-7 2,3,6-TBA	2,3,6-TBA
1746-01-6 2,3,7,8-TCDD (DIOXIN) [SDW, SDW]	2,3,7,8-TETRACHLORODIBENZO-P-DIOXIN (TCDD)
72-54-8 TDE [CAL, CWA, MSL]	DDD
102-71-6 TEA [IAR]	TRIETHANOLAMINE
RR-01521-4 TEAR GAS	TEAR GAS
RR-01521-4 TEAR GAS (IRRITATING SUBSTANCES, LIQUID OR SOLID) N.O.S. [NJL]	TEAR GAS

Chemical Name	Reference Name
34014-18-1 TEBUTHIURON	TEBUTHIURONBUTENOIC ACID, 1-METHYLETHYL ESTER]
34014-18-1 TEBUTHIURON [N-[5-(1,1-DIMETHYLETHYL)-1, 3,4-THIADIAZOL-2-YL)-N,N'-DIMETHYLUREA] [313]	TEBUTHIURONBUTENOIC ACID, 1-METHYLETHYL ESTER]
14119-14-3 TECHNETIUM 93	TECHNETIUM 93
RR-00347-4 TECHNETIUM 93M	TECHNETIUM 93M
14809-55-3 TECHNETIUM 94	TECHNETIUM 94
RR-00346-3 TECHNETIUM 94M	TECHNETIUM 94M
14808-44-7 TECHNETIUM 96	TECHNETIUM 96
RR-00344-1 TECHNETIUM 96M	TECHNETIUM 96M
15759-35-0 TECHNETIUM 97	TECHNETIUM 97
RR-00339-4 TECHNETIUM 97M	TECHNETIUM 97M
32025-58-4 TECHNETIUM 98	TECHNETIUM 98
14133-76-7 TECHNETIUM 99	TECHNETIUM 99
RR-00338-3 TECHNETIUM 99M	TECHNETIUM 99M
14913-92-9 TECHNETIUM 101	TECHNETIUM 101
15701-17-4 TECHNETIUM 104	TECHNETIUM 104
608-73-1 TECHNICAL-GRADE HCH [IAR]	HEXACHLOROCYCLOHEXANE (MIXED ISOMERS)
3689-24-5 TEDP [ALB, ALB, ALB, BC1, BC1, BC1, MAK, MAK, MAK, NIO, NIO, NIO, NIO]	SULFOTEP
3689-24-5 TEDP (SULFOTEP) [OS1, OS1, OS1, OS1]	SULFOTEP
25067-11-2 TEFLON	TEFLON
RR-01640-0 TEFLON DECOMPOSITION PRODUCTS	TEFLON DECOMPOSITION PRODUCTS
7803-68-1 TELLURIC ACID	TELLURIC ACID
13494-80-9 TELLURIUM	TELLURIUM
15125-45-8 TELLURIUM 116	TELLURIUM 116
14304-79-1 TELLURIUM 121	TELLURIUM 121
RR-00335-0 TELLURIUM 121M	TELLURIUM 121M
14304-80-4 TELLURIUM 123	TELLURIUM 123
RR-00332-7 TELLURIUM 123M	TELLURIUM 123M
RR-00330-5 TELLURIUM 125M	TELLURIUM 125M
13981-49-2 TELLURIUM 127	TELLURIUM 127
RR-00329-2 TELLURIUM 127M	TELLURIUM 127M
14269-71-7 TELLURIUM 129	TELLURIUM 129
RR-00326-9 TELLURIUM 129M	TELLURIUM 129M
14683-12-6 TELLURIUM 131	TELLURIUM 131
RR-00324-7 TELLURIUM 131M	TELLURIUM 131M
14234-28-7 TELLURIUM 132	TELLURIUM 132
15759-52-1 TELLURIUM 133	TELLURIUM 133
RR-00323-6 TELLURIUM 133M	TELLURIUM 133M
15701-09-4 TELLURIUM 134	TELLURIUM 134
RR-00614-4 TELLURIUM AND COMPOUNDS [MEX, ALB, ALB, AUS, NIO, NIO, OS1, OS1, TLV, BC1, CEX, GBR]	TELLURIUM COMPOUNDS, N.O.S.
13494-80-9 TELLURIUM AND ITS COMPOUNDS [MAK, MAK]	TELLURIUM
RR-00614-4 TELLURIUM COMPOUNDS [CAL, ONT, QBC]	TELLURIUM COMPOUNDS, N.O.S.
RR-00614-4 TELLURIUM COMPOUNDS, N.O.S.	TELLURIUM COMPOUNDS, N.O.S.
7446-07-3 TELLURIUM DIOXIDE	TELLURIUM DIOXIDE
13494-80-9 TELLURIUM, ELEMENTAL [WHM]	TELLURIUM
7783-80-4 TELLURIUM FLUORIDE (TEF6), (OC-6-11)- [PAL]	TELLURIUM HEXAFLUORIDE

Chemical Name	Reference Name
7783-80-4 TELLURIUM HEXAFLUORIDE	TELLURIUM HEXAFLUORIDE
10026-07-0 TELLURIUM TETRACHLORIDE	TELLURIUM TETRACHLORIDE
846-50-4 TEMAZEPAM	TEMAZEPAM
3383-96-8 TEMEPHOS	TEMEPHOS
3383-96-8 TEMEPHOS (ABATE) [MEX, MEX]	TEMEPHOS
107-49-3 TEPP	TEPP
107-49-3 TEPP (ISO) [GBR, GBR, GBR]	TEPP
5902-51-2 TERBACIL [5-CHLORO-3-(1,1-DIMETHYLETHYL) -6-METHYL-2,4(1H,3H)-PYRIMIDINEDIONE]	TERBACIL [5-CHLORO-3-(1,1-DIMETHYLETHYL)-6-METHYL-2,4(1H,3H)YL)PHOSPHOROTHIOATE
26209-85-8 TERBIUM 147	TERBIUM 147
15065-93-7 TERBIUM 149	TERBIUM 149
15065-95-9 TERBIUM 150	TERBIUM 150
14998-51-7 TERBIUM 151	TERBIUM 151
14981-98-7 TERBIUM 153	TERBIUM 153
15758-64-2 TERBIUM 154	TERBIUM 154
1439-17-4 TERBIUM 155	TERBIUM 155
14391-10-7 TERBIUM 156	TERBIUM 156
RR-00321-4 TERBIUM 156M	TERBIUM 156M
14391-18-5 TERBIUM 157	TERBIUM 157
15759-55-4 TERBIUM 158	TERBIUM 158
13981-29-8 TERBIUM 160	TERBIUM 160
14391-19-6 TERBIUM 161	TERBIUM 161
10042-88-3 TERBIUM CHLORIDE	TERBIUM CHLORIDE
13071-79-9 TERBUFOS	TERBUFOS
5915-41-3 TERBUTHYLAZINE	TERBUTHYLAZINE
886-50-0 TERBUTRYN	TERBUTRYN
133-55-1 TEREPHTHALAMIDE, N,N'-DIMETHYL-N,N'-DINITROSO-	TEREPHTHALAMIDE, N,N'-DIMETHYL-N,N'-DINITROSO-
100-21-0 TEREPHTHALIC ACID	TEREPHTHALIC ACID
1861-32-1 TEREPHTHALIC ACID, TETRACHLORO-, DIMETHYL ESTER	TEREPHTHALIC ACID, TETRACHLORO-, DIMETHYL ESTER
623-26-7 TEREPHTHALONITRILE	TEREPHTHALONITRILE
100-20-9 TEREPHTHALOYL CHLORIDE	TEREPHTHALOYL CHLORIDE
3282-85-7 TERGITOL 7	TERGITOL 7
139-88-8 TERGITOL NO. 4	TERGITOL NO. 4
68956-56-9 TERPENE HYDROCARBONS, N.O.S.	TERPENE HYDROCARBONS, N.O.S.
8001-50-1 TERPENE POLYCHLORINATES (STROBANE6)	TERPENE POLYCHLORINATES (STROBANE6)PHOS)
RR-00912-1 TERPENES AND TERPENOIDS, LIMONENE FRACTION, POLYMER WITH SUBSTITUTED CARBOPOLYCYCLES	TERPENES AND TERPENOIDS, LIMONENE FRACTION, POLYMER WITH SUBANHYDRIDE, ACRYLATE
84-15-1 O-TERPHENYL	O-TERPHENYL
92-06-8 M-TERPHENYL	M-TERPHENYL
92-94-4 P-TERPHENYL	P-TERPHENYL
61788-33-8 TERPHENYL, CHLORINATED [TSC, TSC]	KANECHLOR 500
92-94-4 P-TERPHENYLS [MSL]	P-TERPHENYL
92-94-4 TERPHENYLS [NIO]	P-TERPHENYL
26140-60-3 TERPHENYLS	TERPHENYLS
26140-60-3 TERPHENYLS (SUM OF O-, M-, AND P- ISOMERS) [ONT]	TERPHENYLS
26140-60-3 TERPHENYLS, ALL ISOMERS [GBR]	TERPHENYLS
98-55-5 ALPHA-TERPINEOL	ALPHA-TERPINEOL

512

Chemical Name	Reference Name
8006-39-1 TERPINEOL	TERPINEOL
586-62-9 TERPINOLENE	TERPINOLENE
80-26-2 TERPINYL ACETATE	TERPINYL ACETATE
2593-15-9 TERRAZOLE	TERRAZOLE
25168-15-4 2,4,5-T ESTER [NJL, NJL]	ACETIC ACID, (2,4,5-TRICHLOROPHENOXY)-, ISOOCTYL ESTER
58-22-0 TESTOSTERONE	TESTOSTERONE
58-22-0 TESTOSTERONE AND ITS ESTERS [C65]	TESTOSTERONE
58-20-8 TESTOSTERONE CYPIONATE	TESTOSTERONE CYPIONATE
315-37-7 TESTOSTERONE ENANTHATE	TESTOSTERONE ENANTHATE
RR-00796-5 TESTOSTERONE ESTERS	TESTOSTERONE ESTERS
57-85-2 TESTOSTERONE PROPIONATE	TESTOSTERONE PROPIONATE
13933-32-9 TETRAAMMINEDICHLOROPLATINUM(II)	TETRAAMMINEDICHLOROPLATINUM(II)
2049-95-8 TETRAAMYLBENZENE	TETRAAMYLBENZENE
4067-16-7 3,6,9,12-TETRAZAATETRADECANE-1,14-DI-AMINE	3,6,9,12-TETRAZAATETRADECANE-1,14-DIAMINE
22826-61-5 TETRAAZIDO BENZENE QUINONE	TETRAAZIDO BENZENE QUINONE
92874-42-5 TETRABROMOBISHPENOL-B	TETRABROMOBISHPENOL-BER WITH 2,2'-(OXYBIS (METHYLENE)BIS(2-HYDROXYMETHYL)-1,3-PROPANE-DIOL)-2-PROPENOATE, 2-PROP
79-94-7 TETRABROMOBISPHENOL A	TETRABROMOBISPHENOL A
25327-89-3 TETRABROMOBISPHENOL-A, ALLYL ETHER [OTS]	ALLYL ETHER OF TETRABROMOBISPHENOL-A
21850-44-2 TETRABROMOBISPHENOL-A-BIS-2,3-DIBROMO-PROPYL ETHER	TETRABROMOBISPHENOL-A-BIS-2,3-DIBROMOPROPYL ETHERYLENE)OXY]BIS-
4162-45-2 TETRABROMOBISPHENOL-A-BISETHOXYLATE	TETRABROMOBISPHENOL-A-BISETHOXYLATE
3072-84-2 2,2',6,6'-TETRABROMOBISPHENOL A DIGLY-CIDYL ETHER	2,2',6,6'-TETRABROMOBISPHENOL A DIGLYCIDYL ETHER
55205-38-4 TETRABROMOBISPHENOL-A DIACRYLATE	TETRABROMOBISPHENOL-A DIACRYLATE
3072-84-2 TETRABROMOBISPHENOL A DIGLYCIDYL ETHER, 2,2',6,6'- [OTS]	2,2',6,6'-TETRABROMOBISPHENOL A DIGLYCIDYL ETHER
RR-00799-8 TETRABROMOBISPHENOL-B	TETRABROMOBISPHENOL-BENYLENE) ESTER
488-47-1 TETRABROMOCATECHOL	TETRABROMOCATECHOL
79-27-6 1,1,2,2-TETRABROMOETHANE [AUS, CSR, FLA, GBR, GBR, MAK, MAK, NFP, ONT, ONT]	ACETYLENE TETRABROMIDE
630-16-0 1,1,1,2-TETRABROMOETHANE	1,1,1,2-TETRABROMOETHANE
25167-20-8 TETRABROMOETHANE	TETRABROMOETHANE
632-79-1 TETRABROMOPHTHALICANHYDRIDE [TSC, TSC, TSC]	1,3-ISOBENZOFURANDIONE, 4,5,6,7-TETRABROMO-
311-28-4 TETRABUTYLAMMONIUM IODIDE	TETRABUTYLAMMONIUM IODIDE
13463-39-3 TETRACARBONYLNICKEL [GBR]	NICKEL CARBONYL
128-97-2 1,4,5,8-TETRACARBOXYNAPHTHALENE	1,4,5,8-TETRACARBOXYNAPHTHALENE
30402-14-3 TETRACHLORODIBENZOFURAN	TETRACHLORODIBENZOFURAN
16903-35-8 TETRACHLOROAURIC ACID	TETRACHLOROAURIC ACID
14047-09-7 3,3',4,4'-TETRACHLOROAZOBENZENE	3,3',4,4'-TETRACHLOROAZOBENZENE
21232-47-3 3,3',4,4'-TETRACHLOROAZOXYBENZENE	3,3',4,4'-TETRACHLOROAZOXYBENZENE
95-94-3 1,2,4,5-TETRACHLOROBENZENE	1,2,4,5-TETRACHLOROBENZENE
634-66-2 1,2,3,4-TETRACHLOROBENZENE	1,2,3,4-TETRACHLOROBENZENE
95-94-3 TETRACHLOROBENZENE [RCA]	1,2,4,5-TETRACHLOROBENZENE
959-43-3 1,2,4,5-TETRACHLOROBENZENE	1,2,4,5-TETRACHLOROBENZENE
RR-00143-4 TETRACHLOROBENZENE [NFP, NFP]	TETRACHLOROBENZENES
RR-00143-4 TETRACHLOROBENZENES	TETRACHLOROBENZENES

Chemical Name	Reference Name
15721-02-5 **2,2',5,5'-TETRACHLOROBENZIDINE**	2,2',5,5'-TETRACHLOROBENZIDINE
26914-33-0 **TETRACHLOROBIPHENYL**	TETRACHLOROBIPHENYL
79-95-8 **TETRACHLOROBISPHENOL-A**	TETRACHLOROBISPHENOL-A
118-75-2 **2,3,5,6-TETRACHLORO-2,5-CYCLOHEXADIENE-1,4-DIONE**	2,3,5,6-TETRACHLORO-2,5-CYCLOHEXADIENE-1,4-DIONE
1746-01-6 2,3,7,8-TETRACHLORODIBENZO-P-DIOXIN [ATS, C65, C65, C65, CAA, CAL, CSR, CSR, FLA, MAK]	2,3,7,8-TETRACHLORODIBENZO-P-DIOXIN (TCDD)
1746-01-6 2,3,7,8-TETRACHLORODIBENZO-PARA-DIOXIN [MIN]	2,3,7,8-TETRACHLORODIBENZO-P-DIOXIN (TCDD)
1746-01-6 2,3,7,8-TETRACHLORODIBENZO-P-DIOXIN [MSL, NJL, PQL]	2,3,7,8-TETRACHLORODIBENZO-P-DIOXIN (TCDD)
1746-01-6 **2,3,7,8-TETRACHLORODIBENZO-P-DIOXIN (TCDD)**	2,3,7,8-TETRACHLORODIBENZO-P-DIOXIN (TCDD)
1746-01-6 2,3,7,8-TETRACHLORODIBENZO-PARA-DIOXIN (TCDD) [IAR]	2,3,7,8-TETRACHLORODIBENZO-P-DIOXIN (TCDD)
1746-01-6 2,3,7,8-TETRACHLORODIBENZO-P-DIOXIN (TCDD, DIOXIN) [WHM]	2,3,7,8-TETRACHLORODIBENZO-P-DIOXIN (TCDD)
41903-57-5 **TETRACHLORODIBENZO-P-DIOXIN**	TETRACHLORODIBENZO-P-DIOXIN
RR-00626-8 **TETRACHLORODIBENZO-P-DIOXINS**	TETRACHLORODIBENZO-P-DIOXINS
RR-00626-8 TCDDS (ALL TETRACHLORODIBENZO-P-DIOXINS) [UTS]	TETRACHLORODIBENZO-P-DIOXINS
51207-31-9 **2,3,7,8-TETRACHLORO DIBENZOFURANS**	2,3,7,8-TETRACHLORO DIBENZOFURANS
51207-31-9 TETRACHLORODIBENZOFURAN [ATS]	2,3,7,8-TETRACHLORO DIBENZOFURANS
RR-00508-3 **TETRACHLORODIBENZOFURANS**	TETRACHLORODIBENZOFURANS
RR-00508-3 TCDFS (ALL TETRACHLORODIBENZOFURANS) [UTS]	TETRACHLORODIBENZOFURANS
76-11-9 **1,1,1,2-TETRACHLORO-2,2-DIFLUOROETHANE**	1,1,1,2-TETRACHLORO-2,2-DIFLUOROETHANE
76-11-9 1,1,1,2-TETRACHLORO-2,2-DIFLUOROETHANE (FREON 112) [QBC]	1,1,1,2-TETRACHLORO-2,2-DIFLUOROETHANE
76-11-9 1,1,1,2-TETRACHLORO-2,2-DIFLUORO-ETHANE (FC 112A) [WHM]	1,1,1,2-TETRACHLORO-2,2-DIFLUOROETHANE
76-12-0 **1,1,2,2-TETRACHLORO-1,2-DIFLUOROETHANE**	1,1,2,2-TETRACHLORO-1,2-DIFLUOROETHANE
76-12-0 1,1,2,2-TETRACHLORO-1,2-DIFLUOROETHANE (FREON 113) [QBC]	1,1,2,2-TETRACHLORO-1,2-DIFLUOROETHANE
76-12-0 1,1,2,2-TETRACHLORO-1,2-DIFLUORO-ETHANE (FC 112) [WHM]	1,1,2,2-TETRACHLORO-1,2-DIFLUOROETHANE
76-12-0 1,1,2,2-TETRACHLORO-1,2-DIFLUOROETHANE (FC-112) [CAL]	1,1,2,2-TETRACHLORO-1,2-DIFLUOROETHANE
72-54-8 TETRACHLORODIPHENYLETHANE [CSR, CSR]	DDD
79-34-5 **1,1,2,2-TETRACHLOROETHANE**	1,1,2,2-TETRACHLOROETHANE
79-34-5 TETRACHLOROETHANE, 1,1,2,2- [OTS]	1,1,2,2-TETRACHLOROETHANE
79-34-5 TETRACHLOROETHANE 1,1,2,2 [QBC, QBC]	1,1,2,2-TETRACHLOROETHANE
630-20-6 **1,1,1,2-TETRACHLOROETHANE**	1,1,1,2-TETRACHLOROETHANE
25322-20-7 TETRACHLOROETHANE [ATS]	TETRACHLOROETHANES
25322-20-7 TETRACHLOROETHANE, N.O.S. [RCU]	TETRACHLOROETHANES
25322-20-7 **TETRACHLOROETHANES**	TETRACHLOROETHANES
127-18-4 **TETRACHLOROETHYLENE**	TETRACHLOROETHYLENE
127-18-4 TETRACHLOROETHYLENE (PERCHLOROETHYLENE) [CAA, 313, C65, C65, NTP]	TETRACHLOROETHYLENE
134237-32-4 **TETRACHLOROFLUOROETHANE**	TETRACHLOROFLUOROETHANE
354-11-0 **1,1,1,2-TETRACHLORO-2-FLUOROETHANE (HCFC-121A)**	1,1,1,2-TETRACHLORO-2-FLUOROETHANE (HCFC-121A)

Chemical Name	Reference Name
354-14-3 1,1,2,2-TETRACHLORO-1-FLUOROETHANE (HCFC-121)	1,1,2,2-TETRACHLORO-1-FLUOROETHANE (HCFC-121)
134190-49-1 TETRACHLOROFLUOROPROPANE	TETRACHLOROFLUOROPROPANE
1897-45-6 2,4,5,6-TETRACHLOROISOPHTHALONI-TRILE [PST]	CHLOROTHALONIL
56-23-5 TETRACHLOROMETHANE (CARBON TETRA-CHLORIDE) [CN5]	CARBON TETRACHLORIDE
1335-88-2 TETRACHLORONAPHTHALENE	TETRACHLORONAPHTHALENE
1335-88-2 TETRACHLORONAPHTHALENES, ALL ISO-MERS [GBR, GBR]	TETRACHLORONAPHTHALENE
2438-88-2 2,3,5,6-TETRACHLORO-4-NITROANISOLE	2,3,5,6-TETRACHLORO-4-NITROANISOLE
117-18-0 1,2,4,5-TETRACHLORO-3-NITROBENZENE	1,2,4,5-TETRACHLORO-3-NITROBENZENE
935-95-5 2,3,5,6-TETRACHLOROPHENOL	2,3,5,6-TETRACHLOROPHENOL
58-90-2 2,3,4,6-TETRACHLOROPHENOL	2,3,4,6-TETRACHLOROPHENOL3.BETA.,4.ALPHA.,5.AL-PHA.,6.BETA.)-
58-90-2 TETRACHLOROPHENOL, 2,3,4,6- [CD1, CD1]	2,3,4,6-TETRACHLOROPHENOL
25167-83-3 TETRACHLOROPHENOL	TETRACHLOROPHENOL
53535-27-6 2,3,4,6-TETRACHLOROPHENOL, POTASSIUM SALT [RCU]	POTASSIUM TETRACHLOROPHENATE
25167-83-3 TETRACHLOROPHENOLS [RCA]	TETRACHLOROPHENOL
25567-55-9 2,3,4,6-TETRACHLOROPHENOL, SODIUM SALT	2,3,4,6-TETRACHLOROPHENOL, SODIUM SALT
117-08-8 TETRACHLOROPHTHALIC ANHYDRIDE	TETRACHLOROPHTHALIC ANHYDRIDE
10436-39-2 1,1,2,3-TETRACHLOROPROPENE	1,1,2,3-TETRACHLOROPROPENE
2402-79-1 TETRACHLOROPYRIDINE, 2,3,5,6- [OTS]	PYRIDINE, 2,3,5,6-TETRACHLORO-
29255-31-0 TETRACHLOROTETRAFLUOROPROPANE	TETRACHLOROTETRAFLUOROPROPANE
5216-25-1 P-A,A,A-TETRACHLOROTOLUENE [C65]	4-CHLOROBENZOTRICHLORIDE
5216-25-1 P-ALPHA, ALPHA, ALPHA-TETRACHLORO-TOLUENE [CAB]	4-CHLOROBENZOTRICHLORIDE
134237-37-9 TETRACHLOROTRIFLUOROPROPANE	TETRACHLOROTRIFLUOROPROPANE
134237-39-1 TETRACHLORODIFLUOROPROPANE	TETRACHLORODIFLUOROPROPANE
961-11-5 TETRACHLOROVINPHOS	TETRACHLOROVINPHOS
961-11-5 TETRACHLORVINPHOS [CAL, IAR, PAL]	TETRACHLOROVINPHOSO-1,5,5A,6,9,9A-HEXAHY-DRO-, 3-OXIDE, (3.ALPHA.,5A.BETA.,6.ALPHA.,9.ALPHA., 9A.BETA.)-
961-11-5 TETRACHLORVINPHOS [PHOSPHORIC ACID, 2-CHLORO-1-(2,4,5-TRICHLOROPHENYL) ETHENYL DIMETHYL ESTER] [313]	TETRACHLOROVINPHOS
18919-94-3 TETRACOSAMETHYLCYCLODODECASILOXANE	TETRACOSAMETHYLCYCLODODECASILOXANE
107-53-9 TETRACOSAMETHYLUNDECASILOXANE	TETRACOSAMETHYLUNDECASILOXANE
670-54-2 TETRACYANOMETHYLENE	TETRACYANOMETHYLENE
60-54-8 TETRACYCLINE	TETRACYCLINE
64-75-5 TETRACYCLINE HYDROCHLORIDE	TETRACYCLINE HYDROCHLORIDE
RR-01475-5 TETRACYCLINES	TETRACYCLINES
107-50-6 TETRADECAMETHYLCYCLOHEPTASILOXANE	TETRADECAMETHYLCYCLOHEPTASILOXANE
107-52-8 TETRADECAMETHYLHEXASILOXANE	TETRADECAMETHYLHEXASILOXANE
64036-86-8 TETRADECANE	TETRADECANE
28983-37-1 TERT-TETRADECANETHIOL	TERT-TETRADECANETHIOL
112-72-1 1-TETRADECANOL [OTS]	MYRISTIC ALCOHOL
27196-00-5 TETRADECANOL	TETRADECANOL
1120-36-1 1-TETRADECENE	1-TETRADECENE
38954-75-5 TETRADECYL GLYCIDYL ETHER [EPA, OTS]	OXIRANE, [(TETRADECYLOXY)METHYL]-

Chemical Name	Reference Name
28983-37-1 TERT-TETRADECYL MERCAPTAN [FLA, MSL, NFP, NFP]	TERT-TETRADECANETHIOL
1155-74-4 1-TETRADECYLPYRIDINIUM BROMIDE	1-TETRADECYLPYRIDINIUM BROMIDE
1191-50-0 TETRADECYL SULFATE, SODIUM SALT [PST]	SODIUM TETRADECYL SULFATE
631-41-4 TETRAETHANOL AMMONIUM HYDROXIDE	TETRAETHANOL AMMONIUM HYDROXIDE
122-31-6 TETRAETHOXYPROPANE	TETRAETHOXYPROPANE
56-34-8 TETRAETHYLAMMONIUM CHLORIDE	TETRAETHYLAMMONIUM CHLORIDE
2567-83-1 TETRAETHYLAMMONIUM PERCHLORATE	TETRAETHYLAMMONIUM PERCHLORATE
78-13-7 TETRA (2-ETHYLBUTYL) SILICATE [NFP, NFP]	SILICIC ACID (H4SIO4), TETRAKIS(2-ETHYLBUTYL) ESTER
3689-24-5 TETRAETHYL DITHIONOPYROPHOS-PHATE [ONT, ONT]	SULFOTEP
15108-81-3 TETRAETHYL DITHIOPYROPHOSPHATE	TETRAETHYL DITHIOPYROPHOSPHATE
3689-24-5 TETRAETHYL DITHIOPYROPHOSPHATE (SUL FOTEPP) [CAL]	SULFOTEP
3689-24-5 TETRAETHYLDITHIOPYROPHOSPHATE [EPA]	SULFOTEP
112-60-7 TETRAETHYLENE GLYCOL	TETRAETHYLENE GLYCOL
17831-71-9 TETRAETHYLENE GLYCOL DIACRYLATE	TETRAETHYLENE GLYCOL DIACRYLATE
23783-42-8 TETRAETHYLENE GLYCOL MONOMETHYL ETHER [TSC, TSC, TSC]	2,5,8,11-TETRAOXATRIDECAN-13-OL
36366-93-5 TETRAETHYLENE GLYCOL MONOPHENYL ETHER	TETRAETHYLENE GLYCOL MONOPHENYL ETHER
112-57-2 TETRAETHYLENEPENTAMINE	TETRAETHYLENEPENTAMINE
112-57-2 TETRAETHYLENE PENTAMINE [FLA, MSL, NFP, NFP]	TETRAETHYLENEPENTAMINE
78-10-4 TETRAETHYL ESTER SILICIC ACID [FLA]	ETHYL SILICATE
115-82-2 TETRA (2-ETHYLHEXYL) SILICATE	TETRA (2-ETHYLHEXYL) SILICATE
78-00-2 TETRAETHYL LEAD	TETRAETHYL LEAD
78-00-2 TETRAETHYLLEAD [302]	TETRAETHYL LEAD
78-00-2 TETRAETHYL LEAD, COMPOUNDS [NFP, NFP]	TETRAETHYL LEAD
78-10-4 TETRAETHYL ORTHOSILICATE [GBR, GBR]	ETHYL SILICATE
107-49-3 TETRAETHYL PRYOPHOSPHATE [HM1, HM1, HM1, HM1, HM1]	TEPP
107-49-3 TETRAETHYL PYROPHOSPHATE [CAL, CWA, ONT, ONT]	TEPP
3689-24-5 TETRAETHYL PYROPHOSPHORODITHION-ATE [RCU, RCU]	SULFOTEP
78-10-4 TETRAETHYL SILICATE [HM1, HM1]	ETHYL SILICATE
97-77-8 TETRAETHYLTHIURAM DISULFIDE [CSR, CSR]	DISULFIRAM
97-77-8 TETRAETHYLTHIURAM DISULPHIDE [WHM]	DISULFIRAM
597-64-8 TETRAETHYL TIN	TETRAETHYL TIN
597-64-8 TETRAETHYLTIN [NJL, PAL]	TETRAETHYL TIN
811-97-2 1,1,1,2-TETRAFLUOROETHANE	1,1,1,2-TETRAFLUOROETHANE
116-14-3 TETRAFLUOROETHENE [CTR, TSE]	TETRAFLUOROETHYLENE
116-14-3 TETRAFLUOROETHYLENE	TETRAFLUOROETHYLENE
116-14-3 TETRAFLUOROETHYLENE [ETHENE, TE-TRAFLUORO-] [EP3]	TETRAFLUOROETHYLENE
10036-47-2 TETRAFLUOROHYDRAZINE	TETRAFLUOROHYDRAZINE
75-73-0 TETRAFLUOROMETHANE	TETRAFLUOROMETHANE
1835-49-0 TETRAFLUOROPHTHALONITRILE	TETRAFLUOROPHTHALONITRILE
RR-00194-5 TETRAGLYCIDYLAMINES	TETRAGLYCIDYLAMINES

Chemical Name	Reference Name
100-50-5 1,2,3,6-TETRAHYDROBENZALDEHYDE	1,2,3,6-TETRAHYDROBENZALDEHYDE
1321-16-0 TETRAHYDROBENZALDEHYDE	TETRAHYDROBENZALDEHYDE
1321-16-0 1,2,3,6-TETRAHYDROBENZALDEHYDE [FLA, HM1, HM1, MSL]	TETRAHYDROBENZALDEHYDE
1972-08-3 1-TRANS-DELTA-9-TETRAHYDROCANNABINOL	1-TRANS-DELTA-9-TETRAHYDROCANNABINOL
942-01-8 1,2,3,4-TETRAHYDROCARBAZOLE	1,2,3,4-TETRAHYDROCARBAZOLE
2825-83-4 ENDO-TETRAHYDRODICYLCLOPENTADIENE	ENDO-TETRAHYDRODICYLCLOPENTADIENE
533-74-4 TETRAHYDRO-3,5-DIMETHYL-2H-1,3,5-THIADI-AZINE-2-THIONE	TETRAHYDRO-3,5-DIMETHYL-2H-1,3,5-THIADIAZINE-2-THIONE
109-99-9 TETRAHYDROFURAN	TETRAHYDROFURAN
109-99-9 TETRAHYDROFURANE [QBC]	TETRAHYDROFURAN
97-99-4 TETRAHYDROFURFURYL ALCOHOL	TETRAHYDROFURFURYL ALCOHOL
4795-29-3 TETRAHYDROFURFURYLAMINE	TETRAHYDROFURFURYLAMINE
2455-24-5 TETRAHYDROFURFURYL METHACRYLATE	TETRAHYDROFURFURYL METHACRYLATE
RR-00885-5 TETRAHYDROFURFURYL OLEATE	TETRAHYDROFURFURYL OLEATE
119-64-2 TETRAHYDRONAPHTHALENE	TETRAHYDRONAPHTHALENE
529-33-9 1,2,3,4-TETRAHYDRO-1-NAPHTHOL	1,2,3,4-TETRAHYDRO-1-NAPHTHOL
2954-50-9 1,2,3,4-TETRAHYDRO-2-NAPHTHYLAMINE	1,2,3,4-TETRAHYDRO-2-NAPHTHYLAMINE
RR-01759-4 TETRAHYDRONAPHTHYL HYDROPEROXIDE	TETRAHYDRONAPHTHYL HYDROPEROXIDE
85-43-8 TETRAHYDROPHTHALIC ACID ANHYDRIDE	TETRAHYDROPHTHALIC ACID ANHYDRIDE
85-43-8 TETRAHYDROPHTHALIC ANHYDRIDE [HON, NJL]	TETRAHYDROPHTHALIC ACID ANHYDRIDE
85-40-5 TETRAHYDROPHTHALIMIDE	TETRAHYDROPHTHALIMIDE
100-72-1 TETRAHYDROPYRAN-2-METHANOL	TETRAHYDROPYRAN-2-METHANOL
694-05-3 1,2,3,6-TETRAHYDROPYRIDINE	1,2,3,6-TETRAHYDROPYRIDINE
110-01-0 TETRAHYDROTHIOPHENE	TETRAHYDROTHIOPHENE
131-55-5 2,2',4,4'-TETRAHYDROXYBENZOPHENONE	2,2',4,4'-TETRAHYDROXYBENZOPHENONE
7328-97-4 1,1,2,2-TETRA(P-HYDROXYPHENYL) ETHANE TETRAGLYCIDYL ETHER [EPA]	OXIRANE, 2,2',2'',2'''-[1,2-ETHANEDIYLIDENETETRAKIS-
7328-97-4 1,1,2,2-TETRA(P-HYDROXYPHENYL)-ETHANE TETRAGLYCIDYL ETHER [OTS]	OXIRANE, 2,2',2'',2'''-[1,2-ETHANEDIYLIDENETETRAKIS-
546-68-9 TETRAISOPROPYL TITANATE	TETRAISOPROPYL TITANATE
33125-86-9 TETRAKIS(2-CHLOROETHYL) ETHYLENE DIPHOSPHATE	TETRAKIS(2-CHLOROETHYL) ETHYLENE DIPHOSPHATE
124-64-1 TETRAKIS(HYDROXYMETHYL)PHOSPHONIUM CHLORIDE	TETRAKIS(HYDROXYMETHYL)PHOSPHONIUM CHLORIDE
55566-30-8 TETRAKIS(HYDROXYMETHYL)PHOSPHONIUM SULFATE	TETRAKIS(HYDROXYMETHYL)PHOSPHONIUM SULFATE
RR-01548-5 TETRAKIS(HYDROXYMETHYL) PHOSPHONIUM SALTS	TETRAKIS(HYDROXYMETHYL) PHOSPHONIUM SALTS
65992-66-7 N,N,N',N'-TETRAKIS(OXIRANYLMETHYL)-1,3-CYCLOHEXANEDIMETHANAMINE	N,N,N',N'-TETRAKIS(OXIRANYLMETHYL)-1,3-CYCLO-HEXANEDIMETHANAM
119-64-2 TETRALIN [CSR, PST]	TETRAHYDRONAPHTHALENE
771-29-9 TETRALIN HYDROPEROXIDE	TETRALIN HYDROPEROXIDE
529-34-0 D-TETRALONE	D-TETRALONE
110-18-9 TETRAMETHLYETHYLENEDIAMINE [HON]	1,2-BIS(DIMETHYLAMINO)ETHANE
102-52-3 1,1,3,3-TETRAMETHOXYPROPANE	1,1,3,3-TETRAMETHOXYPROPANE
681-84-5 TETRAMETHOXYSILANE [ONT]	METHYL SILICATE
7696-12-0 TETRAMETHRIN [2,2-DIMETHYL-3-(2-METHYL-1-PROPENYL)CYCLOPROPANECARBOXYLIC ACID (1,3,4,5,6,7-HEXAHYDRO-1,3-DIOXO-2H-ISOINDOL-2-YL)METHYL ESTER] [313]	NEOPYNAMIN

Chemical Name	Reference Name
75-59-2 TETRAMETHYL AMMONIUM HYDROXIDE [NJL, NJL]	TETRAMETHYLAMMONIUM HYDROXIDE
75-59-2 TETRAMETHYLAMMONIUM HYDROXIDE	TETRAMETHYLAMMONIUM HYDROXIDE
RR-01682-0 TETRAMETHYLAMMONIUM SALTS OF ALKYL-BENZENESULFONIC ACID	TETRAMETHYLAMMONIUM SALTS OF ALKYLBENZENE-SULFONIC ACID
95-93-2 1,2,4,5-TETRAMETHYLBENZENE	1,2,4,5-TETRAMETHYLBENZENE
488-23-3 1,2,3,4-TETRAMETHYLBENZENE	1,2,3,4-TETRAMETHYLBENZENE
527-53-7 1,2,3,5-TETRAMETHYLBENZENE	1,2,3,5-TETRAMETHYLBENZENE
RR-00908-5 3,3',5,5'-TETRAMETHYLBIPHENYL-4,4'-DIOL	3,3',5,5'-TETRAMETHYLBIPHENYL-4,4'-DIOL
97-84-7 N,N,N',N'-TETRAMETHYL-1,3-BUTANEDIAMINE	N,N,N',N'-TETRAMETHYL-1,3-BUTANEDIAMINE
5809-08-5 1,1,3,3-TETRAMETHYL BUTYL HYDROPEROX-IDE	1,1,3,3-TETRAMETHYL BUTYL HYDROPEROXIDE
22288-43-3 1,1,3,3-TETRAMETHYLBUTYLPEROXY-2-ETHYL HEXANOATE	1,1,3,3-TETRAMETHYLBUTYLPEROXY-2-ETHYL HEX-ANOATE
9002-93-1 4-(1,1,3,3-TETRAMETHYLBUTYL)PHENOL, ETHOXYLATED [PST]	POLYETHYLENE GLYCOL OCTYLPHENOL ETHER
RR-01153-0 ALPHA-(1,1,3,3-TETRAMETHYLBUTYL)PHE-NOXY-OMEGA-POLYOXYPROPYLENE BLOCK POLYMER WITH POLYOXYETHYLENE	ALPHA-(1,1,3,3-TETRAMETHYLBUTYL)PHENOXY-OMEGA-POLYOXYPROPYLE
2370-88-9 TETRAMETHYLCYCLOTETRASILOXANE	TETRAMETHYLCYCLOTETRASILOXANE
126-86-3 2,4,7,9-TETRAMETHYL-5-DECYNE-4,7-DIOL [PST]	5-DECYNE-4,7-DIOL, 2,4,7,9-TETRAMETHYL-
2627-95-4 TETRAMETHYLDIVINYLDISILOXANE	TETRAMETHYLDIVINYLDISILOXANE
110-60-1 TETRAMETHYLENEDIAMINE	TETRAMETHYLENEDIAMINE
RR-01460-8 TETRAMETHYLENE DIPEROXIDE DICAR-BAMIDE	TETRAMETHYLENE DIPEROXIDE DICARBAMIDE
111-18-2 N,N,N',N'-TETRAMETHYL-1,6-HEXANEDIAMINE	N,N,N',N'-TETRAMETHYL-1,6-HEXANEDIAMINE
75-74-1 TETRAMETHYL LEAD	TETRAMETHYL LEAD
75-74-1 TETRAMETHYLLEAD [302]	TETRAMETHYL LEAD
75-74-1 TETRAMETHYL LEAD [NJL, NJL]	TETRAMETHYL LEAD
75-74-1 TETRAMETHYL LEAD (TML) [EPA, EPA, EPA, WHM]	TETRAMETHYL LEAD
75-74-1 TETRAMETHYL LEAD, COMPOUNDS [NFP, NFP]	TETRAMETHYL LEAD
75-74-1 TETRAMETHYL LEAD [PLUMBANE, TETRAM-ETHYL-] [EP3]	TETRAMETHYL LEAD
51-80-9 TETRAMETHYLMETHYLENEDIAMINE	TETRAMETHYLMETHYLENEDIAMINE
681-84-5 TETRAMETHYL ORTHOSILICATE [GBR, GBR]	METHYL SILICATE
1186-53-4 2,2,3,4-TETRAMETHYL PENTANE [FLA]	PENTANE, 2,2,3,4-TETRAMETHYL-
1186-53-4 2,2,3,4-TETRAMETHYLPENTANE [MSL]	PENTANE, 2,2,3,4-TETRAMETHYL-
1186-53-4 2,2,3,4-TETRAMETHYL PENTANE [NFP, NFP]	PENTANE, 2,2,3,4-TETRAMETHYL-
7154-79-2 2,2,3,3-TETRAMETHYL PENTANE [FLA]	PENTANE, 2,2,3,3-TETRAMETHYL-
7154-79-2 2,2,3,3-TETRAMETHYLPENTANE [MSL]	PENTANE, 2,2,3,3-TETRAMETHYL-
7154-79-2 2,2,3,3-TETRAMETHYL PENTANE [NFP, NFP]	PENTANE, 2,2,3,3-TETRAMETHYL-
75-76-3 TETRAMETHYL SILANE	TETRAMETHYL SILANE
75-76-3 TETRAMETHYLSILANE [SILANE, TETRAM-ETHYL-] [EP3]	TETRAMETHYL SILANE
3333-52-6 TETRAMETHYL SUCCINONITRILE	TETRAMETHYL SUCCINONITRILE
3333-52-6 TETRAMETHYLSUCCINONITRILE [QBC, QBC]	TETRAMETHYL SUCCINONITRILE
3333-52-6 TETRAMETHYL SUCCINONITRILE (DECOM-POSITION PRODUCT OF 2,2'-AZOBISISOBUTY-RONITRILE) [CAL]	TETRAMETHYL SUCCINONITRILE

Chemical Name	Reference Name
3333-52-6 TETRAMETHYLSUCCINONITRILE [NHS, NHS, ONT, ONT]	TETRAMETHYL SUCCINONITRILE
3333-52-6 TETRAMETHYL SUCCINONITRILE (DECOMPOSITION PRODUCT OF 2,2'-AZOBISBUTYRONITRILE) [CEX, CEX]	TETRAMETHYL SUCCINONITRILE
3383-96-8 TETRAMETHYL O,O'-THIO-DI-P-PHENYLENEPHOSPHOROTHIOATE (TEMEPHOS) [CAL]	TEMEPHOS
97-74-5 (ANAD5) TETRAMETHYL THIURAM MONOSULFIDE	(ANAD5) TETRAMETHYL THIURAM MONOSULFIDE
594-27-4 TETRAMETHYL TIN [FLA]	STANNANE, TETRAMETHYL-
594-27-4 TETRAMETHYLTIN [HM1]	STANNANE, TETRAMETHYL-
594-27-4 TETRAMETHYL TIN [MSL, NFP, NFP]	STANNANE, TETRAMETHYL-
632-22-4 TETRAMETHYLUREA	TETRAMETHYLUREA
53014-37-2 TETRANITRO-ANILINE	TETRANITRO-ANILINE
20600-96-8 TETRANITRO DIGLYCERIN	TETRANITRO DIGLYCERIN
509-14-8 TETRANITROMETHANE	TETRANITROMETHANE
509-14-8 TETRANITROMETHANE [METHANE, TETRANITRO-] [EP3]	TETRANITROMETHANERIFLUORO[OXYBIS[METANE]-, T-4-
641-16-7 2,3,4,6-TETRANITROPHENOL	2,3,4,6-TETRANITROPHENOL
RR-01376-3 2,3,4,6-TETRANITROPHENYL METHYL NITRAMINE	2,3,4,6-TETRANITROPHENYL METHYL NITRAMINE
RR-01377-4 2,3,4,6-TETRANITROPHENYLNITRAMINE	2,3,4,6-TETRANITROPHENYLNITRAMINE
RR-01461-9 TETRANITRORESORCINOL	TETRANITRORESORCINOL
RR-01379-6 2,3,5,6-TETRANITROSO-1,4-DINITROBENZENE	2,3,5,6-TETRANITROSO-1,4-DINITROBENZENE
RR-01378-5 2,3,5,6-TETRANITROSO NITROBENZENE	2,3,5,6-TETRANITROSO NITROBENZENE
126505-35-9 2,4,8,10-TETRAOXA-3,9-DIPHOSPHASPIRO [5,5] UNDECANE, 3,9-BIS[2,4,6-TRIS[1,1-DIMETHYLETHYL]PHENOXY]-	2,4,8,10-TETRAOXA-3,9-DIPHOSPHASPIRO[5,5] UNDECANE, 3,9-BIS[
RR-01728-7 3,6,9,13-TETRAOXAHEXADEC-15-ENE-1,11-DIOL, BIS(4-METHYLBENZENESULFONATE)	3,6,9,13-TETRAOXAHEXADEC-15-ENE-1,11-DIOL,BIS(4-METHYLBENZEN
41024-91-3 3,6,9,12-TETRAOXATETRADECANE-1,14-DIOL, BIS(4-METHYLBENZENESULFONATE)	3,6,9,12-TETRAOXATETRADECANE-1,14-DIOL,BIS(4-METHYLBENZENESUENESULFONATE))
RR-01730-1 3,6,9,12-TETRAOXATETRADECANE-1,14-DIOL,7-[(2-PROPENYLOXY)METHYL]-,BIS(4-METHYLBENZENESULFONATE)	3,6,9,12-TETRAOXATETRADECANE-1,14-DIOL,7-[(2-PROPENYLOXY)METZENSULFONATE)
23783-42-8 2,5,8,11-TETRAOXATRIDECAN-13-OL	2,5,8,11-TETRAOXATRIDECAN-13-OL
108-62-3 1,3,5,7-TETRAOXOCANE, 2,4,6,8-TETRAMETHYL- [PAL]	METALDEHYDE
595-90-4 TETRAPHENYL TIN [FLA, MSL, NFP, NFP]	STANNANE, TETRAPHENYL-
757-58-4 TETRAPHOSPHORIC ACID, HEXAETHYL ESTER [TSC, TSC]	HEXAETHYL TETRAPHOSPHATE
7320-34-5 TETRAPOTASSIUM PYROPHOSPHATE [PST]	POTASSIUM PYROPHOSPHATE
3440-75-3 TETRAPROPYL LEAD	TETRAPROPYL LEAD
3087-37-4 TETRAPROPYLORTHOTITANATE	TETRAPROPYLORTHOTITANATE
RR-01154-1 TETRAPROPYL SUCCINIC ACID	TETRAPROPYL SUCCINIC ACIDNE BLOCK POLYMER WITH POLYOXYETHYLENE
69155-42-6 TETRASILOXANE, 1,1,1,3,5,7,7-OCTAMETHYL-3,5-BIS[3-(OXIRANYLMETHOXY)PROPYL]-	TETRASILOXANE, 1,1,1,3,5,7,7-OCTAMETHYL-3,5-BIS[3-(OXIRANYLM
38916-42-6 TETRASODIUM N-(1,2-DICARBOXYETHYL)-N-OCTADECYL SULFOSUCCINAMATE [PST]	DL-ASPARTIC ACID, N-(3-CARBOXY-1-OXO-3-SULFO-PROPYL)-N-OCTADE
64-02-8 TETRASODIUM EDTA	TETRASODIUM EDTA
67401-50-7 TETRASODIUM ETHYLENEDIAMINETETRAACETATE TRIHYDRATE	TETRASODIUM ETHYLENEDIAMINETETRAACETATE TRIHYDRATE

Chemical Name	Reference Name
7722-88-5 TETRASODIUM PYROPHOSPHATE	TETRASODIUM PYROPHOSPHATE
7722-88-5 TETRA SODIUM PYROPHOSPHATE [CAL]	TETRASODIUM PYROPHOSPHATE
36938-50-8 TETRATRIACONTAMETHYLHEXADECASILOXANE	TETRATRIACONTAMETHYLHEXADECASILOXANE
109-27-3 1-TETRAZENE-1-CARBOXIMIDIC ACID, 4-(AMINOMETHYL)-, 2-NITROSOHYDRAZIDE	1-TETRAZENE-1-CARBOXIMIDIC ACID, 4-(AMINOMETHYL)-, 2-NITROSO
70816-59-0 TETRAZINE	TETRAZINE
21732-17-2 TETRAZOL-1-ACETIC ACID	TETRAZOL-1-ACETIC ACID
RR-01462-0 TETRAZOLYL AZIDE	TETRAZOLYL AZIDE
479-45-8 TETRYL	TETRYL
479-45-8 TETRYL (2,4,6-TRINITROPHENYL-METHYL-NITRAMINE) [BC1, BC1, BC1]	TETRYL
479-45-8 TETRYL (2,4,6-TRINITROPHENYL-METHYL-NITRAMINE) [OS1, OS1, OS1, OS1]	TETRYL
479-45-8 TETRYL (2,4,6-TRINITRO-PHENYLMETHYL-NITRAMINE) [ALB, ALB, ALB]	TETRYL
RR-01203-3 TEXTILE MANUFACTURING INDUSTRY	TEXTILE MANUFACTURING INDUSTRY
789-61-7 B-TGDR [MSL]	BETA-THIOGUANIDINE DEOXYRIBOSIDE
50-35-1 THALIDOMIDE	THALIDOMIDE
1314-32-5 THALLIC OXIDE	THALLIC OXIDE
7440-28-0 THALLIUM	THALLIUM
18235-46-6 THALLIUM 194	THALLIUM 194
RR-00320-3 THALLIUM 194M	THALLIUM 194M
26683-69-2 THALLIUM 195	THALLIUM 195
14107-52-9 THALLIUM 197	THALLIUM 197
15743-50-7 THALLIUM 198	THALLIUM 198
RR-00319-0 THALLIUM 198M	THALLIUM 198M
15064-66-1 THALLIUM 199	THALLIUM 199
15720-55-5 THALLIUM 200	THALLIUM 200
15064-65-0 THALLIUM 201	THALLIUM 201
15720-57-7 THALLIUM 202	THALLIUM 202
13968-51-9 THALLIUM 204	THALLIUM 204
7440-28-0 THALLIUM, SOLUBLE COMPOUNDS [MEX]	THALLIUM
563-68-8 THALLIUM(I) ACETATE	THALLIUM(I) ACETATE
563-68-8 THALLIUM ACETATE [WHM]	THALLIUM(I) ACETATE
7789-40-4 THALLIUM BROMIDE	THALLIUM BROMIDE
6533-73-9 THALLIUM(I) CARBONATE [EPA, RCU, RCU, WHM]	THALLOUS CARBONATE
13453-30-0 THALLIUM CHLORATE	THALLIUM CHLORATE
7791-12-0 THALLIUM (I) CHLORIDE [EPA]	THALLOUS CHLORIDE
7791-12-0 THALLIUM CHLORIDE [RCU, RCU]	THALLOUS CHLORIDE
7791-12-0 THALLIUM(I) CHLORIDE [WHM]	THALLOUS CHLORIDE
13473-75-1 THALLIUM (I) CHROMATE	THALLIUM (I) CHROMATE
RR-00575-4 THALLIUM COMPOUNDS	THALLIUM COMPOUNDS
RR-00575-4 THALLIUM COMPOUNDS, N.O.S. [RCU, NJL, HM1, HM1, HM1]	THALLIUM COMPOUNDS
7440-28-0 THALLIUM, ELEMENTAL [WHM]	THALLIUM
7789-27-7 THALLIUM(I) FLUORIDE	THALLIUM(I) FLUORIDE
7790-30-9 THALLIUM(I) IODIDE	THALLIUM(I) IODIDE
2757-18-8 THALLIUM MALONATE [WHM]	THALLOUS MALONATE
10102-45-1 THALLIUM(I) NITRATE	THALLIUM(I) NITRATE

Chemical Name	Reference Name
7722-88-5 TETRASODIUM PYROPHOSPHATE	TETRASODIUM PYROPHOSPHATE
7722-88-5 TETRA SODIUM PYROPHOSPHATE [CAL]	TETRASODIUM PYROPHOSPHATE
36938-50-8 TETRATRIACONTAMETHYLHEXADECASILOXANE	TETRATRIACONTAMETHYLHEXADECASILOXANE
109-27-3 1-TETRAZENE-1-CARBOXIMIDIC ACID, 4-(AMINOMETHYL)-, 2-NITROSOHYDRAZIDE	1-TETRAZENE-1-CARBOXIMIDIC ACID, 4-(AMINOMETHYL)-, 2-NITROSO
70816-59-0 TETRAZINE	TETRAZINE
21732-17-2 TETRAZOL-1-ACETIC ACID	TETRAZOL-1-ACETIC ACID
RR-01462-0 TETRAZOLYL AZIDE	TETRAZOLYL AZIDE
479-45-8 TETRYL	TETRYL
479-45-8 TETRYL (2,4,6-TRINITROPHENYL-METHYL-NITRAMINE) [BC1, BC1, BC1]	TETRYL
479-45-8 TETRYL (2,4,6-TRINITROPHENYL-METHYL-NITRAMINE) [OS1, OS1, OS1, OS1]	TETRYL
479-45-8 TETRYL (2,4,6-TRINITRO-PHENYLMETHYL-NITRAMINE) [ALB, ALB, ALB]	TETRYL
RR-01203-3 TEXTILE MANUFACTURING INDUSTRY	TEXTILE MANUFACTURING INDUSTRY
789-61-7 B-TGDR [MSL]	BETA-THIOGUANIDINE DEOXYRIBOSIDE
50-35-1 THALIDOMIDE	THALIDOMIDE
1314-32-5 THALLIC OXIDE	THALLIC OXIDE
7440-28-0 THALLIUM	THALLIUM
18235-46-6 THALLIUM 194	THALLIUM 194
RR-00320-3 THALLIUM 194M	THALLIUM 194M
26683-69-2 THALLIUM 195	THALLIUM 195
14107-52-9 THALLIUM 197	THALLIUM 197
15743-50-7 THALLIUM 198	THALLIUM 198
RR-00319-0 THALLIUM 198M	THALLIUM 198M
15064-66-1 THALLIUM 199	THALLIUM 199
15720-55-5 THALLIUM 200	THALLIUM 200
15064-65-0 THALLIUM 201	THALLIUM 201
15720-57-7 THALLIUM 202	THALLIUM 202
13968-51-9 THALLIUM 204	THALLIUM 204
7440-28-0 THALLIUM, SOLUBLE COMPOUNDS [MEX]	THALLIUM
563-68-8 THALLIUM(I) ACETATE	THALLIUM(I) ACETATE
563-68-8 THALLIUM ACETATE [WHM]	THALLIUM(I) ACETATE
7789-40-4 THALLIUM BROMIDE	THALLIUM BROMIDE
6533-73-9 THALLIUM(I) CARBONATE [EPA, RCU, RCU, WHM]	THALLOUS CARBONATE
13453-30-0 THALLIUM CHLORATE	THALLIUM CHLORATE
7791-12-0 THALLIUM (I) CHLORIDE [EPA]	THALLOUS CHLORIDE
7791-12-0 THALLIUM CHLORIDE [RCU, RCU]	THALLOUS CHLORIDE
7791-12-0 THALLIUM(I) CHLORIDE [WHM]	THALLOUS CHLORIDE
13473-75-1 THALLIUM (I) CHROMATE	THALLIUM (I) CHROMATE
RR-00575-4 THALLIUM COMPOUNDS	THALLIUM COMPOUNDS
RR-00575-4 THALLIUM COMPOUNDS, N.O.S. [RCU, NJL, HM1, HM1, HM1]	THALLIUM COMPOUNDS
7440-28-0 THALLIUM, ELEMENTAL [WHM]	THALLIUM
7789-27-7 THALLIUM(I) FLUORIDE	THALLIUM(I) FLUORIDE
7790-30-9 THALLIUM(I) IODIDE	THALLIUM(I) IODIDE
2757-18-8 THALLIUM MALONATE [WHM]	THALLOUS MALONATE
10102-45-1 THALLIUM(I) NITRATE	THALLIUM(I) NITRATE

Chemical Name	Reference Name
28249-77-6 THIOBENCARB [CAL]	BOLERO (THIOBENCARB)
28249-77-6 THIOBENCARB [CARBAMIC ACID, DI-ETHYLTHIO-, S-(P-CHLOROBENZYL)] [313]	BOLERO (THIOBENCARB)E]
96-69-5 4,4'-THIOBIS [MIN]	4,4'-THIOBIS(6-TERT-BUTYL-M-CRESOL)
96-69-5 4,4'-THIOBIS(6-TERT-BUTYL-M-CRESOL)	4,4'-THIOBIS(6-TERT-BUTYL-M-CRESOL)
96-69-5 4,4'-THIOBIS(6-TERT, BUTYL-M-CRESOL) [OS1, OS1]	4,4'-THIOBIS(6-TERT-BUTYL-M-CRESOL)
97-18-7 2,2'-THIOBIS-4,6-DICHLORO-PHENOL [NJL]	PHENOL, 2,2'-THIOBIS(4,6-DICHLORO)-
96-69-5 THIOBIS 4,4 6-TERBUTYL-M-CRESOL) [QBC]	4,4'-THIOBIS(6-TERT-BUTYL-M-CRESOL)
102-08-9 THIOCARBANILIDE	THIOCARBANILIDE
2231-57-4 THIOCARBAZIDE	THIOCARBAZIDE
RR-01314-9 THIOCARBONYL TETRACHLORIDE	THIOCARBONYL TETRACHLORIDE
1762-95-4 THIOCYANIC ACID, AMMONIUM SALT [PAL]	AMMONIUM THIOCYANATE
21564-17-0 THIOCYANIC ACID, (2-BENZOTHIAZOLYTHIO) METHYL ESTER	THIOCYANIC ACID, (2-BENZOTHIAZOLYTHIO)METHYL ESTER
21564-17-0 THIOCYANIC ACID, (2-BENZOTHIAZOLYTHIO) METHYL ESTER [MSL]	THIOCYANIC ACID, (2-BENZOTHIAZOLYTHIO)METHYL ESTER
592-87-0 THIOCYANIC ACID, LEAD(2+) SALT [PAL]	LEAD THIOCYANATE
592-85-8 THIOCYANIC ACID, MERCURY(2+) SALT [PAL]	MERCURIC THIOCYANATE
139-65-1 4,4'-THIODIANILINE	4,4'-THIODIANILINE
59669-26-0 THIODICARB	THIODICARB
111-48-8 THIODIGLYCOL [FLA, NFP, NFP, MSL]	ETHANOL, 2,2'-THIOBIS-
2664-63-3 4,4'-THIODIPHENOL	4,4'-THIODIPHENOL
3689-24-5 THIODIPHOSPHORIC ACID ([(HO)2P(S)]2O), TETRAETHYL ESTER [PAL]	SULFOTEP
111-97-7 BETA,BETA'-THIODIPROPIONITRILE	BETA,BETA'-THIODIPROPIONITRILE
111-17-1 THIODIPROPIONIC ACID	THIODIPROPIONIC ACID
39196-18-4 THIOFANOX	THIOFANOXCYL-, TETRASODIUM SALT
60-24-2 THIOGLYCOL [HM1, HM1, NJL, WHM]	2-MERCAPTOETHANOL
68-11-1 THIOGLYCOLIC ACID	THIOGLYCOLIC ACID
789-61-7 BETA-THIOGUANIDINE DEOXYRIBOSIDE	BETA-THIOGUANIDINE DEOXYRIBOSIDE
154-42-7 THIOGUANINE	THIOGUANINE
79-42-5 2-THIOLACTIC ACID	2-THIOLACTIC ACID
79-42-5 THIOLACTIC ACID [HM1, HM1, NJL]	2-THIOLACTIC ACID
RR-00553-8 THIOLS (N-ALKANE MONOTHIOLS)	THIOLS (N-ALKANE MONOTHIOLS)
640-15-3 THIOMETON	THIOMETON
297-97-2 THIONAZIN [302, FLA, MSL, PAL]	O,O-DIETHYL O-PYRAZINYL PHOSPHOROTHIATE
7719-09-7 THIONYL CHLORIDE	THIONYL CHLORIDE
7783-42-8 THIONYL DIFLUORIDE	THIONYL DIFLUORIDE
97-77-8 THIOPEROXYDICARBONIC DIAMIDE ([(H2N)C (S)]2S2), TETRAETHYL- [PAL]	DISULFIRAM
137-26-8 THIOPEROXYDICARBONIC DIAMIDE ([(H2N)C (S)]2S2), TETRAMETHYL- [PAL]	THIRAM
137-26-8 THIOPEROXYDICARBONIC DIAMIDE, TETRAM-ETHYL- [TSC, TSC]	THIRAM
23564-06-9 THIOPHANATE ETHYL [[1,2-PHENYLENEBIS (IMINOCARBONOTHIOYL)]BISCARBAMIC ACID DIETHYL ESTER]	THIOPHANATE ETHYL [[1,2-PHENYLENEBIS(IMINOCAR-BONOTHIOYL)]BIS
23564-05-8 THIOPHANATE-METHYL	THIOPHANATE-METHYLE]
110-02-1 THIOPHENE	THIOPHENE
98-03-3 2-THIOPHENECARBOXALDEHYDE	2-THIOPHENECARBOXALDEHYDE

Chemical Name	Reference Name
7128-64-5 2,2'-(2,5-THIOPHENEDIYL)BIS(5-TERTIARY-BUTYLBENZOXAZOLE)	2,2'-(2,5-THIOPHENEDIYL)BIS(5-TERTIARYBUTYLBEN-ZOXAZOLE)
126-33-0 THIOPHENE, TETRAHYDRO-, 1,1-DIOX-IDE [PAL]	SULFOLANE
108-98-5 THIOPHENOL [302, OTS, T33]	BENZENETHIOL
463-71-8 THIOPHOSGENE	THIOPHOSGENE
3982-91-0 THIOPHOSPHORYL CHLORIDE	THIOPHOSPHORYL CHLORIDE
79-19-6 THIOSEMICARBAZIDE	THIOSEMICARBAZIDE
7783-18-8 THIOSULFURIC ACID (H2S2O3), DIAMMONIUM SALT [PAL]	AMMONIUM THIOSULFATE
52-24-4 THIO-TEPA [FLA]	TRIS(1-AZIRIDINYL)PHOSPHINE SULFIDE
52-24-4 THIOTEPA [IAR]	TRIS(1-AZIRIDINYL)PHOSPHINE SULFIDE
52-24-4 THIO-TEPA [MSL]	TRIS(1-AZIRIDINYL)PHOSPHINE SULFIDE
141-90-2 THIOURACIL	THIOURACILICROTOPHOS)
62-56-6 THIOUREA	THIOUREA
5344-82-1 THIOUREA, (2-CHLOROPHENYL)-	THIOUREA, (2-CHLOROPHENYL)-
614-78-8 THIOUREA, (2-METHYLPHENYL)-	THIOUREA, (2-METHYLPHENYL)-
86-88-4 THIOUREA, 1-NAPHTHALENYL- [PAL]	ANTU
15980-15-1 1,4-THIOXANE [FLA, MSL, NFP, NFP]	1,4-OXATHIANE
13718-26-8 THIRAM	THIRAM
137-26-8 THIRAM (ISO) [GBR, GBR]	THIRAM
137-26-8 THIURAM [RCU, RCU]	THIRAM
7440-29-1 THORIUM	THORIUM
15571-75-2 THORIUM 226	THORIUM 226
15623-47-9 THORIUM 227	THORIUM 227
14274-82-9 THORIUM 228	THORIUM 228
15594-54-4 THORIUM 229	THORIUM 229
14269-63-7 THORIUM 230	THORIUM 230
14932-40-2 THORIUM 231	THORIUM 231
15065-10-8 THORIUM 234	THORIUM 234
1314-20-1 THORIUM DIOXIDE	THORIUM DIOXIDE
13823-29-5 THORIUM NITRATE	THORIUM NITRATE
1314-20-1 THORIUM OXIDE (THO2) [PAL, PAL]	THORIUM DIOXIDE
15832-57-2 THULIUM 162	THULIUM 162
15690-75-2 THULIUM 166	THULIUM 166
13981-30-1 THULIUM 170	THULIUM 170
14333-45-0 THULIUM 171	THULIUM 171
15720-75-9 THULIUM 172	THULIUM 172
14041-46-4 THULIUM 173	THULIUM 173
14041-47-5 THULIUM 175	THULIUM 175
13537-18-3 THULIUM CHLORIDE	THULIUM CHLORIDE
65-71-4 THYMINE	THYMINE
89-83-8 THYMOL	THYMOL
80-59-1 TIGLIC ACID	TIGLIC ACID
7440-31-5 TIN	TIN
15700-33-1 TIN 110	TIN 110
15720-78-2 TIN 111	TIN 111
13966-06-8 TIN 113	TIN 113
RR-00313-4 TIN 117M	TIN 117M
RR-00311-2 TIN 119M	TIN 119M

Chemical Name	Reference Name
14683-06-8 TIN 121	TIN 121
RR-00310-1 TIN 121M	TIN 121M
14683-07-9 TIN 123	TIN 123
RR-00308-7 TIN 123M	TIN 123M
14683-08-0 TIN 125	TIN 125
15832-50-5 TIN 126	TIN 126
15690-89-8 TIN 127	TIN 127
16645-96-8 TIN 128	TIN 128
18282-10-5 TIN OXIDE [MEX, MEX]	STANNIC OXIDE
7772-99-8 TIN(II) CHLORIDE [WHM]	STANNOUS CHLORIDE
10101-75-4 TIN IV CHROMATE	TIN IV CHROMATE
38455-77-5 TIN II CHROMATE	TIN II CHROMATE
RR-00615-5 TIN COMPOUNDS [CAL]	TIN COMPOUNDS, N.O.S.
RR-00043-1 TIN COMPOUNDS, INORGANIC [GBR, GBR]	TIN INORGANIC COMPOUNDS
RR-00615-5 TIN COMPOUNDS, N.O.S.	TIN COMPOUNDS, N.O.S.
RR-00042-0 TIN COMPOUNDS, ORGANIC [GBR, GBR, GBR, MAK, MAK, MAK, MAK]	TIN ORGANIC COMPOUNDS
7440-31-5 TIN, ELEMENTAL [WHM]	TIN
7783-47-3 TIN(II) FLUORIDE [WHM]	STANNOUS FLUORIDE
RR-00043-1 TIN INORGANIC COMPOUNDS	TIN INORGANIC COMPOUNDS
RR-00043-1 TIN, INORGANIC COMPOUNDS [BC1, BC1, BC1]	TIN INORGANIC COMPOUNDS
RR-00043-1 TIN, OXIDE AND INORGANIC COMPOUNDS [QBC]	TIN INORGANIC COMPOUNDS
7440-31-5 TIN (INORGANIC COMPOUNDS EXCEPT OXIDES) [NIO, NIO, NIO]	TIN
7440-31-5 TIN METAL [ONT]	TIN
27344-41-8 TINOPAL CBS	TINOPAL CBS
RR-00042-0 TIN ORGANIC COMPOUNDS	TIN ORGANIC COMPOUNDS
RR-00042-0 TIN, ORGANIC COMPOUNDS [BC1, BC1, MIN]	TIN ORGANIC COMPOUNDS
RR-00042-0 TIN (ORGANIC COMPOUNDS) [NIO, NIO, NIO]	TIN ORGANIC COMPOUNDS
RR-00042-0 TIN, ORGANIC COMPOUNDS [ONT, ONT, QBC, QBC]	TIN ORGANIC COMPOUNDS
18282-10-5 TIN OXIDE [ALB, ALB, MIN]	STANNIC OXIDE
21651-19-4 TIN OXIDE	TIN OXIDE
21651-19-4 TIN OXIDE DUST [ALB]	TIN OXIDE
25324-56-5 TIN PHOSPHIDE	TIN PHOSPHIDE
7646-78-8 TIN TETRACHLORIDE [NJL, NJL]	STANNIC CHLORIDE
25973-55-1 TINUVIN 328 (BENZOTRIAZOLE UV ABSORBER)	TINUVIN 328 (BENZOTRIAZOLE UV ABSORBER)
52829-07-9 TINUVIN 770 (AMINE LIGHT STABILIZER)	TINUVIN 770 (AMINE LIGHT STABILIZER)-[(3-HYDROXYPROPYL)AMINO]-
12056-51-8 TITANATE [TI6O13(2-)], DIPOTASSIUM	TITANATE [TI6O13(2-)], DIPOTASSIUM
7440-32-6 TITANIUM	TITANIUM
15749-33-4 TITANIUM 44	TITANIUM 44
14392-00-8 TITANIUM 45	TITANIUM 45
7550-45-0 TITANIUM CHLORIDE [FLA]	TITANIUM TETRACHLORIDE
7550-45-0 TITANIUM CHLORIDE (TICL4) (T-4)- [PAL]	TITANIUM TETRACHLORIDE
13463-67-7 TITANIUM DIOXIDE	TITANIUM DIOXIDE
11140-68-4 TITANIUM HYDRIDE	TITANIUM HYDRIDE
12137-20-1 TITANIUM OXIDE (TIO)	TITANIUM OXIDE (TIO)
13463-67-7 TITANIUM OXIDE (TIO2) [PAL]	TITANIUM DIOXIDE

Chemical Name	Reference Name
13693-11-3 TITANIUM SULFATE [HM1, HM1, PST]	TITANIUM SULFATE (TI(SIO4)2)
13825-74-6 TITANIUM SULFATE	TITANIUM SULFATE
13693-11-3 TITANIUM SULFATE (TI(SIO4)2)	TITANIUM SULFATE (TI(SIO4)2)
7550-45-0 TITANIUM TETRACHLORIDE	TITANIUM TETRACHLORIDE
7550-45-0 TITANIUM TETRACHLORIDE [TITANIUM CHLORIDE (TICL4) (T-4)-] [EP3]	TITANIUM TETRACHLORIDE
7705-07-9 TITANIUM TRICHLORIDE	TITANIUM TRICHLORIDE
1271-19-8 TITANOCENE DICHLORIDE	TITANOCENE DICHLORIDE
8037-19-2 TOBACCO LEAF, ABSOLUTE [MSL]	OILS, TOBACCO
84961-66-0 TOBACCO DUST	TOBACCO DUST
RR-00031-7 TOBACCO PRODUCTS, SMOKELESS	TOBACCO PRODUCTS, SMOKELESS
RR-00030-6 TOBACCO SMOKE	TOBACCO SMOKE
RR-00031-7 TOBACCO, SMOKELESS PRODUCTS [C65]	TOBACCO PRODUCTS, SMOKELESS
RR-00030-6 TOBACCO SMOKE (PRIMARY) [C65, C65, C65, C65]	TOBACCO SMOKE
49842-07-1 TOBRAMYCIN SULFATE	TOBRAMYCIN SULFATE
58-95-7 D-ALPHA-TOCOPHERYL ACETATE [CSR]	VITAMIN E ACETATE
RR-01508-7 TOE PUFFS, NITROCELLULOSE BASE	TOE PUFFS, NITROCELLULOSE BASE
1156-19-0 TOLAZAMIDE	TOLAZAMIDE
64-77-7 TOLBUTAMIDE	TOLBUTAMIDE
119-93-7 O-TOLIDINE [ALB, AUS, AUS, AUS, MAK, MIN, NHS, NHS, NHS, QBC, QBC, QBC, TLV, TLV, WHM]	3,3'-DIMETHYLBENZIDINE
RR-00520-9 O-TOLIDINE-BASED DYES	O-TOLIDINE-BASED DYES
108-88-3 TOLUENE	TOLUENE
108-88-3 TOLUENE [NFP, NFP]	TOLUENE
108-88-3 TOLUENE (TOLUOL) [QBC, QBC, BC1, BC1, BC1, MAK, MAK, MAK, WHM, ALB, ALB, ALB]	TOLUENE
95-80-7 TOLUENE 2,4-DIAMINE	TOLUENE 2,4-DIAMINE
95-80-7 TOLUENE 2,4 DIAMINE [HON, HON]	TOLUENE 2,4-DIAMINE
91-08-7 TOLUENE 2,6-DIISOCYANATE [BENZENE, 1,3-DIISOCYANATO-2-METHYL-] [EP3]	TOLUENE-2,6-DIISOCYANATE
584-84-9 TOLUENE 2,4-DIISOCYANATE [BENZENE, 2,4-DIISOCYANATO-1-METHYL-] [EP3]	TOLUENE 2,4-DIISOCYANATE
95-80-7 2,4-TOLUENE DIAMINE [CAA]	TOLUENE 2,4-DIAMINE
95-80-7 2,4-TOLUENEDIAMINE [EPA]	TOLUENE 2,4-DIAMINE
95-80-7 TOLUENE-2,4-DIAMINE [MAK, RCA]	TOLUENE 2,4-DIAMINE
95-80-7 TOLUENE 2-4-DIAMINE [WEL, WEL]	TOLUENE 2,4-DIAMINE
95-80-7 TOLUENE-2,4-DIAMINE [WHM]	TOLUENE 2,4-DIAMINE
108-71-4 TOLUENE-3,5-DIAMINE	TOLUENE-3,5-DIAMINE
496-72-0 3,4-TOLUENEDIAMINE [EPA]	DIAMINOTOLUENE
496-72-0 TOLUENE-3,4-DIAMINE [WHM]	DIAMINOTOLUENE
823-40-5 2,6-TOLUENEDIAMINE [EPA]	2,6-DIAMINOTOLUENE
25376-45-8 TOLUENEDIAMINE [RCU, RCU]	DIAMINOTOLUENE (MIXED ISOMERS)
25376-45-8 TOLUENE-AR,AR'-DIAMINE [WHM]	DIAMINOTOLUENE (MIXED ISOMERS)
15481-70-6 2,6-TOLUENEDIAMINE DIHYDROCHLORIDE	2,6-TOLUENEDIAMINE DIHYDROCHLORIDE
6369-59-1 2,5-TOLUENEDIAMINE SULFATE	2,5-TOLUENEDIAMINE SULFATESULFOPHENYL)AZO] PHENYL]AZO][1,1'-BIPHENYL]-4-YL]AZO]-2- HYDROXY-3-METHYL-, DISODIUM SALT
91-08-7 TOLUENE-2,6-DIISOCYANATE	TOLUENE-2,6-DIISOCYANATE

Chemical Name	Reference Name
91-08-7 TOLUENE 2,6-DIISOCYANATE [302]	TOLUENE-2,6-DIISOCYANATE
584-84-9 TOLUENE 2,4-DIISOCYANATE	TOLUENE 2,4-DIISOCYANATE
584-84-9 2,4-TOLUENE DIISOCYANATE [CAA]	TOLUENE 2,4-DIISOCYANATE
584-84-9 TOLUENE DIISOCYANATE [MIN, NTP]	TOLUENE 2,4-DIISOCYANATE
26471-62-5 TOLUENE DIISOCYANATE	TOLUENE DIISOCYANATE
26471-62-5 2,4- & 2,6-TOLUENE DIISOCYANATE [CSR, CSR]	TOLUENE DIISOCYANATE
26471-62-5 TOLUENE DIISOCYANATE (TDI) [ONT, ONT, ONT]	TOLUENE DIISOCYANATE
26471-62-5 TOLUENE-2,4-DIISOCYANATE (TDI) [QBC, QBC, BC1]	TOLUENE DIISOCYANATE
584-84-9 TOLUENE 2,4-DIISOCYANATE(TDI) [FLA]	TOLUENE 2,4-DIISOCYANATE
584-84-9 TOLUENE 2,4-DIISOCYANATE (TDI) [MAK, MAK, MAK]	TOLUENE 2,4-DIISOCYANATE
584-84-9 TOLUENE-2,4-DIISOCYANATE (TDI) [MSL, OS1, OS1, OS1, TLV, TLV]	TOLUENE 2,4-DIISOCYANATE
584-84-9 TOLUENE 2,4-DIISOCYANATE (TDI) [WHM]	TOLUENE 2,4-DIISOCYANATE
26471-62-5 TOLUENE DIISOCYANATE (TDI) [NHS, NHS]	TOLUENE DIISOCYANATE
26471-62-5 TOLUENEDIISOCYANATE (MIXED ISOMERS) [313, NJL, CAN]	TOLUENE DIISOCYANATE
26471-62-5 TOLUENE DIISOCYANATE (MIXED ISOMERS) [EPA]	TOLUENE DIISOCYANATE
26471-62-5 TOLUENE DIISOCYANATES [IAR]	TOLUENE DIISOCYANATE
26471-62-5 TOLUENE DIISOCYANATE [BENZENE, 1,3-DIISOCYANATOMETHYL-] [EP3]	TOLUENE DIISOCYANATE
602-01-7 TOLUENE, 2,3-DINITRO-	TOLUENE, 2,3-DINITRO-
619-15-8 TOLUENE, 2,5-DINITRO-	TOLUENE, 2,5-DINITRO-
70-55-3 P-TOLUENESULFONAMIDE [PST]	BENZENESULFONAMIDE, 4-METHYL-
88-19-7 O-TOLUENESULFONAMIDE	O-TOLUENESULFONAMIDE
RR-00969-8 TOLUENE SULFONAMIDE BISPHENOL A EPOXY ADDUCT	TOLUENE SULFONAMIDE BISPHENOL A EPOXY ADDUCT
104-15-4 P-TOLUENESULFONIC ACID	P-TOLUENESULFONIC ACID
25231-46-3 TOLUENE SULFONIC ACID	TOLUENE SULFONIC ACID
98-59-9 P-TOLUENE SULFONYL CHLORIDE	P-TOLUENE SULFONYL CHLORIDE
98-59-9 PARA-TOLUENESULFONYL CHLORIDE [WEL, WEL]	P-TOLUENE SULFONYL CHLORIDE
98-59-9 TOLUENESULFONYL CHLORIDE [HON]	P-TOLUENE SULFONYL CHLORIDE
25231-46-3 TOLUENESULPHONIC ACID [WHM]	TOLUENE SULFONIC ACID
98-59-9 P-TOLUENESULPHONYL CHLORIDE [GBR]	P-TOLUENE SULFONYL CHLORIDE
95-71-6 TOLUHYDROQUINONE	TOLUHYDROQUINONE
99-94-5 P-TOLUIC ACID	P-TOLUIC ACID
95-53-4 O-TOLUIDINE	O-TOLUIDINE
95-53-4 ORTHO-TOLUIDINE [C65, C65, IAR]	O-TOLUIDINE
95-53-4 TOLUIDINE-O [QBC, QBC, QBC]	O-TOLUIDINE
106-49-0 P-TOLUIDINE	P-TOLUIDINE
106-49-0 PARA-TOLUIDINE [C65]	P-TOLUIDINE
108-44-1 M-TOLUIDINE	M-TOLUIDINE
26915-12-8 TOLUIDINE (ALL ISOMERS) [ALB, ALB, ALB, ALB]	TOLUIDINES
26915-12-8 TOLUIDINE (SUM OF O-, M-, AND P- ISOMERS) [ONT, ONT, ONT]	TOLUIDINES
636-21-5 O-TOLUIDINE HYDROCHLORIDE	O-TOLUIDINE HYDROCHLORIDE

Chemical Name	Reference Name
636-21-5 ORTHO-TOLUIDINE HYDROCHLORIDE [CAL, FLA, C65, C65]	O-TOLUIDINE HYDROCHLORIDE
26915-12-8 TOLUIDINES	TOLUIDINES
136-85-6 TOLUTRIAZOLE	TOLUTRIAZOLE
95-80-7 2,4-TOLUYLENEDIAMINE [HM1, HM1, HM1]	TOLUENE 2,4-DIAMINE
2646-17-5 1-(O-TOLYAZO)-2-NAPHTOL [FLA]	OIL ORANGE SS
91-99-6 M-TOLYDIETHANOLAMINE	M-TOLYDIETHANOLAMINE
91-99-6 META-TOLYDIETHANOLAMINE [NFP, NFP]	M-TOLYDIETHANOLAMINE
140-39-6 P-TOLYL ACETATE	P-TOLYL ACETATE
1334-78-7 TOLYL ALDEHYDE	TOLYL ALDEHYDE
93-69-6 O-TOLYL BIGUANIDE	O-TOLYL BIGUANIDE
91-99-6 2-2'-(M-TOLYLIMIDO) DIETHANOL [FLA]	M-TOLYDIETHANOLAMINE
RR-00875-3 O-TOLYL P-TOLUENE SULFONATE	O-TOLYL P-TOLUENE SULFONATE
29385-43-1 TOLYL TRIAZOLE	TOLYL TRIAZOLE
RR-00759-0 TOTAL TRIHALOMETHANES	TOTAL TRIHALOMETHANES
1330-20-7 TOTAL XYLENES [ATS]	XYLENES (O-, M-, P- ISOMERS)
8001-35-2 TOXAPHENE	TOXAPHENE
8001-35-2 TOXAPHENE (CHLORINATED CAMPHENE) [CAA]	TOXAPHENE
8001-35-2 TOXAPHENE (POLYCHLORINATED CAM-PHENES) [C65, C65, IAR]	TOXAPHENE
RR-01703-8 TOXINS DERIVED FROM FUSARIUM GRAMIN-EARUM, F. CULMORUM AND F. CROOK-WELLENSE	TOXINS DERIVED FROM FUSARIUM GRAMINEARUM, F. CULMORUM AND F.
RR-01704-9 TOXINS DERIVED FROM FUSARIUM SPOROTRICHOIDES	TOXINS DERIVED FROM FUSARIUM SPOROTRICHOIDES CROOKWELLENSE
RR-01706-1 TOXINS DERIVED FROM FUSARIUM MONILI-FORME	TOXINS DERIVED FROM FUSARIUM MONILIFORME
93-72-1 2,4,5-TP ACID [CWA]	SILVEX (2,4,5-TP)
93-72-1 2,4,5-TP ACID (SILVEX) [ATS]	SILVEX (2,4,5-TP)
32534-95-5 2,4,5-TP ACID ESTERS	2,4,5-TP ACID ESTERS
93-72-1 2,4,5-TP ACID; PROPANOIC ACID, 2-(2,4,5-TRICHLOROPHENOXY)- [CAL]	SILVEX (2,4,5-TP)
32534-95-5 2,4,5-TP ESTER; PROPANOIC ACID, 2-(2,4, 5-TRICHLOROPHENOXY)-,ISOOCTYL ES-TER [CAL]	2,4,5-TP ACID ESTERS
32534-95-5 2,4,5-TP ESTERS [CWA, NJL]	2,4,5-TP ACID ESTERS
93-72-1 2,4,5-TP(SILVEX) [RC2, RC2]	SILVEX (2,4,5-TP)
93-72-1 2,4,5-TP (SILVEX) [SDW, SDW]	SILVEX (2,4,5-TP)
66841-25-6 TRALOMETHRIN	TRALOMETHRIN
RR-00778-3 TRANSFORMER OIL	TRANSFORMER OIL
14567-73-8 TREMOLITE	TREMOLITEYL-CCNU)
77536-68-6 TREMOLITE [PAL]	ASBESTOS, TREMOLITE
77536-68-6 TREMOLITE ASBESTOS [MSL]	ASBESTOS, TREMOLITE
299-75-2 TREOSULFAN [C65, CAL]	TREOSULPHAN
299-75-2 TREOSULPHAN	TREOSULPHAN
23523-14-0 TRIACONTAMETHYLCYCLOPENTADECASILOXANE	TRIACONTAMETHYLCYCLOPENTADECASILOXANE
2471-10-5 TRIACONTAMETHYLTETRADECASILOXANE	TRIACONTAMETHYLTETRADECASILOXANE
43121-43-3 TRIADIMEFON [1-(4-CHLOROPHENOXY)-3, 3-DIMETHYL-1-(1H-1,2,4-TRIAZOL-1-YL)-2-BUTANONE]	TRIADIMEFON [1-(4-CHLOROPHENOXY)-3,3-DIMETHYL-1-(1H-1,2,4-TR
541-05-9 TRIALKOXY SILANE	TRIALKOXY SILANE

Chemical Name	Reference Name
2303-17-5 TRIALLATE	TRIALLATEHLORO-2-PROPENYL)ESTER]
102-70-5 TRIALLYLAMINE	TRIALLYLAMINE
1693-71-6 TRIALLYL BORATE	TRIALLYL BORATE
1031-47-6 TRIAMIPHOS	TRIAMIPHOS
396-01-0 TRIAMTERENE	TRIAMTERENE
621-77-2 TRIAMYLAMINE [NFP, NFP]	1-PENTANAMINE, N,N-DIPENTYL-
RR-00886-6 TRIAMYLBENZENE	TRIAMYLBENZENE
621-78-3 TRI-N-AMYL BORATE	TRI-N-AMYL BORATE
621-78-3 TRIAMYL BORATE [NFP, NFP]	TRI-N-AMYL BORATE
RR-01315-0 TRIARYL PHOSPHATES, ISOPROPYLATED	TRIARYL PHOSPHATES, ISOPROPYLATED
RR-01316-1 TRIARYL PHOSPHATES, N.O.S.	TRIARYL PHOSPHATES, N.O.S.
4080-31-3 3,5,7-TRIAZA-1-AZONIATRICYCLODECANE-1-(3-CHLORO-2-PROPENYL)-, CHLORIDE	3,5,7-TRIAZA-1-AZONIATRICYCLODECANE-1-(3-CHLORO-2-PROPENYL)-
RR-01564-5 1,3,5-TRIAZIN-2-AMINE, 4-DIMETHYLAMINO-6-SUBSTITUTED-	1,3,5-TRIAZIN-2-AMINE, 4-DIMETHYLAMINO-6-SUBSTITUTED-
1912-24-9 1,3,5-TRIAZINE-2,4-DIAMINE, 6-CHLORO-N-ETHYL-N'-(1-METHYLETHYL)- [PAL]	ATRAZINE
121-82-4 1,3,5-TRIAZINE, HEXAHYDRO-1,3,5-TRINITRO- [PAL]	CYCLONITE
1014-69-3 S-TRIAZINE, 2-(ISOPROPYLAMINO)-4-(METHYLAMINO)-6-(METHYLTHIO)	S-TRIAZINE, 2-(ISOPROPYLAMINO)-4-(METHYLAMINO)-6-(METHYLTHIO
RR-00506-1 TRIAZINE PESTICIDES, N.O.S.	TRIAZINE PESTICIDES, N.O.S.
29305-12-2 1,3,5-TRIAZINE-2,4,6-TRIAMINE, HYDROBROMIDE	1,3,5-TRIAZINE-2,4,6-TRIAMINE, HYDROBROMIDE
13057-78-8 1,3,5-TRIAZINE-2,4,6-(1H,3H,5H)-TRIONE, 1-CHLORO [PAL]	CHLOROISOCYANURIC ACID
2782-57-2 1,3,5-TRIAZINE-2,4,6-(1H,3H,5H)-TRIONE, 1,3-DICHLORO- [PAL]	DICHLOROISOCYANURIC ACID
2244-21-5 1,3,5-TRIAZINE-2,4,6-(1H,3H,5H)-TRIONE, 1,3-DICHLORO-, POTASSIUM SALT [PAL]	POTASSIUM DICHLOROISOCYANURATE
51580-86-0 1,3,5-TRIAZINE-2,4,6-(1H,3H,5H)-TRIONE, 1,3-DICHLORO-, SODIUM SALT, DIHYDRATE [PAL, PAL]	SODIUM DICHLOROISOCYANURATE DIHYDRATE
87-90-1 1,3,5-TRIAZINE-2,4,6-(1H,3H,5H)-TRIONE, 1,3,5-TRICHLORO- [PAL]	TRICHLORO-S-TRIAZINETRIONE
26603-40-7 1,3,5-TRIAZINE-2,4,6(1H,3H,5H-TRIONE, 1,3,5-TRIS(ISOCYANATOMETHYLPHENYL)-	1,3,5-TRIAZINE-2,4,6(1H,3H,5H-TRIONE, 1,3,5-TRIS(ISOCYANATOM
15875-13-5 1,3,5-TRIAZINE-1,3,5(2H,4H,6H)-TRIPROPANAMINE-N,N,N',N'',N''-HEXAMTHYL-	1,3,5-TRIAZINE-1,3,5(2H,4H,6H)-TRIPROPANAMINE-N,N,N',N',N'',
21087-64-9 1,2,4-TRIAZIN-5(4H)-ONE, 4-AMINO-6-(1,1-DIMETHYLETHYL)-3-(METHYLTHIO)- [PAL]	METRIBUZIN
68-76-8 TRIAZIQUONE [2,5-CYCLOHEXADIENE-1,4-DIONE,2,3,5-TRIS(1-AZIRIDINYL)-] [313]	TRIS(AZIRIDINYL)-P-BENZOQUINONE
24017-47-8 TRIAZOFOS	TRIAZOFOS
28911-01-5 TRIAZOLAM	TRIAZOLAM
61-82-5 1H-1,2,4-TRIAZOL-3-AMINE [PAL, PAL]	AMITROLE
103112-35-2 1H-1,2,4-TRIAZOLE-3-CARBOXYLIC ACID, 1-(2,4-DICHLORPHENYL)-5-(TRICHLOROMETHYL)-ETHYL ESTER	1H-1,2,4-TRIAZOLE-3-CARBOXYLIC ACID, 1-(2,4-DICHLORPHENYL)-5
10101-89-0 TRIBASIC SODIUM PHOSPHATE DODECAHYDRATE	TRIBASIC SODIUM PHOSPHATE DODECAHYDRATE

Chemical Name	Reference Name
101200-48-0　TRIBENURON METHYL [2-(((((4-METHOXY-6-METHYL-1,3,5-TRIAZIN-2-YL)-METHYLAMINO)CARBONYL)AMINO)SULFONYL)-, METHYL ESTER]	TRIBENURON METHYL [2-(((((4-METHOXY-6-METHYL-1,3,5-TRIAZIN-2-YL) -CARBONYL]-AMINO]SULFONYL] BENZOATE]
87-10-5　TRIBORMOSALICYLANILIDE, 3,4',5- [OTS]	3,4',5-TRIBROMOSALICYLANILIDE
57137-10-7　TRIBROMINATED POLYSTYRENE	TRIBROMINATED POLYSTYRENETETRAKIS-, AMMONIUM SALT
626-39-1　1,3,5-TRIBROMOBENZENE	1,3,5-TRIBROMOBENZENE
75-25-2　TRIBROMOMETHANE (BROMOFORM) [UTS]	BROMOFORM
75-25-2　TRIBROMOMETHANE [CSR, CSR, HM1, HM1, HM1, HM1, MAK]	BROMOFORM
118-79-6　2,4,6-TRIBROMOPHENOL [OTS, TSE]	PHENOL, 2,4,6-TRIBROMO-
87-10-5　3,4',5-TRIBROMOSALICYLANILIDE	3,4',5-TRIBROMOSALICYLANILIDE
1116-70-7　TRIBUTYL ALUMINUM	TRIBUTYL ALUMINUM
1116-70-7　TRIBUTYLALUMINUM [FLA, MSL]	TRIBUTYL ALUMINUM
102-82-9　TRIBUTYLAMINE	TRIBUTYLAMINE
688-74-4　TRIBUTYL BORATE	TRIBUTYL BORATE
688-74-4　TRI-N-BUTYL BORATE [FLA, MSL, NFP, NFP]	TRIBUTYL BORATE
77-94-1　TRIBUTYL CITRATE	TRIBUTYL CITRATE
109909-39-9　2,4,6-TRI-SEC-BUTYLPHENOL, ETHOXYLATED, SULFATED, SODIUM SALT	2,4,6-TRI-SEC-BUTYLPHENOL, ETHOXYLATED, SULFATED, SODIUM SAL
126-73-8　TRIBUTYL PHOSPHATE	TRIBUTYL PHOSPHATE
126-73-8　TRIBUTYL PHOSPHATE (TBP) [WHM]	TRIBUTYL PHOSPHATE
998-40-3　TRIBUTYLPHOSPHINE	TRIBUTYLPHOSPHINE
102-85-2　TRIBUTYL PHOSPHITE [FLA, MSL, NFP, NFP]	PHOSPHOROUS ACID, TRIBUTYL ESTER
78-48-8　S,S,S-TRIBUTYL PHOSPHOROTRITHIOATE [CAL]	DEF(DIOXATHION)
150-50-5　S,S,S-TRIBUTYL PHOSPHOROTRITHIOITE	S,S,S-TRIBUTYL PHOSPHOROTRITHIOITE
688-73-3　TRIBUTYLTIN [CAL]	STANNANE, TRI-N-BUTYL-, HYDRIDE
56-36-0　TRIBUTYLTIN ACETATE	TRIBUTYLTIN ACETATE
4342-36-3　TRIBUTYLTIN BENZOATE	TRIBUTYLTIN BENZOATE
1461-22-9　TRIBUTYLTIN CHLORIDE	TRIBUTYLTIN CHLORIDE
RR-01198-3　TRIBUTYLTIN COMPOUNDS	TRIBUTYLTIN COMPOUNDS
1983-10-4　TRIBUTYLTIN FLUORIDE [313, MAK, MAK, MAK]	STANNANE, FLUOROTRIBUTYL-
24124-25-2　TRIBUTYLTIN LINOLEATE	TRIBUTYLTIN LINOLEATE
2155-70-6　TRIBUTYL TIN METHACRYLATE [MAK, MAK, MAK]	TRIBUTYLTIN METHACRYLATE
2155-70-6　TRIBUTYLTIN METHACRYLATE	TRIBUTYLTIN METHACRYLATE
85409-17-2　TRIBUTYLTIN NAPHTHENATE	TRIBUTYLTIN NAPHTHENATE
85409-17-2　TRIBUTYL TIN NAPHTHENATE [MAK, MAK, MAK]	TRIBUTYLTIN NAPHTHENATE
4342-30-7　TRI-N-BUTYLTIN SALICYLATE	TRI-N-BUTYLTIN SALICYLATE
78-48-8　S,S,S-TRIBUTYLTRITHIOPHOSPHATE (DEF) [313]	DEF7-METHANO-1H-INDENE]
78-51-3　TRIBUTYOXYETHYL PHOSPHATE	TRIBUTYOXYETHYL PHOSPHATE
72749-59-8　TRI-C6-12-ALKYL METHYL AMMONIUM CHLORIDE	TRI-C6-12-ALKYL METHYL AMMONIUM CHLORIDEAMETHYLBUTYL)PHENOXY]-1-ANTHRACENYL]AMINO]-2,4,6-TRIMETHYL BENZENESULFONIC ACID DISODIUM SALT
7758-87-4　TRICALCIUM PHOSPHATE [PST]	CALCIUM PHOSPHATE
538-23-8　TRICAPRYLIN	TRICAPRYLIN
5137-55-3　TRICAPRYLYL METHYL AMMONIUM CHLORIDE [PST]	METHYLTRICAPRYLAMMONIUM CHLORIDE

Chemical Name	Reference Name
12079-65-1 TRICARBONYL(ETA-CYCLOPENTADIENYL) MANGANESE [GBR, GBR, GBR]	MANGANESE CYCLOPENTADIENYL TRICARBONYL
12108-13-3 TRICARBONYL(METHYLCYCLOPENTADIENYL) MANGANESE [GBR, GBR, GBR]	METHYLCYCLOPENTADIENYL MANGANESE TRICAR-BONYL
52-68-6 TRICHLORFON	TRICHLORFON
52-68-6 TRICHLORFON (PHOSPHORIC ACID, (2,2,2-TRICHLORO-1-HYDROXYETHYL)-DIMETHYL ESTER] [313]	TRICHLORFON
555-77-1 TRICHLORMETHINE (TRIMUSTINE HY-DROCHLORIDE) [IAR]	TRIS(2-CHLOROETHYL)AMINE
817-09-4 TRICHLORMETHINE (TRIMUSTINE HCL)	TRICHLORMETHINE (TRIMUSTINE HCL)
75-87-6 TRICHLOROACETALDEHYDE [MSL]	CHLORAL
594-65-0 TRICHLOROACETAMIDE	TRICHLOROACETAMIDE
76-03-9 TRICHLOROACETIC ACID	TRICHLOROACETIC ACID
545-06-2 TRICHLOROACETONITRILE	TRICHLOROACETONITRILE
76-02-8 TRICHLOROACETYL CHLORIDE	TRICHLOROACETYL CHLORIDE
87-61-6 1,2,3-TRICHLOROBENZENE	1,2,3-TRICHLOROBENZENE
108-70-3 1,3,5-TRICHLOROBENZENE	1,3,5-TRICHLOROBENZENE
120-82-1 1,2,4-TRICHLOROBENZENE	1,2,4-TRICHLOROBENZENE
120-82-1 TRICHLOROBENZENE, 1,2,4- [OTS]	1,2,4-TRICHLOROBENZENE
120-82-1 TRICHLOROBENZENE 1,2,4- [QBC]	1,2,4-TRICHLOROBENZENE
12002-48-1 TRICHLOROBENZENE	TRICHLOROBENZENE
12002-48-1 TRICHLOROBENZENES [CN4, HM1, HM1, HM1, NJL, RCA]	TRICHLOROBENZENE
50-29-3 1,1,1-TRICHLOROBIS(CHLOROPHENYL) ETHANE [GBR, GBR]	DDT
50-29-3 1,1,1-TRICHLORO-2,2-BIS(P-CHLORPHENYL) ETHANE [ONT]	DDT
2431-50-7 2,3,4-TRICHLOROBUTENE-1 [MAK]	1-BUTENE, 2,3,4-TRICHLORO-
51023-22-4 TRICHLOROBUTENE	TRICHLOROBUTENE
RR-01317-2 TRICHLOROBUTYLENE	TRICHLOROBUTYLENE
3083-25-8 TRICHLOROBUTYLENE OXIDE	TRICHLOROBUTYLENE OXIDE
1558-25-4 TRICHLORO(CHLOROMETHYL)SILANE	TRICHLORO(CHLOROMETHYL)SILANE
1558-25-4 TRICHLORO(CHLOROMETHYL) SILANE [OS3]	TRICHLORO(CHLOROMETHYL)SILANE
27137-85-5 TRICHLORO(DICHLOROPHENYL)SILANE	TRICHLORO(DICHLOROPHENYL)SILANE
27137-85-5 TRICHLORO(DICHLOROPHENYL) SILANE [OS3]	TRICHLORO(DICHLOROPHENYL)SILANE
354-21-2 1,2,2-TRICHLORO-1,1-DIFLUOROETHANE	1,2,2-TRICHLORO-1,1-DIFLUOROETHANE
134237-42-6 TRICHLORODIFLUOROPROPANE	TRICHLORODIFLUOROPROPANE
57321-63-8 TRICHLORO DIPHENYL OXIDE	TRICHLORO DIPHENYL OXIDE
71-55-6 1,1,1-TRICHLOROETHANE	1,1,1-TRICHLOROETHANE
71-55-6 1,1,1-TRICHLOROETHANE (METHYLCHLORO-FORM) [HON, HON]	1,1,1-TRICHLOROETHANE
71-55-6 TRICHLOROETHANE, 1,1,1- [OTS]	1,1,1-TRICHLOROETHANE
79-00-5 1,1,2-TRICHLOROETHANE	1,1,2-TRICHLOROETHANE
79-00-5 TRICHLOROETHANE 1,1,2- [QBC, QBC]	1,1,2-TRICHLOROETHANE
25323-89-1 TRICHLOROETHANE	TRICHLOROETHANE
71-55-6 1,1,1-TRICHLOROETHANE (METHYL CHLORO-FORM) [313, WHM]	1,1,1-TRICHLOROETHANE
79-00-5 TRICHLOROETHANE (1,1,2-) (VINYL TRICHLO-RIDE) [HON, HON]	1,1,2-TRICHLOROETHANE

Chemical Name	Reference Name
71-55-6 1,1,1-TRICHLOROETHANE (METHYL CHLORO-FORM) [ALB, ALB, CN5]	1,1,1-TRICHLOROETHANE
79-01-6 TRICHLOROETHYLENE	TRICHLOROETHYLENE
115-21-9 TRICHLOROETHYLSILANE	TRICHLOROETHYLSILANE
134237-34-6 TRICHLOROFLUOROETHANE	TRICHLOROFLUOROETHANE
75-69-4 TRICHLOROFLUOROMETHANE	TRICHLOROFLUOROMETHANE
75-69-4 TRICHLOROFLUOROMETHANE (CFC-11) [313]	TRICHLOROFLUOROMETHANE
75-69-4 TRICHLOROFLUOROMETHANE (FC 11) [FLA, MIN, MSL, WHM]	TRICHLOROFLUOROMETHANE
75-69-4 TRICHLOROFLUOROMETHANE (FLUO-ROTRICHLOROMETHANE) [ALB, ALB]	TRICHLOROFLUOROMETHANE
134190-51-5 TRICHLOROFLUOROPROPANE	TRICHLOROFLUOROPROPANE
3380-34-5 2,4,4'-TRICHLORO-2-HYDROXY DIPHENYL ETHER [PST]	5-CHLORO-2-(2,4-DICHLOROPHENOXY)PHENOL
87-90-1 TRICHLOROISOCYANURIC ACID [HM1, HM1, NJL, WHM]	TRICHLORO-S-TRIAZINETRIONE
67-66-3 TRICHLOROMETHANE [ONT, ONT]	CHLOROFORM
594-42-3 TRICHLOROMETHANE SULPHURYL CHLORIDE	TRICHLOROMETHANE SULPHURYL CHLORIDE
594-42-3 TRICHLOROMETHANESULFENYL CHLO-RIDE [EPA]	TRICHLOROMETHANE SULPHURYL CHLORIDE
594-42-3 TRICHLORO METHANE SULFENYL CHLO-RIDE [NJL]	TRICHLOROMETHANE SULPHURYL CHLORIDE
594-42-3 TRICHLOROMETHANESULFENYL CHLO-RIDE [ONT]	TRICHLOROMETHANE SULPHURYL CHLORIDE
75-70-7 TRICHLOROMETHANETHIOL	TRICHLOROMETHANETHIOL
67632-66-0 TRICHLOROMETHYL PERCHLORATE	TRICHLOROMETHYL PERCHLORATE
75-69-4 TRICHLOROMONOFLUOROMETHANE [EPA, RCU, RCU]	TRICHLOROFLUOROMETHANE
1321-65-9 TRICHLORONAPHTALENE [QBC, QBC]	TRICHLORONAPHTHALENE
1321-65-9 TRICHLORONAPHTHALENE	TRICHLORONAPHTHALENE
327-98-0 TRICHLORONAT [HM1]	TRICHLORONATE
327-98-0 TRICHLORONATE	TRICHLORONATE
89-69-0 2,4,5-TRICHLORONITROBENZENE	2,4,5-TRICHLORONITROBENZENE
76-06-2 TRICHLORONITROMETHANE [GBR, GBR, ONT, ONT, PST]	CHLOROPICRIN
4259-43-2 1,1,1-TRICHLORO-2,2,3,3,3-PENTAFLUORO-PROPANE	1,1,1-TRICHLORO-2,2,3,3,3-PENTAFLUOROPROPANE
88-06-2 2,4,6-TRICHLOROPHENOL	2,4,6-TRICHLOROPHENOL
88-06-2 TRICHLOROPHENOL, 2,4,6- [CD1, CD1]	2,4,6-TRICHLOROPHENOL
95-95-4 2,4,5-TRICHLOROPHENOL	2,4,5-TRICHLOROPHENOL
609-19-8 3,4,5-TRICHLOROPHENOL	3,4,5-TRICHLOROPHENOL
933-75-5 2,3,6-TRICHLOROPHENOL	2,3,6-TRICHLOROPHENOL
933-78-8 2,3,5-TRICHLOROPHENOL	2,3,5-TRICHLOROPHENOL
15950-66-0 2,3,4-TRICHLOROPHENOL	2,3,4-TRICHLOROPHENOL
25167-82-2 TRICHLOROPHENOL	TRICHLOROPHENOL
25167-82-2 TRICHLOROPHENOL (MIXED) [EP2]	TRICHLOROPHENOL
25167-82-2 TRICHLOROPHENOLS [RCA]	TRICHLOROPHENOL
93-76-5 2,4,5-TRICHLOROPHENOXY ACETIC ACID [FLA]	2,4,5-T
93-76-5 TRICHLOROPHENOXYACETIC ACID, 2,4,5- (2,4, 5-T) [CD1, CD1]	2,4,5-T

Chemical Name	Reference Name
93-76-5 2,4,5-TRICHLOROPHENOXYACETIC ACID [MAK, MAK, MAK]	2,4,5-T
93-76-5 2,4,5-TRICHLOROPHENOXY ACETIC ACID [MSL]	2,4,5-T
93-76-5 2,4,5-(TRICHLOROPHENOXY) ACETIC ACID [NJL, NJL, ONT]	2,4,5-T
93-78-7 2,4,5-(TRICHLOROPHENOXY) ACETIC ACID-ISOPROPYL ESTER	2,4,5-(TRICHLOROPHENOXY) ACETIC ACID-ISOPROPYL ESTER
RR-01622-8 (2,4,5-TRICHLOROPHENOXY)-ACETIC ACID ESTERS	(2,4,5-TRICHLOROPHENOXY)-ACETIC ACID ESTERS
136-25-4 2-(2,4,5-TRICHLOROPHENOXY) ETHYL 2,2-DICHLOPROPIONATE (ERBON)	2-(2,4,5-TRICHLOROPHENOXY) ETHYL 2,2-DICHLOPROPIONATE (ERBON
73826-29-6 TRICHLOROPHENOXYPROPIONIC ACID ESTER [NJL, NJL]	PROPIONIC ACID, 2-(2,4,5-TRICHLOROPHENOXY)-, P-CHLOROPHENACY
93-72-1 TRICHLOROPHENOXYPROPIONIC ACID [NJL]	SILVEX (2,4,5-TP)
6047-17-2 TRICHLOROPHENOXYPROPIONIC ACID ESTER (3-(2,4,5))	TRICHLOROPHENOXYPROPIONIC ACID ESTER (3-(2,4,5))
98-13-5 TRICHLOROPHENYLSILANE [302, EPA, EPA]	PHENYL TRICHLOROSILANE
52-68-6 TRICHLOROPHON [EPA]	TRICHLORFON
96-18-4 1,2,3-TRICHLOROPROPANE	1,2,3-TRICHLOROPROPANE
96-18-4 TRICHLOROPROPANE 1,2,3 [QBC, QBC]	1,2,3-TRICHLOROPROPANE
25735-29-9 TRICHLOROPROPANE	TRICHLOROPROPANE
25735-29-9 TRICHLOROPROPANE, N.O.S. [RCU]	TRICHLOROPROPANE
10025-78-2 TRICHLOROSILANE	TRICHLOROSILANE
10025-78-2 TRICHLOROSILANE [SILANE, TRICHLORO-] [EP3]	TRICHLOROSILANE
134237-38-0 TRICHLOROTETRAFLUOROPROPANE	TRICHLOROTETRAFLUOROPROPANE
87-90-1 TRICHLORO-S-TRIAZINETRIONE	TRICHLORO-S-TRIAZINETRIONE
RR-01541-8 TRICHLOROTRIETHYLAMINE HYDROCHLORIDE	TRICHLOROTRIETHYLAMINE HYDROCHLORIDE
76-13-1 1,1,2-TRICHLORO-1,2,2-TRIFLUOROETHANE	1,1,2-TRICHLORO-1,2,2-TRIFLUOROETHANE
76-13-1 TRICHLORO 1,1,2-, 1,2,2-TRIFLUOROETHANE [QBC, QBC]	1,1,2-TRICHLORO-1,2,2-TRIFLUOROETHANE
76-13-1 1,1,2-TRICHLORO-1,2,2-TRIFLUORO-ETHANE (FC 113) [WHM]	1,1,2-TRICHLORO-1,2,2-TRIFLUOROETHANE
76-13-1 1,1,2-TRICHLORO-1,2,2-TRIFLUOROETHANE (FC-113) [CAL]	1,1,2-TRICHLORO-1,2,2-TRIFLUOROETHANE
76-13-1 1,1,2-TRICHLOROTRIFLUOROETHANE [GBR, GBR]	1,1,2-TRICHLORO-1,2,2-TRIFLUOROETHANE
354-58-5 1,1,1-TRICHLORO-2,2,2-TRIFLUOROETHANE	1,1,1-TRICHLORO-2,2,2-TRIFLUOROETHANE
134237-40-4 TRICHLOROTRIFLUOROPROPANE	TRICHLOROTRIFLUOROPROPANE
RR-01542-9 T2-TRICHOTHECENE	T2-TRICHOTHECENE
57213-69-1 TRICLOPYR, TRIETHYLAMMONIUM SALT	TRICLOPYR, TRIETHYLAMMONIUM SALT
27519-02-4 (Z)-9-TRICOSENE	(Z)-9-TRICOSENE
78-30-8 TRI-O-CRESYL PHOSPHATE [CAL, FLA, MSL, NFP, NFP]	TRIORTHOCRESYL PHOSPHATE
78-30-8 TRI-ORTHO-CRESYL PHOSPHATE [ONT, ONT, WHM]	TRIORTHOCRESYL PHOSPHATE
1330-78-5 TRICRESYL PHOSPHATE	TRICRESYL PHOSPHATE
1330-78-5 TRICRESYLPHOSPHATE [NJL]	TRICRESYL PHOSPHATE
13121-70-5 TRICYCLOHEXYLTIN HYDROXIDE (PLICTRAN) [ALB, ALB, BC1, BC1]	CYHEXATIN
112-70-9 1-TRIDECANOL	1-TRIDECANOL

Chemical Name	Reference Name
112-70-9 TRIDECANOL [NFP, NFP]	1-TRIDECANOL
5116-94-9 TRIDECYL ACID PHOSPHATE	TRIDECYL ACID PHOSPHATE
3076-04-8 TRIDECYL ACRYLATE [NFP, NFP]	2-PROPENOIC ACID, TRIDECYL ESTER
112-70-9 TRIDECYL ALCOHOL [PST]	1-TRIDECANOL
24938-91-8 TRIDECYL ALCOHOL, ETHOXYLATED	TRIDECYL ALCOHOL, ETHOXYLATED
9046-01-9 TRIDECYL ALCOHOL, ETHOXYLATED, PHOSPHATED	TRIDECYL ALCOHOL, ETHOXYLATED, PHOSPHATED-LYPROPYLENE GLYCOL
60883-89-8 TRIDECYLBENZENESULFONIC ACID, DIMETHYLAMINE SALT	TRIDECYLBENZENESULFONIC ACID, DIMETHYLAMINE SALT
60883-90-1 TRIDECYLBENZENESULFONIC ACID, PROPYLAMINE SALT	TRIDECYLBENZENESULFONIC ACID, PROPYLAMINE SALT
2929-86-4 TRIDECYL PHOSPHITE	TRIDECYL PHOSPHITE
4130-35-2 TRI-N-DECYL TRIMELLITATE	TRI-N-DECYL TRIMELLITATE
90-72-2 2,4,6-TRI(DIMETHYLAMINOMETHYL)PHENOL	2,4,6-TRI(DIMETHYLAMINOMETHYL)PHENOL
15468-32-3 TRIDYMITE (SIO2) [PAL]	SILICA, TRIDYMITE
15468-32-3 TRIDYMITE DUST [FLA, MSL]	SILICA, TRIDYMITEOXIME
102-71-6 TRIETHANOLAMINE	TRIETHANOLAMINE
27323-41-7 TRIETHANOLAMINE DODECYLBENZENESULFONATE [EPA]	TRIETHANOLAMINE DODECYLBENZOSULFONATE
27323-41-7 TRIETHANOLAMINE DODECYLBENZENE SULFONATE [MSL]	TRIETHANOLAMINE DODECYLBENZOSULFONATE
27323-41-7 TRIETHANOLAMINE DODECYLBENZOSULFONATE	TRIETHANOLAMINE DODECYLBENZOSULFONATE
139-96-8 TRIETHANOLAMINE LAURYL SULFATE	TRIETHANOLAMINE LAURYL SULFATE
2717-15-9 TRIETHANOLAMINE OLEATE	TRIETHANOLAMINE OLEATE
10017-56-8 TRIETHANOLAMINE PHOSPHATE	TRIETHANOLAMINE PHOSPHATESALT
RR-01562-3 TRIETHANOLAMINE SALTS OF FATTY ACIDS	TRIETHANOLAMINE SALTS OF FATTY ACIDS
4568-28-9 TRIETHANOLAMINE STEARATE	TRIETHANOLAMINE STEARATE
7376-31-0 TRIETHANOLAMINE SULFATE	TRIETHANOLAMINE SULFATE
78-07-9 TRIETHOXYETHYL SILANE	TRIETHOXYETHYL SILANE
998-30-1 TRIETHOXYSILANE	TRIETHOXYSILANE
919-30-2 3-(TRIETHOXYSILYL)PROPYLAMINE	3-(TRIETHOXYSILYL)PROPYLAMINE
1793-90-4 TRIETHYLALLYLGERMANIUM	TRIETHYLALLYLGERMANIUM
97-93-8 TRIETHYLALUMINUM	TRIETHYLALUMINUM
97-93-8 TRIETHYL ALUMINUM [NJL, NJL]	TRIETHYLALUMINUM
121-44-8 TRIETHYLAMINE	TRIETHYLAMINE
78871-22-4 TRIETHYLAMINE CITRATE	TRIETHYLAMINE CITRATE
RR-01155-2 TRIETHYLAMINE NITRILOTRIACETATE	TRIETHYLAMINE NITRILOTRIACETATE
25340-18-5 1,2,4-TRIETHYLBENZENE	1,2,4-TRIETHYLBENZENE
25340-18-5 TRIETHYLBENZENE [HM1]	1,2,4-TRIETHYLBENZENE
97-94-9 TRIETHYLBORANE	TRIETHYLBORANE
77-93-0 TRIETHYL CITRATE	TRIETHYL CITRATE
280-57-9 TRIETHYLENEDIAMINE [WHM]	1,4-DIAZABICYCLO[2,2,2]OCTANE
112-27-6 TRIETHYLENE GLYCOL	TRIETHYLENE GLYCOL
94-28-0 TRIETHYLENE GLYCOL BIS(2-ETHYLHEXANOATE)	TRIETHYLENE GLYCOL BIS(2-ETHYLHEXANOATE)
95-08-9 TRIETHYLENE GLYCOL BIS (2-ETHYLBUTYRATE)	TRIETHYLENE GLYCOL BIS (2-ETHYLBUTYRATE)
111-21-7 TRIETHYLENE GLYCOL DIACETATE [NFP, NFP]	ETHYLENE BIS(OXYETHYLENE) DIACETATE
1680-21-3 TRIETHYLENE GLYCOL DIACRYLATE	TRIETHYLENE GLYCOL DIACRYLATE
1954-28-5 TRIETHYLENE GLYCOL DIGLYCIDYL ETHER	TRIETHYLENE GLYCOL DIGLYCIDYL ETHER

Chemical Name	Reference Name
112-49-2 TRIETHYLENE GLYCOL, DIMETHYL ETHER	TRIETHYLENE GLYCOL, DIMETHYL ETHER
109-16-0 TRIETHYLENE GLYCOL DIMETHACRYLATE	TRIETHYLENE GLYCOL DIMETHACRYLATE
143-22-6 TRIETHYLENE GLYCOL MONOBUTYL ETHER	TRIETHYLENE GLYCOL MONOBUTYL ETHER
143-22-6 TRIETHYLENEGLYCOL MONOBUTYL ETHER [NFP, NFP]	TRIETHYLENE GLYCOL MONOBUTYL ETHER
112-50-5 TRIETHYLENE GLYCOL MONOETHYL ETHER	TRIETHYLENE GLYCOL MONOETHYL ETHER
112-35-6 TRIETHYLENE GLYCOL MONOMETHYL ETHER	TRIETHYLENE GLYCOL MONOMETHYL ETHER
51-18-3 TRIETHYLENEMELAMINE	TRIETHYLENEMELAMINE
112-24-3 TRIETHYLENE TETRAMINE	TRIETHYLENE TETRAMINE
112-24-3 TRIETHYLENETETRAMINE [FLA, MSL, NFP, NFP, PST, WHM]	TRIETHYLENE TETRAMINE
78-40-0 TRIETHYL PHOSPHATE	TRIETHYL PHOSPHATE
122-52-1 TRIETHYL PHOSPHITE	TRIETHYL PHOSPHITE
126-68-1 O,O,O-TRIETHYL PHOSPHOROTHIOATE	O,O,O-TRIETHYL PHOSPHOROTHIOATE
76-05-1 TRIFLUOROACETIC ACID	TRIFLUOROACETIC ACID
407-25-0 TRIFLUOROACETIC ACID ANHYDRIDE	TRIFLUOROACETIC ACID ANHYDRIDE
354-32-5 TRIFLUOROACETYLCHLORIDE	TRIFLUOROACETYLCHLORIDE
354-32-5 TRIFLUOROACETYL CHLORIDE [HM1, HM1, HM1]	TRIFLUOROACETYLCHLORIDE
75-63-8 TRIFLUOROBROMOMETHANE	TRIFLUOROBROMOMETHANE
79-38-9 TRIFLUOROCHLOROETHYLENE [OS3]	CHLOROTRIFLUOROETHYLENE
79-38-9 TRIFLUOROCHLOROETHYLENE [ETHENE, CHLOROTRIFLUORO-] [EP3]	CHLOROTRIFLUOROETHYLENE
79-38-9 TRIFLUOROCHLOROETHYLENE [HM1, HM1, HM1, MSL, NFP, NFP]	CHLOROTRIFLUOROETHYLENE
420-46-2 1,1,1-TRIFLUOROETHANE [TSE]	ETHANE, 1,1,1-TRIFLUORO
27987-06-0 TRIFLUOROETHANE	TRIFLUOROETHANE
75-89-8 2,2,2-TRIFLUOROETHANOL	2,2,2-TRIFLUOROETHANOL
75-46-7 TRIFLUOROMETHANE	TRIFLUOROMETHANE
88-17-5 2-TRIFLUOROMETHYL ANILINE [NJL]	2-(TRIFLUOROMETHYL) ANILINE
88-17-5 2-(TRIFLUOROMETHYL) ANILINE	2-(TRIFLUOROMETHYL) ANILINE
98-16-8 3-TRIFLUOROMETHYLANILINE [HM1, HM1]	BENZENAMINE, 3-(TRIFLUOROMETHYL)-
98-16-8 3-TRIFLUOROMETHYL ANILINE [NJL]	BENZENAMINE, 3-(TRIFLUOROMETHYL)-
98-16-8 3-(TRIFLUOROMETHYL) ANILINE [WHM]	BENZENAMINE, 3-(TRIFLUOROMETHYL)-
1548-13-6 TRIFLOUROMETHYLPHENYLISOCYANATE	TRIFLOUROMETHYLPHENYLISOCYANATE
1548-13-6 3-TRIFLOUROMETHYLPHENYL ISO-CYANATE [HM1]	TRIFLOUROMETHYLPHENYLISOCYANATE
75-63-8 TRIFLUOROMONOBROMOMETHANE [BC1, BC1]	TRIFLUOROBROMOMETHANE
2374-14-3 TRIFLUOROPROPYLMETHYLCYCLOTRISILOXANE	TRIFLUOROPROPYLMETHYLCYCLOTRISILOXANE
1582-09-8 TRIFLURALIN	TRIFLURALIN
1582-09-8 TRIFLURALIN [BENZENEAMINE, 2,6-DINITRO-N, N-DIPROPYL-4-(TRIFLUOROMETHYL)-] [313]	TRIFLURALIN
26644-46-2 TRIFORINE [N,N'-[1,4-PIPERAZINEDIYL-BIS(2,2, 2-TRICHLOROETHYLIDENE)] BISFORMAMIDE]	TRIFORINE [N,N'-[1,4-PIPERAZINEDIYL-BIS(2,2,2-TRICHLOROETHYL
RR-01464-2 TRIFORMOXIME TRINITRATE	TRIFORMOXIME TRINITRATE
112-26-5 TRIGLYCOL DICHLORIDE [FLA, MSL, NFP, NFP]	ETHANE, 1,2-BIS(2-CHLOROETHOXY)-
RR-01615-9 TRIHALOMETHANES	TRIHALOMETHANES
6095-42-7 TRIHEXYL PHOSPHITE	TRIHEXYL PHOSPHITE
100-99-2 TRIISOBUTYL ALUMINUM	TRIISOBUTYL ALUMINUM
100-99-2 TRIISOBUTYLALUMINUM [FLA, MSL, NFP, NFP]	TRIISOBUTYL ALUMINUM

Chemical Name	Reference Name
13195-76-1 TRIISOBUTYL BORATE [FLA, MSL, NFP, NFP]	BORIC ACID (H3BO3), TRIS(2-METHYLPROPYL) ESTER
7756-94-7 TRIISOBUTYLENE	TRIISOBUTYLENE
RR-00510-7 TRIISOCYANATOISOCYANURATE OF ISOPHORONEDIISOCYANATE,SOLUTION	TRIISOCYANATOISOCYANURATE OF ISOPHORONEDI-ISOCYANATE,SOLUTION
122-20-3 TRIISOPROPANOLAMINE	TRIISOPROPANOLAMINE
717-74-8 TRIISOPROPYLBENZENE	TRIISOPROPYLBENZENE
5419-55-6 TRIISOPROPYL BORATE	TRIISOPROPYL BORATE
732-26-3 2,4,6-TRIISOPROPYLPHENOL	2,4,6-TRIISOPROPYLPHENOL
1656-63-9 TRILAURYL TRITHIOPHOSPHITE	TRILAURYL TRITHIOPHOSPHITE
13647-35-3 TRILOSTANE	TRILOSTANE
1317-35-7 TRIMANGANESE TETRAOXIDE [GBR]	MANGANESE TETROXIDE
528-44-9 TRIMELLITIC ACID	TRIMELLITIC ACID
552-30-7 TRIMELLITIC ACID ANHYDRIDE [PST]	TRIMELLITIC ANHYDRIDE
552-30-7 TRIMELLITIC ANHYDRIDE	TRIMELLITIC ANHYDRIDE
127-48-0 TRIMETHADIONE	TRIMETHADIONE
738-70-5 TRIMETHOPRIM	TRIMETHOPRIM
8064-90-2 TRIMETHOPRIM/SULFAMETHOXAZOLE	TRIMETHOPRIM/SULFAMETHOXAZOLE
102-24-9 TRIMETHOXYBOROXINE	TRIMETHOXYBOROXINE
2530-83-8 3-(TRIMETHOXYSILYL)PROPYL GLYCIDYL ETHER [EPA]	GLYCIDOXYPROPYLTRIMETHOXYSILANE
2487-90-3 TRIMETHOXYSILANE	TRIMETHOXYSILANE
1760-24-3 N-[3-(TRIMETHYOXYSILYL)PROPYL]-1,2-ETHANEDIAMINE	N-[3-(TRIMETHYOXYSILYL)PROPYL]-1,2-ETHANEDI-AMINE
3282-30-2 TRIMETHYLACETYL CHLORIDE [HM1, HM1, HM1, NJL, NJL]	PIVALOYL CHLORIDE
75-24-1 TRIMETHYLALUMINIUM	TRIMETHYLALUMINIUM
75-24-1 TRIMETHYL ALUMINIUM [NJL, NJL]	TRIMETHYLALUMINIUM
75-50-3 TRIMETHYLAMINE	TRIMETHYLAMINE
75-50-3 TRIMETHYLAMINE [METHANAMINE, N,N-DIMETHYL-] [EP3]	TRIMETHYLAMINE
90370-29-9 4,4',6-TRIMETHYLANGELICIN PLUS ULTRAVIOLET A RADIATION	4,4',6-TRIMETHYLANGELICIN PLUS ULTRAVIOLET A RADIATION
88-05-1 2,4,6-TRIMETHYLANILINE [IAR, MSL]	ANILINE, 2,4,6-TRIMETHYL-
88-05-1 2,4,6-TRIMETHYL ANILINE [NJL]	ANILINE, 2,4,6-TRIMETHYL-
137-17-7 2,4,5-TRIMETHYLANILINE	2,4,5-TRIMETHYLANILINELCIUM SALT (2:1), (R)-
95-63-6 1,2,4-TRIMETHYLBENZENE [313, CAB, CAN, NFP, NFP]	PSEUDOCUMENE
95-63-6 1,2,4-TRIMETHYL BENZENE [OTV]	PSEUDOCUMENE
95-63-6 1,2,4-TRIMETHYLBENZENE [SDW, WHM]	PSEUDOCUMENE
108-67-8 1,3,5-TRIMETHYLBENZENE	1,3,5-TRIMETHYLBENZENE
108-67-8 TRIMETHYLBENZENE, 1,3,5- [OTS]	1,3,5-TRIMETHYLBENZENE
526-73-8 1,2,3-TRIMETHYLBENZENE	1,2,3-TRIMETHYLBENZENE
25551-13-7 TRIMETHYL BENZENE	TRIMETHYL BENZENE
25551-13-7 TRIMETHYLBENZENE [QBC, WHM]	TRIMETHYL BENZENE
25551-13-7 TRIMETHYLBENZENE (SUM OF ISOMERS) [ONT]	TRIMETHYL BENZENE
25551-13-7 TRIMETHYLBENZENE (ALL ISOMERS) [CAL]	TRIMETHYL BENZENE
25551-13-7 TRIMETHYLBENZENE, ALL ISOMERS [CEX]	TRIMETHYL BENZENE
121-43-7 TRIMETHYL BORATE	TRIMETHYL BORATE
121-43-7 TRIMETHYLBORATE [NJL, NJL]	TRIMETHYL BORATE
464-06-2 2,2,3-TRIMETHYLBUTANE [FLA, MSL, NFP, NFP]	BUTANE, 2,2,3-TRIMETHYL-

Chemical Name	Reference Name
594-56-9 2,3,3-TRIMETHYL-1-BUTENE [FLA, MSL, NFP, NFP]	1-BUTENE, 2,3,3-TRIMETHYL-
75-77-4 TRIMETHYLCHLOROSILANE	TRIMETHYLCHLOROSILANE
75-77-4 TRIMETHYLCHLOROSILANE [SILANE, CHLOROTRIMETHYL-] [EP3]	TRIMETHYLCHLOROSILANE
116-02-9 3,3,5-TRIMETHYLCYCLOHEXANOL	3,3,5-TRIMETHYLCYCLOHEXANOL
RR-00533-4 1,3,5-TRIMETHYLCYCLOHEXANE	1,3,5-TRIMETHYLCYCLOHEXANE
116-02-9 3,3,5-TRIMETHYL-1-CYCLOHEXANOL [FLA, MSL, NFP, NFP]	3,3,5-TRIMETHYLCYCLOHEXANOL
1321-60-4 TRIMETHYLCYCLOHEXANOL	TRIMETHYLCYCLOHEXANOL
2408-37-9 TRIMETHYLCYCLOHEXANONE	TRIMETHYLCYCLOHEXANONE
78-59-1 3,5,5-TRIMETHYL-2-CYCLOHEXEN-1-ONE [FLA]	ISOPHORONE
78-59-1 3,5,5-TRIMETHYLCYCLOHEX-2-ENONE [GBR]	ISOPHORONE
78-59-1 3,5,5-TRIMETHYL-2-CYCLOHEXEN-1-ONE [ONT]	ISOPHORONE
34216-34-7 TRIMETHYLCYCLOHEXYLAMINE	TRIMETHYLCYCLOHEXYLAMINE
52836-31-4 2,2,5-TRIMETHYL-3-DICHLOROACETYL-1,3-OXAZOLIDINE	2,2,5-TRIMETHYL-3-DICHLOROACETYL-1,3-OXAZOLIDINE
504-63-2 TRIMETHYLENE GLYCOL [NFP]	1,3-PROPANEDIOL
142-28-9 TRIMETHYLENE DICHLORIDE [HM1, HM1]	1,3-DICHLOROPROPANE
RR-01465-3 TRIMETHYLENE GLYCOL DIPERCHLORATE	TRIMETHYLENE GLYCOL DIPERCHLORATE
1189-99-7 2,5,5-TRIMETHYLHEPTANE	2,5,5-TRIMETHYLHEPTANE
112-02-7 N,N,N-TRIMETHYL-1-HEXADECANAMINIUM CHLORIDE	N,N,N-TRIMETHYL-1-HEXADECANAMINIUM CHLORIDE
25620-58-0 TRIMETHYLHEXAMETHYLENEDIAMINE	TRIMETHYLHEXAMETHYLENEDIAMINE
25620-58-0 TRIMETHYLHEXAMETHYLENE DI-AMINE [WHM]	TRIMETHYLHEXAMETHYLENEDIAMINE
25620-58-0 TRIMETHYLHEXAMETHYLENEDIAMINES [HM1, HM1]	TRIMETHYLHEXAMETHYLENEDIAMINE
15646-96-5 2,4,4-TRIMETHYLHEXAMETHYLENE DIISO-CYANATE [313]	HEXANE, 1,6-DIISOCYANATO-2,4,4-TRIMETHYL-
16938-22-0 2,2,4-TRIMETHYLHEXAMETHYLENE DIISO-CYANATE [313]	HEXANE, 1,6-DIISOCYANATO-2,2,4-TRIMETHYL-
28679-16-5 TRIMETHYLHEXAMETHYLENE DIISOCYANATE	TRIMETHYLHEXAMETHYLENE DIISOCYANATE
28679-16-5 TRIMETHYLHEXAMETHYLENEDI-ISO-CYANATE [NJL]	TRIMETHYLHEXAMETHYLENE DIISOCYANATE
3522-94-9 2,2,5-TRIMETHYLHEXANE [FLA, MSL, NFP, NFP]	HEXANE, 2,2,5-TRIMETHYL-
3452-97-9 3,5,5-TRIMETHYLHEXANOL [FLA, MSL, NFP, NFP]	1-HEXANOL, 3,5,5-TRIMETHYL-
123-17-1 2,6,8-TRIMETHYL-4-NONANOL [FLA, MSL, NFP, NFP, PST]	4-NONANOL, 2,6,8-TRIMETHYL-
RR-00591-4 2,4,8-TRIMETHYL-6-NONANOL	2,4,8-TRIMETHYL-6-NONANOL
123-18-2 2,6,8-TRIMETHYL-4-NONANONE [NFP, NFP]	ISOBUTYL HEPTYL KETONE
60828-78-6 2,6,8-TRIMETHYL-4-NONYL POLYETHYLENE GLYCOL ETHER	2,6,8-TRIMETHYL-4-NONYL POLYETHYLENE GLYCOL ETHER
77-85-0 TRIMETHYLOLETHANE	TRIMETHYLOLETHANE
77-99-6 TRIMETHYLOLPROPANE [HON, PST]	1,3-PROPANEDIOL, 2-ETHYL-2-(HYDROXYMETHYL)-TER
RR-00230-2 TRIMETHYLOLPROPANE FATTY ACID DI-ACRYLATE	TRIMETHYLOLPROPANE FATTY ACID DIACRYLATE
824-11-3 TRIMETHYLOLPROPANE PHOSPHITE	TRIMETHYLOLPROPANE PHOSPHITE
9040-80-6 TRIMETHYLOLPROPANE, POLYMER WITH TOLUENE DIISOCYANATE AND POLYPROPY-LENE GLYCOL	TRIMETHYLOLPROPANE, POLYMER WITH TOLUENE DIISOCYANATE AND PO

Chemical Name	Reference Name
15625-89-5 TRIMETHYLOLPROPANE TRIACRYLATE	TRIMETHYLOLPROPANE TRIACRYLATE
3290-92-4 TRIMETHYLOLPROPANE TRIMETHACRY-LATE [MIN]	TRIMETHYLOPROPANE TRIMETHACRYLATE
3290-92-4 TRIMETHYLOPROPANE TRIMETHACRYLATE	TRIMETHYLOPROPANE TRIMETHACRYLATE
540-84-1 2,2,4-TRIMETHYLPENTANE [CAA, CAB]	ISOOCTANE
560-21-4 2,3,3-TRIMETHYLPENTANE [FLA, MSL, NFP, NFP]	PENTANE, 2,3,3-TRIMETHYL-
564-02-3 2,2,3-TRIMETHYLPENTANE [FLA, MSL, NFP, NFP]	PENTANE, 2,2,3-TRIMETHYL-
540-84-1 TRIMETHYLPENTANE (2,2,4-) [HON, HON]	ISOOCTANE
144-19-4 2,2,4-TRIMETHYL-1,3-PENTANEDIOL	2,2,4-TRIMETHYL-1,3-PENTANEDIOL
35164-39-7 2,2,4-TRIMETHYL-1,3-PENTANEDIOL BEN-ZOATE ISOBUTYRATE	2,2,4-TRIMETHYL-1,3-PENTANEDIOL BENZOATE ISOBU-TYRATE
6846-50-0 2,2,4-TRIMETHYL-1,3-PENTANEDIOL DIISOBU-TYRATE	2,2,4-TRIMETHYL-1,3-PENTANEDIOL DIISOBUTYRATE
6846-50-0 2,2,4-TRIMETHYL PENTANEDIOL DIISOBU-TYRATE [NFP, NFP]	2,2,4-TRIMETHYL-1,3-PENTANEDIOL DIISOBUTYRATE
6846-50-0 2,2,4-TRIMETHYL-1,3-PENTANEDIOL ES-TER [OTS]	2,2,4-TRIMETHYL-1,3-PENTANEDIOL DIISOBUTYRATE
25265-77-4 2,2,4-TRIMETHYL-1,3-PENTANEDIOL ISOBU-TYRATE [NFP, NFP]	2,2,4-TRIMETHYLPENTANE-1,3-DIOL MONOISOBU-TYRATE
35164-39-7 2,2,4-TRIMETHYLPENTANEDIOL ISOBUTYRATE BENZOATE [NFP, NFP]	2,2,4-TRIMETHYL-1,3-PENTANEDIOL BENZOATE ISOBU-TYRATE
25265-77-4 2,2,4-TRIMETHYLPENTANE-1,3-DIOL MONOISOBUTYRATE	2,2,4-TRIMETHYLPENTANE-1,3-DIOL MONOISOBU-TYRATE
107-39-1 2,4,4-TRIMETHYL-1-PENTENE [FLA, MSL, NFP, NFP]	1-PENTENE, 2,4,4-TRIMETHYL-
107-40-4 2,4,4-TRIMETHYL-2-PENTENE [FLA, MSL, NFP, NFP]	2-PENTENE, 2,4,4-TRIMETHYL-
565-76-4 2,3,4-TRIMETHYL-1-PENTENE [FLA, MSL, NFP, NFP]	1-PENTENE, 2,3,4-TRIMETHYL-
598-96-9 3,4,4-TRIMETHYL-2-PENTENE	3,4,4-TRIMETHYL-2-PENTENE
RR-00757-8 2,4,4-TRIMETHYLPENTYL-2-PEROXYPHENOXY-ACETATE	2,4,4-TRIMETHYLPENTYL-2-PEROXYPHENOXY-AC-ETATE
2655-15-4 2,3,5-TRIMETHYLPHENYL METHYLCARBA-MATE	2,3,5-TRIMETHYLPHENYL METHYLCARBAMATE
512-56-1 TRIMETHYLPHOSPHATE	TRIMETHYLPHOSPHATE
512-56-1 TRIMETHYL PHOSPHATE [MAK, MAK, MSL]	TRIMETHYLPHOSPHATE
121-45-9 TRIMETHYL PHOSPHITE	TRIMETHYL PHOSPHITE
3902-71-4 4,5',8-TRIMETHYLPSORALEN	4,5',8-TRIMETHYLPSORALEN
RR-01181-4 TRIMETHYL SPIROPOLYHETEROCYCLIC NAPHTHALENE COMPOUND	TRIMETHYL SPIROPOLYHETEROCYCLIC NAPHTHA-LENE COMPOUND
139-08-2 TRIMETHYL TETRADECYLPHENYL AMMO-NIUM CHLORIDE [PST]	BENZYLDIMETHYLTETRADECYLAMMONIUM CHLO-RIDE
2489-77-2 TRIMETHYLTHIOUREA	TRIMETHYLTHIOUREA
2782-91-4 TRIMETHYL THIOUREA	TRIMETHYL THIOUREA
1066-45-1 TRIMETHYL TIN CHLORIDE	TRIMETHYL TIN CHLORIDE
1066-45-1 TRIMETHYLTIN CHLORIDE [302, PAL, WHM]	TRIMETHYL TIN CHLORIDE
602-96-0 1,3,5-TRIMETHYL-2,4,6-TRINITROBENZENE	1,3,5-TRIMETHYL-2,4,6-TRINITROBENZENE
RR-01466-4 TRINITROACETIC ACID	TRINITROACETIC ACID
630-72-8 TRINITROACETONITRILE	TRINITROACETONITRILE
RR-01467-5 TRINITROAMINE COBALT	TRINITROAMINE COBALT
26952-42-1 TRINITRO-ANILINE	TRINITRO-ANILINE
606-35-9 TRINITROANISOLE	TRINITROANISOLE

Chemical Name	Reference Name
99-35-4 TRINITROBENZENE	TRINITROBENZENE
99-35-4 1,3,5-TRINITROBENZENE [ATS, EPA]	TRINITROBENZENE
99-35-4 SYM-TRINITROBENZENE [PQL]	TRINITROBENZENE
99-35-4 1,3,5-TRINITROBENZENE [RCU, RCU, TSC, TSE]	TRINITROBENZENE
2508-19-2 TRINITROBENZENESULFONIC ACID	TRINITROBENZENESULFONIC ACID
129-66-8 TRINITROBENZOIC ACID	TRINITROBENZOIC ACID
35860-50-5 TRINITROBENZOIC ACID [HM1, HM1]	TRINITROBENZOIC ACID (VAN)
35860-50-5 TRINITROBENZOIC ACID (VAN)	TRINITROBENZOIC ACID (VAN)
28260-61-9 TRINITROCHLOROBENZENE	TRINITROCHLOROBENZENE
602-99-3 TRINITRO-M-CRESOL	TRINITRO-M-CRESOL
602-99-3 TRINITRO-META-CRESOL [HM1, HM1]	TRINITRO-M-CRESOL
29306-57-8 2,4,6-TRINITRO-1,3-DIAZOBENZENE	2,4,6-TRINITRO-1,3-DIAZOBENZENE
918-54-7 TRINITROETHANOL	TRINITROETHANOL
RR-01468-6 TRINITROETHYLNITRATE	TRINITROETHYLNITRATE
129-79-3 2,4,7-TRINITRO-FLUOREN-9-ONE	2,4,7-TRINITRO-FLUOREN-9-ONE
129-79-3 2,4,7-TRINITROFLUORENONE [MAK]	2,4,7-TRINITRO-FLUOREN-9-ONE
25322-14-9 TRINITROFLUORENONE	TRINITROFLUORENONE
517-25-9 TRINITROMETHANE	TRINITROMETHANE
55810-17-8 TRINITRONAPTHTALENE	TRINITRONAPTHTALENE
4732-14-3 TRINITROPHENETOLE	TRINITROPHENETOLE
88-89-1 TRINITROPHENOL [HM1, HM1]	PICRIC ACID
88-89-1 2,4,6-TRINITROPHENOL [NJL, ONT, ONT, ONT]	PICRIC ACID
RR-01381-0 2,4,6-TRINITROPHENYL GUANIDINE	2,4,6-TRINITROPHENYL GUANIDINE
479-45-8 TRINITROPHENYLMETHYLNITRAMINE [ATS, HM1, HM1]	TETRYL
RR-01382-1 2,4,6-TRINITROPHENYL NITRAMINE	2,4,6-TRINITROPHENYL NITRAMINE
RR-01383-2 2,4,6-TRINITROPHENYL TRIMETHYLOL METHYL NITRAMINE TRINITRATE	2,4,6-TRINITROPHENYL TRIMETHYLOL METHYL NITRAMINE TRINITRATE
82-71-3 TRINITRORESORCINOL	TRINITRORESORCINOL
RR-01384-3 2,4,6-TRINITROSO-3-METHYL NITRAMINOANISOLE	2,4,6-TRINITROSO-3-METHYL NITRAMINOANISOLE
RR-01469-7 TRINITROTETRAMINE COBALT NITRATE	TRINITROTETRAMINE COBALT NITRATE
118-96-7 2,4,6-TRINITROTOLUENE	2,4,6-TRINITROTOLUENE
118-96-7 TRINITROTOLUENE [ATS, FLA, MSL]	2,4,6-TRINITROTOLUENE
118-96-7 TRINITROTOLUENE 2,4,6- [QBC, QBC]	2,4,6-TRINITROTOLUENE
118-96-7 2,4,6-TRINITROTOLUENE (TNT) [TLV, TLV, OS1, OS1, OS1, OS1, AUS, AUS, BC1]	2,4,6-TRINITROTOLUENE
RR-01380-9 2,4,6-TRINITRO-1,3,5-TRIAZIDO BENZENE	2,4,6-TRINITRO-1,3,5-TRIAZIDO BENZENE
RR-01463-1 TRI-(B-NITROXYETHYL) AMMONIUM NITRATE	TRI-(B-NITROXYETHYL) AMMONIUM NITRATE
26523-78-4 TRI[NONYLPHENYL] PHOSPHITE	TRI[NONYLPHENYL] PHOSPHITE
26266-58-0 (Z,Z,Z)-TRI-9-OCTADECENOATE SORBITAN	(Z,Z,Z)-TRI-9-OCTADECENOATE SORBITAN
3028-88-4 TRIOCTYL PHOSPHITE	TRIOCTYL PHOSPHITE
78-30-8 TRIORTHOCRESYL PHOSPHATE	TRIORTHOCRESYL PHOSPHATE
110-88-3 1,3,5-TRIOXANE	1,3,5-TRIOXANE
110-88-3 TRIOXANE [FLA, MSL, NFP, NFP]	1,3,5-TRIOXANE
123-63-7 1,3,5-TRIOXANE, 2,4,6-TRIMETHYL- [PAL]	PARALDEHYDE
154-69-8 TRIPELENNAMINE HYDROCHLORIDE	TRIPELENNAMINE HYDROCHLORIDE
621-77-2 TRIPENTYLAMINE [FLA, MSL]	1-PENTANAMINE, N,N-DIPENTYL-
603-34-9 TRIPHENYL AMINE	TRIPHENYL AMINE

Chemical Name	Reference Name
603-34-9 TRIPHENYLAMINE [CAL, CEX, WHM]	TRIPHENYL AMINE
603-36-1 TRIPHENYL ANTIMONY	TRIPHENYL ANTIMONY
217-59-4 TRIPHENYLENE	TRIPHENYLENE
519-73-3 TRIPHENYLMETHANE	TRIPHENYLMETHANE
115-86-6 TRIPHENYL PHOSPHATE	TRIPHENYL PHOSPHATE
101-02-0 TRIPHENYL PHOSPHITE	TRIPHENYL PHOSPHITE
603-35-0 TRIPHENYLPHOSPHOROUS [NFP, NFP]	PHOSPHINE, TRIPHENYL-
4270-70-6 TRIPHENYLSULFONIUM CHLORIDE	TRIPHENYLSULFONIUM CHLORIDE
900-95-8 TRIPHENYLTIN ACETATE [WHM]	STANNANE, ACETOXYTRIPHENYL-
639-58-7 TRIPHENYLTIN CHLORIDE	TRIPHENYLTIN CHLORIDE
639-58-7 TRIPHENYL TIN CHLORIDE [MSL]	TRIPHENYLTIN CHLORIDE
RR-01199-4 TRIPHENYLTIN COMPOUNDS	TRIPHENYLTIN COMPOUNDS
76-87-9 TRIPHENYLTIN HYDROXIDE	TRIPHENYLTIN HYDROXIDE
76-87-9 TRIPHENYL TIN HYDROXIDE [MSL]	TRIPHENYLTIN HYDROXIDE
7758-29-4 TRIPHOSPHORIC ACID, PENTASODIUM SALT [MSL, PAL]	PENTASODIUM TRIPHOSPHATE
13573-18-7 TRIPHOSPHORIC ACID, SODIUM SALT	TRIPHOSPHORIC ACID, SODIUM SALT
1317-95-9 TRIPOLI [PAL]	SILICA-TRIPOLI
1317-95-9 TRIPOLI DUST [FLA, MSL]	SILICA-TRIPOLI
2399-85-1 TRIPOTASSIUM NITRILOTRIACETATE	TRIPOTASSIUM NITRILOTRIACETATEYL) ESTER
486-12-4 TRIPROLIDINE	TRIPROLIDINE
102-67-0 TRIPROPYL ALUMINUM	TRIPROPYL ALUMINUM
102-69-2 TRIPROPYLAMINE	TRIPROPYLAMINE
13987-01-4 TRIPROPYLENE	TRIPROPYLENE
24800-44-0 TRIPROPYLENE GLYCOL [HON]	PROPANOL, [(1-METHYL-1,2-ETHANEDIYL)BIS(OXY)]BIS-
1638-16-0 TRIPROPYLENE GLYCOL	TRIPROPYLENE GLYCOL
24800-44-0 TRIPROPYLENE GLYCOL [OTS]	PROPANOL, [(1-METHYL-1,2-ETHANEDIYL)BIS(OXY)]BIS-
42978-66-5 TRIPROPYLENE GLYCOL DIACRYLATE	TRIPROPYLENE GLYCOL DIACRYLATE
25498-49-1 TRIPROPYLENE GLYCOL METHYL ETHER [NFP, NFP]	TRIPROPYLENE GLYCOL MONOMETHYL ETHER
20324-33-8 TRIPROPYLENE GLYCOL METHYL ETHER	TRIPROPYLENE GLYCOL METHYL ETHER
25498-49-1 TRIPROPYLENE GLYCOL MONOMETHYL ETHER	TRIPROPYLENE GLYCOL MONOMETHYL ETHER
68-76-8 TRIS(AZIRIDINYL)-P-BENZOQUINONE	TRIS(AZIRIDINYL)-P-BENZOQUINONE
68-76-8 TRIS(AZIRIDINYL)-PARA-BENZO-QUINONE [MIN]	TRIS(AZIRIDINYL)-P-BENZOQUINONE
68-76-8 2,3,5-TRIS(1-AZIRIDINYL)-P-BENZO-QUINONE [NJL]	TRIS(AZIRIDINYL)-P-BENZOQUINONE
68-76-8 TRIS(AZIRIDINYL)-PARA-BENZOQUINONE (TRIAZIQUONE) [C65]	TRIS(AZIRIDINYL)-P-BENZOQUINONE
68-76-8 TRIS(AZIRIDINYL)-PARA-BENZOQUINONE; (TRIAZIQUONE) [CAL]	TRIS(AZIRIDINYL)-P-BENZOQUINONE
68-76-8 TRIS(AZIRIDINYL)-PARA-BENZOQUINONE (TRIAZIQUONE) [IAR]	TRIS(AZIRIDINYL)-P-BENZOQUINONE
52-24-4 TRIS(1-AZIRIDINYL)PHOSPHINE SULFIDE	TRIS(1-AZIRIDINYL)PHOSPHINE SULFIDE
52-24-4 TRIS(1-AZIRIDINYL)PHOSPHINE SULFIDE (THIOTEPA) [C65, C65, NTP, NTP]	TRIS(1-AZIRIDINYL)PHOSPHINE SULFIDE
545-55-1 TRIS(1-AZIRIDINYL) PHOSPHINE OXIDE	TRIS(1-AZIRIDINYL) PHOSPHINE OXIDE
52-24-4 TRIS(AZIRIDINYL)-PHOSPHINE SULFIDE [CSR, CSR]	TRIS(1-AZIRIDINYL)PHOSPHINE SULFIDE

Chemical Name	Reference Name
52-24-4 TRIS(1-AZIRIDINYL)PHOSPHINE SUL-PHIDE [MIN]	TRIS(1-AZIRIDINYL)PHOSPHINE SULFIDE
51-18-3 2,4,6-TRIS(1-AZIRIDINYL)-S-TRIAZINE [CAL, IAR]	TRIETHYLENEMELAMINE
52-24-4 TRIS(1-AZIRIDINYL)-PHOSPHINE SULFIDE [NJL]	TRIS(1-AZIRIDINYL)PHOSPHINE SULFIDE
39409-64-8 TRIS, BIS-BIFLUOROAMINO DIETHOXY PROPANE (TVOPA)	TRIS, BIS-BIFLUOROAMINO DIETHOXY PROPANE (TVOPA)
2718-67-4 TRIS(2-BUTOXYETHYL) PHOSPHITE	TRIS(2-BUTOXYETHYL) PHOSPHITE
555-77-1 TRIS(2-CHLOROETHYL)AMINE	TRIS(2-CHLOROETHYL)AMINE
140-08-9 TRIS(2-CHLOROETHYL)PHOSPHATE [CSR, CSR, IAR, T33]	ETHANOL, 2-CHLORO-, PHOSPHITE (3:1)
115-96-8 TRIS(2-CHLOROETHYL) PHOSPHATE [C65]	ETHANOL, 2-CHLORO-, PHOSPHATE (3:1)
115-96-8 TRIS(2-CHLOROETHYL)PHOSPHATE [OTS]	ETHANOL, 2-CHLORO-, PHOSPHATE (3:1)
38571-73-2 1,2,3-TRIS(CHLOROMETHOXY)PROPANE	1,2,3-TRIS(CHLOROMETHOXY)PROPANE
6145-73-9 TRIS(2-CHLORO-1-PROPYL) PHOSPHATE [T34]	1-PROPANOL, 2-CHLORO-, PHOSPHATE (3:1)
13674-84-5 TRIS(1-CHLORO-2-PROPYL) PHOSPHATE [T33]	2-PROPANOL, 1-CHLORO-, PHOSPHATE (3:1)
RR-01717-4 TRIS(2-CHLORO-1-PROPYL) PHOSPHATE	TRIS(2-CHLORO-1-PROPYL) PHOSPHATE
49690-63-3 TRISDIBROMOPHENYL PHOSPHATE	TRISDIBROMOPHENYL PHOSPHATE
126-72-7 TRIS(DIBROMOPROPYL) PHOSPHATE [WHM]	TRIS(2,3-DIBROMOPROPYL)PHOSPHATE
126-72-7 TRIS(2,3-DIBROMOPROPYL)PHOSPHATE	TRIS(2,3-DIBROMOPROPYL)PHOSPHATE
78-43-3 TRIS(2,3-DICHLOROPROPYL) PHOSPHATE	TRIS(2,3-DICHLOROPROPYL) PHOSPHATE
78-43-3 TRIS(1,3-DICHLORO-2-PROPYL) PHOS-PHATE [T33]	TRIS(2,3-DICHLOROPROPYL) PHOSPHATE
RR-00954-1 TRIS(DISUBSTITUTED ALKYL) HETEROCYCLE	TRIS(DISUBSTITUTED ALKYL) HETEROCYCLEOKETAL
78-42-2 TRIS(2-ETHYLHEXYL)PHOSPHATE	TRIS(2-ETHYLHEXYL)PHOSPHATE
77-86-1 TRIS (HYDROXYMETHYL) AMINOMETHANE	TRIS (HYDROXYMETHYL) AMINOMETHANE
68517-02-2 1,1,1-TRIS(4-HYDROXYPHENYL)PROPANE TRIGLYCIDYL ETHER [EPA]	OXIRANE, 2,2',2"-[PROPYLIDYNETRIS(4,1-PHENYLE-NEOXYMETHYLENE
68517-02-2 TRIS(4-HYDROXYPHENYL)PROPANE-TRIGLY-CIDYL ETHER [OTS]	OXIRANE, 2,2',2"-[PROPYLIDYNETRIS(4,1-PHENYLE-NEOXYMETHYLENE
7422-52-8 TRISILOXANE, 1,1,1,3,5,5,5-HEPTAMETHYL-3-[3-(OXIRANYLMETHOXY)PROPYL]-	TRISILOXANE, 1,1,1,3,5,5,5-HEPTAMETHYL-3-[3-(OXI-RANYLMETHOXY
68937-41-7 TRIS ISOPROPYLATED PHENYL PHOSPHATE	TRIS ISOPROPYLATED PHENYL PHOSPHATE
57-39-6 TRIS(2-METHYL-1-AZIRIDINYL)PHOSPHINE OXIDE	TRIS(2-METHYL-1-AZIRIDINYL)PHOSPHINE OXIDE
150-38-9 TRISODIUM EDTA	TRISODIUM EDTA
150-38-9 TRISODIUM ETHYLENEDIAMINETETRAAC-ETATE TRIHYDRATE (EDTA) [CSR, CSR]	TRISODIUM EDTA
150-38-9 TRISODIUM ETHYLENEDIAMINETETRAAC-ETATE [PST]	TRISODIUM EDTA
139-89-9 TRISODIUM N-(2-HYDROXYETHYL)ETHYLENE-DIAMINETRIACETATE [PST]	N-HYDROXYETHYLETHYLENEDIAMINETRIACETIC ACID TRISODIUM SALT
5064-31-3 TRISODIUM NITRILOTRIACETATE [PST]	NITRILOTRIACETIC ACID TRISODIUM SALT
18662-53-8 TRISODIUM NITRILOTRIACETATE MONOHY-DRATE	TRISODIUM NITRILOTRIACETATE MONOHYDRATE
13721-39-6 TRISODIUM ORTHOVANADATE	TRISODIUM ORTHOVANADATE
7601-54-9 TRISODIUM PHOSPHATE	TRISODIUM PHOSPHATE
13419-59-5 TRISODIUM SULFOSUCCINATE	TRISODIUM SULFOSUCCINATE
RR-01720-9 TRISUBSTITUTED ANTHRACENE	TRISUBSTITUTED ANTHRACENE
RR-01563-4 TRISUBSTITUTED HYDROQUINONE DIESTER	TRISUBSTITUTED HYDROQUINONE DIESTER
RR-00170-7 TRISUBSTITUTED PHENOL	TRISUBSTITUTED PHENOL
638-16-4 TRITHIOCYANURIC ACID	TRITHIOCYANURIC ACID

Chemical Name	Reference Name
10028-17-8 TRITIUM	TRITIUM
78-30-8 TRI-O-TOLYL PHOSPHATE [GBR, GBR]	TRIORTHOCRESYL PHOSPHATE
60864-33-7 TRITON CF 10	TRITON CF 10
RR-01331-0 TRITONAL	TRITONAL
16065-83-1 TRIVALENT CHROMIUM COMPOUNDS [ONT]	CHROMIUM, ION (CR3+)
25155-23-1 TRIXYLENYL PHOSPHATE [HM1]	PHOSPHATE, TRIXYLYL
62450-06-0 TRP-P-1	TRP-P-1
62450-06-0 TRP-P-1 (3-AMINO-1,4-DIMETHYL-5H-PYRIDO(4, 3-B)INDOLE) [IAR]	TRP-P-1
62450-06-0 TRP-P-1 (TRYPTOPHAN-P-1) [C65, C65, CAL]	TRP-P-1
62450-07-1 TRP-P-2	TRP-P-2
62450-07-1 TRP-P-2 (3-AMINO-1-METHYL-5H-PYRIDO(4,3-B)INDOLE) [IAR]	TRP-P-2
62450-07-1 TRP-P-2 (TRYPTOPHAN-P-2) [C65, C65, CAL]	TRP-P-2
72-57-1 TRYPAN BLUE	TRYPAN BLUE
72-57-1 TRYPAN BLUE (COMMERCIAL GRADE) [C65, CAL]	TRYPAN BLUE
72-57-1 TRYPAN BLUE, COMMERICAL GRADE [MSL]	TRYPAN BLUE
73-22-3 L-TRYPTOPHAN	L-TRYPTOPHAN
21259-20-1 T-2 TOXIN	T-2 TOXIN
8001-20-5 TUNG OIL	TUNG OIL
7440-33-7 TUNGSTEN	TUNGSTEN
15749-43-6 TUNGSTEN 176	TUNGSTEN 176
15749-44-7 TUNGSTEN 177	TUNGSTEN 177
15055-23-9 TUNGSTEN 178	TUNGSTEN 178
14683-32-0 TUNGSTEN 179	TUNGSTEN 179
15749-46-9 TUNGSTEN 181	TUNGSTEN 181
14932-41-3 TUNGSTEN 185	TUNGSTEN 185
14983-48-3 TUNGSTEN 187	TUNGSTEN 187
24421-27-0 TUNGSTEN 188	TUNGSTEN 188
RR-00616-6 TUNGSTEN AND COMPOUNDS [GBR, GBR]	TUNGSTEN COMPOUNDS, N.O.S.
69600-03-9 TUNGSTEN ANTIMONATE	TUNGSTEN ANTIMONATE
12070-12-1 TUNGSTEN CARBIDE	TUNGSTEN CARBIDE
RR-00616-6 TUNGSTEN COMPOUNDS [CAL]	TUNGSTEN COMPOUNDS, N.O.S.
RR-00616-6 TUNGSTEN COMPOUNDS, N.O.S.	TUNGSTEN COMPOUNDS, N.O.S.
7440-33-7 TUNGSTEN, ELEMENTAL [WHM]	TUNGSTEN
7783-82-6 TUNGSTEN HEXAFLUORIDE	TUNGSTEN HEXAFLUORIDE
RR-00007-7 TUNGSTEN, INSOLUBLE COMPOUNDS	TUNGSTEN, INSOLUBLE COMPOUNDS
RR-00007-7 TUNGSTEN (INSOLUBLE COMPOUNDS) [QBC, QBC]	TUNGSTEN, INSOLUBLE COMPOUNDS
RR-00008-8 TUNGSTEN, SOLUBLE COMPOUNDS	TUNGSTEN, SOLUBLE COMPOUNDS
RR-00008-8 TUNGSTEN (SOLUBLE COMPOUNDS) [QBC, QBC]	TUNGSTEN, SOLUBLE COMPOUNDS
RR-00007-7 TUNGSTEN, WATER-INSOLUBLE COMPOUNDS [ONT, ONT]	TUNGSTEN, INSOLUBLE COMPOUNDS
RR-00008-8 TUNGSTEN, WATER-SOLUBLE COMPOUNDS [ONT, ONT]	TUNGSTEN, SOLUBLE COMPOUNDS
7783-03-1 TUNGSTIC ACIDS [ONT, ONT]	TUNGSTIC ACID (SOLUBLE TUNGSTEN)
RR-01623-9 TUNGSTIC ACID SALTS	TUNGSTIC ACID SALTS
7783-03-1 TUNGSTIC ACID (SOLUBLE TUNGSTEN)	TUNGSTIC ACID (SOLUBLE TUNGSTEN)
1314-35-8 TUNGSTIC OXIDE	TUNGSTIC OXIDE

Chemical Name	Reference Name
RR-00779-4 **TURBO FUELS**	TURBO FUELS
8002-33-3 TURKEY RED OIL [NFP, NFP]	CASTOR OIL, SULFATED
8024-37-1 **TURMERIC, OLEORESIN (CURCUMIN)**	TURMERIC, OLEORESIN (CURCUMIN)
8006-64-2 **TURPENTINE**	TURPENTINE
9005-90-7 TURPENTINE [PAL, PST]	TURPENTINE (RESIN FROM PINUS SPECIES PARTICULARLY PALUSTRIS)
9005-90-7 **TURPENTINE (RESIN FROM PINUS SPECIES PARTICULARLY PALUSTRIS)**	TURPENTINE (RESIN FROM PINUS SPECIES PARTICULARLY PALUSTRIS)
9005-90-7 TURPENTINE SUBSTITUTE [HM1, HM1, HM1, NJL]	TURPENTINE (RESIN FROM PINUS SPECIES PARTICULARLY PALUSTRIS)
8006-64-2 TURPENTINE (WOOD) [AUS, AUS, AUS]	TURPENTINE
57455-37-5 **ULTRAMARINE BLUE**	ULTRAMARINE BLUE
12713-03-0 UMBER [PST]	RAW UMBER
112-44-7 **UNDECANAL**	UNDECANAL
112-45-8 **10-UNDECANAL**	10-UNDECANAL
110-41-8 **UNDECANAL, 2-METHYL-**	UNDECANAL, 2-METHYL-
1120-21-4 **UNDECANE**	UNDECANE
5332-52-5 **1-UNDECANETHIOL**	1-UNDECANETHIOL
112-42-5 **UNDECANOL**	UNDECANOL
1653-30-1 **2-UNDECANOL**	2-UNDECANOL
112-12-9 **2-UNDECANONE**	2-UNDECANONE
19594-40-2 **4-UNDECANONE, 2-METHYL-**	4-UNDECANONE, 2-METHYL-
143-14-6 **9-UNDECENAL**	9-UNDECENAL
RR-00526-5 **UNDERGROUND HAEMATITE MINING WITH EXPOSURE TO RADON**	UNDERGROUND HAEMATITE MINING WITH EXPOSURE TO RADON
RR-00088-4 **UNLEADED GASOLINE (WHOLLY VAPORIZED)**	UNLEADED GASOLINE (WHOLLY VAPORIZED)
RR-00899-1 **UNSATURATED AMINO ALKYL ESTER SALT**	UNSATURATED AMINO ALKYL ESTER SALT
RR-00900-7 **UNSATURATED AMINO ESTER SALT**	UNSATURATED AMINO ESTER SALT
RR-00998-3 **UNSATURATED ORGANIC COMPOUND**	UNSATURATED ORGANIC COMPOUNDANO-1-METHYL-PROPYL-TERMINATED, POLYMERS WITH BISPHENOL A, EPICHLOROHYDRIN AND 4,4'-(1-METHYLETHYLIDENE) BIS(2,6-DIBROMOPHENOL
RR-00054-4 UNTREATED AND MILDLY REFINED MINERAL OILS [NTP]	UNTREATED AND MILDLY-TREATED OILS
RR-00054-4 **UNTREATED AND MILDLY-TREATED OILS**	UNTREATED AND MILDLY-TREATED OILS
66-75-1 **URACIL MUSTARD**	URACIL MUSTARD
7440-61-1 **URANIUM**	URANIUM
15743-51-8 **URANIUM 230**	URANIUM 230
15700-08-0 **URANIUM 231**	URANIUM 231
14158-29-3 **URANIUM 232**	URANIUM 232
13968-55-3 **URANIUM 233**	URANIUM 233
13966-29-5 **URANIUM 234**	URANIUM 234
15117-96-1 **URANIUM 235**	URANIUM 235
13982-70-2 **URANIUM 236**	URANIUM 236
14269-75-1 **URANIUM 237**	URANIUM 237
13982-01-9 **URANIUM 239**	URANIUM 239
15687-53-3 **URANIUM 240**	URANIUM 240
7440-61-1 URANIUM (NATURAL) [ONT, ONT]	URANIUM
7440-61-1 URANIUM, (NATURAL) SOLUBLE AND INSOLUBLE [MEX, MEX]	URANIUM
541-09-3 URANIUM BIS(ACETATO-O)DIOXO- [PAL]	URANYL ACETATE
10102-06-4 URANIUM BIS(NITRATO-O)DIOXO-, (T-4)- [PAL]	URANYL NITRATE

Chemical Name	Reference Name
36478-76-9 URANIUM BIS(NITRATO-O,O')DIOXO-, (OC-6-11) - [PAL]	URANYL NITRATES-PROPYL ESTER
RR-00617-7 URANIUM COMPOUNDS [MAK, MAK, CAL, ONT, ONT]	URANIUM COMPOUNDS, N.O.S.
RR-00617-7 URANIUM COMPOUNDS, N.O.S.	URANIUM COMPOUNDS, N.O.S.
RR-00617-7 URANIUM COMPOUNDS, NATURAL [GBR, GBR]	URANIUM COMPOUNDS, N.O.S.
7440-61-1 URANIUM, ELEMENTAL [WHM]	URANIUM
7783-81-5 URANIUM HEXAFLUORIDE	URANIUM HEXAFLUORIDE
7440-61-1 URANIUM (INSOLUBLE COMPOUNDS) [NIO, NIO, NIO, NIO]	URANIUM
RR-00045-3 URANIUM INSOLUBLE COMPOUNDS	URANIUM INSOLUBLE COMPOUNDS
7440-61-1 URANIUM (NATURAL) [AUS, AUS]	URANIUM
7440-61-1 URANIUM(NATURAL) [FLA]	URANIUM
7440-61-1 URANIUM (NATURAL) [MSL, TLV, TLV]	URANIUM
7440-61-1 URANIUM (NATURAL), INSOLUBLE [QBC]	URANIUM
RR-00045-3 URANIUM (NATURAL), INSOLUBLE COMPOUNDS [CEX, CEX]	URANIUM INSOLUBLE COMPOUNDS
7440-61-1 URANIUM (NATURAL) SOLUBLE AND INSOLUBLE COMPOUNDS [BC1, BC1]	URANIUM
RR-00044-2 URANIUM (NATURAL), SOLUBLE COMPOUNDS [CEX]	URANIUM SOLUBLE COMPOUNDS
RR-00044-2 URANIUM SOLUBLE COMPOUNDS	URANIUM SOLUBLE COMPOUNDS
RR-00044-2 URANIUM (SOLUBLE COMPOUNDS) [NIO, NIO, NIO, NIO]	URANIUM SOLUBLE COMPOUNDS
541-09-3 URANYL ACETATE	URANYL ACETATE
36478-76-9 URANYL NITRATE	URANYL NITRATES-PROPYL ESTER
13520-83-7 URANYL NITRATE HEXAHYDRATE	URANYL NITRATE HEXAHYDRATE
57-13-6 UREA	UREA
13010-47-4 UREA, N-(2-CHLOROETHYL)-N'-CYCLOHEXYL-N-NITROSO- [PAL, PAL]	1-(2-CHLOROETHYL)-3-CYCLOHEXYL-1-NITROSOUREA
RR-00214-2 UREA, CONDENSATE WITH POLY [OXY(METHYL-1,2-ETHANEDIYL)]-ALPHA-[2-AMINOETHYLETHYL]-X-(2-AMINOETHYLETHOXY)	UREA, CONDENSATE WITH POLY[OXY(METHYL-1,2-ETHANEDIYL)]-
330-54-1 UREA, N'-(3,4-DICHLOROPHENYL)-N,N-DIMETHYL- [PAL]	DIURON
759-73-9 UREA, N-ETHYL-N-NITROSO- [PAL, PAL]	N-NITROSO-N-ETHYLUREA
9011-05-6 UREA-FORMALDEHYDE RESINS [OTS]	UREA, POYLMER WITH FORMALDEHYDE
97380-66-0 UREA-FORMALDEHYDE RESIN [PST, OTS, OTS]	UREA-FORMALDEHYDE-RESINS
97380-66-0 UREA-FORMALDEHYDE-RESINS	UREA-FORMALDEHYDE-RESINS
1129-42-6 UREA, (HEXAHYDRO-6-METHYL-2-OXO-4-PYRIMIDINYL)-	UREA, (HEXAHYDRO-6-METHYL-2-OXO-4-PYRIMIDINYL)
124-43-6 UREA HYDROGEN PEROXIDE [NJL]	UREA PEROXIDE
RR-01338-7 UREA HYDROGEN PEROXIDE	UREA HYDROGEN PEROXIDE
1982-49-6 UREA, 1-(2-METHYLCYCLOHEXYL)-3-PHENYL-	UREA, 1-(2-METHYLCYCLOHEXYL)-3-PHENYL-
13547-17-6 UREA, N,N''-METHYLENEBIS-	UREA, N,N''-METHYLENEBIS-
684-93-5 UREA, N-METHYL-N-NITROSO- [PAL, PAL]	N-NITROSO-N-METHYLUREA
124-47-0 UREA NITRATE	UREA NITRATE
17687-37-5 UREA NITRATE [HM1, HM1]	UREA NITRATE (VAN)
17687-37-5 UREA NITRATE (VAN)	UREA NITRATE (VAN)
124-43-6 UREA PEROXIDE	UREA PEROXIDE
9011-05-6 UREA, POYLMER WITH FORMALDEHYDE	UREA, POYLMER WITH FORMALDEHYDE

Chemical Name	Reference Name
68611-64-3 **UREA, REACTION PRODUCTS WITH FORMALDEHYDE**	UREA, REACTION PRODUCTS WITH FORMALDEHYDE
51-79-6 **URETHANE**	URETHANE
51-79-6 URETHANE (ETHYL CARBAMATE) [313, C65, C65, C65]	URETHANE
RR-00240-4 **URETHANE ACRYLATE**	URETHANE ACRYLATE
69-93-2 **URIC ACID**	URIC ACID
26995-91-5 **UROFOLLITROPIN**	UROFOLLITROPIN
RR-00095-3 **USED ENGINE OILS**	USED ENGINE OILS
64741-49-7 **VACUUM TOWER CONDENSATE (PETROLEUM)**	VACUUM TOWER CONDENSATE (PETROLEUM)
110-62-3 **VALERALDEHYDE**	VALERALDEHYDE
110-62-3 N-VALERALDEHYDE [AUS, ONT, OS1, OTV, TLV]	VALERALDEHYDE
109-52-4 **VALERIC ACID**	VALERIC ACID
109-52-4 N-VALERIC ACID [PST]	VALERIC ACID
108-29-2 **4-VALEROLACTONE**	4-VALEROLACTONE
638-29-9 **VALERYL CHLORIDE**	VALERYL CHLORIDE
2001-95-8 **VALINOMYCIN**	VALINOMYCIN
99-66-1 **VALPROATE**	VALPROATE
99-66-1 VALPROATE (VALPROIC ACID) [C65]	VALPROATE
99-66-1 VALPROIC ACID [MSL]	VALPROATE
19120-62-8 **VANADIC ACID, TRIISOBUTYL ESTER**	VANADIC ACID, TRIISOBUTYL ESTER
1314-62-1 VANADIUM [AUS, MIN, OS1]	VANADIUM PENTOXIDE
7440-62-2 **VANADIUM**	VANADIUM
7440-62-2 VANADIUM (V2O5) [BC1, BC1, BC1]	VANADIUM
14867-38-0 **VANADIUM 47**	VANADIUM 47
14331-97-6 **VANADIUM 48**	VANADIUM 48
14392-01-9 **VANADIUM 49**	VANADIUM 49
1314-62-1 VANADIUM PENTOXIDE [TLV]	VANADIUM PENTOXIDE
7440-62-2 VANADIUM, DUST AND FUME [MEX]	VANADIUM
12604-58-9 VANADIUM ALLOY, BASE, V,C,FE (FER-ROVANADIUM) [PAL]	FERROVANADIUM
11130-21-5 **VANADIUM CARBIDE**	VANADIUM CARBIDE
7632-51-1 VANADIUM CHLORIDE (VCL4), (T-4)- [PAL]	VANADIUM TETRACHLORIDE
RR-00093-1 **VANADIUM COMPOUNDS**	VANADIUM COMPOUNDS
7440-62-2 VANADIUM, ELEMENTAL [WHM]	VANADIUM
7440-62-2 VANADIUM (FUME OR DUST) [MSL, PAL]	VANADIUM
1314-62-1 VANADIUM OXIDE (V2O5) [PAL]	VANADIUM PENTOXIDE
27774-13-6 VANADIUM, OXO[SULFATO(2-)-O]- [PAL]	VANADYL SULFATE
7727-18-6 **VANADIUM OXYTRICHLORIDE**	VANADIUM OXYTRICHLORIDE
1314-62-1 **VANADIUM PENTOXIDE**	VANADIUM PENTOXIDE
1314-62-1 VANADIUM PENTOXIDE, DUST AND FUME [FLA, MSL]	VANADIUM PENTOXIDE
1314-62-1 VANADIUM RESPIRABLE DUST AND FUME [ONT]	VANADIUM PENTOXIDE
7632-51-1 **VANADIUM TETRACHLORIDE**	VANADIUM TETRACHLORIDE
7718-98-1 **VANADIUM TRICHLORIDE**	VANADIUM TRICHLORIDE
1314-34-7 **VANADIUM TRIOXIDE**	VANADIUM TRIOXIDE
27774-13-6 **VANADYL SULFATE**	VANADYL SULFATE
1314-62-1 VANDIUM PENTOXIDE (V2O5) [CEX]	VANADIUM PENTOXIDE
121-33-5 **VANILLIN**	VANILLIN

Chemical Name	Reference Name
128-66-5 VAT YELLOW 4 [IAR]	C.I. VAT YELLOW 4
68956-68-3 VEGETABLE OIL	VEGETABLE OIL
68956-68-3 VEGETABLE OIL MIST [OS1, OS1]	VEGETABLE OIL
8008-89-7 VEGETABLE OIL MISTS	VEGETABLE OIL MISTS
68956-68-3 VEGETABLE OIL MISTS [AUS, ONT, TLV]	VEGETABLE OIL
68956-68-3 VEGETABLE OIL MIST [MEX]	VEGETABLE OIL
120-14-9 VERATRALDEHYDE	VERATRALDEHYDE
1318-00-9 VERMICULITE	VERMICULITE
865-21-4 VINBLASTINE	VINBLASTINE
143-67-9 VINBLASTINE SULFATE	VINBLASTINE SULFATE
143-67-9 VINBLASTINE SULPHATE [IAR]	VINBLASTINE SULFATE
50471-44-8 VINCLOZOLIN [3-(3,5-DICHLOROPHENYL)-5-ETHENYL-5-METHYL-2,4-OXAZOLIDINEDIONE]	VINCLOZOLIN [3-(3,5-DICHLOROPHENYL)-5-ETHENYL-5-METHYL-2,4-OIAZOL-1-YL)-2-BUTANONE]
57-22-7 VINCRISTINE	VINCRISTINE
2068-78-2 VINCRISTINE SULFATE	VINCRISTINE SULFATE
2068-78-2 VINCRISTINE SULPHATE [IAR]	VINCRISTINE SULFATE
75-01-4 VINYL CHLORIDE (CHLOROETHYLENE) [HON, HON]	VINYL CHLORIDE
75-35-4 VINYLIDENE CHLORIDE (1,1-DICHLOROETHYLENE) [HON, HON]	VINYLIDENE CHLORIDE
108-05-4 VINYL ACETATE	VINYL ACETATE
108-05-4 VINYL ACETATE (ACETIC ACID ETHENYL ESTER) [TSC, TSC]	VINYL ACETATE
63181-88-4 VINYL ACETATE - ACRYLIC ESTER COPOLYMERS	VINYL ACETATE - ACRYLIC ESTER COPOLYMERS
25085-41-0 VINYL ACETATE - BUTYL ACRYLATE - ACRYLIC ACID TERPOLYMER	VINYL ACETATE - BUTYL ACRYLATE - ACRYLIC ACID TERPOLYMER
68928-85-8 VINYL ACETATE, CROTONIC ACID, VINYL NEODECANOATE, GLYCIDYL METHACRYLATE POLYMER	VINYL ACETATE, CROTONIC ACID, VINYL NEODECANOATE, GLYCIDYL M
108-05-4 VINYL ACETATE MONOMER [302, MSL]	VINYL ACETATE
108-05-4 VINYL ACETATE MONOMER [ACETIC ACID ETHENYL ESTER] [EP3]	VINYL ACETATE
25086-48-0 VINYL ACETATE-VINYL ALCOHOL CHLORIDE POLYMER	VINYL ACETATE-VINYL ALCOHOL CHLORIDE POLYMER
689-97-4 VINYL ACETYLENE [FLA, MSL, NFP]	1-BUTEN-3-YNE
689-97-4 VINYL ACETYLENE [1-BUTEN-3-YNE] [EP3]	1-BUTEN-3-YNE
3917-15-5 VINYL ALLYL ETHER [FLA, MSL, NFP, NFP]	1-PROPENE, 3-(ETHENYLOXY)-
RR-01156-3 VINYLBENZENE-VEGETABLE OIL COPOLYMER	VINYLBENZENE-VEGETABLE OIL COPOLYMER
30030-25-2 VINYLBENZYL CHLORIDE [FLA, MSL]	BENZENE, (CHLOROMETHYL)ETHENYL-
30030-25-2 VINYLBENZYLCHLORIDE [NFP, NFP]	BENZENE, (CHLOROMETHYL)ETHENYL-
593-60-2 VINYL BROMIDE	VINYL BROMIDE
593-60-2 VINYL BROMIDE (BROMOETHYLENE) [QBC, QBC]	VINYL BROMIDE
111-34-2 VINYL BUTYL ETHER [FLA, MSL, NFP, NFP]	BUTYL VINYL ETHER
123-20-6 VINYL BUTYRATE	VINYL BUTYRATE
75-01-4 VINYL CHLORIDE	VINYL CHLORIDE
75-01-4 VINYL CHLORIDE (CHLOROETHYLENE) [CW2]	VINYL CHLORIDE
75-01-4 VINYL CHLORIDE, MONOMER [AUS, AUS, AUS]	VINYL CHLORIDE
9003-22-9 VINYL CHLORIDE-VINYL ACETATE COPOLYMER	VINYL CHLORIDE-VINYL ACETATE COPOLYMER

Chemical Name	Reference Name
25086-48-0 VINYL CHLORIDE, VINYL ACETATE AND VINYL ALCOHOL COPOLYMER [PST]	VINYL ACETATE-VINYL ALCOHOL CHLORIDE POLYMER
75-01-4 VINYL CHLORIDE [ETHENE, CHLORO-] [EP3]	VINYL CHLORIDE
2549-51-1 VINYL CHLOROACETATE	VINYL CHLOROACETATE
110-75-8 VINYL-2-CHLOROETHYL ETHER [FLA]	2-CHLOROETHYL VINYL ETHER
110-75-8 VINYL 2-CHLOROETHYL ETHER [MSL, NFP, NFP, WHM]	2-CHLOROETHYL VINYL ETHER
14861-06-4 VINYL CROTONATE [FLA, MSL, NFP, NFP]	2-BUTENOIC ACID, ETHENYL ESTER
107-13-1 VINYL CYANIDE (ACRYLONITRILE) [QBC, QBC, QBC]	ACRYLONITRILE
100-40-3 4-VINYL CYLCOHEXENE [FLA]	4-VINYLCYCLOHEXENE
100-40-3 VINYLCYCLOHEXANE, 4- [OTS]	4-VINYLCYCLOHEXENE
100-40-3 4-VINYLCYCLOHEXENE	4-VINYLCYCLOHEXENE
100-40-3 VINYL CYCLOHEXENE [MIN]	4-VINYLCYCLOHEXENE
100-40-3 4-VINYL CYCLOHEXENE [MSL, NFP, NFP, TLV, TLV]	4-VINYLCYCLOHEXENE
100-40-3 VINYLCYCLOHEXENE [WEL]	4-VINYLCYCLOHEXENE
100-40-3 4-VINYLCYCLOHEXENE (CYCLOHEXENE, 4-ETHENYL-) [TSC, TSC, TSC, TSC]	4-VINYLCYCLOHEXENE
106-87-6 4-VINYL-1-CYCLOHEXENE DIEPOXIDE [CAB, CSR, CSR]	VINYL CYCLOHEXENE DIOXIDE
106-87-6 4-VINYL-1-CYCLOHEXENE DIEPOXIDE (VINYL CYCLOHEXENE DIOXIDE) [C65]	VINYL CYCLOHEXENE DIOXIDE
106-87-6 4-VINYL-1-CYCLOHEXENE DIEPOXIDE [NTP]	VINYL CYCLOHEXENE DIOXIDE
106-87-6 VINYL CYCLOHEXENE DIOXIDE	VINYL CYCLOHEXENE DIOXIDE
106-87-6 4-VINYL-1-CYCLOHEXENE DIOXIDE [MAK]	VINYL CYCLOHEXENE DIOXIDE
RR-01002-6 VINYL EPOXY ESTER	VINYL EPOXY ESTER
627-27-0 VINYL ETHYL ALCOHOL	VINYL ETHYL ALCOHOL
109-92-2 VINYL ETHYL ETHER	VINYL ETHYL ETHER
109-92-2 VINYL ETHYL ETHER [ETHENE, ETHOXY-] [EP3]	VINYL ETHYL ETHER
94-04-2 VINYL 2-ETHYLHEXOATE [MSL, NFP, NFP]	HEXANOIC ACID, 2-ETHYL-, ETHENYL ESTER
103-44-6 VINYL 2-ETHYLHEXYL ETHER	VINYL 2-ETHYLHEXYL ETHER
5408-74-2 2-VINYL-5-ETHYLPYRIDINE [FLA, MSL, NFP, NFP]	PYRIDINE, 2-ETHENYL-5-ETHYL-
75-02-5 VINYL FLUORIDE	VINYL FLUORIDE
75-02-5 VINYL FLUORIDE [ETHENE, FLUORO-] [EP3]	VINYL FLUORIDE
RR-00527-6 VINYL HALIDES	VINYL HALIDES
75-35-4 VINYLIDENE CHLORIDE	VINYLIDENE CHLORIDE
75-35-4 VINYLIDENE CHLORIDE (1,1-DICHLOROEHTYLENE) [CAA]	VINYLIDENE CHLORIDE
75-35-4 VINYLIDENE CHLORIDE (1,1-DICHLOROETHYLENE) [OS1]	VINYLIDENE CHLORIDE
9011-06-7 VINYLIDENE CHLORIDE-VINYL CHLORIDE COPOLYMERS	VINYLIDENE CHLORIDE-VINYL CHLORIDE COPOLYMERS
75-35-4 VINYLIDENE CHLORIDE [ETHENE, 1,1-DICHLORO-] [EP3]	VINYLIDENE CHLORIDE
75-38-7 VINYLIDENE FLUORIDE	VINYLIDENE FLUORIDE
75-38-7 VINYLIDENE FLUORIDE [TSE]	VINYLIDENE FLUORIDE
75-38-7 VINYLIDENE FLUORIDE [ETHENE, 1,1-DIFLUORO-] [EP3]	VINYLIDENE FLUORIDE
109-53-5 VINYL ISOBUTYL ETHER	VINYL ISOBUTYL ETHER

Chemical Name	Reference Name
RR-00887-7 VINYL ISOOCTYL ETHER	VINYL ISOOCTYL ETHER
926-65-8 VINYL ISOPROPYL ETHER [FLA, MSL, NFP, NFP]	PROPANE, 2-(ETHENYLOXY)-
1663-35-0 VINYL-2-METHOXYETHYL ETHER [FLA]	ETHENE, (2-METHOXYETHOXY)-
1663-35-0 VINYL 2-METHOXYETHYL ETHER [MSL, NFP, NFP]	ETHENE, (2-METHOXYETHOXY)-
107-25-5 VINYL METHYL ETHER	VINYL METHYL ETHER
107-25-5 VINYL METHYL ETHER [ETHENE, METHOXY-] [EP3]	VINYL METHYL ETHER
RR-01470-0 VINYL NITRATE POLYMER	VINYL NITRATE POLYMER
3048-64-4 VINYLNORBORNENE	VINYLNORBORNENE
RR-00835-5 VINYL OCTADECYL ETHER	VINYL OCTADECYL ETHER
105-38-4 VINYL PROPIONATE [FLA, MSL, NFP, NFP]	PROPANOIC ACID, ETHENYL ESTER
100-43-6 4-VINYLPYRIDINE	4-VINYLPYRIDINE
25086-29-7 VINYLPYRROLIDINONE-STYRENE POLYMER	VINYLPYRROLIDINONE-STYRENE POLYMER
88-12-0 N-VINYL-2-PYRROLIDONE	N-VINYL-2-PYRROLIDONE
88-12-0 1-VINYLPYRROLIDONE [NFP, NFP]	N-VINYL-2-PYRROLIDONE
25013-15-4 VINYL TOLUENE	VINYL TOLUENE
25013-15-4 VINYLTOLUENE [CAL, WHM]	VINYL TOLUENE
25013-15-4 VINYL TOLUENE (MIXTURE OF M- AND P-ISOMERS) [ONT, ONT]	VINYL TOLUENE
79-00-5 VINYL TRICHLORIDE (1,1,2-TRICHLOROETHANE) [C65, C65]	1,1,2-TRICHLOROETHANE
75-94-5 VINYLTRICHLOROSILANE	VINYLTRICHLOROSILANE
75-94-5 VINYL TRICHLOROSILANE [FLA, MSL, NFP, NFP, NJL, NJL]	VINYLTRICHLOROSILANE
1067-53-4 VINYLTRIS(METHOXYETHOXY)SILANE	VINYLTRIS(METHOXYETHOXY)SILANE
5579-85-1 VINYZENE	VINYZENE
68-26-8 VITAMIN A	VITAMIN A
1406-18-4 VITAMIN E	VITAMIN E
58-95-7 VITAMIN E ACETATE	VITAMIN E ACETATE
8032-32-4 VM&P NAPHTHA [MEX, MEX]	LIGROINE
8030-30-6 V.M. & P. NAPHTHA [ONT]	BENZIN
8032-32-4 VM & P NAPHTHA [ALB, ALB, OS1, OS1, TLV]	LIGROINE
8032-32-4 VM&P NAPHTHA [WHM]	LIGROINE
8030-30-6 VM & P (VARNISH MAKERS & PAINTERS) NAPHTHA [CAL]	BENZIN
8030-30-6 VM & P (VARNISH MAKERS AND PAINTERS) NAPHTHA [CEX, CEX]	BENZIN
8030-30-6 VMP NAPHTHA (PAINT AND VARNISH MANU-FACTURING) [QBC]	BENZIN
RR-01157-4 WALNUT SHELLS	WALNUT SHELLS
81-81-2 WARFARIN	WARFARIN
81-81-2 WARFARIN (ISO) [GBR, GBR]	WARFARIN
81-81-2 WARFARIN AND SALTS [313]	WARFARIN
81-81-2 WARFARIN, & SALTS [EPA, RCU, RCU, RCU]	WARFARIN
129-06-6 WARFARIN SODIUM	WARFARIN SODIUM
RR-00530-1 WASTE ANESTHETIC GASES AND VAPORS	WASTE ANESTHETIC GASES AND VAPORS
RR-01600-2 WASTE CRANKCASE OILS	WASTE CRANKCASE OILS
RR-01509-8 WASTE OIL	WASTE OIL
7732-18-5 WATER	WATER

Chemical Name	Reference Name
RR-00803-7 WATER GAS (CARBURETED)	WATER GAS (CARBURETED)
RR-01510-1 WATER REACTIVE SOLID, N.O.S.	WATER REACTIVE SOLID, N.O.S.
RR-01511-2 WAX, LIQUID	WAX, LIQUID
63231-60-7 WAX, MICROCRYSTALLINE [NFP, NFP]	MICROCRYSTALLINE WAX
RR-00838-8 WAX, OXACERITE	WAX, OXACERITE
RR-00838-8 WAX, OZOCERITE [NFP, NFP]	WAX, OXACERITE
8002-74-2 WAX, PARAFFIN [NFP, NFP]	PARAFFIN WAXES AND HYDROCARBON WAXES
RR-00009-9 WELDING FUME [GBR]	WELDING FUMES (NOC)
RR-00009-9 WELDING FUME OR PARTICULATE, NOC [ONT]	WELDING FUMES (NOC)
RR-00009-9 WELDING FUMES [BC1, BC1, IAR, MIN, PAL]	WELDING FUMES (NOC)
RR-00009-9 WELDING FUMES (NOC)	WELDING FUMES (NOC)
RR-00009-9 WELDING FUMES (TOTAL PARTICULATE) [OS1]	WELDING FUMES (NOC)
RR-00009-9 WELDING FUMES (TOTAL OF PARTICLES) [QBC]	WELDING FUMES (NOC)
70983-99-2 WHALE OIL	WHALE OIL
RR-01158-5 WHEAT	WHEAT
8006-95-9 WHEAT GERM OIL	WHEAT GERM OIL
68608-58-2 WHEY	WHEY
1327-53-3 WHITE ARSENIC [HM1, HM1, HM1]	ARSENIC TRIOXIDE
1332-21-4 WHITE ASBESTOS [HM1, HM1, HM1]	ASBESTOS
8012-89-3 WHITE BEESWAX	WHITE BEESWAX
8042-47-5 WHITE MINERAL OIL	WHITE MINERAL OIL
8042-47-5 WHITE MINERAL OIL (PETROLEUM) [PST]	WHITE MINERAL OIL
7723-14-0 WHITE PHOSPHORUS [T33, T34, TSC, TSC]	PHOSPHORUS
12185-10-3 WHITE PHOSPHORUS	WHITE PHOSPHORUS
8052-41-3 WHITE SPIRIT [GBR, GBR]	STODDARD SOLVENT
8052-41-3 WHITE SPIRIT, LOW (15-20%) AROMATIC [HM1]	STODDARD SOLVENT
8052-41-3 WHITE SPIRITS [AUS, AUS]	STODDARD SOLVENT
68917-75-9 WINTERGREEN OIL	WINTERGREEN OIL
13983-17-0 WOLLASTONITE [CAL, IAR]	WOLLASTONITE (CA(SIO3))
13983-17-0 WOLLASTONITE (CA(SIO3))	WOLLASTONITE (CA(SIO3))
13983-17-0 WOLLASTONITE CALCIUM SILICATES [CSR]	WOLLASTONITE (CA(SIO3))
RR-00514-1 WOOD DUST [MAK, MAK]	WOOD DUST, ALL SOFT AND HARD WOODS
RR-00514-1 WOOD DUST (NONALLERGIC) [BC1, BC1, QBC]	WOOD DUST, ALL SOFT AND HARD WOODS
RR-00524-3 WOOD DUST, ALLERGENIC [ALB, ALB]	WOOD DUST, WESTERN RED CEDAR
RR-00524-3 WOOD DUST, ALLERGENIC - WESTERN RED CEDAR [BC1, BC1]	WOOD DUST, WESTERN RED CEDAR
RR-01736-7 WOOD DUST, ALLERGENIC - EXCLUDING WESTERN RED CEDAR	WOOD DUST, ALLERGENIC - EXCLUDING WESTERN RED CEDAR
RR-00514-1 WOOD DUST, ALL SOFT AND HARD WOODS	WOOD DUST, ALL SOFT AND HARD WOODS
RR-00514-1 WOOD DUST, ALL SOFT AND HARD WOODS, EXCEPT WESTERN RED CEDAR [OS1, OS1]	WOOD DUST, ALL SOFT AND HARD WOODS
RR-00016-8 WOOD DUST (BEECH AND OAK) [MAK]	WOOD DUSTS-HARD WOOD
RR-00016-8 WOOD DUST, CERTAIN HARD WOODS [MIN]	WOOD DUSTS-HARD WOOD
RR-00016-8 WOOD DUST (CERTAIN HARDWOODS AS BEACH AND OAK) [AUS, AUS, AUS]	WOOD DUSTS-HARD WOOD
RR-00016-8 WOOD DUST (CERTAIN HARD WOODS AS BEECH AND OAK) [TLV]	WOOD DUSTS-HARD WOOD
RR-00016-8 WOOD DUST (HARD WOOD) [GBR, GBR]	WOOD DUSTS-HARD WOOD

Chemical Name	Reference Name
RR-00514-1 WOOD DUST, NONALLERGENIC [ALB, ALB]	WOOD DUST, ALL SOFT AND HARD WOODS
RR-00524-3 WOOD DUST, RED CEDAR [QBC]	WOOD DUST, WESTERN RED CEDAR
RR-00016-8 WOOD DUSTS - CERTAIN HARD WOODS AS BEECH AND OAK [ONT]	WOOD DUSTS-HARD WOOD
RR-00016-8 WOOD DUSTS-HARD WOOD	WOOD DUSTS-HARD WOOD
RR-00017-9 WOOD DUST (SOFT WOOD) [AUS, AUS, AUS, AUS]	WOOD DUSTS-SOFT WOODS
RR-00017-9 WOOD DUST, SOFT WOOD [TLV, TLV]	WOOD DUSTS-SOFT WOODS
RR-00017-9 WOOD DUST, SOFT WOODS [MIN]	WOOD DUSTS-SOFT WOODS
RR-00017-9 WOOD DUST - SOFT WOOD [ONT, ONT]	WOOD DUSTS-SOFT WOODS
RR-00017-9 WOOD DUSTS-SOFT WOODS	WOOD DUSTS-SOFT WOODS
RR-00524-3 WOOD DUST, WESTERN RED CEDAR	WOOD DUST, WESTERN RED CEDAR
RR-01512-3 WOOD FILLER, LIQUID	WOOD FILLER, LIQUID
RR-00545-8 WOOD INDUSTRIES - FURNITURE & CABINET MAKING	WOOD INDUSTRIES - FURNITURE & CABINET MAKING
RR-01513-4 WOOD PRESERVATIVE, LIQUID	WOOD PRESERVATIVE, LIQUID
RR-01513-4 WOOD PRESERVATIVES (CONTAINING ARSENIC AND CHROMATE) [CAB]	WOOD PRESERVATIVE, LIQUID
11138-66-2 XANTHAN GUM	XANTHAN GUM
140-89-6 XANTHATES [HON]	CARBONODITHIOIC ACID, O-ETHYL ESTER, POTASSIUM SALT
7440-63-3 XENON	XENON
15151-08-3 XENON 120	XENON 120
17913-80-3 XENON 121	XENON 121
15151-09-4 XENON 122	XENON 122
15700-10-4 XENON 123	XENON 123
13994-18-8 XENON 125	XENON 125
13994-19-9 XENON 127	XENON 127
RR-00307-6 XENON 129M	XENON 129M
RR-00288-0 XENON 131M	XENON 131M
14932-42-4 XENON 133	XENON 133
RR-00305-4 XENON 133M	XENON 133M
14995-62-1 XENON 135	XENON 135
RR-00304-3 XENON 135M	XENON 135M
15751-81-2 XENON 138	XENON 138
95-47-6 O-XYLENE	O-XYLENE
106-42-3 P-XYLENE	P-XYLENE
108-38-3 M-XYLENE	M-XYLENE
1330-20-7 XYLENE [CEX, CEX, CEX, FLA, IAR, MAK, MAK, MAK, MIN, MSL, NHS, NHS, PST]	XYLENES (O-, M-, P- ISOMERS)
1330-20-7 XYLENE (O-, M-, P- ISOMERS) [ALB, ALB]	XYLENES (O-, M-, P- ISOMERS)
1330-20-7 XYLENE, ALL ISOMERS [CAL, GBR, GBR, GBR]	XYLENES (O-, M-, P- ISOMERS)
1477-55-0 M-XYLENE-ALPHA, ALPHA'-DIAMINE	M-XYLENE-ALPHA, ALPHA'-DIAMINE
1477-55-0 M-XYLENE A,A'-DIAMINE [AUS, AUS]	M-XYLENE-ALPHA, ALPHA'-DIAMINE
1477-55-0 M-XYLENE-A,A'-DIAMINE [CEX, CEX]	M-XYLENE-ALPHA, ALPHA'-DIAMINE
1477-55-0 M-XYLENE ALPHA, ALPHA'-DIAMINE [OS1, OS1]	M-XYLENE-ALPHA, ALPHA'-DIAMINE
1477-55-0 XYLENE M-, -DIAMINE [QBC, QBC]	M-XYLENE-ALPHA, ALPHA'-DIAMINE
1477-55-0 M-XYLENE ALPHA,ALPHA'-DIAMINE [TLV, TLV]	M-XYLENE-ALPHA, ALPHA'-DIAMINEZYME)
1330-20-7 XYLENE (MIXED) [CWA, EPA]	XYLENES (O-, M-, P- ISOMERS)

Chemical Name	Reference Name
1330-20-7 XYLENE (MIXED ISOMERS) [313, CAN]	XYLENES (O-, M-, P- ISOMERS)
68920-06-9 XYLENE RANGE AROMATIC SOLVENT	XYLENE RANGE AROMATIC SOLVENT
95-47-6 O-XYLENES [CAA]	O-XYLENE
106-42-3 P-XYLENES [CAA]	P-XYLENE
108-38-3 M-XYLENES [CAA]	M-XYLENE
1330-20-7 XYLENES [BEI, CN4, HM1, HM1, HM1, NJL, NJL]	XYLENES (O-, M-, P- ISOMERS)
1330-20-7 XYLENE(S) [RCA]	XYLENES (O-, M-, P- ISOMERS)
1330-20-7 XYLENES (NOS) [HON, HON]	XYLENES (O-, M-, P- ISOMERS)
1330-20-7 XYLENES (TOTAL) [CD1]	XYLENES (O-, M-, P- ISOMERS)
1330-20-7 XYLENES (ISOMERS AND MIXTURE) [CAA]	XYLENES (O-, M-, P- ISOMERS)
1330-20-7 XYLENES (MIXED) [CSR, CSR]	XYLENES (O-, M-, P- ISOMERS)
1330-20-7 XYLENES (O-, M-, P- ISOMERS)	XYLENES (O-, M-, P- ISOMERS)
1330-20-7 XYLENES (TOTAL) [SDW, SDW]	XYLENES (O-, M-, P- ISOMERS)
25321-41-9 XYLENESULFONIC ACID	XYLENESULFONIC ACID
26447-10-9 XYLENESULFONIC ACID, AMMONIUM SALT	XYLENESULFONIC ACID, AMMONIUM SALT
28088-63-3 XYLENESULFONIC ACID, CALCIUM SALT	XYLENESULFONIC ACID, CALCIUM SALT
36729-43-8 XYLENESULFONIC ACID, MAGNESIUM SALT	XYLENESULFONIC ACID, MAGNESIUM SALT
30346-73-7 XYLENESULFONIC ACID, POTASSIUM SALT	XYLENESULFONIC ACID, POTASSIUM SALT
1300-72-7 XYLENESULFONIC ACID, SODIUM SALT [PST]	SODIUM XYLENE SULFONATE
36729-46-1 XYLENESULFONIC ACID, ZINC SALT	XYLENESULFONIC ACID, ZINC SALT
1330-20-7 XYLENE (TOTAL) [PQL]	XYLENES (O-, M-, P- ISOMERS)
95-87-4 2,5-XYLENOL	2,5-XYLENOL
108-68-9 3,5-XYLENOL	3,5-XYLENOL
1300-71-6 XYLENOL	XYLENOL
1300-71-6 XYLENOLS [HM1, HM1, HM1, HM1]	XYLENOL
1300-71-6 XYLENOLS (MIXED) [HON]	XYLENOL
1300-71-6 XYLENOLS, MIXED [PST]	XYLENOL
87-62-7 2,6-XYLIDENE	2,6-XYLIDENE
1300-73-8 XYLIDENE (MIXED ISOMERS) [QBC, QBC, QBC]	XYLIDINE
1300-73-8 XYLIDINE, ALL ISOMERS [GBR, GBR, GBR]	XYLIDINE
87-62-7 O-XYLIDINE [FLA, MSL, NFP, NFP]	2,6-XYLIDENE
95-68-1 2,4-XYLIDINE	2,4-XYLIDINE
95-78-3 2,5-XYLIDINE	2,5-XYLIDINE
1300-73-8 XYLIDINE	XYLIDINE
87-62-7 2,6-XYLIDINE (2,6-DIMETHYLANILINE) [C65]	2,6-XYLIDENE
1300-73-8 XYLIDINE (MIXED ISOMERS) [TLV, TLV, TLV]	XYLIDINE
1300-73-8 XYLIDINES [HM1, HM1, PAL]	XYLIDINE
28258-59-5 XYLYL BROMIDE	XYLYL BROMIDE
35884-77-6 XYLYL BROMIDE [HM1, HM1]	BENZENE, BROMODIMETHYL
RR-01455-1 P-XYLYL DIAZIDE	P-XYLYL DIAZIDE
28347-13-9 XYLYLENE DICHLORIDE	XYLYLENE DICHLORIDE
68876-77-7 YEAST	YEAST
85-84-7 YELLOW AB	YELLOW AB
131-79-3 YELLOW OB	YELLOW OB
50-55-5 YOHIMBAN-16-CARBOXYLIC ACID, 11,17-DIMETHOXY-18-[(3,4,5-TRIMETHOXYBENZOYL)OXY]-, METHYL ESTER, (3.BETA.,16.BETA.,17.ALPHA.,18.BETA.,20.ALPHA.)- [PAL, PAL]	RESERPINE

Chemical Name	Reference Name
24347-38-4 YTTERBIUM 162	YTTERBIUM 162
14834-83-4 YTTERBIUM 166	YTTERBIUM 166
14041-45-3 YTTERBIUM 167	YTTERBIUM 167
14269-78-4 YTTERBIUM 169	YTTERBIUM 169
14041-44-2 YTTERBIUM 175	YTTERBIUM 175
14119-23-4 YTTERBIUM 177	YTTERBIUM 177
29919-07-1 YTTERBIUM 178	YTTERBIUM 178
10361-91-8 YTTERBIUM CHLORIDE	YTTERBIUM CHLORIDE
13768-67-7 YTTERBIUM NITRATE	YTTERBIUM NITRATE
7440-65-5 YTTRIUM	YTTRIUM
14809-53-1 YTTRIUM 86	YTTRIUM 86
RR-00303-2 YTTRIUM 86M	YTTRIUM 86M
14274-68-1 YTTRIUM 87	YTTRIUM 87
13982-36-0 YTTRIUM 88	YTTRIUM 88
10098-91-6 YTTRIUM 90	YTTRIUM 90
RR-00302-1 YTTRIUM 90M	YTTRIUM 90M
14234-24-3 YTTRIUM 91	YTTRIUM 91
RR-00301-0 YTTRIUM 91M	YTTRIUM 91M
15751-59-4 YTTRIUM 92	YTTRIUM 92
14981-70-5 YTTRIUM 93	YTTRIUM 93
15422-72-7 YTTRIUM 94	YTTRIUM 94
15422-71-6 YTTRIUM 95	YTTRIUM 95
7440-65-5 YTTRIUM METAL AND COMPOUNDS [MEX, MEX]	YTTRIUM
7440-65-5 YTTRIUM COMPOUNDS [NIO, NIO]	YTTRIUM
RR-00618-8 YTTRIUM COMPOUNDS	YTTRIUM COMPOUNDS
7440-65-5 YTTRIUM, ELEMENTAL [WHM]	YTTRIUM
7440-65-5 YTTRIUM METAL [ONT]	YTTRIUM
90147-57-2 YUCCA EXTRACTIVES (FROM YUCCA, AGAVACEAE)	YUCCA EXTRACTIVES (FROM YUCCA, AGAVACEAE)
17924-92-4 ZEARALENONE	ZEARALENONE
315-18-4 ZECTRAN [IAR]	MEXACARBATE
9010-66-6 ZEIN	ZEIN
7440-66-6 ZINC	ZINC
14833-23-9 ZINC 62	ZINC 62
14833-26-2 ZINC 63	ZINC 63
13982-39-3 ZINC 65	ZINC 65
13982-23-5 ZINC 69	ZINC 69
RR-00300-9 ZINC 69M	ZINC 69M
RR-00299-3 ZINC 71M	ZINC 71M
15743-55-2 ZINC 72	ZINC 72
7646-85-7 ZINC CHLORIDE FUME [MEX, MEX]	ZINC CHLORIDE
1314-13-2 ZINC OXIDE FUME [MEX, MEX]	ZINC OXIDE
557-34-6 ZINC ACETATE	ZINC ACETATE
14639-97-5 ZINC AMMONIUM CHLORIDE (ZN.CL4.2H4-N)	ZINC AMMONIUM CHLORIDE (ZN.CL4.2H4-N)
14639-98-6 ZINC AMMONIUM CHLORIDE (ZN.CL5.3H4-N)	ZINC AMMONIUM CHLORIDE (ZN.CL5.3H4-N)
52628-25-8 ZINC AMMONIUM CHLORIDE [VAN]	ZINC AMMONIUM CHLORIDE [VAN]
52628-25-8 ZINC AMMONIUM CHLORIDE [NJL]	ZINC AMMONIUM CHLORIDE [VAN]
63885-01-8 ZINC AMMONIUM NITRITE	ZINC AMMONIUM NITRITE
7783-24-6 ZINC AMMONIUM SULFATE	ZINC AMMONIUM SULFATE

Chemical Name	Reference Name
1303-39-5 ZINC ARSENATE	ZINC ARSENATE
10326-24-6 ZINC ARSENITE	ZINC ARSENITE
RR-00517-4 ZINC ASHES	ZINC ASHES
14639-98-6 ZINCATE (3-), PENTACHLORO-, TRIAMMO-NIUM [MSL]	ZINC AMMONIUM CHLORIDE (ZN.CL5.3H4-N)
14639-98-6 ZINCATE(3-), PENTACHLORO-, TRIAMMO-NIUM [PAL]	ZINC AMMONIUM CHLORIDE (ZN.CL5.3H4-N)
14639-97-5 ZINCATE(2-), TETRACHLORO-, DIAMMONIUM, (T-4)- [PAL]	ZINC AMMONIUM CHLORIDE (ZN.CL4.2H4-N)
39413-47-3 ZINC BERYLLIUM SILICATE	ZINC BERYLLIUM SILICATE
14018-95-2 ZINC BICHROMATE [FLA]	ZINC DICHROMATE
4259-15-8 ZINC O,O-BIS(2-ETHYLHEXYL) PHOSPHO-RODITHIOATE	ZINC O,O-BIS(2-ETHYLHEXYL) PHOSPHORODITHIOATE
117-97-5 ZINC BIS(PENTACHLOROPHENOL)	ZINC BIS(PENTACHLOROPHENOL)
1332-07-6 ZINC BORATE	ZINC BORATE
14519-07-4 ZINC BROMATE	ZINC BROMATE
7699-45-8 ZINC BROMIDE	ZINC BROMIDE
7699-45-8 ZINC BROMIDE (ZNBR2) [PAL]	ZINC BROMIDE
3486-35-9 ZINC CARBONATE	ZINC CARBONATE
10361-95-2 ZINC CHLORATE	ZINC CHLORATE
7646-85-7 ZINC CHLORIDE	ZINC CHLORIDE
7646-85-7 ZINC CHLORIDE (ZNCL2) [PAL]	ZINC CHLORIDE
7646-85-7 ZINC CHLORIDE FUME [ALB, ALB, BC1, BC1, FLA, MSL, NIO, NIO, NIO, ONT, ONT, OS1, OS1, OS1, QBC, TLV, TLV]	ZINC CHLORIDE
13530-65-9 ZINC CHROMATE	ZINC CHROMATE
14018-95-2 ZINC CHROMATE [MIN]	ZINC DICHROMATE
15930-94-6 ZINC CHROMATE HYDROXIDE	ZINC CHROMATE HYDROXIDE
37300-23-5 ZINC CHROMATE PIGMENT [MIN]	ZINC YELLOW (ZINC CHROMATE PIGMENT)
RR-00578-7 ZINC COMPOUNDS	ZINC COMPOUNDS
RR-00578-7 ZINC COMPOUNDS, N.O.S. [NJL]	ZINC COMPOUNDS
557-21-1 ZINC CYANIDE	ZINC CYANIDE
557-21-1 ZINC CYANIDE (ZN(CN)2) [PAL]	ZINC CYANIDE
136-23-2 ZINC DIBUTYL DITHIOCARBAMATE	ZINC DIBUTYL DITHIOCARBAMATE
58270-08-9 ZINC, DICHLORO(4,4-DIMETHYL-5-((((METHYL AMINO)CARBONYL)OXY)IMINO)PENTANENI-TRILE)-, (T-4)-	ZINC, DICHLORO(4,4-DIMETHYL-5-((((METHYL AMINO) CARBONYL)OXY)
14018-95-2 ZINC DICHROMATE	ZINC DICHROMATE
557-20-0 ZINC, DIETHYL- [PAL]	DIETHYLZINC
14324-55-1 ZINC DIETHYLDITHIOCARBAMATE	ZINC DIETHYLDITHIOCARBAMATE
557-05-1 ZINC DISTEARATE [GBR, GBR]	ZINC STEARATE
7779-86-4 ZINC DITHIONITE [NJL]	ZINC HYDROSULFITE
136-53-8 ZINC 2-ETHYLHEXOATE	ZINC 2-ETHYLHEXOATE
14634-93-6 ZINC ETHYLPHENYLDITHIOCARBAMATE	ZINC ETHYLPHENYLDITHIOCARBAMATE
7783-49-5 ZINC FLUORIDE	ZINC FLUORIDE
7783-49-5 ZINC FLUORIDE (ZNF2) [PAL]	ZINC FLUORIDE
16871-71-9 ZINC FLUOROSILICATE [HM1, HM1, HM1]	ZINC SILICOFLUORIDE
557-41-5 ZINC FORMATE	ZINC FORMATE
7779-86-4 ZINC HYDROSULFITE	ZINC HYDROSULFITE
20427-58-1 ZINC HYDROXIDE	ZINC HYDROXIDE
12063-19-3 ZINC IRON OXIDE	ZINC IRON OXIDE

Chemical Name	Reference Name
68187-51-9 ZINC IRON YELLOW	ZINC IRON YELLOW
155-04-4 ZINC MERCAPTOBENZOTHIAZOLE	ZINC MERCAPTOBENZOTHIAZOLE
7440-66-6 ZINC (METALLIC) [PST]	ZINC
12001-85-3 ZINC NAPHTHENATE [PST]	NAPHTHENIC ACIDS, ZINC SALTS
7779-88-6 ZINC NITRATE	ZINC NITRATE
557-09-5 ZINC OCTOATE	ZINC OCTOATE
1314-13-2 ZINC OXIDE	ZINC OXIDE
1314-13-2 ZINC OXIDE (ZNO) [PAL]	ZINC OXIDE
1314-13-2 ZINC OXIDE FUME [FLA, MAK, MAK, MSL, NIO, NIO]	ZINC OXIDE
1314-13-2 ZINC OXIDE (FUME) [QBC, QBC]	ZINC OXIDE
23414-72-4 ZINC PERMANGANTE	ZINC PERMANGANTE
1314-22-3 ZINC PEROXIDE	ZINC PEROXIDE
127-82-2 ZINC PHENOLSULFONATE	ZINC PHENOLSULFONATE
51810-70-9 ZINC PHOSPHIDE	ZINC PHOSPHIDE
1314-84-7 ZINC PHOSPHIDE (ZN3P2) [PAL, RCU, RCU, RCU]	ZINC PHOSPHIDE
14332-59-3 ZINC PHOSPHITE	ZINC PHOSPHITE
11103-86-9 ZINC POTASSIUM CHROMATE	ZINC POTASSIUM CHROMATE
14244-62-3 ZINC POTASSIUM CYANIDE	ZINC POTASSIUM CYANIDE
557-28-8 ZINC PROPIONATE	ZINC PROPIONATE
13463-41-7 ZINC PYRITHIONE	ZINC PYRITHIONE
7446-26-6 ZINC PYROPHOSPHATE	ZINC PYROPHOSPHATE
9010-69-9 ZINC RESINATE	ZINC RESINATE
RR-00758-9 ZINC SELENATE	ZINC SELENATE
13597-46-1 ZINC SELENITE	ZINC SELENITE
16871-71-9 ZINC SILICOFLUORIDE	ZINC SILICOFLUORIDE
557-05-1 ZINC STEARATE	ZINC STEARATE
7733-02-0 ZINC SULFATE	ZINC SULFATE
59766-35-7 ZINC SULFATE, BASIC	ZINC SULFATE, BASIC
1314-98-3 ZINC SULFIDE	ZINC SULFIDE
37300-23-5 ZINC YELLOW [CEX]	ZINC YELLOW (ZINC CHROMATE PIGMENT)
37300-23-5 ZINC YELLOW (ZINC CHROMATE PIGMENT)	ZINC YELLOW (ZINC CHROMATE PIGMENT)
12122-67-7 ZINEB	ZINEB
12122-67-7 ZINEB [CARBAMODITHIOIC ACID, 1,2-ETHANEDIYLBIS-, ZINC COMPLEX] [313]	ZINEB3-(2-METHYL-1-PROPENYL)CYCLOPROPANECAR-BOXYLATE]
297-97-2 ZINOPHOS [RCU, RCU]	O,O-DIETHYL O-PYRAZINYL PHOSPHOROTHIATE
137-30-4 ZIRAM	ZIRAM
14940-68-2 ZIRCON	ZIRCON
16923-95-8 ZIRCONATE(2-), HEXAFLUORO-, DIPOTASSIUM, (OC-6-11)- [PAL]	ZIRCONIUM POTASSIUM FLUORIDE
7440-67-7 ZIRCONIUM	ZIRCONIUM
15743-56-3 ZIRCONIUM 86	ZIRCONIUM 86
14681-75-5 ZIRCONIUM 88	ZIRCONIUM 88
13981-27-6 ZIRCONIUM 89	ZIRCONIUM 89
15751-77-6 ZIRCONIUM 93	ZIRCONIUM 93
13967-71-0 ZIRCONIUM 95	ZIRCONIUM 95
14928-30-4 ZIRCONIUM 97	ZIRCONIUM 97
7440-67-7 ZIRCONIUM COMPOUNDS [MEX, MEX]	ZIRCONIUM
17501-44-9 ZIRCONIUM ACETYLACETONATE	ZIRCONIUM ACETYLACETONATE
7440-67-7 ZIRCONIUM AND COMPOUNDS [TLV, TLV]	ZIRCONIUM

Chemical Name	Reference Name
RR-01252-2 ZIRCONIUM[IV], [2,2-BIS[[2-PROPENY-LOXY]METHYL]-1-BUTANOLATO-1,2]TRIS(2-PROPENOATO-O-)	ZIRCONIUM[IV], [2,2-BIS[[2-PROPENYLOXY]METHYL]-1-BUTANOLATO-HACRYLATE, AND SUBSTITUTED SILANE
10026-11-6 ZIRCONIUM CHLORIDE (ZRCL4), (T-4)- [PAL]	ZIRCONIUM TETRACHLORIDE
10119-31-0 ZIRCONIUM CHLORIDE HYDROXIDE	ZIRCONIUM CHLORIDE HYDROXIDE
RR-00624-6 ZIRCONIUM COMPOUNDS [ALB, ALB, AUS, AUS, BC1, BC1, MAK, MAK, MIN, NIO, NIO, NIO, ONT, ONT, OS1, OS1, OS1, QBC, QBC, CAL, CEX, CEX, GBR, GBR]	ZIRCONIUM COMPOUNDS, N.O.S.
RR-00624-6 ZIRCONIUM COMPOUNDS, N.O.S.	ZIRCONIUM COMPOUNDS, N.O.S.
7440-67-7 ZIRCONIUM, ELEMENTAL [WHM]	ZIRCONIUM
22464-99-9 ZIRCONIUM ETHYL HEXOATE	ZIRCONIUM ETHYL HEXOATE
7704-99-6 ZIRCONIUM HYDRIDE	ZIRCONIUM HYDRIDE
11105-16-1 ZIRCONIUM HYDRIDE [HM1, HM1]	ZIRCONIUM HYDRIDE (VAN)
11105-16-1 ZIRCONIUM HYDRIDE (VAN)	ZIRCONIUM HYDRIDE (VAN)
72854-21-8 ZIRCONIUM NAPHTHENATE	ZIRCONIUM NAPHTHENATE
39049-04-2 ZIRCONIUM NEODECANOATE [PST]	NEODECANOIC ACID, ZIRCONIUM SALTATE
13746-89-9 ZIRCONIUM NITRATE	ZIRCONIUM NITRATE
7699-43-6 ZIRCONIUM OXYCHLORIDE	ZIRCONIUM OXYCHLORIDE
63868-82-6 ZIRCONIUM PICRAMATE	ZIRCONIUM PICRAMATE
16923-95-8 ZIRCONIUM POTASSIUM FLUORIDE	ZIRCONIUM POTASSIUM FLUORIDE
10377-98-7 ZIRCONIUM SODIUM LACTATE	ZIRCONIUM SODIUM LACTATE
14644-61-2 ZIRCONIUM SULFATE	ZIRCONIUM SULFATE
10026-11-6 ZIRCONIUM TETRACHLORIDE	ZIRCONIUM TETRACHLORIDE
13826-66-9 ZIRCONYL NITRATE	ZIRCONYL NITRATE

50-00-0 FORMALDEHYDE
FORMALDEHYDE (GAS) [C65, C65, NTP]
FORMALDEHYDE (SOLUTION) [EP3]

50-01-1 GUANIDINE HYDROCHLORIDE

50-06-6 PHENOBARBITAL
PHENOBARBITOL [MIN]

50-07-7 MITOMYCIN C
AZIRINO[2',3';3,4]PYRROLO[1,2-A]INDOLE-4,7-DIONE,6-AMINO-8-[[(AMINOCARBONYL)OXY] METHYL]-1,1A,2,8,8A,8B-HEXAHYDRO-8A- METHOXY-5-MeTHYL-, [1AS-(A.ALPHA.,8.BETA.,8A.ALPHA., 8B.ALPHA.)]- [PAL, PAL]

50-14-6 ERGOCALCIFEROL
1,2-ETHYLIDENE DICHLORIDE [NFP, NFP, PAL]

50-18-0 CYCLOPHOSPHAMIDE
2H-1,3,2-OXAZAPHOSPHORIN-2-AMINE, N,N-BIS(2-CHLOROETHYL) TETRAHYDRO-, 2-OXIDE [PAL, PAL]
CYCLOPHOSPHAMIDE (ANHYDROUS) [C65, C65, C65, C65, C65]

50-21-5 LACTIC ACID

50-24-8 PREDNISOLONE

50-28-2 ESTRADIOL-17B
ESTRA-1,3,5(10)-TRIENE-3,17-DIOL (17.BETA.)- [PAL, PAL]
ESTRADIOL [NJL, NJL]
ESTRADIOL 17B [C65, C65]
OESTRADIOL-17 BETA [FLA]
OESTRADIOL-17B [CAL]

50-29-3 DDT
1,1,1-TRICHLORO-2,2-BIS(P-CHLORPHENYL) ETHANE [ONT]
1,1,1-TRICHLOROBIS (CHLOROPHENYL)ETHANE [GBR, GBR]
4,4'-DDT [CW2, PQL]
4-4'-DDT [MX1]
BENZENE, 1,1'-(2,2,2-TRICHLOROETHYLIDENE)BIS[4-CHLORO- [PAL, PAL]
DDT (1,1,1-TRICHLORO-2,2-BIS (4-CHLOROPHENYL) ETHANE) [CN2]
DDT (1,1,1-TRICHLORO-2,2-BIS(P-CHLORPHENYL) ETHANE) [CAL]
DDT (DICHLORODIPHENYL-TRICHLOROETHANE) [AUS, BC1, BC1, C65, TSC]
DDT (DICHLORODIPHENYL-TRICHLOROSILANE) [TLV]
DDT (ZEIDANE) [QBC, QBC]
DDT AND METABOLITES [CW3]
DDT, P,P'- [ATS]
DICHLORO DIPHENYL TRICHLOROETHANE (D.D.T.) [FLA]
DICHLORODIPHENYL-TRICHLOROETHANE (DDT) [CSR, CSR, OS1, OS1, OS1, OS1]
DICHLOROPHENYL-TRICHLOROETHANE (DDT) + METABOLITES [CD1]
P,P'-DDT [RCA, UTS]

50-31-7 2,3,6-TBA

50-32-8 BENZO(A)PYRENE
BENZ(A)PYRENE [QBC, QBC]
BENZO(A)PYRENE (3,4-BENZO-PYRENE) [CW2]
BENZO(A)PYRENE (3,4-BENZOPY-RENE) [MX1]
BENZO(ALPHA)PYRENE [BC1, BC1]

50-33-9 PHENYLBUTAZONE

50-34-0 PROPANTHELINE BROMIDE

50-35-1 THALIDOMIDE

50-36-2 COCAINE

50-41-9 CLOMIPHENE CITRATE

50-44-2 6-MERCAPTOPURINE

50-53-3 CHLORPROMAZINE (2-CHLORO-10-(3-DIMETHYLAMINOPROPYL) PHENOTHIAZINE)

50-55-5 RESERPINE
YOHIMBAN-16-CARBOXYLIC ACID, 11,17-DIMETHOXY-18-[(3,4, 5-TRIMETHOXYBENZOYL)OXY]-, METHYL ESTER, (3.BETA.,16.BETA., 17.ALPHA.,18.BETA.,20.ALPHA.)- [PAL, PAL]

50-70-4 D-GLUCITOL
SORBITOL [PST]

50-76-0 ACTINOMYCIN D

50-78-2 ACETYLSALICYLIC ACID (AS-PIRIN)
2-(ACETYLOXY) BENZOIC ACID [ONT]
ACETOL (2) [FLA]
ACETYLSALICYLIC ACID [AUS, CAL, MIN, MSL, NJL]
ASPIRIN [C65, C65]
BENZOIC ACID, 2-(ACETYLOXY)- [PAL]
O-ACETYLSALICYLIC ACID [GBR]

50-81-7 L-ASCORBIC ACID

50-99-7 GLUCOSE
DEXTROSE [PST]

51-02-5 PRONETALOL HYDROCHLORIDE

51-03-6 PIPERONYL BUTOXIDE

51-17-2 1-BENZIMIDAZOLE

51-18-3 TRIETHYLENEMELAMINE
2,4,6-TRIS(1-AZIRIDINYL)-S-TRIAZINE [CAL, IAR]

51-21-8 FLUOROURACIL
5-FLUOROURACIL [IAR, NJL, NJL]
FLUOROURACIL (5-FLUOROURACIL) [313]

51-28-5 2,4-DINITROPHENOL
PHENOL, 2,4-DINITRO- [PAL]

51-43-4 1,2-BENZENEDIOL, 4-[1-HY-DROXY-2-(METHYLAMINO) ETHYL]
1,2-BENZENEDIOL,4-[1-HYDROXY-2-(METHYLAMINO)ETHYL]- [MSL]
EPINEPHRINE [EPA, RCU, RCU]

51-52-5 PROPYLTHIOURACIL
4(1H)-PYRIMIDINONE, 2,3-DIHYDRO-6-PROPYL-2-THIOXO- [PAL, PAL]
6-PROPYL-2-THIOURACIL [NJL]

51-75-2 NITROGEN MUSTARD
MECHLORETHAMINE [302, EPA, EPA, PAL]
N-METHYL-BIS(2-CHLOROETHYL) AMINE [MAK, MAK, WHM]
NITROGEN MUSTARD (MECHLORETHAMINE) [C65, C65]
NITROGEN MUSTARD [2-CHLORO-N-(2-CHLOROETHYL)-N-METHYLETHANAMINE] [313]

51-78-5 P-AMINOPHENOL HYDROCHLO-RIDE

51-79-6 URETHANE
CARBAMIC ACID, ETHYL ESTER [EPA, MAK, PAL, PAL]
ETHYL CARBAMATE (URETHANE) [CAA, EP2, RCU, RCU]
URETHANE (ETHYL CARBAMATE) [313, C65, C65, C65]

51-80-9 TETRAMETHYLMETHYLENEDI-AMINE

51-83-2 CARBACHOL CHLORIDE

51-98-9 NORETHISTERONE ACETATE (NORETHINDRONE ACETATE)
NORETHISTERONE ACETATE [CAL]

52-01-7 SPIRONOLACTONE

52-24-4 TRIS(1-AZIRIDINYL)PHOSPHINE SULFIDE
AZIRIDINE, 1,1',1"-PHOSPHINOTH-IOYLIDYNETRIS- [PAL, PAL]
THIO-TEPA [FLA, MSL]
THIOTEPA [IAR]
TRIS(1-AZIRIDINYL)PHOSPHINE SUL-FIDE (THIOTEPA) [C65, C65, NTP, NTP]
TRIS(1-AZIRIDINYL)PHOSPHINE SUL-PHIDE [MIN]
TRIS(1-AZIRIDINYL)-PHOSPHINE SULFIDE [NJL]
TRIS(AZIRIDINYL)-PHOSPHINE SUL-FIDE [CSR, CSR]

52-46-0 APHOLATE

52-51-7 2-BROMO-2-NITRO-1,3-PROPANE-DIOL
2-BROMO-2-NITROPROPANE-1,3-DIOL (BRONOPOL) [313]

52-67-5 D-PENICILLAMINE
PENICILLAMINE [C65]

52-68-6 TRICHLORFON
CHLOROPHOS [MSL]
PHOSPHONIC ACID, (2,2,2-TRICHLORO-1-HYDROXYETHYL)-, DIMETHYLESTER [PAL]
TRICHLORFON (PHOSPHORIC ACID, (2,2,2-TRICHLORO-1-HYDROX-YETHYL)-DIMETHYL ESTER) [313]
TRICHLOROPHON [EPA]

52-76-6 LYNOESTRENOL

52-85-7 FAMPHUR

52-90-4 L-CYSTEINE

53-03-2 PREDNISONE

53-16-7 ESTRONE
ESTRA-1,3,5(10)-TRIEN-17-ONE, 3-HYDROXY- [PAL, PAL]
OESTRONE [FLA, IAR, MIN, MSL]

53-19-0 O,P'-DDD

53-70-3 DIBENZO(A,H)ANTHRACENE
1,2,5,6-DIBENZANTHRACENE (DIBENZO(A,H)ANTHRACENE) [CW2]
DIBENZO(A,H)ANTHRACENE [ATS]
DIBENZ[A,H]ANTHRACENE [IAR]

53-86-1 INDOMETHACIN

53-96-3 2-ACETYLAMINOFLUORENE
ACETAMIDE, N-9H-FLUOREN-2-YL- [PAL, PAL]
ACETAMIDE, N-FLUOREN-2-YL [EP2]
N-FLUOREN-2-YLACETAMIDE [FLA]

54-05-7 CHLOROQUINE

54-11-5 NICOTINE
NICOTINE AND SALTS [EPA]
NICOTINE, & SALTS [RCU, RCU]
PYRIDINE, 3-(1-METHYL-2-PYRRO-LIDINYL)-, (S)- [PAL]

54-17-1 M-PHENYLENEDIAMINE, SUL-FATE SALT

54-21-7 SODIUM SALICYLATE

54-25-1 6-AZAURIDINE

54-31-9 FUROSEMIDE
FUROSEMIDE (FRUSEMIDE) [IAR]

54-62-6 AMINOPTERIN

54-64-8 SODIUM O-(ETHYLMERCU-RITHIO)BENZOATE

54-85-3 ISONICOTINIC ACID HYDRAZINE
ISONICOTINIC ACID HYDRAZINE (ISONIAZID) [CAL, IAR]

54-91-1 PIPOBROMANE

55-18-5 N-NITROSODIETHYLAMINE
ETHANAMINE, N-ETHYL-N-NITROSO-
[PAL, PAL]

55-21-0 BENZAMIDE

55-22-1 ISONICOTINIC ACID

55-31-2 EPINEPHRINE HYDROCHLORIDE

55-38-9 FENTHION
BAYTEX (FENTHION) [ALB, ALB,
ALB, ALB]
FENTHION [O,O-DIMETHYL O-[3-
METHYL-4-(METHYLTHIO)PHENYL]
ESTER, PHOSPHOROTHIOIC ACID]
[313]
PHOSPHOROTHIOIC ACID, O,
O-DIMETHYL O-[3-METHYL-4-
(METHYLTHIO)PHENYL] ESTER
[PAL]

55-63-0 NITROGLYCERIN
1,2,3-PROPANETRIOL, TRINITRATE
[PAL]
GLYCEROL TRINITRATE [GBR,
GBR, GBR]
NITROGLYCERIN (NG) [AUS, AUS,
NJL, NJL, TLV, TLV]
NITROGLYCERINE [EPA, NFP, NFP,
NIO, NIO, NIO, NIO]

55-68-5 PHENYLMERCURIC NITRATE

**55-86-7 NITROGEN MUSTARD HY-
DROCHLORIDE**
ETHANAMINE, 2-CHLORO-N-(2-
CHLOROETHYL)-N-METHYL-,HY-
DROCHLORIDE [PAL, PAL]
NITROGEN MUSTARD HYDROCHLO-
RIDE (MECHLORETHAMINE HY-
DROCHLORIDE) [C65, C65]

55-91-4 ISOFLUORPHATE
DIISOPROPYL FLUOROPHOSPHATE
[EPA]
DIISOPROPYLFLUOROPHOSPHATE
(DFP) [RCU, RCU]

55-98-1 BUSULFAN
1,4-BUTANEDIOL DIMETHANE
SULPHONATE [MIN, MSL]
1,4-BUTANEDIOL DIMETHANESUL-
FONATE [CAB]
1,4-BUTANEDIOL DIMETHANESUL-
FONATE (BUSULFAN) [C65, C65,
CAL]
1,4-BUTANEDIOL DIMETHANE-
SULPHONATE [FLA]
1,4-BUTANEDIOL DIMETHANE-
SULPHONATE (MYLERAN) [IAR]
1,4-BUTANEDIOL DIMETHYLSUL-
FONATE (MYLERAN) [NTP]
1,4-BUTANEDIOL, DIMETHANESUL-
FONATE [PAL, PAL]

56-04-2 METHYLTHIOURACIL
4(1H)-PYRIMIDINONE, 2,3-DIHYDRO-
6-METHYL-2-THIOXO- [PAL, PAL]
6-METHYL-2-THIOURACIL [NJL,
NJL]

56-18-8 3,3'-IMINOBISPROPYLAMINE
3,3'-IMINODIPROPYLAMINE [HM1,
HM1]
BIS(AMINOPROPYL)AMINE [NJL,
NJL]

56-23-5 CARBON TETRACHLORIDE
CARBON TETRACHLORIDE [QBC,
QBC, QBC]
CARBON TETRACHLORIDE (TETRA-
CHLOROMETHANE) [CN2, CW2]
METHANE, TETRACHLORO- [PAL,
PAL]
TETRACHLOROMETHANE (CARBON
TETRACHLORIDE) [CN5]

56-25-7 CANTHARIDIN

**56-34-8 TETRAETHYLAMMONIUM CHLO-
RIDE**

56-35-9 BIS(TRIBUTYLTIN)OXIDE
BIS(TRIBUTYLTIN) OXIDE [313]

56-36-0 TRIBUTYLTIN ACETATE

56-38-2 PARATHION
PARATHION (ISO) [GBR, GBR, GBR]
PARATHION (PHOSPHOROTHIOIC
ACID, O,O-DIETHYL-O-(4-NITRO-
PHENYL)ESTER) [313]
PHOSPHOROTHIOIC ACID, O,O-DI-
ETHYL O-(4-NITROPHENYL) ESTER
[PAL]

56-40-6 GLYCINE

56-49-5 3-METHYLCHOLANTHRENE
BENZ(J)ACEANTHRYLENE, 1,2-DI-
HYRO-3-METHYL- [PAL, PAL]

56-53-1 DIETHYLSTILBESTROL
DIETHYLSTILBESTROL (DES) [C65,
C65, C65]
PHENOL, 4,4'-(1,2-DIETHYL-1,2-
ETHENEDIYL)BIS-, (E)- [PAL, PAL]

56-55-3 BENZ(A)ANTHRACENE
1,2-BENZANTHRACENE (BENZO(A)
ANTHRACENE) [CW2]
BENZO(A)ANTHRACENE [ATS, PQL]
BENZO(A)ANTHRACENE (1,2-BEN-
ZOANTHRACENE) [MX1]
BENZ[A]ANTHRACENE [TLV]

56-57-5 4-NITROQUINOLINE-1-OXIDE
QUINOLINE, 4-NITRO, 1-OXIDE [NJL,
NJL]
QUINOLINE, 4-NITRO-, 1-OXIDE
[PAL]

56-72-4 COUMAPHOS
PHOSPHOROTHIOIC ACID, O-(3-
CHLORO-4-METHYL-2-OXO-2H-1-
BENZOPYRAN-7-YL) O,O-DIETHYL
ESTER [PAL]

56-75-7 CHLORAMPHENICOL
ACETAMIDE, 2,2-DICHLORO-N-[2-
HYDROXY-1-(HYDROXYMETHYL)-2-
(NITROPHENYL)ETHYL]-,[R-(R*,R*)]-
[PAL, PAL]

56-81-5 GLYCERIN
1,2,3-PROPANETRIOL [PAL]
GLYCERIN (MIST) [OS1, OS1]
GLYCERIN MIST [ALB, ALB, AUS,
BC1, MEX, MIN, ONT, QBC, TLV]
GLYCERINE [NFP, NFP]
GLYCEROL [GBR, HON]

56-86-0 ALPHA-GLUTAMIC ACID

**56-93-9 BENZYLTRIMETHYL AMMONIUM
CHLORIDE**

56-95-1 CHLORHEXIDINE DIACETATE

57-06-7 ALLYL ISOTHIOCYANATE
1-PROPENE, 3-ISOTHIOCYANATO-
[PAL]
MUSTARD OIL [MSL, NFP, NFP]

**57-09-0 HEXADECYLTRIMETHYLAMMO-
NIUM BROMIDE**

57-10-3 PALMITIC ACID
HEXADECANOIC ACID [OTS, PST,
TSC]

57-11-4 STEARIC ACID

57-12-5 CYANIDE ANION
CYANIDE [ATS, CD1, CEX, CEX,
NJL, PAL, PQL, QBC, QBC, SDW,
SDW]
CYANIDE (ANION) [MSL]
CYANIDE (AS CN) [BC1, BC1]
CYANIDE (COMPLEXED) [RCA]
CYANIDE COMPOUNDS [CAA]
CYANIDE MIXTURES OR SOLUTIONS
[HM1, HM1, HM1, HM1]
CYANIDE, POTASSIUM AND SODIUM
[ONT, ONT]
CYANIDE, TOTAL [CW2]
CYANIDES [ALB, ALB, ALB, AUS,
AUS, CAN, CN7, CW3, EPA, GBR,
GBR, MAK, MAK, MAK, MEX,
MEX, OS1, OS1, UTS]

57-13-6 UREA

57-14-7 1,1-DIMETHYLHYDRAZINE
1,1-DIMETHYL HYDRAZINE [313,
CAA]
1,1-DIMETHYLHYDRAZINE [NIO,
NIO, NIO, NIO]
1,1-DIMETHYLHYDRAZINE (UDMH)
[C65, RCA]
1,1-DIMETHYLHYDRAZINE [HY-
DRAZINE, 1,1-DIMETHYL-] [EP3]
DIMETHYLHYDRAZINE [302]
DIMETHYLHYDRAZINE, 1,1- [OS3]
DIMETHYLHYDRAZINE, UNSYMMET-
RICAL [HM1, HM1, HM1, HM1]
HYDRAZINE, 1,1-DIMETHYL- [PAL,
PAL]

57-22-7 VINCRISTINE

57-24-9 STRYCHNINE
STRYCHNIDIN-10-ONE [PAL]
STRYCHNINE (2,4A,5,5A,7,8,15,15A,
15B,15C,DECAHYDRO-4,6-METHANO-
6H,14H-INDOLO[3,2,1-IJ]OXEPINO[2,3,
4-DE]PYRROLO[2,3-H]QUINOLIN-14-
ONE) [CN2]
STRYCHNINE AND SALTS [EPA]
STRYCHNINE, & SALTS [RCU, RCU]

57-33-0 PENTOBARBITAL SODIUM

**57-39-6 TRIS(2-METHYL-1-AZIRIDINYL)
PHOSPHINE OXIDE**

57-41-0 PHENYTOIN
2,4-IMIDAZOLIDINEDIONE, 5,5-
DIPHENYL- [PAL, PAL]
DIPHENYLHYDANTOIN (PHENYTOIN)
[C65, C65, CAL, CSR]

57-47-6 PHYSOSTIGMINE

57-50-1 SUCROSE
.ALPHA.-D-GLUCOPYRANOSIDE,
.BETA.-D-FRUCTOFURANOSYL [PAL]
SUGAR [PST]

57-53-4 MEPROBAMATE

57-55-6 1,2-PROPYLENE GLYCOL
1,2-PROPANEDIOL [PAL]
PROPANE-1,2-DIOL [GBR]
PROPYLENE GLYCOL [HON, MIN,
NFP, NFP, PST]

57-56-7 HYDRAZINECARBOXAMIDE
HYDRAZINE CARBOXAMIDE [TSC]

57-57-8 BETA-PROPIOLACTONE
2-OXETANONE [PAL, PAL]
PROPIOLACTONE, BETA [302]

57-63-6 ETHINYLESTRADIOL
19-NORPREGNA-1,3,5(10)-TRIEN-20-
YNE-3,17-DIOL, (17.ALPHA.)- [PAL,
PAL]
ETHINYLOESTRADIOL [CAL, FLA,
MIN, MSL, NJL, NJL]

**57-64-7 PHYSOSTIGMINE SALICYLATE
(1:1)**
PHYSOSTIGMINE, SALICYLATE (1:1)
[302, PAL]
PHYSOSTIGMINE,SALICYLATE [MSL]

57-66-9 PROBENECID

57-68-1 SULFAMETHAZINE

57-74-9 CHLORDANE
4,7-METHANO-1H-INDENE, 1,2,4,5,
6,7,8,8-OCTACHLORO-2,3,3A,4,7,7A-
HEXAHYDRO- [PAL]
CHLORDANE (1,2,4,5,6,7,8,8-OC-
TACHLORO-2,3,3A,4,7,7A-HEXAHY-
DRO-4,7-METHANOINDENE) [CN2]
CHLORDANE (4,7-METHANOINDAN,
1,2,4,5,6,7,8,8-OCTACHLORO-2,3,3A,4,7,
7A-HEXAHYDRO-] [313]
CHLORDANE (ALPHA AND GAMMA
ISOMERS) [UTS]
CHLORDANE (ANALYTICAL GRADE)
[CSR. CSR]
CHLORDANE (TECHNICAL MIXTURE
AND METABOLITES) [CW2, CW3]
CHLORDANE, ALPHA & GAMMA
ISOMERS [RCU, RCU]

57-83-0 PROGESTERONE
PREGN-4-ENE-3,20-DIONE [PAL, PAL]

57-85-2 TESTOSTERONE PROPIONATE

57-88-5 CHOLESTEROL

57-92-1 STREPTOMYCIN

57-97-6 7,12-DIMETHYLBENZ(A)AN-THRACENE
BENZ(A)ANTHRACENE, 7,12-DIMETHYL- [EPA, NJL, PAL, PAL]

58-08-2 CAFFEINE

58-14-0 PYRIMETHAMINE

58-15-1 1-AMINOPYRINE

58-18-4 METHYLTESTOSTERONE
METHYL TESTOSTERONE [CAL]

58-20-8 TESTOSTERONE CYPIONATE

58-22-0 TESTOSTERONE
ANDROST-4-EN-3-ONE, 17-HY-DROXY-,(17.BETA.)- [PAL, PAL]
TESTOSTERONE AND ITS ESTERS [C65]

58-25-3 CHLORDIAZEPOXIDE

58-33-3 PROMETHAZINE HYDROCHLO-RIDE

58-36-6 10,10'-OXYBISPHENOXARSINE
PHENARSAZINE OXIDE [MSL, NJL]
PHENOXARSINE, 10,10'-OXYDI- [302, PAL]

58-55-9 THEOPHYLLINE

58-89-9 LINDANE
CYCLOHEXANE, 1,2,3,4,5,6-HEX-ACHLORO-, (1.ALPHA.,2.ALPHA., 3.BETA.,4.ALPHA.,5.ALPHA.,6.BETA.)- [PAL, PAL]
GAMMA-BHC [HM1, HM1, RCA, UTS]
GAMMA-BHC (ISO) [GBR, GBR, GBR]
GAMMA-BHC (LINDANE) [CW2, EP2]
GAMMA-BHC(LINDANE)-GAMMA [MX1]
GAMMA-HCH (LINDANE) [IAR]
GAMMA-HEXACHLOROCYCLOHEX-ANE [MAK, MAK, MAK]
HEXACHLOROCYCLOHEXANE, GAMMA ISOMER [C65, C65]
HEXACHLOROCYCLOHEXANE, GAMMA- [ATS]
LINDANE (1,2,3,4,5,6-HEXACHLORO-CYCLOHEXANE OR GAMMA-BHC) [CN2]
LINDANE (ALL ISOMERS) [CAA]
LINDANE (GAMMA-HEXACHLORO-CYCLOHEXANE) [NTP]
LINDANE AND OTHER HEX-ACHLOROCYCLOHEXANE ISOMERS [CAL]
LINDANE [CYCLOHEXANE, 1,2,3,4,5, 6-HEXACHLORO-(1ALPHA,2ALPHA, 3BETA,4ALPHA,5ALPHA,6BETA)-] [313]

58-90-2 2,3,4,6-TETRACHLOROPHENOL
PHENOL, 2,3,4,6-TETRACHLORO-[PAL]
TETRACHLOROPHENOL, 2,3,4,6-[CD1, CD1]

58-93-5 HYDROCHLOROTHIAZIDE

58-95-7 VITAMIN E ACETATE
D-ALPHA-TOCOPHERYL ACETATE [CSR]

59-01-8 KANAMYCIN

59-05-2 METHOTREXATE

59-14-3 5-BROMO-2'-DEOXYURIDINE

59-40-5 N-(2-QUINOXALINYL)-SUL-FANILIDE

59-50-7 4-CHLORO-M-CRESOL
P-CHLORO-M-CRESOL [PQL, RCA, RCU, RCU, UTS]
PARA-CHLORO-META-CRESOL [CAL, MX1]
PARACHLOROMETA CRESOL [CW2]
PHENOL, 4-CHLORO-3-METHYL-[PAL]

59-67-6 3-PYRIDINECARBOXYLIC ACID
NICOTINIC ACID [PST]
PYRIDINECARBOXYLIC ACID, 3-[OTS]

59-87-0 NITROFURAZONE
NITROFURAL (NITROFURAZONE) [IAR]

59-88-1 PHENYLHYDRAZINE HY-DROCHLORIDE

59-89-2 N-NITROSOMORPHOLINE
MORPHOLINE, 4-NITROSO- [PAL, PAL]

59-96-1 PHENOXYBENZAMINE
BENZENEMETHANAMINE, N-(2-CHLOROETHYL)-N-(1-METHYL-2-PHENOXYETHYL)- [PAL, PAL]

60-00-4 ETHYLENEDIAMINE TE-TRAACETIC ACID (EDTA)
EDTA [CAL]
ETHYLENEDIAMINE TETRA-ACETIC ACID [NJL]
ETHYLENEDIAMINETETRAACETIC ACID [PST]
ETHYLENEDIAMINETETRAACETIC ACID (EDTA) [CWA]
GLYCINE, N,N'-1,2-ETHANEDIYLBIS [N-(CARBOXYMETHYL)- [PAL]

60-01-5 GLYCERYL TRIBUTYRATE

60-09-3 4-AMINOAZOBENZENE
AMINOAZOBENZENE [WHM]
C.I. SOLVENT YELLOW 1 [NJL]
P-AMINOAZOBENZENE [C65]
PARA-AMINOAZOBENZENE [CAL, IAR, MIN]

60-11-7 4-DIMETHYLAMINOAZOBENZENE
BENZENAMINE, N,N-DIMETHYL-4-(PHENYLAZO)- [PAL, PAL]
DIMETHYL AMINOAZOBENZENE [CAA]
DIMETHYLAMINOAZOBENZENE [EP2]
P-(DIMETHYLAMINO)AZOBENZENE [PQL]
P-DIMETHYLAMINOAZOBENZENE [EPA, FLA, MSL, RCA, UTS]
PARA-DIMETHYLAMINOAZOBEN-ZENE [IAR]

60-12-8 PHENETHYL ALCOHOL

60-13-9 DL-AMPHETAMINE SULFATE

60-15-1 BENZENEETHANAMINE, .ALPHA.-METHYL-

60-23-1 B-MERCAPTOETHYLAMINE HCL

60-24-2 2-MERCAPTOETHANOL
ETHANOL, 2-MERCAPTO- [PAL]
MERCAPTOETHANOL [WEL, WEL]
THIOGLYCOL [HM1, HM1, NJL, WHM]

60-29-7 ETHYL ETHER
DIETHYL ETHER [GBR, GBR, HM1, HM1, HM1, NEU, NJL, NJL, OTS, TSC, TSC, TSC]
DIETHYL ETHER (ETHYL ETHER) [ALB, ALB]
DIETHYL OXIDE (ETHER) [QBC, QBC]
ETHANE, 1,1'-OXYBIS- [PAL]
ETHOXYETHANE [ONT, ONT]
ETHYL ETHER [ETHANE, 1,1'-OXY-BIS-] [EP3]

60-33-3 LINOLEIC ACID

60-34-4 METHYL HYDRAZINE
HYDRAZINE, METHYL- [PAL]
METHYL HYDRAZINE [MEX, MEX, MEX, MEX]
METHYL HYDRAZINE (MONOMETHYL HYDRAZINE) [OS1, OS1, OS1, OS1]
METHYL HYDRAZINE [HYDRAZINE, METHYL-] [EP3]
METHYLHYDRAZINE [C65, HM1, HM1, HM1, HM1, MSL, NFP, NFP, NHS, NHS, NHS, ONT, ONT, WHM]
MONOMETHYL HYDRAZINE [BC1, BC1, MIN]

60-35-5 ACETAMIDE
ACETAMIDE (ETHANAMIDE) [CAI, TSC]

60-41-3 STRYCHNINE, SULFATE
STRYCHNINE SULFATE [302, MSL]

60-51-5 DIMETHOATE

60-54-8 TETRACYCLINE

60-56-0 METHIMAZOLE

60-57-1 DIELDRIN
2,7:3,6-DIMETHANONAPHTH[2,3-B] OXIRENE, 3,4,5,6,9,9-HEXACHLORO-1A,2,2A,3,6,6A,7,7A-OCTAHYDRO-,(1A.ALPHA.,2.BETA.,2A.ALPHA., 3.BETA.,6.BETA.,6A.ALPHA.,7.BETA., 7A.ALPHA.)- [PAL]
DIELDRIN ((1R,4S,4AS,5R,6R,7S,8S,8AR) 1,2,3,4,10,10-HEXACHLORO-1,4,4A,5, 6,7,8,8A-OCTAHYDRO-6,7-EPOXY-1, 4:5,8-DIMETHANONAPHTHALENE) [CN2]
DIELDRIN (ISO) [GBR, GBR, GBR]
DIELDRINE [QBC, QBC]

60-79-7 ERGONOVINE

61-33-6 BENZYLPENICILLIN

61-57-4 NIRIDAZOLE
2-IMIDAZOLIDINONE, 1-(5-NITRO-2-THIAZOLYL)- [PAL, PAL]

61-73-4 METHYLENE BLUE
C.I. BASIC BLUE 9 [PST]

61-76-7 PHENYLEPHRINE HYDROCHLO-RIDE

61-82-5 AMITROLE
1H-1,2,4-TRIAZOL-3-AMINE [PAL, PAL]
3-AMINO 1,2,4- TRIAZOLE (AMIT-ROLE) [QBC, QBC]
3-AMINO-1,2,4-TRIAZOLE (AMIT-ROLE) [ALB, ALB]
3-AMINO-1H-1,2,4-TRIAZOLE [ONT]
AMITROL [MAK]

62-23-7 P-NITROBENZOIC ACID

62-33-9 DISODIUM CALCIUM EDTA
CALCIUM DISODIUM ETHYLENEDI-AMINETETRAACETATE [PST]

62-38-4 PHENYLMERCURIC ACETATE
ACETOXYPHENYLMERCURY [FLA]
MERCURY, (ACETATO-O)PHENYL-[PAL]
PHENYLMERCURY ACETATE [302, EPA, RCU, RCU]

62-44-2 PHENACETIN
ACETAMIDE, N-(4-ETHOXYPHENYL)-[PAL, PAL]
P-ACETOPHENETIDIDE [FLA]

62-49-7 CHOLINE

62-50-0 ETHYL METHANESULFONATE
ETHYL METHANESULPHONATE [CAL, IAR, MIN]
METHANESULFONIC ACID, ETHYL ESTER [NJL, PAL, PAL]

62-51-1 METHACHOLINE CHLORIDE

62-53-3 ANILINE
ANILINE AND HOMOLOGS [NIO, NIO, NIO, NIO, NIO, OS1, OS1, OS1, OS1]

ANILINE AND HOMOLOGUES [ALB, ALB, ALB, AUS, AUS, MEX, MEX, MEX, MIN, ONT, ONT, ONT, ONT, TLV, TLV]
BENZENAMINE [PAL]
BENZENAMINE (ANILINE) [TSC, TSC]

62-55-5 THIOACETAMIDE
ETHANETHIOAMIDE [PAL, PAL]

62-56-6 THIOUREA

62-73-7 DICHLORVOS
DDVP (DICHLORVOS) [ALB, ALB, ALB, C65, C65]
DICHLORVOS (DDVP) [AUS, AUS, BC1, BC1, BC1, CEX, CEX, MAK, MAK, MAK, MAK, OS1, OS1, OS1, OS1]
DICHLORVOS (ISO) [GBR, GBR, GBR]
DICHLORVOS (PHOSPHORIC ACID, 2, 2,-DICHLOROETHENYL DIMETHYL ESTER) [313]
DIMETHYL-O,O-DICHLOROVINYL-2, 2-PHOSPHATE (TECHNICAL) [NFP, NFP]
PHOSPHORIC ACID, 2,2-DICHLOROETHENYL DIMETHYL ESTER [PAL]

62-74-8 SODIUM FLUOROACETATE
ACETIC ACID, FLUORO-, SODIUM SALT [PAL, RCU, RCU, TSC, TSC]
FLUOROACETIC ACID,.SODIUM SALT [EPA]
SODIUM FLUOROACETATE [QBC, QBC, QBC]
SODIUM FLUOROACETATE [TLV, TLV]
SODIUM FLUOROACETATE (1080) [BC1, BC1, BC1]

62-75-9 N-NITROSODIMETHYLAMINE
METHANAMINE, N-METHYL-N-NI-TROSO- [PAL, PAL]
N-NITROSODIMETHYLAMINE (DIMETHYL-NITROSAMINE) [ALB, ALB]
N-NITROSODIMETHYLAMINE (DIMETHYLNITROSOAMINE) [BC1, BC1, BC1]
NITROSODIMETHYLAMINE [302]

63-25-2 CARBARYL
1-NAPHTHALENOL, METHYLCARBA-MATE [PAL]
1-NAPHTHYL N-METHYL CARBA-MATE [ONT]
CARBARYL (ISO) [GBR, GBR]
CARBARYL (SEVIN) [ALB, ALB, BC1, BC1, NIO, NIO, NIO, OS1, OS1, QBC]
CARBARYL [1-NAPHTHALENOL, METHYLCARBAMATE] [313]

63-42-3 BETA-D-LACTOSE

63-68-3 METHIONINE

63-74-1 SULFONAMIDE

63-92-3 PHENOXYBENZAMINE HY-DROCHLORIDE
BENZENEMETHANAMINE, N-(2-CHLOROETHYL)-N-(1-METHYL-2-PHENOXYETHYL)-,HYDROCHLORIDE [PAL, PAL]

63-98-9 PHENACEMIDE

64-00-6 PHENOL, 3-(1-METHYLETHYL)-, METHYLCARBAMATE

64-02-8 TETRASODIUM EDTA
ETHYLENEDIAMINETETRAACETIC ACID, TETRASODIUM SALT [PST]

64-04-0 PHENETHYLAMINE

64-10-8 PHENYLUREA

64-17-5 ETHYL ALCOHOL
ETHANOL [GBR, HM1, HM1, MAK, MAK, MAK, ONT, PAL, PST, TLV, WHM]

ETHYL ALCOHOL (ETHANOL) [ALB, ALB, BC1, OS1, OS1, QBC]

64-18-6 FORMIC ACID

64-19-7 ACETIC ACID
ACETIC ACID, GLACIAL [NFP, NFP]

64-67-5 DIETHYL SULFATE
DIETHYL SULPHATE [IAR]
SULFURIC ACID, DIETHYL ESTER [PAL, PAL, TSC, TSC]

64-69-7 IODACETIC ACID

64-73-3 DEMECLOCYCLINE HY-DROCHLORIDE

64-75-5 TETRACYCLINE HYDROCHLO-RIDE

64-77-7 TOLBUTAMIDE

64-86-8 COLCHICINE

65-30-5 NICOTINE SULFATE

65-45-2 SALICYLAMIDE

65-49-6 4-AMINOSALICYLIC ACID

65-71-4 THYMINE

65-85-0 BENZOIC ACID

65-86-1 OROTIC ACID

66-25-1 HEXALDEHYDE
HEXANAL [FLA, MSL, NFP, NFP, PAL]

66-27-3 METHYL METHANESULFONATE
METHANESULFONIC ACID, METHYL ESTER [NJL, PAL, PAL]
METHYL METHANE SULFONATE [WHM]
METHYL METHANESULPHONATE [IAR, MIN, NTP]

66-71-7 1,10-PHENANTHROLINE

66-75-1 URACIL MUSTARD
2,4(1H,3H)-PYRIMIDINEDIONE, 5-[BIS (2-CHLOROETHYL)AMINO]- [PAL, PAL]
5-(BIS(2-CHLOROETHYL)AMINO) URACIL [NJL]

66-76-2 DICUMAROL

66-77-3 1-NAPHTHALENECARBOXALDE-HYDE

66-81-9 CYCLOHEXIMIDE
2,6-PIPERIDINEDIONE, 4-[2-(3,5-DIMETHYL-2-OXOCYCLOHEXYL)-2-HYDROXYETHYL]-, [1S-[1.ALPHA.(S*).3.ALPHA.,5.BETA.]- [PAL]

67-03-8 THIAZOLIUM, 3-[(4-AMINO-2-METHYL-5-PYRIMIDINYL) METHYL]-5-(2-HYDROXYETHYL) -4-METHYLCHLORIDE, MONOHY-DROCHLORIDE
THIAMIN HYDROCHLORIDE [PST]

67-20-9 NITROFURANTOIN

67-43-6 PENTACARBOXYMETHYL DI-ETHYLENETRIAMINE

67-45-8 FURAZOLIDONE

67-48-1 CHOLINE CHLORIDE

67-56-1 METHYL ALCOHOL
METHANOL [313, BEI, BEI, CAA, CAN, CSR, EPA, GBR, GBR, GBR, HM1, HM1, HM1, HON, HON, MAK, MAK, MAK, MAK, MSL, ONT, ONT, ONT, PAL, TLV, TLV, TLV, UTS]
METHYL ALCOHOL (METHANOL) [ALB, ALB, ALB, BC1, BC1, BC1, QBC, QBC, QBC, WHM]

67-63-0 ISOPROPYL ALCOHOL
2-PROPANOL [PAL]
2-PROPANOL (ISOPROPANOL) [TSC, TSC, TSC]

ISOPROPANOL [ATS, CTR, HM1, HM1, HM1, OTS, TSE, WHM]
PROPAN-2-OL [GBR, GBR, GBR]

67-64-1 ACETONE
2-PROPANONE [PAL]

67-66-3 CHLOROFORM
CHLOROFORM (TRICHLOROMETHANE) [ALB, ALB, BC1, BC1, CW2, OS1, OS1, QBC, QBC]
CHLOROFORM [METHANE, TRICHLORO-] [EP3]
METHANE, TRICHLORO- [PAL, PAL, TSC, TSC]
TRICHLOROMETHANE [ONT, ONT]

67-68-5 DIMETHYL SULFOXIDE

67-71-0 DIMETHYLSULFONE

67-72-1 HEXACHLOROETHANE
ETHANE, HEXACHLORO- [PAL, TSC, TSC, TSC]

68-04-2 SODIUM CITRATE

68-11-1 THIOGLYCOLIC ACID
ACETIC ACID, MERCAPTO- [PAL]
MERCAPTOACETIC ACID [GBR]

68-12-2 DIMETHYLFORMAMIDE
DIMETHYL FORMAMIDE [CAA, OTV, TSC, TSC, TSC]
DIMETHYL FORMAMIDE (DMF) [EPA]
DIMETHYLFORMAMIDE (N,N-) [HON, HON]
FORMAMIDE, N,N-DIMETHYL- [PAL]
N,N-DIMETHYLFORMAMIDE [313, CAL, HM1, HM1, NFP, NFP, ONT, ONT, PST, WHM]
N,N-DIMETHYLFORMAMIDE (DMF) [BEI, BEI]

68-22-4 NORETHISTERONE
19-NORPREGN-4-EN-20-YN-3-ONE, 17-HYDROXY-, (17.ALPHA.)- [PAL, PAL]
NORETHISTERONE (NORETHIN-DRONE) [C65, C65]

68-23-5 NORETHYNODREL

68-26-8 VITAMIN A
RETINOL/RETINYL ESTERS [C65, CAB, CAL]

68-76-8 TRIS(AZIRIDINYL)-P-BENZO-QUINONE
2,3,5-TRIS(1-AZIRIDINYL)-P-BENZO-QUINONE [NJL]
2,5-CYCLOHEXADIENE, 1,4-DIONE, 2, 3,5-TRIS(1-AZIRIDINYL)- [PAL, PAL]
TRIAZIQUONE [2,5-CYCLOHEXA-DIENE-1,4-DIONE,2,3,5-TRIS(1-AZIRIDINYL)-] [313]
TRIS(AZIRIDINYL)-PARA-BENZO-QUINONE [MIN]
TRIS(AZIRIDINYL)-PARA-BENZO-QUINONE (TRIAZIQUONE) [C65, IAR]
TRIS(AZIRIDINYL)-PARA-BENZO-QUINONE; (TRIAZIQUONE) [CAL]

69-05-6 QUINACRINE HYDROCHLORIDE

69-53-4 AMPICILLIN

69-65-8 D-MANNITOL

69-72-7 SALICYLIC ACID

69-79-4 MALTOSE

69-93-2 URIC ACID

70-11-1 PHENACYL BROMIDE

70-18-8 GLUTATHIONE

70-25-7 N-METHYL-N'-NITRO-N-NI-TROSOGUANIDINE
1-METHYL-3-NITRO-1-NITROSO-GUANIDINE [NJL, NJL]
GUANIDINE, N-METHYL-N'-NITRO-N-NITROSO- [EPA, PAL, PAL, RCU, RCU]

MNNG (N-METHYL-N'-NITRO-N-
NITROSOGUANIDINE) [TSC, TSE]
N-METHYL-N'-NITRO-N-NITROSO-
GUANIDINE [CAL]
N-METHYL-N'-NITRO-N-NI-
TROSOGUANIDINE (MNNG) [IAR]
NITROSOGUANIDINE [HM1, HM1,
HM1]

70-30-4 HEXACHLOROPHENE
HEXACHLOROPHENE (HCP) [CAL]

70-49-5 MERCAPTOSUCCINIC ACID

**70-55-3 BENZENESULFONAMIDE, 4-
METHYL-**
P-TOLUENESULFONAMIDE [PST]

70-69-9 PROPIOPHENONE, 4-AMINO-
4-AMINOPROPIOPHENONE [MSL]
PROPIOPHENONE, 4'-AMINO- [PAL]

71-23-8 N-PROPYL ALCOHOL
1-PROPANOL [PAL]
N-PROPANOL [HM1, HM1, PST]
PROPAN-1-OL [GBR, GBR, GBR]
PROPYL ALCOHOL [AUS, AUS,
AUS, BC1, BC1, BC1, MEX, MEX,
MIN, MSL, NFP, NFP, NJL, NJL,
OTV]
PROPYL ALCOHOL (PROPANOL)
[FLA, QBC, QBC, QBC]

71-36-3 N-BUTYL ALCOHOL
1-BUTANOL [NEU, PAL, PST]
BUTAN-1-OL [GBR, GBR]
BUTANOL [HM1, HM1, HM1]
BUTANOL, 1- [OTS]
BUTYL ALCOHOL [NFP, NFP]
N-BUTANOL [RCU, RCU]

71-41-0 AMYL ALCOHOL
1-PENTANOL [PAL]
N-AMYL ALCOHOL [WHM]
PENTYL ALCOHOL [PST]

71-43-2 BENZENE
BENZENE [NFP, NFP]

71-48-7 COBALT(II) ACETATE

71-55-6 1,1,1-TRICHLOROETHANE
1,1,1-TRICHLOROETHANE (METHYL
CHLOROFORM) [313, ALB, ALB,
CN5, WHM]
1,1,1-TRICHLOROETHANE
(METHYLCHLOROFORM) [HON,
HON]
ETHANE, 1,1,1-TRICHLORO- [PAL,
TSC, TSC, TSC, TSC]
METHYL CHLOROFORM [BEI, CAB,
CAL, CEX, CEX, CEX, MEX, MEX,
MIN, NIO, NIO, NIO, NJL, RCU,
RCU, TLV, TLV]
METHYL CHLOROFORM (1,1,1-
TRICHLOROETHANE) [BC1, BC1,
CAA, CAA, OS1, OS1, OS1, QBC,
QBC]
METHYL CHLOROFORM 1,1,1-
TRICHLOROETHANE [EPA]
METHYLCHLOROFORM [HM1, HM1,
HM1, OTV]
TRICHLOROETHANE, 1,1,1- [OTS]

**71-58-9 MEDROXYPROGESTERONE AC-
ETATE**

71-63-6 DIGITOXIN

72-14-0 SULFATHIAZOLE

72-20-8 ENDRIN
2,7:3,6-DIMETHANONAPHTH[2,
3-B]OXIRENE, 3,4,5,6,9,9-HEX-
ACHLORO-1A,2,2A,3,6,6A,7,7A-OC-
TAHYDRO-,(1A.ALPHA.,2.BETA.,
2A.BET A.,2A.BETA.,3.ALPHA.,6.AL-
PHA.,6A.BETA.,7.BETA.,7A.ALPHA.)-
[PAL]
ENDRIN (1,2,3,4,10,10-HEXACHLORO-
6,7-EPOXY-1,4,4A,5,6,7,8,8A-OCTAHY-
DRO-EXO-5,8-DIMETHANONAPHTHA-
LENE) [CN2]
ENDRIN (ISO) [GBR, GBR, GBR]
ENDRIN AND METABOLITES [CW3]

72-33-3 MESTRANOL
19-NORPREGNA-1,3,5(10)-TRIEN-20-
YN-17-OL, 3-METHOXY-, (17.ALPHA.)-
[PAL, PAL]

72-43-5 METHOXYCHLOR
BENZENE, 1,1'-(2,2,2-
TRICHLOROETHYLIDENE)BIS[4-
METHOXY- [PAL]
METHOXYCHLOR (ISO) [GBR]
METHOXYCHLOR [BENZENE, 1,1'-(2,
2,2-TRICHLOROETHYLIDENE)BIS [4-
METHOXY]-] [313]

**72-48-0 1,2-DIHYDROXYAN-
THRAQUINONE**
ALIZARIN [HON]

72-54-8 DDD
1,1-DICHLORO-2,2-BIS-(P-
CHLOROPHENYL)ETHANE [NJL]
4,4'-DDD [PQL]
4,4'-DDD (P,P-TDE) [CW2]
4-4'-DDD (P-P'-TDE) [MX1]
BENZENE, 1,1'-
(2,2-DICHLOROETHYLIDENE)BIS[4-
CHLORO- [PAL]
DDD (DICHLORODIPHENYLDI-
CHLOROETHANE) [C65]
DDD, P,P'- [ATS]
P,P'-DDD [RCA, UTS]
TDE [CAL, CWA, MSL]
TETRACHLORODIPHENYLETHANE
[CSR, CSR]

72-55-9 DDE
1,1-DICHLORO-2,2-BIS(P-
CHLOROPHENYL)ETHYLENE [CAL,
WHM]
4,4'-DDE [PQL]
4,4'-DDE (P,P-DDX) [CW2]
4-4'-DDE (P-P'-DDX) [MX1]
BENZENE, 1,1'-(DICHLOROETHENYLI-
DENE)BIS[4-CHLORO- [PAL]
DDE (DICHLORODIPHENYLDI-
CHLOROETHYLENE) [C65]
DDE, P,P'- [ATS]
P,P'-DDE [MSL, RCA, UTS]
P,P'-DICHLORODIPHE-
NOLDICHLOROETHYLENE [CSR,
CSR]

**72-56-0 DI(P-ETHYLPHENYL)
DICHLOROETHANE**
P,P-ETHYL DDD (PERTHANE) [MSL]
PERTHANE [CAL]

72-57-1 TRYPAN BLUE
2,7-NAPHTHALENEDISULFONIC
ACID, 3,3'-[(3,3'-DIMETHYL-[1,1'-
DIPHENYL]-4,4'-DIYL)BIS(AZO)]BIS[5-
AMINO-4-HYDROXY-, TETRASODIUM
SALT [TSC, TSC]
2,7-NAPHTHALENEDISULFONIC
ACID, 3,3'-[(3,3'-DIMETHYL[1,1'-
BIPHENYL]-4,4'-DIYL)BIS(AZO))BIS[5-
AMINO-4-HYDROXY-, TETRASODIUM
SALT [PAL, PAL]
C.I. DIRECT BLUE 14, TETRASODIUM
SALT [NJL, NJL]
TRYPAN BLUE (COMMERCIAL
GRADE) [C65, CAL]
TRYPAN BLUE, COMMERICAL
GRADE [MSL]

73-22-3 L-TRYPTOPHAN

73-24-5 ADENINE (6-AMINOPURINE)

**74-31-7 N,N'-DIPHENYL-P-PHENYLENEDI-
AMINE**

74-82-8 METHANE

**74-83-9 METHYL BROMIDE (BRO-
MOMETHANE)**
BROMOETHANE [CAN]
BROMOMETHANE [GBR, GBR, GBR,
SDW]
BROMOMETHANE (METHYL BRO-
MIDE) [313, QBC, QBC]
METHANE, BROMO- [PAL, TSC,
TSC, TSC]
METHYL BROMIDE [MX1]

METHYL BROMIDE (BRO-
MOMETHANE) [CAA, CAA, CW2,
HON, HON, UTS]

74-84-0 ETHANE

74-85-1 ETHYLENE
ETHENE [FLA, PAL]
ETHYLENE [ETHENE] [EP3]

74-86-2 ACETYLENE
ACETYLENE [ETHYNE] [EP3]
ETHYNE [PAL]

74-87-3 METHYL CHLORIDE
CHLOROMETHANE [ATS, CAN,
GBR, GBR, OTS, SD4, SDW]
CHLOROMETHANE (METHYL CHLO-
RIDE) [313, UTS]
METHANE, CHLORO- [PAL, TSC,
TSC, TSC, TSC]
METHYL CHLORIDE
(CHLOROMETHANE) [CAA, CW2,
HON, HON, QBC, QBC, QBC]
METHYL CHLORIDE [METHANE,
CHLORO-] [EP3]

74-88-4 METHYL IODIDE
IODOMETHANE [RCA, UTS]
IODOMETHANE (METHYL IODIDE)
[QBC, QBC, QBC]
METHANE, IODO- [PAL, PAL]
METHYL IODIDE (IODOMETHANE)
[CAA]
METHYL IODINE [FLA]

74-89-5 METHYLAMINE
METHANAMINE [PAL]
METHYL AMINE [MIN]
METHYLAMINE [METHANAMINE]
[EP3]
MONOMETHYLAMINE [CWA, EPA]

74-90-8 HYDROGEN CYANIDE
CYANHYDRIC ACID (HYDROGEN
CYANIDE) [QBC, QBC]
HYDROCYANIC ACID [302, CAB,
EP3, HM1, HM1, HM1, HM1,
HM1, NFP, NFP, PAL, RCA, TSC,
TSC]

74-93-1 METHYL MERCAPTAN
1-METHANETHIOL [NHS, NHS]
METHANETHIOL [GBR, HM1, HM1,
HM1, HM1, HM1, ONT, PAL,
RCU, RCU, TSC, TSC, TSE]
METHANETHIOL (METHYL MERCAP-
TAN) [ALB, ALB]
METHYL MERCAPTAN
[METHANETHIOL] [EP3]
METHYLMERCAPTAN [EPA]

74-95-3 METHYLENE BROMIDE
DIBROMOMETHANE [HM1, HM1,
HM1, HM1, HON, MSL, RCA, SD4,
SDW, TSE, UTS, WHM]
METHANE, DIBROMO- [PAL, TSC,
TSC, TSC]

74-96-4 ETHYL BROMIDE
BROMOETHANE [GBR, GBR, IAR,
MAK]
BROMOETHANE (ETHYL BROMIDE)
[CSR, CSR]
BROMOMETHANE (ETHYL BROMIDE)
[QBC, QBC]
ETHANE, BROMO- [PAL]

74-97-5 CHLOROBROMOMETHANE
BROMOCHLOROMETHANE [CAL,
GBR, GBR, HM1, HM1, MAK,
MAK, ONT, ONT, OTS, SDW]
BROMOCHLOROMETHANE
(CHLOROBROMOMETHANE) [ALB,
ALB]
BROMOCHLOROMETHANE/
CHLOROBROMOMETHANE [BC1,
BC1]
CHLOROBROMOMETHANE [QBC]
METHANE, BROMOCHLORO- [PAL,
TSC, TSC, TSC, TSC]

74-98-6 PROPANE

74-99-7 METHYL ACETYLENE
1-PROPYNE [PAL]

METHYL ACETYLENE (PROPYNE)
[BC1, BC1, OS1, OS1, QBC, WHM]
METHYLENE ACETYLENE
(PROPYNE) [ALB, ALB]
PROPYNE [FLA, MSL, NFP, NFP,
ONT, ONT]
PROPYNE [1-PROPYNE] [EP3]

75-00-3 ETHYL CHLORIDE
CHLOROETHANE [ATS, CAN, CSR,
CSR, CW2, EPA, GBR, GBR, IAR,
MAK, MX1, NHS, NHS, OTS, PST,
RCA, SD4, SDW, TSE, UTS]
CHLOROETHANE (ETHYL CHLO-
RIDE) [313, C65]
ETHANE, CHLORO- [PAL, TSC,
TSC, TSC]
ETHYL
CHLORIDE (CHLOROETHANE) [CAA,
HON, HON, QBC]
ETHYL CHLORIDE [ETHANE,
CHLORO-] [EP3]

75-01-4 VINYL CHLORIDE
CHLOROETHENE [PST]
CHLOROETHYLENE (VINYL CHLO-
RIDE) [ALB, ALB, ALB, BC1, BC1,
QBC, QBC, QBC]
ETHENE, CHLORO- [PAL, PAL]
VINYL CHLORIDE (CHLOROETHY-
LENE) [CW2, HON, HON]
VINYL CHLORIDE [ETHENE,
CHLORO-] [EP3]
VINYL CHLORIDE, MONOMER [AUS,
AUS, AUS]

75-02-5 VINYL FLUORIDE
ETHENE, FLUORO- [PAL, TSC, TSC,
TSC]
VINYL FLUORIDE [ETHENE, FLUORO-
] [EP3]

75-03-6 ETHYL IODIDE

75-04-7 ETHYLAMINE
ETHANAMINE [PAL, TSC, TSC]
ETHYLAMINE [ETHANAMINE] [EP3]
MONOETHYLAMINE [CWA, EPA]

75-05-8 ACETONITRILE
METHYL CYANIDE [HM1, HM1,
HM1]

75-07-0 ACETALDEHYDE

75-08-1 ETHYL MERCAPTAN
1-ETHANETHIOL [NHS, NHS]
ETHANETHIOL [GBR, GBR, MAK,
MAK, ONT, PAL]
ETHANETHIOL (ETHYL MERCAPTAN)
[ALB, ALB]
ETHYL MERCAPTAN [QBC]
ETHYL MERCAPTAN [ETHANETHIOL]
[EP3]

75-09-2 METHYLENE CHLORIDE
DICHLOROMETHANE [CAN, CD1,
CN4, GBR, HM1, HM1, HM1,
HM1, IAR, MAK, MAK, MAK,
MAK, ONT, SDW, SDW, SDW]
DICHLOROMETHANE (METHYLENE
CHLORIDE) [313, ALB, ALB, C65,
C65, NTP]
METHANE, DICHLORO- [PAL, PAL,
TSC, TSC, TSC]
METHYLENE CHLORIDE [QBC,
QBC]
METHYLENE CHLORIDE
(DICHLOROMETHANE) [BC1, CAA,
CW2, HON, HON, TLV, TLV]

75-12-7 FORMAMIDE

75-15-0 CARBON DISULFIDE
CARBON BISULPHIDE [HM1, HM1,
HM1, HM1]
CARBON DISULPHIDE [AUS, AUS,
AUS, GBR, GBR]

75-16-1 METHYL MAGNESIUM BROMIDE

75-18-3 DIMETHYL SULFIDE
METHANE, THIOBIS- [PAL]

75-19-4 CYCLOPROPANE

75-20-7 CALCIUM CARBIDE
CALCIUM CARBIDE (CAC2) [PAL]

75-21-8 ETHYLENE OXIDE
ETHYLENE OXIDE (ETHER) [QBC,
QBC]
ETHYLENE OXIDE [OXIRANE] [EP3]
OXIRANE [PAL, PAL, TSC, TSC,
TSC]

75-24-1 TRIMETHYLALUMINIUM
ALUMINUM, TRIMETHYL- [PAL]
TRIMETHYL ALUMINIUM [NJL, NJL]

75-25-2 BROMOFORM
BROMOFORM (TRIBROMOMETHANE)
[313, CW2, QBC, QBC]
METHANE, TRIBROMO- [PAL, TSC,
TSC, TSC]
TRIBROMOMETHANE [CSR, CSR,
HM1, HM1, HM1, HM1, MAK]
TRIBROMOMETHANE (BROMOFORM)
[UTS]

75-26-3 2-BROMOPROPANE

75-27-4 DICHLOROBROMOMETHANE
BROMODICHLOROMETHANE [C65,
C65, CAL, CSR, CSR, IAR, NTP,
PQL, RCA, SD4, SDW, UTS]
METHANE, BROMODICHLORO-
[PAL, TSC, TSC]

75-28-5 ISOBUTANE
ISOBUTANE [PROPANE, 2-METHYL-]
[EP3]
PROPANE, 2-METHYL- [PAL]

75-29-6 2-CHLOROPROPANE
ISOPROPYL CHLORIDE [FLA, HM1,
HM1, HM1, MSL, NFP, NFP]
ISOPROPYL CHLORIDE [PROPANE, 2-
CHLORO-] [EP3]
PROPANE, 2-CHLORO- [PAL, TSC,
TSC]

75-30-9 2-IODOPROPANE

75-31-0 ISOPROPYLAMINE
2-PROPANAMINE [PAL]
ISOPROPYLAMINE [2-PROPANAMINE]
[EP3]

75-33-2 ISOPROPYL MERCAPTAN

75-34-3 1,1-DICHLOROETHANE
1,1-DICHLOROETHANE (ETHYLIDENE
CHLORIDE) [ALB, ALB]
1,1-ETHYLIDENE DICHLORIDE [NFP,
NFP]
DICHLOROETHANE, 1,1- [OTS]
ETHANE, 1,1-DICHLORO- [PAL, TSC,
TSC, TSC]
ETHYLENE DICHLORIDE [RCU,
RCU]
ETHYLIDENE CHLORIDE [CAL]
ETHYLIDENE DICHLORIDE [313,
HM1, HM1, HM1, HM1]
ETHYLIDENE DICHLORIDE (1,1-
DICHLOROETHANE) [CAA, HON]

75-35-4 VINYLIDENE CHLORIDE
1,1 DICHLOROETHYLENE [FLA]
1,1 DICHLOROETHYLENE (VINYLI-
DENE DICHLORIDE) [ALB, ALB]
1,1-DICHLOROETHENE [ATS, ONT,
ONT]
1,1-DICHLOROETHYLENE [CW2,
CW3, CWA, EPA, HM1, HM1,
HM1, HM1, MX1, MX2, OTS, QBC,
RC2, RC2, RCA, RCU, RCU, SDW,
SDW, UTS]
1,1-DICHLOROETHYLENE (VINYLI-
DENE CHLORIDE) [EP2]
ETHENE, 1,1-DICHLORO- [PAL]
VINYLIDENE CHLORIDE (1,1-
DICHLOROEHTYLENE) [CAA]
VINYLIDENE CHLORIDE (1,1-
DICHLOROETHYLENE) [HON, HON,
OS1]
VINYLIDENE CHLORIDE [ETHENE, 1,
1-DICHLORO-] [EP3]

75-36-5 ACETYL CHLORIDE

75-37-6 1,1-DIFLUOROETHANE
DIFLUOROETHANE [HM1, HM1,
MSL]
DIFLUOROETHANE [ETHANE, 1,1-
DIFLUORO-] [EP3]

ETHANE, 1,1-DIFLUORO- [TSC, TSC,
TSC]

75-38-7 VINYLIDENE FLUORIDE
1,1-DIFLUOROETHYLENE [HM1,
HM1, MAK]
ETHENE, 1,1-DIFLUORO- [PAL, TSC,
TSC, TSC]
VINYLIDENE FLUORIDE [TSE]
VINYLIDENE FLUORIDE [ETHENE, 1,
1-DIFLUORO-] [EP3]

75-39-8 ACETALDEHYDE AMMONIA

75-43-4 DICHLOROFLUOROMETHANE
DICHLOROFLUOROMETHANE [AUS,
CEX, FLA, GBR, MAK, MAK,
MSL, NJL, ONT, TLV]
DICHLOROFLUOROMETHANE
(FREON 21) [QBC]
DICHLOROMONOFLUOROMETHANE
(FC 21) [WHM]
DICHLOROMONOFLUOROMETHANE
(FC-21) [CAL]
DICHLOROMONOFLUOROMETHANE
(HCFC-21) [313]
HCFC-21 [CAA]
METHANE, DICHLOROFLUORO-
[PAL]

75-44-5 PHOSGENE
CARBONIC DICHLORIDE [PAL]
CARBONYL CHLORIDE [ONT]
CARBONYL CHLORIDE (PHOSGENE)
[ALB, ALB, BC1, QBC]
PHOSGENE (CARBON DICHLORIDE)
[EP3]
PHOSGENE (CARBONYL CHLORIDE)
[OS1, OS1, WHM]

75-45-6 CHLORODIFLUOROMETHANE
CHLORODIFLUOROMETHANE (FC 22)
[WHM]
CHLORODIFLUOROMETHANE (FC-22)
[CAL]
CHLORODIFLUOROMETHANE
(FREON 22) [QBC]
CHLORODIFLUOROMETHANE
(HCFC-22) [313]
HCFC-22 [CAA]
METHANE, CHLORODIFLUORO-
[PAL, TSC, TSC, TSC]
MONOCHLORODIFLUOROMETHANE
[PST]

75-46-7 TRIFLUOROMETHANE

75-47-8 IODOFORM
IODOFORM (TRIIODOMETHANE)
[QBC]
METHANE, TRIIODO- [PAL]

75-50-3 TRIMETHYLAMINE
METHANAMINE, N,N-DIMETHYL-
[PAL]
TRIMETHYLAMINE [METHANAMINE,
N,N-DIMETHYL-] [EP3]

75-52-5 NITROMETHANE
METHANE, NITRO- [PAL, TSC, TSC,
TSC]

75-54-7 METHYL DICHLOROSILANE
METHYLDICHLOROSILANE [FLA,
MSL, NFP, NFP]
SILANE, DICHLOROMETHYL- [OTS,
PAL]

75-55-8 PROPYLENEIMINE
1,2-PROPYLENEIMINE [EP2, EPA,
RCU, RCU]
1,2-PROPYLENIMINE (2-METHYL
AZIRIDINE) [CAA]
2-METHYLAZIRIDINE [CAB, CAL,
IAR, MIN]
2-METHYLAZIRIDINE (PROPY-
LENEIMINE) [C65, NTP]
AZIRIDINE, 2-METHTL- [PAL, PAL]
PROPYLENE IMINE [AUS, AUS,
AUS, BC1, BC1, FLA, MAK, MAK,
MSL, NIO, NIO, NIO, NIO, NIO,
OS1, OS1, OS1, OS1, QBC, QBC,
QBC, TLV, TLV, TLV]
PROPYLENEIMINE [AZIRIDINE, 2-
METHYL-] [EP3]

75-56-9 PROPYLENE OXIDE
1,2-EPOXYPROPANE [FLA, ONT, ONT]
1,2-EPOXYPROPANE (PROPYLENE OXIDE) [ALB, ALB]
1,2-PROPYLENE OXIDE [CSR, CSR]
OXIRANE, METHYL- [PAL, PAL, TSC, TSC, TSC]
PROPYLENE OXIDE [OXIRANE, METHYL-] [EP3]

75-59-2 TETRAMETHYLAMMONIUM HYDROXIDE
TETRAMETHYL AMMONIUM HYDROXIDE [NJL, NJL]

75-60-5 CACODYLIC ACID
DIMETHYLARSINIC ACID [ATS, HM1, HM1, HM1, HM1]

75-61-6 DIFLUORODIBROMOMETHANE
DIBROMODIFLUOROMETHANE [BC1, BC1, CAL, GBR, GBR, HM1, HM1, MAK, MAK, ONT]
DIBROMODIFLUOROMETHANE (FREON 12B2) [QBC]
METHANE, DIBROMODIFLUORO- [PAL]

75-63-8 TRIFLUOROBROMOMETHANE
BROMOTRIFLUOROMETHANE [CAL, CN2, CN5, GBR, GBR, MAK, MAK, ONT, QBC]
BROMOTRIFLUOROMETHANE (HALON 1301) [313]
HALON 1301 [CAA]
METHANE, BROMOTRIFLUORO- [PAL]
TRIFLUOROMONOBROMOMETHANE [BC1, BC1]

75-64-9 TERT-BUTYLAMINE
2-PROPANAMINE, 2-METHYL- [PAL]

75-65-0 TERT-BUTYL ALCOHOL
2-METHYLPROPAN-2-OL [GBR, GBR]
2-PROPANOL, 2-METHYL- [PAL]
TERT-BUTANOL [PST, TLV, TLV, WHM]

75-66-1 2-METHYL-2-PROPANETHIOL
2-PROPANETHIOL, 2-METHYL- [PAL]

75-68-3 1-CHLORO-1,1-DIFLUOROETHANE
1-CHLORO-1,1-DIFLUOROETHANE (HCFC-142B) [313]
CHLORODIFLUOROETHANE [NJL]
CHLORODIFLUOROETHANE (FC 142) [WHM]
DIFLUORO-1-CHLOROETHANE [FLA, MSL, NFP, NFP]
ETHANE, 1-CHLORO-1,1-DIFLUORO- [PAL, TSC, TSC, TSC]
HCFC-142B [CAA]

75-69-4 TRICHLOROFLUOROMETHANE
CFC-11 [CAA]
FLUOROTRICHLOROMETHANE [BC1, BC1, CAL, NIO, NIO, NIO, OTS, QBC, SD4, SDW]
FLUOROTRICHLOROMETHANE (TRICHLOROFLUOROMETHANE) [OS1, OS1]
METHANE, TRICHLOROFLUORO- [PAL]
TRICHLOROFLUOROMETHANE (CFC-11) [313]
TRICHLOROFLUOROMETHANE (FC 11) [FLA, MIN, MSL, WHM]
TRICHLOROFLUOROMETHANE (FLUOROTRICHLOROMETHANE) [ALB, ALB]
TRICHLOROMONOFLUOROMETHANE [EPA, RCU, RCU]

75-70-7 TRICHLOROMETHANETHIOL

75-71-8 DICHLORODIFLUOROMETHANE
CFC-12 [CAA]
DICHLORODIFLUOROMETHANE (CFC-12) [313]
DICHLORODIFLUOROMETHANE (FC 12) [WHM]
DICHLORODIFLUOROMETHANE (FC-12) [CAL]
DICHLORODIFLUOROMETHANE (FREON-12) [QBC]
METHANE, DICHLORODIFLUORO- [PAL]

75-72-9 CHLOROTRIFLUOROMETHANE
CFC-13 [CAA]
CHLOROTRIFLUOROMETHANE (CFC-13) [313]

75-73-0 TETRAFLUOROMETHANE
CARBON TETRAFLUORIDE [HON]

75-74-1 TETRAMETHYL LEAD
LEAD TETRAMETHYL [QBC, QBC]
PLUMBANE, TETRAMETHYL- [PAL]
TETRAMETHYL LEAD [NJL, NJL]
TETRAMETHYL LEAD (TML) [EPA, EPA, EPA, WHM]
TETRAMETHYL LEAD [PLUMBANE, TETRAMETHYL-] [EP3]
TETRAMETHYL LEAD, COMPOUNDS [NFP, NFP]
TETRAMETHYLLEAD [302]

75-75-2 ALKANE SULFONIC ACID

75-76-3 TETRAMETHYL SILANE
TETRAMETHYLSILANE [SILANE, TETRAMETHYL-] [EP3]

75-77-4 TRIMETHYLCHLOROSILANE
SILANE, CHLOROTRIMETHYL- [EPA, EPA, OTS, PAL]
TRIMETHYLCHLOROSILANE [SILANE, CHLOROTRIMETHYL-] [EP3]

75-78-5 DIMETHYLDICHLOROSILANE
DIMETHYL DICHLOROSILANE [NJL, NJL]
DIMETHYLCHLOROSILANE [FLA]
DIMETHYLDICHLOROSILANE [SILANE, DICHLORODIMETHYL-] [EP3]
SILANE, DICHLORODIMETHYL- [EPA, EPA, OTS, PAL]

75-79-6 METHYL TRICHLOROSILANE
METHYLTRICHLOROSILANE [302, 313, FLA, MSL, NFP, NFP, OS3, WEL, WHM]
METHYLTRICHLOROSILANE [SILANE, TRICHLOROMETHYL-] [EP3]
SILANE, TRICHLOROMETHYL- [EPA, EPA, OTS, PAL]

75-83-2 NEOHEXANE
2,2-DIMETHYLBUTANE [FLA, MSL, NFP, NFP]
BUTANE, 2,2-DIMETHYL- [PAL]

75-84-3 1-PROPANOL, 2,2-DIMETHYL-
TERT-BUTYL CARBINOL [FLA, MSL, NFP, NFP]

75-85-4 2-METHYL-2-BUTANOL
2-BUTANOL, 2-METHYL- [PAL]

75-86-5 ACETONE CYANOHYDRIN
2-METHYLLACTONITRILE [313, RCU, RCU]
PROPANENITRILE, 2-HYDROXY-2-METHYL- [PAL, TSC, TSC]

75-87-6 CHLORAL
ACETALDEHYDE, TRICHLORO [NJL]
ACETALDEHYDE, TRICHLORO- [PAL, TSC, TSC, TSC]
TRICHLOROACETALDEHYDE [MSL]

75-88-7 2-CHLORO-1,1,1-TRIFLUOROETHANE
2-CHLORO-1,1,1-TRIFLUOROETHANE (HCFC-133A) [313]
CHLOROTRIFLUOROETHANE [HM1, HM1]
ETHANE, 2-CHLORO-1,1,1-TRIFLUORO- [TSC]
HCFC-133A [CAA]
HYDROCHLOROFLUOROCARBON (HCFC 133A) [TSE]

75-89-8 2,2,2-TRIFLUOROETHANOL

75-91-2 TERT-BUTYL HYDROPEROXIDE
BUTYL HYDROPEROXIDE (TERTIARY) [OS3]
HYDROPEROXIDE, 1,1-DIMETHYLETHYL [PAL]
HYDROPEROXIDE, 1,1-DIMETHYLETHYL- [OTS]
TERT-BUTYLHYDROPEROXIDE [MAK]

75-94-5 VINYLTRICHLOROSILANE
SILANE, TRICHLOROETHENYL- [PAL]
VINYL TRICHLOROSILANE [FLA, MSL, NFP, NFP, NJL, NJL]

75-95-6 PENTABROMOETHANE

75-98-9 2,2-DIMETHYLPROPANOIC ACID

75-99-0 2,2-DICHLOROPROPIONIC ACID
DALAPON [SDW, SDW]
PROPANOIC ACID, 2,2-DICHLORO- [PAL]

76-01-7 PENTACHLOROETHANE

76-02-8 TRICHLOROACETYL CHLORIDE

76-03-9 TRICHLOROACETIC ACID
ACETIC ACID, TRICHLORO- [PAL]

76-05-1 TRIFLUOROACETIC ACID

76-06-2 CHLOROPICRIN
CHLOROPICRINE (NITROCHLOROMETHANE) [QBC]
METHANE, TRICHLORONITRO- [PAL]
NITROTRICHLOROMETHANE (CHLOROPICRIN) [ALB, ALB]
TRICHLORONITROMETHANE [GBR, GBR, ONT, ONT, PST]

76-09-5 PINACOL

76-11-9 1,1,1,2-TETRACHLORO-2,2-DIFLUOROETHANE
1,1,1,2-TETRACHLORO-2,2-DIFLUORO-ETHANE (FC 112A) [WHM]
1,1,1,2-TETRACHLORO-2,2-DIFLUOROETHANE (FREON 112) [QBC]
ETHANE, 1,1,1,2-TETRACHLORO-2,2-DIFLUORO- [PAL]

76-12-0 1,1,2,2-TETRACHLORO-1,2-DIFLUOROETHANE
1,1,2,2-TETRACHLORO-1,2-DIFLUORO-ETHANE (FC 112) [WHM]
1,1,2,2-TETRACHLORO-1,2-DIFLUOROETHANE (FC-112) [CAL]
1,1,2,2-TETRACHLORO-1,2-DIFLUOROETHANE (FREON 113) [QBC]
1,2-DIFLUORO-1,1,2,2-TETRACHLOROETHANE [CSR]
CFC-112 [CAA]
ETHANE, 1,1,2,2-TETRACHLORO-1,2-DIFLUORO- [PAL]

76-13-1 1,1,2-TRICHLORO-1,2,2-TRIFLUOROETHANE
1,1,2-TRICHLORO-1,2,2-TRIFLUORO-ETHANE (FC 113) [WHM]
1,1,2-TRICHLORO-1,2,2-TRIFLUOROETHANE (FC-113) [CAL]
1,1,2-TRICHLOROTRIFLUOROETHANE [GBR, GBR]
CFC-113 [CAA]
CHLORINATED FLUOROCARBON (CFC-113) [CAB]
ETHANE, 1,1,2-TRICHLORO-1,2,2-TRIFLUORO- [PAL, TSC, TSC]
FREON 113 [313]
TRICHLORO 1,1,2-, 1,2,2-TRIFLUOROETHANE [QBC, QBC]

76-14-2 DICHLOROTETRAFLUOROETHANE
1,2-DICHLORO-1,1,2,2-TETRAFLUOROETHANE [MAK, MAK, ONT, WHM]
1,2-DICHLORO-1,1,2,2-TETRAFLUOROETHANE (FC-114) [CAL]
1,2-DICHLOROTETRAFLUOROETHANE [CEX]
CFC-114 [CAA]
DICHLOROTETRAFLUOROETHANE (CFC-114) [313]
DICHLOROTETRAFLUOROETHANE (FREON 114) [QBC]

76-15-3 CHLOROPENTAFLUOROETHANE
(MONO)CHLOROPENTAFLUO-
ROETHANE (CFC-115) [313]
CFC-115 [CAA]
CHLOROPENTAFLUOROETHANE (FC
115) [WHM]
CHLOROPENTAFLUOROETHANE
(FREON 115) [QBC]
ETHANE, CHLOROPENTAFLUORO-
[PAL]

76-16-4 HEXAFLUOROETHANE

76-19-7 OCTAFLUOROPROPANE

76-22-2 CAMPHOR
BICYCLO(2.2.1)HEPTAN-2-ONE, 1,7,7-
TRIMETHYL- [PAL]
BORNAN-2-ONE [GBR, GBR]
CAMPHOR (SYNTHETIC) [CEX, NIO,
NIO, NIO]
CAMPHOR, SYNTHETIC [ALB, ALB,
AUS, AUS, BC1, BC1, MEX, OS1,
OS1, TLV, TLV]
CAMPHOUR, SYNTHETIC [QBC,
QBC]

76-38-0 METHOXYFLURANE

76-43-7 FLUOXYMESTERONE

76-44-8 HEPTACHLOR
4,7-METHANO-1H-INDENE, 1,4,5,
6,7,8,8-HEPTACHLORO-3A,4,7,7A-
TETRAHYDRO- [PAL]
HEPTACHLOR (AND ITS EPOXIDE)
[RC2, RC2]
HEPTACHLOR + HEPTACHLOR
EPOXIDE [CD1]
HEPTACHLOR AND METABOLITES
[CW3]
HEPTACHLOR[1,4,5,6,7,8,8-HEP-
TACHLORO-3A,4,7,7A-TETRAHYDRO-
4,7-METHANO-1H-INDENE] [313]

76-57-3 CODEINE

76-87-9 TRIPHENYLTIN HYDROXIDE
FENTIN HYDROXIDE [HM1]
TRIPHENYL TIN HYDROXIDE [MSL]

76-93-7 BENZILIC ACID

77-06-5 GIBBERELLIC ACID

77-09-8 PHENOLPHTHALEIN

**77-47-4 HEXACHLOROCYCLOPENTADI-
ENE**
1,3-CYCLOPENTADIENE, 1,2,3,4,5,5-
HEXACHLORO- [PAL, TSC, TSC,
TSC]
HEXACHLOROCYCLO-PENTADIENE
[OS1]

77-58-7 DIBUTYLTIN DILAURATE
STANNANE, DIBUTYLBIS(1-OXODO-
DECYL)OXY]- [TSC, TSC, TSC]

77-65-6 CARBROMAL

77-71-4 5,5-DIMETHYLHYDANTOIN

77-73-6 DICYCLOPENTADIENE
4,7-METHANO-1H-INDENE, 3A,4,7,7A-
TETRAHYDRO- [PAL]

77-74-7 TERT-ISOHEXYL ALCOHOL

77-75-8 1-PENTYN-3-OL, 3-METHYL-
3-METHYL-1-PENTYNOL [NFP, NFP]
METHYLPENTYLNOL [HON]

77-78-1 DIMETHYL SULFATE
DIMETHYL SULPHATE [AUS, AUS,
AUS, IAR]
SULFURIC ACID, DIMETHYL ESTER
[FLA, PAL, PAL, TSC, TSC, TSC]

77-79-2 3-SULFOLENE
SULFOLENE [TSC, TSC, TSC]

77-81-6 TABUN

**77-83-8 OXIRANECARBOXYLIC ACID,
3-METHYL-3-PHENYL-, ETHYL
ESTER**

77-85-0 TRIMETHYLOLETHANE

**77-86-1 TRIS (HYDROXYMETHYL)
AMINOMETHANE**

77-89-4 ACETYL TRIETHYL CITRATE

**77-90-7 1,2,3-PROPANETRICARBOXYLIC
ACID, 2-(ACETYLOXY)-, TRIB-
UTYL ESTER**

77-92-9 CITRIC ACID

77-93-0 TRIETHYL CITRATE

77-94-1 TRIBUTYL CITRATE

**77-99-6 1,3-PROPANEDIOL, 2-ETHYL-2-
(HYDROXYMETHYL)-**
PROPANEDIOL, 2-ETHYL-2-(HY-
DROXYMETHYL)-, 1,3- [OTS]
TRIMETHYLOLPROPANE [HON,
PST]

78-00-2 TETRAETHYL LEAD
LEAD TETRAETHYL [HM1, HM1,
HM1, HM1, QBC, QBC]
PLUMBANE, TETRAETHYL- [PAL]
TETRAETHYL LEAD, COMPOUNDS
[NFP, NFP]
TETRAETHYLLEAD [302]

78-04-6 DIBUTYLTIN MALEATE

78-07-9 TRIETHOXYETHYL SILANE
SILANE, TRIETHOXYETHYL- [PST]

78-08-0 SILANE, ETHENYLTRIETHOXY-

78-10-4 ETHYL SILICATE
SILICIC ACID (H4SIO4), TETRAETHYL
ESTER [PAL]
SILICIC ACID, TETRAETHYL ESTER
[MAK, MAK]
TETRAETHYL ESTER SILICIC ACID
[FLA]
TETRAETHYL ORTHOSILICATE
[GBR, GBR]
TETRAETHYL SILICATE [HM1,
HM1]

**78-11-5 PENTAERYTHRITE TETRANI-
TRATE**
PENTAERYTHRITOL TETRANITRATE
[CSR, CSR]

**78-13-7 SILICIC ACID (H4SIO4), TETRAKIS
(2-ETHYLBUTYL) ESTER**
TETRA (2-ETHYLBUTYL) SILICATE
[NFP, NFP]

78-18-2 CYCLOHEXANONE PEROXIDE
1-HYDROXY-1'-HYDROPEROXY
DICYCLOHEXYL PEROXIDE [NJL]
1-HYDROXY-1'-HYDROPEROXY-
DICYCLOHEXYL PEROXIDE [MAK]

**78-21-7 4-ETHYL-4-HEXADECYLMOR-
PHOLINIUM, ETHYL SULFATE**

78-30-8 TRIORTHOCRESYL PHOSPHATE
PHOSPHORIC ACID, TRIS(2-
METHYLPHENYL) ESTER [PAL, TSC,
TSC]
TRI-O-CRESYL PHOSPHATE [CAL,
FLA, MSL, NFP, NFP]
TRI-O-TOLYL PHOSPHATE [GBR,
GBR]
TRI-ORTHO-CRESYL PHOSPHATE
[ONT, ONT, WHM]

**78-32-0 PHOSPHORIC ACID, TRIS(4-
METHYLPHENYL) ESTER**

**78-33-1 PHE-
NOL, 4-(1,1-DIMETHYLETHYL)-,
PHOSPHATE (3:1)**
PHOSPHATE, TRIS(TERT-
BUTYLPHENYL) [OTS]

78-34-2 DIOXATHION
2,3-P-DIOXANEDITHIOL S,S-BIS(O,O-
DIETHYLPHOSPHORODITHIOATE)
(DIOXATHION) [CAL]
DIOXATHION (DELNAV) [ALB, ALB,
ALB, BC1, BC1, OS1, OS1]
DIOXATHION (DELNAV) (R) [QBC,
QBC]
DIOXATHION (ISO) [GBR, GBR]

PHOSPHORODITHIOIC ACID, S,S'-
1,4-DIOXANE-2,3-DIYL O,O,O',O'-
TETRAETHYL ESTER [PAL]

78-40-0 TRIETHYL PHOSPHATE
PHOSPHORIC ACID, TRIETHYL ESTER
[PST, TSC, TSC, TSC]

**78-42-2 TRIS(2-ETHYLHEXYL)PHOS-
PHATE**
PHOSPHORIC ACID, TRIS(ETHYL-
HEXYL) ESTER [TSC, TSC, TSC]

**78-43-3 TRIS(2,3-DICHLOROPROPYL)
PHOSPHATE**
TRIS(1,3-DICHLORO-2-PROPYL)
PHOSPHATE [T33]

78-44-4 CARISOPRODOL

78-48-8 DEF
S,S,S-TRIBUTYL PHOSPHO-
ROTRITHIOATE [CAL]
S,S,S-TRIBUTYLTRITHIOPHOSPHATE
(DEF) [313]

78-51-3 TRIBUTOXYETHYL PHOSPHATE
ETHANOL, 2-BUTOXY-, PHOSPHATE
(3:1) [TSC, TSC, TSC]

78-53-5 AMITON

78-59-1 ISOPHORONE
2-CYCLOHEXEN-1-ONE, 3,5,5-
TRIMETHYL- [PAL, TSC, TSC, TSC]
3,5,5-TRIMETHYL-2-CYCLOHEXEN-1-
ONE [FLA, ONT]
3,5,5-TRIMETHYLCYCLOHEX-2-
ENONE [GBR]

78-62-6 DIMETHYLDIETHOXYSILANE

**78-63-7 2,5-DIMETHYL-2,5-DI(TERT-
BUTYLPEROXY)HEXANE**

**78-66-0 4-OCTYNE-3,6-DIOL, 3,6-
DIMETHYL-**
3,6-DIMETHYL-4-OCTYNE-3,6-DIOL
[PST]

78-67-1 AZOBISISOBUTYRONITRILE
2,2'-AZOBIS(2-METHYL PROPIONI-
TRILE [FLA]
AZODIISOBUTYRONITRILE [HM1,
HM1, NJL]
PROPANENITRILE, 2,2'-AZOBIS[2-
METHYL- [PAL]

**78-70-6 LINALOOL (EX BOIS DE ROSE;
SYNTHETIC)**

**78-71-7 OXETANE, 3,3-BIS
(CHLOROMETHYL)-**

78-75-1 1,2-DIBROMOPROPANE

78-76-2 2-BROMOBUTANE

78-77-3 1-BROMO-2-METHYLPROPANE

78-78-4 ISOPENTANE
2-METHYLBUTANE [FLA]
BUTANE, 2-METHYL- [PAL]
ISO-PENTANE [MAK, MAK]
ISOPENTANE [BUTANE, 2-METHYL-]
[EP3]

78-79-5 ISOPRENE
1,3-BUTADIENE, 2-METHYL- [PAL]
ISOPRENE [1,3-BUTADIENE, 2-
METHYL-] [EP3]

78-80-8 ISOPROPENYL ACETYLENE
1-BUTEN-3-YNE, 2-METHYL- [PAL]

78-81-9 ISOBUTYLAMINE
1-PROPANAMINE, 2-METHYL- [PAL]
ISO-BUTYLAMINE [CWA, EPA]
ISOBUTYL AMINE [MAK, MAK,
MAK]

78-82-0 ISOBUTYRONITRILE
2-METHYLPROPANENITRILE [FLA]
ISOBUTYRONITRILE [PROPANENI-
TRILE, 2-METHYL-] [EP3]
PROPANENITRILE, 2-METHYL- [PAL]

78-83-1 ISOBUTYL ALCOHOL
1-PROPANOL, 2-METHYL- [PAL, TSC, TSC]
2-METHYLPROPAN-1-OL [GBR, GBR]
ISOBUTANOL [HM1, HM1, HM1, RCA, WHM]

78-84-2 ISOBUTYRALDEHYDE
PROPANAL, 2-METHYL- [OTS, PAL, TSC, TSC, TSC]

78-85-3 METHACRYLALDEHYDE
2-METHYLPROPENAL [MSL, NFP, NFP]
2-PROPENAL, 2-METHYL- [PAL, TSC, TSC, TSC]

78-86-4 BUTANE, 2-CHLORO-
2-CHLOROBUTANE [FLA]
SEC-BUTYL CHLORIDE [MSL, NFP, NFP]

78-87-5 1,2-DICHLOROPROPANE
1,2-DICHLOROPROPANE (PROPYLENE DICHLORIDE) [ALB, ALB, CSR, CSR]
PROPANE, 1,2-DICHLORO- [PAL, TSC, TSC, TSC]
PROPYLENE DICHLORIDE [AUS, AUS, CEX, CEX, FLA, HM1, HM1, HM1, HM1, MEX, MEX, MIN, MSL, NFP, NFP, NIO, NIO, NIO, OS1, OS1, OS1, OTV, RCU, RCU, TLV, TLV, WHM]
PROPYLENE DICHLORIDE (1,2-DICHLOROPROPANE) [BC1, BC1, CAA, HON, HON]

78-88-6 2,3-DICHLOROPROPENE
1-PROPENE, 2,3-DICHLORO- [PAL, TSC, TSC]
2,3-DICHLORO-1-PROPENE [PST]
2,3-DICHLOROPROPYLENE [CSR]

78-89-7 PROPYLENE CHLOROHYDRIN
1-PROPANOL, 2-CHLORO- [PAL]
2-CHLORO-1-PROPANOL [FLA, MSL, NFP, NFP]

78-90-0 PROPYLENE DIAMINE
1,2-PROPANEDIAMINE [FLA, PAL]
1,2-PROPYLENEDIAMINE [HM1, HM1]
PROPYLENEDIAMINE [MSL, NFP, NFP, NJL, NJL]

78-92-2 SEC-BUTYL ALCOHOL
2-BUTANOL [PAL]
BUTAN-2-OL [GBR, GBR]
SEC-BUTANOL [PST, TLV, WHM]

78-93-3 METHYL ETHYL KETONE
2-BUTANONE [ATS, BC1, BC1, CAL, EPA, FDA, MAK, MAK, MAK, NIO, NIO, NIO, ONT, ONT, ONT, ONT, PAL, PST]
2-BUTANONE (METHYL ETHYL KETONE) [ALB, ALB, OS1, OS1, OS1, TSC, TSC, TSC]
BUTAN-2-ONE [GBR, GBR]
ETHYL METHYL KETONE [HM1, HM1, HM1]
METHYL ETHYL KETONE (2-BUTANONE) [CAA, HON, HON]
METHYL ETHYL KETONE (MEK) [AUS, AUS, FLA, MSL, QBC, QBC, RCU, RCU, TLV, TLV, WHM]

78-94-4 METHYL VINYL KETONE
3-BUTEN-2-ONE [FLA, PAL]

78-95-5 CHLOROACETONE
MONOCHLOROACETONE [MSL, WHM]

78-96-6 ISOPROPANOLAMINE
1-AMINO-2-PROPANOL [FLA, MSL, NFP, NFP]
2-PROPANOL, 1-AMINO- [PAL]
MONOISOPROPANOLAMINE [PST]

78-97-7 LACTONITRILE
PROPANENITRILE, 2-HYDROXY- [OTS, PAL, TSC, TSC]

78-98-8 PYRUVIC ALDEHYDE
METHYLGLYOXAL [IAR]

78-99-9 1,1-DICHLOROPROPANE
PROPANE, 1,1-DICHLORO- [PAL, TSC, TSC]

79-00-5 1,1,2-TRICHLOROETHANE
ETHANE, 1,1,2-TRICHLORO- [PAL, TSC, TSC, TSC]
TRICHLOROETHANE (1,1,2-) (VINYL TRICHLORIDE) [HON, HON]
TRICHLOROETHANE 1,1,2- [QBC, QBC]
VINYL TRICHLORIDE (1,1,2-TRICHLOROETHANE) [C65, C65]

79-01-6 TRICHLOROETHYLENE
ACETYLENE TRICHLORIDE [FLA]
ETHENE, TRICHLORO- [PAL]

79-02-7 DICHLOROACETALDEHYDE

79-03-8 PROPIONYL CHLORIDE
PROPANOYL CHLORIDE [PAL]

79-04-9 CHLOROACETYL CHLORIDE
ACETYL CHLORIDE, CHLORO- [PAL]

79-06-1 ACRYLAMIDE
2-PROPENAMIDE [PAL]
2-PROPENAMIDE (ACRYLAMIDE) [TSC, TSC, TSC]

79-07-2 2-CHLOROACETAMIDE

79-08-3 BROMOACETIC ACID

79-09-4 PROPIONIC ACID
PROPANOIC ACID [PAL]

79-10-7 ACRYLIC ACID
2-PROPENOIC ACID [PAL]
ACRYLIC ACID (GLACIAL) [NFP, NFP]

79-11-8 CHLOROACETIC ACID
ACETIC ACID, CHLORO- [PAL]
MONOCHLOROACETIC ACID [CSR, CSR, GBR, GBR, MIN, PST, WEL, WEL, WEL]

79-14-1 HYDROXYACETIC ACID
GLYCOLIC ACID [PST]

79-16-3 N-METHYLACETAMIDE

79-19-6 THIOSEMICARBAZIDE
1-AMINO-2-THIOUREA [WHM]
HYDRAZINE CARBOTHIOAMIDE [RCU, RCU]

79-20-9 METHYL ACETATE
ACETIC ACID, METHYL ESTER [PAL]
METHYL ESTER ACETIC ACID [FLA]

79-21-0 PEROXYACETIC ACID
ETHANEPEROXOIC ACID [PAL]
PERACETIC ACID [302, 313, CAN, EPA, EPA, HON, MAK, MAK, NFP, NFP, OS3]
PERACETIC ACID [ETHANEPEROXIC ACID] [EP3]
PEROXYACETIC ACID [NJL, NJL]

79-22-1 METHYL CHLOROFORMATE
METHYL CHLOROCARBONATE [313, EPA, RCU, RCU, WHM]
METHYL CHLOROFORMATE [CARBONOCHLORIDIC ACID, METHYL ESTER] [EP3]

79-24-3 NITROETHANE
ETHANE, NITRO- [PAL, TSC, TSC, TSC]

79-27-6 ACETYLENE TETRABROMIDE
1,1,2,2-TETRABROMOETHANE [AUS, CSR, FLA, GBR, GBR, MAK, MAK, NFP, ONT, ONT]
ACETYLENE TETRABROMIDE (1,1,2,2-TETRABROMOETHANE) [QBC]
ETHANE, 1,1,2,2-TETRABROMO- [PAL, TSC]
ETHANE, 1,1,2,2-TETRABROMO- [CAI]

79-29-8 2,3-DIMETHYLBUTANE
BUTANE, 2,3-DIMETHYL- [PAL]

79-30-1 ISOBUTYRYL CHLORIDE
ISOBUTYRYLCHLORIDE [NJL]

79-31-2 ISOBUTYRIC ACID
2-METHYLPROPANOIC ACID [TSC]
ISO-BUTYRIC ACID [CWA, EPA, MSL]
METHYLPROPANOIC ACID, 2- [OTS]
PROPANOIC ACID, 2-METHYL- [PAL]

79-34-5 1,1,2,2-TETRACHLOROETHANE
ACETYLENE TETRACHLORIDE [CAL, FLA, HM1, HM1]
ETHANE, 1,1,2,2-TETRACHLORO- [PAL, TSC]
TETRACHLOROETHANE 1,1,2,2 [QBC, QBC]
TETRACHLOROETHANE, 1,1,2,2- [OTS]

79-36-7 DICHLOROACETYL CHLORIDE
ACETYL CHLORIDE [CAL, CWA]
ACETYL CHLORIDE, DICHLORO- [PAL]

79-38-9 CHLOROTRIFLUOROETHYLENE
ETHENE, CHLOROTRIFLUORO- [PAL]
TRIFLUOROCHLOROETHYLENE [HM1, HM1, HM1, MSL, NFP, NFP, OS3]
TRIFLUOROCHLOROETHYLENE [ETHENE, CHLOROTRIFLUORO-] [EP3]

79-39-0 2-METHYLACRYLAMIDE

79-40-3 ETHANEDITHIOAMIDE

79-41-4 METHACRYLIC ACID
2-PROPENOIC ACID, 2-METHYL- [PAL]

79-42-5 2-THIOLACTIC ACID
THIOLACTIC ACID [HM1, HM1, NJL]

79-43-6 DICHLOROACETIC ACID

79-44-7 DIMETHYLCARBAMOYL CHLORIDE
CARBAMIC CHLORIDE, DIMETHYL- [PAL, PAL]
DIMETHYL CARBAMOYL CHLORIDE [AUS, AUS, CAA, TLV]
DIMETHYL CARBAMYL CHLORIDE [BC1, BC1]
DIMETHYLCARBAMOL CHLORIDE [MIN]
DIMETHYLCARBAMOYL CHLORIDE [NJL, NJL]
DIMETHYLCARBAMYL CHLORIDE [313, C65, C65, EPA, QBC, QBC]
N,N-DIMETHYLCARBAMOYL CHLORIDE [WHM]

79-46-9 2-NITROPROPANE
PROPANE, 2-NITRO- [PAL, PAL]

79-57-2 OXYTETRACYCLINE

79-64-1 DIMETHISTERONE

79-69-6 METHYLIONONES (A-)

79-92-5 CAMPHENE

79-94-7 TETRABROMOBISPHENOL A

79-95-8 TETRACHLOROBISPHENOL-A

80-05-7 BISPHENOL A
4,4'-ISOPROPYLIDENEDIPHENOL [313, CAB, MSL, NJL, PAL, PST]
4,4-ISOPROPYLIDENEDIPHENOL [CAN]
PHENOL, 4,4'-(1-METHYLETHYLIDENE)BIS- [TSC, TSC, TSC]

80-07-9 SULFONYL BIS-(4-CHLOROBENZENE)
P,P'-DICHLORODIPHENYL SULFONE [CSR]

80-08-0 DAPSONE
4,4'-DIAMINODIPHENYL SULFONE [TSC, TSC, TSC]
4,4'-SULFONYLDIANILINE (DAPSONE) [CSR, CSR]

80-09-1 4,4'-SULFONYLDIPHENOL
PHENOL, 4,4'-SULFONYLBIS- [TSC, TSC, TSC]

80-10-4 DIPHENYL DICHLOROSILANE
DICHLORODIPHENYLSILANE [FLA]
DIPHENYLDICHLOROSILANE [MSL, NFP, NFP, WHM]
SILANE, DICHLORODIPHENYL- [PAL]

80-11-5 N-METHYL-N-NITROSO-P-TOLUE-NESULFONAMIDE

80-15-9 CUMENE HYDROPEROXIDE
ALPHA, ALPHA-DIMETHYLBENZYL-HYDROPEROXIDE [RCU, RCU]
ALPHA,ALPHA-DIMETHYLBENZYL HYDROPEROXIDE [MAK]
CUMYL HYDROPEROXIDE [DOT]
HYDROPEROXIDE, 1-METHYL-1-PHENYLETHYL [PAL]
HYDROPEROXIDE, 1-METHYL-1-PHENYLETHYL- [EPA, TSC, TSC]

80-17-1 BENZENESULFONIC ACID, HY-DRAZIDE
BENZENESULFOHYDRAZINE [HM1, HM1]

80-26-2 TERPINYL ACETATE

80-39-7 ETHYL P-TOLUENE SULFON-AMIDE
N-ETHYL-P-TOLYLSULFONAMIDE [PST]

80-40-0 ETHYL P-TOLUENE SULFONATE

80-43-3 DICUMYL PEROXIDE

80-46-6 P-TERT-PENTYLPHENOL
P-(TERT-PHENYL) PHENOL [FLA]
PENTAPHEN [MSL, NFP, NFP]
PHENOL, 4-(1,1-DIMETHYLPROPYL)- [PAL]

80-47-7 P-MENTHANE HYDROPEROXIDE

80-48-8 METHYL-P-TOLUENESULFONATE
BENZENESULFONIC ACID, 4-METHYL-, METHYL ESTER [PAL]
METHYL ESTER P-TOLUENE SUL-FONIC ACID [FLA]
METHYL TOLUENE SULFONATE [MSL, NFP, NFP]

80-51-3 DIPHENYLOXIDE-4,4'-DISULFO-HYDRAZIDE
P,P'-OXYBIS(BENZENESULFONYLHY-DRAZIDE) [TSC, TSC, TSC]

80-52-4 P-MENTHANE-1,8-DIAMINE

80-54-6 BENZENEPROPANAL, 4-(1,1-DIMETHYLETHYL)-.ALPHA.-METHYL-

80-56-8 ALPHA-PINENE
BICYCLO(3.1.1)HEPT-2-ENE, 2,6,6-TRIMETHYL- [PAL]

80-59-1 TIGLIC ACID

80-62-6 METHYL METHACRYLATE
2-PROPENOIC ACID, 2-METHYL-, METHYL ESTER [PAL, TSC]
METHACRYLIC ACID, METHYL ES-TER [MAK, MAK, MAK, MAK]

80-63-7 METHYL-2-CHLOROACRYLATE
METHYL 2-CHLOROACRYLATE [302]

81-07-2 SACCHARIN
1,2-BENZISOTHIAZOL-3(2H)-ONE, 1,1-DIOXIDE [PAL, PAL]
SACCHARIN AND SALTS [EPA]
SACCHARIN [1,2-BENZISOTHIAZOL-3 (2H)-ONE,1,1-DIOXIDE] [313]
SACCHARIN, & SALTS [RCU, RCU]

81-11-8 4,4'-DIAMINO-2,2'-STIL-BENEDISULFONIC ACID, DIS-ODIUM SALT
BENZENESULFONIC ACID, 2,2'-(1,2-ETHANEDI- [OTS]

81-15-2 MUSK XYLENE
5-TERT-BUTYL-2,4,6-TRINITRO-M-XYLENE [HM1, HM1]

81-16-3 2-AMINO-1-NAPHTHALENESUL-FONIC ACID
NAPHTHYLAMINE SULFONIC ACID (2,1-) [HON]

81-21-0 2,4-METHANO-2H-INDENO [1,2-B:5,6-B'] BISOXIRENE, OCTAHY-DRO-

81-30-1 1,4,5,8-NAPHTHALENE TE-TRACARBOXYLIC DIANHYDRIDE

81-48-1 C.I. SOLVENT VIOLET 13

81-49-2 1-AMINO-2,4-DIBROMOAN-THRAQUINONE

81-64-1 1,4-DIHYDROXY-9,10-AN-THRACENEDIONE

81-81-2 WARFARIN
2H-1-BENZOPYRAN-2-ONE, 4-HY-DROXY-3-(3-OXO)-1-PHENYLBUTYL)- [PAL]
3-(ALPHA-ACETONYLBENZYL)-4-HYDROXYCOUMARIN [FLA]
WARFARIN (ISO) [GBR, GBR]
WARFARIN AND SALTS [313]
WARFARIN, & SALTS [EPA, RCU, RCU, RCU]

81-82-3 COUMACHLOR

81-84-5 NAPHTHALENEDICARBOXYLIC ANHYDRIDE

81-88-9 C.I. FOOD RED 15
CI FOOD RED 15 [313]
D AND C RED NO. 19 [CAB]
D&C RED NO. 19 [C65]
RHODAMINE B [CAL, IAR, PST, WHM]

82-05-3 BENZANTHRONE

82-22-4 1,1-DIANTHRIMIDE

82-28-0 1-AMINO-2-METHYLAN-THRAQUINONE
9,10-ANTHRACENEDIONE, 1-AMINO-2-METHYL- [PAL, PAL]

82-45-1 1-AMINOANTHRAQUINONE

82-58-6 D-LYSERGIC ACID

82-66-6 DIPHACINONE

82-68-8 PENTACHLORONITROBENZENE
BENZENE, PENTACHLORONITRO- [PAL]
PENTACHLORONITROBENZENE (DCNB) [EPA]
PENTACHLORONITROBENZENE (PCNB) [RCU, RCU]
PENTACHLORONITROBENZENE (QUINTOBENZENE) [CAA]
QUINTOZENE [NJL, NJL]
QUINTOZENE (PENTACHLORONI-TROBENZENE) [CAL, IAR]
QUINTOZENE [PENTACHLORONI-TROBENZENE] [313]

82-71-3 TRINITRORESORCINOL

83-07-8 4-AMINOANTIPYRINE

83-26-1 PINDONE
1H-INDENE-1,3-(2H)-DIONE, 2-(2,2-DIMETHYL-1-OXOPROPYL)- [PAL]
2-(2,2-DIMETHYL-1-OXOPROPYL)-H-INDENE-1,3(2H)-DIONE [ONT]
2-PIVALOYL-1,3-INDANDIONE [FLA]
2-PIVALYL-1,3-INDANDIONE (PIN-DONE) [CAL]
PINDONE [QBC]
PINDONE (2-PIVALYL-1,3-INDAN-DIONE) [OS1, OS1]
PIVAL [MSL, NJL]
PIVAL (2-PIVALOYL-1,3-INDAN-DIONE) [BC1, BC1]
PIVAL (2-PIVALYL-1,3-INDANDIONE) (PINDONE) [ALB, ALB]

83-32-9 ACENAPHTHENE
ACENAPHTHYLENE, 1,2-DIHYDRO- [PAL]

83-34-1 3-METHYLINDOLE

83-44-3 DEOXYCHOLIC ACID

83-63-6 DIACETYLAMINOAZOTOLUENE

83-67-0 THEOBROMINE

83-79-4 ROTENONE (COMMERCIAL)
BARBASCO [PST]
ROTENONE [CSR, CSR, HM1, MAK, MEX, MEX, NIO, NIO, NJL, NJL, OS1, OS1]
ROTENONE (ISO) [GBR, GBR]
ROTENONE, COMMERCIAL [CAL, CEX]
[1]BENZOPYRANO[3,4-B]FURO[2,3-H]
[1]BENZOPYRAN-6(6AH)-ONE,1,212,12A-TETRAHYDRO-8,9-DIMETHOXY-2-(1-METHYLETHENYL)-, [2R-(2.AL-PHA.,6A.ALPHA.,12A.ALPHA.)]- [PAL]

84-15-1 O-TERPHENYL

84-16-2 HEXOESTROL

84-17-3 DIENOESTROL
DEHYDROSTILBESTROL [FLA]
DIENESTROL [C65, CAL]

84-51-5 2-ETHYLANTHRAQUINONE

84-61-7 1,2-BENZENEDICARBOXYLIC ACID, DICYCLOHEXYL ESTER
DICYCLOHEXYL PHTHALATE [GBR]

84-62-8 DIPHENYL PHTHALATE

84-64-0 1,2-BENZENEDICARBOXYLIC ACID, BUTYL CYCLOHEXYL ES-TER

84-65-1 ANTHRAQUINONE

84-66-2 DIETHYL PHTHALATE
1,2-BENZENEDICARBOXYLIC ACID, DIETHYL ESTER [PAL, TSC]

84-69-5 1,2-BENZENEDICARBOXYLIC ACID, BIS(2-METHYLPROPYL) ESTER
DIISOBUTYL PHTHALATE [GBR, NFP, NFP]

84-72-0 1,2-BENZENEDICARBOXYLIC ACID, 2-ETHOXY-2-OXOETHYL ETHYL ESTER
ETHYL PHTHALYL ETHYL GLYCO-LATE [NFP, NFP]

84-74-2 DIBUTYL PHTHALATE
1,2-BENZENEDICARBOXYLIC ACID, DIBUTYL ESTER [PAL, TSC, TSC, TSC]
DI-N-BUTYL PHTHALATE [ATS, CW2, CWA, HM1, HM1, MX1, PQL, RCA, UTS]
DIBUTYL PHTHALATE [TSE]
DIBUTYLPHTHALATE [CAA, NIO, NIO, NIO]

84-75-3 1,2-BENZENEDICARBOXYLIC ACID, DIHEXYL ESTER
DIHEXYL PHTHALATE [TSE]

84-76-4 1,2-BENZENEDICARBOXYLIC ACID, DINONYL ESTER
DINONYL PHTHALATE [GBR]

84-77-5 DECYL PHTHALATE

84-78-6 1,2-BENZENEDICARBOXYLIC ACID, BUTYL OCTYL ESTER

84-80-0 PHYLLOQUINONE
PHYLLOQUINONE (VITAMIN K1) [MSL]

84-83-3 ACETALDEHYDE, (1,3-DIHYDRO-1,3,3-TRIMETHYL-2H-INDOL-2-YLIDENE)

88-89-1 PICRIC ACID
2,4,6-TRINITROPHENOL [NJL, ONT, ONT, ONT]
PHENOL, 2,4,6-TRINITRO- [PAL]
TRINITROPHENOL [HM1, HM1]

88-96-0 PHTHALAMIDE

88-99-3 PHTHALIC ACID

89-13-4 1,2-BENZENEDICARBOXYLIC ACID, 2-ETHYLHEXYL-8-METHYL-NONYLESTER

89-25-8 3-METHYL-1-PHENYL-2-PYRA-ZOLIN-5-ONE
1-PHENYL-3-METHYL-5-PYRA-ZOLONE [CSR, CSR]

89-32-7 1H,3H-BENZO[1,2-C:4,5-C']DIFU-RAN-1,3,5,7-TETRONE

89-52-1 N-ACETYLANTHRANILIC ACID

89-60-1 4-CHLORO-3-NITROTOLUENE

89-61-2 2,5-DICHLORONITROBENZENE
1,4-DICHLORO-2-NITROBENZENE [TSC]
BENZENE, 1,4-DICHLORO-2-NITRO- [OTS]

89-62-3 2-NITRO-P-TOLUIDINE
BENZENAMINE, 4-METHYL-2-NITRO- [PAL]

89-63-4 BENZENAMINE, 4-CHLORO-2-NITRO-
4-CHLORO-2-NITROANILINE [CSR]

89-64-5 4-CHLORO-2-NITROPHENOL

89-65-6 ERYTHORBIC ACID
D-ISOASCORBIC ACID [PST]

89-69-0 2,4,5-TRICHLORONITROBENZENE

89-72-5 O-SEC-BUTYLPHENOL
2-SEC-BUTYLPHENOL [GBR, GBR]
PHENOL, 2-(1-METHYLPROPYL)- [PAL]

89-83-8 THYMOL

89-84-9 4-ACETYLRESORCINOL

89-98-5 BENZALDEHYDE, 2-CHLORO-

90-00-6 O-ETHYLPHENOL

90-02-8 SALICYLALDEHYDE
BENZALDEHYDE, 2-HYDROXY- [TSC, TSC, TSC]

90-03-9 CHLORO(O-HYDROXYPHENYL) MERCURY

90-04-0 O-ANISIDINE
BENZENAMINE, 2-METHOXY- [PAL, PAL]
ORTHO-ANISIDINE [C65, C65, IAR]
ORTHO-METHOXYANILINE [HM1]

90-05-1 GUAIACOL

90-12-0 1-METHYLNAPHTHALENE
NAPHTHALENE, 1-METHYL- [PAL]

90-13-1 NAPHTHALENE, 1-CHLORO-
1-CHLORONAPHTHALENE [NFP, NFP]

90-15-3 ALPHA-NAPHTHOL
1-NAPHTHOL [TSC, TSC, TSC]

90-30-2 N-PHENYL-1-NAPHTHYLAMINE

90-41-5 2-BIPHENYLAMINE
[1,1'-BIPHENYL]-2-AMINE [PAL]

90-42-6 [1,1'-BICYCLOHEXYL]-2-ONE

90-43-7 2-PHENYLPHENOL
O-PHENYL PHENOL [WHM]
O-PHENYLPHENOL [CAN, CSR, CSR, NFP, NFP, NJL]
ORTHO-PHENYLPHENOL [IAR]

90-65-3 PENICILLIC ACID

90-72-2 2,4,6-TRI(DIMETHY-LAMINOMETHYL)PHENOL

90-80-2 DELTA-GLUCONOLACTONE

90-82-4 PSEUDOEPHEDRINE

90-94-8 MICHLER'S KETONE
METHANONE, BIS[4-(DIMETHY-LAMINO0PHENYL]- [PAL, PAL]

91-01-0 BENZHYDROL

91-08-7 TOLUENE-2,6-DIISOCYANATE
BENZENE, 1,3-DIISOCYANATO-2-METHYL- [TSC, TSC]
BENZENE, 1,3-DIISOCYANATO-2-METHYL- (2,6-TOLUENE DIISO-CYANATE) [CAI]
TOLUENE 2,6-DIISOCYANATE [302]
TOLUENE 2,6-DIISOCYANATE [BENZENE, 1,3-DIISOCYANATO-2-METHYL-] [EP3]

91-15-6 O-PHTHALODINITRILE
PHTHALONITRILE [HON]

91-17-8 DECAHYDRONAPHTHALENE
DECALIN [CSR]
NAPHTHALENE, DECAHYDRO- [PAL]

91-20-3 NAPHTHALENE

91-22-5 QUINOLINE

91-23-6 2-NITROANISOLE
O-NITROANISOLE [C65, CSR, HON, NTP]

91-40-7 PHENYL ANTHRANILIC ACID (ALL ISOMERS)

91-44-1 COUMARIN, 7-(DIETHYLAMINO)-4-METHYL-

91-49-6 N-BUTYLACETANILIDE
ACETAMIDE, N-BUTYL-N-PHENYL- [PAL]

91-53-2 ETHOXYQUIN

91-57-6 2-METHYLNAPHTHALENE

91-58-7 BETA-CHLORONAPHTHALENE
2-CHLORONAPHTHALENE [CAL, CW2, NJL, PQL, RCA, UTS]
NAPHTHALENE, 2-CHLORO- [PAL, TSC, TSC]

91-59-8 BETA-NAPHTHYLAMINE
2-NAPHTHALENAMINE [PAL, PAL, RCU, RCU]
2-NAPHTHYLAMINE [AUS, AUS, C65, C65, CAB, EP2, FLA, IAR, MAK, MAK, MSL, NTP, PQL, RCA, UTS]
NAPHTHYLAMINE (2-) [HON]

91-60-1 2-NAPHTHALENETHIOL

91-63-4 QUINALDINE

91-64-5 COUMARIN

91-66-7 N,N-DIETHYLANILINE
BENZENAMINE, N,N-DIETHYL- [PAL]
DIETHYL ANILINE [NJL]

91-80-5 METHAPYRILENE
1,2-ETHANEDIAMINE, N,N-DIETHYL-N'-2-PYRIDINYL-N'-(2-THIENYL-METHYL)- [PAL]

91-84-9 PYRILAMINE

91-92-9 C.I. AZOIC COUPLING COMPONENT
2-NAPHTHALENECARBOXAMIDE, N, N'-(3,3'-DIMETHOXY[1,1'BIPHENYL]-4, 4'-DIYL)BIS[3-HYDROXY- [TSC]

91-93-0 3,3'-DIMETHOXYBENZIDINE-4,4'-DIISOCYANATE

91-94-1 3,3'-DICHLOROBENZIDINE
3,3'-DICHLOROBENZIDINE [QBC, QBC, QBC]
3,3'-DICHLOROBENZIDINE (AND ITS SALTS) [NIO, NIO]
3,3-DICHLOROBENZIDENE [CW2]
3,3-DICHLOROBENZIDINE [313, BC1, BC1, BC1, CAA]
DICHLOROBENZIDINE [CAL, WHM]
[1,1'-BIPHENYL]-4,4'-DIAMINE, 3-3'-DICHLORO- [PAL, PAL]

91-95-2 3,3'-DIAMINOBENZIDINE

91-96-3 C.I. AZOIC COUPLING COMPONENT 5
BUTANAMIDE, N,N'-(3,3'-DIMETHYL [1,1'-BIPHENYL]-4,4'-DIYL)BIS[3-OXO- [TSC]

91-97-4 1,1'-BIPHENYL, 4,4'-DIISO-CYANATO-3,3'-DIMETHYL-
3,3'-DIMETHYL-4,4'-DIPHENYLENE DIISOCYANATE [313]

91-99-6 M-TOLYDIETHANOLAMINE
2-2'-(M-TOLYLIMIDO) DIETHANOL [FLA]
ETHANOL, 2,2'-[(3-METHYLPHENYL) IMINO]BIS- [PAL]
META-TOLYDIETHANOLAMINE [NFP, NFP]

92-04-6 [1,1'-BIPHENYL]-4-OL, 3-CHLORO-
2-CHLORO-4-PHENYLPHENOL [MSL, NFP, NFP, TSC]
3-CHLORO-4-BIPHENYLOL [FLA]

92-06-8 M-TERPHENYL

92-15-9 O-ACETOACETANISIDIDE
ACETOACETYL-O-ANISIDINE [FLA]
BUTANAMIDE, N-(2-METHOXYPHENYL)-3-OXO- [PAL]
O-ACETOACET ANISIDIDE [NFP, NFP]

92-36-4 4-(6-METHYL-2-BENZOTHIA-ZOLYL)-BENZENAMINE

92-43-3 3-PYRAZOLIDINONE, 1-PHENYL-

92-48-8 6-METHYLCOUMARIN
METHYL COUMARIN [CSR]

92-49-9 BENZENAMINE, N-(2-CHLOROETHYL)-N-ETHYL-

92-50-2 ETHANOL, 2-(ETHYLPHENY-LAMINO)-
N-PHENYL-N-ETHYLETHANOLAMINE [FLA, MSL, NFP, NFP]

92-51-3 BICYCLOHEXYL

92-52-4 BIPHENYL
1,1'-BIPHENYL [PAL, TSC, TSC, TSC]
1,1-BIPHENYL [TSE]
BIPHENYL (DIPHENYL) [ALB, ALB, ATS, MEX, MEX]
DIPHENYL [BC1, BC1, HM1, NIO, NIO, NIO, NJL, WHM]
DIPHENYL (BIPHENYL) [OS1, OS1]

92-53-5 4-PHENYLMORPHOLINE
MORPHOLINE, 4-PHENYL- [PAL]

92-59-1 ETHYLBENZYLANILINE
BENZENEMETHANAMINE, N-ETHYL-N-PHENYL- [PAL]
N-ETHYL-N-BENZYLANILINE [HM1, HM1, WHM]

92-66-0 4-BROMODIPHENYL
1,1'-BIPHENYL, 4-BROMO- [PAL, TSC, TSC]
4-BROMO-1,1-BIPHENYL [TSE]

92-67-1 4-AMINOBIPHENYL
4-AMINOBIPHENYL (4-AMIN-ODIPHENYL) [C65, C65]
4-AMINODIPHENYL [ALB, AUS, AUS, AUS, BC1, BC1, BC1, CAL, CEX, CEX, MEX, MIN, NHS, NHS, NHS, NIO, NIO, OS1, TLV, TLV, WHM]
4-AMINODIPHENYLE [QBC, QBC, QBC]
[1,1'-BIPHENYL]-4-AMINE [PAL, PAL]

DIETHYL-2-CHLORO-2-PROPANYL ESTER CARBAMODITHIOIC ACID [FLA]

95-08-9 TRIETHYLENE GLYCOL BIS (2-ETHYLBUTYRATE)

95-13-6 INDENE
1H-INDENE [FLA, PAL]

95-14-7 1,2,3-BENZOTRIAZOLE
1H-BENZOTRIAZOLE [MSL, TSC, TSC, TSC]

95-16-9 BENZOTHIAZOLE

95-19-2 1-(HYDROXYETHYL)-2-(HEPTADE-CYL)IMIDAZOLINE
HEPTADECYL HYDROXYETHYL IMIDAZOLINE [PST]

95-32-9 2-(4-MORPHOLINYLDITHIO)-BEN-ZOTHIAZOLE

95-33-0 N-CYCLOHEXYL-2-BENZOTHIA-ZOLESULFENAMIDE

95-38-5 1H-IMIDAZOLE-1-ETHANOL, 2-(8-HEPTADACENYL)-4,5-DIHYDRO-
2-(8-HEPTADECENYL)-4,5-DIHYDRO-1H-IMIDAZOLE-1-ETHANOL [PST]

95-39-6 2-PROPENOIC ACID, BICYCLO [2.2.1]HEPT-5-EN-2-YL-, METHYL ESTER

95-46-5 O-BROMOTOLUENE
1-BROMO-2-METHYL BENZENE [FLA]
BENZENE, 1-BROMO-2-METHYL [PAL]

95-47-6 O-XYLENE
1,2-DIMETHYL BENZENE [FLA]
BENZENE, 1,2-DIMETHYL- [PAL, TSC, TSC, TSC]
O-XYLENES [CAA]

95-48-7 O-CRESOL
2-METHYL PHENOL [FLA]
CRESOL, O- [302]
CRESOL, ORTHO- [ATS]
O-CRESOL [PQL, RC2, RC2]
O-CRESOL AND CRESYLIC ACID [HON, HON]
ORTHO-CRESOL [CTR]
PHENOL, 2-METHYL- [PAL, TSC, TSC, TSC]

95-49-8 O-CHLOROTOLUENE
1-CHLORO-2-METHYL BENZENE [FLA]
2-CHLOROTOLUENE [GBR, PST]
BENZENE, 1-CHLORO-2-METHYL- [PAL, TSC, TSC, TSC]

95-50-1 O-DICHLOROBENZENE
1,2-DICHLOROBENZENE [313, ATS, CAB, CAN, CN4, CTR, CW2, CW3, FLA, GBR, HM1, HM1, HM1, HM1, MAK, MAK, MAK, MAK, MX1, NJL, TSE]
1,2-DICHLOROBENZENE (O-DICHLOROBENZENE) [CSR, CSR]
BENZENE, 1,2-DICHLORO- [PAL, TSC, TSC, TSC, TSC]
DICHLOROBENZENE, 1,2- [CD1, CD1]
ORTHO-DICHLOROBENZENE [IAR, RCA]

95-51-2 2-CHLOROANILINE
BENZENAMINE, 2-CHLORO- [TSC, TSC, TSC]
O-CHLOROANILINE [CSR]

95-52-3 O-FLUOROTOLUENE

95-53-4 O-TOLUIDINE
BENZENAMINE, 2-METHYL- [PAL, PAL, TSC, TSC, TSC]
O-METHYLANILINE [CAL]
ORTHO-TOLUIDINE [C65, C65, IAR]
TOLUIDINE-O [QBC, QBC, QBC]

95-54-5 O-PHENYLENEDIAMINE
1,2-BENZENEDIAMINE [TSC, TSC]

1,2-PHENYLENEDIAMINE [313]
ORTHO-PHENYLENEDIAMINE [CTR, TSE]
PHENYLENEDIAMINE, ORTHO- [OTS]

95-55-6 O-AMINOPHENOL

95-56-7 O-BROMOPHENOL

95-57-8 2-CHLOROPHENOL
2-CHLOROPHENOL [RCA]
O-CHLOROPHENOL [HON, MSL, NFP, NFP, RCU, RCU, TSC, WHM]
PHENOL, 2-CHLORO- [PAL]

95-63-6 PSEUDOCUMENE
1,2,4-TRIMETHYL BENZENE [OTV]
1,2,4-TRIMETHYLBENZENE [313, CAB, CAN, NFP, NFP, SDW, WHM]
BENZENE, 1,2,4-TRIMETHYL- [TSC, TSC, TSC]

95-65-8 3,4-DIMETHYLPHENOL

95-68-1 2,4-XYLIDINE

95-69-2 4-CHLORO-O-TOLUIDINE
BENZENAMINE, 4-CHLORO-2-METHYL [TSE]
BENZENAMINE, 4-CHLORO-2-METHYL- [PAL, PAL, TSC, TSC]
P-CHLORO-O-TOLUIDINE [313, C65, C65, CAL, NTP]
PARA-CHLORO-ORTHO-TOLUIDINE [IAR, MIN]

95-70-5 2,5-DIAMINOTOLUENE
1,4-BENZENEDIAMINE, 2-METHYL- [TSC, TSC]

95-71-6 TOLUHYDROQUINONE

95-73-8 2,4-DICHLOROTOLUENE

95-74-9 3-CHLORO-P-TOLUIDINE
3-CHLORO-4-METHYL ANILINE [NJL]

95-76-1 3,4-DICHLOROANILINE
3,4-DICHLOROBENZENAMINE [FLA]
BENZENAMINE, 3,4-DICHLORO- [PAL, TSC, TSC, TSC]

95-77-2 3,4-DICHLOROPHENOL

95-78-3 2,5-XYLIDINE

95-79-4 5-CHLORO-O-TOLUIDINE
5-CHLORO-2-METHYL ANILINE [NJL]

95-80-7 TOLUENE 2,4-DIAMINE
1,3-BENZENEDIAMINE, 4-METHYL- [PAL, PAL, TSC, TSC, TSC]
2,4-DIAMINOTOLUENE [NTP]
2,4-DIAMOTOLUENE [313]
2,4-TOLUENE DIAMINE [CAA]
2,4-TOLUENEDIAMINE [EPA]
2,4-TOLUYLENEDIAMINE [HM1, HM1, HM1]
4-METHYL-1,3-BENZENEDIAMINE [FLA]
DIAMINOTOLUENE (MIXED) [EP2]
DIAMINOTOLUENE, 2,4- [OTS]
TOLUENE 2,4 DIAMINE [HON, HON]
TOLUENE 2-4-DIAMINE [WEL, WEL]
TOLUENE-2,4-DIAMINE [MAK, RCA, WHM]

95-81-8 2-CHLORO-5-METHYL ANILINE

95-82-9 BENZENAMINE, 2,5-DICHLORO-

95-83-0 4-CHLORO-O-PHENYLENEDI-AMINE
1,2-BENZENEDIAMINE, 4-CHLORO- [PAL, PAL, TSC, TSC, TSC]
4-CHLORO-1,2-BENZENEDIAMINE [FLA]
4-CHLORO-ORTHO-PHENYLENEDI-AMINE [C65, C65]

95-85-2 2-AMINO-4-CHLOROPHENOL

95-87-4 2,5-XYLENOL

95-88-5 4-CHLORORESORCINOL

95-92-1 ETHYL OXALATE

95-93-2 1,2,4,5-TETRAMETHYLBENZENE

95-94-3 1,2,4,5-TETRACHLOROBENZENE
BENZENE, 1,2,4,5-TETRACHLORO- [EPA, PAL, TSC, TSC, TSC, TSC, TSC]
TETRACHLOROBENZENE [RCA]

95-95-4 2,4,5-TRICHLOROPHENOL
PHENOL, 2,4,5-TRICHLORO- [PAL]

96-05-9 ALLYL METHACRYLATE
2-PROPENOIC ACID, 2-METHYL-, 2-PROPENYL ESTER [TSC, TSC]

96-08-2 7-OXABICYCLO[4.1.0]HEPTANE, 1-METHYL-4-(2-METHYLOXI-RANYL)-

96-09-3 STYRENE OXIDE
OXIRANE, PHENYL- [PAL, TSC, TSC]
PHENYL-OXIRANE [FLA]
STYRENE, OXIDE [MIN]

96-10-6 DIETHYLALUMINUM CHLORIDE
ALUMINUM, CHLORODIETHYL- [PAL]
CHLORODIETHYLALUMINUM [OS3]

96-12-8 1,2-DIBROMO-3-CHLORO-PROPANE
1,2-DIBROMO-3-CHLOROPROPANE (DBCP) [313, C65, C65, C65, MSL, SDW, SDW, SDW, WHM]
DBCP [NJL, NJL]
DBCP (1,2-DIBROMO-3-CHLORO-PROPANE) [CN2]
DIBROMOCHLOROPROPANE [HM1, HM1, HM1, NHS, NHS]
PROPANE, 1,2-DIBROMO-3-CHLORO- [PAL, PAL, TSC, TSC]

96-13-9 2,3-DIBROMO-1-PROPANOL
1-PROPANOL, 2,3-DIBROMO- [TSC, TSC, TSC]

96-14-0 3-METHYLPENTANE
PENTANE, 3-METHYL- [PAL]

96-17-3 BUTANAL, 2-METHYL-
2-METHYLBUTANAL [FLA]
2-METHYLBUTYRALDEHYDE [MSL, NFP, NFP]

96-18-4 1,2,3-TRICHLOROPROPANE
ALLYL TRICHLORIDE [FLA]
PROPANE, 1,2,3-TRICHLORO- [PAL, TSC, TSC]
TRICHLOROPROPANE 1,2,3 [QBC, QBC]

96-20-8 2-AMINO-1-BUTANOL
1-BUTANOL, 2-AMINO- [PAL]

96-22-0 DIETHYL KETONE
3-PENTANONE [ONT, PAL]
PENTAN-3-ONE [GBR, GBR]

96-23-1 1,3-DICHLORO-2-PROPANOL
1,3-DICHLORO-2-PROPANOL [TSE]
1,3-DICHLOROPROPANOL [NJL, TSC, TSC]
1,3-DICHLOROPROPANOL-2 [HM1, HM1]

96-24-2 3-CHLORO-1,2-DIHYDROX-YPROPANE
GLYCEROL ALPHA-MONOCHLORO-HYDRIN [HM1, HM1]
GLYCEROL-ALPHA-MONOCHLORO-HYDRIN [NJL]

96-29-7 METHYL ETHYL KETOXIME
2-BUTANONE, OXIME [TSC, TSC, TSC]

96-32-2 METHYL BROMOACETATE

96-33-3 METHYL ACRYLATE
2-PROPANOIC ACID, METHYL ESTER [FLA]

2-PROPENOIC ACID, METHYL ESTER
[PAL, TSC, TSC]
ACRYLIC ACID, METHYL ESTER
[MAK, MAK, MAK]

96-34-4 **METHYL CHLOROACETATE**
ACETIC ACID, CHLORO-, METHYL
ESTER [PAL]
CHLOROACETIC ACID METHYL
ESTER [MAK, MAK, MAK]
METHYL ESTER CHLOROACETIC
ACID [FLA]

96-37-7 **METHYLCYCLOPENTANE**
CYCLOPENTANE, METHYL- [PAL]
METHYL CYCLOPENTANE [NJL,
NJL]

96-41-3 **CYCLOPENTANOL**

96-45-7 **ETHYLENE THIOUREA**
2-IMIDAZOLIDINETHIONE [FLA,
PAL, PAL]
ETHYLENE THIOUREA (ETU) [CSR,
CSR]
ETHYLENETHIOUREA [CAL, EP2,
EPA, RCU, RCU, SD4]

96-47-9 **METHYLTETRAHYDROFURAN**
2-METHYLTETRAHYDROFURAN
[FLA, MSL, NFP, NFP]
FURAN, TETRAHYDRO-2-METHYL-
[PAL]

96-48-0 **GAMMA-BUTYROLACTONE**
BUTYROLACETONE [HON]
BUTYROLACTONE [NFP, NFP]

96-49-1 **ETHYLENE CARBONATE**
1,3-DIOXOLAN-2-ONE [PAL]

96-50-4 **2-THIAZOLAMINE**

96-53-7 **2-THIAZOLIDINETHIONE**

96-54-8 **1-METHYLPYRROLE**
METHYLPYRROLE [MSL, NFP, NFP]

96-69-5 **4,4'-THIOBIS(6-TERT-BUTYL-M-
CRESOL)**
4,4'-THIOBIS [MIN]
4,4'-THIOBIS(6-TERT, BUTYL-M-
CRESOL) [OS1, OS1]
6,6'-DI-TERT-BUTYL-4,4'-THIODI-M-
CRESOL [GBR, GBR]
PHENOL, 4,4'-THIOBIS[2-(1,1-
DIMETHYLETHYL)-5-METHYL- [PAL]
THIOBIS 4,4 6-TERBUTYL-M-CRESOL)
[QBC]

96-80-0 **N,N-DIISOPROPYL
ETHANOLAMINE**
DIISOPROPYLETHANOLAMINE [NJL]
N,N-DIISOPROPYLETHANOLAMINE
[NFP, NFP]

97-00-7 **1-CHLORO-2,4-DINITROBENZENE**
BENZENE, 1-CHLORO-2,4-DINITRO-
[PAL]

97-02-9 **2,4-DINITROANILINE**
BENZENAMINE, 2,4-DINITRO- [PAL,
TSC, TSC, TSC]

97-05-2 **5-SULFOSALICYLIC ACID**

97-17-6 **DICHLOFENTHION**

97-18-7 **PHENOL, 2,2'-THIOBIS(4,6-
DICHLORO)-**
2,2'-THIOBIS-4,6-DICHLORO-PHENOL
[NJL]
BITHIONOL [MAK]

97-23-4 **DICHLOROPHEN**
DICHLOROPHENE [PST]
DICHLOROPHENE [2,2'-METHYLENE-
BIS(4-CHLOROPHENOL)] [313]
PHENOL, 2,2'-METHYLENEBIS(4-
CHLORO- [TSC, TSC]

97-36-9 **M-ACETOACET XYLIDIDE**
BUTANAMIDE, N-(2,4-
DIMETHYLPHENYL)-3-OXO- [PAL]

97-39-2 **1,3-DI-O-TOLYLGUANIDINE**

97-50-7 **5-CHLORO-2,4-DIMETHOXYANI-
LINE**

97-51-8 **BENZALDEHYDE, 2-HYDROXY-5-
NITRO-**

97-53-0 **EUGENOL**

97-54-1 **ISOEUGENOL**

97-56-3 **O-AMINOAZOTOLUENE**
AMINOAZOTOLUENE [WHM]
BENZENAMINE, 2-METHYL-4-[(2-
METHYLPHENYL)AZO]- [PAL, PAL]
C.I. SOLVENT YELLOW 3 [NJL]
CI SOLVENT YELLOW 3 [313]
ORTHO-AMINOAZOTOLUENE [C65,
C65, IAR, MIN]

97-59-6 **ALLANTOIN**

97-61-0 **2-METHYLPENTANOIC ACID**

97-62-1 **ETHYL ISOBUTYRATE**
PROPANOIC ACID, 2-METHYL-,
ETHYL ESTER [PAL]

97-63-2 **ETHYL METHACRYLATE**
2-PROPENOIC ACID, 2-METHYL-
, ETHYL ESTER [PAL, TSC, TSC,
TSC]
METHACRYLIC ACID, ETHYL ESTER
[WHM]

97-64-3 **ETHYL LACTATE**
PROPANOIC ACID, 2-HYDROXY-,
ETHYL ESTER [PAL]

97-65-4 **BUTANEDIOIC ACID, METHY-
LENE-**

97-72-3 **ISOBUTYRIC ANHYDRIDE**

97-74-5 **(ANAD5) TETRAMETHYL THIU-
RAM MONOSULFIDE**

97-77-8 **DISULFIRAM**
TETRAETHYLTHIURAM DISULFIDE
[CSR, CSR]
TETRAETHYLTHIURAM DISULPHIDE
[WHM]
THIOPEROXYDICARBONIC DIAMIDE
([(H2N)C(S)]2S2), TETRAETHYL-
[PAL]

97-84-7 **N,N,N',N'-TETRAMETHYL-1,3-
BUTANEDIAMINE**

97-85-8 **ISOBUTYL ISOBUTYRATE**

97-86-9 **ISOBUTYL METHACRYLATE**
2-PROPENOIC ACID, 2-METHYL-, 2-
METHYLPROPYL ESTER [TSC, TSC]
ISOBUTYLMETHACRYLATE [NJL]

97-88-1 **BUTYL METHACRYLATE**
2-PROPENOIC ACID, 2-METHYL-
, BUTYL ESTER [PAL, TSC, TSC,
TSC]
N-BUTYL METHACRYLATE [NJL]

97-90-5 **2-PROPENOIC ACID, 2-METHYL-,
1,2-ETHANEDIYL ESTER**

97-93-8 **TRIETHYLALUMINUM**
ALUMINUM, TRIETHYL- [PAL]
TRIETHYL ALUMINUM [NJL, NJL]

97-94-9 **TRIETHYLBORANE**
BORANE, TRIETHYL- [PAL]

97-95-0 **SEC-HEXYL ALCOHOL**
1-BUTANOL, 2-ETHYL- [PAL]
2-ETHYLBUTANOL [HM1, HM1]
2-ETHYLBUTYL ALCOHOL [NFP,
NFP]
ETHYLBUTANOL [NJL]

97-96-1 **2-ETHYLBUTYRALDEHYDE**
BUTANAL, 2-ETHYL- [PAL]
ETHYLBUTYRALDEHYDE [NJL,
NJL]

97-97-2 **DIMETHYL DICHLORACETAL**
DIMETHYL CHLORACETAL [FLA,
NFP, NFP]
ETHANE, 2-CHLORO-1,1-
DIMETHOXY- [PAL]

97-99-4 **TETRAHYDROFURFURYL ALCO-
HOL**
2-FURANMETHANOL, TETRAHYDRO-
[PAL]

98-00-0 **FURFURYL ALCOHOL**
2-FURANMETHANOL [PAL]
FURFURYL ALCOHOL (FURFURAL)
[QBC, QBC, QBC]

98-01-1 **FURFURAL**
2-FURALDEHYDE (FURFURAL) [GBR,
GBR, GBR]
2-FURANCARBOXALDEHYDE [PAL,
TSC, TSC, TSC]
FURFURAL (FURFURAL ALDEHYDE)
[QBC, QBC]

98-03-3 **2-THIOPHENECARBOXALDEHYDE**

98-05-5 **BENZENEARSONIC ACID**
PHENYLARSONIC ACID [WHM]

98-06-6 **TERT-BUTYLBENZENE**
2-METHYL-2-PHENYLPROPANE
[HM1]
BENZENE, (1,1-DIMETHYLETHYL)-
[PAL, TSC, TSC]

98-07-7 **BENZOTRICHLORIDE**
BENZENE, (TRICHLOROMETHYL)-
[PAL, PAL]
BENZENE, TRICHLOROMETHYL-
[TSC, TSC]
BENZOIC TRICHLORIDE (BEN-
ZOTRICHLORIDE) [313]

98-08-8 **BENZOTRIFLUORIDE**
BENZENE, (TRIFLUOROMETHYL)-
[PAL]

98-09-9 **BENZENESULFONYL CHLORIDE**
BENZENE SULFONYL CHLORIDE
[NJL, NJL]

98-10-2 **BENZENESULFONAMIDE**

98-11-3 **PHENOLSULPHONIC ACID**
BENZENESULFONIC ACID [HON]

98-12-4 **CYCLOHEXYLTRICHLOROSI-
LANE**
CYCLOHEXYL TRICHLOROSILANE
[NJL, NJL]
SILANE, TRICHLOROCYCLOHEXYL-
[PAL]

98-13-5 **PHENYL TRICHLOROSILANE**
PHENYL TRICHLORO SILANE [NFP,
NFP]
PHENYLTRICHLOROSILANE [MSL]
SILANE, TRICHLOROPHENYL- [PAL]
TRICHLOROPHENYLSILANE [302,
EPA, EPA]

98-16-8 **BENZENAMINE, 3-(TRIFLUO-
ROMETHYL)-**
3-(TRIFLUOROMETHYL) ANILINE
[WHM]
3-TRIFLUOROMETHYL ANILINE
[NJL]
3-TRIFLUOROMETHYLANILINE
[HM1, HM1]

98-27-1 **P-TERT-BUTYL-O-CRESOL**
PHENOL, 4-(1,1-DIMETHYLETHYL)-2-
METHYL- [PAL]

98-28-2 **4-TERT-BUTYL-2-CHLOROPHE-
NOL**
PHENOL, 2-CHLORO-4-(1,1-
DIMETHYLETHYL)- [PAL]

98-29-3 **4-TERT-BUTYL CATECHOL**
1,2-BENZENEDIOL, 4-(1,1-
DIMETHYLETHYL)- [PAL]
4-TERT BUTYLCATECHOL [WEL,
WEL]
P-TERT-BUTYLCATECHOL [CSR]

98-30-6 **2-AMINO-4-(METHYLSULONYL)
PHENOL**

98-44-2 **2-AMINO-P-BENZENEDISULFONIC
ACID**

98-46-4 **3-NITROBENZOTRIFLUORIDE**
1,3-NITROBENZOTRIFLUORIDE [NFP, NFP]
M-NITROBENZOTRIFLUORIDE [WHM]

98-47-5 **M-NITROBENZENESULFONIC ACID**

98-48-6 **1,3-BENZENEDISULFONIC ACID**
BENZENEDISULFONIC ACID [HON]

98-51-1 **P-TERT-BUTYLTOLUENE**
BENZENE, 1-(1,1-DIMETHYLETHYL)-4-METHYL- [PAL]
P-TERT-BUTYL TOLUENE [MAK, MAK, TLV]
P-TERT-BUTYLTOLUENE [HON]
PARA-TERTIARY-BUTYLTOLUENE [HM1]

98-52-2 **4-TERT-BUTYLCYCLOHEXANOL**

98-54-4 **P-TERT-BUTYL PHENOL**
4-TERT-BUTYLPHENOL [WHM]
P-TERT-BUTYLPHENOL [PST]

98-55-5 **ALPHA-TERPINEOL**

98-56-6 **P-CHLORO-A,A,A-TRIFLUORO-TOLUENE**
BENZENE, 1-CHLORO-4-(TRIFLUORMETHYL)- [OTS]
BENZENE, 1-CHLORO-4-TRIFLUOROMETHYL- [TSC, TSC, TSC]

98-59-9 **P-TOLUENE SULFONYL CHLORIDE**
P-TOLUENESULPHONYL CHLORIDE [GBR]
PARA-TOLUENESULFONYL CHLORIDE [WEL, WEL]
TOLUENESULFONYL CHLORIDE [HON]

98-73-7 **P-TERT-BUTYLBENZOIC ACID**
4-TERT-BUTYL BENZOIC ACID [PST]

98-82-8 **CUMENE**
BENZENE, (1-METHYLETHYL)- [PAL]
CUMENE (1-METHYLETHYLBENZENE) [TSC, TSC, TSC]
ISOPROPYLBENZENE [HM1, HM1, HM1, HM1, SDW, WHM]
ISOPROPYLBENZENE (CUMENE) [QBC, QBC]

98-83-9 **ALPHA-METHYL STYRENE**
2-PHENYLPROPENE [GBR]
A-METHYL STYRENE [AUS, AUS]
ALPHA-METHYLSTYRENE [CAL, CEX, CEX, MSL, NFP, NFP, ONT, ONT, WHM]
BENZENE, (1-METHYLETHENYL)- [TSC, TSC]
ISOPROPENYL BENZENE [NJL]
ISOPROPENYLBENZENE [HM1, HM1, HM1]

98-84-0 **ALPHA-METHYLBENZYLAMINE**
BENZENEMETHANAMINE, ALPHA.-METHYL- [PAL]

98-85-1 **METHYL PHENYL CARBINOL**
METHYLPHENYL CARBINOL [NFP, NFP]

98-86-2 **ACETOPHENONE**
ETHANONE, 1-PHENYL- [PAL]

98-87-3 **BENZAL CHLORIDE**
BENZENE, DICHLOROMETHYL- [TSC, TSC]
BENZYL DICHLORIDE [HON, WHM]
BENZYLIDENE CHLORIDE [HM1, HM1, HM1]

98-88-4 **BENZOYL CHLORIDE**

98-89-5 **CYCLOHEXANE CARBOXYLIC ACID**

98-92-0 **NICOTINAMIDE**

98-94-2 **N,N-DIMETHYLCYCLOHEXY-LAMINE**
DIMETHYL CYCLOHEXYLAMINE [HM1, HM1]
N,N-DIMETHYLCYCLOHEXANAMINE [PST]
N,N-DIMETHYLCYCLOHEXYL AMINE [NJL, NJL]

98-95-3 **NITROBENZENE**
BENZENE, NITRO- [PAL, TSC, TSC, TSC]

98-96-4 **PYRAZINAMIDE**

98-98-6 **PICOLINIC ACID**

99-03-6 **M-AMINOACETOPHENONE**

99-06-9 **M-HYDROXYBENZOIC ACID**

99-08-1 **M-NITROTOLUENE**
BENZENE, 1-METHYL-3-NITRO- [PAL]
NITROTOLUENE [MIN]

99-09-2 **M-NITROANILINE**
BENZENAMINE, 3-NITRO- [TSC, TSC]
NITROANILINE, 3- [OTS]

99-28-5 **2,6-DIBROMO-4-NITROPHENOL**

99-29-6 **BENZENAMINE, 2-BROMO-6-CHLORO-4-NITRO-**

99-30-9 **2,6-DICHLORO-4-NITROANILINE**
BENZENAMINE, 2,6-DICHLORO-4-NITRO- [TSC, TSC, TSC, TSC]
DICHLORAN (2,6-DICHLORO-4-NITROANILINE) [313]

99-35-4 **TRINITROBENZENE**
1,3,5-TRINITROBENZENE [ATS, EPA, RCU, RCU, TSC, TSE]
BENZENE, 1,3,5-TRINITRO- [PAL]
SYM-TRINITROBENZENE [PQL]

99-49-0 **CARVONE**

99-51-4 **M-NITROXYLENE**

99-54-7 **3,4-DICHLORONITROBENZENE**

99-55-8 **5-NITRO-O-TOLUIDINE**
4-NITRO-2-AMINOTOLUENE [MAK]
5-NITRO-ORTHO-TOLUIDINE [IAR]
BENZENAMINE, 2-METHYL-5-NITRO- [EPA, PAL]

99-56-9 **2-AMINO-4-NITROANILINE**
1,2-BENZENEDIAMINE, 4-NITRO- [TSC, TSC]
1,2-DIAMINO-4-NITROBENZENE [IAR]
4-NITRO-O-PHENYLENEDIAMINE [CSR, CSR]

99-57-0 **2-AMINO-4-NITROPHENOL**

99-59-2 **5-NITRO-O-ANISIDINE**
5-NITRO-ORTHO-ANISIDINE [IAR]
BENZENAMINE, 2-METHOXY-5-NITRO- [PAL, PAL]

99-60-5 **2-CHLORO-4-NITROBENZOIC ACID**

99-62-7 **1,3-DIISOPROPYLBENZENE**

99-63-8 **ISOPHTHALOYL CHLORIDE**

99-65-0 **M-DINITROBENZENE**
1,3,-DINITROBENZENE [ATS]
BENZENE, 1,3-DINITRO- [PAL]
DINITROBENZENE (META-) [OS1, OS1]
META-DINITROBENZENE [RCA]

99-66-1 **VALPROATE**
VALPROATE (VALPROIC ACID) [C65]
VALPROIC ACID [MSL]

99-72-9 **BENZENEACETALDEHYDE, 4-METHYL-**

99-76-3 **METHYL P-HYDROXYBENZOATE**

99-80-9 **N-METHYL-N,4-DINITROSOANILINE**

99-87-6 **P-CYMENE**
BENZENE, 1-METHYL-4-(1-METHYLETHYL)- [PAL, TSC]
P-ISOPROPYLTOLUENE [SDW]

99-92-3 **P-AMINOACETOPHENONE**

99-93-4 **P-HYDROXYACETOPHENONE**

99-94-5 **P-TOLUIC ACID**

99-96-7 **P-HYDROXYBENZOIC ACID**

99-98-9 **DIMETHYL-P-PHENYLENEDI-AMINE**

99-99-0 **P-NITROTOLUENE**
BENZENE, 1-METHYL-4-NITRO- [PAL]

100-00-5 **P-NITROCHLOROBENZENE**
1-CHLORO-4-NITROBENZENE [GBR, GBR, GBR, MAK, MAK, MAK, MAK]
4-CHLORONITROBENZENE [CSR]
BENZENE, 1-CHLORO-4-NITRO- [PAL]
P-CHLORONITROBENZENE [HON, ONT]

100-01-6 **P-NITROANILINE**
4-NITROANILINE [GBR, GBR, MAK, MAK]
BENZENAMINE, 4-NITRO- [PAL, TSC, TSC, TSC, TSC]
NITROANILINE (PARA NITROANILINE) [OS3]

100-02-7 **P-NITROPHENOL**
4-NITROPHENOL [313, ATS, CAA, CW2, MX1, MX2, NJL, PST, RCA]
PARATHION DEGRADATION PRODUCT (4-NITROPHENOL) [SD4]
PHENOL, 4-NITRO- [PAL]

100-07-2 **ANISOYL CHLORIDE**

100-10-7 **BENZALDEHYDE, 4-(DIMETHYL-LAMINO)-**

100-14-1 **BENZENE, 1-(CHLOROMETHYL)-4-NITRO-**

100-17-4 **4-NITROANISOLE**
NITROANISOLE [HM1, HM1]
P-NITROANISOLE [HON, NJL]

100-20-9 **TEREPHTHALOYL CHLORIDE**
1,4-BENZENEDICARBONYL DICHLORIDE [PAL]

100-21-0 **TEREPHTHALIC ACID**
PHTHALIC ACID [UTS]

100-25-4 **P-DINITROBENZENE**
1,4-DINITROBENZENE [RCA, UTS]
BENZENE, 1,4-DINITRO- [PAL]
DINITROBENZENE (PARA-) [OS1, OS1]

100-28-7 **BENZENE, 1-ISOCYANATO-4-NI-TRO-**

100-29-8 **4-NITROPHENETOLE**
4-ETHOXYNITROBENZENE [TSC, TSC, TSC]

100-36-7 **DIETHYLETHYLENE DIAMINE**
1,2-ETHANEDIAMINE, N,N-DIETHYL- [PAL]
N,N-DIETHYLETHYLENE-DIAMINE [FLA, NFP, NFP]
N,N-DIETHYLETHYLENEDIAMINE [MSL]

100-37-8 **2-DIETHYLAMINOETHANOL**
2-(DIETHYLAMINO)ETHANOL [CAL, CEX, CEX, ONT, ONT]
DIETHYLAMINOETHANOL [ALB, ALB, ALB, BC1, BC1, MEX, MEX, NJL, QBC, QBC, WHM]
DIETHYLETHANOLAMINE [PST]
ETHANOL, 2-(DIETHYLAMINO)- [PAL]
N,N-DIETHYLETHANOLAMINE [FLA, MSL, NFP, NFP]

100-39-0 **BENZYL BROMIDE**

100-40-3 **4-VINYLCYCLOHEXENE**
4-VINYL CYCLOHEXENE [MSL, NFP, NFP, TLV, TLV]

4-VINYL CYLCOHEXENE [FLA]
4-VINYLCYCLOHEXENE (CYCLO-
HEXENE, 4-ETHENYL-) [TSC, TSC,
TSC, TSC]
CYCLOHEXENE, 4-ETHENYL- [PAL]
VINYL CYCLOHEXENE [MIN]
VINYLCYCLOHEXANE, 4- [OTS]
VINYLCYCLOHEXENE [WEL]

100-41-4 ETHYLBENZENE
BENZENE, ETHYL- [PAL, TSC, TSC,
TSC]
ETHYL BENZENE [ATS, AUS, AUS,
AUS, BC1, BC1, BEI, CAA, FLA,
MIN, MSL, NIO, NIO, NIO, NJL,
NJL, OS1, OS1, OS1, OTV, RCA,
TLV, TLV]

100-42-5 STYRENE
BENZENE, ETHENYL- [PAL, TSC,
TSC]
PHENYLETHYLENE (STYRENE) [ALB,
ALB, ALB]
STYRENE (MONOMER) [CEX, CEX,
CEX]
STYRENE MONOMER [HM1, HM1,
HM1, HM1, NJL, NJL]
STYRENE, MONOMER [AUS, AUS,
CAL, MEX, MEX, MIN, OTV, TLV,
TLV, TLV]
STYRENE, MONOMER (PHENYLETHY-
LENE) [BC1, BC1, BC1, BC1, QBC,
QBC, QBC, QBC]

100-43-6 4-VINYLPYRIDINE

100-44-7 BENZYL CHLORIDE
BENZENE, (CHLOROMETHYL)- [PAL]
BENZENE, CHLOROMETHYL- [TSC,
TSC, TSC]
BENZYL CHLORIDE (CHLORO-
TOLUENE) [QBC]

100-47-0 BENZONITRILE

100-48-1 4-PYRIDINECARBONITRILE

**100-50-5 1,2,3,6-TETRAHYDROBENZALDE-
HYDE**
3-CYCLOHEXENE-1-CARBOXALDE-
HYDE [PAL, TSC, TSC, TSC]

100-51-6 BENZYL ALCOHOL
BENZENEMETHANOL [PAL]

100-52-7 BENZALDEHYDE

100-53-8 BENZYL MERCAPTAN
BENZENEMETHANETHIOL [PAL]

100-54-9 3-PYRIDINECARBONITRILE

100-57-2 PHENYLMERCURIC HYDROXIDE

100-60-7 N-METHYL CYCLOHEXYLAMINE

100-61-8 N-METHYL ANILINE
BENZENAMINE, N-METHYL [PAL]
METHYLANILINE [NJL, QBC, QBC]
MONOMETHYL ANILINE [BC1, BC1,
BC1, NIO, NIO, NIO, NIO, OS1,
OS1, OS1, OS1]
MONOMETHYL ANILINE (N-METHY-
LANILINE) [ALB, ALB, ALB]
N-METHYLANILINE [CAL, CEX,
CEX, GBR, GBR, MAK, MAK,
MAK, WHM]
N-METHYLBENZENEAMINE [ONT,
ONT]

100-63-0 PHENYLHYDRAZINE
HYDRAZINE, PHENYL- [PAL, TSC]

100-64-1 CYCLOHEXANONE OXIME

100-66-3 ANISOLE

100-70-9 2-PYRIDINECARBONITRILE

**100-72-1 TETRAHYDROPYRAN-2-
METHANOL**

100-73-2 ACROLEIN DIMER
2H-PYRAN-2-CARBOXALDEHYDE, 3,
4-DIHYDRO- [PAL]

100-74-3 N-ETHYLMORPHOLINE
4-ETHYLMORPHOLINE [GBR, GBR,
GBR, NFP, NFP]

ETHYLMORPHOLINE [QBC, QBC]
MORPHOLINE, 4-ETHYL- [PAL]
N-ETHYL MORPHOLINE [FLA, MSL]

100-75-4 N-NITROSOPIPERIDINE
PIPERIDINE, 1-NITROSO- [PAL, PAL]

100-79-8 DIOXOLANE

100-88-9 CYCLAMIC ACID

100-97-0 METHENAMINE
HEXAMETHYLENETETRAMINE
[HON, PST]
HEXAMINE [HM1, HM1, NJL]

100-99-2 TRIISOBUTYL ALUMINUM
ALUMINUM, TRIS(2-METHYLPROPYL)
- [PAL]
TRIISOBUTYLALUMINUM [FLA,
MSL, NFP, NFP]

101-02-0 TRIPHENYL PHOSPHITE
PHOSPHOROUS ACID, TRIPHENYL
ESTER [PST]

101-05-3 ANILAZINE
ANILAZINE [4,6-DICHLORO-N-(2-
CHLOROPHENYL)-1,3,5-TRIAZIN-2-
AMINE] [313]

**101-14-4 4,4'-METHYLENEBIS(2-
CHLOROANILINE) (MBOCA)**
2,2'-DICHLORO-4,4'-METHYLENEDI-
ANILINE (MBOCA) [GBR, GBR]
4,4'-METHYLENE BIS (2-CHLOROANI-
LINE) (MBOCA) [OS1, OS1]
4,4'-METHYLENE BIS(2-CHLOROANI-
LINE) [AUS, AUS, AUS, C65, C65,
FLA, MAK, MAK, MSL]
4,4'-METHYLENE BIS(2-CHLOROANI-
LINE) (MOCA) [IAR]
4,4'-METHYLENE BIS(2-CHLOROANI-
LINE) [MOCA] [TLV, TLV, TLV]
4,4'-METHYLENE-BIS-(2-CHLOROANI-
LINE) [ATS]
4,4'-METHYLENEBIS (2-CHLOROANI-
LINE) [CAN, NJL, NJL]
4,4'-METHYLENEBIS (2-CHLOROANI-
LINE) (MBOCA) [ALB, ALB, ALB,
ALB]
4,4'-METHYLENEBIS (2-CHLOROANI-
LINE) (MOCA) [NHS, NHS, NHS]
4,4'-METHYLENEBIS(2-CHLOROANI-
LINE) [BC1, BC1, BC1, CAL, CEX,
CEX, EP2, EPA, MIN, ONT, ONT,
RCA, RCU, RCU, UTS]
4,4-METHYLENE BIS(2-CHLOROANI-
LINE) [CAA]
BENZENAMINE, 4,4'-METHYLENEBIS
[2-CHLORO-](MBOCA) [CAI]
BENZENAMINE, 4-4'-METHYLENEBIS-
(2-CHLORO)- [PAL, PAL]
METHYLENE 4,4-BIS (2-CHLOROANI-
LINE) (3,3-DICHLORO-4,4-DI-
AMINOPHENYLMETHANE) [QBC,
QBC, QBC]

**101-21-3 CIPC (ISOPROPYL N-(3-
CHLOROPHENYL) CARBAMATE)**
CHLOROPROPHAM [IAR]

**101-25-7 DINITROSOPEN-
TAMETHYLENETETRAMINE**
N,N'-DINITROSOPEN-
TAMETHYLENETETRAMINE [PST]
N,N-DINITROSOPEN-
TAMETHYLENETETRAMINE [HM1,
HM1]

101-27-9 BARBAN

**101-39-3 2-PROPENAL, 2-METHYL-3-
PHENYL-**

101-41-7 METHYL PHENYLACETATE

101-42-8 3-PHENYL-1,1-DIMETHYLUREA
FENURON [CAL]

**101-43-9 2-PROPENOIC ACID, 2-METHYL-,
CYCLOHEXYL ESTER**

**101-50-8 4-AMINOAZOBENZENE-3,4'-
DISULFONIC ACID**

101-54-2 P-AMINODIPHENYLAMINE
1,4-BENZENEDIAMINE, N-PHENYL-
[OTS]
N-PHENYL-P-PHENYLENEDIAMINE
[CSR, CSR]

101-55-3 BENZENE, 1-BROMO-4-PHENOXY
4-BROMOPHENYL PHENYL ETHER
[CAL, CW2, MX1, MX2, PQL,
RCA, RCU, RCU, TSC, TSE, UTS]

**101-61-1 4,4'-METHYLENEBIS(N,N-
DIMETHYL) BENZENAMINE**
4,4'-METHYLENE BIS(N,N-DIMETHYL)
BENZENAMINE [CAL, FLA]
4,4'-METHYLENE BIS(N,N-DIMETHY-
LANILINE) [MAK]
4,4'-METHYLENEBIS(N,N-DIMETHY-
LANILINE) [NJL]
BENZENAMINE, 4-4'-METHYLENEBIS-
(N,N-DIMETHYL)- [PAL, PAL]

101-67-7 P,P'-DIOCTYLDIPHENYLAMINE

**101-68-8 METHYLENE BISPHENOL ISO-
CYANATE (MDI)**
4,4'-DIPHENYLMETHANE DIISO-
CYANATE [WHM]
4,4'-METHYLENEDIPHENYL DIISO-
CYANATE [IAR]
BENZENE, 1,1'-METHYLENEBIS(4-
ISOCYANATO- [TSC, TSC, TSC]
BENZENE, 1,1'-METHYLENEBIS(4-
ISOCYANATO- [PAL]
DIPHENYLMETHANE DIISOCYANATE
(MDI) [NHS, NHS]
DIPHENYLMETHANE DIISOCYANATE
(METHYLENE BISPHENYLISO-
CYANATE) [ALB, ALB]
DIPHENYLMETHANE-4,4'-DIISO-
CYANATE [HM1, HM1, MAK,
MAK, MAK, MAK]
METHYLENE BIS(PHENYLISO-
CYANATE) [CEX]
METHYLENE BISPHENYL ISO-
CYANATE [MIN, NIO, NIO, NIO,
NJL, OTV]
METHYLENE BISPHENYL ISO-
CYANATE (MBI) [FLA]
METHYLENE BISPHENYL ISO-
CYANATE (MDI) [TLV]
METHYLENE DIPHENYL DIISO-
CYANATE (4,4') (MDI) [HON, HON]
METHYLENE DIPHENYL DIISO-
CYANATE (MDI) [CAA]
METHYLENE-BIS (4-PHENYL ISO-
CYANATE) [QBC]
METHYLENE-BIS(4-PHENYL ISO-
CYANATE) [CAL]
METHYLENEBIS (PHENYLISO-
CYANATE) [CAN]
METHYLENEBIS(PHENYLISO-
CYANATE) (MDI) [313]
METHYLENEDIPHENYL DIISO-
CYANATE, 4,4'- [OTS]

**101-72-4 N-ISOPROPYL-N'-PHENYL-P-
PHENYLENE-DIAMINE**
BENZENEDIAMINE, N-(1-
METHYLETHYL)-N'-PHENYL-, 1,4-
[OTS]

101-77-9 4,4'-METHYLENEDIANILINE
4,4'-DIAMINODIPHENYL METHANE
[HM1, HM1]
4,4'-DIAMINODIPHENYLMETHANE
[MAK, MAK]
4,4'-DIAMINODIPHENYLMETHANE
(MDA) [WHM]
4,4'-METHYLENE DIANILINE [AUS,
AUS, AUS, CEX, CEX, MSL, NJL,
OS1, TLV, TLV, TLV]
4,4'-METHYLENE DIANILINE [FLA,
MIN]
BENZENAMINE, 4-4'-METHYLENEBIS-
[PAL, TSC, TSC, TSC]
METHYLENEDIANILINE [NFP, NFP]

101-80-4 4,4'-DIAMINODIPHENYL ETHER
4,4'-DIAMINODIPHENYL ETHER (4,4'-
OXYDIANILINE) [C65, C65]
4,4'-OXYDIANILINE [CSR, CSR,
MAK, MAK, NTP]
4,4-OXYDIANILINE [FLA]

BENZENAMINE, 4-4'-OXYBIS- [PAL, PAL]

101-81-5 DIPHENYLMETHANE

101-83-7 DICYCLOHEXYLAMINE
CYCLOHEXANAMINE, N-CYCLO-HEXYL- [PAL]

101-84-8 PHENYL ETHER
1,1'-OXYBISBENZENE [ONT, ONT]
BENZENE, 1,1'-OXYBIS- [PAL, TSC, TSC, TSC]
DIPHENYL ETHER [GBR, HM1, MAK, WHM]
DIPHENYL OXIDE [HON, NFP, NFP]
PHENYL ETHER (VAPOR) [ALB, ALB, NIO, NIO]
PHENYL ETHER (VAPOUR) [BC1, BC1]
PHENYL ETHER VAPOR [FLA, MEX, MEX, MSL]
PHENYL ETHER, VAPOR [CAL]
PHENYL ETHER, VAPOUR [QBC, QBC]

101-86-0 OCTANAL, 2-(PHENYLMETHY-LENE)-
HEXYL CINNAMIC ALDEHYDE [NFP, NFP]

101-90-6 DIGLYCIDYL RESORCINOL ETHER
DIGLYCIDYL RESORCINOL ETHER (DGRE) [C65, C65, CSR, CSR]
OXIRANE, 2,2'-[1,3-PHENYLENEBIS (OXYMETHYLENE)]BIS- [TSC, TSC]
RESORCINOL DIGLYCIDYL ETHER [EPA, OTS, WHM]

101-96-2 1,4-BENZENEDIAMINE, N,N'-BIS(1-METHYLPROPYL)-
N,N'-DI-SEC-BUTYL-P-PHENYLENEDI-AMINE [MSL, NFP, NFP]
N,N-DI-SEC-BUTYL-P-PHENYLENEDI-AMINE [FLA]

101-97-3 ETHYL PHENYLACETATE

102-01-2 ACETOACETANILIDE
BUTANAMIDE, 3-OXO-N-PHENYL- [PAL]

102-06-7 1,3-DIPHENYLGUANIDINE

102-08-9 THIOCARBANILIDE

102-13-6 ISOBUTYL PHENYLACETATE

102-24-9 TRIMETHOXYBOROXINE

102-27-2 N-ETHYL-M-TOLUIDINE

102-30-7 DICLORAN

102-36-3 ISOCYANIC ACID, 3,4-DICHLOROPHENYL ESTER
1,2-DICHLORO-4-PHENYL ISO-CYANATE [NJL]
BENZENE, 1,2-DICHLORO-4-ISO-CYANATO- [MSL, TSC, TSC, TSC]

102-50-1 M-CRESIDINE
META-CRESIDINE [IAR]

102-52-3 1,1,3,3-TETRAMETHOXYPROPANE

102-54-5 DICYCLOPENTADIENYL IRON
BIS(ETA-CYCLOPENTADIENYL) IRON [ONT]
FERROCENE [CSR, FLA, GBR, GBR, PAL, WHM]

102-56-7 2,5-DIMETHOXYANILINE
BENZENAMINE, 2,5-DIMETHOXY- [PAL]

102-67-0 TRIPROPYL ALUMINUM
ALUMINUM, TRIPROPYL- [PAL]

102-69-2 TRIPROPYLAMINE
1-PROPANAMINE, N,N-DIPROPYL- [PAL]

102-70-5 TRIALLYLAMINE

102-71-6 TRIETHANOLAMINE
ETHANOL, 2,2',2''-NITRILOBIS- [TSC, TSC, TSC]

ETHANOL, 2,2',2''-NITRILOTRIS- [PAL]
TEA [IAR]

102-76-1 GLYCEROL TRIACETATE
GLYCERYL TRIACETATE [NFP, NFP, PST]

102-77-2 N-(OXYDIETHYLENE)BENZOTH-IAZOLE-2-SULFENAMIDE

102-79-4 N-BUTYLDIETHANOLAMINE
ETHANOL, 2,2'-(BUTYLIMINO)BIS- [PAL]

102-81-8 2-N-DIBUTYLAMINOETHANOL
2-(DIBUTYLAMINO)ETHANOL [CAL, ONT, ONT]
DIBUTYLAMINOETHANOL [HM1, HM1]
DIBYUTYLAMINOETHANOL [NFP, NFP]
ETHANOL, 2-(DIBUTYLAMINO)- [PAL]
N,N-DI-N-BUTYLAMINOETHANOL [WHM]

102-82-9 TRIBUTYLAMINE
1-BUTANAMINE, N,N-DIBUTYL- [PAL]

102-85-2 PHOSPHOROUS ACID, TRIBUTYL ESTER
TRIBUTYL PHOSPHITE [FLA, MSL, NFP, NFP]

102-96-5 BETA-NITROSTYRENE

103-03-7 PHENICARBAZIDE

103-09-3 2-ETHYLHEXYL ACETATE
ACETIC ACID, 2-ETHYLHEXYL ES-TER [PAL]

103-11-7 2-ETHYLHEXYL ACRYLATE
2-ETHYL HEXYL ACRYLATE [PST]
2-PROPENOIC ACID, 2-ETHYLHEXYL ESTER [PAL, TSC, TSC]

103-23-1 BIS (2-ETHYLHEXYL) ADIPATE
DI(2-ETHYLHEXYL) ADIPATE [OTS]
DI(2-METHYLHEXYL)ADIPATE [CAL, CSR, CSR, IAR, SDW, SDW]
DIOCTYL ADIPATE [NFP, NFP]

103-24-2 DI-2-ETHYL HEXYL AZELATE
BIS(2-ETHYLHEXYL) AZELATE [PST]
DIOCTYL AZELATE [NFP, NFP]

103-26-4 METHYL CINNAMATE

103-29-7 1,2-DIPHENYLETHANE (SYM)

103-30-0 TRANS-STILBENE

103-33-3 AZOBENZENE

103-34-4 4,4'-DITHIODIMORPHOLINE

103-36-6 ETHYL CINNAMATE

103-41-3 BENZYL CINNAMATE

103-44-6 VINYL 2-ETHYLHEXYL ETHER
HEPTANE, 3-[(ETHENYLOXY) METHYL]- [PAL]

103-45-7 2-PHYENYLETHYL ACETATE
PHENYLETHYL ACETATE (BETA) [NFP, NFP]

103-50-4 BENZYL ETHER
DIBENZYL ETHER [NFP, NFP]

103-65-1 PROPYL BENZENE
BENZENE, PROPYL- [PAL]
N-PROPYL BENZENE [HM1, HM1]
N-PROPYLBENZENE [SDW, TSE]
PROPYLBENZENE [FLA, MSL, NFP, NFP]
PROPYLBENZENE, N- [OTS]

103-69-5 N-ETHYLANILINE
BENZENAMINE, N-ETHYL- [PAL]
ETHYLANILINE [NFP, NFP, NJL]

103-71-9 PHENYL ISOCYANATE
BENZENE, ISOCYANATO- [TSC, TSC, TSC]

103-73-1 PHENETOLE
ETHOXYBENZENE [NFP, NFP]

103-75-3 2-ETHOXY-3,4-DIHYDRO-2H-PYRAN
2-ETHOXY-3,4-DI-HYDRO-2-PYRAN [PAL]
2-ETHOXY-3,4-DIHYDRO-2-PYRAN [MSL, NFP, NFP]

103-76-4 1-PIPERAZINEETHANOL
1-(2-HYDROXYETHYL) PIPERAZINE [NFP, NFP]

103-80-0 PHENYLACETYL CHLORIDE

103-82-2 PHENYLACETIC ACID

103-83-3 BENZYL DIMETHYLAMINE
BENZYLDIMETHYLAMINE [WHM]

103-84-4 ACETAMIDE, N-PHENYL-
ACETANILIDE [FLA, HON, MSL, NFP, NFP]

103-85-5 PHENYLTHIOUREA
1-PHENYL-2-THIOUREA [CSR, CSR, WHM]

103-89-9 P-ACETOTOLUIDIDE
ACETAMIDE, N-(4-METHYLPHENYL)- [PAL]

103-90-2 ACETOMINOPHEN (4-HYDROXY-ACETANILIDE)
PARACETAMOL (ACETAMINOPHEN) [IAR]

103-95-7 BENZENEPROPANAL, .ALPHA.-METHYL-4-(1-METHYLETHYL)-
CYCLAMEN ALDEHYDE [NFP, NFP]

104-09-6 BENZENEACETALDEHYDE, 4-METHYL-

104-12-1 BENZENE, 1-CHLORO-4-ISO-CYANATO-
P-CHLOROPHENYL ISOCYANATE [313, HM1]

104-15-4 P-TOLUENESULFONIC ACID
4-METHYLBENZENESULFONIC ACID [PST]
BENZENESULFONIC ACID, 4-METHYL- [PAL]
TOLUENE SULFONIC ACIDS [HON]

104-23-4 C.I. FOOD YELLOW 6

104-38-1 HYDROQUINONE DI-(BETA-HY-DROXYETHYL) ETHER

104-45-0 1-METHOXY-4-PROPYLBENZENE

104-46-1 BENZENE, 1-METHOXY-4-(1-PROPENYL)-
P-ANETHOLE [PST]

104-49-4 BENZENE, 1,4-DIISOCYANATO-
1,4-PHENYLENE DIISOCYANATE [313]

104-50-7 4-HYDROXYOCTANOIC ACID LACTONE

104-51-8 BUTYL BENZENE
1-PHENYLBUTANE [HM1]
BENZENE, BUTYL- [PAL, TSC, TSC]
BUTYLBENZENE [FLA, MSL, NFP, NFP]
N-BUTYLBENZENE [SDW]

104-54-1 CINNAMYL ALCOHOL
CINNAMIC ALCOHOL [PST]

104-55-2 CINNAMALDEHYDE
2-PROPENAL, 3-PHENYL- [TSC, TSC, TSC]

104-68-7 DIETHYLENE GLYCOL MONOPHENYL ETHER

104-72-3 DECYLBENZENE
BENZENE, DECYL- [PAL]

104-75-6 2-ETHYLHEXYLAMINE
1-HEXANAMINE, 2-ETHYL- [PAL]
2-ETHYL HEXYLAMINE [NJL]

104-76-7 **2-ETHYLHEXANOL**
1-HEXANOL, 2-ETHYL- [PAL, TSC, TSC, TSC]
2-ETHYL-1-HEXANOL [PST, TSE]
2-ETHYLHEXYL ALCOHOL [WHM]
ETHYLHEXANOL, 2- [OTS]

104-78-9 **3-(DIETHYLAMINO)PROPY-LAMINE**
1,3-PROPANEDIAMINE, N,N-DIETHYL-[PAL]
DIETHYLAMINOPROPYLAMINE [HM1, HM1]

104-83-6 **P-CHLOROBENZYL CHLORIDE**
PARA-CHLOROBENZYL CHLORIDE [HM1]

104-87-0 **BENZALDEHYDE, 4-METHYL-**

104-88-1 **P-CHLOROBENZALDEHYDE**
BENZALDEHYDE, 4-CHLORO- [PAL, TSC, TSC, TSC]

104-89-2 **2-METHYL-5-ETHYLPIPERIDINE**
PIPERIDINE, 5-ETHYL-2-METHYL-[PAL]

104-90-5 **2-METHYL-5-ETHYLPYRIDINE**
2-PICOLINE, 5-ETHYL- [OTS]
PYRIDINE, 5-ETHYL-2-METHYL-[PAL]

104-93-8 **METHYL PARA-CRESOL**

104-94-9 **P-ANISIDINE**
ANILINE, 4-METHOXY- [OTS]
BENZENAMINE, 4-METHOXY- [PAL]
PARA-ANISIDINE [IAR]

105-05-5 **P-DIETHYL BENZENE**
BENZENE, 1,4-DIETHYL- [OTS, PAL]

105-08-8 **1,4-CYCLOHEXANE DIMETHANOL**

105-11-3 **P-BENZOQUINONE DIOXIME**
P-QUINONE DIOXIME [MSL]
PARA-BENZOQUINONE DIOXIME [CAL, IAR]

105-13-5 **4-METHOXYBENZYL ALCOHOL**

105-16-8 **2-(N,N-DIETHYLAMINO)ETHYL METHACRYLATE**
2-PROPENOIC ACID, 2-METHYL-, 2-(DIETHYLAMINO)ETHYL ESTER [TSC, TSC]

105-21-5 **4-HYDROXYHEPTANOIC ACID, GAMMA-LACTONE**

105-28-2 **2-HEPTADECYL-2-IMIDAZOLINE**

105-30-6 **2-METHYL-1-PENTANOL**

105-31-7 **1-HEXYN-3-OL**

105-36-2 **ETHYL BROMOACETATE**

105-37-3 **ETHYL PROPIONATE**
PROPANOIC ACID, ETHYL ESTER [PAL]

105-38-4 **PROPANOIC ACID, ETHENYL ESTER**
VINYL PROPIONATE [FLA, MSL, NFP, NFP]

105-39-5 **ETHYL CHLOROACETATE**
ACETIC ACID, CHLORO-, ETHYL ESTER [PAL]

105-45-3 **METHYLACETOACETATE**
BUTANOIC ACID, 3-OXO-, METHYL ESTER [PAL]
METHYL ACETOACETATE [FLA, MSL, NFP, NFP]

105-46-4 **SEC-BUTYL ACETATE**
ACETIC ACID, 1-METHYLPROPYL ESTER [PAL]

105-48-6 **ISOPROPYL CHLOROACETATE**

105-53-3 **DIETHYL MALONATE**

105-54-4 **ETHYL BUTYRATE**
BUTANOIC ACID, ETHYL ESTER [PAL]

105-55-5 **N,N'-DIETHYLTHIOUREA**

105-56-6 **ETHYL CYANOACETATE**
ACETIC ACID, CYANO-, ETHYL ESTER [PAL]

105-57-7 **ACETAL**
ETHANE, 1,1-DIETHOXY- [PAL]

105-58-8 **DIETHYL CARBONATE**
CARBONIC ACID, DIETHYL ESTER [PAL]

105-59-9 **METHYLDIETHANOLAMINE**
N-METHYLDIETHANOLAMINE [NFP, NFP]

105-60-2 **CAPROLACTAM**
1,6-HEXANOLACTAM [GBR, GBR]
2H-AZEPIN-2-ONE, HEXAHYDRO-[PAL, TSC, TSC]
CAPROLACTAM [QBC, QBC]
CAPROLACTAM DUST AND VAPOR [MSL]
CAPROLACTAM, VAPOUR AND AEROSOL [ONT]
HEXAHYDRO-2H-AZETIN-2-ONE [FLA]
SIGMA-CAPROLACTAM [MAK, MAK]

105-64-6 **ISOPROPYL PEROXYDICARBON-ATE**
DI-ISOPROPYL PEROXYDICARBON-ATE [NJL, NJL]
DIISOPROPYL PEROXYDICARBON-ATE [DOT, FLA, MSL, NFP, NFP, OS3]
PEROXYDICARBONIC ACID, BIS(1-METHYLETHYL) ESTER [PAL]

105-66-8 **BUTANOIC ACID, PROPYL ESTER**
N-PROPYL BUTYRATE [FLA, MSL, NFP, NFP]

105-67-9 **2,4-DIMETHYLPHENOL**
PHENOL, 2,4-DIMETHYL- [PAL]

105-74-8 **LAUROYL PEROXIDE**
DILAUROYL PEROXIDE [DOT, FLA, MAK, MSL, NJL, OS3]
PEROXIDE, BIS(1-OXODODECYL) [PAL]

105-75-9 **DI-N-BUTYL FUMARATE**

105-76-0 **MALEIC ACID, DIBUTYL ESTER**
DIBUTYL MALEATE [NFP, NFP]

105-77-1 **DIBUTYL XANTHOGEN DISUL-FIDE**

105-83-9 **N,N-BIS(3-AMINOPROPYL) METHYLAMINE**

105-86-2 **GERANYL FORMATE**

105-87-3 **GERANYL ACETATE**

105-90-8 **GERANYL PROPIONATE**

105-99-7 **DIBUTYL ADIPATE**
DI-BUTYL ADIPATE [OTS]

106-11-6 **DIGLYCOL MONOSTEARATE**

106-20-7 **DI(2-ETHYLHEXYL)AMINE**
1-HEXANAMINE, 2-ETHYL-N-(2-ETHYLHEXYL)- [PAL]
BIS(2-ETHYLHEXYL) AMINE [NFP, NFP]
BIS(2-ETHYLHEXYL)AMINE [FLA, MSL]

106-22-9 **CITRONELLOL**

106-23-0 **6-OCTENAL, 3,7-DIMETHYL-**

106-24-1 **GERANIOL**

106-26-3 **2,6-OCTADIENAL, 3,7-DIMETHYL-, (Z)-**

106-27-4 **ISOAMYL BUTYRATE**

106-29-6 **GERANYL BUTYRATE**

106-31-0 **BUTYRIC ANHYDRIDE**

106-32-1 **ETHYL CAPRYLATE**
OCTANOIC ACID, ETHYL ESTER [PAL]

106-35-4 **ETHYL BUTYL KETONE**
3-HEPTANONE [ONT, PAL]
ETHYL BUTYL KETONE (3-HEP-TANONE) [ALB, ALB, OS1, OS1, WHM]
ETHYLBUTYL KETONE (3-HEP-TANONE) [BC1, BC1]
ETHYLBUTYLKETONE (3-HEP-TANONE) [QBC]
HEPTAN-3-ONE [GBR, GBR]

106-36-5 **PROPYL PROPIONATE**
PROPANOIC ACID, PROPYL ESTER [PAL]

106-37-6 **P-DIBROMOBENZENE**

106-38-7 **P-BROMOTOLUENE**
BENZENE, 1-BROMO-4-METHYL-[PAL]

106-40-1 **BENZENAMINE, 4-BROMO-**

106-42-3 **P-XYLENE**
BENZENE, 1,4-DIMETHYL- [PAL, TSC, TSC, TSC, TSC]
P-XYLENES [CAA]

106-43-4 **P-CHLOROTOLUENE**
BENZENE, 1-CHLORO-4-METHYL-[TSC, TSC]

106-44-5 **P-CRESOL**
BENZENE, 4-METHYL- [TSC, TSC, TSC]
CRESOL, PARA- [ATS]
P-CRESOL AND CRESYLIC ACID [HON, HON]
PARA-CRESOL [CTR]
PHENOL, 4-METHYL- [PAL]

106-46-7 **P-DICHLOROBENZENE**
1,4-DICHLOROBENZENE [313, ATS, CAN, CN4, CTR, CW2, CW3, GBR, GBR, MAK, MAK, MAK, MX1, NJL, NTP, RC2, RC2, TSE]
1,4-DICHLOROBENZENE (P-DICHLOROBENZENE) [CSR, CSR]
1,4-DICHLOROBENZENE(P) [CAA]
BENZENE, 1,4-DICHLORO- [PAL, PAL, TSC, TSC, TSC]
DICHLOROBENZENE, 1,4- [CD1, CD1]
PARA-DICHLOROBENZENE [IAR, SDW, SDW]

106-47-8 **P-CHLOROANILINE**
4-CHLOROANILINE [TSE]
BENZENAMINE, 4-CHLORO- [PAL, TSC, TSC, TSC]
PARA-CHLOROANILINE [IAR]

106-48-9 **P-CHLOROPHENOL**
4-CHLOROPHENOL [NJL, NJL]
PHENOL, 4-CHLORO- [PAL]

106-49-0 **P-TOLUIDINE**
BENZENAMINE, 4-METHYL- [PAL, TSC, TSC]
PARA-TOLUIDINE [C65]

106-50-3 **P-PHENYLENE DIAMINE**
1,4-BENZENEDIAMINE [PAL]
1,4-BENZENEDIAMINE (P-PHENYLENEDIAMINE) [TSC, TSC, TSC]
P-PHENYLENEDIAMINE [313, AUS, AUS, CAA, CAL, CAN, CEX, CEX, GBR, GBR, MAK, MAK, MAK, MAK, NJL, ONT, ONT, PQL, TLV]
PARA-PHENYLENEDIAMINE [CTR, IAR, TSE]
PHENYLENEDIAMINE, PARA- [OTS]

106-51-4 **QUINONE**
2,5-CYCLOHEXADIENE, 1,4-DIONE-[PAL]
2,5-CYCLOHEXADIENE-1,4-DIONE (QUINONE) [TSC, TSC, TSC]
BENZOQUINONE [HM1, HM1, HM1]
P-BENZOQUINONE [EPA, GBR, GBR, NFP, NFP, NJL, RCU, RCU, WHM]

P-BENZOQUINONE (QUINONE) [ALB, ALB, QBC]
P-QUINONE [CAN]
PARA-QUINONE [IAR]

106-54-7 P-CHLOROTHIOPHENOL

106-55-8 DIMETHYLPIPERAZINE, CIS-
DIMETHYLPIPERAZINE-CIS [NFP, NFP]

106-63-8 ISOBUTYL ACRYLATE
2-PROPENOIC ACID, 2-METHYL-PROPYL ESTER [PAL, TSC, TSC, TSC]

106-67-2 2-ETHYL-4-METHYL-1-PENTANOL
2-ETHYLISOHEXANOL [NFP, NFP]

106-68-3 ETHYL AMYL KETONE
3-OCTANONE [PAL]
ETHYL AMYL KETONE (5-METHYL-3-HEPTANONE) [ALB, ALB]
ETHYL N-AMYL KETONE [WHM]

106-69-4 1,2,6-HEXANETRIOL

106-70-7 METHYL HEXANOATE

106-71-8 2-CYANOETHYL ACRYLATE
2-PROPENOIC ACID, 2-CYANOETHYL ESTER [PAL, TSC, TSC]

106-72-9 5-HEPTENAL, 2,6-DIMETHYL-

106-74-1 ETHYLENE GLYCOL MO-NOETHYL ETHER ACRYLATE
2-PROPENOIC ACID, 2-ETHOXYETHYL ESTER [TSC, TSC]

106-79-6 DIMETHYL SEBACATE

106-83-2 OXIRANEOCTANOIC ACID, 3-OCTYL-, BUTYL ESTER

106-84-3 OXIRANEOCTANOIC ACID, 3-OCTYL-, OCTYL ESTER

106-87-6 VINYL CYCLOHEXENE DIOXIDE
1,2-EPOXY-4-(EPOXYETHYL)-CYCLO-HEXANE [ONT, ONT]
1-EPOXYETHYL-3,4-EPOXYCYCLO-HEXANE [IAR]
4-VINYL-1-CYCLOHEXENE DIEPOX-IDE [CAB, CSR, CSR, NTP]
4-VINYL-1-CYCLOHEXENE DIEPOX-IDE (VINYL CYCLOHEXENE DIOX-IDE) [C65]
4-VINYL-1-CYCLOHEXENE DIOXIDE [MAK]
7-OXABICYCLO[4.1.0]HEPTANE, 3-OXIRANYL- [PAL, TSC, TSC]

106-88-7 1,2-BUTYLENE OXIDE
1,2-EPOXYBUTANE [CAA, CAB, CSR, CSR, IAR]
BUTYLENE OXIDE [PST, WEL]
OXIRANE, ETHYL- [PAL, TSC, TSC, TSC]

106-89-8 EPICHLOROHYDRIN
1-CHLORO-2,3-EPOXY PROPANE (EPICHLOROHYDRIN) [BC1, BC1, BC1, BC1]
1-CHLORO-2,3-EPOXY-PROPANE (EPICHLOROHYDRIN) [ALB, ALB, ALB]
1-CHLORO-2,3-EPOXYPROPANE [ONT, ONT]
1-CHLORO-2,3-EPOXYPROPANE (EPICHLORHYDRIN) [QBC, QBC, QBC]
EPICHLOROHYDRIN (1-CHLORO-2, 3-EPOXYPROPANE) [CAA, HON, HON]
EPICHLOROHYDRIN [OXIRANE, (CHLOROMETHYL)-] [EP3]
N-CHLORO-2,3-EPOXYPROPANE [RCU, RCU]
OXIRANE, (CHLOROMETHYL)- [PAL, PAL]
OXIRANE, CHLOROMETHYL- [TSC, TSC, TSC]

106-90-1 GLYCIDYL ACRYLATE
2-PROPENOIC ACID, OXIRANYL-METHYL ESTER [TSC, TSC]

106-91-2 GLYCIDYL METHACRYLATE
2-PROPENOIC ACID, 2-METHYL-, OXIRANYLMETHYL ESTER [TSC, TSC]

106-92-3 ALLYL GLYCIDYL ETHER
((2-PROPENYLOXY)METHYL)OXI-RANE [FLA, ONT, ONT, ONT]
ALLYL 2,3-EPOXYPROPYL ETHER [GBR, GBR, GBR]
ALLYL GLYCIDYL ETHER (AGE) [AUS, AUS, AUS, BC1, BC1, BC1, MSL, NHS, NHS, QBC, QBC, TLV, TLV, WHM]
OXIRANE, [(2-PROPENYLOXY) METHYL]- [PAL, TSC, TSC]

106-93-4 ETHYLENE DIBROMIDE (1,2-DI-BROMOETHANE)
1,2 DIBROMOETHANE (ETHYLENE DIBROMIDE) [ALB, ALB, ALB, ALB]
1,2-DIBROMOETHANE [ATS, CSR, CSR, HM1, HM1, HM1, HM1, MAK, MAK]
1,2-DIBROMOETHANE (ETHYLENE DIBROMIDE) [313, BC1, BC1, BC1, BC1, GBR, GBR, NTP, QBC, QBC, QBC]
ETHANE, 1,2-DIBROMO- [PAL, PAL]
ETHYLENE DIBROMIDE [TLV, TLV]
ETHYLENE DIBROMIDE (1,2-DI-BROMO-ETHANE) [WHM]
ETHYLENE DIBROMIDE (1,2-DIBRO-MOETHANE) [CN2]
ETHYLENE DIBROMIDE (DIBRO-MOETHANE) [CAA, HON, HON]
ETHYLENE DIBROMIDE (EDB) [SDW, SDW, SDW]

106-94-5 N-PROPYL BROMIDE
PROPANE, 1-BROMO- [PAL]

106-95-6 ALLYL BROMIDE
1-PROPENE, 3-BROMO- [PAL]
3-BROMOPROPENE [HM1, HM1, HM1]

106-96-7 3-BROMOPROPYNE
1-PROPYNE, 3-BROMO- [PAL]
PROPARGYL BROMIDE [302, EPA, EPA, NFP, NFP]

106-97-8 BUTANE
N-BUTANE [MAK, MAK, PST]

106-98-9 1-BUTENE
BUTENE, 1- [OTS]

106-99-0 1,3-BUTADIENE
BUTA-1,3-DIENE [GBR]
BUTADIENE [HM1, HM1, MIN, WHM]
BUTADIENE (1,3-BUTADIENE) [BC1, BC1, OS1, OS1]

107-00-6 ETHYL ACETYLENE
ETHYL ACETYLENE [1-BUTYNE] [EP3]

107-01-7 2-BUTENE
BUTENE, 2- [OTS]

107-02-8 ACROLEIN
2-PROPENAL [PAL, TSC, TSC, TSC]
ACROLEIN (2-PROPENAL) [OS3]
ACROLEIN [2-PROPENAL] [EP3]
ACROLEINE [QBC, QBC]
ACRYLALDEHYDE [GBR, GBR]

107-03-9 1-PROPANETHIOL
N-PROPYL MERCAPTAN [MIN]
PROPYL MERCAPTAN [MSL, NJL]

107-04-0 1-BROMO-2-CHLOROETHANE
1-CHLORO-2-BROMOETHANE [TSE]
ETHANE, 1-BROMO-2-CHLORO- [TSC, TSC]

107-05-1 ALLYL CHLORIDE
1-PROPENE, 3-CHLORO- [PAL]
3-CHLOROPROPYLENE [UTS]
ALLYL CHLORIDE (3-CHLORO-PROPENE) [QBC, QBC, RCA]
PROPENYL CHLORIDE (CIS-; TRANS-) [HM1, HM1, HM1, HM1]

107-06-2 ETHYLENE DICHLORIDE
1,2-DICHLOROETHANE [ATS, BC1, BC1, CAN, CN4, CSR, CSR, CW2, EP2, IAR, MAK, MX1, MX2, NJL, NJL, NTP, ONT, ONT, ONT, QBC, QBC, QBC, RC2, RC2, RCA, SDW, SDW, UTS]
1,2-DICHLOROETHANE (ETHYLENE DICHLORIDE) [313, ALB, ALB]
1,2-DICHLOROETHANE (ETHYLENEDICHLORIDE) (EDC) [HON, HON]
DICHLOROETHANE, 1,2- [CD1]
ETHANE, 1,2-DICHLORO- [PAL, PAL, TSC, TSC]
ETHYLENE CHLORIDE [HM1, HM1, HM1, HM1]
ETHYLENE DICHLORIDE (1,2-DICHLORO-ETHANE) [FLA, MSL, WHM]
ETHYLENE DICHLORIDE (1,2-DICHLOROETHANE) [C65, C65, CAA, CN2]

107-07-3 ETHYLENE CHLOROHYDRIN
2-CHLOROETHANOL [MAK, MAK, MAK, MAK, NFP, NFP, ONT, ONT]
2-CHLOROETHANOL (ETHYLENE CHLOROHYDRIN) [BC1, BC1, BC1, CSR, CSR]
CHLOROETHANOL [302, EPA, EPA]
ETHANOL, 2-CHLORO- [PAL]
ETHYLENE CHLOROHYDRIN (2-CHLOROETHANOL) [ALB, ALB]
ETHYLENE CHLOROHYDRINE [QBC, QBC]

107-08-4 1-IODOPROPANE

107-10-8 1-PROPANAMINE
PROPYLAMINE [FLA, HM1, HM1, HM1, MSL, NFP, NFP, NJL, NJL, WHM]

107-11-9 ALLYLAMINE
2-PROPEN-1-AMINE [PAL]
ALLYL AMINE [NJL, NJL]
ALLYLAMINE [2-PROPEN-1-AMINE] [EP3]

107-12-0 PROPIONITRILE
ETHYL CYANIDE [RCU, RCU, WHM]
ETHYL CYANIDE (PROPIONITRILE) [UTS]
PROPANENITRILE [PAL]
PROPANENITRILE (ETHYL CYANIDE) [RCA]
PROPIONIC NITRILE [MSL, NFP, NFP]
PROPIONITRILE [PROPANENITRILE] [EP3]

107-13-1 ACRYLONITRILE
2-PROPENENITRILE [PAL, PAL]
ACRYLONITRILE (VINYL CYANIDE) [ALB, ALB, ALB, ALB]
ACRYLONITRILE [2-PROPENENI-TRILE] [EP3]
VINYL CYANIDE (ACRYLONITRILE) [QBC, QBC, QBC]

107-14-2 CHLOROACETONITRILE

107-15-3 ETHYLENEDIAMINE
1,2 DIAMINOETHANE (ETHYLENEDI-AMINE) [ALB, ALB]
1,2-DIAMINOETHANE [GBR, ONT]
1,2-ETHANEDIAMINE [PAL]
ETHYLENE DIAMINE [QBC, WHM]
ETHYLENEDIAMINE [1,2-ETHANEDI-AMINE] [EP3]

107-16-4 FORMALDEHYDE CYANOHYDRIN
GLYCOLONITRILE [MIN, NHS, NHS]

107-18-6 ALLYL ALCOHOL
2-PROPEN-1-OL [PAL]
ALLYL ALCOHOL (2-PROPEN-1-OL) [CN2]
ALLYL ALCOHOL [2-PROPEN-1-OL] [EP3]

107-19-7 PROPARGYL ALCOHOL
2-PROPYN-1-OL [PAL, TSC, TSC]
PROP-2-YN-1-OL [GBR, GBR, GBR]

107-20-0 CHLOROACETALDEHYDE
ACETALDEHYDE, CHLORO- [PAL,
TSC, TSC, TSC]

107-21-1 ETHYLENE GLYCOL
1,2-ETHANEDIOL [PAL, PST]
ETHANE-1,2-DIOL [GBR, GBR]
ETHYLENE GLYCOL (VAPOUR, MIST)
[QBC]
ETHYLENE GLYCOL PARTICULATE
AND VAPOR [FLA]
ETHYLENE GLYCOL, VAPOR [ONT]

107-22-2 GLYOXAL
ETHANEDIAL [TSC, TSC, TSC]
GLYOXAL DIHYDRATE [CSR]

107-25-5 VINYL METHYL ETHER
ETHENE, METHOXY- [PAL]
VINYL METHYL ETHER [ETHENE,
METHOXY-] [EP3]

107-27-7 ETHYLMERCURIC CHLORIDE
CHLOROETHYLMERCURY [FLA]
ETHYL MERCURIC CHLORIDE [NJL,
NJL]

107-29-9 ACETALDEHYDE OXIME
ACETALDOXIME [NJL]

**107-30-2 CHLOROMETHYL METHYL
ETHER**
CHLOROMETHYL METHYL ETHER
(TECHNICAL GRADE) [C65, C65,
EP2, IAR]
CHLOROMETHYL METHYL ETHER
[METHANE, CHLOROMETHOXY-]
[EP3]
METHANE, CHLOROMETHOXY-
[PAL, PAL]
METHYL CHLOROMETHYL ETHER
[CAL, CEX, HM1, HM1, HM1,
HM1, MIN, NHS, NHS, NHS, OS1,
WHM]
MONOCHLORODIMETHYL ETHER
[MAK]

107-31-3 METHYL FORMATE
FORMIC ACID, METHYL ESTER
[MAK, MAK, MAK, PAL]
METHYL FORMATE [FORMIC ACID,
METHYL ESTER] [EP3]

107-37-9 ALLYL TRICHLOROSILANE
SILANE, TRICHLORO-2-PROPENYL-
[PAL]

107-39-1 1-PENTENE, 2,4,4-TRIMETHYL-
2,4,4-TRIMETHYL-1-PENTENE [FLA,
MSL, NFP, NFP]

107-40-4 2-PENTENE, 2,4,4-TRIMETHYL-
2,4,4-TRIMETHYL-2-PENTENE [FLA,
MSL, NFP, NFP]

107-41-5 HEXYLENE GLYCOL
2,4-PENTANEDIOL, 2-METHYL-
[PAL]
2-METHYL-2,4-PENTANEDIOL [NFP,
NFP]
2-METHYLPENTANE-2,4-DIOL [GBR,
GBR]

107-44-8 SARIN

107-45-9 TERT-OCTYLAMINE
2-PENTANAMINE, 2,4,4-TRIMETHYL-
[PAL]

107-46-0 HEXAMETHYLDISILOXANE

107-49-3 TEPP
DIPHOSPHORIC ACID, TETRAETHYL
ESTER [EPA, PAL, RCU, RCU]
TEPP (ISO) [GBR, GBR, GBR]
TETRAETHYL PYROPHOSPHATE
[HM1, HM1, HM1, HM1, HM1]
TETRAETHYL PYROPHOSPHATE
[CAL, CWA, ONT, ONT]

**107-50-6 TETRADECAMETHYLCYCLOHEP-
TASILOXANE**

107-51-7 OCTAMETHYLTRISILOXANE

**107-52-8 TETRADECAMETHYLHEXASILOX-
ANE**

**107-53-9 TETRACOSAMETHYLUNDE-
CASILOXANE**

107-54-0 1-HEXYN-3-OL, 3,5-DIMETHYL-
3,5-DIMETHYL-1-HEXYN-3-OL [PST]
DIMETHYL HEXYNOL [NFP, NFP]

107-58-4 N-TERT-BUTYLACRYLAMIDE

**107-64-2 DISTEARYLDIMETHYLAMMO-
NIUM CHLORIDE**
1-OCTADECANAMINIUM, N,N-
DIMETHYL-N-OCTAD [OTS]
DIMETHYL DIOCTADECYL AMMO-
NIUM CHLORIDE [PST]

107-66-4 DIBUTYL PHOSPHATE
DIBUTYL HYDROGEN PHOSPHATE
[GBR, GBR]
PHOSPHORIC ACID, DIBUTYL ESTER
[OTS, PAL, TSC, TSC, TSC]

**107-68-6 2-(METHYLAMINO)ETHANESUL-
FONIC ACID**

**107-70-0 4-METHOXY-4-METHYLPENTAN-
2-ONE**
4-METHOXY-4-METHYL-2-PEN-
TANONE [PST]

107-71-1 TERT-BUTYL PEROXYACETATE
ETHANEPEROXOIC ACID, 1,1-
DIMETHYLETHYL ESTER [PAL]
TERT-BUTYL PERACETATE [MAK,
NFP, NFP]

107-72-2 AMYL TRICHLOROSILANE
AMYLTRICHLOROSILANE [WHM]
SILANE, TRICHLOROPENTYL- [PAL]

**107-75-5 OCTANAL, 7-HYDROXY-3,7-
DIMETHYL-**
HYDROXYCINTRONELLAL [NFP,
NFP]

107-81-3 2-BROMOPENTANE

107-82-4 BROMOMETHYLBUTANE
1-BROMO-3-METHYLBUTANE [HM1,
HM1]

107-83-5 ISOHEXANE
2-METHYLPENTANE [NFP, NFP,
WHM]
PENTANE, 2-METHYL- [PAL]

107-84-6 ISOAMYL CHLORIDE
BUTANE, 1-CHLORO-3-METHYL-
[PAL]

107-87-9 METHYL PROPYL KETONE
2-PENTANONE [BC1, BC1, CAL,
MAK, MAK, NIO, NIO, NIO, ONT,
ONT, PAL]
2-PENTANONE (METHYL PROPYL
KETONE) [OS1, OS1, OS1]
METHYL N-PROPYL KETONE [NHS,
NHS, PST, WHM]
METHYL PROPYL KETONE (2-PEN-
TANONE) [ALB, ALB]
PENTAN-2-ONE [GBR, GBR]

107-88-0 1,3-BUTANEDIOL
1,3-BUTYLENE GLYCOL [HON]
BETA-BUTYLENE GLYCOL [NFP,
NFP]

107-89-1 ALDOL
ACETALDOL [FLA, HON]
BUTANAL, 3-HYDROXY- [PAL]

107-91-5 2-CYANOACETAMIDE

107-92-6 BUTYRIC ACID
BUTANOIC ACID [PAL]
N-BUTYRIC ACID [CWA]

**107-98-2 PROPYLENE GLYCOL
MONOMETHYL ETHER**
1-METHOXY-2-PROPANOL [ONT,
ONT]
1-METHOXY-2-PROPANAL [HM1,
HM1]
1-METHOXYPROPAN-2-OL [GBR,
GBR, GBR]

2-METHOXY-1-METHYLETHANOL
[PST]
2-PROPANOL, 1-METHOXY- [PAL,
TSC, TSC, TSC]
PROPYLENE GLYCOL 1-METHYL
ETHER [MAK, MAK, MAK]
PROPYLENE GLYCOL METHYL
ETHER [FLA, MSL, NFP, NFP]

108-01-0 2-(DIMETHYLAMINO) ETHANOL
DIMETHYLAMINOETHANOL [OTS]
DIMETHYLETHANOLAMINE [HM1,
HM1]
ETHANOL, 2-(DIMETHYLAMINO)-
[PAL]

108-03-2 1-NITROPROPANE
PROPANE, 1-NITRO- [PAL]

108-05-4 VINYL ACETATE
ACETIC ACID ETHENYL ESTER
[PAL]
VINYL ACETATE (ACETIC ACID
ETHENYL ESTER) [TSC, TSC]
VINYL ACETATE MONOMER [302,
MSL]
VINYL ACETATE MONOMER
[ACETIC ACID ETHENYL ESTER]
[EP3]

108-08-7 2,4-DIMETHYLPENTANE
PENTANE, 2,4-DIMETHYL- [PAL]

108-09-8 1,3-DIMETHYLBUTYLAMINE
2-PENTANAMINE, 4-METHYL- [PAL]

108-10-1 METHYL ISOBUTYL KETONE
2-PENTANONE, 4-METHYL- [PAL,
TSC, TSC, TSC]
4-METHYL-2-PENTANONE [ONT,
ONT]
4-METHYLPENTAN-2-ONE [GBR,
GBR, GBR]
HEXONE [MAK, MAK, MIN, NIO,
NIO, NIO]
HEXONE (METHYL ISOBUTYL KE-
TONE) [ALB, ALB, ALB, BC1, BC1,
BC1, OS1, OS1, OS1, QBC, QBC]
METHYL ISOBUTYL KETONE (HEX-
ONE) [CAA, HON, HON]
METHYL ISOBUTYL KETONE (MIBK)
[BEI]
METHYLISOBUTYL KETONE [MSL]

108-11-2 METHYL ISOBUTYL CARBINOL
2-PENTANOL, 4-METHYL [PAL]
4-METHYL-2-PENTANOL [MAK,
MAK, MAK, ONT, ONT, ONT]
4-METHYLPENTAN-2-OL [GBR,
GBR, GBR]
METHYL AMYL ALCOHOL [NJL]
METHYL AMYL ALCOHOL (METHYL
ISOBUTYL CARBINOL) [ALB, ALB,
ALB, QBC, QBC, QBC]
METHYL ISOBUTYL CARBINOL
(MIBC) [WHM]

**108-16-7 N,N-DIMETHYLISO-
PROPANOLAMINE**
2-PROPANOL, 1-(DIMETHYLAMINO)-
[PAL]

108-18-9 DIISOPROPYLAMINE
2-PROPANAMINE, N-(1-
METHYLETHYL)- [PAL]

108-20-3 ISOPROPYL ETHER
2,2'-OXYBIS(PROPANE) [ONT, ONT]
DIISOPROPYL ETHER [GBR, GBR,
HM1, HM1, NJL, NJL, QBC, QBC]
PROPANE, 2,2'-OXYBIS- [PAL]

108-21-4 ISOPROPYL ACETATE
ACETIC ACID, 1-METHYLETHYL
ESTER [PAL]

108-22-5 ISOPROPENYL ACETATE
1-PROPEN-2-OL, ACETATE [PAL]

108-23-6 ISOPROPYL CHLOROFORMATE

108-24-7 ACETIC ANHYDRIDE
ACETIC ACID, ANHYDRIDE [PAL]

108-29-2 4-VALEROLACTONE

108-30-5 SUCCINIC ANHYDRIDE

108-31-6 MALEIC ANHYDRIDE
2,5-FURANDIONE [PAL]
2,5-FURANDIONE (MALEIC ANHY-
DRIDE) [TSC, TSC, TSC]
MALEIC ACID ANHYDRIDE [MAK,
MAK, MAK]

108-32-7 PROPYLENE CARBONATE
4-METHYL-1,3-DIOXOLANE-2-ONE
[PST]

108-38-3 M-XYLENE
BENZENE, 1,3-DIMETHYL- [PAL,
TSC, TSC, TSC]
M-XYLENES [CAA]

108-39-4 M-CRESOL
M-CRESOL AND CRESYLIC ACID
[HON, HON]
META-CRESOL [CTR]
PHENOL, 3-METHYL- [PAL, TSC,
TSC, TSC]

108-41-8 M-CHLOROTOLUENE

108-42-9 M-CHLOROANILINE
BENZENAMINE, 3-CHLORO- [TSC,
TSC]

108-43-0 M-CHLOROPHENOL
3-CHLOROPHENOL [NJL, NJL]

108-44-1 M-TOLUIDINE

108-45-2 M-PHENYLENEDIAMINE
1,3-BENZENEDIAMINE [TSC, TSC]
1,3-PHENYLENEDIAMINE [313]
META-PHENYLENEDIAMINE [CTR,
IAR, TSE]
PHENYLENEDIAMINE, META- [OTS]

108-46-3 RESORCINOL
1,3-BENZENEDIOL [PAL]

108-55-4 GLUTARIC ANHYDRIDE

108-57-6 DIVINYL BENZENE
BENZENE, 1,3-DIETHENYL- [PAL]
DIVINYLBENZENE [GBR, NFP, NFP]
M-DIVINYL BENZENE [MSL]
M-DIVINYLBENZENE [ONT, WHM]

108-59-8 DIMETHYL MALONATE

**108-60-1 BIS(2-CHLOROISOPROPYL)
ETHER**
2,2-DICHLORO ISOPROPYL ETHER
[FLA, MSL, NFP, NFP]
BIS(2-CHLORO-1-METHYLETHYL)
ETHER [CSR, CSR]
BIS(2-CHLORO-1-METHYLETHYL)
ETHER [IAR]
BIS(2-CHLORO-METHYLETHYL)
ETHER [313, PQL]
BIS(2-CHLOROISOPROPYL) ETHER
[UTS]
DICHLOROISOPROPYL ETHER [EPA,
HM1, HM1, HM1, RCU, RCU,
WHM]
PROPANE, 2,2'-OXYBIS(1-CHLORO-
[TSC, TSC, TSC]

108-62-3 METALDEHYDE
1,3,5,7-TETRAOXOCANE, 2,4,6,8-
TETRAMETHYL- [PAL]

**108-65-6 PROPYLENE GLYCOL
MONOMETHYL ETHER ACETATE**
1-METHOXY-2-PROPANOL ACETATE
[PST]
2-PROPANOL, 1-METHOXY-, AC-
ETATE [TSC, TSC, TSC]
PROPYLENE GLYCOL 1-METHYL
ETHER-2-ACETATE [MAK, MAK,
MAK]
PROPYLENE GLYCOL METHYL
ETHER ACETATE (99% PURE) [NFP,
NFP]

108-67-8 1,3,5-TRIMETHYLBENZENE
BENZENE, 1,3,5-TRIMETHYL- [TSC,
TSC, TSC]
MESITYLENE [MSL]
TRIMETHYLBENZENE, 1,3,5- [OTS]

108-68-9 3,5-XYLENOL

108-69-0 3,5-DIMETHYLANILINE

108-70-3 1,3,5-TRICHLOROBENZENE
BENZENE, 1,3,5-TRICHLORO- [TSC,
TSC, TSC]

108-71-4 TOLUENE-3,5-DIAMINE
1,3-BENZENEDIAMINE, 5-METHYL-
[TSC, TSC]

108-73-6 PHLOROGLUCINOL

108-77-0 CYANURIC CHLORIDE
CYANURIC TRICHLORIDE [NJL,
NJL]

108-78-1 MELAMINE

108-80-5 ISOCYANURIC ACID
CYANURIC ACID [PST]

108-82-7 DIISOBUTYL CARBITOL
DIISOBUTYL CARBINOL [NFP, NFP]

108-83-8 DIISOBUTYL KETONE
2,6-DIMETHYL-4-HEPTANONE [ONT]
2,6-DIMETHYL-4-HEPTANONE (DI-
ISOBUTYL KETONE) [ALB, ALB]
2,6-DIMETHYLHEPTAN-4-ONE [GBR]
2,6-DIMETHYLHEPTANONE [NJL]
4-HEPTANONE, 2,6-DIMETHYL-
[PAL]
DIISOBUTYLKETONE [QBC]
HEPTANONE, 2,6-DIMETHYL-, 4-
[OTS]

108-84-9 SEC-HEXYL ACETATE
1,3-DIMETHYLBUTYL ACETATE
[GBR, GBR, NFP, NFP]
HYXYL ACETATE - SEC [QBC]
METHYL AMYL ACETATE [NJL]
METHYLAMYL ACETATE [HM1,
HM1]

108-86-1 BROMOBENZENE
BENZENE, BROMO- [PAL, TSC, TSC,
TSC]

108-87-2 METHYLCYCLOHEXANE
CYCLOHEXANE, METHYL- [PAL]
METHYL CYCLOHEXANE [OS1, OS1]
METHYL CYCLOHEXANONE [HM1,
HM1]

108-88-3 TOLUENE
BENZENE, METHYL- [PAL, TSC,
TSC, TSC]
METHYLBENZENE [ONT, ONT]
TOLUENE [NFP, NFP]
TOLUENE (TOLUOL) [ALB, ALB,
ALB, BC1, BC1, BC1, MAK, MAK,
MAK, QBC, QBC, WHM]

108-89-4 P-PICOLINE
4-PICOLINE [FLA, MSL, NFP, NFP,
WEL, WEL, WEL]
PYRIDINE, 4-METHYL- [OTS, PAL,
TSC, TSC]

108-90-7 CHLOROBENZENE
BENZENE, CHLORO- [PAL, TSC,
TSC, TSC, TSC]
CHLOROBENZENE (MONO) [QBC]
CHLOROBENZENE
(MONOCHLOROBENZENE) [ALB,
ALB, BC1]
MONOCHLOROBENZENE [CD1, CD1,
CTR, SDW, SDW, SDW, TSE]

108-91-8 CYCLOHEXYLAMINE
CYCLOHEXANAMINE [PAL]
CYCLOHEXYLAMINE [CYCLOHEX-
ANAMINE] [EP3]

108-93-0 CYCLOHEXANOL

108-94-1 CYCLOHEXANONE

108-95-2 PHENOL

108-98-5 BENZENETHIOL
BENZENETHIOL [TSC, TSC, TSC]
PHENYL MERCAPTAN [ALB, ALB,
AUS, BC1, CAL, CEX, FLA, HM1,
HM1, HM1, MEX, MSL,
OS1, OTV, TLV, WHM]
PHENYLMERCAPTAN [QBC]
THIOPHENOL [302, OTS, T33]

108-99-6 3-METHYLPYRIDINE
3-METHYL PYRIDINE [HM1]
3-PICOLINE [WEL, WEL, WEL]
PICOLINE (B-) [HON]
PYRIDINE, 3-METHYL- [OTS, TSC,
TSC]

109-00-2 3-HYDROXYPYRIDINE

109-01-3 1-METHYL PIPERAZINE
PIPERAZINE, 1-METHYL- [PAL]

109-02-4 METHYLMORPHOLINE
4-METHYLMORPHOLINE [FLA, MSL,
NFP, NFP]
MORPHOLINE, 4-METHYL- [PAL]

109-06-8 2-PICOLINE
2-METHYLPYRIDINE [313]
O-PICOLINE [WHM]
PYRIDINE, 2-METHYL- [EPA, OTS,
PAL, TSC, TSC]

109-08-0 2-METHYLPYRAZINE
PYRAZINE, METHYL- [PAL]

109-09-1 2-CHLOROPYRIDINE

**109-13-7 TERT-BUTYL PEROXYISOBU-
TYRATE**

**109-16-0 TRIETHYLENE GLYCOL
DIMETHACRYLATE**
2-PROPENOIC ACID, 2-METHYL-
, 1,2-ETHANEDIYLBIS(OXY-2,1-
ETHANEDIYL) ESTER [TSC, TSC]

109-19-3 BUTYL ISOVALERATE

109-21-7 BUTYL BUTYRATE
BUTANOIC ACID, BUTYL ESTER
[PAL]

**109-27-3 1-TETRAZENE-1-CARBOXIMIDIC
ACID, 4-(AMINOMETHYL)-, 2-
NITROSOHYDRAZIDE**
GUANYL NITROSOAMINOGUANYL-
TETRAZENE [HM1]

109-31-9 DI-N-HEXYL AZELATE

109-43-3 DI-N-BUTYL SEBACATE
BUTYL SEBACATE [NFP, NFP]

109-46-6 DIBUTYL THIOUREA
N,N'-DIBUTYL THIOUREA [PST]

**109-47-7 PHOSPHOROUS ACID, DIBUTYL
ESTER**

109-52-4 VALERIC ACID
N-VALERIC ACID [PST]
PENTANOIC ACID [FLA, MSL, NFP,
NFP, PAL]

109-53-5 VINYL ISOBUTYL ETHER
PROPANE, 1-(ETHENYLOXY)-2-
METHYL- [PAL]

**109-55-7 3-(DIMETHYLAMINO)-PROPY-
LAMINE**
1,3-PROPANEDIAMINE, N,N-
DIMETHYL- [PAL]
1-AMINO-3-DIMETHYLAMINO
PROPANE [OTS]

109-56-8 ISOPROPYL ETHANOLAMINE

109-59-1 ISOPROPOXYETHANOL
2-ISOPROPOXYETHANOL [MAK,
MAK, MAK, MAK, ONT, ONT,
OS1]
ETHANOL, 2-(1-METHYLETHOXY)-
[PAL]
ETHYLENE GLYCOL ISOPROPYL
ETHER [FLA, MSL, NFP, NFP,
WHM]

109-60-4 N-PROPYL ACETATE
ACETIC ACID, PROPYL ESTER [PAL]
PROPYL ACETATE [MAK, MAK,
NFP, NFP, WHM]

109-61-5 PROPYLCHLOROFORMATE
N-PROPYL CHLOROFORMATE [HM1,
HM1, HM1, WHM]
PROPYL CHLOROFORMATE [302,
MSL, PAL]

PROPYL CHLOROFORMATE [CARBONOCHLORIDIC ACID, PROPYL-ESTER] [EP3]

109-63-7 BORON TRIFLUORIDE DI-ETHYLETHERATE
BORON TRIFLOURIDE ETHERATE [MSL]
BORON TRIFLUORIDE DIETHYL ETHERATE [NJL, NJL]
BORON TRIFLUORIDE ETHERATE [FLA, NFP, NFP]
BORON, TRIFLUORO[1,1'-OXYBIS [ETHANE]-, (T-4)- [PAL]

109-65-9 BUTYL BROMIDE
BUTANE, 1-BROMO- [PAL]
N-BUTYL BROMIDE [HM1, HM1, WHM]

109-66-0 PENTANE
N-PENTANE [MAK, MAK, NIO, NIO, NIO, PST]
N-PENTANES [HM1, HM1]
PENTANE, ALL ISOMERS [GBR, GBR]

109-67-1 1-PENTENE

109-69-3 BUTYL CHLORIDE
BUTANE, 1-CHLORO- [PAL]
CHLOROBUTANE, 1- [OTS]
N-BUTYL CHLORIDE [CSR, CSR]

109-70-6 1-CHLORO-3-BROMOPROPANE

109-72-8 BUTYL LITHIUM

109-73-9 BUTYLAMINE (N-)
1-BUTANAMINE [PAL]
BUTYLAMINE [ALB, ALB, BC1, BC1, CEX, CEX, MEX, MEX, NFP, NFP, NJL, NJL, QBC, QBC]
N-BUTYLAMINE [CWA, GBR, GBR, MAK, MAK, MAK, ONT, ONT, OTV, TLV, TLV]

109-74-0 BUTYRONITRILE
BUTANENITRILE [PAL]
N-BUTYRONITRILE [MIN, NHS, NHS]

109-75-1 ALLYL CYANIDE

109-76-2 1,3-PROPANEDIAMINE
1,3-DIAMINOPROPANE [WHM]

109-77-3 MALONONITRILE
PROPANEDINITRILE [TSC, TSC, TSC]

109-78-4 ETHYLENE CYANOHYDRIN
PROPANENITRILE, 3-HYDROXY- [PAL]

109-79-5 BUTYL MERCAPTAN
1-BUTANETHIOL [NFP, NFP, NHS, NHS, ONT, PAL]
BUTANETHIOL [MAK, MAK]
BUTANETHIOL (BUTYL MERCAPTAN) [ALB, ALB]
N-BUTYL MERCAPTAN [AUS, CAL, CEX]

109-83-1 N-METHYLETHANOLAMINE
ETHANOL, 2-(METHYLAMINO)- [PAL]

109-86-4 2-METHOXYETHANOL
2-METHOXYETHANOL [QBC, QBC]
2-METHOXYETHANOL (EGME) [TLV, TLV]
2-METHOXYETHANOL (METHYL CELLOSOLVE) [ALB, ALB, ALB, BC1, BC1, BC1]
ETHANOL, 2-METHOXY- [PAL]
ETHYLENE GLYCOL MONOMETHYL ETHER [C65, C65, CAB, CAL, HM1, HM1, NFP, NFP]
ETHYLENE GLYCOL MONOMETHYL ETHER (EGMME) [CSR]
ETHYLENE GLYCOL MONOMETHYL ETHER ACETATE [HON]
ETHYLENE GLYCOL, MONOMETHYL ETHER [MAK, MAK, MAK, MAK]
METHYL CELLOSOLVE [FLA, MIN, MSL, NIO, NIO, NIO]

METHYL CELLOSOLVE (2-METHOXYETHANOL) [OS1, OS1, OS1, OS1, OS1]

109-87-5 METHYLAL
DIMETHOXYMETHANE [GBR, GBR, MAK, ONT]
DIMETHOXYMETHANE (METHYLAL) [ALB, ALB, QBC]
METHANE, DIMETHOXY- [PAL, TSC, TSC]
METHYLAL (DIMETHOXY-METHANE) [OS1, OS1]
METHYLAL (DIMETHOXYMETHANE) [BC1]

109-89-7 DIETHYLAMINE
ETHANAMINE, N-ETHYL- [PAL, TSC, TSC]

109-90-0 ETHYL ISOCYANATE
ETHANE, ISOCYANATO- [TSC, TSC, TSC]

109-92-2 VINYL ETHYL ETHER
ETHENE, ETHOXY- [PAL]
VINYL ETHYL ETHER [ETHENE, ETHOXY-] [EP3]

109-93-3 DIVINYL ETHER
ETHENE, 1,1'-OXYBIS- [PAL]

109-94-4 ETHYL FORMATE
FORMIC ACID, ETHYL ESTER [MAK, MAK, MAK, PAL]

109-95-5 ETHYL NITRITE
ETHYL NITRITE [NITROUS ACID, ETHYL ESTER] [EP3]
NITROUS ACID, ETHYL ESTER [PAL]

109-97-7 1H-PYRROLE
PYRROLE [FLA, MSL, NFP, NFP]

109-99-9 TETRAHYDROFURAN
FURAN, TETRAHYDRO- [PAL]
TETRAHYDROFURANE [QBC]

110-00-9 FURAN

110-01-0 TETRAHYDROTHIOPHENE

110-02-1 THIOPHENE

110-05-4 DI-TERT-BUTYL PEROXIDE
DIBUTYL PEROXIDE (TERTIARY) [OS3]
PEROXIDE, BIS(1,1-DIMETHYLETHYL) [PAL]
TERT-BUTYL PEROXIDE [PST]

110-12-3 METHYL ISOAMYL KETONE
2-HEXANONE, 5-METHYL- [PAL]
5-METHYL-2-HEXANONE [ONT, PST]
5-METHYLHEXAN-2-ONE [GBR, GBR, HM1, HM1, WHM]

110-13-4 2,5-HEXANEDIONE
ACETONYL ACETONE [NFP, NFP]

110-15-6 SUCCINIC ACID

110-16-7 MALEIC ACID
2-BUTENEDIOIC ACID, (Z)- [PAL]

110-17-8 FUMARIC ACID
2-BUTENEDIOIC ACID, (E)- [PAL]

110-18-9 1,2-BIS(DIMETHYLAMINO) ETHANE
1,2-DI(DIMETHYLAMINO) ETHANE [NJL]
1,2-DI-(DIMETHYLAMINO)ETHANE [HM1, HM1]
TETRAMETHLYETHYLENEDIAMINE [HON]

110-19-0 ISOBUTYL ACETATE
ACETIC ACID, 2-METHYLPROPYL ESTER [PAL]
ACETIC ACID, ISOBUTYL ESTER [PST]
ISO-BUTYL ACETATE [CWA, EPA]
ISOBUTYL ACETATE (2-METHYL-PROPYL ACETATE) [CAL]

110-22-5 DIACETYL PEROXIDE
ACETYL PEROXIDE [MSL, NFP, NJL, NJL]
DI-ACETYL PEROXIDE [FLA]
PEROXIDE, DIACETYL [PAL]

110-25-8 OLEOYL SARCOSINE
N-METHYL-N-OLEOYLGLYCINE [PST]

110-26-9 METHYLENE DIACRYLAMIDE

110-27-0 ISOPROPYL MYRISTATE

110-30-5 N,N'-ETHYLENEBIS(STEAR-AMIDE)
OCTADECANAMIDE, N,N'-1,2-ETHANEDIYLBIS- [OTS, PST]

110-38-3 ETHYL DECANOATE

110-40-7 DIETHYL SEBACATE

110-41-8 UNDECANAL, 2-METHYL-

110-43-0 METHYL N-AMYL KETONE
2-HEPTANONE [ONT, PAL, PST]
AMYL METHYL KETONE [HM1, HM1]
HEPTAN-2-ONE [GBR]
METHYL (N-AMYL) KETONE [FLA, NIO, NIO, NIO, NJL]
METHYL AMYL KETONE [NFP, NFP, WHM]
METHYL N-AMYL KETONE (2-HEP-TANONE) [BC1, BC1, QBC]

110-44-1 SORBIC ACID

110-46-3 AMYL NITRITE

110-47-4 3-ISOPROPOXYPROPIONITRILE

110-49-6 2-METHOXYETHYL ACETATE
2-METHOXYETHYL ACETATE [QBC, QBC]
2-METHOXYETHYL ACETATE (EGMEA) [TLV, TLV]
ETHANOL, 2-METHOXY-, ACETATE [PAL]
ETHYLENE GLYCOL METHYL ETHER ACETATE [MIN]
ETHYLENE GLYCOL METHYL ETHER ACETATE (METHYL CELLO-SOLVEACETATE) [ALB, ALB, ALB]
ETHYLENE GLYCOL MONOMETHYL ETHER ACETATE [C65, C65, CAB, CAL, NFP, NFP]
ETHYLENE GLYCOL MONOMETHYL ETHER ACETATE (METHYL CELLO-SOLVEACETATE) [BC1, BC1, BC1]
ETHYLENE GLYCOL, MONOMETHYL ETHER ACETATE [MAK, MAK, MAK, MAK]
METHYL CELLOSOLVE ACETATE [NIO, NIO, NIO]
METHYL CELLOSOLVE ACETATE (2-METHOXYETHYL ACETATE) [OS1, OS1, OS1, OS1, OS1]

110-53-2 1-BROMOPENTANE
AMYL BROMIDE [NFP, NFP]
PENTANE, 1-BROMO- [PAL]

110-54-3 HEXANE
HEXANE (N-HEXANE) [AUS, BC1, BC1, TLV]
HEXANE (N-HIXANE) [QBC]
N-HEXANE [313, BEI, CSR, GBR, MEX, NIO, NIO, NIO, NJL, NJL, ONT, ONT, OS1, OS1, TSC, WHM]

110-56-5 1,4-DICHLOROBUTANE
BUTANE, 1,4-DICHLORO- [PAL]

110-57-6 TRANS-1,4-DICHLOROBUTENE
TRANS-1,4-DICHLORO-2-BUTENE [313, MSL, PQL]

110-58-7 AMYL AMINE
1-PENTANAMINE [PAL]
AMYLAMINE [FLA, MSL, NFP, NFP]

110-60-1 TETRAMETHYLENEDIAMINE

110-61-2 SUCCINONITRILE

110-62-3 **VALERALDEHYDE**
N-VALERALDEHYDE [AUS, ONT, OS1, OTV, TLV]
PENTANAL [PAL, TSC, TSC, TSC]

110-63-4 **1,4-BUTANEDIOL**

110-64-5 **2-BUTENE-1,4-DIOL**

110-65-6 **1,4-BUTYNEDIOL**

110-66-7 **AMYL MERCAPTAN**
1-PENTANETHIOL [NHS, NHS, PAL]

110-67-8 **PROPANENITRILE, 3-METHOXY-**
3-METHOXYPROPIONITRILE [FLA, MSL, NFP, NFP]

110-68-9 **N-METHYLBUTYLAMINE**
1-BUTANAMINE, N-METHYL- [PAL]

110-69-0 **BUTYRALDOXIME**
BUTANAL OXIME [CSR]
BUTANAL OXIME [PAL]
BUTYL ALDEHYDE OXIME [NJL]

110-71-4 **ETHYLENE GLYCOL DIMETHYL ETHER**
1,2-DIMETHOXYETHANE [HM1, HM1]
ETHANE, 1,2-DIMETHOXY- [PAL]

110-73-6 **2-ETHYLAMINOETHANOL**
ETHYL ETHANOLAMINE [PST]
N-ETHYLETHANOLAMINE [NFP, NFP]

110-74-7 **PROPYL FORMATE**
FORMIC ACID, PROPYL ESTER [PAL]

110-75-8 **2-CHLOROETHYL VINYL ETHER**
2-CHLOROETHYL VINYL ETHER (MIXED) [CW2]
2-CHLOROETHYL VINYL ETHER (MIXTURE) [MX1]
ETHENE, (2-CHLOROETHOXY)- [PAL, TSC, TSC]
VINYL 2-CHLOROETHYL ETHER [MSL, NFP, NFP, WHM]
VINYL-2-CHLOROETHYL ETHER [FLA]

110-76-9 **AMINOETHOXYETHANOL**
2-(2-AMINOETHOXY)ETHANOL [HM1, HM1]

110-77-0 **2-(ETHYLTHIO)ETHANOL**

110-78-1 **N-PROPYL ISOCYANATE**
PROPANE, 1-ISOCYANATO- [TSC, TSC, TSC]
PROPYL ISOCYANATE [NJL]

110-80-5 **2-ETHOXYETHANOL**
2-ETHOXYACETYLENE [FLA]
2-ETHOXYETHANOL (EEGE) [BEI]
2-ETHOXYETHANOL (EGEE) [TLV, TLV]
2-ETHOXYETHANOL (GLYCOL MONOETHYL ETHER) [ALB, ALB, ALB]
ETHANOL, 2-ETHOXY- [PAL]
ETHOXYETHANOL, 2- [OTS]
ETHYLENE GLYCOL MONOETHYL ETHER [C65, C65, CAL, EPA, HM1, HM1, HM1, HON, NFP, NFP, RCU, RCU]
ETHYLENE GLYCOL MONOETHYL ETHER (EGMEE) [CSR]
ETHYLENE GLYCOL, MONOETHYL ETHER [MAK, MAK, MAK, MAK]

110-82-7 **CYCLOHEXANE**

110-83-8 **CYCLOHEXENE**

110-85-0 **PIPERAZINE**

110-86-1 **PYRIDINE**

110-87-2 **DIHYDROPYRAN**
2H-PYRAN, 3,4-DIHYDRO- [PAL]

110-88-3 **1,3,5-TRIOXANE**
TRIOXANE [FLA, MSL, NFP, NFP]

110-89-4 **PIPERIDINE**

110-91-8 **MORPHOLINE**

110-94-1 **GLUTARIC ACID**

110-96-3 **DI-ISOBUTYLAMINE**
1-PROPANAMINE, 2-METHYL-N-(2-METHYLPROPYL)- [PAL]
DIISOBUTYLAMINE [FLA, MSL, NFP, NFP]

110-97-4 **DIISOPROPANOLAMINE**
2-PROPANOL, 1,1'-IMINOBIS- [PAL]

110-98-5 **2-PROPANOL, 1,1'-OXYBIS**
DIPROPYLENE GLYCOL [HON, TSC, TSC, TSC]

110-99-6 **DIGLYCOLIC ACID**

111-11-5 **METHYL ESTER OCTANOIC ACID**
OCTANOIC ACID, METHYL ESTER [OTS]

111-12-6 **METHYL HEPTINE CARBONATE**

111-13-7 **2-OCTANONE**
METHYL HEXYL KETONE [NFP, NFP]

111-14-8 **HEPTANOIC ACID**

111-15-9 **2-ETHOXYETHYL ACETATE**
2,2-ETHOXYETHYL ACETATE (CELLOSOLVE ACETATE) [QBC, QBC]
2-ETHOXYETHANOL [FLA]
2-ETHOXYETHYL ACETATE (CELLOSOLVE ACETATE) [BC1, BC1, BC1, OS1, OS1, OS1, OS1, OS1]
2-ETHOXYETHYL ACETATE (EGEEA) [BEI, TLV, TLV]
2-ETHOXYETHYLACETATE [NJL]
CELLOSOLVE ACETATE [PST]
CELLOSOLVE ACETATE (2-ETHOXYETHYLACETATE) [ALB, ALB, ALB]
ETHANOL, 2-ETHOXY-, ACETATE [PAL]
ETHYLENE GLYCOL MONOETHYL ETHER ACETATE [C65, C65, CAB, CAL, HM1, HM1, HON, NFP, NFP]
ETHYLENE GLYCOL, MONOETHYL ETHER ACETATE [MAK, MAK, MAK, MAK]

111-17-1 **THIODIPROPIONIC ACID**

111-18-2 **N,N,N',N'-TETRAMETHYL-1,6-HEXANEDIAMINE**

111-20-6 **DECANEDIOIC ACID**

111-21-7 **ETHYLENE BIS(OXYETHYLENE) DIACETATE**
TRIETHYLENE GLYCOL DIACETATE [NFP, NFP]

111-25-1 **1-BROMOHEXANE**

111-26-2 **N-HEXYLAMINE**
1-HEXANAMINE [PAL]
HEXYLAMINE [FLA, MSL, NFP, NFP]

111-27-3 **N-HEXANOL**
1-HEXANOL [PAL]
HEXYL ALCOHOL [NFP, NFP, PST]
N-HEXYL ALCOHOL [WHM]

111-29-5 **1,5-PENTANEDIOL**

111-30-8 **GLUTARALDEHYDE**
GLUTARALDEHYDE, ACTIVATED OR UNACTIVATED [BC1]
PENTANEDIAL [PAL, TSC, TSC, TSC]

111-31-9 **1-HEXANETHIOL**

111-34-2 **BUTYL VINYL ETHER**
BUTANE, 1-(ETHENYLOXY)- [PAL]
VINYL BUTYL ETHER [FLA, MSL, NFP, NFP]

111-36-4 **N-BUTYL ISOCYANATE**
BUTANE, 1-ISOCYANATO- [PAL, TSC, TSC, TSC]
N-BUTYLISOCYANATE [NJL]

111-40-0 **DIETHYLENE TRIAMINE**
1,2-ETHANEDIAMINE, N-(2-AMINOETHYL)- [PAL, TSC, TSC, TSC]
2,2'-IMINODI(ETHYLAMINE) [GBR, GBR]
DIETHYLENETRIAMINE [CAL, CEX, CEX, ONT, ONT, OTS, PST, QBC, QBC, TSE, WHM]
DIETHYLENETRIAMINE (DTA) [CTR]

111-41-1 **AMINOETHYLETHANOLAMINE**
(2-AMINOETHYL) ETHANOLAMINE [FLA, MSL, NFP, NFP]
AMINOETHYL ETHANOLAMINE [WHM]
ETHANOL, 2-[(2-AMINOETHYL) AMINO]- [PAL]

111-42-2 **DIETHANOLAMINE**
2,2'-IMINODIETHANOL [GBR]
DIETHANOLAMINE (2,2'-IMINODI-ETHANOL) [HON, HON]
DIETHANOLAMINE (ETHANOL, 2,2'-IMINOBIS-) [TSC, TSC, TSC]
ETHANOL, 2,2'-IMINOBIS- [PAL]

111-43-3 **DIPROPYL ETHER**
N-PROPYL ETHER [MSL, NFP, NFP]
PROPANE, 1,1'-OXYBIS- [PAL]
PROPYL ETHER [FLA]

111-44-4 **BIS(2-CHLOROETHYL) ETHER**
1,1'-OXYBIS(2-CHLOROETHANE) [ONT, ONT, ONT]
2,2'-DICHLORODIETHYL ETHER [MAK, MAK, MAK]
2,2'-DICHLOROETHYL ETHER [NFP, NFP]
2,2-DICHLORODIETHYL ETHER [HM1, HM1, HM1, HM1]
BIS(2-CHLOROETHOXY)METHANE [UTS]
DICHLOROETHYL ETHER [302, ALB, ALB, ALB, AUS, AUS, AUS, BC1, BC1, BC1, CEX, CEX, CEX, FLA, MIN, NIO, NIO, NIO, NIO, NIO, OS1, OS1, OS1, OS1, QBC, QBC, QBC, TLV, TLV, TLV, WHM]
DICHLOROETHYL ETHER (BIS(2-CHLOROETHYL)ETHER) [CAA]
DICHLOROETHYLETHER (BIS(2-CHLOROETHYL) ETHER) [HON, HON]
ETHANE, 1,1'-OXYBIS[2-CHLORO- [PAL]

111-46-6 **DIETHYLENE GLYCOL**
2,2'-OXYDIETHANOL [GBR]
ETHANOL, 2,2'-OXYBIS- [PAL]

111-48-8 **ETHANOL, 2,2'-THIOBIS-**
THIODIGLYCOL [FLA, MSL, NFP, NFP]

111-49-9 **HEXAMETHYLENEIMINE**

111-50-2 **ADIPOYL CHLORIDE**
HEXANEDIOYL DICHLORIDE [PAL]

111-54-6 **1,2-ETHANEDIYLBISCAR-BAMODITHIOIC ACID**
CARBAMODITHIOIC ACID, 1,2-ETHANEDIYLBIS, SALTS AND ESTERS [EPA]
ETHYLENEBISDITHIOCARBAMIC ACID, SALTS & ESTERS [RCU, RCU]
ETHYLENEBISDITHIOCARBAMIC ACID, SALTS AND ESTERS [313]

111-55-7 **1,2-ETHANEDIOL, DIACETATE**
ETHYLENE GLYCOL DIACETATE [HON]
GLYCOL DIACETATE [NFP, NFP]

111-57-9 **N-(2-HYDROXYETHYL)OCTADE-CANAMIDE**

111-60-4 **GLYCOL STEARATE**
ETHYLENE GLYCOL MONOS-TEARATE [PST]

111-64-8 **CAPRYLYL CHLORIDE**
OCTANOYL CHLORIDE [PAL]

PROPOXUR [PHENOL, 2-(1-METHYLETHOXY)-, METHYLCARBAMATE] [313]

114-49-8 SCOPOLAMINE HYDROBROMIDE
BENZENEACETIC ACID, .ALPHA.-(HYDROXYMETHYL)-, 9-METHYL-3-OXA-9-AZATRICYCLO[3.3.1.02,4]NON-7-YL ESTER, HYDROBROMIDE, [7(S)-(1.ALPHA.2.BETA.4.BETA.5.ALPHA.7.BETA)]- [TSC, TSC]

114-83-0 ACETIC ACID, 2-PHENYLHYDRAZIDE

115-02-6 AZASERINE
L-SERINE, DIAZOACETATE (ESTER) [PAL, PAL]

115-07-1 PROPYLENE
1-PROPENE [PAL]
PROPYLENE (1-PROPENE) [EP3]
PROPYLENE (PROPENE) [313, MSL]

115-09-3 MERCURYMETHYLCHLORIDE

115-10-6 DIMETHYL ETHER
METHANE, OXYBIS- [PAL]
METHYL ETHER [FLA, MSL, NFP, NFP]
METHYL ETHER [METHANE, OXYBIS-] [EP3]

115-11-7 ISOBUTYLENE
1-PROPENE, 2-METHYL- [PAL]
2-METHYLPROPENE [FLA, MSL, NFP, NFP, OTS]
2-METHYLPROPENE [1-PROPENE, 2-METHYL-] [EP3]
ISOBUTENE [CSR]

115-18-4 3-BUTEN-2-OL, 2-METHYL-

115-19-5 3-METHYL BUTYNOL
2-METHYL-3-BUTYN-2-OL [PST]
3-BUTEN-2-OL, 2-METHYL- [PAL]
3-BUTYN-2-OL, 2-METHYL- [OTS]

115-21-9 TRICHLOROETHYLSILANE
ETHYLTRICHLORO SILANE [NFP, NFP]
ETHYLTRICHLOROSILANE [NJL, NJL]
SILANE, TRICHLOROETHYL- [PAL]

115-25-3 OCTAFLUOROCYCLOBUTANE

115-26-4 DIMEFOX

115-27-5 4,7-METHANOISOBENZOFURAN-1,3-DIONE, 4,5,6,7,8,8-HEXACHLORO-3A,4,7,7A-TETRAHYDRO-

115-28-6 CHLORENDIC ACID
BICYCLO[2.2.1]HEPT-5-ENE-2,3-DICARBOXYLIC ACID, 1,4,5,6,7,7-HEXACHLORO- [TSC, TSC, TSC]

115-29-7 ENDOSULFAN
6,9-METHANO-2,4,3-BENZODIOXATHIEPIN, 6,7,8,9,10,10-HEXACHLORO-1,5,5A,6,9,9A-HEXAHYDRO-, 3-OXIDE [PAL]
ENDOSULFAN (ISO) [GBR, GBR, GBR]
ENDOSULFAN (THIODAN) [BC1, BC1, BC1]
ENDOSULFAN (THIODAN) (R) [QBC, QBC]
ENDOSULFAN AND METABOLITES [CW3]

115-31-1 ISOBORNYL THIOCYANOACETATE

115-32-2 DICOFOL
1,1-BIS(P-CHLOROPHENYL)-2,2,2-TRICHLOROETHANOL (DICOFOL) [CAL]
DICOFOL [BENZENEMETHANOL, 4-CHLORO-ALPHA-(4-CHLOROPHENYL)-ALPHA-(TRICHLORMETHYL)-] [313]
KELTHANE [CWA, PAL]

115-67-3 PARAMETHADIONE

115-71-9 SANTALOL

115-76-4 2,2-DIETHYL-1,3-PROPANEDIOL
1,3-PROPANEDIOL, 2,2-DIETHYL- [PAL]

115-77-5 PENTAERYTHRITOL
1,3-PROPANEDIOL, 2,2-BIS(HYDROXYMETHYL)- [PAL]

115-82-2 TETRA (2-ETHYLHEXYL) SILICATE

115-84-4 2-ETHYL-2-BUTYL-1,3-PROPANEDIOL
1,3-PROPANEDIOL, 2-BUTYL-2-ETHYL- [PAL]

115-86-6 TRIPHENYL PHOSPHATE
PHOSPHORIC ACID, TRIPHENYL ESTER [PAL, TSC, TSC]

115-90-2 FENSULFOTHION
FENSULFOTHION (DANSANIT) [ALB, ALB]
FENSULFOTHION (DASANIT) [BC1, OS1, QBC]
O,O-DIETHYL O-[4-METHYLSULFINYL)PHENYL]PHOSPHOROTHIOATE (FENSULFOTHION) [CAL]
PHOSPHOROTHIOIC ACID, O,O-DIETHYL O-[4-(METHYLSULFINYL)PHENYL] ESTER [PAL]

115-91-3 O,O-DIETHYL O-[4-METHYLSULFINYL)PHENYL]PHOSPHOROTHIOATE (FENSULFOTHION)

115-95-7 LYNALYL ACETATE (EX BOIS DE ROSE; SYNTHETIC)

115-96-8 ETHANOL, 2-CHLORO-, PHOSPHATE (3:1)
ETHANOL, 2-CHLORO-, PHOSPHATE (3:1) (TRIS(2-CHLOROETHYL) PHOSPHATE) [CAI]
TRIS(2-CHLOROETHYL) PHOSPHATE [C65]
TRIS(2-CHLOROETHYL)PHOSPHATE [CSR, CSR, IAR, OTS]

116-02-9 3,3,5-TRIMETHYLCYCLOHEXANOL
3,3,5-TRIMETHYL-1-CYCLOHEXANOL [FLA, MSL, NFP, NFP]
CYCLOHEXANOL, 3,3,5-TRIMETHYL- [PAL]

116-06-3 ALDICARB
PROPANAL, 2-METHYL-2-(METHYLTHIO)-,O-[(METHYLAMINO)CARBONYL]OXIME [PAL]

116-09-6 ACETONE ALCOHOL

116-14-3 TETRAFLUOROETHYLENE
ETHENE, TETRAFLUORO- [PAL, TSC, TSC]
TETRAFLUOROETHENE [CTR, TSE]
TETRAFLUOROETHYLENE [ETHENE, TETRAFLUORO-] [EP3]

116-15-4 HEXAFLUOROPROPYLENE
1-PROPENE, 1,1,2,3,3,3-HEXAFLUORO- [TSC, TSC]
HEXAFLUOROPROPENE [CTR, OTS]

116-16-5 HEXACHLOROACETONE

116-54-1 METHYL DICHLOROACETATE

116-81-4 BROMAMINE ACID

116-90-5 4,4'-DIBENZANTHRONYL

117-08-8 TETRACHLOROPHTHALIC ANHYDRIDE

117-10-2 1,8-DIHYDROXYANTHRAQUINONE
DANTRON (CHRYSAZIN; 1,8-DIHYDROXYANTHRAQUINONE) [C65, C65, IAR]

117-18-0 1,2,4,5-TETRACHLORO-3-NITROBENZENE

117-34-0 DIPHENYLACETIC ACID

117-37-3 ANISINDIONE

117-39-5 NATURAL YELLOW 10
QUERCETIN [CAL, CSR, CSR, IAR]

117-52-2 COUMAFURYL
FUMARIN [NJL]

117-62-4 2-AMINO-1,5-NAPHTHALINEDISULFONIC ACID

117-79-3 2-AMINOANTHRAQUINONE
2-AMINO-ANTHRAQUINONE [FLA]
9,10-ANTHRACENEDIONE, 2-AMINO- [PAL, PAL]

117-80-6 DICHLONE
1,4-NAPHTHALENEDIONE, 2,3-DICHLORO- [PAL]

117-81-7 DI(2-ETHYLHEXYL)PHTHALATE
1,2-BENZENEDICARBOXYLIC ACID, BIS(2-ETHYLHEXYL) ESTER [ONT, ONT, PAL, PAL, TSC, TSC, TSC]
BIS (2-ETHYLHEXYL) PHTHALATE [MX1]
BIS(2-ETHYLHEXYL) PHTHALATE [CAN, CW2, EP2, PST, RCA]
BIS(2-ETHYLHEXYL)PHTHALATE [CN4, EPA, GBR, GBR, NJL, NJL, PQL]
BIS(2-ETHYLHEXYL)PHTHALATE (DEHP) [CAA]
DI(2-ETHYLHEXYL)PHTHALATE (DEHP) [WHM]
DI-2-ETHYLHEXYL PHTHALATE (DEHP) [NHS, NHS, NHS]
DI-SEC OCTYL PHTHALATE [MSL, NIO, NIO, NIO]
DI-SEC OCTYL PHTHALATE (DI-2-ETHYLHEXYL-PHTHALATE) [BC1, BC1, OS1, OS1, OS1]
DI-SEC-OCTYL PHTHALATE [ALB, ALB, AUS, AUS, CEX, CEX, MEX, MEX, TLV, TLV]
DI-SEC-OCTYL-PHTHALATE [MIN]
DIETHYLHEXYL PHTHALATE [RCU, RCU]
OCTYL PHTHALATE, DI-SEC [QBC, QBC, QBC]

117-82-8 BIS(2-METHOXYETHYL)PHTHALATE
DIMETHOXYETHYL PHTHALATE [NFP, NFP]

117-83-9 BIS[2-N-BUTOXYETHYL] PHTHALATE
DIBUTOXY ETHYL PHTHALATE [NFP, NFP]

117-84-0 DI-N-OCTYL PHTHALATE
1,2-BENZENEDICARBOXYLIC ACID, DIOCTYL ESTER [PAL, TSC, TSC]
DIOCTYL PHTHALATE [MSL, NFP, NFP, PST, WHM]

117-97-5 ZINC BIS(PENTACHLOROPHENOL)

118-28-5 5-AMINO-6-ETHOXY-2-NAPHTHALENESULFONIC ACID

118-33-2 6-AMINO-1,3-NAPHTHALENEDISULFONIC ACID

118-52-5 1,3-DICHLORO-5,5-DIMETHYL HYDANTOIN
1,3-DICHLORO-5,5-DIMETHYLHYDANTOIN [CAL, NIO, NIO, WHM]
2,4-IMIDAXOLIDINEDIONE, 1,3-DICHLORO-5,5-DIMETHYL- [PAL]
3-DICHLORO 5,5-DIMETHYL/HYDANTOIN [QBC, QBC]

118-55-8 PHENYL SALICYLATE

118-58-1 BENZYL SALICILATE

118-60-5 2-ETHYLHEXYL SALICYLATE

118-61-6 ETHYL SALICYLATE

118-69-4 2,6-DICHLOROTOLUENE

118-74-1 HEXACHLOROBENZENE
BENZENE, HEXACHLORO- [PAL, PAL]

118-75-2 2,3,5,6-TETRACHLORO-2,5-CYCLOHEXADIENE-1,4-DIONE
CHLORANIL [OTS]

118-79-6 PHENOL, 2,4,6-TRIBROMO-
2,4,6-TRIBROMOPHENOL [OTS, TSE]

118-83-2 1-NITRO-4-CHLOROBENZO-2-TRIFLUORIDE

118-90-1 BENZOIC ACID, 2-METHYL-

118-91-2 O-CHLOROBENZOIC ACID

118-92-3 O-ANTHRANILIC ACID
ANTHRANILIC ACID [FDA, IAR]

118-96-7 2,4,6-TRINITROTOLUENE
1-METHYL-2,4,6-TRINITROBENZENE [ONT, ONT]
2,4,6-TRINITROTOLUENE (TNT) [AUS, AUS, BC1, OS1, OS1, OS1, OS1, TLV, TLV]
BENZENE, 2-METHYL-1,3,5-TRINITRO- [PAL]
TRINITROTOLUENE [ATS, FLA, MSL]
TRINITROTOLUENE 2,4,6- [QBC, QBC]

119-06-2 1,2-BENZENEDICARBOXYLIC ACID, DITRIDECYL ESTER
DITRIDECYL PHTHALATE [NFP, NFP, TSE]

119-07-3 1,2-BENZENEDICARBOXYLIC ACID, DECYL OCTYL ESTER

119-26-6 2,4-DINITROPHENYLHYDRAZINE

119-33-5 4-METHYL-2-NITRO-PHENOL
4-METHYL-2-NITROPHENOL [TSC, TSC, TSC]

119-34-6 4-HYDROXY-3-NITROANILINE
2-NITRO-4-AMINOPHENOL [MAK, MAK]
4-AMINO-2-NITROPHENOL [CSR, CSR, IAR, MSL]

119-36-8 METHYL SALICYLATE
BENZOIC ACID, 2-HYDROXY-, METHYL ESTER [PAL]

119-38-0 ISOPROPYLMETHYLPYRAZOLYL DIMETHYLCARBAMATE

119-42-6 O-CYCLOHEXYLPHENOL
PHENOL, 2-CYCLOHEXYL- [PAL]

119-47-1 2,2'-METHYLENEBIS(4-METHYL-6-TERT-BUTYLPHENOL)
2,2'-METHYLENEBIS(6-TERT-BUTYL-P-CRESOL) [PST]

119-53-9 BENZOIN

119-61-9 BENZOPHENONE

119-64-2 TETRAHYDRONAPHTHALENE
NAPHTHALENE, 1,2,3,4-TETRAHYDRO- [PAL]
TETRALIN [CSR, PST]

119-65-3 ISOQUINOLINE

119-79-9 1-NAPHTHYLAMINE-6-SULFONIC ACID

119-84-6 3,4-DIHYDROCOUMARIN

119-90-4 3,3'-DIMETHOXYBENZIDINE
3,3'-DIMETHOXYBENZIDINE (O-DIANISIDINE) [RCU, RCU]
3,3'-DIMETHOXYBENZIDINE (ORTHO-DIANISIDINE) [C65]
3,3'-DIMETHOXYBENZIDINE (ORTHO-TOLIDINE) [IAR]
O-DIANISIDINE [MAK, NFP, NFP]
[1,1'-BIPHENYL]-4,4'-DIAMINE, 3-3'-DIMETHOXY- [PAL, PAL]

119-93-7 3,3'-DIMETHYLBENZIDINE
3,3'-DIMETHYL BENZIDINE [CAA]

3,3'-DIMETHYLBENZIDINE (O-TOLIDINE) [313]
3,3'-DIMETHYLBENZIDINE (ORTHO-TOLIDINE) [IAR]
3,3'-DIMETHYLBENZIDINE (ORTHO-TOLUIDINE) [C65]
O-TOLIDINE [ALB, AUS, AUS, AUS, MAK, MIN, NHS, NHS, NHS, QBC, QBC, QBC, TLV, TLV, WHM]
[1,1'-BIPHENYL]-4,4'-DIAMINE, 3-3'-DIMETHYL- [PAL, PAL]

119-94-8 ETHYLBENZYLTOLUIDINE

120-07-0 N-PHENYL DIETHANOLAMINE
N-PHENYLDIETHANOLAMINE [NFP, NFP]

120-12-7 ANTHRACENE

120-14-9 VERATRALDEHYDE
BENZALDEHYDE, 3,4-DIMETHOXY- [TSC, TSC, TSC]

120-18-3 2-NAPHTHALENESULFONIC ACID
NAPHTHALENE SULFONIC ACID (B-) [HON]

120-20-7 ANTHRACENE

120-21-8 BENZALDEHYDE, 4-(DIETHYLAMINO)-

120-32-1 O-BENZYL-P-CHLOROPHENOL
2-BENZYL-4-CHLOROPHENOL [PST, TSC, TSC, TSC]

120-36-5 2[2,4-(DICHLOROPHENOXY)]-PROPIONIC ACID
2,4,-DP [313]
2,4-DP (2,4-DICHLORPHENOXY-PROPIONIC ACID) [CAL]
2[2,4-(DICHLOROPHENOXY)]-PROPANOIC ACID [TSE]

120-40-1 LAURIC ACID DIETHANOLAMIDE CONDENSATE
N,N-BIS(2-HYDROXYETHYL)DODE-CANAMIDE [PST]

120-47-8 ETHYL P-HYDROXYBENZOATE

120-51-4 BENZYL BENZOATE

120-55-8 DIETHYLENE GLYCOL DIBENZOATE

120-57-0 1,3-BENZODIOXOLE-5-CARBOXALDEHYDE
PIPERONAL [FDA]

120-58-1 ISOSAFROLE
1,2-METHYLENEDIOXY-4-PROPENYL BENZENE [NJL]
1,3-BENZODIOXOLE, 5-(1-PROPENYL)- [PAL, PAL]

120-61-6 DIMETHYL P-PHTHALATE
DIMETHYL TEREPHTHALATE [CSR, CSR, HON, MIN, NFP, NFP, OTS]

120-62-7 PIPERONYL SULFOXIDE

120-71-8 P-CRESIDINE
5-METHYL-O-ANISIDINE [MAK]
BENZENAMINE, 2-METHOXY-5-METHYL- [PAL, PAL]
PARA-CRESIDINE [C65, C65, IAR]

120-72-9 INDOLE
1H-INDOLE [PST]

120-78-5 2,2'-DIBENZOTHIAZYL DISULFIDE
BENZTHIAZOLE DISULFIDE [OTS]

120-80-9 CATECHOL
1,2-BENZENEDIOL [PAL, PST]
1,2-DIHYDROXYBENZENE [NJL, ONT]
CATECHOL (PYROCATECHOL) [ALB, ALB, BC1, FLA, OS1, OS1, QBC, QBC]
HYDROXYPHENOL, O- [OTS]
O-DIHYDROXYBENZENE [NFP, NFP]
O-HYDROXYPHENOL [EPA, TSC]
PYROCATECHOL [GBR]

120-82-1 1,2,4-TRICHLOROBENZENE
BENZENE, 1,2,4-TRICHLORO- [PAL, TSC, TSC, TSC]
TRICHLOROBENZENE 1,2,4- [QBC]
TRICHLOROBENZENE, 1,2,4- [OTS]

120-83-2 2,4-DICHLOROPHENOL
DICHLOROPHENOL, 2,4- [CD1, CD1]
PHENOL, 2,4-DICHLORO- [PAL]

120-92-3 CYCLOPENTANONE

120-94-5 METHYLPYRROLIDINE
N-METHYLPYRROLIDINE [PST]
PYRROLIDINE, 1-METHYL- [PAL]

121-14-2 2,4-DINITROTOLUENE
BENZENE, 1-METHYL-2,4-DINITRO- [OTS, PAL]
DINITROTOLUENE [ALB, ALB, ALB, AUS, AUS, FLA, MEX, MEX, MEX, MIN]

121-17-5 2-NITRO-1-CHLOROBENZO-4-TRIFLUORIDE
3-NITRO-4-CHLOROBENZOTRIFLUO-RIDE [HM1, HM1, HM1, WHM]

121-19-7 ROXARSONE

121-21-1 PYRETHRIN I
CYCLOPROPANECARBOXYLIC ACID, 2,2-DIMETHYL-3-(2-METHYL-1-PROPENYL)-,2-METHYL-4-OXO-3-(2,4-PENTADIENYL)-2-CYCLOPENTEN-1-YL ESTER, [1R[1.ALPHA[S*0],3.BETA.]- [PAL]
PYRETHRINS [NJL]

121-29-9 PYRETHRIN II
CYCLOPROPANECARBOXYLIC ACID, 3-(3-METHOXY-2-METHYL-3-OXO-1-PROPENYL)-2,2-DIMETHYL-,2-METHYL-4-OXO-3-(2,4-PENTA DI-ENYL)-2-CYCLOPENTEN-1-YL ES-TER, [R-[1.ALPHA.[3*(Z)], 3.BETA.(E)]- [PAL]
PYRETHRINS [NJL]

121-32-4 BENZALDEHYDE, 3-ETHOXY-4-HYDROXY-

121-33-5 VANILLIN
BENZALDEHYDE, 4-HYDROXY-3-METHOXY- [TSC, TSC, TSC]

121-39-1 OXIRANECARBOXYLIC ACID, 3-PHENYL-, ETHYL ESTER

121-43-7 TRIMETHYL BORATE
BORIC ACID (H3BO3), TRIMETHYL ESTER [PAL]
METHYL BORATE [FLA, MSL, NFP, NFP]
TRIMETHYLBORATE [NJL, NJL]

121-44-8 TRIETHYLAMINE
ETHANAMINE, N,N-DIETHYL- [PAL, TSC, TSC]

121-45-9 TRIMETHYL PHOSPHITE
PHOSPHOROUS ACID, TRIMETHYL ESTER [PAL]
PHOSPHORUS ACID, TRIMETHYL ESTER [TSC, TSC]

121-46-0 BICYCLO(2.2.1)HEPTA-2,5-DIENE
2,5-NORBORNADIENE [MSL, NFP, NFP]
2,5-NORBORNADIENE (DICYCLOHEP-TADIENE) [HM1, HM1]
NORBORNADIENE [FLA]

121-47-1 METANILIC ACID
BENZENESULFONIC ACID, 3-AMINO- [TSC, TSC]

121-54-0 BENZETHONIUM CHLORIDE
P-DI-ISOBUTYLPHENOXYETHOXYETHYL DIMETHYL BENZYL AMMONIUM CHLORIDE [PST]

121-57-3 BENZENESULFONIC ACID, 4-AMINO-
SULFANILIC ACID [HON]

121-60-8 4-(ACETYLAMINO) BENZENESUL-
FONYL CHLORIDE

121-66-4 2-AMINO-5-NITROTHIAZOLE

121-69-7 DIMETHYLANILINE
BENZENAMINE, N,N-DIMETHYL-
[PAL]
DIMETHYLANILINE (N,N) [HON,
HON]
DIMETHYLANILINE (N,N-DIMETHY-
LANILINE) [BC1, BC1, BC1, OS1,
OS1, OS1, OS1, OS1, TLV, TLV,
TLV]
DIMETHYLANILINE N- [QBC, QBC,
QBC]
DIMETHYLANILINE, N,N- [OTS]
N,N-DIETHYL ANILINE (N,N-
DIMETHYLANILINE) [CAA]
N,N-DIMETHYLANILINE [313, ALB,
ALB, ALB, AUS, AUS, AUS, CAB,
CAL, CAN, CEX, CEX, CEX, CSR,
CSR, EPA, GBR, GBR, GBR, HM1,
HM1, IAR, MAK, MAK, MAK,
MAK, NFP, NFP, ONT, ONT,
ONT, T33, WHM]

121-73-3 M-NITROCHLOROBENZENE
BENZENE, 1-CHLORO-3-NITRO-
[PAL]
M-CHLORONITROBENZENE [FLA,
HON, MSL]

121-75-5 MALATHION
MALATHION (ISO) [GBR, GBR]

121-79-9 PROPYL GALLATE
GALLIC ACID, PROPYL ESTER [PST]

121-82-4 CYCLONITE
1,3,5-TRIAZINE, HEXAHYDRO-1,3,5-
TRINITRO- [PAL]
CYCLONITE [QBC, QBC]
CYCLOTRIMETHYLENE-
TRINITRAMINE [CAL]
CYCLOTRIMETHYLEN-
ETRINITRAMINE (RDX) [ATS]
HEXAHYDRO-1,3,5-TRINITRO-1,3,5-
TRIAZINE [GBR, GBR, GBR, ONT,
ONT]
RDX [BC1, BC1, BC1]

121-87-9 BENZENAMINE, 2-CHLORO-4-
NITRO-

121-88-0 2-AMINO-5-NITROPHENOL

121-89-1 3'-NITROACETOPHENONE

121-91-5 ISOPHTHALIC ACID

121-92-6 M-NITROBENZOIC ACID

122-00-9 P-METHYLACETOPHENONE
P-METHYL ACETOPHENONE [NFP,
NFP]

122-07-6 ETHANAMINE, 2,2-DIMETHOXY-
N-METHYL-

122-09-8 ALPHA,ALPHA-
DIMETHYLPHENETHYLAMINE
ALPHA,ALPHA-DIMETHYL
PHENETHYLAMINE [NJL]
BENZENEETHANAMINE, ALPHA,
ALPHA-DIMETHYL- [TSC, TSC]

122-10-1 SWAT

122-14-5 FENITROTHION

122-19-0 DIMETHYLBENZYLOCTADECY-
LAMMONIUM CHLORIDE
DIMETHYL OCTADECYL BENZYL
AMMONIUM CHLORIDE [PST]

122-20-3 TRIISOPROPANOLAMINE
2-PROPANOL, 1,1',1"-NITROTRIS-
[PAL]

122-31-6 TETRAETHOXYPROPANE

122-34-9 SIMAZINE

122-37-2 4-ANILINOPHENOL

122-39-4 DIPHENYLAMINE
BENZENAMINE, N-PHENYL- [PAL]
N,N-DIPHENYLAMINE [ONT, WHM]

122-40-7 HEPTANAL, 2-(PHENYLMETHY-
LENE)-

122-42-9 PROPHAM

122-51-0 ETHYL ORTHOFORMATE
ETHANE, 1,1',1"-[METHYLIDYNETRIS
(OXY)]TRIS- [PAL]
ETHYL FORMATE (ORTHO) [MSL,
NFP, NFP]

122-52-1 TRIETHYL PHOSPHITE

122-57-6 BENZYLIDENE ACETONE

122-59-8 PHENOXYACETIC ACID

122-60-1 PHENYL GLYCIDYL ETHER
(PHENOXYMETHYL) OXIRANE [ONT]
OXIRANE, (PHENOXYMETHYL)-
[PAL, TSC, TSC]
PHENYL GLYCIDYL ETHER (PGE)
[AUS, AUS, BC1, BC1, FLA, MSL,
OS1, OS1, QBC, QBC, TLV, TLV,
TLV, WHM]
PHENYL-2,3-EPOXYPROPYL ETHER
[GBR]

122-62-3 DECANOIC ACID, BIS(2-ETHYL-
HEXYL) ESTER

122-66-7 HYDRAZOBENZENE
1,2-DIPHENYL HYDRAZINE [NJL,
NJL]
1,2-DIPHENYLHYDRAZINE [ATS,
CAA, CW2, EP2, EPA, HON, MX1,
MX2, RCU, SD4, UTS]
1,2-DIPHENYLHYDRAZINE (HYDRA-
ZOBENZENE) [313]
DIPHENYLHYDRAZINE [CW3]
HYDRAZINE, 1,2-DIPHENYL- [PAL,
PAL, TSC, TSC]
HYDRAZOBENZENE (1,2-DIPHENYL-
HYDRAZINE) [C65, C65]

122-78-1 BENZENEACETALDEHYDE
PHENYLACETALDEHYDE [NFP,
NFP]

122-79-2 PHENYL ACETATE

122-80-5 P-ACETOAMINOANILINE

122-82-7 BUTANAMIDE, N-(4-
ETHOXYPHENYL)-3-OXO-
ACETOACET-P-PHENETIDE [FLA,
MSL]
ACETOACET-PARA-PHENETIDE
[NFP, NFP]

122-88-3 P-CHLOROPHENOXYACETIC
ACID

122-97-4 3-PHENYL-1-PROPANOL
PHENYLPROPYL ALCOHOL [NFP,
NFP]

122-98-5 N-PHENYL ETHANOLAMINE
2-ANILINOETHANOL [NFP, NFP]
ETHANOL, 2-(PHENYLAMINO)- [PAL]
N-PHENYLETHANOLAMINE [FLA,
MSL]

122-99-6 ETHYLENE GLYCOL
MONOPHENYL ETHER
2-PHENOXYETHANOL [TSC, TSC,
TSC]
ETHYLENE GLYCOL PHENYL ETHER
[NFP, NFP]

123-00-2 N-AMINOPROPYLMORPHOLINE
4-AMINOPROPYL MORPHOLINE
[FLA]
4-MORPHOLINEPROPANAMINE
[PAL]
AMINOPROPYLMORPHOLINE [NJL,
NJL]
N-(3-AMINOPROPYL) MORPHOLINE
[MSL, NFP, NFP]

123-01-3 DODECYL BENZENE (CRUDE)
DODECYL BENZENE (N-) [HON]
DODECYLBENZENE [OTS]

123-04-6 2-ETHYLHEXYL CHLORIDE
HEPTANE, 3-(CHLOROMETHYL)-
[PAL]

123-05-7 2-ETHYLHEXALDEHYDE
2-ETHYLHEXANAL [FLA, MSL,
NFP, NFP]
ETHYL HEXALDEHYDE [NJL]
HEXANAL, 2-ETHYL- [PAL, TSC,
TSC, TSC]

123-07-9 PHENOL, 4-ETHYL-
P-ETHYLPHENOL [FLA, MSL, NFP,
NFP]

123-08-0 BENZALDEHYDE, 4-HYDROXY-

123-11-5 P-ANISALDEHYDE
BENZALDEHYDE, 4-METHOXY-
[TSC, TSC, TSC]

123-15-9 2-METHYL VALERALDEHYDE
2-METHYLVALERALDEHYDE [FLA,
MSL, NFP, NFP]
ALPHA-METHYLVALERALDEHYDE
[HM1, HM1]
METHYL VALERALDEHYDE [NJL,
NJL]
PENTANAL, 2-METHYL- [PAL]

123-17-1 4-NONANOL, 2,6,8-TRIMETHYL-
2,6,8-TRIMETHYL-4-NONANOL [FLA,
MSL, NFP, NFP, PST]

123-18-2 ISOBUTYL HEPTYL KETONE
2,6,8-TRIMETHYL-4-NONANONE
[NFP, NFP]
4-NONANONE, 2,6,8-TRIMETHYL-
[PAL]

123-19-3 DIPROPYL KETONE
4-HEPTANONE [NFP, NFP, ONT,
PAL]

123-20-6 VINYL BUTYRATE
BUTANOIC ACID, ETHENYL ESTER
[PAL]

123-23-9 SUCCINIC ACID PEROXIDE
DISUCCINIC ACID PEROXIDE [DOT]

123-25-1 DIETHYL SUCCINATE

123-26-2 N,N'-ETHYLENEBIS(12-HYDROXY-
OCTADECANAMIDE)

123-28-4 DILAURYL BETA-THIODIPROPI-
ONATE
DILAURYL THIODIPROPIONATE
[PST]

123-29-5 ETHYL NONANOATE

123-30-8 P-AMINOPHENOL
AMINOPHENOL, P- [OTS]

123-31-9 HYDROQUINONE
1,4-BENZENEDIOL (HYDROQUINONE)
[TSC, TSC, TSC]
1,4-DIHYDROXYBENZENE [ONT,
ONT, ONT]
DIHYDROXYBENZENE (HYDRO-
QUINONE) [ALB, ALB]
P-DIHYDROXYBENZENE [NFP, NFP]

123-32-0 2,5-DIMETHYLPYRAZINE

123-33-1 1,2-DIHYDRO-3,6-PYRIDAZINE-
DIONE
MALEIC HYDRAZIDE [EPA, HON,
IAR, RCU, RCU, TSC, TSC, TSE]

123-35-3 MYRCENE

123-38-6 PROPIONALDEHYDE
PROPANAL [NFP, NFP, OTS, PAL,
TSC, TSC, TSC]

123-39-7 N-METHYLFORMAMIDE

123-42-2 DIACETONE ALCOHOL
2-PENTANONE, 4-HYDROXY-4-
METHYL- [PAL]
4-HYDROXY-4-METHYL-2-PEN-
TANONE [CAL, ONT, ONT]
4-HYDROXY-4-METHYL-PENTAN-2-
ONE [GBR, CBR]
DIACETONE ALCOHOL (4-HYDROXY-
4-METHYL-2-PENTANONE) [ALB,
ALB]
DIACETONE ALCOHOL (4-HYDROXY-
4-METHYL-2-PENTANONE) [BC1,
BC1, OS1, OS1]

123-43-3 SULFOACETIC ACID

123-51-3 ISOAMYL ALCOHOL
1-BUTANOL, 3-METHYL- [PAL]
3-METHYLBUTAN-1-OL [GBR, GBR]
ISOAMYL ALCOHOL (PRIMARY AND SECONDARY) [OS1, OS1, OS1]
ISOAMYL ALCOHOL (PRIMARY) [NIO, NIO, NIO]

123-54-6 2,4-PENTANEDIONE
ACETYL ACETONE [FLA]
ACETYLACETONE [NJL]
PENTANE-2,4-DIONE [HM1, HM1]

123-56-8 SUCCINIMIDE

123-61-5 BENZENE, 1,3-DIISOCYANATO-
1,3-PHENYLENE DIISOCYANATE [313]

123-62-6 PROPIONIC ANHYDRIDE
PROPANOIC ACID, ANHYDRIDE [PAL]
PROPANOIC ANHYDRIDE [TSC, TSC, TSC]

123-63-7 PARALDEHYDE
1,3,5-TRIOXANE, 2,4,6-TRIMETHYL- [PAL]

123-66-0 ETHYL HEXANOATE
ETHYL BUTYL ACETATE [NJL]
ETHYL CAPROATE [FLA, MSL, NFP, NFP]
HEXANOIC ACID, ETHYL ESTER [PAL]

123-68-2 ALLYL CAPROATE

123-72-8 BUTYRALDEHYDE
BUTANAL [PAL, TSC, TSC, TSC]
N-BUTYRALDEHYDE [FLA, MSL]

123-73-9 CROTONALDEHYDE
(E)-CROTONALDEHYDE [WHM]
3-METHYLACROLEINE [HM1, HM1]
CROTONALDEHYDE (E) [MSL]
CROTONALDEHYDE, (E)- [302, OS1, OS1]
CROTONALDEHYDE, (E)- [2-BUTENAL, (E)-] [EP3]

123-75-1 PYRROLIDINE

123-76-2 LEVULINIC ACID

123-77-3 AZODICARBONAMIDE
AZODICARBOXAMIDE [PST]
DIAZENEDICARBOXAMIDE [OTS]

123-79-5 ADIPIC ACID, DIOCTYL ESTER

123-81-9 GLYCOL DIMERCAPTOACETATE
ACETIC ACID, MERCAPTO-, 1,2-ETHANEDIYL ESTER [PAL]

123-86-4 N-BUTYL ACETATE
ACETIC ACID, BUTYL ESTER [PAL, PST]
BUTYL ACETATE [EPA, GBR, GBR, MSL, NFP, NFP]
BUTYL ACETATE, N- [OTS]

123-91-1 1,4-DIOXANE
1,4-DIOXANE (1,4-DIETHYLENEOXIDE) [CAA, HON, HON]
1,4-DIOXANE, TECHNICAL GRADE [GBR, GBR, GBR]
DIOXANE [AUS, AUS, HM1, HM1, HM1, MIN, NHS, NHS, NHS, NIO, NIO, NIO, NIO, OTV, TLV, TLV]
DIOXANE (DIETHYLENE DIOXIDE) [OS1, OS1, OS1, OS1]
DIOXANE (TECH. GRADE) [QBC, QBC, QBC]
DIOXANE, TECH. GRADE [ALB, ALB, ALB, BC1, BC1, MEX, MEX, MEX]
P-DIOXANE [CAL, NFP, NFP]
P-DIOXANE, TECH. GRADE [CEX, CEX]

123-92-2 ISOAMYL ACETATE
1-BUTANOL, 3-METHYL-, ACETATE [PAL]
ISO-AMYL ACETATE [CWA, EPA]
ISOPENTYL ACETATE [GBR, GBR]

123-94-4 GLYCERYL STEARATE

123-95-5 BUTYL STEARATE

124-02-7 DIALLYLAMINE

124-03-8 ETHYLHEXADECYLDIMETHYL-AMMONIUM BROMIDE

124-04-9 ADIPIC ACID
HEXANEDIOIC ACID [PAL]

124-07-2 OCTANOIC ACID

124-09-4 HEXAMETHYLENEDIAMINE
1,6-HEXANEDIAMINE [OTS, TLV]
HEXAMETHYLENE DIAMINE [NJL, NJL]

124-10-7 METHYL TETRADECANOATE

124-13-0 CAPRYLALDEHYDE
OCTANAL [TSC, TSC, TSC]

124-16-3 2-PROPANOL, 1-(2-BUTOXYETHOXY)-
1-(BUTOXYETHOXY)-2-PROPANOL [FLA, MSL, NFP, NFP]
BUTOXYETHOXYPROPANOL [PST]

124-17-4 2-(2-BUTOXYETHOXY)ETHYL ACETATE
DIETHYLENE GLYCOL BUTYL ETHER ACETATE [CTR, TSE]
DIETHYLENE GLYCOL MONOBUTYL ETHER ACETATE [HON, NFP, NFP]

124-18-5 DECANE
N-DECANE [HM1, HM1, OTS]

124-19-6 NONANAL

124-22-1 DODECYLAMINE

124-30-1 N-OCTADECYLAMINE

124-38-9 CARBON DIOXIDE
CARBONIC ANHYDRIDE (CARBON DIOXIDE) [QBC, QBC]

124-40-3 DIMETHYLAMINE
DIMETHYLAMINE [METHANAMINE, N-METHYL-] [EP3]
METHANAMINE, N-METHYL- [PAL]

124-41-4 SODIUM METHYLATE
METHANOL, SODIUM SALT [PAL]
SODIUM METHOXIDE [HON]

124-43-6 UREA PEROXIDE
UREA HYDROGEN PEROXIDE [NJL]

124-47-0 UREA NITRATE

124-48-1 CHLORODIBROMOMETHANE
DIBROMOCHLOROMETHANE [PQL, SD4]
METHANE, DIBROMOCHLORO- [PAL, TSC, TSC]

124-63-0 METHANE SULFONYL CHLORIDE
METHANESULFONYL CHLORIDE [HM1]

124-64-1 TETRAKIS(HYDROXYMETHYL) PHOSPHONIUM CHLORIDE

124-65-2 SODIUM CACODYLATE

124-68-5 2-AMINO-2-METHYL-1-PROPANOL
1-PROPANOL, 2-AMINO-2-METHYL- [PAL]

124-73-2 DIBROMOTETRAFLUOROETHANE (HALON 2402)
DIBROMOTETRAFLUOROETHANE [CN2, CN5]
HALON 2402 [CAA]

124-87-8 PICROTOXIN
COCCULUS [HM1, HM1, HM1, NJL]

125-12-2 ISOBORNYL ACETATE

125-33-7 PRIMACLONE

125-84-8 AMINOGLUTETHIMIDE

126-06-7 3-BROMO-1-CHLORO-5,5-DIMETHYL-2,4-IMIDAZOLIDINE-DIONE

126-07-8 GRISEOFULVIN

126-11-4 1,3-PROPANEDIOL, 2-HYDROXYMETHYL-2-NITRO-

126-14-7 SUCROSE OCTAACETATE

126-15-8 4A(4H)-DIBENZOFURANCARBOXALDEHYDE, 1,5A,6,9,9A,9B,-HEXAHYDRO-

126-17-0 SOLASODINE

126-30-7 1,3-PROPANEDIOL, 2,2-DIMETHYL-
NEOPENTYL GLYCOL [NFP, NFP, PST]
PROPANEDIOL, 2,2-DIMETHYL-, 1,3- [OTS]

126-33-0 SULFOLANE
THIOPHENE, TETRAHYDRO-, 1,1-DIOXIDE [PAL]

126-39-6 2-METHYL-2-ETHYL-1,3-DIOXOLANE
1,3-DIOXOLANE, 2-ETHYL-2-METHYL- [PAL]

126-58-9 1,3-PROPANEDIOL, 2,2'-[OXYBIS (METHYLENE)]

126-68-1 O,O,O-TRIETHYL PHOSPHOROTHIOATE

126-71-6 ISOBUTYL PHOSPHATE
PHOSPHORIC ACID, TRIS(2-METHYLPROPYL) ESTER [TSC, TSC, TSC]

126-72-7 TRIS(2,3-DIBROMOPROPYL)PHOSPHATE
1-PROPANOL, 2,3-DIBROMO-, PHOSPHATE (3:1) [PAL, PAL]
TRIS(DIBROMOPROPYL) PHOSPHATE [WHM]

126-73-8 TRIBUTYL PHOSPHATE
BUTYL PHOSPHATE [NFP, NFP]
PHOSPHORIC ACID TRIBUTYL ESTER [PAL]
PHOSPHORIC ACID, TRIBUTYL ESTER [TSC, TSC, TSC]
TRIBUTYL PHOSPHATE (TBP) [WHM]

126-75-0 DEMETON-S

126-80-7 DISILOXANE, 1,1,3,3-TETRAMETHYL-1,3-BIS[3-OXIRANYLMETHOXY)PROPYL]-
1,3-BIS(3-(2,3-EPOXYPROPOXY)-PROPYL)TETRAMETHYLDISILOXANE [OTS]
1,3-BIS[3-(2,3-EPOXYPROPOXY) PROPYL]TETRAMETHYLDISILOXANE [EPA]

126-85-2 NITROGEN MUSTARD N-OXIDE

126-86-3 5-DECYNE-4,7-DIOL, 2,4,7,9-TETRAMETHYL-
2,4,7,9-TETRAMETHYL-5-DECYNE-4,7-DIOL [PST]

126-92-1 2-ETHYLHEXYL SODIUM SULFATE
SODIUM 2-ETHYLHEXYL SULFATE [PST]

126-98-7 METHYLACRYLONITRILE
2-METHYL-2-PROPENENITRILE [ONT, ONT]
2-PROPENENITRILE, 2-METHYL- [PAL]
ALPHA-METHYLACRYLONITRILE [CAL, CEX, CEX]
METHACRYLONITRILE [302, CSR, GBR, GBR, HM1, HM1, HM1, HM1, MSL, NFP, NFP, PQL, RCA, RCU, RCU]
METHYL ACRYLONITRILE [OS3, WHM]

133-90-4 CHLORAMBEN
CHLORAMBEN [BENZOIC ACID, 3-AMINO-2,5-DICHLORO-] [313]
CLORAMBEN [NJL]

134-03-2 SODIUM ASCORBATE

134-20-3 METHYL ANTHRANILATE

134-29-2 O-ANISIDINE HYDROCHLORIDE
BENZENAMINE, 2-METHOXY-, HYDROCHLORIDE [PAL, PAL]
ORTHO-ANISIDINE HYDROCHLORIDE [C65, C65]

134-31-6 8-HYDROXYQUINOLINE SULFATE

134-32-7 ALPHA-NAPHTHYLAMINE
1-NAPHTHALENAMINE [PAL, RCU, RCU]
1-NAPHTHYLAMINE [C65, CAB, EP2, IAR, MSL, NFP, NFP, NJL, NJL, PQL]
NAPHTHYLAMINE (1-) [HON]

134-50-9 9-AMINOACRIDINE HYDROCHLORIDE

134-62-9 N,N-DIETHYL-M-TOLUAMIDE

134-72-5 EPHEDRINE SULFATE

134-81-6 BENZIL

135-01-3 O-DIETHYL BENZENE
BENZENE, 1,2-DIETHYL- [PAL]

135-02-4 O-ANISALDEHYDE
BENZALDEHYDE, 2-METHOXY- [PAL, TSC, TSC, TSC]
O-METHOXYBENZALDEHYDE [FLA, MSL, NFP, NFP]

135-19-3 2-NAPHTHOL
BETA-NAPHTHOL [NFP, NFP]
NAPHTHOL (B-) [HON]

135-20-6 CUPFERRON
BENZENAMINE, N-HYDROXY-N-NITROSO-, AMMONIUM SALT [PAL, PAL]
CUPFERRON [BENZENEAMINE, N-HYDROXY-N-NITROSO, AMMONIUM SALT] [313]

135-23-9 METHAPYRIDINE HYDROCHLORIDE
METHAPYRILENE HYDROCHLORIDE [CSR]

135-88-6 N-PHENYL-BETA-NAPHTHYLAMINE
2-NAPHTHALENAMINE, N-PHENYL- [PAL]
N-PHENYL-2-NAPHTHYLAMINE [CAL, CSR, CSR, IAR, MAK]
PHENYL-BETA-NAPHTHYLAMINE [NHS, NHS, NHS]

135-98-8 SEC-BUTYLBENZENE
2-PHENYLBUTANE [HM1]
BENZENE, (1-METHYLPROPYL)- [PAL, TSC, TSC]

136-23-2 ZINC DIBUTYL DITHIOCARBAMATE
DIBUTYLDITHIOCARBAMIC ACID, ZINC SALT [PST]

136-25-4 2-(2,4,5-TRICHLOROPHENOXY) ETHYL 2,2-DICHLOPROPIONATE (ERBON)

136-32-3 SODIUM-2,4,5-TRICHLOROPHENATE

136-40-3 PHENAZOPYRIDINE HYDROCHLORIDE
2,6-PYRIDINEDIAMINE, 3-(PHENYLAZO)-, MONOHYDROCHLORIDE [PAL, PAL]

136-44-7 GLYCERYL P-AMINOBENZOATE

136-45-8 DIPROPYL ISOCINCHOMERONATE

136-51-6 CALCIUM 2-ETHYLHEXANOATE

136-53-8 ZINC 2-ETHYLHEXOATE

136-60-7 BUTYL BENZOATE
BENZOIC ACID, BUTYL ESTER [PAL]

136-77-6 4-HEXYLRESORCINOL

136-78-7 SESONE
2-(2,4-DICHLOROPHENOXY) ETHANOL HYDROGEN SULFATE SODIUM SALT [ONT]
CRAG HERBICIDE [BC1, BC1, FLA, MSL, NIO, NIO, NIO]
CRAG HERBICIDE (SESONE) [OS1, OS1]
CRAG HERBICIDE (SODIUM 2,4-DICHLOROPHENOXYETHYL SULPHATE) [ALB, ALB]
ETHANOL, 2-(2,4-DICHLOROPHENOXY)-, HYDROGEN SULFATE, SODIUMSALT [PAL]
SESONE (CRAG HERBICIDE) [MEX]
SESONE (CRAG) [QBC]
SODIUM 2-(2,4-DICHLOROPHENOXY) ETHYL SULFATE [CAL]
SODIUM 2-(2,4-DICHLOROPHENOXY) ETHYL SULPHATE [GBR, GBR]

136-81-2 O-AMYL PHENOL
PHENOL, 2-PENTYL- [PAL]

136-85-6 TOLUTRIAZOLE

137-05-3 METHYL 2-CYANOACRYLATE
2-PROPENOIC ACID, 2-CYANO-, METHYL ESTER [PAL, TSC, TSC]
METHYL CYANOACRYLATE [WHM]
METHYL-2-CYANOACRYLATE [MIN]

137-08-6 N-(2,4-DIHYDROXY-3,3-DIMETHYL-1-OXOBUTYL)-.BETA.-ALANINE, CALCIUM SALT (2:1), (R)-

137-09-7 2,4-DIAMINOPHENOL DIHYDROCHLORIDE
DIAMINOPHENOL HYDROCHLORIDE [HON]
PHENOL, 2,4-DIAMINO-, DIHYDROCHLORIDE [TSC, TSC]

137-17-7 2,4,5-TRIMETHYLANILINE

137-20-2 SODIUM N-METHYL-N-OLEOYLTAURINE
N-METHYL-N-OLEOYLTAURINE, SODIUM SALT [PST, TSC, TSC, TSC]

137-26-8 THIRAM
BIS(DIMETHYLTHIOCARBAMOYL) DISULFIDE [CAL]
BIS(DIMETHYLTHIOCARBONYL) DISULFIDE [ONT]
THIOPEROXYDICARBONIC DIAMIDE ([(H2N)C(S)]2S2), TETRAMETHYL- [PAL]
THIOPEROXYDICARBONIC DIAMIDE, TETRAMETHYL- [TSC, TSC]
THIRAM (ISO) [GBR, GBR]
THIURAM [RCU, RCU]

137-29-1 COPPER(II) DIMETHYLDITHIOCARBAMATE

137-30-4 ZIRAM
MILBAN [MSL]

137-32-6 2-METHYL-1-BUTANOL
1-BUTANOL, 2-METHYL- [PAL]

137-40-6 BIOBAN-S
SODIUM PROPIONATE [PST]

137-41-7 POTASSIUM N-METHYLDITHIOCARBAMATE

137-42-8 METAM-SODIUM

137-66-6 ASCORBYL PALMITATE

137-89-3 BIS(2-ETHYLHEXYL) ISOPHTHALATE

138-00-1 PHENOL, 2,4-DIPENTYL-
2,4-DIAMYLPHENOL [FLA, MSL, NFP, NFP]

138-22-7 N-BUTYL LACTATE
BUTYL LACTATE [GBR, NFP, NFP]
PROPANOIC ACID, 2-HYDROXY-, BUTYL ESTER [PAL]

138-31-8 P-ACETAMIDOBENZENESTIBONIC ACID, SODIUM SALT

138-59-0 SHIKIMIC ACID

138-86-3 DIPENTENE
ALPHA-LIMONENE [PST]

138-89-6 P-NITROSODIMETHYLANILINE

138-93-2 DISODIUM CYANODITHIOIMIDOCARBONATE

139-02-6 SODIUM PHENOLATE

139-05-9 SODIUM CYCLAMATE

139-08-2 BENZYLDIMETHYLTETRADECYLAMMONIUM CHLORIDE
TRIMETHYL TETRADECYLPHENYL AMMONIUM CHLORIDE [PST]

139-13-9 NITRILOTRIACETIC ACID (NTA)
AMINOTRIETHANOIC ACID [PST]
GLYCINE, N,N-BIS(CARBOXYMETHYL)- [PAL, PAL]
NITRILOTRIACETIC ACID [313, C65, C65, CAL, CAN, IAR, MIN, NJL, NJL, NTP]

139-25-3 BENZENE, 1,1'-METHYLENEBIS[4-ISOCYANATO-3-METHYL-
3,3'-DIMETHYLDIPHENYLMETHANE-4,4'-DIISOCYANATE [313]

139-33-3 DISODIUM EDTA
DISODIUM ETHYLENEDIAMINETETRAACETATE [PST]

139-44-6 GLYCERYL TRIS(12-HYDROXYSTEARATE)

139-45-7 GLYCERYL TRIPROPIONATE

139-65-1 4,4'-THIODIANILINE
BENZENAMINE, 4-4'-THIOBIS- [PAL, PAL]

139-66-2 PHENYL SULFIDE

139-87-7 N-ETHYLDIETHANOLAMINE
2,2'-(ETHYLIMINO)DIETHANOL [PST]
ETHANOL, 2,2'-(ETHYLIMINO)BIS- [PAL]

139-88-8 TERGITOL NO. 4
7-ETHYL-2-METHYL-4-UNDECANOL-SULFATE, SODIUM SALT [WHM]

139-89-9 N-HYDROXYETHYLETHYLENEDIAMINETRIACETIC ACID TRISODIUM SALT
TRISODIUM N-(2-HYDROXYETHYL) ETHYLENEDIAMINETRIACETATE [PST]

139-91-3 5-(MORPHOLINOMETHYL)-3-[(5-NITRO-FURFURYLIDENE)-AMINO]-2-OXAZOLIDINONE
2-OXAZOLIDINONE, 5-(4-MORPHOLINYLMETHYL)-3-[[(5-NITRO-2-FURANYL)METHYLENE]AMINO]- [PAL, PAL]
5-(MORPHOLINOMETHYL)-3-((5-NITROFURFURYLIDENE)AMINO)-2-OXAZOLIDINONE [MSL]
5-(MORPHOLINOMETHYL)-3-(5-(NITROFURFURYLIDENE)AMINO)-2-OXAZOLIDINONE [FLA]

139-94-6 NITHIAZIDE

139-96-8 TRIETHANOLAMINE LAURYL SULFATE

140-04-5 BUTYL ACETYL RICINOLEATE
9-OCTADECENOIC ACID, 12-(ACETYLOXY)-, BUTYL ESTER. [R-(Z)]- [PAL]

140-08-9 ETHANOL, 2-CHLORO-, PHOSPHITE (3:1)
TRIS(2-CHLOROETHYL)PHOSPHITE [T33]

140-10-3 TRANS-CINNAMIC ACID

140-11-4 BENZYL ACETATE

140-29-4 BENZYL CYANIDE
BENZENEACETONITRILE [PAL]
PHENYLACETONITRILE [HM1,
HM1, NJL, WHM]

140-31-8 1-(2-AMINOETHYL) PIPERAZINE
1-PIPERAZINEETHANAMINE [PAL]
N-(2-AMINOETHYL)PIPERAZINE
[WHM]
N-AMINOETHYLPIPERAZINE [HM1,
HM1, NJL, NJL]

140-39-6 P-TOLYL ACETATE
P-CRESYL ACETATE [NFP, NFP]

**140-41-0 ACETIC ACID TRICHLORO-
, COMPD. WITH 3-(P-
CHLOROPHENYL)-1,**
MONURON-TCA [CAL]

**140-49-8 4-(CHLOROACETYL)AC-
ETANILIDE**

140-56-7 FENAMINOSULF (LESAN)
FORMULATED FENAMINOSULF
[CSR, CSR]

140-57-8 ARAMITE
2-(P-TERT-BUTYLPHENOXY)ISO-
PROPYL-2-CHLOROETHYL SULFITE
[CAL]
SULFUROUS ACID, 2-CHLOROETHYL
2-[4-(1,1-DIMETHYLETHYL)PHE-
NOXY]-1-METHYLETHYL ESTER
[PAL, PAL]

**140-66-9 PHENOL, 4-(1,1,3,3-TETRAM-
ETHYLBUTYL)-**

140-76-1 PYRIDINE, 2-METHYL-5-VINYL-
2-METHYL-5-VINYLPYRIDINE
[WHM]

140-79-4 1,4-DINITROSOPIPERAZINE

**140-80-7 2-AMINO-5-DIETHYLAMINOPEN-
TANE**
2-AMINO-5-DIETHYL AMINOPEN-
TANE [NJL]

140-88-5 ETHYL ACRYLATE
2-PROPENOIC ACID, ETHYL ESTER
[PAL, PAL, TSC, TSC, TSC]
ACRYLIC ACID, ETHYL ESTER
[MAK, MAK, MAK, MAK]

**140-89-6 CARBONODITHIOIC ACID, O-
ETHYL ESTER, POTASSIUM SALT**
POTASSIUM XANTHATE [FLA, MSL,
NFP, NFP]
XANTHATES [HON]

140-95-4 DIMETHYLOLUREA
N,N'-BIS(HYDROXYMETHYL) UREA
[PST]

141-00-4 CADMIUM SUCCINATE

141-02-6 DI(2-ETHYLHEXYL)FUMARATE

141-03-7 DI-N-BUTYL SUCCINATE

141-04-8 DIISOBUTYL ADIPATE

141-05-9 DIETHYL MALEATE

141-20-8 DIGLYCOL LAURATE

141-25-3 RHODINOL

**141-27-5 2,6-OCTADIENAL, 3,7-DIMETHYL-,
(E)-**

141-28-6 DIETHYL ADIPATE

141-31-1 HYDROXYADIPALDEHYDE

141-32-2 BUTYL ACRYLATE
2-PROPENOIC ACID, BUTYL ESTER
[PAL, TSC, TSC]
ACRYLIC ACID, N-BUTYL ESTER
[MAK, MAK, MAK, MAK]
N-BUTYL ACRYLATE [IAR, ONT,
TLV]
N-BUTYLACRYLATE [HON]

**141-37-7 3,4-EPOXY-6-METHYLCYCLO-
HEXYLMETHYL-3,4-EPOXY-6-
METHYLCYCLOHEXANE CAR-
BOXYLATE**

**141-38-8 OXIRANEOCTANOIC ACID, 3-
OCTYL-, 2-ETHYLHEXYL ESTER**

141-43-5 ETHANOLAMINE
2-AMINOETHANOL [GBR, GBR,
MAK, MAK, ONT, ONT, PST]
2-AMINOETHANOL
(ETHANOLAMINE) [ALB, ALB, QBC,
QBC]
ETHANOL, 2-AMINO- [PAL]
MONOETHANOLAMINE [WHM]

141-53-7 SODIUM FORMATE

141-57-1 PROPYL TRICHLOROSILANE
PROPYLTRICHLOROSILANE [FLA,
MSL, NFP, NFP]
SILANE, TRICHLOROPROPYL- [PAL]

141-59-3 TERT-OCTYL MERCAPTAN
2-PENTANETHIOL, 2,4,4-TRIMETHYL-
[PAL]
TERT-OCTYLMERCAPTAN [HM1,
HM1, HM1]

141-62-8 DECAMETHYLTETRASILOXANE

**141-63-9 DODECAMETHYLPENTASILOX-
ANE**

**141-65-1 PHOSPHORIC ACID, BIS(2-ETHYL-
HEXYL)ESTER, SODIUM SALT**

141-66-2 DICROTOPHOS
3-HYDROXY-N,N-DIMETHYL-CIS-
CROTONAMIDE DIMETHYL PHOS-
PHATE (DICROTOPHOS) [CAL]
DICROTOPHOS (BIDRIN) [ALB, ALB,
ALB, BC1, BC1]
DICROTOPHOS (DIDRIN) [QBC,
QBC]
PHOSPHORIC ACID, 3-(DIMETHY-
LAMINO)-1-METHYL-3-OXO-1-
PROPENYLDIMETHYL ESTER, (E)-
[PAL]

141-75-3 BUTYRYL CHLORIDE

141-78-6 ETHYL ACETATE
ACETIC ACID ETHYL ESTER [PAL]
ETHYLACETATE [T33]

141-79-7 MESITYL OXIDE
3-PENTEN-2-ONE, 4-METHYL- [PAL]
4-METHYLPENT-3-EN-2-ONE [GBR,
GBR]
MESITYL OXIDE (ETHER) [QBC]
MESITYL OXIDE (MO) [CTR]

141-82-2 MALONIC ACID

141-84-4 RHODANINE

**141-85-5 BENZENAMINE, 3-CHLORO-, HY-
DROCHLORIDE**

141-90-2 THIOURACIL

141-91-3 2,6-DIMETHYLMORPHOLINE
MORPHOLINE, 2,6-DIMETHYL- [PAL]

141-93-5 M-DIETHYL BENZENE
BENZENE, 1,3-DIETHYL- [PAL]

141-97-9 ETHYL ACETOACETATE
BUTANOIC ACID, 3-OXO-, ETHYL
ESTER [PAL]

142-03-0 BASIC ALUMINUM ACETATE

142-04-1 ANILINE HYDROCHLORIDE
BENZENAMINE, HYDROCHLORIDE
[PAL, TSC, TSC]

142-09-6 HEXYLMETHACRYLATE
2-PROPENOIC ACID, 2-METHYL-,
HEXYL ESTER [TSC, TSC]
HEXYL METHACRYLATE [NFP,
NFP]

**142-15-4 ISETHIONIC ACID, OLEATE,
SODIUM SALT**

**142-16-5 BIS(2-ETHYLHEXYL)-2-BUTENE-
DIOATE**
BIS(2-ETHYLHEXYL) MALEATE
[NFP, NFP]

**142-18-7 DODECANOIC ACID, 2,3-DIHY-
DROXYPROPYL ESTER**

**142-22-3 DIETHYLENE GLYCOL BIS (AL-
LYLCARBONATE)**

142-26-7 N-ACETYL ETHANOLAMINE

142-28-9 1,3-DICHLOROPROPANE
PROPANE, 1,3-DICHLORO- [PAL,
TSC, TSC, TSC]
TRIMETHYLENE DICHLORIDE [HM1,
HM1]

142-29-0 CYCLOPENTENE

142-31-4 SODIUM OCTYL SULFATE
OCTYL SULFATE, SODIUM SALT
[PST]

142-46-1 2,5-DITHIOBIUREA

142-47-2 GLUTAMIC ACID, SODIUM SALT

**142-58-5 N-(2-HYDROXYETHYL)TETRADE-
CANAMIDE**

**142-59-6 ETHYLENE BIS DITHIOCARBA-
MATE**
NABAM [313, HM1]

142-62-1 CAPROIC ACID
HEXANOIC ACID [PAL, WHM]

142-64-3 PIPERAZINE DIHYDROCHLORIDE
PIPERAZINE DIHYDRO-CHLORIDE
[OS1]
PIPERAZINE HYDROCHLORIDE
[WHM]
PIPERAZINE, DIHYDROCHLORIDE
[PAL]

142-68-7 PENTAMETHYLENE OXIDE
2H-PYRAN, TETRAHYDRO- [PAL]

142-71-2 CUPRIC ACETATE
ACETIC ACID, COPPER(2+) SALT
[PAL]
COPPER ACETATE [PST]

142-77-8 BUTYL OLEATE

142-78-9 COCOMONOETHANOLAMIDE
N-(2-HYDROXYETHYL)DODE-
CANAMIDE [PST]

142-82-5 HEPTANE (N-)
HEPTANE [HM1, HM1, MEX, MEX,
MEX, NFP, NFP, OTV]
HEPTANE (N-HEPTANE) [AUS, AUS,
BC1, BC1, FLA, MSL, OS1, OS1,
OS1, TLV, TLV]
N-HEPTANE [CAL, CEX, CEX, GBR,
GBR, NIO, NIO, NIO, NJL, NJL,
ONT, ONT, ONT, ONT, QBC,
QBC, TSC, TSC, TSC]

142-83-6 SORBALDEHYDE
2,4-HEXADIENAL [CSR, FLA, MSL]

142-84-7 DIPROPYLAMINE
1-PROPANAMINE, N-PROPYL- [PAL,
TSC, TSC]

**142-87-0 SULFURIC ACID, MONODECYL
ESTER, SODIUM SALT**
SODIUM DECYL SULFATE [PST]

142-90-5 LAURYL METHACRYLATE
2-PROPENOIC ACID, 2-METHYL-,
DODECYL ESTER [TSC, TSC]

142-91-6 ISOPROPYL PALMITATE

142-92-7 N-HEXYL ACETATE
HEXYL ACETATE [NFP, NFP]
SEC-HEXYL ACETATE [FLA, ONT]

142-96-1 BUTYL ETHER
BUTANE, 1,1'-OXYBIS- [PAL]
DIBUTYL ETHER [FLA, HM1, HM1,
MSL, NFP, NFP]

143-00-0 DIETHANOLAMMONIUM DODE-
CYL SULFATE

143-07-7 LAURIC ACID

143-08-8 1-NONANOL

143-10-2 1-DECANETHIOL

143-13-5 NONYL ACETATE

143-14-6 9-UNDECENAL

143-15-7 LAURYL BROMIDE

143-16-8 DIHEXYLAMINE
1-HEXANAMINE, N-HEXYL- [PAL]

143-18-0 POTASSIUM OLEATE

143-19-1 SODIUM OLEATE

143-22-6 TRIETHYLENE GLYCOL
MONOBUTYL ETHER
TRIETHYLENEGLYCOL MONOBUTYL
ETHER [NFP, NFP]

143-24-8 2,5,8,11,14-PENTAOXAPENTADE-
CANE
DIMETHOXY TETRAGLYCOL [NFP,
NFP]

143-28-2 CIS-9-OCTADECEN-1-OL

143-33-9 SODIUM CYANIDE
SODIUM CYANIDE (NA(CN)) [302,
PAL]

143-50-0 KEPONE
1,3,4-METHENO-2H-CYCLOBUTA[CD]
PENTALEN-2-ONE, 1,1A,3,3A,4,5,5,5A,
5B,6-DECACHLOROOCTAHYDRO-
[PAL, PAL]
CHLORDECONE [CAB, CAL, MIN]
CHLORDECONE (KEPONE) [C65,
C65, C65, CSR]
CHLORODECONE (KEPONE) [IAR]
KEPONE (CHLORDECONE) [NTP]

143-67-9 VINBLASTINE SULFATE
VINBLASTINE SULPHATE [IAR]

143-74-8 PHENOL RED

143-81-7 BUTABARBITAL SODIUM

144-19-4 2,2,4-TRIMETHYL-1,3-PENTANE-
DIOL

144-34-3 METHYL SELENAC

144-48-9 2-IODOACETAMIDE

144-49-0 FLUOROACETIC ACID

144-55-8 SODIUM BICARBONATE

144-62-7 OXALIC ACID
ETHANEDIOIC ACID [PAL]

145-73-3 ENDOTHALL
7-OXABICYCLO[2.2.1]HEPTANE-2,3-
DICARBOXYLIC ACID [TSC, TSC]

146-84-9 SILVER PICRATE

147-14-8 PHTHALOCYANINE BLUE
C.I. PIGMENT BLUE 15 [OTS]
COPPER PHTHALOCYANIDE BLUE
[PST]

147-24-0 DIPHENHYDRAMINE HY-
DROCHLORIDE

147-47-7 1,2-DIHY-
DRO-2,2,4-TRIMETHYLQUINOLIN
(MONOMER)

147-82-0 BENZENAMINE, 2,4,6-TRIBROMO-

147-94-4 CYSTOSINE ARABINOSIDE
CYTARABINE [C65, CAB, MSL]
CYTARBINE [CSR]

148-01-6 DINITOLMIDE
2-METHYL-3,5-DINITROBENZAMIDE
[ONT, ONT]
3,5-DINITRO-O-TOLUAMIDE [CAL,
MIN]
3,5-DINITRO-O-TOLUAMIDE (ZOA-
LENE) [ALB, ALB, BC1, BC1]

BENZAMIDE, 2-METHYL-3,5-DINI-
TRO- [PAL]
DINITOLMIDE [QBC]
DINITOLMIDE (3,5-DINITRO-O-TOLU-
AMIDE) [OS1]

148-18-5 SODIUM DIETHYLDITHIOCARBA-
MATE

148-24-3 8-HYDROXYQUINOLINE

148-79-8 THIABENDAZOLE [2-(4-THIA-
ZOLYL)-1H-BENIMIDAZOLE, HY-
POPHOSPHITE SALT]
THIABENDAZOLE [2-(4-THIAZOLYL)-
1H-BENIMIDAZOLE] [313]

148-82-3 MELPHALAN
L-PHENYLALANINE, 4-[BIS(2-
CHLOROETHYL)AMINO]- [PAL,
PAL]

149-29-1 PATULIN

149-30-4 2-MERCAPTOBENZOTHIAZOLE
2-MERCAPTOBENZOTHIAZOLE (MBT)
[313]
MERCAPTOBENZOTHIAZOLE [TSC,
TSC, TSC]

149-31-5 2-METHYL-1,3-PENTANEDIOL
1,3-PENTANEDIOLE, 2-METHYL-
[PAL]

149-44-0 SODIUM FORMALDEHYDE SUL-
FOXYLATE

149-57-5 2-ETHYLHEXANOIC ACID
2-ETHYLHEXOIC ACID [WHM]
ETHYL HEXANOIC ACID, 2- [OTS]
HEXANOIC ACID, 2-ETHYL- [TSC,
TSC, TSC]

149-73-5 METHYL ORTHOFORMATE

149-74-6 METHYLPHENYLDICHLOROSI-
LANE
DICHLOROMETHYLPHENYLSILANE
[FLA, MSL, PAL]
DICHLOROMETHYPHENYLSILANE
[302, EPA, EPA]

149-91-7 GALLIC ACID

150-13-0 P-AMINOBENZOIC ACID
PARA-AMINOBENZOIC ACID [IAR,
WEL]

150-19-6 M-GUAIACOL

150-25-4 N,N-(2-DIHYDROXYETHYL)
GLYCINE

150-38-9 TRISODIUM EDTA
TRISODIUM ETHYLENEDIAMINETE-
TRAACETATE [PST]
TRISODIUM ETHYLENEDIAMINETE-
TRAACETATE TRIHYDRATE (EDTA)
[CSR, CSR]

150-50-5 S,S,S-TRIBUTYL PHOSPHO-
ROTRITHIOITE
MERPHOS [313]

150-60-7 DIBENZYL DISULFIDE

150-68-5 MONURON

150-69-6 DULCIN

150-76-5 4-METHOXYPHENOL
HYDROQUINONE MONOMETHYL
ETHER [NFP, NFP]
MEQUINOL (INN) (P-METHOXYPHE-
NOL) [GBR]
P-HYDROXYANISOLE [PST]
P-METHOXYPHENOL [T32, WHM]
PHENOL, 4-METHOXY- [PAL]

150-78-7 HYDROQUINONE, DIMETHYL
ETHER
HYDROQUINONE DIMETHYL ETHER
[PST]

151-05-3 DIMETHYLBENZYLCARBINYL
ACETATE

151-13-3 BUTYL RICINOLEATE

151-21-3 SODIUM LAURYL SULFATE
DODECYL SULFATE, SODIUM SALT
[PST]

151-38-2 METHOXYETHYLMERCURIC
ACETATE

151-41-7 LAURYL SULFATE

151-50-8 POTASSIUM CYANIDE
POTASSIUM CYANIDE (K(CN)) [PAL]

151-56-4 ETHYLENEIMINE
AZIRIDINE [EP2, EPA, IAR, PAL,
RCU, RCU]
ETHYLENE IMINE (AZIRIDINE) [CAA]
ETHYLENEIMINE (AZIRIDINE) [313]
ETHYLENEIMINE (AZIRIDINE) [EP3]
ETHYLENIMINE [AUS, AUS, AUS,
BC1, BC1, MAK, MAK, OTV, TLV,
TLV]
ETHYLENIMINE (AZIRIDINE) [HON]

151-67-7 HALOTHANE

152-16-9 OCTAMETHYLPYROPHOSPHO-
RAMIDE
DIPHOSPHORAMIDE, OCTAMETHYL-
[302, FLA, NJL, PAL]
SCHRADAN [CAL, MSL]

154-23-4 2H-1-BENZOPYRAN-3,5,7-TRIOL,
2-(3,4-DIHYDROXYPHENYL)-3,4-
DIHYDRO-, (2R-TRANS)-

154-42-7 THIOGUANINE

154-69-8 TRIPELENNAMINE HYDROCHLO-
RIDE

154-93-8 BISCHLOROETHYL NI-
TROSOUREA
1,3-BIS(2-CHLOROETHYL)-1-NI-
TROSOUREA [CAL]
1,3-BIS(CHLOROETHYL)-1-NI-
TROSOUREA [CSR]
BISCHLOROETHYL NITROSOUREA
(BCNU) [IAR]
BISCHLOROETHYL NITROSOUREA
(BCNU) (CARMUSTINE) [C65, C65]
N,N-BIS(2-CHLOROETHYL)-N-NI-
TROSOUREA [FLA]

155-04-4 ZINC MERCAPTOBENZOTHIA-
ZOLE

155-41-9 3-OXA-9-AZONIATRICYCLO
[3.3.1.02,4]NONANE, 7-(3-HY-
DROXY-1-OXO-2-PHENYL-
PROPOXY)-9,9-DIMETHYL-, BRO-
MIDE, [7(S)-(1-ALPHA, 2-BETA,4-
BETA,5-ALPHA,7-BETA)]-

156-10-5 P-NITROSODIPHENYLAMINE
BENZENAMINE, 4-NITROSO-N-
PHENYL- [PAL, PAL]
PARA-NITROSODIPHENYLAMINE
[IAR]

156-43-4 P-PHENETIDINE
BENZENAMINE, 4-ETHOXY- [OTS,
PAL]

156-51-4 PHENELZINE SULPHATE

156-57-0 MERCAPTAMINE HYDROCHLO-
RIDE

156-59-2 CIS-1,2-DICHLOROETHYLENE
1,2-DICHLOROETHENE, CIS- [ATS]
DICHLOROETHYLENE-CIS [FLA,
MSL]
ETHENE, 1,2-DICHLORO-, (Z)- [PAL]

156-60-5 1,2-TRANS-DICHLOROETHYLENE
1,2 DICHLOROETHYLENE [MEX,
MEX]
1,2-DICHLOROETHENE, TRANS-
[ATS]
1,2-DICHLOROETHYLENE [MX2,
RCU, RCU]
DICHLOROETHYLENE-TRANS [FLA,
MSL]
ETHENE, 1,2-DICHLORO-, (E)- [PAL]

TRANS-1,2-DICHLOROETHYLENE
[CSR, PQL, RCA, SDW, SDW,
SDW]

156-62-7 CALCIUM CYANAMIDE
CYANAMIDE, CALCIUM SALT (1:1)
[PAL]

156-87-6 PROPANOLAMINE
1-PROPANOL, 3-AMINO- [PAL]
3-AMINOPROPANOL [FLA, MSL,
NFP, NFP]

189-55-9 DIBENZO(A,I)PYRENE
1,2:7,8-DIBENZOPYRENE [EP2]
BENZO(RST)PENTAPHENE [313,
PAL, PAL]
DIBENZ(A,I)PYRENE [EPA]

189-64-0 DIBENZO(A,H)PYRENE
DIBENZO(B,DEF)CHRYSENE [PAL,
PAL]

191-07-1 CORONENE

191-24-2 BENZO(GHI)PERYLENE
1,12-BENZOPERYLENE [CAL]
1,12-BENZOPERYLENE (BENZO(GHI)
PERYLENE) [CW2]
BENZO(G)PERYLENE (1,12-BENZOP-
ERYLENE) [MX1]
BENZO(G,H,I)PERYLENE [ATS, UTS]

191-26-4 ANTHANTHRENE

191-30-0 DIBENZO(A,L)PYRENE

192-47-2 DIBENZO[H,RST]PENTAPHENE

192-65-4 DIBENZO(A,E)PYRENE
NAPHTHO(1,2,3,4-DEF)CHRYSENE
[PAL, PAL]

192-97-2 BENZO(E)PYRENE

**193-39-5 INDENO(1,2,3-CD)PYRENE (2,3-O-
PHENYLENEPYRENE)**
INDENO(1,2,3-CD)PYRENE (2,3-O-
PHENYLENE PYRENE) [CW2]

194-59-2 7H-DIBENZO(C,G)CARBAZOLE
DIBENZO(C,G)CARBAZOLE [NTP]

195-19-7 BENZO[C]PHENANTHRENE

198-55-0 PERYLENE

203-12-3 BENZO[GHI]FLUORANTHENE

203-33-8 BENZO(A)FLUORANTHENE

205-82-3 BENZO(J)FLUORANTHENE

205-99-2 BENZO(B)FLUORANTHENE
3,4-BENZOFLUORANTHENE [MX1]
3,4-BENZOFLUORANTHENE (BENZO
(B) FLUORANTHENE) [CW2]
BENZ(E)ACEPHENANTHRYLENE
[PAL, PAL]
BENZO(B)FLUORANTHENE [UTS]

206-44-0 FLUORANTHENE
BENZO(J,K)FLUORENE [EPA, MSL]

207-08-9 BENZO(K)FLUORANTHENE
11,12-BENZOFLUORANTHENE
(BENZO(K)FLUORANTHENE) [CW2]
BENZO(K)FLUORANTHENE (11,12-
BENZOFLUORANTHENE) [MX1]

208-96-8 ACENAPHTHYLENE

215-58-7 DIBENZ[A,C]ANTHRACENE

217-59-4 TRIPHENYLENE

218-01-9 CHRYSENE
BENZO(A)PHENANTHRENE [313]

224-41-9 DIBENZ[A,J]ANTHRACENE

224-42-0 DIBENZ(A,J)ACRIDINE

225-11-6 BENZ[A]ACRIDINE

225-51-4 BENZ(C)ACRIDINE

226-36-8 DIBENZ(A,H)ACRIDINE

238-84-6 BENZO(A)FLUORENE

260-94-6 ACRIDINE

262-12-4 DIBENZO-P-DIOXIN
DIBENZO-PARA-DIOXIN [CN5]

271-89-6 BENZOFURAN

273-53-0 BENZOXAZOLE

280-57-9 1,4-DIAZABICYCLO[2,2,2]OCTANE
TRIETHYLENEDIAMINE [WHM]

283-56-7 BORATRAN

**283-66-9 HEXAMETHYLENE TRIPEROXIDE
DIAMINE**

285-67-6 6-OXABICYCLO[3.1.0]HEXANE

286-20-4 7-OXABICYCLO[4.1.0]HEPTANE
CYCLOHEXENE OXIDE [CSR]

287-23-0 CYCLOBUTANE

287-92-3 CYCLOPENTANE

288-32-4 1-IMIDAZOLE

288-47-1 THIAZOLE

291-64-5 CYCLOHEPTANE

294-62-2 CYCLODODECANE

297-76-7 ETHYNODIOL DIACETATE

297-78-9 ISOBENZAN
ISOBENZEN [302]

**297-97-2 O,O-DIETHYL O-PYRAZINYL
PHOSPHOROTHIATE**
O,O-DIETHYL O-2-PYRAZINYL PHOS-
PHOROTHIOATE [PQL]
THIONAZIN [302, FLA, MSL, PAL]
ZINOPHOS [RCU, RCU]

298-00-0 METHYL PARATHION
PARATHION-METHYL [302, HM1,
HM1, HM1, HM1, HM1]
PARATHION-METHYL (ISO) [GBR,
GBR, GBR]
PHOSPHOROTHIOIC ACID, O,O-
DIMETHYL O-(4-NITROPHENYL) ES-
TER [PAL]

298-02-2 PHORATE
PHORATE (ISO) [GBR, GBR, GBR]
PHORATE (THEMET (R)) [QBC, QBC,
QBC]
PHORATE (THIMET) [ALB, ALB,
ALB, BC1, BC1, BC1]
PHOSPHORODITHIOIC ACID, O,O-
DIETHYL S-[(ETHYLTHIO)METHYL]
ESTER [PAL]

298-03-3 DEMETON-O

298-04-4 DISULFOTON
DISULFOTON [QBC]
DISULFOTON (ISO) [GBR, GBR]
DISYSTON [BC1, BC1, BC1]
PHOSPHORODITHIOIC ACID, O,O-
DIETHYL S-[2-(ETHYLTHIO)ETHYL]
ESTER [PAL]

**298-07-7 DI(2-ETHYLHEXYL)PHOSPHORIC
ACID**
BIS(2-ETHYLHEXYL) PHOSPHORIC
ACID [NFP]
PHOSPHORIC ACID, BIS(2-ETHYL-
HEXYL) ESTER [TSC, TSC, TSC]

298-14-6 POTASSIUM BICARBONATE

298-18-0 DL-DIEPOXYBUTANE

**298-59-9 METHYLPHENIDATE HY-
DROCHLORIDE**

298-81-7 8-METHOXYPSORALEN
8-METHOXYPSORALEN WITH UL-
TRAVIOLET A THERAPY [C65]
METHOXSALEN [CAL]

**299-27-4 GLUCONIC ACID, MONOPOTAS-
SIUM SALT**
D-GLUCONIC ACID, MONOPOTAS-
SIUM SALT [PST]

299-42-3 EPHEDRINE

299-75-2 TREOSULPHAN
1,2,3,4-BUTANETROL, 1,4-
DIMETHANESULFONATE, [S(R*,R*)]
- [PAL, PAL]
TREOSULFAN [C65, CAL]

299-84-3 RONNEL
FENCHLORPHOS (ISO) [GBR]
PHOSPHOROTHIOIC ACID,
O,O-DIMETHYL O-(2,4,5-
TRICHLOROPHENYL)ESTER [PAL]

299-86-5 CRUFOMATE
4-TERT-BUTYL-2-CHLOROPHENYL-
METHYL METHYLPHOSPHORAMI-
DATE [CAL]
CRUFOMATE (R) [QBC]
PHOSPHORAMIDIC ACID, METHYL-,
2-CHLORO-4-(1,1-DIMETHYLETHYL)
PHENYL METHYL ESTER [PAL]

300-62-9 AMPHETAMINE

300-76-5 NALED
DIBROM [BC1, BC1, MAK, MAK]
DIBROM (NALED) [ALB, ALB]
DIMETHYL-1,2-DIBROMO-2,2-
DICHLOROETHYL PHOSPHATE [NIO,
NIO, NIO, NIO, OS1, OS1, OS1]
DIMETHYL-1,2-DIBROMO-2,2-
DICHLOROETHYL PHOSPHATE
(NALED) [NJL]
NALED [QBC, QBC]
NALED (ISO) [GBR, GBR]
O,O-DIMETHYL O-(1,2-DIBROMO-2,
2-DICHLOROETHYL) PHOSPHATE
(NALED) [CAL]
PHOSPHORIC ACID, 1,2-DIBROMO-
2,2-DICHLOROETHYL DIMETHYL
ESTER [PAL]

**300-92-5 ALUMINUM, HYDROXYBIS
(STEARATO)-**
ALUMINUM DISTEARATE [CEX]

301-04-2 LEAD ACETATE
ACETIC ACID, LEAD(2+) SALT [PAL,
PAL]
LEAD(2+) ACETATE [CSR]

301-12-2 OXYDEMETON-METHYL
OXYDEMETON METHYL [MSL]
OXYDEMETON METHYL [S-(2-
(ETHYLSULFINYL)ETHYL) O,O-
DIMETHYL ESTER PHOSPHOROTH-
IOIC ACID) [313]
OXYDEMETONMETHYL [CAL]

302-01-2 HYDRAZINE
HYDRAZINE (ANHYDROUS) [NFP,
NFP]

**302-17-0 1,1-ETHANEDIOL, 2,2,2-
TRICHLORO-**
CHLORAL HYDRATE [CSR]

**302-70-5 NITROGEN MUSTARD N-OXIDE
HYDROCHLORIDE**
ETHANAMINE, 2-CHLORO-N-(2-
CHLOROETHYL)-N-METHYL-, N-
OXIDE,HYDROCHLORIDE [PAL,
PAL]
NITROGEN MUSTARD N-OXIDE
[CAB]

302-79-4 ALL-TRANS RETINOIC ACID

303-34-4 LASIOCARPINE
2-BUTENOIC ACID, 2-METHYL-
, 7-[[2,3-DIHYDROXY-2-(1-
METHOXYETHYL)-3-METHYL-1-
OXOBUTOXY]METHYL]-2,3,5,7A-TE-
TRAHYDRO-1H-PYRROLIZIN-1-YL
ESTER, [1S[1.ALPHA.(Z),7(2S*,3R*),
7A.ALPHA.]- [PAL, PAL]

303-47-9 OCHRATOXIN A

303-81-1 NOVOBIOCIN

305-03-3 CHLORAMBUCIL
BENZENEBUTANOIC ACID, 4-[BIS
(2-CHLOROETHYL)AMINO]- [PAL,
PAL]

**306-37-6 SYM-DIMETHYLHYDRAZINE DI-
HYDROCHLORIDE**

306-83-2 ETHANE, 2,2-DICHLORO-1,1,1-TRIFLUORO-
2,2-DICHLORO-1,1,1-TRIFLUO-ROETHANE [MAK]
2,2-DICHLORO-1,1,1-TRIFLUO-ROETHANE (HCFC-123) [313]
HCFC-123 [CAA]

309-00-2 ALDRIN
1,4:5,8-DIMETHANONAPHTHALENE, 1,2,3,4,10,10-HEXACHLORO-1,4,4A,5, 8,8A-HEXAHYDRO-, (1.ALPHA.,4.AL-PHA.,4A.BETA.,5.ALPHA., 8.ALPHA., 8A.BETA.)- [PAL]
ALDRIN ((1R,4S,4AS,5S,8R,8AR)1,2,3, 4,10,10-HEXACHLORO-1,4,4A,5,8,8A-HEXAHYDRO-1,4:5,8-DIMETHANON-APHTHALENE) [CN2]
ALDRIN (ISO) [GBR, GBR, GBR]
ALDRIN[1,4:5,8-DIMETHANONAPH-THALENE,1,2,3,4,10,10-HEXACHLORO-1,4,4A,5,8,8A,HEXAHYDRO-(1ALPHA, 4ALPHA,4BETA,5ALPHA, 8ALPHA, 8BETA)-] [313]

309-43-3 SECOBARBITAL SODIUM

311-28-4 TETRABUTYLAMMONIUM IODIDE

311-45-5 DIETHYL-P-NITROPHENYL PHOSPHATE
PARAOXON [HM1, HM1, NJL]

311-89-7 PERFLUOROTRIBUTYLAMINE

314-13-6 DIRECT BLUE 53
EVANS BLUE [CAL, IAR]

314-40-9 BROMACIL
2,4(1H,3H)-PYRIMIDINEDIONE, 5-BROMO-6-METHYL-3-(1-METHYL-PROPYL)- [PAL]
BROMACIL (5-BROMO-6-METHYL-3-(1-METHYLPROPYL)-2,4-(1H,3H)-PYRIMIDINEDIONE [313]
BROMACIL (ISO) [GBR, GBR]

314-42-1 5-BROMO-6-METHYL-3-(1-METHYLETHYL)-2,4(1H,3H)-PYRIMIDINEDIONE

315-18-4 MEXACARBATE
PHENOL, 4-(DIMETHYLAMINO)-3,5-DIMETHYL-, METHYLCARBAMATE (ESTER) [PAL]
ZECTRAN [IAR]

315-22-0 MONOCROTALINE
20-NORCROTALANAN-11,15-DIONE, 14,19-DIHYDRO-12,13-DIHYDROXY-, (13.ALPHA.,14.ALPHA.)- [PAL, PAL]

315-37-7 TESTOSTERONE ENANTHATE

316-42-7 EMETINE DIHYDROCHLORIDE
EMETINE HYDROCHLORIDE [CSR, CSR]
EMETINE, DIHYDROCHLORIDE [NJL, PAL]
EMETINE, HYDROCHLORIDE [302]

319-84-6 ALPHA-BHC
ALPHA-BHC-ALPHA [MX1]
ALPHA-HCH [IAR]
ALPHA-HEXACHLOROCYCLOHEX-ANE [313, FLA, MSL, NJL, NTP]
CYCLOHEXANE, 1,2,3,4,5,6-HEX-ACHLORO-, (1.ALPHA.,2.ALPHA., 3.BETA.,4.ALPHA.,5.BETA.,6.BETA.)- [PAL, PAL]
HEXACHLOROCYCLOHEXANE, AL-PHA ISOMER [C65]
HEXAHCLOROCYCLOHEXANE, AL-PHA- [ATS]

319-85-7 BETA-BHC
BETA-BHC-BETA [MX1]
BETA-HCH [IAR]
BETA-HEXACHLOROCYCLOHEXANE [FLA, MSL, NJL, NTP]
CYCLOHEXANE, 1,2,3,4,5,6-HEX-ACHLORO-, (1.ALPHA.,2.BETA.,3.AL-PHA.,4.BETA.,5.ALPHA.,6.BETA.)- [PAL, PAL]

HEXACHLOROCYCLOHEXANE, BETA ISOMER [C65]
HEXACHOLOCYCLOHEXANE, BETA- [ATS]

319-86-8 DELTA-BHC
CYCLOHEXANE, 1,2,3,4,5,6-HEX-ACHLORO-, (1.ALPHA.,2.ALPHA., 3.ALPHA.,4.BETA.,5.ALPHA.,6.BETA.)- [PAL]
DELTA-BHC (PCB-POLYCHLORI-NATED BIPHENYLS) [CW2]
DELTA-BHC-DELTA [MX1]
HEXACHLOROCYCLOHEXANE, DELTA- [ATS]

320-67-2 5-AZACYTIDINE
AZACITIDINE [C65, IAR]
AZCYTIDINE [NTP]

320-72-9 3,5-DICHLOROSALICYLIC ACID

327-98-0 TRICHLORONATE
TRICHLORONAT [HM1]

328-84-7 3,4-DICHLOROBENZOTRIFLUO-RIDE
BENZENE, 1,2-DICHLORO-4-(TRIFLU-OROMETHYL)-3,4-DICHLOROBEN-ZOTRIFLUORIDE [TSC, TSC, TSC, TSC]

329-01-1 BENZENE, 1-ISOCYANATO-3-(TRI-FLUOROMETHYL)-

329-71-5 2,5-DINITROPHENOL
PHENOL, 2,5-DINITRO- [PAL]

330-54-1 DIURON
DIURON (ISO) [GBR]
UREA, N'-(3,4-DICHLOROPHENYL)-N, N-DIMETHYL- [PAL]

330-55-2 LINURON

331-39-5 CAFFEIC ACID

333-20-0 POTASSIUM THIOCYANATE

333-41-5 DIAZINON
DIAZINON (ISO) [GBR, GBR, GBR]
PHOSPHOROTHIOIC ACID, O, O-DIETHYL O-[6-METHYL-2-(1-METHYLETHYL)-4-PYRIMIDINYL] ESTER [PAL]
PHOSPHOROTHIONIC ACID, O, O-DIETHYL O-(6-METHYL-2-(1-METHYLETHYL)-4-PYRIMIDINYL) ESTER [ONT, ONT]

334-48-5 CAPRIC ACID

334-88-3 DIAZOMETHANE
METHANE, DIAZO- [PAL]

342-69-8 6-METHYLMERCAPTOPURINE RIBONUCLEOSIDE

348-51-6 O-CHLOROFLUOROBENZENE

348-54-9 2-FLUOROANILINE

350-30-1 3-CHLORO-4-FLUORONITROBEN-ZENE

352-93-2 DIETHYL SULFIDE

353-36-6 ETHYL FLUORIDE
ETHANE, FLUORO- [PAL]

353-42-4 BORON TRIFLUORIDE COM-POUND WITH METHYL ETHER (1:1)
BORON TRIFLUORIDE COMPOUND WITH METHYL ETHER [MSL]
BORON TRIFLUORIDE COMPOUND WITH METHYL ETHER (1:1) [BORON, TRIFLUORO[OXYBIS[METANE]-, T-4-[EP3]
BORON TRIFLUORIDE DIMETHYL ETHERATE [HM1, HM1, NJL]

353-50-4 CARBONYL FLUORIDE
CARBON OXYFLUORIDE (CARBONIC DIFLUORIDE) [TSC, TSE]
CARBONIC DIFLUORIDE [PAL, RCU, RCU]

353-59-3 CHLORODIFLUOROBROMO-METHANE
BROMOCHLORODIFLUOROMETHANE [CN2, CN5]
BROMOCHLORODIFLUOROMETHANE (HALON 1211) [313]
CHLORODIFLUOROBROMOMETHANE [NJL]
HALON 1211 [CAA]

354-11-0 1,1,1,2-TETRACHLORO-2-FLUO-ROETHANE (HCFC-121A)

354-14-3 1,1,2,2-TETRACHLORO-1-FLUO-ROETHANE (HCFC-121)

354-21-2 1,2,2-TRICHLORO-1,1-DIFLUO-ROETHANE
ETHANE, 1,2,2-TRICHLORODIFLU-ORO- [TSC]
HCFC-122 [CAA]

354-23-4 1,2-DICHLORO-1,1,2-TRIFLUO-ROETHANE (HCFC-123A)

354-25-6 1-CHLORO-1,1,2,2-TETRAFLUO-ROETHANE (HCFC-124A)

354-32-5 TRIFLUOROACETYLCHLORIDE
TRIFLUOROACETYL CHLORIDE [HM1, HM1, HM1]

354-33-6 ETHENE, PENTAFLUORO-
ETHANE, 1,1,1,2,2-PENTAFLUORO [TSE]
ETHANE, 1,1,1,2,2-PENTAFLUORO-[TSC, TSC, TSC]

354-56-3 PENTACHLOROFLUOROETHANE
CFC-111 [CAA]

354-58-5 1,1,1-TRICHLORO-2,2,2-TRIFLUO-ROETHANE

355-42-0 PERFLUORO-N-HEXANE

357-57-3 BRUCINE
STRYCHNIDIN-10-ONE, 2,3-DIMETHOXY- [TSC, TSC]

359-06-8 FLUOROACETYL CHLORIDE

359-11-5 ETHENE, TRIFLUORO-

360-89-4 OCTAFLUOROBUT-2-ENE

366-18-7 2,2'-BIPYRIDINE

366-70-1 PROCARBAZINE HYDROCHLO-RIDE
BENZAMIDE, N-(1-METHYLETHYL)-4-[(2-METHYLHYDRAZINO)METHYL]-, MONOHYDROCHLORIDE [PAL, PAL]

367-47-5 GLYCERALDEHYDE

371-40-4 4-FLUOROANILINE

371-62-0 ETHYLENE FLUOROHYDRIN

372-09-8 ACETIC ACID, CYANO-
CYANOACETIC ACID [FLA, HON, MSL]

373-02-4 NICKEL(II) ACETATE
NICKEL ACETATE [NTP]

379-79-3 ERGOTAMINE TARTRATE

382-21-8 PERFLUOROISOBUTYLENE

384-22-5 O-NITROBENZOTRIFLUORIDE

389-08-2 NALIDIXIC ACID

392-56-3 HEXAFLUOROBENZENE

396-01-0 TRIAMTERENE

402-54-0 P-NITROBENZOTRIFLUORIDE

407-25-0 TRIFLUOROACETIC ACID ANHY-DRIDE

409-02-9 METHYLHEPTENONE

409-21-2 SILICON CARBIDE
SILICON CARBIDE (SIC) [PAL]

420-04-2 CYANAMIDE
HYDROGEN CYANAMIDE [PST]

420-12-2 ETHYLENE SULFIDE
ETHYLENE SULPHIDE [CAL, IAR]

420-46-2 ETHANE, 1,1,1-TRIFLUORO
1,1,1-TRIFLUOROETHANE [TSE]

421-20-5 METHYL FLUOROSULFATE

422-05-9 1,1-DIHYDROPERFLUORO-
PROPANOL

422-44-6 1,2-DICHLORO-1,1,2,3,3-
PENTAFLUOROPROPANE (HCFC-
225BB)

422-48-0 2,3-DICHLORO-1,1,1,2,3-
PENTAFLUOROPROPANE (HCFC-
225BA)

422-56-0 3,3-DICHLORO-1,1,1,2,2-
PENTAFLUOROPROPANE
3,3-DICHLORO-1,1,1,2,2-PENTAFLUO-
ROPROPANE (HCFC-225CA) [313]
HCFC-225CA [CAA]

428-25-1 HALOALKYL EPOXIDE

428-59-1 HEXAFLUOROPROPYLENE OXIDE
OXIRANE, TRIFLUORO(TRIFLUO-
ROMETHYL)- [TSC, TSC, TSC, TSC]

431-03-8 2,3-BUTADIONE
2,3-BUTANEDIONE [FLA, NFP, NFP,
PAL, PST]
BUTANEDIONE [HM1, HM1, NJL]

431-86-7 1,2-DICHLORO-1,1,3,3,3-
PENTAFLUOROPROPANE (HCFC-
225DA)

431-89-0 PROPANE, 1,1,1,2,3,3,3-HEP-
TAFLUORO-

434-07-1 OXYMETHOLONE
ANDROSTAN-3-ONE, 17-HYDROXY-
2-(HYDROXYMETHYLENE)-17-
METHYL-,(5.ALPHA.,17.BETA.)- [PAL,
PAL]

434-13-9 LITHOCHOLIC ACID

435-97-2 PHENPROCOUMON

438-41-5 CHLORDIAZEPOXIDE HY-
DROCHLORIDE

438-67-5 SODIUM ESTRONE SULFATE
CONJUGATED ESTROGENS: SODIUM
ESTRONE SULFATE [PAL, PAL]

439-14-5 DIAZEPAM

441-38-3 ALPHA-BENZOIN OXIME

443-48-1 METRONIDAZOLE
1H-IMIDAZOLE-1-ETHANOL, 2-
METHYL-5-NITRO- [PAL, PAL]

446-86-6 AZATHIOPRINE
1H-PURINE, 6-[(1-METHYL-4-NITRO-
1H-IMIDAZOL-5-YL)THIO]- [PAL,
PAL]

453-18-9 METHYL FLUOROACETATE

455-19-6 BENZALDEHYDE, 4-(TRIFLUO-
ROMETHYL)-

460-19-5 CYANOGEN
CYANOGEN [ETHANEDINITRILE]
[EP3]
ETHANEDINITRILE [PAL]
OXALONITRILE [GBR]

460-35-5 3-CHLORO-1,1,1-TRIFLUORO-
PROPANE (HCFC-253FB)

461-58-5 GUANIDINE, CYANO
DICYANODIAMIDE [PST]

462-06-6 FLUOROBENZENE
BENZENE, FLUORO- [PAL]

462-08-8 3-AMINO PYRIDINE
3-AMINOPYRIDINE [NJL]

462-95-3 DIETHOXYMETHANE

463-49-0 PROPADIENE
PROPADIENE [1,2-PROPADIENE]
[EP3]

463-51-4 KETENE
ETHENONE [PAL]

463-58-1 CARBONYL SULFIDE
CARBON OXIDE SULFIDE (COS)
[PAL]
CARBON OXYSULFIDE [FLA, MSL,
NFP, NFP]
CARBON OXYSULFIDE [CARBON
OXIDE SULFIDE (COS)] [EP3]

463-71-8 THIOPHOSGENE

463-82-1 2,2-DIMETHYLPROPANE
DIMETHYLPROPANE [NJL, NJL]
PROPANE, 2,2-DIMETHYL- [PAL]
TERT-PENTANE [MAK, MAK]

464-06-2 BUTANE, 2,2,3-TRIMETHYL-
2,2,3-TRIMETHYLBUTANE [FLA,
MSL, NFP, NFP]

464-49-3 (+)-CAMPHOR

465-73-6 ISODRIN

469-21-6 DOXYLAMINE

469-61-4 ALPHA-CEDRENE

470-67-7 7-OXABICYCLO[2.2.1]HEPTANE, 1-
METHYL-4-(1-METHYLETHYL)-

470-82-6 1,8-CINEOL

470-90-6 CHLORFENVINPHOS
CHLORFENVINFOS [302]
PHOSPHORIC ACID, 2-CHLORO-
1-(2,4-DICHLOROPHENYL)
ETHENYLDIETHYL ESTER [PAL]

471-34-1 CARBONIC ACID, CALCIUM SALT
(1:1)
CALCIUM CARBONATE [ALB, GBR,
MEX, MEX, PST]
CALCIUM CARBONATE, INCLUDING
MARBLE [ONT]
CALCIUM CARBONATE/MARBLE
[BC1, BC1, QBC]

471-46-5 OXAMIDE

472-41-3 PHENOL, 4-(3,4-DIHYDRO-2,2,4-
TRIMETHYL-2H-1-BENZOPYRAN-
4-YL)-

473-55-2 PINANE

474-25-9 CHENODIOL

479-45-8 TETRYL
BENZENAMINE, N-METHYL-N,2,4,6-
TETRANITRO- [PAL]
N-METHYL-N,2,4,6-TETRANITROANI-
LINE [GBR, GBR, GBR]
NITRAMINE [ONT, ONT]
TETRYL (2,4,6-TRINITRO-PHENYL-
METHYLNITRAMINE) [ALB, ALB,
ALB]
TETRYL (2,4,6-TRINITROPHENYL-
METHYL-NITRAMINE) [OS1, OS1,
OS1, OS1]
TETRYL (2,4,6-TRINITROPHENYL-
METHYLNITRAMINE) [BC1, BC1,
BC1]
TRINITROPHENYLMETHYL-
NITRAMINE [ATS, HM1, HM1]

479-61-8 CHLOROPHYLL

480-22-8 DITHRANOL

480-54-6 RETRORSINE

480-81-9 SENECIPHYLLINE

481-39-0 5-HYDROXY-1,4-NAPH-
THAQUINONE

482-89-3 3H-INDOL-3-ONE, 2-(1,3-DIHYDRO-
3-OXO-2H-

484-20-8 5-METHOXYPSORALEN
5-METHOXYPSORALEN WITH UL-
TRAVIOLET A THERAPY [C65]

485-31-4 BINAPACRYL

486-12-4 TRIPROLIDINE

488-10-8 JASMONE

488-23-3 1,2,3,4-TETRAMETHYLBENZENE

488-41-5 DIBROMOMANNITOL

488-47-1 TETRABROMOCATECHOL

489-01-0 BUTYLATED HYDROXYANISOLE

492-17-1 2,4'-DIPHENYLDIAMINE

492-80-8 AURAMINE
AURAMINE (TECHNICAL GRADE)
[MIN]
C.I. SOLVENT YELLOW (AURAMINE)
[NJL]
CI SOLVENT YELLOW 34 (AU-
RAMINE) [313]

493-52-7 METHYL RED

494-03-1 N,N-BIS(2-CHLOROETHYL)-2-
NAPHTHYLAMINE
2-NAPHTHALENAMINE, N,N-BIS(2-
CHLOROETHYL)- [PAL, PAL]
CHLORNAPHAZINE [EP2, EPA, NJL,
NJL, RCU, RCU]
N,N-BIS(2-CHLOROETHYL)-2-NAPH-
THYLAMINE (CHLORNAPHAZINE)
[C65, IAR]

494-38-2 ACRIDINE ORANGE

495-48-7 AZOXYBENZENE

495-54-5 C.I. SOLVENT ORANGE 3

495-73-8 BENQUINOX

496-03-7 BUTYRALDOL
HEXANAL, 2-ETHYL-3-HYDROXY-
[PAL]

496-10-6 HYDRINDANE

496-11-7 INDAN

496-46-8 GLYCOURIL

496-72-0 DIAMINOTOLUENE
1,2-BENZENEDIAMINE, 4-METHYL-
[TSC, TSC]
3,4-DIAMINOTOLUENE [MSL, NJL]
3,4-TOLUENEDIAMINE [EPA]
TOLUENE-3,4-DIAMINE [WHM]

497-18-7 CARBONIC DIHYDRAZIDE
CABAZIDE [HM1]

497-19-8 SODIUM CARBONATE

499-75-2 CARVACROL

499-83-2 2,6-PYRIDINEDICARBOXYLIC
ACID

500-28-7 CHLORTHION

501-53-1 BENZYL CHLOROFORMATE

502-39-6 METHYL MERCURY DICYANDI-
AMIDE
METHYLHYDRAZINE [HM1, HM1]
METHYLMERCURIC DICYANAMIDE
[302, MSL, PAL]
METHYLMERCURIC DICYANDI-
AMIDE [FLA]

502-44-3 CAPROLACTONE

503-17-3 2-BUTYNE
CROTONYLENE [HM1, HM1, NJL]

503-74-2 ISOPENTANOIC ACID

504-20-1 PHORONE
2,5-HEPTADIEN-4-ONE, 2,6-
DIMETHYL- [PAL]

504-24-5 4-AMINOPYRIDINE
4-PYRIDINAMINE [MSL]
PYRIDINE, 4-AMINO [FLA]
PYRIDINE, 4-AMINO- [302, PAL]

504-29-0 2-AMINOPYRIDINE
2-PYRIDINAMINE [PAL]
2-PYRIDYLAMINE [GBR, GBR]

504-60-9 1-METHYLBUTADIENE
1,3-PENTADIENE [EP3, EPA, PAL]
1,3-PENTADIENE (CIS & TRANS MIXED) [MSL]
1,3-PENTADIENE (CIS AND TRANS MIX) [NFP, NFP]
1,3-PENTADIENE (MIXED CIS & TRANS) [FLA]
PENTADIENE, 1,3- [OTS]

504-63-2 1,3-PROPANEDIOL
TRIMETHYLENE GLYCOL [NFP]

504-88-1 3-NITROPROPIONIC ACID

505-57-7 2-HEXENAL

505-60-2 MUSTARD GAS
2,2'-DICHLOROETHYL SULFIDE [MAK]
DICHLOROETHYL SULFIDE [HM1]
ETHANE, 1,1'-THIOBIS[2-CHLORO- [PAL, PAL]
MUSTARD GAS (BIS(2-CHLOROETHYL)-SULFIDE) [WHM]
MUSTARD GAS (SULPHUR MUSTARD) [IAR]
MUSTARD GAS [ETHANE, 1,1'-THIO-BIS[2-CHLORO-] [313]

506-59-2 DIMETHYLAMINE HYDROCHLO-RIDE

506-61-6 POTASSIUM SILVER CYANIDE
SILVER POTASSIUM CYANIDE [WHM]

506-64-9 SILVER CYANIDE

506-68-3 CYANOGEN BROMIDE
BROMINE CYANIDE [HM1, HM1, HM1, HM1, HM1]
CYANOGEN BROMIDE (CN)BR [RCU, RCU]

506-77-4 CYANOGEN CHLORIDE
CHLORINE CYANIDE [HM1, HM1, HM1, HM1, HM1]

506-78-5 CYANOGEN IODIDE

506-87-6 AMMONIUM CARBONATE
CARBONIC ACID, DIAMMONIUM SALT [PAL]

506-93-4 GUANIDINE MONONITRATE
GUANIDINE NITRATE [NJL, PST]

506-96-7 ACETYL BROMIDE

507-02-8 ACETYL IODIDE

507-09-5 THIOACETIC ACID

507-19-7 2-BROMO-2-METHYLPROPANE

507-20-0 TERT-BUTYL CHLORIDE
PROPANE, 2-CHLORO-2-METHYL- [PAL]

507-55-1 1,3-DICHLORO-1,1,2,2,3-PENTAFLUOROPROPANE
1,3-DICHLORO-1,1,2,2,3-PENTAFLUO-ROPROPANE (HCFC-225CB) [313]
HCFC-225CB [CAA]

507-70-0 BORNEOL
BICYCLO(2.2.1)HEPTAN-2-OL, 1,7,7-TRIMETHYL-, ENDO- [PAL]

509-09-1 SILVER PENTAFLUOROPROPI-ONATE

509-14-8 TETRANITROMETHANE
METHANE, TETRANITRO- [PAL]
TETRANITROMETHANE [METHANE, TETRANITRO-] [EP3]

510-15-6 CHLOROBENZILATE
BENZENEACETIC ACID, 4-CHLORO-.ALPHA.-(4-CHLOROPHENYL)-AL-PHA.-HYDROXY-,ETHYL ESTER [PAL]
CHLOROBENZILATE [BENZE-NEACETIC ACID, 4-CHLORO-AL-PHA.-(4-CHLOROPHENYL)-ALPHA.-HYDROXY-, ETHYL ESTER] [313]
ETHYL 4,4'-DICHLOROBENZILATE [C65, C65, EP2, NJL, NJL]

512-13-0 ALPHA-FENCHYL ALCOHOL

512-42-5 METHYL SULFATE, SODIUM SALT

512-56-1 TRIMETHYLPHOSPHATE
PHOSPHORIC ACID, TRIMETHYL ESTER [OTS]
TRIMETHYL PHOSPHATE [MAK, MAK, MSL]

512-85-6 ASCARIDOLE

513-35-9 2-METHYL-2-BUTENE
2-BUTENE, 2-METHYL- [PAL]

513-36-0 ISOBUTYL CHLORIDE
PROPANE, 1-CHLORO-2-METHYL- [PAL]

513-37-1 DIMETHYLVINYLCHLORIDE (DMVC)
DIMETHYLVINYL CHLORIDE [NTP]
DIMETHYLVINYL CHLORIDE (DMVC) [CSR, CSR]
DIMETHYLVINYLCHLORIDE [C65, C65]

513-38-2 IODO METHYLPROPANE
1-IODO-2-METHYLPROPANE [NJL]

513-42-8 METHALLYL ALCOHOL
2-PROPEN-1-OL, 2-METHYL- [PAL]

513-48-4 SEC-BUTYL IODIDE
2-IODOBUTANE [HM1, HM1, NJL]

513-49-5 SEC-BUTYLAMINE
SEC-BUTYLAMINE (S-) [MSL]

513-53-1 2-BUTANETHIOL

513-77-9 BARIUM CARBONATE

513-78-0 CADMIUM CARBONATE

513-79-1 COBALT CARBONATE

513-85-9 2,3-BUTANEDIOL
BUTYLENE GLYCOL (PSEUDO) [NFP, NFP]

513-86-0 ACETYL METHYL CARBINOL

514-10-3 ABIETIC ACID

514-73-8 DITHIAZANINE IODIDE

515-98-0 AMMONIUM LACTATE

517-16-8 N-(ETHYLMERCURIC)-P-TOLUE-NESULPHONANNILIDE

517-25-9 TRINITROMETHANE

517-92-0 1,8-DIHYDROXY-2,4,5,7-TETRANI-TROANTHRAQUINONE

518-47-8 SODIUM FLUORESCEIN
C.I. ACID YELLOW 73 [PST]

518-75-2 CITRININ

518-82-1 EMODIN

519-44-8 2,4-DINITRORESORCINOL

519-73-3 TRIPHENYLMETHANE

520-18-3 KAEMPFEROL

520-45-6 DEHYDROACETIC ACID

523-44-4 ORANGE I

523-80-8 4,7-DIMETHOXY-5-(2-PROPENYL)-1,3-BENZODIOXOLE

526-73-8 1,2,3-TRIMETHYLBENZENE
BENZENE, 1,2,3-TRIMETHYL- [TSC, TSC, TSC]

526-83-0 TARTARIC ACID

526-95-4 GLUCONIC ACID
D-GLUCONIC ACID [PST]

526-99-8 MOCIC ACID

527-07-1 SODIUM GLUCONATE
GLUCONIC ACID, SODIUM SALT [PST]

527-09-3 CUPRIC GLUCONATE

527-53-7 1,2,3,5-TETRAMETHYLBENZENE

527-60-6 PHENOL, 2,4,6-TRIMETHYL-

528-29-0 O-DINITROBENZENE
1,2-DINITRO BENZOL [NFP, NFP]
BENZENE, 1,2-DINITRO- [PAL]
DINITROBENZENE (ALPHA-) [OS1, OS1, OS1, OS1]

528-44-9 TRIMELLITIC ACID

528-74-5 DICHLOROMETHOTREXATE

528-75-6 ISOAMYL ALCOHOL (SEC-ONDARY)

529-33-9 1,2,3,4-TETRAHYDRO-1-NAPH-THOL

529-34-0 D-TETRALONE

529-65-7 N-ETHYL ACETANILIDE

530-50-7 HYDRAZINE, 1,1-DIPHENYL

531-18-0 HEXAMETHYLOLMELAMINE

531-76-0 MERPHALAN
DL-PHENYLAMINE, 4-[BIS(2-CHLOROETHYL)AMINO]- [PAL, PAL]

531-82-8 N-[4-(5-NITRO-2-FURYL)-2-THIA-ZOLYL]ACETAMIDE
ACETAMIDE, N-[4-(5-NITRO-2-FU-RANYL)-2-THIAZOLYL]- [PAL, PAL]

531-85-1 [1,1'-BIPHENYL]-4,4'-DIAMINE, DIHYDROCHLORIDE
BENZIDINE DIHYDROCHLORIDE [CSR]

531-86-2 BENZIDINE SALT

532-02-5 2-NAPHTHALENESULFONIC ACID, SODIUM SALT

532-27-4 ALPHA-CHLOROACETOPHENONE
2-CHLORO-1-PHENYLETHANONE [ONT]
2-CHLOROACETOPHENONE [313, CAA, CAB, GBR, HON, HON, NJL]
2-CHLOROACETOPHENONE (CN) [CSR, CSR]
A-CHLOROACETOPHENONE [AUS, TSC, TSC, TSC]
A-CHLOROACETOPHENONE (PHENACYL CHLORIDE) [OS1, OS1]
ALPHA-CHLOROACETOPHENONE [MEX]
ALPHA-CHLOROACETOPHENONE (PHENACYL CHLORIDE) [ALB, ALB, BC1]
CHLOROACETOPHENONE [WHM]
ETHANONE, 2-CHLORO-1-PHENYL- [PAL]

532-28-5 MANDELONITRILE
CHLOROACETHOPHENONE (PHENACYL CHLORIDE) [QBC]

532-32-1 SODIUM BENZOATE

532-82-1 CHRYSOIDINE

533-42-6 2-OCTYLDODECANOL

533-51-7 SILVER OXALATE

533-74-4 TETRAHYDRO-3,5-DIMETHYL-2H-1,3,5-THIADIAZINE-2-THIONE
DAZOMET (TETRAHYDRO-3,5-DIMETHYL-2H-1,3,5-THIADIAZINE-2-THIONE) [313]

533-96-0 SODIUM SESQUICARBONATE

534-07-6 BIS(CHLOROMETHYL)KETONE
1,3-DICHLOROACETONE [HM1, HM1, WHM]
2-PROPANONE, 1,3-DICHLORO- [TSC, TSC]

534-15-6 1,1-DIMETHOXYETHANE
ETHANE, 1,1-DIMETHOXY- [TSC, TSC]

534-17-8 CESIUM CARBONATE

534-22-5 2-METHYLFURAN
2-METHYL FURAN [FLA]
FURAN, 2-METHYL- [PAL]
METHYL FURAN [WHM]

534-52-1 4,6-DINITRO-O-CRESOL
2-METHYL-4,6-DINITROPHENOL
[GBR, GBR, GBR]
4,6-DINITRO-O-CRESOL, & SALTS
[RCU, RCU]
4,6-DINITRO-O-CRESOL, AND SALTS
[CAA]
4,6-DINITRO-ORTO-CRESOL [MX1]
DINITRO-O-CRESOL [ALB, ALB,
ALB, AUS, AUS, BC1, BC1, BC1,
FLA, MEX, MEX, MEX, MIN,
MSL, MX2, NIO, NIO, NIO, NIO,
ONT, ONT, OS1, OS1, OS1, OS1,
QBC, QBC, TLV, TLV]
DINITRO-ORTHO-CRESOL [NHS,
NHS]
DINITROCRESOL [302]
DNOC [HM1, HM1, HM1, HM1]
PHENOL, 2-METHYL-4,6-DINITRO-
[PAL]

535-13-7 ETHYL-2-CHLOROPROPIONATE

535-89-7 CRIMIDINE
CASTRIX [NJL]

536-33-4 ETHIONAMIDE

536-59-4 PERILLA ALCOHOL

536-60-7 CUMINIC ALCOHOL

536-90-3 M-ANISIDINE
BENZENAMINE, 3-METHOXY- [OTS]

**538-07-8 ETHYLBIS(2-CHLOROETHYL)
AMINE**

538-23-8 TRICAPRYLIN

538-68-1 BENZENE, PENTYL-
AMYLBENZENE [NFP, NFP]

538-75-0 DICYCLOHEXYLCARBODIIMIDE

538-93-2 ISOBUTYLBENZENE
BENZENE, (2-METHYLPROPYL)-
[PAL]

539-90-2 ISOBUTYL BUTYRATE

540-18-1 AMYL BUTYRATE

540-42-1 ISOBUTYL PROPIONATE

540-49-8 1,2-DIBROMOETHENE
ACETYLENE DIBROMIDE [HM1]

540-54-5 PROPYL CHLORIDE
PROPANE, 1-CHLORO- [PAL, TSC,
TSC]

540-59-0 1,2-DICHLOROETHYLENE
1,2 DICHLOROETHYLENE [QBC]
1,2-DICHLOROETHENE [ONT, ONT]
ACETYLENE DICHLORIDE [CAL,
FLA]
ACETYLENE DICHLORIDE (1,2-
DICHLOROETHYLENE) [ALB, ALB]
CIS & TRANS 1,2-DICHLOROETHY-
LENE [CSR]
SYM-DICHLOROETHYLENE [WHM]
TRANS-ACETYLENE, DICHLORIDE
[PAL]

540-67-0 METHYL ETHYL ETHER
ETHANE, METHOXY- [PAL]
ETHYL METHYL ETHER [HM1,
HM1]

540-69-2 AMMONIUM FORMATE

540-72-7 SODIUM THIOCYANATE

540-73-8 1,2-DIMETHYLHYDRAZINE
DIMETHYLHYDRAZINE, SYMMETRI-
CAL [HM1, HM1, HM1]
HYDRAZINE, 1,2-DIMETHYL- [PAL,
PAL]

540-82-9 ETHYLSULPHURIC ACID
ETHYL SULFURIC ACID [NJL, NJL]

540-84-1 ISOOCTANE
2,2,4-TRIMETHYLPENTANE [CAA,
CAB]
PENTANE, 2,2,4-TRIMETHYL- [PAL,
TSC, TSC]
TRIMETHYLPENTANE (2,2,4-) [HON,
HON]

540-88-5 TERT-BUTYL ACETATE
ACETIC ACID, 1,1-DIMETHYLETHYL
ESTER [PAL]
ACETIC ACID-TERT-BUTYL ESTER
[NJL]

**540-97-6 DODECAMETHYLCYCLOHEXAS-
ILOXANE**

**541-01-5 HEXADECAMETHYLHEP-
TASILOXANE**

**541-02-6 DECAMETHYLCYCLOPEN-
TASILOXANE**

541-05-9 TRIALKOXY SILANE
HEXAMETHYLCYCLOTRISILOXANE
[TSC, TSC]

541-09-3 URANYL ACETATE
URANIUM BIS(ACETATO-O)DIOXO-
[PAL]

541-25-3 LEWISITE

541-31-1 ISOAMYL MERCAPTAN

541-41-3 ETHYL CHLOROFORMATE
CARBONOCHLORIDIC ACID, ETHYL
ESTER [PAL]

**541-47-9 BETA, BETA-DIMETHACRYLIC
ACID**

541-53-7 DITHIOBIURET
2,4-DITHIOBIURET [313]

**541-69-5 1,3-BENZENEDIAMINE, DIHY-
DROCHLORIDE**

**541-70-8 1,3-BENZENEDIAMINE, SULFATE
(1:1)**
META-PHENYLENEDIAMINE, SUL-
FATE SALT [CTR]
SULFATE SALTS OF META-
PHENYLENEDIAMINE [TSE]

541-73-1 1,3-DICHLOROBENZENE
BENZENE, 1,3-DICHLORO- [PAL,
TSC, TSC]
M-DICHLOROBENZENE [CAL, HON,
PQL, RCA, RCU, RCU, SDW, UTS,
WHM]

541-85-5 ETHYL AMYL KETONE
3-HEPTANONE, 5-METHYL- [PAL]
5-METHYL-3-HEPTANONE [NIO,
NIO, NIO, ONT]
5-METHYLHELPTAN-3-ONE [GBR]
ETHYL AMYL KETONE (5-METHYL-3-
HEPTANONE) [OS1, OS1]
ETHYL SEC-AMYL KETONE [CAL,
CEX, FLA, WHM]
ETHYL SEC-AMYL KETONE (4-
METHYL-3-HEPTANONE) [BC1]

542-18-7 CYCLOHEXYL CHLORIDE
CYCLOHEXANE, CHLORO- [PAL]

542-55-2 ISOBUTYL FORMATE
FORMIC ACID, 2-METHYLPROPYL
ESTER [PAL]

542-56-3 ISOBUTYL NITRITE

542-58-5 ETHANOL, 2-CHLORO-, ACETATE
2-CHLOROETHANOL ACETATE
[FLA]
2-CHLOROETHYL ACETATE [MSL,
NFP, NFP]

542-59-6 ETHYLENE GLYCOL ACETATE
ETHYLENE GLYCOL MONOACETATE
[HON, NFP, NFP]

542-62-1 BARIUM CYANIDE
BARIUM CYANIDE (BA(CN)2) [PAL]

542-69-8 1-IODOBUTANE

542-75-6 1,3-DICHLOROPROPENE
1,2-DICHLOROPROPYLENE [MX1,
MX2]
1,2-DICHLOROPROPYLENE (1,3-
DICHLOROPROPENE) [CW2]
1,3-DICHLOROPROPENE [TLV, TLV]
1,3-DICHLOROPROPENE (CIS AND
TRANS ISOMERS) [QBC, QBC, QBC]
1,3-DICHLOROPROPENE (TECHNICAL
GRADE) [IAR, MIN, NTP]
1,3-DICHLOROPROPENE (TELONE II)
[CSR, CSR]
1,3-DICHLOROPROPYLENE [313]
1-PROPENE, 1,3-DICHLORO- [PAL,
PAL, TSC, TSC]
DICHLOROPROPENE [AUS, AUS,
AUS, CEX, CEX, WHM]
TRANS-1,3-DICHLOROPROPENE
[NJL, NJL]

542-76-7 3-CHLOROPROPIONITRILE
PROPIONITRILE, 3-CHLORO [MSL]
PROPIONITRILE, 3-CHLORO- [302,
FLA, PAL]

542-78-9 MALONALDEHYDE

542-88-1 BIS(CHLOROMETHYL) ETHER
BIS (2-CHLOROMETHYL) ETHER
[NJL, NJL]
BIS(CHLOROMETHYL) ETHER [CAB]
BIS-CHLOROMETHYL ETHER [BC1,
BC1, CEX, CEX, NHS, NHS, NHS,
NIO, NIO]
BISCHLOROMETHYL ETHER [MAK]
CHLOROMETHYL BIS ETHER [QBC,
QBC]
CHLOROMETHYL ETHER [302]
CHLOROMETHYL ETHER [METHANE,
OXYBIS[CHLORO-] [EP3]
DICHLOROMETHYL ETHER [EPA,
RCU, RCU]
METHANE, OXYBIS[CHLORO- [PAL,
PAL]
OXYBIS (CHLOROMETHANE) [ONT]

542-90-5 ETHYL THIOCYANATE
ETHYLTHIOCYANATE [302]

542-92-7 CYCLOPENTADIENE
1,3-CYCLOPENTADIENE [FLA, PAL,
WHM]

543-27-1 ISOBUTYL CHLOROFORMATE

543-49-7 2-HEPTANOL

543-59-9 AMYL CHLORIDE
1-CHLOROPENTANE [FLA, MSL]
PENTANE, 1-CHLORO- [PAL]

543-80-6 BARIUM ACETATE

543-90-8 CADMIUM ACETATE
ACETIC ACID, CADMIUM SALT
[PAL]

544-10-5 1-CHLOROHEXANE
HEXANE, 1-CHLORO- [PAL]
NORMAL-HEXYL CHLORIDE [HM1]

544-16-1 BUTYL NITRITE

544-17-2 CALCIUM FORMATE

544-18-3 COBALTOUS FORMATE
FORMIC ACID, COBALT(2+) SALT
[PAL]

544-25-2 CYCLOHEPTATRIENE

**544-31-0 N-(2-HYDROXYETHYL)HEXADE-
CANAMIDE**

544-60-5 AMMONIUM OLEATE

544-63-8 MYRISTIC ACID

544-76-3 HEXADECANE

544-92-3 COPPER CYANIDE

544-97-8 DIMETHYLZINC

545-06-2 TRICHLOROACETONITRILE

545-55-1 TRIS(1-AZIRIDINYL) PHOSPHINE OXIDE
1-AZIRIDINYL PHOSPHINE OXIDE (TRIS) [NJL]

546-56-5 OCTAPHENYLCYCLOTETRASILOXANE

546-67-8 LEAD (IV) ACETATE

546-68-9 TETRAISOPROPYL TITANATE

546-88-3 ACETOHYDROXAMIC ACID

546-93-0 MAGNESIUM CARBONATE
MAGNESITE [ALB, AUS, BC1, BC1, GBR, MEX, MEX, ONT, OS1, OS1, QBC, TLV]

547-58-0 C.I. ACID ORANGE 52

547-63-7 METHYL ISOBUTYRATE

547-64-8 METHYL LACTATE

548-62-9 C.I. BASIC VIOLET 1
HEXAMETHYL-P-ROSANILINE CHLORIDE [CSR]

548-93-6 2-AMINO-3-HYDROXYBENZOIC ACID

551-74-6 MANNOMUSTINE

552-30-7 TRIMELLITIC ANHYDRIDE
5-ISOBENZOFURANCARBOXYLIC ACID, 1,3-DIHYDRO-1,3-DIOXO- [PAL]
BENZENE-1,2,4-TRICARBOXYLIC ACID 1,2-ANHYDRIDE [GBR, GBR]
TRIMELLITIC ACID ANHYDRIDE [PST]

552-79-4 N-METHYLEPHEDRINE

552-89-6 BENZALDEHYDE, 2-NITRO-

553-70-8 BENZOIC ACID, MAGNESIUM SALT

553-97-9 2-METHYL-1,4-BENZOQUINONE

554-00-7 BENZENAMINE, 2,4-DICHLORO-

554-12-1 METHYL PROPIONATE
PROPANOIC ACID, METHYL ESTER [PAL]

554-13-2 LITHIUM CARBONATE

554-76-7 STIBANILIC ACID

554-84-7 M-NITROPHENOL
3-NITROPHENOL [NJL]
PHENOL, 3-NITRO- [PAL]

555-03-3 3-NITROANISOLE

555-10-2 BETA-PHELLANDRENE

555-16-8 P-NITROBENZALDEHYDE

555-30-6 ALPHA-METHYLDOPA

555-31-7 ALUMINUM ISO-PROPOXIDE
2-PROPANOL, ALUMINUM SALT [PST]

555-54-4 MAGNESIUM DIPHENYL

555-77-1 TRIS(2-CHLOROETHYL)AMINE
TRICHLORMETHINE (TRIMUSTINE HYDROCHLORIDE) [IAR]

555-84-0 1-((5-NITROFURFURYLIDENE) AMINO)-2-IMIDAZOLIDINONE
1-((5-NITROFURFURYLIDENE)-AMINO)-2-IMIDAZOLIDINONE [C65, C65]
2-IMIDAZOLIDINONE, 1-[[(5-NITRO-2-FURANYL)METHYLENE]AMINO]- [PAL, PAL]

556-24-1 METHYL ISOVALERATE

556-52-5 GLYCIDOL
2,3-EPOXY-1-PROPANOL [NJL, ONT]
2,3-EPOXY-1-PROPANOL (GLYCIDOL) [ALB, ALB, QBC]
GLYCIDOL (2,3-EPOXY-1-PROPANOL) [BC1, BC1]

OXIRANEMETHANOL [PAL]
OXIRANEMETHANOL (GLYCIDOL) [TSC, TSC]

556-56-9 ALLYL IODIDE

556-57-0 OCTAMETHYLCYCLOTETRASILOXANE

556-61-6 METHYL ISOTHIOCYANATE
METHYL ISOTHIOCYANATE [ISOTHIOCYANATOMETHANE] [313]

556-64-9 METHYL THIOCYANATE
METHYL THIOCYANATE [THIOCYANIC ACID, METHYL ESTER] [EP3]

556-67-2 OCTAMETHYLCYCLOTETRASILOXANE

556-68-3 HEXADECAMETHYLCYCLOOCTASILOXANE

556-69-4 OCTADECAMETHYLOCTASILOXANE

556-70-7 DOCOSAMETHYLDECASILOXANE

556-71-8 OCTADECAMETHYLCYCLONONASILOXANE

556-88-7 1-NITROGUANIDINE
NITROGUANIDINE [HM1, HM1, NJL]

556-89-8 NITRO UREA

557-04-0 MAGNESIUM STEARATE

557-05-1 ZINC STEARATE
OCTADECANOIC ACID, ZINC SALT [PAL]
ZINC DISTEARATE [GBR, GBR]

557-09-5 ZINC OCTOATE
OCTANOIC ACID, ZINC SALT [PST]

557-17-5 METHYL PROPYL ETHER
METHYL N-PROPYL ETHER [NFP, NFP]
METHYL-N-PROPYL ETHER [FLA, MSL]
PROPANE, 1-METHOXY- [PAL]

557-18-6 DIETHYLMAGNESIUM
DIETHYL MAGNESIUM [NJL]

557-19-7 NICKEL CYANIDE

557-20-0 DIETHYLZINC
DIETHYL ZINC [FLA, MSL]
ZINC, DIETHYL- [PAL]

557-21-1 ZINC CYANIDE
ZINC CYANIDE (ZN(CN)2) [PAL]

557-28-8 ZINC PROPIONATE

557-31-3 ALLYL ETHYL ETHER

557-34-6 ZINC ACETATE
ACETIC ACID, ZINC SALT [PAL]

557-40-4 DIALLYL ETHER
1-PROPENE, 3,3'-OXYBIS- [PAL]
ALLYL ETHER [MSL, NFP, NFP]
DIALLYLETHER [FLA, NJL, NJL]

557-41-5 ZINC FORMATE
FORMIC ACID, ZINC SALT [PAL]

557-75-5 ETHENOL

557-98-2 2-CHLOROPROPENE
1-PROPENE, 2-CHLORO- [PAL]
2-CHLORO PROPYLENE [NFP, NFP]
2-CHLOROPROPYLENE [FLA, HM1, HM1, HM1, MSL]
2-CHLOROPROPYLENE [1-PROPENE, 2-CHLORO-] [EP3]

558-13-4 CARBON TETRABROMIDE
CARBON TETRABROMIDE (TETRABROMOMETHANE) [QBC, QBC]
METHANE, TETRABROMO- [PAL]

558-17-8 2-IODO-2-METHYLPROPANE

558-25-8 METHANESULFONYL FLUORIDE

558-30-5 OXIRANE, 2,2-DIMETHYL-

560-21-4 PENTANE, 2,3,3-TRIMETHYL-
2,3,3-TRIMETHYLPENTANE [FLA, MSL, NFP, NFP]

561-27-3 HEROIN

563-04-2 PHOSPHORIC ACID, TRIS(3-METHYLPHENYL) ESTER

563-12-2 ETHION
ETHION (NIALATE) [ALB, ALB, ALB, BC1, BC1]
ETHION (NIALATE) (R) [QBC, QBC]
PHOSPHORODITHIOIC ACID, S, S'-METHYLENE O,O,O',O'-TETRAETHYLESTER [PAL]

563-41-7 SEMICARBAZIDE HYDROCHLORIDE
HYDRAZINECARBOXAMIDE, MONOHYDROCHLORIDE (SEMICARBAZIDE HYDROCHLORIDE) [CAI]
SEMICARBAZIDE HYDROCHLORIDE (HYDRAZINECARBOXAMIDE, MONOHYDROCHLORIDE) [TSC]

563-43-9 ETHYLALUMINUM DICHLORIDE
ALUMINUM, DICHLOROETHYL- [PAL]
ETHYL ALUMINUM DICHLORIDE [NJL, NJL]

563-45-1 3-METHYL-1-BUTENE
1-BUTENE, 3-METHYL- [PAL]

563-46-2 2-METHYL-1-BUTENE (TECHNICAL)
1-BUTENE, 2-METHYL- [PAL]
2-METHYL-1-BUTENE [EP3, HM1, HM1]
2-METHYL-1-BUTENE (TECHNICAL GRADE) [NFP, NFP]

563-47-3 3-CHLORO-2-METHYLPROPENE
1-PROPENE, 3-CHLORO-2-METHYL- [PAL, PAL]
3-CHLORO-2-METHYL-1-PROPENE [313]
METHALLYL CHLORIDE [FLA, MSL, NFP, NFP]
METHYL ALLYL CHLORIDE [HM1, HM1, NJL]

563-54-2 1-PROPENE, 1,2-DICHLORO-

563-58-6 1,1-DICHLOROPROPENE
1-PROPENE, 1,1-DICHLORO- [TSC, TSC]

563-68-8 THALLIUM(I) ACETATE
ACETIC ACID, THALLIUM(I) SALT [MSL, PAL]
THALLIUM ACETATE [WHM]

563-78-0 2,3-DIMETHYL-1-BUTENE
1-BUTENE, 2,3-DIMETHYL- [PAL]

563-79-1 2,3-DIMETHYL-2-BUTENE
2-BUTENE, 2,3-DIMETHYL- [PAL]

563-80-4 METHYL ISOPROPYL KETONE
2-BUTANONE, 3-METHYL- [PAL]
3-METHYL-2-BUTANONE [ONT, WHM]
3-METHYLBUTAN-2-ONE [HM1, HM1]

564-02-3 PENTANE, 2,2,3-TRIMETHYL-
2,2,3-TRIMETHYLPENTANE [FLA, MSL, NFP, NFP]

564-25-0 DOXYCYCLINE

565-59-3 2,3-DIMETHYLPENTANE
PENTANE, 2,3-DIMETHYL- [PAL]

565-76-4 1-PENTENE, 2,3,4-TRIMETHYL-
2,3,4-TRIMETHYL-1-PENTENE [FLA, MSL, NFP, NFP]

567-18-0 NAPHTHOLSULFONIC ACID (1-)

569-57-3 CHLOROTRIANISENE

569-61-9 C.I. BASIC RED 9 MONOHY-DROCHLORIDE
PARAROSANILINE HYDROCHLORIDE [WHM]

569-64-2 MALACHITE GREEN
C.I. BASIC GREEN 4 [313, CAB, CAN, MSL, NJL, PAL]
MALACHITE GREEN CHLORIDE [CSR]

573-56-8 2,6-DINITROPHENOL
PHENOL, 2,6-DINITRO- [PAL]

573-58-0 C.I. DIRECT RED 28
1-NAPHTHALENESULFONIC ACID, 3,3'-[[1,1'-BIPHENYL]-4,4'-DIYLBIS(AZO)]BIS[(4-AMINO-, DISODIUM SALT [TSC]

576-24-9 2,3-DICHLOROPHENOL

576-26-1 2,6-DIMETHYLPHENOL
DIMETHYLPHENOL, 2,6- [OTS]

577-11-7 DIOCTYL SODIUM SULFOSUCCI-NATE
SODIUM BIS(2-ETHYLHEXYL) SULFO-SUCCINATE [PST]

577-19-5 O-NITROBROMOBENZENE

578-54-1 2-ETHYLANILINE
O-ETHYLANILINE [HON]

578-94-9 DIPHENYLAMINECHLORO-AR-SINE
DIPHENYLAMINE CHLOROARSINE [HM1, HM1, HM1]

579-66-8 2,6-DIETHYLANILINE

580-51-8 [1,1'-BIPHENYL]-3-OL

581-89-5 2-NITRONAPHTHALENE

582-61-6 BENZOYL AZIDE

583-03-9 ALPHA-BUTYL-BENZYL ALCO-HOL

583-15-3 MERCURIC BENZOATE
MERCURY BENZOATE [HM1, HM1, HM1]

583-39-1 2-MERCAPTOBENZIMIDAZOLE

583-57-3 1,2-DIMETHYLCYCLOHEXANE

583-59-5 2-METHYLCYCLOHEXANOL

583-60-8 O-METHYLCYCLOHEXANONE
1-METHYLCYCLOHEXAN-2-ONE [MAK, MAK, MAK]
2-METHYLCYCLOHEXANONE [GBR, GBR, GBR, MSL, ONT, ONT, ONT, WHM]
CYCLOHEXANONE, 2-METHYL-[PAL]
METHYLCYCLOHEXANONE-O [QBC, QBC, QBC]

583-63-1 O-BENZOQUINONE

583-78-8 2,5-DICHLOROPHENOL

584-02-1 3-PENTANOL
DIETHYL CARBINOL [WHM]
DIETHYLCARBINOL [NJL, NJL]

584-03-2 1,2-BUTANEDIOL
ALPHA-BUTYLENE GLYCOL [NFP, NFP]
BUTANEDIOL, 1,2- [OTS]

584-08-7 POTASSIUM CARBONATE

584-79-2 ALLETHRIN

584-84-9 TOLUENE 2,4-DIISOCYANATE
2,4-TOLUENE DIISOCYANATE [CAA]
BENZENE, 2,4-DIISOCYANATO-1-METHYL- [PAL, PÁL, TSC, TSC]
BENZENE, 2,4-DIISOCYANATO-1-METHYL- (2,4-TOLUENE DIISO-CYANATE) [CAI]
TOLUENE 2,4-DIISOCYANATE (TDI) [MAK, MAK, MAK, WHM]

TOLUENE 2,4-DIISOCYANATE [BENZENE, 2,4-DIISOCYANATO-1-METHYL-] [EP3]
TOLUENE 2,4-DIISOCYANATE(TDI) [FLA]
TOLUENE DIISOCYANATE [MIN, NTP]
TOLUENE-2,4-DIISOCYANATE (TDI) [MSL, OS1, OS1, OS1, TLV, TLV]

584-94-1 2,3-DIMETHYLHEXANE
HEXANE, 2,3-DIMETHYL- [PAL]

585-07-9 2-PROPENOIC ACID, 2-METHYL-, 1,1-DIMETHYLETHYL ESTER

585-79-5 M-NITROBROMOBENZENE

586-62-9 TERPINOLENE

586-78-7 P-NITROBROMOBENZENE
NITROBROMOBENZENE [NJL]

587-85-9 DIPHENYLMERCURY

587-98-4 METANIL YELLOW
C.I. ACID YELLOW 36, MONOSODIUM SALT [PST]

588-59-0 STILBENE

589-18-4 P-METHYLBENZYL ALCOHOL

589-34-4 3-METHYLHEXANE
HEXANE, 3-METHYL- [PAL]

589-38-8 3-HEXANONE

589-43-5 2,4-DIMETHYLHEXANE
HEXANE, 2,4-DIMETHYL- [PAL]

589-82-2 3-HEPTANOL

589-90-2 1,4-DIMETHYLCYCLOHEXANE
CYCLOHEXANE, 1,4-DIMETHYL-[PAL]

589-91-3 4-METHYLCYCLOHEXANOL

589-98-0 3-OCTANOL

590-00-1 SORBIC ACID, POTASSIUM SALT

590-01-2 BUTYL PROPIONATE
PROPANOIC ACID, BUTYL ESTER [PAL]

590-18-1 2-BUTENE-CIS
2-BUTENE, (Z)- [PAL]

590-21-6 1-CHLOROPROPYLENE
1-CHLOROPROPYLENE [1-PROPENE, 1-CHLORO-] [EP3]
1-PROPENE, 1-CHLORO- [PAL]

590-28-3 POTASSIUM CYANATE

590-36-3 2-METHYL-2-PENTANOL
2-METHYLPENTAN-2-OL [HM1, HM1]

590-67-0 METHYL CYCLOHEXANOLS

590-86-3 ISOPENTALDEHYDE
BUTANAL, 3-METHYL- [OTS, PAL, TSC, TSC, TSC]

590-88-5 1,3-BUTANEDIAMINE

590-96-5 METHYLAZOXYMETHANOL
METHANOL, (METHYL-ONN-AZOXY)-[PAL, PAL]

591-08-2 1-ACETYL-2-THIOUREA
ACETAMIDE, N-(AMINOTHIOX-OMETHYL)- [MSL, PAL, TSC, TSC]

591-21-9 1,3-DIMETHYL CYCLOHEXANE
1,3-DIMETHYLCYCLOHEXANE [NFP, NFP]
CYCLOHEXANE, 1,3-DIMETHYL-[PAL]

591-23-1 3-METHYLCYCLOHEXANOL

591-27-5 M-AMINOPHENOL

591-47-9 4-METHYLCYCLOHEXENE
CYCLOHEXENE, 4-METHYL- [PAL]

591-50-4 IODOBENZENE

591-60-6 BUTYL ACETOACETATE

591-62-8 N-BUTYLURETHANE

591-76-4 ISOHEPTANE
2-METHYLHEXANE [NFP, NFP]
HEXANE, 2-METHYL- [PAL]

591-78-6 2-HEXANONE
2-HEXANONE [BC1, BC1]
2-HEXANONE (METHYL N-BUTYL KETONE) [ALB, ALB, ALB, OS1, OS1]
HEXAN-2-ONE [GBR, GBR]
METHYL BUTYL KETONE [CEX, CEX, FLA, MSL, NFP, NFP]
METHYL N-BUTYL KETONE [AUS, AUS, CAL, NHS, NHS, OTV, TLV, TLV, TSE]
METHYL N-BUTYL KETONE (2-HEX-ANONE) [TSC, TSC]
METHYL-N-BUTYL KETONE [NJL, QBC, QBC]

591-87-7 ALLYL ACETATE
ACETIC ACID, 2-PROPENYL ESTER [PAL]

591-89-9 MERCURIC POTASSIUM CYANIDE

591-97-9 1-CROTYL CHLORIDE
2-BUTENE, 1-CHLORO- [PAL]

592-01-8 CALCIUM CYANIDE
CALCIUM CYANIDE (CA(CN)2) [PAL]

592-04-1 MERCURIC CYANIDE
MERCURY CYANIDE [HM1, HM1, HM1, HM1]
MERCURY CYANIDE (HG(CN)2) [PAL]

592-05-2 LEAD CYANIDE

592-34-7 N-BUTYL CHLOROFORMATE
BUTYL CHLOROFORMATE [NJL, NJL]

592-41-6 1-HEXENE

592-42-7 HEXADIENE

592-43-8 2-HEXENE (MIXED CIS & TRANS ISOMERS)
2-HEXENE (MIXED CIS- AND TRANS-ISOMERS) [PAL]
2-HEXENE(MIXED CIS & TRANS) [MSL]

592-45-0 1,4-HEXADIENE

592-55-2 BROMOETHYL ETHYL ETHER
2-BROMOETHYL ETHYL ETHER [HM1, HM1]

592-62-1 METHYLAZOXYMETHYL AC-ETATE
METHANOL, (METHYL-ONN-AZOXY)-, ACETATE (ESTER) [PAL, PAL]
METHYLAZOXYMETHANOL AC-ETATE [FLA, MSL]

592-76-7 1-HEPTENE

592-84-7 BUTYL FORMATE
FORMIC ACID, BUTYL ESTER [PAL]
N-BUTYL FORMATE [HM1, HM1]

592-85-8 MERCURIC THIOCYANATE
MERCURY THIOCYANATE [NJL]
THIOCYANIC ACID, MERCURY(2+) SALT [PAL]

592-87-0 LEAD THIOCYANATE
THIOCYANIC ACID, LEAD(2+) SALT [PAL]

593-08-8 METHYL UNDECYL KETONE

593-29-3 POTASSIUM STEARATE

593-45-3 OCTADECANE

593-53-3 METHYL FLUORIDE

593-60-2 VINYL BROMIDE
ETHENE, BROMO- [PAL, TSC, TSC, TSC]
VINYL BROMIDE (BROMOETHYLENE) [QBC, QBC]

615-50-9 1,4-BENZENEDIAMINE, 2-
METHYL-, SULFATE (1:1)

615-53-2 N-NITROSO-N-METHY-
LURETHANE
CARBAMIC ACID, METHYLNITROSO-,
ETHYL ESTER [PAL, PAL]
METHYL URETHANE [NJL]
N-METHYL-N-NITROSOURETHANE
[IAR, MIN]
NITROSOMETHYLURETHANE
[WHM]

615-58-7 2,4-DIBROMOPHENOL
PHENOL, 2,4-DIBROMO- [TSC, TSC,
TSC, TSC]

615-65-6 2-CHLORO-4-METHYL ANILINE

615-66-7 3-CHLORO-4-AMINOANILINE

615-67-8 CHLOROHYDROQUINONE

616-02-4 CITRACONIC ANHYDRIDE

616-21-7 1,2-DICHLOROBUTANE
BUTANE, 1,2-DICHLORO- [PAL]

616-23-9 1-PROPANOL, 2,3-DICHLORO-
ALPHA, BETA-GLYCERIN DICHLORO-
HYDRIN [PAL]
ALPHA,BETA-GLYCERIN DICHLORO-
HYDRIN [NFP, NFP]

616-29-5 1,3-DIAMINO-2-PROPANOL
2-PROPANOL, 1,3-DIAMINO- [PAL]

616-38-6 DIMETHYL CARBONATE
CARBONIC ACID, DIMETHYL ESTER
[PAL]
METHYL CARBONATE [FLA, MSL,
NFP, NFP]

616-45-5 2-PYRROLIDINONE
2-PYRROLIDONE [NFP, NFP]

616-47-7 1-METHYLIMIDAZOLE

616-74-0 4,6-DINITRORESORCINOL

617-50-5 ISOPROPYL ISOBUTYRATE

617-51-6 ISOPROPYL LACTATE
PROPANOIC ACID, 2-HYDROXY-, 1-
METHYLETHYL ESTER [PAL]

617-89-0 FURFURYLAMINE
2-FURANMETHANAMINE [PAL]

617-94-7 2,2-DIMETHYLBENZYL ALCOHOL

618-76-8 IOXYNIL

619-15-8 TOLUENE, 2,5-DINITRO-

619-17-0 4-NITROANTHRANILIC ACID

619-97-6 BENZENE DIAZONIUM NITRATE

620-05-3 BENZYL IODIDE

620-14-4 M-ETHYLTOLUENE
M-ETHYLTOLUENE (BENZENE, 1-
ETHYL-3-METHYL-) [TSC, TSC]

621-64-7 N-NITROSODI-N-PROPYLAMINE
1-PROPANAMINE, N-NITROSO-N-
PROPYL- [PAL, PAL]
DI-N-PROPYLNITROSAMINE [EPA,
PQL, RCA, UTS]
DI-N-PROPYLNITROSOAMINE [RCU,
RCU]
N-NITROSO-DI-N-PROPYLAMINE
[CAL]
N-NITROSODIPROPYLAMINE [MX2,
WHM]

621-77-2 1-PENTANAMINE, N,N-DIPENTYL-
TRIAMYLAMINE [NFP, NFP]
TRIPENTYLAMINE [FLA, MSL]

621-78-3 TRI-N-AMYL BORATE
TRIAMYL BORATE [NFP, NFP]

622-08-2 ETHYLENE GLYCOL MONOBEN-
ZYL ETHER
ETHANOL, 2-(PHENYLMETHOXY)-
[PAL]

622-40-2 MORPHOLINE ETHANOL
4-(2-HYDROXYETHYL) MORPHOLINE
[FLA, MSL, NFP, NFP]
4-MORPHOLINEETHANOL [PAL]

622-44-6 PHENYLCARBYLAMINE CHLO-
RIDE

622-45-7 CYCLOHEXYL ACETATE

622-57-1 N-ETHYL-P-TOLUIDINE
ETHYL TOLUIDINE [NJL]

622-58-2 BENZENE, 1-ISOCYANATO-4-
METHYL-

622-86-6 BENZENE, (2-CHLOROETHOXY)-

622-96-8 P-ETHYLTOLUENE

623-17-6 FURFURYL ACETATE

623-25-6 2,2'-DICHLORO-P-XYLENE

623-26-7 TEREPHTHALONITRILE

623-42-7 METHYL BUTYRATE
BUTANOIC ACID, METHYL ESTER
[PAL]

623-51-8 ETHYL MERCAPTOACETATE

623-70-1 ETHYL CROTONATE

623-84-7 1,2-PROPANEDIOL DIACETATE

623-87-0 GLYCEROL-1,3-DINITRATE

623-91-6 FUMARIC ACID, DIETHYL ESTER
2-BUTENEDIOIC ACID (E)-, DIETHYL
ESTER [OTS]
DIETHYL FUMARATE [NFP, NFP]

624-18-0 P-PHENYLENEDIAMINE DIHY-
DROCHLORIDE
1,4-BENZENEDIAMINE, DIHY-
DROCHLORIDE [TSC, TSC]
1,4-PHENYLENEDIAMINE DIHY-
DROCHLORIDE [313]

624-24-8 METHYL VALERATE

624-29-3 1,4-DIMETHYLCYCLOHEXANE-
CIS
CYCLOHEXANE, 1,4-DIMETHYL-, CIS-
[PAL]

624-48-6 DIMETHYL MALEATE

624-54-4 AMYL PROPIONATE

624-61-3 DIBROMOACETYLENE

624-64-6 2-BUTENE, (E)-
2-BUTENE-(E) [FLA]
2-BUTENE-TRANS [MSL, NFP, NFP]
2-BUTENE-TRANS [2-BUTENE, (E)]
[EP3]

624-74-8 DIIODOACETYLENE

624-83-9 METHYL ISOCYANATE
METHANE, ISOCYANATO- [PAL,
TSC, TSC, TSC]
METHYL ISOCYANATE (M.I.C.) [FLA]
METHYL ISOCYANATE [METHANE,
ISOCYANATO-] [EP3]

624-91-9 METHYL NITRITE

624-92-0 DIMETHYLDISULFIDE
DIMETHYL DISULFIDE [HM1, HM1,
TSC]
DISULFIDE, DIMETHYL (DIMETHYL
DISULFIDE) [CAI]
METHYL DISULFIDE [EPA, EPA,
MSL, PAL]

625-16-1 TERT-AMYL ACETATE
2-BUTANOL, 2-METHYL-, ACETATE
[PAL]

625-27-4 2-METHYL-2-PENTENE
2-PENTENE, 2-METHYL- [PAL]

625-30-9 SEC-AMYLAMINE
2-PENTANAMINE [PAL]

625-50-3 N-ETHYLACETAMIDE

625-55-8 ISOPROPYL FORMATE
FORMIC ACID, 1-METHYLETHYL
ESTER [PAL]

625-58-1 ETHYL NITRATE
NITRIC ACID, ETHYL ESTER [PAL]

625-69-4 2,4-PENTANEDIOL

625-76-3 DINITROMETHANE

625-86-5 2,5-DIMETHYLFURAN
FURAN, 2,5-DIMETHYL- [PAL]

626-17-5 M-PHTHALODINITRILE
1,3-BENZENEDICARBONITRILE [PAL]
1,3-DICYANOBENZENE [TSC, TSC,
TSC]

626-23-3 DI-SEC-BUTYLAMINE
2-BUTANAMINE, N-(1-METHYL-
PROPYL)- [PAL]

626-38-0 SEC-AMYL ACETATE
1-METHYLBUTYL ACETATE [GBR]
2-PENTANOL, ACETATE [PAL]

626-39-1 1,3,5-TRIBROMOBENZENE

626-43-7 BENZENAMINE, 3,5-DICHLORO-

626-67-5 1-METHYLPIPERIDINE

626-86-8 ADIPIC ACID MONOMETHYL
ESTER

627-11-2 CHLOROETHYL CHLOROFOR-
MATE
CHLOROETHYLCHLOROFORMATE
[NJL]

627-12-3 N-PROPYL CARBAMATE

627-13-4 N-PROPYL NITRATE
NITRIC ACID, PROPYL ESTER [PAL]
PROPYL NITRATE [NFP, NFP, OS3]

627-19-0 1-PENTYNE

627-20-3 BETA-AMYLENE-CIS
2-PENTENE, (Z)- [EP3, PAL]

627-27-0 VINYL ETHYL ALCOHOL

627-30-5 3-CHLORO-1-PROPANOL
3-CHLOROPROPANOL [NJL]
3-CHLOROPROPANOL-1 [HM1,
HM1]

627-44-1 DIETHYLMERCURY

627-53-2 ETHANE, 1,1'-SELENOBIS-
DIETHYL SELENIDE [FLA, MSL,
NFP]

627-63-4 FUMARYL CHLORIDE

627-83-8 ETHYLENE GLYCOL DISTEARATE

628-28-4 BUTYL METHYL ETHER

628-32-0 ETHYL PROPYL ETHER
PROPANE, 1-ETHOXY- [PAL]

628-37-5 DIETHYL PEROXIDE
PEROXIDE, DIETHYL [PAL]

628-63-7 N-AMYL ACETATE
ACETIC ACID, PENTYL ESTER [PAL]
AMYL ACETATE [EPA, HM1, HM1,
HM1, MAK, NFP, NFP, PST]
AMYL ACETATE, N- [OTS]
PENTYL ACETATE [GBR, GBR,
WHM]

628-68-2 DIETHYLENE GLYCOL DIAC-
ETATE

628-76-2 1,5-DICHLOROPENTANE
PENTANE, 1,5-DICHLORO- [PAL]

628-81-9 ETHYL BUTYL ETHER
BUTANE, 1-ETHOXY- [PAL]
ETHYLBUTYL ETHER [NJL, NJL]

628-86-4 MERCURY FULMINATE
FULMINATING MERCURY [HM1,
HM1, HM1]
FULMINIC ACID, MERCURY(II)SALT
[MSL, PAL]

628-92-2 CYCLOHEPTENE

628-94-4 ADIPAMIDE

628-96-6 ETHYLENE GLYCOL DINITRATE
1,2-ETHANEDIOL, DINITRATE [PAL]
ETHYLENE DINITRATE [GBR, GBR, GBR]
ETHYLENE GLYCOL DINITRATE (EGDN) [NHS, NHS]
ETHYLENE GLYCOL DINITRATE AND/OR NITROGLYCERIN [BC1, BC1]

629-06-1 1-CHLOROHEPTANE
NORMAL-HEPTYL CHLORIDE [HM1]

629-11-8 1,6-HEXANEDIOL
HEXAMETHYLENE GLYCOL [OTS]

629-14-1 ETHYLENE GLYCOL DIETHYL ETHER
ETHANE, 1,2-DIETHOXY- [PAL]
ETHYLENE GLYCOL DIETHYL ETHER (1,2-DIETHOXYETHANE) [HON]

629-15-2 ETHYLENE GLYCOL DIFORMATE
1,2-ETHANEDIOL DIFORMATE [NFP, NFP]

629-20-9 CYCLOOCTATETRAENE

629-38-9 DIETHYLENE GLYCOL METHYL ETHER ACETATE

629-59-4 TETRADECANE

629-73-2 HEXADECYLENE-1

629-82-3 DIOCTYL ETHER

630-08-0 CARBON MONOXIDE

630-10-4 CARBAMIMIDOSELENOIC ACID
SELENOUREA [EPA, RCU, RCU]

630-16-0 1,1,1,2-TETRABROMOETHANE

630-20-6 1,1,1,2-TETRACHLOROETHANE
ETHANE, 1,1,1,2-TETRACHLORO- [PAL, TSC, TSC]

630-56-8 17A-HYDROXYPROGESTERONE CAPROATE

630-60-4 OUABAIN

630-72-8 TRINITROACETONITRILE

630-93-3 DIPHENYLHYDANTOIN, SODIUM SALT
DIPHENYLHYDANTOIN (PHENYTOIN), SODIUM SALT [C65, CAL]
SODIUM SALT OF PHENYTOIN [PAL, PAL]

631-41-4 TETRAETHANOL AMMONIUM HYDROXIDE

631-60-7 MERCUROUS ACETATE

631-61-8 AMMONIUM ACETATE
ACETIC ACID, AMMONIUM SALT [PAL]

631-71-0 ETHYL ABIETATE

632-22-4 TETRAMETHYLUREA

632-79-1 1,3-ISOBENZOFURANDIONE, 4,5,6,7-TETRABROMO-
TETRABROMOPHTHALIC ANHYDRIDE [HON]
TETRABROMOPHTHALICANHYDRIDE [TSC, TSC, TSC]

632-99-5 MAGENTA
MAGENTA I [IAR, IAR]

633-03-4 C.I. BASIC GREEN 1

633-96-5 ACID ORANGE 7
C.I. ACID ORANGE 7, MONOSODIUM SALT [PST]

634-66-2 1,2,3,4-TETRACHLOROBENZENE
BENZENE, 1,2,3,4-TETRACHLORO- [TSC, TSC]

634-90-2 BENZENE, 1,2,3,5-TETRACHLORO-

634-93-5 BENZENAMINE, 2,4,6-TRICHLORO-
TRICHLOROANILINE (2,4,6-) [HON]

635-22-3 BENZENAMINE, 4-CHLORO-3-NITRO-

636-09-9 DIETHYL TEREPHTHALATE

636-21-5 O-TOLUIDINE HYDROCHLORIDE
BENZENAMINE, 2-METHYL-, HYDROCHLORIDE [PAL, PAL]
ORTHO-TOLUIDINE HYDROCHLORIDE [C65, C65, CAL, FLA]

637-07-0 CLOFIBRATE

637-12-7 ALUMINUM STEARATE
ALUMINUM TRISTEARATE [CEX]

637-78-5 ISOPROPYL PROPIONATE

637-92-3 ETHYL TERT-BUTYL ETHER

638-10-8 ETHYL 3,3-DIMETHACRYLATE

638-11-9 ISOPROPYL BUTYRATE

638-16-4 TRITHIOCYANURIC ACID

638-17-5 THIALDINE
4H-1,3,5-DITHIAZINE, DIHYDRO-2,4,6-TRIMETHYL-, (2.ALPHA.,4.ALPHA.,6.ALPHA.)- [PAL]

638-21-1 PHENYLPHOSPHINE
PHOSPHINE, PHENYL- [PAL]

638-29-9 VALERYL CHLORIDE

638-38-0 MANGANESE(II) ACETATE

638-49-3 AMYL FORMATE
FORMIC ACID, PENTYL ESTER [PAL]

638-56-2 BIS(2-(2-CHLOROETHOXY)ETHYL) ETHER

639-58-7 TRIPHENYLTIN CHLORIDE
TRIPHENYL TIN CHLORIDE [MSL]

640-15-3 THIOMETON

640-19-7 FLUOROACETAMIDE
ACETAMIDE, 2-FLUORO [TSE]
ACETAMIDE, 2-FLUORO- [PAL, RCU, RCU, TSC, TSC]
FLUOROACETAMIDE/1081 [CAL, MSL]
FLUOROACETAMINE [FLA]

641-16-7 2,3,4,6-TETRANITROPHENOL

643-22-1 ERYTHROMYCIN STEARATE

643-28-7 N-ISOPROPYLANILINE
2-ISOPROPYL ANILINE [WHM]
O-ISOPROPYLANILINE [MSL]

643-58-3 2-METHYLBIPHENYL
1,1'-BIPHENYL, 2-METHYL- [PAL]

644-08-6 4-METHYL BIPHENYL

644-31-5 ACETYL BENZOYL PEROXIDE

644-64-4 DIMETILAN

644-97-3 PHENYL PHOSPHORUS DICHLORIDE
BENZENE PHOSPHORUS DICHLORIDE [NJL, NJL]

645-55-6 N-NITROANILINE

645-56-7 P-PROPYLPHENOL

645-62-5 2-ETHYL-3-PROPYLACROLEIN
2-HEXENAL, 2-ETHYL- [PAL]

646-04-8 BETA-AMYLENE-TRANS
2-PENTENE, (E)- [EP3, PAL]

646-06-0 1,3-DIOXOLANE
DIOXOLANE [HM1, HM1, MSL, NJL, NJL]

650-51-1 SODIUM TRICHLOROACETATE

657-27-2 L-LYSINE HCL

662-01-1 1,3-DICHLORO-1,1,2,2,3,3,-HEXAFLUOROPROPANE
CFC-216 [CAA]

670-54-2 TETRACYANOMETHYLENE

671-16-9 PROCARBAZINE
BENZAMIDE, N-(1-METHYLETHYL)-4-[(2-METHYLHYDRAZINO)METHYL]- [PAL, PAL]
N-ISOPROPYL-ALPHA-(2-METHYLHYDRAZINO)-P-TOLUAMIDE [NJL]

674-82-8 DIKETENE
2-OXETANONE, 4-METHYLENE- [PAL]

675-14-9 CYANURIC FLUORIDE

676-83-5 METHYL PHOSPHONOUS DICHLORIDE

676-97-1 METHYL PHOSPHONIC DICHLORIDE
METHYLPHOSPHONIC DICHLORIDE [EPA, EPA]

676-98-2 METHYL PHOSPHONOTHIOIC DICHLORIDE

676-99-3 METHYL PHOSPHONIC DIFLUORIDE

677-21-4 1-PROPENE, 3,3,3-TRIFLUORO-

679-85-6 3-CHLORO-1,1,2,2-TETRAFLUOROPROPANE
HCFC-244 [CAA]

680-31-9 HEXAMETHYL PHOSPHORAMIDE
HEXAMETHYLPHOSPHORAMIDE [313, C65, C65, CAA, CAL, IAR, MIN, NTP, TSC, TSC, TSE]
HEXAMETHYLPHOSPHORIC ACID TRIAMDE [MAK]
HMPA (HEXAMETHYLPHOSPHORAMIDE) [QBC, QBC, QBC]
PHOSPHORIC TRIAMIDE, HEXAMETHYL- [PAL, PAL]

681-84-5 METHYL SILICATE
METHYL ORTHOSILICATE [HM1, HM1, HM1]
SILICIC ACID (H4SIO4), TETRAMETHYL ESTER [PAL]
TETRAMETHOXYSILANE [ONT]
TETRAMETHYL ORTHOSILICATE [GBR, GBR]

682-09-7 1-BUTANOL, 2,2-BIS[(2-PROPENYLOXY)METHYL]-

683-10-3 DODECYLBETAINE

683-18-1 DIBUTYLTIN DICHLORIDE

683-53-4 BROMODICHLOROETHANE

684-16-2 HEXAFLUOROACETONE
2-PROPANONE, 1,1,1,3,3,3-HEXAFLUORO- [PAL]

684-93-5 N-NITROSO-N-METHYLUREA
N-METHYL-N-NITROSOUREA [IAR]
N-NITROSO-N-METHYLUREA CARBAMIDE [MSL]
UREA, N-METHYL-N-NITROSO- [PAL, PAL]

685-91-6 ACETAMIDE, N,N-DIETHYL-

688-73-3 STANNANE, TRI-N-BUTYL-, HYDRIDE
TRIBUTYLTIN [CAL]

688-74-4 TRIBUTYL BORATE
BORIC ACID (H3BO3), TRIBUTYL ESTER [PAL]
TRI-N-BUTYL BORATE [FLA, MSL, NFP, NFP]

688-84-6 2-ETHYLHEXYL METHACRYLATE
2-PROPENOIC ACID, 2-METHYL-, 2-ETHYLHEXYL ESTER [TSC, TSC]

689-12-3 2-PROPENOIC ACID, 1-METHYLETHYL ESTER

689-97-4 1-BUTEN-3-YNE
VINYL ACETYLENE [FLA, MSL, NFP]
VINYL ACETYLENE [1-BUTEN-3-YNE] [EP3]

691-37-2 4-METHYL-1-PENTENE
1-PENTENE, 4-METHYL- [PAL]

692-42-2 ARSINE, DIETHYL-
DIETHYLARSINE [MSL]

693-07-2 2-CHLOROETHYL ETHYL SULFIDE

693-13-0 DIISOPROPYLCARBODIIMIDE

693-21-0 DIETHYLENEGLYCOL DINITRATE

693-23-2 DODECANEDIOIC ACID

693-54-9 2-DECANONE

693-65-2 AMYL ETHER

693-98-1 2-METHYLIMIDAZOLE
1H-IMIDAZOLE, 2-METHYL- [PST]

694-05-3 1,2,3,6-TETRAHYDROPYRIDINE

696-28-6 DICHLOROPHENYLARSINE
ARSINE,DICHLOROPHENYL [MSL]
ARSONOUS DICHLORIDE, PHENYL- [TSC, TSC]
PHENYL DICHLOROARSINE [302, FLA, PAL]
PHENYLDICHLOROARSINE [NJL]

696-29-7 ISOPROPYLCYCLOHEXANE

700-12-9 1,2,3,4,5-PENTAMETHYL BENZENE

702-03-4 N-(2-CYANOETHYL)CYCLOHEXYLAMINE
PROPANENITRILE, 3-(CYCLOHEXYLAMINO)- [PAL]

706-14-9 HYDROXYDECANOIC ACID, GAMMA-LACTONE

707-19-7 PROPARGYL ALCOHOL
PENTYL PROPARGYL ALCOHOL [ALB, ALB, ALB]

709-98-8 PROPANIL
PROPANIL [N-(3,4-DICHLOROPHENYL)PROPANAMIDE] [313]

712-48-1 DIPHENYLCHLOROARSINE

712-68-5 2-AMINO-5-(5-NITRO-2-FURYL)-1, 3,4-THIADIAZOLE
1,3,4-THIADIAZOL-2-AMINE, 5-(5-NITRO-2-FURANYL)- [PAL, PAL]

713-46-2 BETA(P-TERT-BUTYLPHENOXY) ETHANOL

717-74-8 TRIISOPROPYLBENZENE

723-46-6 SULFAMETHOXAZOLE

732-11-6 PHOSMET

732-26-3 2,4,6-TRIISOPROPYLPHENOL

738-70-5 TRIMETHOPRIM

741-58-2 BENZENESULFONAMIDE, N-(2-MERCAPTOETHYL)-, S-ESTER WITH O,O-DIISOPROPYLPHOSPHORODITHIOATE

745-65-3 CYCLOPENTANEHEPTANOIC ACID, 3-.ALPHA.-HYDROXY-2-(3-HYDROXY-1-OCTENYL)-5-OXO-

753-53-7 BORON TRIFLUORIDE ACETIC ACID
BORON TRIFLUORIDE ACETIC ACID COMPLEX [NJL]

756-79-6 DIMETHYL METHYLPHOSPHONATE

757-58-4 HEXAETHYL TETRAPHOSPHATE
HEXAETHYLTETRAPHOSPHATE [HM1, HM1, HM1, HM1, HM1]
TETRAPHOSPHORIC ACID, HEXAETHYL ESTER [TSC, TSC]

759-73-9 N-NITROSO-N-ETHYLUREA
N-ETHYL-N-NITROSUREA [IAR]
UREA, N-ETHYL-N-NITROSO- [PAL, PAL]

759-94-4 ETHYL DIPROPYLTHIOCARBAMATE [EPTC]

760-21-4 2-ETHYL-1-BUTENE
PENTANE, 3-METHYLENE- [PAL]

760-23-6 3,4-DICHLORO-1-BUTENE
3,4-DICHLOROBUTENE [TSC, TSC, TSC]

760-93-0 METHACRYLIC ANHYDRIDE

762-04-9 DIETHYL PHOSPHITE

762-12-9 DECANOYL PEROXIDE
DIDECANOYL PEROXIDE [DOT]

762-13-0 DI-N-NONANOYL PEROXIDE

762-16-3 DI-N-OCTANOYL PEROXIDE

763-29-1 2-METHYL-1-PENTENE
1-PENTENE, 2-METHYL- [PAL]

763-69-9 ETHYL 3-ETHOXYPROPANOATE
PROPANOIC ACID, 3-ETHOXY-, ETHYL ESTER [PST]

764-35-2 METHYL PROPYL ACETYLENE
2-HEXYNE [PAL]

764-41-0 1,4-DICHLORO-2-BUTENE
1,4-DICHLOROBUTENE-2 [MAK]
2-BUTENE, 1,4-DICHLORO- [EPA, MSL, PAL]

764-42-1 FUMARONITRILE

765-34-4 GLYCIDYLALDEHYDE
GLYCIDALDEHYDE [C65, CAL, FLA, IAR, MIN, MSL, NJL, NJL, WHM]

766-09-6 1-ETHYL PIPERDINE

768-52-5 N-ISOPROPYLANILINE
BENZEDRINE [NFP, NFP]
BENZENAMINE, N-(1-METHYLETHYL) - [PAL]

770-35-4 2-PROPANOL, 1-PHENOXY-
1-PHENOXY-2-PROPANOL [TSC, TSC, TSC]

771-29-9 TETRALIN HYDROPEROXIDE

772-54-3 BENZENEMETHANAMINE, N,N-DIETHYL-
N-BENZYLDIETHYLAMINE [FLA, MSL, NFP, NFP]

776-74-9 DIPHENYL METHYL BROMIDE
DIPHENYLMETHYL BROMIDE [NJL, NJL]

777-37-7 4-NITRO-1-CHLOROBENZO-2-TRIFLUORIDE
2-CHLORO-5-NITROBENZOTRIFLUORIDE [FLA, MSL, NFP, NFP]
BENZENE, 1-CHLORO-4-NITRO-2-(TRIFLUOROMETHYL)- [PAL]

778-26-7 MERCUROUS NITRATE

786-19-6 CARBOPHENOTHION

789-02-6 O,P'-DDT

789-07-1 2-NITROPYRENE

789-61-7 BETA-THIOGUANIDINE DEOXYRIBOSIDE
B-TGDR [MSL]

793-24-8 1,4-BENZENEDIAMINE, N-(1,3-DIMETHYLBUTYL)-

794-93-4 PANFURAN S
DIHYDROXYMETHYLFURATRIZINE [CAL]
METHANOL,[[6-[2-(5-NITRO-2-FURANYL)ETHENYL]-1,2,4-TRIAZIN-3-YL]IMINO]BIS- [PAL, PAL]
PANFURAN S (CONTAINING DIHYDROXYMETHYLFURATRIZINE) [IAR]

800-24-8 AZIRIDYL BENZOQUINONE

804-36-4 NITROVIN

811-54-1 LEAD FORMATE

811-97-2 1,1,1,2-TETRAFLUOROETHANE
ETHANE, 1,1,1,2-TETRAFLUORO- [TSC, TSC]

812-00-0 PHOSPHORIC ACID, MONOMETHYL ESTER

812-03-3 PROPANE, 1,1,1,2-TETRACHLORO-

812-04-4 1,1-DICHLORO-1,2,2-TRIFLUOROETHANE (HCFC-123B)

813-94-5 CALCIUM CITRATE

814-49-3 DIETHYL CHLOROPHOSPHATE
DIETHYLCHLOROPHOSPHATE [MSL]

814-68-6 ACRYLYL CHLORIDE
ACRYLYL CHLORIDE [2-PROPENOYL CHLORIDE] [EP3]

814-78-8 METHYL ISOPROPENYL KETONE
3-BUTEN-2-ONE, 3-METHYL- [PAL]
METHYL ISOPROPENYL KETONE, INHIBITED [NJL]

814-91-5 COPPER OXALATE (CUC204)
CUPRIC OXALATE [NJL]

814-93-7 LEAD OXALATE

814-94-8 STANNOUS OXALATE

815-82-7 CUPRIC TARTRATE
BUTANEDIOIC ACID, 2,3-DIHYDROXY-[R-(R*,R*)]-, COPPER(2+)SALT (1:1) [PAL]

815-84-9 TARTARIC ACID, LEAD(2+) SALT
LEAD TARTRATE [MSL]

817-09-4 TRICHLORMETHINE (TRIMUSTINE HCL)

818-08-6 DIBUTYLTIN OXIDE

818-61-1 ETHYLENE GLYCOL MONOACRYLATE
2-HYDROXYETHYL ACRYLATE [NFP, NFP]
2-PROPENOIC ACID, 2-HYDROXYETHYL ESTER [PAL, TSC, TSC]
HYDROXYETHYL ACRYLATE [WHM]

821-08-9 DIVINYL ACETYLENE
1,5-HEXADIEN-3-YNE [PAL]

821-55-6 2-NONANONE
METHYL HEPTYL KETONE [NFP, NFP]

822-06-0 HEXAMETHYLENE DIISOCYANATE
1,6-HEXAMETHYLENEDIISOCYANATE [MAK, MAK, MAK]
HEXAMETHYLENE DIISOCYANATE (HDI) [NHS, NHS]
HEXAMETHYLENE DIISOCYANATE, 1,6- [OTS]
HEXAMETHYLENE-1,6-DIISOCYANATE [313, CAA, CAB]
HEXAMETHYLENE-1,6-DIISOCYANATE (HDI) [ONT, ONT, ONT]
HEXAMETHYLENEDIISOCYANATE [WHM]
HEXANE, 1,6-DIISOCYANATO- [TSC, TSC, TSC]

822-16-2 SODIUM STEARATE

822-36-6 4-METHYLIMIDAZOLE

823-40-5 2,6-DIAMINOTOLUENE
1,3-BENZENEDIAMINE, 2-METHYL- [TSC, TSC]
2,6-TOLUENEDIAMINE [EPA]
DIAMINOTOLUENE [PAL]

824-11-3 TRIMETHYLOLPROPANE PHOS-
PHITE

825-51-4 2-HYDROXYDECALIN

826-62-0 BICYCLO [2,2,1]5-HEPTENE-2,3-
DICARBOXYLIC ANHYDRIDE

827-52-1 CYCLOHEXYLBENZENE
BENZENE, CYCLOHEXYL- [PAL]

827-94-1 BENZENAMINE, 2,6-DIBROMO-4-
NITRO-

828-00-2 DIMETHOXANE
1,3-DIOXAN-4-OL, 2,6-DIMETHYL-,
ACETATE [TSC, TSC]
6-ACETOXY-2,4-DIMETHYL-M-DIOX-
ANE [PST]

831-52-7 SODIUM PICRAMATE

832-69-9 1-METHYLPHENANTHRENE

834-12-8 AMETRYN (N-ETHYL-
N'-(1-METHYLETHYL)-6-
(METHYLTHIO)-1,3,5,-TRIAZINE-
2,4-DIAMINE)

834-28-6 PHENFORMIN HYDROCHLORIDE

836-30-6 P-NITRODIPHENYLAMINE
BENZENAMINE, 4-NITRO-N-PHENYL-
[OTS]

838-88-0 4,4'-METHYLENE BIS(2-METHY-
LANILINE)
4,4'-METHYLENEBIS(2-METHYLANI-
LINE) [MIN]
BENZENAMINE, 4-4'-METHYLENEBIS-
(2-METHYL)- [PAL, PAL]

842-07-9 C.I. SOLVENT YELLOW 14
SUDAN I [CAL, IAR]

846-49-1 LORAZEPAM

846-50-4 TEMAZEPAM

849-99-0 DICYCLOHEXYL ADIPATE

860-22-0 1H-INDOLE-5-SULFONIC ACID,2-
(1,3-DIHYDRO-3-OXO-5-SULFO-2H
FD&C BLUE NO. 2 [PST]

865-21-4 VINBLASTINE

868-18-8 SODIUM TARTRATE

868-77-9 2-HYDROXYETHYL METHACRY-
LATE
2-PROPENOIC ACID, 2-METHYL-, 2-
HYDROXYETHYL ESTER [TSC, TSC]

868-85-9 DIMETHYL HYDROGEN PHOS-
PHITE

869-29-4 ALLYLIDENE DIACETATE
2-PROPENE-1,1-DIOL, DIACETATE
[PAL]

870-08-6 DIOCTYLTIN OXIDE

871-27-2 DIETHYLALUMINUM HYDRIDE
ALUMINUM, DIETHYLHYDRO- [PAL]

872-05-9 1-DECENE
DECENE, N- [OTS]

872-10-6 PENTANE, 1,1'-THIOBIS-
DIAMYL SULFIDE [FLA, MSL, NFP,
NFP]

872-50-4 1-METHYL-2-PYRROLIDONE
1-METHYL-2-PYRROLIDONE [MSL,
NFP, NFP]
2-PYRROLIDINONE, 1-METHYL-
[PAL]
METHYLPYRROLIDONE, N- [OTS]
N-METHYL-2-PYRROLIDINONE [313,
PST]
N-METHYLPYRROLIDINONE [TSC,
TSE]

883-40-9 DIAZODIPHENYLMETHANE

886-50-0 TERBUTRYN

892-21-7 3-NITROFLUORANTHENE

900-95-8 STANNANE, ACETOXYTRIPH-
ENYL-
ACETOXYTRIPHENYL STANNONE
[MSL]
FENTIN ACETATE [HM1]
TRIPHENYLTIN ACETATE [WHM]

915-67-3 C.I. ACID RED 27, TRISODIUM
SALT
AMARANTH [IAR]
C.I. ACID RED 27 [PST]

918-37-6 HEXANITROETHANE

918-54-7 TRINITROETHANOL

919-16-4 LITHIUM CITRATE

919-30-2 3-(TRIETHOXYSILYL)PROPY-
LAMINE

919-86-8 DEMETON-S-METHYL
METHYL-S-DEMETON [NJL]

920-46-7 METHACRYLOYL CHLORIDE

920-66-1 1,1,1,3,3,3-HEXAFLUORO-2-
PROPANOL

921-20-0 3-METHOXY-2,4-DIHYDROXYPEN-
TANE

923-26-2 2-PROPENOIC ACID, 2-METHYL-,
2-HYDROXYPROPYL ESTER

924-16-3 N-NITROSODI-N-BUTYLAMINE
1-BUTANAMINE, N-BUTYL-N-NI-
TROSO- [PAL, PAL]
N-NITROSO-DI-N-BUTYLAMINE
[CAL, RCA]
N-NITROSODIBUTYLAMINE [WHM]

924-42-5 N-METHYLOLACRYLAMIDE
N-METHYLOACRYLAMIDE [C65]

925-60-0 2-PROPENOIC ACID, PROPYL
ESTER

926-06-7 ISOPROPYL METHANE
SULPHONATE

926-56-7 4-METHYL-1,3-PENTADIENE
1,3-PENTADIENE, 4-METHYL- [PAL]

926-57-8 1,3-DICHLOROBUTENE-2
2-BUTENE, 1,3-DICHLORO- [PAL]

926-63-6 DIMETHYLPROPYLAMINE
DIMETHYL-N-PROPYLAMINE [HM1,
HM1]

926-64-7 DIMETHYLAMINOACETONITRILE
2-DIMETHYLAMINOACETO-NITRILE
[NJL]
2-DIMETHYLAMINOACETONITRILE
[HM1, HM1]

926-65-8 PROPANE, 2-(ETHENYLOXY)-
VINYL ISOPROPYL ETHER [FLA,
MSL, NFP, NFP]

927-07-1 TERT-BUTYL PEROXYPIVALATE
PROPANEPEROXOIC ACID, 2,2-
DIMETHYL-, 1,1-DIMETHYLETHYL
ESTER [PAL]

927-80-0 ETHYNE, ETHOXY-
ETHOXYACETYLENE [FLA, MSL,
NFP, NFP]

928-45-0 BUTYL NITRATE
NITRIC ACID, BUTYL ESTER [PAL]

928-55-2 1-PROPENE, 1-ETHOXY-
PROPENYL ETHYL ETHER [FLA,
MSL, NFP, NFP]

928-65-4 HEXYLTRICHLOROSILANE
HEXYL TRICHLOROSILANE [NJL,
NJL]

928-96-1 3-HEXEN-1-OL, (Z)-
3-HEXENOL-CIS [NFP, NFP]

929-06-6 2-(2-AMINOETHOXY)ETHANOL
2-(2-AMINOETHOXY)-ETHANOL
[TSC, TSC, TSC]
2-ETHANOL, (2-AMINOETHOXY)
[MSL]

930-22-3 BUTADIENE MONOXIDE
OXIRANE, ETHENYL- [PAL, TSC,
TSC]

930-37-0 OXIRANE, METHOXYMETHYL-
METHYL GLYCIDYL ETHER [EPA,
OTS]

930-55-2 N-NITROSOPYRROLIDINE
PYRROLIDINE, 1-NITROSO- [PAL,
PAL]

930-68-7 2-CYCLOHEXENE-1-ONE
CYCLOHEXENONE [NFP, NFP]

933-48-2 TRIMETHYLCYCLOHEXANOL

933-75-5 2,3,6-TRICHLOROPHENOL
PHENOL, 2,3,6-TRICHLORO- [PAL]

933-78-8 2,3,5-TRICHLOROPHENOL
PHENOL, 2,3,5-TRICHLORO- [PAL]

935-95-5 2,3,5,6-TETRACHLOROPHENOL

937-14-4 3-CHLOROPEROXYBENZOIC ACID

939-48-0 ISOPROPYL BENZOATE

939-97-9 BENZALDEHYDE, 4-(1,1-
DIMETHYLETHYL)-

941-98-0 1-ACETONAPHTHONE

942-01-8 1,2,3,4-TETRAHYDROCARBAZOLE

944-22-9 FONOFOS
DYFONATE [BC1]
DYFONATE (FONOFOS) [ALB, ALB,
ALB]
FONOFOS [QBC, QBC]
PHOSPHONODITHIOIC ACID, ETHYL-,
O-ETHYL S-PHENYL ESTER [PAL]

947-02-4 PHOSFOLAN
PHOSPHOLAN [EPA, EPA]

950-10-7 MEPHOSFOLAN

950-37-8 METHIDATHION
O,O-DIMETHYL PHOSPHO-
RODITHIOATE, S-ESTER WITH 4-
(MERCAPTOMETHYL)-2-METHOXY-
O2-1,3,4-THIADIAZOLIN-5-ONE
[CAL]
PHOSPHORODITHIOIC ACID, S-[(5-
METHOXY-2-OXO-1,3,4-THIADIAZOL-
3(2H)-YL)METHYL] O,O-DIMETHYL
ESTER [PAL]

952-23-8 PROFLAVIN HYDROCHLORIDE
PROFLAVINE [MSL]

957-51-7 DIPHENAMIDE
DIPHENAMID [313]

959-43-3 1,2,4,5-TETRACHLOROBENZENE

959-98-8 ALPHA-ENDOSULFAN
6,9-METHANO-2,4,3-BENZODI-
OXATHIEPIN, 6,7,8,9,10,10-HEX-
ACHLORO-1,5,5A,6,9,9A-HEXAHY-
DRO-, 3-OXIDE, (3.ALPHA.,5A.BETA.,
6.ALPHA.,9.ALPHA.,9A.BETA.)- [PAL]
ALPHA - ENDOSULFAN [MSL]
ALPHA-ENDOSULFAN-ALPHA [MX1]
ENDOSULFAN I [PQL, RCA, UTS]
ENDOSULFAN, ALPHA [ATS]

961-11-5 TETRACHLOROVINPHOS
TETRACHLORVINPHOS [CAL, IAR,
PAL]
TETRACHLORVINPHOS [PHOS-
PHORIC ACID, 2-CHLORO-1-(2,4,
5-TRICHLOROPHENYL)ETHENYL
DIMETHYL ESTER] [313]

968-81-0 ACETOHEXAMIDE

973-21-7 DINOBUTON

982-57-0 CHLORAMPHENICOL SODIUM
SUCCINATE

989-38-8 C.I. BASIC RED 1
RHODAMINE 6G [CAL, CSR, CSR,
IAR]

991-42-4 NORBORMIDE

992-59-6 C.I. DIRECT RED 2, DISODIUM SALT
1-NAPHTHALENESULFONIC ACID, 3,
3'-[(3,3'-DIMETHYL-[1,1'-BIPHENYL]
-4,4'-DIYL)BIS(AZO)]BIS[4-AMINO-,
DISODIUM SALT [TSC]

993-00-0 METHYLCHLOROSILANE
METHYL CHLOROSILANE [NJL]

993-12-4 DIMETHYL THIOPHOSPHORYL CHLORIDE

993-43-1 ETHYLPHOSPHONOTHIOIC-DICHLORIDE
ETHYL PHOSPHONOTHIOIC DICHLORIDE [HM1, HM1, HM1]

994-05-8 TERT-AMYL METHYL ETHER

995-33-5 N-BUTYL-4,4-DI(TERT-BUTYL-PEROXY)VALERATE
N-BUTYL-4,4-DI(TERT-BUTYL-PEROXY) VALERATE [NJL]

998-30-1 TRIETHOXYSILANE

998-40-3 TRIBUTYLPHOSPHINE

999-21-3 DIALLYL MALEATE

999-61-1 2-HYDROXYPROPYL ACRYLATE
HYDROXYPROPYL ACRYLATE [CAL]
PROPYLENE GLYCOL MONOACRYLATE [NFP, NFP, PAL]

999-81-5 CHLORMEQUAT CHLORIDE
2-CHLOROETHYLTRIMETHYLAMMONIUM CHLORIDE [CSR, CSR]

999-97-3 HEXAMETHYLDISILIZANE

1000-82-4 METHYLOLUREA
METHYLOL UREA [OTS]
METHYLOLUREA (HYDROXYMETHYL UREA) [TSC, TSC, TSC]

1002-16-0 AMYL NITRATE
NITRIC ACID, PENTYL ESTER [PAL]

1002-89-7 AMMONIUM STEARATE

1003-78-7 DIMETHYL SULFOLANE

1014-69-3 S-TRIAZINE, 2-(ISOPROPYLAMINO)-4-(METHYLAMINO)-6-(METHYLTHIO)

1024-57-3 HEPTACHLOR EPOXIDE
2,5-METHANO-2H-INDENO[1,
2-B]OXIRENE, 2,3,4,5,6,7,7-HEPTACHLORO-,(1A.ALPHA.,1B.BETA.,
2.ALPHA.,5.ALPHA.,5A.BETA.,6.BETA.,
A.ALPHA.)- [PAL]
HEPTACHLOR EPOXIDE (BHC-HEXACHLOROCYCLOHEXANE) [CW2]

1031-07-8 ENDOSULFAN SULFATE
6,9-METHANO-2,4,3-BENZODIOXATHIEPIN, 6,7,8,9,10,10-HEXACHLORO-1,5,5A,6,9,9A-HEXAHYDRO-, 3,3-DIOXIDE [PAL]

1031-47-6 TRIAMIPHOS

1047-16-1 QUINACRIDONE

1064-48-8 C.I. ACID BLACK 1

1066-30-4 CHROMIC ACETATE
ACETIC ACID, CHROMIUM(3+) SALT [PAL]
CHROMIUM (III) ACETATE [WHM]

1066-33-7 AMMONIUM BICARBONATE
CARBONIC ACID, MONOAMMONIUM SALT [PAL]

1066-45-1 TRIMETHYL TIN CHLORIDE
TRIMETHYLTIN CHLORIDE [302, PAL, WHM]

1067-20-5 3,3-DIETHYLPENTANE
PENTANE, 3,3-DIETHYL- [PAL]

1067-33-0 DIBUTYLTIN DIACETATE

1067-53-4 VINYLTRIS(METHOXYETHOXY) SILANE

1068-27-5 2,5-DIMETHYL-2,5-DI(TERT-BUTYLPEROXY)HEXYNE-3

1068-57-1 ACETHYDRAZIDE

1068-61-7 LEAD METHACRYLATE

1068-87-7 2,4-DIMETHYL-3-ETHYL PENTANE
2,4-DIMETHYL-3-ETHYLPENTANE [NFP, NFP]
PENTANE, 3-ETHYL-2,4-DIMETHYL- [PAL]

1069-55-2 2-PROPENOIC ACID, 2-CYANO-, ISOBUTYL ESTER

1070-03-7 PHOSPHORIC ACID, MONO(2-ETHYLHEXYL) ESTER
2-ETHYLHEXYL DIHYDROGEN PHOSPHATE [PST]

1070-19-5 TERT-BUTOXYCARBONYL AZIDE

1070-78-6 PROPANE, 1,1,1,3-TETRACHLORO-

1071-73-4 3-ACETYL-1-PROPANOL

1071-83-6 GLYCINE, N-(PHOSPHONOMETHYL)-
GLYPHOSATE [CD1, CSR, SDW, SDW]

1072-35-1 STEARIC ACID, LEAD (2+) SALT
OCTADECANOIC ACID, LEAD(2+) SALT [PAL]

1072-43-1 PROPYLENE SULFIDE

1072-52-2 1-AZIRIDINEETHANOL
2-(1-AZIRIDINYL)ETHANOL [CAL, IAR]

1072-85-1 1-FLUORO-2-BROMOBENZENE

1073-06-9 1-FLUORO-3-BROMOBENZENE

1081-77-2 NONYLBENZENE
NONYLBENZENE (BRANCHED) [HON]

1095-66-5 MORPHOLINE OLEATE

1111-78-0 AMMONIUM CARBAMATE
CARBAMIC ACID, MONOAMMONIUM SALT [PAL]

1113-38-8 AMMONIUM OXALATE

1114-71-2 PEBULATE [BUTYLETHYLCARBAMOTHIOIC ACID S-PROPYL ESTER]

1116-54-7 N-NITROSODIETHANOLAMINE
ETHANOL, 2,2'-(NITROSOIMINO)BIS- [PAL, PAL]

1116-70-7 TRIBUTYL ALUMINUM
ALUMINUM, TRIBUTYL- [PAL]
TRIBUTYLALUMINUM [FLA, MSL]

1118-46-3 BUTYLTRICHLOROSTANNANE
MONOBUTYLTIN TRICHLORIDE [WHM]

1118-58-7 2-METHYL-1,3-PENTADIENE
1,3-PENTADIENE, 2-METHYL- [PAL]

1118-92-9 N,N-DIMETHYLCAPRYLAMIDE

1119-16-0 4-METHYL VALERALDEHYDE

1119-34-2 L-ARGININE HCL

1119-49-9 N-BUTYL ACETAMIDE
ACETAMIDE, N-BUTYL- [PAL]

1120-01-0 HEXADECYL SULFATE, SODIUM SALT

1120-04-3 OCTADECYL SULFATE, SODIUM SALT

1120-21-4 UNDECANE
HENDECANE [NFP, NFP]

1120-23-6 2-BETA-BUTOXYETHOXYETHYL CHLORIDE
2,BETA-BUTOXYETHOXYETHYL CHLORIDE [NFP, NFP]

1120-36-1 1-TETRADECENE

1120-46-3 LEAD OLEATE

1120-48-5 1-OCTANAMINE, N-OCTYL-

1120-71-4 1,3-PROPANE SULTONE
1,2-OXATHIOLANE, 2,2-DIOXIDE [PAL, PAL]
1-2-OXATHIOLANE 2,2-DIOXIDE [NJL]
PROPANE SULTONE [313, AUS, AUS, MIN, QBC, QBC, TLV, WHM]

1121-03-5 2,4-BUTANE SULTONE

1121-30-8 OMADINE

1121-60-4 2-PYRIDINECARBOXALDEHYDE

1122-60-7 NITROCYCLOHEXANE
CYCLOHEXANE, NITRO- [PAL]

1124-33-0 PYRIDINE, 4-NITRO-, 1-OXIDE-

1125-27-5 ETHYL PHENYL DICHLOROSILANE

1126-34-7 M-AMINOBENZENESULFONIC ACID, SODIUM SALT

1126-78-9 N-BUTYLANILINE
BENZENAMINE, N-BUTYL- [PAL]
BUTYLANILINE [NJL]
N-(N-BUTYL)ANILINE [WHM]

1126-79-0 BUTYL PHENYL ETHER

1127-76-0 1-ETHYLNAPHTHALENE

1129-41-5 METOLCARB

1129-42-6 UREA, (HEXAHYDRO-6-METHYL-2-OXO-4-PYRIMIDINYL)-

1132-39-4 1,1'-SELENOBISBENZENE

1133-64-8 N'-NITROSOANABASINE

1134-23-2 CYCLOATE

1137-41-3 4-AMINOBENZOPHENONE

1139-30-6 5-OXATRICYCLO[8.2.0.04,6]DODECANE, 4,12,12-TRIMETHYL-9-METHYLENE-, [1R-(1R*,4R*,6R*, 10S*)]-

1155-74-4 1-TETRADECYLPYRIDINIUM BROMIDE

1156-19-0 TOLAZAMIDE

1162-65-8 AFLATOXIN B1

1163-19-5 DECABROMODIPHENYL OXIDE
DECABROMOBIPHENYL ETHER [NJL]
DECABROMODIPHENYL ETHER [OTS, TSC, TSC, TSC, TSC]

1165-39-5 AFLATOXIN G1

1172-18-5 FLURAZEPAM HYDROCHLORIDE

1172-63-0 JASMOLIN II

1185-55-3 METHYLTRIMETHOXYSILANE

1185-57-5 FERRIC AMMONIUM CITRATE
1,2,3-PROPANETRICARBOXYLIC ACID, 2-HYDROXY-, AMMONIUM IRON(3+) SALT [PAL]

1185-81-5 DIBUTYLTIN BIS(LAURYL MERCAPTIDE)
STANNANE, DIBUTYLBIS(DODECYLTHIO)- [TSC, TSC, TSC]

1186-53-4 PENTANE, 2,2,3,4-TETRAMETHYL-
2,2,3,4-TETRAMETHYL PENTANE [FLA, NFP, NFP]
2,2,3,4-TETRAMETHYLPENTANE [MSL]

1187-59-3 N-METHYLACRYLAMIDE

1189-85-1 TERT-BUTYL CHROMATE
CHROMIC ACID (H2CRO4), BIS(1,1-DIMETHYLETHYL) ESTER [PAL]

1189-99-7 **2,5,5-TRIMETHYLHEPTANE**

1191-15-7 **DIISOBUTYLALUMINUM HYDRIDE**
ALUMINUM, HYDROBIS(2-METHYL-PROPYL)- [PAL]

1191-50-0 **SODIUM TETRADECYL SULFATE**
TETRADECYL SULFATE, SODIUM SALT [PST]

1191-80-6 **MERCURY OLEATE**

1194-65-6 **DICHLOBENIL**
BENZONITRILE, 2,6-DICHLORO- [PAL]

1195-42-2 **ISOPROPYL CYCLOHEXYLAMINE**
CYCLOHEXANAMINE, N-(1-METHYLETHYL)- [PAL]
ISOPROPYLCYCLOHEXYLAMINE [NFP, NFP]

1195-79-5 **FENCHONE**

1197-37-1 **1,2-BENZENEDIAMINE, 4-ETHOXY-**

1200-14-2 **BENZALDEHYDE, 4-BUTYL-**

1208-52-2 **BENZENAMINE, 2-[(4-AMINOPHENYL)METHYL]-**

1212-29-9 **N,N'-DICYCLOHEXYLTHIOUREA**

1241-94-7 **PHOSPHORIC ACID, 2-ETHYLHEXYL DIPHENYL ESTER**
ETHYLHEXYL DIPHENYL PHOSPHATE, 2- [OTS]

1254-78-0 **PHENYL DIDECYL PHOSPHITE**

1271-19-8 **TITANOCENE DICHLORIDE**

1271-28-9 **NICKELOCENE**

1300-21-6 **ETHANE DICHLORIDE**
DICHLOROETHANE [ATS]

1300-64-7 **ANISOYL CHLORIDE**

1300-71-6 **XYLENOL**
PHENOL, DIMETHYL- [PAL, TSC, TSC]
XYLENOLS [HM1, HM1, HM1, HM1]
XYLENOLS (MIXED) [HON]
XYLENOLS, MIXED [PST]

1300-72-7 **SODIUM XYLENE SULFONATE**
SODIUM XYLENESULFONATE [CSR]
XYLENESULFONIC ACID, SODIUM SALT [PST]

1300-73-8 **XYLIDINE**
AR,AR-DIMETHYLBENZENEAMINE (SUM OF ALL ISOMERS) [ONT, ONT]
DIMETHYLAMINOBENZENE (XYLIDENE) [ALB, ALB, ALB]
XYLIDENE (MIXED ISOMERS) [QBC, QBC, QBC]
XYLIDINE (MIXED ISOMERS) [TLV, TLV, TLV]
XYLIDINE, ALL ISOMERS [GBR, GBR, GBR]
XYLIDINES [HM1, HM1, PAL]

1302-42-7 **SODIUM ALUMINATE (NAALO2)**
SODIUM ALUMINATE [HM1, HM1, PST]

1302-52-9 **BERYL**
BERYL (AL2BE3(SIO3)6) [PAL, PAL]
BERYL ORE [FLA, MSL, NTP]

1302-74-5 **CORUNDUM**
ALUMINUM OXIDE [ALB, MAK]
ALUMINUM OXIDE (ALUNDUM, CORUNDUM, ALUMINE) [QBC]
EMERY [TLV]

1302-78-9 **BENTONITE**

1303-00-0 **GALLIUM ARSENIDE**

1303-28-2 **ARSENIC PENTOXIDE**
ARSENIC OXIDE (AS2O5) [PAL, PAL]

1303-32-8 **ARSENIC DISULFIDE**

1303-33-9 **ARSENIC TRISULFIDE**
ARSENIC SULFIDE (AS2S3) [PAL]

1303-39-5 **ZINC ARSENATE**

1303-86-2 **BORON OXIDE**
BORIC ANHYDRIDE [WHM]
BORON OXIDE (B2O3) [PAL]
DIBORON TRIOXIDE [GBR, GBR]

1303-96-4 **BORATES, TETRA, SODIUM SALTS, DECAHYDRATE**
BORATES, TETRA, SODIUM SALTS [TLV]
BORATES, TETRA, SODIUM SALTS DECAHYDRATE [AUS]
BORAX [PST]
BORAX (B4NA2O7.10H2O) [PAL]
SODIUM BORATE [MSL]
SODIUM BORATES [NJL]
SODIUM TETRABORATE [WHM]

1304-28-5 **BARIUM OXIDE**

1304-29-6 **BARIUM PEROXIDE**
BARIUM PEROXIDE (BA(O2)) [PAL]

1304-56-9 **BERYLLIUM OXIDE**
BERYLLIUM OXIDE (BEO) [PAL, PAL]

1304-82-1 **BISMUTH TELLURIDE**
BISMUTH TELLURIDE (BI2TE3) [PAL]
BISMUTH TELLURIDE, UNDOPED [OS1, OS1]
DIBISMUTH TRITELLURIDE [GBR, GBR]

1305-62-0 **CALCIUM HYDROXIDE**
CALCIUM HYDROXIDE (CA(OH)2) [PAL]

1305-78-8 **CALCIUM OXIDE**
CALCIUM OXIDE (CAO) [PAL]

1305-79-9 **CALCIUM PEROXIDE**

1305-99-3 **CALCIUM PHOSPHIDE**

1306-05-4 **FLUORAPATITE**

1306-19-0 **CADMIUM OXIDE**
CADMIUM FUME [NIO, NIO, NIO, NIO, OS1, OS1, OS1]
CADMIUM OXIDE (CDO) [PAL, PAL]
CADMIUM OXIDE FUME [ALB, ALB, BC1, BC1, MEX, QBC, QBC]
CADMIUM OXIDE, PRODUCTION [ONT, ONT]

1306-23-6 **CADMIUM SULFIDE**
CADMIUM SULFIDE (CDS) [PAL, PAL]
CADMIUM SULPHIDE [FLA, GBR, HM1, MSL]

1307-96-6 **COBALT(II) OXIDE**

1308-31-2 **CHROMITE**
CHROMITE ORE [MSL]

1308-38-9 **CHROMIUM (III) OXIDE**
CHROME OXIDE (CR2O3) [PST]
CHROMIUM (III) OXIDE(2:3) [NJL]

1309-32-6 **AMMONIUM FLUOSILICATE**

1309-37-1 **IRON OXIDE**
FERRIC OXIDE [IAR, PST, WHM]
IRON OXIDE (FE2O3) [PAL]
IRON OXIDE DUST AND FUME [NIO, NIO]
IRON OXIDE FUME [ALB, ALB, BC1, BC1, CAL, FLA, MEX, MEX, MSL, NJL, OS1, OS1]
IRON OXIDE FUME (FE2O3) [TLV]
IRON TRIOXIDE DUST AND FUME [QBC]
ROUGE [ONT]

1309-42-8 **MAGNESIUM HYDROXIDE**

1309-48-4 **MAGNESIUM OXIDE**
MAGNESIUM OXIDE (MGO) [PAL]
MAGNESIUM OXIDE FUME [ALB, ALB, BC1, FLA, MEX, MSL, NIO, NIO, ONT, ONT, ONT, OS1, OS1, TLV]
MAGNESIUM OXIDE FUMES [QBC]

1309-60-0 **LEAD PEROXIDE**
LEAD DIOXIDE [HM1, HM1, MSL, NJL]

1309-64-4 **ANTIMONY TRIOXIDE**
ANTIMONY OXIDE (ANTIMONY TRIOXIDE) [C65]
ANTIMONY OXIDE (SB2O3) [PAL]
ANTIMONY TRIOXIDE [QBC, QBC]
ANTIMONY TRIOXIDE, HANDLING AND USE [ALB, ALB]

1310-53-8 **GERMANIUM OXIDE**

1310-58-3 **POTASSIUM HYDROXIDE**
POTASSIUM HYDROXIDE (K(OH)) [PAL]

1310-65-2 **LITHIUM HYDROXIDE**

1310-66-3 **LITHIUM HYDROXIDE MONOHYDRATE**
LITHIUM HYDROXIDE [MIN]

1310-73-2 **SODIUM HYDROXIDE**
SODIUM HYDROXIDE (NA(OH)) [PAL]

1310-82-3 **RUBIDIUM HYDROXIDE**

1312-03-4 **MERCURIC SUBSULFATE**

1312-73-8 **POTASSIUM SULFIDE**
POTASSIUM SULFIDE (K2S) [PAL]

1312-76-1 **POTASSIUM SILICATE**

1313-13-9 **MANGANESE DIOXIDE**

1313-27-5 **MOLYBDENUM TRIOXIDE**

1313-60-6 **SODIUM PEROXIDE**
SODIUM PEROXIDE (NA2(O2)) [PAL]

1313-82-2 **SODIUM SULFIDE**

1313-99-1 **NICKEL OXIDE**
NICKEL (II) OXIDE [CSR, ONT, ONT]
NICKEL MONOXIDE [IAR]
NICKEL OXIDE (NIO) [PAL, PAL]
NICKEL(II) OXIDE [WHM]

1314-06-3 **NICKEL TRIOXIDE**
NICKEL (III) OXIDE [ONT, ONT]
NICKEL OXIDE (NI2O3) [PAL, PAL]
NICKEL(III) OXIDE [WHM]

1314-13-2 **ZINC OXIDE**
ZINC OXIDE (FUME) [QBC, QBC]
ZINC OXIDE (ZNO) [PAL]
ZINC OXIDE FUME [FLA, MAK, MAK, MEX, MEX, MSL, NIO, NIO]

1314-18-7 **STRONTIUM PEROXIDE**
STRONTIUM PEROXIDE (SR(O2)) [PAL]

1314-20-1 **THORIUM DIOXIDE**
THORIUM OXIDE (THO2) [PAL, PAL]

1314-22-3 **ZINC PEROXIDE**

1314-24-5 **PHOSPHORUS TRIOXIDE**

1314-27-8 **LEAD TRIOXIDE**

1314-32-5 **THALLIC OXIDE**
THALLIUM(III) OXIDE [WHM]

1314-34-7 **VANADIUM TRIOXIDE**

1314-35-8 **TUNGSTIC OXIDE**

1314-41-6 **LEAD TETRAOXIDE**

1314-56-3 **PHOSPHORUS PENTOXIDE**
PHOSPHORIC ANHYDRIDE [NJL, NJL, WHM]

1314-61-0 **TANTALUM OXIDE**
TANTALUM, OXIDE DUSTS [TLV]

1314-62-1 **VANADIUM PENTOXIDE**
DIVANADIUM PENTAOXIDE [GBR]
VANADIUM [AUS, MIN, OS1]
VANADIUM OXIDE (V2O5) [PAL]
VANADIUM PENTOXIDE [TLV]
VANADIUM PENTOXIDE, DUST AND FUME [FLA, MSL]

VANADIUM RESPIRABLE DUST AND
FUME [ONT]
VANADIUM PENTOXIDE (V2O5) [CEX]

1314-80-3 PHOSPHORUS PENTASULFIDE
DIPHOSPHORUS PENTASULPHIDE
[GBR, GBR]
PHOSPHORUS PENTASULPHIDE
[AUS, AUS]
PHOSPHORUS SULFIDE [RCU, RCU]
PHOSPHORUS SULFIDE (P2S5) [PAL]

1314-84-7 ZINC PHOSPHIDE
ZINC PHOSPHIDE (ZN3P2) [PAL,
RCU, RCU, RCU]

1314-85-8 PHOSPHORUS SESQUISULFIDE
PHOSPHORUS SULFIDE (P4S3) [PAL]

1314-87-0 LEAD SULFIDE (PBS)
LEAD SULFIDE [CSR, CWA, EPA,
MSL, NJL]

1314-91-6 LEAD TELLURIDE

1314-96-1 STRONTIUM SULFIDE

1314-98-3 ZINC SULFIDE

1315-04-4 ANTIMONY PENTASULFIDE
ANTIMONY SULFIDE (SB2S5) [PAL]

1317-33-5 MOLYBDENUM (IV) SULFIDE
MOLYBDENUM DISULFIDE [ONT,
PST]

1317-35-7 MANGANESE TETROXIDE
MANGANESE OXIDE [FLA]
MANGANOUS-MANGANIC OXIDE
[MAK, MAK]
TRIMANGANESE TETRAOXIDE
[GBR]

1317-36-8 LEAD MONOXIDE
LEAD OXIDE [CSR, MSL]

1317-39-1 COPPER(+1) OXIDE
COPPER(I) OXIDE [WHM]
CUPROUS OXIDE [PST]

1317-40-4 COPPER(II) SULFIDE

1317-42-6 COBALT(II) SULFIDE

1317-43-7 NEMALITE/BRUCITE

1317-60-8 HEMATITE
HAEMATITE [IAR]

1317-65-3 CALCIUM CARBONATE
CACIUM CARBONATE (LIMESTONE)
[QBC]
LIMESTONE [ALB, BC1, BC1, GBR,
MEX, ONT, PAL, PST]

1317-70-0 ANATASE (TIO2)

1317-71-1 OLIVINE
OLIVINE SAND [BC1]

1317-80-2 RUTILE (TIO2)

1317-95-9 SILICA-TRIPOLI
SILICA - CRYSTALLINE, TRIPOLI
[TLV]
SILICA, CRYSTALLINE TRIPOLI [OS1,
OS1]
SILICA, CRYSTALLINE, TRIPOLI
[CEX, WHM]
SILICA, TRIPOLI [NJL]
SILICA-CRYSTALLINE, TRIPOLI
[MIN, QBC]
TRIPOLI [PAL]
TRIPOLI DUST [FLA, MSL]

1318-00-9 VERMICULITE

1318-16-7 BAUXITE (AL2O3.XH2O)

1318-93-0 MORTMORILLONITE

1319-46-6 LEAD CARBONATE HYDROXIDE

1319-72-8 ACETIC ACID, (2,4,5-
TRICHLOROPHENOXY)-, COM-
POUND WITH 1-AMINO-2-
PROPANOL(1:1)
2,4,5-T AMINE [NJL]

1319-73-9 BENZENE, ETHENYL-,
MONOMETHYL DERIV.

1319-77-3 CRESOL
CRESOL (ALL ISOMERS) [CEX, CEX,
NIO, NIO, NIO, QBC, QBC]
CRESOL (CRESYLIC ACID) [RCU,
RCU]
CRESOL (MIXED ISOMERS) [313,
CAN, CSR]
CRESOL (SUM OF O-, M-, AND P-
ISOMERS) [ONT, ONT]
CRESOL AND CRESYLIC ACIDS
(MIXED) [HON, HON]
CRESOL ISOMERS [MEX, MEX]
CRESOL, ALL ISOMERS [ALB, ALB,
ALB, AUS, AUS, BC1, BC1, MIN,
OS1, OS1, OS1, OS1, TLV, TLV]
CRESOLS [EPA]
CRESOLS (CRESYLIC ACID) [RCA]
CRESOLS (O-; M-; P-) [HM1, HM1,
HM1, HM1]
CRESOLS, ALL ISOMERS [GBR, GBR]
CRESOLS/CRESYLIC ACID (ISOMERS
AND MIXTURE) [CAA]
CRESYLIC ACID [NJL, PST]
M- OR P- CRESOL [NFP, NFP]
PHENOL, METHYL- [PAL, TSC,
TSC]

1320-01-0 BENZENE, METHYLPENTYL-
AMYL TOLUENE [FLA, MSL, NFP,
NFP]

1320-05-4 P-TERT-AMYLPHENYL METHYL
ETHER

1320-18-9 2,4-D PROPYLENE GLYCOL
BUTYL ETHER ESTER
2,4-D ESTERS [NJL]
ACETIC ACID, (2,4-DICHLOROPHE-
NOXY)
-, 2-BUTOXYMETHYLETHYLESTER
[PAL]
ACETIC ACID, (2,4-DICHLOROPHE-
NOXY)-, 2-BUTOXYMETHYLETHYL
ESTER [MSL]

1320-21-4 BENZENE, DIMETHYL(PENTY-
LOXY)-
AMYL XYLYL ETHER [MSL, NFP,
NFP]
PENTYL XYLYL ETHER [FLA]

1320-27-0 NAPHTHALENE, PENTYL-

1320-37-2 DICHLOROTETRAFLUO-
ROETHANE
CRYOFLUORANE (INN) [GBR, GBR]
ETHANE, DICHLOROTETRAFLUORO-
[PAL]

1320-98-5 4-METHYL-2-PENTANOL

1321-10-4 CHLOROCRESOL
CHLOROCRESOLS [HM1, HM1,
NJL]

1321-12-6 NITROTOLUENE (MIXED ISO-
MERS)
BENZENE, METHYLNITRO- [PAL]
NITROTOLUENE [BC1, BC1, BC1,
EPA, MAK, MAK, MAK, MEX,
MEX, MSL, NJL, NJL]
NITROTOLUENE (ALL ISOMERS)
[HON, QBC, QBC]
NITROTOLUENE (SUM OF M-, O-
, AND P- ISOMERS) [ONT, ONT,
ONT]
NITROTOLUENE, ALL ISOMERS
[GBR, GBR, GBR]
NITROTOLUENES [HM1, HM1,
HM1]

1321-16-0 TETRAHYDROBENZALDEHYDE
1,2,3,6-TETRAHYDROBENZALDE-
HYDE [FLA, HM1, HM1, MSL]
CYCLOHEXENECARBOXALDEHYDE
[PAL]

1321-31-9 PHENETIDINE
PHENETIDINES [HM1, HM1]

1321-38-6 BENZENE,
DIISOCYANATOMETHYL-
BENZENE, DIISOCYANATOMETHYL-
(UNSPECIFIED TOLUENE DIISO-
CYANATE) [CAI]

1321-60-4 TRIMETHYLCYCLOHEXANOL
CYCLOHEXANOL, TRIMETHYL-
[PAL]

1321-64-8 PENTACHLORONAPHTHALENE
NAPHTHALENE, PENTACHLORO-
[PAL, TSC, TSC, TSC]

1321-65-9 TRICHLORONAPHTHALENE
NAPHTHALENE, TRICHLORO- [PAL,
TSC, TSC, TSC]
TRICHLORONAPHTALENE [QBC,
QBC]

1321-74-0 DIVINYL BENZENE
BENZENE, DIETHENYL- [PAL]
DIVINYLBENZENE [CSR]

1321-94-4 METHYLNAPHTHALENE
METHYLATED NAPHTHALENES
[PST]
METHYLNAPHTHALENES [HM1]

1322-90-3 DYPNONE

1322-93-6 SODIUM DIISOPROPYLNAPH-
THALENESULFONATE

1322-97-0 2-(OCTYLPHENOXY)-ETHANOL

1322-98-1 SODIUM DECYLBENZENESUL-
FONATE
DECYLBENZENESULFONIC ACID,
SODIUM SALT [PST]

1323-19-9 SODIUM TRIISOPROPYLNAPH-
THALENESULFONATE

1323-38-2 GLYCERYL MONORICINOLEATE

1323-83-7 DIGYLCERYL STEARATE

1325-82-2 METHYL VIOLET

1327-33-9 ANTIMONY OXIDE

1327-36-2 ALUMINUM SILICATE

1327-43-1 MAGNESIUM ALUMINUM SILI-
CATE
ALUMINUM - MAGNESIUM SILICATE
[PST]

1327-52-2 ARSENIC ACID
ARSENIC ACID H3ASO4 [EPA]

1327-53-3 ARSENIC TRIOXIDE
ARSENIC OXIDE (AS2O3) [PAL, PAL]
ARSENIC TRIOXIDE (PRODUCTION)
[ALB, ALB, ALB, MEX, MEX,
QBC, QBC]
ARSENIC TRIOXIDE PRODUCTION
[BC1, BC1, BC1, MIN]
ARSENOUS OXIDE [302, FLA]
WHITE ARSENIC [HM1, HM1, HM1,
HM1]

1330-16-1 PINENE

1330-20-7 XYLENES (O-, M-, P- ISOMERS)
BENZENE, DIMETHYL- [PAL, TSC,
TSC]
DIMETHYLBENZENE [QBC, QBC]
DIMETHYLBENZENE (SUM OF O-, M-,
AND P- ISOMERS) [ONT, ONT]
TOTAL XYLENES [ATS]
XYLENE [CEX, CEX, CEX, FLA,
IAR, MAK, MAK, MAK, MIN,
MSL, NHS, NHS, PST]
XYLENE (MIXED ISOMERS) [313,
CAN]
XYLENE (MIXED) [CWA, EPA]
XYLENE (O-, M-, P- ISOMERS) [ALB,
ALB]
XYLENE (TOTAL) [PQL]
XYLENE(S) [RCA]
XYLENE, ALL ISOMERS [CAL, GBR,
GBR, GBR]
XYLENES [BEI, CN4, HM1, HM1,
HM1, NJL, NJL]
XYLENES (ISOMERS AND MIXTURE)
[CAA]

XYLENES (MIXED) [CSR, CSR]
XYLENES (NOS) [HON, HON]
XYLENES (TOTAL) [CD1, SDW, SDW]

1330-38-7 C.I. 74180
C.I. DIRECT BLUE 86 [PST]

1330-43-4 BORATES, TETRA, SODIUM SALTS, ANHYDROUS
BORATES, TETRA, SODIUM SALTS ANHYDROUS [AUS, ONT]
BORON SODIUM OXIDE (B4NA2O7) [PAL]
DISODIUM TETRABORATE [GBR]
SODIUM BORATE [WHM]
SODIUM TETRABORATE [PST]

1330-45-6 CHLOROTRIFLUOROETHANE

1330-61-6 2-PROPENOIC ACID, ISODECYL ESTER
ISODECYL ACRYLATE [HM1]

1330-76-3 BIS(1-METHYLHEPTYL)-2-BUTENEDIOATE

1330-78-5 TRICRESYL PHOSPHATE
PHOSPHORIC ACID, TRIS (METHYLPHENYL) ESTER [TSC, TSC]
TRICRESYLPHOSPHATE [NJL]

1330-80-9 9-OCTADECANOIC ACID (Z)-, MONOESTER WITH 1,2-PROPANEDIOL
PROPYLENE GLYCOL MONOOLEATE [PST]

1331-11-9 3-ETHOXYPROPIONIC ACID

1331-17-5 PROPYLENE GLYCOL, ALLYL ETHER

1331-22-2 METHYLCYCLOHEXANONE

1331-23-3 METHYLCYCLOHEXANOL

1331-28-8 CHLOROSTYRENE
O-CHLOROSTYRENE [ALB, ALB]
O-CHLOROSTYRENE [CAL, CEX, CEX, MEX, MEX, QBC, QBC]

1331-43-7 DIETHYLCYCLOHEXANE
CYCLOHEXANE, DIETHYL- [PAL]

1331-47-1 [1,1'-BIPHENYL]-4,4'-DIAMINO, DICHLORO-

1331-61-9 DODECYLBENZENESULFONIC ACID, AMMONIUM SALT

1331-92-6 2-PROPENAL, 3-PHENYL-, MONOPENTYL DERIVATIVE

1332-07-6 ZINC BORATE
BORIC ACID, ZINC SALT [PAL]

1332-09-8 PUMICE

1332-21-4 ASBESTOS
ASBESTIFORM MINERAL(S) [TSC, TSC]
ASBESTOS (ALL FORMS) [BC1]
ASBESTOS DUST [FLA]
ASBESTOS FIBRES, OTHER [ONT, ONT, ONT, ONT]
ASBESTOS, ALL FORMS [TLV, TLV, TLV]
ASBESTOS, OTHER FORMS [MIN]
WHITE ASBESTOS [HM1, HM1, HM1]

1332-37-2 IRON OXIDE
IRON OXIDE, SPENT [HM1, HM1]

1332-58-7 KAOLIN

1332-77-0 POTASSIUM BORATE
POTASSIUM TETRABORATE [PST]

1333-13-7 TERT-BUTYL-M-CRESOL
PHENOL, (1,1-DIMETHYLETHYL)-3-METHYL- [PAL]

1333-39-7 PHENOLSULPHONIC ACID
PHENOLSULFONIC ACIDS (ALL ISOMERS) [HON]

1333-41-1 PICOLINES
PICOLINE [NJL]
PYRIDINE, METHYL- [TSC, TSC]

1333-74-0 HYDROGEN

1333-82-0 CHROMIUM TRIOXIDE (CRO3)
CHROMIUM ANHYDRIDE [MSL]
CHROMIUM OXIDE [NJL, NJL]
CHROMIUM OXIDE (CRO3) [PAL, PAL]
CHROMIUM TRIOXIDE [NTP]
CHROMIUM TRIOXIDE, ANHYDROUS [HM1, HM1]
CHROMIUM(VI) OXIDE [WHM]
CHROMIUM(VI) OXIDE(1:3) [FLA]

1333-83-1 SODIUM BIFLUORIDE
SODIUM FLUORIDE (NA(HF2)) [PAL]
SODIUM HYDROGEN FLUORIDE [HM1, HM1, HM1, NJL, NJL]

1333-86-4 CARBON BLACK
CARBON BLACK - EXTRACTS [CAL]
CARBON BLACKS [IAR]

1334-78-7 TOLYL ALDEHYDE
BENZALDEHYDE, METHYL- [TSC, TSC, TSC]

1335-10-0 PHENYL PROPYL ALDEHYDE

1335-30-4 ALUMINUM SILICATE, HYDRATE
ALUMINUM SILICATE, HYDRATED [PST]

1335-31-5 MERCURIC OXYCYANIDE
MERCURYOXYCYANIDE [HM1, HM1, HM1]

1335-32-6 LEAD SUBACETATE
LEAD ACETATE, BASIC [WHM]
LEAD, BIS(ACETATO)TETRAHYDROXYTRI- [PAL, PAL]

1335-87-1 HEXACHLORONAPHTHALENE
NAPHTHALENE, HEXACHLORO- [PAL, TSC, TSC, TSC]

1335-88-2 TETRACHLORONAPHTHALENE
NAPHTHALENE, TETRACHLORO- [PAL, TSC, TSC, TSC]
TETRACHLORONAPHTHALENES, ALL ISOMERS [GBR, GBR]

1335-94-0 METHYL IONONE

1336-21-6 AMMONIUM HYDROXIDE
AMMONIUM HYDROXIDE ((NH4)(OH)) [PAL]

1336-36-3 POLYCHLORINATED BIPHENYLS
1,1-BIPHENYL, CHLORO DERIVATIVES [PAL, PAL]
POLYCHLORINATED BIPHENYLS (AROCLORS) [CAA]
POLYCHLORINATED BIPHENYLS (PCB'S) [MSL]
POLYCHLORINATED BIPHENYLS (PCB) [WHM]
POLYCHLORINATED BIPHENYLS (PCBS) [313, CW3, EP2, EPA, ONT, ONT, ONT, ONT]
POLYCHLORINATED BIPHENYLS, N.O.S. [RCU]
POLYCHLOROBIPHENYLS [CAL]
TOTAL PCBS (SUM OF ALL PCB ISOMERS, OR ALL AROCLORS) [UTS]

1336-93-2 MANGANESE NAPHTHENATE

1337-76-4 ATTAPULGITE

1338-02-9 COPPER NAPHTHENATE

1338-23-4 METHYL ETHYL KETONE PEROXIDE
2-BUTANONE, PEROXIDE [PAL]
METHYL ETHYL KETONE PEROXIDE (MEKP) [OS1]
METHYL ETHYL KETONE PEROXIDES [ONT]
METHYL ETHYL KETONE PEROXIDES (MEKP) [GBR]

1338-24-5 NAPTHENIC ACID
NAPHTHENIC ACIDS [HM1, HM1, PAL]

1338-39-2 SORBITAN MONOLAURATE

1338-41-6 SORBITAN MONOOCTADE-CANOATE
SORBITAN MONOSTEARATE [PST]

1338-43-8 SORBITAN MONOOLEATE

1341-24-8 CHLOROACETOPHENONE
CHLOROACETO PHENONE [NFP, NFP]
ETHANONE, 1-PHENYL-, MONOCHLORO DERIV. [PAL]

1341-49-7 AMMONIUM BIFLUORIDE
AMMONIUM FLUORIDE ((NH4)(HF2)) [PAL]

1343-88-0 MAGNESIUM SILICATE

1343-90-4 MAGNESIUM SILICATE, HYDRATE

1343-98-2 SILICIC ACID
SILICA, AMORPHOUS PRECIPITATED [QBC]

1344-00-9 SODIUM ALUMINUM SILICATE
SILICIC ACID, ALUMINUM SODIUM SALT (1:1:1) [PST]

1344-08-7 SODIUM SULFIDE

1344-09-8 SODIUM SILICATE

1344-28-1 ALUMINUM OXIDE
A-ALUMINA [MIN]
ALPHA-ALUMINA [MEX, ONT, OS1, OS1, WHM]
ALUMINUM OXIDE [MAK, MAK, MAK]
ALUMINUM OXIDE (AL2O3) [PAL]
ALUNDUM (AL2O3) [BC1, BC1]

1344-40-7 LEAD OXIDE PHOSPHONATE, HEMIHYDRATE
LEAD PHOSPHITE, DIBASIC [HM1, HM1]

1344-43-0 MANGANESE MONOXIDE
MANGANESE FUME [ALB, ALB, MEX, MEX, OS1, QBC, QBC]
MANGANOUS OXIDE [PST]

1344-48-5 MERCURY SULFIDE
MERCURIC SULPHIDE [HM1]

1344-67-8 COPPER CHLORIDE
COPPER(II) CHLORIDE [WHM]

1344-95-2 CALCIUM SILICATE
CALCIUM SILICATE (SYNTHETIC) [TLV]
SILICIC ACID, CALCIUM SALT [PAL]

1345-04-6 ANTIMONY SULFIDE
ANTIMONY TRISULFIDE [TSC, TSC]

1345-25-1 FERROUS OXIDE

1393-03-9 QUILLAJA (SAPONIN)

1395-21-7 SUBTILISINS (PROTEOLYTIC ENZYMES AS 100% PURE ENZYME)
SUBSTILISINS [MIN]
SUBTILISIN BPN [WHM]
SUBTILISINS (100% PURE CRYSTALLINE ENZYME) [QBC]
SUBTILISINS (PROTEOLYTIC ENZYMES AS 100 PERCENT PURE CRYSTALLINE ENZYME) [PAL]
SUBTILISINS (PROTEOLYTIC ENZYMES AS 100% PURE CRYSTALLINE ENZYME) [TLV]

1397-94-0 ANTIMYCIN A
ANTIMYCIN [NJL]

1401-55-4 TANNIC ACID
TANNIC ACID AND TANNINS [CAL]
TANNINS [PST]

1402-68-2 AFLATOXINS
AFLATOXINS, NATURALLY OCCURRING MIXTURES OF [IAR]

1405-10-3 NEOMYCIN SULFATE

1405-87-4   BACITRACIN

1406-16-2   ACTIVATED ERGOSTEROL

1406-18-4   VITAMIN E

1420-04-8   CLONITRALID
BENZAMIDE, 5-CHLORO-N-(2-CHLORO-4-NITROPHENYL)-2-HYDROXY-,COMPD. WITH 2-AMINOETHANOL (1:1) [PAL]

1420-07-1   DINOTERB

1423-46-7   3-CYCLOHEXENE-1-CARBOX-ALDEHYDE, 2,4,6-TRIMETHYL-

1423-60-5   3-BUTYN-2-ONE

1429-50-1   ETHYLENEDIAMINETETRA (METHYLENEPHOSPHONIC) ACID
PHOSPHONIC ACID, (1,2-ETHANEDIYL-BIS(NITRILOBIS (METHYLENE)))TETRAKIS- (EDTMPA) [TSC]

1439-17-4   TERBIUM 155

1455-21-6   1-NONANETHIOL

1455-77-2   GUANAZOLE

1459-93-4   DIMETHYL ISOPHTHALATE
DIMETHYLISOPHTHALATE [NFP, NFP]

1461-22-9   TRIBUTYLTIN CHLORIDE

1464-42-2   SELENOMETHIONINE

1464-53-5   DIEPOXYBUTANE
1,2,3,4-DIEPOXYBUTANE [NJL, NJL]
1,2;3,4-DIEPOXYBUTANE [EP2]
2,2'-BIOXIRANE [EPA, PAL, PAL, TSC, TSC, TSC]

1465-25-4   N-(1-NAPHTHYL)ETHYLENEDI-AMINE DIHYDROCHLORIDE

1467-79-4   DIMETHYLCYANAMIDE
CYANAMIDE, DIMETHYL- [PAL]

1476-23-9   PROPENE, 3-ISOCYANATO-
PROPENE, 3-ISOCYANATO- [TSC, TSC, TSC]

1477-55-0   M-XYLENE-ALPHA, ALPHA'-DI-AMINE
1,3-BENZENEDIMETHANAMINE [PAL, TSC, TSC]
M-XYLENE A,A'-DIAMINE [AUS, AUS]
M-XYLENE ALPHA, ALPHA'-DIAMINE [OS1, OS1]
M-XYLENE ALPHA,ALPHA'-DIAMINE [TLV, TLV]
M-XYLENE-A,A'-DIAMINE [CEX, CEX]
XYLENE M-, -DIAMINE [QBC, QBC]

1487-49-6   METHYL-3-HYDROXYBUTYRATE

1490-04-6   MENTHOL

1498-40-4   ETHYL PHOSPHONOUS DICHLO-RIDE

1498-51-7   ETHYL PHOSPHORODICHLORI-DATE
PHOSPHORODICHLORIDIC ACID, ETHYL ESTER [TSC, TSC, TSC]

1504-74-1   2-PROPENAL, 3-(2-METHOXYPHENYL)-

1511-62-2   METHANE, BROMODIFLUORO-
METHANE, BROMODIFLUORO [TSE]

1548-13-6   TRIFLOUROMETHYLPHENYLISO-CYANATE
3-TRIFLOUROMETHYLPHENYL ISO-CYANATE [HM1]

1552-12-1   1,5-CYCLOOCTADIENE

1558-25-4   TRICHLORO(CHLOROMETHYL) SILANE
.TRICHLORO(CHLOROMETHYL) SILANE [OS3]

1559-35-9   2-(2-ETHYLHEXYLOXY)ETHANOL

1560-06-1   1-PHENYL-2-BUTENE

1560-69-6   COBALT PROPIONATE

1561-49-5   DICYCLOHEXYL PEROXY-DICAR-BONATE

1563-66-2   CARBOFURAN
2,3-DIHYDRO-2,2-DIMETHYL-7-BENZO-FURANYL METHYLCARBA-MATE (CARBOFURAN) [CAL]
7-BENZOFURANOL, 2,3-DIHYDRO-2, 2-DIMETHYL-, METHYLCARBAMATE [PAL]
CARBOFURAN (FURADAN) [ALB, ALB, BC1, OS1, QBC]
CARBOFURAN (ISO) [GBR]

1563-90-2   N,N-DIBUTYLACETAMIDE

1569-01-3   2-PROPANOL, 1-PROPOXY-
N-PROPOXY-2-PROPANOL [PST]

1569-02-4   PROPYLENE GLYCOL ETHYL ETHER

1569-69-3   CYCLOHEXANETHIOL
CYCLOHEXYL MERCAPTAN [HM1, HM1, NJL]

1582-09-8   TRIFLURALIN
TRIFLURALIN [BENZENEAMINE, 2,6-DINITRO-N,N-DIPROPYL-4-(TRIFLUO-ROMETHYL)-] [313]

1589-47-5   2-METHOXY-1-PROPANOL

1589-49-7   3-METHOXYPROPANOL

1592-23-0   CALCIUM STEARATE

1595-04-6   M-BUTYLTOLUENE

1595-05-7   P-BUTYLTOLUENE

1595-11-5   O-BUTYLTOLUENE

1596-84-5   DAMINOZIDE

1600-27-7   MERCURIC ACETATE
MERCURY ACETATE [HM1, HM1, HM1]

1606-67-3   1-AMINO PYRENE

1606-85-5   ETHANOL, 2,2'-[2-BUTYNE-1,4-DIYLBIS(OXY)]BIS-
2,2'-[2-BUTYNE-1,4-DIYL(OXY)] BISETHANOL [PST]

1609-19-4   CHLORODIETHYLSILANE
SILANE, CHLORODIETHYL- [PAL]

1609-47-8   DIETHYL PYROCARBONATE

1609-86-5   TERT-BUTYL ISOCYANATE
TERT-BUTYLISOCYANATE [NJL]

1610-18-0   PROMETON

1615-80-1   1,2-DIETHYLHYDRAZINE
1,2-DIETHYL HYDRAZINE [NJL, NJL]
DIETHYL HYDRAZINE [MIN]
HYDRAZINE, 1,2-DIETHYL- [PAL, PAL]
N,N'-DIETHYLHYDRAZINE [EPA, RCU, RCU]

1619-57-4   DIETHYL ACETOACETATE

1620-21-9   CHLORCYCLIZINE HYDROCHLO-RIDE

1622-32-8   ETHANESULFONYL CHLORIDE, 2-CHLORO-

1623-15-0   PHOSPHORIC ACID, MONOBUTYL ESTER

1623-24-1   ISOPROPYL PHOSPHORIC ACID
ISOPROPYL ACID PHOSPHATE [HM1, HM1, NJL, NJL]
PHOSPHORIC ACID, MONO(1-METHYLETHYL) ESTER [TSC, TSC, TSC]

1629-58-9   ETHYL VINYL KETONE

1633-83-6   1,4-BUTANESULTONE
1,4-BUTANE SULTONE [MAK]

1634-04-4   METHYL TERT-BUTYL ETHER
METHYL-T-BUTYL ETHER [SD4]
METHYL-T-BUTYL-ETHER [ATS]
METHYLTERTBUTYL ETHER [HM1, HM1]
PROPANE, 2-METHOXY-2-METHYL-[TSC, TSC, TSC, TSC]

1634-78-2   MALAOXON

1638-16-0   TRIPROPYLENE GLYCOL

1638-22-8   P-BUTYLPHENOL

1639-09-4   1-HEPTANETHIOL

1639-66-3   DIOCTYL SODIUM SULFOSUCCI-NATE
SODIUM DIOCTYL SULFOSUCCINATE [PST]

1640-89-7   ETHYL CYCLOPENTANE
CYCLOPENTANE, ETHYL- [PAL]
ETHYLCYCLOPENTANE [NFP, NFP]

1642-54-2   DIETHYLCARBAMAZINE CIT-RATE

1643-20-5   LAURYLDIMETHYLAMINE OXIDE
N,N-DIMETHYLDODECYLAMINE OXIDE [PST]

1646-87-3   ALDICARB SULFOXIDE

1646-88-4   ALDICARB SULFONE

1649-08-7   1,2-DICHLORO-1,1-DIFLUO-ROETHANE
1,2-DICHLORO-1,1-DIFLUO-ROETHANE (HCFC-132B) [313]
ETHANE, 1,2-DICHLORO-1,1-DIFLU-ORO- [TSC, TSC]
HCFC-132B [CAA]
HYDROCHLOROFLUOROCARBON (HCFC 132B) [TSE]

1652-63-7   3-(((HEPTADECAFLUOROOCTYL) SULFONYL)AMINO)-N,N,N-TRIMETHYL-1-PROPANAMINIUM IODIDE

1653-19-6   2,3-DICHLOROBUTADIENE-1,3
1,3-BUTADIENE, 2,3-DICHLORO-[PAL]

1653-30-1   2-UNDECANOL

1656-48-0   3,3'-OXYDIPROPIONITRILE

1656-63-9   TRILAURYL TRITHIOPHOSPHITE

1663-35-0   ETHENE, (2-METHOXYETHOXY)-
VINYL 2-METHOXYETHYL ETHER [MSL, NFP, NFP]
VINYL-2-METHOXYETHYL ETHER [FLA]

1663-39-4   2-PROPENOIC ACID, 1,1-DIMETHYLETHYL ESTER

1666-13-3   PHENYL DISELENIDE

1675-54-3   BISPHENOL A DIGLYCIDYL ETHER
DIGLYCIDYL ETHER OF BISPHENOL A [PST]
OXIRANE, 2,2'-[(1-METHYLETHY-LIDINE)BIS(4,1-PHENYLE-NEOXYMETHYLENE)BIS- [TSC, TSC]

1678-91-7   ETHYL CYCLOHEXANE
CYCLOHEXANE, ETHYL- [PAL]
ETHYLCYCLOHEXANE [NFP, NFP]

1678-93-9   BUTYLCYCLOHEXANE

1680-21-3   TRIETHYLENE GLYCOL DIACRY-LATE

1689-82-3   4-HYDROXYAZOBENZENE

1689-84-5   BROMOXYNIL
BROMOXYNIL (3,5-DIBROMO-4-HYDROXYBENZONITRILE) [313]

1689-99-2 **BROMOXYNIL OCTANOATE (OC-TANOIC ACID,2,6-DIBROMO-4-CYANOPHENYL ESTER)**

1693-71-6 **TRIALLYL BORATE**

1694-09-3 **BENZYL VIOLET 4B**
BENZENEMETHANAMINIUM, N-[4-[[4-(DIMETHYLAMINO)PHENYL][4-[ETHYL[(3-SULFOPHENYL)METHYL]AMINO]PHENYL]METHYLENE]-2,5-CYCLOHEXADIEN-1-YLIDENE]-N-ETHYL-SULFO-,HYDROXIDE, INNER SALT, SODIUM SALT [PAL, PAL]
FD&C VIOLET NO. 1 [PST]

1696-20-4 **MORPHOLINE, 4-ACETYL-**
4-ACETYL MORPHOLINE [FLA]
N-ACETYL MORPHOLINE [MSL, NFP, NFP]

1704-62-7 **DIMETHY-LAMINOETHOXYETHANOL**

1705-85-7 **6-METHYLCHRYSENE**

1706-01-0 **3-METHYLFLUORANTHENE**

1712-64-1 **ISOPROPYL NITRATE**

1717-00-6 **ETHANE, 1,1-DICHLORO-1-FLU-ORO-**
1,1-DICHLORO-1-FLUOROETHANE (HCFC-141B) [313]
HCFC-141B [CAA]

1719-53-5 **DIETHYLDICHLOROSILANE**
DIETHYL DICHLOROSILANE [NJL, NJL]

1737-93-5 **3,5-DICHLORO-2,4,6-TRIFLUO-ROPYRIDINE**

1738-25-6 **BETA-DIMETHYLAMINOPROPI-ONITRILE**
3-(DIMETHYLAMINO)-PROPIONI-TRILE [NFP, NFP]

1746-01-6 **2,3,7,8-TETRACHLORODIBENZO-P-DIOXIN (TCDD)**
2,3,7,8-TCDD (DIOXIN) [SDW, SDW]
2,3,7,8-TETRACHLORODIBENZO-P-DIOXIN [ATS, C65, C65, C65, CAA, CAL, CSR, CSR, FLA, MAK, MSL, NJL, PQL]
2,3,7,8-TETRACHLORODIBENZO-P-DIOXIN (TCDD, DIOXIN) [WHM]
2,3,7,8-TETRACHLORODIBENZO-PARA-DIOXIN [MIN]
2,3,7,8-TETRACHLORODIBENZO-PARA-DIOXIN (TCDD) [IAR]
DIBENZO(B,E][1,4]DIOXIN, 2,3,7,8-TETRACHLORO- [PAL, PAL]

1747-60-0 **6-METHOXY-2-BENZOTHIAZO-LAMINE**

1752-30-3 **ACETONE THIOSEMICARBAZIDE**

1758-61-8 **DI(1-HYDROXYCYCLO-HEXYL) PEROXIDE**
DICYCLOHEXYL PEROXIDE [MAK]

1758-73-2 **METHANESULFONIC ACID, AMINOIMINO-**

1760-24-3 **N-[3-(TRIMETHYOXYSILYL)PROPYL]-1,2-ETHANEDIAMINE**

1762-95-4 **AMMONIUM THIOCYANATE**
THIOCYANIC ACID, AMMONIUM SALT [PAL]

1777-84-0 **3-NITRO-P-ACETOPHENETIDE**
3'-NITRO-P-ACETOPHENETIDE [MSL]

1789-58-8 **ETHYL DICHLOROSILANE**
SILANE, DICHLOROETHYL- [PAL]

1793-90-4 **TRIETHYLALLYLGERMANIUM**

1795-48-8 **ISOPROPYL ISOCYANATE**

1809-19-4 **DIBUTYL PHOSPHITE**
PHOSPHONIC ACID, DIBUTYL ESTER [PAL]

1817-73-8 **BENZENAMINE, 2-BROMO-4,6-DINITRO-**

1825-21-4 **PENTACHLOROANISOLE**

1825-30-5 **NAPHTHALENE, 1,5-DICHLORO-**

1825-31-6 **NAPHTHALENE, 1,4-DICHLORO-**

1835-49-0 **TETRAFLUOROPHTHALONITRILE**

1836-75-5 **NITROFEN**
2,4-DICHLOROPHENYL P-NITRO-PHENYL ETHER [CAL]
BENZENE, 2,4-DICHLORO-1-(4-NITRO-PHENOXY)- [PAL, PAL]
NITROFEN (TECHNICAL GRADE) [C65, C65, IAR]
NITROFEN [BENZENE, 2,4-DICHLORO-1-(4-NITROPHENOXY)-] [313]

1838-59-1 **ALLYL FORMATE**

1843-05-6 **2-HYDROXY-4-N-OCTOXYBEN-ZOPHENONE**

1847-55-8 **CIS-9-OCTADECENYL SULFATE, SODIUM SALT**

1854-23-5 **4,4'-(2-ETHYL-2-NITROTRIMETHYLENE)-DIMORPHOLINE**

1854-26-8 **2-IMIDAZOLIDINONE, 4,5-DIHY-DROXY-1,3-BIS**

1861-32-1 **TEREPHTHALIC ACID, TETRA-CHLORO-, DIMETHYL ESTER**
DCPA (AND ITS ACID METABOLITES) [SD4]

1861-40-1 **N-BUTYL-N-ETHYL-2,6-DINITRO-4-(TRIFLUOROMETHYL)BEN-ZENAMINE**
BENFLURALIN (N-BUTYL-N-ETHYL-2,6-DINITRO-4-(TRIFLUOROMETHYL) BENZENAMINE) [313]

1863-63-4 **AMMONIUM BENZOATE**
BENZOIC ACID, AMMONIUM SALT [PAL]

1867-66-9 **CYCLOHEXANONE, 2-(O-CHLOROPHENYL)-2-(METHY-LAMINE)-, HYDROCHLORIDE**

1873-29-6 **ISOBUTYL ISOCYANATE**

1879-09-0 **2,4-DIMETHYL-6-TERT-BUTYLPHENOL**
6-TERT-BUTYL-2,4-XYLENOL [OTS]

1885-14-9 **PHENYLCHLOROFORMATE**
PHENYL CHLOROFORMATE [WHM]

1886-81-3 **DODECYL BENZENESULFONATE**

1888-71-7 **HEXACHLOROPROPENE**
1-PROPENE, 1,1,2,3,3,3-HEX-ACHLORO- [TSC, TSC, TSC]

1897-45-6 **CHLOROTHALONIL**
2,4,5,6-TETRACHLOROISOPH-THALONITRILE [PST]
CHLOROTHALONIL [1,3-BENZENEDI-CARBONITRILE,2,4,5,6-TETRA-CHLORO-] [313]

1910-42-5 **PARAQUAT**
4,4'-BIPYRIDINIUM, 1,1'-DIMETHYL-, DICHLORIDE [PAL]
PARAQUAT [302, CAL, MAK, MAK, MAK, MIN, NIO, NIO, NIO, NIO, NJL]
PARAQUAT (GRAMOXONE) [BC1, BC1]
PARAQUAT DICHLORIDE [313]
PARAQUAT DICHLORIDE (ISO) [GBR]
PARAQUAT RESPIRABLE DUST [MEX]
PARAQUAT RESPIRABLE SIZES [ALB, ALB, QBC]

1912-24-9 **ATRAZINE**
1,3,5-TRIAZINE-2,4-DIAMINE, 6-CHLORO-N-ETHYL-N'-(1-METHYLETHYL)- [PAL]
ATRAZINE (6-CHLORO-N-ETHYL-N'-(1-METHYLETHYL)-1,3,5-TRIAZINE-2,4-DIAMINE) [313]

1918-00-9 **DICAMBA**
BENZOIC ACID, 3,6-DICHLORO-2-METHOXY- [PAL]
DICAMBA (3,6-DICHLORO-2-METHOXYBENZOIC ACID) [313]

1918-02-1 **PICLORAM**
2-PYRIDINCARBOXYLIC ACID, 4-AMINO-3,5,6-TRICHLORO- [PAL]
4-AMINO-3,5,6-TRICHLORO-2-PYRIDINECARBOXYLIC ACID [ONT, ONT]
PICLORAM (ISO) [GBR, GBR]
PICLORAM (TORDON) [ALB, ALB, BC1, BC1]
PICLORAME (TORDON (R)) [QBC]

1918-16-7 **PROPACHLOR**
PROPACHLOR [2-CHLORO-N-(1-METHYLETHYL)-N-PHENYLACETA-MINDE] [313]

1918-18-9 **SWEP**

1928-38-7 **2,4-D METHYL ESTER**
2,4-D ESTERS [NJL]
ACETIC ACID, (2,4-DICHLOROPHE-NOXY)-, METHYL ESTER [MSL, PAL]

1928-43-4 **2,4-D 2-ETHYLHEXYL ESTER**

1928-47-8 **ACETIC ACID, (2,4,5-TRICHLOROPHENOXY)-, 2-ETHYLHEXYL ESTER**
2,4,5-T ESTER [NJL]

1928-61-6 **2,4-D PROPYL ESTER**
2,4-D ESTERS [NJL]
ACETIC ACID, (2,4-DICHLOROPHE-NOXY)-, PROPYL ESTER [MSL, PAL]

1929-73-3 **2,4-D BUTOXYETHYL ESTER**
2,4-D ESTERS [NJL]
ACETIC ACID, (2,4-DICHLOROPHE-NOXY)-, 2-BUTOXYETHYL ESTER [MSL, PAL]

1929-82-4 **NITRAPYRIN**
2-CHLORO-6-(TRICHLOROMETHYL)PYRIDINE (N-SERVE) [ALB, ALB, BC1, BC1]
2-CHLORO-6-(TRICHLOROMETHYL)PYRIDINE [CAL, GBR, GBR, ONT, ONT]
2-CHLORO-6-TRICHLORO-METHYL PYRIDINE [OS1, OS1]
NITRAPYRIN [QBC, QBC]
NITRAPYRIN (2-CHLORO-6-(TRICHLOROMETHYL)PYRIDINE) [313]
PYRIDINE, 2-CHLORO-6-(TRICHLOROMETHYL)- [PAL]

1931-62-0 **TERT-BUTYL MONOPEROXY-MALEATE**

1934-21-0 **FD&C YELLOW 5**
C.I. ACID YELLOW 23, TRISODIUM SALT [PST]

1936-15-8 **ACID ORANGE 10**
C.I. ACID ORANGE 10 [CSR, CSR, PST]
ORANGE G [IAR]

1937-35-5 **C.I. DIRECT RED 13, DISODIUM SALT**

1937-37-7 **C.I. DIRECT BLACK 38**
2,7-NAPHTHALENEDISULFONIC ACID, 4-AMINO-3-[[4'-[(2,4-DI-AMINOPHENYL)AZO] [1,1'-BIPHENYL]-4-YL]AZO]-5-HYDROXY-6- (PHENY-LAZO)-, DISODIUM SALT [PAL, PAL, TSC, TSC]
DIRECT BLACK 38 [CAB, FLA, MSL, NTP]

DIRECT BLACK 38 (TECHNICAL GRADE) [C65, C65, CAL, IAR, MIN]
DIRECT BLACK META [WHM]

1940-42-7 4-BROMO-2,5-DICHLOROPHENOL

1941-79-3 DIPEROXYAZELAIC ACID

1948-33-0 T-BUTYLHYDROQUINONE
TERT-BUTYLHYDROQUINONE [PST]

1954-28-5 TRIETHYLENE GLYCOL DIGLYCIDYL ETHER

1955-45-9 PIVALOLACTONE

1965-29-3 N-(HYDROXYETHYL)DIETHYLENETRIAMINE

1972-08-3 1-TRANS-DELTA-9-TETRAHYDROCANNABINOL

1982-37-2 METHDILAZINE

1982-47-4 CHLOROXURON

1982-49-6 UREA, 1-(2-METHYLCYCLOHEXYL)-3-PHENYL-
SIDURON [CAL]

1982-69-0 SODIUM DICAMBA [3,6-DICHLORO-2-METHOXYBENZOIC ACID, SODIUM SALT]

1983-10-4 STANNANE, FLUOROTRIBUTYL-
TRIBUTYLTIN FLUORIDE [313, MAK, MAK, MAK]

1984-06-1 SODIUM CAPRYLATE

2001-95-8 VALINOMYCIN

2008-39-1 DIMETHYLAMINE (2,4-DICHLOROPHENOXY)ACETATE

2008-46-0 ACETIC ACID, (2,4,5-TRICHLOROPHENOXY)-, COMPOUND WITH N,N-DIETHYLETHANAMINE
2,4,5-T AMINE [NJL]

2016-56-0 ACETIC ACID, DODECYLAMINE SALT

2016-57-1 DECYLAMINE
1-DECANAMINE [PAL]

2031-67-6 METHYLTRIETHOXYSILANE
SILANE, TRIETHOXYMETHYL- [PST]

2032-59-9 AMINOCARB

2032-65-7 METHIOCARB
MERCAPTODIMETHUR [CAL, EPA, HM1, HM1, NJL]
MESUROL [MSL]
PHENOL, 3,5-DIMETHYL-4-(METHYLTHIO)-, METHYLCARBAMATE [PAL]

2036-15-9 DIPROPYLALUMINUM HYDRIDE
ALUMINUM, HYDRODIPROPYL- [PAL]

2038-03-1 4-(2-AMINOETHYL)-MORPHOLINE
4-MORPHOLINEETHANAMINE [PAL]

2039-87-4 O-CHLOROSTYRENE
BENZENE, 1-CHLORO-2-ETHENYL- [PAL]

2040-96-2 PROPYLCYCLOPENTANE

2049-92-5 P-TERT-AMYLANILINE
BENZENAMINE, 4-(1,1-DIMETHYLPROPYL)- [PAL]

2049-95-8 TETRAAMYLBENZENE

2050-04-6 AMYL PHENYL ETHER

2050-08-0 AMYL SALICYLATE

2050-24-0 1-METHYL-3,5-DIETHYLBENZENE

2050-60-4 BUTYL OXALATE
DIBUTYL OXALATE [PST]

2050-69-3 NAPHTHALENE, 1,2-DICHLORO-

2050-72-8 NAPHTHALENE, 1,6-DICHLORO-

2050-73-9 NAPHTHALENE, 1,7-DICHLORO-

2050-74-0 NAPHTHALENE, 1,8-DICHLORO-

2050-75-1 NAPHTHALENE, 2,3-DICHLORO-

2050-92-2 DI-N-AMYLAMINE
1-PENTANAMINE, N-PENTYL- [PAL]
DIAMYLAMINE [FLA, HM1, HM1, HM1, MSL, NFP, NFP, NJL]

2050-99-9 2,8-DIMETHYL-5-NONANONE

2052-07-5 1,1'-BIPHENYL, 2-BROMO-
2-BROMO-1,1-BIPHENYL [TSE]

2058-46-0 OXYTETRACYCLINE HYDROCHLORIDE

2065-70-5 NAPHTHALENE, 2,6-DICHLORO-

2068-78-2 VINCRISTINE SULFATE
VINCRISTINE SULPHATE [IAR]

2074-50-2 PARAQUAT METHOSULFATE
4,4'-BIPYRIDINIUM, 1,1'-DIMETHYL-, BIS(METHYL SULFATE) [PAL]
PARAQUAT BIS(METHYL SULFATE) [MSL]

2074-87-5 CYANOGEN
CYANOGENE [QBC]

2082-79-3 OCTADECYL 3-(3',5'-DI-TERT-BUTYL-4'-HYDROXYPHENYL) PROPIONATE

2082-81-7 1,4-BUTANEDIOL DIMETHACRYLATE
2-PROPENOIC ACID, 2-METHYL-, 1,4-BUTANEDIYL ESTER [TSC, TSC]

2084-18-6 3-METHYL-2-BUTANETHIOL
2-BUTANETHIOL, 3-METHYL- [PAL]

2090-05-3 CALCIUM BENZOATE

2092-56-0 D&C RED NO. 8

2097-19-0 PHENYLSILATRANE

2100-42-7 2,5-DIMETHOXYCHLOROBENZENE
BENZENE, 2-CHLORO-1,4-DIMETHOXY- [PAL]

2104-64-5 EPN
O-ETHYL O-P-NITROPHENYL PHENYLPHOSPHONOTHIOATE [ONT, ONT]
PHOSPHONOTHIOIC ACID, PHENYL-, O-ETHYL O-(4-NITROPHENYL)ESTER [PAL]

2109-64-0 DIBUTYLISOPROPANOLAMINE
2-PROPANOL, 1-(DIBUTYLAMINO)- [PAL]

2113-57-7 1,1'-BIPHENYL, 3-BROMO-
3-BROMO-1,1-BIPHENYL [TSE]

2116-84-9 PHENYL TRIMETHYL SILOXANE

2144-45-8 DIBENZYL PEROXYDICARBONATE

2150-54-1 C.I. DIRECT BLUE 25, TETRASODIUM SALT
2,7-NAPHTHALENEDISULFONIC ACID, 3,3'-[(3,3'-DIMETHYL[1,1'-BIPHENYL]-4,4'-DIYL)BIS(AZO)]BIS-[4,5-DIHYDROXY-, TETRASODIUM SALT [TSC]

2155-70-6 TRIBUTYLTIN METHACRYLATE
TRIBUTYL TIN METHACRYLATE [MAK, MAK, MAK]

2155-71-7 DI-TERT-BUTYLPEROXYPHTHALATE

2156-56-1 SODIUM DICHLOROACETATE

2156-96-9 DECYL ACRYLATE
2-PROPENOIC ACID, DECYL ESTER [PAL, TSC, TSC]

2157-42-8 SILICIC ACID (H6SI2O7), HEXAETHYL ESTER

2160-93-2 TERT-BUTYLDIETHANOLAMINE
ETHANOL, 2,2'-[(1,1-DIMETHYLETHYL)IMINO] BIS- [PAL]

2163-00-0 1,6-DICHLOROHEXANE

2164-07-0 DIPOTASSIUM ENDOTHALL [7-OXABICYCLO(2.2.1)HEPTANE-2,3-DICARBOXYLIC ACID, DIPOTASSIUM SALT]

2164-08-1 LENACIL

2164-17-2 FLUOMETURON
FLUOMETURON [UREA, N,N-DIMETHYL-N'-[3-(TRIFLUOROMETHYL)PHENYL]-] [313]

2167-23-9 2,2-DI(TERT-BUTYLPEROXY)-BUTANE

2168-68-5 BIS(1-AZIRIDINYL)MORPHOLINOPHOSPHINE SULPHIDE

2179-59-1 ALLYL PROPYL DISULFIDE
DISULFIDE, 2-PROPENYL PROPYL [PAL]

2185-92-4 2-BIPHENYLAMINE HYDROCHLORIDE

2190-04-7 OCTADECYLAMINE ACETATE

2191-10-8 OCTANOIC ACID, CADMIUM SALT (2:1)

2198-75-6 NAPHTHALENE, 1,3-DICHLORO-

2198-77-8 NAPHTHALENE, 2,7-DICHLORO-

2207-04-7 1,4-DIMETHYLCYCLOHEXANE-TRANS
CYCLOHEXANE, 1,4-DIMETHYL-, TRANS- [PAL]

2210-28-8 2-PROPENOIC ACID, 2-METHYL-, PROPYL ESTER

2210-79-9 OXIRANE, [(2-METHYLPHENOXY)METHYL]-
CRESYL GLYCIDYL ETHER, O- [OTS]
O-CRESYL GLYCIDYL ETHER [EPA]

2212-67-1 ORDRAM (MOLINATE)
MOLINATE [CAL]
MOLINATE (1H-AZEPINE-1-CARBOTHIOIC ACID, HEXAHYDRO-S-ETHYL ESTER) [313]

2216-33-3 3-METHYLOCTANE
NONANE (ISO) [NFP]
OCTANE, 3-METHYL [PAL]

2216-34-4 4-METHYLOCTANE
NONANE (ISO) [NFP]
OCTANE, 4-METHYL [PAL]

2216-51-5 1-METHANOL

2217-06-3 DIPICRYL SULFIDE

2223-82-7 NEOPENTYL GLYCOL DIACRYLATE
2-PROPENOIC ACID, 2,2-DIMETHYL-1,3-PROPANEDIYL ESTER [TSC, TSC]

2223-93-0 CADMIUM STEARATE

2224-15-9 OXIRANE, 2,2'-[1,2-ETHANEDIYLBIS (OXYMETHYLENE)]BIS-
ETHYLENE GLYCOL DIGLYCIDYL ETHER [EPA, OTS]

2224-44-4 4-(2-NITROBUTYL)MORPHOLINE

2231-57-4 THIOCARBAZIDE

2234-13-1 OCTACHLORONAPHTHALENE
NAPHTHALENE, OCTACHLORO- [PAL, TSC, TSC, TSC]

2235-25-8 ETHYLMERCURIC PHOSPHATE

2235-46-3 N,N-DIETHYLACETOACETAMIDE

2235-54-3　AMMONIUM LAURYL SULFATE
DODECYL SULFATE AMMONIUM
SALT [PST]

2238-07-5　DIGLYCIDYL ETHER (DGE)
2,2'-(OXYBIS(METHYLENE))BISOXI-
RANE [ONT]
BIS(2,3-EPOXYPROPYL) ETHER
[GBR]
DIGLYCIDYL ETHER [302, CAL,
CEX, EPA, EPA, EPA, FLA, MSL,
NIO, NIO, NIO, NIO, NJL, OTS]
DIGLYCIDYL OXIDE (ETHER) [QBC]
DIGLYCIDYL RESORCINOL ETHER
[MIN]
OXIRANE, 2,2'-[OXYBIS(METHYLENE)]
BIS- [PAL, TSC, TSC]

2243-62-1　1,5-NAPHTHALENEDIAMINE

2244-16-8　CARVONE
D-CARVONE [CSR, CSR]

**2244-21-5　POTASSIUM DICHLOROISOCYA-
NURATE**
1,3,5-TRIAZINE-2,4,6-(1H,3H,5H)-
TRIONE, 1,3-DICHLORO-, POTASSIUM
SALT [PAL]
DICHLOROISOCYANURIC ACID,
POTASSIUM SALT [WHM]
POTASSIUM DICHLORO-S TRI-
AZINETRIONE [FLA]
POTASSIUM DICHLORO-S-TRI-
AZINETRIONE [MSL]

2269-22-9　2-BUTANOL, ALUMINUM SALT

2275-18-5　PROTHOATE

2294-47-5　P-DIAZIDOBENZENE

2300-66-5　DIMETHYLAMINE DICAMBA

2303-16-4　DIALLATE
CARBAMOTHIOIC ACID, BIS(1-
METHYLETHYL)-, S-(2,3-DICHLORO-
2-PROPENYL) ESTER [PAL]
DI-ALLATE [HM1, HM1]
DIALLATE [CARBAMOTHIOIC
ACID, BIS(1-METHYLETHYL)-,S-(2,
3-DICHLORO-2-PROPENYL)ESTER]
[313]

2303-17-5　TRIALLATE

**2305-05-7　GAMMA-N-OCTYL-GAMMA-N-
BUTYROLACTONE**

2310-17-0　PHOSALONE

2312-35-8　PROPARGITE
SULFUROUS ACID, 2-[4-(1,1-
DIMETHYLETHYL)PHENOXY]CYCLO-
HEXYL2-PROPYNYL ESTER [PAL]

**2312-76-7　SODIUM DINITRO-ORTHO-
CRESOLATE**
SODIUM DINITRO-O-CRESOLATE
[HM1, HM1, HM1]

2318-18-5　SENKIRKINE

2321-07-5　FLUORESCEIN
C.I. SOLVENT YELLOW 94 [PST]

2338-12-7　5-NITROBENZOTRIAZOL

2353-45-9　FAST GREEN FCF
C.I. FOOD GREEN 3, DISODIUM SALT
[PST]

**2358-84-1　DIETHYLENE GLYCOL
DIMETHACRYLATE**
2-PROPENOIC ACID, 2-METHYL-
, OXYDI-2,1-ETHANEDIYL ESTER
[TSC, TSC]

**2370-88-9　TETRAMETHYLCYCLOTE-
TRASILOXANE**

**2372-21-6　TERT-BUTYL PEROXYISOPROPYL
CARBONATE**

**2374-14-3　TRIFLUOROPROPYLMETHYLCY-
CLOTRISILOXANE**

2379-81-9　C.I. VAT BLACK 27

2385-85-5　MIREX
1,3,4-METHENO-1H-CYCLOBUTA[CD]
PENTALENE, 1,1A,2,2,3,3A,4,5,5,5A,5B,
6-DODECACHLOROOCTAHYDRO-
[PAL, PAL]
DODECACHLOROPENTACYCLO
[5.3.0.0.0.0] DECANE [CN5]
MIREX (DODECACHLOROPENTACY-
CLO [5.3.0.0.0.0] DECANE) [CN8]

**2386-87-0　3,4-EPOXYCYCLOHEXANECAR-
BOXYLIC ACID (3,4-EPOXYCY-
CLOHEXYLMETHYL) ESTER**

**2386-90-5　BIS(2,3-EPOXYCYCLOPENTYL)
ETHER**

2390-63-8　C.I. BASIC VIOLET 11

**2399-85-1　TRIPOTASSIUM NITRILOTRIAC-
ETATE**

**2402-79-1　PYRIDINE, 2,3,5,6-TETRA-
CHLORO-**
TETRACHLOROPYRIDINE, 2,3,5,6-
[OTS]

2404-44-6　OXIRANE, OCTYL-

2408-37-9　TRIMETHYLCYCLOHEXANONE

2409-55-4　4-METHYL-2-T-BUTYLPHENOL

**2421-27-4　2-PROPENOIC ACID, 2-HYDROXY-
BUTYL ESTER**

**2422-91-5　BENZENE, 1,1',1''-METHYLIDYEN-
ETRIS(4-ISOCYANATO-**

**2425-01-6　OXIRANE, 2,2'-[1,
4-PHENYLENEBIS (OXYMETHY-
LENE)]BIS-**
HYDROQUINONE DIGLYCIDYL
ETHER [EPA, OTS]

2425-06-1　CAPTAFOL
1H-ISOINDOLE-1,3(2H)-DIONE, 3A,
4,7,7A-TETRAHYDRO-2-[(1,1,2,2-
TETRACHLOROETHYL)THIO]- [PAL]
CAPTAFOL (DIFOLATAN) [ALB,
ALB, ALB, BC1, BC1, OS1, QBC,
QBC]
CAPTAFOL (ISO) [GBR, GBR]

**2425-79-8　OXIRANE, 2,2'-[1,4-BUTANEDIYL-
BIS(OXYMETHYLENE)]BIS-**
1,4-BUTANEDIOL DIGLYCIDYL
ETHER [EPA]
BUTANEDIOL DIGLYCIDYL ETHER, 1,
4- [OTS]

2425-85-6　C.I. PIGMENT RED 3
CI PIGMENT RED 3 [IAR]

2426-08-6　N-BUTYL GLYCIDYL ETHER (BGE)
(BUTOXYMETHYL)OXIRANE [ONT,
ONT, ONT]
BUTYL GLYCIDYL ETHER (BGE)
[WHM]
BUTYL GLYCIDYL ETHER, N- [OTS]
N-BUTYL GLYCIDYL ETHER [CAL,
CEX, EPA, MAK, MAK, MIN, NIO,
NIO, NIO]
N-BUTYL GLYCIDYL OXIDE (ETHER)
[QBC]
OXIRANE, (BUTOXYMETHYL)- [PAL]
OXIRANE, BUTOXYMETHYL- [TSC,
TSC]

**2426-54-2　2-(DIETHYLAMINO)ETHYL ACRY-
LATE**
2-PROPENOIC ACID, 2-(DIETHY-
LAMINO)ETHYL ESTER [PAL, TSC,
TSC]

**2429-70-1　C.I. DIRECT RED 10, DISODIUM
SALT**

**2429-71-2　C.I. DIRECT BLUE 8, DISODIUM
SALT**
1-NAPHTHALENESULFONIC ACID, 3,
3'-[3,3'-DIMETHOXY-[1,1'-BIPHENYL]-
4,4'-DIYL)BIS(AZO)]BIS[4-HYDROXY-,
DISODIUM SALT [TSC]

**2429-73-4　C.I. DIRECT BLUE 2, TRISODIUM
SALT**
2,7-NAPHTHALENEDISULFONIC
ACID, 5-AMINO-3-[[4'-[(7-AMINO-
1-HYDROXY-3-SULFO-2-NAPH-
THALENYL)-AZO] [1,1'-BIPHENYL]-
4-YL]AZO]-4-HYDROXY-, TRISODIUM
SALT [TSC]

2429-74-5　C.I. DIRECT BLUE 15
2,7-NAPHTHALENEDISULFONIC
ACID, 3,3'-[(3,3'-DIMETHOXY[1,1'-
BIPHENYL]-4,4'-DIYL)BIS(AZO)]BIS[5-
AMINO-4-HYDROXY-, TETRASODIUM
SALT [TSC]
CI DIRECT BLUE 15 [IAR]

**2429-79-0　BENZOIC ACID, 5-[[4'-[1-AMINO-4-
SULFO-2-NAPHTHALENYL]AZO]
[1,1'-BIPHENYL]-4-YL]AZO]-2-
HYDROXY-, DISODIUM SALT**

2429-80-3　C.I. ACID ORANGE 45

**2429-81-4　C.I. DIRECT BROWN 31, TETRA-
SODIUM SALT**
BENZOIC ACID, 5-[[4'-[[2,6-DIAMINO-
3-[[8-HYDROXY-3,6-DISULFO-7-
[(4-SULFO-1-NAPHTHALENYL)
AZO]-2-NAPHTHALENYL]AZO] -
5-METHYLPHENYL]AZO] [1,1'-
BIPHENYL]-4-YL]AZO]-2-HYDROXY-,
TETRASODIUM SALT [TSC]

**2429-82-5　C.I. DIRECT BROWN 2, DISODIUM
SALT**
BENZOIC ACID, 5-[[4'-[7-AMINO-
1-HYDROXY-3-SULFO-2-NAPH-
THALENYL]AZO][1,1'-BIPHENYL]-4-
YL]AZO]-2-HYDROXY-, DISODIUM
SALT [TSC]

**2429-83-6　C.I. DIRECT BLACK 4, DISODIUM
SALT**
2,7-NAPHTHALENEDISULFONIC
ACID, 4-AMINO-3-[[4'-[(2,4-DIAMINO-
5-METHYLPHENYL)AZO] [1,1'-
BIPHENYL]-4-YL]AZO] 5-HYDROXY-
6- (PHENYLAZO)-, DISODIUM SALT
[TSC]

**2429-84-7　C.I. DIRECT RED 1, DISODIUM
SALT**
BENZOIC ACID, 5-[[4'-[2-
AMINO-8-HYDROXY-6-SULFO-1-
NAPHTHALENYL]AZO][1,1'-BIPHENYL]
-4-YL]AZO]-2-HYDROXY-, DISODIUM
SALT [TSC]

2431-50-7　1-BUTENE, 2,3,4-TRICHLORO-
2,3,4-TRICHLOROBUTENE-1 [MAK]
BUTENE, 2,3,4-TRICHLORO-, 1- [OTS]

2432-99-7　11-AMINOUNDECANOIC ACID

2437-29-8　MALACHITE GREEN OXALATE

**2438-88-2　2,3,5,6-TETRACHLORO-4-NI-
TROANISOLE**

**2439-01-2　CHINOMETHIONAT [6-METHYL-1,
3-DITHIOLO[4,5-B]QUINOXALIN-2-
ONE]**

**2439-10-3　DODECYLGUANIDINE MONOAC-
ETATE**
DODINE [DODECYLGUANIDINE
MONOACETATE] [313]

**2440-22-4　2-BENZOTRIAZOLYL-4-
METHYLPHENOL**
2-(2H-BENZOTRIAZOL-2-YL)-4-
METHYLPHENOL [PST]

2443-39-2　CIS-9,10-EPOXYSTEARIC ACID

**2449-49-2　ALPHA-METHYLBENZYL
DIMETHYL AMINE**
BENZENEMETHANAMINE, N,N,.AL-
PHA.-TRIMETHYL- [PAL]

**2454-37-7　BENZENEMETHANOL, 3-AMINO-
.ALPHA.-METHYL-**
(M-AMINOPHENYL) METHYL
CARBINOL [FLA, NFP, NFP]

(M-AMINOPHENYL)METHYL
CARBINOL [MSL]

**2455-24-5 TETRAHYDROFURFURYL
METHACRYLATE**
2-PROPENOIC ACID, 2-METHYL-,
(TETRAHYDRO-2-FURANYL)METHYL
ESTER [TSC, TSC]

2456-28-2 DIDECYL ETHER

**2461-15-6 OXIRANE, [[(2-ETHYLHEXYL)
OXY]METHYL]-**
2-ETHYLHEXYL GLYCIDYL ETHER
[EPA]
ETHYLHEXYL GLYCIDYL ETHER, 2-
[OTS]

**2461-18-9 OXIRANE, [(DODECYLOXY)
METHYL]-**
LAURYL GLYCIDYL ETHER [EPA,
OTS]

2465-27-2 AURAMINE
AURAMINE (TECHNICAL GRADE)
[IAR]
BENZENAMINE, 4-4'-CARBONIMI-
DOYLBIS[N,N-DIMETHYL-,MONOHY-
DROCHLORIDE [PAL, PAL]
C.I. BASIC YELLOW 2 [PST]
C.I. BASIC YELLOW 2, MONOHY-
DROCHLORIDE [NJL, NJL]

**2471-08-1 HEXACOSAMETHYLDODE-
CASILOXANE**

**2471-09-2 OCTACOSAMETHYLTRIDE-
CASILOXANE**

**2471-10-5 TRIACONTAMETHYLTETRADE-
CASILOXANE**

**2471-11-6 DOTRIACONTAMETHYLPEN-
TADECASILOXANE**

2475-33-4 C.I. VAT BROWN 1

2475-45-8 C.I. DISPERSE BLUE 1
DISPERSE BLUE 1 [C65, C65, CAB,
CAL, IAR]

**2475-46-9 1-METHYLAMINO-4-
ETHANOLAMINOAN-
THRAQUINONE**
C.I. DISPERSE BLUE 3 [PST]

2487-90-3 TRIMETHOXYSILANE

2489-77-2 TRIMETHYLTHIOUREA

**2492-26-4 SODIUM MERCAPTOBENZOTHIA-
ZOLE**
2-MERCAPTOBENZOTHIAZOLE,
SODIUM SALT [PST]

**2493-02-9 BENZENE, 1-BROMO-4-ISO-
CYANATO-**

2497-07-6 OXYDISULFOTON

**2499-59-4 2-PROPENOIC ACID, OCTYL ES-
TER**

**2499-95-8 2-PROPENOIC ACID, HEXYL ES-
TER**

**2508-19-2 TRINITROBENZENESULFONIC
ACID**

2517-43-3 3-METHOXYBUTANOL

**2524-03-0 DIMETHYL PHOSPHOROCHLORI-
DOTHIOATE**
DIMETHYL CHLOROTHIOPHOS-
PHATE [313, OTS]
DIMETHYL THIOPHOSPHORYL
CHLORIDE [HM1, HM1, NJL, NJL]
O,O-DIMETHYL PHOSPHOROCHLORI-
DOTHIOATE [WHM]

**2524-04-1 DIETHYLTHIOPHOSPHORYL
CHLORIDE**
DIETHYL CHLOROTHIOPHOSPHATE
[OTS]

2528-36-1 DIBUTYL PHENYL PHOSPHATE
DI(N-BUTYL) PHENYL PHOSPHATE
[OTS]
DIBUTYLPHENYLPHOSPHATE [MIN]
PHOSPHORIC ACID, DIBUTYL
PHENYL ESTER [TSC, TSC]

**2530-83-8 GLYCIDOXYPROPY-
LTRIMETHOXYSILANE**
3-(TRIMETHOXYSILYL)PROPYL GLY-
CIDYL ETHER [EPA]
GLYCIDOXYPROPYLTRIMETHOXYSI-
LANE, GAMMA- [OTS]
SILANE, TRIMETHOXY[3-(OXIRANYL-
METHOXY)PROPYL]- [TSC]

**2530-85-0 METHACRYLIC ACID, 3-
(TRIMETHOXYSILYL)PROPYL ES-
TER**
2-PROPENOIC ACID, 2-METHYL-, 3-
(TRIMETHOXYSILYL)PROPYL ESTER
[TSC, TSC]
GAMMA-METHACRYLOXYPROPYL-
TRIMETHOXY-SILANE [WHM]

**2536-05-2 BENZENE, 1,1'-METHYLENEBIS(2-
ISOCYANATO-**

2540-82-1 FORMOTHION

**2545-59-7 ACETIC ACID, (2,4,5-
TRICHLOROPHENOXY)-, 2-BU-
TOXYETHYL ESTER**
2,4,5-T ESTER [NJL]
2,4,5-T ESTERS [CWA]

2549-51-1 VINYL CHLOROACETATE

**2549-93-1 1,4-BIS(AMINOMETHYL)CYCLO-
HEXANE**

**2550-33-6 TERT-BUTYL PEROXYDIETHYL-
ACETATE**

2551-62-4 SULFUR HEXAFLUORIDE
SULFUR FLUORIDE (SF6), (OC-6-11)-
[PAL]
SULPHUR HEXAFLUORIDE [AUS,
GBR, GBR]

2553-08-4 P-OCTYLPHENYL SALICYLATE

2554-06-5 METHYLVINYLCYCLOSILOXANE

**2556-36-7 CYCLOHEXANE, 1,4-DIISO-
CYANATO-**
1,4-CYCLOHEXANE DIISOCYANATE
[313]

**2567-83-1 TETRAETHYLAMMONIUM PER-
CHLORATE**

2568-90-3 DIBUTOXYMETHANE

2570-26-5 PENTADECYLAMINE

**2571-88-2 N,N-DIMETHYLOCTADECY-
LAMINE OXIDE**

2577-72-2 3,5-DIBROMOSALICYLANILIDE

2580-77-0 2,2'-SULFONYL BIS-ETHANOL

2581-34-2 3-METHYL-4-NITROPHENOL
PHENOL, 3-METHYL-4-NITRO- [OTS]

**2586-57-4 C.I. DIRECT BLUE 22, DISODIUM
SALT**
1,3-NAPHTHALENEDISULFONIC
ACID, 4-AMINO-5-HYDROXY-
6-[[4'-[(2-HYDROXY-1-NAPH-
THALENYL)AZO]-3,3'-DIMETHOXY[1,
1'-BIPHENYL]-4-YL]AZO]-, DISODIUM
SALT [TSC]

**2586-58-5 BENZOIC ACID, 5-[[4'-[[2,6-DI-
AMINO-3-METHYL-5[(SUL-
FOPHENYL)AZO]PHENYL]AZO]
[1,1'-BIPHENYL]-4-YL]AZO]-2-HY-
DROXY-, DISODIUM SALT**

**2586-60-9 C.I. DIRECT VIOLET 1, DISODIUM
SALT**

**2587-90-8 PHOSPHOROTHIOIC ACID, O,O-
DIMETHYL-S-(2-METHYLTHIO)
ETHYL ESTER**
PHOSPHOROTHIOIC ACID, O,O-
DIMETHYL-S-(2-(METHYLTHIO)
ETHYL)ESTER [MSL]
PHOSPHOROTHIOIC ACID, O,O-
DIMETHYL-S-(2-METHYLTHIO)
ETHYLESTER [PAL]

2591-86-8 1-PIPERIDINECARBOXALDEHYDE

2593-15-9 TERRAZOLE

2602-46-2 C.I. DIRECT BLUE 6
2,7-NAPHTHALENEDISULFONIC
ACID, 3,3'-[[1,1'-BIPHENYL]-4,4'-DIYL-
BIS(AZO)]BIS[5-AMINO-4-HYDROXY-,
TETRASODIUM SALT [TSC, TSC]
2,7-NAPHTHALENEDISULFONIC
ACID, 3,3'-[[1,1'-BIPHENYL]4,4'-DIYL)
BIS(AZO)]BIS[5-AMINO-4-HYDROXY-,
TETRASODIUM SALT [PAL, PAL]
DIRECT BLUE 6 [CAB, FLA, MSL,
NTP]
DIRECT BLUE 6 (TECHNICAL
GRADE) [C65, C65, CAL, MIN,
WHM]
DIRECT BLUE 6 (TECHNICAL-
GRADE) [IAR]

**2605-79-0 N,N-DIMETHYLDECYLAMINE
OXIDE**

**2608-48-2 5-(4-NITROPHENYL)-2,4-PENTA-
DIEN-1-AL**

2610-05-1 C.I. DIRECT BLUE 1
1,3-NAPHTHALENEDISULFONIC
ACID, 6,6'-[(3,3'-DIMETHOXY[1,1'-
BIPHENYL]-4,4'-DIYL)BIS[4-
AMINO-5-HYDROXY-, TETRASODIUM
SALT [TSC, TSC]

**2610-11-9 C.I. DIRECT RED 81, DISODIUM
SALT**

**2612-57-9 2,4-DICHLORO-1-PHENYL ISO-
CYANATE**

2614-76-8 2,2-DIHYDROPEROXY PROPANE

**2615-25-0 CIS- AND TRANS-1,4-CYCLOHEX-
ANEDIAMINE**
(TRANS) 1,4-CYCLOHEXANEDIAMINE
[TSE]

**2618-77-1 2,5-DIMETHYL-2,5-DI-(BEN-
ZOYLPEROXY)HEXANE**

**2627-95-4 TETRAMETHYLDIVINYLDISILOX-
ANE**

2631-37-0 PROMECARB

2631-40-5 ISOPROCARB

2634-33-5 1,2-BENZISOTHIAZOLIN-3-ONE

2636-26-2 CYANOPHOS

2642-71-9 AZINPHOS-ETHYL

2646-17-5 OIL ORANGE SS
1-(O-TOLYAZO)-2-NAPHTOL [FLA]
2-NAPHTHALENOL, 1-[(2-
METHYLPHENYL)AZO]- [PAL, PAL]

**2650-18-2 C.I. ACID BLUE 9, DIAMMONIUM
SALT**

2652-13-3 EICOSAMETHYLNONASILOXANE

**2655-15-4 2,3,5-TRIMETHYLPHENYL
METHYLCARBAMATE**

2664-42-8 N,N-DIMETHYLOLEAMIDE

2664-63-3 4,4'-THIODIPHENOL

**2665-30-7 PHOSPHONOTHIOIC ACID,
METHYL-,O-(4-NITROPHENYL) O-
PHENYL ESTER**
PHOSPHONOTHIOIC ACID;METHYL-
,O-(4-NITROPHENYL)O-PHENYL)
ESTER [MSL]

2673-22-5 BIS(TRIDECYL) SODIUM SULFOS-
UCCINATE
SODIUM DITRIDECYLSULFOSUCCI-
NATE [PST]

2682-20-4 2-METHYL-3-ISOTHIAZOLONE
3(2H)-ISOTHIAZOLONE, 2-METHYL-
[PST]

2687-25-4 1,2-BENZENEDIAMINE, 3-
METHYL-

2691-41-0 CYCLOTETRAMETHYLENETET-
RANITRAMINE

2696-92-6 NITROSYL CHLORIDE

2698-41-1 O-CHLOROBENZYLIDENE MAL-
ONITRILE
O-CHLOROBENZALMALONONITRILE
(CS) [CSR, CSR]
O-CHLOROBENZYLIDENE MAL-
ONONITRILE [MEX, MEX, TLV,
TLV]
O-CHLOROBENZYLIDENE MAL-
ONONITRILE (OCBM) [FLA, MSL,
WHM]
PROPANEDINITRILE, [(2-
CHLOROPHENYL)METHYLENE]-
[PAL]

2699-79-8 SULFURYL FLUORIDE
SULFURYL FLUORIDE [VIKANE]
[313]
SULPHURYL DIFLUORIDE [GBR,
GBR]
SULPHURYL FLUORIDE [AUS, AUS]

2702-72-9 2,4-DICHLOROPHENOXYACETIC
ACID SODIUM SALT
2,4-D SODIUM SALT [313]

2703-13-1 PHOSPHONOTHIOIC ACID,
METHYL-,O-ETHYL O-(4-
(METHYLTHIO)PHENYL) ESTER

2716-10-1 BENZENAMINE, 4,4'-
[1,4-PHENYLENEBIS[1-
METHYLETHYLIDENE]BIS[2,6-
DIMETHYL-,

2716-12-3 BENZENAMINE, 4,4'-
[1,3-PHENYLENEBIS[1-
METHYLETHYLIDENE]BIS[2,6-
DIMETHYL-,

2717-15-9 TRIETHANOLAMINE OLEATE
OLEIC ACID, TRIETHANOLAMINE
SALT [PST]

2718-67-4 TRIS(2-BUTOXYETHYL) PHOS-
PHITE

2724-58-5 16-METHYLHEPTADECANOIC
ACID

2757-18-8 THALLOUS MALONATE
THALLIUM MALONATE [WHM]

2757-90-6 AGARITINE

2761-24-2 SILANE, TRIETHOXYPENTYL-

2763-96-4 MUSCIMOL
3(2H)-ISOXAZOLONE, 5-
(AMINOMETHYL)- [TSC, TSC]
5-(AMINOMETHYL)-3-ISOXAZOLOL
[RCU, RCU]

2764-13-8 2-HYDROXYETHYL DIMETHYL
3-OCTADECANAMIDOPROPYL
AMMONIUM NITRATE

2764-72-9 DIQUAT
DIPYRIDO[1,2-A:2',1'-C]PYRAZINEDI-
IUM, 6,7-DIHYDRO- [PAL]
DIQUAT (REGLONE) [BC1, BC1]
DIQUAT DIBROMIDE [CWA]
DIQUAT, DIBROMIDE [NJL]

2778-04-3 ENDOTHION

2778-42-9 BENZENE, 1,3-BIS(1-ISOCYANATO-
1-METHYLETHYL-

2781-10-4 DI-N-BUTYLTIN DI-2-ETHYLHEX-
ANOATE

2782-57-2 DICHLOROISOCYANURIC ACID
1,3,5-TRIAZINE-2,4,6-(1H,3H,5H)-
TRIONE, 1,3-DICHLORO- [PAL]
DICHLORO-S-TRIAZINETRIONE
[FLA, MSL]

2782-91-4 TRIMETHYL THIOUREA

2783-94-0 FD & C YELLOW NO. 6
FOOD YELLOW #3 [PST]
SUNSET YELLOW FCF [IAR]

2784-94-3 HC BLUE 1
HC BLUE NO. 1 [IAR]

2785-87-7 4-PROPYLGUAIACOL

2806-85-1 3-ETHOXYPROPIONALDEHYDE
3-ETHOXYPROPANOL [NFP, NFP]
PROPANAL, 3-ETHOXY- [PAL]

2807-30-9 ETHYLENE GLYCOL MONO-
PROPYL ETHER

2809-21-4 1-HYDROXYETHANE-1,1-DIPHOS-
PHONIC ACID
1-HYDROXYETHYLIDENE-1,1-
DIPHOSPHONIC ACID [PST]

2812-73-9 ETHYL CHLOROTHIOFORMATE

2820-51-1 NICOTINE HYDROCHLORIDE

2825-15-2 1-NITROHYDANTOIN

2825-83-4 ENDO-TETRAHYDRODICYL-
CLOPENTADIENE

2832-19-1 2-CHLORO-N-HYDROXYMETHY-
LACETAMIDE

2832-40-8 C.I. DISPERSE YELLOW 3
DISPERSE YELLOW 3 [IAR]

2835-39-4 ALLYL ISOVALERATE

2836-32-0 GLYCOLIC ACID, SODIUM SALT

2837-89-0 ETHANE, 2-CHLORO-1,1,1,2-TE-
TRAFLUORO-
2-CHLORO-1,1,1,2-TETRAFLUO-
ROETHANE (HCFC-124) [313]
HCFC-124 [CAA]

2842-38-8 ETHANOL, 2-(CYCLOHEXY-
LAMINO)-
N-(2-HYDROXYETHYL) CYCLOHEXY-
LAMINE [FLA, NFP, NFP]
N-(2-HYDROXYETHYL)CYCLOHEXY-
LAMINE [MSL]

2854-16-2 1-AMINO-2-METHYL-2-PROPANOL

2855-13-2 ISOPHORONE DIAMINE
ISOPHORONEDIAMINE [NJL, NJL]

2855-19-8 OXIRANE, DECYL-

2861-02-1 ANTHRAQUINONE BLUE
2,6-ANTHRACENEDISULFONIC ACID,
4,8-DIAMINO-9,10-DIHYDRO-1,5-
DIHYDROXY-9,10-DIOXO-, DISODIUM
SALT [TSC, TSC, TSC]

2867-47-2 N,N-DIMETHYLAMINOETHYL
METHACRYLATE
2-(DIMETHYLAMINO) ETHYL
METHACRYLATE [FLA, MSL, NFP,
NFP]
2-PROPENOIC ACID, 2-METHYL-, 2-
(DIMETHYLAMINO)ETHYL ESTER
[PAL, TSC, TSC]
DIMETHYLAMINOETHYL
METHACRYLATE [HM1, HM1, NJL]

2871-01-4 HC RED NO. 3
HC RED 3 [CSR, CSR]

2885-00-9 1-OCTADECANETHIOL

2893-78-9 DICHLOROISOCYANURIC ACID,
SODIUM SALT
1,3,5-TRIAZINE-2,4,6-(1H,3H,5H)-
TRIONE, 1,3-DICHLORO-, SODIUM
SALT [PAL]

SODIUM DICHLORO ISOCYANURATE
[FLA]
SODIUM DICHLORO-ISOCYANATE
[NJL]
SODIUM DICHLORO-S-TRIAZINETRI-
ONE [MSL]

2893-80-3 C.I. DIRECT BROWN 6, DISODIUM
SALT
BENZOIC ACID, 5-[[4'-[[2,4-DIHY-
DROXY-3-[(4-SULFOPHENYL)AZO]
[1,1'-BIPHENYL]-4-YL]AZO]-2-HY-
DROXY-, DISODIUM SALT [TSC]

2897-60-1 SILANE, DIETHOXYMETHYL[3-
(OXIRANYLMETHOXY)PROPYL]-
3-(METHYLDIETHOXYSILYL)PROPYL
GLYCIDYL ETHER [EPA, OTS]

2905-65-9 BENZOIC ACID, 3-CHLORO-,
METHYL ESTER

2909-38-8 BENZENE, 1-CHLORO-3-ISO-
CYANATO-

2915-53-9 DIOCTYL MALEATE

2915-57-3 BIS(2-ETHYLHEXYL) SUCCINATE

2917-26-2 1-HEXADECANETHIOL

2917-94-4 ETHANESULFONIC ACID, 2-[2-[2-
[4-(1,1,3,3-TETRAMETHYLBUTYL)
PHENOXY]ETHOXY]ETHOXY]-,
SODIUM SALT

2921-88-2 CHLORPYRIFOS
CHLOROPYRIFOS [MIN]
CHLORPYRIFOS (DURSBAN) [BC1,
BC1, BC1]
CHLORPYRIFOS-ETHYL (DURSBAN)
[QBC, QBC]
PHOSPHOROTHIOIC ACID, O,O-
DIETHYL O-(3,5,6-TRICHLORO-2-
PYRIDINYL) ESTER [PAL]

2929-86-4 TRIDECYL PHOSPHITE

2935-44-6 2,5-HEXANEDIOL

2937-50-0 ALLYL CHLOROCARBONATE
ALLYL CHLOROFORMATE [HM1,
HM1, HM1, MSL, NFP, NFP,
WHM]
CARBONOCHLORIDIC ACID, 2-
PROPENYL ESTER [PAL]

2939-80-2 CAPTAFOL, CIS-

2941-64-2 S-ETHYL CHLOROTHIOFORMATE
ETHYL CHLOROTHIOFORMATE
[HM1, HM1]

2944-67-4 OXALIC ACID, AMMONIUM IRON
(3+) SALT (3:3:1)
ETHANEDIOIC ACID, AMMONIUM
IRON (3+) SALT [MSL]
ETHANEDIOIC ACID, AMMONIUM
IRON(3+) SALT (3:3:1) [PAL]
FERRIC AMMONIUM OXALATE
[CWA, EPA]

2949-22-6 ACETIC ACID, ISOCYANATO-,
ETHYL ESTER

2954-50-9 1,2,3,4-TETRAHYDRO-2-NAPH-
THYLAMINE

2955-38-6 PRAZEPAM

2958-09-0 PHOSPHORIC ACID, MONOOC-
TADECYL ESTER

2971-38-2 2,4-D CHLOROCROTYL ESTER
2,4-D ESTERS [NJL]
ACETIC ACID, (2,4-DICHLOROPHE-
NOXY)-, 4-CHLORO-2-BUTENYL ES-
TER [MSL]
ACETIC ACID, (2,4-DICHLOROPHE-
NOXY)-,4-CHLORO-2-BUTENYL ES-
TER [PAL]

2971-90-6 CLOPIDOL
4-PYRIDINOL, 3,5-DICHLORO-2,6-
DIMETHYL- [PAL]
CLOPIDOL (COYDEN) [ALB, ALB,
BC1, BC1]

2973-10-6 DIISOPROPYL SULFATE

2980-64-5 AMMONIUM DINITRO-O-CRESO-LATE

2985-59-3 4-DODECYLOXY-2-HYDROXYBEN-ZOPHENONE

2998-04-1 DIALLYL ADIPATE

2999-74-8 DIMETHYLMAGNESIUM

3006-13-1 N-ETHYL-N,N-DIMETHYL-1-DODE-CAMINIUM ETHYL SULFATE

3006-15-3 SODIUM 1,4-DIHEXYL SULFOSUC-CINATE

3006-82-4 TERT-BUTYL PEROXY(2-ETHYL)-HEXANOATE

3006-86-8 1,1-DI(TERT-BUTYLPEROXY)-CYCLOHEXANE

3006-93-7 PHTHALIC DIMALEIMIDE

3012-65-5 AMMONIUM CITRATE, DIBASIC
1,2,3-PROPANETRICARBOXYLIC ACID, 2-HYDROXY-, DIAMMONIUM SALT [PAL]
AMMONIUM CITRATE DIBASIC [CAL, CWA]

3018-12-0 DICHLOROACETONITRILE

3025-88-5 2,5-DIMETHYL-2,5-DIHYDRO-PEROXYHEXANE
2,5-DIMETHYL-2,5-DIHYDROPEROXY HEXANE [HM1]

3026-63-9 SODIUM TRIDECYL SULFATE

3028-88-4 TRIOCTYL PHOSPHITE

3030-47-5 N-[2-(DIMETHYLAMINO)ETHYL]-N,N',N'-TRIMETHYL-1,2-ETHANE-DIAMINE

3031-74-1 ETHYL HYDROPEROXIDE

3033-62-3 NIAX A 1

3033-77-0 GLYCIDYL TRIMETHYL AMMO-NIUM CHLORIDE

3034-79-5 DI(2-METHYLBENZOYL)PEROX-IDE

3037-72-7 SILANE, (4-AMINOBUTYL)DI-ETHOXYMETHYL-

3039-83-6 ETHANESULFONIC ACID, SODIUM SALT

3048-64-4 VINYLNORBORNENE

3054-95-3 3,3-DIETHOXYPROPENE

3061-36-7 1,4-DIPHENOXYBENZENE

3064-70-8 BIS(TRICHLOROMETHYL)SUL-FONE

3068-00-6 1,2,4-BUTANETRIOL

3068-88-0 BETA-BUTYROLACTONE
2-OXETANONE, 4-METHYL- [PAL, PAL]
B-BUTYROLACTONE [IAR]

3072-84-2 2,2',6,6'-TETRABROMOBISPHE-NOL A DIGLYCIDYL ETHER
TETRABROMOBISPHENOL A DIGLY-CIDYL ETHER, 2,2',6,6'- [OTS]

3074-75-7 2-METHYL-4-ETHYLHEXANE
HEXANE, 4-ETHYL-2-METHYL- [PAL]

3074-77-9 3-METHYL-4-ETHYLHEXANE
HEXANE, 3-ETHYL-4-METHYL- [PAL]

3076-04-8 2-PROPENOIC ACID, TRIDECYL ESTER
TRIDECYL ACRYLATE [NFP, NFP]

3081-14-9 1,4-BENZENEDIAMINE, N,N'-BIS(1,4-DIMETHYLPENTYL)-
N,N'-BIS(1,4-DIMETHYLPENTYL)-P-PHENYLENEDIAMINE [MSL]
N,N'-BIS-(1,4-DIMETHYL-PENTYL)-P-PHENYLENEDIAMINE [FLA]
N,N'-BIS-(1,4-DIMETHYLPENTYL) P-PHENYLENEDIAMINE [NFP, NFP]

3083-25-8 TRICHLOROBUTYLENE OXIDE
OXIRANE, (2,2,2-TRICHLOROETHYL)-[TSC, TSC, TSC, TSC]

3087-37-4 TETRAPROPYLORTHOTITANATE

3089-11-0 HEXAKIS(METHOXYMETHYL) MELAMINE
HEXA(METHOXYMETHYL) MELAMINE [TSC, TSC, TSC]

3091-25-6 MONOOCTYLTIN TRICHLORIDE

3097-08-3 DODECYL SULFATE, MAGNESIUM SALT

3101-60-8 P-TERT-BUTYL PHENYL GLY-CIDYL ETHER
BUTYLPHENYL GLYCIDYL ETHER, P-TERT- [OTS]
OXIRANE, [[4-(1,1-DIMETHYLETHYL) PHENOXY]METHYL]- [TSC]

3118-97-6 C.I. SOLVENT ORANGE 7
SUDAN II [CAL, IAR]

3121-61-7 2-METHOXYETHYL ACRYLATE
ETHYLENE GLYCOL MONOMETHYL ETHER ACRYLATE [TSC, TSC, TSC]

3126-90-7 DIBUTYL ISOPHTHALATE

3129-90-6 ISOTHIOCYANIC ACID

3129-91-7 DICYCLOHEXYLAMMONIUM NITRITE

3130-19-6 BIS[(3,4-EPOXYCYCLOHEXYL) METHYL] ADIPATE

3132-64-7 EPIBROMOHYDRIN
OXIRANE, BROMOMETHYL- [TSC, TSC, TSC]

3132-99-8 BENZALDEHYDE, 3-BROMO-

3147-75-9 2-(2-HYDROXY-5-TERT-OCTYLPHENYL)BENZOTRIAZOLE

3164-29-2 AMMONIUM TARTRATE, DI-AMMONIUM SALT
AMMONIUM TARTRATE [CAL, EPA, NJL]
BUTANEDIOIC ACID, 2,3-DIHY-DROXY-[R-(R*,R*)]-, DIAMMONIUM-SALT [PAL]
TARTARIC ACID, DIAMMONIUM SALT [MSL]

3164-85-0 POTASSIUM 2-ETHYLHEX-ANOATE

3165-93-3 BENZENAMINE, 4-CHLORO-2-METHYL-, HYDROCHLORIDE
4-CHLORO-O-TOLUIDINE HY-DROCHLORIDE [CSR, CSR, HM1, HM1, HM1, MSL, NJL, WHM]
4-CHLORO-O-TOLUIDINE, HY-DROCHLORIDE [EP2]

3173-53-3 CYCLOHEXYL ISOCYANATE
CYCLOHEXANE, ISOCYANATO-[TSC, TSC, TSC]

3173-72-6 NAPHTHALENE, 1,5-DIISO-CYANATO-
1,5-NAPHTHALENE DIISOCYANATE [313, IAR]
1,5-NAPHTHYLENE DIISOCYANATE [MAK, MAK, MAK]

3174-74-1 2H-PYRAN, 3,6-DIHYDRO-

3178-22-1 TERT-BUTYLCYCLOHEXANE

3179-56-4 ACETYL CYCLOHEXANE SUL-FONYL PEROXIDE

3180-09-4 O-BUTYLPHENOL

3184-13-2 L-ORNITHINE HCL

3188-13-4 CHLOROMETHYL ETHYL ETHER
DICHLOROISOPROPYL ETHER [NFP, NFP]

3188-83-8 2-METHYLOL-4,4'-ISOPROPYLI-DENEDIPHENOL DIGLYCIDYL ETHER

3194-55-6 HEXABROMOCYCLODODECANE
CYCLODODECANE, 1,2,5,6,9,10-HEXABROMO- [TSC, TSC, TSC]

3209-22-1 BENZENE, 1,2-DICHLORO-3-NI-TRO-

3214-47-9 C.I. DIRECT YELLOW 50, TETRA-SODIUM SALT

3221-61-2 2-METHYLOCTANE
NONANE (ISO) [NFP]
OCTANE, 2-METHYL [PAL]

3234-28-4 OXIRANE, DODECYL-

3246-20-6 SODIUM DINONYL SULFOSUCCI-NATE

3248-28-0 PROPIONYL PEROXIDE
DIPROPIONYL PEROXIDE [DOT, HM1, NJL]

3248-91-7 MAGENTA III

3251-23-8 CUPRIC NITRATE
COPPER NITRATE [PST]
COPPER(II) NITRATE [WHM]
NITRIC ACID, COPPER(2+) SALT [PAL]

3252-43-5 DIBROMOACETONITRILE

3254-63-5 PHOSPHORIC ACID, DIMETHYL 4-(METHYLTHIO)PHENYL ESTER

3266-23-7 OXIRANE, 2,3-DIMETHYL-
2,3-BUTYLENE OXIDE [NFP, NFP]

3268-49-3 4-THIAPENTANAL
BETA-METHYL MERCAPTOPROPI-ONALDEHYDE [NFP, NFP]
PROPANAL, 3-(METHYLTHIO)- [TSC, TSC, TSC]
THIA-4-PENTANAL [HM1, HM1]

3268-87-9 OCTACHLORODIBENZO-P-DIOXIN

3275-73-8 NICOTINE TARTRATE

3278-22-6 1,1'-[METHYLENE BIS(SULFONYL)]BISETHENE

3282-30-2 PIVALOYL CHLORIDE
TRIMETHYLACETYL CHLORIDE [HM1, HM1, HM1, NJL, NJL]

3282-85-7 TERGITOL 7

3287-06-7 DIPHENYL DECYL PHOSPHITE

3288-58-2 O,O-DIETHYL S-METHYL DITHIO-PHOSPHATE
PHOSPHORODITHIOIC ACID, O,O-DIETHYL-S-METHYL ESTER [TSC, TSC]

3290-92-4 TRIMETHYLOPROPANE TRIMETHACRYLATE
TRIMETHYLOLPROPANE TRIMETHACRYLATE [MIN]

3296-90-0 2,2-BIS(BROMOMETHYL)-1,3-PROPANEDIOL
1,3-PROPANEDIOL, 2,2-BIS(BRO-MOETHYL)-DIBROMONEOPENTYL GLYCOL [TSC, TSC, TSC]

3312-60-5 N-(3-AMINOPROPYL) CYCLO-HEXYLAMINE
1,3-PROPANEDIAMINE, N-CYCLO-HEXYL- [PAL]

3313-92-6 SODIUM PERCARBONATE

3319-31-1 1,2,4-BENZENETRICARBOXYLIC ACID, TRIS (2-ETHYLHEXYL) ESTER

3322-93-8 1,2-DIBROMO-4-(1,2-DIBRO-MOETHYL) CYCLOHEXANE
CYCLOHEXANE, 1,2-DIBROMO-4-(1,2-DIBROMOETHYL)- [TSC, TSC, TSC]

3324-58-1 PICRIC ACID, SODIUM SALT

3332-27-2 N,N-DIMETHYLTETRADECY-LAMINE OXIDE

3333-52-6 TETRAMETHYL SUCCINONI-TRILE
BUTANEDINITRILE, TETRAMETHYL-[PAL]
TETRAMETHYL SUCCINONITRILE (DECOMPOSITION PRODUCT OF 2,2'-AZOBISBUTYRONITRILE) [CAL, CEX, CEX]
TETRAMETHYLSUCCINONITRILE [NHS, NHS, ONT, ONT, QBC, QBC]

3333-67-3 NICKEL CARBONATE
CARBONIC ACID, NICKEL(2+) SALT (1:1) [PAL, PAL]
NICKEL(II) CARBONATE [WHM]

3337-71-1 ASULAM

3351-28-8 1-METHYLCHRYSENE

3351-30-2 4-METHYLCHRYSENE

3351-31-3 3-METHYLCHRYSENE

3351-32-4 2-METHYLCHRYSENE

3352-87-2 N,N-DIETHYLLAURAMIDE

3380-34-5 5-CHLORO-2-(2,4-DICHLOROPHE-NOXY)PHENOL
2,4,4'-TRICHLORO-2-HYDROXY DIPHENYL ETHER [PST]

3383-96-8 TEMEPHOS
ABATE [BC1, BC1]
ABATE (TEMEPHOS) [ALB, ALB, QBC]
PHOSPHOROTHIOIC ACID, O,O'-(THIODI-4,1-PHENYLENE) O,O,O',O'-TETRAMETHYL ESTER [PAL]
TEMEPHOS (ABATE) [MEX, MEX]
TETRAMETHYL O,O'-THIO-DI-P-PHENYLENEPHOSPHOROTHIOATE (TEMEPHOS) [CAL]

3388-04-3 SILANE, TRIMETHOXY[2-(7-OX-ABICYCLO[4.1.0]HEPT-3-YL) ETHYL]

3389-71-7 1,2,3,4,7,7-HEXACHLORONORBOR-NADIENE

3424-82-6 O,P'-DDE

3425-61-4 TERT-AMYL HYDROPEROXIDE

3437-84-1 DI-ISOBUTRYL PEROXIDE

3440-75-3 TETRAPROPYL LEAD

3452-97-9 1-HEXANOL, 3,5,5-TRIMETHYL-
3,5,5-TRIMETHYLHEXANOL [FLA, MSL, NFP, NFP]

3457-61-2 TERT-BUTYL CUMYL PEROXIDE

3458-22-8 IPD
IPD (3,3'-IMINOBIS-1-PROPANOL DIMETHANESULFONATE (ESTER) HYDROCHLORIDE) [CSR, CSR]

3468-11-9 ISOINDOLINE, 1,3-DIIMINO

3468-63-1 D&C ORANGE NO. 17

3476-90-2 C.I. DIRECT BROWN 59, DIS-ODIUM SALT

3479-86-5 DIBROMOBUTANONE

3486-30-4 C.I. ACID BLUE 7, SODIUM SALT

3486-35-9 ZINC CARBONATE
CARBONIC ACID, ZINC SALT (1:1) [PAL]

3497-00-5 PHENYL PHOSPHORUS THIODICHLORIDE

3520-42-1 C.I. ACID RED 52

3522-94-9 HEXANE, 2,2,5-TRIMETHYL-
2,2,5-TRIMETHYLHEXANE [FLA, MSL, NFP, NFP]

3524-68-3 PENTAERYTHRITOL TRIACRY-LATE

3530-19-6 C.I. DIRECT RED 37
1,3-NAPHTHALENEDISULFONIC ACID, 8-[[4'-[(4-ETHOXYPHENYL) AZO][1,1'-BIPHENYL]-4-YL]AZO]-7-HYDROXY-, DISODIUM SALT [TSC]

3531-19-9 BENZENAMINE, 2-CHLORO-4,6-DINITRO-

3542-36-7 DIOCTYLTIN DICHLORIDE

3546-10-9 PHENESTERIN

3547-04-4 DDE

3564-09-8 PONCEAU 3R
2,7-NAPHTHALENEDISULFONIC ACID, 3-HYDROXY-4-[(2,4,5-TRIMETHYLPHENYL)AZO]-, DIS-ODIUM SALT [PAL, PAL]

3564-21-4 PIGMENT RED 28 (PERMANENT RED)
C.I PIGMENT RED 48, DISODIUM SALT [PST]

3565-26-2 4-NITROQUINOLINE-1-OXIDE

3567-65-5 C.I. ACID RED 85
1,3-NAPHTHALENEDISULFONIC ACID, 7-HYDROXY-8-[[4'-[(4-METHYLPHENYL) SULFONYL]OXY] PHENYL]AZO][1,1'-BIPHENYL]-4-YL]AZO]-, DISODIUM SALT [TSC]

3567-66-6 C.I. ACID RED 33, DISODIUM SALT

3567-69-9 C.I. ACID RED 14
C.I. ACID RED 14, DISODIUM SALT [PST]
CARMOISINE [IAR]

3568-29-4 GLYCEROL 1,3-DIGLYCIDYL ETHER

3569-57-1 SULFOXIDE, 3-CHLOROPROPYL OCTYL

3570-75-0 2-(2-FORMYLHYDRAZINO)-4-(5-NITRO-2-FURYL)THIAZOLE
HYDRAZINECARBOXALDEHYDE, 2-[4-(5-NITRO-2-FURANYL)-2-THIA-ZOLYL]- [PAL, PAL]

3570-80-7 FLUORESCEIN MERCURIC AC-ETATE

3586-14-9 BENZENE, 1-METHYL-3-PHE-NOXY-

3586-55-8 ETHYLENE GLYCOL BIS(SEMI-FORMAL)

3610-27-3 METHOXYTRIGLYCOL ACETATE

3613-30-7 OCTANAL, 7-METHOXY-3,7-DIMETHYL-

3614-12-8 N-DODECYL-N-TETRADECYL BETA-ALANINE

3615-21-2 BENZIMIDAZOLE, 4,5-DICHLORO-2-(TRIFLUOROMETHYL)-
4,5-DICHLORO-2-(TRIFLUO-ROMETHYL)-BENZIMIDAZOLE [MSL]

3618-72-2 C.I. DISPERSE BLUE 79:1 AC-ETAMIDE, N-[5-[BIS[2-(ACETY-LOXYL)ETHYL]AMINO]-2-[(2-BROMO-4,6-DINITROPHENYL) AZO]-4-METHOXY- PHENYL]-
ACETAMIDE, N-[5-[BIS[2-(ACETY-LOXY)ETHYL]AMINO]-2-[(2-BROMO-4,6-DINITROPHENYL)AZO]-4-METHOXYPHENYL]- [TSC, TSC, TSC, TSC]
C.I DISPERSE BLUE 79:1 [TSE]

3618-73-3 ACETAMIDE, N-(5-(BIS(2-(ACETY-LOXY)ETHYL)AMINO)-2-((2-CHLORO-4,6-DINITROPHENYL) AZO)-4-METHOXYPHENYL)-

3622-84-2 N-BUTYLBENZENESULFONAMIDE

3626-28-6 C.I. DIRECT GREEN 1, DISODIUM SALT
2,7-NAPHTHALENEDISULFONIC ACID, 4-AMINO-5-HYDROXY-3-[[4'-HYDROXYPHENYL)AZO] [1,1'-BIPHENYL]-4-YL]-AZO]-6-(PHENY-LAZO)-, DISODIUM SALT [TSC]

3648-20-2 1,2-BENZENEDICARBOXYLIC ACID, DIUNDECYL ESTER
DIUNDECYL PHTHALATE [TSE]

3648-21-3 DIHEPTYL PHTHALATE

3653-48-3 METHOXONE-SODIUM SALT ((4-CHLORO-2-METHYLPHENOXY) ACETATE SODIUM SALT)

3655-00-3 3,3'-(DODECYLIMINO)DIPROPI-ONIC ACID, DISODIUM SALT

3658-48-8 DIISOOCTYL PHOSPHITE

3663-23-8 1,2-BENZENEDIAMINE, 4-BUTYL-

3687-31-8 LEAD ARSENATE

3688-53-7 FURYLFURAMIDE (AF-2)
2-(2-FURYL)-3-(5-NITRO-2-FURYL) ACRYLAMIDE [MSL]
2-FURANACETAMIDE, .ALPHA.-[(5-NITRO-2-FURANYL)METHYLENE]- [PAL, PAL]
AF-2 [MIN]
AF-2 (2-(2-FURYL)-3-(5-NITRO-2-FURYL)ACRYLAMIDE) [IAR]
AF-2 [2-(2-FURYL)-3-(5-NITRO-2-FURYL)ACRYLAMIDE] [CAL]
AF-2; [2-(2-FURYL)-3-(5-NITRO-2-FURYL)]ACRYLAMIDE [C65, C65]

3689-24-5 SULFOTEP
SULFOTEP [QBC, QBC]
SULFOTEP (ISO) [GBR, GBR]
SULFOTEP (TEDP) [MEX, MEX, MEX]
SULFOTEPP [FLA, NJL, PQL]
TEDP [ALB, ALB, ALB, BC1, BC1, BC1, MAK, MAK, MAK, NIO, NIO, NIO, NIO]
TEDP (SULFOTEP) [OS1, OS1, OS1, OS1]
TETRAETHYL DITHIONOPYROPHOS-PHATE [ONT, ONT]
TETRAETHYL DITHIOPYROPHOS-PHATE (SULFOTEPP) [CAL]
TETRAETHYL PYROPHOSPHO-RODITHIONATE [RCU, RCU]
TETRAETHYLDITHIOPYROPHOS-PHATE [EPA]
THIODIPHOSPHORIC ACID ([(HO)2P(S)]2O), TETRAETHYL ESTER [PAL]

3691-35-8 CHLOROPHACINONE
CHLOROPHACINONE (ROZOL) [EPA, EPA]

3697-24-3 5-METHYLCHRYSENE
5-METHYL CHRYSENE [MSL]
CHRYSENE, 5-METHYL- [PAL, PAL]

3710-84-7 DIETHYL HYDROXYLAMINE
N,N-DIETHYLHYDROXYLAMINE [PST]

3724-65-0 CROTONIC ACID
2-BUTENOIC ACID [PAL]

3731-51-9 2-AMINO-4-METHYLPYRIDINE

3734-33-6 BITREX
BENZYL DIETHYL 2,6-XYLYLCAR-BAMOYLMETHYL AMMONIUM BEN-ZOATE [PST]

3734-48-3 CHLORDENE

3734-67-6 C.I. ACID RED 1

3734-97-2 AMITON OXALATE

3735-23-7 METHYL PHENKAPTON

3737-55-1 N-METHYL-N-PALMITOYLTAU-
RINE, SODIUM SALT

3758-54-1 STEARAMINOPROPYLDIMETHYL-
.BETA.-HYDROXYETHYLAMMO-
NIUM PHOSPATE
STEARAMIDOPROPYL DIMETHYL
2-HYDROXYETHYL AMMONIUM
DIHYDROGEN PHOSPHATE [PST]

3759-92-0 5-MORPHOLINOMETHYL-3((5-NI-
TROFURFURYLIDENE)AMINO)-2-
OXAZOLIDINONE HYDROCHLO-
RIDE
5-MORPHOLINOMETHYL-3[(5-NITRO-
FURFURYLIDENE)AMINO]-2-OXAZO-
LIDINONE [IAR]

3761-53-3 PONCEAU MX
2,7-NAPHTHALENEDISULFONIC
ACID, 4-[(2,4-DIMETHYLPHENYL)
AZO]-3-HYDROXY-, DISODIUM SALT
[PAL, PAL]
3-HYDROXY-4-(2,4-XYLYLAZO)-2,7-
NAPHTHALENEDISULFONIC ACID,
DISODIUM SALT [WHM]
C.I. FOOD RED 5 [NJL]
CI FOOD RED 5 [313]

3771-19-5 NAFENOPIN
PROPANOIC ACID, 2-METHYL-2-
[4-(1,2,3,4-TETRAHYDRO-1-NAPH-
THALENYL)PHENOXY] [PAL, PAL]

3772-94-9 PENTACHLOROPHENYL LAU-
RATE

3775-90-4 TERT-BUTYLAMINOETHYL
METHACRYLATE
2-PROPENOIC ACID, 2-METHYL-
, 2-[(1,1-DIMETHYLETHYL)AMINO]
ETHYL ESTER [PAL, TSC, TSC]

3778-73-2 ISOPHOSPHAMIDE
IFOSFAMIDE [C65, CAB]

3794-64-7 SILVER HEPTAFLUOROBU-
TYRATE
NAFENOPIN [MIN]

3794-83-0 PHOSPHONIC ACID, (1-HYDROX-
YETHYLIDENE)BIS-, TETRA-
SODIUM SALT

3795-88-8 5-(MORPHOLINOMETHYL)-3-((5-
NITROFURFURYLIDENE)AMINO)-
2-OXAZOLIDINONE

3810-74-0 STREPTOMYCIN SULFATE

3811-04-9 POTASSIUM CHLORATE
CHLORIC ACID, POTASSIUM SALT
[PAL]

3811-49-2 SALITHION

3811-71-0 C.I. DIRECT BROWN 1

3811-73-2 SODIUM PYRIDITHIONE
SODIUM PYRITHIONE [MAK, MAK,
MAK, MAK]

3813-14-7 ACETIC ACID, (2,4,5-
TRICHLOROPHENOXY)-, COM-
POUND WITH 2,2',2"-NITROTRIS
(ETHANOL) (1:1)
2,4,5-T AMINE [NJL]

3817-11-6 N-NITROSO-N-BUTYL-N-(4-HY-
DROXYBUTYL)AMINE

3825-26-1 AMMONIUM PERFLUOROOC-
TANOATE

3844-45-9 C.I. ACID BLUE 9, DISODIUM
SALT
BRILLIANT BLUE FCF [CAL, IAR]

3851-87-4 DI(3,5,5-TRIMETHYLHEXANOYL)
PEROXIDE

3878-19-1 FUBERIDAZOLE

3882-06-2 DICYCLOHEXYLAMMONIUM
NITRATE

3900-04-7 PHOSPHORIC ACID, MONOHEXYL
ESTER

3902-71-4 4,5',8-TRIMETHYLPSORALEN

3913-02-8 2-BUTYLOCTANOL

3917-15-5 1-PROPENE, 3-(ETHENYLOXY)-
VINYL ALLYL ETHER [FLA, MSL,
NFP, NFP]

3921-30-0 MONODECYL ACID PHOSPHATE

3926-62-3 SODIUM CHLOROACETATE
ACETIC ACID, CHLORO-, SODIUM
SALT [OTS]

3953-10-4 2-ETHYLBUTYL ACRYLATE
2-PROPENOIC ACID, 2-ETHYLBUTYL
ESTER [PAL, TSC, TSC]

3956-55-6 ACETAMIDE, N-[5-[BIS(2-(ACETY-
LOXY)ETHYL)AMINO]-2-[(2-
BROMO-4,6-DINITROPHENYL)
AZO]-4-ETHOXYPHENYL]-

3963-95-9 METHACYCLINE HYDROCHLO-
RIDE

3982-91-0 THIOPHOSPHORYL CHLORIDE

3991-73-9 PHOSPHORIC ACID,
MONOOCTYL ESTER

3999-01-7 LINOLEAMIDE

4016-11-9 EPOXY ETHYLOXY PROPANE
1,2-EPOXY-3-ETHOXYPROPANE
[HM1, HM1]
ETHYL GLYCIDYL ETHER [EPA,
OTS]
OXIRANE, ETHOXYMETHYL- [TSC,
TSC]

4016-14-2 ISOPROPYL GLYCIDYL ETHER
(IGE)
((1-METHYLETHOXY)METHYL)-
OXIRANE [ONT, ONT]
2,3-EPOXYPROPYL ISOPROPYL
ETHER [GBR, GBR]
ISOOPROPYL GLYCIDYL ETHER
[OTS]
ISOPROPYL GLYCIDYL ETHER [CAL,
CEX, CEX, EPA, MIN, NIO, NIO,
NIO]
OXIRANE, [(1-METHYLETHOXY)
METHYL]- [PAL, TSC, TSC]

4032-86-4 3,3-DIMETHYLHEPTANE
HEPTANE, 3,3-DIMETHYL- [PAL]

4035-89-6 HEXAMETHYLENE DIISO-
CYANATE BIURET
IMIDODICARBONIC DIAMIDE, N,N'-2-
TRIS(6-ISOCYANATOHEXYL)- [TSC,
TSC, TSC]

4044-65-9 BITOSCANATE

4065-45-6 2-HYDROXY-4-METHOXYBEN-
ZOPHENONE-5-SULFONIC ACID

4067-16-7 3,6,9,12-TETRAZAATETRADE-
CANE-1,14-DIAMINE

4074-43-5 M-BUTYLPHENOL

4075-79-0 4-ACETYLAMINOBIPHENYL

4075-81-4 CALCIUM PROPIONATE

4080-31-3 3,5,7-TRIAZA-1-AZONIATRICY-
CLODECANE-1-(3-CHLORO-2-
PROPENYL)-, CHLORIDE
1-(3-CHLOROALLYL)-3,5,7-TRIAZA-
1-AZONIAADAMANTANE CHLORIDE
[313, PST]

4098-71-9 ISOPHORONE DIISOCYANATE
CYCLOHEXANE, 5-ISOCYANATO-
1-(ISOCYANATOMETHYL)-1,3,3-
TRIMETHYL- [PAL]
CYCLOHEXANE, 5-ISOCYANATO-
1-(ISOCYANATOMETHYL)-1,3,3-
TRIMETHYL- [TSC, TSC, TSC]

ISOPHORONE DIISOCYANATE (IPDI)
[NHS, NHS, ONT, ONT, ONT]
ISOPHORONEDIISOCYANATE [MAK,
MAK, MAK]

4104-14-7 PHOSACETIM
PHOSPHORAMIDOTHIOIC ACID,
(1-IMINOETHYL)-, O,O-BIS(4-
CHLOROPHENYL) ESTER [PAL]

4109-96-0 DICHLOROSILANE
DICHLOROSILANE [SILANE,
DICHLORO-] [EP3]

4128-73-8 4,4'-DIISOCYANATODIPHENYL
ETHER

4130-35-2 TRI-N-DECYL TRIMELLITATE

4151-51-3 PHENOL, 4-ISOCYANATO-, PHOS-
PHOROTHIOATE (3:1) (ESTER)

4162-45-2 TETRABROMOBISPHENOL-A-
BISETHOXYLATE
ETHANOL, 2,2'-[(1-METHYLETHYLI-
DENE)BIS[(2,6-DIBROMO-4,1-PHENY-
LENE)OXY]BIS- [TSC, TSC, TSC,
TSC]

4170-30-3 CROTONALDEHYDE
2-BUTENAL [PAL]
2-BUTENAL (CROTONALDEHYDE)
[TSC, TSC, TSC, TSC, TSC]
2-BUTENAL, INHIBITED [HM1, HM1,
HM1, HM1, HM1]
CROTONALDEHYDE [CSR, ONT]
CROTONALDEHYDE [2-BUTENAL]
[EP3]
CROTONALDEHYDE, INHIBITED
[NJL]

4175-37-5 OCTYL DIPHENYLAMINE
OCTYLDIPHENYLAMINE [PST]

4196-86-5 PENTAERYTHRITOL TETRABEN-
ZOATE

4246-51-9 DIETHYLENE GLYCOL BIS(3-
AMINOPROPYL) ETHER

4253-22-9 DIBUTYLTIN SULFIDE

4253-34-3 SILANETRIOL, METHYL-, TRIAC-
ETATE

4259-15-8 ZINC O,O-BIS(2-ETHYLHEXYL)
PHOSPHORODITHIOATE
PHOSPHORODITHIOIC ACID, O,O-BIS
(2-ETHYLENE- [OTS]

4259-43-2 1,1,1-TRICHLORO-2,2,3,3,3-
PENTAFLUOROPROPANE
CFC-215 [CAA]

4270-70-6 TRIPHENYLSULFONIUM CHLO-
RIDE

4292-92-6 AMYLCYCLOHEXANE

4300-97-4 CHLOROPIVALOYL CHLORIDE

4301-50-2 FLUENETIL

4316-42-1 N-(N-BUTYL)IMIDAZOLE
N-N-BUTYL-IMIDAZOLE [NJL]

4316-73-8 N-METHYLGLYCINE, SODIUM
SALT

4321-69-1 C.I. ACID VIOLET 7, DISODIUM
SALT

4324-38-3 PROPANOIC ACID, 3-ETHOXY-

4335-09-5 C.I. DIRECT GREEN 6, DISODIUM
SALT
2,7-NAPHTHALENEDISULFONIC
ACID, 4-AMINO-5-HYDROXY-6[[4'-
[(4-HYDROXYPHENYL)AZO] [1,1'-
BIPHENYL]-4-YL]-AZO]-3-[(4-NITRO
PHENYL)AZO]-, DISODIUM SALT
[TSC]

4337-75-1 N-LAUROYL-N-METHYLTAURINE,
SODIUM SALT

4342-03-4 DACARBAZINE
1H-IMIDAZOLE-4-CARBOXAMIDE, 5-(3,3-DIMETHYL-1-TRIAZENYL)- [PAL, PAL]

4342-30-7 TRI-N-BUTYLTIN SALICYLATE

4342-36-3 TRIBUTYLTIN BENZOATE

4395-65-7 C.I. SOLVENT BLUE 68

4403-90-1 D AND C GREEN NO. 5
C.I. ACID GREEN 25 [PST]

4418-26-2 SODIUM DEHYDROACETATE

4418-66-0 PHENOL, 2,2'-THIOBIS(4-CHLORO-6-METHYL)-

4420-74-0 3-MERCAPTOPROPY-LTRIMETHOXYSILANE

4421-95-8 SILICIC ACID (H8SI3O10), OC-TAMETHYL ESTER

4435-53-4 BUTOXYL
3-METHOXYBUTYL ACETATE [NFP, NFP]

4439-24-1 ETHYLENE GLYCOL MONOISOBUTYL ETHER
ETHANOL, 2-(2-METHYLPROPOXY)- [PAL]

4452-58-8 SODIUM PERCARBONATE
CARBONOPEROXOIC ACID, DIS-ODIUM SALT [NJL]

4461-41-0 2-CHLOROBUTENE-2
2-BUTENE, 2-CHLORO- [PAL]

4461-48-7 4-METHYL-2-PENTENE
2-PENTENE, 4-METHYL- [PAL]

4461-52-3 METHOXYMETHANOL

4465-94-5 CYTOXAL ALCOHOL

4466-14-2 JASMOLIN I

4468-93-3 ETHYL BUTOXY ETHANOL
2-ETHYL BUTYL GLYCOL [NFP, NFP]

4472-06-4 AZIDODITHIOCARBONIC ACID

4474-24-2 C.I. ACID BLUE 80

4475-95-0 2-AMINO-2-METHYLBUTANENI-TRILE

4482-55-7 FENURON-TCA

4484-72-4 DODECYL TRICHLOROSILANE

4485-12-5 LITHIUM STEARATE

4501-58-0 3-CYCLOPENTENE-1-ACETALDE-HYDE, 2,2,3-TRIMETHYL-

4521-94-2 SILICIC ACID (H8SI3O10), OC-TAETHYL ESTER

4528-34-1 NITROETHYL NITRATE

4548-53-2 C.I. FOOD RED 1
PONCEAU SX [IAR]

4549-40-0 N-NITROSOMETHYLVINYLAMINE
ETHENAMINE, N-METHYL-N-NI-TROSO- [PAL, PAL]

4568-28-9 TRIETHANOLAMINE STEARATE

4645-07-2 SOLVENT YELLOW #72

4654-26-6 DIOCTYL TEREPHTHALATE

4657-93-6 5-AMINOACENAPHTHENE

4676-39-5 3,4-METHYLENEDIOXYPHENYL-2-PROPANONE

4680-78-8 C.I. ACID GREEN 3
GUINEA GREEN B [CAL, IAR]

4682-03-5 DIAZODINITROPHENOL

4685-14-7 PARAQUAT
4,4'-BIPYRIDINIUM, 1,1'-DIMETHYL- [PAL]

4691-65-0 INOSINE-5-MONOPHOSPHORIC ACID

4706-78-9 DODECYL SULFURIC ACID, POTASSIUM SALT

4712-55-4 DIPHENYL PHOSPHITE

4732-14-3 TRINITROPHENETOLE

4759-48-2 ISOTERTINATE
ISOTRETINOIN [C65, CAB]
RETINOIC ACID [MSL]

4784-77-4 1-CROTYL BROMIDE
2-BUTENE, 1-BROMO- [PAL]

4786-20-3 CROTONONITRILE

4795-29-3 TETRAHYDROFURFURYLAMINE

4806-61-5 ETHYL CYCLOBUTANE
CYCLOBUTANE, ETHYL- [PAL]
ETHYLCYCLOBUTANE [NFP, NFP]

4824-78-6 BROMOPHOS-ETHYL

4835-11-4 HEXAMETHYLENEDIAMINE, N,N'-DIBUTYL-
1,6-HEXANEDIAMINE, N,N'-DIBUTYL- [TSC, TSC]
1,6-HEXANEDIAMINE,N,N'-DIBUTYL [MSL]

4891-67-2 ISOPHTHALIC ANHYDRIDE

4904-61-4 1,5,9-CYCLODODECATRIENE

4979-32-2 N,N-DICYCLOHEXYL-2-BENZOTH-IAZOLE SULFENAMIDE

4985-85-7 AMINOPROPY-LDIETHANOLAMINE

5012-62-4 2,6-BIS(1-METHYLHEPTADECYL)-P-CRESOL

5026-74-4 OXIRANEMETHANAMINE, N-[4-(OXIRANYLMETHOXY)PHENYL]-N-(OXIRANYLMETHYL)-
4-(DIGLYCIDYLAMINO)PHENYL GLYCIDYL ETHER [EPA, OTS]

5042-55-7 1,3-BENZENEDIAMINE, 5-NITRO-

5064-31-3 NITRILOTRIACETIC ACID TRISODIUM SALT
NITRILOTRIACETIC ACID (NTA) TRISODIUM [MSL]
TRISODIUM NITRILOTRIACETATE [PST]

5103-71-9 ALPHA CHLORDANE
CIS-CHLORDANE [ATS]

5103-74-2 GAMMA CHLORDANE
TRANS-CHLORDANE [ATS]

5116-94-9 TRIDECYL ACID PHOSPHATE

5124-30-1 METHYLENE BIS(4-CYCLO-HEXYLISOCYANATE)
1,1'-METHYLENE BIS(4-ISOCYANATO-CYCLOHEXANE) [313]
CYCLOHEXANE, 1,1'-METHYLENEBIS (4-ISOCYANATO- [TSC, TSC, TSC]
CYCLOHEXANE, 1,1'-METHYLENEBIS [4-ISOCYANATO- [PAL]
DICYCLOHEXYLMETHANE 4,4'-DIISOCYANATE (HYDROGENATED MDI) [NHS, NHS]
DICYCLOHEXYLMETHANE-4,4'-DIISOCYANATE [CAL]
METHYLENE BIS-(4-CYCLO-HEXYLISOCYANATE) [MIN]

5131-58-8 1,3-BENZENEDIAMINE, 4-NITRO-

5131-60-2 4-CHLORO-M-PHENYLENEDI-AMINE
1,3-BENZENEDIAMINE, 4-CHLORO- [TSC, TSC]

5131-66-8 1-BUTOXY-2-PROPANOL
2-PROPANOL, 1-BUTOXY- [TSC, TSC, TSC]
PROPYLENE GLYCOL MONOBUTYL ETHER [PST]

5137-52-0 P-TERT-AMYLPHENYL ACETATE

5137-55-3 METHYLTRICAPRYLAMMONIUM CHLORIDE
TRICAPRYLYL METHYL AMMONIUM CHLORIDE [PST]

5138-90-9 P-CHLOROBENZENESULFONIC ACID, SODIUM SALT

5141-20-8 LIGHT GREEN SF
LIGHT GREEN SF (C.I. ACID GREEN 5) [NJL]

5160-02-1 D & C RED NO. 9
D & C RED NO. 9 (CI PIGMENT RED 53:1) [IAR]
D&C RED NO. 9 [C65, C65]

5216-25-1 4-CHLOROBENZOTRICHLORIDE
BENZENE, 1-CHLORO-4-(TRICHLOROMETHYL)- [TSC, TSC]
P-A,A,A-TETRACHLOROTOLUENE [C65]
P-ALPHA, ALPHA, ALPHA-TETRA-CHLOROTOLUENE [CAB]

5234-68-4 CARBOXIN (5,6-DIHYDRO-2-METHYL-N-PHENYL-1,4-OXATHIIN-3-CARBOXAMIDE)

5239-06-5 1,3-DIAZOPROPANE

5246-57-1 2-[(3-AMINOPHENYL)SULFONYL] ETHANOL

5255-75-4 OXIRANE, [(4-NITROPHENOXY) METHYL]-
NITROPHENYL GLYCIDYL ETHER, P- [OTS]
P-NITROPHENYL GLYCIDYL ETHER [EPA]

5281-04-9 LITHOL RUBINE
D AND C RED NO 7 [OTS]

5281-13-0 PIPROTAL

5281-76-5 3-METHOXYBUTYRALDEHYDE

5283-66-9 OCTYL TRICHLOROSILANE

5283-67-0 NONYLTRICHLOROSILANE
NONYL TRICHLOROSILANE [NJL, NJL]

5289-74-7 ECDYSTERONE

5307-02-8 2-METHOXY-1,4-BENZENEDI-AMINE
1,4-BENZENEDIAMINE, 2-METHOXY- [TSC, TSC]

5307-14-2 2-NITRO-P-PHENYLENEDIAMINE
1,4-BENZENEDIAMINE, 2-NITRO- [TSC, TSC]
1,4-DIAMINO-2-NITROBENZENE (2-NITRO-PARA-PHENYLENEDIAMINE) [IAR]
4-AMINO-2-NITROANILINE [WHM]

5309-52-4 2-ETHYL-3-PROPYLACRYLIC ACID
2-HEXENOIC ACID, 2-ETHYL- [PAL]

5323-95-5 SODIUM RICINOLEATE

5324-84-5 SODIUM 1-OCTANESULFONATE

5329-14-6 SULFAMIC ACID
SULPHAMIC ACID [NJL]

5332-52-5 1-UNDECANETHIOL

5332-73-0 3-METHOXYPROPYLAMINE
1-PROPANAMINE, 3-METHOXY- [PAL]

5340-36-3 3-METHYL-3-OCTANOL

5344-82-1 THIOUREA, (2-CHLOROPHENYL)-
1-(O-CHLOROPHENYL)THIOUREA [RCU, RCU]

5350-03-8 DODECANOIC ACID, PENTYL ESTER

5384-21-4 PHENOL, 4,4'-METHYLENEBIS(2,6-DIMETHYL)-

5385-75-1 DIBENZO[A,E]FLUORANTHENE

5388-62-5 BENZENAMINE, 4-CHLORO-2,6-DINITRO-

5392-40-5 CITRAL

5392-82-5 1,4-DICHLORO-2-PHENYL ISO-CYANATE

5395-01-7 BETA-HYDROXYETHYLCARBA-MATE

5405-88-9 DIETHYLENE GLYCOL MONOMETHYL ETHER FORMAL

5408-74-2 PYRIDINE, 2-ETHENYL-5-ETHYL-
2-VINYL-5-ETHYLPYRIDINE [FLA, MSL, NFP, NFP]

5411-22-3 BENZPHETAMINE HYDROCHLO-RIDE
BENZPHETAMINE HYDROCHLORIDE [C65]

5413-75-2 C.I. ACID RED 73, DISODIUM SALT

5419-55-6 TRIISOPROPYL BORATE
BORIC ACID (H3BO3), TRIS(1-METHYLETHYL) ESTER [PAL]

5422-17-3 C.I. DIRECT GREEN 8, TRISODIUM SALT

5431-33-4 GLYCIDYL OLEATE

5432-61-1 N-2-(ETHYLHEXYL)-CYCLOHEXY-LAMINE
CYCLOHEXANAMINE, N-(2-ETHYL-HEXYL)- [PAL]

5434-57-1 HEXYL NEOPENTANOATE

5435-64-3 HEXANAL, 3,5,5-TRIMETHYL-

5455-98-1 1H-ISOINDOLE-1,3(2H)-DIONE, 2-(OXIRANYLMETHYL)-

5456-28-0 ETHYL SELENAC

5459-93-8 N-ETHYLCYCLOHEXYLAMINE
CYCLOHEXANAMINE, N-ETHYL-[PAL]

5466-77-3 2-ETHYLHEXYL-P-METHOXYCIN-NAMATE

5470-11-1 HYDROXYLAMINE, HYDROCHLO-RIDE
HYDROXYLAMINE HYDROCHLORIDE (HYDROXYLAMMONIUM CHLORIDE) [TSC]
HYDROXYLAMINE, HYDROCHLO-RIDE (HYDROXYLAMMONIUM CHLO-RIDE) [CAI]

5493-45-8 1,2-CYCLOHEXANEDICAR-BOXYLIC ACID, BIS(OXIRANYL-METHYL) ESTER
DIGLYCIDYL ESTER OF HEXAHY-DRO-PHTHALIC ACID [OTS]
DIGLYCIDYL ESTER OF HEXAHY-DROPHTHALIC ACID [EPA]

5522-43-0 1-NITROPYRENE

5536-61-8 2-PROPENOIC ACID, 2-METHYL-, SODIUM SALT
METHACRYLIC ACID, SODIUM SALT [PST]

5579-85-1 VINYZENE

5590-18-1 1H-ISOINDOL-1-ONE
C.I. PIGMENT YELLOW 110 [PST]

5593-70-4 BUTYL TITANATE

5598-13-0 CHLORPYRIFOS METHYL [O,O-DIMETHYL-O-(3,5,6-TRICHLORO-2-PYRIDYL)PHOSPHOROTHIOATE

5610-59-3 FULMINATING SILVER

5634-39-9 IODINATED GLYCEROL

5637-83-2 CYANURIC TRIAZIDE

5700-49-2 1,2-DIAMINOETHANE, DIHY-DROIODIDE

5707-69-7 DRAZOXOLON

5714-22-7 SULFUR PENTAFLUORIDE
DISULPHUR DECAFLUORIDE [GBR, GBR]
SULPHUR PENTAFLUORIDE [AUS]

5743-26-0 CALCIUM ACETATE, MONOHY-DRATE

5743-97-5 PERHYDROPHENANTHRENE

5756-43-4 2-ETHYLHEXYL ETHER

5760-50-9 METHYL UNDECYLENATE

5780-07-4 MYRISTIC ALDEHYDE
1,3-BENZODIOXOLE-5-CARBOX-ALDEHYDE, 7-METHOXY- [TSC, TSC, TSC]

5798-79-8 BROMOBENZYL CYANIDE

5809-08-5 1,1,3,3-TETRAMETHYL BUTYL HYDROPEROXIDE

5831-88-9 N,N-DIBUTYL STEARAMIDE
N,N-DIBUTYL STEARAMIDE [NFP, NFP]

5836-29-3 COUMATETRALYL

5847-55-2 DIBUTYLTIN DISTEARATE

5858-33-3 C.I. ACID RED 17, DISODIUM SALT

5858-82-2 PIGMENT RED 48, CI NO. 15865
C.I. PIGMENT RED 52, DISODIUM SALT [PST]

5873-54-1 2,4'-DIPHENYLMETHANE DIISO-CYANATE
BENZENE, 1-ISOCYANATO-2-(4-ISOCYANATOPHENYL)METHYL-[TSC, TSC, TSC]

5881-17-4 3-ETHYLOCTANE

5893-66-3 CUPRIC OXALATE

5894-60-0 HEXADECYLTRICHLOROSILANE
SILANE, TRICHLOROHEXADECYL-[PAL]

5902-51-2 TERBACIL [5-CHLORO-3-(1,1-DIMETHYLETHYL)-6-METHYL-2, 4(1H,3H)-PYRIMIDINEDIONE]

5905-52-2 FERROUS LACTATE

5910-79-2 HEPTADECYL SULFATE, SODIUM SALT

5915-41-3 TERBUTHYLAZINE

5949-05-3 6-OCTENAL, 3,7-DIMETHYL-, (S)-

5970-32-1 MERCURY SALICYLATE

5972-73-6 AMMONIUM OXALATE, UNSPECI-FIED HYDRATE
ETHANEDIOIC ACID, MONOAMMO-NIUM SALT, MONOHYDRATE [MSL, PAL]

5978-70-1 2-OCTANOL

5979-28-2 PIGMENT YELLOW 16
C.I. PIGMENT YELLOW 16 [PST]

5988-91-0 OCTANAL, 3,7-DIMETHYL-

5989-27-5 D-LIMONENE

5989-54-8 L-LIMONENE

6009-70-7 AMMONIUM OXALATE, MONO-HYDRATE
AMMONIUM OXALATE [CAL]
DIAMMONIUM OXALATE [MSL]
ETHANEDIOIC ACID, DIAMMONIUM SALT, MONOHYDRATE [PAL]

6028-57-5 ALUMINUM OCTANOATE

6032-29-7 SEC-AMYL ALCOHOL
2-PENTANOL [MSL, PAL]
2-PENTANOLE [FLA]

SEC-AMYL ALCOHOL [NFP, NFP]

6033-05-2 MORPHINAN-3,6-DIOL, 7,8-DIDE-HYDRO-4,5-EPOXY-17-METHYL-(5A,6A)-, (Z)-9-OCTADECENOATE (SALT)

6046-93-1 COPPER DIACETATE MONOHY-DRATE

6047-17-2 TRICHLOROPHENOXYPROPI-ONIC ACID ESTER (3-(2,4,5))
TRICHLOROPHENOXYPROPIONIC ACID ESTER [NJL]

6055-19-2 CYCLOPHOSPHAMIDE C
CYCLOPHOSPHAMIDE (HYDRATED) [C65, C65, C65, C65, C65]

6055-52-3 1,6-HEXANEDIAMINE, DIHY-DROCHLORIDE

6095-42-7 TRIHEXYL PHOSPHITE

6104-58-1 BRILLIANT BLUE G (ACID BLUE 90)

6104-59-2 BRILLIANT BLUE R (ACID BLUE 83)

6106-46-3 BENZENEACETIC ACID, ALPHA-(HYDROXYMETHYL)-, 9-METHYL-3-OXA-9-AZATRICYCLO[3.3.1.02,4] NON-7-YL ESTER, [7(S)-(1-ALPHA, 2- BETA,4-BETA,5-ALPHA,7-BETA)]-, COMPOUND WITH METHYL NITRATE (1:1)
3-OXA-9-AZONIATRICYCLO[3.3.1.0] NONANE, 7-(3-HYDROXY-1-OXO-2-PHENYLPROPOXY)-9,9-DIMETHYL-, [7(S)-(1A,2B,4B,5A,7B)-, NITRATE (SALT) [TSC, TSC]

6106-81-6 BENZENEACETIC ACID, ALPHA-(HYDROXYMETHYL)-, 9-METHYL-3-OXA-9-AZATRICYCLO[3.3.1.02, 4]NON-7-YL ESTER, N-OXIDE, HYDROBROMI DE, [7(S)-(1-ALPHA, 2-BETA,4-BETA,5-ALPHA,7-BETA)]-

6107-56-8 CALCIUM OCTANOATE

6109-97-3 3-AMINO-9-ETHYLCARBAZOLE HCL
3-AMINO-9-ETHYLCARBAZOLE HY-DROCHLORIDE [C65, C65]
3-AMINO-9-ETHYLCARBAZOLE, HYDROCHLORIDE [MSL]

6112-76-1 MERCAPTOPURINE

6117-91-5 CROTONYL ALCOHOL
2-BUTEN-1-OL [PAL]

6144-28-1 DILINOLEIC ACID

6145-73-9 1-PROPANOL, 2-CHLORO-, PHOS-PHATE (3:1)
TRIS(2-CHLORO-1-PROPYL) PHOS-PHATE [T34]

6153-56-6 ETHANEDIOIC ACID, DIHYDRATE

6164-98-3 CHLORDIMEFORM

6175-45-7 2,2-DIETHOXYACETOPHENONE

6178-32-1 OXIRANE, [(4-NONYLPHENOXY) METHYL]-
P-NONYLPHENYL GLYCIDYL ETHER [EPA, OTS]
PROPANE, 1,2-EPOXY-3-(P-NONYLPHENOXY)- [TSC]

6178-49-0 ACETYLATED LANOLIN ALCO-HOL

6197-30-4 2-PROPENOIC ACID, 2-CYANO-3,3-DIPHENYL-, 2-ETHYLHEXYL ESTER
2-ETHYLHEXYL 2-CYANO-3,3-DIPHENYLACRYLATE [PST]

6219-67-6 1,3-BENZENEDIAMINE, 4-METHOXY-, SULFATE

6219-71-2 1,4-BENZENEDIAMINE, 2-CHLORO-, SULFATE

6219-77-8 1,2-BENZENEDIAMINE, 4-NITRO-, DIHYDROCHLORIDE

6227-14-1 C.I. DIRECT VIOLET 9, DISODIUM SALT

6247-34-3 2-ANTHRACENESULFONIC ACID, 4-((4-(ACETYLAMINO)PHENYL) AMINO)-1-AMINO-9,10-DIHYDRO-9,10-DIOXO-

6247-51-4 C.I. DIRECT BROWN 59

6252-76-2 EXT D & C RED NO. 3

6272-74-8 N-LAUROYL ESTER OF CO-LAMINOFORMYLMETHYLPYRI-DINIUM CHLORIDE

6283-25-6 BENZENAMINE, 2-CHLORO-5-NITRO-

6284-84-0 DIMETHYL PIPERAZINE-CIS

6288-89-7 DIGLYCOL CHLOROHYDRIN

6303-21-5 HYPOPHOSPHORUS ACID

6304-39-8 OCTANOIC ACID, HYDRAZIDE

6315-52-2 1,2-ETHANEDIOL BIS(4-METHYL-BENZENESULFONATE)

6317-18-6 METHYLENE BIS(THIOCYANATE)

6341-97-5 2,4-DICHLOROPHENOXYACETIC ACID ESTER

6358-29-8 C.I. DIRECT RED 39, DISODIUM SALT
1,3-NAPHTHALENEDISULFONIC ACID, 8-[[4'-[(4-ETHOXYPHENYL) AZO]-3,3'-DIMETHYL] [1,1'-BIPHENYL]-4-YL]AZO]-7-HYDROXY-, DISODIUM SALT [TSC]

6358-31-2 PIGMENT YELLOW 74
C.I. PIGMENT YELLOW 74 [PST]

6358-53-8 CITRUS RED NO. 2
2-NAPHTHALENOL, 1-[(2,5-DIMETHOXYPHENYL)AZO]- [PAL, PAL]

6358-85-6 DIARYLANILIDE YELLOW

6359-45-1 C.I. BASIC VIOLET 16

6359-82-6 C.I. ACID YELLOW 11, SODIUM SALT

6359-97-3 C.I. ACID YELLOW 34

6359-98-4 C.I. ACID YELLOW 17, DISODIUM SALT

6360-54-9 C.I. DIRECT BROWN 154
BENZOIC ACID, 5-[[4'-[[2,6-DIAMINO-3-METHYL-5-[(4-SULFOPHENYL)AZO] PHENYL]AZO][1,1'-BIPHENYL]-4-YL]AZO]-2- HYDROXY-3-METHYL-, DISODIUM SALT [TSC]

6368-72-5 SUDAN RED 7B

6369-59-1 2,5-TOLUENEDIAMINE SULFATE
1,4-BENZENEDIAMINE, 2-METHYL-, SULFATE [TSC, TSC]

6369-96-6 ACETIC ACID, (2,4,5-TRICHLOROPHENOXY)-, COMPOUND WITH TRIMETHYLAMINE
2,4,5-T AMINE [NJL]

6369-97-7 ACETIC ACID, (2,4,5-TRICHLOROPHENOXY)-, COMPOUND WITH N-METHYL-METHANAMINE
2,4,5-T AMINE [NJL]

6373-07-5 RHODAMINE WT

6373-20-2 C.I. VAT BLUE 22

6373-74-6 C.I. ACID ORANGE 3
CI ACID ORANGE 3 [IAR]

6381-77-7 SODIUM ERYTHORBATE

6382-07-6 2-(P-TERT-AMYLPHENOXY) ETHANOL

6382-13-4 AMYL STEARATE

6386-38-5 BENZENEPROPANOIC ACID, 3, 5-BIS(1,1-DIMETHYLETHYL)-4-HYDROXY-, METHYL ESTER
BENZENEPROPANOIC ACID, 3,5-BIS(1, 1-DIMETHYLETHYL)- [OTS]

6408-78-2 FD&C BLUE NO.2
C.I. ACID BLUE 25 [PST]

6408-80-6 C.I. ACID BLUE 145

6410-41-9 C.I. PIGMENT RED 5

6416-57-5 SUDAN BROWN RR

6416-68-8 C.I. FLUORESCENT BRIGHTENER 46

6419-19-8 AMINOTRI(METHYLENEPHOS-PHONIC ACID)
NITRILOTRIS(METHYLENE)TRISPHO-SPHONIC ACID [PST]
PHOSPHONIC ACID, [NITRILOTRIS (METHYLENE)]TRIS- [OTS]

6422-86-2 1,4-BENZENEDICARBOXYLIC ACID, BIS(2-ETHYLHEXYL) ESTER

6423-43-4 PROPYLENE GLYCOL DINITRATE
1,2-PROPANEDIOL, DINITRATE [PAL]
1,2-PROPYLENE GLYCOL DINITRATE [ONT, ONT, WHM]
PROPYLENE DINITRATE [GBR, GBR, GBR]

6424-76-6 C.I. VAT BLUE 16

6424-85-7 2-ANTHRACENESULFONIC ACID, 4-[[4-(ACETYLAMINO)PHENYL] AMINO]-1-AMINO-9,10-DIHYDRO-9,10-DIOXO-, MONOSODIUM SALT
C.I. ACID BLUE 40 [PST]

6426-62-6 C.I. DIRECT YELLOW 20

6426-67-1 C.I. DIRECT VIOLET 22, TRISODIUM SALT

6427-21-0 METHOXYMETHYL ISOCYANATE

6449-35-0 1-NAPHTHALENEDISUL-FONIC ACID, 3-[[4'-[(6-AMINO-1-HYDROXY-3-SULFO-2-NAPHTHALENYL)AZO]-3,3'-DIMETHOXY[1,1'-BIPHENYL]-4-YL]AZO]-4-HYDROXY-, DISODIUM SALT

6459-94-5 C.I. ACID RED 114, DISODIUM SALT
C.I. ACID RED 114 [313, C65, CSR, CSR]
CI ACID RED 114 [IAR]

6471-49-4 C.I. PIGMENT RED 23

6484-52-2 AMMONIUM NITRATE
NITRIC ACID AMMONIUM SALT [PAL]

6485-40-1 (-)-CARVONE

6505-86-8 NICOTINE SULFATE

6531-86-8 CYCLOHEXYLCYCLOHEXANOL

6533-00-2 NORGESTREL

6533-68-2 SCOPOLAMINE HYDROBROMIDE TRIHYDRATE

6533-73-9 THALLOUS CARBONATE
THALLIUM(I) CARBONATE [EPA, RCU, RCU, WHM]

6542-37-6 1H,3H,5H-OXAZOLO(3,4-C)OXA-ZOLE-7A,(7H)-METHANOL

6543-62-0 STIBAMINE

6606-65-1 2-PROPENOIC ACID, 2-CYANO-, BUTYL ESTER

6607-45-0 BEN-ZENE, (1,2-DICHLOROETHENYL)-
ALPHA,BETA-DICHLOROSTRYENE [FLA, NFP, NFP]
ALPHA,BETA-DICHLOROSTYRENE [MSL]

6625-46-3 C.I. ACID VIOLET 12, DISODIUM SALT

6637-88-3 C.I. DIRECT ORANGE 6, DIS-ODIUM SALT
BENZOIC ACID, 5-[[4'-[(2,6-DIAMINO-3-METHYL-5-SULFOPHENYL)AZO]-3, 3'-DIMETHYL[1,1'-BIPHENYL]-4-YL] AZO]-2-HYDROXY-, DISODIUM SALT [TSC]

6656-03-7 CUPRATE(3-), [U-[7-[[3,3'-DI-HYDROXY-4'-[[1-HYDROXY-6-(PHENYLAMINO)-3-SULFO-2-NAPHTHALENYL][1,1'-BIPHENYL]-4-YL]AZO] -8-HY-DROXY-1,6-NAPHTHALENEDISUL-FONATO(7-)]DI-, TRISODIUM

6706-59-8 L-GULITOL

6731-36-8 1,1-DI(TERT-BUTYLPEROXY)-3, 3,5-TRIMETHYLCYCLOHEXANE PEROXIDE
1,1-DI(TERT-BUTYLPEROXY)-3,3,5-TRIMETHYLCYCLOHEXANE [NJL]

6737-68-4 HYTHERM BLUE E

6739-62-4 C.I. DIRECT BLACK 91, TRISODIUM SALT
BENZOIC ACID, 2-[[2-AMINO-6-[[4'-[(3-CARBOXY-4-HYDROX-YPHENYL)AZO]-3,3'-DIMETHOXY[1,1'-BIPHENYL]-4-YL]AZO]-5-HYDROXY-7-S ULFO-1-NAPHTHALENYL]AZO]-5-NITRO-, TRISODIUM SALT [TSC]

6742-54-7 BENZENE, UNDECYL-

6752-33-6 ETHANOL, 2,2'-(HEXYLAMINO) BIS-,
ETHANOL, 2,2'-(HEXYLAMINO)BIS-[TSE]

6795-23-9 AFLATOXIN M1

6798-76-1 ABIETIC ACID, ZINC SALT

6806-86-6 CHLOROMETHYL (OXIRANE)
CHLOROMETHYL [FLA]

6834-92-0 SODIUM METASILICATE
SILICIC ACID, DISODIUM SALT, PEN-TAHYDRATE [PST]

6838-85-3 LEAD PHTHALATE

6842-15-5 PROPYLENE TETRAMER
DODECYLENE (ALPHA) [NFP, NFP]

6843-97-6 DODECYLDI(AMINOETHYL) GLYCINE

6844-74-2 C.I. ACID RED 101, DISODIUM SALT

6846-50-0 2,2,4-TRIMETHYL-1,3-PENTANE-DIOL DIISOBUTYRATE
2,2,4-TRIMETHYL PENTANEDIOL DIISOBUTYRATE [NFP, NFP]
2,2,4-TRIMETHYL-1,3-PENTANEDIOL ESTER [OTS]
PROPANOIC ACID, 2-METHYL-, 2,2-DIMETHYL-1-(1-METHYLETHYL)-1,3-PROPANEDIYL ESTER [PST]

6865-35-6 BARIUM STEARATE

6870-67-3 JACOBINE

6895-43-8 ETHYL BIXIN

6915-15-7 MALIC ACID

6923-22-4 MONOCROTOPHOS
3-(DIMETHOXYPHOSPHINYLOXY)
-N-METHYL-CIS-CROTONAMIDE
(MONOCROTOPHOS) [CAL]
MONOCROTOPHOS (AZODRIN (R))
[QBC, QBC]
MONOCROTOPHOS (AZODRIN) [ALB,
ALB, BC1, OS1]
PHOSPHORIC ACID, DIMETHYL 1-
METHYL-3-(METHYLAMINO)-3-OXO-
1-PROPENYL ESTER, (E)- [PAL]

6923-52-0 ANTIMONY(III) ACETATE

6950-84-1 1-NAPHTHYLUREA
NAPHTHYLUREA [HM1, HM1]

**6959-47-3 2-CHLOROMETHYLPYRIDINE
HYDROCHLORIDE**

**6959-48-4 3-CHLOROMETHYLPYRIDINE
HYDROCHLORIDE**
3-(CHLOROMETHYL)PYRIDINE HY-
DROCHLORIDE [MSL]

6969-49-9 N-OCTYL SALICYLATE

6975-98-0 2-METHYLDECANE

6988-21-2 DIOXACARB

**7005-72-3 4-CHLOROPHENYL PHENYL
ETHER**
BENZENE, 1-CHLORO-4-PHENOXY-
[PAL]

7008-42-6 ACRONYCINE

**7023-61-2 C.I. PIGMENT RED 48, CALCIUM
SALT**

**7027-11-4 CYCLOHEXANECARBONITRILE,
1,3,3-TRIMETHYL-5-OXO-,**
ISOPHORONE NITRILE [HON]

7047-84-9 ALUMINUM STEARATE

**7057-92-3 PHOSPHORIC ACID, DIDODECYL
ESTER**

7059-24-7 CHROMOMYCIN A3

7085-85-0 ETHYL CYANOACRYLATE
2-PROPENOIC ACID, 2-CYANO-,
ETHYL ESTER [TSC, TSC]

**7090-25-7 1-NAPHTHYL METHYLNITROSO-
CARBONATE**

**7128-64-5 2,2'-(2,5-THIOPHENEDIYL)BIS(5-
TERTIARYBUTYLBENZOXAZOLE)**

7128-91-8 CETYLDIMETHYLAMINE OXIDE

7149-75-9 4-CHLORO-3-METHYL ANILINE

7154-79-2 PENTANE, 2,2,3,3-TETRAMETHYL-
2,2,3,3-TETRAMETHYL PENTANE
[FLA, NFP, NFP]
2,2,3,3-TETRAMETHYLPENTANE
[MSL]

**7166-19-0 BETA-BROMO-BETA-NI-
TROSTYRENE**

**7173-51-5 DIDECYLDIMETHYLAMMONIUM
CHLORIDE**

**7173-62-8 N-CIS-9-OCTADECENYL-1,3-
PROPANEDIAMINE**

7177-48-2 AMPICILLIN TRIHYDRATE

**7195-45-1 1,2-BENZENEDICARBOXYLIC
ACID, BIS(OXIRANYLMETHYL)
ESTER**
DIGLYCIDYL ESTER OF PHTHALIC
ACID [EPA, OTS]

**7203-90-9 1-(4-CHLOROPHENYL)-3,3-
DIMETHYL TRIAZINE**

7209-38-3 BIS(AMINOPROPYL)PIPERAZINE

7220-79-3 METHYLENE BLUE TRIHYDRATE

7220-81-7 AFLATOXIN B2

**7227-91-0 3,3-DIMETHYL-1-PHENYLTRI-
AZENE**

7241-98-7 AFLATOXIN G2

7280-37-7 PIPERAZINE ESTRONE SULFATE
CONJUGATED ESTROGENS: PIPER-
AZINE ESTRONE SULFATE [PAL,
PAL]

7287-19-6 PROMETRYN
PROMETRYN [N,N'-BIS(1-
METHYLETHYL)-6-METHYLTHIO-1,
3,5-TRIAZINE-2,4-DIAMINE] [313]

7292-16-2 PROPAPHOS

7320-34-5 POTASSIUM PYROPHOSPHATE
TETRAPOTASSIUM PYROPHOSPHATE
[PST]

7320-37-8 OXIRANE, TETRADECYL-

**7324-02-9 2-PROPENOIC ACID, 2-CYANO-, 2-
PROPENYL ESTER**

**7328-97-4 OXIRANE, 2,2',2'',2'''-[1,2-
ETHANEDIYLIDENETETRAKIS-(4,
1-PHENYLENEOXYMETHYLENE)]
TETRAKIS-**
1,1,2,2-TETRA(P-HYDROXYPHENYL)
ETHANE TETRAGLYCIDYL ETHER
[EPA]
1,1,2,2-TETRA(P-HYDROXYPHENYL)
-ETHANE TETRAGLYCIDYL ETHER
[OTS]

7332-32-3 METHYLTRITHION

**7332-46-9 ETHANOL, 2-(2-BUTOXYETHOXY)-
, PHOSPHATE (3:1)**

**7336-20-1 4,4'-DIAMINO-2,2'-STIL-
BENEDISULFONIC ACID, DIS-
ODIUM SALT**

**7346-80-7 SODIUM N-METHYL-N-OLEYL-
TAURATE**

7347-29-7 OLEOYL IMIDAZOLINE

**7373-11-7 2-HYDROXYPROPYLAMINE NI-
TRITE**

7376-31-0 TRIETHANOLAMINE SULFATE

7378-99-6 N,N-DIMETHYLOCTYLAMINE

**7379-27-3 ETHYLENEDIAMINETE-
TRAACETIC ACID, POTASSIUM
SALT**
POTASSIUM ETHYLENEDIAMINETE-
TRAACETATE [PST]

7390-81-0 OXIRANE, HEXADECYL-

7397-62-8 BUTYL GLYCOLATE

**7411-49-6 3,3'-DIAMINOBENZIDINE TE-
TRAHYDROCHLORIDE**

7414-83-7 ETIDRONATE DISODIUM

7415-31-8 1,3-DICHLORO-2-BUTENE

7421-93-4 ENDRIN ALDEHYDE
1,2,4-METHENOCYCLOPENTA[CD]
PENTALENE-5-CARBOXALDEHYDE,
2,2A,3,3,4,7-HEXACHLORODEC-
AHYDRO-,(1.ALPHA.,2.BETA.,
2A.BETA., 4.BETA.,4A.BETA.,5.BETA.,
6A.BETA.6B.BETA.,7R*)- [PAL]

**7422-52-8 TRISILOXANE, 1,1,1,3,5,5,5-HEP-
TAMETHYL-3-[3-(OXIRANYL-
METHOXY)PROPYL]-**
3-(BIS(TRIMETHYLSILOXY)METHYL)
PROPYL GLYCIDYL ETHER [OTS]
3-[BIS(TRIMETHYLSILOXY)METHYL]
PROPYL GLYCIDYL ETHER [EPA]

7428-48-0 LEAD STEARATE
OCTADECANOIC ACID, LEAD SALT
[PAL]
STEARIC ACID, LEAD SALT [MSL]

7429-90-5 ALUMINUM
ALUMINUM METAL [CEX, OS1,
OS1]
ALUMINUM METAL AND OXIDE
[ALB, ALB, CAL, MEX, MEX]
ALUMINUM POWDER [PST]
ALUMINUM, ELEMENTAL [WHM]

7439-89-6 IRON
IRON (FE) [PST]

7439-90-9 KRYPTON

7439-92-1 LEAD
LEAD METAL [EPA]
LEAD, ELEMENTAL [WHM]
LEAD, INORGANIC [AUS, AUS]
LEAD, INORGANIC, DUST AND FUME
[MEX, MEX]
LEAD, INORGANIC, DUST AND
FUMES [FLA]

7439-93-2 LITHIUM

7439-95-4 MAGNESIUM

7439-96-5 MANGANESE
MANGANESE AND COMPOUNDS
[ALB, BC1]
MANGANESE COMPOUNDS [NIO,
NIO]
MANGANESE, ELEMENTAL [WHM]

7439-97-6 MERCURY
MERCURY (VAPOR) [MEX]
MERCURY VAPOR [ALB, ALB, NIO,
NIO, NIO, NIO, QBC]
MERCURY, ELEMENTAL [WHM]
MERCURY, METALLIC [ATS]

7439-98-7 MOLYBDENUM
MOLYBDENUM (INSOLUBLE COM-
POUNDS) [BC1, BC1, NIO, NIO]
MOLYBDENUM METAL [ONT]
MOLYBDENUM, ELEMENTAL [WHM]

7440-01-9 NEON

7440-02-0 NICKEL
METALLIC NICKEL [IAR]
NICKEL (METAL) [QBC]
NICKEL METAL [BC1, BC1, BC1,
MEX, ONT, ONT]
NICKEL METAL AND OTHER COM-
POUNDS [NIO, NIO, NIO]
NICKEL, ELEMENTAL [WHM]
NICKEL, METAL, AND SOLUBLE
COMPOUNDS [MIN]

7440-03-1 NIOBIUM

7440-05-3 PALLADIUM

7440-06-4 PLATINUM
PLATINIUM, ELEMENTAL [WHM]
PLATINUM COMPOUNDS [MAK,
MAK]
PLATINUM METAL [GBR, MEX]
PLATINUM, METAL [CAL]

7440-07-5 PLUTONIUM

7440-09-7 POTASSIUM

7440-14-4 RADIUM

7440-16-6 RHODIUM
RHODIUM (METAL FUME AND IN-
SOLUBLE COMPOUNDS) [NIO, NIO]
RHODIUM METAL [ONT, ONT,
ONT]
RHODIUM, ELEMENTAL [WHM]
RHODIUM, METAL [MEX]
RHODIUM, METAL AND INSOLUBLE
COMPOUNDS [ALB, ALB]
RHODIUM, METAL FUME AND DUSTS
[BC1, BC1, FLA, QBC]

7440-17-7 RUBIDIUM

7440-21-3 SILICON
SILICON POWDER [NJL]

7440-22-4 SILVER
SILVER (METAL DUST AND SOLUBLE
COMPOUNDS) [NIO, NIO]
SILVER, ELEMENTAL [WHM]
SILVER, METAL [MEX, ONT]

SILVER, METAL AND SOLUBLE COMPOUNDS [BC1, BC1, OS1, OS1]

7440-23-5 SODIUM

7440-24-6 STRONTIUM

7440-25-7 TANTALUM
TANTALUM (METAL AND OXIDE DUST) [NIO, NIO]
TANTALUM METAL [ONT, ONT]
TANTALUM METAL [WHM]
TANTALUM, METAL [TLV]
TANTALUM, METAL AND OXIDE DUST [OS1, OS1]

7440-28-0 THALLIUM
THALLIUM (SOLUBLE COMPOUNDS) [NIO, NIO, NIO, NIO]
THALLIUM SOLUBLE COMPOUNDS [ALB, ALB, ALB]
THALLIUM, ELEMENTAL [WHM]
THALLIUM, SOLUBLE COMPOUNDS [BC1, BC1, MEX, OS1, OS1, OS1, OS1, QBC, QBC]
THALLIUM, WATER SOLUBLE COMPOUNDS [ONT, ONT]

7440-29-1 THORIUM

7440-31-5 TIN
TIN (INORGANIC COMPOUNDS EXCEPT OXIDES) [NIO, NIO, NIO]
TIN METAL [ONT]
TIN, ELEMENTAL [WHM]

7440-32-6 TITANIUM

7440-33-7 TUNGSTEN
TUNGSTEN, ELEMENTAL [WHM]

7440-36-0 ANTIMONY
ANITMONY AND COMPOUNDS [MIN]
ANTIMONY AND COMPOUNDS [ALB, ALB, AUS, BC1, MEX, NIO, NIO, NIO, OS1, OS1, TLV]
ANTIMONY, ELEMENTAL [WHM]

7440-37-1 ARGON

7440-38-2 ARSENIC
ARSENIC (INORGANIC COMPOUNDS) [NIO, NIO, NIO, NIO]
ARSENIC AND SOLUBLE COMPOUNDS [ALB, ALB, AUS, AUS, MEX]
ARSENIC, ELEMENTAL [WHM]
INORGANIC ARSENIC [MIN]

7440-39-3 BARIUM
BARIUM (SOLUBLE COMPOUNDS) [ALB, ALB]
BARIUM COMPOUNDS [MAK, MAK]
BARIUM, SOLUBLE COMPOUNDS [MEX, MIN, OS1, OS1]

7440-41-7 BERYLLIUM
BERYLLIUM AND BERYLLIUM COMPOUNDS [OS1, OS1, OS1, OS1, OS1]
BERYLLIUM AND COMPOUNDS [AUS, AUS, NIO, NIO, NIO, NIO, TLV, TLV]
BERYLLIUM AND ITS COMPOUNDS [MAK]
BERYLLIUM POWDER [EPA, RCU, RCU]
BERYLLIUM, ELEMENTAL [WHM]

7440-42-8 BORON

7440-43-9 CADMIUM
CADMIUM AND ITS COMPOUNDS [MAK]
CADMIUM DUST [NIO, NIO, NIO, NIO]
CADMIUM DUST AND SALTS [QBC, QBC]
CADMIUM, DUST & SALTS [BC1, BC1]
CADMIUM, DUST & SALTS (AS CD) [ALB, ALB]
CADMIUM, DUSTS AND SALTS [MEX, MEX]
CADMIUM, ELEMENTAL [WHM]

7440-44-0 CARBON
CARBON BLACK [CEX]

GRAPHITE (NATURAL) [NIO]

7440-45-1 CERIUM

7440-46-2 CESIUM

7440-47-3 CHROMIUM
CHROMIUM (METAL) [AUS]
CHROMIUM (TOTAL) [UTS]
CHROMIUM (VI), WATER SOLUBLE FUME [BEI]
CHROMIUM METAL [CEX, IAR, MIN, NIO, NIO, ONT, TLV, TLV]
CHROMIUM, ELEMENTAL [WHM]
CHROMIUM, METAL [MEX]

7440-48-4 COBALT
COBALT AND BIOAVAILABLE
COBALT COMPOUNDS [MAK, MAK]
COBALT DUST AND FUME [ALB, ALB]
COBALT FUME AND DUST [QBC]
COBALT METAL POWDER [C65, IAR]
COBALT METAL, DUST AND FUME [BC1, BC1, BC1]
COBALT METAL, DUST, AND FUME [NIO, NIO, NIO, OS1, OS1]
COBALT, ELEMENTAL [WHM]
COBALT, METAL, DUST AND FUME [MEX]

7440-50-8 COPPER
COPPER, ELEMENTAL [WHM]
COPPER, FUME, DUSTS AND MISTS [MEX, MEX]

7440-55-3 GALLIUM
GALLIUM, ELEMENTAL [WHM]

7440-56-4 GERMANIUM

7440-58-6 HAFNIUM
HAFNIUM AND COMPOUNDS [NIO, NIO]
HAFNIUM, ELEMENTAL [WHM]

7440-59-7 HELIUM

7440-61-1 URANIUM
URANIUM (INSOLUBLE COMPOUNDS) [NIO, NIO, NIO, NIO]
URANIUM (NATURAL) [AUS, AUS, MSL, ONT, ONT, TLV, TLV]
URANIUM (NATURAL) SOLUBLE AND INSOLUBLE COMPOUNDS [BC1, BC1]
URANIUM (NATURAL), INSOLUBLE [QBC]
URANIUM COMPOUNDS [MAK, MAK]
URANIUM(NATURAL) [FLA]
URANIUM, (NATURAL) SOLUBLE AND INSOLUBLE [MEX, MEX]
URANIUM, ELEMENTAL [WHM]

7440-62-2 VANADIUM
METALLIC VANADIUM [NHS, NHS]
VANADIUM (FUME OR DUST) [MSL, PAL]
VANADIUM (V2O5) [BC1, BC1, BC1]
VANADIUM, DUST AND FUME [MEX]
VANADIUM, ELEMENTAL [WHM]

7440-63-3 XENON

7440-65-5 YTTRIUM
YTTRIUM COMPOUNDS [NIO, NIO]
YTTRIUM METAL [ONT]
YTTRIUM METAL AND COMPOUNDS [MEX, MEX]
YTTRIUM, ELEMENTAL [WHM]

7440-66-6 ZINC
ZINC (METALLIC) [PST]

7440-67-7 ZIRCONIUM
ZIRCONIUM AND COMPOUNDS [TLV, TLV]
ZIRCONIUM COMPOUNDS [ALB, ALB, AUS, AUS, BC1, BC1, MAK, MAK, MEX, MEX, MIN, NIO, NIO, NIO, ONT, ONT, OS1, OS1, OS1, QBC, QBC]
ZIRCONIUM, ELEMENTAL [WHM]

7440-69-9 BISMUTH

7440-70-2 CALCIUM

7440-74-6 INDIUM
INDIUM AND COMPOUNDS [ALB, ALB, AUS, BC1, BC1, OS1, QBC, TLV]
INDIUM, ELEMENTAL [WHM]

7446-07-3 TELLURIUM DIOXIDE

7446-08-4 SELENIUM DIOXIDE
SELENIUM OXIDE [NJL]
SELENIUM OXIDE (SEO2) [PAL]
SELENIUM(IV) DIOXIDE [WHM]

7446-09-5 SULFUR DIOXIDE
SULFUR DIOXIDE (ANHYDROUS) [EP3]
SULPHUR DIOXIDE [AUS, AUS, CN3, GBR, GBR]

7446-11-9 SULFUR TRIOXIDE

7446-14-2 LEAD SULFATE
LEAD SULPHATE [NJL, NJL]
SULFURIC ACID, LEAD (2+) SALT [MSL]
SULFURIC ACID, LEAD(2+) SALT (1:1) [PAL]

7446-15-3 LEAD SELENATE

7446-18-6 THALLOUS SULFATE
SULFURIC ACID, DITHALLIUM(1+) SALT [PAL]
THALLIUM (I) SULFATE [RCU, RCU]
THALLIUM SULFATE [HM1, HM1]
THALLIUM SULFATE (2:1) [MSL]
THALLIUM(I) SULFATE [EPA, WHM]

7446-26-6 ZINC PYROPHOSPHATE

7446-27-7 LEAD PHOSPHATE
PHOSPHORIC ACID, LEAD(2+) SALT (2:3) [PAL, PAL]

7446-34-6 SELENIUM SULFIDE
SELENIUM SULFIDE (SES) [PAL, PAL]

7446-70-0 ALUMINUM CHLORIDE
ALUMINUM CHLORIDE (ALCL3) [PAL]

7446-81-3 SODIUM ACRYLATE
2-PROPENOIC ACID, SODIUM SALT [TSC, TSC]

7447-39-4 CUPRIC CHLORIDE
COPPER CHLORIDE (CUCL2) [PAL]

7447-40-7 POTASSIUM CHLORIDE

7447-41-8 LITHIUM CHLORIDE

7460-82-4 2,2-OXYBISETHANE BIS(4-METHYLBENZENESULFONATE)

7460-84-6 GLYCIDYL STEARATE

7481-89-2 2',3'-DIDEOXYCYTIDINE (AIDS INITIATIVE)

7487-79-8 DIETHANOLAMINE LAURATE

7487-88-9 MAGNESIUM SULFATE

7487-94-7 MERCURIC CHLORIDE

7488-51-9 LEAD SELENITE

7488-56-4 SELENIUM SULFIDE
SELENIUM DISULFIDE [EPA, MSL, NJL, PAL]
SELENIUM(IV) DISULFIDE [WHM]

7492-30-0 POTASSIUM RICINOLEATE

7496-02-8 6-NITROCHRYSENE

7521-80-4 BUTYLTRICHLOROSILANE
BUTYL TRICHLOROSILANE [FLA, NFP, NFP, NJL, NJL]
SILANE, BUTYLTRICHLORO- [PAL]

7530-07-6 CAPRYLYL PEROXIDE (N-OCTANOYL PEROXIDE)
CAPRYLYL PEROXIDE [NJL]

7534-94-3 **2-PROPENOIC ACID, 2-METHYL-, 1,7,7-TRIMETHYLBICYCLO[2.2.1] HEPT-2-YL ESTER, EXO-**

7546-30-7 **MERCUROUS CHLORIDE**

7550-45-0 **TITANIUM TETRACHLORIDE**
TITANIUM CHLORIDE [FLA]
TITANIUM CHLORIDE (TICL4) (T-4)- [PAL]
TITANIUM TETRACHLORIDE [TI-TANIUM CHLORIDE (TICL4) (T-4)-] [EP3]

7553-56-2 **IODINE**

7558-79-4 **SODIUM PHOSPHATE DIBASIC**
DISODIUM PHOSPHATE [PST]
PHOSPHORIC ACID, DISODIUM SALT [MSL, PAL]
SODIUM PHOSPHATE, DIBASIC [NJL]

7558-80-7 **SODIUM ACID PHOSPHATE**
MONOSODIUM O-PHOSPHATE [PST]

7568-93-6 **BENZENEMETHANOL,.ALPHA.- (AMINOMETHYL)-**

7570-26-5 **1,2-DINITROETHANE**

7572-29-4 **DICHLOROACETYLENE**
DICHLORO ACETYLENE [OS3]
DICHLOROETHYNE [ONT]
ETHYNE, DICHLORO- [PAL]

7575-62-4 **DISODIUM 4-DODECYL-2,4'-OXY-DIBENZENESULFONATE**

7580-31-6 **2-ETHYLHEXANOIC ACID, NICKEL SALT**

7580-67-8 **LITHIUM HYDRIDE**
LITHIUM HYDRIDE (LIH) [PAL]

7581-97-7 **2,3-DICHLOROBUTANE**
BUTANE, 2,3-DICHLORO- [PAL]

7585-39-9 **CYCLOHEPTAAMYLOSE**

7601-54-9 **TRISODIUM PHOSPHATE**
PHOSPHORIC ACID, TRISODIUM SALT [MSL, PAL]
SODIUM PHOSPHATE, TRIBASIC [NJL]

7601-89-0 **SODIUM PERCHLORATE**
PERCHLORIC ACID, SODIUM SALT [PAL]

7601-90-3 **PERCHLORIC ACID**

7616-94-6 **PERCHLORYL FLUORIDE**

7623-09-8 **2-CHLOROPROPIONYL CHLO-RIDE**

7631-86-9 **SILICA, AMORPHOUS**
AMORPHOUS SILICA [FLA, MSL]
SILICA [CAL, MIN, ONT, ONT, ONT, PAL]
SILICA FLOUR [ALB]
SILICA SAND [BC1]
SILICA, AMORPHOUS (INCLUDING NATURAL DIATOMACEOUS EARTH) [QBC]
SILICA, AMORPHOUS, FUMED [WHM]
SILICA-AMORPHOUS, FUMED SILICA [AUS]
SILICON DIOXIDE [PST]

7631-89-2 **SODIUM ARSENATE**
ARSENIC ACID (H3ASO4), SODIUM SALT [PAL, PAL]
ARSENIC ACID, SODIUM SALT [MSL]

7631-90-5 **SODIUM BISULFITE**
SODIUM BISULPHITE [AUS]
SODIUM DISULFITE [FLA]
SODIUM HYDROGENSULPHITE [GBR]
SULFUROUS ACID, MONOSODIUM SALT [PAL]

7631-95-0 **SODIUM MOLYBDATE**

7631-98-3 **N-DODECYLSARCOSINE, SODIUM SALT**

7631-99-4 **SODIUM NITRATE**
NITRIC ACID SODIUM SALT [PAL]

7632-00-0 **SODIUM NITRITE**
NITROUS ACID, SODIUM SALT [PAL]

7632-04-4 **SODIUM PERBORATE**
PERBORIC ACID, SODIUM SALT [PST]

7632-05-5 **SODIUM PHOSPHATE**

7632-51-1 **VANADIUM TETRACHLORIDE**
VANADIUM CHLORIDE (VCL4), (T-4)- [PAL]

7637-07-2 **BORON TRIFLUORIDE**
BORANE, TRIFLUORO- [PAL]
BORON TRIFLUORIDE [BORANE, TRIFLUORO-] [EP3]

7645-25-2 **LEAD ARSENATE, UNSPECIFIED**
ARSENIC ACID (H3ASO4), LEAD SALT [PAL]
ARSENIC ACID, LEAD SALT [MSL]
LEAD ARSENATE [CN2]

7646-69-7 **SODIUM HYDRIDE**
SODIUM HYDRIDE (NAH) [PAL]

7646-78-8 **STANNIC CHLORIDE**
STANNANE, TETRACHLORO- [PAL]
TIN TETRACHLORIDE [NJL, NJL]

7646-79-9 **COBALTOUS CHLORIDE**
COBALT(II) CHLORIDE [WHM]

7646-85-7 **ZINC CHLORIDE**
ZINC CHLORIDE (ZNCL2) [PAL]
ZINC CHLORIDE FUME [ALB, ALB, BC1, BC1, FLA, MEX, MEX, MSL, NIO, NIO, NIO, ONT, ONT, OS1, OS1, OS1, QBC, TLV, TLV]

7646-93-7 **POTASSIUM HYDROGEN SUL-FATE**

7647-01-0 **HYDROGEN CHLORIDE**
HYDROCHLORIC ACID [313, CAA, CAB, CAN, CWA, EP3, IAR, OS3, PAL]

7647-10-1 **PALLADIUM(II) CHLORIDE**

7647-14-5 **SODIUM CHLORIDE**

7647-15-6 **SODIUM BROMIDE**

7647-17-8 **CESIUM CHLORIDE**

7647-18-9 **ANTIMONY PENTACHLORIDE**
ANTIMONY CHLORIDE (SBCL5) [PAL]

7647-19-0 **PHOSPHORUS PENTAFLUORIDE**

7664-38-2 **PHOSPHORIC ACID**
ORTHOPHOSPHORIC ACID [GBR, GBR]

7664-39-3 **HYDROGEN FLUORIDE**
HYDROFLUORIC ACID [CWA, OS3, PAL]
HYDROGEN FLUORIDE (HYDROFLU-ORIC ACID) [CAA, TSC, TSC]
HYDROGEN FLUORIDE/HYDROFLUORIC ACID [EP3]

7664-41-7 **AMMONIA**
AMMONIA, ANHYDROUS [NFP, NFP]

7664-93-9 **SULFURIC ACID**
SULPHURIC ACID [AUS, AUS, GBR]

7665-72-7 **TERT-BUTYL GLYCIDYL ETHER**
BUTYL GLYCIDYL ETHER, TERT- [OTS]
OXIRANE, [(1,1-DIMETHYLETHOXY) METHYL]- [TSC]

7681-11-0 **POTASSIUM IODIDE**

7681-38-1 **SODIUM BISULFATE**
SODIUM HYDROGEN SULFATE [HM1, HM1, NJL, NJL]

7681-49-4 **SODIUM FLUORIDE**
SODIUM FLUORIDE (NAF) [PAL]

7681-52-9 **SODIUM HYPOCHLORITE**
HYPOCHLOROUS ACID, SODIUM SALT [PAL]

7681-53-0 **SODIUM HYPOPHOSPHITE**

7681-57-4 **SODIUM METABISULFITE**
DISODIUM DISULPHITE [GBR]
DISULFUROUS ACID, DISODIUM SALT [PAL, PST]
SODIUM METABISULPHITE [AUS]

7681-82-5 **SODIUM IODIDE**

7681-93-8 **PIMARICIN**

7688-21-3 **2-HEXENE-CIS**

7696-12-0 **NEOPYNAMIN**
TETRAMETHRIN [2,2-DIMETHYL-3-(2-METHYL-1-PROPENYL)CYCLO-PROPANECARBOXYLIC ACID (1,3,4,5,6,7-HEXAHYDRO-1,3-DIOXO-2H-ISOINDOL-2-YL)METHYL ESTER] [313]

7697-37-2 **NITRIC ACID**

7699-41-4 **FUSED SILICA**
SILICA, AMORPHOUS, FUSED SILICA [MAK, MAK]

7699-43-6 **ZIRCONIUM OXYCHLORIDE**

7699-45-8 **ZINC BROMIDE**
ZINC BROMIDE (ZNBR2) [PAL]

7700-17-6 **CROTOXYPHOS**

7704-34-9 **SULFUR**
SULPHUR [PST]

7704-99-6 **ZIRCONIUM HYDRIDE**

7705-07-9 **TITANIUM TRICHLORIDE**

7705-08-0 **FERRIC CHLORIDE**
IRON CHLORIDE [NJL, NJL]
IRON CHLORIDE (FECL3) [PAL]

7718-54-9 **NICKEL(II) CHLORIDE**
NICKEL CHLORIDE [EP2, EPA, NJL]
NICKEL CHLORIDE (NICL2) [PAL]

7718-98-1 **VANADIUM TRICHLORIDE**

7719-09-7 **THIONYL CHLORIDE**

7719-12-2 **PHOSPHORUS TRICHLORIDE**
PHOSPHOROUS TRICHLORIDE [PAL]
PHOSPHORUS TRICHLORIDE [PHOS-PHOROUS TRICHLORIDE] [EP3]

7720-78-7 **FERROUS SULFATE**
FERROUS SULPHATE [WHM]
SULFURIC ACID, IRON(2+) SALT (1:1) [PAL]

7722-64-7 **POTASSIUM PERMANGANATE**
PERMANGANIC ACID (HMNO4), POTASSIUM SALT [PAL]

7722-71-6 **OLEYL ETHER PHOSPHATE (ACID)**

7722-76-1 **MONOAMMONIUM PHOSPHATE**
AMMONIUM PHOSPHATE (MONOBA-SIC) [PST]

7722-84-1 **HYDROGEN PEROXIDE**
HYDROGEN PEROXIDE (90%) [ONT]
HYDROGEN PEROXIDE (CONC >52%) [PAL]

7722-88-5 **TETRASODIUM PYROPHOSPHATE**
DIPHOSPHORIC ACID, TETRASODIUM SALT [PAL, PST]
TETRA SODIUM PYROPHOSPHATE [CAL]

7723-14-0 **PHOSPHORUS**
PHOSPHORUS (YELLOW OR WHITE) [313, CAN, ONT]
PHOSPHORUS (YELLOW) [AUS, BC1, BC1, FLA, MEX, MEX, MSL, NIO, NIO, NJL, NJL, OS1, OS1, QBC, TLV]
PHOSPHORUS (YELLOW)) [ALB, ALB]

PHOSPHORUS, WHITE OR YELLOW
[HM1, HM1, HM1, HM1]
PHOSPHORUS, YELLOW [CEX, GBR,
GBR, MAK, MAK, MAK]
WHITE PHOSPHORUS [T33, T34,
TSC, TSC]

7726-95-6 BROMINE

7727-15-3 ALUMINUM BROMIDE

7727-18-6 VANADIUM OXYTRICHLORIDE

7727-21-1 POTASSIUM PERSULFATE
DIPOTASSIUM PEROXYDISULPHATE
[GBR]
PEROXYDISULFURIC ACID ([(HO)S(O)
2]2O2), DIPOTASSIUM SALT [PAL]

7727-37-9 NITROGEN
NITROGEN (LIQUIFIED) [FLA, MSL]

7727-43-7 BARIUM SULFATE
BARIUM SULFATE (1:1) [PST]
BARIUM SULPHATE [AUS, GBR]
BARIUM, SULFATE [MIN]

7727-54-0 AMMONIUM PERSULFATE
DIAMMONIUM PEROXYDISULPHATE
[GBR]

7732-18-5 WATER

7733-02-0 ZINC SULFATE
SULFURIC ACID, ZINC SALT (1:1)
[PAL]

7738-94-5 CHROMIC ACID
CHROMIC (VI) ACID (H2CRO4) [OS1,
OS1]
CHROMIC ACID (H2CRO4) [WHM]
CHROMIC ACID AND CHROMATES
[BC1, NIO, NIO, NIO, NIO]

7745-89-3 STARLICIDE
3-CHLORO-P-TOLUIDINE HY-
DROCHLORIDE [CAL]

7756-94-7 TRIISOBUTYLENE

**7757-74-6 DISULFUROUS ACID, DISODIUM
SALT**

7757-79-1 POTASSIUM NITRATE
NITRIC ACID POTASSIUM SALT
[PAL]

7757-82-6 SODIUM SULFATE (SOLUTION)
SODIUM SULFATE [PST]

7757-83-7 SODIUM SULFITE

7758-01-2 POTASSIUM BROMATE
BROMIC ACID, POTASSIUM SALT
[PAL]

7758-02-3 POTASSIUM BROMIDE

7758-09-0 POTASSIUM NITRITE

7758-11-4 DIPOTASSIUM PHOSPHATE
POTASSIUM PHOSPHATE (DIBASIC)
[PST]

**7758-16-9 DIPHOSPHORIC ACID, DISODIUM
SALT**
SODIUM ACID PYROPHOSPHATE
[PST]

7758-19-2 SODIUM CHLORITE
CHLOROUS ACID, SODIUM SALT
[PAL]

7758-29-4 PENTASODIUM TRIPHOSPHATE
SODIUM PHOSPHATE, TRIBASIC
[NJL]
SODIUM TRIPOLYPHOSPHATE [PST]
TRIPHOSPHORIC ACID, PENTA-
SODIUM SALT [MSL, PAL]

7758-87-4 CALCIUM PHOSPHATE
TRICALCIUM PHOSPHATE [PST]

7758-89-6 COPPER(I) CHLORIDE
CUPROUS CHLORIDE [HM1]

7758-94-3 FERROUS CHLORIDE
IRON CHLORIDE (FECL2) [PAL]

7758-95-4 LEAD CHLORIDE
LEAD CHLORIDE (PBCL2) [PAL]

7758-97-6 LEAD CHROMATE
CHROMIC ACID (H2CRO4), LEAD(2+)
SALT (1:1) [PAL, PAL]

7758-98-7 CUPRIC SULFATE
COPPER SULFATE [CWA, PST]
COPPER(II) SULFATE [WHM]
SULFURIC ACID COPPER(2+) SALT
(1:1) [PAL]

**7758-99-8 COPPER (II) SULFATE PENTAHY-
DRATE (1:1:5)**
CUPRIC SULFATE [CSR]

7759-01-5 LEAD TUNGSTEN OXIDE

7761-88-8 SILVER NITRATE
NITRIC ACID SILVER(1+) SALT [PAL]

7765-84-6 SODIUM PHOSPHATE, TRIBASIC

7772-98-7 SODIUM THIOSULFATE

7772-99-8 STANNOUS CHLORIDE
TIN(II) CHLORIDE [WHM]

7773-01-5 MANGANESE(II) CHLORIDE
MANGANESE CHLORIDE [PST]

7773-06-0 AMMONIUM SULFAMATE
AMMONIUM SULFAMATE (AMATE)
[QBC]
AMMONIUM SULFAMATE (AMMATE)
[BC1, BC1, FLA, MSL]
AMMONIUM SULPHAMATE [AUS]
AMMONIUM SULPHAMIDATE [GBR,
GBR]
SULFAMIC ACID, MONOAMMONIUM
SALT [PAL]

7774-29-0 MERCURIC IODIDE
MERCURY IODIDE [HM1]

7775-09-9 SODIUM CHLORATE
CHLORIC ACID, SODIUM SALT [PAL]

7775-11-3 SODIUM CHROMATE
CHROMIC ACID (H2CRO4), DISODIUM
SALT [PAL]

7775-14-6 SODIUM DITHIONITE
DITHIONOUS ACID, DISODIUM SALT
[PAL]
SODIUM HYDROSULFITE [FLA,
MSL]

7775-19-1 SODIUM METABORATE

7775-27-1 SODIUM PERSULFATE
DISODIUM PEROXODISULPHATE
[GBR]

7778-18-9 CALCIUM SULFATE
CALCIUM SULPHATE [AUS]

7778-39-4 ARSENIC ACID
ARSENIC ACID (H3ASO4) [RCU,
RCU]

**7778-43-0 DISODIUM HYDROGEN ARSEN-
ATE**
ARSENIC ACID, DISODIUM SALT
[FLA, MSL]

7778-44-1 CALCIUM ARSENATE
ARSENIC ACID (H3ASO4), CALCIUM
SALT (2:3) [PAL]

7778-50-9 POTASSIUM DICHROMATE
CHROMIC ACID (H2CR2O7),
DIPOTASSIUM SALT [PAL]
POTASSIUM BICHROMATE [CWA,
EP2, EPA]

7778-54-3 CALCIUM HYPOCHLORITE
HYPOCHLOROUS ACID, CALCIUM
SALT [PAL]

7778-66-7 POTASSIUM HYPOCHLORITE

**7778-73-6 POTASSIUM
PENTACHLOROPHENATE**

7778-74-7 POTASSIUM PERCHLORATE
PERCHLORIC ACID, POTASSIUM
SALT [PAL]

**7778-77-0 DIHYDROGEN POTASSIUM PHOS-
PHATE**
POTASSIUM PHOSPHATE, MONOBA-
SIC [PST]

**7778-80-5 SULFURIC ACID DIPOTASSIUM
SALT**
POTASSIUM SULFATE [PST]

**7779-27-3 HEXAHYDRO-1,3,5-TRIETHYL-S-
TRIAZINE**

7779-86-4 ZINC HYDROSULFITE
DITHIONOUS ACID, ZINC SALT (1:1)
[PAL]
ZINC DITHIONITE [NJL]

7779-88-6 ZINC NITRATE
NITRIC ACID, ZINC SALT [PAL]

7782-39-0 DEUTERIUM

7782-41-4 FLUORINE

7782-42-5 GRAPHITE
GRAPHITE (ALL FORMS) [AUS]
GRAPHITE (NATURAL) [NIO, QBC]
GRAPHITE (NATURAL) DUST [FLA,
MSL]
GRAPHITE (SYNTHETIC) [BC1]
GRAPHITE, NATURAL [ALB, ONT,
OS1, OS1]
GRAPHITE-CONTAINING DUSTS
[MAK, MAK]

7782-44-7 OXYGEN
OXYGEN (DISSOLVED) [MX2]
OXYGEN (LIQUID) [FLA, MSL]

7782-49-2 SELENIUM
SELENIUM AND COMPOUNDS [TLV]
SELENIUM COMPOUNDS [ALB, ALB,
AUS, BC1, MAK, MAK, MEX,
MIN, NIO, NIO, OS1, OS1]
SELENIUM, ELEMENTAL [WHM]

7782-50-5 CHLORINE

7782-61-8 FERRIC NITRATE.9H2O

**7782-63-0 FERROUS SULFATE HEPTAHY-
DRATE**
FERROUS SULFATE [PAL]
FERROUS SULFATE (HEPAHYDRATE)
[MSL]

7782-65-2 GERMANIUM TETRAHYDRIDE
GERMANE [GBR, GBR, HM1, HM1,
HM1, PAL]
GERMANIUM HYDRIDE [NJL]

7782-78-7 NITROSYLSULPHURIC ACID
NITROSYLSULFURIC ACID [NJL,
NJL]

7782-79-8 HYDRAZOIC ACID

7782-82-3 SODIUM SELENITE
SELENIOUS ACID, MONOSODIUM
SALT [MSL]

**7782-85-6 DISODIUM ORTHOPHOSPHATE
HEPTAHYDRATE**

7782-86-7 MERCUROUS NITRATE
MERCUROUS NITRATE (MONOHY-
DRATE) [MSL]
NITRIC ACID, MERCURY(1+) SALT,
MONOHYDRATE [PAL]

7782-89-0 LITHIUM AMIDE

7782-92-5 SODIUM AMIDE

7782-99-2 SULFUROUS ACID
SULFUROUS ACID SOLUTION [NJL,
NJL]

7783-00-8 SELENOUS ACID
SELENIOUS ACID [302, EPA, FLA,
WHM]

**7783-03-1 TUNGSTIC ACID (SOLUBLE TUNG-
STEN)**
TUNGSTIC ACIDS [ONT, ONT]

7783-06-4 HYDROGEN SULFIDE
HYDROGEN SULFIDE (H2S) [PAL]

HYDROGEN SULPHIDE [AUS, AUS, GBR, GBR]

7783-07-5 HYDROGEN SELENIDE
HYDROGEN SELENIDE (H2SE) [PAL]

7783-08-6 SELENIC ACID

7783-18-8 AMMONIUM THIOSULFATE
THIOSULFURIC ACID (H2S2O3), DI-AMMONIUM SALT [PAL]

7783-20-2 AMMONIUM SULFATE
AMMONIUM HYDROGEN SULFATE [WHM]
AMMONIUM SULPHATE, SOLUTION [NJL]
SULFURIC ACID DIAMMONIUM SALT [PAL]

7783-24-6 ZINC AMMONIUM SULFATE

7783-28-0 DIAMMONIUM PHOSPHATE

7783-30-4 MERCUROUS IODIDE

7783-33-7 MERCURY POTASSIUM IODIDE

7783-35-9 MERCURIC SULFATE
MERCURIC SULPHATE [HM1, HM1]
SULFURIC ACID, MERCURY(2+) SALT (1:1) [PAL]

7783-36-0 MERCUROUS SULFATE
MERCUROUS SULPHATE [HM1]

7783-40-6 MAGNESIUM FLUORIDE

7783-41-7 OXYGEN DIFLUORIDE
OXYGEN DIFLUORIDE (FLUORINE MONOXIDE) [OS3]
OXYGEN FLUORIDE (OF2) [PAL]

7783-42-8 THIONYL DIFLUORIDE

7783-46-2 LEAD FLUORIDE
LEAD FLUORIDE (PBF2) [PAL]

7783-47-3 STANNOUS FLUORIDE
TIN(II) FLUORIDE [WHM]

7783-48-4 STRONTIUM FLUORIDE

7783-49-5 ZINC FLUORIDE
ZINC FLUORIDE (ZNF2) [PAL]

7783-50-8 FERRIC FLUORIDE
IRON FLUORIDE (FEF3) [PAL]

7783-54-2 NITROGEN TRIFLUORIDE
NITROGEN FLUORIDE (NF3) [PAL]

7783-55-3 PHOSPHORUS TRIFLUORIDE

7783-56-4 ANTIMONY TRIFLUORIDE
STIBINE, TRIFLUORO- [PAL]

7783-60-0 SULFUR TETRAFLUORIDE
SULFUR FLUORIDE (SF4), (T-4)- [PAL]
SULFUR TETRAFLUORIDE [SULFUR FLUORIDE (SF4), (T-4)-] [EP3]
SULPHUR TETRAFLUORIDE [AUS, GBR, GBR]

7783-61-1 SILICON TETRAFLUORIDE

7783-66-6 IODINE PENTAFLUORIDE

7783-70-2 ANTIMONY PENTAFLUORIDE
ANTIMONY FLUORIDE (SBF5) [PAL]
ANTIMONY(V) PENTAFLUORIDE [WHM]

7783-79-1 SELENIUM HEXAFLUORIDE
SELENIUM (HEXAFLUORIDE) [QBC]
SELENIUM FLUORIDE (SEF6), (OC-6-11)- [PAL]

7783-80-4 TELLURIUM HEXAFLUORIDE
TELLURIUM FLUORIDE (TEF6), (OC-6-11)- [PAL]

7783-81-5 URANIUM HEXAFLUORIDE

7783-82-6 TUNGSTEN HEXAFLUORIDE

7783-85-9 FERROUS AMMONIUM SULFATE.6H2O

7783-91-7 SILVER CHLORITE

7784-01-2 SILVER CHROMATE

7784-08-9 SILVER ARSENITE

7784-21-6 ALUMINUM HYDRIDE

7784-25-0 AMMONIUM ALUM

7784-30-7 ALUMINUM PHOSPHATE

7784-33-0 ARSENIC BROMIDE

7784-34-1 ARSENOUS TRICHLORIDE
ARSENIC CHLORIDE [FLA, MSL]
ARSENIC TRICHLORIDE [CWA, EP2, EPA, HM1, HM1, HM1, HM1, HM1, NJL]
ARSENIC(III) TRICHLORIDE [WHM]

7784-37-4 MERCURIC ARSENATE

7784-40-9 LEAD ARSENATE
ARSENIC ACID (H3ASO4), LEAD (2+) SALT (1:1) [PAL]
LEAD (II) ARSENATE [MSL]

7784-41-0 POTASSIUM ARSENATE
ARSENIC ACID (H3ASO4), MONOPOTASSIUM SALT [PAL, PAL]
POTASSIUM DIHYDROGEN ARSENATE [HM1, HM1, HM1, HM1, WHM]

7784-42-1 ARSINE
ARSENIC AND SOLUBLE COMPOUNDS INCLUDING ARSINE [BEI]

7784-44-3 AMMONIUM ARSENATE

7784-45-4 ARSENOUS TRIIODIDE
ARSENIC IODIDE (ASI3) [NJL]

7784-46-5 SODIUM ARSENITE
SODIUM METAARSENITE [HM1, HM1, HM1, HM1]

7785-84-4 METAPHOSPHORIC ACID, TRISODIUM SALT
METAPHOSPHORIC ACID (H3P3O9), TRISODIUM SALT [PAL]

7785-87-7 MANGANESE(II) SULFATE
MANGANESE (II) SULFATE (1:1) [TSC, TSC]
MANGANESE SULFATE [PST]

7785-88-8 SODIUM ALUMINUM PHOSPHATE

7786-17-6 NAUGAUHITE

7786-30-3 MAGNESIUM CHLORIDE

7786-34-7 MEVINPHOS
2-BUTENOIC ACID, 3-[(DIMETHOXYPHOSPHINYL)OXY]-, METHYL ESTER [PAL]
ALPHA-2-CARBOMETHOXY-1-METHYLVINYL DIMETHYL PHOSPHATE (MEVINPHOS) [CAL]
MEVINPHOS (ISO) [GBR, GBR, GBR]
MEVINPHOS (PHOSDRIN) [MEX, MEX, MEX, QBC, QBC, QBC]
PHOSDRIN [NIO, NIO, NIO, NIO, NJL]
PHOSDRIN (MEVINPHOS) [ALB, ALB, ALB, BC1, BC1, BC1, OS1, OS1, OS1, OS1, OS1]

7786-81-4 NICKEL SULFATE
NICKEL(II) SULFATE [WHM]
SULFURIC ACID, NICKEL(2+) SALT (1:1) [PAL]

7787-32-8 BARIUM FLUORIDE

7787-36-2 BARIUM PERMANGANATE
AMMONIUM PERMANGANATE [OS3]

7787-41-9 BARIUM SELENATE

7787-47-5 BERYLLIUM CHLORIDE
BERYLLIUM CHLORIDE (BECL2) [PAL, PAL]

7787-49-7 BERYLLIUM FLUORIDE
BERYLLIUM FLUORIDE (BEF2) [PAL, PAL]

7787-55-5 BERYLLIUM NITRATE TRIHYDRATE
BERYLLIUM NITRATE (HYDRATED) [MSL]
NITRIC ACID, BERYLLIUM SALT, TRIHYDRATE [PAL]

7787-56-6 BERYLLIUM SULFATE TETRAHYDRATE

7787-71-5 BROMINE TRIFLUORIDE
BROMINE FLUORIDE (BRF3) [PAL]

7788-97-8 CHROMIC FLUORIDE

7788-98-9 AMMONIUM CHROMATE
CHROMIC ACID (H2CRO4), DIAMMONIUM SALT [PAL]

7789-00-6 POTASSIUM CHROMATE
CHROMIC ACID (H2CRO4), DIPOTASSIUM SALT [PAL]

7789-02-8 CHROMIUM (III) NITRATE
CHROMIUM NITRATE [HM1, HM1]

7789-04-0 CHROMIUM PHOSPHATE

7789-06-2 STRONTIUM CHROMATE
CHROMIC ACID (H2CRO4), STRONTIUM SALT (1:1) [PAL, PAL]

7789-09-5 AMMONIUM BICHROMATE
AMMONIUM DICHROMATE [FLA, HM1, HM1, HM1, MSL, NJL, WHM]
CHROMIC ACID (H2CR2O7), DIAMMONIUM SALT [PAL]

7789-17-5 CESIUM IODIDE

7789-18-6 CESIUM NITRATE

7789-21-1 FLUOROSULFONIC ACID
FLUOROSULPHONIC ACID [NJL, NJL]

7789-23-3 POTASSIUM FLUORIDE

7789-24-4 LITHIUM FLUORIDE

7789-27-7 THALLIUM(I) FLUORIDE

7789-29-9 POTASSIUM BIFLUORIDE
POTASSIUM HYDROGEN FLUORIDE [NJL, NJL]

7789-30-2 BROMINE PENTAFLUORIDE
BROMINE FLUORIDE (BRF5) [PAL]

7789-36-8 MAGNESIUM BROMATE

7789-38-0 SODIUM BROMATE

7789-40-4 THALLIUM BROMIDE

7789-42-6 CADMIUM BROMIDE
CADMIUM BROMIDE (CDBR2) [PAL]

7789-43-7 COBALTOUS BROMIDE
COBALT BROMIDE (COBR2) [PAL]
COBALT(II) BROMIDE [WHM]

7789-47-1 MERCURIC BROMIDE

7789-59-5 PHOSPHORUS OXYBROMIDE

7789-60-8 PHOSPHORUS TRIBROMIDE

7789-61-9 ANTIMONY TRIBROMIDE
STIBINE, TRIBROMO- [PAL]

7789-69-7 PHOSPHORUS PENTABROMIDE

7789-75-5 CALCIUM FLUORIDE (CAF2)

7789-80-2 CALCIUM IODATE

7789-82-4 CALCIUM MOLYBDATE

7789-99-3 METHYLAMYL ACETATE

7790-29-6 RUBIDIUM IODIDE

7790-30-9 THALLIUM(I) IODIDE

7790-59-2 POTASSIUM SELENATE

7790-62-7 POTASSIUM PYROSULFATE

7790-69-4 LITHIUM NITRATE

7790-80-9 CADMIUM IODIDE

7790-91-2 **CHLORINE TRIFLUORIDE**
CHLORINE FLUORIDE (CLF3) [PAL]

7790-93-4 **CHLORIC ACID**
CHLORINE DIOXIDE HYDRATE
[NJL]

7790-94-5 **CHLOROSULFONIC ACID**
4-CHLORO-O-TOLUIDINE, HY-
DROCHLORIDE [EPA]
CHLOROSULFURIC ACID [FLA,
MSL, PAL]
CHLOROSULPHONIC ACID [GBR]
PARA-CHLORO-ORTHO-TOLUIDINE
HYDROCHLORIDE [IAR]

7790-98-9 **AMMONIUM PERCHLORATE**
PERCHLORIC ACID, AMMONIUM
SALT [PAL]

7790-99-0 **IODINE MONOCHLORIDE**

7791-10-8 **STRONTIUM CHLORATE**

7791-11-9 **RUBIDIUM CHLORIDE**

7791-12-0 **THALLOUS CHLORIDE**
THALLIUM (I) CHLORIDE [EPA]
THALLIUM CHLORIDE [RCU, RCU]
THALLIUM(I) CHLORIDE [WHM]

7791-18-6 **MAGNESIUM CHLORIDE, HEX-
AHYDRATE**

7791-19-7 **MAGNESIUM CHLORIDE.6H20**

7791-20-0 **NICKEL(II) CHLORIDE HEXAHY-
DRATE (1:2:6)**

7791-21-1 **CHLORINE OXIDE (CL2O)**
CHLORINE MONOXIDE [FLA, MSL,
NFP, NFP]
CHLORINE MONOXIDE [CHLORINE
OXIDE] [EP3]

7791-23-3 **SELENIUM OXYCHLORIDE**

7791-25-5 **SULFURYL CHLORIDE**

7791-27-7 **PYROSULFURYL CHLORIDE**
PYROSULPHURYL CHLORIDE [NJL,
NJL]

7795-91-7 **2-(1-METHYLPROPOXY)
ETHANOL**

7803-49-8 **HYDROXYLAMINE**
HYDROXYLAMINE (OXAMMONIUM)
[CAI]
HYDROXYLAMMONIUM (OXAMMO-
NIUM) [TSC]

7803-51-2 **PHOSPHINE**
PHOSPHINE (HYDROGEN PHOS-
PHIDE) [OS3]

7803-52-3 **STIBINE**
ANTIMONY HYDRIDE [ONT]
STIBINE (ANTIMONY HYDRIDE)
[OS3]

7803-54-5 **MAGNESIUM DIAMIDE**

7803-55-6 **AMMONIUM VANADATE**
AMMONIUM METAVANADATE
[HM1, HM1, HM1, NJL, WHM]

7803-57-8 **HYDRAZINE MONOHYDRATE**
HYDRAZINE HYDRATE [NJL]

7803-62-5 **SILANE**
SILANE (SILICON TETRAHYDRIDE)
[QBC]
SILICON TETRAHYDRIDE [AUS,
CEX, OS1, TLV]
SILICON TETRAHYDRIDE (SILANE)
[ALB, ALB, BC1, BC1]
SILICONE TETRAHYDRIDE [MEX]

7803-63-6 **AMMONIUM HYDROGEN SUL-
FATE**
AMMONIUM BISULFATE [PST]

7803-68-1 **TELLURIC ACID**

8000-26-8 **OILS, PINE NEEDLE**

8000-27-9 **CEDAR WOOD OIL**
OIL OF CEDARWOOD [PST]

8000-29-1 **CITRONELLA OIL**
OIL OF CITRONELLA [PST]

8000-68-8 **PARSLEY APIOLE**

8001-20-5 **TUNG OIL**

8001-22-7 **SOYBEAN OIL**
SOY BEAN OIL [NFP, NFP]

8001-23-8 **SAFFLOWER OIL**

8001-25-0 **OLIVE OIL**

8001-26-1 **LINSEED OIL**
LINSEED OIL, RAW [NFP, NFP]

8001-29-4 **COTTONSEED OIL**
COTTONSEED OIL REFINED [NFP,
NFP]

8001-30-7 **CORN OIL**

8001-31-8 **COPRA**
COCONUT OIL [NFP, NFP, PST]

8001-35-2 **TOXAPHENE**
CAMPHECHLOR [302]
CAMPHECLOR [HM1, HM1]
CHLORINATED CAMPHENE [ALB,
ALB, ALB, AUS, AUS, AUS, BC1,
BC1, BC1, CEX, CEX, CEX, MAK,
MAK, MAK, MEX, MEX, MIN,
NIO, NIO, NIO, NIO, OS1, OS1,
OS1, OS1, OS1]
CHLORINATED CAMPHENE
(TOXAPHENE) [QBC, QBC, QBC,
QBC, TLV, TLV, TLV]
TOXAPHENE (CHLORINATED CAM-
PHENE) [CAA]
TOXAPHENE (POLYCHLORINATED
CAMPHENES) [C65, C65, IAR]

8001-50-1 **TERPENE POLYCHLORINATES
(STROBANE6)**

8001-54-5 **N-ALKYLDIMETHYLBENZYL AM-
MONIUM CHLORIDE**

8001-58-9 **CREOSOTE**
COAL TAR CREOSOTE [NJL, NJL]
CREOSOTE (COAL TAR) [HM1,
HM1]
CREOSOTE (COAL) [NTP]
CREOSOTE OIL [NFP, NFP]
CREOSOTE, COAL TAR [MSL]
CREOSOTES [CAL]

8001-69-2 **COD-LIVER OIL**
COD LIVER OIL [NFP, NFP]

8001-75-0 **CERESIN WAX**

8001-78-3 **HYDROGENATED CASTOR OIL**
CASTOR OIL (HYDROGENATED)
[NFP]
CASTOR OIL, HYDROGENATED
[PST]

8001-79-4 **CASTOR OIL**

8001-86-3 **ISANO OIL**
OILS, BOLEKO [PAL]

8001-97-6 **ALOE**
ALOE VERA GEL [PST]

8002-03-7 **PEANUT OIL**

8002-05-9 **PETROLEUM DISTILLATES
(NAPHTHA)**
CRUDE OIL [CN6]
MINERAL OILS [CAB, MIN]
PETROLEUM [FLA, IAR, IAR, PAL]
PETROLEUM CRUDE [MSL]
PETROLEUM DISTILLATES (NAPH-
THA) (RUBBER SOLVENT) [OS1, OS1]
PETROLEUM DISTILLATES, N.O.S.
[NJL]
PETROLEUM, CRUDE [NFP, NFP]

8002-09-3 **PINE OIL**

8002-13-9 **RAPESEED OIL**
RAPE SEED OIL [NFP, NFP]

8002-16-2 **ROSIN OIL**

8002-24-2 **SPERM OIL**
SPERM OIL NO. 2 [PAL]

8002-26-4 **TALL OIL**

8002-29-7 **TAR OILS**

8002-31-1 **COCOA**

8002-33-3 **CASTOR OIL, SULFATED**
SULFATED CASTOR OIL [PST]
TURKEY RED OIL [NFP, NFP]

8002-43-5 **SOYBEAN LECITHIN**
LECITHIN [PST]

8002-48-0 **BARLEY, MALT**

8002-50-4 **MENHADEN OIL**

8002-64-0 **NEATSFOOT OIL**

8002-72-0 **ONIONS, OIL**

8002-74-2 **PARAFFIN WAXES AND HYDRO-
CARBON WAXES**
PARAFFIN WAX [CEX, GBR, GBR,
MIN, PST]
PARAFFIN WAX (FUME) [AUS]
PARAFFIN WAX FUME [ALB, ALB,
BC1, BC1, CAL, FLA, MEX, MEX,
MSL, ONT, OS1, QBC, TLV]
WAX, PARAFFIN [NFP, NFP]

8002-75-3 **PALM OIL**

8003-03-0 **ASPIRIN, PHENACETIN, AND
CAFFIENE**

8003-19-8 **DICHLOROPROPANE-DICHLORO-
PROPENE (MIXTURE)**
1-PROPENE, 1,3-DICHLORO-, MIXT.
WITH 1,2-DICHLOROPROPANE [PAL]

8003-22-3 **D & C YELLOW NO. 11**
C.I. SOLVENT YELLOW 33 [PST]

8003-34-7 **PYRETHRUM**
PYRETHRINS (ISO) [GBR, GBR]
PYRETHRINS AND PYRETHROIDS
[PAL]

8003-69-8 **C.I. DIRECT BLACK 80**

8004-09-9 **CHLOROPICRIN/METHYL BRO-
MIDE**
CHLOROPICRIN AND METHYL BRO-
MIDE MIXTURE [OS3]

8004-13-5 **PHENYL ETHER-BIPHENYL MIX-
TURE VAPOR**
PHENYL ETHER-BIPHENYL MIXTURE
[MIN, PAL]
PHENYL ETHER-BIPHENYL MIXTURE
(VAPOR) [NIO, NIO]
PHENYL ETHER-BIPHENYL MIXTURE,
VAPOR [OS1, OS1]
PHENYL ETHER-DIPHENYL MIXTURE
(VAPOR) [BC1, BC1]
PHENYL ETHER-DIPHENYL MIXTURE
VAPOR [ALB, ALB]

8004-87-3 **METHYL VIOLET**
C.I. BASIC VIOLET 1 [PST]

8004-92-0 **QUINOLINE YELLOW**
FOOD YELLOW #13 [PST]

8005-03-6 **C.I. ACID BLACK 2**

8006-14-2 **NATURAL GAS**
GAS, NATURAL [MSL, NFP]

8006-20-0 **GAS, PRODUCER**
FUEL GASES, PRODUCER GAS [PAL]

8006-28-8 **SODA LIME**

8006-39-1 **TERPINEOL**

8006-54-0 **LANOLIN**

8006-61-9 **GASOLINE**
GASOLINE, LEADED [HM1, HM1,
HM1]
GAZOLINE [QBC, QBC, QBC]
NATURAL GASOLINE [NJL, NJL]

8006-64-2 **TURPENTINE**
TURPENTINE (WOOD) [AUS, AUS,
AUS]

8006-95-9 **WHEAT GERM OIL**

8007-02-1 LEMONGRASS OIL
OIL OF LEMONGRASS [PST]

8007-40-7 OILS, MUSTARD

8007-43-0 SORBITAN, 9-OCTADECENOATE
SORBITAN SESQUIOLEATE [PST]

8007-45-2 COAL TAR
COAL TAR (COAL TAR PITCH) [NJL]
COAL TAR PITCH VOLATILES [CAL, ONT]
COAL TARS [MIN]
COAL-TARS [IAR]
TAR, COAL [PAL, PAL]

8007-56-5 NITROHYDROCHLORIC ACID (AQUA REGIA)
NITROHYDROCHLORIC ACID [HM1, HM1, NJL, NJL]

8007-70-3 OIL OF ANISE

8007-75-8 BERGAMOT OIL

8007-80-5 CINNAMON BARK OIL

8007-93-0 BELLADONNA LEAF

8008-20-6 KEROSENE
FUEL OIL NO. 1 [NFP, NFP]
KEROSENE (DEODORIZED) [PST]
KEROSINE [FLA, MSL]
KEROSINE (PETROLEUM) [PAL]
NAVY FUELS JP-5 [CSR, CSR]

8008-51-3 CAMPHOR OIL
CAMPHOR OIL (LIGHT) [FLA, MSL, NFP, NFP]
OILS, CAMPHOR [PAL]

8008-56-8 LEMON OIL
OIL OF LEMON [PST]

8008-57-9 ORANGE OIL

8008-74-0 SESAME OIL

8008-79-5 SPEARMINT OIL

8008-89-7 VEGETABLE OIL MISTS

8009-03-8 PETROLATUM

8011-48-1 PINE TAR

8012-45-1 PARAFFIN OIL
LUBRICATING OIL, MINERAL [NFP, NFP]

8012-74-6 LONDON PURPLE

8012-89-3 WHITE BEESWAX
BEESWAX [PST]

8012-95-1 OIL MIST, MINERAL
MINERAL OIL [NFP, NFP]
MINERAL OIL MIST [PAL, QBC, QBC]
OIL (MINERAL) MIST [CEX]
OIL MIST [CAL]
OIL MIST (MINERAL) [NIO, NIO]
OIL, MINERAL [WHM]
OIL, MINERAL - MIST [ONT, ONT, ONT, ONT]

8013-07-8 EPOXIDIZED SOYBEAN OIL

8013-17-0 INVERT SUGAR

8013-75-0 FUSEL OIL

8013-76-1 OIL OF BITTER ALMOND

8013-90-9 IONONE

8014-91-3 C.I. DIRECT BROWN
BENZOIC ACID, 3,3'-[(3,7-DISULFO-1,5-NAPHTHALENE-DIYL)BIS[AZO(6-HYDROXY-3,1-PHENYLENE)AZO[6(OR 7)-SULFO-4,1- NAPHTHALENEDIYL]AZO[1,1'-BIPHENYL]-4,4'-DIYLAZO]BIS [6- HYDROXY-, HEXASODIUM SALT [TSC]
C.I. DIRECT BROWN 74 [NJL, NJL]

8014-95-7 SULFURIC ACID
OLEUM [OS3]
OLEUM (FUMING SULFURIC ACID) [FUMING SULFURIC ACID, MIXTURE WITH SULFUR TRIOXIDE] [EP3]
SULFURIC ACID (MIXTURE WITH SULFUR TRIOXIDE) [MSL]
SULFURIC ACID, FUMING [NJL]

8015-86-9 CARNAUBA WAX

8016-11-3 EPOXIDIZED LINSEED OIL

8016-13-5 FISH OIL

8016-28-2 OILS, LARD
LARD OIL [NFP, NFP]

8016-70-4 HYDROGENATED SOYBEAN OIL

8018-01-7 MANCOZEB

8020-83-5 HYDROCARBON OILS, WHITE MINERAL OIL
HYDROCARBON OILS [PAL, PAL]

8021-27-0 ABIES ALBA OIL

8021-39-4 CREOSOTE, WOOD
CREOSOTE (WOOD TAR) [HM1]
CREOSOTE (WOOD) [NTP]

8021-92-9 FUEL GASES, WATER GAS

8022-00-2 METHYL DEMETON
DEMETHON-METHYL [MAK, MAK, MAK]
DEMETON-METHYL, MIXTURE OF O AND S [QBC, QBC]
PHOSPHOROTHIOIC ACID, O-[2-(ETHYLTHIO)ETHYL] O,O-DIMETHYLESTER, MIXT. WITH S-[2-(ETHYLTHIO)ETHYL] O,O-DIMETHYL PHOSPHOROTHIOATE [PAL]

8022-96-6 ESSENTIAL OILS

8023-53-8 DICHLOROBENZALKONIUM CHLORIDE

8023-74-3 MINK OIL

8023-79-8 PALM KERNEL OIL

8024-37-1 TURMERIC, OLEORESIN (CURCUMIN)

8024-43-9 PERFUMES AND ESSENCES, JASMIN
PERFUMES, JASMIN [PST]

8025-81-8 SPIRAMYCIN

8028-66-8 HONEY

8028-89-5 CARAMEL

8029-43-4 CORN SYRUP

8030-30-6 BENZIN
BENZINE [NJL]
NAPHTHA [HM1, HM1, HM1, PAL, PST]
NAPHTHA (COAL TAR) [NIO, NIO, NIO, OS1, OS1]
NAPHTHA (VM&P NAPHTHA) [MIN]
NAPHTHA V.M. & P. [NFP, NFP]
NAPHTHA VM&P [MSL]
RUBBER SOLVENT (NAPHTHA) [TLV]
V.M. & P. NAPHTHA [ONT]
VM & P (VARNISH MAKERS & PAINTERS) NAPHTHA [CAL]
VM & P (VARNISH MAKERS AND PAINTERS) NAPHTHA [CEX, CEX]
VMP NAPHTHA (PAINT AND VARNISH MANUFACTURING) [QBC]

8030-31-7 NAPHTHA 49 DEGREE BE-COAL TAR TYPE
NAPHTHA, COAL TAR [CAL, CEX]

8031-18-3 FULLERS EARTH
ATTAPULGITE/PALYGORSKITE [MAK]

8032-32-4 LIGROINE
ALIPHATIC NAPHTHA [NJL]
MINERAL SPIRITS [NFP, NFP]
VM & P NAPHTHA [ALB, ALB, OS1, OS1, TLV]
VM&P NAPHTHA [MEX, MEX, WHM]

8037-19-2 OILS, TOBACCO
TOBACCO LEAF, ABSOLUTE [MSL]

8042-47-5 WHITE MINERAL OIL
WHITE MINERAL OIL (PETROLEUM) [PST]

8047-67-4 SACCHARATED IRON OXIDE

8047-99-2 N-ETHYLTOLUENESULFON-AMIDE

8048-52-0 ACRIFLAVINIUM CHLORIDE

8049-17-0 FERROSILICON

8049-47-6 PANCREATIN

8049-99-8 MILORGANITE

8050-07-5 GUM THUS

8050-09-7 ROSIN
GUM ROSIN [PST]

8050-13-3 METHYL ESTER OF ROSIN, PARTIALLY HYDROGENATED

8050-33-7 POLYOXYETHYLENE ESTER OF ROSIN

8050-81-5 SIMETHICONE

8050-93-9 IRON DEXTRIN COMPLEX

8052-10-6 TALL OIL ROSIN

8052-35-5 MOLASSES

8052-41-3 STODDARD SOLVENT
CLEANING SOLVENT, STODDARD SOLVENT [NFP, NFP]
WHITE SPIRIT [GBR, GBR]
WHITE SPIRIT, LOW (15-20%) AROMATIC [HM1]
WHITE SPIRITS [AUS, AUS]

8052-42-4 ASPHALT
ASPHALT (PETROLEUM FUMES) [ALB, ALB]
ASPHALT (PETROLEUM) FUMES [BC1, BC1, CAL, CEX, MEX, MEX, MIN, QBC, TLV]
ASPHALT FUMES [FLA, MSL, NHS, NHS, NJL, ONT]
BITUMEN [MAK]
BITUMEN FUMES [AUS]

8052-48-0 FATTY ACIDS, TALLOW, SODIUM SALTS
SODIUM TALLOW SOAP [PST]

8061-51-6 SODIUM LIGNO SULFONATE
LIGNOSULFONIC ACID, SODIUM SALT [PST]

8061-52-7 CALCIUM LIGNOSULFONATE
LIGNOSULFONIC ACID, CALCIUM SALT [PST]

8061-53-8 LIGNOSULFONIC ACID, AMMONIUM SALT

8061-54-9 LIGNOSULFONIC ACID, MAGNESIUM SALT

8062-15-5 LIGNOSULFONATE
LIGNOSULFONIC ACID [PST]

8064-90-2 TRIMETHOPRIM/SULFAMETHOXAZOLE

8065-48-3 DEMETON
DEMETON (SYSTOX) [BC1, BC1, BC1, OS1, OS1, OS1, OS1]
DEMETON, MIXTURE OF O AND S (SYSTOX) (R) [QBC, QBC]
PHOSPHOROTHIOIC ACID, O,O-DIETHYL O-[2-(ETHYLTHIO)ETHYL] ESTER, MIXT. WITH O,O-DIETHYL S-[2-(ETHYLTHIO)ETHYL] PHOSPHOROTHIOATE [PAL]

8066-33-9 PENTOLITE

8068-05-1 LIGNIN SULPHATE

8070-50-6 CARBON DIOXIDE AND ETHYLENE OXIDE MIXTURES

9000-01-5 GUM ARABIC

9000-07-1 **CARRAGEENAN**
CARRAGEENAN (DEGRADED) [MSL]
CARRAGEENAN, DEGRADED [IAR]

9000-11-7 **CARBOXYMETHYL CELLULOSE**
CELLULOSE, CARBOXYMETHYL
ETHER [PST]

9000-28-6 **GUM GHATTI**

9000-29-7 **GUAIAC GUM**

9000-30-0 **GUAR GUM**

9000-36-6 **KARAYA**

9000-40-2 **LOCUST BEAN GUM**

9000-59-3 **SHELLAC**

9000-65-1 **GUM TRAGACANTH**

9000-69-5 **PECTIN**

9000-70-8 **GELATIN**

9000-71-9 **CASEIN**

9001-73-4 **PAPAIN**

9002-18-0 **AGAR**

9002-68-0 **MENOTROPINS**

9002-81-7 **POLY(OXYMETHYLENE)**
POLYOXYMETHYLENE [PST]

9002-84-0 **POLYTETRAFLUOROETHYLENE**
ETHENE, TETRAFLUORO-, HO-
MOPOLYMER [PAL]
POLYTETRAFLUOROETHYLENE
DECOMPOSITION PRODUCTS [BC1]

9002-86-2 **POLYVINYL CHLORIDE**
POLYVINYL CHLORIDE (PVC) [GBR]
POLYVINYL CHLORIDE RESIN [PST]
PVC [MAK]

9002-88-4 **POLYETHYLENE**

9002-89-5 **POLYVINYL ALCOHOL**

9002-91-9 **ACETALDEHYDE, HOMOPOLY-
MER**

9002-92-0 **LAURYL POLYETHYLENE GLY-
COL ETHER**
DODECYL ALCOHOL, ETHOXY-
LATED [PST]
ETHOXYLATED LAURYL ALCOHOL
[WHM]
POLYOXYETHYLENE LAURYL
ETHER [NFP, NFP]

9002-93-1 **POLYETHYLENE GLYCOL
OCTYLPHENOL ETHER**
4-(1,1,3,3-TETRAMETHYLBUTYL)
PHENOL, ETHOXYLATED [PST]

9002-98-6 **POLYETHYLENEIMINE**

9003-01-4 **ACRYLIC RESIN**
ACRYLIC ACID POLYMER [PST]
POLYACRYLIC ACID [IAR]

9003-04-7 **SODIUM POLYACRYLATE**
ACRYLIC ACID POLYMER, SODIUM
SALT [PST]

9003-05-8 **POLYACRYLAMIDES**
POLYACRYLAMIDE [PST]

9003-06-9 **ACRYLAMIDE-ACRYLIC ACID
POLYMER**
ACRYLAMIDE-ACRYLIC ACID RESIN
[PST]

9003-07-0 **POLYPROPYLENE**

9003-11-6 **POLYETHYLENE-POLYPROPY-
LENE GLYCOL**
BLOCK POLYMER OF POLYPROPY-
LENE OXIDE AND POLYETHYLENE
OXIDE [PST]

9003-13-8 **POLYPROPYLENE GLYCOL,
MONOBUTYL ETHER**
BUTOXYPOLYPROPYLENE GLYCOL
[PST]

9003-17-2 **POLYBUTADIENE RESIN**
BUTADIENE RESIN [PST]

9003-22-9 **VINYL CHLORIDE-VINYL AC-
ETATE COPOLYMER**

9003-28-5 **POLYBUTYLENE**

9003-29-6 **POLYBUTENE**

9003-32-1 **ETHYL ACRYLATE POLYMER**

9003-39-8 **POLYVINYL PYRROLIDONE**
POLYVINYLPYRROLIDONE [PST]

9003-49-0 **POLYBUTYL ACRYLATE**
POLYMERIZED BUTYL ACRYLATE
[PST]

9003-53-6 **POLYSTYRENE**
POLYSTYRENE BEADS, EXPANDED,
MIXTURE WITH FLAMMABLE LIQUID
[NJL]
POLYSTYRENE RESIN [PST]

9003-54-7 **2-PROPENENITRILE POLYMER
WITH ETHENYLBENZENE**
STYRENE-ACRYLONITRILE COPOLY-
MERS [IAR]

9003-55-8 **STYRENE-BUTADIENE POLYMER**
BUTADIENE-STYRENE COPOLYMER
[PST]
STYRENE-BUTADIENE COPOLYMERS
[IAR]

9003-56-9 **ABS RESIN**
ACRYLONITRILE-BUTADIENE-
STYRENE COPOLYMERS [IAR]

9003-63-8 **METHACRYLIC ACID, BUTYL
ESTER, POLYMER**
N-BUTYL METHACRYLATE, POLY-
MERIZED [PST]

9003-70-7 **STYRENE-DIVINYLBENZENE
COPOLYMER**
STYRENE-DIVINYL BENZENE
COPOLYMER RESIN MATRIX [PST]

9004-32-4 **SODIUM CARBOXYMETHYL CEL-
LULOSE**
CELLULOSE CARBOXYMETHYL
ETHER, SODIUM SALT [PST]

9004-34-6 **CELLULOSE**
CELLULOSE (PAPER FIBER) [ALB,
BC1, BC1, MEX, MEX, QBC]
CELLULOSE (PAPER FIBRE) [AUS]

9004-36-8 **CELLULOSE ACETATE BU-
TYRATE**
CELLULOSE ACETATE BUTYRATE
[PST]

9004-53-9 **DEXTRIN**

9004-57-3 **ETHYLCELLULOSE**
CELLULOSE, ETHYL ETHER [PST]

9004-58-4 **ETHYL HYDROXYETHYL CELLU-
LOSE**
CELLULOSE, ETHYL 2-HYDROX-
YETHYL ETHER [PST]

9004-62-0 **HYDROXYETHYL CELLULOSE**
CELLULOSE, 2-HYDROXYETHYL
ETHER [PST]

9004-64-2 **HYDROXYPROPYL ETHER OF
CELLULOSE**
CELLULOSE, 2-HYDROXYPROPYL
ETHER [PST]

9004-65-3 **CELLULOSE HYDROXYPROPYL
METHYL ETHER**
2-HYDROXYPROPYL METHYL CEL-
LULOSE [PST]

9004-66-4 **IRON DEXTRAN COMPLEX**
IRON DEXTRAN [CAL, FLA, MSL,
PAL, PAL]
IRON-DEXTRAN COMPLEX [IAR]

9004-67-5 **METHYLCELLULOSE**
METHYL CELLULOSE [PST]

9004-70-0 **NITROCELLULOSE**
CELLULOSE NITRATE [OS3]

CELLULOSE, NITRATE [PAL]
COLLODION [FLA, MSL, NFP, NFP]

9004-73-3 **METHYLPOLYSILOXANE**

9004-74-4 **POLYETHYLENE GLYCOL
METHYL ETHER**

9004-77-7 **POLYETHYLENE GLYCOL,
MONOBUTYL ETHER**
POLYOXYETHYLENE MONOBUTYL
ETHER [PST]

9004-81-3 **POLYETHYLENE GLYCOL MONO-
LAURATE**
POLYOXYETHYLENE MONOLAU-
RATE [PST]

9004-82-4 **DODECYLOXYPOLY(ETHYLE-
NEOXY) ETHYL SULFATE,
SODIUM SALT**

9004-95-9 **POLYOXYETHYLENE MONOHEX-
ADECYL ETHER**

9004-96-0 **POLYETHYLENE GLYCOL
MONOOLEATE**
POLYOXYETHYLENE MONOLEATE
[PST]

9004-98-2 **POLYETHYLENE GLYCOL
MONOOLEYL ETHER**
POLYOXYETHYLENE MONO(CIS-9-
OCTADECENYL) ETHER [PST]

9004-99-3 **POLYETHYLENE GLYCOL
STEARATE**
POLYOXYETHYLENE MONOS-
TEARATE [PST]

9005-00-9 **POLYOXYETHYLENE MONOOC-
TADECYL ETHER**

9005-07-6 **PEG (400) DIOLEATE**
POLYOXYETHYLENE DIOLEATE
[PST]

9005-08-7 **POLYOXYETHYLENE DIS-
TEARATE**

9005-25-8 **STARCH**
CORNSTARCH [PST]

9005-37-2 **PROPYLENE GLYCOL ALGINATE**

9005-38-3 **ALGENIC ACID, SODIUM SALT**
SODIUM ALGINATE [PST]

9005-42-9 **AMMONIUM CASEINATE**

9005-46-3 **SODIUM CASEINATE**

9005-64-5 **POLYSORBATE 20**
POLYOXYETHYLENE SORBITAN
MONOLAURATE [PST]

9005-65-6 **POLYOXYETHYLENE 20 SORBI-
TAN MONOOLEATE**
POLYOXYETHYLENE SORBITAN
MONOLEATE [PST]
POLYSORBATE 80 (GLYCOL) [CSR,
CSR]

9005-66-7 **POLYOXYETHYLENE SORBITAN
MONOPALMITATE**

9005-67-8 **POLYOXYETHYLENE SORBITAN
MONOSTEARATE**

9005-70-3 **POLYOXYETHYLENE SORBITAN
TRIOLEATE**

9005-71-4 **POLYOXYETHYLENE SORBITAN
TRISTEARATE**

9005-84-9 **AMYLODEXTRIN**
STARCH [GBR]

9005-90-7 **TURPENTINE (RESIN FROM PI-
NUS SPECIES PARTICULARLY
PALUSTRIS)**
MINERAL TURPENTINE [AUS, AUS]
TURPENTINE [PAL, PST]
TURPENTINE SUBSTITUTE [HM1,
HM1, HM1, NJL]

9006-03-5 **CHLORINATED RUBBER**

9006-04-6 NATURAL RUBBERS
RUBBER [AUS, GBR, HM1, HM1, NJL, PST]

9006-26-2 2,5-FURADIONE POLYMER WITH ETHENE

9006-42-2 METIRAM

9006-50-2 ALBUMIN EGG
EGG WHITE [PST]

9006-65-9 DIMETHYL SILICONE
DIMETHICONE [TSC]

9007-13-0 CALCIUM RESINATE
CALCIUM SALT OF TALL-OIL ROSIN [PST]

9007-16-3 ACRYLIC ACID-SUCROSE POLYALLYL ETHER POLYMER

9007-48-1 POLYGLYCEROL ESTER OF OLEIC ACID

9008-34-8 MANGANESE RESINATE

9008-75-7 NYLON

9009-86-3 RICIN

9010-66-6 ZEIN

9010-69-9 ZINC RESINATE
RESIN AND ROSIN ACIDS, ZINC SALTS [PST]

9010-77-9 ACRYLIC ACID POLYMER WITH ETHYLENE
ETHYLENE - ACRYLIC ACID COPOLYMER [PST]

9010-79-1 ETHYLENE-PROPYLENE POLY-MER

9010-88-2 ETHYL ACRYLATE-METHYL METHACRYLATE POLYMER
ETHYL ACRYLATE, COPOLYMER WITH METHYL METHACRYLATE [PST]

9010-98-4 NEOPRENE RUBBER
POLYCHLOROPRENE [IAR]

9011-05-6 UREA, POYLMER WITH FORMALDEHYDE
UREA-FORMALDEHYDE RESIN [PST]
UREA-FORMALDEHYDE RESINS [OTS]

9011-06-7 VINYLIDENE CHLORIDE-VINYL CHLORIDE COPOLYMERS

9011-13-6 MALEIC ANHYDRIDE-STYRENE POLYMER
STYRENE-MALEIC ANHYDRIDE RESIN [PST]

9011-14-7 METHYL METHACRYLATE POLY-MER
POLYMETHYL METHACRYLATE [IAR, PST]

9011-16-9 MALEIC ANHYDRIDE - METHYLVINYL ETHER COPOLY-MER

9014-01-1 SUBTILISINS (PROTEOLYTIC ENZYMES)
SUBTILISIN [PST]
SUBTILISINS [CEX]
SUBTILISINS (PROTEOLYTIC EN-ZYMES AS 100% PURE CRYSTALLINE ENZYME) [GBR, GBR, ONT]

9014-85-1 POLYOXYETHYLENE 2,4,7,9-TE-TRAMETHYL-5-DECYNE-4,7-DIOL

9014-90-8 POLYETHYLENE GLYCOL NONYLPHENYL ETHER SODIUM SULFATE
NONYLPHENOL, ETHOXYLATED, SULFATED, SODIUM SALT [PST]

9014-92-0 ALPHA-(DODECYLPHENYL)-OMEGA-HYDROXY-POLY(OXY-1,2-ETHANEDIYL)
POLYOXYETHYLENE DODECYLPHE-NOL [PST]

9014-93-1 POLYOXYETHYLENE DI-NONYLPHENOL

9015-54-7 HYDROLYZED PROTEIN

9015-68-3 ASPARAGINASE

9015-98-9 POLY(OXYMETHYLENE), .AL-PHA.-HYDRO-.OMEGA.-HY-DROXY-

9016-00-6 POLY[OXY(DIMETHYLSILYLENE)]
POLYDIMETHYLSILOXANE [TSC, TSC]

9016-45-9 ETHYLENE OXIDE-NONYLPHE-NOL POLYMER
POLYOXYETHYLENE NONYLPHENOL [PST]

9016-87-9 POLYMETHYLENE POLYPHENY-LENE ISOCYANATE
ISOCYANIC ACID, POLY-METHYLENEPOLYPHENYLENE ES-TER [TSC, TSC]
POLYMERIC DIPHENYLMETHANE DIISOCYANATE [313]
POLYMERIC MDI [MAK]
POLYMETHYLENE POLYPHENYL ISOCYANATE [IAR]
POLYMETHYLENEPOLYPHENYLENE ISOCYANATE [PST]

9017-09-8 POLYURETHANE

9017-33-8 NAPHTHALENESULFONIC ACID, POLYMER WITH FORMALDE-HYDE

9036-19-5 ETHOXYLATED OCTYLPHENOL
POLYETHYLENE GLYCOL OCTYLPHENYL ETHER [WHM]
POLYOXYETHYLENE MONOOCTYLPHENYL ETHER [PST]

9038-29-3 OXIRANE, METHYL-, POLYMER WITH OXIRANE, DECYL ETHER
(DECYLOXY)POLY(OXYETHYLENE) POLY(OXYPROPYLENE) [PST]

9038-95-3 ETHYLENE OXIDE-PROPY-LENE OXIDE COPOLYMER MONOBUTYL ETHER
POLYETHYLENE-POLYPROPYLENE GLYCOLS MONOBUTYL ETHER [PST]

9040-80-6 TRIMETHYLOLPROPANE, POLY-MER WITH TOLUENE DIISO-CYANATE AND POLYPROPYLENE GLYCOL

9046-01-9 TRIDECYL ALCOHOL, ETHOXY-LATED, PHOSPHATED

9049-05-2 ALGIN GUM

9049-76-7 STARCH HYDROXYPROPYL ETHER

9051-57-4 AMMONIUM NONYLPHENYL POLYOXYETHYLENE SULFATE

9056-38-6 NITROSTARCH

9080-17-5 AMMONIUM POLYSULFIDE

9081-17-8 NONYLPHENOL, ETHOXYLATED AND SULFATED

9084-06-4 FORMALDEHYDE-NAPHTHALE-NESULFONIC ACID POLYMER SODIUM SALT
NAPHTHALENESULFONIC ACID, POLYMER WITH FORMALDEHYDE, SOIDUMSALT [PST]

10017-56-8 TRIETHANOLAMINE PHOSPHATE

10020-43-6 ETHYLENE GLYCOL MONOCTYL ETHER

10022-31-8 BARIUM NITRATE
NITRIC ACID, BARIUM SALT [PAL]

10022-70-5 SODIUM HYPOCHLORITE (PEN-TAHYDRATE)
HYPOCHLOROUS ACID, SODIUM SALT, PENTAHYDRATE [PAL]
SODIUM HYPOCHLORITE [EPA, NJL]

10024-93-8 NEODYMIUM CHLORIDE

10024-97-2 NITROUS OXIDE
NITROGEN MONOXIDE [GBR, GBR]
NITROGEN OXIDE [NHS, NHS]

10025-65-7 PLATINOUS CHLORIDE
PLATINIUM(II) CHLORIDE [WHM]

10025-67-9 SULFUR MONOCHLORIDE
DISULPHUR DICHLORIDE [GBR]
SULFUR CHLORIDE [NJL, NJL, PST, WHM]
SULFUR CHLORIDE (S2CL2) [PAL]
SULPHUR MONOCHLORIDE [AUS]

10025-69-1 STANNOUS CHLORIDE DIHY-DRATE

10025-73-7 CHROMIC CHLORIDE
CHROMIC (III) CHLORIDE [MSL]
CHROMIUM(III) CHLORIDE [WHM]

10025-76-0 EUROPIUM CHLORIDE

10025-78-2 TRICHLOROSILANE
SILANE, TRICHLORO- [PAL]
TRICHLOROSILANE [SILANE, TRICHLORO-] [EP3]

10025-85-1 NITROGEN TRICHLORIDE

10025-87-3 PHOSPHORUS OXYCHLORIDE
PHOSPHORUS OXYCHLORIDE [PHOS-PHORYL CHLORIDE] [EP3]
PHOSPHORYL CHLORIDE [PAL]
PHOSPHORYL TRICHLORIDE [GBR, GBR]

10025-91-9 ANTIMONY TRICHLORIDE
ANTIMONY(III) TRICHLORIDE [WHM]
STIBINE, TRICHLORO- [PAL]

10025-97-5 IRIDIUM TETRACHLORIDE

10025-98-6 POTASSIUM PALLADIUM CHLO-RIDE

10025-99-7 POTASSIUM TETRACHLORO-PLATINATE(II)

10026-04-7 SILICON TETRACHLORIDE

10026-06-9 STANNIC CHLORIDE, HYDRATED

10026-07-0 TELLURIUM TETRACHLORIDE

10026-11-6 ZIRCONIUM TETRACHLORIDE
ZIRCONIUM CHLORIDE (ZRCL4), (T-4) - [PAL]

10026-12-7 NIOBIUM CHLORIDE

10026-13-8 PHOSPHORUS PENTACHLORIDE
PHOSPHORANE, PENTACHLORO-[PAL]

10026-17-2 COBALT(II) FLUORIDE

10026-22-9 COBALT NITRATE

10026-24-1 COBALT(II) SULFATE (1:1), HEP-TAHYDRATE
COBALT SULFATE HEPTAHYDRATE [CSR]

10028-15-6 OZONE

10028-17-8 TRITIUM

10028-22-5 FERRIC SULFATE
SULFURIC ACID, IRON(3+) SALT (3:2) [PAL]

10031-13-7 LEAD ARSENITE

10031-16-0 BARIUM DICHROMATE

10031-18-2 MERCURY BROMIDE (HGBR)
  MERCURY BROMIDE [NJL]

10031-43-3 COPPER(II) NITRATE, TRIHYDRATE (1:2:3)

10031-59-1 THALLIUM SULFATE
  SULFURIC ACID, THALLIUM SALT [PAL]
  THALLOUS SULFATE [NJL]

10031-75-1 BENZENE, 1,1'-(DIISOCYANATOMETHYLENE)BIS-

10031-82-0 BENZALDEHYDE, 4-ETHOXY-

10031-87-5 ACETIC ACID, 2-ETHYLBUTYL ESTER
  2-ETHYLBUTYL ACETATE [NFP, NFP]

10034-76-1 CALCIUM SULFATE HEMIHYDRATE

10034-81-8 MAGNESIUM PERCHLORATE
  PERCHLORIC ACID, MAGNESIUM SALT [PAL]

10034-85-2 HYDRIODIC ACID
  HYDRIOTIC ACID [FDA]
  HYDROGEN IODIDE [HM1, HM1, HM1, WHM]

10034-93-2 HYDRAZINE SULFATE
  HYDRAZINE, SULFATE (1:1) [PAL, PAL]

10034-96-5 MANGANESE SULFATE MONOHYDRATE

10034-99-8 MAGNESIUM SULFATE HEPTAHYDRATE

10035-10-6 HYDROGEN BROMIDE
  HYDROBROMIC ACID [HM1, HM1, HM1, PAL]

10036-47-2 TETRAFLUOROHYDRAZINE

10038-98-9 GERMANIUM TETRACHLORIDE

10039-32-4 PHOSPHORIC ACID, DISODIUM SALT, DODECAHYDRATE
  PHOSPHORIC ACID, DISODIUM SALT, DECAHYDRATE [MSL]
  SODIUM PHOSPHATE, DIBASIC [NJL]

10039-54-0 HYDROXYLAMINE SULFATE (2:1)
  HYDROXYLAMINE SULFATE [HM1, HM1, NJL, NJL, PST, WHM]
  HYDROXYLAMINE SULFATE (2:1) (HYDROXYLAMMONIUM) [CAI]

10042-76-9 STRONTIUM NITRATE
  NITRIC ACID, STRONTIUM SALT [PAL]

10042-88-3 TERBIUM CHLORIDE

10043-01-3 ALUMINUM SULFATE
  SULFURIC ACID, ALUMINUM SALT (3:2) [PAL]

10043-35-3 BORIC ACID (H3BO3)
  BORIC ACID [CSR, CSR]

10043-49-9 GOLD 198

10043-52-4 CALCIUM CHLORIDE

10043-66-0 IODINE 131

10043-67-1 POTASSIUM ALUMINUM SULFATE

10043-92-2 RADON

10045-89-3 FERROUS AMMONIUM SULFATE
  SULFURIC ACID, AMMONIUM IRON (2+) SALT (2:2:1) [PAL]

10045-94-0 MERCURIC NITRATE
  NITRIC ACID, MERCURY(2+) SALT [PAL]

10045-97-3 CESIUM 137

10046-00-1 HYDROXYLAMINE SULFATE (1:1)
  HYDROXYLAMINE, SULFATE (1:1) (HYDROXYLAMINE ACID SULFATE) [CAI]

10048-13-2 STERIGMATOCYSTIN
  7H-FURO[3',2':4,5]FURO[2,3-C]XANTHEN-7-ONE, 3A,12C-DIHYDRO-8-HYDROXY-6-METHOXY, (3AR-CIS)- [PAL, PAL]

10048-32-5 PARASORBIC ACID

10048-95-0 DISODIUM HYDROGEN ARSENATE

10049-04-4 CHLORINE DIOXIDE
  CHLORINE DIOXIDE [CHLORINE OXIDE (CLO2)] [EP3]
  CHLORINE OXIDE (CLO2) [PAL]

10049-05-5 CHROMIC(II) CHLORIDE
  CHROMIUM CHLORIDE (CRCL2) [PAL]
  CHROMOUS CHLORIDE [CWA, EPA, NJL]

10049-07-7 RHODIUM TRICHLORIDE
  RHODIUM CHLORIDE [ONT, ONT, ONT]

10058-23-8 MONOPOTASSIUM PEROXYMONOSULFATE

10061-01-5 CIS-1,3-DICHLOROPROPENE
  1,3-DICHLOROPROPENE, CIS- [ATS]

10061-02-6 TRANS-1,3-DICHLOROPROPENE

10094-62-9 ALPHA,DELTA-GLUCOHEPTONIC ACID, SODIUM SALT, DIHYDRATE

10098-91-6 YTTRIUM 90

10098-97-2 STRONTIUM 90

10099-58-8 LANTHANUM CHLORIDE

10099-59-9 LANTHANUM NITRATE

10099-70-4 DIISOPROPYL MALEATE

10099-71-5 DIAMYL MALEATE

10099-74-8 LEAD NITRATE
  NITRIC ACID, LEAD (2+) SALT [MSL]
  NITRIC ACID, LEAD(2+) SALT [PAL]

10099-79-3 LEAD VANADATE

10101-41-4 CALCIUM SULFATE DIHYDRATE
  CALCIUM SULFATE, DIHYDRATE (GYPSUM) [PST]
  CALCIUM SULFATE, INCLUDING PLASTER OF PARIS [ONT]
  GYPSUM [ALB, BCl, BCl, GBR, MEX, OS1]

10101-50-5 SODIUM PERMANGANATE
  PERMANGANIC ACID, SODIUM SALT [PST]

10101-53-8 CHROMIC SULFATE
  CHROMIUM SULFATE [MSL]
  CHROMIUM(III) SULFATE [WHM]
  SULFURIC ACID, CHROMIUM(3+) SALT (3:2) [PAL]

10101-63-0 LEAD IODIDE
  LEAD IODIDE (PBI2) [PAL]

10101-75-4 TIN IV CHROMATE

10101-89-0 TRIBASIC SODIUM PHOSPHATE DODECAHYDRATE
  PHOSPHORIC ACID, TRISODIUM SALT, DODECAHYDRATE [MSL, PAL]
  SODIUM PHOSPHATE, TRIBASIC [NJL]

10101-97-0 NICKEL(II) SULFATE HEXAHYDRATE (1:1:6)
  NICKEL SULFATE HEXAHYDRATE [CSR]

10102-05-3 PALLADIUM DINITRATE

10102-06-4 URANYL NITRATE
  URANIUM BIS(NITRATO-O)DIOXO-, (T-4)- [PAL]

10102-18-8 SODIUM SELENITE

10102-20-2 SODIUM TELLURITE

10102-43-9 NITRIC OXIDE
  NITRIC OXIDE (NO) [QBC]
  NITRIC OXIDE [NITROGEN OXIDE (NO)] [EP3]
  NITROGEN OXIDE (NO) [PAL, RCU, RCU]

10102-44-0 NITROGEN DIOXIDE
  NITROGEN OXIDE (NO2) [PAL]

10102-45-1 THALLIUM(I) NITRATE
  THALLIUM NITRATE [HM1, HM1, NJL]

10102-48-4 LEAD ARSENATE (PB3(ASS04)2)
  ARSENIC ACID (H3ASO4), LEAD(4+) SALT (3:2) [PAL]
  LEAD (IV) ARSENATE [MSL]
  LEAD ARSENATE [FLA, MIN]

10102-49-5 FERRIC ARSENATE

10102-50-8 FERROUS ARSENATE

10102-53-1 METAARSENIC ACID

10103-46-5 CALCIUM PHOSPHATE

10103-50-1 MAGNESIUM ARSENATE

10103-62-5 CALCIUM ARSENATE

10107-99-0 ABIETIC ACID, DIETHYLENE GLYCOL ESTER

10108-56-2 N-BUTYLCYCLOHEXYLAMINE
  CYCLOHEXANAMINE, N-BUTYL- [PAL]

10108-64-2 CADMIUM CHLORIDE
  CADMIUM CHLORIDE (CDCL2) [PAL, PAL]

10108-73-3 CERIUM NITRATE

10112-91-1 MERCURY CHLORIDE

10117-38-1 POTASSIUM SULFITE

10118-76-0 CALCIUM PERMANGANATE

10119-31-0 ZIRCONIUM CHLORIDE HYDROXIDE

10124-36-4 CADMIUM SULFATE
  CADMIUM SULPHATE [FLA, MSL]
  SULFURIC ACID, CADMIUM SALT (1:1) [PAL, PAL]

10124-37-5 CALCIUM NITRATE

10124-41-1 CALCIUM THIOSULFATE

10124-43-3 COBALT SULFATE
  COBALT(II) SULFATE [WHM]
  SULFURIC ACID, COBALT (2+) SALT (1:1) [TSC, TSC]

10124-48-8 MERCURY AMMONIUM CHLORIDE

10124-50-2 POTASSIUM ARSENITE
  ARSONIC ACID, POTASSIUM SALT [PAL]
  POTASSIUM ARSONITE [EPA]

10124-56-8 SODIUM HEXAMETAPHOSPHATE
  METAPHOSPHORIC ACID (H6P6O18), HEXASODIUM SALT [PAL]
  METAPHOSPHORIC ACID, HEXASODIUM SALT [MSL]
  SODIUM PHOSPHATE, TRIBASIC [NJL]
  SODIUM PHOSPHATE, TRIBASIC (SODIUM HEXAMETAPHOSPHATE) [CWA]

10137-69-6 CYCLOHEXENYL TRICHLOROSILANE

10137-74-3 CALCIUM CHLORATE
  CHLORIC ACID, CALCIUM SALT [PAL]

10137-80-1 BENZENAMINE, N-(2-ETHYL-
HEXYL)-
N-2-(ETHYLHEXYL) ANILINE [FLA,
MSL, NFP, NFP]

10138-04-2 FERRIC AMMONIUM SULFATE

10138-52-0 GADOLINIUM TRICHLORIDE

10138-74-6 N-(2-HYDROXYETHYL) PROPY-
LENE DIAMINE
ETHANOL, 2-[(2-AMINO-1-
METHYLETHYL)AMINO]- [PAL]
N-(2-HYDROXYETHYL) PROPYLENE-
DIAMINE [NFP, NFP]

10139-58-9 RHODIUM NITRATE

10140-65-5 PHOSPHORIC ACID, DISODIUM
SALT, HYDRATE
SODIUM PHOSPHATE, DIBASIC [NJL]

10140-87-1 ETHANOL, 1,2-DICHLORO-, AC-
ETATE
1,2-DICHLOROETHYL ACETATE
[MSL]

10141-00-1 CHROMIUM(III) POTASSIUM SUL-
FATE
CHROMIUM POTASSIUM SULFATE
[MSL]

10141-05-6 COBALT(II) NITRATE
COBALTOUS NITRATE [FLA]
NITRIC ACID, COBALT(2+) SALT
[PAL]

10143-60-9 DI-2-ETHYLHEXYL ETHER

10168-81-7 GADOLINIUM NITRATE

10190-55-3 LEAD MOLYBDATE

10192-29-7 AMMONIUM CHLORATE

10192-30-0 AMMONIUM BISULFITE
SULFUROUS ACID, MONOAMMO-
NIUM SALT [PAL]

10196-04-0 AMMONIUM SULFITE
SULFUROUS ACID, DIAMMONIUM
SALT [PAL]

10198-40-0 COBALT 60

10210-68-1 COBALT CARBONYL
COBALT CARBONYL (CO2(CO)8)
[EPA, EPA, WHM]
COBALT TETRACARBONYL [QBC]

10213-74-8 3-(2-ETHYLBUTOXY) PROPIONIC
ACID
PROPANOIC ACID, 3-(2-ETHYLBU-
TOXY)- [PAL]

10213-77-1 2-[2-(2-METHOXYPROPOXY)
PROPOXY]-1-PROPANOL
1-PROPANOL, 2-[2-(2-
METHOXYPROPOXY)PROPOXY]-
[TSC]

10214-40-1 COPPER SELENITE

10222-01-2 2,2-DIBROMO-3-NITRILOPROPI-
ONAMIDE

10232-93-6 METHOXYETHYL HYDROGEN
MALEATE

10233-88-2 GOLD SODIUM THIOSULFATE

10241-05-1 MOLYBDENUM PENTACHLORIDE

10265-92-6 METHAMIDOPHOS
O,S-DIMETHYL PHOSPHORAMIDOTH-
IOATE [CAL]
PHOSPHORAMIDOTHIOIC ACID, O,S-
DIMETHYL ESTER [PAL]

10279-57-9 HYDRATED SILICA
HYDRATED AMORPHOUS SILICA
[PST]

10290-12-7 COPPER ARSENITE

10294-33-4 BORON TRIBROMIDE
BORANE, TRIBROMO- [PAL]

10294-34-5 BORON TRICHLORIDE
BORON TRICHLORIDE [BORANE,
TRICHLORO-] [EP3]

10294-40-3 BARIUM CHROMATE

10294-53-8 IRON (III) DICHROMATE

10294-54-9 CESIUM SULFATE

10294-56-1 PHOSPHOROUS ACID, ORTHO
PHOSPHOROUS ACID [PST]

10303-47-6 CRESYL DIPHENYL PHOSPHATE

10309-79-2 1-METHYL-2-BENZYLHYDRAZINE

10311-84-9 DIALIFOS
DIALIFOR [302, CAL]

10318-26-0 DIBROMODULCITOL

10325-94-7 CADMIUM NITRATE

10326-21-3 MAGNESIUM CHLORATE

10326-24-6 ZINC ARSENITE

10326-27-9 BARIUM CHLORIDE DIHYDRATE

10347-54-3 CYCLOHEXANE, 1,4-BIS(ISO-
CYANATOMETHYL)-
1,4-BIS(METHYLISOCYANATE)CY-
CLOHEXANE [313]

10361-29-2 AMMONIUM CARBONATE
CARBONIC ACID, AMMONIUM SALT
[PAL]

10361-37-2 BARIUM CHLORIDE

10361-79-2 PRASEODYMIUM CHLORIDE

10361-80-5 PRASEODYMIUM NITRATE

10361-82-7 SAMARIUM TRICHLORIDE

10361-83-8 SAMARIUM NITRATE

10361-89-4 PHOSPHORIC ACID, TRISODIUM
SALT, DECAHYDRATE
SODIUM PHOSPHATE, TRIBASIC
[NJL]
SODIUM PHOSPHATE, TRIBASIC,
DECAHYDRATE [CWA]

10361-91-8 YTTERBIUM CHLORIDE

10361-95-2 ZINC CHLORATE
CHLORIC ACID, ZINC SALT [PAL]

10377-48-7 LITHIUM SULFATE

10377-60-3 MAGNESIUM NITRATE
NITRIC ACID, MAGNESIUM SALT
[PAL]

10377-66-9 MANGANESE NITRATE

10377-98-7 ZIRCONIUM SODIUM LACTATE

10380-28-6 COPPER(II) 8-HYDROXYQUINOLI-
NATE
COPPER 8-HYDROXYQUINOLINE
[IAR]

10380-29-7 CUPRIC SULFATE AMMONIATED
COPPER(2+), TETRAAMINE-, SUL-
FATE (1:1), MONOHYDRATE [PAL]
CUPRIC SULFATE, AMMONIATED
[NJL]

10401-50-0 C.I. DIRECT BLUE 218

10415-75-5 MERCUROUS NITRATE
NITRIC ACID, MERCURY(1+) SALT
[PAL]

10421-48-4 FERRIC NITRATE
NITRIC ACID, IRON(3+) SALT [PAL]

10436-39-2 1,1,2,3-TETRACHLOROPROPENE
1-PROPENE, 1,1,2,3-TETRACHLORO-
[TSC, TSC]

10453-86-8 RESMETHRIN
RESMETHRIN [[5-(PHENYLMETHYL)-
3-FURANYL]METHYL 2,2-DIMETHYL-
3-(2-METHYL-1-PROPENYL)CYCLO-
PROPANECARBOXYLATE] [313]

10471-78-0 2-ISOPROPENYL-2-OXAZOLINE

10476-85-4 STRONTIUM CHLORIDE (SRCL2)

10476-95-6 METHACROLEIN DIACETATE
2-METHYL-2-PROPENE-1,1-DIOL
DIACETATE [WHM]

10489-46-0 RHODIUM SULFATE

10543-95-0 HEXAFLUOROACETONE HY-
DRATE

10544-63-5 ETHYL CROTONATE
2-BUTENOIC ACID, ETHYL ESTER
[PAL]

10544-72-6 NITROGEN TETROXIDE
NITROGEN DIOXIDE [EPA]
NITROGEN OXIDE (N2O4) [PAL]

10544-73-7 NITROGEN TRIOXIDE
NITROGEN OXIDE (N2O3) [PAL]

10545-99-0 SULFUR DICHLORIDE
SULFUR CHLORIDE [HM1]
SULFUR CHLORIDE (SCL2) [PAL]

10546-01-7 SULFUR FLUORIDE (SF5)

10586-17-1 2-PROPENOIC ACID, 2-CYANO-, 1-
METHYLETHYL ESTER

10588-01-9 SODIUM DICHROMATE
CHROMIC ACID (H2CR2O7), DIS-
ODIUM SALT [PAL]
SODIUM BICHROMATE [CWA, EP2,
EPA, FLA, MSL]

10595-95-6 N-NITROSOMETHYLETHY-
LAMINE
ETHANAMINE, N-METHYL-N-NI-
TROSO- [PAL, PAL]

10599-90-3 CHLORAMINE
CHLORAMINES [SD4]

11056-06-7 BLEOMYCIN
BLEOMYCINS [CAL, IAR, MIN]

11069-19-5 DICHLOROBUTENE

11071-47-9 ISOOCTENE
ISOOCTENES [HM1, HM1, NFP,
NFP, PAL]

11096-42-7 POLYETHOXY POLYPROPOXY
OLYETHOXY ETHANOL-IODINE
COMPLEX
ETHOXYLATED NONYLPHENOL
COMPLEX WITH IODINE [PST]

11096-82-5 AROCLOR 1260
PCB-1260 (AROCHLOR 1260) [CW2,
MX1]
POLYCHLORINATED BIPHENYLS
(CONTAINING 60% OR MORE CHLO-
RINE) [C65]

11097-69-1 CHLORODIPHENYL (54% CHLO-
RINE)
AROCLOR 1254 [ATS, CSR, CSR,
EP2, EPA, NTP, PAL, PAL, RCA]
AROCLOR 1254 (PCB) [WHM]
CHLORINATED BIPHENYLS (54%
CHLORINE) [MAK, MAK, MAK,
MAK, MAK]
CHLORODIPHENYL (54% CHLORINE)
(PCB) [OS1, OS1, OS1, OS1]
CHLORODIPHENYL (54% CL) [QBC,
QBC, QBC]
CHLORODIPHENYL 54% [MEX,
MEX, MEX]
PCB-1254 (AROCHLOR 1254) [CW2,
MX1]
PCBS (54% CHLORINE) [AUS, AUS,
AUS, AUS, AUS]

11099-02-8 NICKEL OXIDE

11099-06-2 SILICIC ACID, ETHYL ESTER

11100-14-4 AROCHLOR 1268

11103-86-9 ZINC POTASSIUM CHROMATE

11104-28-2 AROCLOR 1221
PCB-1221 (AROCHLOR 1221) [CW2,
MX1]

12125-02-9 **AMMONIUM CHLORIDE**
AMMONIUM CHLORIDE ((NH4)CL)
[PAL]
AMMONIUM CHLORIDE (FUME)
[QBC, QBC]
AMMONIUM CHLORIDE FUME [CAL,
MEX, MEX, ONT, ONT, OS1, OS1,
TLV, TLV]

12125-56-3 **NICKEL HYDROXIDE**

12135-76-1 **AMMONIUM SULFIDE**
AMMONIUM SULFIDE ((NH4)2S)
[PAL]

12136-15-1 **MERCURY NITRIDE**

12136-45-7 **POTASSIUM OXIDE**
POTASSIUM MONOXIDE [HM1,
HM1]

12137-20-1 **TITANIUM OXIDE (TIO)**

12142-88-0 **NICKEL TELLURIDE**

12161-82-9 **BERTRONDITE**

12164-94-2 **AMMONIUM AZIDE**

12165-69-4 **PHOSPHORUS TRISULFIDE**

12167-20-3 **NITROCRESOLS**

12172-67-7 **ACTINOLITE**

12172-73-5 **ASBESTOS, AMOSITE**
AMOSITE [QBC, QBC, QBC]
AMOSITE ASBESTOS [MSL, WHM]
ASBESTIFORM MINERAL(S) [TSC]
ASBESTOS FIBRES, AMOSITE [ONT,
ONT, ONT, ONT]
ASBESTOS, GRUNERITE [PAL, PAL]
ASBESTOS-AMOSITE [TLV, TLV]

12173-47-6 **HECTORITE**

12174-11-7 **ATTAPULGITE**

12179-04-3 **BORATES, TETRA, SODIUM
SALTS, PENTAHYDRATE**
BORATES, TETRA, SODIUM SALTS
PENTAHYDRATE [AUS]

12185-10-3 **WHITE PHOSPHORUS**

12192-57-3 **AUROTHIOGLUCOSE**

12199-37-0 **SMECTITE**
SMECTITE-GROUP MINERALS [PST]

12217-79-7 **9,10-ANTHRACENEDIONE, 1,5-DI-
AMINOCHLORO-4,8-DIHYDROXY-**

12222-04-7 **C.I. DIRECT BLUE 199**

12222-20-7 **C.I. DIRECT BROWN 111**

12225-18-2 **PIGMENT YELLOW 97, CI 11767**
C.I. PIGMENT YELLOW 97 [PST]

12259-21-1 **IRON OXIDE (FE2O3), HYDRATE**

12259-92-6 **AMMONIUM POLYSULFIDE**

12262-58-7 **CYCLOHEXANONE PEROXIDE**

12263-85-3 **METHYLALUMINUM SESQUIBRO-
MIDE**
ALUMINUM, TRIBRO-
MOTRIMETHYLDI- [PAL]
METHYL ALUMINUM SESQUIBRO-
MIDE [NJL]

12266-38-5 **LEAD ANTIMONIDE**

12269-78-2 **PYROPHYLLITE**

12286-12-3 **MAGNESIUM OXIDE SULFATE**

12298-43-0 **HALLOYSITE**

12327-32-1 **SILICON CARBIDE (SI2C3)**

12385-08-9 **BENZENEDIOL**

12397-35-2 **POTASSIUM CARBONYL**

12401-86-4 **SODIUM MONOXIDE**

12415-34-8 **EMERY**

12427-38-2 **MANEB**
MANEB OR MANEB PREPARATIONS
[HM1, HM1, HM1]
MANEB [CARBAMODITHIOIC ACID,
1,2-ETHANEDIYLBIS-, MANGANESE
COMPLEX] [313]

12440-42-5 **STANNIC PHOSPHIDE**

12504-13-1 **STRONTIUM PHOSPHIDE**

12510-42-8 **ERIONITE**
ERIONITE FIBER [TSC, TSE]

12540-13-5 **COPPER ACETYLIDE**

12542-85-7 **METHYL ALUMINUM
SESQUICHLORIDE**
ALUMINUM,
TRICHLOROTRIMETHYLDI- [PAL]
METHYLALUMINUM SESQUICHLO-
RIDE [FLA, MSL, NFP, NFP]

12587-46-1 **ALPHA RADIATION**

12587-47-2 **BETA RADIATION**

12604-53-4 **FERROMANGANESE**

12604-58-9 **FERROVANADIUM**
FERROVANADIUM (DUST) [QBC,
QBC]
FERROVANADIUM DUST [ALB, ALB,
AUS, AUS, BC1, BC1, CAL, FLA,
MEX, MEX, MSL, NIO, NIO, ONT,
ONT, OS1, OS1, OS1, TLV, TLV]
VANADIUM ALLOY, BASE, V,C,FE
(FERROVANADIUM) [PAL]

12612-47-4 **LEAD CHLORIDE (VAN)**
LEAD CHLORIDE [PAL]

12612-55-4 **NICKEL CARBONYL (VAN)**
NICKEL CARBONYL [BC1, HM1,
HM1, HM1, HM1, QBC]

12640-89-0 **SELENIUM OXIDE (VAN)**
SELENIUM OXIDE [HM1, HM1]

12642-23-8 **POLYCHLORINATED TRIPH-
ENYLS**
POLYCHLORINATED TRIPHENYLS
(AROCLOR 5442) [NJL]

12645-31-7 **PHOSPHORIC ACID, 2-ETHYL-
HEXYL ESTER**

12645-49-7 **IRON MANGANESE ZINC OXIDE**

12645-50-0 **IRON NICKEL ZINC OXIDE**

12653-71-3 **MERCURY OXIDE (VAN)**
MERCURY OXIDE [HM1, HM1,
HM1]

12656-43-8 **ALUMINUM CARBIDE**

12656-57-4 **CADMIUM ORANGE PIGMENT**
MOLYBDATE ORANGE [PST]

12656-85-8 **MOLYBDATE ORANGE (LEAD
CHROMATE PIGMENT)**
C.I. PIGMENT RED 104 [PST]

12663-46-6 **CYCLOCHLOROTINE**

12672-29-6 **AROCLOR 1248**
PCB-1248 (AROCHLOR 1248) [CW2,
MX1]

12674-11-2 **AROCLOR 1016**
PCB-1016 (AROCHLOR 1016) [CW2,
MX1]

12680-48-7 **CHROMIUM SODIUM OXIDE**

12713-03-0 **RAW UMBER**
UMBER [PST]

12736-96-8 **SODIUM POTASSIUM ALUMINUM
SILICATE**

12737-18-7 **CALCIUM SILICIDE**

12751-23-4 **PHOSPHORIC ACID, DODECYL
ESTER**

12751-36-9 **PHARMAMEDIA**

12768-78-4 **C.I. ACID GREEN 16**

12768-82-0 **C.I. BASIC ORANGE 15**

12770-50-2 **BERYLLIUM ALUMINUM ALLOY**
BERYLLIUM-ALUMINUM ALLOY
[MSL]

12771-08-3 **SULFUR CHLORIDE**
SULFUR MONOCHLORIDE [EPA,
NJL]

12772-47-3 **PENTOL**
1-PENTOL [HM1, HM1]

12788-84-0 **SODIUM SALT OF N-COCO BETA
AMINO BUTYRIC ACID**

12788-93-1 **BUTYL PHOSPHORIC ACID**
BUTYL ACID PHOSPHATE [HM1,
HM1, NJL, NJL]
N-BUTYL ACID PHOSPHATE [PST]

12789-46-7 **AMYL ACID PHOSPHATE**

13007-92-6 **CHROMIUM CARBONYL**

13010-47-4 **1-(2-CHLOROETHYL)-3-CYCLO-
HEXYL-1-NITROSOUREA**
1-(2-CHLORO ETHYL)-3-CYCLO-
HEXYL-1-NITROSOUREA [MIN]
1-(2-CHLOROETHYL)-3-CYCLO-
HEXYL-1-NITROSOUREA (CCNU)
[IAR, NTP]
1-(2-CHLOROETHYL)-3-CYCLO-
HEXYL-1-NITROSOUREA (CCNU) (LO-
MUSTINE) [C65, C65]
LOMUSTINE [CSR]
UREA, N-(2-CHLOROETHYL)-N'-
CYCLOHEXYL-N-NITROSO- [PAL,
PAL]

13045-94-8 **MEDPHALAN**

13048-33-4 **1,6-HEXANEDIOL DIACRYLATE**
2-PROPENOIC ACID, 1,6-HEX-
ANEDIYL ESTER [TSC, TSC]
HEXANEDIOL DIACRYLATE [WEL]

13052-09-0 **2,5-DIMETHYL-2,5-DI(2-ETHYL-
HEXANOYLPEROXY)-HEXANE**

13057-78-8 **CHLOROISOCYANURIC ACID**
1,3,5-TRIAZINE-2,4,6-(1H,3H,5H)-
TRIONE, 1-CHLORO [PAL]

13071-79-9 **TERBUFOS**

13081-34-0 **POLY(OXYETHYLENE) 10-DODE-
CYLMERCAPTAN**

13092-75-6 **SILVER ACETYLIDE**

13106-47-3 **BERYLLIUM CARBONATE**
CARBONIC ACID, BERYLLIUM SALT
(1:1) [PAL, PAL]

13106-76-8 **AMMONIUM MOLYBDATE**

13114-62-0 **2-NAPHTHYLUREA**

13121-70-5 **CYHEXATIN**
CYHEXATIN [QBC]
CYHEXATIN (ISO) [GBR, GBR]
CYHEXATIN (TRICYCLOHEXYLTIN
HYDROXIDE) [CN8]
STANNANE, TRICYCLOHEXYLHY-
DROXY- [PAL]
TRICYCLOHEXYLTIN HYDROXIDE
(PLICTRAN) [ALB, ALB, BC1, BC1]

13122-18-4 **TERT-BUTYL PEROXY-3,5,5-
TRIMETHYLHEXANOATE**

13138-45-9 **NICKEL NITRATE (2+ SALT)**
NICKEL NITRATE [FLA, HM1, HM1,
NJL]
NICKEL(II) NITRATE [WHM]
NITRIC ACID, NICKEL (2+) SALT
[MSL]
NITRIC ACID, NICKEL(2+) SALT
[PAL]

13146-28-6 **DI-5-(MORPHOLINOMETHYL)
-3((5-NITROFURFURYLIDENE)
AMINO)-2-OXAZOLIDINONE HY-
DROCHLORIDE**

13150-00-0 **SODIUM LAURYL TRIETHOXY
SULFATE**

13171-21-6 **PHOSPHAMIDON**
PHOSPHAMIDON (2-CHLORO-2-DIETHYLCARBAMOYL-1-METHYLVINYL DEMETHYL PHOSPHATE) [CN8]
PHOSPHORIC ACID, 2-CHLORO-3-(DIETHYLAMINO)-1-METHYL-3-OXO-1-PROPENYL DIMETHYL ESTER [PAL]

13194-48-4 **ETHOPROPHOS**
ETHOPROP [MSL]
ETHOPROP [PHOSPHORODITHIOIC ACID, O-ETHYL S,S-DIPROPYL ESTER] [313]
O-ETHYL S,S-DIPROPYL PHOSPHORODITHIOATE (ETHOPROP) [CAL]
PHOSPHORODITHIOIC ACID, O-ETHYL S,S-DIPROPYL ESTER [PAL]

13195-76-1 **BORIC ACID (H3BO3), TRIS(2-METHYLPROPYL) ESTER**
TRIISOBUTYL BORATE [FLA, MSL, NFP, NFP]

13225-10-0 **A-METHYLGLUCOSIDE TETRANITRATE**

13233-32-4 **RADIUM 224**

13236-02-7 **OXIRANE, 2,2',2"-[1,2,3-PROPANETRIYL TRIS (OXYMETHYLENE)]TRIS-**
GLYCEROL TRIGLYCIDYL ETHER [EPA, OTS]

13256-22-9 **N-NITROSOSARCOSINE**
GLYCINE, N-METHYL-N-NITROSO- [PAL, PAL]

13292-46-1 **RIFAMPICIN**

13311-84-7 **FLUTAMIDE**

13319-75-0 **BORON TRIFLUORIDE DIHYDRATE**

13327-32-7 **BERYLLIUM HYDROXIDE**
BERYLLIUM HYDROXIDE (BE(OH)2) [PAL, PAL]

13347-42-7 **4-CHLORO-2-CYCLOPENTYLPHENOL**

13356-08-6 **BIS(TRIS(2-METHYL-2-PHENYLPROPYL)TIN)OXIDE**
FENBUTATIN OXIDE (HEXAKIS(2-METHYL-2-PHENYL-PROPYL)DISTANNOXANE) [313]

13356-20-2 **MONOOCTYLTIN OXIDE**

13360-57-1 **DIMETHYLSULFAMOYL CHLORIDE**

13360-63-9 **ETHYLBUTYLAMINE**
1-BUTANAMINE, N-ETHYL- [PAL]

13361-32-5 **ALLYL CYANOACETATE**

13366-73-9 **PHOTODIELDRIN**

13393-71-0 **SODIUM PENTADECYLSULFATE**

13396-80-0 **9-PHOSPHABICYCLONONANE**
9-PHOSPHABICYCLONONANES (CYCLOOCTADIENE PHOSPHINES) [HM1, HM1]

13397-24-5 **GYPSUM (CA(SO4).2H2O)**
GYPSUM [OS1]

13410-01-0 **SODIUM SELENATE**

13410-72-5 **BENZENAMINE, 4-ETHOXY-N-[(5-NITRO-2-FURANYL)METHYLENE]-**

13414-54-5 **BENZENE, 1-[(2-METHYL-2-PROPENYL)OXY]-2-NITRO-**

13414-55-6 **BENZOFURAN, 2,3-DIHYDRO-2,2-DIMETHYL-7-NITRO-**

13419-59-5 **TRISODIUM SULFOSUCCINATE**

13423-61-5 **CHROMIC ACID (H2CRO4), MAGNESIUM SALT (1:1)**
MAGNESIUM CHROMATE [MSL]
MAGNESIUM DICHROMATE [FLA]

13424-46-9 **LEAD AZIDE**

13426-91-0 **CUPRIETHYLENEDIAMINE**

13444-75-2 **MERCURY (II) CHROMATE**

13444-85-4 **NITROGEN TRIIODIDE**

13446-10-1 **AMMONIUM PERMANGANATE**
PERMANGANIC ACID (HMNO4), AMMONIUM SALT [PAL]

13446-34-9 **MANGANOUS CHLORIDE**

13446-72-5 **RUBIDIUM CHROMATE**

13446-73-6 **RUBIDIUM DICHROMATE**

13450-90-3 **GALLIUM TRICHLORIDE**
GALLIUM(III) CHLORIDE [WHM]

13450-97-0 **STRONTIUM PERCHLORATE**

13453-06-0 **AMMONIUM TELLURATE**

13453-07-1 **GOLD TRICHLORIDE**

13453-30-0 **THALLIUM CHLORATE**

13453-66-2 **LEAD PYROPHOSPHATE**

13453-71-9 **LITHIUM CHLORATE**

13454-78-9 **CESIUM CHROMATE**

13454-96-1 **PLATINUM TETRACHLORIDE**
PLATINUM(IV) TETRACHLORIDE [WHM]

13455-25-9 **COBALT CHROMATE**

13455-36-2 **COBALT(II) PHOSPHATE**

13457-18-6 **PYRAZOPHOS**

13462-90-3 **NICKEL(II) IODIDE**

13463-39-3 **NICKEL CARBONYL**
NICKEL CARBONYL (NICKEL TETRACARBONYL) [OS3]
NICKEL CARBONYL NI(CO)4), (T-4)- [PAL, PAL]
NICKEL TETRACARBONYL [HM1, HM1]
TETRACARBONYLNICKEL [GBR]

13463-40-6 **IRON PENTACARBONYL**
IRON CARBONYL [NFP, NFP]
IRON CARBONYL (FE(CO)5), (TB-5-11)- [PAL]
IRON, PENTACARBONYL- [OS3]
IRON, PENTACARBONYL- [IRON CARBONYL (FE(CO)5), (TB-5-11)-] [EP3]
IRON, PENTACRABONYL [302]
PENTACARBONYLIRON [GBR]

13463-41-7 **ZINC PYRITHIONE**

13463-67-7 **TITANIUM DIOXIDE**
TITANIUM OXIDE (TIO2) [PAL]

13463-98-4 **CALCIUM ABIETATE**

13464-35-2 **POTASSIUM ARSENITE**

13464-37-4 **ARSENOUS ACID, TRISODIUM SALT**

13464-38-5 **ARSENIC ACID, TRISODIUM SALT**
SODIUM ORTHOARSENATE [HM1]

13465-34-4 **MERCURY (I) CHROMATE**

13465-73-1 **BROMOSILANE**

13465-95-7 **BARIUM PERCHLORATE**

13470-50-3 **2-HEPTADECYL-1-METHYL-1-(2-STEAROYLAMIDO)ETHYL-2-IMIDAZOLINIUM METHYL SULFATE**

13472-45-2 **SODIUM TUNGSTATE**

13473-75-1 **THALLIUM (I) CHROMATE**

13473-90-0 **ALUMINUM NITRATE**

13474-88-9 **1,1-DICHLORO-1,2,2,3,3-PENTAFLUOROPROPANE (HCFC-225CC)**

13477-00-4 **BARIUM CHLORATE**
CHLORIC ACID, BARIUM SALT [PAL]

13477-10-6 **BARIUM HYPOCHLORITE**

13477-34-4 **CALCIUM NITRATE.4H20**

13477-36-6 **CALCIUM PERCHLORATE**

13478-00-7 **NICKEL(II) NITRATE, HEXAHYDRATE (1:2:6)**

13478-10-9 **FERROUS IRON (II) CHLORIDE.4H20**

13478-20-1 **PHOSPHORYL FLUORIDE**
FLUOROPHOSPHORIC ACID, ANHYDROUS [HM1, HM1]

13478-50-7 **LEAD THIOSULFATE**

13483-18-6 **1,2-BIS(CHLOROMETHOXY) ETHANE**

13494-80-9 **TELLURIUM**
TELLURIUM AND COMPOUNDS [ALB, ALB, AUS, MEX, NIO, NIO, OS1, OS1, TLV]
TELLURIUM AND ITS COMPOUNDS [MAK, MAK]
TELLURIUM, ELEMENTAL [WHM]

13494-90-1 **GALLIUM(III) NITRATE**

13510-49-1 **BERYLLIUM SULFATE**
BERYLLIUM SULPHATE [FLA, MSL]
SULFURIC ACID, BERYLLIUM SALT (1:1) [PAL, PAL]

13510-89-9 **LEAD ANTIMONATE**

13515-40-7 **C.I. PIGMENT YELLOW 73**

13517-17-4 **SODIUM CHROMATE DECAHYDRATE**

13520-83-7 **URANYL NITRATE HEXAHYDRATE**

13530-65-9 **ZINC CHROMATE**
CHROMIC ACID (H2CRO4), ZINC SALT (1:1) [PAL, PAL]

13530-68-2 **CHROMIC ACID (H2CR2O7)**

13537-18-3 **THULIUM CHLORIDE**

13537-32-1 **MONOFLUOROPHOSPHORIC ACID**
FLUOROPHOSPHORIC ACID [NJL, NJL]

13547-17-6 **UREA, N,N"-METHYLENEBIS-**

13548-38-4 **CHROMIUM NITRATE**

13548-42-0 **COPPER CHROMATE**

13552-44-8 **4,4'-METHYLENEDIANILINE DIHYDROCHLORIDE**
4,4-METHYLENEDIANILINE AND ITS DIHYDROCHLORIDE [PAL, PAL]

13560-99-1 **2,4,5-T SALTS**
2,4,5-T SALT; ACETIC ACID 2,4,5-TRICHLOROPHENOXY-SODIUM SALT [CAL]
ACETIC ACID, (2,4,5-TRICHLOROPHENOXY)-, SODIUM SALT [PAL]

13561-08-5 **OXIRANE, 2,2'-(OXIRANYLMETHOXY)-1,3-PHENYLENE]BIS (METHYLENE)]BIS-**
2,6-DIGLYCIDYLPHENYL GLYCIDYL ETHER [EPA]
DIGLYCIDYLPHENYL GLYCIDYL ETHER, 2,6- [OTS]

13573-15-4 **CALCIUM MANGANESE SILICON**

13573-18-7 **TRIPHOSPHORIC ACID, SODIUM SALT**

13586-68-0 **2-PROPENAL, 3,4-(1,1-DIMETHYLETHYL)PHENYL-2-METHYL-**

13586-82-8 **COBALT OCTOATE**

13593-03-8 QUINALPHOS

13597-46-1 ZINC SELENITE

13597-99-4 BERYLLIUM NITRATE
NITRIC ACID, BERYLLIUM SALT
[PAL]

13598-15-7 BERYLLIUM PHOSPHATE
PHOSPHORIC ACID, BERYLLIUM
SALT (1:1) [PAL, PAL]

13598-26-0 PHOSPHORIC ACID, BERYLLIUM
SALT (2:3)

13598-36-2 PHOSPHOROUS ACID
PHOSPHOROUS ACID, ORTHO [NJL]

13614-98-7 MINOCYCLINE HYDROCHLORIDE

13637-63-3 CHLORINE PENTAFLUORIDE

13637-71-3 NICKEL PERCHLORATE

13637-76-8 LEAD PERCHLORATE

13647-35-3 TRILOSTANE

13654-09-6 DECARBROMOBIPHENYL
1,1'-BIPHENYL, 2,2',3,3',4,4',5,5',6,6'-
DECABROMO- [TSC, TSC]
POLYBROMINATED BIPHENYLS
[NJL]

13674-84-5 2-PROPANOL, 1-CHLORO-, PHOS-
PHATE (3:1)
2-PROPANOL, 1-CHLORO-, PHOS-
PHATE (3:1) [OTS]
TRIS(1-CHLORO-2-PROPYL) PHOS-
PHATE [T33]

13674-87-8 1,3-DICHLORO ISOPROPYL PHOS-
PHATE
2-PROPANOL, 1,3-DICHLORO-, PHOS-
PHATE (3:1) [TSC, TSC, TSC]

13682-73-0 POTASSIUM CUPROCYANIDE

13684-56-5 DESMEDIPHAM

13693-11-3 TITANIUM SULFATE (TI(SIO4)2)
TITANIUM SULFATE [HM1, HM1,
PST]

13701-59-2 BARIUM METABORATE

13717-00-5 MAGNESITE (MG(CO3))

13718-26-8 THIRAM
SODIUM METAVANADATE [WHM]

13718-59-7 BARIUM SELENITE

13721-39-6 TRISODIUM ORTHOVANADATE

13721-43-2 SODIUM HYPOPHOSPHATE

13746-89-9 ZIRCONIUM NITRATE
NITRIC ACID, ZIRCONIUM(4+) SALT
[PAL]

13762-14-6 COBALT(II) MOLYBDATE

13762-51-1 POTASSIUM BOROHYDRIDE

13765-19-0 CALCIUM CHROMATE
CHROMIC ACID (H2CRO4), CALCIUM
SALT (1:1) [PAL, PAL]

13768-00-8 ACTINOLITE, NON-ASBESTIFORM
ACTINOLITE [QBC, QBC]

13768-67-7 YTTERBIUM NITRATE

13769-43-2 POTASSIUM METAVANADATE

13770-61-1 INDIUM NITRATE

13770-96-2 SODIUM ALUMINUM HYDRIDE

13779-41-4 DIFLUOROPHOSPHORIC ACID

13780-03-5 CALCIUM BISULFITE
CALCIUM HYDROGEN SULFITE
[NJL, NJL]

13814-96-5 LEAD FLUOBORATE
BORATE(1-), TETRAFLUORO-, LEAD
(2+) (2:1) [PAL]
LEAD BORON FLUORIDE [MSL]

13814-98-7 STRONTIUM FLUOBORATE

13820-40-1 AMMONIUM CHLOROPALLADITE

13823-29-5 THORIUM NITRATE
NITRIC ACID, THORIUM(4+) SALT
[PAL]

13825-74-6 TITANIUM SULFATE

13825-86-0 CHROMOUS (II) SULFATE

13826-35-2 BENZENEMETHANOL, 3-PHE-
NOXY-

13826-65-8 NITROUS ACID, LEAD (2+) SALT

13826-66-9 ZIRCONYL NITRATE

13826-83-0 AMMONIUM FLUOBORATE
AMMONIUM FLUOROBORATE [NJL]
BORATE(1-), TETRAFLUORO-, AMMO-
NIUM [PAL]

13826-93-2 DIPOTASSIUM TETRABRO-
MOPALLADATE

13838-16-9 ENFLURANE

13840-33-0 LITHIUM HYPOCHLORITE

13843-59-9 AMMONIUM BROMATE

13843-81-7 LITHIUM DICHROMATE

13845-18-6 SODIUM SULFAMATE

13845-35-7 LEAD TELLURITE

13845-36-8 POTASSIUM TRIPOLYPHOS-
PHATE

13846-18-9 CALCIUM DITHIONITE
CALCIUM HYDROSULFITE [NJL]

13863-41-7 BROMINE CHLORIDE

13863-88-2 SILVER AZIDE

13889-92-4 CARBONOCHLORIDOTHIOIC
ACID, S-PROPYL ESTER
PROPYL CHLOROTHIOFORMATE
[FLA, MSL, NFP, NFP]

13909-09-6 1-(2-CHLOROETHYL)-3-(4-
METHYLCYCLOHEXYL)-1-NI-
TROSOUREA
1-(2-CHLORO ETHYL)-3-(4-METHYL-
CYCLOHEXYL)-1-NITROSOUREA
[MIN]
1-(2-CHLOROETHYL)-3-(4-METHYL-
CYCLOHEXYL)-1-NITROSOUREA
(ME-CCNU) [NTP]
1-(2-CHLOROETHYL)-3-(4-METHYL-
CYCLOHEXYL)-1-NITROSOUREA
(METHYL-CCNU) [C65, CAL, IAR]
METHYL CCNU [CSR]

13933-32-9 TETRAAMMINEDICHLOROPLAT-
INUM(II)

13952-84-6 SEC-BUTYLAMINE
(RS)-SEC-BUTYLAMINE [CWA]
2-BUTANAMINE [PAL]

13961-86-9 OLEIC ACID DIETHANOLAMINE
DIETHANOLAMINE OLEATE [PST]

13966-00-2 POTASSIUM 40

13966-02-4 BERYLLIUM 7

13966-05-7 CALCIUM 45

13966-06-8 TIN 113

13966-29-5 URANIUM 234

13966-31-9 MANGANESE 54

13966-32-0 SODIUM 22

13967-48-1 RUTHENIUM 106

13967-63-0 SCANDIUM 46

13967-64-1 DYSPROSIUM 165

13967-70-9 CESIUM 134

13967-71-0 ZIRCONIUM 95

13967-73-2 STRONTIUM 85

13967-74-3 CERIUM 141

13967-76-5 NIOBIUM 95

13967-90-3 BARIUM BROMATE

13968-49-5 CERIUM 137

13968-50-8 ANTIMONY 127

13968-51-9 THALLIUM 204

13968-53-1 RUTHENIUM 103

13968-55-3 URANIUM 233

13968-59-7 NEPTUNIUM 239

13973-87-0 BROMINE AZIDE

13973-88-1 CHLORINE AZIDE

13981-14-1 PROTACTINIUM 233

13981-15-2 CURIUM 244

13981-16-3 PLUTONIUM 238

13981-17-4 CALIFORNIUM 252

13981-25-4 COPPER 64

13981-27-6 ZIRCONIUM 89

13981-28-7 LANTHANUM 140

13981-29-8 TERBIUM 160

13981-30-1 THULIUM 170

13981-37-8 NICKEL 63

13981-38-9 COBALT 58

13981-41-4 BARIUM 133

13981-43-6 CHLORINE 36

13981-49-2 TELLURIUM 127

13981-50-5 COBALT 57

13981-51-6 MERCURY 197

13981-52-7 POLONIUM 210

13981-53-8 RADIUM 225

13981-54-9 AMERICIUM 242

13981-56-1 FLUORINE 18

13981-83-4 COBALT 61

13981-99-2 NICKEL 57

13982-00-8 TANTALUM 182

13982-01-9 URANIUM 239

13982-04-2 SODIUM 24

13982-06-4 COPPER 60

13982-10-0 PLUTONIUM 242

13982-11-1 CALIFORNIUM 250

13982-13-3 RUBIDIUM 87

13982-20-2 GOLD 193

13982-22-4 GALLIUM 72

13982-23-5 ZINC 69

13982-25-7 COBALT 55

13982-30-4 CERIUM 139

13982-36-0 YTTRIUM 88

13982-38-2 BISMUTH 207

13982-39-3 ZINC 65

13982-63-3 RADIUM 226

13982-70-2 URANIUM 236

13982-78-0 MERCURY 203

13983-17-0 WOLLASTONITE (CA(SIO3))
WOLLASTONITE [CAL, IAR]
WOLLASTONITE CALCIUM SILI-
CATES [CSR]

13983-27-2 **KRYPTON 85**

13987-01-4 **TRIPROPYLENE**
1-PROPENE, TRIMER [PAL]

13994-18-8 **XENON 125**

13994-19-9 **XENON 127**

13994-20-2 **NEPTUNIUM 237**

14017-41-5 **COBALTOUS SULFAMATE**
SULFAMIC ACID, COBALT(2+) SALT
(2:1) [PAL]

14018-95-2 **ZINC DICHROMATE**
CHROMIC ACID (H2CR2O7), ZINC
SALT (1:1) [PAL]
ZINC BICHROMATE [FLA]
ZINC CHROMATE [MIN]

14019-91-1 **CALCIUM SELENATE**

14025-15-1 **CUPRATE(2-), ((ETHYLENEDINI-
TRILO)TETRAACETATO)-, DIS-
ODIUM**
DISODIUM CUPRIC ETHYLENEDI-
AMINETETRAACETATE [PST]

14025-21-9 **DISODIUM [(ETHYLENEDINI-
TRILO)TETRAACETATO] ZINC**

14038-43-8 **FERRIC FERROCYANIDE**
C.I. PIGMENT BLUE 27 [PST]

14041-40-8 **EUROPIUM 158**

14041-42-0 **GADOLINIUM 159**

14041-43-1 **ERBIUM 165**

14041-44-2 **YTTERBIUM 175**

14041-45-3 **YTTERBIUM 167**

14041-46-4 **THULIUM 173**

14041-47-5 **THULIUM 175**

14047-09-7 **3,3',4,4'-TETRACHLOROAZOBEN-
ZENE**

14092-95-6 **CALCIUM 41**

14092-99-0 **MANGANESE 52**

14093-03-9 **COBALT 56**

14093-04-0 **IRON 52**

14093-11-9 **HAFNIUM 172**

14093-12-0 **LUTETIUM 172**

14104-85-9 **MAGNESIUM DICHROMATE**

14107-52-9 **THALLIUM 197**

14109-32-1 **CADMIUM 109**

14119-05-2 **PRASEODYMIUM 144**

14119-08-5 **GALLIUM 66**

14119-09-6 **GALLIUM 67**

14119-13-2 **MOLYBDENUM 93**

14119-14-3 **TECHNETIUM 93**

14119-15-4 **MOLYBDENUM 99**

14119-19-8 **CERIUM 143**

14119-21-2 **GADOLINIUM 148**

14119-23-4 **YTTERBIUM 177**

14119-24-5 **OSMIUM 191**

14119-28-9 **LEAD 205**

14119-30-3 **LEAD 209**

14119-32-5 **PLUTONIUM 241**

14119-33-6 **PLUTONIUM 240**

14119-34-7 **PLUTONIUM 244**

14133-75-6 **INDIUM 110**

14133-76-7 **TECHNETIUM 99**

14147-71-8 **METHYLAMINE NITROFORM**

14158-27-1 **STRONTIUM 89**

14158-29-3 **URANIUM 232**

14158-30-6 **IODINE 124**

14158-31-7 **IODINE 125**

14158-32-8 **IODINE 126**

14158-34-0 **CHLORINE 38**

14158-35-1 **IRIDIUM 194**

14163-25-8 **ARGON 41**

14167-18-1 **SALCOMINE**

14187-32-7 **DIBENZO-18-CROWN-6**

14191-64-1 **PRASEODYMIUM 142**

14191-65-2 **RUBIDIUM 89**

14191-71-0 **INDIUM 115**

14191-76-5 **PRASEODYMIUM 139**

14191-78-7 **EUROPIUM 147**

14191-83-4 **MOLYBDENUM 101**

14216-75-2 **NICKEL NITRATE**
NITRIC ACID, NICKEL SALT [MSL,
PAL]

14220-17-8 **NICKEL POTASSIUM CYANIDE**

14221-47-7 **FERRIC AMMONIUM OXALATE**

14222-20-9 **N-METHYLPSEUDOEPHEDRINE**

14228-73-0 **OXIRANE, 2,2'-[1,4-CYCLO-
HEXANEDILBIS (METHYLE-
NEOXYMETHYLENE)]BIS-**
1,4-BIS(GLYCIDYLOXYMETHYL)
CYCLOHEXANE [EPA]
BIS(GLYCIDYLOXYMETHYL) CYCLO-
HEXANE, 1,4- [OTS]
CYCLOHEXANE, 1,4-BIS[2,3-
EPOXYPROPOXY)METHYL]- [TSC]

14233-37-5 **C.I. SOLVENT BLUE 36**

14234-24-3 **YTTRIUM 91**

14234-28-7 **TELLURIUM 132**

14234-29-8 **CESIUM 136**

14234-35-6 **ANTIMONY 125**

14239-68-0 **CADMIUM DIETHYLDITHIOCAR-
BAMATE**

14244-62-3 **ZINC POTASSIUM CYANIDE**

14255-04-0 **LEAD 210**

14258-49-2 **OXALIC ACID, AMMONIUM SALT**
AMMONIUM OXALATE [EPA]
ETHANEDIOIC ACID, AMMONIUM
SALT [MSL, PAL]

14264-31-4 **COPPER SODIUM CYANIDE**
SODIUM CUPROCYANIDE [HM1,
HM1, HM1, NJL]

14265-44-2 **PHOSPHATE**
ORGANIC PHOSPHATE [HM1, HM1,
HM1]
ORGANIC PHOSPHATE COMPOUND,
SOLID (POISON B) [NJL]
PHOSPHATES [MX2]

14265-71-5 **SELENIUM 75**

14265-75-9 **LUTETIUM 177**

14265-84-0 **IRIDIUM 189**

14265-85-1 **ACTINIUM 225**

14269-63-7 **THORIUM 230**

14269-71-7 **TELLURIUM 129**

14269-74-0 **NEODYMIUM 147**

14269-75-1 **URANIUM 237**

14269-78-4 **YTTERBIUM 169**

14274-68-1 **YTTRIUM 87**

14274-82-9 **THORIUM 228**

14276-49-4 **SILICON 31**

14276-61-0 **SCANDIUM 43**

14276-65-4 **GADOLINIUM 153**

14280-31-0 **SAMARIUM 146**

14280-34-3 **DYSPROSIUM 159**

14280-35-4 **EUROPIUM 156**

14280-36-5 **EUROPIUM 157**

14280-38-7 **BISMUTH 201**

14293-73-3 **POTASSIUM HYDROSULFITE**

14304-78-0 **ARSENIC 74**

14304-79-1 **TELLURIUM 121**

14304-80-4 **TELLURIUM 123**

14307-33-6 **CHROMIC ACID (H2CR2O7), CAL-
CIUM SALT (1:1)**
CALCIUM BICHROMATE [FLA]
CALCIUM DICHROMATE [MSL]

14307-35-8 **LITHIUM CHROMATE**
CHROMIC ACID (H2CRO4),
DILITHIUM SALT [PAL]

14307-43-8 **AMMONIUM TARTRATE**
TARTARIC ACID, AMMONIUM SALT
[MSL]

14320-93-5 **GOLD 195**

14323-43-4 **DIAMMINEDICHLOROPALLA-
DIUM**

14324-55-1 **ZINC DIETHYLDITHIOCARBA-
MATE**

14331-79-4 **BISMUTH 210**

14331-83-0 **ACTINIUM 228**

14331-85-2 **PROTACTINIUM 231**

14331-86-3 **SILVER 112**

14331-88-5 **ANTIMONY 129**

14331-90-9 **BROMINE 84**

14331-91-0 **STRONTIUM 91**

14331-95-4 **RUTHENIUM 105**

14331-97-6 **VANADIUM 48**

14332-59-3 **ZINC PHOSPHITE**

14333-33-6 **CARBON 11**

14333-38-1 **BISMUTH 205**

14333-39-2 **SILVER 106**

14333-45-0 **THULIUM 171**

14336-66-4 **CADMIUM 113**

14336-68-6 **CADMIUM 115**

14336-70-0 **NICKEL 59**

14351-50-9 **N,N-DIMETHYLOLEYLAMINE
OXIDE**

14351-66-7 **ABIETIC ACIDS, SODIUM SALT**

14362-31-3 **CHLORCYCLIZINE HYDROCHLO-
RIDE**

14374-79-9 **ANTIMONY 122**

14374-81-3 **GERMANIUM 71**

14378-12-2 **SOAPSTONE**

14378-21-3 **POTASSIUM 42**

14378-22-4 **POTASSIUM 44**

14378-25-7 **BARIUM 139**

14378-26-8 **RHENIUM 188**

14378-53-1 **RHODIUM 101**

14380-61-1 HYPOCHLORITE ION

14380-75-7 PROMETHIUM 147

14390-89-7 BERYLLIUM 10

14391-10-7 TERBIUM 156

14391-11-8 GOLD 199

14391-16-3 EUROPIUM 155

14391-18-5 TERBIUM 157

14391-19-6 TERBIUM 161

14391-20-9 HOLMIUM 161

14391-24-3 LUTETIUM 173

14391-27-6 TANTALUM 179

14391-29-8 RHENIUM 187

14391-31-2 SAMARIUM 155

14391-45-8 ERBIUM 171

14391-61-8 BROMINE 80

14391-66-3 INDIUM 112

14391-68-5 ANTIMONY 120

14391-72-1 IODINE 128

14391-74-3 GALLIUM 70

14391-86-7 SCANDIUM 48

14391-94-7 SCANDIUM 44

14391-96-9 SCANDIUM 47

14391-97-0 SCANDIUM 49

14391-99-2 CALCIUM 47

14392-00-8 TITANIUM 45

14392-01-9 VANADIUM 49

14392-02-0 CHROMIUM 51

14392-03-1 MANGANESE 51

14392-33-7 SAMARIUM 147

14402-89-2 SODIUM NITROFERRICYANIDE

14408-42-5 2-(8-HEPTADECENYL)-4-METHYL-
2-OXAZOLINE-4-METHANOL

14433-76-2 N,N-DIMETHYLCAPRAMIDE

14448-38-5 HYPONITROUS ACID

14452-47-2 LUTETIUM 176

14452-57-4 MAGNESIUM PEROXIDE

14464-46-1 SILICA, CRISTOBALITE
    CRISTOBALITE (SIO2) [PAL]
    CRISTOBALITE DUST [FLA, MSL]
    QUARTZ, CRISTOBALITE [MAK, MAK]
    SILICA - CRYSTALLINE, CRISTO-BALITE [TLV]
    SILICA CRYSTALLINE CRISTOBALITE [OS1, OS1]
    SILICA, CRYSTALLINE, CRISTO-BALITE [CEX, WHM]
    SILICA-CRYSTALLINE, CRISTO-BALITE [MIN]

14475-73-1 ZIRCONIUM SULFATE

14484-64-1 FERBAM
    FERBAM (ISO) [GBR, GBR]
    FERBAM [TRIS(DIMETHYLCARBAMO-DITHIOATO-S,S') IRON [313]
    FERBAME [QBC]
    IRON, TRIS(DIMETHYLCAR-BAMODITHIOATO-S,S')-, (OC-6-11)-[PAL]

14486-19-2 CADMIUM FLUOBORATE

14489-25-9 CHROMOSULFURIC ACID

14519-07-4 ZINC BROMATE

14546-44-2 HYDRAZINE AZIDE

14567-73-8 TREMOLITE
    ASBESTOS, TREMOLITE [BC1, BC1, CSR, CSR, IAR]

14596-10-2 AMERICIUM 241

14596-12-4 IRON 59

14596-37-3 PHOSPHORUS 32

14634-93-6 ZINC ETHYLPHENYLDITHIOCAR-BAMATE

14639-97-5 ZINC AMMONIUM CHLORIDE (ZN.CL4.2H4-N)
    ZINCATE(2-), TETRACHLORO-, DI-AMMONIUM, (T-4)- [PAL]

14639-98-6 ZINC AMMONIUM CHLORIDE (ZN.CL5.3H4-N)
    ZINCATE (3-), PENTACHLORO-, TRI-AMMONIUM [MSL]
    ZINCATE(3-), PENTACHLORO-, TRI-AMMONIUM [PAL]

14644-61-2 ZIRCONIUM SULFATE
    SULFURIC ACID, ZIRCONIUM(4+) SALT (2:1) [PAL]

14666-78-5 DIETHYL PEROXYDICARBONATE

14674-72-7 CALCIUM CHLORITE

14681-52-8 MANGANESE 56

14681-59-5 IRON 55

14681-63-1 NIOBIUM 94

14681-65-3 NIOBIUM 90

14681-74-4 NIOBIUM 88

14681-75-5 ZIRCONIUM 88

14682-66-7 ALUMINUM 26

14683-06-8 TIN 121

14683-07-9 TIN 123

14683-08-0 TIN 125

14683-10-4 ANTIMONY 124

14683-12-6 TELLURIUM 131

14683-16-0 IODINE 132

14683-19-3 PROMETHIUM 148

14683-23-9 EUROPIUM 152

14683-30-8 LUTETIUM 178

14683-32-0 TUNGSTEN 179

14683-36-4 TANTALUM 183

14684-25-4 BENZENE PHOSPHORUS THIODICHLORIDE
    BENZENE PHOSPHOROUS THIODICHLORIDE [NJL, NJL]

14686-13-6 HEPTYLENE-2-TRANS
    2-HEPTENE, (E)- [PAL]

14686-69-2 BROMINE 82

14687-25-3 LEAD 203

14687-40-2 GERMANIUM 75

14687-50-4 BISMUTH 202

14687-59-3 GERMANIUM 77

14687-60-6 SELENIUM 83

14687-61-7 ARSENIC 77

14687-62-8 BROMINE 83

14694-69-0 IRIDIUM 192

14696-82-3 IODINE AZIDE

14708-14-6 NICKEL(II) FLUOBORATE

14709-52-5 CADMIUM 107

14720-53-7 LEAD BORATE

14729-89-6 DISODIUM IRON(II) ETHYLENEDI-AMINETETRAACETATE

14733-03-0 BISMUTH 214

14762-75-5 CARBON 14

14762-78-8 CERIUM 144

14763-77-0 CUPRIC CYANIDE
    COPPER CYANIDE [NJL]

14797-55-8 NITRATE
    NITRATES [MX2]

14797-65-0 NITRITE
    NITRITES [ATS, MX2]

14798-08-4 BARIUM 140

14807-96-6 TALC
    SILICA, TALC, NON-ASBESTOS FORM [FLA]
    TALC (CONTAINING NO ASBESTOS AND LESS THAN 1% QUARTZ) [NIO, NIO]
    TALC (CONTAINING NO ASBESTOS FIBERS) [ONT, TLV]
    TALC (CONTAINING NO ASBESTOS) [OS1, OS1]
    TALC (FIBROUS) [BC1, BC1]
    TALC (MG3H2(SIO3)4) [PAL]
    TALC (NON-ASBESTIFORM) [ALB]
    TALC (NON-FIBROUS) [QBC, QBC]
    TALC (NONASBESTIFORM, RESP. AND FIBROUS) [MIN]
    TALC NOT CONTAINING ASBESTI-FORM FIBRES [IAR]

14808-44-7 TECHNETIUM 96

14808-60-7 QUARTZ
    QUARTZ (SIO2) [PAL]
    QUARTZ DUST [FLA]
    QUARTZ-CONTAINING FINE DUSTS (> 1% QUARTZ) [MAK, MAK]
    SILICA (QUARTZ) [QBC, QBC]
    SILICA - CRYSTALLINE, QUARTZ [TLV]
    SILICA, CRYSTALLINE [GBR, NIO, NIO, NIO]
    SILICA, CRYSTALLINE QUARTZ [OS1, OS1]
    SILICA, CRYSTALLINE, QUARTZ [CEX, WHM]
    SILICA-CRYSTALLINE, QUARTZ [MIN]

14809-44-0 ARSENIC 69

14809-45-1 ARSENIC 70

14809-47-3 BROMINE 75

14809-48-4 RUBIDIUM 79

14809-49-5 STRONTIUM 81

14809-51-9 STRONTIUM 83

14809-53-1 YTTRIUM 86

14809-55-3 TECHNETIUM 94

14809-68-8 KRYPTON 87

14833-09-1 CHROMIUM 48

14833-23-9 ZINC 62

14833-26-2 ZINC 63

14833-32-0 SILVER 102

14833-35-3 INDIUM 109

14833-49-9 NICKEL 65

14834-67-4 IODINE 133

14834-68-5 IODINE 135

14834-69-6 LANTHANUM 137

14834-71-0 IRIDIUM 187

14834-72-1 PROMETHIUM 143

14834-73-2 PROMETHIUM 144

14834-74-3 PROMETHIUM 146

14834-83-4 YTTERBIUM 166

14838-15-4 PHENYLPROPANOLAMINE

14857-34-2 DIMETHYLETHOXYSILOXANE

14859-67-7 RADON 222

14860-53-8 PHOSPHONIC ACID, (1-HY-
DROXYETHYLIDINE)BIS-, TE-
TRAPOTASSIUMSALT

14861-06-4 2-BUTENOIC ACID, ETHENYL
ESTER
   VINYL CROTONATE [FLA, MSL,
   NFP, NFP]

14866-68-3 CHLORATE

14867-38-0 VANADIUM 47

14867-54-0 GADOLINIUM 152

14877-64-6 NEODYMIUM 141

14877-67-9 SAMARIUM 141

14900-21-1 HAFNIUM 181

14900-25-5 BERKELIUM 249

14901-07-6 IONONE, BETA (BETA-IONONE)

14901-08-7 CYCASIN
   .BETA.-D-GLUCOPYRANOSIDE,
   (METHYL-ONN-AZOXY)METHYL
   [PAL, PAL]

14903-02-7 POTASSIUM 43

14907-88-1 EUROPIUM 146

14907-89-2 EUROPIUM 149

14913-49-6 BISMUTH 212

14913-89-4 RHODIUM 105

14913-92-9 TECHNETIUM 101

14914-02-4 IODINE 130

14914-12-6 LUTETIUM 174

14914-27-3 IODINE 134

14914-66-0 INDIUM 117

14914-68-2 ANTIMONY 119

14914-75-1 BARIUM 131

14914-76-2 CESIUM 131

14922-51-1 HAFNIUM 170

14922-70-4 PLATINUM 188

14928-14-4 SILVER 105

14928-29-1 STRONTIUM 92

14928-30-4 ZIRCONIUM 97

14928-36-0 RUBIDIUM 88

14928-39-3 PLUTONIUM 235

14930-96-2 CYTOCHALASIN B

14932-40-2 THORIUM 231

14932-41-3 TUNGSTEN 185

14932-42-4 XENON 133

14932-53-7 RUBIDIUM 86

14932-64-0 NICKEL 56

14937-16-7 GADOLINIUM 149

14937-17-8 GADOLINIUM 151

14940-68-2 ZIRCON

14952-27-3 PROMETHIUM 141

14952-31-9 GADOLINIUM 147

14952-32-0 GADOLINIUM 146

14952-40-0 ACTINIUM 227

14967-67-0 ERBIUM 161

14967-68-1 PALLADIUM 103

14967-69-2 SILVER 103

14977-61-8 CHROMYL CHLORIDE
   CHROMIUM OXYCHLORIDE [HM1,
   HM1, HM1, NJL, NJL]
   CHROMIUM, DICHLORODIOXO-, (T-4)
   - [PAL]

14981-64-7 PALLADIUM 109

14981-70-5 YTTRIUM 93

14981-79-4 PRASEODYMIUM 143

14981-86-3 EUROPIUM 145

14981-91-0 IRIDIUM 190

14981-97-6 DYSPROSIUM 157

14981-98-7 TERBIUM 153

14982-00-4 DYSPROSIUM 155

14983-46-1 RHENIUM 184

14983-48-3 TUNGSTEN 187

14983-72-3 KRYPTON 77

14986-84-6 SODIUM PHOSPHATE TRIBISINE

14993-35-2 OSMIUM 180

14993-36-3 OSMIUM 182

14993-39-6 PLATINUM 186

14993-64-7 OSMIUM 181

14993-65-8 RHENIUM 181

14993-75-0 AMERICIUM 243

14995-61-0 KRYPTON 88

14995-62-1 XENON 135

14998-27-7 CHLORITE

14998-51-7 TERBIUM 151

14998-63-1 RHENIUM 186

14999-33-8 MANGANESE 53

15005-90-0 CHROMOSULFURIC ACID

15034-49-8 GERMANIUM 69

15034-51-2 GALLIUM 73

15046-84-1 IODINE 129

15047-05-9 CESIUM 129

15055-11-5 CERIUM 134

15055-23-9 TUNGSTEN 178

15055-30-8 PLATINUM 189

15064-65-0 THALLIUM 201

15064-66-1 THALLIUM 199

15064-97-8 MERCURY 194

15065-02-8 SAMARIUM 145

15065-10-8 THORIUM 234

15065-93-7 TERBIUM 149

15065-95-9 TERBIUM 150

15066-92-9 CESIUM 130

15066-93-0 LANTHANUM 132

15067-28-4 LEAD 214

15086-94-9 EOSIN

15092-71-4 MAGNESIUM 28

15092-72-5 SILICON 32

15092-94-1 LEAD 212

15096-52-3 SODIUM ALUMINUM FLUORIDE
   ALUMINUM SODIUM FLUORIDE
   [WHM]

15100-28-4 PROTACTINIUM 234

15108-81-3 TETRAETHYL DITHIOPYROPHOS-
PHATE

15116-79-7 SILVER 104

15116-82-2 MERCURY 193

15116-90-2 NEPTUNIUM 234

15116-95-7 AMERICIUM 240

15117-45-0 CALIFORNIUM 246

15117-48-3 PLUTONIUM 239

15117-53-0 SULFUR 35

15117-96-1 URANIUM 235

15123-69-0 COPPER SELENATE

15125-02-7 RUTHENIUM 94

15125-45-8 TELLURIUM 116

15125-66-3 PRASEODYMIUM 137

15125-75-4 HOLMIUM 155

15128-03-7 COPPER 61

15139-70-5 CADMIUM 117

15142-96-8 PHOSPHORIC ACID, [1,2-
ETHANEDIYLBIS[NITRILOBIS
(METHYLENE)]TETRAKIS-, HEX-
ASODIUM SALT

15151-08-3 XENON 120

15151-09-4 XENON 122

15159-40-7 N-CHLOROFORMYL MORPHO-
LINE

15191-85-2 SILICIC ACID (H4SIO4), BERYL-
LIUM SALT (1:2)
   BERYLLIUM SILICATE [FLA, MSL]

15194-98-6 CALCIUM ARSENITE (2:1)

15195-06-9 STRONTIUM ARSENITE
   STRONTIUM ORTHOARSENITE
   [HM1, HM1, HM1]

15229-36-4 BARIUM 126

15237-97-5 CALIFORNIUM 249

15245-44-0 1,3-BENZENEDIOL, 2,4,6-TRINI-
TRO-, LEAD SALT
   LEAD TRINITRORESORCINATE
   [MSL]

15262-20-1 RADIUM 228

15263-52-2 CARTAP HYDROCHLORIDE

15271-41-7 BICYCLO[2.2.1]HEPTANE-2-
CARBONITRILE, 5-CHLORO-6-
((((METHYLAMINO)CARBONYL)
OXY)IMINO)-,(1ST-(1-ALPHA, 2-
BETA, 4-ALPHA, 5-ALPHA, 6E))-
   2-EXO-CHLORO-5-ENDO-CYANO-2-
   NORBORNANONE O-(METHYLCAR-
   BAMOLY)OXIME [MSL]
   BICYCLO[2.2.1]HEPTANE-2-CARBONI-
   TRILE, 5-CHLORO-6- [NJL]

15356-70-4 DL-MENTHOL

15375-84-5 DISODIUM MANGANESE
ETHYLENEDIAMINETETRAAC-
ETATE

15385-58-7 MERCURY BROMIDE
   MERCUROUS BROMIDE [HM1]
   MERCURY (I) BROMIDE (1:1) [NJL]

15411-90-2 CURIUM 240

15411-91-3 CURIUM 241

15411-92-4 PLUTONIUM 236

15411-93-5 PLUTONIUM 237

15414-89-8 1,3-DIMETHYL-5,5-DIMETHYLHY-
DANTOIN

15422-57-8 SELENIUM 73

15422-58-9 **SELENIUM 81**

15422-59-0 **ARSENIC 73**

15422-71-6 **YTTRIUM 95**

15422-72-7 **YTTRIUM 94**

15467-20-6 **NITRILOTRIACETIC ACID DIS-ODIUM SALT**

15468-32-3 **SILICA, TRIDYMITE**
QUARTZ, TRIDYMITE [MAK, MAK]
SILICA - CRYSTALLINE, TRIDYMITE [TLV]
SILICA, CRYSTALLINE TRIDYMITE [OS1, OS1]
SILICA, CRYSTALLINE, TRIDYMITE [CEX, WHM]
SILICA-CRYSTALLINE, TRIDYMITE [MIN, QBC]
TRIDYMITE (SIO2) [PAL]
TRIDYMITE DUST [FLA, MSL]

15475-56-6 **METHOTREXATE SODIUM**

15478-11-2 **KRYPTON 79**

15480-34-9 **IODINE 120**

15481-70-6 **2,6-TOLUENEDIAMINE DIHY-DROCHLORIDE**

15501-74-3 **SEPIOLITE**

15503-86-3 **ISATIDINE**

15510-73-3 **CURIUM 242**

15520-11-3 **DI(4-TERT-BUTYLCYCLOHEXYL)-PEROXYDICARBONATE**

15521-65-0 **NICKEL DIMETHYLDITHIOCAR-BAMATE**

15545-48-9 **N'-(3-CHLORO-4-METHYLPHENYL)-N,N-DIMETHYLUREA**

15545-97-8 **2,2'-AZODI-(2,4-DIMETHYL-4-METHOXYVALERONITRILE)**

15546-11-9 **DI-N-BUTYLTIN BIS(METHYL MALEATE)**
DIBUTYLTIN METHYL MALEATE [WHM]

15546-12-0 **BIS((2-ETHYLHEXYLOXY)MALE-OLOXY) DIBUTYLSTANNANE**

15546-16-4 **DI-N-BUTYLTIN DI(MONOBUTYL) MALEATE**

15571-58-1 **DIOCTYLTIN BIS(2-ETHYLHEXYL THIOGLYCOLATE)**

15571-75-2 **THORIUM 226**

15575-20-9 **ARSENIC 76**

15585-10-1 **EUROPIUM 154**

15585-26-9 **CHLORINE 39**

15594-54-4 **THORIUM 229**

15598-34-2 **PYRIDINE PERCHLORATE**

15621-76-8 **CURIUM 245**

15623-45-7 **RADIUM 223**

15623-47-9 **THORIUM 227**

15625-89-5 **TRIMETHYLOLPROPANE TRI-ACRYLATE**

15630-89-4 **SODIUM PERCARBONATE**
SODIUM PERCARBONATE COM-POUND WITH HYDROGEN PEROXIDE [NJL]

15646-96-5 **HEXANE, 1,6-DIISOCYANATO-2,4,4-TRIMETHYL-**
2,4,4-TRIMETHYLHEXAMETHYLENE DIISOCYANATE [313]

15662-33-6 **RYANODINE**

15663-27-1 **CISPLATIN**
CIS-DIAMMINEDICHLOROPLATINUM (II) [WHM]
CIS-PLATINOUS DIAMMINE DICHLO-RIDE [FLA]
PLATINUM, DIAMMINEDICHLORO-, (SP-4-2)- [PAL, PAL]

15678-91-8 **KRYPTON 81**

15684-18-1 **CIS-DIAMMINEDICHLOROPALLA-DIUM(II)**

15687-53-3 **URANIUM 240**

15690-69-4 **PALLADIUM 100**

15690-75-2 **THULIUM 166**

15690-77-4 **MOLYBDENUM 90**

15690-82-1 **NEODYMIUM 151**

15690-84-3 **NEPTUNIUM 240**

15690-89-8 **TIN 127**

15696-43-2 **LEAD OCTOATE**
LEAD CAPRYLATE [MSL]

15699-18-0 **NICKEL AMMONIUM SULFATE**
NICKEL(II) AMMONIUM SULFATE [WHM]
SULFURIC ACID, AMMONIUM NICKEL(2+) SALT (2:2:1) [PAL]

15700-08-0 **URANIUM 231**

15700-10-4 **XENON 123**

15700-33-1 **TIN 110**

15700-34-2 **NEODYMIUM 138**

15700-36-4 **NEPTUNIUM 236**

15700-37-5 **NEPTUNIUM 235**

15700-40-0 **NIOBIUM 89**

15700-41-1 **NIOBIUM 98**

15700-49-9 **HOLMIUM 162**

15701-07-2 **CURIUM 249**

15701-09-4 **TELLURIUM 134**

15701-12-9 **SAMARIUM 142**

15701-15-2 **STRONTIUM 80**

15701-16-3 **TANTALUM 186**

15701-17-4 **TECHNETIUM 104**

15701-21-0 **TANTALUM 184**

15701-22-1 **TANTALUM 185**

15706-36-2 **PLATINUM 191**

15706-37-3 **PLUTONIUM 243**

15706-41-9 **POTASSIUM 45**

15706-44-2 **PROMETHIUM 145**

15706-50-0 **RHODIUM 107**

15706-54-4 **PLATINUM 199**

15708-41-5 **SODIUM IRON(III) ETHYLENETE-TRAACETATE**

15715-02-3 **BERKELIUM 246**

15715-04-5 **LANTHANUM 131**

15715-05-6 **LUTETIUM 169**

15715-08-9 **IODINE 123**

15715-94-3 **SAMARIUM 151**

15718-71-5 **ETHYLENE DIAMINE DIPER-CHLORATE**

15720-26-0 **BROMINE 74**

15720-29-3 **CALIFORNIUM 253**

15720-35-1 **CESIUM 127**

15720-45-3 **POLONIUM 207**

15720-47-5 **PROMETHIUM 150**

15720-55-5 **THALLIUM 200**

15720-57-7 **THALLIUM 202**

15720-75-9 **THULIUM 172**

15720-78-2 **TIN 111**

15721-02-5 **2,2',5,5'-TETRACHLOROBENZI-DINE**

15726-30-4 **CESIUM 135**

15735-70-3 **PLATINUM 193**

15735-74-7 **PLATINUM 197**

15739-80-7 **LEAD SULFATE**
LEAD SULPHATE [NJL]
SULFURIC ACID, LEAD SALT [MSL]

15741-25-0 **BARIUM 128**

15741-29-4 **BARIUM 141**

15741-32-9 **LUTETIUM 170**

15743-50-7 **THALLIUM 198**

15743-51-8 **URANIUM 230**

15743-55-2 **ZINC 72**

15743-56-3 **ZIRCONIUM 86**

15743-84-7 **RADIUM 227**

15749-33-4 **TITANIUM 44**

15749-43-6 **TUNGSTEN 176**

15749-44-7 **TUNGSTEN 177**

15749-46-9 **TUNGSTEN 181**

15749-54-9 **PALLADIUM 101**

15749-66-3 **PHOSPHORUS 33**

15749-81-2 **NEODYMIUM 149**

15749-97-0 **HOLMIUM 164**

15750-02-4 **HOLMIUM 159**

15750-04-6 **HOLMIUM 167**

15750-13-7 **HAFNIUM 175**

15750-15-9 **INDIUM 111**

15750-23-9 **FERMIUM 254**

15750-24-0 **FERMIUM 255**

15750-26-2 **FERMIUM 257**

15751-59-4 **YTTRIUM 92**

15751-77-6 **ZIRCONIUM 93**

15751-81-2 **XENON 138**

15752-22-4 **IRIDIUM 188**

15752-27-9 **LUTETIUM 171**

15752-34-8 **MENDELEVIUM 257**

15752-38-2 **BERKELIUM 247**

15752-86-0 **LEAD 202**

15755-17-6 **IODINE 121**

15755-18-7 **ANTIMONY 117**

15755-27-8 **ANTIMONY 116**

15755-33-6 **ARSENIC 72**

15755-35-8 **ARSENIC 78**

15755-39-2 **ASTATINE 211**

15755-53-0 **BERKELIUM 250**

15755-89-2 **LUTETIUM 179**

15755-98-3 **ACTINIUM 224**

15756-15-7 **MERCURY 195**

15756-26-0 **AMERICIUM 244**

15756-29-3 ANTIMONY 131

15756-32-8 ANTIMONY 126

15756-34-0 ANTIMONY 128

15756-35-1 ANTIMONY 130

15756-76-0 GERMANIUM 67

15756-77-1 GERMANIUM 68

15756-83-9 GERMANIUM 78

15756-84-0 GERMANIUM 66

15756-89-5 GOLD 194

15756-90-8 FERMIUM 252

15756-98-6 FRANCIUM 223

15757-14-9 GALLIUM 68

15757-23-0 HAFNIUM 173

15757-86-5 COPPER 67

15757-87-6 CURIUM 243

15757-90-1 CURIUM 246

15757-94-5 CERIUM 135

15758-03-9 CESIUM 132

15758-14-2 CHROMIUM 49

15758-24-4 CALIFORNIUM 248

15758-27-7 CESIUM 125

15758-29-9 CESIUM 138

15758-32-4 CURIUM 247

15758-33-5 CURIUM 248

15758-35-7 RUTHENIUM 97

15758-45-9 SELENIUM 79

15758-54-0 TANTALUM 174

15758-55-1 TANTALUM 176

15758-64-2 TERBIUM 154

15759-26-9 TANTALUM 172

15759-27-0 TANTALUM 177

15759-28-1 TANTALUM 175

15759-29-2 TANTALUM 180

15759-35-0 TECHNETIUM 97

15759-52-1 TELLURIUM 133

15759-55-4 TERBIUM 158

15759-70-3 SAMARIUM 156

15760-04-0 SILVER 111

15760-07-3 SILVER 115

15765-19-2 CALIFORNIUM 251

15765-23-8 PRASEODYMIUM 145

15765-24-9 PRASEODYMIUM 147

15765-31-8 PROMETHIUM 149

15765-38-5 BROMINE 76

15765-39-6 BROMINE 77

15765-78-3 RHENIUM 189

15765-79-4 RHODIUM 99

15765-80-7 RHODIUM 100

15765-82-9 RHODIUM 102

15765-86-3 RUBIDIUM 84

15766-00-4 SAMARIUM 153

15766-03-7 PROMETHIUM 151

15766-06-0 PROTACTINIUM 232

15766-09-3 PROTACTINIUM 228

15766-10-6 PROTACTINIUM 230

15766-25-3 NEPTUNIUM 238

15766-33-3 NICKEL 66

15766-50-4 OSMIUM 185

15766-57-1 OSMIUM 194

15776-16-6 AMERICIUM 246

15776-19-9 BISMUTH 206

15776-20-2 BISMUTH 213

15790-07-5 C.I. PIGMENT YELLOW 104

15792-67-3 FD&C BLUE NO. 1 (ALUMINUM SALT)

15816-77-0 LEAD 211

15816-85-0 LANTHANUM 135

15816-87-2 LANTHANUM 138

15816-88-3 LANTHANUM 141

15816-89-4 LANTHANUM 142

15816-99-6 IRIDIUM 195

15825-70-4 MANNITOL HEXANITRATE
MANNITOL HEXANITRATE (NITRO-MANNITE) [HM1, HM1]

15827-56-2 (CIS) 1,4-CYLCOHEXANEDIAMINE

15829-53-5 MERCUROUS OXIDE

15832-32-3 NIOBIUM 96

15832-34-5 HOLMIUM 157

15832-40-3 HAFNIUM 183

15832-46-9 NEPTUNIUM 233

15832-50-5 TIN 126

15832-57-2 THULIUM 162

15840-01-4 DYSPROSIUM 166

15840-02-5 EINSTEINIUM 253

15840-03-6 EINSTEINIUM 254

15840-13-8 ERBIUM 169

15840-14-9 ERBIUM 172

15840-15-0 EUROPIUM 148

15840-16-1 EUROPIUM 150

15843-02-4 NICKEL FORMATE

15845-52-0 DIBASIC LEAD PHOSPHATE
LEAD HYDROGEN PHOSPHATE [MSL]
LEAD PHOSPHATE, DIBASIC [NJL]

15869-86-0 4-ETHYLOCTANE

15869-92-8 3,4-DIMETHYLOCTANE

15872-73-8 PHENOL, 2,4-DIAMINO-6-METHYL-

15875-13-5 1,3,5-TRIAZINE-1,3,5(2H,4H,6H)-TRIPROPANAMINE-N,N,N',N',N", N"-HEXAMTHYL-

15905-86-9 NITRIC ACID, URANIUM SALT

15922-78-8 SODIUM OMADINE

15930-94-6 ZINC CHROMATE HYDROXIDE

15950-66-0 2,3,4-TRICHLOROPHENOL
PHENOL, 2,3,4-TRICHLORO- [PAL]

15956-58-8 2-ETHYLHEXANOIC ACID, MANGANESE SALT

15965-99-8 OXIRANE, [(HEXADECYLOXY) METHYL]-
HEXADECYL GLYCIDYL ETHER [EPA, OTS]

15972-60-8 ALACHLOR
ALACHLOR (2-CHLORO-2',6'-DI-ETHYL-N-METHOXYMETHYL AC-ETANILIDE) [CN8]

15980-15-1 1,4-OXATHIANE
1,4-THIOXANE [FLA, MSL, NFP, NFP]

16044-16-9 CALIFORNIUM 244

16057-77-5 OSMIUM 193

16065-83-1 CHROMIUM, ION (CR3+)
CHROMIUM (III) [SD4]
TRIVALENT CHROMIUM COMPOUNDS [ONT]

16066-38-9 DI-N-PROPYL PEROXYDICARBONATE

16071-86-6 C.I. DIRECT BROWN 95
CUPRATE(2-), [5-[[4'-[[2,6-DIHYDROXY-3-[(2-HYDROXY-5-SULFOPHENYL)AZO]PHENYL]AZO][1,1'-BIPHENYL]-4-YL]AZO]-2-HYDROXY BENZOATO(4-)]-, DISODIUM [TSC]
CUPRATE(2-), [5-[[4'-[[2,6-DIHYDROXY-3-[(2-HYDROXY-5-SULFOPHENYL)AZO]PHENYL]AZO] 1,1'-BIPHENYL]4-YL]AZO-2-HYDROXY BENZOATO(4-)]-, DISODIUM [PAL, PAL]
DIRECT BROWN 95 [CAL, FLA, MSL]
DIRECT BROWN 95 (TECH. GRADE) [CAB]
DIRECT BROWN 95 (TECHNICAL GRADE) [C65, C65, IAR, MIN, WHM]

16091-18-2 DIOCTYLTIN MALEATE

16110-89-7 BENZENESULFONIC ACID, 4-[(4, 6-DICHLORO-1,3,5-TRIAZIN-2-YL) AMINO]-

16111-62-9 DI(2-ETHYLHEXYL)PEROXY-DICARBONATE

16143-79-6 CUPRATE(4-), [U-[[6,6'-[3,3'-DI-HYDROXY[1,1'-BIPHENYL]-4,4'-DIYL]BIS(AZO)BIS[4-AMINO-5-HYDROXY-1,3-NAPHTHALENE DISULFONATO](8-)]DI-, TETRASODIUM

16215-49-9 BUTYL PEROXYDICARBONATE
DI-N-BUTYL PEROXYDICARBONATE [DOT]
N-BUTYL PEROXYDICARBONATE [HM1]

16219-75-3 ETHYLIDENE NORBORNENE
5-ETHYLIDENE-2-NORBORNENE [ONT, WHM]
BICYCLO(2.2.1)HEPT-2-ENE, 5-ETHYLIDENE- [PAL]

16222-66-5 THALLIUM(III) SULFATE

16245-77-5 1,4-BENZENEDIAMINE, SULFATE (1:1)
PARA-PHENYLENEDIAMINE, SULFATE SALT [CTR]
SULFATE SALTS OF PARA-PHENYLENEDIAMINE [TSE]

16245-97-9 OXIRANE, [(OCTADECYLOXY) METHYL]-
N-OCTADECYL GLYCIDYL ETHER [EPA]
OCTADECYL GLYCIDYL ETHER, N-[OTS]

16337-84-1 CARBONIC ACID, NICKEL SALT

16368-97-1 PHOSPHORIC ACID, BIS(2-ETHYLHEXYL) PHENYL ESTER

16389-88-1 DOLOMITE [CAMG(CO3)2]
DOLOMITE [PST]

16415-43-3 AMERICIUM 245

16423-68-0 FLUORESCEIN, 2',4',5',7'-TE-TRAIODO, DISODIUM SALT

16484-86-9 DIETHYL GLYCOL

16501-01-2 METHOXY ETHYL PHTHALATE

16532-79-9 4-BROMOBENZYLCYANIDE
BENZENE ACETONITRILE, 4-BROMO-
[TSC, TSC]

16543-55-8 N-NITROSONORNICOTINE
N'-NITROSONORNICOTINE [IAR,
WHM]
PYRIDINE, 3-(1-NITROSO-2-PYRRO-
LIDINYL)-, (S)- [PAL, PAL]

16568-02-8 GYROMITRIN
GYROMITRIN (ACETALDEHYDE
METHYLFORMYLHYDRAZONE) [C65,
C65, CAL]
HYDRAZINECARBOXALDEHYDE,
ETHYLIDENEMETHYL- [PAL, PAL]

16588-67-3 3-[N-ETHYL-4-[[6-(METHYLSUL-
FONYL)-2-BENZOTHIAZOLYL]
AZO]-M-TOLUIDINO]-PROPIONI-
TRILE

16645-96-8 TIN 128

16645-99-1 LEAD 200

16646-00-7 LEAD 198

16652-07-6 BERKELIUM 245

16652-10-1 AMERICIUM 239

16680-47-0 SODIUM EQUILIN SULFATE
CONJUGATED ESTROGENS: SODIUM
EQUILIN SULFATE [PAL, PAL]

16685-55-5 ARSENIC 71

16715-83-6 2-PROPENOIC ACID, 2-METHYL-,
2-[BIS(1-METHYLETHYL)AMINO]
ETHYL ESTER

16721-80-5 SODIUM HYDROSULFIDE
SODIUM HYDROSULPHIDE [WHM]
SODIUM SULFIDE [FLA]
SODIUM SULFIDE (NA(SH)) [PAL]

16729-61-6 LANTHANUM 143

16729-74-1 POLONIUM 203

16729-76-3 POLONIUM 205

16731-55-8 POTASSIUM METABISULFITE

16752-77-5 METHOMYL
ETHANIMIDOTHIOIC ACID, N-
[[(METHYLAMINO)CARBONYL]OXY]-,
METHYL ESTER [PAL]
METHOMYL (ISO) [GBR, GBR]
METHOMYL (LANNATE) [ALB, ALB,
ALB, BC1, BC1, OS1]
METHOMYL (LANNATE) (R) [QBC]
S-METHYL N-[(METHYLCAR-
BAMOYL)OXY]-THIOACETAMIDATE
(METHOMYL) [CAL]

16842-03-8 COBALT HYDROCARBONYL
COBALT, TETRACARBONYLHYDRO-
[PAL]

16845-29-7 RUTHENIUM CHLORIDE HY-
DROXIDE

16853-85-3 LITHIUM ALUMINUM HYDRIDE
ALUMINATE(1-), TETRAHYDRO-,
LITHIUM,(T-4)- [PAL]
LITHIUM TETRAHYDROALUMINATE
[FLA, MSL]

16871-71-9 ZINC SILICOFLUORIDE
SILICATE(2-), HEXAFLUORO-, ZINC
(1:1) [PAL]
ZINC FLUOROSILICATE [HM1, HM1,
HM1]

16871-90-2 POTASSIUM SILICOFLUORIDE
POTASSIUM FLUOROSILICATE [HM1,
HM1]

16872-11-0 FLUOROBORIC ACID

16887-00-6 CHLORIDE

16893-85-9 SODIUM SILICOFLUORIDE
SODIUM FLUOROSILICATE [NJL]
SODIUM SILICA FLUORIDE [MSL]

16901-76-1 THALLIUM NITRATE

16903-35-8 TETRACHLOROAURIC ACID

16919-19-0 AMMONIUM SILICOFLUORIDE
AMMONIUM FLUOROSILICATE [PST]
SILICATE(2-), HEXAFLUORO-, DI-
AMMONIUM [PAL]

16919-58-7 AMMONIUM CHLOROPLATINATE

16919-73-6 DIPOTASSIUM HEX-
ACHLOROPALLADATE

16920-93-7 DIPOTASSIUM HEXABROMO-
PLATINATE

16921-30-5 POTASSIUM HEXACHLOROPLATI-
NATE(IV)

16922-44-4 GALLIUM 65

16923-95-8 ZIRCONIUM POTASSIUM FLUO-
RIDE
ZIRCONATE(2-), HEXAFLUORO-,
DIPOTASSIUM, (OC-6-11)- [PAL]

16924-00-8 POTASSIUM HEPTAFLUOROTAN-
TALATE

16938-22-0 HEXANE, 1,6-DIISOCYANATO-2,2,
4-TRIMETHYL-
2,2,4-TRIMETHYLHEXAMETHYLENE
DIISOCYANATE [313]

16940-66-2 SODIUM BOROHYDRIDE

16940-81-1 HEXAFLUOROPHOSPHORIC ACID

16941-12-1 CHLOROPLATINIC ACID

16949-15-8 LITHIUM BOROHYDRIDE

16961-83-4 FLUOROSILICIC ACID
HYDROFLUOSILICIC ACID [MSL]
SILICOFLUORIC ACID [NJL]

16962-07-5 ALUMINUM BOROHYDRIDE

16984-48-8 FLUORIDES
FLUORIDE [ALB, ALB, ATS, BC1,
CAL, CD1, CEX, GBR, QBC, RCA,
SDW, SDW, SDW]

16996-40-0 LEAD 2-ETHYLHEXOATE

16996-51-3 LEAD LINOLEATE

17014-71-0 POTASSIUM PEROXIDE
POTASSIUM PEROXIDE (K2(O2))
[PAL]

17026-81-2 3-AMINO-4-ETHOXYAC-
ETANILIDE

17056-36-9 RUBIDIUM 83

17068-78-9 ANTHOPHYLLITE, NON-ASBESTI-
FORM
ANTHOPHYLLITE [PAL, PAL]
ASBESTIFORM MINERAL(S) [TSC]

17086-28-1 DOXYCYCLINE MONOHYDRATE

17099-81-9 FERRIC EDTA
ETHYLENEDIAMINETETRAACETIC
ACID, FE(III) CHELATE [PST]

17109-49-8 EDIFENPHOS

17125-80-3 BARIUM SILICOFLUORIDE

17230-88-5 DANAZOL

17237-93-3 CARBONIC ACID, NICKEL(2+)
SALT (2:1)

17239-85-9 BISMUTH 200

17239-87-1 LEAD 201

17256-39-2 2-PROPANAMINE, 1-CHLORO-N,N-
DIMETHYL-, HYDROCHLORIDE

17268-47-2 3-DIMETHYLAMINO-N,N-
DIMETHYLPROPIONAMIDE

17354-14-2 1,4-BIS(BUTYLAMINO)-9,10-AN-
THRAQUINONE
C.I. SOLVENT BLUE 35 [PST]

17372-87-1 ACID RED 87
FLUORESCEIN, 2',4',5',7'-TETRA-
BROMO-, DISODIUM SALT [PST]

17418-58-5 9,10-ANTHRACENEDIONE, 1-
AMINO-4-HYDROXY-2-PHENOXY-

17421-79-3 SODIUM ETHYLENEDIAMINETE-
TRAACETATE

17462-58-7 SEC-BUTYL CHLOROFORMATE

17476-04-9 LITHIUM ALUMINUM TRI-TERT-
BUTOXY-HYDRIDE

17501-44-9 ZIRCONIUM ACETYLACETONATE

17518-47-7 2-(1-NAPHTHYL)THIOACETAMIDE

17557-23-2 OXIRANE, 2,2'-[(2,2-DIMETHYL-1,
3-PROPANEDIYL)BIS(OXYMETHY-
LENE)]BIS-
NEOPENTYL GLYCOL DIGLYCIDYL
ETHER [EPA, OTS]

17557-67-4 6-(METHYLSULFONYL)-2-BEN-
ZOTHIAZOLAMINE

17572-97-3 ETHYLENEDIAMINETE-
TRAACETIC ACID, TRIPOTAS-
SIUM SALT

17601-96-6 2-AMINO-4-[(2-HYDROXYETHYL)
SULFONYL]PHENOL

17617-23-1 FLURAZEPAM

17620-10-9 ANTIMONY 115

17637-99-9 PALLADIUM 107

17639-93-9 METHYL-2-CHLOROPROPIONATE

17687-37-5 UREA NITRATE (VAN)
UREA NITRATE [HM1, HM1]

17688-68-5 4-PHENYLTHIOMORPHOLINE, 1,
1-DIOXIDE-

17702-41-9 DECABORANE
DECABORANE (14) [302]
DECABORANE(14) [PAL]

17702-57-7 FORMPARANATE

17741-62-7 4-[4-[(2,6-DICHLORO-4-NITRO-
PHENYL)AZO]PHENYL]THIOMOR-
PHOLINE,1,1-DIOXIDE-

17754-90-4 BENZALDEHYDE, 4-(DIETHY-
LAMINO)-2-HYDROXY-

17804-35-2 BENOMYL
BENOMYL [MEX, MEX]
BENOMYL (ISO) [GBR, GBR]
CARBAMIC ACID, [1-[(BUTYLAMINO)
CARBONYL]-1H-BENZIMIDAZOL-2-
YL]-, METHYL ESTER [PAL]

17831-71-9 TETRAETHYLENE GLYCOL DI-
ACRYLATE

17861-62-0 NICKEL NITRITE

17913-80-3 XENON 121

17924-92-4 ZEARALENONE

17963-04-1 SILANE, ETHOXYDIMETHYL[3-
(OXIRANYL-METHOXY)PROPYL]-
3-(DIMETHYLETHOXYSILYL)PROPYL
GLYCIDYL ETHER [EPA, OTS]

17977-09-2 2-PROPENOIC ACID, 2,2-DINITRO-
PROPYL ESTER

18015-76-4 C.I. BASIC GREEN 4, OXALATE

18130-74-0 BIFLUORIDE, N.O.S.

18233-96-0 AMERICIUM 238

18235-46-6 THALLIUM 194

18266-52-9 1,4-BENZENEDIAMINE, 2-NITRO-, DIHYDROCHLORIDE

18268-34-3 RUBIDIUM 81

18268-70-7 HEXANOIC ACID, 2-ETHYL-, DI-ESTER WITH TETRAETHYLENE GLYCOL

18282-10-5 STANNIC OXIDE
TIN OXIDE [ALB, ALB, MEX, MEX, MIN]

18307-23-8 SEPIOLITE

18312-04-4 OCTANOIC ACID, ZIRCONIUM SALT

18312-31-7 STEARYL OCTANOATE

18351-85-4 PHOSPHORIC ACID, MONO(2-METHYLPHENYL) ESTER

18378-89-7 PLICAMYCIN

18396-20-8 FERMIUM 253

18411-36-4 NEODYMIUM 139

18414-36-3 DIBENZYLDICHLOROSILANE
DIBENZYL DICHLOROSILANE [NJL, NJL]

18432-28-5 TETRAZOLYL AZIDE

18435-45-5 NONADECANE

18454-12-1 LEAD CHROMATE OXIDE
LEAD CHROMATE [FLA]
LEAD CHROMATE OXIDE (PB2(CRO4)O) [PAL, PAL]

18472-51-0 CHLOROHEXIDINE DIGLU-CONATE

18495-30-2 PROPANE, 1,1,2,3-TETRACHLORO-

18496-04-3 NIOBIUM 97

18496-25-8 SULFIDE
SULPHIDE (AS H2S) [CD1]
SULPHIDES [MX2]

18540-29-9 CHROMIUM (VI)
CHROMIUM, HEXAVALENT [ATS]
CHROMIUM, ION (CR6+) [PAL]
HEXAVALENT CHROMIUM [MX2, RCA]

18633-25-5 OXIRANE, TRIDECYL-

18662-53-8 TRISODIUM NITRILOTRIAC-ETATE MONOHYDRATE
NITRILOTRIACETIC ACID TRISODIUM MONOHYDRATE [CSR, CSR]
NITRILOTRIACETIC ACID, TRISODIUM SALT MONOHYDRATE [C65, C65, CAB]
NTA TRISODIUM SALT.H2O [MSL]

18760-44-6 3-(DECYLOXY)TETRAHYDROTH-IOPHENE 1,1-DIOXIDE

18766-38-6 DOCOSAMETHYLCYLCLOUNDE-CASILOXANE

18772-36-6 EICOSAMETHYLCYCLODE-CASILOXANE

18784-52-6 PLUTONIUM 245

18810-58-7 BARIUM AZIDE

18844-04-7 HEXATRIACONTAMETHYLHEP-TADECASILOXANE

18853-08-2 RHENIUM 178

18853-09-3 RHENIUM 177

18868-43-4 MOLYBDENUM OXIDE
MOLYBDENUM DIOXIDE [ONT]

18879-37-3 BARIUM 142

18883-66-4 STREPTOZOTOCIN
D-GLUCOSE, 2-DEOXY-2-[[(METHYL-NITROSOAMINO)CARBONYL]AMINO]- [PAL, PAL]
STREPTOZOCIN [NJL, NJL]

18912-80-6 ETHANOL, 2-[2-(2-METHYL-PROPOXY)ETHOXY]-
DIETHYLENE GLYCOL MONOISOBUTYL ETHER [NFP, NFP]

18917-93-6 MAGNESIUM LACTATE

18919-94-3 TETRACOSAMETHYLCYCLODO-DECASILOXANE

18972-56-0 MAGNESIUM SILICOFLUORIDE
MAGNESIUM FLUOROSILICATE [HM1, HM1]

19010-66-3 LEAD DIMETHYLDITHIOCARBA-MATE
LEAD BIS (DIMETHYLDITHIOCARBA-MATE) [MSL]

19044-88-3 ORYZALIN [4-(DIPROPYLAMINO)-3,5-DINITROBENZENESULFON-AMIDE]

19074-24-9 BIS(2-ETHYLHEXYL) DODECANE-DIOATE

19120-62-8 VANADIC ACID, TRIISOBUTYL ESTER

19168-23-1 AMMONIUM CHLOROPALLA-DATE

19201-36-6 2-BUTENEDIOIC ACID (Z-)-, MONO [2-[(1-OXO-2-PROPENYL)OXY] ETHYL]-ESTER

19249-03-7 ETHANOL,2,2'-[OXYBIS(2,1-ETHANEDIYLOXY)]BIS-,BIS(4-METHYLBENZENESULFONATE)

19287-45-7 DIBORANE
DIBORAN [QBC]
DIBORANE(6) [PAL]

19355-69-2 2-AMINO-2-METHYLPROPANENI-TRILE

19372-44-2 CALCIUM, BIS(2,4-PENTANEDION-ATO-O,O')

19381-50-1 C.I. ACID GREEN 1

19408-74-3 1,2,3,7,8,9-HEXACHLORODIBENZO-P-DIOXIN

19472-24-3 ORTHO-BROMOBENZYL CYANIDE

19529-38-5 DIETHYLENETRIAMINEPEN-TAACETIC ACID, DISODIUM IRON (III) SALT

19594-40-2 4-UNDECANONE, 2-METHYL-

19624-22-7 PENTABORANE
PENTABORANE(9) [PAL]

19660-16-3 2-PROPENOIC ACID, 2,3-DIBRO-MOPROPYL ESTER

19666-30-9 OXADIAZON
OXYDIAZON [3-[2,4-DICHLORO-5-(1-METHYLETHOXY)PHENYL]-5-(1,1-DIMETHYLETHYL)-1,3,4-OXADIAZOL-2(3H)-ONE] [313]

19721-22-3 1-PROPANOL, 3-MERCAPTO

19780-11-1 DODECENYLSUCCINIC ANHY-DRIDE

19783-14-3 LEAD HYDROXIDE

19869-93-3 SELENIUM 70

19910-65-7 DI-SEC-BUTYL PEROXYDICAR-BONATE

20018-09-1 1-(DIIODOMETHYL)SULFONYL-4-METHYL BENZENE

20062-22-0 HEXANITROSTILBINE
HEXANITROSTILBENE [NJL]

20091-45-6 GOLD 200

20103-09-7 1,4-BENZENEDIAMINE, 2,5-DICHLORO-

20217-01-0 OXIRANE, [(2,4-DIBROMOPHE-NOXY)METHYL]-
2,4-DIBROMOPHENYL GLYCIDYL ETHER [EPA]
DIBROMOPHENYL GLYCIDYL ETHER, 2,4- [OTS]

20265-96-7 P-CHLOROANILINE HYDROCHLO-RIDE

20265-97-8 P-ANISIDINE HYDROCHLORIDE

20268-51-3 7-NITROBENZO[A]ANTHRACENE

20282-70-6 [1,1'-BIPHENYL]-4,4'-BIS(DIAZO-NIUM), 3,3'-DIMETHOXY-

20324-32-7 DIPROPYLENE GLYCOL METHYL ETHER
2-PROPANOL, 1-(2-METHOXY-1-METHYLETHOXY)- [TSC, TSC, TSC]

20324-33-8 TRIPROPYLENE GLYCOL METHYL ETHER

20325-40-0 3,3'-DIMETHOXYBENZIDINE DI-HYDROCHLORIDE
3,3'-DIMETHOXYBENZIDINE DIHY-DROCHLORIDE (O-DIANISIDINE DI-HYDROCHLORIDE) [313]
3,3'-DIMETHOXYBENZIDINE DIHY-DROCHLORIDE (ORTHO-DIANISIDINE DIHYDROCHLORIDE) [C65]

20354-26-1 METHAZOLE [2-(3,4-DICHLOROPHENYL)-4-METHYL-1,2,4-OXADIAZOLIDINE-3,5-DIONE]

20379-10-6 ACTINIUM 226

20403-41-2 LEAD MYRISTATE

20427-58-1 ZINC HYDROXIDE

20427-59-2 COPPER HYDROXIDE

20526-58-3 SODIUM SULFOSUCCINATE

20539-63-3 3-NITROPERYLENE

20600-96-8 TETRANITRO DIGLYCERIN

20601-76-7 ASTATINE 207

20624-25-3 DIETHYLCARBAMODITHIOIC ACID SODIUM SALT TRIHYDRATE

20667-12-3 SILVER OXIDE

20727-33-7 DIOCTYL SODIUM SULFOSUCCI-NATE

20770-41-6 POTASSIUM PHOSPHIDE

20816-12-0 OSMIUM TETROXIDE
OSMIUM OXIDE (OSO4), (T-4)- [PAL]
OSMIUM TETRAOXIDE [GBR, GBR]

20820-44-4 NITRO ISOBUTANE TRIOL TRINI-TRATE

20824-56-0 DIAMMONIUM ETHYLENEDI-AMINETETRAACETATE

20830-75-5 DIGOXIN

20830-81-3 DAUNOMYCIN
5,12-NAPHTHACENEDIONE, 8-ACETYL-10-[(3-AMINO-2,3,6-TRIDEOXY-.ALPHA.-L-LYXO-HEX-OPYRANOSYL)OXY]-7,8,9,10-TE-TRAHYDRO-6,8,11-TRIHYDROXY-8-(HYDROXYACETYL)-1-METHOXY-, HYDROCHLORIDE, (8S-CIS)- [PAL, PAL]

20837-86-9 LEAD CYANAMIDE

20859-73-8 ALUMINUM PHOSPHIDE
ALUMINUM PHOSPHIDE (ALP) [PAL]

20941-65-5 ETHYL TELLURAC

21041-93-0 COBALT HYDROXIDE

21087-64-9 METRIBUZIN
1,2,4-TRIAZIN-5(4H)-ONE, 4-
AMINO-6-(1,1-DIMETHYLETHYL)-3-
(METHYLTHIO)- [PAL]
METRIBUZIN (SENCOR) [ALB, ALB]

21129-18-0 MANGANESE PROPIONATE

21160-95-2 METHYLAMMONIUM N-
METHYLDITHIOCARBAMATE

21232-47-3 3,3',4,4'-TETRACHLOROAZOXY-
BENZENE

21245-02-3 4-(DIMETHYLAMINO)BENZOIC
ACID, 2-ETHYLHEXYL ESTER

21259-20-1 T-2 TOXIN

21351-79-1 CESIUM HYDROXIDE
CAESIUM HYDROXIDE [AUS, GBR]
CESIUM HYDROXIDE (CS(OH)) [PAL]

21368-68-3 DL-CAMPHOR

21416-87-5 ICRF-159

21429-43-6 ACETAMIDE, N-[5-[BIS[2-(ACETY-
LOXY)ETHYL]AMINO]-2-[(2-
CHLORO-4,6-DINITROPHENYL)
AZO]-4-ETHOXYPHENYL]-

21459-71-2 RHENIUM 182

21548-32-3 FOSTHIETAN

21564-17-0 THIOCYANIC ACID, (2-BENZOTH-
IAZOLYTHIO)METHYL ESTER
THIOCYANIC ACID, (2-BENZOTHIA-
ZOLYTHIO) METHYL ESTER [MSL]

21609-90-5 LEPTOPHOS
LEPTOPHOS (O-(4-BROMO-2,5-
DICHLOROPHENYL) O-METHYL-
PHENYLPHOSPHONOTHIOATE [CN8]

21645-51-2 ALUMINUM HYDROXIDE

21651-19-4 TIN OXIDE
TIN OXIDE DUST [ALB]

21652-27-7 OLEYL BASED IMIDAZOLINE
(Z)-2-(8-HEPTADECENYL)-2-IMIDA-
ZOLINE-1-ETHANOL [PST]

21679-31-2 CHROMIUM ACETYLACETONATE

21725-46-2 CYANAZINE
BLADEX [NJL]

21732-17-2 TETRAZOL-1-ACETIC ACID

21739-91-3 CYTEMBENA

21850-44-2 TETRABROMOBISPHENOL-A-BIS-
2,3-DIBROMOPROPYL ETHER

21884-44-6 LUTEOSKYRIN

21908-53-2 MERCURIC OXIDE

21923-23-9 CHLORTHIOPHOS

21982-43-4 2-PROPENOIC ACID, 2-CYANO-,
ETHOXY ETHYL ESTER

21985-87-5 PENTANITROANILINE

22092-38-2 OXIRANE, PENTADECYL-

22095-52-9 NEODYMIUM 136

22095-53-0 PRASEODYMIUM 136

22095-76-7 CALIFORNIUM 254

22095-77-8 TANTALUM 173

22128-62-7 CHLOROMETHYLCHLOROFOR-
MATE
CHLOROMETHYL CHLOROFORMATE
[WHM]

22205-45-4 COPPER(I) SULFIDE

22224-92-6 FENAMINPHOS
ETHYL 3-METHYL-4-(METHYLTHIO)
-PHENYL (1-METHYLETHYL) PHOS-
PHORAMIDATE [CAL]
FENAMIPHOS [ONT, ONT]
FENAMIPHOS (NEMACUR) [ALB,
ALB, ALB]
PHOSPHORAMIDIC ACID, (1-
METHYLETHYL)-, ETHYL-3-METHYL-
4-(METHYLTHIO)PHENYL ESTER
[PAL]

22288-43-3 1,1,3,3-TETRAMETHYLBUTYLPER-
OXY-2-ETHYL HEXANOATE

22349-59-3 1,4-DIMETHYLPHENANTHRENE

22397-33-7 3,3,6,6,9,9-HEXAMETHYL-1,2,4,5-
TETRAOXOCYCLONONANE

22398-80-7 INDIUM PHOSPHIDE

22421-59-6 2,6-DIBROMO-4-METHYLPHENYL
GYCIDYL ETHER
DIBROMO-4-METHYLPHENYL GLY-
CIDYL ETHER, 2,6- [OTS]

22464-99-9 ZIRCONIUM ETHYL HEXOATE
2-ETHYLHEXANOIC ACID, ZIRCO-
NIUM SALT [PST]

22481-48-7 RADON 220

22506-53-2 3,9-DINITROFLUORANTHENE

22541-79-3 CHROMIUM (II)
CHROMIUM, ION (CR2+) [PAL]
DIVALENT CHROMIUM COMPOUNDS
[ONT]

22571-95-5 SYMPHYTINE

22750-93-2 ETHYL PERCHLORATE

22751-24-2 M-NITROBENZENE DIAZONIUM
PERCHLORATE

22781-23-3 BENDIOCARB
BENDIOCARB [2,2-DIMETHYL-1,3-
BENZODIOXOL-4-OL METHYLCAR-
BAMATE] [313]

22826-61-5 TETRAAZIDO BENZENE QUINONE

22830-45-1 IRON (II) GLUCONATE

22966-79-6 ESTRADIOL MUSTARD
OESTRADIOL MUSTARD [CAL, IAR]

22967-92-6 METHYLMERCURY
METHYL MERCURY [C65, MAK,
MAK, MAK, MAK]

22975-76-4 4,4'-DIMETHYLANGELICIN PLUS
ULTRAVIOLET A RADIATION

23023-91-8 2-PROPENOIC ACID, 2-CYANO-, 2,
2,2-TRIFLUOROMETHYL ESTER

23054-60-6 N-(2-HYDROXYPROPYL)OC-
TANAMIDE

23054-61-7 N-(2-HYDROXYPROPYL)DE-
CANAMIDE

23092-17-3 HALAZEPAM

23103-98-2 PIRIMICARB

23135-22-0 OXAMYL
OXAMYL (VYDATE) [SDW, SDW]

23209-59-8 CALCIUM SODIUM METAPHOS-
PHATE

23214-92-8 ADRIAMYCIN
5,12-NAPHTHACENEDIONE, 10-[(3-
AMINO-2,3,6-TRIDEOXY-.ALPHA.-
L-LYXO-HEXOPYRANOSYL)OXY]-7,
8,9,10-TETRAHYDRO-6,8,11-TRIHY-
DROXYACETYL)-1-METHOXY, (8S-
CIS)- [PAL, PAL]
ADRIAMYCIN (DOXORUBICIN HY-
DROCHLORIDE) [C65]

23238-59-7 GOLD 201

23246-96-0 RIDDELLINE

23255-69-8 FUSARENON-X
FUSARENONE X [IAR]

23255-93-8 HYCANTHONE MESYLATE

23282-20-4 NIVALENOL

23305-64-8 2-[2-(PROPOXYETHOXY)ETHOXY]
-ETHANOL

23315-89-1 GADOLINIUM 145

23383-59-7 CUMYL PEROXYPIVALATE

23414-72-4 ZINC PERMANGANTE

23422-53-9 FORMETANATE
FORMETANATE HYDROCHLORIDE
[302, EPA, EPA, MSL]

23436-19-3 2-PROPANOL, 1-(2-METHYL-
PROPOXY)-

23474-91-1 TERT-BUTYL PEROXYCROTO-
NATE

23505-41-1 PIRIMPHOS-ETHYL
PIRIMFOS-ETHYL [MSL]

23523-12-8 HEXATRIACONTAMETHYLCY-
CLOOCTADECASILOXANE

23523-14-0 TRIACONTAMETHYLCYCLOPEN-
TADECASILOXANE

23537-16-8 RUGULOSIN

23541-50-6 DAUNORUBICIN HYDROCHLO-
RIDE

23560-59-0 HEPTENOPHOS

23564-05-8 THIOPHANATE-METHYL

23564-06-9 THIOPHANATE ETHYL [[1,2-
PHENYLENEBIS(IMINOCAR-
BONOTHIOYL)]BISCARBAMIC
ACID DIETHYL ESTER]

23606-32-8 SILVER AMMONIUM NITRATE

23696-28-8 OLAQUINDOX (N-(2-HYDROX-
YETHYL)-3-METHYL-2-QUINOXA-
LINECARBOXAMIDE 1,4-DIOXIDE)

23732-94-7 HEXACOSAMETHYLCY-
CLOTRIDECASILOXANE

23745-86-0 POTASSIUM FLUOROACETATE

23746-34-1 POTASSIUM BIS(2-HYDROX-
YETHYL)DITHIOCARBAMATE

23783-42-8 2,5,8,11-TETRAOXATRIDECAN-13-
OL
TETRAETHYLENE GLYCOL
MONOMETHYL ETHER [TSC, TSC,
TSC]

23950-58-5 PRONAMIDE
BENZAMIDE, 3,5-DICHLORO-N-(1,1-
DIMETHYL-2-PROPYNYL)- [PAL]

24017-47-8 TRIAZOFOS

24072-75-1 5,6-DICHLORO-2-BENZOTHIAZO-
LAMINE

24124-25-2 TRIBUTYLTIN LINOLEATE

24267-56-9 IODINE-131

24347-38-4 YTTERBIUM 162

24360-05-2 HEXAMETHYLENETETRAMINE
HYDROGEN CHLORIDE

24382-04-5 MALONALDEHYDE, SODIUM
SALT

24383-94-6 BISMUTH 203

24390-14-5 DOXYCYCLINE HYCLATE

24421-27-0 TUNGSTEN 188

24442-57-7 ETHANOL, 1,2-DIBROMO-, AC-
ETATE

24447-13-0 IRIDIUM 186

24468-13-1 2-ETHYLHEXYL CHLOROFOR-MATE
ETHYL HEXYLCHLOROFORMATE [NJL, NJL]

24613-89-6 CHROMIUM(III) CHROMATE
CHROMIUM CHROMATE [MSL]

24634-61-5 POTASSIUM SORBATE

24800-44-0 PROPANOL, [(1-METHYL-1,2-ETHANEDIYL)BIS(OXY)]BIS-
TRIPROPYLENE GLYCOL [HON, OTS]

24884-69-3 2-NITRO-2-METHYLPROPANOL NITRATE

24934-91-6 CHLORMEPHOS

24937-78-8 ETHYLENE-VINYL ACETATE POLYMER
ETHYLENE, POLYMER WITH VINYL ACETATE [PST]

24938-04-3 POLYETHYLENE TEREPHTHA-LATE - POLYETHYLENE ISOPH-THALATE FILM

24938-91-8 TRIDECYL ALCOHOL, ETHOXY-LATED
POLYOXYETHYLENE TRIDECYL ALCOHOL [PST]

24968-79-4 ACRYLIC ACID METHYL ESTER, POLYMER WITH ACRYLONI-TRILE AND 1,3-BUTADIENE

24991-31-9 POLYVINYL BUTYRATE

25013-15-4 VINYL TOLUENE
BENZENE, ETHENYLMETHYL- [PAL]
METHYL STYRENE (ALL ISOMERS) [MAK, MAK]
METHYLSTYRENES [HM1, HM1, HM1]
METHYLSTYRENES, ALL ISOMERS [GBR, GBR]
VINYL TOLUENE (MIXTURE OF M- AND P- ISOMERS) [ONT, ONT]
VINYLTOLUENE [CAL, WHM]

25013-16-5 BUTYLATED HYDROXYANISOLE
BUTYLATED HYDROXYANISOLE (BHA) [IAR, WHM]

25014-41-9 ACRYLONITRILE POLYMER
POLYACRYLONITRILE [PST]

25035-68-1 ETHYL ACRYLATE-METHACRYLIC ACID-STYRENE POLYMER
METHACRYLIC ACID, POLYMER WITH STYRENE AND ETHYL ACRY-LATE [PST]

25035-69-2 METHYL METHACRY-LATE-BUTYL ACRYLATE-METHACRYLIC ACID POLYMER
METHACRYLIC ACID, POLYMER WITH BUTYL ACRYLATE AND METHYL METHACRYLATE [PST]

25038-54-4 NYLON 6
POLYAMIDE RESINS [PST]

25038-59-9 MYLAR

25057-89-0 BENTAZON

25067-11-2 TEFLON

25068-38-6 BISPHENOL A-EPICHLOROHY-DRIN POLYMER
BISPHENOL A - EPICHLOROHYDRIN CONDENSATE [PST]

25085-02-3 ACRYLAMIDE-SODIUM ACRY-LATE POLYMER
SODIUM ACRYLATE, POLYMER WITH ACRYLAMIDE [PST]

25085-34-1 STYRENE ACRYLATE COPOLY-MER
STYRENE ACRYLIC ACID COPOLY-MER [PST]

25085-41-0 VINYL ACETATE - BUTYL ACRY-LATE - ACRYLIC ACID TERPOLY-MER

25085-99-8 OXIRANE,2,2'-4-BUTYLI-DENEBISPHENYLE-NEOXYMETHYLENE (DGEBA)
OXIRANE, 2,2'-[(1-METHYLETHY-LIDINE)BIS(4,1-PHENYLE-NEOXYMETHYLENE)]BIS-, HO-MOPOLYMER [TSC, TSC]

25086-29-7 VINYLPYRROLIDINONE-STYRENE POLYMER

25086-48-0 VINYL ACETATE-VINYL ALCO-HOL CHLORIDE POLYMER
VINYL CHLORIDE, VINYL ACETATE AND VINYL ALCOHOL COPOLYMER [PST]

25086-89-9 POLYVINYLPYRROLIDONE-VINYL ACETATE COPOLYMER

25103-52-0 ISOOCTANOIC ACID

25103-58-6 TERT-DODECYL MERCAPTAN
TERT-DODECANTHIOL [PAL]

25104-37-4 POLY(VINYL ETHYL ETHER)

25109-57-3 1,2-DIBROMO-3-BUTANONE
1,2-DIBROMOBUTAN-3-ONE [HM1, HM1]

25134-21-8 MEMTETRAHYDROPHTHALIC ANHYDRIDE
4,7-METHANOISOBENZOFURAN-1,3-DIONE, 3A,4,7,7A-TETRAHY-DROMETHYL- [TSC, TSC]

25136-55-4 DIMETHYLDIOXANE
1,4-DIOXANE, DIMETHYL- [PAL]
DIMETHYLDIOXANES [HM1, HM1]

25152-84-5 2,4-DECADIENAL

25154-42-1 CHLOROBUTANE
CHLOROBUTANES [HM1, HM1]

25154-52-3 NONYLPHENOL
NONYL PHENOL [WHM]
PHENOL, NONYL- [PAL]

25154-54-5 DINITROBENZENE (MIXED)
BENZENE, DINITRO- [PAL]
DINITROBENZENE [MAK, MAK, MEX, MEX, MEX, NJL]
DINITROBENZENE (ALL ISOMERS) [BC1, BC1, BC1, NIO, NIO, NIO, NIO, OS1, OS1, QBC, QBC]
DINITROBENZENE (SUM OF M-, O-, AND P- ISOMERS) [ONT, ONT]
DINITROBENZENE, ALL ISOMERS [ALB, ALB, ALB, GBR, GBR, GBR]
DINITROBENZENE, N.O.S. [RCU]
DINITROBENZENES [HM1, HM1, HM1]
DINITROBENZENES (NOS) [HON]

25154-55-6 NITROPHENOL (MIXED ISOMERS)
NITROPHENOL (MIXED) [CWA, EPA, MSL]
NITROPHENOL, MIXED ISOMERS [NJL]
NITROPHENOLS [CW3, PAL]

25155-15-1 CYMENE
CYMENES [HM1, HM1]

25155-23-1 PHOSPHATE, TRIXYLYL
PHENOL, DIMETHYL-, PHOSPHATE (3:1) [TSC, TSC]
TRIXYLENYL PHOSPHATE [HM1]

25155-30-0 SODIUM DODECYLBENZENESUL-FONATE
BENZENESULFONIC ACID, DODE-CYL-, SODIUM SALT [PAL]
DODECYLBENZENESULFONIC ACID, SODIUM SALT [PST]
SODIUM DODECYLBENZENE SUL-FONATE [MSL, NJL]
SODIUM DODECYLBENZENE-SUL-FONATE [CAL]

25156-49-4 C.I. DIRECT BLACK 4

25167-20-8 TETRABROMOETHANE

25167-32-2 BENZENESULFONIC ACID, OXY-BIS[DODECYL-,DISODIUM SALT]
DISODIUM 2,2'-OXYBIS(4-DODECYL-BENZENESULFONATE) [PST]

25167-67-3 BUTYLENE
BUTENE [EP3]
BUTENES [PAL]

25167-70-8 DIISOBUTYLENE
DI-ISOBUTYLENE [NJL, NJL]
PENTENE, 2,4,4-TRIMETHYL- [PAL]

25167-82-2 TRICHLOROPHENOL
PHENOL, TRICHLORO- [PAL]
TRICHLOROPHENOL (MIXED) [EP2]
TRICHLOROPHENOLS [RCA]

25167-83-3 TETRACHLOROPHENOL
TETRACHLOROPHENOLS [RCA]

25167-93-5 CHLORONITROBENZENE
BENZENE, CHLORONITRO- [PAL]
NITROCHLOROBENZENE [FLA, MSL, NFP, NFP]

25168-04-1 NITROXYLOL
NITROXYLENE [HON]

25168-05-2 CHLOROTOLUENES
BENZENE, CHLOROMETHYL- [PAL]
CHLOROTOLUENE [ATS, NFP, NFP]

25168-06-3 ISOPROPYL PHENOL

25168-15-4 ACETIC ACID, (2,4,5-TRICHLOROPHENOXY)-, ISOOCTYL ESTER
2,4,5-T ESTER [NJL]
2,4,5-T ESTERS [PAL]

25168-21-2 DIBUTYLTIN BIS (ISOOCTYL MALEATE)-2-BUTENOIC ACID, 4,4'-[(DIBUTYLSTANNYLENE)BIS (OXY)]BIS[4-OXO-, DIISOOCTYL ESTER, (Z,Z)-
2-BUTENOIC ACID, 4,4'-[(DIBUTYL-STANNYLENE)BIS(OXY)]BIS(4-OXO-, DIISOOCTYL ESTER (Z,Z)- [TSC, TSC, TSC]

25168-24-5 DIBUTYLTIN BIS(ISOOCTYL MER-CAPTOACETATE)
ACETIC ACID, 2,2'-[(DIBUTYLSTAN-NYLENE)BIS(THIO)]BIS-, DIISOOCTYL ESTER [TSC, TSC]
BIS(2-ETHYLHEXYLOXYCARBONYL-METHYL-THIO)DIBUTYLSTANNANE [WHM]

25168-26-7 2,4-D ISOOCTYL ESTER
2,4-D ESTER (2,4-D ISOOCTYL ESTER) [CWA]
2,4-D ESTERS [NJL]
ACETIC ACID, (2,4-DICHLOROPHE-NOXY)-, ISOOCTYL ESTER [MSL, PAL]

25180-19-2 C.I. DIRECT BLUE 2

25180-27-2 C.I. DIRECT BLUE 25

25180-39-6 C.I. DIRECT BROWN 6

25180-41-0 C.I. DIRECT BROWN 31

25180-45-4 C.I. DIRECT GREEN 1

25180-46-5 C.I. DIRECT GREEN 6

25180-47-6 C.I. DIRECT GREEN 8

25188-24-3 C.I. DIRECT RED 1

25188-29-8 C.I. DIRECT RED 10

25188-30-1 C.I. DIRECT RED 13

25188-42-5 C.I. DIRECT RED 81

25188-44-7 C.I. DIRECT VIOLET 1

25214-69-1 HYDROLYZED POLYACRYLONI-TRILE

25231-21-4 POLYPROPYLENE STEARYL ETHER

25231-46-3 **TOLUENE SULFONIC ACID**
TOLUENESULPHONIC ACID [WHM]

25251-51-8 **3-TERT-BUTYL PEROXY-3-PHENYLPHTHALIDE**

25255-06-5 **C.I. DIRECT BROWN 2**

25260-60-0 **1,4-DIACETOXYBUT-2-ENE (1,4)**

25265-68-3 **METHYL TETRAHYDROFURAN**

25265-71-8 **DIPROPYLENE GLYCOL**
PROPANOL, OXYBIS- [PAL]

25265-76-3 **PHENYLENEDIAMINES**
PHENYLENEDIAMINE [QBC, QBC, RCU]
PHENYLENEDIAMINES (BENZENEDI-AMINES) [TSC]

25265-77-4 **2,2,4-TRIMETHYLPENTANE-1,3-DIOL MONOISOBUTYRATE**
2,2,4-TRIMETHYL-1,3-PENTANEDIOL ISOBUTYRATE [NFP, NFP]
PROPANOIC ACID, 2-METHYL-, MO-NOESTER [OTS]

25311-71-1 **ISOFENPHOS [2-[[ETHOXYL[(1-METHYLETHYL)AMINO]PHOS-PHINOTHIOYL]OXY]BENZOIC ACID 1-METHYLETHYL ESTER]**

25316-40-9 **5,12-NAPHTHACENEDIONE, 10-[(3-AMINO-2,3,6-TRIDEOXY-.ALPHA.-L-LYXO-HEXOPYRANOSYL)OXY]-7,8,9,10-TETRAHYDRO-6,8,11-TRI-HYDROXY-8-(HYDROXYACETYL)-1-METHOXY-, HYDROCHLORIDE, (8S-CIS)-**

25321-09-9 **DIISOPROPYL BENZENE**
DIISOPROPYL BENZENES [HM1]

25321-14-6 **DINITROTOLUENE (MIXED ISO-MERS)**
BENZENE, METHYLDINITRO- [PAL]
DINITROTOLUENE [ATS, BC1, BC1, BC1, CW3, EPA, MSL, NIO, NIO, NIO, NIO, NIO, NJL, OS1, OS1, OS1, OS1, QBC, QBC, TLV, TLV, TLV]
DINITROTOLUENE (MIXED) [EP2]
DINITROTOLUENE (SUM OF ALL ISOMERS) [ONT, ONT]
DINITROTOLUENES [MAK, MAK, NHS, NHS, NHS]

25321-22-6 **DICHLOROBENZENE (MIXED ISOMERS)**
BENZENE, DICHLORO- [PAL]
DICHLOROBENZENE [ATS]
DICHLOROBENZENE (META; OR-THO; PARA) [HM1, HM1]
DICHLOROBENZENE (MIXED) [EPA, MSL]
DICHLOROBENZENE, N.O.S. [RCU]
DICHLOROBENZENES [MX2, RCA]
DICHLOROBENZENES, MIXED ISO-MERS [NJL]

25321-41-9 **XYLENESULFONIC ACID**

25322-01-4 **NITROPROPANE**
NITROPROPANES [HM1, HM1]

25322-14-9 **TRINITROFLUORENONE**

25322-20-7 **TETRACHLOROETHANES**
TETRACHLOROETHANE [ATS]
TETRACHLOROETHANE, N.O.S. [RCU]

25322-68-3 **POLYETHYLENE GLYCOL**
POLYETHYLENE GLYCOLS [MIN, NFP, NFP]

25322-69-4 **POLYPROPYLENE GLYCOL**
POLYPROPYLENE GLYCOLS [MIN, NFP, NFP]

25323-30-2 **DICHLOROETHYLENES**
DICHLOROETHYLENE [HM1, HM1]
DICHLOROETHYLENE, N.O.S. [RCU]

25323-89-1 **TRICHLOROETHANE**

25324-56-5 **TIN PHOSPHIDE**
STANNIC PHOSPHIDE [NJL]

25327-89-3 **ALLYL ETHER OF TETRABROMO-BISPHENOL-A**
BENZENE, 1,1'-(1-METHYLETHYLI-DENE)BIS[3,5-DIBROMO-4-(2-PROPENYLOXY)- [TSC, TSC, TSC, TSC]
TETRABROMOBISPHENOL-A, ALLYL ETHER [OTS]

25329-82-2 **C.I. DIRECT VIOLET 22**

25338-51-6 **TERT-BUTYL STYRENE**
TERT-BUTYLSTYRENE [NFP, NFP, PAL]

25338-55-0 **DIMETHYLAMINOMETHYLPHE-NOL**

25339-17-7 **ISODECANOL**
ISODECANOL, MIXED ISOMERS [NFP, NFP]
ISODECYL ALCOHOL [PST]

25339-56-4 **HEPTENE**
HEPTYLENE [FLA, MSL, NFP, NFP]
N-HEPTENE [HM1, HM1]

25340-17-4 **DIETHYLBENZENE**

25340-18-5 **1,2,4-TRIETHYLBENZENE**
TRIETHYLBENZENE [HM1]

25360-09-2 **TERT-HEXADECANETHIOL**

25360-10-5 **TERT-NONYL MERCAPTAN**
TERT-NONANETHIOL [PAL]

25376-45-8 **DIAMINOTOLUENE (MIXED ISO-MERS)**
BENZENEDIAMINE, AR-METHYL-[TSC, TSC]
DIAMINOTOLUENE [MSL]
DIAMINOTOLUENE (MIXED) [C65, CAL]
DIAMINOTOLUENE, MIXED ISOMERS [NJL]
TOLUENE-AR,AR'-DIAMINE [WHM]
TOLUENEDIAMINE [RCU, RCU]

25377-21-3 **DI-TERT-BUTYL-P-CRESOL**
2,6-DITERTIARY-BUTYL-PARA-CRESOL [GBR]

25377-72-4 **AMYLENE, NORMAL**
N-AMYLENE [HM1, HM1]

25377-83-7 **OCTENE (MIXED ISOMERS)**
2-OCTENE (MIXED CIS- AND TRANS-ISOMERS) [NFP, NFP]

25402-06-6 **CINERIN I**

25417-20-3 **DIBUTYLNAPHTHALENESUL-FONIC ACID, SODIUM SALT**

25429-29-2 **KANECHLOR 500**

25455-73-6 **SILVER OXIDE (AG2O2)**

25481-93-0 **BORATES, TETRA, SODIUM SALTS, PENTAHYDRATE**

25495-90-3 **HEXANE, CHLORO-**

25496-08-6 **FLUOROTOLUENE**
FLUOROTOLUENES [HM1, HM1]

25496-72-4 **GLYCEROL MONOOLEATE**
GLYCERYL MONOOLEATE [PST]

25497-29-4 **CHLORODIFLUOROETHANES**

25498-49-1 **TRIPROPYLENE GLYCOL MONOMETHYL ETHER**
(2-(2-METHOXYMETHYLETHOXY) METHYLETHOXY) PROPANOL [PST]
PROPANOL, [2-(2-METHOXYMETHYLETHOXY) METHYLETHOXY- [TSC, TSC, TSC]
TRIPROPYLENE GLYCOL METHYL ETHER [NFP, NFP]

25512-65-6 **DIHYDRO-2H-PYRAN**
2H-PYRAN, DIHYDRO- [PAL]

25550-14-5 **BENZENE, ETHYLMETHYL-**

25550-55-4 **DINITROSOBENZENE**

25550-58-7 **DINITROPHENOL**
DINITROPHENOL, ALL ISOMERS [WHM]
PHENOL, DINITRO- [PAL]

25550-98-5 **PHOSPHOROUS ACID, (DIISODE-CYL)PHENYL ETHER**
DIISODECYL PHENYL PHOSPHITE [TSE]
DIISODECYLPHENYL PHOSPHITE [PST]

25551-13-7 **TRIMETHYL BENZENE**
BENZENE, TRIMETHYL- [PAL, TSC, TSC, TSC]
TRIMETHYLBENZENE [QBC, WHM]
TRIMETHYLBENZENE (ALL ISOMERS) [CAL]
TRIMETHYLBENZENE (SUM OF ISO-MERS) [ONT]
TRIMETHYLBENZENE, ALL ISOMERS [CEX]

25551-14-8 **AZODI-(1,1'-HEXAHYDROBEN-ZONITRILE)**
1,1'-AZODI-(HEXAHYDROBENZONI-TRILE) [HM1, HM1]

25551-28-4 **NAPHTHALENE DIISOCYANATE**
NAPHTHALENE DIISOCYANATE (NDI) [NHS, NHS]

25567-55-9 **2,3,4,6-TETRACHLOROPHENOL, SODIUM SALT**

25567-67-3 **DINITROCHLOROBENZENE**
BENZENE, CHLORODINITRO- [PAL]
DINITROCHLOROBENZENES [HM1]

25567-68-4 **CHLORONITROTOLUENE**
CHLORONITROTOLUENES [HM1, HM1, HM1]

25584-83-2 **2-PROPENOIC ACID, MONOESTER WITH 1,2-PROPANEDIOL**

25586-43-0 **NAPHTHALENE, CHLORO-**
CHLORONAPHTHALENE [MX1]
CHLORONAPHTHALENES [MX2]
CHLORONAPHTHATLENE [HON]

25618-55-7 **POLYGLYCERINE**

25620-58-0 **TRIMETHYLHEXAMETHYLENE-DIAMINE**
TRIMETHYLHEXAMETHYLENE DI-AMINE [WHM]
TRIMETHYLHEXAMETHYLENEDI-AMINES [HM1, HM1]

25637-84-7 **GLYCERYL DIOLEATE**

25637-99-4 **CYCLODODECANE, HEXABROMO-**

25638-17-9 **BUTYLNAPHTHALENESULFONIC ACID, SODIUM SALT**

25639-42-3 **METHYLCYCLOHEXANOL**
CYCLOHEXANOL, METHYL- [PAL]
METHYL CYCLOHEXANOL [NJL]
METHYLCYCLOHEXANOL (META-AND PARA-ISOMER MIXTURE) [CEX]

25640-78-2 **ISOPROPYL BIPHENYL**
2-ISOPROPYLBIPHENYL [NFP, NFP]

25641-53-6 **SODIUM DINITRO-O-CRESOLATE**

25659-31-8 **LEAD IODATE**

25667-42-9 **POLYETHER-SULPHONE**

25704-18-1 **SODIUM POLYSTYRENE SUL-FONATE**

25704-81-8 **C.I. VAT GREEN 2**

25719-52-2 **DODECYL 2-METHYLACRYLATE POLYMER**

25719-60-2 **BETA-PINENE POLYMER**

25721-38-4 **LEAD PICRATE**

25722-70-7 **POLYGLYCIDOL**

26635-64-3 **ISOOCTANE (VAN)**
ISOOCTANE [PAL]

26636-01-1 **DIBUTYLTIN S,S'-BIS(ISOOCTYL MERCAPTOACETATE)**
ACETIC ACID, 2,2'-[(DIMETHYLSTAN-NYLENE)BIS(THIO)]BIS-, DIISOOCTYL ESTER [TSC, TSC]

26636-39-5 **EICOSYLOXYPOLY(ETHYLE-NEOXY) ETHANOL**

26636-40-8 **POLYOXYETHYLENE DOCOSYL ETHER**

26638-19-7 **DICHLOROPROPANE**
DICHLOROPROPANE, N.O.S. [RCU]
DICHLOROPROPANES [CAL]

26644-46-2 **TRIFORINE [N,N'-[1,4-PIPERAZINEDIYL-BIS(2,2,2-TRICHLOROETHYLIDENE)] BIS-FORMAMIDE]**

26645-10-3 **DIETHYLGOLD BROMIDE**

26658-19-5 **SORBITAN, TRIOCTADECANOATE**

26658-42-4 **EPICHLOROHYDRIN, POLYMER WITH TETRAETHYLENEPEN-TAMINE**

26675-46-7 **ISOFLURANE**

26683-69-2 **THALLIUM 195**

26747-90-0 **1,3-DIAZETIDINE-2,4-DIONE, 1,3-BIS(3-ISO-CYANATOMETHYLPHENYL)-**

26748-41-4 **TERT-BUTYL PEROXYNEODE-CANOATE**

26748-47-0 **CUMYL PEROXYNEODECANOATE**

26750-50-5 **1,1'-[OXYBIS(METHYLENESUL-FONYL)] BISETHENE**

26760-64-5 **METHYL BUTENE**
METHYLBUTENE [NJL, NJL]

26761-40-0 **DIISODECYL PHTHALATE**
1,2-BENZENEDICARBOXYLIC ACID, DIISODECYL ESTER [TSC, TSC, TSC, TSE]

26761-45-5 **NEODECANOIC ACID, OXIRANYL-METHYL ESTER**
GLYCIDYL ESTER OF NEODECANOIC ACID [EPA, OTS]
NEODECANOIC ACID, 2,3-EPOXYPROPYL ESTER [TSC, TSC]

26762-93-6 **DIISOPROPYLBENZENE HY-DROPEROXIDE**

26780-96-1 **1,2-DIHYDRO-2,2,4-TRIMETHYLQUINOLINE POLY-MER**
1,2-DIHYDRO-2,2,4-TRIMETHYLQUINO-LINE (POLYMER) [CSR]

26782-43-4 **HYDROXYSENKIRKINE**

26885-07-4 **N-EICOSANOYL-N-METHYLTAU-RINE, SODIUM SALT**

26896-20-8 **NEODECANOIC ACID**

26914-02-3 **IODO PROPANE**
IODOPROPANES [HM1, HM1]

26914-33-0 **TETRACHLOROBIPHENYL**

26915-12-8 **TOLUIDINES**
TOLUIDINE (ALL ISOMERS) [ALB, ALB, ALB, ALB]
TOLUIDINE (SUM OF O-, M-, AND P-ISOMERS) [ONT, ONT, ONT]

26915-70-8 **ALPHA-PHOSPHONO-OMEGA-(TRIDECYLOXY) POLY(OXY-1,2-ETHANEDIYL)**

26952-21-6 **ISOOCTYL ALCOHOL**
ISOOCTANOL [PAL]

ISOOCTYL ALCOHOL (ISOOCTANOL) [WHM]
ISOOCTYL ALCOHOL (MIXED ISO-MERS) [GBR]

26952-23-8 **DICHLOROPROPENE**
1-PROPENE, DICHLORO- [PAL, TSC, TSC]
DICHLOROPROPANE, N.O.S. [RCU]
DICHLOROPROPENES [CAL]

26952-42-1 **TRINITRO-ANILINE**

26967-76-0 **PHOSPHATE, TRIS(ISOPROPY-LPHENYL)**

26982-03-6 **CHLOROPHENOL (MIXED ISO-MERS)**
CHLOROPHENOLS [313, CAB, CAL, HM1, HM1, HM1, IAR, PAL, PAL]
CHLOROPHENOLS, N.O.S. [NJL]

26995-91-5 **UROFOLLITROPIN**

27107-89-7 **MONOOCTYLTIN TRIS(2-ETHYL-HEXYL THIOGLYCOLATE)**

27134-26-5 **CHLOROANILINES**

27134-27-6 **DICHLOROANILINE**
DICHLOROANILINE (MIXED ISO-MERS) [HON]
DICHLOROANILINE, ALL ISOMERS [WHM]

27136-73-8 **SUBSTITUTED IMIDAZOLE**
1-HYDROXYETHYL-2-HEPTADE-CENYL GLYOXALIDINE [PST]

27137-85-5 **TRICHLORO(DICHLOROPHENYL) SILANE**
DICHLOROPHENYL TRICHLOROSI-LANE [NJL, NJL]
DICHLOROPHENYLTRICHLOROSI-LANE [HM1, HM1, HM1, WHM]
TRICHLORO(DICHLOROPHENYL) SILANE [OS3]

27138-57-4 **DIHYDROXYBENZOIC ACID (RE-SORCYLIC ACID)**

27152-57-4 **CALCIUM ARSENITE (2:3)**

27156-03-2 **DICHLORODIFLUOROETHYLENE**

27176-87-0 **DODECYLBENZENESULFONIC ACID**
BENZENESULFONIC ACID, DODE-CYL- [PAL]
DODECYLBENZENE SULFONIC ACID [NJL, NJL]

27178-16-1 **DIISODECYL ADIPATE**

27178-87-6 **DIMETHYLNAPHTHALENESUL-FONIC ACID, SODIUM SALT**

27193-28-8 **OCTYLPHENOL**

27193-86-8 **DODECYLPHENOL**
DODECYL PHENOL [NFP, NFP]

27193-93-7 **NONYLNAPHTHALENE**

27196-00-5 **TETRADECANOL**

27208-37-3 **CYCLOPENTA[C,D]PYRENE**

27213-90-7 **DIISOBUTYLNAPHTHALENESUL-FONIC ACID, SODIUM SALT**

27215-10-7 **DIISOOCTYL ACID PHOSPHATE**
PHOSPHORIC ACID, DIISOOCTYL ESTER [TSC, TSC, TSC]

27215-95-8 **NONENE**

27236-46-0 **ISOHEXENE**

27253-28-7 **LEAD NEODECANOATE**

27253-30-1 **LITHIUM NEODECANOATE**

27253-31-2 **COBALT NEODECANOATE**

27253-33-4 **CALCIUM NEODECANOATE**

27254-36-0 **NITRONAPHTHALENE**

27277-00-5 **2-AMINO-4,5-DIHYDRO-6-METHYL-4-PROPYL-5-TRIAZOLO-(1,5-C)-PYRAMIDIN-5-ONE**

27304-13-8 **OXYCHLORDANE**

27306-79-2 **POLYOXYETHYLENE MONOTE-TRADECYL ETHER**

27314-13-2 **NORFLURAZON [4-CHLORO-5-METHYLAMINO]-2-[3-(TRIFLUO-ROMETHYL)PHENYL]-3(2H)-PYRI-DAZINONE]**

27323-18-8 **CHLOROBIPHENYL**

27323-41-7 **TRIETHANOLAMINE DODECYL-BENZOSULFONATE**
BENZENESULFONIC ACID, DO-DECYL-, COMPD. WITH 2,2',2"-NI-TRILOTRIS[ETHANOL](1:1) [PAL]
DODECYLBENZENESULFONIC ACID, TRIETHANOLAMINE SALT [PST]
TRIETHANOLAMINE DODECYLBEN-ZENE SULFONATE [MSL]
TRIETHANOLAMINE DODECYLBEN-ZENESULFONATE [EPA]

27344-41-8 **TINOPAL CBS**
DISODIUM 4,4'-BIS(2-SULFOSTYRYL) BIPHENYL [WHM]

27354-18-3 **C.I. SOLVENT RED 169**

27417-39-6 **1H-PYRROLE, METHYL-**

27458-92-0 **ISOTRIDECYL ALCOHOL**

27458-93-1 **ISOOCTADECANOL**

27478-34-8 **DINITRONAPHTHALENES**

27486-00-6 **LEAD 199**

27497-51-4 **CHLORODIFLUOROETHANES**

27519-02-4 **(Z)-9-TRICOSENE**

27554-26-3 **DIISOOCTYL PHTHALATE**
1,2-BENZENEDICARBOXYLIC ACID, DIISOOCTYL ESTER [TSC]

27598-85-2 **AMINOPHENOL**
AMINOPHENOLS [HM1, HM1, NJL]

27742-26-3 **IRIDIUM 184**

27753-52-2 **1,1'-BIPHENYL, NONABROMO-**
NONABROMO-1,1'-BIPHENYL [TSE]

27757-95-5 **C.I. PIGMENT RED 52**

27774-13-6 **VANADYL SULFATE**
VANADIUM, OXO[SULFATO(2-)-O]-[PAL]

27816-23-5 **2-PROPENOIC ACID, 2-CYANO-, 2-METHOXYETHYL ESTER**

27858-07-7 **OCTABROMOBIPHENYL**
1,1'-BIPHENYL, OCTABROMO- [TSC, TSC]

27923-56-4 **DIISOPROPYLPHENOLS**

27939-60-2 **3-CYCLOHEXENE-1-CARBOX-ALDEHYDE, DIMETHYL-**

27978-54-7 **HYDRAZINE PERCHLORATE**

27987-06-0 **TRIFLUOROETHANE**

28005-74-5 **2,2'-[(2-METHYLPHENYL)IMINO] BISETHANOL**

28057-48-9 **D-TRANS-ALLETHRIN [D-TRANS-CHRYSANTHEMIC ACID OF D-ALLETHRONE]**

28088-63-3 **XYLENESULFONIC ACID, CAL-CIUM SALT**

28108-99-8 **PHOSPHATE, ISOPROPYLPHENYL DIPHENYL**
ISOPROPYLPHENYL DIPHENYL PHOSPHATE [OTS]
PHOSPHORIC ACID, (1-METHYLETHYL)PHENYL DIPHENYL ESTER [TSC]

28165-71-1 DICHLOROPHENOXYACETIC ACID ESTER
2,6-DICHLOROPHENOXYACETIC ACID ESTER [NJL]

28178-42-9 BENZENE, 2-ISOCYANATO-1,3-BIS (1-METHYLETHYL)-

28182-81-2 HEXAMETHYLENE DIISO-CYANATE HOMOPOLYMER

28188-24-1 PENTAERYTHRITOL TRIS-TEARATE

28249-77-6 BOLERO (THIOBENCARB)
THIOBENCARB [CAL]
THIOBENCARB [CARBAMIC ACID, DI-ETHYLTHIO-, S-(P-CHLOROBENZYL)] [313]

28258-59-5 XYLYL BROMIDE

28260-61-9 TRINITROCHLOROBENZENE

28299-41-4 BENZENE, 1,1'-OXYBIS[METHYL-

28300-74-5 ANTIMONY POTASSIUM TAR-TRATE
ANTIMONATE(2-), BIS[.MU.-[2,3-DIHY-DROXYBUTANEDIOATO(4-)-O1,O2:O3, O4]DI-, DIPOTASSIUM, TRIHYDRATE, STEREOISOMER [PAL]

28314-03-6 ACETAMIDE, N-9H-FLUOREN-1-YL-

28324-52-9 PINANE HYDROPEROXIDE
PINANYL HYDROPEROXIDE [DOT]

28330-26-9 MALEIC HYDRAZIDE SODIUM SALT

28347-13-9 XYLYLENE DICHLORIDE
BENZENE, BIS(CHLOROMETHYL)- [MSL]

28348-53-0 SODIUM CUMENESULFONATE

28348-64-3 ISOPROPYLNAPTHALENESUL-FONIC ACID, SODIUM SALT

28348-65-4 ISOBUTYLNAPTHALENESUL-FONIC ACID, SODIUM SALT

28407-37-6 C.I. DIRECT BLUE 218

28434-00-6 2,2-DIMETHYL-3-(2-METHYL-1-PROPENYL)CYCLOPROPANECAR-BOXYLIC ACID, 2-METHYL-4-OXO-3-(2-PROPENYL)-2-CY-CLOPENTEN-1-YL ESTER

28434-86-8 3,3'-DICHLORO-4,4'-DIAMIN-ODIPHENYL ETHER
BENZENAMINE, 4-4'-OXYBIS(2-CHLORO)- [PAL, PAL]

28479-22-3 3-CHLORO-4-METHYLPHENYL ISOCYANATE
3-CHLORO-4-METHYL PHENYL ISO-CYANATE [NJL]

28519-02-0 DISODIUM DODECYLDIPHENYL ETHER DISULFONATE

28522-15-8 KRYPTON 74

28522-17-0 KRYPTON 76

28553-12-0 DIISONONYL PHTHALATE
1,2-BENZENEDICARBOXYLIC ACID, DIISONONYL ESTER [TSC]
DIISONONYLPHTHALATE [PST]

28554-00-9 CHLOROPROPIONIC ACID

28556-81-2 BENZENE, 2-ISOCYANATO-1,3-DIMETHYL-

28602-27-9 BENZALDEHYDE, (DIMETHY-LAMINO)-

28604-91-3 2,2'-AZODI-(2,4-DIMETHYL-VALERONITRILE)

28629-66-5 PHOSPHORODITHIOIC ACID, O,O-DIISOOCTYL

28631-66-5 C.I. ACID BLUE 22

28652-72-4 PHENYLTOLUENE
PHENYLTOLUENE O [NFP, NFP]

28675-17-4 DODECYLANILINE

28677-93-2 1-PROPANOL, METHOXY-
METHOXY-1-PROPANOL [TSC, TSC, TSC]

28679-16-5 TRIMETHYLHEXAMETHYLENE DIISOCYANATE
TRIMETHYLHEXAMETHYLENEDI-ISOCYANATE [NJL]

28772-56-7 BROMADIOLONE

28805-86-9 BUTYL PHENOL
BUTYLPHENOLS [HM1, HM1, HM1]

28911-01-5 TRIAZOLAM

28981-97-7 ALPRAZOLAM

28983-37-1 TERT-TETRADECANETHIOL
TERT-TETRADECYL MERCAPTAN [FLA, MSL, NFP, NFP]

28983-56-4 C.I. ACID BLUE 93

28984-85-2 NITROBIPHENYL
1,1'-BIPHENYL, NITRO- [PAL]

28984-89-6 1,1'-BIPHENYL, PHENOXY-

28987-17-9 BARIUM NONYLPHENATE

29027-17-6 2-CHLORO-3-METHYL ANILINE

29054-43-1 IRIDIUM 185

29054-62-4 IRIDIUM 182

29061-61-8 DODECYLBENZENESULFONIC ACID, DIISOPROPYLAMINE SALT

29061-63-0 DODECYLBENZENESULFONIC ACID, TRIETHYLAMINE SALT

29069-24-7 PREDNIMUSTINE

29116-98-1 SORBITAN DIOLEATE

29135-62-4 HEXANITROOXANILIDE

29169-69-5 SODIUM N-OLEYL TAURINE

29171-20-8 6-OCTEN-1-YN-3-OL, 3,7-DIMETHYL-

29191-52-4 ANISIDINE (O-, P- ISOMERS)
ANISIDINE [FLA]
ANISIDINE (O- AND P- ISOMERS) [MSL]
ANISIDINE (O-P ISOMERS) [MIN]
ANISIDINE (ORTHO AND PARA ISO-MERS) [CEX, CEX]
ANISIDINE (SUM OF O-, P- ISOMERS) [ONT, ONT]
ANISIDINES [HM1, HM1]
BENZENAMINE, AR-METHOXY- [PAL]

29232-93-7 PIRIMIPHOS METHYL [O-(2-(DIETHYLAMINO)-6-METHYL-4-PYRIMIDINYL)-O, O-DIMETHYLPHOSPHOROTH-IOATE]

29255-31-0 TETRACHLOROTETRAFLUORO-PROPANE
CFC-214 [CAA]

29291-35-8 N-NITROSOFOLIC ACID

29305-12-2 1,3,5-TRIAZINE-2,4,6-TRIAMINE, HYDROBROMIDE

29306-57-8 2,4,6-TRINITRO-1,3-DIAZOBEN-ZENE

29385-43-1 TOLYL TRIAZOLE
1H-BENZOTRIAZOLE, METHYL-[TSC, TSC, TSC]
METHYL-1H-BENZOTRIAZOLE [PST]

29387-84-6 1,2-PROPANEDIOL, MONOISO-PROPYL ETHER
3-ISOPROPOXY-1-PROPANOL [PST]

29387-86-8 PROPYLENE GLYCOL MONOBUTYL ETHER

29473-77-6 LEAD SEBACATE

29492-78-2 AMERICIUM 237

29492-85-1 HAFNIUM 182

29590-42-9 2-PROPENOIC ACID, ISOOCTYL ESTER
ISO-OCTYL ACRYLATE [OTS]

29595-25-3 AMMONIUM DINITRO-O-CRESO-LATE

29665-18-7 MENDELEVIUM 258

29687-28-3 HAFNIUM 184

29687-31-8 PLATINUM 200

29687-52-3 NEPTUNIUM 232

29689-14-3 CHROMIUM CARBONATE

29710-25-6 OCTYL 12-HYDROXYSTEARATE

29710-31-4 CETYL OCTANOATE

29733-86-6 BUTENEDIOL

29756-38-5 BROMOPENTANE

29757-24-2 N-NITROANILINE
NITROANILINES (O-; M-; P-) [HM1, HM1]
O-NITROANILINE [NJL, NJL]

29761-21-5 PHOSPHATE, ISODECYL DIPHENYL
ISODECYL DIPHENYL PHOSPHATE [HM1, OTS]
PHOSPHORIC ACID, ISODECYL DIPHENYL ESTER [TSC, TSC]

29790-52-1 NICOTINE SALICYLATE

29836-26-8 BETA-D-GLUCOPYRANOSIDE, 1-O-OCTYL

29887-08-9 CALCO OIL BLUE

29901-97-1 PROTACTINIUM 227

29903-04-6 DI-(1-NAPHTHOYL) PEROXIDE

29911-28-2 2-PROPANOL, 1-(2-BUTOXY-1-METHYLETHOXY)-
DIPROPYLENE GLYCOL BUTYL ETHER [TSC, TSC, TSC]

29919-07-1 YTTERBIUM 178

29965-97-7 CYCLOOCTADIENE
CYCLOOCTADIENES [HM1, HM1]

30007-47-7 5-BROMO-5-NITRO-1,3-DIOXANE

30025-38-8 DIPROPYLENE GLYCOL MO-NOETHYL ETHER

30026-92-7 TERT-BUTYL ISOPROPYL BEN-ZENE HYDROPEROXIDE

30030-25-2 BENZENE, (CHLOROMETHYL) ETHENYL-
VINYLBENZYL CHLORIDE [FLA, MSL]
VINYLBENZYLCHLORIDE [NFP, NFP]

30125-47-4 C.I. PIGMENT YELLOW 138

30136-13-1 PROPYLENE GLYCOL MONO-PROPYL ETHER

30174-58-4 TERT-DECANETHIOL
TERT-DECYLMERCAPTAN [FLA, MSL, NFP, NFP]

30232-11-2 METHYLCYCLOHEXYL ACETATE

30236-29-4 SUCROSE OCTANITRATE

30310-80-6 N-NITROSOHYDROXYPROLINE

30325-89-4 PARAFORMALDEHYDE

30346-73-7 XYLENESULFONIC ACID, POTAS-SIUM SALT

30399-84-9 **ISOOCTADECANOIC ACID**

30402-14-3 **TETRACHLORODIBENZOFURAN**

30402-15-4 **PENTACHLORODIBENZOFURAN**

30516-87-1 **3'-AZIDO-3'-DEOXYTHYMIDINE (AIDS INITIATIVE)**

30525-89-4 **PARAFORMALDEHYDE**
PARAFORMALDEHYDE (POLYMER) [MSL]

30526-22-8 **POTASSIUM TOLUENESUL-FONATE**

30526-26-2 **NONYLPHENOL PHOSPHATE**

30553-04-9 **NAPHTHYLTHIOUREA**

30554-72-4 **CYCLOHEXANE, TETRABRO-MODICHLORO-**

30554-73-5 **CYCLOHEXANE, TRIBRO-MOCHLORO-**
CYCLOHEXANE, TRIBRO-MOTRICHLORO- [TSC, TSC, TSC]

30560-19-1 **ACEPHATE**
ACEPHATE (ACETYLPHOSPHO-RAMIDEOTHIOIC ACID O,S-DIMETHYL ESTER [313]

30586-10-8 **DICHLOROPENTANE**
DICHLOROPENTANES [NFP, NFP, PAL]

30586-18-6 **PENTAMETHYLHEPTANE**

30618-84-9 **GLYCERYL MONOTHIOGLYCO-LATE**

30674-80-7 **METHACRYLOYLOXYETHYL ISOCYANATE**
2-PROPENOIC ACID, 2-METHYL-2-ISOCYANATOETHYL ESTER [TSC, TSC, TSC]

30704-63-3 **POLYOXYETHYLENE P-TERT-BUTYLPHENOL - FORMALDE-HYDE RESIN**

30714-78-4 **ETHYL BUTYL CARBONATE**
CARBONIC ACID, BUTYL ETHYL ESTER [PAL]

30905-38-5 **CADMIUM 104**

30938-41-1 **POLYMER OF VINYL ACETATE, N-BUTYL ACRYLATE, VINYL CHLORIDE, AND ACRYLIC ACID**

30968-45-7 **HYDROGENATED METHYL ABI-ETE**

30989-40-3 **CURIUM 238**

31206-94-7 **2,5,8,10,13-PENTAOXAHEXADEC-15-ENOIC ACID,9,14-DIOXO-2-[(1-OXO-2-PROPENYL)OXY]ETHYL ESTER**

31212-28-9 **NITROBENZENESULPHONIC ACID**
NITROBENZENESULFONIC ACID [NJL, NJL]

31218-83-4 **PROPETAMPHOS**
PROPETAMPHOS [3-[[(ETHYLAMINO) METHOXYPHOSPHINOTHIOYL]OXY]-2-BUTENOIC ACID, 1-METHYLETHYL ESTER] [313]

31242-93-0 **CHLORINATED DIPHENYL OXIDE**
CHLORINATED DIPHENYL OXIDE (ETHER) [QBC]
HEXACHLORO DIPHENYL OXIDE [NFP, PAL]

31394-54-4 **ISOHEPTANE**

31512-74-0 **POLY[OXYETHYLENE (DIMETHYLIMINO) ETHYLENE (DIMETHYLIMINO) ETHYLENE DICHLORIDE]**

31566-31-1 **GLYCERYL MONOSTEARATE**

31711-50-9 **BUTYL NAPHTHALENE**
BUTYLNAPHTHALENE [PST]

31800-88-1 **OCTYLOXYPOLY(ETHYLENEOXY) ETHYL PHOSPHATE**

31807-55-3 **ISODODECANE**

31866-76-9 **1-OXYETHYL-2-STEARIC IMIDA-ZOLINE**

31906-04-4 **3-CYCLOHEXENE-1-CARBOX-ALDEHYDE, 4-(4-HYDROXY-4-METHYLPENTYL)-**

32020-21-6 **IRON 60**

32025-58-4 **TECHNETIUM 98**

32052-51-0 **ISOCYANIC ACID, TRIMETHYL-CYCLOHEXYL ESTER**

32241-08-0 **NAPHTHALENE, HEPTACHLORO-**

32280-46-9 **N,N-DIETHYL-1,3-BUTANEDI-AMINE**
1,3-BUTANEDIAMINE, N,N'-DIETHYL- [PAL]

32315-10-9 **METHANOL, TRICHLORO-, CAR-BONATE (2:1)**

32360-05-7 **2-PROPENOIC ACID, 2-METHYL-, OCTADECYL ESTER**

32458-06-3 **BUTYL METHACRYLATE, METHYL METHACRYLATE, 2-HY-DROXYETHYL METHACRYLATE AND STYRENE COPOLYMER**

32492-61-8 **POLYOXYETHYLENE ISOPROPY-LIDENEDIPHENOL**

32534-81-9 **PENTABROMODIPHENYLOXIDE**
PENTABROMODIPHENYL ETHER [OTS, TSC, TSC, TSC, TSC]
PENTABROMODIPHENYL OXIDE [TSE]

32534-95-5 **2,4,5-TP ACID ESTERS**
2,4,5-T ACID ESTERS [MSL]
2,4,5-TP ESTER; PROPANOIC ACID, 2-(2,4,5-TRICHLOROPHENOXY)-, ISOOCTYL ESTER [CAL]
2,4,5-TP ESTERS [CWA, NJL]
PROPANOIC ACID, 2-(2,4,5-TRICHLOROPHENOXY)-, ISOOCTYL ESTER [PAL]

32536-52-0 **OCTABROMODIPHENYLOXIDE**
OCTABROMODIPHENYL ETHER [OTS, TSC, TSC, TSC, TSC]

32568-89-1 **2,4-IMIDAZOLIDINEDIONE, 5, 5-DIMETHYL-3-[2-(OXIRANYL-METHOXY)PROPYL]-1-(OXI-RANYLMETHYL)-**
2,4-IMIDAZOLIDINEDIONE, 5, 5-DIMETHYL-3-[2-(OXIRANYL-METHOXY)PROPYL]-1-(OXIRANYL-METHYL)- [TSC]
3-(2-GLYCIDYLOXYPROPYL)-1-GLY-CIDOL-5,5-DIMETHYL-HYDANTOIN [EPA, OTS]

32588-76-4 **ETHYLENE BIS(TETRABROMOPH-THALIMIDE)**

32612-48-9 **POLYETHYLENE GLYCOL MONO-LAURYL ETHER SULFATE AMMO-NIUM SALT**
AMMONIUM DODECYL POLY-OXYETHYLENE SULFATE [PST]
SODIUM LAURYL ETHER SULFATE [WHM]

32749-94-3 **PENTANAL, 2,3-DIMETHYL-**
2,3-DIMETHYL PENTALDEHYDE [MSL]
2,3-DIMETHYL VALERALDEHYDE [FLA]

32809-16-8 **PROCYMIDONE**

33089-61-1 **AMITRAZ**

33125-86-9 **TETRAKIS(2-CHLOROETHYL) ETHYLENE DIPHOSPHATE**
PHOSPHORIC ACID, 1,2-ETHANEDIYL TETRAKIS (2-CHLOROETHYL)ESTER [TSC, TSC, TSC]

33213-65-9 **BETA-ENDOSULFAN**
6,9-METHANO-2,4,3-BENZODI-OXATHIEPIN, 6,7,8,9,10,10-HEX-ACHLORO-1,5,5A,6,9,9A-HEXAHY-DRO-, 3-OXIDE, (3.ALPHA.,5A.ALPHA., 6.BETA.,9.BETA.,9A.ALPHA.)- [PAL]
BETA - ENDOSULFAN [MSL]
BETA-ENDOSULFAN-BETA [MX1]
ENDOSULFAN II [PQL, RCA, UTS]
ENDOSULFAN, BETA- [ATS]

33229-34-4 **HC BLUE 2**
HC BLUE NO. 2 [IAR]

33419-42-0 **ETOPOSIDE**

33453-96-2 **A,A'-DI-(NITROXY) METHYLETHER**

33543-31-6 **2-METHYLFLUORANTHENE**

33568-99-9 **DIOCTYLTIN BIS(ISOOCTYL MALEATE)**

33791-58-1 **2-PROPENOIC ACID, 3A,4,5,6,7,7A-HEXAHYDRO-4,7-METHANO-1H-INDENYL ESTER**

33857-26-0 **2,7-DICHLORODIBENZO-P-DIOXIN**

34014-18-1 **TEBUTHIURON**
TEBUTHIURON [N-[5-(1,1-DIMETHYLETHYL)-1,3,4-THIADIA-ZOL-2-YL)-N,N'-DIMETHYLUREA] [313]

34018-47-8 **PLUTONIUM 234**

34077-87-7 **DICHLOROTRIFLUOROETHANE**

34099-73-5 **ETHYL BORATE**

34216-34-7 **TRIMETHYLCYCLOHEXYLAMINE**

34256-82-1 **ACETOCHLOR**

34274-30-1 **PHOSPHORIC ACID, [1,2-ETHANEDIYLBIS[NITRILO-BIS(METHYLENE)]]TETRAKIS-, POTASSIUM SALT**
PHOSPHONIC ACID, [1,2-ETHANEDIYLBIS[NITRILOBIS(METHY-LENE)]TETRAKIS-, POTASSIUM SALT [TSC]

34314-63-1 **3-OXAURACIL**

34364-42-6 **PHOSPHORIC ACID, (1-METHYL-1-PHENYLETHYL)PHENYL DIPHENYLESTER**

34465-46-8 **HEXACHLORODIBENZODIOXIN**
HEXACHLORODIBENZO-P-DIOXIN [ATS]

34590-94-8 **DIPROPYLENE GLYCOL MONOMETHYL ETHER**
DIPROPYLENE GLYCOL METHYL ETHER [BC1, BC1, BC1, CEX, CEX, CEX, MIN, MSL, NIO, NIO, NIO, OS1, OS1, OS1, TLV, TLV, TLV, WHM]
DIPROPYLENE GLYCOL, METHYL ETHER [AUS, AUS, AUS]
DIPROPYLENE GLYCOL, MONOMETHYL ETHER [MAK, MAK]
PROPANOL, (2-METHOXYMETHYLETHOXY)- [PAL]

34621-99-3 **1,2-ETHANEDIYL TETRAKIS(2-CHLORO-1-METHYLETHYLENE) PHOSPHATE**

34651-95-1 **MONO-(TRICHLORO)TETRA (MONO-POTASSIUM DICHLORO) -PENTA-S-TRIAZINE-TRIONE**
MONO-(TRICHLORO)TETRA (MONOPOTASSIUM DICHLORO)-PENTA-S-TRIAZINE-TRIONE [PAL]

34689-46-8 CRESYLIC ACID SODIUM SALT

34731-32-3 CARBAMODITHIOIC ACID, 1,2-ETHANEDIYL ESTER

34893-92-0 1,3-DICHLORO-5-PHENYL ISO-CYANATE
BENZENE, 1,3-DICHLORO-5-ISO-CYANATO- [TSC, TSC, TSC]

35029-96-0 LEAD (II) METHYLTHIOLATE

35087-77-5 D-GLUCONIC ACID, POTASSIUM SALT

35089-00-0 BERYLLIUM PHOSPHATE
PHOSPHORIC ACID, BERYLLIUM SALT [PAL, PAL]

35164-39-7 2,2,4-TRIMETHYL-1,3-PENTANE-DIOL BENZOATE ISOBUTYRATE
2,2,4-TRIMETHYLPENTANEDIOL ISOBUTYRATE BENZOATE [NFP, NFP]

35243-89-1 OXIRANE, [(1,2-DIBROMO-PROPOXY)METHYL]-
1,2-DIBROMOPROPYL GLYCIDYL ETHER [EPA]
DIBROMOPROPYL GLYCIDYL ETHER, 1,2- [OTS]

35365-94-7 N,N-DIETHYLETHANAMINE PHOSPHATE

35367-38-5 DIFLUBENZURON

35400-43-2 SULPROFOS
BOLSTAR (SULPROFOS) [ALB, ALB]
PHOSPHORODITHIOIC ACID, O-ETHYL O-[4-(METHYLTHIO)PHENYL] S-PROPYL ESTER [PAL]
SULPROFOS [O-ETHYL O-[4-(METHYLTHIO)PHENYL]PHOSPHO-RODITHIOIC ACID S-PROPYL ESTER] [313]
SULPROPHOS [HM1]

35554-44-0 IMAZALIL [1-[2-(2,4-DICHLOROPHENYL)-2-(2-PROPENYLOXY)ETHYL]-1H-IMI-DAZOLE]

35691-65-7 1,2-DIBROMO-2,4-DICYANOBU-TANE
1-BROMO-1-(BROMOMETHYL)-1,3-PROPANEDICARBONITRILE [313]
2-BROMO-2-(BROMOETHYL)PEN-TANEDINITRILE [PST]

35822-46-9 1,2,3,4,6,7,8-HEPTACHLORODIBENZO-P-DIOXIN

35860-31-2 HEXANITRODIPHENYLAMINE

35860-50-5 TRINITROBENZOIC ACID (VAN)
TRINITROBENZOIC ACID [HM1, HM1]

35860-51-6 DINITRORESORCINOL

35884-77-6 BENZENE, BROMODIMETHYL
XYLYL BROMIDE [HM1, HM1]

36088-22-9 1,2,3,7,8-PENTACHLORODIBENZO-P-DIOXIN
PENTACHLORODIBENZO-P-DIOXIN [ATS]

36311-34-9 ISOHEXADECANOL

36355-01-8 HEXABROMOBIPHENYL
1,1'-BIPHENYL, HEXABROMO- [TSC, TSC]
HEXABROMO-1,1-BIPHENYL [TSE]
POLYBROMINATED BIPHENYLS [IAR, MIN, NTP]

36366-93-5 TETRAETHYLENE GLYCOL MONOPHENYL ETHER

36393-56-3 NORPSEUDOEPHEDRINE

36445-71-3 DECYL PHENOXYBENZENEDISUL-FONIC ACID, DISODIUM SALT

36478-76-9 URANYL NITRATE
URANIUM BIS(NITRATO-O,O')DIOXO-, (OC-6-11)- [PAL]

36483-57-5 1-PROPANOL, 2,2-DIMETHYL-, TRIBROMO DERIVATIVE

36653-82-4 CETYL ALCOHOL

36724-43-3 2,2'-[OXYBIS(METHYLENESUL-FONYL)] BISETHANOL

36729-43-8 XYLENESULFONIC ACID, MAGNE-SIUM SALT

36729-46-1 XYLENESULFONIC ACID, ZINC SALT

36768-62-4 4-AMINO-2,2,6,6-TETRAM-ETHYLPIPERIDINE

36791-04-5 RIBAVIRIN

36840-25-2 FRANCIUM 222

36938-50-8 TETRATRIACONTAMETHYLHEX-ADECASILOXANE

36938-52-0 OCTATRIACONTAMETHYLOC-TADECASILOXANE

37187-22-7 ACETYL ACETONE PEROXIDE

37189-83-6 LINOLEIC ACID DIMER-DI-ETHYLENETRIAMINE POLYMER

37199-81-8 MALEIC ANHYDRIDE-DIISOBUTY-LENE POLYMER, SODIUM SALT
MALEIC ANHYDRIDE, POLYMER WITH 2,4,4-TRIMETHYLPENTENE, SODIUM SALT [PST]

37206-20-5 METHYL ISOBUTYL KETONE PEROXIDE

37211-05-5 NICKEL CHLORIDE

37220-82-9 OLEIC ACID ESTER WITH 1,2,3-PROPANETRIOL
GLYCERINE OLEATE [PST]

37248-34-3 POTASSIUM SULFIDE (VAN)
POTASSIUM SULFIDE [PAL]

37251-69-7 POLYOXYETHYLENE POLY-OXYPROPYLENE NONYLPHENOL

37264-96-3 COBALT CARBONYLS
COBALT CARBONYL [CEX, PAL]

37271-20-8 POLYETHOXYAMINE HK

37273-91-9 METALDEHYDE (VAN)
METALDEHYDE [HM1, HM1, PAL]

37280-82-3 POLYOXYETHYLENE POLY-OXYPROPYLENE PHOSPHATE

37281-78-0 POLYOXYPROPYLENE OLEATE BUTYL ETHER

37286-64-9 POLYPROPYLENE GLYCOL METHYL ETHER

37293-14-4 BISMUTH TELLURIDE
BISMUTH TELLURIDE (SE DOPED) [ALB, ALB]
BISMUTH TELLURIDE, SE DOPED [OS1]
BISMUTH TELLURIDE, SELENIUM-DOPED [BC1, BC1]

37299-86-8 RHODAMINE WT

37300-23-5 ZINC YELLOW (ZINC CHROMATE PIGMENT)
ZINC CHROMATE PIGMENT [MIN]
ZINC YELLOW [CEX]

37310-83-1 OLEYL ETHER PHOSPHATE (NEU-TRAL)

37311-02-7 METHYL OXIRANE POLYMER WITH OXIRANE, MONOOCTYL ETHER

37319-17-8 ELMIRON (SODIUM PENTOSAN-POLYSULFATE)

37320-91-5 MERCURY IODIDE (VAN)
MERCURY IODIDE [HM1, HM1]

37324-23-5 AROCHLOR 1262

37332-31-3 CAPRYLIC/CAPRIC TRIGLYC-ERIDE

37340-60-6 NONYLPHENOXYPOLY(ETHYLE-NEOXY) ETHYL PHOSPHATE, SODIUM SALT

37449-19-7 MANGANESE ISOOCTANOATE

37452-11-2 DIMETHYLAMINE DODECYLBEN-ZENESULFONATE

37523-33-4 POP NONYLPHENOL - FORMALDEHYDE RESIN

37677-14-8 3-CYCLOHEXENE-1-CARBOX-ALDEHYDE, 4-(4-METHYL-3-PEN-TENYL)-

37764-25-3 N,N-DIALLYL-2,3-DICHLOROAC-ETAMIDE

37853-59-1 FIREMASTER 680
1,2-BIS(TRIBROMOPHENOXY)-ETHANE [TSE]
BENZENE, 1,1'-[1,2-ETHANEDIYLBIS (OXY)]BIS[2,4,6-TRIBROMO- [TSC, TSC, TSC, TSC]
ETHANE, 1,2-BIS(2,4,6-TRIBROMOPHE-NOXY)- [OTS]

37853-61-5 BISMETHYLETHER OF TETRA-BROMOBISPHENOL-A

37860-51-8 ETHANOL,2,2'-[OXYBIS(2,1-ETHANEDIYLOXY)]BIS,BIS(4-METHYLBENZENESULFONATE))

37871-00-4 HEPTACHLOROBENZO-P-DIOXIN

37971-36-1 2-PHOSPHONO-1,2,4-BUTANETRI-CARBOXYLIC ACID
2-PHOSPHONOBUTANE-1,2,4-TRICAR-BOXYLIC ACID [PST]
BUTANETRICARBOXYLIC ACID, 1,2,4- [OTS]

38051-10-4 2,2-BIS(CHLOROMETHYL)-1,3-PROPANEDIYL TETRAKIS(2-CHLOROETHYL) PHOSPHATE

38232-63-2 MERCUROUS AZIDE

38252-74-3 N-NITROSO-N-BUTYL-N-(3-CAR-BOXYPROPYL)AMINE

38304-52-8 2,4-IMIDAZOLIDINEDIONE, 3, 3'-[2-(OXIRANYLMETHOXY) -1,3-PROPANEDIYL]BIS[5,5-DIMETHYL-1-(OXIRANYL-METHYL)-
1,3-BIS(5,5-DIMETHYL-1-GLYCIDYL-HYDANTOIN-3-YL)-2-GLYCIDYL [OTS]
1,3-BIS(5,5-DIMETHYL-1-GLYCIDYL-HYDANTOIN-3-YL)-2-GLYCIDY-LOXYPROPANE [EPA]

38455-77-5 TIN II CHROMATE

38483-28-2 METHYLENE GLYCOL DINI-TRATE

38565-52-5 OXIRANE, (2,2,3,3,4,4,5,5,6,6,7,7,7-TRIDECAFLUOROHEPTYL)-

38571-73-2 1,2,3-TRIS(CHLOROMETHOXY) PROPANE

38661-72-2 CYCLOHEXANE, 1,3-BIS(ISO-CYANATOMETHYL)-
1,3-BIS(METHYLISOCYANATE)CY-CLOHEXANE [313]

38727-55-8 DIETHATYL ETHYL

38888-98-1 1,1-DIPHENYLETHANE (UNS)

38916-42-6 **DL-ASPARTIC ACID, N-(3-CAR-BOXY-1-OXO-3-SULFOPROPYL)-N-OCTADECYL-, TETRASODIUM SALT**
TETRASODIUM N-(1,2-DICAR-BOXYETHYL)-N-OCTADECYL SUL-FOSUCCINAMATE [PST]

38954-75-5 **OXIRANE, [(TETRADECYLOXY) METHYL]-**
TETRADECYL GLYCIDYL ETHER [EPA, OTS]

38998-75-3 **1,2,3,4,7,8,9-HEPTACHLORO DIBENZOFURAN**
HEPTACHLORODIBENZOFURAN [ATS]

39001-02-0 **OCTACHLORODIBENZOFURAN**

39049-04-2 **NEODECANOIC ACID, ZIRCO-NIUM SALT**
ZIRCONIUM NEODECANOATE [PST]

39156-41-7 **2,4-DIAMINOANISOLE SULFATE**
1,3-BENZENEDIAMINE, 4-METHOXY-, SULFATE (1:1) [PAL, PAL, TSC, TSC]
2,4-DIAMINOANISOLE SULPHATE [FLA]

39196-18-4 **THIOFANOX**

39227-28-6 **1,2,3, 4,7,8-HEXACHLORODIBENZO-P-DIOXIN**

39236-46-9 **IMIDAZOLINIDYL UREA**

39300-45-3 **DINOCAP**

39300-88-4 **TARA GUM**

39354-45-5 **DISODIUM SALT OF LAURYL AL-COHOL POLYETHYLENE GLYCOL ETHER SULFOSUCCINATE**

39377-49-6 **COPPER CYANIDE (VAN)**
COPPER CYANIDE [HM1, HM1]

39382-31-5 **METHYLPENTADIENES**

39390-00-6 **LEAD CHLORIDE SILICATE**

39393-37-8 **SANTICIZER 711**

39404-03-0 **MAGNESIUM SILICIDE**

39409-64-8 **TRIS, BIS-BIFLUOROAMINO DI-ETHOXY PROPANE (TVOPA)**

39413-47-3 **ZINC BERYLLIUM SILICATE**
BERYLLIUM ZINC SILICATE [NTP]
SILICIC ACID, BERYLLIUM ZINC SALT [PAL, PAL]

39421-75-5 **HYDROXYPROPYL GUAR GUM**

39464-64-7 **DINONYLPHENOL, ETHOXY-LATED, PHOSPHATED**

39515-41-8 **FENPROPATHRIN**
FENPROPATHRIN [2,2,3,3-TETRAM-ETHYLCYCLOPROPANE CAR-BOXYLIC ACID CYANO(3-PHENOXY-PHENYL) METHYL ESTER] [313]

39515-51-0 **BENZALDEHYDE, 3-PHENOXY-**

39638-32-9 **BIS(2-CHLOROISOPROPYL)ETHER**
PROPANE, 2,2'-OXYBIS[2-CHLORO- [PAL]

39817-09-9 **OXIRANE, 2,2'-[METHYLENEBIS (PHENYLENEOXYMETHYLENE)] BIS-**
2,2'-[METHYLENEBIS(PHENYLE-NEOXYMETHYL)]BISOXIRANE [TSC]

39831-55-5 **AMIKACIN SULFATE**

39920-37-1 **1,3-DICHLORO-2-PHENYL ISO-CYANATE**

39974-35-1 **1-NITRO-2-CHLOROBENZO-3-TRIFLUORIDE**

40058-87-5 **ISOPROPYL-2-CHLOROPROPI-ONATE**

40321-76-4 **1,2,3,7,8-PENTACHLORODIBENZO-P-DIOXIN**

40372-66-5 **2-PHOSPHONO-1,2,4-BUTANET-RICARBOXYLIC ACID, SODIUM SALT**
1,2,4-BUTANETRICARBOXYLIC ACID, 2-PHOSPHONO-, SODIUM SALT [PST]

40487-42-1 **PENDIMETHALIN [N-(1-ETHYL-PROPYL)-3,4-DIMETHYL-2,6-DINI-TROBENZENEAMINE]**

41024-91-3 **3,6,9,12-TETRAOXATETRADE-CANE-1,14-DIOL,BIS(4-METHYL-BENZENESULFONATE)**

41098-56-0 **1,4-BENZENEDISULFONIC ACID, 2,2'-[1,2-ETHENEDIYLBIS[(3-SULFO-4,1-PHENYLENE)IMINO [6-(DIETHYLAMINO)-1,3,5-TRI-AZINE-4,2-DIYL]IMINO]BIS-, HEX-ASODIUM SALT**

41123-59-5 **1,1'-[METHYLENEBIS(SULFONYL)] BIS-2-CHLOROETHANE**

41123-69-7 **2,2'-[METHYLENEBIS(SULFONYL)] BISETHANOL**

41195-90-8 **1,2-DICHLORO-3-PHENYL ISO-CYANATE**

41198-08-7 **CURACRON**
PROFENOFOS [O-(4-BROMO-2-CHLOROPHENYL)-O-ETHYL-S-PROPYL PHOSPHOROTHIOATE] [313]

41291-34-3 **ETHYLENE(5,6-DIBROMONOR-BORNANE-2,3-DICARBOXIMIDE)**

41372-08-1 **METHYLDOPA SESQUIHYDRATE**

41444-43-3 **SEC-PENTAMINE**
SEC-PENTANAMINE [PAL]

41444-55-7 **DECYL GLUCOSIDE**

41453-50-3 **LEAD B-RESORCYLATE**

41468-25-1 **PYRIDOXAL-5'-PHOSPHATE**

41556-26-7 **BIS (1,2,2,6,6-PENTAMETHYL-4-PIPERIDINYL) SEBACATE**
BIS (1,2,2,6,6-PENTAMETHYL-4-PIPERIDINYL) DECANEDIOATE [PST]

41575-94-4 **CARBOPLATIN**

41587-36-4 **CHLORONITROANILINE**
CHLORONITROANILINE, ALL ISO-MERS [WHM]
CHLORONITROANILINES [HM1, HM1, HM1]

41683-62-9 **1,2-DICHLOROMETHOXYETHANE**

41687-30-3 **2-[(3-NITROPHENYL)SULFONYL] ETHANOL**

41766-75-0 **3,3'-DIMETHYLBENZIDINE DIHY-DROFLUORIDE (O-TOLUIDINE DI-HYDROFLUORIDE)**
3,3'-DIMETHYLBENZIDINE DIHY-DROFLUORIDE (O-TOLIDINE DIHY-DROFLUORIDE) [313]

41903-57-5 **TETRACHLORODIBENZO-P-DIOXIN**

41928-09-0 **POLY(OXYETHYLENE) METHYLENEBIS(OCTYLPHENOL)**

42296-74-2 **HEXADIENE**

42350-99-2 **2-CHLORO-4,6-DI-TERT-AMYLPHENOL**
PHENOL, 2-CHLORO-4,6-BIS(1,1-DIMETHYLPROPYL)- [PAL]

42389-30-0 **1,2-BENZENEDIAMINE, 5-CHLORO-3-NITRO-**

42397-64-8 **1,6-DINITROPYRENE**

42397-65-9 **1,8-DINITROPYRENE**

42404-50-2 **2-PROPENOIC ACID, 2-[[[[[1,3, 3-TRIMETHYL-5-[[[2-[(1-OXO-2-PROPENYL)OXY]ETHOXY]CAR-BONYL]AMINO]CYCLOHEXYL] METHYL]AMINO] CARBONYL] OXY]ETHYL ESTER**

42479-87-8 **4-TERT-BUTYL-2-PHENYLPHE-NOL**

42504-46-1 **ISOPROPANOLAMINE DODECYL-BENZENE SULFONATE**
ISOPROPANOLAMINE DODECYL-BENZENESULFONATE [CAL]
ISOPROPANOLAMINE DODECYLBEN-ZENESULFONATE [CWA, EPA]

42509-80-8 **PHOSPHOROTHIOIC ACID, O-(5-CHLORO-1-(1-METHYLETHYL) -1H-1,2,4-TRIAZOL-3-YL) O,O-DIETHYL ESTER**
CGA-12223 [MSL]

42557-13-1 **POLY[OXY[METHYL(3,3,3-TRI-FLUOROPROPYL)SILYLENE], AL-PHA-(TRIMETHYLSILYL)-OMEGA-[(TRIMETHYLSILYL)OXY]-**

42589-07-1 **2,4,5-T AMINE**

42808-36-6 **9-HYDROXYOCTADECANOIC ACID, BUTYL ESTER, HYDROGEN SULFATE, SODIUM SALT**

42874-03-3 **OXYFLUORFEN**

42874-96-4 **2-CHLORO-1-(3-METHYLPHE-NOXY)-4-(TRIFLUOROMETHYL) BENZENE**

42978-66-5 **TRIPROPYLENE GLYCOL DI-ACRYLATE**
2-PROPANOIC ACID, (1-METHYL-1, 2-ETHANEDIYL)BIS[OXY(METHYL-2, 1-ETHANEDIYL)]ESTER [TSC, TSC, TSC]

43121-43-3 **TRIADIMEFON [1-(4-CHLOROPHE-NOXY)-3,3-DIMETHYL-1-(1H-1,2,4-TRIAZOL-1-YL)-2-BUTANONE]**

43133-95-5 **METHYLPENTANE**

49690-63-3 **TRISDIBROMOPHENYL PHOS-PHATE**

49842-07-1 **TOBRAMYCIN SULFATE**

50373-55-2 **METHYL PHENYL CARBINYL ACETATE**

50471-44-8 **VINCLOZOLIN [3-(3,5-DICHLOROPHENYL)-5-ETHENYL-5-METHYL-2,4-OXAZOLIDINE-DIONE]**

50475-76-8 **LITHIUM ACETYLIDE ETHYLENE-DIAMINE**

50594-77-9 **PHENOL, 3-[2-CHLORO-4-(TRIFLU-OROMETHYL)PHENOXY]-, AC-ETATE**

50696-42-9 **DIAMYL NAPHTHALENE**

50769-39-6 **BUTYLPOLYETHOXYETHANOL ESTERS OF PHOSPHORIC ACID**

50782-69-9 **PHOSPHONOTHIOIC ACID, METHYL-, S-[2-[BIS(1-METHYLETHYL)AMINOETHYL) O-ETHYL ESTER**

50789-44-1 **BENZENEMETHANOL, 3-PHE-NOXY-, ACETATE**

50810-25-8 **ALUMINUM SILICON (AL SI5)**

50825-29-1 **LEAD NAPHTHALATE**

50892-23-4 PEROXISOME PROJECT (WY-14643)

50922-29-7 BASIC ZINC CHROMATE

51004-61-6 AF 2(FOAMING AGENT)

51023-22-4 TRICHLOROBUTENE

51142-18-8 BACTOPEPTONE

51200-87-4 4,4-DIMETHYLOXAZOLIDINE

51207-31-9 2,3,7,8-TETRACHLORO DIBENZO-FURANS
TETRACHLORODIBENZOFURAN [ATS]

51218-45-2 METOLACHLOR

51235-04-2 HEXAZINONE

51240-95-0 NEODECANEPEROXOIC ACID, 1,1, 3,3-TETRAMETHYLBUTYL ESTER

51274-00-1 IRON OXIDE YELLOW
C.I. PIGMENT YELLOW 42 [PST]

51317-24-9 LEAD NITRORESORCINATE

51338-27-3 HOELON
DICLOFOP-METHYL [CD1]
DICLOFOP-METHYL [2-[4-(2,4-DICHLOROPHENOXY)PHENOXY] PROPANOIC ACID, METHYL ESTER] [313]

51344-60-6 ETHOXYLATED ABIETYLAMINE

51344-62-8 DEHYDROABIETYLAMINE-ETHY-LENE OXIDE ADDUCT
DEHYDROABIETYLAMINE-ETHY-LENE OXIDE CONDENSATE [PST]

51363-64-5 PHOSPHATE, DIISODECYL PHENYL
PHOSPHORIC ACID, DIISODECYL PHENYL ESTER [TSC, TSC]

51481-10-8 DEOXYNIVALENOL

51481-61-9 CIMETIDINE

51580-86-0 SODIUM DICHLOROISOCYANU-RATE DIHYDRATE
1,3,5-TRIAZINE-2,4,6-(1H,3H,5H)-TRIONE, 1,3-DICHLORO-, SODIUM SALT, DIHYDRATE [PAL]
SODIUM DICHLORO-S-TRIAZINE TRIONEDIHYDRATE [MSL]

51609-41-7 P-NONYLPHENOL, ETHOXY-LATED AND PHOSPHATED

51617-79-9 ALPHA-(OCTADECYLPHENYL)-OMEGA-HYDROXYPOLY(OXY-1,2-ETHANEDIYL)

51630-58-1 FENVALERATE
FENVALERATE [4-CHLORO-ALPHA-(1-METHYLETHYL)BENZENEACETIC ACID CYANO(3-PHENOXYPHENYL) METHYL ESTER] [313]

51632-16-7 BENZENE, 1-(BROMOMETHYL)-3-PHENOXY-

51787-44-1 BENZ(C)ACRIDINE, 7,8,9,11-TE-TRAMETHYL-
BENZ(C)ACRIDINE [FLA]

51810-70-9 ZINC PHOSPHIDE

51811-79-1 NONYLPHENOL, ETHOXYLATED AND PHOSPHATED

51845-86-4 BORIC ACID, ETHYL ESTER
ETHYL BORATE [HM1, HM1]

52061-60-6 P-METHANE HYDROPEROXIDE
PARAMETHANE HYDROPEROXIDE [NJL]

52181-51-8 CHLOROBENZOTRIFLUORIDE
CHLOROBENZOTRIFLUORIDES [HM1, HM1]

52217-48-8 CITRIC ACID, TRIS(DIMETHY-LAMINE) SALT

52218-35-6 2-[(6-AMINO-2-NAPHTHALENYL) SULFONYL] ETHANOL

52277-29-9 PHENOLSULFONIC ACID - FORMALDEHYDE - UREA CON-DENSATE, SODIUM SALT

52277-33-5 POLY(OXY-1,4-BUTANEDIYL), ALPHA-(1-OXO-2-PROPENYL)-X-[(1-OXO-2-PROPENYL)OXY]-
POLY(OXY-1,4-BUTANEDIYL), AL-PHA-(1-OXO-2-(HYDROXYETHYL) TALLOWAMINE OXIDE, PHOSPHATE [TSC, TSC]

52299-20-4 2-[(HYDROXYMETHYL)AMINO]-2-METHYL PROPANOL

52315-07-8 CYPERMETHRIN

52326-66-6 DISTEARYL PEROXYDI-CARBON-ATE
DISTEARYL PEROXY DI-CARBONATE [NJL]

52470-25-4 GUANIDINE NITRATE (VAN)
GUANIDINE NITRATE [HM1, HM1]

52475-86-2 3-CYCLOHEXENE-1-CARBOX-ALDEHYDE, 1-METHYL-4-(4-METHYL-3-PENTENYL)-

52495-71-3 POLY(OXY-1,2-ETHANEDIYL), ALPHA-HYDRO-3-(OXIRANYL-METHYL)-1,3-PROPANEDIOL (3:1)
POLY(OXY-1,2-ETHANEDIYL), ALPHA-HYDRO-3-(OXIRANYL-METHOXY)-,ETHER WITH 2-ETHYL-2-(HYDROXYMETHYL)-1,3-PROPANE-DIOL (3:1) [TSC, TSC]

52503-15-8 NONYLPHENOL, ETHOXYLATED, PHOSPHATED, POTASSIUM SALT

52628-25-8 ZINC AMMONIUM CHLORIDE [VAN]
AMMONIUM ZINC CHLORIDE [PAL]
ZINC AMMONIUM CHLORIDE [NJL]

52645-53-1 PERMETHRIN
PERMETHRIN [3-(2,2-DICHLOROETHENYL)-2,2-DIMETHYL-CYCLOPROPANECARBOXYLIC ACID, (3-PHENOXYPHENYL)METHYL ES-TER] [313]

52652-59-2 LEAD STEARATE DIBASIC
LEAD STEARATE [EPA]
STEARIC ACID, LEAD SALT, DIBASIC [MSL]

52673-60-6 POLYPROPYLENE GLYCOL AL-PHA-METHYL GLUCOSIDE ETHER (4:1)

52740-16-6 CALCIUM ARSENITE
ARSONIC ACID, CALCIUM SALT (1:1) [PAL]

52783-44-5 HEPTADECANOL

52821-24-6 1H-BENZ(DE)ISOQUINOLINE-1,3(2H)-DIONE, 2-(3-HYDROX-YPROPYL)-6-[(3-HYDROX-YPROPYL)AMINO]-

52829-07-9 TINUVIN 770 (AMINE LIGHT STA-BILIZER)
BIS(2,2,6,6-TETRAMETHYL-4-PIPERIDINYL)SEBACETATE [PST]

52836-31-4 2,2,5-TRIMETHYL-3-DICHLOROACETYL-1,3-OXAZO-LIDINE

52896-87-4 4-ISOPROPYLHEPTANE

52907-07-0 ETHYLENE BIS(5,6-DI-BROMONORBORNANE-2,3-DICAR-BOXIMIDE)

52918-63-5 DELTAMETHRIN

53014-37-2 TETRANITRO-ANILINE

53061-10-2 1,1'-[OXYBIS(METHYLENESUL-FONYL)] BIS-2-CHLOROETHANE

53124-00-8 STARCH, HYDROGEN PHOS-PHATE, 2-HYDROXYPROPY-LETHER

53220-22-7 DIMYRISTYL PEROXYDI-CAR-BONATE
DIMYRISTYL PEROXYDICARBONATE [NJL]

53231-79-1 PLATINIUM(II) SULFATE

53404-12-9 ARSENIC ACID, LEAD (4+) SALT

53404-19-6 BROMACIL, LITHIUM SALT [2, 4-(1H,3H)-PYRIMIDINEDIONE, 5-BROMO-6-METHYL-3-(1-METHYL-PROPYL), LITHIUM SALT]

53404-37-8 2,4-D 2-ETHYL-4-METHYLPENTYL ESTER

53404-44-7 DIPOTASSIUM 12-SULFATO-9-OCTADECENOATE

53404-49-2 ETHYLENE GLYCOL ETHER OF PINENE

53404-60-7 DAZOMET, SODIUM SALT [TE-TRAHYDRO-3,5-DIMETHYL-2H-1,3, 5-THIADIAZINE-2-THIONE, ION(1-) , SODIUM]

53408-91-6 MERCURY THIOCYANATE (VAN)
MERCURY THIOCYANATE [HM1, HM1]

53422-49-4 AZIDOETHYL NITRATE

53461-82-8 OXYDI-2,1-ETHANEDIYL TE-TRAKIS(2-CHLOROETHYL) PHOS-PHATE

53467-00-8 SODIUM DODECYL DIPHENYL OXIDE SULFONATE

53467-11-1 2,4-D ESTERS
2,4-DICHLOROPHENOXYACETIC ACID, ESTER [NJL]
POLY(OXYMETHYL-1,2-ETANEDIYL)),A-[(2,4-DICHLOROPHENOXY) ACETYL]-OMEGA-BUTOXY- [MSL]
POLY[OXY(METHYL-1,2-ETHANEDIYL)],.ALPHA.-[(2,4-DICHLOROPHENOXY)ACETYL]-.OMEGA.-BUTOXY- [PAL]

53469-21-9 AROCLOR 1242
CHLORINATED BIPHENYLS (42% CHLORINE) [MAK, MAK, MAK, MAK, MAK]
CHLORODIPHENYL (42% CHLORINE) [ALB, ALB, ALB, BC1, BC1, BC1, CEX, CEX, FLA, MIN, MSL, NIO, NIO, NIO, NIO, TLV, TLV]
CHLORODIPHENYL (42% CHLORINE) (PCB) [OS1, OS1, OS1, OS1]
CHLORODIPHENYL (42% CHLORINE) PCB [WHM]
CHLORODIPHENYL (42% CL) [QBC, QBC, QBC]
CHLORODIPHENYL 42% [MEX, MEX, MEX]
PCB-1242 (AROCHLOR 1242) [CW2, MX1]
PCBS (42% CHLORINE) [AUS, AUS, AUS, AUS, AUS]

53494-70-5 ENDRIN KETONE

53496-15-4 ACETIC ACID, SEC-PENTYL ES-TER

53535-27-6 POTASSIUM TETRA-CHLOROPHENATE
2,3,4,6-TETRACHLOROPHENOL, POTASSIUM SALT [RCU]

53558-25-1 PYRIMINIL

53799-46-5 CROCIDOLITE (FE5NA2(SIO3)8)

53894-28-3 PHENOL, [[(2-AMINOETHYL) AMINO]METHYL]-

53973-98-1 POLYGEENAN

53980-88-4 2-CYCLOHEXENE-1-OCTANOIC ACID, 5(OR 6)-CARBOXY-4-HEXYL-

53988-05-9 CALCIUM ISONONANOATE

54116-08-4 SODIUM TRIDECYLPOLY (OXYETHYLENE) SULFATE

54140-64-6 POLY[OXY(METHYL-1,2-ETHANEDIYL)], ALPHA,ALPHA'-[(1-METHYLETHYLIDENE)DI-4,1-PHENYLENE]BIS[3-(OXIRANYL-METHOXY)-
POLY[OXY-1,2-ETHANEDIYL), ALPHA, ALPHA'-[(1-METHYLETHYLIDENE)DI-4,1-PHENYLENE]BIS[3-(OXIRANYL-METHOXY)- [TSC, TSC]

54150-69-5 2,4-DIMETHOXYANILINE HY-DROCHLORIDE

54208-63-8 OXIRANE, 2,2'-[METHYLENEBIS(2,1-PHENYLENEOXYMETHYLENE)] BIS-
BISPHENOL F DIGLYCIDYL ETHER [EPA, OTS]

54350-48-0 ETRETINATE

54363-49-4 METHYLPENTADIENE

54532-97-7 2-BUTYLBIPHENYL

54566-73-3 BORON OXIDE

54579-28-1 C.I. DIRECT ORANGE 1

54590-52-2 ISOPROPANOLAMINE DODECYL-BENZENESULFONATE (1:1)
BENZENESULFONIC ACID, 4-DODE-CYL-, COMPD. WITH 1-AMINO-2-PROPANOL (1:1) [PAL]
ISOPROPANOLAMINE DODECYLBEN-ZENESULFONATE [NJL]

54612-36-1 NONYLPHENYLPOLYOXYETHY-LENE SULFOSUCCINATE

54693-46-8 DIACETONE ALCOHOL PEROX-IDE

54749-90-5 CHLOROZOTOCIN

54846-79-6 POLYOXYETHYLENE SORBITAN HEPTAOLEATE

54849-38-6 MONOMETHYLTIN TRIS (ISOOOCTYL MERCAPTOAC-ETATE)
ACETIC ACID, 2,2',2"-[(METHYLSTAN-NYLIDYNE)TRIS(THIO)]TRIS-, TRI-ISOOCTYL ESTER [TSC, TSC]

54965-24-1 TAMOXIFEN CITRATE

54972-97-3 1-PENTANOL, METHYL-

55069-68-6 POLYETHYLENE GLYCOL, OLEIC ACID DIESTER

55205-38-4 TETRABROMOBISPHENOL-A DI-ACRYLATE
2-PROPENOIC ACID, (1-METHYLETHYLIDENE)BIS(2,6-DI-BROMO-4,1-PHENYLENE) ESTER [TSC, TSC, TSC, TSC]

55290-64-7 DIMETHIPIN [2,3-DIHYDRO-5,6-DIMETHYL-1,4-DITHIIN-1,1,4,4-TETRAOXIDE]

55299-18-8 DIPHENYLHYDRAZINE

55406-53-6 3-IODO-2-PROPYNYL BUTYLCAR-BAMATE
3-IODO-2-PROPYNYL BUTYL CARBA-MATE [PST]

55488-87-4 FERRIC AMMONIUM OXALATE, UNSPECIFIED HYDRATE
ETHANEDIOIC ACID, AMMONIUM IRON SALT [MSL]
FERRIC AMMONIUM OXALATE [PAL]

55557-01-2 N-NITROSOGUVACINE

55557-02-3 N-NITROSOGUVACOLINE

55566-30-8 TETRAKIS(HYDROXYMETHYL) PHOSPHONIUM SULFATE

55673-89-7 1,2,3,4,7,8,9-HEXACHLORODIBEN-ZOFURAN

55684-94-1 HEXACHLORODIBENZOFURAN

55720-99-5 CHLORINATED DIPHENYL OXIDE
CHLORINATED BIPHENYL OXIDE [MAK, MAK]

55738-54-0 TRANS-2-((DIMETHYLAMINO) METHYLIMINO)-5-(2-(5-NITRO-2-FURYL)VINYL)-1,3,4-OXADIAZOLE
METHANIMIDAMIDE, N,N-DIMETHYL-N'-[5-[2-(5-NITRO-2-FU-RANYL)ETHENYL]-1,3,4-OXADIAZOL-2-YL]-,(E)- [PAL, PAL]

55810-17-8 TRINITRONAPHTHALENE

55845-06-2 FORMALDEHYDE, POLYMER WITH NONYLPHENOL AND OXI-RANE

56073-10-0 BRODIFACOUM

56093-45-9 SELENIUM SULFIDE

56189-09-4 DIBASIC LEAD STEARATE
LEAD, BIS(OCTADECANOATO)DIOX-ODI- [PAL]
STEARIC ACID, LEAD SALT, DIBASIC [MSL]

56320-22-0 ARSENIC DISULFIDE

56343-50-1 N,N-DIMETHYL DICHLOROAC-ETAMIDE

56391-57-2 NETILMICIN SULFATE

56709-13-8 1H,3H,5H-OXAZOLO(3,4-C)OXA-ZOLE, POLY(OXYMETHYLENE) DERIVATIVE

56797-01-4 CERIUM 2-ETHYLHEXOATE

56802-99-4 CHLORINATED TRISODIUM PHOSPHATE

56803-37-3 TERT-BUTYLPHENYL DIPHENYL PHOSPHATE
PHOSPHATE, TERT-BUTYLPHENYL DIPHENYL [OTS]
PHOSPHORIC ACID, (1,1-DIMETHYLETHYL) PHENYL DIPHENYL ESTER [TSC, TSC]

56832-73-6 BENZOFLUORANTHENE

56863-02-6 DIETHANOLAMIDE OF LINOLEIC ACID

56894-91-8 1,4-BIS(CHLOROMETHOXYMETHYL) BENZENE

57011-27-5 PHOSPHORIC ACID, [1,2-ETHANEDIYLBIS[NITRILOBIS (METHYLENE)]TETRAKIS-, AM-MONIUM SALT

57018-52-7 PROPYLENE GLYCOL T-BUTYL ETHER
2-PROPANOL, 1-(1,1-DIMETHYLETHOXY)- [TSC, TSC, TSC]

57109-90-7 CLORAZEPATE DIPOTASSIUM

57117-31-4 2,3,4,7,8-PENTACHLORO DIBEN-ZOFURANS

57117-41-6 1,2,3,7,8-PENTACHLORO DIBEN-ZOFURAN

57117-44-9 1,2,3,6,7,8-HEXACHLORO DIBEN-ZOFURAN

57137-10-7 TRIBROMINATED POLYSTYRENE

57171-56-9 POLYOXYETHYLENE SORBITOL HEXAOLEATE

57213-69-1 TRICLOPYR, TRIETHYLAMMO-NIUM SALT

57308-10-8 CALCIUM HYDRIDE

57321-63-8 TRICHLORO DIPHENYL OXIDE

57407-26-8 EMERY

57451-03-3 NONYLPHENOL, ETHOXYLATED, SULFATED, TRIETHANOLAMINE SALT

57455-37-5 ULTRAMARINE BLUE
C.I. PIGMENT BLUE 29 [PST]

57485-31-1 ALUMINUM SILICON

57608-40-9 AMMONIUM NITRATE PHOS-PHATE

57653-85-7 1,2,3, 6,7,8-HEXACHLORODIBENZO-P-DIOXIN

57835-92-4 4-NITROPYRENE

57837-19-1 METALAXYL

58145-14-5 DIETHYLAMINO ETHANOLAMINE

58164-88-8 ANTIMONY LACTATE

58270-08-9 ZINC, DICHLORO(4,4-DIMETHYL-5-((((METHYL AMINO)CARBONYL) OXY)IMINO)PENTANENITRILE)-, (T-4)-

58449-37-9 ISONONANOYL PEROXIDE

58500-38-2 SILICIC ACID, BERYLLIUM SALT

59080-40-9 2,2',4,4',5,5'-HEXABROMO-1,1'-BIPHENYL

59139-23-0 NONYLPHENOL, ETHOXY-LATED, PHOSPHATED, MO-NOETHANOLAMINE SALT

59227-89-3 N-DODECYLCAPROLACTAM

59355-75-8 MAPP
METHYL ACETYLENE - PROPADIENE MIXTURE [MEX, MEX]
METHYL ACETYLENE - PROPADIENE MIXTURE (MAPP) [ALB, ALB]
METHYL ACETYLENE-PROPADIENE MIXTURE [CEX, CEX, HM1, HM1, MIN, NIO, NIO, NIO]
METHYL ACETYLENE-PROPADIENE MIXTURE (MAPP) [AUS, AUS, BC1, BC1, FLA, ONT, ONT, OS1, OS1, OS1, TLV, TLV]
METHYL ACETYLENEPROPADIENE MIXTURE [CAL]
METHYLACETYLENE-PROPADIENE (MAPP) [QBC, QBC]

59467-96-8 MIDAZOLAM HYDROCHLORIDE

59536-65-1 FIREMASTER BP-6
FIREMASTER BP 6 [PAL, PAL]
POLYBROMINATED BIPHENYLS [FLA, NJL]
POLYBROMINATED BIPHENYLS (PBB'S) [MSL]

59669-26-0 THIODICARB

59716-87-9 2-AMINO-5-(5-NITRO-2-FURYL)-1, 3,4-THIADIAZOLE

59720-42-2 1H,3H,5H-OXAZOLO(3,4-C)OXA-ZOLE, METHANOL DERIVATIVE

59766-31-3 POTASSIUM TITANATE (K2TI8O17)

59766-35-7 ZINC SULFATE, BASIC

59789-51-4 1H-PYROLE-2,5-DIONE, 1-(2,4,6-TRIBROMOPHENYL)-

59820-43-8 HC YELLOW 4
HC YELLOW NO. 4 [IAR]

59865-13-3 CYCLOSPORIN A
CYCLOSPORIN [NTP]

59917-23-6 HYDROXYL AMINE IODIDE

60102-37-6 PETASITENINE

60153-49-3 3-METHYLNITROSOAMINOPRO-
PIONITRILE
3-(N-NITROSOMETHYLAMINO)PROPI-
ONITRILE [C65, CAL, IAR, MIN]
3-METHYLNITROSAMINO PROPIONI-
TRILE [MSL]
3-METHYLNITROSAMINOPROPIONI-
TRILE [PAL, PAL]

60168-88-9 FENARIMOL [.ALPHA.-(2-
CHLOROPHENYL)-.ALPHA.-4-
CHLOROPHENYL)-5-PYRIM-
IDINEMETHANOL]

60207-90-1 PROPICONAZOLE [1-[2-(2,4-
DICHLOROPHENYL)-4-PROPYL-
1,3-DIOXOLAN-2-YL]-METHYL-1H-
1,2,4,-TRIAZOLE]

60209-82-7 ISODECYL NEOPENTANOATE

60381-61-5 2-ETHYLHEXYL ACRYLATE,
POLYMER WITH VINYL TOLUENE

60501-41-9 OXIRANE, [(9-OCTADECENY-
LOXY)METHYL]-, (Z)-
OLEYL GLYCIDYL ETHER [EPA,
OTS]

60568-05-0 FURMECYCLOX

60616-74-2 MAGNESIUM HYDRIDE

60646-36-8 ARSENIC TRICHLORIDE

60676-86-0 SILICA, FUSED
SILICA - AMORPHOUS, SILICA, FUSED
[TLV]
SILICA, AMORPHOUS, FUSED [WHM]
SILICA, AMORPHOUS, QUARTZ
GLASS [MAK, MAK]
SILICA, CRYSTALLINE, SILICA,
FUSED [CEX]
SILICA, FUSED CRYSTALLINE [QBC]
SILICA, FUSED, DUST [FLA, MSL]
SILICA, VITREOUS [PST]
SILICA-CRYSTALLINE, SILICA [MIN]

60816-37-7 DODECYLBENZENESULFONIC
ACID, 1,3-PROPANEDIAMINE
SALT

60816-39-9 DODECYLBENZENESULFONIC
ACID, N,N-DIMETHYL-1,3-
PROPANEDIAMINESALT

60816-63-9 ETHYLENEDIAMINETE-
TRAACETIC ACID, TRIETHY-
LAMINE SALT

60828-78-6 2,6,8-TRIMETHYL-4-NONYL
POLYETHYLENE GLYCOL ETHER

60851-34-5 2,3,4,6,7,8-HEXACHLORO DIBEN-
ZOFURANS

60857-97-8 REACTION PRODUCT OF HY-
DROXYETHYL ACRYLATE AND
METHYL OXIRANE

60864-33-7 TRITON CF 10
BENZYL ETHER OF OCTYLPHE-
NOXYPOLYETHOXY ETHANOL [PST]

60874-89-7 POLY(OXYETHYLENE)
METHYLENEBIS(DIAMYLPHE-
NOL)

60874-90-0 SODIUM ISOPROPYL ISOBUTYL
NAPHTHALENESULFONATE

60883-84-3 SODIUM NONYLMETHYLNAPH-
THALENESULFONATE

60883-89-8 TRIDECYLBENZENESULFONIC
ACID, DIMETHYLAMINE SALT

60883-90-1 TRIDECYLBENZENESULFONIC
ACID, PROPYLAMINE SALT

61105-31-5 CROCIDOLITE (FE2MG3NA2(SIO3)
8)

61204-26-0 BISMUTH CHROMATE

61262-53-1 ETHYLENE BIS(PENTABRO-
MOPHENOXIDE)

61386-02-5 BENZOIC ACID, 3,3'-
METHYLENEBIS[6-AMINO-DI-2-
PROPENYL] ESTER

61524-98-9 ETHOXYLATED HYDROABIETHYL
ALCOHOL

61578-04-9 OXIRANE, [[4-(1-METHYL-1-
PHENYLETHYL)PHENOXY]
METHYL]-
CUMYLPHENYL GLYCIDYL ETHER,
P- [OTS]
P-CUMYLPHENYL GLYCIDYL ETHER
[EPA]

61702-44-1 2-CHLORO-P-PHENYLENEDI-
AMINE SULFATE

61702-73-6 1H-IMIDAZOLIUM, 1,1-BIS(CAR-
BOXYMETHYL)-4,5-DIHYDRO-
2-UNDECYL-, HYDROXIDE, DIS-
ODIUM SALT

61702-81-6 1,2-BENZENEDICARBOXYLIC
ACID, HEXYL ISODECYL ESTER

61702-88-3 BENZENE, 1,1'-OXYBIS[(1,1,3,3-
TETRAMETHYLBUTYL)-

61703-05-7 C.I. DIRECT BLACK 114

61725-89-1 OXIRANE METHYL-, POLYMER
WITH OXIRANE, TRIDECYL
ETHER

61788-32-7 HYDROGENATED TERPHENYLS

61788-33-8 KANECHLOR 500
TERPHENYL, CHLORINATED [TSC,
TSC]

61788-47-4 COCONUT OIL FATTY ACIDS

61788-48-5 ACETYLATED LANOLIN

61788-59-8 FATTY ACIDS, COCO, METHYL
ESTERS

61788-60-1 METHYL ESTERS OF COTTON-
SEED OIL

61788-72-5 OCTYL EPOXYTALLATES
EPOXIDIZED OCTYL TALLATE [PST]

61788-76-9 CHLOROALKANES
ALKANES, CHLORO- [TSC]
CHLORINATED PARAFFINS (C10 -
C13) [HM1]

61788-89-4 FATTY ACIDS, C18-UNSATU-
RATED, DIMERS

61788-90-7 N,N-DIMETHYL (COCONUT OIL
ALKYL) AMINE OXIDE
COCO ALKYLDIMETHYLAMINES, N-
OXIDES [PST]

61788-93-0 COCO ALKYLDIMETHYLAMINES

61789-01-3 FATTY ACIDS, TALL OIL, EPOXI-
DIZED, 2-ETHYLHEXYL ESTERS
TALL OIL, EPOXIDIZED, 2-ETHYL-
HEXYL ESTERS [PST]

61789-18-2 COCO ALKYLTRIMETHYL QUA-
TERNARY AMMONIUM CHLO-
RIDES

61789-28-4 CREOSOTE OIL
CREOSOTE, OIL [MSL]

61789-30-8 COCONUT FATTY ACIDS, POTAS-
SIUM SALT
POTASSIUM COCONUT OIL SOAP
[PST]

61789-31-9 POTASSIUM COCONUT FATTY
OIL SOAP
COCONUT OIL SOAP [PST]

61789-32-0 FATTY ACIDS, COCO, 2-SUL-
FOETHYL ESTERS, SODIUM
SALTS

61789-36-4 CALCIUM NAPHTHENATE

61789-40-0 COCOAMIDO(PROPYLBETAINE)
N-(COCO ALKYL) AMIDO PROPYL
DIMETHYL BETAINE [PST]

61789-51-3 COBALT NAPHTHENATE
COBALT NAPHTHA [NFP, NFP]
COBALT NAPHTHENATES [HM1,
HM1]

61789-52-4 COBALT TALLATE

61789-65-9 ALUMINUM RESINATE

61789-72-8 DIMETHYL BENZYL HYDRO-
GENATED TALLOW AMMONIUM
CHLORIDE

61789-73-9 BENZYL BIS(HYDROGENATED
TALLOW ALKYL) METHYL AM-
MONIUM CHLORIDE

61789-75-1 BENZYL DIMETHYL (TALLOW
ALKYL) AMMONIUM CHLORIDE

61789-76-2 DIOCO ALKYLAMINE

61789-77-3 DI(COCO ALKYL) DIMETHYL
AMMONIUM CHLORIDE

61789-81-9 BIS(HYDROGENATED TALLOW
ALKYL)DIMETHYL QUATER-
NARY COMPOUNDS, METHYL
SULFATES

61789-86-4 CALCIUM PETROLEUM SUL-
FONATE
CALCIUM SALTS OF PETROLEUM
SULFONIC ACIDS [PST]

61789-91-1 JAJOBA BEAN OIL

61789-97-7 TALLOW OIL
TALLOW [NFP, NFP, PST]

61789-98-8 CORK

61789-99-9 LARD

61790-12-3 TALL OIL FATTY ACIDS
FATTY ACIDS, TALL-OIL [PST]

61790-14-5 LEAD NAPHTHENATE

61790-50-9 RESIN ACIDS AND ROSIN ACIDS,
POTASSIUM SALT
POTASSIUM SALT OF WOOD ROSIN
ACIDS [PST]

61790-51-0 RESIN ACIDS AND ROSIN ACIDS,
SODIUM SALTS
SODIUM SALT OF HYDROCARBON
INSOLUBLE FRACTION OF ROSIN
[PST]

61790-53-2 DIATOMACEOUS EARTH
CELITE [PST]
DIATOMACEOUS EARTH, NATURAL
[GBR]
SILICA - AMORPHOUS, DIATOMA-
CEOUS EARTH (UNCALCINED) [TLV,
TLV]
SILICA, AMORPHOUS, DIATOMA-
CEOUS EARTH [CEX]
SILICA, AMORPHOUS, DIATOMA-
CEOUS EARTH (UNCALCINED)
[MAK, MAK]
SILICA, AMORPHOUS, DIATOMA-
CEOUS EARTH, CONTAINING LESS
THAN 1% CRYSTALLINE SILICA [OS1,
OS1]
SILICA-AMORPHOUS, DIATOMA-
CEOUS EARTH (UNCALCINED) [AUS]

61790-59-8 HYDROGENATED TALLOW
ALKYL AMINE ACETATE

61790-63-4 COCONUT OIL ACID, DI-ETHANOLAMINE SALT
FATTY ACIDS, COCO, COMPOUNDS WITH DIETHANOLAMINE [PST]

61790-81-6 LANOLIN, ETHOXYLATED
ETHOXYLATED LANOLIN [PST]

61790-85-0 ETHOXYLATED N-(TALLOW ALKYL)TRIMETHYLENE DI-AMINES

61790-86-1 POLYOXYETHYLENE SORBITAN MONOTALLATE

61790-88-3 SORBITAN, TALL-OIL FATTY ACID TRIESTERS, ETHOXYLATED

61790-90-7 ETHOXYLATED SORBITOL HEX-AESTER OF TALL-OIL ACIDS

61791-00-2 TALL OIL FATTY ACIDS, ETHOXYLATED
FATTY ACIDS, TALL-OIL, ETHOXLATED [PST]

61791-10-4 QUATERNARY AMMONIUM COMPOUNDS, COCO ALKYL-BIS(HYDROXYETHYL)METHYL, ETHOXYLATED, CHLORIDES

61791-12-6 ETHOXYLATED CASTOR OIL
CASTOR OIL, POLYOXYETHYLATED [PST]

61791-14-8 ETHOXYLATED COCONUT OIL ALKYL AMINE
ETHOXYLATED COCO ALKYL AMINES [PST]

61791-23-9 POLYOXYETHYLENE SOYA ACID ESTERS

61791-24-0 AMINES, SOYA ALKYL, ETHOXY-LATED
POLYOXYETHYLENE SOYA ALKYL AMINES [PST]

61791-26-2 TALLOW AMINE, ETHOXYLATED
ETHOXYLATED TALLOWAMINE [PST]

61791-28-4 ALCOHOLS, TALLOW, ETHOXY-LATED

61791-34-2 N-(SOYA ALKYL)-N-ETHYLMOR-PHOLINIUM ETHYLSULFATE

61791-41-1 SODIUM N-(TALL-OIL ALKYL)-N-METHYLTAURINE

61791-47-7 BIS(2-HYDROXYETHYL) CO-COAMINE OXIDE

61791-53-5 TALLOW DIAMINE
N-TALLOW ALKYLTRIMETHYLENE-DIAMINE OLEATES [PST]

61792-07-2 ACETIC ACID, (2,4,5-TRICHLOROPHENOXY)-, 1-METHYL PROPYL ESTER
2,4,5-T ESTER [NJL]

61827-84-7 METHYL OXIRANE POLYMER WITH OXIRANE, OCTYL ETHER

61849-72-7 POLYPROPYLENE GLYCOL BETA-METHYL GLUCOSIDE ETHER (4:1)

61886-60-0 1,2-BENZENEDICARBOXYLIC ACID, ISODECYL TRIDECYL ES-TER

61969-42-4 C.I. SOLVENT BLUE 53

62015-39-8 MISOPROSTOL

62073-57-8 DIMETHYLOL UREA - FORMALDEHYDE - MONOMETHYLOL UREA POLY-MER

62207-76-5 COBALT, ((2,2'-(1,2-ETHANEDIYL-BIS(NITRILOMETHYLIDYNE))BIS (6-FLUOROPHENOLATO))(2)-
COBALT(II), N,N'-ETHYLENEBIS(3-FLUOROSALICYLIDENEIMINATO)-[WHM]
COBALT(II),N,N'-ETHYLENEBIS (3-FLUOROSALICYLIDENEIMINATO)-[MSL]

62449-33-6 2-(8-HEPTADECENYL)-4,5-DIHY-DRO-1H-IMIDAZOLE-1-ETHANOL, MONOHYDROCHLORIDE, (Z)-

62450-06-0 TRP-P-1
5H-PYRIDO[4,3-B]INDOL-3-AMINE, 1, 4-DIMETHYL- [PAL, PAL]
TRP-P-1 (3-AMINO-1,4-DIMETHYL-5H-PYRIDO(4,3-B)INDOLE) [IAR]
TRP-P-1 (TRYPTOPHAN-P-1) [C65, C65, CAL]

62450-07-1 TRP-P-2
5H-PYRIDO[4,3-B]INDOL-3-AMINE, 1-METHYL- [PAL, PAL]
TRP-P-2 (3-AMINO-1-METHYL-5H-PYRIDO(4,3-B)INDOLE) [IAR]
TRP-P-2 (TRYPTOPHAN-P-2) [C65, C65, CAL]

62476-59-9 ACIFLUORFEN
ACIFLUORFEN, SODIUM SALT [5-(2-CHLORO-4-(TRIFLUOROMETHYL) PHENOXY)-2-NITRO-BENZOIC ACID, SODIUM SALT] [313]

62654-17-5 1,4-BENZENEDIAMINE, ETHANE-DIOATE (1:1)

62765-43-9 NIAX CATALYST ESN

62765-93-9 NIAX CATALYST ESN

63041-90-7 6-NITROBENZO(A)PYRENE

63089-86-1 POLYOXYETHYLENE SORBITOL TETRAOLEATE

63134-33-8 4-[[4-(PHENYLMETHOXY) PHENYL]SULFONYL] PHENOL

63148-56-1 SILOXANES AND SILICONES, METHYL 3,3,3-TRIFLUORO-PROPYL

63148-62-9 POLY(DIMETHYLSILOXANE)
DIMETHYL SILICONES AND SILOX-ANE [TSC, TSC]
SILICONES AND SILOXANES, DIMETHYL [PST]

63148-65-2 POLYVINYL BUTYRAL
POLYVINYL BUTYRAL RESIN [PST]

63148-69-6 SOYA ALKYL RESIN

63181-88-4 VINYL ACETATE - ACRYLIC ES-TER COPOLYMERS

63182-08-1 SODIUM VINYLBENZENESUL-FONATE, POLYMER WITH DI-VINYLBENZENE

63231-60-7 MICROCRYSTALLINE WAX
WAX, MICROCRYSTALLINE [NFP, NFP]

63231-67-4 SILICA GEL
SILICA, AMORPHOUS GEL [QBC]
SILICA-AMORPHOUS, SILICA GEL [AUS]

63283-80-7 PROPANE, 2,2'-OXYBIS [DICHLORO-

63393-89-5 COUMARONE-INDENE RESIN

63393-93-1 ISOPROPYL LANOLIN

63449-39-8 CHLORINATED HYDROCARBONS (CHORINATED PARAFFINS)
CHLORINATED PARAFFIN WAXES [CN4]
CHLORINATED WAX [PST]
CHLOROPARRAFFINS [MAK]
PARAFFIN WAXES AND HYDROCAR-BON WAXES, CHLORINATED [TSC]

63597-41-1 OCTADIENE

63734-62-3 BENZOIC ACID, 3-[2-CHLORO-4-(TRIFLUOROMETHYL)PHENOXY]-

63868-82-6 ZIRCONIUM PICRAMATE

63885-01-8 ZINC AMMONIUM NITRITE

63918-97-8 LEAD STYPHNATE

63937-14-4 MERCURY GLUCONATE

63938-10-3 CHLOROTETRAFLUOROETHANE

63989-69-5 FERRIC ARSENITE

64036-86-8 TETRADECANE

64037-54-3 3,4-DICHLOROBUTENE-1
1-BUTENE, 3,4-DICHLORO-,(.+-.)-1 [PAL]

64047-28-5 BIS(DI-(BETA-CHLOROETHYL) SULFIDE)PALLADOUS CHLORIDE

64051-23-6 2-BUTOXYETHYL DIHYDROGEN-PHOSPHATE, DIETHYLAMINE SALT

64070-14-0 O-ANISIDINE ANTIMONYL TAR-TRATE

64082-35-5 LITHIUM FERROSILICON

64083-59-6 C.I. DIRECT ORANGE 8

64091-90-3 3-(N-NITROSOMETHYLAMINO)-4-(3-PYRIDYL)-1-BUTANAL (NNA)

64091-91-4 4-(N-NITROSOMETHYLAMINO)-1-(3-PYRIDYL)-1-BUTANONE
4-(METHYLNITROSAMINO)-1-(3-PYRIDYL)-1-BUTANONE [MSL]
4-(N-NITROSOMETHYLAMINO)-1-(3-PYRIDYL)-1-BUTANONE (NNK) [CAL, IAR, NTP]

64093-79-4 CHROMOSULFURIC ACID
NEOCHROMIUM [NJL]

64175-94-6 CALCIUM CHLORIDE HYDROX-IDE HYPOCHLORITE, DIHY-DRATE

64365-06-6 ISOPARAFFINIC PETROLEUM HYDROCARBONS

64382-04-3 PHTHALIC ACID, POLYMER WITH 1,3-DIHYDRO-1,3-DIOXO-5-ISOBENZOFURANCARBOXYLIC ACID AND TRIMETHYLOL-PROPANE, LINOLEATE

64399-38-8 2-(DIETHYLAMINO)ETHYL METHACRYLATE, POLYMER WITH DODECYL METHACRY-LATE, STYRENE, HEXADEXYL METHACRYLATE AND TETRADE-CYL METHACRYLATE

64503-07-7 BENZYL DIBROMOACETATE

64630-63-3 2-PROPENOIC ACID, 7-OXABICY-CLO[4.1.0]HEPT-3-YLMETHYL ES-TER

64665-10-7 SODIUM TRIBUTYLNAPHTHALE-NESULFONATE

64741-44-2 DISTILLATES (PETROLEUM), STRAIGHT-RUN MIDDLE
GAS OIL [HM1, HM1, NFP, NFP, NJL]

64741-49-7 VACUUM TOWER CONDENSATE (PETROLEUM)
MINERAL OIL, PETROLEUM CON-DENSATES, VACUUM TOWER [MSL]

64741-50-0 PETROLEUM DISTILLATES, LIGHT PARAFFINIC
DISTILLATES (PETROLEUM), LIGHT PARAFFINIC [PST]
MINERAL OIL, PETROLEUM DISTIL-LATES, LIGHT PARAFFINIC [MSL]

64741-51-1 PETROLEUM DISTILLATES, HEAVY PARAFFINIC
DISTILLATES (PETROLEUM), HEAVY PARAFFINIC [PST]
MINERAL OIL, PETROLEUM DISTILLATES, HEAVY PARAFFINIC [MSL]

64741-52-2 LIGHT NAPHTHENIC OIL
DISTILLATES (PETROLEUM), LIGHT NAPHTHENIC [PST]
MINERAL OIL, PETROLEUM DISTILLATES, LIGHT NAPHTHENIC [MSL]

64741-53-3 PETROLEUM DISTILLATES, HEAVY NAPHTHENIC
DISTILLATES (PETROLEUM), HEAVY NAPTHENIC [PST]
MINERAL OIL, PETROLEUM DISTILLATES, HEAVY NAPHTHENIC [MSL]

64741-58-8 LIGHT VACUUM GAS OIL (PETROLEUM)
LIGHT VACUUM DISTILLATES [IAR]

64741-59-9 PETROLEUM DISTILLATES, LIGHT CATALYTIC CRACKED
LIGHT CATALYTICALLY CRACKED DISTILLATES [IAR]

64741-61-3 PETROLEUM DISTILLATES, HEAVY CATALYTIC CRACKED
HEAVY CATALYTICALLY CRACKED DISTILLATES [IAR]

64741-67-9 HEAVY AROMATIC BOTTOMS

64741-71-5 RESIDUAL HEAVY POLYMER (PETROLEUM)
RESIDUAL (HEAVY) FUEL OILS [C65, IAR]

64741-89-5 PETROLEUM DISTILLATES, SOLVENT-REFINED LIGHT PARAFFINIC
MINERAL OIL, PETROLEUM DISTILLATES, SOLVENT-REFINED LIGHT-PARAFFINIC [MSL]

64741-97-5 DISTILLATES (PETROLEUM), SOLVENT REFINED LIGHT NAPHTHENIC
MINERAL OIL, PETROLEUM DISTILLATES, SOLVENT-REFINED LIGHT-NAPHTHENIC [MSL]

64742-03-6 LIGHT DISTILLATE SOLVENT EXTRACT (PETROLEUM)
MINERAL OIL, PETROLEUM EXTRACTS, LIGHT NAPHTHENIC DISTILLATE SOLVENT [MSL]

64742-04-7 PARAFFINIC DISTILLATE SOLVENT EXTRACT
MINERAL OIL, PETROLEUM EXTRACTS, HEAVY PARAFFINIC DISTILLATESOLVENT [MSL]

64742-05-8 LIGHT PARAFFINIC DISTILLATE SOLVENT EXTRACT (PETROLEUM)
MINERAL OIL, PETROLEUM EXTRACTS, LIGHT PARAFFINIC DISTILLATESOLVENT [MSL]

64742-06-9 EXTRACTS (PETROLEUM), MIDDLE DISTILLATE SOLVENT
MINERAL SEAL OIL TYPICAL [NFP, NFP]

64742-10-5 RESIDUAL OIL SOVENT EXTRACT (PETROLEUM)
MINERAL OIL, PETROLEUM EXTRACTS, RESIDUAL OIL SOLVENT [MSL]

64742-11-6 HEAVY NAPHTHENIC DISTILLATE SOLVENT EXTRACT
MINERAL OIL, PETROLEUM EXTRACTS, HEAVY NAPHTHENIC DISTILLATE SOLVENT [MSL]

64742-14-9 ACID TREATED LIGHT DISTILLATE (PETROLEUM)
DISTILLATES (PETROLEUM), ACID TREATED LIGHT [PST]

64742-16-1 PETROLEUM RESINS

64742-17-2 ACID TREATED RESIDUAL OIL (PETROLEUM)
MINERAL OIL, PETROLEUM RESIDUAL OILS, ACID-TREATED [MSL]

64742-18-3 PETROLEUM DISTILLATES, ACID TREATED, HEAVY NAPHTHENIC
MINERAL OIL, PETROLEUM RESIDUAL OILS, ACID-TREATED HEAVY-NAPHTHENIC [MSL]

64742-19-4 PETROLEUM DISTILLATES, ACID TREATED, LIGHT NAPHTHENIC
MINERAL OIL, PETROLEUM RESIDUAL OILS, ACID-TREATED LIGHT-NAPHTHENIC [MSL]

64742-20-7 PETROLEUM DISTILLATES, ACID TREATED, HEAVY PARAFFINIC
MINERAL OIL, PETROLEUM DISTILLATES, ACID-TREATED HEAVY-PARAFFINIC [MSL]

64742-21-8 PETROLEUM DISTILLATES, ACID TREATED, LIGHT PARAFFINIC
MINERAL OIL, PETROLEUM DISTILLATES, ACID-TREATED LIGHT-PARAFFINIC [MSL]

64742-24-1 SLUDGE ACID
ACID SLUDGE [NJL]

64742-42-3 HYDROCARBON WAXES (PETROLEUM), CLAY-TREATED MICROCRYSTALLINE
PETROLEUM WAX, CLAY-TREATED, MICROCRYSTALLINE [PST]

64742-53-6 PETROLEUM DISTILLATES, HYDROTREATED LIGHT NAPHTHENIC
MINERAL OIL, PETROLEUM DISTILLATES, HYDROTREATED LIGHT-NAPHTHENIC [MSL]

64742-55-8 PETROLEUM DISTILLATES, HYDROTREATED LIGHT PARAFFINIC
MINERAL OIL, PETROLEUM DISTILLATES, HYDROTREATED LIGHT-PARAFFINIC [MSL]

64742-56-9 PETROLEUM DISTILLATES, SOLVENT DEWAXED LIGHT PARAFFINIC
MINERAL OIL, PETROLEUM DISTILLATES, SOLVENT-DEWAXED LIGHT-PARAFFINIC [MSL]

64742-61-6 SLACK WAX (PETROLEUM)
SLACK WAX [PST]

64742-64-9 DISTILLATES (PETROLEUM), SOLVENT-DEWAXED LIGHT NAPHTHENIC
MINERAL OIL, PETROLEUM DISTILLATES, SOLVENT-DEWAXED LIGHT-NAPHTHENIC [MSL]

64742-68-3 NAPHTHENIC OILS (PETROLEUM), CATALYTIC DEWAXED HEAVY
MINERAL OIL, PETROLEUM NAPHTHENIC OILS, CATALYTIC DEWAXEDHEAVY [MSL]

64742-69-4 NAPHTHENIC OILS (PETROLEUM), CATALYTIC DEWAXED LIGHT
MINERAL OIL, PETROLEUM NAPHTHENIC OILS, CATALYTIC DEWAXEDLIGHT [MSL]

64742-70-7 PARAFFIN OILS (PETROLEUM), CATALYTIC DEWAXED HEAVY
MINERAL OIL, PETROLEUM PARAFFIN OILS, CATALYTIC DEWAXED HEAVY [MSL]

64742-71-8 PARAFFIN OILS, CATALYTIC DEWAXED LIGHT
MINERAL OIL, PETROLEUM PARAFFIN OILS, CATALYTIC DEWAXED LIGHT [MSL]

64742-88-7 SOLVENT NAPHTHA (PETROLEUM), MEDIUM ALIPHATIC

64742-89-8 SOLVENT NAPHTHA (PETROLEUM), LIGHT ALIPHATIC
NAPHTHA, SOLVENT [NJL]

64742-94-5 NAPHTHA (PETROLEUM), HEAVY AROMATIC
SOLVENT NAPHTHA (PETROLEUM), HEAVY AROMATIC [PST]

64742-95-6 PETROLEUM NAPHTHA, LIGHT AROMATIC
SOLVENT NAPHTHA (PETROLEUM), LIGHT AROMATIC [PST, TSC, TSC]

64742-96-7 SOLVENT NAPHTHA (PETROLEUM), HEAVY ALIPHATIC

64754-90-1 CHLORINATED POLYETHYLENE

64902-72-3 CHLORSULFURON [2-CHLORO-N-[[4-METHOXY-6-METHYL-1,3, 5-TRIAZIN-2-YL) AMINO]CARBONYL]BENZENESULFONAMIDE

64969-34-2 3,3'-DICHLOROBENZIDINE SULFATE

64992-16-1 ETHANAMINIUM, 2-[[2-CYANO-3-[4-(DIETHYLAMINO)PHENYL]-1-OXO-2-PROPENYL]OXY]-N,N,N-TRIMETHYL-, CHLORIDE

65071-95-6 TALL OIL, ETHOXYLATED

65122-06-7 C.I. BASIC RED 14

65646-68-6 RETINOID PROJECT 2 (4-HYDROXYPHENYL)

65652-41-7 PHOSPHATE, BIS(TERT-BUTYLPHENYL) PHENYL

65879-44-9 PHENOL, 2,4-DIAMINO-6-METHYL-, HYDROCHLORIDE

65992-66-7 N,N,N',N'-TETRAKIS(OXIRANYLMETHYL)-1,3-CYCLOHEX-ANEDIMETHANAMINE

65996-68-1 FLUE GASSES, FERROUS METAL, BLAST FURNACE
GAS, BLAST FURNACE [MSL, NFP]

65996-77-2 COKE (COAL)

65996-79-4 NAPHTHA (COAL TAR)
COAL TAR NAPHTHA [NJL]

65996-89-6 TAR, COAL, HIGH-TEMP
HIGH-TEMPERATURE COAL TARS [NTP]

65996-90-9 TAR, COAL, LOW-TEMP

65996-91-0 COAL TAR LIGHT OIL
COAL TAR OIL [MAK]
DISTILLATES (COAL TAR), UPPER [PAL]

65996-92-1 COAL TAR DISTILLATE
COAL TAR DISTILLATES, FLAMMABLE [HM1, HM1]

65996-93-2 COAL TAR PITCHES
COAL TAR PITCH [ATS, MSL, NFP, NFP, NHS, NHS, NHS]
COAL TAR PITCH VOLATILES [ALB, ALB, ALB, AUS, AUS, CEX, MEX, MEX, OSI, OSI, TLV, TLV, WHM]
COAL TAR PITCH VOLATILES (BENZENE SOLUBLE FRACTION) [QBC, QBC]
COAL TAR PITCH VOLATILES (BENZENE-SOLUBLE FRACTION) [NIO, NIO, NIO, NIO]
COAL-TAR PITCHES [IAR]
PARTICULATE POLYCYCLIC AROMATIC HYDROCARBONS [BC1]

65997-05-9 ROSIN POLYMERS
ROSIN, PARTIALLY DIMERIZED [PST]

65997-06-0 ROSIN, PARTIALLY HYDRO-GENATED

65997-15-1 PORTLAND CEMENT
CEMENT, PORTLAND, CHEMICALS [PAL]

65997-17-3 GLASS, OXIDE
GLASS [MIN]

66070-87-9 POLYGLYCERYL PHTHALATE ESTER OF COCONUT OIL FATTY ACID

66104-24-3 BERYLLIUM CARBONATE

66108-37-0 PHOSPHORIC ACID, 2,2-BIS (BROMOMETHYL)-3-CHLORO-PROPYL BIS[2-CHLORO-1-(CHLOROMETHYL)ETHYL] ESTER

66215-27-8 CYROMAZINE

66327-54-6 3-CYCLOHEXENE-1-CARBOX-ALDEHYDE, 1-METHYL-4-(4-METHYLPENTYL)-

66422-95-5 ETHANOL, 2-(2,4-DIAMINOPHE-NOXY)-, DIHYDROCHLORIDE

66441-23-4 FENOXAPROP ETHYL [2-(4-((6-CHLORO-2-BENZOXAZOLYLEN OXY)PHENOXY) PROPANOIC ACID, ETHYL ESTER]

66507-71-9 2,2',2"-NITRILOTRISETHANOL POLYMER WITH 1,4-DICHLORO-2-BUTENE AND N,N,N',N'-TETRAM-ETHYL-2-BUTENE-1,4-DIAMINE

66733-21-9 ERIONITE
ERIONITE FIBER [TSC]

66841-25-6 TRALOMETHRIN

67254-79-9 MIXED FATTY ACIDS

67401-50-7 TETRASODIUM ETHYLENEDI-AMINETETRAACETATE TRIHY-DRATE

67485-29-4 HYDRAMETHYLNON [TETRAHY-DRO-5,5-DIMETHYL-2(1H)-PYRIMIDINONE[3-[4-(TRIFLU-OROMETHYL)PHENYL]-1-[2-[4-(TRIFLUOROMETHYL)PHENYL] ETHENYL]-2-PROPENYLIDENE] HYDRAZONE]

67562-39-4 1,2,3, 4,6,7,8-HEPTACHLORODIBENZO-FURAN

67632-66-0 TRICHLOROMETHYL PERCHLO-RATE

67700-49-6 ROSIN POLYMER WITH P-TERT-BUTYLPHENOL, FORMALDEHYDE AND GLYCEROL

67701-05-7 FATTY ACIDS, C8-18 AND C18-UNSATURATED
FATTY ACIDS CONTAINING LAURIC, MYRISTIC, PALMITIC, STEARIC, OLEIC, AND/OR LINOLEIC [PST]

67708-83-2 DIBROMOCHLOROPROPANE

67711-86-8 LEAD SILICATE SULFATE

67711-90-4 FLUE DUST, POISONOUS

67730-10-3 GLU-P-2
2-AMINODIPYRIDO(1,2-A:3',2'-D) IMIDAZOLE [MSL]
GLU-P-2 (2-AMINOPYRIDO[1,2-AL-PHA:3',2'-D]IMIDAZOLE) [C65, C65, CAL]
GLU-P-2 (2-AMINOPYRIDO[1,2-AL-PHA:3',2'-D]IMIDAZOLE) [IAR]

67730-11-4 GLU-P-1
2-AMINO-6-METHYLDIPYRIDO[1,2-A:3',2'-D]IMIDAZOLE [MSL]
GLU-P-1 (2-AMINO-6-METHYLDIPYRIDO[1,2-ALPHA:3',2'-D] IMIDAZOLE) [C65, C65, CAL, IAR]

67762-90-7 DIMETHYL SILICONE POLYMER WITH SILICA
DIMETHYL SILICONES AND SILOX-ANES, REACTION PRODUCTS WITH SILICA [TSC, TSC]

67762-94-1 DIMETHYLMETHYLVINYLSILOX-ANE

67774-32-7 FIREMASTER FF-1
HEXABROMOBIPHENYL (FIREMAS-TER FF-1) [NTP]
POLYBROMINATED BIPHENYL MIX-TURE (FIREMASTER FF-1) [CSR, CSR]
POLYBROMINATED BIPHENYLS [ATS, NJL]
POLYBROMINATED BIPHENYLS (2, 4,5,2',4',5'-HEXABROMO BIPHENYL) [WHM]

67785-61-9 SODIUM POLYMETHACRYLATE

67785-93-7 POTASSIUM SULFATORICI-NOLEATE

67786-03-2 OXIRANE, 2,2'-[[[(2-OXIRANYL-METHOXY)PHENYL]METHYLENE] BIS(4,1-PHENYLENEOXYMETHY-LENE)]BIS-
[BIS(4-GLYCIDYLOXYPHENYL)]-(2-GLYCIDYL-OXYPHENYL)METHANE [OTS]
[BIS(4-GLYCIDYLOXYPHENYL)]-(2-GLYCIDYLOXYPHENYL)METHANE [EPA]

67801-06-3 1,3-BENZENEDIAMINE, 4-ETHOXY-, DIHYDROCHLORIDE

67859-56-7 2,3-DIHYDROXYPROPYL-3'-(HEXYLTHIO)PROPIONATE

67859-60-3 BOROXIN, TRIS((2-ETHYLHEXYL) OXY)-

67860-04-2 OXIRANE, HEPTADECYL-

67874-81-1 CEDROL METHYL ETHER

67880-17-5 HYDRAZINE DICARBONIC ACID DIAZIDE

67891-79-6 HEAVY AROMATIC DISTILLATE (PETROLEUM)

67891-80-9 LIGHT AROMATIC DISTILLATE (PETROLEUM)

67922-57-0 NONYLPHENOL, ETHOXYLATED, PHOSPHATED, MAGNESIUM SALT

67924-23-6 COBALTATE (6-), [[[1,2-ETHANEDIYLBIS[NITRILOBIS (METHYLENE)]TETRAKIS[PHOS-PHONATO](8-)-N,N',O,O",O"", O"""]-, PENTAPOTASSIUM HY-DROGEN, (OC-6-21)-

67952-66-3 DODECYLBENZENESULFONIC ACID, ETHYLENEDIAMINE SALT

67969-67-9 COBALTATE (6-), [[[1,2-ETHANEDIYLBIS[NITRILOBIS (METHYLENE)]TETRAKIS[PHOS-PHONATO](8-)-N,N',O,O',O"", O"""]-, PENTASODIUM HYDRO-GEN, (OC-6-21)-

67989-28-0 SOYBEAN OIL, POLYMER WITH PENTAERYTHRITOL, TOLUENEDIISOCYANATE AND TUNG OIL

67989-89-3 CUPRATE(6-), [[[1,2-ETHANEDIYL-BIS[NITRILOBIS(METHYLENE)] TETRAKIS[PHOSPHONATO](8-)N, N',O,O",O"",O"""]-, PENTAPOTAS-SIUM HYDROGEN, (OC-6-21)-

67990-27-6 C.I. SOLVENT YELLOW 107

67993-50-4 DODECYLPHENOXYBENZENE DISULFONIC ACID

68006-83-7 2-AMINO-3-METHYL-9H-PYRIDO (2,3,-B)INDOLE (METHYL A-AL-PHA-C)
2-AMINO-3-METHYL-9H-PYRIDO(2,3,-B)INDOLE [MSL]
ME-A-ALPHA-C (2-AMINO-3-METHYL-9H-PYRIDO[2,3-B]INDOLE) [C65, C65, CAL]
MEA-A-C [MIN]
MEA-ALPHA-C (2-AMINO-3-METHYL-9H-PYRIDO[2,3-B]INDOLE) [IAR]

68015-98-5 1,3-BENZENEDIAMINE, 4-ETHOXY-, SULFATE (1:1)

68025-39-8 COBALTATE (6-), [[[1,2-ETHANEDIYLBIS[NITRILOBIS (METHYLENE)]TETRAKIS[PHOS-PHONATO](6-)-N,N',O,O",O"", O"""]-, PENTAAMMONIUM HY-DROGEN, (OC-6-21)-

68037-05-8 AMMONIUM C6-10-ALKYL POLY-OXYETHYLENE SULFATE

68037-59-2 DIMETHYLHYDROPOLYSILOX-ANE

68037-64-9 SILOXANES AND SILICONES, DIMETHYL, METHYL HY-DROGEN, REACTION PROD-UCTS WITH POLYETHYLENE-POYPROPYLENE GLYCOL MONOACETATE ALLYL ETHER

68037-74-1 DIMETHYLPOLYSILOXANES

68071-54-5 CASTOR OIL, DEHYDRATED, POLYMER WITH P-TERT-BUTYL-BENZOIC ACID, GLYCEROL AND PHTHALIC ANHYDRIDE

68081-81-2 BENZENESULFONIC ACID, C10-16-ALKYL DERIVATIVES, SODIUM SALTS

68081-84-5 OXIRANE, MONO[(C10-16-ALKY-LOXY)METHYL] DERIVATIVES
ALKYL (C10 - C16) GLYCIDYL ETHER [EPA]
ALKYL (C10-C16) GLYCIDYL ETHER [OTS]
OXIRANE, MONO((C10-C16-ALKY-LOXY)METHYL]DERIVATIVES [TSC, TSC]

68083-14-7 POLYDIMETHYLDIPHENYL SILOXANE COPOLYMER
DIMETHYLDIPHENYLSILOXANE [TSC, TSC]

68084-55-9 DODECYLBENZENESULFONIC ACID, N-(2-AMINOETHYL) ETHANOLAMINE SALT

68085-85-8 CYHALOTHRIN [3-(2-CHLORO-3,3, 3-TRIFLUORO-1-PROPENYL)-2,2-DIMETHYLCYCLOPROPANECAR-BOXYLIC ACID CYANO(3-PHE-NOXYPHENYL) METHYL ESTER]

68092-47-7 BENZOIC ACID, 3-METHYL-, BAR-IUM SALT

68122-64-5 SOYBEAN OIL POLYMERIZED

68122-86-1 IMIDAZOLIUM COMPOUNDS, 4,5-DIHYDRO-1-METHYL-2-NORTAL-LOWALKYL-1-(2-TALLOW AMI-DOETHYL), METHYL SULFATES
TALLOW IMIDAZOLINIUM METHYL SULFATE [PST]

68130-43-8 OXO ALCOHOL STILL BOTTOMS, SULFATED, SODIUM SALT

68131-12-4 MEAT MEAL

68131-39-5 ALCOHOLS, C-12 - C-15, ETHOXY-LATED
ALCOHOL C12 - C15, POLY(1-3) ETHOXYLATE [HM1]

68131-40-8 ALCOHOLS, C11-15-SECONDARY, ETHOXYLATED

68132-21-8 PERILLA OIL

68132-78-5 ETHOXYLATED TALLOW AMINE HYDROCHLORIDE

68133-37-9 DIETHANOLAMINE ETHYLENEDI-AMINETETRAACETATE

68134-06-5 1,3-DIMETHYLBUTYL GLYCIDYL ETHER
DIMETHYLBUTYL GLYCIDYL ETHER, 1,3- [OTS]

68134-07-6 6-METHYLHEPTYL GLYCIDYL ETHER
METHYLHEPTYL GLYCIDYL ETHER, 6- [OTS]

68139-30-0 PROPANAMINIUM, N-(3-AMINO-PROPYL)-2-HYDROXY-N,N-DIMETHYL-3-SULFO-, N-COCO ACYL DERIVATIVES, HYDROX-IDES, INNER SALTS

68139-89-9 MALEIC ANHYDRIDE, COM-POUND WITH TALL-OIL FATTY ACIDS

68140-00-1 COCONUT MONOETHANOLAMIDE

68140-76-1 1,1'-(METHYLIMINO)BIS(3-CHLORO-2-PROPANOL), POLY-MER WITH N,N,N',N'-TETRAM-ETHYL-1,2-ETHANEDIAMINE

68152-55-6 ETHYLENE GLYCOL POLYMER WITH FUMARIC ACID & ROSIN

68153-33-3 DI-C10-16-ALKYL DIMETHYL AMMONIUM CHLORIDE

68153-35-5 ETHANAMINIUM, 2-AMINO-N-(2-AMINOETHYL)-N-(2-HYDROX-YETHYL)-N-METHYL-, N,N'-DI-TALLOW ACYL DERIVATIVES, METHYL SULFATES (SALTS)

68153-88-8 LINSEED OIL, TUNG OIL, 4-(T-BUTYL)PHENOL, BISPHENOL A, FORMALDEHYDE POLYMER

68153-99-1 AMINES, N-TALLOW ALKYLTRIMETHYLENEDI-, DI-OLEATES

68154-36-9 FATTY ACIDS, COCO, MO-NOESTERS WITH SORBITAN

68155-01-1 POLYOXYETHYLENE CETYL AND OLEYL ALCOHOLS

68155-06-6 N,N-BIS(2-HYDROXYETHYL)-C12-18-ALKYLAMINE

68155-33-9 AMINES, C14-18-ALKYL, ETHOXY-LATED

68187-51-9 ZINC IRON YELLOW
C.I. PIGMENT YELLOW 119 [PST]

68187-69-9 QUATERNARY AMMONIUM COM-POUNDS, (HYDROGENATED TALLOW ALKYL) BIS(HYDROX-YETHYL) METHYL, ETHOXY-LATED, CHLORIDES

68187-71-3 CALCIUM SALTS OF TALL-OIL FATTY ACIDS

68187-76-8 SULFONATED CASTOR OIL, SODIUM SALT
CASTOR OIL, SULFATED, SODIUM SALT [PST]

68187-84-8 CASTOR OIL, OXIDIZED

68187-85-9 FATTY ACIDS, TALL-OIL, ESTERS WITH ETHYLENE GLYCOL

68188-27-2 FATTY ACIDS, TALL-OIL, ESTERS WITH PENTAERYTHRITOL

68188-96-5 PHOSPHORIC ACID, [1,2-ETHANEDIYLBIS[NITRILOBIS (METHYLENE)]TETRAKIS-, TE-TRAPOTASSIUM SALT

68213-23-0 ALCOHOLS, C12-18, ETHOXY-LATED

68213-57-0 OILS, MENHADEN, POLYMER-IZED

68239-06-5 CYCLOHEXANE, 2-HEPTYL-3,4-BIS(9-ISOCYANATONONYL)-1-PENTYL-

68239-42-9 ETHOXYLATED METHYL GLUCO-SIDE

68239-80-5 1,3-BENZENEDIAMINE, 4-CHLORO-, SULFATE (1:1)

68239-82-7 1,2-BENZENEDIAMINE, 4-NITRO-, SULFATE (1:1)

68239-83-8 1,4-BENZENEDIAMINE, 2-NITRO-, SULFATE (1:1)

68298-14-6 METHYL EPOXYSTEARATE AND TETRAETHYLENE PENTAMINE (TEPA), REACTION PRODUCTS

68298-46-4 7-AMINO-2,2-DIMETHYL-2,3-DIHY-DROBENZOFURAN
7-BENZOFURANAMINE, 2,3-DIHY-DRO-2,2-DIMETHYL- [TSC, TSC, TSC]

68308-34-9 SHALE OILS
SHALE OIL [NJL]
SHALE-OILS [C65]

68308-36-1 SOYBEAN MEAL

68308-53-2 SOYA FATTY ACIDS

68308-74-7 N,N-DIMETHYL TALL-OIL FATTY AMIDES

68309-98-8 CADMATE (6-), [[[1,2-ETHANEDIYLBIS[NITRILOBIS (METHYLENE)]TETRAKIS[PHOS-PHONATO](8-)-N,N',O,O'',O''', O'''']-, PENTAPOTASSIUM HY-DROGEN, (OC-6-21)-

68332-64-9 P-TERT-BUTYLPHENOL, ETHOXY-LATED, PHOSPHATED

68333-69-7 ROSIN, MALEATED, POLYMER WITH PENTAERYTHRITOL
PENTAERYTHRITOL ESTER OF MALEIC ANHYDRIDE - MODIFIED WOOD ROSIN [PST]

68334-00-9 HYDROGENATED COTTONSEED OIL

68334-28-1 OILS, VEGETABLE, HYDRO-GENATED

68334-30-5 DIESEL OIL (PETROLEUM)
DIESEL FUEL [NJL]
DIESEL FUEL (MARINE) [CAB]
DIESEL FUEL MARINE [CSR]
DIESEL FUEL OIL [NFP, NFP, PAL]
MARINE DIESEL FUEL [IAR]

68359-37-5 CYFLUTHRIN [3-(2,2-DICHLOROETHENYL)-2,2-DIMETHYLCYCLOPROPANECAR-BOXYLIC ACID, CYANO(4-FLU-ORO-3-PHENOXYPHENYL) METHYL ESTER]

68389-88-8 POLY(OXY-1,2-ETHANEDIYL), ALPHA-[2-[BIS(2-AMINOETHYL) METHYLAMMONIO]ETHYL]-OMEGA-HYDROXY-, N,N'-DICOCO ACYL DERIVATIVES, METHYL SULFATES (SALTS)

68389-89-9 POLY(OXY-1,2-ETHANEDIYL), ALPHA-[2-[BIS(2-AMINOETHYL) METHYLAMMONIO]ETHYL-OMEGA]-HYDROXY-, N,N'-BIS(HY-DROGENATED TALLOW ACYL) DERIVATIVES, METHYL SUL-FATES (SALTS)

68390-66-9 2-IMIDAZOLINIUM, 1-(CAR-BOXYMETHYL)-1-(2-HYDROX-YETHYL)-2-COCOYLNOR, HY-DROXIDE, MONOSODIUM SALT

68391-01-5 ALKYLBENZYLDIMETHYLAMMO-NIUM CHLORIDE
QUATERNARY AMMONIUM COMPOUNDS, BENZYL-C12-18-ALKYLDIMETHYL,CHLORIDES [PST]

68400-67-9 POLYURETHANE

68409-75-6 BONE MEAL

68410-69-5 POLY(OXY-1,2-ETHANEDIYL), ALPHA-[2-[BIS(2-AMINOETHYL) METHYLAMMONIO]ETHYL]-OMEGA-HYDROXY-, N,N'-DI-TALLOW ACYL DERIVATIVES, METHYL SULFATES (SALTS)
N,N'-DITALLOW N''-METHYL N''-POLYETHOXYAMIDO AMMONIUM METHOSULFATE [PST]

68411-30-3 BENZENESULFONIC ACID, C10-13-ALKYL DERIVATIVES SODIUM SALT
BENZENESULFONIC ACID, LINEAR ALKYL, SODIUM SALT [WHM]

68412-54-4 POLYETHYLENE GLYCOL BRANCHED NONYLPHENYL ETHER
POLYOXYETHYLENE BRANCHED-C9-ALKYLPHENOL [PST]

68413-04-7 METHYL, TALLOW DIETHYLEN-ETRIAMINE CONDENSATE, PROPOXYLATED,METHYL SUL-FATE
POLY[OXY(METHYL-1,2-ETHANEDIYL)], ALPHA-[2-[BIS(2-AMINOETHYL)METHYLAMMONIO] METHYLETHYL]-OMEGA-HYDROXY-, N,N'-DITALLOW ACYL DERIVATIVES, METHYL SULFATES (SALTS) [TSC, TSC]

68422-69-5 SULFATED BUTYL OLEATE

68424-10-2 COTTONSEED MEAL

68424-85-1 ALKYL (C12-16)DIMETHYLBENZY-LAMMONIUM CHLORIDE
ALKYL(C12-16) DIMETHYL BENZYL AMMONIUM CHLORIDE [PST]

68424-94-2 (COCO ALKYL)DIMETHYL BE-TAINES

68425-13-8 LATEX, POLYMER OF 2-METHYL-1,3-BUTADIENE

68425-31-0 GASOLINE, NATURAL
GASOLINE (NATURAL GAS), NATURAL [PAL]
NATURAL GASOLINE [HM1, HM1]

68425-44-5 AMIDES, COCO, N-(HYDROXYETHYL), ETHOXYLATED

68439-45-2 ALCOHOLS, C6-12, ETHOXYLATED

68439-49-6 ALCOHOLS, C16-18, ETHOXYLATED

68439-50-9 ALCOHOLS, C12-14, ETHOXYLATED

68439-56-5 SODIUM ALPHA-OLEFINSULFONATE

68439-72-5 POLYOXYETHYLENE C8-18 AND C18 UNSATURATED, ALKYLAMINES

68441-17-8 ETHENE, HOMOPOLYMER, OXIDIZED
OXIDIZED POLYETHYLENE [PST]

68441-84-9 SODIUM SALT OF CRESOL SULFONIC ACID CONDENSED WITH UREA FORMALDEHYDE

68442-82-0 CALCIUM CARBONATE DIMETHYLHEXANOATE

68442-85-3 CELLULOSE, REGENERATED

68442-99-9 MANGANESE BORON NEODECANOATE

68443-05-0 9-OCTADECENOIC ACID (Z)-, SULFONATED, SODIUM SALT
SODIUM SULFONATED OLEATE [PST]

68457-75-0 ISOBUTYLATED, STYRENEATED CRESOLS

68458-49-1 POLYPHOSPHORIC ACIDS, ESTERS WITH POLYETHYLENEGLYCOL-NONYLPHENYL ETHER

68458-88-8 LANOLIN, ETHOXYLATED PROPOXYLATED

68459-31-4 C9-19 FATTY ACID ESTER PHTHALIC ALKYL

68459-98-3 1,2-BENZENEDIAMINE, 4-CHLORO-, SULFATE (1:1)

68476-25-5 FELDSPAR

68476-26-6 FUEL GASES
FUEL OIL [HM1, HM1]
GAS, OIL GAS [NFP]
OIL GAS [NJL]

68476-30-2 FUEL OIL NO. 2

68476-31-3 FUEL OIL NO. 4

68476-33-5 FUEL OIL

68476-37-9 GLUE (AS DEPOLYMERIZED ANIMAL COLLAGEN)

68476-40-4 HYDROCARBON PROPELLANT

68476-78-8 MOLASSES
CANE SYRUP [PST]

68476-82-4 PEANUTS

68476-85-7 L.P.G. (LIQUIFIED PETROLEUM GAS)
L.P.G. [CEX, MIN, NIO, NIO, NIO]
LIQUEFIED PETROLEUM GAS [NJL]
LIQUEFIED PETROLEUM GAS (L.P.G.) [MSL]
LIQUEFIED PETROLEUM GAS (LPG) [GBR, GBR]
PETROLEUM GASES, LIQUEFIED [PAL]

68476-86-8 PETROLEUM GASES, LIQUIFIED, SWEETENED
PETROLEUM GASES, LIQUIFIED [HM1, HM1]

68476-87-9 PETROLEUM REFINING RESIDUES, POLYMERIZED
CRACKED RESIDUES DERIVED FROM THE REFINING OF CRUDE OIL [IAR]

68476-95-9 SHALE

68477-30-5 DISTILLATES (PETROLEUM), CATALYTIC REFORMER FRACTIONATORRESIDUE, INTERMEDIATE BOILING
AROMATIC PETROLEUM DERIVATIVE SOLVENT [PST]

68477-31-6 PETROLEUM DISTILLATES, CATALYTIC, REFORMER FRACTIONATOR RESIDUE, LOW-BOILING
AROMATIC PETROLEUM HYDROCARBON SOLVENT [PST]

68478-54-6 CALCIUM 2-ETHYLHEXANOATE/ISONONANOATE COMPLEXES

68511-39-7 ETHOXYLATED LINEAR C12-15-SEC-ALCOHOL SULFATE

68513-95-1 SOY FLOUR

68514-61-4 MILK, HYDROLYZED

68514-76-1 CITRUS PULP, ORANGE

68515-42-4 PHTHALIC ACID, DIALKYL (C7-C11) ESTER
1,2-BENZENEDICARBOXYLIC ACID, DI-C(7-11)-BRANCHED AND LINEAR-ALKYL ESTERS [TSE]
1,2-BENZENEDICAROBOXIC ACID, DI-C(7-11)-BRANCHED AND LINEAR-ALKYL ESTERS [DI(HEPTYL-, NONYL-, UNDECYL) PHTHALATE (MIXED ISOMERS)] [TSC]

68515-44-6 PHTHALIC ACID, DIALKYL (C7) ESTER
DIHEPTYL PHTHALATE (BRANCHED AND LINEAR ISOMERS) [TSC]

68515-45-7 PHTHALIC ACID, DIALKYL (C9) ESTER
DINONYL PHTHALATE (BRANCHED AND LINEAR ISOMERS) [TSC]

68515-47-9 DITRIDECYL PHTHALATE (MIXED ISOMERS)
1,2-BENZENEDICAROBOXIC ACID, DI-C(11-14)-BRANCHED ALKYL ESTERS, C(13)-RICH [DITRIDCEYL PHTHALATE (MIXED ISOMERS)] [TSC, TSC]

68515-49-1 DIISODECYL PHTHALATE (MIXED ISOMERS)
1,2-BENZENEDICAROBOXIC ACID, DI-C(9-11)-BRANCHED ALKYL ESTERS, C(10-RICH [(DIISODECYL PHTHALATE (MIXED ISOMERS)] [TSC, TSC]

68515-50-4 DIHEXYL PHTHALATE (MIXED ISOMERS)
1,2-BENZENEDICAROBOXIC ACID, DIHEXYL ESTER, BRANCHED AND LINEAR [DIHEXYL PHTHALATE (MIXED ISOMERS)] [TSC, TSC]

68517-02-2 OXIRANE, 2,2',2''-[PROPYLIDYNETRIS(4,1-PHENYLENEOXYMETHYLENE)]TRIS-
1,1,1-TRIS(4-HYDROXYPHENYL) PROPANE TRIGLYCIDYL ETHER [EPA]
TRIS(4-HYDROXYPHENYL)PROPANE-TRIGLYCIDYL ETHER [OTS]

68526-82-9 OCTYL ALCOHOL BOTTOMS

68526-90-9 DECYL ALCOHOL BOTTOMS

68526-94-3 ALCOHOLS, C12-20, ETHOXYLATED

68527-23-1 NAPHTHA (PETROLEUM), LIGHT STEAM-CRACKED AROMATIC

68527-99-1 1H-IMIDAZOLIUM, 1,3-BIS(CARBOXYMETHYL)-4,5-DIHYDRO-1-(2-HYDROXYETHYL)-2-UNDECYL-, DIHYDROXIDE, DISODIUM SALT

68551-07-5 OXO ALCOHOL STILL BOTTOMS (C8-18 ALCOHOLS)

68551-13-3 POLYETHOXYLATED POLYPROPOXYLATED C12-15 ALCOHOLS

68551-17-7 ALKANES, C10-13-ISO-
ALKANES [MIN]

68551-42-8 FATTY ACIDS, C6-19-BRANCHED, MANGANESE SALTS

68553-00-4 FUEL OIL NO. 6

68553-81-1 RICE BRAN OIL

68554-06-3 POLY(OXY-1,2-ETHANEDIYL), ALPHA-[3-[BIS(2-AMINOETHYL) METHYLAMMONIO]-2-HYDROXYPROPYL]-OMEGA-HYDROXY-, N-COCO ACYL DERIVATIVES, METHYL SULFATES (SALTS)

68554-37-0 PENTAERYTHRITOL, PHTHALIC ANHYDRIDE AND TALL-OIL POLYMER WITH ROSIN

68554-65-4 SILOXANES AND SILICONES, DIMETHYL, POLYMERS WITH METHYL SILSESQUIOXANES AND POLYETHYLENE-POYPROPYLENE GLYCOL MONOBUTYL ETHER

68555-09-9 LINEAR ALKYLBENZENE (C-20 TO C-48)

68584-15-6 1,3-BENZENEDICARBOXYLIC ACID, POLYMER WITH 5-AMINO-1,3,5-TRIMETHYLCYCLOHEXYLAMINE, MODIFIED

68584-22-5 BENZENESULFONIC ACID, C10-16-ALKYL DERIVATIVES

68584-23-6 BENZENESULFONIC ACID, C10-16-ALKYL DERIVATIVES, CALCIUM SALTS

68584-24-7 BENZENESULFONIC ACID, C10-16-ALKYL DERIVATIVES, COMPOUNDS WITH 2-PROPANAMINE

68584-25-8 BENZENESULFONIC ACID, C10-16-ALKYL DERIVATIVES, COMPOUNDS WITH TRIETHANOLAMINE

68584-26-9 BENZENESULFONIC ACID, C10-16-ALKYL DERIVATIVES, MAGNESIUM SALTS

68584-27-0 BENZENESULFONIC ACID, C10-16-ALKYL DERIVATIVES, POTASSIUM SALTS

68585-34-2 ALPHA-ALKYL(C10-16)-OMEGA-HYDROXYPOLY(OXYETHYLENE) SULFATE,SODIUM SALT

68585-36-4 ALKYL(C10-14) OXYPOLY (ETHYLENEOXY)ETHYL PHOSPHATE

68603-42-9 COCONUT DIETHANOLAMIDE
AMIDES, COCO, N,N-BIS(2-HYDROXYETHYL) [PST]
COCONUT OIL ACID DIETHANOLAMIDE CONDENSATE [CSR]

68607-27-2 QUATERNARY AMMONIUM COMPOUNDS, (HYDROGENATED TALLOW ALKYL) BIS(HYDROXYETHYL) METHYL, CHLORIDES

68608-26-4 SULFONIC ACIDS, PETROLEUM, SODIUM SALTS
SODIUM SALTS OF PETROLEUM SULFONIC ACIDS [PST]

68608-58-2 WHEY

68609-86-9 MANGANESE, 2-ETHYLHEXANOATE NAPHTHENATE COMPLEXES

68609-96-1 ALKYL (C8-C10) GLYCIDYL ETHER
OXIRANE, MONO[C(8-10)-ALKYLOXY) METHYL]DERIVATIVES [TSC]

68609-97-2 ALKYL (C12-C14) GLYCIDYL ETHER
OXIRANE, MONO[(C(12-14)-ALKYLOXY)METHYL]DERIVATIVES [TSC]

68610-92-4 CELLULOSE, OMEGA-ETHER WITH ETHOXYLATED 2-HYDROXY-3-(TRIMETHYLAMMONIO)PROPANOL, CHLORIDE

68611-14-3 LIGNOSULFONIC ACID, ETHOXYLATED, SODIUM SALTS

68611-55-2 MIXED C10-16-ALKYL SULFATES

68611-64-3 UREA, REACTION PRODUCTS WITH FORMALDEHYDE
UREA-FORMALDEHYDE RESIN [OTS]

68630-89-7 6-CARBOXY-4-HEXYL-2-CYCLOHEXENE-1-OCTANOIC ACID, MONOPOTASSIUM SALT

68630-92-2 1H-IMIDAZOLE-1-PROPANOIC ACID, 3-[2-(2-CARBOXYETHOXY) ETHYL]-2-HEPTYL-2,3-DIHYDRO-, DISODIUM SALT

68630-96-6 1H-IMIDAZOLIUM, 1-(2-CARBOXYETHYL)-4,5-DIHYDRO-3-(2-HYDROXYETHYL)-2-ISOHEPTADECYL-, HYDROXIDE, INNER SALT

68648-12-4 POLYOXYPROPYLENE DITALLOIL ESTER

68648-20-4 SORBITOL, TALL-OIL FATTY ACID SESQUIESTERS, ETHOXYLATED

68648-38-4 POLYOXYETHYLENE LANOLIN ALCOHOL

68648-98-6 BENZENESULFONIC ACID, MONO-C9-17-BRANCHED ALKYL DERIVATIVES

68649-00-3 BENZENESULFONIC ACID, MONO-C9-17-BRANCHED ALKYL DERIVATIVES, ISOPROPYLAMINE SALTS

68649-45-6 CARBOXYPOLYMETHYLENE RESIN

68650-09-9 FATTY ACIDS, TALL-OIL, MIXED ESTERS WITH GLYCEROL AND POLYETHYLENE GLYCOL

68650-39-5 IMIDAZOLIUM COMPOUNDS, 1-[2-(CARBOXYMETHOXY)ETHYL]-1-CARBOXYMETHYL)-4,5-DIHYDRO-2-NORCOCO ALKYL, HYDROXIDES, SODIUM SALTS

68755-33-9 POLYETHOXYLATED PRIMARY AMINE (C14-18)

68783-08-4 C7-C35 REFINED PETROLEUM OIL
REFINED PETROLEUM PRODUCTS [CN6]

68783-78-8 DITALLOW DIMETHYL AMMONIUM CHLORIDE

68808-54-8 3-AMINO-1,4-DIMETHYL-5H-PYRIDO (4,3-B) INDOLE ACETATE

68833-55-6 MERCURY ACETYLIDE

68848-64-6 LITHIUM SILICON

68855-54-9 SILICA, AMORPHOUS, DIATOMACEOUS EARTH
DIATOMACEOUS EARTH, UNCALCINED [ONT]
SILICA AMORPHOUS DIATOMACEOUS EARTH (UNCALCINED) [PAL]
SILICA, AMORPHOUS [ALB]
SILICA, AMORPHOUS, DIATOMACEOUS EARTH (CALCINED) [MAK, MAK]
SILICA-AMORPHOUS DIATOMACEUS EARTH [MIN]

68876-77-7 YEAST

68877-34-9 N-(NONYLOXYPROPYL)-1,3-PROPANEDIAMINE

68891-29-2 ALPHA-ALKYL(C8-C10)-OMEGA-HYDROXYPOLY(OXYETHYLENE) AMMONIUMSULFATE

68901-17-7 PHOSPHONIC ACID, [1,2-ETHANEDIYLBIS[NITRILOBIS (METHYLENE)]TETRAKIS-, OCTAAMMONIUM SALT

68908-35-0 POLYACRYLONITRILE, OXIDIZED

68908-63-4 ALPHA-ALKYL(C12-C15)-OMEGA-HYDROXYPOLYOXYETHYLENE

68909-59-1 PHOSPHORIC ACID MONO(C8-C10)ALKYL SODIUM SALTS

68911-49-9 DRIED BLOOD

68915-32-2 QUASSIA, EXTRACT

68916-96-1 MATE

68917-09-9 OCOTEA CYMBARUM OIL
OCOTEA OIL [PST]

68917-75-9 WINTERGREEN OIL

68920-06-9 XYLENE RANGE AROMATIC SOLVENT

68920-70-7 ALKANES, C6-18, CHLORO-
ALKANES, C(6-18), CHLORO- [TSC]

68928-85-8 VINYL ACETATE, CROTONIC ACID, VINYL NEODECANOATE, GLYCIDYL METHACRYLATE POLYMER

68937-41-7 TRIS ISOPROPYLATED PHENYL PHOSPHATE
PHENOL ISOPROPYLATED PHOSPHATE [OTS]

68937-54-2 SILOXANES AND SILICONES, DIMETHYL, 3-HYDROXYPROPYL METHYL, ETHOXYLATED

68937-99-5 SUNFLOWER SEEDS

68938-15-8 FATTY ACIDS, COCO, HYDROGENATED

68953-36-6 TALL OIL FATTY ACIDS, REACTION PRODUCT WITH TETRAETHYLENE PENTAMINE

68955-41-9 ALKANES, C(10-18)-BROMOCHLORO-

68955-55-5 ALKYL(C12-14) DIMETHYL AMINE OXIDE

68955-64-6 HEXANEDINITRILE, HYDROGENATED, HIGH-BOILING FRACTION, PHOSPHONOMETHYLATED

68956-56-9 TERPENE HYDROCARBONS, N.O.S.

68956-68-3 VEGETABLE OIL
VEGETABLE OIL MIST [MEX, OSI, OS1]
VEGETABLE OIL MISTS [AUS, ONT, TLV]

68958-86-1 NICKELATE(6-), [[[1,2-ETHANEDIYLBIS[NITRILOBIS (METHYLENE)]TETRAKIS[PHOSPHONATO](8-)], PENTAAMMONIUM HYDROGEN, (OC-6-21)-

68958-87-2 NICKELATE(6-), [[[1,2-ETHANEDIYLBIS[NITRILOBIS (METHYLENE)]TETRAKIS[PHOSPHONATO](8-)], PENTAPOTASSIUM HYDROGEN, (OC-6-21)-

68958-88-3 NICKELATE(6-), [[[1,2-ETHANEDIYLBIS[NITRILOBIS (METHYLENE)]TETRAKIS[PHOSPHONATO](8-)], PENTASODIUM HYDROGEN, (OC-6-21)-

68959-23-9 OXIRANE, 2,2',2"-[1,2,6-HEXANETRIYLTRIS-(OXYMETHYLENE)]TRIS-
1,2,6-HEXANETRIOL TRIGLYCIDYL ETHER [EPA]
HEXANETRIOL TRIGLYCIDYL ETHER, 1,2,6- [OTS]

68966-84-7 1,3-BENZENEDIAMINE, AR-ETHYL-AR-METHYL-

68975-47-3 ISOHEPTENE

68987-80-4 OXIRANE, MONO[C(6-12)-ALKYLOXY)METHYL]DERIVATIVES
ALKYL (C6 - C12) GLYCIDYL ETHER [EPA]
ALKYL (C8-C12) GLYCIDYL ETHER [OTS]

68987-86-0 ISOPROPYLATED CRESOL

68988-56-7 POLYTRIMETHYLHYDROSILYLSILICONE

68988-76-1 9-OCTADECENOIC ACID (Z)-, SULFONATED

68991-42-4 PAPRIKA

69009-90-1 DIISOPROPYL BIPHENYL
DIISOPROPYLBIPHENYL [PST]

69011-04-7 PHOSPHORIC ACID, BUTYL ESTER, MANGANESE(2+) SALT

69011-36-5 POLYOXYETHYLENE TRIMETHYLDECYL ALCOHOL

69011-71-8 ALUMINUM DROSS

69012-64-2 SILICA FUME (AMORPHOUS)
SILICA - AMORPHOUS, FUME [TLV]

69013-18-9 ALCOHOLS, C8-18, ETHOXYLATED PROPOXYLATED
POLYETHOXYLATED POLYPROPOXYLATED C8-C18 ALCOHOLS [PST]

69013-19-0 ALCOHOLS, C8-22, ETHOXYLATED

69029-39-6 POLYOXYETHYLENE POLYOXYPROPYLENE DI-SEC-BUTYLPHENOL

69155-42-6 TETRASILOXANE, 1,1,1,3,5,7,7-OCTAMETHYL-3,5-BIS[3-(OXIRANYLMETHOXY)PROPYL]-
1,1,1,3,5,7,7-OCTAMETHYL-3,5-BIS(6,7-EPOXY-4-OXAHEPTYL)- [OTS]

1,1,1,3,5,7,7'-OCTAMETHYL-3,5-BIS (6,7-EPOXY-4-OXAHEPTYL) TE-TRASILOXANE [EPA]

69227-21-0 ALCOHOLS, C12-18, ETHOXY-LATED PROPOXYLATED

69227-22-1 ALCOHOLS, C10-16, ETHOXY-LATED PROPOXYLATED
ALPHA-ALKYL(C10-16)-OMEGA-HY-DROXY-POLYOXYETHYLENE POLY-OXYPROPYLENE POLYOXYETHY-LENE [PST]

69409-94-5 FLUVALINATE [N-[2-CHLORO-4-(TRIFLUOROMETHYL)PHENYL] -DL-VALINE(+)-CYANO (3-PHE-NOXYPHENYL)METHYL ESTER

69430-24-6 CYCLOSILANES, DIMETHYL-CYCLOPOLYDIMETHYLSILOXANE [TSC, TSC]

69523-06-4 FERROCERIUM

69600-03-9 TUNGSTEN ANTIMONATE

69655-05-6 2',3'-DIDEOXYINOSINE (AIDS INI-TIATIVE)

69669-25-6 POTASSIUM SALTS OF FATTY ACIDS (C12-C20)

69806-50-4 FLUAZIFOP-BUTYL [2-[4-[[5-(TRI-FLUOROMETHYL)-2-PYRIDINYL] OXY]-PHENOXY]PROPANOIC ACID, BUTYL ESTER

69834-19-1 BENZENE, 1,1'-OXYBIS[DODECYL-

69867-71-6 PENTYL PHENYL ACID PHOS-PHATE

69882-11-7 PHENOL, 2,4(OR 2,6)-DIBROMO-, HOMOPOLYMER

69898-00-6 ALPHA-OLEFINS

69943-75-5 POLY(OXY-1,2-ETHANEDIYL), AL-PHA-[4-OXIRANYLMETHOXY) BENZOYL]-OMEGA-[[(4-OXI-RANYLMETHOXY)BENZOYL]OXY] -

70042-58-9 TERT-BUTYLCYCLOHEXYL CHLOROFORMATE
TERT-BUTYLCYCLOHEXYLCHLORO-FORMATE [NJL]

70124-77-5 FLUCYTHRINATE

70131-50-9 BENTONITE, ACID-LEACHED CLAY [PST]

70131-67-8 SILOXANES AND SILICONES, DI-ME, HYDROXY-TERMINATED

70191-75-2 DECYL PHENOXYBENZENEDISUL-FONIC ACID

70206-24-5 ALKYL IMIDAZOLINIUM METHYL SULFATE (DERIVED FROM OLEIC ACID)

70247-86-8 POLYOXYETHYLENE C4-18-ALKYL ESTER OF PHOSPHORIC ACID, MONOETHANOLAMINE SALTS

70476-82-3 MITOXANTRONE HYDROCHLO-RIDE

70592-78-8 PETROLEUM DISTILLATES, VAC-UUM
HEAVY VACUUM DISTILLATES [IAR]

70592-80-2 ALKYL(C10-16) DIMETHYLAMINE OXIDE

70632-06-3 POLY(OXY-1,2-ETHANEDIYL) , ALPHA-(CARBOXYMETHYL) -OMEGA-HYDROXY-, C12-15-ALKYL ETHERS, SODIUM SALTS

70648-26-9 1,2,3,4,7,8-HEXACHLORO DIBEN-ZOFURAN

70657-70-4 2-METHOXYPROPYL-1-ACETATE

70750-53-7 POLYTERPENE RESINS, SYN-THETIC

70776-03-3 NAPHTHALENE, CHLORO DERIVATIVES
CHLORINATED NAPHTHALENE [CW3]
CHLORINATED NAPHTHALENE, N.O.S. [RCU]
CHLORINATED NAPHTHALENES [CAL]

70816-59-0 TETRAZINE

70903-62-7 O,P-DINONYLPHENOL, ETHOXY-LATED, PHOSPHATED, SODIUM SALT

70904-61-9 AMIDOSULFOSUCCINATE

70914-09-9 POLY(OXY-1,2-ETHANEDIYL), ALPHA-[2-[BIS(2-AMINOETHYL) METHYLAMMONIO]ETHYL]-OMEGA-HYDROXY-, N,N'-DI-C(14-18) ACYL DERIVAT IVES, METHYL SULFATES (SALTS)

70955-37-2 DODECYL AND HIGHER ALIPHATIC KETONES

70983-99-2 WHALE OIL

71012-10-7 OLEIC ACID, 2-(2-(2-(2-HYDROX-YETHOXY)ETHOXY)ETHOXY) ETHYL ESTER

71033-08-4 OXIRANE, 2,2'-[(1-METHYLETHY-LIDINE)BIS[4,1-PHENYLE-NEOXY[1-(BUTOXYMETHYL)-2, 1-ETHANEDIYL]OXYMETHYLENE] BIS-
2,2-BIS(P-GLYCIDYLOXY-3-BU-TOXYPROPYLOXY)-PHENYL) PROPANE [OTS]
2,2-BIS(P-2-GLYCIDYLOXY-3-BU-TOXYPROPYLOXY)-PHENYL) PROPANE [EPA]

71113-21-8 [[[3-(DIMETHYLAMINO)PROPYL] IMINO]BIS(METHYLENE)]BIS-PHOSPHONIC ACID, MONOHY-DROCHLORIDE

71121-36-3 ISOCYANATOBENZOTRIFLUO-RIDE
ISOCYANATOBENZO-TRIFLUORIDE [NJL]
ISOCYANATOBENZOTRIFLUORIDES [HM1, HM1, HM1, WHM]

71243-46-4 POLYOXYETHYLENE LINEAR PRIMARY C8-16 ALCOHOLS

71267-22-6 N'-NITROSOANATABINE

71328-89-7 AROCLOR 1240

71526-07-3 1-OXA-4-AZASPIRO(4,5)DECANE, 4-(DICHLOROACETYL)-
1-OXA-4-AZASPIRO[4,5]DECANE, 4-(DICHLOROACETYL)- [TSE]

71549-82-1 POLYOXYETHYLENE C12-13-ALKYL ESTER OF PHOSPHORIC ACID, MONOETHANOLAMINE SALTS

71751-41-2 ABAMECTIN [AVERMECTIN B1]

71786-60-2 N,N-BIS(2-HYDROXYETHYL)-C12-18-ALKYLAMINE

71808-64-5 SILANE, [(3-CHLOROPROPYL) DIMETHOXY[3-(OXIRANYL-METHOXY)PROPYL]-
3-(CHLOROPROPYL)DIMETHOXY-[3-(OXIRANYLMETHOXY)PROPYL] SILANE [EPA]

DIMETHOXYSILANE, (3-GLYCIDOXY-PROPYL)(3-CHLOROPROPYL)- [OTS]

71819-49-3 C.I. SOLVENT BLUE 98

71819-51-7 C.I. SOLVENT RED 164

71873-51-3 C.I. ACID YELLOW 218

71927-89-4 C.I. ACID GREEN 28

72067-21-1 O,P-DINONYLPHENOL, ETHOXY-LATED, PHOSPHATED, POTAS-SIUM SALT

72139-15-2 C.I. REACTIVE RED 56

72152-45-5 C.I. REACTIVE GREEN 12

72152-54-6 C.I. ACID BLUE 182

72178-02-0 FOMESAFEN
FOMESAFEN [5-(2-CHLORO-4-TRI-FLUOROMETHYL)PHENOXY)-N-METHYLSULFONYL)-2-NITROBEN-ZAMIDE] [313]

72214-01-8 2-ETHYLHEXYL SULFATE

72243-90-4 3-[[4-AMINO-9,10-DIHYDRO-9,10-DIOXO-3-[SULFO-4-(1,1,3,3-TE-TRAMETHYLBUTYL)PHENOXY]-1-ANTHRACENYL]AMINO]-2,4,6-TRIMETHYL BENZENESULFONIC ACID DISODIUM SALT

72252-48-3 BENZOIC ACID, 3-[2-CHLORO-4-(TRIFLUOROMETHYL)PHENOXY]-, POTASSIUM SALT

72254-58-1 3-AMINO-1-METHYL-5H-PYRIDO (4,3-B) INDOLE ACETATE

72319-24-5 OXIRANE, 2,2'-[(1-METHYLETHY-LIDINE)BIS[4,1-PHENYLE-NEOXY-3,1-PROPANEDIY-LOXY-4,1-PHENYLENE(1-METHYLETHYLIDENE)-4,1-PHENYLENEOXYMETHYLENE] BIS-
2,2'-[(1-METHYLETHYLIDENE)BIS(4, 1-PHENYLENEOXY-3,1-PROPANE-DIOXY [OTS]
2,2'-[(1-METHYLETHYLIDINE) BIS[4,1-PHENYLENEOXY-3,1-PROPANEDIYLOXY-4,1-PHENY-LENE(1-METHYLETHYLIDENE)-4,1-PHENYLENEOXYMETHYLENE]BIS (OXIRANE) [EPA]

72490-01-8 FENOXYCARB [2-(4-PHE-NOXYPHENOXY)ETHYL]CAR-BAMIC ACID ETHYL ESTER

72749-59-8 TRI-C6-12-ALKYL METHYL AM-MONIUM CHLORIDE

72854-21-8 ZIRCONIUM NAPHTHENATE

72918-21-9 1,2,3,7,8,9-HEXACHLORO DIBEN-ZOFURAN

73090-68-3 TERT-BUTYL TETRALIN
NAPHTHALENE, (1,1-DIMETHYLETHYL)-1,2,3,4-TETRAHY-DRO- [PAL]

73090-69-4 CHLORO-4-TERT-AMYLPHENOL
PHENOL, CHLORO-4-(1,1-DIMETHYL-PROPYL)- [PAL]

73138-28-0 DIMETHYLDIOCTADECYLAMMO-NIUM BENTONITE

73398-64-8 DI(C8-18)ALKYL DIMETHYL AM-MONIUM CHLORIDE

73455-30-8 DIMETHYLAMINE ETHYLENEDI-AMINETETRAACETATE

73459-03-7 5-METHYLANGELICIN

73506-32-8 HYDRAZINE SELENATE

73506-91-9 DICHLOROBROMOETHANE

73513-30-1 METHYLPENTALDEHYDE
PENTANAL, METHYL- [PAL]

73597-26-9 2-PROPENOIC ACID, 2-METHYL-, 2-[[[[[5-ISOCYANATO-1,3,3-TRIMETHYLCYCLOHEXYL]METHYL]AMINO]CARBONYL]OXY]ETHYL ESTER

73680-58-7 ALUMINUM FLUOROSULFATE, HYDRATE

73826-29-6 PROPIONIC ACID, 2-(2,4,5-TRICHLOROPHENOXY)-, P-CHLOROPHENACYL ESTER
TRICHLOROPHENOXYPROPIONIC ACID ESTER [NJL]

74007-80-0 OXYBIS(DIBUTYL(2,4,5-TRICHLOROPHENOXY)TIN)

74051-80-2 SETHOXYDIM [2-[1-(ETHOXY-IMINO)BUTYL]-5-[2-(ETHYLTHIO)PROPYL]-3-HYDROXY-2-CYCLO-HEXEN-1-ONE]

74204-30-1 C.I. REACTIVE BLUE 52

74222-97-2 SULFFOMETURON METHYL
SULFOMETURON METHYL [TLV, TLV]

74398-71-3 1,2,3-PROPANETRIYL ESTER OF 12-(OXIRANYLMETHOXY)-9-OC-TADECANOIC ACID

74499-22-2 METHYL TALLATE

74811-65-7 CROSCARMELLOSE SODIUM

74849-88-0 DICYANOETHYL DIETHYLENE-TRIAMINE

75150-13-9 2,4-DIBROMO-6-METHYLPHENYL GYCIDYL ETHER

75321-20-9 1,3-DINITROPYRENE

75330-75-5 LOVASTATIN

75364-04-4 SILICIC ACID (H6SI2O7), COBALT (2+) MAGNESIUM SALT (1:2:1)

75405-06-0 PENTANENITRILE, 3-AMINO-

75768-93-3 C.I. DIRECT RED 81, TRI-ETHANOLAMINE SALT

75790-84-0 BENZENE, 2-ISOCYANATO-4-[(4-ISOCYANATO PHENYL)METHYL]-1-METHYL-
4-METHYLDIPHENYLMETHANE-3,4-DIISOCYANATE [313]

75790-87-3 BENZENE, 1-ISOCYANATO-2-[(4-ISOCYANATOPHENYL)THIO]-
2,4'-DIISOCYANATODIPHENYL SUL-FIDE [313]

75980-60-8 DIPHENYL-2,4,6-TRIMETHYLBEN-ZOYL PHOSPHINE OXIDE

76180-96-6 IQ
2-AMINO-3-METHYLIMIDAZO(4,5-F)QUINOLINE [MSL]
IQ (2-AMINO-3-METHYLIMIDAZO[4,5-F]QUINOLINE) [C65, C65, IAR]
IQ; (2-AMINO-3-METHYLIMIDAZO[4,5-F]QUINOLINE) [CAL]

76543-88-9 INTERFERON A (AIDS INITIA-TIVE)

76578-14-8 QUIZALOFOP-ETHYL [2-[4-[(6-CHLORO-2-QUINOXALINYL)OXY]PHENOXY]PROPANOIC ACID ETHYL ESTER]

76649-15-5 2-CHLORO-1-METHYLETHYL BIS (2-CHLOROPROPYL) PHOSPHATE

77094-11-2 MEIQ (2-AMINO-3,4-DIMETHYLIMIDAZO[4,5-F]QUINO-LINE)

77500-04-0 MEIQX (2-AMINO-3,8-DIMETHYLIMIDAZO[4,5-F]QUINOXALINE)

77500-13-1 BIS(2-HYDROXYETHYL)-3-(DECY-LOXY)PROPYLAMINE OXIDE

77501-63-4 LACTOFEN
LACTOFEN [5-(2-CHLORO-4-(TRIFLU-OROMETHYL)PHENOXY)-2-NITRO-2-ETHOXY-1-METHYL-2-OXOETHYL ESTER] [313]

77536-66-4 ASBESTOS, ACTINOLITE
ACTINOLITE ASBESTOS [MSL]

77536-67-5 ASBESTOS, ANTHOPHYLITE
ANTHOPHYLITE [QBC, QBC]
ANTHOPHYLITE ASBESTOS [IAR]
ANTHOPHYLLITE ASBESTOS [MSL]

77536-68-6 ASBESTOS, TREMOLITE
TREMOLITE [PAL]
TREMOLITE ASBESTOS [MSL]

77851-17-3 BENZENE, (1-METHYLETHYL)(2-PHENYLETHYL)-

78330-30-0 SODIUM SALTS OF ALPHA-SULFO-OMEGA-HYDROXYPOLY (OXY-1,2-ETHANEDIYL) C11-14 ISOALKYL ETHERS, (C13-RICH)

78491-02-8 DIAZOLIDINYL UREA

78812-39-2 CARBAMIC ACID, MANGANESE SALT

78871-22-4 TRIETHYLAMINE CITRATE

79171-09-8 ETHYLAMINE, DISTILLATION PRODUCTS

79217-60-0 CYCLOSPORINE
CICLOSPORIN [IAR]

79637-74-4 2-PROPENOIC ACID, OCTAHY-DRO-4,7-METHANO-1H-INDENYL ESTER

79660-25-6 ACETAMIDE, 2,2-DICHLORO-N-(1,3-DIOXOLAN-2-YLMETHYL)-N-2-PROPENYL-

79748-81-5 FUSARIN C

79771-08-7 ISOPROPYLAMINE, DISTILLA-TION RESIDUES
ISOPROPYLAMINE DISTILLATION RESIDUES [TSE]

79771-09-8 ETHYLAMINE, DISTILLATION RESIDUES
ETHYLAMINE DISTILLATION RESIDUES [TSE]

79869-58-2 PROPANETHIOL
PROPANETHIOLS [HM1, HM1]

80466-34-8 2,4-HEXADIENAL

80508-23-2 PYRIDINE, 3-(1-NITROSO-2-PYRROLIDINYL)
N'-NITROSONORNICOTINE [MSL]

80584-98-1 DEHYDROABIETYLAMINE, ETHOXYLATED

81228-87-7 CYCLOBUTYL CHLOROFORMATE
CYCLOBUTYLCHLOROFORMATE [NJL, NJL]

81457-65-0 COPPER, [29H, 31H-PHTHALO-CYANINATO(2-)-N29,N30,N31,N32-, [[3-(METHYLETHOXY)-, PROPY-LAMINOSULFONYL DERIVATIVES

81624-04-6 HEPTENE

82065-80-3 DIISOTRIDECYL PEROXYDICAR-BONATE, TECHNICALLY PURE
DIISOTRIDECYL PEROXYDICARBON-ATE [DOT]

82428-30-6 2-PROPENOIC ACID, 2-METHYL-, 7-OXABICYCLO[4.1.0]HEPT-3-YLMETHYL ESTER

82657-04-3 BIFENTHRIN

83463-62-1 BROMOCHLOROACETONITRILE

83968-18-7 DIALKYL 79 PHTHALATE

84002-64-2 A-METHYLGLYCEROL TRINI-TRATE

84540-57-8 PROPYLENE GLYCOL MONOMETHYL ETHER ACETATE

84812-04-4 1,4,7,10,13,16-HEXAOXACYCLOOC-TADECANE,2-[(2-PROPENYLOXY) METHYL]

84852-15-3 4-NONYLPHENOL, BRANCHED
NONYLPHENOL, 4-BRANCHED [OTS]

84961-66-0 TOBACCO DUST

85081-53-4 DODECENYLSUCCINIC ACID, MONOTRIDECYL ESTER

85095-61-0 N,N'-(METHYLENEDI-4,1-PHENY-LENE)BIS[2,2-DIMETHYL-
HYDRAZINECARBOXAMIDE, N,N'-(METHYLENE-4,1-PHENYLENE)BIS[2,2'-DIMETHYL- [TSC, TSC]
HYDRAZINECARBOXYMIDE, N,N'-(METHYLENEDI-4,1-PHENYLENE)BIS[2,2-DIMETHYL] [TSE]

85409-17-2 TRIBUTYLTIN NAPHTHENATE
TRIBUTYL TIN NAPHTHENATE [MAK, MAK, MAK]

85502-23-4 3-(N-NITROSOMETHYLAMINO)PROPIONALDEHDYE

85712-26-1 POTASSIUM N,N-BIS (HYDROX-YETHYL) COCOAMINE OXIDE PHOSPHATE
POTASSIUM N,N-BIS (HYDROX-YETHYL) COCOAMINE OXIDE PHOS-PHATE [TSE]

85712-27-2 POTASSIUM N,N'-BIS(HYDROX-YETHYL) TALLOWAMINE OXIDE PHOSPHATE

85763-69-5 IRON, C3-13-CARBOXYLATE NAPHTHENATE COMPLEXES

86220-42-0 NAFRELIN ACETATE

86290-81-5 GASOLINE, MOTOR FUEL
GASOLINE [PAL]
MOTOR FUEL, N.O.S. [NJL]
PETROL (GASOLINE) [AUS, AUS]
PETROL, LEADED [HM1]

87320-05-6 2-PROPENOIC ACID, [2-[1,1-DIMETHYL-2-[(1-OXO-2-PROPENYL)OXY]ETHYL]-5-ETHYL-1,3-DIOXAN-5-YL)METHYL ESTER

87625-62-5 PTAQUILOSIDE

87903-39-7 LEAD HYDROXYSALICYLATE

88497-56-7 BENZENE, ETHENYL-, HO-MOPOLYMER, BROMINATED

88671-89-0 MYCLOBUTANIL [.AL-PHA.-BUTYL-.ALPHA.-(4-CHLOROPHENYL)-1H-1,2,4-TRI-AZOLE-1-PROPANETRILE]

88917-22-0 PROPANOL 1 (OR 2)-2-METHOXYMETHYLETHOXY, AC-ETATE
DIPROPYLENE GLYCOL MONOMETHYL ETHER ACETATE [TSC, TSC, TSC]

89460-82-2 METHYL NITRAMINE

89800-10-2 POLYOXY(1-OXO-1,6-HEX-ANEDIYL), ALPHA-HYDRO-OMEGA-HYDROXY-ESTER WITH 2,2'-(OXYBIS(METHYLENE) BIS(2-HYDROXYMETHYL)-1,3-PROPANEDIOL)-2-PROPENOATE, 2-PROP

90147-57-2 YUCCA EXTRACTIVES (FROM YUCCA, AGAVACEAE)

90370-29-9 4,4',6-TRIMETHYLANGELICIN PLUS ULTRAVIOLET A RADIA-TION

90454-18-5 DICHLORO-1,1,2-TRIFLUO-ROETHANE

90884-29-0 PHENOL, 4,4'-(OXYBIS(2,1-ETHANEDIYLTHIO))BIS-

90982-32-4 CHLORIMURON ETHYL [ETHYL-2-[[[(4-CHLORO-6-METHOXYPYRIMIDIN-2-YL) -CAR-BONYL]-AMINO]SULFONYL]BEN-ZOATE]

91031-95-7 MANNITAN, COCONUT OIL ES-TER

91681-63-9 4H-BENZOPYRAN-4-ONE, 6-(2,3-DIHYDROXY-3-METHYLBUTYL) -3-(2,4-DIHYDROXYPHENYL)-7-HYDROXY-

92874-42-5 TETRABROMOBISHPENOL-B

93589-69-6 PHENOL, 4,4'-[METHYLENEBIS (OXY-2,1-ETHANEDIYLTHIO)]BIS-

93763-70-3 PERLITE
PERLITE DUST [AUS, PAL]
PERLITE, EXPANDED [PST]

93918-16-2 MANGANESE NEONONOATE

94088-85-4 DOXYCYCLINE CALCIUM

94313-89-0 AMMONIUM DIISODECYL SULFO-SUCCINATE

96478-09-0 2-PROPENOIC ACID, 2-METHYL-, 2-[3-(2H-BENZOTRIAZOL-2-YL)-4-HYDROXYPHENYL]ETHYL ESTER

96915-49-0 POLYOXY(1-OXO-1,6-HEX-ANEDIYL), ALPHA-HYDRO-OMEGA-HYDROXY-ESTER WITH 3-HYDROXY-2,2-DIMETHYL-PROPYL 3-HYDROXY-2,2-DIMETHYLPROPANOATE (2:1), DI-2-[PROPANOATE

96915-50-3 POLYOXY[OXY(1-OXO-1,6 HEX-ANEDIYL)], ALPHA-(1-OXO-2-PROPENYL)-OMEGA-((TETRAHY-DRO-2-FURANYL)METHOXY)-

96949-21-2 RHAMSAN GUM

97043-91-9 ALCOHOLS, C9-16, ETHOXY-LATED

97380-66-0 UREA-FORMALDEHYDE-RESINS
UREA-FORMALDEHYDE RESIN [OTS]

97553-00-9 COD OIL, SULFONATED

98730-04-2 2H-1,4-BENZOXAZINE, 4-(DICHLOROACETYL)-3,4-DIHY-DRO-3-METHYL-

100545-50-4 1-DECANAMINE, N-DECYL-N-METHYL-N-OXIDE

101200-48-0 TRIBENURON METHYL [2-((((((4-METHOXY-6-METHYL-1,3,5-TRI-AZIN-2-YL)-METHYLAMINO)CAR-BONYL)AMINO)SULFONYL)-, METHYL ESTER]

102093-68-5 4-METHYL-2,6-BIS(METHYLTHIO)-

103112-35-2 1H-1,2,4-TRIAZOLE-3-CARBOXYLIC ACID, 1-(2, 4-DICHLORPHENYL)-5-(TRICHLOROMETHYL)-, ETHYL ESTER

104503-68-6 3,6,9,12,15,18,21-HEPTAOXATE-TRATRIAOCTANOIC ACID, SODIUM SALT

104810-48-2 POLY(OXY-1,2-ETHANEDIYL), AL-PHA-(3-(3-(2H-BENZOTRIAZOL-2-YL)-5-(1,1-DIMETHYLETHYL) -4-HYDROXYPHENYL)-1-OXO-PROPYL)-OMEGA- HYDROXY-

104983-85-9 1,3-BENZENEDIAMINE, 2-METHYL-4,6-BIS(METHYLTHIO)-

105554-30-1 HEXAMETHYLOL BENZENE HEX-ANITRATE

105650-23-5 PHIP (2-AMINO-1-METHYL-6-PHENYLIMIDAZO-[4,5-B]PYRI-DINE)

105735-71-5 3,7-DINITROFLUORANTHENE

105839-17-6 CASTOR OIL, EPOXIDIZED

105859-97-0 LIGNIN, ALKALI, REACTION PRODUCTS WITH DISODIUM SUL-FITE ANDFORMALDEHYDE

106151-63-7 P-NONYLPHENOL, ETHOXY-LATED, PHOSPHATED, DIPOTAS-SIUM SALT

106158-22-9 POLY(OXY-1,2-ETHANEDIYL), .ALPHA.-HYDRO-.W.-HYDROXY-, ETHERWITH 2-ETHYL-2-(HY-DROXYMETHYL)-1,3-PROPANE-DIOL (3:1) DI-2-PROPENOATE, METHYL ETHER

106264-79-3 MIXTURE OF 1,3-BENZENE-DIAMINE, 2-METHYL-4,6-BIS (METHYLTHIO)AND 1,3-BEN-ZENEDIAMINE, 4-METHYL-2,6-BIS (METHYLTHIO)

106790-31-2 4-ISOCYANATO-N,N-BIS(4-ISOCYANATOPHENYL)-2,5-DIMETHOXYBENZENEAMINE

108171-26-2 CHLORINATED PARAFFINS: C12, 60% CHLORINE
CHLORINATED PARAFFINS [C65, CAL]
CHLORINATED PARAFFINS (C12, 60% CHLORINE) [NTP]
CHLORINATED PARAFFINS OF AV-ERAGE CARBON-CHAIN LENGTH C12 AND AVERAGE DEGREE OF CHLORI-NATION 60% [IAR]

108171-27-3 CHLORINATED PARAFFINS: C23, 43% CHLORINE

109909-39-9 2,4,6-TRI-SEC-BUTYLPHENOL, ETHOXYLATED, SULFATED, SODIUM SALT

110843-98-6 1,5-DIOXASPIRO [5.5] UNDE-CANE-3,3-DICARBOXYLIC ACID BIS (1,2,2,6,6-PENTAMETHYL-4-PIPERIDINYL) ESTER

111381-89-6 DI(HEPTYL-, NONYL-) PHTHA-LATE (BRANCHED AND LINEAR ISOMERS)

111381-90-9 DI(HEPTYL-, UNDECYL-) PHTHA-LATE (BRANCHED AND LINEAR ISOMERS)

111381-91-0 DI(NONYL-, UNDECYL-) PHTHA-LATE (BRANCHED AND LINEAR ISOMERS)

111512-56-2 1,1-DICHLORO-1,2,3,3,3-PENTAFLUOROPROPANE (HCFC-225EB)

111984-09-9 3,3'-DIMETHOXYBENZIDINE HY-DROCHLORIDE (O-DIANISIDINE HYDROCHLORIDE)

112926-00-8 SILICA, AMORPHOUS, PRECIPI-TATED AND GEL
PRECIPITATED SILICA [ONT, PAL]
SILICA - AMORPHOUS, PRECIPI-TATED SILICA AND SILICA GEL [TLV]
SILICA-AMORPHOUS, PRECIPITATED SILICA [AUS]

113221-69-5 MALEIC ANHYDRIDE, POLYMER WITH ETHYL ACRYLATE AND VINYL ACETATE, HYDROLYZED

114719-15-2 ETHANOL,2,2'-[[1-[(2-PROPENYLOXY)METHYL]-1,2-ETHANEDIYL]BIS(OXY)]BIS-,BIS (4-METHYLBENZENESULFONATE)

114719-19-6 1,2-PROPANEDIOL,3-(2-PROPENY-LOXY)-,BIS(4-METHYLBENZENE-SULFONATE)

115361-68-7 DIMETHYLMETHYL 3,3,3-TRIFLU-OROPROPYL SILOXANE

115965-75-8 2-PROPENOIC ACID, 2-(2-OXO-3-OXAZOLIDINYL) ETHYL ESTER

116355-83-0 FUMONISIN B1

116355-84-1 FUMONISIN B2

116712-07-3 NORPOR 10

119239-21-3 CYCLOHEXYL METHACRYLATE, 2-HYDROXYLETHYL METHACRY-LATE, ISODECYL METHACRY-LATE, 2-MORPHOLINOETHYL METHACRYLATE COPOLYMER

121158-58-5 DODECYL PHENOL (BRANCHED)

122436-67-3 M-CRESOL, POLYMER WITH FORMALDEHYDE AND SUL-FANILIC ACID

124028-99-5 2-PROPANOL,1-[2-[2-[[(4-METHYLPHENYL)SULFONYL] OXY]ETHOXY]ETHOXY]-3-(2-PROPENYLOXY)-,4-METHYLBEN-ZENESULFONATE

124029-00-1 ETHANOL,2-[1-[[2-[2-[[(4-METHYLPHENYL)SULFONYL] OXY]ETHOXY]ETHOXY]METHYL] -2-(PROPENYLOXY)ETHOXY]-,4-METHYLBENZENESULFONATE

124213-39-4 2-PROPANOL,1-[2-[[(4-METHYLPHENYL)SULFONYL] OXY]ETHOXY]-3-(2-PROPENY-LOXY)-4-METHYLBENZENESUL-FONATE

124737-31-1 BENZENEDIAZONIUM, 4-(DIMETHYLAMINO)-, SALT WITH 2-HYDROXY-5-SULFOBENZOIC ACID (1:1)

124993-63-1 HYDRAZINE, [4-[1-METHYLB-UTOXY]PHENYL], MONOHY-DROCHLORIDE

125304-11-2 POLY(OXY-1,2-ETHANEDIYL) , ALPHA-(1-OXO-2-PROPENYL) -OMEGA-HYDROXY-, C(10-16)-ALKYL ETHERS

125426-39-3 2-CHLORO-1,1,1,2,3,3,3-HEP-TAFLUOROPROPANE
CFC-217 [CAA]

125972-19-2 N,N-DIMETHYLISOOCTADE-CANAMINE, N-OXIDE

WOOD DUST, CERTAIN HARD WOODS [MIN]
WOOD DUSTS - CERTAIN HARD WOODS AS BEECH AND OAK [ONT]

RR-00017-9 WOOD DUSTS-SOFT WOODS
WOOD DUST (SOFT WOOD) [AUS, AUS, AUS, AUS]
WOOD DUST - SOFT WOOD [ONT, ONT]
WOOD DUST, SOFT WOOD [TLV, TLV]
WOOD DUST, SOFT WOODS [MIN]

RR-00018-0 PROGESTERONE AND OESTRO-GENS

RR-00019-1 ALUMINUM, PYRO POWDERS
ALUMINUM PYRO POWDERS [CAL, PAL, QBC]

RR-00020-4 ALUMINUM, WELDING FUMES
ALUMINUM WELDING FUMES [PAL]

RR-00021-5 ALUMINUM, SOLUBLE SALTS
ALUMINUM SALTS, SOLUBLE [GBR]
ALUMINUM SOLUBLE SALTS [PAL]
ALUMINUM, WATER-SOLUBLE SALTS, N.O.S. [WHM]

RR-00022-6 ALUMINUM, ALKYLS (NOC)
ALKYLALUMINUMS [OS3]
ALUMINUM ALKYL COMPOUNDS [GBR, WHM]
ALUMINUM ALKYLS [NJL, PAL, QBC]
ALUMINUM, ALKYLS [CAL, OS1]

RR-00023-7 CHROMIUM (II) COMPOUNDS
CHROMIUM (II) COMPOUNDS, N.O.S. [WHM]
CHROMIUM VI COMPOUNDS, WATER SOLUBLE [QBC]
CHROMIUM, SOLUBLE CHROMOUS SALTS [BC1]

RR-00024-8 CHROMIUM (III) COMPOUNDS
CHROMIUM (III) COMPOUNDS, N.O.S. [WHM]
CHROMIUM COMPOUNDS, TRIVA-LENT [IAR]
CHROMIUM III COMPOUNDS [ALB, ALB]
CHROMIUM, SOLUBLE CHROMIC SALTS [BC1]

RR-00025-9 CHROMIUM (VI) COMPOUNDS-WATER SOLUBLE

RR-00026-0 CHROMIUM (VI) COMPOUNDS
CHROMIUM (HEXAVALENT COM-POUNDS) [C65, C65]
CHROMIUM (VI) COMPOUNDS, N.O.S. [WHM]
CHROMIUM COMPOUNDS, HEXAVA-LENT [PAL, PAL]
HEXAVALENT CHROMIUM COM-POUNDS [IAR, ONT, ONT]

RR-00027-1 BETEL QUID WITH TOBACCO

RR-00028-2 SOOTS
SOOT [PAL, PAL]

RR-00029-3 TALC CONTAINING ASBESTOS FIBERS
TALC (CONTAINING ASBESTOS FIBERS) [MIN, TLV]
TALC (FIBROUS) [ALB, ALB, QBC]
TALC CONTAINING ASBESTIFORM FIBERS [C65, IAR]

RR-00030-6 TOBACCO SMOKE
ENVIRONMENTAL TOBACCO SMOKE [CAB]
TOBACCO SMOKE (PRIMARY) [C65, C65, C65, C65]

RR-00031-7 TOBACCO PRODUCTS, SMOKE-LESS
TOBACCO, SMOKELESS PRODUCTS [C65]

RR-00032-8 BRACKEN FERN
BRACKERN FERN [MSL]

RR-00033-9 PROGESTINS

RR-00034-0 NORGESTREL AND OESTROGENS

RR-00035-1 ARSENIC ORGANIC COMPOUNDS
ARSENIC, ORGANIC COMPOUNDS [MIN]

RR-00036-2 MOLYBDENUM SOLUBLE COM-POUNDS
MOLYBDENUM (SOLUBLE COM-POUNDS) [NIO, NIO]
MOLYBDENUM, SOLUBLE COM-POUNDS [QBC]

RR-00037-3 MOLYBDENUM INSOLUBLE COM-POUNDS
MOLYBDENUM, INSOLUBLE COM-POUNDS [QBC]

RR-00038-4 NICKEL SOLUBLE COMPOUNDS
NICKEL AND CERTAIN NICKEL COMPOUNDS [C65]
NICKEL, SOLUBLE COMPOUNDS [BC1, BC1, BC1, QBC]
NICKEL, WATER-SOLUBLE COM-POUNDS [ONT]
NICKEL, WATER-SOLUBLE INOR-GANIC COMPOUNDS, N.O.S. [WHM]

RR-00039-5 NORETHISTERONE AND OESTRO-GENS

RR-00040-8 RHODIUM SOLUBLE COM-POUNDS
RHODIUM (SOLUBLE COMPOUNDS) [NIO, NIO]
RHODIUM WATER-SOLUBLE COM-POUNDS [ONT, ONT, ONT]
RHODIUM, SOLUBLE SALTS [BC1, BC1, MEX, QBC]

RR-00041-9 SILVER SOLUBLE COMPOUNDS
SILVER, WATER SOLUBLE COM-POUNDS [ONT]
SILVER, WATER-SOLUBLE COM-POUNDS, N.O.S. [WHM]

RR-00042-0 TIN ORGANIC COMPOUNDS
TIN (ORGANIC COMPOUNDS) [NIO, NIO, NIO]
TIN COMPOUNDS, ORGANIC [GBR, GBR, GBR, MAK, MAK, MAK, MAK]
TIN, ORGANIC COMPOUNDS [BC1, BC1, MIN, ONT, ONT, QBC, QBC]

RR-00043-1 TIN INORGANIC COMPOUNDS
TIN COMPOUNDS, INORGANIC [GBR, GBR]
TIN, INORGANIC COMPOUNDS [BC1, BC1, BC1]
TIN, OXIDE AND INORGANIC COM-POUNDS [QBC]

RR-00044-2 URANIUM SOLUBLE COMPOUNDS
URANIUM (NATURAL), SOLUBLE COMPOUNDS [CEX]
URANIUM (SOLUBLE COMPOUNDS) [NIO, NIO, NIO, NIO]

RR-00045-3 URANIUM INSOLUBLE COM-POUNDS
URANIUM (NATURAL), INSOLUBLE COMPOUNDS [CEX, CEX]

RR-00046-4 PLATINUM SOLUBLE SALTS
PLATINUM (SOLUBLE SALTS) [NIO, NIO, QBC]
PLATINUM SALTS, SOLUBLE [GBR, GBR]
PLATINUM, SOLUBLE SALTS [BC1]
PLATINUM, WATER-SOLUBLE COM-POUNDS [ONT]
PLATINUM, WATER-SOLUBLE SALTS, N.O.S. [WHM]

RR-00047-5 RHODIUM INSOLUBLE COM-POUNDS
RHODIUM, WATER INSOLUBLE COM-POUNDS [ONT, ONT, ONT]

RR-00048-6 THALLIUM, WATER-SOLUBLE COMPOUNDS, N.O.S.

RR-00049-7 BARIUM SOLUBLE COMPOUNDS
BARIUM (SOLUBLE COMPOUNDS) [NIO, NIO, NIO, QBC]

BARIUM COMPOUNDS, SOLUBLE [GBR]
BARIUM, SOLUBLE COMPOUNDS [BC1, CAL]
BARIUM, WATER-SOLUBLE COM-POUNDS, N.O.S. [WHM]

RR-00051-1 OESTROGENS, NONSTEROIDAL
NONSTEROIDAL OESTROGENS [IAR]

RR-00052-2 OESTROGENS, STEROIDAL
STEROIDAL OESTROGENS [IAR]

RR-00053-3 MOPP
MOPP AND OTHER COMBINED CHEMOTHERAPY INCLUDING ALKY-LATING AGENTS [IAR]
MOPP(COMBINED THERAPY W/NITROGEN MUSTARD,VIN-CRISTINE,PROCAR. [PAL, PAL]

RR-00054-4 UNTREATED AND MILDLY-TREATED OILS
UNTREATED AND MILDLY REFINED MINERAL OILS [NTP]

RR-00055-5 ANALGESIC MIXTURES CON-TAINING PHENACETIN
ANALGESIC MIXTURE CONTAINING PHENACETIN [MIN]

RR-00056-6 8-METHOXYPSORALEN PLUS ULTRAVIOLET RADIATION
8-METHOXYPSORALEN (METHOXSALEN) PLUS ULTRAVIO-LET RADIATION [IAR]
METHOXSALEN WITH ULTRA-VIO-LET [MIN]
METHOXSALEN WITH ULTRA-VIO-LET A THERAPY [PAL, PAL]
METHOXSALEN WITH ULTRA-VIO-LET A THERAPY (PUVA) [NTP]

RR-00057-7 ANDROGENIC (ANABOLIC) STEROIDS
ANABOLIC STEROIDS [C65, C65]
ANABOLIC STEROIDS (ANDROGENIC STEROIDS) [CAL]

RR-00059-9 CHLOROPHENOXY HERBICIDES

RR-00060-2 CARBON BLACK EXTRACTS
CARBON-BLACK EXTRACTS [C65, IAR]
SOLVENT (BENZENE) EXTRACTS OF MOST CARBON BLACKS [PAL, PAL]

RR-00061-3 BITUMENS, EXTRACTS OF STEAM-REFINED AND AIR-RE-FINED
BITUMENS [IAR, IAR, MIN]

RR-00062-4 O-DIANISIDINE-BASED DYES

RR-00063-5 AMYL CHLORIDES (MIXED)
AMYL CHLORIDES [HM1, HM1]

RR-00064-6 FOAMING AGENTS

RR-00065-7 INORGANIC ARSENIC
ARSENIC (INORGANIC ARSENIC COMPOUNDS) [C65]
ARSENIC, INORGANIC [NHS, NHS, NHS]
ARSENIC, INORGANIC COMPOUNDS [CAB, NJL]

RR-00066-8 RADIONUCLIDES
RADIONUCLIDES (INCLUDING RADON) [CAA]

RR-00067-9 GLYCOL ETHERS

RR-00068-0 ISOPROPYL ALCOHOL MANU-FACTURE (STRONG-ACID PRO-CESS)

RR-00069-1 ORGANORHODIUM COMPLEX
ORGANORHODIUM COMPLEX (PMN-82-147) [PAL]

RR-00070-4 CHROMITE ORE PROCESSING
CHROMITE ORE PROCESSING (CHROMATE) [CEX, PAL, TLV, TLV]
CHROMITE ORE PROCESSING (CHROMATES) [QBC, QBC]

RR-00169-4   1,1-DIMETHYLPROPYL PEROX-
             YESTER

RR-00170-7   TRISUBSTITUTED PHENOL

RR-00171-8   HALOALKYL SUBSTITUTED
             CYCLIC ETHERS

RR-00172-9   BHUSA

RR-00173-0   BIPYRIDILIUM PESTICIDES,
             N.O.S.

RR-00174-1   ALKYLATED DIPHENYL OXIDE

RR-00175-2   PHOSPHONIUM SALT

RR-00177-4   BISULFITES
             BISULFITES, INORGANIC, AQUEOUS
             SOLUTION, N.O.S. [HM1, HM1, NJL]

RR-00178-5   ALKYLENE GLYCOL TEREPH-
             THALATE AND SUBSTITUTED
             BENZOATEESTERS

RR-00179-6   ALKYLATED DIARYLAMINE,
             SULFURIZED

RR-00180-9   BORATE AND CHLORATE MIX-
             TURES
             CHLORATE AND BORATE MIXTURES
             [HM1, HM1, NJL]

RR-00181-0   BORDEAUX ARSENITE, LIQUID
             OR SOLID
             BORDEAUX ARSENITE [NJL]

RR-00183-2   BROMATES, INORGANIC, N.O.S.

RR-00185-4   T-BUTYL PEROXYDIETHYL-AC-
             ETATE WITH T-BUTYL PEROXY-
             BENZOATE

RR-00186-5   CARBAMATE PESTICIDES, N.O.S.

RR-00188-7   POLYMER OF ALKANEPOLYOL
             AND POLYALKYLPOLYISO-
             CYANATOCARBOMONOCYCLE,
             ACETONE OXIME-BLOCKED

RR-00189-8   POLYMER OF ALKENOIC ACID,
             SUBSTITUTED ALKYLACRYLATE,
             SODIUMSALT

RR-00190-1   PERFLUOROALKYL EPOXIDE

RR-00191-2   DIALKENYLAMIDE

RR-00192-3   ALKYLPHENOXY-
             POLYALKOXYAMINE

RR-00193-4   CAUSTIC ALKALI LIQUIDS, N.O.S.

RR-00194-5   TETRAGLYCIDYLAMINES

RR-00195-6   CELLULOID

RR-00196-7   DISUBSTITUTED ALKYL TRI-
             AZINES

RR-00197-8   SUBSTITUTED PHOSPHATE ES-
             TER

RR-00198-9   6-NITRO-2(3H)-BENZOOXA-
             ZOLONE

RR-00199-0   3-ALKYL-2-(2-ANILINO)VINYLTHI-
             AZOLINIUM SALT

RR-00200-6   CHLORATE AND MAGNESIUM
             CHLORIDE MIXTURES

RR-00201-7   CHLORATES, INORGANIC, N.O.S.

RR-00202-8   CHLORINATED ETHANES

RR-00204-0   CHLORINATED PHENOLS

RR-00205-1   CHLORITE, INORGANIC, N.O.S.
             CHLORITES, INORGANIC, N.O.S.
             [HM1, HM1]

RR-00206-2   CHLOROALKYL ETHERS
             CHLOROALKYL ETHERS, N.O.S.
             [RCU]

RR-00207-3   CHLOROFORMATE, N.O.S.
             CHLOROFORMATE COMPOUNDS,
             N.O.S. [WHM]

             CHLOROFORMATES, N.O.S. [HM1,
             HM1, HM1]

RR-00208-4   CHLOROSILANES, N.O.S.

RR-00209-5   DIPHENYL ETHER/BIPHENYL
             MIXTURE

RR-00210-8   SUBSTITUTED AROMATIC

RR-00211-9   N,N'-BIS[2-[2-(3-ALKYL)THIAZO-
             LINE]VINYL]-1,4-PHENYLENEDI-
             AMINE METHYL SULFATE, DOU-
             BLE SALT

RR-00212-0   BROMINATED AROMATIC COM-
             POUND

RR-00213-1   SODIUM SALT OF AN ALKY-
             LATED, SULFONATED AROMATIC

RR-00214-2   UREA, CONDENSATE WITH POLY
             [OXY(METHYL-1,2-ETHANEDIYL)]
             -ALPHA-[2-AMINOETHYLETHYL]-
             X-(2-AMINOETHYLETHOXY)

RR-00215-3   ALKEHYLDICARBOXYLIC
             ACIDS, POLYMERS WITH ALKA-
             NEPOLYOL ANDTDI, ALKANOL
             BLOCKED, ACRYLATE

RR-00216-4   COAL GAS
             GAS, COAL GAS [NFP]

RR-00217-5   CAPROLACTONE, POLYMER
             WITH HEXAMETHYLENE DI-
             ISOCYANATE, HYDROXYALKYL
             ACRYLATE ESTER, REACTION
             PRODUCTS WITH SUBSTITUTED
             ALKANOIC ACID AND METAL
             HETEROMONOCYCLE

RR-00218-6   METHYLENEBIS(4-ISOCYANA-
             TOBENZENE), POLYMER WITH
             POLYCAPROLACTONE TRIOL
             AND ALKOXYLATED ALKA-
             NEPOLYOL, HYDROXYALKYL
             METHACRYLATE ESTER

RR-00219-7   COBALT RESINATE, PRECIPI-
             TATED

RR-00220-0   COMBUSTIBLE LIQUID, N.O.S.

RR-00221-1   ALKENOIC ACID, TRISUBSTI-
             TUTED BENZYL-DISUBSTITUTED
             PHENYLESTER

RR-00222-2   ALKENOIC ACID, TRISUBSTI-
             TUTED PHENYLALKYL DISUBSTI-
             TUTEDPHENYL ESTER

RR-00223-3   HYDROXYALKYL METHACRY-
             LATE, ALKYL ESTER

RR-00224-4   COMPRESSED OR LIQUIFIED
             GASES, N.O.S.

RR-00225-5   CARBAMIC ACID, (TRIALKY-
             LOXYSILYALKYL)-SUBSTITUTED
             ACRYLATEESTER

RR-00226-6   POLYALKYLPOLYSILAZANE, BIS
             (SUBSTITUTED ACRYLATE)

RR-00227-7   POLYMER OF ALKYL CAR-
             BOMONOCYCLE DIISOCYANATE
             WITH ALKANEPOLYOL POLY-
             ACRYLATES

RR-00228-8   COPPER BASED PESTICIDES,
             N.O.S.

RR-00229-9   SUBSTITUTED OXIDE-ALKYLENE
             POLYMER, METHACRYLATE

RR-00230-2   TRIMETHYLOLPROPANE FATTY
             ACID DIACRYLATE

RR-00231-3   BISPHENOL A, EPICHLOROHY-
             DRIN, METHYLENEBIS(SUBSTI-
             TUTEDCARBOMONOCYCLE),
             POLYALKYLENE GLYCOL, ALKA-
             NOL, METHACRYLATE POLYMER

RR-00232-4   REACTION PRODUCT OF A
             MONOALKYL SUCCINIC ANHY-
             DRIDE WITH ANX-HYDROXY
             METHACRYLATE

RR-00233-5   CORROSIVE LIQUIDS, N.O.S.

RR-00234-6   2-PROPENOIC ACID, 1-[HYDROX-
             YMETHYL]PROPYL ESTER

RR-00235-7   POLY(OXY-1,2-ETHANEDIYL)
             , ALPHA-(2-METHYL-1-OXO-2-
             PROPENYL)-X-HYDROXY-, C10-
             16-ALKYL ETHERS

RR-00236-8   CORROSIVE SOLIDS, N.O.S.

RR-00237-9   SUBSTITUTED ACRYLATED
             ALKOXYLATED ALIPHATIC
             POLYOL

RR-00238-0   ACID MODIFIED ACRYLATED
             EPOXIDE

RR-00239-1   ALIPHATIC DIURETHANE ACRY-
             LATE ESTER

RR-00240-4   URETHANE ACRYLATE

RR-00241-5   AMINO ACRYLATE MONOMER

RR-00242-6   POLYSUBSTITUTED POLYOL

RR-00243-7   ALKOXYLATED ALKANE
             POLYOL, POLYACRYLATE ESTER

RR-00244-8   COUMARIN DERIVATIVE PESTI-
             CIDES, N.O.S.

RR-00245-9   POLYOL CARBOXYLATE ESTER

RR-00246-0   1,3-BENZENEDIAMINE, 4-(1,1-
             DIMETHYLETHYL)-AR-METHYL

RR-00249-3   DERIVATIVE OF TETRA-
             CHLOROETHYLENE

RR-00250-6   1,4-DI-(2-T-BUTYLPEROXY ISO-
             PROPYL)BENZENE

RR-00252-8   SUBSTITUTED 2-PHENOXYPYRI-
             DINE

RR-00253-9   2,2-DI(T-BUTYLPEROXY)
             PROPANE

RR-00254-0   SUBSTITUTED METHYLPYRIDINE

RR-00255-1   DICHLOROTRIAZINETRIONE AND
             ITS SALTS

RR-00256-2   2,2-DI(4,4-DI-T-BUTYL-PEROXY
             CYCLOHEXYL)-PROPANE 42%

RR-00257-3   DI-2,4-DICHLOROBENZOYL PER-
             OXIDE

RR-00258-4   DICARBOXYLIC ACID MO-
             NOESTER

RR-00259-5   1,2-BENZENEDIAMINE, 4-
             ETHOXY-, SULFATE

RR-00260-8   SUBSTITUTED POLYGLYCIDYL
             BENZENEAMINE

RR-00261-9   2-PROPYLBIPHENYL

RR-00262-0   DINITROPHENATES ALKALI MET-
             ALS, DRY OR WETTED

RR-00263-1   DINITROPHENOLATES

RR-00264-2   DISINFECTANTS, N.O.S.

RR-00265-3   DISUBSTITUTED DIAMINO
             ANISOLE

RR-00266-4   GASOLINE ENGINE EXHAUST
             (CONDENSATES/EXTRACTS)

RR-00405-7 MERCURY 197M

RR-00406-8 MERCURY 195M

RR-00407-9 MERCURY 193M

RR-00408-0 MANGANESE 52M

RR-00409-1 LUTETIUM 178M

RR-00410-4 LUTETIUM 177M

RR-00411-5 PARATHION AND COMPRESSED GAS MIXTURE

RR-00413-7 PERCHLORATES, INORGANIC, N.O.S.

RR-00414-8 DIALKYLNITROSAMINES

RR-00415-9 LUTETIUM 176M

RR-00416-0 PERMANGANATES, INORGANIC, N.O.S.

RR-00417-1 PEROXIDES, INORGANIC, N.O.S.

RR-00418-2 PESTICIDES, N.O.S.
PESTICIDES AND THEIR BY-PRODUCTS [CN7]

RR-00419-3 LUTETIUM 174M

RR-00420-6 LEAD 202M

RR-00421-7 LEAD 195M

RR-00422-8 PETROLEUM DISTILLATES, N.O.S.

RR-00423-9 PETROLEUM OIL

RR-00424-0 PHENOXY PESTICIDES, N.O.S.

RR-00425-1 KRYPTON 85M

RR-00426-2 KRYPTON 83M

RR-00427-3 IRIDIUM 195M

RR-00428-4 PHENYLMERCURIC COMPOUND, SOLID, N.O.S.
PHENYLMERCURIC COMPOUNDS, N.O.S. [HM1, HM1, HM1]

RR-00429-5 PHENYL UREA PESTICIDES, N.O.S.

RR-00430-8 IRIDIUM 194M

RR-00431-9 IRIDIUM 192M

RR-00432-0 IRIDIUM 190M

RR-00433-1 PHTHALATE ESTERS

RR-00434-2 PHTHALIMIDE DERIVATIVE PESTICIDES, N.O.S.

RR-00435-3 IODINE 132M

RR-00436-4 IODINE 120M

RR-00437-5 INDIUM 119M

RR-00438-6 INDIUM 117M

RR-00439-7 INDIUM 116M

RR-00440-0 POISONOUS LIQUIDS, N.O.S.

RR-00441-1 INDIUM 115M

RR-00442-2 INDIUM 114M

RR-00443-3 INDIUM 113M

RR-00444-4 HOLMIUM 166M

RR-00445-5 POISONOUS SOLIDS, N.O.S.

RR-00446-6 HOLMIUM 164M

RR-00447-7 HOLMIUM 162M

RR-00448-8 HAFNIUM 182M

RR-00449-9 HAFNIUM 180M

RR-00450-2 POLYNUCLEAR AROMATIC HYDROCARBONS

RR-00451-3 HAFNIUM 179M

RR-00452-4 POTASSIUM NITRATE AND SODIUM NITRITE, MIXTURE
POTASSIUM NITRATE AND SODIUM NITRITE MIXTURES [HM1, HM1]

RR-00453-5 POTASSIUM SALTS OF NITRO-AROMATIC DERIVATIVES, EXPLOSIVE
POTASSIUM SALTS OF AROMATIC NITRO-DERIVATIVES [HM1, HM1]

RR-00454-6 PYROPHORIC LIQUIDS, N.O.S.

RR-00455-7 PYROPHORIC METALS OR ALLOYS, N.O.S.

RR-00456-8 PYROPHORIC SOLIDS, N.O.S.

RR-00457-9 PYROSULFURIC ACID

RR-00458-0 HAFNIUM 178M

RR-00459-1 RADIOACTIVE MATERIALS, N.O.S.
HIGH-LEVEL RADIOACTIVE MATTER (INCLUDING RADIOACTIVE WASTES) [CN6]

RR-00460-4 HAFNIUM 177M

RR-00461-5 GOLD 200M

RR-00462-6 GOLD 198M

RR-00463-7 EUROPIUM 152M

RR-00464-8 EINSTEINIUM 254M

RR-00465-9 COBALT 62M

RR-00466-0 COBALT 60M

RR-00467-1 COBALT 58M

RR-00468-2 CESIUM 135M

RR-00469-3 CESIUM 134M

RR-00470-6 CERIUM 137M

RR-00471-7 CADMIUM 117M

RR-00472-8 REFRIGERANT GASES, N.O.S.

RR-00473-9 CADMIUM 115M

RR-00474-0 CADMIUM 113M

RR-00475-1 RESIN OIL

RR-00476-2 RESIN SOLUTION

RR-00477-3 BROMINE 80M

RR-00478-4 BROMINE 74M

RR-00479-5 RODENTICIDES, LIQUID OR SOLID, N.O.S.

RR-00480-8 BISMUTH 210M

RR-00481-9 BARIUM 135M

RR-00482-0 BARIUM 133M

RR-00483-1 BARIUM 131M

RR-00484-2 ANTIMONY 126M

RR-00485-3 SELENATES, N.O.S.

RR-00486-4 ANTIMONY 124M

RR-00487-5 ANTIMONY 118M

RR-00488-6 ANTIMONY 116M

RR-00489-7 BROMINATED FLUOROCARBONS

RR-00490-0 SODIUM FLUORIDES

RR-00491-1 SODIUM SALTS OF NITRO-AROMATIC DERIVATIVES, N.O.S.
SODIUM SALTS OF AROMATIC NITRO-DERIVATIVES, N.O.S. [HM1, HM1]

RR-00492-2 AMERICIUM 246M

RR-00493-3 AMERICIUM 244M

RR-00495-5 AMERICIUM 242M

RR-00496-6 SUBSTITUTED NITROPHENOL PESTICIDES, N.O.S.

RR-00497-7 PHOSPHORODITHIOIC AND PHOSPHOTHIOIC ACID ESTERS

RR-00498-8 PHENOLIC COMPOUNDS
PHENOLIC COMPOUNDS (4AAP) [CAL]

RR-00499-9 CHLOROPHENOXY DERIVATIVE ACIDS, ESTER, ETHERS, AMINES AND OTHER SALTS

RR-00500-5 CYANIDE (SALTS)
CYANIDE SALTS [NHS, NHS]
CYANIDES (SOLUBLE CYANIDE SALTS), NOT OTHERWISE SPECIFIED [RCU, RCU]

RR-00502-7 C9 AROMATIC HYDROCARBON FRACTION

RR-00503-8 2-AMINO-2,3-DIMETHYLBUTYRONITRILE

RR-00504-9 SELSUN

RR-00505-0 HEXACHLORODIBENZOFURANS
HXCDFS (ALL HEXACHLORODIBENZOFURANS) [UTS]

RR-00506-1 TRIAZINE PESTICIDES, N.O.S.

RR-00507-2 PENTACHLORODIBENZOFURANS
PECDFS (ALL PENTACHLORODIBENZOFURANS) [UTS]

RR-00508-3 TETRACHLORODIBENZOFURANS
TCDFS (ALL TETRACHLORODIBENZOFURANS) [UTS]

RR-00509-4 HEXACHLORODIBENZO-P-DIOXINS
HXCDDS (ALL HEXACHLORODIBENZO-P-DIOXINS) [UTS]

RR-00510-7 TRIISOCYANATOISOCYANURATE OF ISOPHORONEDIISOCYANATE, SOLUTION

RR-00511-8 PENTACHLORODIBENZO-P-DIOXINS
PECDDS (ALL PENTACHLORODIBENZO-P-DIOXINS) [UTS]

RR-00512-9 PENTACHLOROPHENOL DERIVATIVES

RR-00513-0 AMYL MALEATE

RR-00514-1 WOOD DUST, ALL SOFT AND HARD WOODS
WOOD DUST [MAK, MAK]
WOOD DUST (NONALLERGIC) [BC1, BC1, QBC]
WOOD DUST, ALL SOFT AND HARD WOODS, EXCEPT WESTERN RED CEDAR [OS1, OS1]
WOOD DUST, NONALLERGENIC [ALB, ALB]

RR-00515-2 BROWN COAL TAR

RR-00516-3 TARS (POLYCYCLIC AROMATIC HYDROCARBONS)

RR-00517-4 ZINC ASHES

RR-00518-5 SULFATE
SULPHATE [CD1]
SULPHATES [MX2]

RR-00519-6 3-HEPTENE (MIXED CIS- AND TRANS- ISOMERS)
3-HEPTENE (MIXED CIS AND TRANS) [NFP, NFP]

RR-00520-9 O-TOLIDINE-BASED DYES

RR-00521-0 IRON SALTS (SOLUBLE)
IRON SALTS [GBR, GBR, PAL]
IRON SALTS, (WATER-SOLUBLE) [ONT]
IRON SALTS, SOLUBLE [BC1, BC1, CAL, CEX, QBC, TLV]

RR-00607-5 NITROXYLENE, ALL ISOMERS
            NITROXYLENES (O-; M-; P-) [HM1,
            HM1, HM1]

RR-00608-6 OSMIUM, TETRAAMINETETRA-
            CHLORO-

RR-00609-7 PERCHLORATE COMPOUNDS,
            N.O.S.

RR-00610-0 PERSULFATE COMPOUNDS
            PERSULFATES (ALKALI METALS)
            [ALB, ALB]
            PERSULFATES, ALKALI METAL
            [ONT, ONT]

RR-00611-1 RHODIUM COMPOUNDS, N.O.S.
            RHODIUM COMPOUNDS [CAL]

RR-00612-2 SELENIUM COMPOUNDS
            SELENIUM AND COMPOUNDS [GBR]
            SELENIUM COMPOUNDS, N.O.S. [NJL,
            RCU, WHM]

RR-00613-3 SILICOFLUORIDE COMPOUNDS,
            N.O.S.
            SILICOFLUORIDES, N.O.S. [NJL]

RR-00614-4 TELLURIUM COMPOUNDS, N.O.S.
            TELLURIUM AND COMPOUNDS
            [BC1, CEX, GBR]
            TELLURIUM COMPOUNDS [CAL,
            ONT, QBC]

RR-00615-5 TIN COMPOUNDS, N.O.S.
            TIN COMPOUNDS [CAL]

RR-00616-6 TUNGSTEN COMPOUNDS, N.O.S.
            TUNGSTEN AND COMPOUNDS [GBR,
            GBR]
            TUNGSTEN COMPOUNDS [CAL]

RR-00617-7 URANIUM COMPOUNDS, N.O.S.
            URANIUM COMPOUNDS [CAL, ONT,
            ONT]
            URANIUM COMPOUNDS, NATURAL
            [GBR, GBR]

RR-00618-8 YTTRIUM COMPOUNDS

RR-00619-9 K125-HAZARDOUS WASTES

RR-00620-2 AMYL LACTATE

RR-00621-3 ALUMINUM MAGNESIUM PHOS-
            PHIDE
            MAGNESIUM ALUMINUM PHOSPHIDE
            [HM1, HM1, NJL]

RR-00622-4 7-ETHYL-2-METHYL-4-HENDE-
            CANOL

RR-00623-5 N-ETHYLBENZYLTOLUIDINE, ALL
            ISOMERS
            N-ETHYLBENZYLTOLUIDINES [HM1,
            HM1]

RR-00624-6 ZIRCONIUM COMPOUNDS, N.O.S.
            ZIRCONIUM COMPOUNDS [CAL,
            CEX, CEX, GBR, GBR]

RR-00625-7 ARSENIC COMPOUNDS, N.O.S.
            ARSENIC AND ARSENIC COM-
            POUNDS [IAR]
            ARSENIC AND COMPOUNDS [GBR]
            ARSENIC COMPOUNDS [313, BC1,
            CAL, CN4, CN7, CW3, QBC]
            ARSENIC COMPOUNDS (INORGANIC
            INCLUDING ARSINE) [CAA]

RR-00626-8 TETRACHLORODIBENZO-P-DIOX-
            INS
            TCDDS (ALL TETRA-
            CHLORODIBENZO-P-DIOXINS) [UTS]

RR-00627-9 1,1-DIPHENYLPROPANE

RR-00628-0 DDT METABOLITES

RR-00629-1 2-CHLORO-4-TERT-AMYLPHENYL
            METHYL ETHER

RR-00630-4 LEAD COMPOUNDS
            LEAD AND COMPOUNDS [GBR]
            LEAD COMPOUNDS, N.O.S. [HM1,
            HM1, HM1, NJL, RCU]

RR-00631-5 D001-IGNITABLE UNLISTED HAZ-
            ARDOUS WASTES

RR-00632-6 D002-CORROSIVE UNLISTED HAZ-
            ARDOUS WASTES

RR-00633-7 D003-REACTIVE UNLISTED HAZ-
            ARDOUS WASTES

RR-00634-8 CHROMIUM COMPOUNDS
            CHROMIUM COMPOUNDS, N.O.S.
            [NJL, RCU]

RR-00635-9 F001-HAZARDOUS WASTES

RR-00636-0 F002-HAZARDOUS WASTES

RR-00637-1 F003-HAZARDOUS WASTES

RR-00638-2 F004-HAZARDOUS WASTES

RR-00639-3 F005-HAZARDOUS WASTES

RR-00640-6 F006-HAZARDOUS WASTES

RR-00641-7 F007-HAZARDOUS WASTES

RR-00642-8 F008-HAZARDOUS WASTES

RR-00643-9 F009-HAZARDOUS WASTES

RR-00644-0 F010-HAZARDOUS WASTES

RR-00645-1 F011-HAZARDOUS WASTES

RR-00646-2 F012-HAZARDOUS WASTES

RR-00647-3 F019-HAZARDOUS WASTES

RR-00648-4 F020-HAZARDOUS WASTES

RR-00649-5 F021-HAZARDOUS WASTES

RR-00650-8 F022-HAZARDOUS WASTES

RR-00651-9 F023-HAZARDOUS WASTES

RR-00652-0 F024-HAZARDOUS WASTES

RR-00653-1 F026-HAZARDOUS WASTES

RR-00654-2 F027-HAZARDOUS WASTES

RR-00655-3 F028-HAZARDOUS WASTES

RR-00656-4 K001-HAZARDOUS WASTES

RR-00657-5 K002-HAZARDOUS WASTES

RR-00658-6 K003-HAZARDOUS WASTES

RR-00659-7 K004-HAZARDOUS WASTES

RR-00660-0 K005-HAZARDOUS WASTES

RR-00661-1 K006-HAZARDOUS WASTES

RR-00662-2 K007-HAZARDOUS WASTES

RR-00663-3 K008-HAZARDOUS WASTES

RR-00664-4 K009-HAZARDOUS WASTES

RR-00665-5 K010-HAZARDOUS WASTES

RR-00666-6 K011-HAZARDOUS WASTES

RR-00667-7 K013-HAZARDOUS WASTES

RR-00668-8 K014-HAZARDOUS WASTES

RR-00669-9 K015-HAZARDOUS WASTES

RR-00670-2 K016-HAZARDOUS WASTES

RR-00671-3 K017-HAZARDOUS WASTES

RR-00672-4 K018-HAZARDOUS WASTES

RR-00673-5 K019-HAZARDOUS WASTES

RR-00674-6 K020-HAZARDOUS WASTES

RR-00675-7 K021-HAZARDOUS WASTES

RR-00676-8 K022-HAZARDOUS WASTES

RR-00677-9 K023-HAZARDOUS WASTES

RR-00678-0 K024-HAZARDOUS WASTES

RR-00679-1 K025-HAZARDOUS WASTES

RR-00680-4 K026-HAZARDOUS WASTES

RR-00681-5 K027-HAZARDOUS WASTES

RR-00682-6 K028-HAZARDOUS WASTES

RR-00683-7 K029-HAZARDOUS WASTES

RR-00684-8 K030-HAZARDOUS WASTES

RR-00685-9 K031-HAZARDOUS WASTES

RR-00686-0 K032-HAZARDOUS WASTES

RR-00687-1 K033-HAZARDOUS WASTES

RR-00688-2 K034-HAZARDOUS WASTES

RR-00689-3 K035-HAZARDOUS WASTES

RR-00690-6 K036-HAZARDOUS WASTES

RR-00691-7 K037-HAZARDOUS WASTES

RR-00692-8 K038-HAZARDOUS WASTES

RR-00693-9 K039-HAZARDOUS WASTES

RR-00694-0 K040-HAZARDOUS WASTES

RR-00695-1 K041-HAZARDOUS WASTES

RR-00696-2 K042-HAZARDOUS WASTES

RR-00697-3 K043-HAZARDOUS WASTES

RR-00698-4 K044-HAZARDOUS WASTES

RR-00699-5 K045-HAZARDOUS WASTES

RR-00700-1 K046-HAZARDOUS WASTES

RR-00701-2 K047-HAZARDOUS WASTES

RR-00702-3 K048-HAZARDOUS WASTES

RR-00703-4 K049-HAZARDOUS WASTES

RR-00704-5 K050-HAZARDOUS WASTES

RR-00705-6 K051-HAZARDOUS WASTES

RR-00706-7 K052-HAZARDOUS WASTES

RR-00707-8 K060-HAZARDOUS WASTES

RR-00708-9 K061-HAZARDOUS WASTES

RR-00709-0 K062-HAZARDOUS WASTES

RR-00710-3 K064-HAZARDOUS WASTES

RR-00711-4 K065-HAZARDOUS WASTES

RR-00712-5 K066-HAZARDOUS WASTES

RR-00713-6 K069-HAZARDOUS WASTES

RR-00714-7 K071-HAZARDOUS WASTES

RR-00715-8 K073-HAZARDOUS WASTES

RR-00716-9 K083-HAZARDOUS WASTES

RR-00717-0 K084-HAZARDOUS WASTES

RR-00718-1 K085-HAZARDOUS WASTES

RR-00719-2 K086-HAZARDOUS WASTES

RR-00720-5 K087-HAZARDOUS WASTES

RR-00721-6 K088-HAZARDOUS WASTES

RR-00722-7 K090-HAZARDOUS WASTES

RR-00723-8 K091-HAZARDOUS WASTES

RR-00724-9 K093-HAZARDOUS WASTES

RR-00725-0 K094-HAZARDOUS WASTES

RR-00726-1 K095-HAZARDOUS WASTES

RR-00727-2 K096-HAZARDOUS WASTES

RR-00728-3 K097-HAZARDOUS WASTES

RR-00729-4 K098-HAZARDOUS WASTES

RR-00730-7 K099-HAZARDOUS WASTES

RR-00731-8 K100-HAZARDOUS WASTES

RR-00732-9 K101-HAZARDOUS WASTES

RR-00733-0 K102-HAZARDOUS WASTES

RR-00734-1 K103-HAZARDOUS WASTES

RR-00735-2 K104-HAZARDOUS WASTES

RR-00856-0 ETHYLENE GLYCOL ETHYL-BUTYL ETHER

RR-00857-1 ETHYLENE GLYCOL ETHYL-HEXYL ETHER

RR-00858-2 ETHYLENE GLYCOL MONOMETHYL ETHER ACETAL

RR-00859-3 ETHYLENE GLYCOL MONOMETHYL ETHER FORMAL

RR-00860-6 FUEL OIL NO. 5

RR-00861-7 GASOLINE (CASINGHEAD)
CASINGHEAD GASOLINE [NJL]

RR-00862-8 GLUCOSE PENTAPROPIONATE

RR-00863-9 ISOBUTYLCYCLOHEXANE

RR-00864-0 ISODECALDEHYDE

RR-00865-1 ISOOCTYL NITRATE

RR-00866-2 ISOPROPYL BICYCLOHEXYL

RR-00867-3 METHYL DIHYDROABIETATE

RR-00868-4 METHYL HEPTADECYL KETONE

RR-00869-5 METHYL PENTADECYL KETONE

RR-00870-8 METHYLENE DIISOCYANATE

RR-00871-9 N,N-BIS(1-METHYLHEPTYL) ETHYLENEDIAMINE

RR-00872-0 N,N-DIBUTYLTOLUENESULFON-AMIDE

RR-00873-1 N,N-DIETHYLSTEARAMIDE

RR-00874-2 N-(2-PHENOXYETHYL) ANILINE

RR-00875-3 O-TOLYL P-TOLUENE SUL-FONATE

RR-00876-4 P-TERT-AMYLPHENYL BUTYL ETHER

RR-00877-5 PETROLEUM SULFONATE

RR-00878-6 PHENYL DI-O-XENYL PHOS-PHATE

RR-00879-7 PINE PITCH

RR-00880-0 POPPY SEED OIL

RR-00881-1 PROPYLCYCLOHEXANE

RR-00882-2 SEC-BUTYLCYCLOHEXANE

RR-00883-3 STRAW OIL

RR-00884-4 TERT-BUTYLDECALIN

RR-00885-5 TETRAHYDROFURFURYL OLEATE

RR-00886-6 TRIAMYLBENZENE

RR-00887-7 VINYL ISOOCTYL ETHER

RR-00888-8 K141-HAZARDOUS WASTES

RR-00889-9 K142-HAZARDOUS WASTES

RR-00890-2 K143-HAZARDOUS WASTES

RR-00891-3 K144-HAZARDOUS WASTES

RR-00892-4 K145-HAZARDOUS WASTES

RR-00893-5 K147-HAZARDOUS WASTES

RR-00894-6 K148-HAZARDOUS WASTES

RR-00895-7 K149-HAZARDOUS WASTES

RR-00896-8 K150-HAZARDOUS WASTES

RR-00897-9 K151-HAZARDOUS WASTES

RR-00898-0 SUBSTITUTED NITRILE

RR-00899-1 UNSATURATED AMINO ALKYL ESTER SALT

RR-00900-7 UNSATURATED AMINO ESTER SALT

RR-00901-8 PHOSPHORIC ACID, 1,2-ETHANEDIYL TETRAKIS(2-CHLORO-1-METHYLETHYL) ESTER

RR-00902-9 SUBSTITUTED DIALKYL OXA-ZOLONE

RR-00903-0 BENZENE, SUBSTITUTED, ALKYL ACRYLATE DERIVATIVE

RR-00904-1 METHYLENEBISTRISUBSTITUTED ANILINE

RR-00905-2 ACRYLAMIDE, POLYMER WITH SUBSTITUTED ALKYLACRY-LAMIDE SALT

RR-00906-3 DISULFONIC ACID ROSIN AMINE SALT OF A BENZIDINE DERIVA-TIVE

RR-00907-4 EPOXIDIZED POLYBUTENE

RR-00908-5 3,3',5,5'-TETRAMETHYL-BIPHENYL-4,4'-DIOL

RR-00909-6 2,5-DIMERCAPTO-1,3,4-THIADI-AZOLE, ALKYL POLYCARBOXY-LATE

RR-00910-9 2-PROPENAMIDE, N-[3-(DIMETHY-LAMINO)PROPYL]-

RR-00911-0 POLYMER OF SUBSTITUTED ALKYLPHENOL FORMALDEHYDE AND PHTHALICANHYDRIDE, ACRYLATE

RR-00912-1 TERPENES AND TERPENOIDS, LIMONENE FRACTION, POLY-MER WITH SUBSTITUTED CAR-BOPOLYCYCLES

RR-00913-2 BIS(SUBSTITUTED)CARBOMONO-CYCLIC AZOCARBOMONOCYLI-COL

RR-00914-3 FATTY ACID, AMINE SALT

RR-00915-4 2-OXEPANONE, POLYMER WITH 4,4'-(1-METHYLETHYLI-DENE)BISPHENOLAND 2,2-[(1-METHYLETHYLIDENE)BIS(4,1-PHENYLENEOXYMETHYLENE)] BISOXIRANE, GRAFT

RR-00916-5 POLYALKYLENEPOLYOL ALKY-LAMINE

RR-00917-6 METAL SALT OF A COMPLEX INORGANIC OXYACID

RR-00918-7 N,N'-1,2-ETHANEDIYLBIS-, POLY-MER WITH 2,4,6-TRICHLORO-1, 3,5-TRIAZINE, REACTION PROD-UCTS WITH N-BUTYL-2,2,6,6-TE-TRAMETHYL-4-PIPERIDINAMINE

RR-00919-8 2-CHLORO-N-METHYL-N-SUBSTI-TUTED ACETAMIDE

RR-00920-1 SUBSTITUTED ALIPHATIC ACID HALIDE

RR-00921-2 DISUBSTITUTED NITROBENZENE

RR-00922-3 SUBSTITUTED AMINOBENZOIC ACID ESTER

RR-00923-4 ALKYL ESTER

RR-00924-5 TERT-AMYL PEROXY ALKYLENE ESTER

RR-00925-6 ALKYL (HETEROCYCLICYL) PHENYLAZOHETERO MONO-CYCLIC POLYONE

RR-00926-7 SUBSTITUTED TRIAZINE ISOCYA-NURATE

RR-00927-8 ALKYL (HETEROCYCLICYL) PHENYLAZOHETERO MONO-CYCLIC POLYONE, [(ALKYLIMI-DAZOLYL)METHYL] DERIVATIVE

RR-00928-9 MODIFIED ACRYLIC ESTER

RR-00929-0 HALONITROBENZOIC ACID, SUB-STITUTED

RR-00930-3 ALKYL-BISOXYALKYL (SUBSTITUTED-1,1-DIMETHYLETHYLPHENYL) BEN-ZOTRIAZOLE

RR-00931-4 METALATED ALKYLPHENOL COPOLYMER

RR-00932-5 NITROPHENOXYALKANOIC ACID SUBSTITUTED THIAZINO HY-DRAZIDE

RR-00933-6 POLY(SUBSTITUTED TRIAZINYL) PIPERAZINE

RR-00934-7 4,4'-(9H-FLUOREN-9-YLIDENE)BIS

RR-00935-8 DIGLYCIDYL ETHER OF DISUB-STITUTED CARBOPOLYCYCLE

RR-00936-9 NITRATE POLYETHER POLYOL

RR-00937-0 SULPHUR-BRIDGED SUBSTI-TUTED PHENOLS

RR-00938-1 ALKANOIC ACID, BUTANEDIOL AND CYCLOHEXANEALKANOL POLYMER

RR-00939-2 DISTILLATES (PETROLEUM), C (3-6), POLYMERS WITH STYRENE AND MIXED TERPENES

RR-00940-5 PHOSPHORYLATED OXO-HETEROMONOCYCLE POLY-OXYETHYLENE ALKYL ETHER

RR-00941-6 PHOSPHORYLATED CAPRO-LACTONE, ALKYLOXO-HETEROMONO-CYCLE AND POLYALKYLENE POLYOL ALKYL ETHER

RR-00942-7 SUBSTITUTED ALKYL HALIDE

RR-00943-8 PERHALO ALKOXY ETHER

RR-00944-9 SUBSTITUTED HYDROXYLAMINE

RR-00945-0 ALKYL PEROXY-2-ETHYL HEX-ANOATE

RR-00946-1 ALKYL ALKENOATE, AZOBIS-

RR-00947-2 BISPHENOL A, EPICHLOROHY-DRIN, POLYALKYLENEPOLYOL AND POLYISOCYANATO DERIVA-TIVE

RR-00948-3 REACTION PRODUCT OF ALKA-NEDIOL AND EPICHLOROHYDRIN

RR-00949-4 POLYMER OF POLYETHYLENEPOLYAMINE AND ALKANEDIOL DIGLYCIDYL ETHER

RR-00950-7 SILANE, (1,1-DIMETHYLETHOXY) DIMETHOXY (2-METHYLPROPYL)

RR-00952-9 SUBSTITUTED TRIPHENYL-METHANE

RR-00953-0 BIS(2,2,6,6-TETRAM-ETHYLPIPERIDINYL) ESTER OF CYCLOALKYL SPIROKETAL

RR-00954-1 TRIS(DISUBSTITUTED ALKYL) HETEROCYCLE

RR-00955-2 ISOCYANATE TERMINATED POLYOLS

RR-01032-2 ALKYL(C8-10) POLYETHOXY-POLYPROPOXYBENZENE ETHER

RR-01033-3 ALPHA-ALKYL(C8-C14)-OMEGA-HYDROXYPOLY(OXYPROPY-LENE) BLOCK COPOLYMER WITH POLYOXYETHYLENE

RR-01034-4 ALMOND HULLS

RR-01035-5 ALUMINUM - MAGNESIUM STEARATE

RR-01036-6 AMMONIUM DODECYL ALCO-HOL POLYOXYETHYLENE PHOS-PHATE

RR-01037-7 APPLE POMACE

RR-01038-8 BANCROFT CLAY

RR-01039-9 BARIUM ZINC METHYLBEN-ZOATE-2-ETHYLHEXANOATE

RR-01040-2 BENZENESULFONIC ACID, C8-24-ALKYL DERIVATIVES

RR-01041-3 BENZENESULFONIC ACID, C8-24-ALKYL DERIVATIVES, AMMO-NIUM, MAGNESIUM, POTASSIUM OR ZINC SALT

RR-01042-4 BORAX PENTAHYDRATE

RR-01043-5 BRAN

RR-01044-6 BREAD CRUMBS

RR-01045-7 BUTOXYTRIETHYLENE GLYCOL PHOSPHATE

RR-01046-8 CALCIUM ALKYL(C8-24)BEN-ZENESULFONATE

RR-01047-9 CALCIUM AND SODIUM SALTS OF SUGAR DERIVED ACIDS

RR-01048-0 CANARY SEED

RR-01049-1 CARDBOARD

RR-01050-4 CARROTS

RR-01051-5 CASTOR OIL, MALEIC ANHY-DRIDE, AND POLYETHYLENE GLYCOL COPOLYMER

RR-01052-6 CAT FOOD

RR-01053-7 CHEESE

RR-01054-8 COCO SHELL FLOUR

RR-01055-9 COCOAMINO PROPIONIC ACID

RR-01056-0 COCOAMPHOCARBOXYGLYCI-NATE, DISODIUM SALT

RR-01057-1 COCONUT OIL, POLYMER WITH ISOPHTHALIC ACID, TRIMELIT-TIC ANHYDRIDE AND TRIMETHY-LOLPROPANE

RR-01058-2 COFFEE GROUNDS
COFFEE [IAR]

RR-01059-3 CONDENSATION PRODUCT OF SORBITOL EPICHLOROHYDRIN WITH THE OLEIC ACID DIAMIDE OF DIETHYLENETRIAMINE

RR-01060-6 CORN

RR-01061-7 CRACKED OATS AND WHEAT

RR-01062-8 CROSS LINKED POLYMER OF SEBACRYL CHLORIDE & POLY-METHYLENEPOLYPHENYL ISO-CYANATE, WITH ETHYLENEDI-AMINE AND DIETHYLENETRIA MINE

RR-01063-9 DI ACETYL TARTARIC ESTERS OF MONO AND DIGLYCERIDES OF EDIBLE FATTY ACIDS

RR-01064-0 DIACETYL TARTARIC ACID ES-TERS OF MONO AND DIGLYC-ERIDES OF EDIBLE FATS

RR-01065-1 DICYCLOPENTADIENE, POLY-MER WITH (MIXED STYRENE AND ALPHA-METHYLSTYRENE), (MIXED INDENE AND METHYL INDENE), AND VINYL TOLUENE

RR-01066-2 DIMETHYL DIALKYL AMMONIUM CHLORIDE POWDER

RR-01067-3 O,P-DINONYLPHENOL, ETHOXY-LATED, PHOSPHATED, AMMO-NIUM SALT

RR-01068-4 O,P-DINONYLPHENOL, ETHOXY-LATED, PHOSPHATED, CALCIUM SALT

RR-01069-5 O,P-DINONYLPHENOL, ETHOXY-LATED, PHOSPHATED, ZINC SALT

RR-01070-8 DIPROPYLENETRIAMINE AMIDE OF TALL-OIL FATTY ACID RE-ACTED WITH TALL-OIL FATTY ACID-POLYETHYLENE GLYCOL ESTER AND MALEIC ANHYDRIDE

RR-01071-9 DISODIUM DODECYLIMIDA-ZOLINIUM DICARBOXYLATE

RR-01072-0 DODECYL DIMETHYL BENZYL AMMONIUM NAPHTHENATE

RR-01073-1 DODECYLBENZENESULFONIC ACID, 1,1,2,3-TETRAMETHYL-BUTYLAMINE SALT

RR-01074-2 DOUGLAS FIR BARK

RR-01075-3 EPOXIDIZED TALL-OIL FATTY ACIDS WITH TALL-OIL ROSIN

RR-01076-4 ETHOXYLATED SORBITAN POLYSORBATE

RR-01077-5 ETHOXYLATED SORBITOL PEN-TAESTER OF TALL-OIL ACIDS

RR-01078-6 FATTY ACIDS, COCO, REAC-TION PRODUCTS WITH 2-[(2-AMINOETHYL)AMINO] ETHANOL, ALKYLATION PRODUCTS WITH METHYL ACRYLATE, SODIUM SALTS

RR-01079-7 FATTY ACIDS, DEHYDRATED CASTOR-OIL, POLYMERS WITH MALEIC ANHYDRIDE, TRI-ETHANOLAMINE SALTS

RR-01080-0 FERTILIZER

RR-01081-1 FISH MEAL

RR-01082-2 FLOUR

RR-01083-3 FRIANITE

RR-01084-4 GLYCINE OF TALL-OIL FATTY ACIDS

RR-01085-5 GRAPES (POMACE)

RR-01086-6 GRAVEL

RR-01087-7 2-IMIDAZOLINIUM, 1-(CAR-BOXYMETHYL)-1-(2-HYDROX-YETHYL)-2-COCOYLNOR, CHLO-RIDE, MONOSODIUM SALT

RR-01088-8 IMIDAZOLIUM COMPOUNDS, 2-(C16-18 ALKYL)-1-(2-(C16-18 AMIDO)ETHYL)-4,5-DIHYDRO-1-METHYL, METHYL SULFATES

RR-01089-9 IMIDAZOLIUM COMPOUNDS, 2-HEPTADECYL-4,5-DIHYDRO-1-METHYL-1-(2-TALLOW AMI-DOETHYL), METHYL SULFATES

RR-01090-2 ISOPROPYLAMINE SALT OF OLEOYLISOPROPANOLAMIDE DERIVATIVE OFSULFOSUCCINIC ACID

RR-01091-3 ISOPROPYLAMINE SALT OF STEARYLISOPROPANOLAMIDE DERIVATIVE OF SULFOSUCCINIC ACID

RR-01092-4 4,4'-ISOPROPYLIDENE DIPHENOL ALKYL (C12-15) PHOSPHITES

RR-01093-5 LIME GREEN PIGMENT

RR-01094-6 MANURE

RR-01095-7 MILLET SEED

RR-01096-8 MILO

RR-01097-9 MIXED FATTY AND ROSIN ACIDS

RR-01098-0 MONOETHANOLAMINE ALKYL (C8-C24) BENZENE SULFONATE

RR-01099-1 NAPHTHALENESULFONIC ACID/PARAMETALDEHYDE CON-DENSATE, AMMONIUM SALT

RR-01100-7 NAPHTHENIC ACID SOAP OF N-ALKYL (C16-C18)TRIMETHYLENE-DIAMINE

RR-01101-8 NICKEL COMPLEX OF DIMETHY-LAMINO METHYL OCTYL PHE-NOL

RR-01102-9 NITROGEN FIXING BACTERIA

RR-01103-0 P-NONYLPHENOL, ETHOXY-LATED, PHOSPHATED, CALCIUM SALT

RR-01104-1 OATS AND OATMEAL

RR-01105-2 OLEIC ACID ESTER OF TETRA (HYDROXYETHYL)ETHYLENEDI-AMINE

RR-01106-3 OLIGOESTER DERIVFED BY CON-DENSATION OF ADIPIC ACID, PH-THALIC ANHYDRIDE, ETHYLENE GLYCOL, N-OCTYL ALCOHOL AND N-DECYL ALCOHOL

RR-01107-4 OYSTER SHELLS

RR-01108-5 PAPER

RR-01109-6 PEANUT BUTTER

RR-01110-9 PEANUT SHELLS

RR-01111-0 PEAT MOSS

RR-01112-1 PECAN SHELL FLOUR

RR-01113-2 POE SORBITAN DITALLATE

RR-01114-3 POLY(OXY(METHYL-1,2-ETHANEDIYL)), ALPHA-(2-(DI-ETHYLMETHYLAMMONIO) ETHYL-OMEGA-HYDROXY-, CHLORIDE

RR-01115-4 POLY-OXO ALUMINUM STEARATE

RR-01116-5 POLYMER OF N-BUTYL ACRY-LATE, METHYL METHACRY-LATE, METHACRYLIC ACID AND AMINOPROPYL METHACRYLATE

RR-01117-6 POLYMERIZED FATTY ACID PHOSPHATE (100% C12-C18)

RR-01118-7 POLYOXYALKYLENE GLYCOL

RR-01119-8 POLYOXYALKYLENE SILOXANE

RR-01120-1 POLYOXYETHYLENE AMYLPHE-NOL - FORMALDEHYDE RESIN

RR-01225-9 FATTY ACID AMINE CONDEN-
SATE, POLYCARBOXYLIC ACID
SALTS

RR-01226-0 COCO ACID TRIAMINE CONDEN-
SATE, POLYCARBOXYLIC ACID
SALTS

RR-01227-1 2-BUTENEDIOIC ACID (Z), MONO
(2-((1-OXO-PROPENYLOXY)
ETHYL)ESTER

RR-01228-2 ISOPROPYLIDENE, BIS(1,1-
DIMETHYLPROPYL) DERIVATIVE

RR-01229-3 ALKANAMINIUM, POLYALKYL-
[(2-METHYL-1-OXO-2-PROPENYL)
OXY]SALT, POLYMER WITH
ACRYLAMIDE AND SUBSTITUTED
ALKYL METHACRYLATE

RR-01230-6 1-PROPANOL, 3,3'-OXYBIS[2,3-BIS
(BROMOMETHYL)]-

RR-01231-7 PERFLUOROALKYL AROMATIC
CARBAMATE MODIFIED ALKYL
METHACRYLATE COPOLYMER

RR-01232-8 CARBOPOLYCYCLICOL
AZOALKYLAMINOALKYLCAR-
BOMONOCYCLIC ESTER, HALO-
GEN ACID SALTS

RR-01233-9 ALKYLBENZENESULFONIC ACIDS
AND SODIUM SALTS

RR-01234-0 ALPHA-OLEFIN SULFONATE,
SODIUM SALTS

RR-01235-1 FORMALDEHYDE, CONDEN-
SATED POLYOXYETHYLENE
FATTY ACID, ESTER WITH STYRE-
NATED PHENOL, ETHYLENE OX-
IDE ADDUCT

RR-01236-2 FATTY ACID, ESTER WITH
STYRENATED PHENOL, ETHY-
LENE OXIDE ADDUCT

RR-01237-3 BENZENAMINE, 4-(1-METHYLBU-
TOXY)-, HYDROCHLORIDE

RR-01238-4 BENZENE, 1-(1-METHYLBUTOXY)-
4-NITRO-,

RR-01239-5 DIMETRIDAZOLE

RR-01240-8 POLYMETHYLCARBOMONOCY-
CLE, REACTION PRODUCT WITH
2-HYDROXYETHYL ACRYLATE

RR-01241-9 ETHOXYBENZOTHIAZOLE DISUL-
FIDE

RR-01242-0 ETHANEDIMIDIC ACID

RR-01243-1 SUBSTITUTED TRIAZOLE

RR-01244-2 SULFONAMIDE

RR-01245-3 DIALKYLDITHIOPHOSPHORIC
ACID, ALIPHATIC AMINE SALT

RR-01246-4 HALOGENATED BIPHENYL GLY-
CIDYL ETHERS

RR-01247-5 FLUORENE SUBSTITUTED ARO-
MATIC AMINE

RR-01248-6 ALKYLCARBAMIC ACID,
ALKYNYL ESTER

RR-01249-7 REACTION PRODUCTS OF SUB-
STITUTED HYDROXYLALKA-
NES AND POLYALKYLPOLYISO-
CYANATOCARBOMONOCYCLE

RR-01250-0 TALL OIL FATTY ACIDS, RE-
ACTION PRODUCTS WITH
POLYAMINES, ALKYL SUBSTI-
TUTED

RR-01251-1 BUTYL ACRYLATE, POLYMER
WITH SUBSTITUTED METHYL
STYRENE, METHACRYLATE, AND
SUBSTITUTED SILANE

RR-01252-2 ZIRCONIUM[IV], [2,2-BIS[[2-
PROPENYLOXY]METHYL]-
1-BUTANOLATO-1,2]TRIS(2-
PROPENOATO-O-)

RR-01253-3 BROMINATED TRIAZINE DERIVA-
TIVE

RR-01254-4 POLYMER OF ALKANEDIOIC
ACID, METHYLENEBISCAR-
BOMONOCYCLIC DIISOCYANATE,
AND ALKYLENE GLYCOLS, HY-
DROXYALKYL ACRYLATE ESTER

RR-01255-5 EPOXIDIZED COPOLYMER OF
PHENOL AND SUBSTITUTED PHE-
NOL

RR-01256-6 RECOVERED METAL HYDROXIDE

RR-01257-7 POLYMER OF DISODIUM
MALEATE, ALLYL ETHER, AND
ETHYLENE OXIDE

RR-01258-8 ACRYLATE SUBSTITUTED SILOX-
ANES AND SILICONES

RR-01259-9 REACTION PRODUCTS OF PHE-
NOLIC PENTAERYTHRITOL TE-
TRAESTER WITH FATTY ACID
OILS AND ESTERS, AND GLYC-
ERIDE TRIESTERS

RR-01260-2 PENTAERYTHRITOL, MIXED ES-
TERS WITH CARBOXYLIC ACIDS

RR-01261-3 POLYALKYLENE GLYCOL SUB-
STITUTED ACETATE

RR-01262-4 MODIFIED HYDROCARBON RESIN

RR-01263-5 REACTION PRODUCT OF
ETHOXYLATED FATTY ACID OILS
AND A PHENOLIC PENTAERY-
THRITOL TETRAESTER

RR-01264-6 ALKALI METAL NITRITES

RR-01271-5 ALCOHOL C-13 - C-15 POLY(1-3)
ETHOXYLATE

RR-01272-6 ALCOHOL C-6 - C-17 (SEC-
ONDARY)POLY(3-6) ETHOXYLATE

RR-01273-7 BIPHENYL PHENYL ETHER AND
DIPHENYL OXIDE, MIXTURES

RR-01274-8 BROMOALLYLENE

RR-01275-9 CALCIUM ARSENATE AND CAL-
CIUM ARSENITE, MIXTURES,
SOLID

RR-01276-0 CHLOROPHENATES
CHLOROPHENATES, N.O.S. [HM1,
HM1, HM1]

RR-01277-1 ALPHA-CHLOROPROPYLENE

RR-01278-2 DICHLOROANILINES

RR-01279-3 DICHLOROETHER

RR-01280-6 DICHLOROETHYL OXIDE

RR-01281-7 DICHLOROPHENOLS

RR-01282-8 2,4-DICHLOROPHENOXYACETIC
ACID DIETHANOLAMINE SALT

RR-01283-9 2,4-DICHLOROPHENOXYACETIC
ACID DIMETHYLAMINE SALT

RR-01284-0 2,4-DICHLOROPHENOXYACETIC
ACID TRIISOPROPYLAMINE SALT

RR-01285-1 DIMETHYLPHENOLS

RR-01286-2 DIPHENYL OXIDE AND BIPHENYL
PHENYL ETHER MIXTURES

RR-01287-3 ETHYL FLUID

RR-01288-4 ETHYLENE DIBROMIDE AND
METHYL BROMIDE MIXTURES,
LIQUID

RR-01289-5 HYDROXYDIMETHYLBENZENES

RR-01290-8 ISOFENPHOS

RR-01291-9 ISOXATHION

RR-01292-0 LEAD ARSENITES

RR-01293-1 MERCARBAM

RR-01294-2 MERCURY BISULPHATES

RR-01295-3 MERCURIC GLUCONATE

RR-01296-4 MERCUROUS BISULPHATE

RR-01297-5 MERCURY ACETATES

RR-01298-6 MERCURY POTASSIUM CYANIDE

RR-01299-7 MERCURY SULFATES

RR-01300-3 METHYL BROMIDE AND ETHY-
LENE DIBROMIDE MIXTURES,
LIQUID

RR-01301-4 METHYLCHLOROBENZENES

RR-01302-5 METHYLNITROPHENOLS

RR-01303-6 METHYLVINYLBENZENES

RR-01304-7 NAPHTHALENE

RR-01305-8 NITROBENZOTRIFLUORIDES

RR-01306-9 ORTHOARSENIC ACID

RR-01307-0 PENTANETHIOLS

RR-01308-1 PHENTHOATE

RR-01309-2 PIRIMIPHOS-ETHYL

RR-01310-5 POLYHALOGENATED BIPHENYLS

RR-01311-6 PROPYLIDENE DICHLORIDE

RR-01312-7 SILVER ORTHOARSENITE

RR-01313-8 SODIUM COPPER CYANIDE

RR-01314-9 THIOCARBONYL TETRACHLO-
RIDE

RR-01315-0 TRIARYL PHOSPHATES, ISO-
PROPYLATED

RR-01316-1 TRIARYL PHOSPHATES, N.O.S.

RR-01317-2 TRICHLOROBUTYLENE

RR-01318-3 CHLOROBENZYLCHLORIDES

RR-01319-4 5-ETHYL-2-PICOLINE

RR-01322-9 ADHESIVES

RR-01323-0 AEROSOLS

RR-01325-2 AIR

RR-01327-4 GUANYL
NITROSOAMINOGUANYLIDENE
HYDRAZINE

RR-01328-5 GUANYL
NITROSOAMINOGUANYLTE-
TRAZENE

RR-01329-6 HEXOLITE

RR-01330-9 OCTOLITE

RR-01331-0 TRITONAL

RR-01332-1 5-MERCAPTOTETRAZOL-1-
ACETIC ACID

RR-01333-2 DINITROGLYCOLURIL

RR-01334-3 NITROTRIAZOLONE

RR-01335-4 OCTYL ALDEHYDES

RR-01336-5 PHOSPHORUS, AMORPHOUS

RR-01337-6 POTASSIUM, METAL ALLOYS

RR-01466-4 TRINITROACETIC ACID

RR-01467-5 TRINITROAMINE COBALT

RR-01468-6 TRINITROETHYLNITRATE

RR-01469-7 TRINITROTETRAMINE COBALT NITRATE

RR-01470-0 VINYL NITRATE POLYMER

RR-01471-1 AMINOGLYCOSIDES

RR-01472-2 ANGIOTENSIN CONVERTING ENZYME (ACE)
ANGIOTENSIN CONVERTING ENZYME (ACE) INHIBITORS [C65]

RR-01473-3 BARBITURATES

RR-01474-4 BENZODIAZEPINES

RR-01475-5 TETRACYCLINES

RR-01476-6 DIOXINS, POLYHALOGENATED DIBENZO-P-

RR-01477-7 FURANS, POLYHALOGENATED DIBENZO-

RR-01500-9 LACQUER

RR-01501-0 MINERAL FIBERS
MINERAL FIBRES [CN4]

RR-01502-1 MINING REAGENT, LIQUID

RR-01503-2 PAPER, UNSATURATED OIL TREATED, INCOMPLETELY DRY

RR-01504-3 POLISH

RR-01505-4 RARE GASES, MIXTURES

RR-01506-5 REDUCING LIQUID

RR-01507-6 RESIN FAST BLACK WP

RR-01508-7 TOE PUFFS, NITROCELLULOSE BASE

RR-01509-8 WASTE OIL

RR-01510-1 WATER REACTIVE SOLID, N.O.S.

RR-01511-2 WAX, LIQUID

RR-01512-3 WOOD FILLER, LIQUID

RR-01513-4 WOOD PRESERVATIVE, LIQUID
WOOD PRESERVATIVES (CONTAINING ARSENIC AND CHROMATE) [CAB]

RR-01514-5 AEROSOL DISPENSERS

RR-01516-7 DI-ISOTRIDECYL PEROXYDICARBONATE

RR-01517-8 LATEX, PAINTS

RR-01518-9 PAINT

RR-01519-0 PERFUMERY PRODUCTS, WITH FLAMMABLE SOLVENT

RR-01520-3 PICRITE

RR-01521-4 TEAR GAS
TEAR GAS (IRRITATING SUBSTANCES, LIQUID OR SOLID) N.O.S. [NJL]

RR-01522-5 TERT-OCTYL HYDROPEROXIDE

RR-01523-6 POLYCYCLIC AROMATIC HYDROCARBONS
POLYCYCLIC AROMATIC HYDROCARBONS (BENZENE-SOLUBLE FRACTION) [MEX, MEX]

RR-01524-7 ACRYLIC FIBRES

RR-01525-8 ANAESTHETICS, VOLATILE

RR-01526-9 ANGELICIN PLUS ULTRAVIOLET A RADIATION

RR-01527-0 BENZO[C]FLUORENE

RR-01528-1 3-CARBETHOXYPSORALEN

RR-01529-2 CARRAGEENAN, NATIVE

RR-01530-5 CHLORDANE/HEPTACHLOR

RR-01531-6 CHLORINATED DIBENZODIOXINS

RR-01532-7 PARA-DIMETHYLAMINOAZOBENZENEDIAZO SODIUM SULPHONATE

RR-01533-8 4,5'-DIMETHYLANGELICIN PLUS ULTRAVIOLET A RADIATION

RR-01534-9 IRON SORBITOL-CITRIC ACID COMPLEX

RR-01535-0 5-METHYLANGELICIN PLUS ULTRAVIOLET A RADIATION

RR-01536-1 7-METHYLPYRIDO[3,4-C]PSORALEN

RR-01537-2 MODACRYLIC FIBRES

RR-01538-3 POLYVINYL ACETATE

RR-01539-4 PROFLAVINE SALTS

RR-01540-7 PYRIDO[3,4-C]PSORALEN

RR-01541-8 TRICHLOROTRIETHYLAMINE HYDROCHLORIDE

RR-01542-9 T2-TRICHOTHECENE

RR-01543-0 N-NITROSOPROLINE

RR-01544-1 OESTROGEN-PROGESTIN REPLACEMENT THERAPY

RR-01545-2 GLASS FILAMENTS

RR-01546-3 PETROLEUM SOLVENTS

RR-01547-4 HYPOCHLORITE SALTS

RR-01548-5 TETRAKIS(HYDROXYMETHYL) PHOSPHONIUM SALTS

RR-01549-6 JET FUEL

RR-01550-9 DISTILLATE (LIGHT) FUELS

RR-01551-0 METABISULFITES

RR-01552-1 SULFITES

RR-01553-2 DIISOPROPYL SULFATE

RR-01554-3 CHROMATES
CHROMATES (WATER INSOLUBLE CHROMIUM (IV) COMPOUNDS [ALB] CHROMIUM VI COMPOUNDS, CERTAIN INSOLUBLE SALTS [QBC]

RR-01555-4 CHLOROBIPHENYLS

RR-01556-5 OXIDANTS (OZONE)

RR-01557-6 SUSPENDED PARTICULATE MATTER

RR-01558-7 SUBSTITUTED ETHYL ALKENAMIDE

RR-01559-8 POLYALKYLENE POLYAMINE

RR-01560-1 1H-BENZOTRIAZOLE, 5-(PENTYLOXY)-, SODIUM SALT

RR-01561-2 1H-BENZOTRIAZOLE, 5-(PENTYLOXY)-, POTASSIUM SALT

RR-01562-3 TRIETHANOLAMINE SALTS OF FATTY ACIDS

RR-01563-4 TRISUBSTITUTED HYDROQUINONE DIESTER

RR-01564-5 1,3,5-TRIAZIN-2-AMINE, 4-DIMETHYLAMINO-6-SUBSTITUTED-

RR-01565-6 POLYAMINOPOLYACID

RR-01566-7 BISPHENOL DERIVATIVE

RR-01567-8 QUATERNARY AMMONIUM SALT OF FLUORINATED ALKYLARYL AMIDE

RR-01568-9 OXOSUBSTITUTED AMINOALKANOIC ACID DERIVATIVE

RR-01569-0 SUBSTITUTED DIACRYLATE

RR-01570-3 ALKYLATED DIPHENYLS

RR-01571-4 DIMETHYL-3-SUBSTITUTED HETEROMONOCYCLE

RR-01572-5 DIMETHYL-3-SUBSTITUTED HETEROMONOCYCLE AMINE

RR-01573-6 METHYLPOLYCHLORO ALIPHATIC KETONE

RR-01574-7 POLYAMINE DITHIOCARBAMATE

RR-01575-8 ACRYLATES OF ALIPHATIC POLYOL

RR-01576-9 BIS(1,2,2,6,6-PENTAMETHYL-4-PIPERIDIN-4-OL) ESTER OF CYCLOALIPHATIC SPIROKETAL

RR-01577-0 TERT-BUTYLARSINE

RR-01600-2 WASTE CRANKCASE OILS

RR-01601-3 CREOSOTE IMPREGNATED WASTE MATERIALS

RR-01602-4 FUEL CONTAINING TOXIC SUBSTANCES THAT ARE DANGEROUS GOODS

RR-01603-5 ORGANOHALOGEN COMPOUNDS

RR-01604-6 PERSISTENT SYNTHETIC MATERIALS (INCLUDING PERSISTENT PLASTICS)

RR-01605-7 PETROLEUM DISTILLATE RESIDUES

RR-01606-8 CRUDE OIL WASTES

RR-01607-9 BIOLOGICAL WARFARE SUBSTANCES

RR-01608-0 CHEMICAL WARFARE SUBSTANCES

RR-01609-1 ORGANOSILICON COMPOUNDS

RR-01610-4 2,3,4,5-BIS(2-BUTYLENE)TETRAHYDRO-2-FURFURAL

RR-01611-5 POLYCHLORINATED TERPHENYLS

RR-01613-7 CHLORINATED WASTEWATER EFFLUENTS

RR-01614-8 ALDRIN AND DIELDRIN

RR-01615-9 TRIHALOMETHANES

RR-01620-6 DICHROMATES

RR-01621-7 MOLYBDATES

RR-01622-8 (2,4,5-TRICHLOROPHENOXY)-ACETIC ACID ESTERS

RR-01623-9 TUNGSTIC ACID SALTS

RR-01624-0 CHLORINATED DIPHENYL OXIDES

RR-01632-0 3'-AZIDO-3'-DEOXYTHYMIDINE/2',3'-DIDEOXYCYTIDINE (AIDS INITIATIVE)

RR-01633-1 3'-AZIDO-3'-DEOXYTHYMIDINE + 2',3'-DIDEOXYINOSINE (AIDS INITIATIVE)

RR-01634-2 HALOGENATED ETHANES CLASS STUDY

RR-01636-4 HALOGENATED ETHANES

RR-01637-5 PHOSPHOTUNGSTIC ACIDS

RR-01638-6 CADMIUM OXIDE PRODUCTION
CADMIUM OXIDE PRODUCTION (MEASURED AS CADMIUM) [MEX]

RR-01639-7 ANTIMONY TRIOXIDE PRODUCTION

RR-01640-0 TEFLON DECOMPOSITION PRODUCTS

RR-01641-1 HALOGENATED PHOSPHATE ESTERS

RR-01642-2 AMINOETHYLETHYLENE UREA METHACRYLAMIDE

RR-01643-3 PROPANOIC ACID, 2,2-DIMETHYL-, ETHENYL ESTER

RR-01644-4 HEXANEDIOIC ACID, DIETHENYL ESTER

RR-01645-5 SUBSTITUTED ACRYLAMIDE

RR-01646-6 2-PROPENOIC ACID, 2-[[(1-METHYLETHOXY) CARBONYL] AMINO]ETHYLESTER

RR-01647-7 2,5,8,10,13-PENTAOXAHEXADEC-15-ENOIC ACID, 9,14-DIOXO-2-[(1-OXO-2-PROPENYL)OXY]ETHYL ESTER

RR-01648-8 HEXANOIC ACID, 2-ETHYL-, ETHENYL ESTER

RR-01649-9 SUBSTITUTED DICHLOROBEN-ZOTHIAZOLES

RR-01650-2 ACETAMIDE, N-[4-(PENTYLOXY) PHENYL]- ACETAMIDE, N-[2-NITRO-4-(PENTYLOXY)PHENYL]-, AND ACETAMIDE, N-[2-AMINO-4-(PENTYLOXY)PHENYL]

RR-01651-3 NEONONANOIC ACID, ETHENYL ESTER

RR-01652-4 ALKYLPHENOXYPOLY (OXYETHYLENE SULFURIC ACID ESTER, SUBSTITUTED AMINE SALT

RR-01653-5 1,2-CYCLOHEXANEDICAR-BOXYLIC ACID, 2,2-BIS[[[2-[(OXI-RANYLMETHOXY)CARBONYL] CYCLOHEXYLCARBONYL]OXY] METHYL]-1,3-PROPANEDIYL BIS (OXIRANYLMETHYL) ESTER

RR-01654-6 PHENYL(DISUBSTITUTEDPOLY-CYCLIC)

RR-01655-7 DIPEROXY KETAL

RR-01656-8 MANNICH-BASED ADDUCT

RR-01657-9 POLYSUBSTITUTED PHENYLA-ZOPOLYSUBSTITUTED PHENYL DYE

RR-01658-0 PHOSPHONOCARBOXYLATE SALTS

RR-01659-1 ALKYL PHOSPHONATE AMMO-NIUM SALTS

RR-01660-4 AROMATIC SULFONIC ACID COMPOUND WITH AMINE

RR-01661-5 ORGANOTIN CATALYSTS

RR-01662-6 SILICONE ESTER POLYACRY-LATE

RR-01663-7 BENZOIC ACID, 2-(3-PHENYL-BUTYLIDENE)AMINO-, METHYL ESTER

RR-01664-8 CAPPED ALIPHATIC ISO-CYANATE

RR-01665-9 ETHOXYLATED SUBSTITUTED NAPHTHOL

RR-01666-0 PHOSPHORAMIDE

RR-01667-1 HALOGENATED PHENYL ALKANE

RR-01668-2 AROMATIC AMINO ETHER

RR-01669-3 ACRYLATED EPOXY PHENOLIC RESIN

RR-01671-7 SUBSTITUTED PHENYL AZO SUB-STITUED BENZENEDIAZONIUM SALT

RR-01672-8 ACRYLAMIDE-SUBSTITUTED EPOXY

RR-01673-9 NICKEL SALT OF AN ORGANO COMPOUND CONTAINING NI-TROGEN

RR-01674-0 BUTOXY-SUBSTITUTED ETHER ALKANE

RR-01675-1 DISUBSTITUTED DIPHENYLSUL-FONE

RR-01676-2 SUBSTITUTED NITROBENZENE

RR-01677-3 POLYOXYALKYLENE SUBSTI-TUTED AROMATIC AZO COL-ORANT

RR-01679-5 CYCLIC PHOSPHAZENE, METHACRYLATE DERIVATIVE

RR-01680-8 ALIPHATIC DIFUNCTIONAL ACRYLIC ACID ESTER

RR-01681-9 ALIPHATIC DICARBOXYLIC ACID SALT

RR-01682-0 TETRAMETHYLAMMONIUM SALTS OF ALKYLBENZENESUL-FONIC ACID

RR-01683-1 2-PROPENOIC ACID, REAC-TION PRODUCT WITH 2-OXE-PROPANONE AND ALKYL TRIOL

RR-01684-2 2-PROPENOIC ACID, DOCOSYL ESTER

RR-01685-3 2-PROPENOIC ACID, C-18-26 AND C>20 ALKYL ESTERS

RR-01686-4 AROMATIC AMINE POLYOLS

RR-01687-5 NITROBENZOIC ACID OCTYL ESTER

RR-01688-6 ALKYLATED SULFONATED DIPHENYL OXIDE, ALKALI AND AMINE SALTS

RR-01689-7 ALKYLENEDIOALKYL ETHER

RR-01690-0 ALKENYL ETHER OF ALKA-NETRIOL POLYMER

RR-01691-1 1,4-BIS(3-HYDROXY-4-BEN-ZOYLPHENOXY)BUTANE

RR-01692-2 POLYESTER POLYURETHANE ACRYLATE

RR-01693-3 SUBSTITUTED BENZENEDIAZO-NIUM

RR-01699-9 MANGANESE METHYLCY-CLOPENTADIENYL TRICAR-BONYL

RR-01701-6 HAIR COLORANTS, PERSONAL USE OF

RR-01702-7 HAIRDRESSER OR BARBER, OC-CUPATIONAL EXPOSURES AS

RR-01703-8 TOXINS DERIVED FROM FUSAR-IUM GRAMINEARUM, F. CULMO-RUM AND F. CROOKWELLENSE

RR-01704-9 TOXINS DERIVED FROM FUSAR-IUM SPOROTRICHOIDES

RR-01705-0 PICKLED VEGETABLES, TRADI-TIONAL ASIAN

RR-01706-1 TOXINS DERIVED FROM FUSAR-IUM MONILIFORME

RR-01707-2 SALTED FISH, CHINESE STYLE

RR-01708-3 HYDROBROMOFLUOROCARBONS HYDROBROMOFLUOROCARBONS (EXCEPT HBFC-22B1) [CAA]

RR-01709-4 STRYCHNINE SALTS

RR-01710-7 SULFONES

RR-01711-8 CYANOACRYLATES

RR-01712-9 ALKYL-, BROMO-, CHLORO-, HY-DROXYMETHYL DIARYL ETHERS

RR-01713-0 SILOXANES

RR-01714-1 CHLOROALKYL PHOSPHATES

RR-01715-2 PROPYLENE GLYCOL ETHERS AND ESTERS

RR-01716-3 METHYL ETHYLENE GLYCOL ETHERS AND ESTERS

RR-01717-4 TRIS(2-CHLORO-1-PROPYL) PHOSPHATE

RR-01718-5 POLYPIPERIDINOL-ACRYLATE METHACRYLATE

RR-01719-6 2-CHLORO-4,6-BIS(SUBSTITUTED) -1,3,5-TRIAZINE,DIHYDROCHLO-RIDE

RR-01720-9 TRISUBSTITUTED ANTHRACENE

RR-01721-0 CARBOXYLIC ACID GLYCIDYL ESTERS

RR-01722-1 LECITHINS, PHOSPHOLIPASE A2-HYDROLYZED

RR-01723-2 POLYEPOXYPOLYOL

RR-01724-3 SUBSTITUTED BENZENEDICAR-BOXYLIC ACID ESTER

RR-01725-4 ORGANOTIN LITHIUM COM-POUNDS

RR-01726-5 ALKYLSULFONIUM SALT

RR-01727-6 SUBSTITUTED QUINOLINE

RR-01728-7 3,6,9,13-TETRAOXAHEXADEC-15-ENE-1,11-DIOL,BIS(4-METHYL-BENZENESULFONATE)

RR-01729-8 3,6,9,12,16-PENTAOXANONADEC-18-ENE-1,14-DIOL,BIS(4-METHYL-BENZENSULFONATE)

RR-01730-1 3,6,9,12-TETRAOXATETRADE-CANE-1,14-DIOL,7-[(2-PROPENY-LOXY)METHYL]-,BIS(4-METHYL-BENZENESULFONATE)

RR-01731-2 PHOSPHATED POLYARYLPHE-NOL ETHOXYLATE, POTASSIUM SALT

RR-01732-3 2-PROPENOIC ACID 3-(TRIMETHOXYSILYL) PROPYL ES-TER

RR-01733-4 ALIPHATIC ETHER

RR-01734-5 SUBSTITUTED 2-NITROBENZENE-SULFONAMIDE

RR-01735-6 SUBSTITUTED 2-AMINOBENZENE-SULFONAMIDE

RR-01736-7 WOOD DUST, ALLERGENIC - EX-CLUDING WESTERN RED CEDAR

RR-01737-8 CHLOROBENZENES, N.O.S.

RR-01738-9 GASOLINE ENGINE EXHAUST

RR-01739-0 MINERAL FIBERS (OTHER THAN MANMADE)

RR-01740-3 POLYCYCLIC AROMATIC HYDROCARBON DERIVATIVES

RR-01741-4 RESIDUAL (HEAVY) FUEL OILS

RR-01742-5 ESTROGENS, NON-STEROIDAL

RR-01743-6 ESTROGENS, STEROIDAL

RR-01744-7 CONJUGATED ESTROGENS

RR-01745-8 NITRILOTRIACETIC ACID (SALTS)

RR-01746-9 ALKYL ANTHRAQUINONES

RR-01747-0 AMINOPHENOL SULFONIC ACID

RR-01749-2 TERT-AMYL PEROXY-2-EHTYL-HEXANOATE

RR-01750-5 TERT-AMYLPEROXY-3,5,5-TRIMETHYLHEXANOATE

RR-01751-6 TERT-BUTYLPEROXY STEARYL-CARBONATE

RR-01752-7 DI-(2-TERT-BUTYLPEROXYISO-PROPYL)-BENZENE

RR-01753-8 DI-4-CHLOROBENZOYL PEROXIDE

RR-01754-9 2,5-DIMETHYL-2,5-DI-(2-ETHYL-HEXANOYLPEROXY) HEXANE

RR-01755-0 2,5-DIMETHYL-2,5-DI-(3,5,5-TRI-METHYLHEXANOYLPEROXY) HEXANE

RR-01756-1 DIPEROXY DODECANE DIACID

RR-01757-2 DI-(2 PHENOXYETHYL) PEROXY-DICARBONATE

RR-01758-3 METHYLCYCLOHEXANONE PEROXIDE

RR-01759-4 TETRAHYDRONAPHTHYL HYDROPEROXIDE

RR-01760-7 CELLULOSE NITRATE, WET WITH ALCOHOL

RR-01761-8 PENT-ACETATE

RR-01762-9 BENZOL DILUENT

RR-01763-0 PETROLEUM ETHER

RR-01765-2 ACRYLATE ESTERS

RR-01766-3 C9 AROMATIC HYDROCARBON FRACTION

RR-01767-4 CHLOROFLUOROALKANES, FULLY HALOGENATED

RR-01768-5 GROUTS

RR-01769-6 HEXAVALENT CHROMIUM CHEMICALS

RR-01770-9 NITRATE COMPOUNDS

RR-01771-0 POLYCHLORINATED ALKANES

RR-01772-1 ANTHRANILIC ACID SALTS

RR-01773-2 EPHEDRINE SALTS, OPTICAL ISOMERS, AND SALTS OF OPTICAL ISOMERS

RR-01774-3 ERGONOVINE SALTS

RR-01775-4 ERGOTAMINE SALTS

RR-01776-5 N-ACETYLANTHRANILIC ACID SALTS

RR-01777-6 NORPSEUDOEPHEDRINE SALTS, OPTICAL ISOMERS, AND SALTS OF OPTICAL ISOMERS

RR-01778-7 PHENYLACETIC ACID SALTS

RR-01779-8 PHENYLPROPANOLAMINE SALTS, OPTICAL ISOMERS, AND SALTS OF OPTICAL ISOMERS

RR-01780-1 PIPERIDINE SALTS

RR-01781-2 PSEUDOEPHEDRINE SALTS, OPTICAL ISOMERS, AND SALTS OF OPTICALISOMERS

RR-01782-3 METHYLAMINE SALTS

RR-01783-4 ETHYLAMINE SALTS

RR-01784-5 D-LYSERGIC ACID SALTS, OPTICAL ISOMERS, AND SALTS OF OPTICALISOMERS

RR-01785-6 N-METHYLEPHEDRINE SALTS, OPTICAL ISOMERS, AND SALTS OF OPTICAL ISOMERS

RR-01786-7 N-ETHYLEPHEDRINE SALTS, OPTICAL ISOMERS, AND SALTS OF OPTICAL ISOMERS

RR-01787-8 N-METHYLPSEUDOEPHEDRINE SALTS, OPTICAL ISOMERS, AND SALTS OFOPTICAL ISOMERS

RR-01788-9 N-ETHYLPSEUDOEPHEDRINE SALTS, OPTICAL ISOMERS, AND SALTS OFOPTICAL ISOMERS

RR-01789-0 N-ETHYLEPHEDRINE

RR-01790-3 N-ETHYLPSEUDOEPHEDRINE

RR-01791-4 CHROMIUM (VI) COMPOUNDS (CERTAIN WATER INSOLUBLE FORMS)

RR-01792-5 REFRACTORY FIBRES (CERAMIC OR OTHERS)

RR-01805-3 PARA-ARAMIDE FIBRES (KEVLAR, TWARON)

RR-01806-4 POLYOLEFINES FIBRES

RR-01807-5 ORGANIC SYNTHETIC FIBRES (CARBON AND GRAPHITE FIBRES)

RR-01808-6 FIBROUS GLASS, MICROFIBRES

RR-01809-7 HEXANE (OTHER ISOMERS) HEXANE (ISOMERS OTHER THAN "N") [QBC, QBC]

RR-01810-0 POLYNUCLEAR AROMATIC HYDROCARBIDES